Yacht Have

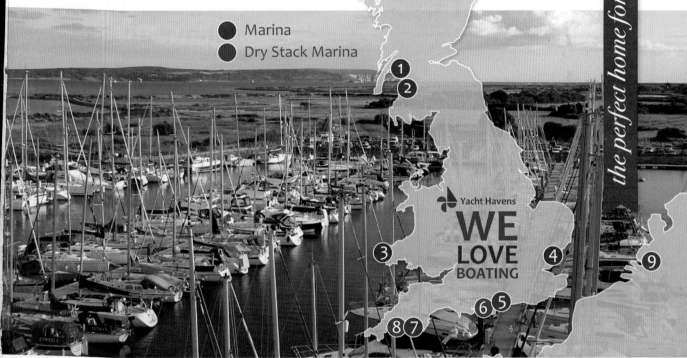

● Marina
● Dry Stack Marina

Yacht Havens
WE
LOVE
BOATING

the perfect home for y

9 Prime Marina Locations... *with the most helpful staff*

Whether you're looking for an annual berth, a safe winter haven or a visitor berth in a new cruising area, you'll find first-rate facilities and a warm welcome from our friendly, experienced team at whichever Yacht Havens location you choose.

Attractive berthing rates and flexible contracts to save you money, superb modern facilities for your enjoyment, free Wi-Fi at every berth for your convenience, and the freedom to exchange berths with other Yacht Havens.

We can provide you with the best boating experience across the UK, and now the Netherlands. Just ask us what we can do for you.

1 Largs Yacht Haven
Largs, Ayrshire
Tel: 01475 675333

2 Troon Yacht Haven
Troon, Ayrshire
Tel: 01292 315553

3 Neyland Yacht Haven
Neyland, Pembrokeshire
Tel: 01646 601601

4 Fambridge Yacht Haven
River Crouch, Essex
Tel: 01621 740370

5 Lymington Yacht Haven
Lymington, Hampshire
Tel: 01590 677071

6 Haven Quay
Lymington, Hampshire
Tel: 01590 677072

7 Plymouth Yacht Haven
Plymouth, Devon
Tel: 01752 404231

8 Yacht Haven Quay
Plymouth, Devon
Tel: 01752 481190

NEW!
9 Jachthaven Biesbosch
Drimmelen, Netherlands
Tel: +31(0)162 68 22 49

CALL US TODAY OR VISIT www.yachthavens.com

REEDS
NAUTICAL
ALMANAC
2016

ADLARD COLES NAUTICAL

EDITORS **Perrin Towler and Mark Fishwick**

Free updates are available at www.reedsalmanacs.co.uk

IMPORTANT SAFETY NOTE AND LEGAL DISCLAIMER

This Almanac provides basic navigational data for planning and executing passages. The tidal prediction data has been reproduced by permission of national hydrographic offices. Chartlets illustrate items in the text, orientate the user and highlight key features; they should not be relied on for navigational purposes and must always be used in conjunction with a current, corrected navigational chart. Any waypoint or position listed in this Almanac must first be plotted on the appropriate chart to assess its accuracy, safety in the prevailing circumstances and relevance to the Skipper's intentions.

Navigational guidance or suggestions are based on the accumulated experience of editors, agents, harbour masters and users. They are generic and in compiling a passage plan or pilotage notebook all other available publications and information should be consulted. They take no account of the characteristics of individual vessels nor the actual or forecast meteorological conditions, sea or tidal state. These need to be checked with appropriate local authorities for the intended area of operation prior to departure.

While every care has been taken in compiling the Almanac, it is a human endeavour with many contributors. Despite rigorous checking there may be inadvertent errors or inaccuracies, and omissions resulting from the time of notification in relation to the publication date. To the extent that the editors or publishers become aware of these, corrections will be published on the website www.reedsalmanacs.co.uk (requires registration). Readers should therefore regularly check the website between January and June for any such corrections. Data in this Almanac is corrected up to Weekly Edition 25/2015 of Admiralty Notices to Mariners.

The publishers, editors and their agents accept no responsibility for any errors or omissions, or for any accident, loss or damage (including without limitation any indirect, consequential, special or exemplary damages) arising from the use or misuse of, or reliance upon, the information contained in this Almanac.

The use of any data in this Almanac is entirely at the discretion of the Skipper or other individual with responsibility for the command, conduct or navigation of the vessel in which it is relied upon.

Reeds Nautical Almanac

An imprint of Bloomsbury Publishing Plc
50 Bedford Square, London, WC1B 3DP
1385 Broadway, New York, NY 10018, USA

www.bloomsbury.com

Tel: +44 (0)207 631 5600
Fax: +44 (0)207 631 5800
info@reedsalmanacs.co.uk
editor.britishisles@reedsalmanacs.co.uk
editor.continental@reedsalmanacs.co.uk
www.reedsalmanacs.co.uk

Almanac manager Chris Stevens

Reeds Nautical Almanac 2016 – ISBN 978 1 4729 1938 0

Reeds Looseleaf Almanac 2016 – ISBN 978 1 4729 1942 7

Reeds Looseleaf Update Pk 2016 – ISBN 978 1 4729 1943 4

British Library Cataloguing-in-Publication Data
A catalogue record for this book is available from the British Library.

Library of Congress Cataloguing-in-Publication data has been applied for.

REEDS, ADLARD COLES NAUTICAL and the Buoy logo are trademarks of Bloomsbury Publishing Plc

First published 2015

Bloomsbury Publishing Plc makes every effort to ensure that the papers used in the manufacture of our books are natural, recyclable products made from wood grown in well-managed forests. Our manufacturing processes conform to the environmental regulations of the country of origin.

To find out more about our authors and books visit www.bloomsbury.com. Here you will find extracts, author interviews, details of forthcoming events and the option to sign up for our newsletters.

Printed by Bell & Bain Ltd.

ADVERTISEMENT SALES

Enquiries about advertising space should be addressed to:

adlardcoles@bloomsbury.com

Come what may!

PANTAENIUS
Sail & Motor Yacht Insurance

Germany · Great Britain* · Monaco · Denmark · Austria · Spain · Sweden · USA · Australia
Plymouth · Phone +44 17 52 22 36 56

pantaenius.co.uk

Chapter Contents

CHAPTER CONTENTS.. vii

NAVIGATIONAL CONTENTS ix

Introduction   1
Editorial • Using the almanac • Improvements • Agents wanted • Corrections • Acknowledgements and Permissions

Chapter 1   Reference data   5
Symbols and Abbreviations • Useful addresses (postal and e-mail) and websites • 2016 Calendar • Eclipse notes • Conversion tables • Time • Vocabularies

Chapter 2   Regulations   21
Cruising formalities • HM Revenue and Customs • Foreign Customs • VAT • International Regulations for Preventing Collisions at Sea (COLREGs)

Chapter 3   Navigation   33
Passage planning and SOLAS V • Conduct of the plan • Light characteristics • IALA buoyage • Satellite navigation • Loran • Electronic charts • Waypoints • Chart plotters • Radar and AIS • Racons and ramarks • Navigational warnings • Hazards • Offshore energy installations • Navigational tables: Distance of horizon • Distance, speed and time • Distance off by sextant • True bearing of sun at sunrise/set • Sun's declination • True bearing of sun • Distance off rising/dipping lights • Sunrise/set times • Moonrise/set times • Phases of the Moon

Chapter 4   Tides   53
Introduction • Definitions • Calculating times/heights of tide • Overhead clearances • Computer predictions • Tidal streams • Meteorological effects • Tidal stream rates

Chapter 5   Communications   63
International code of signals • Radio communication • Global communications • Navigational warnings • Gunfacts and subfacts • Coastal radio stations (CRS) • Coastguard contact details for UK • Coast Guard contact details for Republic of Ireland, Denmark, Germany, The Netherlands, Belgium, France, Spain, Portugal, Gibraltar

Chapter 6   Weather   81
Beaufort wind scale • Sea states • Heavy weather warnings • GMDSS weather • NAVTEX • HM Coastguard broadcasts • HF facsimile • BBC broadcasts • Internet weather resources • Sources of foreign weather information

Chapter 7   Safety   105
Introduction • Safety equipment • GMDSS • Definitions of emergencies • Emergency signals • MAYDAY • PAN PAN • SÉCURITÉ • Sound & visual distress signals • Helicopter rescue • Abandon ship • UK SAR • HM Coastguard • NCI • RNLI

Chapter 8   First aid   117
Objectives • Medical care abroad • Medical advice by R/T • Emergency resuscitation • Choking • Drowning • Hypothermia • Shock • Acute illness • Stroke • Bites/stings • Bleeding • Burns • Chest and crush injuries • Cuts • Dental • Eyes • Fish hooks • Fractures • Poison • Sprains • First aid kits • Drugs • Emergency childbirth • Injections • Physiological measurements • Observation form

Chapter 9   Harbours, coasts and tides   129
Map of areas • Area information • Harbour information • Environmental guidance • Distance tables: English Channel, Irish Sea, North Sea, Bay of Biscay • Tidal coefficients
See page ix for details

General index   1041

Gazetteer (index of place names)   1044

Navigational Contents

Chapter 9, Harbours, coasts and tides   **129**

Map of areas • Harbour details and tide tables • Tidal streams • Lights, buoys and waypoints • Passage information • Ferries • Distance tables • Special notes

Area 1   **South West England, including Isles of Scilly** 137
     Isles of Scilly to Anvil Point

Area 2   Central Southern England 191
     Anvil Point to Selsey Bill

Area 3   **South East England** 245
     Selsey Bill to North Foreland

Area 4   **East England** 269
     North Foreland to Great Yarmouth

Area 5   **North East England** 329
     Winterton to Berwick-upon-Tweed

Area 6   **South East Scotland** 363
     Eyemouth to Rattray Head

Area 7   North East Scotland 389
     Rattray Head to Cape Wrath, incl Orkney and Shetland Islands

Area 8   North West Scotland 427
     Cape Wrath to Crinan Canal

Area 9   **South West Scotland** 465
     Crinan Canal to Mull of Galloway

Area 10   **North West England, Isle of Man and North Wales** 495
     Mull of Galloway to Bardsey Island

Area 11   **South Wales and Bristol Channel** 525
     Bardsey Island to Land's End

Area 12   **South Ireland** 563
     Malahide clockwise to Liscannor Bay

Area 13   North Ireland 605
     Liscannor Bay clockwise to Lambay Island

Area 14   **West Denmark** 643
     Skagen to Romø

Area 15   Germany 659
     Danish border to Emden

Area 16   **The Netherlands and Belgium** 695
     Delfzijl to Nieuwpoort

Area 17   North France 743
     Dunkerque to Cap de la Hague

Area 18   **Central North France** 791
     Cap de la Hague to St Quay-Portrieux

Area 19   **Channel Islands** 815
     Alderney to Jersey

Area 20   North Brittany 841
     Paimpol to Pointe de Penmarc'h

Area 21   South Brittany 881
     Loctudy to St Nazaire

Area 22   **South Biscay** 911
     River Loire to Spanish border

Area 23   **North and North West Spain** 947
     Fuenterrabía to Bayona

Area 24   **Portugal** 981
     Viana do Castelo to Vila Real de Santo Antonio

Area 25   South West Spain, Gibraltar and Morocco 1003
     Ayamonte to Europa Point, Tangier and Ceuta

Area 26   **Portuguese Islands** 1021
     Azores: Flores, Faial, Pico, São Jorge, Graciosa, Terceira, Santa Maria and São Miguel. **Madeira:** Porto Santo and Madeira

TAKE THE EASY ROUTE
with eSeries powered by LightHouse™ II

ⓔ Series

- **Completely redesigned** for clarity and simplicity, eSeries with LightHouse II makes everything easier
- **More Powerful** with smarter touch controls and all new features
- **Point and Steer** with Evolution's 9-axis autopilot control
- **Connect anywhere** onboard with Wi-Fi and Raymarine mobile apps
- **Network** with any LightHouse powered MFD

For more information, go to
www.raymarine.com

Raymarine®

xxix

Introduction

0.1   **Introduction**   2

Editorial

0.2   **Using the Almanac**   2

Numbering system • Contents and Index

0.3   **Help improve the Almanacs**   2

Improvements and corrections • Harbour Agents • Agents wanted

0.4   **Corrections and updates**   2

Sources of corrections • Online updates

0.5   **Acknowledgements & permissions**   2

Acknowledgements • Permissions • Disclaimer

0.1 INTRODUCTION: USING THE ALMANAC

0.1.1 Editorial

Welcome to the 2016 edition of the Almanac. It includes numerous developments that have occurred since the last edition as well as fully updated safety and navigational information – in all 45,000 changes each year. See 0.4.2 below for details of free online monthly updates.

The economic recovery is enabling marinas and harbours throughout Europe to both modernise and expand. This together with a reduction in fishing fleets, notably in W France and Spain, has improved berth availability to visitors. Numerous chartlets have been revised to reflect this, and facilities have been extensively reviewed. These are now represented by symbols to differentiate them from the text.

Technological innovation continues with the widespread introduction of LED navigational marks in the UK and Ireland. The maximum range of lights is gradually being reduced to 18M and sound signals are being removed from the majority of locations. The onus is increasingly falling on individual vessels to maintain a safe offing or make good a landfall. Virtual Aids to Navigation are now in service. These are only displayed on AIS and hence the reliance on electronic equipment is growing. This is increasingly challenging for small craft where battery capacity is limited.

The emphasis on Offshore Energy Installations is shifting towards tidal turbines and trials are underway from the Solent to the Shetlands. The maintenance of wind farms has led to increased traffic around many harbours such as Ramsgate and Hartlepool as well as commercial ports.

In light of these developments the Passage information provided in each area has been revised. Harbour information reflects new capacity including Stromness and Knight's Town, Valentia, a 1600 berth extension to Port de Minimes Marina, La Rochelle, and the re-build of Port Vauban at St Malo. The Cowes breakwater is finished and work is expected on a new marina behind the Shrape Breakwater in E Cowes. The Queensferry crossing will be completed this year and congestion in the Forth eased.

The reference chapters have now been re-written to reflect the new UK Coastguard structure and communications network. Contact details have been amended throughout. Tide tables have been augmented by additional sets of tidal coefficients along the Channel and Western coasts of France. These more accurately reflect the local differences which affect tidal streams and the operation of locks and sills.

Every effort has been made to maintain the accuracy and coherence of the Almanac so that information remains clear, concise, crisp and readily to hand in any eventuality.

0.2 USING THE ALMANAC

0.2.1 Numbering system

The nine chapters contain numbered sections. In Chapter 4, for example, sections are numbered 4.1 to 4.9. Within section 4.1 the key paragraphs are numbered 4.1.1 and 4.1.2.

Diagrams are numbered: Fig 4(1) to Fig 4(9). Tables are similarly numbered.

Each main harbour entry in Chapter 9 is shown with the chapter number, the Area number and a sequential number within that Area. Thus Plymouth is shown as 9.1.15.

0.2.2 Contents and Index

Main subject headings and page numbers are listed on the contents page at the start of each section. A General Index and Gazetteer are at the back of the Almanac.

0.3 HELP IMPROVE THE ALMANACS

0.3.1 Improvements and corrections

Corrections – there will inevitably be some – and suggestions, however minor, for improvements are always welcome. All, especially those based on passages and port visits, will be carefully considered. Please send these directly to the relevant Editor (see below) or to info@reedsalmanacs.co.uk. Otherwise, a note to Adlard Coles Nautical (see page v) will be forwarded as necessary. Please address all queries regarding Reeds Online to contactus@reedsnauticalalmanac.co.uk

Perrin Towler – editor.britishisles@reedsalmanacs.co.uk covers Areas 1 to 13 (*the UK and Ireland*) and is responsible for Chapters 3 (*Navigation*), 4 (*Tides*), 7 (*Safety*) and 8 (*First Aid*).

Mark Fishwick – editor.continental@reedsalmanacs.co.uk covers Areas 14 to 26 (*Skagen to Gibraltar*) *and* is responsible for Chapters 1 (*Reference Data*), 2 (*Regulations*), 5 (*Communications*) and 6 (*Weather*).

0.3.2 Harbour Agents

Our Harbour Agents provide invaluable local information not always available from official sources. Vacancies are shown below. If you would like to earn a free copy of the Almanac every year, please apply to the relevant Editor.

0.3.3 Agents wanted

We are keen to recruit Agents for the following areas:

UK: London and Thames Estuary.
Continental: River Seine/Paris, Bénodet/Port-La-Forêt/Concarneau, Île de Ré/Ars-en-Ré/St Martin-de-Ré.

0.4 CORRECTIONS AND UPDATES

0.4.1 Sources of corrections

This Almanac is corrected to Weekly Edition 25/2015 of Admiralty Notices to Mariners. Where possible, corrections from other sources (Agents, HMs etc) received by 1 June 2015 have also been incorporated.

0.4.2 Online updates

It is important that this Almanac is kept up to date to maintain its accuracy. For free monthly updates from January to June, please register at www.reedsnauticalalmanac.co.uk

Tick the boxes to show which updates have been applied:

Jan ☐   Feb ☐   Mar ☐   Apr ☐   May ☐   Jun ☐

0.5 ACKNOWLEDGEMENTS & PERMISSIONS

The Editors thank the many authorities which have kindly provided information essential to the preparation of this Almanac. As always, Frank Singleton has supported Reeds with information and advice about weather.

The contents include material obtained with the permission of the UK Hydrographic Office, HM Stationery Office and: Trinity House, the Northern Lighthouse Board, HM Nautical Almanac Office, HM Revenue and Customs, the Meteorological Office, the Maritime and Coastguard Agency, the ABP, BBC, IBA, NCI, PLA and RNLI.

Important contributions have also been made by the following authorities in the Republic of Ireland and continental Europe: The Commissioners of Irish Lights, Deutsche Gesellschaft zur Rettung Schiffbrüchiger (DgzRS), Koninklijke Nederlandse Redding Maatschappij (KNRM).

In particular we thank Harbourmasters, Marina Managers, our Harbour Agents and numerous individual yachtsmen who have contributed new ideas and suggestions based on their invaluable local knowledge.

0.5.1 Permissions

Information from Admiralty Charts, the Admiralty Lists of Lights, Admiralty Tide Tables, Admiralty Sailing Directions and the Admiralty Lists of Radio Signals is reproduced with the permission of the UK Hydrographic Office (Licence No GB DQ – 001 – Adlard Coles) and the Controller of HMSO.

Extracts from the International Code of Signals, 1969 and Meteorological Office Weather Services for Shipping are published by permission of the Controller of HMSO.

Phases of the Moon and Sun/Moon rising and setting times are included by permission of HM Nautical Almanac Office.

Tidal stream diagrams are reproduced from Admiralty charts and tidal stream atlases by permission of the UKHO and the Controller of HMSO.

Tidal stream data in 9.1.5, 9.12.3 and 9.13.3 is printed by kind permission of the Royal Cruising Club and the Irish Cruising Club respectively.

UK and Foreign tidal predictions and curves are supplied by the UK Hydrographic Office, with the permission (and Authorisation Numbers) of the following authorities to use the Standard Port predictions stated:

- Danmarks Meteorologiske Institut, Denmark: Esbjerg.
- SHOM, France: Dunkerque, Boulogne, Dieppe, Le Havre, Cherbourg, St Malo, Brest, Pte de Grave; and tidal coefficients for Calais, St Malo, Brest and Pte de Grave.
- BSH, Germany: Helgoland, Wilhelmshaven and Cuxhaven (11123/2008-07).
- Rijkswaterstaat, The Netherlands: Vlissingen and Hoek van Holland.
- Vlaamse Hydrografie, Belgium: Zeebrugge.
- Marinho Instituto Hidrográfico, Portugal: Lisboa.
- Instituto Hidrográfico de la Marina, Spain: La Coruña.

0.5.2 Disclaimer

Neither the UK nor any other National Hydrographic Office has verified the information in this product and none accepts liability for the accuracy of reproduction or any modifications made thereafter. No National HO warrants that this product satisfies national or international regulations regarding the use of the appropriate products for navigation.

1

Reference Data

| 1.1 | Symbols and abbreviations | 6 |

Admiralty chart symbols • Symbols and Abbreviations

| 1.2 | Postal and e-mail addresses with websites | 9 |

Listed by countries

| 1.3 | Calendar 2016 and ephemeris | 10 |

Public holidays • Eclipses

| 1.4 | Conversion tables | 11 |

Conversion factors • Feet to metres and vice versa

| 1.5 | Time | 12 |

Zone, UT, Standard, DST • Tide tables • Radio time signals

| 1.6 | Vocabularies | 12 |

English • Deutsch • Nederlands • Français • Español • Português

Fig 1(1) Admiralty chart symbols

Symbol	Description	Symbol	Description	Symbol	Description
Pyl ⦵28⦵ Pyl	Power transmission line with pylons and safe overhead clearance as defined by the responsible authority	⊥	Church, chapel		Drying contour LW line, Chart Datum
20	Vertical clearance above HAT		Radio mast, television mast		Blue ribbon or differing blue tints may be shown
			Monument (including column, pillar, obelisk, statue)		Anchoring prohibited
			Chimney		
～～～	Submarine cable		Flare stack (on land)		Marine Farm
-·-·-·-	Buried pipeline		Tanks		Wreck, depth unknown, danger to navigation
～～～	Overfalls, tide rips and races		Crane		Wreck, depth unknown, no danger to navigation
3 kn →	Flood stream Mean spring rate		Bird sanctuary	4 Wk 6	Wreck, depth obtained by sounding
3 Kn →	Ebb stream Mean spring rate	C G	Coastguard Station	4 Wk 6	Wreck, swept by wire to the depth shown
◇D	Designated position of tabulated tidal streams	⋏ ⌒	Woods in general		Wreck showing any part at level of chart datum
(a) (b)	Examples: (a) fixed beacon; (b) buoy	⚲	Withy - starboard hand		Rock which covers and uncovers, height above chart datum
	Limit of safety zone around offshore installation	⚱	Withy - port hand		Rock awash at level of chart datum
⚲	Major light	◁▷	Floodlit		Dangerous underwater rock of unknown depth
	Port hand lt bcn (PHM)	⋘	Marsh		Dangerous underwater rock of known depth
⊙ V-AIS	Virtual AIS	⤳	Kelp	⚓	Recommended anchorage
⊖	Fishing harbour	⊙)))	Fog signal	✦	Rescue station, lifeboat station, rocket station
Ⓥ	Visitor's berth	⛵	Yacht berths, no facilities	Ⓛ	Yacht harbour, Marina
	Fuel station (Petrol, Diesel)	Ⓥ	Visitor's buoy	⚓	Harbourmaster's Office
◣	Public slipway		Mooring buoy	⊖	Customs office
⊤	Water tap	▢	Laundrette	⊕	Hospital
	Public landing, steps, ladder	◀	Yacht Club, Sailing Club	✉	Post office

1.1 SYMBOLS AND ABBREVIATIONS

The following more common Symbols and Abbreviations feature in Reeds Almanacs and in some Admiralty charts and publications. See also Fig 1(1). Admiralty NP 5011 is the authority for Symbols and Abbreviations on printed charts. Those symbols in the tinted box immediately below are frequently used within the text describing each port.

Symbol	Meaning
�особ	Alongside berth
©	Automatic telling machine (ATM), cashpoint
®	Bank
	Bottled gas available, Calor Gas, Camping Gaz, or Kosangas
	Boatyard
	Boat hoist (+tons)
	Crane (+ tons)
	Chandlery
	Customs
	Diesel (supply by hose)
	Diesel (in cans)
	Electrical repairs
©	Electronic repairs
	Fresh water supply
	Food shop/supermarket
H	Hospital
≠	In transit with, ie ldg marks/lts
@	Internet café/access
	Landing place/steps/ladder
LB, ◆	Lifeboat, inshore lifeboat
	Light float, minor
	Lt float, major; Light vessel or Lanby
	Licensed bar, Public house, Inn
©	National Coastwatch Institution (NCI)
	Marine engineering repairs
	Mooring buoy
	Petrol (supply by hose)
	Petrol (in cans)
⊠	Post Office
	Pump out facility
×	Restaurant/cafe
	Rigger
	Sailmaker
	Shipwright (esp wooden hulls)
	Shower
	Shore power (electrical)
	Slipway
ⓥ	Visitors' berths
	Visitors' mooring
I	Wind turbine
	Weather forecast available
	Yacht Club/(YC)/Sailing Club(SC)

Abbr	Meaning
abm	Abeam
ABP	Associated British Ports
ACA	Admiralty Chart, AC Agent
ACN	Adlard Coles Nautical (Publisher)
Aff Mar	Affaires Maritimes
AIS	Automatic Identification System
aka	Also known as
ALL	Admiralty List of Lights
ALRS	Admiralty List of Radio Signals
Al	Alternating light
ANWB	Association of road & waterway users (Dutch)
ATT	Admiralty Tide Tables
ATT	Atterisage (landfall/SWM) buoy
Auto	Météo Répondeur Automatique
B.	Bay, Black
BE	Belgian chart
Bk	Broken (nature of seabed)
Bkwtr	Breakwater
BMS	Bulletin Météorologique Spécial (Strong wind/Gale warning)
	Bn, bcn(s) Beacon, beacon(s)
BSH	German Hydrographic Office/chart(s)
BST	British Summer Time (= DST)
Bu	Blue
By(s)	Buoy, buoys
C.	Cape, Cabo, Cap
c	Coarse (sand; nature of seabed)
ca	Cable (approx 185m long)
Cas	Castle
CD	Chart datum (tidal)
CEVNI	Code Européen de Voies de la Navigation Intérieure (inland waterway signs etc)
cf	Compare, cross-refer to
CG	Coast Guard, HM Coastguard (in the UK)
CGOC	Coast Guard Operations Centre (UK)
chan	Channel (navigational)
Ch	Channel (VHF)
Ch, ⊕	Church
Chy	Chimney
Co	Coral (nature of seabed)
Col	Column, pillar, obelisk
CPA	Closest Point of Approach
CROSS	Centre Régional Opérationnel de Surveillance et Sauvetage (= MRCC)
CRS	Coast Radio Station
CRT	Canal and River Trust
C/S	COSPAS/SARSAT (satellite)
Cy	Clay (nature of seabed)
Dec	Declination (of the Sun) abbr: December
Defib	Automated External Defibrillator
dest	Destroyed
DF, D/F	Radio Direction Finding
DG	De-gaussing (range)
DGPS	Differential GPS
Dia	Diaphone (fog signal)
Dir Lt	Directional light
discont	Discontinued
DLR	Dockland Light Railway
Dn(s)	Dolphin(s)
DR	Dead Reckoning or Dries (secondary tides)
DSC	Digital Selective Calling
DST	Daylight Saving Time
DW	Deep Water (route)
DYC	Dutch Yacht Chart(s)
DZ	Danger Zone (buoy)
E	East
ECM	East cardinal mark (buoy/beacon)
ED	Existence doubtful. European Datum
EEA	European Economic Area
Elev	Elevation
Ent	Entrance, entry, enter
EP, △	Estimated position
ETA	Estimated Time of Arrival
ETD	Estimated Time of Departure
F	Fixed light
f	Fine (eg sand; nature of seabed)
F&A	Fore and aft (berth/mooring)
FFL	Fixed and Flashing light
Fl	Flashing light
FM	Frequency Modulation
Foc	Free of charge
Fog Det lt	Fog Detector light
Freq, Fx	Frequency
FS	Flagstaff, Flagpole
ft	Foot, feet
Ft,	Fort
FV	Fishing vessel
G	Gravel (nature of seabed), Green
GC	Great Circle
GDOP	Geometric Dilution of Precision (GPS)

GHA	Greenwich Hour Angle
GLA	General Lighthouse Authority
GMDSS	Global Maritime Distress & Safety System
grt	Gross Registered Tonnage
Gy	Grey
H, h, Hrs	Hour(s)
H−, H+	Minutes before, after the whole hour
H24	Continuous
HAT	Highest Astronomical Tide
HF	High Frequency
HFP	High Focal Plane (buoy)
HIE	Highlands & Islands Enterprise
HJ	Day service only, sunrise to sunset
HM	Harbour Master
HMRC	HM Revenue & Customs
HMSO	Her Majesty's Stationery Office
HN	Night service only, sunset to sunrise
HO	Office hours, Hydrographic Office
(hor)	Horizontally disposed (lights)
hPa	Hectopascal (= 1millibar)
HT	High Tension (overhead electricity line)
HW	High Water
HX	No fixed hours
IALA	International Association of Marine Aids to Navigation and Lighthouse Authorities
iaw	In accordance with
IDM	Isolated Danger Mark (buoy/beacon)
IHO	International Hydrographic Organisation
IMO	International Maritime Organisation
INMARSAT	International Maritime Satellite Organisation
intens	Intensified (light sector)
IPTS	International Port Traffic Signals
IQ	Interrupted quick flashing light
IRPCS	International Regulations for the Prevention of Collisions at Sea
Is, I	Island, Islet
ISAF	International Sailing Federation
Iso	Isophase light
ITU	International Telecommunications Union
ITZ	Inshore Traffic Zone (TSS)
IUQ	Interrupted ultra quick flashing light
IVQ	Interrupted very quick flashing light
kn	Knot(s)
kW	Kilowatts
L	Lake, Loch, Lough
Lat	Latitude
LAT	Lowest Astronomical Tide
Lanby, ⌐	Large automatic navigational buoy
Ldg	Leading (light)
LF	Low frequency
L Fl	Long flash
LH	Left hand
LNG	Liquefied Natural Gas
LNTM	Local Notice To Mariners
LOA	Length overall
Long, lng	Longitude
LPG	Liquefied Petroleum Gas
LT	Local time
Lt(s), ✪ ✩	Light(s)
M	Moorings. Nautical (sea) mile(s). Mud
m	Metre(s)
Mag	Magnetic. Magnitude (of star)
Mb, mb	Millibar (= 1 hectopascal, hPa)
MCA	Maritime and Coastguard Agency
Met/Météo	Meteorology/Météorologie (weather)
MHWN	Mean High Water Neaps
MHWS	Mean High Water Springs
MHz	Megahertz
ML	Mean Level (tidal)
MLWN	Mean Low Water Neaps
MLWS	Mean Low Water Springs

MMSI	Maritime Mobile Service Identity
Mo	Morse
Mon	Monument. Abbrev: Monday
MRCC	Maritime Rescue Co-ordination Centre
MRSC	Maritime Rescue Sub-Centre (not in the UK)
MSI	Maritime Safety Information
N	North
Navi	Navicarte (French charts)
NB	Nota Bene. Notice Board
NCM	North Cardinal Mark (buoy/beacon)
ND	No Data (secondary tides)
NGS	Naval Gunfire Support (buoy)
NM	Notice(s) to Mariners
NMOC	National Maritime Operations Centre (HMCG)
np	Neap tides
NP	Naval Publication (plus number)
NRT	Net registered tonnage
NT	National Trust (land/property)
Obscd	Obscured
Obstn	Obstruction
Oc	Occulting light
ODAS	Ocean Data Acquisition System (buoy)
Or	Orange (see also Y)
OT	Other times
P	Pebbles
(P)	Preliminary (NM)
PA	Position approximate
Pax	Passenger(s)
PC	Portuguese chart
PD	Position doubtful
PHM	Port-hand Mark (buoy/beacon)
PLA	Port of London Authority
Pos	Position
Prog	Prognosis (weather charts)
prom	Prominent
PSSA	Particularly Sensitive Sea Area
Pt(e).	Point(e)
Pta	Punta (point)
Q	Quick flashing light
QHM	Queen's Harbour Master
qv	Refer to (quod vide)
R	Red. River.
Racon	Radar transponder beacon
Ramark	Radar beacon
RCD	Recreational Craft Directive
RG	Emergency RDF station
RH	Right hand
Rk, Rky	Rock, Rocky (nature of seabed)
RNLI	Royal National Lifeboat Institution
ROI	Republic of Ireland
R/T	Radiotelephony
Ru	Ruins
RYA	Royal Yachting Association
S	South, Sand (nature of seabed)
S, St, Ste	Saint(s)
SAMU	Service d'Aide Médicale Urgente (ambulance)
SAR	Search and Rescue
SBM	Single buoy mooring
SC	Sailing Club. Spanish chart
SCM	South Cardinal Mark (buoy/beacon)
SD	Sailing Directions, Semi-diameter (of sun) Sounding of doubtful depth
sf	Stiff (nature of seabed)
Sh	Shells (nature of seabed). Shoal
SHM	Starboard-hand Mark (buoy/beacon); Simplified Harmonic Method (tides)
SHOM	Service Hydrographique et Océanographique de la Marine (FrenchHO/Chart)
Si	Silt (nature of seabed)
SIGNI	Signalisation de la Navigation Intérieure
SMS	Short Message Service (mobile texting)
so	Soft (eg mud; nature of seabed)

SOLAS	Safety of Life at Sea (IMO Convention)	UQ	Ultra Quick flashing light
Sp	Spire	UT	Universal Time (= approx GMT)
sp	Spring tides	Var	Variation (magnetic)
SPM	Special Mark (buoy/beacon)	(vert)	Vertically disposed (lights)
SR	Sunrise	Vi	Violet
SRR	Search and Rescue Region	vis	Visibility, visible
SS	Sunset. Signal Station	VLCC	Very large crude carrier (Oil tanker)
SSB	Single Sideband (radio)	VNF	Voie Navigable de France (canals)
St	Stones (nature of seabed)	VQ	Very Quick flashing light
Stbd	Starboard	VTS	Vessel Traffic Service
subm	Submerged	W	West. White
SWM	Safe Water Mark (buoy/beacon)	WCM	West Cardinal Mark (buoy/beacon)
sy	Sticky (eg mud; nature of seabed)	Wd	Weed (nature of seabed)
(T), (Temp)	Temporary	wef	With effect from
tbc	To be confirmed	WGS	World Geodetic System (GPS datum)
tbn	To be notified	wi-fi	Wireless Fidelity (internet access)
TD	Temporarily Discontinued (fog signal)	WIG	Wing in ground effect (craft)
TE	Temporarily Extinguished (light)	WIP	Work in progress
tfn	Till further notice	Wk, ⌐ ⊕	Wreck (see also Fig 1(1))
Tr, twr	Tower	WMO	World Meteorological Organisation
T/R	Traffic Report (route notification)	WPT, ⊕	Waypoint
TSS	Traffic Separation Scheme	WZ	Code for UK coastal navigation warning
uncov	Uncovers	Y	Yellow, Amber, Orange

1.2   POSTAL AND E-MAIL ADDRESSES WITH WEBSITES

Alderney Visitor Information Centre
51 Victoria Street, St Anne, Alderney GY9 3TA. ☎ 01481 823737. www.visitalderney.com

BBC Radio
Broadcasting House, London W1A 1AA. ☎ 020 7580 4468. www.bbc.co.uk

Canal and River Trust
First Floor North, Station House, 500 Elder Gate, Milton Keynes, MK9 1BB. ☎ 0303 040 4040. www.canalrivertrust. org.gov customer.services@canalrivertrust.org.gov

Cruising Association (CA)
CA House, 1 Northey St, Limehouse Basin, London E14 8BT. ☎ 020 7537 2828. 🖷 020 7537 2266 www.cruising.org.uk office@cruising.org.uk

ECMWF (European Centre for Medium-range Weather Forecasts) The Director-General, Shinfield Park, Reading RG2 9AX, UK. ☎ 0118 949 9000. www.ecmwf.int   DG@ecmwf.int

Guernsey Tourist Information Centre
North Esplanade, St Peter Port, Guernsey GY1 3AN. ☎ 01481 723552. www.visitguernsey.com enquiries@visitguernsey. com

HM Revenue and Customs (HMRC)
Helpline ☎ 0845 010 9000; www.hmrc.gov.uk

Hydrographic Office, The United Kingdom
Admiralty Way, Taunton, Somerset TA1 2DN.
☎ 01823 337900. 🖷 01823 337900. www.ukho.gov.uk customerservices@ukho.gov.uk

International Maritime Organisation (IMO)
4 Albert Embankment, London SE1 7SR. ☎ 020 7735 7611. 🖷 020 7587 3210. www.imo.org info@imo.org

Jersey Visitor Services Centre
Liberation Place, St Helier, Jersey JE1 1BB. ☎ 01534 448877. 🖷 01534 448899. www.jersey.com info@jersey.com

Lloyd's
One Lime Street, London EC3M 7HA. ☎ 020 73271000; 🖷 020 7327 6827. www.lloyds.com enquiries@lloyds.com

Marine Accident Investigation Branch (MAIB)
Mountbatten House, Grosvenor Square, Southampton SO15 2JU. ☎ 023 8039 5500. 🖷 023 8039 5500. www.maib.gov.uk maib@dft.gsi.gov.uk

Maritime and Coastguard Agency (MCA)
Spring Place, 105 Commercial Rd, Southampton SO15 1EG. ☎ 023 8032 9100; Info 0870 6006 505. 🖷 023 8032 9105. www.dft.gov.uk/mca   infoline@mcga.gov.uk

Meteorological Office
FitzRoy Road, Exeter EX1 3PB. ☎ 0370 900 0100. 🖷 0370 900 5050, ☎ +44 1392 885680. 🖷 +44 1392 885681 from abroad. www.metoffice.gov.uk enquiries@metoffice.gov.uk

National Coastwatch Institution (NCI)
☎ 0300 111 1202. 1,Barras St, Liskeard, Cornwall PL14 6AD www.nci.org.uk info@nci.org.uk

Northern Lighthouse Board
84 George St, Edinburgh EH2 3DA. ☎ 0131 4733100. 🖷 0131 220 2093. www.nlb.org.uk enquiries@nlb.org.uk

Registry of Shipping and Seamen/ Small Ships Register
Anchor Court, Keen Road,Cardiff CF24 5JW; ☎ 029 2044 8800; www.ukshipregister.co.uk rss@mcga.gov.uk
Part I (Pleasure vessels) ☎ 02920 448 866.part1_registry@ mcga.gov.uk SSR ☎ 029 2044 8813.ssr_registry@mcga.gov.uk

Royal Institute of Navigation (RIN)
1 Kensington Gore, London SW7 2AT. ☎ 020 7591 3134. 🖷 020 7591 3131. www.rin.org.uk admin@rin.org.uk

Royal National Lifeboat Institution (RNLI)
West Quay Road, Poole, Dorset BH15 1HZ. ☎ 0845 122 6999. www.rnli.org.uk info@rnli.org

Royal Yachting Association (RYA)
RYA House, Ensign Way, Hamble, SO31 4YA.
☎ 023 8060 4100. 🖷 023 8060 4299. www.rya.org.uk info@rya.org.uk

Royal Yachting Association (Scotland)
Caledonia House, South Gyle, Edinburgh EH12 9DQ.
☎ 0131 317 7388. 🖷 0844 556 9549. www.ryascotland.org.uk admin@ryascotland.org.uk

Scottish Canals
Canal House, Applecross St, Glasgow G4 9SP. ☎ 0141 332 6936. www.scottishcanals.co.uk enquiries@scottishcanals. co.uk

Trinity House, Corporation of
Tower Hill, London EC3N 4DH. ☎ 020 7481 6900. 🖷 020 7480 7662. www.trinityhouse.co.uk enquiries@thls.org

BELGIUM

Belgian Tourist Information
1a Cavendish Sq, London W1G 0LD. ☎ 020 7307 7738
www.visitflanders.co.uk info@visitflanders.co.uk

British Embassy (Consular Services)
Avenue d'Auderghem 10, 1040 Brussels. ☎ +32 2287 6211
consularsection.brussels@fco.gov.uk

Vlaamse Hydrografie
Administratief Centrum, Vrijhavenstraat 3, B-8400 Oostende.
☎ +32 59 554 211. ▨ +32 59 507 037
www.vlaamsehydrografie.be kust@vlaanderen.be.

DENMARK

British Embassy
Kastelsvej 36-40, DK-2100 København Ø.
☎ +45 35 44 52 00. Consular. copenhagen@fco.gov.uk

Danish Hydrographic Office
Kort-og Matrikelstyrelsen, Rentemestervej 8, DK-2400
København. ☎ +45 35 87 50 50. www.kms.dk kms@kms.dk

Søfartsstyrelsen (Danish Maritime Authority)
Søsportens Sikkerhedsråd (Safety for yachts)
Vermundsgade 38C, DK-2100 København Ø.
☎ +45 39 27 15 15. ▨ +45 39 17 44 01
www.dma.dk dma@dma.dk

FRANCE

British Embassy (Consular Services)
16 rue d'Anjou, 75008 Paris.
☎ +33 1 44 51 31 00. ▨ +33 1 44 51 31 27.

Service Hydrographique et Océanographique de la Marine
13 rue du Chatellier, CS 92803, 29228 Brest Cedex 2.
☎ +33 2 56 31 23 12. ▨ +33 2 98 22 08 99. www.shom.fr
bps@shom.fr

Société Nationale de Sauvetage en Mer (SNSM)
31, cité d'Antin, 75009 Paris. ☎ +33 1 56 02 64 64. ▨ +33 1
56 89 30 01. www.snsm.org

GERMANY

British Embassy (Consular Services)
Wilhelmstrasse 70/71, 10117 Berlin. ☎ +49 30 2045 7-0.
▨ +49 30 204 570. www.ukingermany.fco.gov.uk
ukingermany@fco.gov.uk

Bundesamt für Seeschiffahrt und Hydrographie (BSH).
(German HO) Bernhard-Nocht-Str 78, 20359 Hamburg.
☎ +49 40 31900. ▨ +49 40 3190 50 00. www.bsh.de info@
bsh.de

Deutscher Gesellschaft zur Rettung Schiffbrüchiger (DGzRS
= Rescue service). PO Box 10 63 40, 28063 Bremen.
☎ +49 421 53707-0. ▨ +49 421 53707-690. www.dgzrs.de
info@seenotretter.de

German National Tourist Board
Beethovenstrasse 69, D-60325 Frankfurt/Main.
☎ +49 0 6997 4640. ▨ +49 0 6975 1903.
www.germany-tourism.de info@d-z-t.com

IRELAND, REPUBLIC OF

British Embassy
29 Merrion Road, Ballsbridge, Dublin 4. ☎ +353 1 205 3700.
▨ +353 1 205 3779 www.britishembassy.ie
consular.dubli@fco.gov.uk

Irish Coast Guard (IRCG)
Department of Transport, Leeson Lane, Dublin 2, RoI.
☎ +353 1 678 3455 www.ircg.ie admin@irishcoastguard.ie

Irish Lights, Commissioners of
Harbour Road, Dun Laoghaire, Co Dublin, RoI. ☎ +353 1
2715400; ▨ +353 1 2715564. www.cil.ie info@cil.ie

Irish Tourist Board
103 Wigmore St, London W1U 1QS. ☎ 0207 518 0800 / 0800
313 4000. www.discoverireland.com/gb

MOROCCO

British Embassy
28 Avenue S.A.R. Sidi Mohammed, Rabat 10105 (BP 45), ☎
+212 537 63 33 33 Rabat.Consular@fco.gov.uk
www.ukinmorocco.fco.gov.uk

Moroccan National Tourist Office
Angle Rue Oued Al Makhazine, Rue Zalaga-BP., 19-Agdal,
Rabat ☎ +212 537 27 83 00 / 19 www.visitmorocco.com

NETHERLANDS

British Embassy (Consular Services)
Koningslaan 44, 1075 AE Amsterdam.
☎ +31 20 676 4343. www.ukinnl.fco.gov.uk ukinnl@fco.gov.uk

Dienst der Hydrografie
Van der Burchlaan 31, PO Box 90701, 2509 LS Den Haag.
☎ +31 70 316 2800. ▨ +31 70 316 2843. www.hydro.nl
info@hydro.nl

Stichting Classificatie Waterrecreatiebedrijven
PO Box 93345, 2509 AH Den Haag. ☎ +31 70 328 3807.

PORTUGAL

British Embassy (Consular Services)
Rua de São Bernardo 33, 1249-082 Lisboa. ☎ +351 21 359
4082 ▨ +351 21 3924 153. www.ukinportugal.fco.gov.uk
portugal.consulate@fco.gov.uk

Hydrographic Office
Instituto Hidrografico, Rua das Trinas 49, 1296 Lisboa.
☎ +351 21 3914 000. ▨ +351 21 3914 199. www.
hidrografico.pt mail@hidrografico.pt

SPAIN

British Embassy (Consular Services)
Torre Espacio, Paseo de la Castellana 259, 28046 Madrid.
☎ +34 917 146 300 or from inside Spain: 902 109 356. ▨ +34
917 146 301 info.consulate@fco.gov.uk

Instituto Hidrografico de la Marina
Plaza de San Severiano No 3, 11007 Cadiz. ☎ +34 956 54 52
05. www.armada.mde.es/ihm orpflota@fn.mde.es

Spanish Tourist Office
6th floor, 64 North Row, London W1K 7DE. ☎ 020 7317
2011. www.spain.info info.londres@tourspain.es

1.3 2016 CALENDAR AND EPHEMERIS

1.3.1 UK public holidays 2016

Jan 1 New Year
Jan 2 (Scotland only, falls on Saturday)
Jan 4 (Scotland only, in lieu of Jan 2)
Mar 17 (St Patrick, N Ireland only)
Mar 25 (Good Friday)
Mar 28 (Easter Monday, not in Scotland)
May 2 (Early Spring)
May 30 (Spring)

Jul 12 (Battle of the Boyne, N Ireland only)
Aug 1 (Summer, Scotland only)
Aug 29 (Summer, not in Scotland)
Dec 25 (Christmas Day, falls on Sunday)
Dec 26 (Boxing Day)
Dec 27 in lieu of Dec 25

Note: Clauses in the *Banking and Financial Dealings Act*
allow the Government to alter dates at short notice

1.3.2 Eclipses visible from the British Isles & Europe

Sadly, no eclipses of the Sun or Moon will be visible in 2016.

1.4 CONVERSION TABLES

1.4.1 Conversion factors

To convert	Multiply by	To convert	Multiply by
Area			
sq in to sq mm	645·16	sq mm to sq in	0·00155
sq ft to sq m	0·0929	sq m to sq ft	10·76
Length (See 1.4.2 for feet to metres, and vice versa)			
in to mm	25·4	mm to in	0·0394
yds to m	0·914	m to yds	1·094
fathoms to m	1·8288	m to fathoms	0·5468
nautical miles (M) to kilometres	1·852	kilometres to nautical miles	0·539957
nautical miles to statute miles	1·1515	statute miles to nautical miles	0·8684
(1 cable equals 0·1 M, approx 185m)			
Velocity (See also 6.1.1 Beaufort scale)			
ft/sec to m/sec	0·3048	m/sec to ft/sec	3·281
ft/sec to miles/hr	0·682	miles/hr to ft/sec	1·467
ft/sec to knots	0·592	knots to ft/sec	1·689
ft/min to m/sec	0·0051	m/sec to ft/min	196·8
knots to miles/hr	1·1515	miles/hr to knots	0·868
knots to km/hr	1·852	km/hr to knots	0·54
knots to m/sec	0·5144	m/sec to knots	1·944
Mass			
oz to grams	28·35	grams to oz	0·0353
lb to kg	0·4536	kg to lb	2·205
tons to tonnes (1000 kg)	1·016	tonnes to tons (2240 lb)	0·9842
Pressure			
inches of mercury to millibars	33·86	millibars to inches of mercury	0·0295
lb/sq in to kg/sq cm	0·0703	kg/sq cm to lb/sq in	14·22
lb/sq in to atmospheres	0·068	atmospheres to lb/sq in	14·7
Volume			
cu in to cu cm (cc or ml)	16·387	cu cm (cc or ml) to cu in	0·061
cu ft to cu metres	0·0283	cu metres to cu ft	35·31
cu ft to galls	6·25	galls to cu ft	0·16
cu ft to litres	28·33	litres to cu ft	0·035
Capacity			
pints to litres	0·568	litres to pints	1·76
Imp galls to litres	4·546	litres to Imp galls	0·22
Imp galls to US galls	1·2	US galls to Imp galls	0·833
Torque			
lbf ft to Nm	1·3558	Nm to lbf ft	0·7376
kgf m to Nm	9·8066	Nm to kgf m	0·102
lbf ft to kgf m	0·1383	kgf m to lbf ft	7·233

1.4.2 Feet to metres, metres to feet

Explanation: The central columns of figures in bold type can be referred to the left to convert metres into feet, or to the right to convert feet into metres, eg, five lines down: 5 metres = 16·40 feet, or 5 feet = 1·52 metres. Alternatively multiply feet by 0·3048 for metres, or multiply metres by 3·2808 for feet.

Feet		Metres	Feet		Metres	Feet		Metres	Feet		Metres
3·28	**1**	0·30	45·93	**14**	4·27	88·58	**27**	8·23	131·23	**40**	12·19
6·56	**2**	0·61	49·21	**15**	4·57	91·86	**28**	8·53	134·51	**41**	12·50
9·84	**3**	0·91	52·49	**16**	4·88	95·14	**29**	8·84	137·80	**42**	12·80
13·12	**4**	1·22	55·77	**17**	5·18	98·43	**30**	9·14	141·08	**43**	13·11
16·40	**5**	1·52	59·06	**18**	5·49	101·71	**31**	9·45	144·36	**44**	13·41
19·69	**6**	1·83	62·34	**19**	5·79	104·99	**32**	9·75	147·64	**45**	13·72
22·97	**7**	2·13	65·62	**20**	6·10	108·27	**33**	10·06	150·92	**46**	14·02
26·25	**8**	2·44	68·90	**21**	6·40	111·55	**34**	10·36	154·20	**47**	14·33
29·53	**9**	2·74	72·18	**22**	6·71	114·83	**35**	10·67	157·48	**48**	14·63
32·81	**10**	3·05	75·46	**23**	7·01	118·11	**36**	10·97	160·76	**49**	14·94
36·09	**11**	3·35	78·74	**24**	7·32	121·39	**37**	11·28	164·04	**50**	15·24
39·37	**12**	3·66	82·02	**25**	7·62	124·67	**38**	11·58	167·32	**51**	15·54
42·65	**13**	3·96	85·30	**26**	7·92	127·95	**39**	11·89	170·60	**52**	15·84

1.5 TIME

Zone Time For timekeeping purposes the world is divided into 24 Time Zones, each 15° of Longitude wide (like orange segments). Zone 0 (7½°E to 7½°W) straddles 0° longitude (the Greenwich Meridian). In the marine world zones E of it (ahead in time, or earlier) are labelled Zone –1, –2 and so on to –11. Zones to the W (later in time) are +1, +2 etc to +11. *Caution: land-based time-zone charts often refer to zones the other way around – be careful which one you are using.*

1.5.1 Universal Time (UT) is a datum to which broadcast times, times of sunrise/set, scientific data etc are referred. UT is also the name of the Standard Time (see below) usually kept in countries wholly or partly within Zone 0.

1.5.2 Standard Time, also called legal time, is that kept in different countries **but is modified by Daylight Saving Time in the summer** (see 1.5.3).

For all practical purposes, Greenwich Mean Time (GMT) is the same as UT.

Fig 1(2) Zone Time and Standard Times for Europe

UT is the Standard Time in the UK, including the Channel Islands and Northern Ireland (Zone 0); also the Irish Republic, Portugal and Morocco (Zone 1/2) and Madeira (Zone 2).

UT –1 is the Standard Time in Denmark and Germany (Zone –1); also The Netherlands, Belgium, France, Spain and Gibraltar (Zone 0). Subtract 1 hour from the local Standard Time to give UT, or add 1 hour to UT to give Standard Time in these places.

UT +1 is the Standard Time in the Azores (Zone +2). Add 1 hour to the local Standard Time to give UT, or subtract 1 hour from UT to give Standard Time in these places.

See also Special Notes for each country in Chapter 9.

1.5.3 Daylight Saving Time (DST), or BST in the UK, is an artificial adjustment of 1 hour ahead of Standard Time made during the summer months in most European and many other countries.

In EU countries DST is kept from the last Sunday in March until the last Sunday in October.

1.5.4 Tide Tables Tidal predictions are referred to the Standard Time kept by the port, which is shown to the top left of its tide tables, along with instructions for converting the predicted times to UT; or to local summer time.

In the summer months (no tinted background) add 1 hour to the printed times to convert to local DST.

RADIO TIME SIGNALS

1.5.5 BBC Radio time signals in the UK are broadcast in clock time (ie, in BST during summer; UT in the winter).

	BBC Radio 1	BBC Radio 2	BBC Radio 3
Mainland	97·6–99·8 MHz	88–90·2 MHz	90·2–92·4 MHz
Channel Is	97·1 MHz	89·6 MHz	91·1 MHz
Mon-Thu	0630, 0900, 1000	0700, 0800, 1700	0700, 0800
Fri	0630, 0900, 1000, 1900	0700, 0800, 1700	0700, 0800
Sat	No info	0700, 0800	0700, 0800
Sun	No info	0800, 0900	0700

BBC Radio 4 time signals are broadcast on AM (long wave) 198 kHz every hour of every day except 0000 and Mon-Fri: 1800, 2200; Sat: 1500, 1800; Sun: 1000, 1100, 1800, 2200. See 6.10.1 for AM (medium wave) and FM frequencies.

1.5.6 The BBC World Service provides time signals on the following frequencies (kHz) at the times shown below:

A:	198	B:	648	C:	1296	D:	3955
E:	6195	F:	7150	G:	7230	H:	7325
I:	9410	J:	9750	K:	9760	L:	9915
M:	12095	N:	15070	O:	15340	P:	17640
Q:	17705						

Time	A	B	C	D	E	F	G	H	I	J	K	L	M	N	O	P	Q
0000	♦	♦						♦				♦	♦	♦			
0200	♦	♦	♦		♦			♦	♦			♦	♦				
0300	♦	♦	♦		♦			♦	♦			♦	♦				
0400	♦				♦	♦		♦				♦					
0500	♦							♦	♦			♦	♦				
0600		♦	♦		♦			♦					♦	♦			
0700	♦							♦			♦		♦	♦		♦	
0800	♦						♦				♦		♦	♦		♦	♦
0900	♦						♦	♦	♦	♦	♦		♦	♦	♦	♦	♦
1100	♦									♦	♦	♦		♦	♦	♦	
1200	♦									♦	♦	♦		♦	♦	♦	
1300	♦									♦	♦			♦	♦	♦	
1500	♦			♦						♦	♦			♦	♦	♦	♦
1600	♦			♦							♦			♦	♦		♦
1700	♦			♦							♦			♦	♦		
1800	♦			♦							♦			♦	♦		
1900	♦			♦							♦			♦	♦		
2000	♦			♦							♦		♦	♦			
2200	♦	♦						♦	♦			♦	♦		♦	♦	♦
2300	♦	♦						♦	♦			♦	♦	♦	♦	♦	

1.5.7 Worldwide time signals

Anthorn (54°54'N 03°16'W), UK (MSF) broadcasts UT signals continuously on 60 kHz. This station can be received widely in N and W Europe.

Fort Collins, Colorado (WWV) and Kekaha, Hawaii (WWVH) broadcast UT signals continuously on frequencies 2·5, 5, 10, 15 and 20 (not Kekaha) MHz. The station is identified by voice (English) at every H and H+30. The hour is signalled by a 1500 Hz tone and minutes and seconds by 1200 Hz tones. WWV also broadcasts Atlantic weather in two parts at H+08 and H+09 and weather in the Eastern N. Pacific at H+10.

1.6 VOCABULARIES

English, German, Dutch, French, Spanish and Portuguese vocabularies of useful words are listed, for easy reference, under broad subject headings on the following pages.

English	Deutsch	Nederlands	Français	Español	Português
ON ARRIVAL	**BEI ANKUNFT**	**BIJ AANKOMST**	**À L'ARRIVÉE**	**AL LLEGAR**	**CHEGADA**
Beam	Schiffsbreite	Breedte	Largeur	Manga	Boca
Certificate of competence	Befähigungszeugnis	Getuigschrift	Permis bateau	Título	Certificado de competência
Crew list	Besatzungsliste	Bemanningslijst	Liste d'équipage	Lista de tripulación	Lista da tripulação
Customs	Zoll	Douane	Douanes	Aduana	Alfândega
Draught	Tiefgang	Diepgang	Tirant d'eau	Calado	Calado
Insurance certificate	Versicherungspolice	Verzekeringsbewijs	Attestation d'assurance	Certificado de seguro	Certificado de seguro
Length over all (LOA)	Länge über Alles (LüA)	Lengte over alles (LOA)	Longeur hors-tout	Eslora total	Comprimento fora a fora
Passport	Reisepaß	Paspoort	Passeport	Pasaporte	Passaporte
Port of registry	Heimathafen	Thuishaven	Quartier maritime	Puerto de matrícula	Porto de registo
Port police	Polizei	Haven politie	Police du port	Policía de puerto	Polícia marítima
Pratique	Frei Verkehrserlaubnis	Inklaren	Libre-pratique	Libre prática	Livre prática
Registration certificate	Registrierungszertifikat	Zeebrief	Acte de francisation	Certificado de Registro	Certificado de Registro
Ship's logbook	Logbuch	Logboek	Livre de bord	Cuaderno de bitácora	Diário de bordo
Ship's papers	Schiffspapiere	Scheepspapieren	Papiers de bord	Documentación del barco	Documentação do barco
Surveyor	Gutachter	Opzichter, expert	Expert maritime	Inspector	Inspector
VAT certificate	Mehrwertsteuerbescheinigung	BTW-dokument	Attestation de TVA	Certificado IVA	Certificado de IVA
ASHORE	**AN LAND**	**AAN LAND**	**À TERRE**	**A TIERRA**	**A TERRA**
Airport	Flughafen	Vliegveld	Aéroport	Aeropuerto	Aeroporto
Bakery	Bäckerei	Bakkerij	Boulangerie, patisserie	Panadería	Padaria, Pastelaria
Bank	Bank	Bank	Banque	Banco	Banco
Bus station	Busbahnhof	Busstation	Gare routière	Estación de autobuses	Estação do autocarro
Butcher	Fleischerei	Slagerij	Boucherie	Carnicería	Talho
Chandlery	Yachtausrüster	Scheepswinkel	Shipchandler	Efectos navales	Aprestos
Chemist	Apotheke	Apotheek	Pharmacie	Farmacia	Farmácia
Dentist	Zahnarzt	Tandarts	Dentiste	Dentista	Dentista
Engineer	Motorenservice	Monteur	Mécanicien	Mecánico	Engenheiro
Exchange (money)	Wechselstelle	Geldwisselkantoor	Bureau de change	Cambio	Cambio
Garage	Autowerkstatt	Garage	Garage	Garage	Oficina mecânica
Hospital	Krankenhaus	Ziekenhuis	Hôpital	Hospital	Hospital
Ironmongery	Eisenwarenhandlung	IJzerhandelaar	Quincaillerie	Quincalleria	Loja de ferragens
Launderette	Waschsalon	Wasserette	Laverie	Lavandería	Lavanderia
Market	Markt	Markt	Marché	Mercado	Mercado
Post office	Postamt	Postkantoor	Bureau de poste, PTT	Correos	Correio, CTT
Railway station	Bahnhof	Station	Gare de chemin de fer	Estación de ferrocanil	Estação de comboios
Sailmaker	Segelmacher	Zeilmaker	Voilier	Velero, veleria	Veleiro
Shops	Geschäfte	Winkels	Boutiques	Tiendas	Lojas
Stamps, postal	Marken	Postzegels	Timbres	Sellos	Selos
Supermarket	Supermarkt	Supermarkt	Supermarché	Supermercado	Supermercado
HARBOUR	**HAFEN**	**HAVEN**	**PORT**	**PUERTO**	**PORTO**
Anchoring, anchorage	Ankern, Ankerplatz	Ankeren, ankerplaats	Mouiller l'ancre	Fondeadero	Fundeadouro
Boathoist	Bootskran	Botenlift	Travelift	Travelift	Pórtico elevador
Boatyard	Bootswerft	Jachtwerf	Chantier naval	Astilleros	Estaleiro naval
Breakwater, mole	Außenmole	Pier	Brise-lame	Escollera	Quebra-mar, molhe
Bridge, pedestrian	Schlengel	Loopbrug	Passerelle	Pasarela	Ponte para peões
Bridge, swing	Drehbrücke	Draaibrug	Pont tournant	Puente giratorio	Ponte giratória
Cable (electrical)	Kabel	Kabel	Câble	Cable	Cabo (electrico)
Commercial port	Handelshafen	Commerciële haven	Port de commerce	Puerto comercial	Porto comercial
Crane	Kran	Kraan	Grue	Grua	Grua
Drying port	Trockenfallender Hafen	Droogvallende haven	Port d'échouage	Puerto seco	Porto secar
Dues, berthing tariff	Hafengebühren	Havengeld	Tarifs portuaire	derechos de puerto	Taxas
Ferry	Fähre	Veerboot	Ferry, bac, transbordeur	Ferry	ferry
Fishing harbour	Fischereihafen	Vissershaven	Port de pêche	Puerto de pesca	Porto de pesca

English	Deutsch	Nederlands	Français	Español	Português
Fuel berth	Tankstelle	Bunkerplaats	Ponton à carburants	Surtidor	Estação para combustível
Harbour entrance	Hafeneinfahrt	Haveningang	Entrée du port	Entrada de puerto	Entrada
Harbour guide	Hafenhandbuch	Haven gids	Guide du port	Guia del puerto	Guia de Porto
Harbour master	Hafenmeister	Havenmeester	Capitaine du port	Capitán del puerto	Capitão marítimo
Height, headroom	Höhe	Hoogte	Hauteur	Alturas	Altura
Jetty	Steg	Steiger	Jetée	Malecón	Cais
Landing place	Anlegeplatz	Plaats om aan land te gaan	Débarcadère	Desembarcadero	Desembarcadouro
Lock	Schleuse	Sluis	Ecluse	Esclusa	Eclusa
Marina, yacht harbour	Yachthafen	Jachthaven	Port de plaisance	Puerto deportivo	Doca de recreio
Mast crane	Mastenkran	Mastenkraan	Grue	Grúa	Guindaste (mastro)
Mooring (buoy)	Anlegen	Meerplaats	Coffre d'amarrage	Fondeadero	Bóia de atracação
Permitted	Erlaubt	Toegestaan	Autorisé	Permitido	Autorizado
Pier	Pier, Mole	Pier	Appontement/quai	Muelle	Molhe
Prohibited	Verboten	Verboden	Interdit	Prohibido	Proibido
Slip, slipway	Slip, Slipbahn	Helling	Cale	Varadero	Rampa
Water, fresh	Süßwasser	Zoet water	Eau douce	Agua dulce	Água doce
Yacht club	Yachtclub	Jachtclub	Club nautique	Club náutico	Clube náutico

NAVIGATION / NAVIGATION / NAVIGATIE / NAVIGATION / NAVIGACIÓN / NAVEGAÇÃO

English	Deutsch	Nederlands	Français	Español	Português
Abeam	Querab	Dwarsscheeps	Par le travers	Por el través	Por de través
Ahead	Voraus	Voor	Par l'avant	Avante	À proa
Astern	Achteraus	Achter	Par l'arrière	Atrás	À popa
Buoy	Tonne	Boei/ton	Bouée	Boya	Bóia
Binoculars	Fernglas	Verrekijker	Jumelles	Prismáticos	Binóculos
Channel (navigational)	Kanal	Kanaal/geul	Chenal	Canal	Canal
Chart	Seekarte	Zeekaart	Carte	Carta náutica	Carta hidrográfica

Compass / Kompaß / Kompas / Compas / Compás / Agulha

English	Deutsch	Nederlands	Français	Español	Português
Bearing	Peilung	Peiling	Relèvement	Marcación	Azimute
Compass course	Kompaßkurs	Kompaskoers	Cap du compas	Rumbo de aguja	Rumo do agulha
Course True	Wahrer Kurs	Ware Koers	Cap vrai	Rumbo verdadero	Rumo verdadeiro
Deviation	Ablenkung	Deviatie	Déviation	Desvio	Desvios
East	Ost	Oost	Est	Este	Este
Handbearing compass	Handpeilkompass	Handpeil kompas	Compas de relèvement	Compás de marcaciones	Agulha portátil
North	Nord	Noord	Nord	Norte	Norte
South	Süd	Zuid	Sud	Sur	Sul
Variation	Missweisung	Variatie	Déclinaison magnétique	Variación	Variação
West	West	West	Ouest	Oeste	Oeste
Current	Strömung	Stroom	Courant	Corriente	Corrente
Dead reckoning	Koppelnavigation	Gegist bestek	Navigation à l'estime	Estima	Navegação estimada
Degree	Grad	Graad	Degré	Grado	Grau
Depth	Wassertiefe	Diepte	Profondeur	Profundidad	Profundidade
Distance	Entfernung	Afstand, verheid	Distance	Distancia	Distância
Downstream	Flußabwärts	Stroomafwaarts	En aval	Río abajo	Jusante
Dredged (channel)	Baggerrinne	Gebaggerd (geul)	Dragué	Dragado	Dragado
Echosounder	Echolot	Dieptemeter	Sondeur	Sonda	Sondador acústica
Estimated position	Gegißte Position	Gegiste positie	Point estimé	Posición estimada	Posição estimada
Firing range	Schießgebiet	Schietoefeningen	Zone de tir	Zona de tiro	Zone de tiro
Hazard	Hindernis	Gevaar	Danger	Peligro	Perigo
Latitude	Geographische Breite	Breedte	Latitude	Latitud	Latitude
Leeway	Abdrift	Drift	Dérive	Sotavento	Abatimento

Lights list / Leuchtfeuerverzeichnis / Lichtenlijst / Liste des feux / Libro de faros / Lista de faróis

English	Deutsch	Nederlands	Français	Español	Português
Leading lights	Richtfeuer	Geleidelichten	Feux d'alignement	Luz de enfilación	Luz de enfiamento
Lighthouse	Leuchtturm	Vuurtoren	Phare	Faro	Farol
Lightship	Feuerschiff	Lichtschip	Bâteau-feu	Buque faro	Barco-farol

English	Deutsch	Nederlands	Français	Español	Português
Log	Log	Log	Loch	Corredera	Odómetro
Longitude	Geographische Länge	Lengte	Longitude	Longitud	Longitude
Nautical mile	Seemeile	Zeemijl	Mille nautique	Milla marina	Milha marítima
Numbers	**Nummern**	**Nummers**	**Nombres**	**Numeros**	**Números**
One	Ein	Een	Un	Uno	Um/uma
Two	Zwei	Twee	Deux	Dos	Dois/duas
Three	Drei	Drie	Trois	Tres	Três
Four	Vier	Vier	Quatre	Cuatro	Quatro
Five	Fünf	Vijf	Cinq	Cinco	Cinco
Six	Sechs	Zes	Six	Seis	Seis
Seven	Sieben	Zeven	Sept	Siete	Sete
Eight	Acht	Acht	Huit	Ocho	Oito
Nine	Neun	Negen	Neuf	Nueve	Nove
Ten	Zehn	Tien	Dix	Diez	Dez
Pilot (guide)	Lotse	Loods, Gids	Pilote	Práctico	Roteiro da costa, Pilôto
Port (side)	Backbord	Bakboord	Bâbord	Babor	Bombordo
Prohibited area	Sperrgebiet	Verboden gebied	Zone interdite	Zona de prohibida	Zonaproibida
River mouth	Flußmündung	Riviermond	Embouchure	Desembocadura	Foz
Starboard	Steuerbord	Stuurboord	Tribord	Estribor	Estibordo
Tide	**Tide, Gezeiten**	**Getijde**	**Marée**	**Marea**	**Maré**
Chart datum (CD)	Kartennull	Reductievlak	Zero des cartes	Datum	Zero hidrográfico
Covers	Deckt	Onderlopend	Couvre	Cubierto	Cobre
Dries	Trockenfallend	Droogvallend	Découvre	Descubierto	Fica em sêco
Ebb	Ebbe	Eb	Jusant	Marea menguante	Corrente evasante
Flood	Flut	Vloed	Flot	Flujo de marea	Corrente de enchente
High water (HW)	Hochwasser	Hoogwater	Pleine mer (PM)	Pleamar	Praia mar
Low water (LW)	Niedrigwasser	Laagwater	Basse mer (BM)	Bajamar	Baixa-mar
Neaps	Nipptide	Doodtij	Morte-eau (ME)	Marea muerta	Mares mortas
Springs	Springtide	Springtij, springvloed	Vive-eau (VE)	Marea viva	Mares vivas
Tidal streams	Gezeitenströme	Getijstromen	Courants de marée	Corrientes de marea	Correntes de maré
Tide tables	Tidenkalender	Getijdetafel	Annuaire des marées	Anuario de mareas	Tabela de marés
Underwater	Unterwasser	Onderwater	Immergé	Debajo del agua	Submerso
Upstream	Flußaufwärts	Stroomopwaarts	En amont	Aguasarriba	Montante
Waypoint	Wegepunkt	Waypoint/Route punt	Point de destination	Waypoint	Ponto de chegada
Wreck	Wrack	Wrak	Epave	Naufragio	Naufrágio
METEOROLOGY	**METEOROLOGIE**	**METEOROLOGIE**	**MÉTÉO**	**METEOROLOGÍA**	**METEOROLOGIA**
Area	Gebiet	Gebied	Zone	Zona	Área
Clouds, cloudy	Wolken	Wolken	Nuages, nuageux	Nublado	Nublado
Cold	Kalt	Koud	Froid	Frio	Frio
Direction	Richtung	Richting	Direction	Dirección	Indicação
Forecast	Vorhersage	Vooruitzicht	Prévision	Previsión	Previsão
Fresh, cool, chilly	Frisch	Fris	Frais	Fresco	Fresco, frio
Local	Örtlich	Plaatselijk	Local	Local	Local
Outlook	Aussichten	Vooruitzicht	Evolution probable	Evolución probable	Prospecto
Precipitation	**Niederschlag**	**Neerslag**	**Précipitation**	**Precipitación**	**Precipitação**
Drizzle	Niesel	Motregen	Bruine	Llovizna	Chuvisco
Frequent	Häufig	Veelvuldig	Fréquent	Frecuente	Frequente
Hail	Hagel	Hagel	Grêle	Granizo	Granizo
Heavy	Schwer	Zwaar	Abondant	Abunante	Forte
Isolated	Vereinzelt	Geïsoleerd	Isolé	Aislado	Isolado
Lightning	Blitze	Bliksem	Eclair de foudre	Relampago	Relâmpago
Rain	Regen	Regen	Pluie	Iluvia	Chuva
Scattered	Vereinzelt	Verspreid	Esparses	Difuso	Esporádico

English	Deutsch	Nederlands	Français	Español	Português
Shower	Schauer	Bui	Averse	Chubasco	Aguaceiro
Sleet	Schneeregen	Natte sneeuw	Neige fondue	Aguanieve	Geada miúda
Snow	Schnee	Sneeuw	Neige	Nieve	Neve
Thunder	Donner	Donder	Tonnerre	Trueno	Trovão
Thunderstorm	Gewitter	Onweer	Orage	Tormenta	Trovoada
Pressure systems	**Luftdrucksysteme**	**Druk**	**Pression**	**Presión**	**Pressão**
Air mass	Luftmasse	Luchtmassa	Masse d'air	Masa de aire	Massa de ar
Cold front	Kaltfront	Koufront	Front froid	Frente frio	Frente fria
Decrease	Abrahme	Afnemen	Affaiblissement	Disminuir	Enfraquecer
Deep	Tief	Diep	Profond	Profundo	Profundo
Deepening	Vertiefend	Verdiepend	Se creusant	Ahondamiento	Agravar-se
Disturbance	Störung	Storing	Perturbation	Perturbación	Perturbação
Extending	Ausdehnung	Uitstrekkend	S'étendant	Extendiendose	Alastrando
Falling	Fallend	Dalend	Baissant	Bajando	Descendo
Filling	Auffüllend	Vullend	Se comblant	Rellenandose	Encher-se
Front	Front	Front	Front	Frente	Frente
Gradient	Druckgefälle	Gradiënt	Gradient	Gradiente	Gradiente
High (anticyclone)	Hoch	Hoog, Hogedrukgebied	Anticyclone	Alta presión, Anticiclón	Anticiclone
Isobar	Isobare	Isobaar	Isobare	Isobara	Isobaro
Low (depression)	Tief, Sturmtief	Laag, Depressie	Dépression	Baja presión, Depresión	Depressão
Moving	Bewegend	Bewegend	Se déplacant	Movimiento, moviendose	Afastar-se
Occlusion, occluded	Okklusion	Occlusie	Occlus	Oclusión	Oclusão
Quickly	Schnell	Snel	Rapidement	Rápidamente	Rápidamente
Ridge	Hochdruckbrücke	Rug	Dorsale, crête	Cresta	Crista
Rising	Ansteigend	Stijgend	Montant	Subiendo	Subindo
Slowly	Langsam	Langzaam	Lentement	Lentamente	Lentamente
Steady	Gleichbleibend	Vast	Stable	Fijo	Constante
Trough	Trog, Tiefausläufer	Trog	Thalweg	Seno	Linha de baixa pressão
Warm front	Warmfront	Warmtefront	Front chaud	Frente calido	Frente quente
Sea/sea state	**See/Seegang**	**Zee**	**Mer**	**Mar**	**Mar**
Breakers	Brecher	Brekende zee	Lames déferlantes	Rompientes	Arrebentação
Calm	Flaute	Kalmte	Calme	Calma	Calma
Choppy	Kabbelig	Kort en steil	Croisée, hachée, clapot	Picada	Mareta
Overfalls (tide race)	Stromkabbelung	Stroomkabbeling	Clapot, brisants	Escarceos, remolino	Bailadeiras
Rough	Rauh	Ruw	Forte	Fuerte marejada	Agitado
Seaway	Seegang	Zeegang	Haute mer	Alta mar	Mare alta
Short, steep	Kurz, steil	Kort, steil	Vagues courtes, creuse	Mar corta, gruesa	Mar curta, cavado
Slight	Leicht	Licht	Legère, peu agitée	Marejadilla	Ligeiramente
Swell	Schwell	Deining	Houle	Mar de fondo	Ondulação
Sun	Sonne	Zon	Soleil	Sol	Sol
Sunrise	Sonnenaufgang	Zonsopgang	Lever du soleil	Orto	Nascer do sol
Sunset	Sonnenuntergang	Zonsondergang	Coucher du soleil	Ocaso	Pôr do sol
Synoptic situation	Wetterlage	Synoptische toestand	Situation générale	Situación sinoptica	Situação sinóptica
Visibility	**Sichtweite**	**Zicht**	**Visibilité**	**Visibilidad**	**Visibilidade**
Dispersing	Auflösend	Oplossend	Se dispersant	Disipación	Limpando
Extensive	Ausgedehnt	Uitgebreid	Etendu	Extenso	Extenso
Fog	Nebel	Mist	Brouillard	Niebla	Nevoeiro
Fog bank	Nebelbank	Mistbank	Banc de brume	Banco de niebla	Banco do nevoeiro
Good	Gut	Goed	Bon	Bueno	Bom/boa
Haze	Diesig	Nevelig	Brume	Calina	Cerração
Mist	Dunst	Nevel	Brume légere	Neblina	Neblina
Moderate	Mäßig	Matig	Médiocre, reduite	Moderardo	Moderada
Poor	Schlecht	Slecht	Mauvais	Malo	Fraca má
Weather	Wetter	Weerraport	Temps	Tiempo	Boletim meteorológico

Wind / Beaufort / direction

English	Deutsch (Wind)	Nederlands (Wind)	Français (Vent)	Español (Viento)	Português (Vento)
F0, Calm	Windstille	Stil	Calme	Calma	Calma
F1, Light airs	Leiser Zug	Flauw en stil	Très légère brise	Ventolina	Aragem
F2, Light breeze	Leichte Brise	Flauwe koelte	légère brise	Flojito	Fraco
F3, Gentle breeze	Schwache Brise	Lichte koelte	Petite brise	Flojo	Bonançoso
F4, Moderate breeze	Mäßige Brise	Matige koelte	Jolie brise	Bonancible	Moderado
F5, Fresh breeze	Frische Brise	Frisse bries	Bonne brise	Fresquito	Frésco
F6, Strong breeze	Starker Wind	Stijve bries	Vent frais	Fresco	Muito frêsco
F7, Near gale	Steifer Wind	Harde wind	Grand frais	Freschachón	Forte
F8, Gale	Stürmischer Wind	Stormachtig	Coup de vent	Duro	Muito forte
F9, Strong gale	Sturm	Storm	Fort coup de vent	Muy duro	Tempestuoso
F10, Storm	Schwerer Sturm	Zware storm	Tempête	Temporal	Temporal
Backing	Rückdrehender Wind	Krimpende wind	Vent revenant	Rolar el viento	Sentido retrógrado
Gale warning	Sturmwarnung	Stormwaarschuwing	Avis de coup de vent	Aviso de viento duro	Aviso de vento multo forte
Gust	Windstoß	Windvlaag	Rafale	Ráfaga, racha	Rajada
Increasing	Zunehmend	Toenemend	Augmentant	Aumentar	Aumentando
Land breeze	Landbrise	Landbries	Brise de terre	Brisa de tierra	Brisa da terra
Moderating	Abnehmend	Afnemend	Décroissant	Disminuyendo	Decrescendo
Sea breeze	Seebrise	Zeebries	Brise de mer	Brisa de mar	Brisa do mar
Squall	Bö	Windbui	Grain	Turbonada	Borrasca
Variable	Umlaufend	Veranderlijk	Variable	Variable	Variável
Veering	Rechtdrehend	Ruimende wind	Virant	Cambiar	Sentido horário

Safety / Emergency

English (SAFETY/EMERGENCY)	Deutsch (SICHERHEIT/NOTFALL)	Nederlands (VEILIGHEID/NOODGEVAL)	Français (SECURITE/URGENCE)	Español (SEGURIDAD/EMERGENCIAS)	Português (SEGURANÇA/EMERGÊNCIAS)
Aground	Aufgelaufen	Vastgelopen	Echoué	Encallado	Encalhar
Capsize	Kentern	Omslaan	Sancir, chavirer	Volcó	Virar-se
Coastguard	Küstenwache	Kustwacht	Gendarmerie maritime	Guarda costas	Policia Marítima
Collision	Kollision	Aanvaring	Abordage	Colisión, abordaje	Colisão abalroamento
Dismasted	Mastbruch	Mastbreuk	Démâté	Desarbolar	Partir o mastro
Distress	Seenot	Nood	Détresse	Socorro	Distress
Distress flares	Signalraketen	Noodvuurwerk	Fusées de détresse	Bengalas	Fachos de socorro
Dragging anchor	Schlierender Anker	Krabbend anker	Chasser sur l'ancre	Ancla garreando	Âncora agarrar
EPIRB	Seenotfunkboje	EPIRB	Balise de détresse	Baliza	EPIRB
Fire extinguisher	Feuerlöscher	Brandblusser	Extincteur	Extintor	Extintor
Harness, safety	Sicherheitsgurt	Harnas	Harnais	Arnés de seguridad	Cinto de segurança
Help!	Hilfe!	Help!	Au secours!	¡Socorro!	Socorro!
Lifeboat	Rettungsboot	Reddingsboot	Canot de sauvetage	Lancha de salvamento	Barco salva-vidas
Lifejacket	Schwimmweste	Reddingsvest	Gilet de sauvetage	Chaleco salvavidas	Colete de salvação
Liferaft	Rettungsinsel	Reddingsvlot	Radeau de sauvetage	Balsa salvavidas	Jangada de salvação
Man overboard	Mann über Bord	Man over boord	Homme à la mer	Hombre al agua	Homem ao mar
Seacock	Seeventil	Afsluiter	Vanne	Grifos de fondo	Torneira de segurança
Seaworthy	Seetüchtig	Zeewaardig	Marin	Marinero	Condição de navegar
Sinking	Sinken	Zinkend	En train de couler	Hundiendose	Afundate
Tow line	Schleppleine	Sleeplijn	Filin de remorque	Cabo remolque	Cabo de rebocador

Medical

English (MEDICAL)	Deutsch (MEDIZINISCHES)	Nederlands (MEDISCHE)	Français (MEDICALE)	Español (MEDICO)	Português (MEDICAS)
Antibiotic	Antibiotika	Antibiotica	Antibiotique	Antibiotico	Antibiótico
Bandage	Verband	Verband	Pansement	Vendas	Ligadura
Bleeding	Blutend	Bloeden	Saignement	Sangrando	Sangrar
Burns	Verbrennung	Brandwonden	Brûlures	Quemadura	Queimadura
Concussion	Gehirnerschütterung	Hersenschudding	Traumatisme crânien	Conmoción	Traumatismo
Dehydration	Austrocknung	Uitdroging	Déshydratation	Deshidratación	Desidratação
Dentist	Zahnarzt	Tandarts	Dentiste	Dentista	Dentista
Doctor	Doktor	Dokter/arts	Médecin	Médico	Médico
Drown(ing)	Ertrinken	Verdrinken, (verdrinking)	Noyade, (se noyant)	Ahogarse	Afogar-se

English	Deutsch	Nederlands	Français	Español	Português
Exhaustion	Erschöpfung	Uitputting	Epuisement	Agotamiento	Exausto
Fever	Fieber	Koorts	Fièvre	Fiebre	Febre
First aid	Erste Hilfe	Eerste hulp	Premier secours	Primeros auxillos	Primeiros socorros
Fracture	Fraktur	Breuk	Fracture	Fractura	Fractura
Headache	Kopfschmerz	Hoofdpijn	Mal à la tête	Dolor de cabeza	Dor de cabeça
Heart attack	Herzanfall	Hartaanval	Crise cardiaque	Ataque corazón	Ataque cardíaco
Hospital	Krankenhaus	Ziekenhuis	Hôpital	Hospital	Hospital
Hypothermia	Unterkühlung	Onderkoeling	Hypothermie	Hypotermia	Hipotermia
Illness	Krankheit	Ziekte	Maladie	Enfermo	Doença
Injury	Verletzung	Verwonding	Blessure	Lesión	Lesão
Pain	Schmerz	Pijn	Douleur	Dolor	Dôr
Painkiller(s)	Schmerztabletten	Pijnstillers	Analgésique	analgesico	Analgésico
Poisoning	Vergiftung	Vergiftiging	Empoisonnement	Envenenamiento	Envenenamento
Pulse	Puls	Hartslag	Pouls	Pulso	Pulso
Rest	Ruhen	Rust	Repos	Reposo	Descanso
Seasickness	Seekrankheit	Zeeziekte	Mal de mer	Mareo	Enjoar
Shock	Schock	Shock	Choc	Choque	Choque
Splint	Schiene	Spalk	Attelle	Tablilla	Colocar em talas
Sticking plaster	Heftpflaster	Pleister	Pansement adhésif	Esparadrapo	Adesivo
Stomach upset	Bauchbeschwerden	Maag van streek	Indigestion	Mal de estómago	Cólicas
Stretcher	Trage	Draagbaar	Brancard	Camilla	Maca
Sunburn	Sonnenbrand	Zonnebrand	Coup de soleil	Quemadura del sol	Queimadura de sol
Swelling	Schwellung	Zwelling	Enflure	Hinchazón	Inchaço
Toothache	Zahnschmerzen	Tandpijn	Rage de dents	Dolor de muelas	Dôr dos dentes
Unconscious	Bewußtlos	Bewusteloos	Inconscient	Inconsciente	Inconsciente
Wound	Wunde	Wond	Blessure	Herida	Ferida

THE BOAT | DAS BOOT | DE BOOT | LE BATEAU | EL BARCO | DO BARCO

Rigging and Sails	Rigg und Segel	Tuigage en Zeilen	Gréement et Voiles	Jarcias y Velas	Massame e Velas
Babystay	Babystag	Baby stag	Bas-étai	Babystay	Babystay
Backstay	Achterstag	Achterstag	Pataras	Estay de popa	Brandal fixo da pôpa
Batten	Segellatte	Zeillat	Latte	Sables	Régua
Beating	Kreuzen	Kruisen	Au prés	Ciñendo	Bolinar
Boom	Baum	Giek	Bôme	Botavara	Retranca
Clevis pin	Schäkelbolzen	Borstbout	Axe à anneau brisé	Pasador de seguridad	Perno com troço de argola
Foresail	Vorsegel	Voorzeil	Voile avant/foc	Foque	Pana de proa
Forestay	Vorstag	Voorstag	Etai	Estay de proa	Estai real
Genoa	Genua	Genua	Génois	Génova	Genoa
Halyard	Fall	Val	Drisse	Driza	Adriça
Jib	Fock	Fok	Foc	Foque	Fiocco
Mainsail	Großsegel	Grootzeil	Grand' voile	Mayor	Vela grande
Mast	Mast	Mast	Mât	Mast	Mastro
Mast track	Mastschiene	Mastrail	Rail	Carril	Calha do mastro
Mizzen	Besan	Bezaan	Artimon	Mesana	Mezena
Reaching	Raumschots	Ruime wind	Au portant	Viento a través	A um largo
Rod kicker	Rod kicker	Giekophouder	Hale-bas rigide	Contra rígido	Rod kicker
Roller reefing foresail	Rollfock	Rolfock	Génois à enrouleur	Enrollador de génova	Genoa de enrolar
Running	Vorm Wind	Voor de wind	Vent arrière	Viento a favor	Vento à pôpa
Sheet	Schot	Schoot	Ecoute	Escota	Escota
Shrouds	Wanten	Wanten	Haubans	Obenques	Enxarcia
Spinnaker	Spinnaker	Spinnaker	Spi	Spi	Spinnaker
Spinnaker boom	Spinnakerbaum	Spinnaker boom	Tangon de spi	Tangon	Pau de spinnaker
Spreader, crosstree	Saling	Zaling	Barre de flèche	Crucetas	Vaus
Staysail	Stagsegel	Stagzeil	Trinquette	Trinqueta	Estai

English	Deutsch	Nederlands	Français	Español	Português
Storm jib	Sturmfok	Stormfok	Tourmentin	Tormentín	Estai de tempo
Topping lift	Dirk (for the boom)	Kraanlijn, Dirk	Balancine	Amantillo	Amantilho
Trysail	Trysegel	Stormzeil, Trysail	Voile de cape	Vela de capa	Cachapana
Upwind	Am Wind	Aan de wind/te loevert	Au vent	Vienta en contra	Bolinando
Whipping twine	Takelgarn	Takelgaren	Fil à surlier	Piolilla	Cordão de pequeña bitola
Working jib	Arbeitsfock	Werkfok	Foc de route	Foque	Estai
On deck	**An Deck**	**Aan dek**	**Sur le pont**	**En cubierta**	**No convés**
Aft	Achtern, achteraus	Achter, achteruit	En arrière	Atrás	À ré
Ahead	Voraus	Vooruit	Par l'avant	Avante	À proa
Anchor	Anker	Anker	Ancre	Ancla	Âncora
Anchor chain	Ankerkette	Ankerketting	Chaîne d'ancre	Cadena	Corrente
Anchor warp	Ankerleine	Ankerlijn	Orin	Cabo de fondeo	Espia da âncora
Astern	achtern	Achteruit, slaan	Par l'arrière	Hacia atrás	À popa
Block	Block	Blok	Poulie	Motón	Moitão
Bow	Bug	Boeg	Etrave	Proa	Proa
Bridgedeck	Brückendeck	Brugdek	Bridgedeck	Cubierta de puente	Pavimento da ponte
Cockpit	Cockpit	Cockpit	Cockpit	Bañera	Cockpit
Deck	Deck	Dek	Pont	Cubierta	Convés
Dinghy, Tender	Jolle, Beiboot	Bijboot	Annexe	Chinchorro, Bote	Escaler, Côco
Fender	Fender	Stootwil	Défense	Defensa	Defensa
Inflatable (dinghy)	Schlauchboot	Opblaasbare boot	Gonflable	Bote hinchable	Inflável
Ladder	Leiter	Ladder, Trap	Echelle	Escala	Escada
Pulpit	Bugkorb	Preekstoel	Balcon avant	Púlpito	Guarda proeiro
Pushpit	Heckkorb	Hekrailing	Balcon arrière	Balcón de popa	Varandim
Railing	Reling	Railing	Filières	Guardamencebos	Balaustrada
Rope	Tauwerk	Touw	Cordage	Cabo, Cuerda, Cabulleria	Cabo, Corda
Rudder	Ruder	Roer	Safran, Gouvernail	Pala de Timón	Leme
Shackle	Schäkel	Sluiting	Manille	Grillete	Manilha
Ship	Schiff	Schip	Navire	Buque	Navio
Stanchion	Seerelingsstütze	Scepter	Chandelier	Candelero	Balaustre
Stern	Heck	Spiegel/hek	Arrière	Popa	Popa
Superstructure	Aufbau	Opbouw	Superstructure	Superestructura	Super-estrutura
Tiller	Pinne	Helmstok	Barre franche	Caña	Cana de leme
Toe rail	Fußleiste	Voetrail	Rail de fargue	Regala	Amurada inferior
Topsides	Rumpfseiten	Bovenwaterschip	Oeuvres mortes	Obra muerta	Costada
Wheel, steering	Steuerrad	Stuurwiel	Barre à roue	Rueda	Roda do leme
Winch	Winsch	Lier	Winch	Winche	Molinete
Windlass, capstan	Ankerwinsch	Ankerlier, Kaapstander	Guindeau	Molinete	Molinete da âncora
Yacht, sailing	Segelyacht	Zeilboot	Voilier	Velero	Veleiro
Below deck	**Unter Deck**	**Onderdeks**	**Sous le pont**	**Bajo cubierta**	**Abaixo do convés**
Bilge	Bilge	Bilge	Galbord	Sentina	Esgôto
Bilge keel	Kimmkiel	Kimkiel	Bi-quilles	Quillas	Quilha dupla
Bilge pump	Bilgepumpe	Bilge pomp	Pompe de cale	Bomba de achique	Bomba de esgoto
Cabin	Kajüte	Kajuit	Cabine	Cabina	Cabine
Calorifier	Wärmespender	Verwarming	Echangeur de chaleur	Calentador	Calorifico
Companionway	Niedergang	Opgang	Descente	Entrada cámera	Escotilha de passagem
Deckhead	Kajütdach	Kajuitdak	Tête de pont	Techo de cabina	Cabeço do convés
Fin keel	Kurzkiel	Vinkiel	Quille courte	Quilla de aleta	Patilhão
Fresh water tank	Frischwassertank	Drinkwatertank	Réservoir d'eau douce	Tanque de agua dulce	Tanque de aguada
Galley	pantry/Kombüse	Kombuis	Cuisine	Cocina	Cozinha
Gas alarm	Gasmelder	Gasalarm	Alarme anti-gaz	Alarma de gas	Alarme de gás
Holding tank	Schmutzwassertank	Vuilwatertank	Réservoir à eaux usées	Tanque aguas negras	Deposito de esgotos
Hull	Rumpf	Romp	Coque	Carena	Casco

English	Deutsch	Nederlands	Français	Español	Português
Keel	Kiel	Kiel	Quille	Quilla	Quilha
Long keel	Langk el	Langkieler	Quille longue	Quilla corrida	Quilha corrida
Oven	Backo˜en	Oven	Four	Horno	Forno
Skeg	Skeg	Scheg	Talon	Skeg	Skeg
Underwater hull	Unterwasserschiff	Onderwaterschip	Oeuvres vives	Obra viva	Obras vivas
ELECTRICS	**ELEKTROTECHNIK**	**ELECTRICITEIT**	**ELECTRIQUES**	**ELECTRICOS**	**ELECTRICO**
Alternator	Wechselstromgenerator	Generator	Alternateur	Alternador	Alternador
Battery	Batterie	Accu, batterij	Batterie/accumulateur	Batería	Bateria
Battery charger	Batterieladegerät	Acculader	Chargeur de batterie	Cargador debatería	Carregador debateria
Bulb	Glühb rne	Lamp	Ampoule	Bombilla	Lâmpada
Circuit breaker	Ausschalter	Automaat	Disjoncteur	Disyuntor	Interruptor
Coil	Spule	Spoel	Bobine	Bobina	Bobine
Dynamo	Lichtm aschine	Dynamo	Dynamo	Dinamo	Dínamo
Electrical wiring	Elektrik	Elektrische bedrading	Réseau électrique	Circuito eléctrico	Circuito electrico
Fuse	Sicherung	(Hoofd)Zekering	Fusible	Fusible	Fusivel
Invertor	Wechselrichter	Inverter	Convertisseur	Inverter	Inversor
Regulator (voltage)	Spann.ngsregler	Spanningsregelaar	Régulateur de tension	Regulador	Regulador
Shore power	Landanschluß	Walstroom	Courant de secteur	Corriente de tierra	Corrente alterna
Solar panel	Solarzellen	Zonnepaneel	Paneau solaire	Placa solar	Painel solar
Soldering iron	Lötkolben	Soldeerbout	fer à souder	Soldador eléctrico	ferro de soldar
Switch	Schalter	Schakelaar	Interrupteur	Interruptor	Interruptor
THE ENGINE	**DER MOTOR**	**DE MOTOR**	**LE MOTEUR**	**EL MOTOR**	**DO MOTOR**
Air filter	Luftfil er	Luchtfilter	Filtre à air	Filtro a aire	Filtro de ar
Carburettor	Vergaser	Carburateur	Carburateur	Carburador	Carburador
Cooling water	Kühlwasser	Koelwater	Eau de refroidissement	Agua refrigerado	Agua fresca
Diesel	Diesel	Dieselolie	Gazole	Gas-oil	Gasóleo
Drive belt	Treibriemen	Aandrijfriem	Courroie d'entrainement	Correa de transmissión	Correia
Engine mount	Motorfundament	Motorsteun	Support moteur	Bancada del motor	Fixe do motor
Engine oil	Motoröl	Motorolie	Huile de moteur	Aceite de motor	Óleo de motor
Exhaust pipe	Auspuff	Uitlaat	Tuyau d'échappement	Tubo de escape	Tubo de escape
Fuel filter	Kraftstoffilter	Brandstoffilter	Filtre de carburant	Filtro de combustible	Filtro de gasóleo
Fuel tank	Tank, Brennstofftank	Brandstof tank	Réservoir à carburant	Tanque de combustible	Tanque de combustivel
Gasket	Dichtung	Pakking	Joint	Empaquetadura junta	Junta
Gearbox	Getriebekasten	Keerkoppeling	Boîte de vitesses	Caja de cambio	Caixa de velocidades
Grease	Fett	Vet	Graisse	Grasa	Massa
Head gasket	Zylinderkopfdichtung	Koppakking	Joint de culasse	Junta de culata	Junta do cabeça
Injector	Einspritzdüse	Injector, Verstuiver	Injecteur	Injector	Injector
Jubilee clip	Schlauchklemme	Slangklem	Collier de serrage	Abrazadera	Abraçadeira ajustáavel
Methylated spirits	Brennspiritus	Spiritus/Methylalcohol	Alcool à brûler	Alcohol desnaturalizada	Alcool metilico
Oil filter	Ölfilte-	Oliefilter	Filtre à huile	Filtro de aceite	Filtro de óleo
Outboard motor	Außenbordmotor	Buitenboordmotor	Moteur hors-bord	Motor fuerabordas	Motor fora de borda
Paraffin	petroleum	Petroleum, Kerosine	Kérosène, Pétrole	Petróleo, Keroseno	Petróleo
P-bracket	Wellenlagerbock	Schroefas uithouder	Support d'hélice	Casquillo	Polée em P
Petrol	Benzin	Benzine	Essence	Gasolina	Gasolina
Propeller	Propeller, Schraube	Schroef	Hélice	Hélice	Hélice
Propeller shaft	Schraubenwelle	Schroefas	Arbre d'hélice	Eje de la hélice	Veio do hélice
Spark plug	Zündkerze	Bougie	Bougie	Bujía	Vela
Starter	Anlasser	Startmotor	Démarreur	Motor de arranque	Motor de arranque
Stern gland	Stopfbuchse	Schroefasdoorvoer	Presse étoupe	Bocina	Bucim
Throttle	Gas	Gashendel	Accélérateur	Acelerador	Acelerador
Two-stroke	Zweitakt	Tweetakt	A deux temps	Dos tiempos	De dois tempos
Water pump	Wasserpumpe	Waterpomp	Pompe à eau	Bomba de agua	Bomba de água

2

Regulations

2.1   **Cruising formalities**   **22**
Documents including ICC • Registration • Insurance

2.2   **HM Revenue & Customs**   **22**
The EU • Duty-free stores • Departure/Arrival • Immigration

2.3   **Foreign Customs**   **23**
The Schengen Treaty • Red diesel

2.4   **EU VAT on vessels**   **23**
VAT • Temporary importation

2.5   **e-Borders**   **23**

2.6   **Travelling with pets**   **23**

2.7   **International Regulations for Preventing Collisions at Sea 24**
Rules 2, 5–7, 10 and 12–19 verbatim • Navigation lights and shapes •
Learning the Collision Regulations

2.1 CRUISING FORMALITIES

Before your annual cruise, especially beyond UK waters, check that yacht and personal documents are in-date.

Yacht documents

Registration certificate (2.1.1). Marine insurance valid for the intended cruising area, including adequate third-party cover (2.1.2). Proof of VAT status (2.4). Ship's radio licence (5.2.3). Ship's Log, itinerary and a crew list. All must be **originals,** keep them in transparent files for easy inspection; failure to produce them can them can potentially incur fines abroad.

Personal documents

Valid passports and European health insurance card (8.2). Radio Operator's certificate of competence (5.2.3).

International Certificate of Competence

An International Certificate of Competence (ICC) (Pleasure Craft), valid for 5 years, is required in most European countries, especially if inland waterways are part of your cruising. plans. It can be endorsed for power or sail, inland or coastal waters, or all four.

The inland waterways endorsement requires a short, written test on CEVNI (ColRegs and buoyage in European inland waterways), but this does not apply to UK inland waterways. If evidence of competence (eg Day Skipper or above) is not held, a practical test must be passed before the ICC is issued. E-mail certification@rya.org.uk or call ☎ 0845 345 0376.

Health regulations

Comply with health regulations (eg report any infectious disease); check if vaccination certificates are needed.

2.1.1 Registration

The Merchant Shipping Act 1995 and related Merchant Shipping (Registration of Ships) Regulations govern the registration of British ships in the UK.

The Register of British Ships is in 4 parts, covering:

Part I	Merchant ships and pleasure vessels.
Part II	Fishing vessels.
Part III	Small ships (Small Ships Register).
Part IV	Bareboat charter ships.

- Registration under Part I is a relatively complex and expensive business (£124), since a yacht has to be surveyed and follows the same procedure as a large merchant vessel. It costs £49 to renew for 5 years. The certificate establishes the ship's nationality and tonnage. Details of registered ownership and mortgages can be obtained from the RSS at Cardiff.

- Part III, The Small Ships Register (SSR), is for owners who want a simple registration, to prove the yacht's nationality and meet registration requirement for a privileged ensign, but it registers neither 'Title' nor mortgages. A small ship is defined as < 24m LOA. The cost is £25 for five years, and measurement only requires taking the boat's LOA.

For further details contact: Registry of Shipping and Seamen/Small Ships Register, Anchor Court, Keen Road, Cardiff CF24 5JW; SSR ☎ 029 2044 8813; rss@mcga.gov.uk

Application forms and the online SSR registration service can be found at www.ukshipregister.com.

2.1.2 Insurance

Every cruising boat should be adequately insured against loss or damage. The insured value should be the cost of replacing the boat and all equipment. Third-party cover for up to £3,000,000 is also needed. Spanish translation needed for Spanish waters. The cruising area must be adjusted and paid for if cruising beyond the usual limits. Failure to disclose all relevant facts may invalidate a policy.

2.2 HM REVENUE & CUSTOMS (HMRC)

2.2.1 The European Union (EU)

EU yachtsmen can move freely within the EU, provided Customs duty, VAT, or any other Customs charges, have been paid in an EU country. Most nations make random checks on yachts and you may be asked for evidence that VAT, or its equivalent, has been paid on your vessel. See 2.3.2 for notes on using UK-bought red diesel overseas.

EU members are listed in Notice No 8. The Channel Islands and the Canary Islands do not operate a VAT system under EU rules, and are therefore treated as outside the EU.

2.2.2 Customs Notice No 8 (Dec 2002)

This Notice 'Sailing your pleasure craft to and from the UK' (with Update 5 of April 2012) is the UK's interpretation of what the law says about pleasure craft and compliance with HM Revenue & Customs requirements, as summarised below; it is a good idea to have a copy aboard.

This Notice and further information may be obtained from the National Advice Service, ☎ 0845 010 9000 or +44 208 929 0152, which is open M-F, 0800-2000. Or visit HM Revenue & Customs website at: www.hmrc.gov.uk

Yachtsmen are warned that a boat may be searched at any time. Penalties are severe for non-declaration of prohibited or restricted goods, and the carriage and non-declaration of prohibited drugs and firearms. If goods are smuggled the vessel may be seized and the persons concerned may be liable to prosecution, a heavy fine and/or prison sentence.

2.2.3 Duty-free stores

Duty-free stores may be allowed on vessels going south of Brest or north of the N bank of the R. Eider (Denmark), by prior application to a Customs office. Duty-free stores cannot be taken to the Republic of Ireland nor to the Channel Islands. Contact the National Advice Service for details of how to embark stores and of the conditions to be satisfied. For more information read Notice 69A Duty free ship's stores.

2.2.4 C1331 Pleasure craft departing the UK
To another EU country

No report is needed, unless a Customs Officer asks for one.

To a country which is outside the EU:

You must notify HM Customs of your departure on Form C1331, Part 1 'Leaving the UK', available from most YCs, marinas, Customs offices, the National Advice Service or via the HMR&C website.

Before you expect to leave the UK, return the completed Part 1 to a Customs Officer, put it in a Customs post box or take/post it to a Customs office. C1331 is valid for 48 hrs from the stated time of departure. If departure is not notified, delay, inconvenience and possible prosecution may result on return. Keep Part 2 'Arriving in the UK' on board for use on your return. If your voyage is abandoned, Part 2, marked 'Voyage abandoned', should be returned to where the original form was sent.

2.2.5 Arrival from an EU country

If arriving directly from another EU country there is no need to fly flag 'Q', complete any paperwork, or contact Customs. You must, however, contact the Customs 'Yachtline' (see 2.2.8) if you have goods to declare, or have non-EU nationals on board. You must also declare animals or birds (see 2.6); prohibited or restricted goods, eg controlled drugs, firearms, radio transmitters not accepted in the UK; counterfeit goods; duty-free stores; or the boat itself if duty and VAT are owed on it. Further details are in *Notice No 1*.

2.2.6 Arrival from a non-EU country

If arriving directly from a non-EU country (including the Channel Islands), yachts are subject to Customs control. When inside the 12-mile limit, ie UK Territorial Waters, fly flag 'Q' where most easily seen (lowering it when formalities are complete); and fill in Part 2 of Form C1331.

On arrival, call the Customs 'Yachtline' (see 2.2.8) and inform them if: VAT has not been paid on the vessel; goods in excess of your allowances as detailed in Notice 1, or duty-free stores are on board; you have prohibited or restricted goods; a crew has a notifiable illness or needs immigration clearance; repairs or modifications (not running repairs) have been made since leaving the EU.

Comply with any instructions from the Customs Officer; failure to do so may incur a penalty. Do not land anybody or goods or transfer them to another vessel until a Customs Officer says you may.

2.2.7 Immigration

In most yachting centres the Customs Officer also acts as the Immigration Officer. The skipper is responsible for ensuring that any non-EU national aboard gets an Immigration Officer's permission to enter the UK from any country except the Isle of Man, the Channel Islands or ROI.

2.2.8 Customs telephone numbers

The National Yachtline ☎ 0845 723 1110 must only be used by yachtsmen wishing to clear Customs on return to the UK from a non-EU country (2.2.6). Use the National Advice Service ☎ 0845 010 9000 for all other Customs queries and requests for forms and notices.

2.2.9 Drug smuggling

Preventing drug smuggling is a key role for HMR&C. Public support is very important. If you see or become aware of suspicious activity, call ☎ 0800 595000. This is a 24-hour, 7/7 free and anonymous hotline. There may be a reward.

2.3 FOREIGN CUSTOMS

Other EU countries should apply the regulations outlined in 2.2.1 to 2.2.6 above, and any Customs formalities are likely to be minimal. If boarded by Customs, stay polite and helpful at all times. Before departure, skippers are recommended to check the procedures in force in their destination country.

2.3.1 The Schengen Treaty

In 1985 France, Germany, Luxembourg, Netherlands and Belgium signed a Convention which abolished internal border controls, establishing a single external border, around what is sometimes called 'Schengen-land'. Today all 28 EU members, except the UK, Ireland, Romania, Bulgaria, Croatia and Cyprus, are signatories to Schengen and the decisions taken since 1985 are now part of EU law.

Those travelling from the UK to a Schengen country, or vice versa, are crossing the external border and may be subject to additional checks. Thus, Belgium and the Netherlands require a vessel to report on arrival *even if coming from another EU country*. The RYA has forms available for download from their website (www.rya.org.uk/infoadvice/boatingabroad/Pages/customsformalities.aspx), which are simply a crew list (DoB, place of birth, passport No, nationality) and brief itinerary (last/next ports, dates etc).

In practice these forms are not always requested, but if you have them aboard you may avoid fines which can legally be imposed for not observing Schengen/EU law. On arrival check with the HM whether any Customs or Immigration forms need to be completed.

2.3.2 Red diesel

Problems with UK pleasure craft using red (dyed green in ROI) reduced duty diesel in other EU countries have mostly been resolved using the following guidelines. Carry on board all receipts for red/green diesel (dated, and marked 'UK or ROI duty paid' if possible); keep diesel in the main fuel tank(s) only, *do not carry any red/green diesel in cans*; log engine hours. However, Belgium re-introduced their previous ban on dyed diesel in early 2015 and with it the renewed risk of fines. No reports of fines had been received by June 2015, but caution is advised. Check for changes on the RYA website www.rya.org.uk or the free online monthly updates at www.reedsnauticalalmanacs.co.uk (January 2016 -June 2016).

2.4 EU VAT ON VESSELS

An EU resident can move a yacht between member states without restriction, providing VAT has been paid. Documentary evidence supporting a vessel's VAT status should be carried at all times. A leaflet *UK guide for Yachts,* available on the Customs website, gives further advice; see also *Notice 8*, section 6.

2.4.1 Temporary Importation (TI)

TI is a complex process likely to affect few yachtsmen. The maximum stay in the EU is 18 months. See *Notice 8*, Section 5 and *Notice 308*, available on the Customs website or from the National Advice Service. On arrival in the UK fill in Form C108 as well as the usual Form C1331.

2.5 e-BORDERS

The UK Border Force (UKBF) plan to introduce an onerous reporting procedure for private pleasure craft embarking on cross-border voyages to and from other EU states remains on hold for the immediate future due to the compexity of finding a practical solution for implementing it. The existing procedures **HMRC C1331 Pleasure craft departing the UK (2.2.4, 2.2.5 2.2.6 and 2.2.7)** therefore remain unchanged.

2.6 TRAVELLING WITH PETS

Pets can travel to France from the UK in private leisure craft providing they comply with French regulations, see http://www.ambafrance-uk.org/Importing-animals-into-France,2780. Travelling with pets from the EU (and certain non-EU countries) to the UK in private leisure craft is **not** permitted; they can only travel with '**an approved transport company on an authorised route**', see https://www.gov.uk/take-pet-abroad/overview Before travel, all pets must be microchipped, with up to date vaccinations, Pet passport and any other paperwork in order (varies according to country). Dogs must also have tapeworm medication administered **by a vet** <120hrs and >24 hrs before entering the UK.

2.7 INTERNATIONAL REGULATIONS FOR PREVENTING COLLISIONS AT SEA

The 1972 International Regulations for Preventing Collisions at Sea (IRPCS) are also referred to as the Colregs or Rule of the Road. They should be read as an entity, with the explanatory notes and associated diagrams below. See also *Learning the Rule of the Road* (Basil Mosenthal/ACN), the RYA booklet G2 and *The Mariner's Handbook* NP 100.

Because the exact wording of certain Rules is very important, Rules 2, 3 (in part), 5–10, 12–19 and 25 are quoted verbatim in *blue italics* and then briefly discussed – especially those of specific interest to yachtsmen. This does not imply that this chapter should be consulted in the heat of the moment – since all mariners must have a sound working knowledge of the Rules – but an aide-memoire is sometimes handy.

2.7.1 Part A: General

Rule 1 (Application) is mainly introductory and therefore not quoted verbatim, but note that many harbours have their own local rules – which must be obeyed.

Rule 2 (Responsibility).

(a) Nothing in these Rules shall exonerate any vessel, or the owner, master or crew thereof, from the consequences of any neglect to comply with these Rules or of the neglect of any precaution which may be required by the ordinary practice of seamen, or by the special circumstances of the case.

(b) In construing and applying these Rules due regard shall be had to all dangers of navigation and collision and to all special circumstances, including the limitations of the vessels involved, which may make a departure from these Rules necessary to avoid immediate danger.

The Rules must be interpreted in a seamanlike way. No vessel has a 'right of way' over another regardless of special circumstances – eg other vessels under way or at anchor, shallow water, poor visibility, TSS, fishing fleets, etc – or the handling characteristics of the vessels concerned in the prevailing conditions. Sometimes vessels may need to depart from the Rules to avoid immediate danger (2b).

Rule 3 (Definitions). The following are selected extracts:

(c) 'Sailing vessel' means any vessel under sail provided that propelling machinery, if fitted, is not being used.

(f) 'Vessel not under command' means one which through some exceptional circumstance is unable to manoeuvre as required by these Rules and is therefore unable to keep out of the way of another vessel.

(g) 'Vessel restricted in her ability to manoeuvre' means one which from the nature of her work is thus restricted ... (6 examples are given)

(h) 'Vessel constrained by her draught' means a power-driven vessel which, because of her draught in relation to the available depth and width of navigable water is severely restricted in her ability to deviate from her course.

(i) 'Underway' means that a vessel is not at anchor, or made fast to the shore, or aground. (Note: 'Making way' means progressing through the water, rather than drifting.)

(l) 'Restricted visibility' means any condition in which visibility is restricted by fog, mist, falling snow, heavy rainstorms, sandstorms or any other similar causes.

(m) A 'Wing-in-Ground (WIG) craft' means a multimodal craft which, in its main operational mode, flies in close proximity to the surface by utilising surface-effect action.

2.7.2 Part B: Steering and Sailing Rules
Section I – In any condition of visibility

Rule 4 (Application) is self-explanatory. In complying with Rules 4–18 ask yourself, and answer, the following three questions and take action if so required:

> • Is there a risk of collision?
> • If there is, am I the give-way vessel?
> • If I am, what action must I take?

This is not part of the IRPCS, but it concentrates the mind.

Rule 5 (Look-out).

Every vessel shall at all times maintain a proper look-out by sight and hearing as well as by all available means appropriate in the prevailing circumstances and conditions so as to make a full appraisal of the situation and of the risk of collision.

Look-out means not only by eyes and ears, but also by radar and VHF, especially at night or in low visibility. Do not neglect blind arcs, eg look below the genoa and directly to windward. Rule 5 is arguably the most important Rule of all.

Rule 6 (Safe speed).

Every vessel shall at all times proceed at a safe speed so that she can take proper and effective action to avoid collision and be stopped within a distance appropriate to the prevailing circumstances and conditions so as to make a full appraisal of the situation and of the risk of collision. In determining a safe speed the following factors are among those taken into account:

(a) By all vessels:
  (i) the state of visibility;
  (ii) the traffic density including concentrations of fishing vessels or any other vessels;
  (iii) the manoeuvrability of the vessel with special reference to stopping distance and turning ability in the prevailing conditions;
  (iv) at night the presence of background light such as from shore lights or from back scatter of her own lights;
  (v) the state of wind, sea and current, and the proximity of navigational hazards;
  (vi) the draught in relation to the available depth of water;

(b) Additionally, by vessels with operational radar:
  (i) the characteristics, efficiency and limitations of the radar equipment;
  (ii) any constraints imposed by the radar range scale in use;
  (iii) the effect on radar detection of the sea state, weather and other sources of interference;
  (iv) the possibility that small vessels, ice and other floating objects may not be detected by radar at an adequate range;
  (v) the number, location and movement of vessels detected by radar;
  (vi) the more exact assessment of the visibility that may be possible when radar is used to determine the range of vessels or other objects in the vicinity.

The conditions which determine what is a safe speed are clearly listed. Excessive speed gives less time to assess the situation and take avoiding action, and produces a worse collision if such action fails. Even 4 knots may be too fast in a crowded river.

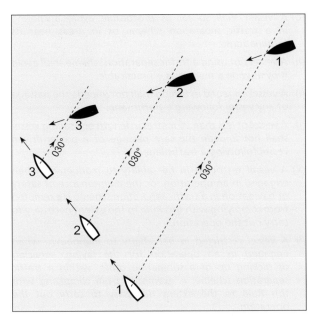

Fig 2(1) Rule 7. The bearing of black from white is steady on 030°. Well before reaching position 2, white should have altered to starboard by at least 45° to pass astern of black.

Rule 7 (Risk of collision).

(a) *Every vessel shall use all available means appropriate to the prevailing circumstances and conditions to determine if risk of collision exists. If there is any doubt such risk shall be deemed to exist.*

(b) *Proper use shall be made of radar equipment if fitted and operational, including long-range scanning to obtain early warning of risk of collision and radar plotting or equivalent systematic observation of detected objects.*

(c) *Assumptions shall not be made on the basis of scanty information, especially scanty radar information.*

(d) *In determining if risk of collision exists the following considerations shall be among those taken into account:*
   (i) *such risk shall be deemed to exist if the compass bearing of an approaching vessel does not appreciably change;*
   (ii) *such risk may sometimes exist even when an appreciable bearing change is evident, particularly when approaching a very large vessel or a tow or when approaching a vessel at close range.*

A vessel should take a series of compass bearings on a closing ship. Unless the bearings change appreciably, a risk of collision exists; see Fig 2(1). If in doubt, assume that there is a risk. Radar, properly used, offers early warning of risk of collision, but careful systematic plotting of contacts is needed to obtain maximum collision-avoidance data.

Rule 8 (Action to avoid collision).

(a) *Any action taken to avoid collision shall be taken in accordance with the Rules of this Part and shall, if the circumstances of the case admit, be positive, made in ample time and with due regard to the observance of good seamanship.*

(b) *Any alteration of course and/or speed to avoid collision shall, if the circumstances of the case admit, be large enough to be readily apparent to another vessel observing visually or by radar; a succession of small alterations of course and/or speed should be avoided.*

(c) *If there is sufficient sea room, alteration of course alone may be the most effective action to avoid a close-quarters situation provided that it is made in good time, is substantial and does not result in another close-quarters situation.*

(d) *Action taken to avoid collision with another vessel shall be such as to result in passing at a safe distance. The effectiveness of the action shall be carefully checked until the other vessel is finally past and clear.*

(e) *If necessary to avoid collision or allow more time to assess the situation, a vessel shall slacken her speed or take all way off by stopping or reversing her means of propulsion.*

(f) (i) *A vessel which by any of these Rules is required not to impede the passage or safe passage of another vessel shall, when required by the circumstances of the case, take early action to allow sufficient sea room for the safe passage of the other vessel.*
   (ii) *A vessel required not to impede the passage or safe passage of another vessel is not relieved of this obligation if approaching the other vessel so as to involve risk of collision and shall, when taking action, have full regard to the action which may be required by the Rules of this Part.*
   (iii) *A vessel the passage of which is not to be impeded remains fully obliged to comply with the Rules of this Part when the two vessels are approaching one another so as to involve risk of collision.*

Any such actions must be positive, seamanlike and taken early. Large alterations of course and/or speed are more obvious to the other vessel, especially at night or on radar. Turn away (even through 180°), slow down, stop (or go astern). In avoiding one vessel, watch out for others.

Rule 9 (Narrow channels).

(a) *A vessel proceeding along the course of a narrow channel or fairway shall keep as near to the outer limit of the channel or fairway which lies on her starboard side as is safe and practicable.*

(b) *A vessel of less than 20 metres in length or a sailing vessel shall not impede the passage of a vessel which can safely navigate only within a narrow channel or fairway.*

(c) *A vessel engaged in fishing shall not impede the passage of any other vessel navigating within a narrow channel or fairway.*

(d) *A vessel shall not cross a narrow channel or fairway if such crossing impedes the passage of a vessel which can safely navigate only within such channel or fairway. The latter vessel may use the sound signal prescribed in Rule 34 (d) if in doubt as to the intention of the crossing vessel.*

(e) (i) *In a narrow channel or fairway when overtaking can take place only if the vessel to be overtaken has to take action to permit safe passing, the vessel intending to overtake shall indicate her intention by sounding the appropriate signal prescribed in Rule 34 (c) (i). The vessel to be overtaken shall, if in agreement, sound the appropriate signal prescribed in Rule 34 (c) (ii) and take steps to permit safe passing. If in doubt she may sound the signals prescribed in Rule 34 (d).*
   (ii) *This Rule does not relieve the overtaking vessel of her obligation under Rule 13.*

(f) *A vessel nearing a bend or an area of a narrow channel or fairway where other vessels may be obscured by an* intervening obstruction shall navigate with particular alertness and caution and shall sound the appropriate signal prescribed in Rule 34 (e).

(g) Any vessel shall, if the circumstances of the case admit, avoid anchoring in a narrow channel.

Rule 9 (a) requires a vessel ... to keep as near to the starboard side of a narrow channel as is safe and practicable. Yachts and other small craft can often keep outside the channel altogether, thus ensuring even greater safety.

Rule 10 (Traffic Separation Schemes).

(a) This Rule applies to traffic separation schemes adopted by the Organization (IMO) and does not relieve any vessel of her obligation under any other Rule.

(b) A vessel using a traffic separation scheme shall:
- (i) proceed in the appropriate traffic lane in the general direction of traffic flow for that lane;
- (ii) so far as practicable keep clear of a traffic separation line or separation zone;
- (iii) normally join or leave a traffic lane at the termination of the lane, but when joining or leaving from either side shall do so at as small an angle to the general direction of traffic flow as practicable.

(c) A vessel shall so far as practicable avoid crossing traffic lanes, but if obliged to do so shall cross on a heading as nearly as practicable at right angles to the general direction of traffic flow.

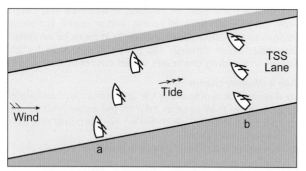

Fig 2(2) Rule 10c. A yacht crossing a TSS lane shall head at right angles to the lane axis, as in (a), regardless of the course made good as a result of wind or tidal streams. Yacht (b) is not heading at right angles and is therefore contravening Rule 10(c).

(d) (i) A vessel shall not use an inshore traffic zone when she can safely use the appropriate traffic lane within the adjacent traffic separation scheme. However, vessels of less than 20 metres in length, sailing vessels and vessels engaged in fishing may use the inshore traffic zone.

- (ii) Notwithstanding sub-paragraph (d) (i) a vessel may use an inshore traffic zone when en route to or from a port, offshore installation or structure, pilot station or any other place situated within the traffic zone, or to avoid immediate danger.

(e) A vessel other than a crossing vessel or a vessel joining or leaving a lane shall not normally enter a separation zone or cross a separation line except:
- (i) in cases of emergency to avoid immediate danger;
- (ii) to engage in fishing within a separation zone.

(f) A vessel navigating in areas near the terminations of traffic separation schemes shall do so with particular caution.

(g) A vessel shall so far as practicable avoid anchoring in a traffic separation scheme or in areas near its terminations.

(h) A vessel not using a traffic separation scheme shall avoid it by as wide a margin as is practicable.

(i) A vessel engaged in fishing shall not impede the passage of any vessel following a traffic lane.

(j) A vessel of less than 20 metres in length or a sailing vessel shall not impede the safe passage of a power-driven vessel following a traffic lane.

(k) A vessel restricted in her ability to manoeuvre when engaged in an operation for the maintenance of safety of navigation in a traffic separation scheme is exempted from complying with this Rule to the extent necessary to carry out the operation.

(l) A vessel restricted in her ability to manoeuvre when engaged in an operation for the laying, servicing or picking up of a submarine cable, within a traffic separation scheme, is exempted from complying with this Rule to the extent necessary to carry out the operation.

TSS are essential to the safety of larger vessels and are shown in Chapter 9 and on most charts. Whilst inconvenient for yachtsmen, TSS shall be avoided where possible, or accepted as another element of passage planning. All vessels, including yachts, must comply with Rule 10.

Fig 2(2) illustrates how to cross a TSS (10c). Note well that craft <20m LOA, and any sailing yacht, shall not impede a power vessel using a traffic lane (10j). However, when two vessels meet or converge in a TSS with a risk of collision, Rule 10 does not modify any other provisions of the IRPCS.

Yachts should use inshore traffic zones (ITZ). It is unusual and certainly undesirable for yachts to use TSS lanes, due to the substantial speed difference between yachts and large vessels. Yachts using TSS lanes may be told by the monitoring radar station (eg Dover CG, CROSS Corsen) to leave asap.

2.7.3 Section II – Vessels in sight of one another

Rule 11 (Application) is self explanatory.

Rule 12 (Sailing vessels).

(a) When two sailing vessels are approaching one another, so as to involve risk of collision, one of them shall keep out of the way of the other as follows:
- (i) when each has the wind on a different side, the vessel which has the wind on the port side shall keep out of the way of the other;
- (ii) when both have the wind on the same side, the vessel which is to windward shall keep out of the way of the vessel which is to leeward;
- (iii) if a vessel with the wind on the port side sees a vessel to windward and cannot determine with certainty whether the other vessel has the wind on the port or on the starboard side, she shall keep out of the way of the other.

(b) For the purposes of this Rule the windward side shall be deemed to be the side opposite to that on which the mainsail is carried or, in the case of a square-rigged vessel, the side opposite to that on which the largest fore-and-aft sail is carried.

Rule 12 does not apply if either yacht is motor sailing. When two yachts under sail are at risk of collision, (a) (i) and (ii) are clear.

(a) (iii) applies where it is unclear which tack a windward yacht is on. Fig 2(3) illustrates the practical application of Rule 12 in these three cases.

Boats on opposite tacks – yacht on port tack gives way

Both yachts on the same tack – windward boat gives way

Port tack boat gives way

Fig 2(3) Rule 12. A port tack yacht keeps clear. If both yachts are on the same tack the windward boat keeps clear. If in doubt, port tack always keeps clear.

Rule 12 (a) (iii) might cause doubt in 2 other situations:

* The windward side is deemed, Rule 12 (b), to be the side opposite to that on which the mainsail is set or would be set, if sailing under a headsail alone. When sailing under spinnaker alone, windward would be the side on which its boom is set.
* Heaving-to, ideally, set mainsail to port on starboard tack with jib backed to windward. Others (except those with special privileges) will then become the give-way.vessel.

Rule 13 (Overtaking).

(a) *Notwithstanding anything contained in the Rules of Part B, Sections I & II any vessel overtaking any other shall keep out of the way of the vessel being overtaken.*

(b) *A vessel shall be deemed to be overtaking when coming up with another vessel from a direction more than 22½° abaft her beam, that is, in such a position with reference to the vessel she is overtaking, that at night she would be able to see only the sternlight of that vessel but neither of her sidelights.*

(c) *When a vessel is in any doubt as to whether she is overtaking another, she shall assume that this is the case and act accordingly.*

(d) *Any subsequent alteration of the bearing between the two vessels shall not make the overtaking vessel a crossing vessel within the meaning of these Rules or relieve her of the duty of keeping clear of the overtaken vessel until she is finally past and clear.*

An overtaking vessel, whether power or sail, shall keep clear of the vessel being overtaken. It is courteous, but not in the IRPCS, for an overtaking yacht under sail to pass to leeward of the yacht being overtaken.

Rule 14 (Head-on situation).

(a) *When two power-driven vessels are meeting on reciprocal or nearly reciprocal courses so as to involve risk of collision each shall alter her course to starboard so that each shall pass on the port side of the other.*

(b) *Such a situation shall be deemed to exist when a vessel sees the other ahead or nearly ahead and by night she could see the masthead lights of the other in a line or nearly in a line and/or both sidelights and by day she observes the corresponding aspect of the other vessel.*

(c) *When a vessel is in any doubt as to whether such a situation exists she shall assume that it does exist and act accordingly.*

When two power-driven vessels approach head-on, or nearly so, with risk of collision, each must alter course to starboard, to pass port to port. The alteration should be large enough as to be obvious to the other vessel, sounding one short blast, if in audible range, to make intentions clear.

Rule 15 (Crossing situation).

When two power-driven vessels are crossing so as to involve risk of collision, the vessel which has the other on her own starboard side shall keep out of the way and shall, if the circumstances of the case admit, avoid crossing ahead of the other vessel.

The give-way vessel would normally alter to starboard to pass astern of the other. A yacht with limited power and speed should, if possible, avoid crossing ahead of the other. Exceptionally, an alteration to port may be justified (eg shoal water to starboard), in which case a large alteration may be needed to avoid crossing ahead of the other.

Rule 16 (Action by give-way vessel).

Every vessel which is directed to keep out of the way of another vessel shall, so far as possible, take early and substantial action to keep well clear.

'Early … substantial … well clear' are the key words, ie don't dither or get into a close-quarters situation.

Rule 17 (Action by stand-on vessel).

(a) (i) *Where one of two vessels is to keep out of the way, the other shall keep her course and speed.*

(ii) *The latter vessel may however take action to avoid collision by her manoeuvre alone, as soon as it becomes apparent to her that the vessel required to keep out of the way is not taking appropriate action in compliance with these Rules.*

(b) *When, from any cause, the vessel required to keep her course and speed finds herself so close that collision cannot be avoided by the action of the give-way vessel alone, she shall take such action as will best aid to avoid collision.*

(c) *A power-driven vessel which takes action in a crossing situation in accordance with sub-paragraph (a) (ii) of this Rule to avoid collision with another power-driven vessel* shall, if the circumstances of the case admit, not alter course to port for a vessel on her own port side.

(d) This Rule does not relieve the give-way vessel of her obligation to keep out of the way.

This Rule describes a deteriorating sequence of events (1–3 below) and the actions that a stand-on vessel may or shall (ie must) take if it appears that the give-way vessel is failing to keep clear.

1. At least in the initial stages, the stand-on vessel should maintain her course/speed (17a. i) for fear of inducing a collision by needless avoiding action of her own.

2. The stand-on vessel may manoeuvre to avoid collision (17a. ii) if it appears that the give-way vessel is failing to keep clear. In so manoeuvring a power-driven vessel shall, if possible, not alter course to port for a vessel on her own port side (17c). Usually she would alter substantially to starboard, to reduce the risk of both vessels turning towards each other.

3. When the stand-on vessel finds herself so close that collision cannot be avoided by the action of the give-way vessel alone, the stand-on vessel *shall* take such action (17b), as will best aid to avoid collision. This is likely to be drastic, eg Full astern and/or helm hard over.

Rule 18 (Responsibilities between vessels).

Except where Rules 9, 10 and 13 otherwise require:

(a) A power-driven vessel underway shall keep out of the way of:
   (i)   a vessel not under command;
   (ii) a vessel restricted in her ability to manoeuvre;
   (iii) a vessel engaged in fishing;
   (iv) a sailing vessel.

(b) A sailing vessel underway shall keep out of the way of:
   (i)   a vessel not under command;
   (ii) a vessel restricted in her ability to manoeuvre;
   (iii) a vessel engaged in fishing.

(c) A vessel engaged in fishing when underway shall, so far as possible, keep out of the way of:
   (i) a vessel not under command;
   (ii) a vessel restricted in her ability to manoeuvre.

(d) (i) Any vessel other than a vessel not under command or a vessel restricted in her ability to manoeuvre shall, if the circumstances of the case admit, avoid impeding the safe passage of a vessel constrained by her draught, exhibiting the signals in Rule 28.
   (ii) A vessel constrained by her draught shall navigate with particular caution having full regard to her special condition.

(e) A seaplane on the water shall, in general, keep well clear of all vessels and avoid impeding their navigation. In circumstances, however, where risk of collision exists, she shall comply with the Rules of this Part.

(f) (i) A WIG craft shall, when taking off, landing and in flight near the surface, keep well clear of all other vessels and avoid impeding their navigation.
   (ii) A WIG craft operating on the water surface shall comply with the Rules of this Part as a power-driven vessel.

Rule 18 states priorities according to manoeuvrability. The hierarchy is clear and logical: (b) applies to yachts. New para (f) includes Wing in Ground (WIG) craft, defined in Rule 3 (m). These air-cushion vehicles are more aircraft than vessel.

2.7.4   Section III – Vessels in restricted visibility
Rule 19 (Restricted visibility).

(a) This Rule applies to vessels not in sight of one another when navigating in or near an area of restricted visibility.

(b) Every vessel shall proceed at a safe speed adapted to the prevailing circumstances and conditions of restricted visibility. A power-driven vessel shall have her engines ready for immediate manoeuvre.

(c) Every vessel shall have due regard to the prevailing circumstances and conditions of restricted visibility when complying with the Rules of Section 1 of this Part.

(d) A vessel which detects by radar alone the presence of another vessel shall determine if a close-quarters situation is developing and/or risk of collision exists. If so, she shall take avoiding action in ample time, provided that when such action consists of an alteration of course, so far as possible the following shall be avoided:
   (i) an alteration of course to port for a vessel forward of the beam, other than for a vessel being overtaken;
   (ii) an alteration of course towards a vessel abeam or abaft the beam.

(e) Except where it has been determined that a risk of collision does not exist, every vessel which hears apparently forward of her beam the fog signal of another vessel, or which cannot avoid a close-quarters situation with another vessel forward of her beam, shall reduce her speed to the minimum at which she can be kept on her course. She shall if necessary take all her way off and in any event navigate with extreme caution until danger of collision is over.

This rule can cause difficulty due to misunderstanding and/or the negative nature of para (d). Stand-on and give-way vessels, as in Section II of the Rules, no longer exist in fog. Every vessel is simply trying to avoid other unseen vessels whose presence, ie fog signal, is only heard or detected by radar. Safe speed (b) is commonsense, but is sometimes ignored by commercial ships. Slowing down to maintain steerage way for a fog signal forward of the beam (e) has limited application to an already slow-moving yacht.

Radar can save your life, but only if it is efficiently used and correctly interpreted by an experienced operator. Rule 7c warns against making assumptions based on scanty information, especially radar information. AIS, if fitted and switched on in both vessels, may resolve such assumptions.

If a vessel is detected by radar alone (d), decide whether a collision risk exists and/or a close-quarters situation is developing. If so, take early action to avoid collision.

(d) (i) and (ii) tell you what you should *avoid* doing. Try to envisage the situations which these two clauses address. It may help to transpose their words into more positive terms:

> • Turn to starboard in all cases except for a vessel on or abaft your starboard beam, or for a vessel being overtaken.

Safety dictates that you should: sound the appropriate fog signal, keep a good look-out, deploy an efficient radar reflector and use radar if fitted. Consider also using the engine to assist manoeuvrability, especially in light winds. In thick fog it is best to anchor in shallow water; not easy if you are crossing a TSS in mid-Channel. Heaving to in the central separation zone to await better visibility might be an option, but see Rule 10 (b) and (e).

2.7.5 Part C: Lights and shapes

Many of the Rules (20–31) in Part C are semi-technical, not relevant to yachts or can be studied in slow time. The verbatim text is therefore only included for Rule 25 (Lights and shapes for sailing vessels underway).

Rule 20 (Application).

Part C Rules must be complied with in all weathers. All lights and shapes must comply with Annex I. The required lights must be shown from sunset to sunrise, and by day in low visibility. The required shapes must be displayed by day.

Rule 21 (Definitions).

Six types of navigation lights are defined, see Fig 2(4). A vessel <20m LOA may combine her side lights into 1 lantern on the fore and aft centreline, Rule 21 (b). A masthead lt, aka steaming lt, is rarely at the masthead; see Annex 1, para 2.

Rule 22 (Visibility of lights).

This Rule gives the *minimum* visibility range of lights shown by vessels of different LOAs. If <12m: masthead, stern and all-round 2M; side 1M. 12–50m: masthead 5M (but 3M if <20m); stern, side and all-round 2M. See Fig 2(4).

Rule 23 (Power-driven vessels underway).

(a) specifies a masthead light forward and a second, higher light aft if >50m LOA; plus sidelights and sternlight.
(b) refers to the all-round Fl Y light on an air-cushion vessel.
(c) a WIG craft must show a high-intensity all-round Fl R light.
(d) (i) A power-driven vessel <12m LOA may combine her masthead and sternlights into one all-round Ⓦ light.

Rule 24 (Towing and pushing).

This Rule is lengthy and detailed and need not concern yacht owners. However all mariners should know the lights shown by tugs and their tows; see Fig 2(5).

Rule 25 (Sailing vessels underway and vessels under oars).

(a) A sailing vessel underway shall exhibit:
   (i) sidelights; (ii) a sternlight.
(b) In a sailing vessel <20m LOA, the lights prescribed in paragraph (a) of this Rule may be combined in one lantern at or near the top of the mast where it can best be seen.
(c) A sailing vessel underway may, in addition to the lights prescribed at (a) of this Rule, exhibit at or near the top of the mast where they can best be seen two all-round lights, in a vertical line, the upper being red and the lower green, but these lights shall not be exhibited in conjunction with the combined lantern permitted by paragraph (b) of this Rule.
(d) (i) A sailing vessel <7m LOA shall, if practicable, exhibit the lights prescribed in paragraphs (a) or (b) of this Rule, but if she does not, she shall have ready at hand an electric torch or lighted lantern showing a white light which shall be exhibited in sufficient time to prevent collision.
   (ii) A vessel under oars may exhibit the lights prescribed in this Rule for sailing vessels, but if she does not, she shall have ready at hand an electric torch or lighted lantern showing a white light which shall be exhibited in sufficient time to prevent collision.
(e) A vessel proceeding under sail when also being propelled by machinery shall exhibit forward where it can best be seen a conical shape, apex downwards.

The tricolour light at (b) shall not be switched on at the same time as the normal side and stern lights, and must never be used when under power. It gives max brilliance for minimum battery drain, but note that LED lights easily out-perform conventional lights in power consumption and brilliance, but cost much more. In congested waters or where there are shore lights in the background, the normal side and stern lights are deemed much easier to see from the bridge of a large ship than a masthead tricolour.

Rule 26 (Fishing vessels).

All mariners should know the lights shown by fishing vessels; see Fig 2(5).

Rule 27 (Vessels not under command, or restricted in their ability to manoeuvre).

All mariners should know the lights shown by such vessels. They do not apply to vessels <12m LOA. If engaged in diving operations exhibit a rigid replica of flag 'A' at least 1m high.

Rule 28 (Vessels constrained by their draught).

All mariners should know the lights shown by such vessels; see Fig 2(5).

Rule 29 (Pilot vessels).

All mariners should know the lights shown by such vessels; see Fig 2(5).

Rule 30 (Vessels anchored or aground).

Yachts at ⚓, like other vessels, *shall* show an ⚓ light or ball; see Fig 2(5). This stipulation warns other mariners and has safety and insurance implications. Abroad, especially in the Netherlands and Germany, the ball requirement is often strictly enforced. Rules 30e and 30f grant 2 minor exemptions for yachts <7m LOA and <12m LOA.

Rule 31 (Seaplanes).

This Rule applies to WIG craft as well as seaplanes.

2.7.6 Part D: Sound and light signals

Rule 32 (Definitions).

> A whistle in a yacht effectively means a foghorn.
> · A short blast on a foghorn lasts about 1 second.
> – A prolonged blast lasts four to six seconds.

Rule 33 (Equipment for sound signals).

Vessels >12m LOA must have a whistle; plus a bell if >20m. A boat <12m LOA may make an efficient sound signal by other means, judged by audibility within the enclosed bridge of a large ship and against conflicting background noises.

Rule 34 (Manoeuvring and warning signals).

Power-driven vessels in sight of each other:
The following sound signals (short or long blasts on a foghorn) may also be supplemented by light flashes:

·	I am altering course to starboard.
··	I am altering course to port.
···	I am operating astern propulsion.
·····	I do not understand your intentions/actions or I doubt if sufficient action is being taken to avoid collision.

In a narrow channel:

––·	I intend to overtake on your starboard side.
––··	I intend to overtake you on your port side.
–·–·	I agree with your overtaking signal.
–	Warning by vessel nearing a bend where other vessels may not be visible.

Fig 2(4) Navigation lights

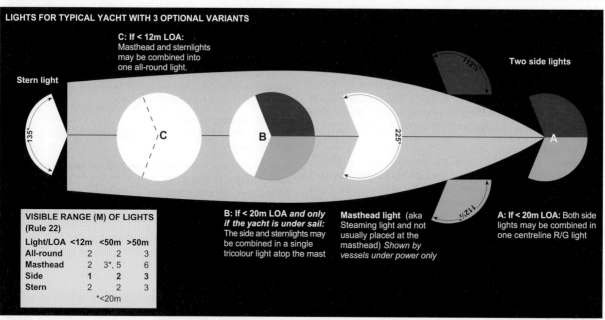

LIGHTS FOR TYPICAL YACHT WITH 3 OPTIONAL VARIANTS

C: If < 12m LOA: Masthead and sternlights may be combined into one all-round light.

Stern light

Two side lights

VISIBLE RANGE (M) OF LIGHTS (Rule 22)			
Light/LOA	<12m	<50m	>50m
All-round	2	2	3
Masthead	2	3*, 5	6
Side	1	2	3
Stern	2	2	3
		*<20m	

B: If < 20m LOA *and only if the yacht is under sail:* The side and sternlights may be combined in a single tricolour light atop the mast

Masthead light (aka Steaming light and not usually placed at the masthead) *Shown by vessels under power only*

A: If < 20m LOA: Both side lights may be combined in one centreline R/G light

PLAN VIEWS OF LIGHTS FOR SAILING VESSELS UNDERWAY AND UNDER SAIL ONLY
Note: If motor-sailing, the lights appropriate for a power-driven vessel must be shown, as below

Sailing vessel <7m shows, if practicable, side and sternlights; if not, a white light to prevent collision. Rule 25 d i

Sailing vessel <20m may show:
Either tricolour light atop the mast. Rule 25 b

Or combined centreline side lights plus stern light. Rule 25 a

Sailing vessel >20m Separate sidelights and sternlight. Rule 25 a

Sailing vessel under sail may show (in addition to other lights) two all-round lights, red over green, near the top of the mast, **but never** at the same time as a tricolour light atop the mast. Rule 25 c

PLAN VIEWS OF LIGHTS FOR POWER-DRIVEN VESSELS UNDERWAY AND SAILING CRAFT UNDER POWER

<7m and <7 knots (all-round light and, if practicable, side lights). Rule 23 d ii

<12m (combined masthead & sternlight). Rule 23 d i

<20m (sidelights combined in one centreline light). Rule 21 b

>20m showing masthead, stern and separate sidelights. Rule 21 a, b, c

>50m with two masthead lights, the aft one higher. Rule 23 a i & ii

Fig 2(5) Principal navigation lights and shapes. (Note: All vessels seen from starboard side)

Vessels being towed and towing

Vessel towed shows sidelights (forward) and sternlight

Tug shows two masthead lights, sidelights, sternlight, yellow towing light

Towing by day — Length of tow more than 200m

Towing vessel and tow display diamond shapes. By night, the towing vessel shows three masthead lights instead of two as for shorter tows

Motor sailing

Cone point down, forward. At night the lights of a power-driven vessel underway

Vessel fishing

All-round red light over all-round white, plus sidelights and sternlight when making way

Fishing/Trawling

A shape consisting of two cones point to point in a vertical line one above the other

Vessel trawling

All-round green light over all-round white, plus sidelights and sternlight when making way

Vessel restricted in her ability to manoeuvre

All-round red, white, red lights vertically, plus normal steaming lights when making way

Three shapes in a vertical line: ball, diamond, ball

Not under command

Two all-round red lights, plus sidelights and sternlight when making way

Two balls vertically

Dredger

All round red, white, red lights vertically, plus two all-round red lights (or two balls) on foul side, and two all-round green (or two diamonds) on clear side

Divers down

Letter 'A' International Code

Constrained by draught

Three all-round red lights in a vertical line, plus normal steaming lights. By day — a cylinder

Pilot boat

All-round white light over all-round red, plus sidelights and sternlight when underway, or anchor light

Vessel at anchor

All-round white light; if over 50m, a second light aft and lower

Ball forward

Vessel aground

Anchor light(s), plus two all-round red lights in a vertical line

Three balls in a vertical line

Rule 35 (Sound signals in restricted visibility).

–	Power-driven vessel making way.
– –	Power-driven vessel underway, but stopped and not making way.
– • •	Vessel not under command; restricted in her ability to manoeuvre; constrained by her draught; or engaged in fishing, towing or pushing; **or a sailing vessel**.
– • • •	Vessel being towed, or if more than one vessel is towed, the last vessel in the tow.
• • • •	Pilot vessel engaged on pilotage duties.

The maximum interval between sound signals for vessels underway in restricted visibility is two minutes, but more frequently if other craft are near.

For flag and sound signals which have a special meaning under IRPCS, see the asterisked items in Fig 5(1).

	Bell rung rapidly for about 5 seconds, every minute.
=	Vessel at anchor.
	After the above signal, gong rung rapidly for about 5 seconds every minute; the bell being sounded in the fore part of the vessel and the gong aft
=	Vessel of 100m or more in length at anchor.
• – •	Vessel at anchor (optional extra sound signal).
	Bell rung rapidly for about 5 seconds, with three separate and distinct strokes of the bell before and after the rapid ringing
=	Vessel aground.

Rule 36 (Signals to attract attention).
These signals, whether sound or light, must be such as not to be mistaken for other signals in these Rules or an aid to navigation. Avoid using high intensity strobe lights.

Rule 37 (Distress signals).
See Annex IV and Chapter 7, Fig 7(1).

2.7.7   Annexes I to IV
Annex I: vertical/horizontal positioning and technical details of lights and shapes in various vessels.

Annex II: lights which may be shown by fishing vessels working close together.

Annex III: technical details of frequencies, audibility and directional properties of sound signals.

Annex IV: distress signals; see also Chapter 7, Fig 7(1).

2.7.8   Learning the Collision Regulations
Most of this chapter deals with the vitally important Rules of the Collision Regulations (ColRegs). To many people learning these is a chore; something to be swotted up, often parrot-fashion, so as to pass an exam – and then be forgotten. This is not learning, it is simply a memory test without understanding of the subject.

The ColRegs are written by seamen for seamen, unambiguously and in language which leaves little room for misunderstanding. Every word or sentence reflects a real life situation which has occurred at sea. They are steeped in practicality and hard won experience. Above all they exist to prevent collisions and hence save lives.

If we can adopt a better way of learning then what seemed like a chore may turn out to be a pleasure. Start from the Known and move with growing interest towards the Unknown. Grasp the big picture, in this case a site map below, showing how the ColRegs consist of five Parts (A-E), sub-divided, as required, into Sections; thence into the 38 individual Rules; finally 4 Annexes with technical minutiae.

Now for the details: As you study each Rule envisage the scenario which is being described. Ask yourself 'Do I fully understand this Rule?' 'What would I do in this situation?' This is the best way of gaining a sound working knowledge of the Rules – and disposing of that parrot.

PARTS			RULES
A: GENERAL			1–3
B: STEERING	I	In any visibility	4–10
& SAILING	II	In sight of each other	11–18
	III	In restricted visibility	19
C: LIGHTS & SHAPES			20–31
			Annexes I/II
D: SOUND & LIGHT SIGNALS			32–37
			Annexes III/IV
E: EXEMPTIONS			38

3

Navigation

3.1 Passage planning and SOLAS V   34
Passage planning form • Planning considerations

3.2 Conduct of the plan   35
Estimating position • Navigational wrinkles

3.3 Lights   36
Light characteristics • Light sectors and arcs of visibility

3.4 IALA buoyage (Region A)   38

3.5 Satellite navigation   39
GPS • EGNOS • GLONASS • Galileo • Horizontal datums

3.6 LORAN-C, NELS and eLORAN   40

3.7 Electronic charts   40
Raster charts • Vector charts

3.8 Waypoints   40
Waypoint navigation • Loading waypoints

3.9 Chartplotters   41

3.10 Radar and AIS   41
Display modes • Radar as an aid to navigation • Collision avoidance • Radar in fog • AIS

3.11 Racons and Ramarks   43

3.12 Navigational warnings   43

3.13 Hazards   43
Submarines • Gunfacts • Mines

3.14 Offshore energy installations   44
Gas and oil rigs • Wind farms • Wave and tidal installations

3.15 Navigational tables   44
Notes on Tables 3(1) to 3(6) • Checking a compass for accuracy

3.16 Sunrise/set times   49
Notes on table for Lats 56°N, 48°N and 40°N

3.17 Moonrise/set times   51
Notes on table for Lats 56°N, 48°N and 40°N

3.18 Phases of the Moon   52

Fig 3(1) Passage planning form

DATE:............................ FROM: .... TO: .... DIST: ....M

ALTERNATIVE DESTINATION(S): ....

WEATHER FORECAST: ....
....

FORECASTS AVAILABLE DURING PASSAGE: ....
....

TIDES

DATE:.. | DATE:.. | DATE:..

PLACE:.. | PLACE:.. | PLACE:..

HW .... | HW .... | HW ....

LW .... | LW .... | LW ....

HW .... | HW .... | HW ....

LW .... | LW .... | LW ....

COEFFICIENT: ....

HEIGHT OF TIDE AT: ....
........................hrs ....m ....hrs ....m ....hrs ....m

DEPTH CONSTRAINTS: ....

TIDAL STREAMS AT: ....

TURNS .... AT .... TOTAL SET (FM .... TO ....): ....° .... M

TURNS .... AT .... TOTAL SET (FM .... TO ....): ....° .... M

NET TIDAL STREAM FOR PASSAGE: ....° .... M

ESTIMATED TIME: ....hrs ETD: .... ETA: ....

SUN/MOON SUNRISE: .... SUNSET: ....

MOONRISE: .... MOONSET: .... PHASE: ....

WAYPOINTS | NO | NAME | TRACK/DISTANCE (TO NEXT WAYPOINT)
| .... | .... | .... / ....
| .... | .... | .... / ....
| .... | .... | .... / ....
| .... | .... | .... / ....
| .... | .... | .... / ....

DANGERS CLEARING BEARINGS/RANGES/DEPTHS
....
....
....

LIGHTS/MARKS EXPECTED ....
....
....

COMMUNICATIONS PORT/MARINA .... VHF .... ☎ ....
PORT/MARINA .... VHF .... ☎ ....

NOTES (CHARTS PREPARED & PAGE NUMBERS OF RELEVANT PILOTS/ALMANACS/ETC): ....
....
....

3.1 PASSAGE PLANNING AND SOLAS V

All passages by any vessel that goes to sea *must* be planned. 'Going to sea' is defined as proceeding beyond sheltered waters. Even in very familiar waters every passage, however short, should be properly planned. Before you set sail you need to determine where you are going, how to get there and what factors may influence the plan.

Full passage planning requirements may be found in Chapter V of the International Convention for Safety of Life at Sea (SOLAS), but more digestible guidance for small craft is in the MCA's Pleasure Craft Information Pack at: **www.dft.gov.uk/mca/pleasure_craft_information_packdec07-2.pdf.**

Although the passage plan does not have to be recorded on paper, in the event of legal action a written plan is clear proof that the required planning has been completed; it can also be referred to during the passage. A suggested passage planning form is at Fig 3(1) opposite. When completed this would constitute, with due consideration of the points at 3.1.1 below, a reasonable passage plan. The blank form may be photocopied and modified to suit individual needs.

Although spot checks on small craft are unlikely, the MCA could, following an accident or incident, take action under the Merchant Shipping Act if it could be proved that the skipper did not have a reasonable passage plan.

3.1.1 Passage planning considerations

All passage plans should at least consider the following:

- **Weather.** Before setting out check the weather forecast and know how to get regular updates during the passage.
- **Tides.** Check tidal predictions and determine if there are any limiting depths at your port of departure, during the passage and at the port of arrival (and at alternative ports, if applicable). Tidal streams will almost certainly affect the plan.
- **Vessel.** Confirm she is suitable for the intended trip, is properly equipped, and has sufficient fuel, water and food on board.
- **Crew.** Take into account the size of crew, their experience, expertise and stamina. Cold, tiredness and seasickness can be debilitating – and skippers are not immune.
- **Navigation.** Make sure you are aware of all navigational dangers by consulting up to date charts, pilot books and this Almanac. Never *rely* on GPS for fixing your position.
- **Contingency plan.** Consider bolt holes which can be entered *safely* in an emergency, if the weather deteriorates or mutiny threatens.
- **Information ashore.** Make sure someone ashore knows your plans, when they should become concerned and what action to take if necessary. Be sure to join the Coastguard Voluntary Safety Identification Scheme (see 7.11.2).

'The winds and waves are always on the side of the ablest navigators' (Edward Gibbon).

3.1.2 Passage planning form

The following notes amplify some of the items on the form:

- The height of tide affects the depth, and tidal streams may hinder sensible progress – off headlands or in narrow passages, for example. On longer passages, which cross main tidal streams, determine the net

effect of the streams and calculate a course to steer – more efficient than trying to maintain a track.
- Note which harbours and marinas have restricted times of access due to bars, sills or locks. These may affect your ETA and, probably, your ETD.
- Prepare a detailed pilotage plan for entry/exit of any unfamiliar harbour; a sketch is invaluable.
- Squinting into the setting sun can make pilotage very difficult. At night, light from a full moon can help enormously.
- Know when you expect to see lights at night – a good check on progress. See Table 3(6).
- Look up and note relevant VHF channels and/or phone numbers for ports of departure and arrival.

3.2 CONDUCT OF THE PLAN

Take every opportunity to fix your position by all available means – visual bearings and electronic aids. Although GPS normally gives accurate positional information 24 hours a day, the system is very vulnerable to external interference. Onboard, the equipment depends on a steady power supply and is susceptible to moisture or accidental damage. It is therefore essential to maintain a written record of positions, headings and distances run at regular intervals. From this an updated position may be established if a fix is unobtainable.

A dead reckoning (DR) plot must be started as soon as there is any doubt about your ability to fix your position.

3.2.1 Estimating position

A well adjusted compass, an accurate timepiece and a log to measure distance run through the water are the essential tools of dead reckoning (DR) navigation. A DR plot starts from your last known or estimated position. After a certain interval, depending on your proximity to dangers, the time, log reading, distance run and course steered are noted. These are then plotted on the chart to obtain an updated position.

This will be more accurate if leeway is applied to the course steered. For example, a yacht steering 090° with a northerly wind might make 5° of leeway; 095° would therefore be the course plotted. To refine this position still further, include the tidal stream rate and direction experienced since the last plotted position. This results in an estimated position (EP), which is the most accurate position in the absence of a fix.

By using conventional symbols on the chart (Fig 3(2)), others can immediately appreciate what they see and check the accuracy of the navigator's work. A sensible navigator always welcomes a second opinion.

Fig 3(2) DR/EP plot

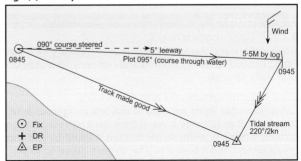

Navigation

The DR or EP should always be extrapolated to ensure that the vessel will not run into danger. When plotting an EP include possible inaccuracies in the log (for distance run), the tidal stream and the course made good. Leeway, for example, is usually only an inspired guess. From this, instead of a range and bearing from your last position, you will construct a 'lozenge of probability'. Always assume your position is on that part of the lozenge which is closest to danger.

3.2.2 Navigational wrinkles
Distance, speed and time equations (see also Table 3(2)):

Distance (M)	=	Speed (kn) x time (hours)
Speed	=	Distance ÷ time
Time	=	Distance ÷ Speed

Tidal streams. Look at the stream flowing past fishing floats, buoys or any other fixed objects to better estimate the *actual* direction and rate – often significantly different from those obtained from tidal stream atlases or the chart.

Running fix. This is particularly useful at night when only one suitable light is visible. Take a bearing and plot the position line (PL) from the mark; note the time. When the angle has changed by about 45°, take a second bearing and plot the PL. Transfer the first PL by the distance and direction the boat has travelled *over the ground* in the interval between the two bearings. Your position at the time of the second bearing is where the first PL crosses the second.

Pilotage. A detailed sketch of the pilotage passage in a notebook saves constant trips to the chart table. Be very wary of having charts in the cockpit where they can get wet or blow away.

Blind pilotage. In 6 minutes (1/10 hour) you will cover 1/10 of your speed in miles. So at 6 knots you will cover 0·6 miles (6 cables); at 12 knots you will cover 1·2 miles. In 3 minutes you will cover half that distance; in 12 minutes, twice the distance. Remember to allow for any tidal stream.

Distance off track is easily calculated by knowing that 1° at 6 miles subtends an arc of 1 cable; 5° at 60 miles = 5 miles.

A Dutchman's log can be used when the ship's log has failed. Time how long it takes for an object dropped at the bows to pass the stern. Speed (kn) = (0·592 x LOA (ft)) ÷ T (seconds). Thus for an LOA of 34ft and a time of 5 seconds, the speed will be 34 x 0·592 ÷ 5 = 4 knots. A table can be drawn up for your boat for various speeds.

Distances. Although the term Nautical Mile (NM) is commonly used, the strictly correct unit for distance at sea is the Sea Mile. This is the length of one minute of latitude at any particular place. It varies from about 1843m at the equator to about 1862m at the poles. A NM is defined as 1852m.

3.3 LIGHTS

3.3.1 Light characteristics
Abbreviations and characteristics of marine lights are shown in Fig 3(5) and in NP 5011 (Symbols and Abbreviations used on Admiralty Charts).

3.3.2 Light sectors and arcs of visibility
Coloured sector lights are used to indicate dangerous areas (usually red) or to guide a vessel through a narrow passage (usually white). A white sector is often flanked by red and green sectors: red is seen if too far to port; green to stbd.

The limits of light sectors, arcs of visibility and the alignment of directional and leading lights are shown **as seen from your vessel**. All bearings are True, clockwise from north, and are shown for selected lights in Chapter 9 – either in the Lights, buoys and waypoints sections or in individual harbour entries under Lights and Marks.

Fig 3(3) shows a simple sectored light listed as *Oc WR 4s 37m* **W17M**, *R13M; 151°-W-115°-R-151°*.

Fig 3(3) A simple sectored light

The light, showing white and red sectors, occults every 4s, is 37 metres above the level of MHWS, its white sector has a nominal range of 17 miles (ranges greater than 15 miles are in **bold**) and the red light's range is 13 miles. The white sector is visible when it bears 151° clockwise to 115°, an arc of 324°; the red sector is visible when it bears 115° to 151°, an arc of 36°.

In Fig 3(4) is a slightly more complex sectored light listed as: *Q WRG 9m 10M; 015°-G-058°-W-065°-R-103°-G-143·5°-W-146·5°-R-015°*.

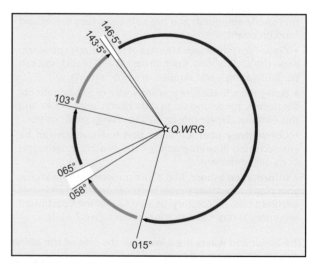

Fig 3(4) A more complex sectored light

This quick flashing light has white, red and green sectors, is 9m above MHWS and has a nominal range of 10M. By inspection there are two sets of WRG sectors providing directional guidance to vessels approaching from SW and NW. In each case a narrow (7° and 3°) white sector is flanked by a red sector to port and a green sector to starboard.

Fig 3(5) Light characteristics

CLASS OF LIGHT		International abbreviations	National abbreviations	Illustration Period shown ⊢————————⊣
FIXED		F		
OCCULTING *(duration of light longer than dark)*				
Single-occulting		Oc	Occ	
Group-occulting	eg		Oc(2)	
Composite group-occulting	eg	Oc(2+3)	Gp Occ(2+3)	
ISOPHASE *(light and dark equal duration)* Iso				
FLASHING *(duration of light shorter than dark)*				
Single-flashing		Fl		
Long-flashing *(flash 2s or longer)*			L Fl	
Group-flashing	eg	Fl(3)	Gp Fl(3)	
Composite group-flashing	eg	Fl(2+1)	Gp Fl(2+1)	
QUICK *(50 to 79 flashes per minute, usually 50 or 60)*				
Continuous quick		Q	Qk Fl	
Group quick	eg	Q(3)	Qk Fl(3)	
Interrupted quick		IQ	Int Qk Fl	
VERY QUICK *(80 to 159 flashes per minute, usually 100 or 120)*				
Continuous very quick		VQ	V Qk Fl	
Group very quick	eg	VQ(3)	V Qk Fl(3)	
Interrupted very quick		IVQ	Int V Qk Fl	
ULTRA QUICK *(160 or more flashes per minute, usually 240 to 300)*				
Continuous ultra quick		UQ		
Interrupted ultra quick		IUQ		
MORSE CODE	eg	Mo (K)		
FIXED AND FLASHING		F Fl		
ALTERNATING	eg	Al. WR	Alt. WR	

COLOUR	International abbreviations	NOMINAL RANGE in miles		International abbreviations
White	W *(may be omitted)*	Light with single range	eg	**15M**
Red	R	Light with two different ranges	eg	**15/10M**
Green	G	Light with three or more ranges	eg	**15-7M**
Blue	Bu	**PERIOD** is given in seconds	eg	**90s**
Violet	Vi	**DISPOSITION** Horizontally disposed		**(hor)**
Yellow	Y	Vertically disposed		**(vert)**
Orange	Y	Synchronised with (an)other light(s)		**(sync)**
Amber	Y	**ELEVATION** is given in metres (m) above the level of MHWS		

3.4 IALA BUOYAGE (REGION A)

IALA buoyage system – Regions

Region A, shown here, includes Europe; Region B covers the Americas and the Pacific.

Note that different rules (the CEVNI Code) apply on Inland Waterways in Europe.

3.4.1 Cardinal marks (buoys and beacons) indicate where the best navigable water lies relative to a danger or hazard and, specifically, that:

- The deepest water is on the named side of the mark
- The named side is the safe side on which to pass

They may also be used to emphasise a bend or fork in a channel, or to mark the end of a shoal.

Cardinal marks are referenced to the points of the compass and named after the quadrant in which they are placed relative to the danger or hazard they mark. The four quadrants (N, E, S and W) are defined by the True bearings NW-NE, NE-SE, SE-SW and SW-NW from the danger or hazard. It follows that you should normally pass north of a NCM, east of an ECM and so on.

Topmarks: 2 black cones in various unique combinations.
Colour: yellow & black bands (black always relates to the points of the cones), as shown in the diagram.
Lights: always white, Q or VQ, as defined in Fig 3(5). The number of flashes relates to the 'clock code': a NCM at 12

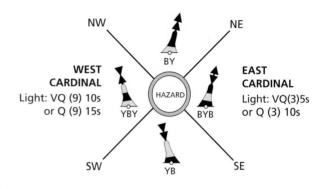

NORTH CARDINAL
Light: VQ or Q

WEST CARDINAL
Light: VQ (9) 10s or Q (9) 15s

EAST CARDINAL
Light: VQ(3)5s or Q (3) 10s

SOUTH CARDINAL
Light: VQ (6) + L Fl 10s or Q (6) + L Fl 15s

o'clock flashes continuously; an ECM at 3 o'clock flashes groups of 3; a SCM flashes groups of 6 + a long flash (at least 2 secs duration); and a WCM flashes groups of 9.

3.4.2 Lateral marks indicate the sides of a well-defined channel according to the direction of buoyage (3.4.8). If PHM or SHM marks do not rely on shape for identification, they carry the appropriate topmarks where practicable. In Region A:

- PHM marks are red. Shape: can, spar or pillar. Topmark, if fitted: red can. Even numbering. Light: red; any rhythm except (2+1).

- SHM marks are green (exceptionally, black). Shape: conical, spar or pillar. Topmark, if fitted: green cone, point up. Odd numbering. Light: green; any rhythm except (2+1).

R   R

Port hand marks (PHM)

Navigable channel

Direction of buoyage

G   G

Starboard hand marks (SHM)

3.4.3 Preferred channel marks At a channel junction the preferred channel may be shown by a lateral mark with red or green bands. The alternate channel (to the same destination) lies on the other side of this mark. Topmarks are those of a PHM or SHM. Think of a PCP (see below) as a SHM modified by a R band, and the PCS as a PHM modified by a G band.

Preferred channel to port (PCP)
SHM mark, green with red band. Light (if fitted): Fl (2+1) G.

Preferred channel to starboard (PCS)
PHM mark, red with green band. Light (if fitted): Fl (2+1) R.

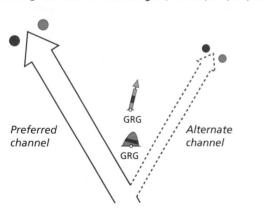

Preferred channel

GRG

GRG

Alternate channel

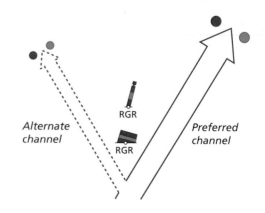

Alternate channel

RGR

RGR

Preferred channel

3.4.4 Isolated danger marks (IDM) are placed on or above an isolated danger of limited extent, eg a rock or a wreck, which has navigable water all around it.

BRB   BRB

Colour: black with red band(s). Topmark: 2 black balls (vert). Light, if any, is a group of two white flashes, Fl (2).

3.4.5 Safe water marks (SWM) indicate that there is navigable water all round the mark. They are used as centreline, mid-channel or landfall marks.

RW   RW

Shape: spherical, pillar or spar. Colour: red & white vertical stripes. Topmark: 1 R ball, but only on pillar or spar buoys. Light, if any, is white, occulting, isophase, L Fl 10s or Morse code 'A' (· −)

3.4.6 Special marks are not of navigational significance but indicate a special area usually depicted on the chart, eg spoil grounds, naval exercise areas, water-ski and bathing areas. Other features include yacht racing marks, cables or pipelines, outfalls, ODAS buoys and TSS marks (where conventional channel marks might cause confusion).

Y   Y   Y   Y

Colour: yellow. Shape: optional, but not conflicting with lateral or safe water marks. If can or conical buoys are used they indicate the side on which to pass. Topmark if fitted: yellow X. Light if fitted: yellow, any rhythm not used for Cardinal, IDM and SWM marks.

3.4.7 New dangers Emergency wreck-marking buoys are used before other marks can be deployed. They are pillar or spar shaped with blue & yellow vert stripes; WRECK is painted vert in black on yellow; optional topmark yellow +; light Al Bu/Y 3s 4M; Racon 'D' (−··) and/or AIS are optional.

YBu   YBu

Newly discovered hazards – wrecks, obstructions, rocks or shoals – are marked by IALA cardinal or lateral buoys, often 4 or more where the risk to other vessels is high. For emphasis one of the marks may be duplicated until the hazard has been adequately publicised. A Racon coded 'D' (−··) and/or an AIS transducer may also be fitted.

3.4.8 Conventional direction of buoyage
This is defined as either: the **Local** direction when entering estuaries and large waterways from seaward.
Or elsewhere as the **General** direction of buoyage which in principle is clockwise around continents.
Around the UK the general direction is N up the W coast and the Irish Sea, E up the English Channel, thence N up the North Sea (necessitated by the clockwise direction around mainland Europe).
If there is doubt, the direction is marked on charts by a broad arrowhead as shown in 3.4.2. Off Orfordness, Suffolk, for example, the local direction into the Thames Estuary is SW but the general direction is N.

3.5   SATELLITE NAVIGATION

3.5.1   GPS

The NAVSTAR Global Positioning System (GPS) is owned and operated by the United States Department of Defense (DoD). It provides world-wide, 24 hours a day, precise position, velocity and time information to anyone with a suitable receiver. The system consists of at least 24 operational satellites, orbiting 10,900M (20,200km) above the earth, in 6 orbital planes inclined at 55° to the equator. Positional accuracy is within 33m for 95% of the time; for the remaining 5% of the time errors of up to 100m are theoretically possible. Signal propagation delays can cause errors of up to 5m; other system errors are up to 2·5m.

GPS accuracy can be enhanced by Differential GPS (DGPS). An additional MF diferential signal in the 283.5-315 Khz band is transmitted from terrestrial beacons to provide overlapping coverage out to 50M from the coast. The accuracy obtainable from the Standard Positioning Service (as opposed to the Precise Positioning Service for military use) is generally quite adequate for yachting purposes. However, errors may occur because older surveys were not conducted to the same accuracy as is now possible with GPS. The source diagrams on Admiralty charts show survey dates. Also, some charts are not yet referred to the WGS84 datum, and corrections must be made before plotting positions obtained from GPS (see 3.5.5).

When choosing one of the many GPS receivers available, consider size (handheld, portable or fixed installation); check that it has multi-channels (12+ is the norm) and whether it can receive EGNOS. Compatibility with Galileo (see 3.5.4) may also influence your choice. Much depends on where it is to be used, the degree of accuracy required and budget.

3.5.2   EGNOS

EGNOS is a Wide Area Augmentation System (WAAS). It consists of 3 geostationary satellites and a network of 34 ground stations for ranging and integrity monitoring, plus 4 master control centres and 6 up-link stations. The 'Wide Area' is bounded approximately by Iceland, Murmansk, Greece and Marrakesh. Many receivers now display a 'WAAS operating' indication. Measured performance (99% accuracy of 1–2m) far exceeds expectations. EGNOS is due to cease operating when Galileo is fully established.

3.5.3   GPS integrity monitoring

Urgent information is given in MSI broadcasts by HM CG on VHF or MF, and by Navtex message category J (see Ch 6).

The US CG issues daily status information which may be found on www.navcen.uscg.gov. The US Government states that, although GPS is primarily for military use, it is committed to provide the best possible service to civil and commercial users worldwide – both in times of peace and conflict.

3.5.4   Other satellite-based systems

The *Global Navigation Satellite System* (GLONASS) operated by the Russian Federation is similar to GPS with 27 satellites in orbit. It is not readily available to non-commercial users.

Galileo is an independent European satellite system, compatible with GPS but promising greater accuracy. Six satellites have been launched. Initial capability is expected mid-decade, with full operation projected by 2020.

More details are in ALRS Vol 2, NP 282.

3.5.5 Horizontal datums

Spheroids are mathematical models of the earths surface from which horizontal datums are derived for specifying position on the Earth's surface. GPS uses the World Geodetic System 84 (WGS84). With the exception of much of the coast of Ireland and the Western Scotland, Admiralty charts of UK waters have now been converted to the ETRS 89 datum, as have most charts of NW Europe. This is WGS84 compatible. Many charts of Iberia are still to ED 50 but progress is being made. Chartlets in this Almanac are referenced to WGS84.

GPS fixes can be plotted directly onto charts which are referenced to WGS84, but on other charts the coordinates read from the GPS receiver must be converted to the datum of the chart in use. The datum used is always printed on the chart, normally under 'SATELLITE-DERIVED POSITIONS'. This gives the latitude and longitude corrections to be applied to satellite positions before plotting them on the chart. The difference between WGS84 and other horizontal datums varies with location and can vary from tens of metres, up to a maximum of 7 miles in the Pacific. There are 2 options:

- Set the receiver to WGS84 and manually apply the adjustment given on the chart before plotting the position. Take care to apply the adjustment accurately; there is a worked example on each chart.

- Set your receiver to the datum of the chart in use. The corrections will automatically be applied by the receiver's software. This method is convenient but not necessarily very accurate due to the random nature of the differences and software limitations.

The first option is advised by UKHO. Always check the datum of your chart. Do not assume that an adjacent chart will use the same datum. See ALRS Vol 2 for further information.

3.6 LORAN-C, NELS AND eLORAN

3.6.1 LORAN-C (LOng RAnge Navigation) is a low frequency electronic position fixing system, with a master transmitting station and two or more slave stations making up each chain; 17 such chains are operational. Stations transmit at a specified Group Repetition Interval (GRI) which is unique to each chain. Typical groundwave ranges are 800M to 1200M; accuracy is about 100⁺m, decreasing with range.

3.6.2 NELS (NW European Loran-C System) covers an area broadly defined by Biscay, Denmark, North Cape, Svalbard, Jan Mayen Island, Iceland and Ireland. Operational chains, with their GRIs, are: Ejde (Faeroe Island)(9007), Bø (Norway) (7001), Sylt (Germany)(7499) and Lessay (France)(6731).

3.6.3 eLORAN is an enhanced Loran service from Anthorn, Cumbria. The station was declared fully operational in January 2008 and will transmit Eurofix DGNSS, differential Loran correction messages and time signals. Accuracy is 10–20m. Use of eLoran is actively encouraged by the UK government and the General Lighthouse Authorities as an independent terrestrial back-up to satellite-based systems.

3.7 ELECTRONIC CHARTS

Electronic charts fall into two main types: raster and vector. They are produced as a cartridge-type chip for use in chartplotters, as CDs for PCs and laptops or can be downloaded. Coverage is virtually worldwide. It is advisable to ascertain what updating service is provided; for many it is a simple part-exchange of old for new for a reasonable fee.

ARCS (Admiralty Raster Chart Service) has a licence system with update discs supplied weekly, quarterly or annually.

Electronic charts are excellent value for money but, depending as they do on a reliable power source and electronic equipment, they should be used to complement rather than replace paper charts.

See *The Electronic Yachtmaster* (DVD) by Tom Cunliffe (www. tomcunliffe.com).

3.7.1 Raster charts

Raster charts are produced by scanning the paper version of the chart. They are selective but exact electronic copies of originals. ARCS charts are Admiralty raster charts; Maptech is the US equivalent.

Adjacent charts can be automatically linked (known as quilting) to prevent 'sailing off the edge' of the previous one.

Users should be aware that zooming in on a raster chart does not necessarily produce more detail, it simply enlarges the image. Detailed larger scale charts of the area are included in the folio for this purpose.

3.7.2 Vector charts

Vector charts are produced by re-digitising paper charts. They are used in most dedicated chartplotters and build the information up in layers, such as seabed detail, depth contours, buoyage, and topographic features. Specific layers can be de-selected to increase the clarity of the plotter display.

The data is seamless, which usually enables the user to buy one cartridge (or CD) covering a specific area such as the Central English Channel rather than a folio of electronic charts. It provides total cartographic coverage right down to harbour details. Zooming in provides large scale coverage with more detail, rather than a simple enlargement of the chart image. Zooming out provides a smaller scale overview of the area.

Vector charts are 'intelligent' and interact with their host systems which allows the user to interrogate chart symbols or be warned of impending dangers. Blue Charts, C-Map, and Navionics are examples of vector charts.

3.8 WAYPOINTS

3.8.1 Waypoint navigation

- Waypoints (WPT or ⊕) are any geographical positions, usually associated with electronic navigation, and normally expressed as a latitude and longitude. They may be used as departure or arrival points, route turning points, or any other chosen positions to aid navigation.

- In Chapter 9 the waypoint for each harbour under the Navigation sub-heading is a safe position from which to begin the approach. The WPT below the title is usually the harbour entrance or final WPT.

- Section 4 of each Area (Lights, buoys and waypoints) lists the positions of selected lights and buoys which may be used as waypoints, but see 3.8.2 below. The Lat/Long, taken from a large scale chart, is usually stated to 0'·01 of a minute.

- Discrepancies in position may result from different horizontal datums, incomplete data or survey data deemed inadequate by modern standards. Charts may also contain very small integral errors.

- 'Track' means track over the ground, not heading. For example, if required to approach a harbour (or waypoint) on a track of 090°, the waypoint should be kept on a constant bearing of 090°. Any significant deviation from that track could stand you into danger.

3.8.2 Loading waypoints

It is very easy to make mistakes when loading waypoints. The following measures may prevent errors:

- Try to work without interruption.
- While planning your route, plot intended waypoints on the chart(s) before loading them into electronic memories. This ensures that the track(s) between waypoints will not lead you into danger, across shoals, rocks or even over dry land.
- Published waypoints (including those in this Almanac) should be verified by first plotting them on the chart. It helps to label each waypoint with a number or name.
- Undetected errors in lat/long values can cause major navigational problems, usually at the most awkward moment.
- Ensure that positions have been accurately extracted from the paper chart.
- Use the correct format for entering data. Check in the equipment's handbook.
- Check that tracks and distances between waypoints, as computed by the equipment, agree with those measured directly off the paper chart.
- Check longitude for E or W, especially when near the Greenwich Meridian.
- Ask another crew member to double-check your work.

3.9 CHARTPLOTTERS

Chartplotters, linked to GPS receivers, fall into three main categories: handheld/portable, fixed and PC-based. Their greatest benefit is that they continuously display the boat's position on an electronic chart; many also give tidal data. The downside is that they may make you navigationally lazy, to the extent that a total GPS failure could leave some navigators literally lost. See 3.2 and 3.2.1.

Some plotters can be interfaced with an autopilot, engine data, instrument data and even video/live-cam coverage of remote onboard areas. They may display depth and radar overlays on the same screen as the chart. There are many available options, and choice depends on what you want and what you wish to spend. However:

- Although a chartplotter greatly assists the navigator, it is only an *aid* to navigation. It is excellent for pilotage in narrow waters, but is no substitute for a good lookout.
- Be particularly wary if the GPS is coupled with an autopilot – not recommended in confined waters.
- GPS can easily be interrupted; it may even fail completely. It is therefore essential to maintain a written log of the boat's positions, courses and speeds. From this a reliable and reasonably accurate DR/EP plot may be drawn up.

3.9.1 Handheld and portable chartplotters

Modern handheld plotters offer electronic chart displays with all the usual GPS features. Although they are capable of chart overview, they have very limited screen size. They are ideal as an independent battery powered back-up to the boat's primary GPS receiver and plotter.

Portable plotters have a screen at least 3" x 4" (75 x 103mm), about the smallest practical size to avoid excessive zooming. They are ideal for use in smaller craft or as a cockpit display and back-up in larger craft. They can be removed in harbour for security and used ashore for passage planning.

3.9.2 Fixed plotters

Fixed plotters are permanently mounted, at the chart table, on the pedestal or at the wheel on a bridge or fly-bridge, with optional repeaters elsewhere. Many plotters have split or multiple screens and may, with professional installation, allow much onboard data to be displayed. Increasingly they perform additional functions with dedicated power supplies, ruggedised build and water resistance.

3.9.3 PC based plotters

PCs and laptops are used at sea with navigational software and electronic charts linked to a GPS receiver to provide real-time navigation. Software add-ons include tidal stream data, forecasts, communication links, autopilot controls and, for the racing yachtsman, tactical features such as tacking and gybing angles, instrument displays, countdown timing and startline overview. If not fitted with an integral or USB GPS receiver, they require linking to the vessel's GPS NMEA output. Mounting, shock resistance, configuration and a reliable power supply may pose problems. Laptops have many other uses onboard and can, of course, be taken ashore for security and passage planning.

3.9.4 Tablets and mobile phones

Tablets and mobile phones are increasingly being used as handheld and portable plotters. Electronic chart packages are cheap, with the GPS receiver normally integrated in the device which is readily carried everywhere. They can, with appropriate applications, be used effectively. However, their ease of use and familiarity can lead to poor practice. All advice as to origin and accuracy of data applies. Battery life can be a problem as can reliability in the marine environment, and they should not be relied on as the sole navigational aid.

3.10 RADAR AND AIS

3.10.1 Radar

Radar (**Ra**dio **D**irection **a**nd **R**ange) is used for both collision avoidance and navigation. To realise its full potential a sound understanding of its operation and limitations is required. The manufacturer's instruction book should be read carefully; time spent practising with the set to obtain optimum performance in different conditions will build confidence. Many controls have automatic default settings, but the best results can often be obtained by manual tuning. Correct interpretation of the displayed information is fundamental to success.

Small craft radars transmit very short concentrated pulses of microwaves at a frequency of about 9500 MHz (X-band) and 3cm wavelength. Each pulse lasts less than a microsecond (pulse length) and the number of pulses transmitted per second (pulse repetition frequency or PRF) is normally between 800 and 3000. The reflected pulses are displayed on a screen.

3.10.2 Radar display modes

Typical small craft radars can be interfaced with a flux gate compass to give three heading modes:

- **North-up:** North is at the top of the display, the same as the chart. This mode helps general orientation.

- **Head-up:** Ship's current heading is at the top of the display. As the heading changes, the picture rotates; all bearings of radar returns are relative to the ship's head.

- **Course-up:** The selected course is at the top of the display. If a new course is selected, the picture rotates. As heading alters, the heading marker moves in unison. *This is arguably the best mode for use in collision avoidance.*

In *relative motion* mode the ship is usually at the centre of the display, and the land appears to move in the opposite direction to the vessel. The display can be offset so that a larger area of the screen is ahead, astern or abeam of the vessel. In *true motion* mode the vessel moves across the screen, and the land stays still. This may seem ideal, but there are disadvantages: it is not obvious if a contact is closing on a steady bearing; precise speed input is essential; and the image must be reset when close to the edge of the screen.

Some radar sets have split screens for simultaneous displays of, for example, radar picture, chartplotter and echosounder data. The ability to overlay the radar picture on the electronic chart can be invaluable when trying to distinguish charted objects from other contacts.

3.10.3 Radar as an aid to navigation

When used properly radar is a very useful aid to navigation, but it does have limitations:

- Radar cannot see behind objects or around corners; it may not detect small objects and it may not differentiate between two targets that are close together.

- Objects with flat surfaces, such as buildings or cliffs, reflect better than those with curved or sloping surfaces. A low coastline which is below the radar horizon will not be detected, but returns from more distant high ground may be received, giving a false impression of the range of the nearest land.

- Radar beams do not discriminate bearing accurately: 3 ranges will give a better fix than 3 radar bearings.

- The distance of the radar horizon is slightly greater than that of the visible horizon under average atmospheric conditions. It is given by $R = 2.2 \times \sqrt{H^s}$ where R = Range (M) and H^s scanner height (m). Most small craft radars have a maximum range of 16 to 24M depending upon power output and scanner and target heights.

3.10.4 Radar for collision avoidance

To determine if a risk of collision exists individual contacts need to be tracked in order to, assess their courses, speeds and closest point of approach (CPA). Practice makes perfect; you should be able to track, plot and assess up to 3 contacts simultaneously. Modern digital radar plotters automate this. In relative motion mode another vessel's course and speed are not obvious from the radar display; it needs to be worked out by resolving the relative velocity triangle.

- A vessel moving in the same direction and at the same speed as you is stationary relative to your boat.

- A risk of collision exists if another contact is on a steady bearing with its range decreasing. If the bearing changes appreciably, assess whether the vessel will cross ahead or astern of you.

- To decide the correct action to take, plot the contact at regular intervals on a plotting sheet and determine its course, speed, CPA and time to CPA.

- Plotting intervals depend on the contact's rate of closure. Keep plotting until all risk of collision is over.

3.10.5 Radar in fog

When navigating in poor visibility a thorough working knowledge of Rule 19 of the Colregs is absolutely essential. Rule 19 is unusual in that:

- It differs markedly from Rules 11–18 (Conduct of vessels in sight of one another) which clearly distinguish between stand-on and give-way vessels. Rule 19 contains no such distinction; every vessel is responsible for keeping clear of others.

- It states what avoiding actions should **not** be taken, leaving you to infer what positive options are open to you. These boil down to:

- Alter to *starboard* for a vessel forward of the beam (unless you are overtaking) and alter *away* from a vessel abeam or abaft the beam.

- In practice this means that you should always alter your course to starboard **except** for a vessel on your own starboard quarter – or if you are overtaking.

- Some commercial vessels greatly exceed the safe speed stipulated by the Colregs. Small craft skippers should be aware of this when deciding what avoiding action to take in restricted visibility.

GRP yachts produce a poor radar return; an effective radar reflector is essential to enable detection by other vessels.

3.10.6 AIS

Automatic Identification System (AIS) transmits information to other ships and shore authorities. SOLAS requires it to be fitted to vessels over 300 GRT. In addition to vessel tracking AIS data is widely used to augment collision avoidance by presenting position, course, speed and identity information continuously in near real-time.

Each ship's course and speed vector is shown by a symbol on an AIS screen or overlaid on radar, chart plotter or PC. This data may also appear in a text box as heading, COG & SOG, range, CPA, position, ship's name (derived from its MMSI), and her status – under power or sail, anchored, constrained by draught, restricted in ability to manoeuvre, not under command.

Increasingly lights, buoys and other aids to navigation (AtoN) are now fitted with AIS. On Admiralty charts, these are shown by a magenta circle around the A2N and the notation 'AIS'. Information available depends on the display system fitted onboard. Not all (esp early) receivers are A2N capable.

Caveats: Many small vessels are not fitted with AIS, and some may not have it switched on; some only display 3 lines of text, not a plot; in busy areas only the strongest signals may be shown; AIS may distract a bridge watchkeeper or lookout from his visual and radar watch; GPS/electronic failures invalidate AIS and it does not yet feature in the Colregs.

AIS is not mandatory for leisure craft, but Class B transmitters/transceivers for non-SOLAS vessels are available. Receivers, either standalone, or integrated in a plotter or computer, will provide accurate and continuous display of other ships' courses and speeds (over the ground) and identity. In areas of traffic concentration Class B transmitters may be excluded from the network if there are no available time slots.

Note: AIS is to some extent dependent on user inputs, and it is not a radar despite what some advertisements may imply.

3.11 RACONS AND RAMARKS

3.11.1 Racons

A Racon is a transponder beacon often fitted to light vessels, lighthouses and important buoys. It appears on charts as a magenta circle marked *Racon*. When triggered by a vessel's radar transmission it emits a distinctive coded return, much more easily recognised than a normal radar return.

Fig 3(6) Radar beacon responses

A Racon's return or 'flash' appears on the radar display as a line or narrow sector radiating outwards from a point slightly beyond the Racon's actual position. Thus a vessel's distance from the Racon, measured from where the flash begins, is slightly more than the actual distance. This is due to the slight response delay in the Racon apparatus. The coded identification flash is usually a Morse Code letter whose dashes/dots form a bold tail; see Fig 3(6). Racons fitted to temporary 'new danger' buoys are coded 'D', have a 10M range and are 'frequency agile'.

Racon's maximum range is typically 10M but may be greater. In practice, range depends on the power and elevation of both the Racon and vessel's radar. Thus a 2–3kW radar with a scanner height of 5m might not receive a buoy-mounted Racon, height also 5m, until only 6M from it.

Abnormal radio propagation may cause a spurious flash to be seen far beyond the quoted range, appearing along the correct bearing but at random ranges. A flash can only be relied upon if it seems to be stable and is within its quoted range. At short ranges a Racon can cause unwanted interference on the radar which may be reduced by adjusting the rain clutter control on the radar receiver.

3.11.2 Racon details

Racons are shown under 'Lights, buoys and waypoints' for each Area in Chapter 9 of this Almanac. Details given are:

- The Morse identification signal.
- Approximate maximum reception range (M).
- The angular sector (if other than 360°) within which the Racon signal can be received.

Most Racons respond through 360° and scan the 3cm (X-band) frequency range. Frequency agile Racons respond automatically to both 3cm and 10cm (S-band) emissions. But the agile response is switched off for a few seconds every minute, so that the Racon response does not obscure any radar echoes which are needed.

3.11.3 Ramarks

Ramarks are similar to Racons but are not triggered by a ship's radar. Their continuous transmission forms a line of Morse characters which radiates from the centre of the radar display to its edge, and cannot therefore give range. There are very few, if any, in the area covered by this Almanac.

3.12 NAVIGATIONAL WARNINGS

Navigational warnings are promulgated for 16 NavAreas (I–XVI) worldwide. An Area co-ordinator oversees each NavArea and handles inputs from national and other authorities. The 3 types of navigational warnings are:

NavArea warnings which deal with Oceanic matters and are transmitted by SafetyNet and Navtex at scheduled times in English and other languages.

Coastal warnings, up to 250M offshore, are issued by a national authority and broadcast in English or the local language by Navtex, SafetyNet or VHF Coast radio stations. These warnings cover navaids, lights, buoys, rocks, wrecks, naval exercises, SAR operations, cable laying, piracy, etc.

Local warnings supplement the above for inshore waters and within port limits. They are issued by port, pilotage or CG authorities in English and/or the local language.

Full details for all the Areas covered by this Almanac, including broadcast times, are in Chapter 5.

3.13 HAZARDS

3.13.1 Submarines

Fishing vessels, and occasionally yachts, have been snagged or hit by submarines operating just below the surface. The risk is greatest in the western English Channel, Irish Sea, Firth of Clyde, North Channel and West Scotland, especially at night. Submarine exercise areas are shown by name and/or number in 9.8.23 and 9.9.16.

Listen to and take heed of Subfacts (see Chapter 5). If possible avoid areas of known submarine activity, but when in an exercise area the best advice is:

- Keep clear of any vessel flying Code flags 'NE2' which indicate that submarines are in the vicinity.
- Run the engine and operate the echo sounder, which may indicate your presence to the submarine.
- At night show navigation lights low down, on pulpit and stern. They are more likely to be seen by a submarine at periscope depth than a tricolour at the masthead.

3.13.2 Gunfacts

Gunfacts are warnings of intended naval gunnery and missile practice firings, and underwater explosions. They do not restrict the passage of any vessel, and the onus for safety is on the firing warship. For further details see Chapter 5.

3.13.3 Mines

In the very unlikely event of coming across a mine at sea:

- Keep well clear, and do not prod it with a boathook.
- Do not try to tow it to port. Financial rewards for such action have been discontinued.
- Inform the Coastguard on Ch 16 immediately, and make an 'All Ships' Sécurité broadcast so that ships in the vicinity are aware of the situation.

3.14 OFFSHORE ENERGY INSTALLATIONS

In addition to well-established offshore oil and gas platforms, Offshore Renewable Energy Installations (OREIs) may be encountered across the area covered by the Almanac. OREIs include wind farms, operational and under construction; wave and tidal/current generators are also being trialled inshore and offshore posing new challenges to navigation. There is no right of access to any OREI and any approach should only be made in controlled circumstances by trained personnel. The visibility of offshore wind, wave and tidal energy installations depends on the device type; some are totally submerged while others protrude slightly above the sea surface. The marking of offshore wave and tidal energy installations is based on IALA's recommendations.

3.14.1 Gas and oil rigs

Gas and oil field structures are found mainly in the North Sea, Liverpool Bay and Morecambe Bay. There are also numerous installations off the west coasts of Denmark, Germany and the Netherlands. With careful passage planning and study of the relevant charts they can usually be passed at a safe distance without undue detours. Most are protected by Safety Zones, usually 500m radius, in which it is an offence for unauthorised vessels to enter, anchor or fish. Small craft must *stay well clear of all such installations at all times.*

3.14.2 Wind farms

Wind farms are proliferating, mostly in the shallow waters over coastal banks. Around the UK they are especially prevalent in the Thames Estuary, from the north Norfolk coast to Flamborough Head, in Liverpool Bay and Morecambe Bay. A large wind farm is planned SW of the Isle of Wight. The turbines give early visual warning of where the intervening navigable channels lie, thus they may assist rather than hinder navigation. Spacing between turbines is generally at least 500m, and the lowest point of the rotor sweep is at least 22m above MHWS. The number of wind turbines worldwide is forecast to rise steeply, and they will tend to move into deeper waters (20–25m) and further offshore.

More information is in the relevant Areas of this Almanac and at: www.mcga.gov.uk; www.bwea.com/offshore; and www.thecrownestate.co.uk.

3.14.3 Effects on navaids

Trials show that wind farms have no adverse effects on GPS, magnetic compasses and VHF. Predictably, however, radar performance is affected by blind or shadow areas, and by large or spurious echoes. This will inevitably lead to increasingly confused radar pictures in areas of multiple wind farms; small craft may be 'lost' on other vessels' radars. Mariners should report any unusual effects found in or near wind farms to the MCA and/or the RYA.

3.14.4 Navigating within a wind farm

It is often impracticable to avoid wind farms without making unacceptable detours. Unlike gas and oil installations, the majority of existing wind farms have Safety Zones. During construction a 500m safety zone, which is applicable to all unauthorised craft, usually exists around the works area.

Mariners should exercise caution navigating in close proximity to wind farms being aware that underwater obstructions may exist. Particular care should be taken passing structures with vessels in attendance and in any event avoid closing within 50m of individual turbines or associated structures.

When planning a voyage, mariners should assess hazards and associated risks in proximity to wind farms and turbines:

- spacing – wind farm turbines are usually spaced 500m or more apart depending on the size of the turbine
- depth of water – most wind turbines operate in shallow water, possibly restricting navigable channels
- seabed changes – wind farm structures may affect the depth of water through silting/scouring of the seabed making adjacent depth information unreliable
- tidal streams – wind farm structures may obstruct tidal streams locally, creating eddies nearby
- small craft – vessels involved in turbine maintenance and safety duties may be encountered or obscured within or around a wind farm
- shore marks – in coastal areas shore marks may also be obscured by wind farm structures
- electrical transformer stations – can be found in or adjacent to larger wind farms
- rotor effects – can change the flow of the wind
- no anchoring is possible due to submarine cables.

3.14.5 Wave and tidal installations

There are also many devices that may have unique and very different designs. They vary from attenuators and overtopping devices on or near the surface to submerged turbines. It is important to identify the hazard and its means of attachment to the seabed. Some may be seabed mounted; others may be tethered by floating flexible moorings allowing considerable lateral movement.

Great care should be taken in the vicinity of these installations, which should whenever possible be kept well clear.

3.15 NAVIGATIONAL TABLES

Table 3 (1) – Distance of horizon

Enter with height (m/ft) of eye, and extract the theoretical distance (M) of horizon. Actual distance could be reduced by abnormal refraction. Maximum height of eye is desirable; climbing onto the boom may double your height of eye.

TABLE 3(1) DISTANCE OF HORIZON					
Height of eye		Horizon distance	Height of eye		Horizon distance
metres	feet	M	metres	feet	M
1	3·3	2·1	21	68·9	9·5
2	6·6	2·9	22	72·2	9·8
3	9·8	3·6	23	75·5	10·0
4	13·1	4·1	24	78·7	10·2
5	16·4	4·7	25	82·0	10·4
6	19·7	5·1	26	85·3	10·6
7	23·0	5·5	27	88·6	10·8
8	26·2	5·9	28	91·9	11·0
9	29·6	6·2	29	95·1	11·2
10	32·8	6·6	30	98·4	11·4
11	36·1	6·9	31	101·7	11·6
12	39·4	7·2	32	105·0	11·8
13	42·7	7·5	33	108·3	12·0
14	45·9	7·8	34	111·6	12·1
15	49·2	8·1	35	114·8	12·3
16	52·5	8·3	36	118·1	12·5
17	55·8	8·6	37	121·4	12·7
18	59·1	8·8	38	124·7	12·8
19	62·3	9·1	39	128·0	13·0
20	65·6	9·3	40	131·2	13·2

Table 3(2) – Distance, speed and time

Enter with time (minutes or decimals of an hour, if using a calculator) and speed (kn) to determine distance run (M). Or extract speed or time as a function of the other two variables; for example speed over a measured mile. For speeds >20kn, use multiples of lesser speeds.

TABLE 3(2)   DISTANCE, SPEED AND TIME

Decimal of hr	Mins	3.0	4.0	5.0	6.0	7.0	8.0	9.0	10.0	11.0	12.0	13.0	14.0	15.0	16.0	17.0	18.0	19.0	20.0	Mins	Decimal of hr
·0167	1		0·1	0·1	0·1	0·1	0·1	0·2	0·2	0·2	0·2	0·2	0·2	0·3	0·3	0·3	0·3	0·3	0·3	1	·0167
·0333	2	0·1	0·1	0·2	0·2	0·2	0·3	0·3	0·3	0·4	0·4	0·4	0·5	0·5	0·5	0·6	0·6	0·6	0·7	2	·0333
·0500	3	0·1	0·2	0·2	0·3	0·3	0·4	0·4	0·5	0·5	0·6	0·6	0·7	0·7	0·8	0·8	0·9	0·9	1·0	3	·0500
·0667	4	0·2	0·3	0·3	0·4	0·5	0·5	0·6	0·7	0·7	0·8	0·9	0·9	1·0	1·1	1·1	1·2	1·3	1·3	4	·0667
·0833	5	0·2	0·3	0·4	0·5	0·6	0·7	0·7	0·8	0·9	1·0	1·1	1·2	1·2	1·3	1·4	1·5	1·6	1·7	5	·0833
·1000	6	0·3	0·4	0·5	0·6	0·7	0·8	0·9	1·0	1·1	1·2	1·3	1·4	1·5	1·6	1·7	1·8	1·9	2·0	6	·1000
·1167	7	0·4	0·5	0·6	0·7	0·8	0·9	1·1	1·2	1·3	1·4	1·5	1·6	1·8	1·9	2·0	2·1	2·2	2·3	7	·1167
·1333	8	0·4	0·5	0·7	0·8	0·9	1·1	1·2	1·3	1·5	1·6	1·7	1·9	2·0	2·1	2·3	2·4	2·5	2·7	8	·1333
·1500	9	0·4	0·6	0·7	0·9	1·0	1·2	1·3	1·5	1·6	1·8	1·9	2·1	2·2	2·4	2·5	2·7	2·8	3·0	9	·1500
·1667	10	0·5	0·7	0·8	1·0	1·2	1·3	1·5	1·7	1·8	2·0	2·2	2·3	2·5	2·7	2·8	3·0	3·2	3·3	10	·1667
·1833	11	0·5	0·7	0·9	1·1	1·3	1·5	1·6	1·8	2·0	2·2	2·4	2·6	2·7	2·9	3·1	3·3	3·5	3·7	11	·1833
·2000	12	0·6	0·8	1·0	1·2	1·4	1·6	1·8	2·0	2·2	2·4	2·6	2·8	3·0	3·2	3·4	3·6	3·8	4·0	12	·2000
·2167	13	0·6	0·9	1·1	1·3	1·5	1·7	2·0	2·2	2·4	2·6	2·8	3·0	3·2	3·5	3·7	3·9	4·1	4·3	13	·2167
·2333	14	0·7	0·9	1·2	1·4	1·6	1·9	2·1	2·3	2·6	2·8	3·0	3·3	3·5	3·7	4·0	4·2	4·4	4·7	14	·2333
·2500	15	0·7	1·0	1·2	1·5	1·8	2·0	2·2	2·5	2·7	3·0	3·2	3·5	3·7	4·0	4·2	4·5	4·7	5·0	15	·2500
·2667	16	0·8	1·1	1·3	1·6	1·9	2·1	2·4	2·7	2·9	3·2	3·5	3·7	4·0	4·3	4·5	4·8	5·1	5·3	16	·2667
·2833	17	0·8	1·1	1·4	1·7	2·0	2·3	2·5	2·8	3·1	3·4	3·7	4·0	4·2	4·5	4·8	5·1	5·4	5·7	17	·2833
·3000	18	0·9	1·2	1·5	1·8	2·1	2·4	2·7	3·0	3·3	3·6	3·9	4·2	4·5	4·8	5·1	5·4	5·7	6·0	18	·3000
·3167	19	1·0	1·3	1·6	1·9	2·1	2·5	2·9	3·2	3·5	3·8	4·1	4·4	4·8	5·1	5·4	5·7	6·0	6·3	19	·3167
·3333	20	1·0	1·3	1·7	2·0	2·3	2·7	3·0	3·3	3·7	4·0	4·3	4·7	5·0	5·3	5·7	6·0	6·3	6·7	20	·3333
·3500	21	1·0	1·4	1·7	2·1	2·4	2·8	3·1	3·5	3·8	4·2	4·5	4·9	5·2	5·6	5·9	6·3	6·6	7·0	21	·3500
·3667	22	1·1	1·5	1·8	2·2	2·6	2·9	3·3	3·7	4·0	4·4	4·8	5·1	5·5	5·9	6·2	6·6	7·0	7·3	22	·3667
·3833	23	1·1	1·5	1·9	2·3	2·7	3·1	3·4	3·8	4·2	4·6	5·0	5·4	5·7	6·1	6·5	6·9	7·3	7·7	23	·3833
·4000	24	1·2	1·6	2·0	2·4	2·8	3·2	3·6	4·0	4·4	4·8	5·2	5·6	6·0	6·4	6·8	7·2	7·6	8·0	24	·4000
·4167	25	1·3	1·7	2·1	2·5	2·9	3·3	3·8	4·2	4·6	5·0	5·4	5·8	6·3	6·7	7·1	7·5	7·9	8·3	25	·4167
·4333	26	1·3	1·7	2·2	2·6	3·0	3·5	3·9	4·3	4·8	5·2	5·6	6·1	6·5	6·9	7·4	7·8	8·2	8·7	26	·4333
·4500	27	1·3	1·8	2·2	2·7	3·1	3·6	4·0	4·5	4·9	5·4	5·8	6·3	6·7	7·2	7·6	8·1	8·5	9·0	27	·4500
·4667	28	1·4	1·9	2·3	2·8	3·3	3·7	4·2	4·7	5·1	5·6	6·1	6·5	7·0	7·5	7·9	8·4	8·9	9·3	28	·4667
·4833	29	1·5	1·9	2·4	2·9	3·4	3·9	4·3	4·8	5·3	5·8	6·3	6·8	7·2	7·7	8·2	8·7	9·2	9·7	29	·4833
·5000	30	1·5	2·0	2·5	3·0	3·5	4·0	4·5	5·0	5·5	6·0	6·5	7·0	7·5	8·0	8·5	9·0	9·5	10·0	30	·5000
·5167	31	1·6	2·1	2·6	3·1	3·6	4·1	4·7	5·2	5·7	6·2	6·7	7·2	7·8	8·3	8·8	9·3	9·8	10·3	31	·5167
·5333	32	1·6	2·1	2·7	3·2	3·7	4·3	4·8	5·3	5·9	6·4	6·9	7·5	8·0	8·5	9·1	9·6	10·1	10·7	32	·5333
·5500	33	1·6	2·2	2·7	3·3	3·8	4·4	4·9	5·5	6·0	6·6	7·1	7·7	8·2	8·8	9·3	9·9	10·4	11·0	33	·5500
·5667	34	1·7	2·3	2·8	3·4	4·0	4·5	5·1	5·7	6·2	6·8	7·4	7·9	8·5	9·1	9·6	10·2	10·8	11·3	34	·5667
·5833	35	1·7	2·3	2·9	3·5	4·1	4·7	5·2	5·8	6·4	7·0	7·6	8·2	8·7	9·3	9·9	10·5	11·1	11·7	35	·5833
·6000	36	1·8	2·4	3·0	3·6	4·2	4·8	5·4	6·0	6·6	7·2	7·8	8·4	9·0	9·6	10·2	10·8	11·4	12·0	36	·6000
·6117	37	1·8	2·4	3·1	3·7	4·3	4·9	5·5	6·1	6·7	7·3	8·0	8·6	9·2	9·8	10·4	11·0	11·6	12·2	37	·6117
·6333	38	1·9	2·5	3·2	3·8	4·4	5·1	5·7	6·3	7·0	7·6	8·2	8·9	9·5	10·1	10·8	11·4	12·0	12·7	38	·6333
·6500	39	1·9	2·6	3·2	3·9	4·5	5·2	5·8	6·5	7·1	7·8	8·4	9·1	9·7	10·4	11·0	11·7	12·3	13·0	39	·6500
·6667	40	2·0	2·7	3·3	4·0	4·7	5·3	6·0	6·7	7·3	8·0	8·7	9·3	10·0	10·7	11·3	12·0	12·7	13·3	40	·6667
·6833	41	2·0	2·7	3·4	4·1	4·8	5·5	6·1	6·8	7·5	8·2	8·9	9·6	10·2	10·9	11·6	12·3	13·0	13·7	41	·6833
·7000	42	2·1	2·8	3·5	4·2	4·9	5·6	6·3	7·0	7·7	8·4	9·1	9·8	10·5	11·2	11·9	12·6	13·3	14·0	42	·7000
·7167	43	2·2	2·9	3·6	4·3	5·0	5·7	6·5	7·2	7·9	8·6	9·3	10·0	10·8	11·5	12·2	12·9	13·6	14·3	43	·7167
·7333	44	2·2	2·9	3·7	4·4	5·1	5·9	6·6	7·3	8·1	8·8	9·5	10·3	11·0	11·7	12·5	13·2	13·9	14·7	44	·7333
·7500	45	2·2	3·0	3·7	4·5	5·2	6·0	6·7	7·5	8·2	9·0	9·7	10·5	11·2	12·0	12·7	13·5	14·2	15·0	45	·7500
·7667	46	2·3	3·1	3·8	4·6	5·4	6·1	6·9	7·7	8·4	9·2	10·0	10·7	11·5	12·3	13·0	13·8	14·6	15·3	46	·7667
·7833	47	2·3	3·1	3·9	4·7	5·5	6·3	7·0	7·8	8·6	9·4	10·2	11·0	11·7	12·5	13·3	14·1	14·9	15·7	47	·7833
·8000	48	2·4	3·2	4·0	4·8	5·6	6·4	7·2	8·0	8·8	9·6	10·4	11·2	12·0	12·8	13·6	14·4	15·2	16·0	48	·8000
·8167	49	2·5	3·3	4·1	4·9	5·7	6·5	7·4	8·2	9·0	9·8	10·6	11·4	12·2	13·1	13·9	14·7	15·5	16·3	49	·8167
·8333	50	2·5	3·3	4·2	5·0	5·8	6·7	7·5	8·3	9·2	10·0	10·8	11·7	12·5	13·3	14·2	15·0	15·8	16·7	50	·8333
·8500	51	2·5	3·4	4·2	5·1	5·9	6·8	7·6	8·5	9·3	10·2	11·0	11·9	12·7	13·6	14·4	15·3	16·1	17·0	51	·8500
·8667	52	2·6	3·5	4·3	5·2	6·1	6·9	7·8	8·7	9·5	10·4	11·3	12·1	13·0	13·9	14·7	15·6	16·5	17·3	52	·8667
·8833	53	2·6	3·5	4·4	5·3	6·2	7·1	7·9	8·8	9·7	10·6	11·5	12·4	13·2	14·1	15·0	15·9	16·8	17·7	53	·8833
·9000	54	2·7	3·6	4·5	5·4	6·3	7·2	8·1	9·0	9·9	10·8	11·7	12·6	13·5	14·4	15·3	16·2	17·1	18·0	54	·9000
·9167	55	2·8	3·7	4·6	5·5	6·4	7·3	8·3	9·2	10·1	11·0	11·9	12·8	13·8	14·7	15·6	16·5	17·4	18·3	55	·9167
·9333	56	2·8	3·7	4·7	5·6	6·5	7·5	8·4	9·3	10·3	11·2	12·1	13·1	14·0	14·9	15·9	16·8	17·7	18·7	56	·9333
·9500	57	2·8	3·8	4·7	5·7	6·6	7·6	8·5	9·5	10·4	11·4	12·3	13·3	14·2	15·2	16·1	17·1	18·0	19·0	57	·9500
·9667	58	2·9	3·9	4·8	5·8	6·8	7·7	8·7	9·7	10·6	11·6	12·6	13·5	14·5	15·5	16·4	17·4	18·4	19·3	58	·9667
·9833	59	2·9	3·9	4·9	5·9	6·9	7·9	8·8	9·8	10·8	11·8	12·8	13·8	14·7	15·7	16·7	17·7	18·7	19·7	59	·9833
1·0000	60	3·0	4·0	5·0	6·0	7·0	8·0	9·0	10·0	11·0	12·0	13·0	14·0	15·0	16·0	17·0	18·0	19·0	20·0	60	1·0000

TIME — SPEED IN KNOTS — TIME

Navigation

Table 3(3) – Distance off by vertical sextant angle (VSA)
From height (m/ft) of the object read across to the VSA obtained (corrected for index error). At the top of the table read off the distance (M) of the object.

If the base of, say, a lighthouse is below the horizon, the VSA will over-read the distance. For very precise ranges, add the amount by which sea level is below MHWS to the height of the object (above MHWS) – before entering the table.

TABLE 3(3) DISTANCE OFF BY VERTICAL SEXTANT ANGLE

HEIGHT OF OBJECT ft	m	DISTANCE OF OBJECT (NAUTICAL MILES)															
		0·2 ° ′	0·4 ° ′	0·6 ° ′	0·8 ° ′	1·0 ° ′	1·2 ° ′	1·4 ° ′	1·6 ° ′	1·8 ° ′	2·0 ° ′	2·4 ° ′	2·8 ° ′	3·2 ° ′	3·6 ° ′	4·0 ° ′	5·0 ° ′
33	10	1 33	0 46	0 31	0 23	0 19	0 15	0 13	0 12	0 10							
39	12	1 51	0 56	0 37	0 28	0 22	0 19	0 16	0 14	0 12	0 11	0 10					
46	14	2 10	1 05	0 43	0 32	0 26	0 22	0 19	0 16	0 14	0 13	0 11					
53	16	2 28	1 14	0 49	0 37	0 30	0 25	0 21	0 19	0 16	0 15	0 12	0 11				
59	18	2 47	1 24	0 56	0 42	0 33	0 28	0 24	0 21	0 19	0 17	0 14	0 12	0 10			
66	20	3 05	1 33	1 02	0 46	0 37	0 31	0 27	0 23	0 21	0 19	0 15	0 13	0 12	0 10		
72	22	3 24	1 42	1 08	0 51	0 41	0 34	0 29	0 26	0 23	0 20	0 17	0 15	0 13	0 11	0 10	
79	24	3 42	1 51	1 14	0 56	0 45	0 37	0 32	0 28	0 25	0 22	0 19	0 16	0 14	0 12	0 11	
85	26	4 01	2 01	1 20	1 00	0 48	0 40	0 34	0 30	0 27	0 24	0 20	0 17	0 15	0 13	0 12	
92	28	4 19	2 10	1 27	1 05	0 52	0 43	0 37	0 32	0 29	0 26	0 22	0 19	0 16	0 14	0 13	0 10
98	30	4 38	2 19	1 33	1 10	0 56	0 46	0 40	0 35	0 31	0 28	0 23	0 20	0 17	0 15	0 14	0 11
105	32	4 56	2 28	1 39	1 14	0 59	0 49	0 42	0 37	0 33	0 30	0 25	0 21	0 19	0 16	0 15	0 12
112	34	5 15	2 38	1 45	1 19	1 03	0 53	0 45	0 39	0 35	0 31	0 26	0 23	0 20	0 17	0 16	0 13
118	36	5 33	2 47	1 51	1 24	1 07	0 56	0 48	0 42	0 37	0 33	0 28	0 24	0 21	0 19	0 17	0 13
125	38	5 41	2 56	1 58	1 28	1 11	0 59	0 50	0 44	0 39	0 35	0 29	0 25	0 22	0 20	0 18	0 14
131	40	6 10	3 05	2 04	1 33	1 14	1 02	0 53	0 46	0 41	0 37	0 31	0 27	0 23	0 21	0 19	0 15
138	42	6 28	3 15	2 10	1 37	1 18	1 05	0 56	0 49	0 43	0 40	0 32	0 28	0 24	0 22	0 19	0 16
144	44	6 46	3 24	2 16	1 42	1 22	1 08	0 58	0 51	0 45	0 41	0 34	0 29	0 25	0 23	0 20	0 16
151	46	7 05	3 33	2 22	1 47	1 25	1 11	1 01	0 53	0 47	0 43	0 36	0 30	0 27	0 24	0 21	0 17
157	48	7 23	3 42	2 28	1 51	1 29	1 14	1 04	0 56	0 49	0 45	0 37	0 32	0 28	0 25	0 22	0 18
164	50	7 41	3 52	2 35	1 56	1 33	1 17	1 06	0 58	0 52	0 46	0 39	0 33	0 29	0 26	0 23	0 19
171	52	7 59	4 01	2 41	2 01	1 36	1 20	1 09	1 00	0 54	0 48	0 40	0 34	0 30	0 27	0 24	0 19
177	54	8 18	4 10	2 47	2 05	1 40	1 23	1 12	1 03	0 56	0 50	0 42	0 36	0 31	0 28	0 25	0 20
184	56	8 36	4 19	2 53	2 10	1 44	1 27	1 14	1 05	0 58	0 52	0 43	0 37	0 32	0 29	0 26	0 21
190	58	8 54	4 29	2 59	2 15	1 48	1 30	1 17	1 07	1 00	0 54	0 45	0 38	0 34	0 30	0 27	0 21
197	60	9 12	4 38	3 05	2 19	1 51	1 33	1 20	1 10	1 02	0 56	0 46	0 40	0 35	0 31	0 28	0 22
203	62	9 30	4 47	3 12	2 24	1 55	1 36	1 22	1 12	1 04	0 58	0 48	0 41	0 36	0 32	0 29	0 23
210	64	9 48	4 56	3 18	2 28	1 59	1 39	1 25	1 14	1 06	0 59	0 49	0 42	0 37	0 33	0 30	0 24
217	66	10 06	5 05	3 24	2 33	2 02	1 42	1 27	1 17	1 08	1 01	0 51	0 44	0 38	0 34	0 31	0 25
223	68	10 24	5 15	3 30	2 38	2 06	1 45	1 30	1 19	1 10	1 03	0 53	0 45	0 39	0 35	0 32	0 25
230	70	10 42	5 24	3 36	2 42	2 09	1 48	1 33	1 21	1 12	1 05	0 54	0 46	0 41	0 36	0 32	0 26
236	72	11 00	5 33	3 42	2 47	2 14	1 51	1 35	1 24	1 14	1 07	0 56	0 48	0 42	0 37	0 33	0 27
246	75	11 27	5 47	3 52	2 54	2 19	1 56	1 39	1 27	1 17	1 10	0 58	0 50	0 44	0 39	0 35	0 28
256	78	11 54	6 01	4 01	3 01	2 24	2 01	1 43	1 30	1 20	1 12	1 00	0 52	0 45	0 40	0 36	0 29
266	81	12 20	6 14	4 10	3 08	2 30	2 05	1 47	1 34	1 23	1 15	1 03	0 54	0 47	0 42	0 38	0 30
276	84	12 47	6 28	4 19	3 15	2 36	2 10	1 51	1 37	1 27	1 18	1 05	0 56	0 49	0 43	0 39	0 31
289	88	13 22	6 46	4 32	3 24	2 43	2 16	1 57	1 42	1 31	1 22	1 08	0 58	0 51	0 45	0 41	0 33
302	92	13 57	7 05	4 44	3 33	2 51	2 22	2 02	1 47	1 35	1 25	1 11	1 01	0 53	0 47	0 43	0 34
315	96	14 32	7 23	4 56	3 42	2 58	2 28	2 07	1 51	1 39	1 29	1 14	1 04	0 56	0 49	0 45	0 36
328	100	15 07	7 41	5 09	3 52	3 05	2 35	2 13	1 56	1 43	1 33	1 17	1 06	0 58	0 52	0 46	0 37
341	104	15 41	7 59	5 21	4 01	3 13	2 41	2 18	2 01	1 47	1 36	1 20	1 09	1 00	0 54	0 48	0 39
358	109	16 24	8 22	5 36	4 12	3 22	2 48	2 24	2 06	1 52	1 41	1 24	1 12	1 03	0 56	0 51	0 40
374	114	17 06	8 45	5 51	4 24	3 31	2 56	2 31	2 12	1 58	1 46	1 28	1 16	1 06	0 59	0 53	0 42
394	120	17 57	9 12	6 10	4 38	3 42	3 05	2 39	2 19	2 04	1 51	1 33	1 20	1 10	1 02	0 56	0 45
427	130	19 20	9 57	6 40	5 01	4 01	3 21	2 52	2 31	2 14	2 01	1 41	1 26	1 15	1 07	1 00	0 48
459	140	20 42	10 42	7 11	5 24	4 19	3 36	3 05	2 42	2 24	2 10	1 48	1 33	1 21	1 12	1 05	0 52
492	150	22 03	11 27	7 41	5 47	4 38	3 52	3 19	2 54	2 35	2 19	1 56	1 39	1 27	1 17	1 10	0 56
574	175	25 17	13 17	8 57	6 44	5 24	4 30	3 52	3 23	3 00	2 42	2 15	1 56	1 41	1 30	1 21	1 05
656	200	28 22	15 07	10 12	7 41	6 10	5 09	4 25	3 52	3 26	3 05	2 35	2 13	1 56	1 43	1 33	1 14
738	225		16 54	11 27	8 38	6 56	5 47	4 58	4 21	3 52	3 29	2 54	2 29	2 10	1 56	1 44	1 24
820	250		18 39	12 41	9 35	7 41	6 25	5 30	4 49	4 17	3 52	3 13	2 46	2 25	2 09	1 56	1 33
902	275		20 22	13 54	10 31	8 27	7 03	6 03	5 18	4 43	4 15	3 32	3 02	2 39	2 22	2 08	1 42
984	300		22 03	15 07	11 27	9 12	7 41	6 36	5 47	5 09	4 38	3 52	3 19	2 54	2 35	2 19	1 51
1148	350			17 29	13 17	10 42	8 57	7 41	6 44	6 00	5 24	4 30	3 52	3 23	3 00	2 42	2 10
1312	400			19 48	15 07	12 11	10 12	8 46	7 41	6 51	6 10	5 09	4 25	3 52	3 26	3 05	2 28

Tables 3(4) & 3(5) – Checking a compass for accuracy

The accuracy of a boat's magnetic compass can be checked against the true bearing of the sun at sunrise or sunset. The sun's declination and the boat's approximate latitude are all that are needed; no sextant or sight reduction tables.

Since bearings probably cannot be taken with the steering compass, steer the boat directly towards or away from the rising or setting sun, aligning it with the mast.

Due to refraction, which is about 34' when observing bodies on the horizon, the sun's bearing should be taken when its lower limb is just over half a diameter above the horizon.

Table 3(4) – Sun's declination

Enter with the appropriate date band and read across to the centre column.

Extract and record declination (bold type), noting whether it is North or South.

Table 3(5) – True bearing of sun at sunrise and sunset

Enter with the approximate latitude and declination, obtained from Table 3(4). Extract the tabulated bearing.

The True bearing is measured from North if declination is North, or from South if declination is South; towards the East if rising or towards the West if setting.

For example on 26 Aug, declination 10°N and latitude 45°N: the tabulated bearing is 75·8°, say 76°. If the sun is rising the True bearing will be N76°E, ie 076°T.

Apply local magnetic variation to 076°T and compare the result with the compass reading to determine deviation on the course steered.

3 examples of different Dec N/S and whether rising/setting:

Lat	Dec	SR/SS	Tab brg	True bearing	
45°	10°N	Setting	76°	N76°W	= 284°
45°	10°S	Rising	76°	S76°E	= 104°
45°	10°S	Setting	76°	S76°W	= 256°

Table 3(4) Sun's declination for 2016			
SOUTH		**DECLINATION**	**NORTH**
Dec 06 - Jan 06		23°	Jun 04 - Jul 07
Jan 07 - Jan 13	Nov 29 - Dec 05	22°	May 28 - Jun 03	Jul 08 - Jul 14
Jan 14 - Jan 18	Nov 24 - Nov 28	21°	May 22 - May 27	Jul 15 - Jul 20
Jan 19 - Jan 23	Nov 19 - Nov 23	20°	May 18 - May 21	Jul 21 - Jul 24
Jan 24 - Jan 27	Nov 15 - Nov 18	19°	May 13 - May 17	Jul 25 - Jul 29
Jan 28 - Jan 30	Nov 11 - Nov 14	18°	May 09 - May 12	Jul 30 - Aug 02
Jan 31 - Feb 03	Nov 08 - Nov 10	17°	May 06 - May 08	Aug 03 - Aug 05
Feb 04 - Feb 06	Nov 04 - Nov 07	16°	May 02 - May 05	Aug 06 - Aug 09
Feb 07 - Feb 09	Nov 01 - Nov 03	15°	Apr 29 - May 01	Aug 10 - Aug 12
Feb 10 - Feb 12	Oct 29 - Oct 31	14°	Apr 26 - Apr 28	Aug 13 - Aug 16
Feb 13 - Feb 15	Oct 26 - Oct 28	13°	Apr 23 - Apr 25	Aug 17 - Aug 19
Feb 16 - Feb 18	Oct 23 - Oct 25	12°	Apr 20 - Apr 22	Aug 20 - Aug 22
Feb 19 - Feb 21	Oct 20 - Oct 22	11°	Apr 17 - Apr 19	Aug 23 - Aug 25
Feb 22 - Feb 24	Oct 17 - Oct 19	10°	Apr 14 - Apr 16	Aug 26 - Aug 27
Feb 25 - Feb 26	Oct 15 - Oct 16	9°	Apr 11 - Apr 13	Aug 28 - Aug 30
Feb 27 - Feb 29	Oct 12 - Oct 14	8°	Apr 09 - Apr 10	Aug 31 - Sep 02
Mar 01 - Mar 03	Oct 09 - Oct 11	7°	Apr 06 - Apr 08	Sep 03 - Sep 05
Mar 04 - Mar 05	Oct 07 - Oct 08	6°	Apr 03 - Apr 05	Sep 06 - Sep 07
Mar 06 - Mar 08	Oct 04 - Oct 06	5°	Apr 01 - Apr 02	Sep 08 - Sep 10
Mar 09 - Mar 10	Oct 02 - Oct 03	4°	Mar 29 - Mar 31	Sep 11 - Sep 13
Mar 11 - Mar 13	Sep 29 - Oct 01	3°	Mar 27 - Mar 28	Sep 14 - Sep 15
Mar 14 - Mar 15	Sep 26 - Sep 28	2°	Mar 24 - Mar 26	Sep 16 - Sep 18
Mar 16 - Mar 18	Sep 24 - Sep 25	1°	Mar 21 - Mar 23	Sep 19 - Sep 20
Mar 19 - Mar 20	Sep 21 - Sep 23	0°	Mar 19 - Mar 20	Sep 21 - Sep 23

TABLE 3(5)   TRUE BEARING OF SUN AT SUNRISE AND SUNSET

DECLINATION

LAT	0°	1°	2°	3°	4°	5°	6°	7°	8°	9°	10°	11°	LAT
30°	90	88·8	87·7	86·5	85·4	84·2	83·1	81·9	80·7	79·6	78·4	77·3	30°
31°	90	88·8	87·7	86·5	85·3	84·2	83·0	81·9	80·6	79·5	78·3	77·1	31°
32°	90	88·8	87·6	86·5	85·3	84·1	82·9	81·7	80·5	79·4	78·2	77·0	32°
33°	90	88·8	87·6	86·4	85·2	84·0	82·8	81·6	80·4	79·2	78·0	76·8	33°
34°	90	88·8	87·6	86·4	85·2	84·0	82·7	81·5	80·3	79·1	77·9	76·7	34°
35°	90	88·8	87·5	86·3	85·1	83·9	82·7	81·4	80·2	79·0	77·8	76·5	35°
36°	90	88·8	87·5	86·3	85·0	83·8	82·6	81·3	80·1	78·8	77·6	76·3	36°
37°	90	88·7	87·5	86·2	85·0	83·7	82·5	81·2	80·0	78·7	77·4	76·2	37°
38°	90	88·7	87·5	86·2	84·9	83·6	82·4	81·1	79·8	78·5	77·3	76·0	38°
39°	90	88·7	87·4	86·1	84·8	83·6	82·3	81·0	79·7	78·4	77·1	75·8	39°
40°	90	88·7	87·4	86·1	84·8	83·5	82·1	80·8	79·5	78·2	76·9	75·6	40°
41°	90	88·7	87·3	86·0	84·7	83·4	82·0	80·7	79·4	78·0	76·7	75·3	41°
42°	90	88·6	87·3	86·0	84·6	83·3	81·9	80·6	79·2	77·8	76·5	75·1	42°
43°	90	88·6	87·3	85·9	84·5	83·1	81·8	80·4	79·0	77·6	76·3	74·9	43°
44°	90	88·6	87·2	85·8	84·4	83·0	81·6	80·2	78·8	77·4	76·0	74·6	44°
45°	90	88·6	87·2	85·7	84·3	82·9	81·5	80·1	78·6	77·2	75·8	74·3	45°
46°	90	88·6	87·1	85·7	84·2	82·8	81·3	79·9	78·4	77·0	75·5	74·0	46°
47°	90	88·5	87·1	85·6	84·1	82·6	81·2	79·7	78·2	76·7	75·2	73·7	47°
48°	90	88·5	87·0	85·5	84·0	82·5	81·0	79·5	78·0	76·5	75·0	73·4	48°
49°	90	88·5	86·9	85·4	83·9	82·4	80·8	79·3	77·7	76·2	74·6	73·1	49°
50°	90	88·4	86·9	85·3	83·8	82·2	80·6	79·1	77·5	75·9	74·3	72·7	50°
51°	90	88·4	86·8	85·2	83·6	82·0	80·4	78·8	77·2	75·6	74·0	72·4	51°
52°	90	88·4	86·7	85·1	83·5	81·9	80·2	78·6	76·9	75·3	73·6	71·9	52°
53°	90	88·3	86·7	85·0	83·3	81·7	80·0	78·3	76·6	74·9	73·2	71·5	53°
54°	90	88·3	86·6	84·9	83·2	81·5	79·8	78·0	76·3	74·6	72·8	71·1	54°
55°	90	88·2	86·5	84·8	83·0	81·3	79·5	77·7	76·0	74·2	72·4	70·6	55°
56°	90	88·2	86·4	84·6	82·8	81·0	79·2	77·4	75·6	73·8	71·9	70·0	56°
57°	90	88·2	86·3	84·5	82·6	80·8	78·9	77·0	75·2	73·3	71·4	69·5	57°
58°	90	88·1	86·2	84·3	82·4	80·5	78·6	76·7	74·8	72·8	70·9	68·9	58°
59°	90	88·1	86·1	84·2	82·2	80·3	78·3	76·3	74·3	72·3	70·3	68·3	59°
60°	90	88·0	86·0	84·0	82·0	80·0	77·9	75·9	73·8	71·8	69·7	67·6	60°

DECLINATION

LAT	12°	13°	14°	15°	16°	17°	18°	19°	20°	21°	22°	23°	LAT
30°	76·1	74·9	73·8	72·6	71·4	70·3	69·1	67·9	66·7	65·5	64·4	63·2	30°
31°	76·0	74·8	73·6	72·4	71·2	70·0	68·9	67·7	66·5	65·3	64·1	62·9	31°
32°	75·8	74·6	73·4	72·2	71·0	69·8	68·6	67·4	66·2	65·0	63·8	62·6	32°
33°	75·6	74·4	73·2	72·1	70·8	69·6	68·4	67·1	65·9	64·7	63·5	62·2	33°
34°	75·5	74·2	73·0	71·8	70·6	69·3	68·1	66·9	65·6	64·4	63·1	61·9	34°
35°	75·3	74·1	72·8	71·6	70·3	69·1	67·8	66·6	65·3	64·1	62·8	61·5	35°
36°	75·1	73·8	72·6	71·3	70·1	68·8	67·5	66·3	65·0	63·7	62·4	61·1	36°
37°	74·9	73·6	72·4	71·1	69·8	68·5	67·2	65·9	64·6	63·3	62·0	60·7	37°
38°	74·7	73·4	72·0	70·8	69·5	68·2	66·9	65·6	64·3	62·9	61·6	60·3	38°
39°	74·5	73·2	71·9	70·5	69·2	67·9	66·6	65·2	63·9	62·5	61·2	59·8	39°
40°	74·2	72·9	71·6	70·2	68·9	67·6	66·2	64·8	63·5	62·1	60·7	59·3	40°
41°	74·0	72·7	71·3	69·9	68·6	67·2	65·8	64·4	63·0	61·6	60·2	58·8	41°
42°	73·7	72·4	71·0	69·6	68·2	66·8	65·4	64·0	62·6	61·2	59·7	58·3	42°
43°	73·5	72·1	70·7	69·3	67·9	66·4	65·0	63·6	62·1	60·7	59·2	57·7	43°
44°	73·2	71·8	70·3	68·9	67·5	66·0	64·6	63·1	61·6	60·1	58·6	57·1	44°
45°	72·9	71·4	70·0	68·5	67·0	65·6	64·1	62·6	61·1	59·5	58·0	56·4	45°
46°	72·6	71·1	69·6	68·1	66·6	65·1	63·6	62·0	60·5	58·9	57·4	55·8	46°
47°	72·2	70·7	69·2	67·7	66·2	64·6	63·1	61·5	59·9	58·3	56·7	55·0	47°
48°	71·9	70·3	68·8	67·2	65·7	64·1	62·5	60·9	59·3	57·6	55·9	54·3	48°
49°	71·5	69·9	68·4	66·8	65·1	63·5	61·9	60·2	58·6	56·9	55·2	53·4	49°
50°	71·1	69·5	67·9	66·2	64·6	62·9	61·3	59·6	57·8	56·1	54·3	52·6	50°
51°	70·7	69·1	67·4	65·7	64·0	62·3	60·6	58·8	57·1	55·3	53·5	51·6	51°
52°	70·3	68·6	66·9	65·1	63·4	61·6	59·9	58·1	56·3	54·4	52·5	50·6	52°
53°	69·8	68·1	66·3	64·5	62·7	60·9	59·1	57·3	55·4	53·5	51·5	49·5	53°
54°	69·3	67·5	65·7	63·9	62·0	60·2	58·3	56·4	54·4	52·4	50·4	48·3	54°
55°	68·7	66·9	65·1	63·2	61·3	59·4	57·4	55·4	53·4	51·3	49·2	47·1	55°
56°	68·2	66·3	64·4	62·4	60·5	58·5	56·5	54·4	52·3	50·1	47·9	45·7	56°
57°	67·6	65·6	63·6	61·6	59·6	57·5	55·4	53·3	51·1	48·9	46·5	44·2	57°
58°	66·9	64·9	62·8	60·8	58·7	56·5	54·3	52·1	49·8	47·4	45·0	42·5	58°
59°	66·2	64·1	62·0	59·8	57·6	55·4	53·1	50·8	48·4	45·9	43·3	40·7	59°
60°	65·4	63·3	61·1	58·8	56·5	54·2	51·8	49·4	46·8	44·2	41·5	38·6	60°

Table 3(6) – Distance off lights when rising or dipping
Enter with height of eye and height of light to extract the range (M) of a light when it first rises above, or finally dips below, the horizon. This range, with a bearing, gives a fix.

TABLE 3(6) LIGHTS – DISTANCE OFF (M) WHEN RISING OR DIPPING

HEIGHT OF LIGHT		HEIGHT OF EYE									
metres	feet	metres	1	2	3	4	5	6	7	8	9	10
metres	feet	feet	3	7	10	13	16	20	23	26	30	33
10	33		8·7	9·5	10·2	10·8	11·3	11·7	12·1	12·5	12·8	13·2
12	39		9·3	10·1	10·8	11·4	11·9	12·3	12·7	13·1	13·4	13·8
14	46		9·9	10·7	11·4	12·0	12·5	12·9	13·3	13·7	14·0	14·4
16	53		10·4	11·2	11·9	12·5	13·0	13·4	13·8	14·2	14·5	14·9
18	59		10·9	11·7	12·4	13·0	13·5	13·9	14·3	14·7	15·0	15·4
20	66		11·4	12·2	12·9	13·5	14·0	14·4	14·8	15·2	15·5	15·9
22	72		11·9	12·7	13·4	14·0	14·5	14·9	15·3	15·7	16·0	16·4
24	79		12·3	13·1	13·8	14·4	14·9	15·3	15·7	16·1	16·4	17·0
26	85		12·7	13·5	14·2	14·8	15·3	15·7	16·1	16·5	16·8	17·2
28	92		13·1	13·9	14·6	15·2	15·7	16·1	16·5	16·9	17·2	17·6
30	98		13·5	14·3	15·0	15·6	16·1	16·5	16·9	17·3	17·6	18·0
32	105		13·9	14·7	15·4	16·0	16·5	16·9	17·3	17·7	18·0	18·4
34	112		14·2	15·0	15·7	16·3	16·8	17·2	17·6	18·0	18·3	18·7
36	118		14·6	15·4	16·1	16·7	17·2	17·6	18·0	18·4	18·7	19·1
38	125		14·9	15·7	16·4	17·0	17·5	17·9	18·3	18·7	19·0	19·4
40	131		15·3	16·1	16·8	17·4	17·9	18·3	18·7	19·1	19·4	19·8
42	138		15·6	16·4	17·1	17·7	18·2	18·6	19·0	19·4	19·7	20·1
44	144		15·9	16·7	17·4	18·0	18·5	18·9	19·3	19·7	20·0	20·4
46	151		16·2	17·0	17·7	18·3	18·8	19·2	19·6	20·0	20·3	20·7
48	157		16·5	17·3	18·0	18·6	19·1	19·5	19·9	20·3	20·6	21·0
50	164		16·8	17·6	18·3	18·9	19·4	19·8	20·2	20·6	20·9	21·3
55	180		17·5	18·3	19·0	19·6	20·1	20·5	20·9	21·3	21·6	22·0
60	197		18·2	19·0	19·7	20·3	20·8	21·2	21·6	22·0	22·3	22·7
65	213		18·9	19·7	20·4	21·0	21·5	21·9	22·3	22·7	23·0	23·4
70	230		19·5	20·3	21·0	21·6	22·1	22·5	22·9	23·2	23·6	24·0
80	262		20·7	21·5	22·2	22·8	23·3	23·7	24·1	24·5	24·8	25·2

3.16 SUNRISE/SET TIMES 2016

Showing times of Sunrise (SR) and Sunset (SS) for every 3rd day, the table below is much simpler than in earlier editions.

This does not imply any lack of accuracy since the times of Sunrise and Sunset never change by more than 8 minutes (and often by only 1–3 minutes) between the given dates.

The tables are based on Longitude 0°, so longitude corrections are required: add 4 minutes of time for every degree West of Greenwich; subtract if East.

Latitudes 56°N, 48°N and 40°N are chosen so as to split into 3 equal bands the 24° span of latitude from Shetland 60°N to Gibraltar 36°N. Be sure to enter the correct table.

Times are in UT – add 1 hour in non-shaded months to convert to BST

LATITUDE 56°N

	Rise	Set	Rise	Set	Rise	Set	Rise	Set	Rise	Set	Rise	Set
	JANUARY		FEBRUARY		MARCH		APRIL		MAY		JUNE	
1	08 31	15 36	07 56	16 32	06 51	17 35	05 30	18 39	04 15	19 40	03 22	20 35
4	08 30	15 39	07 50	16 38	06 43	17 41	05 22	18 45	04 09	19 46	03 19	20 39
7	08 29	15 44	07 44	16 45	06 35	17 48	05 14	18 51	04 02	19 52	03 16	20 42
10	08 27	15 48	07 38	16 52	06 28	17 54	05 06	18 57	03 56	19 58	03 15	20 45
13	08 24	15 53	07 31	16 58	06 20	18 00	04 59	19 04	03 50	20 04	03 14	20 47
16	08 21	15 59	07 24	17 05	06 12	18 06	04 51	19 10	03 45	20 09	03 13	20 49
19	08 17	16 05	07 17	17 11	06 04	18 13	04 44	19 16	03 40	20 15	03 13	20 50
22	08 13	16 11	07 10	17 18	05 56	18 19	04 36	19 22	03 35	20 20	03 13	20 51
25	08 08	16 17	07 03	17 24	05 48	18 25	04 29	19 28	03 30	20 25	03 15	20 51
28	08 03	16 23	06 56	17 31	05 40	18 31	04 22	19 34	03 26	20 29	03 16	20 50
31	07 58	16 30			05 32	18 37			03 23	20 34		

	JULY		AUGUST		SEPTEMBER		OCTOBER		NOVEMBER		DECEMBER	
1	03 19	20 49	04 05	20 07	05 05	18 53	06 04	17 34	07 08	16 19	08 07	15 31
4	03 21	20 47	04 10	20 01	05 11	18 46	06 10	17 26	07 14	16 12	08 12	15 28
7	03 25	20 45	04 16	19 54	05 17	18 38	06 16	17 19	07 20	16 06	08 16	15 27
10	03 28	20 42	04 22	19 47	05 22	18 30	06 22	17 11	07 27	16 00	08 20	15 25
13	03 32	20 38	04 28	19 41	05 28	18 22	06 28	17 03	07 33	15 55	08 24	15 25
16	03 37	20 34	04 34	19 33	05 34	18 14	06 34	16 56	07 39	15 50	08 26	15 25
19	03 42	20 30	04 39	19 26	05 40	18 06	06 40	16 49	07 45	15 45	08 29	15 26
22	03 47	20 25	04 45	19 19	05 46	17 58	06 47	16 41	07 51	15 41	08 30	15 27
25	03 52	20 20	04 51	19 11	05 52	17 50	06 53	16 34	07 57	15 37	08 31	15 29
28	03 57	20 15	04 57	19 04	05 58	17 42	06 59	16 27	08 02	15 33	08 32	15 32
31	04 03	20 09	05 03	18 56			07 06	16 21			08 31	15 35

Times are in UT – add 1 hour in non-shaded months to convert to BST

LATITUDE 48°N

	JANUARY Rise	Set	FEBRUARY Rise	Set	MARCH Rise	Set	APRIL Rise	Set	MAY Rise	Set	JUNE Rise	Set
1	07 50	16 17	07 29	16 59	06 40	17 45	05 38	18 31	04 41	19 14	04 05	19 52
4	07 50	16 20	07 24	17 04	06 35	17 49	05 32	18 35	04 37	19 18	04 03	19 54
7	07 49	16 23	07 20	17 09	06 29	17 54	05 26	18 39	04 32	19 22	04 01	19 57
10	07 48	16 27	07 16	17 13	06 23	17 59	05 20	18 44	04 28	19 26	04 01	19 59
13	07 47	16 31	07 11	17 18	06 17	18 03	05 14	18 48	04 24	19 30	04 00	20 01
16	07 45	16 35	07 06	17 23	06 11	18 07	05 08	18 52	04 20	19 34	04 00	20 02
19	07 42	16 39	07 01	17 28	06 04	18 12	05 02	18 57	04 16	19 38	04 00	20 03
22	07 40	16 44	06 55	17 33	05 58	18 16	04 57	19 01	04 13	19 41	04 01	20 03
25	07 37	16 48	06 50	17 37	05 52	18 21	04 52	19 05	04 10	19 45	04 02	20 04
28	07 33	16 53	06 44	17 42	05 46	18 25	04 46	19 09	04 07	19 48	04 03	20 03
31	07 30	16 57			05 40	18 29			04 05	19 51		

	JULY Rise	Set	AUGUST Rise	Set	SEPTEMBER Rise	Set	OCTOBER Rise	Set	NOVEMBER Rise	Set	DECEMBER Rise	Set
1	04 05	20 03	04 37	19 35	05 18	18 40	06 00	17 39	06 45	16 41	07 29	16 09
4	04 07	20 02	04 40	19 31	05 22	18 34	06 04	17 33	06 50	16 37	07 33	16 08
7	04 09	20 00	04 44	19 26	05 27	18 28	06 08	17 26	06 55	16 32	07 36	16 07
10	04 12	19 59	04 48	19 21	05 31	18 22	06 12	17 21	06 59	16 28	07 39	16 07
13	04 14	19 57	04 52	19 16	05 35	18 16	06 17	17 15	07 04	16 24	07 42	16 07
16	04 18	19 54	04 57	19 11	05 39	18 10	06 21	17 09	07 08	16 21	07 44	16 07
19	04 21	19 51	05 01	19 05	05 43	18 03	06 26	17 03	07 13	16 18	07 46	16 08
22	04 24	19 48	05 05	19 00	05 47	17 57	06 30	16 58	07 17	16 15	07 48	16 10
25	04 28	19 44	05 09	18 54	05 51	17 51	06 35	16 53	07 21	16 13	07 49	16 12
28	04 31	19 41	05 13	18 48	05 55	17 45	06 39	16 48	07 25	16 11	07 50	16 14
31	04 35	19 37	05 17	18 42			06 44	16 43			07 50	16 17

LATITUDE 40°N

	JANUARY Rise	Set	FEBRUARY Rise	Set	MARCH Rise	Set	APRIL Rise	Set	MAY Rise	Set	JUNE Rise	Set
1	07 22	16 45	07 09	17 18	06 33	17 52	05 44	18 25	05 00	18 55	04 33	19 23
4	07 22	16 48	07 06	17 22	06 28	17 55	05 39	18 28	04 56	18 58	04 32	19 25
7	07 22	16 50	07 03	17 26	06 24	17 59	05 34	18 31	04 53	19 01	04 31	19 27
10	07 22	16 53	07 00	17 29	06 19	18 02	05 29	18 34	04 49	19 04	04 31	19 28
13	07 21	16 56	06 56	17 33	06 14	18 05	05 25	18 37	04 46	19 07	04 31	19 30
16	07 20	17 00	06 52	17 36	06 09	18 08	05 20	18 40	04 44	19 10	04 31	19 31
19	07 18	17 03	06 49	17 40	06 05	18 11	05 16	18 43	04 41	19 12	04 31	19 32
22	07 17	17 06	06 44	17 43	06 00	18 14	05 12	18 46	04 39	19 15	04 32	19 33
25	07 15	17 10	06 40	17 47	05 55	18 17	05 07	18 49	04 37	19 18	04 32	19 33
28	07 13	17 14	06 36	17 50	05 50	18 21	05 04	18 52	04 35	19 20	04 34	19 33
31	07 10	17 17			05 45	18 24			04 34	19 22		

	JULY Rise	Set	AUGUST Rise	Set	SEPTEMBER Rise	Set	OCTOBER Rise	Set	NOVEMBER Rise	Set	DECEMBER Rise	Set
1	04 35	19 33	04 59	19 13	05 28	18 31	05 57	17 42	06 29	16 57	07 03	16 35
4	04 37	19 32	05 01	19 10	05 31	18 26	05 59	17 37	06 33	16 54	07 06	16 35
7	04 38	19 31	05 04	19 07	05 34	18 21	06 03	17 32	06 36	16 51	07 09	16 35
10	04 40	19 30	05 07	19 03	05 36	18 16	06 06	17 28	06 40	16 48	07 11	16 35
13	04 43	19 29	05 10	18 59	05 39	18 12	06 09	17 23	06 43	16 45	07 13	16 35
16	04 45	19 27	05 13	18 55	05 42	18 07	06 12	17 19	06 47	16 43	07 16	16 36
19	04 47	19 25	05 16	18 51	05 45	18 02	06 15	17 14	06 50	16 41	07 17	16 37
22	04 50	19 23	05 19	18 46	05 48	17 57	06 18	17 10	06 53	16 39	07 19	16 39
25	04 52	19 20	05 21	18 42	05 51	17 52	06 22	17 06	06 57	16 37	07 20	16 40
28	04 55	19 17	05 24	18 37	05 54	17 47	06 25	17 02	07 00	16 36	07 21	16 42
31	04 58	19 14	05 27	18 33			06 28	16 58			07 22	16 45

3.17   MOONRISE/SET TIMES 2016

The table below gives the times of Moonrise (MR) and Moonset (MS) for every 3rd day; interpolation is necessary for other days. The aim is simply to indicate whether the night in question will be brightly moonlit, partially moonlit or pitch black – depending, of course, on cloud cover.

The table is based on Longitude 0°. To correct for longitude, add 4 minutes of time for every degree West; subtract if East.

Latitudes 56°N, 48°N and 40°N are chosen so as to split into three equal bands the 24° span of latitude from Shetland 60°N to Gibraltar 36°N. Be sure to enter the correct table.

** Indicates that the phenomenon does not occur.

The phase of the Moon, ie New Moon, 1st Quarter, Full Moon and Last Quarter, are shown on Tide tables and in detail on the penultimate page of Chapter 4 (Tides).

Times are in UT – add 1 hour in non-shaded months to convert to BST

LATITUDE 56°N

	Rise	Set	Rise	Set	Rise	Set	Rise	Set	Rise	Set	Rise	Set
	JANUARY		FEBRUARY		MARCH		APRIL		MAY		JUNE	
1	** **	11 25	01 04	10 55	00 58	09 55	02 26	11 01	02 12	12 19	01 57	15 24
4	02 15	12 27	04 13	12 39	03 47	12 14	04 11	14 43	03 30	16 26	03 29	19 30
7	05 30	14 02	06 42	15 43	05 45	15 52	05 31	19 00	04 58	20 40	06 07	22 27
10	08 09	16 53	08 16	19 46	07 07	20 07	07 07	23 06	07 24	23 50	09 37	23 59
13	09 48	20 45	09 32	23 53	08 35	** **	09 40	01 09	10 45	01 04	13 03	00 40
16	11 01	** **	11 13	02 24	10 50	02 22	12 58	03 00	14 09	02 14	16 20	01 40
19	12 31	03 22	13 55	05 14	13 59	04 28	16 19	04 07	17 27	03 13	19 28	03 06
22	15 01	06 32	17 17	06 51	17 21	05 41	19 36	05 07	20 37	04 31	21 48	05 36
25	18 23	08 23	20 39	07 55	20 40	06 40	22 42	06 30	23 08	06 44	23 16	09 09
28	21 47	09 30	23 55	08 58	23 50	07 56	00 25	08 51	00 17	10 04	00 03	13 06
31	** **	10 31			01 40	10 02			01 33	14 00		

	Rise	Set	Rise	Set	Rise	Set	Rise	Set	Rise	Set	Rise	Set
	JULY		AUGUST		SEPTEMBER		OCTOBER		NOVEMBER		DECEMBER	
1	01 26	17 07	02 37	18 55	05 02	18 54	06 17	18 00	08 31	17 38	09 21	17 32
4	03 48	20 18	06 07	20 29	08 30	19 55	09 36	19 07	11 23	19 36	11 25	20 29
7	07 16	22 02	09 38	21 31	11 47	21 06	12 35	20 53	13 21	22 41	12 45	** **
10	10 47	23 06	12 56	22 37	14 43	23 01	14 47	23 48	14 40	01 11	13 55	02 51
13	14 06	** **	16 00	** **	16 51	01 00	16 15	02 19	15 57	05 23	15 42	07 02
16	17 16	01 05	18 21	02 13	18 16	04 49	17 32	06 32	17 59	09 31	18 52	10 07
19	19 47	03 25	19 50	05 55	19 34	09 02	19 22	10 42	21 16	12 12	22 34	11 40
22	21 22	06 56	21 05	10 00	21 30	12 57	22 21	13 35	** **	13 34	00 53	12 40
25	22 34	10 53	22 45	13 56	** **	15 33	00 41	15 05	03 05	14 33	04 12	13 45
28	** **	14 52	00 28	16 52	02 50	16 59	04 08	16 06	06 23	15 40	07 15	15 28
31	01 37	18 09	03 51	18 31			07 27	17 10			09 29	18 19

LATITUDE 48°N

	Rise	Set	Rise	Set	Rise	Set	Rise	Set	Rise	Set	Rise	Set
	JANUARY		FEBRUARY		MARCH		APRIL		MAY		JUNE	
1	** **	11 25	00 46	11 15	00 32	10 23	01 55	11 31	01 53	12 35	02 03	15 14
4	02 00	12 44	03 42	13 11	03 15	12 46	03 56	14 56	03 31	16 20	03 55	19 00
7	05 00	14 33	06 14	16 10	05 26	16 09	05 37	18 50	05 21	20 13	06 39	21 57
10	07 39	17 22	08 07	19 51	07 09	20 01	07 33	22 37	07 56	23 19	09 54	23 46
13	09 35	20 55	09 45	23 36	08 57	23 45	10 11	00 38	11 05	00 42	13 01	00 39
16	11 09	** **	11 43	01 56	11 22	01 51	13 15	02 41	14 11	02 09	16 01	01 56
19	12 58	02 57	14 24	04 44	14 21	04 04	16 18	04 04	17 11	03 25	18 55	03 37
22	15 33	06 00	17 31	06 35	17 25	05 34	19 18	05 21	20 07	04 59	21 20	06 07
25	18 40	08 04	20 35	07 56	20 26	06 51	22 11	06 59	22 38	07 17	23 05	09 24
28	21 46	09 28	23 33	09 16	23 21	08 22	** **	09 22	** **	10 24	00 06	12 59
31	** **	10 46			01 07	10 34			01 31	13 58		

	Rise	Set	Rise	Set	Rise	Set	Rise	Set	Rise	Set	Rise	Set
	JULY		AUGUST		SEPTEMBER		OCTOBER		NOVEMBER		DECEMBER	
1	01 49	16 40	03 09	18 26	05 15	18 45	06 14	18 06	08 06	18 04	08 47	18 05
4	04 21	19 47	06 24	20 15	08 24	20 03	09 15	19 30	10 49	20 09	10 58	20 54
7	07 36	21 46	09 35	21 36	11 24	21 30	12 02	21 26	12 56	23 04	12 36	** **
10	10 48	23 08	12 36	22 58	14 10	23 34	14 20	** **	14 34	01 21	14 07	02 42
13	13 49	** **	15 28	00 10	16 26	01 28	16 06	02 32	16 13	05 09	16 13	06 32
16	16 45	01 34	17 53	02 43	18 10	04 59	17 44	06 22	18 32	08 59	19 21	09 36
19	19 17	03 57	19 41	06 08	19 50	08 49	19 53	10 12	21 42	11 45	22 44	11 27
22	21 09	07 13	21 16	09 51	22 01	12 26	22 50	13 05	** **	13 25	00 50	12 46
25	22 41	10 49	23 14	13 28	** **	15 05	00 58	14 52	03 00	14 41	03 51	14 07
28	** **	14 27	01 01	16 22	03 06	16 47	04 06	16 10	06 00	16 05	06 42	16 02
31	02 09	17 37	04 10	18 16			07 07	17 31			09 00	18 47

Times are in UT – add 1 hour in non-shaded months to convert to BST

LATITUDE 40°N

	Rise JANUARY	Set	Rise FEBRUARY	Set	Rise MARCH	Set	Rise APRIL	Set	Rise MAY	Set	Rise JUNE	Set
1	23 57	11 26	00 32	11 30	00 13	10 42	01 34	11 52	01 40	12 47	02 07	15 07
4	01 48	12 57	03 20	13 34	02 53	13 08	03 44	15 06	03 33	16 16	04 13	18 39
7	04 39	14 55	05 55	16 28	05 12	16 20	05 41	18 43	05 38	19 54	07 01	21 37
10	07 18	17 42	08 01	19 55	07 11	19 56	07 51	22 16	08 19	22 58	10 06	23 36
13	09 26	21 02	09 54	23 24	09 14	23 27	10 33	00 15	11 20	00 26	13 00	00 38
16	11 15	** **	12 03	01 36	11 44	01 28	13 28	02 26	14 12	02 06	15 47	02 08
19	13 17	02 39	14 45	04 23	14 36	03 48	16 18	04 02	16 59	03 34	18 33	03 58
22	15 54	05 38	17 41	06 24	17 28	05 29	19 05	05 32	19 45	05 18	21 00	06 29
25	18 53	07 50	20 32	07 57	20 15	06 59	21 49	07 19	22 16	07 39	22 58	09 34
28	21 45	09 27	23 18	09 29	23 00	08 40	** **	09 44	** **	10 37	00 08	12 54
31	** **	10 57			00 45	10 57			01 30	13 57		

	Rise JULY	Set	Rise AUGUST	Set	Rise SEPTEMBER	Set	Rise OCTOBER	Set	Rise NOVEMBER	Set	Rise DECEMBER	Set
1	02 05	16 21	03 31	18 05	05 25	18 38	06 11	18 11	07 49	18 22	08 23	18 29
4	04 44	19 25	06 36	20 06	08 20	20 09	09 00	19 46	10 26	20 32	10 39	21 12
7	07 50	21 34	09 34	21 39	11 07	21 47	11 40	21 49	12 39	23 20	12 29	** **
10	10 49	23 09	12 22	23 14	13 47	23 57	14 01	** **	14 30	01 28	14 16	02 36
13	13 37	00 09	15 06	00 33	16 08	01 48	15 59	02 42	16 24	05 00	16 35	06 11
16	16 23	01 54	17 34	03 04	18 06	05 06	17 54	06 15	18 55	08 37	19 41	09 15
19	18 56	04 20	19 35	06 18	20 02	08 39	20 15	09 51	22 00	11 26	22 51	11 18
22	20 59	07 25	21 25	09 45	22 23	12 04	23 10	12 44	00 03	13 17	00 48	12 50
25	22 47	10 46	23 35	13 08	00 17	14 46	01 11	14 42	02 56	14 47	03 36	14 24
28	00 04	14 09	01 23	16 00	03 17	16 39	04 05	16 13	05 43	16 23	06 19	16 25
31	02 32	17 14	04 24	18 04			06 54	17 47			08 40	19 06

3.18 PHASES OF THE MOON 2016 which principally determine Springs, Neaps and tide times (UT)

New Moon ●	d	h	m	First Quarter ◐	d	h	m	Full Moon ○	d	h	m	Last Quarter ◑	d	h	m
Jan	10	01	31	Jan	16	23	26	Jan	24	01	46	Feb	1	03	28
Feb	8	14	39	Feb	15	07	46	Feb	22	18	20	Mar	1	23	11
Mar	9	01	54	Mar	15	17	03	Mar	23	12	01	Mar	31	15	17
Apr	7	11	24	Apr	14	03	59	Apr	22	05	24	Apr	30	03	29
May	6	19	30	May	13	17	02	May	21	21	14	May	29	12	12
June	5	03	00	June	12	08	10	June	20	11	02	June	27	18	19
July	4	11	01	July	12	00	52	July	19	22	57	July	26	23	00
Aug	2	20	45	Aug	10	18	21	Aug	18	09	27	Aug	25	03	41
Sept	1	09	03	Sept	9	11	49	Sept	16	19	05	Sept	23	09	56
Oct	1	00	11	Oct	9	04	33	Oct	16	04	23	Oct	22	19	14
Oct	30	17	38	Nov	7	19	51	Nov	14	13	52	Nov	21	08	33
Nov	29	12	18	Dec	7	09	03	Dec	14	00	06	Dec	21	01	56
Dec	29	06	53												

4

Tides

4.1   Introduction   54

Tide tables • Spanish Secondary Ports

4.2   Definitions   54

CD • LAT • HAT • OD (Newlyn) • Charted depth • Drying height •
Vertical clearances • Elevation of lights • Height of tide • Rise and Fall
of tide • Duration • Interval • Springs and Neaps • MHWS, MHWN •
MLWS, MLWN • MSL/ML • Range • Standard and Secondary Ports •
Tidal Coefficients

4.3   Calculating times and heights of High and Low Water   55

Standard and Secondary Ports: times of HW and LW • Secondary Ports:
heights of HW and LW • Graphical interpolation

4.4   Calculating intermediate times and heights of tide   56

Standard and Secondary Ports • Factors • The Rule of Twelfths • Co-Tidal
and Co-Range charts

4.5   Calculating clearances below overhead objects   58

4.6   Tidal prediction by computer   58

EasyTide   • TotalTide • Commercial software

4.7   Tidal streams   59

Tidal stream atlases • Tidal stream diamonds • Flood streams around UK •
Calculating rates • Tidal streams in rivers

4.8   Meteorological effects   61

Wind • Barometric pressure • Storm warning services

Graph for calculating tidal stream rates   62

4.1 INTRODUCTION

This chapter provides background information on the tidal data in Chapter 9 of this Almanac, where times and heights of HW and LW are shown for Standard Ports, and time and height differences for their associated Secondary Ports. Tides are predicted for average meteorological conditions. In the UK the average pressure is about 1013mb. A difference of 34mb can cause the tide to rise (lower pressure) or fall (higher pressure) by about 0·3m. See 4.8 for more details.

4.1.1   Admiralty Tide Tables (ATT) are the source for all tidal data in this Almanac; they are published in 4 volumes:

- Vol 1 (NP 201) UK and Ireland (including Channel ports from Hoek van Holland to Brest);
- Vol 2 (NP 202) Europe (excluding UK, Ireland and Channel ports), Mediterranean and the Atlantic;
- Vol 3 (NP 203) Indian Ocean and South China Sea;
- Vol 4 (NP 204) Pacific Ocean.

4.1.2   Spanish Secondary Ports referenced to Lisboa
In Areas 23 and 25, some Spanish secondary ports have Lisboa as Standard Port. Time differences for these ports, when applied to the printed times of HW and LW for Lisboa (UT), automatically give HW and LW times in the Zone Time for Spain (ie UT –1), *not* Portugal (UT). No other corrections are required, except for DST when applicable.

4.2 DEFINITIONS

Chart Datum (CD)
CD is the reference level from which heights of tide are predicted and charted depths are measured. In the UK it normally approximates to LAT, and the tide will not frequently fall below it. The actual depth of water in any particular position is the charted depth plus the height of tide.

Lowest Astronomical Tide (LAT)
LAT is the lowest level which can be predicted under average meteorological, and any combination of astronomical, conditions. This level will not be reached every year. Storm surges can cause even lower levels to be reached.

Highest Astronomical Tide (HAT)
HAT is the highest level which can be predicted to occur under average meteorological conditions and under any combination of astronomical conditions, except storm surges. It is the level above which vertical clearances under bridges and power lines are measured; see 4.5.

Ordnance Datum (Newlyn)
Ordnance Datum (Newlyn) is the datum of the land levelling system on mainland England, Scotland and Wales, and to which all features on UK land maps are referred. The difference between Ordnance Datum (Newlyn) and CD is shown at the foot of each page of tide tables in this Almanac. Differences between CD and foreign land levelling datums are similarly quoted.

Charted depth
Charted depths are printed on charts in metres and decimetres (0·1m) and show the depth of water below CD. (Not to be confused with a sounding which is the actual depth of water (charted depth + height of tide) in a particular position.)

Drying height
A drying height is the height above CD of any feature which at times is covered by water. The figures, in metres and decimetres, are underlined on the chart. The depth of water over a drying feature is the height of tide minus the drying height. If the result is negative, then the feature is uncovered at that time.

Vertical clearances under bridges and power lines
These are measured above HAT. Some older charts may still show clearances above MHWS; see 4.5.

Elevation of lights
The charted height of a light (its elevation) is measured above MHWS.

Height of tide
The height of the tide is the vertical distance of the sea level above (or very occasionally below) CD. Predicted heights are given in metres and decimetres.

Rise/Fall of tide
The Rise of the tide is the amount the tide has risen since the earlier Low Water. The Fall of a tide is the amount the tide has fallen since the last High Water.

Duration
Duration is the time between LW and the next HW, normally slightly more than six hours. It can be used to find the approximate time of LW when only the time of HW is known.

Interval
The interval is a period of time quoted in hours and minutes before (–) or after (+) HW. Intervals are printed in hourly increments (–6hrs to +6hrs) along the bottom of each tidal curve diagram in Chapter 9.

Spring tides
Spring tides occur roughly every 16 days, near to Full ○ and New ● Moon, when the tide-raising forces of Sun and Moon are at a maximum. See 3.18, for phases of the Moon.

Neap tides
Neaps occur roughly every 16 days, near the Moon's first ◑ and last ◐ quarters, when the tide-raising forces of Sun and Moon are at a minimum. See 3.18 for phases of the Moon.

Mean High Water and Low Water Springs/Neaps
MHWS and MHWN are the means of predicted HW heights of Sp or Np tides over a period of 18·6 years. Similarly, MLWS and MLWN are the means of LW heights for Sp and Np tides respectively. Mean tide level (MTL) is the mean of the above values.

Mean Sea Level (MSL or ML)
This is the average level of the sea's surface over a long period, preferably 18·6 years.

Range
The range of a tide is the difference between the heights of successive HWs and LWs. Spring range is the difference between MHWS and MLWS, and Neap range is the difference between MHWN and MLWN.

Standard Ports have tidal characteristics observed over a long period and are suitable as a reference for secondary ports on the adjacent coasts.

Secondary Ports have similar tidal characteristics to those of their Standard Port. Time and height differences are applied to the Standard Port predictions. 'Secondary' does not imply lesser importance.

Tidal Coefficients indicate the range of a tide. Daily values are listed and explained in 9.0.9.

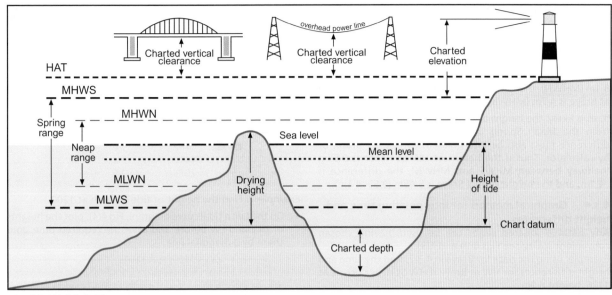

Fig 4(1) Tidal definitions

4.3 CALCULATING TIMES AND HEIGHTS OF HIGH AND LOW WATER

4.3.1 Standard Ports

The Standard Ports for which daily predictions are given in this Almanac are listed below by geographic Areas. Not all ports shown in the ATTs are included, but adjustments have been made to Secondary Port differences where necessary.

1	Falmouth, Plymouth, Dartmouth, Portland.
2	Poole, Southampton, Portsmouth, Chichester.
3	Shoreham, Dover.
4	Margate, Sheerness, London Bridge, Walton-on-the-Naze, Harwich, Lowestoft.
5	Immingham, R Tees, Tyne (North Shields).
6	Leith, Aberdeen.
7	Invergordon, Wick, Lerwick.
8	Stornoway, Ullapool, Oban.
9	Greenock.
10	Liverpool, Holyhead.
11	Milford Haven, Bristol (Avonmouth).
12	Dublin, Cobh, Tarbert Island.
13	Galway, River Foyle, Galway.
14	Esbjerg.
15	Helgoland, Cuxhaven and Wilhelmshaven.
16	Hoek van Holland, Vlissingen, Zeebrugge.
17	Dunkerque, Dieppe, Le Havre, Cherbourg.
18	St Malo.
19	St Peter Port, St Helier.
20	Brest.
22	Pointe de Grave.
23	La Coruña.
24	Lisboa.
25	Gibraltar.
26	Ponta Delgada, Casablanca.

Predicted times and heights of HW and LW are tabulated for each Standard Port. Note that these are only predictions and take no account of the effects of wind and barometric pressure (see 4.8). See 1.5 for Zone times and Daylight Saving Time (DST).

4.3.2 Secondary Ports – times of HW and LW

Each Secondary Port listed in Chapter 9 has a data block for calculating times of HW and LW. The following example is for Braye (Alderney):

TIDES –0400 Dover; ML 3·5; Duration 0545
Standard Port ST HELIER (⟶)

Times				Height (metres)			
High Water		Low Water		MHWS	MHWN	MLWN	MLWS
0300	0900	0200	0900	11·0	8·1	4·0	1·4
1500	2100	1400	2100				
Differences BRAYE							
+0050	+0040	+0025	+0105	–4·8	–3·4	–1·5	–0·5

In this example –0400 Dover means that, on average, HW Braye is 4 hours 00 minutes before HW Dover (the Range and times of HW Dover are in Area 3 and on the bookmark).

ML (or MSL) is defined in 4.2.

Duration 0545 means that LW Braye occurs 5 hours and 45 minutes before the next HW.

The arrow (⟶) after the Standard Port's name points to where the tide tables are in the book.

The most accurate method of prediction uses Standard Port times and Secondary Port time differences as in the block. When HW at St Helier occurs at 0300 and 1500, the difference is +0050, and thus HW at Braye occurs at 0350 and 1550. When HW at St Helier occurs at 0900 and 2100, the difference is +0040, and HW at Braye occurs at 0940 and 2140.

If, as will usually be the case, HW St Helier occurs at some other time, then the difference for Braye must be found by interpolation, either by eye, by the graphical method or by calculator.

So when HW St Helier occurs at 1200 (midway between 0900 and 1500), the difference is +0045 (midway between +0040 and +0050), and therefore HW Braye is 1245.

The same method is used for calculating the times of LW. Times thus obtained are in the Secondary Port's Zone Time. For calculating heights of HW and LW see 4.3.3.

4.3.3 Secondary Ports – heights of HW and LW

The Secondary Port data block also contains height differences which are applied to the heights of HW and LW at the Standard Port. When the height of HW at St Helier is 11·0m (MHWS), the difference is –4·8m, so the height of HW Braye is 6·2m (MHWS). When the height of HW St Helier is 8·1m (MHWN), the difference is –3·4m, and the height of HW at Braye is 4·7m (MHWN).

If, as is likely, the height of tide at the Standard Port differs from the Mean Spring or Neap level, then the height difference must also be interpolated: by eye, by graph or by calculator. Thus if the height of HW St Helier is 9·55m (halfway between MHWS and MHWN), the difference is –4·1m, and the height of HW Braye is 5·45m (9·55–4·1m).

4.3.4 Graphical method for interpolating time and height differences

Any suitable squared paper can be used, having chosen convenient scales; see Figs 4(2) and 4(3).
Example: using the data for Braye in 4.3.2, find the time and height differences for HW at Braye if HW St Helier occurs at 1126, height 8·9m.

Time difference

On the horizontal axis of Fig 4(2) select a scale for the time at St Helier covering 0900 to 1500 (for which the relevant time differences for Braye are known). On the vertical axis the scale must cover +0040 to +0050, the time differences given for 0900 and 1500.

Plot point A, the time difference (+0040) for HW St Helier at 0900; and point B, the time difference (+0050) for HW St Helier at 1500. Join AB. Enter the graph at time 1126 (HW St Helier); intersect AB at C then go horizontally to read +0044 on the vertical axis. On that morning HW Braye is 44 minutes after HW St Helier, ie 1210.

Height difference

In Fig 4(3) the horizontal axis covers the height of HW at St Helier (ie 8·1 to 11·0m) and the vertical axis shows the relevant height differences (–3·4 to –4·8m).

Plot point D, the height difference (–3·4) at Neaps when the height of HW St Helier is 8·1m; and E, the height difference (–4·8m) at Springs when the height of HW St Helier is 11·0m. Join DE.

Enter the graph at 8·9m (the height of HW St Helier that morning) and mark F where that height meets DE. From F follow the horizontal line to read off the corresponding height difference: –3·8m. So that morning the height of HW Braye is 5·1m.

Fig 4(2)

4.4 CALCULATING INTERMEDIATE TIMES AND HEIGHTS OF TIDE

4.4.1 Standard Ports

Intermediate heights and times are best calculated from the Mean Spring and Neap Curves for Standard Ports in Chapter 9. Examples below are for Leith, on a day when the predictions are:

		UT	Ht (m)
22		0202	5·3
		0752	1·0
	TU	1417	5·4
		2025	0·5

Example 1: Find the height of tide at Leith at 1200.

• On the Leith tidal curve diagram, Fig 4(4), plot the heights of HW and LW before and after the required time. Join them by a diagonal line.

Fig 4(3)

• Enter the HW time and other times as necessary in the boxes below the curves.

• From the required time, go vertically to the curves. The Spring curve is red, and the Neap curve (where it differs) is blue. Interpolate between the curves by comparing the actual range, in this example 4·4m, with the Mean Ranges printed beside the curves. Never extrapolate. Here the Spring curve applies.

• Go horizontally to the diagonal line first plotted, thence vertically to the height scale, to extract 4·2m.

Example 2: Find the time in the afternoon when the height of tide has fallen to 3·7m.

• On Fig 4(5) plot the heights of HW and LW above and below the required height of tide. Join them by a sloping line.

• In the boxes below the curves enter HW time and other hourly times to cover the required timescale.

• From the required height, drop vertically to the diagonal line and thence horizontally to the curves. Interpolate between them as in Example 1; do not extrapolate. In this example the actual range is 4·9m, so the Spring curve applies.

• Drop vertically to the timescale, and read off the time required: 1637.

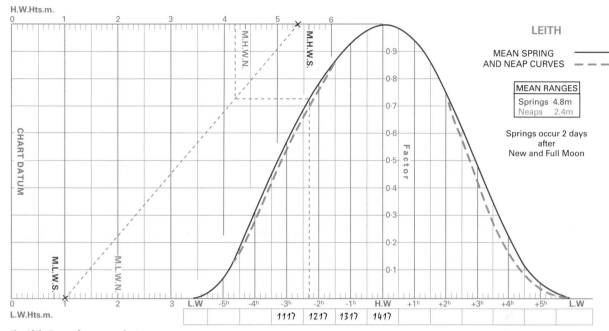

Fig 4(4) Curve for example 1

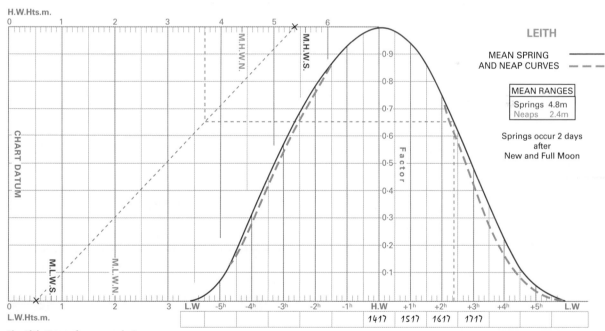

Fig 4(5) Curve for example 2

4.4.2 Secondary Ports

On coasts where the tidal curves for adjacent Standard Ports change little in shape, and where the duration of rise or fall at the Secondary Port is similar to its relevant Standard Port (ie where HW and LW time differences are nearly the same), intermediate times and heights are calculated from the tidal curves for the Standard Port in a similar manner to 4.4.1. The curves are entered with the times and heights of HW and LW at the Secondary Port, calculated as in 4.3.2 and 4.3.3.

Interpolate by eye between the curves, using the range at the Standard Port as argument. Do not extrapolate – use the Spring curve for ranges greater than Springs, and the Neap curve for ranges less than Neaps. With a large change in duration between Springs and Neaps the results may have a slight error, greater near LW.

For places between Christchurch and Selsey Bill (where the tidal regime is complex) special curves are given in 9.2.9.

4.4.3 The use of factors

Factors (in green on tidal curves) are another method of tidal calculation, but the Admiralty tidal prediction form (see 9.2.9) no longer contains the relevant fields. A factor of 1 = HW, and 0 = LW. Tidal curve diagrams show the factor of the range attained at times before and after HW. Thus the factor represents the percentage of the mean range (for the day in question) which has been reached at any particular time. Simple equations used are:

	Range x Factor = Rise
or	Factor = Rise ÷ Range

In determining or using the factor it may be necessary to interpolate between the Spring and Neap curves as described in 4.4.2.

Factors are particularly useful when calculating hourly height predictions for ports with special tidal problems (9.2.9).

4.4.4 The Rule of Twelfths

The Rule of Twelfths estimates by mental arithmetic the height of the tide between HW and LW. It assumes that the duration of rise or fall is 6 hours, the curve is symmetrical and approximates to a sine curve. *Thus the rule is invalid in the Solent, for example, where these conditions do not apply, nor should it be used if accuracy is critical.*

The rule states that from one LW to the next HW, and vice versa, the tide rises or falls by:

1/12th of its range in the 1st hour
2/12ths of its range in the 2nd hour
3/12ths of its range in the 3rd hour
3/12ths of its range in the 4th hour
2/12ths of its range in the 5th hour
1/12th of its range in the 6th hour

4.4.5 Co-Tidal and Co-Range charts

These charts are used to predict tidal times and heights for an offshore position, as distinct from port or coastal locations which are covered by tide tables. There are Admiralty charts for the following offshore areas of NW Europe:

5057	Dungeness to Hoek van Holland
5058	British Isles and adjacent waters
5059	Southern North Sea

They depict Co-Range and Co-Tidal lines, as defined below, with detailed instructions for use.

A Co-Range line joins points of equal Mean Sp (or Np) Range which are simply the difference in level between MHWS and MLWS (or MHWN and MLWN).

A Co-Tidal line joins points of equal Mean HW (or LW) Time Interval. This is defined as the mean time interval between the passage of the Moon over the Greenwich Meridian and the time of the next HW (or LW) at the place concerned.

To find times and heights of tide at an offshore location, say in the Thames Estuary, needs some pre-planning especially if intending to navigate through a shallow gap. The *Tidal Stream Atlas for the Thames Estuary* (NP 249) also contains Co-Tidal and Co-Range charts. These are more clearly arranged and described than charts 5057–5059, but prior study is still necessary. The calculations require predictions for Sheerness, Walton-on-the-Naze or Margate, as relevant to the vessel's position.

4.5 CALCULATING CLEARANCES BELOW OVERHEAD OBJECTS

A diagram often helps when calculating vertical clearances below bridges, power cables etc. Fig 4(6) shows the relationship to CD. The height of such objects as shown on the chart is usually measured above HAT, so the actual clearance will almost always be more. The height of HAT above CD is given at the foot of each page of the tide tables. Most Admiralty charts now show clearances above HAT, but check the **Heights** block below the chart Title.

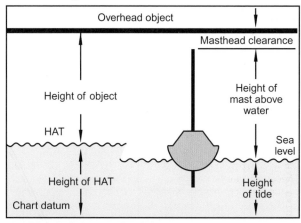

Fig 4(6) Calculating masthead clearance

Insert the dimensions into the following formula, carefully observing the conventions for brackets:

> Masthead clearance = (Height of object above HAT + height of HAT above CD) minus (height of tide at the time + height of the masthead above waterline)

4.6 TIDAL PREDICTION BY COMPUTER

Tidal prediction by computer is simple, fast and accurate, and is therefore arguably better than by manual means. The UKHO offers three tidal prediction programmes.

4.6.1 EasyTide

EasyTide is a free, on-line tidal prediction service for the leisure sailor. It provides tidal predictions for the current day, and the next 6 consecutive days. Times and heights of HW and LW for the required port are presented in both tabular and graphical formats. Over 6,000 standard and secondary ports are available worldwide. To open EasyTide visit www.ukho.gov.uk/easytide.

4.6.2 TotalTide

TotalTide (DP 550) is claimed to be the world's most comprehensive tidal prediction system. It is aimed primarily at commercial shipping and satisfies SOLAS Regulations and the UK Merchant Shipping Safety of Navigation Regulations 2002. It may thus be carried in lieu of paper tide tables.

TotalTide gives fast and accurate tidal predictions for over 7,000 ports and tidal stream data for more than 3,000 locations worldwide. The CD contains a free calculation programme. There are 10 Area Data Sets (ADS), with world-wide coverage. Areas 1-4 (see below) are in one ADS.

Access is via a permit system. Annual updates are available, and are essential to satisfy safety regulations, although TotalTide will still run without them but less accurately.

The ten ADS cover:

1–4.	Europe, Mediterranean and northern waters
5.	Red Sea, the Gulf and Indian Ocean (north)
6.	Singapore to Japan and Philippines
7.	Australia and Borneo
8.	Pacific Ocean, New Zealand and America (west)
9.	North America (east coast) and Caribbean
10.	South Atlantic and Indian Ocean (southern part)

TotalTide displays 7 days of times and heights of HW and LW in both tabular and graphical formats. Other useful facilities include: a display of heights at specified times and time intervals; a continuous plot of height against time; indications of periods of daylight and twilight, moon phases, springs/neaps; the option to insert the yacht's draft; calculations of under-keel and overhead clearances.

TotalTide will run on most versions of Microsoft Windows, but check your system is compatible before buying.

4.6.3 Commercial software

Various commercial firms sell tidal prediction software for use on computers or calculators. Such software is mostly based on NP 159 and can often predict many years ahead. It remains essential to install annual updates.

Many chart plotters incorporate a tidal prediction facility. These are embedded in the software, and their accuracy depends on age and edition. Following a lay-up, at the start of the season or if the power has been disconnected it is necessary to check the date–time to which the plotter is set. If on an unfamiliar boat it is prudent to confirm the prediction against another known data source.

Tablet computer and smart phone applications should be selected on the basis of reputation, accuracy and reliability. A secondary method of prediction should be carried on board to cover eventualities such as equipment or power failure. Do not rely solely upon mobile signal dependent options.

4.7   TIDAL STREAMS

Tidal streams are the horizontal movement of water caused by the vertical rise and fall of the tide; see Fig 4(8). They normally change direction about every six hours. They are quite different from ocean currents, such as the Gulf Stream, which run indefinitely in the same direction.

The direction (set) of a tidal stream is always that towards which it is running. The speed (rate) of tidal streams is important to yachtsmen: about 2kn is common, 6–8kn is not unusual in some areas at springs; 16kn has been recorded in the Pentland Firth.

4.7.1   Tidal stream atlases

Admiralty Tidal Stream Atlases, listed below, show the rate and set of tidal streams in the more important areas around the UK and NW Europe.

The set is shown by arrows graded in weight and, where possible, in length to indicate the rate. Thus $\longrightarrow$ is a weak stream and $\longrightarrow$ a strong stream. The figures against the arrows give the mean Np and Sp rates in tenths of a knot; thus 19,34 indicates a mean Np rate of 1·9kn and a mean Sp rate of 3·4kn. The comma between 19 and 34 is roughly where the observations were made. Inshore eddies are rarely shown in detail due to limitations of scale.

Tidal stream chartlets in this Almanac are derived from the following Admiralty Tidal Stream Atlases.

NP	Title
209	Orkney and Shetland Islands
218	North Coast of Ireland, West Coast of Scotland
219	Portsmouth Harbour and Approaches
220	Rosyth Harbour and Approaches
221	Plymouth Harbour and Approaches
222	Firth of Clyde and Approaches
233	Dover Strait
249	Thames Estuary (with co-tidal charts)
250	The English Channel
251	North Sea, Southern Part
252	North Sea, North Western Part
253	North Sea, Eastern Part
254	The West Country, Falmouth to Teignmouth
255	Falmouth to Padstow, including the Isles of Scilly
256	Irish Sea and Bristol Channel
257	Approaches to Portland
258	Bristol Channel (Lundy to Avonmouth)
259	Irish Sea, Eastern part
263	Lyme Bay
264	The Channel Islands and adjacent Coast of France
265	France, West Coast
337	The Solent and adjacent waters

Useful information is also given in Admiralty Sailing Directions: times of slack water, when the tide turns, overfalls, races etc.

Along open coasts the tidal stream does not necessarily turn at HW and LW; it often occurs at about half tide. The tidal stream usually turns earlier inshore than offshore. In larger bays the tide often sets inshore towards the coast.

The Yachtsman's Manual of Tides by Michael Reeve-Fowkes (ACN) covers all theoretical and practical aspects of tides. It includes four regional **Tidal Atlases**: *Western Channel; Central Channel and the Solent; Southern North Sea and Eastern Channel; Channel Ports and Approaches.* These are also available separately. The atlases are referenced to HW Cherbourg and include annual Cherbourg tide tables (which are downloadable free from www.adlardcoles.com).

Many cruising guides and pilot books also represent tidal stream information for specific areas and can be used in conjunction with the above publications.

Tidal streams referred to HW Aberdeen

Hours	◇ Direction of streams (degrees)	Geographical Position — Rates at spring tides (knots) / Rates at neap tides (knots)			◇ A 59°13'.0N 2°43'.0W			◇ B 59°12'.0N 2°52'.1W		
Before High Water 6				−6	204	5·0	2·0	148	2·5	1·0
5				−5	205	5·2	2·1	149	4·7	1·9
4				−4	207	4·0	1·5	137	3·7	1·5
3				−3	205	2·5	1·0	129	2·2	0·9
2				−2		0·0	0·0	118	1·1	0·5
1				−1	026	2·9	1·1	065	0·6	0·2
High Water				0	023	4·1	1·6	347	1·1	0·5
After High Water 1				+1	024	4·5	1·8	330	2·4	1·0
2				+2	025	4·0	1·6	320	4·2	1·7
3				+3	026	3·0	1·2	310	4·0	1·6
4				+4	034	1·2	0·5	309	2·4	1·0
5				+5	209	1·7	0·7	328	1·3	0·5
6				+6	204	4·8	2·0	145	1·2	0·5

Fig 4(7) Tidal stream diamonds

4.7.2 Tidal stream diamonds

Lettered diamonds (◇) on many medium scale charts refer to tables on the chart giving the most accurate tidal stream set and rate (for springs and neaps) at hourly intervals before and after HW at a convenient Standard Port. See Fig 4(7).

The tidal stream diamonds above are extracted from the table on AC 2250 which covers the eastern part of the Orkney Islands. The streams are referred to HW Aberdeen.

Diamond 'A' is located in the 1M wide Sound of Eday (between the islands of Eday and Sanday).

By inspection the first slack water is sharply defined at HW −2. The stream then sets NNE until a much less obvious slack water occurs between HW +4 and +5. The stream turns at about HW +4½, but rates are still between 1·2 and 1·7kn at springs (0·5 and 0·7kn at neaps).

The SSW-going stream reaches a max spring rate of 5·2kn at HW −5; neap rates are less than half these values.

There is considerable detail in these tables. Careful study enables accurate planning and timing for the transit of narrow waters and best use of the tidal streams.

4.7.3 Flood stream around UK

The main flood stream sets ENE up the English Channel, NE into the Bristol Channel and N up the W coasts of Ireland and Scotland (Fig 4(8)). The ebb stream is the reverse.

Two variations are worth noting:

The flood flows NE-N through St George's Channel. However the flood also sets SE through the North Channel and S into the Irish Sea. Both streams meet off the Isle of Man.

Off the E coasts of Scotland and England the stream sets S all the way from Orkney to the Thames Estuary, finally meeting the N-going stream which has curled round North Foreland and up towards London.

Fig 4(8) The main flood stream around the British Isles

4.7.4   Calculating tidal stream rates

The tidal stream rate at any time may be calculated, assuming that it varies with the range of the tide at Dover. In tidal stream atlases, and on the tidal stream chartlets in this Almanac, the rates are shown in tenths of a knot. Thus '05,27' translates as 0·5 knots at Neaps and 2·7 knots at Springs.

Example: Calculate the tidal stream rate off the north tip of Skye at 0420 UT on a day when the tide at Dover is:

UT	Ht (m)	
0328	1·4	By inspection the mean Range
0819	6·3	at Dover is:
1602	1·1	$\frac{(4\cdot9 + 5\cdot2 + 5\cdot3)}{3} = 5\cdot1m$
2054	6·4	

In 9.8.3 the tidal stream chartlet for HW Dover −4 shows '08,18': a mean Neap rate of 0·8 knots and a Spring rate of 1·8 knots.

On Fig 4(9) overleaf, from the horizontal Rates axis mark 08 on the horizontal blue Neaps line; likewise 18 on the Springs line. Join these two marks with a diagonal. From the range 5·1 on the vertical axis go horizontally to cut the diagonal just drawn. From this point go vertically to the Rates axis, top or bottom, and read off the predicted rate of 15 (1·5 knots).

4.7.5   Tidal streams in rivers

Tidal streams in rivers are influenced by the local topography of the river bed as well as solar and lunar forces. At or near Springs, in a river which is obstructed, for example by a bar, the time of HW gets later going up the river. The time of LW also gets later, but more rapidly so the duration of the flood becomes shorter, and duration of ebb becomes longer. At the entrance the flood stream starts at an interval after LW which increases with the degree of obstruction of the channel; this interval between local LW and the start of the flood increases with the distance upriver. The ebb begins soon after local HW along the length of the river. Hence the duration of flood is less than that of the ebb and the difference increases with distance upriver.

The flood stream is normally stronger than the ebb, and runs harder during the first half of the rise of tide. At Neaps the flood and ebb both start soon after local LW and HW respectively, and their durations and rates are about equal.

4.8   METEOROLOGICAL EFFECTS

Prolonged strong winds and unusually high/low pressure significantly affect tidal heights. Early or late times of HW or LW are principally caused by the wind. The effects of wind or pressure individually may not be great, but the combined effect, which is more likely, may be much greater.

4.8.1   Wind

Wind >Force 5 raises the sea level on a lee shore and lowers it on a windward, but there are wide variations due to topography. Strong on/offshore winds may affect the predicted times of HW or LW by up to 1 hour.

Strong winds blowing parallel to a coast tend to set up long waves, which travel along the coast. Sea level is raised at the crest of these waves (positive surge) and lowered in the troughs (negative surge). A storm surge is an unusually severe positive surge. Under exceptional conditions this can raise the height of HW by >1m; a negative surge can lower the height of LW by the same amount – clearly more serious for yachtsmen and other mariners.

The southern North Sea and the Thames Estuary are prone to such surges. In Jan 1953 a deep depression and an exceptionally severe and prolonged northerly storm caused a positive storm surge which raised sea level by almost 3m on the UK's E coast and by even more on the Dutch coast – with heavy loss of life.

4.8.2   Barometric pressure

Over a period of time and across a wide area sea levels are lowered by high pressure and raised by low. The former is of more practical concern to mariners. Mean sea level pressures are quoted in Admiralty Pilots: 1017mb, for example, along the UK south coast in July; 1014mb in January. 1013mb at Wick in July; 1007mb in January.

Intense minor depressions, line squalls or other abrupt changes in the weather can cause wave oscillations, a phenomenon known as a *seiche*. The wave period can vary from a few minutes up to two hours, with heights up to a metre – usually less, rarely more. Certain harbours, eg Wick and Fishguard, due to their shape or size, are particularly prone to *seiches* especially in winter.

4.8.3   Storm Warning Services

These warn of possible coastal flooding caused by abnormal meteorological conditions. A negative storm surge causes abnormally low tidal levels in the Dover Strait, Thames Estuary and Southern North Sea. 6–12 hrs warning is given, Sept–April, when tidal levels at Dover, Sheerness or Lowestoft are forecast to be >1m below predicted levels.

Warnings are broadcast by Navtex, HM Coastguard and the Channel Navigation Information Service (CNIS) on VHF and MF as normally used for navigation warnings.

Fig 4(9) Graph for calculating tidal stream rates

5

Communications

5.1 The International Code of Signals   **64**
Using the International code

5.2 Operating a radio   **65**
Frequency bands • Regulations • Licences • R/T procedures • Prowords

5.3 Radio communications systems   **65**
VHF radio • MF radio • Mobile phones • VOIP • wi-fi/Wi-MAX

5.4 Global communication systems   **66**
HF radio (SSB) • Satellite communications (Satcoms)

5.5 Navigational warnings   **68**

5.6 Gunfacts and subfacts   **68**

5.7 Coast radio stations (CRS)   **68**
Functions • Track reports • Link calls

5.8 United Kingdom (excluding the Channel Islands)   **69**

5.9 The Channel Islands   **70**

5.10 Republic of Ireland   **70**

5.11 West Denmark   **71**

5.12 Germany   **72**

5.13 The Netherlands   **73**

5.14 Belgium   **74**

5.15 France   **74**

5.16 Spain   **76**

5.17 Mainland Portugal   **77**

5.18 Portuguese Islands, Gibraltar & Morocco   **77**

5.19 Vessel Traffic Services (VTS)   **78**

5.20 International Port Traffic Signals (IPTS)   **78**

5.21 Flag etiquette   **79**
Ensigns • Burgees • Salutes • Dressing ship • Ensigns and flags

5.1 THE INTERNATIONAL CODE OF SIGNALS

Marine communication is rooted in the International Code of Signals (2005 edition published by IMO) which provides for safety of navigation and of persons, especially where there are language problems. The present Code came into force in 1969 and is available in nine languages: English, French, German, Greek, Italian, Japanese, Norwegian, Russian and Spanish. Ships, aircraft and shore stations can communicate in these languages without knowing a foreign tongue, provided they have the relevant Code. The English language edition is published by HMSO.

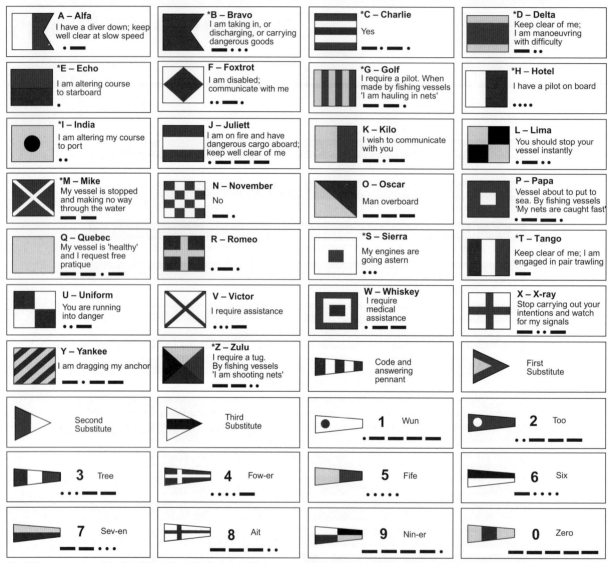

Fig 5(1) International Code of Signals: code flags, single letter meanings, phonetic alphabet (NATO/ITU) and Morse code symbols

5.1.1 Using the International Code

The Code can be used to convey information by: voice, using R/T or loud-hailer; Morse by flashing light and sound signals; alphabetical flags and numeral pendants; or by hand flags.

Signals consist of: single-letter signals which are urgent or much used; two-letter signals in the General Section; and three-letter signals starting with the letter 'M' in the Medical Section. Fig 5(1) above shows the Code flags for letters and numerals, the alphanumeric phonetic alphabet, Morse code for letters and numerals, and the meanings of important or common single-letter signals. **Signals marked with * when made by sound may only be used in compliance with COLREGs Rules 34 and 35.**

You may also see or hear the following important two-letter signals (there are many others):

NC	I am in distress and require immediate assistance.
NE2	You should proceed with great caution; submarines are exercising in this area.
RY	You should proceed at slow speed when passing me (or vessels making signals).
YG	You appear to be contravening the rules of a Traffic Separation Scheme.

5.2   OPERATING A RADIO

5.2.1   Frequency bands

For convenience radio frequencies with similar ranges and transmission characteristics are grouped into different 'bands', such as MF, HF, VHF and UHF (meaning 'medium', 'high', 'very high' and 'ultra-high' frequencies, respectively). Their ranges and characteristics are described below.

5.2.2   Regulations

The regulations for using R/T communications are in the *Handbook for Marine Radio Communication* (Lloyd's of London Press). They are lengthy and form part of the syllabus and examination for certificates of competence.

Some of the more important stipulations are:

Operators must not divulge the contents of messages heard; distress calls have priority; Coast radio stations, or MRCCs as appropriate, control communications in their respective areas. Check that the channel is free before transmitting; unnecessary or superfluous messages are prohibited, as is bad language. In harbour a yacht may not use inter-ship channels except for safety; a radio log must be used to record all messages sent and received.

Communications between MRCCs and second parties are tape recorded, as are intercepted communications with a third party. The following statement is included at the request of the MCA:

'Radio & telephone calls to/from MRCCs are recorded for public safety, prevention/detection of crime and to maintain the operational standards of HM Coastguard'.

5.2.3   Radio licences

- The **vessel** requires a Ship Radio Licence for any VHF, UHF, MF, HF, satellite or EPIRB equipment on board. It is valid 'for life' and is issued free of charge by: Ofcom, Riverside House, 2a Southwark Bridge Rd, London SE1 9HA. ☎ 020 7981 3040; 🖷 020 7981 3235. www.ofcom.org.uk.

  This allows the use of international maritime channels, Ch M for communications between yachts, marinas and clubs, and M2 for race control.

- The **person** in charge of a set requires a Certificate of Competence (Pt 1) and Authority to Operate (Pt 2). This will usually be the Short Range Certificate (SRC), incorporating VHF DSC procedures. The syllabus and conduct of examinations are the RYA's responsibility; see RYA booklet G26. The former Restricted Certificate of Competence (VHF only) is no longer issued, but remains valid for VHF-only radios.

  The Long Range Certificate (LRC) covers MF, HF, Satellite communications and EPIRBs, as does the professional seafarers' General Operator's Certificate (GOC).

- Citizen's Band (CB) and Amateur (Ham) Radio can be useful means of communication for other than safety matters, but only supplement maritime radio on VHF, MF or HF. CB no longer requires a licence in the UK, but Amateur Radio licences are issued by: Ofcom, see contact details above.

- ATIS (Automatic Transmission Identification System) is compulsory on European inland waters. An ATIS capable VHF radio must be used. Ofcom issues a Notice of Variation (NOV) to your exisiting VHF licence on request to licencingcentre@ofcom.org.uk

5.2.4   R/T procedures

Communications between a vessel and a Coast radio station are controlled by the latter, except for Distress, Urgency or Safety messages. Between two vessels, the vessel which is called nominates the working channel.

Before calling, decide exactly what you wish to say; it may help to write the message down. Speak slowly and distinctly. Repeat (or spell phonetically) names or important words; see below for the use and meanings of common prowords.

Your position should be given as Lat/Long or the vessel's bearing and distance *from* a charted object, eg 'My position 225° Isle of May 4M' means you are 4M SW of the Isle of May (*not* 4M NE). Use the 360° True bearing notation and the 24 hours clock (0001–2359), specifying UT or LT.

5.2.5   Prowords

These simplify and expedite communications:

ACKNOWLEDGE	Have you received and understood?
CONFIRM	My version is … is that correct?
CORRECTION	An error has been made; the correct version is …
I SAY AGAIN	I repeat … (eg vital words)
I SPELL	What follows is spelled phonetically.
OUT	End of working.
OVER	I have completed this part of my message, and am inviting you to reply.
RECEIVED	Receipt acknowledged.
SAY AGAIN	Repeat your message (or part of it, ie: All after/before …).
STATION CALLING	Used when a station is unsure of the identity of the calling station.

5.3   RADIO COMMUNICATIONS SYSTEMS

5.3.1   VHF radio

VHF radio (Marine band 156·00–174·00 MHz) is used by most vessels. Range is slightly better than the line of sight between aerials, typically about 20M between yachts and up to 65M to a shore station depending on aerial heights. It always pays to fit a good aerial, as high as possible.

VHF sets may be **Simplex**, ie transmit and receive on the same frequency, so only one person can talk at a time. However, most modern sets transmit and receive simultaneously on different frequencies (full Duplex) so conversation is normal, but two aerials are needed.

Marine VHF frequencies are known by their international channel number (Ch), as shown below.

Channels are grouped according to three main purposes, but some have more than one purpose.

***Public correspondence:* (via Coast radio stations).**
Ch 26, 27, 25, 24, 23, 28, 04, 01, 03, 02, 07, 05, 84, 87, 86, 83, 85, 88, 61, 64, 65, 62, 66, 63, 60, 82, 78, 81.
All channels can be used for Duplex.

Inter-ship:
Ch 06*, 08*, 10, 13, 09, 72*, 73, 69, 77*, 15, 17.
These are all Simplex channels. * for use in UK.

Port Operations:
Simplex: Ch 12, 14, 11, 13, 09, 68, 71, 74, 69, 73, 17, 15.
Duplex: Ch 20, 22, 18, 19, 21, 05, 07, 02, 03, 01, 04, 78, 82, 79, 81, 80, 60, 63, 66, 62, 65, 64, 61, 84.

The following channels have one specific purpose:

Ch 0 (156·00 MHz): SAR ops, not available to yachts.

Ch 10 (156·50 MHz), **23** (161·750 MHz), **84** (161·825 MHz) and **86** (161·925 MHz): MSI broadcasts. The optimum channel number is stated on Ch 16 in the announcement prior to the broadcast itself.

Ch 13 (156·650 MHz): Inter-ship communications relating to safety of navigation; a possible channel for calling a merchant ship if no contact on Ch 16.

Ch 16 (156·80 MHz): Distress, Safety and calling. Ch 16, in parallel with DSC Ch 70, will be monitored by ships, CG rescue centres (and, in some areas, any remaining Coast Radio Stations) for Distress and Safety until further notice. Yachts should monitor Ch 16. After an initial call, stations concerned **must** switch to a working channel, except for Distress and Safety matters.

Ch 67 (156·375 MHz): Small craft safety channel used by all UK CG centres, accessed via Ch 16.

Note: For routine (non-emergency) traffic contact NMOC Fareham on Ch 67 using call-sign Solent Coastguard.,

Ch 70 (156·525 MHz): Digital Selective Calling for Distress and Safety purposes under GMDSS.

Ch 80 (157·025 MHz): Primary working channel between yachts and UK marinas.

Ch M (157·85 MHz): Secondary working channel, formerly known as Ch 37, but no longer.

Ch M2 (161·425 MHz): for race control, with Ch M as stand-by. YCs may apply to use Ch M2.

5.3.2   MF radio
MF radios operate in the 1605–4200 kHz wavebands giving typical ranges of 100–300M using the Single Sideband (SSB) mode of transmission. Double sideband (DSB) mode transmissions are prohibited except for emergency transmissions on the marine Distress frequency (2182 kHz).

MF transmissions tend to follow the earth's curvature, which makes them suitable for direction-finding equipment. The GMDSS DSC Distress alerting frequency is 2187·5 kHz.

5.3.3   Mobile phones (GSM/GPRS/3G/4G broadband)
Mobile phones are not suitable for emergency situations, except in extremis (see 7.6.6), but permit low cost messaging as well as roaming when in different countries. Many smartphones now incorporate GPS and mapping functionality as well as wi-fi (see 5.3.5).

Ordinary GSM (Global System for Mobile Communications) wireless telephony allows a PC to connect wirelessly to the Internet at a speed of 9.6kbps (thousand bits per second); slow, but sufficient for sending and receiving basic emails and browsing text-only websites; GSM is more likely to be available than faster, enhanced services (see below), particularly in less populated areas.

Enhancements: many data services may be available, depending upon where you sail and if you have a compatible handset. These include Global Packet Radio Service (GPRS), boosting GSM download rates to around 36–48kbps. Wireless Broadband services such as 3G delivered increasingly better speeds, theoretically up to 14 mbps. The latest Generation, 4G, is about 5-7 times faster than 3G, with a maximum potential speed of around 80 mbps, providing rapid mobile broadband internet access for large downloads like films,

mobile TV, connections to laptops with wireless modems, smartphones and other mobile devices.

The connection is 'always on'; charges are according to the amount of data transmitted, not the duration of the call. Availability of each service from base stations varies and range offshore cannot be relied upon, especially off sparsely populated coasts where only a limited service is provided.

5.3.4   Voice over IP (VOIP)
Internet Protocol, or IP, is the 'language' that drives the Internet, and VOIP (Voice over IP) is a system where voice communication is encoded and transmitted very efficiently as computer data. Free to join VOIP services such as Skype provide free voice communication between 'members', and also (for a small charge) with normal land-line telephones. VOIP will work over mobile broadband, but network charges will be high if roaming, so wired broadband connection is required, often available on the marina pontoons.

5.3.5   Wi-fi/Wi-MAX
Basic wireless telephony services such as GSM handle voice communications; 3G/4G wireless broadband service adds data capability that may be used with an on board PC (or Macintosh) to send and receive email, browse websites and download updates and chart corrections, in exactly the same way as from a home PC.

Wi-fi, on the other hand, is an increasingly common wireless system, independent of the phone operators' networks, that provides broadband Internet access via 'hotspots', or 'wireless access points', often located at marinas. Even with special antennae, range is limited and this prevents wi-fi being useful to yachts while at sea, but speeds of up to 10,000kbps in the marina are typical and are perfectly suited to normal Internet use like web browsing and downloading Notices to Mariners or Almanac updates.

The next big thing, Wi-MAX, is an emerging technology that promises Internet access at up to 40mbps over much greater ranges than wi-fi. Services are appearing but its onset has been slower than anticipated.

5.4   GLOBAL COMMUNICATION SYSTEMS
Once out of range of VHF/MF or wireless telephony/broadband service, the yachtsman's communications options are limited to MF/HF radio and satellite systems (Satcoms).

A yachtsman embarking on an extended offshore venture would usually choose a mix of equipment for sensible reasons of redundancy, this will allow a choice of listening and transmitting, via terrestrial and satellite radio systems, to meet his needs at various times and in different circumstances. A typical setup might include a fixed or handheld satcoms transceiver, a HF SSB transceiver and receiver. For data capability, these would be interfaced to an on-board PC or Mac.

5.4.1   HF/single sideband radio (HF-SSB)
HF SSB radios use frequencies in the 4, 8, 12, 16 and 22 MHz bands (short wave), provide worldwide coverage and usually include MF frequencies as well.

Despite rapid growth in marine satellite communications, HF SSB remains a popular choice amongst long-distance cruisers, providing a cost-free voice (and limited email) capability for cruisers, sometimes operating over vast distances. Operators have the Long Range Certificate (LRC; see 5.2.3) or General Operators Certificate (GOC). To use Amateur (ham) bands (giving more frequencies and higher power, therefore range) operators must take the ham examination.

Using HF SSB radio for email requires a radio modem, often proprietary to the supplier. Though slow and requiring some skill to operate effectively, the almost-nil operating costs appeal to many and SSB radio has a strong following amongst blue-water cruisers. Established suppliers include Sailmail and Globe Wireless.

HF SSB radio is also extensively used for receipt of weatherfax images (see Chapter 6), although a receive-only SSB radio (with an adequate, grounded antenna installation) may be used rather than a full transceiver. Though declining in popularity, several useful weatherfax transmitting stations remain, including Northwood, UK and Offenbach, Germany.

5.4.2   Satellite Communications (Satcoms)

Satellite communications systems operate over **Ultra High Frequency (UHF)** radio using digital technology that makes them easier and more reliable to use than SSB radio; they operate with either a dedicated ship installation or a standalone handset.

Ongoing costs usually include a monthly service fee and usage charges that will be related to either the amount of satellite time used, or the volume of data transmitted and received.

For two-way voice communications, the 'Ship Station' (aka 'Mobile Earth Station') transmits to a visible satellite that is simultaneously in sight of a 'Land Earth Station', eg Goonhilly, Cornwall. From there, the call is routed to its destination through the normal terrestrial telephone network.

The satellite 'constellations' have different architectures. Inmarsat for example has several geostationary (GEO) satellites, each positioned over the equator. Because they are comparatively high up 19,400M (36,000km), each satellite has a large signal 'footprint', overlapping the next one and thus world-wide coverage is provided (although not in the polar regions above about 70°N and 70°S).

	Wired Broadband	Wi-fi	Cellular GSM	Cellular GPRS	Cellular 3G	Wi-Max
Range offshore[1] (NM)	0	0-1	0-15	0-15	0-15	0-30
Voice communication	Y[2]	Y[2]	Y	Y	Y	Y[2]
Text messages	Y	Y	Y	Y	Y	Y
Light email traffic	Y	Y	Y[3]	Y	Y	Y
Text weather forecasts	Y	Y	Y[3]	Y	Y	Y
Heavy email traffic	Y	Y	✗	Y[3]	Y	Y
Graphical weather forecasts	Y	Y	✗	Y[3]	Y	Y
Full web browsing	Y	Y	✗	Y[3]	Y	Y

[1] typical, if in coverage area
[2] using VOIP
[3] Likely to be expensive

Other systems employ many more Low Earth Orbit (LEO) satellites orbiting the Earth about 540M (1,000km) up.

The smaller, low-data rate, handheld voice terminals incorporate an omni-directional antenna that works best with an unobstructed view of the satellite. Fixed installations use an external gyro-stabilised antenna (to keep it pointing at the satellite as the boat moves); more powerful systems that support higher data rates employ antenna radomes that are really too large for installation aboard a 10–15m yacht.

The GMDSS provides automatic distress, urgency and safety communications, with some satcom systems (eg Inmarsat C) providing a red button that alerts a Coastguard Operations Centre (CGOC) when pressed.

There are several service providers, each offering an array of capabilities. Table 5(1) below provides a useful summary of systems that might be used aboard a 10–20m yacht.

This is a fast moving field: check with manufacturers/retailers for up-to-date specifications and prices.

TABLE 5(1)   Satellite Communications (Satcoms)

System	Antenna	Satellites	Coverage	GMDSS	Phone (voice)	Fax	SMS text	Position tracking	Data rate
INMARSAT C	Omnidirectional	GEO	Global excepting polar regions	Yes	No	Yes	Yes	Yes (GPS)	Very low, uneconomic for email
INMARSAT Mini-M	Gyrostabilised 40cms diameter	GEO	Global excepting polar regions	No	Yes	Yes	No	No	2.4kbps
INMARSAT D+	Omnidirectional	GEO	Global excepting polar regions	No	No	No	No	Yes (GPS)	Very low, uneconomic for email
Fleet 33	Gyrostabilised 40cms diameter	GEO	Global excepting polar regions	No	Yes	Yes	Yes	No	9.6kbps uncompressed
Iridium	Omnidirectional	66 LEO	Global	No	Yes	Out only	Yes	No	Data kit required 2.4kbps (higher with compression)
Globalstar	Omnidirectional	40 LEO	Global excepting polar regions	No	Yes	Yes	Yes	To 10km	9.6kbps uncompressed, 38.6kbps compressed, 56k with data kit
Thuraya	Omnidirectional	1 GEO	Europe, N Africa & Middle East, India	No	Yes	Yes	Yes	Yes (GPS)	9.6kbps (144kbps possible with land *DSL receiver)

*DSL (Digital Subscriber Line) allows broadband Internet over normal copper telephone lines. It is used in the wireless/Satcoms arena to denote 'broadband-like speeds' over the wireless link and usually means speeds of 5 to 24mbps which is relatively fast for Satcoms.

5.5 NAVIGATIONAL WARNINGS

Navigational warnings are promulgated for 16 NavAreas (I–XVI) worldwide. An Area co-ordinator oversees each NavArea and handles inputs from national and other authorities. The 3 types of navigational warnings are:

NavArea warnings which deal with Oceanic matters and are transmitted over satcoms by SafetyNet and via NAVTEX at scheduled times in English and other languages.

Coastal warnings, up to 250M offshore, are issued by a national authority and broadcast in English or the local langauge by NAVTEX, SafetyNet or VHF Coast radio stations. These warnings cover navaids, lights, buoys, rocks, wrecks, naval exercises, SAR operations, cable laying, piracy etc.

Local warnings supplement the above for inshore waters and within port limits. They are issued by port, pilotage or CG authorities in English and/or the local language.

Details of country by country navwarning provision are given in the relevant country listing.

5.6 GUNFACTS AND SUBFACTS

5.6.1 Gunfacts

Gunfacts are warnings of intended naval firing practice, broadcast on Ch 16. They include gunnery and missile firings and underwater explosions. (For the latter only, broadcasts are made 1 hour, 30 mins and immediately before detonation.) They do not restrict the passage of any vessel. The onus for safety is on the firing warship. The broadcasts include:

- Time (LT) and approximate location of firings, with a declared safe distance in nautical miles.
- Whether illuminants, ie flares etc, are to be fired.

Gunfacts (S Coast) are issued from Plymouth for English Channel exercise areas (9.1.16). See Subfacts (below) for details.

Gunfacts (Ship) are issued by a nominated warship and cover activity in any UK areas. An initial announcement is made on VHF 16.

5.6.2 Subfacts

These are warnings of planned or known submarine activity.

Subfacts (South Coast) give details for the English Channel; see 9.1.16. They are broadcast on NAVTEX (see Chapter 6) and, after an announcement on Ch 16, by Falmouth CGOC every 4 hours from 0140 UT on VHF. The VHF channels are shown on Fig 6(4).

Subfacts (Clyde) give details for the W coast of Scotland (see 9.8.22 and 9.9.16). They are broadcast on NAVTEX (see Chapter 6) and, after announcements on Ch 16, by Stornoway CGOC every 4 hours from 0410 UT; and by Belfast CGOC every 4 hours from 0410 UT. The VHF channels are shown on Fig 6(4).

NAVTEX also transmits Subfacts and Gunfacts twice daily: Niton (E) at 0440 and 1640 UT; Portpatrick (O) at 0620 and 1820 UT.

5.7 COAST RADIO STATIONS (CRS)

In recent years many European nations have closed, or plan to close, their coast radio stations due in part to the popularity of mobile phones. **The UK, France and the Netherlands have no CRS.**

Ireland (but see 5.10.4), Denmark, Belgium, Spain and Portugal still operate CRS. In Germany a commercial company, DP07 Seefunk, provides a limited link call service. Details of national CRS are given below in the country by country listing.

5.7.1 Functions

CRS functions include:

- Control of ship-shore communications; see 5.2.4.
- Providing link calls; see 5.7.3. **Note: The Coastguard, or foreign equivalent, do not handle link calls.**
- Broadcasting at scheduled times traffic lists (messages awaiting ships at sea); navigational warnings; weather bulletins and gale warnings (Chapter 6). Note: Do not transmit on a designated broadcast channel when a broadcast is scheduled.
- Handling Distress and Urgency calls in some countries.

5.7.2 Track Reports

A Track Report (TR) may be passed to a CRS or Coastguard centre, as appropriate, stating a yacht's point of departure, destination, ETA and number of people onboard. This is an obvious safety measure, especially in more remote regions, should a yacht subsequently become overdue. On arrival be sure to inform the local CRS or CG and ask that your TR be cancelled with the original CRS/CG. See also Chapter 7.

5.7.3 Making a VHF link call

- Within range (up to 40M), select a 'clear' channel, ie no transmissions. A busy channel will have carrier noise, speech or the engaged signal (a series of pips).
- Call the CRS on a working channel related to the position of the yacht. Calls should state the calling channel and last at least 6 seconds in order to activate the response equipment at the CRS.

For example:

Bilbao Radio, this is Yacht Seabird, Yacht Seabird, Golf Oscar Romeo India Fower, Channel 27, Over.

A four-tone signal or pips indicate a temporary delay, but you will be answered when an operator is free.

- When the call is accepted you will hear pips, indicating that the channel has been engaged. Wait for the operator to speak. If no response, do not change channel since you may lose your turn.

  If the pips are not heard, the station's transmitter has not been activated or may be out of range. Try another station or call again when closer.

- The operator will request: Vessel's call sign (eg Golf Oscar Romeo India Fower), name, accounting code (see 5.7.4), type of message (eg telegram, telephone call) and the telephone number required (and in the case of a personal call the name of the person called).

5.7.4 Paying for a link call

Calls are accounted for worldwide by quoting an 'Accounting Authority Indicator Code' (AAIC). This must previously have been arranged with an ITU-recognised authority such as Cable and Wireless, who use GB 02 as their AAIC. All CRS hold the complete ITU listing of ship stations and their AAICs.

5.7.5 MF link calls

Calls should be initiated on 2182 kHz or the nominated working frequency. The Coast station will allocate a working channel, and cue you into the system.

5.8   UNITED KINGDOM (EXCLUDING THE CHANNEL ISLANDS)

5.8.1   HM Coastguard Communications

Contact details for UK Coastguard Operations Centres are shown below:

EASTERN REGION

NATIONAL MARITIME OPERATIONS CENTRE (NMOC)   50°51'·50N 01°14'·90W   DSC MMSI, 002320011
☎ 02392 552100   🖷 02392 554131/01305 760451. Kites Croft, Fareham PO14 4LW.
Area: South Coast: Topsham (River Exe) to Beachy Head (Note: in this area make initial contact using call sign *Solent Coastguard* on Ch 67 (H24) to avoid congestion on Ch 16). East coast: Reculver Towers to Southwold.

DOVER COASTGUARD   50°08'N 01°20'E   DSC MMSI 002320010   ☎ 01304 210008   🖷 01304 225762
Langdon Battery, Swingate, Dover CT15 5NA.   Area: Beachy Head to Reculver Towers
Operates Channel Navigation Information Service (CNIS); see 9.3.13.

LONDON COASTGUARD   51°30'N 00°03'E   MMSI 002320063   ☎ 0208 3127380 🖷 0208 3098196. Thames Barrier
Navigation Centre, Unit 28, 34 Bowater Road, Woolwich, London SE18 5TF.
Area: River Thames from Shell Haven Pt (N bank) and Egypt Bay (S bank), up-river to Teddington Lock

†HUMBER COASTGUARD   54°06'N 00°11'W   MMSI 002320007   ☎ 01262 672317   🖷 01262 400779
Lime Kiln Lane, Bridlington, Yorks E Riding YO15 2LX.   Area: Southwold to the Scottish border.

SCOTLAND & NORTHERN IRELAND REGION

†ABERDEEN COASTGUARD   57°08'N 02°05'W   MMSI 002320004   ☎ 01224 592334   🖷 01224 575920
Marine House, Blaikies Quay, Aberdeen AB11 5PB.   Area: English border to Cape Wrath, incl Pentland Firth.

†SHETLAND COASTGUARD   60°09'N 01°08'W   MMSI 002320001   ☎ 01595 692976   🖷 01595 693634
Knab Road, Lerwick ZE1 0AX.   Area: Orkney, Fair Isle and Shetland.

†*STORNOWAY COASTGUARD   58°12'N 06°22'W   MMSI 002320024   ☎ 01851 702013   🖷 01851 706796
Battery Point, Stornoway, Isle of Lewis H51 2RT.   Area: Cape Wrath to Mull, Western Isles and St Kilda.

†*BELFAST COASTGUARD   54°40'N 05°40'W   MMSI 002320021   ☎ 02891 463933   🖷 02891 469854
Bregenz House, Quay St, Bangor, Co Down BT20 5ED.
Area: Carlingford Lough to Lough Foyle and Firth of Clyde, inc islands.

WESTERN REGION

†HOLYHEAD COASTGUARD   53°19'N 04°38'W   MMSI 002320018   ☎ 01407 762051. 🖷 01407 761613
Prince of Wales Rd, Holyhead, Anglesey LL65 1ET.   Area: Mull of Galloway to Friog (1·6M S of Barmouth).

†MILFORD HAVEN COASTGUARD   51°42'N 05°03'W   MMSI 002320017   ☎ 01646 690909. 🖷 01646 697287
Gorsewood Drive, Hakin, Milford Haven, SA73 2HD.   Area: Friog to Marsland Mouth (near Bude).

†*FALMOUTH COASTGUARD   50°09'N 05°03'W   MMSI 002320014   ☎ 01326 317575. 🖷 01326 315610
Pendennis Point, Castle Drive, Falmouth TR11 4WZ.
Area: Marsland Mouth (near Bude) to Topsham (River Exe) falmouthcoastguard@mcga.gov.uk

NOTES: † Monitors DSC MF 2187.5 kHz.
        * Broadcasts Gunfacts/Subfacts; see 5.6.
The 3 Regions above are Search and Rescue Regions.

CAUTION: The HM Coastguard modernisation programme should be completed by the end of 2015. However, information on this page still remains subject to potential change and readers are advised to check the free monthly updates from January 2016 to June 2016 available for download at www.reedsnauticalalmanacs.co.uk.

THE NATIONAL COASTWATCH INSTITUTION (NCI) operates on dedicated VHF Ch 65. There are 50 operational stations (2015) around the English and Welsh coast. Manned by volunteers keeping a visual watch during daylight hours throughout the year, they are able to provide, on request, details of actual weather conditions and sea state at their individual coastal locations, (see Safety, 7.12), radio checks, and local knowledge, from visitors' moorings to hazards.

Fig 5(2) UK Coastal (WZ) Navwarning areas 1–17. The area(s) for which each CGOC is responsible are in brackets

5.8.2 Navwarnings (NavArea I Co-ordinator)

After an announcement on Ch 16, HMCG broadcasts coastal navigational warnings (WZ) on receipt and every 12 hours at the same times and on the same frequencies as the Group A MSI broadcasts. See Table 6(2) for details. WZs are also broadcast by NAVTEX at the times shown at in Table 6(1), and by SafetyNET at 1730 UT daily for areas not covered by NAVTEX. See also weekly *Notices to Mariners* which include a list of warnings in force.

Urgent warnings are broadcast as soon as possible after receipt on Ch 16 and on 2182kHz. New dangers to navigation should be notified to other craft and the nearest MRCC by a *Sécurité* call. More details at www.mcga.gov.uk (see Safety Information/Navigational Safety).

5.9 THE CHANNEL ISLANDS

5.9.1 Coastguard communications

Guernsey and Jersey Coastguard stations direct SAR operations in the North and South of the CI area respectively, they also provide communications on VHF and DSC.

Close liaison is maintained with French SAR authorities and a distress situation may be controlled by whichever is more

appropriate. For example a British yacht in difficulty in French waters may be handled by St Peter Port or Jersey so as to avoid language problems; and vice versa for a French yacht.

GUERNSEY COASTGUARD (CRS)
49°27'N 02°32'W   DSC MMSI 002320064
☎ 01481 720672   📠 01481 714177   VHF Ch 16, 20.
Area: The Channel Islands Northern area.

JERSEY COASTGUARD (CRS)
49°10'·90N 02°06'·80W   DSC MMSI 002320060
☎ 01534 447705/04   📠 01534 499089   VHF Ch 16, 82.
Area: The Channel Islands Southern area.

5.9.2 Emergency VHF D/F

In the Channel Islands a yacht in distress should call on Ch 16; expect to be switched to Ch 20 for Guernsey or Ch 82 for Jersey. The emergency D/F sites, *from* which the true bearing of the yacht in distress is given, are: VHF D/F stations, h RG.

Site	Lat/Long	Controlled via
Guernsey	49°26'·27N 02°35'·77W	*Guernsey Coastguard*
Jersey	49°10'·85N 02°14'·30W	*Jersey Coastguard*

5.9.3 Coast Radio Stations

Guernsey Coastguard and Jersey Coastguard are operated by the States of Guernsey and Jersey respectively. They handle Distress and Urgency situations as well as providing link calls (in emergencies only).

Guernsey Coastguard
VHF 16, **20** (primary working Ch, call direct for safety traffic), 16, small craft distress/emergency. (Ch 62, link calls for emergencies only.)
DSC Ch 70: **MMSI** 002320064.
Traffic lists: Vessels are called individually on Ch 16.

Jersey Coastguard
VHF 16, **82**, primary working Ch, call direct for safety traffic, 16, small craft distress/emergency. (Ch 25, no continuous listening watch, link calls for emergencies only.)
DSC Ch 70: **MMSI** 002320060.
Navigational warnings, weather bulletins and traffic lists are broadcast on Ch 82 at times in Chapter 6.

5.10 REPUBLIC OF IRELAND

5.10.1 Coast Guard communications

DUBLIN (MRCC)
53°20'N 06°15W   DSC MMSI 002500300 (+2187·5 kHz)
☎ +353 1 662 0922/3   📠 +353 1 662 0795
Area: Carlingford Lough to Youghal.

VALENTIA (MRSC)
51°56'N 10°21'W   DSC MMSI 002500200 (+2187·5 kHz)
☎ +353 669 476 109   📠 +353 669 476 289
Area: Youghal to Slyne Head.

MALIN HEAD (MRSC)
55°22'N 07°20W   DSC MMSI 002500100 (+2187·5 kHz)
☎ +353 74 9370103   📠 +353 74 9370221
Area: Slyne Head to Lough Foyle

Co-ordinates SAR operations around ROI via Dublin MRCC, Malin Head and Valentia MRSCs and remote sites. It may liaise with the UK and France during any rescue operation

Fig 5(3) Irish coast radio stations, CG centres and boundaries

5.10.4 Coast Radio Stations

The main CRS at Malin Head, Dublin and Valentia are co-located with Coast Guard centres. Remote CRS sites, as listed below, ensure full coastal coverage in the three regions (NW, SE and SW Ireland). Callsign is station name and Radio.

All monitor Ch 16 (H24). Broadcasts are made on a working channel after pre-warning on Ch 16 and 2182 kHz. Ch 67 is only used for Safety messages. See Chapter 6 for times and details of gale warnings and weather bulletins.

Commercial link calls, ship to shore, are not available, but Medico link calls can be made on both VHF and MF. Call CRS on a working channel; only call Ch 16 if in difficulty or emergency.

Ships should pass their intended voyage or position to Malin Head Coast Guard, Dublin Coast Guard or Valentia Coast Guard, as appropriate. Update voyage details by TR.

NW Ireland (clockwise from the west)

Clifden Coast Guard Radio	26
Belmullet Coast Guard Radio	83
Donegal Bay Coast Guard Radio	02
Glen Head Coast Guard Radio	24
MALIN HEAD Coast Guard RADIO	23, 85
MF: Tx 1677 kHz, Rx 2102	

SE Ireland

Carlingford Coast Guard Radio	04
DUBLIN Coast Guard RADIO	(No MF) 83
Wicklow Head Coast Guard Radio	02
Rosslare Coast Guard Radio	23
Mine Head Coast Guard Radio	83

SW Ireland

Cork Coast Guard Radio	26
Mizen Coast Guard Head Radio	04
Bantry Coast Guard Radio	23
VALENTIA Coast Guard RADIO	24
MF: Tx 1752 kHz, Rx 2096	
Shannon Coast Guard Radio	28
Galway Coast Guard Radio	04

within 100M of the Irish coast. Dept of Transport, Leeson Lane, Dublin 2. ☎ (01) 6620922; admin@irishcoastguard.ie The Irish EPIRB Registry is also at this address.

The MRCC/MRSCs are co-located with the Coast radio stations of the same name and manned by the same staff. All stations keep watch H24 on VHF Ch 16 and DSC Ch 70. If ashore dial 999 or 112 in an emergency and ask for Marine Rescue.

5.10.2 SAR resources

The Irish Coast Guard provides some 50 units around the coast and is on call H24. The RNLI maintains four stations around the coast and operates 42 lifeboats. Additionally, six community-run inshore rescue boats are available.

Helicopters, based at Dublin, Waterford, Shannon and Sligo, can respond within 15 to 45 minutes and operate to a radius of 200M. They are equipped with infrared search equipment and can uplift 30 survivors.

Military and civilian aircraft and vessels, together with the Garda (Police) and lighthouse service, can also be involved.

Some stations provide specialist cliff climbing services. They are manned by volunteers, who are trained in first aid and equipped with inflatables, breeches buoys, cliff ladders etc. Their ☎ numbers (the Leader's residence) are given, where appropriate, under each port.

5.10.3 Navwarnings

Navwarnings are broadcast by Malin Hd and Valentia NAVTEX 518 kHz, see Table 6(1) and by MRCCs/MRSCs every 4 hours on VHF: Dublin and Malin Head from 0033 UT; Valentia from 0233 UT.

5.11 WEST DENMARK

5.11.1 Coast Guard communications

The national SAR agency is:

JRCC Admiral Danish Fleet, PO Box 1483, DK-8220 Brabrand, Denmark. ☎ +45 894 33099; ⊠ +45 728 50384.

The SAR coordinator is:

JRCC Denmark, ☎ +45 728 50000 switchboard; ⊠ +45 728 50384; jrcc@sok.dk

MMSI 002191000

MF DSC 2187·5kHz TX/RX (Routine national traffic TX 1624·5kHz, RX 2159·5kHz; Routine international TX 2177·0kHz, RX 2189·5kHz).

JRCC Denmark has no direct communications with vessels in distress, but operates via two Maritime Operations Centres (MOC's) and Coast Radio Station (CRS) Lyngby Radio.

MOC North (MRSC) ☎ +45 992 21520; ⊠ +45 992 21538, mocn-orum@mil.dk, covers the W coast of Denmark.

5.11.2 SAR resources

There are at least 12 lifeboats stationed at the major harbours along this west facing coast. They double up as Pilot boats.

Fig 5(4) Danish Coast Guard and Coast Radio Stations

5.11.3 Navwarnings
Navwarnings are broadcast by Lyngby Radio on VHF and MF 1734 kHz, 1738 kHz every 4 hours from 0133 UT.

5.11.4 Firing practice areas
There are 4 such areas on the W coast as in Fig 5(4) and listed below. Firing times are broadcast daily by Danmarks Radio 1 at 1645 UT after the weather. Times are also available from the Range office by e or Ch 16.

P	Tranum & Blokhus ☎ 982 35088 or call *Tranum*.
Q	Nymindegab e 765 23334 or call *Nymindegab*.
R	Oksbøl ☎ 765 41213, call *Oksbøl* or *www. forsvaret.dk/oksbl*
S	Rømø E ☎ 737 55219 or call *Fly Rømø*; Rømø W ☎ 745 41340.

5.11.5 Coast Radio Stations
All Danish VHF/MF CRS are remotely controlled by **Lyngby** and use the callsign *Lyngby Radio*.

The stations listed below monitor Ch 16 and DSC Ch 70, H24. Call on working frequencies, to keep Ch 16 clear. MF stations do not monitor 2182 kHz. Traffic lists are broadcast on all VHF channels every odd H+05.

VHF and MF CRS
Call Ch 16/2182kHz callsign *Lyngby Radio*

Skagen VHF Ch 04, 28; MF Tx 1758kHz, Rx 2102kHz.
Hirtshals VHF Ch 66.
Hanstholm VHF Ch 01.
Bovbjerg VHF Ch 02; MF Tx 1767kHz, Rx 2111kHz.
Blåvand VHF 23; MF Tx 1734kHz, Rx 2078kHz.

HF RT Lyngby Radio, in the course of normal working in the 4, 8, 12 & 16 MHz bands, *receives* calls from ships on the frequencies in *black italic*; and transmits on the paired frequencies in green. Ships may be directed to other paired frequencies as listed in ALRS Vol 1(1).

4116	4408	*16411*	17293
4137	4429	*16423*	17305
8216	8740	*16462*	17344
8246	8770	*22045*	22741
12257	13104	*22051*	22747
12269	13116	*22105*	22801
12296	13143	*25070*	26145
16408	17290	*25073*	26148

5.12 GERMANY

5.12.1 Coast Guard communications
The national SAR agency is:

> *Deutsche Gesellschaft zur Rettung Schiffbrüchiger* (DGzRS), German Maritime Search & Rescue Service. Werderstrasse 2, D-28199 Bremen. info@seenotretter.de ☎ +49 421 536870 (emergency), 421 53707-0 (ops); 📠 +49 421 53707-690.

DGzRS is responsible for coordinating SAR operations, and operates 60 rescue units.

Bremen MRCC (☎ + 49 421 536870; MMSI 002111240), using callsign *Bremen Rescue,* maintains an H24 watch on MF 2182kHz and Ch 16 VHF, also DSC MF 2187·5kHz and VHF Ch 70 via remote Coast radio stations at:

Blumenthal	Norderney	Borkum
Stade	Cuxhaven	Wangerooge
Helgoland	Westerhever	Kampen

Bremen MRCC guarantees assistance, on request, to foreign MRCCs co-ordinating SAR measures for German vessels in foreign waters.

5.12.2 SAR resources
There are 20 offshore lifeboats, LOA 20–46m, and 20 smaller <11m lifeboats, based at Borkum, Juist, Norderney, Norddeich, Baltrum, Langeoog, Neuharlingersiel, Wangerooge, Horumersiel, Hooksiel, Wilhelmshaven, Fedderwardersiel, Bremerhaven, Cuxhaven, Brunsbüttel, Helgoland, Büsum Eiderdamm, Nordstrand, , Amrum, Hörnum and List. There are also many inshore lifeboats.

Fig 5(5) German VTS Centres and Coastal Radio Stations

5.12.3 Navwarnings

Navwarnings *(Nautische Warnnachricht)* are broadcast in English by Pinneberg NAVTEX 518 kHz (see Table 6(1)) every 4 hours from 0300 UT. Bremen MRCC broadcasts vital Navwarnings for the North Sea Coast on VHF 16 on receipt of a warning and every H and H+30 until cancelled.

5.12.4 Coast Radio Stations

DP07 – Seefunk, Estedeich 84, 21129 Hamburg. ☎+49 040 23855782. ✉+49 040 74134242. info@dp07.com www.dp07.com   MMSI 002113100

This company operates seven CRS on the North Sea coast and five on the Baltic coast as shown on Fig 5(5) and listed below. The Coast Guard CRS monitor Ch 16 and DSC Ch 70.

Traffic Lists: at 0745, 0945, 1245, 1645 and 1945LT. Callsign is the name of the remote station.

Nordfriesland	26	**Baltic coast**	
Elbe-Weser	24	Kiel	23
Hamburg	83	Flensburg	27
Bremen	25		
Accumersiel	28		
Borkum	61		

5.13   THE NETHERLANDS

5.13.1 Coast Guard communications
The national SAR agency is:

SAR Commission Maritime, Ministry of Infrastructure and the Environment, PO Box 5807, 2280 HV The Hague, The Netherlands.

The Netherlands CG at Den Helder, co-located with the Navy HQ, coordinates SAR operations as the Dutch JRCC for A1 and A2 Sea Areas. (JRCC = Joint Rescue Coordination Centre – marine & aeronautical.) Callsign is *Netherlands Coast Guard*, but *Den Helder Rescue* during SAR operations.

The JRCC keeps a listening watch H24 on DSC Ch 70, and MF DSC 2187·5 kHz (but not on 2182 kHz); MMSI 002442000.

Coast Guard Operations can be contacted H24 via:

In emergency:
☎ +31 9000 111.

Operational telephone number:
☎+31 223 542300. ✉+31 223 658358; ccc@kustwacht.nl
If using a mobile phone, call +31 9000 111, especially if the International emergency number 112 is subject to delays.

Admin/info (HO):
☎ +31 223 658300. ✉ +31 223 658303. info@kustwacht.nl
PO Box 10.000/MPC 10A, 1780 CA Den Helder.

Remote CG stations are shown on Fig 5(6). Working channels are VHF 23 and 83.

5.13.2 Medical advice
Call initially on Ch 16, DSC Ch 70 or 2187·5 kHz (MMSI 002442000). Working chans are VHF Ch 23 & 83 or MF 2824 kHz (transmit), 2520 kHz (receive).

5.13.3 SAR resources
The Dutch Lifeboat Ass'n (KNRM) manages 45 all-weather lifeboat stations and 13 inshore lifeboat stations along the coast. The 75 lifeboats include 13m LOA water-jet, rigid inflatables capable of 36 kn. Helicopters, fixed wing aircraft and ships of the RNN can be called upon; also Air Force helicopters at Leeuwarden. The area of activity extends across the Dutch Continental Shelf and into the Waddenzee, IJsselmeer and estuaries of Zuid Holland and Zeeland.

Fig 5(6) Netherlands Coast Guard and Radio stations

5.13.4 Navwarnings

Navwarnings are broadcast by Netherlands Coast Guard on NAVTEX 518 kHz (see Table 6(1)) every 4 hours from 0230 UT, and in English on VHF Ch 23 or Ch 83 and MF 3673 kHz every 4 hours from 0333 UT.

5.14 BELGIUM

5.14.1 Coast Guard communications

The Belgian Coast Guard coordinates SAR operations from Oostende MRCC, callsign *Coast Guard Oostende*. The MRCC and *Oostende Radio* (CRS, see below) both keep listening watch H24 on Ch 16, 67, 2182 kHz and DSC Ch 70 and 2187·5 kHz.

Coast Guard stations

> **MRCC OOSTENDE**
> ☎ +32 59 701000   📠 +32 59 703605
> MMSI 002059981
> mrcc@mrcc.be
>
> **RCC BRUSSELS** (Point of contact for COSPAS/SARSAT)
> ☎ +32 (0)2 751 4615   📠 +32 (0)2 7524201
> rcc@mil.be

5.14.2 SAR resources

Offshore and inshore lifeboats are based at Nieuwpoort, Oostende and Blankenberge.

The Belgian Air Force provides helicopters from Koksijde near the French border. The Belgian Navy also participates in SAR operations as required, ☎ +32 50 558324   📠 +32 50 558319 mik@mil.be

5.14.3 Navwarnings

Navwarnings are broadcast by Oostende on NAVTEX 518 kHz (see Chapter 6, Table 6(1)), and in English every 4 hrs on VHF 27 and MF 2761 kHz on receipt, every H+03 and H+33, and every 4 hours from 0233 UT. Local messages are broadcast on 490kHz. Navwarnings are broadcast by Antwerpen for the Schelde on VHF 24 on receipt and every H+55.

5.14.4 Coast Radio Stations

> **OOSTENDE RADIO**
> ☎ +32 50 558241   📠 50 558748   rmd@mil.be
> MMSI 002050480 Ch 16, DSC Ch 70 and 2187·5 kHz (H24)
> MF: Tx 2484, Rx 3178 kHz. VHF Ch 27, 63, 78, 85
> Zeebrugge   Ch 27, 63
> Middelkerke (Oostende)   Ch 27
> De Panne (French border)   Ch 78

> **Antwerpen Radio** (remote control by Oostende CRS)
> Ch 16, DSC Ch 70; MMSI 002050485. Working channels:
> Antwerp Ch 07, 27; Ghent Ch 81; rest of Belgium Ch 24.

5.15 FRANCE

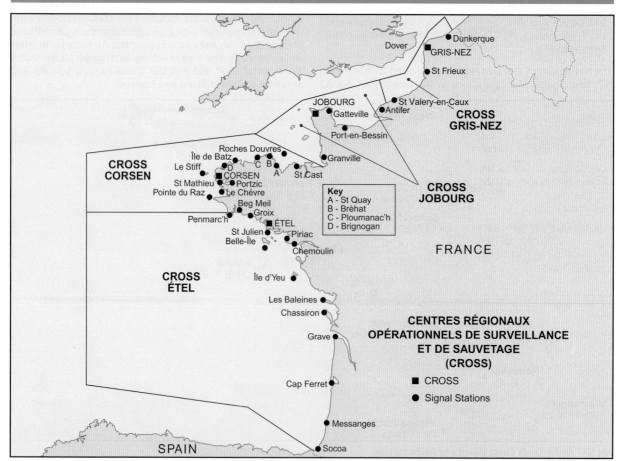

Fig 5(7) CROSS centres and signal stations (for remote stations see 6(12)

5.15.1 French CROSS infrastructure

CROSS (Centres Régionaux Opérationnels de Surveillance et de Sauvetage) provide an all-weather presence H24 along the French coast in liaison with foreign CGs. CROSS is an MRCC. The main functions include:

- Co-ordinating SAR operations
- Navigational surveillance
- Broadcasting navigational warnings (5.5)
- Broadcasting weather information (6.20)
- Anti-pollution control
- Marine and fishery surveillance

All centres monitor VHF Ch 16 and DSC Ch 70 and co-ordinate SAR on Ch 10, 13, 15, **67**, **68**, 73; DSC Ch 70, Corsen MRCC additionally monitors MF DSC (2187·5 kHz). They can be contacted by phone 112; via the National Gendarmerie, Affaires Maritimes or a Semaphore station. If a nautical emergency *is witnessed from the land*, phone direct on 196. CROSS also monitors the TSS in the Dover Strait, off Casquets and off Ouessant using, for example, callsign *Ushant Traffic*.

For medical advice at sea call the appropriate CROSS which will contact the French TMAS (Telemedical Maritime Assistance Service) and co-ordinate an evacuation if necessary. ⚠ Note: SAMU (Ambulance) will respond more quickly to a medical emergency in a harbour or marina than calling a doctor. Simply dial 15.

5.15.2 CROSS locations/contact details

Any CROSS can be telephoned simply by dialling **112**.

CROSS Gris-Nez 50°52'N 01°35'E
MMSI 002275100 ☎ 03·21·87·21·87 📠 03·21·87·78·55 gris-nez@mrccfr.eu Area: Belgian border–Cap d'Antifer. CROSS Gris-Nez responds to COSPAS/SARSAT alerts.

CROSS Jobourg 49°41'N 01°54'W
MMSI 002275200 ☎ 02·33·52·16·16 📠 02·33·52·78·23. jobourg@mrccfr.eu Area: Cap d'Antifer–Mt St Michel

CROSS Corsen 48°24'N 04°47'W
MMSI 002275300 ☎ 02·98·89·31·31 📠 02·98·89·65·75 corsen@mrccfr.eu Area: Mt St Michel–Pointe de Penmarc'h

CROSS Étel 47°39'N 03°12'W
MMSI 002275000 ☎ 02·97·55·35·35 📠 02·97·55·49·34 etel@mrccfr.eu Area: Pointe de Penmarc'h to the Spanish border.

5.15.3 Lifeboats

Société National de Sauvetage en Mer (SNSM; lifeboat service) comes under CROSS. At sea, all vessels must contact CROSS to call upon the services of SNSM, they should not be contacted directly in an emergency. A hefty charge may be levied for a false call-out.

5.15.4 Navwarnings (NavArea II Co-ordinator)

Navwarnings are of two types: **Long-range** for NavArea II issued by SHOM and broadcast by SafetyNet.

AvurNavs include Coastal and Local warnings. They are issued by Cherbourg for the coast from the Belgian border to Mont St Michel, and by Brest from Mont St Michel to the Spanish border. They are broadcast by CROSS (Corsen) NAVTEX in

English every 4 hours from 0000 UT, and in French every 4 hours from 0040 UT. CROSS also broadcasts Navwarnings on VHF 79/80 after initial call on VHF 16 (Jobourg rarely).

5.15.5 Semaphore (signal) stations

These keep visual, radar and radio watch (VHF Ch 16) around the coast and are manned by the French Navy. They relay emergency calls to CROSS and are equipped with VHF DF. They show gale warning signals, repeat forecasts and offer local weather reports. They are listed below.

Semaphore (Signal) stations

Dunkerque	03·28·66·86·18
Boulogne	03·21·31·32·10
Ault	03·22·60·47·33
Dieppe	02·35·84·23·82
Fécamp	02·35·28·00·91
La Hève	02·35·46·07·81
Villerville	02·31·88·11·13
Port-en-Bessin	02·31·21·81·51
St Vaast	02·33·54·44·50
Barfleur	02·33·54·04·37
Le Homet	02·33·92·60·08
La Hague	02·33·52·71·07
Barneville-Carteret	02·33·53·85·08
Granville Le Roc	02·33·50·05·85
* St Cast	02·96·41·85·30
St Quay-Portrieux	02·96.70.42.18
* Bréhat	02·96·20·00·12
Ploumanac'h	02·96·91·46·51
* Batz	02·98·61·76·06
Brignogan	02·98·83·50·84
Ouessant Stiff	02·98·48·81·50
St-Mathieu	02·98·89·01·59
Portzic (Ch 08)	02·98·22·21·47
* Toulinguet	02·98·27·90·02
* Cap-de-la-Chèvre	02·98·27·09·55
Pointe-du-Raz	02·98·70·66·57
Penmarc'h	02·98·58·61·00
* Beg Meil	02·98·94·98·92
Étel Mât Fenoux (HW-3 to HW+2)	02·97·55·35·39
Beg Melen (Groix)	02·97·86·80·13
* Talut (Belle-Île)	02·97·31·85·07
St Julien	02·97·50·09·35
* Piriac-sur-Mer	02·40·23·59·87
Chemoulin	02·40·91·99·00
St Sauveur (Yeu)	02·51·58·31·01
* Les Baleines (Ré)	05·46·29·42·06
Chassiron (Oléron)	05·46·47·85·43
Pointe-de-Grave	05·56·09·60·03
Cap Ferret	05·56·60·60·03
* Messanges	05·58·48·94·10
Socoa	05·59·47·18·54

* Operate sunrise to sunset only. Remainder are H24.

5.15.6 Emergency VHF DF stations – France

A vessel in emergency can obtain its true bearing *from* a DF station. DF stations are remotely controlled by a semaphore (signal) station, a †Naval lookout station or a *CROSS, all of which keep primary watch on Ch 16.

To obtain a true bearing from asterisked DF stations, call CROSS on Ch 16, 11 or 67. Also available are seven continuously scanned frequencies from among Ch 1–29, 36, 39, 48, 50, 52, 55, 56 and 60–88.

VHF DF stations are listed below (NE to NW then S):

Dunkerque	51°03'·40N 02°20'·40E
*Gris-Nez	50°52'·20N 01°35'·01E
Boulogne	50°44'·00N 01°36'·00E
Ault	50°06'·50N 01°27'·50E
Dieppe	49°56'·00N 01°05'·20E
Fécamp	49°46'·10N 00°22'·20E
La Hève	49°30'·60N 00°04'·20E
Villerville	49°23'·20N 00°06'·50E
Port-en-Bessin	49°21'·10N 00°46'·30W
St Vaast	49°34'·50N 01°16'·50W
Barfleur	49°41'·90N 01°15'·90W
Levy	49°41'·70N 01°28'·20W
†Homet	49°39'·50N 01°37'·90W
*Jobourg	49°41'·50N 01°54'·50W
La Hague	49°43'·60N 01°56'·30W
Carteret (Barneville)	49°22'·40N 01°48'·30W
Le Roc (Granville)	48°50'·10N 01°36'·90W
Grouin (Cancale) (HJ)	48°42'·60N 01°50'·60W
St Cast	48°38'·60N 02°14'·70W
St Quay-Portrieux (HJ)	48°39'·30N 02°49'·50W
Bréhat	48°51'·30N 03°00'·10W
Ploumanac'h (HJ)	48°49'·50N 03°28'·20W
Batz	48°44'·80N 04°00'·60W
Brignogan (HJ)	48°40'·60N 04°19'·70W
Creac'h (Ouessant)	48°27'·60N 05°07'·70W
*Creac'h (Ouessant)	48°27'·60N 05°07'·80W
†St Mathieu	48°19'·80N 04°46'·20W
Toulinguet (HJ)	48°16'·80N 04°37'·50W
Cap de la Chèvre (HJ)	48°10'·20N 04°33'·00W
Pointe du Raz	48°02'·30N 04°43'·80W
Penmarc'h	47°47'·90N 04°22'·40W
Beg-Meil (HJ)	47°51'·30N 03°58'·40W
Beg Melen	47°39'·20N 03°30'·10W
*Etel	47°39'·80N 03°12'·00W
St Julien (HJ)	47°29'·70N 03°07'·50W
Taillefer	47°21'·80N 03°09'·00W
Le Talut	47°17'·70N 03°13'·00W
Piriac	47°22'·50N 02°33'·40W
Chemoulin	47°14'·10N 02°17'·80W
St Sauveur	46°41'·70N 02°18'·80W
Les Baleines (HJ)	46°14'·60N 01°33'·70W
Chassiron	46°02'·80N 01°24'·50W
La Coubre	45°41'·90N 01°13'·40W
Pointe de Grave	45°34'·30N 01°03'·90W
Cap Ferret	44°37'·50N 01°15'·00W
Messanges (HJ)	43°48'·80N 01°23'·90W
Socoa	43°23'·30N 01°41'·10W

5.16 SPAIN

5.16.1 Coast Guard communications

The Society for Maritime Rescue and Safety (Sociedad de Salvamento y Seguridad Maritima – SASEMAR) is the national agency for SAR operations (and the prevention of pollution); akin to MCA in the UK.

MRCC Madrid coordinates SAR operations via 3 MRCCs (Coruña, Gijon and Finisterre) on the N coast and Tarifa MRCC on the SW coast – as listed below.

All facilities monitor (H24) VHF Ch 16, MF 2182 kHz and DSC Ch 70, 2187·5 kHz. They broadcast weather as shown in Chapter 6 and Nav warnings, Chapter 3; they do not handle link calls.

North and North West Spain

In N and NW Spain CG Centres do not keep continuous watch on Ch 16, so call on a working channel; see Fig 5(8).

Madrid MRCC   MMSI, 002241022.
☎ 917 559 132   📠 915 261 440. cncs@sasemar.es

Bilbao MRCC
43°21'N 03°02'W   MMSI 002240996
☎ 944 839 411   📠 944 839 161   bilbao@sasemar.es

Santander MRSC
43°28'N 03°43'W   MMSI 002241009
☎ 942 213 030   📠 942 213 638   santander@sasemar.es

Gijón MRCC
43°34'N 05°42'W   MMSI 002240997
☎ 985 326 050   📠 985 320 908   gijon@sasemar.es

Finisterre MRCC
42°42'N 08°59'W   MMSI 002240993
☎ 981 767 320   📠 981 767 740   finister@sasemar.es

Coruña MRSC
43°22'N 08°23'W   MMSI 002240992
☎ 981 209 541 📠 981 209 518   coruna@sasemar.es

Vigo MRSC
42°10'N 08°41'W   MMSI 002240998
☎ 986 222 230   📠 986 228 957   vigo@sasemar.es

South West Spain

Tarifa MRCC coordinates SAR in SW Spain and the Gibraltar Strait, thus accepts Distress, Urgency and Safety traffic only.

Tarifa MRCC
36°01'N 05°35'W   MMSI 002240994
☎ 956 684 740   📠 956 680 606   tarifa@sasemar.es

Huelva MRSC
37°13'N 07°07'W   MMSI 002241012
☎ 959 243 000   📠 959 242 103   algeciras@sasemar.es
huelva@sasemar.es

Cádiz MRSC
36°32'N 06°18'W   MMSI 002241011
☎ 956 214 253   📠 956 226 091   cadiz@sasemar.es

Algeciras MRSC (controlled by Malaga MRCC)
36°08'N 05°26'W   MMSI 002241001
☎ 956 580 930   📠 956 585 402   algeciras@sasemar.es

5.16.2   Navwarnings (NavArea III Co-ordinator)

Long-range Navwarnings for NavArea III are broadcast by SafetyNet. Coastal warnings are broadcast in English and Spanish by Coast radio stations on NAVTEX 518 kHz (see Table 6(1)) and on VHF; monitor VHF Ch 16 or MF 2182 kHz.

5.16.3   Coast Radio Stations (CRS)

CRS monitor (H24) Ch 16, 2182kHz and DSC (VHF & MF) call Ch 16, then switch to a working channel. Traffic lists not broadcast on VHF, but passed on a working VHF channel; or on MF at every odd H +33, except 0133 and 2133.

North Spain (Callsign for all: *Bilbao Radio*)
Remote stations controlled by Bilbao Comms Centre are:

Pasajes		Ch 27
Machichaco	*MF: Tx 1677 & 2182; Rx 2102 & 2182.* No VHF or DF	
Bilbao		Ch 26
Santander		Ch 24

North West Spain (Callsign for all: *Coruña Radio*)
Remote stations controlled by Coruña Comms Centre are:

Cabo Ortegal		Ch 02
Coruña	*MF: Tx 1707 & 2182; Rx 2132 & 2182.*	Ch 26
Finisterre	*MF: Tx 1698 & 2182; Rx 2123 & 2182.*	Ch 22
Vigo		Ch 20
La Guardia		Ch 82
Cabo Peñas		Ch 27
Navia		Ch 62

South West Spain
Remote stations controlled by Malaga Comms Centre include:

Chipiona Radio	*MF: Tx 1656; Rx 2081.* No VHF	
Cádiz Radio		Ch 28
Tarifa Radio	*MF: Tx 1704; Rx 2129.* Ch 83	

5.17 MAINLAND PORTUGAL

5.17.1 Coast Guard communications

The Portuguese Navy coordinates SAR in two regions, Lisboa and Santa Maria (Azores), via MRCCs at Lisboa, Ponta Delgada (Azores) and Funchal MRSC (Madeira). A network of CRS maintains an H24 listening watch on all distress frequencies; see Fig 5(8).

Their HQ (Ministério da Defensa Nacional – Marinha) is at:

Estado-Maior da Armada, Rua do Arsenal, 1149-001 Lisboa
☎ 21 098 4450 (HO); 21 3217666 (H24) 📠 21 440 1954

Lisboa MRCC
38°41'N 09°19'W   MMSI 002630100
☎ +351 21 4401919 (Emergency), 21 4401955, 21 4411646; 📠 21 4401954
mrcc.lisboa@marinha.pt   comar.dir@marinha.pt;
Full DSC Ch70 2187.5 kHz coverage not complete but 7 stations remotely controlled by Lisboa MRCC now operational, DSC Ch70/ Voice, VHF Ch16/ 2182kHz.

5.17.2 Coast Radio Stations (CRS)
Lisboa Radio controls all remote stations.
☎ +351 214 190424 📠 +351 214 199900
lisboa.radio@telecom.pt

All CRS, Ch 16 (H24). Call name, (see Fig 5(8)) plus 'Radio'.

5.18 PORTUGUESE ISLANDS

5.18.1 Coast Guard communications

(CENCOMARACORES) & Delgada MRCC
37°45'N 25°38'W MMSI 002040100 VHF Ch 16 2182kHz
Planned DSC Ch 70; 2182 kHz ☎ +351 296 281777
(Emergency); 205227 (Ops); 📠 296 205239
mrcc.delgada@marinha.pt

Horta MRSC
38°32'N 28°38'W   MMSI 002040200   VHF Ch 11
2182 kHz

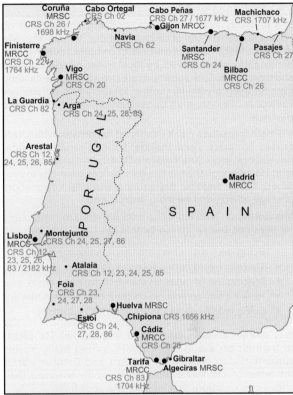

Fig 5(8) Spanish and Portuguese CG radio stations

Funchal MRSC (CENCOMARMADEIRA)
38°38'N 16°54'W   MMSI 002550100   VHF Ch 11
2182 kHz ☎ +351 291 213112 (Emergency)

5.18.2 Coast Radio Stations (CRS)
All CRS monitor Ch 16 (H24) and are remotely controlled by Lisboa Radio. The following MF frequencies and VHF channels are used:

Flores Radio		23
Faial Radio	MF: Tx 2741 & Rx 2237kHz.	25 (H24), 26, 28
Pico Radio		24 (H24), 26, 28
São Miguel Radio	MF: Tx & Rx 2182kHz.	23 (H24), 26, 27
Porto Santo Radio		23, 28 (H24)

GIBRALTAR

SAR communications

Gibraltar VTS
36°09'N 5°22'W   MMSI 002361001   VHF & MF DSC
☎ +350 200 462 54 📠 +350 200 770 11
ops@port.gov.gi

MOROCCO

Coast Guard communications

Rabat MRCC
33°49'N 07°09'W   MMSI 002424133   VHF & MF DSC
☎ +212 537 625897   emergency +212 537 625877
mrcc.rabat@mpm.gov.ma

Agadir MRSC
30°25'N 09°38'W   MMSI 002424136   VHF & MF DSC;
Ch 16 ☎ +212 528 842964   mrsc.agadir@mpm.gov.ma

5.19 VESSEL TRAFFIC SERVICES (VTS)

VTS primarily assist commercial shipping and depend on correct sequential use of communications. VTS usually exist in areas of high traffic density with complex routeing and/or navigational constraints, eg shoal water, narrow channels and tight bends. Such conditions are stressful and demanding even for the professional mariner. Small craft must therefore keep clear of big ships, eg by navigating just outside buoyed channels.

VTS both formalises, and is shorthand for, services which have existed for years. These include: Position reporting; radar surveillance; procedural control of traffic (not to be confused with TSS, although the two may exist in close proximity to each other); traffic, weather, navigational and tidal information; pilotage advice; and berthing instructions.

Monitoring of VHF radio channels is therefore compulsory for big ships, but yachts can benefit by:

- Being aware of what commercial ships are doing
- Thereby being better able to avoid them
- Making use of radar assistance if the need arises
- Hearing any constraints on their own movements
- Being readily contactable by the VTS authority.

It is rarely necessary (and often undesirable) for yachts to transmit on a busy VTS channel; indeed many marine authorities discourage this. In Dutch and German waters the letter of the law is often more rigidly enforced than elsewhere; violations may incur on-the-spot fines.

VTS diagrams appear in the relevant geographic area of Chapter 9, with detailed notes in the harbour text.

5.20 INTERNATIONAL PORT TRAFFIC SIGNALS (IPTS)

The International system (IPTS) is now widely used on the Continent, but its universal adoption may take some years. Obsolescent national signals may occasionally be seen in a few minor ports. In some major ports, eg Calais, these important signals are the primary means of controlling traffic, thereby much reducing or even eliminating VHF radio 'chatter'.

- The main movement message always comprises three lights, vertically disposed. No additional light shall be added to the column carrying the main message.

  Thus the main message is always recognisable as IPTS, and not as lights of navigational significance. IPTS may also be used to control traffic at locks and bridges.

- Red lights indicate *Do not proceed*.

- Green lights indicate *Proceed, subject to the conditions stipulated*. Red and green lights are never displayed together, to avoid confusion.

- Signals may be omni-directional, ie seen by all vessels simultaneously; or directional, ie seen only by vessels entering or by vessels leaving harbour.

- Signal 1 *Serious Emergency* must show at least 60 flashes/minute. All other signals may be fixed or slow occulting (useful when background glare is a problem). A mixture of fixed and occulting lights must not be used.

- Signal No 5 assumes that VHF, signal lamp, loud-hailer, auxiliary signal or other means of communication will specifically inform a vessel that she may proceed.

- Exemption signals. A single yellow light, displayed to the left of the column carrying main messages Nos 2 or

No	Lights		Main message
1		Flashing	Serious emergency – all vessels to stop or divert according to instructions
2		Fixed or Slow Occulting	Vessels shall not proceed (*Note:* Some ports may use an exemption signal, as in 2a below)
3			Vessels may proceed. One-way traffic
4			Vessels may proceed. Two-way traffic
5			A vessel may proceed only when she has received specific orders to do so. (*Note:* Some ports may use an exemption signal, as in 5a below)
	Exemption signals and messages		
2a		Fixed or Slow Occulting	Vessels shall not proceed, except that vessels which navigate outside the main channel need not comply with the main message
5a			A vessel may proceed when she has received specific orders to do so, except that vessels which navigate outside the main channel need not comply with the main message
	Auxiliary signals and messages		
	White and/or yellow lights, displayed with the main lights		Local meanings, as promulgated in local port orders

5 and level with the upper light, means *Vessels which can safely navigate outside the main channel need not comply with the main message*. This signal, shown as Nos 2a and 5a, is obviously important to small craft.

- Signals which are auxiliary to the main message may be devised by local authorities. Such auxiliary signals should show only white and/or yellow lights, and be displayed to the right of the column carrying the main message. Ports with complex entrances and much traffic may need many auxiliary signals, which will have to be documented. Smaller harbours with less traffic may only need the basic signals, such as Nos 2 and 3.

5.21   FLAG ETIQUETTE

Flags are still used to express identity – by national ensigns, club burgees, etc. Brief guidance on their use is given below and in *Reeds Maritime Flag Handbook* (ACN).

5.21.1   Ensign

A yacht's ensign is the national maritime flag corresponding to the nationality of her owner; Fig 5(9). Thus a British yacht should wear the Red Ensign unless she qualifies for a Special ensign (5.21.2). The national ensign should be kept clean and in good repair. It should normally be worn at the stern, or if that's not possible the nearest position should be used, eg at the peak in a gaffer, at the mizzen masthead in a ketch or yawl, or about two-thirds up the leech of the mainsail. In harbour or at anchor the proper position is at the stern.

In British harbours by custom the ensign is hoisted at 0800 (0900 between 1 Nov and 14 Feb) or as soon after that time as people come on board; and lowered when the crew leaves the boat for the evening, at sunset or at 2100LT, whichever occurs first.

Abroad the ensign (and burgee) are increasingly left flying at all times, to identify nationality – even at night in harbour, unless the boat is unattended.

At sea the ensign must be worn when meeting other vessels, when entering or leaving foreign ports, or when approaching forts, Signal and CG stations – ie virtually at all times.

When racing the ensign should not be worn after the five minute gun. It should be hoisted on finishing or retiring.

5.21.2   Special ensigns

Members of certain clubs may apply for a permit to wear a special ensign (eg Blue Ensign, defaced Blue Ensign, or defaced Red Ensign). The yacht must either be registered under Part I of the Merchant Shipping Act 1995 and of at least 2 tons gross tonnage, or be registered on the Small Ships Register and of at least 7m LOA. The owner(s) must be British subjects, and the yacht must not be used for any professional, business or commercial purpose. Full details can be obtained from Secretaries of the clubs concerned. The permit must be carried on board.

A special ensign must only be worn when the owner is on board or ashore in the vicinity, and only when the yacht is flying the burgee (or a Flag Officer's flag) of the club concerned. When the yacht is sold, or the owner ceases to be a member of the club, the permit must be returned to the Secretary of the club.

5.21.3   Courtesy ensign

It is customary when abroad to fly a small maritime ensign of the country concerned at the starboard spreader. *A courtesy ensign must not be worn below any other flag on the same halyard.* Thus a club burgee, if usually flown at the starboard spreader, must be shifted to the port spreader when abroad, permitting the courtesy ensign to be close up at the starboard spreader. The correct courtesy flag for a foreign yacht in British waters is the Red Ensign. British yachts do not need to fly a courtesy flag in the Channel Islands.

5.21.4   Burgee

A burgee shows that a yacht is in the charge of a member of the club indicated, and does not necessarily indicate ownership. It should be flown at the masthead.

If this is impossible due to wind sensors, radio aerials etc, the burgee may be flown at the starboard spreader, but this should be avoided unless strictly necessary. A long bamboo cane can readily be hoisted to the masthead as a staff, keeping the burgee clear above sensors and aerials.

A yacht should not fly more than one burgee. A burgee is not flown when a yacht is racing. If the yacht is on loan, or is chartered, it is correct to use the burgee of the skipper or charterer – not that of the absent owner. Normal practice has been to lower the burgee at night, at the same time as the ensign, but nowadays many owners leave the burgee flying if they are on board or ashore in the vicinity.

5.21.5   Choice of burgee

An owner who is not a Flag Officer, and who belongs to more than one club, should normally fly the burgee (and if authorised the special ensign) of the senior club in the harbour where the yacht is lying. An exception may be if another club is staging a regatta or similar function.

5.21.6   Flag Officer's flag

Clubs authorise their Flag Officers to fly special swallow-tailed flags, with the same design as the club burgee and in place of it. The flags of a vice-commodore and a rear-commodore carry one and two balls respectively. A Flag Officer's flag is flown day and night while he is on board, or ashore nearby. A Flag Officer should fly his flag with the Red Ensign (or special ensign, where authorised) in preference to the burgee of some other club.

It is customary for a Flag Officer to be saluted (not more than once a day) by a yacht flying the burgee of that club.

5.21.7   House flag

An owner may fly his personal flag when he is on board in harbour. It must not conflict with the design of an existing flag. A house flag is normally rectangular, and is flown at the crosstrees in a sloop or cutter, at the mizzen masthead in a ketch or yawl, or at the foremast head in a schooner.

5.21.8   St George's Flag

This is predominantly a land flag, commonly misused afloat. It may only be flown at sea by admirals on duty and members of the Association of Dunkirk Little Ships (in conjunction with their burgee).

5.21.9   Q Flag

Vessels requiring customs' clearance should fly the yellow Q flag. This declares that 'my vessel is healthy and I request free pratique' ie to clear in and come ashore. This is not required for vessels from, and sailing within, the EU.

5.21.10   Salutes

Yachts should salute all Royal Yachts, and all warships of whatever nationality. A salute is made by dipping the ensign (only). The vessel saluted responds by dipping her ensign, and then re-hoisting it, whereupon the vessel saluting re-hoists hers.

5.21.11   Dressing ship

Dressing overall (as opposed to dressing with masthead flags) is normally only done in harbour to mark national festivals. The international code flags and pennants (Fig 5(1)) are flown in the following order from stem to masthead(s) to stern:

E, Q, p 3, G, p 8, Z, p 4, W, p 6, P, p 1, I, Code, T, Y, B, X, 1st Sub, H, 3rd Sub, (masthead) D, F, 2nd Sub, U, A, O, M, R, p 2, J, p 0, N, p 9, K, p 7, V, p 5, L, C, S. (Total = 40).

Fig 5(9) Ensigns and flags

UK WHITE ENSIGN

UK BLUE ENSIGN

UK RED ENSIGN

AUSTRALIA

AUSTRIA

BASQUE FLAG

BELGIUM

CANADA

CYPRUS

DENMARK

FINLAND

FRANCE

GERMANY

GREECE

GUERNSEY

JERSEY

IRELAND

ISRAEL

ITALY

LIBERIA

MALTA

MONACO

MOROCCO

NETHERLANDS

NEW ZEALAND

NORWAY

PANAMA

POLAND

PORTUGAL

SOUTH AFRICA

SPAIN

SWEDEN

SWITZERLAND

TUNISIA

TURKEY

USA

6

Weather

6.1   Wind/sea state scales and terminology   82
Beaufort wind scale • Douglas Sea State Scale • Cloud types (pictures) •
Terms used in bulletins • Geostrophic scales • Unit conversion

6.2   Heavy weather warnings   85
International storm warning signals

6.3   Sources of weather information   86

6.4   Using weather forecasts   86

6.5   GMDSS – Maritime Safety Information (MSI)   86

6.6   NAVTEX   86
Messages • UK 518 and 490 kHz stations • Coverage abroad

6.7   High seas bulletins   89

6.8   MSI Broadcasts by HM Coastguard   89
Shipping and Inshore forecasts • Remote CG transmitters

6.9   HF radio facsimile broadcasts   91
GMDSS (Offenbach) • Royal Navy (Northwood)

6.10   Met Office/BBC weather forecasts   92
UK Forecast areas • Shipping Forecast • Frequencies • Schedule

6.11   Internet weather resources   94

6.12   Other weather resources   96
Newspapers • Television • Volmet

6.13–6.23 Weather information abroad   96
6.13 Channel Islands – 96 • *6.14* Republic of Ireland – 96 •
6.15 West Denmark – 97 • *6.16* Germany – 98 •
6.17 The Netherlands – 99 • *6.18* Belgium – 100 •
6.19 France – 100 • *6.20* N and NW Spain – 102 •
6.21 Portugal, Azores and Madeira – 103 •
6.22 SW Spain – 103 • *6.23* Gibraltar – 104

Weather

6.1 WIND/SEA STATE SCALES AND TERMINOLOGY

6.1.1 Beaufort wind scale

Force	Wind speed			Description	Deep sea criteria	Probable mean (max) wave ht (m)
	(knots)	(km/h)	(m/sec)			
0	<1	0–2	0–0.5	Calm	Calm, glassy, like a mirror	—
1	1–3	2–6	0.3–1.5	Light air	Calm. Ripples like scales are formed	0.1 (0.1)
2	4–6	7–11	1.6–3.3	Light breeze	Small wavelets, still short but more pronounced, not breaking	0.1 (0.3)
3	7–10	13–19	3.4–5.4	Gentle breeze	Large wavelets, crests begin to break; glassy foam, perhaps a few white horses	0.6 (1)
4	11–16	20–30	5.5–7.9	Moderate breeze	Small waves becoming longer; fairly frequent white horses	1 (1.5)
5	17–21	31–39	8.0–10.7	Fresh breeze	Moderate waves, taking a more pronounced form; many white horses, perhaps some spray	2 (2.5)
6	22–27	41–50	10.8–13.8	Strong breeze	Large waves forming; white foam crests are extensive everywhere; probably some spray	3 (4)
7	28–33	52–61	13.9–17.1	Near gale	Sea heaps up; white foam from breaking waves begins to be blown in streaks	4 (5.5)
8	34–40	63–74	17.2–20.7	Gale	Moderately high waves of greater length; edge of crests break into spindrift; foam blown in well-marked streaks	5.5 (7.5)
9	41–47	76–87	20.8–24.4	Strong gale	High waves with tumbling crests; dense streaks of foam; spray may affect visibility	7 (10)
10	48–55	89–102	24.5–28.4	Storm	Very high waves with long overhanging crests; dense streaks of foam make the surface of sea white. Heavy, shock-like tumbling sea; visibility affected	9 (12.5)
11	56–63	104–117	28.5–32.6	Violent storm	Exceptionally high waves; sea completely covered with long white patches of foam; edges of wave crests blown into froth. Visibility affected	11.5 (16)
12	64 plus	118 plus	32.7 plus	Hurricane	Air filled with foam and spray; sea completely white with driving spray; visibility seriously affected	14 (—)

Notes

- The actual force exerted on a sail is proportional to the square of the windspeed and directly to the density. Thus a wind of Force 6 (say 25 knots) has more than three times the 'weight' of a wind of Force 4 (say 14 knots) – which is why even a small increase in windspeed, or gusts, have a significant effect on a sailing yacht. Temperature has a smaller effect; at 5°C the force of the wind is about 6% higher than at 20°C.

- Wave height depends upon the distance travelled (fetch), the length of time the wind has been blowing and any currents. The probable mean wave heights (max. in brackets) shown in the table above are for open sea, away from land. In enclosed waters, or near to land, with an offshore wind, wave heights will be smaller but often steeper – particularly with wind against tide.

Swell

The length and height in metres of swell waves are described and quantified as follows:

Length of Swell (m)	
Short	0–100
Average	100–200
Long	over 200

Height of Swell (m)	
Low	0–2
Moderate	2–4
Heavy	over 4

6.1.2 Douglas Sea State Scale (wind force/wave height):

Wind force	Wave ht (m)	Douglas	Sea State
0	-	0	Calm
1	0.1	1	Calm
2	0.2	2	Smooth
3	0.6	3	Slight
4	1.0	3-4	Slight-mod
5	2.0	4	Moderate
6	3.0	5	Rough
7	4.0	5-6	Rough-V.Rough
8	5.5	6-7	V.Rough-High
9	7.0	7	High
10	9.0	8	Very High
11	11.5	8	Very High
12	14+	9	Phenomenal

Fig 6(1) Typical sea states associated with various Beaufort wind forces

Photography:
Force 0, 11, 12 © *Crown*; Force 1, 2, 7, © *GJ Simpson*; Force 3, 4, 5, 6, © *IG MacNeil*; Force 8, © *WAE Smith*; Force 9, © *JP Laycock*; Force 10, © *G Allen*

Force 0   0–1kn   Calm   Wave Ht 0m

Force 1   1–3kn   Light Air   Wave Ht 0.1m

Force 2   4–6kn   Light Breeze   Wave Ht 0.2m

Force 3   7–10kn   Gentle Breeze   Wave Ht 0.6m

Force 4   11–16kn Moderate Breeze   Wave Ht 1m

Force 5   17–21kn   Fresh Breeze   Wave Ht 2m

Force 6   22–27kn   Strong Breeze   Wave Ht 3m

Force 7   28–33kn   Near Gale   Wave Ht 4m

Force 8   34–40kn   Gale   Wave Ht 5.5m

Force 9   41–47kn   Severe Gale   Wave Ht 7m

Force 10   48–55kn   Storm   Wave Ht 9m

Force 11   56–63kn   Violent Storm   Wave Ht 11.5m

Force 12   64+kn   Hurricane   Wave Ht 14m

6.1.3   Cloud types

Cirrus (Ci) Very high, silky-white clouds (mare's tails) consisting of ice-particles, associated with strong winds aloft. Often indicates the approach of a frontal system and deteriorating weather conditions.   © *RK Pilsbury*

Cirrocumulus (Cc) Very high small white puffs in ripples, sometimes forming banks; the classic Mackerel sky. Thicker than Cirrus and formed by turbulence. Not usually associated with any particular weather.   © *RK Pilsbury*

Cirrostratus (Cs) A thin high whitish veil which does not blur the outline of the sun or moon; the only cloud to show a halo. If it thickens, it can indicate an approaching front or depression.   © *CS Broomfield*

Altocumulus (Ac) Essentially a greyish-white layer cloud with gaps between rounded globules of cumulus. If forming in bands, it may indicate an approaching front.
© *CS Broomfield*

Altostratus (As) A medium level, sheet of blue-grey cloud like a layer of fog. The sun or moon may shine through it, without halo. Usually indicates that frontal rain can be expected in a few hours, turning to Ns.   © *S Cornford*

Nimbostratus (Ns) A low, dark grey, rain-bearing cloud mass obscuring the whole sky. Often follows Altostratus and the passing of a front. Steady rain, sleet or snow will set in, the base covering low hills.   © *RK Pilsbury*

Stratus (St) Thin, grey, fog-like sheet of cloud as low as 1000 or even 500 feet. Dense Stratus can produce drizzle and poor visibility. It is usually associated with warm sector conditions.
© *Crown*

Stratocumulus (Sc) Dense, lumpy grey cloud below 5000 feet; persistent in winter with high pressure. If it thickens to form a solid grey layer across the sky, expect rain at any time.
© *GA Watts*

Cumulus (Cu) Fair weather cloud with vertical development due to rising warm air, especially over coasts. Plenty of blue sky but risk of gusty showers due to instability. Disappears by early evening.   © *RK Pilsbury*

Cumulonimbus (Cb) Large, dramatic convection cloud rising to great heights with huge tops, some anvil-shaped. Heavy rain and thundery squalls are likely. Down-currents produce sudden strong gusts, hazardous to yachts.   © *Crown*

6.1.4 Terms used in weather bulletins

Speed of movement of pressure systems

Slowly	< 15 knots
Steadily	15–25 knots
Rather quickly	25–35 knots
Rapidly	35–45 knots
Very rapidly	> 45 knots

Visibility

Good	> 5 nautical miles
Moderate	2–5 nautical miles
Poor	1000 metres–2 nautical miles
Fog	Less than 1000 metres

Barometric pressure changes (tendency)

Rising or falling slowly:
Pressure change of 0·1 to 1·5 hPa/mb in the preceding 3 hours.

Rising or falling:
Pressure change of 1·6 to 3·5 hPa/mb in the preceding 3 hours.

Rising or falling quickly:
Pressure change of 3·6 to 6 hPa/mb in the preceding 3 hours.

Rising or falling very rapidly:
Pressure change of more than 6 hPa/mb in the preceding 3 hours.

Now rising (or falling):
Pressure has been falling (rising) or steady in the preceding 3 hours, but at the time of observation was definitely rising (falling).

6.1.5 Geostrophic scales

When interpreting a synoptic chart it often helps to be able to estimate the Beaufort wind force; similarly the speed in knots at which warm, cold and occluded fronts are moving – hence when a warm front or a post-cold front clearance may arrive.

On UK charts there is a scale shown. To use, set your dividers between adjacent isobars, then read off from the left (high) end of the scale the gradient windspeed in knots. Be careful to use the scale at the appropriate latitude. Warm fronts move at about 2/3 of this speed and cold or occluded fronts move at about 3/4. Surface winds can be at least 3/4 of the gradient speed in cold, unstable air and between 1/2 and 2/3 in warm, stable air coming from a southerly point.

Fig 6(2) Example (not to scale) of geostrophic wind scale found on UK synoptic charts

6.1.6 Unit conversion

Weather administrations may use a variety of units. To convert between them, refer to the following table:

Atmospheric pressure

Hectopascals (hPa) = Millibars (mb).
Hectopascals (hPa) X 0·295 = inches of mercury (InHg).
Hectopascals (hPa) X 0·75 = mm of mercury (mmHg).
Inches of mercury (InHg) X 33·8639 = Hectopascals (hPa).
mm of mercury (mmHg) X 1·33322 = Hectopascals (hPa).

Temperature

Multiply °C by 1·8 then add 32 to obtain °F.
Subtract 32 from °F, then divide by 1·8 to obtain °C.

6.2 HEAVY WEATHER WARNINGS

Gale warnings

A *Gale* warning means that average winds of at least F8 (34–40kn) or gusts up to 43–51kn are possible (i.e. may occur) somewhere within the area, but not necessarily over the whole area.

Severe Gale means winds of at least F9 (41–47kn) or gusts reaching 52–60kn.

Storm means winds of F10 (48–55kn) or gusts of 61–68kn.

Violent Storm means winds of F11 (56–63kn) or gusts of 69kn or more.

Hurricane Force means winds of F12 (64kn or more).

Gale warnings remain in force until amended or cancelled ('gales now ceased'). If a gale persists for more than 24 hours the warning is re-issued.

Timing of gale warnings

Imminent:	Within 6 hrs from time of issue
Soon:	6–12 hrs from time of issue
Later:	More than 12 hrs from time of issue

Note: when there is some uncertainty, UK forecasters may say 'Perhaps gale 8 later' but without issuing a gale warning. However within the *Imminent* and *Soon* periods, a warning must be issued even if the forecaster says 'perhaps...'

Strong wind warnings

Strong wind warnings will only be issued if the wind is forecast to reach F6 or more and was not predicted when the last forecast was issued. They will be broadcast on receipt and will be included in the repetition forecasts but will be incorporated into the next new inshore forecast if still in force.

Wind

Wind direction: The direction from which the wind is blowing.

Winds becoming cyclonic: Indicates that there will be considerable changes in wind direction across the path of a depression within the forecast area.

Veering: A clockwise change in the wind direction, eg SW to W.

Backing: An anti-clockwise change in the wind direction, eg W to SW.

6.2.1 International storm warning signals

France and the Netherlands use the international system of visual storm warning signals shown here, shapes by day, lights at night. Portugal and Gibraltar use national variants.

Day	Night	Meaning
●	Ⓦ ●	Strong wind (force 6–7)
▲	● ●	NW gale
▲▲	● Ⓦ	NE gale
▼	Ⓦ Ⓦ	SW gale
▼▼	Ⓦ ●	SE gale
✚	● ● ●	Hurricane force12, any direction
▬ }	Rectangular flags of any colour	Wind veering
▬ }		Wind backing

6.3 SOURCES OF WEATHER INFORMATION

There is no shortage of weather information available to today's yachtsman, indeed it could be argued that there is almost too much choice, with 'official' and 'unofficial' forecasts of varying quality provided by both private and government organisations over a variety of media, some entirely free to receive, some not.

For the first 24 to 48 hours, forecasts from national maritime authorities should always be consulted using forecasts from different sources if available. It can then be more useful to compare successive forecasts from the same source.

6.4 USING WEATHER FORECASTS

Weather forecasting has advanced greatly over the past 30 years or so and is still improving, due to increasing computer power, improved measuring equipment and advanced data analysis techniques. This is particularly apparent in the ability to forecast large scale patterns up to five or six days ahead. There are also improvements, but less obvious, in short term detailed prediction of marine weather.

Single observer forecasting, oft-practised in days past, is less relevant in Europe, now there is a well developed meteorological infrastructure. Watching the sky and reading the barometer are still useful but more for verifying the 'official' forecast than for creating your own. Instead, the sailor is better advised to monitor every weather broadcast, paying attention to trends and changes in the forecast. Consistency from one forecast to the next and between forecasts from different sources builds confidence in them; the degree of inconsistency implies the degree of uncertainty. Daily monitoring of forecasts is important over periods of around five days. Monitoring forecasts on a six hourly basis is essential to keep ahead of small but important changes that can occur over such timescales.

It is most productive, and sensible, to use your own observations and experience to interpret forecasts to your local area. Useful observations will include:

- The barometer, logging carefully whether it is steady, rising or falling and at what hourly rate; see 6.1.4.
- Actual wind direction and speed, noting any tendency to back or veer, increase or moderate; see 6.2.
- The appearance of the sky, especially cloud types, their estimated heights and direction of travel; see 6.1.3.
- When precipitation starts and ends may also provide valuable clues to frontal movement.

> The pocket sized *Reeds Weather Handbook*, by Frank Singleton, is a very useful guide to the use of weather forecasts.

Compare your forecast with what actually happened. Analyse errors, and remember that practice makes perfect.

6.5 GMDSS – MARITIME SAFETY INFORMATION

As well as the better known distress reporting and rescue aspects of the Global Maritime Distress and Safety System (GMDSS) (see 7.3), weather information is grouped with navigational safety information as the second major portion of GMDSS and is collectively termed Maritime Safety Information (MSI).

Member nations of the International Maritime Organization (IMO) are responsible for ensuring that GMDSS MSI is broadcast over communications systems covering inshore, offshore and deep oceanic regions.

GMDSS MSI is the primary source of weather information to yachtsmen worldwide and is disseminated by national maritime administrations (eg MCA) over systems such as SafetyNET, NAVTEX, Marine VHF, MF and HF radio. Its content may also be rebroadcast by national broadcast (eg BBC) and Coast Radio Stations (CRS) and private organisations.

The sources of this information are national Meteorological Offices. Contact details for the UK Met Office are given in 1.2, other nations' are as follows:

> Republic of Ireland 6.14; Denmark 6.15; Germany 6.16; The Netherlands 6.17; Belgium 6.18; France 6.19; Spain 6.20 and 6.22; Portugal 6.21; Gibraltar 6.23.

6.6 NAVTEX

NAVTEX is the prime method of disseminating MSI from the fairway buoy to at least 200 miles offshore. A dedicated aerial and receiver with an LCD screen (or integrated printer) are required. The user selects which stations and message categories are recorded for automatic display or printing. Two frequencies are used: On the international frequency, 518kHz, messages are always available in English with excellent coverage of Europe. Interference between stations is **minimised** by scheduling time slots and by **limiting transmission power**; see Fig 6(3). NAVTEX information applies only to the geographical area for which each station is responsible.

On the national frequency 490kHz (shown in purple throughout this chapter) the UK issues inshore waters forecasts and coastal station actuals. Elsewhere it is used mainly for transmissions in the national language. 490khz stations have different identification letters from 518kHz stations.

NAVTEX is convenient, and particularly useful when preoccupied handling your vessel as you will not miss potentially important information.

6.6.1 NAVTEX Messages

Each message is prefixed by a four-character group: The **first character** is the code letter of the transmitting station (eg E for Niton). The **second character** is the message category, see 6.6.2. The **third and fourth** are message serial numbers, running from 01 to 99 and then re-starting at 01. The serial number 00 denotes urgent messages which are always displayed.

NAVTEX printers reject corrupt or already displayed messages; soft copy sets show all valid messages. Weather messages, and some other message types, are dated/timed. All NAVTEX messages begin with ZCZC, ending with NNNN.

Note: Navareas and Metareas have the same boundaries; in this chapter they are referred to as Metareas, but in Chapter 3 as Navareas. See also Fig 6(5) for Shipping forecast areas and Fig 6(4) for Inshore waters forecast boundaries.

6.6.2 NAVTEX Message categories

A*	Navigational warnings
B*	Meteorological warnings
C*	Ice reports
D*	SAR info and Piracy attack warnings
E	Weather forecasts
F	Pilot service
G	AIS
H	LORAN
I	Spare
J	SATNAV
K	Other electronic Navaids
L	Navwarnings additional to **A**
M-U	Spare
V-Y	Special services – as allocated
Z	No messages on hand

* On SOLAS sets these may not be rejected. Some leisure sets may be set to reject all message types except D.

Fig 6(3) NAVTEX stations/areas – UK and W Europe

6.6.3 UK/ROI 518 kHz broadcasts

Below (see next page for overseas), times (UT) of transmissions are in chronological order to facilitate the rapid finding of the next transmission. Weather messages are in **bold**. These are forecasts for 24 hours plus a brief 24 hour outlook. Times of an extended outlook are in italics; this is for 3 to 5 days beyond the shipping forecast 24 hour outlook. The 24 hour outlook and the extended outlook make NAVTEX arguably a more useful source of information than Radio 4. Times of an extended outlook (2 or 3 days beyond the shipping forecast period) are in *italics*.

E	–	**Niton** Thames clockwise to Fastnet, excluding Trafalgar.	*0040*	0440	**0840**	1240	1640	**2040**
G	–	**Cullercoats** Fair Isle clockwise to Thames, excluding N & S Utsire, Fisher and German Bight.	*0100*	0500	**0900**	1300	1700	**2100**
O	–	**Portpatrick** Lundy clockwise to SE Iceland.	*0220*	0620	1020	1420	**1820**	2220
Q	–	**Malin Head** Sea areas, Shannon, Rockall, Malin & Bailey		0640			1840	

6.6.4 UK/ROI 490 kHz stations

These provide four 24 hour forecasts and outlooks daily for UK inshore waters (to 12M offshore) at times in bold. Reports of actual weather at the places listed below are broadcast at times in *italics*. To receive these reports select 490 kHz and message category 'V' (6.6.2) on your NAVTEX receiver. Times (UT) of transmissions are listed in chronological order.

Actual Met data includes: Sea level pressure (mb), wind direction and speed (kn), weather, visibility (M), air and sea temperatures (°C), dewpoint temperature (°C) and mean wave height (m).

A – Malin Head	0000	0400	0800	1200	1600	2000

Lough Foyle to Carlingford Lough, Mull of Galloway to Cape Wrath, the Minch.

C – Portpatrick	0020	*0420*	0820	1220	*1620*	2020

Land's End to Cape Wrath, the Minch, Lough Foyle to Carlingford Lough
N Rona, Stornoway, S Uist, Lusa (Skye bridge), Tiree, Macrihanish, Belfast, Malin Head, Belmullet, St Bees Head, Ronaldsway, Crosby, Valley, Aberporth, Roches Point, Valentia, St Mawgan.

I – Niton	0120	0520	*0920*	1320	1720	*2120*

Gibraltar Point to St David's Head and the Channel Islands
Sandettie Lt V, Greenwich Lt V, Solent, Hurn airport, Guernsey airport, Jersey airport, Portland, Channel Lt V, Plymouth, Culdrose, Seven Stones Lt V, St Mawgan and Roches Point (Cork).

U – Cullercoats	*0320*	0720	1120	*1520*	1920	2320

Cape Wrath to North Foreland and Shetland
Sandettie Lt V, Manston, Shoeburyness, Weybourne, Donna Nook or Bridlington, Boulmer, Leuchars, Aberdeen, Lossiemouth, Wick, Kirkwall, Lerwick, Foula, K7 Met buoy, Sule Skerry.

6.6.5 NAVTEX coverage abroad

Selected NAVTEX stations in Metareas I and II, with their identity codes, are listed in chronological order of transmissions. Times of weather messages are shown in **bold**. Gale warnings are usually transmitted 4 hourly.

Note: these overseas stations all have UK sea areas in their broadcasts: Rogaland, Hamburg, Oostende, IJmuiden, Valentia, Malin Head. Though MRCC Tórshavn's ('D', Faroe Islands) does not, its forecasts may be useful (www.mrcc.fo)

TABLE 6(1) **NAVTEX broadcast schedule for Western Europe**

METAREA I (Co-ordinator – UK) — Transmission times (UT)

B –	Oostende, Belgium (Note 7)	0010	0410	0810	**1210**	1610	2010
D –	Tórshavn, Faroe Islands	0030	0430	0830	1230	1530	2030
K –	Niton (Note 1)	**0140**	0540	0940	**1340**	1740	2140
L –	Rogaland, Norway	0150	0550	0950	1350	1750	2150
L –	Pinneberg, Hamburg (Note 5)	0150	0550	0950	1350	1750	2150
M –	Jeløya, Norway	**0200**	0600	1000	**1400**	1800	2200
P –	Netherlands CG, Den Helder	**0230**	0630	1030	**1430**	1830	2230
Q –	Malin Head, Ireland (Note 3)	**0240**	**0640**	**1040**	1440	**1840**	**2240**
S –	Pinneberg, Hamburg (Note 6)	**0300**	**0700**	1100	**1500**	**1900**	**2300**
T –	Oostende, Belgium (Note 4)	**0310**	**0710**	1110	1510	**1910**	2310
V –	Oostende, Belgium (Note 2)	0330	0730	1130	1530	1930	2330
W –	Valentia, Ireland (Note 3)	0340	**0740**	**1140**	1540	**1940**	**2340**

Notes 1 In English, only Nav warnings for coast from Cap Gris Nez to Île de Bréhat. 2 Nav warnings only.
3 See 6.6 and Fig 6(5) for the Atlantic areas covered by 'Q' at 1040, 2240 and by 'W' at 1140, 2340.
4 Forecasts and strong wind warnings for Thames and Dover, plus nav info for the Belgian coast.
5 In German, weather and Nav warnings for German Bight (also W and S Baltic) at times in bold above.
6 Weather broadcasts every 4 hours for the North Sea. 7 Weather & Nav warnings, in Dutch occ English.

METAREA II (Co-ordinator – France) — Transmission times (UT)

A –	Corsen, Le Stiff, France	**0000**	0400	0800	**1200**	1600	2000
D –	Coruña, Spain	0030	0430	0830	1230	1630	**2030**
E –	Corsen, Le Stiff, France (in French)	0040	0440	0840	1240	1640	2040
F –	São Miguel, Azores, Portugal	**0050**	**0450**	**0850**	1250	**1650**	2050
G –	Monsanto, Portugal (in Portuguese)	0100	0500	0900	1300	1700	2100
M –	Porto Santo, Madeira, Portugal (in Portuguese)	0100	0500	0900	1300	1700	2100
G –	Tarifa, Spain	0100	0500	**0900**	1300	1700	**2100**
I –	Las Palmas, Islas Canarias, Spain	0120	0520	**0920**	1320	**1720**	2120
J –	São Miguel, Azores, Portugal (in Portuguese)	0130	0530	0930	1330	1730	2130
P –	Porto Santo, Madeira, Portugal	0230	0630	1030	1430	1830	2230
R –	Monsanto, Portugal	**0250**	**0650**	**1050**	**1450**	**1850**	**2250**
T –	Tarifa, Spain (in Spanish)	0310	0710	1110	1510	1910	2310
W –	Coruña, Spain (in Spanish)	0340	0740	**1140**	1540	**1940**	2340

6.7 GMDSS HIGH SEAS BULLETINS

In the N Atlantic part of MetArea I, dotted outline in Fig 6(3), weather data is available from:

- SafetyNET which provides MSI to vessels in A3 sea areas. Goonhilly Coast Earth Station (CES) transmits Met data at 0930 & 2130 UT via the **AOR-E** and **AOR-W** satellites.

- NAVTEX. Malin Head (Q) and Valentia (W) cover the E Northern and E Central sections, as well as Sole, Shannon, Rockall and Bailey.

6.8 MSI BROADCASTS BY HM COASTGUARD

MRCC routinely broadcast Maritime Safety Information (MSI) over VHF every 3 hours at the UTC times in Table 6(2) below.

Each broadcast contains one of 3 different Groups of MSI:

> **Group A**
> The full broadcast contains the Shipping forecast, Inshore waters forecast (for local Areas as in Table 6(1)), Gale and strong wind warnings.
>
> A Fisherman's 3 day forecast is broadcast (1 Oct–31 Mar) by MRCC which are MF equipped (see 6.8.3), except Tiree.
>
> Times of 'A' broadcasts in **bold type**.

> **Group B**
> Contains a new Inshore waters forecast, new outlook, and Gale warnings.'B' broadcast times in plain type.
>
> **Group C**
> A repeat of the Inshore waters forecast and Gale warnings (as per the previous Group A or B) plus any new Strong wind warnings. 'C' broadcast times are *italicised*.

Notes: Fig 6(4) shows the boundaries of the 17 areas which are covered by Inshore waters forecasts. In the forecast these areas are referred to not by their numbers, but by their defining places/headlands, eg Whitby to Gibraltar Point.

Subfacts & Gunfacts are broadcast by the following CGOC's Falmouth, Aberdeen (occas), Stornoway and Belfast. See 5.5, 5.6, Navigation (WZ) warnings, Subfacts & Gunfacts.

6.8.1 Actual weather

On request MRCC may report the actual weather in their vicinity. HM CG stresses that they are not qualified Met observers and such reports may be no more than a look out of the window or originate from passing ships and yachts.

The National Coastwatch Institution (NCI) operates on dedicated VHF Ch 65. There are 50 operational stations (2015) around the English and Welsh coast. Manned by volunteers keeping a visual watch during daylight hours throughout the year, they are able to provide, on request, details of actual weather conditions and sea state at their individual coastal locations (see Safety, 7.12).

TABLE 6(2) VHF broadcasts of shipping and inshore waters forecasts by HM Coastguard

Coastguard	Shipping forecast areas	Inshore waters forecast areas	Broadcast Groups and times UTC							
South Coast			B	C	A	C	B	C	A	C
Falmouth CGOC	Portland, Plymouth, Sole, Shannon Lundy, Fastnet, Biscay, Fitzroy	8, 9	0110	*0410*	**0710**	*1010*	1310	*1610*	**1910**	*2210*
Fareham NMOC	Portland, Wight	6, 7	0130	*0430*	**0730**	*1030*	1330	*1630*	**1930**	*2230*
Dover CGOC	Thames, Dover, Wight	5, 6	0110	*0410*	**0710**	*1010*	1310	*1610*	**1910**	*2210*
East Coast			B	C	A	C	B	C	A	C
Fareham NMOC	Thames, Humber, German Bight	5, 6	0110	*0410*	**0710**	*1010*	1310	*1610*	**1910**	*2210*
Humber CGOC	Forties, Forth, Tyne, Dogger, Fisher, Humber, German Bight	3, 4	0150	*0450*	**0750**	*1050*	1350	*1650*	**1950**	*2250*
Aberdeen CGOC	Cromarty, Forties, Forth, Fair Isle Viking, S Utsire, Fisher	1, 2	0130	*0430*	**0730**	*1030*	1330	*1630*	**1930**	*2230*
Shetland CGOC	Viking, Fair Isle, Faroes, N Utsire	18	0110	*0410*	**0710**	*1010*	1310	*1610*	**1910**	*2210*
West Coast			B	C	A	C	B	C	A	C
Stornoway CGOC	Rockall, Malin, Hebrides, Bailey, Fair Is, Faroes, SE Iceland	15,16,17	0110	*0410*	**0710**	*1010*	1310	*1610*	**1910**	*2210*
Belfast CGOC	Irish Sea, Rockall, Malin	13,14,	0210	*0510*	**0810**	*1110*	1410	*1710*	**2010**	*2310*
Holyhead CGOC	Irish Sea	11,12	0130	*0450*	**0730**	*1050*	1330	*1650*	**1930**	*2250*
Milford Haven CGOC	Lundy, Fastnet, Irish Sea, Shannon	9,10	0150	*0450*	**0750**	*1050*	1350	*1650*	**1950**	*2250*

CAUTION: The HM Coastguard modernisation programme should completed by the end of 2015. However, information on this and the next two pages (Fig 6(4) and 6.8.2 Remote MSI transmitters) still remains subject to potential change and readers are advised to check at www.mcga,gov.uk (search for: Maritime safety information leaflet), and also the free monthly updates from January 2016 to June 2016 available for download at www.reedsnauticalalmanacs.co.uk.

Fig 6(4) Remote Coastguard transmitters and Inshore waters forecast areas used by the UK Coastguard network

6.8.2 Remote MSI transmitters

Coastguard MSI broadcasts are transmitted via remote aerial sites giving optimum coverage. Figure 6(4) opposite shows their locations and VHF broadcast channels (either Ch 10, 23, 84 or 86), after a prior announcement on Ch 16. Broadcasts are made every 3 hours from the start times (UTC) shown below. **MF frequencies** are also used, see 6.8.3.

INSHORE WATERS FORECAST AREAS/MSI TRANSMITTERS/START TIMES

1 Cape Wrath to Rattray Head inc Orkney, 0130

A1	Durness (Loch Eriboll)
B1	Wideford Hill (Kirkwall, Orkney)
C1	Noss Head (Wick)
D1	Rosemarkie (Cromarty)
E1	Windy Head Hill (2M W Fraserburgh)

2 Rattray Head to Berwick-upon-Tweed, 0130

A2	Greg Ness (MF: see 6.8.3)
B2	Inverbervie
C2	Fife Ness
D2	Craigkelly (Burntisland)
E2	St Abbs Head

3 Berwick-upon-Tweed to Whitby, 0150

A3	Newton
B3	Cullercoats (Blyth) (MF: see 6.8.3)
C3	Boulby (10M N of Whitby)

4 Whitby to Gibraltar Point, 0150

A4	Ravenscar
B4	Flamborough Head
C4	Easington (Spurn Head)
D4	Skegness

5 Gibraltar Point to North Foreland. 0110

A5	Guy's Head (Wisbech)
B5	Langham (Blakeney)
C5	Trimingham (Cromer)
D5	Caister (Great Yarmouth)
E5	Lowestoft
F5	Bawdsey (R Deben)
G5	Walton-on-the-Naze
H5	Bradwell (R Blackwater)
I5	Shoeburyness

6 North Foreland to Selsey Bill, 0110

A6	North Foreland)
B6	Langdon (Dover)
C6	Fairlight (Hastings)
D6	Newhaven

7 Selsey Bill to Lyme Regis, 0130

A7	Boniface Down (Ventnor IoW)
B7	Needles
C7	St Alban's Head
D7	Grove Pt (Portland Bill)
E7	Beer Head (Lyme Bay)

8 Lyme Regis to Land's End inc the Isles of Scilly, 0110

A8	Berry Head (Torbay)
B8	Dartmouth
C8	East Prawle (Salcombe)
D8	Rame Head
E8	Fowey
F8	Falmouth
G8	Lizard
H8	Scillies (MF: see 6.8.3)

9 Land's End to St David's Head, inc the Bristol Channel, 0150

A9	Trevose Head
B9	Hartland Point
C9	Combe Martin (N Devon)
D9	Severn Bridges
E9	St Hilary (Barry)
F9	Monkstone (Tenby)
G9	St Ann's (Milford Haven)

10 St David's Head to Great Orme Head inc St George's Channel, 0150

A10	Dinas Head (Fishguard)
B10	Blaenplwyf (Aberystwyth)
C10	South Stack (Holyhead)
D10	Great Orme

11 Great Orme Head to the Mull of Galloway, 0130

A11	Moel-y-Parc(N Wales)
B11	Langthwaite (Lancaster)
C11	Caldbeck (Carlisle)

12 Isle of Man, 0130

A12	Snaefell (Isle of Man)

13 Lough Foyle to Carlingford Lough, 0210

A13	Slievemartin (Rostrevor)
B13	Orlock Head (Bangor)
C13	Black Mountain (Belfast
D13	West Torr (Fair Head)
E13	Limvady (Lough Foyle)
F13	Navar (Lower Lough Erne)

14 Mull of Galloway to Mull of Kintyre incl the Firth of Clyde and North Channel, 0210

A14	Rhu Staffnish (Kintyre)
B14	Lawhill (Ardrossan)
C14	Clyde (Greenock)

15 Mull of Kintyre to Arnamurchan Point, 0210

A15	Kilchiaran (SW Islay)
B15	South Knapdale (Loch Fyne)
C15	Torosay (E Mull)
D15	Glengorm (N Mull)
E15	Tiree (MF: see 6.8.3)

16 The Minch, 0110

A16	Arisaig (S of Mallaig)
B16	Drumfearn (SE Skye)
C16	Skriag (Portree, Skye)
D16	Rodel (S Harris)
E16	Melvaig (Loch Ewe)
F16	Portnaguran (E Lewis))

17 Arnamurchan Point to Cape Wrath, 0110

A17	Barra
B17	Clettreval (N Uist)
C17	Forsnaval (W Lewis)
D17	Butt of Lewis (MF: see 6.8.3)

18 Shetland Isles and 60 nautical mile radius, 0110

A18	Fitfull Head (Sumburgh)
B18	Lerwick (MF: see 6.8.3)
C18	Collafirth
D18	Saxa Vord (Unst)

6.8.3 MF MSI broadcasts

For the time-being MSI is broadcast via MF from the following Radio Sites (with an initial announcement on 2182 kHz):

MF Radio Site	Group A	Inshore waters gale warnings*
Tiree (1883 kHz)	0810, 2010	0210, 1410
Shetland (1770 kHz)	0710, 1910	0110, 1310
Butt of Lewis (1743 kHz)	0710, 1910	0110, 1310
Scillies (1880 kHz)	0710, 1910	0110, 1310
Gregness (2226 kHz)	0730, 1930	0130, 1330
Cullercoats (1925 kHz)	0750, 1950	0150, 1350

*Only made if there are gale warnings in force.

6.9 HF RADIO FACSIMILE BROADCASTS

Radio-Fax machines (or HF/SSB receiver and a computer) able to receive images, eg weather charts, are useful on board. Isobaric charts, sea and swell charts, satellite cloud images, sea temperature charts and wind charts are all of special interest to the blue water yachtsman. See ALRS, Vol 3.

6.9.1 DWD Offenbach

This reliable GMDSS related broadcast, is from the German weather bureau DWD (www.dwd.de). Transmitted from Pinneburg, DDH and DDK on 3855 kHz (10 kW), 7880 and 13881 kHz (both at 20 kW) it includes actual and forecast surface charts and sea state forecast, for up to 108 hours. Broadcast at 1110 UTC daily it can be downloaded from www.dwd.de/sendeplan

6.9.2 RN Northwood

Frequencies for the **RN, Northwood**, North Atlantic schedule are: 2618·5 kHz (2000–0600), 4610 & 8040 (H24), 11086·5 (0600–2000). The full schedule is broadcast at 0100 & 1300 UT; gale warnings: 0348, 0548, 1148, 1548, 1748, 2348.

6.10 SHIPPING FORECAST AREAS

Fig 6(5) UK forecast areas, SafetyNet area I/II boundary

SHIPPING FORECAST RECORD TIME/DAY/DATE ....

GENERAL SYNOPSIS   At 0048, 0520, 1201, 1754 UT/BST

System	Present position	Movement	Forecast position	At

Gales	SEA AREA FORECAST	Wind (At first)	(Later)	Weather	Visibility
	VIKING				
	NORTH UTSIRE				
	SOUTH UTSIRE				
	FORTIES				
	CROMARTY				
	FORTH				
	TYNE				
	DOGGER				
	FISHER				
	GERMAN BIGHT				
	HUMBER				
	THAMES				
	DOVER				
	WIGHT				
	PORTLAND				
	PLYMOUTH				
	BISCAY				
	TRAFALGAR (0048)				
	FITZROY				
	SOLE				
	LUNDY				
	FASTNET				
	IRISH SEA				
	SHANNON				
	ROCKALL				
	MALIN				
	HEBRIDES				
	BAILEY				
	FAIR ISLE				
	FAROES				
	S E ICELAND				

COASTAL REPORTS at .... UT BST	Wind Direction	Force	Weather	Visibility	Pressure	Change
Tiree auto (T)						
Stornoway (St)						
Lerwick (L)						
Leuchars (Le)						
Bridlington (B)						
Sandettié auto (S)						

COASTAL REPORTS	Wind Direction	Force	Weather	Visibility	Pressure	Change
Greenwich auto (G)						
Jersey (J)						
Channel auto (C)						
Scilly auto (Sc)						
Valentia (V)						
Ronaldsway (R)						
Malin Head (M)						

Fig 6(6) UK shipping forecast pro forma

6.10.1 BBC Radio 4 shipping and other forecasts

Although not part of the GMDSS, for many the shipping forecast and its associated reports, as broadcast by BBC Radio 4, remain the principal source of weather data. NAVTEX broadcasts a more complete service, see 6.6. The main benefit of the shipping forecast is that it is updated 6 hourly. HM Coastguard broadcasts repeats as detailed in 6.8.

LT	Bands	Contents summary
0048	LW, MW, FM	Shipping, and 18 inshore waters forecasts. Long list of station reports
0520	LW, MW, FM	Shipping, and 18 inshore waters forecasts. Short list of station reports
1201	LW only	Shipping forecast
1754	LW, FM (Sat/Sun)	Shipping forecast

Sea area Trafalgar is only included in the 0048 forecast.

Frequencies

LW	198 kHz	
MW	Tyneside 603 kHz. London & N Ireland 720 kHz Redruth 756 kHz. Plymouth & Enniskillen 774 kHz Aberdeen 1449 kHz. Carlisle 1485 kHz	
FM	**England** 92·4–94·6 MHz	
	Scotland 91·3–96·1 MHz	103·5–104·9 MHz
	Wales 92·8–96·1 MHz	103·5–104·9 MHz
	N Ireland 93·2–96·0 MHz	103·5–104·6 MHz
	Channel Is 94·8 MHz	

Contents of the shipping forecast

Time of issue; summary of gale warnings in force at that time; a general synopsis of weather systems and their expected development and movement over the next 24 hours; sea area forecasts for the same 24 hours, including wind direction/force, weather and visibility in each; and an outlook for the following 24 hours.

Gale warnings for all affected areas are broadcast at the earliest break in Radio 4 programmes after receipt, as well as after the next news bulletin.

Weather reports from coastal stations follow the 0048 and 0520 shipping forecasts; see Figs 6(5) and 6(6). They include wind direction and force, present weather, visibility, and sea-level pressure and tendency, if available.

Shipping forecasts cover large sea areas. The Inshore waters forecast (below) can provide more specific information to mariners on coastal passages.

6.10.2 BBC Radio 4 Inshore waters forecast

A forecast for inshore waters (up to 12M offshore) in 17 areas around the UK and N Ireland, valid for 24 hrs, is broadcast after the 0048 and 0520 coastal station reports. It includes forecasts of wind direction and force, weather, visibility, and sea state. The 17 inshore areas (Fig 6(4)) are defined by the following places and headlands, clockwise around the UK:

Cape Wrath, Orkney, Rattray Hd, Berwick-upon-Tweed, Whitby, Gibraltar Point, N Foreland, Selsey Bill, Lyme Regis, Land's End, St David's Head, Great Ormes Head,

Isle of Man, Mull of Galloway, Carlingford Lough, Lough Foyle, Mull of Kintyre, Ardnamurchan Pt and Shetland.

Reports of actual weather. The list of stations changes from time to time. Currently the following station reports are broadcast after the 0048 shipping forecast:

Tiree*, Stornoway, Lerwick, *Wick*, Aberdeen, Leuchars, *Boulmer*, Bridlington, Sandettie LV*, Greenwich LV*, *St Catherines Point*, Jersey, Channel LV*, Scilly*, *Milford Haven, Aberporth, Valley, Liverpool Crosby*, Valentia, Ronaldsway, Malin Head, *Macrihanish*.

After the 0520 shipping forecast, stations shown in italics are omitted. Those shown with an asterix are described as automatic weather stations. At these the current weather, such as rain, may not be reported.

Current observations can also be obtained from the internet, see www.xcweather.co.uk/ or www.metoffice.gov.uk/.

6.10.3 BBC general (land) forecasts

Land area forecasts are broadcast by BBC Radio 4 on the frequencies in 6.10.1. They can often include an outlook up to 48 hours beyond the shipping forecast, and more details of frontal systems and coastal weather.

Land area forecasts – Wind strength

Wind descriptions, with Beaufort scale equivalents, are:

Calm:	0	Fresh:	5
Light:	1–3	Strong:	6–7
Moderate:	4	Gale:	8

Land area forecasts – Visibility

The following definitions apply to land forecasts:

Mist:	Visibility between 2000m and 1000m
Fog:	Visibility less than 1000m
Dense fog:	Less than 50m

6.11 INTERNET WEATHER FORECASTS

The Internet is not an operational system but does provide a useful back-up and an alternative source for GMDSS services as well as information otherwise not available. Website addresses can change without notice; many services are automated and lacking quality control. There are no amendment procedures.

The UK Met Office website www.metoffice.gov.uk/weather/marine/ has texts of all marine forecasts, warnings and weather actuals from coastal stations, light vessels and data buoys. Additionally, the BBC website, http://www.bbc.co.uk/weather/coast_and_sea/ has 'Coastal forecasts' which are output from the Met Office high resolution weather prediction model. These give objective, 30 hour forecasts, at 6 hour intervals for locations around the UK.

The Met Office has a number of Apps, some of use for sailors (see: www.metoffice.gov.uk/services/public/free/mobile).

National meteorological services provide the most reliable forecasts for their own waters. Texts of GMDSS forecasts can be downloaded from text only web pages or by using free email text retrieval services. (See www.weather.mailasail.com/Franks-Weather/Home for details.) Some private firms also provide forecast services, e.g. www.weatheronline.co.uk

Useful, particularly for planning purposes, are GRIB files. These are automatically created from numerical weather prediction models, usually the US Global Forecast System (GFS). They can be obtained free by email from Saildocs (www.saildocs.com)

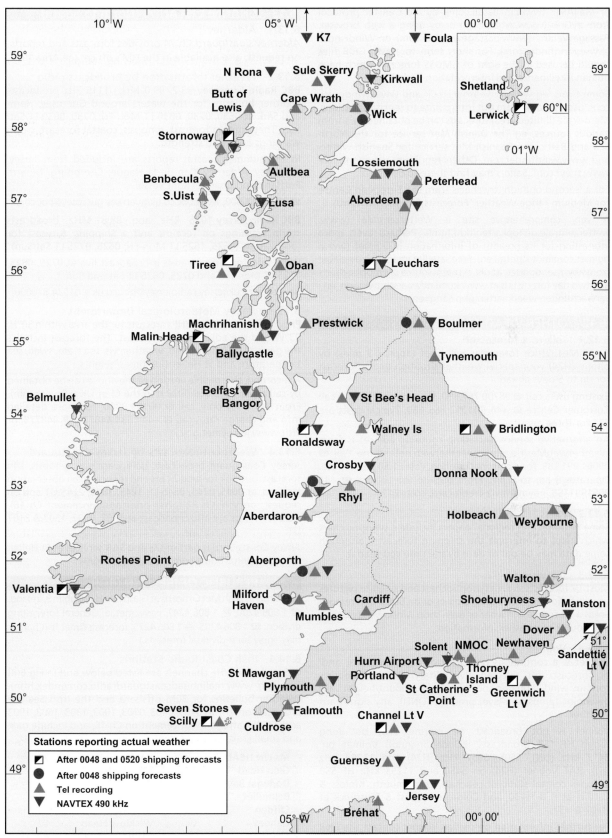

Fig 6(7) Stations reporting actual weather via BBC Radio 4, telephone recordings and NAVTEX

or MailASail (www.mailasail.com), by file transfer protocol from zyGrib (www.zygrib.org/) or by using a web browser, PassageWeather(www.passageweather.com) or WindFinder (www.windfinder.com). For short term forecasts, GRIB files should be used in the light of GMDSS forecasts which have human intelligence and interpretation added.

Some firms, eg www.xcweather.co.uk and www.windguru.com, use the same data but interpolate to specific locations. Meso-scale (limited area) forecasts can be obtained free from various sources, eg The Danish Met service for the North Sea and Baltic, the Spanish Met service for Spanish waters and www.windfinder.com. Others require prepayment, like www.theyr.com. Sailors may find these to be useful.

For a 'second opinion' try charts from the European Centre for Medium-range Weather Forecasts (www.ecmwf.int/).

A very comprehensive site is Wetterzentrale (www.wetterzentrale.de/topkarten/tknf.html). Perhaps even more interesting for its breadth of information is Ogimet (www.ogimet.com/index.phtml.en). Free satellite imagery is available from www.sat.dundee.ac.uk/. A useful source of synoptic charts and two day forecasts is at www.knmi.nl/waarschuwingen_en_verwachtingen/weerkaarten.php?lang=en

6.12   OTHER WEATHER RESOURCES

6.12.1   Talk to a forecaster
A UK Met Office forecaster will brief (approx 3 mins) by phone (H24) your special weather situation and a forecast for up to 5 days ahead.

Existing users call ☎ 08700 767 890. New users in the UK call Customer Centre ☎ +44 (0)1392 885 680. Typical costs are £21 for three minutes.

An alternative service (including routeing advice) is from WeatherwebMarine, www.weatherweb.net/marine Call ☎ 09061 991 189. for a personal briefing, cost £1.50 per minute. Operates 8 am to 6 pm. From outside the UK call ☎ +44 9061991189, payment via credit card, costs £12.

6.12.2   Press
Some national and regional papers include a synoptic chart which helps to interpret the shipping forecast – although some data may be out of date by the time you read it.

6.12.3   Television
Most TV forecasts show a synoptic chart and satellite pictures usually following news programmes or on BBC 24. In remote areas a TV forecast in a bar or even shop window may be the only source of weather information.

6.12.4   Volmet
Volmet is a continuous broadcast of actual weather and/or forecasts for selected civil/military airfields. Reports contain wind direction/spd, vis, cloud amount/height, air/dewpoint temps, sea level pressure (QNH), any significant weather.

Volmet is not broadcast on marine VHF, but long range HF SSB broadcasts include Shannon Volmet on: 3413 kHz (HN), 5505, 8957 kHz (H24) and 13264 kHz (HJ). RAF Volmet (H24) on 5450 and 11253 kHz HF SSB includes coastal stations: Leuchars, Lossiemouth, Kinloss & Cardiff at H & H+30; and Culdrose, Belfast & Prestwick at H+07 & +37.

For further information on civilian Volmet contact the Civil Aviation Authority. For military Volmet contact No 1 AIDU, RAF Northolt, West End Road, Ruislip, Middlesex HA4 6NG.

6.13   CHANNEL ISLANDS (North to South)

6.13.1   Alderney
Alderney Coastguard Ch 74 provides forecasts and reports on request; also available in the HM's office; see Area 19.

6.13.2   Weather information by broadcast radio
BBC Radio Guernsey 93·2, 99·0 MHz; 1116 kHz broadcasts weather bulletins for the waters around Guernsey, Herm and Sark at 0630, 0730, 0830 LT Mon-Fri; 0730, 0830 LT Sat/Sun. They contain forecast, synopsis, coastal forecast, storm warnings and wind strength.

In the summer coastal reports are included from Jersey, Guernsey, Alderney, Cap de la Hague, Cherbourg, Dinard, Portland and Channel light vessel.

☎ 01481 200600. www.bbc.co.uk/guernsey guernsey@bbc.co.uk

BBC Radio Jersey 1026 kHz and 88·8 MHz broadcasts storm warnings on receipt; and a Shipping forecast for local waters (0625, 1625 LT Mon-Fri; 0625, 0725 LT Sat/Sun).

Wind info for Jersey waters is broadcast live at 0725, 0825, 1725 LT Mon-Fri; and 0725, 0825 LT Sat and Sun.

www.bbc.co.uk/jersey radiojersey@bbc.co.uk e 01534 870000.

6.13.3   Jersey Meteorological Department
Provides 6 hourly 24hr VHF forecasts for the area within 50°N, 03°W and the adjacent French coast. The forecast includes the general situation, wind, weather, vis, sea state, swell, sea temperature and St Helier tide times/heights.

A recorded, chargeable version of the above can be obtained by calling ☎ 0900 669 0022 if in the CI or UK (60p per min). From Guernsey only, call ☎ 06969 8800. For more detailed info and charge rates call ☎ +44 1534 448770, 🖷 448778; or visit www.jerseymet.gov.je

6.13.4   Weather broadcasts by Jersey Coastguard
Jersey Coastguard broadcasts gale warnings, synopsis, 24h forecast, outlook for next 24 hrs and reports from observation stations, at 0645, 0745, 0845 LT; 1245, 1845, 2245 UT and on request; on VHF Ch 82 (after prior announcement Ch 16). Gale warnings are also broadcast at 0307, 0907, 1507 & 2107 UT.

Jersey Coastguard is part of CG and SAR services at St Helier. www.jersey-harbours.com e 01534 447705/04.

6.14   REPUBLIC OF IRELAND (ROI)
Met Éireann (Irish Met Office) is at Glasnevin Hill, Dublin 9, Ireland. ☎ 1 806 4200, 🖷 1 806 4247, www.met.ie General forecasting division: ☎ 1 806 4255, 🖷 1 806 4275, forecasts@met.ie (Charges may apply for individual forecasts.)

6.14.1   Irish Coast Radio Stations
CRS and their VHF channels are listed below and in Fig 6(8) (see also www.malinheadcoastguardradio.com/index.htm). Weather bulletins for 30M offshore and the Irish Sea are broadcast on VHF at 0103, 0403, 0703, 1003, 1303, 1603, 1903 and 2203LT after announcement on Ch 16, and include gale and small craft warnings, synopsis and a 24-hour forecast.

MALIN HEAD	23	Bantry	23
Glen Head	24	Mizen Head	04
Donegal Bay	02	Cork	26
Belmullet	83	Mine Head	83
Clifden	26	Rosslare	23
Galway	04	Wicklow Head	02
Shannon	28	DUBLIN	83
VALENTIA	24	Carlingford	04

Gale warnings are broadcast on receipt and at 0033, 0633, 1233 and 1833 UT, after an announcement on Ch 16. **MF** Valentia Radio broadcasts forecasts for Shannon and Fastnet on 1752 kHz at 0833 & 2033 UT, and on request. Gale warnings are broadcast on receipt and at 0303, 0903, 1503 and 2103 (UT) after an announcement on 2182 kHz. Weather is not broadcast on MF by Malin Head or Dublin.

6.14.2   NAVTEX

Malin Head MRCC broadcasts (ID 'A', see Fig 6(3)) an inshore forecast at 0000, 0400, 0800, 1200, 1600 and 2000 for North of Ireland, West of Scotland and North of Scotland. Also worth receiving from Portpatrick.

6.14.3   Weather by telephone

The latest sea area forecasts and gale warnings are available H24 from Weatherdial (www.weatherdial.ie) as recorded messages. Dial ☎ 1550 123 plus the following suffixes: 850 Munster; 851 Leinster; 852 Connaught; 853 Ulster; 854 Dublin (plus winds in Dublin Bay and HW times); 855 Coastal waters and Irish Sea.

6.14.4   Weather by fax

Similar information, plus isobaric, swell and wave charts and any Small craft warnings (>F6 up to 10M offshore; Apr–Sep inc) is available H24 by Weatherdial 🖷 1570 131 838 (from within ROI only). From the menu below select the required 4-digit product code (see code 0400 for full listing):

- 0015: Latest analysis chart
- 0016: Forecast valid for next 24 hrs
- 0017: Forecast valid for next 36 hrs
- 0018: Forecast valid for next 48 hrs
- 0021: Forecasts for coastal waters and Irish Sea
- 0031, 0032, 0033, 0034: Forecast (days 1–4) for sea and swell wave heights and periods
- 5-day forecasts (plain language, farming/national) 0001: Munster. 0002: Leinster. 0003: Connaught. 0004: Ulster. 0005: Dublin.

Sea Planners provide graphic forecasts <5 days (updated at 0430 d'ly) of expected winds/waves at 7 offshore positions:

0041: 53°N 05° 30'W.	0045: 54°N 11°W.
0042: 51°N 06°W.	0046: 55°N 10°W.
0043: 51°N 10°30'W. 0044: 53°N 12°W.	0047: 56°N 08°W.

6.14.5   Radio Telefís Éireann (RTE) Radio 1

RTE Radio 1 broadcasts weather bulletins daily at 0602, 1253, & 2355LT on 252kHz (LW) Summerhill (15M W of Dublin airport) and FM (88·2–90·0MHz).

Bulletins contain a situation, forecast and coastal reports for Irish Sea and coastal waters. The forecast includes: wind, weather, vis, swell (if higher than 4m) and a 24 hrs outlook.

Gale warnings are included in hourly news bulletins on FM & MF.

Coastal reports include wind, weather, visibility, pressure, with pressure tendency over the last 3 hrs described as:

Steady	=	0–0·4 hPa change
Rising/falling slowly	=	0·5–1·9
Rising/falling	=	2·0–3·4
Rising/falling rapidly	=	3·5–5·9
Rising/falling very rapidly	=	> 6·0

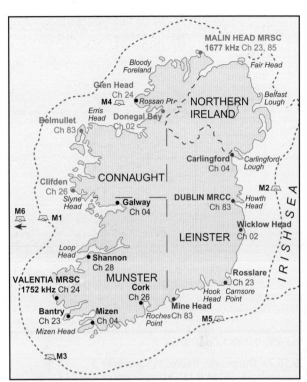

Fig 6(8) shows the CRS, Met buoys M1–M6, provinces and headlands named in forecasts

FOREIGN LANGUAGE WEATHER INFORMATION

It's worth learning some basic meteorological terms in other languages (see 1.6). Especially so in France where CROSS, for the most part, take great care to read forecasts on marine VHF reasonably slowly and with good, clear diction. It is a good idea to use the Internet before you go, to become familiar with the local forecast structure, terms used and the places they refer to. Do not bank on being able to download an English language forecast in the area!

6.15   WEST DENMARK

6.15.1   Gale warnings and forecasts are broadcast on receipt, on request and at 0133, 0533, 0933, 1333, 1733, 2133 UT, in Danish/English by remote CRS, callsign *Lyngby Radio:*

Areas, Fig 6(9)	**VHF Channels**
2 South Baltic	02, 04
3 West Baltic	01, 02, 03, 04, 61, 85
4 The Belts & the Sound	01, 02, 03, 04, 05, 07, 61, 65, 83, 85
5 Kattegat	03, 04, 05, 07, 64, 65, 66, 83
6 Skagerrack	01, 02, 04, 64, 66
8 Fisher	01, 02, 23, 66
9 German Bight	02, 23

Areas 2–5 are north of the Kiel Canal and east of 10°E.

MF: Blåvand 1734 kHz, Skagen 1758 kHz, Skamlebæk 1704 kHz and Rønne 2586 kHz broadcast gale warnings for all areas in Danish/English on receipt.

KEY:
2 S Baltic
3 W Baltic
4 The Belts and the Sound
5 Kattegat
6 Skagerrak
8 Fisher

KEY:
9 German Bight
13 Forties
14 Dogger
15 Humber

Fig 6(9) Danish forecast areas

6.15.2 Danmarks Radio, Programme 1
Kalundborg (55°44'N 11°E) broadcasts on LW 243 kHz:
• Gale warnings, weather reports and forecast for areas 2-6, 8 and 9 at 0545, 0845, 1145, 1745LT.
• A 5 day forecast for areas 2-6, 8, 9 and 13–15 at 0545, 0845 & 1745LT.

6.15.3 Danmarks Meteorologiske Institut (DM)
Provides marine forecasts in **English** on www.dmi.dk/eng/index/forecasts.htm.

Stockholm Radio offers two PDF downloads showing VHF weather broadcast channels in Denmark: www.stockholmradio.se Click on Kustradio / VHF

6.16 GERMANY

6.16.1 Deutscher Wetterdienst (DWD)
DWD (German weather service) provides Met info which is updated twice daily; more often for weather reports and text forecasts. SEEWIS (Sea Weather Information System) allows data to be accessed by telephone/modem and fed into an onboard computer.

DWD ☎ + 49 (0) 69 8062 6201. 🖷 + 49 (0) 69 8062 6193. www.dwd.de   seeschifffahrt@dwd.de

6.16.2 Traffic Centres
Traffic Centres, Fig 6(10), broadcast local storm warnings, weather bulletins, visibility in German or English on request.

Traffic Centre	VHF Ch	Every
German Bight Traffic	80	H+00
Cuxhaven-Elbe Traffic	71 (outer Elbe)	H+35
Brunsbüttel-Elbe Traffic	68 (lower Elbe)	H+05
Kiel Kanal II (E-bound)	02	H+15 & H+45
Kiel Kanal III (W-bound)	03	H+20 & H+50
Bremerhaven-Weser Traffic	02, 04, 05, 07, 21, 22, 82	H+20
Bremen-Weser Traffic	19, 78, 81	H+30
Hunte Traffic	63	H+30
Jade Traffic	20, 63	H+10
Ems Traffic	15, 18, 20, 21	H+50

6.16.3 Coast Radio Stations
DP07 (Seefunk) operates commercial CRS, Fig 6(10), at:

Nordfriesland (Sylt)	Ch 26
Elbe-Weser	Ch 01, 24
Hamburg (Control centre)	Ch 83
Bremen	Ch 25
Accumersiel	Ch 28
Borkum	Ch 61
Helgoland	Ch 86

DP07 broadcasts (only in German): gale and strong wind warnings on receipt. At 0745Ⓐ, 0945, 1245, 1645 and 1945Ⓐ

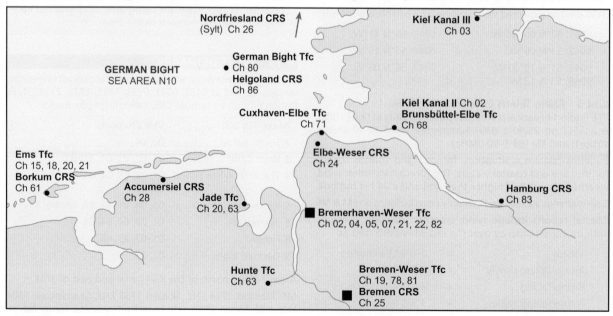

Nordfriesland CRS (Sylt) Ch 26

Kiel Kanal III Ch 03

GERMAN BIGHT SEA AREA N10

German Bight Tfc Ch 80
Helgoland CRS Ch 86

Kiel Kanal II Ch 02
Brunsbüttel-Elbe Tfc Ch 68

Cuxhaven-Elbe Tfc Ch 71

Ems Tfc Ch 15, 18, 20, 21
Borkum CRS Ch 61

Elbe-Weser CRS Ch 24

Hamburg CRS Ch 83

Accumersiel CRS Ch 28
Jade Tfc Ch 20, 63

Bremerhaven-Weser Tfc Ch 02, 04, 05, 07, 21, 22, 82

Hunte Tfc Ch 63

Bremen-Weser Tfc Ch 19, 78, 81
Bremen CRS Ch 25

Fig 6(10) German MSI broadcasts

LT for Fisher, German Bight and Humber, DP07 broadcasts: a synopsis, 12-hour forecast, 24 hrs outlook and coastal station reports. ⒶSummer only. Also a 4–5 day outlook for the North Sea (and Baltic) at 0945 and 1645.

6.16.4   NAVTEX
Pinneberg (S) in **English** on 518 kHz and Pinneberg (L) in German on 490 kHz transmit Weather bulletins (and Nav warnings) for German Bight (area N10) at the times listed in Table 6(1).

6.16.5   Radio broadcasting
NDR 1 Welle Nord (FM)
A summary, outlook and wind forecast for the German Bight is broadcast after the news (1006, 1300, 1906 LT) by **Sylt 92.7** MHz; **Helgoland** 92.5 MHz; **Hamburg** 92.3MHz; **Flensburg** 87.7MHz; **Heide** 87.9MHz; **Keil** 99.7MHz.

Radio Bremen (MW and FM)
Warnings of extreme weather conditions in German Bight, with potential for corresponding hazards, are broadcast after the news by: **Bremerhaven** 936 kHz; 89·3, 92·1, 95·4 & 100·8 MHz; and by **Bremen** 88·3, 93·8, 96·7 & 101·2 MHz.

6.16.6   HF Radio
Offenbach (Main) Programme 1 broadcasts in **English** on 4583, 7646 and 10100·8 kHz at 0005, 0020, 0305, 0320, 0535, 0550, 0835, 0850, 1135, 1150, 1435, 1450, 1735, 1750, 2035 and

2050 UT: Weather situation, 12 hour forecast and a further 12 hours outlook for N Sea, Baltic and coastal areas.

At 0355, 1530 UT: North Sea weather and 5 day forecasts in **English**. Further broadcast schedule: www.dwd.de/sendeplan.

6.16.7   Recorded telephone forecasts for yachtsmen
For wind forecast and outlook (1/4–30/9; only within Germany) call ☎ 0900 111 6050 plus two digits as follows:

00	Number for pleasure craft
54	Denmark
55	Netherlands, IJsselmeer, Maas and Schelde
59	Windline latest station reports

For year-round telephone weather synopsis, forecast and outlook, call ☎ 0900 111 6920 plus two digits as follows:

00	General maritime weather
21	North Sea and Baltic
22	German Bight, Fisher and SW North Sea

Call ☎ +49 (0)69 8062 6101 (H24) for the latest wind warnings (> F6) for individual North Sea areas. If no warning is in force, a wind forecast for the German Bight and W/S Baltic is given.

6.17   THE NETHERLANDS

Fig 6(11)   Netherlands MSI broadcasts

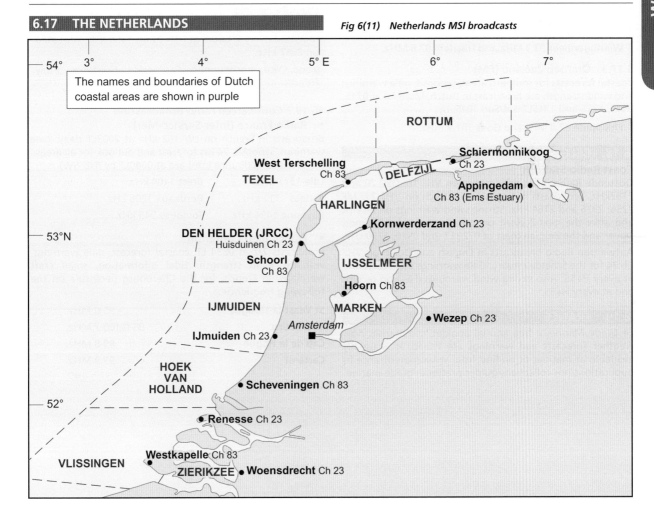

The names and boundaries of Dutch coastal areas are shown in purple

6.17.1 Netherlands Coastguard
VHF MSI broadcasts – Fig 6(11), NE to SW
Forecasts for 7 areas to 30M offshore and 3 inland waterways (IJsselmeer, Marken and Zierikzee) are broadcast in **English** and Dutch at 0805, 1305, 1905, 2305 LT on VHF as shown below, **without** prior announcement on Ch 16 or 70.

> **VHF Ch 23**: Schiermonnikoog, Kornwerderzand, Wezep, Huisduinen (Den Helder), IJmuiden, Renesse, Woensdrecht.
>
> **VHF Ch 83**: Appingedam, West Terschelling, Hoorn, Schoorl, Scheveningen, Westkapelle.
> All stations monitor Ch 16.

Gale warnings are broadcast 0333, 0733, 1133, 1533, 1933 and 2333UT and on receipt.
Ouddorp CG (51°48'N 03°52'E) broadcasts weather info Ch 74, 25 at H+30 and gives radar assistance on request H24.

MF weather broadcasts
Forecasts for Viking, Forties, Dogger, Fisher, German Bight, Humber, Thames and Dover are broadcast by Scheveningen in **English** at 0940 and 2140UT on 3673 kHz. Gale warnings for these areas are broadcast in **English** at 0333, 0733, 1133, 1533, 1933 and 2333UT and on receipt.

6.17.2 Radio Noord-Holland (FM)
Coastal forecasts for northern areas, gale warnings and wind strength are broadcast in Dutch, Mon–Fri at 0730, 0838, 1005, 1230 and 1705LT; Sat/Sun 1005, by:

Wieringermeer 93.9 MHz and **Haarlem** 97.6 MHz.

6.17.3 Omroep Zeeland (FM)
Coastal forecasts for southern areas, synopsis, gale warnings and wind strength are broadcast in Dutch, Mon–Fri at 0715, 0915, 1215 and 1715LT; Sat/Sun 1015, by:

Philippine 97.8 MHz and **Goes** 101.9 MHz.

6.18 BELGIUM

Coast Radio Stations
Oostende Radio, after prior notice on VHF 16, DSC 70 and 2182kHz, broadcasts in **English** and Dutch on VHF 27, MF 2256, 2376 and 2761 kHz: Strong wind warnings on receipt and after the next 2 silent periods. Forecasts for Thames, Dover and the Belgian coast at 0720 LT and 0820, 1720 UT.

Antwerpen Radio broadcasts in **English** and Dutch on VHF Ch 24 for the Schelde estuary: Gale warnings on receipt and at every H+55. Also strong wind warnings (F6+) on receipt and at every H+55.

6.19 FRANCE
Le Guide Marine, a free annual booklet, summarises how weather forecasts and warnings are issued. It is usually available at marinas or on-line from www.meteofrance.fr/publications/nos collections/guides-pratiques/guide-marine.

6.19.1 CROSS VHF and MF broadcasts
CROSS broadcasts Met bulletins on VHF in French, after an announcement on Ch 16. In the English Channel broadcasts can be given in **English**, on request Ch 16. Remote stations, VHF channels and local times are shown in Fig 6(12) on page 101. There are continuous broadcasts on VHF Ch 63 from Étel for Pointe de Penmarc'h to L'Anse de L'Aiguillon and from Chassiron for L'Anse de L'Aiguillon to the Spanish border.

A useful source of texts for all Météo France marine forecasts is www.meteo-marine.com/meteo-marine

Broadcasts include: Gale warnings, general situation, 24 hrs forecast (actual weather, wind, sea state and vis) and trends for coastal waters (20M offshore).

Gale warnings (*Bulletins Météorologique Spéciaux* or BMS). are broadcast on VHF at H+03. Warnings and forecasts on MF frequencies are as shown below.

> **CROSS GRIS-NEZ**
> Humber, Thames, Dover Wight. Forecasts at 0833 and 2033. Warnings at HH+03, 1650 kHz after announcement on 2667 kHz.
>
> Thames, Dover, Wight, Portland, Plymouth. 3-day forecast at 1633 on 1650 kHz
>
> **CROSS CORSEN**
> Atlantic and English Channel. Forecasts at 0815 and 2015 repeated every 2 hours at HH+03 on 1650 kHz, repeated on 2667 kHz.
>
> Iroise, Yeu, Rochebonne, Cantabrico, Finisterre. 3-day forecast at 1615, on 1650 kHz.

16.19.2 Commercial radio broadcasting
- **Radio France (Inter-Service-Mer)**
Broadcasts in French on LW 162 kHz at 2003LT daily: gale warnings, synopsis, 24 hrs forecast and outlook for all areas. MF broadcasts (France Info) are at 0640LT by (NE–SW):

Lille 1377 kHz.	**Brest** 1404 kHz.
Rennes 711 kHz.	**Bordeaux** 1206 kHz.
Bayonne 1494 kHz.	**Toulouse** 945 kHz.

- **France Bleu Contentin**
Broadcasts 7/7 at 0834 LT: coastal forecast, gale warnings, visibility, wind strength, tidal information, small craft warnings, in French, for the Cherbourg peninsula on the following frequencies:

St Vaast-la-Hougue	95·0 MHz.
Cherbourg	95·0/100·7 MHz.
Cap de la Hague	99·8 MHz.
Carteret	99·9 MHz.

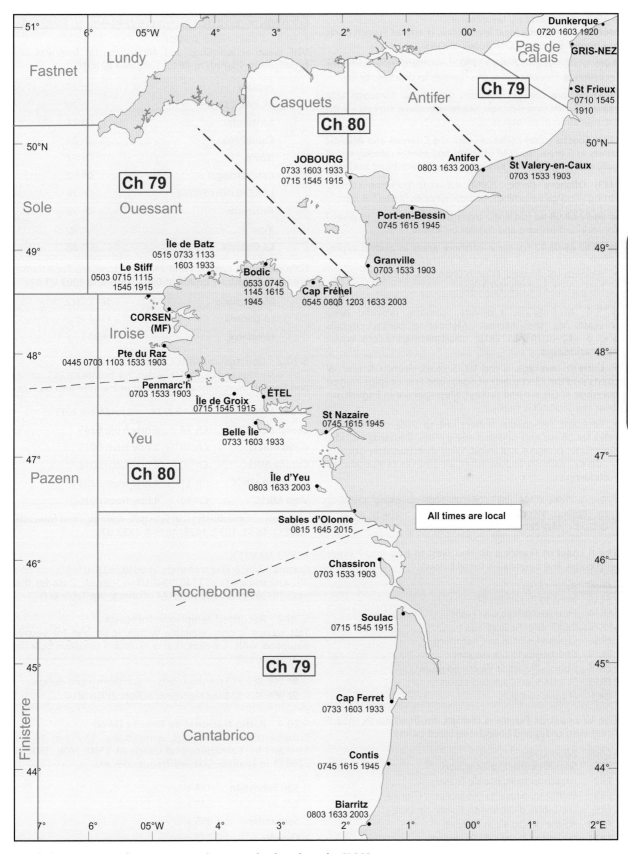

Fig 6(12) Météo France forecast areas and VHF weather broadcasts by CROSS

6.19.3   Forecasts by telephone

Météo is the ☎ of a local Met Office, if any, at French ports in Chapter 9, under COMMUNICATIONS.

Auto: Dial 08·92·68·32·50 (or 3250 in mainland France only) for recorded ☎ forecasts, plus the number in brackets below:

(331) **Inshore** (*rivage*) bulletins give 7-day forecasts (2M seaward), tide times, signals, sea temperature, surf etc – with an indication of reliability.

(332) **Coastal** (*côte*) bulletins cover the Channel and Atlantic coasts in 5 areas, to 20M offshore and contain strong wind/gale warnings, general synopsis, 7-day forecast and outlook.

(333) **Offshore** (*large*, 200M seaward) bulletins contain strong wind/gale warnings, general synopsis, 7-day forecast.

☎ 08·92·68·08·dd (dd is the *département* number, as given) for localised inshore and coastal bulletins.

☎ 08·92·68·08·77 for same offshore areas as above; saying 'STOP' as it is named, selects 1 of 3 offshore areas:
- Western UK, Ireland, English Channel and Bay of Biscay;
- Eastern UK and the North Sea.

6.19.4   Metmarine, a private company, provides Area forecasts via the Internet. Metmarine contact details are: e +44 (0)207 6177818; info@metmarine.com www.metmarine.com

- Current coverage is the UK S coast, French N and W coasts and the Med coasts of Spain and France (plus limited coverage of Greece and Turkey). Messages are in **English**; no prior registration is needed.

- Forecasts for coastal waters out to 20M offshore are valid for 36 hrs and updated every 6 hrs. Contents include: Time/date; pressure and trend; wind direction/strength and tendency; visibility and weather; air and sea temperatures. For example:

> *Today from 0700 BST 10/9/08. 1008mb rising slowly to 1009mb. Wind NE2, backing NNE2. Visibility good, drizzle.   Max temp 23°C. Sea temp 17°C.*

The N coast of France is divided, west to east, into 7 Areas with codes and boundaries listed below):

Code	Boundaries
FR13	Brest to Ploumanac'h
FR14	Ploumanac'h to Cap Fréhel
FR15	Cap Fréhel to Granville
FR16	Channel Islands to Cherbourg
FR17	Cherbourg to Courseulles
FR18	Courseulles to St Valéry-en-Caux
FR19	St Valéry-en-Caux to Étaples
UK1	Étaples to Calais

The W coast of France is divided, south to north, into 7 Areas with codes and boundaries listed below:

Code	Boundaries
FR6	Spanish border to Arcachon
FR7	Arcachon to Pointe de Grave
FR8	Pointe de Grave to Les Sables d'Olonne
FR9	Les Sables d'Olonne to Pointe de Croisic
FR10	Pointe de Croisic to Lorient
FR11	Lorient to Pointe de Penmarc'h
FR12	Pointe de Penmarc'h to Brest

6.20   NORTH AND NORTH WEST SPAIN

6.20.1   Coast radio stations

VHF weather warnings and 48 hrs coastal forecasts are broadcast in Spanish at 0840, 1240 and 2010 UT by:

Pasajes	Ch 27
BILBAO CENTRE	Ch 26
Santander	Ch 24
Cabo Peñas	Ch 27
Navia	Ch 62
Cabo Ortegal	Ch 02
LA CORUÑA CENTRE	Ch 26
Finisterre	Ch 22
Vigo	Ch 20
La Guardia	Ch 82

Gale warnings, synopsis and 24/48 hrs forecasts for Atlantic areas are broadcast on MF at 0703, 1303 and 1903 UT by:

Machichaco	1677 kHz
La Coruña	1707 kHz
Finisterre	1698 kHz

6.20.2   Coastguard MRCC/MRSC

Gale warnings and coastal forecasts are broadcast in Spanish and **English** on receipt and as listed below (see also 6.22.2):

Bilbao MRCC*	Chs 10 & 74	2 hrly from 0033
Santander MRSC	Ch 74	2 hrly from 0245
Gijón MRCC	Ch 10	2 hrly from 0015
Coruña MRSC	Ch 10	2 hrly from 0015
Finisterre MRCC	Ch 11	4 hrly from 0233
Vigo MRSC	Ch 10	4 hrly from 0015

*****Bilbao** also broadcasts High Seas gale warnings and forecasts at 0233, 0633, 1033, 1433, 1833 & 2233 UT.

6.20.3 NAVTEX

Coruna (D) 518 kHz transmits at 0830, 2030 UT. Coruna (W) 490 kHz transmits at 1140, 2340 UT in Spanish. Data for the next 18–36 hrs is valid to 450M offshore; see Table 6(1).

6.20.4   Recorded telephone forecasts

This service is only available within Spain and for vessels equipped with Autolink. For a recorded weather bulletin in Spanish, call:

☎ 906 365 372 for the coasts of Cantábrico and Galicia.
☎ 906 365 374 for High Seas bulletins; Fig 6(14).

6.20.4   Radio Nacional de España (MW)

Broadcasts storm warnings, synopsis and 12 hrs or 18 hrs forecasts for Cantábrico and Galicia at 1100, 1400, 1800 & 2200 LT in Spanish. Stations/frequencies are:

San Sebastián	774 kHz
Bilbao	639 kHz
Santander	855 kHz
Oviedo	729 kHz
La Coruña	639 kHz

Fig 6(13) North and North West Spain

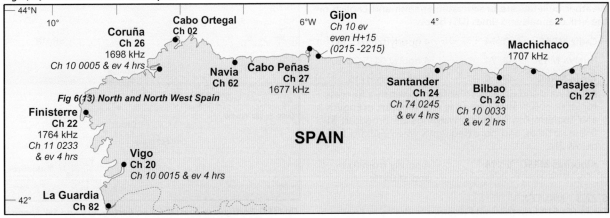

6.21 PORTUGAL, AZORES AND MADEIRA

6.21.1 Radionaval Portugal weather broadcasts

Broadcasts in Portuguese and **English** are on Ch 11 at the times (UT) listed below. They contain: Storm, gale and poor visibility warnings; synopsis and 24 hrs forecasts for three coastal zones out to 20M offshore.

Fig 6(14) Portuguese coastal zones, inset shows forecast areas for MetArea II: High Seas and Coastal/Offshore Areas

Leixões Coastal waters of N and Central zones.	0705, 1905
Cencomar (also on MF 2657 kHz) Coastal waters of all 3 zones.	0905, 2105
Faro Coastal waters of Central and S Zones.	0805, 2005
Horta Azores archipelago (MF 2657 kHz)	0935, 2135 UT
Cencomarmadeira (Madeira) for 20M radius of Madeira and Porto Santo (MF 2657 kHz) VHF Ch 11	0735, 1935 1030, 1630 LT

6.21.2 Radiofusão Portuguesa

Broadcasts weather bulletins for coastal waters in Portuguese at 1100 UT. Stations and frequencies are:

Porto, Elvas	720 kHz
Viseu, Lisboa 1	666 kHz
Montemer (Coimbra), **Miranda do Douro**	630 kHz
Faro	97·6 MHz, 720 kHz

6.21.3 NAVTEX

Transmitters: Monsanto/Lisbon(R),(G), São Miguel/Azores (F),(J), and Porto Santo/Madeira – (P),(M), broadcast weather forecasts in **English** for the next 24H on 518 kHz, every four hours. (R) starts 0250, (F) 0050, (P) 0230 inc. nav warnings within 20M of the coast. These are repeated in Portuguese on 490kHz, (G) starts 0100, (J) 0130, (M) 0100.

6.22 SOUTH WEST SPAIN

6.22.1 Coast radio stations

The following CRS broadcast gale warnings, synopsis, 24 hrs and 48 hrs forecasts for Atlantic and Mediterranean areas, in Spanish, at the times (UT) and on the VHF and MF frequencies shown below (gale warnings broadcast on receipt):

Cádiz	*Ch 28*	*0833*	*1133*	*2003*
Tarifa	*Ch 83*	*0833*	*1133*	*2003*
	1656 kHz	*0733*	*1233*	*1933*
Malaga	*Ch 26*	*0833*	*1133*	*2003*

6.22.2 Coastguard MRCC/MRSC

Weather bulletins are broadcast in Spanish and **English** on the VHF channels and times (UT) below:

Cadiz MRSC	Ch 74	4 hourly from 0315
Huelva MRSC	Ch 74	4 hourly 0415-2015
Tarifa MRCC	Ch 10, 67	Every even H+15.

Actual wind and visibility at Tarifa, followed by a forecast for Strait of Gibraltar, Cádiz Bay and Alborán, in English and Spanish. Fog (visibility) warnings are broadcast at the same times, and more frequently when visibility falls below 2M.

Algeciras MRSC	Ch 74	4 hourly from 0315 and at 0515.

6.22.3 NAVTEX

Tarifa (G) 518 kHz transmits weather bulletins in English at 0900 and 2100 UT. These include: Gale warnings, general synopsis and development and a forecast, valid for 18 hrs from 0900 or for 36 hrs from 2100, for the N Atlantic and W Mediterranean within 450M of the coast.

Tarifa (T) 490 kHz transmits at 1110, 2310 UT in Spanish. Data for the next 18–36 hrs is valid to 450M offshore; see 6.6.6.

6.22.4 Recorded telephone forecasts

The service is only available within Spain and for Autolink-equipped vessels. For a coastal waters bulletin in Spanish call:

- ☎ 906 365 373 for Atlantic Andalucia and the Canaries.
- ☎ 906 365 374 for High Seas bulletin for Atlantic areas.

6.23 GIBRALTAR

6.23.1 Gibraltar Broadcasting Corporation (GBC)

GBC (Gibraltar Radio) broadcasts in English: General synopsis, situation, wind direction and force, visibility and sea state, radius 5M from Gibraltar. Frequencies are 1458 kHz, 91·3 MHz, 92·6 MHz and 100·5 MHz. Times (UT):

Mon–Fri:	0530, 0630, 0730, 1030, 1230
Sat:	0530, 0630, 0730, 1030
Sun:	0630, 0730, 1030

Fig 6(15) Portuguese coastal zones, Portuguese and SW Spanish weather broadcast stations

6.23.2 British Forces Broadcasting Service (BFBS)

Gale warnings for the Gibraltar area are broadcast on receipt by BFBS 1 and 2. All broadcasts are in English for waters within 5M of Gib and contain: Shipping forecast, wind, weather, visibility, sea state, swell, HW and LW times.

BFBS 1

Frequencies and times (Local): 93·5, 97·8* MHz FM.

Mon–Fri:	0745, 0845, 1005, 1605
Sat:	0845, 0945, 1202
Sun:	0845, 0945, 1202, 1602

* This frequency is reported to have greater range.

BFBS 2

Frequencies and time (Local): 89·4, 99·5 MHz FM.

Mon–Fri: 1200

6.23.3 Other sources

www.weatheronline.co.uk gives Gib marine weather. Metmarine forecast (Tarifa to Malaga) text MET SP1 to 80818. RAF Volmet HF 5450, 11253 kHz (H24). NAVTEX Tarifa (G), see 6.22.3.

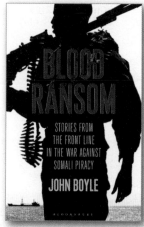

7

Safety

| 7.1 | **Introduction** | 107 |
| | Emergency procedures | |

| 7.2 | **Safety equipment** | 107 |
| | Table of recommended safety equipment (Page 106) | |

| 7.3 | **GMDSS** | 107 |
| | Sub-systems • EPIRBs • Sea areas • Functions | |

| 7.4 | **Definitions of emergency** | 108 |
| | Distress – MAYDAY • Urgency – PAN PAN • Safety – SÉCURITÉ | |

| 7.5 | **Distress signals by radio** | 108 |
| | DSC alerts • MAYDAY procedures | |

| 7.6 | **Urgency and Safety signals** | 109 |
| | PAN PAN • SÉCURITÉ • MSI • Medical advice • Mobile telephones | |

| 7.7 | **Sound and visual distress signals** | 110 |
| | Flares • Distress and life-saving signals • SAR aircraft signals | |

| 7.8 | **Helicopter rescue** | 113 |
| | Communications • Preparations • Double lift • Hi-line | |

| 7.9 | **Abandon ship** | 113 |

| 7.10 | **Search and rescue (SAR) – UK** | 114 |
| | Organisation • Raising the alarm | |

| 7.11 | **HM Coastguard** | 114 |
| | Organisation • Responsibilities • Safety Identification Scheme (CG66) | |

| 7.12 | **National Coastwatch Institution** | 115 |

| 7.13 | **Royal National Lifeboat Institution** | 115 |

TABLE 7(1) This is a precis of recommended safety equipment which should be carried, based on the categories in 7.2.1. A bullet (•) recommends that the item be carried, but the number, method or contents is left to the judgement of the skipper.

		CATEGORY	A	B	C	D
1	**Propulsion**					
1.1	A storm trisail or deep reef to reduce the mainsail area by at least 40% (sailing vessels)		•	•	•	
1.2	For engine starting, a battery isolated from all other electrical systems or hand cranking		•	•	•	
2	**Anchors**, with appropriate lengths/diameter of warp and/or chain		2+	2	2	1
3	**Bailing and pumping**					
3.1	Buckets of 9–14 ltrs capacity with lanyard and strongly secured handle		2	2	2	
3.2	Hand bilge pumps discharging overboard and operable with all hatches closed		2	2	1	
3.3	Softwood bungs attached adjacent to all through-hull fittings		•	•	•	•
4	**Detection**					
4.1	Radar reflector (if practicable) with as large a radar cross-section as is reasonable		1	1	1	1
4.2	Fixed navigation lights complying with IRPCS; motoring cone and anchor ball		•	•	•	
4.3	Foghorn		•	•	•	
4.4	Powerful torch		•	•	•	•
5	**Pyrotechnics (all in date)**					
5.1	Hand-held red flares)		6	4	4	2
5.2	Buoyant orange smoke signals) to SOLAS standards		2	2		
5.3	Red parachute rockets)		12	4	2	
5.4	Hand-held orange smoke signals				2	2
5.5	Hand-held white flares ('ship scarers')		4	4	4	
6	**Fire fighting**					
6.1	Fire blanket (BS EN 1869) for vessels with cooking equipment		•	•	•	•
6.2	Multi-purpose extinguishers (minimum 5A/34B to BS EN 3 (craft with galley *or* with engine fuel)		3	2	1	1
6.3	Additionally, for craft with *both* galley *and* engine fuel, fire extinguishers (as above)		1	1	1	1
6.4	Additionally, for craft with engines over 25hp, a fixed automatic or semi-automatic fire fighting system to discharge into the engine space		•	•	•	•
7	**Personal safety (per crew member)**					
7.1	Warm clothing, oilskins, seaboots and hat		•	•	•	•
7.2	Lifejacket or buoyancy aid (BS EN 395) 100 Newtons					1
7.3	Lifejacket (BS EN 396) 150 Newtons, with crotch straps, spray hood and light		1	1	1	
7.4	Safety harness (may be reduced by 50% for yachts with enclosed wheelhouse)		1	1	1	
7.5	Immersion suit		1			
7.6	Jackstays and cockpit clip-on strong points		•	•	•	
8	**Liferaft**					
8.1	Liferaft, designed solely for saving life, sufficient to carry all on board		1	1	1	
8.2	Emergency grab bag (see text for contents)		1	1	1	
9	**Man overboard recovery**					
9.1	Horseshoe lifebelts fitted with drogue and automatic light		2	2	1	
9.2	Buoyant sling on floating line (may replace 1 horseshoe lifebelt if 2 are carried)		1	1		
9.3	Buoyant heaving line, at least 30m long, with quoit		1	1	1	
9.4	Boarding ladder capable of rapid and secure attachment		1	1	1	
9.5	Dan buoy with a large flag		1	1		
10	**Radio**					
10.1	Radio able to receive shipping forecasts on 198 kHz and local radio station forecasts		1	1	1	1
10.2	VHF (preferably DSC) radio and emergency aerial with prepared deck mounting		1	1	1	
10.3	Waterproof hand-held VHF radio		1	1	1	1
10.4	Navtex receiver		1	1	1	
10.5	Marine band HF/SSB radio and/or global SatCom system		1			
10.6	EPIRB (406 MHz)		1			
10.7	SART		1			
11	**Navigation**					
11.1	Paper charts, tide tables and navigational publications of the cruising area and adjacent areas		•	•	•	
11.2	Steering compass, able to be lit at night.		1	1	1	1
11.3	Hand-bearing compass		•	•	•	
11.4	Navigational drawing instruments		•	•	•	
11.5	Barometer		1	1	1	
11.6	Echosounder and lead line		1	1	1	
11.7	Radio navigation system (eg GPS)		1	1	1	
11.8	Watch or clock		1	1	1	1
11.9	Distance measuring log		1	1	1	
11.10	Binoculars		1	1	1	
11.11	Sextant and associated reduction tables		1			
12	**First aid kit and manual** (see Chapter 8)		1	1	1	1
13	**General equipment**					
13.1	Emergency tiller (for wheel-steered vessels)		1	1	1	1
13.2	Towing warp		1	1	1	
13.3	Bosun's chair (sit harness BS EN 813 1997)		1	1	1	
13.4	Mooring warps and fenders		•	•	•	•
13.5	Waterproof torch		•	•	•	•
13.6	Rigid or inflatable tender		3	2	2	1
13.7	Spares and repair tools for engine & electrics, and bosun's bag of shackles, twine etc		•	•	•	•
13.8	Emergency water supply, isolated from main tanks		•	•	•	
13.9	Emergency repair materials		•	•	•	

7.1 INTRODUCTION

The safety of the crew and boat must be every skipper's number one priority. While all onboard must take a share of the responsibility for their individual safety, the skipper makes the final decisions. He or she must ensure the boat is sound and properly equipped, and the crew is well briefed and capable of undertaking the planned trip. In extremis it is the skipper who decides when to call for help. This chapter includes advice on safety equipment, emergency procedures and the Search and Rescue (SAR) organisations in the areas covered by this Almanac.

7.1.1 Emergency procedures

Knowing when to call for help and the correct procedures to be used may be a matter of life or death. It is a SOLAS requirement to have a table of life-saving signals readily available, but an emergency at sea is no time to start reading. The definitions of Distress and Urgency must be thoroughly understood; an unnecessary MAYDAY call may divert scarce resources from a more pressing emergency. Unless you or your vessel is in *grave and imminent danger* and you require *immediate assistance* a PAN PAN or even a routine call may be appropriate. Of course, you should not flinch from making an appropriate emergency call when you believe it to be justified.

7.2 SAFETY EQUIPMENT

Table 7(1) is a recommended list of equipment which should be carried onboard depending on Category (see below).

Although non-commercial leisure craft of less than 13·7m LOA are not obliged to carry any specific life saving or fire fighting equipment, SOLAS requires *all* vessels to: plan each passage (see Chapter 3); carry a radar reflector if practicable; have access to a table of life-saving signals (see Figs 7(1) and 7(2)); report serious hazards to navigation (if not already reported); and respond to distress signals.

7.2.1 Recommended safety equipment for seagoing vessels of less than 13·7m LOA

The minimum safety equipment to be carried depends on LOA, the area in which you are cruising and the risk of being caught out in heavy weather. Based on these parameters yachts are categorised as follows:

A Ocean Ocean passages of any length. LOA is likely to be longer than 10m.

B Offshore Yachts which make offshore passages of 50–500M around the UK and NW Europe. LOA is likely to be 8–13.7m.

C Inshore Coastal cruises by day/night, within 10M of land and 4 hours of a port of refuge. LOA is likely to be less than 8m.

D Sheltered In estuaries, inshore or inland waters, day only. Within 1 hour of a port of refuge. LOA is likely to be less than 6m.

7.3 GMDSS

The Global Maritime Distress and Safety System (GMDSS) came into force in 1999. Most seagoing vessels over 300 tons are required by SOLAS to comply with GMDSS, but it is not compulsory for yachts. However, it is important that the principles of the system are understood, and you should at least consider fitting compliant equipment depending on your cruising area. What follows is a brief summary of the system. Full details may be found in ALRS Vol 5 (NP 285).

7.3.1 Sub-systems

GMDSS comprises 'sub-systems' which are coordinated through Maritime Rescue Coordination Centres (MRCC) to ensure safety at sea. The main sub-systems are:

- **Digital Selective Calling (DSC)** uses terrestrial communications for making initial contact and, in a distress situation, provides the vessel's identity, nature of distress and position (entered manually or automatically if linked with the GPS). In all DSC messages every vessel and relevant shore station has a 9-digit Maritime Mobile Service Identity (MMSI) which is in effect an automatic electronic callsign. Dedicated frequencies are: VHF Ch 70, MF 2187·5 kHz. See also 7.5.1.

- **Satellite Communication System** Inmarsat is currently the only provider of GMDSS communications and services. Coverage is between 76°N and 76°S. A **Satellite Ship Earth Station (SES)** enables a ship to communicate via a satellite system.

- **Maritime Safety Information (MSI)** includes the broadcast of navigation and meteorological warnings, weather forecasts and other urgent or safety-related messages. See Chapter 6.

- **Emergency Position Indicating Radio Beacons (EPIRB)** Hand-held or float-free, they transmit on 406 MHz. (Note that 121·5 MHz, 243·0 MHz and L-band EPIRBS are no longer used.) Most EPIRBs have a built-in GPS and about 48 hours of battery life. They communicate via the Cospas-Sarsat (C/S) network of geostationary and polar orbit satellites. Although C/S will relay an EPIRB signal to earth with no delay, the position of a non-GPS EPIRB may take several hours to determine. The cost of an EPIRB ranges from less than £300 to more than £1000 for float-free, GPS models.

- **Personal Locator Beacons (PLB)** operate on the same principle. They must also be registered but they do not have a vessel-specific MMSI. A PLB with GPS costs about £250.

- **Search And Rescue Transponders (SART)** are radar transceivers which operate on 9 GHz and respond to 3 cm (X-band) radars, like a small portable Racon (see Chapter 3). They are primarily intended for use in liferafts to help searching ships and aircraft find survivors. The transmitted signal shows on a radar screen as 12 dots radiating out from the SART's position. A SART costs about £500.

7.3.2 EPIRB registration

An EPIRB transmits data which is uniquely coded to identify the individual beacon and, because it may be transmitting from anywhere in the world, it must be registered. The UK registration centre is:

The EPIRB Registry, The Maritime and Coastguard Agency, Falmouth CGOC, Pendennis Point, Castle Drive, Falmouth, Cornwall TR11 4WZ; ☎ 01326 211569; email: epirb @mcga.gov.uk.

It is just as important that any changes are immediately notified.

7.3.3 EPIRB false alerts

False alerts caused by inadvertent or incorrect use of EPIRBs put a significant burden on SAR Centres and may coincide with an actual distress situation.

All crew members should be aware of the proper use of the particular EPIRB onboard. Make sure that testing is adequately supervised; that the EPIRB is correctly installed and maintained; and that it is not activated if assistance is already available.

If an EPIRB is activated accidentally, make every effort to advise the nearest MRCC of the false alert as soon as possible.

7.3.4   Sea areas
The type of equipment carried by a vessel depends on her operating area. The four GMDSS Areas are:

A1   An area within R/T coverage of at least one VHF Coastguard or Coast radio station in which continuous VHF alerting is available via DSC. Range: 20–50M from the CG/CRS.

A2   An area, excluding sea area A1, within R/T coverage of at least one MF CG/CRS in which continuous DSC alerting is available. Range: approx 50–250M from the CG/CRS.

A3   An area between 76°N and 76°S, excluding sea areas A1 and A2, within coverage of HF or an Inmarsat satellite in which continuous alerting is available.

A4   An area outside sea areas A1, A2 and A3, ie the polar regions, within coverage of HF.

In each area certain types of radio equipment must be carried by GMDSS vessels: in A1, VHF DSC; A2, VHF and MF DSC; A3, VHF, MF and HF or SatCom; A4 VHF, MF and HF. A Navtex receiver must also be carried in GMDSS vessels, regardless of their operating area.

Most UK yachts will operate in A1 areas where a DSC VHF radio and a Navtex receiver will meet GMDSS requirements. VHF radios without DSC may still be used for sending distress calls; Coastguard stations will continue to maintain a loudspeaker watch on Ch 16 for the foreseeable future. However, DSC is rapidly becoming the preferred method of making initial contact for routine, urgency and, particularly, distress messages.

7.3.5   Functions
Regardless of the sea areas in which they operate, vessels fully complying with GMDSS must be able to perform certain functions:

- Transmit ship-to-shore distress alerts by two independent means;
- Receive shore-to-ship distress alerts;
- Transmit and receive ship-to-ship distress alerts;
- Transmit signals for locating incidents;
- Transmit and receive communications for SAR co-ordination;
- Transmit and receive Maritime Safety Information (MSI).

7.4   DEFINITIONS OF EMERGENCY
7.4.1   Distress – MAYDAY
Distress applies to any situation where a *vessel or person is in grave and imminent danger and requires immediate assistance*. The radio prefix used with a distress message is MAYDAY. A distress call has priority over all other transmissions.

7.4.2   Urgency – PAN PAN
Urgency applies to a *very urgent message concerning the safety of a vessel or person*. For example, for a vessel disabled but not in immediate danger, or for an urgent medical problem. The radio prefix associated with an urgency message is PAN PAN. An urgency call has priority over all other transmissions except distress.

7.4.3   Safety – SÉCURITÉ
A safety message is usually associated with a notification of a navigational hazard or meteorological warning. The radio prefix used with a safety message is SÉCURITÉ (pronounced SAY-CURE-E-TAY).

7.5   DISTRESS SIGNALS BY RADIO
A distress call may only be made with the skipper's authority, and only if the *boat or a person* is in *grave and imminent danger*. The initial call may be made by DSC or voice. A distress call may also be made on behalf of another vessel which is unable to make her own call. A man overboard will usually justify a distress call.

7.5.1   Distress alert by DSC
Depending on the layout and switches fitted to your particular VHF DSC radio, a distress alert is typically sent as follows:

- Briefly press the red, guarded Distress button. The set will automatically switch to Ch 70. Press *again* for 5 seconds to transmit a basic distress alert with position and time. The set then reverts to Ch 16.
- If time permits, select the nature of the distress from the menu (eg Collision, Fire, Flooding) then press the Distress button for 5 seconds to send a full distress alert.

A CG/CRS should automatically send a distress acknowledgement on Ch 70 before replying on Ch 16. Ships in range should reply directly on Ch 16.

If a distress acknowledgement is not received from a CG/CRS, the distress alert will automatically be repeated by the radio every four minutes.

When a DSC distress acknowledgement has been received, or after about 15 seconds, the vessel in distress should transmit a MAYDAY message by voice on Ch 16. This follows the format on the next page but should include the **MMSI** and **callsign** after the vessel's name to ensure that the voice and digital messages correlate.

7.5.2   False alerts
If a distress alert is inadvertently transmitted, an All Stations voice message on VHF Ch 16 cancelling the false alert must be sent at once. The message should include the name of the vessel, MMSI number, position, and the words 'Cancel my Distress Alert of (date/time (GMT/UT)'.

7.5.3   MAYDAY call by voice
Brief your crew so they are all able to send a distress message; the MAYDAY message format below should be displayed near the radio. A MAYDAY call by voice should usually be sent on VHF Ch 16 or MF 2182 kHz, but any frequency may be used if help would thus be obtained more quickly. A distress call has priority over all other transmissions. If you hear a MAYDAY call, immediately cease all transmissions that may interfere with the call, listen on the frequency concerned and note down the details.

Before making the call check that the main battery switch and the radio are both ON, and that HIGH POWER (25 watts) and Ch 16 (or 2182 kHz for MF) are selected.

Press and hold down the transmit button, and say slowly and distinctly:

> • **MAYDAY MAYDAY MAYDAY**
> • **THIS IS** ....
> (name of boat, spoken three times)
> • **MAYDAY** ....
> (name of boat spoken once)
> • **CALLSIGN / MMSI Number** ....
> (Following a DSC alert)
> • **MY POSITION IS** ....
> (latitude and longitude, true bearing and
>   distance *from* a known point, or general location)
> • **Nature of distress** ....
> (sinking, on fire etc)
> • **Help required** ....
> (immediate assistance)
> • **Number of persons on board** ....
> • **Any other important, helpful information** ....
> (you are taking to the liferaft; distress rockets are
>   being fired etc)
> • **OVER**

On completion of the distress message, release the transmit button and listen. The boat's position is of vital importance and should be repeated if time allows. If an acknowledgement is not received, check the set and repeat the distress call.

7.5.4   MAYDAY acknowledgement
In coastal waters an immediate acknowledgement should be expected, as follows:

> **MAYDAY** ....
> (name of vessel sending the distress message, spoken three times)
> **THIS IS** ....
> (name of station acknowledging, spoken three times)
> **RECEIVED MAYDAY**

If you hear a distress message, write down the details and, if you can help, acknowledge accordingly but only after giving an opportunity for the nearest Coastguard station or more suitable vessel to do so.

7.5.5   MAYDAY relay
If you hear a distress message from a vessel, and it is not acknowledged, you should pass on the message as follows:

> **MAYDAY RELAY** ....
> (spoken three times)
> **THIS IS** ....
> (name of vessel re-transmitting the distress message, spoken three times), followed by the intercepted message.

7.5.6   Control of MAYDAY traffic
A MAYDAY call imposes general radio silence until the vessel concerned or some other authority (eg the nearest Coastguard station) cancels the distress. If necessary the station controlling distress traffic may impose radio silence as follows:

> **SEELONCE MAYDAY**, followed by its name or other identification, on the distress frequency.

If some other station nearby believes it necessary to do likewise, it may transmit:

> **SEELONCE DISTRESS**, followed by its name or other identification.

7.5.7   Relaxing radio silence
When complete radio silence is no longer necessary on the frequency being used for distress traffic, the controlling station may relax radio silence, indicating that restricted working may be resumed, as follows:

> **MAYDAY**
> **ALL STATIONS, ALL STATIONS, ALL STATIONS**
> **THIS IS** ....
> (name or callsign)
> Time ....
> Name of the vessel in distress ....
> **PRUDONCE**

Normal working on a distress frequency may then be resumed having listened carefully, as always, before transmitting. Subsequent calls from the casualty should be prefixed by the Urgency signal (7.6.1).

If distress working continues on other frequencies these will be identified. For example: PRUDONCE on 2182 kHz, but SEELONCE on VHF Ch 16.

7.5.8   Cancelling radio silence
When the problem is resolved, the distress call must be cancelled by the coordinating station using the prowords SEELONCE FEENEE as follows:

> **MAYDAY**
> **ALL STATIONS, ALL STATIONS, ALL STATIONS**
> **THIS IS** ....
> (name or callsign)
> The time ....
> The name of the vessel in distress ....
> **SEELONCE FEENEE**

7.6   URGENCY AND SAFETY SIGNALS

7.6.1   PAN PAN – Urgency signal
The radio Urgency prefix, consisting of the words PAN PAN spoken three times, indicates that a vessel or station has a very *urgent message concerning the safety of a ship or person*. It may be used when urgent medical advice is needed.

This is an example of an urgency call:

> **PAN PAN, PAN PAN, PAN PAN**
> **ALL STATIONS, ALL STATIONS, ALL STATIONS**
> **THIS IS YACHT SEABIRD, SEABIRD, SEABIRD**
> In position two nine zero degrees Needles lighthouse two miles
> Dismasted and propeller fouled
> Anchor dragging and drifting east north east towards Shingles Bank
> Require urgent tow
> **OVER**

PAN PAN messages take priority over all traffic except distress and are sent on VHF Ch 16 or MF 2182 kHz. They should be cancelled when the urgency is over.

If the message is long (eg a medical call) or communications traffic is heavy, it may be passed on a working frequency after an initial call on Ch 16 or 2182 kHz. At the end of the initial call you should indicate that you are switching to a working frequency.

If you hear an urgency call react in the same way as for a distress call.

7.6.2   SÉCURITÉ – Safety prefix

The word SÉCURITÉ (pronounced SAY-CURE-E-TAY), spoken three times, indicates that the station is about to transmit *an important navigational or meteorological warning*. Such messages usually originate from a CG Centre or a Coast Radio Station, and are transmitted on a working channel after an announcement on the distress/calling channel (VHF Ch 16 or MF 2182 kHz).

Safety messages are usually addressed to 'All Stations'. An example of a Sécurité message is:

> **SÉCURITÉ, SÉCURITÉ, SÉCURITÉ** ....
>
> **THIS IS** ....
> (CG Centre or Coast Radio Station callsign, spoken three times)
>
> **ALL STATIONS** ....
> (spoken three times) followed by instructions to change channel, then the message.

7.6.3   Maritime Safety Information (MSI)

MSI consists of navigational, weather and safety messages. See also Chapters 5 and 6.

GMDSS transmits MSI in English by two independent but complementary means:

• **Navtex** on MF (518 kHz and 490 kHz) has a range of about 300 miles offshore (see 6.6).

• **SafetyNet** uses Inmarsat satellites to cover beyond MF range.

In coastal areas MSI may be broadcast by CG/CRS on VHF.

7.6.4   Medical advice by R/T

Medical advice and help can be obtained via any UK and most European CG MRCC; such calls are free. The initial call should be by an urgency message (PAN PAN x 3) on VHF Ch 16, by DSC on VHF Ch 70 or MF 2187·5 kHz. The Coastguard is always the first point of contact. See also Chapter 8 for further medical advice.

7.6.5   Survival times

Average sea temperatures around the UK range from +5° to +15°C. If you fall overboard your chances of survival in certain conditions, unless quickly recovered, are not good.

> • From 10°–16°C survival is probable up to about 6 hours if wearing a lifejacket.
>
> • At 10°C survival could be 4–8 hours if wearing a survival or wet suit, but otherwise only an hour or so.
>
> • Below 10°C only 50% of people will survive for more than ONE hour; a survival or wet suit might extend this time.

7.6.6   Mobile telephones

Most yachts will have at least one mobile telephone onboard but *their use in a distress or urgency situation is strongly discouraged* as the existing dedicated and well-established maritime distress communications system would be by-passed. Mobile telephones have limited coverage (typically no more than about 10 miles offshore); they do not 'broadcast' so other vessels which might be able to render assistance cannot monitor the situation; and they are extremely vulnerable to getting wet.

However, when all other means of communication have failed, the use of mobile telephones have saved many lives, sometimes in bizarre circumstances.

7.7   SOUND AND VISUAL DISTRESS SIGNALS

Annex IV of the IRPCS lists in full the recognised distress signals. Some are depicted in Fig 7(1); those below, with notes on their use, are most relevant to yachts and small craft.

> • Continuous sounding of any fog signalling apparatus.
>   *To avoid confusion, this is best done by repeatedly sending the letters SOS in Morse (··· ─ ─ ─ ···).*
>
> • An SOS signal made by any method.
>   *For a small craft the most likely methods are by fog horn or by flashing light.*
>
> • The International Code signal 'NC'.
>   *This can be made by hoisting the international code flags 'NC'. See Chapter 5, Fig 5(1).*
>
> • A square flag with a ball, or anything resembling a ball, above or below it.
>   *This is simply made from any square flag and a round fender or anchor ball.*
>
> • Slow and repeated raising/lowering of arms outstretched to each side.
>   *Raise and lower the arms together, above and below the horizontal; this is usually obvious inside a mile.*
>
> • A rocket parachute red flare or a hand-held red flare.
>   *See 7.7.1 below.*
>
> • An orange smoke signal.

By day orange smoke signals are more effective than flares, although the smoke disperses quickly in a strong wind.

If help is needed but the boat is not in immediate danger, the proper signal is 'V', meaning 'I require assistance'. This can be sent by light or sound (···─), or by hoisting flag 'V'.

7.7.1   Use of flares and other pyrotechnics

Fig 7(1) shows how flares are best used to attract attention and to communicate with SAR teams once help has arrived. SOLAS requires this information to be on all sea-going vessels; all crew members should know where to find it.

Flares serve two purposes:

• to raise the alarm, and
• to pinpoint the boat's position.

Within about three miles from land a red hand flare will do both; it is the most effective distress signal at night. At greater distances a red parachute rocket (which projects a suspended flare higher than 1,000ft (300m), burning for more than 40 seconds) is used to raise the alarm, but use red hand flares to pinpoint the boat's position.

If possible wear a protective glove and hold flares firmly, downwind of yourself. Rockets turn into wind: in normal conditions fire them vertically, but in strong winds aim about 15° downwind. Do not aim them into wind, or they will not gain altitude. If there is low cloud, fire rockets at 45° downwind, so that the flare burns under the cloud. Initially, fire two flares in succession – the first to attract attention, the second to confirm.

Fig 7(1) Distress and life-saving signals

(1) Signals to be used by Ships, Aircraft or Persons in Distress

Fire rocket vertically

If windy fire 15° downwind

In low cloud fire at 45°

NEVER fire a parachute flare if a helicopter is approaching

WIND

NEVER fire into the wind

Handheld pinpoint flare shows exactly where you are - use inshore or in sight of other vessels

Orange smoke for use by day - especially in bright conditions

For use at night or in dull conditions to pinpoint position

Long range signal for use offshore

Floating orange smoke

Red hand flare

Red parachute rocket

Buoyant smoke

White collision warning flare

White pinpoint

Handheld orange smoke

Orange hand smoke

(2) Surface to Air Signals

Message	International Code of Signals		ICAO Visual Signals
'I require assistance'	'V'	⊠ (· · · —)	V
'I require medical assistance'	'W'	▣ (· — —)	X
'No' or 'negative'	'N'	▦ (— ·)	N
'Yes' or 'affirmative'	'C'	▤ (— · — ·)	Y
'Proceed in this direction'			↑

(3) Air to Surface replies

'Message understood'

Drop a message   Rocking wings   Flash landing or navigation lights twice

or 'T' (—)
or 'R' (· — ·)

Morse code signal by light

'Message not understood - repeat'

Straight and level flight

Circling

or 'RPT'
(· — · · — · —)

Morse code signal by light

(4) Air to Surface Direction Signals

Sequence of 3 manœuvres meaning proceed in this direction

(1) Circle vessel at least once

(2) Cross low ahead of vessel, rocking wings

(3) Overfly vessel and head in required direction

Your assistance is no longer required

Cross low astern of vessel rocking wings

Note: as an alternative to rocking wings, the aircraft engine pitch or volume may be varied

(5) Surface to Air replies

'Message understood - I will comply'

Change course or direction

or

'T' (—)   Morse Code signal by light

or

    Code and answering pennant 'Close up'

'I am unable to comply'

'N' (— ·)   Morse Code signal by light

or

    International flag 'N'

Fig 7(2)   Life-saving signals (Shore-to-Ship) (see 7.7.4)

Signals (1) to (4) may be indicated by a white flare, a rocket showing white stars on bursting, or an explosive sound signal.

(1) Landing signals for the guidance of small boats with crews or persons in distress

	Other signals	*Meaning*
Vertical motion of a white flag (or white light or flare by night) or of the arms	International Code letter **'K'** (— · —) by light or sound	**'This is the best place to land'** (An indication of direction may be given by a steady white light or flare at a lower level)

Horizontal motion of a white flag (or white light or flare by night) or of the arms extended horizontally	International Code letter **'S'** (· · ·) by light or sound	**'Landing here is highly dangerous'**

Horizontal motion of a white flag followed by 2. placing the white flag in the ground and 3. by carrying another white flag in the direction to be indicated. By night white lights or flares are used instead of white flags	1. Signalling the code letter **'S'** (· · ·), followed by the code letter **'R'** (· — ·) if the better landing place is more to the right in the direction of the approach, or 2. by the code letter **'L'** (· — · ·) if the better landing place is more to the left in the direction of approach	**'Landing here is highly dangerous. A more favourable location for landing is in the direction indicated'**

(2) Signals to be made in connection with the use of shore apparatus for life-saving

Signal	*Meaning*	**Signal**	*Meaning*
Vertical motion of a white flag (or white light or flare by night) or of the arms	**In general:** 'affirmative' Specifically: 'rocket line is held - tail block is made fast - hawser is made fast - man is in the breeches buoy - haul away'	**Horizontal** motion of a white flag (or white light or flare by night) or of the arms	**In general:** 'negative' Specifically: 'slack away - stop hauling'

(3) Replies from life-saving stations etc. to distress signals made by ships or persons

Pyrotechnic signals
Orange smoke signal

White star rocket - three single signals fired at intervals of about one minute

'You are seen - assistance will be given as soon as possible'

(4) Signals to be used to warn a ship which is standing into danger

International Code signals **'U'**

or **'NF'**

'You are standing into danger'

International Code signals **'U'** by light or sound · · —

(5) Signals used by Sub-Aqua divers

 'I am OK'

 'I need assistance'

Note: White flares are not distress signals but are used to indicate your presence to another vessel or acknowledge the sighting of a red flare. On night passages have at least four onboard and have at least one readily to hand from the cockpit. When using them shield your eyes to protect night vision.

7.7.2 Signals used by SAR aircraft

A searching aircraft normally flies at about 3,000–5,000ft, or below cloud, firing a green Very light every five or ten minutes and at each turning point.

If a green flare is seen, a vessel in distress should take the following action:

- Wait for the green flare to die out.
- Fire one red flare.
- Fire another red flare some 20 seconds later. (The aircraft aligns itself on the transit of both flares.)
- Fire a third red flare when the aircraft is overhead, or if it appears to be going badly off course.

Other signals are shown at Figs 7(1) and 7(2).

7.7.3 Directing signals by aircraft

Fig 7(1) shows how an aircraft directs a vessel towards a ship in distress; how it tells the vessel that its assistance is no longer required; and how the vessel may reply.

7.7.4 Shore-to-ship: visual signals

If no radio contact is possible, some of the signals shown in Fig 7(2) may be used to a vessel in distress close to the shore.

7.8 HELICOPTER RESCUE

SAR helicopters in the UK are operated by HM Coastguard at Sumburgh, Stornoway, Portland and Lee-on-Solent; the Royal Navy at Prestwick and Culdrose; the RAF at Lossiemouth, Boulmer, Valley, Leconfield, Wattisham and Chivenor.

Sea King SAR helicopters operated by the RN and RAF have a radius of action of 180–240M and can lift up to 18 survivors. The AW139 helicopters operated by HMCG have a radius of action of 180M and can lift 9 survivors or 2 stretchers. Both types are fully night capable.

7.8.1 Communications

SAR helicopters are generally fitted with VHF, FM & AM, UHF and HF SSB radios and can communicate with lifeboats on VHF FM, using callsign *Rescue nn*. Communications between ship and helicopter should normally be on VHF Ch 16 or 67; 2182 kHz SSB may also be available. If contact is difficult communication can often be achieved through a Nimrod aircraft, if on scene, or through a lifeboat or HM Coastguard.

When the helicopter is sighted by a vessel in distress, a flare (fired away from, not at, the helicopter!), an orange smoke signal, dye marker or an signal lamp (not pointed directly at the helicopter) will assist recognition – very important if other vessels are in the vicinity. Spray dodgers with the yacht's name or sail number are useful aids to identification.

7.8.2 On the yacht

The helicopter will pass advice before it gets so close that communications become difficult because of the noise of its rotors. Listen carefully and, if possible, do exactly as requested. You may be asked to keep the sails up in order to stabilise the yacht. If they are lowered, securely lash them and any loose gear. If possible remove items such as danbuoys or long aerials from the cockpit area.

If a crewman descends from the helicopter he will take charge. Obey his instructions promptly. *Never secure the winch wire to the yacht*, and beware that it may carry a dangerous static charge if it is not dipped (earthed) in the sea before handling.

7.8.3 Double lift

Survivors may be lifted by double lift in a strop, accompanied by the crewman in a canvas seat, or it may be necessary to position yourself in the strop. Put your head and shoulders through the strop so that the padded part is in the small of the back and the toggle is in front of the face. Pull the toggle down, as close to the chest as possible.

When ready, give a thumbs-up sign with an extended arm, then place both arms close down by the side of the body to ensure that the strop does not slide up. Resist the temptation to hang on to the strop. On reaching the helicopter do exactly as instructed by the crew. Injured persons can be lifted in a special stretcher provided by the helicopter.

7.8.4 Hi-line

Often a 'Hi-line' technique using a 'Heaving-in' line is used. This is a rope tail, attached to the helicopter winch wire by a weak link and weighted at its lower end. When it is lowered to the yacht do not make it fast but coil it down carefully. The helicopter pays out the winch wire and then moves to one side of the yacht and descends while the yacht takes in the slack (keeping it outboard and clear of all obstructions) until the winch hook and strop are on board.

A helicopter crewman may be lowered with the strop. When ready to lift, the helicopter ascends and takes in the wire. Pay out the tail, keeping enough weight on it to keep it taut until the end is reached, then cast it off well clear of the yacht. But if a further lift is to be made the tail should be retained on board (again, not made fast) to facilitate recovery of the strop for the next lift.

When leaving a helicopter ashore, beware of the tail rotor which can be difficult to see. Obey all instructions given by the helicopter crew.

7.9 ABANDON SHIP

Do not abandon a vessel unless she is definitely sinking. A vessel is easier to find than a liferaft and provides better shelter. While she is still afloat use her resources (eg the radio for distress communications) and put extra equipment in the liferaft or the dinghy, which should also be taken if possible. Make all preparations.

The aim should be to board the liferaft without getting wet. Before boarding and cutting it adrift:

- Send a MAYDAY call or DSC distress alert, saying that the vessel is being abandoned. Include your position and the number of people on board.
- Dress warmly with sweaters under foul weather gear. Wear life jackets and take extra clothes.
- Fill any available containers with fresh water to about ¾ full so that they will float.
- Collect additional food, tins and tin-opener.

- Collect navigational gear, EPIRB, SART, handheld VHF, torch, extra flares, bucket, length of line, first-aid kit, knife, fenders and any other equipment which might aid your survival.

Some items should already be in a survival bag – essential when a vessel is abandoned in a hurry. See *The Grab Bag Book* (ACN).

Launch the liferaft on the lee side of the vessel, having first checked the painter is secured to a strongpoint. Pull in the painter, which may be up to 10m long, and give a sharp tug to activate the CO_2 bottle. If the liferaft inflates upside down, try to right it from the vessel. Keep the liferaft close alongside while boarding; release or cut the painter before the vessel sinks.

Once in the liferaft, plan for the worst: a long wait for rescue. Protection, location, water, food are the priorities – in that order. Always believe that you will survive.

- Keep the inside of the raft as dry as possible. Huddle together for warmth. Close the openings as necessary, but keep a good lookout for ships and aircraft.
- Stream the drogue if necessary for stability, or so as to stay near the original position.
- Take anti-seasickness pills.
- Don't drink any water in the first 24 hours, then ration fresh water to ½ litre per person per day. Do not drink sea water or urine. Collect rain water.
- Use flares sparingly.

7.10   SEARCH AND RESCUE (SAR) – UK

The Maritime & Coastguard Agency (MCA) is responsible for developing and maintaining a high level of marine safety, minimising loss of life amongst sefarers and protecting the marine environment. Its responsibility for emergency response is met by HM Coastguard, which co-ordinates civil maritime SAR in the UK Search and Rescue Region (SRR).

HM Coastguard (HM CG) manages the UK maritime SAR helicopter service and shore rescues. It coordinates the response of ambulance, fire and police services to maritime incidents in and around the UK. The UK CG possesses some boats, but lifeboats and their crews are provided by non-governmental charities, primarily the RNLI. UK CG enlists the assistance of merchant vessels and warships in the response to SAR events beyond coastal waters.

7.10.1   Raising the alarm

If you see an incident which merits an emergency response at sea or along the shoreline, dial 999 and ask for the Coastguard. You will be asked for details of the incident and may be asked to remain contactable to provide further information.

If you see a distress signal or receive a distress message at sea and are in the immediate vicinity you are obliged to respond with all speed unless or until you are specifically released. Even if unable to give effective aid, your presence may provide reassurance or act as a beacon for other help.

When alerted, HM CG summons the most appropriate assets: they may direct vessels in the vicinity of the distress, request the launch of a lifeboat or scramble a helicopter and alert other vessels by radio, digital or satellite communications. For shore incidents, including cliff and mud rescue, it will activate its rescue teams and liaise with the other services.

7.11   HM COASTGUARD (HM CG)

The EU Integrated Maritime Policy promotes the wellbeing of citizens by ensuring their security and safety as well as responsible economic practice and environmental protection. HM CG contributes to this through dedicated surveillance infrastructure, managed by a resilient command, control and communications network. Vessel monitoring allows greater awareness and understanding of maritime activity in order to to speed response in rescue co-ordination, intervention against illegal acts and pollution prevention.

Nine networked Coastguard Operation Centres CGOC co-ordinated by the Maritime Operations Centre (MOC) at Fareham control nominated areas of responsibility (AOR). Each has the ability to control operations across all AOR, sharing data and access to communcations nationwide.

CG locations and boundaries are shown in Chapter 6, Fig 6(4), and in Chapter 9 on the maps at the start of Areas 1–11.

The Dover CG monitors the TSS in the Dover Strait by radar and AIS, and supports the traffic safety through the Channel Navigation Information Service (CNIS).

The CG manage vessel traffic messages (VTM) for merchant vesels. MSI broadcasts disseminate essential safety information. Details of messages are shown in Chapter 6.

The CG Rescue Service consists of regular officers supported by 3,500 Auxiliary Coastguards who are on call to cover emergencies around the coastline. These rescue teams, respond to, assess and carry out searches and cliff and mud rescues (depending on location). They liaise with the RNLI.

E-mail: CG centre name before coastguard@mcga.gov.uk e.g. Milford Haven is: milfordhavencoastguard@mcga.gov.uk. Radio and phone calls to and from CG Centres are recorded, but HM CG does *not* provide or accept telephone link calls.

Appropriate CG (or continental equivalent) telephone numbers are listed for each port entry.

7.11.1   HM CG routine services

HM Coastguard routinely supports maritime activity by:

- Organising and conducting search and rescue operations at sea or on the UK's cliffs or foreshore.
- Keeping an H24 radio watch on VHF and GMDSS.
- Controlling the UK maritime SAR helicopter service.
- Providing medical advice from a doctor via telephone link call; and arranging medical assistance. Call initially on VHF / MF DSC or on Ch 16 or 2182 kHz and expect to be switched to a working Channel/frequency.
- Broadcasting Maritime Safety Information (MSI) messages at 4 hourly intervals on VHF and MF. These include WZ, local Subfacts and Gunfacts in exercise areas as well as local inshore forecasts and warnings related to navigation, weather, tidal surges and the failure of navigation aids.
- Vessel traffic monitoring and surveillance of UK Territorial Waters and Exclusive Economic Zone by AIS and VTM.
- Recording Traffic Reports (TRs) from yachts and other small craft either by VHF or telephone. When passing a TR include your vessel's name, callsign, MMSI number, ETD, ETA and number of persons onboard. The boat's safe arrival should be passed to the nearest CGOC, for relay to the departure CG. Also, be sure to report any change of plan to the nearest CG while on passage. *Overdue action is only taken if concern for the safety of the vessel is reported to HM CG, by a shore contact, for example.*

7.11.2 CG66 Voluntary Safety Identification Scheme

Owners/skippers may register details of their vessel with HM CG for use in emergency free of charge. If in difficulty, missing or overdue, the CG66 provides accurate information on the vessel including a description to expedite assistance.

The CG66 may be completed on the MCA's website (www. mcga.gov.uk and search for 'CG66') or it is available from MRCCs, RNLI lifeboat stations and possibly your sailing club or marina. Details are kept for two years then destroyed if not renewed.

7.12 NATIONAL COASTWATCH INSTITUTION

The National Coastwatch Institution (NCI) - www.nci.org.uk is a voluntary organisation keeping a visual watch along the UK's coast assisting the preservation and protection of life at sea. There are 50 operational stations (2015) manned by over 2000 volunteers. They are marked on the Area maps by ©. All NCI stations monitor VHF Ch 16 and provide actual weather, sea state, hazards and local information on request on their dedicated channel **VHF Ch 65**. Many are equipped with AIS Ⓐ, DSC Ⓓ and radar Ⓡ. Contact details:

Area 1		
St Mary's	01720	422641
Gwennap Head (Land's End) Ⓐ Ⓓ Ⓡ	01736	871351
Penzance Ⓐ Ⓓ	01736	367063
Bass Point (Lizard) Ⓐ Ⓓ Ⓡ	01326	290212
Nare Point (Helford River) Ⓐ Ⓓ Ⓡ	01326	231113
Portscatho (E of Fal; summer w/e only)	01872	580180
Charlestown Ⓡ	01726	817068
Polruan (Fowey) Ⓐ Ⓡ	01726	870291
Rame Head (Plymouth) Ⓐ Ⓓ Ⓡ	01752	823706
Prawle Point (Salcombe) Ⓐ Ⓓ Ⓡ	01548	511259
Froward Point (Dartmouth) Ⓐ Ⓓ Ⓡ	07976	505649
Torbay Ⓐ Ⓓ Ⓡ	01803	411145
Teignmouth Ⓐ Ⓓ Ⓡ	01626	772377
Exmouth Ⓓ	01395	222492
Lyme Bay Ⓓ	07745	756872
Portland Bill Ⓐ Ⓓ Ⓡ	01305	860178
St Alban's Head Ⓐ Ⓓ Ⓡ	01929	439220
Swanage Ⓐ Ⓓ Ⓡ	01929	422596

Area 2		
Peveril Pt (Swanage) Ⓐ Ⓓ Ⓡ	01929	422596
Needles Ⓐ	01983	754231
Calshot Ⓐ Ⓓ Ⓡ	02380	893562
Lee on Solent Ⓐ Ⓓ	02392	556758
Gosport	02392	765194

Area 3		
Shoreham Ⓐ Ⓓ Ⓡ	01273	463292
Newhaven Ⓐ Ⓡ	01273	516464
Folkestone (Copt Pt) Ⓐ Ⓓ	01303	227132

Area 4		
Herne Bay (w/ends only) Ⓐ Ⓓ Ⓡ	01227	744454
Whitstable (w/ends only) Ⓓ	07864	646209
Holehaven (Canvey Island) Ⓐ Ⓓ	01268	696971
Southend (w/ends only) Ⓐ Ⓓ	07815	945210
Felixstowe Ⓓ	01394	670808
Gorleston (Great Yarmouth) Ⓐ Ⓓ Ⓡ	01493	440384
Caister Ⓐ Ⓓ	07527	977613

Area 5		
Mundesley (Norfolk) Ⓐ Ⓓ Ⓡ	01263	722399
Runton Ⓐ Ⓓ Ⓡ	01263	513725
Wells-next-the-sea (Norfolk) Ⓐ Ⓓ Ⓡ	01328	710587
Mablethorpe Ⓐ Ⓓ	07958	038564
Sunderland Ⓐ Ⓡ	01915	672579

Area 10		
Rossall Pt (Fleetwood) Ⓐ Ⓓ Ⓡ	01253	681378
Point Lynas (being established 2015)		
Porth Dinllean Ⓓ	07773	362005
Rhoscolyn Ⓐ	07887	955496

Area 11		
Morfa Bychan (Porthmadog) Ⓓ	07521	446050
Wooltack Point (St Brides Bay) Ⓐ Ⓓ	01646	636802
Worms Head (Gower) Ⓓ	01792	390167
Nells Point (Barry) Ⓐ Ⓓ	01446	420746
Boscastle (North Cornwall) Ⓐ Ⓓ Ⓡ	01840	250965
Stepper Point (Padstow) Ⓐ Ⓓ Ⓡ	01841	530039
St Agnes Head (N Cornwall) Ⓐ Ⓡ	01872	552073
St Ives Ⓐ Ⓡ	01736	799398
Cape Cornwall (Land's End) Ⓐ Ⓓ Ⓡ	01736	787890

7.13 ROYAL NATIONAL LIFEBOAT INSTITUTION

The RNLI is a registered charity which provides a 24-hour lifeboat service up to 100M off the UK and Irish coasts. There are more than 230 lifeboat stations. Of these, four are on the River Thames, and another four are inland water stations. The RNLI also operates over 100 lifeguard units. All new lifeboats are capable of at least 25 knots. When launched on service, lifeboats >10m keep watch on VHF and MF DSC as well as VHF Ch 16. They can also use alternative frequencies to contact other vessels, SAR aircraft, HM CG, Coast Radio Stations or other SAR agencies. All lifeboats show a quick-flashing blue light when on operational service.

The RNLI actively promotes safety at sea by providing a free comprehensive safety service (eg, the RNLI Sea Safety Check) to members and the general public including advice, publications and demonstrations. The RNLI aims to save lives and prevent accidents by helping people be prepared through water safety awareness. The *RNLI Skippers' Handbook* covers emergencies, seamanship, weather, first aid, navigation and engines. For more details of how the RNLI can help you to be safer at sea, or to support the work of the RNLI, call the RNLI on 0845 122 6999, visit www.rnli.org.uk.

Safety

Come what may!

8

First Aid

8.1–2	**First aid objectives – Medical care abroad**	**118**
8.3	**Medical advice by R/T**	**118**
8.4	**ABC – Emergency resuscitation**	**118**
	Airway • Breathing • Circulation • External chest compression	
8.5–6	**Choking – Drowning**	**120**
8.7–8	**Hypothermia – Shock**	**120**
8.9	**Acute (sudden) illness**	**121**
	Abdominal pain • Allergies • Children • Constipation • Convulsions • Diabetes • Diarrhoea • Fever • Heart attack • Heat stroke • Seasickness	
8.10	**Stroke**	**122**
8.11–12	**Bites and stings – Fish hook injury**	**122**
8.13–14	**Bleeding (open wound) – Bleeding (internal)**	**122**
8.15–16	**Burns and scalds – Chest injury**	**122**
8.17–18	**Crush injuries – Cuts and wounds**	**122**
8.19–20	**Dental problems – Eye injury**	**123**
8.21	**Fractures and dislocations**	**123**
	Specific fractures • Skull, Nose, Cheek, Jaw, Neck, Spine, Ribs, Upper and Lower limbs • Compound fractures	
8.22–24	**Frostbite – Head problems – Nail injuries**	**124**
8.25–27	**Nosebleed – Poisoning – Strains and sprains**	**125**
8.28–30	**First aid kit – Drugs – Emergency childbirth**	**125**
8.31–32	**Injections – Books on first aid**	**126**
8.33	**Normal physiological measurements**	**126**
8.34	**Observation form**	**127**

First Aid

8.1 FIRST AID OBJECTIVES

The objectives of First Aid at sea are to:

- **Preserve life**
- **Prevent further damage**
- **Relieve pain and distress**
- **Deliver a live casualty ashore**

With any casualty be calm, reassuring and methodical. Examine the whole casualty at the earliest opportunity.

Completion of a First Aid Course is strongly recommended. In the UK, RYA approved, one-day, basic training is available through sailing clubs and schools. The St Johns Ambulance and British Red Cross provide a variety of recognised courses.

8.2 MEDICAL CARE ABROAD

For cruising in European waters it is advisable for each crew member to carry the European Health Insurance Card (EHIC) which entitles you to reduced-cost, sometimes free, medical treatment that becomes necessary while you are in the EU, Iceland, Liechtenstein and Norway. It is free and is available by phone, by post or online; see: *www.dh.gov.uk/en/Policyandguidance/Healthadvicefortravellers/index.htm*.

A useful booklet *Health Advice for Travellers,* which gives details of how to claim in the 28 participating countries, is obtainable from Post Offices. A further 30+ countries have reciprocal health care agreements with the UK, but others, including Turkey, the Americas, Canada, Caribbean, India, Middle East, Africa, Asia, Thailand, Japan, Hong Kong and the Pacific region (except Australia and New Zealand), do not. Private health insurance is strongly advised in these areas.

Skippers should be aware of crew members' medical conditions and any ongoing medication, which may seriously alter the implications of an injury or change treatment. If cruising to distant destinations sufficient medication must be carried and additional sterile needles and syringes may be advisable, if you or a crewman are diabetic, for example. Seek advice from your doctor or local hospital before leaving. Syringes and needles are attractive to intravenous drug abusers so should be securely locked away.

Yachtsmen cruising abroad may naturally be concerned about the risk of being given infected blood. Normally it is not possible to carry blood or plasma in a yacht. If a blood transfusion is essential, try to ensure that the blood used has been screened against HIV and Hepatitis B.

8.3 MEDICAL ADVICE BY R/T

Medical advice can be obtained almost anywhere in European waters (and elsewhere) by making an all-stations 'PAN PAN' call or a DSC Urgency Alert to the Coastguard, or to a Coast Radio Station (CRS) in those countries where CRS still exist. You will be connected to a suitable medical authority – usually a doctor or the nearest hospital.

The Urgency signal 'PAN PAN' is always advised, especially abroad, because it is internationally understood and cuts through most language problems; see also 7.5.

Urgent help needed is shown in bold italic type against the more serious medical problems in the following pages and this implies a PAN PAN call. You should also recognise that as a layman you are not qualified to judge how serious the casualty's condition is – so get the best possible advice and/or help as quickly as possible.

Be ready to give a detailed summary of the patient's symptoms – pulse rate, breathing rate, temperature, skin colour, conscious state (with reference to pupil size, responses to verbal command and to firm pinching); site and description of any pain, site and type of injury, amount of blood lost etc. See the Observation Form at 8.34.

If medical help is needed by way of a doctor coming aboard, or if a serious casualty has to be landed, the arrangements will be made by the Coastguard.

If you are sure that the situation is not urgent, you may wish to forewarn the port authority so that a doctor or paramedic can meet you on arrival. Such a call could be made in adequate time on the harbour's working channel.

8.4 ABC – EMERGENCY RESUSCITATION

Advice is obtainable at: *www.resus.org.uk/pages/guide.htm*
The immediate procedure for any collapsed or apparently unconscious person is:

> Assess whether or not the casualty is conscious. Carefully shake his shoulders and ask loudly 'What's happened?' or 'Are you all right?' or give a command such as 'Open your eyes'. An unconscious casualty will not respond.

A = Airway

Be aware of the risk of neck or spinal injury following a fall or head injury. If suspected, try to keep the head and neck still and in line, then lift the chin to maintain a clear airway rather than tilting the head. However, the priority must be to ensure the casualty is indeed breathing. Remove any visible obstruction from the casualty's mouth (leave well-fitting dentures in place). Listen at the mouth for breathing. Tilt the head backwards, using head tilt and chin lift to maintain a clear airway; Fig 8(1). Look, listen and feel for *no more than 10 seconds* to determine if the casualty is breathing normally. If in any doubt, act as if it is *not* normal.

Fig 8(1) Ensuring a clear airway

If breathing, place casualty in recovery position; Fig 8(2). Check the area is clear of danger.

Fig 8(2) Recovery position

B = Breathing and C = Circulation

If not breathing and the airway is clear, start chest compressions; Fig 8(3). This situation is called **cardiac arrest**.

The casualty will be unconscious and may appear very pale, grey or bluish in colour. An artificial circulation will have to be provided by chest compression. If the circulation stops, the breathing will also stop. Casualties with cardiac arrest will need both rescue breaths and chest compression, a combination known as Cardio Pulmonary Resuscitation (**CPR**).

Chest compressions

To start external chest compression, lay the casualty on a hard, flat surface if possible. Kneel beside casualty. The point at which pressure will be applied is the centre of the chest. (It is now considered a waste of valuable time for an amateur first aider to search for the point of pressure by the 'rib margin' method.)

Place the heel of your first hand on top of the other hand and interlock your fingers. Depress breastbone 5–6cm (2–2½in) then release 100–120 times per minute; Fig 8(3).

Fig 8(3) Chest compression

With either one or two operators give 30 chest compressions and continue cycles of 2 breaths to 30 compressions; see Fig 8(5) and Fig 8(6). Use a compression rate of 100 per minute. If trained, chest compression should be combined with rescue breaths; after 30 compressions, give 2 effective rescue breaths (2 breaths should not take more than 5 secs). *Do not stop.*

Rescue breaths

Kneel beside the casualty, maintain head tilt and chin lift, and pinch the nostrils. Take a deep breath and blow two full breaths into patient's mouth, each breath should be 1 second in duration. Watch for rise and fall of chest; Fig 8(4).

Fig 8(4) Rescue breaths

Action plan for the resuscitation of adults
Casualty is unconscious but is breathing normally:

- ***Urgent help needed***
- Turn casualty into the recovery position
- Check for continued breathing

Fig 8(5) One operator

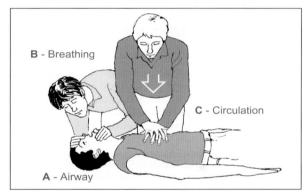

Fig 8(6) Two operators

Casualty is unconscious and not breathing:

- ***Urgent help needed***
- Start chest compressions
- After 30 chest compressions, give 2 rescue breaths

Continue with chest compressions and rescue breaths in a ratio of 30:2. If you are untrained, or are unwilling to give rescue breaths, give chest compressions only. Stop to re-check casualty only if he/she starts to show signs of regaining consciousness; coughing, opening eyes, speaking or moving purposefully *and* starts to breathe normally.

If breathing restarts, place casualty in the recovery position.

Otherwise, ***do not interrupt resuscitation***.

Action plan for the resuscitation of children

Blow into both mouth and nose if necessary, but resuscitate at same rate as an adult. For external chest compression use gentle compression with one hand only, or just fingers for a baby; use a compression rate of up to 100 per minute, one breath every 5 compressions. In the case of a baby give up to 5 initial rescue breaths. Use a faster breathing rate (one inflation every 3 seconds) and smaller breaths.

Problems during resuscitation

It is unlikely that the casualty's pulse will return spontaneously without other more advanced techniques (especially *defibrillation*) so do not waste time by stopping CPR to check the circulation. Only stop and re-check if the casualty shows signs of life (movement or breathing). Otherwise, carry on until either the emergency services arrive, another rescuer can take over, or you are too exhausted to keep going.

Drowning (8.6) and Hypothermia (8.7)

Drowning and hypothermia casualties may exhibit all the signs of apparent death, yet may totally recover. Only abandon resuscitation if really necessary, and after thorough and repeated attempts have been made to warm the victim.

8.5 CHOKING

If blockage by some object is suspected, turn the casualty on his/her side and give up to 5 sharp back slaps with the flat of the hand between the shoulder blades. Check mouth and remove any obstruction.

If unsuccessful, wrap both your arms around the victim's waist from behind, and give 5 sharp upward thrusts with both fists into the abdomen above the navel but below the ribs so as to initiate coughing. Clear object from mouth.

In an unconscious adult administer abdominal thrusts with the victim lying on his back. Attempt mouth to mouth breathing. Do not give up.

Infants and small children should just be given 5 forceful blows on the back before proceeding to chest thrusts which are performed in the same way as chest compressions but should be sharper and performed at a slower rate, with each thrust trying to relieve the obstruction. Continue if no relief, with the sequence of 5 rescue breaths, 5 back blows and 5 chest thrusts.

8.6 DROWNING

The resuscitation of an apparently drowned person may be complicated by two factors:

- An acute (sudden) illness (8.9 and 8.10) – a stroke, accident or blow to the head, for example – may have precipitated the fall into the water.
- The time spent in the water may have produced marked hypothermia. The water around the UK is rarely warmer than 15°C (60°F).

During rescue and after resuscitation try to keep the head low so that vomit is not inhaled and water is able to drain from the mouth.

Treatment

A = Airway Clear airway: seaweed, dentures etc. See 8.4.

B = Breathing If not breathing start mouth to mouth ventilation as soon as possible and in the water if practicable. See 8.4.

C = Circulation If pulse is absent, start chest compression as soon as aboard. See 8.4.

- If stomach is bulging, turn casualty on to side to empty water, or he may vomit large quantities of water which could be inhaled.
- Prevent cooling. Remove wet clothes; wrap casualty in blankets to warm him/her.
- Continue resuscitation until the casualty revives or death is certain. Hypothermia may mimic death. Do not abandon resuscitation until the casualty has been warmed or signs of death persist despite attempts at warming.
- Once revived, put in the recovery position, Fig 8(2).
- Any person rescued from drowning may collapse in the next 24 hours as the lungs react to inhaled water. *Urgent help needed*.

8.7 HYPOTHERMIA

Lowered body temperature may follow immersion in the sea or prolonged exposure on deck.

Symptoms include: unreasonable behaviour followed by apathy and confusion; unsteady gait, stumbling; slurring of speech; pale, cold skin; slow, weak pulse; slow breathing; shivering. Hypothermia may lead to collapse, unconsciousness and, ultimately, death.

Treatment

A = Airway control; put in recovery position.

B = Breathing If not breathing, start mouth to mouth ventilation.

C = Circulation Be prepared to use chest compressions. Remove wet clothing. Avoid wind chill. Dry and wrap in blankets or sleeping bag plus warm hat and cover, if available, in foil survival bag. *Urgent help needed*. Give hot sweet drinks if conscious.

Do not give alcohol, rub the skin or place very hot objects against skin.

8.8 SHOCK

Shock can result from almost any accident or medical emergency and, depending upon the cause, may range in severity from a simple faint to near death. Shock occurs when the delivery of oxygen to the tissues is impaired because of inadequate or inefficient circulation of the blood. Possible causes include:

- Loss of blood: internal or external bleeding.
- Loss of fluid: diarrhoea, peritonitis, burns.
- Heart failure: heart attack.
- Lung failure: drowning.
- Brain failure: stroke, head injury.
- Illness: diabetes.

Signs and symptoms

Thirst, apathy, nausea, restlessness. Pale, cold, clammy skin, sweating. Rapid, weak pulse. Rapid, shallow breathing. Dull, sunken eyes, bluish lips; it can lead to collapse.

Treatment

- *ABC – airway, breathing, circulation* see 8.4.
- Control bleeding, if present.
- Lay the casualty flat; elevate legs to 20°.

Exceptions:

i   Bleeding from mouth – use recovery position.

ii Unconscious – use recovery position.

iii Chest injury – sitting may be preferred.

- Splint any fractures; avoid movement.

- Keep reasonably warm; don't overheat.
- Relieve pain – give painkillers.

Exceptions:

i   Head injury with impaired consciousness.

ii Cases with severe breathing difficulty.

- Reassure the casualty.
- Do not let the casualty eat, drink, smoke or move unnecessarily. If he complains of thirst, moisten the lips with a little water.

Exception

Fluids may be life saving in cases of dehydration (diarrhoea, vomiting, severe burns, for example). Give half a cup of water at 15 minute intervals. Add a pinch of salt and a little sugar. Never give alcohol. Avoid fluids if severe abdominal pain or internal injury. *Never* attempt to give fluids by mouth to an unconscious person.

Collapse and signs of shock after an accident when external blood loss is slight or absent must suggest internal bleeding. Clues may be few. The casualty may cough up or vomit blood, or pass blood in urine or from bowel. He/she may complain of worsening pain in abdomen or chest. *Urgent help needed.*

Medical illnesses, such as diabetes, severe infections or heart disease, may produce shock without giving many clues as to the cause. *Urgent help needed.*

8.9   ACUTE (SUDDEN) ILLNESS

Before a long passage be aware of any crew medical problems and record medication being taken. Seek medical advice on likely symptoms and treatment. Unless forewarned, diagnosis may be very difficult once at sea.

- *Abdominal pain (minor)*
  i   Upper abdomen, intermittent, burning, no tenderness, otherwise well. May follow large alcohol intake. Eased by milk or antacid; bland meals; no alcohol.

  ii More generalised, cramping or colicky pain, no tenderness, may have diarrhoea or vomiting. May be gastroenteritis or food poisoning. Take oral fluid with a pinch of salt added. Avoid dehydration. Seek advice.

- *Abdominal pain (major)* Severe abdominal pain, usually constant and generalised. Abdomen may be rigid or very tender to touch, fever may be present, rapid pulse rate, generally unwell, nausea and vomiting. Make the casualty comfortable, give pain relief (injection if possible). Give nothing to eat or drink. *Urgent help needed.*

- *Allergies* Mild cases may just have a rash which responds to calamine lotion and antihistamine tablets. Severe cases may collapse with breathing difficulty and require emergency ABC resuscitation. Seek advice.

- *Children* may become ill with alarming rapidity. Ear and throat infections are especially common. Children are also more susceptible to the effects of dehydration; if ill get them to drink copious fluids. Reduce drug dosage to a proportion of adult dose based on weight (average adult 70kg/155lb). Seek advice.

- *Constipation* Common at sea. Prevent by eating fruit, vegetables, bran and, if necessary, anti-constipation medication (eg senna preparations).

- *Convulsion* Casualty may be a known epileptic. **Do not** put anything in the mouth. Prevent injury. Place in recovery position; protect airway (he/she may still look very blue). After fit, allow him/her to sleep. *Urgent help needed.*

- *Diabetes* A diabetic may become unconscious if his/her blood sugar is too high or too low. For **hyper**glycaemia (too much sugar) insulin is needed. **Hypo**glycaemia (too little sugar) may be caused by too much insulin, unusual stress or exercise, or too little food. In either case, if rousable, first give sweets, sugar, soft drinks. If recovery not rapid, *Urgent help needed*.

- *Diarrhoea* Can be serious, especially in young children if much fluid is lost. Stop food, give plenty of fluid. Water is sufficient in most cases or, alternatively, add salt (1 teaspoon/litre) and sugar (4–5 teaspoons/litre) to water. Lomotil or Imodium tablets are very effective in adults.

- *Fever* May be associated with anything from common cold, appendicitis, heat stroke to an infected toe. Except for major abdominal problems, prescribe copious fluids, paracetamol or aspirin (not in the case of children) and antibiotics if infection is present. Seek advice.

- *Heart attack* Severe central 'crushing' chest pain; may spread to shoulders, neck or arms. Sweating, then bluish lips, then collapse. Breathing and heart may stop. Give one aspirin tablet 300mg, to be chewed not swallowed whole. Note that some casualties (diabetics, for example) may not experience the chest pains. *Urgent help needed.*

  i   Early symptoms; rest, reassure.

  ii If unconscious: recovery position; observe breathing and pulse.

  iii If breathing stops or pulse is absent, commence mouth to mouth ventilation and chest compression immediately and do not stop. See 8.4.

- *Heat stroke* Cool casualty by spraying with cold water or wrap the casualty in a cold wet sheet until their temperature falls to 38°C under the tongue; encourage drinking (one teaspoon of salt/half litre of water). If casualty stops sweating, has a rapid pounding pulse and is becoming unconscious, *urgent help needed*.

- *Seasickness* is basically an inner ear disturbance caused by motion. The condition is aggravated by fear, anxiety, fatigue and boredom. It may manifest itself as lethargy, dizziness or headache as well as nausea and vomiting. Avoid strong food tastes and too much alcohol. Take small amounts of fluid and food frequently if you feel ill. Avoid fatigue; adequate sleep will often relieve the sick feeling. Keep warm. Stay on deck and concentrate on some task if possible – taking the helm can work wonders. Sick crew on deck must be secured by lifeline. Turn in if all else fails. Vomiting may cause serious loss of fluid.

No one remedy suits everybody. Try the alternatives until you find one that is effective with minimal side effects; most tend to cause a dry mouth and some tiredness. Tablets available over the counter include: Avomine (promethazine) and Stugeron (cinnarizine). Take the first tablet some hours before going to sea, and then regularly for as long as necessary. Take a tablet just before going to sleep if possible. Various preparations can be applied behind the ear or worn as a wrist band. Prolonged seasickness is serious – seek advice.

8.10 STROKE

Stroke presents in different ways. However, if a person develops **sudden** onset weakness on one side of the body, problems with speech &/or visual problems: *think of stroke*. Recognise a stroke using the FAST test:

- Face – Has one side of face fallen; can they smile?
- Arms – Can both arms be raised and kept up?
- Speech – Is it slurred or muddled?
- Time – To call for help.

If there is difficulty with *any one* of these suspect, a stroke.

Place in recovery position and check airway.

Urgent help needed – specialist treatment within 4½ hours may reverse the effects of a stroke. If you are more than 4½ hours from help, and the casualty is conscious and able to swallow, give 300mg of asprin, even if he/she is improving.

ACCIDENTS AND INJURIES

8.11 BITES AND STINGS

Injected poison from bites and stings usually only causes local swelling and discomfort, but some individuals may react severely. For insect stings, resuscitate if collapse occurs; otherwise give rest, painkillers, antihistamines (eg chlorpheniramine).

In warmer water, sea snakes and various sea stingers can inject deadly poison: prevent drowning, resuscitate if necessary. If sting is caused by jelly fish or Portuguese Man O'War etc, pour vinegar onto sting to reduce further poison release.

If the victim becomes weak and breathless, lightly compress the limb above the wound with a rolled up roller bandage to delay the spread of poison. **Do not** apply a tourniquet; this practice is out of date due to the danger of losing a limb. Start resuscitation. *Urgent help needed*.

Many large cities maintain a 24-hour poison information centre. Use the radio for advice.

8.12 FISH HOOK INJURY

Push the hook round until the point and barb appear. Cut off the point and barb and withdraw the hook. Dress the holes and give an antibiotic.

8.13 BLEEDING – OPEN WOUND

Bleeding is often dramatic, but is almost always controllable.

Treatment

- Apply firm continuous direct pressure; bandage on a large pad. If bleeding continues, bandage more pads on top of initial pads; then press directly over wound for at least 10 minutes (blood takes this long to clot).
- Elevate if wound is on a limb.
- **Do not** apply a tourniquet. This practice is out of date and risks losing a limb.

8.14 BLEEDING – INTERNAL (CLOSED INJURY)

Follows fractured bones, crush injuries, or rupture of organs such as the liver or spleen. Treat for shock which may appear rapidly. *Urgent help needed*.

8.15 BURNS AND SCALDS

Treatment

- Move the victim into fresh air to avoid inhaling smoke.
- *ABC – Airway, Breathing, Circulation;* see 8.4.
- Stop further injury: dip the whole of the burnt part into cold water for 10–15 minutes. Seawater is excellent but may be very painful.
- Remove loose clothing only; do not pull off clothing stuck to the skin.
- Cover with sterile dressing. If skin is broken or blistered, use sterile paraffin gauze beneath the dressing. Separate burnt fingers with paraffin gauze. Never use adhesive dressings.
- Do not prick blisters or apply ointments.
- Elevate burnt limb and immobilise.
- Give strong painkillers.
- Treat for shock: give frequent and copious drinks of water.
- Start giving antibiotics for major burns.
- If burns are extensive or deep, *urgent help needed*.

Sunburn can be very severe. Best treated by keeping the skin cool by fanning or careful sponging with chilled water; give painkillers. For prevention keep well covered and only use sunscreen preparations with high protection factor, greater than SPF 20.

8.16 CHEST INJURY

May result in fractured ribs. These are very painful, and breathing may be uncomfortable and shallow. The fractured ribs may puncture the lung or, if a number of ribs are each broken in two places (eg after crush injury), then this part of the chest may move independently of the rest of the chest and seriously impair breathing.

Treatment

- *ABC – Airway, Breathing, Circulation;* see 8.4.
- Casualty may be more comfortable sitting up.
- Tape a plastic cover over any hole on 3 sides (top and both sides) if air is sucking in and out.
- Support any unstable chest segment with your hand.
- For fractured ribs prescribe rest and strong painkillers if necessary. Very painful.

Avoid tight strapping lest it restricts breathing even further. *Urgent help needed* for any case with impaired breathing.

8.17 CRUSH INJURIES

These injuries involve skin, underlying muscle and tendon and sometimes bone as well. Common in hands and fingers (from winches and anchor chains) and elsewhere when tissue is sandwiched. Bleeding may be slight, but bruising and swelling considerable and extremely painful. Keep affected part as cold as possible; elevate if a hand or foot and give painkillers. *Urgent help needed*.

8.18 CUTS AND WOUNDS

Often dramatic but only potentially serious if nerves, tendons or blood vessels are severed.

Treatment

- Clean thoroughly with antiseptic. Remove dirt or other foreign bodies from the wound.
- Small clean cuts can be closed using Steristrips. Skin must be dry. Use as many Steristrips as necessary to keep the skin edges together. Leave for at least 5 days.
- Larger deep cuts may require special suture techniques (stitches); apply a dressing and seek help. Do not try amateur surgery at sea.
- Ragged lacerations or very dirty wounds – do not attempt to close these; dead tissues may have to be trimmed away to prevent infection. Clean as well as possible, sprinkle antibiotic powder in wound and apply a dressing. Seek help.

If in doubt a wound is best left open and lightly covered to keep it clean and dry.

Fingers and toes
Blood may collect under the nail following an injury. Release the blood by piercing the nail with a red hot needle or paper clip. It will not hurt!

8.19   DENTAL PROBLEMS

Dental pain usually seems worse at sea; prevention is better than cure. It is advisable to see a dentist about 4 weeks before departing on a long voyage to allow time for any necessary treatment. If X-Rays are taken, be sure to keep them with you.

Consider carrying a dental mirror, tongue spatula, pen torch, cotton wool rolls, tweezers and a ready-mixed temporary filling. Dentanurse is an emergency treatment pack which enables an amateur to make basic temporary repairs, eg replacing crowns, lost fillings. It contains zinc oxide and Eugenol.

Treatment

- Throbbing toothache, made worse by hot or cold or when bitten on. If a cavity, clean out and apply temporary filling. Take an anti-inflammatory painkiller.
- Dull toothache, tender to bite on; gum swollen or red with possible discharge. Treat as above but also take an antibiotic.
- Broken tooth or filling. Cover exposed surfaces with zinc oxide paste.
- Teeth which have been knocked out should be placed in a clean container with milk or moist gauze and taken to a dentist at the first opportunity for re-implantation. This can be attempted onboard. If the tooth is not re-implanted within 24 hours (preferably within 1 hour), chance of success is poor.
- Bleeding gums. Clean teeth more thoroughly. If accompanied by foul odour and metallic taste, use regular hot salt water rinses and antibiotics.
- Pain around wisdom tooth. Toothbrush to clean area; use hot salt water rinses; take antibiotics and painkillers.

8.20   EYE PROBLEMS

All eye injuries or illnesses are potentially serious. Never put old or previously opened ointment or drops into an eye; serious infection could result.

Treatment

- Foreign object. Flush the eye with clean water, pull the bottom lid out to inspect, remove object with a clean tissue. For objects under upper eyelid, ask casualty to grasp lashes and pull the upper lid over the lower lid. A proper eye-bath is very effective. Blinking under water may also remove the object. After removal of object, insert sterile antibiotic ointment inside pulled out lower lid. Cover with pad.
- Corrosive fluid. Flush continuously with water for 15 minutes. Give painkillers and chloramphenicol ointment; cover with pad. *Seek help as soon as possible*.
- Conjunctivitis. A sticky, weeping eye with yellow discharge. Chloramphenicol ointment 4 times a day.

8.21   FRACTURES AND DISLOCATIONS

Fracture is a broken bone. Dislocation is a displaced joint. Both result from major trauma and will produce pain (which is worse on attempted movement), localised swelling, abnormal shape, and a grating feeling on movement (when a fracture is present). Blood vessels or nerves around the fracture or dislocation may also be damaged resulting in a cold, pale, or numb limb below the site of the injury.

Fractures of large bones such as the femur (upper leg) will result in major internal bleeding and may cause shock. When complications occur, *Urgent help needed*.

Early application of a splint and elevation of the injured limb where possible will reduce pain and minimise complications. Treat for shock and pain.

Specific fractures and dislocations

Cheek   Caused by a direct blow. Rarely serious but requires specialist care.

Jaw   Beware of associated brain or spinal injury. Remove blood and teeth fragments; leave loose teeth in place; protect broken teeth (see 8.19). Ensure airway is clear. Start regular antiseptic mouth washes and antibiotics. Support jaw with bandage over top of the head. Give only fluids by mouth.

Neck   May result from a direct blow, a fall or a whiplash type injury. If conscious, casualty may complain of pain, tingling, numbness or weakness in limbs below the injury. *Mishandling may damage the spinal cord, resulting in paralysis or death*. Avoid movement and support the head. Immobilise by wrapping a folded towel around the neck. If movement is necessary then lift the victim as one rigid piece, never allowing the neck to bend. *Urgent help needed*.

Nose   Control bleeding by pinching (8.25).

Ribs   See chest injury (8.16). Often very painful. Strapping is not advised.

Skull   See head injury (8.23).

Spine   Fracture of the spine may occur below the neck, also resulting in *paralysis or death*. Mishandling of the victim may greatly worsen the damage. Avoid movement if possible. Lift the casualty without allowing the spine to sag. *Urgent help needed*.

Upper Limb

Collar bone (clavicle). Support arm in sling, Fig 8(7).

- **Dislocated shoulder**. If this has happened before, the casualty may remedy the dislocation himself; otherwise do not attempt to remedy it in case a fracture exists.
- **Upper arm (humerus)**. Support the arm with a collar and cuff inside the shirt as illustrated in Fig 8(8), ie tie a clove hitch around the wrist and loop the ends behind the neck.
- **Forearm and wrist**. Splint (eg with battens or wood). Do not bandage tightly. Elevate or support in a sling.
- **Fingers**. Elevate hand and, unless badly crushed, leave unbandaged; keep moving. If very wobbly, bandage to adjacent finger.

Fig 8(7) Sling

Fig 8(8) Collar and cuff

Lower limb

- **Thigh**. Shock may be considerable. Splint by strapping to other leg with padding between the legs. Gently straighten the casualty's lower leg. If necessary apply traction at the ankle to help straighten the leg. Do not bandage too tightly.
- **Knee**. Twisting injuries or falls damage the ligaments and cartilages of the knee. Very painful and swollen. Treat as for fracture.

- **Lower leg**. Pad very well. Splint using oar, broom handle or similar pieces of wood.
- **Ankle**. Fracture or severe sprain may be indistinguishable. Immobilise in neutral position with foot at right angles. Elevate the limb.

To be really effective a splint must be rigid and extend to the joints above and below the fracture. This is not always possible. The splint must be very well padded.

If the limb beyond the bandage or splint becomes swollen or discoloured, the bandage must be loosened to improve circulation. If you improvise a splint it is essential not to enclose the whole limb and risk cutting off the circulation (see 8.13, reference non-use of tourniquet).

As a general rule, if a fracture is suspected then seek advice.

Compound fractures

If a deep wound overlies the fracture, or the bone ends are visible, do not try to close the wound or replace the bone ends. Clean thoroughly with antiseptic and cover with a sterile dressing. *Seek help*. Start antibiotics if help is delayed.

8.22   FROSTBITE

Usually affects the extremities: toes, fingers, ears or nose. The affected part may be very painful, numb, stiff and discoloured. Warm gently (on someone else's back, for example).

Immersion in water less than 43°C (110°F); any higher temperature will cause more damage. Do not rub the affected part with anything.

8.23   HEAD INJURY

A blow to the head, with or without fracture, may result in immediate unconsciousness or more delayed effects.

Treatment

- Immediate unconsciousness, but quick recovery with slight drowsiness or headache. Prescribe rest and watch carefully.
- Immediate unconsciousness, no sign of recovery. Put in recovery position (beware of associated spine injury). Check airway. Observe the following and record every 10 minutes: pulse rate, breathing rate, pupil size (both sides), responses to verbal command, response to firm pinching. *Urgent help needed*.
- Delayed deterioration (either not unconscious immediately, or apparently recovering then worsening). Increasing drowsiness, change in mental state and eventually unconsciousness. Treat as above. *Urgent help needed*.

Scalp wounds may bleed profusely. Control with very firm pressure; cut away hair, and close using Steristrips if no fracture beneath. *If in doubt seek help*. Avoid giving drugs after head injury.

8.24   NAIL INJURIES

Nails can be torn or lifted off the underlying flesh and bleeding can be impressive. Do not try to remove the damaged nail which will support the underlying tissue or bone if it is also damaged. Treat as for any bleeding wound with a pad and a bandage wrapped around the finger.

Blood may collect under the nail following a blow. Release the blood by piercing the nail with a red hot needle or paper clip. It will not hurt!

8.25 NOSEBLEED

Lean forwards and pinch the soft part of the nose firmly for at least 10 minutes to allow the blood to clot. Do not blow nose or try to remove clot.

If bleeding continues repeat the pressure for longer than 10 minutes. If still bleeding after 30 minutes insert as much 50mm (2in) gauze bandage (moistened with water) into the nostrils as you can, using forceps to feed the bandage in. *Urgent help needed.*

8.26 POISONING

Poison may reach the body when swallowed, inhaled or injected through the skin (for bites and stings see 8.11).

Treatment

- *ABC – Airway, Breathing, Circulation;* see 8.4.
- Recovery position if unconscious.
- *Seek help*.

Swallowed poison
The poison container may give instructions or suggest antidote(s). For corrosive or petroleum products (acids, alkalis, bleach, detergent, petrol) *do not induce vomiting*. Administer copious fluids (eg milk).

For other substances (pills, medicines) *do not induce vomiting* – it is often ineffective and may cause further harm to the casualty. If collapsed or unconscious **urgent help needed**.

Inhaled poison
Poison may be inhaled from sources such as carbon monoxide or other exhaust fumes, bottled gas which has leaked into the bilge, or fire extinguisher gas. Carbon monoxide inhalation produces cherry red lips and skin. Move into fresh air at once. If breathing is absent, start resuscitation. **Urgent help needed**.

8.27 STRAINS AND SPRAINS

Torn ligaments, pulled muscles and other injuries. Rest the injured part; elevate if possible; apply ice packs (wrapped in a towel) if possible; administer painkillers. If in doubt, treat as a fracture and immobilise.

8.28 FIRST AID KIT

A made-up Offshore First Aid kit (below) costs about £50; a more comprehensive pre-packed kit might exceed £180. Prescriptions marked * are needed for three of the drugs listed in 8.29. Out of date drugs are potentially dangerous and should be destroyed. Special preparations are available for children. Stow the following suggested items in a readily accessible, clearly marked waterproof container:

Triangular bandage x 2 (doubles as bandage or sling)

Crepe bandage 75mm x 2

Gauze bandage 50mm x 2

Elastoplast 75mm x 1

Band Aids (or similar) various shapes and sizes

Wound dressings, 1 large, 1 medium

Sterile non-adhesive dressing (Melolin) x 5

Steristrips x 5 packs

Cotton wool

Scissors and forceps, good quality stainless steel

Safety pins

Thermometer

Disposable gloves

Antiseptic solution (eg Savlon)

Eye wash

Antifungal powder or cream (athlete's foot)

Insect repellent (DEET, diethyltoluamide)

Individual choice of anti-seasickness tablets

Sunscreen with high protection factor

Moisture cream (for cracked hands and lips)

Medication – see 8.29 for basic suggestions. The list will vary with voyage area, duration and crew embarked

Additional items for extended cruising
Vaccinations (possibly start 6 months before departure)

Syringes 2ml x 2 (if carrying injections)

Dental kit – see 8.19.

8.29 DRUGS AND MEDICATION

Drug	Type/Use	Dose
Paracetamol 500mg tablets	painkiller	1–2 tablets every 4 hours
Aspirin	blood thinner	300mg single dose in case of heart attack or stroke
Ibuprofen	anti-inflammatory	400mg every 8 hours (avoid if history of asthma or stomach ulcer)
*Dihydrocodeine 30mg tablets	strong painkiller	1–2 tablets every 4 hours
Cetirizine	antihistamine	10mg once a day
Aludrox	indigestion	1–2 before meals
Loperamide 2mg capsules	diarrhoea	Take 2 capsules initially followed by 1 after each loose stool, up to a maximum of 8 per day
Senokot tablets	constipation	2–4 tablets daily
*Amoxycillin 250mg	antibiotic	250–500mg every 8 hours (beware penicillin allergy)
*Erythromycin 250mg	antibiotic	For penicillin-allergic adults 4 tablets daily
Cinnarizine 15mg tablets	seasickness	2 before voyage, then 1 every 8 hours
*Scopolamine patches	seasickness	1 patch behind ear 5-6 hrs before voyage; replace after 72 hrs if necessary
Chloramphenicol ointment 1%	eye infection	3-4 times per day

* Only available on prescription.

8.30 EMERGENCY CHILDBIRTH

A yacht is no place to give birth. Unless remaining within about one hour of harbour, a woman who is more than 32 weeks' pregnant should not even consider going to sea in a small boat. Before 30 weeks into the pregnancy, there should be no problem embarking on short offshore passages.

Although giving birth may be an entirely natural process, it is potentially fraught with danger – for both mother and baby – and should ideally be supervised by a midwife or other suitably qualified person. Labour may start with backache, regular pains in the abdomen, a show of blood-stained mucus and a gush of water from the birth canal. Such a situation clearly demands a PAN PAN call followed by rapid assessment and transfer to hospital by lifeboat or helicopter.

Should help not be available, the best advice is to stay calm, unhurried and let nature take its course.

8.31 INJECTIONS

A doctor's prescription is required for injections. Stringent regulations apply to injectable painkiller drugs, which are probably only warranted for long passages. It is safest to inject into the muscle on the outer part of the mid-thigh. Clean the area, then plunge the needle swiftly an inch or so through the skin, pull back on the plunger to ensure that a blood vessel has not been entered, then slowly complete the injection.

8.32 BOOKS ON FIRST AID

The books below give detailed advice. At least one should be on board and the crew should know where it is stowed:

First Aid at Sea by Douglas Justins & Colin Berry (ACN, £10.99). Recommended by the RORC.

Doctor On Board by Jürgen Hauert (ACN, £12.99). Step by step photos and advice.

8.33 NORMAL PHYSIOLOGICAL MEASUREMENTS

Pulse rate	Adults 60–80/minute Children up to 100/minute
Breathing	12–15/minute
Temperature	36·7°C (98·4°F)

8.34 OBSERVATION FORM

The information recorded by you on this form will be invaluable in helping doctors or paramedics to diagnose the problem and arrange the best possible treatment for your casualty. This is particularly important if there may be a significant time lapse between requesting medical help and the casualty reaching hospital.

- Keep photocopies of this form in your First Aid kit so as to preserve the original.

- Whilst awaiting help, record your observations by ticking or annotating the various boxes at 10 minute intervals; trends can be important and will help doctors detect any improvement or deterioration in the casualty's condition.

- Continue recording observations until help arrives.

- If within radio range of shore attempt to pass the observations via the Coastguard to a medical authority; or ask a ship to relay.

- Before the casualty is taken off the yacht ensure that this form and personal documents are secured to him/her.

DATE .... CASUALTY'S NAME .... AGE .... M/F							
Times of observations @ 10 minute intervals:		**10**	**20**	**30**	**40**	**50**	**60**
EYES Observe for reactions whilst testing other responses	Open spontaneously						
	Open when spoken to						
	Open to painful stimulus						
	Nil response						
MOVEMENT Apply painful stimulus: Pinch ear lobe or skin on back of hand	Obeys commands						
	Responds						
	Nil response						
SPEECH Speak clearly and directly, close to the casualty's ear	Responds sensibly to queries						
	Seems confused						
	Uses inappropriate words						
	Incomprehensible sounds						
	Nil response						
PULSE (Beats per minute) Take adult's pulse at wrist or neck. Note rate and whether beats are: weak (w); strong (s); regular (reg) or irregular (irreg)	Over 110						
	101-110						
	91-100						
	81-90						
	71-80						
	61-70						
	Below 61						
BREATHING (Breaths per minute) Note rate and whether breathing is: quiet (q); noisy (n); easy (e); or difficult (d)	Over 40						
	31-40						
	21-30						
	11-20						
	Below 11						

9

Harbours, Coasts and Tides

Map of Areas • Area information • Harbour information • Environmental guidance • Distance tables: English Channel, Irish Sea, North Sea, Bay of Biscay • Tidal coefficients

9.1	South West England	137
9.2	Central South England	191
9.3	South East England	245
9.4	East England	269
9.5	North East England	329
9.6	South East Scotland	363
9.7	North East Scotland, Orkney and Shetland	389
9.8	North West Scotland	427
9.9	South West Scotland	465
9.10	NW England, Isle of Man and N Wales	495
9.11	South Wales and Bristol Channel	525
9.12	South Ireland	563
9.13	North Ireland	605
9.14	West Denmark	643
9.15	Germany	659
9.16	Netherlands and Belgium	695
9.17	North France	743
9.18	Central North France	791
9.19	Channel Islands	815
9.20	North Brittany	841
9.21	South Brittany	881
9.22	South Biscay	911
9.23	North and North West Spain	947
9.24	Portugal	981
9.25	South West Spain, Gibraltar and Morocco	1003
9.26	Portuguese Islands, Azores and Madeira	1021

Area	Page	Area	Page
1	137	**14**	643
2	191	**15**	659
3	245	**16**	695
4	269	**17**	743
5	329	**18**	791
6	363	**19**	815
7	389	**20**	841
8	427	**21**	881
9	465	**22**	911
10	495	**23**	947
11	525	**24**	981
12	563	**25**	1003
13	605	**26**	1021

9.0.2 AREA INFORMATION

The 26 geographic Areas are arranged as follows:

An Area Map which includes harbours, principal lights, TSS, MRCCs, NCI stations, airports, main ferry routes, magnetic variation and a distance table. Wind farms and other offshore energy installations are not routinely shown.

Tidal stream chartlets showing hourly rates and set.

Lights, buoys and waypoints (LBW) listing positions and characteristics of selected lights and other marks, their daytime appearance, fog signals and Racons. Arcs of visibility and alignment of sector/leading lights are true bearings as seen from seaward. Lights are white unless otherwise stated; any colours are shown between the bearings of the relevant arcs. AIS is widely fitted to navigational marks, but not normally shown in LBW Use of 'Virtual AIS' aids to navigation is increasing. See the relevant official charts/publications for details.

Passage Information (PI) is at Section 5 in most Areas (eg 9.1.5 in Area 1) with further information geographically arranged between the harbour entries, as shown on the Area's Contents page.

Special notes giving data specific to a country or area.

Harbour information (see below).

9.0.3 HARBOUR INFORMATION

Each harbour entry is arranged as follows:

HARBOUR NAME followed by the County or Unitary Council (or foreign equivalent) and the lat/long of the harbour entrance, or equivalent, for use as the final waypoint.

Harbour ratings (✿ ۵ ✿), inevitably subjective, which grade a port based on the following criteria:

Ease of access:
✿✿✿ *Can be entered in almost any weather from most directions and at all states of tide, by day or night.*
✿✿ *Accessible in strong winds from most directions; possible tidal or pilotage constraints.*
✿ *Only accessible in calm, settled conditions by day with little or no swell; possible bar and difficult pilotage.*

Facilities available:
۵۵۵ *Good facilities for vessel and crew.*
۵۵ *Most domestic needs catered for, but limited boatyard facilities.*
۵ *Possibly some domestic facilities, but little else.*

Ambience:
✿✿✿ *An attractive place; well worth visiting.*
✿✿ *Average for this part of the coast.*
✿ *Holds no particular attraction.*

CHARTS show Admiralty (AC), Imray, and foreign charts, all smallest scale first. Admiralty Leisure Folios (56XX), which cover most of the UK, Channel Islands and Ireland, and Imray 2000 series folios (2X00) are also shown.

TIDES include a time difference (usually on Dover in the UK), ML, Duration and the harbour's Standard Port. Time and height differences for nearby Secondary Ports are also shown. Tidal coefficients are tabled at 9.0.9.

SHELTER assesses how protected a harbour is from wind, sea, surge and swell. It warns of any access difficulties and advises on safe berths and anchorages.

NAVIGATION gives guidance on the approach and entry, and shows the position of the approach waypoint with its bearing and distance to the harbour entrance or next significant feature. Some waypoints may not be shown on the chartlet. Access times are only stated where a lock, gate, sill or other obstruction restricts entry. Otherwise the minimum charted depth of water in the approaches, where it is less than 2m, is usually shown, but always consult up to date official charts.

Chartlets are based on official charts augmented with local information. Due to their scale, they may not cover the whole area referred to in the text nor do they show every depth, mark, light or feature.

The chartlets are not intended to be used for navigation; positions taken from them should not be used as waypoints in chart plotters. The publisher and editors disclaim any responsibility for resultant accidents or damage if they are so used. The largest scale official chart, properly corrected, should always be used.

Drying areas and an indicative 5m depth contour are shown as: Dries ▨ <5m ▨ >5m

Wrecks around the UK which are of archaeological or historic interest are protected by law. Sites are listed under harbour entries or in Passage Information. Unauthorised interference, including anchoring and diving on these sites, may lead to a substantial fine.

LIGHTS AND MARKS describes, in more detail than is shown on the chartlets, any unusual characteristics of marks, their appearance by day and features not listed elsewhere.

COMMUNICATIONS shows the telephone area code followed by local telephone and VHF contact details for: coastguard, weather, police, doctor/medical, harbourmaster/office, other. Marina contact details are not usually duplicated if they are shown under the marina entry. International telephone calls from/to the UK are described in Special Notes, as are national numbers for emergency services: normally 112 in the EU; 999 in the UK. Radio callsigns, if not obvious, are in *italics*.

FACILITIES symbols describe berthing options and facilities in harbours, marinas and yacht clubs (see the free **Reeds Marina Guide** for detailed marina plans in the UK, Channel Islands and Ireland). Water, electricity, showers and toilets are available in marinas unless otherwise stated. Most yacht clubs welcome visiting crews who belong to a recognised club and arrive by sea. Any rail and air links are also shown.

The overnight cost of a visitor's alongside berth (⌓), *correct with regard to information supplied at the time of going to press (Summer 2015)*, is the average charge per metre LOA (unless otherwise stated) during high season, usually June to Sept. It includes VAT, harbour dues and, where possible, any tourist taxes (per head). The cost of pile moorings, ۩s or ⚓s, where these are the norm, may also be given. Shore electricity is usually free abroad but extra in the UK.

The number of ❶ berths is a marina's estimate of how many visitors may be accommodated at any one time. It is always advisable to call the marina beforehand.

TransEurope Marinas (www.transeuropemarinas.com) is an expanding grouping of independent marinas in the UK and abroad. Many hold Blue Flags and 4 or 5 Gold Anchor Awards; it is a condition of membership that they are well-equipped and maintain high standards. They operate a discounted reciprocal berthing scheme and are shown by the symbol ⊕.

Rover Tickets, offering good discounts, are available for many berthing facilities in Scotland and the offlying islands.

Distance tables

9.0.4 ENVIRONMENTAL GUIDANCE

- You must comply with regulations for navigation and conduct in Marine Nature Reserves, Marine Conservation Zones (MCZ), National Water Parks etc.
- In principle, avoid ditching rubbish at sea; keep it on board then dispose of it in shoreside refuse bins.
- Readily degradable foodstuffs may be ditched at sea when more than 3 miles offshore (more than 12 miles in the English Channel).
- Other foodstuffs and materials which are not readily degradable should never be thrown overboard.
- Only flush the heads overboard when well offshore, and do not discharge holding tanks until at least 3 miles off. Many countries now require holding tanks to be fitted; pump-out facilities (⚓) are shown in the text.
- Avoid discharging 'grey' water into a marina, anchorage or moorings area. Note any harbour/marina regulations.
- Deposit used engine oil and oily waste ashore at a recognised facility. Do not allow an automatic bilge pump to discharge oily bilge water overboard.
- Dispose of toxic waste (some antifoulings, old batteries etc) at an approved disposal facility.
- Minimise noise, wash and disturbance.
- Respect wild birds, plants, fish and marine animals. Avoid protected nesting sites and breeding colonies.
- Do not anchor or dry out on vulnerable seabed species, eg soft corals and eel grass.

9.0.5 DISTANCES (M) ACROSS THE ENGLISH CHANNEL

France/CI \ England	Longships	Falmouth	Fowey	Plymouth bkwtr	Salcombe	Dartmouth	Torbay	Exmouth	Weymouth	Poole Hbr Ent	Needles Lt Ho	Nab Tower	Littlehampton	Shoreham	Brighton	Newhaven	Eastbourne	Folkestone	Dover	Ramsgate
Le Conquet	114	118	134	142	139	152	158	168	181	199	203	236	256	260	264	269	281	313	318	342
L'Aberwrac'h	101	103	114	120	122	133	142	151	171	191	197	215	236	246	247	248	258	297	301	327
Roscoff	110	97	101	97	91	100	107	117	130	144	149	165	184	193	197	200	211	246	252	298
Trébeurden	118	105	107	100	95	103	110	122	128	148	148	168	186	197	201	206	218	257	262	288
Tréguier	131	114	112	102	93	99	104	114	118	132	136	153	170	182	187	192	211	252	253	277
Lézardrieux	139	124	119	107	97	102	105	115	117	129	133	150	167	179	184	189	208	249	250	274
St Quay-Portrieux	151	134	128	120	109	114	115	123	128	133	136	151	174	180	185	188	200	238	242	270
St Cast Le Guildo	164	145	139	127	114	117	120	128	124	129	132	147	170	177	181	184	198	234	239	264
St Malo	172	152	145	132	118	121	124	132	125	130	133	148	170	176	182	185	195	235	240	264
St Helier	156	133	124	110	95	97	100	109	103	108	110	125	145	155	160	163	175	211	218	241
St Peter Port	140	115	105	89	75	71	73	81	81	87	90	105	126	135	139	142	155	190	196	227
Braye (Alderney)	149	129	107	90	74	72	71	74	63	68	71	86	108	115	120	123	135	173	178	202
Cherbourg	173	144	130	111	98	94	93	95	67	64	63	72	89	99	102	107	120	153	158	185
St Vaast	194	165	151	134	119	117	119	127	86	78	72	69	84	91	94	97	104	140	147	178
Ouistreham	231	203	190	173	158	155	159	164	121	111	100	88	93	94	94	93	96	132	133	160
Deauville	238	210	198	181	164	163	160	160	127	114	103	89	90	89	87	87	89	124	126	152
Le Havre	239	210	197	180	164	162	165	159	125	112	98	84	84	83	82	82	83	119	120	146
Fécamp	246	218	206	188	173	169	165	164	127	111	98	81	72	69	66	63	63	98	98	122
Dieppe	273	245	231	214	200	194	191	189	151	133	121	97	86	76	72	67	62	78	80	100
Boulogne	294	264	247	231	216	204	202	199	161	143	132	101	86	77	73	65	54	28	28	51
Calais	307	276	261	244	229	218	214	211	174	155	143	114	99	91	86	78	64	29	24	30

NOTES ON DISTANCE TABLES 9.0.5–9.0.8

Approximate distances in nautical miles are by the most direct practicable route, taking into account the requirement to cross Traffic Separation Schemes (or the recommendation to cross main shipping lanes) on a heading at right angles to the general direction of traffic flow. They do not take into account any diversions to avoid exclusion zones around oil or gas installations.

Skippers are reminded of the need to consult up to date official charts, navigational publications and weather forecasts to choose a ground track that avoids dangers (especially lee shores), adding a margin of safety that is appropriate for the conditions likely to be expected during the voyage.

9.0.6 DISTANCES (M) ACROSS THE IRISH SEA

Scotland England Wales / Ireland	Port Ellen (Islay)	Campbeltown	Troon	Portpatrick	Mull of Galloway	Kirkcudbright	Maryport	Fleetwood	Pt of Ayre (IOM)	Port St Mary (IOM)	Liverpool	Holyhead	Pwllheli	Fishguard	Milford Haven	Swansea	Portishead	Ilfracombe	Padstow	Longships
Tory Island	75	107	132	119	134	170	185	215	156	167	238	207	260	279	307	360	405	372	372	399
Malin Head	45	76	101	88	103	139	154	184	125	136	207	176	229	248	276	329	374	341	341	368
Lough Foyle	33	61	86	73	88	124	139	169	110	121	192	161	214	233	261	314	359	320	326	353
Portrush	31	50	76	64	80	116	131	161	102	113	184	153	206	225	253	306	351	308	318	345
Carnlough	42	35	57	32	45	81	96	126	67	78	149	115	168	187	215	268	313	265	280	307
Larne	51	39	58	24	37	72	88	118	58	70	141	106	159	178	206	259	304	254	271	298
Belfast Lough	64	48	65	26	34	69	85	115	55	66	138	101	154	173	201	254	299	249	266	293
Bangor	63	48	64	22	30	65	81	111	51	62	134	97	150	169	197	250	295	245	262	289
Strangford Loch	89	72	84	36	30	63	76	97	41	37	107	69	121	141	167	219	264	214	231	258
Carlingford Loch	117	100	112	64	60	90	103	112	70	51	118	67	111	124	149	202	247	197	214	241
Dun Laoghaire	153	136	148	100	93	119	126	120	93	69	119	56	82	94	109	162	207	157	174	201
Wicklow	170	153	165	117	108	133	140	127	108	83	123	56	67	71	90	143	188	138	155	182
Arklow	182	165	177	129	120	144	149	133	117	93	131	64	71	65	79	132	178	128	144	167
Rosslare	215	202	208	161	154	179	180	164	152	125	156	90	83	55	58	109	156	110	119	137
Tuskar Rock	216	203	209	162	155	179	182	165	152	126	152	91	82	48	51	105	149	103	112	130
Dunmore East	250	237	243	196	189	213	216	199	186	160	189	127	116	79	76	130	176	124	127	136
Youghal	281	268	274	227	220	244	247	230	217	191	220	158	147	110	103	156	199	148	139	138
Cork Harbour	300	287	293	246	239	263	266	249	236	210	239	177	166	131	118	170	215	163	151	144
Baltimore	346	333	339	292	285	309	312	295	282	256	285	223	212	172	160	209	253	198	178	161
Fastnet Rock	354	341	347	300	293	317	320	303	290	264	293	231	220	181	169	216	259	207	185	170

9.0.7 DISTANCES (M) ACROSS THE NORTH SEA

Norway to France / UK	Bergen	Stavanger	Lindesnes	Skagen	Esjberg	Sylt (List)	Brunsbüttel	Helgoland	Bremerhaven	Willhelmshaven	Delfzijl	Den Helder	IJmuiden	Scheveningen	Roompotsluis	Vlissingen	Zeebrugge	Oostende	Nieuwpoort	Dunkerque
Lerwick	210	226	288	403	428	442	517	470	510	500	493	486	497	505	551	550	552	555	562	588
Kirkwall	278	275	323	438	439	452	516	467	507	497	481	460	473	481	515	514	516	519	526	545
Wick	292	283	323	437	428	440	498	449	489	479	458	433	444	451	485	484	486	489	496	514
Inverness	356	339	381	485	461	462	529	479	519	509	487	460	471	478	513	512	514	517	524	542
Fraserburgh	288	266	296	410	383	384	451	404	444	434	412	385	396	403	430	429	431	434	441	456
Aberdeen	308	279	298	411	371	378	433	382	432	412	386	353	363	369	401	400	402	405	412	426
Dundee	362	329	339	451	394	401	448	396	436	426	395	352	359	364	390	389	385	388	395	412
Port Edgar	391	355	362	472	409	413	457	405	445	435	401	355	361	366	391	390	386	389	396	413
Berwick	374	325	320	431	356	361	408	355	395	385	355	310	315	320	342	341	337	340	347	364
Hartlepool	409	353	340	440	340	331	367	312	352	342	302	241	243	247	266	265	261	264	271	288
Grimsby	463	395	362	452	324	318	342	291	332	325	288	187	182	185	199	198	190	191	201	198
Kings Lynn	485	416	379	466	330	333	343	292	344	336	283	184	183	183	197	195	187	188	198	195
Lowestoft	508	431	380	453	308	300	295	262	284	271	218	118	104	98	95	99	87	87	89	106
Harwich	540	461	410	483	330	331	320	287	309	296	243	147	126	114	94	100	84	77	80	80
Brightlingsea	558	479	428	501	348	349	338	305	327	314	261	165	144	105	108	106	92	88	86	87
Burnham/Crouch	567	488	437	510	357	358	347	314	336	323	270	174	151	112	109	115	99	92	93	95
London Bridge	620	543	490	560	400	408	395	361	382	374	320	222	199	149	153	149	134	125	126	114
Sheerness	580	503	450	520	360	367	353	319	340	334	280	180	157	109	113	109	94	85	86	74
Ramsgate	575	498	446	516	368	346	339	305	323	315	262	161	144	121	89	85	77	65	58	42
Dover	588	511	459	529	378	359	352	328	336	328	275	174	155	132	101	92	79	65	58	44

NOTES

This Table applies to Areas 3–7 and 14–17, each of which also contains its own internal Distance Table.

9.0.8 DISTANCES (M) ACROSS THE BAY OF BISCAY

West France \ N & NW Spain	Bayona	Vigo	Sanxenxo	Vilagarcia	Muros	Cabo Finisterre	Camariñas	La Coruña	Ria de Cedeira	Cabo Ortegal	Ria de Vivero	Ria de Ribadeo	Cabo Peñas	Gijon	San Vicente de la B	Santander	Bilbao	Cabo Machichaco	Zumaya	Fuenterrabia
Le Conquet	440	435	432	431	409	387	373	348	323	305	308	306	286	297	300	301	315	308	322	330
Lampaul (Ushant)	436	431	428	427	405	383	369	346	321	303	307	306	290	295	306	310	323	317	331	337
Camaret	444	439	436	435	413	391	377	352	327	309	309	306	287	302	300	304	316	308	322	329
Brest (Moulin B)	450	445	442	441	419	397	383	360	335	317	315	314	294	311	307	310	321	315	329	336
Douarnenez	441	436	433	432	410	388	374	350	325	307	309	307	287	295	300	302	316	308	323	329
Audierne	425	420	417	416	394	372	358	331	306	288	291	289	268	272	279	281	290	284	298	304
Port-la-Forêt	439	434	431	430	408	386	372	340	315	297	298	292	270	272	276	271	278	273	285	290
Lorient entrance	443	438	435	434	412	390	376	345	320	302	304	295	266	268	264	261	264	258	270	274
Belle Ile (Le Palais)	439	434	431	430	408	386	372	341	316	298	296	287	255	258	249	244	246	238	250	254
La Trinité	450	445	442	441	419	397	383	352	327	309	311	300	267	270	260	256	258	249	260	247
Piriac	453	448	445	444	422	400	386	353	328	310	311	301	266	265	255	246	246	239	248	252
Pornichet	455	450	447	446	424	402	388	355	330	312	309	298	262	264	249	242	239	229	242	247
L'Herbaudière	461	456	453	452	430	408	394	363	338	320	293	291	253	264	238	245	236	220	228	233
Sables d'Olonne	451	446	443	442	420	398	384	351	326	308	300	284	243	243	217	209	199	187	195	199
La Rochelle	469	464	461	460	438	416	402	363	338	320	311	293	248	249	216	208	193	179	187	192
Royan/Port-Médoc	460	455	452	451	429	407	493	353	328	310	307	287	237	236	198	191	172	158	166	146
Arcachon	470	465	462	461	439	417	403	337	312	294	289	266	214	207	167	150	115	121	112	90
Capbreton	430	425	422	421	399	377	363	323	298	280	272	245	191	186	131	117	78	59	45	23
Anglet/Bayonne	430	425	422	421	399	377	363	324	299	281	267	242	188	186	126	117	77	54	35	16
Hendaye	423	418	415	414	392	370	356	315	290	272	258	232	179	172	115	105	66	44	23	0·6

NOTES

1. This Table can be read in conjunction with those in West France (9.20.2, 9.21.2, 9.22.2) and North Spain (9.23.2).

2. Yachts leaving Ireland for Longships use 9.0.6, if applicable; then use 9.0.5 for distances across the English Channel.

3. **Landes firing range** in SE Biscay (see 9.22.26 and AC 1104) extends 106M S from Pte de la Négade to Capbreton and up to 45M offshore. Distances are based on routes passing west of Landes range, except that from Royan/Port-Médoc* and Arcachon* the distances to Fuenterrabia and Zumaya are by the inshore route E of the range.

If this route is not available, add 25M to the tabulated distance from Royan/Port-Médoc to Fuenterrabia; add nil for Zumaya. From Arcachon add 44M and 16M to Fuenterrabia and Zumaya respectively, using the 8M wide, safe corridor extending 270°/39M from ATT-ARC buoy through the range. This corridor is also used in calculating distances from Arcachon to all headlands and destinations west of Zumaya.

Wish you were here?

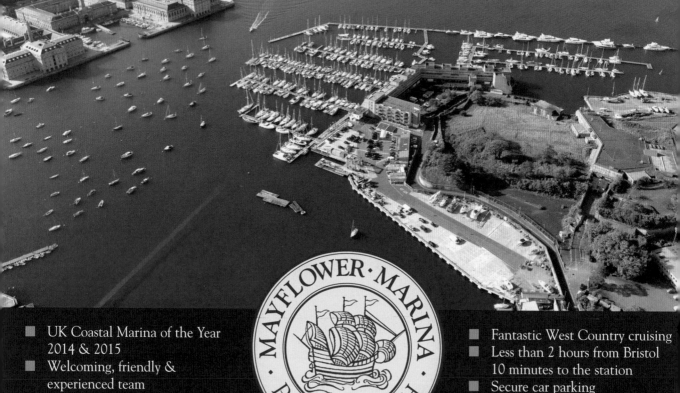

South West England

Isles of Scilly to Anvil Point

9.1.1	Contents	137
9.1.2	Area map	138
9.1.3	Tidal stream charts	140
9.1.4	Lights, buoys and waypoints	142
9.1.5	Passage information	144

Isles of Scilly, Land's End – 144 • Land's End to Falmouth – 148 • Falmouth to Fowey – 152 • Fowey to Plymouth – 156 • Plymouth to Start Point – 158 • Start Point to Teignmouth – 168 • Lyme Bay – 175 • Crossing the English Channel (western part), Portland Race, Portland Bill to Anvil Point – 179

9.1.6	Isles of Scilly	145
9.1.7	St Mary's (Scilly)	147
9.1.8	Newlyn	147
	Mousehole	
9.1.9	Penzance	148
	St Michael's Mount • Porthleven • Mullion Cove • Cadgwith • Coverack	
9.1.10	Helford River	149
9.1.11	Falmouth & River Fal, Standard Port, tide tables/curves	150
	Portscatho • Gorran Haven • Portmellon • Polkerris	
9.1.12	Mevagissey	156
	Dodman Pt gunnery range • Charlestown • Par	
9.1.13	Fowey	157
9.1.14	Looe	158
	Polperro	
9.1.15	Plymouth, tide tables/curves	159
9.1.16	Naval exercise areas	165
	River Lynher • River Tamar	
9.1.17	River Yealm	166
	River Erme • River Avon • Hope Cove	
9.1.18	Salcombe	167
	Kingsbridge	
9.1.19	Dartmouth, tide tables/curves	168
	Dittisham • Totnes	
9.1.20	Brixham	173
	Paignton	
9.1.21	Torquay	174
9.1.22	Teignmouth	174
9.1.23	River Exe	175
	Axmouth/Beer	
9.1.24	Lyme Regis	178
9.1.25	West Bay (Bridport)	178
9.1.26	Portland tidal stream chartlets	180
9.1.27	Portland Harbour, tide tables/curves	185
9.1.28	Weymouth	187
9.1.29	Lulworth ranges	188
	Church Ope Cove • Ringstead Bay • Durdle Door • Lulworth Cove • Worbarrow Bay • Chapman's Pool	

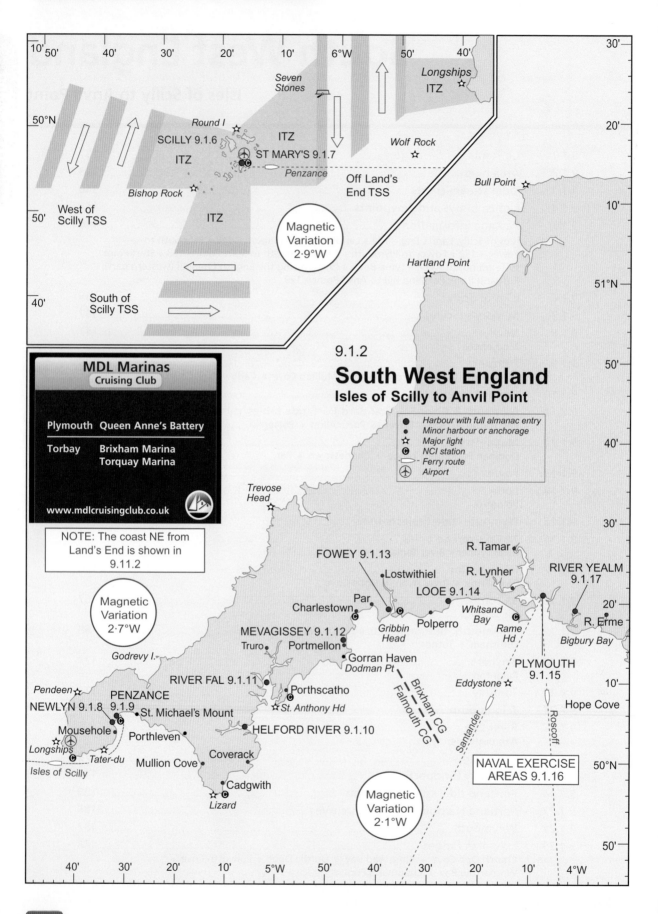

10' 50' 40' 30' 20' 10' 6°W 50' 40' 30'

50°N

Seven Stones

Longships ITZ

Round I

SCILLY 9.1.6

ITZ

ST MARY'S 9.1.7

Wolf Rock

ITZ

Penzance

Bishop Rock

Off Land's End TSS

Bull Point

West of Scilly TSS

ITZ

Magnetic Variation 2·9°W

Hartland Point

50' South of Scilly TSS

51°N

9.1.2

South West England
Isles of Scilly to Anvil Point

●	Harbour with full almanac entry
•	Minor harbour or anchorage
☆	Major light
ⓒ	NCI station
⊂⊃	Ferry route
✈	Airport

MDL Marinas
Cruising Club

Plymouth   Queen Anne's Battery

Torbay   Brixham Marina
           Torquay Marina

www.mdlcruisingclub.co.uk

NOTE: The coast NE from Land's End is shown in 9.11.2

Magnetic Variation 2·7°W

Trevose Head

R. Tamar

FOWEY 9.1.13
Lostwithiel
R. Lynher
RIVER YEALM 9.1.17

Par
LOOE 9.1.14
Charlestown
Whitsand Bay
Polperro
R. Erme
Gribbin Head
Rame Hd
Bigbury Bay

MEVAGISSEY 9.1.12
Truro   _Portmellon_
Gorran Haven
Dodman Pt
PLYMOUTH 9.1.15

Godrevy I.

RIVER FAL 9.1.11
Eddystone ☆
Porthscatho
Hope Cove

Pendeen ☆
PENZANCE
NEWLYN 9.1.8 9.1.9
St. Michael's Mount
St. Anthony Hd

Mousehole
Porthleven
HELFORD RIVER 9.1.10

Longships
Tater-du
Coverack
NAVAL EXERCISE AREAS 9.1.16

Isles of Scilly
Mullion Cove
Santander
Roscoff
Brixham CG
Falmouth CG

Cadgwith
Lizard

Magnetic Variation 2·1°W

40' 30' 20' 10' 5°W 50' 40' 30' 20' 10' 4°W

DISTANCE TABLE
Approximate distances in nautical miles are by the most direct route while avoiding dangers and allowing for TSS. Places in *italics* are in adjoining areas; places in **bold** are in 9.0.5, Cross-Channel Distances. See also 9.2.2.

		1	2	3	4	5	6	7	8	9	10	11	12	13	14	15	16	17	18	19	20	21	22	23
1	Longships	1																						
2	Scilly (Crow Sd)	22	2																					
3	**Newlyn**	14	34	3																				
4	Penzance	15	35	1	4																			
5	Lizard Point	23	42	16	16	5																		
6	Helford River	36	57	29	29	13	6																	
7	**Falmouth**	39	60	32	32	16	3	7																
8	Mevagissey	52	69	46	46	28	19	17	8															
9	Fowey	57	76	49	49	34	24	22	7	9														
10	Looe	63	80	57	57	39	31	29	16	11	10													
11	**Plymouth Bkwtr**	70	92	64	64	49	41	39	25	22	11	11												
12	R Yealm (ent)	72	94	66	66	51	43	41	27	24	13	4	12											
13	**Salcombe**	81	102	74	74	59	52	50	40	36	29	22	19	13										
14	Start Point	86	103	80	80	63	57	55	45	40	33	24	21	7	14									
15	Dartmouth	95	116	88	88	72	65	63	54	48	42	35	32	14	9	15								
16	Torbay	101	118	96	96	78	72	70	62	55	50	39	36	24	15	11	16							
17	Teignmouth	109	126	104	104	86	80	78	70	63	58	47	44	32	23	19	8	17						
18	**Exmouth**	113	131	107	107	90	84	82	73	67	61	51	48	33	27	24	12	6	18					
19	Lyme Regis	126	144	120	120	104	98	96	86	81	74	63	60	48	41	35	30	25	20	19				
20	West Bay	130	148	124	124	108	102	100	90	86	78	67	64	52	45	39	34	30	25	7	20			
21	Portland Bill	135	151	128	128	112	106	104	93	89	81	73	70	55	49	45	42	39	36	22	17	21		
22	Weymouth	143	159	136	136	120	114	112	111	97	89	81	79	63	57	53	50	47	44	30	25	8	22	
23	St Alban's Hd	151	167	144	144	128	122	120	109	105	97	89	86	71	65	61	58	55	52	38	33	16	21	23

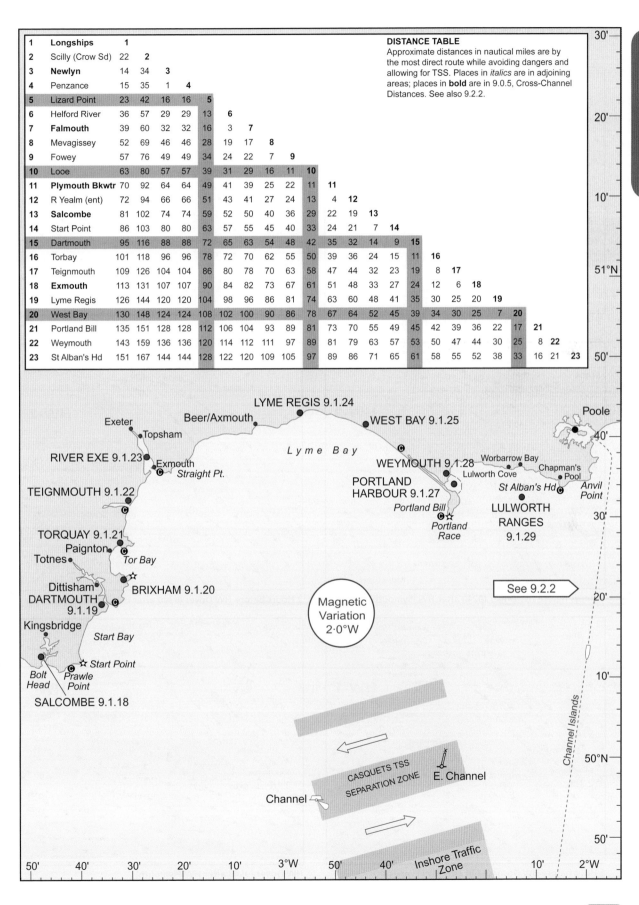

Poole

Exeter
Topsham

LYME REGIS 9.1.24
Beer/Axmouth
WEST BAY 9.1.25

RIVER EXE 9.1.23
Exmouth
Straight Pt.

Lyme Bay

Worbarrow Bay
WEYMOUTH 9.1.28
Lulworth Cove
Chapman's Pool

TEIGNMOUTH 9.1.22

PORTLAND HARBOUR 9.1.27
St Alban's Hd
Anvil Point

Portland Bill
LULWORTH RANGES 9.1.29

TORQUAY 9.1.21
Paignton
Totnes
Tor Bay
Portland Race

See 9.2.2

Dittisham
BRIXHAM 9.1.20
DARTMOUTH 9.1.19

Magnetic Variation 2·0°W

Kingsbridge

Start Bay

Bolt Head
☆ Start Point
Prawle Point

SALCOMBE 9.1.18

CASQUETS TSS SEPARATION ZONE
E. Channel

Channel

Channel Islands

Inshore Traffic Zone

9.1.3 AREA 1 TIDAL STREAMS

5 Hours before HW Dover (0040 after HW Plymouth)

4 Hours before HW Dover (0140 after HW Plymouth)

3 Hours before HW Dover (0240 after HW Plymouth)

2 Hours before HW Dover (0340 after HW Plymouth)

1 Hour before HW Dover (0440 after HW Plymouth)

HW Dover (0540 after HW Plymouth)

Eastward 9.2.3   Portland 9.1.26   Isle of Wight 9.2.16   Northward 9.11.3   Southward 9.20.3   Channel Is 9.19.3

1 Hour after HW Dover (0545 before HW Plymouth)

2 Hours after HW Dover (0445 before HW Plymouth)

3 Hours after HW Dover (0345 before HW Plymouth)

4 Hours after HW Dover (0245 before HW Plymouth)

5 Hours after HW Dover (0145 before HW Plymouth)

6 Hours after HW Dover (0045 before HW Plymouth)

141

9.1.4 LIGHTS, BUOYS AND WAYPOINTS

Bold print = light with a nominal range of 15M or more. CAPITALS = place or feature. *CAPITAL ITALICS* = light-vessel, light float or Lanby. *Italics* = Fog signal. ***Bold italics*** = Racon. Many marks/buoys are fitted with AIS, <u>MMSI No</u>; see relevant charts.

ISLES OF SCILLY TO LAND'S END

Bishop Rock ☆ Fl (2) 15s 44m **20M**; part obsc 204°-211°, obsc 211°-233° and 236°-259°; Gy ○ twr with helo platform; ***Racon T, 18M, 254°-215°***; 49°52'·37N 06°26'·74W; <u>992351137</u>.

Round Rk ⌕ 49°53'·10N 06°25'·19W.

Old Wreck ⌕ VQ; 49°54'·26N 06°22'·81W.

ST AGNES and ST MARY'S

Peninnis Hd ⚲ Fl 20s 36m 9M; 231°-117° but part obsc 048°-083° within 5M; W ○ twr on B frame, B cupola; 49°54'·28N 06°18'·21W.

Gugh ⌕ Fl (5) Y 20s; 49°53'·52N 06°18'·73W.

Spanish Ledge ⌕ Q (3) 10s; *Bell;* 49°53'·94N 06°18'·86W.

Woolpack ⌕ Q (6) + L Fl 15s; 49°54'·40N 06°19'·37W.

Bartholomew Ledges ⌕ QR 12m; 49°54'·37N 06°19'·89W.

N Bartholomew ⌕ Fl R 5s; 49°54'·49N 06°19'·99W.

Bacon Ledge ⌕ Fl (4) R 5s; 49°55'·22N 06°19'·26W.

Ldg lts 097·3°: Front, Iso RW (vert) 2s; W △, 49°55'·12N 06°18'·50W. Rear, Oc WR (vert) 10s; Or X on W bcn.

St Mary's Pool pier ⚲ Fl WRG 2s 5m 4M; 070°-R-100°-W-130°-G-070°; 49°55'·11N 06°19'·00W.

Crow Rock ⌕ Fl (2) 10s; 49°56'·26N 06°18'·49W.

Hats ⌕ VQ (6) + L Fl 10s; 49°56'·21N 06°17'·14W.

AROUND TRESCO, BRYHER and ST MARTIN'S

Tresco Flats, Hulman ⌕ Fl G 4s, 49°56'·29N 06°20'·30W.

Little Rag Ledge ⌕ Fl (2) R 5s, 49°56'·44N 06°20'·43W.

Bryher, Bar ⌕ Q (3) 10s, 49°57'·37N 06°20'·84W.

Bryher, Church Quay ⌕ Q (3) 10s, 49°57'·18N 06°20'·97W.

Spencers Ledge ⌕ Q (6) + L Fl 15s; 49°54'·78N 06°22'·06W.

Steeple Rock ⌕ Q (9) 15s; 49°55'·46N 06°24'·24W.

Round Island ☆ Fl 10s 55m **18M**, also shown in reduced vis; 021°-288°; W ○ twr; *Horn (4) 60s;* 49°58'·74N 06°19'·39W.

St Martin's, Higher Town quay ⚲ Fl R 5s, 49°57'·45N 06°16'·84W.

SCILLY to LAND'S END

Seven Stones Lt V ⚓ Fl (3) 30s 12m **15M**; R hull; *Horn (3) 60s; Racon O, 15M;* 50°03'·63N 06°04'·32W; <u>992351023</u>.

Wolf Rock ☆ Fl 15s 34m **16M**; H24; *Horn 30s;* ***Racon T, 10M***; 49°56'·72N 05°48'·55W; <u>992351128</u>.

Longships ☆ Fl (2) WR 10s 35m **W15M**, R11M; 189°-R-327°-W-189°; also shown in reduced vis; Gy ○ twr with helicopter platform; *Horn 10s;* 50°04'·01N 05°44'·81W.

Carn Base ⌕ Q (9) 15s; 50°01'·48N 05°46'·18W.

Runnel Stone ⌕ Q (6) + L Fl 15s; *Whis;* 50°01'·18N 05°40'·36W.

LAND'S END TO PLYMOUTH

Tater-du ☆ Fl (3) 15s 34m **20M**; 241°-072°; W ○ twr. FR 31m 13M, 060°-072° over Runnel Stone; 50°03'·14N 05°34'·67W.

NEWLYN

Low Lee ⌕ Q (3) 10s; 50°05'·56N 05°31'·38W.

S Pier ⚲ Fl 5s 10m 9M; W ○ twr; 253°-336°; 50°06'·18N 05°32'·57W.

N Pier ⚲ F WG 4m 2M; 238°-G-248°, 50°06'·18N 05°32'·62W.

PENZANCE

S Pier ☆ Fl WR 5s 11m **W17M**, R12M; 159°-R (unintens)-224°-R-268°-W-344·5°-R-shore; 50°07'·06N 05°31'·68W.

Mountamopus ⌕ Q (6) + L Fl 15s; 50°04'·62N 05°26'·25W.

Lizard ☆ Fl 3s 70m **26M**; 250°-120°, partly visible 235°-250°; W 8-sided twr; *Horn 30s;* 49°57'·61N 05°12'·13W.

Manacle ⌕ Q (3) 10s; *Bell;* 50°02'·81N 05°01'·91W.

FALMOUTH

St Anthony Head ☆ Iso WR 15s 22m, **W16M**, R14M, H24; 295°-W-004°-R (over Manacles)-022°-W-172°; W 8-sided twr; *Horn 30s;* 50°08'·46N 05°00'·96W.

Black Rock Fl (2) 10s 3M; B IDM bcn twr; 50°08'·72N 05°02'·00W.

Black Rock ⌕ Fl R 2·5s; 50°08'·68N 05°01'·74W.

Castle ▲ Fl G 2·5s; 50°08'·99N 05°01'·62W.

St Mawes ⌕ Q (6) + L Fl 15s; 50°09'·10N 05°01'·42W.

The Governor ⌕ VQ (3) 5s; 50°09'·15N 05°02'·40W.

West Narrows ⌕ Fl (2) R 10s; 50°09'·39N 05°02'·07W.

East Narrows ▲ Fl (2) G 10s; 50°09'·43N 05°01'·90W.

The Vilt ▲ Fl (4) G 15s; 50°09'·99N 05°02'·28W.

Northbank ⌕ Fl R 4s; 50°10'·34N 05°02'·26W.

St Just ⌕ QR; 50°10'·44N 05°01'·72W.

Mylor appr chan ▲ Fl G 6s; 50°10'·79N 05°02'·70W. ⌕ Fl R 5s.

Messack ▲ Fl G 15s; 50°11'·31N 05°02'·22W.

Carrick ▲ Fl (2) G 10s; 50°11'·59N 05°02'·74W.

Pill ▲ Fl (3) G 15s; 50°12'·05N 05°02'·40W.

Turnerware Bar ▲ Fl G 5s; 50°12'·40N 05°02'·15W.

Dock Basin Dir ⚲ 266°: WRG 2s 15m, 10M; 258°-Iso G-226°-Al WG-264°-Iso W-268°-Al WR-270°-Iso R-274°; 50°09'·38N 05°03'·29W.

N Arm ⚲ QR 5m 3M; 50°09'·42N 05°03'·20W.

Inner Hbr Dir ⚲ 233°: WRG 3s 5m, 9M; 226°-Iso G-231°-Iso W-236°-Iso R-241°; 50°09'·20N 05°03'·94W.

Falmouth Haven Marina ⚲ 2 FR (vert); 50°09'·27N 05°03'·91W.

Falmouth Marina ⌕ VQ (3) 5s; 50°09'·91N 05°04'·99W.

DODMAN POINT and MEVAGISSEY

Naval gunnery targets SSE of Dodman Point:

'A' ⌕ Fl Y 10s; 50°08'·53N 04°46'·37W.

'B' ⌕ Fl Y 5s; 50°10'·30N 04°45'·00W.

'C' ⌕ Fl Y 2s; 50°10'·40N 04°47'·51W.

Gwineas ⌕ Q (3) 10s; *Bell;* 50°14'·48N 04°45'·40W.

Mevagissey, Victoria Pier ⚲ Fl (2) 10s 9m 12M; *Dia 30s;* 50°16'·15N 04°46'·92W.

FOWEY

Cannis Rock ⌕ Q (6) + L Fl 15s; *Bell;* 50°18'·38N 04°39'·95W.

Fowey ⚲ L Fl WR 5s 28m W11M, R9M; 284°-R-295°-W-028°-R-054°; W 8-sided twr, R lantern; 50°19'·63N 04°38'·83W.

St Catherine's Pt ⚲ Fl R 2·5s 15m 2M; vis 150°-295°; 50°19'·69N 04°38'·66W.

Lamp Rock ⌕ Fl G 5s 3m 2M; vis 357°-214°; 50°19'·70N 04°38'·41W.

Whitehouse Pt ⚲ Iso WRG 3s 11m W11M, R/G8M; 017°-G-022°-W-032°-R-037°; R col; 50°19'·98N 04°38'·28W.

POLPERRO, LOOE, EDDYSTONE and WHITSAND BAY

Udder Rock ⌕ VQ (6) + L Fl 10s; *Bell;* 50°18'·93N 04°33'·85W.

POLPERRO, W pier ⚲ FW 4m 4M; FR when hbr closed in bad weather; 50°19'·86N 04°30'·96W.

Spy House Pt ⚲ Iso WR 6s 30m 7M; W288°-060°, R060°-288°; 50°19'·81N 04°30'·69W.

LOOE, Ranneys ⌕ Q (6) + L Fl 15s; 50°19'·85N 04°26'·37W.

Mid Main ⌕ Q (3) 10s 2M; 50°20'·56N 04°26'·94W.

Banjo Pier ☆ Oc WR 3s 8m **W15M**, R12M; 207°-R267°- W-313°-R-332°; 50°21'·06N 04°27'·06W.

White Rock ⚲ Fl R 3s 5m 2M; 50°21'·03N 04°27'·09W.

Eddystone ☆ Fl (2) 10s 41m **17M**. Same twr, Iso R 10s 28m 8M; vis 110°-133° over Hand Deeps; Gy twr, helicopter platform; *Horn 30s;* ***Racon T, 10M***; 50°10'·84N 04°15'·94W; <u>992351125</u>.

Hand Deeps ⌕ Q (9) 15s; 50°12'·68N 04°21'·10W.

PLYMOUTH

PLYMOUTH SOUND, WESTERN CHANNEL

Draystone ⌕ Fl (2) R 5s; 50°18'·85N 04°11'·07W.

Knap ▲ Fl G 5s; 50°19'·56N 04°10'·02W.

Plymouth bkwtr W head, ⚲ Fl WR 10s 19m W12M, R9M; 262°-W-208°-R-262°; W ○ twr. Same twr, Iso 4s 12m 10M; vis 033°-037°; *Horn 15s;* 50°20'·07N 04°09'·52W.

Maker ⚲ Fl (2) WRG 10s 29m, W11M, R/G6M; 270°-W-004°-R-050°; W twr, R stripe; 50°20'·51N 04°10'·87W.

Queens Ground ⌕ Fl (2) R 10s; 50°20'·29N 04°10'·08W.

New Ground ⌕ Fl R 2s; 50°20'·47N 04°09'·43W.

Melampus ⌕ Fl R 4s; 50°21'·15N 04°08'·72W.

PLYMOUTH SOUND, EASTERN CHANNEL

Wembury Pt ⚲ Oc Y 10s 45m; occas; 50°19'·01N 04°06'·63W.

West Tinker ⚓ VQ (9) 10s; 50°19'·25N 04°08'·64W.
East Tinker ⚓ Q (3) 10s; 50°19'·20N 04°08'·30W.
Whidbey ⚓ Oc (2) WRG 10s 29m, W8M, R/G6M; H24; 000°-G-137·5°-W-139·5°-R-159°; Or and W col; 50°19'·53N 04°07'·27W.
The Breakwater, E head ⚓ L Fl WR 10s 9m W8M, R6M; 190°-R-353°-W-001°-R-018°-W-190°; 50°20'·01N 04°08'·24W.
Staddon Pt ⚓ Oc WRG 10s 15m W8M, R/G5M; H24. 348°-G-038°-W-050°-R-090°; W structure, R bands; 50°20'·17N 04°07'·54W.
Withyhedge Dir ⚓ 070° (for W Chan): WRG 13m W13M, R/G5M; H24; 060°-FG-065°-Al WG (W phase increasing with brg) -069°-FW-071°-Al WR (R phase increasing with brg)-075°-F R 080°; W ▽, orange stripe on col. Same col, Fl (2) Bu 5s; vis 120°-160°; 50°20'·75N 04°07'·44W.

SMEATON PASS (W of Mount Batten and S of The Hoe)
Ldg lts 349°. Front, Mallard Shoal ⚓ Q WRG 5m W10M, R/G3M; W △, Or bands; 233°-G-043°- R-067°- G-087°-W-099°-R-108° (ldg sector); 50°21'·60N 04°08'·33W. Rear, 396m from front, Hoe ⚓ Oc G 1·3s 11m 3M, 310°-040°; W ▽, Or bands; 50°21'·81N 04°08'·39W.
S Mallard ⚓ VQ (6) + L Fl 10s; 50°21'·51N 04°08'·30W.
W Mallard ⚓ QG; 50°21'·57N 04°08'·36W.
S Winter ⚓ Q (6) + L Fl 15s; 50°21'·40N 04°08'·55W.
NE Winter ⚓ QR; 50°21'·54N 04°08'·50W.
NW Winter ⚓ VQ (9) 10s; 50°21'·55N 04°08'·70W.

ENTRANCE TO THE CATTEWATER
QAB (Queen Anne's Battery) ldg lts ⚓ 048·5°. Front, FR; Or/W bcn; 50°21'·84N 04°07'·84W. Rear, Oc R 8s 14m 3M; 139m NE. Fishers Nose ⚓ Fl (3) R 10s 6m 4M; 50°21'·80N 04°08'·01W. Also F Bu ≠ 026·5° with F Bu 50°22'·00N 04°07'·86W, for Cobbler Chan.

DRAKE CHANNEL, THE BRIDGE and THE NARROWS
Ravenness Dir ⚓ 225°: WRG 11m, W13M, R/G5M; H24. 217°-FG-221°-Al WG-224° (W phase increasing with brg)-FW-226°-Al WR-229° (R phase increasing with bearing)-FR-237°; W ▽, orange stripe on col. In fog, 160°-FW-305°; power failure QY; 50°21'·14N 04°10'·07W.
Asia ⚓ Fl (2) R 5s; 50°21'·47N 04°08'·85W.
St Nicholas ⚓ QR; 50°21'·55N 04°09'·20W.
N Drakes Is ⚓ Fl R 4s; 50°21'·52N 04°09'·38W.
E Vanguard ⚓ QG; 50°21'·47N 04°09'·70W.
W Vanguard ⚓ Fl G 3s; 50°21'·49N 04°09'·98W.
Devils Point ⚓ QG 5m 3M; Fl 5s in fog; 50°21'·59N 04°10'·04W.
Battery ⚓ Fl R 2s; 50°21'·52N 04°10'·21W.

The Bridge Channel
No 1, ⚓ QG 4m; 50°21'·03N 04°09'·53W. No 2, ⚓ QR 4m. No 3, ⚓ Fl (3) G 10s 4m. No 4, ⚓ Fl (4) R 10s 4m; 50°21'·09N 04°09'·63W.
Mount Wise, Dir ⚓ 343°: WRG 7m, W13M, R/G5M; H24. 331°-FG-338°-Al WG-342° (W phase increasing with brg)-FW-344°-Al WR-348° (R phase increasing with bearing)-FR-351°. In fog, 341·5°-FW-344·5°; 50°21'·96N 04°10'·33W.
Ocean Court Dir Q WRG 15m, W11M, R/G3M; 010°-G-080°-W-090°-R-100°; 50°21'·85N 04°10'·11W.

PLYMOUTH TO START POINT

RIVER YEALM Sand bar ⚓ Fl R 5s; 50°18'·59N 04°04'·12W.

SALCOMBE
Sandhill Pt Dir ⚓ 000°: Fl WRG 2s 27m W/R/G 8M; 337·5°-G-357·5°-W-002·5°-R-012·5°; R/W ⚓ on W mast, rear daymark; 50°13'·77N 03°46'·67W. Front daymark, Pound Stone R/W ⚓.
Bass Rk ⚓ Fl R 5s; 50°13'·47N 03°46'·71W.
Wolf Rk ⚓ Fl G 5s; 50°13'·53N 03°46'·58W.
Blackstone Rk ⚓; 50°13'·61N 03°46'·51W.
Ldg lts 042·5°, front Fl 2s 5m 8M, 50°14'·53N 03°45'·31W; rear Fl 5s 45m 8M.

Start Pt ☆ Fl (3) 10s 62m **25M**; 184°-068°. Same twr: FR 55m 9M; 210°-255° over Skerries Bank; *Horn 60s;* 50°13'·34N 03°38'·54W.

START POINT TO PORTLAND BILL
DARTMOUTH
Kingswear Dir ⚓ 328°: Iso WRG 3s 9m 8M; 318°-G-325°-W-331°-

R-340°; W ○ twr; 50°20'·82N 03°34'·09W.
Mewstone ⚓ VQ (6) + L Fl 10s; 50°19'·92N 03°31'·89W.
West Rock ⚓ Q (6) + L Fl 15s; 50°19'·86N 03°32'·47W.
Homestone ⚓ QR; 50°19'·61N 03°33'·55W.
Castle Ledge ⚓ Fl G 5s; 50°19'·99N 03°33'·11W.
Checkstone ⚓ Fl (2) R 5s; 50°20'·45N 03°33'·81W.
Dir ⚓ 104·5°: FW 5m 9M; vis 102°-107°; 50°20'·65N 03°33'·80W.

BRIXHAM
Berry Head ☆ Fl (2) 15s 58m **19M**; vis 100°-023°; W twr; 50°23'·98N 03°29'·01W. R lts on radio mast 5·7M NW, inland of Paignton.
Victoria bkwtr ⚓ Oc R 15s 9m 6M; W twr; 50°24'·33N 03°30'·78W.
No. 1 ⚓ Fl G; 50°24'·30N 03°30'·89W.
No. 2 ⚓ Fl R; 50°24'·32N 03°30'·83W.

PAIGNTON and TORQUAY
⚓ QG (May-Sep); 50°27'·42N 03°31'·80W, 85m off Haldon Pier.
Haldon Pier (E) ⚓ QG 9m 6M; 50°27'·43N 03°31'·73W.
Princess Pier (W) ⚓ QR 9m 6M; 50°27'·46N 03°31'·73W.

TEIGNMOUTH
Outfall ⚓ Fl Y 5s; 50°31'·97N 03°27'·77W, 288°/1·3M to hbr ent.
Bar ⚓ Fl G 2s; 50°32'·44N 03°29'·25W.
Trng wall, middle ⚓ Oc R 6s 4m 3M; 50°32'·33N 03°29'·93W.
The Point ⚓ Oc G 6s 3M & FG (vert); 50°32'·42N 03°30'·05W.

RIVER EXE to SIDMOUTH and AXMOUTH
Exe ⚓ Mo(A) 10s; 50°35'·86N 03°23'·79W.
No. 1 ⚓ 50°36'·05N 03°23'·84W.
No. 2 ⚓ 50°36'·02N 03°23'·94W.
Exmouth Dir ⚓ 305°: WRG 6m, 6M; 299°-Iso G-304°-Iso W-306°-Iso R-311°; W col, 50°36'·99N 03°25'·34W.
No. 10 ⚓ Fl R 3s; 50°36'·73N 03°24'·77W.
No. 12 Warren Pt ⚓ 50°36'·91N 03°25'·41W.
Sidmouth ⚓ Fl R 5s 5m 2M; 50°40'·48'N 03°14'·43W.
Axmouth jetty ⚓ Fl G 4s 7m 2M; 50°42'·12N 03°03'·29W.

LYME REGIS
Outfall ⚓ Q (6) + L Fl 15s; 50°43'·17N 02°55'·66W.
Ldg lts 284°: Front, Victoria Pier ⚓ Oc WR 8s 6m, W9M, R7M; 284°-R-104°-W-284°; Bu col; 50°43'·19N 02°56'·17W. Rear, FG 8m 9M.

WEST BAY (BRIDPORT)
W pier root, Dir ⚓ 336°: F WRG 5m 4M; 165°-G-331°-W-341°-R-165°; 50°42'·62N 02°45'·89W.
W pier outer limit ⚓ Iso R 2s 5m 4M; 50°42'·51N 02°45'·83W.
E pier outer limit ⚓ Iso G 2s 5m 4M; 50°42'·53N 02°45'·80W.

PORTLAND BILL TO ANVIL POINT
Portland Bill lt ho ☆ Fl (4) 20s 43m **25M**. vis 221°-244° (gradual change from 1 Fl to 4 Fl); 244°-117° (shows 4 Fl); 117°-141° (gradual change from 4 Fl to 1 Fl). W ○ twr; *Dia 30s;* 50°30'·85N 02°27'·39W. Same twr, FR 19m 13M; 265°-291° over Shambles.
W Shambles ⚓ Q (9) 15s; *Bell;* 50°29'·78N 02°24'·41W.
E Shambles ⚓ Q (3) 10s; *Bell;* 50°31'·26N 02°20'·08W.

PORTLAND HARBOUR
Outer Bkwtr Fort Head (N end) ⚓ QR 14m 5M; 013°-268°; 50°35'·11N 02°24'·87W.
NE Bkwtr (A Hd) ⚓ Fl 2·5s 22m 10M; 50°35'·16N 02°25'·07W.
NE Bkwtr (B Hd) ⚓ Oc R 15s 11m 5M; 50°35'·65N 02°25'·88W.
N Arm (C Hd) ⚓ Oc G 10s 11m 5M; 50°35'·78N 02°25'·95W.

WEYMOUTH
Ldg lts 239·6°: both FR 5/7m 7M; Front 50°36'·46N 02°26'·87W, S Pier hd ⚓ Q 10m 9M; 50°36'·58N 02°26'·49W. IPTS 190m SW.

LULWORTH RANGE TO ANVIL POINT
Targets:	DZ 'A' ⚓, Fl Y 2s,	50°33'·34N 02°06'·52W.
off	DZ 'B' ⚓, Fl Y 10s,	50°32'·11N 02°05'·92W.
St Alban's Hd	DZ 'C' ⚓, Fl Y 5s,	50°32'·76N 02°04'·56W.

Anvil Pt ⚓ Fl 10s 45m 9M; vis 237°-076° (H24); W ○ twr and dwelling; 50°35'·51N 01°57'·60W. Measured mile close west.

9.1.5 PASSAGE INFORMATION

More passage information is threaded between the harbours in this area. Notes for crossing the western English Channel are after 9.1.25; see 9.0.5 for distances. Admiralty Leisure Folios: 5602 Falmouth to Teignmouth, 5603 Falmouth to Hartland Point, includes the Isles of Scilly. **Bibliography:** *West Country Cruising Companion* (Fernhurst/Fishwick); *Channel Havens* (ACN/Endean); *Shell Channel Pilot* (Imray/Cunliffe); *Channel Pilot* (Admiralty NP27).

ISLES OF SCILLY TO LAND'S END

The 48 islands lie 21-31M WSW of Land's End, with many rocky outcrops and offlying dangers; see AC 34, 883. Care is needed particularly in poor visibility. No one ⚓ gives shelter from all winds and swell, so be ready to move at short notice. Follow the approach transits on AC 34 because the tidal streams are difficult to predict with accuracy. They run harder off Points and over rocks, where overfalls may occur.

The N/S lanes of the Land's End TSS (AC 1148) are roughly defined by Seven Stones rocks to the west and Wolf Rock to the south east. The **Seven Stones** (lt ship, fog sig, Racon, AIS) lie 16M W of Land's End; many of them dry, with ledges in between. **Wolf Rock** (lt ho, fog sig, Racon, AIS) 8M SSW of Land's End, is in contrast steep-to. Beware commercial vessels turning N near Wolf Rock to enter the N-bound lane of the TSS; and those leaving the S-bound lane.

▶*Between Scilly and Land's End (AC 1148) streams are rotatory, clockwise, setting W from HW Dover; N from HW + 2; NE from HW + 4; E from HW –6; SSE from HW – 4; and SW from HW – 2. Sp rates are about 1kn. From Mounts Bay to Scilly leave the Runnel Stone at HW Dover –2; a fair W-going tide lasts for only 3 hrs, with cross tides setting SW then NW to N. Consider arriving at dawn. For the return passage streams are a little less critical.◀*

LAND'S END

The peninsula (AC 1148, 1149, 777 and 2345) is always a critical tidal gate and often a dangerous lee shore. There are many inshore and offlying rocks but no ports of refuge. The main features are: Gwennap Head, with Runnel Stone buoy 1M to the S; Land's End and Longships reef 1M to the W; Cape Cornwall and The Brisons; Pendeen and The Wra. Further E see Area 11.

Passage between **Gwennap Head** and the Runnel Stone (0·5m; 7ca S) is not advised even in calm weather at HW, owing to rocks and uncharted wrecks closer inshore. These dangers are in the R sectors of Longships and Tater-du lts. From Gwennap Hd to Land's End, 2M NW, rocks extend up to 1½ca offshore, and depths are irregular causing a rough seas in strong winds against the tide.

4 cables S of Land's End Armed Knight, a jagged 27m high rock, overlooks **Longships**. This is an extensive and very dangerous reef made up of Carn Bras, on which the lt ho stands, and other rky islets. About 5ca to the E and NE are Kettle's Bottom 5m and Shark's Fin 0·9m, both isolated and very dangerous drying rocks. The ½M wide passage between Land's End and Kettle's Bottom is safe in calm, settled weather, but never at night. To clear Kettle's Bottom and Shark's Fin, keep the Brisons High summit just open W of Low summit brg 001°. In adverse weather navigate in the 3M wide ITZ, well W of Longships.

▶*Local streams exceed 4kn at Sp and are unpredictable.◀*

1M NNE of Land's End, after rounding Cowloe Rks and Little Bo 3·4m and Bo Cowloe 6m, the transit 150° of two bns on the cliffs E of Sennen Cove leads to an ⚓ in about 2m on the S side of **Whitesand Bay**, only safe in fair weather and offshore winds; avoid all underwater cables. Sennen ⌖ twr (110m) is conspic, almost on the 150° transit.

Cape Cornwall (conspic ruined chy) is about 3M further N. It overlooks The Brisons, two rocky islets (27 and 22m, High and Low summits) about 5ca to SW. There is no safe passage inside The Brisons. The Vyneck rock 1·8m is 3 cables to the NW.

3M NNE is **Pendeen Head** (lt ho) with the Wra (Three Stone Oar), drying rocks, close N. Overfalls and a race extend up to 1½M W and SW of Pendeen; avoid except in calm weather and at slack water. A conspic TV mast is 1.5M S of Pendeen.

TIDAL STRATEGY FOR ROUNDING LAND'S END

The tidal stream chartlets and notes below are reproduced by kind permission of the Royal Cruising Club Pilotage Foundation, *Yachting Monthly* magazine in which they were first published and the author Hugh Davies.

▶*The chartlets referenced to HW Dover, illustrate tidal streams and inshore currents. Streams run hard around Land's End, setting N/S and E/W past it. It is truly a tidal gate, and one which favours a N-bound passage – with careful timing nearly 9½ hrs of fair tide can be carried, from HWD –3 to HWD +5. Use currents close inshore running counter to the tidal streams.*

Example N-bound: At HWD+1 the N-going flood starts off Gwennap and does not turn NE along the N Cornish coast until HWD+3, but as early as HWD–3 an inshore current is beginning to run N'ly. So, N-bound, use this by arriving off Runnel Stone at HWD–2 and then keep within ¼M of the shore. If abeam Brisons at HWD, the tide and current should serve for the next 6 or 7 hrs to make good St Ives, or even Newquay and Padstow.

Example S-bound: If S-bound from St Ives to Newlyn, aim to reach the Runnel Stone by HWD+5, ie with 2 hrs of E-going tide in hand for the remaining 9M to Newlyn. This would entail leaving St Ives 5 hrs earlier, to make the 20M passage; buck a foul tide for the first 3 hrs but use an inshore S-going current, keeping as close inshore as prudent, but having to move offshore to clear the Wra and the Brisons. This timing would suit passages from S Wales and the Bristol Channel, going inshore of Longships if the weather suits.

From Ireland, ie Cork or further W, the inshore passage would not benefit. But plan to be off the Runnel Stone at HWD+5 if bound for Newlyn; or at HWD+3 if bound for Helford/Falmouth, with the W-going stream slackening and 5 hrs of fair tide to cover the remaining 20M past the Lizard.◀

9.1.6 ISLES OF SCILLY

The Isles of Scilly consist of 48 islands and numerous rocky outcrops, covering an area approx 10M by 7M and lying 21–31M WSW of Land's End. Only St Mary's, St Martin's, Tresco, Bryher, St Agnes and Gugh are inhabited. The islands belong to the Duchy of Cornwall. There is a LB and CG Sector Base at St Mary's.

CHARTS AC 1148, 34, 883, 5603; Imray C10, C7, 2400

TIDES Standard Port is Plymouth. Differences for St Mary's are given in 9.1.7.

▶*Tidal heights, times, and streams are irregular. See the 12 hourly tidal stream chartlets below and AC 34 for 5 tidal stream diamonds, all referred to HW Plymouth.*◀

SHELTER The Isles of Scilly are exposed to Atlantic swell and wind. Weather can be unpredictable and fast-changing. It is not a place for inexperienced navigators or poorly equipped yachts. Normal yacht ⚓s may drag on fine sand, even with plenty of scope, but holding is mostly good. That said, the islands are most attractive. The following are some of the many ⚓s, anti-clockwise:

ST MARY'S Hugh Town Harbour.
Porth Cressa (S of Hugh Town). Beware of dangers on each side of entrance and submarine cables. Good ⚓ (2m) in W/NW'lies, but exposed to swell from SE to SW. 14 ⚓s £20/night www. porthcressamoorings.co.uk see 9.1.7 on p. 147
Watermill Cove (NE corner of St Mary's). Excellent shelter in winds S to NW. ⚓ in approx 5m.

ST MARTIN'S, W end. Tean Sound needs careful pilotage, but attractive ⚓ in better weather. More suitable for shoal draught boats which can ⚓ or take the ground out of main tidal stream in channel. 4 ⚓s £20/night (Star Castle Hotel ☎07730 475559). Hotel re-opened as Karma St Martin's ☎422368.

ST HELEN'S POOL (S of St Helen's Is). Ent via St Helen's Gap to ⚓ in 1·5m–7m. Secure, but may be swell near HW; see chartlet.

OLD GRIMSBY SOUND. Old Grimsby Hbr comprises Green Porth & Raven's Porth, divided by a quay, both dry 2·3m. Beware cable in Green Porth. ⚓s 1½ca NE of quay in 2·5m; access more difficult than New Grimsby. Well sheltered in SW'lies but open to swell if wind veers N of W. Facilities: 7 ⚓s (R or Y) £15, ⚓ (quay) ⚓ hotel.

NEW GRIMSBY SOUND Appr (line G) between Tresco and Bryher, or with adequate rise of tide from the S across Tresco Flats. Good shelter except in NW'lies. Popular ⚓ between Hangman Is and the quay in 1·5–4·5m. Beware cables. 22 ⚓s, R or Y, £15 for <18m LOA, £30 for >18m; £5 to ⚓. Tresco HM ☎07778 601237. VHF Ch 08 for ✕ ⊐ ⚓ ⚒ ⊠. Ferry to St Mary's.

ST AGNES/GUGH The Cove, well sheltered from W and N winds, except when the sandbar between the islands covers near HWS with a strong NW wind. Beware cables.
Porth Conger, N of the sandbar, sheltered in S'lies except when the bar is covered. Facilities: ⚓ (two quays), ferry to St Mary's; in Middle Town (¾M), ⚒ ⊠ ✕ ⊐.

NAVIGATION See 9.1.2 for TSS to the E, S and W. If unable to identify approach ldg lines/marks, it is best to lie off. Many chans between islands have dangerous shallows, often with rky ledges. Some of the most used ldg lines, as on chartlet, include:

Line A. ⊕ 49°53'·55N 06°18'·00W. *North Carn of Mincarlo ≠ W extremity of Great Minalto 307·1°*. This is the usual approach to St Mary's Road, via St Mary's Sound. From the E or SE avoid Gilstone Rk (dries 4m) 3ca E of Peninnis Hd. Spanish Ledges, off Gugh, are marked by an ECM buoy; thence past Woolpack SCM bn, Bartholomew Ledges lt bn and N Bartholomew PHM buoy Fl R 5s to ent St Mary's Road on 040·5° (see Line B).

Line B. ⊕ 49°54'·69N 06°20'·33W. *St Martin's daymark ≠ the top of Creeb 040·5°*. This clears Woodcock Ledge, breaks in bad weather.

Line C. ⊕ 49°52'·06N 06°21'·24W. *Summit of Castle Bryher ≠ gap between the summits of Great Smith 350·5°*. Via Smith Sound, between the drying rocks off St Agnes and Annet.

Line D. ⊕ 49°52'·40N 06°27'·80W. *Summit of Great Ganilly just open north of Bant's Carn 059°*. From the SW enter Broad Sound between Bishop Rk lt ho and Flemming's Ledge, 7ca to the N; pass between Round Rk NCM and Gunner SCM buoys to Old Wreck NCM By; beware Jeffrey Rk, close to port. Ldg marks are more than 7M off and at first not easy to see. See also Line C, Smith Sound.

Line E. ⊕ 49°54'·34N 06°27'·80W. *St Agnes Old lt ho ≠ Carn Irish 099·7°*. From WNW, passing close N of Crim Rocks and Gunners.

Line F. ⊕ 49°56'·34N 06°26'·46W. *St Agnes Old lt ho ≠ Tins Walbert bcn 127°*. The NW Passage is about 7ca wide with good ldg marks, but beware cross tide. A lit WCM marks Steeple Rk (0·1m). Intercept Line D for St Mary's Road.

Line G. ⊕ 49°59'·00N 06°21'·96W. *Star Castle Hotel ≠ W side of Hangman Is 157°*. From the N, leads into New Grimsby Sound between Bryher and Tresco. Crossing Tresco Flats need adequate rise of tide (plan for max drying ht of 1·7m) and moderate vis.

Line H. ⊕ 49°55'·66N 06°14'·38W. *Summit of Samson Hill ≠ the NE extremity of Innisidgen 284·5°*. From the E or NE: Crow Snd can be rough in strong SE'lies, but, with sufficient rise of tide, is not difficult. From the NE pass close to Menawethan and Biggal Rk, avoiding Trinity Rk and the Ridge, which break in bad weather. Hats SCM buoy marks a shoal with an old boiler, (0·6m) on it. Maintain 284·5° between Bar Pt and Crow Bar (0·7m), pass N of Crow Rk IDM lt bn (for best water), then turn SSW for St Mary's.

Historic Wrecks are at 49°52'·2N 06°26'·5W Tearing Ledge, 2ca SE of Bishop Rk Lt; and at 49°54'·30N 06°19'·89W, Bartholomew Ledges, 5ca N of Gugh.

LIGHTS AND MARKS See 9.1.4 for lts, including Seven Stones lt float, Wolf Rock lt ho and Longships lt ho. A working knowledge of the following daymarks and conspic features will greatly aid pilotage (from NE to SW):

St Martin's E end: Conical bcn twr (56m) with RW bands.

Round Is: roughly conical shaped, with W lt ho 19/55m high.

Tresco: Abbey & FS, best seen from S. Cromwell's Castle and Hangman Is from the NW.

Bryher: Watch Hill, stone bn (43m); rounded Samson Hill.

St Mary's: TV & radio masts (R lts) at N end, are visible from all over Scilly. Crow Rk IDM bn, 11m on rk drying 4·6m, at N.

St Agnes: Old lt ho, ○ W tr, visible from all directions.

Bishop Rk lt ho: Grey ○ tr, 44m; helo pad above lamp.

9.1.7 ST MARY'S

Isles of Scilly 49°55'·14N 06°18'·71W ✿✿✿✿✿✿✿✿✿

CHARTS AC 2665, 34, 883, 5603; Imray C10, C7, 2400

TIDES –0630 Dover; ML 3·2; Duration 0600

Standard Port PLYMOUTH (→)

Times				Height (metres)			
High Water		Low Water		MHWS	MHWN	MLWN	MLWS
0000	0600	0000	0600	5·5	4·4	2·2	0·8
1200	1800	1200	1800				
Differences ST MARY'S							
–0052	–0103	–0048	–0045	+0·2	–0·1	–0·2	–0·1

SHELTER Good in St Mary's Hbr, except in W/NW gales when 14 Y ⚓s or ⚓ in Porth Cressa may be more comfortable. Strictly no ⚓ in the approaches and harbour limits (Newman rock to Newford Is).

NAVIGATION WPT 49°53'·96N 06°18'·83W (abeam Spanish Ledge ECM lt buoy) 307°/1·2M via St Mary's Sound to transit line B (040°). The 097·3° transit leads S of Bacon Ledge (0·3m) marked by PHM lt buoy. A charted 151° transit leads into the Pool between Bacon Ledge and the Cow and Calf (drying 0·6m and 1·8m). Pilotage is compulsory for yachts >30m LOA. Hbr speed limit 3kn. NB: Do not impede the ferry *Scillonian III* which arrives about 1200 and sails at 1630 Mon-Fri; Sat times vary with month. Also the blue-hulled cargo ship *Gry Maritha* thrice weekly.

LIGHTS AND MARKS See chartlet and 9.1.4. 097·3° ldg marks: W bcns; front, white △; rear, orange x; lts as chartlet. The R and G sectors of the Pierhead lt are not on the conventional sides of the W sector. Buzza Hill twr and power stn chy (48m) are conspic.

COMMUNICATIONS (Code 01720) CGOC (01326) 317575; Ⓗ 422392; Dr 422628; Police 101; Tourist Info 422536. Pilot 422078. HM 422768(H24 ansafone); emergency out of hrs 07789 273626. *St Mary's Hbr* and Pilot Ch 14, 16 (0800-1700LT). *Falmouth CG* covers Scilly and the TSS/ITZ off Land's End on Ch 16, 23.

FACILITIES hm@stmarys-harbour.co.uk www.stmarys-harbour.co.uk 38 Y ⚓s lie in 6 trots close E of the ♦ in 1·5-2·7m; 28 ⚓s are for <12m LOA and 10 for <18m. Fees by LOA: £18<8m<£24·00<12m>£36; multihulls +50%. 3 drying ⚓ against hbr wall with tender landing pontoon; ⚓ at inner drying berth H24. 🔧🔩(Sibleys ☎422431) at middle berth. ➘📰⛽🔧🔌🖥️⚠ @. HM holds mail if addressed c/o HM, St Mary's, Isles of Scilly TR21 0HU.

Hugh Town Essential shops, 🏪 ACA ✉ Ⓑ 🖥️ 🛒 ✕ 🏧.

Porth Cressa 14 Y ⚓ £20<15m Apr-Oct. candpcarss@tiscali.co.uk ☎423758 mob 07825 477695. ⚓ 📰(Sibleys) wi-fi. ⚓ clear of the marked power cable area.

Mainland access: Ferry leaves Penzance 0915 Mon-Fri (not Sun) and 1630 St Mary's (Sat varies); booking ☎0845 710 5555. ✈ to St Just, Newquay, Exeter. Booking www.islesofscilly-travel.co.uk ☎(01736) 334220.

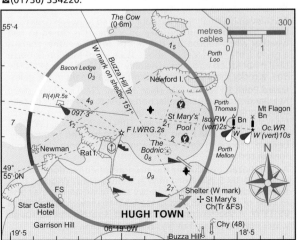

9.1.8 NEWLYN

Cornwall 50°06'·19N 05°32'·58W ✿✿✿✿✿✿✿✿

CHARTS AC 777, 2345, 5603; Imray C10, C7, 2400

TIDES –0635 Dover; ML 3·2; Duration 0555

Standard Port PLYMOUTH (→)

Times				Height (metres)			
High Water		Low Water		MHWS	MHWN	MLWN	MLWS
0000	0600	0000	0600	5·5	4·4	2·2	0·8
1200	1800	1200	1800				
Differences NEWLYN							
–0053	–0108	–0035	–0036	+0·1	0·0	–0·2	0·0

SHELTER Good, but in SE'lies heavy swell enters hbr; access H24. FVs take priority. Good ⚓ in Gwavas Lake in offshore winds.

NAVIGATION WPT 50°06'·19N 05°31'·80W, 270°/0·5M to S pier. From NE, beware The Gear and Dog Rk 3½ca NE of hbr ent; from S, avoid Low Lee (1·1m) ECM buoy Q(3)10s and Carn Base (1·8m).

LIGHTS AND MARKS See chartlet and 9.1.4. Do not confuse S pier hd lt, Fl 5s, with Penzance pierhead lt, Fl WR 5s, 033°/1M. N pierhd lt: G sector (238°-248°) clears The Gear; W sector over hbr.

COMMUNICATIONS (Code 01736) CGOC (01326) 317575, Police 101, Ⓗ, Dr: as Penzance. HM www.newlynharbour.co.uk 362523 non-HO 361017. *Newlyn Hbr* Ch 09 12 16 (M-F 0800-1700, Sat 0800-1200LT).

FACILITIES SW of Mary Williams Pier finger pontoons are offered to visiting yachts only for overnight stays. Preference is given to FVs; yachtsmen may be asked to move and must pre-check with HM. Fish landing results in dog ban on pontoons/harbour property. NE pontoon: 40⚓ <15m (2m); 4 SW pontoon: 40⚓ <10m (¾m–CD) Charges: £12/night<8m>£18/night<12m>£27.00.

Services ➘ ⚓ 📰🔌🔧🔨🖥️ Ⓔ ⚠ 🏗(6t) 🛢.

Town ✉ Ⓑ(am only) 🖥️ 🛒 ✕ 🏧, bus to Penzance ⇌ ✈.

ADJACENT HARBOUR

MOUSEHOLE, Cornwall, 50°04'·97N 05°32'·26W. ✿✿✿✿✿✿. AC 2345, 5603. HW +0550 on Dover; Tides as Newlyn; ML 3·2m; Duration 0600. Best appr from S, midway between St Clements Is and bkwtr. N pier lt, 2 FG (vert) = hbr open, summer; 3 FR (vert) = hbr closed (in winter). Ent 11m wide; hbr dries approx 2·4m. Shelter good except in NE and S or SE winds; protected by St Clements Is from E'lies. Ent is closed with timber baulks from Nov-Apr. HM ☎(01736) 731897. Per night £10, multi hull £12. Facilities limited: ➘ ⚓ 🍺. Buses to Penzance.

9.1.9 PENZANCE

Cornwall 50°07′·09N 05°31′·68W ✿✿♒♒♒♒✿✿✿✿

CHARTS AC 777, 2345, 5603; Imray C10, C7, 2400

TIDES –0635 Dover; ML 3·2; Duration 0550

Standard Port PLYMOUTH (→)

Times				Height (metres)			
High Water		Low Water		MHWS	MHWN	MLWN	MLWS
0000	0600	0000	0600	5·5	4·4	2·2	0·8
1200	1800	1200	1800				

Differences PENZANCE are the same as for Newlyn (9.1.8)
PORTHLEVEN

–0045	–0105	–0030	–0025	0·0	–0·1	–0·2	0·0

LIZARD POINT

–0045	–0100	–0030	–0030	–0·2	–0·2	–0·3	–0·2

SHELTER Excellent in the wet dock. ⚓/dry out close N of wet dock or ⚓ E of hbr, but Mounts Bay is unsafe ⚓ in S or SE winds, which, if strong, also render the hbr ent dangerous.

NAVIGATION WPT 50°06′·77N 05°31′·06W, 306°/0·5M to S pier hd. Approach dries. Beware The Gear Rk 4·5ca S. Western Cressar (4ca NE of ent) and Ryeman Rks are marked by unlit SCM bns. Hbr speed limit 5kn. Dock gate opens 2 hrs before local HW until 1 hr after, 12 waiting buoys are sited S of S pier.

LIGHTS AND MARKS Hbr lts often hard to see against shore lts. No ldg lts/marks. IPTS 2 & 3, shown on N knuckle near dock gate, indicates when it is open/shut, but may not be shown for yachts.

8 unlit Y spar buoy racing marks are laid, Apr-Sep, at: 50°07′·30N 05°31′·48W; 50°07′·11N 05°30′·33W; 50°06′·92N 05°29′·00W; 50°06′·63N 05°32′·03W; 50°06′·55N 05°31′·45W; 50°06′·33N 05°32′·48W; 50°06′·25N 05°30′·30W; 50°05′·70N 05°29′·35W.

COMMUNICATIONS (Code 01736) CGOC (01326) 317575; Police 101; Dr 363866; Ⓗ 362382. HM 366113, in emergency 07967 240660; VHF Ch 09, **12**, 16 (HW –2 to HW +1, and office hrs).

FACILITIES Wet dock 50Ⓥ<10m £1·80. Registration required. ▷⚓🛢⚓✂🔧ⒸⒶ⛟(3t) ⛽🔌(dry dock).
Penzance SC ☎364989, ⚓⚓✕🗑.
Town 🏪✉️Ⓑ🛒🍺🔃✕🗑@⇌✈.

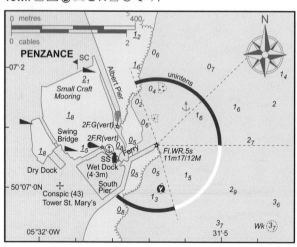

MINOR HARBOURS AROUND MOUNTS BAY

ST MICHAEL'S MOUNT, Cornwall, **50°07′·17N 05°28′·67W**. AC 2345, 777, 5603. HW +0550 on Dover; Tides as Penzance; ML 3·2m; Duration 0550. Beware: Hogus Rks (5·5) 250m NW of hbr; Outer Penzeath Rk (0·4) 5ca W of hbr; and Maltman Rk (0·9) 1ca SSW of the Mount. Shelter good from N to SE, but only in fair wx. Unlit hbr dries 2·1; approx 3·3m at MHWS and 1·4m at MHWN. Dry out against W pier or ⚓ W of the hbr in 2-3m on sand, space for 6 boats. HM mob ☎07870 400282 or 07917 583284. Facilities: ⚓✕ café. Marazion, 3ca to N: ✉️🛒✕🗑 Mounts Bay SC.

PORTHLEVEN, Cornwall, **50°04′·92N 05°19′·13W**. ✿✿♒♒✿. AC 777, 2345, 5603. HW –0635 on Dover; ML 3·1m; Duration 0545. Hbr dries 2m above the old LB ho but has approx 2·3m in centre of ent. It is very exposed to W and SW. Beware rks round pier hd and Deazle Rks to W. Lt on S pier, FG 10m 4M, = hbr open. Inside hbr FG, 033°-067°, shown as required to vessels entering. Visitors berth on the quay 2m, E side of inner hbr. HM ☎(01326) 574270. Facilities: **Inner Hbr** ⚓ £8, ⚓⚓✂Ⓒ🗑. Village ✉️Ⓑ🛒🗑✕🗑.

MULLION COVE, Cornwall, **50°00′·90N, 05°15′·54W**. ✿✿♒✿✿✿. AC 777, 2345, 5603. Lizard HW –0630 on Dover, –0050 and –0·2m on HW Plymouth; ML 3·0m; Duration 0545. Porth Mellin hbr dries about 2·4m and is open to SW'lies. Due to lack of space, visiting boats may only stay in hbr briefly to load/unload. No lts. There is a slip on E side of hbr. ⚓ in Mullion Cove where shown on the chart and especially in lee of N end of Mullion Island in approx 4-5m, but best to ask local advice. NT owns the island and the hbr. HM ☎(01326) 240222. Mullion village (1M) 🛒🗑.

Historic Wrecks are located at:
50°03′·44N 05°17′·16W (*St Anthony*), 2M SE of Porthleven.
50°02′·37N 05°16′·46W (*Schiedam*), 1·5M N of Mullion Is.
49°58′·54N 05°14′·51W, Rill Cove, 1·4M NW of Lizard Pt.
49°57′·49N 05°12′·98W (*Royal Anne*) The Stags, Lizard Pt.

LAND'S END TO FALMOUTH

Close E of Gwennap ⚓ off Porthcurno, owing to cables (AC 777). Western approach to Mounts Bay is marked by Tater-du Light. The Bucks (3·3m) are 2ca ESE of Tater-du, and Gull Rock (24m) is 9ca NE of close off the E point of Lamorna Cove. Little Heaver (dries) is 100m SW of Gull Rk, and Kemyel Rock (dries) is 1¾ca ENE.

Mousehole is a small drying harbour, sheltered from W and N, but exposed to E or S winds, when entrance may be shut. Approach from SW side of St Clement's Is. In W winds there is good ⚓ off the harbour.

Low Lee, (dangerous rock 1·1m) 4ca NE of Penlee Pt is marked by ECM lit buoy. Carn Base (rk 1·8m) lies 3ca NNW of Low Lee.

Newlyn is the sole harbour in Mount's Bay safe to approach in strong onshore winds, but only near to HW. East towards **Penzance** beware the unmarked Dog Rock (1·1m) and The Gear (1·9m,IDM lt bcn).

From Penzance to St Michael's Mount the head of the bay is shoal, drying 4ca off in places. Dangerous rocks include Cressar, Long, Hogus and Outer Penzeath. Keep south of a limiting line joining Penzance Market Ho Dome (55m) with pierheads of **St Michael's Mount** harbour at 089° to clear these dangers. The tiny hbr dries 2·1m, but is well sheltered, with ⚓ about 1ca W of entrance.

Two dangerous rocks, Guthen and Maltman (0·9m), lie 2ca W and S of St Michael's Mount respectively. 1M SE is The Greeb (7m), with rocks between it and shore. The Bears (dry) lie 1¾ca E of The Greeb. The Stone (dries) is 5ca S of Cudden Pt, while offshore is Mountamopus shoal marked by a SCM lit buoy. Welloe Rk (0·8m) lies 5ca SW of Trewavas Head.

Porthleven is a small tidal harbour, entered between Great and Little Trigg Rocks and the pier on S side. It is closed in bad weather when approach is dangerous. In fair weather there is a good ⚓ off Porth Mellin, about 1½ca NE of Mullion Is; Porth Mellin hbr (dries) is for temp use only. 2·5M W of Lizard Pt is The Boa, a rocky shoal (11m) on which the sea breaks in SW gales.

Lizard Point (Fl 3s70m 26M Horn 30s; AC 2345) is a bold, steep headland. Close inshore reflection of the lt may clearly be seen under some conditions. Dangerous rocks, mostly drying, lie 5-7ca from the lt ho in its SW quadrant. Either side of local LW these rocks are visible. 1M E and 1·5M NE of Lizard lt ho respectively, beware Vrogue (1·8m) and Craggan (1·5m) both dangerous sunken rocks.

A race, hazardous during the strength of either flood or ebb extends 2–3M S; it is worst in W'ly winds against W-going tide. Race conditions with short, heavy seas in westerlies may also exist SE of the Lizard. Pass at least 3M to seaward via 49°55′N 05°13′W. A shorter inner route which is rarely free of rough water may be taken, but go no further N than 49°56′·8N.

▶*In the race slack water occurs at about HW Plymouth –4 and the stream sets NE from HWP –3. The next slack water is at HWP +2½ and the stream sets SW from HWP +3; max springs rates reach 2-3kn. Outside the race the NE-going Channel flood starts at approx HWP –3¾ and the SW-going at HWP +3. Study NP 255 tidal stream atlas, Falmouth to Padstow inc Scilly.*◀

To await slack water, ⚓ (W to E) at: Kynance Cove, Housel Bay (3ca ENE of the lt ho and SSE of a conspic hotel), Parn Voose Cove (49°58'·4N 05°11'·1W), Cadgwith or Coverack.

N of Black Head, rocks extend at least 1ca offshore; a rock drying 1·6m lies off Chynhalls Pt (AC 1267). Coverack is a good ⚓ in W'lies. WSW of Lowland Pt drying rocks extend 2½ca offshore.

The Manacles (dry), 7½ca E and SE of Manacle Pt, are marked by ECM lt buoy and are in R sector of St Anthony Hd lt.

▶*Here the stream sets NE from HW Plymouth –0345, and SW from HW+ 0200, Sp rates 1·25kn.*◀

There are no offshore dangers on courses NNW to Nare Pt and Helford River or N to Falmouth.

MINOR HARBOURS NORTH EAST OF THE LIZARD

CADGWITH, Cornwall, **49°59'·22N 05°10'·68W**. ❀❀❀✿☆☆☆. AC 154, 2345, 5603. HW –0625 on Dover; –0030 on Plymouth; –0·2m on Devonport; ML 3·0m. Use Differences Lizard Pt under 9.1.9. Hbr dries; it is divided by The Todden, a rky outcrop. Beware the extension of this, rks called The Mare; also beware Boa Rk to ESE which cover at quarter tide. ⚓ off The Mare in about 2–3m, but not recommended in on-shore winds. There are no lts. Many local FVs operate from here and are hauled up on the shingle beach. Facilities: ✉ ✕ 🖥 shops at Ruan Minor (1M).

COVERACK, Cornwall, **50°01'·44N 05°05'·66W**. ❀❀❀✿☆☆☆. AC 154, 147, 5603. HW –0620 on Dover; ML 3·0m; Duration 0550. From the S beware the Guthens, off Chynhalls Pt; from the N, Davas and other rks off Lowland Pt, and to the NE Manacle Rks (ECM lt buoy). There are no lts. The tiny hbr dries and is full of small FVs. It has 3·3m at MHWS and 2·2m at MHWN. In good weather and off-shore winds it is better to ⚓ outside. HM ☎(01326) 280679. Facilities: ⚓ (hotel) ⛽ 🔧 from garage (2M uphill) ✉ 🛒.

9.1.10 HELFORD RIVER

Cornwall **50°05'·79N 05°06'·06W** (Ent) ❀❀❀✿✿✿

CHARTS AC 154, 147, 5603; Imray C10, C6, 2400

TIDES –0615 Dover; ML 3·0; Duration 0550

Reference Port PLYMOUTH (→)

Times				Height (metres)			
High Water		Low Water		MHWS	MHWN	MLWN	MLWS
0000	0600	0000	0600	5·5	4·4	2·2	0·8
1200	1800	1200	1800				
Differences HELFORD RIVER (Entrance)							
–0030	–0035	–0015	–0010	–0·2	–0·2	–0·3	–0·2
COVERACK							
–0030	–0050	–0020	–0015	–0·2	–0·2	–0·3	–0·2

SHELTER Excellent, except in E'lies when an uncomfortable swell reaches Frenchman's Creek. Access by day; no lights.

NAVIGATION WPT 50°05'·86N 05°04'·84W, 270°/1·9M to The Pool. From N beware August Rock marked by SHM buoy, Fl G 5s. From SE keep well off Nare Pt and Dennis Hd. At the entrance to Gillan harbour an ECM buoy marks Car Croc, a rock drying 1m. A NCM buoy marks the Voose, a rky reef E of Bosahan Pt.

Beware an extensive drying mud bank N and W of Bar SHM buoy, which is partially obscured by ⚓s. Speed limit 6kn in the river. The channel up to Gweek is buoyed but dries W of 05°10'·35W.

LIGHTS AND MARKS Marks as chartlet, all unlit.

COMMUNICATIONS (Code 01326) CGOC 317575; ⊞ 572151; Police 101. Moorings Officer: moorings@helford-river.com ☎250749; water taxi ☎250749, Ch M. Ferry also acts as a water taxi, on request HO. Helford River SC Ch **80** M.

FACILITIES Local info at www.helfordrivermoorings.co.uk. In The Pool 25 G ⚓s, marked 'Visitors' with G pick-up buoys, are randomly numbered (see diagram); <10m £15, <11m £17, <12m £18; 7th night free. No prior reservations. Rafting essential in high season.

Yachts may ⚓ in Durgan Bay but not inshore at the E end due to eelgrass beds (see chartlet). ⚓/taking the ground is permitted W of Groyne Pt, but not S of Porth Navas Creek owing to oyster beds. Ditch rubbish ashore at the Helford River SC or the Ferryboat Inn (N bank).

N bank Helford Passage ⛴ ⚓. Porth Navas YC☎340065, limited ⚓ ⚓ ✕ 🖥; www.portnavasyachtclub.co.uk.

S bank: Gillan Creek ⚓ ⚓ ⛽. **Helford River SC**☎231460 ⛴ ⚓ 🔧 ✕ 🖥. **Helford village** ✉ bus to Falmouth ⇌.

Gweek Quay ☎221657 🖶221685, info@gweek-quay.com ⚓ ⚓ 🔧 ✕ 🔧 🖧 🛢 🏗(100t) ⛽ ✉.

Visitors' Moorings
Diagram only

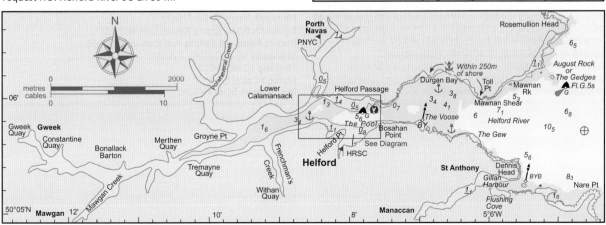

9.1.11 FALMOUTH & RIVER FAL

Cornwall **50°08'·61N 05°01'·48W** (Ent) ❀❀❀❀♨♨♨❁❁❁

CHARTS AC 154, 32, 18, 5602; Imray C6, Y58

TIDES −0610 Dover; ML 3·0; Duration 0550

SHELTER Excellent, depending on wind direction; see Facilities. Access H24 in any weather, although the entrance, 1M wide and deep, is rough in fresh on-shore winds over the ebb.

NAVIGATION WPT 50°08'·00N 05°01'·74W, 000°/7ca to Black Rock PHM lt buoy, which marks the edge of the deep water E channel. At night this is advised, although by day small vessels may pass either side of Black Rk (2·1m), IDM Fl(2) 10s, close W of mid-entrance. A Wave Energy Test Area is situated 155° St Anthony Hd 1.5 NM; a Y SPM Fl(5)Y 20s marks its SW corner.

Falmouth is a deep water port with sizeable commercial docks. Do not impede ships up to 90,000 tons nor ⚓ in prohib areas. Outside the buoyed chans large areas are quite shallow; care is needed below half tide.

Speed limits: 8kn in Falmouth inner hbr (ie W of the docks), the Penryn River and, N of Turnaware Bar, in the R Fal/Truro and offlying creeks.

5kn in St Mawes and the Percuil River, Mylor, Restronguet and St Just Creeks and Portscatho (50°10'·83N 04°58'·32W), 2 pages on.

LIGHTS AND MARKS See chartlet and 9.1.4. R sector of St Anthony Hd lt (H24) covers the Manacles Rocks.

COMMUNICATIONS (Code 01326). CGOC 317575. Police 101. Dr Falmouth 212120, Penryn 372502. Ⓗ Falmouth 434700. Ⓗ Truro (A&E) (01872) 250000.

Falmouth Hbr Radio Ch **12** 14 (M-F 0800-1700). HM Truro and launch Ch 12. Falmouth Haven Marina and St Mawes Hbr: Ch 12. Pendennis & Falmouth marinas, Royal Cornwall YC **80/M**. Mylor Yacht Hbr Ch 80/M, 37/M1, Malpas Marine, St Mawes SC: Ch M (HO).

FACILITIES from seaward, clockwise to St Just:

Port Pendennis Marina 50°09'·15N 05°03'·70W. 45+20Ⓥ in 3-4·5m in outer harbour (access H24) www.portpendennis.com ☎211211 £3·25; max LOA 70m, short stay <12m £15 ❁s. Marina village access HW±3 via ent gate/tfc lts. YC, ✕ Ⓓ tennis court.

Falmouth Haven Marina (FHM) 50°09'·22N 05°03'·89W. ☎310990/1; Ch 12. Access H24, min depth 1.4m on 235° ldg line, 2 orange △s leaving 4 Y buoys to port. Dir bn WRG 3s on N Quay 233°. 100 ⌷, price standard /Jul-Aug: £2·60/£2·80; short stay (<2 hrs) £1/m. Ⓥ ⚓ up to 200m E of FHM in 2-5m (see FHC below). 🚢 🛥 Ⓓ wi-fi.

Falmouth Harbour Commissioners (FHC) www.falmouthport.co.uk info@falmouthport.co.uk; 21 G ❁s off Prince of Wales pier, prices ❁/⚓: £1·90/£1·30.

Falmouth Marina 50°09'·90N 05°05'·02W up the Penryn River. www.premiermarinas.com ☎316620. Access H24, min depth 2m. In near appr, pass close N of outer pontoon, 2FR (vert), leaving ECM lt bn (hard to see) close to stbd. Ignore unlit PHM & SHM buoys, close NE of ECM bcn, which mark the Penryn R. Berth on J hammerhead/fuel berth, then as directed. A disued pipeline drying 1·8m crosses the marina E/W; adjacent depth gauge shows depth over it. 300+20Ⓥ, £3 (min £24); £1·54 for 4 hrs (min £12·30). Services: 🅑 Ⓓ 🛥(H24) 🚢 ⚒ 🔧 ⒼⒹ ⚓ ▣(30t) 🅑(25t) 🔧 🛒 Ⓒ ✕ 🗒.

Town 🚢 ⚓ ⚑ ACA ✉ Ⓑ 🛒 ✕ 🗒 ⇌ ✈ (Newquay).

Fuel Falmouth & Mylor marinas, Falmouth Haven Marina.

Water taxis Falmouth Water Taxi www.falmouthwatertaxi.co.uk ☎07522 446659, VHF Ch 12 (after 09:30).

Penryn HM Penryn River ☎373352. Half tide access via drying buoyed channel to Freeman's Wharf 50°10'·00N 05°05'·75W www. freemanswharf.com ☎377509. 45 ½ tide drying ⌷, £2/ft/wk +VAT. ⚓ 🔧 ✕ Ⓖ ⚒ ⚓ Ⓔ Ⓓ (500m), ACA, 🍴 🗒 (20t).

Mylor Yacht Hbr ⚓ 50°10'·74N 05°03'·14W, NW bank of Carrick Road. Good shelter in W'lies but exposed to E'lies. PHM and SHM lit buoys mark ent to the fairway to the Yacht Hbr, dredged 2m; access H24. Marina www.mylor.com enquiries@mylor.com ☎372121. 140 ⌷ + 40 Ⓥ, £3·50; 15 ❁, £2·00. 🅑 🛥 🚢 ⚒ ✕ ⒼⒹ Ⓔ ⚓ ▭ 🗒 (35t), 🗒 (4t), 🅑 🛒 ✕ 🗒.

Mylor Pool has many moorings in 1m to 2·2m. Mylor Creek dries or is shoal, <1m. Mylor Creek Boatyard welcomes visitors who can take the ground www.mylorcreekboatyard.com.

St Mawes Hbr 50°09'·49N 05°00'·84W. Keep S of St Mawes SCM buoy marking Lugo Rk, 0·6m. HM www.stmawesharbour.co.uk ☎270553. 13 G ❁s to SW of the hbr toward Carricknath Pt. £16<10m>£21<12m>£32<15m. ▭ ▭ ⚓ 🚢 ⚒ ✕ ⒼⒹ ⚓ 🍴. ⚓ above Castle Pt well sheltered from all but SW winds, £0·50. Beware numerous moorings and oyster beds, especially in Percuil River, the upper reaches of which dry above the village.

St Just, good ⚓ off the creek, except in strong W/SW'lies.

FACILITIES UP-RIVER TO TRURO See the chartlet overleaf.
Restronguet Creek 50°11'·53N 05°03'·56W, dries completely. Beware Carick Carlys Rk 0·8m, 3ca E of ent, marked by NCM & SCM posts. Moorings fill the pool (12·4m) so only space to ⚓ is outside. Pandora Inn ☎372678, pontoon dries 0·9m, Ⓓ 🛒 ✕ 🗒.

River Fal/Truro River At N end of Carrick Roads beware strong tides and rips off Turnaware Point and Bar. *King Harry chain ferry has right of way over all passing vessels* except those with a pilot aboard. She shows the lts/marks of a vessel with limited manoeuvrability. Fl Y lts indicate her direction of travel.

HM Truro www.portoftruro.co.uk ☎(01872) 272130 administers 4Ⓥ pontoons, E of Channals Creek, N of Ruan Creek, off Woodbury Pt and at Malpas (max draught 1m): £15/day all LOA; £3 for 2 hrs ⚓ £6.

A pontoon, W bank just S of King Harry ferry, is for tourist access to Trelissick House (⚓ Fl R 5s). Tenders may berth on the inshore side; yachts may berth outboard briefly for ⚓ only.

Good ⚓s, £5/day, off Tolcarne Creek, Channals Creek, Tolverne, Ruan Creek, Church Creek, Mopus Reach and Malpas.

Malpas Marine ☎ (01872) 271260. No facilities for visiting yachts 8ca NW of Malpas a flood barrier gate is usually open, 2 FR/FG (vert). Gate is shut HW ±2¼, and 3 Fl R lts shown, if HW >5·6m is forecast due to storm surge. Smugglers Cottage at Tolverne, www.smugglerscottage.co.uk, ⚓ £15<9m>£18<12m>£22 per day, ⚓ at pontoon.

Truro 50°15'·72N 05°02'·85W. HW Truro, sp & nps, is approx HW Falmouth −0022 and −2·0m. Drying ⌷ (£10 all LOA) close SW of HM's office ⒹⒹ ⚓ 🏴. **City** ⇌ all facilities.

YACHT CLUBS
Royal Cornwall YC (visitors most welcome) ☎311105/312126 (Secretary), ▭ ⚓ ✕ 🗒.
Falmouth Town SC ☎313662.
Falmouth Watersports Association ☎211223.
Port of Falmouth Sailing Association ☎211555.
St Mawes SC ☎270686 🗒 (visitors most welcome).
Flushing SC ☎374043.
Mylor YC ☎374391, 🗒.
Restronguet SC ☎374536.

H.W.Hts.m

FALMOUTH

MEAN SPRING AND NEAP CURVES

MEAN RANGES	
Springs	4.6m
Neaps	2.3m

CHART DATUM

M.H.W.N. M.H.W.S.

M.L.W.S. M.L.W.N. M.L.W.N.

L.W.Hts.m

L.W -5ʰ -4ʰ -3ʰ -2ʰ -1ʰ H.W. +1ʰ +2ʰ +3ʰ +4ʰ +5ʰ +6ʰ L.W

Factor

FALMOUTH TO FOWEY

Porthscatho, 3M NNE of St Anthony Hd, is a safe ‡ in W'lies. Gull Rock (38m high) and The Whelps (dry 4.6m) lie 6ca E/SE of Nare Hd and may be passed inshore. In Veryan Bay beware Lath Rk (2.1m), 1.6M NE of Gull Rock.

On E side of Veryan Bay, Dodman Pt is a 110m rounded shoulder, with a conspic stone cross. Depths are irregular for 1M S, with heavy overfalls in strong winds over sp tide, when it is best to pass 2M off. 3 SPM lt buoys (naval gunnery targets) lie 2.3–4.8M SSE of Dodman Pt.

2.1M NE of Dodman Pt is Gwineas Rk (8m high) and Yaw Rk (0.9m), marked by ECM lt buoy. Passage inside Gwineas Rk is possible, but not advised in strong onshore winds or poor vis. **Gorran Haven**, 1M WSW of Gwineas Rk, a sandy cove with L-shaped pier which dries at sp, is a good ‡ in offshore winds. So too are **Portmellon, Mevagissey Bay** and **St Austell Bay**.

MINOR HARBOURS AND ANCHORAGES FROM ST ANTHONY HEAD TO MEVAGISSEY

PORTSCATHO, Cornwall, **50°10'.84N 04°58'.32W**. AC 1267, 154, 5602. HW −0600 on Dover, HW −0025 and −0.2m on Plymouth; ML 3.0m; Duration 0550. Small drying hbr, but in settled weather and off-shore winds pick up a ⌀ or ‡ outside moorings in good holding. No lts. HM ☎(01872) 580243. Facilities: ⚓ 🏪 🏪 ✉ Ⓑ 1000-1230 Mon, Wed, Fri 🍴 ✗ ▱.

GORRAN HAVEN, Cornwall, **50°14'.49N 04°47'.16W**. AC 1267, 148, 5602. HW −0600 on Dover, HW −0010 and −0.1m on Plymouth. Shelter good with flat sand beach for drying out in off-shore wind; good ‡ 100 to 500m E of harbour. Not suitable ‡ when wind is in E. Beware Gwineas Rk and Yaw Rk marked by ECM lt buoy. Beware pot markers on appr. Fin keelers without legs should not ‡ closer than 300m from hbr wall where depth is 1.8m at MLWS. Facilities: 🛒 ✗ ▱.

PORTMELLON, Cornwall, **50°15'.74N 04°46'.98W**. AC1267, 148, 5602. HW −0600 on Dover, HW −0010 and −0.1m on Plymouth; ML 3.1m; Duration 0600. Shelter good but only suitable as a temp ‡ in settled weather and offshore winds. Facilities: ✗ ▱ (summer).

POLKERRIS, Cornwall, **50°20'.15N 04°41'.40W**. AC 148. Tides as Charlestown/Par. Very good ‡ in NE/SE winds. Facilities: ✗ ▱.

STANDARD TIME (UT)
For Summer Time add ONE hour in **non-shaded areas**

FALMOUTH LAT 50°09'N LONG 5°03'W
TIMES AND HEIGHTS OF HIGH AND LOW WATERS

Dates in red are **SPRINGS**
Dates in blue are NEAPS

YEAR 2016

JANUARY

Time	m	Time	m
1 0334	1.4	**16** 0337	0.9
0919	4.6	0933	4.8
F 1602	1.5	SA 1608	1.0
2149	4.2	◑ 2204	4.4
2 0415	1.7	**17** 0430	1.2
1007	4.3	1030	4.6
SA 1649	1.7	SU 1705	1.2
◑ 2244	4.0	2309	4.3
3 0510	1.9	**18** 0534	1.4
1104	4.2	1138	4.4
SU 1751	1.8	M 1814	1.4
2349	4.0		
4 0620	1.9	**19** 0025	4.2
1210	4.1	0648	1.5
M 1901	1.8	TU 1257	4.3
		1934	1.4
5 0104	4.0	**20** 0145	4.3
0731	1.8	0814	1.4
TU 1325	4.1	W 1413	4.3
2006	1.6	2054	1.2
6 0215	4.2	**21** 0251	4.5
0834	1.6	0930	1.1
W 1433	4.3	TH 1516	4.5
2103	1.4	2159	1.0
7 0309	4.5	**22** 0345	4.7
0929	1.3	1030	0.8
TH 1525	4.5	F 1608	4.7
2152	1.1	2253	0.7
8 0352	4.7	**23** 0431	4.9
1017	1.1	1121	0.6
F 1610	4.7	SA 1654	4.8
2239	0.9	2341	0.5
9 0433	4.9	**24** 0514	5.1
1104	0.8	1207	0.4
SA 1652	4.9	SU 1736	4.9
2323	0.7	○	
10 0513	5.1	**25** 0024	0.5
1148	0.6	0553	5.2
SU 1735	4.9	M 1247	0.4
●		1814	4.9
11 0007	0.6	**26** 0102	0.5
0553	5.2	0631	5.2
M 1231	0.5	TU 1323	0.4
1817	5.0	1851	4.9
12 0049	0.5	**27** 0135	0.6
0635	5.3	0707	5.2
TU 1313	0.4	W 1355	0.6
1900	5.0	1927	4.9
13 0129	0.5	**28** 0204	0.7
0718	5.3	0741	5.1
W 1354	0.5	TH 1423	0.8
1942	4.9	2001	4.8
14 0210	0.6	**29** 0230	1.0
0800	5.2	0814	4.9
TH 1435	0.6	F 1448	1.0
2024	4.8	2034	4.6
15 0251	0.7	**30** 0254	1.2
0844	5.0	0847	4.7
F 1519	0.7	SA 1512	1.3
2110	4.6	2109	4.4
		31 0321	1.4
		0924	4.5
		SU 1544	1.5
		2152	4.2

FEBRUARY

Time	m	Time	m
1 0401	1.7	**16** 0504	1.3
1012	4.2	1105	4.2
M 1632	1.7	TU 1739	1.5
◑ 2248	4.0	2351	4.1
2 0500	1.9	**17** 0617	1.6
1113	4.1	1235	4.0
TU 1745	1.9	W 1901	1.6
2355	4.0		
3 0626	2.0	**18** 0123	4.1
1222	4.0	0754	1.6
W 1911	1.8	TH 1400	4.1
		2039	1.5
4 0108	4.1	**19** 0235	4.3
0747	1.8	0920	1.3
TH 1338	4.1	F 1504	4.3
2022	1.6	2148	1.1
5 0222	4.3	**20** 0329	4.6
0853	1.5	1018	0.9
F 1447	4.3	SA 1554	4.5
2121	1.3	2240	0.8
6 0318	4.6	**21** 0414	4.8
0950	1.1	1107	0.6
SA 1541	4.6	SU 1637	4.7
2215	1.0	2326	0.6
7 0404	4.9	**22** 0454	5.0
1042	0.8	1149	0.5
SU 1628	4.8	M 1716	4.9
2304	0.7	○	
8 0449	5.1	**23** 0006	0.5
1131	0.5	0532	5.1
M 1714	5.0	TU 1228	0.4
● 2352	0.4	1753	5.0
9 0534	5.3	**24** 0041	0.5
1217	0.3	0609	5.2
TU 1800	5.1	W 1301	0.4
		1829	5.0
10 0036	0.3	**25** 0112	0.5
0618	5.4	0645	5.1
W 1301	0.2	TH 1329	0.5
1844	5.2	1904	5.0
11 0119	0.2	**26** 0138	0.6
0702	5.4	0718	5.1
TH 1342	0.1	F 1352	0.7
1926	5.1	1936	4.9
12 0158	0.2	**27** 0159	0.8
0745	5.4	0748	4.9
F 1422	0.2	SA 1412	0.9
2007	5.0	2005	4.7
13 0237	0.4	**28** 0219	1.0
0827	5.2	0816	4.7
SA 1502	0.5	SU 1434	1.1
2047	4.8	2031	4.5
14 0319	0.6	**29** 0245	1.2
0910	4.9	0844	4.5
SU 1544	0.8	M 1503	1.3
2132	4.6	2101	4.3
15 0406	1.0		
1000	4.6		
M 1635	1.2		
◑ 2230	4.3		

MARCH

Time	m	Time	m
1 0320	1.4	**16** 0441	1.3
0922	4.3	1037	4.1
TU 1541	1.6	W 1711	1.5
◑ 2152	4.1	2317	4.1
2 0408	1.7	**17** 0552	1.6
1025	4.1	1217	3.9
W 1639	1.8	TH 1831	1.7
2303	4.0		
3 0523	1.9	**18** 0101	4.0
1138	3.9	0734	1.6
TH 1812	1.9	F 1345	3.9
		2021	1.6
4 0017	4.0	**19** 0214	4.2
0701	1.9	0904	1.3
F 1255	4.0	SA 1447	4.2
1942	1.8	2130	1.3
5 0133	4.2	**20** 0308	4.5
0819	1.5	0959	1.0
SA 1410	4.2	SU 1535	4.4
2051	1.4	2220	0.9
6 0240	4.5	**21** 0352	4.7
0923	1.1	1044	0.7
SU 1512	4.6	M 1614	4.7
2150	1.0	2303	0.7
7 0334	4.9	**22** 0431	4.9
1019	0.7	1125	0.5
M 1603	4.8	TU 1652	4.9
2244	0.6	2341	0.6
8 0423	5.2	**23** 0509	5.0
1111	0.4	1201	0.4
TU 1651	5.1	W 1728	5.0
2334	0.3	○	
9 0511	5.4	**24** 0015	0.5
1159	0.1	0545	5.1
W 1738	5.2	TH 1233	0.5
●		1804	5.0
10 0020	0.1	**25** 0045	0.5
0558	5.5	0621	5.1
TH 1244	0.0	F 1300	0.6
1824	5.3	1839	5.0
11 0103	0.0	**26** 0110	0.6
0644	5.5	0654	5.0
F 1326	-0.1	SA 1321	0.7
1907	5.3	1910	4.9
12 0143	0.0	**27** 0130	0.8
0728	5.4	0723	4.9
SA 1405	0.1	SU 1341	0.8
1947	5.2	1936	4.8
13 0222	0.2	**28** 0151	0.9
0809	5.2	0747	4.7
SU 1443	0.3	M 1405	1.0
2025	5.0	1957	4.6
14 0301	0.5	**29** 0218	1.1
0849	4.9	0809	4.5
M 1523	0.7	TU 1434	1.2
2106	4.7	2016	4.5
15 0346	0.9	**30** 0252	1.3
0934	4.5	0832	4.3
TU 1610	1.2	W 1510	1.4
◑ 2158	4.4	2058	4.3
		31 0337	1.5
		0944	4.1
		TH 1601	1.7
		◑ 2221	4.1

APRIL

Time	m	Time	m
1 0444	1.8	**16** 0030	4.0
1105	4.0	0656	1.6
F 1726	1.9	SA 1320	3.9
2339	4.1	1937	1.7
2 0622	1.8	**17** 0145	4.1
1221	4.0	0830	1.4
SA 1904	1.8	SU 1420	4.1
		2057	1.4
3 0053	4.2	**18** 0239	4.4
0745	1.5	0927	1.1
SU 1336	4.2	M 1507	4.4
2020	1.4	2148	1.1
4 0204	4.5	**19** 0324	4.6
0854	1.1	1012	0.9
M 1442	4.6	TU 1547	4.6
2123	1.0	2231	0.9
5 0304	4.9	**20** 0404	4.8
0953	0.7	1052	0.7
TU 1536	4.9	W 1624	4.8
2220	0.6	2309	0.7
6 0357	5.2	**21** 0442	4.9
1047	0.3	1128	0.6
W 1626	5.1	TH 1702	4.9
2312	0.3	2343	0.7
7 0448	5.4	**22** 0520	4.9
1137	0.1	1200	0.6
TH 1715	5.3	F 1738	5.0
●		○	
8 0000	0.0	**23** 0014	0.7
0537	5.5	0556	4.9
F 1224	-0.1	SA 1228	0.7
1801	5.4	1813	5.0
9 0045	-0.1	**24** 0041	0.7
0624	5.5	0629	4.9
SA 1307	-0.1	SU 1253	0.7
1846	5.4	1844	4.9
10 0127	-0.1	**25** 0105	0.8
0710	5.4	0659	4.8
SU 1347	0.1	M 1317	0.8
1927	5.3	1910	4.8
11 0206	0.1	**26** 0130	0.9
0752	5.2	0725	4.7
M 1426	0.4	TU 1344	1.1
2006	5.1	1934	4.7
12 0246	0.5	**27** 0200	1.0
0832	4.8	0750	4.5
TU 1506	0.7	W 1415	1.1
2046	4.8	2000	4.6
13 0330	0.9	**28** 0236	1.2
0916	4.5	0821	4.4
W 1552	1.2	TH 1453	1.3
2135	4.4	2044	4.4
14 0423	1.3	**29** 0320	1.4
1015	4.1	0923	4.2
TH 1650	1.6	F 1543	1.6
◑ 2244	4.1	2155	4.3
15 0530	1.6	**30** 0424	1.6
1154	3.8	1039	4.1
F 1803	1.8	SA 1659	1.7
		◑ 2309	4.3

Chart Datum is 2·91 metres below Ordnance Datum (Newlyn). HAT is 5·7 metres above Chart Datum.

》》 **FREE** monthly updates. Register at 《
www.reedsnauticalalmanac.co.uk

153

Falmouth tides

FALMOUTH   LAT 50°09'N   LONG 5°03'W
TIMES AND HEIGHTS OF HIGH AND LOW WATERS

Dates in red are SPRINGS
Dates in blue are NEAPS

YEAR **2016**

MAY

Day	Time	m	Time	m	Time	m	Time	m
1 SU	0550	1.6	1152	4.1	1828	1.7		
16 M	0100	4.1	0727	1.5	1341	4.0	1955	1.6
2 M	0022	4.4	0711	1.4	1305	4.3	1945	1.4
17 TU	0201	4.2	0855	1.3	1432	4.3	2057	1.4
3 TU	0132	4.6	0822	1.1	1412	4.6	2053	1.0
18 W	0250	4.4	0926	1.1	1515	4.5	2146	1.1
4 W	0236	4.9	0925	0.7	1510	4.9	2153	0.7
19 TH	0333	4.6	1010	1.0	1555	4.7	2229	1.0
5 TH	0332	5.1	1022	0.4	1602	5.1	2248	0.3
20 F	0414	4.7	1049	0.8	1634	4.8	2307	0.9
6 F ●	0425	5.3	1114	0.2	1652	5.3	2339	0.1
21 SA ○	0453	4.8	1125	0.8	1711	4.9	2342	0.8
7 SA	0516	5.4	1203	0.0	1740	5.4		
22 SU	0530	4.8	1157	0.8	1747	4.9		
8 SU	0027	0.0	0606	5.4	1249	0.0	1826	5.4
23 M	0014	0.8	0604	4.8	1228	0.8	1818	4.9
9 M	0111	0.0	0653	5.3	1332	0.2	1910	5.3
24 TU	0044	0.8	0637	4.8	1258	0.9	1849	4.9
10 TU	0153	0.2	0736	5.1	1412	0.4	1950	5.1
25 W	0115	0.8	0709	4.7	1330	0.9	1919	4.8
11 W	0235	0.5	0818	4.8	1453	0.8	2031	4.9
26 TH	0150	0.9	0742	4.6	1405	1.1	1954	4.7
12 TH	0318	0.8	0901	4.5	1538	1.1	2115	4.6
27 F	0228	1.0	0820	4.5	1446	1.2	2039	4.6
13 F ◗	0408	1.2	0953	4.2	1630	1.5	2212	4.3
28 SA	0314	1.2	0912	4.3	1536	1.4	2137	4.5
14 SA	0506	1.5	1110	3.9	1732	1.7	2336	4.1
29 SU ◗	0413	1.3	1016	4.3	1642	1.5	2244	4.5
15 SU	0613	1.6	1237	3.9	1841	1.7		
30 M	0524	1.4	1126	4.2	1756	1.5	2353	4.5
31 TU	0638	1.3	1236	4.4	1910	1.3		

JUNE

Day	Time	m	Time	m	Time	m	Time	m
1 W	0103	4.6	0749	1.1	1344	4.6	2020	1.1
16 TH	0207	4.2	0831	1.4	1438	4.3	2055	1.4
2 TH	0210	4.8	0855	0.8	1446	4.8	2125	0.8
17 F	0258	4.4	0922	1.2	1524	4.5	2145	1.2
3 F	0311	5.0	0956	0.6	1541	5.0	2225	0.5
18 SA	0343	4.6	1008	1.1	1605	4.7	2230	1.1
4 SA	0406	5.1	1053	0.4	1633	5.2	2320	0.3
19 SU	0424	4.7	1050	1.0	1644	4.8	2311	0.9
5 SU ●	0459	5.2	1145	0.3	1722	5.3		
20 M ○	0503	4.8	1129	0.9	1720	4.9	2350	0.9
6 M	0011	0.2	0550	5.2	1234	0.2	1810	5.3
21 TU	0541	4.8	1207	0.8	1756	5.0		
7 TU	0059	0.2	0638	5.1	1319	0.3	1854	5.3
22 W	0027	0.8	0618	4.8	1244	0.8	1831	5.0
8 W	0142	0.3	0722	5.0	1400	0.5	1935	5.1
23 TH	0105	0.8	0656	4.8	1321	0.8	1908	5.0
9 TH	0224	0.5	0803	4.8	1440	0.7	2014	5.0
24 F	0143	0.8	0734	4.7	1400	0.9	1947	4.9
10 F	0305	0.8	0843	4.6	1521	1.0	2055	4.7
25 SA	0224	0.8	0814	4.7	1441	1.0	2030	4.9
11 SA	0348	1.1	0926	4.3	1606	1.3	2141	4.5
26 SU	0308	0.9	0900	4.6	1528	1.1	2121	4.8
12 SU ◗	0436	1.3	1018	4.1	1657	1.6	2236	4.2
27 M ◗	0400	1.1	0955	4.4	1623	1.2	2220	4.6
13 M	0530	1.5	1125	4.0	1754	1.7	2345	4.1
28 TU	0500	1.2	1059	4.4	1728	1.3	2326	4.6
14 TU	0629	1.6	1240	4.0	1856	1.7		
29 W	0606	1.2	1208	4.4	1837	1.3		
15 W	0103	4.1	0731	1.6	1345	4.1	1958	1.6
30 TH	0037	4.5	0716	1.2	1321	4.5	1950	1.2

JULY

Day	Time	m	Time	m	Time	m	Time	m
1 F	0149	4.6	0828	1.0	1428	4.7	2102	1.0
16 SA	0214	4.2	0838	1.5	1449	4.4	2106	1.5
2 SA	0255	4.7	0936	0.8	1526	4.9	2208	0.8
17 SU	0310	4.4	0931	1.3	1536	4.6	2157	1.2
3 SU	0353	4.9	1037	0.6	1619	5.1	2307	0.5
18 M	0356	4.6	1019	1.1	1617	4.8	2244	1.0
4 M ●	0446	5.0	1132	0.5	1708	5.2		
19 TU ○	0438	4.7	1105	0.9	1655	5.0	2329	0.8
5 TU	0000	0.4	0535	5.0	1222	0.4	1754	5.2
20 W	0519	4.9	1149	0.8	1734	5.1		
6 W	0048	0.3	0622	5.0	1307	0.4	1837	5.3
21 TH	0012	0.7	0600	4.9	1231	0.7	1814	5.2
7 TH	0130	0.3	0704	5.0	1347	0.5	1917	5.2
22 F	0054	0.6	0642	5.0	1312	0.6	1855	5.2
8 F	0209	0.5	0743	4.9	1424	0.7	1954	5.1
23 SA	0135	0.5	0723	4.9	1352	0.6	1936	5.2
9 SA	0246	0.7	0819	4.7	1459	0.9	2031	4.9
24 SU	0215	0.6	0803	4.9	1432	0.7	2018	5.1
10 SU	0322	1.0	0856	4.5	1535	1.2	2108	4.7
25 M	0257	0.7	0845	4.8	1514	0.9	2104	5.0
11 M	0359	1.3	0937	4.3	1615	1.5	2151	4.4
26 TU ◗	0342	0.9	0934	4.6	1603	1.1	2157	4.7
12 TU ◗	0442	1.5	1025	4.2	1702	1.7	2242	4.2
27 W	0435	1.1	1032	4.5	1701	1.3	2300	4.5
13 W	0534	1.7	1124	4.1	1801	1.9	2343	4.1
28 TH	0537	1.3	1143	4.4	1809	1.4		
14 TH	0635	1.8	1234	4.0	1905	1.9		
29 F	0015	4.4	0648	1.4	1303	4.4	1927	1.4
15 F	0057	4.1	0738	1.7	1350	4.2	2008	1.7
30 SA	0136	4.4	0809	1.3	1417	4.5	2050	1.2
31 SU	0247	4.5	0925	1.1	1517	4.8	2200	1.0

AUGUST

Day	Time	m	Time	m	Time	m	Time	m
1 M	0345	4.7	1028	0.9	1608	5.0	2258	0.7
16 TU	0328	4.6	0953	1.2	1548	4.9	2220	1.0
2 TU ●	0435	4.9	1121	0.6	1655	5.1	2349	0.5
17 W	0413	4.8	1043	0.9	1629	5.1	2309	0.8
3 W	0521	5.0	1209	0.5	1738	5.2		
18 TH ○	0456	5.0	1131	0.7	1711	5.3	2355	0.5
4 TH	0034	0.4	0603	5.0	1251	0.5	1817	5.3
19 F	0540	5.1	1216	0.5	1755	5.4		
5 F	0114	0.4	0642	5.0	1328	0.5	1855	5.2
20 SA	0039	0.4	0624	5.2	1259	0.4	1838	5.4
6 SA	0149	0.5	0718	5.0	1401	0.7	1930	5.1
21 SU	0121	0.3	0707	5.2	1339	0.4	1921	5.4
7 SU	0220	0.7	0752	4.9	1431	0.9	2003	5.0
22 M	0201	0.3	0748	5.1	1418	0.5	2003	5.3
8 M	0249	1.0	0826	4.7	1458	1.2	2036	4.8
23 TU	0241	0.5	0828	5.0	1458	0.7	2046	5.1
9 TU	0316	1.2	0900	4.5	1525	1.4	2111	4.6
24 W	0322	0.8	0912	4.8	1543	1.0	2135	4.8
10 W ◗	0345	1.5	0939	4.3	1557	1.7	2155	4.3
25 TH ◗	0411	1.1	1006	4.5	1637	1.3	2237	4.4
11 TH	0425	1.8	1029	4.2	1649	1.9	2251	4.1
26 F	0510	1.4	1119	4.3	1746	1.6		
12 F	0529	1.9	1131	4.1	1807	2.0	2358	4.0
27 SA	0001	4.2	0624	1.6	1250	4.3	1914	1.6
13 SA	0647	2.0	1244	4.1	1925	1.9		
28 SU	0131	4.2	0759	1.6	1407	4.4	2049	1.4
14 SU	0115	4.1	0758	1.8	1405	4.3	2031	1.7
29 M	0241	4.4	0920	1.3	1507	4.7	2154	1.1
15 M	0234	4.3	0859	1.5	1504	4.6	2128	1.4
30 TU	0336	4.6	1018	1.0	1555	4.9	2247	0.7
31 W	0421	4.8	1107	0.7	1638	5.1	2333	0.5

Chart Datum is 2·91 metres below Ordnance Datum (Newlyn). HAT is 5·7 metres above Chart Datum.

STANDARD TIME (UT)
For Summer Time add ONE hour in **non-shaded** areas

FALMOUTH LAT 50°09′N LONG 5°03′W
TIMES AND HEIGHTS OF HIGH AND LOW WATERS

Dates in red are **SPRINGS**
Dates in blue are NEAPS

YEAR **2016**

SW England

SEPTEMBER

Time	m		Time	m
1 0502	5.0		**16** 0431	5.1
1151	0.5		1111	0.6
TH 1717	5.2		F 1648	5.4
●			○ 2335	0.4
2 0014	0.4		**17** 0517	5.3
0539	5.1		1157	0.4
F 1230	0.5		SA 1734	5.5
1754	5.3			
3 0051	0.4		**18** 0020	0.2
0616	5.1		0602	5.4
SA 1304	0.6		SU 1241	0.2
1830	5.3		1819	5.6
4 0122	0.6		**19** 0103	0.1
0651	5.1		0646	5.4
SU 1334	0.7		M 1322	0.2
1904	5.2		1904	5.5
5 0149	0.7		**20** 0143	0.2
0724	5.0		0728	5.3
M 1359	0.9		TU 1402	0.3
1935	5.0		1947	5.4
6 0212	1.0		**21** 0222	0.4
0756	4.9		0809	5.2
TU 1420	1.1		W 1441	0.6
2005	4.8		2029	5.1
7 0232	1.2		**22** 0303	0.8
0825	4.7		0851	4.9
W 1440	1.4		TH 1525	1.0
2036	4.6		2116	4.7
8 0255	1.4		**23** 0349	1.2
0859	4.5		0942	4.6
TH 1508	1.6		F 1618	1.4
2115	4.4		☽ 2219	4.3
9 0328	1.7		**24** 0448	1.6
0944	4.3		1056	4.3
F 1550	1.9		SA 1728	1.7
☽ 2211	4.1		2353	4.1
10 0420	2.0		**25** 0605	1.8
1046	4.1		1234	4.2
SA 1702	2.1		SU 1906	1.7
2319	4.0			
11 0551	2.1		**26** 0123	4.1
1157	4.1		0750	1.7
SU 1842	2.1		M 1352	4.4
			2040	1.5
12 0034	4.0		**27** 0230	4.3
0721	2.0		0907	1.4
M 1313	4.3		TU 1450	4.6
1959	1.8		2139	1.1
13 0155	4.3		**28** 0320	4.6
0830	1.6		1001	1.1
TU 1424	4.6		W 1536	4.9
2102	1.4		2227	0.8
14 0258	4.6		**29** 0401	4.8
0928	1.3		1046	0.8
W 1517	4.9		TH 1616	5.1
2156	1.0		2309	0.6
15 0346	4.9		**30** 0438	5.0
1021	0.9		1126	0.6
TH 1603	5.2		F 1653	5.2
2247	0.7		2347	0.5

OCTOBER

Time	m		Time	m
1 0513	5.1		**16** 0452	5.4
1203	0.6		1136	0.3
SA 1729	5.2		SU 1712	5.6
●			○ 2359	0.1
2 0021	0.5		**17** 0539	5.5
0548	5.1		1221	0.1
SU 1235	0.6		M 1800	5.6
1804	5.2			
3 0051	0.6		**18** 0043	0.1
0623	5.1		0625	5.5
M 1304	0.7		TU 1305	0.1
1837	5.1		1846	5.5
4 0116	0.8		**19** 0125	0.2
0656	5.1		0709	5.4
TU 1327	0.9		W 1346	0.2
1908	5.0		1930	5.3
5 0137	1.0		**20** 0205	0.4
0727	5.0		0750	5.3
W 1347	1.1		TH 1427	0.5
1937	4.8		2014	5.0
6 0157	1.2		**21** 0246	0.8
0754	4.8		0833	5.0
TH 1408	1.3		F 1511	0.9
2006	4.6		2101	4.6
7 0222	1.4		**22** 0332	1.2
0824	4.6		0922	4.7
F 1437	1.5		SA 1603	1.3
2041	4.4		☽ 2201	4.3
8 0255	1.6		**23** 0428	1.6
0906	4.4		1030	4.3
SA 1517	1.7		SU 1710	1.6
2137	4.2		2335	4.1
9 0340	1.9		**24** 0541	1.8
1009	4.2		1206	4.2
SU 1617	2.0		M 1839	1.7
☽ 2248	4.0			
10 0459	2.1		**25** 0101	4.0
1120	4.2		0716	1.8
M 1759	2.0		TU 1325	4.3
			2011	1.5
11 0002	4.0		**26** 0206	4.2
0641	2.0		0838	1.5
TU 1233	4.3		W 1423	4.5
1925	1.8		2111	1.2
12 0118	4.3		**27** 0255	4.5
0758	1.7		0932	1.2
W 1344	4.6		TH 1509	4.7
2032	1.4		2157	0.9
13 0225	4.6		**28** 0335	4.7
0901	1.3		1016	1.0
TH 1444	4.9		F 1549	4.9
2130	1.0		2238	0.8
14 0318	4.9		**29** 0411	4.9
0956	0.9		1055	0.8
F 1535	5.2		SA 1626	5.0
2222	0.6		2314	0.7
15 0405	5.2		**30** 0446	5.0
1047	0.5		1131	0.7
SA 1623	5.4		SU 1703	5.1
2312	0.3		● 2348	0.7
			31 0522	5.1
			1204	0.7
			M 1739	5.1

NOVEMBER

Time	m		Time	m
1 0018	0.7		**16** 0024	0.2
0557	5.1		0605	5.5
TU 1234	0.8		W 1249	0.1
1812	5.0		1829	5.4
2 0045	0.8		**17** 0109	0.2
0631	5.1		0651	5.4
W 1259	0.9		TH 1333	0.2
1844	4.9		1915	5.2
3 0108	1.0		**18** 0151	0.4
0701	5.0		0734	5.3
TH 1322	1.0		F 1415	0.5
1914	4.8		2000	5.0
4 0133	1.1		**19** 0233	0.7
0730	4.9		0817	5.1
F 1348	1.1		SA 1459	0.8
1945	4.6		2045	4.6
5 0201	1.3		**20** 0317	1.1
0800	4.7		0902	4.8
SA 1419	1.3		SU 1548	1.2
2020	4.4		2137	4.3
6 0235	1.5		**21** 0409	1.5
0840	4.5		0957	4.5
SU 1459	1.5		M 1646	1.5
2111	4.2		☽ 2250	4.1
7 0319	1.7		**22** 0510	1.7
0939	4.4		1113	4.2
M 1554	1.8		TU 1754	1.7
☽ 2219	4.1			
8 0426	1.9		**23** 0018	4.0
1047	4.3		0620	1.8
TU 1719	1.9		W 1238	4.2
2330	4.1		1912	1.6
9 0559	1.9		**24** 0126	4.1
1157	4.4		0738	1.7
W 1846	1.7		TH 1343	4.3
			2023	1.5
10 0042	4.3		**25** 0219	4.3
0720	1.7		0845	1.5
TH 1307	4.6		F 1435	4.5
1958	1.4		2116	1.2
11 0150	4.5		**26** 0303	4.5
0828	1.3		0935	1.2
F 1411	4.8		SA 1519	4.7
2100	1.0		2159	1.0
12 0248	4.7		**27** 0342	4.7
0928	0.9		1017	1.0
SA 1508	5.1		SU 1559	4.8
2156	0.6		2237	0.9
13 0340	5.1		**28** 0420	4.9
1022	0.6		1056	0.9
SU 1600	5.3		M 1637	4.9
2248	0.4		2314	0.8
14 0429	5.3		**29** 0457	5.0
1114	0.3		1133	0.8
M 1651	5.4		TU 1715	4.9
○ 2338	0.2		● 2347	0.8
15 0518	5.4		**30** 0534	5.1
1203	0.2		1206	0.8
TU 1741	5.5		W 1751	4.9

DECEMBER

Time	m		Time	m
1 0018	0.8		**16** 0056	0.3
0609	5.1		0635	5.4
TH 1236	0.9		F 1322	0.2
1824	4.9		1901	5.1
2 0047	0.9		**17** 0140	0.4
0641	5.0		0718	5.3
F 1305	0.9		SA 1405	0.4
1857	4.8		1943	4.9
3 0116	1.0		**18** 0221	0.6
0712	4.9		0800	5.2
SA 1335	1.0		SU 1447	0.6
1930	4.7		2025	4.7
4 0148	1.1		**19** 0301	0.9
0745	4.8		0840	4.9
SU 1409	1.1		M 1529	0.9
2005	4.5		2108	4.5
5 0224	1.3		**20** 0344	1.2
0823	4.7		0924	4.7
M 1449	1.3		TU 1615	1.3
2050	4.4		2157	4.2
6 0307	1.5		**21** 0433	1.5
0913	4.6		1016	4.4
TU 1539	1.4		W 1708	1.5
2149	4.2		☽ 2258	4.0
7 0404	1.6		**22** 0530	1.7
1016	4.5		1119	4.2
W 1646	1.6		TH 1808	1.7
☽ 2257	4.2			
8 0518	1.7		**23** 0013	4.0
1122	4.4		0633	1.8
TH 1804	1.5		F 1236	4.1
			1914	1.7
9 0006	4.2		**24** 0127	4.1
0637	1.6		0740	1.7
F 1231	4.5		SA 1348	4.2
1918	1.4		2018	1.5
10 0115	4.4		**25** 0224	4.3
0751	1.4		0843	1.6
SA 1339	4.7		SU 1444	4.3
2027	1.1		2113	1.3
11 0219	4.7		**26** 0312	4.5
0858	1.1		0936	1.3
SU 1442	4.9		M 1531	4.5
2129	0.8		2159	1.1
12 0317	4.9		**27** 0355	4.7
0959	0.7		1022	1.1
M 1540	5.1		TU 1613	4.7
2226	0.5		2241	1.0
13 0410	5.2		**28** 0435	4.9
1054	0.5		1103	1.0
TU 1634	5.2		W 1653	4.8
2319	0.4		2319	0.9
14 0500	5.3		**29** 0513	5.0
1147	0.3		1141	0.9
W 1725	5.3		TH 1731	4.8
○			● 2355	0.8
15 0009	0.3		**30** 0549	5.0
0549	5.4		1216	0.8
TH 1236	0.2		F 1807	4.8
1814	5.2			
			31 0030	0.8
			0623	5.0
			SA 1251	0.8
			1842	4.8

Chart Datum is 2·91 metres below Ordnance Datum (Newlyn). HAT is 5·7 metres above Chart Datum.

》》 **FREE** monthly updates. Register at 《
www.reedsnauticalalmanac.co.uk 《

9.1.12 MEVAGISSEY

Cornwall **50°16'·16N 04°46'·93W** ※❀⚓⚓❀❀❀

CHARTS AC 1267, 148, 147, 5602; Imray C10, C6, 2400

TIDES –0600 Dover; ML 3·1; Duration 0600

Standard Port PLYMOUTH (→)

Times				Height (metres)			
High Water		Low Water		MHWS	MHWN	MLWN	MLWS
0000	0600	0000	0600	5·5	4·4	2·2	0·8
1200	1800	1200	1800				
Differences MEVAGISSEY							
–0015	–0020	–0010	–0005	–0·1	–0·1	–0·2	–0·1

SHELTER Exposed only to E'lies; if >F3 go to Fowey. Appr in strong SE'lies is dangerous. Access all tides.

NAVIGATION WPT 50°16'·16N 04°45'·98W, 270°/6ca to Victoria pier head. Beware rky ledges off N Quay. Speed limit 3kn.

LIGHTS AND MARKS As chartlet and 9.1.4. S Pier lt ho is conspic.

COMMUNICATIONS (Code 01726) CGOC (01326) 317575; Police 101; Dr 843701. HM 843305.
Ch 16 14 (Summer 0900-2100. Winter 0900-1700); call HM for berth.

FACILITIES www.mevagisseyharbour.co.uk meva.harbour@talk21.com
Outer Hbr Victoria Pier is heavily used by fishing boats Ⓥ may berth only with prior permission in about 2m, £14 all LOAs, short stay free. Keep clear of 🛢 hose & pleasure boats at inner end. Fishing boats use pier as early/late temporary berth to catch tide for entry/exit.

Bilge keelers/cats can dry out on sandy beach, SE side of W Quay on request to HM. There are 2 trots of 3 cylindrical ⚓s off N Pier: moor fore & aft, rafting possible.

No ⚓ inside the hbr due to overcrowding and many FVs. ⚓ off is only advised in settled weather with no E in the wind.

Inner Hbr (dries 1·5m) reserved for FVs, unless taking on ⚓. ⚓ 🔧 ⚓(1t) ⚓. 🚻 ⚓ on W Quay; key held at HM Office.

Village 🛢⊠🚂📮Ⓘ Ice ✕⊡@, Ⓑ (Jun-Sep1000-1430, Oct-May1000-1300), ⇌ (bus to St Austell), ✈ Newquay.

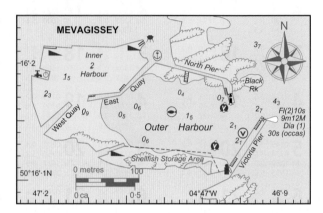

GUNNERY RANGE OFF DODMAN PT AND GRIBBIN HEAD

Naval gunnery practice takes place to seaward of Dodman Point and Gribbin Head, under the control of Flag Officer Sea Training (FOST), HMS Drake, Plymouth PL2 2BG. For info ☎(01752) 557550 (H24) or call *FOST OPS* on VHF Ch 74, which is monitored by all warships in the exercise areas S of Plymouth.

Firing is by day only, approx 1-2 times per week, in a 2 hrs block, although actual firing only lasts about 15 mins. Planned firings are broadcast in Gunfacts and Navtex. Advance details are also printed in local newspapers and are available from HMs at Fowey, Looe, Polperro and Mevagissey. Firings are not planned for 2 weeks at Christmas and 4 weeks in August.

Warships, from a position some 2·5 to 9M SSE of Gribbin Hd, fire WSW at 3 target buoys: **A** Fl Y 10s 50°08'·53N 04°46'·37W; **B** Fl Y 5s 50°10'·31N 04°45'·00W; and **C** Fl Y 2s 50°10'·41N 04°47'·51W, about 3·7M SSE of Dodman Pt.

If safety criteria cannot be met, eg owing to vessels in the range area, firing will not take place. A helicopter provides range safety surveillance. A range safety boat will advise other craft of firings and may suggest a course alteration to clear the area. Yachts are legally entitled to transit the range area without undue delay.

MINOR HARBOURS BETWEEN MEVAGISSEY AND FOWEY

CHARLESTOWN, Cornwall, **50°19'·84N 04°45'·35W**. AC 1267, 148, 31. HW –0555 on Dover, –0010 on Plymouth; HW –0·1m on Plymouth; ML 3·1m; Duration 0605. HM ☎01726 70241, VHF Ch 14, 16 (HW –2, only when vessel expected). Still a china clay port, also home to square riggers for filming/publicity.
Enter the inner hbr via entry gate as agreed with HM, but only in W'lies. Ent dries and should only be attempted by day and in offshore winds with calm weather. Gate fee £35, but free if within Square Sail's schedule. Hbr is shut in SE'lies. Waiting buoy 2ca S of hbr. Bkwtrs 2FG & 2FR(vert) 5m 1M. Ent sig: ● (night) = hbr shut. Facilities: www.square-sail.com; ⚌ £3.12/m, ⚓ ⚓ ⚓ (or pre-arranged tanker) ✕ 🔧 ⚓ ⚓(8t) ⊠ ✕ ⊡.

PAR, Cornwall, **50°20'·61N 04°42'·06W**. AC 1267, 148, 31. HW –0555 on Dover. The harbour dries 1·2m and should only be used in an emergency for yachts. There are plans for a marina, but possibly not for several years. Beware Killyvarder Rock (dries 2·4m) 3ca SE of ent, marked by unlit SHM bn. Only enter by day, in calm weather with offshore winds. Facilities in village.

FOWEY TO PLYMOUTH

(AC 1267, 148, 1613) Gribbin Head has a conspic daymark, a 25m high □ twr with R & W bands. In bad weather the sea breaks on rocks around the Head. Cannis Rock (4·3m) is 2½ca SE, marked by a lit SCM.

Fowey is a fine natural harbour, accessible in most conditions although exposed to strong SSW'lies. 3M E of Fowey and 5ca offshore at the East end of Lantivet Bay is Udder Rock (0·6m) marked by a lit SCM. 1M ENE of this buoy Larrick Rock (4·3m) is 1½ca off Nealand Pt.

Polperro harbour dries, but the inlet is a good ⚓ in offshore winds. Beware Polca Rock (1m) roughly in the approach. East of Polperro shoals lie 2½ca off Downend Point. A measured Mile lies between Polperro and **Looe**. The channel between Looe Island and Hannafore Point nearly dries; the charted Boat passage is not advised. The Ranneys (dry, marked by SCM lt buoy) are reefs extending 2½ca E and SE of Looe Island where there are overfalls in bad weather. In Whitsand Bay a buoyed artificial reef (the former *HMS Scylla*) lies 1·4M NW of Rame Head.

Eddystone rocks (AC 1613) lie 8M SSW of Rame Head. NE Rock (0·9m) and shoals are 280m NE. Close NW of the light house is the stump of the old lt ho. The sea can break on Hand Deeps (7m), rocks 3·4M NW of Eddystone, marked by a lit WCM.

9.1.13 FOWEY

Cornwall **50°19'·65N 04°38'·54W** ❄❄❄❄🌢🌢🌢❁❁❁

CHARTS AC 1267, 148, 31, 5602; Imray C10, C6, 2400

TIDES –0540 Dover; ML 2·9; Duration 0605

Standard Port PLYMOUTH (→)

Times				Height (metres)			
High Water		Low Water		MHWS	MHWN	MLWN	MLWS
0000	0600	0000	0600	5·5	4·4	2·2	0·8
1200	1800	1200	1800				
Differences FOWEY							
–0010	–0015	–0010	–0005	–0·1	–0·1	–0·2	–0·2
LOSTWITHIEL							
+0005	–0010	DR	DR	–4·1	–4·1	DR	DR
PAR							
–0010	–0015	–0010	–0005	–0·4	–0·4	–0·4	–0·2

SHELTER Good at Mixtow in all winds. Lower hbr exposed to S-SW, S and SW'ly gales can cause heavy swell and confused seas, especially on the ebb. Entry H24 at any tide in almost any conditions.

NAVIGATION WPT 50°19'·33N 04°38'·80W, 027°/7ca through hbr ent to Whitehouse Pt Dir lt in W sector. Appr in W sector of Fowey lt ho. 3M E of ent beware Udder Rk marked by SCM lt buoy. From SW beware Cannis Rk (4ca SE of Gribbin Hd) marked by SCM lt buoy. Entering hbr, keep well clear of Punch Cross Rks to stbd. Fowey is a busy commercial clay port with ship movements day and night. Speed limit 6kn. Give way to the Bodinnick–Caffa-Mill ferry.

Unmarked, mostly drying chan is navigable up to Golant (1M N of Wiseman's Pt), but moorings restrict ⚓ space. Access on the tide by shoal draught to Lerryn (1·6M) and Lostwithiel (3M) (18m power cables and 5·3m rail bridge).

LIGHTS AND MARKS See chartlet and 9.1.4. An unlit RW twr 33m on Gribbin Hd (1·3M WSW of hbr ent) is conspicuous from seaward, as is a white house 3ca E of hbr ent. Fowey light house is conspicuous. The W sector (022°-032°) of Whitehouse Pt dir lt leads through the 200m wide harbour entrance. At the entrance, St Catherine's Point is a lamp box; Lamp Rock is on a G bcn. At Whitehouse Pt a conspic radar and CCTV mast monitors traffic in the entrance.

COMMUNICATIONS (Code 01726) CGOC (01326) 317575; NCI Polruan 870291; Police 101; Ⓗ 832241; Dr 08444 992767; HM 832471.

Fowey Hbr Radio Ch **12** 16 (HO). Hbr Patrol (0900-2000LT) Ch 12. Water taxi (07774 906730) Ch 06. Pilots & Tugs Ch 09, 12.

FACILITIES from seaward:
Polruan Quay (E bank) short stay (2 hrs) pontoon, ⚓ ⚓ (3t). **Toms BY** ☎870232, ⚓ ⚓ ⬭⬜🏠⚓ 🗲 ⚓ ⬭ (50t) 🅿.

FOWEY All ⚓s are marked 'FHC VISITORS'. Berth/ moor as directed by Hbr Patrol. **V** pontoons are in situ May-Oct. Average overnight fees on ⚓/ pontoon: £2·00/m, but £2·40/m on Mixtow Pill. Reductions for 3 or 7 days, except Jul/Aug, 30 min free wi-fi included.

Pont Pill, on the E side, offers double-berth fore and aft ⚓s and ⬭ in 2m on two 36m floating pontoons; there is also a refuse barge.

Albert Quay HM's Office, ⚓ ⚓. The 'T' shaped landing pontoon is for short stay (2 hrs). Off the E bank is a trot of blue swinging ⚓s (1-50) and another 36m floating pontoon (*Underhills*).

Berrills BY short stay (2 hrs) landing pontoon, ♿ access, 250m N of Albert Quay, ⚓ 🛢 oil disposal.

Midway between Bodinnick and Mixtow Pill a sheltered double-sided pontoon (*Gridiron*) is off the E bank with 100m of **V** on its W side.

Mixtow Pill (5ca upriver) is a quieter 135m, shore-linked pontoon in 2·2m, **V** on S side. ⬭ ⚓ 🅿 🏠 🖀(8.4t) by arrangement. A ⚓ is 30m N of Wiseman's Point.

Royal Fowey YC ☎833573, ⚓⬭✕⬜ ☎832245 Cafe.
Fowey Gallants SC ☎832335, ⬭✕⬜.
Services M, 🛢🅿 ACA, Ⓔ ⚓ ⬭ ⚓ 🖀 ⚓ ⬭ (7 ton).
Town ✕⬜✉Ⓑ ⇄ (bus to Par), ✈ (Newquay).

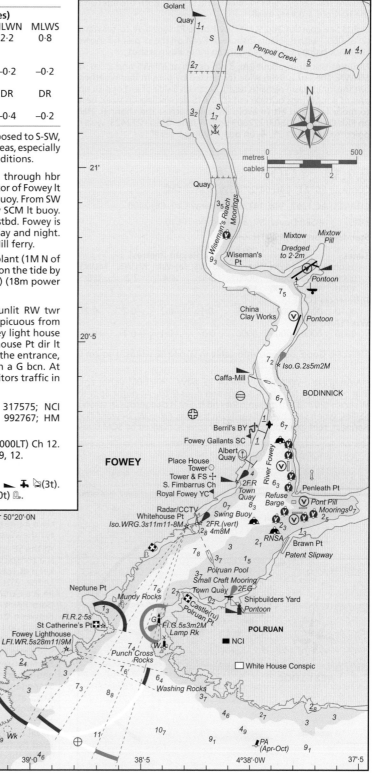

MINOR HARBOUR BETWEEN FOWEY AND LOOE

POLPERRO, Cornwall, **50°19´·78N 04°30´·79W**. AC 1267, 148, 5602. HW −0554 on Dover; HW −0007 and −0·2m on Plymouth; ML 3·1m; Duration 0610.

Shelter Good.

Navigation WPT 50°19´·74N 04°30´·72W, 310°/2ca to inner basin. Beware The Raney, rocks W of ent, and rks to E. Both the inlet and the harbour are very small.

Lights See chartlet and 9.1.4. Spy House Pt shows W to seaward (288°-060°) with R sectors inshore. A FW lt on the W pier hd shows FR (black ball by day) when gate to inner Basin is shut.

Facilities Hbr dries about 2m; 3·3m depth at MHWS and 2·5m at MHWN. The ent is 9·8m wide (protected by gate in bad weather, but the inner basin remains tidal). Berth on E side of ent or pick up one of 4 moorings outside ent (£10-£15 depending on size); best to moor fore and aft to avoid swinging across the channel. HM on Fish Quay, ☎(01503) 272809 or 'Girl Jane' 01503 272423; www. polperro.org ⌂£1.25. ⚓ on quays. **Village** ✉ ✕ 🗐.

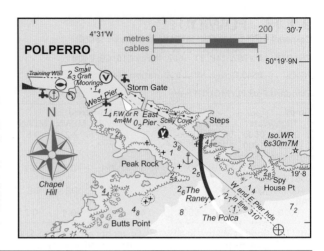

9.1.14   LOOE

Cornwall **50°21´·04N 04°27´·03W** ❄☀⚓⚓🌸🌸

CHARTS   AC 1267, 148, 147, 5602; Imray C10, C6, 2400

TIDES   −0538 Dover; ML 3·0; Duration 0610

Standard Port PLYMOUTH (→)

Times				Height (metres)			
High Water		Low Water		MHWS	MHWN	MLWN	MLWS
0000	0600	0000	0600	5·5	4·4	2·2	0·8
1200	1800	1200	1800				
Differences LOOE							
−0010	−0010	−0005	−0005	−0·1	−0·2	−0·2	−0·2
WHITSAND BAY							
0000	0000	0000	0000	0·0	+0·1	−0·1	+0·2

SHELTER   Good, but uncomfortable in strong SE winds. ⚓ in 2m E of the pier hd; no ⚓ in hbr.

NAVIGATION   WPT 50°20´·68N 04°25´·60W, 290°/1·0M to hbr ent. The outer chan has rky outcrops and dries 0·7m. At night appr in W sector (267°-313°) of Banjo Pier hd lt. Ent is dangerous in strong SE'lies, when seas break heavily on the bar. From W, beware The Ranneys, reef extending 500m SE of Looe Is, marked by SCM buoy

Q (6) + L Fl 15s. 3ca NE of hbr ent, avoid the Limmicks, rks extending 1½ca offshore. Do not attempt the rky passage between Looe Is and the mainland except with local knowledge and at HW. At sp, ebb tide runs up to 5kn. Speed limit 5 mph (sic) within the harbour.

LIGHTS AND MARKS   See chartlet and 9.1.4. Looe Island (aka St George's Is) is conspic (45m), 8ca S of the ent. Mid Main ECM lt bcn is off Hannafore Pt, halfway between pier hd and Looe Is. Siren (2) 30s (fishing) at Nailzee Pt. No lts inside hbr.

COMMUNICATIONS   (Code 01503) CGOC (01326) 317575; Police 101; Dr 263195. HM: Ch 16 (occas).

FACILITIES ❶ berth on W Quay, marked VISITORS, £12.00 all LOA, dries about 3·3m, sloping gently to firm, level sand; rafting is possible and frequently practised.

HM www.looecornwall.com ☎262839, mob 07918 7289550 OH.

W Quay ⚓ ⚓ 📷 🗑 🛢 ✕(Wood) 🔧 🔲 Ⓔ.

E Quay ⚓ For FVs, but access HW ±3 for 🛢 🗐(3t).

Looe SC www.looesailingclub.co.uk ☎262559 🔌 ✕ 🗐.

Town ✉ Ⓑ 🗑 🗐 ✕ @ ⇌.

PLYMOUTH TO START POINT

Rame Head, on the W side of the entrance to **Plymouth Sound**, is a conspicuous conical headland, with a small chapel on top; rocks extend about 1ca off and wind-over-tide overfalls may be met 1·5M to seaward. Approaching Plymouth from the West, clear Rame Head and Penlee Point by about 8ca, then steer NNE for the Western end of the Breakwater.

At the SE entrance to Plymouth Sound, Great Mewstone (57m) is a conspicuous rocky islet 4ca off Wembury Pt. From the East keep at least 1M offshore until clear of the drying Mewstone Ledge, 2½ca SW of Great Mewstone. The Slimers, which dry, lie 2ca E of Mewstone. E and W Ebb Rocks (awash) lie 2½ca off Gara Point (AC 30). Wembury Bay leads to the **River Yealm**. Between Gara Point and Stoke Point, 2·5M to the E, dangers extend up to 4ca offshore.

In Bigbury Bay beware Wells Rock (1m) and other dangers 5ca S of Erme Head. From Bolt Tail to Bolt Head keep 5ca offshore to clear Greystone Ledge, sunken rocks near Ham Stone (11m), and Gregory Rocks (2m) 5ca SE of Ham Stone. The Little Mew Stone (3m high) and Mew Stone lie directly below Bolt Head. Keep at least 3ca SE of the Mewstones before turning N for **Salcombe**. Overfalls occur off Prawle Point, 2·6M E of Bolt Head.

Naval exercise areas from the Isles of Scilly to Start Point are used by submarines and warships, especially near Plymouth. Yachts should try to stay clear.

9.1.15 PLYMOUTH

Devon **50°20'·04N 04°10'·07W** (W Chan) 🌸🌸🌸❀🌀🌀🌀🏵🏵🏵
50°20'·04N 04°08'·07W (E Chan)

CHARTS AC 1267, 1613, 5602, 1900, 30, 1902, 1901, 1967, 871; Imray C10, C14, C6, 2400

TIDES –0540 Dover; ML 3·3; Duration 0610
Standard Port PLYMOUTH (→)

Times				Height (metres)			
High Water		Low Water		MHWS	MHWN	MLWN	MLWS
0000	0600	0000	0600	5·5	4·4	2·2	0·8
1200	1800	1200	1800				
Differences BOVISAND PIER							
–0010	–0010	0008	–0009	–0·1	0·0	+0·2	+0·2
TURNCHAPEL (Cattewater)							
0000	0000	+0010	–0015	0·0	+0·1	+0·2	+0·1
JUPITER POINT (R Lynher)							
+0010	+0005	0000	–0005	0·0	0·0	+0·1	0·0
ST GERMANS (R Lynher)							
0000	0000	+0020	+0020	–0·3	–0·1	0·0	+0·2
SALTASH (R Tamar)							
0000	+0010	0000	–0005	+0·1	+0·1	+0·1	+0·1
LOPWELL (R Tavy)							
ND	ND	DR	DR	–2·6	–2·7	DR	DR
CARGREEN (R Tamar)							
0000	+0010	+0020	+0020	0·0	0·0	–0·1	0·0
COTEHELE QUAY (R Tamar)							
0000	+0020	+0045	+0045	–0·9	–0·9	–0·8	–0·4

NOTE: Winds from SE to W increase the flood and retard the ebb; vice versa in winds from the NW to E.

In Plymouth Sound N all traffic N of the Breakwater conforms to IRPC Rule 9 for 'Narrow channels. There is extensive CCTV coverage of all the enclosed waters.

Speed limits: 10kn* N of Plymouth Bkwtr; 8kn in Cattewater; 4kn N of a line from Fisher's Nose to S side of QAB; 5kn in Sutton Hbr. *Vessels <15m LOA are exempt from the 10kn limit when more than 400m from the shore or in the access lane for water/jet skiers (Fisher's Nose to the W end of Mount Batten Bkwtr and from Royal Plymouth Corinthian YC to W Mallard buoy).

SHELTER Good or better in 4 major and some minor marinas; see Facilities. Around the Sound ⚓s, sheltered according to the wind: Cawsand Bay, Barn Pool (below Mt Edgcumbe), N of Drake's Island, below The Hoe and in Jennycliff Bay. Also good shelter W of Cremyll and off the Hamoaze in the R Lynher and R Tamar above Saltash.

Plymouth is a Naval Base (Devonport) and a commercial, ferry and fishing port. The whole harbour is under the QHM's jurisdiction, but Millbay and the Cattewater are controlled by ABP and Cattewater Commissioners (Capt Charlesworth).

NAVIGATION From the west: WPT 50°18'·81N 04°10'·90W (abeam Draystone PHM buoy), 035°/1·5M to W Bkwtr lt. From the east: WPT 50°18'·81N 04°08'·00W, 000°/1·2M to abeam E Bkwtr lt. The Sound can be entered via the W (main) or E Chans which are well lit/buoyed, but in strong W'lies the E Chan can be a hazardous lee shore for yachts. It is vital to keep well clear of the unlit Shag Stone. There are 3·2m patches NE of E Tinker ECM lt buoy. Yachts need not keep to the deep water chans. They must not impede the Torpoint chain ferries at 50°22'·50N 04°11'·30W. In Cattewater vessels <20m LOA keep clear of vessels >20m LOA.

The Bridge (short cut SW of Drake's Island) is marked by 2 PHM and 2 SHM lt bns, both with tide gauges showing height of tide above CD; least charted depth is 1·3m. There are drying rocky patches either side of the marked channel which is aligned 327°/147° with the LH (blue roof) of 3 conspic high-rise blocks (5ca NW of Mayflower marina). In the chan spring tides reach 346°/1·6kn at HW −4 and 150°/2·4kn at HW +4.

Historic Wrecks are at: 50°21'·73N 04°07'·70W (N of Mt Batten), 50°19'·00N 04°11'·64W and 50°18'·61N 04°12'·05W.

NAVAL ACTIVITY Call Devonport Ops ☎563777 Ext 2182/3. Naval Ops, ☎501182 (H24). www.qhmplymouth.org.uk contains much info of interest to all mariners.

Warships have right of way in the main and DW channels; obey MOD Police. Keep >50m off berthed warships and 100m off berthed submarines.

Yachts should avoid impeding the safe passage of larger vessels.

Submarines: Do not pass within 200m or cross astern within 800m of any submarine under way. Submarines may secure to a buoy close N of the Breakwater, showing a Fl Y anti-collision light.

Diving: Keep clear of Bovisand Pier, the Breakwater Fort and Ravenness Pt when diving signals (Flag A) are displayed.

Ferries Roscoff: 12/week; 5 Hrs; Brittany Ferries (www.brittany-ferries.co.uk); Santander: weekly; 19½ Hrs; Brittany Ferries.

LIGHTS AND MARKS Principal daymarks: conical Rame Head to the W; Great Mew Stone & Staddon Heights to the E; The Breakwater; on The Hoe: Smeaton Tower (R/W bands) & the Naval War Memorial; and Ocean Court (a white bldg) overlooking The Narrows and Mayflower marina.

See chartlets and 9.1.4 for the many lts, some hard to see against shore lts. Dir WRG lts defining the main chans are shown H24** from: Whidbey (138·5°), Staddon Pt (044°), Withyhedge (070°), W Hoe bn (315°), Western King (271°), Millbay** (048·5°), Ravenness (225°), Mount Wise (343°), and Ocean Court** (085°). ** Not H24.

Notes: In fog W lts may, on request to Port Control, be shown from: Mallard (front ldg lt) Fl 5s; West Hoe bn F; Eastern King Fl 5s; Ravenness Fl (2) 15s; Mount Wise F; Ocean Court Fl 5s. Major lts in The Sound show QY if mains power fails. N of The Bkwtr, four large mooring buoys (C, D, E & F) have Fl Y lts.

Wind strength warning flags (R & W vert stripes) are flown at QAB and Mayflower marinas and at The Camber (HO only) to warn of excessive winds as follows:

> 1 wind flag = Force 5–7 (17–27kn).
> 2 wind flags = > Force 7 (>27kn).

COMMUNICATIONS (Code 01752) CGOC (01326) 317575; Police 101; Dr 663138; ⊞ 668080. QHM 836952; DQHM 836485; Port surveyor 836962; Longroom Port Control 836528; *Longroom Port Control* (H24) monitor Ch 13, 14 or 16 underway. Flagstaff Port Control 552413 *Flag* Ch 13, 16.

Cattewater HM 665934; ABP Millbay 662191; Mayflower & QAB marinas and Torpoint Yacht Hbr: Ch **80** M. Plymouth Yacht Haven: Ch **80**. *Sutton Lock*, for opening and marina: Ch **12** (H24). *Cattewater Hbr* Ch 14 (Mon-Fri, 0900-1700LT). *Millbay Docks* Ch 12 14 (only during ferry ops).

FACILITIES Marinas and Rivers from seaward (SE-NW):

Plymouth Yacht Haven

Plymouth Yacht Haven www.yacht havens.com ☎404231, mobile 07721 498422 (o/night); access H24, channel dredged 2·25m. ⚓ 450⌂ inc ❻ £3.25 inc ⚡, 🚾🏧⛽⚓(H24) 🔋⚒🔧🗑🛒🔲(75t) mobile ⚓(4t) 🔩🚻🛒✕🖥 wi-fi.

Turnchapel Wharf ryan@turnchapelwharf.com commercial/large yacht. Max draught 5m max LOA50m, ⚓ at all states of tide.

Yacht Haven Quay 50°21´·87N 04°06´·61W ☎481190. Access H24 to pontoon in 1·2m, 2 FG (vert); ⚒🔧 C; dry berths/storage. Ferry ½hrly from Mount Batten–Barbican, 5 mins: daily 0700-2300, Ch **M** or ☎408590 mob 07930 838614.

Queen Anne's Battery (QAB) www.queenannesbattery.co.uk ☎671142, mobile 07740 806039. ⚓ 235⌂+40❻ short stay £8·45, £3·50<12.5m>£4·35<18m>£4·85<18·1m inc ⚡; 🚾🏧🔋(0830-1830, 7/7) 🔋⚒🔧🗑🛒🔲(25t) 🔩🚻Ⓒ🖥. Extensive facilities.

Sutton Harbour 50°21´·98N 04°07´·96W. West Pier and Guy's Quay marinas are mainly for locals (no shwrs).

Lock operates H24, free. IPTS: sigs 1, 2 & 3. Call *Sutton Lock* Ch 12 for entry and allocation of a berth; secure to floating pontoons either side of lock. The lock retains CD +3·5m in the hbr; when rise of tide >3m, approx HW ±3, free-flow is in operation.

Sutton Hbr Marina www.suttonharbourmarina.com ☎204702, ⚓ 467⌂ inc ❻ £3.50 inc ⚡, 🚾🏧⛽🔋⚒🔲(25t) 🖥 all maintenance/repairs. Nearby 🔩🛒.

King Point Marina www.kingpointmarina.co.uk ☎424297; 171⌂ inc ❻ (development continues), poa approx £3, 🚾⚡🏧⛽🔋 ice, wi-fi(weak).

Mill Bay Village Marina ☎226785, Ch **M**. No ❻.

Mayflower Marina ④ www.mayflowermarina.co.uk ☎556633 (HO) mob 07840 116853. 27m max LOA with 3·5m min depth, but 1·5m near ⛽ and fuel berth. Limit speed to maintain steerage. ⚓ 396⌂ inc 40❻ £3·15 (inc ⚡) wi-fi, £5 then £3/hr<4 hrs. 🔋⚒🏧🔲 ✕ 🔧🗑🛒🔩🔲(33t) ⚓(1½ t) Divers 🔩🛒✕🖥 ☎500008.

Southdown Marina www.southdownmarina.co.uk ☎823084. Approach dries 2m. 🛟s. ⚓ 35⌂ in 2m inc❻ £20 all LOA, or at drying quay, 🏧🔋⚓(20t) 🖥.

Torpoint Yacht Hbr www.torpointyachtharbour.co.uk ☎813658. Access H24, 2m. 100 🛟s 70⌂ pre-call for a ❻ berth £2.50, 🔋⚒🔲 ✕⚓🗑 free water taxi.

YACHT CLUBS
Royal Western Yacht Club of England ☎660077, ▲✕🖥.
Royal Plymouth Corinthian YC ☎664327, VHF ChM, ▲✕🖥⚓.
Plym YC ☎404991. **RNSA** ☎567854. **Mayflower SC** ☎ 492566.
Torpoint Mosquito SC ☎812508, ✕🖥 visitors welcome.
Saltash SC ☎845988. **Tamar River SC** ☎362741.

STANDARD TIME (UT)
For Summer Time add ONE hour in **non-shaded areas**

PLYMOUTH   LAT 50°22'N   LONG 4°11'W
TIMES AND HEIGHTS OF HIGH AND LOW WATERS

Dates in **red** are **SPRINGS**
Dates in blue are **NEAPS**

YEAR 2016

JANUARY

Time	m	Time	m
1 0346	1.8	**16** 0348	1.3
0943	4.9	1005	5.2
F 1612	1.9	SA 1619	1.4
2211	4.6	◗ 2233	4.9
2 0427	2.1	**17** 0438	1.6
1028	4.7	1059	5.0
SA 1658	2.1	SU 1714	1.6
◗ 2302	4.4	2333	4.7
3 0520	2.3	**18** 0541	1.8
1122	4.5	1206	4.8
SU 1756	2.2	M 1824	1.9
4 0007	4.4	**19** 0048	4.6
0626	2.4	0702	2.0
M 1232	4.5	TU 1324	4.7
1902	2.2	1948	1.9
5 0124	4.4	**20** 0206	4.7
0735	2.3	0830	1.9
TU 1351	4.5	W 1439	4.7
2007	2.1	2106	1.7
6 0234	4.6	**21** 0314	4.9
0839	2.1	0941	1.6
W 1458	4.7	TH 1544	4.9
2107	1.9	2208	1.4
7 0331	4.8	**22** 0412	5.1
0936	1.8	1038	1.3
TH 1554	4.9	F 1640	5.1
2200	1.6	2301	1.1
8 0421	5.1	**23** 0503	5.3
1027	1.5	1128	1.0
F 1643	5.0	SA 1729	5.2
2249	1.3	2348	0.9
9 0506	5.3	**24** 0548	5.5
1115	1.2	1213	0.8
SA 1729	5.2	SU 1813	5.3
2334	1.1	○	
10 0549	5.4	**25** 0030	0.8
1200	1.0	0628	5.5
SU 1813	5.3	M 1255	0.7
●		1852	5.3
11 0018	1.0	**26** 0109	0.8
0630	5.5	0705	5.5
M 1244	0.9	TU 1332	0.8
1854	5.3	1927	5.3
12 0101	0.9	**27** 0144	0.9
0712	5.6	0738	5.5
TU 1327	0.8	W 1406	0.9
1936	5.4	1958	5.2
13 0142	0.9	**28** 0215	1.1
0753	5.6	0809	5.4
W 1408	0.8	TH 1435	1.1
2017	5.3	2028	5.1
14 0223	1.0	**29** 0243	1.3
0835	5.5	0841	5.2
TH 1450	0.9	F 1503	1.4
2059	5.2	2100	4.9
15 0304	1.1	**30** 0310	1.5
0918	5.4	0914	5.0
F 1532	1.1	SA 1529	1.6
2143	5.1	2135	4.7
		31 0337	1.8
		0950	4.8
		SU 1558	1.9
		2215	4.6

FEBRUARY

Time	m	Time	m
1 0412	2.0	**16** 0512	1.7
1033	4.6	1140	4.6
M 1640	2.1	TU 1748	1.9
◗ 2304	4.4		
2 0511	2.3	**17** 0018	4.5
1129	4.4	0626	2.0
TU 1753	2.3	W 1304	4.4
		1914	2.1
3 0010	4.3	**18** 0143	4.5
0638	2.3	0810	2.0
W 1242	4.3	TH 1424	4.5
1915	2.3	2049	1.9
4 0134	4.4	**19** 0256	4.7
0754	2.2	0930	1.7
TH 1411	4.4	F 1531	4.6
2026	2.0	2156	1.6
5 0252	4.6	**20** 0356	4.9
0901	1.9	1026	1.3
F 1523	4.6	SA 1627	4.9
2129	1.7	2248	1.2
6 0351	4.9	**21** 0446	5.2
1000	1.5	1114	1.0
SA 1619	4.9	SU 1713	5.1
2225	1.4	2332	0.9
7 0441	5.2	**22** 0529	5.4
1054	1.2	1157	0.8
SU 1708	5.1	M 1754	5.2
2317	1.1	○	
8 0528	5.4	**23** 0013	0.8
1144	0.8	0607	5.5
M 1755	5.3	TU 1236	0.7
●		1829	5.3
9 0005	0.8	**24** 0050	0.7
0612	5.6	0642	5.5
TU 1232	0.6	W 1311	0.7
1839	5.4	1902	5.3
10 0050	0.6	**25** 0123	0.8
0657	5.7	0714	5.4
W 1316	0.5	TH 1342	0.8
1922	5.5	1932	5.2
11 0133	0.5	**26** 0150	0.9
0741	5.7	0745	5.3
TH 1358	0.4	F 1408	1.0
2005	5.5	2002	5.2
12 0213	0.5	**27** 0215	1.1
0823	5.7	0815	5.2
F 1438	0.5	SA 1429	1.2
2045	5.4	2032	5.0
13 0253	0.7	**28** 0236	1.3
0905	5.5	0846	5.0
SA 1518	0.8	SU 1449	1.4
2126	5.3	2102	4.9
14 0333	1.0	**29** 0258	1.5
0949	5.3	0917	4.8
SU 1559	1.1	M 1513	1.6
2210	5.0	2134	4.7
15 0417	1.3		
1037	4.9		
M 1647	1.5		
◗ 2303	4.7		

MARCH

Time	m	Time	m
1 0327	1.8	**16** 0450	1.7
0953	4.6	1119	4.5
TU 1547	1.9	W 1721	1.9
◗ 2216	4.5	2349	4.4
2 0410	2.0	**17** 0559	2.0
1044	4.4	1247	4.3
W 1640	2.2	TH 1842	2.2
2315	4.3		
3 0530	2.3	**18** 0121	4.4
1151	4.2	0748	2.1
TH 1821	2.3	F 1407	4.3
		2029	2.1
4 0032	4.3	**19** 0235	4.5
0713	2.2	0911	1.8
F 1319	4.3	SA 1513	4.5
1949	2.1	2136	1.7
5 0204	4.5	**20** 0334	4.8
0828	1.9	1007	1.4
SA 1449	4.5	SU 1606	4.8
2100	1.8	2227	1.3
6 0317	4.8	**21** 0422	5.1
0934	1.5	1052	1.0
SU 1552	4.8	M 1650	5.0
2202	1.4	2310	1.0
7 0413	5.1	**22** 0504	5.2
1032	1.0	1134	0.8
M 1644	5.1	TU 1728	5.2
2257	0.9	2350	0.8
8 0503	5.4	**23** 0542	5.3
1125	0.7	1211	0.7
TU 1732	5.4	W 1802	5.3
2347	0.6	○	
9 0551	5.6	**24** 0025	0.7
1213	0.4	0616	5.4
W 1819	5.5	TH 1245	0.7
●		1834	5.3
10 0033	0.3	**25** 0056	0.8
0638	5.8	0648	5.4
TH 1258	0.2	F 1314	0.8
1903	5.6	1905	5.3
11 0117	0.2	**26** 0123	0.9
0723	5.8	0720	5.3
F 1341	0.2	SA 1338	1.0
1946	5.7	1935	5.2
12 0158	0.3	**27** 0146	1.0
0807	5.8	0751	5.2
SA 1421	0.3	SU 1358	1.1
2027	5.6	2005	5.1
13 0237	0.5	**28** 0207	1.2
0849	5.6	0820	5.0
SU 1500	0.6	M 1419	1.3
2107	5.4	2033	5.0
14 0317	0.8	**29** 0230	1.4
0931	5.2	0849	4.8
M 1540	1.0	TU 1443	1.5
2147	5.1	2102	4.8
15 0359	1.2	**30** 0259	1.6
1017	4.9	0924	4.6
TU 1625	1.5	W 1517	1.8
◗ 2235	4.7	2141	4.6
		31 0341	1.8
		1013	4.4
		TH 1605	2.0
		◗ 2238	4.4

APRIL

Time	m	Time	m
1 0447	2.1	**16** 0053	4.4
1119	4.2	0709	2.1
F 1730	2.2	SA 1342	4.3
2350	4.4	1948	2.1
2 0635	2.1	**17** 0205	4.5
1239	4.3	0838	1.8
SA 1914	2.1	SU 1444	4.5
		2103	1.8
3 0116	4.5	**18** 0302	4.7
0758	1.8	0935	1.5
SU 1412	4.5	M 1535	4.7
2031	1.8	2155	1.5
4 0240	4.8	**19** 0351	4.9
0907	1.4	1022	1.2
M 1522	4.8	TU 1618	4.9
2136	1.3	2240	1.2
5 0343	5.1	**20** 0433	5.1
1007	1.0	1103	1.0
TU 1617	5.1	W 1656	5.1
2233	0.9	2319	1.0
6 0437	5.4	**21** 0511	5.2
1101	0.6	1140	0.9
W 1707	5.4	TH 1732	5.2
2325	0.5	2355	0.9
7 0527	5.7	**22** 0547	5.3
1150	0.3	1213	0.9
TH 1755	5.6	F 1806	5.3
●		○	
8 0012	0.3	**23** 0026	0.9
0615	5.8	0622	5.3
F 1237	0.1	SA 1242	0.9
1841	5.7	1839	5.3
9 0058	0.1	**24** 0054	0.9
0703	5.8	0656	5.2
SA 1321	0.1	SU 1308	1.0
1925	5.7	1911	5.2
10 0140	0.2	**25** 0120	1.0
0748	5.7	0728	5.1
SU 1402	0.3	M 1332	1.1
2007	5.6	1941	5.2
11 0221	0.4	**26** 0145	1.2
0832	5.5	0759	5.0
M 1442	0.6	TU 1356	1.3
2047	5.4	2010	5.1
12 0301	0.7	**27** 0212	1.3
0915	5.2	0830	4.8
TU 1522	1.0	W 1425	1.5
2126	5.1	2040	4.9
13 0344	1.2	**28** 0244	1.5
1001	4.8	0906	4.7
W 1606	1.5	TH 1500	1.7
2210	4.8	2120	4.8
14 0434	1.6	**29** 0327	1.7
1102	4.4	0954	4.5
TH 1700	1.9	F 1549	1.9
◗ 2319	4.5	2214	4.6
15 0537	2.0	**30** 0429	1.9
1226	4.2	1056	4.4
F 1810	2.2	SA 1703	2.1
		◗ 2321	4.5

Chart Datum is 3·22 metres below Ordnance Datum (Newlyn). HAT is 5·9 metres above Chart Datum.

〉〉**FREE** monthly updates. Register at 〈
www.reedsnauticalalmanac.co.uk

SW England

STANDARD TIME (UT)
For Summer Time add ONE
hour in **non-shaded areas**

PLYMOUTH   LAT 50°22'N   LONG 4°11'W
TIMES AND HEIGHTS OF HIGH AND LOW WATERS

YEAR **2016**

MAY

Time	m		Time	m
1 0601	1.9	**16**	0122	4.4
1211	4.4		0740	1.9
SU 1839	2.0	M	1403	4.4
			2007	2.0
2 0040	4.6	**17**	0221	4.6
0725	1.7		0848	1.7
M 1336	4.6	TU	1454	4.6
1959	1.8		2109	1.8
3 0203	4.8	**18**	0312	4.7
0837	1.4		0939	1.5
TU 1448	4.8	W	1540	4.8
2108	1.4		2159	1.5
4 0311	5.1	**19**	0357	4.9
0939	1.0		1024	1.3
W 1547	5.2	TH	1621	5.0
2207	1.0		2241	1.3
5 0409	5.4	**20**	0438	5.0
1035	0.6		1103	1.1
TH 1640	5.4	F	1700	5.1
2301	0.6		2319	1.1
6 0503	5.6	**21**	0518	5.1
1127	0.4		1138	1.1
F 1730	5.6	SA	1738	5.2
● 2350	0.4	○	2354	1.1
7 0554	5.7	**22**	0557	5.2
1214	0.3		1210	1.1
SA 1818	5.7	SU	1814	5.3
8 0037	0.3	**23**	0026	1.0
0642	5.7		0634	5.1
SU 1300	0.4	M	1241	1.1
1903	5.7		1849	5.3
9 0122	0.3	**24**	0058	1.1
0730	5.6		0710	5.1
M 1343	0.4	TU	1312	1.2
1947	5.6		1922	5.2
10 0205	0.5	**25**	0129	1.1
0815	5.4		0743	5.0
TU 1425	0.7	W	1342	1.2
2028	5.4		1953	5.2
11 0247	0.8	**26**	0202	1.2
0900	5.1		0817	4.9
W 1506	1.1	TH	1416	1.4
2108	5.2		2027	5.1
12 0330	1.1	**27**	0239	1.3
0945	4.8		0856	4.8
TH 1550	1.5	F	1455	1.5
2148	4.9		2108	5.0
13 0417	1.5	**28**	0323	1.5
1040	4.5		0943	4.7
F 1639	1.8	SA	1543	1.7
◑ 2241	4.6		2159	4.8
14 0512	1.8	**29**	0419	1.7
1152	4.3		1039	4.6
SA 1738	2.1	SU	1646	1.9
		◑	2259	4.8
15 0007	4.4	**30**	0532	1.7
0619	2.0		1147	4.5
SU 1303	4.3	M	1805	1.9
1849	2.1			
		31	0011	4.7
			0651	1.7
		TU	1303	4.6
			1925	1.7

JUNE

Time	m		Time	m
1 0130	4.8	**16**	0223	4.5
0805	1.4		0840	1.8
W 1416	4.8	TH	1456	4.7
2038	1.5		2105	1.8
2 0242	5.0	**17**	0317	4.7
0950	1.1		0934	1.6
TH 1519	5.1	F	1544	4.9
2141	1.1		2157	1.6
3 0344	5.2	**18**	0405	4.8
1010	0.9		1020	1.4
F 1615	5.3	SA	1628	5.0
2238	0.8		2242	1.4
4 0441	5.4	**19**	0450	5.0
1105	0.6		1102	1.3
SA 1707	5.5	SU	1711	5.2
2331	0.6		2323	1.2
5 0534	5.5	**20**	0533	5.1
1155	0.5		1141	1.2
SU 1757	5.6	M	1751	5.2
●		○		
6 0019	0.5	**21**	0002	1.1
0624	5.5		0614	5.1
M 1242	0.5	TU	1220	1.1
1844	5.6		1829	5.3
7 0106	0.5	**22**	0041	1.0
0713	5.4		0654	5.1
TU 1327	0.6	W	1257	1.1
1928	5.6		1906	5.3
8 0150	0.6	**23**	0118	1.0
0759	5.3		0732	5.1
W 1409	0.8	TH	1334	1.1
2010	5.4		1942	5.3
9 0232	0.8	**24**	0156	1.1
0843	5.1		0809	5.0
TH 1450	1.0	F	1411	1.2
2049	5.2		2019	5.2
10 0314	1.1	**25**	0235	1.1
0924	4.9		0849	5.0
F 1531	1.4	SA	1451	1.3
2125	5.0		2100	5.2
11 0356	1.4	**26**	0318	1.2
1006	4.6		0933	4.9
SA 1614	1.7	SU	1536	1.4
2204	4.8		2147	5.1
12 0441	1.7	**27**	0407	1.4
1055	4.4		1024	4.8
SU 1702	1.9	M	1629	1.6
◑ 2254	4.6	◐	2241	4.9
13 0534	1.9	**28**	0506	1.5
1157	4.3		1123	4.7
M 1758	2.1	TU	1733	1.7
			2345	4.8
14 0005	4.4	**29**	0616	1.6
0633	2.0		1233	4.7
TU 1304	4.4	W	1850	1.8
1900	2.1			
15 0122	4.4	**30**	0101	4.8
0738	2.0		0733	1.6
W 1403	4.5	TH	1347	4.8
2006	2.0		2009	1.6

JULY

Time	m		Time	m
1 0217	4.9	**16**	0234	4.5
0845	1.4		0844	1.9
F 1454	5.0	SA	1507	4.7
2119	1.4		2113	1.8
2 0324	5.0	**17**	0333	4.7
0950	1.2		0940	1.6
SA 1555	5.2	SU	1559	4.9
2220	1.1		2207	1.5
3 0424	5.2	**18**	0424	4.8
1047	0.9		1030	1.4
SU 1649	5.4	M	1645	5.1
2315	0.8		2256	1.3
4 0518	5.3	**19**	0511	5.0
1139	0.8		1117	1.2
M 1739	5.5	TU	1729	5.3
●		○ 2341	1.1
5 0005	0.7	**20**	0555	5.1
0609	5.3		1201	1.1
TU 1227	0.7	W	1811	5.4
1826	5.6			
6 0052	0.6	**21**	0025	0.9
0656	5.3		0637	5.2
W 1312	0.7	TH	1244	1.0
1910	5.5		1851	5.4
7 0136	0.6	**22**	0107	0.8
0740	5.2		0719	5.2
TH 1353	0.8	F	1324	0.9
1950	5.5		1930	5.5
8 0216	0.8	**23**	0148	0.8
0820	5.1		0759	5.2
F 1431	1.0	SA	1404	0.9
2025	5.3		2010	5.4
9 0253	1.0	**24**	0227	0.8
0855	5.0		0839	5.2
SA 1507	1.2	SU	1443	1.0
2057	5.1		2051	5.4
10 0329	1.3	**25**	0308	1.0
0928	4.8		0921	5.1
SU 1543	1.5	M	1524	1.2
2130	4.9		2134	5.2
11 0405	1.6	**26**	0351	1.2
1004	4.6		1006	4.9
M 1622	1.8	TU	1610	1.4
2209	4.7		2223	5.0
12 0446	1.8	**27**	0441	1.4
1048	4.5		1100	4.8
TU 1708	2.0	W	1706	1.6
◑ 2257	4.5		2322	4.8
13 0537	2.0	**28**	0543	1.7
1145	4.4		1206	4.6
W 1805	2.2	TH	1818	1.8
2359	4.4			
14 0638	2.1	**29**	0037	4.6
1256	4.4		0702	1.8
TH 1910	2.2	F	1324	4.6
			1944	1.8
15 0119	4.4	**30**	0159	4.6
0742	2.0		0825	1.7
F 1407	4.5	SA	1437	4.8
2014	2.0		2104	1.6
		31	0311	4.8
			0936	1.5
		SU	1540	5.0
			2209	1.3

AUGUST

Time	m		Time	m
1 0412	5.0	**16**	0400	4.8
1035	1.2		1003	1.5
M 1635	5.2	TU	1620	5.1
2304	1.0		2231	1.3
2 0506	5.1	**17**	0448	5.0
1126	0.9		1055	1.2
TU 1724	5.4	W	1706	5.3
● 2352	0.7		2321	1.0
3 0554	5.2	**18**	0534	5.2
1212	0.7		1143	0.9
W 1809	5.5	TH	1750	5.5
		○		
4 0037	0.6	**19**	0007	0.7
0637	5.3		0618	5.3
TH 1255	0.7	F	1228	0.8
1849	5.5		1832	5.6
5 0118	0.6	**20**	0052	0.6
0717	5.3		0701	5.4
F 1333	0.7	SA	1310	0.6
1925	5.5		1915	5.7
6 0155	0.7	**21**	0133	0.5
0751	5.2		0742	5.4
SA 1408	0.9	SU	1351	0.6
1957	5.4		1956	5.6
7 0228	0.9	**22**	0213	0.6
0821	5.1		0823	5.4
SU 1439	1.1	M	1430	0.7
2027	5.2		2038	5.5
8 0257	1.2	**23**	0252	0.8
0852	4.9		0904	5.3
M 1508	1.4	TU	1509	1.0
2058	5.0		2119	5.3
9 0324	1.5	**24**	0332	1.1
0925	4.8		0946	5.1
TU 1537	1.7	W	1551	1.3
2133	4.8		2205	5.0
10 0353	1.8	**25**	0418	1.4
1003	4.6		1036	4.8
W 1610	2.0	TH	1643	1.6
◑ 2212	4.6	◑	2301	4.7
11 0430	2.0	**26**	0516	1.8
1049	4.4		1142	4.6
TH 1701	2.2	F	1752	2.0
2303	4.4			
12 0534	2.2	**27**	0020	4.5
1150	4.3		0636	2.0
F 1818	2.3	SA	1306	4.5
			1929	2.0
13 0011	4.3	**28**	0150	4.4
0653	2.3		0813	2.0
SA 1310	4.3	SU	1424	4.7
1931	2.2		2056	1.8
14 0146	4.3	**29**	0303	4.6
0803	2.1		0927	1.7
SU 1430	4.5	M	1528	4.9
2038	2.0		2159	1.4
15 0303	4.5	**30**	0403	4.9
0907	1.8		1023	1.3
M 1530	4.8	TU	1621	5.2
2138	1.6		2250	1.0
		31	0452	5.1
			1111	1.0
		W	1707	5.4
			2336	0.8

Chart Datum is 3·22 metres below Ordnance Datum (Newlyn). HAT is 5·9 metres above Chart Datum.

STANDARD TIME (UT)
For Summer Time add ONE hour in **non-shaded areas**

PLYMOUTH   LAT 50°22'N   LONG 4°11'W
TIMES AND HEIGHTS OF HIGH AND LOW WATERS

Dates in red are **SPRINGS**
Dates in blue are **NEAPS**

YEAR 2016

SEPTEMBER

Day	Time	m		Day	Time	m
1 TH ●	0535	5.2		16 F ○	0510	5.3
	1154	0.8			1121	0.8
	1748	5.5			1726	5.6
					2346	0.6
2 F	0017	0.6		17 SA	0555	5.5
	0614	5.3			1208	0.6
	1234	0.7			1811	5.8
	1824	5.5				
3 SA	0055	0.6		18 SU	0031	0.4
	0648	5.3			0638	5.6
	1309	0.7			1252	0.5
	1858	5.5			1855	5.8
4 SU	0129	0.7		19 M	0114	0.4
	0719	5.3			0721	5.7
	1341	0.9			1333	0.5
	1928	5.4			1939	5.8
5 M	0157	0.9		20 TU	0155	0.5
	0749	5.2			0803	5.6
	1407	1.1			1413	0.6
	1958	5.3			2021	5.6
6 TU	0221	1.2		21 W	0234	0.7
	0819	5.1			0844	5.5
	1431	1.3			1453	0.9
	2029	5.1			2103	5.4
7 W	0241	1.4		22 TH	0314	1.1
	0850	4.9			0926	5.2
	1452	1.6			1535	1.3
	2100	4.9			2148	5.0
8 TH	0301	1.7		23 F ◐	0358	1.5
	0924	4.7			1014	4.9
	1515	1.9			1625	1.7
	2135	4.6			2244	4.6
9 F ◑	0329	2.0		24 SA	0454	1.9
	1005	4.5			1120	4.6
	1551	2.1			1733	2.1
	2222	4.4				
10 SA	0415	2.2		25 SU	0009	4.4
	1059	4.4			0613	2.2
	1707	2.4			1250	4.5
	2324	4.2			1915	2.2
11 SU	0558	2.4		26 M	0140	4.4
	1211	4.3			0800	2.2
	1853	2.4			1408	4.6
					2043	1.9
12 M	0050	4.2		27 TU	0251	4.6
	0728	2.3			0912	1.8
	1344	4.5			1511	4.9
	2007	2.1			2142	1.5
13 TU	0231	4.5		28 W	0347	4.9
	0838	2.0			1004	1.4
	1458	4.8			1602	5.2
	2110	1.7			2230	1.1
14 W	0333	4.8		29 TH	0432	5.1
	0938	1.6			1050	1.1
	1552	5.1			1645	5.4
	2207	1.3			2313	0.9
15 TH	0423	5.1		30 F	0511	5.3
	1032	1.2			1130	0.9
	1640	5.4			1723	5.5
	2258	0.9			2352	0.8

OCTOBER

Day	Time	m		Day	Time	m
1 SA ●	0547	5.4		16 SU ○	0529	5.6
	1208	0.8			1145	0.6
	1758	5.5			1748	5.8
2 SU	0027	0.8		17 M	0008	0.4
	0618	5.4			0618	5.8
	1242	0.8			1231	0.4
	1830	5.5			1834	5.9
3 M	0059	0.9		18 TU	0053	0.4
	0649	5.4			0659	5.8
	1311	1.0			1314	0.4
	1901	5.4			1920	5.8
4 TU	0124	1.0		19 W	0135	0.5
	0719	5.3			0743	5.8
	1336	1.1			1356	0.6
	1932	5.3			2004	5.6
5 W	0146	1.2		20 TH	0216	0.7
	0750	5.2			0825	5.6
	1357	1.3			1438	0.9
	2002	5.1			2048	5.4
6 TH	0205	1.4		21 F	0257	1.1
	0820	5.1			0908	5.3
	1418	1.6			1521	1.3
	2032	4.9			2135	5.0
7 F	0225	1.7		22 SA ◐	0342	1.6
	0852	4.9			0955	5.0
	1442	1.8			1611	1.7
	2106	4.7			2231	4.6
8 SA	0254	1.9		23 SU	0436	2.0
	0929	4.7			1059	4.7
	1517	2.0			1715	2.1
	2151	4.5			2354	4.4
9 SU ◑	0337	2.2		24 M	0548	2.3
	1021	4.5			1226	4.6
	1614	2.3			1847	2.2
	2251	4.3				
10 M	0450	2.5		25 TU	0118	4.4
	1128	4.4			0727	2.3
	1811	2.4			1342	4.6
					2014	2.0
11 TU	0008	4.3		26 W	0225	4.6
	0650	2.4			0842	2.0
	1251	4.5			1443	4.8
	1934	2.1			2114	1.7
12 W	0148	4.5		27 TH	0319	4.8
	0807	2.1			0936	1.6
	1417	4.8			1533	5.1
	2041	1.7			2201	1.4
13 TH	0301	4.8		28 F	0404	5.1
	0911	1.6			1021	1.3
	1519	5.1			1616	5.2
	2140	1.3			2243	1.1
14 F	0354	5.2		29 SA	0442	5.3
	1006	1.2			1101	1.1
	1611	5.4			1655	5.4
	2233	0.9			2322	1.0
15 SA	0443	5.4		30 SU ●	0517	5.4
	1058	0.8			1138	1.0
	1700	5.7			1730	5.4
	2322	0.6			2356	1.0
				31 M	0550	5.5
					1212	1.0
					1804	5.4

NOVEMBER

Day	Time	m		Day	Time	m
1 TU	0027	1.0		16 W	0033	0.5
	0622	5.5			0639	5.9
	1241	1.1			1257	0.5
	1837	5.4			1903	5.8
2 W	0053	1.1		17 TH	0118	0.6
	0655	5.4			0725	5.8
	1308	1.2			1342	0.6
	1910	5.3			1950	5.6
3 TH	0116	1.3		18 F	0201	0.8
	0726	5.3			0809	5.7
	1332	1.4			1425	0.9
	1941	5.1			2036	5.3
4 F	0139	1.5		19 SA	0243	1.1
	0757	5.2			0853	5.4
	1356	1.5			1510	1.2
	2012	5.0			2122	5.0
5 SA	0204	1.6		20 SU	0328	1.5
	0828	5.0			0939	5.1
	1424	1.7			1557	1.6
	2045	4.8			2215	4.7
6 SU	0236	1.8		21 M ◐	0417	1.9
	0905	4.9			1033	4.9
	1501	1.9			1652	1.9
	2129	4.6			2323	4.5
7 M ◑	0318	2.1		22 TU	0516	2.2
	0954	4.7			1146	4.7
	1555	2.2			1800	2.1
	2226	4.5				
8 TU	0421	2.3		23 W	0039	4.4
	1056	4.6			0630	2.3
	1724	2.3			1301	4.6
	2336	4.4			1921	2.1
9 W	0603	2.4		24 TH	0145	4.5
	1209	4.6			0751	2.2
	1857	2.1			1403	4.7
					2030	1.9
10 TH	0059	4.5		25 F	0240	4.7
	0730	2.1			0855	1.9
	1330	4.8			1456	4.9
	2008	1.8			2123	1.7
11 F	0220	4.8		26 SA	0327	4.9
	0839	1.7			0945	1.7
	1442	5.1			1542	5.0
	2110	1.4			2208	1.5
12 SA	0322	5.2		27 SU	0409	5.1
	0939	1.3			1028	1.4
	1541	5.4			1624	5.2
	2206	1.0			2248	1.3
13 SU	0415	5.4		28 M	0447	5.3
	1033	0.9			1107	1.3
	1635	5.6			1703	5.3
	2258	0.7			2324	1.2
14 M ○	0504	5.7		29 TU ●	0524	5.4
	1123	0.7			1142	1.2
	1726	5.8			1741	5.3
	2346	0.5			2357	1.2
15 TU	0552	5.8		30 W	0600	5.4
	1211	0.5			1214	1.2
	1815	5.8			1817	5.3

DECEMBER

Day	Time	m		Day	Time	m
1 TH	0026	1.2		16 F	0103	0.6
	0635	5.4			0709	5.8
	1245	1.3			1329	0.6
	1852	5.2			1937	5.5
2 F	0055	1.3		17 SA	0147	0.8
	0709	5.4			0755	5.7
	1315	1.3			1414	0.7
	1926	5.1			2022	5.3
3 SA	0124	1.4		18 SU	0230	1.0
	0741	5.3			0838	5.5
	1346	1.4			1456	1.1
	1959	5.0			2106	5.1
4 SU	0154	1.5		19 M	0311	1.3
	0813	5.2			0919	5.2
	1418	1.6			1539	1.4
	2033	4.9			2149	4.9
5 M	0229	1.7		20 TU	0354	1.6
	0850	5.1			1000	5.0
	1456	1.7			1623	1.7
	2115	4.8			2234	4.6
6 TU	0310	1.9		21 W ◐	0440	2.0
	0935	4.9			1047	4.8
	1545	1.9			1713	2.0
	2206	4.6			2331	4.5
7 W ◑	0404	2.0		22 TH	0534	2.2
	1031	4.8			1149	4.6
	1649	2.0			1812	2.1
	2308	4.6				
8 TH	0517	2.2		23 F	0041	4.4
	1136	4.8			0638	2.3
	1811	2.0			1303	4.5
					1920	2.2
9 F	0020	4.6		24 SA	0147	4.5
	0645	2.1			0750	2.2
	1250	4.8			1407	4.6
	1930	1.8			2028	2.0
10 SA	0138	4.8		25 SU	0243	4.7
	0804	1.8			0855	2.0
	1406	5.0			1503	4.7
	2039	1.5			2124	1.8
11 SU	0248	5.0		26 M	0333	4.9
	0911	1.5			0948	1.8
	1513	5.2			1552	4.9
	2141	1.2			2211	1.6
12 M	0348	5.3		27 TU	0418	5.1
	1010	1.2			1033	1.6
	1613	5.4			1637	5.0
	2236	0.9			2252	1.4
13 TU	0443	5.6		28 W	0500	5.2
	1104	0.9			1114	1.4
	1707	5.6			1720	5.1
	2328	0.7			2330	1.3
14 W ○	0533	5.7		29 TH ●	0540	5.4
	1155	0.7			1152	1.3
	1759	5.6			1800	5.2
15 TH	0017	0.6		30 F	0006	1.2
	0622	5.8			0618	5.4
	1244	0.6			1228	1.2
	1849	5.6			1839	5.2
				31 SA	0041	1.2
					0655	5.4
					1304	1.2
					1916	5.2

Chart Datum is 3·22 metres below Ordnance Datum (Newlyn). HAT is 5·9 metres above Chart Datum.

>> **FREE** monthly updates. Register at <<
www.reedsnauticalalmanac.co.uk

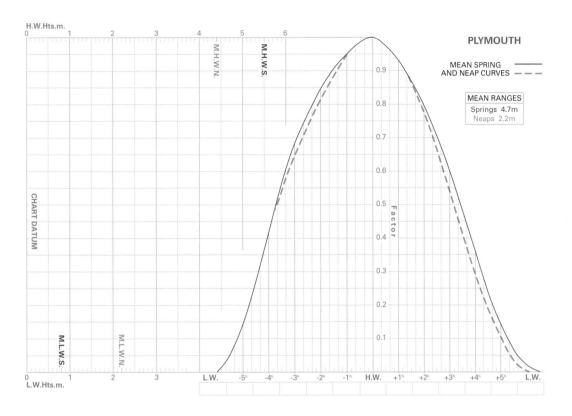

PLYMOUTH

MEAN SPRING
AND NEAP CURVES

MEAN RANGES
Springs 4.7m
Neaps 2.2m

9.1.16 NAVAL EXERCISE AREAS (SUBFACTS AND GUNFACTS)

Submarines and warships frequently exercise south of Plymouth and in the western part of the English Channel. Details of submarine activity and naval gunnery and missile firings are broadcast daily by HM Coastguard. See Chapter 3 for advice when sailing in areas of known submarine activity, and Chapter 6 for timings of daily broadcasts. Although the onus for safety is on the relevant naval vessels, if possible it is best to avoid areas where exercises are taking place.

LYNHER (or ST GERMANS) RIVER AC 871. This river flows into The Hamoaze about 0·8M SSW of the Tamar Bridge. On the tide it is navigable for some 4M inland. The channel, entered at Lynher PHM lt buoy 50°23′·78N 04°12′·83W, is marked by 2 more lt buoys in the first mile to Sandacre Pt and carries 2-5m up to Ince Castle. Caution: underwater cables/gaspipe, as charted. Thereafter it carries less than 1m or dries, except at Dandy Hole, a pool with 3-5m, 50°22′·54N 04°16′·29W. Here the navigable chan bends NW then dries completely; it is marked by small R and G posts.

⚓s, amid local moorings, are: off Sand Acre Bay (N bank, beware foul ground); in 2·5-5m at the ent to Forder Lake (N bank opposite Jupiter Pt, where the pontoons and Y moorings are for naval use only); SE of Ince Pt and Castle in about 3m; and at Dandy Hole. St Germans Quay is private, but temp ⌇, M may be pre-arranged with Quay SC ☎(01503) 250370. Facilities: ✉ ≣ ⌕(½M).

RIVER TAMAR AC 871. The river is navigable on the tide for 12M via Calstock and Morwellham to Gunnislake weir. **Jubilee Green** (pontoon in 3m, max LOA 10m) is on W bank close N of the Royal Albert (30m) & Tamar (35m) bridges. Pwr cables (21m) cross 0·4M S of Cargreen (0°26′·55N 04°12′·25W) sited on the W bank 0·7M N of the Tavy with a quay and many local moorings/⚓ in 2·5-5m. Overhead cables 1M N have 16m clearance en-route Weir Quay. **Weir Quay BY** ☎(01822) 840474, ⚓s ⌇ ▲ ▮ ⚒ ⚹ ⌇ ⊞(20t) ⌇(12t) ⌕. **Weir Quay SC** ☎(01822) 840960, ⚓. Upstream the river S-bends, narrows and partly dries with channel above Halton Quay from 0·1m to >2·5m. N of Cotehele Quay it turns 90° stbd to Calstock, ⚓ in 2m E of viaduct (24m); ≣ ✉ ≷. **Calstock BY** ☎(01822) 832502. ⚓ ⚹ SH, ⌇(8t), ⊞(10t)

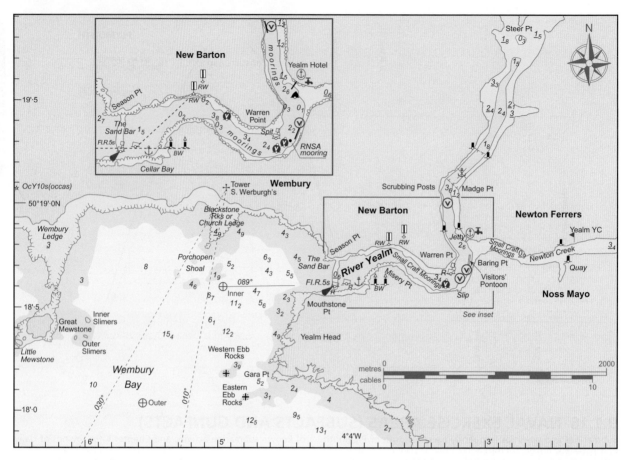

9.1.17 RIVER YEALM

Devon **50°18′·58N 04°04′·13W** (Ent) ❀❀⚓⚓🏠🏠🏠

CHARTS AC 1613, 1900, 30, 5602; Imray C10, C6, C14, 2400

TIDES –0522 Dover; ML 3·2; Duration 0615

Standard Port PLYMOUTH (←—)

Times				Height (metres)			
High Water		Low Water		MHWS	MHWN	MLWN	MLWS
0000	0600	0000	0600	5·5	4·4	2·2	0·8
1200	1800	1200	1800				
Differences RIVER YEALM ENTRANCE							
+0006	+0006	+0002	+0002	–0·1	–0·1	–0·1	–0·1

NOTE: Strong SW winds hold up the ebb and raise levels, as does the river if in spate.

SHELTER Very good in river. Entry is easy except in strong SW-W winds; ⚓ in Cellar Bay is open from SW to NW. Space in the hbr is very limited, and in wind-over-tide conditions, when moored yachts lie across the stream, larger vessels may find turning difficult. Vessels >18m LOA may not enter without the HM's permission.

NAVIGATION Outer WPT 50°18′·03N 04°05′·55W, 033°/7ca to the Inner WPT 50°18′·59N 04°04′·98W; then 089°/5½ca to the Sand Bar. Keep St Werburgh's ✠ twr between 010° and 030° to clear the Slimers and the W & E Ebb Rocks, then turn onto 089°; ldg bns (both W △ with B stripe). ≠ clears Mouthstone Ledge, but not the sand bar which extends SW. Two PHM buoys mark the S end of the Sand Bar and must be left to port on entry; the seaward buoy is lit, Fl R 5s. After passing the second (E'ly, unlit) buoy turn onto 047° ldg bns (both W with R stripe), on the N shore, then follow line of the river. There is only 1m at MLWS between the Sand Bar and Misery Point. Leave Spit PHM buoy to port and Office SHM to starboard. No ⚓ in river; speed limit 6kn.

LIGHTS AND MARKS Great Mewstone (57m) is conspic 1·5M to W of river ent; no passage between it and the coast.

COMMUNICATIONS (Code 01752) CGOC (01326) 317575; Dr 880567; Police 101; HM 872533; Water taxi ☎07817 132757.

FACILITIES 1⚓ off Misery Pt, 3⚓ off Warren Pt and ⓥ pontoons in The Pool and 3ca up-river. Pontoon, ⚓/⚓ £1.70. 🚾/🚿 at HM. 🛒🔧. **Yealm YC** ☎872291, 🍴 Bistro, 🍺.
Newton Ferrers 🛒🛢🔧✉️🍴✕🍺.
Noss Mayo ⌕ at Pope's Quay (dries), 🛒🛢🔧✕🍺.
Bridgend (Newton Creek) ⌕ (dries), 🔧🅿️.
🏧🏧🛎 at Yealmpton (3M).

ADJACENT ANCHORAGES IN BIGBURY BAY
RIVER ERME, Devon, **50°18′·15N 03°57′·67W.** AC1613. HW –0525 on Dover; +0015 and –0·6m on HW Plymouth. Temp day ⚓ in 3m at mouth of drying river, open to SW. Enter only in offshore winds and settled weather. Beware Wells Rk (1m) 1M SE of ent. Appr from SW, clear of Edwards Rk. Ent between Battisborough Is and W. Mary's Rk (dries 1·1m) keeping to the W. No facilities. Two Historic Wrecks are at 50°18′·15N 03°57′·41W and 50°18′·41N 03°57′·19W on W side of the ent.

RIVER AVON, Devon, **50°16′·63N 03°53′·33W.** AC1613. Tides as R Erme, above. Appr close E of conspic Burgh Is (⚓) & Murray's Rk, marked by bcn. Enter drying river HW –1, only in offshore winds and settled weather. Narrow chan hugs cliffy NW shore, then S-bends SE and N off Bantham. Local knowledge or a dinghy recce near LW would assist. Streams run hard, but able to dry out in good shelter clear of moorings. HM ☎(01548) 561196. ✉️🛒 🍺 at Bantham. Aveton Gifford accessible by dinghy, 2·5M.

HOPE COVE, Devon, **50°14′·65N 03°51′·78W.** AC 1613. Tides as R Erme; ML 2·6m; Duration 0615. Popular day ⚓ in centre of cove, but poor holding and only in offshore winds. Appr with old LB ho in SE corner of the cove brg 130° and ⚓ WSW of the pier head, clear of moorings. Basse Rock extends 30m around the pier head. Off the S shore Goody Rock (dries 2·5m) is cleared by brgs of 060° on the pier head and 120° on the old LB ho. No lights. Facilities: very limited in village, good at Salcombe (4M bus).

9.1.18 SALCOMBE

Devon 50°13'·17N 03°46'·67W (The Bar) ❀❀❀❀❀❀❀❀

CHARTS AC 1613, 1634, 28, 5602; Imray C10, C6, C5, 2300

TIDES –0523 Dover; ML 3·1; Duration 0615

Standard Port PLYMOUTH (←)

Times				Height (metres)			
High Water		Low Water		MHWS	MHWN	MLWN	MLWS
0100	0600	0100	0600	5·5	4·4	2·2	0·8
1300	1800	1300	1800				
Differences SALCOMBE							
0000	+0010	+0005	–0005	–0·2	–0·3	–0·1	–0·1
START POINT							
+0015	+0015	+0005	+0010	–0·1	–0·2	+0·1	+0·2

SHELTER Good, but fresh S'lies can cause uncomfortable swell off the town. Better shelter in The Bag at ♥ pontoon. ‡s at Sunny Cove; off SE shore from YC start line to fuel barge (where poor holding has been reported); 200m W and 300m SSE of Salt Stone SHM bcn 50°15'·19N 03°45'·55W.

NAVIGATION WPT 50°12'·43N 03°46'·67W, 000°/1·3M to Sandhill Pt lt. In normal conditions The Bar (least depth 1m) is not a problem, *but it should not be crossed on an ebb tide with strong onshore winds or swell when dangerous breakers can occur;* if in doubt, call

HM Ch 14 before approaching. The deepest water is close inshore to the west off the leading line. The spring flood reaches 2·5kn. Speed limit 8kn, marked in season by 3 Y SPM between Cadmus and Chapple Rks, but 6kn between Marine Hotel/racing start line and the Saltstone (approx 50°15'·2N 03°45'·5W); minimise wash; radar checks in force. In Jul/Aug within the hbr cruising yachts must motor, not sail. When dinghy racing in progress (Sat/Sun & regatta weeks: Fl Y lt at YC and warning on Ch 14) yachts should use a fairway marked by SPM buoys close to the NW bank and straddling the start line. Two Historic Wrecks lie 1M WNW of Prawle Pt at 50°12'·74N 03°44'·40W & 50°12'·73N 03°44'·75W.

LIGHTS AND MARKS Outer 000° ldg marks, both R/W bcns (a conspic gabled house close N is more easily seen): front, Poundstone; rear, Sandhill Pt Dir lt 000°, stay in the W sector (357·5°-002·5°). Close W of ldg line beware Bass Rk (0·8m) marked by PHM lt buoy; and, close E, Wolf Rk (0·6m), marked by SHM lt buoy. When past Wolf Rk, leave Black Stone Rk (5m), G/W bcn, well to stbd and pick up inner ldg lts 042·5°.

COMMUNICATIONS (Code 01548) CGOC (01326) 317575; Police 101; Dr 842284. HM ☎843791: May to mid-Sep: 0900-1645 Mon-Thu; 0900-1615 Fri-Sun; *Salcombe Hbr Launch* (May to mid-Sep: daily 0715 -2300; rest of year: Mon-Fri 0900-1600) Ch 14. *Harbour taxi* Ch 12. *Fuel Barge* Ch 06. ICC HQ, *Egremont* Ch M.

FACILITIES www.salcombeharbour.co.uk www.southhams.gov.uk

Numerous ⚓s either side of the fairway off the town. The Bag (50°14'·58N 03°45'·56W): pontoon for 150♥. 21 marked ⚓s (orange <8m LOA, yellow >8m). Pontoon & ⚓ £1·80 (inc hbr dues). ‡ £0·90/m. Short stay pontoon (½ hr max, 0700-1900, for ⚓/stores) in 0·9m is E of HM's office for adjacent 🚻/🚿. Public ⛽ 🅿(12t) in Batson Creek (Fishermens Quay). ⚓ on pontoon in The Bag. 🗑 at Whitestrand Quay.
Fuel Barge 🛢 ⚓ ☎07801 798862, summer 7/7 0830-1700; winter Mon-Fri 0830-1700.
Salcombe YC ☎842872 / 842593, ⚓ ✕ 🖴.
Island CC ☎531776 in *Egremont*, N of Ox Pt.
Services ✕ ⚒ 🚿 🅔 ⛽ ⚓ ACA.
Town ✉ Ⓑ 🅿 ✕ 🖴 🚈 (bus to Totnes).

ADJACENT HARBOUR, 3M north
KINGSBRIDGE, 50°16'·88N 03°46'·52W. AC 28. HW = HW Salcombe +0005. The 3M chan, which dries approx 2·4m in the upper reaches, is marked beyond Salt Stone SHM perch by R/W PHM poles with R can topmarks. 6ca N of Salt Stone a secondary chan marked by PHM buoys gradually diverges E into Balcombe Creek (dries). Ferry berths at private pontoon at New Quay, 3ca S of drying Kingsbridge basin and on outside of the new mooring pontoon. Max LOA 11m.
♥ boats < 5·5m ⛴ on pontoon on west wall. ♥ 5·5m < ⛴ < 9m inside middle of pontoon. ♥ >9m ⛴ on the wall N of the pontoon (drying 3·4m to soft mud). Best to pre-check berth availability with Salcombe HM; fees and Hbr dues are payable.
Services ⛽ ⚓.
Town ✉ Ⓑ 🛒 🅿 ✕ 🖴 ℍ852349 (A&E).

START POINT TO TEIGNMOUTH

(AC 1613, 1634, 3315) Start Point (Fl(3) 10s 62m25M, *Horn 60s*), 3·3M ENE of Prawle Point, is a long headland with a distinctive cock's comb spine, a W lt ho on the Point, and conspicuous radio masts to the WNW. Black Stone rock (6m high) is visible 2½ca SSE of the lt ho, with Cherrick Rocks (1m) 1ca further S; other drying rocks lie closer inshore. A race may extend 1M to the S and 1·6M to the E. In fair weather any overfalls can be avoided by passing close to seaward of rks; there is no clear-cut inshore passage as such. In bad weather keep at least 2M off.

▶ *The NE-going flood begins at HW Plymouth −2; max (3·1kn Sp) at HWP −1. The SW-going ebb begins at HWP +5; max (2·2kn Sp) at HWP +6. But at HWP +4 it is possible to round Start Point close inshore using a back eddy. On both the flood and the ebb back eddies form between Start Point and Hallsands, 1M NW. Inshore the tide turns 30 minutes earlier.*◀

Skerries Bank (least depth 2·1m), on which the sea breaks in bad weather, lies from 9ca to 4·1M NE of Start Pt (AC 1634). In off-shore winds there is a good ⚓ in 3m 1ca off Hallsands (1M NW of Start). A fish farm lies about 3½ ca off. Between **Dartmouth** and **Brixham** rocks extend 5ca offshore. Berry Head (Fl 2 15s), a bold, flat-topped headland (55m), gives a good radar return.

▶ *Here the stream turns N at HW Plymouth −3, and S at HWP +3; max Sp rates 3kn.*◀

In **Torbay** (AC 26) the more obvious dangers are steep-to, but beware the Sunker (awash at CD) 100m SW of Ore Stone, and Morris Rogue (0·8m) 5ca W of Thatcher Rock.

In W'lies Hope Cove, Anstey's Cove (beware the Three Brothers, rks drying 0·6m, S side) and Babbacombe Bay are good ⚓s on sand. There are no offlying dangers from Long Quarry Point for 4M N to **Teignmouth** whose ent is dangerous in onshore winds.

▶ *Off Teignmouth Bar the NNE-going stream begins at HW Plymouth −1½ and the SSW-going at HWP +5¼. In the ent the flood begins at HWP −5½ and the ebb at HWP +¾. The Spring ebb reaches 4–5kn off The Point.*◀

9.1.19 DARTMOUTH

Devon 50°20'·66N 03°33'·96W ✿✿✿✿⚓⚓⚓☆☆☆

CHARTS AC 1613, 1634, 2253, 5602; Imray C10, C5, 2300

TIDES −0510 Dover; ML 2·8; Duration 0630
DARTMOUTH (→). The differences below are referred to Plymouth as Standard Port, not to Dartmouth.

Times				Height (metres)			
High Water		Low Water		MHWS	MHWN	MLWN	MLWS
0100	0600	0100	0600	5·5	4·4	2·2	0·8
1300	1800	1300	1800				
Differences GREENWAY QUAY (DITTISHAM)							
+0030	+0045	+0025	+0005	−0·6	−0·6	−0·2	−0·2
STOKE GABRIEL (DUNCANNON)							
+0035	+0040	+0020	+0030	−0·9	−0·8	−0·4	−0·3
TOTNES							
+0030	+0040	+0115	+0030	−2·0	−2·1	DR	DR

SHELTER Excellent shelter inside the harbour and up-river. Three marinas and many other ⚓ and ⚓ options. River navigable to Totnes depending on tide and draught.

NAVIGATION WPT 50°19'·53N 03°32'·83W, 328°/1·5M in the white sector of Kingswear Dir lt. Bayard's Cove Dir lt leads 293° to abeam

Royal Dart YC where the main fairway opens. There is no bar and hbr access is H24, but ent can be difficult in strong SE to SW winds. Speed limit 6kn from Castle Ledge buoy up-river to 1M below Totnes. *Caution: The Lower and Higher car ferries S and N of Dartmouth have right of way; give way early.*

LIGHTS AND MARKS as on the chartlet and/or 9.1.4. E of the ent, on Inner Froward Pt (167m) is a conspic daymark, obelisk (24·5m). Lateral and cardinal lt buoys mark all dangers to seaward of conspic Dartmouth Castle. Within hbr, all jetty/ pontoon lts to the W are 2FR (vert); and 2FG (vert) to the E.

COMMUNICATIONS (Code 01803) CGOC (01326) 317575; Police 101; Dr 832212; ⊞ 832255. HM 832337, 835220 (out of hours emergency); *Dartnav* VHF Ch 11 (summer 0730-2100). Darthaven, Dart and Noss marinas Ch 80. Fuel barge Ch 06. Water taxis: *Yacht taxi* (DHNA) Ch 69, 07970 346571 (summer 0800-2300, winter 1000-1600). *Greenway ferry* Ch 10 or 844010 (to Dittisham).

FACILITIES **Dart Harbour and Navigation Authority** (DHNA), www.dartharbour.org ☎832337, 450⚓+90⓿.
Town jetty: £19<10m>£23·50<12m>£29<15m>£34<18m.
S Embankment: £1·60; other pontoons, ⚓s and ⚓ £1·55/m; all inc Hbr dues (£0·75) & VAT. ⚓ ⚓ 🚾/⚓ to N of Boat Float.

DHNA ⓿ pontoons are marked by blue flags and ⚓s are blue with black 'V' or 'Visitors' in yellow and black. Check availability with HM or call DHNA. The most likely ⓿ berths from S to N are:

W bank: pontoon (max 9m LOA) off Dartmouth YC (May-Sep). Town pontoon W side only, but E side 1700-0845. N'ly of 2 pontoons just S of Dart marina (26'/8m max LOA).

E of fairway: The ⓿ pontoon by Kingswear rail station is shared by DHNA and Darthaven marina; the former are for 2 hrs max. The 3 ⓿ pontoons are N of the fuel barge.

Only space to ⚓ is E of fairway and Nos 3A to 5 buoys, with HM's agreement. These large unlit mooring buoys (plus Nos 3, 5A and 6) in mid-stream are for commercial vessels/FVs; do not ⚓ over their ground chains, as shown on AC 2253.

Darthaven Marina www.darthaven.co.uk ☎752242, 250⚓+20 ⓿ £2·50+VAT+hbr dues. ⚓ ⚓ ⚓ ⚓ ⚓ ⚓(35t) ⚓ wi-fi.

Dart Marina www.dartmarinayachtharbour.com ☎837161. 100⚓ +10⓿ £2·50+VAT+hbr dues. ⚓ ✕ wi-fi.

Noss Marina www.nossmarina.co.uk ☎839087 mob 07920425452. ⚓180⚓ £2·98 + dues, ⚓ ⚓ ⚓ ⚓(10t) wi-fi.

YACHT CLUBS (visitors welcome): **Royal Regatta**, last week Aug. **Royal Dart YC** ☎752272, short stay pontoon, ⚓ ✕ ⚓ ☎752880. **Dartmouth YC** ☎832305, ⚓ ⚓ ⚓ ✕ ⚓.

SERVICES Fuel Barge next to No 6 buoy, Ch 06, ⚓ ⚓; ☎07801 798861 summer 7/7 0800-1800; winter Mon/Wed/Fri 1200-1700, Sat/Sun 1000-1700.

Creekside BY (Old Mill Creek) ☎832649, ⚓ ⚓ dry dock, ⚓(customer only) ⚓ ⚓ ✕ ⚓ ⚓(14t) ⚓.

Dartside Quay (Galmpton Creek): ☎845445 ⚓⚓⚓⚓⚓⚓✕⚓ ⚓ ⚓ ⚓(65t/16t), ⚓ wi-fi.

Town www.dartmouth-tourism.org.uk, ⚓ 🚾/⚓ ⚓ (0800-2000) ⚓⚓⊠⚓⚓✕⚓⚓ ☎555872 (steam train in season to Paignton); bus to Totnes/Paignton, ✈ Exeter.

UP-RIVER TO DITTISHAM AND TOTNES
R Dart is navigable by day (unlit above Stoney Ground) on the flood to Totnes bridge, 5·5M above Dittisham. Use AC 2253 and DHNA annual handbook. HW Totnes = HW Dartmouth +0015. Speed limit 6kn to S end of Home Reach, then 'Dead Slow' (no wash). Leave Anchor Stone to port. From Dittisham brgs of 020° / 310° on successive Boat Houses lead between Lower Back and Flat Owers; or keep E of the latter. Up-river the 11 unlit buoys are numbered 1-11 in sequence, rather than odd/even convention for lateral buoys.

Dittisham ⚓s/⚓s/**Berths:** For ⚓s above Anchor Stone, call *DartNav* Ch 11. Many ⚓s (lt blue with black V) between Anchor Stone & Dittisham; no ⚓. 4 ⚓s off Stoke Gabriel by SHM bcn, QG. 🚾/⚓ in village park to N (5mins walk) of Ferry Boat Inn.

Totnes All berths dry. W bank: Baltic Wharf ☎867922, ⚓ (⓿ pre-call) ⚓ ⚓ ⚓ ⚓(16t) ⚓(35t) ⚓ ⚓.

E bank: (**V** call DHNA) for 5⌣ at Steamer Quay, keep clear of ferry berths. Limited ⌣ on soft mud in the W Arm N of Steam Packet Inn ☎863880, ⚓⬡▣✕⬡. **Totnes** All amenities; mainline ⇌.

STANDARD TIME (UT)
For Summer Time add ONE hour in **non-shaded areas**

DARTMOUTH LAT 50°21'N LONG 3°35'W
TIMES AND HEIGHTS OF HIGH AND LOW WATERS

Dates in red are SPRINGS
Dates in blue are NEAPS

YEAR **2016**

JANUARY

#	Time m	Time m	Time m	Time m
1	0343 1.6	1003 4.3	F 1609 1.7	2230 4.0
16	0345 1.1	1024 4.6	SA 1616 1.2	◑ 2251 4.3
2	0424 1.9	1047 4.1	SA 1654 1.9	◐ 2320 3.8
17	0434 1.4	1117 4.4	SU 1710 1.4	2350 4.1
3	0516 2.1	1139 3.9	SU 1751 2.0	
18	0536 1.6	1222 4.2	M 1819 1.7	
4	0023 3.8	0621 2.2	M 1247 3.9	1858 2.0
19	0103 4.0	0658 1.8	TU 1340 4.1	1944 1.7
5	0140 3.8	0731 2.1	TU 1408 3.9	2003 1.9
20	0223 4.1	0827 1.7	W 1457 4.1	2103 1.5
6	0252 4.0	0836 1.9	W 1517 4.1	2104 1.7
21	0333 4.3	0939 1.4	TH 1604 4.3	2206 1.2
7	0351 4.2	0934 1.6	TH 1615 4.3	2158 1.4
22	0433 4.5	1037 1.1	F 1702 4.5	2300 0.9
8	0443 4.5	1025 1.3	F 1705 4.4	2248 1.1
23	0526 4.7	1127 0.8	SA 1753 4.6	2347 0.7
9	0529 4.7	1114 1.0	SA 1753 4.6	2333 0.9
24	0613 4.9	1212 0.6	SU 1838 4.7	○
10	0614 4.8	1159 0.8	SU 1838 4.7	●
25	0030 0.6	0653 4.9	M 1255 0.5	1916 4.7
11	0017 0.8	0654 4.9	M 1244 0.7	1918 4.7
26	0109 0.6	0729 4.9	TU 1331 0.6	1950 4.7
12	0101 0.7	0736 5.0	TU 1327 0.6	1959 4.8
27	0143 0.7	0801 4.9	W 1405 0.7	2020 4.6
13	0141 0.7	0815 5.0	W 1407 0.6	2039 4.7
28	0214 0.9	0831 4.8	TH 1433 0.9	2050 4.5
14	0222 0.8	0856 4.9	TH 1448 0.7	2120 4.6
29	0241 1.1	0902 4.6	F 1501 1.2	2121 4.3
15	0302 0.9	0938 4.8	F 1529 0.9	2203 4.5
30	0308 1.3	0935 4.4	SA 1527 1.4	2155 4.1
31	0334 1.6	1009 4.2	SU 1555 1.7	2234 3.8

FEBRUARY

#	Time m	Time m	Time m	Time m
1	0409 1.8	1051 4.0	M 1636 1.9	◑ 2322 3.8
16	0508 1.5	1157 4.0	TU 1743 1.7	
2	0507 2.1	1146 3.8	TU 1748 2.1	
17	0034 3.9	0621 1.8	W 1319 3.8	1910 1.9
3	0026 3.7	0634 2.1	W 1257 3.7	1911 2.1
18	0159 3.9	0806 1.8	TH 1442 3.9	2046 1.7
4	0150 3.8	0750 2.0	TH 1428 3.8	2022 1.8
19	0315 4.1	0928 1.5	F 1551 4.0	2154 1.4
5	0311 4.0	0858 1.7	F 1543 4.0	2126 1.5
20	0417 4.3	1024 1.1	SA 1649 4.3	2247 1.0
6	0412 4.3	0958 1.3	SA 1641 4.3	2223 1.2
21	0509 4.6	1113 0.8	SU 1736 4.5	2331 0.7
7	0503 4.6	1053 1.0	SU 1731 4.5	2316 0.9
22	0553 4.8	1156 0.6	M 1819 4.6	○
8	0552 4.8	1143 0.6	M 1820 4.7	●
23	0012 0.6	0632 4.9	TU 1236 0.5	1854 4.7
9	0004 0.6	0637 5.0	TU 1232 0.4	1903 4.8
24	0050 0.5	0706 4.9	W 1311 0.5	1926 4.7
10	0050 0.4	0721 5.1	W 1316 0.3	1945 4.9
25	0123 0.6	0738 4.8	TH 1341 0.6	1955 4.6
11	0132 0.3	0804 5.1	TH 1357 0.2	2027 4.9
26	0149 0.7	0807 4.7	F 1407 0.8	2024 4.6
12	0212 0.3	0845 5.1	F 1436 0.3	2106 4.8
27	0214 0.9	0837 4.6	SA 1428 1.0	2053 4.4
13	0251 0.5	0926 4.9	SA 1516 0.6	2146 4.7
28	0234 1.1	0907 4.4	SU 1447 1.2	2123 4.3
14	0330 0.8	1008 4.7	SU 1556 0.9	2229 4.4
29	0256 1.3	0937 4.2	M 1511 1.4	2154 4.1
15	0414 1.1	1055 4.3	M 1643 1.3	◑ 2321 4.1

MARCH

#	Time m	Time m	Time m	Time m
1	0325 1.6	1012 4.0	TU 1544 1.7	◐ 2235 3.9
16	0446 1.5	1136 3.9	W 1717 1.7	
2	0407 1.8	1102 3.8	W 1636 2.0	2332 3.7
17	0005 3.8	0554 1.8	TH 1302 3.7	1838 2.0
3	0525 2.1	1207 3.6	TH 1816 2.1	
18	0137 3.8	0744 1.9	F 1424 3.7	2025 1.9
4	0047 3.7	0709 2.0	F 1335 3.7	1945 1.9
19	0253 3.9	0908 1.6	SA 1532 3.9	2134 1.5
5	0221 3.9	0824 1.7	SA 1508 3.9	2057 1.6
20	0354 4.2	1005 1.2	SU 1627 4.2	2225 1.1
6	0337 4.2	0932 1.3	SU 1613 4.2	2200 1.2
21	0444 4.5	1051 0.8	M 1713 4.4	2309 0.8
7	0434 4.5	1031 0.8	M 1706 4.5	2256 0.7
22	0527 4.6	1133 0.6	TU 1752 4.6	2349 0.6
8	0526 4.8	1124 0.5	TU 1754 4.6	2346 0.4
23	0606 4.7	1210 0.5	W 1827 4.7	○
9	0616 5.0	1212 0.2	W 1844 4.9	●
24	0024 0.5	0641 4.8	TH 1245 0.5	1858 4.7
10	0033 0.1	0702 5.2	TH 1258 0.0	1927 5.0
25	0056 0.6	0712 4.8	F 1314 0.6	1929 4.7
11	0117 0.0	0746 5.2	F 1340 0.0	2008 5.1
26	0123 0.7	0743 4.7	SA 1337 0.8	1958 4.6
12	0157 0.1	0829 5.2	SA 1420 0.1	2049 5.0
27	0145 0.8	0813 4.6	SU 1357 0.9	2027 4.5
13	0235 0.3	0910 5.0	SU 1458 0.4	2128 4.8
28	0206 1.0	0842 4.4	M 1418 1.1	2054 4.4
14	0315 0.6	0951 4.6	M 1537 0.8	2206 4.5
29	0228 1.2	0910 4.2	TU 1441 1.3	2123 4.2
15	0356 1.0	1021 4.3	TU 1622 1.3	◑ 2253 4.1
30	0257 1.4	0944 4.0	W 1515 1.6	2201 4.0
31	0338 1.6	1032 3.8	TH 1602 1.8	◐ 2256 3.8

APRIL

#	Time m	Time m	Time m	Time m
1	0443 1.9	1136 3.6	F 1725 2.0	
16	0108 3.8	0705 1.9	SA 1358 3.7	1944 1.9
2	0006 3.8	0631 1.9	SA 1254 3.7	1910 1.9
17	0222 3.9	0835 1.6	SU 1502 3.9	2100 1.6
3	0132 3.9	0754 1.6	SU 1429 3.9	2028 1.6
18	0321 4.1	0933 1.3	M 1555 4.1	2153 1.3
4	0258 4.2	0904 1.2	M 1542 4.2	2134 1.1
19	0412 4.3	1020 1.0	TU 1640 4.3	2239 1.0
5	0403 4.5	1005 0.8	TU 1639 4.5	2232 0.7
20	0455 4.5	1102 0.8	W 1719 4.5	2318 0.8
6	0459 4.8	1100 0.4	W 1730 4.8	2324 0.3
21	0534 4.6	1139 0.7	TH 1756 4.6	2354 0.7
7	0551 5.1	1149 0.1	TH 1820 5.0	●
22	0612 4.7	1212 0.7	F 1831 4.7	○
8	0011 0.1	0640 5.2	F 1237 -0.1	1905 5.1
23	0025 0.7	0647 4.7	SA 1242 0.7	1903 4.7
9	0058 -0.1	0727 5.2	SA 1321 -0.1	1948 5.1
24	0054 0.7	0720 4.6	SU 1308 0.8	1935 4.6
10	0139 0.0	0810 5.1	SU 1401 0.1	2029 5.0
25	0120 0.8	0751 4.5	M 1331 0.9	2004 4.6
11	0220 0.2	0853 4.9	M 1440 0.4	2108 4.8
26	0144 1.0	0821 4.4	TU 1355 1.1	2032 4.5
12	0259 0.5	0935 4.6	TU 1520 0.8	2146 4.5
27	0211 1.1	0851 4.2	W 1424 1.3	2101 4.3
13	0341 1.0	1020 4.2	W 1603 1.3	2229 4.2
28	0242 1.3	0927 4.1	TH 1458 1.5	2140 4.2
14	0430 1.4	1120 3.8	TH 1656 1.7	◑ 2336 3.9
29	0325 1.5	1013 3.9	F 1546 1.7	2233 4.0
15	0532 1.8	1242 3.6	F 1805 2.0	
30	0426 1.7	1114 3.8	SA 1659 1.9	◐ 2338 3.9

Chart Datum is 2·62 metres below Ordnance Datum (Newlyn). HAT is 5·3 metres above Chart Datum.

STANDARD TIME (UT)
For Summer Time add ONE hour in **non-shaded areas**

DARTMOUTH   LAT 50°21'N   LONG 3°35W

TIMES AND HEIGHTS OF HIGH AND LOW WATERS

Dates in red are **SPRINGS**
Dates in blue are NEAPS

YEAR 2016

SW England

MAY

	Time	m		Time	m
1 SU	0556	1.7	**16** M	0138	3.8
	1227	3.8		0736	1.7
	1835	1.8		1420	3.8
				2003	1.8
2 M	0055	4.0	**17** TU	0239	4.0
	0721	1.5		0845	1.5
	1352	4.0		1513	4.0
	1955	1.6		2106	1.6
3 TU	0220	4.2	**18** W	0331	4.1
	0834	1.2		0937	1.3
	1507	4.2		1600	4.2
	2105	1.2		2157	1.3
4 W	0330	4.5	**19** TH	0418	4.3
	0937	0.8		1022	1.1
	1608	4.6		1643	4.4
	2205	0.8		2240	1.1
5 TH	0430	4.8	**20** F	0500	4.4
	1034	0.4		1102	0.9
	1702	4.8		1723	4.5
	2300	0.4		2318	0.9
6 F ●	0526	5.0	**21** SA ○	0542	4.5
	1126	0.2		1137	0.9
	1754	5.0		1802	4.6
	2349	0.2		2353	0.9
7 SA	0619	5.1	**22** SU	0622	4.6
	1213	0.1		1209	0.9
	1843	5.1		1839	4.7
8 SU	0037	0.1	**23** M	0025	0.8
	0706	5.1		0658	4.5
	1300	0.1		1241	0.9
	1927	5.1		1913	4.7
9 M	0122	0.1	**24** TU	0058	0.9
	0753	5.0		0734	4.5
	1342	0.2		1312	1.0
	2009	5.0		1945	4.6
10 TU	0204	0.3	**25** W	0129	0.9
	0837	4.8		0806	4.4
	1424	0.5		1341	1.0
	2050	4.8		2015	4.6
11 W	0245	0.6	**26** TH	0201	1.0
	0921	4.5		0839	4.3
	1504	0.9		1415	1.2
	2129	4.6		2049	4.5
12 TH	0327	0.9	**27** F	0237	1.1
	1004	4.2		0917	4.2
	1547	1.3		1453	1.3
	2207	4.3		2129	4.4
13 F ◑	0414	1.3	**28** SA	0321	1.3
	1058	3.9		1003	4.1
	1635	1.6		1540	1.5
	2259	4.1		2218	4.2
14 SA	0508	1.6	**29** SU ◑	0416	1.5
	1208	3.7		1057	4.0
	1733	1.9		1642	1.7
				2317	4.2
15 SU	0023	3.8	**30** M	0527	1.5
	0614	1.8		1203	3.9
	1318	3.7		1800	1.7
	1845	1.9			
			31 TU	0027	4.1
				0647	1.5
				1318	4.0
				1921	1.5

JUNE

	Time	m		Time	m
1 W	0146	4.2	**16** TH	0241	3.9
	0801	1.2		0837	1.6
	1434	4.2		1515	4.1
	2035	1.3		2102	1.6
2 TH	0300	4.4	**17** F	0337	4.1
	0908	0.9		0932	1.4
	1539	4.5		1604	4.3
	2139	0.9		2155	1.4
3 F	0404	4.6	**18** SA	0426	4.2
	1008	0.7		1018	1.2
	1637	4.7		1650	4.4
	2237	0.6		2241	1.4
4 SA	0503	4.8	**19** SU	0513	4.4
	1104	0.4		1101	1.1
	1730	4.9		1734	4.6
	2330	0.4		2322	1.0
5 SU ●	0558	4.9	**20** M ○	0557	4.5
	1154	0.3		1140	1.0
	1822	5.0		1816	4.6
6 M	0018	0.3	**21** TU	0001	0.9
	0649	4.9		0639	4.5
	1242	0.3		1219	0.9
	1908	5.0		1854	4.7
7 TU	0106	0.3	**22** W	0041	0.8
	0737	4.8		0718	4.5
	1327	0.4		1257	0.9
	1951	5.0		1930	4.7
8 W	0149	0.4	**23** TH	0118	0.8
	0821	4.7		0755	4.5
	1408	0.6		1333	0.9
	2032	4.8		2005	4.7
9 TH	0230	0.6	**24** F	0155	0.9
	0904	4.5		0831	4.4
	1448	0.8		1410	1.0
	2110	4.6		2041	4.6
10 F	0312	0.9	**25** SA	0233	0.9
	0944	4.3		0910	4.4
	1528	1.2		1449	1.1
	2145	4.4		2121	4.6
11 SA	0353	1.2	**26** SU	0316	1.0
	1025	4.0		0953	4.3
	1611	1.5		1533	1.2
	2223	4.2		2206	4.5
12 SU ◑	0437	1.5	**27** M ◑	0404	1.2
	1113	3.8		1043	4.2
	1658	1.7		1626	1.4
	2312	4.0		2259	4.3
13 M	0529	1.7	**28** TU	0502	1.3
	1213	3.7		1140	4.1
	1753	1.9		1728	1.5
14 TU	0021	3.8	**29** W	0001	4.2
	0629	1.8		0611	1.4
	1319	3.8		1248	4.1
	1856	1.9		1846	1.6
15 W	0138	3.8	**30** TH	0116	4.2
	0734	1.6		0729	1.4
	1420	3.9		1404	4.2
	2002	1.8		2005	1.4

JULY

	Time	m		Time	m
1 F	0235	4.3	**16** SA	0252	3.9
	0842	1.2		0841	1.7
	1513	4.4		1526	4.1
	2116	1.2		2110	1.6
2 SA	0344	4.4	**17** SU	0353	4.1
	0948	1.0		0938	1.4
	1616	4.6		1620	4.3
	2218	0.9		2205	1.3
3 SU	0446	4.6	**18** M	0446	4.2
	1046	0.7		1029	1.2
	1712	4.8		1708	4.5
	2314	0.6		2255	1.1
4 M ●	0542	4.7	**19** TU ○	0534	4.4
	1138	0.6		1116	1.0
	1803	4.9		1753	4.7
				2340	0.9
5 TU	0004	0.5	**20** W	0620	4.5
	0634	4.7		1200	0.9
	1226	0.5		1836	4.8
	1851	5.0			
6 W	0052	0.4	**21** TH	0024	0.7
	0720	4.7		0701	4.6
	1312	0.5		1244	0.8
	1934	4.9		1915	4.8
7 TH	0135	0.4	**22** F	0107	0.6
	0803	4.6		0742	4.6
	1352	0.6		1324	0.7
	2012	4.9		1953	4.9
8 F	0215	0.6	**23** SA	0147	0.6
	0842	4.5		0821	4.6
	1429	0.8		1403	0.7
	2047	4.7		2032	4.8
9 SA	0251	0.8	**24** SU	0226	0.6
	0916	4.4		0900	4.6
	1505	1.0		1441	0.8
	2118	4.5		2112	4.8
10 SU	0327	1.1	**25** M	0306	0.8
	0948	4.2		0941	4.5
	1540	1.3		1522	1.0
	2150	4.3		2154	4.6
11 M	0402	1.4	**26** TU	0348	1.0
	1023	4.0		1025	4.3
	1619	1.6		1607	1.2
	2228	4.1		2242	4.4
12 TU ◑	0442	1.6	**27** W	0437	1.2
	1106	3.9		1118	4.2
	1704	1.8		1702	1.4
	2315	3.9		2339	4.2
13 W	0532	1.8	**28** TH	0538	1.5
	1201	3.8		1222	4.0
	1800	2.0		1813	1.6
14 TH	0015	3.8	**29** F	0052	4.0
	0634	1.9		0658	1.6
	1311	3.8		1340	4.0
	1906	2.0		1940	1.6
15 F	0135	3.8	**30** SA	0216	4.0
	0738	1.8		0821	1.5
	1424	3.9		1455	4.2
	2010	1.8		2101	1.4
			31 SU	0330	4.2
				0934	1.3
				1600	4.4
				2207	1.1

AUGUST

	Time	m		Time	m
1 M	0433	4.4	**16** TU	0421	4.2
	1034	1.0		1001	1.3
	1657	4.6		1642	4.5
	2303	0.8		2230	1.1
2 TU ●	0529	4.5	**17** W	0511	4.4
	1125	0.7		1054	1.0
	1748	4.8		1729	4.7
	2351	0.5		2320	0.8
3 W	0619	4.6	**18** TH	0558	4.6
	1211	0.5		1142	0.7
	1834	4.9		1815	4.9
				○	
4 TH	0037	0.4	**19** F	0006	0.5
	0701	4.7		0643	4.7
	1255	0.5		1227	0.6
	1913	4.9		1856	5.0
5 F	0118	0.4	**20** SA	0052	0.4
	0740	4.7		0725	4.8
	1332	0.5		1310	0.4
	1948	4.9		1938	5.1
6 SA	0154	0.5	**21** SU	0132	0.3
	0813	4.6		0805	4.8
	1407	0.7		1350	0.4
	2019	4.8		2018	5.0
7 SU	0227	0.7	**22** M	0212	0.4
	0843	4.5		0845	4.8
	1437	0.9		1428	0.5
	2049	4.6		2059	4.9
8 M	0255	1.0	**23** TU	0250	0.6
	0913	4.3		0925	4.7
	1506	1.2		1507	0.8
	2119	4.4		2139	4.7
9 TU	0322	1.3	**24** W	0329	0.9
	0945	4.2		1005	4.5
	1534	1.5		1548	1.1
	2153	4.2		2224	4.4
10 W ◑	0350	1.6	**25** TH ◑	0415	1.2
	1022	4.0		1054	4.2
	1607	1.8		1639	1.4
	2231	4.0		2319	4.1
11 TH	0426	1.8	**26** F	0512	1.6
	1107	3.8		1159	4.0
	1657	2.0		1747	1.8
	2321	3.8			
12 F	0529	2.0	**27** SA	0036	3.9
	1206	3.7		0632	1.8
	1813	2.1		1321	3.9
				1925	1.8
13 SA	0027	3.7	**28** SU	0207	3.8
	0649	2.1		0809	1.8
	1325	3.7		1442	4.1
	1927	2.0		2053	1.6
14 SU	0203	3.7	**29** M	0322	4.0
	0759	1.9		0924	1.5
	1448	3.9		1548	4.3
	2035	1.8		2157	1.2
15 M	0322	3.9	**30** TU	0424	4.3
	0904	1.6		1021	1.1
	1550	4.2		1643	4.6
	2136	1.4		2249	0.8
			31 W	0515	4.5
				1110	0.8
				1730	4.8
				2335	0.6

Chart Datum is 2·62 metres below Ordnance Datum (Newlyn). HAT is 5·3 metres above Chart Datum.

》》 FREE monthly updates. Register at 《
www.reedsnauticalalmanac.co.uk 《

171

DARTMOUTH LAT 50°21'N LONG 3°35'W
TIMES AND HEIGHTS OF HIGH AND LOW WATERS

STANDARD TIME (UT)
For Summer Time add ONE hour in **non-shaded areas**

Dates in red are SPRINGS
Dates in blue are NEAPS

YEAR 2016

SEPTEMBER

Time	m		Time	m
1 0559	4.6	**16**	0533	4.7
1153	0.6		1120	0.6
TH 1813	4.9		F 1750	5.0
●			○ 2345	0.4
2 0016	0.4	**17**	0620	4.9
0639	4.7		1207	0.4
F 1234	0.5		SA 1836	5.2
1849	4.9			
3 0055	0.4	**18**	0031	0.2
0712	4.7		0702	5.0
SA 1309	0.5		SU 1252	0.3
1922	4.9		1919	5.0
4 0129	0.5	**19**	0114	0.2
0742	4.7		0744	5.1
SU 1340	0.7		M 1332	0.3
1951	4.8		2002	5.2
5 0156	0.7	**20**	0154	0.3
0811	4.6		0825	5.0
M 1406	0.9		TU 1412	0.4
2020	4.7		2043	5.0
6 0220	1.0	**21**	0232	0.5
0841	4.5		0905	4.9
TU 1429	1.1		W 1451	0.7
2051	4.5		2124	4.8
7 0239	1.2	**22**	0312	0.9
0911	4.3		0946	4.6
W 1450	1.4		TH 1532	1.1
2121	4.3		2207	4.4
8 0259	1.5	**23**	0355	1.3
0944	4.1		1033	4.3
TH 1513	1.7		F 1622	1.5
2155	4.0		◔ 2302	4.0
9 0327	1.8	**24**	0450	1.7
1024	3.9		1137	4.0
F 1548	1.9		SA 1728	1.9
◑ 2241	3.8			
10 0412	2.0	**25**	0025	3.8
1117	3.8		0608	2.0
SA 1703	2.2		SU 1305	3.9
2341	3.6		1911	2.0
11 0553	2.2	**26**	0156	3.8
1227	3.7		0756	2.0
SU 1849	2.2		M 1425	4.0
			2040	1.7
12 0105	3.6	**27**	0310	4.0
0724	2.1		0909	1.6
M 1400	3.9		TU 1530	4.3
2003	1.9		2140	1.3
13 0249	3.9	**28**	0408	4.3
0835	1.8		1002	1.2
TU 1517	4.2		W 1623	4.6
2107	1.5		2229	0.9
14 0353	4.2	**29**	0454	4.5
0936	1.4		1049	0.9
W 1613	4.5		TH 1708	4.8
2205	1.1		2312	0.7
15 0445	4.5	**30**	0534	4.7
1031	1.0		1129	0.7
TH 1702	4.8		F 1747	4.9
2257	0.7		2351	0.6

OCTOBER

Time	m		Time	m
1 0612	4.8	**16**	0553	5.0
1207	0.6		1144	0.4
SA 1823	4.9		SU 1813	5.2
●			○	
2 0026	0.6	**17**	0007	0.2
0643	4.8		0639	5.2
SU 1242	0.6		M 1231	0.2
1854	4.9		1858	5.3
3 0059	0.7	**18**	0053	0.2
0713	4.8		0723	5.2
M 1311	0.8		TU 1314	0.2
1925	4.8		1943	5.2
4 0124	0.8	**19**	0134	0.3
0742	4.7		0806	5.2
TU 1335	0.9		W 1355	0.4
1955	4.7		2026	5.0
5 0145	1.0	**20**	0215	0.5
0812	4.6		0847	5.0
W 1356	1.1		TH 1436	0.7
2024	4.5		2109	4.8
6 0204	1.2	**21**	0255	0.9
0842	4.5		0929	4.7
TH 1417	1.4		F 1519	1.1
2053	4.3		2155	4.4
7 0224	1.5	**22**	0339	1.4
0913	4.3		1014	4.4
F 1440	1.6		SA 1608	1.5
2127	4.1		◔ 2249	4.0
8 0252	1.7	**23**	0432	1.8
0949	4.1		1117	4.1
SA 1515	1.8		SU 1711	1.9
2210	3.9			
9 0334	2.0	**24**	0010	3.8
1040	3.9		0543	2.1
SU 1611	2.1		M 1242	4.0
◑ 2309	3.7		1843	2.0
10 0446	2.3	**25**	0134	3.8
1145	3.8		0723	2.1
M 1806	2.2		TU 1358	4.0
			2010	1.8
11 0024	3.7	**26**	0243	4.0
0646	2.2		0839	1.8
TU 1306	3.9		W 1501	4.2
1930	1.9		2111	1.5
12 0205	3.9	**27**	0339	4.2
0803	1.9		0934	1.4
W 1435	4.2		TH 1553	4.5
2038	1.5		2159	1.2
13 0320	4.2	**28**	0425	4.5
0908	1.4		1019	1.1
TH 1539	4.5		F 1638	4.6
2138	1.1		2242	0.9
14 0415	4.6	**29**	0504	4.7
1004	1.0		1100	0.9
F 1632	4.8		SA 1718	4.8
2232	0.7		2321	0.8
15 0505	4.8	**30**	0541	4.8
1057	0.6		1137	0.8
SA 1723	5.1		SU 1754	4.8
2321	0.4		● 2355	0.8
		31	0615	4.9
			1211	0.8
			M 1829	4.8

NOVEMBER

Time	m		Time	m
1 0026	0.8	**16**	0033	0.3
0647	4.9		0703	5.3
TU 1241	0.9		W 1257	0.3
1901	4.8		1927	5.2
2 0053	0.9	**17**	0118	0.4
0719	4.8		0748	5.2
W 1308	1.0		TH 1341	0.4
1934	4.7		2012	5.0
3 0116	1.1	**18**	0200	0.6
0749	4.7		0831	5.1
TH 1331	1.2		F 1424	0.7
2004	4.5		2057	4.7
4 0138	1.3	**19**	0241	0.9
0819	4.6		0914	4.8
F 1355	1.3		SA 1508	1.0
2034	4.4		2142	4.4
5 0203	1.4	**20**	0326	1.3
0850	4.4		0959	4.5
SA 1423	1.5		SU 1554	1.4
2106	4.2		2234	4.1
6 0234	1.6	**21**	0414	1.7
0926	4.3		1051	4.3
SU 1459	1.7		M 1648	1.7
2149	4.0		◑ 2340	3.9
7 0316	1.9	**22**	0512	2.0
1013	4.1		1202	4.1
M 1552	2.0		TU 1755	1.9
◑ 2245	3.9			
8 0418	2.1	**23**	0054	3.8
1114	4.0		0626	2.1
TU 1720	2.1		W 1316	4.0
2353	3.8		1917	1.9
9 0558	2.2	**24**	0202	3.9
1225	4.0		0747	2.0
W 1853	1.9		TH 1420	4.1
			2027	1.7
10 0114	3.9	**25**	0258	4.1
0726	1.9		0852	1.7
TH 1346	4.2		F 1515	4.3
2004	1.6		2120	1.5
11 0238	4.2	**26**	0347	4.3
0836	1.5		0943	1.5
F 1500	4.5		SA 1602	4.4
2107	1.2		2206	1.3
12 0342	4.6	**27**	0430	4.5
0937	1.1		1026	1.2
SA 1601	4.8		SU 1646	4.6
2204	0.8		2247	1.1
13 0437	4.8	**28**	0510	4.7
1032	0.7		1106	1.1
SU 1657	5.0		M 1726	4.7
2257	0.5		2323	1.0
14 0527	5.1	**29**	0548	4.8
1122	0.5		1141	1.0
M 1750	5.2		TU 1805	4.7
○ 2345	0.3		● 2356	1.0
15 0617	5.2	**30**	0625	4.8
1210	0.3		1213	1.0
TU 1840	5.2		W 1842	4.7

DECEMBER

Time	m		Time	m
1 0025	1.0	**16**	0103	0.4
0659	4.8		0733	5.2
TH 1245	1.1		F 1329	0.4
1916	4.6		2000	4.9
2 0055	1.1	**17**	0146	0.6
0733	4.8		0817	5.1
F 1315	1.1		SA 1413	0.6
1949	4.5		2044	4.7
3 0124	1.2	**18**	0228	0.8
0804	4.7		0859	4.9
SA 1345	1.2		SU 1454	0.9
2021	4.4		2127	4.5
4 0153	1.3	**19**	0309	1.1
0835	4.6		0939	4.7
SU 1417	1.4		M 1536	1.2
2054	4.3		2208	4.3
5 0228	1.5	**20**	0351	1.4
0911	4.5		1019	4.4
M 1454	1.5		TU 1620	1.5
2135	4.2		2252	4.0
6 0308	1.7	**21**	0436	1.8
0955	4.3		1105	4.2
TU 1542	1.7		W 1709	1.8
2225	4.0		◑ 2348	3.9
7 0401	1.8	**22**	0529	2.0
1049	4.2		1205	4.0
W 1645	1.8		TH 1807	1.9
◑ 2326	4.0			
8 0513	2.0	**23**	0056	3.8
1153	4.2		0634	2.1
TH 1806	1.8		F 1318	3.9
			1916	2.0
9 0036	4.0	**24**	0204	3.9
0641	1.9		0746	2.0
F 1305	4.2		SA 1424	4.0
1926	1.6		2024	1.8
10 0154	4.2	**25**	0301	4.1
0800	1.6		0852	1.8
SA 1423	4.4		SU 1522	4.1
2036	1.3		2121	1.6
11 0307	4.4	**26**	0353	4.3
0908	1.3		0946	1.6
SU 1532	4.6		M 1613	4.3
2139	1.0		2209	1.4
12 0409	4.7	**27**	0440	4.5
1008	1.0		1032	1.4
M 1634	4.8		TU 1659	4.4
2235	0.7		2251	1.2
13 0505	5.0	**28**	0523	4.6
1103	0.7		1113	1.2
TU 1730	5.0		W 1744	4.5
2327	0.5		2329	1.1
14 0557	5.1	**29**	0604	4.8
1154	0.5		1151	1.1
W 1824	5.0		TH 1825	4.6
○			●	
15 0016	0.4	**30**	0005	1.0
0647	5.2		0643	4.8
TH 1244	0.4		F 1227	1.0
1913	5.0		1903	4.6
		31	0041	1.0
			0719	4.8
			SA 1304	1.0
			1939	4.6

Chart Datum is 2·62 metres below Ordnance Datum (Newlyn). HAT is 5·3 metres above Chart Datum.

》》FREE monthly updates. Register at 《
www.reedsnauticalalmanac.co.uk

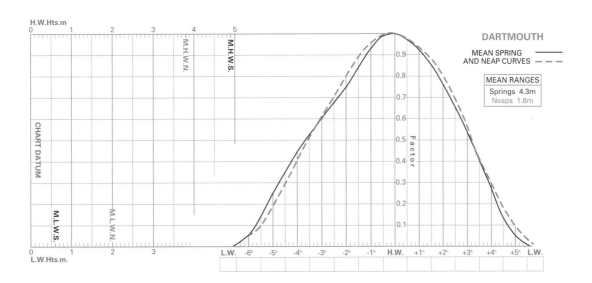

DARTMOUTH

MEAN SPRING ——————
AND NEAP CURVES – – – – –

| MEAN RANGES |
| Springs 4.3m |
| Neaps 1.8m |

9.1.20 BRIXHAM

Devon 50°24′·31N 03°30′·85W
✿✿✿⚓⚓⚓✿✿

CHARTS AC *3315*, 1613, 1634, 5602, 26;
Imray C10, C5, 2300

TIDES –0505 Dover; ML 2·9; Duration 0635.

Use **TORQUAY** Differences on **PLYMOUTH**.

SHELTER Very good in marina; good at YC pontoon in SW of hbr, but outer hbr is dangerous in strong NW'lies. ⚓ NW of fairway or in Fishcombe and Elberry Coves W of the hbr; beware water skiers.

NAVIGATION WPT 50°24′·70N 03° 31′·09W, (3ca off chartlet), 159°/0·4M to harbour entrance. Beware fish farm 5Ca WNW of harbour; easy access H24. Inshore around Torbay controlled areas (May-Sep, mainly for swimmers) are marked by unlit Y SPM buoys; boats may enter with caution, speed limit 5kn.

LIGHTS AND MARKS Berry Hd, a conspic headland and good radar return, is 1·25M ESE of entrance. The fairway is marked by two pairs of lit lateral buoys. See chartlet and 9.1.4 for lt details.

COMMUNICATIONS (Code 01803) CGOC (01326) 317575; Police 101; Dr 855897; ⊞ 882153. HM 853321; Pilot 882214; Ch 14 (May-Sep 0800-1800; Oct-Apr 0900-1700, Mon-Fri). Marina Ch 80.

Water Taxi: *Shuttle* Ch M.

FACILITIES Brixham Marina
brixham@mdlmarinas.co.uk ☎882929, ⌿
500⌒ inc ⓥ £3·40<12·5m>£4·20<18m, Short stay<4 hrs £8·00.
Events pontoon (no ⚡) £2·32 < 12·5m>£2·90.
⚓(0900-1800, Apr-Oct) ▣ ▭(50t).

Brixham YC ☎853332, ⓥ pontoon, ⚓⌿⌿⚓✕▭.
Hbr Office ⚓ each side of fairway, ⌿⌿⚓⚓▭(4t).
Town pontoon ⌒ £1·90, ⚡ no WC/⚓.
Town ⚓✉Ⓑ⚓⚡Ⓔ⚓ ACA ✕▭ bus to Paignton, ⚑ ✈ Exeter.

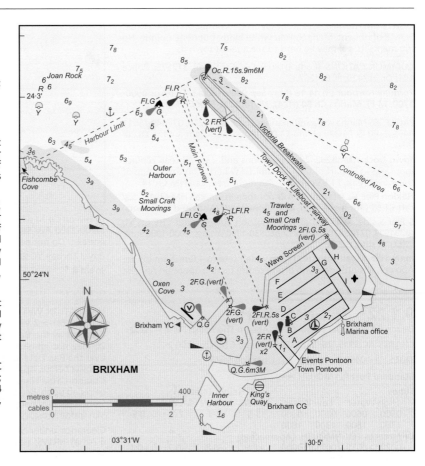

MINOR HARBOUR 2·3M NW OF BRIXHAM

PAIGNTON, Devon, **50°25′·96N 03°33′·36W** 500m S of Paignton pier. AC 1613, 26. HW +0035 and –0·6m on Plymouth; Duration 0640; ML 2·9m. Hbr dries 1·3m, only suitable for max LOA 8·2m. A heavy swell enters in E'lies. Drying rks extend 180m ENE from E pier, ⚡ QR, to Black Rk, ECM twr ⚡ Q (3) 10s 5m 3M. In the approach chan **keep to port.** HM (summer only) ☎(01803) 557812, VHF Ch 14.
Paignton SC ☎525817. Facilities: ⚓⚓✕⚓⚓ ACA.

9.1.21 TORQUAY

Devon 50°27'·45N 03°31'·73W ✿✿✿❀♨♨♨✿✿

CHARTS AC 3315, 1613, 5602, 26; Imray C5

TIDES –0500 Dover; ML 2·9; Duration 0640

Standard Port PLYMOUTH (←—)

Times				Height (metres)			
High Water		Low Water		MHWS	MHWN	MLWN	MLWS
0100	0600	0100	0600	5·5	4·4	2·2	0·8
1300	1800	1300	1800				
Differences TORQUAY							
+0025	+0045	+0010	0000	–0·6	–0·7	–0·2	–0·1

NOTE: There is often a stand of about 1 hour at HW

SHELTER Good, but some swell in hbr with strong SE'lies. ⚓s NW of Hope's Nose at Hope Cove, Anstey's Cove & Babbacombe Bay are well sheltered in W'lies. No ⚓ in hbr.

NAVIGATION WPT 50°27'·03N 03°31'·57W, 339°/0·40M to round SHM buoy 80m WSW of Haldon pier. Semi-blind ent; keep a good lookout. Access at all tides, but in strong SE winds backwash may make the narrow ent difficult. Hbr speed limit 5 knots.

Controlled areas (mainly for swimmers, May-Sep) close inshore around Torbay are marked by unlit Y SPM buoys; boats may enter with caution, speed limit 5 knots. Watch out for canoeists.

LIGHTS AND MARKS Ore Stone (32m rk) is conspic off Hope's Nose, 2·3M E of hbr ent. Many conspic white bldgs, but none unique. No ldg marks/lts. Lts may be hard to see against town lts.

COMMUNICATIONS (Code 01803) CGOC (01326) 317575; Police 101; Dr 212429; Ⓗ 614567.

Torquay Harbour Ch **14** 16 (May-Sep 0800-1800LT; Oct-Apr 0900-1700, M-F). Marina Ch **80** (H24), M. *Torquay Fuel* Ch M.

FACILITIES Marina www.mdlmarinas.co.uk ☎200210, 07764 175611 440◻+60 ❤ £3.32<12·5m>£4.09<18m>£4.60<24m; <4 hrs £8.00; ◻ ♿.

Town Dock HM ☎292429. 4 pontoons and a wavebreak replace moorings in SE part of the hbr. Limited ❤ berths on the W side of the wavebreak, £1.55/m/night, 4th night free. Call HM Ch 14.

Haldon Pier ◣ at root, 96m ❤ pontoon in 2m, call HM Ch 14: ◻/ raft £1.55 (4th night free), ⚓ ⬚▷(long lead). Also used for events.

S Pier S side: 🛢🛟 LPG at fuel pontoon, *Riviera Fuel* VHF Ch M; ☎294509, 07786 370324 (Apr-Sept 0830-1900 W/days, 1000-1900 Sun); ◻(6t). N side: ◻⚓.

Inner Hbr A wall with cill (11·6m wide), retains 1·0m–2·8m inside. Above it a footbridge with lifting centre section opens on request

to HM on *Ch 14* when the flapgate is down, approx HW –3½ to +3 (0700-2300 Apr-Sep, 0900-1700 Oct-Mar). IPTS sigs 2 & 4 control ent/exit. Do not pass under the bridge when it is down. ◻, ⚓.

Royal Torbay YC ☎292006, ✕ 🍴.

Town 🛢✕🔧 📱Ⓔ⛽🛢 ACA ✉ Ⓑ 🛒 ◻ ✕ 🍴 @ ⇌ ✈(Exeter).

9.1.22 TEIGNMOUTH

Devon 50°32'·37N 03°30'·00W (Abeam The Point) ✿✿♨♨✿✿

CHARTS AC 3315, 26, 5601, 5602; Imray C10, C5, 2300

TIDES –0450 Dover; ML 2·7; Duration 0625

Standard Port PLYMOUTH (←—)

Times				Height (metres)			
High Water		Low Water		MHWS	MHWN	MLWN	MLWS
0100	0600	0100	0600	5·5	4·4	2·2	0·8
1300	1800	1300	1800				
Differences TEIGNMOUTH (Approaches)							
+0020	+0050	+0025	0000	–0·9	–0·8	–0·2	–0·1
TEIGNMOUTH (New Quay)							
+0025	+0055	+0040	+0005	–0·8	–0·8	–0·2	+0·1

Tidal streams. At the hbr ent the flood starts HW Plymouth –5½ and the ebb at HWP +¾. Be aware that from about HW –3, as The Salty covers, the flood sets NW across it and a small back eddy flows S past the ❤ pontoons. HW slack occurs at approx HWP +½.

SHELTER Good, but entry hazardous in strong onshore winds (NE-S) when surf forms on the bar.

NAVIGATION WPT 50°32'·37N 03°29'·15W, 265°/5ca to trng wall lt (Oc R 6s) ≠ 2 white ◻s on seawall beyond. The E-W chan is dredged as the Bar, Pole and Spratt Sands shift frequently; AC 26 may not immediately show such shifts. Depths reduce to CD over the Bar. Detailed pilotage and hbr notes are in the excellent website below. Appr chan is not well buoyed, contact HM for latest situation. Small buoys laid by Pilots near the appr chan, should not be relied upon. Beware rks off The Ness; and variable extent of The Salty, a bank of hard gravel/sand. Max speed 6kn over the ground; 5kn inside Y buoys off The Ness & Teignmouth beaches.

Clearance under Shaldon bridge is 2·9m at MHWS and approx 6·7m at MLWS; near its N end there is a 9m wide drawbridge section. Avoid a Historic wreck site (50°32'·95N 03°29'·24W; just off chartlet), close inshore ENE of Ch twr.

LIGHTS AND MARKS The Ness, a 50m high red sandstone headland, and church tower are both conspic from afar. Close NE of the latter, just off N edge of chartlet, Teign Corinthian YC bldg (cream colour) is also conspic. Ignore two FR (NNE of The Point); they are not ldg lts. N of The Point, two F Bu lts align 023°, but are not ldg lts. A Y can buoy, Fl Y 5s, at 50°31'·97N 03°27'·78W marks the seaward end of outfall, 103° The Ness 1·3M.

COMMUNICATIONS (Code 01626) CGOC (01326) 317575; Police

101; Dr 774355; Ⓗ 772161. HM 773165 (or 07796 178456 in emergency or out of hrs).

Hbr Ch **12** 16 (Mon-Fri 0900-1400).

FACILITIES www.teignmouth-harbour. com; info@teignmouth-harbour.com. In suitable weather up to 10 boats can raft on 2 detached 20m Ⓥ pontoons (Jubilee & Trafalgar) at 50°32'·62N 03°29'·96W in 2·5m; £1.10/m. Third pontoon (Newfoundland) is for local boats only. No ⚓ in hbr due to many moorings and strong tidal streams. In fair weather ⚓ approx 1·5ca SE of The Ness or Teignmouth Pier.

E Quay Polly Steps ⚓ (up to 10m). **Teign Corinthian YC** ☎772734, Ⓓ£8 Ⓓ.

Services ⚓ ⚓ ⚓ ⚒ ⚒ ⚓ Ⓔ ⚓ ⚓(8t) ⚓.

Town ⚓ ⚓ (1M), ⚓ ⚓ ⚓ Ⓑ ⚓ ⚓ ✕ ⚓ ⚓ ✈(Exeter).

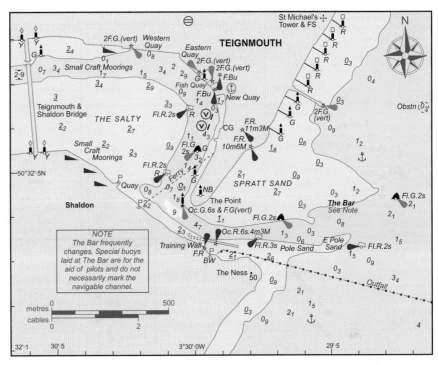

LYME BAY: TEIGNMOUTH TO PORTLAND BILL

Lyme Bay curves 65M in a great arc N, E and SE from Start Pt to Portland Bill (AC 3315).

▶*Tides are weak, rarely more than 0·75kn. From west to east the tidal curve becomes progressively more distorted, especially on the rising tide. The rise is relatively fast for the 1st hr after LW; then slackens noticeably for the next 1½ hrs, before resuming the rapid rate of rise. There is often a stand at HW, not very noticeable at Start Pt but lasting about 1½ hrs at Lyme Regis.*◀

From Teignmouth to Dawlish rocks extend 1ca offshore. Beware Dawlish Rock (2·1m) about 5ca off North end of town. Pole Sand (dries up to 3·5m) and Warren Sand on the West side of the **River Exe** estuary are liable to shift. A firing range at Straight Point has a danger zone extending 1·5M east marked by 2 Y lit DZ buoys.

3½ M SE of Sidmouth on the 20m contour a mussel farm is marked by 2 pairs of SPM (Fl Y 5s). Another mussel farm, similarly marked lies 4¾ M further S in 25m of water.

Between Torbay and Portland there is no harbour accessible in strong onshore winds, and yachtsmen must take care not to be caught on a lee shore. Marked by 4 cardinal buoys, the wreck of the *EMSSTROM* (16·5m) lies 2½M E of Hope's Nose. There are no dangers offshore. In offshore winds there is a good ⚓ NE of Beer Hd, the western-most chalk cliff in England. Golden Cap (186m and conspic) is 3·5M E of **Lyme Regis**. High Ground (3m) and Pollock (4m) are rocky patches 7ca offshore, respectively 1·5M W and 6ca SW of **West Bay** (Bridport).

From 6M SE of Bridport Chesil Beach runs SE for about 9M to the N end of the Isle of Portland. From a distance this peninsula looks like an island, with its distinctive wedge-shaped profile sloping down from 145m at the N to sea level at the Bill. Although mostly steep-to, an 18m high, white obelisk on the tip of the Bill warns of a rocky ledge extending about 50m S of the Bill. If the highest window on the lighthouse is seen above the top of the obelisk, you are clear of the outermost rock, but still very close inshore.

If heading up-Channel from Start Point, Dartmouth or Torbay time your departure to pass Portland Bill with a fair tide at all costs, especially at springs. If late, a temporary ⚓ can be found close inshore at Chesil Cove, abeam the highest part of Portland.

9.1.23 RIVER EXE

Devon 50°36'·94N 03°25'·40W (Abeam Exmouth) ❀❀⚓⚓✿✿

CHARTS AC 3315, 2290, 5601; Imray C10, C5, 2300

TIDES –0445 Dover; ML 2·1; Duration 0625

Standard Port PLYMOUTH (←)

Times				Height (metres)			
High Water		Low Water		MHWS	MHWN	MLWN	MLWS
0100	0600	0100	0600	5·5	4·4	2·2	0·8
1300	1800	1300	1800				
Differences EXMOUTH (Approaches)							
+0030	+0050	+0015	+0005	–0·9	–1·0	–0·5	–0·3
EXMOUTH DOCK							
+0035	+0055	+0050	+0020	–1·5	–1·6	–0·9	–0·5
STARCROSS							
+0040	+0100	+0055	+0025	–1·4	–1·5	–0·8	–0·1
TURF LOCK							
+0045	+0100	+0034	ND	–1·6	–1·6	–1·2	–0·4
TOPSHAM							
+0045	+0105	ND	ND	–1·5	–1·6	ND	ND

NOTE: In the appr chan the sp ebb reaches 3·3kn; with wind against tide a confused, breaking sea quickly builds. Off Warren Pt the flood stream runs at 3-4kn and the ebb can exceed 4½kn when the banks uncover.

SHELTER Good upstream of No 13 buoy and in Dock.

NAVIGATION WPT 50°35'·92N 03°23'·75W, Exe SWM buoy, 335°/6ca to No 7 SHM buoy. Up to date AC 2290 is essential. Leave WPT at approx LW+2 (or when there is sufficient rise of the tide) when hazards can be seen and some shelter obtained, or at HW–1 or HW–2 to reach Exmouth or Topsham, respectively, at slack water.

The approach channel, with a least depth of 0·4m, is well marked with PHM/SHM buoys, some of which are lit. Night entry is not advised; the alignment of the ldg lts is questionable. After No 10 PHM buoy do not cut the corner round Warren Point; turn to the SW after passing No 12 PHM buoy. Beware of the flood stream sweeping NW past the marina entrance into moorings and shallow banks.

The estuary bed is sand/mud, free of rocks. Follow the curve of the channel rather than straight lines between the buoys; some bends are marked on the outside only. *Continued overleaf*

The estuary is an international conservation area.

LIGHTS AND MARKS See chartlet and 9.1.4. A caravan site is conspic close W of Straight Pt with red cliffs to W and NNE. Exmouth ⊕ tr and FS are conspic. At Exmouth the 305° ldg bcns are hard to see by day. Following buoys in the upper reaches are lit:

No 22 PHM: Ting Tong   – Fl R 2s.
No 24 PHM: Exe   – Fl R 4s.
No 41 SHM: Turf Green   – Fl G 2s.
No 45 SHM: Topsham reach   – Fl G 4s.

A jetski/PWC area is marked by yellow buoys to NE of SWM.

Caution: Firing range N of Straight Pt has a danger area to ESE, marked by 2 DZ SPM lt buoys (not the sewer outfall SPM buoy, Fl Y 5s, shown on chartlet). R flags are flown when the range is active (likely times 0800-1600, Mon-Fri); call *Straight Pt Range* VHF Ch 08 16. From the E, check also with safety launch.

COMMUNICATIONS (Code 01395) CGOC (01326) 317575; Police 101; Dr 273001; ⊞ 279684. Council River Team (01392) 274306; Port of Exeter Ch 12 16 (Mon-Fri: 0730-1730LT). Exmouth Marina/ bridge Ch 14. *Water taxi* Ch M.

FACILITIES

EXMOUTH Marina, ☎269314, mostly berth holders, but at least 4 visitors berths in 2.0m depth. Pontoon in ent chan is used by ferries/water taxi and for 🛢. Call Ch 14 for berth availability. Strong cross tides at ent. Footbridge lifts on request Ch 14 0800-1700; stays open HN. 172◯ inc some ◍ in 2m £3·80 (15t)

Exe SC ☎264607, ✕ ◻.

Town 🛢🛒🏚🛠📠 Ⓔ ⚠ 🗓 ACA ✉ Ⓑ 🛒 🔲 ✕ ◻ 🚆. ⚓£12·00, lie near No 15 SHM buoy, off Turf Lock and Topsham SC; call water taxi Ch M ☎07970 918418 for advice.

STARCROSS (01626) Village 🛒 ◻ ✉ Ⓑ 🚆. **Starcross Garage** ☎890225, 🛢 🛒 🛠 Gas. **Starcross Fishing & Cruising Club** 2◍ (☎891996 (evenings) for availability). **Starcross YC** ☎890470.

TURF LOCK, 50°39'·88N 03°28'·06W, operates daily in working hours by prior arrangement with the lock keeper ☎274306.

Exeter Ship Canal (3·0m depth). Contact HM ☎ (below) or Ch 12 for non-tidal pontoon berths for ◍ in Turf Basin, £15·00 for min of 2 days (also laying up).

TOPSHAM (01392) Here the river carries about 0·4m. Options are ◯ (pontoon in ¾m at LW) at Trouts BY or ⚓ in appr's, dry out at Town Quay in mud on rough wall (☎07801 203338 – £9·60 /day) or find a mooring.

Topsham SC ☎877524, 2◍s ◣ ⚓ 🛢 ◻.

Trouts BY www.trouts boatyard.co.uk ☎873044. MLWS max depth 0·75m; 🛢 ◯ £16·50<6m>£22 (+ £20 key deposit); short stay £12. 🛢🛢🛢🛠🔲🗓.

Retreat BY www.retreatboatyard.co.uk ☎874720, appr channel dries approx 1m, 🛢 poss ◯🛢🛢🛠🔲🗓(10½t) 🗓.

Town 🛢⚠🗓 ACA ✉ Ⓑ 🛒 🔲 ✕ ◻ 🚆.

EXETER www.exeter.gov.uk river.canal@exeter.gov.uk Waterways, Civic Centre, Paris St, Exeter EX1 1JN; hours 0730-1530 M-F, Sat/Sun as required. The HM is responsible for the whole River Exe and Exeter Canal. www.exe-estuary.org makes interesting reading. The City Basin accepts visitors and lay-ups by prior arrangement.

City all facilities; 🛠 🗓 ACA @ 🚆 ✈.

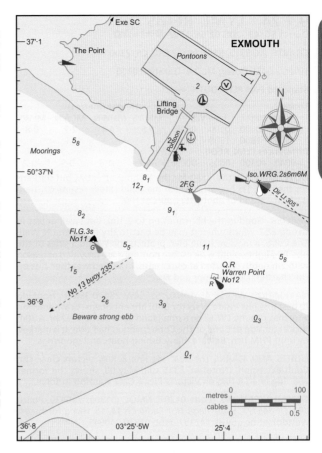

TWO MINOR HARBOURS WEST OF LYME REGIS

BEER, Devon, **50°41'·64N 03°05'·22W**. Beer Roads is an ⚓ in 2-5m on sand, shingle and gravel sheltered by Beer Head from prevailing W'lies, but open to E-SW winds. Beer Head is 142m high, topped by a caravan site. Its chalk cliffs are white, becoming red sandstone to the E towards Seaton. A conspic radio/tel mast is above the centre of the beach. Church tower is conspic. Do not use the historic approach of the beach light ≠ church tower light. All beach lights are purely for illumination, not for navigation. Land on the open beach, as in centuries past; some traditional local boats are still hauled up the beach on skids.

Beer & Seaton are adjacent towns N and NNE of Beer Head. **Beer** 🗓 ✉ 🛒 ✕ ◻. **Seaton** 🛢 🛒 shops ✕ ◻.

2·5M WSW of Beer Head and 8 cables offshore are the remains of the *MSC Napoli* wreck (50°40'·38N 03°09'·58W), marked by E, S and W cardinal buoys and a No Entry zone, radius 3 cables.

AXMOUTH, Devon, **50°42'·13N 03°03'·29W**, 1·85M ENE of Beer Head. AC 3315. HW –0455 on Dover; +0045 and –1·1m on Plymouth; ML 2·3m; Duration 0640. MHWS 4·3m, MHWN 3·2m.

A small drying hbr at the mouth of the R Axe for boats max draught 1·2m, LOA 8·5m (longer by prior arrangement), able to dry out. Appr chan to bar (dries 0·5m) often shifts about a NNE/SSW axis. A dinghy recce (best near LW) to view the access channel and a prior brief by HM/YC are advised. Enter only in settled weather from HW –2½ to +1½ nps; HW –2½ or –½ sp. After 7m wide ent and the pierhead SHM beacon, Fl G 4s 7m 2M, turn 90° to port inside; hug the hbr wall to stbd until into the mooring basin; moor as directed. The low bridge (2m clearance) just N of moorings is the head of navigation.

HM ☎(01297) 22180/07939 044109. **Axe YC** ☎20043: 🛢 ◣ pontoon, 🚩(6t) ◻. Facilities: 🛢 🛠 🗓. Excellent pilotage notes

at www.axeyachtclub.co.uk are essential reading for first time visitors. **Seaton**, as above for victuals etc.

The chartlet is reproduced by kind permission of the Axe Yacht Club whose Commodore and others revised these notes.

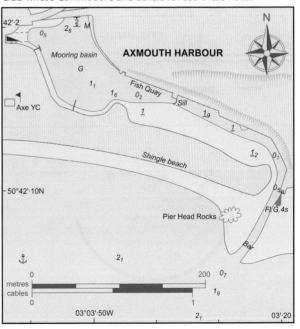

9.1.24 LYME REGIS

Dorset 50°43'·20N 02°56'·19W ❀❀❀❀❀❀

CHARTS AC 3315, 5601; Imray C10, C5, 2300

TIDES −0455 Dover; ML 2·4; Duration 0700

Standard Port PLYMOUTH (←)

Times				Height (metres)			
High Water		Low Water		MHWS	MHWN	MLWN	MLWS
0100	0600	0100	0600	5·5	4·4	2·2	0·8
1300	1800	1300	1800				
Differences LYME REGIS							
+0040	+0100	+0005	−0005	−1·2	−1·3	−0·5	−0·2

NOTE: Rise is relatively fast for the 1st hour after LW, but slackens for the next 1½ hrs, after which the rapid rate is resumed. There is often a stand of about 1½ hours at HW.

SHELTER Good in the hbr (dries up to 2·1m), but swell enters in strong E/SE winds when it may be best to dry inside the N Wall. The Cobb, a massive stone pier, protects the W and S sides of the hbr. Two lightweight walk ashore pontoons are provided in season with 2m depth reported at outer ends. In settled weather ⚓ as on the chartlet, clear of appr and ⚓s.

NAVIGATION WPT 50°43'·10N 02°55'·64W, 284°/0·35M to front ldg lt; best line just inside W sector until hbr ent opens. Craft drawing >1m may ground LW ±2 on a small sand bar outside the hbr. A 70m rock extension to E end of The Cobb covers at half tide; it is marked by unlit PHM bcn. Beware many fishing floats and moorings.

LIGHTS AND MARKS The 284° ldg line & R/W sectors clear The Cobb extension. (Note: AC 3315 still, May 08, shows the former 296° ldg line.) R flag on Victoria Pier = Gale warning in force.

COMMUNICATIONS (Code 01297) NMOC (01305) 760439; Police 101; Dr 445777. *Lyme Regis Hbr Radio* Ch **14**, 16. HM g.forshaw@ westdorset-dc.gov.uk 442137, mob 07870 240645.

FACILITIES www.dorsetforyou/lymeregisharbour.com Dry out against Victoria Pier (0·3–1·3m) on clean, hard sand, max LOA 11m. ⌓ 7–11m £14.00, >11m £20.00. A lightweight, 50m long pontoon has been installed against Victoria Pier.

9 Y ⚓s lie ENE of hbr ent in 1·5-2m, £7.00. ⚓ ⚓ 🏠 📞 ⛴ (mobile) ✕ 🔧 🗓 Ⓔ 🏪 (ACA Axminster, 5M). SC ☎442373, 🔧 ⚓ ✕ 🖥.

Power Boat Club ☎07768 725959, ✕ 🖥.

Town 🍴 ✉ Ⓑ 🏪 🗓 ✕ 🖥 bus to Axminster ⇌ ✈ (Exeter).

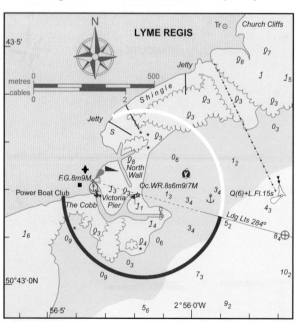

9.1.25 WEST BAY (BRIDPORT)

Dorset 50°42'·51N 02°45'·81W ❀❀❀❀❀❀

CHARTS AC 3315, 5601; Imray C10, C5, 2300

TIDES −0500 Dover; ML 2·3; Duration 0650

Standard Port PLYMOUTH (←)

Times				Height (metres)			
High Water		Low Water		MHWS	MHWN	MLWN	MLWS
0100	0600	0100	0600	5·5	4·4	2·2	0·8
1300	1800	1300	1800				
Differences BRIDPORT (West Bay)							
+0025	+0040	0000	0000	−1·4	−1·4	−0·6	−0·1
CHESIL BEACH							
+0040	+0055	−0005	+0010	−1·6	−1·5	−0·5	0·0
CHESIL COVE							
+0035	+0050	−0010	+0005	−1·5	−1·6	−0·5	−0·2

NOTE: Rise is relatively fast for first hr after LW; it then slackens for the next 1½ hrs, after which the rapid rise is resumed. There is often a stand of about 1½ hrs at HW.

SHELTER Access in W'lies and shelter in the outer hbr are improved by the new W pier, but the ent is exposed to SE'lies and swell/surge may affect the outer hbr in any wind direction.

NAVIGATION WPT 50°42'·05N 02°45'·49W, 336°/0·50M to ent in W sector of Dir lt. High Ground shoal 3m is 1·4M W of the ent. 3·7 and 4·6m shoals are SW & SE of hbr. SPM outfall buoy, Fl Y 5s, is 5ca SSW of ent.

LIGHTS AND MARKS See chartlet and 9.1.4 for lts. High cliffs are conspic close ESE of the hbr.

COMMUNICATIONS (Code 01308) NMOC (01305) 760439; Police 101; Dr 421109. *Bridport Radio* Ch **11** 16. HM 423222, mob 07870 240636

FACILITIES, www. dorsetforyou/westbayharbour.com. **Outer hbr** is silted up to 0·5m below CD. Ⓥ pontoon is in front of the slipway, close W of ≮ Fl R 8s. ⌓ £15/night. ⚓ 🔌 🏠 📞 ✕ 🔧.

Inner hbr dries, except for a pool scoured 2·7m by sluice water, and is full of FVs and local boats. The entry sill dries 0·1m.

Bridport (1½M N of the hbr) 🏪 ✉ Ⓑ 🏪 🗓 ✕ 🖥 ⇌ (bus to Axminster) ✈ (Exeter).

CROSSING THE ENGLISH CHANNEL, WESTERN PART

These notes cover cross-Channel passages between harbours between Scilly and Tor Bay to French harbours between Ouessant and Île de Bréhat. See AC 2655, 2656; 9.2.5 for the Central English Channel; and 9.0.6 for cross-Channel distances. Passage planning advice and SOLAS V requirements are in Chapter 3.

The Western Channel differs from the central and eastern parts; it is wider and its French harbours are less easy to approach.

Distance is a major ingredient in any passage plan. The shortest is 88M (Salcombe to Bréhat or Trébeurden); the longest 108M (Fowey to L'Aberwrac'h). Salcombe and Dartmouth are favourite departure points, for the Channel Islands and points West. Falmouth is often preferred for rounding Ushant. Taking an average distance of 100M:

Time at 5 knots is 20 hours, or up to 30 hours if the wind is southerly. Therefore it is important to consider:

- **Crew** stamina and experience. A short-handed crew (husband & wife, for example) may cope quite easily with a 20-hour passage, but 30 hours beating to windward might prove too much.

- **Weather**. A window of at least 24 hrs (preferably 48 hrs) is needed during which the maximum forecast wind is F5 (depending on direction); wind not N'ly (ie onshore along the Brittany coast); with a low chance of fog on that coast or in mid-Channel.

- **ETD and ETA**. Some night sailing is inevitable, but in Jul/Aug there is only about 6 hours of darkness. To avoid leaving, arriving or crossing shipping lanes in the dark, a departure at about 0900 allows you to cross the lanes in daylight and arrive after dawn.

Shipping lanes. Whilst not having to cross a TSS, lines linking the Casquets TSS and Ushant TSS will indicate where the main west-bound and east-bound traffic will be likely to be encountered. Be alert for shipping not using this route, and try to avoid close quarters situations. Do not expect big ships to give way to you; they will probably be travelling fast (24 knots is not uncommon, even in poor visibility) and may have difficulty detecting you if it is at all rough.

Contingencies. North Brittany is not the easiest of coasts owing to an abundance of offshore rocks which demand precise landfalls and careful pilotage. Fresh onshore winds complicate the task, as does fog. There is no all-weather, all-tide, port of refuge along this coast, so the forecast weather window is crucial. If things are not going as planned, it is no disgrace to turn back or make for an alternative destination. The Channel Islands are a possible diversion, or consider heaving-to for a rest or to await an improvement in the weather.

PORTLAND RACE

South of the Bill lies Portland Race (AC 2255) in which severe and very dangerous sea states occur. Even in settled weather it should be carefully avoided by small craft, although at neaps it may be barely perceptible.

The Race occurs at the confluence of two strong tidal streams which at springs run S down each side of the Isle of Portland for almost 10 out of 12 hours. These streams meet the main E-W stream of the Channel, producing large eddies on either side of Portland Bill and a highly confused sea state with heavy overfalls in the Race. The irregular contours of the sea-bed, which shoals rapidly from depths of about 100m some 2M south of the Bill to as little as 10m on Portland Ledge 1M further N, greatly contribute to the violence of the Race. Portland Ledge strongly deflects the flow of water upwards, so that on the flood the Race lies SE of the Bill and SW on the ebb. Conditions deteriorate with wind-against-tide, especially at springs. In an E'ly gale against the flood stream the Race may spread eastward to The Shambles bank. The Race normally extends about 2M S of the Bill, but further S in bad weather.

The Race can be avoided by passing to seaward of it, ie 3–5M S of the Bill; or by using the inshore passage if conditions suit.

The **inshore passage** offers relatively smooth water between 1ca and 3ca off the Bill (depending on wind). Do not use it in onshore winds >F4/5, nor at night under power due to lobster pot floats and especially not at springs with wind against tide. From W or E, start close inshore at least 2M N of the Bill to utilise the S-going stream; hug the Bill to avoid being set into the Race.

▶ *The tidal stream chartlets at 9.1.26 which merit careful study show the approx hourly positions of the Race. They are referred to HW Plymouth, to HW Portland for those bound from/to Weymouth, and to HW Dover if passing S of the Race. See also the smaller scale chartlets at 9.1.3 and 9.2.3 and tidal diamonds 'A' on AC 2610 or 'R' on AC 2255.* ◀

Timing is vital to catch 'slackish' water in the inshore passage and to enjoy a fair tide thereafter. The windows are as follows:

▶ **Westbound**: *HW Dover –1 to HW +2; HW Plymouth +5 to HW –5; or HW Portland +4 to HW –6. Timing is easier if starting from Portland Harbour, Weymouth or Lulworth Cove.*

Eastbound: *HW Dover +5 to HW –4; HW Plymouth –2 to HW +2; or HW Portland –3 to HW +1.* ◀

PORTLAND BILL TO ANVIL POINT

The buoyed Shambles bank lies 2–5M E of Portland Bill and is best avoided, as is the gap between it and the Race (AC 2610). In bad weather the sea breaks heavily on it. E of **Weymouth** rocky ledges extend 3ca offshore as far as Lulworth Cove, which provides a reasonable ⚓ in fine, settled weather and offshore winds; as do Worbarrow Bay and Chapman's Pool further E.

A firing range extends 5M offshore between Lulworth Cove and St Alban's Head. Yachts should avoid the range when it is active or pass through as quickly as possible. Beware Kimmeridge Ledges which extend over 5ca seaward. Three rarely used yellow naval target buoys (DZ A, B and C) lie on St Alban's Ledge.

St Alban's Head (107m) is bold and steep-to.

A sometimes vicious race forms over **St Alban's Ledge**, a rocky dorsal ridge (least depth 8·5m) which extends 4M SW from St Alban's Hd.

▶ *1M S of St Alban's Head the E-going flood starts at HW Plymouth –1, and the W-going ebb at HWP +0525; inshore max Sp rates 4¾kn; see also diamond 'U' on AC 2610. Slack water lasts barely 30 mins. The race moves E on the flood and W on the ebb, which is more dangerous since overfalls extend 2½M further SW than on the flood. A back eddy sets almost continuously SSE down the W side of the Head.* ◀

In settled weather and at neaps the race may be barely perceptible and can safely be crossed. Otherwise avoid it by passing to seaward via 50°31'·40N 02°07'·80W (thus clearing Lulworth Range); or by using the inshore passage very close inshore below St Alban's Head. This is at most 5ca wide and avoids the worst of the overfalls, but expect your decks to get wet. In onshore gales stay offshore. The light on St Alban's Hd Iso R 2s is only shown when the ranges are active.

The rugged, cliffy coastline between St Alban's Hd and Anvil Point (Fl 10s 45m 9M) is steep-to quite close inshore. Measured mile beacons stand either side of Anvil Pt lt ho; the track is 083·5°.

9.1.26 PORTLAND TIDAL STREAMS

5 Hours before HW Plymouth (Devonport)

4 Hours before HW Plymouth (Devonport)

3 Hours before HW Plymouth (Devonport)

2 Hours before HW Plymouth (Devonport)

1 Hour before HW Plymouth (Devonport)

HW Plymouth (Devonport)

General Area 1: 9.1.3 and 9.2.3

HW+1 (HW Portland) (0440 before HW Dover)
1 Hour after HW Plymouth (Devonport)

HW+2 (0100 after HW Portland) (0340 before HW Dover)
2 Hours after HW Plymouth (Devonport)

HW+3 (0200 after HW Portland) (0240 before HW Dover)
3 Hours after HW Plymouth (Devonport)

HW+4 (0300 after HW Portland) (0140 before HW Dover)
4 Hours after HW Plymouth (Devonport)

HW+5 (0400 after HW Portland) (0040 before HW Dover)
5 Hours after HW Plymouth (Devonport)

HW+6 (0500 after HW Portland) (0020 after HW Dover)
6 Hours after HW Plymouth (Devonport)

PORTLAND — LAT 50°34'N LONG 2°26'W
TIMES AND HEIGHTS OF HIGH AND LOW WATERS

STANDARD TIME (UT) — For Summer Time add ONE hour in **non-shaded areas**

Dates in red are SPRINGS · Dates in blue are NEAPS

YEAR 2016

JANUARY

Day							
1 F	0306 0.7	1037 1.7	1541 0.6	2305 1.5			
16 SA	0330 0.4	1050 1.8	1608 0.4	☽ 2320 1.7			
2 SA	0334 0.8	1115 1.6	1613 0.7	◑ 2349 1.4			
17 SU	0419 0.6	1147 1.7	1701 0.5				
3 SU	0420 0.9	1202 1.4	1702 0.7				
18 M	0027 1.5	0519 0.7	1307 1.5	1806 0.6			
4 M	0053 1.4	0527 0.9	1310 1.4	1916 0.7			
19 TU	0152 1.5	0634 0.8	1429 1.5	1928 0.7			
5 TU	0218 1.4	0747 0.9	1434 1.4	2010 0.7			
20 W	0303 1.6	0836 0.8	1538 1.5	2049 0.6			
6 W	0328 1.6	0839 0.8	1545 1.5	2057 0.6			
21 TH	0406 1.7	0934 0.7	1643 1.7	2144 0.5			
7 TH	0425 1.8	0926 0.7	1646 1.6	2143 0.5			
22 F	0503 1.9	1022 0.6	1742 1.8	2233 0.4			
8 F	0516 1.9	1013 0.6	1740 1.8	2228 0.5			
23 SA	0554 2.1	1106 0.5	1833 1.9	2317 0.3			
9 SA	0602 2.0	1059 0.5	1828 1.9	2314 0.4			
24 SU	0642 2.2	1145 0.3	1920 2.0	○ 2357 0.3			
10 SU	0645 2.1	1143 0.4	● SU 1911 2.0	2358 0.3			
25 M	0725 2.2	1222 0.3	2002 2.0				
11 M	0725 2.2	1226 0.3	1950 2.0				
26 TU	0034 0.2	0804 2.2	1258 0.2	2035 2.0			
12 TU	0039 0.3	0803 2.2	1307 0.2	2028 2.0			
27 W	0110 0.2	0837 2.1	1334 0.3	2103 1.9			
13 W	0121 0.2	0841 2.2	1350 0.2	2105 2.0			
28 TH	0144 0.3	0908 2.0	1408 0.3	2130 1.8			
14 TH	0203 0.3	0921 2.1	1434 0.2	2146 1.9			
29 F	0214 0.4	0938 1.9	1436 0.4	2157 1.7			
15 F	0246 0.3	1003 2.0	1519 0.3	2230 1.8			
30 SA	0239 0.5	1007 1.7	1459 0.5	2225 1.6			
31 SU	0304 0.6	1036 1.6	1528 0.5	2257 1.5			

FEBRUARY

Day				
1 M	0341 0.6	1111 1.5	1611 0.6	◑ 2339 1.4
16 TU	0443 0.7	1239 1.4	1731 0.7	
2 TU	0435 0.7	1200 1.4	1709 0.6	
17 W	0118 1.4	0603 0.8	1410 1.3	1902 0.7
3 W	0042 1.4	0545 0.8	1314 1.3	1816 0.7
18 TH	0241 1.4	0833 0.8	1523 1.4	2027 0.7
4 TH	0224 1.4	0812 0.8	1456 1.4	1948 0.7
19 F	0348 1.6	0918 0.7	1628 1.5	2124 0.6
5 F	0340 1.6	0904 0.7	1610 1.5	2116 0.6
20 SA	0447 1.8	1003 0.6	1724 1.7	2213 0.5
6 SA	0441 1.8	0953 0.5	1712 1.7	2209 0.5
21 SU	0538 1.9	1046 0.4	1814 1.9	2258 0.3
7 SU	0535 1.9	1042 0.4	1807 1.8	2258 0.3
22 M	0625 2.1	1126 0.3	1900 2.0	○ 2338 0.2
8 M	0625 2.1	1128 0.3	● 1857 2.0	2344 0.2
23 TU	0708 2.1	1203 0.2	1941 2.0	
9 TU	0712 2.2	1212 0.1	1945 2.1	
24 W	0015 0.1	0747 2.2	1238 0.1	2016 2.0
10 W	0027 0.1	0758 2.3	1254 0.1	2028 2.1
25 TH	0050 0.1	0818 2.1	1311 0.2	2041 2.0
11 TH	0109 0.1	0839 2.3	1336 0.0	2103 2.1
26 F	0122 0.2	0846 2.0	1340 0.2	2105 1.9
12 F	0150 0.1	0917 2.2	1417 0.0	2136 2.0
27 SA	0149 0.3	0913 1.9	1405 0.3	2128 1.7
13 SA	0230 0.2	0954 2.1	1459 0.1	2212 1.9
28 SU	0213 0.3	0937 1.7	1427 0.3	2149 1.6
14 SU	0310 0.3	1035 1.9	1542 0.3	2254 1.7
29 M	0237 0.3	1000 1.6	1454 0.4	2214 1.5
15 M	0352 0.5	1123 1.7	1630 0.5	◑ 2347 1.5

MARCH

Day				
1 TU	0308 0.5	1033 1.5	1531 0.5	◑ 2252 1.5
16 W	0416 0.6	1215 1.4	1700 0.7	
2 W	0355 0.6	1120 1.4	1627 0.6	2348 1.4
17 TH	0030 1.4	0543 0.8	1353 1.3	1834 0.8
3 TH	0505 0.7	1225 1.3	1737 0.6	
18 F	0220 1.4	0740 0.8	1508 1.3	1953 0.8
4 F	0109 1.3	0623 0.7	1415 1.3	1856 0.7
19 SA	0329 1.5	0843 0.7	1609 1.5	2052 0.7
5 SA	0301 1.5	0847 0.6	1541 1.4	2053 0.6
20 SU	0427 1.7	0931 0.5	1702 1.7	2143 0.5
6 SU	0410 1.7	0934 0.5	1646 1.6	2148 0.4
21 M	0517 1.8	1015 0.4	1750 1.8	2230 0.3
7 M	0509 1.9	1022 0.3	1744 1.8	2238 0.3
22 TU	0603 2.0	1057 0.2	1834 2.0	2313 0.2
8 TU	0602 2.1	1109 0.2	1837 2.0	2326 0.1
23 W	0645 2.1	1137 0.2	1915 2.0	○ 2352 0.1
9 W	0653 2.2	1153 0.0	● 1929 2.1	
24 TH	0722 2.1	1213 0.1	1949 2.1	
10 TH	0009 0.0	0744 2.3	1235 -0.1	2016 2.2
25 F	0026 0.1	0754 2.1	1244 0.1	2016 2.0
11 F	0051 0.0	0831 2.4	1316 -0.1	2054 2.2
26 SA	0057 0.2	0822 2.0	1312 0.2	2041 1.9
12 SA	0131 0.0	0911 2.3	1356 0.0	2125 2.1
27 SU	0124 0.2	0847 1.9	1336 0.3	2102 1.8
13 SU	0210 0.1	0945 2.1	1436 0.1	2155 2.0
28 M	0148 0.3	0909 1.7	1400 0.3	2118 1.7
14 M	0248 0.2	1021 1.9	1516 0.3	2231 1.8
29 TU	0214 0.3	0931 1.6	1426 0.3	2142 1.6
15 TU	0327 0.4	1006 1.6	1601 0.5	◑ 2315 1.6
30 W	0243 0.4	1006 1.5	1459 0.4	2222 1.5
31 TH	0324 0.5	1056 1.4	1550 0.5	◑ 2318 1.4

APRIL

Day				
1 F	0435 0.6	1200 1.3	1707 0.6	
16 SA	0148 1.4	0650 0.8	1446 1.3	1911 0.8
2 SA	0031 1.4	0817 0.7	1342 1.3	1829 0.7
17 SU	0303 1.5	0755 0.8	1544 1.5	2012 0.7
3 SU	0225 1.4	0827 0.6	1515 1.4	2022 0.6
18 M	0400 1.6	0848 0.5	1508 1.3	2105 0.6
4 M	0341 1.6	0911 0.4	1619 1.6	2124 0.4
19 TU	0449 1.7	0936 0.4	1721 1.8	2155 0.4
5 TU	0442 1.8	0958 0.3	1718 1.8	2215 0.3
20 W	0534 1.9	1022 0.3	1804 1.9	2241 0.3
6 W	0537 2.0	1045 0.1	1813 2.0	2303 0.2
21 TH	0616 2.0	1105 0.2	1844 2.0	2323 0.2
7 TH	0630 2.2	1130 0.0	● 1904 2.2	2348 0.0
22 F	0653 2.0	1143 0.2	○ 1919 2.1	
8 F	0722 2.3	1213 -0.1	1952 2.3	
23 SA	0000 0.2	0727 2.0	1215 0.2	1950 2.0
9 SA	0030 0.0	0813 2.4	1254 -0.1	2035 2.3
24 SU	0031 0.2	0757 2.0	1243 0.2	2017 2.0
10 SU	0110 0.0	0857 2.3	1333 0.0	2110 2.2
25 M	0059 0.2	0823 1.9	1310 0.3	2039 1.9
11 M	0149 0.1	0934 2.1	1412 0.2	2141 2.1
26 TU	0127 0.3	0845 1.8	1338 0.3	2057 1.7
12 TU	0227 0.3	1009 1.9	1452 0.4	2215 1.9
27 W	0157 0.3	0912 1.7	1408 0.3	2124 1.7
13 W	0309 0.4	1051 1.6	1537 0.6	2256 1.6
28 TH	0231 0.4	0951 1.5	1443 0.4	2206 1.6
14 TH	0401 0.6	1150 1.4	1636 0.8	◑ 2352 1.5
29 F	0315 0.4	1042 1.4	1533 0.5	2301 1.5
15 F	0526 0.8	1329 1.3	1759 0.9	
30 SA	0421 0.5	1145 1.3	1648 0.6 ◑	

Chart Datum is 0·93 metres below Ordnance Datum (Newlyn). HAT is 2·5 metres above Chart Datum.

》》 FREE monthly updates. Register at 《 www.reedsnauticalalmanac.co.uk

STANDARD TIME (UT)
For Summer Time add ONE hour in **non-shaded areas**

PORTLAND LAT 50°34'N LONG 2°26'W

TIMES AND HEIGHTS OF HIGH AND LOW WATERS

Dates in red are **SPRINGS**
Dates in blue are NEAPS

YEAR 2016

MAY

Time	m	Time	m
1 0010	1.4	**16** 0214	1.4
0538	0.6	0711	0.7
SU 1315	1.3	M 1511	1.4
1807	0.7	1932	0.8
2 0147	1.5	**17** 0319	1.5
0711	0.5	0807	0.6
M 1446	1.5	TU 1601	1.6
1937	0.6	2027	0.7
3 0308	1.6	**18** 0411	1.6
0839	0.4	0857	0.5
TU 1551	1.7	W 1647	1.7
2052	0.5	2118	0.5
4 0411	1.8	**19** 0458	1.7
0930	0.3	0945	0.4
W 1649	1.9	TH 1730	1.9
2148	0.4	2207	0.4
5 0509	2.0	**20** 0541	1.8
1019	0.2	1031	0.3
TH 1744	2.1	F 1810	2.0
2238	0.2	2253	0.4
6 0604	2.1	**21** 0622	1.9
1106	0.1	1112	0.3
F 1835	2.2	SA 1848	2.1
● 2326	0.2	○ 2333	0.3
7 0657	2.3	**22** 0659	1.9
1150	0.0	1147	0.3
SA 1924	2.3	SU 1924	2.1
8 0009	0.1	**23** 0007	0.3
0750	2.3	0733	1.9
SU 1231	0.0	M 1218	0.3
2010	2.3	1956	2.0
9 0050	0.1	**24** 0039	0.3
0838	2.2	0802	1.9
M 1311	0.1	TU 1250	0.3
2051	2.2	2022	2.0
10 0130	0.2	**25** 0111	0.3
0919	2.1	0829	1.8
TU 1351	0.4	W 1324	0.3
2127	2.1	2046	1.9
11 0211	0.3	**26** 0147	0.3
0957	1.9	0902	1.7
W 1432	0.4	TH 1400	0.3
2202	1.9	2119	1.8
12 0255	0.5	**27** 0228	0.3
1036	1.7	0943	1.6
TH 1517	0.6	F 1441	0.4
2241	1.7	2200	1.7
13 0349	0.6	**28** 0315	0.4
1122	1.5	1032	1.5
F 1612	0.7	SA 1532	0.5
◑ 2327	1.6	2251	1.6
14 0457	0.7	**29** 0413	0.5
1235	1.3	1131	1.4
SA 1721	0.8	SU 1635	0.6
		◑ 2353	1.5
15 0030	1.4	**30** 0518	0.5
0608	0.7	1248	1.4
SU 1410	1.3	M 1743	0.6
1830	0.9		

JUNE

Time	m	Time	m
1 0234	1.6	**16** 0312	1.5
0751	0.4	0822	0.6
W 1520	1.7	TH 1604	1.6
2014	0.6	2045	0.7
2 0340	1.7	**17** 0411	1.6
0859	0.3	0911	0.5
TH 1619	1.8	F 1651	1.8
2120	0.5	2136	0.6
3 0441	1.9	**18** 0503	1.7
0952	0.3	0958	0.4
F 1715	2.0	SA 1735	1.9
2215	0.4	2224	0.5
4 0540	2.0	**19** 0550	1.8
1042	0.2	1042	0.4
SA 1808	2.2	SU 1818	2.0
2305	0.3	2308	0.4
5 0635	2.1	**20** 0633	1.9
1129	0.1	1122	0.3
SU 1858	2.3	M 1859	2.1
● 2352	0.2	○ 2348	0.4
6 0730	2.1	**21** 0713	1.9
1212	0.1	1159	0.3
M 1946	2.3	TU 1937	2.1
7 0034	0.2	**22** 0024	0.3
0821	2.1	0748	1.9
TU 1253	0.2	W 1236	0.3
2031	2.3	2010	2.1
8 0116	0.2	**23** 0102	0.3
0904	2.0	0819	1.9
W 1333	0.3	TH 1315	0.3
2111	2.2	2041	2.0
9 0158	0.3	**24** 0141	0.3
0941	1.9	0855	1.8
TH 1414	0.4	F 1355	0.3
2147	2.0	2115	1.9
10 0241	0.4	**25** 0224	0.3
1016	1.7	0935	1.7
F 1457	0.5	SA 1438	0.3
2223	1.8	2155	1.9
11 0329	0.5	**26** 0310	0.3
1053	1.6	1021	1.6
SA 1543	0.7	SU 1525	0.4
2301	1.7	2240	1.8
12 0423	0.6	**27** 0400	0.3
1137	1.4	1114	1.6
SU 1638	0.8	M 1618	0.5
◑ 2345	1.5	◑ 2335	1.6
13 0525	0.7	**28** 0456	0.4
1237	1.3	1220	1.5
M 1745	0.8	TU 1717	0.6
14 0041	1.4	**29** 0044	1.6
0630	0.7	0559	0.5
TU 1407	1.4	W 1341	1.5
1853	0.8	1825	0.7
15 0158	1.4	**30** 0205	1.6
0729	0.6	0711	0.5
W 1513	1.5	TH 1452	1.6
1952	0.8	1941	0.7

JULY

Time	m	Time	m
1 0315	1.6	**16** 0316	1.4
0829	0.5	0839	0.6
F 1553	1.8	SA 1608	1.6
2059	0.6	2110	0.6
2 0421	1.7	**17** 0423	1.5
0932	0.4	0928	0.5
SA 1652	1.9	SU 1700	1.8
2202	0.5	2200	0.5
3 0523	1.8	**18** 0520	1.7
1025	0.3	1015	0.4
SU 1747	2.1	M 1749	1.9
2255	0.4	2248	0.5
4 0621	2.0	**19** 0609	1.8
1113	0.2	1101	0.4
M 1838	2.2	TU 1834	2.1
● 2342	0.3	○ 2332	0.4
5 0715	2.0	**20** 0655	1.9
1156	0.2	1143	0.3
TU 1927	2.3	W 1918	2.1
6 0023	0.2	**21** 0012	0.3
0806	2.1	0737	1.9
W 1237	0.2	TH 1224	0.2
2012	2.3	1959	2.2
7 0103	0.2	**22** 0052	0.2
0849	2.0	0814	1.9
TH 1316	0.2	F 1305	0.2
2053	2.2	2035	2.1
8 0142	0.2	**23** 0132	0.1
0922	1.9	0849	1.9
F 1355	0.3	SA 1346	0.2
2127	2.1	2108	2.1
9 0222	0.3	**24** 0213	0.1
0952	1.8	0925	1.9
SA 1434	0.4	SU 1427	0.2
2200	1.9	2145	2.0
10 0302	0.4	**25** 0256	0.2
1022	1.6	1006	1.8
SU 1511	0.5	M 1510	0.3
2232	1.7	2226	1.9
11 0340	0.5	**26** 0341	0.3
1057	1.5	1053	1.7
M 1545	0.7	TU 1557	0.4
2307	1.6	◑ 2315	1.7
12 0414	0.6	**27** 0431	0.4
1137	1.4	1150	1.6
TU 1621	0.8	W 1650	0.6
◑ 2348	1.5		
13 0450	0.7	**28** 0018	1.5
1230	1.4	0530	0.5
W 1715	0.8	TH 1310	1.6
		1756	0.7
14 0040	1.4	**29** 0145	1.5
0640	0.7	0642	0.6
TH 1347	1.4	F 1430	1.5
1917	0.8	1922	0.8
15 0156	1.4	**30** 0302	1.5
0747	0.7	0815	0.6
F 1506	1.5	SA 1536	1.7
2017	0.7	2106	0.7
		31 0410	1.6
		0920	0.5
		SU 1637	1.8
		2200	0.6

AUGUST

Time	m	Time	m
1 0512	1.7	**16** 0452	1.6
1012	0.4	0952	0.5
M 1732	2.0	TU 1720	1.9
2248	0.4	2228	0.4
2 0608	1.9	**17** 0546	1.8
1059	0.3	1040	0.4
TU 1822	2.1	W 1810	2.1
● 2331	0.3	2313	0.3
3 0659	2.0	**18** 0635	1.9
1141	0.2	1126	0.3
W 1909	2.2	TH 1857	2.2
		○ 2356	0.2
4 0010	0.2	**19** 0722	2.0
0747	2.1	1209	0.2
TH 1220	0.2	F 1942	2.3
1953	2.3		
5 0047	0.2	**20** 0036	0.1
0828	2.0	0806	2.1
F 1258	0.2	SA 1250	0.1
2032	2.2	2025	2.3
6 0123	0.2	**21** 0116	0.0
0859	2.0	0841	2.1
SA 1334	0.2	SU 1330	0.1
2103	2.1	2100	2.2
7 0158	0.2	**22** 0156	0.1
0924	1.9	0913	2.0
SU 1408	0.3	M 1410	0.2
2133	1.9	2133	2.1
8 0230	0.3	**23** 0236	0.1
0950	1.7	0948	1.9
M 1438	0.4	TU 1451	0.3
2201	1.8	2211	1.9
9 0256	0.4	**24** 0318	0.3
1018	1.6	1029	1.8
TU 1503	0.6	W 1533	0.4
2231	1.6	2256	1.7
10 0318	0.5	**25** 0404	0.5
1050	1.5	1120	1.6
W 1532	0.6	TH 1623	0.6
◑ 2304	1.5	◑ 2358	1.5
11 0352	0.6	**26** 0501	0.6
1131	1.4	1241	1.5
TH 1619	0.7	F 1732	0.7
2347	1.4		
12 0444	0.7	**27** 0135	1.4
1227	1.4	0622	0.8
F 1724	0.8	SA 1414	1.5
		1947	0.8
13 0048	1.3	**28** 0255	1.4
0550	0.7	0803	0.7
SA 1359	1.4	SU 1523	1.6
1959	0.8	2055	0.7
14 0225	1.3	**29** 0402	1.5
0810	0.7	0902	0.6
SU 1523	1.5	M 1623	1.8
2051	0.7	2143	0.6
15 0347	1.4	**30** 0500	1.7
0903	0.6	0953	0.5
M 1626	1.7	TU 1716	2.0
2140	0.5	2228	0.4
		31 0551	1.9
		1039	0.4
		W 1805	2.1
		2311	0.3

Chart Datum is 0·93 metres below Ordnance Datum (Newlyn). HAT is 2·5 metres above Chart Datum.

FREE monthly updates. Register at
www.reedsnauticalalmanac.co.uk

183

STANDARD TIME (UT)
For Summer Time add ONE hour in **non-shaded areas**

PORTLAND   LAT 50°34'N   LONG 2°26'W
TIMES AND HEIGHTS OF HIGH AND LOW WATERS

Dates in **red** are **SPRINGS**
Dates in **blue** are **NEAPS**

YEAR 2016

SEPTEMBER
Time m

1 0638 2.0 / 1122 0.2 / TH 1850 2.2 / ● 2350 0.2	**16** 0612 2.0 / 1104 0.3 / F 1833 2.2 / ○ 2333 0.1
2 0722 2.1 / 1200 0.2 / F 1931 2.2	**17** 0700 2.1 / 1149 0.2 / SA 1920 2.3
3 0025 0.1 / 0801 2.1 / SA 1236 0.1 / 2007 2.2	**18** 0015 0.1 / 0746 2.2 / SU 1230 0.1 / 2006 2.4
4 0058 0.1 / 0830 2.1 / SU 1310 0.2 / 2036 2.1	**19** 0055 0.0 / 0824 2.3 / M 1310 0.1 / 2046 2.3
5 0129 0.2 / 0852 1.9 / M 1341 0.3 / 2103 2.0	**20** 0134 0.1 / 0855 2.2 / TU 1350 0.1 / 2120 2.2
6 0156 0.3 / 0916 1.8 / TU 1407 0.4 / 2129 1.8	**21** 0213 0.2 / 0927 2.1 / W 1429 0.3 / 2156 2.0
7 0217 0.4 / 0941 1.7 / W 1429 0.5 / 2155 1.6	**22** 0253 0.4 / 1006 1.9 / TH 1510 0.5 / 2241 1.7
8 0237 0.5 / 1006 1.6 / TH 1456 0.6 / 2223 1.5	**23** 0336 0.6 / 1054 1.7 / F 1600 0.7 / ◗ 2347 1.5
9 0308 0.6 / 1040 1.5 / F 1537 0.7 / ◖ 2304 1.4	**24** 0433 0.8 / 1208 1.5 / SA 1721 0.8
10 0357 0.7 / 1132 1.4 / SA 1644 0.8	**25** 0126 1.4 / 0607 0.9 / SU 1356 1.5 / 1915 0.8
11 0005 1.3 / 0509 0.8 / SU 1250 1.4 / 1951 0.8	**26** 0245 1.4 / 0733 0.9 / M 1508 1.6 / 2021 0.7
12 0145 1.3 / 0628 0.8 / M 1443 1.5 / 2034 0.7	**27** 0348 1.6 / 0834 0.8 / TU 1606 1.8 / 2112 0.6
13 0319 1.4 / 0840 0.7 / TU 1553 1.7 / 2119 0.5	**28** 0441 1.8 / 0925 0.6 / W 1657 1.9 / 2158 0.4
14 0425 1.6 / 0929 0.5 / W 1651 1.9 / 2204 0.4	**29** 0529 1.9 / 1012 0.5 / TH 1743 2.1 / 2241 0.3
15 0521 1.8 / 1017 0.4 / TH 1744 2.1 / 2249 0.3	**30** 0613 2.1 / 1056 0.3 / F 1826 2.2 / 2322 0.2

OCTOBER
Time m

1 0654 2.2 / 1135 0.2 / SA 1906 2.2 / ● 2358 0.2	**16** 0632 2.3 / 1124 0.2 / SU 1854 2.3 / ○ 2350 0.1
2 0730 2.2 / 1211 0.2 / SU 1939 2.2	**17** 0716 2.4 / 1208 0.2 / M 1940 2.4
3 0030 0.2 / 0758 2.1 / M 1243 0.2 / 2007 2.1	**18** 0031 0.1 / 0756 2.4 / TU 1248 0.1 / 2025 2.3
4 0059 0.3 / 0822 2.0 / TU 1313 0.3 / 2034 2.0	**19** 0111 0.1 / 0832 2.3 / W 1328 0.2 / 2105 2.2
5 0123 0.4 / 0845 1.9 / W 1338 0.4 / 2059 1.8	**20** 0150 0.3 / 0908 2.2 / TH 1409 0.3 / 2146 2.0
6 0144 0.4 / 0905 1.8 / TH 1401 0.5 / 2121 1.7	**21** 0229 0.5 / 0949 2.0 / F 1452 0.5 / 2233 1.8
7 0205 0.5 / 0925 1.7 / F 1427 0.5 / 2149 1.6	**22** 0313 0.7 / 1036 1.8 / SA 1546 0.7 / ◗ 2335 1.6
8 0231 0.6 / 0959 1.6 / SA 1502 0.6 / 2234 1.4	**23** 0411 0.9 / 1138 1.6 / SU 1708 0.8
9 0306 0.7 / 1053 1.5 / SU 1610 0.7 / ◖ 2337 1.3	**24** 0106 1.4 / 0536 1.0 / M 1328 1.5 / 1830 0.9
10 0431 0.8 / 1206 1.4 / M 1943 0.8	**25** 0225 1.5 / 0652 1.0 / TU 1443 1.6 / 1935 0.8
11 0110 1.3 / 0558 0.9 / TU 1403 1.4 / 2017 0.7	**26** 0327 1.6 / 0755 0.9 / W 1541 1.7 / 2030 0.7
12 0253 1.5 / 0808 0.8 / W 1521 1.6 / 2053 0.5	**27** 0417 1.8 / 0848 0.8 / TH 1631 1.8 / 2118 0.5
13 0358 1.7 / 0901 0.6 / TH 1621 1.8 / 2137 0.4	**28** 0502 1.9 / 0937 0.6 / F 1717 2.0 / 2203 0.4
14 0454 1.9 / 0950 0.5 / F 1715 2.0 / 2221 0.3	**29** 0544 2.1 / 1023 0.5 / SA 1759 2.1 / 2246 0.3
15 0545 2.1 / 1038 0.3 / SA 1806 2.2 / 2306 0.2	**30** 0624 2.2 / 1105 0.4 / SU 1837 2.1 / ● 2326 0.3
	31 0658 2.2 / 1143 0.3 / M 1910 2.1

NOVEMBER
Time m

1 0000 0.3 / 0728 2.2 / TU 1218 0.3 / 1940 2.1	**16** 0008 0.2 / 0730 2.4 / W 1228 0.2 / 2005 2.3
2 0029 0.3 / 0755 2.1 / W 1247 0.4 / 2009 2.0	**17** 0049 0.2 / 0813 2.4 / TH 1310 0.3 / 2053 2.2
3 0055 0.4 / 0819 2.0 / TH 1314 0.4 / 2034 1.8	**18** 0130 0.3 / 0855 2.3 / F 1353 0.4 / 2137 2.0
4 0119 0.5 / 0840 1.9 / F 1341 0.5 / 2058 1.7	**19** 0211 0.5 / 0938 2.1 / SA 1439 0.5 / 2221 1.8
5 0145 0.5 / 0902 1.8 / SA 1412 0.5 / 2130 1.6	**20** 0255 0.7 / 1023 1.9 / SU 1532 0.7 / 2310 1.6
6 0212 0.6 / 0937 1.7 / SU 1450 0.6 / 2215 1.5	**21** 0348 0.9 / 1112 1.7 / M 1638 0.8 / ◖
7 0245 0.7 / 1029 1.6 / M 1551 0.7 / ◗ 2315 1.4	**22** 0021 1.5 / 0456 1.0 / TU 1220 1.6 / 1746 0.8
8 0358 0.8 / 1137 1.5 / TU 1708 0.7	**23** 0148 1.5 / 0607 1.0 / W 1359 1.5 / 1849 0.8
9 0036 1.4 / 0530 0.9 / W 1310 1.5 / 2008 0.7	**24** 0253 1.5 / 0711 1.0 / TH 1504 1.6 / 1946 0.7
10 0219 1.5 / 0656 0.8 / TH 1443 1.6 / 2013 0.6	**25** 0344 1.7 / 0808 0.9 / F 1556 1.7 / 2037 0.6
11 0326 1.7 / 0821 0.7 / F 1546 1.8 / 2103 0.4	**26** 0431 1.8 / 0859 0.7 / SA 1644 1.8 / 2125 0.5
12 0421 1.9 / 0919 0.6 / SA 1643 2.0 / 2151 0.3	**27** 0513 2.0 / 0947 0.6 / SU 1728 1.9 / 2211 0.5
13 0513 2.1 / 1010 0.4 / SU 1736 2.1 / 2238 0.2	**28** 0552 2.1 / 1034 0.5 / M 1807 2.0 / 2254 0.4
14 0601 2.3 / 1059 0.3 / M 1827 2.3 / ○ 2325 0.2	**29** 0628 2.2 / 1117 0.5 / TU 1844 2.0 / ● 2332 0.4
15 0647 2.4 / 1145 0.3 / TU 1916 2.3	**30** 0702 2.2 / 1154 0.4 / W 1918 2.0

DECEMBER
Time m

1 0004 0.4 / 0733 2.1 / TH 1226 0.4 / 1950 1.9	**16** 0033 0.3 / 0759 2.4 / F 1257 0.3 / 2043 2.2
2 0033 0.4 / 0802 2.1 / F 1257 0.4 / 2018 1.9	**17** 0115 0.3 / 0844 2.3 / SA 1340 0.3 / 2125 2.0
3 0104 0.5 / 0827 2.0 / SA 1330 0.4 / 2047 1.8	**18** 0156 0.4 / 0927 2.2 / SU 1425 0.4 / 2203 1.8
4 0137 0.6 / 0855 1.9 / SU 1407 0.5 / 2121 1.7	**19** 0238 0.6 / 1007 2.0 / M 1511 0.5 / 2241 1.7
5 0211 0.6 / 0931 1.8 / M 1449 0.5 / 2203 1.6	**20** 0322 0.7 / 1046 1.8 / TU 1603 0.6 / 2323 1.6
6 0250 0.6 / 1017 1.7 / TU 1542 0.6 / 2256 1.5	**21** 0412 0.8 / 1130 1.6 / W 1701 0.7 / ◖
7 0347 0.7 / 1115 1.6 / W 1645 0.6 / ●	**22** 0017 1.4 / 0516 0.9 / TH 1225 1.5 / 1804 0.8
8 0003 1.4 / 0503 0.8 / TH 1229 1.5 / 2028 0.7	**23** 0146 1.4 / 0628 1.0 / F 1345 1.4 / 1905 0.7
9 0133 1.4 / 0617 0.8 / F 1359 1.6 / 1907 0.6	**24** 0258 1.5 / 0731 0.9 / SA 1503 1.5 / 2000 0.7
10 0248 1.6 / 0733 0.8 / SA 1510 1.7 / 2021 0.5	**25** 0351 1.6 / 0827 0.8 / SU 1601 1.6 / 2051 0.6
11 0347 1.8 / 0845 0.7 / SU 1611 1.8 / 2121 0.4	**26** 0437 1.8 / 0917 0.7 / M 1653 1.7 / 2139 0.5
12 0443 2.0 / 0944 0.6 / M 1710 2.0 / 2214 0.3	**27** 0519 1.9 / 1007 0.6 / TU 1738 1.8 / 2226 0.5
13 0535 2.2 / 1038 0.5 / TU 1806 2.1 / 2303 0.3	**28** 0600 2.0 / 1053 0.5 / W 1821 1.9 / 2308 0.4
14 0624 2.3 / 1128 0.4 / W 1900 2.2 / ○ 2350 0.2	**29** 0639 2.1 / 1135 0.5 / TH 1901 1.9 / ● 2345 0.4
15 0712 2.4 / 1214 0.3 / TH 1952 2.2	**30** 0716 2.1 / 1212 0.4 / F 1937 1.9
	31 0020 0.4 / 0750 2.1 / SA 1246 0.4 / 2010 1.9

Chart Datum is 0·93 metres below Ordnance Datum (Newlyn). HAT is 2·5 metres above Chart Datum.

9.1.27 PORTLAND HARBOUR

Dorset 50°35'·72N 02°25'·91W (N Ship Chan) ❀❀❀❀♜♜♜❀

CHARTS AC 2610, 2255, 2268, 5601; Imray C10, C12, C5, C4, 2300

TIDES –0430 Dover; ML 1·0
Standard Port PORTLAND (→)

Times				Height (metres)			
High Water		Low Water		MHWS	MHWN	MLWN	MLWS
0100	0700	0100	0700	2·1	1·4	0·8	0·1
1300	1900	1300	1900				
LULWORTH COVE and MUPE BAY (Worbarrow Bay)							
+0005	+0015	–0005	0000	+0·1	+0·1	+0·2	+0·1

Double LWs occur between Portland and Lulworth Cove. Predictions are for the 1st LW. The 2nd LW occurs 3–4 hrs later.

SHELTER The large harbour is exposed to winds from the E and NE, but in westerlies the moorings on the W and S sides are reasonably sheltered by Chesil Beach and the Isle of Portland. The marina is protected by substantial rock breakwaters. ⚓ on W side of hbr in about 3m, as on chartlet, or further N off Castle Cove SC.

NAVIGATION If approaching from the W inshore of Portland Race study the tidal stream diagrams carefully and note the advice in Passage Information.

WPT 50°35'·97N 02°25'·24W, 240°/0·5M to North Ship Channel (NSC). This is the recommended route to the marina via the Fairway, keeping clear of commercial shipping using East Ship Channel (ESC). Small craft (LOA<20m) may use ESC, but must monitor Ch 74 and obey IPTS lights shown on 'C' and Fort Heads. Small craft crossing ESC should remain well clear of the Heads (>0.5M) so that they may be seen by commercial vesels navigating within the channel. Vessels >20m must obtain permission from Portland Harbour Radio before entering or leaving the harbour.

South Ship Channel (SSC) is permanently closed and blocked by overhead wires. 7ca SE of SSC is a noise range, ⚓ marked by 4 Fl Y buoys. Beware of high speed ferries to/from Weymouth. There is a rocky reef extending about 1·5ca offshore NE of Castle Cove SC, and shoal water E of Small Mouth.

A 12kn speed limit applies throughout the harbour and in Weymouth Bay, but vessels under 10m LOA are exempt. However, a 6kn speed limit applies to *all* vessels in the areas depicted and within 150m of any breakwater or harbour premises.

Approach to Weymouth & Portland National Sailing Academy (WPNSA) and Portland Marina. The Fairway remains the recommended route for small craft (<20m LOA) transiting the

harbour and engines must be used (if fitted) when proceeding between the marina/WPNSA and NSC. From NSC head SW to the SWM buoy (see chartlet) then S to the marina entrance, leaving the 2 PHM buoys to port. Obey traffic signals on 'C' Head and on the marina breakwater:

● Flashing = Entrance closed; await instructions (Ch 74 for
● Harbour Control; Ch 80 for Marina)
●

● Fixed = Commercial vessel/sailing fleet departing; do not
● impede
●

● Fixed = Commercial vessel/sailing fleet arriving; do not
● impede
●

LIGHTS AND MARKS The precipitous N face of the Verne, near the N end of the Isle of Portland, is conspic from afar. 'A' Hd lt ho is a conspic W twr, Fl 2·5s 22m 10M.

COMMUNICATIONS (Code 01305) NMOC 760439; Police 101; Dorchester Ⓗ 824055; Ⓗ (A&E) 820341. Port Control ☎824044.

Monitor Ch **74** *Portland Hbr Radio* for commercial traffic. *Portland Marina* Ch 80.

FACILITIES **Portland Port,** www.portland-port.co.uk owns the harbour which is a busy commercial port with an international bunkering station. Port and Harbour Dues (£0.70/m/day) apply to all vessels in Portland Inner Harbour; they are rigorously collected.

Castle Cove SC ☎783708, ⚓ ⬎ ⚓.

WPNSA The 3 pontoons for keel boats are protected by a rock breakwater. admin@wpnsa.org.uk www.wpnsa.org.uk ☎866000. Temporary pontoon berths may be available for 1 or 2 nights by prior arrangement.

RNSA moorings adjacent to fairway and WPNSA breakwater are private moorings for members occupied Apr-Oct.

Portland Marina The entrance, 50°34'·44N 02°27'·32W, is between the NW end of the marina breakwater and the WPNSA breakwater. www.portlandmarina.co.uk ☎08454 30 2012; 600⌂+Ⓥ berths on pontoons R, S & T £2.95 inc ⏚ for <14m LOA, short stay <4hrs £8, ⬎ ▣ ⚓ ⚓(inc bio-diesel) ⬎ ⚓ ✕ ⚓ (⎯) Ⓔ ⚓ ⬎(50/320t) ⚓ Diver ✕ ⬜ Helipad.

Castletown is SE of the marina. ✉ Ⓑ ✕ ⬜ 501 bus to Weymouth ⇌ ✈ (Hurn).

Chartlet overleaf

PORTLAND HARBOUR *continued*

9.1.28 WEYMOUTH

Dorset 50°36'·57N 02°26'·58W ✿✿✿✿⚓⚓⚓ ✿✿✿

CHARTS AC 2610, 2255, 2268, 2172, 5601; Imray C10, C12, C5, C4, 2300

TIDES –0438 Dover; ML 1·1

Standard Port PORTLAND (⟵)

Use Portland predictions and Note. Mean ranges are small: 0·6m at np and 2·0m at sp. NOTE: A LW stand lasts about 4 hrs at sp and 1 hr at nps. Due to an eddy, the tidal stream in Weymouth Roads is W-going at all times except for 2 hrs, HW –0510 to HW –0310.

SHELTER Good, but swell enters the outer harbour and The Cove in strong E winds. It is feasible to ⚓ in Weymouth Bay NNE of the harbour entrance in about 3·5m, about 4ca off the beach to seaward of the buoyed bathing area.

NAVIGATION WPT 50°36'·69N 02°26'·23W, 240°/2ca to South Pierhead. The harbour entrance is 1M NW of North Ship Channel into Portland Harbour.

If bound to/from the E, check whether Lulworth Firing Ranges are active. See 9.1.29 for further information.

A DG Range, marked by 3 SPM buoys (one Fl Y 2s), lies 2·5ca SE of South Pier; ⚓. See ACs 2255 and 2268 for other restrictions on ⚓ in Weymouth Bay.

Comply with traffic signals displayed from a RW pole near the root of South Pier (see below). Speed limit is 'Dead Slow'. A rowing boat ferry crosses the fairway just E of the LB (see chartlet) and has right of way.

LIGHTS AND MARKS Traffic signals, shown *vertically* on S Pier:

No signals = Proceed with caution;

⬤⬤⬤ = Large vessel leaving, do not obstruct the pierhead area;

⬤⬤⬤ = Large vessel is entering, do not leave harbour;

⬤⬤ over ⬤ = Harbour closed, do not enter the pierheads area;

⬤○⬤ = Only proceed on receipt of instructions;

⬤⬤⬤ (flashing) = All vessels await instructions.

If in any doubt, call *Weymouth Harbour* on Ch 12.

Conspic ⌖ spire, 6ca NNW of the harbour entrance, is a useful daymark. Pierhead lights may be hard to see against shore lights. Weymouth Eye, a 53m twr on the hard NW of N Pier, is floodlit. Together with Portland 'A' Head lt ho, a conspic W twr 1·7M SE of hbr ent, it is readily visible day and night.

Ldg lts 239·6°, 2 FR (H24), are 500m inside the pierheads; daymarks (same position) are R open ◊s on W poles; they are not visible until the entrance is opened.

A LIFTING BRIDGE allows masted craft access to the marina.

Lifting times (LT):

15 Apr-15 Sep, 0800, 1000, 1200, 1400, 1600, 1800, 2000; **plus Jun, Jul, Aug,** 2100.

16 Sep-14 Apr: 0800, 1000, 1200, 1400, 1600, 1800.

NOTE: Minimum 1 hr notice is required for lifts 16 Sep-14 Apr ☎838423 (ansafone), 789357 or Ch 12.

Be visible from bridge 5 mins before lift time; listen on Ch 12 for any broadcasts. Outbound vessels usually have priority. Waiting pontoons are on the S bank, either side of the bridge. 3FR or 3FG (vert) on both sides of the bridge are traffic lights and must be obeyed at all times, even when the bridge is closed. Clearance when closed is 2·75m above HAT.

COMMUNICATIONS (Code 01305) NMOC 760439; Dr 774411; Ⓗ Dorchester 251150; Police 101. HM 838423.

Weymouth Harbour and *Weymouth Town Bridge* (at opening times): Ch 12 (0800-2000 in summer and when vessel due). *Weymouth Marina* Ch 80. Ch 60 for ⚓.

FACILITIES (http://harbour.weymouth.gov.uk) **Outer Harbour** ⌓ The Cove (S side) on ❶ pontoons <10m LOA> Custom House Quay. Quays from The Cove to lift bridge are for FVs and tourist boats. ⌓£2·50 inc ⚡, ⚡£2·50/day discount for >3 nights (not Jul/Aug); <4 hrs 6m>£8<12m>£11. ⚓ Ⓢ (coin operated).

Inner Harbour No ❶ berths on council pontoons W of the bridge.

Weymouth Marina www.weymouth-marina.co.uk ☎767576, dredged 2·5m, 290⌓ inc ❶ £2·85 inc ⚡ £8 for <14m short stay <4 hrs: call Ch 80 for berth.

Fuel ⛽ from jetty W of LB station or by bowser (see HM or Harbour Guide for details); no petrol in harbour.

Weymouth SC ☎785481, ⌂. **Royal Dorset YC** ☎786258, ✕ ⌂.

Town ⌂ Ⓓ ⚓ ⌘ ⚓ ⚒ ⚒ Ⓔ ⚑ ⛪ Diver ⚓ ✉ Ⓑ ⛟ ✕ ⌂ @ ⇌ ✈ (Bournemouth).

Ferries St Malo: 6/week; 8¼ Hrs.

9.1.29 LULWORTH RANGES

LULWORTH FIRING RANGES comprise an inner (D026) and an outer (D026B) sea danger area; the former is more likely to impact on yachts, so to speak. See chartlet above and AC 2610.

The inner area, shaded pink, extends 5·5M offshore. It runs from just E of Lulworth Cove coastwise to Kimmeridge Bay, thence seaward clockwise and back to just E of Lulworth. The Lat/Longs of boundary corners are shown.

The outer area extends 12M offshore and adjoins the seaward boundary of the inner as shown. It is rarely used.

INFORMATION Firing times and recorded range activity are available H24 primarily from Ansafone ☎(01929) 404819; or secondarily, during office hours from Range Control ☎(01929) 404712.

Times can also be obtained from Range Control and Range Safety Boats (Ch 08), Portland CG (Ch 16), the St Alban's Head National Coastwatch station ☎(01929) 439220; and from local HMs, marinas, YCs and newspapers. Annual firing weekends and No Firing periods are given at www.reedsalmanacs.co.uk in the Almanac's January Update.

NAVAL FIRING Warships may use the inner and outer areas, firing eastward from Adamant Shoal (50°33'N 02°19'W) at the 3 DZ target buoys (up to 3M SW of St Alban's Head), which should be avoided by at least 1M. Warships fly red flags and other vessels/ helicopters may patrol the area.

ARMY FIRING takes place on the inner range most weekdays from 0930-1700 (1230 on Fri), often on Tues and Thurs nights for 3-4 hrs and for up to six weekends per year. There is NO firing in Aug and on Public Holidays. When firing is in progress red flags (at night Iso R 2s) are flown from St Alban's Head and Bindon Hill. However some red flags are flown whether or not firing is taking place; these mark the boundary of the range's land area.

REGULATIONS When the ranges are active the Range Safety boats will intercept yachts in the range and request them (Ch 08) to clear the range as quickly as possible.

However all the land danger area and the inner sea danger area are subject to The Lulworth Ranges Byelaws 1978 operative from 10 Nov 1978 – Statutory Instruments 1978 No 1663. A key passage states: *The Byelaws shall not apply to any vessel in the ordinary course of navigation, not being used for fishing, in the Sea Area and remaining in the Sea Area no longer than is reasonably necessary to pass through the Sea Area.*

Yachts should however make every reasonable effort to keep clear when firing is in progress. If on passage between Weymouth and Anvil Point, a track via 50°30'N 02°10'W just clips the SW corner of the inner range, avoids St Alban's Race and is only 3·3M longer than a direct track. See AC 2610 or 5601.4.

The Range Safety boats, which are capable of 30kn, are based in Portland Harbour.

SW England

ANCHORAGES BETWEEN PORTLAND BILL AND ANVIL PT

Helpful to read *Inshore along the Dorset Coast* (Peter Bruce).

CHURCH OPE COVE, Dorset, **50°32´·26N 02°25´·64W.** AC 2255, 2268, 5601.8. Tidal data as for Portland. A small cove on the E side of the Isle of Portland, about midway between the Bill and Portland Hbr. It is used by divers & completely open to the E, but could serve as a temporary ⚓ in about 3m off the shingle beach, to await a fair tide around the Bill.

RINGSTEAD BAY, Dorset, **50°37´·83N 02°20´·48W.** AC 2610, 5601. Tides as for Weymouth, 4M to WSW. Tempy ⚓ in 3-5m toward the E end of the bay. Ringstead Ledges, drying, define the W end of the bay. Rks on the E side restrict the effective width to about 3ca; easiest appr is from SE.

DURDLE DOOR, Dorset, **50°37´·27N 02°16´·58W.** AC 2610, 5601. Tides as for Lulworth Cove, 1M E. Durdle Door is a conspic rock archway. Close E of it Man o' War Cove offers ⚓ for shoal draught in settled weather. To the W, ⚓ may be found – with considerable caution – inside The Bull, Blind Cow, The Cow and The Calf which form part of a rocky reef.

LULWORTH COVE, Dorset, **50°37´·00N 02°14´·82W.** AC 2172, 5601. HW –0449 on Dover. Tides; ML 1·2m. Good shelter in fair weather and offshore winds, but ocasionally a squally katabatic wind may blow down from the surrounding cliffs at night. Heavy swell enters the Cove in S/SW winds; if strong the ⚓ becomes untenable. Enter the Cove slightly E of centre. 8kn speed limit. ⚓ in NE quadrant in 2·5m. Holding is poor. Local ⛽ village and ⛵ are on W side. Facilities: ⛽ ⚓ taps in car park and at slip, ✉ ✕ ⌂ museum.

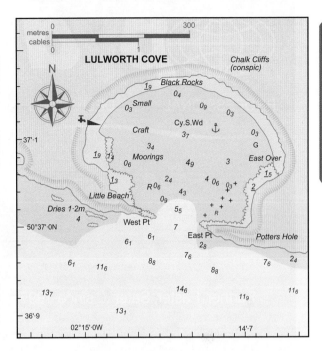

WORBARROW BAY, Dorset, **50°37´·03N 02°12´·08W.** AC 2172, 5601. Tides as Lulworth Cove/Mupe Bay. Worbarrow is a 1½M wide bay, close E of Lulworth Cove. It is easily identified from seaward by the V-shaped gap in the hills at Arish Mell, centre of bay just E of Bindon Hill. Bindon Hill also has a white chalk scar due to cliff falls. Caution: Mupe Rks at W end and other rks 1ca off NW side. ⚓s in about 3m sheltered from W or E winds at appropriate end. The bay lies within Lulworth Ranges; landing prohibited at Arish Mell. No lights/facilities. Tyneham village, deserted since WW2, is close to the E.

CHAPMAN'S POOL, Dorset, **50°35´·53N 02°03´·93W.** AC 2172, 5601. Tidal data: interpolate between Mupe Bay and Swanage. Chapman's Pool, like Brandy Bay and Kimmeridge Bay, is picturesque and comfortable when the wind is off-shore. ⚓ in depths of about 3m in centre of bay (to avoid tidal swirl). From here to St Alban's Hd the stream sets SSE almost continuously due to a back eddy. No lts. Facilities: village shop and 'Square & Compass' ⌂ ☎01929 439229, are at Worth Matravers (1.5M walk).

Central South England

Anvil Point to Selsey Bill

9.2.1	Contents	191
9.2.2	Area map, Nab Deep Water Channel and Distance table	192
9.2.3	Channel tidal stream charts	194
9.2.4	Lights, buoys and waypoints	196
9.2.5	Passage information	198
	Crossing the English Channel (central part) – 198 • Swanage to Christchurch, Western approaches to the Solent, the Solent, Isle of Wight south coast – 209 • Eastern approaches to the Solent – 237	
9.2.6	Swanage	199
9.2.7	Studland Bay	200
9.2.8	Poole Harbour, tide tables/curves	200
	Wareham	
9.2.9	Christchurch	210
9.2.10	Keyhaven	210
9.2.11	Needles Channel	211
	Alum Bay • Totland Bay • Freshwater Bay	
9.2.12	Yarmouth	212
9.2.13	Lymington	213
9.2.14	Newtown River	214
9.2.15	Beaulieu River	215
9.2.16	Isle of Wight and Solent tidal stream charts	216
9.2.17	Special tidal curves from Christchurch to Selsey Bill	218
9.2.18	The Central Solent	220
9.2.19	Cowes/River Medina	221
	Folly Reach • Newport	
9.2.20	Southampton, tide tables/curves	223
9.2.21	River Hamble	228
	Ashlett Creek • Hill Head	
9.2.22	Wootton Creek	230
9.2.23	Ryde	230
9.2.24	Bembridge	231
9.2.25	Portsmouth tide tables/curves	232
9.2.26	Langstone Harbour	238
9.2.27	Chichester Harbour, tide tables/curves	239

Central S England

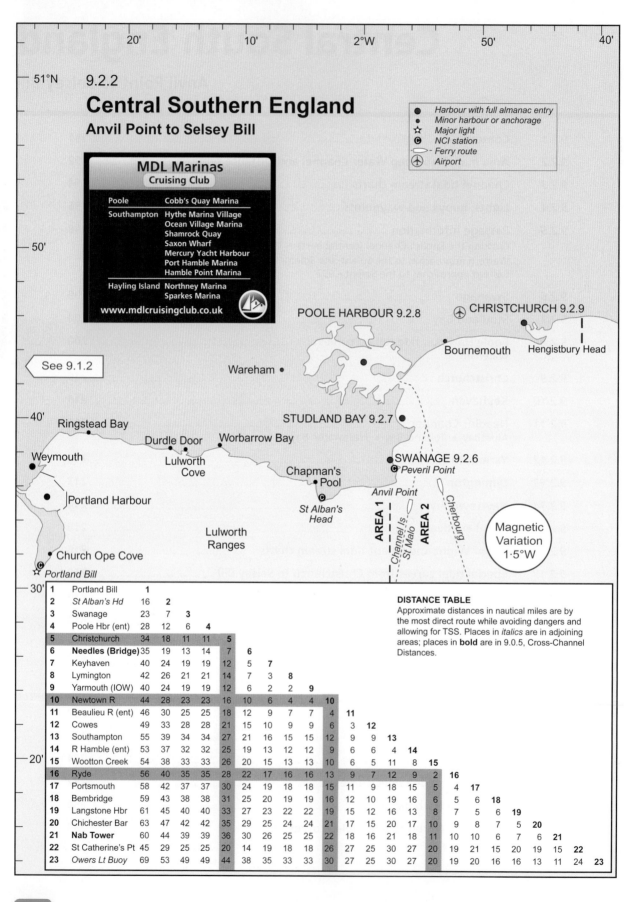

9.2.2
Central Southern England
Anvil Point to Selsey Bill

Legend:
- ● Harbour with full almanac entry
- • Minor harbour or anchorage
- ☆ Major light
- Ⓒ NCI station
- Ferry route
- ✈ Airport

MDL Marinas — Cruising Club

Poole	Cobb's Quay Marina
Southampton	Hythe Marina Village
	Ocean Village Marina
	Shamrock Quay
	Saxon Wharf
	Mercury Yacht Harbour
	Port Hamble Marina
	Hamble Point Marina
Hayling Island	Northney Marina
	Sparkes Marina

www.mdlcruisingclub.co.uk

See 9.1.2

POOLE HARBOUR 9.2.8   CHRISTCHURCH 9.2.9

Bournemouth   Hengistbury Head

Wareham

STUDLAND BAY 9.2.7

Ringstead Bay   Durdle Door   Worbarrow Bay

Weymouth   Lulworth Cove   Chapman's Pool

SWANAGE 9.2.6   Peveril Point

Portland Harbour   St Alban's Head   Anvil Point

Lulworth Ranges

AREA 1   Channel Is St Malo   AREA 2   Cherbourg

Church Ope Cove

Portland Bill

Magnetic Variation 1·5°W

DISTANCE TABLE

Approximate distances in nautical miles are by the most direct route while avoiding dangers and allowing for TSS. Places in *italics* are in adjoining areas; places in **bold** are in 9.0.5, Cross-Channel Distances.

		1	2	3	4	5	6	7	8	9	10	11	12	13	14	15	16	17	18	19	20	21	22	23
1	Portland Bill	1																						
2	*St Alban's Hd*	16	2																					
3	Swanage	23	7	3																				
4	Poole Hbr (ent)	28	12	6	4																			
5	Christchurch	34	18	11	11	5																		
6	**Needles (Bridge)**	35	19	13	14	7	6																	
7	Keyhaven	40	24	19	19	12	5	7																
8	Lymington	42	26	21	21	14	7	3	8															
9	Yarmouth (IOW)	40	24	19	19	12	6	2	2	9														
10	Newtown R	44	28	23	23	16	10	6	4	4	10													
11	Beaulieu R (ent)	46	30	25	25	18	12	9	7	7	4	11												
12	Cowes	49	33	28	28	21	15	10	9	9	6	3	12											
13	Southampton	55	39	34	34	27	21	16	15	15	12	9	9	13										
14	R Hamble (ent)	53	37	32	32	25	19	13	12	12	9	6	6	4	14									
15	Wootton Creek	54	38	33	33	26	20	15	13	13	10	6	5	11	8	15								
16	Ryde	56	40	35	35	28	22	17	16	16	13	9	7	12	9	2	16							
17	Portsmouth	58	42	37	37	30	24	19	18	18	15	11	9	18	15	5	4	17						
18	Bembridge	59	43	38	38	31	25	20	19	19	16	12	10	19	16	6	5	6	18					
19	Langstone Hbr	61	45	40	40	33	27	23	22	22	19	15	12	16	13	8	7	5	6	19				
20	Chichester Bar	63	47	42	42	35	29	25	24	24	21	17	15	20	17	10	9	8	7	5	20			
21	**Nab Tower**	60	44	39	39	36	30	26	25	25	22	18	16	21	18	11	10	10	6	7	6	21		
22	*St Catherine's Pt*	45	29	25	25	20	14	19	18	18	26	27	25	30	27	20	19	21	15	20	19	15	22	
23	*Owers Lt Buoy*	69	53	49	49	44	38	35	33	33	30	27	25	30	27	20	19	20	16	16	13	11	24	23

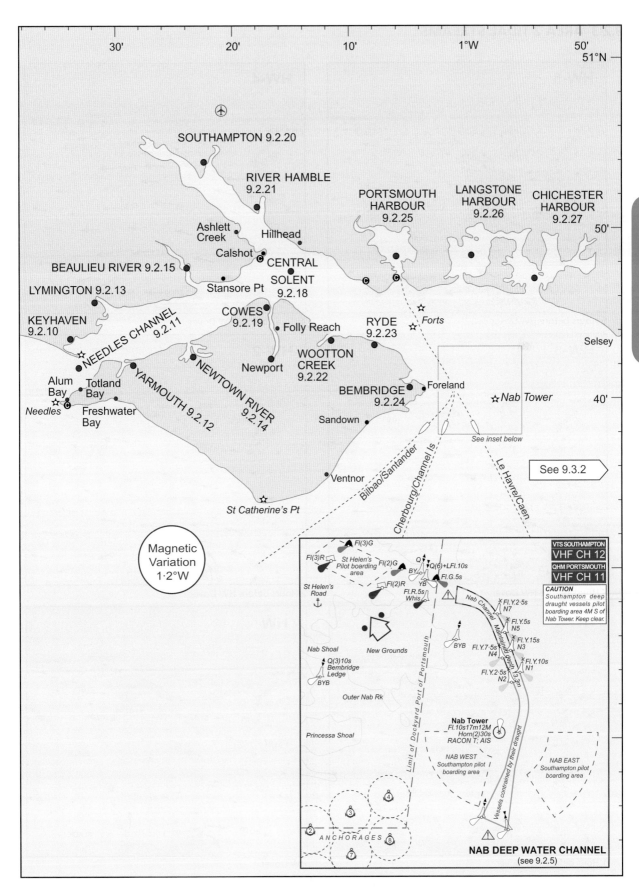

Central S England

51°N
30'   20'   10'   1°W   50'

✈
SOUTHAMPTON 9.2.20

RIVER HAMBLE
9.2.21

PORTSMOUTH
HARBOUR
9.2.25

LANGSTONE
HARBOUR
9.2.26

CHICHESTER
HARBOUR
9.2.27

Ashlett
Creek   Hillhead

Calshot
BEAULIEU RIVER 9.2.15   © CENTRAL
SOLENT
9.2.18

50'

LYMINGTON 9.2.13   Stansore Pt

©   ©

KEYHAVEN
9.2.10

NEEDLES CHANNEL
9.2.11

COWES
9.2.19   Folly Reach   RYDE
9.2.23

☆
Forts

Selsey

☆

Newport   WOOTTON
CREEK
9.2.22

Alum
Bay   Totland
Bay   NEWTOWN RIVER
9.2.14

YARMOUTH 9.2.12

BEMBRIDGE
9.2.24   Foreland

☆ Nab Tower   40'

©
Needles   Freshwater
Bay

Sandown

See inset below

Ventnor

St Catherine's Pt

Magnetic
Variation
1·2°W

Bilbao/Santander

Cherbourg/Channel Is

Le Havre/Caen

See 9.3.2

Fl(3)G

Fl(3)R   St Helen's   Fl(2)G   Q   Q(6)+LFl.10s
Pilot boarding   BY   Fl.G.5s
area   YB

St Helen's   Fl(2)R
Road   Fl.R.5s
⚓   Whis

VTS SOUTHAMPTON
VHF CH 12
QHM PORTSMOUTH
VHF CH 11

CAUTION
Southampton deep
draught vessels pilot
boarding area 4M S of
Nab Tower. Keep clear.

Nab Channel   Fl.Y.2·5s
N7
Fl.Y.5s
N5
Fl.Y.15s
N3

BYB   Fl.Y.7·5s
N4   Fl.Y.10s
N1

Nab Shoal   New Grounds

Q(3)10s
Bembridge
Ledge
BYB

Outer Nab Rk

Princessa Shoal

Nab Tower
Fl.10s17m12M
Horn(2)30s
RACON T; AIS

NAB WEST
Southampton pilot
boarding area

NAB EAST
Southampton pilot
boarding area

Fl.Y.2·5s
N2   13·3m

Limit of Dockyard Port of Portsmouth

Vessels constrained by their draught

4

3

2   8

A N C H O R A G E S

7

NAB DEEP WATER CHANNEL
(see 9.2.5)

9.2.3 AREA 2 TIDAL STREAMS

HW-5

5 Hours before HW Dover

HW-4

4 Hours before HW Dover

HW-3

3 Hours before HW Dover

HW-2

2 Hours before HW Dover

HW-1

1 Hour before HW Dover

HW

HW Dover

Westward 9.1.3   Portland 9.1.26   Isle of Wight 9.2.16   Eastward 9.3.3   Southward 9.18.3   Channel Is 9.19.3

1 Hour after HW Dover

2 Hours after HW Dover

3 Hours after HW Dover

4 Hours after HW Dover

5 Hours after HW Dover

6 Hours after HW Dover

9.2.4 LIGHTS, BUOYS AND WAYPOINTS

Bold print = light with a nominal range of 15M or more. CAPITALS = place or feature. *CAPITAL ITALICS* = light-vessel, light float or Lanby. *Italics* = Fog signal. ***Bold italics*** = Racon. Some marks/buoys are fitted with AIS (<u>MMSI No</u>); see relevant charts.

SWANAGE TO ISLE OF WIGHT

SWANAGE
Pier Hd ⚡ 2 FR (vert) 6m 3M; 50°36'·56N 01°56'·95W.
Peveril Ledge ⚓ QR; 50°36'·41N 01°56'·10W.

POOLE BAR and SWASH CHANNEL
Poole Bar (No. 1) ⚑ QG; *Bell*; 50°39·29N 01°55'·14W.
(Historic wreck) ⚓ Fl Y 5s; 50°39'·70N 01°54'·86W.
South Hook (No. 11) ⌁ 50°39'·70N 01°55'·20W.
No. 2 ⚓ Fl R 2s; 50°39'·23N 01°55'·24W.
No. 3 ⚑ Fl G 3s; 50°39'·76N 01°55'·49W.
No. 4 ⚓ Fl R 2s; 50°39'·72N 01°55'·60W.
Training Bank ⌁ 2 FR (vert); 50°39'·82N 01°55'·86W.
No. 6 ⚓ Fl R 4s; 50°40'·14N 01°55'·91W.
No. 5 ⚑Fl G 5s; 50°40'·19N 01°55'·81W.
Hook Sands (No. 7) ⚑ Fl G 3s; 50°40'·50N 01°56'·16W.
Channel (No. 8) ⚓ Fl R 2s; 50°40'·45N 01°56'·27W.
No. 10 ⚓ Fl R 4s; 50°40'·84N 01°56'·86W.
Swash (No. 9) ⌁ Q (9) 15s; 50°40'·88N 01°56'·70W.

EAST LOOE CHANNEL
East Hook ⚓ 50°40'·58W 01°55'·23W.
East Looe 1 ⚑ Fl G 5s; 50°41'·09N 01°55'·82W.
East Looe 2 ⚓ Fl R 4s; 50°41'·07N 01°55'·83W.
East Looe 3 (Limit 10 knots) ⚑ Fl G 3s; 50°41'·11N 01°56'·16W.
East Looe 4 (Limit 10 knots) ⚓ Fl R 2s; 50°41'·09N 01°56'·17W.
North Hook ⚓ Fl (2) R 5s; 50°41'·01N 01°56'·44W.

BROWNSEA ROADS
No. 12 ⚓ Q R; 50°40'·94N 01°57'·17W.
No.14 ⚓ Fl R 2s; 50°41'·03N 01°57'·32W.
N Haven ⌁ Q (9) 15s 5m; 50°41'·15N 01°57'·17W.
Brownsea ⌁ Q (3) 10s; 50°41'·14N 01°57'·39W.
Brownsea Island Dir lt 299°F WRG; 296·5°-G-297·8°-AltWG-298·8°-W-299·2°-AltWR-300·2°-R-301·5°; 50°41'·16N 01°57'·67W (only shown for commercial vessels); 2FR(vert); 301·5°-296·5° (H24).

MIDDLE SHIP CHANNEL
Bell (No. 15) ⌁ Q (6) + L Fl 15s; 50°41'·36N 01°57'·12W.
No. 16 ⚓ VQ R; 50°41'·43N 01°57'·25W.
No. 17 ⚑ Fl G 3s; 50°41'·68N 01°57'·02W.
Aunt Betty (No. 22) ⌁ Q (3)10s; 50°41'·97N 01°57'·25W.
Diver (No. 25) ⌁ Q (9) 15s; 50°42'·29N 01°58'·32W.

NORTH CHANNEL
Salterns Marina Outer Bkwtr Hd ⚡ 2 FR (vert) 2M; Tfc sigs; 50°42'·23N 01°57'·10W.
Parkstone YC platform ⚡ Q 8m 1M; hut on dolphin; 50°42'·37N 01°58'·08W.
Stakes (No. 29) ⌁ Q (6) + L Fl 15s; 50°42'·43N 01°59'·00W.

POOLE BAY
Bournemouth Rocks ⚓ 50°42'·32N 01°53'·40W.
Lightwave ⚓ 50°41'·50N 01°51'·68W.
Christchurch Ledge ⚓ 50°41'·57N 01°41'·55W (Apr-Oct).
⌁ (x2) Fl (5) Y 20s; 50°37'·98N 01°43'·02W; (265° 2.6M from Needles F'wy).

WESTERN APPROACHES TO THE SOLENT

NEEDLES CHANNEL
Needles Fairway ⌁ L Fl 10s; *Whis*; 50°38'·24N 01°38'·98W.
SW Shingles ⌁Fl R 2·5s; 50°39'·29N 01°37'·52W.
Bridge ⌁ VQ (9) 10s; ***Racon (T) 10M***; 50°39'·63N 01°36'·88W.

NEEDLES ☆ 50°39'·73N 01°35'·50W; Oc (2) WRG 20s 24m **W17M**, R14M, R13M G14M; ○ Twr, R band and lantern; vis: shore-R-300°-W-083°-R (unintens)-212°-W-217°-G-224° (H24). *Horn (2) 30s.*
Shingles Elbow ⚓ Fl (2) R 5s; 50°40'·37N 01°36'·05W.
Mid Shingles ⚓ Fl (3) R 10s; 50°41'·21N 01°34'·66W.
Warden ⚑ Fl G 2·5s; *Bell*; 50°41'·48N 01°33'·55W.
NE Shingles ⌁ Q (3) 10s; 50°41'·96N 01°33'·41W.

Hurst Point ☆ 50°42'·48N 01°33'·03W; FL (4) WR 15s 23m W13M, R11M; W ○ Twr; vis:080°-W(unintens)-104°, 234°-W-244°-R-250°-W-053°. Same structure, Iso WRG 4s 19m **W21M, R18M, G17M**; vis: 038·8°-G-040·8°-W-041·8°-R- 043·8°; By day W7M, R5M, G5M.

NORTH CHANNEL
North Head ⚑ Fl (3) G 10s; 50°42'·69N 01°35'·52W.

THE WESTERN SOLENT

Note: Numerous yellow yacht racing buoys are laid throughout the Solent (seasonal, Mar-Dec). Most, but not all, are lit Fl Y 4s.

SOLENT MARKS
Sconce ⌁ Q; *Bell*; 50°42'·53N 01°31'·43W.
Black Rock ⚑ Fl G 5s; 50°42'·57N 01°30'·59W.
Lymington Bank ⚓ Fl (2) R 5s; *Bell*; 50°43'·10N 01°30'·85W.
Solent Bank ⚓ Fl (3) R 10s; 50°44'·23N 01°27'·37W.
Hamstead Ledge ⚑ Fl (2) G 5s; 50°43'·87N 01°26'18W.
Newtown River ⌁ Q (9) 15s; 50°43'·75N 01°24'·96W.
W Lepe ⚓ Fl R 5s; 50°45'·24N 01°24'·09W.
Salt Mead ⚑ Fl (3) G 10s; 50°44'·51N 01°23'·04W.
Gurnard Ledge ⚑ Fl (4) G 15s; 50°45'·51N 01°20'·59W.
E Lepe ⚓ Fl (2) R 5s; *Bell*; 50°45'·93N 01°21'·07W.
Lepe Spit ⌁ Q (6) + L Fl 15s; 50°46'·78N 01°20'·64W.
Gurnard ⌁ Q; 50°46'·22N 01°18'·84W.

YARMOUTH
East Fairway ⚓ Fl R 2s; 50°42'·62N 01°29'·95W.
Poole Belle ⚓ Fl Y 5s; 50°42'·54N 01°30'·17W.
Pier Head, centre, ⚡ 2 FR (vert) 2M; G col. High intensity FW (occas); 50°42'·51N 01°29'·97W.
Ldg Lts 187·6° Front FG 5m 2M; 50°42'·36N 01°30'·06 W. Rear, 63m from front, FG 9m 2M; both W ◊.

LYMINGTON
Ldg Lts 319·5°, Or posts. Front, FR 12m 8M; 50°45'·19N 01°31'·65W. vis: 309·5°-329·5°. Rear, FR 17m 8M.
Jack in the Basket ⚓ Fl R 2s 9m; 50°44'·27N 01°30'·57W.
No. 1 ⌁ Fl G 2s 2m 3M; G △ on pile; 50°44'·41N 01°30'·48W.
No. 2 (Cross Boom) ⌁ Fl R 2s 4m 3M; 50°44'·36N 01°30'·58W.
Yacht Haven ldg lts 244°. Front FY 4m; R △; 50°45'·09N 01°31'·53W. Rear, 22m from front, FY 6m; R ▽.

BEAULIEU RIVER
Millennium Dir lt 334°. ⚡ Oc WRG 4s 13m W4M, R3M, G3M; vis: 318°-G-330°-W-337°-R-348°; 50°47'·12N 01°21'·90W.
Beaulieu Spit, E end ⌁ Fl R 5s 3M; R dolphin, W band; 50°46'·85N 01°21'·76W.
No. 1 ⌁ 50°46'·91N 01°21'·70W.
No 2 ⌁ 50°46'·92N 01°21'·78W.

COWES
Prince Consort ⌁ VQ; 50°46'·41N 01°17'·56W.
Trinity House ⚑ Fl Y 5s; 50°46'·31N 01°17'·75W.
No. 1 ⚑ Fl G 3s; 50°46'·07N 01°18'·03W.
No. 2 ⚓ Q R; 50°46'·07N 01°17'·87W.
No 2A ⚓ LFl R 10s; 50°45'·92N 01°17'·70W.
Tide gauge ⌁ L Fl R 5s; 50°45'·78N 01°17'·66W.
E. Cowes Bkwtr Hd ⚡ 2F R; 50°45'·88N 01°17'·52W.
⚑ Fl G 3s; 50°45'·91N 01°17'·55W.
⚓ Fl R 3s; 50°45'·89N 01°17'·53W.
Shrape Mud ⚑ Fl G 5s; 50°46'·02N 01°17'·38W.
⚓ Fl R 5s; 50°46'·00N 01°17'·37W.

Boat Chan Ent ◣ Fl G 2s; 50°46'·08N 01°17'·05W.
  ◤ Q R; 50°46'·06N 01°17'·04W.
Shrape Bn ⌕ LFl R 10S 3m 3M; 50°46'·09N 01°16'·90W.

CENTRAL SOLENT AND SOUTHAMPTON WATER

Note: Numerous yellow yacht racing buoys are laid throughout the Solent (seasonal, Mar-Dec). Most, but not all, are lit Fl Y 4s.

SOLENT MARKS

Lepe Spit ⌕ Q (6) + L Fl 15s; 50°46'·78N 01°20'·64W.
NE Gurnard ◿Fl (3) R 10s; 50°47'·06N 01°19'·42W.
South Bramble ◣ Fl G 2·5s; 50°46'·98N 01°17'·72W.
W Bramble ⌕ VQ (9) 10s; *Bell*; **Racon (T) 3M**; 50°47'·20N 01°18'·65W.
Thorn Knoll ◣Fl G 5s; 50°47'·50N 01°18'·44W.
Bourne Gap ◿ Fl R 3s; 50°47'·83N 01°18'·34W.
West Knoll ◿ Fl Y 2·5s; 50°47'·43N 01°17'·84W.
North Thorn ◣ QG; 50°47'·91N 01°17'·84W.
Stanswood Outfall ⌕ Iso R 10s 6m 5M; 4 FR Lts; 50°48'·26N 01°18'·82W.

CALSHOT REACH

East Knoll ◣ 50°47'·96N 01°16'·83W.
CALSHOT SPIT ◔ Fl 5s 5m 10M; R hull, Lt Twr amidships; *Horn (2) 60s*; 50°48'·35N 01°17'·64W.
Calshot ⌕ VQ; *Bell*; 50°48'·44N 01°17'·03W.
Castle Point ◿ IQ R 10s; 50°48'·71N 01°17'·67W.
Reach ◣ Fl (3) G 10s; 50°49'·05N 01°17'·65W .
Black Jack ◿ Fl (2) R 4s; 50°49'·13N 01°18'·09W.
Hook ⌕ QG; *Horn (1) 15s*; 50°49'·52N 01°18'·30W.
Coronation ◣ Fl Y 5s; 50°49'·55N 01°17'·62W.
Bald Head ◣ Fl G 2·5s; 50°49'·80N 01°18'·06W.

RIVER HAMBLE

Hamble Pt ⌕ Q (6) + L Fl 15s; 50°50'·15N 01°18'·66W.
Hamble Common Dir lt 351·7°, Oc (2) WRG 12s 5m W4M; R4M; G4M; vis: 348·7°-G-350·7°-W-352·7°-R-354·7°; 50°51'·00N 01°18'·84W.
Sailing Club Dir lt 028·9° ⚓ Iso WRG 6s 5m W4M, R4M, G4M: vis: 025·9°-G-027·9°-W-029·9°-R-031·9°; 50°51'·10N 01°18'·34W.

SOUTHAMPTON WATER

Fawley Marine Terminal. SE end ⚓ 2 FR (vert) 9m 10M; 50°50'·06N 01°19'·42W.
Greenland ◣ Iso G 2s; 50°51'·11N 01°20'·38W.
Cadland ◿ Fl R 3s; 50°51'·02N 01°20'·54W.
Lains Lake ◿ Fl (2) R 4s; 50°51'·59N 01°21'·65W.
Hound ◣ Fl (3) G 10s; 50°51'·68N 01°21'·52W.
Netley ◣ Fl G 3s; 50°52'·03N 01°21'·81W.
Deans Elbow ◿ Oc R 4s; 50°52'·16N 01°22'·76W.
NW Netley ◣ Fl G 7s; 50°52'·31N 01°22'·73W.
Weston Shelf ◣ Fl (3) G 15s; 50°52'·71N 01°23'·26W.

HYTHE

Hythe Pier Hd ⚓ 2 FR (vert) 12m 5M; 50°52'·49N 01°23'·61W.
Hythe Marina Ent ⌕ Q (3) 10s; 50°52'·63N 01°23'·88W.
Hythe Knock ◿ Fl R 3s; 50°52'·83N 01°23'·81W.

SOUTHAMPTON and RIVER ITCHEN

Swinging Ground No. 1 ◣ Oc G 4s; 50°53'·00N 01°23'·44W.
E side. No. 1 ⌕ QG; 50°53'·15N 01°23'·40W.
No. 2 ⌕ Fl G 5s 2M; 50°53'·29N 01°23'·38W.
No. 3 ⌕ Fl G 7s; 50°53'·48N 01°23'·28W.
No. 4 ⌕ QG 4m 2M; 50°53'·62N 01°23'·16W.

SOUTHAMPTON and RIVER TEST

Queen Elizabeth II Terminal, S end ⚓ 4 FG (vert) 16m 3M; 50°53'·00N 01°23'·71W.
Gymp ◿ QR; 50°53'·17N 01°24'·30W.
Town Quay Ldg Lts 329°, both F 12/22m 3/2M.
Gymp Elbow ◿ Oc R 4s; 50°53'·50N 01°24'·68W.
Dibden Bay ⌕ Q; 50°53'·70N 01°24'·92W.

THE EASTERN SOLENT

Note: Numerous yellow yacht racing buoys are laid throughout the Solent (seasonal, Mar-Dec). Most, but not all, are lit Fl Y 4s.

SOLENT MARKS

West Ryde Middle ⌕ Q (9) 15s; 50°46'·48N 01°15'·79W.
Norris ◿ Fl (3) R 10s; 50°45'·97N 01°15'·51W.
North Ryde Middle ◿ Fl (4) R 20s; 50°46'·61N 01°14'·31W.
South Ryde Middle ◣ Fl G 5s; 50°46'·13N 01°14'·16W.
Peel Bank ◿ Fl (2) R 5s; 50°45'·49N 01°13'·35W.
SE Ryde Middle ⌕ VQ (6)+L Fl 10s; 50°45'·93N 01°12'·10W.
NE Ryde Middle ◿ Fl (2) R 10s; 50°46'·21N 01°11'·88W.
Mother Bank ◿ Fl R 3s; 50°45'·49N 01°11'·21W.
Browndown ◣ Fl G 15s; 50°46'·57N 01°10'·95W.
Fort Gilkicker ⚓ Oc G 10s 7M; 50°46'·43N 01°08'·47W.
N Sturbridge ⌕ VQ; 50°45'·33N 01°08'·23W.
Ryde Sands ⌕ Fl R 10s; 50°44'·56N 01°07'·26W.
Ryde Sands ⌕ L Fl R 12s; 50°44'·16N 01°05'·99W.
No Man's Land Fort ⚓ Iso R 2s 21m 8M; 50°44'·40N 01°05'·70W.
Horse Sand Fort ⚓ Iso G 2s 21m 8M; 50°45'·01N 01°04'·34W.
Saddle ◣ VQ (3) G 10s; 50°45'·20N 01°04'·98W.

NORTH CHANNEL and HILLHEAD

Calshot ⌕ VQ; *Bell (1) 30s*; 50°48'·44N 01°17'·03W.
Hillhead ◿ Fl R 2·5s; 50°48'·07N 01°16'·00W.
E Bramble ⌕ VQ (3) 5s; 50°47'·23N 01°13'·64W.

WOOTTON CREEK

Wootton Beacon ⌕ Q 1M; (NB); 50°44'·53N 01°12'·13W.
Dir lt. Oc WRG 10s vis: 220·8°-G-224·3°-W-225·8°-R-230·8°; 50°44'·03N 01°12'·86W.

RYDE

Ryde Pier ⚓, NW corner, N and E corner marked by 2 FR (vert). In fog FY from N corner, vis: 045°-165°, 200°-320°; 50°44'·34N 01°09'·72W.
Leisure Hbr E side ⚓ 2 FR (vert) 7m 1M. FY 6m shown when depth of water in Hbr greater than 1m; 2 FY 6m when depth exceeds 1·5m; 50°43'·99N 01°09'·29W.

PORTSMOUTH APPROACHES

Horse Sand ◣ Fl G 2·5s; 50°45'·53N 01°05'·27W.
Outer Spit ⌕ Q (6) + L Fl 15s; 50°45'·58N 01°05'·50W.
Mary Rose ◿ Fl Y 5s; 50°45'·80N 01°06'·20W.
Boyne ◣ Fl G 5s; 50°46'·15N 01°05'·26W.
Spit Refuge ◿ Fl R 5s; 50°46'·15N 01°05'·46W.
Spit Sand Fort ⚓ Fl R 5s; 18m 7M. 50°46'·24N 01°05'·94W.
Castle (NB) ◿ Fl (2) G 6s; 50°46'·45N 01°05'·38W.
Southsea Castle N corner ⚓ Iso 2s 16m 11M, W stone Twr, B band; vis: 337°-071°; 50°46'·69N 01°05'·33W.
Southsea Castle Dir lt 001·5° WRG 11m W13M, R5M, G5M; same structure; vis: 351·5°-FG-357·5°-Al WG (W phase incr with brg), 000°-FW-003°-AlWR(R phase incr with brg), 005·5°-FR-011·5°; 50°46'·69N 01°05'·33W.
Ridge ◿ Fl (2) R 6s; 50°46'·44N 01°05'·65W.
No. 1 Bar (NB) ◣ Fl (3) G 10s; 50°46'·77N 01°05'·81W.
No. 2 ◿ Fl (3) R 10s; 50°46'·69N 01°05'·97W.
No. 3 ◣ QG; 50°47'·08N 01°06'·24W.
No. 4 ◿ QR; 50°47'·01N 01°06'·36W.
BC Outer ⌕ Oc R 15s; 50°47'·32N 01°06'·68W.

PORTSMOUTH HARBOUR

Fort Blockhouse ⚓ Dir lt 320°; WRG 6m W13M, R5M, G5M; vis: 310°- Oc G-316°-Al WG(W phase incr with brg), 318·5°-Oc-321·5°- Al WR (R phase incr with brg), 324°-Oc R-330°. 2 FR (vert) 20m E; 50°47'·37N 01°06'·74W.
Ballast ⌕ Fl R 2·5s; 50°47'·62N 01°06'·83W.
Hbr Ent Dir lt (NB) (Fuel Jetty) ⚓ WRG 2m 1M; vis: 322·5°-Iso G 2s -330°-Al WG-332·5°-Iso 2s(main chan) -335°-Al WR-337·5°-Iso R 2s-345° (Small Boat Chan) H24; 50°47'·85N 01°06'·98W.

EASTERN APPROACHES to THE SOLENT
Outer Nab 1 ⚓ VQ (9) 10s; 50°38'·18N 00°56'·88W.
Outer Nab 2 ⚓ VQ (3) 5s; 50°38'·43N 00°57'·70W.
Nab Tower ⚓ Fl 10s 17m 12M; *Horn (2) 30s*; RACON T;
50°40'·08N 00°57'·15W; <u>992351136</u>.
N 2 ⚓ Fl Y 2·5s. 6M; 50°41'·03N 00°56'·74W.
N 1 ⚓ Fl Y (4)10s; 50°41'·26N 00°56'·52W.
N 4 ⚓ Fl Y 7·5s; 50°41'·86N 00°57'·24W.
N 3 ⚓ Fl (3) Y 15s; 50°41'·63N 00°56'·74W.
N 5 ⚓ Fl Y 5s; 50°41'·99N 00°56'·97W.
N 7 ⚓ Fl Y 2·5s; 50°42'·35N 00°57'·20W.
New Grounds ⚓ VQ (3) 5s; 50°41'·84N 00°58'·49W.
Nab End ⚓ Fl R 5s; *Whis*; 50°42'·63N 00°59'·49W.
Dean Tail ▲ Fl G 5s; 50°42'·99N 00°59'·17W.
Dean Tail S ⚓ Q (6) + L Fl 10s; 50°43'·04N 00°59'·57W.
Dean Tail N ⚓ Q; 50°43'·13N 00°59'·57W.
Horse Tail ▲ Fl (2) G 10s; 50°43'·23N 01°00'·23W.
Nab East ⚓ Fl (2) R 10s; 50°42'·86N 01°00'·80W.
Dean Elbow ▲ Fl (3) G 15s; 50°43'·69N 01°01'·88W.
St Helens ⚓ Fl (3) R 15s; 50°43'·36N 01°02'·41W.
Horse Elbow ▲ QG; 50°44'·26N 01°03'·88W.
Cambrian Wreck ⚓ 50°44'·43N 01°03'·43W.
Warner ⚓ QR; *Whis*; 50°43'·87N 01°03'·99W.

BEMBRIDGE
St Helen's Fort ☆ (IOW) Fl (3) 10s 16m 8M; large ○ stone
structure; 50°42'·30N 01°05'·05W.
Bembridge Tide Gauge ⚓ Fl Y 2s 1M; 50°42'·46N 01°05'·02W.

SOUTH EAST COAST of the ISLE OF WIGHT
St Catherine's Point ☆ 50°34'·54N 01°17'·87W; Fl 5s 41m **25M**;
vis: 257°-117°; FR 35m 13M (same Twr) vis: 099°-116°.
Ventnor Haven W Bwtr ⚓ 2 FR (vert) 3M; 50°35'·50N 01°12'·30W.
Sandown Pier Hd ⚓ 2 FR (vert) 7m 2M; 50°39'·05N 01°09'·18W.
W Princessa ⚓ Q (9) 15s; 50°40'·16N 01°03'·65W.
Bembridge Ledge ⚓ Q (3) 10s; 50°41'·15N 01°02'·81W

LANGSTONE and APPROACHES
Eastney Pt Fraser Trials Range ⚓ FR, Oc (2) Y 10s, and FY Lts
(occas) when firing taking place; 50°47'·19N 01°02'·22W.
Winner ⚓; 50°45'·10N 01°00'·10W.
Roway Wk ⚓ Fl (2) 5s; 50°46'·11N 01°02'·28W.
Langstone Fairway ⚓ L Fl 10s; 50°46'·32N 01°01'·36W.
Eastney Pt Outfall ⚓ QR 2m 2M; 50°47'·23N 01°01'·68W.
East Milton ⚓ Fl (4) R 10s; 50°48'·16N 01°01'·76W.
NW Sinah ▲ Fl G 5s; 50°48'·14N 01°01'·58W.

CHICHESTER ENTRANCE
West Pole (tripod) ⚓ Fl R 5s 14m 7M; 50°45'·45N 00°56'·59W.
Bar ⚓ Fl(2) R 10s 10m 4M; 50°45'·92N 00°56'·46W.
Eastoke ⚓ QR; 50°46'·68N 00°56'·11W.
West Winner ⚓ QG; Tide gauge. 50°46'·88N 00°55'·98W.

EMSWORTH CHANNEL
Fishery ⚓ Q (6) + L Fl 15s; 50°47'·38N 00°56'·07W.
NW Pilsey ▲ Fl G 5s; 50°47'·50N 00°56'·20W.
Verner ⚓ Fl R 10s; 50°48'·20N 00°56'·63W.
Marker Pt ⚓ Fl (2) G 10s 8m; 50°48'·91N 00°56'·72W.
Emsworth ⚓ Q (6) + L Fl 15s; tide gauge. 50°49'·66N 00°56'·76W.

THORNEY CHANNEL
Camber ⚓ Q (6) + L Fl 15s; 50°47'·87N 00°54'·06W.
Pilsey ⚓ Fl (2) R 10s; 50°47'·98N 00°54'·24W.
Thorney ⚓ Fl G 5s; 50°48'·20N 00°54'·28W.

CHICHESTER CHANNEL
NW Winner ▲ Fl G 10s; 50°47'·19N 00°55'·92W.
N Winner ▲ Fl (2) G 10s; 50°47'·31N 00°55'·83W.
Mid Winner ▲ Fl (3) G 10s; 50°47'·38N 00°55'·69W.
Stocker ⚓ Fl (3) R 10s; 50°47'·45N 00°55'·52W.
Copyhold ⚓ Fl (4) R 10s; 50°47'·50N 00°54'·93W.
East Head Spit ▲ Fl (4) G 10s; 50°47'·45N 00°54'·82W.
Snowhill ▲ Fl G 5s; 50°47'·52N 00°54'·34W.
Sandhead ⚓ Fl R 10s; 50°47'·67N 00°54'·25W.
Chalkdock ⚓ Fl (2) G 10s; 50°48'·49N 00°53'·30W.

9.2.5 PASSAGE INFORMATION

More passage information is threaded between the harbours
in this area. See 9.0.5 for distances across the central part of the
English Channel. Admiralty Leisure Folio 5601 covers Exmouth
to Christchurch, and 5600 covers The Solent & Approaches.
Bibliography: *Channel Havens* (ACN/Endean); *Shell Channel
Pilot* (Imray/Cunliffe); *Channel Pilot* (Admiralty NP27).

CROSSING THE ENGLISH CHANNEL – CENTRAL PART

(AC 2656) These notes cover the crossing (about 13 hrs)
Weymouth/Poole/Solent to Channel Is/Cherbourg/Le Havre
and should help in compiling a cross-Channel passage plan.
Distances are tabulated in 9.0.5.

Departure and destination factors to be considered include:

Portland/Weymouth: Good angle on the wind in SW/lies but
passage extended by 7/8½M respectively to clear the E end of
The Shambles Bank. There is no advantage to closing Portland
Bill and Portland Race which should be avoided. Portland Bill
Light loom provides a good headmark on night N'ly crossing.
Check the tidal streams and monitor course made good.

Poole: Leave on the ebb. At springs with a S/SE wind beware
short steep seas in the Swash Channel. Studland Bay or
Swanage Roads are useful departure ⚓s. Anvil Pt Light and the
Ferries provide useful visual cues on N'ly crossing. Off Handfast,
Peveril and Anvil Pts overfalls occur with wind against tide.

Solent ports: Decide whether to leave to the East via Nab Tower,
using the tide to gain a lift on the ebb past Bembridge Ledge
and across St Catherine's Deep or west via the Needles.

▶The latter usually requires a fair tide through Hurst Narrows
(HW Portsmouth –0100 to +0430) which dictates the ETD. ◀

The Easterly option has no tidal gate, but is longer if heading
to Cherbourg or the Channel Islands and gives a less favourable
angle on a SW'ly wind. It can offer better shelter initially in the
lee of the IoW, but often has an uncomfortable chop especially
on a return passage. Loom of St Catherine's and the Needles
are useful guides at night, but can be deceptively far away.

Braye (Alderney) is accessible at all times, but allowance must
be made to avoid being swept W to the Casquets at springs.
Plan and monitor the track so as to approach from well up-tide.
Braye Hbr is unsafe in fresh or strong E/NE'lies. From Braye you
may make Guernsey, having negotiated the Alderney Race, or
the Swinge. A direct passage is achievable off the Casquets or
in benign conditions through Ortac Channel.

Cherbourg is accessible H24. Especially at springs, aim well up-
tide and approching the peninsula allow a large offset angle
to maintain track. Harbours on the W side of the Cotentin
(**Diélette** to **Granville**) are tidally constrained and exposed to
the W but are options in E'lies if conditions in the Alderney
Race permit on a W-going ebb tide or at slack water.

To the E, **Barfleur** and **St Vaast** are sheltered from the prevailing
W'lies, but are tidally constrained. Barfleur dries and yachts off
St Vaast until the marina gates open. Further to the E **Le Havre**
is a port of refuge although approach requires care.

Thorough planning is a requirement for a safe and efficient
crossing. Refer to 3.1, Passage Planning.

- Study the meteorological situation several days before departure so that advantageous wind and weather windows may be predicted and bad weather avoided.

- High pressure may provide quiescent conditions, but is often accompanied by reduced visibility owing to fog. Even if radar equipped and competent in its use, be wary of crossing if fog or poor visibility is forecast.

- For a yacht, consider the forecast wind direction, likely shifts and sea breezes which may affect the angle on the wind. Prevailing winds are SW/W, except in the spring when NE/E winds are equally likely. Check the tidal stream and the effect that this will have on Course to Steer. It is best to place the tide on the lee bow in order to gain the maximum benefit. It is very advantageous to get well to windward and up-tide of the destination. For example, from the Needles it may be beneficial initially to proceed W'wards, working the tides to advantage.

- For motor boats, windows with the wind in the N provide the best down sea conditions, maximising speed and reducing slamming.

- Choose the route, departure points and landfalls so that passage time out of sight of identifiable marks is minimised. This reduces the risk of navigational errors, anxiety and fatigue. A landfall at night or dawn/dusk is frequently easier due to the visibility of navigational marks on clear nights. The loom of a powerful lighthouse is often visible on the clouds at a distance greater than its nominal range.

- Consider tidal constraints at the points of departure and destination, and any tidal gates en route such as Hurst Narrows or the French coast between Cap Barfleur and Cap de la Hague.

▶ *Calculate the hourly direction/rate of tidal streams expected during the crossing so as to lay off the total drift angle required to make good the desired track. Rarely do 6 hrs of E-going tide cancel out 6 hrs of W-going (or vice versa); streams off the French coast are usually stronger. Note times and areas of races/overfalls and keep well clear.* ◀

- Consider actions to be taken if fog sets in as the risk of collision is much increased. GPS may provide a position, but navigating 'blind' by radar demands intense attention. Other small boats and yachts may not be detected.

- AIS will only display vessels or marks which transmit.

- Cross on a heading as near 90° as practicable to the Traffic Routeing between the Casquets TSS and the Greenwich Meridian. Consider motoring to expedite such a crossing.

- Keep a very sharp lookout particularly when in the Traffic Routeing area in mid-Channel. Ensure the crew is aware of the directions from which to expect approaching traffic. It is commonsense and good seamanship to alter course early in order to keep well clear of large commercial ships. These may not alter for a small craft despite Rule 18 (a) (iv).

- Make use of additional navigational information such as soundings, noting when crossing distinctive seabed contours; the rising or dipping ranges of major lights; if equipped with radar, use clearly identifiable targets.

- Plan a harbour of refuge in case of fog, bad weather, accident or gear failure. For example, if unable to make Cherbourg in a strong SSW'ly and E-going tide, consider bearing away for St Vaast in the lee of the peninsula. Alternatively, in the event of engine failure identify a sheltered ⚓. In extremis heave to and stay at sea notifying the appropriate CG or safety service.

9.2.6 SWANAGE

Dorset 50°36'·70N 01°56'·55W ❀❀❀⚓⚓🌢🌢

CHARTS AC 5601, 2615, 2172; Imray C10, C12, C4, 2300

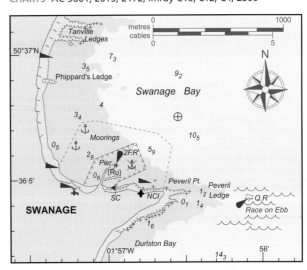

TIDES HW Sp −0235 & +0125, Np −0515 & +0120 on Dover; ML 1·5

Standard Port POOLE HARBOUR (→)

Times				Height (metres)			
High Water		Low Water		MHWS	MHWN	MLWN	MLWS
—	—	0500	1100	2·2	1·7	1·2	0·6
—	—	1700	2300				
Differences SWANAGE							
—	—	−0045	−0050	−0·1	+0·1	+0·2	+0·2

NOTE: From Swanage to Christchurch double HWs occur except at neaps. HW differences refer to the higher HW and are approximate.

SHELTER Large, well sheltered bay with just 1·6m range at springs. Good ⚓ in all winds except from NE to SE; Poole is closest port of refuge. ⌔ may be possible on the pier which is owned by a trust and used extensively in the summer by tourist boats, ferries and other commercial concerns. Seek advice at the gate on arrival. Gate closes at dusk Oct to Apr.

NAVIGATION WPT 50°36'·70N 01°56'·55W, 240°/0.3M to Pier Hd. Tidal races off Peveril Point to the south and Handfast Point to the north can be lively in strong onshore winds and an ebb tide. Do not cut inside Peveril Ledge buoy.

Beware of ruins of old pier south of the main pier; Tanville and Phippards Ledges in west side of bay both dry. At night, main hazard is other boats which may be difficult to see against background lights.

LIGHTS AND MARKS 2 FR (vert) on the Pier, difficult to see due to back scatter of street lights. Peveril Ledge PHM buoy, Q.R, but can be hard to pick out by day.

COMMUNICATIONS (Code 01929) NMOC (01305) 760439; Police 101; Dr 422231; ⊞ 422202; Pier 427058. No VHF.

FACILITIES Pier Possible⌔; Swanage SC ☎422987 ⅙ ⎘ 🛆 Boat Park (Peveril Pt) ⎘ ⎅ ⚓ Services: Diving. Town ⌂ ⌂ (1½M) 🛒 ✉ Ⓑ ✕ 🖵 ⇌ (bus to Wareham and Bournemouth), ✈ (Hurn).

9.2.7 STUDLAND BAY

Dorset 50°38'·80N 01°55'·50W ✺✺✺🌢✿✿✿

CHARTS AC 5601, 2175, 2172; Imray C4

TIDES Approx as for Swanage.

SHELTER Good in winds from S, W and NW, but exposed to N and E. Best ⚓ in about 3m, 4ca WNW of Handfast Pt, but beware poor holding due to weed. Nearest port of refuge is Poole.

NAVIGATION No dangers except very close inshore; N end of bay is shallow. Beware unmarked Redend Rocks (reported to dry 0·5m) ESE of Redend Point. Sp limit 5kn in buoyed areas off beach. Beware of swimmers. PWCs not permitted to land on nor launch from beach. ⚓ in the area between the 6 small yellow buoys in the S end of the bay.

LIGHTS AND MARKS See chartlet. Handfast Point and Old Harry Rock are conspicuous on approach to the bay.

COMMUNICATIONS None.

FACILITIES Village: 🔧 at tap behind beach cafe, 🏧 🛢 ✉ 🚮 ✗ 🍴 hotel. No marine facilities.

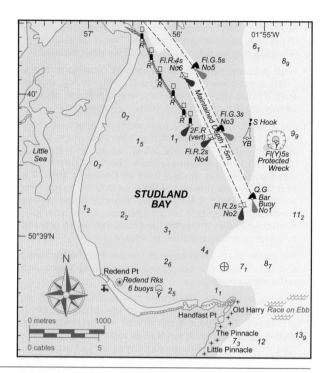

9.2.8 POOLE HARBOUR

Dorset 50°40'·93N 01°56'·96W (Ent) ✺✺✺🌢🌢🌢✿✿✿

CHARTS AC 5601, 2175, 2611; Imray C10, C12, C4, Y23, 2300

TIDES Town Quay ML 1·6. The tide is above ML from about LW+2 to next LW–2.

Standard Port POOLE HARBOUR (→)

Times				Height (metres)			
High Water		Low Water		MHWS	MHWN	MLWN	MLWS
—	—	0500	1100	2·2	1·7	1·2	0·6
—	—	1700	2300				
Differences POOLE HARBOUR ENTRANCE							
—	—	–0025	–0010	0·0	0·0	0·0	0·0
POTTERY PIER							
—	—	+0010	+0010	–0·2	0·0	+0·1	+0·2
CLEAVEL POINT							
—	—	–0005	–0005	–0·1	–0·2	0·0	–0·1
WAREHAM (River Frome)							
—	—	+0130	+0045	0·0	0·0	0·0	+0·3

Daily predictions of the times and heights of HW/LW are for the Standard Port of **POOLE HARBOUR** (near the Ro-Ro terminal). Double HWs occur except at neaps. The height of the 2nd HW is always about 1·8m; the height of the 1st HW varies from springs to neaps. The listed time of HW refers to the highest of the two.

Strong and continuous winds from E to SW may raise sea levels by as much as 0·2m; W to NE winds may lower levels by 0·1m. Barometric pressure can affect the tide by as much as 0·3m. At Wareham the height of LW does not usually fall below 0·7m.

SHELTER An excellent natural harbour with a narrow entrance. Accessible in all conditions, the entrance can become very rough, especially on the ebb in E/SE gales. Yachts may berth at the marinas listed under Facilities. The Town Quay, very exposed in strong E/SE winds, is used by yachts>15m and by arrangement with the Boat Haven. At busy periods yachts may be rafted several deep or allocated a berth in the Port of Poole Marina. Good ⚓s may be found wherever sheltered from the wind and clear of channels, moorings and shellfish beds, especially clear of the buoyed fairways in South Deep and off Pottery Pier (W end of Brownsea Island) or off Shipstal Point, about 1·3M west of Pottery Pier. All are within a Quiet Area (see chartlet and speed limits).

NAVIGATION WPT Poole Bar (No 1 SHM) Buoy, QG, 50°39'·29N 01°55'·14W, 328°/1·95M to Haven Hotel. In very strong SE-S winds the Bar is dangerous, especially on the ebb. Beware cross-Channel high speed ferries which operate through the area. Monitor VHF Ch 14 for information about commercial traffic and MOD activity. In Studland Bay and close to training bank beware lobster pots. From Poole Bar to Shell Bay a **Boat Channel**, suitable for craft <3m draught, parallels the W side of the Swash Channel, close E of the Training Bank, and should be used whenever possible.

East Looe Channel The well lit and buoyed channel, least depth 1·3m, is the recommended approach from the E when height of tide allows. It is liable to shift and the buoys moved accordingly. Check LNTMs from www.phc.co.uk; there may be less water than charted. The groynes to the N are marked by SHM beacons, the two most W'ly are lit, 2 FG (vert).

Within the harbour the two principal channels (Middle Ship and North) lead up to Poole Town Quay. Outside of these there are extensive shoal or drying areas.

Middle Ship Chan is dredged to 7·5m depth and 105m width for ferries and commercial shipping to/from the Continental Freight and Ferry Terminal. Vessels designated 'controlled movements' have moving exclusion zones which must be avoided. Small craft should use the **Boat Channel** which runs parallel to the main channel between the PHM buoys and the unlit beacons further S which mark the edge of the bank. Depth is 2·0m in this channel, but 1·5m closer to the stakes. Caution: When large ferries pass, a temporary but significant reduction in depth may be experienced; it is then prudent to keep close to the PHM buoys.

North Channel is no longer routinely dredged but remains usable for leisure craft; the best water is on the outside of channel bends.

Lulworth gunnery range. Prior to proceeding W all craft should check for activity, as shown in the HM's office and in the Updates to this Almanac.

LIGHTS AND MARKS See chartlet and 9.2.4 for main buoys, beacons and lights.
Poole Bridge traffic lights, shown from bridge tower:

● (3 vert Fl) = Emergency stop. Do not proceed;

● (3 vert) = Do not proceed past the signal;

○ + 3 ● vert = Vessels may proceed with caution;

● (3 vert) = Vessels may proceed.

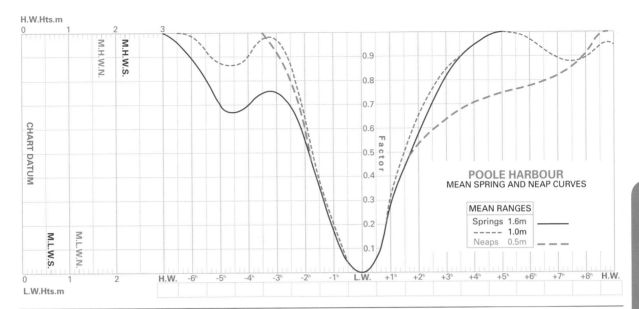

POOLE HARBOUR
MEAN SPRING AND NEAP CURVES

MEAN RANGES	
Springs	1.6m
	1.0m
Neaps	0.5m

POOLE BRIDGES SUMMER SCHEDULE:

Poole Bridge and the new Twin Sails Bridge co-ordinate openings with *scheduled times* at which vessels are advised to be on station. The timings are biased to outbound traffic am, and inbound pm. The second bridge will open asap after the first, which may open again to release traffic left between the closed bridges. Daily:

Twin Sail: 0530, 0630, 0730, 0930, 1030, 1130, 1230, 1430.

Poole Br: *ASAP thereafter.*

Poole Br: 1530,1630, 1830, 1930, 2130, 2230, 2330.

Twin Sail: *ASAP thereafter*.

+ Sat/Sun: Twin sail: 0830, 1330; Poole Br: 1730, 2030.

Pleasure craft may pass when bridges open for commercial traffic. Timings at www.phc.co.uk. Monitor Ch 12 for more information.

COMMUNICATIONS (Code 01202) NMOC (01305) 760439; Police 101; ⊞ 665511. HM 440233; *Poole Hbr Control* 440230, Poole Bridge 674115.

VHF Ch **14** 16 (H24), *Poole Bridge* Ch12. Poole Quay Boat Haven, Cobbs Quay (call *CQ Base*) Ch 80; Salterns Marina Ch M 80; Parkstone Haven, Poole YC Haven (*Pike*), Dorset Yacht Co Ch M.

FACILITIES **Marinas** from seaward (harbour dues are usually included, but best to check):

Salterns Marina marina@salterns.co.uk ☎709971. max draught 2·5m 300⌒ few❶, £3·50/m, short stay £1/m, ⓘ 🛢 🛠(H24) 🔥 ✕ ✎ ⓑ Ⓔ ⛟(5t) ⛟(45t) ⚓ 🗑 ✉ ✕ 🏠. Appr from North Channel near NC 7 SHM buoy, Fl G 3s.

Parkstone YC Haven ☎743610 (Parkstone YC). Access from North Channel near NC 11 SHM buoy, Fl G 3s. Approach chan, dredged 2·5m, is marked by PHM and SHM unlit buoys. Ldg daymarks 006°, both Y ◇s. Some ❶ berths, £4·00; dredged 2m.

Poole Quay Boat Haven poolequayboathaven.co.uk ☎649488. max draught 3·0m, 100❶ £4·07 ⛟ £3/day; short stay <4hrs £6·50<10m, £12·50<20m. ⚓ ⓘ ⚓ 🛠 ⛟. Office hrs 0700-2200 Apr-Sep.

Town Quay (£2·78) **& Port of Poole Marina** ☎660120 by arrangement.

Lake Yard Marina www.lakeyard.com ☎674531. Ent marked by 2FR (vert) and 2FG (vert). ⚓ £1·30, 56⌒+❶ if space £3·20, ⛟(5t) ⛟(50t) Club: 🏠; Water taxi, weekends Apr-Oct.

Sunseeker International Marina ☎685335, 50⌒ 🛢 🔥 ✎ ⓑ⛟(30t) ⛟(36t) ⚓ 🗑 ✕ 🏠.

Cobbs Quay Marina ☎674299. 850⌒ some ❶ £3·25, 🛥 ⓘ 🛢 🛢 🛢 LPG 🔥 ✎ ⓑ Ⓔ ⚓ ⛟(10t) ⚓ ✕ 🏠.

Swinging moorings in Poole Harbour £15, contact PHC through the Boat Haven.

Public Landing Places: Steps at Poole Quay. **Public** 🛥: Baiter Harbourside Park, car/trailer parking – machine payment.

Services A complete range of marine services is available; consult marina/HM for exact locations. **Fuel** Corrals (S side of Poole Quay adjacent bridge) D.

Town: All domestic facilities, ⇌ ✈ (Bournemouth 12M and Southampton by direct rail link). **Ferries:**

Cherbourg: 2-3/day; 4½ Hrs (HSS 2¼ hrs); Brittany (www. brittany-ferries.co.uk);

St Malo/Jersey/Guernsey: 2/day HSS; 2¾–4½ hrs; Condor (www. condorferries.co.uk).

YACHT CLUBS

Royal Motor YC ☎707227, ⚓ ✕ 🏠.
Parkstone YC ☎743610 (Parkstone Haven).
Poole YC ☎672687.

POOLE HARBOUR *continued*

QUIET AREA

South of a line between Patchins Point and South Haven Point is a Quiet Area with an advisory 6kn speed limit. Keep noise and wash to a minimum.

WAREHAM, Dorset, **50°41´·03N 02°06´·56W**. AC 5601, 2611. HW −0030 (Np), +0320 (Sp) on Dover. LWs do not fall below 0·7m except under very low river flow conditions. **Shelter** Very good. Access via narrow and winding chan and R Frome (min depth approx 0·1m), but well marked by buoys and posts, lit to N of Gigger's Island; keep to the outside of all bends. Passage is unlit beyond Gigger's Island. There is a water-ski area between Gold Pt on the N side of the Arne peninsula and WH11/WH12 buoys. Beware prohib ⚓s (salmon holes) marked on the chart. Max draught 1·2m to Wareham Quay. Enforced 4kn speed limit in river.

Facilities: Ridge Wharf Yacht Centre www.ridgewharf.co.uk ☎(01929) 552650. ½M upstream of R Frome ent. 165⬱+♥ £2·70, ⬧ ⬟ ⚓ ⊞ ⛽ ▮ ⚓ ⚒ ⛏ ⬚ △ ◪(18t) ⬪.

Redcliffe YC ☎551227, Ch 80, Visitors welcome (½M below bridge). **Wareham Quay** ⬱ (free) ⚓ ✗. **Town** ▦ ▦ ▮ ✉ Ⓑ ⬚ ✗ ⬚ ⇌.

Central S England

POOLE

See separate chartlet

POOLE BRIDGE OPENING TIMES
The bridge opens daily in winter, only if vessels are waiting, at the following times:

0530, 0630, 0730, 0930, 1030, 1230, 1430, 1530, 1630, 1830, 1930, 2130 & 2345 (also at 1730 on Sat, Sun and Bank Holidays).

Pleasure craft may pass when bridge is opened for commercial traffic. Monitor Ch 12 for further information

BOAT CHANNEL
A Boat Chan for pleasure craft and FVs up to 1.5m draft lies SW of the Middle Ship Channel

BOAT CHANNEL
A Boat Chan for pleasure craft and FVs up to 3m draft lies on the SW side of the Swash Chan

HISTORIC WRECKS VICINITY S HOOK
These sites are protected from unauthorised interference. See Admiralty Annual Notice to Mariners No 16

203

HEIGHTS OF TIDE AT POOLE (TOWN QUAY)

The four curves below are an alternative to those shown on page 201. Although their accuracy should be regarded as approximate, they enable a speedy estimate to be made of the Height of Tide at hourly intervals after the time of LW at Poole Town Quay. The curves are drawn for LW heights above CD of: 0·3m, 0·6m (MLWS), 0·9m and 1·2m (MLWN). The small range of tide (neaps 0·4m; springs 1·5m) is immediately apparent, as is the HW stand at neaps and the double HWs at springs.

Note: All references to are to LW because at Poole the times and heights of LW are more sharply defined than those of HW. HW times and heights are complicated by a stand of tide at HW at neaps and by double HWs at springs. The times of the two HWs cannot therefore be readily predicted with great accuracy.

Procedure:
- Extract time and height of the preceding LW from the Poole Harbour tide tables (⟶).
- Using the curve whose LW height is closest to the predicted height, note the time of LW and enter the curve at the number of hours after LW for the time required.
- Extract the estimated height of tide (above CD).
- For a more exact estimate, interpolate between the curves if appropriate.

POOLE BRIDGE OPENING TIMES

The bridge opens daily in winter, only if vessels are waiting, at the following times:

0530, 0630, 0730, 0930, 1030, 1230, 1430, 1530, 1630, 1830, 1930, 2130 & 2345 (also at 1730 on Sat, Sun and Bank Holidays).

Pleasure craft may pass when bridge is opened for commercial traffic. Monitor Ch 12 for further information

Poole Quay Boat Haven

SAFETY RECOMMENDATIONS

- Check LNTMs (www.phc.co.uk) to keep abreast of changes.
- Look out for large ferries/commercial ships which are limited in manoeuvrability and speed. **Stay out of exclusion zones.** Keep well clear and make your intentions obvious.
- Use designated small boat channels whenever possible.
- Be alert in the harbour entrance where the stream runs hard; it is a particularly hazardous area with great potential for an accident – **the chain ferry has right of way.**
- Keep to the speed limits and be particularly aware of the wash you are creating. This can cause injury to swimmers and other boaters or damage to other craft and property.

- Proceed slowly in the vicinity Town Quay, marinas and moorings to minimise wash that can damage moored craft.
- Anticipate sailing craft manoeuvres, give them room to tack.
- Poole Harbour Guide is available from the Harbour Master and at www.phc.co.uk. Areas are designated for personal watercraft, waterskiing, wind & kite surfing. 'Quiet Zones' exclude PWC and have speed limits of 6kts.
- Before sailing check the tidal situation and comply with Passage planning requirements.
- Bird sensitive areas exist in most bays and lakes in the harbour. Virtually all intertidal mud areas are nominated Sites of Special Scientific Interest (SSSI).

Central S England

Poole Harbour tides

Note – Sea level is above mean tide level from 2 hrs after LW to 2 hrs before the next LW. HW occurs between 5 hrs after LW and 3 hrs before the next LW, the time shown is approximate and should be checked for suitability.

STANDARD TIME (UT)
For Summer Time add ONE hour in **non-shaded areas**

POOLE HARBOUR LAT 50°43'N LONG 1°59'W
HEIGHTS OF HIGH WATER AND TIMES AND HEIGHTS OF LOW WATERS

Dates in **red** are SPRINGS
Dates in **blue** are NEAPS

YEAR 2016

JANUARY

Day	Time m	Day	Time m
1 F	0832 1.2 / 1133 1.8 / 2049 1.0	16 SA	0115 1.9 / 0835 0.9 / 1322 1.9 / ☽ 2059 0.8
2 SA	0001 1.6 / 0918 1.3 / 1213 1.7 / ◑ 2137 1.1	17 SU	0204 1.8 / 0938 1.0 / 1410 1.8 / 2206 0.9
3 SU	0049 1.6 / 1012 1.3 / 1302 1.6 / 2235 1.2	18 M	0259 1.7 / 1051 1.1 / 1503 1.7 / 2320 0.9
4 M	0203 1.5 / 1119 1.3 / 1408 1.5 / 2345 1.2	19 TU	0803 2.0 / 1203 1.0 / 1609 1.7
5 TU	0852 1.8 / 1235 1.2 / 1556 1.5	20 W	0029 0.9 / 0649 1.9 / 1307 0.9 / 1901 1.8
6 W	0054 1.2 / 0644 1.8 / 1330 1.1 / 1714 1.6	21 TH	0131 0.9 / 0726 2.0 / 1404 0.7 / 1945 1.9
7 TH	0147 1.1 / 0548 1.9 / 1418 0.9 / 1921 1.7	22 F	0227 0.8 / 0805 2.0 / 1456 0.6 / 2026 1.9
8 F	0235 0.9 / 0627 2.0 / 1501 0.8 / 2004 1.9	23 SA	0317 0.7 / 0841 2.1 / 1543 0.5 / 2103 2.0
9 SA	0320 0.8 / 0722 2.1 / 1545 0.6 / 2050 2.0	24 SU	0403 0.6 / 0913 2.2 / 1627 0.4 / ○ 2137 2.1
10 SU	0404 0.7 / 0854 2.2 / 1627 0.5 / ● 2135 2.1	25 M	0446 0.6 / 0943 2.2 / 1708 0.4 / 2209 2.1
11 M	0447 0.6 / 0944 2.3 / 1709 0.4 / 2219 2.1	26 TU	0527 0.6 / 1013 2.2 / 1747 0.4 / 2242 2.1
12 TU	0529 0.6 / 1029 2.3 / 1750 0.4 / 2302 2.1	27 W	0606 0.7 / 1047 2.2 / 1824 0.5 / 2318 2.0
13 W	0612 0.6 / 1112 2.3 / 1832 0.4 / 2345 2.1	28 TH	0642 0.8 / 1121 2.0 / 1857 0.6 / 2358 1.9
14 TH	0656 0.7 / 1155 2.2 / 1916 0.5	29 F	0715 0.9 / 1044 1.9 / 1931 0.8 / 2259 1.7
15 F	0029 2.0 / 0743 0.8 / 1238 2.1 / 2004 0.6	30 SA	0752 1.0 / 1102 1.8 / 2009 0.9 / 2322 1.7
		31 SU	0833 1.0 / 1135 1.7 / 2053 1.0 / 2358 1.6

FEBRUARY

Day	Time m	Day	Time m
1 M	0921 1.1 / 1214 1.6 / 2145 1.1 ☽	16 TU	0214 1.7 / 1024 1.0 / 1433 1.6 / 2258 1.1
2 TU	0045 1.6 / 1018 1.2 / 1304 1.6 / 2246 1.2	17 W	0307 1.6 / 1141 1.0 / 1537 1.6
3 W	0159 1.5 / 1124 1.2 / 1425 1.5 / 2354 1.2	18 TH	0014 1.1 / 0836 1.9 / 1251 0.9 / 1901 1.8
4 TH	0409 1.6 / 1235 1.1 / 1629 1.5	19 F	0120 1.0 / 0719 1.8 / 1350 0.8 / 1940 1.8
5 F	0105 1.1 / 0506 1.7 / 1342 1.0 / 1917 1.7	20 SA	0216 0.9 / 0755 1.9 / 1442 0.6 / 2018 1.8
6 SA	0207 1.0 / 0559 1.9 / 1437 0.8 / 1957 1.8	21 SU	0305 0.8 / 0829 2.0 / 1528 0.5 / 2050 1.9
7 SU	0259 0.8 / 0801 2.0 / 1524 0.5 / 2041 2.0	22 M	0349 0.6 / 0855 2.1 / 1610 0.4 / ○ 2116 2.0
8 M	0346 0.6 / 0850 2.2 / 1609 0.3 / ● 2124 2.1	23 TU	0430 0.6 / 0915 2.1 / 1649 0.4 / 2140 2.1
9 TU	0430 0.5 / 0936 2.3 / 1652 0.2 / 2206 2.2	24 W	0507 0.6 / 0945 2.2 / 1725 0.4 / 2211 2.1
10 W	0513 0.4 / 1018 2.4 / 1734 0.2 / 2246 2.3	25 TH	0542 0.6 / 1022 2.2 / 1758 0.5 / 2248 2.1
11 TH	0555 0.4 / 1100 2.4 / 1815 0.2 / 2327 2.2	26 F	0613 0.7 / 1059 2.1 / 1827 0.6 / 2325 1.9
12 F	0638 0.5 / 1141 2.3 / 1858 0.3	27 SA	0643 0.7 / 1127 1.9 / 1858 0.7 / 2240 1.8
13 SA	0007 2.1 / 0723 0.6 / 1221 2.2 / 1944 0.5	28 SU	0717 0.8 / 1036 1.8 / 1934 0.8 / 2248 1.7
14 SU	0049 1.9 / 0812 0.8 / 1302 2.0 / 2035 0.7	29 M	0756 0.9 / 1058 1.8 / 2016 0.9 / 2312 1.7
15 M	0130 1.8 / 0910 0.9 / 1344 1.8 / ◐ 2138 0.9		

MARCH

Day	Time m	Day	Time m
1 TU	0842 1.0 / 1130 1.7 / 2106 1.1 / ◐ 2348 1.7	16 W	0116 1.7 / 0959 1.0 / 1414 1.6 / 2241 1.2
2 W	0937 1.0 / 1214 1.6 / 2207 1.2	17 TH	0218 1.6 / 1119 1.0 / 1959 1.9 / 2358 1.2
3 TH	0040 1.6 / 1042 1.1 / 1318 1.5 / 2316 1.2	18 F	0327 1.6 / 1229 1.0 / 2125 1.9
4 F	0250 1.6 / 1154 1.1 / 2057 1.7	19 SA	0103 1.1 / 0426 1.6 / 1328 0.8 / 1926 1.8
5 SA	0031 1.2 / 0440 1.7 / 1308 1.0 / 1909 1.7	20 SU	0157 1.0 / 0518 1.7 / 1419 0.7 / 2000 1.8
6 SU	0142 1.0 / 0543 1.8 / 1411 0.7 / 1944 1.8	21 M	0245 0.8 / 0614 1.8 / 1505 0.6 / 2030 1.9
7 M	0237 0.8 / 0750 2.0 / 1501 0.5 / 2026 2.0	22 TU	0328 0.7 / 0826 2.0 / 1547 0.5 / 2044 2.0
8 TU	0326 0.6 / 0837 2.2 / 1548 0.3 / 2107 2.2	23 W	0408 0.6 / 0833 2.1 / 1626 0.5 / ○ 2101 2.1
9 W	0411 0.4 / 0921 2.4 / 1631 0.1 / ● 2147 2.3	24 TH	0444 0.6 / 0914 2.1 / 1700 0.5 / 2138 2.2
10 TH	0453 0.2 / 1003 2.5 / 1713 0.1 / 2228 2.4	25 F	0516 0.6 / 0955 2.2 / 1730 0.6 / 2217 2.1
11 F	0535 0.2 / 1044 2.5 / 1755 0.1 / 2307 2.3	26 SA	0545 0.6 / 1034 2.1 / 1758 0.7 / 2254 2.0
12 SA	0618 0.3 / 1125 2.4 / 1838 0.3 / 2346 2.2	27 SU	0614 0.7 / 1108 1.9 / 1829 0.7 / 2241 1.9
13 SU	0702 0.5 / 1205 2.2 / 1923 0.5	28 M	0647 0.7 / 1015 1.8 / 1904 0.8 / 2217 1.8
14 M	0024 2.0 / 0750 0.7 / 1244 2.0 / 2014 0.8	29 TU	0724 0.8 / 1027 1.8 / 1945 1.0 / 2237 1.8
15 TU	0057 1.8 / 1324 1.8 / ◐ 2118 1.0	30 W	0808 0.9 / 2033 1.1 / 2316 1.8
		31 TH	0900 1.0 / 1146 1.7 / 2133 1.2 ◑

APRIL

Day	Time m	Day	Time m
1 F	0007 1.7 / 1007 1.1 / 1246 1.6 / 2246 1.3	16 SA	0245 1.6 / 1158 1.0 / 2059 2.0
2 SA	0125 1.6 / 1121 1.1 / 2020 1.8	17 SU	0034 1.2 / 0353 1.7 / 1257 0.9 / 1903 1.8
3 SU	0003 1.2 / 0420 1.7 / 1236 0.9 / 1849 1.8	18 M	0129 1.0 / 0448 1.7 / 1350 0.8 / 1935 1.8
4 M	0114 1.0 / 0641 1.8 / 1341 0.7 / 1924 1.9	19 TU	0217 0.9 / 0542 1.8 / 1437 0.7 / 1827 1.9
5 TU	0212 0.8 / 0730 2.0 / 1434 0.5 / 2005 2.1	20 W	0301 0.8 / 0641 1.9 / 1519 0.7 / 1927 2.0
6 W	0301 0.5 / 0818 2.2 / 1522 0.3 / 2046 2.3	21 TH	0342 0.7 / 0749 2.0 / 1559 0.6 / 2019 2.1
7 TH	0348 0.3 / 0902 2.4 / 1607 0.1 / ● 2126 2.4	22 F	0418 0.7 / 0843 2.1 / 1633 0.7 / ○ 2104 2.2
8 F	0431 0.2 / 0945 2.5 / 1650 0.1 / 2207 2.5	23 SA	0449 0.6 / 0928 2.1 / 1702 0.7 / 2145 2.2
9 SA	0514 0.2 / 1027 2.5 / 1733 0.2 / 2247 2.4	24 SU	0518 0.6 / 1009 2.1 / 1732 0.8 / 2222 2.1
10 SU	0557 0.3 / 1108 2.4 / 1817 0.4 / 2326 2.3	25 M	0548 0.7 / 1045 2.0 / 1804 0.8 / 2232 2.0
11 M	0642 0.4 / 1149 2.2 / 1904 0.6	26 TU	0622 0.7 / 1019 1.9 / 1840 0.9 / 2204 1.9
12 TU	0002 2.1 / 0731 0.6 / 1231 2.0 / 1958 0.9	27 W	0659 0.8 / 1008 1.8 / 1920 1.0 / 2217 1.9
13 W	0025 1.9 / 0827 0.8 / 1316 1.8 / 2102 1.1	28 TH	0742 0.8 / 1041 1.8 / 2002 1.1 / 2258 1.8
14 TH	0017 1.7 / 0935 1.0 / 1838 2.0 / ◐ 2219 1.2	29 F	0833 0.9 / 1129 1.7 / 2108 1.2 / 2351 1.8
15 F	0114 1.6 / 1050 1.0 / 1946 2.0 / 2331 1.2	30 SA	0937 1.0 / 1232 1.6 / 2221 1.2 ◐

Chart Datum: 1·40 metres below Ordnance Datum (Newlyn). HAT is 2·6 metres above Chart Datum.

》》 **FREE** monthly updates. Register at 《
www.reedsnauticalalmanac.co.uk

Note - Sea level is above mean tide level from 2 hrs after LW to 2 hrs before the next LW. HW occurs between 5 hrs after LW and 3 hrs before the next LW, the time shown is approximate and should be checked for suitability.

STANDARD TIME (UT)
For Summer Time add ONE hour in **non-shaded areas**

POOLE HARBOUR LAT 50°43'N LONG 1°59'W

HEIGHTS OF HIGH WATER AND TIMES AND HEIGHTS OF LOW WATERS

Dates in red are **SPRINGS**
Dates in blue are NEAPS

YEAR 2016

Central S England

MAY

Time	m	Time	m
1 0106 1.7 / 1052 1.0 / SU 1956 1.9 / 2336 1.2		**16** 0309 1.6 / 1221 1.0 / M 1841 1.8	
2 0356 1.7 / 1204 0.9 / M 1824 1.8		**17** 0054 1.1 / 0415 1.6 / TU 1315 1.0 / 1910 1.8	
3 0044 1.0 / 0616 1.8 / TU 1308 0.7 / 1900 2.0		**18** 0145 1.0 / 0511 1.7 / W 1405 0.9 / 1800 1.9	
4 0143 0.8 / 0708 2.0 / W 1405 0.5 / 1942 2.1		**19** 0231 0.9 / 0607 1.8 / TH 1449 0.9 / 1847 2.0	
5 0235 0.6 / 0757 2.2 / TH 1455 0.4 / 2024 2.3		**20** 0312 0.8 / 0710 1.9 / F 1529 0.8 / 1938 2.1	
6 0323 0.4 / 0843 2.3 / F 1543 0.3 / ● 2105 2.4		**21** 0349 0.8 / 0812 2.0 / SA 1604 0.8 / ○ 2028 2.2	
7 0409 0.3 / 0926 2.4 / SA 1628 0.2 / 2146 2.5		**22** 0422 0.7 / 0902 2.1 / SU 1636 0.8 / 2112 2.2	
8 0453 0.2 / 1009 2.4 / SU 1713 0.3 / 2227 2.4		**23** 0453 0.7 / 0946 2.1 / M 1709 0.8 / 2151 2.2	
9 0538 0.3 / 1051 2.3 / M 1759 0.5 / 2306 2.3		**24** 0527 0.7 / 1025 2.0 / TU 1744 0.9 / 2219 2.1	
10 0624 0.4 / 1134 2.2 / TU 1848 0.7 / 2343 2.1		**25** 0603 0.7 / 1101 1.9 / W 1822 0.9 / 2227 2.0	
11 0714 0.6 / 1217 2.0 / W 1941 0.9		**26** 0642 0.7 / 1035 1.8 / TH 1905 1.0 / 2216 1.9	
12 0007 1.9 / 0808 0.8 / TH 1306 1.8 / 2042 1.1		**27** 0725 0.8 / 1036 1.8 / F 1953 1.1 / 2253 1.9	
13 0907 0.9 / 1822 2.1 / F 2148 1.2 / ◑		**28** 0815 0.8 / 1122 1.7 / SA 2051 1.1 / 2345 1.8	
14 0032 1.7 / 1014 1.0 / SA 1922 2.1 / 2256 1.2		**29** 0915 0.9 / 1225 1.7 / SU 2158 1.2 / ◑	
15 0133 1.6 / 1120 1.0 / SU 2025 2.0 / 2358 1.2		**30** 0058 1.8 / 1024 0.9 / M 1935 2.0 / 2308 1.1	
		31 0327 1.8 / 1134 0.9 / TU 1802 1.9	

JUNE

Time	m	Time	m
1 0014 1.0 / 0435 1.8 / W 1238 0.8 / 1839 2.0		**16** 0109 1.2 / 0429 1.6 / TH 1329 1.1 / 1850 1.9	
2 0114 0.8 / 0649 2.0 / TH 1336 0.6 / 1921 2.1		**17** 0157 1.1 / 0528 1.7 / F 1415 1.1 / 1816 1.9	
3 0209 0.7 / 0739 2.1 / F 1430 0.5 / 2004 2.2		**18** 0240 1.0 / 0624 1.8 / SA 1456 1.0 / 1852 2.0	
4 0300 0.5 / 0826 2.2 / SA 1520 0.5 / 2047 2.3		**19** 0319 0.9 / 0741 1.9 / SU 1534 0.9 / 1940 2.1	
5 0349 0.4 / 0910 2.3 / SU 1609 0.4 / ● 2128 2.4		**20** 0355 0.8 / 0840 2.0 / M 1612 0.9 / ○ 2038 2.2	
6 0435 0.3 / 0953 2.3 / M 1656 0.5 / 2208 2.4		**21** 0432 0.7 / 0927 2.0 / TU 1650 0.8 / 2129 2.2	
7 0521 0.3 / 1035 2.3 / TU 1743 0.6 / 2247 2.3		**22** 0510 0.6 / 1011 2.1 / W 1729 0.8 / 2214 2.2	
8 0608 0.4 / 1117 2.2 / W 1832 0.7 / 2324 2.2		**23** 0549 0.6 / 1054 2.1 / TH 1809 0.9 / 2257 2.1	
9 0655 0.5 / 1158 2.0 / TH 1922 0.9 / 2356 2.0		**24** 0629 0.6 / 1136 2.0 / F 1852 0.9 / 2339 2.1	
10 0744 0.7 / 1242 1.9 / F 2014 1.1 / 2328 1.9		**25** 0712 0.7 / 1221 1.9 / SA 1939 1.0	
11 0835 0.8 / 1336 1.7 / SA 2110 1.2		**26** 0023 2.0 / 0759 0.7 / SU 1308 1.9 / 2032 1.0	
12 0000 1.8 / 0930 1.0 / SU 1240 1.7 / ◑ 2211 1.2		**27** 0111 1.9 / 0853 0.8 / M 1359 1.8 / ◑ 2133 1.1	
13 0043 1.7 / 1032 1.1 / M 1943 2.0 / 2315 1.3		**28** 0204 1.9 / 0956 0.9 / TU 1455 1.8 / 2240 1.1	
14 0138 1.6 / 1137 1.2 / TU 2042 2.0		**29** 0300 1.8 / 1105 0.9 / W 1749 1.9 / 2347 1.0	
15 0014 1.2 / 0308 1.6 / W 1236 1.2 / 1903 1.8		**30** 0402 1.8 / 1211 0.9 / TH 1825 2.0	

JULY

Time	m	Time	m
1 0049 0.9 / 0637 1.9 / F 1312 0.8 / 1907 2.1		**16** 0115 1.2 / 0438 1.6 / SA 1334 1.2 / 1839 1.9	
2 0148 0.8 / 0727 2.0 / SA 1410 0.7 / 1950 2.2		**17** 0204 1.1 / 0539 1.7 / SU 1423 1.1 / 1808 2.0	
3 0242 0.6 / 0814 2.1 / SU 1504 0.7 / 2032 2.3		**18** 0248 0.9 / 0738 1.8 / M 1507 1.0 / 1859 2.1	
4 0333 0.5 / 0857 2.2 / M 1555 0.6 / ● 2112 2.3		**19** 0331 0.8 / 0829 1.9 / TU 1550 0.9 / ○ 2026 2.2	
5 0421 0.4 / 0939 2.2 / TU 1642 0.6 / 2151 2.4		**20** 0412 0.6 / 0916 2.1 / W 1632 0.8 / 2121 2.3	
6 0506 0.3 / 1019 2.2 / W 1728 0.6 / 2228 2.3		**21** 0453 0.5 / 1000 2.2 / TH 1713 0.7 / 2208 2.3	
7 0551 0.4 / 1057 2.2 / TH 1814 0.7 / 2302 2.3		**22** 0533 0.5 / 1042 2.2 / F 1754 0.7 / 2251 2.3	
8 0635 0.5 / 1134 2.1 / F 1858 0.8 / 2334 2.1		**23** 0614 0.5 / 1123 2.2 / SA 1836 0.7 / 2334 2.3	
9 0718 0.6 / 1211 2.0 / SA 1942 1.0 / 2331 2.0		**24** 0655 0.5 / 1204 2.1 / SU 1920 0.8	
10 0800 0.8 / 1249 1.8 / SU 2026 1.1 / 2332 1.9		**25** 0016 2.2 / 0740 0.6 / M 1247 2.0 / 2009 0.9	
11 0842 1.0 / 1159 1.7 / M 2112 1.2		**26** 0059 2.0 / 0829 0.7 / TU 1332 1.9 / ◑ 2106 1.0	
12 0006 1.8 / 0927 1.1 / TU 1237 1.7 / ◑ 2203 1.3		**27** 0146 1.9 / 0928 0.9 / W 1421 1.8 / 2214 1.1	
13 0049 1.7 / 1018 1.2 / W 1330 1.7 / 2305 1.3		**28** 0237 1.8 / 1039 1.0 / TH 1521 1.8 / 2324 1.1	
14 0145 1.6 / 1122 1.3 / TH 2040 1.9		**29** 0338 1.7 / 1151 1.0 / F 1821 1.9	
15 0016 1.3 / 0307 1.6 / F 1236 1.3 / 2147 1.9		**30** 0031 1.0 / 0633 1.8 / SA 1257 1.0 / 1900 2.0	
		31 0134 0.8 / 0721 1.9 / SU 1358 0.9 / 1941 2.1	

AUGUST

Time	m	Time	m
1 0230 0.7 / 0806 2.0 / M 1453 0.8 / 2021 2.2		**16** 0219 0.9 / 0738 1.8 / TU 1443 1.0 / 1843 2.0	
2 0321 0.5 / 0847 2.1 / TU 1543 0.7 / ● 2057 2.3		**17** 0307 0.7 / 0821 2.0 / W 1529 0.8 / 2025 2.2	
3 0407 0.4 / 0925 2.2 / W 1629 0.6 / 2131 2.3		**18** 0351 0.5 / 0904 2.1 / TH 1613 0.7 / ○ 2112 2.4	
4 0450 0.4 / 0959 2.2 / TH 1711 0.6 / 2204 2.4		**19** 0433 0.4 / 0945 2.3 / F 1654 0.6 / 2156 2.5	
5 0531 0.4 / 1031 2.2 / F 1752 0.7 / 2236 2.3		**20** 0514 0.3 / 1025 2.3 / SA 1735 0.5 / 2238 2.5	
6 0611 0.5 / 1104 2.2 / SA 1831 0.8 / 2309 2.2		**21** 0554 0.3 / 1105 2.3 / SU 1816 0.6 / 2319 2.4	
7 0648 0.6 / 1139 2.0 / SU 1908 0.9 / 2338 2.0		**22** 0635 0.4 / 1145 2.3 / M 1858 0.7	
8 0723 0.8 / 1213 1.9 / M 1943 1.0 / 2305 1.9		**23** 0001 2.3 / 0718 0.5 / TU 1226 2.1 / 1945 0.8	
9 0758 0.9 / 1125 1.8 / TU 2022 1.1 / 2331 1.8		**24** 0043 2.1 / 0806 0.7 / W 1308 2.0 / 2040 1.0	
10 0839 1.1 / 1156 1.7 / W 2108 1.2 / ◐		**25** 0128 1.9 / 0903 1.0 / TH 1353 1.9 / ◐ 2150 1.1	
11 0008 1.7 / 0927 1.2 / TH 1238 1.7 / 2202 1.3		**26** 0221 1.8 / 1020 1.1 / F 1450 1.8 / 2308 1.1	
12 0056 1.6 / 1024 1.3 / F 1337 1.7 / 2304 1.3		**27** 0739 1.9 / 1139 1.2 / SA 2003 2.0	
13 0203 1.6 / 1130 1.3 / SA 1520 1.7		**28** 0019 1.0 / 0633 1.8 / SU 1249 1.1 / 1851 2.0	
14 0014 1.2 / 0400 1.6 / SU 1244 1.3 / 1833 1.8		**29** 0122 0.9 / 0716 1.9 / M 1349 1.0 / 1929 2.0	
15 0125 1.1 / 0705 1.7 / M 1351 1.2 / 1744 1.9		**30** 0216 0.7 / 0757 1.9 / TU 1441 0.9 / 2007 2.1	
		31 0305 0.6 / 0834 2.0 / W 1528 0.7 / 2040 2.2	

Chart Datum: 1·40 metres below Ordnance Datum (Newlyn). HAT is 2·6 metres above Chart Datum.

FREE monthly updates. Register at
www.reedsnauticalalmanac.co.uk

207

Note - Sea level is above mean tide level from 2 hrs after LW to 2 hrs before the next LW. HW occurs between 5 hrs after LW and 3 hrs before the next LW, the time shown is approximate and should be checked for suitability.

STANDARD TIME (UT)
For Summer Time add ONE hour in **non-shaded areas**

POOLE HARBOUR LAT 50°43'N LONG 1°59'W
HEIGHTS OF HIGH WATER AND TIMES AND HEIGHTS OF LOW WATERS

Dates in red are **SPRINGS**
Dates in blue are **NEAPS**

YEAR 2016

SEPTEMBER

Day	Time	m		Day	Time	m
1	0349	0.4		**16**	0327	0.4
	0906	2.1			0845	2.2
	TH 1611	0.7			F 1550	0.6
	● 2106	2.3			○ 2056	2.4
2	0431	0.4		**17**	0410	0.3
	0932	2.2			0926	2.3
	F 1651	0.6			SA 1632	0.4
	2133	2.3			2139	2.6
3	0509	0.4		**18**	0451	0.2
	1000	2.3			1006	2.5
	SA 1728	0.7			SU 1713	0.4
	2206	2.3			2221	2.6
4	0545	0.5		**19**	0532	0.2
	1032	2.2			1046	2.5
	SU 1802	0.8			M 1754	0.4
	2242	2.2			2303	2.5
5	0617	0.7		**20**	0613	0.3
	1108	2.1			1125	2.4
	M 1833	0.8			TU 1837	0.6
	2317	2.1			2345	2.4
6	0647	0.8		**21**	0657	0.6
	1141	2.0			1205	2.2
	TU 1905	0.9			W 1923	0.7
	2241	1.9				
7	0720	0.9		**22**	0027	2.2
	1055	1.9			0745	0.8
	W 1942	1.0			TH 1244	2.0
	2255	1.8			2018	0.9
8	0800	1.1		**23**	0115	1.9
	1116	1.8			0845	1.1
	TH 2026	1.1			F 1325	1.9
	2324	1.7			◐ 2130	1.1
9	0847	1.2		**24**	0218	1.8
	1147	1.8			1007	1.2
	F 2120	1.2			SA 1420	1.8
	◐				2252	1.1
10	0002	1.7		**25**	0726	2.0
	0946	1.3			1208	1.2
	SA 1234	1.7			SU 1800	1.9
	2223	1.3				
11	0106	1.6		**26**	0002	1.0
	1053	1.4			0623	1.9
	SU 1409	1.7			M 1235	1.2
	2333	1.2			1831	1.9
12	0836	1.8		**27**	0102	0.9
	1208	1.3			0700	1.9
	M 1817	1.8			TU 1332	1.1
					1910	1.9
13	0047	1.1		**28**	0155	0.8
	0657	1.8			0738	1.9
	TU 1322	1.2			W 1422	0.9
	1846	1.9			1946	2.0
14	0150	0.9		**29**	0242	0.6
	0725	1.9			0813	2.0
	W 1418	1.0			TH 1507	0.8
	1925	2.1			2015	2.1
15	0240	0.7		**30**	0326	0.5
	0805	2.0			0839	2.1
	TH 1505	0.8			F 1549	0.7
	2012	2.2			2029	2.2

OCTOBER

Day	Time	m		Day	Time	m
1	0407	0.5		**16**	0344	0.3
	0854	2.2			0904	2.4
	SA 1627	0.7			SU 1608	0.4
	● 2057	2.3			○ 2121	2.6
2	0444	0.5		**17**	0427	0.2
	0922	2.3			0945	2.5
	SU 1702	0.7			M 1650	0.3
	2136	2.3			2204	2.6
3	0517	0.6		**18**	0509	0.2
	0958	2.3			1025	2.5
	M 1733	0.7			TU 1733	0.3
	2216	2.2			2246	2.5
4	0546	0.7		**19**	0552	0.4
	1036	2.2			1105	2.4
	TU 1801	0.8			W 1817	0.5
	2253	2.1			2329	2.4
5	0615	0.8		**20**	0638	0.6
	1107	2.0			1145	2.3
	W 1831	0.9			TH 1904	0.7
	2228	1.9				
6	0647	1.0		**21**	0014	2.1
	1028	1.9			0729	0.9
	TH 1907	0.9			F 1223	2.1
	2221	1.8			2000	0.9
7	0726	1.1		**22**	0104	1.9
	1034	1.9			0831	1.1
	F 1949	1.0			SA 1252	1.9
	2241	1.8			◐ 2109	1.0
8	0812	1.2		**23**	0609	2.1
	1058	1.8			0951	1.3
	SA 2041	1.1			SU 1326	1.8
	2320	1.7			2227	1.1
9	0910	1.3		**24**	0716	2.1
	1141	1.8			1106	1.3
	SU 2146	1.2			M 1447	1.7
	◐				2335	1.0
10	0015	1.6		**25**	0604	1.9
	1022	1.4			1209	1.2
	M 1254	1.7			TU 1806	1.8
	2259	1.2				
11	0751	1.9		**26**	0034	0.9
	1139	1.4			0637	1.9
	TU 1756	1.8			W 1306	1.1
					1639	1.8
12	0012	1.1		**27**	0127	0.8
	0637	1.8			0714	1.9
	W 1251	1.2			TH 1356	1.0
	1827	1.9			1730	1.8
13	0117	0.9		**28**	0215	0.8
	0704	1.9			0746	2.0
	TH 1349	1.0			F 1441	0.9
	1907	2.0			1827	2.0
14	0210	0.6		**29**	0259	0.7
	0742	2.1			0802	2.1
	F 1439	0.7			SA 1523	0.8
	1953	2.2			1931	2.1
15	0258	0.4		**30**	0340	0.7
	0823	2.3			0757	2.2
	SA 1524	0.5			SU 1602	0.7
	2038	2.4			● 2022	2.2
				31	0417	0.7
					0842	2.3
					M 1636	0.7
					2107	2.2

NOVEMBER

Day	Time	m		Day	Time	m
1	0449	0.8		**16**	0449	0.3
	0925	2.3			1006	2.5
	TU 1705	0.7			W 1714	0.3
	2150	2.2			2231	2.4
2	0519	0.8		**17**	0534	0.5
	1004	2.2			1047	2.4
	W 1734	0.8			TH 1800	0.4
	2230	2.1			2315	2.3
3	0548	0.9		**18**	0622	0.7
	1031	2.1			1127	2.3
	TH 1805	0.8			F 1848	0.6
	2302	1.9				
4	0622	1.0		**19**	0000	2.1
	1011	2.0			0714	0.9
	F 1841	0.9			SA 1205	2.1
	2158	1.8			1941	0.7
5	0701	1.1		**20**	0049	1.9
	1005	1.9			0813	1.1
	SA 1921	0.9			SU 1153	1.9
	2216	1.8			2043	0.9
6	0745	1.2		**21**	0556	2.1
	1033	1.9			0921	1.2
	SU 2010	1.0			M 1225	1.8
	2257	1.7			◑ 2151	1.0
7	0840	1.3		**22**	0655	2.1
	1120	1.8			1031	1.3
	M 2111	1.1			TU 1330	1.7
	◑ 2353	1.6			2259	1.0
8	0951	1.3		**23**	0759	2.1
	1224	1.7			1136	1.2
	TU 2225	1.1			W 1454	1.7
9	0726	2.0		**24**	0000	1.9
	1107	1.3			0619	1.9
	W 1731	1.8			TH 1233	1.2
	2338	1.0			1558	1.7
10	0616	1.9		**25**	0055	1.0
	1218	1.2			0653	1.9
	TH 1804	1.9			F 1326	1.1
					1655	1.7
11	0042	0.9		**26**	0145	0.9
	0642	2.0			0723	1.9
	F 1318	1.0			SA 1413	1.0
	1847	2.0			1752	1.8
12	0139	0.7		**27**	0231	0.9
	0720	2.1			0627	2.0
	SA 1410	0.7			SU 1456	0.7
	1933	2.2			1852	1.9
13	0230	0.5		**28**	0313	0.8
	0801	2.3			0714	2.1
	SU 1458	0.5			M 1535	0.8
	2019	2.3			1950	2.0
14	0318	0.3		**29**	0350	0.8
	0843	2.4			0803	2.2
	M 1545	0.3			TU 1610	0.8
	○ 2103	2.4			● 2040	2.1
15	0404	0.3		**30**	0424	0.8
	0925	2.5			0850	2.2
	TU 1629	0.3			W 1641	0.7
	2147	2.5			2126	2.1

DECEMBER

Day	Time	m		Day	Time	m
1	0455	0.9		**16**	0519	0.5
	0932	2.2			1031	2.4
	TH 1712	0.7			F 1744	0.3
	2209	2.0			2259	2.2
2	0528	0.9		**17**	0607	0.6
	1004	2.1			1110	2.3
	F 1746	0.7			SA 1832	0.4
	2250	1.9			2342	2.1
3	0604	1.0		**18**	0657	0.8
	1012	2.0			1146	2.1
	SA 1823	0.8			SU 1921	0.6
	2327	1.8				
4	0644	1.0		**19**	0025	1.9
	1006	2.0			0749	1.0
	SU 1903	0.8			M 1206	2.0
	2206	1.8			2012	0.8
5	0728	1.1		**20**	0114	1.8
	1024	1.9			0844	1.1
	M 1949	0.8			TU 1150	1.8
	2245	1.7			2108	0.9
6	0818	1.2		**21**	0624	2.1
	1109	1.9			0945	1.2
	TU 2043	0.9			W 1229	1.7
	2337	1.7			◑ 2211	1.1
7	0920	1.2		**22**	0718	2.0
	1206	1.8			1052	1.3
	W 2150	1.0			TH 1323	1.6
	◑				2318	1.1
8	0056	1.6		**23**	0818	2.0
	1032	1.2			1156	1.2
	TH 1503	1.7			F 1456	1.6
	2301	0.9				
9	0802	2.0		**24**	0019	1.1
	1143	1.1			0648	1.8
	F 1614	1.8			SA 1253	1.2
					1616	1.6
10	0008	0.8		**25**	0114	1.1
	0624	1.9			0714	1.8
	SA 1246	1.2			SU 1344	1.1
	1828	1.9			1720	1.7
11	0109	0.7		**26**	0202	1.0
	0701	2.1			0708	1.9
	SU 1343	0.8			M 1429	1.0
	1916	2.0			1818	1.7
12	0204	0.6		**27**	0245	1.0
	0743	2.2			0639	2.0
	M 1435	0.6			TU 1509	0.9
	2003	2.2			1920	1.8
13	0255	0.5		**28**	0325	0.9
	0826	2.3			0722	2.1
	TU 1524	0.4			W 1546	0.8
	2049	2.3			2018	1.9
14	0344	0.4		**29**	0401	0.9
	0908	2.4			0815	2.1
	W 1612	0.3			TH 1620	0.7
	○ 2133	2.3			● 2108	2.0
15	0432	0.4		**30**	0436	0.8
	0950	2.4			0909	2.2
	TH 1658	0.3			F 1656	0.6
	2216	2.3			2154	2.0
				31	0513	0.8
					0955	2.2
					SA 1732	0.6
					2237	2.0

Chart Datum: 1·40 metres below Ordnance Datum (Newlyn). HAT is 2·6 metres above Chart Datum.

SWANAGE TO CHRISTCHURCH BAY

Studland Bay (AC 2172) is a good ⚓ especially in winds from the S and W. Poole Bay offers good sailing in waters sheltered from the W and N, with no off-lying dangers.

▶ *1M NE of Durlston Head, Peveril Ledge runs 2½ca seaward, causing a significant race which extends nearly 1M eastwards, particularly on W-going stream against a SW wind. Proceeding towards the excellent shelter of Poole Harbour, overfalls may be met off Ballard Pt and Old Harry on the W-going stream.*◀

(AC 2615) There is deep water quite close inshore between St Alban's Hd and Anvil Pt.

▶ *Tidal streams are weak North of a line between Handfast Point and Hengistbury Head and within Christchurch Bay. The tide runs hard over the ledge at springs, and there may be overfalls*◀

Hengistbury Hd is a dark reddish headland S of Christchurch Hbr with a groyne extending 1ca S and Beerpan Rks a further 100m E of groyne. Beware lobster pots in this area. Christchurch Ledge extends 2·75M SE from Hengistbury Hd. There is a passage between Christchurch Ledge and Dolphin Bank, some 1½M further SE. A wreck hazardous to surface navigation (PA) has been reported in this chan (2013).

WESTERN APPROACHES TO THE SOLENT

(AC 2035) The Needles are distinctive rocks at the W end of the Isle of Wight. The adjacent chalk cliffs of High Down are conspic from afar, but the Lt Ho may not be seen by day until relatively close. Goose Rk, and dries, is about 50m WNW of the Lt Ho, 100-150m WSW of which is a drying wreck. When rounding the Needles an offing of 1½ca will clear these. The NW side of Needles Chan is defined by the Shingles bank, parts of which dry and on which the sea breaks violently in the least swell. The SE side of the bank is fairly steep-to; the NW side shelves more gradually.

▶ *The ENE-going flood runs from HW Portsmouth +0500 until HW –0130, Sp rates 3.1kn at The Bridge, a reef which runs 8ca W of the Lt ho and is marked by WCM, and 3.9kn at Hurst. The WSW-going ebb runs from HW –0100 until HW +0430, sp rates 4.4kn at Hurst, 3.4kn WSW across the Shingles and at The Bridge. The Needles Chan is well marked and in fair weather presents no significant problems, but in a SW 4 or above over the ebb, breaking seas occur near The Bridge and SW Shingles. In bad weather broken water and overfalls extend along The Bridge. S to W gales against the ebb raise very dangerous breaking seas in the Needles Chan and approaches. The sea state can be at its worst shortly after LW when the flood has just begun. There is then no wind-over-tide situation, but a substantial swell is raised as a result of the recently turned stream.* ◀

In such conditions use the E route to the Solent, S of the IoW and via Nab Tower; or find shelter at Poole or Studland.

In strong winds the North Channel, N of the Shingles, is preferable to the Needles Channel. The two join S of Hurst Pt, where overfalls and tide rips may be met. Beware The Trap, a very shallow gravel shoal spit, 150m SE of Hurst Castle.

In E winds Alum Bay, with its coloured cliffs close NE of the Needles, is an attractive daytime ⚓ but beware Long Rk (dries) in the middle of the bay, and Five Fingers Rock 1½ca SW of Hatherwood Pt on N side. Totland Bay is good ⚓ in settled weather, but avoid Warden Ledge.

THE SOLENT

(AC 2035, 2036, 2037) There are few dangers in mid-chan. The most significant is Bramble Bank (dries) between Cowes and Calshot. The main fairway (buoyed) passes S and W of the Brambles, but yachts can use the North Chan to the NE of the Brambles at any state of tide.

▶ *Tidal streams are strong at Sp, see 9.2.16.* ◀

A Precautionary Area between Cowes and Calshot provides priority and safety for large commercial shipping. There are many yacht racing buoys *(seasonal, Mar–Dec)* in the Solent and Southampton Water. Most are fitted with a Fl Y 4s light.

Several inshore spits, banks, rocks and ledges include: Pennington and Lymington Spits on the N shore; Black Rk 4ca W of entrance to Yarmouth ; Hamstead Ledge 8ca W of entrance to Newtown River and Saltmead Ledge 1·5M to E; Gurnard Ledge 1·5M W of Cowes; Lepe Middle and Beaulieu Spit, S and W of the ent to Beaulieu R; the shoals off Stone Pt, marked by Lepe Spit SCM buoy; Shrape Mud, which extends N from the breakwater of Cowes Hbr and along to Old Castle Pt; the shoals and isolated rks which fringe the island shore from Old Castle Pt to Ryde, including either side of the ent to Wootton Creek; and Calshot Spit which extends almost to the deep water chan into Southampton Water.

Southampton Water is a commercial waterway with large tankers, container ships, frequent Ro-Ro and High Speed ferries. Yachts should monitor VHF Ch 12 (Southampton VTS) for shipping movements. Between the Fawley Marine Terminal and the BP jetty on the E side the fairway is narrow with little room for yachts to take refuge. To the N there is adequate water for yachts close outboard of the main buoyed channel; the banks are of gently shelving soft mud, apart from foul ground between Hythe and Marchwood. At night, unlit marks and large mooring buoys may be hard to see against the many shore lights. Except in strong N'lies, Southampton Water and the R Test and Itchen provide sheltered sailing. The R Hamble is convenient, but crowded.

Depending on the wind direction, there are many good ⚓s: in W winds on E side of Hurst, as close inshore as depth permits, NE of Hurst lt; in S winds, or in good weather, W of Yarmouth Harbour entrance, as near shore as possible; in winds between W and N in Stanswood Bay, about 1M NE of Stansore Pt. Just N of Calshot Spit there is shelter from SW and W, while Osborne Bay, 2M E of Cowes, is sheltered from winds between S and W; in E winds Gurnard Bay, to the W of Cowes, is preferable. Stokes Bay is well sheltered from the N, and at the E end of the IoW there is good ⚓ in Priory Bay off Bembridge in winds from S, SW or W; but clear out if wind goes into E.

Shoal-draught boats can explore the estuaries at the top of the tide: Ashlett Creek between Fawley and Calshot, Eling up the R Test, and the upper reaches of the R Medina.

ISLE OF WIGHT – SOUTH COAST

(AC 2045) From the Needles eastward to Freshwater Bay the cliffs can be approached to within 1ca, but beyond the E end of the chalk cliffs there are ledges off Brook and Atherfield which require at least 5ca offing.

▶ *The E-going stream sets towards these dangers. 4M SSW of the Needles the stream turns NNE at HW Portsmouth +0500, and W at HW –0030, Sp rate 2kn. It is safe to pass 2ca off St Catherine's lt ho (conspic), but a race occurs off the point and can be very dangerous at or near Sp against a strong wind, particularly SE of the point on a W-going stream. St Catherine's should then be rounded at least 2M off. 1·25M SE of the point the stream turns E at HW Portsmouth +0520, and W at HW –0055, sp rate 3·75kn.* ◀

Rocks extend about 2½ca either side of Dunnose where a race occurs. In Sandown Bay ⚓ off Shanklin or Sandown where the tidal streams are weak inshore.

▶ *Off the centre of the Bay they turn NE x E at HW Portsmouth +0500, and SW x W at HW –0100, sp rates 2 kn.* ◀

The Yarborough Monument is conspicuous above Culver Cliff. Whitecliff Bay provides ⚓ in winds between W and N. From here to Foreland (Bembridge Pt) the coast is fringed by a ledge of drying rocks extending up to 3ca offshore, and it is advisable to keep to seaward of Bembridge Ledge ECM lt buoy.

9.2.9 CHRISTCHURCH

Dorset 50°43'·53N 01°44'·33W ❀⚓♨❁❁❁

CHARTS AC 5601, 2035, 2172; Imray C10, C12, C4, 2300, 2200

TIDES HW Sp −0210, Np, −0140 Dover; ML 1·2

Standard Port PORTSMOUTH (→); Tidal Curve see 9.2.17 (→)

Times				Height (metres)			
High Water		Low Water		MHWS	MHWN	MLWN	MLWS
0000	0600	0500	1100	4·7	3·8	1·9	0·8
1200	1800	1700	2300				
Differences BOURNEMOUTH							
−0240	+0055	−0050	−0030	−2·6	−2·0	−0·6	−0·2
CHRISTCHURCH (Entrance)							
−0230	+0030	−0035	−0035	−2·9	−2·4	−1·2	−0·2
CHRISTCHURCH (Quay)							
−0210	+0100	+0100	+0055	−2·9	−2·4	−1·0	0·0
CHRISTCHURCH (Tuckton bridge)							
−0205	+0110	+0110	+0105	−3·0	−2·5	−1·0	+0·1

NOTE: Double HWs occur, except near neaps; predictions are for the higher HW. Near neaps there is a stand; predictions are for mid-stand. Tidal levels are for inside the bar; outside, the tide is about 0·6m lower at springs. Water flow in Rivers Avon and Stour cause considerable variations from predicted heights. Tuckton LWs do not fall below 0·7m except under very low river flow conditions.

SHELTER Good in lee of Hengistbury Hd; elsewhere exposed to SW winds. R Stour navigable at HW up to Tuckton, and the R Avon up to the first bridge. Both are well sheltered. Most ⚓s in the hbr dry.

NAVIGATION WPT 50°43'·53N 01°43'·58W, 270°/0·5M to NE end of Mudeford Quay. Outer approach is marked with PHM and SHM buoys; the bar and chan are liable to shift and may not be as shown on the chart. Beware groynes S of Hengistbury Hd, Beerpan, Yarranton/Clarendon Rks. The ebb stream reaches 4-5kn in The Run; best to enter or leave at HW/stand. Night entry is not recommended without local knowledge. Chan inside hbr is narrow and mostly shallow (approx 0·3m) with sand/mud. No ⚓ in chan. No berthing at ferry jetty by Mudeford sandbank (beware frequent ferries between Mudeford Quay, The Run and Christchurch). Hbr speed limit 4kn. If fishing from any craft, a local licence is obligatory.

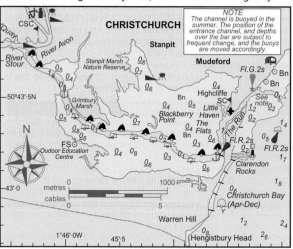

LIGHTS AND MARKS See chartlet and 9.2.4. Unlit chan buoys in hbr and apps are locally laid Apr-Oct inc; info from ☎483250.

COMMUNICATIONS (Code 01202) NMOC (01305) 760439; Police 101; Ⓗ 303626; Casualty 704167.

FACILITIES Rossiter Yachts ☎483250, ⚓£28; **Christchurch SC (CSC)** ☎483150, ⛽£10 limited ⚓ £20, monohulls <10m only.
Services ⚓⚒⚓ ▮▲🛢⚙⚓⚓⚒⚓⚙🛒 Ⓔ 🅂(10t) ⚓ ACA.
Town ✉ Ⓑ ⇌ ✈ (Hurn).

9.2.10 KEYHAVEN

Hampshire 50°42'·85N 01°33'·26W ❀⚓♨❁❁

CHARTS AC 5600, 2035, 2021; Imray C4, C3, 2300, 2200

TIDES −0020, +0105 Dover; ML 2·0

Standard Port PORTSMOUTH (→); Tidal Curve see 9.2.17 (→)

Times				Height (metres)			
High Water		Low Water		MHWS	MHWN	MLWN	MLWS
0000	0600	0500	1100	4·7	3·8	1·9	0·8
1200	1800	1700	2300				
Differences HURST POINT							
−0115	−0005	−0030	−0025	−2·0	−1·5	−0·5	−0·1
TOTLAND BAY							
−0130	−0045	−0035	−0045	−2·2	−1·7	−0·4	−0·1
FRESHWATER BAY							
−0210	+0025	−0040	−0020	−2·1	−1·5	−0·4	0·0

NOTE: Double HWs occur at or near sp; predictions are for the first HW. Off springs there is a stand of about 2 hrs; predictions are then for mid-stand.

SHELTER Reasonable, but all moorings and ⚓s are exposed to winds across the marshland. The 'U' bend carries up to 3m but there is only about 0·5m in the entrance. Keyhaven Lake and Mount Lake all but dry. The river gets very congested during the sailing season.

NAVIGATION WPT 50°42'·70N 01°32'·79W is on 308° leading line marked by two B beacons in transit. Follow transit for about 3ca until through the pair of unlit R and G buoys then start turning to port, giving North Point a wide berth. The channel is marked with unlit G buoys. Entry is difficult on the ebb and should not be attempted in strong E winds. Entrance is possible, with local knowledge, via Hawkers Lake (unmarked), approx 0·3m less water than main channel. Beware lobster pots. When approaching from the W beware the Shingles Bank over which seas break and which partly dries. At Hurst Narrows give The Trap a wide berth.

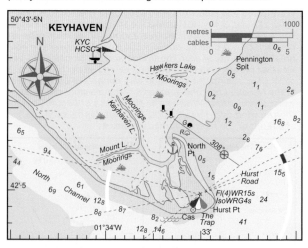

LIGHTS AND MARKS See 9.2.4, chartlet above and 9.2.11 for sectors of Hurst Point light.

COMMUNICATIONS (Code 01590) NMOC (02392) 552100; Police 101; Dr 643022; Ⓗ 677011; *River Warden* VHF Ch ⚓ 645695.

FACILITIES Quay ⚓⚒; **Keyhaven YC** ☎642165, ⚓⛽£12 ⚓🛢⚓. **Hurst Castle SC** ☎(01590) 645589, ⚓⚒⚓. **New Forest District Council** (R.Warden), ⚓⚒.
W Solent Boat Builders ☎642080, ⚓ ✕ ⚒ Ⓔ 🅂(9t) ⚓.
Village ✕ 🛒. Milford-on-Sea (1M): ▮▲🛢🛒✉ Ⓑ ✕ 🛒 ⇌ (bus to New Milton), ✈ (Hurn).

9.2.11 NEEDLES CHANNEL

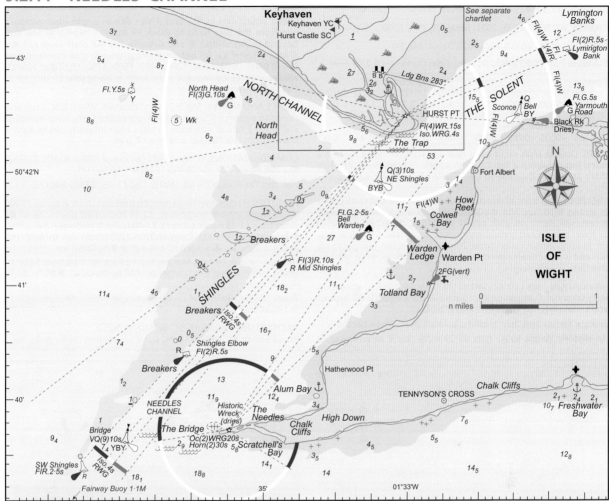

- The NW side of the Needles Channel is defined by the Shingles, a bank parts of which dry, and on which seas breaks in the least swell. The SE side of the bank is fairly steep-to, the NW side is less defined. Dredgers work on the Pot Bank, 4ca S of Bridge WCM.

- On the ebb the stream sets very strongly (3-4kn) WSW across the Shingles. The Needles Channel is well lit and buoyed and in fair weather presents no significant problems. Be aware that even a SW Force 4 against the ebb will raise breaking seas near Bridge and SW Shingles buoys. Contact National Coastwatch Needles on VHF Ch 65 for local conditions.

- In strong winds the North Channel, N of the Shingles, is preferable to the Needles Channel. The two join S off Hurst Point where overfalls and tide rips may be met. Beware The Trap, a shoal spit extending 150m SE of Hurst Castle.

- In bad weather broken water and overfalls extend along The Bridge, a reef which runs 8ca W of the lt ho. Lt's W extremity is marked by Bridge WCM lt buoy, but in calm weather it is feasible to pass E of Bridge buoy to avoid the worst of the ebb stream.

- Gales from S-W raise very dangerous breaking seas against the ebb in the Needles Chan, at points only 250m wide. The sea state can be at its worst shortly after LW when the flood has just begun. Although wind is with tide a substantial swell is raised as a result of turbulence in the recently turned stream. In such conditions consider using North Channel; the E route to the Solent, S of the IoW and via Nab Tower; or find shelter at Poole or Studland.

The Needles are distinctive rocks at the W end of the Isle of Wight (see AC 5600, 2035, 2021). By day the chalk cliffs of High Down are conspic from afar but the lt ho may not be seen by day until relatively close. Goose Rk, dries, is about 50m WNW of the lt ho, 100-150m WSW of which is a drying wreck.

ANCHORAGES BETWEEN THE NEEDLES AND YARMOUTH

ALUM BAY, 50°40'.10N 01°34'.33W. AC 5600, 2021. Tides as for Totland Bay. Very good shelter in E and S winds, but squally in gales. Distinctive white cliffs to S and multi-coloured cliffs and chairlift to E. Appr from due W of chairlift to clear Five Fingers Rk, to the N and Long Rk, a reef drying 0·9m at its E end, to the S. ⚓ in about 4m.

TOTLAND BAY, 50°40'.98N 01°32'.86W. AC 5600, 2035. ML 1·9m. Good shelter in E'lies in wide shelving bay between Warden Ledge (rks 4ca offshore) to the N, and Hatherwood Pt to the SW. Appr W of Warden SHM buoy Fl G 2·5s to ⚓ out of the stream in 2m between pier (2FG vert) and old LB house; good holding. Colwell Bay, to the N between Warden Pt and Fort Albert, is generally rocky and shallow.

ANCHORAGE EAST OF THE NEEDLES, SOUTH IOW

FRESHWATER BAY, 50°40'.07N 01°30'.61W. AC 5600, 2021. ML 1·6m. Good shelter from the N, open to the S. The bay is 3·2M E of Needles Lt ho and 1·2M E of Tennyson's Cross. Conspic marks: Redoubt Fort on W side; a hotel on N side; Stag and Mermaid Rks to the E. The bay is shallow, with rocky drying ledges ¾ca either side and a rock (dries 0·1m) almost in the centre. Best to ⚓ in about 2m just outside the bay. ✉ ☕ ✕ ☐.

9.2.12 YARMOUTH

Isle of Wight **50°42′·42N 01°30′·05W** ✿✿✿≋≋≋✿✿✿

CHARTS AC 5600, 2035, 2021; Imray C3, C15, 2200

TIDES Sp −0050, +0150, Np +0020 Dover; ML 2·0

Standard Port PORTSMOUTH (→); Tidal Curve see 9.2.17 (→)

Times				Height (metres)			
High Water		Low Water		MHWS	MHWN	MLWN	MLWS
0000	0600	0500	1100	4·7	3·8	1·9	0·8
1200	1800	1700	2300				
Differences YARMOUTH							
−0105	+0005	−0025	−0030	−1·7	−1·2	−0·3	0·0

NOTE: Double HWs occur at or near sp; at other times there is a stand lasting about 2 hrs. Differences refer to the first HW when there are two; otherwise to the middle of the stand. See 9.2.17.

SHELTER Good from all directions of wind and sea, but swell enters in strong N/NE'lies. Hbr dredged 2m from ent to swing bridge; access H24. Berth on walk ashore pontoons, detached pontoons or town quay. Craft should give notice of arrival. S Quay is only for loading/embarkation (max 20 mins). In season hbr gets very full and is sometimes closed to visitors. 38 Or ✪s outside hbr (see chartlet). ⚓ well clear.

NAVIGATION WPT 50°42′·58N 01°30′·01W, 188°/2ca to abeam car ferry terminal. Dangers on approach are Black Rock (SHM lt buoy) and shoal water to the N of the E/W bkwtr.

Beware ferries and their wash/turbulence even when berthed.

An **Historic Wreck** 50°42′·55N 01°29′·67W, 2ca E of pierhead; is marked by a Y SPM. Royal Solent moorings lie between the two.

Caution: strong ebb in the ent and parts of the harbour at Sp.

Speed limits: 4kn in hbr and R Yar; 6kn in approaches within a line from the pier head to Black Rk buoy. ⚓ prohib in hbr and beyond swing bridge (R and G traffic lts). All craft must call *Yar Bridge* on VHF Ch 68 before approaching. Bridge opens during BST at 0800, 0900, 1000, 1200, 1400, 1600, 1730, 1830 and 2000LT; winter months by request only. River is navigable by dinghy at HW to Freshwater.

LIGHTS AND MARKS See 9.2.4 and chartlet. When hbr is full and closed to visitors a R flag is flown from end of Ferry Jetty; 'Harbour Full' signs E side of entrance. In fog a high intensity white light is shown from the Pier Hd and inner E pier.

COMMUNICATIONS (Code 01983 (whole of IOW)) NMOC (02392) 552100; Police 101; Dr 08448 151428. HM 760321.

HM and *Yar Bridge* Ch **68**. Water Taxi Ch 15, 07969 840173.

FACILITIES Harbour www.yarmouth-harbour.co.uk Approx 250✔, £3·39 (walk ashore pontoons), £2·11 (detached pontoons or ✪) short stay: £1·39 (walk ashore), £1·00 (other); midweek discounts. ▲▲ (winter 0700-1900, summer 0700-2200 (times vary through the year)) ⬡◻ ⚓ ⚓ ┤▷(£3·50-£4/visit) ⊹ ⬟ ⬟ (5t) ice.

Hayles ⌂ in SW corner; Marine Services are located near Salterns Quay, 500m above the bridge, or ½M by road. ◗ ⬟ ✕ ⬟ ⬟ ⬟ ⬟. Divers.

Town ⬡⬡✉ⓒ(PO) ⬟ ✕ ⬟. Ferries to Lymington, ⇌ (Lymington), ✈ (Hurn/Southampton).

YACHT CLUBS Yarmouth SC ☎760270, ◗ ⬟.
Royal Solent YC ☎760256, ◗ ◗ ✕ ⬟; visiting crews welcome.

9.2.13 LYMINGTON

Hampshire 50°45′·13N 01°31′·40W ❋❋❋⚓⚓⚓❀❀❀

CHARTS AC 5600, 2035, 2021; Imray C3, C15, 2200

TIDES Sp −0040, +0100, Np +0020 Dover; ML 2·0

Standard Port PORTSMOUTH (→); Tidal Curve see 9.2.17 (→)

Times				Height (metres)			
High Water		Low Water		MHWS	MHWN	MLWN	MLWS
0000	0600	0500	1100	4·7	3·8	1·9	0·8
1200	1800	1700	2300				
Differences LYMINGTON							
−0110	+0005	−0020	−0020	−1·6	−1·2	−0·4	−0·1

SEE NOTE under Tides at 9.2.12.

Lymington Harbour Commissioners www.lymington harbour.co.uk ☎672014; HM patrol for berthing advice; wi-fi on river berths.
Dan Bran Pontoon up to 40🅥 £18–£80 ⬭+£2 ⚓ and use of LTSC. Long walk ashore pontoon suitable for rallies and other events.
Town Quay public pontoon, many ⚡ ⬡ £15-£20/craft (9·5m-10·5m LOA), £8·20 for short stay; >15m LOA by arrangement with HM; ⚓ ⬢ (see HM) ⚓ ⬢ ⬢ 🚻/♿.
Scrubbing grids (5) See HM for details.
Berthon Lymington Marina www.berthon.co.uk ☎647405. 250+100🅥 £3·80, ⬭ 🛢⬢⬢⬢🔧 🗜 Ⓔ ⚠ 🏴(75t) ⬡(20t) 🅛 wi-fi.
Town all facilities, ⇌ Ferries to IoW, ✈ (Hurn or Southampton).

YACHT CLUBS Royal Lymington YC ☎672677, ⬢ ✕ 🍴.
Lymington Town SC ☎674514, ⬢ (Sailability access hoist) ✕ 🍴.

SHELTER Excellent above Harper's Post/Wave Screen.

NAVIGATION WPT 50°44′·0N 01°30′·05W, 320°/5ca to abeam No 1 and Cross Boom bns. Strong cross-tide at river entrance. Min depth 2.2m in mid-chan until above ferry terminal, then 1·8m to Town Quay. Speed limits (strictly enforced): 6kn N of Jack in the Basket; 4kn above Wave Screen. ⚓ in the river is prohibited.

> All craft must keep clear of IoW ferries in the river (IRPC Rule 9). Do not follow close astern, or overtake, as they make the final approach to their berth and be aware of wash. They sound one short blast before departure. Ferries may pass each other on N-S transit abreast Pylewell Boom

LIGHTS AND MARKS Conspic YC starting platform is close E of first (No1) SHM bn, Fl G 2s; Jack-in-the-Basket PHM bn, Fl R 2s, is 1·6ca SW of platform. Ldg Lts 319·5°, both FR 12/17m 8M on dayglow Or posts are difficult to see by day. The narrow channel is well marked by lit bns and one lit PHM buoy, Fl.R.2s to port, Fl.G.2s to starboard. Two BW posts, Dir FW 007·5°, and two RW posts, Dir FW 187·5°, are used by ferries passing in the river. Two FY ldg lts 244° mark the ent into Lymington Yacht Haven.

COMMUNICATIONS (Code 01590) NMOC (02392) 552100. Dr 0845 4647; 🏥663000; Police 101; HM 672014; Marinas VHF Ch **80** M (OH).

FACILITIES **Lymington Yacht Haven** www.yachthavens.com ☎677071, 2.5m depth, all tides access. 475⬡+100🅥, £3·25, ⬢ 🅟 🛢 (mobile) ▮▮⚓✕ 🔧 ⬢ 🅛 ⬡ 🏴(50t) ⬡(4t).

9.2.14 NEWTOWN RIVER

Isle of Wight **50°43′·45N 01°24′·66W** ❀❀❀◊✿✿✿

CHARTS AC 5600, 2035, 2036, 2021; Imray C3, C15, 2200

TIDES Sp −0108, Np +0058, Dover; ML 2·3
Standard Port PORTSMOUTH (→); Tidal Curve see 9.2.17 (→)

Times				Height (metres)			
High Water		Low Water		MHWS	MHWN	MLWN	MLWS
0000	0600	0500	1100	4·7	3·8	1·9	0·8
1200	1800	1700	2300				
Differences YARMOUTH/NEWTOWN ENTRANCE							
−0105	+0005	−0025	−0030	−1·7	−1·2	−0·3	0·0

NOTE: Double HWs occur at or near springs; at other times there is a stand which lasts about 2 hrs. Differences refer to the first HW when there are two. At other times they refer to the middle of the stand. See 9.2.17.

SHELTER 3½M E of Yarmouth, Newtown gives good shelter, but is exposed to N'ly winds. There are 5 W ⚓s in Clamerkin Lake and 20 W ⚓s in the main arm leading to Shalfleet Quay. R buoys are private; all are numbered; check with HM.

Do not ⚓ beyond yellow buoys indicating 'Anchorage Limit' due to oyster beds. Fin keel boats can stay afloat from ent to Hamstead landing or to yellow buoys indicating 'Anchorage Limit'. If no room in river, good ⚓ in 3-5m W of ent, beware rky ledges SSE of Hamstead Ledge SHM lt buoy, and possible underwater obstns.

Public landing on E side of river at Newtown Quay by conspic black boathouse and where marked to avoid controlled areas. Eastern peninsula out to Fishhouse Pt is a nature reserve; no landing.

NAVIGATION WPT 50°43′·85N 01°25′·21W, 130°/0·42M to front leading beacon. Without local knowledge, best time to enter is from about HW −4, on the flood while the mud flats are still visible. From W, make good Hamstead Ledge SHM buoy, thence E to pick up leading marks.

From E, keep N of Newtown Gravel Banks where fresh W/SW winds over a spring ebb can raise steep breaking seas. Leave WCM buoy, Q(9) 15s, to port.

Min depth to inside the entrance is about 2m. 5kn speed limit in harbour is strictly enforced. Near junction to Causeway Lake depth is only 0·9m and beyond this water quickly shoals.

At ent to Clamerkin Lake (1·2–1·8m) keep to SE to avoid gravel spit off W shore, marked by two SHMs; the rest of chan is marked by occasional buoys. Beware many oyster beds in Western Haven and Clamerkin Lake.

Rifle Range at top of Clamerkin Lake and in Spur Lake; R flags flown during firing. High voltage power line across Clamerkin at 50°42′·81N 01°22′·66W has clearance of only 8·8m and no shore markings.

LIGHTS AND MARKS See 9.2.4 and chartlet. Conspic TV mast (152m) bearing about 150° (3·3M from hbr ent) provides initial approach track. In season many yachts at anchor or on the buoys inside the hbr are likely to be evident. The ldg bns, 130°, are on Fishhouse Pt on NE side of entrance. Front: RW pile with Y-shaped topmark; rear: R pile with W spherical topmark. Once inside, there are no lights.

COMMUNICATIONS (Code 01983 (for whole IOW)) NMOC (02392) 552100; Police 101; Dr 08448 151428; HM 531424; Taxi 884353.

FACILITIES Average mooring charges overnight £19, short/day stay £9·50. ⚓ no formal charge applies but donations to National Trust are welcomed.
Lower Hamstead Landing ⚓.
Newtown Quay ⚓ ⚓.
Newtown ✉ Ⓑ (Yarmouth or Newport), ⇌ (bus to Yarmouth, ferry to Lymington), ✈ (Bournemouth or Southampton).
Shalfleet Quay ⚓⚓⚓ ◯(drying), dry waste facilities.
Shalfleet Village ▤ ✕ ⌂.

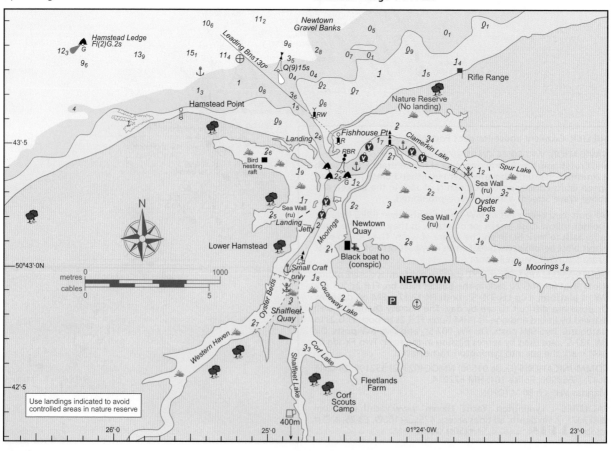

9.2.15 BEAULIEU RIVER

Hampshire 50°46'·89N 01°21'·72W (Ent) ❄❄❄❄❄❄❄ ❀❀❀

CHARTS AC 5600, 2036, 2021; Imray C3, C15, 2200

TIDES −0100 and +0140 Dover; ML 2·4

Standard Port PORTSMOUTH (→); Tidal Curve see 9.2.17 (→)

Times				Height (metres)			
High Water		Low Water		MHWS	MHWN	MLWN	MLWS
0000	0600	0500	1100	4·7	3·8	1·9	0·8
1200	1800	1700	2300				
BUCKLERS HARD							
−0040	−0010	+0010	−0010	−1·0	−0·8	−0·2	−0·3
STANSORE POINT							
−0030	−0010	−0005	−0015	−0·7	−0·5	−0·3	−0·3

NOTE: Double HWs occur at or near springs; the 2nd HW is approx 1¾ hrs after the 1st. On other occasions there is a stand which lasts about 2 hrs. Differences refer to the first HW when there are two, or to the middle of the stand. See 9.2.17.

SHELTER Very good above Needs Ore Pt in all winds. ⚓ possible in reach between Inchmery Ho and Needs Ore Pt, but for facilities proceed to Bucklers Hard Yacht Hbr (3·5M from entrance). No ⚓ E of Inchmery Ho or 1M either side of Bucklers Hard.

NAVIGATION WPT 50°46'·60N 01°21'·40W, close E of seasonal ⬙ (Mar-Oct) in 4m, 324°/3·3ca to abeam Beaulieu Spit Dn. Least depth between WPT and No 2 beacon approx 0·9m. Speed limit of 5kn applies throughout the river. If possible proceed under power in the river.

Ldg marks 324° at entrance should be carefully aligned due to shoal water either side of ldg line, although there is possibly more water to NE of line. Beaulieu Spit Dn should be left approx 40m to port. At night keep in W sector of Millennium Bn (Lepe) until No 4 PHM Bn abeam.

Many of the landing stages and slips shown on the chartlet and AC 2021 are privately owned and must not be used without permission. The river is navigable for about 2M above Bucklers Hard, but the uppermost reaches are best explored by dinghy. There are limited alongside drying berths at Beaulieu extending from the weir not more than 20m downstream.

LIGHTS AND MARKS See 9.2.4 and chartlet. Front ldg mark is red No 2 Bn, 0·7ca N of Beaulieu Spit Dn; the rear is W edge of Lepe Ho. The Old CG cottages and Boat House are conspic, approx 320m E of Lepe Ho. The river is clearly marked by R and G bns and withies/perches. SHM bns 5, 9, 19 and 21 are all Fl G 4s; PHM bns 8, 12 and 20 are Fl R 4s.

COMMUNICATIONS (Code 01590) NMOC (02392) 552100; Police 101; Dr 612451 or (02380) 845955; Ⓗ 0845 4647.

FACILITIES **Bucklers Hard/HM,** www.beaulieuriver.co.uk ☎616200, 115⚓+40Ⓥ, min charge £3·40, short stay £7·50. Detached pontoons & Ⓥ s £2/m, short stay from £6, ⚓ in river £5·50. ♿ⓘ🛒(0800-1800) ⛽⚒ ⚓🛢(35t), scrubbing grid (£25) ⚓🅿🛒🍴🚻.

Royal Southampton YC (Gins Club) ☎616213, www.rsyc.org.uk; Ch 77(Sea Echo); ⚓(temp) ⚓⚓🍴🚻.

Village ✉ Ⓑ Ⓒ 🛒 (☎616293), 🍴 🚻(Beaulieu) ⚇ (bus to Brockenhurst), ✈ (Hurn or Southampton).

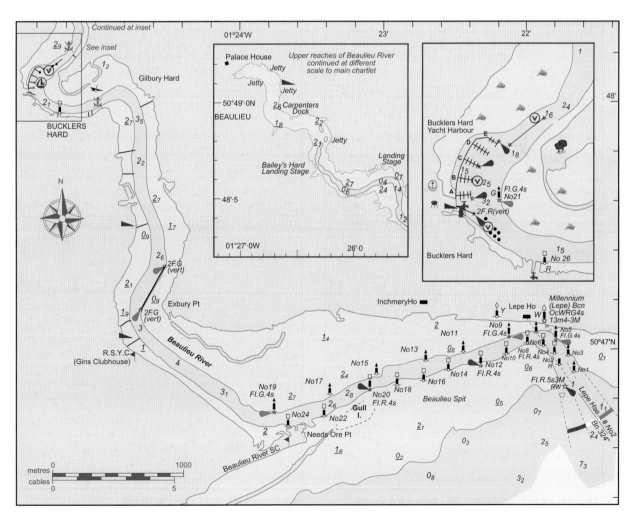

9.2.16 ISLE OF WIGHT AND SOLENT TIDAL STREAMS

5 Hours before HW Portsmouth (0445 before HW Dover)

4 Hours before HW Portsmouth (0345 before HW Dover)

3 Hours before HW Portsmouth (0245 before HW Dover)

2 Hours before HW Portsmouth (0145 before HW Dover)

1 Hour before HW Portsmouth (0045 before HW Dover)

HW Portsmouth (0015 after HW Dover)

General Area 2: 9.2.3

1 Hour after HW Portsmouth (0115 after HW Dover)

2 Hours after HW Portsmouth (0215 after HW Dover)

3 Hours after HW Portsmouth (0315 after HW Dover)

4 Hours after HW Portsmouth (0415 after HW Dover)

5 Hours after HW Portsmouth (0515 after HW Dover)

6 Hours after HW Portsmouth (0615 after HW Dover)

9.2.17 SPECIAL TIDAL CURVES FROM CHRISTCHURCH TO SELSEY BILL

Due to the complex tidal patterns between Christchurch and Selsey Bill, **special curves** for each Secondary Port are used to obtain the most accurate values of times and heights. Because their LW points are more sharply defined than HW, *all times are referenced to LW.* HW times on the tables are approximate.

Critical curve. The spring and neap curves at harbours from Christchurch to Yarmouth differ considerably in shape and duration so a third, 'critical', curve is provided for the Portsmouth range at which the heights of the two HWs are equal for the port concerned. Interpolate between this critical curve and either the spring or neap curve as appropriate.

Note: Whilst the critical curve extends throughout the tidal range, the spring and neap curves stop at the higher HW. For example, with a Portsmouth range of 3·8m (near springs), at 7hrs after LW Lymington the factor should be referenced to the *next* LW. Had the Portsmouth range been 2·0m (near neaps), it should be referenced to the *previous* LW.

Example: Find the height of tide at Christchurch at 1420 on a day when the tidal predictions for Portsmouth are:

18
SA

0110	4·6
0613	1·1
1318	4·6
1833	1·0

Standard Port PORTSMOUTH

Times				Height (metres)			
High Water		Low Water		MHWS	MHWN	MLWN	MLWS
0000	0600	0500	1100	4·7	3·8	1·9	0·8
1200	1800	1700	2300				
Differences CHRISTCHURCH ENTRANCE							
−0230	+0030	−0035	−0035	−2·9	−2·4	−1·2	−0·2

- On the tidal prediction form complete fields 2-5, 8-10 (by interpolation) and thus 13-15.
- On the Christchurch tidal graph, plot the HW & LW heights (fields 14 & 15) and join them with a diagonal line.
- Time required (1420) is 3hr 38min before LW Christchurch (field 13); from this value go vertically towards the curves.
- Interpolate for the Portsmouth range (3·6m) between the spring curve (3·9m) and the critical curve (2·6m).
- Go horizontally to the diagonal; thence vertically to read off the height of tide at 1420: 1·6m.

STANDARD PORT ..Portsmouth..TIME/HEIGHT REQUIRED..1420..

SECONDARY PORT ..Christchurch..DATE..18 Nov..TIME ZONE..0(UT)

	TIME		HEIGHT		
	HW	LW	HW	LW	RANGE
STANDARD PORT	1	2 1833	3 4·6	4 1·0	5 3·6
Seasonal change	Standard Port		6	6	
DIFFERENCES	7	8 −0035	9 −2·8	10 −0·3	
Seasonal change	Standard Port		11	11	
SECONDARY PORT	12	13 1758	14 1·8	15 0·7	
Duration	16				

STANDARD PORT ....TIME/HEIGHT REQUIRED...............

SECONDARY PORT ....DATE................. TIME ZONE...............

	TIME		HEIGHT		
	HW	LW	HW	LW	RANGE
STANDARD PORT	1	2	3	4	5
Seasonal change	Standard Port		6	6	
DIFFERENCES	7	8	9	10	
Seasonal change	Standard Port		11	11	
SECONDARY PORT	12	13	14	15	
	16				

Note*** At Tuckton LWs do not fall below 0·7m except under very low river flow conditions.

Note: From Christchurch to Yarmouth height differences always refer to the higher HW which should be used to obtain the range at the Secondary Port. HW time differences also refer to the higher HW, but are not required for this calculation.

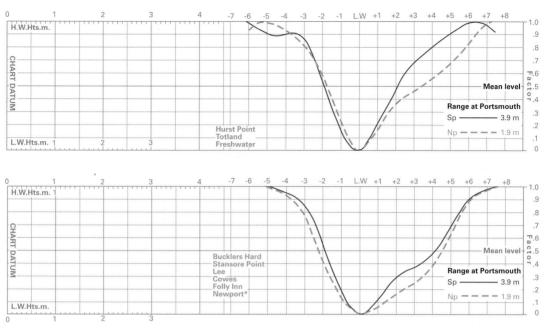

Note* Due to the constriction of the R Medina, Newport requires special treatment since the hbr dries 1·4m. The calculation should be made using the LW time and height differences for Cowes, and the HW height differences for Newport. Any calculated heights which fall below 1·4m should be treated as 1·4m.

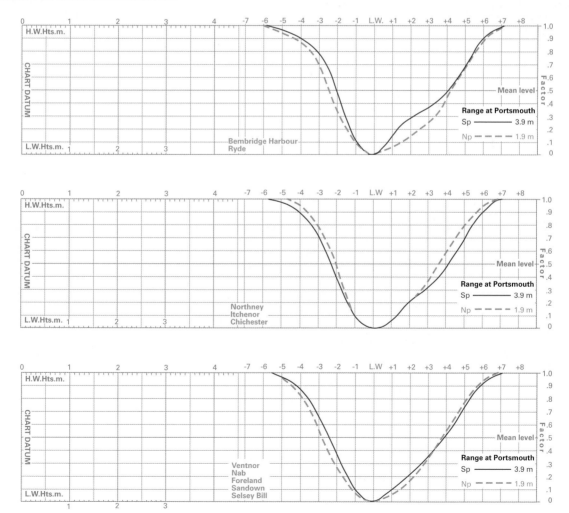

9.2.18   THE CENTRAL SOLENT

CHARTS AC 5600, 2036.

Reference and websites *Solent Cruising Companion,* Aslett/Wiley; *Solent Hazards,* Peter Bruce; *Solent Year Book* SCRA; *Yachtsman's Guide* ABP Southampton www.southamptonvts.co.uk for detailed live port info including: shipping movements, VTS radar displays (Southampton to S of Nab Tr), webcams, tide & weather. See www.bramblemet.co.uk for live weather & tidal info at the Bramble Bk.

Vessel Traffic Service (VTS) Southampton VTS operates on VHF Ch **12** 14, advising Solent shipping between the Needles and Nab Tower including Southampton Water. VTS monitors and coordinates the safe passage of commercial ships. It includes a radar service on request. **QHM Portsmouth,** VHF Ch **11**, controls Portsmouth Harbour and its approaches N of a line from Gilkicker Pt to Outer Spit Buoy.

All small craft in the Solent are strongly advised to monitor VHF Ch 12 Southampton, and/or Ch 11 QHM Portsmouth.

Small Craft and Commercial Shipping Always consider the restricted field of vision from large ships at close quarters and their limited ability to manoeuvre at slow speeds with minimal under keel clearance in restricted waters. Smaller ships keeping clear of these vessels will be encountered outside the channel. Maintain a proper lookout, especially astern. If crossing the main fairway, try to do so at 90° to it. Southampton Hbr byelaw 2003/10 refers. Be aware of the frequent cross-Solent High Speed and Ro-Ro ferries.

The **Precautionary Area** is the potentially dangerous part of the Solent in which large vessels make tight turns into and out of the Thorn Channel leading to Southampton Water. It extends from just NW of Cowes to the Hook buoy N of Calshot Castle (see chartlet and AC 2036). This can be a very busy area. Be particularly alert when between East Lepe and Ryde Middle. Inbound ships pass Prince Consort NCM turning to *port* towards Gurnard NCM, before starting their critical turn to starboard into the Thorn Channel which is, for large ships, very narrow.

Clear Channel Vessels are over 220m LOA and require a clear and unimpeded passage ahead when transiting the Precautionary Area. Vessels may enter the Precautionary Area maintaining a safe distance astern of the 'clear channel vessel'.

Moving Prohibited Zone (MPZ) Any vessel over 150m LOA when in the Precautionary Area is enclosed by an MPZ which extends 1000m ahead of the vessel and 100m on either beam. Small craft under 20m LOA must remain outside the MPZ and use seaman-like anticipation of its route and sea-room required when turning.

Escort Towage for VLCCs From S of the Nab Tr VLCCs over 60,000dwt bound to and from Fawley Marine Terminal will have an escort tug secured to the stern by tow wire. Outwards, the escort tug may leave a tanker at Prince Consort. Use extreme caution when passing round the stern of a large tanker and do not pass between it and the escort tug.

Local Signals Clear Channel & MPZ vessels display a black cylinder by day or 3 all-round ● lts (vert) by night. They are normally preceded by a Harbour Patrol Launch (Ch 12 *Callsign SP*) exhibiting a flashing blue light. Southampton patrol launches have 'HARBOUR MASTER' painted on their superstructure. At night a fixed ● all-round lt is shown above the W masthead light.

Coastguard The National Maritime Operations Centre (NMOC) at Fareham, ☎(02392) 552100, coordinates SAR activities from Topsham to mid-channel; E to the Greenwich Lt V; thence N to Beachy Hd. Owing to the concentration of small craft in its area the NMOC keeps watch on VHF Ch 67/16. Uniquely, an initial call to *Solent Coastguard* should be made on Ch 67 for routine traffic and radio checks, although radio checks on a working channel with other stations – marinas, yachts etc – are preferred and encouraged.

Solent Yacht Racing Buoys Numerous yellow yacht racing buoys are laid throughout the Solent (*seasonal, Mar-Dec*). Most, but not all, are lit Fl Y 4s.

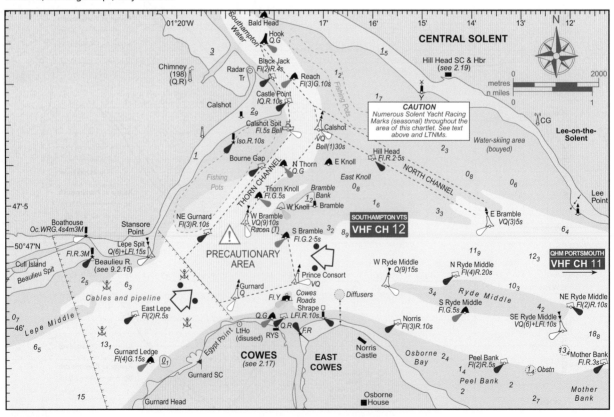

9.2.19 COWES/RIVER MEDINA

Isle of Wight 50°46'·08N 01°17'·95W ✦✦✦✦✿✿✿✿✿✿✿✿

CHARTS AC 5600, 2036, 2793; Imray C3, C15, 2200

TIDES +0029 Dover; ML 2·7

Standard Port PORTSMOUTH (→); Tidal Curve see 9.2.17 (←)

Times				Height (metres)			
High Water		Low Water		MHWS	MHWN	MLWN	MLWS
0000	0600	0500	1100	4·7	3·8	1·9	0·8
1200	1800	1700	2300				
Differences COWES							
–0015	+0015	0000	–0020	–0·5	–0·3	–0·1	0·0
FOLLY INN							
–0015	+0015	0000	–0020	–0·6	–0·4	–0·1	+0·2
NEWPORT							
ND	ND	ND	ND	–0·6	–0·4	+0·1	+0·8

NOTE: Double HWs occur at/near Sp. At other times there is a stand of about 2 hrs. Times are for the middle of the stand. See 9.2.17.

SHELTER Excellent now that outer hbr is protected from N/NE winds by outer bkwtr. ⚓ prohib in hbr. 10 large ⬤s off The Green; pontoons S of chain ferry on both sides of river and in Folly Reach and several marinas. Good ⚓ in Osborne Bay, 2M E, sheltered from SE to W but exposed to wash from large ships; no landing.

NAVIGATION WPT 50°46'·35N 01°18'·0W, 173°/2·7ca to mid chan between Nos 1 and 2 buoys. Shrape Mud, on the E side of the hbr, extends to Old Castle Pt. A small craft channel crosses from Shrape Bn to the fairway, passing N of Shrape Bwtr. Depth is 0.2m below CD, but in the approach is close to CD and it should only be used if craft's draught is less than the existing height of tide. R Medina is navigable to Newport, but dries S of Folly Inn.

- Yachts should use engines. Sp limit 6kn enforced in hbr.
- Strong tidal streams: HW±3 Outer Bwtr creates tidal shadow to the W and HW-3 to HW sets vessels towards W shore in Inner Fairway; no sailing through or ⚓ amongst moorings. Beware ebb (4·5kn at Sp) in vicinity of chain ferry.
- Beware frequent ferry movements, High-Speed to W Cowes, large Ro-Ro ferries to E Cowes, and commercial shipping.
- Ro-Ro ferries may use thrust when manoeuvring at E Cowes causing considerable turbulence across the river. Hi-Speed ferries turn short round to stbd when appr Jubilee Pontoon.
- All vessels < 20m should give way to the chain ferry which displays a Y Fl lt when about to depart and under way; it runs daily 0530-0030, Sun 0635-0030.

LIGHTS AND MARKS See 9.2.4 and chartlet. Small craft chan is marked by Shrape Bn LFl R 10s 3m 3M, and paired PHM and SHM: inbound Fl R 2s/Fl G 2s, Fl R 5s/Fl G 5s (Outer Bwtr) Fl R 3s/Fl G 3s (E Bwtr) with 3 lit Y SPM and R Bn, L Fl R 5s where it joins fairway.

COMMUNICATIONS (Code 01983 (for all IOW)) NMOC (02392) 552100; ⚓ (02380) 228844; Police 101; ⊞ 524081; Dr 294902; Cowes HM 293952; Folly HM (for ferry and berthing) 07884 400046. Monitor *Cowes Hbr Radio* VHF Ch **69** for hbr launches and HM's ⬤s. Yachts >20m LOA should advise arr/dep, and call *Chain Ferry* Ch 69 if passing. Marinas Ch 80. Harbour taxi 07855 767918; *Hbr Taxi* Ch 77 or Sally Water Taxi 07831 331717/299033; *Water Taxi* Ch 06. Casualties: (Ch 16/69) for ambulance at S end of Trinity Landing.

FACILITIES www.cowes.co.uk / www.cowesharbourcommission.co.uk from seaward: **RYS Haven** (strictly members only).

Trinity Landing, on The Parade, short stay on inner side only, pick-up/landing on outer side, call HM Ch 69.

Cowes Yacht Haven www.cowesyachthaven.com ☎299975; 35◖+ 200 ⓥ, £3·00, &🔌🗙⚓🖨Ⓔ🏧🛒(35t)🅿(12t)✕.

Shepard's Wharf Marina, www.shepards.co.uk ☎297821, ◣ 30◖+100ⓥ £2·50, &🔌⛽🗙🖨Ⓔ♨🛒✕🅿 wi-fi.

UKSA ☎294941, info@uksa.org, 5 ◖ ⓥ, £3·30, &✕🅿.

East Cowes Marina www.eastcowesmarina.co.uk ☎293983. 230 ◖+150ⓥ £3·00, &🔌🍴🛒✕🅿 wi-fi.

Town All marine services and facilities.

⚓ Town Quay, Trinity Landing, and Whitegates public pontoon.
Fuel: Lallows 🛒, Cowes Hbr Fuels pontoon off Souters 🛒.
Scrubbing berths Town Quay, UKSA, Sealift, Folly Inn.
Ferries Red Funnel www.redfunnel.co.uk to/from Southampton: ☎(02380) 334010 Car/pax – E Cowes; foot pax – W Cowes. **Chain Ferry** ☎293041, W/E Cowes.

YACHT CLUBS Royal Yacht Squadron ☎292191; **Island SC** ☎296621; **Royal Corinthian YC** ☎293581; **Royal London YC** ☎299727; **Cowes Corinthian YC** ☎296333; **RORC** ☎295144 (Cowes Week); **E Cowes SC** www.eastcowessc.co.uk ☎531687. **Cowes Combined Clubs** ccc@cowesweek.co.uk ☎295744.

RIVER MEDINA, FOLLY REACH TO NEWPORT
FOLLY REACH, 50°44'·03N 01°16'·99W. Above Medham min depth 0·6m to S Folly bn. ◖ on ⓥ pontoons off W bank (£1·25). *Folly Launch* Ch 72/☎07884 400046; Folly Waterbus ☎07974 864627.
Folly Inn ☎297171, &♨◣ scrubbing berth ◖ (pontoon) for 9ⓥ(max LOA 40') £1·80.
Island Harbour Marina (5ca S of Folly Inn) island-harbour.co.uk ☎539994, ◣ 96◖+20ⓥ £3·00, &🔌🛒🗙(50t)⚓✕🅿. Call *Island Harbour Control* on Ch 80 on entering Folly Reach. Excellent shelter; approach via marked channel (dries approx 0·1m) with waiting pontoon to stbd, withies to port. Access HW±3 (draught 1·5m) via lock (width 7·9m) 0800-1730, or by arrangement.

NEWPORT, 50°42'·21N 01°17'·43W. HM ☎525994/ 07970 009589. VHF Ch 69. Tides: Special treatment is required as the hbr dries 1·4m. Use the LW time and height differences for Cowes, and the HW height differences for Newport. Above Island Hbr Marina the chan to Newport (1·2M) dries, but from HW Portsmouth –1½ to +2 it carries 2m or more. S from Folly Inn, the Hbr Authority is IoW Council. Sp limit 6kn to Seaclose, S of Newport Rowing Club (NRC); thence 4kn to Newport. Buoyed and partially lit chan favours W bank. Power lines have 33m clearance. 192° ldg marks/lts are W ◇ bns (lit) on E bank.

Odessa Boatyard ☎524337, odessaboatyard@hotmail.co.uk 75◖+5ⓥs on pontoons downstream on W side £1·50, &⚓🛗🔩 🛒 slipway (35t) 🅿(6t) 🔒.

Newport Yacht Hbr ☎525994, 50ⓥs pontoons on the E/SE sides of the basin have 1·4m at HW ±2 £1·75. Bilge keelers lie alongside pontoons (soft mud); fin keelers against quay wall (firm level bottom). Fender boards can be supplied. ⚓ 🛗 🛒 🅿(8t) Classic Boat Centre. **Town** ◣ 🍴🍴🗙🖨.

Chartlets overleaf

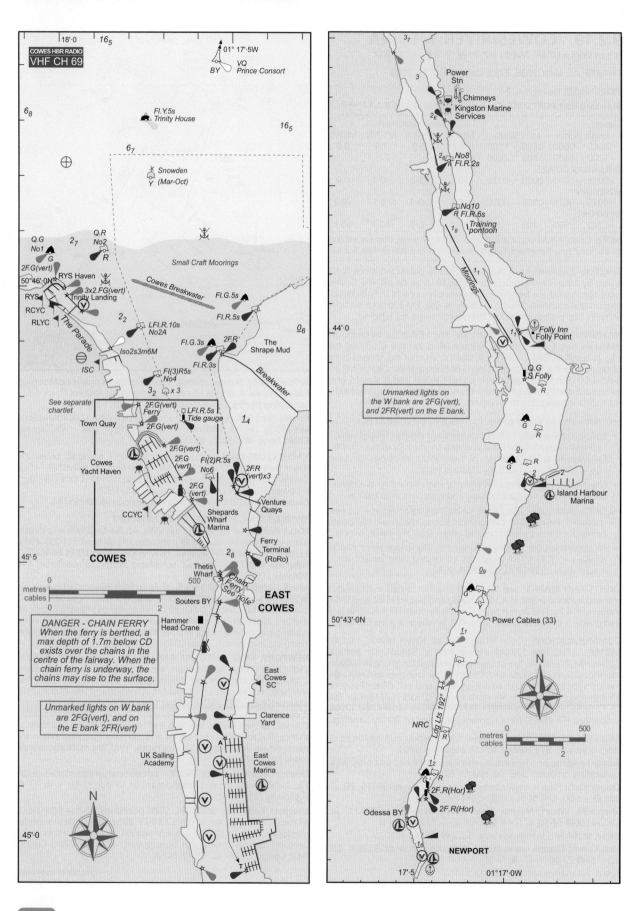

COWES HBR RADIO
VHF CH 69

01° 17'·5W
VQ
BY Prince Consort

Fl.Y.5s
Trinity House

6_8

16_5

6_7

Snowden
(Mar-Oct)
Y

Small Craft Moorings

Q.G
No1
G

2_7

Q.R
No2
R

2F.G(vert)
50°46'·0N
RYS Haven
3x2.FG(vert)
Trinity Landing
RYS
RCYC
RLYC

The Parade

2_2

Cowes Breakwater

Fl.G.5s

Fl.R.5s

0_6

The Shrape Mud

Iso2s3m6M

ISC

LFl.R.10s
No2A

Fl.G.3s

2FR

Fl.R.3s

Breakwater

Fl(3)R.5s
No4

3_2 x 3

See separate
chartlet

Town Quay

2F.G(vert)
Ferry

LFl.R.5s
Tide gauge

1_4

2F.G(vert)

Cowes
Yacht Haven

2F.G(vert)

2F.G
(vert)

Fl(2)R.5s
No6

2F.R
(vert)x3

CCYC

2F.G
(vert)

3

Shepards
Wharf
Marina

Venture
Quays

COWES

2_8

Thetis
Wharf

Chain
Ferry
See note

Souters BY

EAST
COWES

metres
cables

0   500

0   2

DANGER - CHAIN FERRY
When the ferry is berthed, a
max depth of 1.7m below CD
exists over the chains in the
centre of the fairway. When the
chain ferry is underway, the
chains may rise to the surface.

Hammer
Head Crane

East
Cowes
SC

Clarence
Yard

Unmarked lights on W bank
are 2FG(vert), and on
the E bank 2FR(vert)

UK Sailing
Academy

East
Cowes
Marina

N

45'·5

45'·0

3_7

Power
Stn
Chimneys
Kingston Marine
Services

2_6

2_6

No8
Fl.R.2s

No10
R Fl.R.6s

Training
pontoon

1_8

1_1

Moorings

1_1

Folly Inn
Folly Point

Q.G
S.Folly
R

44'·0

Unmarked lights on
the W bank are 2FG(vert),
and 2FR(vert) on the E bank.

G
R

0_1
G
R

2

Island Harbour
Marina

0_9

G
R

50°43'·0N

Power Cables (33)

1_1

N

metres
cables

0   500

0   2

Ldg Lts 192°

NRC
R

1_2
G
R
2F.R(Hor)

Odessa BY

2F.R(Hor)

1_6

NEWPORT

17'·5

01°17'·0W

Note Double HWs occur at Southampton. The predictions are for the first HW.

STANDARD TIME (UT)
For Summer Time add ONE hour in **non-shaded areas**

SOUTHAMPTON   LAT 50°53'N   LONG 1°24'W
TIMES AND HEIGHTS OF HIGH AND LOW WATERS

Dates in red are **SPRINGS**
Dates in blue are **NEAPS**

YEAR 2016

Central S England

JANUARY

Time	m	Time	m
1 0307	4.1	**16** 0309	4.5
0849	1.7	0843	1.2
F 1516	3.9	SA 1532	4.3
2110	1.7	◑ 2108	1.2
2 0357	4.0	**17** 0416	4.3
0938	1.9	0943	1.5
SA 1612	3.8	SU 1618	4.1
◐ 2204	1.9	2214	1.5
3 0527	4.0	**18** 0502	4.2
1037	2.1	1057	1.7
SU 1711	3.7	M 1732	4.0
2307	2.1	2333	1.6
4 0548	3.9	**19** 0618	4.1
1142	2.1	1222	1.7
M 1821	3.7	TU 1852	4.0
5 0011	2.0	**20** 0054	1.6
0655	3.9	0805	1.2
TU 1246	2.1	W 1332	1.5
1928	3.7	2000	4.1
6 0112	1.9	**21** 0157	1.5
0755	4.0	0825	4.3
W 1343	1.8	TH 1428	1.2
2029	3.9	2121	4.3
7 0205	1.7	**22** 0250	1.3
0851	4.2	0914	4.4
TH 1432	1.5	F 1517	1.0
2120	4.1	2207	4.3
8 0253	1.5	**23** 0338	1.0
0933	4.3	1020	4.5
F 1517	1.2	SA 1603	0.7
2202	4.3	2250	4.4
9 0337	1.2	**24** 0423	0.8
1012	4.5	1100	4.5
SA 1600	0.9	SU 1646	0.5
2237	4.4	○ 2333	4.4
10 0420	1.0	**25** 0506	0.7
1104	4.6	1140	4.5
SU 1643	0.7	M 1728	0.5
● 2327	4.6	1808	0.6
11 0503	0.8	**26** 0014	4.4
1138	4.7	0548	0.7
M 1725	0.6	TU 1218	4.4
		1808	0.6
12 0002	4.7	**27** 0052	4.4
0544	0.7	0627	0.8
TU 1216	4.7	W 1255	4.4
1805	0.5	1845	0.7
13 0042	4.7	**28** 0127	4.4
0626	0.7	0702	1.0
W 1258	4.7	TH 1332	4.3
1847	0.5	1918	1.0
14 0126	4.7	**29** 0203	4.3
0708	0.8	0733	1.2
TH 1344	4.7	F 1356	4.2
1930	0.7	1947	1.2
15 0214	4.6	**30** 0241	4.2
0753	1.0	0802	1.4
F 1434	4.5	SA 1431	4.0
2016	0.9	2017	1.4
		31 0257	4.1
		0837	1.6
		SU 1515	3.9
		2056	1.6

FEBRUARY

Time	m	Time	m
1 0340	3.9	**16** 0443	4.1
0925	1.8	1025	1.6
M 1558	3.7	TU 1717	3.9
◑ 2151	1.9	2305	1.7
2 0429	3.8	**17** 0637	4.0
1030	2.0	1159	1.7
TU 1656	3.6	W 1832	3.9
2304	2.1		
3 0531	3.6	**18** 0039	1.8
1144	2.0	0703	3.9
W 1815	3.5	TH 1317	1.5
		1941	3.9
4 0019	2.1	**19** 0144	1.6
0721	3.8	0804	4.0
TH 1256	1.9	F 1413	1.4
2009	3.8	2117	4.0
5 0130	1.9	**20** 0236	1.4
0826	4.0	0927	1.2
F 1401	1.6	SA 1501	1.1
2059	4.0	2200	4.2
6 0229	1.6	**21** 0322	1.1
0913	4.2	1006	4.3
SA 1455	1.2	SU 1545	0.8
2143	4.3	2238	4.3
7 0319	1.2	**22** 0406	0.8
1011	4.4	1043	4.3
SU 1543	0.8	M 1627	0.5
2215	4.4	○ 2315	4.4
8 0404	0.8	**23** 0447	0.6
1043	4.6	1120	4.4
M 1627	0.5	TU 1708	0.5
● 2307	4.6	2353	4.4
9 0448	0.6	**24** 0527	0.6
1119	4.8	1156	4.4
TU 1710	0.3	W 1747	0.5
2343	4.7		
10 0530	0.4	**25** 0026	4.4
1158	4.8	0605	0.7
W 1751	0.2	TH 1231	4.3
		1822	0.7
11 0023	4.8	**26** 0056	4.3
0611	0.3	0636	0.9
TH 1241	4.8	F 1304	4.3
1831	0.2	1850	0.9
12 0107	4.8	**27** 0126	4.3
0653	0.4	0659	1.0
F 1325	4.8	SA 1336	4.3
1913	0.4	1910	1.0
13 0154	4.7	**28** 0156	4.3
0735	0.6	0723	1.1
SA 1415	4.6	SU 1410	4.2
1957	0.6	1937	1.1
14 0248	4.5	**29** 0232	4.2
0821	0.9	0756	1.2
SU 1512	4.4	M 1450	4.1
2046	1.0	2015	1.3
15 0358	4.3		
0915	1.3		
M 1601	4.1		
◑ 2145	1.4		

MARCH

Time	m	Time	m
1 0313	4.1	**16** 0424	4.0
0838	1.4	0958	1.6
TU 1521	3.9	W 1703	3.9
◑ 2101	1.6	2241	1.8
2 0343	3.8	**17** 0531	3.9
0932	1.7	1128	1.8
W 1631	3.7	TH 1809	3.8
2204	1.9		
3 0503	3.7	**18** 0013	1.9
1049	1.9	0634	3.7
TH 1711	3.5	F 1248	1.7
2332	2.1	1915	3.8
4 0546	3.5	**19** 0121	1.8
1216	1.9	0737	3.8
F 2018	3.8	SA 1348	1.5
		2016	3.9
5 0058	1.7	**20** 0214	1.5
0834	3.9	0835	3.9
SA 1334	1.6	SU 1436	1.3
2110	4.0	2106	4.1
6 0208	1.6	**21** 0300	1.2
0915	4.1	0950	4.1
SU 1434	1.2	M 1521	0.9
2119	4.3	2225	4.2
7 0300	1.1	**22** 0343	0.9
0946	4.4	1021	4.2
M 1523	0.7	TU 1603	0.6
2151	4.5	2251	4.3
8 0346	0.7	**23** 0424	0.7
1019	4.6	1057	4.3
TU 1608	0.4	W 1643	0.5
2243	4.7	○ 2325	4.4
9 0429	0.4	**24** 0503	0.6
1057	4.8	1118	4.3
W 1651	0.1	TH 1721	0.6
● 2321	4.8	2357	4.4
10 0511	0.1	**25** 0539	0.7
1137	4.9	1207	4.3
TH 1732	0.0	F 1755	0.7
11 0002	4.9	**26** 0027	4.4
0552	0.1	0608	0.8
F 1221	4.9	SA 1237	4.3
1812	0.0	1820	0.9
12 0046	4.8	**27** 0054	4.4
0633	0.2	0627	0.9
SA 1306	4.8	SU 1307	4.3
1854	0.2	1839	1.0
13 0133	4.7	**28** 0123	4.3
0715	0.4	0652	0.9
SU 1355	4.6	M 1340	4.3
1937	0.5	1908	1.0
14 0226	4.5	**29** 0158	4.3
0800	0.7	0726	1.0
M 1456	4.4	TU 1420	4.2
2024	0.9	1945	1.2
15 0309	4.2	**30** 0239	4.2
0852	1.2	0807	1.2
TU 1545	4.1	W 1505	4.1
◑ 2122	1.4	2030	1.5
		31 0328	4.0
		0857	1.5
		TH 1559	3.9
		◑ 2127	1.8

APRIL

Time	m	Time	m
1 0406	3.7	**16** 0558	3.7
1006	1.8	1205	1.7
F 1706	3.7	SA 1839	3.8
2254	2.0		
2 0546	3.6	**17** 0042	1.8
1140	1.9	0701	3.6
SA 1803	3.6	SU 1309	1.6
		1939	3.8
3 0026	1.9	**18** 0141	1.6
0658	3.7	0758	3.7
SU 1302	1.6	M 1403	1.4
2045	4.1	2034	4.0
4 0141	1.6	**19** 0230	1.3
0849	4.1	0850	3.9
M 1407	1.2	TU 1450	1.1
2118	4.3	2116	4.1
5 0236	1.1	**20** 0315	1.0
0919	4.4	0956	4.1
TU 1458	0.7	W 1533	0.9
2144	4.6	2221	4.3
6 0323	0.7	**21** 0357	0.8
0954	4.6	1031	4.2
W 1544	0.4	TH 1615	0.8
2219	4.8	2234	4.3
7 0407	0.3	**22** 0435	0.8
1034	4.8	1049	4.2
TH 1628	0.1	F 1652	0.8
● 2258	4.9	○ 2309	4.4
8 0449	0.1	**23** 0510	0.8
1116	4.9	1145	4.3
F 1709	0.1	SA 1725	0.9
2340	4.9		
9 0531	0.0	**24** 0000	4.4
1200	4.9	0538	0.9
SA 1751	0.0	SU 1215	4.3
		1750	1.0
10 0024	4.9	**25** 0027	4.4
0613	0.1	0600	0.9
SU 1247	4.8	M 1243	4.3
1833	0.3	1814	1.0
11 0111	4.7	**26** 0056	4.4
0656	0.3	0628	0.9
M 1338	4.6	TU 1317	4.3
1918	0.6	1845	1.1
12 0206	4.5	**27** 0132	4.4
0741	0.7	0704	0.9
TU 1446	4.4	W 1356	4.3
2006	1.0	1924	1.2
13 0323	4.2	**28** 0214	4.3
0832	1.1	0745	1.1
W 1530	4.1	TH 1443	4.2
2103	1.5	2009	1.4
14 0359	4.0	**29** 0303	4.1
0935	1.5	0835	1.3
TH 1638	4.0	F 1538	4.0
◑ 2215	1.8	2105	1.7
15 0501	3.8	**30** 0401	3.9
1051	1.7	0939	1.6
F 1737	3.8	SA 1643	3.9
2333	1.9	◑ 2225	1.9

Chart Datum: 2·74 metres below Ordnance Datum (Newlyn). HAT is 5·0 metres above Chart Datum.

》 **FREE** monthly updates. Register at 《
www.reedsnauticalalmanac.co.uk

223

Note Double HWs occur at Southampton. The predictions are for the first HW.

STANDARD TIME (UT)
For Summer Time add ONE hour in **non-shaded areas**

SOUTHAMPTON LAT 50°53′N LONG 1°24′W
TIMES AND HEIGHTS OF HIGH AND LOW WATERS

Dates in red are **SPRINGS**
Dates in blue are **NEAPS**

YEAR 2016

MAY

Time	m	Time	m
1 SU 0451 3.7 / 1107 1.7 / 1736 3.8 / 2352 1.8		**16** M 0641 3.6 / 1224 1.7 / 1855 3.8	
2 M 0608 3.8 / 1226 1.5 / 1913 4.1		**17** TU 0101 1.8 / 0747 3.7 / 1324 1.6 / 1951 3.9	
3 TU 0106 1.6 / 0734 4.0 / 1333 1.2 / 2012 4.3		**18** W 0155 1.5 / 0812 3.8 / 1415 1.4 / 2042 4.1	
4 W 0206 1.1 / 0851 4.3 / 1429 0.8 / 2119 4.6		**19** TH 0243 1.3 / 0927 4.0 / 1501 1.2 / 2147 4.3	
5 TH 0257 0.7 / 0930 4.6 / 1518 0.5 / 2155 4.8		**20** F 0325 1.1 / 1007 4.1 / 1543 1.1 / 2205 4.3	
6 F 0343 0.4 / 1012 4.7 / 1604 0.3 / 2236 4.9 ●		**21** SA 0405 1.0 / 1028 4.2 / 1621 1.0 / 2243 4.4 ○	
7 SA 0427 0.2 / 1055 4.8 / 1647 0.2 / 2319 4.9		**22** SU 0440 0.9 / 1127 4.3 / 1656 1.0 / 2317 4.4	
8 SU 0511 0.1 / 1141 4.8 / 1731 0.2		**23** M 0510 0.9 / 1158 4.3 / 1726 1.1 / 2350 4.4	
9 M 0004 4.8 / 0554 0.2 / 1230 4.7 / 1815 0.4		**24** TU 0540 0.9 / 1226 4.4 / 1756 1.1	
10 TU 0052 4.6 / 0638 0.6 / 1323 4.5 / 1901 0.7		**25** W 0036 4.4 / 0612 0.9 / 1259 4.4 / 1830 1.1	
11 W 0145 4.4 / 0724 0.7 / 1436 4.4 / 1949 1.0		**26** TH 0112 4.4 / 0649 0.9 / 1340 4.4 / 1910 1.2	
12 TH 0250 4.2 / 0814 1.0 / 1507 4.2 / 2044 1.4		**27** F 0155 4.3 / 0732 1.0 / 1426 4.3 / 1957 1.3	
13 F 0327 4.0 / 0911 1.4 / 1609 4.1 / 2146 1.7 ◐		**28** SA 0244 4.2 / 0821 1.2 / 1520 4.2 / 2052 1.5	
14 SA 0425 3.8 / 1015 1.6 / 1702 3.9 / 2252 1.8		**29** SU 0341 4.1 / 0921 1.4 / 1621 4.1 / 2201 1.7 ◐	
15 SU 0523 3.7 / 1120 1.7 / 1826 3.9 / 2358 1.9		**30** M 0426 3.9 / 1035 1.5 / 1736 4.1 / 2317 1.7	
		31 TU 0538 3.9 / 1150 1.5 / 1825 4.1	

JUNE

Time	m	Time	m
1 W 0030 1.5 / 0654 4.0 / 1258 1.3 / 1936 4.3		**16** TH 0114 1.8 / 0804 3.8 / 1335 1.7 / 2028 4.0	
2 TH 0135 1.2 / 0802 4.3 / 1359 1.0 / 2056 4.6		**17** F 0206 1.6 / 0901 3.9 / 1425 1.5 / 2057 4.1	
3 F 0231 0.9 / 0910 4.5 / 1453 0.8 / 2135 4.7		**18** SA 0252 1.4 / 0948 4.1 / 1510 1.4 / 2140 4.3	
4 SA 0321 0.6 / 0955 4.6 / 1542 0.6 / 2217 4.8		**19** SU 0333 1.2 / 1007 4.2 / 1551 1.2 / 2221 4.3	
5 SU 0409 0.4 / 1040 4.6 / 1629 0.5 / 2302 4.7 ●		**20** M 0411 1.0 / 1047 4.3 / 1630 1.1 / 2257 4.4 ○	
6 M 0454 0.3 / 1128 4.6 / 1715 0.5 / 2348 4.7		**21** TU 0448 0.9 / 1143 4.4 / 1707 1.1 / 2332 4.4	
7 TU 0539 0.3 / 1217 4.6 / 1800 0.6		**22** W 0524 0.9 / 1211 4.4 / 1743 1.0	
8 W 0034 4.6 / 0624 0.5 / 1310 4.5 / 1846 0.8		**23** TH 0019 4.5 / 0601 0.8 / 1244 4.5 / 1821 1.0	
9 TH 0124 4.4 / 0709 0.7 / 1343 4.3 / 1933 1.0		**24** F 0056 4.5 / 0640 0.8 / 1324 4.5 / 1902 1.0	
10 F 0217 4.2 / 0755 0.9 / 1439 4.2 / 2021 1.3		**25** SA 0139 4.5 / 0722 0.9 / 1409 4.5 / 1947 1.1	
11 SA 0250 4.0 / 0844 1.2 / 1534 4.1 / 2114 1.5		**26** SU 0227 4.4 / 0809 1.0 / 1501 4.4 / 2037 1.3	
12 SU 0347 3.9 / 0938 1.5 / 1628 4.0 / 2211 1.7 ◐		**27** M 0321 4.2 / 0902 1.2 / 1600 4.3 / 2136 1.4 ◐	
13 M 0441 3.7 / 1036 1.7 / 1718 3.9 / 2311 1.9		**28** TU 0422 4.1 / 1005 1.3 / 1708 4.2 / 2245 1.5	
14 TU 0536 3.6 / 1137 1.8 / 1811 3.9		**29** W 0513 4.0 / 1116 1.4 / 1755 4.2 / 2358 1.5	
15 W 0013 1.9 / 0701 3.7 / 1237 1.8 / 1933 3.9		**30** TH 0623 4.0 / 1228 1.4 / 1907 4.3	

JULY

Time	m	Time	m
1 F 0109 1.4 / 0739 4.1 / 1336 1.3 / 2043 4.5		**16** SA 0121 1.8 / 0807 3.8 / 1346 1.8 / 2028 4.0	
2 SA 0211 1.1 / 0836 4.3 / 1435 1.1 / 2104 4.5		**17** SU 0215 1.6 / 0901 4.0 / 1438 1.6 / 2116 4.2	
3 SU 0305 0.9 / 0927 4.4 / 1527 0.9 / 2152 4.6		**18** M 0303 1.3 / 0947 4.2 / 1525 1.4 / 2201 4.3	
4 M 0354 0.6 / 1033 4.5 / 1616 0.7 / 2250 4.6 ●		**19** TU 0347 1.1 / 1027 4.3 / 1608 1.1 / 2255 4.5 ○	
5 TU 0441 0.5 / 1120 4.5 / 1702 0.7 / 2334 4.6		**20** W 0429 0.9 / 1102 4.4 / 1649 1.0 / 2327 4.5	
6 W 0525 0.4 / 1207 4.5 / 1747 0.7		**21** TH 0510 0.7 / 1137 4.5 / 1729 0.8	
7 TH 0017 4.5 / 0609 0.5 / 1254 4.4 / 1831 0.8		**22** F 0000 4.6 / 0550 0.6 / 1226 4.6 / 1810 0.8	
8 F 0101 4.4 / 0652 0.6 / 1345 4.4 / 1913 0.9		**23** SA 0039 4.6 / 0628 0.6 / 1306 4.6 / 1849 0.8	
9 SA 0146 4.2 / 0733 0.9 / 1407 4.3 / 1956 1.2		**24** SU 0121 4.6 / 0709 0.6 / 1350 4.6 / 1932 0.9	
10 SU 0235 4.1 / 0815 1.1 / 1451 4.2 / 2039 1.4		**25** M 0207 4.5 / 0753 0.8 / 1440 4.5 / 2018 1.0	
11 M 0304 3.9 / 0859 1.4 / 1542 4.1 / 2126 1.7		**26** TU 0300 4.4 / 0841 1.0 / 1537 4.4 / 2112 1.3 ◐	
12 TU 0357 3.8 / 0949 1.7 / 1632 4.0 / 2219 1.8 ◐		**27** W 0400 4.2 / 0938 1.3 / 1649 4.3 / 2216 1.5	
13 W 0450 3.7 / 1045 1.9 / 1758 3.9 / 2318 2.0		**28** TH 0451 4.0 / 1048 1.5 / 1733 4.2 / 2333 1.6	
14 TH 0552 3.6 / 1145 2.0 / 1826 3.8		**29** F 0606 3.9 / 1207 1.6 / 1939 4.2	
15 F 0019 1.9 / 0704 3.7 / 1247 1.9 / 1932 3.9		**30** SA 0053 1.6 / 0727 4.0 / 1322 1.6 / 2031 4.3	
		31 SU 0200 1.3 / 0830 4.2 / 1424 1.4 / 2051 4.3	

AUGUST

Time	m	Time	m
1 M 0254 1.1 / 0945 4.3 / 1516 1.1 / 2137 4.4		**16** TU 0238 1.4 / 0927 4.2 / 1502 1.4 / 2158 4.4	
2 TU 0342 0.8 / 1028 4.4 / 1603 0.9 / 2238 4.5 ●		**17** W 0326 1.1 / 1008 4.2 / 1548 1.1 / 2231 4.5	
3 W 0427 0.6 / 1111 4.5 / 1648 0.7 / 2319 4.5		**18** TH 0410 0.7 / 1042 4.6 / 1631 0.8 / 2303 4.7 ○	
4 TH 0510 0.5 / 1153 4.5 / 1731 0.7 / 2359 4.4		**19** F 0452 0.5 / 1129 4.7 / 1712 0.6 / 2339 4.8	
5 F 0551 0.5 / 1235 4.4 / 1812 0.7		**20** SA 0532 0.4 / 1205 4.8 / 1752 0.5	
6 SA 0037 4.4 / 0631 0.6 / 1315 4.4 / 1851 0.9		**21** SU 0018 4.8 / 0612 0.3 / 1245 4.8 / 1832 0.5	
7 SU 0117 4.3 / 0709 0.8 / 1355 4.3 / 1927 1.1		**22** M 0101 4.8 / 0652 0.4 / 1329 4.7 / 1913 0.6	
8 M 0157 4.2 / 0743 1.1 / 1437 4.3 / 2000 1.3		**23** TU 0147 4.7 / 0733 0.6 / 1418 4.6 / 1957 0.8	
9 TU 0220 4.0 / 0816 1.4 / 1522 4.2 / 2034 1.6		**24** W 0239 4.5 / 0819 1.0 / 1517 4.4 / 2048 1.2	
10 W 0302 3.9 / 0852 1.6 / 1533 4.0 / 2118 1.8 ◐		**25** TH 0324 4.2 / 0914 1.4 / 1602 4.2 / 2151 1.6 ◐	
11 TH 0345 3.7 / 0943 1.9 / 1620 3.8 / 2217 2.0		**26** F 0435 4.0 / 1026 1.7 / 1724 4.1 / 2317 1.8	
12 F 0440 3.6 / 1050 2.1 / 1719 3.7 / 2326 2.1		**27** SA 0559 3.9 / 1157 1.9 / 1833 4.2	
13 SA 0600 3.5 / 1200 2.1 / 1855 3.8		**28** SU 0043 1.7 / 0711 3.9 / 1313 1.8 / 1941 4.1	
14 SU 0034 2.0 / 0745 3.7 / 1309 2.0 / 2002 3.9		**29** M 0147 1.5 / 0817 4.1 / 1412 1.5 / 2036 4.2	
15 M 0142 1.8 / 0843 4.0 / 1411 1.8 / 2055 4.1		**30** TU 0239 1.2 / 0944 4.3 / 1501 1.3 / 2145 4.3	
		31 W 0325 0.9 / 1019 4.4 / 1546 1.0 / 2223 4.4	

Chart Datum: 2·74 metres below Ordnance Datum (Newlyn). HAT is 5·0 metres above Chart Datum.

FREE monthly updates. Register at www.reedsnauticalalmanac.co.uk

Note Double HWs occur at Southampton. The predictions are for the first HW.

STANDARD TIME (UT)
For Summer Time add ONE hour in **non-shaded areas**

SOUTHAMPTON   LAT 50°53'N   LONG 1°24'W
TIMES AND HEIGHTS OF HIGH AND LOW WATERS

Dates in red are **SPRINGS**
Dates in blue are **NEAPS**

YEAR 2016

Central S England

SEPTEMBER

Time	m		Time	m
1 0408	0.6		**16** 0347	0.7
1056	4.5		1030	4.7
TH 1629	0.7		F 1608	0.7
● 2300	4.4		○ 2238	4.8
2 0449	0.5		**17** 0429	0.4
1135	4.5		1103	4.9
F 1710	0.7		SA 1650	0.4
2338	4.4		2316	4.9
3 0529	0.5		**18** 0510	0.2
1212	4.5		1142	4.9
SA 1749	0.7		SU 1731	0.3
			2357	5.0
4 0013	4.4		**19** 0550	0.2
0607	0.7		1223	4.9
SU 1245	4.4		M 1811	0.4
1826	0.9			
5 0048	4.3		**20** 0040	4.9
0642	0.9		0631	0.4
M 1316	4.4		TU 1307	4.8
1856	1.1		1853	0.5
6 0122	4.3		**21** 0127	4.8
0708	1.1		0713	0.6
TU 1347	4.3		W 1357	4.7
1919	1.3		1936	0.8
7 0157	4.2		**22** 0221	4.5
0730	1.3		0759	1.0
W 1421	4.2		TH 1459	4.4
1945	1.4		2026	1.2
8 0235	4.1		**23** 0336	4.3
0800	1.5		0854	1.5
TH 1441	4.1		F 1547	4.2
2022	1.6		◑ 2131	1.6
9 0300	3.9		**24** 0428	4.0
0843	1.8		1011	1.9
F 1528	3.9		SA 1706	4.0
◑ 2113	1.9		2302	1.8
10 0415	3.8		**25** 0542	3.9
0944	2.1		1142	2.0
SA 1620	3.7		SU 1811	3.9
2233	2.1			
11 0450	3.6		**26** 0021	1.8
1117	2.3		0646	3.9
SU 1725	3.6		M 1252	1.9
2357	2.1		1910	3.9
12 0718	3.7		**27** 0123	1.6
1237	2.2		0755	4.0
M 1933	3.8		TU 1349	1.7
			2011	4.0
13 0112	1.9		**28** 0214	1.4
0819	4.0		0850	4.2
TU 1346	1.9		W 1438	1.4
2058	4.1		2138	4.2
14 0212	1.5		**29** 0300	1.1
0904	4.3		0931	4.3
W 1439	1.4		TH 1522	1.1
2133	4.4		2205	4.3
15 0302	1.1		**30** 0343	0.8
0942	4.5		1039	4.5
TH 1525	1.0		F 1605	0.8
2204	4.6		2239	4.4

OCTOBER

Time	m		Time	m
1 0424	0.7		**16** 0404	0.4
1112	4.5		1038	5.0
SA 1645	0.7		SU 1626	0.4
● 2315	4.4		○ 2253	5.0
2 0504	0.7		**17** 0446	0.2
1120	4.5		1117	5.0
SU 1724	0.8		M 1708	0.3
2335	4.4		2335	5.0
3 0541	0.8		**18** 0528	0.2
1215	4.5		1200	5.0
M 1758	1.0		TU 1750	0.3
4 0022	4.4		**19** 0020	4.9
0612	1.0		0610	0.4
TU 1242	4.4		W 1245	4.9
1823	1.1		1833	0.5
5 0052	4.4		**20** 0109	4.8
0632	1.2		0654	0.7
W 1311	4.4		TH 1336	4.7
1841	1.2		1917	0.8
6 0125	4.3		**21** 0205	4.5
0654	1.3		0741	1.1
TH 1343	4.4		F 1440	4.4
1909	1.3		2007	1.2
7 0201	4.3		**22** 0339	4.3
0727	1.4		0837	1.5
F 1421	4.2		SA 1525	4.2
1946	1.5		◑ 2110	1.6
8 0245	4.1		**23** 0413	4.1
0808	1.7		0950	1.9
SA 1508	4.1		SU 1638	4.0
2033	1.7		2231	1.8
9 0338	3.9		**24** 0513	4.0
0900	2.0		1110	2.0
SU 1545	3.8		M 1739	3.9
◑ 2138	2.1		2345	1.9
10 0443	3.8		**25** 0613	3.9
1031	2.3		1218	2.0
M 1723	3.7		TU 1836	3.8
2319	2.1			
11 0531	3.7		**26** 0047	1.8
1204	2.2		0716	4.0
TU 1818	3.7		W 1317	1.8
			1936	3.9
12 0038	1.9		**27** 0141	1.6
0827	4.1		0814	4.1
W 1315	1.9		TH 1408	1.5
1954	4.1		2030	4.0
13 0142	1.5		**28** 0230	1.3
0909	4.4		0902	4.3
TH 1411	1.5		F 1454	1.2
2106	4.4		2152	4.2
14 0234	1.1		**29** 0314	1.1
0908	4.6		0938	4.4
F 1459	1.0		SA 1537	1.0
2136	4.6		2218	4.3
15 0320	0.7		**30** 0356	0.9
1002	4.8		1043	4.5
SA 1543	0.6		SU 1618	0.9
2213	4.9		● 2233	4.4
			31 0436	0.9
			1056	4.5
			M 1656	0.9
			2329	4.4

NOVEMBER

Time	m		Time	m
1 0513	1.0		**16** 0507	0.4
1148	4.5		1140	5.0
TU 1729	1.0		W 1731	0.3
2 0001	4.4		**17** 0003	4.9
0542	1.2		0552	0.5
W 1216	4.5		TH 1227	4.9
1753	1.1		1815	0.5
3 0030	4.4		**18** 0053	4.7
0603	1.3		0637	0.8
TH 1244	4.5		F 1317	4.7
1814	1.2		1901	0.8
4 0101	4.4		**19** 0151	4.6
0629	1.3		0726	1.1
F 1316	4.4		SA 1416	4.4
1845	1.2		1951	1.1
5 0137	4.4		**20** 0236	4.3
0704	1.4		0819	1.4
SA 1354	4.3		SU 1528	4.2
1923	1.3		2047	1.4
6 0221	4.3		**21** 0342	4.2
0746	1.6		0921	1.7
SU 1440	4.2		M 1601	4.0
2009	1.6		◑ 2154	1.7
7 0313	4.1		**22** 0444	4.1
0837	1.9		1030	1.9
M 1535	4.0		TU 1701	3.9
◑ 2106	1.8		2302	1.8
8 0414	4.0		**23** 0612	4.0
0950	2.1		1136	2.0
TU 1621	3.8		W 1824	3.8
2235	2.0			
9 0500	3.9		**24** 0005	1.8
1123	2.1		0633	4.0
W 1731	3.8		TH 1239	1.9
2358	1.9		1856	3.8
10 0627	4.0		**25** 0104	1.7
1237	1.9		0730	4.0
TH 1902	4.0		F 1334	1.7
			1952	3.9
11 0105	1.6		**26** 0156	1.5
0752	4.3		0824	4.2
F 1338	1.5		SA 1424	1.4
2006	4.3		2041	4.0
12 0202	1.2		**27** 0243	1.3
0835	4.6		0907	4.3
SA 1430	1.1		SU 1508	1.2
2110	4.6		2159	4.2
13 0252	0.8		**28** 0327	1.2
0935	4.8		1014	4.4
SU 1517	0.7		M 1550	1.1
2149	4.8		2236	4.3
14 0338	0.5		**29** 0407	1.1
1014	5.0		1032	4.5
M 1603	0.4		TU 1628	1.0
○ 2232	4.9		● 2317	4.4
15 0423	0.4		**30** 0445	1.2
1055	5.0		1108	4.5
TU 1647	0.3		W 1702	1.1
2316	5.0		2351	4.4

DECEMBER

Time	m		Time	m
1 0516	1.2		**16** 0537	0.6
1139	4.5		1211	4.8
TH 1729	1.1		F 1802	0.4
2 0017	4.4		**17** 0040	4.7
0543	1.3		0624	0.7
F 1225	4.5		SA 1259	4.6
1757	1.1		1847	0.6
3 0045	4.4		**18** 0134	4.5
0613	1.3		0710	0.9
SA 1256	4.5		SU 1350	4.4
1829	1.1		1933	0.9
4 0120	4.4		**19** 0239	4.4
0649	1.3		0757	1.2
SU 1335	4.4		M 1448	4.2
1909	1.2		2021	1.2
5 0203	4.4		**20** 0307	4.2
0731	1.4		0850	1.5
M 1420	4.4		TU 1525	4.0
1953	1.3		2116	1.5
6 0252	4.3		**21** 0406	4.1
0820	1.6		0947	1.8
TU 1511	4.2		W 1619	3.9
2046	1.5		◑ 2215	1.7
7 0348	4.2		**22** 0459	4.0
0921	1.8		1050	1.9
W 1612	4.0		TH 1718	3.8
◑ 2154	1.7		2319	1.9
8 0453	4.1		**23** 0553	3.9
1039	1.9		1156	2.0
TH 1701	3.9		F 1817	3.7
2314	1.7			
9 0543	4.1		**24** 0022	1.9
1156	1.8		0715	4.0
F 1812	4.0		SA 1258	1.9
			1946	3.8
10 0027	1.6		**25** 0121	1.8
0659	4.3		0814	4.0
SA 1305	1.5		SU 1353	1.7
1928	4.2		2011	3.9
11 0131	1.3		**26** 0212	1.6
0807	4.5		0906	4.2
SU 1404	1.2		M 1440	1.5
2028	4.4		2107	4.0
12 0226	1.0		**27** 0258	1.5
0855	4.7		0925	4.3
M 1455	0.8		TU 1523	1.3
2132	4.6		2152	4.2
13 0317	0.8		**28** 0341	1.3
0956	4.9		1011	4.4
TU 1544	0.6		W 1602	1.1
2217	4.7		2237	4.3
14 0405	0.6		**29** 0420	1.2
1039	4.9		1050	4.5
W 1631	0.4		TH 1639	1.0
○ 2303	4.8		● 2312	4.4
15 0451	0.5		**30** 0455	1.2
1125	4.9		1122	4.5
TH 1716	0.4		F 1713	1.0
2350	4.7			
			31 0006	4.4
			0529	1.1
			SA 1154	4.5
			1746	0.9

Chart Datum: 2·74 metres below Ordnance Datum (Newlyn). HAT is 5·0 metres above Chart Datum.

》 **FREE** monthly updates. Register at 《
www.reedsnauticalalmanac.co.uk

225

SOUTHAMPTON
MEAN SPRING AND NEAP CURVES

MEAN RANGES
Springs 4.0m
Neaps 1.9m

9.2.20 SOUTHAMPTON

Hampshire 50°52'·65N 01°23'·37W ✿✿✿✿◊◊◊✿✿

CHARTS AC 5600, 2036, 2041; Imray C3, C15, 2200

TIDES HW (1st) −0001 Dover; ML 2·9

Standard Port SOUTHAMPTON (←—)

Times				Height (metres)			
High Water		Low Water		MHWS	MHWN	MLWN	MLWS
0400	1100	0000	0600	4·5	3·7	1·8	0·5
1600	2300	1200	1800				
Differences REDBRIDGE							
−0020	+0005	0000	−0005	−0·1	−0·1	−0·1	−0·1

NOTE: At springs there are two separate HWs about 2 hrs apart; at neaps there is a long stand. Predictions are for the first HW when there are two, otherwise for the middle of the stand. NE gales and a high barometer may lower sea level by 0·6m.

SHELTER Good in most winds, although a choppy sea builds in SE winds >F4, when it may be best to shelter in marinas. ⓥ berths available in Hythe, Town Quay, Ocean Village, Shamrock Quay and Kemp's Marinas. No specific yacht anchorages but temporary ⚓ is permitted (subject to HM) off club moorings at Netley, Hythe, Weston and Marchwood in about 2m. Keep clear of all channels and Hythe Pier. Public moorings for larger yachts opposite Royal Pier near Gymp Elbow PHM buoy in 4m (contact HM); nearest landing is at Town Quay Marina.

NAVIGATION WPT 50°50'·0N 01°18'·6W, just S of Hamble Point SCM, to Weston Shelf SHM buoy 312°/4M. Main channels are well marked.

- For a Yachtsman's Guide see www.southamptonvts.co.uk.
- Yachts should keep just outside the buoyed lit fairway and when crossing it should do so at 90°, abeam Fawley chy, at Cadland and Greenland buoys, abeam Hythe and abeam Town Quay.

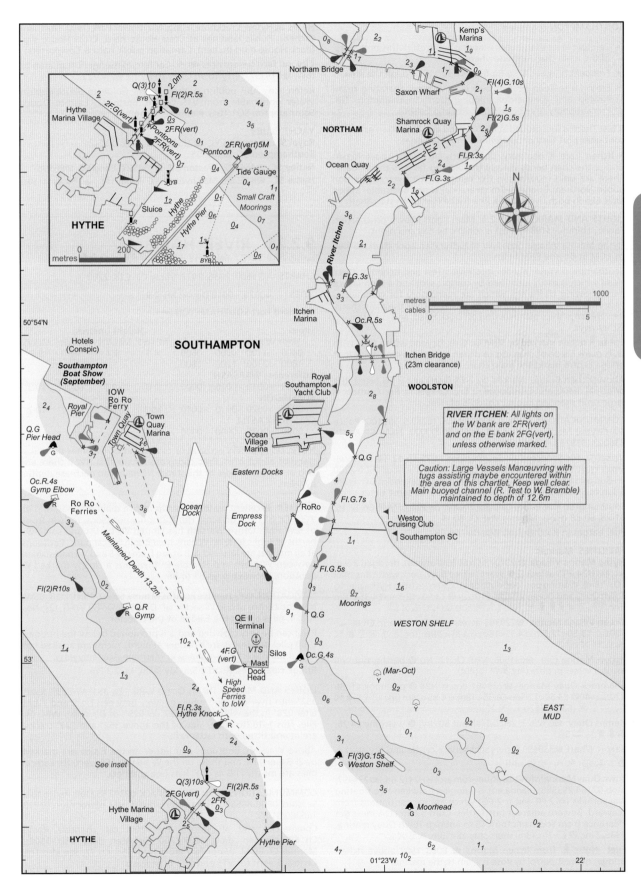

Central S England

SOUTHAMPTON *continued*

- Caution: several large unlit mooring buoys off Hythe, both sides of the main chan, and elsewhere.
- Frequent hi-speed & Ro-Ro ferries operate through the area.
- Essential to keep clear of very large tankers operating from Fawley and commercial shipping from Southampton. NB Precautionary Area between Cowes and Calshot.

R Test There is foul ground at Marchwood and Royal Pier; extensive container port further upstream. Eling Chan dries.

R Itchen Care is necessary, particularly at night. Above Itchen Bridge the chan bends sharply to port and favours the W bank. There are unlit moorings in the centre of the river. Navigation above Northam Bridge (4·7m clearance) is not advisable. Speed limit 6kn in both rivers N of a line Hythe Pier to Weston Shelf.

LIGHTS AND MARKS See 9.2.4. Main lights shown on chartlet. Fawley chimney (198m, R lts) is conspic by day and night.

Hythe Marina Village, close NW of Hythe Pier: appr chan marked by ECM lit Bn and PHM lit Bn.

Southampton Water divides into Rivers Test and Itchen at Dock Head which is easily identified by conspic silos and a high lattice mast showing traffic sigs which are mandatory for commercial vessels, but may be disregarded by yachts outside the main chans. Beware large ships manoeuvring off Dock Head, and craft leaving either river.

Ent to R Itchen marked by SHM Oc G 4s, beyond which piles with G lts mark E side of chan ldg to Itchen bridge (24·4m); a FW lt at bridge centre marks the main chan.

Above Itchen Bridge, marked by 2 FR (vert) and 2 FG (vert), the principal marks are: Crosshouse lit Bn; Chapel lit Bn. **Caution:** 8 large unlit mooring buoys in middle of river.

COMMUNICATIONS (Code 02380) NMOC (02392) 552100; Police 101; Hythe Medical Centre ☎845955; ⊞ 777222. HM ABP & VTS 330022 (339733 outside HO).

Vessel Traffic Services Call: *Southampton VTS* Ch **12** 14 16 (H24). Small craft traffic info broadcast on Ch 12 on the hour 0600-2200 Fri-Sun and Bank Holidays from Easter to last weekend in Oct. From 1 Jun to 30 Sep broadcasts are every day at the same times.

Southampton Hbr Patrol Call: *Southampton Patrol* VHF Ch **12** 16, 01-28, 60-88 (H24). Marinas VHF Ch **80** M.

Fuel barges in R Itchen, call *Wyefuel* Ch 08.

FACILITIES Marinas:
Hythe Marina Village ☎ 207073 Ch 80 (call ahead), dredged 2·5m. Access H24 IPTS at lock (2 ● over ○ = free flow) (lock width 8m; max LOA 18m). Waiting pontoon outside lock. Ferries from Hythe pier to Town Quay and Ocean Village. 210⌷+♥ £3·50<12·5m>£4·35<18m (inc ⌁), ⬟ ⚓ ▮ ♨ ✕ ⊡ ⬚(30t) ⌷(12t) ⬕ ✕ ⬚.

Ocean Village Marina ☎229385 access H24 ♥ pre-book Ch 80 ⬛ 450⌷ £3·50<12·5m>£4·35<18m>£4·85<24m (inc ⌁), ⬟ ⬕ ▮ ♨ ✎ ⬕ ⬛ ✕ ⬚.

Itchen Marina (50) ☎631500, VHF Ch 12. No ♥ berths, but will assist a vessel in difficulty, ▮ ⬚ ⬚(40t).

Shamrock Quay Marina ☎229461 access H24 ♥ pre-book Ch 80 220⌷+40♥ £3·50<12·5m>£4·35<18m>£4·85<24m (inc ⌁), ⬕ ▮ ♨ ✎ ⬕ ⬕ ⬡ (63t) ⌷(12t) ⬕ ⬚ ✕ ⬚.

Kemps Quay ☎632323, 260⌷ (all but 50 dry) ♥ welcome £4·26, ⚓ ▮ ▮ ✎ ⬚(5t)

Saxon Wharf ☎339490 Superyacht service centre and dry-stack. ♥ ⬕ ⚓ ⌁ ♨ ✎ ⬕ ⬕ wi-fi.

Town Quay Marina info@townquay.com access H24 (2·6m) ☎234397, mob 07764 293588; marina ent is a dogleg between two floating wavebreaks (☆ 2 FR and ☆ 2 FG) which appear continuous from seaward. Beware adjacent fast ferries. Craft >20m LOA must get clearance from Southampton VTS to ent/dep Town Quay marina. 133⌷ inc ♥ £3·20<13m, short stay £8<13m, ⬕ ⬕ ✕ ⬚

Fuel Note: ▮ from Itchen Marina ☎ 8063 1500 above Itchen Bridge, nearest petrol by hose is from Hythe marina.

Hards at Hythe, Cracknore, Eling, Mayflower Park (Test), Northam (Itchen). Public landings at Cross House hard, Cross House slip, Block House hard (Itchen), Carnation public hard & Cowporters.

City All facilities services ⇌ ✈, Car ferry/passenger catamaran to IoW, ☎019192; ferry Town Quay to Hythe ☎840722.

Reference ABP publish a *Yachtsman's Guide to Southampton Water* obtainable from VTS Centre, Berth 37, Eastern Docks, Southampton SO1 1GG. Please send sae.

YACHT CLUBS
Royal Southampton YC ☎223352, ⚓ ⚓ ⬕ ⬚ ✕ ⬚.
Southampton SC ☎446575.
Netley SC ☎454272; **Weston SC** ☎452527; **Eling SC** ☎863987.
Hythe SC ☎846563; **Marchwood YC** ☎666141, ⚓ ↖ ⚓ ⬚(10t) ⬚.

9.2.21 RIVER HAMBLE
Hampshire **50°51'·0N 01°18'·50W** ❂❂❂⬙⬙⬙ ❀❀❀

CHARTS AC 5600, 2036, 2022; Imray C3, C15, 2200

TIDES +0020, –0010 Dover; ML 2·9

Standard Port SOUTHAMPTON (←)

Times				Height (metres)			
High Water		Low Water		MHWS	MHWN	MLWN	MLWS
0400	1100	0000	0600	4·5	3·7	1·8	0·5
1600	2300	1200	1800				
Differences WARSASH							
+0020	+0010	+0010	0000	0·0	+0·1	+0·1	+0·3
BURSLEDON							
+0020	+0020	+0010	+0010	+0·1	+0·1	+0·2	+0·2
CALSHOT CASTLE							
0000	+0025	0000	0000	0·0	0·0	+0·2	+0·3

NOTE: Double HWs occur at or near sp; at other times there is a stand of about 2 hrs. Predictions are for the first HW if there are two or for the middle of the stand. NE gales can decrease depths by 0·6m.

SHELTER Excellent, with ♥s berths at five main marinas, and at YCs, SCs, BYs and on some Hbr Authority pontoons. Jetty in front of B/W HM's Office Warsash has limited ♥s berths and some ♥s berths at Public Jetty on W bank near Bugle Inn car park. Pontoons at all marinas are lettered A, B et seq from the S end. In mid-stream between Warsash and Hamble Pt Marina is a clearly marked ♥ pontoon, between piles B1 to B6.

NAVIGATION WPT Hamble Point SCM buoy, 50°50'·15N 01°18'·66W, is in the centre of the W sector of Dir 352°, Oc (2) WRG 12s; Nos 1, 3, 5, piles are on the E edge of the G sector.

River can be very crowded and ⚓ is prohibited below the bridges. Unlit piles and buoys are a danger at night. Yachts are advised not to use spinnakers above Warsash Maritime Academy jetty.

Bridge clearances: Road 3·5m; Rly 5·5m; M27 3·8m.

LIGHTS AND MARKS Dir ☆, Oc (2) WRG 12s, 351°-W-353°, leads 352° into river ent. and when midway between No 5 and 7 bns alter 028° in the white sector, 027°-W-029°, of Dir ☆, Iso WRG 6s. Piles Nos 1–10 (to either side of the above 352° and 028° tracks) are lit and fitted with radar refl.

Above Warsash, pontoons and jetties on the E side are marked by 2FG (vert) lts, and those on the W side by 2FR (vert) lts. Lateral piles are mostly Fl G 4s or Fl R 4s (see chartlet).

COMMUNICATIONS (Code 02380) NMOC 02392 552100; Police 101; Seastart 0800 885500. HM 01489 576387, HM patrol mob 07718 146380/81/99.

Commercial and all vessels over 20m call: *Hamble Radio* Ch **68** (Apr-Sep: daily 0600-2200. Oct-Mar: daily 0800-1800). Marinas Ch **80** M. Water Taxi ☎07827157155, Ch 77. See also The Central Solent for VTS and info broadcasts.

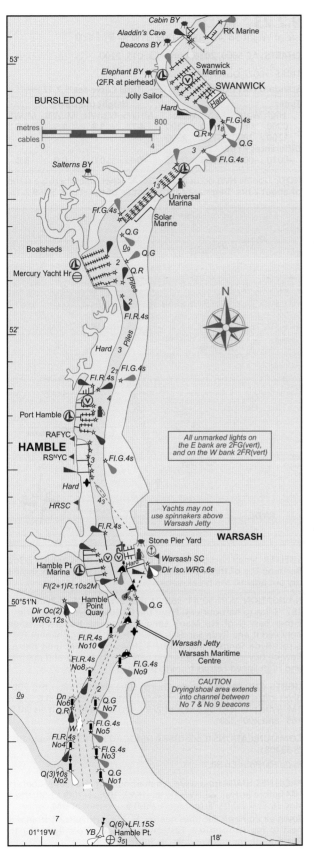

FACILITIES Marinas from seaward:

Hamble Pt Marina ☎452464, access H24, pre-book Ch 80. 220⊂ £3·40, short stay £8·20; ♠ ⚒ ⚓ Ⓔ ⊞ ⛟(65t) ⛴(7t) ⚓ ⛒ ⛁.

Stone Pier Yard ☎01489 579589, access H24, 56⊂ £3·00, <4hrs £8; ⚓ ⊞ ⚒ ⚓ ⚓ ⛟(10t) ⚓. Nearby: ✉ Ⓒ ⛒.

Port Hamble Marina ☎452741 access H24 pre-book Ch 80. ⚓ 310⊂+Ⓥ £3·40; ⛁ ♠ ♠ ♠ Ⓔ ⚒ ⛟(60t) ⛴(7t) ⚓ ⛁.

Mercury Yacht Hbr ☎455994, access H24, pre-book Ch 80. 346⊂+Ⓥ £3·40, short stay £8·20; ⚓ ⛁ ♠ ♠ ♠ ⚓ Ⓔ ⚓ ⛟(20t) ⚓ ⛁.

Universal Marina ⚓ ☎01489 574272, access H24, pre-book Ch 80. 249⊂+Ⓥ £3·20, short stay £8·00; ⛁ ♠ ⊞ ⚒ ⚓ ⛟(80t) ⛁/café.

Swanwick Marina ☎01489 884081 (H24), 380⊂+Ⓥ £3·55, £1·50 <4hrs (min £5); ⛁ ♠ ♠ ♠ ⚒ ⚓ ⚓ ⛟(60t) ⛴(12t) ⚓ ✉ ⛒ @.

The Hbr Authority jetty in front of the conspic B/W HM's Office at Warsash has limited ⊂ (£2), also Ⓥ pontoons in mid-stream (both £1·50), ⚓ toilets ashore. Public Jetty on W bank near the Bugle Inn car park (£1·75). See www.hants.gov.uk/hamble harbour.

YACHT CLUBS
Hamble River SC ☎452070
RAFYC ☎452208, ⚓ ✕ ⛁
Royal Southern YC ☎450300
Warsash SC ☎01489 583575.

SERVICES Wide range of marine services available; consult marina/ HM for locations. *Hamble River Guide* available from HM.
Hards At Warsash, Hamble and Swanwick.
Slips at Warsash, Hamble, Bursledon & Lower Swanwick.
Maintenance piles at Warsash by HM's slipway; on W bank upriver of public jetty near Bugle; by slip opposite Swanwick Marina.
✉ (Hamble, Bursledon, Warsash and Lower Swanwick); (Hamble, Bursledon, Sarisbury Green, Swanwick, Warsash).
⚍ (Hamble and Bursledon); ✈ (Southampton).

ADJACENT HARBOURS
ASHLETT CREEK, Hants, 50°50′·01N 01°19′·49W. AC 5600, 2038, 2022. Tides approx as Calshot Castle (opposite). Small drying (2·1m) inlet across from R Hamble; best for shoal-draught vessels. Appr at HW close to Fawley Marine Terminal. Unlit chan, marked by 3 PHM buoys, then in sequence a pair of PHM and SHM buoys, a PHM beacon and SHM buoy (No 7). Then follow PHM and SHM beacons. The single dog leg is indicated by an arrow on a PHM. A final SHM buoy leads to the drying sailing club pontoon or the drying quay. Ldg bns are hard to find; local knowledge desirable. Berth at drying quay. Facilities: ♠ ⚓ ⚓ Hard ⛁. **Ashlett SC.**

HILL HEAD, Hants, 50°49′·08N 01°14′·54W. AC 5600, 2036, 2022. HW +0030 on Dover; use LEE-ON-SOLENT differences. Short term ⚓ for small craft at mouth of R Meon. Bar dries ¼M offshore. Ent dries 1·2m at MLWS. Appr on 030° towards Hill Head SC ho (W, conspic); Small hbr to W inside ent (very narrow due to silting) where small yachts can lie in soft mud alongside wall. Facilities: **Hill Head SC** ☎(01329) 664843. **Hill Head village** ⛽ ⛏ ⚓ ✉ ⛒.

9.2.22 WOOTTON CREEK
Isle of Wight 50°44'·09N 01°12'·77W ✿✿◊◊✿✿✿

CHARTS AC 5600, 2036, 2022; Imray C3, C15, 2200

TIDES +0023 Dover; ML 2·8. Use RYDE differences.

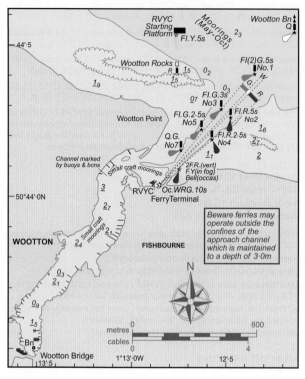

SHELTER Good except in strong N or E winds. Creek dries above the ferry terminal. ABs on RVYC pontoons (dry); No ⚓ in the fairway. Speed limit 5kn.

NAVIGATION WPT Wootton NCM Bn, Q, 50°44'·53N 01°12'·13W, 222°/6·3ca to ferry slip. An obstruction drying 1·4m verges Peel Bank 8 Ca NNE of the Starting Platform. Beware seasonal moorings (May-Oct) laid between the platform & No 1 bn. Large car ferries may operate outside the confines of the approach channel when proceeding to and from Fishbourne Ferry Terminal.

LIGHTS AND MARKS Entrance to approach channel due S of SE Ryde Middle SCM and 1·75M W of Ryde Pier. From the W, pass N of the Starting Platform & No 1 bn. The channel is marked by four SHM bns and two PHMs, all lit. Keep in W sector of Dir lt, Oc WRG 10s, 221°-G-224°-W-225½°-R-230½°. By ferry terminal, turn onto ldg marks on W shore △ ▽, which form a ◊ when in transit 270°.

COMMUNICATIONS (Code 01983) NMOC (02392) 552100; Police 101; Dr 562955.

FACILITIES Fishbourne, **Royal Victoria YC** ☎882325, ⚓ ⟲£1·50/m (shortstay £5·00), ⚓ ⚓ ✕ ⌂. **Village** Wootton Bridge ⛽ ⛽ ✉ Ⓑ ☵ ✕ ⌂ (Ryde), ⇌ (ferry to Portsmouth), ✈ (Southampton).

9.2.23 RYDE
Isle of Wight 50°43'·98N 01°09'·31W ✿✿◊◊✿✿

CHARTS AC 5600, 2036; Imray C3, C15, 2200

TIDES +0022 Dover; ML 2·8m

Standard Port PORTSMOUTH (→); Tidal Curve see 9.2.17 (←)

Times				Height (metres)			
High Water		Low Water		MHWS	MHWN	MLWN	MLWS
0000	0600	0500	1100	4·7	3·8	1·9	0·8
1200	1800	1700	2300				
Differences RYDE							
−0010	−0010	−0005	−0005	−0·1	0·0	0·0	0·0

NOTE: Double HWs occur at/near Sp. At other times there is a stand of about 2 hrs. Times are for the middle of the stand. See 9.2.17.

SHELTER Small harbour 300m E of Ryde Pier; dries approx 2·3m. Berth on E'ly of three pontoons; long and fin keel yachts should dry out against the breakwater.

NAVIGATION WPT 50°44'·36N 01°09'·13W (2.7ca E of Ryde Pier), 197°/3·9ca to harbour ent. From the E beware Ryde Sand (dries). Either pass N of No Man's Land Fort or use inshore passage between it and Ryde Sands Bcns (lit PHMs). Drying channel 197° across Ryde Sands is marked by 3 SHM and 3 PHM unlit buoys. Beware hovercraft manoeuvring between Ryde pier and marina, and High Speed Ferries from/to pierhead.

LIGHTS AND MARKS Ryde Church spire (Holy Trinity) brg 200° gives initial appr. Harbour entrance lights are 2 FR and Fl G 3s 7m 1M. Ryde pier is lit by 3 sets of 2FR (vert) and a FY fog lt vis 045°-165° and 200°-320°.

COMMUNICATIONS (Code 01983) NMOC (02392) 552100; Police 101; Ⓗ 524081.

Ryde Harbour Ch 80.

FACILITIES **HM/Marina** www.rydeharbour.com ☎613879. Contact HM before arrival. ⚓ HW–2½ to +HW2, 100⟲+70❷ £1.50, short stay £6. ⛽ ⚓ ✉ ⛽ ⛽ (garage 1·25M) ⚓. Internet at HM Office.
Town all domestic facilities nearby ☵ ✕ ⌂. Hovercraft from slip next to hbr to Southsea. Fast cat (passenger) from Ryde Pier to Portsmouth for mainland ⇌; ✈ Southampton.

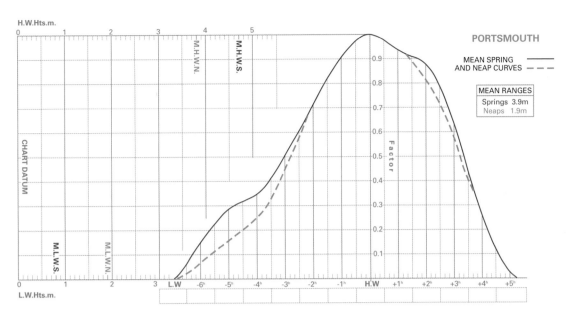

PORTSMOUTH

MEAN SPRING AND NEAP CURVES

MEAN RANGES
Springs 3.9m
Neaps 1.9m

9.2.24 BEMBRIDGE

Isle of Wight 50°41'·62N 01°06'·40W ✿♤♤♤✿✿✿

CHARTS AC 5600, 2037, 2022; Imray C9, C3, C15, 2200

TIDES +0020 Dover

Standard Port PORTSMOUTH (⟶); Tidal Curve see 9.2.17 (⟵)

Times				Height (metres)			
High Water		Low Water		MHWS	MHWN	MLWN	MLWS
0000	0600	0500	1100	4·7	3·8	1·9	0·8
1200	1800	1700	2300				
Differences BEMBRIDGE APPROACHES							
–0010	–0005	0000	+0005	+0·1	+0·1	0·0	+0·1
BEMBRIDGE HARBOUR							
+0020	0000	+0100	+0020	–1·5	–1·4	–1·3	–1·0
VENTNOR							
–0025	–0030	–0025	–0030	–0·8	–0·6	–0·2	+0·2
SANDOWN							
0000	+0005	+0010	+0025	–0·6	–0·5	–0·2	0·0

NOTE: Double HWs occur at/near Sp. At other times there is a stand of about 2 hrs. Times are for the middle of the stand. **See 9.2.17.**

SHELTER Good, but approach difficult in strong N/NE winds. ♥ berths rafting on long pontoons at Duver Marina dredged to 2m, or dry out, with ⚓s fore and aft, on sandy beach to port just inside harbour ent. No ⚓ in chan and hbr, but Priory Bay is sheltered ⚓ in winds from S to WNW (dries inshore but ½M off gives 1·5m).

NAVIGATION WPT tide gauge, Fl Y 2s, 50°42'·46N 01°05'·02W, approx 2ca E of ent to well-buoyed but unlit channel. Carefully check tide gauge which indicates depth over the bar, between Nos 2A and 3 buoys, which almost dries. Beware the gravel banks between St Helen's Fort, Nodes Pt and N to Seaview. Speed limit 6kn.

LIGHTS AND MARKS St Helen's Fort Fl (3) 10s 16m 8M; no ⚓ within 1ca of Fort. Conspic W seamark on shore where chan turns S. Beware many unlit Y racing marks off Bembridge and Seaview (Mar-Oct).

COMMUNICATIONS (Code 01983) NMOC (02392) 552100; Police 101; Dr 872614; HM 872828; Hbr staff Ch 80; *Water Taxi* Ch 80.

FACILITIES **Duver Marina** ☎872828. +🅿 ⚓ 140🛥🅥 £2.60, ♿🚿 ⚓ ⛽ ⚒⚓ 🔧📠 🅿 ℞. **Drying out area £6 flat rate. St Helen's Quay** ⚓ 🅿. **Bembridge Marina** permanent berthholders. **Brading Haven YC** ☎872289, ⚓ ✕ 🅿. **Bembridge SC** ☎872686.

Town 🛒 🛒 🚮 ✕ 🅿 ✉ Ⓑ ⇌ (Ryde), ✈ (So'ton).

HARBOUR ON THE SOUTH COAST OF THE ISLE OF WIGHT
VENTNOR HAVEN, 50°35'·53N 01°12'·50W. AC 2045. Bkwtrs to protect FVs. Very exposed to E to SSE. 8🅾 (in <F4; seasonal), ☎07976 009260 or VHF Ch 17 for availability. ⚓ (quay) 🛒 🚮. **Town** All domestic facilities, some up a steep hill.

STANDARD TIME (UT)
For Summer Time add ONE hour in **non-shaded areas**

PORTSMOUTH LAT 50°48′N LONG 1°07′W
TIMES AND HEIGHTS OF HIGH AND LOW WATERS

Dates in **red** are **SPRINGS**
Dates in blue are NEAPS

YEAR 2016

JANUARY

Day	Time	m	Time	m		Day	Time	m	Time	m
1	0350	4.3	F 1556	4.0		16	0352	4.6	SA 1607	4.3
	0853	1.7	2112	1.7			0852	1.2	◐ 2118	1.2
2	0433	4.1	SA 1645	3.9		17	0452	4.4	SU 1712	4.2
	0945	1.9	◑ 2209	1.9			0956	1.4	2228	1.4
3	0522	4.0	SU 1747	3.7		18	0553	4.3	M 1818	4.1
	1059	2.0	2326	2.0			1118	1.6	2352	1.6
4	0626	3.9	M 1913	3.7		19	0657	4.3	TU 1926	4.1
	1210	2.0					1235	1.6		
5	0030	1.9	TU 1307	1.9		20	0101	1.5	W 1338	1.4
	0746	4.0	2025	3.8			0801	4.3	2030	4.2
6	0124	1.8	W 1356	1.7		21	0202	1.4	TH 1433	1.2
	0847	4.1	2121	4.0			0859	4.4	2131	4.3
7	0213	1.7	TH 1441	1.4		22	0255	1.3	F 1523	1.0
	0936	4.3	2207	4.2			0951	4.5	2224	4.4
8	0259	1.4	F 1525	1.2		23	0344	1.1	SA 1609	0.8
	1016	4.4	2246	4.4			1037	4.6	2310	4.5
9	0344	1.2	SA 1608	0.9		24	0430	1.0	SU 1653	0.7
	1052	4.6	2321	4.6			1119	4.6	○ 2353	4.6
10	0427	1.0	SU 1650	0.8		25	0513	0.9	M 1735	0.7
	1127	4.7	● 2356	4.7			1201	4.6		
11	0510	0.9	M 1732	0.6		26	0035	4.6	TU 1243	4.6
	1203	4.7					0553	0.9	1813	0.7
12	0033	4.7	TU 1242	4.7		27	0117	4.6	W 1323	4.5
	0550	0.8	1812	0.6			0631	0.9	1848	0.8
13	0115	4.7	W 1323	4.7		28	0157	4.5	TH 1402	4.4
	0632	0.8	1853	0.6			0705	1.1	1920	1.0
14	0201	4.7	TH 1410	4.6		29	0233	4.5	F 1437	4.3
	0714	0.8	1936	0.7			0736	1.2	1951	1.2
15	0254	4.7	F 1504	4.5		30	0306	4.4	SA 1513	4.2
	0800	1.0	2023	0.9			0808	1.4	2025	1.4
						31	0342	4.2	SU 1554	4.0
							0846	1.6	2106	1.6

FEBRUARY

Day	Time	m	Time	m		Day	Time	m	Time	m
1	0425	4.1	M 1644	3.8		16	0527	4.2	TU 1757	4.0
	0933	1.8	◑ 2158	1.8			1049	1.5	2329	1.6
2	0516	3.9	TU 1745	3.7		17	0634	4.1	W 1909	3.9
	1036	1.9	2312	2.0			1213	1.6		
3	0621	3.8	W 1932	3.6		18	0045	1.7	TH 1321	1.5
	1204	2.0					0742	4.1	2020	4.0
4	0040	2.0	TH 1318	1.8		19	0149	1.6	F 1419	1.3
	0803	3.9	2052	3.8			0845	4.1	2127	4.2
5	0143	1.8	F 1414	1.5		20	0244	1.4	SA 1509	1.1
	0906	4.1	2141	4.1			0938	4.3	2218	4.3
6	0237	1.5	SA 1503	1.2		21	0331	1.2	SU 1554	0.9
	0950	4.3	2221	4.3			1022	4.4	2255	4.4
7	0325	1.2	SU 1549	0.9		22	0415	1.0	M 1635	0.7
	1028	4.5	2257	4.5			1102	4.5	○ 2333	4.5
8	0410	0.9	M 1633	0.6		23	0456	0.8	TU 1715	0.6
	1105	4.7	● 2334	4.7			1142	4.5		
9	0454	0.7	TU 1716	0.4		24	0011	4.6	W 1222	4.5
	1143	4.8					0533	0.8	1751	0.7
10	0014	4.8	W 1224	4.8		25	0050	4.6	TH 1301	4.5
	0535	0.5	1757	0.3			0608	0.8	1823	0.8
11	0056	4.9	TH 1306	4.8		26	0126	4.5	F 1336	4.4
	0617	0.5	1838	0.3			0638	0.9	1852	0.9
12	0142	4.8	F 1353	4.8		27	0158	4.5	SA 1406	4.4
	0659	0.5	1920	0.5			0704	1.0	1919	1.0
13	0233	4.8	SA 1446	4.6		28	0227	4.4	SU 1438	4.3
	0743	0.7	2004	0.7			0733	1.1	1949	1.2
14	0329	4.6	SU 1546	4.4		29	0301	4.3	M 1517	4.1
	0831	1.0	2054	1.0			0807	1.3	2026	1.4
15	0427	4.5	M 1650	4.2						
	0929	1.3	◑ 2200	1.4						

MARCH

Day	Time	m	Time	m		Day	Time	m	Time	m
1	0343	4.1	TU 1604	3.9		16	0502	4.2	W 1739	4.0
	0849	1.5	◑ 2111	1.6			1024	1.5	2308	1.7
2	0433	3.9	W 1700	3.7		17	0610	3.9	TH 1851	3.9
	0942	1.7	2211	1.9			1148	1.7		
3	0533	3.7	TH 1814	3.6		18	0024	1.8	F 1257	1.6
	1055	1.9	2347	2.0			0721	3.9	2006	3.9
4	0655	3.7	F 2014	3.8		19	0129	1.7	SA 1356	1.5
	1241	1.9					0828	3.9	2144	4.1
5	0118	1.9	SA 1348	1.6		20	0224	1.5	SU 1447	1.2
	0831	3.9	2110	4.0			0924	4.1	2245	4.3
6	0216	1.5	SU 1441	1.2		21	0311	1.2	M 1531	1.0
	0920	4.2	2151	4.3			1003	4.2	2235	4.4
7	0306	1.2	M 1528	0.8		22	0353	1.0	TU 1612	0.8
	1001	4.5	2230	4.6			1041	4.4	2309	4.5
8	0351	0.8	TU 1613	0.5		23	0433	0.8	W 1650	0.7
	1040	4.7	2309	4.7			1120	4.4	○ 2345	4.5
9	0435	0.5	W 1656	0.3		24	0509	0.7	TH 1725	0.7
	1121	4.8	● 2350	4.9			1159	4.5		
10	0517	0.3	TH 1738	0.2		25	0022	4.5	F 1238	4.5
	1202	4.9					0542	0.8	1756	0.8
11	0032	4.9	F 1246	4.9		26	0057	4.5	SA 1311	4.5
	0558	0.3	1819	0.2			0610	0.9	1824	0.9
12	0118	4.9	SA 1333	4.9		27	0126	4.5	SU 1340	4.4
	0640	0.3	1900	0.4			0636	0.9	1851	1.0
13	0208	4.8	SU 1427	4.7		28	0154	4.4	M 1414	4.4
	0723	0.5	1944	0.6			0704	1.0	1920	1.1
14	0303	4.7	M 1528	4.5		29	0227	4.3	TU 1448	4.2
	0809	0.8	2034	1.0			0736	1.1	1955	1.3
15	0402	4.4	TU 1633	4.3		30	0307	4.2	W 1534	4.1
	0909	1.1	◑ 2138	1.4			0816	1.3	2039	1.5
						31	0357	4.0	TH 1630	3.9
							0906	1.5	◑ 2135	1.8

APRIL

Day	Time	m	Time	m		Day	Time	m	Time	m
1	0458	3.8	F 1741	3.8		16	0653	3.8	SA 1943	4.0
	1013	1.8	2258	2.0			1224	1.7		
2	0615	3.7	SA 1919	3.8		17	0058	1.8	SU 1324	1.6
	1202	1.8					0802	3.8	2123	4.1
3	0051	1.8	SU 1320	1.5		18	0156	1.6	M 1416	1.4
	0747	3.9	2032	4.1			0902	4.0	2215	4.3
4	0152	1.5	M 1415	1.2		19	0243	1.4	TU 1502	1.2
	0846	4.2	2119	4.4			0940	4.1	2208	4.4
5	0242	1.1	TU 1503	0.8		20	0326	1.1	W 1543	1.0
	0931	4.5	2201	4.7			1018	4.3	2241	4.5
6	0328	0.7	W 1549	0.5		21	0405	1.0	TH 1621	0.9
	1014	4.7	2242	4.9			1057	4.4	2318	4.5
7	0412	0.4	TH 1633	0.3		22	0441	0.9	F 1656	0.9
	1056	4.9	● 2324	5.0			1137	4.4	○ 2355	4.5
8	0455	0.3	F 1716	0.2		23	0513	0.9	SA 1728	1.0
	1140	4.9					1215	4.5		
9	0008	5.0	SA 1226	5.0		24	0028	4.5	SU 1249	4.5
	0537	0.2	1758	0.2			0542	0.9	1757	1.0
10	0054	5.0	SU 1315	4.9		25	0058	4.5	M 1318	4.5
	0620	0.3	1841	0.4			0611	0.9	1826	1.1
11	0143	4.8	M 1410	4.8		26	0127	4.5	TU 1350	4.4
	0704	0.5	1926	0.7			0640	1.0	1858	1.1
12	0238	4.7	TU 1515	4.6		27	0200	4.4	W 1427	4.3
	0750	0.8	2016	1.1			0714	1.0	1934	1.3
13	0336	4.4	W 1618	4.3		28	0240	4.2	TH 1514	4.2
	0845	1.2	2119	1.5			0754	1.2	2018	1.4
14	0436	4.1	TH 1719	4.1		29	0330	4.1	F 1611	4.0
	0959	1.5	◑ 2241	1.8			0843	1.4	2114	1.7
15	0542	3.9	F 1827	4.0		30	0432	3.9	SA 1719	4.0
	1117	1.7	2353	1.9			0947	1.6	◑ 2230	1.8

Chart Datum: 2·73 metres below Ordnance Datum (Newlyn). HAT is 5·1 metres above Chart Datum.

》》 **FREE** monthly updates. Register at 《
www.reedsnauticalalmanac.co.uk 《

STANDARD TIME (UT)
For Summer Time add ONE hour in **non-shaded areas**

PORTSMOUTH LAT 50°48'N LONG 1°07'W
TIMES AND HEIGHTS OF HIGH AND LOW WATERS

Dates in red are **SPRINGS**
Dates in blue are NEAPS

YEAR 2016

Central S England

MAY

Time	m		Time	m
1 0546	3.8	**16**	0018	1.8
1118	1.7		0720	3.8
SU 1838	4.0	M	1245	1.7
			2004	4.1
2 0013	1.8	**17**	0118	1.7
0705	3.9		0820	3.9
M 1246	1.5	TU	1339	1.6
1951	4.2		2053	4.2
3 0121	1.5	**18**	0209	1.5
0810	4.2		0908	4.0
TU 1344	1.2	W	1426	1.4
2045	4.5		2133	4.3
4 0214	1.1	**19**	0253	1.3
0900	4.5		0952	4.2
W 1435	0.9	TH	1508	1.3
2131	4.7		2213	4.4
5 0302	0.8	**20**	0333	1.2
0947	4.7		1034	4.3
TH 1523	0.6	F	1548	1.2
2215	4.9		2252	4.5
6 0348	0.5	**21**	0410	1.1
1032	4.8		1116	4.4
F 1609	0.4	SA	1625	1.1
● 2300	5.0	○	2329	4.5
7 0433	0.4	**22**	0444	1.0
1119	4.9		1155	4.5
SA 1654	0.4	SU	1701	1.1
2345	5.0			
8 0517	0.3	**23**	0003	4.5
1207	4.9		0517	1.0
SU 1738	0.4	M	1230	4.5
			1734	1.1
9 0032	4.9	**24**	0035	4.5
0601	0.4		0550	0.9
M 1300	4.8	TU	1301	4.5
1823	0.6		1807	1.1
10 0121	4.8	**25**	0106	4.5
0646	0.6		0623	1.0
TU 1358	4.7	W	1335	4.5
1909	0.8		1842	1.1
11 0214	4.6	**26**	0141	4.4
0733	0.8		0659	1.0
W 1505	4.6	TH	1414	4.4
1959	1.1		1921	1.2
12 0311	4.4	**27**	0222	4.3
0825	1.1		0741	1.1
TH 1601	4.4	F	1502	4.3
2058	1.5		2006	1.3
13 0407	4.2	**28**	0311	4.2
0929	1.4		0829	1.2
F 1654	4.2	SA	1558	4.2
◐ 2207	1.7		2100	1.5
14 0507	3.9	**29**	0412	4.1
1039	1.6		0928	1.4
SA 1753	4.1	SU	1701	4.2
2314	1.8	◐	2208	1.6
15 0613	3.8	**30**	0521	4.0
1144	1.7		1043	1.5
SU 1900	4.0	M	1808	4.2
			2331	1.6
		31	0632	4.1
			1206	1.4
		TU	1914	4.3

JUNE

Time	m		Time	m
1 0044	1.4	**16**	0128	1.7
0736	4.2		0834	3.9
W 1312	1.2	TH	1346	1.6
2012	4.5		2059	4.2
2 0144	1.4	**17**	0215	1.5
0832	4.4		0926	4.1
TH 1407	1.0	F	1432	1.5
2104	4.7		2145	4.3
3 0236	0.9	**18**	0258	1.4
0924	4.6		1013	4.2
F 1458	0.8	SA	1515	1.4
2152	4.8		2227	4.4
4 0326	0.7	**19**	0338	1.2
1014	4.7		1056	4.4
SA 1547	0.7	SU	1556	1.3
2240	4.9		2305	4.5
5 0414	0.6	**20**	0417	1.1
1104	4.8		1135	4.5
SU 1635	0.6	M	1636	1.2
● 2327	4.9	○	2340	4.5
6 0500	0.5	**21**	0456	1.0
1155	4.8		1211	4.5
M 1722	0.6	TU	1715	1.1
7 0014	4.8	**22**	0013	4.5
0545	0.5		0533	0.9
TU 1248	4.8	W	1245	4.5
1807	0.7		1752	1.0
8 0103	4.7	**23**	0048	4.5
0631	0.6		0610	0.9
W 1347	4.7	TH	1321	4.6
1853	0.9		1830	1.0
9 0153	4.6	**24**	0125	4.5
0716	0.8		0649	0.9
TH 1448	4.6	F	1402	4.5
1941	1.1		1910	1.1
10 0245	4.4	**25**	0206	4.5
0804	1.1		0730	0.9
F 1536	4.5	SA	1450	4.5
2032	1.4		1954	1.1
11 0337	4.2	**26**	0255	4.4
0856	1.3		0816	1.0
SA 1622	4.3	SU	1543	4.4
2128	1.6		2045	1.3
12 0430	4.0	**27**	0354	4.3
0956	1.6		0910	1.2
SU 1711	4.2	M	1641	4.4
◐ 2230	1.8	◐	2145	1.4
13 0527	3.9	**28**	0458	4.2
1058	1.7		1015	1.3
M 1807	4.0	TU	1742	4.3
2333	1.9		2257	1.5
14 0632	3.8	**29**	0604	4.1
1158	1.8		1132	1.4
TU 1909	4.0	W	1845	4.4
15 0033	1.8	**30**	0012	1.5
0736	3.8		0709	4.2
W 1255	1.7	TH	1243	1.4
2007	4.1		1946	4.4

JULY

Time	m		Time	m
1 0118	1.3	**16**	0135	1.7
0811	4.3		0901	3.9
F 1344	1.2	SA	1357	1.7
2043	4.6		2118	4.2
2 0216	1.1	**17**	0224	1.5
0909	4.4		0951	4.1
SA 1440	1.1	SU	1446	1.6
2136	4.7		2203	4.3
3 0309	0.9	**18**	0310	1.3
1003	4.6		1035	4.3
SU 1532	1.0	M	1532	1.4
2226	4.7		2242	4.4
4 0359	0.7	**19**	0354	1.1
1055	4.6		1113	4.5
M 1621	0.8	TU	1615	1.2
● 2313	4.8	○	2317	4.5
5 0447	0.6	**20**	0436	0.9
1145	4.7		1148	4.6
TU 1708	0.8	W	1657	1.0
2359	4.7		2352	4.6
6 0532	0.6	**21**	0517	0.7
1236	4.6		1223	4.6
W 1753	0.8	TH	1737	0.9
7 0044	4.7	**22**	0028	4.6
0615	0.7		0556	0.8
TH 1327	4.7	F	1302	4.7
1837	0.9		1816	0.8
8 0131	4.6	**23**	0107	4.6
0658	0.8		0635	0.6
F 1418	4.6	SA	1344	4.7
1920	1.1		1856	0.8
9 0218	4.4	**24**	0148	4.6
0739	1.0		0716	0.7
SA 1503	4.5	SU	1431	4.7
2002	1.3		1939	0.9
10 0305	4.3	**25**	0237	4.5
0821	1.2		0759	0.8
SU 1544	4.4	M	1524	4.6
2046	1.5		2026	1.1
11 0350	4.1	**26**	0334	4.4
0906	1.5		0849	1.0
M 1625	4.2	TU	1619	4.5
2135	1.7	◐	2121	1.3
12 0437	3.9	**27**	0437	4.3
0958	1.7		0948	1.3
TU 1709	4.1	W	1718	4.4
◐ 2233	1.8		2230	1.5
13 0532	3.8	**28**	0542	4.2
1102	1.9		1106	1.5
W 1803	4.0	TH	1821	4.3
2338	1.9		2349	1.5
14 0645	3.7	**29**	0650	4.1
1206	1.9		1224	1.6
TH 1916	3.9	F	1927	4.3
15 0040	1.9	**30**	0101	1.5
0759	3.8		0758	4.2
F 1305	1.9	SA	1330	1.5
2024	4.0		2030	4.4
		31	0203	1.3
			0902	4.3
		SU	1429	1.3
			2125	4.5

AUGUST

Time	m		Time	m
1 0258	1.1	**16**	0245	1.4
0959	4.4		1010	4.3
M 1522	1.2	TU	1510	1.4
2214	4.6		2216	4.6
2 0347	0.9	**17**	0332	1.1
1048	4.6		1046	4.5
TU 1610	1.0	W	1555	1.1
● 2259	4.6		2252	4.6
3 0433	0.7	**18**	0415	0.8
1133	4.6		1121	4.7
W 1655	0.9	TH	1637	0.9
2342	4.6	○	2328	4.7
4 0516	0.6	**19**	0457	0.8
1217	4.7		1158	4.8
TH 1737	0.8	F	1718	0.7
5 0024	4.6	**20**	0005	4.8
0557	0.7		0537	0.5
F 1301	4.7	SA	1238	4.8
1817	0.9		1758	0.6
6 0108	4.5	**21**	0045	4.8
0636	0.8		0617	0.4
SA 1345	4.6	SU	1320	4.8
1855	1.0		1838	0.6
7 0150	4.6	**22**	0127	4.8
0712	1.0		0657	0.5
SU 1425	4.5	M	1406	4.8
1930	1.2		1920	0.7
8 0230	4.3	**23**	0215	4.7
0745	1.2		0740	0.7
M 1502	4.4	TU	1459	4.7
2003	1.4		2005	0.9
9 0308	4.2	**24**	0313	4.5
0818	1.4		0827	1.0
TU 1536	4.3	W	1556	4.6
2038	1.6		2058	1.2
10 0346	4.1	**25**	0417	4.3
0857	1.6		0925	1.4
W 1614	4.1	TH	1655	4.4
◐ 2122	1.8	◐	2208	1.5
11 0431	3.9	**26**	0524	4.2
0947	1.9		1051	1.7
TH 1659	4.0	F	1800	4.2
2221	1.9		2335	1.7
12 0527	3.7	**27**	0636	4.0
1100	2.0		1213	1.8
F 1757	3.8	SA	1911	4.1
2343	2.0			
13 0711	3.7	**28**	0048	1.6
1225	2.1		0751	4.1
SA 1940	3.8	SU	1320	1.7
			2018	4.2
14 0057	1.9	**29**	0151	1.5
0834	3.8		0901	4.2
SU 1328	1.9	M	1419	1.5
2050	4.0		2115	4.3
15 0155	1.7	**30**	0244	1.2
0928	4.1		0958	4.4
M 1422	1.7	TU	1509	1.3
2138	4.2		2201	4.5
		31	0332	1.0
			1038	4.6
		W	1554	1.0
			2242	4.6

Chart Datum: 2·73 metres below Ordnance Datum (Newlyn). HAT is 5·1 metres above Chart Datum.

》》 FREE monthly updates. Register at 《
www.reedsnauticalalmanac.co.uk 《

233

STANDARD TIME (UT)
For Summer Time add ONE hour in **non-shaded areas**

PORTSMOUTH   LAT 50°48'N   LONG 1°07'W
TIMES AND HEIGHTS OF HIGH AND LOW WATERS

Dates in red are SPRINGS
Dates in blue are NEAPS

YEAR 2016

SEPTEMBER

Day	Time m	Time m	Time m	Time m
1 TH ●	0415 0.8	1115 4.6	1637 0.9	2322 4.6
2 F	0456 0.7	1154 4.7	1717 0.8	
3 SA	0002 4.6	0535 0.7	1234 4.7	1754 0.9
4 SU	0043 4.6	0610 0.8	1312 4.6	1828 1.0
5 M	0122 4.5	0642 1.0	1348 4.6	1857 1.1
6 TU	0156 4.4	0710 1.2	1419 4.5	1924 1.3
7 W	0228 4.3	0738 1.3	1451 4.4	1955 1.4
8 TH	0303 4.2	0812 1.5	1529 4.2	2034 1.6
9 F ◐	0348 4.0	0855 1.8	1615 4.0	2125 1.9
10 SA	0442 3.8	0954 2.1	1711 3.8	2238 2.1
11 SU	0551 3.7	1144 2.2	1824 3.7	
12 M	0023 2.0	0800 3.8	1304 2.0	2016 3.9
13 TU	0129 1.8	0859 4.1	1359 1.8	2108 4.2
14 W	0221 1.4	0941 4.4	1447 1.4	2147 4.4
15 TH	0307 1.1	1017 4.6	1532 1.1	2223 4.7
16 ○	0351 0.7	1053 4.8	1614 0.8	2301 4.8
17 SA	0434 0.5	1130 4.9	1656 0.5	2340 4.9
18 SU	0515 0.4	1210 5.0	1737 0.5	
19 M	0020 5.0	0556 0.4	1253 5.0	1817 0.5
20 TU	0105 4.9	0636 0.6	1339 4.9	1859 0.6
21 W	0154 4.8	0719 0.7	1432 4.8	1944 0.9
22 TH	0253 4.6	0806 1.1	1531 4.6	2036 1.2
23 F ◐	0401 4.4	0906 1.5	1634 4.3	2150 1.6
24 SA	0510 4.2	1039 1.8	1741 4.1	2319 1.7
25 SU	0622 4.1	1157 1.9	1852 4.0	
26 M	0028 1.7	0741 4.1	1302 1.8	2003 4.1
27 TU	0130 1.6	0921 4.3	1400 1.6	2106 4.2
28 W	0223 1.4	1025 4.5	1449 1.4	2147 4.4
29 TH	0309 1.1	1107 4.6	1533 1.1	2223 4.5
30 F	0352 0.9	1054 4.6	1614 1.0	2301 4.6

OCTOBER

Day	Time m	Time m	Time m	Time m
1 SA ●	0432 0.8	1128 4.7	1653 0.9	2340 4.6
2 SU	0509 0.8	1205 4.7	1728 0.9	
3 M	0019 4.6	0542 0.9	1241 4.7	1759 1.0
4 TU	0055 4.6	0612 1.1	1314 4.6	1826 1.1
5 W	0126 4.5	0638 1.2	1342 4.5	1852 1.2
6 TH	0155 4.4	0706 1.3	1413 4.4	1921 1.3
7 F	0230 4.3	0738 1.5	1451 4.3	1958 1.5
8 SA	0314 4.2	0818 1.7	1538 4.1	2044 1.8
9 SU ◐	0409 4.0	0911 2.0	1635 3.9	2148 2.0
10 M	0516 3.8	1037 2.2	1745 3.8	2345 2.1
11 TU	0652 3.8	1237 2.1	1922 3.9	
12 W	0059 1.8	0820 4.1	1332 1.8	2030 4.2
13 TH	0152 1.4	0906 4.4	1420 1.4	2113 4.5
14 F	0240 1.1	0944 4.7	1505 1.0	2153 4.7
15 SA	0325 0.7	1022 4.9	1549 0.7	2233 4.9
16 SU ○	0408 0.5	1102 5.1	1632 0.5	2315 5.0
17 M	0451 0.4	1143 5.1	1714 0.4	2358 5.1
18 TU	0533 0.4	1226 5.1	1757 0.4	
19 W	0044 5.0	0616 0.5	1313 5.0	1839 0.6
20 TH	0136 4.9	0700 0.8	1406 4.8	1925 0.9
21 F	0237 4.7	0748 1.2	1506 4.6	2017 1.2
22 SA ◐	0348 4.5	0849 1.6	1611 4.3	2129 1.6
23 SU	0454 4.3	1017 1.9	1717 4.1	2253 1.8
24 M	0602 4.1	1131 1.9	1827 4.0	
25 TU	0000 1.8	0721 4.1	1234 1.9	1940 4.0
26 W	0100 1.7	0908 4.3	1332 1.7	2136 4.1
27 TH	0154 1.5	1003 4.4	1422 1.5	2224 4.3
28 F	0241 1.3	1042 4.5	1506 1.3	2202 4.4
29 SA	0323 1.1	1028 4.6	1547 1.1	2239 4.5
30 SU ●	0403 1.0	1102 4.7	1625 1.0	2318 4.6
31 M	0440 1.0	1138 4.7	1700 1.0	2357 4.6

NOVEMBER

Day	Time m	Time m	Time m	Time m
1 TU	0514 1.1	1213 4.7	1731 1.1	
2 W	0033 4.6	0544 1.2	1245 4.6	1759 1.1
3 TH	0104 4.6	0612 1.3	1313 4.6	1826 1.2
4 F	0132 4.5	0641 1.4	1343 4.5	1857 1.3
5 SA	0206 4.4	0714 1.5	1420 4.3	1933 1.4
6 SU	0250 4.3	0754 1.7	1507 4.2	2018 1.6
7 M ◐	0344 4.1	0845 1.9	1604 4.0	2116 1.8
8 TU	0450 4.0	0955 2.1	1713 3.9	2240 1.9
9 W	0605 4.0	1152 2.0	1832 3.9	
10 TH	0020 1.8	0724 4.2	1258 1.8	1943 4.2
11 F	0119 1.5	0823 4.5	1350 1.4	2037 4.4
12 SA	0209 1.1	0909 4.7	1437 1.1	2123 4.7
13 SU	0257 0.8	0952 4.9	1523 0.8	2208 4.9
14 M ○	0343 0.6	1036 5.1	1608 0.5	2253 5.0
15 TU	0428 0.5	1120 5.1	1653 0.5	2340 5.0
16 W	0513 0.5	1205 5.1	1738 0.5	
17 TH	0029 5.0	0557 0.6	1253 5.0	1822 0.6
18 F	0122 4.9	0643 0.9	1344 4.8	1909 0.8
19 SA	0226 4.7	0733 1.2	1442 4.6	2000 1.2
20 SU	0335 4.6	0830 1.5	1544 4.4	2102 1.5
21 M ◐	0433 4.4	0942 1.8	1646 4.1	2216 1.7
22 TU	0532 4.2	1054 1.9	1750 3.9	2323 1.8
23 W	0638 4.1	1159 1.9	1857 3.9	
24 TH	0023 1.8	0752 4.2	1258 1.8	2001 4.0
25 F	0119 1.7	0911 4.3	1351 1.7	2054 4.1
26 SA	0208 1.5	0925 4.4	1436 1.5	2137 4.3
27 SU	0252 1.4	0959 4.5	1518 1.3	2218 4.4
28 M	0333 1.3	1037 4.6	1556 1.2	2300 4.5
29 TU ●	0411 1.2	1115 4.6	1632 1.1	2340 4.6
30 W	0447 1.2	1150 4.6	1706 1.1	

DECEMBER

Day	Time m	Time m	Time m	Time m
1 TH	0016 4.6	0520 1.2	1223 4.6	1737 1.1
2 F	0048 4.6	0552 1.3	1252 4.6	1809 1.1
3 SA	0118 4.5	0624 1.3	1323 4.5	1842 1.1
4 SU	0152 4.5	0659 1.4	1359 4.4	1918 1.2
5 M	0233 4.4	0740 1.5	1443 4.3	2001 1.3
6 TU	0325 4.3	0828 1.6	1537 4.1	2054 1.5
7 W ◐	0426 4.2	0928 1.8	1643 4.0	2159 1.7
8 TH	0533 4.2	1049 1.9	1755 4.0	2326 1.7
9 F	0641 4.3	1215 1.7	1904 4.1	
10 SA	0041 1.5	0744 4.4	1318 1.4	2004 4.3
11 SU	0139 1.3	0838 4.7	1411 1.1	2058 4.6
12 M	0231 0.9	0928 4.8	1501 0.9	2149 4.7
13 TU	0321 0.8	1016 5.0	1549 0.7	2238 4.8
14 W ○	0409 0.7	1102 5.0	1636 0.5	2328 4.9
15 TH	0457 0.7	1149 5.0	1723 0.5	
16 F	0019 4.9	0543 0.7	1237 4.9	1808 0.6
17 SA	0113 4.8	0630 0.9	1326 4.8	1854 0.7
18 SU	0213 4.7	0717 1.1	1419 4.6	1941 1.0
19 M	0314 4.6	0807 1.3	1515 4.4	2032 1.3
20 TU	0404 4.5	0903 1.6	1610 4.2	2131 1.5
21 W ◐	0453 4.3	1008 1.8	1707 4.0	2237 1.7
22 TH	0547 4.1	1115 1.9	1808 3.8	2341 1.8
23 F	0647 4.1	1218 1.9	1913 3.8	
24 SA	0039 1.8	0747 4.1	1316 1.8	2014 3.9
25 SU	0132 1.8	0842 4.2	1405 1.6	2109 4.0
26 M	0220 1.6	0930 4.3	1449 1.5	2158 4.2
27 TU	0304 1.5	1014 4.4	1529 1.3	2243 4.4
28 W	0345 1.4	1054 4.5	1608 1.2	2323 4.5
29 TH ●	0424 1.3	1130 4.6	1645 1.0	
30 F	0000 4.5	0501 1.2	1203 4.6	1720 1.0
31 SA	0033 4.6	0536 1.1	1235 4.6	1755 0.9

Chart Datum: 2·73 metres below Ordnance Datum (Newlyn). HAT is 5·1 metres above Chart Datum.

》》 **FREE** monthly updates. Register at 《
www.reedsnauticalalmanac.co.uk 《

9.2.25   PORTSMOUTH
Hampshire **50°47'·38N 01°06'·67W** (Entrance) ❀❀❀❀◊◊◊❀❀❀

CHARTS AC 5600, 2037, 2625, 2631, 2629, 2628; Imray C9, C3, C15, 2200

TIDES +0020 Dover; ML 2·8

Standard Port PORTSMOUTH (←—)

Times				Height (metres)			
High Water		Low Water		MHWS	MHWN	MLWN	MLWS
0500	1000	0000	0600	4·7	3·8	1·9	0·8
1700	2200	1200	1800				

Differences LEE-ON-THE-SOLENT

−0005	+0005	−0015	−0010	−0·2	−0·1	+0·1	+0·2

NOTE: Strong winds from NE to SE, coupled with a high barometer, may lower levels by 1m and delay times of HW and LW by 1hr; the opposite may occur in strong W'lies with low pressure.

SHELTER Excellent. This very large harbour affords shelter in some area from any wind. There are marinas at Gosport, Fareham and Portchester, plus several yacht pontoons/jetties and many swinging moorings. On the Portsmouth side, The Camber offers good shelter, but this is a busy commercial dock and often full; beware Isle of Wight car ferries berthing near the entrance.

Portsmouth is a major Naval Base and Dockyard Port; all vessels come under the authority of the Queen's Harbour Master (QHM). If over 20m LOA, QHM's permission (VHF Ch 11) must be obtained before entering, leaving or moving in the harbour. ⚓ is prohibited in the harbour and fishing only allowed clear of main channels.

NAVIGATION WPT No 4 Bar buoy, QR, 50°47'·01N 01°06'·36W, 330°/4·5ca to entrance. Beware strong tidal streams in the narrow entrance which is invariably very busy. Commercial shipping, cross-Channel ferries, warships, High Speed Ferries and small craft all operate within this area.

Approaches: From the W, yachts can use the Swashway Chan (to NW of Spit Sand Fort) which carries about 2m. The transit of the War Memorial and RH edge of block of flats (049·4°) indicates the deepest water, but need not be followed exactly except at LWS. The Inner Swashway (Round Tr brg approx 035°) carries only 0·3m; local knowledge required. Approaching inshore from the E, the submerged barrier, which extends from Southsea to Horse Sand Fort, should only be crossed via the unlit Inshore Boat Passage (0·9m) 1ca off the beach, marked by R & G piles; or via the lit Main Passage (min depth 1·2m), 7ca further S, marked by a G pile and a concrete dolphin.

- **The Small Boat Channel (SBC)** is *mandatory* for craft under 20m LOA when entering or leaving harbour. It runs just outside the W edge of the main dredged channel from abeam No 4 Bar buoy to Ballast Bcn PHM. A depth gauge, showing height above chart datum, is on BC4.

- Craft may not enter harbour on the E side of the main chan and must not enter the SBC on its E side; craft may enter or leave the SBC anywhere on its W side (beware Hamilton Bank, some of which dries).

- Engines (if fitted) must be used when in the SBC between No 4 Bar buoy and Ballast Bcn, *which should always be left to port*.

- If crossing to/from Gunwharf Quays or The Camber obtain approval from QHM (Ch 11) and then do so at 90° to the main channel N of Ballast Bcn.

- At night the SBC is covered by the Oc R sector (324°-330°) of the Dir WRG lt on Fort Blockhouse until close to the hbr ent. Thereafter the Iso R 2s sector (337·5°-345°) of the Dir WRG lt on the dolphin E of Gosport Marina leads 341° through the entrance to Ballast Bcn.

- If joining or leaving the SBC via the Inner Swashway, BC Outer Bn *must* be left to port.

- The SBC may be used even if the main channel is closed.

- Gosport Ferry plies between Gosport Pier and The Hard. It runs every 15mins with a continuous service at peak times.

Speed limit is 10kn (through the water) within the harbour and within 1000 yds of the shore in any part of the Dockyard Port.

Exclusion zones: Do not approach within 50m of any MoD vessel or establishment, or within 100m of any submarine. Yellow buoys (see chartlet above) are sometimes deployed to delineate the 50m zone. Some ships are protected by a zone of 250m while underway. They are escorted by armed vessels; listen to QHM on VHF 11 or 13 for details. If you fail to respond to warnings it will be assumed you have hostile intentions.

Historic wrecks are: *Mary Rose* (sank 1545) at 50°45'·8N 01°06'·2W (5ca SSW of Spit Sand Fort), and *Invincible* (sank 1758) *at* 50°44'·36N 01°02'·32W (1·4M ESE of Horse Sand Fort). *Mary Rose* is marked by SPM buoys, Fl Y 5s.

Navigational piles should not be approached too closely as many are on steep-to mud banks; those lit in Portchester Lake may be difficult to see at night due to background lighting.

FOG ROUTINE
- Broadcast on VHF Ch **11** and **13**, Fog Routine comes into force when QHM considers that visibility is so low that normal shipping movements would be dangerous.

- Small craft may continue at the skipper's discretion, but with great caution, keeping well clear of the main and approach channels. Monitor VHF Ch **11** at all times.

- Be aware that the presence of radar echoes from small vessels within the main channel can cause much doubt and difficulty to the master of a large vessel.

- For their own safety, small craft are strongly advised not to proceed when fog routine is in force.

- 3 W (vert) lights are displayed in fog at the centre of Gosport Ferry pier.

LIGHTS AND MARKS See 9.2.4 and chartlet. From S and E of the IoW Nab Tower is conspic about 9·5M SE of the harbour entrance. The entrance to the Eastern Solent is between Horse Sand Fort and No Man's Land Fort. Spit Sand Fort lies 160°/1·2M from the harbour entrance. All forts are conspic round stone structures.

St Jude's ✠ spire and Southsea Castle light house in transit 003° lead between Outer Spit SCM buoy and Horse Sand SHM buoy. At night keep in the W sector (000°-003°) of Southsea Castle Dir WRG lt, which also shows Iso 2s 16m 11M, vis 337°-071° (94°).

FIRING RANGE **Tipner Rifle Range**, N of Whale Island, danger area extends 2,500m from firing range. R flag or ● lt on Tipner Range FS indicates firing is in progress; yachts should clear the range danger area or transit it as quickly as possible.

COMMUNICATIONS (Code Portsmouth/Gosport 02392) NMOC 552100; Police 101; ⊞ 286000; QHM Hbr Control (H24) 723694; Commercial Docks & Camber Berthing Offices ☎297395.

Yachts should monitor Ch **11** (*QHM*) for traffic and navigational information. *Portsmouth Hbr Radio* (Commercial Port) Ch 14 (H24). *Haslar Marina* and *Port Solent* Ch 80 M (H24). *Gosport Marina* call Ch 80 M (HO). Fareham Marine Ch M (summer 0900-1730). Gunwharf Quay Ch 80. **Naval activities** to the S/SE of Portsmouth and IOW may be advised by Solent CG Ch 67 or ☎552100; or Naval Ops ☎722008. Naval vessels use Ch 13. The positions and times of naval firings and underwater explosions are broadcast daily at 0800 and 1400LT Ch 06; preceded by a Securité call on Ch 16. Warnings of underwater explosions will also be broadcast on Ch 16 at 1 hour, at 30 mins and just before the detonation.

FACILITIES

Haslar Marina berths@haslarmarina.co.uk ☎601201. Access H24 580⌴+ 50**Ⓥ** at L/M pontoons, inboard of green light vessel, £3·10 (max 60m LOA); short stay £8. ⓑ⚓◣ ▮ ✎ Ⓔ ⌂ ⌴ ✕ @ wi-fi. A floating boat lift is at the end of 'F' pontoon. Other Haslar Creek moorings/pontoons are private; Haslar bridge (1·8m clearance).

Gunwharf Quays www.gunwharf-quays.com ☎836732 ⚓⍨ ▷⊡ ⌇⑂ **Ⓥ** £3·58 (max 80m LOA), short stay £1·13/m, booking advisable (or try VHF 80 for short notice requirements). £5 landing fee/vessel for drop offs to Gunwharf Quays shopping centre. Note Small Boat Channel regulations.

Town Quay (Camber) ☎833166, berth on wall, <12m £17·10, short stay (3hrs) £7·90, pay 'KB Boat Park'; no ⚓/⍨.

Gosport Marina www.premiermarinas.com ☎524811. 471⌴ inc **Ⓥ** if berths available, £3·15, short stay £1·41/m. ⓑ⊡♂▮◣✎(14t) dry stack storage for motor boats. ⌂ ☲ ✕ ⌂.

Endeavour Quay www.endeavourquay.co.uk ☎584200, fully equipped ⍨ (some **Ⓥ** berths) ✕ ✎ ⚓ ⍽ ⌇(180t) ◣(35t) ⌂ wi-fi.

Royal Clarence Marina ⚓ www.royalclarencemarina.org. ☎523523, 150⌴, **Ⓥ** space varies, up to 50m, < 12m £2·75 (winter £27·00), special rates for yacht rallies. ⓑ⍨⊡✎ ⌇ ✕ wi-fi.

Port Solent portsolent@premiermarinas.com ☎210765. Lock 9m wide. 900⌴ inc **Ⓥ**, £3·05 (min fee £24·40) £1·41/m <12m (min £10). ⓑ⊡♂◣▮◣✎✕Ⓔ⚓⌇(40t)⌂☲ ✕ ⌂.

Portchester Lake is marked by lit and unlit piles. Unusually, PHMs are numbered 57 to 74 (from seaward), and SHMs 95 to 75.

Beware unlit naval buoys at the S end. Do not delay crossing Tipner Range, S limit marked by piles 63/87 and N by 70/78. Portchester Castle is conspic 5ca SSW of marina. Call marina Ch 80 when inbound passing pile 78. Pile B, Fl (4) G 10s, marks the waiting pontoon (see chartlet). Access H24 via chan dredged 1·5m to lock (43m x 9·1m); enter on 3 ● (vert) or on loudspeaker instructions.

WicorMarine Yacht Haven (Portchester) www.wicormarine.co.uk ☎(01329) 237112, some **Ⓥ** on pontoons, £1·50/m (rallies welcome), ⚓⚓⍨▮◣✕✎⊡Ⓔ⌇(12t) ◣(7t) ⌂ ☲ ✕.

Portsmouth Marine Engineering ☎(01329) 232854, ⌴(drying) **Ⓥ** £15/night, £40/week, ⍨ ◣(18t).

Upper Quay Marina (dries) upperquaymarine1@gmail.com ☎469855 mob 07712189444, 50⌴+**Ⓥ** by arrangement, ⚓ ⍨.

TOWN CENTRES

PORTSMOUTH ✕✎⊡Ⓔ⚓⌂ ACA ⌂⊠Ⓑ⇌✈ (Southampton).
Ferries:
Brittany Ferries (www.brittany-ferries.co.uk):-
Bilbao: 2/week; 24 or 32 hrs; Santander: 2/week; 24 hrs;
Caen: 3/day; 6 hrs; St Malo: daily; 8½ hrs;
Cherbourg HSS: up to 3/day; 3 hrs.
Le Havre HSS: up to 2/day (summer only); 3¼ hrs.
Condor Ferries (www.condorferries.co.uk):
Guernsey/Jersey: 6/week; 6½–12½ hrs.
IoW: frequent car and passenger services to Ryde/Fishbourne.
Water Taxis between Gunwharf Quays, Historic Dockyard and RN Submarine Museum (Haslar Jetty, Gosport).

GOSPORT ✕✎⊡Ⓔ⌂⌂ ACA ⊠Ⓑ⇌(Portsmouth) ✈ (S'ton).

FAREHAM (01329) ⊠Ⓑ☲⌂⇌✈(Southampton). Fareham Lake is well marked, partially lit up to Bedenham Pier and unlit thereafter. Chan dries 0·9m in final 5ca to Town Quay.

YACHT CLUBS

Royal Naval Sailing Association ☎521100.
Royal Naval & Royal Albert YC ☎825924, ⌂.
Portsmouth SC ☎ 820596.
Portchester SC ☎376375, detached pontoon⚓ (close No71 bn).
Hardway SC secretary@hardwaysailingclub.co.uk ⌇▬⌴✕⌂.
Gosport CC ☎(01329) 47014 or (0860) 966390.
Fareham Sailing & Motor Boat Club ☎(01392) 280738.

EASTERN APPROACHES TO THE SOLENT

(AC 2045, 2036) Nab Tower (Fl 10s, *Horn(2) 30s*), a conspic steel and concrete structure 4·5M E of Foreland, marks Nab Shoal for larger vessels but is of no direct significance to yachts. The Nab Channel (maintained depth 13.3m) runs northwards to the East of the tower. It is indicated by Y SPM as far as New Grounds. Keep well clear of shipping manoeuvring in the Pilot Boarding Areas and using the Nab Deep Water Channel.

NW of Nab Tr the E approach to the Solent via Spithead presents few problems and is far safer in SW/W gales than the Needles Channel. A good lookout should be kept for ferry traffic and vessels anchoring in St Helen's Roads.

The main shipping chan is well buoyed and easy to follow, but there is plenty of water for yachts over New Grounds and Warner Shoal when approaching No Man's Land Fort (Iso R 2s)and Horse Sand Fort (Iso G 2s). A submerged barrier lies N of the latter; the barrier to SW of the former has been demolished and therefore yachts may now pass safely SW of No Man's Land Fort; the Ryde Sand Bcns (both PHM and lit) mark the shallows to the SW. Ryde Sand dries extensively and is a trap for the unwary.

Care should be taken to avoid Hamilton Bank to the South of the small boat channel close East of Portsmouth Harbour entrance.

Langstone and Chichester Harbours have offlying sands which are dangerous in strong S'ly winds. E and W Winner flank the entrance to Langstone Hbr. E and W Pole Sands, drying 1m, lie either side of the appr channel to Chichester Hbr. Bracklesham Bay is shallow, with a distinct inshore set at certain states of the tide. SE along a low-lying coast is Selsey Bill with extensive offshore rocks and shoals. Pass these to seaward via Owers SCM lt buoy or the Looe Chan, which in suitable conditions is much used by yachts on passage to/from points E of the Solent. Although lit, the many lobster pot markers demand caution at night; it is dangerous in onshore gales as searoom is limited by extensive shoals on which the sea breaks.

Looe Channel: ▶ *the E-going flood runs from HW Portsmouth +0430 (HW Dover +0500) until HW Portsmouth −0130 (HW Dover −0100), at sp reaching 2.4kn near the Boulder and Street light buoys which mark its narrow western end; they may be hard to see. Max Neap rate is 1.2kn.*
The W-going ebb runs from HW Portsmouth −0130 (HW Dover −0100) until HW Portsmouth +0430 (HW Dover +0500), at Springs reaching 2.6kn near Boulder and Street. Max Neap rate is 1.3kn.◀

Boulder SHM lt buoy is at the W ent to this chan, about 6M SE of Chichester Bar Bn. Medmery Bank, 3·7m, is 1M WNW of Boulder.

9.2.26 LANGSTONE HARBOUR

Hampshire 50°47'·23N 01°01'·54W (Ent) ✿✿✿✿✿✿✿✿

CHARTS AC 5600, 2037, 3418; Imray C3, C9, Y33, 2200

TIDES +0022 Dover; ML 2·9

Standard Port PORTSMOUTH (←—)

Times				Height (metres)			
High Water		Low Water		MHWS	MHWN	MLWN	MLWS
0500	1000	0000	0600	4·7	3·8	1·9	0·8
1700	2200	1200	1800				
Differences LANGSTONE							
0000	–0015	0000	–0010	+0·1	+0·1	0·0	0·0
NAB TOWER							
+0015	0000	+0015	+0015	–0·2	0·0	+0·2	0·0
SELSEY BILL							
+0010	–0010	+0035	+0020	+0·5	+0·3	–0·1	–0·2

SHELTER Very good in marina (2·4m) to W inside ent, access over tidal flap 1·6m at CD; waiting pontoon. Ent is 7m wide. Buoys on W side of harbour entrance administered by Eastney Cruising Association, max LOA 9m. Or ⚓ out of the fairway in Russell's Lake or Langstone Chan (water ski area); or see HM (E side of ent). Hbr speed limit 10kn; 5kn in Southsea Marina channel.

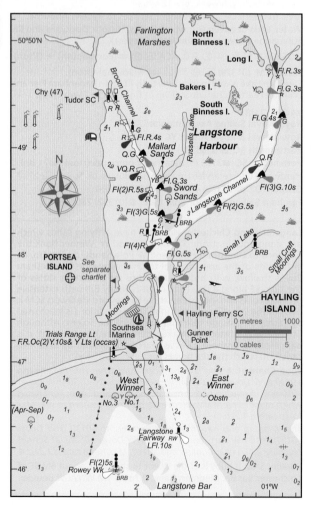

NAVIGATION WPT 50°46'·31N 01°01'·36W, Langstone Fairway SWM Bcn 348°/9·4ca to QR light at ent (Eastney Pt).

Bar has about 1·2m. Entrance channel lies between East and West Winner drying banks, which afford some protection. Appr is straightforward in most conditions, best from HW –3 to +1, but avoid entry against the ebb, esp at sp and in strong onshore winds. In strong S/SE winds do not attempt entry.

LIGHTS AND MARKS Ldg marks (concrete dolphins), or Fairway beacon in line with conspic chy, leads 344° just clear of East Winner. The entrance itself deepens and favours the W side. The narrow approach channel to Southsea Marina is marked by 5 SHM piles, only the first of which is lit, Fl G. There are 9 PHM piles; see chartlet for details.

COMMUNICATIONS (Code 02392) NMOC 552100; Police 101; Ⓗ 286000; HM 463419.

Harbour VHF Ch **12** 16 (0900-1700 daily). Marina Ch **80** M.

FACILITIES **Southsea Marina** www.premiermarinas.com ☎82 2719 Access channel dredged to 0·5m. Marina entrance, width 7m, has R/G lights indicating if the cill is up or down. 320◌, £3·05 (min fee £24·40); free 🔌 for 2 nights. £1.38/m for <4hrs (min fee £11·28). **Services:** ⛽🔧♣⌂🛢⚓🔨(25t) ⛟(18t) ⛽🛒✕⛴.

Hayling Pontoon (E side of ent), www.langstoneharbour.org.uk ⛵ (H24 with pre-payment tag available from HO) 🚻. Vacant mooring buoys in harbour may be available, £13·60/night.

Langstone SC ☎ 9248 4577 ⚓⛵🔨⛴.

Eastney Cruising Association (ECA) ☎9273 4103, 6 buoys for craft under 9m by prior arrangement, ⛽✕⛴. Visitors welcome – access to ECA by intercom at road barrier or ☎827396 (bar).

Tudor SC (Hilsea) ☎9266 2002 ⚓⛵🔨⛴.

Hayling Ferry SC; Locks SC ☎9282 9833.

Towns: ✉ (Eastney, Hayling), Ⓑ (Havant, Hayling, Emsworth), ⇌ (bus to Havant), ✈ (Southampton).

9.2.27 CHICHESTER HARBOUR

W. Sussex **50°46'·86N 00°56'·06W** ❀❀♦♦♦❀❀❀

CHARTS AC 5600, 2045, 3418; Imray C9, C3, Y33, 2200

TIDES +0027 Dover; ML 2·8

Standard Port CHICHESTER (→)

Times				Height (metres)			
High Water		Low Water		MHWS	MHWN	MLWN	MLWS
0500	1000	0000	0600	4·9	4·0	1·9	0·9
1700	2200	1200	1800				
Differences NORTHNEY							
+0020	+0010	0000	+0005	0·0	−0·2	−0·2	−0·4
BOSHAM							
+0010	+0005	ND	ND	0·0	−0·1	ND	ND
ITCHENOR							
+0005	0000	−0010	+0005	−0·1	−0·2	−0·2	−0·3
DELL QUAY							
+0015	+0010	ND	ND	0·0	−0·1	ND	ND

SHELTER Excellent in all five main chans: Emsworth, Thorney, Chichester, Bosham, Itchenor Reach and Fishbourne. There are six yacht habours and marinas (see Facilities); also about 50 ⚓s at Emsworth and Itchenor. ⚓s off E Head (uncomfortable in NE'lies); in Thorney Chan off Pilsey Is; and in Chichester Channel off Chalkdock Pt. Harbour speed limit of 8kn; max fine £2,500.

NAVIGATION WPT 50°45'·32N 00°56'·60W, 013°/1·6M to entrance. Best entry is HW −3 to +1, to avoid confused seas on the ebb, especially in onshore winds >F5. Leave West Pole Bn close to port; the channel Northward is effectively only about 200m wide.

- Bar is periodically dredged to 1.5m below CD. After severe gales the depth can vary markedly and it is prudent to assume a least depth of 0.5m below CD. Be very wary of entering in S'ly winds >F6. Depths may vary ±0·75m after gales.

- Chichester Harbour is very popular for recreational boat and dinghy racing. With 10,000 resident vessels the navigation channels can be very busy at weekends. Anticipate sailing craft manoeuvres and give them room to tack as necessary.

- ⚓ N of E Head remains the most popular spot for visiting yachts. Vessels should ⚓ S of a line between East Head Spit and Snowhill SHM to remain clear of channel.

APPROACHES: From the W, Horse Sand Fort astern, bearing 265°, leads to the West Pole Bn. Leave W Pole 30-80m to port and alter 013° as Eastoke Pt opens E of W Pole and Bar bns. Pass approx 50m off Bar Bn. For latest survey see www.conservancy.co.uk or LNTM.

From the E/SE, via Looe Chan, keep W for 2M then alter NW toward the West Pole Bn keeping Nab Tr astern brg 184° to clear shoals in Bracklesham Bay. Beware the two old concrete targets on and just south of East Pole Sand. Note: Historic Wrecks lie 137°/1·2M and 095°/3·2M from West Pole Bn.

ENTRANCE Pass between Eastoke PHM buoy (QR) and W Winner SHM bn (QG and tide gauge). Three SHM lit buoys mark the edge of the Winner shoal (dries), to starboard of the ent. Near Fishery SCM lit buoy, depths may change and buoys are moved accordingly. Here the channel divides: N towards Emsworth and ENE towards Chichester. Stocker's Sands (dries 1·9m), is marked by 2 lit PHM buoys. East Head lit SHM marks W end of recommended ⚓.

EMSWORTH CHANNEL is straight, broad, deep and well marked/lit in the 2·5M reach to Emsworth SCM bn, Q (6) + L Fl 15s, where Sweare Deep forks NW to Northney. Pass close to Hayling Is SC pontoon entering/leaving this channel to avoid shallow patch just S of bn.

THORNEY CHANNEL Strangers should go up at half-flood. Ent is at Camber SCM bn, Q (6) + L Fl 15s; pass between Pilsey and Thorney Lt bns, thereafter channel is marked by perches. Above Stanbury Pt chan splits, Prinsted Channel to port and Nutbourne Channel to stbd; both dry at N ends. Good ⚓ off Pilsey Island.

CHICHESTER CHANNEL to Itchenor Reach and Bosham Channel. From East Head SHM buoy pass between Snowhill SHM buoy and Sandhead PHM buoy, then head about 030° for 1M to leave Chaldock NCM Bn to starboard. The transit shown on AC 3418 need not be identified nor followed exactly. 6 ca E of Chaldock Bn the channel divides: turn N into Bosham Channel or continue ESE into Itchenor Reach for Birdham Pool and Chichester Marina,

beyond which is the drying Fishbourne Channel to Dell Quay. ⚓ in Itchenor Reach and Bosham Channel.

LIGHTS AND MARKS Approach marked by West Pole PHM tripod bn at 50°45'·45N 00°56'·59W; Bar PHM bn is 6ca N. E side of ent channel is marked by W Winner SHM lit pile, with tide gauge; and by 3 SHM buoys: NW Winner, N Winner, Mid Winner. All channels within the harbour are well marked by day. Emsworth Channel is well lit; Thorney Channel is partly lit; Bosham Channel is unlit. Itchenor Reach is unlit except for Birdham Pool and Chichester Marina entrance bns. Two unlit SHM lie inshore NE of E Head to indicate edge of shoal water in ⚓.

COMMUNICATIONS (Code 01243) NMOC (02392) 552100; Police 101; ⊞ 787970. Chichester Hbr Office www.conservancy.co.uk ☎512301; Emsworth Hbr Office 376422. www.chimet.co.uk ⚓ at Bar.

Chichester Hbr Radio VHF Ch 14 16 (Apr-Sep: 0830-1700; Sat 0900-1300) or Chichester Hbr Patrol (w/e Apr-Oct). Northney Marina Ch 80 M. Chichester Marina Ch 80.

FACILITIES

HAYLING ISLAND Sparkes Marina ☎02392 463572, mob 07770 365610. Access all tides via chan dredged 2m; from close N of Sandy Pt appr on ≠ 277° of 2 bns; thence alter S, past 3 PH bns to marina. Pontoons have 1·6m. 150⚓+20❶ £3·50 (inc ⚡), £8 < 4hrs. **Services:** ⚓ 🛢 🖉 ♠ ⚓ LPG, ⚒ ⚒ 🅱 (16t) ✕.

Northney Marina northney@mdlmarinas.co.uk ☎02392 466321, access via chan 1m. ➴ 228⚓+30❶ £3·32, £8 <4hrs. 🖉 ♠ ⚒ ⚒ 🅱 🖥(35t) 🅻 wi-fi.

EMSWORTH CHANNEL Jetty (dries 1·6m) abreast millpond wall between Emsworth SC and Slipper SC to N, 50m long, ✫ 2FR (vert). Free for <2 hrs stay, ♁. ➴ at South St, Kings St, Slipper Mill; contact the Warden ☎376422. Ferry £1 to ⚓ Ch 14 *Emsworth Mobile.*

Emsworth Yacht Hbr ☎377727, access over 2·4m cill retaining 1·5m. 200⚓+20❶ £2·75, ⚓ ♠ 🖉 ⚒ ⚒ 🅱 🖥(50t) 🖥(40t) 🅻 ACA.

THORNEY CHANNEL Thornham Marina, ☎375335. Appr chan dries, 81⚓+2 ❶ £22·46/craft, berths dry suiting multihull or lift keel vessel. **Services:** ➴ ⚒ ⚒ 🖵 🖥(12t) 🖥(10t) ✕ 🍴.

CHICHESTER CHANNEL/ITCHENOR REACH Hard available at all stages of the tide. There are berths for approx 40 yachts at Itchenor on both buoys and pontoons. £11·80/night. Ferry/water taxi VHF Ch 08, ☎07970 378350, to Bosham and Birdham. ⚓ ➴ (£3·60) ♁ ⚓ 🖙 🖚 🖛 ⚒ ⚒ 🅱 🅔 🖚(hbr office) ✕ 🍴(Ship Inn).

BOSHAM CHANNEL For moorings (200+) contact the Quaymaster ☎573336. ⚓ prohib in chan which mostly dries.

Bosham Quay Hard **Services:** ⚓ ♁ ⚓ 🛋 🛆. EC Wed.

CHICHESTER LAKE Birdham Pool Marina ⚓ ☎512310. Enter drying channel (dredged 2015) HW±3 (depth>1m) at Birdham SHM bn (depth gauge) access via lock (5m wide). Visitors should call 07831 466815 in advance. VHF Ch 80. ➴ 230⚓+10❶ £2·60, 🖉 ♠ 🖉 ♠ ⚒ ⚒ 🅱 🅔 🅻 🛆 🖥(3t). ✕ 🍴 (Chichester YC ☎512916).

Chichester Marina ☎512731, the well marked channel has a minimum depth of of ½m (dredged in Mar 2011). Enter at CM SHM lit pile, with depth gauge. A waiting pontoon is outside the lock (7m wide).

Lock signals: ⬤ = Wait; ⬤ = Enter.

Q (top of tower) = both gates open (free flow).

Call lock-keeper on Ch 80, ☎512731, H24 to book exit outwith freeflow. 1000⚓+50 ❶ £3·03, £1·28/m for <4hrs, 🖉 ♠ 🖉 ♠ LPG, 🖉 ⚒ ⚒ 🅱 🅔 🖵 🖥(65t) 🛆 🅻 ACA ✕ 🍴.

FISHBOURNE CHANNEL Dell Quay: Possible drying berth against the quay, apply to Hbr Office. ⚓ ⚓ ➴ ⚒ 🖵.

YACHT CLUBS

Bosham SC ☎572341; **Chichester YC** ☎512918, 🖉 ✕ 🍴; **Chichester Cruiser and Racing Club** ☎371731; **Dell Quay SC** ☎785080; **Emsworth SC** ☎373065; **Emsworth Slipper SC** ☎372523; **Hayling Island SC** ☎(02392) 463768; **Itchenor SC** ☎512400; **Mengham Rithe SC** ☎(02392) 463337; **Thorney Island SC** ☎371731.

Cobnor Activities Centre Trust (at Cobnor Pt) attracts many young people, inc disabled 🦽 afloat. ☎01243 572791.

CHICHESTER HARBOUR

MEAN SPRING AND NEAP CURVES

MEAN RANGES
Springs 4.0m
Neaps 2.1m

H.W.Hts.m.
CHART DATUM
M.H.W.N. M.H.W.S.
M.L.W.S. M.L.W.N.
L.W.Hts.m.

Factor
0.9 0.8 0.7 0.6 0.5 0.4 0.3 0.2 0.1

H.W. -5ʰ -4ʰ -3ʰ -2ʰ -1ʰ L.W. +1ʰ +2ʰ +3ʰ +4ʰ +5ʰ +6ʰ H.W.

CHICHESTER HARBOUR LAT 50°47'N LONG 0°56'W

TIMES AND HEIGHTS OF HIGH AND LOW WATERS

STANDARD TIME (UT)
For Summer Time add ONE hour in **non-shaded areas**

Dates in red are **SPRINGS**
Dates in blue are **NEAPS**

YEAR 2016

JANUARY

Time m	Time m
1 F 0351 4.4 / 0904 1.8 / 1603 4.2 / 2132 1.7	**16** SA 0352 4.7 / 0914 1.2 / 1612 4.5 / ◑ 2145 1.2
2 SA 0438 4.2 / 1001 2.0 / 1655 4.0 / ◐ 2232 1.9	**17** SU 0450 4.5 / 1021 1.4 / 1714 4.3 / 2254 1.4
3 SU 0532 4.1 / 1116 2.1 / 1756 3.9 / 2343 2.0	**18** M 0550 4.4 / 1132 1.5 / 1821 4.2
4 M 0636 4.0 / 1239 2.1 / 1911 3.9	**19** 0004 1.6 / TU 0656 4.3 / 1243 1.6 / 1935 4.2
5 TU 0055 2.0 / 0745 4.1 / 1348 2.0 / 2019 4.0	**20** W 0113 1.6 / 0808 4.4 / 1350 1.5 / 2053 4.3
6 W 0155 1.9 / 0844 4.2 / 1432 1.8 / 2113 4.2	**21** TH 0215 1.5 / 0914 4.5 / 1445 1.3 / 2158 4.5
7 TH 0235 1.8 / 0930 4.4 / 1502 1.6 / 2157 4.4	**22** F 0307 1.3 / 1009 4.7 / 1532 1.1 / 2252 4.6
8 F 0310 1.6 / 1010 4.6 / 1534 1.4 / 2237 4.5	**23** SA 0351 1.2 / 1053 4.7 / 1615 0.9 / 2338 4.7
9 SA 0348 1.4 / 1046 4.7 / 1612 1.1 / 2316 4.6	**24** SU 0434 1.0 / 1131 4.8 / 1656 0.8 / ○
10 SU 0428 1.2 / 1125 4.8 / 1652 1.0 / ● 2356 4.7	**25** M 0014 4.8 / 0514 1.0 / 1210 4.8 / 1737 0.8
11 M 0511 1.0 / 1205 4.8 / 1735 0.8	**26** TU 0046 4.8 / 0555 1.0 / 1249 4.8 / 1817 0.8
12 TU 0039 4.8 / 0555 0.9 / 1248 4.9 / 1819 0.7	**27** W 0121 4.8 / 0634 1.0 / 1327 4.7 / 1855 0.9
13 W 0123 4.9 / 0641 0.9 / 1332 4.9 / 1905 0.7	**28** TH 0157 4.7 / 0711 1.1 / 1405 4.6 / 1932 1.1
14 TH 0209 4.9 / 0728 0.9 / 1420 4.8 / 1953 0.8	**29** F 0233 4.6 / 0747 1.3 / 1442 4.5 / 2007 1.3
15 F 0258 4.8 / 0818 1.0 / 1513 4.7 / 2045 1.0	**30** SA 0310 4.5 / 0823 1.5 / 1520 4.3 / 2043 1.5
	31 SU 0349 4.3 / 0901 1.7 / 1604 4.1 / 2124 1.8

FEBRUARY

Time m	Time m
1 M 0435 4.1 / 0951 2.0 / 1658 3.9 / ◑ 2218 2.0	**16** TU 0521 4.3 / 1106 1.5 / 1755 4.1 / 2343 1.6
2 TU 0532 3.9 / 1102 2.1 / 1805 3.8 / 2335 2.1	**17** W 0627 4.1 / 1223 1.6 / 1911 4.0
3 W 0645 3.9 / 1241 2.1 / 1933 3.8	**18** TH 0057 1.7 / 0744 4.1 / 1334 1.6 / 2050 4.1
4 TH 0101 2.1 / 0802 4.0 / 1355 1.9 / 2041 4.0	**19** F 0202 1.6 / 0914 4.2 / 1430 1.4 / 2200 4.3
5 F 0204 1.9 / 0859 4.2 / 1433 1.6 / 2132 4.2	**20** SA 0252 1.4 / 1016 4.4 / 1516 1.1 / 2251 4.5
6 SA 0248 1.6 / 0944 4.4 / 1512 1.3 / 2214 4.5	**21** SU 0335 1.2 / 1101 4.6 / 1557 0.9 / 2332 4.7
7 SU 0330 1.3 / 1024 4.7 / 1553 1.0 / 2255 4.7	**22** M 0416 1.0 / 1129 4.7 / 1637 0.8 / ○
8 M 0412 1.0 / 1104 4.8 / 1636 0.7 / ● 2336 4.9	**23** TU 0000 4.8 / 0455 0.9 / 1157 4.7 / 1716 0.7
9 TU 0456 0.8 / 1146 5.0 / 1720 0.5	**24** W 0025 4.8 / 0534 0.8 / 1230 4.8 / 1754 0.8
10 W 0019 5.0 / 0540 0.6 / 1231 5.0 / 1804 0.4	**25** TH 0056 4.8 / 0611 0.8 / 1306 4.7 / 1830 0.7
11 TH 0104 5.1 / 0626 0.5 / 1316 5.0 / 1849 0.4	**26** F 0129 4.8 / 0645 0.9 / 1340 4.7 / 1903 0.9
12 F 0149 5.1 / 0711 0.6 / 1403 5.0 / 1935 0.5	**27** SA 0201 4.7 / 0717 1.0 / 1412 4.5 / 1933 1.1
13 SA 0236 5.0 / 0758 0.7 / 1453 4.8 / 2023 0.8	**28** SU 0232 4.5 / 0747 1.2 / 1444 4.4 / 2003 1.3
14 SU 0326 4.8 / 0849 1.0 / 1549 4.6 / 2118 1.1	**29** M 0305 4.3 / 0818 1.4 / 1523 4.1 / 2036 1.5
15 M 0421 4.6 / 0950 1.3 / 1649 4.3 / ◑ 2226 1.4	

MARCH

Time m	Time m
1 TU 0346 4.1 / 0857 1.7 / 1613 3.9 / ◑ 2121 1.8	**16** W 0453 4.2 / 1034 1.5 / 1730 4.1 / 2316 1.7
2 W 0440 3.9 / 0953 1.9 / 1715 3.7 / 2230 2.0	**17** TH 0557 4.0 / 1155 1.7 / 1843 3.9
3 TH 0546 3.8 / 1126 2.0 / 1831 3.7	**18** F 0033 1.8 / 0714 3.8 / 1306 1.6 / 2037 3.9
4 F 0011 2.1 / 0706 3.8 / 1259 1.9 / 1959 3.9	**19** SA 0137 1.7 / 0908 4.0 / 1404 1.5 / 2147 4.2
5 SA 0130 1.9 / 0821 4.0 / 1359 1.5 / 2101 4.2	**20** SU 0229 1.5 / 1007 4.2 / 1451 1.2 / 2235 4.4
6 SU 0223 1.5 / 0915 4.3 / 1446 1.2 / 2148 4.5	**21** M 0313 1.2 / 1049 4.4 / 1533 1.0 / 2311 4.6
7 M 0309 1.1 / 1000 4.6 / 1531 0.8 / 2231 4.8	**22** TU 0353 1.0 / 1116 4.5 / 1613 0.8 / 2335 4.7
8 TU 0353 0.8 / 1042 4.9 / 1615 0.5 / 2313 5.0	**23** W 0432 0.8 / 1139 4.6 / 1652 0.7 / ○
9 W 0437 0.5 / 1126 5.0 / 1659 0.3 / ● 2357 5.1	**24** TH 0000 4.8 / 0510 0.7 / 1211 4.7 / 1729 0.7
10 TH 0521 0.3 / 1212 5.1 / 1744 0.2	**25** F 0031 4.8 / 0546 0.7 / 1244 4.7 / 1803 0.8
11 F 0041 5.2 / 0605 0.3 / 1258 5.1 / 1828 0.3	**26** SA 0102 4.7 / 0618 0.9 / 1316 4.6 / 1834 0.9
12 SA 0127 5.2 / 0650 0.3 / 1346 5.1 / 1913 0.4	**27** SU 0132 4.7 / 0647 0.9 / 1347 4.5 / 1902 1.0
13 SU 0213 5.0 / 0735 0.5 / 1436 4.9 / 1959 0.7	**28** M 0200 4.5 / 0716 1.0 / 1418 4.4 / 1931 1.2
14 M 0302 4.8 / 0822 0.8 / 1529 4.6 / 2049 1.0	**29** TU 0231 4.4 / 0746 1.2 / 1455 4.2 / 2004 1.4
15 TU 0355 4.5 / 0917 1.2 / 1627 4.3 / ◑ 2154 1.4	**30** W 0310 4.2 / 0824 1.4 / 1542 4.0 / 2047 1.6
	31 TH 0402 3.9 / 0915 1.6 / 1642 3.9 / ◑ 2150 1.9

APRIL

Time m	Time m
1 F 0508 3.8 / 1038 1.8 / 1751 3.8 / 2333 2.0	**16** SA 0637 3.8 / 1228 1.7 / 1942 3.9
2 SA 0620 3.8 / 1214 1.7 / 1909 3.9	**17** SU 0103 1.8 / 0835 3.8 / 1328 1.6 / 2117 4.1
3 SU 0054 1.8 / 0737 4.0 / 1322 1.5 / 2023 4.2	**18** M 0159 1.6 / 0940 4.0 / 1420 1.4 / 2204 4.3
4 M 0154 1.4 / 0843 4.3 / 1416 1.1 / 2118 4.5	**19** TU 0246 1.3 / 1019 4.2 / 1505 1.2 / 2235 4.5
5 TU 0244 1.0 / 0934 4.6 / 1505 0.8 / 2204 4.8	**20** W 0328 1.1 / 1045 4.4 / 1547 1.0 / 2300 4.6
6 W 0330 0.7 / 1020 4.9 / 1550 0.5 / 2248 5.1	**21** TH 0408 0.9 / 1114 4.5 / 1626 0.9 / 2330 4.7
7 TH 0414 0.4 / 1105 5.0 / 1635 0.3 / ● 2333 5.2	**22** F 0445 0.9 / 1147 4.6 / 1702 0.9 / ○
8 F 0458 0.3 / 1153 5.1 / 1720 0.3	**23** SA 0003 4.7 / 0519 0.8 / 1222 4.6 / 1734 0.9
9 SA 0018 5.2 / 0542 0.3 / 1242 5.1 / 1805 0.3	**24** SU 0035 4.7 / 0550 0.9 / 1254 4.6 / 1804 1.0
10 SU 0105 5.2 / 0627 0.3 / 1331 5.0 / 1849 0.5	**25** M 0105 4.6 / 0619 0.9 / 1326 4.5 / 1834 1.1
11 M 0152 5.0 / 0711 0.5 / 1421 4.9 / 1935 0.8	**26** TU 0134 4.5 / 0649 1.0 / 1359 4.4 / 1906 1.2
12 TU 0240 4.8 / 0757 0.8 / 1513 4.7 / 2023 1.1	**27** W 0207 4.4 / 0723 1.1 / 1436 4.3 / 1942 1.3
13 W 0332 4.5 / 0849 1.1 / 1608 4.4 / 2123 1.4	**28** TH 0246 4.2 / 0803 1.3 / 1522 4.2 / 2027 1.5
14 TH 0427 4.2 / 0958 1.5 / 1706 4.1 / ◑ 2242 1.7	**29** F 0337 4.1 / 0854 1.4 / 1619 4.1 / 2130 1.7
15 F 0527 3.9 / 1118 1.7 / 1811 3.9 / 2358 1.8	**30** SA 0440 3.9 / 1010 1.6 / 1722 4.0 / ◑ 2300 1.8

Chart Datum: 2·74 metres below Ordnance Datum (Newlyn). HAT is 5·3 metres above Chart Datum.

FREE monthly updates. Register at
www.reedsnauticalalmanac.co.uk

STANDARD TIME (UT)
For Summer Time add ONE hour in **non-shaded areas**

CHICHESTER HARBOUR LAT 50°47'N LONG 0°56'W
TIMES AND HEIGHTS OF HIGH AND LOW WATERS

Dates in red are **SPRINGS**
Dates in blue are NEAPS

YEAR 2016

Central S England

MAY

Day	Time m	Time m	Time m	Time m
1 SU	0548 3.9	1135 1.6	1831 4.1	
16 M	0025 1.8	0710 3.8	1250 1.7	1959 4.0
2 M	0017 1.6	0658 4.0	1244 1.4	1943 4.3
17 TU	0125 1.7	0835 3.9	1346 1.6	2105 4.2
3 TU	0121 1.4	0809 4.3	1344 1.1	2045 4.6
18 W	0218 1.5	0928 4.1	1436 1.4	2146 4.3
4 W	0215 1.0	0907 4.6	1436 0.9	2136 4.9
19 TH	0303 1.3	1007 4.2	1520 1.3	2223 4.5
5 TH	0304 0.7	0957 4.8	1525 0.6	2223 5.0
20 F	0343 1.2	1044 4.4	1558 1.2	2258 4.5
6 F ●	0351 0.5	1045 5.0	1611 0.5	2308 5.1
21 SA ○	0419 1.1	1121 4.5	1632 1.2	2333 4.6
7 SA	0436 0.4	1134 5.0	1657 0.5	2355 5.1
22 SU	0451 1.0	1157 4.5	1704 1.2	
8 SU	0521 0.4	1226 5.0	1743 0.6	
23 M	0006 4.6	0521 1.0	1232 4.5	1735 1.2
9 M	0044 5.0	0605 0.5	1318 5.0	1828 0.7
24 TU	0039 4.5	0553 1.0	1307 4.5	1809 1.2
10 TU	0132 4.9	0651 0.7	1408 4.7	1914 0.9
25 W	0112 4.5	0628 1.0	1343 4.5	1847 1.2
11 W	0221 4.7	0737 0.9	1458 4.7	2002 1.2
26 TH	0149 4.4	0707 1.1	1422 4.4	1928 1.3
12 TH	0310 4.5	0827 1.1	1548 4.5	2057 1.5
27 F	0230 4.3	0750 1.1	1508 4.4	2016 1.4
13 F ◐	0402 4.2	0927 1.4	1640 4.1	2206 1.7
28 SA	0320 4.2	0843 1.3	1601 4.3	2117 1.5
14 SA	0456 4.0	1039 1.6	1735 4.1	2319 1.8
29 SU ◐	0419 4.1	0950 1.4	1659 4.3	2233 1.6
15 SU	0557 3.8	1147 1.7	1840 4.0	
30 M	0521 4.1	1104 1.4	1801 4.3	2344 1.5
31 TU	0627 4.2	1211 1.4	1907 4.4	

JUNE

Day	Time m	Time m	Time m	Time m
1 W	0048 1.3	0737 4.3	1313 1.2	2013 4.6
16 TH	0149 1.7	0837 3.9	1406 1.7	2101 4.2
2 TH	0148 1.1	0842 4.5	1410 1.0	2110 4.8
17 F	0238 1.5	0929 4.1	1452 1.6	2147 4.3
3 F	0241 0.9	0937 4.7	1503 0.9	2159 4.9
18 SA	0318 1.4	1013 4.3	1528 1.5	2226 4.4
4 SA	0330 0.7	1028 4.8	1551 0.8	2246 5.0
19 SU	0350 1.4	1053 4.4	1600 1.4	2303 4.5
5 SU ●	0417 0.6	1119 4.9	1638 0.7	2334 5.0
20 M ○	0421 1.2	1131 4.4	1634 1.3	2338 4.5
6 M	0502 0.6	1212 4.9	1724 0.8	
21 TU	0455 1.1	1208 4.5	1711 1.3	
7 TU	0023 4.9	0548 0.7	1305 4.9	1810 0.9
22 W	0014 4.5	0532 1.0	1247 4.5	1750 1.2
8 W	0112 4.8	0633 0.8	1353 4.8	1856 1.1
23 TH	0052 4.5	0612 1.0	1326 4.6	1832 1.2
9 TH	0159 4.6	0719 0.9	1439 4.7	1943 1.2
24 F	0131 4.5	0655 0.9	1407 4.6	1917 1.2
10 F	0246 4.5	0807 1.1	1523 4.5	2033 1.4
25 SA	0214 4.5	0740 1.0	1452 4.5	2005 1.2
11 SA	0334 4.3	0859 1.4	1609 4.4	2130 1.6
26 SU	0302 4.4	0831 1.1	1542 4.5	2102 1.3
12 SU ◐	0423 4.1	1000 1.6	1658 4.2	2236 1.8
27 M ◐	0358 4.3	0931 1.2	1637 4.4	2208 1.4
13 M	0516 3.9	1105 1.7	1751 4.0	2344 1.9
28 TU	0457 4.3	1038 1.3	1735 4.4	2316 1.4
14 TU	0616 3.8	1209 1.8	1855 4.0	
29 W	0600 4.2	1145 1.4	1837 4.4	
15 W	0049 1.8	0729 3.8	1311 1.8	2004 4.1
30 TH	0022 1.4	0708 4.3	1249 1.3	1945 4.5

JULY

Day	Time m	Time m	Time m	Time m
1 F	0126 1.2	0820 4.4	1351 1.2	2048 4.6
16 SA	0210 1.7	0854 4.0	1419 1.8	2113 4.2
2 SA	0224 1.1	0922 4.5	1447 1.1	2141 4.7
17 SU	0247 1.6	0944 4.2	1455 1.6	2156 4.3
3 SU	0315 0.9	1016 4.7	1537 1.0	2229 4.8
18 M	0318 1.4	1025 4.3	1531 1.4	2233 4.4
4 M ●	0402 0.8	1108 4.7	1623 0.9	2316 4.8
19 TU ○	0353 1.2	1104 4.4	1609 1.3	2310 4.5
5 TU	0447 0.7	1159 4.8	1709 0.9	
20 W	0432 1.0	1142 4.5	1650 1.1	2348 4.6
6 W	0004 4.8	0532 0.8	1248 4.8	1753 1.0
21 TH	0513 0.9	1222 4.6	1733 1.0	
7 TH	0050 4.7	0616 0.8	1330 4.7	1837 1.1
22 F	0028 4.7	0556 0.8	1303 4.7	1817 0.9
8 F	0135 4.6	0700 0.9	1411 4.7	1920 1.2
23 SA	0111 4.7	0640 0.7	1346 4.7	1902 0.9
9 SA	0218 4.5	0744 1.1	1452 4.6	2004 1.4
24 SU	0154 4.7	0726 0.7	1430 4.7	1949 1.0
10 SU	0302 4.4	0814 0.9	1534 4.4	2050 1.6
25 M	0242 4.6	0814 0.9	1519 4.6	2041 1.1
11 M	0346 4.2	0916 1.5	1618 4.3	2143 1.8
26 TU ◐	0335 4.5	0909 1.1	1612 4.5	2142 1.3
12 TU ◐	0433 4.0	1012 1.7	1705 4.1	2248 1.9
27 W	0434 4.3	1014 1.3	1709 4.4	2252 1.4
13 W	0525 3.9	1115 1.9	1759 4.0	
28 TH	0536 4.2	1123 1.4	1811 4.3	
14 TH	0002 1.9	0629 3.8	1224 1.9	1908 3.9
29 F	0002 1.5	0645 4.1	1233 1.5	1920 4.3
15 F	0115 1.9	0749 3.8	1329 1.9	2018 4.0
30 SA	0110 1.4	0803 4.2	1338 1.4	2031 4.4
31 SU	0211 1.2	0917 4.4	1435 1.3	2132 4.5

AUGUST

Day	Time m	Time m	Time m	Time m
1 M	0302 1.0	1017 4.5	1524 1.1	2222 4.6
16 TU	0248 1.3	0957 4.3	1506 1.4	2204 4.4
2 TU ●	0348 0.9	1107 4.7	1609 1.0	2305 4.7
17 W	0328 1.0	1035 4.5	1548 1.1	2242 4.6
3 W	0431 0.8	1149 4.7	1651 0.9	2347 4.7
18 TH ○	0410 0.8	1114 4.7	1630 0.9	2322 4.7
4 TH	0514 0.7	1226 4.7	1733 0.9	
19 F	0453 0.6	1155 4.8	1714 0.7	
5 F	0028 4.7	0555 0.7	1303 4.7	1814 1.0
20 SA	0004 4.8	0536 0.5	1237 4.9	1757 0.6
6 SA	0109 4.6	0636 0.8	1340 4.7	1854 1.1
21 SU	0048 4.9	0621 0.5	1321 4.9	1842 0.6
7 SU	0149 4.6	0715 1.0	1418 4.6	1932 1.2
22 M	0133 4.9	0705 0.5	1406 4.9	1928 0.7
8 M	0228 4.4	0752 1.2	1456 4.5	2008 1.4
23 TU	0221 4.8	0752 0.7	1454 4.8	2016 0.9
9 TU	0307 4.3	0829 1.4	1535 4.3	2047 1.6
24 W	0314 4.6	0843 1.0	1547 4.6	2113 1.2
10 W ◐	0348 4.1	0909 1.6	1616 4.1	2134 1.8
25 TH ◐	0413 4.4	0947 1.3	1645 4.4	2226 1.4
11 TH	0435 3.9	1002 1.9	1704 3.9	2241 2.0
26 F	0516 4.2	1104 1.5	1747 4.2	2343 1.6
12 F	0532 3.8	1113 2.0	1804 3.8	
27 SA	0625 4.0	1217 1.6	1857 4.1	
13 SA	0006 2.0	0647 3.7	1233 2.0	1924 3.9
28 SU	0053 1.5	0750 4.0	1324 1.6	2020 4.3
14 SU	0122 1.9	0815 3.8	1337 1.9	2036 4.0
29 M	0154 1.4	0923 4.2	1420 1.4	2139 4.3
15 M	0208 1.6	0913 4.1	1424 1.6	2124 4.2
30 TU	0245 1.2	1021 4.5	1507 1.2	2229 4.5
31 W	0329 0.9	1105 4.6	1550 1.0	2302 4.6

Chart Datum: 2·74 metres below Ordnance Datum (Newlyn). HAT is 5·3 metres above Chart Datum.

》 FREE monthly updates. Register at 《
www.reedsnauticalalmanac.co.uk

243

STANDARD TIME (UT)
For Summer Time add ONE hour in **non-shaded areas**

CHICHESTER HARBOUR LAT 50°47'N LONG 0°56'W
TIMES AND HEIGHTS OF HIGH AND LOW WATERS

Dates in red are **SPRINGS**
Dates in blue are **NEAPS**

YEAR 2016

SEPTEMBER

Day	Time m	Day	Time m
1	0411 0.8 / 1137 4.7 / TH 1631 0.9 / ● 2332 4.7	**16**	0346 0.7 / 1045 4.9 / F 1608 0.7 / ○ 2256 4.9
2	0452 0.7 / 1203 4.8 / F 1711 0.8	**17**	0429 0.4 / 1126 5.0 / SA 1652 0.5 / 2339 5.0
3	0007 4.7 / 0531 0.7 / SA 1236 4.8 / 1749 0.8	**18**	0513 0.3 / 1210 5.1 / SU 1735 0.4
4	0045 4.7 / 0609 0.7 / SU 1311 4.7 / 1825 0.9	**19**	0025 5.0 / 0557 0.4 / M 1255 5.1 / 1819 0.5
5	0122 4.6 / 0644 0.9 / M 1345 4.6 / 1858 1.1	**20**	0112 5.0 / 0643 0.4 / TU 1341 5.0 / 1904 0.6
6	0156 4.5 / 0716 1.1 / TU 1418 4.5 / 1929 1.3	**21**	0202 4.9 / 0727 0.7 / W 1430 4.8 / 1950 0.8
7	0230 4.4 / 0746 1.2 / W 1450 4.3 / 2001 1.5	**22**	0256 4.7 / 0816 1.0 / TH 1525 4.6 / 2043 1.2
8	0306 4.2 / 0820 1.5 / TH 1529 4.1 / 2040 1.7	**23**	0356 4.4 / 0917 1.4 / F 1624 4.3 / ◐ 2156 1.5
9	0353 4.0 / 0903 1.8 / F 1618 3.9 / ◐ 2134 1.9	**24**	0459 4.2 / 1040 1.6 / SA 1726 4.1 / 2319 1.6
10	0450 3.8 / 1011 2.0 / SA 1718 3.8 / 2303 2.0	**25**	0606 4.0 / 1156 1.7 / SU 1835 4.0
11	0556 3.7 / 1146 2.1 / SU 1827 3.8	**26**	0029 1.6 / 0734 4.0 / M 1302 1.7 / 2017 4.0
12	0030 1.9 / 0719 3.8 / M 1301 1.9 / 1947 3.9	**27**	0130 1.5 / 0916 4.2 / TU 1357 1.5 / 2139 4.2
13	0130 1.6 / 0835 4.1 / TU 1355 1.6 / 2048 4.2	**28**	0221 1.3 / 1010 4.5 / W 1445 1.3 / 2225 4.4
14	0218 1.3 / 0924 4.4 / W 1442 1.3 / 2133 4.5	**29**	0306 1.1 / 1050 4.6 / TH 1528 1.1 / 2256 4.5
15	0303 0.9 / 1005 4.6 / TH 1525 0.9 / 2215 4.7	**30**	0347 0.9 / 1116 4.7 / F 1608 0.9 / 2315 4.6

OCTOBER

Day	Time m	Day	Time m
1	0427 0.8 / 1138 4.8 / SA 1647 0.8 / ● 2347 4.7	**16**	0405 0.5 / 1100 5.2 / SU 1629 0.5 / ○ 2317 5.1
2	0506 0.7 / 1209 4.8 / SU 1724 0.8	**17**	0449 0.4 / 1144 5.2 / M 1712 0.4
3	0021 4.7 / 0541 0.8 / M 1242 4.8 / 1757 0.9	**18**	0005 5.1 / 0533 0.4 / TU 1231 5.2 / 1756 0.5
4	0056 4.6 / 0613 0.9 / TU 1314 4.7 / 1828 1.0	**19**	0055 5.1 / 0619 0.6 / W 1319 5.1 / 1841 0.6
5	0128 4.5 / 0643 1.1 / W 1342 4.5 / 1857 1.2	**20**	0148 4.9 / 0705 0.8 / TH 1410 4.9 / 1928 0.9
6	0159 4.4 / 0713 1.3 / TH 1412 4.4 / 1928 1.3	**21**	0244 4.7 / 0754 1.1 / F 1506 4.6 / 2020 1.2
7	0234 4.3 / 0745 1.5 / F 1449 4.2 / 2004 1.5	**22**	0343 4.5 / 0852 1.4 / SA 1605 4.4 / ◐ 2127 1.5
8	0320 4.1 / 0826 1.7 / SA 1540 4.0 / 2052 1.8	**23**	0444 4.3 / 1011 1.7 / SU 1705 4.1 / 2249 1.7
9	0418 3.9 / 0926 2.0 / SU 1644 3.8 / ◐ 2212 2.0	**24**	0546 4.1 / 1127 1.8 / M 1811 4.0 / 2358 1.7
10	0524 3.8 / 1106 2.1 / M 1751 3.8 / 2347 1.9	**25**	0702 4.1 / 1232 1.8 / TU 1941 3.9
11	0635 3.9 / 1226 1.9 / TU 1903 3.9	**26**	0058 1.7 / 0849 4.2 / W 1329 1.7 / 2117 4.1
12	0053 1.7 / 0749 4.1 / W 1325 1.6 / 2010 4.2	**27**	0152 1.5 / 0944 4.4 / TH 1420 1.5 / 2202 4.3
13	0147 1.3 / 0848 4.5 / TH 1415 1.2 / 2104 4.5	**28**	0240 1.3 / 1021 4.6 / F 1505 1.2 / 2230 4.5
14	0235 1.0 / 0935 4.8 / F 1501 0.9 / 2149 4.8	**29**	0323 1.1 / 1044 4.7 / SA 1546 1.1 / 2253 4.6
15	0321 0.7 / 1017 5.0 / SA 1545 0.6 / 2232 5.0	**30**	0403 1.0 / 1116 4.7 / SU 1624 1.0 / ● 2325 4.6
		31	0441 1.0 / 1142 4.8 / M 1700 1.0 / 2359 4.6

NOVEMBER

Day	Time m	Day	Time m
1	0515 1.0 / 1214 4.7 / TU 1732 1.0	**16**	0514 0.6 / 1210 5.2 / W 1739 0.6
2	0033 4.6 / 0545 1.1 / W 1245 4.7 / 1802 1.1	**17**	0042 5.0 / 0600 0.7 / TH 1301 5.1 / 1825 0.7
3	0105 4.6 / 0616 1.2 / TH 1314 4.6 / 1832 1.2	**18**	0138 4.9 / 0648 0.9 / F 1353 4.9 / 1912 0.9
4	0137 4.5 / 0647 1.3 / F 1345 4.5 / 1905 1.3	**19**	0234 4.8 / 0737 1.2 / SA 1448 4.7 / 2004 1.2
5	0213 4.4 / 0723 1.5 / SA 1422 4.3 / 1942 1.4	**20**	0330 4.6 / 0832 1.4 / SU 1545 4.4 / 2103 1.4
6	0258 4.2 / 0804 1.7 / SU 1511 4.1 / 2029 1.6	**21**	0425 4.4 / 0939 1.7 / M 1642 4.2 / ◗ 2215 1.7
7	0355 4.1 / 0900 1.9 / M 1616 4.0 / ◗ 2137 1.8	**22**	0520 4.3 / 1052 1.9 / TU 1741 4.0 / 2324 1.8
8	0458 4.0 / 1027 2.0 / TU 1722 4.0 / 2306 1.8	**23**	0620 4.1 / 1158 1.9 / W 1847 3.9
9	0604 4.1 / 1148 1.9 / W 1830 4.1	**24**	0025 1.8 / 0733 4.1 / TH 1259 1.8 / 2010 4.0
10	0016 1.6 / 0711 4.3 / TH 1252 1.6 / 1937 4.3	**25**	0123 1.7 / 0849 4.3 / F 1355 1.7 / 2114 4.2
11	0114 1.4 / 0815 4.6 / F 1346 1.3 / 2036 4.6	**26**	0215 1.6 / 0932 4.4 / SA 1444 1.5 / 2151 4.3
12	0207 1.1 / 0907 4.9 / SA 1436 1.0 / 2126 4.9	**27**	0301 1.4 / 1006 4.6 / SU 1526 1.3 / 2226 4.5
13	0256 0.8 / 0952 5.1 / SU 1522 0.7 / 2212 5.0	**28**	0342 1.3 / 1041 4.7 / M 1605 1.2 / 2302 4.6
14	0343 0.7 / 1036 5.2 / M 1608 0.6 / ○ 2259 5.1	**29**	0418 1.2 / 1116 4.7 / TU 1639 1.2 / ● 2339 4.6
15	0428 0.6 / 1122 5.2 / TU 1653 0.5 / 2348 5.1	**30**	0451 1.3 / 1149 4.7 / W 1710 1.2

DECEMBER

Day	Time m	Day	Time m
1	0014 4.6 / 0522 1.3 / TH 1222 4.7 / 1741 1.2	**16**	0033 5.0 / 0547 0.9 / F 1245 5.0 / 1812 0.8
2	0048 4.6 / 0545 1.3 / F 1253 4.6 / 1814 1.2	**17**	0129 4.9 / 0635 1.0 / SA 1336 4.9 / 1900 0.9
3	0122 4.5 / 0630 1.4 / SA 1327 4.5 / 1851 1.2	**18**	0221 4.8 / 0722 1.2 / SU 1427 4.7 / 1948 1.1
4	0159 4.5 / 0709 1.4 / SU 1405 4.4 / 1930 1.3	**19**	0309 4.7 / 0812 1.4 / M 1518 4.5 / 2040 1.3
5	0242 4.4 / 0752 1.5 / M 1451 4.3 / 2017 1.4	**20**	0356 4.5 / 0907 1.6 / TU 1603 4.3 / 2138 1.6
6	0335 4.3 / 0844 1.7 / TU 1551 4.2 / 2115 1.6	**21**	0445 4.4 / 1011 1.8 / W 1703 4.1 / ◗ 2244 1.7
7	0434 4.3 / 0954 1.8 / W 1655 4.1 / ◗ 2230 1.7	**22**	0538 4.2 / 1120 1.9 / TH 1802 4.0 / 2349 1.9
8	0536 4.3 / 1111 1.8 / TH 1800 4.2 / 2340 1.6	**23**	0637 4.1 / 1227 1.9 / F 1908 3.9
9	0640 4.4 / 1218 1.6 / F 1907 4.3	**24**	0053 1.9 / 0743 4.1 / SA 1331 1.9 / 2017 4.0
10	0043 1.5 / 0745 4.6 / SA 1319 1.4 / 2012 4.5	**25**	0153 1.8 / 0844 4.3 / SU 1426 1.7 / 2114 4.2
11	0142 1.3 / 0843 4.8 / SU 1415 1.1 / 2108 4.8	**26**	0243 1.7 / 0933 4.4 / M 1511 1.5 / 2200 4.3
12	0236 1.1 / 0932 5.0 / M 1505 0.9 / 2158 4.9	**27**	0325 1.6 / 1015 4.5 / TU 1548 1.4 / 2241 4.4
13	0326 0.9 / 1019 5.1 / TU 1553 0.8 / 2247 5.0	**28**	0359 1.5 / 1052 4.6 / W 1620 1.3 / 2319 4.5
14	0414 0.8 / 1105 5.2 / W 1639 0.7 / ○ 2338 5.0	**29**	0430 1.4 / 1128 4.6 / TH 1651 1.2 / ● 2356 4.6
15	0501 0.8 / 1154 5.1 / TH 1726 0.7	**30**	0504 1.3 / 1201 4.7 / F 1725 1.1
		31	0031 4.6 / 0540 1.3 / SA 1236 4.7 / 1801 1.1

Chart Datum: 2·74 metres below Ordnance Datum (Newlyn). HAT is 5·3 metres above Chart Datum.

FREE monthly updates. Register at www.reedsnauticalalmanac.co.uk

South East England

Selsey Bill to North Foreland

9.3.1	Contents	245
9.3.2	Area map and Distance table	246
9.3.3	Tidal stream charts	248
9.3.4	Lights, buoys and waypoints	250
9.3.5	Passage information	251

The Eastern Channel, Selsey Bill and the Owers, Owers to Beachy Head – 251 • Crossing the Eastern English Channel, Beachy Head to Dungeness, Dungeness to North Foreland, Lydd Firing Ranges, Hythe Firing Ranges – 260

9.3.6	Littlehampton	252
9.3.7	Shoreham, tide tables/curves	253
9.3.8	Brighton	257
9.3.9	Newhaven	258
9.3.10	Eastbourne Hastings	259
9.3.11	Rye	261
9.3.12	Folkestone	262
9.3.13	Dover, tide tables/curves	263
9.3.14	Dover Strait	267
9.3.15	Ramsgate Sandwich	268

SE England

9.3.2
South East England
Selsey Bill to North Foreland

		1	2	3	4	5	6	7	8	9	10	11	12	13	14	15	16
1	*Portland Bill*	**1**															
2	Nab Tower	60	**2**														
3	Boulder Lt Buoy	65	5	**3**													
4	Owers Lt Buoy	69	10	6	**4**												
5	**Littlehampton**	78	19	13	12	**5**											
6	**Shoreham**	90	32	24	21	13	**6**										
7	**Brighton**	93	35	28	24	17	5	**7**									
8	**Newhaven**	97	40	34	29	24	12	7	**8**								
9	Beachy Head	104	46	41	36	30	20	14	8	**9**							
10	Eastbourne	111	51	45	40	34	24	19	12	7	**10**						
11	**Rye**	129	72	67	62	56	46	41	34	25	23	**11**					
12	Dungeness	134	76	71	66	60	50	44	38	30	26	9	**12**				
13	**Folkestone**	152	92	84	81	76	65	60	53	43	40	23	13	**13**			
14	Dover	157	97	89	86	81	70	65	58	48	45	28	18	5	**14**		
15	Ramsgate	172	112	104	101	96	85	80	73	63	60	43	33	20	15	**15**	
16	*N Foreland*	175	115	107	104	99	88	83	76	66	63	46	36	23	18	3	**16**

DISTANCE TABLE
Approximate distances in nautical miles are by the most direct route while avoiding dangers and allowing for TSS. Places in *italics* are in adjoining areas; places in **bold** are in 9.0.5, Cross-Channel Distances; places underlined are in 9.0.7, Distances across the North Sea.

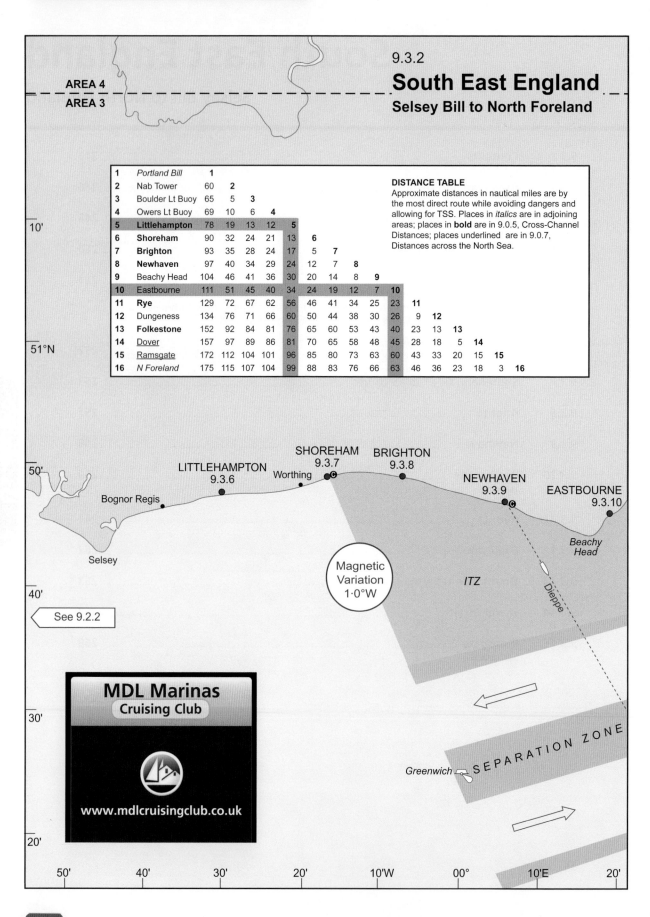

AREA 4
AREA 3

10'
51°N
50'
40'
30'
20'

LITTLEHAMPTON
9.3.6

SHOREHAM
9.3.7

BRIGHTON
9.3.8

NEWHAVEN
9.3.9

EASTBOURNE
9.3.10

Worthing

Bognor Regis

Selsey

Beachy Head

Magnetic Variation 1·0°W

ITZ

Dieppe

See 9.2.2

Greenwich

SEPARATION ZONE

MDL Marinas
Cruising Club
www.mdlcruisingclub.co.uk

50' 40' 30' 20' 10'W 00° 10'E 20'

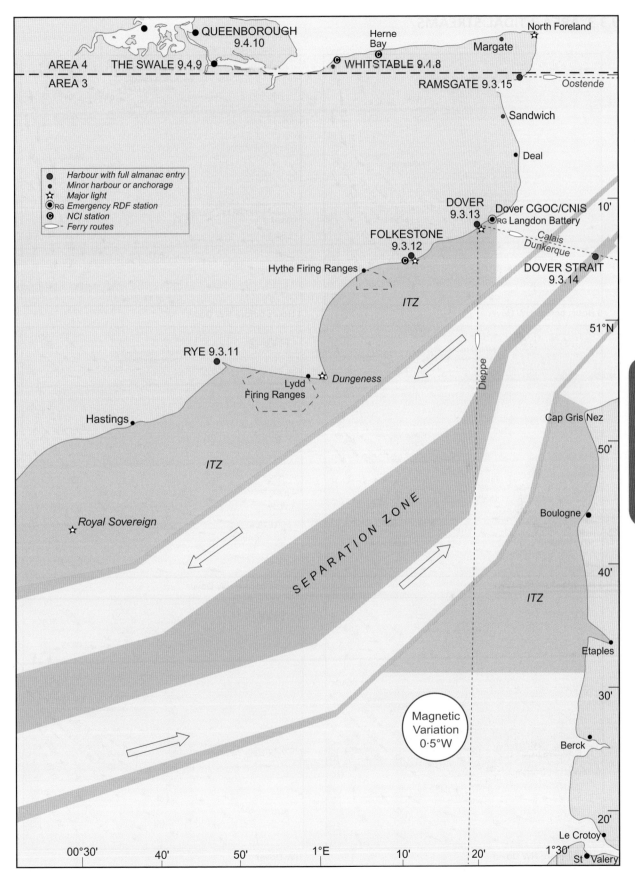

QUEENBOROUGH 9.4.10

Herne Bay

North Foreland

Margate

AREA 4   THE SWALE 9.4.9   WHITSTABLE 9.4.8

AREA 3   RAMSGATE 9.3.15   Oostende

Sandwich

Deal

- Harbour with full almanac entry
- Minor harbour or anchorage
☆ Major light
Ⓡ RG Emergency RDF station
Ⓒ NCI station
⬭ Ferry routes

DOVER 9.3.13   Dover CGOC/CNIS   10'
RG Langdon Battery

FOLKESTONE 9.3.12   Calais
Dunkerque

Hythe Firing Ranges   DOVER STRAIT 9.3.14

ITZ

51°N

RYE 9.3.11

Lydd Firing Ranges   Dungeness

Hastings   Cap Gris Nez   50'

ITZ

☆ Royal Sovereign   Boulogne

40'

S E P A R A T I O N   Z O N E

Dieppe

ITZ   Etaples

30'

Magnetic Variation 0·5°W

Berck

20'

Le Crotoy

00°30'   40'   50'   1°E   10'   20'   1°30'   St Valery

9.3.3 AREA 3 TIDAL STREAMS

5 Hours before HW Dover

4 Hours before HW Dover

3 Hours before HW Dover

2 Hours before HW Dover

1 Hour before HW Dover

HW Dover

Westward 9.2.3   Southward 9.17.3   Northward 9.4.3   Thames Estuary 9.4.6   Eastward 9.16.3

1 Hour after HW Dover

2 Hours after HW Dover

3 Hours after HW Dover

4 Hours after HW Dover

5 Hours after HW Dover

6 Hours after HW Dover

SE England

9.3.4 LIGHTS, BUOYS AND WAYPOINTS

Bold print = light with a nominal range of 15M or more. CAPITALS = place or feature. *CAPITAL ITALICS* = light-vessel, light float or Lanby. *Italics* = Fog signal. ***Bold italics*** = Racon. Some marks/buoys are fitted with AIS (MMSI No); see relevant charts.

OWERS TO BEACHY HEAD

SELSEY BILL and THE OWERS
S Pullar ∫ VQ (6) + L Fl 10s; 50°38'·84N 00°49'·29W.
Pullar ∫ Q (9) 15s; 50°40'·47N 00°50'·09W.
Boulder ⌐ Fl G 2·5s; 50°41'·56N 00°49'·09W.
Street ⌐ QR; 50°41'·69N 00°48'·89W.
Mixon Bn ∫ Q(6) + L Fl 15s; 50°42'·38N 00°46'·32E.
Owers ∫ Q (6) + L Fl 15s; *Bell*; ***Racon (O) 10M***; 50°38'·59N 00°41'·09W.
E Borough Hd ∫ Q (3) 10s *Bell*; 50°41'·54N 00°39'·09W.

LITTLEHAMPTON
West Pier Hd ∫ QR 7m 6M; 50°47'·88N 00°32'·46W.
Training Wall Hd ∫ QG 10m 2M; 50°47'·87N 00°32'·38W.
Ldg lts 346°. Front, E Pier Hd ∡ FG 6m 7M; B col. Rear, 64m from front, Oc W 7·5s 9m10M; W twr; vis: 290°-W- 042°; 50°48'·09N 00°32'·51W.
Outfall ⌐ Fl Y 5s; 50°46'·27N 00°30'·53W.
Littlehampton ∡ Fl (5) Y 20s 9m 5M; 50°46'·19N 00°29'·54W.

WORTHING and SHOREHAM
Met Mast ∡ Mo(U) 15s 16m 10M; Aero FR (103m), *Horn Mo(U) 15s;* 50°41'·29N 00°20'·59W.
Outfall ∫ Fl R 2·5s 3m; 50°48'·38N 00°20'·34W.
Express ∫ Fl Y 5s; (Apr-Oct); 50°47'·28N 00°17'·09W.
W Bkwtr Head ∡ Fl R 5s 7m 7M; 50°49'·49N 00°14'·89W.
Ldg Lts 355°. Middle Pier Front, Oc 5s 8m 10M; W watch-house, R base; tidal Lts, tfc sigs; *Horn 20s.* **Rear**, 192m from front, Fl 10s 13m **15M**; Gy twr vis: 283°-103°; 50°49'·75N 00°14'·88W.
Outfall ∫ 50°49'·47N 00°14'·39W.
Shoreham Outfall ∫ Q (6) + L Fl 15s; 50°47'·88N 00°13'·72W.

BRIGHTON and BRIGHTON MARINA
Black Rk Ledge ∫ Fl Y 4s; 50°48'·07N 00°06'·46W.
W Bkwtr Hd ∡ QR 10m 7M; W ○ structure, R bands; *Horn (2) 30s;* 50°48'·50N 00°06'·38W.
E Bkwtr Hd ∡ QG 8m 7M and Fl (4) WR 20s 16m W10M, R8M; W pillar, G bands; vis: 260°-R-295°-W-100°; 50°48'·47N 00°06'·37W.
Saltdean Outfall ⌐ Fl Y 5s; 50°46'·72N 00°02'·13W.

NEWHAVEN
Bkwtr Head ∡ Oc (2) 10s 17m 12M; 50°46'·56N 00°03'·50E.
E Pier Hd ∡ Iso G 10s 12m 6M; W twr; 50°46'·81N 00°03'·59E.

OFFSHORE MARKS
CS 1 ∫ Fl Y 2·5s; *Whis;* 50°33'·69N 00°03'·92W.
GREENWICH ⌐ 50°24'·54N 00°00'·10E; Fl 5s 12m **15M**; Riding light FW; R hull; ***Racon (M) 10M***; *Horn 30s* .
CS 2 ∫ Fl Y 5s; 50°39'·14N 00°32'·60E.
CS 3 ∫ Fl Y 10s; 50°52'·04N 01°02'·18E.

BEACHY HEAD TO DUNGENESS

Beachy Head ∡ 50°44'·03N 00°14'·49E; Fl (2) 20s 31m 8M; W round twr, R band and lantern; vis: 248°-101°; (H24).
Royal Sovereign ∡ Fl 20s 28m 12M; W ○ twr, R band on W cabin on col; *Horn (2) 30s;* 50°43'·45N 00°26'·09E.
Royal Sovereign ⌐ QR; 50°44'·23N 00°25'·84E.

EASTBOURNE and SOVEREIGN HARBOUR
SH ∫ L Fl 10s; 50°47'·40N 00°20'·71E.
Martello Tower ∡ Fl (3) 15s 12m 7M.; 50°47'·24N 00°19'·83E.
Dir lt 258°Fl WRG 5s 4m 1M; vis: 252·5°-G-256·5°-W-259·5°-R-262·5°; 50°47'·28N 00°19'·71E.
S Bkwtr Hd ∡ Fl (4) R 12s 3m 6M; 50°47'·30N 00°20'·03E.
St Leonard's Outfall ⌐ Fl Y 5s; 50°49'·31N 00°31'·95E.

HASTINGS
Ldg Lts 356·3°. Front, FR 14m 4M; 50°51'·29N 00°35'·38E. Rear, West Hill, 357m from front, FR 55m 4M; W twr.

RYE
Rye Fairway, L Fl 10s; 50°54'·04N 00°48'·04E.
W Groyne Hd No. 2 ∡ Fl R 5s 7m 6M; 50°55'·58N 00°46'·55E.
E Arm Hd No. 1 ∫ Q (9) 15s 7m 5M; G △; *Horn 7s;* 50°55'·73N 00°46'·46E.
Dungeness Outfall ∫ Q (6) + L Fl 15s; 50°54'·45N 00°58'·21E.
Dungeness ☆ 50°54'·81N 00°58'·56E; Fl 10s 40m **21M**; B ○ twr, W bands and lantern, floodlit; Part obsc 078°-shore; (H24). F RG 37m 10M (same twr); vis: 057°-R-073°-G-078°-196°-R-216°; *Horn (3) 60s;* FR Lts shown between 2·4M and 5·2M WNW when firing taking place. QR on radio mast 1·2M NW.

DUNGENESS TO NORTH FORELAND

FOLKESTONE
Hythe Flats Outfall ⌐ Fl Y 5s; 51°02'·52N 01°05'·32E.
Breakwater Head ☆ Fl (2) 10s 14m **22M**; 51°04'·56N 01°11'·69E.

DOVER
Admiralty Pier Extension Head ☆ 51°06'·69N 01°19'·66E; Fl 7·5s 21m **20M**; W twr; vis: 096°-090°, obsc in The Downs by S Foreland inshore of 226°; *Horn 10s;* Int Port Tfc sigs.
S Bkwtr W Hd ☆ 51°06'·78N 01°19'·80E; Oc R 30s 21m **18M**; W twr.
Knuckle ☆ 51°07'·04N 01°20'·49E; Fl (4) WR 10s 15m **W15M**, R13M; W twr; vis: 059°-R-239°-W-059°.
N Head ∡ Fl R 2·5s 11m 5M; 51°07'·20N 01°20'·61E.
Eastern Arm Hd ∡ Fl G 5s 12m 5M; *Horn (2) 30s;* Int port tfc sigs; 51°07'·31N 01°20'·59E.

DOVER STRAIT
Bullock Bank ∫ VQ; 50°46'·94N 01°07'·60E.
Ridens SE ∫ VQ (3) 5s; 50°43'·47N 01°18'·87E.
Colbart SW ∫ VQ (6) + L Fl 10s; 50°48'·86N 01°16'·30E.
South Varne ∫ Q (6) + L Fl 15s; 50°55'·64N 01°17'·30E.
Mid Varne ∫ VQ(9)10s; 50°58'·94N 01°19'·88E.
East Varne ∫ VQ(3)5s; 50°58'·22N 01°20'·90E.
Colbart N ∫ VQ; 50°57'·45N 01°23'·29E.
Varne NW ∫ Q; 51°00'·80N 01°22'·70E.
Varne NE ∫ Q (3) 10s; 50°59'·80N 01°22'·70E.
VARNE ⌐ 51°01'·29N 01°23'·90E; Fl R 5s 12m **15M**; ***Racon (T) 10M;*** *Horn 30s;* 992351038.
CS 4 ∫ Fl (4) Y 15s; 51°08'·62N 01°33'·92E.
MPC ∫ Fl Y 2·5s; ***Racon (O) 10M***; 51°06'·12N 01°38'·20E; 992351122.
SW Goodwin ∫ Q (6) + L Fl 15s; 51°08'·50N 01°28'·88E; 992351036.
S Goodwin ⌐ Fl (4) R 15s; 51°10'·60N 01°32'·26E.
SE Goodwin ⌐ Fl (3) R 10s; 51°12'·99N 01°34'·45E.
E GOODWIN ⌐ 51°13'·26N 01°36'·37E; Fl 15s 12m **23M**; R hull with lt twr amidships; ***Racon (T) 10M;*** *Horn 30s;* 992351035.
E Goodwin ∫ Q (3) 10s; 51°15'·67N 01°35'·69E.
NE Goodwin ∫ Q (3) 10s; ***Racon (M) 10M.*** 51°20'·31N 01°34'·16E.

DEAL and THE DOWNS
Trinity Bay VQ (9) 10s; 51°11'·60N 01°29'·00E.
Deal Bank ⌐ QR; 51°12'·92N 01°25'·57E.
Goodwin Fork ∫ Q (6) + L Fl 15s; *Bell;* 51°14'·38N 01°26'·70E.
Downs ⌐ Fl (2) R 5s; *Bell;* 51°14'·50N 01°26'·22E.

GULL STREAM
W Goodwin ⌐ Fl G 5s; 51°15'·61N 01°27'·38E.
S Brake ⌐ Fl (3) R 10s; 51°15'·77N 01°26'·82E.
NW Goodwin ∫ Q (9) 10s; 51°16'·73N 01°28'·60E.
Brake ⌐ Fl (4) R 15s; 51°16'·98N 01°28'·19E.
N Goodwin ⌐ Fl G 2·5s; 51°18'·12N 01°30'·35E.
Gull Stream ⌐ QR; 51°18'·26N 01°29'·69E.

Gull ⚓ VQ (3) 5s; 51°19'·57N 01°31'·30E.
Goodwin Knoll ▲ Fl (2) G 5s; 51°19'·57N 01°32'·20E.

RAMSGATE CHANNEL
B2 ▲ Fl (2) G 5s; 51°18'·26N 01°23'·93E.
W Quern ⚓Q (9) 15s; 51°18'·98N 01°25'·39E.

RAMSGATE
RA ⚓ Q(6) + L Fl 15s; 51°19'·60N 01°30'·13E.
E Brake ⚓ Fl R 5s; 51°19'·47N 01°29'·20E.
No. 1 ▲QG; 51°19'·56N 01°27'·29E.
No. 2 ⚓ Fl (4) R 10s; 51°19'·46N 01°27'·28E.
No. 3 ▲ Fl G 2·5s; 51°19'·56N 01°26'·61E.
No. 4 ⚓ QR; 51°19'·46N 01°26'·60E.
N Quern ⚓ Q; 51°19'·41N 01°26'·11E.
No. 5 ⚓ Q (6) + L Fl 15s; 51°19'·56N 01°25'·91E.
No. 6 ⚓ QR; 51°19'·46N 01°25'·91E.
South Bkwtr Hd ⚓ VQ R 10m 5M; 51°19'·46N 01°25'·41E.
N Bkwtr Hd ⚓ QG 10m 5M; 51°19'·56N 01°25'·47E.
Western Marine terminal Dir lt 270° ⚓, Oc WRG 10s 10m 5M;
B △, Or stripe; vis: 259°-G-269°-W- 271°-R-281°; 51°19'·51N

01°24'·85E. Rear 493m from front Oc 5s 17m 5M; B ▽, Or stripe;
vis: 263°-278°.

BROADSTAIRS and NORTH FORELAND
Broadstairs Knoll ⚓ Fl R 2·5s; 51°20'·88N 01°29'·48E.
Pier SE End ⚓ 2 FR (vert) 7m 4M; 51°21'·50N 01°26'·74E.
Elbow ⚓ Q; 51°23'·23N 01°31'·59E.

North Foreland ☆ 51°22'·49N 01°26'·70E; Fl (5) WR 20s 57m
W19M, R16M, R15M; W 8-sided twr; vis: shore-W-150°-R(**16M**)-
181°-R(**15M**)-200°-W-011°; H24; 992351020.

OFFSHORE MARKS
SANDETTIE ⚓ 51°09'·36N 01°47'·12E; Fl 5s 12m **15M**; R hull
with lt twr amidships; *Racon; Horn 30s*; 992351029.

F1 ⚓ Fl (4) Y 15s; 51°11'·21N 01°44'·91E.
South Falls ⚓ Q (6) + L Fl 15s; 51°13'·84N 01°43'·93E.
Sandettie W ▲ Fl (3) G 12s; 51°15'·09N 01°54'·47E.
Mid Falls ⚓ Fl (3) R 10s; 51°18'·63N 01°46'·99E.
Inter Bank⚓ Fl Y 5s; *Bell; Racon (M) 10M*; 51°16'·47N 01°52'·23E.
F2 ⚓ Fl (4) Y 15s; 51°20'·41N 01°56'·19E.

9.3.5 PASSAGE INFORMATION

More passage information is threaded between the harbours
in this area. See 9.0.5 for distances across the eastern part of the
English Channel. Admiralty Leisure Folio 5605 covers Chichester
to Ramsgate including the Dover Strait. **Bibliography:** *Shell
Channel Pilot* (Imray/Cunliffe); *Channel Pilot and Dover Strait
Pilot* (Admiralty NP27 & 28).

THE EASTERN CHANNEL

One of the greatest densities of commercial shipping occurs
in this area. In such waters the greatest danger is a collision
with a larger vessel, especially in poor visibility. Even when
coastal cruising it is essential to know the limits of the TSS
and ITZ. For example, the SW-bound TSS lane from the Dover
Strait passes only 3.5M off Dungeness. Radar surveillance of
the Dover Strait is maintained at all times by the Channel
Navigation Information Service (CNIS). In addition to the
many large ships using the traffic lanes, ferries cross between
English and continental ports; fishing vessels operate both
inshore and offshore; in the summer months many yachts may
be encountered; and static dangers such as lobster pots and
fishing nets are concentrated in certain places.

▶ *Although the rates of tidal streams vary with locality, they
are greatest in the narrower parts of the Channel and off
major headlands. In the Dover Strait spring rates can reach 4kn,
but elsewhere in open water they seldom exceed 2kn.* ◀

N'ly winds, which may give smooth water and pleasant sailing
off the English coast, can cause rough seas on the French coast.
In strong S'lies the English coast between the Isle of Wight and
Dover is very exposed and shelter is hard to find. The Dover
Strait has a funnelling effect, which can increase the wind
speed and cause very rough seas.

ON PASSAGE UP CHANNEL

▶ *There are three tidal gates; the Looe, Beachy Head and
Dungeness. These form a tidal sequence which can enable a
reasonably fast yacht to carry a fair tide from Selsey Bill to Dover.
Based on a mean speed over the ground of 7kn, transit the Looe
at about slack water, HW Dover +0500 (HW Portsmouth +0430);
Beachy Head will be passed at HW Dover +0100; Dungeness at
HW D +0300 and Dover at HW +0530, only stemming the first
of the ebb in the last hour. The down-Channel passage is less
efficient and a stop at Sovereign Harbour or Brighton may be
considered.* ◀

SELSEY BILL AND THE OWERS

(AC 1652) Selsey Bill is a low headland off which lie several
drying rocks and shoals. Just W and SW of Selsey, The Streets
extend 1·25M seaward. 1·25M SSW of the Bill are The Grounds
(or Malt Owers) and The Dries. Inshore1M East of The Dries
1·25M South of the lifeboat house on E side of Selsey Bill is
The Mixon, a group of rocks with a red lit bn at the E end. The
Owers extend 3M to the S and 5M to the SE of the Bill. Looe
Chan is immediately S of these dangers marked by buoys at the
W end, where it is narrowest, between Brake (or Cross) Ledge
to the N and Boulder Bk to the South. It is navigable in good
visibility and suitable weather. Beware of lobster pots.

▶ *The E-going stream begins at HW Portsmouth +0430, and the
W-going at HW Portsmouth –0135, spring rate 2·5kn.*◀

In poor visibility, bad weather or at night keep S of the Owers SCM
lt buoy, 7M SE of Selsey Bill, marking SE end of the Outer Owers.

▶ *Much of the Outer Owers is less than 3m, and large parts virtually
dry. A combination of tidal streams and strong winds produces
heavy breaking seas and overfalls over a large area.*◀

OWERS TO BEACHY HEAD

(AC 1652) The coast from Selsey Bill to Brighton is low, faced
by a shingle beach and with few off-lying dangers. Bognor Rks
(dry in places) extend 1·75M ESE from a point 1M W of the
pier, and Bognor Spit extends another 0·5M further E and S.
Middleton ledge, with rocks running 8ca offshore, about 2M E
of Bognor pier, with depths of less than 1m. Shelley Rks lie 5ca
S of Middleton ledge, with depths of less than 1m.

Winter Knoll, 2·5M SSW of Littlehampton, has depths of 2·1m.
Kingston Rks, depth 2m, lie about 3·25M ESE of Littlehampton.
An unlit outfall bn is 3ca off Goring-by-sea (2M W of Worthing
pier). Grass Banks, an extensive shoal with a wreck swept to
1·2m at W end, lie about 1M S of Worthing pier. Elbow shoal,
with depth of 2·5m, lies E of Grass Banks. A meteorological mast
103m high (Mo(U)W 15s, Horn Mo(U)) is 7M S of Worthing.

Off Shoreham, Church Rks, with depth of 0·3m, lie 1·5M W of
the harbour entrance and 2½ca offshore. Jenny Rks, with depth
0·9m, are 1·25M E of the entrance, 3ca offshore. A diffuser is
marked by a SCM and a wreck swept to 3.3m by a Y SPM.

From Brighton to Beachy Head the high chalk cliffs of the S
Downs are conspicuous. With the exception of outfalls there
are no dangers more than 3ca offshore, until Birling Gap,
where Head Ledge (dries) extends about 4ca S of the cliff foot.
Beachy Head lt ho (FL(2) 20s 31m 8M) is built on this. Beware
of fishing gear markers close inshore.

9.3.6 LITTLEHAMPTON

W. Sussex 50°47′·87N 00°32′·43W ❄♨♨♧♧

CHARTS AC 5605, 1652, 1991; Imray C12, C9, 2100

TIDES +0015 Dover; ML 2·8
Standard Port SHOREHAM (→)

Times				Height (metres)			
High Water		Low Water		MHWS	MHWN	MLWN	MLWS
0500	1000	0000	0600	6·3	4·8	1·9	0·6
1700	2200	1200	1800				
Differences LITTLEHAMPTON (ENT)							
+0010	0000	−0005	−0010	−0·4	−0·4	−0·2	−0·2
ARUNDEL							
ND	+0120	ND	ND	−3·1	−2·8	ND	ND
PAGHAM							
+0015	0000	−0015	−0025	−0·7	−0·5	−0·1	−0·1
BOGNOR REGIS							
+0010	−0005	−0005	−0020	−0·6	−0·5	−0·2	−0·1

NOTE: Tidal hts in hbr are affected by flow down R Arun. Tide seldom falls lower than 0·9m above CD.

SHELTER Good, but ent is hazardous in strong SE winds which cause swell up the hbr. The bar is rough in SW'lies. Visitors can berth at Town Quay, where marked, in front of HM's office.

NAVIGATION WPT 50°47′·53N 00°32′·30W, 346°/0·6M to front ldg lt. Bar 5ca offshore.

- Approach and entrance dry up to 0·9m.
- Shoal bank on West side of channel South of East Pier.
- The ebb runs fast (3–5 kn) at springs.
- From HW−1 to HW+4 a strong W-going tidal stream sets across the ent; keep to E side. Speed limit 6½kn from 15m S of W Pier –Old Town Bridge in Arundel. NO WASH passing moored craft.

On E side of ent chan a training wall which covers at half-tide is marked by 7 poles and lit bn at S end. The W pier is a long, prominent structure of wood piles; beware shoal ground within its arm. A tide gauge on end shows height of tide above CD. To obtain depth on the bar subtract 0·8m from indicated depth.

River Arun. ⚓ and trawling is prohibited where submarine pipelines run across the channel from a position close N of the Lt ho to the first yacht berths, and SE of a retractable footbridge (3·6m clearance MHWS; 9·4m above CD). Sited 3ca above Town Quay, this gives access for masted craft to Littlehampton Marina. It is opened by request to HM before 1630 previous day. The River Arun is navigable on the tide by small, unmasted craft for 24M.

LIGHTS AND MARKS High-rise bldg (38m) is conspic 0·4M NNE of hbr ent. A pile with small platform and ✰, Fl Y (5) 20s 5M, is 2·5M SE of hbr ent at 50°46′·1N 00°29′·5W.

Ldg lts 346°: Front FG on B column; Rear, lt ho Oc W 7·5s at root of E breakwater, 290°-W- 042°. Craft approaching from the W should keep to seaward of the 2m contour until W Pier is positively identified. The Fl G 3s lt at Norfolk Wharf leads craft upstream, once inside hbr ent. The outer ends of some Yacht Club mooring pontoons on the W bank are lit QR.

When Pilot boat with P1 at the bow displays the Pilot flag 'H' or ○ over ● lts, all boats keep clear of entrance; large ship moving.

Footbridge sigs, from high mast to port:
Fl ● = open
Fl ● = bridge moving
○ = closed

Bridge's retractable centre section (22m wide) has 2 FR (vert) to port and 2 FG (vert) to stbd at both upstream and downstream ends.

COMMUNICATIONS (Code 01903) NMOC (02392) 552100; Police 101; Dr 714113. HM www.littlehampton.org.uk 721215, VHF Ch 71 16 (0900-1700LT); Pilots Ch 71 16 when vessel due. Bridge Ch 71. Marinas Ch 80 M (HO).

FACILITIES Town Quay, pontoon connected to shore £20 <8m>£23<11m>£32, ♨ ⛽ ▭ ᵂᶜ ☞ in harbour office. **Services:** ⚓ (pay Hbr dues).

Littlehampton Yacht Club ☎713990, ♨♨▭ www.littlehamptonyachtclub.co.uk.

The Shipyard: www.shipyardlittlehampton. co.uk ☎713327, ⚓ 6▭♥ £1·00, ♨ ⛽ ✂ ✕(Wood) ♨ ᾱ(50t).

Arun YC ☎716016, ♨ ⚓ 92▭+10♥ £14·00/yacht (dries), ⛽♨ ⛽ ✕ ▭.

Marina www.littlehamptonmarina.co.uk ☎713553, ⚓ 120▭ £2·00, ⛽♨♨✂✂ ▭(12t) ♨ ✕ ▭.

Ship and Anchor Marina ☎(01243) 551262, 2M up-river at Ford, ⚓ 50▭+♥♨♨✂♨♨ ✕ ▭.

Town ▭▭ ACA ✉ Ⓑ ♨ ✕ ▭ ⇌ ✈ (Gatwick).

LITTLEHAMPTON

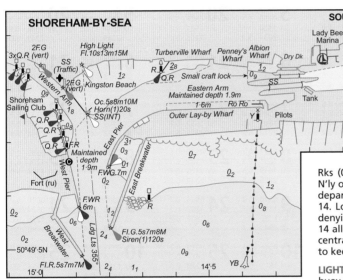

9.3.7 SHOREHAM

W. Sussex 50°49'·53N 00°14'·85W ✿❀⚓⚓✿✿

CHARTS AC 5605, 1652, 2044; Imray C12, C9, 2100

TIDES +0009 Dover; ML 3·3; Duration 0605

Standard Port SHOREHAM (→)

Times				Height (metres)			
High Water		Low Water		MHWS	MHWN	MLWN	MLWS
0500	1000	0000	0600	6·3	4·8	1·9	0·6
1700	2200	1200	1800				
Differences WORTHING							
+0010	0000	−0005	−0010	−0·1	−0·2	0·0	0·0

SHELTER Excellent, once through the lock and into The Canal. The shallow water (dredged 1·9m) at the ent can be very rough in strong on-shore winds and dangerous in onshore gales. Lady Bee and Aldrington (E end of The Canal) marinas, least depth 2m, both welcome visitors. They are managed by the Hbr Authority. Visitors are advised not to use the drying Western Arm if possible. Hbr speed limit = 4kn.

NAVIGATION WPT 50°49'·23N 00°14'·81W, 355°/0·52M to front ldg lt. From E beware Jenny Rks (0·9m) and, from the W, Church

Rks (0·3m). The Eastern Arm leads to Prince George Lock, the N'ly of two locks, which opens H24 @ H+30 for arrivals and H for departures; also, if not busy, at other times on request VHF Ch 14. Lock ent is 5·5m wide, pontoon inside on S wall; sill at CD, denying access only at about LWS±1. Lock will advise depth on Ch 14 allowing a safety margin of 0·2m. Waiting space on S side of central lead-in pier. Commercial ships use Prince Philip lock; yachts to keep clear. A Ro-Ro ramp and terminal are on S side of lock.

LIGHTS AND MARKS Radio mast is conspic 170m N of High lt. SCM lit buoy, marks outfall diffusers, 157°/1·8M from hbr ent. Chimney 100m high, silver with blk top. 3F.R(hor) at top and 3F.R(hor) mid height. SYC yacht race signal mast on Shoreham Beach 00°16'·06W.

Ldg lts 355°: front Oc 5s 8m 10M; rear High lt, Fl 10s 13m 15M.

Traffic Sigs IPTS (Sigs 2 and 5, Oc) are shown from Middle Pier. Note : ● Fl lt exempts small craft.
Oc R 3s (from LB ho, directed at E or W Arms) = No exit.

Lock Sigs (Comply strictly to avoid turbulence):
3 ● (vert) = do not approach lock.
● ○ ● (vert) = clear to approach lock.

COMMUNICATIONS NMOC (02392) 552100; Police 101; Ⓗ (01903) 205111; (Code 01273) Dr 461101. HM 598100; Locks 592366; HM and lock call *Shoreham Hbr Radio* VHF Ch **14** 16 (H24). Lock advises Marina of arrivals, 0830-1800 Mon-Sat; 1000-1400 Sun.

FACILITIES 🛥 🛶 on N side of W Arm HW±3; ACA.

Lady Bee Marina ☎596880 (0900-1700), mob 07802 848915 (best to book), access as lock times, ⚓110🛥 +10 Ⓥ £25/craft<12m inc ⌁ 🏧 🏧 ⚒ ⚓ 🅿 ⚓ ⚓ ✕. FV at Nicholson's Marina ☎598100.

Sussex YC ☎464868, welcomes visitors, on a drying ½ tide pontoon in the Western Arm (limited Ⓥ) but with only 1🛥 in the canal, so prior notice advised; also, ⚓ ✕ 🍺 .

Town ✉ Ⓑ 🚮 🅾 ⇌ ✈ (Gatwick).

SHOREHAM LAT 50°50'N LONG 0°15'W
TIMES AND HEIGHTS OF HIGH AND LOW WATERS

STANDARD TIME (UT)
For Summer Time add ONE hour in **non-shaded areas**

Dates in red are SPRINGS
Dates in blue are NEAPS

YEAR 2016

JANUARY

Day	Time m	Time m	Time m	Time m
1	0325 5.5	0929 1.6	F 1539 5.2	2153 1.6
16	0312 6.0	0935 1.1	SA 1535 5.8	☽ 2203 1.1
2	0412 5.3	1019 1.8	SA 1633 4.9	☽ 2245 1.9
17	0408 5.8	1037 1.3	SU 1636 5.5	2307 1.4
3	0508 5.0	1117 2.0	SU 1736 4.8	2345 2.0
18	0514 5.5	1148 1.5	M 1751 5.2	
4	0610 4.9	1224 2.0	M 1840 4.8	
19	0021 1.5	0632 5.4	TU 1303 1.5	1915 5.2
5	0056 2.0	0711 5.0	TU 1340 1.9	1941 4.9
20	0136 1.5	0751 5.4	W 1412 1.4	2031 5.4
6	0209 1.9	0809 5.3	W 1442 1.7	2037 5.2
21	0241 1.4	0858 5.7	TH 1512 1.1	2133 5.7
7	0304 1.6	0900 5.5	TH 1530 1.4	2126 5.5
22	0337 1.1	0954 5.9	F 1604 0.9	2225 6.0
8	0350 1.4	0945 5.8	F 1612 1.1	2211 5.8
23	0426 0.9	1043 6.1	SA 1650 0.7	2312 6.2
9	0431 1.2	1027 6.1	SA 1653 0.9	2253 6.1
24	0511 0.8	1127 6.2	SU 1734 0.6	○ 2354 6.3
10	0510 1.0	1108 6.2	SU 1732 0.8	● 2334 6.2
25	0552 0.8	1208 6.3	M 1814 0.6	
11	0549 0.9	1148 6.3	M 1812 0.7	
26	0033 6.3	0631 0.8	TU 1245 6.2	1851 0.7
12	0014 6.3	0629 0.8	TU 1230 6.4	1852 0.6
27	0110 6.2	0706 0.9	W 1319 6.1	1926 0.8
13	0056 6.4	0711 0.7	W 1312 6.3	1935 0.6
28	0142 6.1	0740 1.0	TH 1349 5.9	2000 0.9
14	0139 6.3	0755 0.8	TH 1355 6.2	2019 0.7
29	0212 5.9	0815 1.1	F 1418 5.7	2035 1.1
15	0223 6.2	0842 0.9	F 1442 6.0	2108 0.8
30	0239 5.7	0852 1.3	SA 1448 5.5	2113 1.4
31	0309 5.4	0933 1.6	SU 1525 5.2	2155 1.6

FEBRUARY

Day	Time m	Time m	Time m	Time m
1	0350 5.2	1022 1.8	M 1615 4.9	◐ 2247 1.9
16	0445 5.4	1119 1.5	TU 1722 5.1	2355 1.6
2	0452 4.9	1123 2.0	TU 1740 4.7	2354 2.1
17	0601 5.1	1240 1.7	W 1851 5.0	
3	0620 4.8	1235 2.1	W 1857 4.7	
18	0117 1.7	0731 5.1	TH 1356 1.6	2018 5.1
4	0112 2.1	0729 4.9	TH 1354 1.9	2001 5.0
19	0227 1.6	0848 5.3	F 1459 1.3	2123 5.5
5	0227 1.8	0828 5.3	F 1458 1.5	2058 5.4
20	0325 1.3	0945 5.7	SA 1550 1.0	2214 5.8
6	0323 1.5	0920 5.6	SA 1547 1.2	2148 5.7
21	0413 1.0	1032 5.9	SU 1635 0.8	2258 6.1
7	0408 1.1	1006 6.0	SU 1630 0.8	2233 6.1
22	0455 0.8	1114 6.1	M 1715 0.7	○ 2337 6.2
8	0450 0.9	1050 6.3	M 1712 0.6	● 2317 6.3
23	0534 0.7	1152 6.2	TU 1753 0.6	
9	0531 0.7	1134 6.5	TU 1754 0.4	
24	0014 6.3	0609 0.7	W 1226 6.2	1827 0.6
10	0000 6.5	0613 0.5	W 1217 6.6	1836 0.3
25	0047 6.2	0642 0.7	TH 1256 6.1	1900 0.7
11	0043 6.6	0656 0.4	TH 1300 6.6	1920 0.3
26	0114 6.1	0714 0.8	F 1320 6.0	1932 0.8
12	0126 6.6	0741 0.4	F 1343 6.5	2004 0.4
27	0136 6.0	0746 0.9	SA 1344 5.8	2004 1.0
13	0208 6.5	0826 0.6	SA 1428 6.3	2051 0.6
28	0158 5.8	0817 1.1	SU 1411 5.7	2035 1.2
14	0254 6.2	0915 0.8	SU 1516 6.0	2141 0.9
29	0227 5.6	0850 1.3	M 1444 5.4	2107 1.5
15	0344 5.9	1011 1.1	M 1612 5.5	◐ 2240 1.3

MARCH

Day	Time m	Time m	Time m	Time m
1	0303 5.3	0929 1.6	TU 1526 5.1	◐ 2150 1.8
16	0421 5.4	1052 1.5	W 1701 5.1	2331 1.7
2	0350 5.0	1022 1.9	W 1623 4.8	2255 2.1
17	0534 5.0	1218 1.8	TH 1828 4.9	
3	0501 4.7	1139 2.0	TH 1811 4.6	
18	0059 1.9	0709 4.8	F 1338 1.7	2001 5.0
4	0023 2.1	0650 4.7	F 1307 1.9	1928 4.8
19	0211 1.7	0832 5.1	SA 1440 1.4	2106 5.4
5	0149 1.9	0758 5.1	SA 1424 1.6	2031 5.3
20	0307 1.4	0928 5.4	SU 1531 1.1	2155 5.7
6	0254 1.5	0855 5.5	SU 1519 1.2	2124 5.7
21	0354 1.1	1013 5.8	M 1614 0.9	2236 6.0
7	0343 1.1	0945 6.0	M 1605 0.8	2212 6.2
22	0434 0.9	1053 6.1	TU 1652 0.8	2314 6.1
8	0427 0.7	1031 6.3	TU 1649 0.5	2258 6.4
23	0510 0.8	1130 6.1	W 1728 0.7	○ 2349 6.2
9	0510 0.4	1117 6.5	W 1732 0.3	● 2343 6.7
24	0545 0.7	1202 6.1	TH 1802 0.7	
10	0554 0.3	1202 6.7	TH 1817 0.1	
25	0019 6.2	0617 0.7	F 1229 6.0	1834 0.7
11	0026 6.8	0638 0.2	F 1246 6.7	1901 0.1
26	0043 6.1	0648 0.8	SA 1252 6.0	1905 0.8
12	0109 6.8	0723 0.2	SA 1330 6.6	1946 0.2
27	0103 6.0	0719 0.8	SU 1316 5.9	1935 0.9
13	0152 6.6	0808 0.4	SU 1413 6.4	2031 0.5
28	0127 5.9	0747 1.0	M 1344 5.8	2003 1.1
14	0235 6.3	0855 0.7	M 1500 6.0	2120 0.9
29	0157 5.7	0817 1.1	TU 1416 5.6	2034 1.3
15	0323 5.9	0948 1.1	TU 1554 5.6	◐ 2217 1.3
30	0232 5.5	0854 1.4	W 1456 5.3	2117 1.6
31	0316 5.1	0944 1.7	TH 1549 5.0	◐ 2216 1.9

APRIL

Day	Time m	Time m	Time m	Time m
1	0417 4.8	1053 1.9	F 1716 4.7	2340 2.1
16	0034 1.9	0635 4.7	SA 1311 1.7	1930 5.0
2	0608 4.7	1223 1.9	SA 1856 4.9	
17	0145 1.7	0803 4.9	SU 1413 1.5	2036 5.3
3	0111 1.9	0727 5.0	SU 1347 1.6	2002 5.3
18	0241 1.5	0901 5.2	M 1503 1.3	2126 5.6
4	0222 1.4	0828 5.5	M 1448 1.1	2058 5.8
19	0327 1.2	0947 5.5	TU 1546 1.1	2208 5.8
5	0315 1.0	0921 5.9	TU 1538 0.7	2149 6.2
20	0407 1.0	1027 5.8	W 1624 0.9	2245 6.0
6	0402 0.6	1010 6.3	W 1623 0.4	2236 6.5
21	0444 0.9	1102 5.9	TH 1700 0.8	2318 6.1
7	0447 0.3	1057 6.6	TH 1709 0.2	● 2322 6.7
22	0518 0.8	1133 6.0	F 1735 0.8	○ 2346 6.1
8	0532 0.2	1144 6.7	F 1755 0.1	
23	0552 0.8	1200 6.0	SA 1808 0.8	
9	0007 6.8	0618 0.1	SA 1231 6.7	1840 0.1
24	0010 6.0	0624 0.8	SU 1226 6.0	1840 0.9
10	0051 6.8	0704 0.2	SU 1316 6.6	1926 0.3
25	0035 6.0	0655 0.8	M 1254 5.9	1910 1.0
11	0135 6.6	0750 0.3	M 1400 6.4	2012 0.5
26	0103 5.9	0724 0.9	TU 1324 5.8	1939 1.1
12	0219 6.3	0837 0.6	TU 1447 6.0	2101 0.9
27	0135 5.8	0755 1.1	W 1358 5.7	2013 1.3
13	0306 5.8	0927 1.0	W 1540 5.6	2155 1.3
28	0211 5.6	0833 1.2	TH 1438 5.5	2057 1.5
14	0401 5.3	1028 1.5	TH 1642 5.2	◑ 2306 1.7
29	0256 5.3	0922 1.5	F 1530 5.2	2155 1.7
15	0508 4.9	1151 1.7	F 1757 4.9	
30	0354 5.0	1026 1.7	SA 1643 5.0	◑ 2310 1.8

Chart Datum: 3·27 metres below Ordnance Datum (Newlyn). HAT is 6·9 metres above Chart Datum.

»» FREE monthly updates. Register at ««
www.reedsnauticalalmanac.co.uk

SHOREHAM LAT 50°50'N LONG 0°15'W

TIMES AND HEIGHTS OF HIGH AND LOW WATERS

STANDARD TIME (UT)
For Summer Time add ONE hour in **non-shaded areas**

Dates in red are **SPRINGS**
Dates in blue are NEAPS

YEAR 2016

MAY

Time	m	Time	m
1 SU 0523 / 1146 / 1819	4.8 / 1.7 / 5.1	**16** M 0108 / 0712 / 1335 / 1951	1.8 / 4.8 / 1.7 / 5.1
2 M 0035 / 0653 / 1309 / 1930	1.7 / 5.0 / 1.5 / 5.4	**17** TU 0206 / 0829 / 1427 / 2046	1.6 / 5.0 / 1.5 / 5.4
3 TU 0148 / 0759 / 1415 / 2030	1.4 / 5.4 / 1.1 / 5.8	**18** W 0254 / 0911 / 1513 / 2131	1.4 / 5.3 / 1.3 / 5.6
4 W 0246 / 0855 / 1509 / 2123	1.0 / 5.9 / 0.7 / 6.2	**19** TH 0336 / 0952 / 1553 / 2209	1.2 / 5.5 / 1.2 / 5.8
5 TH 0336 / 0947 / 1558 / 2212	0.6 / 6.2 / 0.5 / 6.5	**20** F 0415 / 1027 / 1631 / 2242	1.0 / 5.7 / 1.0 / 5.9
6 F 0424 / 1037 / 1646 / ●2300	0.3 / 6.5 / 0.3 / 6.7	**21** SA 0451 / 1059 / 1708 / ○2312	0.9 / 5.8 / 1.0 / 5.9
7 SA 0511 / 1126 / 1734 / 2347	0.2 / 6.6 / 0.2 / 6.7	**22** SU 0527 / 1131 / 1744 / 2342	0.9 / 5.9 / 1.0 / 6.0
8 SU 0559 / 1214 / 1821	0.2 / 6.6 / 0.3	**23** M 0601 / 1203 / 1818	0.8 / 5.9 / 1.0
9 M 0033 / 0645 / 1302 / 1908	6.7 / 0.2 / 6.6 / 0.4	**24** TU 0012 / 0634 / 1235 / 1851	6.0 / 0.9 / 5.9 / 1.0
10 TU 0119 / 0732 / 1348 / 1955	6.5 / 0.4 / 6.4 / 0.6	**25** W 0045 / 0706 / 1309 / 1923	5.9 / 0.9 / 5.9 / 1.1
11 W 0204 / 0819 / 1435 / 2042	6.2 / 0.7 / 6.1 / 0.9	**26** TH 0119 / 0740 / 1345 / 2000	5.8 / 1.0 / 5.8 / 1.2
12 TH 0250 / 0908 / 1524 / 2134	5.8 / 1.0 / 5.7 / 1.3	**27** F 0158 / 0820 / 1428 / 2044	5.7 / 1.1 / 5.7 / 1.3
13 F 0342 / 1003 / 1619 / ◐2235	5.4 / 1.4 / 5.4 / 1.6	**28** SA 0244 / 0908 / 1518 / 2139	5.5 / 1.2 / 5.5 / 1.5
14 SA 0440 / 1111 / 1721 / 2355	5.0 / 1.6 / 5.1 / 1.8	**29** SU 0339 / 1007 / 1621 / ◐2246	5.3 / 1.4 / 5.3 / 1.6
15 SU 0548 / 1230 / 1835	4.8 / 1.8 / 5.0	**30** M 0450 / 1117 / 1739	5.1 / 1.5 / 5.3
		31 TU 0002 / 0614 / 1235 / 1854	1.5 / 5.2 / 1.4 / 5.5

JUNE

Time	m	Time	m
1 W 0115 / 0727 / 1344 / 2000	1.3 / 5.4 / 1.1 / 5.8	**16** TH 0215 / 0812 / 1436 / 2038	1.6 / 5.0 / 1.6 / 5.3
2 TH 0218 / 0829 / 1443 / 2058	1.0 / 5.7 / 0.9 / 6.1	**17** F 0302 / 0904 / 1522 / 2124	1.4 / 5.2 / 1.4 / 5.5
3 F 0313 / 0926 / 1537 / 2151	0.7 / 6.0 / 0.6 / 6.3	**18** SA 0345 / 0947 / 1603 / 2203	1.2 / 5.5 / 1.3 / 5.7
4 SA 0404 / 1020 / 1627 / 2242	0.5 / 6.3 / 0.5 / 6.5	**19** SU 0424 / 1026 / 1643 / 2241	1.1 / 5.7 / 1.1 / 5.8
5 SU 0453 / 1111 / 1717 / ●2331	0.4 / 6.4 / 0.4 / 6.5	**20** M 0502 / 1107 / 1721 / ○2317	1.0 / 5.8 / 1.0 / 5.9
6 M 0542 / 1201 / 1805	0.3 / 6.5 / 0.4	**21** TU 0540 / 1141 / 1758 / 2353	0.9 / 5.9 / 1.0 / 6.0
7 TU 0018 / 0629 / 1249 / 1851	6.5 / 0.4 / 6.5 / 0.5	**22** W 0616 / 1218 / 1834	0.8 / 6.0 / 1.0
8 W 0104 / 0716 / 1335 / 1937	6.4 / 0.5 / 6.3 / 0.7	**23** TH 0029 / 0652 / 1255 / 1910	6.0 / 0.8 / 6.0 / 1.0
9 TH 0149 / 0802 / 1419 / 2023	6.1 / 0.7 / 6.1 / 0.9	**24** F 0107 / 0729 / 1334 / 1949	6.0 / 0.8 / 6.0 / 1.0
10 F 0233 / 0847 / 1504 / 2109	5.8 / 0.9 / 5.9 / 1.2	**25** SA 0148 / 0809 / 1417 / 2033	5.9 / 0.9 / 5.9 / 1.1
11 SA 0319 / 0933 / 1551 / 2158	5.5 / 1.2 / 5.6 / 1.5	**26** SU 0233 / 0855 / 1505 / 2124	5.8 / 1.0 / 5.8 / 1.2
12 SU 0408 / 1024 / 1642 / ◐2255	5.2 / 1.5 / 5.3 / 1.7	**27** M 0324 / 0949 / 1600 / ◐2224	5.6 / 1.1 / 5.7 / 1.3
13 M 0503 / 1124 / 1738	4.9 / 1.7 / 5.1	**28** TU 0425 / 1053 / 1704 / 2334	5.4 / 1.3 / 5.5 / 1.4
14 TU 0006 / 0603 / 1236 / 1838	1.8 / 4.8 / 1.8 / 5.0	**29** W 0537 / 1205 / 1818	5.3 / 1.5 / 5.5
15 W 0118 / 0708 / 1342 / 1942	1.8 / 4.8 / 1.8 / 5.1	**30** TH 0046 / 0655 / 1318 / 1931	1.3 / 5.3 / 1.3 / 5.6

JULY

Time	m	Time	m
1 F 0154 / 0807 / 1423 / 2037	1.2 / 5.5 / 1.1 / 5.8	**16** SA 0224 / 0817 / 1448 / 2042	1.7 / 5.0 / 1.7 / 5.3
2 SA 0254 / 0911 / 1521 / 2136	0.9 / 5.8 / 0.9 / 6.1	**17** SU 0315 / 0911 / 1536 / 2131	1.5 / 5.3 / 1.5 / 5.6
3 SU 0349 / 1008 / 1610 / 2229	0.7 / 6.0 / 0.7 / 6.2	**18** M 0358 / 0958 / 1619 / 2214	1.2 / 5.6 / 1.2 / 5.8
4 M 0439 / 1100 / 1703 / ●2319	0.6 / 6.2 / 0.6 / 6.3	**19** TU 0439 / 1040 / 1659 / ○2255	1.0 / 5.9 / 1.1 / 6.0
5 TU 0528 / 1149 / 1750	0.5 / 6.4 / 0.6	**20** W 0518 / 1121 / 1738 / 2334	0.8 / 6.0 / 0.9 / 6.1
6 W 0006 / 0614 / 1236 / 1835	6.3 / 0.5 / 6.4 / 0.7	**21** TH 0557 / 1201 / 1816	0.7 / 6.2 / 0.8
7 TH 0050 / 0658 / 1319 / 1918	6.2 / 0.6 / 6.3 / 0.8	**22** F 0014 / 0635 / 1241 / 1855	6.2 / 0.7 / 6.3 / 0.8
8 F 0132 / 0740 / 1400 / 2000	6.1 / 0.7 / 6.2 / 0.8	**23** SA 0054 / 0715 / 1321 / 1936	6.2 / 0.6 / 6.3 / 0.8
9 SA 0211 / 0820 / 1439 / 2039	5.9 / 0.9 / 6.0 / 1.1	**24** SU 0135 / 0756 / 1403 / 2020	6.2 / 0.7 / 6.2 / 0.8
10 SU 0250 / 0859 / 1518 / 2120	5.6 / 1.1 / 5.7 / 1.3	**25** M 0219 / 0841 / 1448 / 2107	6.1 / 0.8 / 6.1 / 0.9
11 M 0331 / 0941 / 1601 / 2204	5.3 / 1.4 / 5.4 / 1.6	**26** TU 0307 / 0931 / 1538 / ◐2202	5.9 / 0.9 / 5.9 / 1.1
12 TU 0418 / 1027 / 1649 / ◐2256	5.1 / 1.6 / 5.2 / 1.8	**27** W 0402 / 1030 / 1637 / 2308	5.6 / 1.2 / 5.6 / 1.3
13 W 0513 / 1122 / 1745 / 2358	4.8 / 1.9 / 5.0 / 1.9	**28** TH 0509 / 1140 / 1748	5.3 / 1.4 / 5.4
14 TH 0614 / 1228 / 1846	4.7 / 2.0 / 4.9	**29** F 0022 / 0630 / 1257 / 1908	1.5 / 5.2 / 1.5 / 5.4
15 F 0113 / 0717 / 1346 / 1946	1.9 / 4.8 / 1.9 / 5.1	**30** SA 0137 / 0753 / 1409 / 2024	1.4 / 5.3 / 1.4 / 5.5
		31 SU 0242 / 0903 / 1510 / 2127	1.2 / 5.6 / 1.2 / 5.8

AUGUST

Time	m	Time	m
1 M 0338 / 1001 / 1603 / 2221	0.9 / 5.9 / 0.9 / 6.1	**16** TU 0333 / 0932 / 1554 / 2150	1.3 / 5.6 / 1.3 / 5.8
2 TU 0428 / 1051 / 1651 / ●2309	0.7 / 6.2 / 0.8 / 6.2	**17** W 0415 / 1017 / 1635 / 2233	1.0 / 6.0 / 1.0 / 6.1
3 W 0514 / 1137 / 1735 / 2353	0.6 / 6.3 / 0.7 / 6.3	**18** TH 0455 / 1100 / 1715 / ○2315	0.8 / 6.2 / 0.8 / 6.3
4 TH 0557 / 1219 / 1817	0.6 / 6.4 / 0.7	**19** F 0535 / 1141 / 1756 / 2357	0.6 / 6.4 / 0.7 / 6.4
5 F 0033 / 0637 / 1259 / 1856	6.2 / 0.6 / 6.3 / 0.8	**20** SA 0615 / 1222 / 1837	0.5 / 6.5 / 0.6
6 SA 0111 / 0714 / 1335 / 1932	6.1 / 0.7 / 6.2 / 0.9	**21** SU 0038 / 0656 / 1304 / 1919	6.5 / 0.4 / 6.5 / 0.6
7 SU 0145 / 0750 / 1409 / 2006	6.0 / 0.9 / 6.0 / 1.0	**22** M 0120 / 0739 / 1346 / 2003	6.4 / 0.5 / 6.5 / 0.6
8 M 0216 / 0824 / 1440 / 2042	5.8 / 1.0 / 5.8 / 1.2	**23** TU 0203 / 0824 / 1429 / 2049	6.3 / 0.6 / 6.3 / 0.6
9 TU 0247 / 0901 / 1510 / 2121	5.5 / 1.3 / 5.5 / 1.5	**24** W 0249 / 0912 / 1517 / 2141	6.0 / 0.9 / 6.0 / 1.1
10 W 0322 / 0942 / 1547 / ◐2207	5.2 / 1.6 / 5.2 / 1.7	**25** TH 0342 / 1009 / 1614 / ◐2244	5.7 / 1.2 / 5.6 / 1.4
11 TH 0409 / 1031 / 1643 / 2303	4.9 / 1.9 / 5.0 / 1.9	**26** F 0448 / 1119 / 1725	5.3 / 1.6 / 5.3
12 F 0523 / 1133 / 1758	4.7 / 2.1 / 4.8	**27** SA 0002 / 0613 / 1242 / 1852	1.7 / 5.0 / 1.7 / 5.1
13 SA 0012 / 0635 / 1249 / 1907	2.1 / 4.7 / 2.1 / 4.9	**28** SU 0124 / 0745 / 1358 / 2017	1.6 / 5.1 / 1.6 / 5.3
14 SU 0134 / 0742 / 1411 / 2009	2.0 / 4.9 / 1.9 / 5.1	**29** M 0232 / 0857 / 1500 / 2120	1.4 / 5.5 / 1.4 / 5.6
15 M 0244 / 0850 / 1509 / 2103	1.6 / 5.2 / 1.6 / 5.5	**30** TU 0327 / 0952 / 1551 / 2211	1.1 / 5.8 / 1.1 / 6.0
		31 W 0414 / 1038 / 1635 / 2255	0.9 / 6.2 / 0.9 / 6.2

Chart Datum: 3·27 metres below Ordnance Datum (Newlyn). HAT is 6·9 metres above Chart Datum.

SE England

》》FREE monthly updates. Register at 《
www.reedsnauticalalmanac.co.uk

255

STANDARD TIME (UT)
For Summer Time add ONE hour in **non-shaded areas**

SHOREHAM LAT 50°50′N LONG 0°15′W
TIMES AND HEIGHTS OF HIGH AND LOW WATERS

Dates in red are **SPRINGS**
Dates in blue are NEAPS

YEAR 2016

SEPTEMBER

Day	Time m	Time m	Time m	Time m
1 TH ●	0456 0.7	1120 6.3	1716 0.8	2335 6.3
16 F ○	0429 0.7	1036 6.4	1650 0.7	2254 6.5
2 F	0535 0.7	1158 6.4	1754 0.7	
17 SA	0510 0.5	1119 6.6	1732 0.5	2337 6.6
3 SA	0013 6.2	0611 0.7	1234 6.3	1829 0.8
18 SU	0552 0.4	1201 6.7	1815 0.4	
4 SU	0046 6.2	0646 0.8	1306 6.2	1902 0.9
19 M	0020 6.7	0634 0.3	1244 6.7	1858 0.4
5 M	0114 6.0	0718 0.9	1333 6.1	1934 1.0
20 TU	0103 6.6	0719 0.4	1326 6.6	1943 0.5
6 TU	0139 5.9	0750 1.0	1355 5.9	2007 1.1
21 W	0146 6.4	0805 0.6	1410 6.4	2030 0.8
7 W	0204 5.7	0823 1.2	1420 5.7	2041 1.4
22 TH	0233 6.1	0853 0.9	1458 6.0	2122 1.1
8 TH	0234 5.4	0859 1.5	1453 5.4	2120 1.7
23 F ◐	0326 5.7	0949 1.3	1555 5.6	2223 1.5
9 F ◑	0313 5.1	0942 1.9	1537 5.0	2211 2.0
24 SA	0433 5.3	1100 1.7	1707 5.1	2344 1.8
10 SA	0408 4.8	1043 2.1	1649 4.7	2322 2.2
25 SU	0558 5.0	1227 1.9	1838 5.0	
11 SU	0555 4.6	1204 2.3	1831 4.7	
26 M	0109 1.8	0732 5.1	1344 1.8	2005 5.2
12 M	0047 2.1	0710 4.8	1331 2.1	1939 5.0
27 TU	0217 1.6	0842 5.5	1445 1.5	2106 5.5
13 TU	0209 1.8	0812 5.2	1440 1.7	2036 5.4
28 W	0311 1.3	0933 5.8	1533 1.2	2153 5.9
14 W	0304 1.4	0906 5.7	1528 1.3	2125 5.9
29 TH	0355 1.0	1017 6.1	1615 0.9	2235 6.1
15 TH	0348 1.0	0952 6.1	1610 0.9	2210 6.2
30 F	0434 0.8	1056 6.3	1653 0.8	2313 6.2

OCTOBER

Day	Time m	Time m	Time m	Time m
1 SA ●	0510 0.8	1132 6.4	1729 0.8	2348 6.2
16 SU ○	0444 0.5	1055 6.8	1707 0.4	2316 6.7
2 SU	0545 0.8	1205 6.3	1803 0.8	
17 M	0528 0.3	1139 6.9	1752 0.3	
3 M	0017 6.2	0617 0.8	1233 6.2	1834 0.9
18 TU	0001 6.8	0612 0.3	1223 6.8	1837 0.3
4 TU	0042 6.1	0649 0.9	1255 6.1	1905 1.0
19 W	0047 6.7	0659 0.4	1307 6.7	1924 0.5
5 W	0106 6.0	0720 1.1	1318 5.9	1936 1.1
20 TH	0132 6.5	0746 0.7	1352 6.4	2012 0.8
6 TH	0132 5.8	0751 1.3	1344 5.8	2007 1.3
21 F	0220 6.2	0835 1.0	1441 6.0	2103 1.1
7 F	0202 5.6	0823 1.5	1417 5.5	2042 1.6
22 SA ◑	0314 5.8	0931 1.4	1538 5.5	2203 1.5
8 SA	0238 5.3	0903 1.8	1458 5.2	2129 1.9
23 SU	0418 5.4	1039 1.8	1646 5.1	2321 1.8
9 SU ◑	0327 5.0	1001 2.1	1556 4.8	2235 2.1
24 M	0535 5.1	1205 2.0	1811 4.9	
10 M	0455 4.7	1121 2.3	1751 4.7	
25 TU	0044 1.9	0704 5.1	1321 1.8	1938 5.1
11 TU	0002 2.1	0635 4.9	1250 2.1	1907 5.0
26 W	0151 1.7	0814 5.4	1420 1.6	2040 5.4
12 W	0127 1.9	0741 5.3	1404 1.7	2007 5.4
27 TH	0244 1.4	0905 5.7	1508 1.3	2128 5.7
13 TH	0230 1.4	0837 5.7	1456 1.3	2059 5.8
28 F	0329 1.2	0949 6.0	1550 1.1	2209 5.9
14 F	0318 1.0	0925 6.2	1541 0.9	2146 6.3
29 SA	0408 1.0	1027 6.2	1627 0.9	2246 6.1
15 SA	0401 0.7	1010 6.5	1624 0.6	2231 6.6
30 SU ●	0444 0.9	1103 6.2	1703 0.9	2319 6.1
31 M	0519 0.9	1133 6.2	1737 0.9	2348 6.1

NOVEMBER

Day	Time m	Time m	Time m	Time m
1 TU	0552 0.9	1159 6.2	1810 0.9	
16 W	0554 0.4	1204 6.8	1820 0.4	
2 W	0013 6.1	0624 1.0	1224 6.1	1841 1.0
17 TH	0032 6.7	0641 0.5	1251 6.6	1907 0.5
3 TH	0040 6.0	0655 1.1	1250 6.0	1912 1.1
18 F	0120 6.5	0729 0.7	1338 6.4	1956 0.7
4 F	0109 5.9	0725 1.3	1319 5.8	1942 1.3
19 SA	0209 6.3	0818 1.0	1426 6.0	2046 1.0
5 SA	0139 5.8	0758 1.5	1353 5.6	2017 1.5
20 SU	0300 5.9	0911 1.3	1520 5.6	2141 1.4
6 SU	0216 5.5	0838 1.7	1434 5.3	2102 1.7
21 M ◐	0357 5.6	1012 1.7	1620 5.2	2247 1.7
7 M ◑	0304 5.3	0932 1.9	1527 5.0	2201 1.9
22 TU	0501 5.3	1128 1.9	1729 5.0	
8 TU	0410 5.0	1043 2.1	1649 4.8	2317 2.0
23 W	0004 1.9	0614 5.2	1244 1.9	1850 5.0
9 W	0552 5.0	1206 2.0	1829 5.0	
24 TH	0113 1.8	0730 5.3	1345 1.7	2001 5.1
10 TH	0040 1.8	0705 5.3	1322 1.7	1934 5.4
25 F	0209 1.7	0828 5.5	1436 1.5	2054 5.4
11 F	0150 1.5	0805 5.7	1421 1.3	2030 5.8
26 SA	0257 1.5	0915 5.7	1521 1.3	2138 5.6
12 SA	0245 1.1	0857 6.2	1512 0.9	2121 6.2
27 SU	0340 1.3	0955 5.9	1601 1.1	2216 5.8
13 SU	0334 0.8	0945 6.5	1559 0.6	2209 6.5
28 M	0418 1.2	1030 6.0	1638 1.0	2249 6.0
14 M ○	0420 0.5	1031 6.7	1645 0.4	2257 6.7
29 TU ●	0454 1.1	1101 6.1	1714 1.0	2320 6.0
15 TU	0507 0.4	1118 6.8	1732 0.3	2345 6.7
30 W	0529 1.1	1130 6.1	1748 0.9	2350 6.1

DECEMBER

Day	Time m	Time m	Time m	Time m
1 TH	0604 1.1	1159 6.1	1822 1.0	
16 F	0021 6.6	0626 0.6	1238 6.6	1853 0.5
2 F	0020 6.0	0636 1.1	1230 6.0	1854 1.0
17 SA	0110 6.5	0716 0.7	1325 6.4	1940 0.6
3 SA	0052 6.0	0707 1.2	1302 5.9	1926 1.1
18 SU	0156 6.3	0801 0.9	1411 6.1	2027 0.9
4 SU	0125 5.9	0741 1.3	1337 5.8	2000 1.3
19 M	0243 6.1	0849 1.2	1458 5.8	2115 1.2
5 M	0203 5.7	0820 1.5	1418 5.6	2043 1.4
20 TU	0331 5.8	0939 1.5	1548 5.4	2205 1.5
6 TU	0247 5.6	0909 1.6	1508 5.3	2135 1.6
21 W ◐	0422 5.5	1035 1.7	1643 5.1	2303 1.7
7 W ◑	0343 5.4	1010 1.8	1610 5.1	2239 1.7
22 TH	0518 5.2	1144 1.9	1743 4.9	
8 TH	0456 5.3	1124 1.8	1734 5.1	2356 1.7
23 F	0014 1.9	0619 5.1	1258 1.9	1850 4.9
9 F	0620 5.4	1240 1.6	1855 5.3	
24 SA	0124 1.9	0727 5.2	1358 1.8	2002 5.0
10 SA	0110 1.5	0729 5.6	1347 1.3	1959 5.6
25 SU	0221 1.8	0828 5.3	1450 1.6	2058 5.4
11 SU	0214 1.2	0828 6.0	1445 1.0	2057 6.0
26 M	0310 1.6	0916 5.6	1534 1.4	2141 5.5
12 M	0310 0.9	0922 6.3	1538 0.7	2150 6.3
27 TU	0353 1.4	0956 5.8	1615 1.2	2219 5.7
13 TU	0401 0.7	1012 6.6	1628 0.5	2242 6.5
28 W	0432 1.2	1031 5.9	1653 1.0	2254 5.9
14 W ○	0450 0.6	1102 6.7	1717 0.4	2332 6.6
29 TH ●	0509 1.1	1106 6.0	1729 1.0	2329 6.0
15 TH	0539 0.5	1150 6.7	1805 0.4	
30 F	0545 1.1	1139 6.1	1805 0.9	
31 SA	0003 6.1	0619 1.1	1214 6.1	1839 0.9

Chart Datum: 3·27 metres below Ordnance Datum (Newlyn). HAT is 6·9 metres above Chart Datum.

》》FREE monthly updates. Register at 〈 www.reedsnauticalalmanac.co.uk 〈

9.3.8 BRIGHTON
E. Sussex **50°48'·53N 00°06'·38W** ✿✿✿✿▲▲▲✿✿✿

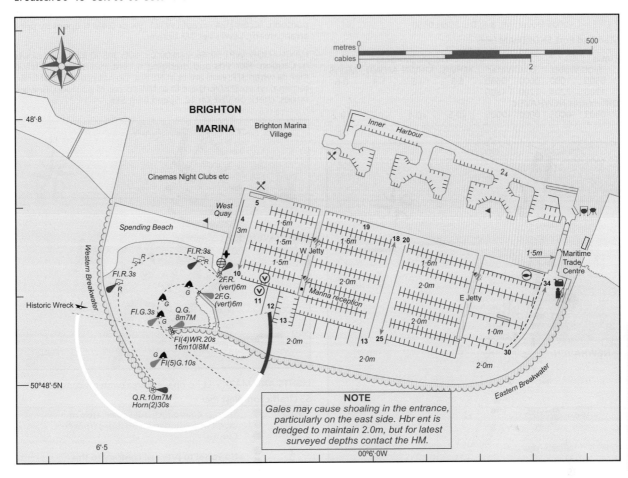

NOTE
Gales may cause shoaling in the entrance, particularly on the east side. Hbr ent is dredged to maintain 2.0m, but for latest surveyed depths contact the HM.

CHARTS AC 5605, 1652, 1991; Imray C12, C31, C9, 2100

TIDES +0004 Dover; ML 3·5; Duration 0605

Standard Port SHOREHAM (◄—)

Times				Height (metres)			
High Water		Low Water		MHWS	MHWN	MLWN	MLWS
0500	1000	0000	0600	6·3	4·8	1·9	0·6
1700	2200	1200	1800				
Differences BRIGHTON MARINA							
–0002	–0002	0000	0000	+0·2	+0·2	+0·1	+0·1

SHELTER Good in the marina under all conditions, but in strong S'ly winds confused seas can make the final appr very rough. Speed limit 5kn.

NAVIGATION WPT 50°48'·23N 00°06'·39W, 000°/0·26M to W bkwtr lt.

- Ent chan dredged 2·0m, but after gales shoaling occurs especially on E side; craft drawing >1·5m should keep to the W side of chan until past the second SHM buoy.
- In heavy weather, best appr is from SSE to avoid worst of the backlash from bkwtrs; beware shallow water E of ent in R sector of lt Fl (4) WR 20s.

W-going stream starts at Brighton HW–1½ and E-going at HW+4½. Inshore the sp rate reaches approx 1·5kn. A Historic Wreck is at 50°48'·6N 00°06'·49W, immediately W of the marina's W bkwtr.

LIGHTS AND MARKS The marina is at the E end of the town, where white cliffs extend eastward. Daymark: conspic white hospital block, brg 334° leads to ent. Six Y spar lt buoys (up to 2M offshore) used as racing buoys:

1. 50°48'·06N 00°06'·41W	2. 50°47'·61N 00°08'·43W
3. 50°46'·61N 00°07'·00W	4. 50°47'·00N 00°15'·33W
5. 50°48'·40N 00°19'·40W	6. 50°47'·40N 00°05'·00W

A sewer outfall can buoy, Fl Y 5s, is 1·1M off the coast. Navigational lts may be hard to see against shore glare:

E bkwtr Fl (4) WR 20s (intens) 16m 10/8M; vis 260°-R-295°-W-100°. E bkwtr hd QG 8m 7M. W bkwtr hd, tr R/W bands, QR 10m 7M; Horn (2) 30s. Inner Hbr lock controlled by normal R/G lts, 0800-1800LT.

COMMUNICATIONS (Code 01273) NMOC (02392) 552100; Police 101; ⊞ 696955; Dr 686863; HM 819919; Call: *Brighton Control* VHF Ch **M** 80 16 (H24).

FACILITIES **Brighton Marina** www. premiermarinas.com ☎819919; Inner Hbr has least depth of 2·4m. 1600⌾ inc ❹ £2·41, £1·00 <4hrs, ⬤◎♁⚓⚔◇💻Ⓔ🛈♺◪(60t) ◭(35t) Divers ◭ ACA ✕ ◻. Fuel pontoon 🛢🛢🛢 (H24).

Brighton Marina YC ☎818711, ✕ ◻.

Bus service from marina; timetable info ☎674881. Electric railway runs from marina to Palace Pier, Mar-Oct.

Town ✉ Ⓑ 🛒 ✕ ◻ ⇌ ✈ (Gatwick).

9.3.9 NEWHAVEN

E. Sussex **50°46'·84N 00°03'·53E** ✺✺✺✺♦♦♦☆☆

CHARTS AC 5605, 1652, 2154; Imray C12, C31, C9, 2100

TIDES +0004 Dover; ML 3·6; Duration 0550

Standard Port SHOREHAM (←)

Times				Height (metres)			
High Water		Low Water		MHWS	MHWN	MLWN	MLWS
0500	1000	0000	0600	6·3	4·8	1·9	0·6
1700	2200	1200	1800				
Differences NEWHAVEN							
−0003	−0005	0000	+0005	+0·5	+0·4	+0·2	+0·2

SHELTER Good in all weathers, but in strong on-shore winds there is often a dangerous sea at the ent. Appr from the SW, to pass 50m off bkwtr hd to avoid heavy breaking seas on E side of dredged chan. At marina (mostly dredged to 1m), berth on inside of ❶ pontoon, access H24 except LWS±1. Sw br 0.4M N opens by prior arrangement. Lewes lies 7M upriver.

NAVIGATION WPT 50°46'·24N 00°03'·60E, 348°/0·32M to W bkwtr lt. Caution: Hbr silts and dredging is continuous. Do not leave marina when 3FR(vert) are lit at NE ent. Do not proceed S of RoRo terminal when 3FR(vert) are lit at NW corner of pontoon. Beware ferries; check on VHF Ch 12. Speed limit 5kn.

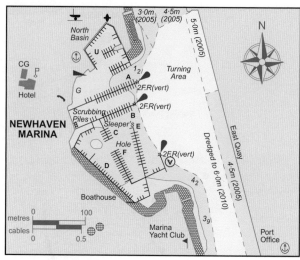

LIGHTS AND MARKS Lt ho on west breakwater is conspic.

SS (Traffic) on West Pier

Fl	●	Serious emergency. All vessels stop or
Fl	●	divert according to instructions from Port
Fl	●	Control
F	●	No vessel to proceed contrary to this
F	●	signal
F	●	
F	●	
F	●	Small vessels may proceed. Two way traffic
F	○	
F	●	Proceed only when instructed by Port
F	○	Control. All other vessels keep clear
F	●	

COMMUNICATIONS (Code 01273) NMOC (02392) 552100; Police 101; Dr 515076; ⊞696955; HM / Port Control vts@newhavenport authority.co.uk ☎612926 (H24), fax 612878; 'Newhaven Radio' Ch 12 (H24). Swing bridge opening Ch 12. Marina Ch **80** M (0800-1700).

FACILITIES Marina ☎513881; ⚓ (☎612612 launch/recovery £17.50) 280⛵+20❶ £2.25, ⚒🅿🔧(HO,1ca N of marina entrance) ⬛🗡🔩 📠 📺(18t) 🛒(10t) 🛢🧺 ✗ ⛽.

Newhaven and Seaford SC ☎(01323) 890077, ⚓ ⚓.

Town: 📠 Ⓔ ⚠ ACA ✉ Ⓑ 🔧 ✗ 🗒 ⇌ ✈ (Gatwick); **Ferries:** Dieppe; 2-3/day; 4 Hrs; Transmanche Ferries (transmancheferries.co.uk).

9.3.10 EASTBOURNE

E. Sussex **50°47´·34N 00°19´·90E** ❀❀◊◊◊❀❀

CHARTS AC 536, 5605; Imray 12, C31, C8, 2100

TIDES –0005 Dover; ML 3·8; Duration 0540

Standard Port SHOREHAM (←)

Times				Height (metres)			
High Water		Low Water		MHWS	MHWN	MLWN	MLWS
0500	1000	0000	0600	6·3	4·8	1·9	0·6
1700	2200	1200	1800				
Differences EASTBOURNE							
–0010	–0005	+0015	+0020	+1·1	+0·6	+0·2	+0·1

SHELTER Very good, but in strong SE'lies the entrance may be hazardous and should be avoided at HW±1½. Access via buoyed channel (2m) and twin locks into inner basin (4m). The channel is prone to shoaling after gales and is dredged regularly. If uncertain of up-to-date situation, contact HM before entry.

NAVIGATION WPT 50°47´·37N 00°20´·81E, SWM buoy 'SH', L Fl 10s, 259°/0·45M to hbr ent. There are shoals to the NE in Pevensey Bay and from 2·5M SE toward Royal Sovereign lt (tide rips). From Beachy Hd, keep 0·75M offshore to clear Holywell Bank. 3kn speed limit in approach chan and Sovereign Harbour.

LIGHTS AND MARKS See 9.3.4 and chartlet. From E, by day R roofs are conspic. Dir lt, Fl WRG 5s leads 258° through appr channel. Large wreck on N side marked by 2 lit SHM. Eastbourne pier, 2 FR, is 2M S of hbr ent; unlit PHM buoy is approx 5ca S.

COMMUNICATIONS (Code 01323) CGOC (01304) 210008; Police 101; Dr470370; ⊞ 417400. HM 470099. **VHF Ch 17 call** *Sovereign Hbr* at SWM.

SAFETY RECOMMENDATIONS

- Monitor VHF Ch **17** (H24) for nav info, lock status and berthing.
- Always follow IPTS (Sigs 2, 3 and 5) for locks entry.
- At busy periods obey instructions of Lock Keeper or Assistant on lock-side or directions by VHF.
- Locks close every H and H+30.
- Always approach locks with fenders (set about 1m above the waterline) and mooring lines rigged on both sides of craft. Secure to the pontoons as far to the front of the lock as possible. Set springs to reduce surging, particularly at LW.
- Switch off engines and radar whilst in lock.
- When secured alongside, assist other craft berthing.

FACILITIES Sovereign Harbour www.premiermarinas.com ☎470099 860⌿ £2·41, ⛽⊙⚓🛥🚿(H24) ⛽🍴(50t) ✕🔧⚒.
Sovereign Harbour YC ☎470888, ✕🍴 visitors welcome.
Retail Park with S/market and all domestic facilities.
Town (2½M) all needs, ⇌ ✈ (Gatwick).

ADJACENT ANCHORAGE

HASTINGS, E Sussex, **50°50´·88N 00°35´·50E.** AC 536.
Tides, see RYE; ML 3·8m; Duration 0530. Strictly a settled weather ⚓ or emergency shelter. The stone breakwater is in disrepair and only protects FVs. Beware dangerous wreck 3ca SE of pier hd. Ldg lts 356°, both FR 14/55m 4M: front on W metal column; rear 357m from front, on 5-sided W tr on West Hill. Pier under refit, now lit Q R 1s. W bkwtr hd Fl R 2·5s 5m 4M; Fl G 5s 2m, 30m from head of No3 Groyne (E bkwtr). A Historic Wreck (*Amsterdam*) is about 2M W of pier, close inshore at 50°50´·7N 00°31´·65E. Facilities: ACA (St Leonards). Few marine services, but domestic facilities at Hastings and St Leonards. YC ☎(01424) 420656.

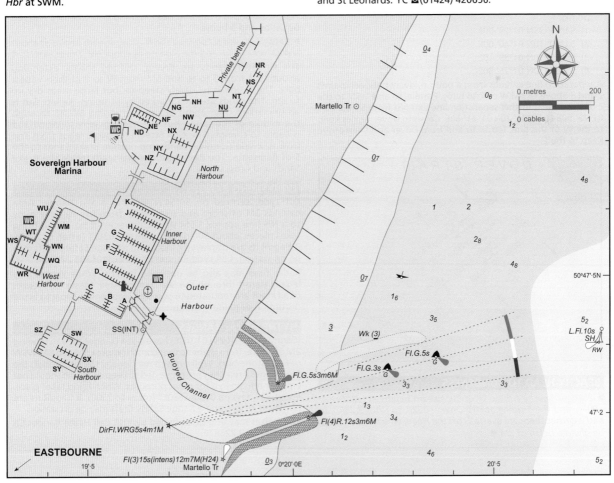

CROSSING THE EASTERN ENGLISH CHANNEL

This section applies to crossings, ranging from the short (4–5 hrs) Dover Strait route to Boulogne or Calais to the moderate (10–15 hrs) ones from Brighton, Newhaven or Eastbourne to Dieppe and adjacent French ports. Distance tables are at 9.0.5. Routes cross the Dover Strait TSS where the SW-bound lane lies only 7M S of Beachy Head and 4.5M SE of Dover.

- Dover to Calais or Boulogne is only about 25M but crosses the most congested part of Dover Strait TSS.

- Plan to use the tidal streams to minimise crossing time and to be up-tide and to windward of the destination port when clear of the TSS.

- Always cross the TSS on a heading of 90° to the traffic flow, and minimise the time spent in the TSS. Be prepared to motor.

- When crossing the TSS, listen to the VHF broadcasts of navigational and traffic information made by CNIS. These include details of vessels which appear to be contravening Rule 10.

- Keep a very sharp lookout for ships in the traffic lanes and cross-Channel ferries. Ensure that the crew is aware of the directions from which to expect traffic in the TSS.

- Unless equipped with radar, and you are competent in its operation, do not attempt the crossing if fog or poor visibility is forecast. Even with radar, navigate with extreme caution.

- A Voluntary Separation Scheme (VSS) is in use to help ferries safely cross the Dover TSS. It is not a formal separation scheme and is not shown on official charts, but small craft should be aware of its existence. Most cross-Channel ferries can be expected to be using the VSS but high speed ferries are not included in the scheme. The VSS is bounded by the following points:

  1) 51°05′·35N 01°28′·00E
  2) 51°00′·10N 01°40′·00E
  3) 50°59′·30N 01°39′·10E
  4) 51°04′·70N 01°26′·80E

A separation line extends from a point midway between points 2 and 3 above to 1M W of CA6 buoy. Parallel to and 1·5M to the E of this zone a further separation line extends from 51°06′·40N 01°29′·90E to 51°01′·20N 01°41′·60E. Calais to Dover ferries keep to the W of this line; ferries to and from Dover and Dunkerque keep to the E.

BEACHY HEAD TO DUNGENESS

▶ *2M S of Beachy Hd the W-going stream begins at HW Dover +0030, and the E-going at HW Dover –0520, sp rates 2·25kn. In bad weather there are overfalls off the Head, which should then be given a berth of 2M.* ◀

(AC 536) Royal Sovereign lt tr is 7·4M E of Beachy Head. The extensive Royal Sovereign shoals lie from 3M NW of the tr to 1·5M N of it, and have a minimum depth of 3·5m.

▶ *There are strong eddies over the shoals at Sp, and the sea breaks on them in bad weather.* ◀

On the direct course from the Royal Sovereign to Dungeness there are no dangers. Along the coast in Pevensey and Rye Bays there are drying rocky ledges or shoals extending 5ca offshore in places. These include Boulder Bank near Wish Tr, S of Eastbourne; Oyster Reef off Cooden; Bexhill Reef off Bexhill-on-Sea; Bopeep Rocks off St Leonards; and the shoals at the mouth of R Rother, at the entrance to Rye. There are also banks 2–3M offshore, on which the sea builds in bad weather.

Avoid the firing range danger area between Rye and Dungeness (lt, fog sig, RC). The nuclear power station is conspic at S extremity of the low-lying spit. The point is steep-to on SE side.

▶ *2M SE of Dungeness the NE-going flood starts at HW Dover –0100, max Sp rate 1.9kn. The SW-going ebb starts at HW Dover +0430, max Sp rate 2.1kn.* ◀

DUNGENESS TO NORTH FORELAND

(AC 1892, 1828) From Dungeness to Folkestone the coast forms a bay. Beware Roar bank, depth 2·7m, E of New Romney: otherwise there are no off-lying dangers apart from Hythe firing range. Good ⌁ off Sandgate in offshore winds.

▶ *Off Folkestone the E-going stream starts at HW Dover –0155, Sp rate 2kn; the W-going at HW Dover +0320, sp rate 1·5kn.* ◀

Passing Dover and S Foreland keep 1M offshore. Do not pass too close to Dover as the many ferries leave at speed and there can be considerable backwash. It is advisable to call Dover Port Control (Ch 74) to identify yourself before passing the harbour entrances. 8M S of Dover in the TSS is the Varne, a shoal 7M long, with least depth 3·3m and heavy seas in bad weather, marked by Lanby and 5 buoys.

▶ *Between S and N Foreland the N-going stream begins at about HW Dover –0150, and the S-going at about HW Dover +0415.* ◀

The Goodwin Sands are drying, shifting shoals, extending about 10M from S to N, and 5M from W to E at their widest part. The E side is relatively steep-to, but large areas dry up to 3m. The sands are well marked by lit buoys. Kellett Gut is an unmarked chan about 5ca wide, running SW/NE through the middle of the sands, but it is not regularly surveyed and is liable to change. The Gull Stream (buoyed) leads from The Downs, inside Goodwin Sands and outside Brake Sands to the S of Ramsgate. The Ramsgate channel leads inside the Brake Sands and Cross Ledge.

LYDD FIRING RANGES

Off Lydd, centred on 50°54′N 00°53′E, a Sea Danger Area extends 3M offshore and stretches E from Rye Fairway buoy to a N/S line approx 1·5M W of Dungeness lt ho. When firing takes places, about 300 days p.a. 0830–1630LT (often to 2300), R flags/R lts are displayed ashore and a Range Safety Craft may be on station. Call Lydd Ranges Ch 73 or ☎01303 225518.

Radar fixes may also be obtained by VHF. While vessels may legally transit through the Sea Danger Area, Masters of vessels of all kinds are requested to pass S of Stephenson Shoal to avoid disruption of military firings.

HYTHE FIRING RANGES

(Centred on 51°02′N 01°03′E) have a Sea Danger Area extending 2M offshore, from Hythe to Dymchurch (approx 5M and 8M WSW of Folkestone hbr). While vessels may legally transit through the Sea Danger Area, they are requested to keep clear, and should not enter or remain in it for other purposes. When firing takes place, about 300 days pa 0830–1630LT (often to 2300), R flags/R lts are displayed ashore and a Range Safety Craft may be on station.

Radar fixes may also be obtained by VHF. Call *Hythe Ranges* Ch 73 or ☎01303 225879.

9.3.11 RYE

E. Sussex 50°55'·60N 00°46'·58E ✿✿✿✿✿✿✿✿✿

CHARTS AC 5605, 536, 1991; Imray C12, C31, C8, 2100

TIDES ML 2·0; Duration 3·25hrs sp, 5hrs nps

Standard Port DOVER (→)

Times				Height (metres)			
High Water		Low Water		MHWS	MHWN	MLWN	MLWS
0000	0600	0100	0700	6·8	5·3	2·1	0·8
1200	1800	1300	1900				
Differences RYE (approaches)							
+0005	–0010	ND	ND	+1·0	+0·7	ND	ND
RYE HARBOUR							
+0005	–0010	DR	DR	–1·4	–1·7	DR	DR
HASTINGS							
0000	–0010	–0030	–0030	+0·8	+0·5	+0·1	–0·1

SHELTER Very good in R Rother which dries completely to soft mud. Rye Bay is exposed to prevailing SW'lies with little shelter, when there is good ⚓ in lee of Dungeness (6M to E). In N'lies ⚓ 5ca N of the Rye Fairway buoy.

Rye Hbr is a small village, ¾M inside ent on W bank, used by commercial shipping. Berth initially on Admiralty Jetty (E bank) and see HM for ⌒ or M. No ⚓. Max speed 6kn.

Rye Town (a Cinque Port) is 2M up river. Enter via Rock Channel for ⓥ ⌒ along NE side of Strand Quay.

NAVIGATION WPT Rye Fairway SWM lt buoy, 50°54'·04N 00°48'·02E, 329°/1·8M to W Arm tripod lt, 50°55'·55N 00°46'·65E. For details of Lydd and Hythe Firing Ranges see Passage Information (←).

- Bar dries 2·75m about 2ca offshore and needs care when wind >F6 from SE to SW. Also shoals E and W of ent with ground swell or surf; narrow ent (42m) and chan (30m).
- Enter HW –2 to HW +2; flood runs 4·5kn (max HW –2 to HW –1).
- Depth of water over the bar can be judged by day from horizontal timbers at base of West Arm tripod structure (approx 2ca N of the bar) these are set at 1·5, 3 and 4·5m above CD.

Note: A Historic Wreck (*Anne*) is about 4M WSW of Rye, close inshore at 50°53'·42N 00°41'·91E.

LIGHTS AND MARKS W Arm lt Fl R 5s 7m 6M, wooden tripod, radar reflector. E Arm hd, Q (9) 15s 7m 5M; Horn 7s, G △. On E Pier a floodlit 'Welcome to Rye' sign may be helpful. Rock Chan entrance marked by a QR and QG lt buoy.

IPTS (Sigs 2 & 5 only) are shown to seaward (3M) from HM's office and up-river (1M) from HM's office.

COMMUNICATIONS (Code 01797) CGOC (01304) 210008; Police 101; Dr 222031; Ⓗ 224499. HM 225225 Ch 14 (0900-1700LT, HW±2 or when vessel due). Call HM and monitor Ch 14 before arrival/ departure.

FACILITIES (from seaward) **Admiralty Jetty** ⚓▬▬⚓
Rye Hbr ▬ HW±3 (26t launch £10) ✕ ✎ Ⓑ Ⓔ ⌂(3t/15t) Ⓛ ACA
Rye Hbr SC (Sec'y) ☎223376.

Rye Town, Strand Quay ⌒, £14.75<8m>£17.74<10m>£23.57 wood fender posts against steel piled wall with numbered ladders. ⓥ must stop at timber staging on E bank and report to HM Office. ⚓ Ⓛ Ⓣ Ⓣ ⚓ ⋯.

Town ✉ Ⓑ ⌗ ⇌ ✈ (Lydd).

SE England

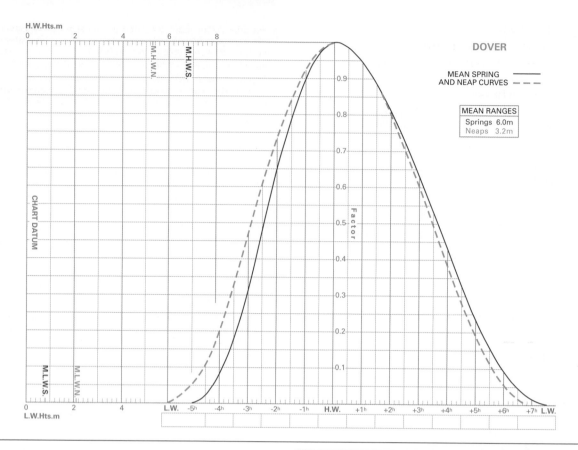

DOVER

MEAN SPRING ——————
AND NEAP CURVES – – – –

MEAN RANGES
Springs 6.0m
Neaps 3.2m

9.3.12   FOLKESTONE

Kent **51°04'·59N 01°11'·67E** ❁❁❁◊◊✿✿

CHARTS AC 5605, 1892, 1991; Imray C12, C8, 2100

TIDES –0010 Dover; ML 3·9; Duration 0500
Standard Port DOVER (→)

Times				Height (metres)			
High Water		Low Water		MHWS	MHWN	MLWN	MLWS
0000	0600	0100	0700	6·8	5·3	2·1	0·8
1200	1800	1300	1900				
Differences FOLKESTONE							
–0020	–0005	–0010	–0010	+0·4	+0·4	0·0	–0·1
DUNGENESS							
–0010	–0015	–0020	–0010	+1·0	+0·6	+0·4	+0·1

SHELTER Good except in strong E-S winds when seas break at the harbour entrance.

NAVIGATION WPT 51°04'·33N 01°11'·89E, 330°/0·26M to bkwtr hd lt. Depth gauge on head of E Pier. Beware drying Mole Hd Rks and Copt Rks to stbd of the entrance; from/to the NE, keep well clear of the latter due to extended sewer outfall pipe. Inner Hbr, dries, has many FVs and local shoal draught boats. For Lydd firing ranges see Passage Information.

LIGHTS AND MARKS Hotel block is conspicuous at W end of Inner Hbr.

COMMUNICATIONS (Code 01303) CGOC (01304) 210008; Police 101. Folkestone Hbr Co Ltd/HM 254597, mob 07401 627563 (0600–1800); *Folkestone Port Control* Ch 15.

FACILITIES Berth on **South Quay** (dries); £20/craft for � (fender board needed) or ⚓, ▬ (free), ⊤.

Folkestone Yacht & Motor Boat Club ☎251574, F&A▲▬ ⌂⊤⛵⊔.

Town ⓣⓣ(100m) ✉ Ⓑ ☕ ✗ ⌂ ≋ ✈ (Lydd).

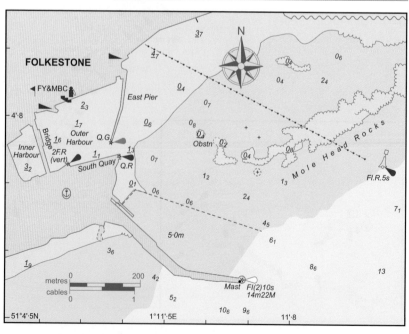

9.3.13 DOVER

Kent   **51°06′·74N 01°19′·73E** (W ent) ❀❀❀⚓⚓⚓❀❀
     51°07′·25N 01°20′·61E (E ent)

CHARTS AC 5605, 1892, 1828, 1698; Imray C30, C12, C8, 2100

TIDES 0000 Dover; ML 3·7; Duration 0505

Standard Port DOVER (→)

Times				Height (metres)			
High Water		Low Water		MHWS	MHWN	MLWN	MLWS
0000	0600	0100	0700	6·8	5·3	2·1	0·8
1200	1800	1300	1900				
Differences DEAL							
+0012	+0010	+0004	+0002	−0·5	−0·3	0·0	+0·1

SHELTER The Outer Hbr ⚓ is tenable in offshore winds, but exposed to winds from NE through S to SW; in gales a heavy sea builds up. Small craft must not be left unattended at ⚓ in Outer Hbr. Very good in marina.

NAVIGATION Advise Port Control of intentions when 2M off. WPT from SW, 51°06′·18N 01°19′·67E, 000°/0·5M to Admiralty Pier lt ho. WPT from NE, 51°07′·30N 01°21′·41E, 270°/0·5M to S end Eastern Arm. Beware lumpy seas/overfalls outside the bkwtrs and the frequent ferries and high speed craft using both ents. Strong tides across ents and high walls make ent under sail slow and difficult; use of engine is very strongly recommended.

Note: A Historic wreck is adjacent to the Eastern Arm bkwtr at 51°07′·6N 01°20′·7E (see chartlet). There are 4 more Historic wrecks on the Goodwin Sands.

- **Specific permission** to ent/leave the hbr via E or W ent must first be obtained from Port Control on Ch **74**; advise if you have no engine. Comply with any VHF instructions from Port Control or hbr patrol launch. Clearance for small/slow craft is not normally given until within 200m of ent. Visitors are welcomed and usually escorted by hbr launch to the marina.

- All vessels with a draught > 2.5m wishing to visit the marina must inform Dover Port Control, Ch **74**, prior to entry.

- **If no VHF:** Stay safe distance clear of either entrance, call Port Control +44(0)1304 206063 or make visual contact (Aldis lamp) with Port Control (E Arm) who may direct Hbr launch to assist. Q W lt from Port Control twr or patrol launch = keep clear of ent you are approaching.

LIGHTS AND MARKS As per 9.3.4 and chartlet. IPTS (for all vessels) are shown for the E ent by Port Control (conspic twr) at S end of E Arm and for W entrance, from the head of Admiralty Pier.

COMMUNICATIONS (Code 01304) CGOC 210008; Police 101; ⊞ (01233) 633331. HM 240400 ext 4520, Duty HM mob 07836 262713; Port Control 240400 ext 5530.

Call *Dover Port Control* (also Hbr launch) Ch 74, 12, 16 for clearance to enter. Request clearance to leave hbr as soon as clear of the marina. *Dover Marina* Ch 80, only within marina. *Dover Coastguard* Ch 16, maintains TSS surveillance.

Chan Nav Info Service (CNIS) broadcasts tfc/ nav/weather/tidal info Ch 11 at H+40; also, if vis < 2M, at H+55.

FACILITIES Dover Harbour Marina ④ ☎241663; on arrival berth at reception pontoon to visit marina office or await gate opening. 3 berthing options are available in the marina:

- Tidal hbr, E of waiting pontoon (1·5m), access H24. 107⌇ on 3 pontoons (C, B, A) in 2·5m, £2·70. Mostly for single night stop.
- Granville Dock, gate open approx HW–3½ to +4½, 133⌇, £2·40.
- Wellington Dock. Gate and swing bridge open HW ±2 nps and approx HW–1½ to HW+1½ or 2½ sp, depending on range. In the final appr, especially near LW, stay in deep water as defined by the W sector (324°-333°) of the F WR lt, 2 unlit SHM poles and a G conical buoy. IPTS are shown, plus a small Fl ● lt 5 mins before Wellington bridge is swung open. 160⌇ £2·20.

Services: www.doverport.co.uk ⚓ 370⌇ inc ▼ ⇥ 🗑 ⛽(H24) 🔧 LPG 🔋🗡 🔩 🛒 Ⓔ ⚓ 🏗(50t) 🛢 🅿 ACA, wi-fi.

Royal Cinque Ports YC www.rcpyc.co.uk ☎206262, ⚓🗜.

White Cliffs M and YC www.wcmyc.co.uk ☎211666.

Town: ⌂ ☏(H24) ✉ Ⓑ 🛒 ✕ 🖃 ⇌.

Ferries: Services to Calais (passenger/vehicle), Dunkerque (vehicle).

STANDARD TIME (UT)
For Summer Time add ONE hour in **non-shaded areas**

DOVER LAT 51°07'N LONG 1°19'E
TIMES AND HEIGHTS OF HIGH AND LOW WATERS

Dates in red are **SPRINGS**
Dates in blue are **NEAPS**

YEAR 2016

JANUARY

Day	Time	m	Time	m	Time	m	Time	m
1 F	0313	6.1	1019	1.7	1541	5.7	2226	2.0
16 SA	0258	6.6	1027	1.2	1526	6.2	2244 ☽	1.4
2 SA	0401	5.8	1101	1.9	1637	5.4	2314 ☽	2.2
17 SU	0356	6.3	1120	1.4	1632	5.9	2342	1.6
3 SU	0459	5.5	1159	2.1	1742	5.3		
18 M	0504	6.0	1224	1.6	1753	5.7		
4 M	0023	2.4	0604	5.4	1310	2.1	1848	5.3
19 TU	0052	1.8	0624	5.8	1334	1.7	1915	5.6
5 TU	0142	2.3	0708	5.5	1417	2.0	1947	5.5
20 W	0205	1.8	0744	5.8	1447	1.7	2025	5.8
6 W	0249	2.1	0804	5.6	1516	1.8	2037	5.7
21 TH	0319	1.7	0852	6.0	1608	1.5	2125	6.1
7 TH	0345	1.8	0852	5.9	1609	1.5	2120	6.0
22 F	0436	1.4	0956	6.2	1719	1.2	2215	6.3
8 F	0436	1.5	0934	6.1	1658	1.3	2200	6.3
23 SA	0538	1.1	1038	6.4	1811	1.1	2257	6.5
9 SA	0522	1.3	1015	6.4	1744	1.1	2239	6.5
24 SU	0628	0.9	1119	6.5	1855	1.0	2336 ○	6.7
10 SU	0607	1.1	1055	6.5	1828 ●	1.0	2318	6.7
25 M	0711	0.9	1157	6.6	1933	1.0		
11 M	0651	1.0	1135	6.7	1911	0.9	2357	6.8
26 TU	0014	6.8	0749	0.9	1233	6.5	2004	1.0
12 TU	0734	0.9	1215	6.7	1952	0.9		
27 W	0051	6.7	0821	0.9	1309	6.4	2031	1.1
13 W	0038	6.9	0816	0.8	1257	6.7	2032	0.9
28 TH	0127	6.6	0849	1.1	1343	6.3	2053	1.3
14 TH	0121	6.9	0858	0.9	1342	6.6	2112	1.0
29 F	0159	6.5	0913	1.2	1415	6.1	2117	1.4
15 F	0208	6.8	0941	1.0	1431	6.4	2155	1.2
30 SA	0229	6.2	0940	1.4	1445	6.0	2148	1.6
31 SU	0300	6.0	1014	1.6	1520	5.6	2227	1.9

FEBRUARY

Day	Time	m	Time	m	Time	m	Time	m
1 M	0341	5.7	1057	1.9	1614	5.3	2316 ☽	2.2
16 TU	0435	6.0	1157	1.7	1718	5.6		
2 TU	0448	5.4	1159	2.1	1747	5.1		
17 W	0024	1.9	0554	5.6	1308	1.9	1846	5.4
3 W	0032	2.4	0614	5.3	1324	2.2	1901	5.2
18 TH	0141	2.0	0727	5.5	1426	1.9	2010	5.5
4 TH	0201	2.3	0724	5.4	1437	2.0	2002	5.5
19 F	0301	1.8	0847	5.7	1600	1.7	2116	5.8
5 F	0309	2.0	0822	5.7	1538	1.7	2053	5.8
20 SA	0428	1.5	0947	6.0	1710	1.4	2206	6.2
6 SA	0407	1.6	0911	6.0	1633	1.4	2138	6.2
21 SU	0530	1.2	1032	6.3	1800	1.1	2245	6.4
7 SU	0459	1.3	0956	6.3	1724	1.1	2220	6.5
22 M	0617	0.9	1108	6.4	1841	1.0	2321 ○	6.6
8 M	0549	1.0	1039	6.6	1813	0.9	2301 ●	6.8
23 TU	0656	0.8	1141	6.5	1915	1.0	2356	6.7
9 TU	0638	0.8	1120	6.8	1859	0.8	2342	7.0
24 W	0729	0.8	1214	6.5	1942	1.0		
10 W	0725	0.6	1201	6.9	1943	0.7		
25 TH	0030	6.7	0756	0.9	1246	6.5	2003	1.0
11 TH	0024	7.1	0808	0.5	1244	6.9	2023	0.6
26 F	0102	6.6	0819	1.0	1314	6.4	2023	1.1
12 F	0107	7.1	0849	0.5	1327	6.8	2101	0.7
27 SA	0127	6.5	0841	1.1	1337	6.2	2048	1.2
13 SA	0151	7.0	0928	0.7	1413	6.7	2141	0.8
28 SU	0147	6.3	0907	1.2	1358	6.1	2118	1.4
14 SU	0239	6.8	1010	0.9	1504	6.3	2225	1.1
29 M	0212	6.2	0939	1.4	1427	5.9	2153	1.6
15 M	0332	6.4	1058	1.3	1604	5.9	2318 ☽	1.5

MARCH

Day	Time	m	Time	m	Time	m	Time	m
1 TU	0248	5.9	1016	1.7	1508	5.6	2234 ☽	1.9
16 W	0413	5.9	1134	1.7	1651	5.6		
2 W	0336	5.6	1104	2.0	1606	5.3	2331	2.2
17 TH	0003	1.9	0528	5.5	1247	2.1	1814	5.3
3 TH	0502	5.2	1220	2.3	1817	5.1		
18 F	0121	2.0	0709	5.4	1408	2.1	1949	5.4
4 F	0110	2.3	0650	5.2	1359	2.2	1929	5.3
19 SA	0245	1.9	0838	5.6	1541	1.8	2059	5.8
5 SA	0234	2.1	0756	5.6	1508	1.8	2026	5.7
20 SU	0412	1.6	0934	5.9	1649	1.5	2147	6.1
6 SU	0337	1.6	0850	6.0	1607	1.4	2115	6.2
21 M	0511	1.0	1015	6.2	1737	1.2	2225	6.4
7 M	0434	1.2	0937	6.4	1701	1.1	2200	6.6
22 TU	0556	1.0	1049	6.4	1816	1.1	2300	6.6
8 TU	0529	0.9	1021	6.7	1754	0.8	2242	6.9
23 W	0632	0.9	1120	6.5	1848	1.0	2334 ○	6.6
9 W	0621	0.6	1103	6.9	1843 ●	0.6	2324	7.1
24 TH	0702	0.9	1150	6.5	1913	1.0		
10 TH	0710	0.4	1144	7.0	1928	0.5		
25 F	0006	6.6	0725	0.9	1220	6.5	1932	1.0
11 F	0006	7.2	0754	0.3	1227	7.1	2008	0.4
26 SA	0035	6.5	0747	1.0	1246	6.4	1955	1.0
12 SA	0049	7.2	0812	1.0	1309	7.0	2046	0.5
27 SU	0055	6.4	0812	1.0	1304	6.3	2022	1.1
13 SU	0133	7.1	0912	0.5	1355	6.8	2125	0.7
28 M	0113	6.4	0840	1.1	1325	6.3	2053	1.3
14 M	0220	6.8	0952	0.8	1445	6.5	2208	1.0
29 TU	0140	6.3	0911	1.3	1356	6.2	2126	1.5
15 TU	0312	6.4	1037	1.3	1543	6.0	2258 ☽	1.5
30 W	0215	6.1	0946	1.6	1437	5.9		
31 TH	0301	5.7	1030	1.9	1530	5.5	2257 ☽	2.1

APRIL

Day	Time	m	Time	m	Time	m	Time	m
1 F	0407	5.3	1131	2.2	1716	5.2		
16 SA	0101	2.0	0634	5.3	1343	2.1	1909	5.4
2 SA	0020	2.2	0618	5.3	1317	2.2	1855	5.4
17 SU	0219	1.9	0810	5.5	1501	1.9	2026	5.7
3 SU	0158	2.0	0729	5.6	1435	1.9	1957	5.8
18 M	0335	1.6	0906	5.8	1609	1.6	2116	6.0
4 M	0306	1.6	0826	6.0	1537	1.5	2050	6.2
19 TU	0435	1.3	0947	6.0	1659	1.4	2157	6.3
5 TU	0406	1.2	0916	6.4	1634	1.1	2137	6.6
20 W	0521	1.1	1021	6.2	1740	1.2	2234	6.4
6 W	0504	0.8	1001	6.7	1729	0.8	2221	6.9
21 TH	0557	1.0	1052	6.4	1812	1.1	2307	6.5
7 TH	0600	0.6	1044	6.9	1821 ●	0.6	2304	7.2
22 F	0626	1.0	1123	6.4	1838	1.1	2338 ○	6.5
8 F	0651	0.4	1126	7.1	1908	0.4	2347	7.2
23 SA	0652	1.0	1152	6.4	1903	1.1		
9 SA	0737	0.3	1208	7.1	1951	0.4		
24 SU	0005	6.4	0718	1.0	1218	6.4	1930	1.1
10 SU	0030	7.2	0818	0.3	1253	7.0	2030	0.4
25 M	0027	6.4	0747	1.0	1239	6.4	2001	1.1
11 M	0116	7.1	0857	0.5	1339	6.8	2111	0.7
26 TU	0049	6.3	0818	1.1	1303	6.4	2033	1.2
12 TU	0203	6.8	0937	0.9	1429	6.5	2154	1.0
27 W	0117	6.3	0850	1.3	1336	6.3	2107	1.4
13 W	0255	6.4	1021	1.3	1525	6.1	2244	1.5
28 TH	0154	6.1	0925	1.5	1418	6.1	2147	1.6
14 TH	0354	5.9	1111	1.8	1627	5.7	2347 ☽	1.8
29 F	0241	5.9	1008	1.8	1512	5.8	2237	1.9
15 F	0503	5.5	1226	2.1	1739	5.4		
30 SA	0345	5.5	1105	2.0	1634	5.5	2348 ☽	2.0

Chart Datum: 3·67 metres below Ordnance Datum (Newlyn). HAT is 7·4 metres above Chart Datum.

STANDARD TIME (UT)
For Summer Time add ONE hour in **non-shaded areas**

DOVER   LAT 51°07′N   LONG 1°19′E
TIMES AND HEIGHTS OF HIGH AND LOW WATERS

Dates in red are **SPRINGS**
Dates in blue are NEAPS

YEAR **2016**

MAY

Day	Time m	Time m	Time m	Time m
1 SU	0540 5.4	1233 2.1	1816 5.5	
16 M	0141 1.9	0716 5.4	1414 2.0	1935 5.6
2 M	0121 1.9	0659 5.6	1359 1.8	1924 5.8
17 TU	0245 1.7	0821 5.6	1515 1.8	2035 5.8
3 TU	0232 1.6	0759 6.0	1503 1.5	2021 6.2
18 W	0342 1.5	0908 5.8	1609 1.6	2122 6.1
4 W	0334 1.2	0851 6.3	1602 1.1	2111 6.6
19 TH	0431 1.4	0946 6.0	1654 1.4	2201 6.2
5 TH	0435 0.9	0939 6.6	1701 0.9	2159 6.9
20 F	0512 1.2	1020 6.2	1731 1.3	2235 6.3
6 F ●	0535 0.6	1024 6.8	1756 0.6	2245 7.1
21 SA ○	0547 1.2	1052 6.3	1803 1.2	2306 6.3
7 SA	0630 0.5	1109 7.0	1847 0.5	2330 7.1
22 SU	0620 1.1	1123 6.4	1836 1.1	2336 6.3
8 SU	0719 0.4	1153 7.0	1934 0.4	
23 M	0653 1.1	1152 6.4	1910 1.1	
9 M	0015 7.1	0802 0.4	1238 7.0	2017 0.5
24 TU	0003 6.3	0727 1.1	1220 6.4	1944 1.1
10 TU	0101 6.9	0844 0.6	1325 6.6	2059 0.7
25 W	0032 6.3	0801 1.1	1250 6.4	2020 1.2
11 W	0149 6.7	0925 0.9	1414 6.6	2143 1.0
26 TH	0104 6.3	0836 1.3	1326 6.4	2056 1.3
12 TH	0240 6.3	1008 1.3	1506 6.3	2231 1.4
27 F	0143 6.2	0912 1.4	1409 6.2	2137 1.5
13 F ◐	0335 5.9	1058 1.7	1602 5.9	2328 1.7
28 SA	0232 6.0	0955 1.6	1503 6.0	2225 1.6
14 SA	0435 5.6	1159 2.0	1703 5.6	
29 SU ◑	0332 5.8	1048 1.8	1611 5.8	2329 1.7
15 SU	0033 1.9	0547 5.4	1307 2.1	1816 5.5
30 M	0455 5.6	1200 1.8	1731 5.8	
31 TU	0047 1.7	0621 5.7	1321 1.8	1847 5.9

JUNE

Day	Time m	Time m	Time m	Time m
1 W	0158 1.5	0729 5.9	1428 1.5	1951 6.2
16 TH	0248 1.8	0818 5.6	1517 1.8	2036 5.8
2 TH	0302 1.2	0826 6.2	1531 1.3	2047 6.5
17 F	0340 1.6	0905 5.8	1608 1.6	2121 6.0
3 F	0406 1.0	0919 6.4	1633 1.0	2140 6.7
18 SA	0427 1.4	0945 6.0	1652 1.4	2159 6.1
4 SA	0512 0.8	1008 6.6	1734 0.8	2230 6.8
19 SU	0511 1.3	1021 6.2	1733 1.3	2234 6.2
5 SU ●	0611 0.7	1056 6.8	1829 0.7	2318 6.9
20 M ○	0552 1.2	1056 6.3	1812 1.1	2309 6.3
6 M	0703 0.6	1141 6.9	1919 0.6	
21 TU	0632 1.1	1129 6.4	1852 1.1	2343 6.3
7 TU	0004 6.9	0749 0.6	1226 6.9	2005 0.6
22 W	0711 1.1	1203 6.5	1931 1.1	
8 W	0050 6.8	0832 0.8	1311 6.8	2049 0.7
23 TH	0018 6.4	0749 1.1	1239 6.5	2010 1.1
9 TH	0135 6.6	0913 1.0	1357 6.6	2131 1.0
24 F	0055 6.4	0827 1.1	1318 6.5	2050 1.1
10 F	0222 6.3	0953 1.3	1444 6.4	2214 1.3
25 SA	0136 6.3	0905 1.2	1401 6.5	2131 1.2
11 SA	0312 6.0	1034 1.6	1534 6.1	2301 1.5
26 SU	0222 6.2	0946 1.3	1451 6.4	2216 1.3
12 SU ◑	0405 5.7	1121 1.9	1627 5.8	2353 1.8
27 M ◑	0316 6.1	1034 1.5	1548 6.2	2310 1.5
13 M	0504 5.5	1218 2.1	1727 5.6	
28 TU	0422 5.9	1133 1.6	1655 6.0	
14 TU	0051 1.9	0610 5.4	1320 2.1	1834 5.5
29 W	0016 1.5	0540 5.8	1245 1.7	1809 6.0
15 W	0151 1.9	0720 5.4	1421 2.0	1941 5.6
30 TH	0126 1.5	0659 5.8	1356 1.6	1923 6.0

JULY

Day	Time m	Time m	Time m	Time m
1 F	0233 1.4	0805 6.0	1503 1.4	2029 6.2
16 SA	0257 1.8	0823 5.6	1527 1.8	2040 5.7
2 SA	0342 1.3	0905 6.2	1611 1.2	2128 6.4
17 SU	0351 1.6	0910 5.9	1619 1.6	2125 6.0
3 SU	0454 1.1	0959 6.4	1718 1.0	2222 6.6
18 M	0441 1.4	0952 6.1	1706 1.3	2206 6.2
4 M ●	0558 0.9	1047 6.6	1817 0.8	2311 6.7
19 TU ○	0527 1.2	1030 6.3	1751 1.2	2246 6.3
5 TU	0651 0.8	1132 6.7	1908 0.7	2356 6.7
20 W	0612 1.1	1108 6.5	1835 1.0	2324 6.5
6 W	0738 0.8	1214 6.8	1954 0.7	
21 TH	0655 1.0	1145 6.6	1918 0.9	
7 TH	0038 6.6	0819 0.9	1256 6.8	2036 0.8
22 F	0002 6.5	0737 1.0	1224 6.7	2000 0.9
8 F	0119 6.5	0856 1.0	1338 6.7	2114 0.9
23 SA	0041 6.6	0816 1.0	1304 6.8	2041 0.9
9 SA	0201 6.3	0929 1.2	1420 6.5	2149 1.2
24 SU	0122 6.6	0855 1.0	1347 6.8	2120 0.9
10 SU	0244 6.1	1000 1.5	1503 6.3	2223 1.4
25 M	0207 6.5	0934 1.1	1433 6.6	2202 1.1
11 M	0329 5.9	1030 1.7	1549 6.0	2300 1.7
26 TU ◑	0257 6.3	1018 1.3	1526 6.4	2250 1.3
12 TU ◑	0420 5.6	1109 2.0	1640 5.7	2347 1.9
27 W	0355 6.0	1109 1.5	1627 6.2	2348 1.5
13 W	0518 5.4	1206 2.2	1739 5.5	
28 TH	0508 5.8	1214 1.7	1740 5.9	
14 TH	0050 2.1	0623 5.3	1320 2.2	1844 5.4
29 F	0057 1.7	0633 5.7	1328 1.8	1903 5.8
15 F	0156 2.0	0727 5.4	1428 2.1	1947 5.5
30 SA	0210 1.7	0750 5.8	1442 1.7	2019 6.0
31 SU	0326 1.5	0856 6.0	1557 1.5	2124 6.2

AUGUST

Day	Time m	Time m	Time m	Time m
1 M	0447 1.3	0952 6.3	1710 1.2	2219 6.4
16 TU	0415 1.5	0924 6.1	1641 1.4	2142 6.2
2 TU ●	0550 1.1	1039 6.5	1808 0.9	2305 6.6
17 W	0504 1.3	1006 6.4	1729 1.1	2224 6.5
3 W	0640 0.9	1120 6.7	1857 0.8	2345 6.6
18 TH ○	0552 1.1	1045 6.7	1816 0.9	2304 6.7
4 TH	0724 0.9	1159 6.8	1939 0.7	
19 F	0637 0.9	1125 6.9	1902 0.8	2343 6.8
5 F	0022 6.6	0800 0.9	1238 6.8	2016 0.8
20 SA	0721 0.9	1204 7.0	1945 0.7	
6 SA	0059 6.5	0832 1.0	1316 6.8	2048 0.9
21 SU	0022 6.8	0800 0.8	1245 7.1	2026 0.7
7 SU	0135 6.4	0857 1.2	1353 6.6	2114 1.1
22 M	0104 6.8	0839 0.8	1328 7.0	2104 0.7
8 M	0211 6.2	0918 1.4	1428 6.4	2138 1.3
23 TU	0147 6.7	0917 0.9	1413 6.8	2144 0.9
9 TU	0247 6.0	0942 1.6	1503 6.1	2206 1.6
24 W	0236 6.5	0959 1.1	1504 6.6	2228 1.2
10 W ◐	0327 5.7	1015 1.8	1544 5.8	2244 1.9
25 TH ◑	0333 6.1	1047 1.5	1604 6.2	2323 1.6
11 TH	0420 5.4	1100 2.1	1643 5.5	2338 2.1
26 F	0444 5.8	1150 1.8	1719 5.8	
12 F	0533 5.2	1207 2.4	1756 5.3	
27 SA	0033 1.9	0611 5.5	1307 2.0	1849 5.6
13 SA	0102 2.3	0644 5.2	1340 2.4	1906 5.3
28 SU	0154 2.0	0737 5.6	1428 1.9	2015 5.8
14 SU	0219 2.1	0746 5.4	1451 2.1	2007 5.6
29 M	0324 1.8	0848 5.9	1555 1.6	2123 6.1
15 M	0321 1.8	0839 5.8	1549 1.7	2057 5.9
30 TU	0444 1.5	0943 6.2	1705 1.2	2213 6.3
31 W	0539 1.2	1026 6.5	1757 1.0	2254 6.5

Chart Datum: 3·67 metres below Ordnance Datum (Newlyn). HAT is 7·4 metres above Chart Datum.

SE England

》》FREE monthly updates. Register at 《
www.reedsnauticalalmanac.co.uk 《

265

STANDARD TIME (UT)
For Summer Time add ONE hour in **non-shaded areas**

DOVER   LAT 51°07'N   LONG 1°19'E
TIMES AND HEIGHTS OF HIGH AND LOW WATERS

Dates in red are **SPRINGS**
Dates in blue are **NEAPS**

YEAR **2016**

SEPTEMBER

Time	m		Time	m
1 0624	1.0	**16**	0528	1.1
1104	6.7		1021	6.9
TH 1841	0.8	F	1754	0.9
● 2328	6.6	○	2241	6.9
2 0703	1.0	**17**	0615	0.9
1140	6.8		1101	7.1
F 1918	0.8	SA	1841	0.7
			2321	7.0
3 0001	6.6	**18**	0659	0.7
0734	1.0		1142	7.2
SA 1216	6.8	SU	1926	0.6
1949	0.9			
4 0034	6.6	**19**	0001	7.1
0800	1.1		0740	0.7
SU 1251	6.8	M	1224	7.3
2014	1.0		2006	0.6
5 0106	6.5	**20**	0043	7.0
0819	1.2		0819	0.7
M 1322	6.6	TU	1307	7.2
2034	1.2		2045	0.7
6 0135	6.3	**21**	0127	6.9
0838	1.3		0859	0.8
TU 1348	6.4	W	1353	6.9
2056	1.3		2124	0.9
7 0158	6.1	**22**	0217	6.6
0905	1.5		0941	1.1
W 1410	6.2	TH	1445	6.6
2125	1.5		2208	1.3
8 0222	5.9	**23**	0314	6.2
0938	1.7		1029	1.5
TH 1440	5.9	F	1546	6.1
2200	1.8	◐	2302	1.8
9 0257	5.6	**24**	0424	5.8
1018	2.0		1131	1.9
F 1526	5.6	SA	1701	5.7
◑ 2245	2.1			
10 0357	5.3	**25**	0014	2.1
1110	2.3		0546	5.5
SA 1707	1.6	SU	1251	2.1
2352	2.4		1834	5.5
11 0604	5.2	**26**	0141	2.2
1244	2.5		0718	5.6
SU 1834	5.2	M	1419	2.0
			2008	5.7
12 0139	2.4	**27**	0317	1.9
0713	5.4		0832	5.9
M 1416	2.3	TU	1547	1.7
1938	5.5		2112	6.0
13 0251	2.0	**28**	0427	1.6
0809	5.7		0925	6.2
TU 1519	1.8	W	1649	1.3
2032	5.9		2157	6.3
14 0348	1.6	**29**	0517	1.3
0857	6.2		1006	6.5
W 1614	1.4	TH	1736	1.0
2118	6.3		2233	6.5
15 0439	1.3	**30**	0559	1.1
0940	6.5		1042	6.7
TH 1705	1.1	F	1816	0.9
2200	6.6		2305	6.6

OCTOBER

Time	m		Time	m
1 0634	1.1	**16**	0549	0.8
1116	6.8		1037	7.2
SA 1849	1.0	SU	1817	0.6
● 2335	6.7	○	2258	7.1
2 0702	1.1	**17**	0636	0.7
1150	6.8		1119	7.3
SU 1915	1.0	M	1904	0.5
			2340	7.2
3 0006	6.6	**18**	0719	0.6
0723	1.2		1202	7.3
M 1223	6.7	TU	1947	0.5
1936	1.1			
4 0036	6.6	**19**	0023	7.1
0743	1.2		0801	0.7
TU 1250	6.6	W	1247	7.2
1958	1.2		2027	0.7
5 0100	6.4	**20**	0109	7.0
0807	1.3		0843	0.8
W 1309	6.4	TH	1334	6.9
2023	1.4		2108	1.0
6 0117	6.3	**21**	0200	6.7
0836	1.4		0926	1.1
TH 1329	6.3	F	1428	6.5
2053	1.5		2152	1.4
7 0142	6.2	**22**	0257	6.3
0909	1.6		1016	1.5
F 1400	6.0	SA	1529	6.1
2128	1.8	◑	2246	1.8
8 0218	5.9	**23**	0402	5.9
0947	1.9		1117	1.9
SA 1442	5.7	SU	1640	5.7
2209	2.1		2356	2.2
9 0307	5.6	**24**	0515	5.6
1035	2.3		1234	2.1
SU 1544	5.3	M	1807	5.5
◑ 2304	2.4			
10 0512	5.2	**25**	0119	2.3
1146	2.5		0641	5.6
M 1801	5.2	TU	1358	2.0
			1943	5.6
11 0046	2.5	**26**	0244	2.1
0636	5.4		0801	5.8
TU 1336	2.3	W	1517	1.7
1909	5.5		2046	5.9
12 0215	2.2	**27**	0353	1.7
0737	5.7		0855	6.1
W 1446	1.9	TH	1618	1.4
2005	5.9		2130	6.2
13 0317	1.8	**28**	0444	1.5
0827	6.2		0937	6.4
TH 1543	1.5	F	1704	1.2
2052	6.3		2205	6.4
14 0410	1.4	**29**	0525	1.3
0912	6.6		1015	6.6
F 1636	1.1	SA	1742	1.1
2136	6.7		2237	6.5
15 0501	1.1	**30**	0559	1.2
0955	7.0		1049	6.7
SA 1728	0.8	SU	1813	1.1
2217	7.0	●	2308	6.6
		31	0627	1.2
			1122	6.7
		M	1838	1.1
			2339	6.6

NOVEMBER

Time	m		Time	m
1 0650	1.2	**16**	0702	0.6
1153	6.6		1144	7.2
TU 1903	1.2	W	1931	0.6
2 0007	6.6	**17**	0008	7.1
0714	1.2		0747	0.7
W 1219	6.5	TH	1231	7.1
1929	1.2		2014	0.8
3 0030	6.5	**18**	0056	7.0
0743	1.3		0832	0.8
TH 1239	6.4	F	1320	6.8
1958	1.3		2057	1.0
4 0050	6.4	**19**	0146	6.8
0815	1.4		0918	1.1
F 1303	6.3	SA	1412	6.5
2030	1.5		2141	1.4
5 0118	6.3	**20**	0239	6.5
0848	1.6		1007	1.4
SA 1335	6.1	SU	1510	6.1
2105	1.7		2231	1.8
6 0155	6.1	**21**	0337	6.1
0926	1.8		1103	1.7
SU 1417	5.9	M	1613	5.8
2145	2.0	◑	2332	2.1
7 0244	5.8	**22**	0440	5.8
1011	2.1		1208	2.0
M 1513	5.6	TU	1724	5.5
◑ 2235	2.2			
8 0354	5.5	**23**	0041	2.3
1112	2.2		0551	5.7
TU 1713	5.4	W	1318	2.0
2347	2.4		1851	5.5
9 0549	5.5	**24**	0153	2.2
1247	2.2		0710	5.7
W 1834	5.5	TH	1427	1.9
			2001	5.6
10 0129	2.2	**25**	0300	2.0
0659	5.8		0814	5.9
TH 1406	1.9	F	1528	1.7
1934	5.9		2052	5.9
11 0238	1.8	**26**	0357	1.8
0754	6.2		0903	6.1
F 1508	1.5	SA	1619	1.5
2025	6.3		2132	6.1
12 0336	1.5	**27**	0443	1.5
0844	6.6		0944	6.3
SA 1605	1.1	SU	1700	1.3
2111	6.6		2207	6.3
13 0431	1.1	**28**	0521	1.4
0929	6.9		1020	6.4
SU 1700	0.9	M	1734	1.3
2155	6.9		2240	6.4
14 0523	0.9	**29**	0552	1.3
1014	7.1		1054	6.5
M 1754	0.7	TU	1805	1.2
○ 2238	7.1	●	2312	6.5
15 0614	0.7	**30**	0622	1.2
1059	7.2		1124	6.4
TU 1844	0.6	W	1836	1.2
2323	7.1		2342	6.5

DECEMBER

Time	m		Time	m
1 0653	1.2	**16**	0739	0.7
1152	6.4		1221	6.9
TH 1908	1.2	F	2007	0.8
2 0008	6.5	**17**	0045	7.0
0726	1.2		0826	0.8
F 1219	6.3	SA	1308	6.7
1941	1.3		2049	1.0
3 0035	6.5	**18**	0132	6.8
0800	1.3		0910	0.9
SA 1248	6.3	SU	1356	6.5
2015	1.4		2130	1.2
4 0105	6.4	**19**	0220	6.6
0836	1.4		0954	1.2
SU 1321	6.2	M	1447	6.2
2050	1.5		2212	1.5
5 0143	6.3	**20**	0310	6.3
0913	1.6		1040	1.5
M 1403	6.1	TU	1540	5.9
2129	1.7		2257	1.8
6 0230	6.1	**21**	0404	6.0
0956	1.8		1130	1.8
TU 1455	5.8	W	1639	5.6
2215	1.9	◑	2351	2.1
7 0329	5.9	**22**	0503	5.7
1050	1.9		1227	2.0
W 1605	5.6	TH	1746	5.4
◑ 2314	2.1			
8 0446	5.8	**23**	0054	2.3
1202	1.9		0610	5.6
TH 1742	5.6	F	1329	2.0
			1858	5.4
9 0035	2.1	**24**	0159	2.2
0611	5.8		0720	5.6
F 1322	1.8	SA	1429	1.9
1857	5.8		2002	5.5
10 0154	1.9	**25**	0301	2.1
0719	6.1		0821	5.8
SA 1429	1.5	SU	1525	1.8
1956	6.1		2053	5.8
11 0259	1.6	**26**	0355	1.8
0816	6.4		0910	5.9
SU 1532	1.3	M	1614	1.6
2048	6.4		2135	6.0
12 0400	1.3	**27**	0441	1.6
0908	6.7		0950	6.1
M 1634	1.0	TU	1657	1.4
2138	6.6		2212	6.2
13 0500	1.0	**28**	0521	1.4
0958	6.9		1025	6.2
TU 1735	0.8	W	1736	1.3
2226	6.8		2247	6.3
14 0557	0.8	**29**	0558	1.3
1047	7.0		1059	6.3
W 1830	0.7	TH	1814	1.2
○ 2313	7.0	●	2319	6.4
15 0650	0.7	**30**	0635	1.2
1134	7.0		1131	6.4
TH 1921	0.7	F	1852	1.2
2359	7.0		2350	6.5
		31	0712	1.1
			1204	6.4
		SA	1929	1.2

Chart Datum: 3·67 metres below Ordnance Datum (Newlyn). HAT is 7·4 metres above Chart Datum.

FREE monthly updates. Register at
www.reedsnauticalalmanac.co.uk

9.3.14 DOVER STRAIT

For orientation only – due to scale not all lights and buoys are shown

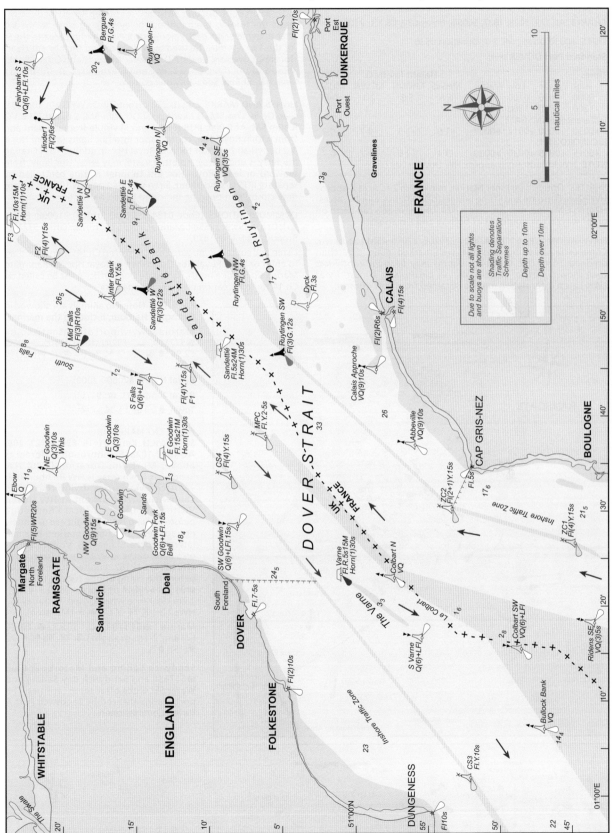

SE England

9.3.15 RAMSGATE

Kent 51°19'·51N 01°25'·50E ❀❀❀♁♁♁✿✿✿

CHARTS AC 323, 1828, 1827, 5605/6; Imray C30, C8, C1, 2100

TIDES +0030 Dover; ML 2·7; Duration 0530

Standard Port DOVER (⟵)

Times				Height (metres)			
High Water		Low Water		MHWS	MHWN	MLWN	MLWS
0000	0600	0100	0700	6·8	5·3	2·1	0·8
1200	1800	1300	1900				
Differences RAMSGATE							
+0030	+0030	+0017	+0007	−1·6	−1·3	−0·7	−0·2
RICHBOROUGH							
+0015	+0015	+0030	+0030	−3·4	−2·6	−1·7	−0·7

NOTE: HW Broadstairs is approx HW Dover +0037.

SHELTER Options: (a) Inner Marina, min depth 2m. Access approx HW ±2 via flap gate and lifting bridge; (b) W Marina, min 2m, access H24; (c) E Marina, min 2m, access H24. Larger vessels can berth on outer wavebreak pontoons of both W and E marinas.

NAVIGATION WPT 51°19'·43N 01°27'·70E, 270°/1·45M to S bkwtr. Commercial shipping uses the well-marked main E-W chan dredged 7·5m. *Due to silting, depths may be significantly less than shown; parts of Eastern Marina, particularly, almost dry at LWS.* Latest information may be obtained from Port Control.

For ent/dep yachts must use the Recommended Yacht Track on the S side of the main buoyed chan. Ent/dep under power, or advise Port Control if unable to motor. Ent/dep Royal Hbr directly; cross the turning basin without delay **keeping close to the W Pier to avoid shoal patch alongside E Pier.** Holding area to the S of the S bkwtr must be used by yachts to keep the hbr ent clear for freight vessels. Beware Dike Bank to the N and Quern Bank close S of the chan. Cross Ledge and Brake shoals are further S. Speed limit 5kn. See www.rma.eu.com.

LIGHTS AND MARKS Ldg lts 270°: front Dir Oc WRG 10s 10m 5M; rear, Oc 5s 17m 5M. N bkwtr hd = QG 10m 5M; S bkwtr hd = VQ R 10m 5M. At E Pier, **IPTS** (Sigs 2 and 3) visible from seaward and from within Royal Hbr, control appr into hbr limits (abeam Nos 1 & 2 buoys) and ent/exit to/from Royal Hbr. In addition a Fl Orange lt = ferry is under way; no other vessel may enter Hbr limits from seaward or leave Royal Hbr. Ent to inner marina controlled by separate IPTS to stbd of ent. Siren sounded approx 10 mins before gate closes; non-opening indicated by red ball or light.

COMMUNICATIONS (Code 01843) CGOC (01304) 210008; Police 101; Dr 852853; Ⓗ 225544. Marina Office 572110, Hbr Office 572100; Broadstairs HM 861879.

Listen and contact *Ramsgate Port Control* on Ch 14 when intending to enter or leave Royal Hbr. Only when in Royal Hbr call *Ramsgate Marina* Ch 80 for a berth. Ramsgate Dock Office must be called on Ch 14 for information on Inner Marina Lock.

FACILITIES Note: No animals are allowed in Ramsgate Harbour, including the marinas.

Marina www.portoframsgate.co.uk
☎572100, ⚓ £32/craft<5t, 510⚓+300Ⓥ £2·96, ⚓⛽🔌🅿🔧⚒🛒Ⓔ🔺🏁(40t)🔩 ACA.

Royal Hbr ⚓🏁(0600-2200).

Royal Temple YC ☎591766, 🍴.

Town ⚓✉Ⓑ🛒✕🍴⇌.

ADJACENT HARBOUR
SANDWICH, Kent, **51°16'·83N 01°21'·20E**. AC 1827 1828. Richborough differences above; ML 1·4m; Duration 0520. HW Sandwich Town quay is HW Richborough +1. Access via narrow drying channel to Sandwich; arrive off ent at HW Dover. Visitors should seek local knowledge before arriving by day; night ent definitely not advised. The chan is marked by small lateral buoys and beacons. Visitors' berths on S bank of the River Stour at Town Quay ☎(01304) 612162. Limited turning room before the swing bridge (opens 1H notice ☎01304 620644 or Mobile 0860 378792), 1·7m clearance when closed. Facilities: ⚓ HW±2 ✉Ⓑ🛒✕🍴⇌✈ (Manston).

Marina ☎613783, ⚓ 50⚓+Ⓥ £2·20 (<18m, 2·1m draught) ⚓🗑🔧⚒🔺🏁(15t) ⚓🔩.

Sandwich Sailing and Motorboat Club ☎617650 and **Sandwich Bay Sailing and Water Ski Clubs** offer some facilities. The port is administered by Sandwich Port & Haven Commissioners.

East England

North Foreland to Great Yarmouth

9.4.1	Contents	269
9.4.2	Area map and Distance table	270
9.4.3	Tidal stream charts	272
9.4.4	Lights, buoys and waypoints	274
9.4.5	Passage information	277
	Thames Estuary, North Foreland to London Bridge, Thames/Orwell to Belgium or The Netherlands – 277 • Shoeburyness to River Colne – 301 • River Colne to Harwich, Harwich to Orford Ness, Orford Ness to Great Yarmouth – 309	
9.4.6	Thames Estuary tidal stream charts	278
9.4.7	Thames Estuary (Margate, tide tables/curves)	280
9.4.8	Whitstable	285
	Margate • Herne Bay	
9.4.9	The Swale	286
9.4.10	Queenborough	287
9.4.11	River Medway (Sheerness, tide tables/curves)	288
9.4.12	River Thames (London Bridge, tide tables/curves)	293
	Holehaven • Gravesend • Thames tidal barrier • Gallions Point Marina • South Dock Marina • Poplar Dock Marina • Limehouse Basin Marina • St Katharine Haven • Chelsea Harbour • Brentford Dock Marina	
9.4.13	Southend-on-Sea	301
	Leigh-on-Sea	
9.4.14	River Roach/Havengore	302
9.4.15	Burnham-on-Crouch, tide tables/curves	303
9.4.16	River Blackwater	307
9.4.17	River Colne	308
9.4.18	Walton Backwaters (Walton-on-the-Naze, tide tables/curves)	310
9.4.19	River Stour	315
9.4.20	River Orwell (Harwich, tide tables/curves)	316
9.4.21	River Deben	320
9.4.22	River Ore/Alde	321
9.4.23	Southwold	322
9.4.24	Lowestoft, tide tables/curves	323
9.4.25	Great Yarmouth	327
	Norfolk Broads	

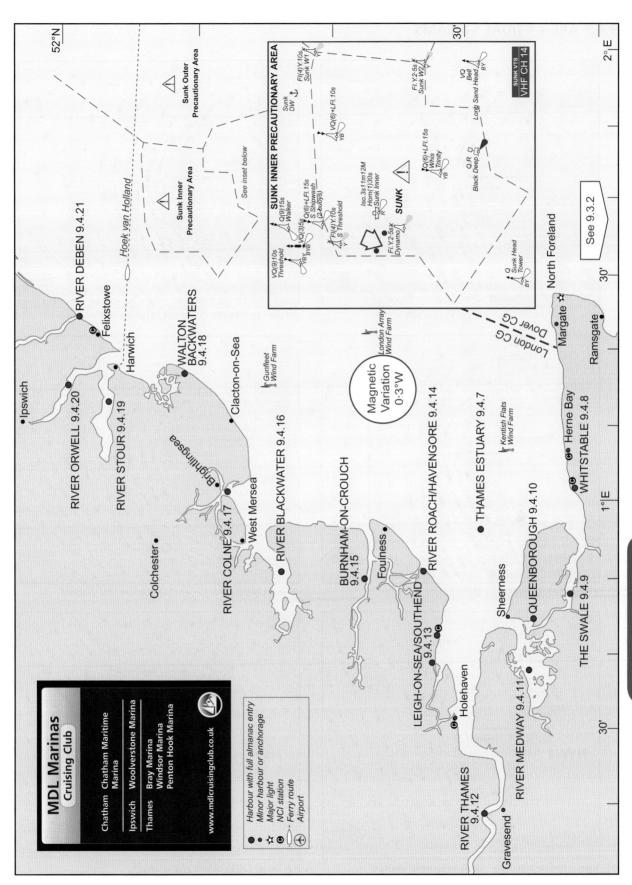

9.4.3 AREA 4 TIDAL STREAMS

5 Hours before HW Dover (0530 before HW Walton)

4 Hours before HW Dover (0430 before HW Walton)

3 Hours before HW Dover (0330 before HW Walton)

2 Hours before HW Dover (0230 before HW Walton)

1 Hour before HW Dover (0130 before HW Walton)

HW Dover (0030 before HW Walton)

Southward 9.3.3   Thames Estuary 9.4.6   Northward 9.5.3   Eastward 9.16.3

1 Hour after HW Dover (0030 after HW Walton)

2 Hours after HW Dover (0130 after HW Walton)

3 Hours after HW Dover (0230 after HW Walton)

4 Hours after HW Dover (0330 after HW Walton)

5 Hours after HW Dover (0430 after HW Walton)

6 Hours after HW Dover (0530 after HW Walton)

9.4.4 LIGHTS, BUOYS AND WAYPOINTS

Bold print = light with a nominal range of 15M or more. CAPITALS = place or feature. *CAPITAL ITALICS* = light-vessel, light float or Lanby. *Italics* = Fog signal. ***Bold italics*** = Racon. Some marks/buoys are fitted with AIS (<u>MMSI No</u>); see relevant charts.

IMPORTANT NOTE. Regular changes are made to Thames Estuary buoyage. Check Notices to Mariners for the latest information.

THAMES ESTUARY – SOUTHERN

(Direction of buoyage generally East to West)

APPROACHES to THAMES ESTUARY
Foxtrot 3 � 51°24'·15N 02°00'·38E; Fl 10s 12m **15M**; *Racon (T) 10M*; *Horn 10s*; <u>992351033</u>.
Falls Hd ⚲ Q; 51°28'·23N 01°49'·89E.
Drill Stone ⚲ Q (3) 10s; 51°25'·88N 01°42'·89E.
Thanet N ⚲ VQ; 51°28'·28N 01°38'·06E.
NE Spit ⚲ VQ (3) *Racon (T) 10M*; 51°27'·93N 01°29'·89E; <u>992351037</u>
East Margate ⚲ Fl R 2·5s; 51°27'·03N 01°26'·40E.
Elbow ⚲ Q; 51°23'·23N 01°31'·59E.
Foreness Pt Outfall ⚲ Fl R 5s; 51°24'·61N 01°26'·02E.
Longnose ⚲ 51°24'·15N 01°26'·08E.
Longnose Spit ⚲ Fl R 2·5s 5m 2M; 51°23'·93N 01°25'·68E.

MARGATE and GORE CHANNEL
SE Margate ⚲ Q (3) 10s; 51°24'·05N 01°20'·40E.
S Margate ⚲ Fl G 2·5s; 51°23'·83N 01°16'·65E.
Copperas ⚲ QG; 51°23'·81N 01°11'·18E.
Reculver ⚲ QR; 51°23'·63N 01°12'·56E.

HERNE BAY
Beltinge Bay Bn ⚲ Fl Y 5s; 51°22'·73N 01°08'·63E.
Landing Stage ⚲ Q 18m 4M, (isolated); 51°22'·91N 01°06'·89E.
N Pier Hd ⚲ 2 FR (vert); 51°22'·43N 01°07'·27E.

WHITSTABLE
Whitstable Street ⚲ ; 51°24'·00N 01°01'·54E. (See The Swale.)
Oyster ⚲ Fl (2) R 10s; 51°22'·14N 01°01'·16E.
W Quay Dn ⚲ Fl G 5s 2m.

THE SWALE
Whitstable Street ⚲ Fl R 2s; 51°24'·00N 01°01'·54E.
Columbine ⚲ Fl G 2s; 51°24'·26N 01°01'·34E.
Columbine Spit ⚲ Fl (3) G 10s; 51°23'·86N 01°00'·03E.
Ham Gat ⚲ Q G; 51°23'·08N 00°58'·32E.
Pollard Spit ⚲ Q R; 51°22'·98N 00°58'·57E.
Sand End ⚲ Fl G 5s; 51°21'·43N 00°55'·90E.
Receptive Point ⚲ Fl G 10s; 51°20'·86N 00°54'·41E.
Queenborough Spit ⚲ Q (3) 10s; 51°25'·81N 00°43'·93E.
South Oaze ⚲ Fl R 2s; 51°21'·34N 00°56'·01E.

QUEENS CHANNEL and FOUR FATHOMS CHANNEL
E Margate ⚲ Fl R 2·5s; 51°27'·03N 01°26'·40E.
Spaniard ⚲ Q (3) 10s; 51°26'·23N 01°04'·00E.
Spile ⚲ Fl G 2·5s; 51°26'·43N 00°55'·70E.

PRINCES CHANNEL
Tongue Sand E ⚲ VQ (3) 5s; 51°29'·48N 01°22'·21E.
Tongue Sand N ⚲ Q; 51°29'·68N 01°22'·03E.
Princes Outer ⚲ VQ (6) + L Fl 10s; 51°28'·89N 01°20'·43E.
Princes North ⚲ Q G; 51°29'·25N 01°18'·35E.
Princes South ⚲ Q R; 51°28'·74N 01°18'·26E.
Princes No.1 ⚲ Fl (4) G 15s; 51°29'·23N 01°16'·02E.
Princes No.2 ⚲ Fl (2) R 5s; 51°28'·81N 01°13'·08E.
Princes No.3 ⚲ Fl (2) G 5s; 51°29'·33N 01°13'·10E.
Princes No.4 ⚲ Fl (3) R 10s; 51°28'·83N 01°09'·90E.
Princes No.5 ⚲ Fl (3) G 10s; 51°29'·39N 01°10'·00E.
Princes Mid ⚲ Fl Y 5s; 51°29'·19N 01°09'·00E.
Shivering Sand Twr N ⚲ Q; 51°30'·01N 01°04'·76E.
Shivering Sand Twr S ⚲ Q (6)+L Fl 15s; *Bell*; 51°29'·75N 01°04'·83E.
Princes No.8 ⚲ Fl (2) R 5s; 51°29'·14N 01°03'·00E.
Princes Inner ⚲ Fl Y 2·5s; 51°29'·59N 01°03'·47E.

FOULGER'S GAT and KNOB CHANNEL
N Edinburgh Channel is not buoyed. Foulger's Gat, lies within the boundary of the London Array Wind Farm.
Long Sand Outer ⚲ Iso 2s; 51°34'·61N 01°28'·34E.
Long Sand Middle ⚲ Iso 5s; 51°35'·60N 01°26'·45E.
Long Sand Inner ⚲ Iso 2s; 51°38'·78N 01°25'·44E.
SE Knob ⚲ Fl G 5s; 51°30'·89N 01°06'·41E.
Knob ⚲ Iso 5s; *Whis*; 51°30'·69N 01°04'·28E.

OAZE DEEP
Oaze Deep ⚲ Fl (2) G 5s; 51°29'·18N 00°56'·77E.
Red Sand Trs N ⚲ Fl (3) R 10s; *Bell*; 51°28'·73N 00°59'·32E.
N Oaze ⚲ QR; 51°30'·03N 00°57'·65E.
Oaze ⚲ Fl (4) Y 10s; 51°29'·06N 00°56'·93E.
W Oaze ⚲ Iso 5s; 51°29'·06N 00°55'·43E.
Oaze Bank ⚲ Q G 5s; 51°33'·24N 00°46'·32E.
Cant ⚲ (unlit); 51°27'·77N 00°53'·36E.
East Cant ⚲ QR; 51°28'·53N 00°55'·60E.

MEDWAY, SHEERNESS
Medway ⚲ Mo (A) 6s; 51°28'·83N 00°52'·81E.
No. 1 ⚲ Fl G 2·5s; 51°28'·55N 00°50'·50E.
No. 2 ⚲ Q; 51°28'·33N 00°50'·52E.
No. 7 ⚲ Fl G 10s; 51°27'·91N 00°47'·52E.
No. 9 ⚲ Fl G 5s; 51°27'·74N 00°46'·61E.
No. 11 ⚲ Fl (3) G 10s; 51°27'·51N 00°45'·80E.
Grain Hard ⚲ Fl G 5s; 51°26'·98N 00°44'·17E.
Isle of Grain ⚲ Q 20m 13M; RW ◇ on R twr; 51°26'·70N 00°43'·38E.
Queenborough Spit ⚲ Q (3) 10s; 51°25'·81N 00°43'·93E.

RIVER THAMES

SEA REACH, NORE and YANTLET
Sea Reach 1 (N) ⚲ Fl Y 2·5s; *Racon (T) 10M*; 51°29'·59N 00°52'·71E.
Sea Reach 1 (S) ⚲ Fl R 2·5s; 51°29'·39N 00°52'·45E.
Sea Reach 2 (N) ⚲ Fl R 2·5s; 51°29'·49N 00°49'·73E.
Sea Reach 2 (S) ⚲ Fl R 2·5s; 51°29'·30N 00°49'·75E.
Sea Reach 3 (N) ⚲ Q G (sync); 51°29'·41N 00°46'·80E.
Sea Reach 3 (S) ⚲ Q R (sync); 51°29'·22N 00°46'·71E.
Sea Reach 4 (N) ⚲ Fl (2) G 5s; 51°29'·69N 00°44'·19E.
Sea Reach 4 (S) ⚲ Fl (2) R 5s; 51°29'·50N 00°44'·12E.
Sea Reach 5 (N) ⚲ VQ G; 51°30'·04N 00°41'·47E.
Sea Reach 5 (S) ⚲ VQ R; 51°29'·85N 00°41'·43E.
Sea Reach 6 (N) ⚲ Fl G 5s; 51°30'·13N 00°39'·87E.
Sea Reach 6 (S) ⚲ Fl R 5s; 51°29'·93N 00°39'·84E.
Sea Reach 7 (N) ⚲ Fl Y 2·5s; 51°30'·21N 00°36'·94E.
Sea Reach 7 (S) ⚲ Fl R 2·5s; *Racon (T) 10M*; 51°30'·01N 00°36'·91E.
Nore Swatch ⚲ Fl (4) R 15s; 51°28'·28N 00°45'·55E.
Mid Swatch ⚲ Fl G 5s; 51°28'·68N 00°44'·16E.
W Nore Sand ⚲ Fl (3) R 10s; 51°29'·48N 00°40'·87E.
East Blyth ⚲ Q; 51°29'·73N 00°37'·81E.
Mid Blyth ⚲ Q; 51°30'·08N 00°32'·38E.

LEIGH-ON-SEA and SOUTHEND-ON-SEA
Leigh ⚲; 51°31'·07N 00°42'·56E.
Southend Pier E End ⚲ 2 FG (vert) 7m; *Horn Mo (N) 30s, Bell (1)*

LOWER HOPE REACH
West Blyth ⚲ Q R; 51°29'·57N 00°28'·39E.
Mucking 1 ⚲ Q G; 51°29'·70N 00°37'·28E.
Lower Hope ⚲ Fl R 5s; 51°29'·32N 00°28'·02E.
Mucking 3 ⚲ Fl G 2·5s; 51°29'·33N 00°27'·68E.
Mucking 5 ⚲ Fl (3) G 10s; 51°28'·76N 00°27'·19E.
Mucking 7 ⚲ Fl G 5s; 51°28'·03N 00°26'·77E.
Bell Ovens ⚲ Q G; 51°27'·50N 00°26'·35E.
Higham ⚲ Fl (2) R 5s; 51°27'·40N 00°26'·85E.
Tilbury ⚲ Q(6) + L Fl 15s; 51°27'·16N 00°25'·50E.

GRAVESEND
Shornmead ⚲ Fl (2) WR 10s 12m, W11/7M, R11M; vis 070°-W-084°-R(Intens)-089°-W(Intens)-094°-W-250°; 51°26'·92N 00°26'·24E.
Northfleet Upper ☆ Oc WRG 10s 30m **W16M**, R12M, G12M; vis:126°-R-149°-W-159°-G-268°-W-279°; 51°26'·93N 00°20'·06E.

THAMES TIDAL BARRIER
Spans B, C, D, E, F, G are navigable. Spans C – F are for large ships. Eastbound small craft/yachts use Span B (51°29'·73N 00°02'·23E) Westbound small craft/yachts use Span G (51°29'·91N 00°02'·21E). **Red X** indicates span closed **Green** → indicates span open. W lights show either side of open spans in poor visibility.

KENTISH KNOCK
Kentish Knock ⚓ Q (3) 10s; 51°38'·08N 01°40·43E.
S Knock ⚓ Q (6) + L Fl 15s; *Bell*; 51°34'·13N 01°34'·29E.

KNOCK JOHN CHANNEL
No. 7 ▲ Fl (4) G 15s; 51°32'·03N 01°06'·40E.
No. 5 ▲ Fl (3) G 10s; 51°32'·49N 01°07'·75E.
No. 4 ⬟ QR 10s; 51°32'·40 N 01°08'·08E.
No. 2 ⬟ Fl (3) R 10s; 51°33'·03N 01°09'·63E.
No. 3 ⚓ Q (6) + L Fl 15s; 51°33'·29N 01°09'·83E.
No. 1 ▲ Fl G 5s; 51°33'·75N 01°10'·72E.
Knock John ⬟ Fl (2) R 5s; 51°33'·61N 01°11'·37E.

BLACK DEEP
No. 12 ⬟ Fl (4) R 15s; 51°33'·83N 01°13'·50E.
No. 11 ▲ Fl (3) G 10s; 51°34'·33N 01°13'·40E.
No. 10 ⬟ Fl (3) R 10s; 51°34'·74N 01°15'·60E.
No. 9 ⚓ Q (6) + L Fl 15s; 51°35'·13N 01°15'·09E.
No. 8 ⚓ Q (9) 15s; 51°36'·36N 01°20'·43E.
No. 7 ▲ QG. 51°37'·08N 01°17'·69E.
No. 6 ⬟ Fl R 2·5s; 51°38'·53N 01°24'·41E.
No. 5 ⚓ VQ (3) 5s; 51°39'·53N 01°23'·00E.
No. 4 ⬟ Fl (2) R 5s; 51°41'·42N 01°28'·49E.
Long Sand Bcn ⚐ ; 51°41'·48N 01°29'·49E.
No. 3 ▲ Fl (3) G 15s; 51°42'·39N 01°26'·66E.
No. 1 ▲ Fl G 5s; 51°44'·03N 01°28'·09E.
No. 2 ⬟ Fl (4) R 15s; 51°45'·63N 01°32'·20E.
SHM ⚓ Fl Y 2·5s; *Racon (T) 10M*; 51°29'·59N 00°52'·71E.

SUNK
Sunk Head Tower ⚓ Q; *Whis*; 51°46'·63N 01°30'·51E.
Black Deep ⬟ QR; 51°48'·10N 01°36'·60E.
Long Sand Head ⚓ VQ; *Whis*; 51°47'·90N 01°39'·42E.
SUNK CENTRE ⬡ 51°50'·11N 01°46'·02E; Fl (2) 20s 12m **16M**; *Racon (C) 10M*; *Horn (2) 60s*; 992351094.
Dynamo ⚓ Fl Y 2·5s; 51°29'·59N 00°52'·71E.
Trinity ⚓ Q (6) + L Fl 15s; 51°49'·03N 01°36'·39E.
SUNK INNER ⬡ 51°51'·17N 01°34'·40E; Iso 3s 11m 12M; *Racon (C) 10M*; *Horn 30s*; 992351027.

FISHERMANS GAT
Outer Fisherman ⚓ Q (3) 10s; 51°34'·02N 01°25'·10E.
Fisherman No. 1 ▲ Fl G 2·5s (sync); 51°34'·50N 01°23'·52E.
Fisherman No. 2 ⬟ Fl R 2·5s (sync); 51°34'·30N 01°23'·50E.
Fisherman No. 3 ▲ Fl G 5s; 51°34'·78N 01°22'·65E.
Fisherman No. 4 ⬟ Fl (2) R 5s; 51°34'·77N 01°22'·08E.
Fisherman No. 5 ▲ Fl G 5s; 51°35'·25N 01°21'·84E.
Fisherman No. 6 ⬟ Fl (3) R 10s; 51°35'·08N 01°21'·56E.
Inner Fisherman ⬟ Q R; 51°36'·10N 01°19'·97E.

BARROW DEEP
SW Barrow ⚓ Q(6) + L Fl 15s; *Bell*; 51°32'·29N 01°00'·31E.
Alice Wk ⚓ Fl (2) 5s; 51°32'·33N 01°04'·22E.
Barrow No. 14 ⬟ Fl R 2·5s; 51°31'·83N 01°00'·43E.
Barrow No. 13 ▲ Fl (2) G 5s; 51°32'·82N 01°03'·07E.
Barrow No. 12 ▲ Fl (2) R 5s; 51°33'·24N 01°05'·94E.
Barrow No.11 ▲ Fl (3) G 10s; 51°34'·08N 01°06'·70E.
Barrow No. 9 ⚓ VQ (3) 5s; 51°35'·34N 01°10'·30E.
Barrow No. 8 ⬟ Fl (2) R 5s; 51°35'·03N 01°11'·40E.
Barrow No. 6 ⬟ Fl (4) R 15s; 51°37'·30N 01°14'·69E.
Barrow No. 5 ▲ Fl G 10s; 51°40'·03N 01°16'·20E.
Barrow No. 4 ⚓ VQ (9) 10s; 51°39'·88N 01°17'·48E.
Barrow No. 3 ⚓ Q (3) 10s; *Racon (M)10M*; 51°42'·02N 01°20'·24E.
Barrow No. 2 ⬟ Fl (2) R 5s; 51°41'·98N 01°22'·89E.

WEST SWIN and MIDDLE DEEP
Blacktail (W) ⚐ ; 51°31'·46N 00°55'·19E.
Maplin ⚓ Q G (sync with W Swin); *Bell*; 51°33'·66N 01°01'·40E.
W Swin ⚓ Q R (sync with Maplin); 51°33'·40N 01°01'·97E.
Maplin Edge ▲ Fl G 2.5s; 51°35'·33N 01°03'·64E.
Maplin Bank ⬟ Fl (3) R 10s; 51°35'·50N 01°04'·70E.

EAST SWIN and KING'S CHANNEL
NE Maplin ▲ Fl G 5s; *Bell*; 51°37'·43N 01°04'·90E.
W Hook Middle ⬟ 51°39'·18N 01°07'·97E.
S Whitaker ▲ Fl (2) G 10s; 51°40'·17N 01°09'·11E.
N Middle ⚓ Q; 51°41'·35N 01°12'·61E.
W Sunk ⚓ Q (9) 15s; 51°44'·33N 01°25'·80E.
Gunfleet Spit ⚓ Q (6) + L Fl 15s; *Bell*; 51°45'·33N 01°21'·70E.

WHITAKER CHANNEL and RIVER CROUCH
Whitaker ⚓ Q (3) 10s; 51°41'·43N 01°10'·51E.
Inner Whitaker ⚓ VQ (6) + L Fl 10s; 51°40'·76N 01°08'·40E.
Swin Spitway ⚓ Iso 10s; *Bell*; 51°41'·95N 01°08'·35E.
Whitaker 1 ▲ FlG 5s (sync Whit 2), 51° 40'.69N 01° 06'.67E.
Whitaker 2 ⬟ FlR 5s (sync Whit 1), 51° 40'.41N 01° 06'.78E.
Whitaker 3 ▲ Fl(2) G 5s (sync Whit 4) 51° 40'.41N 01° 04'.74E.
Whitaker 4 ⬟ Fl(2) R 5s (sync Whit 4) 51° 40'.16N 01° 04'.91E.
Whitaker 5 ▲ Fl(3) G 10s (sync Whit 6) 51° 40'.03N 01° 03'.22E.
Whitaker 6 ⬟ Fl(3) R 10s(sync Whit 5) 51° 39'.77N 01° 03'.43E.
Whitaker 7 ▲ Fl(4) G 10s(sync Whit 8) 51° 39'.54N 01° 02'.00E.
Whitaker 8 ⬟ Fl(4) R 10s(sync Whit 7), 51° 39'.35N 01° 02'.00E.
Swallowtail 1 ⬟ Fl Y 5s, 51° 41'.38N 01° 08'.20E.
Swallowtail 2 ⬟ Fl Y 10s, 51° 41'.19N 01° 06'.28E.
Swallowtail 3 ⬟ Fl Y 15s, 51° 40.84N 01° 04.39E.
Swallowtail 4 ⬟ (2) Y 10s, 51° 40'.52N 01° 03'.47E.
Buxey Edge ▲ Fl G 10s, 51° 40'.65N 01° 03'.48E.
Swallowtail ⚓ VQ (9) 10s, 51° 40'.04N 01° 02'.65E.
Sunken Buxey ⚓ VQ; 51°39'·59N 01°00'·77E.
Buxey No. 1 ⚓ VQ (6) + L Fl 10s; 51°39'·18N 01°01'·13E.
Buxey No. 2 ⚓ Q; 51°39'·08N 01°00'·23E.
Outer Crouch 1 ▲ Fl G 5s (sync Cro 2) 51° 38'.71N 00° 59'.00E.
Outer Crouch 2 ⬟ Fl R 5s (sync Cro 1) 51° 38'.62N 00° 59'.20E.
Outer Crouch 3 ▲ Fl G 10s (sync Cro 4) 51° 38'.10N 00° 57'.83E.
Outer Crouch 4 ⬟ Fl R 10s (sync Cro 3) 51° 37'.95N 00° 58'.00E.
Crouch ⚓ Q.Fl, 51° 37.650N 00° 56.582E.
Inner Crouch ⚓ L Fl 10s, 51° 37'.19N 00° 55'.09E.
Branklet (RGR) **Pref Chan to Stbd,** Comp grp Fl (R) 2+1 10s, 51° 36'.99N 00° 52'.10E.

GOLDMER GAT and WALLET
NE Gunfleet ⚓ Q (3) 10s; 51°49'·93N 01°27'·79E.
Wallet No. 2 ⬟ Fl R 5s; 51°48'·88N 01°22'·99E.
Wallet No. 4 ⬟ Fl (4) R 10s; 51°46'·53N 01°17'·23E.
Wallet Spitway ⚓ L Fl 10s; *Bell*; 51°42'·86N 01°07'·30E.
Knoll ⚓ Q; 51°43'·88N 01°05'·07E.
Eagle ▲ QG; 51°44'·13N 01°03'·82E.
N Eagle ⚓ Q; 51°44'·71N 01°04'·32E.
NW Knoll ⬟ Fl (2) R 5s; 51°44'·35N 01°02'·17E.
Colne Bar ▲ Fl (2) G 5s; 51°44'·61N 01°02'·57E.
Bench Head ▲ Fl (3) G 10s; 51°44'·69N 01°01'·10E.

RIVER BLACKWATER
The Nass ⚓ VQ (3) 5s 6m 2M; 51°45'·83N 00°54'·83E.
Thirslet ▲ Fl (3) G 10s; 51°43'·73N 00°50'·39E.
No. 1 ▲ ; 51°43'·44N 00°48'·02E.

RIVER COLNE and BRIGHTLINGSEA
Inner Bench Head No. 2 ⬟ Fl (2) R 5s; 51°45'·96N 01°01'·74E.
Colne Pt No. 1 ▲ Fl G 3s; 51°46'·01N 01°01'·92E.
No. 8 ⬟ Fl R 3s; 51°46'·90N 01°01'·30E.
No. 9 ▲ Fl G 3s; 51°47'·36N 01°01'·07E.

Ldg lts 041°. Front, FR 7m 4M; W □, R stripe on post; vis: 020°-080°; 51°48'·39N 01°01'·20E. Rear, 50m from front, FR 10m 4M; W □, R stripe on post. FR lts are shown on 7 masts between 1·5M and 3M NW when firing occurs.

WALTON BACKWATERS
Pye End ⊿ L Fl 10s; 51°55'·03N 01°17'·90E.
No. 2 ⌓ Fl (2) 5s; 51°54'·62N 01°16'·80E.
Crab Knoll No. 3 ▲ Fl G 5s; 51°54'·41N 01°16'·41E.
No. 16 ⌓ 51°53'·17N 01°15'·52E.

HARWICH APPROACHES
(Direction of buoyage is North to South)

MEDUSA CHANNEL
Medusa ▲ Fl G 5s; 51°51'·23N 01°20'·35E.
Stone Banks ⌓ FlR 5s; 51°53'·19N 01°19'·23E.
Pennyhole ⌔ ; 51°53'·55N 01°18'·00E (Mar–Sep).

CORK SAND and ROUGH SHOALS
S Cork ⌕ Q (6) + L Fl 15s; 51°51'·33N 01°24'·09E.
SE Roughs Tower ⌕ Q (3) 10s; 51°53'·64N 01°28'·94E.
NW Roughs Tower ⌕ VQ (9) 10s; 51°53'·81N 01°28'·77E.
Cork Sand ⌕ Fl (3) R 10s; 51°55'·51N 01°25'·42E.

HARWICH CHANNEL
S Threshold ⌕ Fl (4) Y 10s; 51°52'·20N 01°33'·14E.
S Shipwash ⌕⌕ 2 By(s) Q (6) + L Fl 15s; 51°52'·71N 01°33'·97E.
Outer Tidal Bn ⌕ Mo (U) 15s 2m 3M; 51°52'·85N 01°32'·34E.
E Fort Massac ⌕ VQ (3) 5s; 51°53'·36N 01°32'·79E.
W Fort Massac ⌕ VQ (9) 10s; 51°53'·36N 01°32'·49E.
Walker ⌕ Q (9)15s; 51°53'·79N 01°33'·90E.
N Threshold ⌕ Fl Y 5s; 51°54'·49N 01°33'·47E.
SW Shipwash ▲ Fl Y 2·5s; 51°54'·75N 01°34'·21E.
Haven ⌕ Mo (A) 5s; 51°55'·76N 01°32'·56E.
W Shipwash ⌓ Fl (2) R 10s; 51°57'·13N 01°35'·89E.
NW Shipwash ⌓ Fl R 5s; 51°58'·98N 01°37'·01E.
Harwich App (HA) ⌕ Iso 5s; 51°56'·75N 01°30'·66E.
Cross ⌕ Fl (3) Y 10s; 51°56'·23N 01°30'·48E.
Harwich Chan No. 1 ⌓ Fl Y 2·5s; *Racon (T) 10M*; 51°56'·13N 01°27'·06E.
Harwich Chan No. 3 ⌓ Fl (3) Y 10s; 51°56'·04N 01°25'·54E.
Harwich Chan No. 5 ⌓ Fl (5) Y 10s; 51°55'·96N 01°24'·01E.
Harwich Chan No. 7 ⌓ Fl (3) Y 10s; 51°55'·87N 01°22'·49E.
S Bawdsey ⌕ Q (6) + L Fl 15s; *Whis*; 51°57'·23N 01°30'·19E.
Washington ▲ QG; 51°56'·52N 01°26'·59E.
Felixstowe Ledge ▲ Fl (3) G 10s; 51°56'·30N 01°23'·72E.
Wadgate Ledge ⌕ Fl (4) G 15s; 51°56'·16N 01°21'·99E.
Platters ⌕ Q (6) + L Fl 15s; 51°55'·64N 01°20'·97E.
Rolling Ground ▲ QG; 51°55'·55N 01°19'·75E.
Beach End ▲ Fl (2) G 5s; 51°55'·62N 01°19'·21E.
Cork Sand Yacht Bn ⌕ VQ 2M; 51°55'·21N 01°25'·20E.
Rough ⌕ VQ; 51°55'·19N 01°31'·00E.
Pitching Ground ⌓ Fl (4) R 15s; 51°55'·43N 01°21'·05E.
Inner Ridge ⌓ QR; 51°55'·38N 01°20'·20E.
Deane ⌓ L Fl R 6s; 51°55'·36N 01°19'·28E.
Landguard ⌕ Q; 51°55'·45N 01°18'·84E.

RIVERS STOUR AND ORWELL
RIVER STOUR and HARWICH
Shotley Spit ⌕ Q (6) + L Fl 15s; 51°57'·21N 01°17'·69E.
Shotley Marina Lock E side Dir lt 339·5° 3m 1M (uses Moiré pattern); Or structure; 51°57'·46N 01°16'·60E.
Guard ⌓ Fl R 5s; *Bell*; 51°57'·07N 01°17'·86E.

RIVER ORWELL and IPSWICH
Suffolk Yacht Harbour. Ldg lts Front Iso Y 1M; 51°59'·73N 01°16'·09E. Rear Oc Y 4s 1M.

HARWICH TO ORFORD NESS
FELIXSTOWE, R DEBEN and WOODBRIDGE HAVEN
Woodbridge Haven ⌓ Mo(A)15s; 51°58'·20N 01°23'·85E.
Deben ⌔ ; 51°59'·30N 01°23'·53E.

RIVERS ORE and ALDE
Orford Haven ⌓ L Fl 10s; *Bell*. 52°01'·85N 01°28'·20E.

OFFSHORE MARKS
S Galloper ⌕ Q (6) L Fl 15s; *Racon (T) 10M*; 51°43'·98N 01°56'·43E.
N Galloper ⌕ Q; 51°49'·84N 01°59'·99E.
S Inner Gabbard ⌕ Q (6) + L Fl 15s. 51°49'·92N 01°51'·89E.
N Inner Gabbard ⌕ Q; 51°59'·20N 01°56'·00E.
Outer Gabbard ⌕ Q (3) 10s; *Racon (O) 10M*; 51°57'·83N 02°04'·19E.
NHR-SE ▲ Fl G 5s; 51°45'·39N 02°39'·89E.

SHIPWASH and BAWDSEY BANK
E Shipwash ⌕ VQ (3) 5s; 51°57'·08N 01°37'·89E.
NW Shipwash ⌓ Fl R 5s; 51°58'·98N 01°37'·01E.
N Shipwash ⌕ Q 7M; *Racon (M) 10M*; *Whis*; 52°01'·73N 01°38'·27E; 992351069.
S Bawdsey ⌕ Q (6) + L Fl 15s; *Whis*; 51°57'·23N 01°30'·22E.
Mid Bawdsey ▲ Fl (3) G 10s; 51°58'·88N 01°33'·59E.
NE Bawdsey ▲ Fl G 10s; 52°01'·73N 01°36'·09E.

CUTLER and WHITING BANKS
Cutler ▲ QG; 51°58'·51N 01°27'·48E.
SW Whiting ⌕ Q (6) + L Fl 10s; 52°00'·96N 01°30'·69E.
Whiting Hook ⌓ Fl R 10s; 52°02'·98N 01°31'·82E.
NE Whiting ⌕ Q (3) 10s; 52°03'·61N 01°33'·32E.

ORFORD NESS TO WINTERTON
(Direction of buoyage is South to North)
Orford Ness Lt Ho (dis) 52°05'·04N 01°34'·45E; W ○ twr, R bands.
Aldeburgh Ridge ⌓ QR; 52°06'·49N 01°36'·95E.

SOUTHWOLD
Southwold ☆ 52°19'·63N 01°40'·89E; Fl 10s 37m **W24M**; vis: 204°-W-032·5°, 992351019.

LOWESTOFT and APPR VIA STANFORD CHANNEL
E Barnard ⌕ Q (3) 10s; 52°25'·14N 01°46'·38E.
Newcome Sand ⌕ QR; 52°26'·28N 01°46'·97E.
S Holm ⌕ VQ (6) + L Fl 10s; 52°26'·85N 01°47'·15E.
Stanford ⌓ Fl R 2·5s; 52°27'·35N 01°46'·67E.
SW Holm ▲ Fl (2) G 5s; 52°27'·87N 01°46'·99E.
Kirkley ⌕ Oc WRG 10s 17m, W8M, R6M, G6M, vis: 210°-G-224°-W-229°-R-313°; 52°27'·71N 01°44'·54E.
Outer Hbr S Pier Hd ⌕ Oc R 5s 12m 6M; *Horn (4) 60s*; Tfc sigs; 52°28'·29N 01°45'·36E.
N Newcome ⌓ Fl (4) R 15s; 52°28'·39N 01°46'·37E.
Lowestoft ☆ 52°29'·22N 01°45'·35; Fl 15s 37m **23M**; W twr; part obscd 347°- shore.

LOWESTOFT NORTH ROAD and CORTON ROAD
Lowestoft Ness SE ⌕ Q (6) + L Fl 15s; 52°28'·84N 01°46'·25E.
Lowestoft Ness N ⌕ VQ (3) 5s; *Bell*; 52°28'·89N 01°46'·23E.
W Holm ▲ Fl (3) G 10s; 52°29'·80N 01°47'·09E.
NW Holm ▲ Fl (4) G 15s; 52°31'·93N 01°46'·70E.

GREAT YARMOUTH APPROACH via HOLM CHANNEL
E Newcome ⌓ Fl (2) R 5s; 52°28'·51N 01°49'·21E.
Holm Approach ⌕ Q (3) 10s; 52°30'·88N 01°50'·22E.
Holm Sand ⌕ Q (9) 15s; 52°33'·18N 01°46'·54E.
S Corton ⌕ Q (6) + L Fl 15s; *Bell*; 52°32'·94N 01°49'·12E.
NE Holm ⌓ Fl R 2·5s; 52°32'·69N 01°48'·48E.
Mid Corton ▲ Fl G 2·5s; 52°33'·62N 01°48'·01E.
N Holm ⌕ Q. 52°33'·93N 01°47'·23E.
W Corton ▲ Fl (3) G 10s; 52°34'·12N 01°47'·50E.

GREAT YARMOUTH and GORLESTON
Gorleston South Pier Hd ⌕ Fl R 3s 11m 11M; vis: 235°-340°; 52°34'·33N 01°44'·28E.
N Pier Hd ⌕ QG 8m 6M; vis: 176°-078°; *Horn(3) 60s*; 52°34'·38N 01°44'·38E.
Outer Hbr S Pier ⌕ Q R 17m 3M; 52°34'·61N 01°44'·73E.
Outer Hbr N Pier ⌕ Q G 17m 3M; 52°34'·70N 01°44'·74E.

9.4.5 PASSAGE INFORMATION

More passage information is threaded between the harbours in this area. See 9.0.7 for distances across the North Sea. Admiralty Leisure Folios: 5606 Thames Estuary from Ramsgate to Tower Bridge, London; 5607 Northern Thames Estuary to Orford Ness; 5614 Orford Ness to Whitby. See also eastcoastsailing.co.uk.
Bibliography: *East Coast Rivers* (Fernhurst/Harber); *East Coast Pilot* (Imray/Jarman,Holness,Cooper); *Dover Strait Pilot* and *North Sea (West) Pilot* (Admiralty NP28 and 54); *North Sea Passage Pilot* (Imray/Navin).

THE THAMES ESTUARY

(AC 1183, 1975, 1607, 1606, 1609) The sandbanks shift constantly in the Thames Estuary; charted depths cannot be relied upon, and an accurate echo sounder is vital. Up to date charts showing the latest buoyage changes and a detailed tidal stream atlas are essential. The commercial deep water route from the Sunk Inner precautionary area to the Yantlet runs through Black Deep and Knock John Channel. It is well marked, as are Barrow Deep, Princes Channel and Fisherman's Gat. The lesser channels and swatchways, which are convenient for yachtsmen, particularly when crossing the estuary, are mainly unmarked and unlit. They should be used with great caution. Good visibility is needed to pick out buoys and marks, and to avoid shipping. Care should be exercised when plotting electronic waypoints. Many former navigation marks and towers have been removed or allowed to disintegrate. Those that are still standing may be in very shallow water.

Newcomers to the estuary are advised to choose a period of neap tides when minimum depths are greater and tidal streams weaker. If a crossing at springs is unavoidable, keep in deep water even if that involves a longer passage.

Wind strength and direction needs careful consideration for each of the course changes during a typical crossing to avoid becoming trapped between shoals, down-wind and down-stream of safe water.

▶ *Study the tides carefully so as to work the streams to best advantage and to ensure sufficient depth at the times and places where you expect to be, or might be later. In principle it is best to make most of the crossing on a rising tide. However, a typical passage from N Foreland to the Blackwater will take 7-9 hours, so some adverse stream is almost inevitable. The stream at springs runs at 3kn in places, mostly along the channels but sometimes across the intervening banks. A short, steep sea may be raised with the wind against tide.* ◀

CROSSING THE THAMES ESTUARY

(AC 1183, 1975, 1607, 1606, 1609) *Crossing the Thames Estuary* (Imray) gives 32 suggested routes. Making N from N Foreland to Orford Ness or beyond it may be better to keep to seaward of the main banks, via Kentish Knock and Long Sand Head buoys, thence to N Shipwash lt buoy, 14M further N.

Bound NW from North Foreland it is approximately 35M to the mouths of the Rivers Crouch, Blackwater or Colne. One route is through the Princes Channel, thence N of the Oaze Deep precautionary area and S of the Knob and West Barrow banks to the West Swin, before turning NE into Middle Deep and the East Swin. The Wallet Spitway then leads NW to the Colne and Blackwater. Many routes may be followed, depending on wind direction, tidal conditions and confidence in electronic aids in the absence of marks. Passage over Sunk Sand and Knock John bank is best avoided, particularly in winds >F5.

Southbound from the Colne or Blackwater, leave 1-2 hours before HW and make best speed to skirt or cross the shoals while there is sufficient depth of water. The latter part of the passage will be in generally deeper water of Edinburgh Channel, Fisherman's or Foulger's Gat or (keeping clear of the turbines in the London Array), with a favourable S-going stream just starting off North Foreland.

From the NE follow Goldmer Gat and the Wallet (for Rivers Colne/Blackwater) or King's Channel and Whittaker Channel for the River Crouch.

London VTS has radar coverage from Greenwich to a line between the Naze and Margate. Keep a listening watch on VHF Ch 69 to monitor shipping activity. If really necessary, *London VTS* may be able to give navigational help to yachts.

NORTH FORELAND TO LONDON BRIDGE

North Foreland has a conspic lt ho (AC 1828), with buoys offshore.
▶ *From HW Dover –0120 to +0045 the stream runs N from The Downs and W into Thames Estuary. From HWD +0045 to +0440 the N-going stream from The Downs meets the E-going stream from Thames Estuary, which in strong winds causes a bad sea. From HWD –0450 to –0120 the streams turn W into Thames Estuary and S towards The Downs. If bound for London, round N Foreland against the late ebb in order to carry a fair tide from Sheerness onward.* ◀

The most direct route from North Foreland to the Thames and Medway is via South, Gore and Copperas Channels, then Overland Passage, Four Fathoms Channel and The Cant. Minimum depth is about 2m but may be significantly less off Reculver and leaving Copperas Channel. It is not well marked. Beware of Kentish Flats wind farm (30 turbines) with submarine cables to Herne Bay. An alternative, deeper route is east of Margate Sand and the Tongue, via the Princes Channel to the Oaze Deep. The N & S Edinburgh Channels are unmarked but navigable with caution.

▶ *W-going streams begin at approx HW Sheerness –0600 and E-going at HW Sheerness +0030.* ◀

Margate or Whitstable afford little shelter for yachts. The Swale provides an interesting inside route S of the Isle of Sheppey with access to Sheerness and the R Medway. If sailing from N Foreland to the Thames, Queenborough offers the first easily accessible, all-tide, deep-water shelter. In the east part of The Swale, Harty Ferry provides a sheltered ⚓, but not in strong winds from N or E. The Medway Channel is the main approach to Sheerness from The Warp.

CROSSING FROM THAMES/ORWELL TO BELGIUM OR THE NETHERLANDS

(ACs 1408, 1406, 1610, 1630, 1872, 1183) Up to date charts are *essential*. Thanet Wind Farm is now established 6Nm ENE of N Foreland and bounded by cardinal buoys. The southern North Sea is an extremely busy area with several large and complex Traffic Separation Schemes and associated Precautionary Areas which should be avoided if at all possible. Where practicable, navigate outside the schemes using any available ITZs, and always cross a TSS on a heading at 90° to the traffic flow in accordance with the Collision Regulations. From the Thames to Zeebrugge, the most direct route crosses the Nord Hinder South TSS in the vicinity of F3 light float. This is a focal point for dense crossing traffic. The very busy area around West Hinder must then be negotiated. A longer but much safer route (or if bound for Dunkerque, Nieuwpoort or Oostende) via S Falls, Sandettié and Dyck minimises time spent in the TSS and provides the option of breaking the passage in Ramsgate.

From the Orwell (or adjacent rivers) to the Netherlands, it may be best to pass north of the Sunk Inner and Sunk Outer TSS/Precautionary Areas before setting course for Hoek van Holland. From S Shipwash make for N Inner Gabbard before setting course for NHR-S to cross the North Hinder TSS to NHR-SE, then pass S of the Maas-West TSS leaving MW1, MW3 and MW5 buoys to port before shaping up for MV-N buoy.

▶ *Care must be taken throughout with the tidal streams, which may be setting across the yacht's track.* ◀

The whole area is relatively shallow, and in bad weather seas are steep and short. (See also Area 16.)

9.4.6 THAMES ESTUARY TIDAL STREAMS

Due to very strong rates of tidal streams in some areas, eddies may occur. Where possible, some indication of these is shown, but in many areas there is insufficient information or eddies are unstable.

5 Hours before HW Sheerness (0335 before HW Dover)

4 Hours before HW Sheerness (0235 before HW Dover)

3 Hours before HW Sheerness (0135 before HW Dover)

2 Hours before HW Sheerness (0035 before HW Dover)

1 Hour before HW Sheerness (0025 after HW Dover)

HW Sheerness (0125 after HW Dover)

Due to very strong rates of tidal streams in some areas, eddies may occur. Where possible, some indication of these is shown, but in many areas there is insufficient information or eddies are unstable.

1 Hour after HW Sheerness (0225 after HW Dover)

2 Hours after HW Sheerness (0325 after HW Dover)

3 Hours after HW Sheerness (0425 after HW Dover)

4 Hours after HW Sheerness (0525 after HW Dover)

5 Hours after HW Sheerness (0600 before HW Dover)

6 Hours after HW Sheerness (0500 before HW Dover)

E England

9.4.7 THAMES ESTUARY

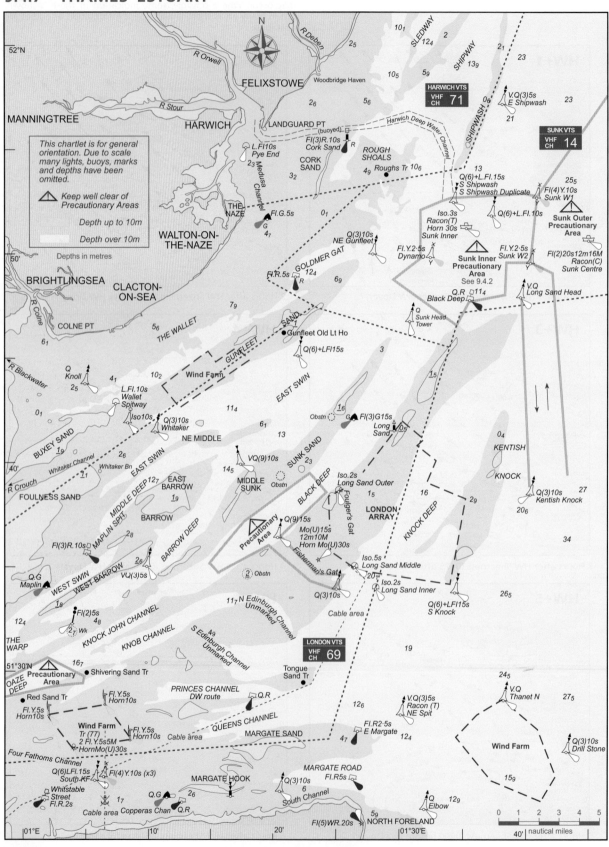

This chartlet is for general orientation. Due to scale many lights, buoys, marks and depths have been omitted.

⚠ Keep well clear of Precautionary Areas

Depth up to 10m

Depth over 10m

Depths in metres

STANDARD TIME (UT)
For Summer Time add ONE hour in **non-shaded areas**

MARGATE   LAT 51°23'N   LONG 1°23'E
TIMES AND HEIGHTS OF HIGH AND LOW WATERS

Dates in red are **SPRINGS**
Dates in blue are **NEAPS**

YEAR 2016

JANUARY

Time	m		Time	m
1 0357	4.3	**16**	0357	4.6
1023	0.9		1029	0.6
F 1631	4.1	SA	1639	4.4
2225	1.4	◑	2239	1.0
2 0441	4.2	**17**	0449	4.5
1109	1.1		1125	0.7
SA 1718	4.0	SU	1737	4.3
◐ 2317	1.5		2341	1.1
3 0534	4.0	**18**	0552	4.4
1207	1.2		1232	0.8
SU 1823	3.9	M	1850	4.2
4 0023	1.6	**19**	0057	1.2
0650	3.9		0709	4.3
M 1321	1.3	TU	1345	0.9
1939	3.9		2005	4.2
5 0158	1.6	**20**	0216	1.2
0804	3.9		0824	4.3
TU 1430	1.2	W	1456	1.0
2041	4.0		2116	4.3
6 0305	1.4	**21**	0329	1.0
0905	4.0		0939	4.4
W 1526	1.1	TH	1601	1.0
2136	4.2		2220	4.4
7 0359	1.2	**22**	0437	0.9
0959	4.2		1043	4.5
TH 1615	1.0	F	1658	0.9
2224	4.4		2313	4.5
8 0447	1.0	**23**	0534	0.7
1047	4.4		1134	4.6
F 1659	0.9	SA	1745	0.9
2306	4.5		2357	4.6
9 0531	0.8	**24**	0621	0.5
1129	4.5		1219	4.7
SA 1739	0.9	SU	1825	0.8
2346	4.6	○		
10 0612	0.7	**25**	0035	4.7
1211	4.6		0701	0.6
SU 1817	0.8	M	1259	4.7
●			1901	0.8
11 0025	4.7	**26**	0110	4.7
0651	0.6		0738	0.5
M 1253	4.7	TU	1336	4.7
1855	0.7		1934	0.8
12 0106	4.8	**27**	0145	4.7
0732	0.5		0811	0.5
TU 1338	4.8	W	1411	4.6
1935	0.7		2007	0.9
13 0147	4.8	**28**	0220	4.7
0813	0.4		0841	0.6
W 1422	4.8	TH	1445	4.5
2018	0.7		2039	1.0
14 0230	4.8	**29**	0254	4.6
0856	0.4		0912	0.7
TH 1507	4.7	F	1517	4.4
2102	0.8		2112	1.0
15 0312	4.7	**30**	0327	4.5
0941	0.5		0944	0.8
F 1551	4.6	SA	1551	4.3
2148	0.9		2148	1.2
		31	0403	4.3
			1021	0.9
		SU	1629	4.1
			2230	1.3

FEBRUARY

Time	m		Time	m
1 0444	4.1	**16**	0529	4.4
1106	1.1		1159	0.9
M 1715	3.9	TU	1821	4.1
◐ 2322	1.5			
2 0536	3.9	**17**	0031	1.2
1205	1.3		0647	4.2
TU 1813	3.8	W	1318	1.1
			1941	4.0
3 0033	1.6	**18**	0156	1.2
0645	3.8		0811	4.2
W 1323	1.4	TH	1439	1.2
1939	3.8		2059	4.1
4 0208	1.5	**19**	0321	1.1
0814	3.9		0932	4.3
TH 1442	1.3	F	1553	1.1
2055	4.0		2208	4.3
5 0321	1.3	**20**	0434	0.9
0924	4.0		1048	4.4
F 1543	1.1	SA	1652	1.0
2154	4.2		2303	4.5
6 0418	1.0	**21**	0527	0.7
1022	4.3		1128	4.5
SA 1634	1.0	SU	1736	0.9
2244	4.4		2346	4.6
7 0508	0.7	**22**	0609	0.6
1111	4.5		1209	4.6
SU 1719	0.9	M	1810	0.9
2328	4.6	○		
8 0554	0.6	**23**	0021	4.6
1155	4.6		0643	0.5
M 1801	0.8	TU	1243	4.6
●			1841	0.8
9 0009	4.7	**24**	0052	4.7
0636	0.5		0713	0.5
TU 1240	4.8	W	1313	4.6
1842	0.7		1913	0.8
10 0051	4.9	**25**	0122	4.7
0718	0.3		0742	0.5
W 1324	4.9	TH	1344	4.6
1923	0.6		1945	0.8
11 0134	4.9	**26**	0155	4.7
0759	0.3		0811	0.5
TH 1409	4.9	F	1415	4.5
2005	0.6		2015	0.8
12 0216	4.9	**27**	0228	4.6
0840	0.3		0839	0.6
F 1453	4.8	SA	1447	4.5
2047	0.6		2046	0.9
13 0258	4.9	**28**	0300	4.5
0921	0.4		0908	0.7
SA 1535	4.7	SU	1519	4.3
2130	0.7		2117	1.0
14 0341	4.8	**29**	0332	4.4
1004	0.5		0940	0.9
SU 1618	4.5	M	1553	4.2
2218	0.9		2153	1.1
15 0429	4.6			
1055	0.7			
M 1710	4.3			
◐ 2315	1.0			

MARCH

Time	m		Time	m
1 0408	4.2	**16**	0513	4.4
1019	1.1		1130	1.1
TU 1633	4.0	W	1754	4.1
◐ 2238	1.3			
2 0454	4.0	**17**	0010	1.1
1111	1.3		0630	4.1
W 1724	3.9	TH	1253	1.3
2340	1.4		1917	4.0
3 0554	3.8	**18**	0138	1.2
1224	1.4		0757	4.1
TH 1833	3.8	F	1421	1.3
			2036	4.1
4 0110	1.5	**19**	0308	1.0
0718	3.8		0919	4.2
F 1354	1.4	SA	1538	1.2
2007	3.9		2147	4.2
5 0240	1.3	**20**	0418	0.8
0847	4.0		1033	4.4
SA 1509	1.2	SU	1635	1.1
2120	4.1		2243	4.4
6 0347	1.0	**21**	0508	0.7
0955	4.2		1112	4.5
SU 1607	1.0	M	1716	1.0
2217	4.4		2326	4.5
7 0442	0.7	**22**	0546	0.6
1049	4.5		1150	4.5
M 1657	0.8	TU	1748	0.9
2305	4.6		2359	4.6
8 0531	0.5	**23**	0615	0.5
1136	4.7		1220	4.5
TU 1741	0.7	W	1819	0.8
2349	4.8	○		
9 0615	0.4	**24**	0027	4.7
1221	4.8		0642	0.5
W 1824	0.6	TH	1246	4.6
●			1850	0.7
10 0031	4.9	**25**	0057	4.7
0657	0.2		0711	0.5
TH 1306	4.9	F	1314	4.6
1907	0.5		1922	0.7
11 0116	5.0	**26**	0129	4.7
0739	0.2		0740	0.5
F 1351	4.9	SA	1346	4.6
1949	0.5		1953	0.7
12 0200	5.1	**27**	0203	4.6
0819	0.2		0808	0.6
SA 1434	4.9	SU	1419	4.5
2032	0.5		2023	0.8
13 0243	5.0	**28**	0236	4.5
0858	0.3		0837	0.7
SU 1515	4.7	M	1451	4.4
2114	0.6		2053	0.9
14 0326	4.9	**29**	0307	4.4
0939	0.5		0908	0.8
M 1556	4.5	TU	1522	4.3
2200	0.8		2127	1.0
15 0414	4.6	**30**	0340	4.2
1027	0.8		0944	1.0
TU 1646	4.3	W	1559	4.1
◐ 2256	0.9		2208	1.1
		31	0423	4.1
			1032	1.2
		TH	1648	4.0
		◑	2306	1.2

APRIL

Time	m		Time	m
1 0522	3.9	**16**	0114	1.1
1141	1.4		0733	4.1
F 1753	3.9	SA	1352	1.4
			2003	4.0
2 0029	1.3	**17**	0237	1.0
0639	3.9		0850	4.1
SA 1310	1.4	SU	1506	1.3
1920	3.9		2111	4.2
3 0201	1.2	**18**	0344	0.8
0818	4.0		0953	4.3
SU 1434	1.2	M	1603	1.1
2042	4.1		2208	4.3
4 0314	0.9	**19**	0433	0.7
0924	4.3		1042	4.4
M 1538	1.0	TU	1645	1.0
2144	4.3		2253	4.4
5 0412	0.6	**20**	0509	0.7
1022	4.5		1120	4.4
TU 1630	0.8	W	1720	0.9
2236	4.6		2327	4.5
6 0503	0.4	**21**	0540	0.6
1112	4.7		1149	4.5
W 1717	0.6	TH	1753	0.8
2323	4.8		2358	4.6
7 0549	0.3	**22**	0610	0.6
1158	4.8		1214	4.5
TH 1803	0.5	F	1827	0.7
●		○		
8 0007	4.9	**23**	0029	4.6
0632	0.2		0641	0.5
F 1244	4.8	SA	1245	4.6
1847	0.5		1900	0.7
9 0054	5.1	**24**	0103	4.7
0715	0.2		0711	0.6
SA 1329	4.9	SU	1318	4.6
1933	0.4		1932	0.7
10 0141	5.1	**25**	0139	4.6
0756	0.3		0740	0.6
SU 1413	4.9	M	1353	4.6
2017	0.4		2003	0.7
11 0227	5.0	**26**	0214	4.5
0836	0.4		0810	0.7
M 1455	4.7	TU	1427	4.5
2100	0.5		2035	0.8
12 0313	4.9	**27**	0248	4.4
0917	0.6		0844	0.8
TU 1537	4.5	W	1500	4.3
2147	0.7		2111	0.8
13 0401	4.6	**28**	0323	4.3
1003	0.9		0922	0.9
W 1625	4.3	TH	1537	4.2
2241	0.8		2153	0.9
14 0458	4.4	**29**	0406	4.2
1102	1.2		1010	1.1
TH 1728	4.1	F	1625	4.1
◑ 2351	1.0		2248	1.0
15 0612	4.1	**30**	0501	4.1
1223	1.4		1113	1.2
F 1848	4.0	SA	1725	4.0
		◑		

Chart Datum: 2·5 metres below Ordnance Datum (Newlyn). HAT is 5·2 metres above Chart Datum.

E England

》FREE monthly updates. Register at 《
www.reedsnauticalalmanac.co.uk

281

STANDARD TIME (UT)
For Summer Time add ONE hour in **non-shaded areas**

MARGATE   LAT 51°23'N   LONG 1°23'E
TIMES AND HEIGHTS OF HIGH AND LOW WATERS

Dates in red are SPRINGS
Dates in blue are NEAPS

YEAR 2016

MAY

Day	Time	m	Day	Time	m
1 SU	0002	1.1	16 M	0153	1.0
	0612	4.0		0806	4.1
	1234	1.3		1421	1.4
	1841	4.0		2024	4.1
2 M	0127	1.0	17 TU	0254	0.9
	0736	4.1		0906	4.2
	1358	1.2		1519	1.2
	2003	4.1		2121	4.2
3 TU	0241	0.8	18 W	0344	0.8
	0851	4.3		0958	4.3
	1506	1.0		1607	1.1
	2109	4.4		2210	4.3
4 W	0341	0.6	19 TH	0426	0.8
	0952	4.5		1039	4.3
	1602	0.8		1648	0.9
	2205	4.6		2252	4.4
5 TH	0433	0.4	20 F	0504	0.7
	1045	4.7		1114	4.4
	1652	0.6		1727	0.8
	2255	4.8		2328	4.5
6 F ●	0521	0.3	21 SA	0540	0.7
	1133	4.8		1145	4.5
	1741	0.5		1804	0.7
	2344	4.9	○		
7 SA	0607	0.3	22 SU	0003	4.6
	1220	4.8		0613	0.6
	1829	0.4		1218	4.6
				1839	0.7
8 SU	0034	5.0	23 M	0039	4.6
	0652	0.3		0645	0.6
	1307	4.9		1254	4.6
	1917	0.4		1913	0.7
9 M	0124	5.0	24 TU	0117	4.6
	0735	0.4		0717	0.7
	1352	4.8		1331	4.6
	2004	0.4		1947	0.6
10 TU	0212	5.0	25 W	0155	4.6
	0817	0.5		0751	0.7
	1436	4.7		1409	4.5
	2050	0.5		2022	0.7
11 W	0259	4.8	26 TH	0234	4.5
	0858	0.7		0828	0.8
	1518	4.6		1446	4.4
	2136	0.6		2101	0.7
12 TH	0347	4.6	27 F	0313	4.4
	0942	1.0		0909	0.9
	1604	4.4		1525	4.4
	2227	0.7		2145	0.8
13 F ☽	0440	4.3	28 SA	0356	4.3
	1035	1.2		0956	1.0
	1700	4.2		1610	4.3
	2328	0.9		2237	0.8
14 SA	0545	4.1	29 SU	0447	4.2
	1143	1.4		1053	1.1
	1811	4.1		1704	4.2
			☽	2341	0.9
15 SU	0040	1.0	30 M	0550	4.1
	0659	4.0		1202	1.2
	1309	1.5		1810	4.2
	1921	4.0			
			31 TU	0055	0.8
				0705	4.2
				1322	1.1
				1925	4.2

JUNE

Day	Time	m	Day	Time	m
1 W	0208	0.7	16 TH	0258	1.0
	0818	4.3		0909	4.1
	1433	1.0		1528	1.2
	2035	4.4		2128	4.2
2 TH	0310	0.6	17 F	0347	0.9
	0923	4.5		0958	4.2
	1534	0.8		1616	1.0
	2136	4.6		2217	4.3
3 F	0406	0.5	18 SA	0432	0.8
	1020	4.6		1041	4.4
	1629	0.7		1701	0.9
	2232	4.7		2300	4.4
4 SA	0457	0.4	19 SU	0513	0.8
	1112	4.7		1119	4.5
	1722	0.6		1742	0.8
	2326	4.9		2340	4.5
5 SU	0546	0.4	20 M	0550	0.8
	1200	4.8		1155	4.5
	1815	0.5		1820	0.7
●			○		
6 M	0018	4.9	21 TU	0018	4.6
	0633	0.5		0624	0.7
	1247	4.8		1233	4.6
	1905	0.4		1856	0.6
7 TU	0109	5.0	22 W	0057	4.6
	0719	0.5		0659	0.7
	1333	4.8		1312	4.6
	1954	0.4		1933	0.6
8 W	0157	4.9	23 TH	0138	4.6
	0802	0.7		0735	0.7
	1417	4.7		1352	4.6
	2040	0.4		2011	0.6
9 TH	0244	4.8	24 F	0220	4.6
	0842	0.8		0815	0.7
	1459	4.6		1432	4.6
	2124	0.5		2052	0.6
10 F	0329	4.6	25 SA	0303	4.5
	0922	1.0		0857	0.8
	1542	4.5		1513	4.5
	2208	0.7		2136	0.6
11 SA	0416	4.4	26 SU	0346	4.4
	1006	1.1		0943	0.9
	1628	4.3		1556	4.5
	2257	0.8		2224	0.7
12 SU ☽	0509	4.2	27 M ☽	0433	4.3
	1057	1.3		1035	1.0
	1725	4.2		1645	4.4
	2355	1.0		2320	0.7
13 M	0613	4.0	28 TU	0528	4.3
	1205	1.4		1135	1.1
	1833	4.1		1743	4.4
14 TU	0101	1.0	29 W	0026	0.7
	0717	4.0		0635	4.2
	1327	1.5		1248	1.1
	1936	4.0		1852	4.4
15 W	0203	1.0	30 TH	0136	0.7
	0816	4.0		0749	4.3
	1433	1.4		1403	1.1
	2034	4.1		2005	4.4

JULY

Day	Time	m	Day	Time	m
1 F	0243	0.7	16 SA	0310	1.1
	0858	4.4		0919	4.1
	1510	0.9		1543	1.2
	2113	4.5		2143	4.2
2 SA	0344	0.7	17 SU	0401	1.0
	1000	4.5		1010	4.3
	1612	0.8		1634	1.0
	2217	4.6		2234	4.3
3 SU	0440	0.7	18 M	0447	0.9
	1056	4.6		1055	4.4
	1711	0.6		1720	0.8
	2315	4.7		2318	4.5
4 M	0532	0.6	19 TU	0528	0.9
	1146	4.7		1135	4.5
	1806	0.5		1801	0.7
●			○	2359	4.6
5 TU	0007	4.8	20 W	0606	0.7
	0620	0.7		1214	4.6
	1233	4.7		1840	0.6
	1856	0.4			
6 W	0056	4.8	21 TH	0039	4.7
	0704	0.7		0643	0.7
	1316	4.8		1254	4.7
	1943	0.4		1919	0.5
7 TH	0142	4.8	22 F	0122	4.7
	0745	0.8		0721	0.7
	1358	4.8		1335	4.8
	2025	0.4		1958	0.5
8 F	0225	4.7	23 SA	0206	4.7
	0823	0.8		0802	0.7
	1438	4.7		1416	4.8
	2104	0.5		2039	0.4
9 SA	0306	4.6	24 SU	0249	4.7
	0858	1.0		0844	0.7
	1516	4.6		1457	4.7
	2140	0.6		2121	0.5
10 SU	0345	4.4	25 M	0331	4.6
	0935	1.1		0927	0.8
	1554	4.5		1538	4.7
	2218	0.8		2205	0.6
11 M	0425	4.2	26 TU	0415	4.5
	1016	1.2		1014	0.9
	1635	4.3		1624	4.6
	2301	0.9	☽	2255	0.7
12 TU	0510	4.1	27 W	0504	4.3
	1104	1.4		1109	1.1
	1725	4.1		1718	4.5
☽	2355	1.1		2355	0.8
13 W	0610	3.9	28 TH	0606	4.2
	1206	1.5		1218	1.1
	1832	4.0		1825	4.4
14 TH	0103	1.2	29 F	0107	0.9
	0720	3.9		0723	4.2
	1336	1.5		1338	1.2
	1943	4.0		1944	4.3
15 F	0211	1.2	30 SA	0221	1.0
	0823	4.0		0838	4.2
	1446	1.4		1454	1.1
	2046	4.1		2100	4.4
			31 SU	0329	0.9
				0947	4.4
				1605	0.9
				2212	4.5

AUGUST

Day	Time	m	Day	Time	m
1 M	0431	0.9	16 TU	0422	1.1
	1047	4.5		1029	4.4
	1708	0.7		1654	0.9
	2311	4.7		2256	4.5
2 TU	0524	0.8	17 W	0506	0.9
	1137	4.6		1113	4.6
	1800	0.6		1739	0.7
●				2339	4.6
3 W	0000	4.7	18 TH	0547	0.8
	0608	0.8		1153	4.7
	1221	4.7		1820	0.6
	1845	0.5	○		
4 TH	0044	4.8	19 F	0020	4.8
	0648	0.7		0626	0.7
	1259	4.8		1233	4.8
	1925	0.4		1900	0.5
5 F	0124	4.7	20 SA	0103	4.8
	0724	0.8		0705	0.7
	1336	4.8		1314	4.9
	2002	0.5		1940	0.4
6 SA	0202	4.7	21 SU	0146	4.9
	0759	0.9		0746	0.7
	1412	4.8		1356	5.0
	2035	0.6		2020	0.4
7 SU	0237	4.6	22 M	0230	4.9
	0833	0.9		0827	0.7
	1446	4.7		1437	4.9
	2106	0.7		2059	0.4
8 M	0310	4.5	23 TU	0311	4.7
	0906	1.0		0909	0.8
	1520	4.6		1519	4.8
	2137	0.8		2140	0.6
9 TU	0343	4.3	24 W	0352	4.5
	0940	1.1		0954	0.9
	1556	4.4		1604	4.7
	2212	0.9		2226	0.8
10 W	0419	4.2	25 TH	0439	4.4
	1020	1.3		1046	1.0
	1636	4.3		1657	4.6
☽	2255	1.1	☽	2325	1.0
11 TH	0503	4.0	26 F	0541	4.2
	1109	1.4		1155	1.2
	1725	4.1		1807	4.4
	2351	1.3			
12 F	0559	3.9	27 SA	0042	1.2
	1216	1.6		0702	4.1
	1831	3.9		1319	1.2
				1931	4.3
13 SA	0108	1.4	28 SU	0205	1.2
	0723	3.8		0822	4.2
	1353	1.5		1444	1.1
	1956	3.9		2055	4.3
14 SU	0228	1.4	29 M	0322	1.2
	0838	4.0		0936	4.3
	1507	1.3		1602	0.9
	2108	4.1		2209	4.5
15 M	0330	1.2	30 TU	0425	1.0
	0938	4.2		1037	4.5
	1605	1.1		1702	0.7
	2207	4.3		2306	4.6
			31 W	0514	1.0
				1126	4.6
				1749	0.6
				2351	4.7

Chart Datum: 2·5 metres below Ordnance Datum (Newlyn). HAT is 5·2 metres above Chart Datum.

STANDARD TIME (UT)
For Summer Time add ONE hour in **non-shaded areas**

MARGATE LAT 51°23'N LONG 1°23'E
TIMES AND HEIGHTS OF HIGH AND LOW WATERS

Dates in red are **SPRINGS**
Dates in blue are NEAPS

YEAR 2016

SEPTEMBER

Day	Time	m	Time	m	Time	m	Time	m
1 TH ●	0553	0.9	1205	4.7	1827	0.5		
2 F	0028	4.7	0627	0.9	1239	4.8	1900	0.5
3 SA	0101	4.7	0701	0.8	1310	4.8	1931	0.5
4 SU	0131	4.7	0734	0.8	1342	4.8	2001	0.6
5 M	0202	4.6	0807	0.9	1416	4.7	2030	0.7
6 TU	0234	4.5	0838	1.0	1449	4.6	2059	0.8
7 W	0306	4.4	0909	1.1	1523	4.5	2129	1.0
8 TH	0340	4.3	0943	1.2	1559	4.4	2206	1.2
9 F ◑	0419	4.1	1025	1.3	1643	4.1	2255	1.4
10 SA	0508	3.9	1123	1.5	1739	3.9		
11 SU	0004	1.6	0614	3.8	1249	1.6	1857	3.9
12 M	0137	1.6	0747	3.9	1423	1.4	2027	4.0
13 TU	0255	1.4	0901	4.1	1531	1.4	2136	4.3
14 W	0353	1.1	0958	4.4	1625	0.9	2229	4.5
15 TH	0441	1.0	1046	4.6	1712	0.7	2314	4.7
16 F ○	0523	0.8	1127	4.8	1754	0.5	2357	4.8
17 SA	0605	0.7	1207	4.9	1835	0.4		
18 SU	0039	4.9	0645	0.6	1249	5.0	1915	0.4
19 M	0122	5.0	0727	0.6	1332	5.1	1955	0.4
20 TU	0205	4.9	0809	0.6	1416	5.0	2034	0.5
21 W	0247	4.8	0851	0.7	1500	4.9	2114	0.6
22 TH	0329	4.6	0936	0.8	1547	4.8	2159	0.9
23 F ◑	0416	4.4	1028	1.0	1642	4.5	2258	1.1
24 SA	0518	4.2	1138	1.1	1754	4.3		
25 SU	0019	1.4	0641	4.1	1304	1.2	1921	4.2
26 M	0150	1.4	0804	4.1	1433	1.1	2047	4.3
27 TU	0309	1.3	0917	4.3	1548	0.9	2157	4.5
28 W	0411	1.2	1019	4.5	1644	0.7	2251	4.6
29 TH	0457	1.0	1106	4.6	1727	0.6	2333	4.7
30 F	0532	1.0	1143	4.7	1800	0.6		

OCTOBER

Day	Time	m	Time	m	Time	m	Time	m
1 SA ●	0005	4.7	0604	0.9	1213	4.7	1828	0.6
2 SU	0031	4.7	0637	0.8	1241	4.8	1858	0.6
3 M	0057	4.7	0710	0.8	1313	4.8	1928	0.4
4 TU	0128	4.7	0742	0.8	1347	4.7	1956	0.7
5 W	0201	4.6	0812	0.9	1421	4.6	2024	0.9
6 TH	0234	4.5	0842	1.0	1454	4.5	2054	1.0
7 F	0306	4.4	0914	1.1	1528	4.3	2128	1.2
8 SA	0342	4.2	0953	1.2	1609	4.2	2213	1.4
9 SU ◑	0428	4.0	1046	1.4	1703	4.0	2316	1.6
10 M	0528	3.9	1201	1.4	1814	3.9		
11 TU	0041	1.6	0650	3.9	1335	1.4	1942	4.0
12 W	0212	1.5	0817	4.0	1452	1.1	2059	4.3
13 TH	0318	1.2	0921	4.3	1550	0.8	2157	4.6
14 F	0410	1.0	1012	4.6	1640	0.6	2245	4.8
15 SA	0456	0.8	1057	4.8	1725	0.5	2329	4.9
16 SU ○	0540	0.7	1139	5.0	1807	0.4		
17 M	0012	5.0	0623	0.6	1223	5.1	1848	0.4
18 TU	0056	5.0	0708	0.5	1310	5.1	1930	0.4
19 W	0141	4.9	0752	0.5	1358	5.1	2010	0.5
20 TH	0224	4.8	0836	0.6	1446	5.0	2052	0.7
21 F	0308	4.7	0923	0.7	1535	4.8	2138	1.0
22 SA ◑	0356	4.4	1016	0.9	1630	4.5	2234	1.3
23 SU	0456	4.2	1123	1.0	1740	4.3	2352	1.5
24 M	0616	4.1	1244	1.1	1904	4.2		
25 TU	0123	1.5	0735	4.1	1408	1.0	2023	4.3
26 W	0241	1.4	0846	4.2	1518	0.9	2130	4.4
27 TH	0343	1.3	0947	4.4	1613	0.8	2223	4.6
28 F	0429	1.1	1035	4.5	1653	0.7	2304	4.6
29 SA	0506	1.0	1113	4.6	1725	0.7	2334	4.6
30 SU ●	0539	0.9	1143	4.7	1756	0.7	2359	4.7
31 M	0613	0.8	1213	4.7	1827	0.7		

NOVEMBER

Day	Time	m	Time	m	Time	m	Time	m
1 TU	0026	4.7	0647	0.8	1246	4.7	1857	0.7
2 W	0058	4.7	0720	0.8	1321	4.7	1926	0.8
3 TH	0133	4.7	0750	0.8	1357	4.6	1955	0.9
4 F	0207	4.6	0821	0.9	1431	4.5	2027	1.0
5 SA	0240	4.4	0854	0.9	1506	4.3	2103	1.1
6 SU	0313	4.3	0933	1.0	1546	4.2	2146	1.3
7 M ◑	0357	4.2	1023	1.1	1637	4.1	2243	1.4
8 TU	0454	4.0	1128	1.2	1740	4.0	2357	1.5
9 W	0603	4.0	1250	1.2	1859	4.1		
10 TH	0123	1.5	0727	4.1	1409	1.0	2018	4.3
11 F	0238	1.2	0839	4.3	1513	0.8	2122	4.5
12 SA	0337	1.0	0936	4.6	1606	0.6	2215	4.7
13 SU	0428	0.8	1027	4.8	1654	0.5	2303	4.8
14 M ○	0516	0.7	1114	4.9	1740	0.4	2348	4.9
15 TU	0603	0.6	1202	5.1	1825	0.5		
16 W	0033	5.0	0651	0.5	1253	5.1	1909	0.5
17 TH	0120	4.9	0739	0.5	1344	5.1	1952	0.6
18 F	0206	4.8	0826	0.5	1433	5.0	2035	0.8
19 SA	0251	4.7	0914	0.6	1523	4.8	2120	1.0
20 SU	0338	4.5	1005	0.7	1616	4.5	2211	1.3
21 M ◑	0432	4.3	1104	0.9	1719	4.3	2316	1.5
22 TU	0542	4.1	1214	1.0	1833	4.2		
23 W	0040	1.6	0656	4.1	1328	1.0	1943	4.2
24 TH	0158	1.5	0802	4.1	1433	1.0	2047	4.3
25 F	0301	1.4	0902	4.2	1527	0.9	2141	4.4
26 SA	0352	1.2	0955	4.4	1611	0.9	2225	4.5
27 SU	0435	1.1	1038	4.5	1650	0.8	2301	4.5
28 M	0514	0.9	1115	4.5	1726	0.8	2331	4.6
29 TU ●	0551	0.8	1149	4.6	1800	0.8		
30 W	0001	4.6	0626	0.8	1224	4.6	1832	0.8

DECEMBER

Day	Time	m	Time	m	Time	m	Time	m
1 TH	0035	4.7	0700	0.7	1300	4.6	1902	0.8
2 F	0110	4.7	0732	0.7	1337	4.6	1933	0.9
3 SA	0147	4.6	0805	0.7	1414	4.5	2008	0.9
4 SU	0222	4.5	0841	0.8	1452	4.4	2046	1.0
5 M	0258	4.4	0921	0.8	1532	4.3	2129	1.1
6 TU	0338	4.3	1007	0.9	1617	4.2	2220	1.3
7 W ◑	0428	4.2	1103	1.0	1712	4.2	2322	1.4
8 TH	0528	4.2	1211	1.0	1819	4.2		
9 F	0036	1.4	0639	4.2	1327	0.9	1937	4.3
10 SA	0156	1.2	0756	4.3	1435	0.8	2047	4.4
11 SU	0303	1.0	0903	4.5	1535	0.7	2148	4.6
12 M	0401	0.9	1002	4.7	1628	0.6	2241	4.7
13 TU	0455	0.7	1057	4.8	1718	0.6	2331	4.8
14 W ○	0548	0.6	1149	4.9	1807	0.6		
15 TH	0018	4.9	0640	0.5	1242	5.0	1854	0.6
16 F	0105	4.9	0730	0.4	1332	5.0	1940	0.7
17 SA	0151	4.8	0818	0.4	1421	4.9	2022	0.8
18 SU	0236	4.7	0904	0.5	1509	4.7	2103	1.0
19 M	0320	4.6	0949	0.6	1556	4.5	2146	1.2
20 TU	0405	4.4	1036	0.8	1647	4.3	2235	1.4
21 W ◑	0459	4.2	1131	0.9	1748	4.1	2335	1.5
22 TH	0607	4.1	1234	1.1	1854	4.0		
23 F	0058	1.6	0713	4.0	1340	1.1	1955	4.1
24 SA	0211	1.5	0814	4.1	1438	1.1	2052	4.1
25 SU	0311	1.4	0912	4.1	1530	1.1	2144	4.3
26 M	0403	1.2	1004	4.3	1617	1.0	2229	4.4
27 TU	0449	1.0	1050	4.4	1659	1.0	2307	4.5
28 W	0531	0.9	1129	4.5	1737	0.9	2342	4.6
29 TH ●	0608	0.8	1205	4.5	1812	0.9		
30 F	0016	4.6	0643	0.7	1242	4.6	1843	0.8
31 SA	0053	4.7	0716	0.6	1320	4.6	1917	0.8

Chart Datum: 2·5 metres below Ordnance Datum (Newlyn). HAT is 5·2 metres above Chart Datum.

E England

》》 **FREE** monthly updates. Register at 《 www.reedsnauticalalmanac.co.uk

283

9.4.8 WHITSTABLE

Kent **51°21′·86N 01°01′·46E** ❀❀◊◊❀❀

CHARTS AC 5606, 1607, 2571; Imray C1, 2100

TIDES +0135 Dover; ML 3·0; Duration 0605

Standard Port MARGATE (←)

Times				Height (metres)			
High Water		Low Water		MHWS	MHWN	MLWN	MLWS
0100	0700	0100	0700	4·8	3·9	1·4	0·5
1300	1900	1300	1900				
Differences HERNE BAY							
+0022	+0020	+0019	+0017	+0·6	+0·3	+0·2	+0·1
WHITSTABLE APPROACHES							
+0042	+0029	+0025	+0050	+0·6	+0·6	+0·1	0·0

SHELTER Good, except in strong winds from NNW to NE. Harbour dries up to 0·4m. Yacht berths are limited since priority is given to commercial shipping. Fender board needed against piled quays or seek a mooring to NW of harbour (controlled by YC).

NAVIGATION WPT 51°22′·65N 01°01′·10E, 345°/0·83M to W Quay dolphin. From E keep well seaward of Whitstable Street, a hard drying sandspit, which extends 1M N from the coast; shoals a further 1M to seaward are marked by Whitstable Street NCM lt buoy. From W avoid Columbine and Pollard Spits.

Approach, not before half flood, via Oyster PHM lt buoy to harbour entrance. Entry best at HW±1.

LIGHTS AND MARKS Off head of W Quay on a dolphin, ⚡ Fl G 5s, covers the approaches.

COMMUNICATIONS (Code 01227) NMOC 02392 552100; Police 101; Dr 594400. HM 274086.

Whitstable Harbour Radio VHF Ch **09** 12 16 (Mon–Fri: 0830–1700 LT. Other times: HW –3 to HW+1). Tidal info on request.

FACILITIES www.canterbury.gov.uk HM ☎274086, ⌑ £10 ⚓ ⚒.
Whitstable YC ☎272942, ⚓ ◣ HW ±3 ⚓⚒× ⌑.
Services: ⚓ ⚒ Ⓔ ⌂ ⟁ ⟁ ACA.
Town: 🏤 ✉ Ⓑ ⟟ 🛒 × ⌑ ⇌ ✈ Lydd/Manston.

MINOR HARBOURS WEST OF NORTH FORELAND

MARGATE, Kent, **51°23′·43N 01°22′·65E**. AC 5606, 1 827, 1828, 323; Imray Y7, C1; Stanfords 5, 8. HW+0045 on Dover; ML 2·6; Duration 0610; Standard Port (←). Small hbr drying 3m, inside bkwtr (Stone Pier) FR 17m 4M; exposed to NW'lies. Appr: from E, via Longnose NCM buoy, keeping about 5ca offshore; from N, via Margate PHM lt buoy; from W via Gore Chan and S Chan to SE Margate ECM. Facilities: Margate YC ☎(01843) 292602 ⌑.

Town 10 ◣ HW±2, book at Foreshore Office ☎(01843) 577529, 🏤 🏤 ✉ Ⓑ 🛒 × ⌑ ⇌ ✈ Manston.

HERNE BAY, Kent, **51°22′·40N 01°07′·22E**. AC 5606, 1607. Close E of Pavilion (old pier), 400m long bkwtr 'Neptunes Arm' gives well sheltered drying ⚓ for craft <8m. Exposed to strong NE'lies at HW springs when Thames Barrier closed as seas may then top the bkwtr. 3 ◣ (HW±2½ to HW±3). Foreshore Manager ☎(01227) 266719. Herne Bay SC ☎(01227) 375650. Lts: QW 8m 4M is 6ca offshore (former pier hd); bkwtr hd 2FR (vert); pier hd 2FG (vert); R bn on B dolphin, Fl Y 5s, is approx 1M ENE of bkwtr hd. Reculvers twrs are conspic 3M to the E.

Town, 🏤 🏤 Ⓔ ✉ Ⓑ 🖳 🛒 × ⌑ ⇌ ✈ Lydd/Manston.

9.4.9 THE SWALE
Kent ✵ ✵ ⚓ ⚓ ⚓ ❀ ❀

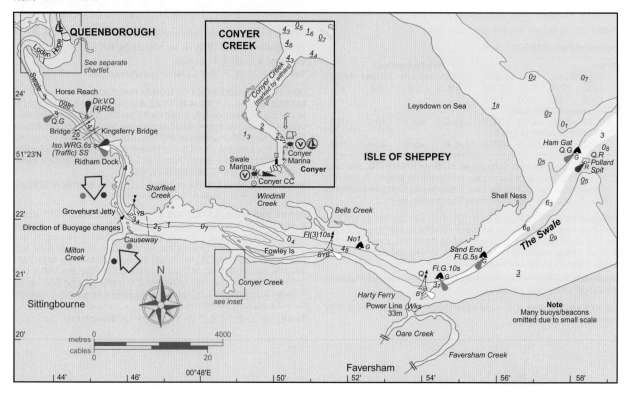

CHARTS AC 5606, 2482, 2571, 2572, 1834, 3683; Imray Y18, C1, 2100

TIDES Queenborough +0130 Dover; Harty Ferry +0120 Dover; ML (Harty Ferry) 3·0; Duration 0610. Faversham HW differences are –0·2m on Sheerness; no other data.

Standard Port SHEERNESS (→)

Times				Height (metres)			
High Water		Low Water		MHWS	MHWN	MLWN	MLWS
0200	0800	0200	0700	5·8	4·7	1·5	0·6
1400	2000	1400	1900				
Differences R. SWALE (Grovehurst Jetty)							
–0007	0000	0000	+0016	0·0	0·0	0·0	–0·1

Grovehurst Jetty is close N of the ent to Milton Creek.

SHELTER Excellent in the Swale, the 14M chan between the Isle of Sheppey and the N Kent coast, from Shell Ness in the E to Queenborough in the W. Yachts can enter the drying creeks of Faversham, Oare, Conyer, and Milton. Beware wrecks at ent to Faversham Creeks. Many moorings line the chan from Faversham to Conyer Creeks. See Queenborough for all-tide access.

NAVIGATION E ent WPT: Whitstable Street lit PHM, 51°24'·00N 01°01'·54E, at ent to well marked buoyed chan. Speed limit 8kn. The middle section from 1·5M E of Conyer Creek to 0·5M E of Milton Creek is narrowed by drying mudbanks and carries least depths of 0·4m. At Milton Creek direction of buoyage changes. There are numerous oyster beds in the area. Kingsferry Bridge opens for masted craft on request, but subject to railway trains; temp ⚓s off SW bank. The power lines crossing SE of the br have a clearance of 30m. **W ent** is marked by Queenborough Spit ECM buoy, Q (3) 10s, 1M S of Garrison Pt, at 51°25'·81N 00°43'·93E.

LIGHTS AND MARKS See chartlet and 9.4.4. In W Swale lights intended for large coasters using the narrow chan are:

No 5 Bn Oc WRG 6s; vis: 161°-G-166°-W-167°-R-172°. Round Loden Hope bend: two Q WG and one Q WRG on bns; keep in G sectors. See Queenborough chartlet. Horse Reach ldg lts 113°: front QG 7m 5M; rear Fl G 3s 10m 6M. Dir lt 098°, VQ (4) R 5s 6m 5M. Kingsferry Bridge: Dir WRG, 142°-G-147°-W-148°-R-153° 9m. Lts on bridge: two x 2 FG (vert) on SW buttresses; two x 2 FR (vert) on NE. Road bridge has a vertical clearance of 28m.

Kingsferry Bridge traffic sigs:

No lts	= Bridge down (3·35m MHWS).
Al Q ●/◐	= Centre span lifting.
F ●	= Bridge open (29m MHWS).
Q ●	= Centre span lowering. Keep clear.
Q ◌	= Bridge out of action.

Request bridge opening on VHF Ch 10 (or Ch 74 if no response).

COMMUNICATIONS (Code 01795) NMOC 02392 552100; Police 101; Dr or Ⓗ via Medway Navigation Service 663025. HM (Medway Ports Ltd) 596593.

Medway VTS VHF Ch **74** 16 22 (H24); Kingsferry Bridge Ch 10 (H24).

FACILITIES **FAVERSHAM:** Town ⛽ ✉ Ⓑ 🍴 ✕ ⌑ ⇌ ✈(Gatwick).

OARE CREEK ⚓ ◡ ✕ ✎ 🖼 ◨(8t) ⚓.

Hollow Shore Cruising Club ⌑.

CONYER CREEK: Swale Marina ☎521562. ⛽ ◣ ◡(dredged 2m) £14/craft, ⚓ ⊡ ◎ 🏴 🚿 ⚓ ✕ ✎ ◨(30t) ⚓(30t); **Conyer CC.**

Conyer Marina ☎521711, ◣ ◡£10, 🏴 ✕ ✎ ⚓ Ⓔ ⚓ 🍴 🛒 ⚓.

MILTON CREEK (Sittingbourne): **Crown Quay** ⚓ ⚓.

Town ✉ Ⓑ 🍴 ✕ ⌑ ⇌ ✈ (Gatwick); also the Dolphin Yard Sailing Barge Museum.

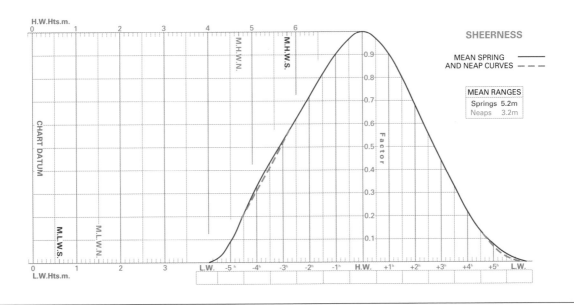

SHEERNESS

MEAN SPRING
AND NEAP CURVES

MEAN RANGES
Springs 5.2m
Neaps 3.2m

9.4.10 QUEENBOROUGH

Kent (Isle of Sheppey) **51°25'·04N 00°44'·19E** ✿✿✿✿♒♒♒✿✿

CHARTS AC 5606, 1834, 2572, 3683; Imray C1, Y18, 2100

TIDES Use 9.4.11 Sheerness, 2M to the N. +0130 Dover; ML 3·0; Duration 0610

SHELTER Good, except near HW in strong N'ly winds. The first deep-water refuge W of N Foreland, accessible at all tides from Garrison Pt (9.4.11); or from the Swale (9.4.9) on the tide. An all-tide pontoon/jetty (5m depth at end) on E bank is for landing/short stay only; both sides of the jetty are foul. 2 Y Øs (raft up to 4 craft <10m each) lie N of the jetty and a pair Gy Øs (raft up to 6 craft <10m each) lie to the S; also single visitors Y Øs Nos 14-26 for craft <10m, with HM's permission craft >10m berth on W side of Concrete Lighter. When Øs are full the hbr controller will offer spare buoys. Smaller R buoys (numbered) are for locals. ⚓ is discouraged due to commercial traffic. Speed limit 8kn.

NAVIGATION WPT: see R Medway for appr via Garrison Pt. Enter the river at Queenborough Spit ECM buoy, Q (3) 10s, 51°25'·75N 00°43'·93E. The chan narrows between drying banks and moorings. See The Swale if approaching from the E.

LIGHTS AND MARKS Lights as chartlet. Note: No 5 Bn Dir Oc WRG 6s, vis 161°-G-166°-W-167°-R-172°, on river bend covers the appr chan. All-tide landing 2 FR (vert). Concrete lighter Fl G 3s.

COMMUNICATIONS (Code 01795) NMOC 02392 552100; Police 101; Dr 583828; Ⓗ (01634) 830000 (Gillingham). HM 662051.

Monitor *Medway VTS* VHF Ch **74** for tfc info. Call Ch 08 *Sheppey One* (Q'boro HM) for berths, also water taxi at weekends only (£1.20 to landing).

FACILITIES HM/Hbr Controller, ✿ and ⌂ on concrete lighter 9/day or £48/week. Tokens for all-tide-landing gate may be purchased at the YC, local grocery or pubs. ⚓ on all-tide landing jetty.

Queenborough YC Wed, Fri, Sat evenings, Sat, Sun lunchtimes, ♠ ♿ 🚿 📶 ✕ 🍴.

The Creek ☎07974 349018, ⚓ (HW±1½) Scrubbing grid ⚓ 🔌 📷.

Services: ⚓ ✕ ✎ 🖧 🍴 🏪(10t).

Town 🏧 🏨 ✉ 🚂 ✕ 🍴 ⇌ ✈ (Gatwick).

Chartlet labels

River Medway
Queenborough
Spit Q(3)10s
BYB
Or
No1
No5
Queenborough Spit
Swale Ness
Deadmans Island
The Lappel
Queenborough
Point
Q.R.
Fl.R4s
No1
No2
No3
No4
(13 Y + 2 Gy buoys)
2.F.R (vert) x2
All-tide landing
QUEENBOROUGH
(Note: up to 6 boats allowed on each Ⓥ)
Concrete Lighter
Craft Ⓥ >10m
Moorings
Fl.R.3s
Fl.G.3s
Queenborough YC
Q.WG
Bn
Q.WG
Fl.R.3s
Loden Hope
West Point
2FR (vert)
Ⓥ (2 Gy buoys)
Q
FlG2·5s
Q.G.
G
Fl.R.2·5s
Q.WRG
Q.R.
R
West Swale
Long Pt
Oc.WRG.6s

metres / cables
0 ... 600 / 0 ... 3

51°25'·0N
25'·5

9.4.11 RIVER MEDWAY (SHEERNESS TO ROCHESTER)

Kent **51°27′·03N 00°44′·50E** (Off Garrison Pt) ✳✳✳✳◊◊◊✿✿✿

CHARTS AC 5606, 1835, 1834, 1185, 2482, 3683; IMRAY C1, Y18, 2100

TIDES +0130 Dover; ML 3·1; Duration 0610

Standard Port SHEERNESS (→)

Times				Height (metres)			
High Water		Low Water		MHWS	MHWN	MLWN	MLWS
0200	0800	0200	0700	5·8	4·7	1·5	0·6
1400	2000	1400	1900				
Differences BEE NESS							
+0002	+0002	0000	+0005	+0·2	+0·1	0·0	0·0
BARTLETT CREEK							
+0016	+0008	ND	ND	+0·1	0·0	ND	ND
DARNETT NESS							
+0004	+0004	0000	+0010	+0·2	+0·1	0·0	−0·1
CHATHAM (Lock Approaches)							
+0010	+0012	+0012	+0018	+0·3	+0·1	−0·1	−0·2
UPNOR							
+0015	+0015	+0015	+0025	+0·2	+0·2	−0·1	−0·1
ROCHESTER (STROOD PIER)							
+0018	+0018	+0018	+0028	+0·2	+0·2	−0·2	−0·3
WOULDHAM							
+0030	+0025	+0035	+0120	−0·2	−0·3	−1·0	−0·3
NEW HYTHE							
+0035	+0035	+0220	+0240	−1·6	−1·7	−1·2	−0·3
ALLINGTON LOCK							
+0050	+0035	ND	ND	−2·1	−2·2	−1·3	−0·4

SHELTER There are 4 marinas down-river of Rochester Bridge and 5 above. Sheerness is solely a commercial hbr. See Queenborough for access to/from The Swale. Lower reaches of the Medway are exposed to strong NE winds, but Stangate and Half Acre Creeks are secure in all winds and give access to lesser creeks. There are good ⚓s in Sharfleet Creek; with sufficient rise of the tide it is possible to go via the 'back-door' into Half Acre Creek. Speed limit is 6kn W of Folly Pt (Hoo Island).

NAVIGATION WPT 51°28′·84N 00°52′·78E, SWM buoy, 253°/4·5M to No 11 SHM buoy, then follow recommended route 240°/1M to Grain Hard SHM buoy. The wreck of the *Richard Montgomery* is visible 2M NE of estuary entrance and a Military Wreck depth 2_2 in Kethole Reach marked by E Bulwark SHM. There is a huge area to explore, although much of it dries to mud. Some minor creeks are buoyed. The river is well buoyed/marked up to Rochester and tidal up to Allington Lock (21·6M). Above Rochester bridge the river shoals appreciably and in the upper reaches there is only about 1m at LW. A new bridge is being built at Holborough Reach.

Bridge Clearances (HAT), going up-river:	
Rochester	5·4m
Medway (M2)	16·2m
Holborough Reach	5·0m
New Hythe (footbridge)	10·7m
Aylesford (pedestrian)	2·3m
Aylesford (road)	2·7m
Maidstone bypass (M20)	8·9m

LIGHTS AND MARKS See 9.4.4 and chartlet for details of most lts. NB: not all buoys are shown due to small scale. Isle of Grain lt Q 20m 13M. Power stn chy (242m) Oc and FR lts. Tfc Sigs: Powerful lt, Fl 7s, at Garrison Pt means large vessel under way: if shown up river = inbound; if to seaward = outbound.

COMMUNICATIONS (Codes 01634 Medway) NMOC 02392 552100; Police 101; Dr via Medway Navigation Service (01795) 663025; Sheerness HM (01795) 596593.

Medway VTS VHF Ch **74** 16 (H24). Monitor Ch 74 underway and Ch 16 at ⚓. Radar assistance is available on request Ch 22. Ch **80** M for marinas: Gillingham, Hoo (H24), Chatham Maritime (H24), Medway Bridge (0900–1700LT) and Port Medway (0800–2000LT).

YACHT CLUBS Sheppey YC ☎(01795) 663052;
Lower Halstow YC ☎(01227) 458554;
Medway Cruising Club (Gillingham) ☎856489, ⚓🔧⚓🛖.
Hoo Ness YC ☎250052, ⚓🔧⚓✕🛖.
Hundred of Hoo SC ☎250102;
Medway Motor Cruising Club ☎827194;
Medway Motor YC ☎(01622) 737647;
Medway YC (Upnor) ☎718399; **Upnor SC** ☎718043;
Royal Engineers YC ☎844555; **RNSA Medway** ☎200970;

GILLINGHAM MARINA — River Medway

Rochester CC ☎841350, ⚓🔧⚓🛖✕🛖; **Strood YC** ☎718261, ⚓🔧⚓🛖(1½t)🛖.

FACILITIES (☎ code 01634 unless otherwise stated). All moorings run by YCs or marinas. Slips at Commodores Hard, and Gillingham.

Marinas (From seaward to Rochester Bridge)

Gillingham Marina cross-tide, esp. the ebb, off the lock ent are reduced by a timber baffle at 90° to the stream (close W of lock). An angled pontoon deflects the stream and is also the fuel berth; the outboard end is lit by 2FR (vert).⚓ (252+12Ⓥs) ☎280022, pre-book. £2·90 locked basin (access via lock (width 6m) HW±4½), £2·00 tidal basin HW±2, 🛖🚻🔧✕🔧🛒Ⓔ🛖(65t)🛖(1t)🛒🛖.

Medway Pier Marine ☎851113, 🔧⚓🛖🚻Ⓔ🛖(6t).

PORT WERBURGH

Port Werburgh (120+ residential 🛏 only), ☎252107 access to W basin HW±2½; HW±3 to E basin (via sill 1m above CD); an unlit WCM buoy marks chan ent; 4 waiting buoys in river; 🚻🛖.

Chatham Maritime Marina – see inset on previous page (←) ☎899200, (300🛏), access via lock H24 and sill 1.3m below CD (1.5m at MLWS), £3.20. C(15T), ⚓🛖🛒🛖🚻🔧🛖🛖.

Victory Marina ☎07785 971797, Ⓥ⚓🛖.

Marinas (Up-river from Rochester Bridge)

Medway Bridge Marina ☎843576, 🔧 160🛏+4Ⓥ £1.20, 🚻🛖🚻✕🔧🛖Ⓔ🛖(3t)🛖(10t)🛒🛖✕🛖.

Port Medway Marina ☎720033, 50🛏 inc Ⓥ✕🛖(16t)🛖.

Cuxton Marina ☎721941, 🔧 150🛏+Ⓥ✕🔧🛖Ⓔ🛖(12t)🛖.

Elmhaven Marina ☎240489, 60🛏✕🔧🛖🛖.

Allington Lock operates HW–3 to +2, ☎(01622) 752864.

Allington Marina ☎(01622) 752057, above the lock; 🔧 120🛏⚓🚻🛖🛖✕🔧🛖🛖(10t)🛖.

Towns: All facilities ✉🚂🛖✕⇌✈ (Gatwick).

SHEERNESS LAT 51°27'N LONG 0°45'E
TIMES AND HEIGHTS OF HIGH AND LOW WATERS

STANDARD TIME (UT)
For Summer Time add ONE hour in **non-shaded areas**

Dates in **red** are **SPRINGS**
Dates in blue are **NEAPS**

YEAR 2016

JANUARY

Day	Time	m	Day	Time	m
1 F	0434 / 1039 / 1707 / 2241	5.2 / 1.1 / 5.1 / 1.5	**16** SA	0429 / 1055 / 1709 / 2302 ◗	5.6 / 0.7 / 5.5 / 1.1
2 SA	0517 / 1120 / 1754 / 2329 ◗	5.0 / 1.2 / 4.9 / 1.6	**17** SU	0521 / 1142 / 1807 / 2357	5.5 / 0.9 / 5.3 / 1.2
3 SU	0609 / 1214 / 1849	4.8 / 1.4 / 4.7	**18** M	0623 / 1244 / 1914	5.3 / 1.0 / 5.1
4 M	0031 / 0713 / 1329 / 1953	1.8 / 4.7 / 1.5 / 4.7	**19** TU	0108 / 0736 / 1402 / 2026	1.3 / 5.1 / 1.1 / 5.1
5 TU	0155 / 0823 / 1446 / 2059	1.8 / 4.7 / 1.4 / 4.8	**20** W	0229 / 0853 / 1519 / 2137	1.3 / 5.1 / 1.1 / 5.2
6 W	0311 / 0929 / 1546 / 2159	1.6 / 4.8 / 1.3 / 5.1	**21** TH	0347 / 1005 / 1630 / 2242	1.2 / 5.3 / 1.1 / 5.4
7 TH	0410 / 1027 / 1638 / 2252	1.3 / 5.1 / 1.1 / 5.3	**22** F	0500 / 1108 / 1731 / 2337	1.0 / 5.5 / 1.0 / 5.5
8 F	0501 / 1117 / 1724 / 2338	1.1 / 5.3 / 1.0 / 5.5	**23** SA	0601 / 1202 / 1819	0.8 / 5.6 / 0.9
9 SA	0549 / 1202 / 1808	0.9 / 5.5 / 0.9	**24** SU	0024 / 0650 / 1249 / 1859 ○	5.7 / 0.6 / 5.8 / 0.9
10 SU ●	0021 / 0635 / 1244 / 1850	5.7 / 0.8 / 5.7 / 0.8	**25** M	0107 / 0732 / 1330 / 1936	5.7 / 0.5 / 5.8 / 0.8
11 M	0101 / 0720 / 1326 / 1933	5.8 / 0.7 / 5.8 / 0.7	**26** TU	0146 / 0810 / 1409 / 2009	5.7 / 0.5 / 5.8 / 0.9
12 TU	0141 / 0806 / 1407 / 2016	5.8 / 0.5 / 5.9 / 0.7	**27** W	0222 / 0845 / 1445 / 2040	5.7 / 0.6 / 5.7 / 0.9
13 W	0220 / 0851 / 1449 / 2058	5.9 / 0.5 / 5.9 / 0.7	**28** TH	0255 / 0914 / 1519 / 2107	5.6 / 0.7 / 5.6 / 1.0
14 TH	0301 / 0934 / 1533 / 2137	5.8 / 0.5 / 5.8 / 0.8	**29** F	0327 / 0940 / 1552 / 2135	5.5 / 0.8 / 5.4 / 1.1
15 F	0343 / 1014 / 1619 / 2218	5.8 / 0.6 / 5.7 / 0.9	**30** SA	0400 / 1007 / 1626 / 2206	5.4 / 0.9 / 5.2 / 1.2
			31 SU	0435 / 1039 / 1704 / 2243	5.2 / 1.1 / 5.0 / 1.4

FEBRUARY

Day	Time	m	Day	Time	m
1 M	0516 / 1119 / 1749 / 2331 ◗	5.0 / 1.3 / 4.8 / 1.5	**16** TU	0600 / 1212 / 1846	5.3 / 1.1 / 5.1
2 TU	0609 / 1214 / 1848	4.7 / 1.5 / 4.7	**17** W	0039 / 0713 / 1331 / 2000	1.3 / 5.1 / 1.3 / 4.9
3 W	0035 / 0719 / 1335 / 2000	1.7 / 4.6 / 1.6 / 4.7	**18** TH	0208 / 0836 / 1456 / 2118	1.3 / 5.0 / 1.3 / 5.0
4 TH	0210 / 0839 / 1501 / 2114	1.7 / 4.6 / 1.5 / 4.8	**19** F	0335 / 0955 / 1614 / 2228	1.2 / 5.2 / 1.2 / 5.2
5 F	0330 / 0950 / 1604 / 2219	1.4 / 4.9 / 1.2 / 5.1	**20** SA	0454 / 1100 / 1717 / 2324	1.0 / 5.4 / 1.1 / 5.4
6 SA	0432 / 1050 / 1658 / 2313	1.2 / 5.2 / 1.0 / 5.4	**21** SU	0552 / 1152 / 1805	0.8 / 5.6 / 1.0
7 SU	0527 / 1141 / 1748	0.9 / 5.5 / 0.9	**22** M	0011 / 0636 / 1235 / 1842 ○	5.6 / 0.6 / 5.7 / 0.9
8 M ●	0000 / 0619 / 1227 / 1834	5.6 / 0.7 / 5.8 / 0.7	**23** TU	0051 / 0714 / 1313 / 1916	5.7 / 0.6 / 5.8 / 0.8
9 TU	0044 / 0708 / 1310 / 1920	5.7 / 0.5 / 6.0 / 0.6	**24** W	0126 / 0748 / 1348 / 1947	5.7 / 0.5 / 5.8 / 0.6
10 W	0125 / 0755 / 1352 / 2004	5.9 / 0.4 / 6.1 / 0.5	**25** TH	0159 / 0819 / 1419 / 2017	5.7 / 0.5 / 5.7 / 0.8
11 TH	0205 / 0840 / 1434 / 2046	6.0 / 0.2 / 6.1 / 0.5	**26** F	0229 / 0847 / 1449 / 2045	5.7 / 0.6 / 5.7 / 0.8
12 F	0245 / 0922 / 1516 / 2126	6.0 / 0.3 / 6.0 / 0.6	**27** SA	0259 / 0913 / 1519 / 2111	5.6 / 0.7 / 5.6 / 0.9
13 SA	0327 / 1001 / 1600 / 2203	6.0 / 0.4 / 5.8 / 0.7	**28** SU	0329 / 0938 / 1550 / 2137	5.5 / 0.8 / 5.4 / 1.0
14 SU	0411 / 1038 / 1647 / 2243	6.0 / 0.6 / 5.6 / 0.9	**29** M	0401 / 1006 / 1624 / 2207	5.4 / 1.0 / 5.2 / 1.2
15 M	0501 / 1118 / 1741 / 2332 ◗	5.6 / 0.8 / 5.3 / 1.1			

MARCH

Day	Time	m	Day	Time	m
1 TU	0438 / 1039 / 1703 / 2247 ◗	5.1 / 1.2 / 5.0 / 1.3	**16** W	0541 / 1146 / 1819	5.3 / 1.2 / 5.0
2 W	0524 / 1124 / 1754 / 2342	4.9 / 1.4 / 4.8 / 1.5	**17** TH	0018 / 0654 / 1303 / 1935	1.2 / 5.0 / 1.4 / 4.8
3 TH	0625 / 1232 / 1904	4.7 / 1.6 / 4.6	**18** F	0151 / 0818 / 1433 / 2056	1.3 / 5.0 / 1.5 / 4.9
4 F	0106 / 0748 / 1412 / 2029	1.6 / 4.6 / 1.6 / 4.7	**19** SA	0321 / 0939 / 1553 / 2208	1.2 / 5.1 / 1.4 / 5.1
5 SA	0250 / 0913 / 1530 / 2144	1.5 / 4.8 / 1.3 / 5.0	**20** SU	0437 / 1043 / 1657 / 2304	0.9 / 5.4 / 1.2 / 5.4
6 SU	0402 / 1021 / 1630 / 2245	1.2 / 5.2 / 1.1 / 5.3	**21** M	0532 / 1133 / 1743 / 2350	0.8 / 5.6 / 1.0 / 5.5
7 M	0503 / 1117 / 1725 / 2336	0.9 / 5.6 / 0.9 / 5.6	**22** TU	0614 / 1214 / 1819	0.7 / 5.7 / 0.9
8 TU	0600 / 1204 / 1815	0.6 / 5.9 / 0.7	**23** W	0028 / 0648 / 1250 / 1851 ○	5.6 / 0.6 / 5.7 / 0.8
9 W ●	0021 / 0651 / 1249 / 1903	5.9 / 0.4 / 6.1 / 0.5	**24** TH	0102 / 0719 / 1322 / 1922	5.7 / 0.6 / 5.7 / 0.8
10 TH	0103 / 0738 / 1332 / 1948	6.0 / 0.2 / 6.2 / 0.4	**25** F	0133 / 0748 / 1351 / 1952	5.7 / 0.6 / 5.7 / 0.7
11 F	0145 / 0822 / 1414 / 2031	6.2 / 0.1 / 6.2 / 0.4	**26** SA	0202 / 0817 / 1419 / 2022	5.7 / 0.6 / 5.7 / 0.8
12 SA	0226 / 0904 / 1456 / 2111	6.2 / 0.2 / 6.1 / 0.5	**27** SU	0232 / 0845 / 1448 / 2049	5.7 / 0.7 / 5.7 / 0.8
13 SU	0309 / 0941 / 1539 / 2149	6.1 / 0.3 / 5.9 / 0.6	**28** M	0302 / 0912 / 1518 / 2116	5.6 / 0.8 / 5.5 / 0.9
14 M	0353 / 1017 / 1625 / 2227	5.6 / 0.6 / 5.6 / 0.8	**29** TU	0334 / 0939 / 1551 / 2143	5.4 / 1.0 / 5.4 / 1.1
15 TU	0443 / 1055 / 1716 / 2314 ◗	5.7 / 0.9 / 5.3 / 1.0	**30** W	0410 / 1008 / 1630 / 2218	5.2 / 1.2 / 5.2 / 1.2
			31 TH	0454 / 1050 / 1717 / 2309 ◗	5.0 / 1.4 / 5.0 / 1.3

APRIL

Day	Time	m	Day	Time	m
1 F	0551 / 1152 / 1821	4.8 / 1.5 / 4.7	**16** SA	0131 / 0753 / 1401 / 2026	1.2 / 5.0 / 1.6 / 4.9
2 SA	0025 / 0708 / 1326 / 1945	1.5 / 4.7 / 1.6 / 4.7	**17** SU	0255 / 0911 / 1518 / 2138	1.1 / 5.1 / 1.5 / 5.0
3 SU	0210 / 0836 / 1454 / 2107	1.4 / 4.9 / 1.4 / 5.0	**18** M	0405 / 1015 / 1622 / 2235	1.0 / 5.3 / 1.3 / 5.3
4 M	0329 / 0950 / 1559 / 2213	1.1 / 5.3 / 1.1 / 5.3	**19** TU	0500 / 1106 / 1710 / 2321	0.8 / 5.5 / 1.1 / 5.4
5 TU	0435 / 1049 / 1657 / 2307	0.8 / 5.7 / 0.9 / 5.6	**20** W	0541 / 1147 / 1748	0.8 / 5.6 / 1.0
6 W	0535 / 1139 / 1751 / 2355	0.6 / 5.9 / 0.7 / 5.9	**21** TH	0000 / 0614 / 1222 / 1822	5.5 / 0.7 / 5.6 / 0.9
7 TH ●	0629 / 1225 / 1842	0.4 / 6.1 / 0.5	**22** F	0034 / 0646 / 1252 / 1855 ○	5.6 / 0.7 / 5.7 / 0.8
8 F	0039 / 0716 / 1310 / 1929	6.1 / 0.2 / 6.2 / 0.4	**23** SA	0105 / 0717 / 1321 / 1927	5.7 / 0.6 / 5.7 / 0.7
9 SA	0123 / 0801 / 1353 / 2013	6.2 / 0.2 / 6.2 / 0.3	**24** SU	0136 / 0748 / 1351 / 1959	5.7 / 0.7 / 5.7 / 0.7
10 SU	0207 / 0842 / 1435 / 2055	6.3 / 0.2 / 6.1 / 0.4	**25** M	0207 / 0819 / 1421 / 2030	5.7 / 0.7 / 5.7 / 0.8
11 M	0251 / 0920 / 1519 / 2135	6.2 / 0.4 / 5.9 / 0.5	**26** TU	0239 / 0849 / 1453 / 2100	5.6 / 0.8 / 5.6 / 0.9
12 TU	0337 / 0957 / 1604 / 2215	6.0 / 0.7 / 5.7 / 0.7	**27** W	0313 / 0919 / 1527 / 2130	5.5 / 0.9 / 5.5 / 1.0
13 W	0428 / 1035 / 1654 / 2301	5.7 / 1.0 / 5.3 / 0.9	**28** TH	0351 / 0950 / 1606 / 2205	5.3 / 1.1 / 5.3 / 1.1
14 TH ◑	0526 / 1123 / 1754	5.4 / 1.3 / 5.0	**29** F	0436 / 1031 / 1653 / 2254	5.2 / 1.3 / 5.1 / 1.2
15 F	0002 / 0635 / 1234 / 1907	1.1 / 5.1 / 1.5 / 4.8	**30** SA ◗	0531 / 1130 / 1753	5.0 / 1.4 / 4.9

Chart Datum: 2·90 metres below Ordnance Datum (Newlyn). HAT is 6·3 metres above Chart Datum.

FREE monthly updates. Register at
www.reedsnauticalalmanac.co.uk

SHEERNESS   LAT 51°27'N   LONG 0°45'E
TIMES AND HEIGHTS OF HIGH AND LOW WATERS

STANDARD TIME (UT)
For Summer Time add ONE hour in **non-shaded areas**

Dates in red are **SPRINGS**
Dates in blue are NEAPS

YEAR 2016

MAY

Time	m	Time	m
1 0003	1.3	**16** 0214	1.2
0641	5.0	0830	5.0
SU 1251	1.5	M 1432	1.6
1909	4.9	2055	4.9
2 0134	1.2	**17** 0319	1.1
0803	5.1	0935	5.1
M 1416	1.4	TU 1535	1.4
2030	5.0	2156	5.1
3 0256	1.0	**18** 0413	1.0
0917	5.4	1028	5.3
TU 1524	1.1	W 1628	1.2
2138	5.3	2246	5.3
4 0404	0.8	**19** 0458	0.9
1019	5.7	1112	5.4
W 1626	0.9	TH 1712	1.1
2236	5.6	2327	5.4
5 0507	0.6	**20** 0537	0.8
1113	5.9	1149	5.5
TH 1724	0.7	F 1751	0.9
2328	5.9		
6 0603	0.4	**21** 0004	5.5
1201	6.1	0612	0.8
F 1819	0.5	SA 1222	5.6
		1828	0.8
7 0016	6.1	**22** 0039	5.6
0653	0.3	0646	0.8
SA 1248	6.1	SU 1254	5.7
1909	0.4	1903	0.8
8 0103	6.2	**23** 0112	5.6
0738	0.3	0720	0.7
SU 1332	6.2	M 1327	5.7
1956	0.3	1939	0.7
9 0150	6.2	**24** 0146	5.6
0821	0.4	0755	0.8
M 1416	6.1	TU 1400	5.7
2041	0.3	2015	0.7
10 0236	6.1	**25** 0221	5.6
0900	0.5	0830	0.8
TU 1501	5.9	W 1434	5.6
2123	0.5	2051	0.8
11 0324	5.9	**26** 0258	5.6
0938	0.8	0905	1.0
W 1546	5.6	TH 1510	5.5
2205	0.6	2127	0.9
12 0414	5.7	**27** 0338	5.5
1016	1.1	0941	1.1
TH 1634	5.4	F 1550	5.4
2249	0.9	2205	1.0
13 0508	5.4	**28** 0423	5.4
1100	1.3	1022	1.2
F 1729	5.1	SA 1636	5.3
◑ 2344	1.1	2251	1.0
14 0610	5.1	**29** 0516	5.2
1159	1.5	1115	1.3
SA 1834	4.9	SU 1732	5.1
		◑ 2350	1.1
15 0058	1.2	**30** 0620	5.2
0718	5.0	1221	1.4
SU 1317	1.6	M 1839	5.1
1945	4.8		
		31 0105	1.0
		0733	5.2
		TU 1338	1.3
		1954	5.2

JUNE

Time	m	Time	m
1 0223	0.9	**16** 0322	1.2
0845	5.4	0940	5.0
W 1450	1.2	TH 1542	1.4
2105	5.3	2202	5.0
2 0333	0.8	**17** 0413	1.1
0950	5.6	1030	5.2
TH 1555	1.0	F 1634	1.2
2208	5.6	2251	5.2
3 0438	0.7	**18** 0458	1.0
1047	5.8	1114	5.4
F 1658	0.8	SA 1720	1.1
2305	5.8	2334	5.4
4 0538	0.6	**19** 0540	0.9
1140	5.9	1153	5.5
SA 1758	0.6	SU 1802	0.9
2357	5.9		
5 0631	0.5	**20** 0014	5.5
1228	6.0	0619	0.9
SU 1853	0.5	M 1231	5.6
●		○ 1842	0.8
6 0047	6.0	**21** 0052	5.6
0718	0.5	0657	0.8
M 1315	6.0	TU 1307	5.7
1942	0.4	1922	0.8
7 0136	6.1	**22** 0130	5.7
0801	0.6	0736	0.8
TU 1401	6.0	W 1343	5.7
2029	0.4	2003	0.7
8 0224	6.0	**23** 0207	5.7
0842	0.7	0815	0.8
W 1445	5.8	TH 1420	5.7
2113	0.6	2045	0.7
9 0311	5.9	**24** 0246	5.7
0920	0.9	0855	0.9
TH 1529	5.7	F 1458	5.7
2154	0.6	2126	0.7
10 0358	5.7	**25** 0328	5.7
0956	1.1	0934	1.0
F 1614	5.5	SA 1538	5.6
2234	0.8	2206	0.8
11 0446	5.4	**26** 0412	5.6
1034	1.3	1014	1.1
SA 1701	5.2	SU 1623	5.5
2318	1.0	2248	0.8
12 0537	5.2	**27** 0502	5.4
1120	1.5	1100	1.2
SU 1754	5.0	M 1713	5.4
◐		◐ 2337	0.9
13 0011	1.1	**28** 0559	5.3
0634	5.0	1155	1.3
M 1219	1.6	TU 1813	5.3
1854	4.9		
14 0118	1.2	**29** 0037	1.0
0736	4.9	0704	5.3
TU 1333	1.7	W 1303	1.3
2000	4.8	1923	5.2
15 0224	1.2	**30** 0151	1.0
0840	4.9	0815	5.3
W 1443	1.6	TH 1418	1.2
2105	4.9	2036	5.3

JULY

Time	m	Time	m
1 0305	0.9	**16** 0329	1.3
0923	5.4	0943	5.0
F 1530	1.1	SA 1556	1.4
2145	5.4	2214	5.0
2 0414	0.9	**17** 0423	1.2
1026	5.6	1039	5.3
SA 1639	0.9	SU 1649	1.2
2248	5.6	2306	5.3
3 0518	0.8	**18** 0511	1.0
1123	5.7	1126	5.5
SU 1745	0.7	M 1738	1.0
2345	5.8	2351	5.5
4 0613	0.7	**19** 0555	1.0
1215	5.8	1209	5.6
M 1842	0.6	TU 1824	0.9
●		○	
5 0037	5.9	**20** 0033	5.6
0701	0.7	0637	0.9
TU 1303	5.9	W 1249	5.7
1931	0.4	1907	0.7
6 0126	6.0	**21** 0113	5.8
0744	0.7	0719	0.8
W 1347	5.9	TH 1328	5.8
2017	0.4	1952	0.6
7 0212	6.1	**22** 0153	5.9
0824	0.8	0801	0.8
TH 1429	5.8	F 1406	5.8
2059	0.5	2036	0.5
8 0255	5.9	**23** 0233	5.9
0842	1.0	0844	0.8
F 1510	5.7	SA 1444	5.8
2137	0.6	2119	0.5
9 0337	5.7	**24** 0314	5.9
0934	1.1	0924	0.8
SA 1549	5.6	SU 1524	5.8
2211	0.7	2159	0.6
10 0418	5.5	**25** 0357	5.8
1005	1.2	1002	0.9
SU 1629	5.4	M 1607	5.7
2244	0.9	2238	0.7
11 0459	5.3	**26** 0443	5.6
1040	1.4	1043	1.1
M 1710	5.2	TU 1654	5.6
2319	1.1	◐ 2319	0.8
12 0544	5.1	**27** 0536	5.4
1122	1.5	1130	1.2
TU 1759	5.0	W 1749	5.4
◐			
13 0006	1.3	**28** 0011	1.0
0635	4.9	0637	5.3
W 1219	1.7	TH 1233	1.3
1857	4.8	1856	5.3
14 0112	1.4	**29** 0122	1.1
0735	4.8	0748	5.2
TH 1335	1.7	F 1352	1.3
2004	4.7	2014	5.2
15 0226	1.4	**30** 0242	1.2
0840	4.8	0901	5.2
F 1453	1.6	SA 1513	1.2
2112	4.8	2130	5.3
		31 0356	1.1
		1011	5.4
		SU 1630	1.0
		2240	5.5

AUGUST

Time	m	Time	m
1 0504	1.0	**16** 0443	1.2
1112	5.6	1058	5.4
M 1738	0.8	TU 1714	1.0
2339	5.7	2327	5.5
2 0600	0.9	**17** 0532	1.0
1204	5.7	1145	5.6
TU 1833	0.6	W 1804	0.8
●			
3 0029	5.8	**18** 0012	5.7
0646	0.9	0617	0.9
W 1250	5.8	TH 1228	5.8
1919	0.5	○ 1850	0.7
4 0115	5.9	**19** 0054	5.9
0727	0.9	0701	0.8
TH 1332	5.9	F 1308	5.9
2001	0.5	1936	0.5
5 0156	5.9	**20** 0134	6.0
0803	0.9	0745	0.7
F 1410	5.9	SA 1347	5.9
2038	0.5	2021	0.4
6 0234	5.9	**21** 0215	6.1
0837	0.9	0828	0.7
SA 1446	5.8	SU 1426	6.1
2112	0.6	2103	0.4
7 0311	5.7	**22** 0256	6.1
0907	1.0	0909	0.7
SU 1521	5.7	M 1506	6.0
2141	0.7	2143	0.5
8 0345	5.6	**23** 0338	5.9
0935	1.1	0947	0.8
M 1554	5.5	TU 1548	5.9
2207	0.9	2220	0.5
9 0419	5.4	**24** 0422	5.7
1003	1.3	1024	1.0
TU 1629	5.3	W 1634	5.7
2235	1.1	2257	0.9
10 0455	5.2	**25** 0512	5.5
1037	1.4	1109	1.1
W 1708	5.1	TH 1728	5.5
◐ 2312	1.3	◐ 2345	1.1
11 0538	5.0	**26** 0612	5.2
1121	1.6	1208	1.3
TH 1757	4.8	F 1837	5.2
12 0002	1.5	**27** 0055	1.3
0631	4.8	0724	5.1
F 1222	1.7	SA 1333	1.4
1902	4.6	1958	5.1
13 0117	1.6	**28** 0223	1.4
0740	4.7	0843	5.1
SA 1353	1.8	SU 1503	1.3
2020	4.6	2121	5.2
14 0243	1.6	**29** 0343	1.3
0855	4.8	0958	5.3
SU 1516	1.6	M 1624	1.0
2135	4.9	2232	5.5
15 0349	1.4	**30** 0453	1.2
1003	5.1	1100	5.6
M 1619	1.3	TU 1730	0.8
2236	5.2	2330	5.7
		31 0547	1.1
		1150	5.7
		W 1820	0.7

Chart Datum: 2·90 metres below Ordnance Datum (Newlyn). HAT is 6·3 metres above Chart Datum.

E England

>> **FREE** monthly updates. Register at <<
www.reedsnauticalalmanac.co.uk

STANDARD TIME (UT)
For Summer Time add ONE hour in **non-shaded areas**

SHEERNESS   LAT 51°27'N   LONG 0°45'E
TIMES AND HEIGHTS OF HIGH AND LOW WATERS

Dates in red are **SPRINGS**
Dates in blue are **NEAPS**

YEAR 2016

SEPTEMBER

Time	m		Time	m
1 TH 0017	5.8	**16** F 0553	0.9
0629	1.0	1202	5.9
1234	5.8	1829	0.6 ○
1901	0.6 ●		
2 F 0058	5.9	**17** SA 0029	6.1
0705	0.9	0640	0.7
1312	5.9	1243	6.0
1937	0.6	1915	0.4
3 SA 0135	5.9	**18** SU 0111	6.2
0738	0.9	0725	0.6
1346	5.9	1324	6.2
2010	0.6	1959	0.3
4 SU 0208	5.8	**19** M 0152	6.2
0810	0.9	0809	0.6
1419	5.8	1404	6.2
2040	0.6	2042	0.3
5 M 0240	5.8	**20** TU 0234	6.2
0839	1.0	0850	0.6
1450	5.7	1446	6.2
2107	0.8	2121	0.5
6 TU 0310	5.7	**21** W 0316	6.0
0906	1.1	0929	0.7
1520	5.6	1529	6.0
2132	0.9	2158	0.7
7 W 0340	5.5	**22** TH 0400	5.8
0931	1.2	1008	0.9
1552	5.4	1616	5.8
2158	1.1	2235	1.0
8 TH 0413	5.3	**23** F 0449	5.5
0959	1.3	1052	1.1
1628	5.2	1712	5.5
2230	1.3	2322	1.3 ◑
9 F 0450	5.1	**24** SA 0548	5.2
1037	1.5	1152	1.3
1711	4.9	1821	5.2
2312	1.5 ◑		
10 SA 0538	4.9	**25** SU 0031	1.5
1129	1.7	0701	5.0
1808	4.7	1320	1.4
		1944	5.1
11 SU 0015	1.7	**26** M 0202	1.6
0642	4.7	0823	5.0
1247	1.8	1452	1.3
1927	4.6	2107	5.2
12 M 0151	1.8	**27** TU 0324	1.5
0806	4.7	0939	5.2
1433	1.7	1611	1.0
2053	4.8	2217	5.5
13 TU 0312	1.5	**28** W 0433	1.3
0924	5.0	1041	5.5
1546	1.3	1712	0.8
2203	5.2	2312	5.7
14 W 0412	1.3	**29** TH 0525	1.1
1026	5.4	1130	5.7
1645	1.0	1758	0.7
2258	5.6	2357	5.8
15 TH 0505	1.0	**30** F 0605	1.0
1117	5.7	1211	5.8
1739	0.8	1835	0.7
2346	5.9		

OCTOBER

Time	m		Time	m
1 SA 0035	5.8	**16** SU 0002	6.1
0638	1.0	0614	0.7
1247	5.8	1216	6.1
1906	0.7 ●	1849	0.4 ○
2 SU 0108	5.8	**17** M 0046	6.2
0709	0.9	0702	0.6
1319	5.8	1259	6.2
1936	0.7	1935	0.4
3 M 0139	5.8	**18** TU 0129	6.3
0741	0.9	0748	0.5
1350	5.8	1342	6.3
2005	0.7	2018	0.4
4 TU 0208	5.8	**19** W 0211	6.2
0810	0.9	0832	0.6
1420	5.8	1427	6.2
2033	0.8	2058	0.5
5 W 0236	5.7	**20** TH 0255	6.0
0838	1.0	0913	0.6
1450	5.7	1512	6.1
2059	0.9	2136	0.8
6 TH 0306	5.6	**21** F 0339	5.8
0904	1.1	0955	0.8
1522	5.5	1602	5.8
2126	1.1	2214	1.1
7 F 0338	5.4	**22** SA 0429	5.5
0931	1.2	1040	1.0
1557	5.3	1658	5.5
2155	1.3	2300	1.4 ◑
8 SA 0414	5.2	**23** SU 0527	5.2
1004	1.4	1139	1.2
1638	5.1	1805	5.2
2234	1.5		
9 SU 0458	5.0	**24** M 0005	1.6
1052	1.5	0637	5.0
1731	4.9	1303	1.3
2331	1.7 ◑	1922	5.1
10 M 0556	4.8	**25** TU 0132	1.7
1201	1.7	0755	5.0
1842	4.7	1429	1.2
		2041	5.2
11 TU 0055	1.8	**26** W 0252	1.6
0715	4.7	0910	5.1
1342	1.6	1543	1.1
2009	4.8	2150	5.4
12 W 0229	1.6	**27** TH 0359	1.4
0841	4.9	1012	5.4
1507	1.3	1641	0.9
2125	5.2	2245	5.6
13 TH 0336	1.3	**28** F 0452	1.2
0949	5.3	1102	5.5
1611	1.0	1726	0.8
2225	5.6	2329	5.7
14 F 0432	1.1	**29** SA 0533	1.1
1044	5.6	1143	5.6
1708	0.8	1801	0.8
2316	5.9		
15 SA 0524	0.9	**30** SU 0007	5.7
1132	5.9	0607	1.0
1801	0.6	1219	5.7
		1831	0.8 ●
		31 M 0039	5.8
		0640	0.9
		1252	5.7
		1901	0.8

NOVEMBER

Time	m		Time	m
1 TU 0109	5.8	**16** W 0107	6.2
0712	0.9	0729	0.5
1322	5.8	1324	6.3
1931	0.8	1955	0.5
2 W 0138	5.8	**17** TH 0152	6.1
0744	0.9	0816	0.5
1353	5.7	1411	6.2
2001	0.8	2037	0.6
3 TH 0207	5.7	**18** F 0237	6.0
0814	0.9	0901	0.5
1425	5.7	1459	6.1
2031	0.9	2116	0.8
4 F 0238	5.7	**19** SA 0322	5.8
0844	1.0	0945	0.7
1459	5.5	1549	5.8
2101	1.1	2155	1.1
5 SA 0310	5.5	**20** SU 0411	5.5
0913	1.1	1030	1.0
1535	5.4	1643	5.5
2132	1.3	2239	1.3
6 SU 0347	5.3	**21** M 0505	5.3
0946	1.2	1123	1.1
1616	5.2	1744	5.3
2209	1.4	2333	1.6 ◑
7 M 0430	5.2	**22** TU 0607	5.0
1030	1.4	1232	1.2
1706	5.0	1851	5.1
2300	1.6 ◑		
8 TU 0524	5.0	**23** W 0046	1.7
1131	1.5	0717	4.9
1810	4.9	1349	1.3
		2002	5.1
9 W 0012	1.7	**24** TH 0205	1.7
0633	4.9	0828	5.0
1255	1.4	1457	1.2
1928	5.0	2110	5.2
10 TH 0139	1.6	**25** F 0312	1.6
0755	5.0	0933	5.1
1423	1.3	1555	1.1
2045	5.2	2208	5.3
11 F 0254	1.4	**26** SA 0408	1.4
0908	5.3	1027	5.3
1533	1.0	1643	1.0
2150	5.6	2256	5.4
12 SA 0355	1.1	**27** SU 0455	1.2
1008	5.6	1113	5.4
1634	0.8	1722	0.9
2245	5.9	2335	5.5
13 SU 0453	0.9	**28** M 0535	1.1
1101	5.8	1150	5.5
1731	0.6	1757	0.9
2335	6.0		
14 M 0548	0.7	**29** TU 0009	5.6
1150	6.1	0611	1.0
1823	0.5	1226	5.6
○		1829	0.9 ●
15 TU 0022	6.2	**30** W 0042	5.7
0640	0.6	0647	0.9
1238	6.2	1259	5.6
1911	0.4	1902	0.8

DECEMBER

Time	m		Time	m
1 TH 0113	5.7	**16** F 0137	6.0
0721	0.8	0805	0.4
1332	5.7	1400	6.1
1935	0.9	2020	0.7
2 F 0144	5.7	**17** SA 0222	5.9
0755	0.8	0851	0.4
1406	5.6	1448	6.0
2009	0.9	2100	0.8
3 SA 0217	5.7	**18** SU 0307	5.8
0830	0.9	0935	0.6
1441	5.6	1536	5.8
2042	1.0	2137	1.0
4 SU 0251	5.6	**19** M 0353	5.6
0904	1.0	1016	0.7
1519	5.5	1624	5.6
2117	1.1	2215	1.2
5 M 0328	5.4	**20** TU 0440	5.4
0940	1.1	1059	0.9
1600	5.4	1715	5.3
2154	1.3	2257	1.4
6 TU 0410	5.3	**21** W 0531	5.2
1020	1.1	1148	1.1
1648	5.2	1810	5.1
2239	1.4	2350	1.6 ◑
7 W 0500	5.2	**22** TH 0629	5.0
1111	1.2	1249	1.3
1744	5.1	1911	4.9
2338	1.5		
8 TH 0600	5.1	**23** F 0059	1.7
1216	1.2	0734	4.8
1852	5.1	1358	1.3
		2015	4.9
9 F 0050	1.5	**24** SA 0214	1.7
0712	5.1	0842	4.8
1337	1.2	1500	1.3
2006	5.2	2119	4.9
10 SA 0209	1.4	**25** SU 0319	1.6
0827	5.2	0944	5.0
1453	1.0	1554	1.2
2115	5.4	2215	5.1
11 SU 0318	1.2	**26** M 0415	1.4
0934	5.5	1037	5.1
1601	0.8	1642	1.1
2216	5.7	2301	5.3
12 M 0422	1.0	**27** TU 0504	1.2
1034	5.7	1122	5.3
1703	0.7	1724	1.0
2311	5.8	2341	5.4
13 TU 0524	0.8	**28** W 0547	1.0
1129	5.9	1202	5.4
1800	0.6	1803	0.9
14 W 0002	6.0	**29** TH 0018	5.6
0622	0.6	0625	0.9
1221	6.0	1240	5.5
1850	0.6 ○	1839	0.9 ●
15 TH 0050	6.0	**30** F 0053	5.6
0715	0.5	0703	0.8
1311	6.1	1315	5.6
1937	0.6	1915	0.9
		31 SA 0128	5.7
		0742	0.8
		1351	5.7
		1952	0.8

Chart Datum: 2·90 metres below Ordnance Datum (Newlyn). HAT is 6·3 metres above Chart Datum.

》》 **FREE** monthly updates. Register at 《
www.reedsnauticalalmanac.co.uk 《

9.4.12 RIVER THAMES

CHARTS AC 5606, 1185,1186, 2151, 3337, 2484, 3319. Imray C1, C2, Y18, 2100

TIDES The river is tidal up to Teddington. +0252 Dover; ML 3·6; Duration 0555.

Standard Port LONDON BRIDGE (→)

Times				Height (metres)			
High Water		Low Water		MHWS	MHWN	MLWN	MLWS
0300	0900	0400	1100	7·1	5·9	1·3	0·5
1500	2100	1600	2300				
Differences TILBURY							
−0055	−0040	−0050	−0115	−0·7	−0·5	+0·1	0·0
WOOLWICH (GALLIONS POINT)							
−0020	−0020	−0035	−0045	−0·1	0·0	+0·2	0·0
ALBERT BRIDGE							
+0025	+0020	+0105	+0110	−0·9	−0·8	−0·7	−0·4
HAMMERSMITH BRIDGE							
+0040	+0035	+0205	+0155	−1·4	−1·3	−1·0	−0·5
KEW BRIDGE							
+0055	+0050	+0255	+0235	−1·8	−1·8	−1·2	−0·5
TEDDINGTON (RICHMOND) LOCK							
+0105	+0055	+0325	+0305	−2·2	−2·7	−0·9	−0·5

Tide Boards, showing height of tide above CD, are at Lower Pool (Met Police Boatyard), King's Reach (Temple Stairs), Nine Elms Reach (Cringle Wharf) and Battersea Reach (Plantation Wharf). Above Putney the height of LW may be below CD if the water flow over Teddington Weir is low; warnings are broadcast by London VTS. Thames Barrier closure will alter predicted water levels greatly.

SHELTER Very good in marinas. PLA Drawdocks, arrowed, are drying inlets offering emergency refuge, but subject to wash.

NAVIGATION WPT 51°29′·62N 00°44′·17E, Sea Reach No 4 buoy, 276°/6·8M to abm Holehaven Creek and 37·8M to Tower Bridge. Pleasure craft should always keep clear of commercial vessels. To clear tanker berths off Holehaven, Coryton and Thames Haven yachts should navigate following small craft crossing routes:

Inward from north: Keep close to SHM buoys & bns. At W Leigh Middle ensure the fairway is clear, then cross to the S side of Yantlet Channel. Make for E Blyth buoy before turning onto the inward track, so as to clear outward vessels passing close to the PHM buoys. It is safe to pass S of Tanker, Cliffe Fleet, West Blyth and Lower Hope buoys. When safe, cross to the N side in Lower Hope Reach as quickly as possible.

Outward to north: as above in reverse, but cross to the N side between Sea Reach Nos 4 & 5 buoys.

Inward from the south: Keep well south of Yantlet Channel, crossing to the N side in Lower Hope Reach as described above.

Sea Reach, keep well S of the main chan to clear tankers turning off Canvey Is and Coryton. Keep at least 60m clear of oil/gas jetties and give a wide berth to London Gateway container terminal. Warning lights are displayed at Cliffe and Canvey Is when large ships are manoeuvring.

Lower Hope Reach, hold the NW bank until Ovens SHM; long groynes extend from the N bank for the next 2M. Tilbury Landing Stage is used by the Gravesend ferry and cruise liners.

Gravesend Reach, beware 6 groynes off the N bank, WSW of Coalhouse Pt, which project almost into the fairway. Outer ends marked by SHM beacons, Fl G 2·5s. *No inside passage exists.*

Above **Gravesend** keep to stbd side of channel – do not cut corners.

Northfleet Hope, beware ships/tugs turning into Tilbury Docks; container berths and a grain terminal are close up-river. A light, Iso 6s, on Tilbury Cargo Jetty (51°27′·06N 00°20′·95E) is shown when vessels are manoeuvring off Tilbury Dock locks or for a berth in Northfleet Hope. *Proceed with caution.*

Above **Cherry Garden Pier** vessels > 40m LOA and tugs/tows always have priority. Very heavy traffic/frequent River Ferries from/to Greenwich, Tower of London and Westminster. Police and PLA launches are always willing to assist.

Speed limits. Minimise wash. 8kn speed limit applies at Canvey Is, Coryton and Thurrock when gas tankers are berthed, in all creeks and above Wandsworth Bridge. Voluntary 12kn speed limit to Wandsworth Bridge.

Exclusion Zones. These are in place off the Houses of Parliament on the North Embankment and government offices below Vauxhall Br on the S side of the river.

> The Port of London Authority (PLA), London River House, Royal Pier Road, Gravesend, Kent DA12 2BG, ☎01474 562200, publish (download free from www.pla.co.uk) *Recreational Users' Guide; General Directions; Permanent Notices to Mariners; Port of London River Byelaws; Tide Tables* and a *Port Handbook*.

LIGHTS AND MARKS See 9.4.4 and chartlets. Glare from shore lts can make navigation by night difficult or even hazardous. Some ♠s show F.Bu lights; High Speed Craft show Fl.Y lights.

COMMUNICATIONS Port of London Authority: www.pla.co.uk HM: lower district 01474 562200, upper district 020 7743 7912; NMOC 02392 552100; Port Controller Gravesend 01474 560311, Police Marine Support Unit 020 7275 4421; Thames Barrier Navigation Centre 020 8855 0315; Tower Bridge 020 7940 3984; Richmond lock 020 8940 0634; Teddington lock 020 8940 8723.

London VTS works from centres at Gravesend and the Thames Barrier, dividing the estuary/river into three sectors. Routine MSI broadcasts are made at the times shown (in brackets) below.

Sector 1: Outer limits – Sea Reach No 4 buoy. Ch **69** (H +15, H +45)

Sector 2: Sea Reach 4 – Crayfordness. Ch **68** (H and H +30)

Sector 3: Crayfordness – Teddington. Ch **14** (H +15 and H +45)

PLA launches *Thames Patrol* Ch 69, 68, 14. Police launches *Thames Police* Ch 14. **Do not** use tug Channels 8, 10, 13, 36, 72, 77. **In emergency** call *London Coastguard* Ch 16, 67, co-located with London VTS at the Thames Barrier Navigation Control and covering from Canvey Island to Teddington. *Thames Coastguard* covers the area to seaward of Canvey Island. The RNLI operates four LB stations at Gravesend, Victoria Embankment (Waterloo Br), Chiswick and Teddington.

HARBOURS IN SEA REACH and GRAVESEND REACH

HOLEHAVEN CREEK, Essex, **51°30′·57N 00°33′·16E.** AC 2484, 1186. HW +0140 on Dover; use differences for Coryton (at 9.4.13); ML 3·0m; Duration 0610. Shelter is good, but beware swell from passing traffic. Keep to Canvey Is side on ent. Temp ⚓ in lee of Chainrock jetty or call PLA ☎01474 562462 for possible mooring. 5ca N of ent an overhead oil pipe (2 FY horiz) crosses chan with 11m clearance, plus. 2 FG (vert) on all jetty heads, plus Chainrock SHM buoy Fl G 5s. ⚓ from Lobster Smack pub, ☎ ☎ from Canvey Village (1M); all other facilities on Canvey Is.

GRAVESEND, Kent, **51°26′·61N 00°22′·92E.** AC 1186, 2151. HW +0150 on Dover; ML 3·3m; Duration 0610. Use Tilbury diffs. ⚓ E of Club's Jetty close to S shore, but remote from town.

Gravesend SC www.gravesendsailingclub.co.uk ☎07538 326623.

PLA ⚓s are upriver, £25/night. **London River Moorings** ☎01474 535700 All tide ⚓s £25/night, half tide ⚓. New all tide landing stage available.

Embankment Marina www.theembankmentmarina.co.uk ☎01474 535700; £25/night. Lock into Canal Basin HW −1 to HW Tilbury by arrangement: 1 Apr-31 Oct, 0800-2000; Winter 1000-1600. Lock closes at local HW. Out of hours openings up to HW by arrangement with lock keeper; facilities ⚓ ⚓ ⚓ ⚓ ⚓ ⚓ ⚓ Ⓔ.

Town Tourist info ☎(01474) 337600, ☎ ☎ ⚓ ⚓ Ⓔ Ⓔ ⚓(70t) at Denton Wharf, ⚓ ✕.

Thurrock YC 51°28′·28N 00°19′·47E at Grays, ☎01375 373720. 1⚓ ☎ ☎ (2M), ⚓ ✕ (occas). M-F 1000-1500; Thu 2000-2300.

CANVEY ISLAND TO CRAYFORDNESS (AC 1185, 1186, 2484, 2151)

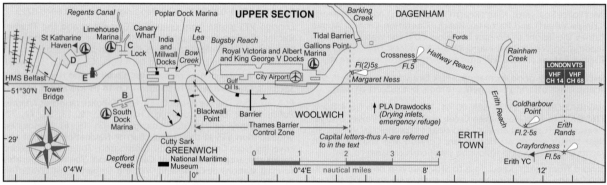

CRAYFORDNESS TO TOWER BRIDGE (AC 2484, 2151, 3337)

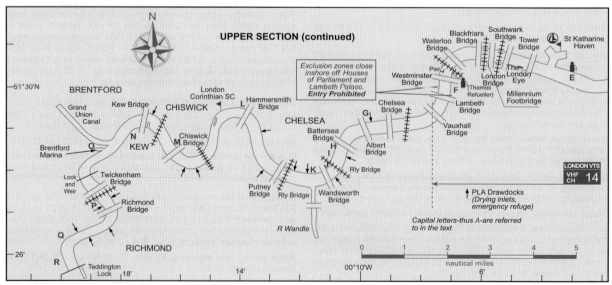

TOWER BRIDGE TO TEDDINGTON (AC 3319)

OBSTRUCTIONS

THAMES TIDAL BARRIER
51°29'·91N 00°02'·21E
(Span G), AC 5606, 2484, 3337. At Woolwich Reach are 9 piers between which gates can be rotated upwards from the river bed to form a barrier. Piers are numbered 1-9 from N to S; Spans are lettered A-K from S to N (see diagram). A, H, J & K are not navigable. C-F, with depth 5·8m and 61m wide, are for larger ships. Spans B and G, with 1·25m, are for small craft:
W-bound via G and E-bound via B (51°29'·73N 00°02'·24E).
On N side of river (Spans E and F) a cross-tide component is reported.

Small craft should not navigate between Thames Refinery Jetty and Gulf Oil Island, unless intending to transit the Barrier. Yachts must transit the barrier under power, not sail. When all spans are closed, keep 200m clear to avoid turbulence. It is dangerous to transit a span closed to navigation as the gate may be semi-raised.

In reduced visibility, open spans are marked by high intensity lts. *London VTS* broadcasts information on Ch **14** at H + 15 / H + 45.

CONTROL AND COMMUNICATIONS The Thames Barrier Navigation Centre controls all traffic in a Zone from Margaret Ness (51°30'·5N 00°05'·56E) to Blackwall Point (51°30'·3N 00°00'·2E), using the callsign *London VTS* on VHF Ch **14**, 22, 16.

Inbound call *London VTS* to obtain clearance to proceed through the Barrier when passing Margaret Ness (51°30'·54N 00°05'·50E).

Outbound call *London VTS* to obtain clearance to proceed through the Barrier when passing Blackwall Point (51°30'·30N 00°00'·18E).

Non-VHF craft should, if possible, pre-notify the Thames Barrier Control ☎020 8855 0315; then observe all visual signals, proceed with caution keeping clear of larger vessels and use spans B or G as appropriate. Call before and after transiting the barrier.
There are fixed warning noticeboards (CEVNI) at: Barking Creek (N bank) Blackwall Stairs (N bank) Blackwall Point (S bank).

TESTING The Barrier is completely closed for testing once a month for about 3hrs, LW ±1½. Visit www.pla.co.uk (navigation safety) or call ☎020 8305 4188 for details. In Sept/Oct an annual closure lasts about 10 hrs, LW to LW.

THAMES GATEWAY CABLE CAR Joining the O2 Millennium Dome site at Greenwich with the Excel Centre at Victoria Docks, the Cable Car crosses the river between 2 pylons giving an air draught of 60m.

RICHMOND WEIR & TEDDINGTON LOCK At **Richmond** a half-tide lock and a weir with overhead sluice gates give access up-river; fee payable. 3 R discs (3 ● lts at night) in a ▽ show when the sluice gates are down (closed) and maintaining at least 1·72m between Richmond and Teddington bridges. The half-tide lock, on the Surrey bank, must then be used. At other times (approx HW ±2) two ● lights at each arch = weir open to navigation. Pass freely below the 3 central arches.

Teddington lock marks the end of the tidal Thames (semi-tidal above Richmond). Red/green lights indicate which of the launch or barge locks is in operation.

BRIDGES

Craft approaching a bridge against the tide give way to those approaching with the tide. All bridges above Tower Bridge are fixed. A ▽ of R discs (● lts) below a bridge span = this arch closed; a bundle of straw = reduced headroom. Clearances (m) above HAT are as follows:

Tower Bridge fog horn sounds a 10s blast every 20s when bascules are open for shipping; standby signal is a bell 30s. On bridge arches between Tower and Wandsworth bridges high intensity white signal lts are electronically triggered by one or more large vessels approaching a bridge. Iso W 4s means a single large vessel, VQ a second large vessel, nearing the same bridge. Other vessels keep clear of that arch. NB air-draught boards at Harrods Depository and Chiswick Eyot for Hammersmith Bridge.

Dartford (QEII)	53	Albert	4.5
Tower (closed)	8·0	Battersea road	5·0
Tower (open)	42	Battersea rail	5·5
London	8·2	Wandsworth	5·3
Cannon St	6·7	Fulham	6·3
Southwark	6·8	Putney	4·8
Millennium	8·4	Hammersmith	3·1
Blackfriars rail	6·8	Barnes	4·9
Blackfriars road	6·5	Chiswick	6·1
Waterloo	8·1	Kew rail	5·0
Charing Cross	6·5	Kew road	4·7
Westminster	4·8	Richmond foot	5·0
Lambeth	5·9	Twickenham	8·5*
Vauxhall	5·2	Richmond rail	8·0*
Victoria Railway	5·5	Richmond road	7·9*
Chelsea	6·1	* Above maintained water level

FACILITIES

Marinas See overleaf for Gallions Point and Poplar Dock. Other marinas and YCs (Bold red letters in brackets appear on River Thames chartlets):

(A) Greenwich YC ☎0844 736 5846; VHF Ch M. ⚓ ⚙ ✎.

(B) South Dock Marina ☎020 7252 2244 (details overleaf).

(C) Limehouse Marina ☎020 7308 9930 (details overleaf).

(D) St Katharine Haven ☎020 7264 5312 (details overleaf).

West India Dock ☎020 7517 5500; group bookings (min 6) only.

Hermitage Moorings (3·5ca downstream of Tower bridge) ☎020 7481 2122; VHF Ch M/80. 2❶ on pontoons (max 40m LOA) £30/£40 per night, ⚓ ⎓ ⛵ ☈/⚑.

(E) Fuel Barge ☎020 7481 1774; VHF Ch 14 *Heiko*. ⬥ ⚓ M-F 0630-1430. W/ends Apr-Oct, HW±2, between 0600 & 1800.

(F) Westminster Petroleum Ltd Fuel barge moored off Houses of Parliament. Open Mon-Fri (and Sat from Easter to mid-Oct). Call *Thames Refueller* VHF Ch 14, ☎07831 110681, ⬥ ⚓ dry stores.

(G) Cadogan 020 8748 2715, give 24 hrs notice. Chiswick 020 8742 2713. Dove 020 8748 2715. St Katharine 020 7488 0555. Putney 020 7378 1211. 020 7930 2062 for Kew Pier and Richmond Lndg Stage.

(H) Chelsea Yacht & Boat Co ☎020 7352 1427, ⚓ ⚓.

(I) Chelsea Harbour ☎020 7225 9100 (details overleaf).

(J) Imperial Wharf Marina pier@imperialwharfmarina.co.uk. ☎07932 603284, 270m pontoon, 2m at MLWS, £4/m/night, ⚓ ⎓, visitors welcome.

(K) Hurlingham YC ☎020 8788 5547, ⚓.

Chiswick Pier ☎020 8742 2713, all tide access, max draught 1·4m £0.95m, long pontoon, 2FG (vert), ⚓ ⎓ ✗. Visitors welcome.

(L) Chiswick Quay Marina ☎020 8994 8743, Access HW±2 via lock, ⚓ 50◯ ⚓ ✗ ✎ ▣ ⛽ ⚒.

(M) Dove Marina.

(N) Kew Marina ☎020 8940 8364, ⚓ ⬥ ⚓ ⚓ ⚒.

(O) Brentford Dock ☎020 8232 8941, (details overleaf).

(P) Richmond Slipway ⚓ ⚓ ⬥ ⚓ ✗ ✎ ▣ ⛽ ⚒.

(Q) Eel Pie Island ⚓ ⚓ ⎓ ⚓ ✗ ✎ ▣ ⛽(6t) ⚒.

(R) Swan Island Hbr ⚓ ⚓ ◯ ⚓ ⬥ ⚓ ✗ ✎ ▣ ⛽(30t) ⚒.

Piers where landing may be possible, by prior arrangement:
Call London River Services ☎020 7941 2400 for: Greenwich, Tower Millennium, Embankment, Bankside, Festival, Westminster, Millbank, Blackfriars and Waterloo.

E England

MARINAS ON THE TIDAL THAMES, from seaward to Bentford
AC 2484, 3337, 3319

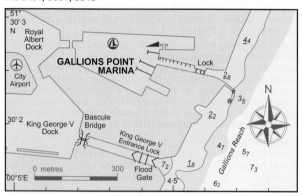

GALLIONS POINT MARINA, **51°30'·30N 00°04'·66E** sited in the entry basin for Royal Albert Dock. www.gallionspoint marina.co.uk ☎ 020 7476 7054. Ch M/80, when vessel expected. 2 x ✩s 2FG (vert) on river pier. 8m depth in basin. Access via lock (width 7·6m) HW±5; Locking £5 each way; H24 security, ⬡ from £12/yacht, ⬡ WC ⬡ ⬡ ⬡ ⬡ ⬡ slip. DLR from N Woolwich to central London, until 0030. Woolwich ferry & foot tunnel 15 mins walk. ✈ (City) is adjacent.

POPLAR DOCK MARINA, **51°30'·07N 00°00'·50W** (lock entrance). ☎020 7308 9930, *but no visitors berths*. Tides: India & Millwall Docks ent., 4·5M below Tower Bridge. Canary Wharf twrs (244m) are conspic 4ca W of marina. Enter via West India Entrance Lock (spanned by Docklands 'Blue Bridge'). Lock (300m x 24m) opens 0700-1700LT at HW for outbound and HW-1 to HW+1 for arrivals, foc, but OT £20. Least width 12·1m into marina. VHF Ch 13 (H24).

Facilities: ⬡ ⬡ ⬡ ⬡ ⬡ ⬡ ⬡; Dr ☎020 7237 1078; Ⓗ 020 7955 5000; Police 101. ⇌ Blackwall DLR, Canary Wharf Tube, ✈ City.

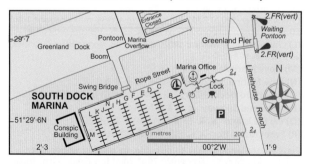

SOUTH DOCK MARINA (B), **51°29'·65N 00°01'·97W**. ☎020 7252 2244 Mob 0795 662 10710 (OH 0700-1900 Mon-Fri / 0800-1700 w/e) Tides as for Surrey Dock Greenland Ent. 1·1M above Greenwich,

2·5M below Tower Bridge. Baltic Quay building at SW end of marina is conspic with five arched rooftops. Waiting pontoon at Greenland Pier. Approx access via lock (width 6·4m) HW-2 to HW+1½ for 2m draught. 250 + Ⓥ, £22·62< 9·5m>£24·89<15m>£33·89<20m>£39·6 0.VHF Ch **M** 80; Dr ☎237 1078; Ⓗ; ⬡ ⬡ ☎020 7955 5000; Police 101.
Facilities: ⬡ ⬡ ⬡ ⬡ ⬡ ⬡(20t) ⬡ ⬡ ⬡ ⬡ ⇌ Surrey Quays and Canada Water tube, ✈ City Airport.

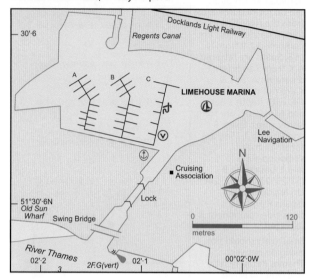

LIMEHOUSE MARINA (C), **51°30'·57N 00°02'·27W** www.bwml.co.uk ☎020 7308 9930. Entry HW±3 via swing bridge/lock (width 8m) 0800-1800LT daily Apr-Sep; 0800-1600 Oct-Mar; other times by prior arrangement (24hrs notice) to BWML Lock control ☎020 7308 9930, beware cross-tide at lock cut, mid flood/ebb. Waiting pontoon in lock entrance is accessible outside LW±2. Call VHF Ch 80 *Limehouse Marina*.

Facilities: 90⬡ from £28·00/craft Ⓥ H24 security, ⬡ ⬡ ⬡ ⬡ ⬡ ⇌ DLR; also entry to Regents Canal and Lee Navigation. Visitors welcome at CA clubhouse.

ST KATHARINE HAVEN (D), **51°30'·36N 00°04'·35W** www.skdocks. co.uk ☎020 77264 5312 HW +0245 on Dover. Tides as London Bridge. Beware of cross tide at mid-flood/ebb. Good shelter under all conditions. Tower Bridge and the Tower of London are uniquely conspic, close upriver. Six Y waiting buoys are close downstream of entrance. Berthing on St Katharine Pier, 30m upriver, is strictly by prior arrangement and only for shoal draught vessels with permission. Pleasure launches berth on S side of pier.

St Katharine Haven Ch **80** M. Lock (41m x 12·5m with 2 small lifting bridges), access HW –2 to HW +1½, season 0600-2030, winter 0800-1800LT; other times by prior arrangement. R/G traffic lights at entr.

Facilities: 160⬡ inc Ⓥ in 3 basins £4·50+⬡– usually Centre Basin or East Dock; ⬡ ⬡ ⬡ ⬡ ⬡ @.

LONDON BRIDGE

MEAN SPRING
AND NEAP CURVES ─────

MEAN RANGES	
Springs	6.6m
Neaps	4.6m

CHELSEA HARBOUR MARINA (I), London, **51°28′·48N 00°10′·82W.** Tides: see Albert Bridge. Good shelter in all conditions, 5M above Tower Bridge, reached via 14 fixed bridges. Battersea railway bridge is 120m upstream. Belvedere Tower (80m high with tide ball) is conspic, next to the lock and bascule bridge. Max sizes for lock: LOA 24m, beam 5·5m, draught 1·8m. Lock opens when there is 2·5m water above sill; tide gauge outside; R/G traffic lights. Limited waiting berths and shore access on Chelsea Hbr Pier (1·7m) close upriver. **Marina** Mobile ☎07770 542783. VHF Ch 80. 50⌣+10♥ £25<12m>£30, 🖔 ⛟ 🎣 WC 🗑 Ⓑ 🛒 ✕ 🖵.

BRENTFORD DOCK MARINA (O), 51°28′·92N 00°18′·05W, 1100m beyond Kew bridge. AC 3319. Tides, see Kew Bridge. 60⌣ inc ♥, £20 for all LOA. Mobile ☎07970 143 987. No VHF. Access approx HW±2½ via lock 4·8m x 9·8m (longer LOA during freeflow); 15m waiting pontoon. **Facilities:** ⬷ ⛟ 🗑 🛢 🔨 ⚓ 🛒 🛒 ✕ 🖵; Gunnersbury & Ealing tubes. Ent to Grand Union Canal via lock is nearby: ⚓ ⌣.

E England

STANDARD TIME (UT)
For Summer Time add ONE hour in **non-shaded areas**

LONDON BRIDGE LAT 51°30'N LONG 0°05'W
TIMES AND HEIGHTS OF HIGH AND LOW WATERS

Dates in **red** are **SPRINGS**
Dates in blue are NEAPS

YEAR 2016

JANUARY

Time	m		Time	m
1 0547	6.3	**16** 0003	0.9
1210	1.0	0551	6.8
F 1816	6.2	SA 1229	0.6
		☾ 1826	6.7
2 0015	1.5	**17** 0047	1.1
0624	6.0	0641	6.6
SA 1248	1.1	SU 1313	0.8
☾ 1900	6.0	1924	6.5
3 0057	1.6	**18** 0140	1.3
0712	5.8	0743	6.4
SU 1333	1.3	M 1413	1.0
1958	5.8	2032	6.3
4 0148	1.8	**19** 0248	1.4
0829	5.6	0859	6.3
M 1437	1.5	TU 1526	1.1
2108	5.8	2142	6.3
5 0301	1.9	**20** 0405	1.4
0946	5.7	1017	6.3
TU 1602	1.5	W 1641	1.2
2214	5.9	2255	6.3
6 0438	1.7	**21** 0526	1.2
1049	6.0	1125	6.5
W 1706	1.3	TH 1804	1.1
2316	6.2		
7 0543	1.4	**22** 0005	6.5
1146	6.3	0641	0.9
TH 1804	1.1	F 1229	6.7
		1909	1.0
8 0011	6.5	**23** 0102	6.7
0641	1.1	0742	0.6
F 1236	6.6	SA 1323	6.9
1857	1.0	2002	0.9
9 0100	6.7	**24** 0150	6.8
0735	0.9	0833	0.5
SA 1321	6.8	SU 1409	7.0
1948	0.9	○ 2047	0.9
10 0145	6.8	**25** 0233	6.9
0826	0.7	0918	0.4
SU 1404	7.0	M 1451	7.1
● 2036	0.9	2127	0.9
11 0227	6.9	**26** 0311	7.0
0914	0.6	0957	0.5
M 1446	7.2	TU 1529	7.1
2122	0.8	2201	1.0
12 0308	7.0	**27** 0345	6.9
0959	0.4	1029	0.5
TU 1528	7.3	W 1604	7.0
2205	0.8	2226	1.0
13 0348	7.0	**28** 0417	6.9
1041	0.3	1052	0.6
W 1610	7.3	TH 1636	6.8
2245	0.7	2249	1.0
14 0427	7.0	**29** 0447	6.7
1119	0.4	1114	0.7
TH 1652	7.1	F 1707	6.6
2323	0.8	2317	1.1
15 0508	6.9	**30** 0517	6.5
1153	0.5	1141	0.8
F 1737	6.9	SA 1738	6.4
		2347	1.2
		31 0549	6.3
		1209	0.9
		SU 1813	6.2

FEBRUARY

Time	m		Time	m
1 0019	1.4	**16** 0117	1.1
0626	6.1	0719	6.5
M 1242	1.1	TU 1341	1.0
☽ 1854	6.0	2002	6.2
2 0057	1.5	**17** 0217	1.3
0713	5.8	0831	6.3
TU 1322	1.4	W 1452	1.3
1950	5.7	2115	6.0
3 0147	1.7	**18** 0335	1.4
0821	5.6	0950	6.2
W 1423	1.6	TH 1611	1.4
2115	5.7	2235	6.1
4 0303	1.8	**19** 0459	1.3
0957	5.7	1110	6.3
TH 1612	1.6	F 1741	1.3
2231	5.9	2351	6.3
5 0457	1.5	**20** 0625	0.9
1107	6.1	1218	6.6
F 1725	1.3	SA 1851	1.0
2338	6.3		
6 0606	1.2	**21** 0049	6.6
1206	6.5	0726	0.6
SA 1827	1.1	SU 1312	6.8
		1943	0.9
7 0035	6.6	**22** 0136	6.8
0709	0.9	0814	0.5
SU 1258	6.8	M 1356	7.0
1928	1.0	○ 2027	0.9
8 0125	6.9	**23** 0216	6.9
0809	0.6	0856	0.4
M 1346	7.1	TU 1435	7.0
● 2025	0.8	2106	0.9
9 0210	7.0	**24** 0251	6.9
0902	0.4	0932	0.5
TU 1430	7.3	W 1509	7.0
2115	0.7	2139	0.8
10 0253	7.2	**25** 0323	7.0
0949	0.2	1001	0.5
W 1514	7.4	TH 1539	7.0
2200	0.6	2205	0.9
11 0334	7.3	**26** 0352	7.0
1032	0.1	1024	0.5
TH 1556	7.4	F 1607	6.9
2241	0.5	2229	0.9
12 0414	7.3	**27** 0420	6.9
1110	0.1	1047	0.6
F 1638	7.3	SA 1635	6.8
2319	0.6	2256	0.9
13 0454	7.2	**28** 0450	6.8
1143	0.2	1112	0.6
SA 1721	7.0	SU 1705	6.6
2354	0.7	2323	1.0
14 0535	7.0	**29** 0520	6.6
1214	0.5	1137	0.8
SU 1806	6.7	M 1738	6.4
		2349	1.1
15 0032	0.9		
0622	6.8		
M 1251	0.7		
☽ 1858	6.4		

MARCH

Time	m		Time	m
1 0555	6.3	**16** 0055	0.9
1203	1.0	0658	6.6
TU 1815	6.1	W 1311	1.1
☽		1933	6.1
2 0019	1.3	**17** 0150	1.2
0637	6.1	0808	6.3
W 1238	1.2	TH 1420	1.5
1902	5.9	2048	5.9
3 0102	1.4	**18** 0306	1.3
0731	5.8	0927	6.1
TH 1328	1.5	F 1543	1.5
2009	5.7	2211	5.9
4 0205	1.6	**19** 0430	1.2
0856	5.7	1050	6.3
F 1454	1.7	SA 1713	1.4
2145	5.7	2329	6.2
5 0404	1.6	**20** 0602	0.9
1026	6.0	1159	6.6
SA 1645	1.4	SU 1826	1.1
2303	6.1		
6 0528	1.2	**21** 0028	6.6
1135	6.5	0702	0.6
SU 1755	1.1	M 1252	6.8
		1918	0.9
7 0006	6.6	**22** 0114	6.8
0640	0.8	0748	0.5
M 1233	6.9	TU 1335	6.9
1905	0.9	2001	0.9
8 0100	6.9	**23** 0152	6.9
0748	0.5	0826	0.5
TU 1323	7.2	W 1411	6.9
2007	0.7	○ 2039	0.8
9 0147	7.2	**24** 0226	6.9
0843	0.3	0859	0.5
W 1410	7.4	TH 1443	7.0
● 2100	0.6	2112	0.8
10 0231	7.4	**25** 0256	7.0
0931	0.1	0928	0.5
TH 1454	7.5	F 1511	7.0
2146	0.4	2141	0.8
11 0313	7.5	**26** 0324	7.1
1014	0.0	0953	0.5
F 1537	7.5	SA 1537	7.0
2228	0.3	2208	0.7
12 0355	7.5	**27** 0353	7.0
1052	0.0	1018	0.5
SA 1619	7.4	SU 1606	6.9
2306	0.4	2235	0.8
13 0435	7.5	**28** 0423	6.9
1124	0.2	1045	0.6
SU 1701	7.1	M 1637	6.7
2340	0.5	2302	0.9
14 0518	7.2	**29** 0455	6.7
1152	0.5	1109	0.8
M 1745	6.7	TU 1710	6.5
		2326	1.0
15 0014	0.7	**30** 0530	6.5
0604	6.9	1135	0.9
TU 1226	0.8	W 1746	6.2
☽ 1834	6.4	2353	1.0
		31 0611	6.3
		1208	1.1
		TH 1831	6.0
		☾	

APRIL

Time	m		Time	m
1 0033	1.1	**16** 0237	1.2
0703	6.0	0900	6.1
F 1256	1.3	SA 1511	1.6
1931	5.8	2138	5.9
2 0131	1.3	**17** 0356	1.2
0816	5.9	1018	6.2
SA 1410	1.6	SU 1633	1.4
2101	5.8	2255	6.1
3 0312	1.4	**18** 0515	1.0
0949	6.1	1129	6.5
SU 1609	1.4	M 1747	1.2
2226	6.1	2356	6.4
4 0451	1.1	**19** 0622	0.7
1102	6.5	1223	6.7
M 1723	1.1	TU 1843	1.0
2334	6.6		
5 0606	0.7	**20** 0043	6.7
1203	6.9	0709	0.6
TU 1836	0.8	W 1306	6.8
		1928	0.9
6 0031	7.0	**21** 0123	6.8
0721	0.4	0747	0.6
W 1257	7.2	TH 1342	6.8
1944	0.7	2007	0.8
7 0121	7.2	**22** 0157	6.9
0819	0.2	0822	0.6
TH 1346	7.4	F 1414	6.9
● 2039	0.5	○ 2043	0.8
8 0207	7.5	**23** 0228	7.0
0908	0.1	0853	0.6
F 1432	7.5	SA 1442	6.9
2127	0.3	2115	0.7
9 0250	7.6	**24** 0258	7.1
0952	0.1	0923	0.5
SA 1516	7.5	SU 1510	6.9
2210	0.2	2146	0.7
10 0334	7.7	**25** 0328	7.1
1030	0.1	0951	0.6
SU 1559	7.3	M 1541	6.9
2249	0.2	2216	0.7
11 0417	7.6	**26** 0400	7.0
1102	0.3	1020	0.6
M 1642	7.1	TU 1614	6.7
2323	0.4	2244	0.8
12 0501	7.3	**27** 0434	6.8
1130	0.6	1047	0.8
TU 1726	6.7	W 1648	6.5
2356	0.6	2309	0.8
13 0548	7.0	**28** 0511	6.6
1202	0.9	1116	0.9
W 1813	6.4	TH 1726	6.3
		2338	0.9
14 0034	0.8	**29** 0554	6.4
0641	6.6	1151	1.1
TH 1246	1.2	F 1811	6.1
☾ 1908	6.1		
15 0125	1.1	**30** 0017	1.0
0746	6.3	0645	6.2
F 1349	1.5	SA 1239	1.3
2019	5.9	☽ 1908	5.9

Chart Datum: 3·20 metres below Ordnance Datum (Newlyn). HAT is 7·7 metres above Chart Datum.

FREE monthly updates. Register at
www.reedsnauticalalmanac.co.uk

STANDARD TIME (UT)
For Summer Time add ONE hour in **non-shaded areas**

LONDON BRIDGE   LAT 51°30'N   LONG 0°05'W
TIMES AND HEIGHTS OF HIGH AND LOW WATERS

Dates in red are **SPRINGS**
Dates in blue are NEAPS

YEAR 2016

MAY

Time	m	Time	m
1 0113	1.1	**16** 0317	1.1
0753	6.1	0935	6.1
SU 1351	1.5	M 1546	1.5
2027	5.9	2207	6.0
2 0241	1.2	**17** 0422	1.0
0918	6.2	1043	6.2
M 1535	1.4	TU 1655	1.3
2152	6.2	2311	6.2
3 0418	0.9	**18** 0522	0.9
1030	6.6	1142	6.4
TU 1651	1.1	W 1756	1.1
2301	6.6		
4 0530	0.7	**19** 0004	6.5
1134	6.9	0615	0.8
W 1805	0.8	TH 1230	6.6
		1847	1.0
5 0001	6.9	**20** 0048	6.7
0648	0.5	0702	0.7
TH 1231	7.2	F 1309	6.7
1916	0.7	1932	0.8
6 0054	7.2	**21** 0126	6.8
0751	0.3	0744	0.7
F 1322	7.4	SA 1344	6.8
● 2014	0.5	○ 2013	0.7
7 0142	7.4	**22** 0201	6.9
0842	0.2	0822	0.7
SA 1410	7.3	SU 1416	6.8
2105	0.3	2051	0.7
8 0229	7.6	**23** 0234	7.0
0928	0.3	0857	0.7
SU 1455	7.3	M 1448	6.9
2150	0.2	2127	0.6
9 0314	7.6	**24** 0306	7.0
1008	0.4	0929	0.7
M 1540	7.2	TU 1522	6.8
2232	0.2	2201	0.6
10 0359	7.5	**25** 0341	7.0
1042	0.5	1001	0.7
TU 1624	7.0	W 1557	6.7
2308	0.3	2232	0.6
11 0445	7.3	**26** 0418	6.9
1111	0.8	1032	0.8
W 1708	6.7	TH 1634	6.6
2342	0.5	2301	0.7
12 0532	7.0	**27** 0458	6.7
1143	1.0	1106	0.9
TH 1754	6.4	F 1713	6.4
		2332	0.8
13 0017	0.7	**28** 0542	6.6
0623	6.6	1145	1.0
F 1224	1.2	SA 1758	6.3
◑ 1844	6.1		
14 0103	0.9	**29** 0012	0.8
0721	6.3	0632	6.4
SA 1317	1.5	SU 1234	1.2
1946	5.9	◐ 1852	6.2
15 0205	1.1	**30** 0104	0.9
0827	6.1	0735	6.3
SU 1429	1.6	M 1340	1.4
2057	5.9	2001	6.1
		31 0221	1.0
		0851	6.4
		TU 1505	1.3
		2120	6.3

JUNE

Time	m	Time	m
1 0347	0.9	**16** 0432	1.0
1001	6.6	1049	6.1
W 1622	1.1	TH 1706	1.3
2230	6.6	2318	6.3
2 0458	0.7	**17** 0528	0.9
1106	6.8	1145	6.4
TH 1736	0.9	F 1804	1.1
2333	6.9		
3 0613	0.6	**18** 0010	6.5
1206	7.0	0621	0.8
F 1850	0.7	SA 1234	6.6
		1857	0.9
4 0030	7.1	**19** 0056	6.7
0723	0.5	0711	0.8
SA 1301	7.1	SU 1316	6.7
1951	0.5	1945	0.8
5 0122	7.3	**20** 0136	6.8
0807	0.5	0756	0.8
SU 1351	7.1	M 1355	6.8
● 2045	0.3	○ 2029	0.7
6 0211	7.4	**21** 0213	6.9
0906	0.5	0837	0.8
M 1439	7.2	TU 1432	6.8
2133	0.3	2111	0.6
7 0259	7.5	**22** 0250	7.0
0949	0.6	0915	0.7
TU 1525	7.1	W 1509	6.8
2217	0.2	2151	0.5
8 0345	7.4	**23** 0327	7.0
1031	0.7	0952	0.7
W 1609	7.0	TH 1546	6.9
2256	0.3	2229	0.5
9 0431	7.2	**24** 0406	7.0
1057	0.9	1029	0.7
TH 1652	6.8	F 1624	6.7
2330	0.5	2304	0.5
10 0516	6.9	**25** 0447	6.9
1127	1.0	1107	0.8
F 1734	6.5	SA 1703	6.6
		2336	0.6
11 0002	0.7	**26** 0530	6.8
0602	6.6	1146	0.9
SA 1202	1.2	SU 1746	6.5
1818	6.3		
12 0039	0.8	**27** 0013	0.7
0651	6.3	0618	6.6
SU 1245	1.4	M 1232	1.1
◑ 1909	6.0	◐ 1835	6.4
13 0127	1.0	**28** 0059	0.8
0738	6.1	0715	6.5
M 1338	1.6	TU 1328	1.2
2010	5.9	1935	6.3
14 0227	1.1	**29** 0200	0.9
0847	6.0	0824	6.4
TU 1447	1.6	W 1438	1.3
2116	5.9	2050	6.3
15 0332	1.1	**30** 0316	0.9
0948	6.0	0933	6.4
W 1602	1.5	TH 1554	1.2
2219	6.0	2202	6.5

JULY

Time	m	Time	m
1 0429	0.9	**16** 0446	1.2
1040	6.5	1059	6.1
F 1709	1.0	SA 1723	1.3
2308	6.7	2332	6.3
2 0544	0.8	**17** 0545	1.1
1146	6.7	1158	6.4
SA 1826	0.8	SU 1822	1.0
3 0011	6.9	**18** 0025	6.6
0658	0.7	0640	0.9
SU 1246	6.8	M 1249	6.6
1931	0.5	1917	0.8
4 0108	7.1	**19** 0111	6.8
0757	0.7	0733	0.9
M 1339	6.9	TU 1334	6.8
● 2028	0.3	○ 2009	0.7
5 0159	7.2	**20** 0154	6.9
0848	0.7	0821	0.9
TU 1427	7.0	W 1415	6.9
2118	0.2	2057	0.5
6 0247	7.3	**21** 0234	7.1
0933	0.7	0907	0.8
W 1512	7.1	TH 1455	7.0
2203	0.2	2143	0.4
7 0332	7.3	**22** 0314	7.2
1012	0.8	0950	0.7
TH 1554	7.0	F 1534	7.0
2243	0.3	2225	0.3
8 0415	7.2	**23** 0354	7.2
1045	0.9	1031	0.7
F 1633	6.9	SA 1612	7.1
2316	0.4	2304	0.3
9 0456	7.0	**24** 0434	7.1
1112	1.0	1109	0.7
SA 1711	6.7	SU 1650	6.9
2343	0.6	2338	0.4
10 0535	6.7	**25** 0516	7.0
1140	1.1	1146	0.8
SU 1748	6.4	M 1731	6.8
11 0010	0.7	**26** 0010	0.5
0614	6.4	0600	6.7
M 1213	1.3	TU 1225	1.0
1826	6.2	◐ 1815	6.7
12 0044	0.9	**27** 0047	0.7
0658	6.1	0652	6.5
TU 1252	1.4	W 1312	1.1
◑ 1914	6.0	1910	6.5
13 0127	1.1	**28** 0136	0.9
0749	5.9	0756	6.3
W 1340	1.6	TH 1412	1.3
2018	5.8	2020	6.4
14 0225	1.3	**29** 0246	1.1
0852	5.8	0907	6.2
TH 1448	1.7	F 1527	1.3
2127	5.8	2136	6.4
15 0340	1.3	**30** 0402	1.1
0956	5.9	1018	6.3
F 1616	1.6	SA 1645	1.1
2231	6.0	2249	6.5
		31 0521	1.1
		1131	6.4
		SU 1806	0.9
		2359	6.7

AUGUST

Time	m	Time	m
1 0641	0.9	**16** 0610	1.1
1236	6.7	1221	6.5
M 1917	0.6	TU 1849	0.8
2 0059	7.0	**17** 0044	6.8
0741	0.8	0709	0.9
TU 1329	6.9	W 1310	6.8
● 2013	0.3	1948	0.6
3 0150	7.1	**18** 0131	7.0
0831	0.7	0804	0.9
W 1415	7.0	TH 1354	7.1
2102	0.2	○ 2040	0.4
4 0235	7.2	**19** 0214	7.2
0916	0.8	0855	0.8
TH 1457	7.1	F 1435	7.1
2146	0.2	2128	0.3
5 0317	7.2	**20** 0255	7.3
0955	0.8	0941	0.7
F 1535	7.1	SA 1515	7.3
2223	0.3	2212	0.2
6 0355	7.2	**21** 0336	7.4
1028	0.9	1023	0.6
SA 1610	7.0	SU 1554	7.3
2253	0.4	2251	0.1
7 0430	7.1	**22** 0416	7.3
1052	1.0	1101	0.6
SU 1643	6.9	M 1632	7.2
2315	0.6	2326	0.3
8 0503	6.8	**23** 0457	7.1
1115	1.0	1137	0.7
M 1714	6.7	TU 1712	7.1
2337	0.7	2355	0.5
9 0534	6.5	**24** 0540	6.8
1143	1.1	1212	0.9
TU 1746	6.4	W 1756	6.9
10 0004	0.9	**25** 0027	0.7
0607	6.2	0627	6.5
W 1214	1.3	TH 1253	1.1
◑ 1821	6.2	◐ 1847	6.6
11 0035	1.1	**26** 0110	1.0
0646	6.0	0727	6.2
TH 1250	1.5	F 1347	1.2
1906	5.9	1954	6.4
12 0114	1.4	**27** 0216	1.3
0740	5.7	0841	6.0
F 1337	1.7	SA 1501	1.3
2015	5.7	2115	6.3
13 0212	1.6	**28** 0338	1.4
0859	5.6	0959	6.1
SA 1454	1.8	SU 1623	1.2
2142	5.7	2234	6.4
14 0356	1.6	**29** 0503	1.3
1014	5.8	1118	6.3
SU 1640	1.5	M 1751	0.9
2252	6.0	2348	6.7
15 0508	1.3	**30** 0625	1.0
1123	6.2	1224	6.5
M 1747	1.1	TU 1903	0.5
2353	6.4		
		31 0048	7.0
		0723	0.8
		W 1316	6.9
		1956	0.3

Chart Datum: 3·20 metres below Ordnance Datum (Newlyn). HAT is 7·7 metres above Chart Datum.

E England

》》FREE monthly updates. Register at 《
www.reedsnauticalalmanac.co.uk

299

STANDARD TIME (UT)
For Summer Time add ONE hour in **non-shaded areas**

LONDON BRIDGE   LAT 51°30'N   LONG 0°05'W
TIMES AND HEIGHTS OF HIGH AND LOW WATERS

Dates in **red** are SPRINGS
Dates in blue are NEAPS

YEAR **2016**

SEPTEMBER

Day	Time	m	Time	m	Day	Time	m	Time	m
1 TH	0137	7.1	0812	0.8	16 F	0103	7.1	0742	0.8
	1359	7.0	●2042	0.3		1326	7.1	○2017	0.4
2 F	0218	7.2	0854	0.8	17 SA	0148	7.3	0835	0.7
	1437	7.1	2121	0.3		1410	7.3	2107	0.2
3 SA	0256	7.2	0932	0.8	18 SU	0232	7.5	0933	0.8
	1511	7.1	2155	0.4		1451	7.4	2151	0.1
4 SU	0329	7.1	1003	0.9	19 M	0314	7.5	1007	0.5
	1542	7.1	2221	0.5		1531	7.5	2231	0.2
5 M	0359	7.0	1027	0.9	20 TU	0355	7.4	1046	0.6
	1612	7.0	2241	0.6		1611	7.5	2305	0.3
6 TU	0427	6.8	1049	0.9	21 W	0436	7.1	1122	0.6
	1641	6.8	2303	0.7		1652	7.3	2334	0.6
7 W	0456	6.6	1115	1.0	22 TH	0518	6.8	1156	0.7
	1711	6.6	2328	0.9		1737	7.0		
8 TH	0526	6.4	1142	1.2	23 F	0004	0.9	0605	6.4
	1744	6.3	2353	1.1		1235	0.9	◑1828	6.7
9 F	0601	6.1	1209	1.3	24 SA	0046	1.2	0702	6.1
	1823	6.1	◐			1325	1.2	1935	6.4
10 SA	0023	1.3	0644	5.8	25 SU	0150	1.5	0816	5.9
	1246	1.5	1913	5.8		1438	1.3	2055	6.2
11 SU	0108	1.6	0745	5.6	26 M	0314	1.6	0939	6.0
	1343	1.7	2032	5.6		1601	1.2	2216	6.3
12 M	0224	1.8	0923	5.6	27 TU	0442	1.4	1059	6.2
	1546	1.7	2207	5.9		1732	0.9	2331	6.7
13 TU	0427	1.6	1043	6.0	28 W	0602	1.1	1204	6.6
	1709	1.2	2316	6.3		1841	0.6		
14 W	0536	1.2	1147	6.5	29 TH	0030	7.0	0659	0.9
	1816	0.8				1254	6.9	1931	0.4
15 TH	0013	6.8	0640	1.0	30 F	0116	7.1	0746	0.8
	1240	6.8	1921	0.6		1336	7.0	2013	0.4

OCTOBER

Day	Time	m	Time	m	Day	Time	m	Time	m
1 SA	0155	7.1	0827	0.8	16 SU	0121	7.4	0810	0.7
	1411	7.0	●2049	0.5		1342	7.4	○2041	0.3
2 SU	0229	7.1	0903	0.8	17 M	0206	7.5	0901	0.5
	1443	7.1	2119	0.5		1426	7.6	2127	0.3
3 M	0259	7.1	0933	0.8	18 TU	0251	7.5	0947	0.4
	1512	7.1	2143	0.6		1509	7.6	2208	0.3
4 TU	0326	7.0	0959	0.9	19 W	0334	7.4	1028	0.4
	1541	7.1	2206	0.6		1551	7.6	2243	0.5
5 W	0354	6.9	1024	0.8	20 TH	0417	7.1	1105	0.5
	1611	6.9	2231	0.7		1635	7.4	2313	0.7
6 TH	0423	6.7	1050	0.9	21 F	0500	6.8	1140	0.6
	1642	6.7	2256	0.9		1722	7.1	2343	1.0
7 F	0454	6.5	1115	1.1	22 SA	0546	6.4	1217	0.8
	1715	6.5	2320	1.1		1814	6.7	◑	
8 SA	0527	6.2	1139	1.2	23 SU	0024	1.3	0641	6.1
	1753	6.2	2348	1.3		1306	1.1	1918	6.4
9 SU	0608	6.0	1212	1.3	24 M	0123	1.6	0751	5.9
	1840	6.0	◐			1415	1.2	2032	6.2
10 M	0030	1.5	0702	5.7	25 TU	0245	1.7	0910	5.9
	1303	1.5	1944	5.8		1534	1.2	2148	6.3
11 TU	0132	1.8	0822	5.6	26 W	0408	1.6	1027	6.1
	1424	1.6	2119	5.9		1653	1.0	2302	6.5
12 W	0337	1.8	0958	5.9	27 TH	0526	1.3	1133	6.4
	1627	1.3	2236	6.3		1804	0.7		
13 TH	0456	1.3	1108	6.4	28 F	0002	6.8	0626	1.0
	1736	0.9	2338	6.8		1225	6.7	1855	0.6
14 F	0604	1.0	1205	6.8	29 SA	0049	6.9	0713	0.9
	1846	0.6				1306	6.8	1935	0.6
15 SA	0032	7.2	0712	0.8	30 SU	0127	6.9	0754	0.8
	1256	7.2	1948	0.4		1342	6.9	●2010	0.6
					31 M	0200	7.0	0831	0.8
						1414	7.0	2040	0.7

NOVEMBER

Day	Time	m	Time	m	Day	Time	m	Time	m
1 TU	0229	7.0	0903	0.8	16 W	0230	7.4	0927	0.4
	1444	7.1	2108	0.7		1449	7.6	2146	0.5
2 W	0257	7.0	0933	0.8	17 TH	0315	7.3	1011	0.3
	1514	7.1	2135	0.7		1535	7.6	2224	0.6
3 TH	0326	6.9	1001	0.8	18 F	0400	7.1	1051	0.4
	1545	7.0	2203	0.8		1621	7.4	2256	0.9
4 F	0357	6.8	1029	0.9	19 SA	0445	6.8	1127	0.5
	1617	6.8	2230	0.9		1709	7.1	2328	1.1
5 SA	0429	6.6	1054	1.0	20 SU	0531	6.5	1204	0.7
	1653	6.6	2257	1.1		1800	6.8		
6 SU	0504	6.4	1120	1.0	21 M	0006	1.3	0621	6.2
	1732	6.4	2328	1.2		1248	1.0	◑1857	6.4
7 M	0544	6.2	1153	1.1	22 TU	0056	1.6	0722	6.0
	1819	6.2	●			1347	1.1	2002	6.2
8 TU	0009	1.4	0634	5.9	23 W	0205	1.7	0832	5.9
	1241	1.2	1917	6.1		1457	1.2	2110	6.2
9 W	0107	1.6	0741	5.8	24 TH	0323	1.7	0943	6.0
	1349	1.4	2037	6.1		1603	1.1	2218	6.3
10 TH	0245	1.7	0912	5.9	25 F	0435	1.5	1050	6.2
	1541	1.1	2157	6.4		1705	1.0	2322	6.4
11 F	0416	1.4	1028	6.3	26 SA	0539	1.3	1146	6.4
	1656	0.9	2302	6.8		1801	0.9		
12 SA	0528	1.1	1130	6.8	27 SU	0013	6.6	0632	1.1
	1805	0.6				1232	6.6	1848	0.8
13 SU	0001	7.1	0639	0.9	28 M	0055	6.7	0718	1.0
	1225	7.1	1915	0.5		1312	6.8	1930	0.8
14 M	0054	7.3	0744	0.7	29 TU	0131	6.8	0759	0.8
	1315	7.4	○2013	0.4		1347	6.9	●2007	0.8
15 TU	0143	7.4	0838	0.5	30 W	0203	6.9	0836	0.8
	1402	7.6	2102	0.4		1420	7.0	2041	0.8

DECEMBER

Day	Time	m	Time	m	Day	Time	m	Time	m
1 TH	0233	6.9	0911	0.8	16 F	0301	7.2	0957	0.3
	1451	7.0	2112	0.8		1522	7.5	2209	0.8
2 F	0305	6.8	0944	0.8	17 SA	0347	7.1	1040	0.3
	1524	7.0	2142	0.9		1609	7.4	2245	0.9
3 SA	0338	6.8	1015	0.8	18 SU	0432	6.9	1118	0.4
	1559	6.9	2212	0.9		1655	7.1	2317	1.1
4 SU	0413	6.7	1043	0.8	19 M	0515	6.7	1152	0.6
	1636	6.8	2244	1.0		1742	6.8	2349	1.3
5 M	0449	6.5	1111	0.9	20 TU	0559	6.4	1227	0.8
	1717	6.6	2319	1.1		1831	6.5		
6 TU	0529	6.4	1145	0.9	21 W	0027	1.4	0647	6.1
	1802	6.5				1310	1.0	◑1923	6.2
7 W	0001	1.3	0616	6.2	22 TH	0115	1.6	0745	5.9
	1230	1.0	◑1855	6.3		1406	1.2	2022	6.0
8 TH	0054	1.5	0713	6.1	23 F	0218	1.8	0850	5.8
	1328	1.1	2003	6.2		1509	1.2	2123	6.0
9 F	0207	1.6	0829	6.1	24 SA	0336	1.8	0955	5.9
	1454	1.1	2121	6.4		1611	1.2	2226	6.0
10 SA	0336	1.5	0950	6.3	25 SU	0444	1.6	1058	6.1
	1617	1.0	2229	6.6		1709	1.1	2327	6.2
11 SU	0454	1.3	1057	6.6	26 M	0545	1.3	1154	6.3
	1728	0.8	2332	6.8		1804	1.0		
12 M	0610	1.0	1158	7.0	27 TU	0019	6.4	0640	1.1
	1843	0.7				1241	6.6	1855	0.9
13 TU	0030	7.0	0719	0.7	28 W	0102	6.6	0729	0.9
	1253	7.2	1947	0.6		1322	6.7	1941	0.9
14 W	0123	7.1	0817	0.5	29 TH	0141	6.7	0813	0.8
	1344	7.4	○2040	0.6		1359	6.8	●2022	0.9
15 TH	0213	7.2	0909	0.3	30 F	0217	6.8	0854	0.8
	1433	7.5	2127	0.6		1434	6.9	2058	0.9
					31 SA	0251	6.8	0932	0.7
						1509	7.0	2132	0.9

Chart Datum: 3·20 metres below Ordnance Datum (Newlyn). HAT is 7·7 metres above Chart Datum.

FREE monthly updates. Register at
www.reedsnauticalalmanac.co.uk

9.4.13 SOUTHEND-ON-SEA/LEIGH-ON-SEA

Essex 51°31'·07N 00°42'·57E ✿✿✿✿✿✿

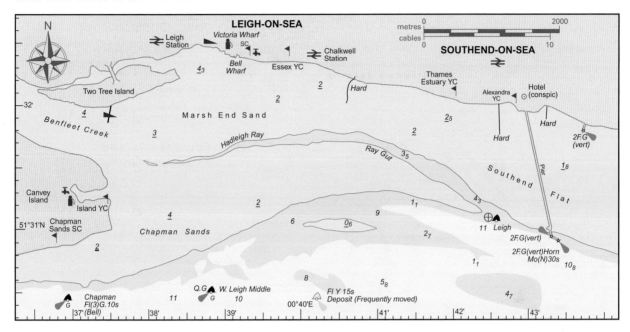

CHARTS AC 5606, 1183, 1185; Imray C1, C2, Y18, 2100

TIDES +0125 Dover; ML 3·0; Duration 0610

Standard Port SHEERNESS (←—)

Times				Height (metres)			
High Water		Low Water		MHWS	MHWN	MLWN	MLWS
0200	0800	0200	0700	5·8	4·7	1·5	0·6
1400	2000	1400	1900				
Differences SOUTHEND-ON-SEA							
−0005	−0005	−0005	−0005	+0·1	0·0	−0·1	−0·1
CORYTON							
+0005	+0010	+0010	+0010	+0·4	+0·3	0·0	0·0

SHELTER The whole area dries soon after half ebb, except Ray Gut (0·4–4·8m) which leads to Leigh Creek and Hadleigh Ray, either side of Two Tree Island, thence to Benfleet Creek where the limit of W navigation is the Tidal Barrier just above Benfleet YC; all are buoyed, but echo-sounder is essential. Craft can take the ground alongside Bell Wharf or Victoria Wharf. It is also possible to secure at the end of Southend Pier to collect stores, ⚓. NOTE: Southend-on-Sea and Leigh-on-Sea are both part of the lower PLA Area and an 8kn speed limit is enforced in inshore areas.

NAVIGATION WPT 51°31'·07N 00°42'·57E, Leigh SHM buoy, at ent to Ray Gut; this SHM buoy can be left close to port on entering Ray Gut, since there is now more water NE of it than to the SW. Appr from Shoeburyness, keep outside the W Shoebury SHM buoy, Fl G 2·5s.

Beware some 3000 small boat moorings 1M either side of Southend Pier. Speed limit in Canvey Island/Hadleigh Ray areas is 8kn.

LIGHTS AND MARKS See chartlet.

COMMUNICATIONS (Code 01702) NMOC 02392 552100; Essex Police Marine Section (01268) 775533; Police 101; Dr 225500; ℍ 348911. HM 215620; HM Leigh-on-Sea 710561.

Port Control London: *London VTS* VHF Ch 68 inward from Sea Reach No4, *London VTS* VHF Ch 69 seaward from Sea Reach No 4.

FACILITIES
SOUTHEND-ON-SEA: Southend Pier ☎215620, ⌕ (craft to be attended), £20<40'>£40<50'>£50/night (short stay free) ⚓⛽. **Alexandra YC** ☎340363, ⚓⌂; **Thorpe Bay YC** ☎587563, ⚓⚓✕⌂; **Thames Estuary YC** ☎345967; **Halfway YC** ☎582025, book 1⚓⚓; **Town** Ⓔ ⌂ ACA ✉ Ⓑ 🛒 ✕ ⌂ ≠ ✈.

LEIGH-ON-SEA: Essex YC ☎478404, ⚓⌂; **Leigh on Sea SC** ☎476788, ⚓⌂; **Bell Wharf** ⌕ 24hrs free then £11/night; **Victoria Wharf** ⌕ 24hrs free then £11/night, ⚓⚓⛽ (Two Tree Is HW±2½). **Town:** ✕🔧⚓🚒⛽🛒⌂.

CANVEY ISLAND: (Code 01268) Access via drying channels. **Island YC** ☎510360, ⌕ (max 9m), dries; **Benfleet YC** (on S side of Benfleet Creek on Canvey Island) ☎792278, ⚓⛽⚓🔧 (by day), **Services** ⚓✕🔧⚓🚒🛒🔧⌂⌂.

SHOEBURYNESS TO RIVER COLNE

(Charts 1185, 1975) Maplin and Foulness Sands extend nearly 6M NE from Foulness Pt, the extremity being marked by Whitaker bn. On N side of Whitaker chan leading to R. Crouch and R Roach lies Buxey Sand, inshore of which is the Ray Sand chan (dries), a convenient short cut between R. Crouch and R. Blackwater with sufficient rise of tide.

To seaward of Buxey Sand and the Spitway, Gunfleet Sand extends 10M NE, marked by buoys and dries in places. Gunfleet Sands Wind Farm, lies along the shoal between Swin Spitway and Gunfleet Old Lt Ho. 50 turbines are marked by Fl Y lts and R Aero obstn lts, the northernmost sounds a Horn every 10s in fog.

▶ A conspic disused lt tr stands on SE side of Gunfleet Sand, about 6M SSE of the Naze tr, and here the SW-going (flood) stream begins about HW Sheerness +0600, and the NE-going stream at about HW Sheerness −0030, sp rates 2kn. ◀

The Rivers Blackwater and Colne share a common estuary which is approached from the NE via NE Gunfleet lt buoy; thence along Goldmer Gat and the Wallet towards Knoll and Eagle lt buoys. For the Colne turn NNW via Colne Bar buoy towards Inner Bench Hd buoy keeping in mid-chan. For R Blackwater, head WNW for NW Knoll and Bench Hd buoys. From the S or SE, make for the Whitaker ECM buoy, thence through the Spitway, via Swin Spitway and Wallet Spitway buoys to reach Knoll buoy and deeper water.

9.4.14 RIVER ROACH/HAVENGORE

Essex 51°36'·98N 00°52'·14E (Branklet SPM buoy), R Roach ❁❁❁◊❁❁❁. 51°33'·62N 00°50'·52E, Havengore Bridge ❁◊❁

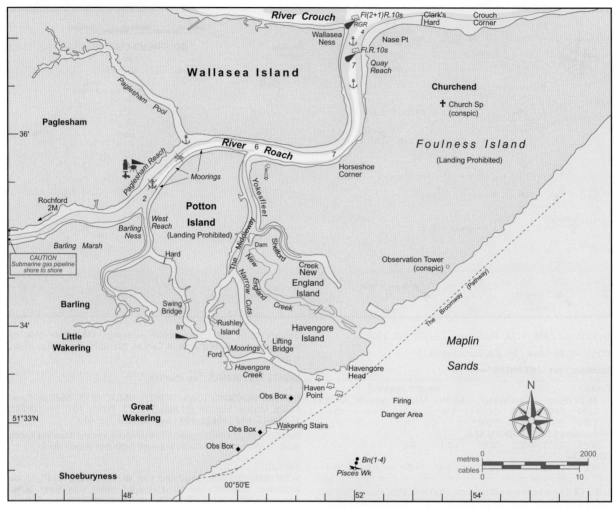

CHARTS AC 5607, 1185, 3750; Imray C1, Y17, 2100, 2000

TIDES +0110 Dover; ML 2·7; Duration 0615

Standard Port WALTON-ON-THE-NAZE (→)

Times				Height (metres)			
High Water		Low Water		MHWS	MHWN	MLWN	MLWS
0000	0600	0500	1100	4·2	3·4	1·1	0·4
1200	1800	1700	2300				
Differences ROCHFORD							
+0050	+0040	DR	DR	−0·8	−1·1	DR	DR

SHELTER Good. The Roach gives sheltered sailing and access to a network of secluded creeks, including Havengore (the 'backdoor' from the Crouch to the Thames Estuary). No ⌁ available in the area. ⚓s behind sea-walls can be found for all winds at: Quay Reach (often more protected than the Crouch), Paglesham Reach, West Reach, Barling Reach and Yokes Fleet. An ⚓ light is essential due to freighters H24. Speed limit 8kn. Crouch Hbr Authority controls R Roach and Crouch, out to Foulness Pt.

NAVIGATION Normal access H24 to the Roach is from R Crouch; the ent between Branklet buoy and Nase Pt is narrowed by mudbanks. Unlit buoys up-river to Barling Ness, above which few boats go. To exit at Havengore Creek, appr via Middleway and Narrow Cuts to reach the bridge before HW.

Entry via **Havengore Creek** is possible in good weather, with great care and adequate rise of tide (max draught 1·5m at HW sp). Shoeburyness Range is usually active Mon-Fri 0600-1700LT; give 24hrs notice to Range Officer by ☎. Subsequent clearance on VHF by Havengore lifting bridge (☎HW±2, HJ); no passage unless bridge raised. Least water is over the shifting bar, just inside creek ent. From the S, cross Maplin Sands at HW −1 from S Shoebury SHM buoy, leaving Pisces wreck (conspic, 1M from ent) to port.

LIGHTS AND MARKS Unlit, but night entry to R Roach may be possible.

COMMUNICATIONS (Code 01702 Southend) NMOC 02392 552100; Dr 218678. Crouch HM (01621) 783602; Range Officer 383211; Havengore Br 383436; Swing Br to Potton Island 219491.

VHF Ch 72 16 is worked by Range Officer (*Shoe Base*) (HO); Radar Control (*Shoe Radar*) (HO); & Bridge keeper (*Shoe Bridge*) (HW±2 by day). Radar guidance may be available.

FACILITIES Paglesham (East End) ⚓ ⛽ ⚓ ⚒ ⚓ ⬚ (at ⁞) ⁞ ⬚.
Gt Wakering: ⚓ ⚓ ⚓ ⚓ ⚒ ⚓ ⬚ ⬚ ⬚ (from ⁞ at Rochford) Wakering YC ☎530926, ⚓⚓⬚. **Towns** Gt Wakering & Rochford; ▦ ✕ ⬚ ⬚ (Great Wakering and Barling); most facilities, ⒷB and ⇌ in Rochford and Shoeburyness, ✈ (Southend).

9.4.15 BURNHAM-ON-CROUCH

Essex 51°37'·50N 00°48'·23E (Yacht Hbr) ✤✤✤✦✦✦✦✿✿

CHARTS AC 5607, 1183, 1975, 3750; Imray C1, Y17, 2100, 2000

TIDES +0115 Dover; ML 2·5; Duration 0610

Standard Port WALTON-ON-THE-NAZE (→)

Times				Height (metres)			
High Water		Low Water		MHWS	MHWN	MLWN	MLWS
0000	0600	0500	1100	4·2	3·4	1·1	0·4
1200	1800	1700	2300				
Differences WHITAKER BEACON							
+0022	+0024	+0033	+0027	+0·6	+0·5	+0·2	+0·1
HOLLIWELL POINT							
+0034	+0037	+0100	+0037	+1·1	+0·9	+0·3	+0·1
NORTH FAMBRIDGE							
+0115	+0050	+0130	+0100	+1·1	+0·8	0·0	−0·1
HULLBRIDGE							
+0115	+0050	+0135	+0105	+1·1	+0·8	0·0	−0·1
BATTLESBRIDGE							
+0120	+0110	DR	DR	−1·8	−2·0	DR	DR

SHELTER River is exposed to most winds. Cliff Reach (off W edge of lower chartlet) is sheltered from SW'lies. ‡ prohib in fairway but possible just E or W of the moorings.

NAVIGATION Appr from East Swin or the Wallet via Swin Spitway SWM / Whitaker ECM, into the well marked Whitaker Channel.

Near Sunken Buxey seas can be hazardous with strong wind over tide. Ray Sand Chan (dries 1·7m), usable on the tide by shoal draught boats as a short cut from/to the Blackwater. Shoeburyness Artillery ranges lie E and S of Foulness Pt, clear of fairway. The buoyed chan leads either side of shoal patch 0·7 marked by Sunken Buxey NCM and Buxey No1 SCM. R Crouch is navigable to Battlesbridge, 10M beyond Burnham. No landing on Foulness or Bridgemarsh Is. Wallasea Wetlands are being developed; keep well clear of ships using unloading pontoon at Ringwood Pt. ⚓ in swinging area between Horse Shoal and No 3 SHM buoy. Work is expected to be complete by 2019.

LIGHTS AND MARKS There are few landmarks to assist entry, but Whitaker Chan, Swallow Tail Bank and the river are well lit/buoyed to 0·5M W of Essex Marina. From Buxey No2 steer 240° in buoyed channel past Foulness Pt, N of Crouch NCM into the river. There is a 1·3M unlit gap between Inner Crouch SWM lit buoy, and the entrance to R Roach, which is marked by Pinto PHM and Branklet PCM en-route Horse Shoal NCM.

COMMUNICATIONS (Code 01621 Maldon) NMOC 02392 552100; Police 101; Dr 782054. HM 783602.

VHF Ch 80 for: Crouch HM Launch (0900-1700LT, w/e); Essex Marina; Burnham Yacht Harbour; W Wick Marina (1000-1700), also Ch M.

FACILITIES There are five marinas or yacht hbrs. Speed limit 8kn.

BURNHAM, Burnham Yacht Hbr ☎782150, access H24, some ⚓s, ⬛ 350 £2·50 ⚡+£2, 🔧🛢♨✗⚒🖥🚻⚓(30t) 🅿✗💧.
Royal Corinthian YC ☎782105, ⚓🛥🍽✚✗💧.
Royal Burnham YC ☎782044, 🛥✗💧.
Crouch YC ☎782252, 🛥⚓🛥🛒🏴🛢♨✗⚒🖥⚓🚻(15t) 🅿 ACA ✗💧. **Town** ✉🅑🛒🛍✗🚃✈ (Southend).

ALTHORNE, Bridgemarsh Marina ☎740414 ⬛ £20/craft/night 🛢♨⚓🚻🛢.

WALLASEA ISLAND, Essex Marina ☎(01702) 258531, ⚓ 400 £2·13, ⬛🛢⛽LPG🛢✗⚒🖥🔧⚓🚻(40t)⚓(13t) 🅿✗💧. Ferry to Burnham Town Hard at w/ends in season, ☎07704 060482. **Essex YC**. ACA.

FAMBRIDGE, Fambridge Yacht Haven ④ www.yachthavens. com ☎740370; Min 1m depth in approach, Stow Cr marked by bn Q(4) y 8s/F lts in line 356°. ⚓s £1·50, ⬛ (£6-£15) 170⬛+Ⓥ £2·50, short stay <3 hrs<15m £5, 🔧🏴♨✗⚒🖥(£6-£15)🚻⚓(25t) ⚓(5t) 🛒 YC 💧.

Fambridge Yacht Station, Ⓥ pontoon £5/night, ⚒⚡.

Brandy Hole Yacht Stn ☎(01702) 230248, Access via drying channel ⚓🛥🛥 120⬛🛢🏴✗⚒🚻🛢.

STANDARD TIME (UT)
For Summer Time add ONE hour in **non-shaded areas**

BURNHAM-ON-CROUCH LAT 51°37′N LONG 0°48′E
TIMES AND HEIGHTS OF HIGH AND LOW WATERS

Dates in red are SPRINGS
Dates in blue are NEAPS

YEAR 2016

JANUARY

Day	Time m	Time m	Time m	Time m	Day	Time m	Time m	Time m	Time m
1 F	0420 4.4	1105 0.7	1701 4.3	2303 1.1	16 SA	0430 4.9	1105 0.4	1713 4.9	◑2313 0.8
2 SA	0503 4.3	1148 0.8	1747 4.2	◐2352 1.2	17 SU	0520 4.8	1158 0.6	1808 4.7	
3 SU	0554 4.1	1238 0.9	1842 4.2		18 M	0011 0.9	0618 4.6	1301 0.7	1911 4.5
4 M	0047 1.3	0657 4.0	1333 0.9	1945 4.2	19 TU	0120 1.0	0728 4.4	1415 0.7	2021 4.5
5 TU	0149 1.3	0813 4.0	1433 0.9	2051 4.3	20 W	0242 0.9	0847 4.5	1532 0.6	2131 4.6
6 W	0255 1.2	0924 4.2	1533 0.8	2151 4.6	21 TH	0401 0.6	0957 4.7	1636 0.5	2232 4.9
7 TH	0401 1.0	1023 4.5	1630 0.7	2245 4.8	22 F	0504 0.3	1058 4.9	1730 0.3	2326 5.1
8 F	0500 0.8	1115 4.7	1722 0.6	2334 5.0	23 SA	0558 0.1	1152 5.2	1819 0.2	
9 SA	0553 0.5	1202 5.0	1810 0.5		24 SU	0016 5.2	0647 -0.1	1241 5.2	○1903 0.2
10 SU	0019 5.2	0640 0.6	1247 5.1	●1854 0.4	25 M	0101 5.2	0732 -0.1	1326 5.2	1943 0.3
11 M	0102 5.3	0725 0.2	1330 5.3	1936 0.4	26 TU	0142 5.1	0814 -0.1	1406 5.0	2017 0.5
12 TU	0142 5.3	0808 0.2	1412 5.3	2016 0.5	27 W	0216 5.0	0850 0.1	1440 4.9	2045 0.6
13 W	0222 5.3	0851 0.2	1454 5.3	2057 0.5	28 TH	0243 4.8	0920 0.3	1510 4.7	2110 0.7
14 TH	0302 5.2	0933 0.2	1538 5.2	2138 0.6	29 F	0310 4.7	0948 0.4	1541 4.6	2140 0.8
15 F	0344 5.1	1017 0.3	1624 5.1	2223 0.7	30 SA	0343 4.7	1019 0.5	1617 4.5	2216 0.9
					31 SU	0422 4.6	1058 0.7	1659 4.5	2258 0.9

FEBRUARY

Day	Time m	Time m	Time m	Time m	Day	Time m	Time m	Time m	Time m
1 M	0507 4.4	1143 0.8	1747 4.4	◐2347 1.1	16 TU	0557 4.7	1229 0.7	1845 4.4	
2 TU	0559 4.2	1236 0.9	1843 4.3		17 W	0048 0.8	0705 4.4	1344 0.8	1957 4.3
3 W	0045 1.2	0703 4.1	1339 1.0	1951 4.2	18 TH	0214 0.8	0829 4.3	1511 0.9	2113 4.4
4 TH	0158 1.2	0827 4.0	1450 0.9	2107 4.3	19 F	0343 0.6	0945 4.5	1620 0.6	2217 4.6
5 F	0320 1.1	0946 4.2	1601 0.8	2213 4.6	20 SA	0449 0.3	1047 4.8	1716 0.4	2313 4.9
6 SA	0435 0.8	1048 4.5	1703 0.6	2310 4.8	21 SU	0543 0.1	1140 5.0	1803 0.3	
7 SU	0536 0.5	1141 4.9	1755 0.5	2359 5.1	22 M	0002 5.1	0630 -0.1	1228 5.1	○1846 0.2
8 M	0627 0.2	1229 5.2	1842 0.3 ●		23 TU	0046 5.1	0713 -0.1	1310 5.1	1923 0.3
9 TU	0044 5.3	0713 0.0	1314 5.4	1925 0.2	24 W	0124 5.1	0750 0.0	1346 4.9	1954 0.4
10 W	0127 5.4	0756 -0.1	1357 5.5	2006 0.2	25 TH	0155 4.9	0821 0.2	1414 4.8	2016 0.5
11 TH	0208 5.4	0838 -0.1	1439 5.5	2046 0.2	26 F	0218 4.8	0844 0.5	1438 4.7	2037 0.5
12 F	0248 5.4	0919 -0.1	1521 5.4	2126 0.3	27 SA	0242 4.8	0907 0.4	1505 4.7	2105 0.5
13 SA	0329 5.3	1001 0.1	1606 5.2	2207 0.4	28 SU	0313 4.9	0937 0.4	1540 4.7	2139 0.6
14 SU	0413 5.2	1044 0.2	1652 5.0	2253 0.5	29 M	0351 4.8	1014 0.5	1621 4.7	2217 0.6
15 M	0501 4.9	1132 0.4	1744 4.7	◐2344 0.7					

MARCH

Day	Time m	Time m	Time m	Time m	Day	Time m	Time m	Time m	Time m
1 TU	0434 4.7	1055 0.6	1706 4.6	◐2301 0.8	16 W	0539 4.7	1200 0.7	1819 4.3	
2 W	0523 4.5	1144 0.8	1758 4.4	2353 0.9	17 TH	0025 0.6	0648 4.4	1312 0.9	1933 4.2
3 TH	0619 4.2	1244 1.0	1859 4.2		18 F	0148 0.7	0813 4.3	1444 0.9	2052 4.2
4 F	0059 1.1	0731 4.0	1403 1.1	2017 4.1	19 SA	0320 0.6	0929 4.4	1557 0.7	2158 4.5
5 SA	0235 1.1	0906 4.1	1530 1.0	2139 4.3	20 SU	0426 0.3	1030 4.7	1653 0.5	2253 4.7
6 SU	0407 0.9	1020 4.4	1641 0.7	2243 4.6	21 M	0519 0.1	1121 4.9	1740 0.3	2341 4.9
7 M	0513 0.5	1117 4.7	1736 0.5	2336 5.0	22 TU	0605 0.0	1207 5.0	1822 0.3	
8 TU	0605 0.2	1206 5.2	1823 0.2		23 W	0024 5.0	0645 0.0	1247 5.0	○1857 0.3
9 W	0022 5.2	0651 -0.1	1252 5.4	●1906 0.1	24 TH	0102 5.0	0719 0.1	1321 4.9	1926 0.4
10 TH	0106 5.5	0735 -0.2	1336 5.6	1948 0.0	25 F	0131 4.9	0745 0.3	1346 4.8	1947 0.4
11 F	0148 5.6	0817 -0.3	1418 5.6	2028 0.0	26 SA	0154 4.8	0806 0.4	1408 4.8	2008 0.4
12 SA	0230 5.6	0857 -0.2	1501 5.5	2108 0.0	27 SU	0218 4.9	0831 0.4	1435 4.9	2037 0.4
13 SU	0312 5.5	0938 -0.1	1544 5.3	2150 0.1	28 M	0249 4.9	0902 0.4	1509 4.9	2111 0.4
14 M	0356 5.3	1020 0.2	1630 5.0	2234 0.2	29 TU	0326 4.9	0938 0.5	1550 4.9	2149 0.4
15 TU	0444 5.0	1101 0.4	1720 4.7	◐2324 0.4	30 W	0409 4.8	1019 0.6	1635 4.8	2233 0.6
					31 TH	0457 4.6	1106 0.7	1724 4.5	◐2323 0.7

APRIL

Day	Time m	Time m	Time m	Time m	Day	Time m	Time m	Time m	Time m
1 F	0551 4.4	1202 0.9	1820 4.3		16 SA	0125 0.6	0751 4.3	1405 1.0	2024 4.1
2 SA	0023 0.9	0656 4.2	1316 1.1	1929 4.1	17 SU	0247 0.5	0904 4.4	1522 0.9	2130 4.3
3 SU	0152 1.0	0824 4.1	1453 1.1	2057 4.2	18 M	0353 0.4	1003 4.6	1620 0.7	2225 4.6
4 M	0334 0.8	0946 4.4	1611 0.8	2210 4.5	19 TU	0445 0.3	1054 4.8	1708 0.6	2313 4.8
5 TU	0444 0.5	1047 4.8	1709 0.5	2306 4.8	20 W	0530 0.2	1138 4.9	1750 0.5	2356 4.9
6 W	0537 0.2	1139 5.2	1758 0.3	2355 5.2	21 TH	0609 0.2	1217 4.9	1826 0.4	
7 TH	0625 -0.1	1226 5.5	1843 0.0 ●		22 F	0033 4.9	0641 0.3	1251 4.9	○1855 0.5
8 F	0041 5.5	0709 -0.2	1311 5.6	1926 -0.1	23 SA	0105 4.8	0707 0.4	1317 4.8	1919 0.4
9 SA	0126 5.6	0751 -0.3	1355 5.6	2008 -0.2	24 SU	0131 4.8	0732 0.4	1341 4.9	1944 0.4
10 SU	0210 5.7	0833 -0.2	1438 5.5	2051 -0.1	25 M	0157 4.8	0800 0.4	1410 4.9	2014 0.4
11 M	0254 5.6	0914 0.0	1522 5.3	2134 -0.1	26 TU	0229 4.9	0833 0.4	1445 5.0	2049 0.4
12 TU	0340 5.4	0957 0.2	1607 5.0	2220 0.1	27 W	0308 4.9	0910 0.4	1525 5.0	2130 0.4
13 W	0429 5.1	1044 0.5	1657 4.7	2311 0.3	28 TH	0351 4.9	0953 0.6	1610 4.8	2215 0.5
14 TH	0524 4.7	1137 0.7	1754 4.3	◐	29 F	0439 4.7	1041 0.8	1659 4.6	2306 0.7
15 F	0011 0.5	0631 4.4	1243 1.0	1905 4.1	30 SA	0532 4.5	1136 0.9	1752 4.4	◐

Chart Datum: 2·35 metres below Ordnance Datum (Newlyn). HAT is 5·8 metres above Chart Datum.

》》 FREE monthly updates. Register at 《
www.reedsnauticalalmanac.co.uk

STANDARD TIME (UT)
For Summer Time add ONE hour in **non-shaded areas**

BURNHAM-ON-CROUCH   LAT 51°37'N   LONG 0°48'E

TIMES AND HEIGHTS OF HIGH AND LOW WATERS

Dates in red are **SPRINGS**
Dates in blue are NEAPS

YEAR 2016

MAY

Time	m	Time	m
1 0005	0.8	**16** 0205	0.5
0633	4.4	0827	4.3
SU 1243	1.1	M 1430	1.1
1854	4.2	2051	4.2
2 0125	0.9	**17** 0308	0.5
0750	4.3	0926	4.4
M 1411	1.1	TU 1533	1.0
2013	4.2	2148	4.4
3 0258	0.8	**18** 0402	0.5
0910	4.5	1017	4.6
TU 1534	0.9	W 1626	0.8
2131	4.4	2237	4.5
4 0411	0.5	**19** 0448	0.5
1015	4.8	1102	4.7
W 1637	0.7	TH 1711	0.7
2233	4.8	2322	4.7
5 0507	0.2	**20** 0528	0.4
1110	5.1	1142	4.8
TH 1731	0.4	F 1750	0.6
2327	5.1		
6 0557	0.0	**21** 0001	4.7
1159	5.4	0603	0.4
F 1819	0.1	SA 1218	4.8
●		○ 1824	0.6
7 0016	5.4	**22** 0038	4.8
0643	-0.1	0635	0.5
SA 1247	5.5	SU 1250	4.9
1905	-0.1	1856	0.5
8 0103	5.6	**23** 0110	4.8
0728	-0.2	0705	0.5
SU 1332	5.5	M 1320	4.9
1951	-0.1	1927	0.5
9 0150	5.6	**24** 0141	4.8
0811	-0.1	0737	0.5
M 1417	5.4	TU 1351	5.0
2036	-0.2	2000	0.4
10 0236	5.5	**25** 0216	4.9
0854	0.1	0812	0.5
TU 1501	5.3	W 1427	5.0
2122	-0.1	2037	0.4
11 0323	5.3	**26** 0254	4.9
0938	0.3	0852	0.5
W 1546	5.0	TH 1507	5.0
2209	0.0	2118	0.4
12 0413	5.0	**27** 0338	4.9
1024	0.5	0935	0.6
TH 1634	4.7	F 1551	4.9
2300	0.2	2204	0.5
13 0507	4.7	**28** 0425	4.8
1115	0.8	1023	0.8
F 1727	4.4	SA 1638	4.7
◑ 2356	0.4	2255	0.6
14 0608	4.4	**29** 0517	4.6
1212	1.0	1116	0.9
SA 1830	4.2	SU 1730	4.6
		◑ 2354	0.7
15 0058	0.5	**30** 0615	4.5
0717	4.3	1218	1.0
SU 1319	1.1	M 1829	4.4
1943	4.1		
		31 0106	0.7
		0722	4.4
		TU 1334	1.1
		1938	4.3

JUNE

Time	m	Time	m
1 0226	0.7	**16** 0308	0.7
0836	4.5	0929	4.3
W 1455	1.0	TH 1529	1.0
2055	4.4	2155	4.3
2 0339	0.5	**17** 0400	0.6
0943	4.7	1019	4.5
TH 1606	0.7	F 1624	0.7
2202	4.7	2244	4.4
3 0440	0.3	**18** 0447	0.6
1042	5.0	1105	4.6
F 1706	0.5	SA 1713	0.8
2301	5.0	2330	4.6
4 0534	0.1	**19** 0531	0.5
1135	5.2	1147	4.8
SA 1759	0.2	SU 1758	0.6
2354	5.3		
5 0623	0.0	**20** 0011	4.7
1224	5.4	0611	0.5
SU 1849	0.0	M 1226	4.9
●		○ 1838	0.5
6 0044	5.5	**21** 0051	4.8
0710	0.0	0649	0.5
M 1312	5.4	TU 1303	5.0
1937	-0.1	1917	0.4
7 0133	5.5	**22** 0128	4.9
0755	0.0	0726	0.5
TU 1357	5.3	W 1338	5.0
2024	-0.2	1956	0.4
8 0220	5.4	**23** 0205	5.0
0838	0.2	0803	0.5
W 1441	5.2	TH 1415	5.0
2111	-0.1	2035	0.4
9 0307	5.2	**24** 0245	5.0
0921	0.4	0842	0.5
TH 1525	5.0	F 1454	5.0
2157	0.0	2116	0.4
10 0354	5.0	**25** 0327	5.0
1004	0.6	0924	0.6
F 1609	4.7	SA 1536	5.0
2244	0.2	2200	0.4
11 0442	4.7	**26** 0413	4.9
1049	0.8	1010	0.7
SA 1654	4.5	SU 1621	4.8
2332	0.3	2248	0.5
12 0534	4.4	**27** 0502	4.8
1137	1.0	1059	0.8
SU 1746	4.2	M 1711	4.7
◑		◑ 2342	0.6
13 0023	0.5	**28** 0556	4.6
0630	4.2	1155	0.9
M 1230	1.1	TU 1805	4.5
1846	4.1		
14 0117	0.6	**29** 0044	0.7
0732	4.1	0657	4.5
TU 1327	1.2	W 1302	1.0
1954	4.0	1909	4.4
15 0213	0.7	**30** 0157	0.7
0834	4.1	0805	4.5
W 1428	1.1	TH 1420	1.0
2059	4.1	2025	4.4

JULY

Time	m	Time	m
1 0313	0.6	**16** 0310	0.8
0915	4.6	0932	4.2
F 1539	0.8	SA 1534	1.0
2138	4.6	2206	4.2
2 0421	0.4	**17** 0410	0.8
1018	4.8	1028	4.5
SA 1647	0.5	SU 1639	0.8
2241	4.9	2259	4.5
3 0518	0.3	**18** 0505	0.6
1115	5.0	1118	4.7
SU 1745	0.2	M 1734	0.6
2338	5.2	2347	4.7
4 0610	0.1	**19** 0554	0.5
1207	5.2	1204	4.9
M 1838	0.0	TU 1823	0.4
●		○	
5 0030	5.3	**20** 0032	4.9
0657	0.1	0638	0.6
TU 1255	5.3	W 1246	5.0
1926	-0.2	1908	0.3
6 0119	5.4	**21** 0114	5.1
0742	0.1	0719	0.4
W 1341	5.2	TH 1325	5.1
2013	-0.2	1950	0.2
7 0205	5.3	**22** 0154	5.2
0824	0.2	0758	0.4
TH 1423	5.1	F 1404	5.1
2057	-0.2	2030	0.2
8 0249	5.1	**23** 0234	5.2
0903	0.4	0837	0.4
F 1502	4.9	SA 1442	5.1
2139	0.0	2111	0.2
9 0330	4.9	**24** 0315	5.2
0940	0.5	0916	0.4
SA 1539	4.7	SU 1522	5.1
2218	0.1	2152	0.2
10 0411	4.6	**25** 0358	5.1
1016	0.7	0957	0.5
SU 1616	4.5	M 1605	5.0
2257	0.4	2235	0.4
11 0451	4.4	**26** 0444	4.9
1054	0.9	1042	0.6
M 1656	4.4	TU 1652	4.8
2338	0.6	◑ 2322	0.5
12 0535	4.2	**27** 0535	4.7
1137	1.0	1132	0.7
TU 1742	4.2	W 1744	4.6
◑			
13 0023	0.7	**28** 0017	0.7
0625	4.1	0631	4.5
W 1227	1.1	TH 1233	0.8
1838	4.0	1845	4.4
14 0114	0.8	**29** 0127	0.8
0723	4.0	0738	4.4
TH 1323	1.1	F 1349	0.9
1950	3.9	2002	4.3
15 0210	0.9	**30** 0251	0.8
0829	4.1	0902	4.4
F 1426	1.1	SA 1520	0.8
2103	4.0	2122	4.5
		31 0407	0.6
		1001	4.6
		SU 1635	0.5
		2229	4.7

AUGUST

Time	m	Time	m
1 0507	0.4	**16** 0441	0.7
1100	4.8	1050	4.6
M 1734	0.1	TU 1712	0.6
2327	5.0	2324	4.7
2 0558	0.2	**17** 0535	0.5
1153	5.0	1141	4.9
TU 1826	-0.1	W 1805	0.3
●			
3 0019	5.2	**18** 0011	5.0
0645	0.1	0622	0.4
W 1241	5.2	TH 1226	5.1
1913	-0.3	○ 1851	0.1
4 0106	5.3	**19** 0055	5.2
0728	0.1	0705	0.2
TH 1325	5.2	F 1308	5.2
1956	-0.3	1934	-0.1
5 0149	5.2	**20** 0137	5.4
0806	0.2	0745	0.2
F 1404	5.1	SA 1347	5.3
2036	-0.2	2015	-0.1
6 0228	5.0	**21** 0217	5.4
0840	0.3	0824	0.2
SA 1437	4.9	SU 1426	5.3
2111	0.0	2054	0.0
7 0302	4.8	**22** 0258	5.4
0909	0.5	0902	0.2
SU 1507	4.8	M 1506	5.3
2142	0.2	2134	0.1
8 0334	4.7	**23** 0340	5.2
0936	0.6	0942	0.3
M 1537	4.6	TU 1548	5.1
2212	0.4	2214	0.2
9 0407	4.5	**24** 0425	5.0
1009	0.7	1024	0.5
TU 1612	4.5	W 1633	5.0
2246	0.6	2258	0.4
10 0445	4.4	**25** 0513	4.7
1047	0.8	1112	0.6
W 1653	4.4	TH 1725	4.7
◑ 2327	0.7	◑ 2349	0.7
11 0528	4.3	**26** 0608	4.5
1131	0.9	1209	0.7
TH 1742	4.2	F 1826	4.4
12 0016	0.9	**27** 0056	0.9
0619	4.2	0715	4.2
F 1224	1.0	SA 1326	4.2
1840	4.0	1946	4.2
13 0114	1.0	**28** 0230	0.9
0722	4.1	0835	4.2
SA 1328	1.1	SU 1507	0.7
2000	3.9	2112	4.4
14 0222	1.0	**29** 0353	0.7
0840	4.1	0947	4.5
SU 1447	1.1	M 1622	0.4
2125	4.0	2220	4.7
15 0335	0.9	**30** 0452	0.5
0952	4.3	1047	4.8
M 1608	0.9	TU 1719	0.1
2230	4.4	2316	5.0
		31 0543	0.2
		1139	5.0
		W 1809	-0.2

Chart Datum: 2·35 metres below Ordnance Datum (Newlyn). HAT is 5·8 metres above Chart Datum.

》》 **FREE** monthly updates. Register at 《 www.reedsnauticalalmanac.co.uk

305

E England

BURNHAM-ON-CROUCH LAT 51°37'N LONG 0°48'E
TIMES AND HEIGHTS OF HIGH AND LOW WATERS

STANDARD TIME (UT)
For Summer Time add ONE hour in **non-shaded areas**

Dates in **red** are **SPRINGS**
Dates in blue are NEAPS

YEAR **2016**

SEPTEMBER

#	Time	m	#	Time	m
1 TH ●	0006 0627 1225 1853	5.2 0.1 5.2 -0.3	**16** F ○	0600 1201 1827	0.3 5.2 0.0
2 F	0051 0707 1307 1933	5.2 0.1 5.2 -0.2	**17** SA	0031 0643 1245 1910	5.4 0.2 5.4 -0.2
3 SA	0130 0743 1342 2008	5.1 0.2 5.1 -0.1	**18** SU	0114 0725 1326 1951	5.5 0.1 5.5 -0.2
4 SU	0204 0811 1411 2035	5.0 0.4 4.9 0.1	**19** M	0156 0805 1407 2031	5.6 0.1 5.5 -0.1
5 M	0231 0833 1435 2059	4.8 0.5 4.8 0.3	**20** TU	0237 0844 1448 2111	5.5 0.1 5.5 0.0
6 TU	0256 0858 1502 2125	4.7 0.5 4.8 0.4	**21** W	0320 0924 1531 2151	5.3 0.2 5.3 0.2
7 W	0326 0928 1536 2158	4.7 0.6 4.7 0.5	**22** TH	0404 1008 1617 2234	5.1 0.3 5.1 0.5
8 TH	0403 1004 1617 2237	4.6 0.7 4.6 0.7	**23** F ◐	0451 1056 1710 2326	4.8 0.5 4.8 0.7
9 F ◑	0446 1046 1703 2323	4.5 0.8 4.4 0.8	**24** SA	0546 1155 1813	4.4 0.6 4.4
10 SA	0535 1136 1757	4.4 0.9 4.2	**25** SU	0032 0655 1314 1935	1.0 4.2 0.8 4.3
11 SU	0019 0631 1239 1904	1.0 4.2 1.0 4.0	**26** M	0206 0818 1450 2059	1.0 4.2 0.6 4.4
12 M	0133 0745 1403 2039	1.1 4.1 1.1 4.0	**27** TU	0330 0931 1602 2204	0.8 4.5 0.3 4.7
13 TU	0258 0912 1536 2158	1.1 4.2 0.9 4.3	**28** W	0430 1029 1657 2258	0.6 4.8 0.2 5.0
14 W	0414 1020 1646 2256	0.8 4.5 0.6 4.7	**29** TH	0519 1119 1744 2346	0.3 5.0 -0.1 5.2
15 TH	0511 1114 1740 2346	0.6 4.9 0.2 5.1	**30** F	0603 1203 1826	0.2 5.2 -0.2

OCTOBER

#	Time	m	#	Time	m
1 SA ●	0028 0641 1244 1903	5.2 0.2 5.2 0.0	**16** SU ○	0003 0617 1218 1843	5.4 0.2 5.5 -0.1
2 SU	0105 0714 1318 1933	5.1 0.3 5.1 0.1	**17** M	0048 0701 1302 1926	5.6 0.1 5.6 -0.2
3 M	0136 0739 1344 1956	4.9 0.4 4.9 0.3	**18** TU	0132 0743 1346 2007	5.6 0.0 5.7 -0.1
4 TU	0159 0759 1407 2017	4.8 0.5 4.9 0.4	**19** W	0215 0826 1430 2048	5.5 0.0 5.6 0.1
5 W	0222 0823 1434 2044	4.8 0.5 4.9 0.5	**20** TH	0258 0909 1515 2130	5.4 0.1 5.5 0.3
6 TH	0252 0854 1508 2118	4.8 0.5 4.8 0.5	**21** F	0343 0955 1604 2216	5.1 0.2 5.2 0.5
7 F	0329 0931 1549 2158	4.8 0.6 4.7 0.7	**22** SA ◑	0431 1047 1658 2310	4.8 0.4 4.9 0.8
8 SA	0413 1013 1636 2244	4.7 0.7 4.6 0.8	**23** SU	0527 1147 1801	4.5 0.6 4.5
9 SU ◑	0501 1104 1728 2339	4.6 0.8 4.4 1.0	**24** M	0014 0634 1301 1919	1.0 4.3 0.6 4.4
10 M	0555 1205 1830	4.4 1.0 4.2	**25** TU	0136 0754 1423 2036	1.1 4.2 0.6 4.5
11 TU	0049 0700 1326 1952	1.2 4.2 1.1 4.1	**26** W	0256 0905 1532 2139	0.9 4.5 0.4 4.7
12 W	0218 0825 1502 2119	1.2 4.2 0.9 4.3	**27** TH	0358 1002 1626 2232	0.7 4.7 0.2 4.9
13 TH	0340 0943 1615 2223	1.0 4.5 0.6 4.7	**28** F	0448 1052 1713 2318	0.5 5.0 0.1 5.1
14 F	0441 1042 1711 2315	0.7 4.8 0.3 5.1	**29** SA	0532 1137 1753 2359	0.4 5.1 0.1 5.1
15 SA	0532 1132 1759	0.4 5.2 0.0	**30** SU ●	0610 1216 1828	0.4 5.1 0.2
			31 M	0035 0643 1251 1856	5.0 0.4 5.0 0.4

NOVEMBER

#	Time	m	#	Time	m
1 TU	0106 0708 1320 1919	4.9 0.5 4.9 0.5	**16** W	0108 0725 1326 1946	5.6 0.0 5.8 0.0
2 W	0129 0730 1344 1943	4.9 0.5 4.9 0.5	**17** TH	0153 0811 1413 2030	5.5 0.1 5.7 0.2
3 TH	0153 0756 1412 2012	4.9 0.5 4.9 0.6	**18** F	0238 0858 1500 2115	5.4 0.1 5.5 0.4
4 F	0224 0828 1447 2048	5.0 0.5 4.9 0.6	**19** SA	0323 0947 1550 2202	5.1 0.1 5.3 0.6
5 SA	0303 0907 1529 2130	4.9 0.6 4.9 0.7	**20** SU	0411 1039 1643 2254	4.8 0.3 5.0 0.8
6 SU	0346 0951 1616 2217	4.9 0.7 4.7 0.8	**21** M ◑	0504 1136 1743 2353	4.6 0.4 4.7 1.0
7 M	0435 1042 1708 2312	4.7 0.8 4.6 1.0	**22** TU	0606 1232 1851	4.3 0.5 4.5
8 TU	0528 1142 1807	4.5 0.9 4.4	**23** W	0059 0718 1347 2001	1.1 4.2 0.5 4.5
9 W	0017 0627 1259 1917	1.2 4.3 1.0 4.3	**24** TH	0210 0828 1451 2103	1.1 4.3 0.5 4.6
10 TH	0138 0741 1426 2037	1.2 4.3 0.9 4.5	**25** F	0315 0928 1547 2157	0.9 4.5 0.4 4.7
11 F	0300 0901 1541 2146	1.1 4.4 0.7 4.8	**26** SA	0410 1019 1634 2244	0.8 4.7 0.4 4.9
12 SA	0407 1006 1640 2243	0.8 4.8 0.4 5.1	**27** SU	0456 1105 1716 2326	0.7 4.9 0.5 4.9
13 SU	0503 1101 1731 2334	0.5 5.2 0.1 5.4	**28** M	0537 1147 1752	0.6 4.9 0.5
14 M ○	0553 1151 1818	0.3 5.5 0.0	**29** TU ●	0004 0612 1225 1823	5.0 0.6 4.9 0.5
15 TU	0021 0639 1239 1903	5.6 0.1 5.7 0.1	**30** W	0037 0643 1258 1851	5.0 0.6 4.9 0.6

DECEMBER

#	Time	m	#	Time	m
1 TH	0105 0710 1327 1919	5.0 0.6 4.9 0.6	**16** F	0134 0801 1358 2017	5.5 -0.1 5.6 0.2
2 F	0133 0740 1358 1951	5.0 0.5 4.9 0.6	**17** SA	0220 0849 1446 2102	5.4 -0.1 5.5 0.4
3 SA	0205 0814 1434 2028	5.0 0.5 4.9 0.7	**18** SU	0304 0937 1534 2147	5.2 0.0 5.3 0.6
4 SU	0243 0853 1515 2111	5.0 0.6 4.9 0.7	**19** M	0350 1025 1623 2233	4.9 0.1 5.0 0.8
5 M	0326 0938 1601 2158	4.9 0.7 4.9 0.9	**20** TU	0437 1115 1716 2323	4.7 0.3 4.7 1.0
6 TU	0413 1028 1652 2251	4.8 0.7 4.8 1.0	**21** W ◐	0529 1207 1812	4.4 0.5 4.5
7 W ◐	0504 1126 1747 2350	4.6 0.8 4.6 1.1	**22** TH	0016 0628 1302 1913	1.1 4.2 0.6 4.3
8 TH	0600 1233 1850	4.5 0.9 4.5	**23** F	0114 0735 1359 2015	1.2 4.1 0.7 4.3
9 F	0101 0705 1351 2001	1.2 4.4 0.9 4.6	**24** SA	0216 0841 1456 2112	1.2 4.2 0.7 4.4
10 SA	0220 0820 1506 2111	1.1 4.4 0.7 4.7	**25** SU	0318 0939 1548 2204	1.1 4.3 0.7 4.6
11 SU	0333 0932 1611 2213	0.9 4.7 0.5 5.0	**26** M	0414 1031 1636 2250	0.9 4.5 0.7 4.7
12 M	0437 1033 1707 2308	0.6 5.0 0.3 5.2	**27** TU	0502 1117 1718 2333	0.8 4.7 0.7 4.9
13 TU	0532 1128 1758 2359	0.4 5.3 0.2 5.4	**28** W	0546 1159 1757	0.7 4.8 0.6
14 W ○	0624 1220 1846	0.1 5.6 0.1	**29** TH ●	0011 0625 1238 1833	5.0 0.6 4.9 0.6
15 TH	0047 0713 1310 1932	5.5 -0.1 5.7 0.1	**30** F	0047 0701 1314 1907	5.0 0.6 4.9 0.6
			31 SA	0120 0736 1348 1942	5.1 0.5 5.0 0.6

Chart Datum: 2·35 metres below Ordnance Datum (Newlyn). HAT is 5·8 metres above Chart Datum.

9.4.16 RIVER BLACKWATER

Essex 51°45'·33N 00°54'·90E ✿✿✿✿◊◊◊◊✿✿✿

CHARTS AC 5607, 1183, 1975, 3741; Imray C1, Y17, 2000

TIDES Maldon +0130 Dover; ML 2·8; Duration 0620)

Standard Port WALTON-ON-THE-NAZE (→)

Times				Height (metres)			
High Water		Low Water		MHWS	MHWN	MLWN	MLWS
0000	0600	0500	1100	4·2	3·4	1·1	0·4
1200	1800	1700	2300				
Differences SUNK HEAD							
0000	+0002	–0002	+0002	–0·3	–0·3	–0·1	–0·1
WEST MERSEA							
+0035	+0015	+0055	+0010	+0·9	+0·4	+0·1	+0·1
BRADWELL							
+0035	+0023	+0047	+0004	+1·0	+0·8	+0·2	0·0
OSEA ISLAND							
+0057	+0045	+0050	+0007	+1·1	+0·9	+0·1	0·0
MALDON							
+0107	+0055	ND	ND	–1·3	–1·1	ND	ND

SHELTER Good, as appropriate to wind. ⚓ in creeks restricted by oyster beds and many moorings.

NAVIGATION WPT Knoll NCM, Q, 51°43'·88N 01°05'·07E, 287°/6·7M to Nass bn, ECM VQ (3) 5s. Speed limit 8kn W of Osea Is.

WEST MERSEA Min depth 0·6m in approach to Nass Bn. Avoid oyster beds between Cobmarsh and Packing Marsh Is and in Salcott Chan. ⚓ in Mersea Quarters in 5-6m (exposed). There is a pontoon for landing (limited waiting); also pile moorings in Ray Channel.

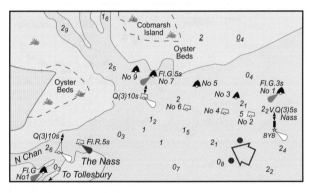

TOLLESBURY FLEET Proceed via S Chan up Woodrolfe Creek. A tide gauge shows depth over marina ent sill (approx 2·4m at MHWS and 1·4m at MHWN). Speed limits: Tollesbury Fleet S Chan 8kn; Woodrolfe Creek upper reaches 4kn.

BRADWELL No dangers. Creek only suitable for small craft and gets very crowded. Ent has SCM bn Q with tide gauge in creek; leave to **STBD** on entry. 4 PHM buoys, 3 SHM withies and 2 B/W △ ldg bns mark the channel which doglegs past a SHM buoy to marina entrance. Min depth 0·2m in approach channel.

MALDON From S of Osea Is, 'The Doctor', No 3 SHM buoy on with Blackwater SC lt, Iso G 5s, lead 300° approx up the chan; or No 3 and No 8 buoys in line at 305°.

LIGHTS AND MARKS See chartlets and 9.4.4. The channel to Maldon has some lit marks.

COMMUNICATIONS (Codes 01621 Maldon; 01206 Colchester/W Mersea) NMOC 02392 552100; Police 101; Dr 854118.

Bradwell and Tollesbury Marinas Ch **80** M (HO); Blackwater Marina Ch ⚓ 0900-2300; Heybridge Lock Ch 80 HW –2+1; W Mersea Clark and Carter launch: call CC1.

FACILITIES

WEST MERSEA (01206) **W Mersea YC**382947, launch service call *YC1* on Ch M or ☎07752 309435, ⚓ ✕ 🏠. **Town** ⚓ 🏪🏤🏦⚓ ⛽ 🔧 ✕ 🛒 ⚓ (bus to Colchester, Ⓗ01206-853535), ✈ (Southend/Stansted).

TOLLESBURY Marina ✉ marina@woodrolfe.com ☎(01621) 869202, Sill to marina dries 1·9m. ⚓ 220⚓+20☻ £2·00, ⚓ 🛢 LPG ⚓ ✕ 🔧 ⚓ 🔧(20t) ⚓(5t) ⚓. **Tollesbury Cruising Club** ☎869561, ⚓ ✕ 🏠 **Village** 🏪 ⛽ ✕ 🛒 ✉ Ⓑ (Tues, Thurs 1000-1430), ⚓ (bus to Witham), ✈ (Southend or Cambridge).

BRADWELL (01621). **Bradwell Marina** ☎776235. Min depth 0·2m in appr chan. 300⚓ some ☻ £2·00, short stay £3/hr. ⚓ 🛢⚓🔧✕🔧 ⚓ 🔧(45t) ⚓ ✕ 🏠. **Bradwell Quay YC** ☎890173, ⚓⚓ ⚓ ⚓ 🏠. **Town** ✉ ⚓ (bus/taxi to Southminster), ✈ (Southend).

MAYLANDSEA (01621). **Blackwater Marina** ☎740264, 150⚓ ⚓ £5 in chan, ⚓ 230⚓ all dry £15/craft, 🛢⚓✕🔧 🏠. Taxi to Southminster ⚓.

MALDON (01621). **Maldon Quay** Beyond No 8 buoy, the chan which shifts and carries 0·2m, is lit and buoyed. Access near HW; pontoons dry to soft mud. HM/River bailiff ☎856487, Mob 07818 013723; ⚓ ⚓ ⚓ £15, ⚓ 🛢⚓. **Maldon Little Ship Club** ☎854139, 🏠. **Services:** ⚓🔧 ⚓ Ⓔ ⚓ ⚓ ACA. **Town** ✉ Ⓑ ⚓ (bus to Chelmsford, Ⓗ(01245) 440761), ✈ (Southend, Cambridge or Stansted).

HEYBRIDGE BASIN (01621). **Lock** ☎853506, opens HW –1 to HW approx. A SHM buoy opposite the lock marks the deep water ent. Access to Chelmer & Blackwater Canal (not navigable). **Blackwater SC** ☎853923, ⚓⚓. **Services:** at CRS Marine ☎854684, Mobile 07850 543873 (pontoon outside lock), ⚓⚓⚓⚓🔧⚓ ⚓ ⚓. Bus to Heybridge/Maldon.

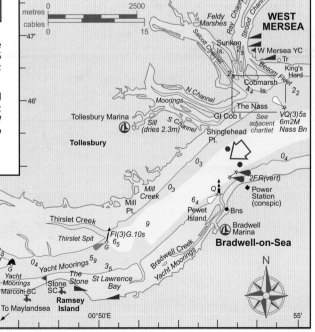

9.4.17 RIVER COLNE ✽✽♨♨♨✿✿

Essex **51°47'·98N 01°00'·60E** (Brightlingsea) ✽✽♨♨♨✿✿

CHARTS AC 5607, 1183, 1975, 3741; Imray C1, Y17, 2000

TIDES +0050 Dover; ML 2·5; Duration 0615

Standard Port WALTON-ON-THE-NAZE (→)

Times				Height (metres)			
High Water		Low Water		MHWS	MHWN	MLWN	MLWS
0000	0600	0500	1100	4·2	3·4	1·1	0·4
1200	1800	1700	2300				
Differences BRIGHTLINGSEA							
+0025	+0021	+0046	+0004	+0·8	+0·4	+0·1	0·0
COLCHESTER							
+0035	+0025	DR	DR	0·0	−0·3	DR	DR
CLACTON-ON-SEA							
+0012	+0010	+0025	+0008	+0·3	+0·1	+0·1	+0·1

SHELTER Suitable shelter can be found from most winds. Brightlingsea outer hbr is exposed to W'lies. In the creek S of Cindery Island are moorings (as shown) and long pontoons in about 1·5m, with possible ⌒ for Ⓥs. ⚓ prohib in Brightlingsea Hbr, but there are ⚓s to the NW of Mersea Stone Pt and in Pyefleet

Chan, E of Pewit Island. R Colne is navigable for 4·5m draught to Wivenhoe and Rowhedge, where the river dries; and to The Hythe, Colchester (3m draught).

NAVIGATION WPT Colne Bar By, SHM Fl (2) G 5s, 51°44'·61N 01°02'·57E, 340°/3·6M to Mersea Stone. Extensive mud and sand banks flank the entrance channel. At HW sand barges regularly visit Ballast Quay below the tidal barrier at Wivenhoe. The ent to Brightlingsea Creek is narrow at LW and carries about 1m.

Tidal barrier 2ca below Wivenhoe church is usually open (30m wide) allowing unrestricted passage; keep to stbd, max speed 5kn. 3 F R (vert), shown on N Pier, visible up/downstream indicate barrier gates are shut; see also LIGHTS AND MARKS.

Speed limits in approaches and up-river (by buoy no): No13–15 = 8kn; No15–18 = no spd limit; No18–31 = 8kn; No 31 – Colchester = 5kn; Brightlingsea Harbour = 4kn.

LIGHTS AND MARKS Bateman's Tr (conspic) by Westmarsh Pt Fl(3) 20s. Brightlingsea ldg lts/marks 041°; FR 7/10m 4M, W□ on R striped posts; adjusted to suit channel. Pass Spit SCM, and follow lateral marks to NCM (Q) where chan is divided by Cindery Is. Pyefleet Chan and other creeks are unlit.

Lit lateral buoys up to Wivenhoe. Passage through flood barrier marked by 2FR/FG (vert) each side and there are bns, QR/QG, up/downstream on the banks.

COMMUNICATIONS (Brightlingsea 01206) NMOC 02392 552100; Police 101; Dr 302522; HM (Brightlingsea) 302200, mob 07952 734814.

Brightlingsea Harbour Radio VHF Ch 68. *Waterside Marina* Ch 80.

River Colne
Caution - Buoyage liable to be moved. Consult AC3741 for position of oyster beds between Mersea Island and Brightlingsea

FACILITIES

BRIGHTLINGSEA:

Town Hard ☎302200, ⚓⚓⛽♿🅿. Pontoon berths £17<26ft, £19<36ft, £22<50ft, (inc water taxi) short stay £5; at fuel jetty.

Waterside Marina ☎308709, sill 1m above CD; ⌒£2·00 inc 🚿.

Colne YC ☎302594, ⚓⛽✕🖥 wi-fi.

Brightlingsea SC ⚓🅿.

Services 🅿 🅿(HO) ♿✗🔧🔌Ⓔ⛽🅰 🅿(mobile) ⓛ ACA.

Town ✉ Ⓑ 🛒 ✕ 🅿 bus (Wivenhoe or Colchester for ⇄), ✈ (Southend or Stansted).

WIVENHOE: Wivenhoe SC ☎822132. ⌒ (drying Ⓥ berths may be available on the inside pontoons – first come first served, no charge).

Village 🅿✉🛒🅿⇄.

ROWHEDGE: Quay ⌒(drying) some Ⓥ.

RIVER COLNE TO HARWICH

(Chart AC 1975, 1593) 4M SW of the Naze twr at Hollands Haven a conspic radar twr (67m, unlit) is an excellent daymark. From the S, approach Walton and Harwich via the Medusa chan about 1M E of the twr. Beware pot buoys and drift nets off the Naze. At N end of this chan, 1M off Dovercourt, Pye End buoy marks chan SSW to Walton Backwaters. Harwich and Landguard Pt are close to the N. Making Harwich from the SE beware the drying Cork Sand, which lies N/S.

Sunk Inner SWM buoy, 11M E of The Naze, marks the outer apps to Harwich, an extensive and well sheltered hbr accessible at all times (chart AC 2693).

- Harwich, Felixstowe Docks and Ipswich are referred to collectively as the Haven Ports.
- Small craft should give plenty of sea room to shipping manoeuvring to board/disembark Pilots in the Sunk Inner Precautionary Area, see Thames Estuary chartlet.
- Harwich Deep Water chan runs N from 1·5M NNW of Sunk Inner between Rough and Shipwash shoals, then W past the Cork Sand PHM. Constantly used by commercial shipping, small craft should use the recommended track for yachts.

Approaching from NE and 2M off the ent to R. Deben, beware Cutler shoal, with least depth of 1·2m, marked by SHM buoy on E side; Wadgate Ledge and the Platters are about 1·5M ENE of Landguard Point; beware pot buoys off Cobbolds Point.

▶ S of Landguard Point the W-going (flood) stream begins at HW Harwich +0600, and the E-going stream at HW Harwich, sp rates about 1·5kn. Note: HW Harwich is never more than 7 mins after HW Walton; LW times are about 10 mins earlier. ◄

HARWICH TO ORFORD NESS

(Chart AC 2052) Shipwash shoal, buoyed and with a drying patch, runs NNE from 9M E of Felixstowe to 4M SSE of Orford Ness. Inshore of this is Shipway Chan, then Bawdsey Bank, buoyed with depths of 2m, on which the sea breaks in E'ly swell.

The Sledway Chan lies between Bawdsey Bank and Whiting Bank (buoyed) which is close SW of Orford Ness, and has depths less than 1m. Hollesley Chan, about 1M wide, runs inshore W and N of this bank. In the SW part of Hollesley B is the ent to Orford Haven and the R Ore/Alde.

▶ There are overfalls S of Orford Ness on both the ebb and flood streams. 2M E of Orford Ness the SW-going stream begins at HW Harwich +0605, sp rate 2·5kn; the NE-going stream begins at HW Harwich –0010, sp rate 3kn. ◄

Note: The direction of local buoyage becomes S to N off Orford Ness (52°05'N).

ORFORD NESS TO GREAT YARMOUTH

(Chart AC 1543) N of Orford Ness seas break on Aldeburgh Ridge (1·3m), but the coast is clear of offlying dangers past Aldeburgh and Southwold, as far as Benacre Ness, 5M S of Lowestoft. Sizewell power station is a conspic 3 bldg 1·5M N of Thorpe Ness. Keep 1·5M offshore to avoid fishing floats.

▶ Lowestoft is best approached from both S and E by the buoyed/lit Stanford chan, passing E of Newcome Sand and SW of Holm Sand; beware possible strong set across hbr ent. ◄

From the N, approach through Cockle Gatway, Caister Road, Yarmouth Road, passing Great Yarmouth (beware of prohibited area N of harbour entrance); then proceed S through Gorleston, Corton and Lowestoft North Roads (buoyed).

▶ 1M E of hbr ent, the S-going stream begins at HW Dover –0600, and the N-going at HW Dover, sp rates 2·6kn. ◄

Scroby Sands Wind Farm consists of 30 turbines centred on 52°39'·00N 01°47'·00E. Each turbine is 61m high, with 80m diameter blades and clearance height of 18m and six of the perimeter ones are lit. Vessels to keep well clear and not enter the area.

In the approaches to Great Yarmouth from seaward the banks are continually changing; use the buoyed chans which, from N and S, are those described in the preceding paragraph. But from the E the shortest approach is via Corton ECM lt buoy and the Holm Channel leading into Gorleston Road.

▶ The sea often breaks on North Scroby, Middle Scroby and Caister Shoal (all of which dry), and there are heavy tide rips over parts of Corton and South Scroby Sands, Middle and South Cross Sands, and Winterton Overfalls. ◄

▶ 1M NE of ent to Gt Yarmouth the S-going stream begins at HW Dover –0600, and the N-going at HW Dover –0015, sp rates 2·3kn. Breydon Water (tidal) affects streams in the Haven; after heavy rain the out-going stream at Brush Quay may exceed 5kn. ◄

About 12M NE of Great Yarmouth lie Newarp Banks, on which the sea breaks in bad weather.

WALTON ON THE NAZE

MEAN SPRING ——————
AND NEAP CURVES - - - - -

MEAN RANGES
Springs 3.8m
Neaps 2.3m

9.4.18 WALTON BACKWATERS

Essex 51°54'·57N 01°16'·79E (No 2 PHM buoy) 🌊🌊💧💧💧🌸🌸

CHARTS AC 5607, 2052, 2695; Imray C1, Y16, 2000

TIDES +0030 Dover; ML 2·2; Duration 0615. Walton is a Standard Port. Predictions are for Walton Pier, ie to seaward. Time differences for Bramble Creek (N of Hamford Water) are +10, –7, –5, +10 mins; height differences are all +0.3m.

SHELTER Good in all weather, but ent not advised if a big sea is running from the NE. Good ⚓s in Hamford Water (clear of Oakley Creek) and in N end of Walton Chan, 2ca S of Stone Pt on E side.

NAVIGATION From S, appr via Medusa Chan; from N and E via the Harwich recomended yacht track (9.4.19). WPT Pye End SWM buoy, L Fl 10s, 51°55'·05N 01°17'·89E, 234°/1·1M to Crab Knoll SHM at chan ent. NB this stretch carries only 0·9m. Beware lobster pots off the Naze and Pye Sands and oyster beds in the Backwaters. Buoys may be moved to suit channel changes.

At narrow ent to Walton Chan leave NCM buoy to stbd, and 3 PHM buoys close to port. Fm No.9 SHM expect significant shoaling, ebb>4kn and strong back eddies ivo Stone Pt. Best water to E side, with narrowest/shallowest water at No 16 PHM; stay in channel. Beware tidal sheer and vessels with strong tide under them.

LIGHTS AND MARKS Naze Tr (49m) is conspic 3M S of Pye End buoy. 2M NNE at Felixstowe, cranes and Y flood lts are conspic D/N. No.9 SHM Fl G 5s, No.10 PHM Q R.

COMMUNICATIONS (Code 01255) CGOC 020 8312 7380; Police 101; Ⓗ 421145; HM 851899.

Titchmarsh Marina Ch 80, 0800-2000 in season.

FACILITIES
Titchmarsh Marina ☎672185, pontoons on R Twizzle. Access over sill 1·3m (tide gauge at entrance) berths dredged to approx 2m. ⚓ 420⛵ inc Ⓥ, £17 <8m, £19 <10m, £20 <12m, £22 <13m, £26m; short stay (<3 hrs) £5/hr. ⛽ LPG ⚓🔧🛠 ⛽📠(10t) 🚽(35t) ⚓🛠 × 🚿.
Walton & Frinton YC, ☎678161, ×🚿. **Yacht Basin** ☎675873; appr dries. ⚓🛥 60⛵ (long stay) 🛠🔧🔌🛠🔧🛠🚿🛠(½t) wi-fi.
Town 🏪🛒✉🍴× 🛒 ⇌ ✈ (Southend/Cambridge/Stansted).

STANDARD TIME (UT)
For Summer Time add ONE hour in **non-shaded areas**

WALTON-ON-THE-NAZE   LAT 51°51'N   LONG 1°17'E
TIMES AND HEIGHTS OF HIGH AND LOW WATERS

Dates in red are **SPRINGS**
Dates in blue are **NEAPS**

YEAR 2016

JANUARY

Time	m	Time	m
1 0345	3.8	**16** 0343	4.1
0956	0.7	1004	0.4
F 1616	3.7	SA 1615	4.0
2154	1.2	☽ 2216	0.9
2 0432	3.7	**17** 0434	4.0
1041	0.8	1102	0.5
SA 1708	3.5	SU 1714	3.8
☽ 2246	1.3	2320	1.0
3 0527	3.5	**18** 0535	3.9
1142	0.9	1207	0.6
SU 1806	3.4	M 1822	3.7
4 0001	1.4	**19** 0031	1.0
0628	3.4	0649	3.8
M 1254	1.0	TU 1317	0.7
1907	3.4	1937	3.7
5 0131	1.3	**20** 0145	0.9
0733	3.4	0808	3.8
TU 1359	1.0	W 1426	0.7
2010	3.5	2051	3.8
6 0237	1.2	**21** 0258	0.8
0836	3.6	0917	3.9
W 1457	0.9	TH 1533	0.8
2109	3.6	2154	3.9
7 0332	1.0	**22** 0406	0.7
0931	3.7	1015	4.0
TH 1548	0.8	F 1631	0.8
2200	3.8	2247	4.1
8 0419	0.8	**23** 0503	0.5
1019	3.9	1104	4.1
F 1631	0.8	SA 1719	0.8
2245	4.0	2332	4.2
9 0501	0.6	**24** 0551	0.4
1103	4.0	1149	4.2
SA 1710	0.7	SU 1801	0.7
2327	4.1	○	
10 0541	0.5	**25** 0014	4.2
1145	4.2	0632	0.3
SU 1747	0.7	M 1230	4.2
●		1838	0.7
11 0010	4.2	**26** 0053	4.3
0621	0.4	0710	0.3
M 1228	4.3	TU 1309	4.2
1825	0.6	1913	0.6
12 0052	4.3	**27** 0130	4.2
0702	0.3	0745	0.3
TU 1311	4.4	W 1347	4.1
1905	0.6	1945	0.8
13 0134	4.3	**28** 0204	4.2
0744	0.2	0816	0.4
W 1352	4.4	TH 1423	4.0
1948	0.6	2013	0.8
14 0216	4.3	**29** 0237	4.1
0828	0.3	0845	0.5
TH 1438	4.3	F 1458	3.9
2033	0.7	2042	0.9
15 0258	4.2	**30** 0311	4.0
0913	0.5	0915	0.6
F 1524	4.2	SA 1534	3.8
2122	0.8	2116	1.0
		31 0348	3.8
		0950	0.7
		SU 1613	3.6
		☽ 2157	1.1

FEBRUARY

Time	m	Time	m
1 0431	3.6	**16** 0511	3.9
1035	0.9	1137	0.7
M 1702	3.5	TU 1755	3.6
☽ 2249	1.2		
2 0526	3.5	**17** 0003	0.9
1137	1.0	0626	3.7
TU 1804	3.4	W 1250	0.9
		1914	3.5
3 0008	1.3	**18** 0124	0.9
0635	3.4	0751	3.6
W 1306	1.1	TH 1409	1.0
1913	3.4	2035	3.6
4 0146	1.2	**19** 0250	0.8
0749	3.4	0906	3.8
TH 1417	1.1	F 1525	0.9
2024	3.5	2143	3.8
5 0253	1.0	**20** 0404	0.7
0808	3.8	1007	3.9
F 1516	0.9	SA 1623	0.9
2128	3.7	2235	4.0
6 0349	0.8	**21** 0457	0.5
0954	3.8	1055	4.0
SA 1606	0.8	SU 1707	0.8
2221	3.9	2319	4.1
7 0437	0.6	**22** 0539	0.4
1042	4.0	1136	4.1
SU 1650	0.7	M 1744	0.7
2308	4.1	○ 2358	4.2
8 0522	0.4	**23** 0614	0.3
1127	4.2	1213	4.2
M 1731	0.6	TU 1818	0.7
● 2352	4.3		
9 0605	0.3	**24** 0033	4.2
1211	4.4	0646	0.3
TU 1812	0.5	W 1249	4.2
		1850	0.7
10 0035	4.4	**25** 0106	4.2
0647	0.2	0716	0.3
W 1256	4.5	TH 1323	4.1
1853	0.5	1921	0.7
11 0118	4.5	**26** 0137	4.2
0730	0.1	0745	0.3
TH 1339	4.5	F 1355	4.1
1935	0.5	1948	0.7
12 0200	4.5	**27** 0208	4.1
0812	0.1	0811	0.4
F 1423	4.4	SA 1426	4.0
2019	0.5	2015	0.7
13 0242	4.4	**28** 0239	4.0
0854	0.2	0838	0.5
SA 1508	4.3	SU 1457	3.9
2104	0.6	2046	0.8
14 0325	4.3	**29** 0312	3.9
0940	0.3	0908	0.6
SU 1556	4.0	M 1531	3.7
2155	0.7	2122	0.9
15 0413	4.1		
1033	0.5		
M 1650	3.8		
☽ 2253	0.8		

MARCH

Time	m	Time	m
1 0348	3.7	**16** 0451	3.9
0945	0.8	1108	0.9
TU 1612	3.6	W 1731	3.6
☽ 2205	1.0	2340	0.8
2 0433	3.6	**17** 0607	3.6
1035	1.0	1223	1.0
W 1705	3.4	TH 1850	3.5
2305	1.2		
3 0535	3.4	**18** 0105	0.9
1154	1.2	0733	3.6
TH 1818	3.3	F 1330	1.1
		2014	3.5
4 0049	1.4	**19** 0239	0.8
0658	3.3	0851	3.7
F 1336	1.2	SA 1510	1.0
1939	3.4	2123	3.7
5 0215	1.0	**20** 0351	0.6
0822	3.5	0951	3.9
SA 1444	1.0	SU 1607	0.9
2054	3.6	2215	3.9
6 0317	0.8	**21** 0440	0.5
0927	3.7	1038	4.0
SU 1539	0.8	M 1648	0.8
2154	3.8	2258	4.0
7 0411	0.5	**22** 0517	0.4
1019	4.0	1116	4.1
M 1626	0.7	TU 1722	0.7
2244	4.1	2334	4.1
8 0459	0.3	**23** 0548	0.4
1106	4.2	1151	4.1
TU 1711	0.6	W 1754	0.7
2330	4.3	○	
9 0544	0.2	**24** 0007	4.2
1151	4.4	0616	0.3
W 1754	0.5	TH 1224	4.1
●		1826	0.6
10 0013	4.5	**25** 0039	4.2
0627	0.1	0645	0.3
TH 1236	4.5	F 1256	4.1
1836	0.4	1856	0.6
11 0057	4.6	**26** 0109	4.2
0709	0.0	0713	0.4
F 1320	4.5	SA 1326	4.1
1920	0.4	1924	0.6
12 0140	4.6	**27** 0140	4.2
0751	0.1	0739	0.4
SA 1404	4.4	SU 1356	4.0
2003	0.4	1952	0.6
13 0222	4.6	**28** 0211	4.1
0831	0.2	0805	0.5
SU 1449	4.3	M 1427	3.9
2047	0.5	2021	0.7
14 0306	4.4	**29** 0243	4.0
0915	0.4	0834	0.6
M 1535	4.0	TU 1500	3.8
2136	0.6	2055	0.7
15 0353	4.2	**30** 0318	3.8
1005	0.6	0909	0.8
TU 1628	3.8	W 1539	3.7
☽ 2231	0.7	2135	0.8
		31 0401	3.7
		0955	1.0
		TH 1628	3.5
		☽ 2229	1.0

APRIL

Time	m	Time	m
1 0457	3.5	**16** 0041	0.8
1105	1.1	0706	3.6
F 1735	3.4	SA 1319	1.2
2357	1.1	1942	3.5
2 0612	3.4	**17** 0211	0.8
1250	1.2	0823	3.7
SA 1856	3.4	SU 1440	1.1
		2052	3.6
3 0136	0.9	**18** 0322	0.7
0743	3.5	0924	3.8
SU 1409	1.0	M 1538	1.0
2018	3.5	2146	3.8
4 0244	0.7	**19** 0411	0.6
0856	3.8	1011	3.9
M 1508	0.8	TU 1621	0.9
2123	3.8	2229	3.9
5 0341	0.4	**20** 0447	0.5
0953	4.0	1049	4.0
TU 1559	0.7	W 1656	0.9
2216	4.1	2305	4.0
6 0431	0.3	**21** 0516	0.5
1042	4.3	1124	4.0
W 1647	0.5	TH 1729	0.7
2304	4.3	2338	4.1
7 0519	0.1	**22** 0545	0.4
1128	4.4	1156	4.1
TH 1733	0.4	F 1801	0.6
● 2349	4.5	○	
8 0604	0.1	**23** 0009	4.2
1213	4.5	0614	0.4
F 1818	0.4	SA 1228	4.1
		1832	0.5
9 0033	4.6	**24** 0042	4.2
0647	0.1	0643	0.4
SA 1259	4.5	SU 1259	4.1
1903	0.3	1902	0.5
10 0118	4.7	**25** 0114	4.2
0729	0.1	0710	0.5
SU 1344	4.4	M 1331	4.1
1948	0.3	1932	0.6
11 0202	4.6	**26** 0147	4.1
0810	0.3	0737	0.6
M 1429	4.3	TU 1403	4.0
2033	0.4	2003	0.6
12 0247	4.4	**27** 0221	4.0
0853	0.5	0808	0.7
TU 1516	4.1	W 1438	3.9
2120	0.5	2037	0.6
13 0335	4.2	**28** 0258	3.9
0941	0.7	0846	0.8
W 1607	3.8	TH 1517	3.8
2214	0.6	2117	0.7
14 0433	3.9	**29** 0341	3.8
1040	1.0	0933	0.9
TH 1708	3.6	F 1606	3.6
☽ 2318	0.7	2210	0.8
15 0546	3.7	**30** 0435	3.7
1153	1.1	1040	1.0
F 1822	3.5	SA 1707	3.5
		☽ 2330	0.9

Chart Datum: 2·16 metres below Ordnance Datum (Newlyn). HAT is 4·7 metres above Chart Datum.

E England

WALTON-ON-THE-NAZE
LAT 51°51′N   LONG 1°17′E

TIMES AND HEIGHTS OF HIGH AND LOW WATERS

STANDARD TIME (UT)
For Summer Time add ONE hour in **non-shaded areas**

Dates in red are **SPRINGS**
Dates in blue are **NEAPS**

YEAR 2016

MAY

Day	Time m	Time m	Time m	Time m
1 SU	0544 3.6	1211 1.1	1822 3.5	
16 M	0123 0.8	0740 3.6	1351 1.2	2007 3.6
2 M	0101 0.8	0707 3.6	1332 1.0	1941 3.6
17 TU	0233 0.8	0844 3.7	1456 1.1	2105 3.7
3 TU	0211 0.6	0822 3.8	1435 0.8	2050 3.8
18 W	0327 0.7	0935 3.8	1546 0.9	2151 3.8
4 W	0310 0.4	0924 4.1	1531 0.7	2147 4.1
19 TH	0407 0.6	1017 3.9	1626 0.8	2231 3.9
5 TH	0403 0.3	1017 4.2	1622 0.5	2238 4.3
20 F	0441 0.6	1135 4.0	1702 0.7	2306 4.0
6 F ●	0453 0.2	1105 4.4	1712 0.4	2325 4.5
21 SA ○	0514 0.6	1128 4.0	1737 0.6	2341 4.1
7 SA	0540 0.2	1152 4.4	1801 0.3	
22 SU	0547 0.5	1202 4.1	1811 0.5	
8 SU	0011 4.6	0626 0.2	1239 4.5	1849 0.3
23 M	0016 4.2	0617 0.5	1236 4.1	1843 0.5
9 M	0057 4.6	0710 0.3	1325 4.4	1936 0.3
24 TU	0052 4.2	0647 0.6	1311 4.1	1916 0.5
10 TU	0143 4.6	0753 0.5	1411 4.3	2022 0.3
25 W	0128 4.1	0718 0.6	1347 4.0	1951 0.5
11 W	0230 4.4	0835 0.6	1458 4.1	2108 0.4
26 TH	0204 4.1	0753 0.7	1424 4.0	2028 0.6
12 TH	0319 4.2	0921 0.8	1548 3.9	2158 0.5
27 F	0244 4.0	0834 0.8	1504 3.9	2110 0.6
13 F ◑	0415 3.9	1012 1.0	1644 3.7	2255 0.7
28 SA	0328 3.9	0923 0.9	1551 3.8	2203 0.7
14 SA	0519 3.7	1115 1.2	1748 3.6	
29 SU ◐	0420 3.8	1024 1.0	1648 3.7	2312 0.7
15 SU	0006 0.8	0629 3.6	1234 1.2	1857 3.5
30 M	0523 3.8	1139 1.0	1754 3.7	
31 TU	0029 0.6	0637 3.8	1256 1.0	1907 3.8

JUNE

Day	Time m	Time m	Time m	Time m
1 W	0139 0.5	0750 3.9	1403 0.9	2018 3.9
16 TH	0230 0.8	0849 3.7	1503 1.1	2107 3.7
2 TH	0240 0.4	0856 4.0	1503 0.7	2120 4.1
17 F	0322 0.8	0939 3.8	1553 0.9	2154 3.8
3 F	0336 0.4	0954 4.2	1600 0.6	2214 4.3
18 SA	0407 0.7	1022 3.9	1636 0.8	2236 4.0
4 SA	0430 0.3	1046 4.3	1654 0.5	2304 4.4
19 SU	0448 0.7	1101 4.0	1715 0.7	2316 4.1
5 SU ●	0521 0.3	1135 4.4	1747 0.3	2352 4.5
20 M ○	0524 0.7	1139 4.1	1751 0.6	2355 4.1
6 M	0609 0.4	1222 4.4	1838 0.3	
21 TU	0557 0.7	1217 4.1	1827 0.5	
7 TU	0040 4.6	0655 0.5	1309 4.4	1926 0.3
22 W	0033 4.2	0630 0.6	1256 4.2	1903 0.5
8 W	0127 4.5	0738 0.6	1355 4.3	2012 0.3
23 TH	0113 4.2	0705 0.6	1335 4.2	1942 0.4
9 TH	0214 4.4	0820 0.7	1441 4.2	2056 0.4
24 F	0152 4.2	0743 0.7	1414 4.1	2022 0.4
10 F	0302 4.2	0900 0.9	1526 4.0	2140 0.5
25 SA	0233 4.1	0826 0.7	1455 4.0	2105 0.5
11 SA	0352 3.9	0943 1.0	1615 3.9	2227 0.6
26 SU	0317 4.1	0913 0.8	1539 4.0	2154 0.5
12 SU ◐	0447 3.8	1033 1.1	1708 3.7	2322 0.7
27 M ◐	0406 4.0	1008 0.9	1629 3.9	2253 0.6
13 M	0545 3.6	1137 1.2	1806 3.6	
28 TU	0503 3.9	1112 1.0	1728 3.9	
14 TU	0025 0.8	0646 3.6	1254 1.3	1908 3.6
29 W	0000 0.6	0610 3.8	1223 1.0	1836 3.9
15 W	0130 0.8	0749 3.6	1403 1.2	2011 3.6
30 TH	0109 0.6	0722 3.8	1334 0.9	1949 3.9

JULY

Day	Time m	Time m	Time m	Time m
1 F	0213 0.6	0833 3.9	1440 0.8	2058 4.0
16 SA	0241 1.0	0855 3.6	1517 1.0	2118 3.7
2 SA	0315 0.6	0937 4.0	1543 0.7	2157 4.2
17 SU	0336 0.9	0949 3.8	1609 0.8	2208 3.9
3 SU	0413 0.6	1033 4.2	1643 0.5	2251 4.3
18 M	0424 0.8	1035 4.0	1652 0.7	2252 4.0
4 M ●	0507 0.5	1123 4.3	1738 0.4	2340 4.4
19 TU ○	0503 0.8	1117 4.1	1732 0.6	2334 4.1
5 TU	0556 0.6	1210 4.3	1829 0.3	
20 W	0539 0.7	1158 4.2	1810 0.5	
6 W	0026 4.4	0641 0.6	1255 4.3	1915 0.3
21 TH	0015 4.2	0615 0.7	1240 4.3	1849 0.4
7 TH	0113 4.4	0723 0.7	1339 4.3	1958 0.3
22 F	0057 4.2	0652 0.6	1320 4.3	1929 0.3
8 F	0157 4.3	0802 0.8	1421 4.2	2038 0.4
23 SA	0138 4.3	0732 0.6	1401 4.3	2010 0.3
9 SA	0241 4.2	0837 0.9	1501 4.1	2114 0.5
24 SU	0220 4.3	0814 0.7	1441 4.3	2052 0.4
10 SU	0324 4.0	0912 1.0	1542 4.0	2151 0.6
25 M	0303 4.2	0859 0.8	1522 4.2	2137 0.4
11 M	0409 3.8	0950 1.1	1626 3.8	2233 0.7
26 TU ◐	0349 4.1	0949 0.9	1609 4.1	2229 0.5
12 TU ◐	0458 3.7	1038 1.2	1716 3.7	2327 0.8
27 W	0442 3.9	1047 0.9	1703 4.0	2332 0.6
13 W	0552 3.5	1143 1.3	1813 3.6	
28 TH	0545 3.8	1155 1.0	1810 3.9	
14 TH	0032 1.0	0650 3.5	1306 1.3	1915 3.5
29 F	0041 0.7	0657 3.7	1309 1.0	1927 3.8
15 F	0139 1.0	0753 3.5	1416 1.2	2019 3.6
30 SA	0151 0.8	0814 3.8	1422 0.9	2043 3.9
31 SU	0300 0.8	0925 3.9	1534 0.7	2147 4.1

AUGUST

Day	Time m	Time m	Time m	Time m
1 M	0404 0.8	1023 4.1	1639 0.6	2242 4.2
16 TU	0358 0.9	1008 3.9	1626 0.7	2229 4.0
2 TU ●	0458 0.7	1113 4.2	1732 0.4	2330 4.3
17 W	0441 0.8	1053 4.1	1709 0.6	2313 4.2
3 W	0544 0.7	1158 4.3	1818 0.3	
18 TH ○	0519 0.7	1136 4.2	1749 0.4	2345 4.3
4 TH	0014 4.4	0625 0.7	1240 4.3	1859 0.3
19 F	0557 0.7	1218 4.4	1830 0.3	
5 F	0056 4.3	0703 0.7	1319 4.3	1937 0.3
20 SA	0037 4.4	0636 0.6	1259 4.5	1910 0.3
6 SA	0136 4.3	0739 0.8	1356 4.3	2010 0.4
21 SU	0119 4.5	0716 0.6	1340 4.5	1950 0.3
7 SU	0214 4.2	0811 0.8	1431 4.2	2041 0.5
22 M	0202 4.4	0758 0.6	1421 4.5	2031 0.3
8 M	0251 4.0	0841 0.9	1505 4.1	2111 0.6
23 TU	0244 4.3	0842 0.7	1502 4.4	2113 0.4
9 TU	0328 3.9	0912 1.0	1542 3.9	2144 0.7
24 W	0329 4.1	0929 0.8	1547 4.2	2203 0.6
10 W ◑	0408 3.7	0950 1.1	1625 3.8	2226 0.9
25 TH ◑	0420 3.9	1025 0.9	1640 4.1	2304 0.8
11 TH	0455 3.6	1040 1.2	1718 3.6	2326 1.1
26 F	0521 3.7	1131 0.9	1748 3.9	
12 F	0553 3.4	1154 1.3	1823 3.5	
27 SA	0017 0.9	0636 3.6	1249 1.0	1912 3.8
13 SA	0049 1.2	0659 3.4	1327 1.3	1935 3.5
28 SU	0135 1.0	0800 3.6	1412 0.9	2033 3.9
14 SU	0203 1.2	0810 3.5	1438 1.1	2043 3.6
29 M	0252 1.0	0914 3.8	1532 0.7	2139 4.0
15 M	0306 1.0	0915 3.7	1537 0.9	2140 3.8
30 TU	0358 0.9	1012 4.0	1633 0.6	2233 4.2
31 W	0448 0.9	1100 4.2	1721 0.4	2318 4.3

Chart Datum: 2·16 metres below Ordnance Datum (Newlyn). HAT is 4·7 metres above Chart Datum.

WALTON-ON-THE-NAZE   LAT 51°51'N   LONG 1°17'E

TIMES AND HEIGHTS OF HIGH AND LOW WATERS

STANDARD TIME (UT)
For Summer Time add ONE hour in **non-shaded areas**

Dates in red are **SPRINGS**
Dates in blue are NEAPS

YEAR 2016

SEPTEMBER

Time	m		Time	m
1 0529	0.8	**16** 0454	0.7
1141	4.3	1109	4.3
TH 1800	0.4	F 1723	0.4
● 2357	4.3	○ 2330	4.4
2 0605	0.7	**17** 0535	0.6
1218	4.3	1151	4.4
F 1835	0.4	SA 1805	0.3
3 0034	4.3	**18** 0013	4.5
0640	0.7	0615	0.6
SA 1253	4.3	SU 1234	4.6
1907	0.4	1846	0.2
4 0110	4.3	**19** 0057	4.6
0713	0.7	0657	0.5
SU 1326	4.3	M 1316	4.6
1937	0.4	1926	0.2
5 0144	4.2	**20** 0140	4.5
0744	0.8	0740	0.5
M 1358	4.3	TU 1358	4.6
2005	0.5	2006	0.3
6 0216	4.1	**21** 0223	4.4
0811	0.8	0824	0.6
TU 1430	4.1	W 1441	4.5
2031	0.6	2048	0.5
7 0248	3.9	**22** 0308	4.2
0840	0.9	0911	0.7
W 1503	4.0	TH 1526	4.3
2059	0.8	2137	0.7
8 0320	3.8	**23** 0358	3.9
0914	1.0	1006	0.8
TH 1539	3.8	F 1620	4.1
2133	0.9	◓ 2238	1.0
9 0359	3.6	**24** 0459	3.7
0955	1.1	1112	0.9
F 1622	3.6	SA 1731	3.8
◐ 2220	1.2	2353	1.1
10 0449	3.5	**25** 0616	3.6
1052	1.2	1233	0.9
SA 1722	3.5	SU 1858	3.7
2336	1.3		
11 0601	3.4	**26** 0117	1.2
1229	1.3	0741	3.6
SU 1845	3.4	M 1401	0.9
		2019	3.9
12 0121	1.3	**27** 0238	1.1
0721	3.4	0856	3.8
M 1357	1.2	TU 1520	0.7
2006	3.5	2125	4.0
13 0231	1.2	**28** 0343	1.0
0836	3.6	0953	4.0
TU 1501	0.9	W 1617	0.6
2110	3.8	2217	4.2
14 0327	1.0	**29** 0430	0.9
0936	3.8	1039	4.1
W 1554	0.7	TH 1700	0.5
2201	4.0	2258	4.2
15 0412	0.8	**30** 0507	0.8
1025	4.1	1118	4.2
TH 1640	0.5	F 1735	0.5
2247	4.3	2335	4.3

OCTOBER

Time	m		Time	m
1 0541	0.8	**16** 0510	0.6
1152	4.3	1123	4.5
SA 1805	0.4	SU 1737	0.2
●		○ 2348	4.6
2 0008	4.3	**17** 0554	0.5
0614	0.7	1207	4.7
SU 1224	4.3	M 1820	0.2
1834	0.5		
3 0041	4.3	**18** 0033	4.6
0647	0.7	0639	0.4
M 1255	4.3	TU 1252	4.7
1903	0.6	1902	0.3
4 0113	4.2	**19** 0118	4.6
0717	0.7	0724	0.4
TU 1326	4.3	W 1336	4.7
1930	0.6	1944	0.4
5 0143	4.1	**20** 0203	4.4
0745	0.7	0810	0.5
W 1358	4.2	TH 1421	4.5
1954	0.7	2027	0.5
6 0212	4.0	**21** 0249	4.2
0813	0.8	0858	0.5
TH 1430	4.0	F 1509	4.3
2020	0.8	2116	0.9
7 0244	3.9	**22** 0339	3.9
0844	0.9	0951	0.6
F 1504	3.9	SA 1604	4.0
2053	1.0	◐ 2214	1.1
8 0320	3.8	**23** 0439	3.7
0922	1.0	1055	0.8
SA 1544	3.7	SU 1715	3.8
2135	1.1	2326	1.3
9 0405	3.6	**24** 0552	3.6
1011	1.1	1213	0.8
SU 1635	3.6	M 1836	3.7
◐ 2235	1.3		
10 0506	3.4	**25** 0049	1.3
1130	1.2	0713	3.6
M 1747	3.5	TU 1337	0.8
		1954	3.8
11 0022	1.4	**26** 0209	1.3
0628	3.4	0826	3.7
TU 1313	1.1	W 1453	0.7
1920	3.5	2100	4.0
12 0149	1.3	**27** 0316	1.1
0751	3.5	0925	3.9
W 1422	0.9	TH 1549	0.6
2032	3.8	2151	4.1
13 0250	1.0	**28** 0404	1.0
0858	3.8	1011	4.0
TH 1518	0.6	F 1631	0.6
2129	4.1	2232	4.1
14 0340	0.8	**29** 0442	0.9
0952	4.1	1049	4.1
F 1607	0.4	SA 1703	0.6
2218	4.3	2308	4.2
15 0425	0.7	**30** 0516	0.8
1039	4.3	1123	4.2
SA 1653	0.3	SU 1732	0.6
2303	4.5	● 2341	4.2
		31 0549	0.7
		1154	4.2
		M 1802	0.5

NOVEMBER

Time	m		Time	m
1 0012	4.2	**16** 0012	4.6
0622	0.6	0623	0.4
TU 1226	4.3	W 1231	4.7
1831	0.6	1843	0.4
2 0043	4.2	**17** 0059	4.5
0653	0.6	0712	0.3
W 1258	4.2	TH 1317	4.7
1858	0.6	1928	0.5
3 0114	4.2	**18** 0145	4.4
0722	0.7	0759	0.4
TH 1331	4.2	F 1405	4.5
1924	0.7	2012	0.7
4 0145	4.1	**19** 0232	4.2
0751	0.7	0848	0.4
F 1404	4.1	SA 1454	4.3
1952	0.8	2059	0.9
5 0218	4.0	**20** 0322	4.0
0823	0.7	0938	0.5
SA 1439	3.9	SU 1548	4.1
2026	0.9	2150	1.1
6 0254	3.9	**21** 0417	3.8
0900	0.8	1035	0.6
SU 1519	3.8	M 1652	3.8
2108	1.1	◑ 2251	1.3
7 0338	3.7	**22** 0521	3.7
0947	0.9	1143	0.7
M 1609	3.7	TU 1802	3.7
◑ 2204	1.2		
8 0433	3.6	**23** 0006	1.4
1054	1.0	0632	3.6
TU 1711	3.6	W 1256	0.8
2327	1.3	1914	3.7
9 0544	3.5	**24** 0124	1.3
1226	0.9	0744	3.6
W 1831	3.6	TH 1407	0.8
		2022	3.8
10 0059	1.3	**25** 0234	1.2
0704	3.6	0846	3.7
TH 1341	0.8	F 1508	0.7
1950	3.8	2117	3.9
11 0209	1.1	**26** 0330	1.1
0817	3.8	0936	3.9
F 1441	0.6	SA 1553	0.7
2054	4.1	2201	4.0
12 0305	0.9	**27** 0413	0.9
0917	4.0	1017	4.0
SA 1534	0.4	SU 1628	0.7
2148	4.3	2239	4.1
13 0357	0.7	**28** 0450	0.8
1009	4.3	1052	4.0
SU 1624	0.3	M 1701	0.7
2238	4.4	2314	4.1
14 0446	0.6	**29** 0526	0.7
1057	4.5	1126	4.1
M 1712	0.3	TU 1734	0.6
○ 2325	4.5	● 2347	4.2
15 0535	0.4	**30** 0600	0.6
1144	4.6	1201	4.2
TU 1758	0.3	W 1806	0.7

DECEMBER

Time	m		Time	m
1 0019	4.2	**16** 0044	4.5
0632	0.6	0702	0.3
TH 1236	4.2	F 1303	4.6
1834	0.7	1916	0.6
2 0053	4.2	**17** 0131	4.4
0703	0.6	0751	0.3
F 1310	4.2	SA 1351	4.5
1902	0.7	2000	0.7
3 0126	4.1	**18** 0217	4.3
0736	0.6	0837	0.3
SA 1345	4.1	SU 1439	4.3
1934	0.8	2042	0.9
4 0201	4.0	**19** 0303	4.1
0810	0.6	0922	0.4
SU 1422	4.0	M 1529	4.1
2010	0.9	2125	1.0
5 0239	4.0	**20** 0351	3.9
0849	0.6	1009	0.5
M 1503	3.9	TU 1622	3.9
2053	1.0	2212	1.2
6 0321	3.9	**21** 0444	3.8
0934	0.7	1101	0.7
TU 1550	3.9	W 1720	3.7
2146	1.1	◑ 2310	1.3
7 0411	3.8	**22** 0542	3.6
1031	0.8	1201	0.8
W 1646	3.8	TH 1822	3.6
◑ 2253	1.2		
8 0511	3.7	**23** 0024	1.4
1145	0.8	0645	3.5
TH 1754	3.8	F 1305	0.9
		1927	3.6
9 0012	1.2	**24** 0138	1.3
0622	3.7	0752	3.5
F 1300	0.7	SA 1407	0.9
1909	3.8	2031	3.6
10 0128	1.1	**25** 0244	1.2
0737	3.8	0852	3.6
SA 1405	0.6	SU 1504	0.9
2020	4.0	2124	3.8
11 0233	0.9	**26** 0339	1.0
0845	4.0	0941	3.8
SU 1504	0.5	M 1553	0.8
2122	4.1	2209	3.9
12 0331	0.7	**27** 0425	0.8
0944	4.2	1023	3.9
M 1559	0.4	TU 1635	0.8
2217	4.3	2248	4.0
13 0426	0.6	**28** 0504	0.7
1036	4.4	1102	4.0
TU 1651	0.4	W 1712	0.7
2308	4.4	2324	4.1
14 0520	0.5	**29** 0540	0.6
1126	4.5	1140	4.1
W 1742	0.4	TH 1746	0.7
○ 2356	4.4	●	
15 0612	0.3	**30** 0001	4.1
1214	4.6	0613	0.5
TH 1830	0.5	F 1218	4.1
		1816	0.7
		31 0037	4.2
		0647	0.5
		SA 1254	4.2
		1848	0.7

Chart Datum: 2·16 metres below Ordnance Datum (Newlyn). HAT is 4·7 metres above Chart Datum.

E England

》》 FREE monthly updates. Register at 《
www.reedsnauticalalmanac.co.uk
313

9.4.19 & 20   RIVERS STOUR AND ORWELL

9.4.19 RIVER STOUR

Essex/Suffolk **51°57'·06N 01°17'·77E** (Guard PHM buoy)

❀❀❀♨♨♨❀❀❀

CHARTS AC 5607, 2052, 2693, 1491, 1594; Imray C29, C1, Y16, 2000

TIDES Harwich +0050 Dover; ML 2·1; Duration 0630

Standard Port HARWICH (→)

Times				Height (metres)			
High Water		Low Water		MHWS	MHWN	MLWN	MLWS
0000	0600	0000	0600	4·0	3·4	1·1	0·4
1200	1800	1200	1800				
Differences MISTLEY							
+0025	+0015	+0005	0000	+0·2	+0·1	−0·1	−0·1

SHELTER Good at Shotley Marina; access (H24) via chan dredged 2m, outer limits lit, to lock. Enter only on F.G tfc lt. ☒ also at Halfpenny Pier (H24) and Manningtree (dries). No yachts at Parkeston Quay. ⚓s off Wrabness Pt (untenable in some wind against tide conditions), Holbrook Creek and Stutton Ness.

NAVIGATION WPT 51°55'·33N 01°25'·50E, on recommended yacht track, S of Cork Sand Bn (see also River Orwell page).

- Keep clear of commercial shipping and new container berth at Felixstowe. Outside the hbr, yachts should cross the DW chan at 90° between Rolling Ground and Platters buoys. Stay clear of the DW chan (Y lit SPMs) using Recommended Yacht Track, running S and W of DW chan to beyond Harwich.

- **Caution:** Bkwtr, ESE of Blackman's Hd, covers at HW; marked by PHM lit Bn, only 5ca W of main chan. Beware of passing W of Harwich Shelf ECM buoy without sufficient height of tide. The Guard Shoal (0·8m), about 2ca S of Guard PHM, lies close to the Recommended Yacht Track. R Stour is well marked; sp limit 8kn. Beware 'The Horse' 4ca NE and drying bank 1½ca NW of Wrabness Pt. Mistley Quay local knowledge advised for the narrow, tortuous chan to Manningtree.

Special local sound signals
Commercial vessels may use these additional signals:

Four short and rapid blasts followed by one short blast	} = I am turning short around to stbd.
Four short and rapid blasts followed by two short blasts	} = I am turning short around to port.
One prolonged blast	= I am leaving dock, quay or ⚓.

LIGHTS AND MARKS The R Stour to Mistley is lit. Shotley Marina: a Dir lt at lock indicates the dredged chan (2·0m) by Inogen (or Moiré) visual marker lt which is a square, ambered display; a vert B line indicates on the appr centre line 339°. If off the centre line, arrows indicate the direction to steer to regain it.

COMMUNICATIONS (Code 01255) NMOC 02392 552100; Police 101; Dr 201299; Ⓗ 201200. Harwich HM 243030; Hbr Ops 243000.

Harwich VTS Ch **71** 11 20 (H24). Yachts should monitor Ch 71 for traffic info, but not transmit. ⚓, tidal and (possibly) help in poor vis may be available on request. The Hbr Patrol launch listens on Ch 11 (summer weekends, 0800–1800). Hbr Radar Ch 20. Shotley Marina Ch **80** (lock master). Monitor Ch 14 for an information service by Dover CG for the Sunk TSS.

FACILITIES HARWICH Halfpenny Pier Available Apr–Oct, no reservations. Free 0900-1600; overnight: £1·00/m<20m>£30, max 72hrs, ⚓ ⚓.

Town ⚓ ⚓ ⚓ ⚓ ⚓ ⚓ ⚓ ⚓ ⚓ ⚓ ⚓ ⚓ ⚓. **Ferries:** Hook of Holland: 2/day; 6¼ Hrs; Stena (www.stenaline.co.uk).

SHOTLEY (01473) **Shotley Marina** www.shotleymarina.com ☎ 788982 (H24), access H24 via lock (width 7m). 350⚓, Ⓥs welcome £2·50, ⚓ ⚓ ⚓ (H24) ⚓ ⚓ ⚓ Ⓔ ⚓ ⚓ ⚓(40t) ⚓ ⚓ ⚓ ⚓ ⚓ Ferries to Harwich.

Shotley SC ☎787500, ⚓ ⚓ ⚓.

WRABNESS: ⚓ ⚓ ⚓.

MANNINGTREE, ☒ (drying), ⚓. Stour SC ☎(01206) 393924, M (drying but free), ⚓ ⚓ ⚓. **Town** ⚓ ⚓.

E England

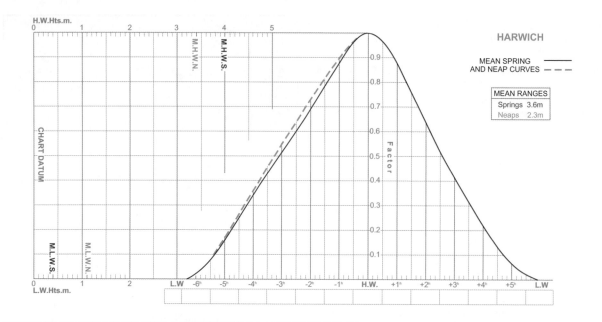

9.4.20 RIVER ORWELL ❀❀❀❀❀❀❀❀❀

Suffolk 51°57'·06N 01°17'·77E (Guard PHM by)

CHARTS AC 5607, 2052, 2693, 1491; Imray C29, C1, Y16, 2000

TIDES Pin Mill +0100 Dover; Ipswich +0115 Dover; ML 2·4; Duration 0555

Standard Port HARWICH (→)

Times				Height (metres)			
High Water		Low Water		MHWS	MHWN	MLWN	MLWS
0000	0600	0000	0600	4·0	3·4	1·1	0·4
1200	1800	1200	1800				
Differences IPSWICH							
+0015	+0025	0000	+0010	+0·2	0·0	−0·1	−0·1

SHELTER Good. Ent and river well marked, but many unlit moorings line both banks. ♩s above Shotley Pt on W side, or off Pinmill. No yacht facilities at Felixstowe. ♥'s (pre-book) at Suffolk Yacht Hbr, Woolverstone Marina, Fox's Marina (Ipswich), Neptune Marina, Ipswich Haven Marina, the latter two accessed via Prince Philip Lock; *call Ch 68 at Orwell Bridge.*

NAVIGATION Approach/entry from seaward as for R Stour. WPT Shotley Spit SCM buoy, Q (6)+L Fl 15s, 51°57'·21N 01°17'·68E, at river ent. A *Yachting Guide to Harwich Harbour and its Rivers* has much useful info; it can be obtained free from Harwich Haven Authority, The Quay, Harwich CO12 3HH; ☎(01255) 243030.

- Keep clear of the many large ships, ferries from/to Harwich, Felixstowe and Ipswich, especially container ships turning between Trinity container terminal, Shotley Spit and Guard buoys.

- Major harbour works planned for 2015 to cater for larger container ships. Advise referral to NM. Ch 71 should be monitored whilst in Harwich Harbour but only used in emergency. Call sign Harwich VTS. Harwich HM 01255 243030. Refer also to 9.4.19 River Stour.

- 6kn speed limit in R Orwell. Ipswich Dock ent closed when tide >4.6m above CD. Ipswich New Cut entrance, *Velocity Control Structure is raised from seabed when required with the top of it just below sea level;* 3 FR (vert) = no passage.

LIGHTS AND MARKS Suffolk Yacht Hbr appr marked by SWM then four bns and ldg lts: front Iso Y; rear Oc Y 4s.

A14 bridge lts: Centre FY (clearance 38m)
No 9 Pier 2 FR (vert) } shown up and
No 10 Pier 2 FG (vert) } downstream.
● and ● tfc lts control ent to Ipswich Dock (H24).
Ipswich New Cut: 3 FR (vert) when closed to navigation.

COMMUNICATIONS (Code 01473) NMOC 02392 552100; Police 101; ℍ 712233; Orwell Navigation Service 211066, Ipswich HM 211771, Ipswich Lock Control 213526.

Ipswich Port Radio and *Ipswich Lock Control* VHF Ch 68 (H24). Once above Shotley Pt or anywhere in the area, monitor Ch 68 continuously in fog. Suffolk Yacht Hbr, Woolverstone Marina, Fox's Marina: Ch 80 M. Neptune Marina: Ch ⚓ 80 (0800-1730LT), Ipswich Haven Marina Ch M 80.

FACILITIES
LEVINGTON Suffolk Yacht Hbr (SYH) info@syharbour.co.uk ☎659240, Ch 80. Access H24. ⚓ (launch £25) 550♨+♥ £2·64, ♿ ⚑ ▣ ♠ ♦(0815-1750) LPG ♠ ⚒ ⚓ (E) ⚠ ♦(10/60t) ♦(15t) ⚙ ☕.
Haven Ports YC ☎659658, × ▢.
Town ✉ Ⓑ (Felixstowe), ⇌ (bus to Ipswich), ✈ (Cambridge).

PIN MILL King's Boatyard ☎780258; M (£10 proceeds to RNLI, 07714 260568), ⚓ ⚒ ⚓ ▣ ♠ LPG ⚒ ⚓ (E) ⚠ ♦(6t) ☕(1M).
Haven Ports YC ☎659658 × ▢.
Pin Mill SC ☎780271.

WOOLVERSTONE Marina ☎780206 (H24), Mob 07803 968209.
235♨, ♥ £2·88 inc ⚡⊳, ♿ ⚓ ▣ ♦(HO) ♠ ⚓ (E) ⚠ ♨ ♦(20t) ⚙ ☕ ×.
Royal Harwich YC www.royalharwichyachtclub.co.uk ☎780319 ♨ £2·11, × ▢.
Town Chelmondiston: ✉ ☕.

IPSWICH, Fox's Marina ⚓ ☎ 689111, Ch 80. 75♨+♥ £2·59, ♿ ♠ ♦(0800-1730) ♠ ⚒ ⚓ (E) ⚠ ♨ ♦(44/70t) ♦(7t) ⚙ ACA ☕ wi-fi.
Marina YC ☎684111 ▢ ×.
Orwell YC ☎602288 ⚓ ⚓ ☕ ▢

Prince Philip Lock near city centre gives H24 access to:
Neptune Marina ☎215204, 150♨+♥ £2·64, ♦(0800-1700).
Ipswich Haven Marina ☎236644 ⚓ 320♨+♥ £2·47, ♦(0800-1900) ♠ ▣ ⚒ (E) ♨ ♦(70t) ♦ × ▢ wi-fi.
Town ⊞ ⊠ ✉ Ⓑ ⇌ ✈ (Cambridge/Norwich).

STANDARD TIME (UT)
For Summer Time add ONE hour in **non-shaded areas**

HARWICH LAT 51°57'N LONG 1°17'E
TIMES AND HEIGHTS OF HIGH AND LOW WATERS

Dates in **red** are **SPRINGS**
Dates in blue are NEAPS

YEAR 2016

JANUARY

Day	Time	m	Day	Time	m
1 F	0342 / 0941 / 1611 / 2143	3.7 / 0.7 / 3.5 / 1.1	**16** SA ☽	0339 / 0937 / 1621 / 2152	3.8 / 0.4 / 3.7 / 0.8
2 SA ☽	0428 / 1032 / 1659 / 2240	3.5 / 0.8 / 3.4 / 1.2	**17** SU	0432 / 1032 / 1720 / 2251	3.7 / 0.6 / 3.6 / 0.9
3 SU	0522 / 1134 / 1753 / 2351	3.4 / 0.9 / 3.4 / 1.3	**18** M	0538 / 1144 / 1827	3.5 / 0.7 / 3.4
4 M	0622 / 1238 / 1852	3.3 / 1.0 / 3.3	**19** TU	0009 / 0651 / 1306 / 1937	1.0 / 3.5 / 0.8 / 3.4
5 TU	0101 / 0726 / 1337 / 1957	1.3 / 3.3 / 0.9 / 3.4	**20** W	0132 / 0804 / 1416 / 2045	1.0 / 3.5 / 0.8 / 3.5
6 W	0203 / 0831 / 1432 / 2101	1.2 / 3.4 / 0.9 / 3.5	**21** TH	0243 / 0912 / 1516 / 2147	0.9 / 3.6 / 0.7 / 3.7
7 TH	0259 / 0930 / 1523 / 2156	1.0 / 3.5 / 0.8 / 3.7	**22** F	0344 / 1011 / 1608 / 2241	0.7 / 3.8 / 0.7 / 3.9
8 F	0349 / 1021 / 1610 / 2244	0.8 / 3.7 / 0.7 / 3.8	**23** SA	0437 / 1104 / 1654 / 2329	0.5 / 3.9 / 0.6 / 4.0
9 SA	0436 / 1107 / 1654 / 2327	0.7 / 3.8 / 0.6 / 3.9	**24** SU ○	0525 / 1152 / 1737	0.4 / 4.0 / 0.6
10 SU ●	0520 / 1150 / 1736	0.6 / 3.9 / 0.6	**25** M	0013 / 0608 / 1236 / 1816	4.0 / 0.3 / 4.0 / 0.7
11 M	0008 / 0602 / 1232 / 1816	4.0 / 0.5 / 4.0 / 0.6	**26** TU	0054 / 0649 / 1316 / 1851	4.0 / 0.3 / 4.0 / 0.7
12 TU	0049 / 0644 / 1314 / 1855	4.0 / 0.4 / 4.0 / 0.6	**27** W	0131 / 0725 / 1352 / 1922	4.0 / 0.4 / 3.9 / 0.7
13 W	0129 / 0725 / 1357 / 1935	4.0 / 0.3 / 4.1 / 0.6	**28** TH	0203 / 0756 / 1424 / 1953	3.9 / 0.5 / 3.8 / 0.8
14 TH	0210 / 0806 / 1442 / 2017	4.0 / 0.3 / 4.0 / 0.6	**29** F	0236 / 0826 / 1457 / 2027	3.9 / 0.5 / 3.7 / 0.8
15 F	0253 / 0850 / 1529 / 2102	3.9 / 0.3 / 3.9 / 0.7	**30** SA	0311 / 0901 / 1534 / 2106	3.8 / 0.6 / 3.7 / 0.9
			31 SU	0352 / 0943 / 1617 / 2152	3.7 / 0.7 / 3.6 / 1.0

FEBRUARY

Day	Time	m	Day	Time	m
1 M ☽	0439 / 1035 / 1706 / 2248	3.5 / 0.9 / 3.4 / 1.2	**16** TU	0511 / 1114 / 1756 / 2342	3.6 / 0.8 / 3.3 / 1.0
2 TU	0534 / 1142 / 1803	3.3 / 1.0 / 3.3	**17** W	0627 / 1242 / 1913	3.4 / 0.9 / 3.3
3 W	0003 / 0635 / 1253 / 1904	1.2 / 3.2 / 1.0 / 3.3	**18** TH	0112 / 0748 / 1357 / 2026	1.0 / 3.4 / 1.0 / 3.3
4 TH	0118 / 0741 / 1358 / 2010	1.2 / 3.2 / 1.0 / 3.3	**19** F	0229 / 0859 / 1459 / 2130	0.9 / 3.5 / 0.9 / 3.5
5 F	0224 / 0804 / 1456 / 2118	1.1 / 3.3 / 0.9 / 3.5	**20** SA	0333 / 0959 / 1553 / 2224	0.7 / 3.7 / 0.7 / 3.8
6 SA	0324 / 0954 / 1549 / 2217	0.9 / 3.6 / 0.7 / 3.7	**21** SU	0426 / 1051 / 1639 / 2312	0.5 / 3.9 / 0.7 / 3.9
7 SU	0416 / 1046 / 1636 / 2307	0.7 / 3.8 / 0.6 / 3.9	**22** M ○	0511 / 1137 / 1720 / 2355	0.4 / 4.0 / 0.6 / 4.0
8 M ●	0503 / 1133 / 1719 / 2352	0.5 / 4.0 / 0.5 / 4.0	**23** TU	0552 / 1218 / 1757	0.3 / 4.0 / 0.6
9 TU	0547 / 1217 / 1800	0.3 / 4.1 / 0.5	**24** W	0035 / 0628 / 1256 / 1830	4.0 / 0.4 / 3.9 / 0.7
10 W	0034 / 0628 / 1301 / 1839	4.1 / 0.2 / 4.2 / 0.4	**25** TH	0110 / 0659 / 1329 / 1858	3.9 / 0.4 / 3.9 / 0.7
11 TH	0116 / 0709 / 1343 / 1919	4.1 / 0.1 / 4.2 / 0.4	**26** F	0140 / 0725 / 1356 / 1926	3.9 / 0.4 / 3.8 / 0.7
12 F	0157 / 0749 / 1427 / 2001	4.1 / 0.1 / 4.1 / 0.4	**27** SA	0209 / 0752 / 1425 / 1959	3.9 / 0.5 / 3.8 / 0.7
13 SA	0239 / 0831 / 1511 / 2045	4.1 / 0.2 / 4.0 / 0.4	**28** SU	0241 / 0826 / 1459 / 2035	3.8 / 0.5 / 3.8 / 0.7
14 SU	0322 / 0916 / 1558 / 2132	4.0 / 0.3 / 3.8 / 0.6	**29** M	0318 / 0903 / 1540 / 2115	3.7 / 0.6 / 3.7 / 0.8
15 M ☽	0411 / 1007 / 1652 / 2227	3.8 / 0.5 / 3.6 / 0.8			

MARCH

Day	Time	m	Day	Time	m
1 TU ☽	0401 / 0946 / 1627 / 2201	3.6 / 0.8 / 3.5 / 1.0	**16** W ☽	0451 / 1049 / 1727 / 2325	3.6 / 0.9 / 3.3 / 0.9
2 W	0452 / 1039 / 1723 / 2303	3.4 / 1.0 / 3.3 / 1.1	**17** TH	0608 / 1219 / 1848	3.4 / 1.1 / 3.2
3 TH	0554 / 1204 / 1825	3.2 / 1.1 / 3.2	**18** F	0057 / 0732 / 1336 / 2004	0.9 / 3.3 / 1.1 / 3.3
4 F	0032 / 0701 / 1326 / 1931	1.2 / 3.2 / 1.1 / 3.2	**19** SA	0212 / 0842 / 1438 / 2107	0.8 / 3.5 / 1.0 / 3.5
5 SA	0152 / 0814 / 1431 / 2042	1.1 / 3.3 / 1.0 / 3.4	**20** SU	0314 / 0940 / 1532 / 2201	0.6 / 3.7 / 0.8 / 3.7
6 SU	0259 / 0926 / 1526 / 2149	0.8 / 3.5 / 0.8 / 3.6	**21** M	0406 / 1030 / 1619 / 2249	0.5 / 3.9 / 0.7 / 3.9
7 M	0355 / 1024 / 1615 / 2243	0.6 / 3.8 / 0.6 / 3.8	**22** TU	0450 / 1115 / 1659 / 2332	0.4 / 4.0 / 0.7 / 4.0
8 TU	0443 / 1112 / 1658 / 2330	0.4 / 4.0 / 0.6 / 4.0	**23** W ○	0528 / 1156 / 1735	0.4 / 4.0 / 0.6
9 W ●	0527 / 1158 / 1740	0.2 / 4.2 / 0.3	**24** TH	0011 / 0601 / 1233 / 1805	4.0 / 0.4 / 3.9 / 0.6
10 TH	0014 / 0608 / 1242 / 1820	4.2 / 0.0 / 4.3 / 0.3	**25** F	0045 / 0653 / 1303 / 1832	3.9 / 0.5 / 3.8 / 0.6
11 F	0057 / 0648 / 1324 / 1901	4.2 / 0.0 / 4.3 / 0.2	**26** SA	0114 / 0653 / 1328 / 1902	3.8 / 0.5 / 3.8 / 0.6
12 SA	0138 / 0729 / 1407 / 1943	4.0 / 0.0 / 4.2 / 0.2	**27** SU	0142 / 0722 / 1355 / 1934	3.8 / 0.5 / 3.8 / 0.6
13 SU	0221 / 0810 / 1450 / 2027	4.2 / 0.1 / 4.0 / 0.3	**28** M	0213 / 0755 / 1441 / 2009	3.8 / 0.5 / 3.8 / 0.6
14 M	0304 / 0854 / 1535 / 2114	4.1 / 0.3 / 3.8 / 0.5	**29** TU ☽	0248 / 0830 / 1507 / 2047	3.8 / 0.6 / 3.7 / 0.7
15 TU	0353 / 0944 / 1625 / 2209	3.9 / 0.6 / 3.6 / 0.7	**30** W	0329 / 0909 / 1552 / 2130	3.6 / 0.8 / 3.6 / 0.8
			31 TH ☽	0418 / 0955 / 1646 / 2224	3.5 / 1.0 / 3.4 / 1.0

APRIL

Day	Time	m	Day	Time	m
1 F	0519 / 1058 / 1750 / 2340	3.3 / 1.1 / 3.2 / 1.1	**16** SA	0036 / 0708 / 1309 / 1934	0.8 / 3.4 / 1.2 / 3.3
2 SA	0628 / 1248 / 1858	3.2 / 1.2 / 3.2	**17** SU	0146 / 0815 / 1411 / 2036	0.8 / 3.5 / 1.1 / 3.4
3 SU	0119 / 0742 / 1402 / 2009	1.0 / 3.3 / 1.0 / 3.3	**18** M	0246 / 0912 / 1505 / 2131	0.6 / 3.7 / 0.9 / 3.6
4 M	0230 / 0856 / 1500 / 2117	0.8 / 3.5 / 0.8 / 3.6	**19** TU	0337 / 1003 / 1553 / 2220	0.5 / 3.9 / 0.8 / 3.8
5 TU	0329 / 0957 / 1550 / 2214	0.5 / 3.8 / 0.6 / 3.8	**20** W	0421 / 1048 / 1634 / 2304	0.5 / 3.9 / 0.7 / 3.9
6 W	0419 / 1048 / 1635 / 2303	0.3 / 4.1 / 0.4 / 4.0	**21** TH	0457 / 1129 / 1709 / 2343	0.5 / 4.0 / 0.6 / 3.9
7 TH ●	0503 / 1134 / 1718 / 2350	0.1 / 4.2 / 0.3 / 4.2	**22** F ○	0528 / 1205 / 1739	0.5 / 3.9 / 0.6
8 F	0545 / 1219 / 1800	0.0 / 4.3 / 0.2	**23** SA	0018 / 0555 / 1235 / 1808	3.8 / 0.5 / 3.8 / 0.6
9 SA	0034 / 0626 / 1303 / 1843	4.3 / 0.0 / 4.3 / 0.1	**24** SU	0048 / 0625 / 1300 / 1840	3.8 / 0.5 / 3.8 / 0.6
10 SU	0118 / 0707 / 1346 / 1926	4.3 / 0.0 / 4.2 / 0.2	**25** M	0117 / 0656 / 1327 / 1913	3.8 / 0.6 / 3.8 / 0.6
11 M	0202 / 0750 / 1429 / 2011	4.3 / 0.2 / 4.1 / 0.3	**26** TU	0149 / 0729 / 1401 / 1949	3.8 / 0.6 / 3.8 / 0.6
12 TU	0248 / 0834 / 1514 / 2059	4.1 / 0.4 / 3.8 / 0.4	**27** W	0225 / 0804 / 1440 / 2027	3.8 / 0.6 / 3.8 / 0.6
13 W	0337 / 0923 / 1602 / 2155	3.9 / 0.7 / 3.6 / 0.6	**28** TH	0306 / 0843 / 1524 / 2110	3.7 / 0.8 / 3.6 / 0.7
14 TH	0435 / 1026 / 1702 / 2312	3.6 / 1.0 / 3.3 / 0.8	**29** F	0355 / 0929 / 1616 / 2202	3.6 / 0.8 / 3.4 / 0.8
15 F	0549 / 1153 / 1820	3.4 / 1.2 / 3.2	**30** SA ☽	0453 / 1026 / 1718 / 2308	3.4 / 1.1 / 3.3 / 0.9

Chart Datum: 2·02 metres below Ordnance Datum (Newlyn). HAT is 4·4 metres above Chart Datum.

E England

》》**FREE** monthly updates. Register at 《
www.reedsnauticalalmanac.co.uk《

STANDARD TIME (UT)
For Summer Time add ONE hour in **non-shaded areas**

HARWICH LAT 51°57′N LONG 1°17′E
TIMES AND HEIGHTS OF HIGH AND LOW WATERS

Dates in red are **SPRINGS**
Dates in blue are **NEAPS**

YEAR 2016

MAY

Time	m		Time	m
1 0602	3.3	**16**	0112	0.7
1152	1.2		0738	3.4
SU 1828	3.3		M 1337	1.2
			1958	3.4
2 0043	0.8	**17**	0209	0.7
0714	3.4		0837	3.6
M 1326	1.0		TU 1432	1.0
1938	3.4		2055	3.5
3 0158	0.7	**18**	0300	0.6
0824	3.6		0929	3.7
TU 1428	0.9		W 1522	0.9
2044	3.6		2146	3.7
4 0259	0.5	**19**	0344	0.6
0927	3.8		1017	3.8
W 1522	0.6		TH 1604	0.8
2144	3.8		2232	3.8
5 0352	0.3	**20**	0421	0.6
1021	4.1		1059	3.9
TH 1611	0.4		F 1639	0.7
2237	4.0		2313	3.8
6 0439	0.1	**21**	0454	0.6
1110	4.2		1135	3.9
F 1657	0.3		SA 1712	0.6
● 2326	4.2		○ 2350	3.8
7 0523	0.1	**22**	0527	0.6
1156	4.3		1206	3.8
SA 1742	0.2		SU 1746	0.6
8 0013	4.3	**23**	0023	3.8
0605	0.1		0601	0.6
SU 1242	4.3		M 1235	3.8
1826	0.1		1822	0.6
9 0100	4.3	**24**	0055	3.8
0648	0.2		0636	0.6
M 1326	4.2		TU 1306	3.8
1912	0.2		1858	0.5
10 0146	4.3	**25**	0130	3.8
0731	0.3		0711	0.6
TU 1410	4.1		W 1341	3.8
1958	0.2		1935	0.5
11 0233	4.1	**26**	0208	3.8
0816	0.5		0746	0.7
W 1455	3.9		TH 1420	3.8
2048	0.4		2014	0.5
12 0323	3.9	**27**	0250	3.8
0904	0.8		0826	0.8
TH 1543	3.7		F 1504	3.7
2143	0.5		2057	0.6
13 0417	3.7	**28**	0339	3.7
1000	1.0		0912	0.9
F 1638	3.5		SA 1553	3.6
◗ 2252	0.7		2147	0.6
14 0522	3.5	**29**	0435	3.6
1116	1.2		1006	1.0
SA 1744	3.3		SU 1652	3.4
			◗ 2248	0.7
15 0006	0.7	**30**	0539	3.5
0632	3.4		1114	1.0
SU 1233	1.2		M 1800	3.4
1854	3.3			
		31	0007	0.7
			0647	3.5
			TU 1242	1.0
			1909	3.5

JUNE

Time	m		Time	m
1 0125	0.6	**16**	0213	0.8
0755	3.6		0848	3.5
W 1353	0.9		TH 1442	1.0
2015	3.6		2106	3.5
2 0229	0.5	**17**	0300	0.7
0858	3.8		0939	3.6
TH 1453	0.7		F 1527	0.9
2116	3.8		2157	3.6
3 0326	0.4	**18**	0342	0.7
0956	4.0		1025	3.7
F 1548	0.5		SA 1608	0.8
2213	4.0		2242	3.7
4 0416	0.3	**19**	0422	0.7
1048	4.1		1105	3.8
SA 1638	0.4		SU 1647	0.7
2306	4.1		2323	3.8
5 0503	0.3	**20**	0502	0.6
1137	4.2		1141	3.9
SU 1727	0.2		M 1727	0.6
● 2356	4.2		○	
6 0548	0.3	**21**	0001	3.8
1224	4.2		0541	0.6
M 1814	0.2		TU 1216	3.9
			1806	0.5
7 0044	4.3	**22**	0038	3.8
0632	0.4		0619	0.6
TU 1310	4.2		W 1250	3.9
1900	0.2		1846	0.5
8 0132	4.2	**23**	0115	3.9
0715	0.5		0657	0.7
W 1354	4.1		TH 1327	3.9
1947	0.2		1925	0.5
9 0219	4.1	**24**	0155	3.9
0759	0.6		0734	0.7
TH 1438	3.9		F 1406	3.8
2035	0.3		2005	0.4
10 0306	4.0	**25**	0238	3.9
0843	0.8		0814	0.7
F 1522	3.8		SA 1449	3.8
2126	0.5		2047	0.4
11 0354	3.8	**26**	0325	3.8
0930	1.0		0858	0.8
SA 1609	3.6		SU 1535	3.7
2221	0.6		2134	0.5
12 0446	3.6	**27**	0418	3.7
1026	1.2		0948	0.8
SU 1702	3.5		M 1629	3.6
◗ 2323	0.7		◗ 2228	0.5
13 0544	3.4	**28**	0517	3.6
1137	1.2		1047	0.9
M 1802	3.4		TU 1733	3.5
			2335	0.6
14 0026	0.8	**29**	0622	3.6
0647	3.4		1201	1.0
TU 1248	1.2		W 1841	3.5
1906	3.3			
15 0123	0.8	**30**	0052	0.6
0750	3.4		0727	3.6
W 1349	1.1		TH 1320	0.9
2009	3.4		1949	3.6

JULY

Time	m		Time	m
1 0202	0.6	**16**	0216	0.9
0833	3.7		0849	3.5
F 1428	0.8		SA 1446	1.0
2054	3.7		2118	3.5
2 0304	0.6	**17**	0308	0.8
0934	3.8		0946	3.6
SA 1529	0.6		SU 1537	0.9
2156	3.9		2212	3.6
3 0359	0.5	**18**	0356	0.8
1031	4.0		1034	3.8
SU 1625	0.5		M 1624	0.7
2252	4.0		2258	3.8
4 0448	0.5	**19**	0441	0.7
1122	4.1		1117	3.9
M 1716	0.3		TU 1708	0.6
● 2344	4.1		○ 2341	3.9
5 0534	0.5	**20**	0523	0.6
1210	4.1		1157	3.9
TU 1804	0.3		W 1751	0.5
6 0032	4.2	**21**	0021	3.9
0617	0.5		0603	0.6
W 1256	4.1		TH 1237	4.0
1850	0.2		1832	0.4
7 0119	4.2	**22**	0102	4.0
0659	0.6		0642	0.6
TH 1338	4.1		F 1315	4.0
1934	0.3		1912	0.3
8 0202	4.1	**23**	0142	4.0
0739	0.7		0720	0.6
F 1419	4.0		SA 1354	4.0
2018	0.3		1952	0.3
9 0244	3.9	**24**	0225	4.0
0818	0.8		0800	0.6
SA 1458	3.9		SU 1434	4.0
2059	0.4		2032	0.3
10 0325	3.8	**25**	0309	3.9
0855	0.9		0842	0.7
SU 1537	3.8		M 1517	3.9
2140	0.6		2116	0.4
11 0407	3.6	**26**	0358	3.8
0936	1.0		0929	0.7
M 1620	3.6		TU 1606	3.8
2226	0.7		◗ 2205	0.5
12 0451	3.5	**27**	0453	3.7
1028	1.2		1023	0.9
TU 1710	3.5		W 1705	3.7
◗ 2322	0.8		2306	0.6
13 0541	3.4	**28**	0555	3.5
1133	1.2		1130	1.0
W 1806	3.4		TH 1815	3.5
14 0022	0.9	**29**	0025	0.6
0637	3.3		0702	3.5
TH 1244	1.2		F 1253	1.0
1908	3.3		1928	3.5
15 0121	0.9	**30**	0141	0.8
0741	3.3		0812	3.5
F 1348	1.2		SA 1410	0.9
2015	3.3		2040	3.6
		31	0247	0.8
			0918	3.7
			SU 1517	0.7
			2145	3.8

AUGUST

Time	m		Time	m
1 0345	0.7	**16**	0332	0.9
1017	3.9		1004	3.7
M 1616	0.5		TU 1601	0.7
2242	4.0		2234	3.8
2 0435	0.7	**17**	0420	0.7
1109	4.0		1052	3.9
TU 1706	0.4		W 1648	0.6
● 2333	4.1		2320	4.0
3 0520	0.6	**18**	0503	0.6
1156	4.1		1136	4.0
W 1753	0.3		TH 1732	0.4
			○	
4 0019	4.2	**19**	0002	4.1
0601	0.6		0543	0.6
TH 1239	4.1		F 1218	4.1
1835	0.3		1813	0.3
5 0103	4.1	**20**	0044	4.1
0641	0.7		0623	0.5
F 1319	4.1		SA 1257	4.1
1916	0.3		1852	0.2
6 0142	4.0	**21**	0125	4.2
0717	0.7		0701	0.5
SA 1355	4.0		SU 1336	4.1
1952	0.4		1931	0.2
7 0218	3.9	**22**	0206	4.1
0749	0.8		0741	0.5
SU 1429	3.9		M 1415	4.1
2024	0.5		2011	0.2
8 0252	3.8	**23**	0249	4.0
0820	0.9		0824	0.5
M 1503	3.9		TU 1457	4.1
2055	0.6		2053	0.3
9 0327	3.7	**24**	0335	3.9
0857	0.9		0909	0.6
TU 1541	3.7		W 1543	3.9
2133	0.7		2141	0.5
10 0406	3.6	**25**	0426	3.7
0940	1.0		1001	0.8
W 1625	3.6		TH 1633	3.7
◗ 2221	0.9		◖ 2239	0.8
11 0452	3.5	**26**	0527	3.5
1034	1.2		1107	1.0
TH 1718	3.4		F 1752	3.5
2325	1.0			
12 0546	3.4	**27**	0002	1.0
1146	1.3		0640	3.4
F 1818	3.3		SA 1236	1.1
			1914	3.4
13 0035	1.1	**28**	0125	1.0
0645	3.3		0756	3.4
SA 1301	1.3		SU 1357	1.0
1923	3.3		2030	3.6
14 0140	1.1	**29**	0232	1.0
0751	3.4		0903	3.6
SU 1408	1.1		M 1507	0.8
2035	3.4		2134	3.8
15 0239	1.0	**30**	0330	0.9
0903	3.5		1001	3.8
M 1509	1.0		TU 1604	0.6
2142	3.6		2229	4.0
		31	0419	0.8
			1051	4.0
			W 1653	0.4
			2317	4.1

Chart Datum: 2·02 metres below Ordnance Datum (Newlyn). HAT is 4·4 metres above Chart Datum.

STANDARD TIME (UT)
For Summer Time add ONE hour in **non-shaded areas**

HARWICH LAT 51°57′N LONG 1°17′E
TIMES AND HEIGHTS OF HIGH AND LOW WATERS

Dates in red are **SPRINGS**
Dates in blue are NEAPS

YEAR 2016

SEPTEMBER

Time	m		Time	m
1 0502	0.7		**16** 0440	0.6
1136	4.1		1109	4.0
TH 1735	0.3		F 1708	0.3
●			○ 2338	4.2
2 0001	4.2		**17** 0521	0.5
0542	0.7		1152	4.2
F 1219	4.1		SA 1749	0.2
1814	0.3			
3 0042	4.1		**18** 0020	4.3
0619	0.7		0600	0.4
SA 1256	4.1		SU 1233	4.3
1850	0.4		1828	0.1
4 0118	4.0		**19** 0102	4.3
0651	0.8		0640	0.4
SU 1329	4.0		M 1314	4.3
1919	0.5		1907	0.2
5 0149	3.9		**20** 0144	4.3
0718	0.8		0722	0.4
M 1358	3.9		TU 1355	4.3
1945	0.6		1948	0.2
6 0217	3.8		**21** 0226	4.1
0748	0.8		0805	0.5
TU 1429	3.9		W 1437	4.2
2014	0.6		2030	0.4
7 0248	3.8		**22** 0310	3.9
0823	0.8		0851	0.6
W 1504	3.8		TH 1524	4.0
2050	0.7		2117	0.6
8 0325	3.7		**23** 0359	3.7
0903	0.9		0943	0.8
TH 1545	3.7		F 1620	3.7
2131	0.9		◗ 2214	0.9
9 0410	3.6		**24** 0500	3.4
0949	1.1		1051	1.0
F 1635	3.5		SA 1735	3.5
◗ 2223	1.1		2341	1.2
10 0503	3.4		**25** 0620	3.3
1051	1.2		1223	1.0
SA 1735	3.3		SU 1902	3.4
2346	1.3			
11 0604	3.3		**26** 0106	1.2
1218	1.3		0737	3.4
SU 1842	3.2		M 1343	0.9
			2015	3.6
12 0107	1.2		**27** 0212	1.1
0710	3.3		0841	3.6
M 1335	1.2		TU 1449	0.7
1954	3.3		2115	3.8
13 0211	1.1		**28** 0309	1.0
0821	3.4		0937	3.8
TU 1440	1.0		W 1544	0.6
2109	3.5		2208	4.0
14 0307	1.0		**29** 0358	0.9
0929	3.6		1027	4.0
W 1536	0.7		TH 1631	0.4
2206	3.8		2255	4.1
15 0356	0.8		**30** 0441	0.8
1023	3.9		1112	4.1
TH 1624	0.5		F 1711	0.4
2254	4.0		2338	4.2

OCTOBER

Time	m		Time	m
1 0520	0.7		**16** 0456	0.5
1153	4.1		1124	4.2
SA 1747	0.4		SU 1722	0.2
●			○ 2355	4.4
2 0016	4.1		**17** 0538	0.4
0554	0.7		1208	4.3
SU 1229	4.0		M 1803	0.1
1818	0.5			
3 0050	4.0		**18** 0038	4.4
0623	0.7		0620	0.3
M 1300	4.0		TU 1252	4.4
1843	0.6		1844	0.2
4 0117	3.9		**19** 0121	4.3
0650	0.7		0703	0.3
TU 1328	3.9		W 1335	4.4
1909	0.6		1925	0.3
5 0142	3.9		**20** 0204	4.2
0720	0.7		0748	0.4
W 1357	3.9		TH 1420	4.2
1940	0.7		2009	0.5
6 0212	3.9		**21** 0249	4.0
0755	0.8		0835	0.5
TH 1431	3.8		F 1509	4.0
2014	0.8		2056	0.8
7 0248	3.8		**22** 0337	3.7
0833	0.9		0929	0.7
F 1510	3.7		SA 1606	3.8
2052	0.9		◗ 2153	1.1
8 0331	3.7		**23** 0436	3.5
0915	1.0		1039	0.9
SA 1557	3.6		SU 1718	3.5
2136	1.1		2315	1.3
9 0422	3.5		**24** 0554	3.3
1007	1.1		1205	0.9
SU 1655	3.4		M 1840	3.5
◗ 2234	1.3			
10 0525	3.3		**25** 0039	1.3
1126	1.2		0709	3.4
M 1804	3.3		TU 1319	0.9
			1949	3.6
11 0026	1.4		**26** 0145	1.3
0633	3.3		0812	3.5
TU 1300	1.1		W 1421	0.7
1917	3.3		2048	3.8
12 0139	1.2		**27** 0242	1.1
0743	3.4		0908	3.7
W 1407	0.9		TH 1515	0.6
2031	3.6		2140	4.0
13 0237	1.0		**28** 0332	0.9
0851	3.6		0958	3.9
TH 1505	0.7		F 1601	0.5
2133	3.8		2227	4.1
14 0327	0.8		**29** 0416	0.8
0949	3.8		1043	4.0
F 1555	0.5		SA 1641	0.5
2224	4.1		2310	4.1
15 0413	0.6		**30** 0454	0.8
1038	4.1		1125	4.0
SA 1640	0.3		SU 1716	0.6
2311	4.3		● 2348	4.1
			31 0528	0.7
			1201	4.0
			M 1745	0.6

NOVEMBER

Time	m		Time	m
1 0021	4.0		**16** 0016	4.4
0556	0.7		0601	0.3
TU 1233	3.9		W 1233	4.4
1811	0.7		1823	0.3
2 0047	3.9		**17** 0101	4.3
0625	0.7		0647	0.3
W 1301	3.9		TH 1319	4.4
1840	0.7		1907	0.4
3 0111	3.9		**18** 0146	4.2
0658	0.7		0734	0.3
TH 1330	3.9		F 1407	4.3
1912	0.7		1951	0.6
4 0142	3.9		**19** 0231	4.0
0733	0.7		0823	0.4
F 1405	3.9		SA 1456	4.1
1946	0.8		2038	0.8
5 0218	3.9		**20** 0319	3.8
0810	0.8		0916	0.6
SA 1444	3.8		SU 1550	3.8
2024	0.9		2130	1.1
6 0259	3.7		**21** 0413	3.6
0851	0.9		1021	0.7
SU 1529	3.7		M 1653	3.6
2106	1.0		◗ 2238	1.3
7 0347	3.6		**22** 0519	3.4
0940	1.0		1137	0.8
M 1624	3.5		TU 1805	3.5
◗ 2158	1.2			
8 0446	3.4		**23** 0001	1.4
1042	1.0		0630	3.4
TU 1730	3.4		W 1247	0.8
2309	1.3		1912	3.5
9 0556	3.3		**24** 0111	1.3
1213	1.0		0734	3.4
W 1842	3.4		TH 1346	0.8
			2013	3.6
10 0054	1.3		**25** 0209	1.2
0707	3.4		0832	3.6
TH 1330	0.9		F 1439	0.7
1953	3.6		2107	3.7
11 0200	1.1		**26** 0301	1.0
0814	3.6		0925	3.7
F 1431	0.7		SA 1526	0.7
2057	3.8		2156	3.9
12 0255	0.9		**27** 0347	0.9
0910	3.8		1012	3.8
SA 1524	0.5		SU 1607	0.7
2153	4.1		2240	3.9
13 0345	0.7		**28** 0427	0.8
1008	4.0		1055	3.9
SU 1613	0.3		M 1642	0.7
2243	4.2		2319	4.0
14 0432	0.5		**29** 0501	0.7
1058	4.2		1134	3.9
M 1658	0.2		TU 1714	0.7
○ 2330	4.3		● 2353	3.9
15 0517	0.4		**30** 0533	0.7
1146	4.3		1208	3.9
TU 1741	0.2		W 1746	0.7

DECEMBER

Time	m		Time	m
1 0021	3.9		**16** 0045	4.2
0606	0.7		0635	0.3
TH 1239	3.8		F 1306	4.3
1819	0.7		1851	0.5
2 0049	3.9		**17** 0131	4.2
0641	0.7		0723	0.3
F 1310	3.9		SA 1354	4.2
1853	0.8		1935	0.6
3 0121	3.9		**18** 0216	4.0
0718	0.7		0811	0.3
SA 1346	3.9		SU 1441	4.1
1927	0.8		2019	0.8
4 0157	3.9		**19** 0300	3.9
0755	0.7		0900	0.5
SU 1425	3.8		M 1530	3.9
2004	0.9		2104	1.0
5 0237	3.8		**20** 0347	3.7
0835	0.7		0953	0.6
M 1510	3.8		TU 1621	3.7
2046	0.9		2154	1.2
6 0322	3.7		**21** 0438	3.5
0921	0.8		1055	0.7
TU 1602	3.6		W 1717	3.5
2135	1.0		◗ 2259	1.3
7 0415	3.5		**22** 0537	3.4
1014	0.8		1201	0.8
W 1702	3.5		TH 1820	3.4
◗ 2233	1.1			
8 0519	3.4		**23** 0019	1.3
1121	0.8		0642	3.4
TH 1809	3.5		F 1303	0.9
2348	1.2		1925	3.4
9 0631	3.4		**24** 0127	1.3
1244	0.8		0747	3.4
F 1918	3.6		SA 1357	0.9
			2025	3.5
10 0113	1.1		**25** 0225	1.1
0739	3.5		0846	3.5
SA 1354	0.7		SU 1447	0.8
2023	3.7		2120	3.6
11 0220	0.9		**26** 0315	1.0
0843	3.7		0939	3.6
SU 1454	0.6		M 1530	0.8
2124	3.9		2208	3.7
12 0318	0.7		**27** 0357	0.9
0942	3.9		1027	3.7
M 1548	0.4		TU 1610	0.7
2219	4.1		2251	3.8
13 0410	0.6		**28** 0435	0.8
1037	4.1		1109	3.8
TU 1637	0.4		W 1648	0.7
2310	4.2		2328	3.9
14 0500	0.4		**29** 0512	0.7
1128	4.2		1147	3.8
W 1723	0.4		TH 1726	0.7
2358	4.2		●	
15 0548	0.3		**30** 0001	3.9
1218	4.3		0550	0.6
TH 1808	0.4		F 1222	3.8
			1802	0.7
			31 0034	3.9
			0628	0.6
			SA 1256	3.8
			1838	0.7

Chart Datum: 2·02 metres below Ordnance Datum (Newlyn). HAT is 4·4 metres above Chart Datum.

E England

》 **FREE** monthly updates. Register at 《
www.reedsnauticalalmanac.co.uk

319

9.4.21 RIVER DEBEN

Suffolk **51°59'·38N 01°23'·58E** (Felixstowe Ferry) ✲♦♦♦♦✿✿✿

CHARTS AC 5607, 2052, 2693; Imray C29, C1, C28, Y16, 2000

TIDES Woodbridge Haven +0025 Dover; Woodbridge +0105 Dover; ML 1·9; Duration 0635

Standard Port WALTON-ON-THE-NAZE (←—)

Times				Height (metres)			
High Water		Low Water		MHWS	MHWN	MLWN	MLWS
0100	0700	0100	0700	4·2	3·4	1·1	0·4
1300	1900	1300	1900				
Differences FELIXSTOWE PIER (51° 57'N 1° 21'E)							
−0005	−0007	−0018	−0020	−0·5	−0·4	0·0	0·0
BAWDSEY							
−0016	−0020	−0030	−0032	−0·8	−0·6	−0·1	−0·1
WOODBRIDGE HAVEN (Entrance)							
0000	−0005	−0020	−0025	−0·5	−0·5	−0·1	+0·1
WOODBRIDGE (Town)							
+0045	+0025	+0025	−0020	−0·2	−0·3	−0·2	0·0

SHELTER Good in Tide Mill Yacht Hbr (TMYH) Woodbridge. ⚓s upriver, clear of moorings: N of Horse Sand; off Ramsholt, The Rocks, Waldringfield, The Tips, Methersgate, Kyson Pt and Woodbridge.

NAVIGATION WPT 51°58'·20N 01°23'·85E Woodbridge Haven SWM buoy, thence NNW'ly past W Knoll PHM to Mid Knoll SHM and then head north up-river to the entrance hugging the sea wall/W bank until close to Deben PHM. Chartlet shows the ent at latest survey (2013), showing west'ly movement of the Knolls. Depths can change significantly and buoyage is often adjusted Visitors may obtain updates at www.debenestuarypilot.co.uk.

Cross the shifting shingle bar HW−3 to HW+2 depending on draught. When Horse Sand is covered there is approx 2m at the bar. Best to enter after half-flood, and leave on the flood. The ent is only 1ca wide and in strong on-shore winds gets dangerously choppy; channel is well buoyed/marked. Keep to the W shore until PHM opposite the SC, then move E of Horse Sand. No commercial traffic. 8kn max speed N of Green Reach.

LIGHTS AND MARKS See 9.4.4 and chartlets.

COMMUNICATIONS (Code 01473) NMOC 02392 552100; Police 101.

HM mob 07803 476621, Asst HM mob 07860 191768.
Pilotage for river entr, call *Odd Times* Ch 08 or HM on mobile.

FACILITIES
FELIXSTOWE FERRY (01394), ⚓⚓⚓✕✕🖊🏠🛢✕🛒; Felixstowe Ferry SC ☎283785, 200⚓ ━(£10) 🚿🚻 ☎282173.

RAMSHOLT HM ☎07930 304061, ⚓⚓✕🛒.

WALDRINGFIELD AC 2693. Craft >11m should ⚓ at the Rocks (NW of Prettyman's Pt) or upriver at The Tips. Possibly ⚓ for <11m by arrangement with HM waldringfieldhm @ btinternet.com ☎736291 mob 07925 081062; 🚻 may have ⚓, ☎736260 VHF 80; ━. Sailing Club www.waldringfieldsc.com 🛒 (Wed and Sat evenings only).

WOODBRIDGE (01394) **Tide Mill Yacht Hbr**, 200+ⓥ (max 24m) £2. info@tidemillyachtharbour.co.uk; ☎385745. Ent by No 26 PHM. Depth over sill, dries 1·5m, is 1·6m at MHWN and 2·5m MHWS, with very accurate tide gauge and 6 waiting buoys. ⚓⚓⚓🖊⚓ 🖊✕🛒🖊 Ⓔ ⚒🛒(18t)🚿🛢 ACA.

Woodbridge Cruising Club, ☎386737.

Deben YC www.debenyachtclub.co.uk.

Town ⚓🏠🖊✉️Ⓑ🚻🛒✕🛒≈✈ (Cambridge, Norwich).

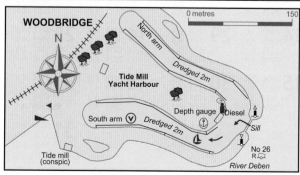

9.4.22 RIVER ORE/ALDE

Suffolk **52°02'·13N 01°27'·49E** (Ent) ⚓🌊🌊🌊🌊✿✿✿

CHARTS AC 5607, 1543, 2052, 2693, 2695; Imray C29, C28, 2000

TIDES Ent. +0015 Dover Slaughden Quay +0155 Dover; ML1·6; Duration 0620

Standard Port WALTON-ON-THE-NAZE (⟵)

Times				Height (metres)			
High Water		Low Water		MHWS	MHWN	MLWN	MLWS
0100	0700	0100	0700	4·2	3·4	1·1	0·4
1300	1900	1300	1900				
Differences ORFORD HAVEN BAR							
−0026	−0030	−0036	−0038	−1·0	−0·8	−0·1	0·0
ORFORD QUAY							
+0040	+0040	+0055	+0055	−1·4	−1·1	0·0	+0·2
SLAUGHDEN QUAY							
+0105	+0105	+0125	+0125	−1·3	−0·8	−0·1	+0·2
IKEN CLIFF							
+0130	+0130	+0155	+0155	−1·3	−1·0	0·0	+0·2

SHELTER Good shelter within the river, but the entrance should not be attempted in strong E/ESE onshore winds and rough seas or at night. The only safe ⚓ is in Short Gull; ⚓ in mooring areas. Landing on Havergate Island bird sanctuary is prohibited. ⚓ at Orford have small pick-up buoys marked V. Possible use of private mooring via Upson's BY at Slaughden.

NAVIGATION WPT Orford Haven SWM buoy, 52°01'·85N 01°28'·21E, is moved as required. For latest position call London CG ☎020 8312 7380. Proceed NW'ly to Oxley PHM shaping a more NNW'ly course to leave Weir SHM and North Shoal to starboard.

Chartlet is based on surveys carried out in 2013, but depths can change significantly. North Shoal has fragmented and the entr moved to the N. For regularly updated local information see *eastcoastsailing.co.uk*, call Small Craft Deliveries: ☎(01394) 387672, sales@scd-charts.co.uk; or local marinas or chandlers.

- The bar (approx 0·5m) shifts after onshore gales and is dangerous in rough or confused seas caused by tidal streams offshore running against those within the shingle banks. Sp ebb runs up to 6kn.
- Without local info do not enter before half flood or at night. For a first visit, appr at about LW+2½ in settled conditions and at nps.
- Beware Horse Shoal close WNW of N Weir Pt and shoals S & SW of Dove Pt (SW tip of Havergate Island).

R Ore (re-named R Alde between Orford and Slaughden Quay) is navigable up to Snape. The upper reaches are shallow and winding, and marked by withies with red and green topmarks. Shellfish beds on the E bank centred on 52°08'·0N 01°35'·6E.

LIGHTS AND MARKS As chartlet. Ent and river are unlit. Oxley and Weir buoys are seasonal Apr-Oct and the Or bn on the shore of Oxley Marshes is not maintained. Shingle Street, about 2ca S of ent, is identified by Martello tr 'AA', CG Stn, terrace houses and DF aerial. Up-river, Orford Ch and Castle are conspic; also Martello Tr 'CC', 3ca S of Slaughden Quay.

COMMUNICATIONS (Codes 01394 Orford; 01728 Aldeburgh) NMOC 02392 552100; Police 101; Orford Dr 450315(HO); Aldeburgh Dr 452027(HO). HM 07528 092635; Small Craft Deliveries (pilotage info) 382655.

Chantry on Ch 08.

FACILITIES

ORFORD: Orford Quay for ⚓ phone HM for availability (max 1 hour with competent crew); M (marked 'V'; call HM. No rafting) £8/ night collected by launch 'Chantry'; 🔧⚓☎ Orford SC (OSC) visitors may use 🚿 (£6 deposit, key held in Quay Office), ✕ 🍴 Internet (White Lion Hotel Aldeburgh).

Village (¼M) 🏪 🏪(HO) ⛽ ⊠ 🛒 ✕ 🍴 ⇌ (twice daily bus to Woodbridge).

ALDEBURGH: Slaughden Quay 🔧⚓⚓🏗(20t).

Aldeburgh YC (AYC) ☎452562 ♿.

Slaughden SC (SSC). **Services:** ⚓ (via Upson's BY if any vacant) £5, ✕⚓🏗⛽🔧⚓🚿⛽.

Town (¾M) 🏪 🚿 ✕ 🍴 ⊠ Ⓑ ⇌ (bus to Wickham Market), ✈ (Norwich).

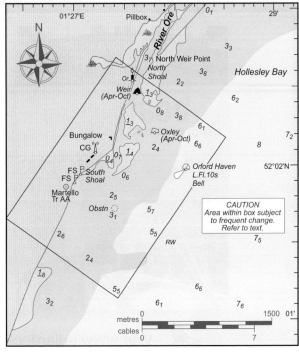

9.4.23 SOUTHWOLD

Suffolk 52°18'·78N 01°40'·54E ✿❋⚓⚓⚓✿✿✿

CHARTS AC 5614, 1543, 2695; Imray C29, C28

TIDES −0105 Dover; ML 1·5; Duration 0620

Standard Port LOWESTOFT (→)

Times				Height (metres)			
High Water		Low Water		MHWS	MHWN	MLWN	MLWS
0300	0900	0200	0800	2·4	2·1	1·0	0·5
1500	2100	1400	2000				
Differences SOUTHWOLD							
+0105	+0105	+0055	+0055	0·0	0·0	−0·1	0·0
MINSMERE SLUICE							
+0110	+0110	+0110	+0110	0·0	−0·1	−0·2	−0·2
ALDEBURGH (seaward)							
+0130	+0130	+0115	+0120	+0·3	+0·2	−0·1	−0·2
ORFORD NESS							
+0135	+0135	+0135	+0125	+0·4	+0·6	−0·1	0·0

NOTE: HW time differences (above) for Southwold apply up the hbr. At the ent mean HW is HW Lowestoft +0035.

SHELTER Good, but the ent is dangerous in strong winds from N through E to S. Visitors berth (must book ahead) on pontoon by HM office or a staging 6ca from the ent, on N bank near to the Harbour Inn. If rafted, shore lines are essential due to current. Some larger vessels moor on N wall of Dock and may encroach channel.

NAVIGATION WPT 52°18'·09N 01°41'·69E, 315°/1M to N Pier lt.

- Enter on the flood as the ebb runs up to 6kn. Shoals are unpredictable up to ¼M offshore; a sand and shingle bar, extent/depth variable, lies off the ent and a shoal builds inside N Pier. Obtain details of appr chans from HM before entering.
- Enter between piers in midstream. When chan widens keep close to The Knuckle (2 FG vert), turn stbd towards LB House; keep within 3m of quay wall until it ends, then resume midstream.
- **Caution** Rowing ferry (which has right of way at all times) 3ca and unlit low footbridge 7.5ca upstream of ent.

LIGHTS AND MARKS See 9.4.4 and chartlet. Walberswick ✠ in line with N Pier lt ≠ 268°. Hbr ent opens on 300°. 3 FR (vert) at N pier ≠ port closed. Lt ho, W ○ tr, Fl 10s 24M, in Southwold town, 0·86M NNE of hbr ent.

COMMUNICATIONS (Code 01502) CGOC (01262) 672317; Police 101; Dr 722326; Ⓗ 723333. HM 724712.
Southwold Port Radio Ch **12** 16 (as reqd). HM for fuel 12.

FACILITIES **Hbr** southwold harbour@waveney.gov.uk ⚓ HW±3 (Hbr dues £7·70) ⚓ £16·50 <20ft, £19·50 <30ft, £24·00 <40ft (+⚓) ⚓ ⚓ at new berth also by bowser 100 l/min, ⚓ ⚓ ⚓ ⚓(20t) ⚓.
Southwold SC; **Town** (¾M) ⚓ ✉ Ⓑ ⚓ ✕ ⚓ (bus to Brampton/ Darsham), ✈ (Norwich).

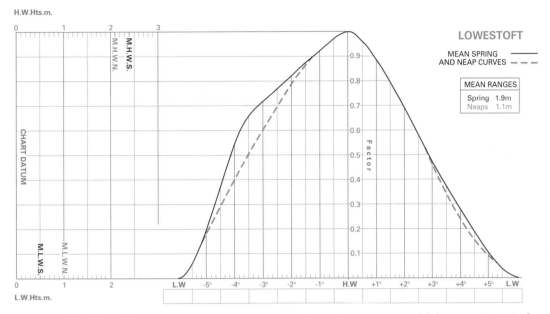

H.W.Hts.m.

LOWESTOFT

MEAN SPRING
AND NEAP CURVES

MEAN RANGES	
Spring	1.9m
Neaps	1.1m

L.W.Hts.m.

9.4.24 LOWESTOFT

Suffolk 52°28'·31N 01°45'·39E ❀❀❀✿✿✿✿✿✿✿

CHARTS AC 1543, 1535, 5614; Imray C29, C28

TIDES –0133 Dover; ML 1·6; Duration 0620

SHELTER Good; accessible H24. Wind over tide, especially on the ebb, can make the entrance lively. Fairway is dredged to 4·7m. RN&S YC yacht basin in the SW corner of Outer Hbr with 2m to 2·5m; no berthing on N side of S Pier. Lowestoft Haven Marina has facilities 1.5M upriver with additional berths in Hamilton Dock.

NAVIGATION Keep watch on Ch 14; req permission from *Harbour Control* to enter/leave harbour and pass (either way) between Outer Harbour and Waveney/Hamilton Docks to deconflict with helicopter movements. Sands continually shift and buoys are moved to suit. Beware shoals and drying areas esp S of hbr ent; do not cross banks in bad weather, nor at mid flood/ebb. Speed limit in harbour is 4 kn.

From S, WPT is E Barnard ECM buoy, Q (3) 10s, 52°25'·15N 01°46'·36E; thence via Stanford Chan E of Newcome Sand QR, Stanford Fl R 2.5s and N Newcome Fl (4) R 15s, all PHM buoys. S Holm SCM, VQ (6)+L Fl 10s, and SW Holm SHM, Fl (2) G 5s, buoys mark the seaward side of this channel.

From E, WPT is Holm Approach ECM buoy, Q (3) 10s, 52°30'·88N 01°50'·22E, then via Holm Chan (buoyed) into Corton Road. Or approach direct to S Holm SCM buoy for Stanford Channel.

From N, appr via Yarmouth, Gorleston and Corton Roads.

Bridge to Inner Hbr (Lake Lothing and Lowestoft CC) lifts at the following times (20 mins notice required): daily 0300, 0500, 0700, 0945, 1115, 1430, 1600, 1800; W/Es + 🚩s only 1900, 2100 and 2400. Small craft may pass under the bridge (clearance 2·2m) at any time but call on Ch 14. Craft are to await bridge opening on pontoon at E end of Trawl Dock. **Vessels may only transit bridge chan when G lights shown on N side of appr** (showing Br locked in position).

LIGHTS AND MARKS Lowestoft lt ho, W twr, is 1M N of entrance.

Traffic Signals shown on E arm of RN&SYC Yacht Basin to exit, and tip of S Pier for entry/exit:

3 FR (vert) = Do not proceed
GWG (vert) = proceed only as/when instructed

Bridge Sigs (on N bank each side of bridge):
● = bridge operating, keep 150m clear.
◉ = vessels may pass through bridge channel.

COMMUNICATIONS (Code 01502) CGOC (01262) 672317; Police 101; Ⓗ 01493 452452. HM & Bridge Control 572286; Mutford Bridge and Lock 531778 (+Ansafone, checked daily at 0830, 1300 & 1730); Oulton Broad Yacht Stn 574946; Pilot 572286 ext 243.

Lowestoft Hbr Control (ABP) VHF Ch **14** 16 11 (H24). *Lowestoft Haven Marina* Ch **80**, 37. *Oulton Broad YS and Mutford Lock Control* Ch 73. Pilot Ch 14. RN & SYC Ch 14, 80.

FACILITIES

Royal Norfolk & Suffolk YC (www.rnsyc.net) admin@rnsyc.org.uk, ☎566726, ⬤£2·30 inc YC facilities, × ⬛.

Lowestoft Cruising Club www.lowestoftcruisingclub.co.uk, ☎07913 391950, ⬤(max 13m) £2·00, ⬆, ⚓ ▬ (emergency).

Lowestoft Haven Marina ⬤ www. lowestofthavenmarina.co.uk ☎580300, 140⬤ £2·21 (+ 47 ⬤ in Hamilton Dock, £2·37) ⬧ ⬛ @ ⬤⬛⚒⚓ ⬛ Ⓔ ⬛(70t) ⬛ ACA.

Town ⬛ Ⓑ ⬛ × ⬛ ⇌ ✈ (Norwich).

Entry to the Broads: See Gt Yarmouth & www.norfolkbroads.com. Passage to Oulton Broad from Lake Lothing is via two bridges and Mutford Lock (width 5·9m). The lock is available 0800-1200 and 1300-1730 daily; 0800-1200 and 1300-1930 at weekends/Bank Holidays; 0900-1200 only 1 Nov to 31 Mar. Charge is £11/day. Booking 24 hours in advance is recommended, ☎(01502) 531778 or Ch 73. Oulton Broad Yacht Station ☎(01502) 574946. From Oulton Broad, access into the R Waveney is via Oulton Dyke.

STANDARD TIME (UT)
For Summer Time add ONE hour in **non-shaded areas**

LOWESTOFT LAT 52°28'N LONG 1°45'E
TIMES AND HEIGHTS OF HIGH AND LOW WATERS

Dates in red are **SPRINGS**
Dates in blue are **NEAPS**

YEAR 2016

JANUARY

Time	m	Time	m
1 0125 / 0800 / F 1428 / 1941	2.4 / 0.8 / 2.1 / 1.2	**16** 0117 / 0756 / SA 1403 / ☽ 1953	2.6 / 0.6 / 2.3 / 1.0
2 0211 / 0849 / SA 1538 / ☽ 2028	2.4 / 0.9 / 2.1 / 1.3	**17** 0207 / 0852 / SU 1512 / 2050	2.5 / 0.7 / 2.2 / 1.1
3 0304 / 0955 / SU 1643 / 2132	2.3 / 1.0 / 2.1 / 1.4	**18** 0309 / 1001 / M 1639 / 2208	2.4 / 0.8 / 2.2 / 1.1
4 0414 / 1104 / M 1737 / 2320	2.2 / 1.0 / 2.2 / 1.4	**19** 0431 / 1114 / TU 1742 / 2334	2.4 / 0.8 / 2.3 / 1.1
5 0523 / 1158 / TU 1826	2.2 / 1.0 / 2.2	**20** 0546 / 1219 / W 1839	2.4 / 0.9 / 2.3
6 0021 / 0621 / W 1244 / 1910	1.2 / 2.2 / 1.0 / 2.3	**21** 0044 / 0654 / TH 1320 / 1932	0.9 / 2.4 / 0.9 / 2.4
7 0109 / 0715 / TH 1325 / 1950	1.1 / 2.2 / 0.9 / 2.4	**22** 0150 / 0800 / F 1416 / 2021	0.8 / 2.4 / 0.9 / 2.5
8 0154 / 0805 / F 1406 / 2028	1.0 / 2.3 / 0.9 / 2.4	**23** 0247 / 0855 / SA 1505 / 2105	0.7 / 2.4 / 0.8 / 2.5
9 0238 / 0850 / SA 1448 / 2106	0.8 / 2.4 / 0.8 / 2.5	**24** 0335 / 0942 / SU 1547 / ○ 2146	0.5 / 2.4 / 0.8 / 2.6
10 0323 / 0933 / SU 1531 / ● 2145	0.7 / 2.4 / 0.7 / 2.6	**25** 0418 / 1024 / M 1625 / 2226	0.5 / 2.4 / 0.8 / 2.6
11 0409 / 1015 / M 1615 / 2225	0.6 / 2.5 / 0.7 / 2.6	**26** 0458 / 1105 / TU 1659 / 2304	0.5 / 2.4 / 0.8 / 2.6
12 0454 / 1057 / TU 1658 / 2306	0.5 / 2.5 / 0.7 / 2.6	**27** 0535 / 1142 / W 1730 / 2341	0.5 / 2.3 / 0.9 / 2.6
13 0539 / 1140 / W 1740 / 2347	0.5 / 2.5 / 0.8 / 2.7	**28** 0610 / 1217 / TH 1759	0.6 / 2.3 / 0.9
14 0623 / 1224 / TH 1822	0.5 / 2.4 / 0.8	**29** 0016 / 0643 / F 1252 / 1829	2.5 / 0.7 / 2.2 / 1.0
15 0031 / 0708 / F 1311 / 1905	2.6 / 0.5 / 2.3 / 0.9	**30** 0053 / 0717 / SA 1328 / 1905	2.5 / 0.8 / 2.2 / 1.1
		31 0134 / 0753 / SU 1409 / 1948	2.4 / 0.9 / 2.1 / 1.1

FEBRUARY

Time	m	Time	m
1 0220 / 0837 / M 1505 / ☽ 2039	2.3 / 1.0 / 2.1 / 1.2	**16** 0253 / 0930 / TU 1601 / 2146	2.4 / 0.9 / 2.2 / 1.0
2 0318 / 0938 / TU 1630 / 2150	2.2 / 1.1 / 2.1 / 1.3	**17** 0424 / 1053 / W 1714 / 2320	2.3 / 1.0 / 2.2 / 1.0
3 0441 / 1105 / W 1735 / 2335	2.1 / 1.1 / 2.1 / 1.2	**18** 0545 / 1206 / TH 1818	2.3 / 1.0 / 2.2
4 0552 / 1206 / TH 1829	2.1 / 1.1 / 2.2	**19** 0035 / 0702 / F 1313 / 1918	0.9 / 2.3 / 1.0 / 2.3
5 0036 / 0652 / F 1257 / 1917	1.1 / 2.2 / 1.0 / 2.3	**20** 0142 / 0805 / SA 1409 / 2008	0.8 / 2.3 / 0.9 / 2.4
6 0127 / 0747 / SA 1343 / 2002	0.9 / 2.2 / 0.9 / 2.4	**21** 0236 / 0851 / SU 1454 / 2050	0.6 / 2.4 / 0.9 / 2.5
7 0216 / 0834 / SU 1429 / 2043	0.8 / 2.3 / 0.8 / 2.5	**22** 0320 / 0931 / M 1532 / ○ 2129	0.5 / 2.4 / 0.8 / 2.5
8 0305 / 0917 / M 1515 / ● 2124	0.6 / 2.4 / 0.7 / 2.6	**23** 0359 / 1007 / TU 1606 / 2205	0.5 / 2.4 / 0.8 / 2.6
9 0353 / 0959 / TU 1600 / 2205	0.5 / 2.5 / 0.6 / 2.7	**24** 0435 / 1041 / W 1637 / 2241	0.5 / 2.4 / 0.8 / 2.6
10 0439 / 1040 / W 1644 / 2247	0.4 / 2.5 / 0.6 / 2.7	**25** 0509 / 1113 / TH 1706 / 2315	0.5 / 2.4 / 0.8 / 2.6
11 0523 / 1121 / TH 1726 / 2329	0.3 / 2.5 / 0.6 / 2.7	**26** 0540 / 1143 / F 1733 / 2349	0.6 / 2.3 / 0.8 / 2.5
12 0606 / 1204 / F 1807	0.3 / 2.4 / 0.7	**27** 0609 / 1213 / SA 1802	0.6 / 2.3 / 0.8
13 0012 / 0649 / SA 1249 / 1848	2.7 / 0.4 / 2.4 / 0.8	**28** 0024 / 0638 / SU 1247 / 1836	2.4 / 0.7 / 2.2 / 0.9
14 0058 / 0733 / SU 1337 / 1934	2.6 / 0.5 / 2.3 / 0.8	**29** 0102 / 0711 / M 1325 / 1916	2.4 / 0.7 / 2.2 / 1.0
15 0149 / 0824 / M 1436 / ☽ 2029	2.5 / 0.7 / 2.2 / 0.9		

MARCH

Time	m	Time	m
1 0146 / 0750 / TU 1411 / ☽ 2004	2.3 / 1.0 / 2.1 / 1.1	**16** 0248 / 0900 / W 1521 / 2134	2.3 / 1.0 / 2.2 / 0.9
2 0237 / 0840 / W 1509 / 2104	2.2 / 1.1 / 2.1 / 1.2	**17** 0424 / 1033 / TH 1642 / 2307	2.2 / 1.1 / 2.1 / 0.9
3 0353 / 0951 / TH 1636 / 2245	2.1 / 1.2 / 2.1 / 1.2	**18** 0544 / 1152 / F 1751	2.2 / 1.1 / 2.2
4 0526 / 1128 / F 1747	2.1 / 1.2 / 2.1	**19** 0020 / 0700 / SA 1300 / 1857	0.9 / 2.3 / 1.1 / 2.2
5 0004 / 0631 / SA 1229 / 1842	1.0 / 2.2 / 1.1 / 2.2	**20** 0124 / 0757 / SU 1354 / 1949	0.7 / 2.3 / 1.0 / 2.3
6 0100 / 0727 / SU 1320 / 1931	0.9 / 2.2 / 1.0 / 2.3	**21** 0215 / 0838 / M 1436 / 2030	0.6 / 2.3 / 0.9 / 2.4
7 0152 / 0815 / M 1409 / 2016	0.7 / 2.3 / 0.8 / 2.4	**22** 0257 / 0913 / TU 1512 / 2107	0.6 / 2.4 / 0.8 / 2.4
8 0243 / 0857 / TU 1456 / 2100	0.5 / 2.4 / 0.7 / 2.6	**23** 0334 / 0945 / W 1543 / ○ 2141	0.5 / 2.4 / 0.8 / 2.5
9 0332 / 0938 / W 1543 / ● 2142	0.4 / 2.5 / 0.6 / 2.7	**24** 0407 / 1015 / TH 1613 / 2215	0.5 / 2.4 / 0.7 / 2.5
10 0418 / 1019 / TH 1627 / 2226	0.2 / 2.5 / 0.5 / 2.8	**25** 0438 / 1043 / F 1641 / 2249	0.5 / 2.4 / 0.7 / 2.5
11 0503 / 1100 / F 1709 / 2309	0.2 / 2.5 / 0.5 / 2.8	**26** 0507 / 1111 / SA 1709 / 2323	0.6 / 2.4 / 0.7 / 2.4
12 0545 / 1142 / SA 1751 / 2354	0.2 / 2.5 / 0.5 / 2.7	**27** 0535 / 1141 / SU 1739 / 2358	0.7 / 2.3 / 0.8 / 2.4
13 0627 / 1225 / SU 1834	0.4 / 2.4 / 0.6	**28** 0604 / 1214 / M 1813	0.7 / 2.3 / 0.8
14 0042 / 0710 / M 1313 / 1920	2.6 / 0.6 / 2.3 / 0.7	**29** 0035 / 0636 / TU 1252 / 1851	2.3 / 0.8 / 2.3 / 0.9
15 0136 / 0758 / TU 1407 / ☽ 2015	2.5 / 0.8 / 2.2 / 0.8	**30** 0117 / 0715 / W 1335 / 1937	2.2 / 0.9 / 2.2 / 0.9
		31 0207 / 0802 / TH 1427 / ☽ 2035	2.1 / 1.1 / 2.1 / 1.0

APRIL

Time	m	Time	m
1 0313 / 0904 / F 1532 / 2201	2.1 / 1.2 / 2.1 / 1.0	**16** 0529 / 1126 / SA 1716 / 2353	2.2 / 1.2 / 2.2 / 0.7
2 0459 / 1041 / SA 1658 / 2331	2.1 / 1.2 / 2.1 / 0.9	**17** 0641 / 1232 / SU 1823	2.3 / 1.2 / 2.2
3 0607 / 1157 / SU 1804	2.2 / 1.1 / 2.2	**18** 0053 / 0735 / M 1327 / 1920	0.7 / 2.3 / 1.1 / 2.2
4 0030 / 0703 / M 1253 / 1857	0.8 / 2.3 / 1.0 / 2.3	**19** 0144 / 0816 / TU 1410 / 2003	0.7 / 2.3 / 1.0 / 2.3
5 0125 / 0751 / TU 1344 / 1946	0.6 / 2.4 / 0.8 / 2.4	**20** 0226 / 0850 / W 1445 / 2040	0.6 / 2.3 / 0.9 / 2.3
6 0217 / 0834 / W 1434 / 2032	0.4 / 2.5 / 0.7 / 2.5	**21** 0302 / 0919 / TH 1517 / 2115	0.6 / 2.4 / 0.8 / 2.4
7 0307 / 0915 / TH 1522 / ● 2118	0.3 / 2.5 / 0.6 / 2.7	**22** 0335 / 0946 / F 1547 / ○ 2150	0.6 / 2.4 / 0.7 / 2.4
8 0355 / 0956 / F 1608 / 2204	0.2 / 2.6 / 0.5 / 2.7	**23** 0405 / 1013 / SA 1617 / 2224	0.6 / 2.4 / 0.7 / 2.4
9 0439 / 1038 / SA 1652 / 2251	0.2 / 2.6 / 0.4 / 2.7	**24** 0434 / 1042 / SU 1648 / 2300	0.6 / 2.4 / 0.7 / 2.4
10 0522 / 1120 / SU 1736 / 2339	0.3 / 2.5 / 0.4 / 2.7	**25** 0504 / 1114 / M 1720 / 2336	0.7 / 2.4 / 0.7 / 2.3
11 0605 / 1204 / M 1821	0.4 / 2.5 / 0.5	**26** 0535 / 1148 / TU 1756	0.8 / 2.4 / 0.7
12 0030 / 0647 / TU 1252 / 1908	2.6 / 0.6 / 2.4 / 0.6	**27** 0014 / 0609 / W 1226 / 1834	2.3 / 0.8 / 2.3 / 0.8
13 0129 / 0734 / W 1344 / 2005	2.4 / 0.9 / 2.3 / 0.7	**28** 0056 / 0649 / TH 1308 / 1919	2.2 / 0.9 / 2.3 / 0.8
14 0245 / 0832 / TH 1448 / ☽ 2120	2.3 / 1.1 / 2.2 / 0.8	**29** 0145 / 0736 / F 1358 / 2015	2.1 / 1.1 / 2.2 / 0.9
15 0413 / 1003 / F 1605 / 2245	2.2 / 1.2 / 2.2 / 0.8	**30** 0246 / 0834 / SA 1456 / ☽ 2131	2.1 / 1.2 / 2.2 / 0.9

Chart Datum: 1·50 metres below Ordnance Datum (Newlyn). HAT is 2·9 metres above Chart Datum.

>> **FREE** monthly updates. Register at
www.reedsnauticalalmanac.co.uk <<

STANDARD TIME (UT)
For Summer Time add ONE hour in **non-shaded areas**

LOWESTOFT LAT 52°28'N LONG 1°45'E
TIMES AND HEIGHTS OF HIGH AND LOW WATERS

Dates in red are **SPRINGS**
Dates in blue are **NEAPS**

YEAR 2016

MAY

Date	Time m	Time m	Time m	Time m
1 SU	0428 2.1	0952 1.2	1606 2.2	2257 0.8
16 M	0606 2.2	1152 1.3	1738 2.2	
2 M	0540 2.2	1119 1.1	1723 2.2	
17 TU	0015 0.7	0703 2.2	1248 1.2	1836 2.2
3 TU	0000 0.7	0635 2.3	1222 1.0	1822 2.3
18 W	0105 0.7	0747 2.3	1335 1.1	1927 2.3
4 W	0056 0.5	0724 2.4	1316 0.9	1915 2.4
19 TH	0148 0.7	0822 2.3	1413 1.0	2009 2.3
5 TH	0149 0.4	0808 2.5	1409 0.7	2006 2.5
20 F	0226 0.7	0851 2.3	1447 0.9	2048 2.3
6 F	0241 0.3	0851 2.5	1500 0.6	● 2056 2.6
21 SA	0300 0.7	0918 2.4	1520 0.8	○ 2125 2.3
7 SA	0331 0.3	0934 2.6	1549 0.5	2146 2.7
22 SU	0332 0.7	0945 2.4	1554 0.7	2202 2.3
8 SU	0417 0.3	1017 2.6	1636 0.4	2236 2.7
23 M	0404 0.7	1016 2.5	1628 0.7	2239 2.3
9 M	0501 0.4	1101 2.6	1723 0.4	2328 2.6
24 TU	0437 0.7	1051 2.5	1705 0.7	2317 2.3
10 TU	0544 0.6	1146 2.5	1809 0.4	
25 W	0513 0.8	1127 2.4	1744 0.7	2357 2.3
11 W	0021 2.5	0626 0.7	1233 2.5	1857 0.5
26 TH	0550 0.8	1206 2.4	1824 0.7	
12 TH	0120 2.4	0711 0.9	1323 2.4	1952 0.6
27 F	0039 2.2	0630 0.9	1248 2.4	1909 0.7
13 F	0232 2.3	0802 1.1	1418 2.2	◑ 2058 0.7
28 SA	0128 2.2	0716 1.0	1335 2.3	2002 0.8
14 SA	0350 2.2	0913 1.3	1524 2.2	2213 0.7
29 SU	0224 2.2	0810 1.1	1428 2.3	◑ 2107 0.8
15 SU	0500 2.2	1044 1.3	1634 2.2	2318 0.7
30 M	0345 2.2	0915 1.2	1529 2.3	2223 0.7
31 TU	0508 2.2	1037 1.1	1642 2.3	2330 0.6

JUNE

Date	Time m	Time m	Time m	Time m
1 W	0606 2.3	1148 1.1	1751 2.4	
16 TH	0024 0.9	0708 2.2	1254 1.2	1844 2.2
2 TH	0028 0.6	0656 2.4	1249 0.9	1848 2.4
17 F	0109 0.9	0748 2.3	1338 1.1	1936 2.2
3 F	0124 0.5	0744 2.4	1346 0.8	1944 2.4
18 SA	0149 0.8	0820 2.3	1417 0.9	2022 2.2
4 SA	0218 0.5	0829 2.5	1442 0.6	2038 2.6
19 SU	0225 0.8	0849 2.4	1454 0.8	2103 2.3
5 SU	0309 0.5	0851 2.6	1534 0.5	● 2133 2.6
20 M	0301 0.8	0920 2.5	1532 0.7	○ 2143 2.3
6 M	0357 0.5	0959 2.6	1624 0.4	2226 2.6
21 TU	0338 0.8	0954 2.5	1612 0.7	2222 2.3
7 TU	0442 0.6	1044 2.6	1711 0.4	2318 2.5
22 W	0416 0.7	1031 2.5	1653 0.6	2301 2.4
8 W	0525 0.7	1129 2.5	1757 0.4	
23 TH	0455 0.8	1109 2.5	1734 0.6	2342 2.3
9 TH	0010 2.4	0606 0.8	1214 2.5	1843 0.4
24 F	0535 0.8	1149 2.5	1817 0.6	
10 F	0104 2.3	0647 0.9	1300 2.5	1932 0.5
25 SA	0024 2.3	0616 0.9	1231 2.5	1900 0.6
11 SA	0203 2.2	0730 1.1	1348 2.4	2026 0.6
26 SU	0110 2.3	0700 1.0	1315 2.5	1948 0.6
12 SU	0313 2.2	0819 1.2	1440 2.3	◑ 2130 0.7
27 M	0201 2.2	0748 1.0	1404 2.5	◑ 2044 0.7
13 M	0420 2.2	0926 1.3	1541 2.2	2236 0.8
28 TU	0305 2.2	0845 1.1	1459 2.4	2151 0.7
14 TU	0521 2.2	1058 1.3	1648 2.2	2334 0.8
29 W	0433 2.2	0955 1.1	1609 2.4	2301 0.7
15 W	0618 2.2	1202 1.3	1748 2.2	
30 TH	0537 2.3	1117 1.1	1725 2.4	

JULY

Date	Time m	Time m	Time m	Time m
1 F	0003 0.7	0630 2.4	1226 1.0	1829 2.4
16 SA	0031 1.0	0700 2.3	1302 1.1	1906 2.2
2 SA	0103 0.7	0721 2.4	1329 0.8	1930 2.5
17 SU	0114 1.0	0742 2.3	1346 1.0	2000 2.2
3 SU	0159 0.7	0810 2.5	1429 0.7	2030 2.5
18 M	0155 0.9	0819 2.4	1409 0.9	2045 2.3
4 M	0253 0.7	0857 2.6	1524 0.5	● 2127 2.5
19 TU	0235 0.9	0856 2.5	1512 0.8	○ 2126 2.3
5 TU	0342 0.7	0943 2.6	1613 0.4	2218 2.5
20 W	0316 0.8	0933 2.5	1555 0.6	2205 2.4
6 W	0426 0.7	1028 2.7	1659 0.4	2306 2.5
21 TH	0358 0.8	1011 2.6	1639 0.6	2244 2.4
7 TH	0507 0.8	1111 2.7	1743 0.4	2353 2.4
22 F	0440 0.7	1050 2.6	1722 0.5	2324 2.4
8 F	0545 0.8	1153 2.6	1825 0.4	
23 SA	0522 0.8	1130 2.7	1804 0.5	
9 SA	0038 2.3	0621 0.9	1234 2.6	1906 0.5
24 SU	0005 2.4	0602 0.8	1212 2.7	1846 0.5
10 SU	0125 2.2	0657 1.0	1315 2.5	1949 0.7
25 M	0049 2.4	0643 0.9	1255 2.6	1930 0.6
11 M	0216 2.2	0735 1.1	1359 2.4	2036 0.8
26 TU	0136 2.3	0728 1.0	1341 2.6	◑ 2019 0.7
12 TU	0320 2.1	0819 1.3	1449 2.3	◑ 2134 0.9
27 W	0232 2.3	0820 1.0	1435 2.5	2119 0.7
13 W	0424 2.1	0916 1.3	1553 2.2	2243 1.0
28 TH	0351 2.2	0925 1.1	1547 2.4	2234 0.8
14 TH	0520 2.1	1102 1.3	1705 2.2	2342 1.0
29 F	0507 2.3	1053 1.1	1710 2.4	2344 0.9
15 F	0612 2.2	1211 1.3	1807 2.2	
30 SA	0607 2.3	1211 1.0	1821 2.4	
31 SU	0048 0.9	0702 2.4	1319 0.8	1930 2.4

AUGUST

Date	Time m	Time m	Time m	Time m
1 M	0149 0.9	0755 2.5	1422 0.7	2031 2.4
16 TU	0128 1.0	0747 2.4	1403 0.9	2025 2.3
2 TU	0243 0.9	0843 2.6	1515 0.5	● 2122 2.5
17 W	0212 0.9	0828 2.5	1449 0.7	2105 2.4
3 W	0329 0.8	0927 2.6	1601 0.4	2207 2.5
18 TH	0256 0.8	0908 2.6	1535 0.6	○ 2144 2.5
4 TH	0410 0.8	1009 2.7	1643 0.4	2249 2.5
19 F	0340 0.8	0948 2.7	1619 0.5	2223 2.5
5 F	0448 0.8	1049 2.7	1722 0.5	2329 2.4
20 SA	0424 0.7	1028 2.8	1703 0.4	2302 2.6
6 SA	0522 0.9	1128 2.7	1759 0.5	
21 SU	0505 0.7	1109 2.8	1745 0.4	2343 2.5
7 SU	0007 2.4	0553 0.9	1206 2.6	1834 0.6
22 M	0546 0.7	1150 2.8	1826 0.4	
8 M	0044 2.3	0624 0.9	1243 2.5	1908 0.7
23 TU	0025 2.5	0626 0.8	1234 2.7	1908 0.5
9 TU	0120 2.2	0658 1.1	1322 2.5	1944 0.8
24 W	0111 2.4	0710 0.9	1322 2.6	1954 0.7
10 W	0200 2.2	0737 1.2	1407 2.4	◑ 2025 1.0
25 TH	0203 2.3	0801 1.0	1418 2.4	◑ 2050 0.9
11 TH	0252 2.2	0826 1.3	1503 2.2	2119 1.1
26 F	0312 2.3	0906 1.0	1538 2.4	2209 1.0
12 F	0409 2.2	0931 1.3	1623 2.2	2244 1.2
27 SA	0436 2.3	1041 1.0	1709 2.4	2330 1.1
13 SA	0516 2.2	1123 1.2	1737 2.2	2352 1.2
28 SU	0543 2.3	1201 0.9	1825 2.4	
14 SU	0611 2.3	1226 1.2	1841 2.2	
29 M	0038 1.1	0643 2.4	1311 0.8	1937 2.4
15 M	0042 1.1	0702 2.3	1317 1.0	1939 2.3
30 TU	0141 1.0	0739 2.5	1411 0.7	2030 2.5
31 W	0232 1.0	0826 2.6	1500 0.6	2112 2.5

Chart Datum: 1·50 metres below Ordnance Datum (Newlyn). HAT is 2·9 metres above Chart Datum.

E England

》》 **FREE** monthly updates. Register at 《 www.reedsnauticalalmanac.co.uk

325

STANDARD TIME (UT)
For Summer Time add ONE hour in **non-shaded areas**

LOWESTOFT   LAT 52°28'N   LONG 1°45'E
TIMES AND HEIGHTS OF HIGH AND LOW WATERS

Dates in red are SPRINGS
Dates in blue are NEAPS

YEAR 2016

SEPTEMBER

Day	Time	m	Time	m	Time	m	Time	m
1 TH	0314	0.9	0908	2.6	1542	0.5	● 2150	2.5
2 F	0351	0.9	0947	2.7	1620	0.5	2226	2.5
3 SA	0425	0.8	1025	2.7	1656	0.5	2301	2.5
4 SU	0456	0.9	1101	2.7	1729	0.6	2333	2.4
5 M	0526	0.9	1136	2.6	1759	0.7		
6 TU	0004	2.4	0554	0.9	1212	2.6	1828	0.8
7 W	0037	2.3	0627	1.0	1250	2.5	1859	0.9
8 TH	0114	2.3	0705	1.1	1332	2.4	1936	1.0
9 F	0158	2.3	0750	1.2	1423	2.3	◐ 2021	1.2
10 SA	0252	2.2	0847	1.3	1536	2.2	2123	1.3
11 SU	0411	2.2	1022	1.3	1710	2.2	2304	1.3
12 M	0525	1.2	1150	1.2	1817	2.2		
13 TU	0010	1.2	0622	2.3	1245	1.0	1913	2.3
14 W	0100	1.1	0712	2.4	1334	0.8	2000	2.4
15 TH	0147	1.0	0757	2.5	1422	0.7	2041	2.5
16 F	0233	0.9	0839	2.6	1509	0.5	○ 2119	2.6
17 SA	0319	0.8	0921	2.7	1555	0.4	2158	2.6
18 SU	0403	0.7	1003	2.8	1639	0.3	2238	2.6
19 M	0446	0.7	1046	2.9	1721	0.3	2319	2.6
20 TU	0528	0.7	1130	2.8	1803	0.4		
21 W	0002	2.6	0610	0.7	1216	2.8	1844	0.6
22 TH	0047	2.5	0656	0.8	1307	2.6	1930	0.8
23 F	0138	2.4	0749	0.9	1410	2.5	◐ 2025	1.0
24 SA	0242	2.3	0857	1.0	1542	2.4	2146	1.2
25 SU	0404	2.3	1031	1.0	1709	2.4	2315	1.2
26 M	0516	2.3	1148	0.9	1826	2.4		
27 TU	0025	1.2	0620	2.4	1254	0.8	1931	2.4
28 W	0126	1.1	0718	2.4	1351	0.7	2017	2.5
29 TH	0214	1.1	0805	2.5	1437	0.6	2054	2.5
30 F	0253	1.0	0844	2.6	1516	0.6	2128	2.5

OCTOBER

Day	Time	m	Time	m	Time	m	Time	m
1 SA	0327	0.9	0922	2.6	1552	0.6	● 2200	2.5
2 SU	0400	0.9	0958	2.7	1625	0.6	2230	2.5
3 M	0430	0.8	1034	2.6	1655	0.7	2259	2.5
4 TU	0459	0.9	1109	2.6	1723	0.7	2329	2.5
5 W	0529	0.9	1144	2.5	1751	0.8		
6 TH	0001	2.4	0601	0.9	1222	2.4	1821	0.9
7 F	0038	2.4	0638	1.0	1303	2.3	1858	1.1
8 SA	0120	2.4	0722	1.1	1351	2.2	1942	1.2
9 SU	0210	2.3	0816	1.2	1454	2.2	◐ 2038	1.3
10 M	0310	2.3	0933	1.2	1640	2.2	2158	1.4
11 TU	0430	2.3	1110	1.1	1749	2.3	2330	1.3
12 W	0540	2.3	1210	0.9	1844	2.4		
13 TH	0028	1.2	0634	2.4	1302	0.8	1931	2.4
14 F	0118	1.0	0723	2.5	1352	0.6	2013	2.5
15 SA	0207	0.9	0808	2.6	1441	0.5	2053	2.6
16 SU	0255	0.8	0853	2.8	1528	0.4	○ 2133	2.7
17 M	0341	0.7	0939	2.8	1613	0.3	2214	2.7
18 TU	0427	0.6	1025	2.9	1657	0.4	2256	2.7
19 W	0512	0.6	1112	2.8	1740	0.5	2340	2.6
20 TH	0557	0.6	1202	2.7	1823	0.7		
21 F	0026	2.6	0645	0.7	1258	2.6	1908	0.9
22 SA	0117	2.5	0739	0.8	1406	2.4	◐ 2001	1.1
23 SU	0216	2.4	0847	0.8	1538	2.4	2116	1.3
24 M	0329	2.4	1012	0.9	1657	2.4	2248	1.3
25 TU	0444	2.3	1124	0.8	1809	2.4	2358	1.3
26 W	0550	2.4	1226	0.8	1910	2.4		
27 TH	0058	1.2	0649	2.4	1321	0.7	1955	2.4
28 F	0147	1.1	0738	2.4	1406	0.7	2031	2.5
29 SA	0226	1.0	0818	2.5	1445	0.7	2103	2.5
30 SU	0301	1.0	0856	2.5	● 1519	0.7	2132	2.5
31 M	0333	0.9	0932	2.6	1551	0.7	2200	2.5

NOVEMBER

Day	Time	m	Time	m	Time	m	Time	m
1 TU	0404	0.8	1008	2.5	1620	0.7	2228	2.5
2 W	0435	0.8	1044	2.5	1649	0.8	2259	2.5
3 TH	0507	0.8	1120	2.4	1719	0.9	2332	2.5
4 F	0541	0.9	1157	2.4	1751	0.9		
5 SA	0009	2.5	0618	0.9	1238	2.3	1829	1.0
6 SU	0050	2.4	0701	1.0	1325	2.3	1913	1.2
7 M	0137	2.4	0753	1.0	1421	2.2	◐ 2006	1.3
8 TU	0232	2.3	0900	1.0	1551	2.2	2112	1.3
9 W	0336	2.3	1026	1.0	1715	2.3	2241	1.3
10 TH	0452	2.3	1134	0.8	1812	2.4	2351	1.2
11 F	0556	2.4	1229	0.7	1900	2.5		
12 SA	0047	1.1	0649	2.5	1321	0.6	1944	2.5
13 SU	0139	0.9	0739	2.6	1412	0.5	2027	2.6
14 M	0230	0.8	0828	2.7	1502	0.4	○ 2109	2.7
15 TU	0321	0.6	0918	2.8	1550	0.4	2152	2.7
16 W	0410	0.6	1008	2.8	1635	0.5	2236	2.7
17 TH	0458	0.5	1100	2.7	1720	0.6	2321	2.7
18 F	0546	0.5	1153	2.6	1804	0.8		
19 SA	0008	2.6	0635	0.6	1250	2.5	1848	1.0
20 SU	0058	2.5	0728	0.6	1355	2.4	1937	1.1
21 M	0152	2.5	0830	0.7	1516	2.3	◑ 2037	1.3
22 TU	0253	2.4	0943	0.8	1631	2.3	2203	1.4
23 W	0404	2.3	1051	0.8	1738	2.3	2319	1.4
24 TH	0511	2.3	1151	0.8	1838	2.3		
25 F	0020	1.3	0611	2.3	1244	0.8	1927	2.4
26 SA	0112	1.2	0704	2.4	1330	0.8	2005	2.4
27 SU	0155	1.1	0750	2.4	1410	0.8	2037	2.4
28 M	0232	1.0	0831	2.4	1445	0.8	2105	2.4
29 TU	0307	0.9	0909	2.5	● 1518	0.8	2132	2.5
30 W	0340	0.8	0946	2.4	1548	0.8	2201	2.5

DECEMBER

Day	Time	m	Time	m	Time	m	Time	m
1 TH	0414	0.8	1023	2.4	1620	0.8	2234	2.6
2 F	0449	0.8	1100	2.4	1653	0.9	2309	2.5
3 SA	0526	0.8	1139	2.4	1729	0.9	2347	2.5
4 SU	0605	0.8	1219	2.3	1808	1.0		
5 M	0027	2.5	0647	0.9	1304	2.3	1851	1.1
6 TU	0112	2.4	0736	0.9	1354	2.2	1940	1.2
7 W	0202	2.4	0833	0.9	1457	2.2	◐ 2038	1.2
8 TH	0257	2.4	0945	0.9	1633	2.2	2150	1.3
9 F	0404	2.3	1057	0.8	1737	2.3	2310	1.2
10 SA	0518	2.4	1158	0.7	1829	2.4		
11 SU	0015	1.1	0619	2.5	1254	0.6	1917	2.5
12 M	0114	0.9	0714	2.6	1347	0.5	2003	2.5
13 TU	0210	0.8	0809	2.6	1440	0.6	2048	2.6
14 W	0305	0.7	0903	2.7	1530	0.6	○ 2133	2.7
15 TH	0358	0.5	0958	2.7	1618	0.6	2219	2.7
16 F	0447	0.4	1051	2.6	1703	0.7	2305	2.7
17 SA	0535	0.4	1143	2.6	1746	0.8	2351	2.7
18 SU	0623	0.5	1235	2.5	1828	0.9		
19 M	0037	2.6	0711	0.5	1332	2.3	1910	1.1
20 TU	0125	2.5	0803	0.6	1438	2.2	1956	1.2
21 W	0216	2.4	0902	0.8	1552	2.2	◑ 2052	1.3
22 TH	0315	2.3	1009	0.9	1656	2.2	2219	1.4
23 F	0424	2.3	1111	0.9	1755	2.2	2335	1.3
24 SA	0529	2.3	1205	0.9	1849	2.3		
25 SU	0033	1.3	0627	2.2	1254	0.9	1933	2.3
26 M	0123	1.1	0722	2.3	1336	0.9	2009	2.4
27 TU	0205	1.0	0811	2.3	1413	0.9	2038	2.4
28 W	0243	0.9	0853	2.3	1448	0.8	2107	2.5
29 TH	0319	0.8	0930	2.4	● 1522	0.8	2139	2.5
30 F	0357	0.8	1007	2.4	1558	0.8	2214	2.6
31 SA	0435	0.7	1044	2.4	1635	0.8	2251	2.6

Chart Datum: 1·50 metres below Ordnance Datum (Newlyn). HAT is 2·9 metres above Chart Datum.

》》FREE monthly updates. Register at 《
www.reedsnauticalalmanac.co.uk 《

9.4.25 GREAT YARMOUTH

Norfolk 52°34'·36N 01°44'·39E ❁❁⌂✿✿

CHARTS AC 5614, 1543, 1535, 1534; Imray C29, C28

TIDES –0210 Dover; ML 1·5; Duration 0620

Standard Port LOWESTOFT (←)

Times				Height (metres)			
High Water		Low Water		MHWS	MHWN	MLWN	MLWS
0300	0900	0200	0800	2·4	2·1	1·0	0·5
1500	2100	1400	2000				
Differences GORLESTON (To be used for Great Yarmouth)							
–0035	–0035	–0030	–0030	0·0	0·0	0·0	0·0
CAISTER-ON-SEA							
–0120	–0120	–0100	–0100	0·0	–0·1	0·0	0·0
WINTERTON-ON-SEA							
–0225	–0215	–0135	–0135	+0·8	+0·5	+0·2	+0·1

• Rise of tide occurs mainly during 3.5 hours after LW. From HW Lowestoft –3 until HW the level is usually within 0·3m of predicted HW. Flood tide runs until about HW +15 and ebb until about LW +25.

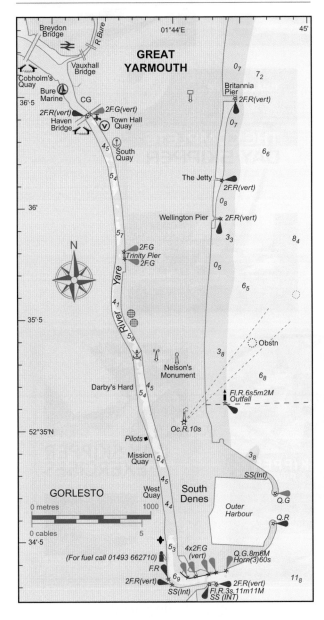

SHELTER Excellent on Town Hall Quay, close S of Haven Bridge; Yachts and small craft are not allowed in the outer harbour and ⚓ prohibited in the commercial hbrs, R Yar and its approaches.

NAVIGATION WPT 52°34'·43N 01°45'·56E, 264°/1M to front ldg lt.

• Access is H24, subject to clearance, but small craft must not attempt ent in strong SE winds which cause dangerous seas, especially on the ebb. Except at local slack water, which occurs at HW+1½ and LW+1¾, tidal streams at the ent are strong.

• On the flood, the stream eddies NW past S pier, thence up-river; a QY tidal lt on S pier warns of this D/N. Beware being set onto the N pier.

• Temporary shoaling may occur in the ent during strong E'lies, with depths 1m less than those charted.

• Beware strong tidal streams that sweep through the Haven Bridge. Proceed slowly past Bure Marine.

• Outer Harbour is for commercial vessels only.

LIGHTS AND MARKS See 9.4.4 and Chartlet. Gorleston Pier lt, Fl R 3s, is on R brick building (W lower half floodlit). Entrance and bend marked by five 2FG(vert) and seven 2FR(vert).

TRAFFIC SIGNALS Port operations advise on Ch 12, this should always be heeded. **Inbound:** now IALA sigs on S pier: 3 Fl ● = hbr closed; 3F ● = do not proceed; 3 F ● = vessels may proceed, oneway; ● ○ ● = proceed only when told to; **Outbound:** 3 ● (vert) = no vessel to go down river south of LB shed. Haven and Breydon bridges: 3 ● (vert) = passage prohib.

COMMUNICATIONS (Code 01493) CGOC (01262) 672317; Police 101; HM 335511; Breydon Bridge 651275.

Yarmouth Radio Ch **12** (H24). *Breydon Bridge* Ch **12** (OH).

FACILITIES Town Hall Quay (50m stretch) ⌑ £15/yacht, then £7 for successive days.
Burgh Castle Marina (top of Breydon Water 5M) ☎780331, access HW ±4 for 1m draught ⚓ 90+10Ⓥ £16, ▯ ⚓ ▮ ▮ ⚒ ⚲ ⊞ diving ⚓ ACA ⚒ ✕ ⌑.
Goodchild Marine Services (top of Breydon Water 5M) ☎782301, access 6ft at LW. ⚓ ☏ (27⌑+6Ⓥ £12.50), ⚓ ⚓ ▮ ⚲ ⚲ ⊞ ⌂ ⌂(32t, including mast stepping).
Town ⌂ ⌂ ⚓ ACA ⊠ Ⓑ ⚒ ✕ ⌑ ⇌ ✈ (Norwich).

NORFOLK BROADS: The Broads comprise about 120 miles of navigable rivers and lakes in Norfolk and Suffolk. The main rivers (Bure, Yare and Waveney) are tidal, flowing into the sea at Great Yarmouth. The N Broads have a 2·3m headroom limit. Br clearances restrict cruising to R Yare (Great Yarmouth to Norwich, but note that 3M E of Norwich, Postwick viaduct on S bypass has 10·7m HW clearance) and River Waveney (Lowestoft to Beccles). The Broads may be entered also at Lowestoft (9.4.24). Broads Navigation Authority ☎(01603) 610734.

Entry to the Broads: Pass up R Yare at slack LW, under Haven Bridge (1·8m HAT) thence to Breydon Water via Breydon Bridge (4·0m) or to R Bure. Both bridges lift in co-ordination to pass small craft in groups. All br lifts on request to Hbr Office (01493) 335503 during office hrs. Weekend and bank holiday lifts to be requested working afternoon before. Call the Bridge Officer on VHF Ch 12. R Bure has two fixed bridges (2·3m MHWS).

Tidal data on the rivers and Broads is based on the time of LW at Yarmouth Yacht Stn (mouth of R Bure), which is LW Gorleston +0100 (see TIDES). Add the differences below to time of LW Yarmouth Yacht Stn to get local LW times:

R Bure		R Waveney		R Yare	
Acle Bridge	+0230	Berney Arms	+0100	Reedham	+0130
Horning	+0300	St Olaves	+0130	Cantley	+0200
Potter Heigham	+0300	Oulton Broad	+0300	Norwich	+0300

LW at Breydon (mouth of R Yare) is LW Yarmouth Yacht Stn +0100. Tide starts to flood on Breydon Water whilst still ebbing from R Bure. Max draught is 1·8m; 2m with care. Tidal range 0·6m to 1·8m.

Licences (compulsory) are obtainable from: The Broads Authority, 18 Colegate, Norwich NR3 1BQ, ☎01603-610734; the Info Centre, Yarmouth Yacht Stn or the River Inspectors. *Hamilton's Guide to the Broads* is recommended. www.hamilton publications.com.

E England

North East England

Winterton to Berwick-upon-Tweed

9.5.1	Contents	329
9.5.2	Area map and Distance table	330
9.5.3	Tidal stream charts	332
9.5.4	Lights, buoys and waypoints	334
9.5.5	Passage information	336

North Sea passages, Oil and Gas Installations – 336 • North Norfolk Coast – 336, The Wash – 337 • The Wash to the River Humber, River Humber – 340 • River Humber to Hartlepool, Hartlepool to Coquet Island – 346 • Hartlepool to Southern Netherlands, Hartlepool to the German Bight – 353 • Coquet Island to Farne Islands, Farne Islands, Holy Island to Berwick – 361

9.5.6	Blakeney	336
9.5.7	Wells-next-the-Sea	337
	Burnham Overy Staithe • Brancaster Staithe	
9.5.8	King's Lynn	338
9.5.9	Wisbech	338
9.5.10	Boston	339
	River Welland • Wainfleet	
9.5.11	River Humber (Immingham, tide tables/curves)	344
	Grimsby • Hull • Goole • South Ferriby • Brough Haven • Naburn	
9.5.12	Bridlington	347
	Filey	
9.5.13	Scarborough	347
9.5.14	Whitby	348
	Runswick Bay • River Tees, Middlesbrough	
9.5.15	Hartlepool (River Tees, tide tables/curves)	353
9.5.16	Seaham	354
9.5.17	Sunderland	354
9.5.18	River Tyne/North Shields, tide tables/curves	355
	Cullercoats	
9.5.19	Blyth	359
	Newbiggin	
9.5.20	Amble	360
	Boulmer • Craster • Newton Haven and Beadnell Bay • Seahouses (North Sunderland) • Farne Islands	
9.5.21	Holy Island	361
9.5.22	Berwick-upon-Tweed	362

NE England

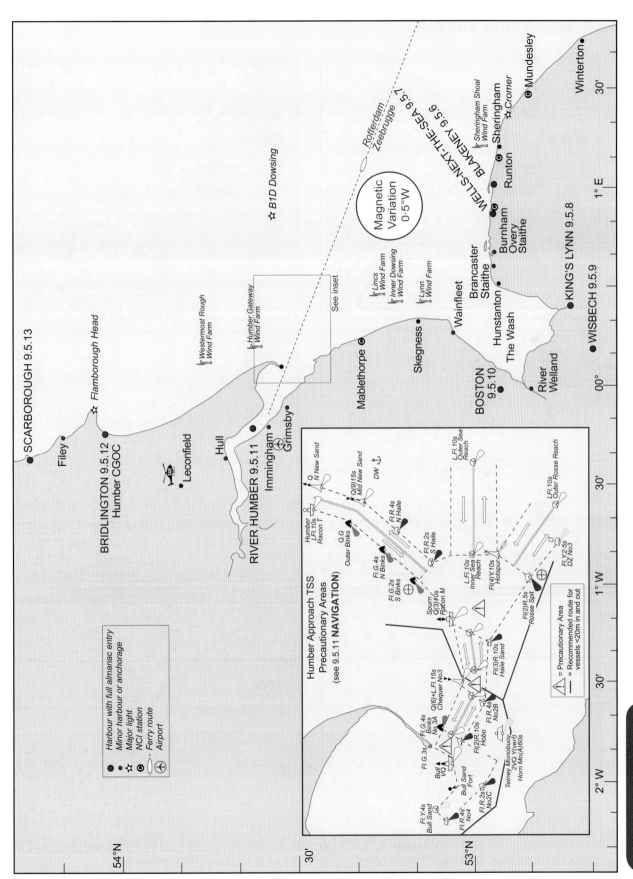

Magnetic Variation 0·5°W

☆ B1D Dowsing

WELLS-NEXT-THE-SEA 9.5.7

BLAKENEY 9.5.6

Rotterdam Zeebrugge

Sheringham Shoal Wind Farm

Mundesley
⊙ Cromer
Sheringham ★
Runton
◉ Winterton

KING'S LYNN 9.5.8

Burnham Overy Staithe

Brancaster Staithe
Hunstanton
Wainfleet
The Wash

WISBECH 9.5.9

Lincs Wind Farm
Inner Dowsing Wind Farm
Lynn Wind Farm

Westermost Rough Wind Farm

Humber Gateway Wind Farm

See inset

River Welland

BOSTON 9.5.10

Skegness

Mablethorpe ⊙

SCARBOROUGH 9.5.13
Filey

Flamborough Head ☆

BRIDLINGTON 9.5.12
Humber CGOC

Leconfield ✈

Hull

RIVER HUMBER 9.5.11

Immingham
Grimsby ⊙

Humber Approach TSS
Precautionary Areas
(see 9.5.11 **NAVIGATION**)

Q N New Sand

Q(9)15s Mid New Sand

DW ⚓

Humber LFl.10s Racon T

Fl.R.4s N Haile
Fl.R.2s S Haile

Q.G Outer Binks

Fl.G.4s N Binks

Fl.G.2s S Binks

Spurn Q(3)10s Racon M

LFl.10s Inner Sea Reach

LFl.10s Outer Sea Reach

LFl.10s Outer Rosse Reach

Fl(4)Y.10s Hotspur

Fl.Y.2·5s DZ No3

Fl(2)R.5s Rosse Spit

Fl(3)R.10s Haile Sand

Q(6)+LFl.15s Chequer No3

Fl.R.4s No2B

Fl(2)R.10s Hobo

Fl.G.4s Binks No3A

Fl.G.3s

Bull VQ

Bull Sand Fort

Tetney Monobuoy
2WQ.Y(vert)
Horn Mo(A)/60s

Fl.R.2s No2C

Fl.Y.4s Bull Sand
No4

Fl.R.4s No4

△ = Precautionary Area
│ = Recommended route for vessels <20m in and out

Harbour with full almanac entry
Minor harbour or anchorage
Major light
NCI station
Airport
Ferry route

54°N
30'
53°N

2°W   1°W   30'   00'   30'   1°E   30'

9.5.3 AREA 5 TIDAL STREAMS

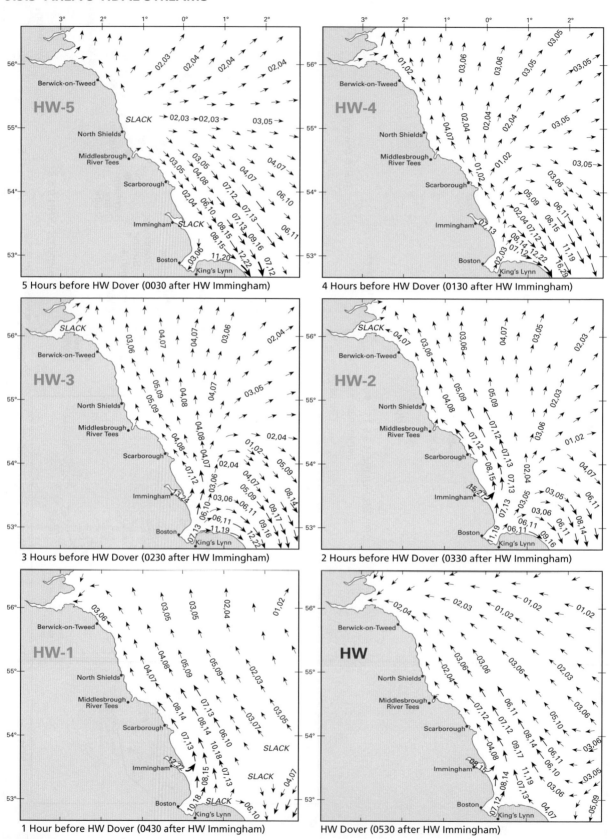

5 Hours before HW Dover (0030 after HW Immingham)

4 Hours before HW Dover (0130 after HW Immingham)

3 Hours before HW Dover (0230 after HW Immingham)

2 Hours before HW Dover (0330 after HW Immingham)

1 Hour before HW Dover (0430 after HW Immingham)

HW Dover (0530 after HW Immingham)

Northward 9.6.3   Southward 9.4.3

1 Hour after HW Dover (0555 before HW Immingham)

2 Hours after HW Dover (0455 before HW Immingham)

3 Hours after HW Dover (0355 before HW Immingham)

4 Hours after HW Dover (0255 before HW Immingham)

5 Hours after HW Dover (0155 before HW Immingham)

6 Hours after HW Dover (0055 before HW Immingham)

NE England

9.5.4 LIGHTS, BUOYS AND WAYPOINTS

Bold print = light with a nominal range of 15M or more. CAPITALS = place or feature. *CAPITAL ITALICS* = light-vessel, light float or Lanby. *Italics* = Fog signal. ***Bold italics*** = Racon. Some marks/buoys are fitted with AIS (<u>MMSI No</u>); see relevant charts.

GREAT YARMOUTH TO THE WASH

(Direction of buoyage ⇕ South to North)

YARMOUTH and CAISTER ROADS/COCKLE GATEWAY
SW Scroby ◣ Fl G 2·5s; 52°35'·13N 01°46'·69E.
Scroby Elbow ◣ Fl (2) G 5s; *Bell*; 52°36'·55N 01°46'·26E.
Yarmouth Outfall ◥ Q R; 52°37'·58N 01°45'·70E.
Mid Caister ◥ Fl (2) R 5s; *Bell*; 52°38'·99N 01°45'·66E.
NW Scroby ◣ Fl (3) G 10s; 52°40'·36N 01°46'·31E.
N Caister ◥ Fl (3) R 10s; 52°40'·77N 01°45'·65E.
Hemsby ◥ Fl R 2·5s; 52°41'·80N 01°46'·00E.
N Scroby ⚏ VQ; 52°41'·39N 01°46'·47E.
Cockle ⚏ VQ (3) 5s; *Bell*; 52°44'·03N 01°43'·59E.

OFFSHORE ROUTE
Cross Sand ⚏ L Fl 10s 6m 5M; ***Racon (T) 10M***; 52°37'·03N 01°59'·14E.
E Cross Sand ◥ Fl (4) R 15s; 52°38'·55N 01°53'·55E.
NE Cross Sand ⚏ VQ (3) 5s; 52°44'·22N 01° 53'·80E.
Smith's Knoll ⚏ Q (6) + L Fl 15s 7M; ***Racon (T) 10M***; 52°43'·52N 02°17'·89E.
S Winterton Ridge ⚏ Q (6) + L Fl 15s; 52°47'·21N 02°03'·44E.
E Hammond Knoll ⚏ Q (3) 10s; 52°52'·32N 01°58'·64E.
Hammond Knoll ⚏ Q (9) 15s; 52°49'·68N 01°57'·54E.
Newarp ⚏ L Fl 10s 7M; ***Racon (O) 10M***; 52°48'·37N 01°55'·69E; <u>992351118</u>.
S Haisbro ⚏ Q (6) + L Fl 15s; *Bell*; 52°50'·82N 01°48'·29E.
Mid Haisbro ◣ Fl (2) G 5s; 52°54'·22N 01°41'·59E.
N Haisbro ⚏ Q; ***Racon (T) 10M***; 53°00'·22N 01°32'·29E; <u>992351031</u>.
Happisburgh ☆ Fl (3) 30s 41m 14M; 52°49'·21N 01°32'·18E.

(Direction of buoyage ⇕ East to West)

CROMER
Cromer ☆ 52°55'·45N 01°19'·01E; Fl 5s 84m **21M**; W 8-sided twr; vis: 102°-307° H24; ***Racon (O) 25M***; <u>992351015</u>.
Tayjack Wk ◥ Fl R 2·5s; 52°57'·61N 01°15'·37E.
E Sheringham ⚏ Q (3) 10s; 53°02'·21N 01°14'·84E.
W Sheringham ⚏ Q (9) 15s; 53°02'·95N 01°06'·72E.

BLAKENEY
Blakeney Fairway (SWM) ◔ ; 52°59'·31N 00°57'·83E (approx).
Blakeney Overfalls ◥ Fl (2) R 5s; *Bell*; 53°03'·01N 01°01'·37E.

WELLS-NEXT-THE-SEA/BRANCASTER STAITHE
Wells WCM ⚏ Q(9) 15s; 52°59'·64N 00°51'·00E.
Bridgirdle ◥ Fl R 2·5s; 53°01'·73N 00°43'·95E.

APPROACHES TO THE WASH
S Race ⚏ Q (6) + L Fl 15s; *Bell*; 53°07'·81N 00°57'·34E.
E Docking ◥ Fl R 2·5s; 53°09'·82N 00°50'·39E.
N Race ◣ Fl G 5s; *Bell*; 53°14'·98N 00°43'·87E.
N Docking ⚏ Q; 53°14'·82N 00°41'·49E.
Scott Patch ⚏ VQ (3) 5s; 53°11'·12N 00°36'·39E.
S Inner Dowsing ⚏ Q (6) + L Fl 15s; *Bell*; 53°12'·12N 00°33'·69E.
Boygrift Tower ⸬ Fl (2) 10s 12m 5M; 53°17'·63N 00°19'·24E.
Burnham Flats ⚏ Q (9) 15s; *Bell*; 53°07'·53N 00°34'·89E.
Lincs WF E ⚏ Q (3) 10s; 53°07'·61N 00°31'·06E.

THE WASH
Met Mast ⌀ Mo(U) 15s 12m 10M Mast (87); *Horn Mo(U) 30s*; 53°18'·82N 00°44'·85E.
West Ridge ⚏ Q (9) 15s; 53°19'·06N 00°44'·47E.

N Well ⚏ L Fl 10s; *Whistle*; ***Racon (T) 10M***; 53°03'·02N 00°27'·87E.
Roaring Middle L Fl 10s 7m 8M; 52°58'·64N 00°21'·08E.

CORK HOLE/KING'S LYNN
Sunk ⚏ Q (9) 15s; 52°56'·29N 00°23'·40E.
Seal Sand ⚏ Q; *Bell*; 52°56'·00N 00°20'·00E.

WISBECH CHANNEL/RIVER NENE
Beacons are moved as required.

FREEMAN CHANNEL
Boston Roads ◔ L Fl 10s; 52°57'·69N 00°16'·96E.
Boston No. 1 ◣ Fl G 3s; 52°57'·88N 00°15'·16E.
Alpha ◥ Fl R 3s; 52°57'·65N 00°14'·99E.
No. 3 ◣ Fl G 6s; 52°57'·93N 00°15'·42E.
No. 5 ◣ Fl G 3s; 52°58'·51N 00°12'·72E.
Freeman Inner ⚏ Q (9) 15s; 52°58'·59N 00°11'·36E.
Delta ◥ Fl R 6s; 52°58'·38N 00°11'·25E.

BOSTON LOWER ROAD
Echo ◥ Fl R 3s; 52°58'·38N 00°10'·13E.
Foxtrot ◥ Fl R 3s; 52°57'·60N 00°08'·94E.
Boston No. 9 ◣ Fl G 3s; 52°57'·58N 00°08'·36E.
Black Buoy ◥ Fl (2) R 6s; 52°56'·82N 00°07'·74E.
Tabs Head ⚏ Q WG 4m 1M; R □ on W mast; vis: W shore - 251°- G - shore; 52°56'·00N 00°04'·91E.

BOSTON, NEW CUT AND RIVER WITHAM
Ent N side, Dollypeg ⚏ QG 4m 1M; B △ on Bn; 52°56'·13N 00°05'·03E.
New Cut ⚏ Fl G 3s; △ on pile; 52°55'·98N 00°04'·67E.
New Cut Ldg Lts 240°. Front, No. 1, F 5m 5M; 52°55'·85N 00°04'·40E. Rear, 90m from front, F 8m 5M.

WELLAND CUT/RIVER WELLAND
SE side ⚐ Iso R 2s; NW side Iso G 2s. Lts QR (to port) and QG (to stbd) mark the chan upstream; 52°55'·72E 00°04'·68E.

(Direction of buoyage ⇕ North to South)

BOSTON DEEP/WAINFLEET ROADS
Scullridge ◣ ; 52°59'·76N 00°13'·86E.
Off Ingoldmells Point ⚐ Fl Y 5s 22m 5M; Mast; *Mo (U) 30s*; 53°12'·49N 00°25'·85E.

THE WASH TO THE RIVER HUMBER

(Direction of buoyage ⇕ South to North)

Dudgeon ⚏ Q (9) 15s 7M; ***Racon (O) 10M***; 53°16'·62N 01°16'·90E.
E Dudgeon ⚏ Q (3) 10s; 53°19'·72N 00°58'·69E.
W Ridge ⚏ Q (9) 15s; 53°19'·04N 00°44'·50E.
Inner Dowsing ⚏ Q (3) 10s 7M, ***Racon (T) 10M***; Horn 60s; 53°19'·10N 00°34'·80E.
Mid Outer Dowsing ◣ Fl (3) G 10s; 53°24'·82N 01°07'·79E.
N Outer Dowsing ⚏ Q; 53°33'·52N 00°59'·59E; ***Racon (T) 10M***; <u>992351031</u>.
B.1D Platform Dowsing ⬡ 53°33'·68N 00°52'·63E; Fl (2) 10s 28m **22M**; Morse (U) R 15s 28m 3M; *Horn (2) 60s*.

RIVER HUMBER APPROACHES
Protector ◥ Fl R 2·5s; 53°24'·84N 00°25'·12E.
DZ No. 4 ◿ Fl Y 5s; 53°27'·15N 00°19'·06E.
DZ No. 3 ◿ Fl Y 2·5s 53°29'·30N 00°19'·21E.
Rosse Spit ◥ Fl (2) R 5s 53°30'·56N 00°16'·60E.
Outer Sand ⚏ Q (3) 10s; 53°36'·41N 00°29'·39E; <u>992351119</u>.
Haile Sand No. 2 ◥ Fl (3) R 10s; 53°32'·42N 00°13'·18E.
Humber lt float L Fl G 10s 5M; ***Racon (T) 7M***; 53°38'·70N 00°21'·24E.
Humber Gateway Met Mast ⬡ Mo(U) 15s 12m 10M; 53°38'·24N 00°15'·73E; *Horn Mo(U) 30s*; Y framework mast.
N Binks ◣ Fl G 4s; 53°36'·01N 00°18'·28E.
S Binks ◣ Fl G 2s 53°34'·74N 00°16'·55E.

SPURN Q (3) 10s 10m 8M; *Racon (M) 5M*; 53°33'·49N 00°14'·20E.
SE CHEQUER VQ (6)+L Fl 10s 6m 6M; 53°33'·27N 00°12'·33E.
Chequer No. 3 ↙ Q (6) + L Fl 15s; 53°33'·07N 00°10'·63E.
No 2B ↘ Fl R 4s; 53°32'·33N 00°09'·10E.
TETNEY ⬱ 2 VQ Y (vert); *Horn Mo (A)60s*; QY on 290m floating hose; 53°32'·35N 00°06'·76E.

RIVER HUMBER/GRIMSBY/HULL
Binks No. 3A ◤ Fl G 4s; 53°33'·92N 00°07'·43E.
Spurn Pt ↯ Fl G 3s 11m 5M; 53°34'·37N 00°06'·47E.
BULL ↙ VQ 8m 6M; 53°33'·54N 00°05'·70E.
Bull Sand ↘ Q R ; 53°34'·45N 00°03'·69E.
North Fort ↙ Q; 53°33'·80N 00°04'·19E.
South Fort ↙ Q (6) + L Fl 15s; 53°33'·65N 00°03'·96E.
Haile Sand Fort ↙ Fl R 5s 21m 3M; 53°32'·07N 00°01'·99E.
Haile Chan No. 4 ↘ Fl R 4s; 53°33'·64N 00°02'·84E.
Middle No. 7 VQ (6) + L Fl 10s; 53°35'·80N 00°01'·50E.
Grimsby Royal Dock ent E side ↙ Fl (2) R 6s 10m 8M; Dn; 53°35'·08N 00°04'·04W.
Killingholme Lts in line 292°. Front, Iso R 2s 10m 14M. 53°38'·78N 00°12'·96W. Rear, 215m from front, F WRG 22m 3M; vis: 289·5°-G-290·5°-G/W-291·5°-W-292·5°-R/W-293·5° -R-294·5° (H24).

RIVER HUMBER TO WHITBY
Canada & Giorgios Wreck ↙ VQ (3) 5s; 53°42'·37N 00°07'·16E.
Albrough intake ↙ Q R 17m 2M; 53°49'·01N 00°03'·38W.

BRIDLINGTON
SW Smithic ↙ Q (9) 15s; 54°02'·41N 00°09'·21W.
N Pier Hd ↙ Fl 2s 12m 9M; *Horn 60s*; (Tidal Lts) Fl R or Fl G; 54°04'·77N 00°11'·19W.
N Smithic ↙ VQ; *Bell*; 54°06'·22N 00°03'·90W.
Flamborough Hd ☆ 54°06'·98N 00°04'·96W; Fl (4) 15s 65m **24M**; W ○ twr; *Horn (2) 90s*.

FILEY/SCARBOROUGH/WHITBY
Filey Brigg ↙ Q (3) 10s; *Bell*; 54°12'·74N 00°14'·60E.
Scarborough E Pier Hd ↙ QG 8m 3M; 54°16'·88N 00°23'·36W.
Scarborough Pier ↙ Iso 5s 17m 9M; W ○ twr; vis: 219°- 039° (tide sigs); *Dia 60s*; 54°16'·91N 00°23'·40W.
Whitby ↙ Q; *Bell*; 54°30'·33N 00°36'·58W.
Whitby High ☆ 54°28'·67N 00°34'·10W; Fl WR 5s 73m **18M**, R16M; W 8-sided twr & dwellings; vis:128°-R-143°-W- 319°.
Whitby E Pier ↙ Fl(2) R 4s 16m 5M; 54°29'·57N 00°36'·71W; Masonry Twr.
Whitby E Pier Ext Hd ↙ Q R 12m 5M; 54°29'·65N 00°36'·74W.
Whitby W Pier ↙ Fl(2) G 4s 24m 5M; 54°29'·57N 00°36'·78W; Masonry Twr.
Whitby W Pier Ext Hd ↙ Q G 12m 5M; 54°29'·65N 00°36'·80W.

WHITBY TO THE RIVER TYNE
RUNSWICK/REDCAR
Salt Scar ↙ 54°38'·12N 01°00'·12W VQ; *Bell*.
Luff Way Ldg Lts 197°. Front, on Esplanade, FR 8m 7M; vis: 182°-212°; 54°37'·10N 01°03'·71W. Rear, 115m from front, FR 12m 7M; vis: 182°-212°.
High Stone. Lade Way Ldg Lts 247°. Front, Oc R 2·5s 9m 7M; 54°37'·15N 01°03'·92W. Rear, 43m from front, Oc R 2·5s 11m 7M; vis: 232°-262°.

TEES APPROACHES/HARTLEPOOL
Tees Approach ↙ L FL 10s 8M; *Racon (B)*; *Horn*; 54°40'·60N 01°04'·00W; AIS.
Bkwtr Hd S Gare ☆ 54°38'·85N 01°08'·25W; Fl WR 12s 16m **W20M, R17M**; W ○ twr; vis: 020°-W-240°-R-357°; Sig Stn.
Ldg Lts 210·1° Front, FR 18m 13M; 54°37'·22N 01°10'·20W.
Rear, 560m from front, FR 20m **16M**.
Longscar ↙ Q (3) 10s; *Bell*; 54°40'·86N 01°09'·89W.

The Heugh ☆ 54°41'·79N 01°10'·56W; Fl (2) 10s 19m **19M**; W twr.
Hartlepool Marina Lock Dir Lt 308° Dir Fl WRG2s 6m 3M; vis: 305·5°-G-307°-W-309°-R-310·5°; 54°41'·45N 01°11'·92W.

SEAHAM/SUNDERLAND
Seaham N Pier Hd ↙ Fl G 10s 12m 5M; W col, B bands; 54°50'·26N 01°19'·26W.
Sunderland Roker Pier Hd ☆ 54°55'·28N 01°21'·15W; Fl 5s 25m **18M**; W □ twr, 3 R bands and cupola: vis: 211°- 357°; *Siren 20s*.
Old N Pier Hd ↙ QG 12m 8M; metal column; 54°55'·13N 01°21'·61W.
DZ ↘ 54°57'·04N 01°18'·90W and ↘ 54°58'·61N 01°19'·90W; both Fl Y 2·5s.

TYNE ENTRANCE/NORTH SHIELDS
Ent North Pier Hd ☆ 55°00'·88N 01°24'·18W; Fl (3) 10s 26m **26M**; Gy □ twr, W lantern; *Horn 10s*.
Herd Groyne Hd Dir Ldg Lt 249°, Oc 10s 14m **19M**; vis: 246·5°-W-251·5°; R pile structure, R&W lantern. Same structure Oc RG 10s 13m, 11M; vis:224°-G-246·5°, 251·5°-R-277°; sync with Dir lt FR (unintens) 080°-224°. *Bell (1) 5s*; 55°00'·49N 01°25'·44W.

RIVER TYNE TO BERWICK-UPON-TWEED
CULLERCOATS and BLYTH
Cullercoats Ldg Lts 256°. Front, FR 27m 3M; 55°02'·06N 01°25'·91W. Rear, 38m from front, FR 35m 3M.
Blyth Ldg Lts 324°. Front ↯, F Bu 11m 10M; 55°07'·42N 01°29'·82W. Rear ↯, 180m from front, F Bu 17m 10M. Both Or ↯ on twr.
Blyth E Pier Hd ☆ 55°06'·98N 01°29'·21W; Fl (4) 10s 19m **21M**, W twr; same structure FR 13m 13M, vis:152°-249°; Horn (3) 30s.
Met Mast ↙; Mo(U)15s 17m 10M; 55°08'·78N 01°25'·25W; Aero 3 FR (107m), *Horn 30s*.

COQUET ISLAND, AMBLE and WARKWORTH
Coquet ☆ 55°20'·03N 01°32'·39W; Fl (3) WR 20s 25m **W19M, R15M**; W□twr, turreted parapet, lower half Gy; vis: 330°-R-140°-W-163°-R-180°-W-330°; sector boundaries are indeterminate and may appear as Alt WR; *Horn 30s*.
Amble N Pier Head ↙ Fl G 6s 12m 6M; 55°20'·39N 01°34'·25W.

SEAHOUSES, BAMBURGH and FARNE ISLANDS
N Sunderland ↘ Fl R 2·5s; 55°34'·62N 01°37'·12W.
Bamburgh Black Rocks Point ☆ 55°36'·99N 01°43'·45W; Oc(2) WRG 8s 12m **W14M**, R11M, G11M; W bldg; vis: 122°-G-165°-W-175°-R-191°-W- 238°-R- 275°-W- 289°-G-300°.
Inner Farne ↙ Fl (2) WR 15s 27m W10M, R7M; W ○ twr; vis: 119°- R - 280° - W - 119°; 55°36'·92N 01°39'·35W.
Longstone ☆ W side 55°38'·63N 01°36'·66W; Fl 20s 23m **24M** (H24); R twr, W band; 992351127.
Swedman ◤ Fl G 2·5s; 55°37'·65N 01°41'·63W.

HOLY ISLAND
Ridge ↙ Q (3) 10s; 55°39'·70N 01°45'·97W.
Plough Rock ↙ Q (9) 15s; 55°40'·24N 01°46'·00W.
Old Law E Bn ↙ (Guile Pt) Oc WRG 6s 9m 4M; vis: 180·5°-G-258·5°-W-261·5°-R-300°, 55°39'·49N 01°47'·59W; stone obelisk.
Heugh ↙ Oc WRG 6s 24m 5M; vis: 135°-G-308°-W-311-R- shore; 55°40'·09N 01°47'·99W.
Plough Seat ↘ QR; 55°40'·37N 01°44'·97W.
Goldstone ◢ QG; 55°40'·25N 01°43'·64W.

BERWICK-UPON-TWEED
Bkwtr Hd ↙ Fl 5s 15m 6M; vis: 201°-009°, (obscured 155°-201°); W ○ twr, R cupola and base; (FG (same twr) 8m 1M; vis 009°-G-155°; 55°45'·88N 01°59'·06W.

9.5.5 PASSAGE INFORMATION

More passage information is threaded between the harbours in this area. Admiralty Leisure Folios: 5614 Orford Ness to Whitby, 5615 Whitby to Berwick-upon-Tweed. **Bibliography:** *Forth, Tyne, Dogger, Humber* – Blakeney to St Abbs – (Imray/Irving) covers the southern part of this less frequented cruising ground. Northwards, use Royal Northumberland *YC's Sailing Directions Humber to Rattray Head. The North Sea (West) Pilot* (Admiralty NP54) covers the whole area.

NORTH SEA PASSAGES

See 9.0.7 for distances across the North Sea. There is further Passage Information in Area 4 for the southern North Sea, in Area 6 for crossing to Norway and the Baltic, and in Area 16 for crossings from the Frisian Islands, German Bight, Belgium and the Netherlands.

OIL AND GAS INSTALLATIONS

See 3.14. Any craft going offshore in the N Sea is likely to encounter oil or gas installations. These are shown on Admiralty charts, where scale permits; the position of mobile rigs is updated in weekly NMs. Safety zones of radius 500m are established round all permanent platforms, mobile exploration rigs, and tanker loading moorings, as described in the Annual Summary of Admiralty Notices to Mariners No 20. Some of these platforms are close together or inter-linked. Unauthorised vessels, including yachts, must not enter these zones except in emergency or due to stress of weather.

Platforms show a main lt, Fl Mo (U) 15s 15M. In addition secondary lts, Fl Mo (U) R 15s 2M, synchronised with the main lt, may mark projections at each corner of the platform if not marked by a W lt. The fog signal is Horn Mo (U) 30s. See the Admiralty List of Lights and Fog Signals, Vol A.

NORTH NORFOLK COAST

(AC 1503, 108) The coast of N Norfolk is unfriendly in bad weather, with no harbour accessible when there is any N in the wind. The hbrs all dry, and seas soon build up in the entrances or over the bars, some of which are dangerous even in a moderate breeze and an ebb tide. But in settled weather and moderate offshore winds it is a peaceful area to explore, particularly for boats which can take the ground. At Blakeney and Wells chans shift almost every year, so local knowledge is essential and may best be acquired in advance from the HM.

Haisborough Sand (buoyed) lies parallel to and 8M off the Norfolk coast at Happisburgh lt ho, with depths of less than 1m in many places, and drying 0·4m near the mid-point. The shoal is steep-to, on its NE side in particular, and there are tidal eddies. Even a moderate sea or swell breaks on the shallower parts. There are dangerous wks near the S end. Haisborough Tail and Hammond Knoll (with wk depth 1·6m) lie to the E of S end of Haisborough Sand. Newarp lt F is 5M SE. Similar banks lie parallel to and up to 60M off the coast.

▶ *The streams follow the generally NW/SE direction of the coast and offshore chans. But in the outer chans the stream is somewhat rotatory: when changing from SE-going to NW-going it sets SW, and when changing from NW-going to SE-going it sets NE, across the shoals. Close S of Haisborough Sand the SE-going stream begins at HW Immingham –0030; the NW-going at HW Immingham +0515, sp rates up to 2·5kn. It is possible to carry a fair tide from Gt Yarmouth to the Wash.* ◀

If proceeding direct from Cromer to the Humber, pass S of Sheringham Shoal (buoyed) and the windfarm close North, to the NE of Blakeney Overfalls (min depth 1m) and Docking Shoal, and then to SW of Race Bank, so as to fetch Inner Dowsing lt tr (lt, fog sig). Thence pass E of Protector Overfalls, and steer for Rosse Spit buoy at SE ent to R Humber.

▶ *At Sheringham Shoal the ESE-going stream begins at HW Immingham –0225, and the WNW-going at +0430.* ◀

9.5.6 BLAKENEY

Norfolk **52°59'·19N 00°57'·90E** (ent shifts) ✿⚓⚓✿✿✿

CHARTS AC 1190, 108; Imray C29, C28, Y9

TIDES –0445 Dover; ML Cromer 2·8; Duration 0530; Zone 0 (UT)

Standard Port IMMINGHAM (➞)

Times				Height (metres)			
High Water		Low Water		MHWS	MHWN	MLWN	MLWS
0100	0700	0100	0700	7·3	5·8	2·6	0·9
1300	1900	1300	1900				
Differences BLAKENEY BAR (approx 52°59'N 00°59'E)							
+0035	+0025	+0030	+0040	–1·6	–1·3	ND	ND
BLAKENEY (approx 52°57'N 01°01'E)							
+0115	+0055	ND	ND	–3·9	–3·8	ND	ND
CROMER							
+0044	+0032	+0108	+0059	–2·3	–1·8	–0·7	0·0

SHELTER Very good but limited access over shallow bar. In fresh on-shore winds conditions at the entrance deteriorate quickly, especially on the ebb. Moorings in The Pit area dry out.

NAVIGATION WPT 53°00'·00N 00°58'·20E, approx 225°/1M to Fairway RW buoy which is moved as required.

- Hjordis lit IDM buoy marks large dangerous wreck, close E of entrance.
- The bar is shallow, shifts often and the channel is marked by unlit SHM and PHM buoys, relaid each spring.
- Strangers are best advised to follow local boats in. Beware mussel lays, drying, off Blakeney Spit. Speed limit 8kn.

LIGHTS AND MARKS Conspic marks: Blakeney and Langham churches; a chimney on the house on Blakeney Point neck; TV mast (R lts) approx 2M S of entrance.

COMMUNICATIONS (Code 01263) CGOC (01262) 672317; Police 101; Dr 740314.

FACILITIES ⚓ ⛴ ⌖ ♠ ⚓ ⚓ ⤓ ♀ ⛴(15t); **Services** ⚒ ⚓ 🛢 ⛵ ⚓ (Stratton Long) ☎740362. **Village** 🍴 🍴 🛢 ✉ Ⓑ🍺 ✕ ⚑ ⇌ (Sheringham) ✈(Norwich).

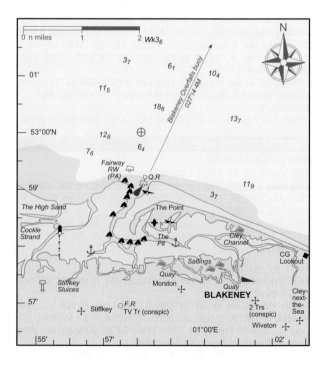

9.5.7 WELLS-NEXT-THE-SEA

Norfolk **52°59′·30N 00°49′·75E** (ent shifts) ✺❋♦♦♦♦☆☆☆

CHARTS AC 1190, 108; Imray C29, C28, Y9

TIDES –0445 Dover; ML 1·2 Duration 0540

Standard Port IMMINGHAM (⟶)

Times				Height (metres)			
High Water		Low Water		MHWS	MHWN	MLWN	MLWS
0100	0700	0100	0700	7·3	5·8	2·6	0·9
1300	1900	1300	1900				
Differences WELLS BAR (approx 52°59′N 00°49′E)							
+0020	+0020	+0020	+0020	–1·3	–1·0	ND	ND
WELLS-NEXT-THE-SEA (approx 52°58′N 00°51′E)							
+0035	+0045	+0340	+0310	–3·8	–3·8	CD+	CD+
BURNHAM OVERY STAITHE							
+0045	+0055	ND	ND	–5·0	–4·9	ND	ND
HUNSTANTON							
+0010	+0020	+0105	+0025	+0·1	–0·2	–0·1	0·0

NOTE: LW time differences at Wells are for the end of a LW stand which lasts about 4 hrs at sp and about 5 hrs at nps.

SHELTER Good, but entrance difficult in strong on-shore winds. Outer Harbour is mainly for commercial wind farm tenders and survey vessels, but with HM prior permission, vessels >2m draught may shelter in adverse conditions. Limited access (max draught 3m at springs); best attempted on the flood. Quay berths/pontoons at W end of Town Quay with 2m+ at LW.

NAVIGATION WPT Wells, WCM 52°59′·65N 00°50′·36E. Ent varies in depth and position; buoys are altered to suit. Initially keep to W side of chan to counter E-going tide; and to E side of chan from No 12 PHM lt buoy to Quay.

Obtain HM's advice, Hbr Launch often available to escort visitors into hbr & up to Quay. Spd limits: No 6 buoy to Quay 8kn; above this 5kn. See www.wellsharbour.co.uk for latest harbour chart.

LIGHTS AND MARKS From WCM Q (9) 15s the channel is well marked by lateral buoys, all Fl.3s, R to port, G to stbd. The lifeboat house (W with R roof) at N end of harbour is conspicuous.

COMMUNICATIONS (Code 01328) CGOC (01262) 672317; Police 101; Dr 710741; ⊞ 710097. HM 711646, mob 07775 507284.

Call '*Wells Harbour*' Ch 12 16 before entering, listens HW–2 to HW+2 and when vessels expected.

FACILITIES Main Quay ⌷ on pontoons or quay, £20-£24/night; ⌂ᴰ, ᴵᴼ, ᵂᶜ, ▯, ♣, ⚒ (9 ton), ⚓ (25 ton mobile), ⌂, ⌂, ✕ ⌂.

E Quay ⌐ (£11/craft/day), ⚓ ➘; Wells SC ☎710622, ⌐ ⌂;

Services Ⓔ ACA, ⚠ LB.

Town: ⌂ ⌂ (Hbr fuelling stn HW–2 to HW+3; supply up to 14,000 ltrs), Gas, ▯ ▯ R, ⌂ ⊠ Ⓑ ⇌ (bus to Norwich/King's Lynn), ✈ (Norwich).

ADJACENT HARBOURS

BURNHAM OVERY STAITHE, Norfolk, **52°58′·95N 00°46′·55E** ✺♦☆☆☆. AC 1190, 108. HW –0420 on Dover. See 9.5.8. Small drying hbr; ent chan has 0·3m MLWS. ⚓ off the Staithe only suitable in good weather. No lts. Scolt Hd is conspic to W and Gun Hill to E; Scolt Hd Island is conspic 3M long sandbank which affords some shelter. Chan varies constantly and buoys are moved to suit. Local knowledge advisable. Facilities: **Burnham Overy Staithe SC** ☎(01328) 738348 ⚓➘. Services: ⚓ ⌐ ♣ ✕ ✎ ⌂. **Burnham Market** EC Wed; ⌂ ⌂ ⚒ ✕ ⌂.

BRANCASTER STAITHE, Norfolk, **52°59′·02N 00°37′·65E** ✺♦☆☆☆. AC 1190, 108. HW –0425 on Dover; as Burnham. Small drying hbr, dangerous to enter except by day in settled weather. Spd limit 6kn. Appr from due N. Conspic golf club house with lt, Fl 5s 8m 3M, is 0·5M S of chan ent and Fairway buoy. Beware wk shown on chart. Sandbanks shift; buoys changed to suit. Scolt Hd conspic to E. Local knowledge or Pilot advised. Possible ⚓ in The Hole. HM ☎(01485) 210638. Facilities: **Brancaster Staithe SC** ☎210249 ✕ ⌂; **Hunstanton SC** ☎(01485) 534705. Services: ♣ ✕ ✎ ⌂ ⌂ ⌂ ⌂. ⌂ ⌂ ⚒ ✕ ⌂.

THE WASH

(AC 108,1200) The Wash is formed by the estuaries of the rivers Great Ouse, Nene, Welland and Witham; it is an area of shifting sands, most of which dry. Important features are the strong tidal streams, the low-lying shore, and the often poor vis. Watch the echo sounder carefully, because buoys may have been moved to accommodate changes in the channel. ▶*Near North Well, the in-going stream begins at HW Immingham –0430, and the out-going at HW Immingham +0130, sp rates about 2kn. The in-going stream is usually stronger than the out-going, but its duration is less. Prolonged NE winds cause an in-going current, which can increase the rate and duration of the in-going stream and raise the water level at the head of the estuary. Do not attempt entry to the rivers too early on the flood, which runs hard in the rivers.* ◀

North Well SWM lt buoy and Roaring Middle lt F are the keys to entering the Wash from N or E. But from the E it is also possible to appr via a shallow route N of Stiffkey Overfalls and Bridgirdle PHM buoy; thence via Sledway and Woolpack PHM lt buoy to North Well and into Lynn Deeps. Near north end of Lynn Deeps there are overfalls over Lynn Knock at sp tides. Approach King's Lynn via the buoyed/lit Bulldog chan. For R Nene (Wisbech) follow the Wisbech Chan. Boston and R Welland are reached via Freeman Channel, westward from Roaring Middle; or via Boston Deep, all lit.

The NW shore of The Wash is fronted by mudflats extending 2–3M offshore and drying more than 4m; a bombing range is marked by Y bns and buoys. Wainfleet Swatchway should only be used in good vis; the buoyed chan shifts constantly, and several shoals (charted depths unreliable) off Gibraltar Pt obstruct access to Boston Deep. In an emergency Wainfleet offers good shelter (dries).

Wisbech Cut provides access inland to the Fens Middle Levels, which are navigable by narrow boat and river cruisers. Dog in a Doublet Lock above Wisbech is the practical limit of sea going vessels.

The Gt Ouse is navigable as far as Ely 24M upstream. There is a lock half-way at Denver Sluice ☎(01366) 382340/VHF Ch 73, and low bridges beyond.

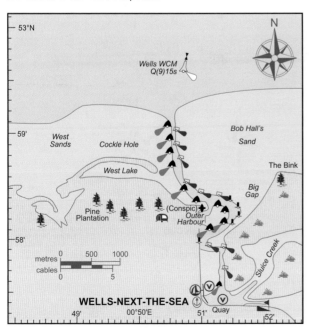

WELLS-NEXT-THE-SEA

9.5.8 KING'S LYNN

Norfolk 52°49'·72N 00°21'·14E (W Stones Bn) ✿❀☽✿✿

CHARTS AC 1190, 108, 1200, 5614. Imray C29, Y9

TIDES –0443 Dover, +0030 Immingham; ML 3·6; Duration 0340

Standard Port IMMINGHAM (→)

Times				Height (metres)			
High Water		Low Water		MHWS	MHWN	MLWN	MLWS
0100	0700	0100	0700	7·3	5·8	2·6	0·9
1300	1900	1300	1900				
Differences KING'S LYNN							
+0030	+0030	+0305	+0140	–0·5	–0·8	–0·8	+0·1
WEST STONES							
+0025	+0025	+0115	+0040	–0·3	–0·4	–0·3	+0·2

SHELTER Port is well sheltered; entry only recommended HW–3½. A busy commercial and fishing port. 3 Visitors pontoons on S quay. River moorings available from local clubs; keep clear of FV moorings.

NAVIGATION WPT Sunk WCM 52° 56'.29N 00° 23'.40E follow Bulldog chan buoyed/lit apprs to King's Lynn. Upstream to Denver Sluice (ent to inland routes) 6 bridges span river, min clearance 9·15m less ht of tide above dock sill at King's Lynn. Allow 1½h for passage. The sandbanks regularly shift and extend several miles into the Wash; buoyage is altered accordingly. Contact HM a week beforehand for info or see www.sailthewash.com.

LIGHTS AND MARKS See 9.5.4 and chartlet.

COMMUNICATIONS Code (01553) CGOC (01262) 672317; Police 101; HM 0773411 Docks 691555.

Kings Lynn Hbr Radio Ch **14** 11 (M-F: 0800-1730 LT. Other times: HW –4 to HW+1.) (ABP) Ch **14** 16 11 (HW–2½ to HW).

FACILITIES ⌷ £42 (<9m/48hrs), ⚓ ▲ ⚒ ⚓ ⑂ Ⓔ ⬡(32t) ⛽.

Town ⑂ 🏦 ✉ Ⓑ 🛒 ✕ 🗲 ✈ (Norwich).

Ely Marina 24M up the Great Ouse River, ☎(01353) 664622, ⚓ ⬛
⚓ ♿ ▲ ⚒ ⚓ ⑂ ⬡(10t) ⛽.

9.5.9 WISBECH

Cambridgeshire 52°40'·02N 00°09'·55E ✿❀☽✿✿

CHARTS AC 1190, 108, 1200, 5614. Imray C29, Y9

TIDES –0443 Dover, +0030 Immingham; ML 3·6; Duration 0340

Standard Port IMMINGHAM (→)

Times				Height (metres)			
High Water		Low Water		MHWS	MHWN	MLWN	MLWS
0100	0700	0100	0700	7·3	5·8	2·6	0·9
1300	1900	1300	1900				
OUTER WESTMARK KNOCK							
+0010	+0015	+0040	+0020	–0·2	–0·5	–0·6	–0·4
WISBECH CUT							
+0020	+0010	+0120	+0055	–0·3	–0·7	–0·4	ND
WISBECH							
+0055	+0040	DR	DR	–0·2	–0·6	DR	DR

SHELTER Excellent in river with Wisbech Yacht Harbour pontoon berths and facilities immediately below Freedom Br in Wisbech. Vessels of 4·8m draught can reach Wisbech at sp (3·4m at nps).

NAVIGATION WPT Roaring Middle Lt Flt 52° 58'.64N 00° 21'.08E 220°/7M to Wisbech No 1 SHM Fl G 5s. Call HM at RAF No 4 ECM (⚓ for waiting) to check commercial traffic and Cross Keys Sw Br. Red flags and lights shown when Holbeach Firing Range in use. Chan to R Nene well marked with lit buoys/bns. Best ent HW–3. Given 24H notice of visit Cross Keys Sw Br in river opens by request. Waiting pontoon 0.5M below bridge thence 6M to Wisbech. Departure best at HW; if draught >3m, clear Kerr NCM by HW+3. For pilotage notes ☎ Hbr Office or visit www.fenland.gov.uk.

LIGHTS AND MARKS See 9.5.4 and chartlet.

COMMUNICATIONS Code (01945) CGOC (01262) 672317; Police 101; Dr 582133; Ⓗ 585781; HM 588059 (24H); Cross Keys Swing Bridge ☎01406 350364.

HM *Wisbech Harbour* Ch **09** call HM and Cross Keys Sw Br HW±3.

FACILITIES **Yacht Hbr** (102⌷+15Ⓥ) £1, min charge £5, ⚓ ⬡ ⚒ ⛽.
♿(barge). **Town** ⑂ 🏦 🛒 ✕ ⚓.

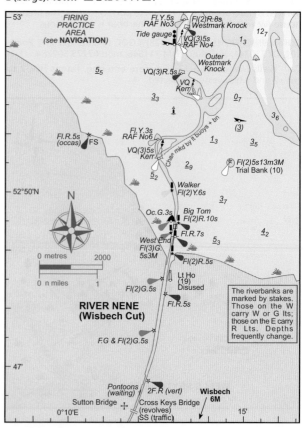

9.5.10 BOSTON

Lincs **52°56'·00N 00°04'·92E** (Tabs Head bn)   🌸🌸⚓⚓🌸🌸

CHARTS AC 108, 1200, 5614; Imray C29, Y9

TIDES –0415 Dover; ML 3·3; Duration Flood 0500, Ebb 0700

Standard Port IMMINGHAM (→)

Times				Height (metres)			
High Water		Low Water		MHWS	MHWN	MLWN	MLWS
0100	0700	0100	0700	7·3	5·8	2·6	0·9
1300	1900	1300	1900				
Differences BOSTON							
0000	+0010	+0140	+0050	–0·5	–1·0	–0·9	–0·5
TABS HEAD (WELLAND RIVER)							
0000	+0005	+0125	+0020	+0·2	–0·2	–0·2	–0·2
SKEGNESS							
+0010	+0015	+0030	+0020	–0·4	–0·5	–0·1	0·0
INNER DOWSING LIGHT TOWER							
0000	0000	+0010	+0010	–0·9	–0·7	–0·1	+0·3

SHELTER Very good. Except in emergency, berthing in the Dock is prohib without HM permission. Yachts secure just above Dock ent and see HM. The port is administered by Port of Boston Ltd.

Yachts capable of lowering masts can pass through the Grand Sluice (***24 hrs notice required***) into fresh water. Sluice 16m x 9m and opens twice a tide at approx HW±2. It leads into the R Witham Navigation which goes 31M to Lincoln. Marina is to stbd immediately beyond the sluice. British Waterways have 50 berths, with ⚓ and ⛽, beyond Grand Sluice.

NAVIGATION WPT Boston Rds SWM lt buoy, 52°57'·61N 00°16'·29E, 282°/0·75M to Freeman Chan ent. Thence Bar Chan is well marked, but liable to change. SW of Clay Hole a new chan has formed which dries 0·6m at entr; it is marked by SHM lt buoys Nos 11N, 13N, and 15. Tabs Head marks the ent to the river; it should be passed not earlier than HW–3 to enable the Grand Sluice to be reached before the start of the ebb. Beware rocky bottom within 100m of the Sluice. On reaching Boston Dock, masts should be lowered to negotiate swing bridge (cannot always be opened) and three fixed bridges. Chan through town is un-navigable at LW.

LIGHTS AND MARKS St Boltoph's ch tr (the Boston Stump) is conspic from afar. New Cut and R Witham are marked by bns with topmarks. FW lts mark ldg lines: three pairs going upstream and six pairs going downstream.

COMMUNICATIONS (Code 01205) CGOC (01262) 672317; Police 101; Ⓗ 364801. Dock office 365571 (H24 & emergency); HM 362328; Grand Sluice Control 364864 (not always manned but has answerphone) mob 07712 010920.

All vessels between No 9 buoy and Grand Sluice must listen Ch 12 (also advisable to maintain watch while in local tidal waters). Call: *Boston Port Control* VHF Ch 12 (HW –2½ to HW +1½ when commercial shipping is moving). Ch 11 used by dock staff as required. *Grand Sluice* Ch 74.

FACILITIES Boston Marina ☎364420, (48⛵+Ⓥ) £7/night any size inc ⛽; ⚓ ⚓ ACA, ⚓ in dock (see HM); **Grand Sluice** 🚻/🚿; **BWB** moorings: 1st night free, then £5·00. **Services:** 🔋 ⚒ ⚓ Ⓔ. **Town** Thurs; ✉ Ⓑ 🛒 ✕ 🛏 ⇌ ✈(Humberside).

ADJACENT HARBOURS

RIVER WELLAND, Lincolnshire, **52°56'·00N 00°04'·92E** (Tabs Head bn). AC 1190, 1200. At Welland Cut HW –0440 on Dover; ML 0·3m; Duration 0520. See 9.5.10. At Tabs Head bn HW ±3, ent Welland Cut which is defined by training walls and lt bns. Beware sp flood of up to 5kn. Berth 6M up at **Fosdyke Yacht Haven** ☎01205 260240. (50⛵ + 6Ⓥ 2m depth); ⛽ ⚓ 🚻/🚿 🔋 by prior arrangement, ⚒ ⚓ (50t) – only one between R. Humber & Lowestoft ✕ 🛏.

WAINFLEET, Lincolnshire, **53°04'·79N 00°19'·89E** (chan ent). AC 108. Skegness HW +0500 on Dover. See 9.5.10 (Skegness). ML 4·0m; Duration 0600. Shelter good. Recommended max boat length 10m, draught 1·5m. Lit PHM indicates entrance to drying channel close S of Gibraltar Point. Chan through saltings marked by buoys and bns with lateral topmarks but no lts. Enter HW ±1½. Facilities: ⚓ ⛵ (larger boats at fishing jetties, smaller at YC), ⚓ at YC ⚓. **Town** All facilities at Skegness (3½ miles).

Skegness YC skegnessyachtclub.co.uk ☎0779 941856 (Commodore) on the R Steeping inside Gibraltar Pt has a membership of about 65 with a small clubhouse. ⚓⚓⚓ at ⚓ ⛵ ⚓ 🚻 🚿.

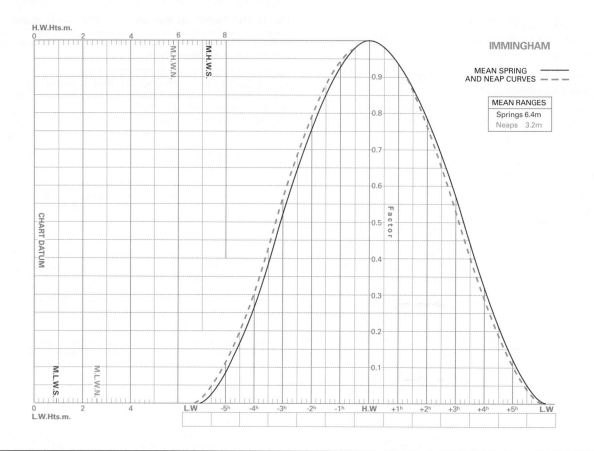

IMMINGHAM

MEAN SPRING
AND NEAP CURVES

MEAN RANGES

Springs 6.4m

Neaps 3.2m

THE WASH TO THE RIVER HUMBER

(AC 108, 107) Inner Dowsing is a narrow N/S sandbank with a least depth of 1·2m, 8M offshore between Skegness and Mablethorpe. There are overfalls off the W side of the bank at the N end. Inner Dowsing ECM (Q(3) 10s, Racon T) is to the NE of the bank, S Inner Dowsing and Scott Patch indicate its southernmost extent. Inshore of the bank lie three wind farms: Lynn (27 turbines), Inner Dowsing (27 turbines) and Lincs Offshore (75 Turbines) some lit Fl Y 5s and some by R Aero lts. The extremities of the wind farms are indicated by a series of cardinal marks while under construction; when completed selected turbines are marked by lights and fog signals.

In the outer approaches to The Wash and R. Humber there are many offlying banks, but few of them are of direct danger to yachts. The sea however breaks on some of them in bad weather, when they should be avoided. Fishing vessels may be encountered, and there are many oil/gas installations offshore.

RIVER HUMBER

(AC 104, 1188, 3497) The Humber is formed by the Ouse and the Trent, which meet 13M above Kingston-upon-Hull. The river is commercially important and gives access to these rivers and inland waterways; it also drains most of Yorkshire and the Midlands. Where the Humber estuary reaches the sea between Northcoates Pt and Spurn Hd it is 4M wide. A VTS scheme is in operation to regulate commercial shipping in the Humber, Ouse and Trent and provide full radar surveillance. Yachts are advised to monitor the appropriate Humber VTS frequency. Advice is at www.humber.com/Yachting_and_Leisure/Pleasure_Craft_Navigation/

Approaching from the S, a yacht should make good Rosse Spit and then Haile Sand No 2, both PHM lt buoys, before altering westward leading SW of Bull Channel.

If bound to/from the N, avoid The Binks, a shoal (dries 1·6m in places) extending 3M E from Spurn Hd, with a rough sea when wind is against tide. Depths offshore are irregular and subject to frequent change; it is best to cross the main channel at No 3 Chequer following the small vessel route and follow a line N of SE Chequer to close Spurn before heading NNE to pass inshore of the Humber Gateway wind farm (72 turbines). In poor weather extend offshore to S Binks SHM.

Haile Sand and Bull Sand Forts are both conspic to the SW of Spurn Head; beyond them it is advisable to keep just outside one of the buoyed chans, since shoals are liable to change. Hawke Chan (later Sunk) is the main dredged chan to the N. Haile Chan favours the S side and Grimsby. Bull Chan takes a middle course before merging with Haile Chan. There are good yachting facilities at Grimsby and Hull. ⌓ inside Spurn Head.

▶ *Streams are strong, even fierce at Sp; local info suggests that they are stronger than shown in Tidal Streams based upon NP 251 (Admiralty Tidal Stream Atlas). 5ca S of Spurn Hd the flood sets NW from about HW Immingham −0520, Sp rate 3·5kn; the ebb sets SE from about HW Immingham, Sp rate 4kn. The worst seas are experienced in NW gales against a strong flood tide. 10M E of Spurn Hd the tidal streams are not affected by the river; relative to HW Immingham, the S-going stream begins at −0455, and the N-going at +0130. Nearer the entrance the direction of the S-going stream becomes more W'ly, and that of the N-going stream more E'ly.* ◀

IMMINGHAM   LAT 53°38'N   LONG 0°11'W
TIMES AND HEIGHTS OF HIGH AND LOW WATERS

STANDARD TIME (UT)
For Summer Time add ONE hour in **non-shaded areas**

Dates in red are **SPRINGS**
Dates in blue are **NEAPS**

YEAR 2016

JANUARY

Day	Time	m	Day	Time	m
1 F	0411 / 1019 / 1610 / 2221	1.9 / 6.1 / 2.4 / 6.2	16 SA	0415 / 1018 / 1628 / 2228	1.3 / 6.5 / 1.8 / 6.8
2 SA	0456 / 1111 / 1659 / 2320	2.2 / 5.8 / 2.7 / 5.9	17 SU	0509 / 1120 / 1727 / 2333	1.6 / 6.3 / 2.1 / 6.5
3 SU	0550 / 1215 / 1800	2.4 / 5.7 / 2.8	18 M	0613 / 1232 / 1839	1.9 / 6.1 / 2.3
4 M	0031 / 0652 / 1322 / 1908	5.8 / 2.5 / 5.7 / 2.9	19 TU	0050 / 0726 / 1346 / 1959	6.3 / 2.1 / 6.1 / 2.3
5 TU	0141 / 0758 / 1424 / 2020	5.8 / 2.4 / 5.9 / 2.7	20 W	0209 / 0840 / 1455 / 2114	6.3 / 2.0 / 6.2 / 2.1
6 W	0244 / 0900 / 1519 / 2126	5.9 / 2.2 / 6.2 / 2.4	21 TH	0321 / 0944 / 1556 / 2217	6.4 / 1.9 / 6.5 / 1.7
7 TH	0338 / 0955 / 1606 / 2221	6.2 / 2.0 / 6.5 / 2.0	22 F	0424 / 1040 / 1647 / 2312	6.6 / 1.6 / 6.8 / 1.3
8 F	0426 / 1043 / 1649 / 2310	6.4 / 1.7 / 6.7 / 1.7	23 SA	0518 / 1130 / 1733	6.8 / 1.5 / 7.0
9 SA	0511 / 1128 / 1730 / 2356	6.7 / 1.5 / 7.0 / 1.4	24 SU	0002 / 0605 / 1215 / 1814 ○	1.1 / 7.0 / 1.3 / 7.2
10 SU ●	0555 / 1211 / 1810	6.9 / 1.4 / 7.2	25 M	0048 / 0646 / 1256 / 1853	1.0 / 7.0 / 1.3 / 7.3
11 M	0040 / 0637 / 1253 / 1849	1.2 / 7.0 / 1.2 / 7.3	26 TU	0129 / 0724 / 1333 / 1930	0.9 / 7.0 / 1.3 / 7.3
12 TU	0123 / 0719 / 1334 / 1927	1.0 / 7.1 / 1.2 / 7.4	27 W	0205 / 0759 / 1405 / 2003	1.0 / 6.9 / 1.4 / 7.2
13 W	0204 / 0800 / 1414 / 2006	0.9 / 7.1 / 1.2 / 7.4	28 TH	0237 / 0830 / 1434 / 2033	1.2 / 6.7 / 1.6 / 7.0
14 TH	0246 / 0842 / 1455 / 2048	1.0 / 7.0 / 1.3 / 7.3	29 F	0306 / 0859 / 1503 / 2104	1.5 / 6.5 / 1.8 / 6.7
15 F	0328 / 0927 / 1538 / 2134	1.1 / 6.8 / 1.5 / 7.1	30 SA	0337 / 0931 / 1535 / 2139	1.7 / 6.3 / 2.0 / 6.4
			31 SU	0411 / 1010 / 1613 / 2221	2.0 / 6.1 / 2.3 / 6.1

FEBRUARY

Day	Time	m	Day	Time	m
1 M	0455 / 1100 / 1703 / 2316 ◑	2.3 / 5.8 / 2.6 / 5.8	16 TU	0541 / 1156 / 1810	2.0 / 6.0 / 2.3
2 TU	0553 / 1209 / 1809	2.5 / 5.6 / 2.8	17 W	0028 / 0658 / 1315 / 1937	6.1 / 2.4 / 5.9 / 2.4
3 W	0035 / 0700 / 1329 / 1922	5.6 / 2.6 / 5.6 / 2.8	18 TH	0156 / 0821 / 1431 / 2101	6.0 / 2.4 / 6.0 / 2.2
4 TH	0200 / 0812 / 1438 / 2037	5.7 / 2.5 / 5.9 / 2.6	19 F	0315 / 0930 / 1538 / 2206	6.1 / 2.2 / 6.3 / 1.8
5 F	0308 / 0918 / 1535 / 2147	5.9 / 2.4 / 6.2 / 2.2	20 SA	0421 / 1026 / 1632 / 2259	6.4 / 1.9 / 6.6 / 1.4
6 SA	0404 / 1015 / 1624 / 2246	6.3 / 1.9 / 6.6 / 1.6	21 SU	0512 / 1114 / 1716 / 2346	6.7 / 1.6 / 6.9 / 1.1
7 SU	0453 / 1106 / 1709 / 2337	6.6 / 1.6 / 6.9 / 1.3	22 M ○	0553 / 1157 / 1756	6.8 / 1.4 / 7.1
8 M ●	0539 / 1153 / 1751	6.9 / 1.3 / 7.2	23 TU	0028 / 0629 / 1237 / 1833	1.0 / 6.9 / 1.3 / 7.2
9 TU	0025 / 0623 / 1238 / 1832	1.1 / 7.1 / 1.1 / 7.5	24 W	0107 / 0702 / 1312 / 1908	0.9 / 6.9 / 1.3 / 7.2
10 W	0110 / 0705 / 1321 / 1912	0.7 / 7.3 / 0.9 / 7.7	25 TH	0141 / 0732 / 1344 / 1939	1.0 / 6.9 / 1.3 / 7.1
11 TH	0153 / 0745 / 1402 / 1953	0.6 / 7.3 / 0.9 / 7.7	26 F	0211 / 0759 / 1411 / 2007	1.1 / 6.8 / 1.4 / 7.0
12 F	0233 / 0825 / 1443 / 2034	0.6 / 7.3 / 0.9 / 7.6	27 SA	0238 / 0825 / 1437 / 2036	1.3 / 6.7 / 1.5 / 6.8
13 SA	0314 / 0907 / 1524 / 2118	0.8 / 7.1 / 1.2 / 7.4	28 SU	0304 / 0854 / 1504 / 2106	1.6 / 6.5 / 1.7 / 6.6
14 SU	0355 / 0953 / 1608 / 2209	1.1 / 6.7 / 1.5 / 7.0	29 M	0332 / 0927 / 1536 / 2140	1.8 / 6.3 / 2.0 / 6.2
15 M	0442 / 1047 / 1701 / 2310 ◑	1.6 / 6.3 / 1.9 / 6.5			

MARCH

Day	Time	m	Day	Time	m
1 TU	0407 / 1007 / 1619 / 2224 ◑	2.1 / 6.0 / 2.3 / 5.9	16 W	0512 / 1124 / 1747	2.2 / 6.0 / 2.2
2 W	0458 / 1100 / 1720 / 2328	2.5 / 5.7 / 2.6 / 5.6	17 TH	0014 / 0629 / 1248 / 1919	5.9 / 2.6 / 5.7 / 2.4
3 TH	0609 / 1224 / 1837	2.7 / 5.5 / 2.7	18 F	0144 / 0800 / 1407 / 2045	5.8 / 2.7 / 5.8 / 2.2
4 F	0114 / 0728 / 1357 / 1957	5.5 / 2.7 / 5.7 / 2.5	19 SA	0305 / 0911 / 1516 / 2147	6.0 / 2.4 / 6.1 / 1.8
5 SA	0241 / 0843 / 1504 / 2114	5.8 / 2.4 / 6.0 / 2.1	20 SU	0409 / 1006 / 1610 / 2238	6.3 / 2.0 / 6.5 / 1.4
6 SU	0344 / 0948 / 1557 / 2221	6.2 / 2.0 / 6.5 / 1.6	21 M	0455 / 1052 / 1654 / 2322	6.6 / 1.7 / 6.8 / 1.2
7 M	0435 / 1043 / 1644 / 2316	6.6 / 1.6 / 6.9 / 1.1	22 TU	0532 / 1134 / 1733	6.7 / 1.5 / 7.0
8 TU	0521 / 1132 / 1728	7.0 / 1.2 / 7.3	23 W ○	0002 / 0604 / 1213 / 1808	1.0 / 6.9 / 1.3 / 7.1
9 W ●	0005 / 0604 / 1220 / 1811	0.5 / 7.3 / 0.9 / 7.6	24 TH	0040 / 0635 / 1249 / 1842	1.0 / 6.9 / 1.2 / 7.1
10 TH	0051 / 0646 / 1304 / 1853	0.5 / 7.5 / 0.7 / 7.8	25 F	0113 / 0702 / 1320 / 1913	1.0 / 6.9 / 1.3 / 7.0
11 F	0134 / 0726 / 1347 / 1936	0.3 / 7.5 / 0.6 / 7.9	26 SA	0143 / 0729 / 1348 / 1942	1.1 / 6.9 / 1.3 / 6.9
12 SA	0215 / 0805 / 1428 / 2018	0.4 / 7.5 / 0.6 / 7.8	27 SU	0210 / 0755 / 1413 / 2011	1.3 / 6.8 / 1.4 / 6.8
13 SU	0255 / 0845 / 1508 / 2103	0.6 / 7.2 / 0.9 / 7.4	28 M	0235 / 0824 / 1440 / 2040	1.5 / 6.7 / 1.6 / 6.6
14 M	0334 / 0928 / 1551 / 2152	1.1 / 6.9 / 1.3 / 6.9	29 TU	0300 / 0854 / 1510 / 2112	1.7 / 6.5 / 1.8 / 6.3
15 TU	0418 / 1019 / 1641 / 2253 ◑	1.6 / 6.4 / 1.8 / 6.4	30 W	0332 / 0930 / 1550 / 2154	2.0 / 6.2 / 2.1 / 6.0
			31 TH	0419 / 1018 / 1647 / 2252 ◑	2.4 / 5.9 / 2.3 / 5.7

APRIL

Day	Time	m	Day	Time	m
1 F	0527 / 1127 / 1804	2.6 / 5.6 / 2.5	16 SA	0124 / 0728 / 1338 / 2017	5.7 / 2.8 / 5.8 / 2.1
2 SA	0028 / 0649 / 1309 / 1924	5.5 / 2.7 / 5.7 / 2.4	17 SU	0238 / 0842 / 1445 / 2118	5.9 / 2.6 / 6.0 / 1.9
3 SU	0213 / 0809 / 1428 / 2043	5.7 / 2.5 / 6.0 / 2.2	18 M	0339 / 0937 / 1540 / 2207	6.2 / 2.2 / 6.3 / 1.6
4 M	0319 / 0918 / 1526 / 2152	6.2 / 2.0 / 6.5 / 1.5	19 TU	0425 / 1024 / 1625 / 2250	6.4 / 1.9 / 6.6 / 1.3
5 TU	0412 / 1016 / 1616 / 2249	6.7 / 1.6 / 7.0 / 1.0	20 W	0502 / 1106 / 1705 / 2330	6.6 / 1.6 / 6.8 / 1.2
6 W	0459 / 1108 / 1703 / 2340	7.0 / 1.1 / 7.4 / 0.6	21 TH	0535 / 1145 / 1741	6.8 / 1.4 / 6.8
7 TH ●	0543 / 1157 / 1749	7.3 / 0.8 / 7.7	22 F ○	0007 / 0604 / 1222 / 1814	1.1 / 6.9 / 1.3 / 6.9
8 F	0027 / 0624 / 1244 / 1834	0.4 / 7.5 / 0.5 / 7.8	23 SA	0043 / 0632 / 1255 / 1846	1.2 / 6.9 / 1.3 / 6.9
9 SA	0112 / 0704 / 1329 / 1918	0.3 / 7.6 / 0.4 / 7.9	24 SU	0115 / 0701 / 1325 / 1919	1.2 / 6.9 / 1.3 / 6.8
10 SU	0154 / 0744 / 1412 / 2003	0.5 / 7.5 / 0.5 / 7.7	25 M	0144 / 0731 / 1353 / 1950	1.3 / 6.9 / 1.3 / 6.7
11 M	0235 / 0824 / 1454 / 2049	0.7 / 7.3 / 0.8 / 7.4	26 TU	0210 / 0800 / 1422 / 2022	1.5 / 6.8 / 1.5 / 6.6
12 TU	0314 / 0907 / 1538 / 2139	1.1 / 6.9 / 1.2 / 6.9	27 W	0238 / 0831 / 1454 / 2056	1.7 / 6.6 / 1.7 / 6.4
13 W	0356 / 0955 / 1627 / 2240	1.7 / 6.5 / 1.7 / 6.3	28 TH	0312 / 0907 / 1535 / 2139	1.9 / 6.4 / 1.9 / 6.1
14 TH ◑	0447 / 1058 / 1731	2.3 / 6.0 / 2.1	29 F	0357 / 0954 / 1630 / 2236	2.2 / 6.1 / 2.1 / 5.8
15 F	0000 / 0557 / 1219 / 1856	5.9 / 2.7 / 5.8 / 2.3	30 SA ◐	0500 / 1057 / 1741	2.5 / 5.9 / 2.2

Chart Datum: 3·90 metres below Ordnance Datum (Newlyn). HAT is 8·0 metres above Chart Datum.

NE England

IMMINGHAM LAT 53°38'N LONG 0°11'W
TIMES AND HEIGHTS OF HIGH AND LOW WATERS

STANDARD TIME (UT)
For Summer Time add ONE hour in **non-shaded areas**

Dates in red are **SPRINGS**
Dates in blue are NEAPS

YEAR 2016

MAY

Time	m		Time	m
1 0000	5.7	**16** 0156	5.8
0617	2.6	0756	2.7
SU 1222	5.9	M 1406	5.9
1857	2.1	2036	2.0
2 0141	5.9	**17** 0256	6.0
0735	2.4	0859	2.4
M 1348	6.1	TU 1503	6.1
2012	1.8	2128	1.8
3 0249	6.2	**18** 0345	6.2
0846	2.0	0949	2.1
TU 1452	6.5	W 1551	6.3
2120	1.4	2213	1.6
4 0344	6.7	**19** 0426	6.5
0948	1.6	1033	1.8
W 1547	6.9	TH 1633	6.5
2220	1.0	2254	1.5
5 0433	7.0	**20** 0501	6.6
1043	1.2	1114	1.6
TH 1638	7.3	F 1711	6.6
2313	0.7	2334	1.4
6 0519	7.3	**21** 0533	6.8
1135	0.8	1153	1.5
F 1728	7.6 ●	SA 1746	6.7 ○
7 0003	0.5	**22** 0012	1.3
0602	7.5	0604	6.9
SA 1225	0.6	SU 1230	1.4
1816	7.7	1822	6.7
8 0050	0.5	**23** 0048	1.3
0644	7.5	0637	6.9
SU 1312	0.5	M 1305	1.4
1904	7.7	1858	6.7
9 0134	0.6	**24** 0120	1.4
0726	7.5	0711	6.9
M 1357	0.5	TU 1337	1.4
1951	7.5	1935	6.7
10 0216	0.8	**25** 0151	1.5
0807	7.3	0743	6.9
TU 1442	0.7	W 1410	1.4
2038	7.2	2011	6.6
11 0256	1.2	**26** 0223	1.6
0850	7.0	0816	6.8
W 1526	1.1	TH 1446	1.5
2128	6.8	2048	6.5
12 0338	1.7	**27** 0300	1.8
0936	6.6	0854	6.6
TH 1615	1.5	F 1528	1.6
2227	6.3	2132	6.3
13 0424	2.2	**28** 0345	2.0
1033	6.2	0940	6.4
F 1711	1.9 ◑	SA 1621	1.8
2337	5.9	2227	6.1
14 0522	2.6	**29** 0441	2.2
1145	5.9	1038	6.2
SA 1820	2.1	SU 1723	1.8
		2339	5.9 ◑
15 0050	5.8	**30** 0550	2.3
0637	2.8	1150	6.2
SU 1300	5.8	M 1832	1.8
1934	2.2		
		31 0104	6.0
		0703	2.3
		TU 1310	6.3
		1943	1.7

JUNE

Time	m		Time	m
1 0215	6.3	**16** 0258	6.0
0814	2.0	0904	2.4
W 1420	6.5	TH 1511	6.1
2051	1.4	2130	1.9
2 0315	6.6	**17** 0345	6.2
0920	1.7	0956	2.1
TH 1521	6.8	F 1558	6.2
2153	1.2	2217	1.8
3 0408	6.9	**18** 0425	6.5
1020	1.3	1042	1.8
F 1617	7.1	SA 1641	6.4
2249	1.0	2301	1.6
4 0457	7.1	**19** 0503	6.7
1115	1.0	1126	1.6
SA 1711	7.3	SU 1721	6.6
2341	0.8	2343	1.5
5 0543	7.3	**20** 0539	6.8
1207	0.7	1207	1.5
SU 1803	7.4 ●	M 1802	6.7 ○
6 0030	0.8	**21** 0023	1.4
0627	7.4	0616	6.9
M 1258	0.6	TU 1247	1.4
1853	7.4	1842	6.7
7 0116	0.8	**22** 0101	1.4
0710	7.4	0654	7.0
TU 1345	0.6	W 1325	1.3
1942	7.3	1922	6.8
8 0159	1.0	**23** 0136	1.4
0752	7.3	0730	7.0
W 1430	0.8	TH 1403	1.2
2028	7.1	2001	6.8
9 0239	1.3	**24** 0213	1.5
0834	7.1	0806	7.0
TH 1514	1.0	F 1442	1.3
2116	6.8	2041	6.7
10 0319	1.7	**25** 0251	1.6
0918	6.8	0845	6.9
F 1558	1.4	SA 1523	1.3
2206	6.4	2124	6.5
11 0359	2.0	**26** 0334	1.7
1006	6.4	0929	6.8
SA 1645	1.7	SU 1611	1.4
2302	6.1	2215	6.4
12 0445	2.4	**27** 0425	1.9
1104	6.1	1021	6.6
SU 1736	2.0 ◐	M 1705	1.6
		2316	6.2 ◑
13 0003	5.8	**28** 0524	2.1
0540	2.6	1125	6.5
M 1210	5.9	TU 1808	1.7
1834	2.2		
14 0105	5.7	**29** 0028	6.1
0646	2.7	0633	2.2
TU 1316	5.8	W 1239	6.4
1937	2.2	1915	1.7
15 0204	5.8	**30** 0141	6.2
0759	2.7	0746	2.1
W 1417	5.9	TH 1353	6.5
2037	2.1	2025	1.6

JULY

Time	m		Time	m
1 0247	6.4	**16** 0302	6.0
0857	1.9	0911	2.4
F 1501	6.6	SA 1524	6.0
2131	1.5	2140	2.1
2 0346	6.7	**17** 0351	6.3
1002	1.5	1010	2.1
SA 1604	6.8	SU 1614	6.3
2230	1.3	2231	1.8
3 0439	6.9	**18** 0435	6.6
1100	1.2	1100	1.8
SU 1702	7.0	M 1700	6.5
2324	1.2	2318	1.6
4 0528	7.1	**19** 0517	6.8
1155	0.9	1147	1.5
M 1756	7.1 ●	TU 1744	6.7 ○
5 0014	1.1	**20** 0002	1.5
0613	7.3	0557	7.0
TU 1246	0.7	W 1231	1.3
1846	7.2	1826	6.9
6 0100	1.1	**21** 0043	1.3
0656	7.3	0636	7.2
W 1333	0.7	TH 1314	1.1
1932	7.2	1908	7.0
7 0143	1.2	**22** 0123	1.3
0737	7.3	0715	7.3
TH 1416	0.8	F 1354	1.0
2015	7.0	1948	7.0
8 0221	1.3	**23** 0201	1.2
0817	7.2	0753	7.3
F 1457	1.0	SA 1433	0.9
2055	6.8	2028	7.0
9 0257	1.6	**24** 0240	1.3
0856	7.0	0832	7.3
SA 1535	1.3	SU 1513	1.0
2135	6.5	2110	6.9
10 0331	1.9	**25** 0321	1.4
0935	6.7	0915	7.2
SU 1611	1.6	M 1556	1.2
2217	6.2	2155	6.7
11 0408	2.2	**26** 0407	1.6
1018	6.3	1004	6.9
M 1651	1.9	TU 1644	1.4
2303	6.0	2249	6.4 ◑
12 0451	2.4	**27** 0500	1.9
1111	6.0	1102	6.6
TU 1738	2.2 ◑	W 1741	1.7
2359	5.8	2355	6.2
13 0544	2.7	**28** 0605	2.2
1214	5.8	1214	6.4
W 1834	2.4	TH 1850	1.9
14 0101	5.7	**29** 0109	6.1
0647	2.8	0722	2.2
TH 1322	5.7	F 1333	6.3
1936	2.4	2005	2.0
15 0204	5.8	**30** 0222	6.2
0759	2.7	0841	2.1
F 1427	5.8	SA 1450	6.4
2042	2.3	2115	1.9
		31 0328	6.4
		0950	1.7
		SU 1559	6.6
		2216	1.7

AUGUST

Time	m		Time	m
1 0425	6.7	**16** 0409	6.5
1050	1.4	1036	1.8
M 1659	6.8	TU 1641	6.5
2310	1.5	2254	1.7
2 0514	7.0	**17** 0453	6.9
1143	1.0	1113	1.4
TU 1751	7.0	W 1726	6.8
2359	1.3 ●	2341	1.5
3 0558	7.2	**18** 0535	7.1
1232	0.8	1212	1.1
W 1836	7.1	TH 1809	7.0 ○
4 0043	1.2	**19** 0024	1.3
0639	7.3	0616	7.4
TH 1317	0.7	F 1256	0.9
1916	7.1	1850	7.2
5 0124	1.2	**20** 0106	1.1
0718	7.3	0656	7.6
F 1357	0.8	SA 1338	0.7
1953	7.0	1930	7.3
6 0200	1.3	**21** 0146	1.1
0755	7.3	0735	7.6
SA 1433	1.0	SU 1417	0.7
2027	6.9	2009	7.3
7 0232	1.5	**22** 0225	1.0
0829	7.1	0815	7.6
SU 1505	1.2	M 1456	0.9
2059	6.7	2049	7.1
8 0302	1.7	**23** 0305	1.2
0902	6.8	0858	7.4
M 1535	1.6	TU 1536	1.1
2129	6.4	2131	6.9
9 0332	1.9	**24** 0348	1.5
0936	6.5	0945	7.1
TU 1608	1.9	W 1620	1.5
2205	6.2	2221	6.5
10 0407	2.2	**25** 0438	1.8
1017	6.2	1042	6.7
W 1647	2.2	TH 1714	1.9
2250	5.9 ◑	2324	6.2
11 0453	2.5	**26** 0542	2.2
1111	5.8	1155	6.2
TH 1738	2.5	F 1825	2.3
2354	5.7		
12 0553	2.8	**27** 0042	6.0
1225	5.6	0704	2.4
F 1842	2.6	SA 1322	6.0
		1948	2.4
13 0111	5.7	**28** 0201	6.0
0704	2.8	0831	2.2
SA 1345	5.6	SU 1445	6.1
1953	2.6	2104	2.3
14 0221	5.8	**29** 0312	6.3
0822	2.6	0941	1.8
SU 1455	5.8	M 1557	6.4
2104	2.4	2204	2.0
15 0320	6.2	**30** 0410	6.7
0937	2.3	1038	1.4
M 1552	6.2	TU 1654	6.7
2204	2.0	2255	1.7
		31 0457	7.0
		1127	1.1
		W 1739	6.9
		2341	1.4

Chart Datum: 3·90 metres below Ordnance Datum (Newlyn). HAT is 8·0 metres above Chart Datum.

FREE monthly updates. Register at
www.reedsnauticalalmanac.co.uk

STANDARD TIME (UT)	IMMINGHAM LAT 53°38'N LONG 0°11'W	Dates in red are SPRINGS
For Summer Time add ONE hour in **non-shaded areas**	TIMES AND HEIGHTS OF HIGH AND LOW WATERS	Dates in blue are NEAPS

YEAR 2016

SEPTEMBER

Time	m	Time	m
1 0538 1212 TH 1817 ●	7.2 0.9 7.1	**16** 0509 1147 F 1747 ○	7.3 1.0 7.2
2 0023 0617 F 1254 1852	1.3 7.3 0.8 7.1	**17** 0001 0551 SA 1232 1827	1.2 7.6 0.7 7.4
3 0102 0654 SA 1331 1925	1.3 7.3 0.9 7.0	**18** 0044 0632 SU 1315 1907	0.9 7.8 0.5 7.5
4 0135 0728 SU 1404 1954	1.3 7.3 1.1 7.0	**19** 0126 0714 M 1355 1945	0.8 7.9 0.5 7.5
5 0205 0800 M 1432 2021	1.4 7.1 1.3 6.8	**20** 0207 0756 TU 1435 2025	0.8 7.8 0.7 7.4
6 0232 0829 TU 1459 2048	1.6 6.9 1.6 6.7	**21** 0248 0840 W 1514 2106	1.0 7.6 1.1 7.1
7 0258 0900 W 1526 2119	1.8 6.6 1.8 6.4	**22** 0330 0928 TH 1556 2154	1.3 7.1 1.6 6.6
8 0329 0934 TH 1558 2157	2.1 6.3 2.2 6.1	**23** 0419 1026 F 1647 ◑ 2255	1.8 6.6 2.1 6.2
9 0409 1018 F 1645 ◑ 2249	2.4 5.9 2.5 5.8	**24** 0523 1142 SA 1800	2.2 6.1 2.6
10 0506 1126 SA 1750	2.7 5.6 2.8	**25** 0016 0649 SU 1313 1930	5.9 2.4 5.9 2.7
11 0013 0619 SU 1306 1908	5.6 2.8 5.5 2.8	**26** 0139 0818 M 1436 2047	6.0 2.2 6.1 2.5
12 0142 0740 M 1426 2027	5.7 2.7 5.7 2.6	**27** 0251 0925 TU 1546 2145	6.2 1.8 6.4 2.2
13 0248 0902 TU 1528 2134	6.1 2.3 6.1 2.2	**28** 0348 1018 W 1637 2234	6.6 1.5 6.7 1.8
14 0340 1007 W 1619 2228	6.5 1.8 6.6 1.8	**29** 0434 1104 TH 1716 2317	6.9 1.2 6.9 1.5
15 0426 1100 TH 1704 2315	6.9 1.3 6.9 1.5	**30** 0514 1146 F 1751 2357	7.1 1.0 7.0 1.4

OCTOBER

Time	m	Time	m
1 0552 1224 SA 1823 ●	7.2 1.0 7.1	**16** 0525 1204 SU 1802 ○	7.7 0.6 7.5
2 0035 0627 SU 1300 1853	1.3 7.3 1.0 7.1	**17** 0020 0609 M 1249 1842	0.8 7.9 0.5 7.6
3 0108 0700 M 1331 1921	1.3 7.2 1.2 7.0	**18** 0105 0653 TU 1331 1922	0.7 7.9 0.5 7.6
4 0137 0731 TU 1359 1947	1.4 7.1 1.4 7.0	**19** 0149 0738 W 1412 2002	0.7 7.8 0.8 7.5
5 0203 0800 W 1425 2014	1.5 6.9 1.6 6.8	**20** 0231 0824 TH 1452 2044	0.9 7.6 1.2 7.2
6 0229 0830 TH 1450 2043	1.7 6.7 1.8 6.6	**21** 0315 0914 F 1534 2131	1.2 7.1 1.7 6.8
7 0258 0902 F 1519 2117	1.9 6.4 2.1 6.3	**22** 0404 1012 SA 1623 ◑ 2230	1.7 6.4 2.2 6.3
8 0335 0942 SA 1601 2201	2.2 6.0 2.5 6.0	**23** 0507 1129 SU 1731 2348	2.1 6.1 2.7 6.0
9 0428 1039 SU 1703 ◑ 2307	2.5 5.7 2.8 5.7	**24** 0630 1254 M 1859	2.3 5.9 2.9
10 0542 1221 M 1824	2.7 5.5 2.9	**25** 0110 0752 TU 1410 2017	6.0 2.2 6.0 2.7
11 0053 0702 TU 1354 1946	5.7 2.6 5.7 2.8	**26** 0220 0857 W 1516 2117	6.2 1.9 6.3 2.4
12 0210 0822 W 1459 2057	6.0 2.3 6.2 2.4	**27** 0318 0949 TH 1606 2206	6.5 1.6 6.6 2.0
13 0307 0931 TH 1552 2155	6.5 1.8 6.6 1.9	**28** 0406 1033 F 1645 2249	6.8 1.4 6.8 1.7
14 0355 1027 F 1638 2246	6.9 1.3 7.0 1.5	**29** 0447 1114 SA 1720 2329	6.9 1.3 7.0 1.5
15 0440 1117 SA 1721 2334	7.3 0.9 7.3 1.1	**30** 0524 1151 SU 1752 ●	7.1 1.2 7.0
		31 0006 0559 M 1227 1821	1.4 7.1 1.2 7.1

NOVEMBER

Time	m	Time	m
1 0041 0632 TU 1300 1850	1.4 7.1 1.3 7.1	**16** 0045 0636 W 1309 1902	0.7 7.8 0.7 7.6
2 0111 0705 W 1329 1919	1.4 7.0 1.4 7.0	**17** 0132 0724 TH 1352 1944	0.7 7.7 0.9 7.5
3 0139 0737 TH 1356 1947	1.5 6.9 1.6 6.9	**18** 0218 0812 F 1433 2026	0.8 7.5 1.2 7.3
4 0206 0809 F 1422 2017	1.7 6.7 1.8 6.8	**19** 0303 0902 SA 1515 2112	1.1 7.1 1.7 6.9
5 0237 0841 SA 1453 2049	1.8 6.4 2.0 6.5	**20** 0352 0959 SU 1601 2206	1.5 6.6 2.2 6.5
6 0314 0920 SU 1533 2130	2.0 6.2 2.4 6.2	**21** 0448 1107 M 1656 ◑ 2315	1.9 6.2 2.6 6.2
7 0403 1013 M 1628 ◑ 2227	2.3 5.9 2.7 6.0	**22** 0556 1221 TU 1809	2.2 5.9 2.9
8 0510 1134 TU 1742 2349	2.5 5.7 2.8 5.9	**23** 0031 0711 W 1329 1929	6.0 2.2 5.9 2.9
9 0627 1312 W 1903	2.4 5.8 2.7	**24** 0140 0816 TH 1431 2038	6.1 2.1 6.1 2.6
10 0120 0742 TH 1422 2016	6.0 2.1 6.1 2.4	**25** 0241 0911 F 1524 2131	6.2 1.9 6.3 2.3
11 0227 0852 F 1519 2120	6.4 1.7 6.6 2.0	**26** 0333 0957 SA 1609 2217	6.5 1.7 6.6 2.0
12 0322 0952 SA 1608 2216	6.9 1.3 7.0 1.5	**27** 0417 1038 SU 1647 2259	6.6 1.6 6.8 1.7
13 0412 1045 SU 1654 2308	7.3 1.0 7.3 1.2	**28** 0457 1118 M 1721 2338	6.8 1.5 6.9 1.6
14 0500 1135 M 1738 ○ 2357	7.6 0.8 7.5 0.9	**29** 0533 1155 TU 1753 ●	6.8 1.4 7.0
15 0549 1223 TU 1820	7.8 0.7 7.6	**30** 0014 0608 W 1231 1824	1.5 6.9 1.4 7.1

DECEMBER

Time	m	Time	m
1 0049 0643 TH 1303 1857	1.5 6.9 1.5 7.1	**16** 0119 0715 F 1335 1929	0.7 7.5 1.0 7.5
2 0120 0718 F 1333 1929	1.5 6.8 1.6 7.0	**17** 0206 0803 SA 1417 2012	0.7 7.3 1.2 7.4
3 0151 0753 SA 1403 2000	1.6 6.7 1.7 6.9	**18** 0251 0850 SU 1457 2055	1.0 7.0 1.6 7.1
4 0224 0829 SU 1436 2032	1.7 6.5 1.9 6.7	**19** 0336 0940 M 1537 2142	1.3 6.7 1.9 6.7
5 0301 0908 M 1515 2112	1.8 6.3 2.1 6.5	**20** 0422 1035 TU 1621 2236	1.6 6.3 2.2 6.4
6 0347 0956 TU 1604 2202	1.9 6.1 2.3 6.3	**21** 0513 1135 W 1712 ◑ 2341	2.0 6.0 2.7 6.1
7 0444 1059 W 1707 ◑ 2307	2.1 5.9 2.5 6.2	**22** 0611 1237 TH 1815	2.3 5.8 2.9
8 0552 1221 TH 1821	2.1 5.9 2.6	**23** 0049 0715 F 1339 1930	5.9 2.4 5.8 2.9
9 0025 0704 F 1339 1935	6.2 2.0 6.1 2.4	**24** 0155 0819 SA 1436 2044	6.0 2.3 6.0 2.7
10 0143 0814 SA 1443 2045	6.4 1.8 6.4 2.1	**25** 0254 0914 SU 1528 2141	6.1 2.1 6.2 2.3
11 0249 0919 SU 1539 2148	6.7 1.5 6.8 1.7	**26** 0345 1001 M 1612 2228	6.3 1.9 6.5 2.0
12 0347 1018 M 1629 2245	7.1 1.2 7.1 1.3	**27** 0430 1045 TU 1651 2311	6.4 1.7 6.7 1.8
13 0442 1111 TU 1717 2339	7.3 1.0 7.3 1.0	**28** 0509 1126 W 1727 2351	6.6 1.6 6.9 1.6
14 0534 1202 W 1803 ○	7.3 0.9 7.5	**29** 0547 1205 TH 1803 ●	6.7 1.5 7.0
15 0030 0625 TH 1250 1846	0.8 7.6 0.9 7.6	**30** 0030 0625 F 1242 1839	1.5 6.8 1.6 7.1
		31 0107 0703 SA 1316 1913	1.4 6.8 1.5 7.1

Chart Datum: 3·90 metres below Ordnance Datum (Newlyn). HAT is 8·0 metres above Chart Datum.

NE England

9.5.11 RIVER HUMBER

S bank: NE and N Lincolnshire. N bank: E Riding of Yorks and City of Kingston-upon-Hull.
Hull marina: **53°44'·24N 00°20'·15W** ❀❀◌◌◌◌ ✿✿✿ ✿✿

CHARTS AC 1190, 107, 104, 1188, 3496, 3497, 5614; Imray C29; ABP

TIDES –0510 Immingham, –0452 Hull, Dover; ML 4·1; Duration 0555

Standard Port IMMINGHAM (←)

Times				Height (metres)			
High Water		Low Water		MHWS	MHWN	MLWN	MLWS
0100	0700	0100	0700	7·3	5·8	2·6	0·9
1300	1900	1300	1900				
Differences BULL SAND FORT							
–0020	–0030	–0035	–0015	–0·4	–0·3	+0·1	+0·2
GRIMSBY							
–0012	–0014	–0015	–0013	–0·3	–0·2	+0·1	+0·4
HULL (ALBERT DOCK)							
+0019	+0019	+0033	+0027	+0·3	+0·1	–0·1	–0·2
HUMBER BRIDGE							
+0027	+0022	+0049	+0039	–0·1	–0·4	–0·7	–0·6
BURTON STATHER (R Trent)*							
+0105	+0050	+0240	+0205	–2·0	–2·7	–2·2	–1·1
KEADBY (R Trent)*							
+0135	+0120	+0425	+0410	–2·8	–3·3	–2·3	–0·9
BLACKTOFT (R Ouse)†							
+0100	+0055	+0325	+0255	–1·6	–1·8	–2·2	–1·1
GOOLE (R Ouse)†							
+0130	+0115	+0355	+0350	–1·6	–2·1	–1·9	–0·6

* Normal river level at Burton Stather is about 0·1m below CD, and at Keadby 0·1m to 0·2m below CD.

† Heights of LW can increase by up to 0·3m at Blacktoft and 0·6m at Goole when river in spate. HW hts are little affected.

SHELTER R Humber is the estuary of the rivers Ouse and Trent. The estuary has strong tidal streams: a strong NW'ly against a spring flood of 3-4kn causes a short steep sea, as does a SE'ly over the ebb stream. Off Hull Marina it can be very choppy with fresh winds over spring tidal streams.

Anchorages Inside Spurn Hd only with winds from NE to ESE, or close off Haile Sand Fort in fair weather. In S to W winds there is a good ⚓ off the SW bank 8ca above N Killingholme Oil jetty, well out of main chan. In N'lies ⚓ off Hawkin's Pt, N of S9 buoy.

Marinas at Hull, S Ferriby and the docks at Goole are all entered by lock access HW±3. At Grimsby lock best to await levels at HW±2. Otherwise £10 lock fee. Gates may shut at extreme HW Sp. Immingham should only be used in emergency. S Ferriby should not be attempted without up-to-date ABP charts for the ever-changing buoyed chan above Hull. Waiting pontoon at S Ferriby.

Do not attempt entry to Winteringham or Brough Havens without contacting Humber Yawl Club for details of approach channel and mooring availability. Both dry to soft mud. Winteringham is prone to heavy silting, but is dredged.

NAVIGATION Note TSS on chartlet 9.5.2 and on AC 109, guidance at www.humber.com/Yachting_and_Leisure/Pleasure_Craft_Navigation
From S, WPT 53°30'·50N 00°16'·50E, 1ca SW of Rosse Spit PHM buoy, Fl (2) R 5s. Thence pass S of Haile Sand No 2, then via No 2B, Tetney monobuoy and No 2C into Haile Channel.

From N, WPT 53°34'·74N 00°16'·55E, S Binks SHM buoy, Fl G 2s. Thence passing N of Spurn Lt Float, SE Chequer to Chequer No 3 crossing to S of Bull Channel at this point. This avoids strong tidal streams off Spurn Hd. Best arrival at LW. Sp tides are fierce: 4·4kn ebb off Spurn Head and Immingham. There is a big ship ⚓ S of Spurn Hd. Keep clear of large commercial vessels using Hawke (8·4m) and Sunk (8·8m) Chans; these are marked by S1-S9 SHM buoys, all Fl G 1·5s (S8 is a SHM bn, Fl G 1·5s with tide gauge); and by P2-P9 PHM buoys, all Fl R 1·5s. Foul Holme Chan (SHM buoys FH1 to FH9) off Immingham is recommended for small craft.

For **Grimsby**, the lock into Fish Docks is 255°/1·35M from Lower Burcom No 6 Lt Float, Fl R 4s. The jetty off the W entrance to Royal Dock has been extended further into Grimsby Road. Many offshore energy service cats use the port travelling at up to 28kts.

Off **Kingston-upon-Hull** a tidal eddy and streams can be rotatory: the flood makes W up Hull Roads for ¾hr whilst the ebb is already running down-river over Skitter Sand on the opposite bank (reaches 2½kn at sp). Humber Bridge (conspic) has 29m clearance.

LIGHTS AND MARKS The Humber is well marked. At Grimsby a conspic tr (94m with 'minaret' on top) is 300m W of Fish Dock lock. IPTS at Grimsby, Immingham, Killingholme, Hull and Goole. A number of jetties with F lts have additional FL lts shown in fog.

COMMUNICATIONS (Codes: Grimsby 01472; Hull 01482) Police 101; Humber HM 01482 327171; see also: www.humber.com

Advise *VTS Humber* on Ch 14 (or ☎01482 212191) position and intentions on appr from seaward of Clee Ness lt float; listen on Ch 14 / 12 when W of the float and Ch 15 upstream fm the Humber Bridge to Gainsborough (R Trent) and Goole (R Ouse). ⚓, nav & tidal information is broadcast on Ch 12/14 every odd H+03; more detailed info, including height of tide, is available on request.

GRIMSBY Port Director 327171; CGOC (01262) 672317.

HULL Lock ☎330508; CGOC (01262) 672317; ⊖ 782107.

Other VHF stns: *Grimsby Docks Radio* Ch **74** (H24) 18 79 call Humber Cruising for marina staff or *Fishdock Island* for lock keeper. Immingham *Docks Radio* Ch 19 68 (H24). R Hull Port Ops Service call *Drypool Radio* Ch 22 (Mon-Fri HW–2 to HW+1; Sat 0900-1100 LT). *Hull Marina* Ch M **80** (H24); *Albert Dock Radio* Ch 09. *Ferriby Sluice* (R Ancholme Lock) Ch 74. Humber YC, Ch M (if racing). *Goole Docks Radio* Ch 14 (H24) 09 19. Boothferry Bridge Ch 09 (H24). Selby Railway and Toll Bridges Ch 09. Br Waterways locks Ch 74. *Brough and Winteringham Havens, Humber Yawl Club* Ch M.

FACILITIES

Humber Mouth YC ☎(01472) 812063, M drying at Tetney Haven. Humber Cruising Association.

GRIMSBY NE Lincs (01472) **Fish Docks** Lock access (R/G tfc lts) HW±3 £10 locking fee for ⓥ, both gates open for free flow HW±2, enter free of charge with authorisation by *Fish Dock Island*, Ch **74**. Gates may close at extreme HW springs to prevent flooding.

Meridian Quay Marina www.hcagrimsby.co.uk ☎(01472) 268424, 200 ⌵+30 ⓥ, £3·00 inc ⌁, ▯, ♠⚒⚒ ⌁ ▱ 🖥 (35ton) ⌷ wi-fi.

Grimsby and Cleethorpes YC ☎(01482) 356678; enter via ABP Royal Dock, avoiding the extended piers enclosing tidal basin to N and W, limited ⌵; ♠ ⚓ ▯ ✕ ⌷. **Town** all facilities.

Hull Marina ⊕ www.bwml.co.uk/hull_marina ☎(01482) 609960; (312+20ⓥ £2·85), access HW±3 via lock (width 8·5m) 150m to the E. 1 Apr-31 Oct H24, 1 Nov-31 Mar 0700-2100 – otherwise 24hrs notice. **No** shelter from W through S to E when waiting outside lock in river especially in strong W winds or if marina shut. The pontoon in basin for craft awaiting lock in dries, *do not settle into the mud.* Marina office open 0900-1700 daily; ⌁ ⌁ ♠ ♠ ♂ ⚒ ⌁ △ 🖥(50t) ⌂(2t) ⌁ **ACA. City** All facilities.

Ferries: Rotterdam/Zeebrugge daily; 10 hrs; www.poferries.com.

South Ferriby Marina N Lincs. southferribymarina.com ☎(01652) 635620; access HW±3, 100+20ⓥ £10/craft<10m<+£1/m max LOA 12m. Lock fee £7. VHF 80; ⚓ ⌁ ♠ ⌁ ⌂(20t). **Village** ▱ ⌷.

S Ferriby Lock ☎01652 635219. River Authority ⓥ berths £8/48hrs. Lock Fee £7 (70ft LOA, 18ft beam, 6ft draught); ♂. VHF 74: 0730-1600.

Humber Yawl Club has two facilities: **Brough Haven** E Riding, ☎(01482) 667224, entry HW±1½; Ch 37/M1 limited ⌵; ♠ ⚓ ⌷. **Winteringham Haven** N Lincs ☎(01724) 734452. Limited access, dredging in progress; Ch 37/M1. 🖥(10t) ⌧.

GOOLE Take the flood up the River Humber, follow the buoyed channel past Hull, Brough and when approaching Trent Falls Apex alter course to stbd and follow the River Ouse to Goole. Situated off the Aire and Calder Navigation and accessed via the commercial lock for Goole Docks (Ch 14) are two havens, limited ⓥ.

Goole Boathouse ☎(01405) 763985 ⌵£5<64ft, max draught 1·5m.

Viking Marina www.vikingmarine.com ☎(01405) 765737, ⌵ £8·00/boat/nt, <80' LOA, 20' beam, 9' draught; ♠ ⌁ ♠⚒ ⚒ 🖥(18t) ⌷.

NABURN follow R Ouse, 4M S of York and 80M above Spurn Pt; **Naburn Marina** ☎(01904) 621021; 300+50ⓥ max draught 1m £3·00 (min £21) ⚓ ⌁ (£2·00) 🚾/📶 ♠ ♠ ♂ ✕ ⚒ 🖥(16t) ⌁ ✕.

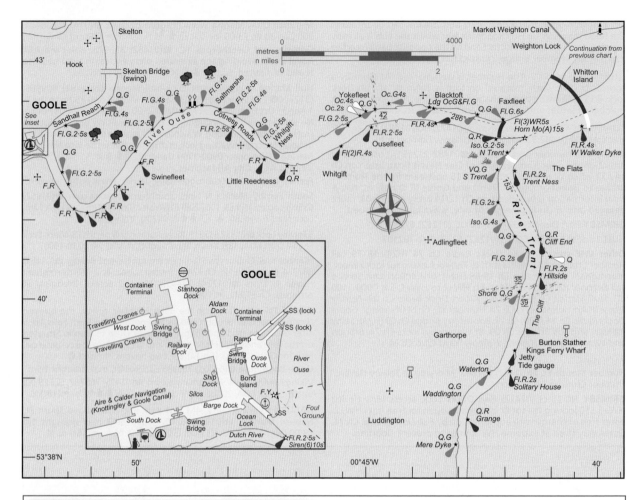

RIVER HUMBER TO HARTLEPOOL

(AC 107, 121, 129, 134, 5615) NE of the Humber Estuary clear of the main channel lies the Humber Gateway Wind Farm marked by 5 cardinal buoys and a 79m high met mast. Westermost Rough Wind Farm, marked by 4 cardinal buoys, lies 3M off Withernsea on the inshore route to Bridlington. A wreck hazardous to surface navigation lies in 8m, 2M off Tunstall. Navigation is aided by the disused light house and prominent church spires/towers along the coast. A small firing practice area lies inshore off the beach at Royston. Bridlington Bay (chart 1882) is clear of dangers apart from Smithic Shoals (marked by N and S cardinal lt buoys), lying S of Flamborough Hd about 2½M off Bridlington; seas break on these shoals in strong N or E winds even at HW.

Flamborough Head (lt, fog sig, RG) is a steep, W cliff with conspic light house on summit. The light may be obscured by cliffs when close inshore. An old lt ho, also conspic, is 2½ca WNW.

▶ Tides run hard around the Head which, in strong winds against a sp tide, is best avoided by 2M. ◀

From here the coast runs NW, with no offshore dangers until Filey Brigg where rocky ledges extend 5ca ESE, marked by a lit ECM. ⚓ in Filey B in N or offshore winds. NW of Filey Brigg beware Old Horse Rks and foul ground 5ca offshore; maintain this offing past Scarborough to Whitby High Lt. Off Whitby beware Whitby Rk and The Scar (dries in places) to the E of hbr, and Upgang Rks (dries in places) 1M to WNW; swell breaks heavily on all these rocks.

From Whitby to Hartlepool dangers lie no more than 1M offshore. Runswick Bay (AC 1612), 5M NW of Whitby, provides ⚓ in winds from S and W but is dangerous in onshore winds. 2½M further NW the little harbour of Staithes is suitable for yachts which can take the ground, but only in good weather and offshore winds. Redcliff, dark red and 205m high, is a conspic feature of this

coast which, along to Hunt Cliff, is prone to landslides and is fringed with rky ledges which dry for about 3ca off. There is a conspic radio mast 4ca SSE of Redcliff. Off Redcar and Coatham beware Salt Scar and West Scar, drying rky ledges lying 1– 8ca offshore. Other ledges lie close SE and S of Salt Scar which has NCM lt buoy. Between R Tees and Hartlepool beware Long Scar, detached rky ledge (dries 2m) with extremity marked by ECM lt buoy. Tees and Hartlepool Bays are exposed to strong E/SE winds. The R Tees and Middlesbrough are highly industrialised. At Hartlepool there is a centre specialising in the maintenance and restoration of Tall Ships.

HARTLEPOOL TO COQUET ISLAND

(AC 134, 152, 156, 5615) From The Heugh an offing of 1M clears all dangers until past Seaham and approaching Sunderland, where White Stones, rky shoals with depth 1·8m, lie 1·75M SSE of Roker Pier lt ho, and Hendon Rk, depth 0·9m, lies 1·25M SE of the lt ho. 1M N of Sunderland is Whitburn Steel, a rky ledge with less than 2m over it; a dangerous wreck (buoyed) lies 1ca SE of it. The coast N of Tynemouth is foul, and on passage to Blyth it should be given an offing of 1M. St Mary's Island (with disused conspic lt ho) is 3·5M N of Tynemouth, joined to the mainland by a causeway. The tiny drying harbour of Seaton Sluice, 1M NW of St Mary's Island, is accessible only in offshore winds via a narrow entrance.

Proceeding N from Blyth, keep well seaward of The Sow and Pigs rks, and set course to clear Newbiggin Pt and Beacon Pt by about 1M. Newbiggin church spire is prominent from N and S, and NW of Beacon Pt are conspic chys of aluminium smelter and power stn. 2M NNW of Beacon Pt is Snab Pt where rks extend 3ca seaward. Further offshore Cresswell Skeres, rky patches with depth 3m, lie about 1·5M NNE of Snab Pt.

9.5.12 BRIDLINGTON

E Riding of Yorkshire **54°04'·78N 00°11'·21W** ❄❄⚓⚓✿✿

CHARTS AC 1191, 1190, 129, 121, 1882, 5614; Imray C29

TIDES +0553 Dover; ML 3·6; Duration 0610

Standard Port RIVER TEES (→)

Times				Height (metres)			
High Water		Low Water		MHWS	MHWN	MLWN	MLWS
0000	0600	0000	0600	5·5	4·3	2·0	0·9
1200	1800	1200	1800				
Differences BRIDLINGTON							
+0100	+0050	+0055	+0050	+0·6	+0·4	+0·3	+0·2
FILEY BAY							
+0042	+0042	+0047	+0034	+0·3	+0·6	+0·4	+0·1

SHELTER Good, except in E, SE and S winds. Hbr dries completely to soft black mud. Berth on S pier or near HM's Office as directed. A marina is planned but agreement not yet reached.

NAVIGATION WPT SW Smithic WCM, Q (9) 15s, 54°02'·41N 00°09'·21W, 333°/2·6M to ent. Close-in appr is with N pier hd lt on brg 002° to keep W of drying patch (The Canch). Beware bar, 1m at MLWN, could dry out at MLWS.

LIGHTS AND MARKS Hbr is 4M WSW of Flamborough Hd lt, Fl (4) 15s 65m 24M. Y racing marks are laid in the Bay, Apr-Oct. Tidal sigs, by day from S pier: R flag = >2·7m in hbr; No flag = < 2·7m. At night from N pier: Fl ● = >2·7m in hbr; Fl ● = < 2·7m.

COMMUNICATIONS (Code 01262) CGOC 672317; Police 101; Ⓗ 606666. HM 670148/9, mob 0860 275150. Call on VHF Ch 16, then Ch **12** for working.

FACILITIES S Pier � £2.80/m²/week or £20/yacht <3days, ⚓ ⚓ see HM, ▬ ⚓(☎500227), ⚓(3t) ⚓(75t); **Royal Yorks YC** ☎672041 ⚓⚓ ✕ 🏠. **Town** 🏪⚒⚓🏧⚓✉Ⓑ🍴✕🏠⚓✈(Humberside).

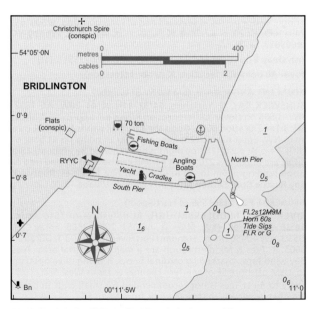

ADJACENT ANCHORAGE (7M SE of Scarborough)

FILEY, N Yorkshire, 54°12'·80N 00°16'·20W, AC 1882, 129. HW +0532 on Dover; ML 3·5m; Duration 0605. See 9.5.12. Good ⚓ in winds from S to NNE in 4 – 5m on hard sand. Lt on cliff above CG Stn, G metal column, FR 31m 1M vis 272°-308°. Filey Brigg, a natural bkwtr, is marked by ECM buoy, Q(3)10s, Bell. Beware Old Horse Rks, 2M WNW of Filey Brigg, foul ground extending ½M offshore. An unmarked Historic Wreck (see 9.0.3) lies at 54°11'·51N 00°13'·48W. Facilities: ☎ ✉ Ⓑ 🍴 ✕ 🏠 Ⓗ ☎(01723) 68111 ⚓.

9.5.13 SCARBOROUGH

N. Yorkshire **54°16'·88N 00°23'·36W** ❄❄⚓⚓⚓✿✿✿

CHARTS AC 1191, 129, 1612, 5614; Imray C29

TIDES +0527 Dover; ML 3·5; Duration 0615

Standard Port RIVER TEES (→)

Times				Height (metres)			
High Water		Low Water		MHWS	MHWN	MLWN	MLWS
0000	0600	0000	0600	5·5	4·3	2·0	0·9
1200	1800	1200	1800				
Differences SCARBOROUGH							
+0040	+0040	+0030	+0030	+0·2	+0·3	+0·3	0·0

SHELTER Good in Outer Harbour (dries) for local residents, access via narrow (10m) entrance by East pier, but not in strong E/SE'lies. During winter months access to Outer Harbour is via the drawbridge only.

Ⓥ � on SE side of pontoons in Inner Harbour, dredged to 2m.

NAVIGATION WPT 54°16'·50N 00°22'·00W, 302°/0·83M to E pier lt. Appr from the E to avoid Ramsdale Scar, rky shoal 0·9m. Keep careful watch for salmon nets E & SE of ent. Beware rks extending approx 20m SW of E pier head. Give Castle Headland close N of hbr a wide berth due to coastal defences. Min depths in approach and entrance to Inner Harbour approx 0·1m.

LIGHTS AND MARKS Lt ho (conspic), Iso 5s, Dia 60s, on Vincent Pier is shown by night or B ● by day when there is more than 3·7m over bar in entrance. A Y SPM Fl Y 4s, 113° E pier 1·6NM marks the end of an outfall; another, N of the harbour 0·7NM NE of Scalby Ness, is marked by a PHM Fl R 5s. Scarborough YC sets lit Y SPM. Other lts as shown on chartlet.

COMMUNICATIONS (Code 01723) CGOC (01262) 672317; Police 101; Ⓗ 368111. Port Control 373877, Call *Scarborough Port Control* VHF Ch **12** 16 (H24). Watchkeeper will offer guidance to approaching visitors.

FACILITIES HM www.scarboroughbc.gov.uk ☎373530 (HO); � £12/night, inner harbour pontoon berth £26/night <10m LOA, ⚓ ⚓ ⚒ ⚓ ⚓(4t) ▬; **Scarborough YC** ☎373821 ⚓ ▬ � ⚓ ⚒ 🏧 ⚓.

Services ⚒ 🏧 ✕ 🏧 Ⓔ.

Town ▬ (£11·50) 🏪🏪✉Ⓑ🍴✕🏠⚓✈(Humberside).

9.5.14 WHITBY

N. Yorkshire 54°29'·65N 00°36'·78W ✿✿✿❀↓↓↓↓✿✿✿

CHARTS AC 129, 134, 1612, 5614, 5615; Imray C29, C24

TIDES +0500 Dover; ML 3·3; Duration 0605

Standard Port RIVER TEES (→)

Times				Height (metres)			
High Water		Low Water		MHWS	MHWN	MLWN	MLWS
0000	0600	0000	0600	5·5	4·3	2·0	0·9
1200	1800	1200	1800				
Differences WHITBY							
+0020	+0020	+0018	+0017	+0·1	+0·1	+0·2	+0·1

SHELTER Good, except in lower hbr in strong NW to NE winds.

NAVIGATION WPT 54°30'·21N 00°36'·96W, 169°/0·57M to ent. The harbour can be entered safely under most conditions except during N-NE gales when the approach is dangerous due to breaking seas and entry should not be attempted.

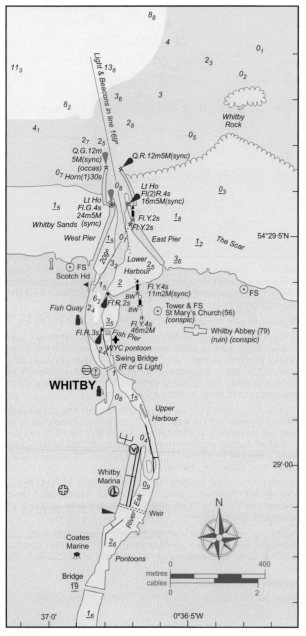

- From the SE beware Whitby Rk; leave Whitby NCM lit buoy to port.
- Beware strong set to E from HW −2 to HW, when nearing piers.

Vessels >37m LOA must embark pilot; via HM. All vessels must proceed at a safe speed within the Harbour and sailing vessels are to use auxiliary power if so equipped.

Min depth in entrance approx 1·0m. Bridge opens on request HW±2 every H and H+30. Additional openings at w/e and BH: May-Sep 0930 and1730; Aug 1300 (irrespective of tides). FG lts = open; FR lts = shut. Wait on pontoon at Fish Pier if necessary. If alongside Fish Quay or FVs, craft must not be left unattended. Priority for Swing Bridge passage:
1. Commercial vessels with pilot embarked (keep well clear);
2. Large vessels; 3. Vessels inbound fm seaward; 4. When both bridge sections open 2-way traffic in single file keeping to stbd. Establish comms with Whitby Bridge Ch **11** prior to transit.

LIGHTS AND MARKS See Lights, buoys & waypoints and chartlet. Ruins of Whitby Abbey are to the E of the entrance. Whitby High Lt ho (white 8-sided tower) is 2M ESE of harbour entrance, which is marked on either side by 5m high R and G wooden towers respectively lit with QR and QG lts. The E Pier Lt Ho Fl (2)R 4s on a 16m masonry tower has been re-commissioned. This together with The W Pier Lt Ho Fl(2)G 4s 24m, are synchronised with lts on the pier extensions. Leading lines for entering hbr:

2 bns (W △, and W ○ with B stripe) seen between the outer pierheads (Q Fl R and Q Fl G), lead 169° into hbr. At night the Fwd bn sync Fl Y 4s, aligns with a lt sync Fl Y 4s (46m) on a lamp post. Maintain this line until bns on E pier (2 sync Fl Y 2s) are abeam. Thence keep these same bns in transit astern bearing 029°.

COMMUNICATIONS (Code 01947) CGOC (01262) 672317; Police 101; Dr 820888. HM www.yorkshireports.co.uk 602354/602272.

Hbr and Marina VHF Ch **11** 16 12 (H24). Whitby Bridge Ch **11** 16 06 (listens on Ch 16 HW−2 to HW+2).

FACILITIES Whitby Marina (dredged 1.5m) is 2ca beyond swing bridge; visitor berths at seaward end of long pontoon. ☎600165. 240+10Ⓥ, £2.20; ⬡ ▬ 🗑 ⛽ ⚒ ⚓ 🛒 🔧 ⛽ 🚿 ⚓. ACA. **Whitby YC** ☎603623, ▲ ⚓ 🚻.

Fish Quay ⬢(in commercial quantities only, ☎602255) 🗝(1t).

Town All domestic facilities, ✕ ⬜ ⇌ ✈(Teesside).

ADJACENT ANCHORAGE (5M WNW of Whitby)

RUNSWICK BAY, N. Yorkshire, **54°32'·11N 00°44'·20W.** AC 1612. HW +0505 on Dover: Differences on R Tees are approx as Whitby; ML 3·1m; Duration 0605. Good shelter in all winds from SSE thru W to NW. Enter bay at 225° keeping clear of many rks at base of cliffs. Two W posts (2FY by night when required by lifeboat) 18m apart are ldg marks 270° to LB ho and can be used to lead into ⚓. Good holding in 6m to 9m in middle of bay. Facilities: **Runswick Bay Rescue Boat Station** ☎(01947) 840965. **Village** 🛒✕⬜.

ADJACENT PORT (3M SSE of Hartlepool)

RIVER TEES / MIDDLESBROUGH, Middlesbrough/Stockton, **54°38'·94N 01°08'·48W.** ✿✿✿❀✿. AC 152, 2567, 2566; Imray C29; OS 93. HW +0450 Dover; Standard Port River Tees. ML 3·1; Duration 0605. R. Tees & Middlesbrough are a major industrial area. Tees Bay wind farm, marked by cardinal buoys, is being developed up to 1½M offshore SE of the Tees channel as far as West Scar.

Entry to River Tees is not recommended for small craft in heavy weather, especially in strong winds from NE to SE.

Tees Fairway SWM buoy, Iso 4s 8m 8M, Horn 5s, Racon, is at 54°40'·93N 01°06'·38W, 030°/2·4M from S Gare bkwtr. The channel is well buoyed from the Fairway buoy to beyond Middlesbrough. Ldg lts 210°, both FR on framework trs. At Old CG stn a Q lt, or 3 ● (vert) = no entry without HM's consent. Call: *Tees Port Control* VHF Ch **14** 22 16 12 (H24). Monitor Ch 14; also info Ch 14 22. *Tees Barrage Radio* Ch M (37). HM ☎(01642) 277201.

South Gare Marine Club ☎491039(occas) ▲ ▬ ⚓; **Castlegate Marine Club** ☎583299 ▲ ▬ ⚓ 🔧 🔧; **Tees Motor Boat Club** ▲; Services: ⬡ 🔧 Ⓔ ⚓. ACA. **City** All domestic facilities, ⇌ ✈. Hartlepool Marina (9.5.14) lies 3M to NNW with all yacht facilities.

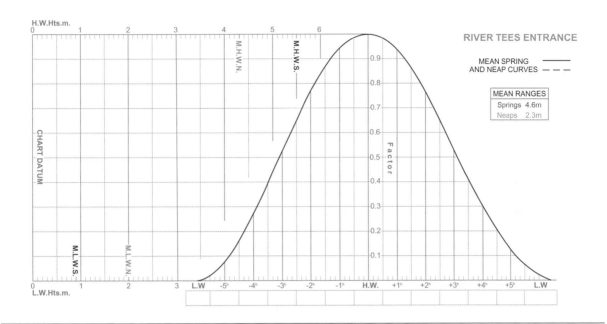

RIVER TEES ENTRANCE

MEAN SPRING
AND NEAP CURVES

MEAN RANGES
Springs 4.6m
Neaps 2.3m

NE England

STANDARD TIME (UT)
For Summer Time add ONE hour in **non-shaded areas**

RIVER TEES   LAT 54°38'N   LONG 1°09'W
TIMES AND HEIGHTS OF HIGH AND LOW WATERS

Dates in red are **SPRINGS**
Dates in blue are **NEAPS**

YEAR 2016

JANUARY

#	Time	m	#	Time	m
1	0201	1.5	**16**	0156	1.0
	0812	4.5		0811	4.9
F	1403	2.0	SA	1413	1.5
	2022	4.7	◑	2018	5.1
2	0249	1.7	**17**	0252	1.2
	0904	4.4		0910	4.7
SA	1458	2.2	SU	1515	1.7
◑	2118	4.5		2122	4.9
3	0347	1.9	**18**	0357	1.5
	1002	4.3		1016	4.6
SU	1610	2.3	M	1628	1.8
	2222	4.3		2233	4.7
4	0454	2.0	**19**	0511	1.6
	1105	4.3		1126	4.6
M	1725	2.3	TU	1746	1.8
	2329	4.3		2349	4.7
5	0600	2.0	**20**	0629	1.6
	1207	4.4		1238	4.7
TU	1831	2.1	W	1902	1.6
6	0032	4.4	**21**	0103	4.8
	0657	1.8		0735	1.5
W	1302	4.6	TH	1341	4.9
	1925	1.9		2004	1.3
7	0127	4.6	**22**	0206	5.0
	0746	1.6		0830	1.3
TH	1348	4.8	F	1433	5.1
	2011	1.6		2057	1.1
8	0214	4.8	**23**	0258	5.1
	0829	1.4		0916	1.2
F	1430	5.0	SA	1518	5.3
	2054	1.3		2143	0.9
9	0257	5.0	**24**	0342	5.2
	0910	1.2		0958	1.1
SA	1510	5.2	SU	1559	5.4
	2135	1.1	○	2224	0.8
10	0338	5.1	**25**	0423	5.3
	0950	1.1		1036	1.1
SU	1548	5.4	M	1636	5.4
●	2216	0.9		2302	0.7
11	0418	5.3	**26**	0501	5.2
	1030	1.0		1111	1.2
M	1627	5.5	TU	1710	5.4
	2256	0.7		2337	0.8
12	0500	5.3	**27**	0537	5.2
	1110	1.0		1143	1.2
TU	1707	5.6	W	1745	5.4
	2338	0.6			
13	0543	5.3	**28**	0011	0.9
	1151	1.0		0612	5.0
W	1749	5.6	TH	1215	1.4
				1820	5.2
14	0021	0.7	**29**	0045	1.1
	0628	5.2		0648	4.9
TH	1235	1.1	F	1246	1.5
	1833	5.5		1858	5.0
15	0106	0.8	**30**	0119	1.3
	0717	5.1		0727	4.7
F	1321	1.3	SA	1320	1.7
	1922	5.3		1939	4.8
			31	0156	1.6
				0811	4.5
			SU	1401	1.9
				2027	4.5

FEBRUARY

#	Time	m	#	Time	m
1	0241	1.8	**16**	0328	1.6
	0902	4.3		0946	4.5
M	1454	2.2	TU	1601	1.8
◑	2124	4.3		2212	4.5
2	0341	2.0	**17**	0446	1.8
	1002	4.2		1101	4.4
TU	1609	2.3	W	1728	1.8
	2231	4.2		2335	4.4
3	0457	2.1	**18**	0615	1.9
	1109	4.2		1222	4.5
W	1734	2.2	TH	1854	1.7
	2344	4.2			
4	0613	2.0	**19**	0059	4.6
	1215	4.3		0726	1.7
TH	1847	2.0	F	1330	4.7
				1958	1.4
5	0051	4.4	**20**	0201	4.8
	0714	1.8		0820	1.5
F	1314	4.6	SA	1422	5.0
	1944	1.7		2048	1.1
6	0148	4.6	**21**	0249	5.0
	0805	1.5		0903	1.3
SA	1403	4.9	SU	1505	5.2
	2033	1.3		2130	0.9
7	0235	4.9	**22**	0329	5.1
	0851	1.3		0942	1.2
SU	1447	5.2	M	1543	5.3
	2117	1.0	○	2206	0.8
8	0319	5.1	**23**	0405	5.2
	0934	1.0		1016	1.1
M	1529	5.4	TU	1617	5.4
●	2200	0.7		2239	0.7
9	0401	5.4	**24**	0439	5.2
	1016	0.8		1047	1.1
TU	1609	5.6	W	1648	5.4
	2242	0.4		2311	0.8
10	0443	5.5	**25**	0510	5.1
	1057	0.7		1117	1.1
W	1650	5.8	TH	1719	5.3
	2324	0.3		2341	0.9
11	0526	5.5	**26**	0541	5.1
	1137	0.7		1145	1.2
TH	1732	5.8	F	1751	5.2
12	0005	0.3	**27**	0010	1.0
	0609	5.5		0613	5.0
F	1219	0.8	SA	1215	1.3
	1815	5.7		1825	5.1
13	0048	0.5	**28**	0041	1.2
	0656	5.3		0648	4.8
SA	1302	1.0	SU	1247	1.4
	1903	5.5		1903	4.9
14	0134	0.8	**29**	0115	1.4
	0745	5.0		0728	4.6
SU	1350	1.2	M	1325	1.7
	1957	5.2		1946	4.6
15	0225	1.2			
	0841	4.7			
M	1448	1.5			
◑	2059	4.8			

MARCH

#	Time	m	#	Time	m
1	0156	1.7	**16**	0302	1.7
	0815	4.4		0919	4.5
TU	1410	1.9	W	1540	1.7
◑	2038	4.4		2155	4.4
2	0247	2.0	**17**	0424	2.0
	0912	4.2		1036	4.3
W	1510	2.1	TH	1712	1.8
	2143	4.2		2323	4.3
3	0400	2.2	**18**	0558	2.0
	1018	4.1		1201	4.4
TH	1638	2.2	F	1840	1.6
	2258	4.1			
4	0528	2.1	**19**	0047	4.4
	1129	4.2		0709	1.9
F	1806	2.0	SA	1311	4.6
				1942	1.4
5	0014	4.3	**20**	0146	4.6
	0642	1.9		0801	1.6
SA	1236	4.4	SU	1402	4.8
	1913	1.6		2029	1.1
6	0118	4.6	**21**	0231	4.8
	0739	1.6		0843	1.4
SU	1333	4.8	M	1444	5.0
	2007	1.2		2107	0.9
7	0210	4.9	**22**	0308	5.0
	0828	1.2		0919	1.2
M	1421	5.2	TU	1521	5.2
	2055	0.8		2141	0.8
8	0256	5.2	**23**	0342	5.1
	0913	0.9		0951	1.1
TU	1505	5.5	W	1553	5.3
	2140	0.5	○	2212	0.8
9	0339	5.5	**24**	0412	5.1
	0956	0.7		1021	1.0
W	1547	5.7	TH	1624	5.3
●	2222	0.2		2241	0.8
10	0422	5.6	**25**	0441	5.2
	1038	0.5		1050	1.0
TH	1630	5.9	F	1653	5.2
	2304	0.1		2310	0.9
11	0505	5.7	**26**	0509	5.1
	1119	0.5		1118	1.0
F	1713	5.9	SA	1723	5.2
	2346	0.2		2338	1.0
12	0548	5.6	**27**	0540	5.1
	1200	0.5		1148	1.1
SA	1758	5.8	SU	1756	5.0
13	0028	0.4	**28**	0009	1.1
	0633	5.4		0614	4.9
SU	1244	0.7	M	1221	1.3
	1846	5.5		1832	4.9
14	0112	0.8	**29**	0044	1.3
	0721	5.1		0653	4.7
M	1331	1.0	TU	1257	1.5
	1939	5.2		1915	4.6
15	0201	1.2	**30**	0123	1.6
	0815	4.8		0738	4.5
TU	1427	1.4	W	1340	1.7
◑	2041	4.7		2005	4.4
			31	0211	1.9
				0832	4.3
			TH	1436	1.9
			◑	2108	4.2

APRIL

#	Time	m	#	Time	m
1	0317	2.1	**16**	0526	2.1
	0937	4.2		1128	4.3
F	1556	2.0	SA	1809	1.6
	2222	4.1			
2	0447	2.1	**17**	0019	4.3
	1049	4.2		0638	2.0
SA	1726	1.9	SU	1239	4.5
	2338	4.3		1910	1.4
3	0606	1.9	**18**	0118	4.5
	1158	4.4		0731	1.8
SU	1839	1.5	M	1332	4.7
				1957	1.2
4	0046	4.6	**19**	0203	4.7
	0708	1.6		0813	1.5
M	1259	4.8	TU	1415	4.9
	1937	1.1		2036	1.1
5	0141	4.9	**20**	0240	4.9
	0800	1.2		0850	1.3
TU	1351	5.2	W	1452	5.0
	2028	0.7		2110	1.0
6	0229	5.3	**21**	0313	5.0
	0847	0.9		0923	1.2
W	1438	5.5	TH	1526	5.1
	2114	0.4		2141	0.9
7	0314	5.5	**22**	0343	5.1
	0933	0.6		0954	1.1
TH	1523	5.8	F	1557	5.1
●	2159	0.2	○	2212	0.9
8	0358	5.7	**23**	0412	5.1
	1016	0.4		1024	1.0
F	1609	5.9	SA	1628	5.1
	2243	0.1		2241	0.9
9	0442	5.7	**24**	0440	5.1
	1100	0.3		1054	1.0
SA	1654	5.9	SU	1659	5.1
	2325	0.2		2311	1.0
10	0526	5.6	**25**	0512	5.1
	1143	0.4		1125	1.1
SU	1742	5.7	M	1732	5.0
				2343	1.1
11	0008	0.5	**26**	0547	5.0
	0610	5.4		1200	1.2
M	1228	0.6	TU	1810	4.9
	1831	5.5			
12	0053	0.8	**27**	0019	1.3
	0658	5.1		0625	4.8
TU	1316	0.9	W	1238	1.3
	1925	5.1		1852	4.7
13	0141	1.3	**28**	0059	1.6
	0751	4.8		0709	4.7
W	1412	1.3	TH	1322	1.5
	2026	4.7		1943	4.5
14	0241	1.8	**29**	0146	1.8
	0853	4.5		0801	4.5
TH	1522	1.6	F	1417	1.7
◑	2138	4.4		2043	4.3
15	0358	2.1	**30**	0249	2.0
	1007	4.3		0903	4.4
F	1646	1.7	SA	1529	1.7
	2259	4.2	◑	2153	4.3

Chart Datum: 2·85 metres below Ordnance Datum (Newlyn). HAT is 6·1 metres above Chart Datum.

STANDARD TIME (UT)
For Summer Time add ONE hour in **non-shaded areas**

RIVER TEES LAT 54°38'N LONG 1°09'W

TIMES AND HEIGHTS OF HIGH AND LOW WATERS

Dates in red are **SPRINGS**
Dates in blue are **NEAPS**

YEAR 2016

MAY

	Time	m		Time	m
1 SU	0411	2.1	**16** M	0552	2.1
	1013	4.4		1154	4.4
	1650	1.6		1826	1.5
	2306	4.4			
2 M	0529	1.9	**17** TU	0035	4.4
	1122	4.6		0650	1.9
	1802	1.4		1252	4.6
				1916	1.4
3 TU	0013	4.6	**18** W	0125	4.6
	0633	1.6		0737	1.7
	1225	4.9		1339	4.7
	1903	1.0		1958	1.3
4 W	0111	4.9	**19** TH	0206	4.8
	0729	1.2		0817	1.5
	1321	5.2		1420	4.9
	1958	0.7		2036	1.2
5 TH	0202	5.2	**20** F	0241	4.9
	0820	0.9		0853	1.3
	1412	5.5		1457	5.0
	2048	0.4		2111	1.1
6 F ●	0250	5.5	**21** SA ○	0314	5.0
	0909	0.6		0927	1.2
	1501	5.7		1532	5.0
	2136	0.3		2143	1.0
7 SA	0335	5.6	**22** SU	0344	5.1
	0956	0.4		1000	1.1
	1550	5.8		1605	5.0
	2222	0.3		2215	1.0
8 SU	0421	5.6	**23** M	0415	5.1
	1042	0.4		1034	1.0
	1639	5.8		1639	5.0
	2307	0.4		2248	1.1
9 M	0506	5.6	**24** TU	0449	5.1
	1128	0.4		1108	1.0
	1728	5.6		1714	5.0
	2351	0.6		2323	1.1
10 TU	0551	5.4	**25** W	0525	5.1
	1214	0.6		1145	1.1
	1819	5.4		1754	4.9
11 W	0035	1.0	**26** TH	0001	1.3
	0638	5.2		0604	5.0
	1304	0.8		1225	1.2
	1912	5.0		1837	4.8
12 TH	0124	1.4	**27** F	0042	1.5
	0729	4.9		0647	4.9
	1358	1.1		1310	1.3
	2009	4.7		1927	4.7
13 F ◑	0218	1.8	**28** SA	0130	1.7
	0827	4.6		0738	4.7
	1459	1.4		1404	1.4
	2113	4.4		2024	4.5
14 SA	0325	2.0	**29** SU ◑	0228	1.8
	0933	4.4		0836	4.6
	1610	1.6		1508	1.5
	2223	4.3		2128	4.5
15 SU	0441	2.2	**30** M	0339	1.9
	1045	4.4		0942	4.6
	1723	1.6		1619	1.4
	2334	4.3		2236	4.5
			31 TU	0453	1.8
				1050	4.7
				1728	1.3
				2341	4.7

JUNE

	Time	m		Time	m
1 W	0559	1.6	**16** TH	0038	4.4
	1155	4.9		0655	1.9
	1832	1.1		1259	4.5
				1918	1.6
2 TH	0042	4.9	**17** F	0126	4.6
	0659	1.3		0742	1.7
	1255	5.1		1347	4.7
	1931	0.8		2002	1.4
3 F	0137	5.1	**18** SA	0208	4.8
	0756	1.0		0824	1.5
	1352	5.4		1430	4.8
	2025	0.7		2041	1.3
4 SA	0228	5.3	**19** SU	0245	4.9
	0849	0.8		0903	1.3
	1445	5.5		1509	4.9
	2117	0.6		2118	1.2
5 SU ●	0317	5.5	**20** M ○	0320	5.1
	0940	0.7		0940	1.1
	1537	5.6		1546	5.0
	2205	0.6		2154	1.1
6 M	0403	5.5	**21** TU	0355	5.2
	1028	0.5		1016	1.0
	1627	5.6		1622	5.1
	2251	0.6		2231	1.1
7 TU	0449	5.5	**22** W	0430	5.2
	1116	0.5		1054	0.9
	1716	5.5		1700	5.1
	2336	0.8		2309	1.1
8 W	0534	5.4	**23** TH	0507	5.2
	1202	0.7		1133	0.9
	1805	5.3		1740	5.1
				2348	1.2
9 TH	0020	1.1	**24** F	0547	5.2
	0619	5.3		1214	1.0
	1249	0.8		1824	5.0
	1854	5.1			
10 F	0104	1.4	**25** SA	0030	1.3
	0706	5.1		0629	5.1
	1338	1.0		1300	1.1
	1945	4.8		1912	4.9
11 SA	0151	1.7	**26** SU	0116	1.4
	0756	4.8		0717	5.0
	1429	1.3		1349	1.1
	2039	4.5		2005	4.8
12 SU ◑	0245	1.9	**27** M ◑	0208	1.6
	0853	4.6		0812	4.9
	1526	1.5		1446	1.2
	2137	4.3		2103	4.7
13 M	0348	2.1	**28** TU	0310	1.7
	0956	4.4		0914	4.8
	1628	1.7		1550	1.3
	2239	4.3		2207	4.6
14 TU	0456	2.2	**29** W	0419	1.7
	1101	4.4		1021	4.8
	1732	1.7		1657	1.3
	2341	4.3		2312	4.7
15 W	0600	2.1	**30** TH	0529	1.6
	1203	4.4		1130	4.9
	1829	1.7		1805	1.3

JULY

	Time	m		Time	m
1 F	0017	4.8	**16** SA	0043	4.4
	0635	1.4		0707	1.9
	1237	5.0		1314	4.5
	1910	1.1		1929	1.7
2 SA	0119	5.0	**17** SU	0134	4.6
	0738	1.2		0756	1.6
	1339	5.2		1403	4.7
	2010	1.0		2015	1.5
3 SU	0214	5.2	**18** M	0218	4.9
	0836	0.9		0840	1.4
	1436	5.3		1446	4.9
	2103	0.9		2057	1.3
4 M ●	0304	5.4	**19** TU	0257	5.1
	0929	0.7		0921	1.2
	1528	5.4		1526	5.1
	2152	0.9	○	2136	1.2
5 TU	0351	5.5	**20** W	0335	5.2
	1018	0.6		1000	1.0
	1617	5.4		1605	5.2
	2238	0.9		2216	1.1
6 W	0435	5.5	**21** TH	0412	5.4
	1103	0.6		1040	0.8
	1703	5.4		1644	5.3
	2320	1.0		2255	1.0
7 TH	0517	5.5	**22** F	0450	5.4
	1147	0.6		1120	0.7
	1748	5.3		1725	5.3
				2335	1.0
8 F	0000	1.1	**23** SA	0529	5.5
	0557	5.4		1201	0.7
	1229	0.8		1808	5.3
	1831	5.1			
9 SA	0040	1.3	**24** SU	0015	1.1
	0639	5.2		0611	5.4
	1310	1.0		1245	0.7
	1914	4.9		1853	5.2
10 SU	0118	1.5	**25** M	0059	1.2
	0722	5.0		0656	5.3
	1353	1.2		1331	0.9
	1959	4.7		1942	5.0
11 M	0200	1.8	**26** TU	0147	1.4
	0810	4.8		0748	5.2
	1439	1.5		1423	1.1
	2048	4.5	◑	2037	4.8
12 TU ◑	0249	2.0	**27** W	0243	1.6
	0904	4.5		0849	5.0
	1531	1.8		1523	1.3
	2142	4.3		2139	4.7
13 W	0351	2.2	**28** TH	0350	1.7
	1006	4.4		0958	4.8
	1632	1.9		1631	1.5
	2242	4.2		2247	4.6
14 TH	0502	2.2	**29** F	0505	1.7
	1112	4.3		1112	4.7
	1737	1.9		1746	1.6
	2345	4.3		2358	4.6
15 F	0610	2.1	**30** SA	0622	1.6
	1216	4.4		1227	4.8
	1838	1.8		1859	1.5
			31 SU	0106	4.8
				0731	1.3
				1336	5.0
				2001	1.3

AUGUST

	Time	m		Time	m
1 M	0205	5.1	**16** TU	0149	4.8
	0830	1.1		0817	1.4
	1433	5.2		1423	4.9
	2054	1.2		2035	1.4
2 TU ●	0255	5.3	**17** W	0232	5.1
	0921	0.8		0900	1.1
	1522	5.3		1504	5.1
	2140	1.1		2117	1.2
3 W	0339	5.4	**18** TH	0312	5.3
	1006	0.7		0941	0.8
	1606	5.4		1544	5.3
	2222	1.0	○	2158	1.0
4 TH	0419	5.5	**19** F	0351	5.5
	1047	0.6		1022	0.6
	1646	5.4		1624	5.5
	2300	1.0		2238	0.8
5 F	0457	5.5	**20** SA	0430	5.7
	1125	0.6		1102	0.5
	1725	5.3		1705	5.5
	2336	1.1		2317	0.8
6 SA	0533	5.4	**21** SU	0509	5.7
	1201	0.8		1143	0.4
	1802	5.2		1747	5.5
				2357	0.8
7 SU	0009	1.2	**22** M	0551	5.7
	0609	5.3		1224	0.5
	1237	1.0		1831	5.4
	1838	5.0			
8 M	0042	1.4	**23** TU	0039	0.9
	0647	5.1		0636	5.6
	1312	1.2		1309	0.8
	1917	4.8		1918	5.2
9 TU	0115	1.6	**24** W	0125	1.2
	0728	4.9		0728	5.3
	1349	1.5		1358	1.1
	1959	4.6		2012	4.9
10 W	0154	1.9	**25** TH	0219	1.4
	0815	4.6		0828	5.0
	1432	1.8		1457	1.5
◑	2047	4.4	◑	2113	4.7
11 TH	0244	2.1	**26** F	0327	1.7
	0912	4.4		0939	4.7
	1529	2.0		1610	1.8
	2144	4.3		2224	4.5
12 F	0356	2.3	**27** SA	0450	1.8
	1019	4.2		1101	4.6
	1641	2.1		1735	1.9
	2250	4.2		2343	4.5
13 SA	0520	2.2	**28** SU	0617	1.7
	1132	4.2		1226	4.7
	1756	2.1		1854	1.8
	2358	4.3			
14 SU	0633	2.0	**29** M	0058	4.7
	1240	4.4		0727	1.4
	1859	1.9		1334	4.9
				1953	1.5
15 M	0059	4.5	**30** TU	0156	5.0
	0729	1.7		0822	1.1
	1336	4.6		1426	5.1
	1950	1.6		2042	1.4
			31 W	0243	5.2
				0908	0.9
				1510	5.2
				2124	1.2

Chart Datum: 2·85 metres below Ordnance Datum (Newlyn). HAT is 6·1 metres above Chart Datum.

》》 FREE monthly updates. Register at 《
www.reedsnauticalalmanac.co.uk 《

351

NE England

STANDARD TIME (UT)
For Summer Time add ONE hour in **non-shaded areas**

RIVER TEES LAT 54°38'N LONG 1°09'W
TIMES AND HEIGHTS OF HIGH AND LOW WATERS

Dates in red are SPRINGS
Dates in blue are NEAPS

YEAR 2016

SEPTEMBER
Time m   Time m

1 0324 5.4 / 0948 0.7 / TH 1548 5.3 / ● 2201 1.1   **16** 0245 5.4 / 0917 0.7 / F 1519 5.5 / ○ 2135 0.9

2 0400 5.5 / 1024 0.7 / F 1624 5.3 / 2235 1.1   **17** 0326 5.7 / 0959 0.4 / SA 1600 5.7 / 2216 0.7

3 0433 5.5 / 1058 0.8 / SA 1657 5.3 / 2307 1.1   **18** 0406 5.9 / 1040 0.3 / SU 1641 5.7 / 2257 0.6

4 0506 5.4 / 1129 0.9 / SU 1729 5.2 / 2337 1.2   **19** 0448 5.9 / 1121 0.3 / M 1723 5.7 / 2337 0.6

5 0538 5.3 / 1200 1.0 / M 1802 5.1   **20** 0531 5.9 / 1203 0.5 / TU 1808 5.5

6 0007 1.3 / 0612 5.2 / TU 1231 1.2 / 1836 5.0   **21** 0020 0.8 / 0618 5.7 / W 1247 0.8 / 1854 5.3

7 0038 1.5 / 0650 4.9 / W 1305 1.5 / 1915 4.8   **22** 0107 1.0 / 0711 5.3 / TH 1336 1.2 / 1947 5.0

8 0114 1.7 / 0734 4.7 / TH 1344 1.8 / 2001 4.5   **23** 0201 1.4 / 0812 4.9 / F 1434 1.6 / ◗ 2049 4.7

9 0157 2.0 / 0825 4.4 / F 1434 2.1 / ◖ 2055 4.3   **24** 0311 1.7 / 0926 4.6 / SA 1551 2.0 / 2202 4.5

10 0257 2.2 / 0930 4.2 / SA 1545 2.3 / 2200 4.2   **25** 0438 1.8 / 1052 4.5 / SU 1723 2.1 / 2326 4.5

11 0428 2.3 / 1047 4.2 / SU 1713 2.3 / 2312 4.2   **26** 0608 1.7 / 1220 4.6 / M 1841 1.9

12 0555 2.1 / 1204 4.3 / M 1826 2.1   **27** 0042 4.7 / 0715 1.4 / TU 1323 4.8 / 1938 1.7

13 0020 4.5 / 0659 1.8 / TU 1306 4.6 / 1922 1.8   **28** 0139 5.0 / 0806 1.1 / W 1411 5.0 / 2023 1.5

14 0117 4.8 / 0750 1.4 / W 1355 4.9 / 2009 1.4   **29** 0223 5.2 / 0848 1.0 / TH 1451 5.2 / 2102 1.3

15 0203 5.1 / 0835 1.0 / TH 1438 5.2 / 2053 1.1   **30** 0302 5.3 / 0924 0.9 / F 1526 5.3 / 2136 1.2

OCTOBER
Time m   Time m

1 0336 5.4 / 0956 0.8 / SA 1558 5.3 / ● 2208 1.1   **16** 0300 5.8 / 0933 0.4 / SU 1534 5.7 / ○ 2153 0.6

2 0408 5.4 / 1027 0.9 / SU 1628 5.3 / 2238 1.1   **17** 0343 5.9 / 1016 0.3 / M 1617 5.8 / 2236 0.5

3 0438 5.4 / 1057 1.0 / M 1656 5.3 / 2307 1.1   **18** 0428 6.0 / 1059 0.3 / TU 1701 5.8 / 2319 0.5

4 0509 5.3 / 1125 1.1 / TU 1727 5.2 / 2337 1.2   **19** 0515 5.9 / 1142 0.5 / W 1746 5.6

5 0542 5.1 / 1155 1.3 / W 1801 5.1   **20** 0004 0.7 / 0604 5.6 / TH 1227 0.9 / 1833 5.4

6 0009 1.4 / 0618 5.0 / TH 1229 1.5 / 1839 4.9   **21** 0052 0.9 / 0658 5.3 / F 1316 1.3 / 1925 5.1

7 0044 1.6 / 0700 4.7 / F 1307 1.8 / 1923 4.7   **22** 0147 1.2 / 0759 4.9 / SA 1415 1.8 / ◗ 2026 4.8

8 0126 1.8 / 0750 4.5 / SA 1354 2.1 / 2015 4.4   **23** 0256 1.6 / 0911 4.6 / SU 1530 2.1 / 2137 4.5

9 0219 2.1 / 0851 4.3 / SU 1458 2.3 / ◖ 2117 4.3   **24** 0418 1.7 / 1034 4.4 / M 1657 2.2 / 2258 4.5

10 0338 2.2 / 1005 4.2 / M 1628 2.4 / 2228 4.3   **25** 0543 1.6 / 1156 4.5 / TU 1814 2.1

11 0511 2.1 / 1123 4.3 / TU 1748 2.2 / 2338 4.5   **26** 0013 4.6 / 0648 1.4 / W 1259 4.7 / 1911 1.9

12 0622 1.7 / 1230 4.6 / W 1849 1.8   **27** 0111 4.8 / 0738 1.3 / TH 1346 4.9 / 1956 1.7

13 0039 4.8 / 0717 1.3 / TH 1323 5.0 / 1939 1.5   **28** 0156 5.0 / 0819 1.1 / F 1424 5.1 / 2035 1.5

14 0130 5.2 / 0805 0.9 / F 1409 5.3 / 2025 1.1   **29** 0235 5.2 / 0854 1.0 / SA 1459 5.2 / 2109 1.3

15 0216 5.5 / 0850 0.6 / SA 1452 5.6 / 2109 0.8   **30** 0310 5.3 / 0941 1.0 / SU 1530 5.3 / ● 2141 1.2

31 0343 5.3 / 0957 1.0 / M 1559 5.3 / 2212 1.1

NOVEMBER
Time m   Time m

1 0414 5.3 / 1026 1.1 / TU 1627 5.3 / 2242 1.1   **16** 0412 5.9 / 1041 0.5 / W 1641 5.8 / 2304 0.5

2 0445 5.2 / 1055 1.2 / W 1658 5.3 / 2313 1.2   **17** 0502 5.8 / 1126 0.7 / TH 1727 5.7 / 2351 0.6

3 0518 5.1 / 1127 1.3 / TH 1732 5.2 / 2346 1.3   **18** 0553 5.6 / 1211 1.0 / F 1815 5.5

4 0555 4.9 / 1201 1.5 / F 1810 5.0   **19** 0041 0.8 / 0647 5.3 / SA 1300 1.4 / 1905 5.2

5 0022 1.5 / 0636 4.8 / SA 1240 1.7 / 1852 4.8   **20** 0134 1.1 / 0745 4.9 / SU 1355 1.8 / 2001 4.9

6 0104 1.7 / 0724 4.6 / SU 1325 2.0 / 1942 4.6   **21** 0235 1.4 / 0848 4.6 / M 1459 2.1 / ◖ 2105 4.7

7 0155 1.8 / 0822 4.4 / M 1423 2.2 / ◗ 2040 4.5   **22** 0344 1.6 / 0959 4.4 / TU 1614 2.3 / 2216 4.5

8 0302 2.0 / 0929 4.3 / TU 1542 2.3 / 2146 4.4   **23** 0500 1.7 / 1113 4.4 / W 1730 2.2 / 2328 4.5

9 0424 1.9 / 1042 4.4 / W 1703 2.2 / 2255 4.6   **24** 0608 1.6 / 1218 4.5 / TH 1833 2.1

10 0538 1.6 / 1150 4.6 / TH 1809 1.9 / 2359 4.8   **25** 0031 4.7 / 0701 1.5 / F 1310 4.7 / 1923 1.9

11 0640 1.3 / 1248 4.9 / F 1905 1.5   **26** 0122 4.8 / 0745 1.4 / SA 1353 4.9 / 2005 1.7

12 0055 5.1 / 0733 1.0 / SA 1338 5.3 / 1956 1.2   **27** 0205 4.9 / 0823 1.3 / SU 1429 5.0 / 2042 1.5

13 0147 5.5 / 0822 0.7 / SU 1425 5.5 / 2044 0.9   **28** 0244 5.0 / 0858 1.2 / M 1503 5.2 / 2117 1.3

14 0236 5.7 / 0909 0.5 / M 1511 5.7 / ○ 2131 0.6   **29** 0320 5.1 / 0930 1.2 / TU 1533 5.2 / ● 2150 1.2

15 0324 5.8 / 0955 0.4 / TU 1556 5.8 / 2218 0.5   **30** 0354 5.1 / 1001 1.2 / W 1603 5.3 / 2222 1.2

DECEMBER
Time m   Time m

1 0427 5.1 / 1033 1.2 / TH 1635 5.3 / 2255 1.2   **16** 0452 5.6 / 1113 0.8 / F 1712 5.7 / 2341 0.5

2 0501 5.1 / 1106 1.3 / F 1710 5.2 / 2330 1.2   **17** 0542 5.5 / 1158 1.0 / SA 1758 5.5

3 0537 5.0 / 1142 1.4 / SA 1747 5.1   **18** 0028 0.7 / 0632 5.3 / SU 1243 1.3 / 1844 5.3

4 0007 1.3 / 0618 4.9 / SU 1221 1.6 / 1828 5.0   **19** 0116 0.9 / 0723 5.0 / M 1330 1.6 / 1934 5.1

5 0049 1.4 / 0704 4.7 / M 1304 1.8 / 1914 4.8   **20** 0207 1.2 / 0817 4.7 / TU 1421 1.9 / 2028 4.8

6 0136 1.5 / 0757 4.6 / TU 1356 2.0 / 2007 4.7   **21** 0302 1.5 / 0915 4.5 / W 1521 2.2 / ◖ 2128 4.6

7 0234 1.6 / 0858 4.5 / W 1501 2.1 / ◗ 2108 4.7   **22** 0404 1.7 / 1017 4.4 / TH 1630 2.3 / 2233 4.5

8 0343 1.7 / 1004 4.5 / TH 1617 2.1 / 2215 4.7   **23** 0511 1.8 / 1122 4.3 / F 1741 2.2 / 2340 4.4

9 0455 1.6 / 1111 4.6 / F 1727 1.9 / 2321 4.8   **24** 0614 1.8 / 1222 4.5 / SA 1842 2.1

10 0601 1.4 / 1213 4.9 / SA 1831 1.6   **25** 0041 4.5 / 0706 1.7 / SU 1315 4.6 / 1932 1.9

11 0024 5.0 / 0702 1.1 / SU 1310 5.1 / 1929 1.3   **26** 0134 4.7 / 0751 1.6 / M 1358 4.8 / 2015 1.6

12 0123 5.3 / 0758 0.9 / M 1402 5.4 / 2023 1.0   **27** 0219 4.8 / 0831 1.5 / TU 1437 5.0 / 2054 1.4

13 0218 5.5 / 0850 0.7 / TU 1452 5.6 / 2115 0.8   **28** 0259 4.9 / 0907 1.4 / W 1511 5.1 / 2131 1.3

14 0311 5.6 / 0939 0.7 / W 1539 5.7 / ○ 2205 0.6   **29** 0336 5.0 / 0942 1.3 / TH 1544 5.2 / ● 2206 1.2

15 0402 5.7 / 1027 0.7 / TH 1626 5.7 / 2254 0.5   **30** 0412 5.1 / 1016 1.2 / F 1617 5.3 / 2241 1.1

31 0447 5.1 / 1051 1.2 / SA 1652 5.3 / 2317 1.0

Chart Datum: 2·85 metres below Ordnance Datum (Newlyn). HAT is 6·1 metres above Chart Datum.

》》**FREE** monthly updates. Register at 《
www.reedsnauticalalmanac.co.uk 《

9.5.15 HARTLEPOOL

Hartlepool 54°41'·26N 01°11'·90W (West Hbr ent) ❄❄⚓⚓⚓✿✿

CHARTS AC 152, 2567, 2566, 5615; Imray C24

TIDES +0437 Dover; ML 3·0; Duration 0600

Standard Port RIVER TEES (←—)

Times				Height (metres)			
High Water		Low Water		MHWS	MHWN	MLWN	MLWS
0000	0600	0000	0600	5·5	4·3	2·0	0·9
1200	1800	1200	1800				
Differences HARTLEPOOL							
−0004	−0004	−0006	−0006	−0·1	−0·1	−0·2	−0·1
MIDDLESBROUGH							
0000	+0002	0000	−0003	+0·1	+0·2	+0·1	−0·1

SHELTER Excellent in marina (5m) 4Ca SW of port. Strong E/SE winds raise broken water and swell in the bay, making ent channel hazardous, but possible. In such conditions, ask marina's advice.

NAVIGATION WPT 54°40'·86N 01°09'·90W, Longscar ECM buoy, 295°/1·06M to W Hbr ent. (For Victoria Hbr, 308°/0·65M to Nos 1/2 buoys.) From S, beware Longscar Rks, 4ca WSW of WPT. Tees Fairway SWM buoy, (54°40'·94N 01°06'·47W) is 2M E of Longscar ECM buoy and may assist the landfall. Tees Bay Wind Farm, consists of 27 turbines (extremities Fl Y 5s 13m 5M) up to 1½M offshore SE of the Tees channel as far as West Scar. Speed limit 4kn in W Hbr.

LIGHTS AND MARKS Tees Fairway Iso 4s 8m 8M, Horn(1) 5s. Longscar ECM buoy Q(3), bell, lies off the entrance. Steetly Works chimney, 1·4M NW of marina ent, and St Hilda's Church tower on The Heugh, are both prominent. Entry to commercial port on ldg lt Dir Iso WRG 324° following clearly buoyed channel. Dir lt Fl WRG leads 308° to marina lock, between W Hbr outer piers, Oc R and Oc G, and inner piers, FR and FG; bright street lts on S pier.

Lock sigs: ● = Proceed; ● = Wait; ●● = Lock closed.

Dir lt, Iso WRG, leads 325° via lit buoyed chan to Victoria Hbr.

COMMUNICATIONS (Code 01429) CGOC (01262) 672317; Police 101; Dr 272679; Tees Port Authority (01642) 277205; Marina Lock Ch ⚓ 80. Marina Ch M, 80. *Tees Port Control* info Ch 14, 22 (H24). *Hartlepool Dock Radio* Ch 12, only for ship docking.

FACILITIES

Hartlepool Marina www.hartlepool-marina.com ☎865744 Access via channel dredged to CD and lock(max beam 8.5m), over tidal cill 0.8m below CD, (HW± 5), (H24). 500+100♥ £2.16, ⚒🖬🛋(H24) ⬛LPG ⛽⚒🔧📠🕐⚠🛒🖃 (40/300t) 🗓 (13t).

Victoria Hbr (commercial dock, not normally for yachts), access H24. Call *Tees Port Control* Ch 14 for short-stay.

Tees & Hartlepool YC ☎233423 ⚓ 🖵.

Town 🏦🏧✉🅑🍴(H24) ✗🖵⇌✈(Teesside).

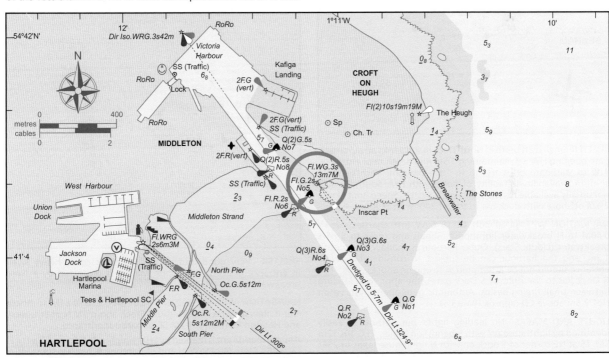

HARTLEPOOL TO SOUTHERN NETHERLANDS

(AC 2182A, 1191, 1190, 1503, 1408, 1610, 3371, 110) From abeam Whitby the passage can, theoretically, be made on one direct course, but this would conflict with oil/gas activities and platforms including Rough and Amethyst fields off Humber, Hewett off Cromer and very extensive fields further offshore. Commercial, offshore support and fishing vessels may be met S of Flamborough Hd and particularly off NE Norfolk where it is advisable to follow an inshore track avoiding the growing number of wind farms.

After passing Flamborough Hd, N Outer Dowsing and Dudgeon cardinals pass inside Haisborough Shoal, either:

Proceed SSE inshore of Cross Sand to Outer Gabbard departing SE to cross Nord Hinder TSS at right angles before heading for Roompotsluis via Middelbank and subsequent buoyed channel.

Or, set course ESE from the vicinity of Smith's Knoll, so as to cross the N/S deep-water traffic routes to the E altering SE towards Hoek van Holland, keeping N of Maas Approaches TSS.

HARTLEPOOL TO THE GERMAN BIGHT

(AC 2182A, 1191, 266, 1405) Taking departure eastward from abeam Whitby High lt, skirt the SW Patch off Dogger Bank, clearing S of Katy Gas Field passing E Friesland windfarm and then N of German Bight W Approach TSS. Thence head for the Elbe or Helgoland. The latter may also serve as a convenient haven in order to adjust the passage for Elbe tides and streams, without greatly increasing passage distance.

► *Tidal streams are less than 1kn away from the coast and run E/W along much of the route.* ◄

9.5.16 SEAHAM

Durham **54°50'·24N 01°19'·28W** ❀⚓✿

CHARTS AC 152, 1627, 5615; Imray C24

TIDES +0435 Dover; ML 3·0; Duration 0600

Standard Port RIVER TEES (←——)

Times				Height (metres)			
High Water		Low Water		MHWS	MHWN	MLWN	MLWS
0000	0600	0000	0600	5·5	4·3	2·0	0·9
1200	1800	1200	1800				
Differences SEAHAM							
–0015	–0015	–0015	–0015	–0·3	–0·2	0·0	–0·2

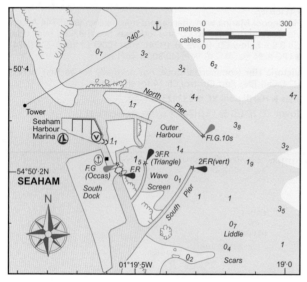

SHELTER Small craft berth in North Dock marina with an automatic gate opening HW ± 3 maintaining 1.5m min depth near entrance. 4 Ⓥ to starboard on entry up to 1·9m draught. Yachts should confirm availability/suitability of berth with HM. South Dock is not usually available to non-commercial vessels. Speed limit 5kn. ⚓ 2½ca offshore with clock twr in transit 240° with St John's ✠ twr.

NAVIGATION WPT 54°50'·36N 01°18'·60W (off chartlet), 256°/ 0·40M to N breakwater lighthouse. Shoals and rocks to S of S breakwater (Liddle Scars). Outer harbour is shallow, min depth 1.2m, and the harbour can be subject to considerable swell; entrance should not be attempted in strong on-shore winds. The N Dock lies to the right of S Dock gates, the appr chan leading between small drying basins. Automatic gate opens/shuts when HoT 2.5m above CD (approx HW ±3½).

LIGHTS AND MARKS No leading lights, but harbour is easily identified by lighthouse (W with B bands) on N pier, Fl G 10s 12m 5M. FS at NE corner of S dock on with N lighthouse leads in 256° clear of Tangle Rocks. 3FR lts on wave screen are in form of a △.

Traffic sigs at S Dock: ● = Vessels enter; ● = Vessels leave.

Traffic sigs at N Dock gate: 3 ● (vert) = shut : 3 ● (vert) = open.

COMMUNICATIONS (Code 0191) CGOC (01262) 672317; Police 101; Dr 5812332. HM 07801 215236.

VHF Ch **12** 16 06 0800–1700 (Mon-Fri), *Seaham Marina* Ch 80.

FACILITIES N Dock Marina ☎581 8998; 78 ⌒ inc 4Ⓥ up to 10m. **S Dock** (Seaham Hbr Dock Co) ☎516 1700; ⚓ ⌂(40t).

Town (½M) 🕭 🕭 🖂 ⚒ and ⛽ available out of town (about 5M). 🖂 Ⓑ 🛒 ✕ ⌸ ⇌ ✈(Teesside or Newcastle).

9.5.17 SUNDERLAND

Tyne and Wear **54°55'·23N 01°21'·15W** ❀❀❀⚓⚓✿✿

CHARTS AC 152, 1627, 5615; Imray C24

TIDES +0430 Dover; ML 2·9; Duration 0600; Zone 0 (UT)

Standard Port RIVER TEES (←——)

Times				Height (metres)			
High Water		Low Water		MHWS	MHWN	MLWN	MLWS
0000	0600	0000	0600	5·5	4·3	2·0	0·9
1200	1800	1200	1800				
Differences SUNDERLAND							
–0017	–0017	–0016	–0016	–0·2	–0·1	0·0	0·0

SHELTER Very good, but strong Easterlies cause heavy swell in entrance and outer harbour. Marina is protected by a fixed breakwater; access H24.

NAVIGATION WPT 54°55'·21N 01°20'·10W, 278°/0·61M to Roker Pier lt. Beware wreck at Whitburn Steel, about 1M N of entrance, and Hendon Rk (0·9m), 1·2M SE of harbour entrance.

Appr to marina ent is marked by SHM dolphin, Fl G 5s, and E jetty, Fl (2) G 10s. Close (4m) SSW of end of pontoon ▮ beware substantial obst'n (1·6m) in surrounding depth 2·1m, see chartlet.

LIGHTS AND MARKS 3 Fl ● at Pilot Stn (Old N Pier) = danger in harbour; no entry/departure.

COMMUNICATIONS (Code 0191) CGOC (01262) 672317; ⊖ (0191) 257 9441; Police 101; Ⓗ 565 6256. HM 567 2626 (HO), 567 0161 (OT).

Sunderland Marina Ch ⚓ 80. Port VHF Ch **14** (H24); tide and visibility reports on request.

FACILITIES Sunderland Marina 120 pontoon berths in 2·1m, max LOA 13m; and 110 ⚓. ☎ 5144721. Ⓥ (if available) £2.05/m; ⚓ £10/ craft, ♿ ⚓ ▮(H24) ⛽.

Sunderland YC ☎ 567 5133, ▬(dinghy) ⌒⚓⌸.

Wear Boating Association ☎567 5313, ⌒.

Services ⚒ 🕭 ⛽ ⛽.

Town 🕭 ▮ 🖂 Ⓑ 🛒 ⎕ ✕ ⌸ ⇌ ✈(Newcastle).

9.5.18 R TYNE/NORTH SHIELDS

Tyne and Wear **55°00'·89N 01°24'·10W** ❀❀♨♨♨❀❀

CHARTS AC 152, 1191, 1934, 5615; Imray C24

TIDES +0430 Dover; ML 3·0; Duration 0604

Standard Port NORTH SHIELDS (→)

Times				Height (metres)			
High Water		Low Water		MHWS	MHWN	MLWN	MLWS
0200	0800	0100	0800	5·0	3·9	1·8	0·7
1400	2000	1300	2000				
Differences NEWCASTLE-UPON-TYNE							
+0003	+0003	+0008	+0008	+0·3	+0·2	+0·1	+0·1

SHELTER Good in all weather. Access H24, but in strong E and NE winds appr may be difficult for smaller craft due to much backwash off the piers. Confused seas can build at the ent in severe weather, and a large steep swell can break dangerously when meeting an ebb tide up to half a mile inside the harbour. In emergency or as a refuge yachts may berth on Fish Quay; contact HM.

NAVIGATION WPT 55°01'·00N 01°23'·00W, 250°/0·7M to hbr ent following recommended route in Ldg lt Oc sector. Outbound keep S of hbr ent centreline. From S, no dangers. From N, beware Bellhues Rk (approx 1M N of hbr and ¾M off shore); give N pier a wide berth. Dredged chan in Lower Hbr is buoyed. 6kn speed limit in the river west of Herd Groyne Lt Ho. In river proceed under power keeping to the starboard side. Tacking across the channel is prohibited. The 7 bridges at Newcastle have least clearance above HAT of 25m, or 4·5m when Millennium Bridge is closed.

LIGHTS AND MARKS See 9.5.4 and chartlet. W Lt LFl 10s shows from centre span of the Millennium Bridge when fully open.

COMMUNICATIONS (Code 0191) CGOC (01262) 672317; Met 2326453; Police 101; Ⓗ (Tynemouth) 2596660; Ⓗ (Newcastle) 2325131. HM 2570407; Tyne VTS 2572080. *Tyne VTS* Ch **12**, 11 (H24). Marinas: Royal Quays Ch 80. St Peter's Ch 80 M.

FACILITIES Notes: A conservancy fee (£10) may be levied by the Port Authority on all visiting craft. Craft waiting to enter the marinas should do so outside the fairway.

Royal Quays Marina ⊕ 54°59'·79N 01°26'·84W ☎2728282 2M upriver from pierheads in former Albert Edward Dock; 302 � inc Ⓥ, £2·30 7·9m depth. Lock (42.5m x 7·5m) departures H and H+30; arrivals H+15 and H+45, or on request at quiet times. Waiting pontoon outside lock. ⚓ 🅿 ♦ ⬆(H24) ♦ ⬚⊡ 🛒 🖤(30t) 🗐 🛒.

St Peter's Marina 54°57'·94N 01°34'·35W, 8M upriver, and 1M E of city. ☎2654472; 140+20Ⓥ, £2.27. Access approx HW±3½ over sill, dries 0·8m, which retains 2·5m within. Traffic lights at ent. ♦ on pontoon outside ent in 2m. **Services** ⚓ 🅿 🗡🔧 🗐 ACA, Ⓔ.

Millennium Bridge in city centre, limited pontoon berthing below the bridge; pre-book ☎2211363.

Newcastle City Marina enquiries@newcastlecitymarina.co.uk ☎2211348, mob 07435 788426; Boats <26m: price on sliding scale 5m £10 → £30 >14m, short stay (< 5hrs) free. ⬚⊡ ⚓.

City. All amenities, ⇌ (Newcastle/S Shields), ✈ (Newcastle). **Ferries:** Ijmuiden; daily; 16 hrs; DFDS (www.dfdsseaways.co.uk).

ADJACENT HARBOUR

CULLERCOATS, 55°02'·08N 01°25'·81W. AC 1191. +0430 Dover. Tides as 9.5.17. Small drying hbr 1·6M N of R Tyne ent. Appr on ldg line 256°, two bns (FR lts), between drying rks. An occasional fair weather ⚓ or dry against S pier. Facilities at Tynemouth.

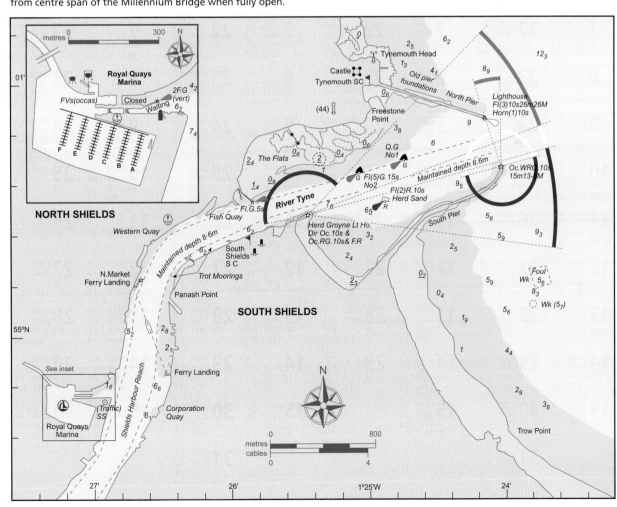

STANDARD TIME (UT)
For Summer Time add ONE hour in **non-shaded areas**

RIVER TYNE/NORTH SHIELDS   LAT 55°00'N   LONG 1°26'W
TIMES AND HEIGHTS OF HIGH AND LOW WATERS

Dates in red are **SPRINGS**
Dates in blue are **NEAPS**

YEAR 2016

JANUARY

Day	Time m	Time m	Time m	Time m	Day	Time m	Time m	Time m	Time m
1	0146 1.4	0759 4.3	F 1348 1.9	2008 4.4	16	0144 1.0	0753 4.6	SA 1357 1.5	2002 4.9
2	0235 1.7	0852 4.1	SA 1443 2.1	2105 4.2	17	0241 1.2	0855 4.5	SU 1500 1.6	2108 4.6
3	0334 1.8	0950 4.0	SU 1554 2.2	2209 4.1	18	0348 1.4	1003 4.3	M 1616 1.7	2222 4.5
4	0440 1.9	1055 4.0	M 1711 2.2	2318 4.1	19	0502 1.5	1116 4.3	TU 1736 1.7	2340 4.5
5	0546 1.9	1200 4.1	TU 1820 2.1		20	0616 1.5	1226 4.5	W 1850 1.5	
6	0024 4.2	0645 1.7	W 1256 4.3	1916 1.8	21	0052 4.6	0722 1.4	TH 1327 4.7	1953 1.3
7	0120 4.3	0735 1.6	TH 1342 4.5	2003 1.6	22	0153 4.7	0817 1.3	F 1417 4.9	2046 1.0
8	0205 4.5	0818 1.4	F 1422 4.7	2044 1.3	23	0243 4.9	0904 1.2	SA 1501 5.0	2131 0.8
9	0246 4.7	0859 1.2	SA 1500 4.9	2124 1.1	24	0327 4.9	0945 1.1	SU 1541 5.1	2213 0.7
10	0325 4.9	0938 1.2	SU 1536 5.1	2204 0.9	25	0408 5.0	1022 1.1	M 1618 5.2	2250 0.7
11	0404 5.0	1018 1.0	M 1613 5.2	2245 0.7	26	0446 4.9	1056 1.1	TU 1654 5.1	2325 0.8
12	0444 5.1	1057 0.9	TU 1652 5.3	2327 0.6	27	0522 4.9	1127 1.2	W 1729 5.0	2358 0.9
13	0526 5.1	1138 1.0	W 1733 5.3		28	0558 4.7	1158 1.3	TH 1805 4.9	
14	0010 0.7	0610 5.0	TH 1220 1.1	1817 5.2	29	0031 1.0	0634 4.6	F 1231 1.4	1842 4.7
15	0055 0.8	0659 5.0	F 1305 1.3	1906 5.0	30	0106 1.2	0712 4.4	SA 1306 1.6	1923 4.5
					31	0146 1.5	0756 4.2	SU 1348 1.8	2010 4.3

FEBRUARY

Day	Time m	Time m	Time m	Time m	Day	Time m	Time m	Time m	Time m
1	0233 1.7	0848 4.0	M 1443 2.1	2108 4.1	16	0319 1.5	0934 4.3	TU 1550 1.7	2203 4.3
2	0334 1.9	0949 4.0	TU 1559 2.2	2217 3.9	17	0438 1.7	1051 4.2	W 1718 1.7	2328 4.3
3	0448 2.0	1100 3.9	W 1724 2.1	2335 4.0	18	0602 1.7	1209 4.3	TH 1841 1.5	
4	0601 1.9	1211 4.1	TH 1838 1.9		19	0046 4.4	0714 1.6	F 1315 4.5	1946 1.3
5	0045 4.1	0704 1.7	F 1310 4.3	1935 1.6	20	0148 4.5	0808 1.4	SA 1407 4.7	2036 1.0
6	0140 4.4	0755 1.5	SA 1357 4.6	2022 1.3	21	0235 4.7	0852 1.3	SU 1449 4.9	2118 0.8
7	0225 4.7	0839 1.2	SU 1437 4.9	2106 0.9	22	0315 4.8	0929 1.1	M 1526 5.0	2155 0.7
8	0306 4.9	0921 1.0	M 1516 5.1	2148 0.6	23	0351 4.9	1003 1.0	TU 1600 5.1	2228 0.7
9	0346 5.1	1002 0.8	TU 1554 5.3	2230 0.4	24	0424 4.9	1034 1.0	W 1632 5.1	2300 0.7
10	0427 5.2	1042 0.7	W 1634 5.4	2312 0.3	25	0455 4.9	1103 1.0	TH 1704 5.0	2329 0.8
11	0509 5.2	1123 0.7	TH 1716 5.5	2354 0.4	26	0527 4.8	1132 1.1	F 1736 4.9	2359 0.9
12	0552 5.1	1204 0.8	F 1800 5.4		27	0559 4.7	1201 1.2	SA 1810 4.8	
13	0037 0.5	0637 5.0	SA 1247 1.0	1848 5.2	28	0030 1.1	0633 4.5	SU 1234 1.4	1846 4.6
14	0123 0.8	0728 4.7	SU 1336 1.2	1942 4.9	29	0105 1.3	0711 4.3	M 1310 1.6	1928 4.3
15	0215 1.1	0826 4.5	M 1434 1.5	2047 4.6					

MARCH

Day	Time m	Time m	Time m	Time m	Day	Time m	Time m	Time m	Time m
1	0145 1.6	0755 4.2	TU 1356 1.8	2019 4.1	16	0255 1.6	0908 4.2	W 1532 1.6	2149 4.2
2	0237 1.9	0851 4.1	W 1459 2.0	2126 3.9	17	0417 1.9	1026 4.1	TH 1703 1.6	2315 4.1
3	0350 2.0	1002 3.9	TH 1628 2.1	2247 3.9	18	0547 1.9	1148 4.2	F 1827 1.5	
4	0517 2.0	1121 3.9	F 1755 1.9		19	0034 4.2	0659 1.7	SA 1257 4.3	1930 1.3
5	0009 4.0	0631 1.8	SA 1232 4.2	1903 1.6	20	0134 4.4	0752 1.5	SU 1349 4.6	2018 1.1
6	0112 4.3	0729 1.5	SU 1326 4.5	1956 1.2	21	0218 4.6	0833 1.3	M 1430 4.7	2057 0.9
7	0201 4.6	0816 1.2	M 1411 4.8	2043 0.8	22	0255 4.7	0908 1.2	TU 1506 4.9	2130 0.8
8	0244 4.9	0900 0.9	TU 1452 5.1	2127 0.4	23	0328 4.8	0940 1.0	W 1538 5.0	2201 0.7
9	0325 5.1	0942 0.6	W 1532 5.4	2210 0.2	24	0358 4.9	1009 0.9	TH 1608 5.0	2231 0.7
10	0406 5.3	1023 0.5	TH 1614 5.5	2252 0.1	25	0427 4.9	1038 0.9	F 1639 5.0	2300 0.8
11	0447 5.3	1105 0.4	F 1657 5.6	2334 0.2	26	0457 4.8	1107 1.0	SA 1710 4.9	2329 0.9
12	0530 5.2	1147 0.5	SA 1742 5.4		27	0528 4.7	1137 1.1	SU 1743 4.7	2359 1.1
13	0017 0.4	0615 5.0	SU 1230 0.7	1831 5.2	28	0600 4.6	1208 1.2	M 1819 4.6	
14	0102 0.8	0704 4.8	M 1318 1.0	1926 4.9	29	0032 1.3	0635 4.5	TU 1244 1.4	1858 4.4
15	0152 1.2	0800 4.5	TU 1416 1.3	2032 4.5	30	0110 1.5	0717 4.3	W 1327 1.6	1947 4.1
					31	0157 1.8	0808 4.1	TH 1425 1.8	2051 4.0

APRIL

Day	Time m	Time m	Time m	Time m	Day	Time m	Time m	Time m	Time m
1	0305 2.0	0916 3.9	F 1545 1.9	2210 3.9	16	0518 2.0	1117 4.1	SA 1800 1.5	
2	0434 2.0	1035 4.0	SA 1714 1.7	2331 4.0	17	0007 4.1	0631 1.8	SU 1227 4.2	1901 1.3
3	0555 1.8	1150 4.2	SU 1828 1.4		18	0106 4.3	0724 1.6	M 1321 4.4	1948 1.2
4	0039 4.3	0657 1.5	M 1252 4.5	1925 1.0	19	0151 4.5	0806 1.4	TU 1403 4.6	2026 1.0
5	0132 4.7	0749 1.2	TU 1341 4.8	2015 0.7	20	0227 4.6	0841 1.2	W 1439 4.7	2059 0.9
6	0218 5.0	0835 0.8	W 1426 5.2	2102 0.4	21	0300 4.7	0913 1.1	TH 1512 4.8	2131 0.8
7	0300 5.2	0919 0.6	TH 1509 5.4	2146 0.2	22	0330 4.8	0944 1.0	F 1544 4.9	2201 0.8
8	0342 5.4	1003 0.4	F 1553 5.5	2230 0.1	23	0400 4.8	1015 0.9	SA 1615 4.8	2232 0.8
9	0425 5.4	1046 0.3	SA 1639 5.5	2314 0.2	24	0430 4.8	1045 0.9	SU 1648 4.8	2302 0.9
10	0508 5.3	1130 0.4	SU 1726 5.4	2357 0.5	25	0501 4.8	1116 1.0	M 1722 4.7	2334 1.1
11	0554 5.1	1215 0.6	M 1818 5.1		26	0534 4.7	1149 1.1	TU 1758 4.6	
12	0041 0.9	0642 4.8	TU 1305 0.9	1914 4.8	27	0007 1.3	0609 4.6	W 1227 1.2	1838 4.4
13	0131 1.3	0738 4.5	W 1402 1.2	2019 4.4	28	0046 1.5	0650 4.4	TH 1310 1.4	1927 4.2
14	0232 1.7	0843 4.3	TH 1515 1.4	2131 4.2	29	0132 1.7	0740 4.3	F 1405 1.5	2027 4.1
15	0351 1.9	0957 4.1	F 1639 1.5	2251 4.1	30	0235 1.9	0843 4.1	SA 1516 1.6	2140 4.0

Chart Datum: 2·60 metres below Ordnance Datum (Newlyn). HAT is 5·7 metres above Chart Datum.

STANDARD TIME (UT)
For Summer Time add ONE hour in **non-shaded areas**

RIVER TYNE/NORTH SHIELDS   LAT 55°00'N   LONG 1°26'W
TIMES AND HEIGHTS OF HIGH AND LOW WATERS

Dates in red are **SPRINGS**
Dates in blue are **NEAPS**

YEAR 2016

MAY

Day	Time	m	Time	m	Time	m	Time	m
1 SU	0356	1.9	0956	4.1	1638	1.5	2256	4.1
2 M	0517	1.8	1110	4.3	1751	1.3		
3 TU	0005	4.4	0622	1.5	1216	4.5	1852	1.0
4 W	0102	4.7	0718	1.2	1311	4.8	1946	0.7
5 TH	0151	5.0	0809	0.9	1400	5.1	2036	0.4
6 F ●	0236	5.2	0857	0.6	1448	5.3	2123	0.3
7 SA	0320	5.3	0943	0.4	1535	5.4	2209	0.3
8 SU	0404	5.3	1030	0.3	1621	5.4	2254	0.4
9 M	0448	5.3	1116	0.4	1713	5.3	2338	0.7
10 TU	0534	5.1	1203	0.5	1805	5.0		
11 W	0023	1.0	0623	4.9	1253	0.8	1901	4.7
12 TH	0111	1.3	0716	4.6	1347	1.0	2001	4.4
13 F ◑	0206	1.7	0816	4.4	1450	1.3	2105	4.2
14 SA	0315	1.9	0923	4.2	1602	1.5	2214	4.1
15 SU	0433	2.0	1034	4.1	1715	1.5	2324	4.1
16 M	0546	1.9	1143	4.2	1818	1.4		
17 TU	0025	4.2	0644	1.8	1242	4.3	1908	1.3
18 W	0114	4.3	0730	1.6	1329	4.4	1949	1.2
19 TH	0154	4.5	0809	1.4	1409	4.5	2025	1.1
20 F	0230	4.6	0845	1.2	1445	4.6	2100	1.0
21 SA ○	0303	4.7	0919	1.1	1520	4.7	2133	1.0
22 SU	0334	4.8	0952	1.0	1554	4.7	2206	1.0
23 M	0406	4.8	1025	0.9	1628	4.7	2239	1.0
24 TU	0439	4.8	1100	0.9	1704	4.7	2313	1.1
25 W	0513	4.8	1136	1.0	1742	4.6	2349	1.2
26 TH	0550	4.7	1215	1.1	1824	4.5		
27 F	0029	1.4	0631	4.6	1300	1.2	1911	4.4
28 SA	0116	1.5	0720	4.5	1351	1.3	2008	4.3
29 SU ◑	0212	1.7	0818	4.4	1454	1.3	2114	4.2
30 M	0322	1.8	0924	4.4	1605	1.3	2224	4.3
31 TU	0438	1.7	1035	4.4	1717	1.2	2332	4.4

JUNE

Day	Time	m	Time	m	Time	m	Time	m
1 W	0548	1.5	1142	4.6	1821	1.0		
2 TH	0033	4.6	0649	1.2	1244	4.8	1919	0.8
3 F	0127	4.9	0745	1.0	1340	5.0	2013	0.6
4 SA	0215	5.1	0838	0.7	1432	5.2	2104	0.5
5 SU ●	0301	5.2	0925	0.5	1522	5.3	2151	0.5
6 M	0346	5.2	1017	0.4	1611	5.3	2237	0.6
7 TU	0431	5.2	1104	0.4	1701	5.1	2321	0.8
8 W	0517	5.1	1151	0.5	1752	5.0		
9 TH	0005	1.1	0604	5.0	1238	0.7	1842	4.7
10 F	0049	1.3	0653	4.8	1325	1.0	1935	4.5
11 SA	0135	1.6	0745	4.6	1417	1.2	2030	4.3
12 SU ◑	0228	1.8	0842	4.3	1514	1.4	2128	4.1
13 M	0332	2.0	0943	4.2	1617	1.6	2229	4.0
14 TU	0443	2.0	1048	4.1	1719	1.6	2332	4.1
15 W	0550	2.0	1152	4.1	1817	1.6		
16 TH	0029	4.2	0647	1.8	1249	4.2	1907	1.5
17 F	0117	4.3	0735	1.6	1337	4.4	1951	1.3
18 SA	0159	4.5	0817	1.4	1419	4.5	2031	1.2
19 SU	0236	4.7	0855	1.2	1457	4.6	2108	1.1
20 M ○	0311	4.8	0932	1.1	1534	4.7	2144	1.1
21 TU	0345	4.9	1008	0.9	1610	4.8	2220	1.0
22 W	0419	4.9	1045	0.8	1648	4.8	2257	1.0
23 TH	0455	4.9	1124	0.8	1727	4.8	2335	1.1
24 F	0533	4.9	1204	0.8	1809	4.7		
25 SA	0016	1.2	0614	4.9	1248	0.9	1855	4.6
26 SU	0100	1.3	0701	4.8	1336	1.0	1947	4.5
27 M ◑	0151	1.5	0755	4.7	1432	1.1	2047	4.4
28 TU	0252	1.6	0856	4.6	1536	1.2	2153	4.4
29 W	0403	1.6	1005	4.5	1646	1.2	2301	4.4
30 TH	0516	1.5	1116	4.6	1755	1.1		

JULY

Day	Time	m	Time	m	Time	m	Time	m
1 F	0007	4.5	0625	1.4	1225	4.7	1858	1.0
2 SA	0107	4.7	0728	1.1	1327	4.8	1957	0.9
3 SU	0200	4.9	0825	0.9	1422	5.0	2050	0.8
4 M ●	0248	5.1	0918	0.6	1513	5.1	2138	0.8
5 TU	0333	5.2	1006	0.5	1602	5.1	2222	0.8
6 W	0417	5.2	1052	0.5	1648	5.1	2304	0.9
7 TH	0500	5.2	1135	0.5	1733	4.9	2343	1.1
8 F	0542	5.1	1216	0.7	1817	4.8		
9 SA	0021	1.3	0625	4.9	1256	0.9	1901	4.6
10 SU	0059	1.5	0709	4.7	1338	1.1	1947	4.4
11 M	0141	1.7	0757	4.5	1424	1.4	2037	4.2
12 TU ◑	0230	1.9	0850	4.3	1515	1.6	2131	4.0
13 W	0333	2.0	0950	4.1	1617	1.7	2232	4.0
14 TH	0446	2.1	1056	4.0	1722	1.8	2337	4.0
15 F	0558	2.0	1205	4.1	1824	1.7		
16 SA	0037	4.2	0659	1.8	1305	4.2	1918	1.6
17 SU	0127	4.4	0749	1.5	1353	4.4	2004	1.4
18 M	0210	4.6	0832	1.3	1435	4.6	2045	1.2
19 TU ○	0247	4.8	0911	1.1	1514	4.7	2124	1.1
20 W	0323	4.9	0950	0.9	1552	4.9	2202	1.0
21 TH	0359	5.1	1029	0.7	1630	5.0	2241	0.9
22 F	0436	5.1	1109	0.6	1709	5.0	2320	0.9
23 SA	0515	5.2	1150	0.6	1751	4.9		
24 SU	0000	1.0	0556	5.1	1233	0.6	1835	4.8
25 M	0043	1.1	0641	5.1	1318	0.8	1924	4.7
26 TU ◑	0130	1.3	0733	4.9	1409	1.0	2020	4.5
27 W	0225	1.4	0832	4.7	1509	1.2	2124	4.4
28 TH	0334	1.6	0942	4.6	1620	1.4	2235	4.4
29 F	0453	1.6	1058	4.5	1735	1.4	2347	4.4
30 SA	0610	1.5	1215	4.5	1846	1.4		
31 SU	0053	4.6	0720	1.2	1322	4.7	1948	1.2

AUGUST

Day	Time	m	Time	m	Time	m	Time	m
1 M	0150	4.8	0819	1.0	1418	4.8	2040	1.1
2 TU ●	0238	5.0	0909	0.7	1506	5.0	2125	1.0
3 W	0321	5.1	0954	0.6	1550	5.0	2206	0.9
4 TH	0401	5.2	1035	0.5	1630	5.0	2243	1.0
5 F	0439	5.2	1113	0.6	1709	4.9	2318	1.0
6 SA	0517	5.1	1148	0.7	1746	4.8	2351	1.2
7 SU	0554	5.0	1222	0.9	1824	4.7		
8 M	0023	1.3	0631	4.8	1257	1.1	1902	4.5
9 TU	0059	1.5	0712	4.6	1335	1.3	1945	4.3
10 W ◑	0139	1.7	0759	4.4	1419	1.6	2034	4.1
11 TH	0231	2.0	0855	4.1	1515	1.8	2133	4.0
12 F	0341	2.1	1002	4.0	1626	2.0	2240	4.0
13 SA	0505	2.1	1118	3.9	1741	1.9	2352	4.1
14 SU	0621	1.9	1230	4.1	1846	1.8		
15 M	0053	4.3	0720	1.6	1327	4.3	1938	1.6
16 TU	0141	4.5	0807	1.3	1412	4.6	2023	1.3
17 W	0222	4.8	0849	1.0	1451	4.8	2103	1.1
18 TH ○	0259	5.0	0929	0.7	1530	5.0	2143	0.9
19 F	0336	5.2	1009	0.5	1608	5.2	2222	0.8
20 SA	0414	5.4	1050	0.4	1648	5.2	2301	0.7
21 SU	0454	5.4	1131	0.4	1729	5.2	2341	0.8
22 M	0536	5.4	1212	0.5	1813	5.1		
23 TU	0023	0.9	0621	5.3	1257	0.7	1900	4.9
24 W	0110	1.1	0713	5.0	1345	1.0	1954	4.6
25 TH ◑	0204	1.3	0813	4.7	1444	1.3	2058	4.4
26 F	0313	1.6	0926	4.5	1559	1.6	2213	4.3
27 SA	0438	1.7	1049	4.3	1722	1.7	2332	4.3
28 SU	0604	1.5	1212	4.4	1840	1.6		
29 M	0043	4.5	0715	1.3	1319	4.6	1941	1.5
30 TU	0140	4.8	0811	1.0	1412	4.8	2029	1.3
31 W	0226	5.0	0857	0.8	1455	4.9	2110	1.1

Chart Datum: 2·60 metres below Ordnance Datum (Newlyn). HAT is 5·7 metres above Chart Datum.

NE England

≫ **FREE** monthly updates. Register at ≪
www.reedsnauticalalmanac.co.uk

357

STANDARD TIME (UT)
For Summer Time add ONE hour in **non-shaded areas**

RIVER TYNE/NORTH SHIELDS   LAT 55°00'N   LONG 1°26'W
TIMES AND HEIGHTS OF HIGH AND LOW WATERS

Dates in red are **SPRINGS**
Dates in blue are **NEAPS**

YEAR 2016

SEPTEMBER

Day	Time	m	Day	Time	m
1 TH ●	0306 / 0937 / 1533 / 2146	5.1 / 0.6 / 5.0 / 1.0	16 F ○	0232 / 0905 / 1505 / 2120	5.1 / 0.6 / 5.2 / 0.8
2 F	0341 / 1013 / 1608 / 2220	5.2 / 0.6 / 5.0 / 1.0	17 SA	0311 / 0946 / 1544 / 2200	5.4 / 0.4 / 5.3 / 0.6
3 SA	0416 / 1046 / 1641 / 2251	5.2 / 0.6 / 5.0 / 1.0	18 SU	0350 / 1027 / 1624 / 2241	5.6 / 0.2 / 5.4 / 0.6
4 SU	0449 / 1117 / 1713 / 2321	5.1 / 0.7 / 4.9 / 1.1	19 M	0432 / 1109 / 1705 / 2322	5.6 / 0.3 / 5.4 / 0.6
5 M	0522 / 1147 / 1746 / 2351	5.0 / 0.9 / 4.8 / 1.2	20 TU	0516 / 1151 / 1749	5.5 / 0.4 / 5.2
6 TU	0557 / 1218 / 1821	4.9 / 1.1 / 4.6	21 W	0005 / 0603 / 1235 / 1836	0.8 / 5.3 / 0.7 / 5.0
7 W	0023 / 0634 / 1252 / 1858	1.4 / 4.7 / 1.3 / 4.5	22 TH	0053 / 0656 / 1324 / 1930	1.0 / 5.0 / 1.1 / 4.7
8 TH	0100 / 0717 / 1331 / 1943	1.6 / 4.4 / 1.6 / 4.3	23 F ◐	0148 / 0800 / 1423 / 2035	1.3 / 4.7 / 1.5 / 4.5
9 F ◐	0145 / 0808 / 1421 / 2038	1.8 / 4.2 / 1.9 / 4.1	24 SA	0259 / 0916 / 1541 / 2152	1.5 / 4.4 / 1.9 / 4.3
10 SA	0247 / 0914 / 1531 / 2146	2.1 / 4.0 / 2.1 / 4.0	25 SU	0428 / 1041 / 1711 / 2314	1.6 / 4.3 / 1.9 / 4.3
11 SU	0412 / 1033 / 1656 / 2303	2.1 / 3.9 / 2.1 / 4.0	26 M	0554 / 1204 / 1829	1.5 / 4.4 / 1.8
12 M	0540 / 1153 / 1812	2.0 / 4.0 / 2.0	27 TU	0027 / 0703 / 1309 / 1927	4.5 / 1.3 / 4.5 / 1.6
13 TU	0014 / 0647 / 1256 / 1910	4.2 / 1.7 / 4.3 / 1.7	28 W	0124 / 0755 / 1357 / 2012	4.7 / 1.1 / 4.7 / 1.4
14 W	0109 / 0738 / 1344 / 1957	4.5 / 1.3 / 4.6 / 1.4	29 TH	0208 / 0837 / 1436 / 2050	4.9 / 0.9 / 4.9 / 1.2
15 TH	0153 / 0823 / 1426 / 2039	4.8 / 0.9 / 4.9 / 1.1	30 F	0246 / 0913 / 1510 / 2123	5.0 / 0.8 / 5.0 / 1.1

OCTOBER

Day	Time	m	Day	Time	m
1 SA ●	0319 / 0945 / 1542 / 2154	5.1 / 0.7 / 5.0 / 1.0	16 SU ○	0245 / 0921 / 1519 / 2137	5.5 / 0.4 / 5.4 / 0.6
2 SU	0351 / 1015 / 1612 / 2224	5.1 / 0.8 / 5.0 / 1.0	17 M	0327 / 1004 / 1600 / 2220	5.6 / 0.3 / 5.5 / 0.5
3 M	0422 / 1045 / 1642 / 2253	5.1 / 0.8 / 5.0 / 1.1	18 TU	0412 / 1047 / 1642 / 2304	5.7 / 0.3 / 5.5 / 0.5
4 TU	0455 / 1114 / 1713 / 2323	5.0 / 1.0 / 4.9 / 1.2	19 W	0458 / 1130 / 1727 / 2350	5.6 / 0.5 / 5.3 / 0.7
5 W	0528 / 1144 / 1745 / 2355	4.9 / 1.2 / 4.8 / 1.3	20 TH	0549 / 1215 / 1815	5.3 / 0.9 / 5.1
6 TH	0604 / 1216 / 1821	4.7 / 1.4 / 4.6	21 F	0039 / 0644 / 1305 / 1909	0.9 / 5.0 / 1.3 / 4.8
7 F	0030 / 0644 / 1253 / 1902	1.5 / 4.5 / 1.6 / 4.4	22 SA ◐	0136 / 0748 / 1403 / 2013	1.2 / 4.7 / 1.7 / 4.5
8 SA	0113 / 0733 / 1339 / 1952	1.7 / 4.2 / 1.9 / 4.2	23 SU	0246 / 0902 / 1520 / 2128	1.5 / 4.4 / 2.0 / 4.3
9 SU ◐	0209 / 0835 / 1443 / 2057	1.9 / 4.0 / 2.1 / 4.1	24 M	0410 / 1022 / 1648 / 2248	1.6 / 4.3 / 2.1 / 4.3
10 M	0326 / 0952 / 1610 / 2214	2.0 / 3.9 / 2.2 / 4.1	25 TU	0532 / 1141 / 1804	1.5 / 4.3 / 2.0
11 TU	0455 / 1112 / 1733 / 2329	1.9 / 4.1 / 2.1 / 4.2	26 W	0001 / 0638 / 1245 / 1902	4.4 / 1.4 / 4.5 / 1.8
12 W	0608 / 1220 / 1837	1.6 / 4.3 / 1.8	27 TH	0059 / 0729 / 1333 / 1947	4.6 / 1.2 / 4.6 / 1.6
13 TH	0030 / 0704 / 1341 / 1927	4.5 / 1.3 / 4.7 / 1.4	28 F	0144 / 0810 / 1411 / 2024	4.8 / 1.1 / 4.8 / 1.4
14 F	0120 / 0752 / 1357 / 2012	4.9 / 0.9 / 5.0 / 1.1	29 SA	0221 / 0844 / 1444 / 2057	4.9 / 1.0 / 4.9 / 1.2
15 SA	0203 / 0837 / 1438 / 2055	5.2 / 0.6 / 5.3 / 0.8	30 SU ●	0255 / 0915 / 1515 / 2129	5.0 / 1.0 / 5.0 / 1.1
			31 M	0327 / 0946 / 1545 / 2200	5.0 / 0.9 / 5.0 / 1.1

NOVEMBER

Day	Time	m	Day	Time	m
1 TU	0359 / 1016 / 1615 / 2230	5.0 / 1.0 / 5.0 / 1.1	16 W	0356 / 1028 / 1622 / 2250	5.6 / 0.5 / 5.5 / 0.5
2 W	0432 / 1046 / 1646 / 2301	5.0 / 1.1 / 5.0 / 1.2	17 TH	0445 / 1113 / 1708 / 2338	5.5 / 0.7 / 5.3 / 0.6
3 TH	0506 / 1117 / 1718 / 2334	4.8 / 1.2 / 4.9 / 1.3	18 F	0537 / 1158 / 1756	5.3 / 1.0 / 5.2
4 F	0542 / 1149 / 1753	4.7 / 1.4 / 4.7	19 SA	0028 / 0632 / 1247 / 1849	0.8 / 5.0 / 1.4 / 4.9
5 SA	0010 / 0621 / 1226 / 1832	1.4 / 4.5 / 1.6 / 4.6	20 SU	0123 / 0733 / 1341 / 1949	1.1 / 4.7 / 1.7 / 4.7
6 SU	0052 / 0707 / 1310 / 1919	1.6 / 4.3 / 1.9 / 4.4	21 M ◐	0226 / 0839 / 1447 / 2056	1.3 / 4.4 / 2.0 / 4.5
7 M ◐	0144 / 0804 / 1407 / 2018	1.7 / 4.2 / 2.1 / 4.3	22 TU	0338 / 0948 / 1606 / 2208	1.5 / 4.2 / 2.1 / 4.3
8 TU	0251 / 0915 / 1524 / 2129	1.8 / 4.1 / 2.2 / 4.2	23 W	0451 / 1100 / 1721 / 2320	1.6 / 4.2 / 2.1 / 4.3
9 W	0410 / 1030 / 1647 / 2242	1.8 / 4.2 / 2.1 / 4.3	24 TH	0558 / 1206 / 1825	1.6 / 4.3 / 2.0
10 TH	0525 / 1140 / 1757 / 2349	1.6 / 4.4 / 1.8 / 4.6	25 F	0022 / 0652 / 1258 / 1915	4.4 / 1.5 / 4.5 / 1.8
11 F	0627 / 1238 / 1853	1.3 / 4.7 / 1.5	26 SA	0113 / 0735 / 1341 / 1955	4.6 / 1.4 / 4.6 / 1.6
12 SA	0045 / 0720 / 1328 / 1943	4.9 / 0.9 / 5.0 / 1.2	27 SU	0155 / 0812 / 1417 / 2032	4.7 / 1.3 / 4.8 / 1.4
13 SU	0136 / 0810 / 1413 / 2031	5.2 / 0.7 / 5.2 / 0.9	28 M	0232 / 0846 / 1450 / 2106	4.8 / 1.2 / 4.9 / 1.3
14 M ○	0222 / 0857 / 1456 / 2117	5.4 / 0.5 / 5.4 / 0.6	29 TU ●	0307 / 0919 / 1522 / 2139	4.9 / 1.1 / 5.0 / 1.2
15 TU	0309 / 0942 / 1539 / 2204	5.6 / 0.4 / 5.5 / 0.5	30 W	0341 / 0951 / 1553 / 2212	4.9 / 1.1 / 5.0 / 1.1

DECEMBER

Day	Time	m	Day	Time	m
1 TH	0414 / 1023 / 1624 / 2245	4.9 / 1.2 / 5.0 / 1.1	16 F	0436 / 1059 / 1653 / 2328	5.4 / 0.8 / 5.4 / 0.6
2 F	0449 / 1056 / 1657 / 2319	4.8 / 1.2 / 4.9 / 1.2	17 SA	0526 / 1143 / 1740	5.2 / 1.1 / 5.2
3 SA	0525 / 1130 / 1732 / 2357	4.7 / 1.4 / 4.9 / 1.2	18 SU	0016 / 0617 / 1228 / 1828	0.7 / 5.0 / 1.3 / 5.0
4 SU	0604 / 1207 / 1811	4.6 / 1.5 / 4.8	19 M	0105 / 0710 / 1313 / 1920	0.9 / 4.7 / 1.6 / 4.8
5 M	0037 / 0647 / 1249 / 1854	1.4 / 4.5 / 1.7 / 4.6	20 TU	0156 / 0806 / 1404 / 2017	1.2 / 4.5 / 1.9 / 4.6
6 TU	0125 / 0739 / 1340 / 1947	1.5 / 4.4 / 1.9 / 4.5	21 W ◐	0252 / 0904 / 1506 / 2119	1.5 / 4.2 / 2.1 / 4.4
7 W ◐	0222 / 0840 / 1443 / 2049	1.6 / 4.3 / 2.0 / 4.4	22 TH	0355 / 1006 / 1618 / 2225	1.7 / 4.1 / 2.2 / 4.2
8 TH	0330 / 0949 / 1559 / 2159	1.6 / 4.3 / 2.0 / 4.4	23 F	0500 / 1112 / 1730 / 2333	1.8 / 4.1 / 2.1 / 4.2
9 F	0443 / 1059 / 1714 / 2308	1.5 / 4.4 / 1.8 / 4.6	24 SA	0602 / 1213 / 1833	1.7 / 4.2 / 2.0
10 SA	0550 / 1203 / 1819	1.3 / 4.6 / 1.6	25 SU	0034 / 0655 / 1306 / 1924	4.3 / 1.7 / 4.4 / 1.8
11 SU	0013 / 0650 / 1300 / 1917	4.8 / 1.1 / 4.8 / 1.3	26 M	0126 / 0740 / 1349 / 2007	4.4 / 1.5 / 4.5 / 1.6
12 M	0112 / 0746 / 1351 / 2011	5.0 / 0.9 / 5.1 / 1.0	27 TU	0209 / 0820 / 1427 / 2046	4.6 / 1.4 / 4.7 / 1.4
13 TU	0205 / 0837 / 1437 / 2102	5.2 / 0.7 / 5.3 / 0.7	28 W	0248 / 0857 / 1501 / 2122	4.7 / 1.3 / 4.8 / 1.2
14 W ○	0256 / 0926 / 1522 / 2152	5.4 / 0.7 / 5.4 / 0.6	29 TH ●	0324 / 0932 / 1534 / 2156	4.8 / 1.2 / 4.9 / 1.1
15 TH	0346 / 1013 / 1607 / 2240	5.4 / 0.7 / 5.4 / 0.5	30 F	0359 / 1006 / 1607 / 2231	4.8 / 1.2 / 5.0 / 1.0
			31 SA	0434 / 1041 / 1640 / 2307	4.8 / 1.2 / 5.0 / 1.0

Chart Datum: 2·60 metres below Ordnance Datum (Newlyn). HAT is 5·7 metres above Chart Datum.

》》 FREE monthly updates. Register at 《
www.reedsnauticalalmanac.co.uk

9.5.19   BLYTH

Northumberland **55°06'·98N 01° 29'·27W** ❀❀❀☆☆❀❀

CHARTS AC 152, 156, 1626, 5615; Imray C24

TIDES +0430 Dover; ML 2·8; Duration 0558

Standard Port NORTH SHIELDS (←)

Times				Height (metres)			
High Water		Low Water		MHWS	MHWN	MLWN	MLWS
0200	0800	0100	0800	5·0	3·9	1·8	0·7
1400	2000	1300	2000				
Differences BLYTH							
+0005	−0007	−0001	+0009	0·0	0·0	−0·1	+0·1

SHELTER Very good. Access H24 but at LW in strong SE winds, seas break across entrance.

NAVIGATION WPT 55°06'·57N 01°28'·74W, 324°/0·5M to harbour entrance. From N, beware The Pigs, The Sow and Seaton Sea Rks. A meteorological mast is 3M offshore and 2 wind turbines on North Spit (1·2M N of hbr ent) centred on 55°08'·16N 01°29'·41W. No dangers from S. See RNYC Sailing Directions, from RNYC, House Yacht Tyne, S Hbr, Blyth NE24 3PB (website below).

LIGHTS AND MARKS Blyth E Pier Hd, W twr on S end. Outer ldg lts 324°, Or ◊ on framework trs. Inner ldg lts 338°, front W 6-sided tr; rear W △ on mast. See chartlet and 9.5.4 for lt details.

COMMUNICATIONS (Code 01670) CGOC (01262) 672317; Police 101. HM 352678, Call *Blyth Hbr Control* Ch **12** 11 (H24) for clearance to enter/ leave.

FACILITIES R Northumberland YC www.rnyc.org.uk ☎353636. In SE part of South Hbr, E side of Middle Jetty, ♥ berth on N side of RNYC pontoons in 4·4m least depth. 123 ⌓ £1·50, ⌦ ▮ (20t) ⌐(1½t for trailer sailers) @ ⌗ ⚒ (www.millermarine.co.uk).

South Hbr ☎352678, ⊥ ⌐(50t).

Town ⌦ ⌦ ▮ ✉ Ⓑ 🛒 ✕ ⌂ bus to Newcastle ⇌ ✈.

ADJACENT ANCHORAGE
NEWBIGGIN, Northumberland, **55°10'·75N 01°30'·10W**. AC 156. Tides approx as for Blyth, 3·5M to the S. Temp, fair weather ⚓ in about 4m in centre of the bay, off the conspic sculptures on the breakwater; sheltered from SW to N winds. Caution: offlying rky ledges to N and S. Conspic church on N side of bay; bkwtr lt Fl G 10s 4M. Facilities: **SC** (dinghies). **Town** 🛒 ✕ ⌂.

9.5.20 AMBLE

Northumberland 55°20'·37N 01°34'·25W ❄❄◊◊◊◊❀❀

CHARTS AC 156, 1627, 5615; Imray C24

TIDES +0412 Dover; ML 3·1; Duration 0606

Standard Port NORTH SHIELDS (←—)

Times				Height (metres)			
High Water		Low Water		MHWS	MHWN	MLWN	MLWS
0200	0800	0100	0800	5·0	3·9	1·8	0·7
1400	2000	1300	2000				
Differences AMBLE							
–0013	–0013	–0016	–0020	0·0	0·0	+0·1	+0·1
COQUET ISLAND							
–0010	–0010	–0020	–0020	+0·1	+0·1	0·0	+0·1

SHELTER The Hbr (alias Warkworth Hbr) is safe in all winds. But ent is dangerous in strong N to E winds or in swell, when heavy seas break on Pan Bush shoal and on the bar at hbr ent, where least depth is 0·8m. ⚓ in Coquet Road or NNE of North Pier.

NAVIGATION WPT 55°21'·00N 01°33'·10W, 225° to hbr ent, 0·9M. Ent recommended from NE, passing N and W of Pan Bush. The S-going stream sets strongly across ent. Once inside the N bkwtr, beware drying banks to stbd, ie on N side of channel; keep close (15m) to S quays. 4kn speed limit in hbr.

- In NE'ly gales broken water can extend to Coquet Island, keep E of island and go to Blyth where appr/ent may be safer.
- Coquet Chan (min depth 0·3m) is not buoyed and is only advised with caution, by day, in good vis/weather, slight sea state and with adequate rise of tide, ie HW–2.

LIGHTS AND MARKS Coquet Island lt ho, (conspic) W☐tr, turreted parapet, lower half grey; horn 30s. See chartlet and 9.5.4.

COMMUNICATIONS (Code 01665) HM 710306; CGOC (01262) 672317; local CG 710575; Police 101; Ⓗ (01670) 521212.
Amble Marina Ch **80** (7/7, 0900-1700). HM Ch 16, works Ch 14 (Mon-Fri 0900-1700 LT). Coquet YC Ch M (occas).

FACILITIES Amble Marina marina@ amble.co.uk ☎712168, is about 5ca from hbr ent. Tidal gauge shows depth over sill, 0·8m above CD; access between PHM buoy and old jetty, both unlit. Pontoons are 'A' to 'F' from sill; Ⓥ berth on 'B' initially. 200 ⌁+40 Ⓥ, £2·50. 🛢🖥⚓🛁♨✂✕ 🖥Ⓔ⚠🛒🔧🏴(50t) ⛴🛒✕☐ Hbr 🛟⌁.
Coquet YC ☎711179, ⚓ ⛴ ☐ ♠⚓.
Town ✉🛒✕☐🚂 (Alnmouth), ✈ (Newcastle).

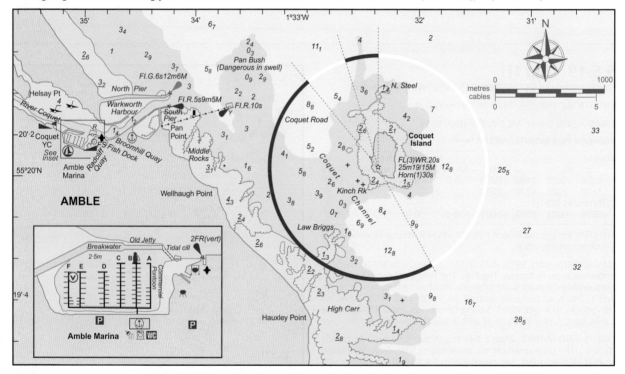

HBRS AND ANCHORAGES BETWEEN AMBLE AND HOLY ISLAND

BOULMER, Northumberland, **55°25'·00N 01°33'·90W**. AC 156. Tides approx as for Amble, 4·6M to the S. A small haven almost enclosed by N and S Rheins, rky ledges either side of the narrow (30m) ent, Marmouth; only advised in settled offshore weather. 2 FW bns (lit when req'd by LB or FVs) lead approx 257° through the ent, leaving close to stbd a bn on N Rheins. ⚓ just inside in about 1·5m or dry out on sand at the N end. Few facilities: ✉☐ in village. Alnwick is 4M inland.

CRASTER, Northumberland, **55°28'·40N 01°35'·30W**. AC 156. Tidal differences: interpolate between Amble (9.5.20) and N Sunderland (9.5.21). Strictly a fair weather ⚓ in offshore winds, 1M S of the conspic Dunstanburgh Castle (ru). The ent, 40m wide, is N of Muckle Carr and S of Little Carr which partly covers and has a bn on its S end. ⚓ in about 3·5m just inshore of these 2 rocky outcrops; or berth at the E pier on rk/sand inside the tiny drying hbr. Facilities: 🛒 ✕ ☐.

NEWTON HAVEN and BEADNELL BAY, Northumberland, **55°30'·90N 01° 36'·70W**. AC 156. HW +0342 on Dover; Tidal differences: interpolate between Amble (9.5.20) and N Sunderland (9.5.21). ML 2·6m; Duration 0625. A safe ⚓ in winds from NNW to SE via S but susceptible to swell. Ent to S of Newton PHM buoy and Newton Pt. Beware Fills Rks. ⚓ between Fills Rks and Low Newton by the Sea in 4/5m. A very attractive ⚓ with no lts, marks or facilities except a pub. Further ⚓ S of Beadnell Pt (1M N of Newton Pt) in 4–6m; small, fishing hbr whose wall is newly rebuilt; Beadnell SC. Village 0·5M.

SEAHOUSES (N Sunderland), Northumberland, **55°35'·04N 01°38'·91W**. AC 1612. HW +0340 on Dover; Duration 0618. See 9.5.21. Good shelter except in on-shore winds when swell makes outer hbr berths (0·7m) very uncomfortable and dangerous. Inner hbr has excellent berths but usually full of FVs. Beware The Tumblers (rks) to the W of ent and rks protruding NE from bkwtr hd Fl R 2·5s 6m NW pier hd FG 11m 3M; vis 159°–294°, W tr; traffic sigs; Siren 90s when vessels expected. Good facilities. If dangerous to enter, ● is shown over the FG lt (or R flag over a Bu flag) on NW pier hd.

FARNE ISLANDS, Northumberland, **55°37′·15N 01°39′·37W**. AC 160, 156, 111. HW +0345 on Dover; ML 2·6m; Duration 0630. See 9.5.21. The islands are a NT nature reserve in a beautiful area; they should only be attempted in good weather. Landing is only allowed on Farne Is, Staple Is and Longstone. In the inner group, ⚓ in The Kettle on the NE side of Inner Farne; near the Bridges (connecting Knocks Reef to West Wideopen); or to the S of West Wideopen. In the outer group, ⚓ in Pinnacle Haven (between Staple Is/Brownsman). Beware turbulence over Knivestone and Whirl Rks and eddy S of Longstone during NW tidal streams. Lts and marks: Black Rocks Pt, Oc (2) WRG 8s 12m 14/11M;122°-G-165°-W-175°-R-191°-W-238°-R-275°-W-289°-G-300°. Bamburgh Castle is conspic 6ca to the SE. Farne Is lt ho at SW Pt, Fl (2) WR 15s 27m 8/6M; W ○ tr; 119°-R-280°-W-119°. Longstone Fl 20s 23m

24M, R tr with W band (conspic), temp'ly replaced by Light vessel Fl(2) 20s15M (2015). Caution: reefs extend about 7ca seaward. No facilities.

NATIONAL NATURE RESERVE (Holy Island) A National Nature Reserve (NNR) extends from Budle Bay (**55°37′N 01°45′W**, close to Black Rocks Point lt ho) along the coast to Cheswick Black Rocks, 3M SE of Berwick-upon-Tweed. The NNR extends seaward from the HW shoreline to the drying line; it includes Holy Island and the adjacent islets.

Visiting craft are asked to respect two constraints:

- Landing is prohibited on the small island of Black Law (55°39′·68N 01°47′·50W) from April to August inclusive.
- Boats should not be landed or recovered anywhere in the NNR except at the designated and buoyed watersports zone on the SE side of Budle Bay.

COQUET ISLAND TO FARNE ISLANDS

(AC 156) Coquet Is (lt, fog sig) lies about 5ca offshore at SE end of Alnmouth B, and nearly 1M NNE of Hauxley Pt, off which dangerous rks extend 6ca offshore, drying 1·9m. On passage, normally pass 1M E of Coquet Is in the W sector of the lt. Coquet chan may be used in good vis by day; but it is only 2ca wide, not buoyed, has least depth of 0·9m near the centre. In S or W winds, there are good ⚓s in Coquet Road, W and NW of the Island.

▶The stream runs strongly: S-going from HW Tyne −0515 and N-going from HW Tyne +0045. ◀

Amble (Warkworth) Hbr ent is about 1M W of Coquet Is, and 1·5M SE of Warkworth Castle (conspic). 4ca NE and ENE of ent is Pan Bush, rky shoal with least depth of 0·9m on which dangerous seas can build in any swell. The bar has varying depths, down to less than 1m. The entrance is dangerous in strong winds from N/E when broken water may extend to Coquet Is. Once inside, the hbr is safe.

Between Coquet Is and the Farne Is, 19M to N, keep at least 1M offshore to avoid various dangers. To seaward, Craster Skeres lie 5M E of Castle Pt, and Dicky Shad and Newton Skere lie 1·75M and 4·5M E of Beadnell Pt; these are three rky banks on which the sea breaks heavily in bad weather.

FARNE ISLANDS

(AC 111, 160) The coast between N Sunderland Pt (Snook) and Holy Island, 8M NW, has fine hill (Cheviots) scenery fronted by dunes and sandy beaches. The Farne Is and offlying shoals extend 4·5M offshore, and are a mini-cruising ground well worth visiting in good weather. The islands are a bird sanctuary, owned and operated by the National Trust, with large colonies of sea birds and grey seals. The R Northumberland YC's Sailing Directions (see 9.5.19) are useful; AC 111 is essential.

Inner Sound separates the islands from the mainland. In good conditions it is a better N/S route than keeping outside the whole group; but the stream runs at 3kn at sp, and with strong wind against tide there is rough water. If course is set outside the Farne Islands, pass 1M E of Longstone (lit) to clear Crumstone Rk 1M

to S, and Knivestone (dries 3·6m) and Whirl Rks (depth 0·6m) respectively 5 and 6ca NE of Longstone lt ho. The sea breaks on these rocks.

The islands, rks and shoals are divided by Staple Sound, running NW/SE, into an inner and outer group. The former comprises Inner Farne, W and E Wideopens and Knock's Reef. Inner Farne (lt) is the innermost Is; close NE there is ⚓ called The Kettle, sheltered except from NW, but ⚓ out of stream close to The Bridges connecting Knock's Reef and W Wideopen. 1M NW of Inner Farne Is and separated by Farne Sound, which runs NE/SW, lies the Megstone, a rk 5m high. Beware Swedman reef (dries 0·5m), marked by SHM buoy 4ca WSW of Megstone.

The outer group of Islands comprises Staple and Brownsman Islands, N and S Wamses, the Harcars and Longstone. There is occas ⚓ between Staple and Brownsman Is.

▶ Piper Gut and Crafords Gut may be negotiated in calm weather and near HW, stemming the S-going stream. ◀

HOLY ISLAND TO BERWICK

(AC 1612, 111, 160) ▶ Near the Farne Is and Holy Is the SE-going stream begins at HW Tyne −0430, and the NW-going at HW Tyne +0130. Sp rates are about 2.5kn in Inner Sound, 4kn in Staple Sound and about 3·5kn 1M NE of Longstone, decreasing to seaward. There is an eddy S of Longstone on NW-going stream. ◀

Holy Is (or Lindisfarne) lies 6M WNW of Longstone, and is linked to mainland by a causeway covered at HW. There is a good ⚓ on S side (AC 1612) with conspic daymarks and dir lts. The castle and a W obelisk at Emmanuel Head are also conspic.

▶ The stream runs strongly in and out of hbr, W-going from HW Tyne +0510, and E-going from HW Tyne −0045. E of Holy Is, Goldstone chan runs N/S between Goldstone Rk (dries) SHM buoy on E side and Plough Seat Reef and Plough Rk (both dry) on W side, with PHM buoy. ◀

Berwick Bay has some offlying shoals. Berwick-upon-Tweed is easily visible against the low shoreline, which rises again to high cliffs further north. The hbr entrance is restricted by a shallow bar, dangerous in onshore winds.

9.5.21 HOLY ISLAND

Northumberland **55°39′·57N 01°46′·81W** ❋❋♦♢♢♢

TIDES +0344 Dover; ML No data; Duration 0630

Standard Port NORTH SHIELDS (←—)

Times				Height (metres)			
High Water		Low Water		MHWS	MHWN	MLWN	MLWS
0200	0800	0100	0800	5·0	3·9	1·8	0·7
1400	2000	1300	2000				
Differences HOLY ISLAND							
−0043	−0039	−0105	−0110	−0·2	−0·2	−0·3	−0·1
NORTH SUNDERLAND (Seahouses)							
−0048	−0044	−0058	−0102	−0·2	−0·2	−0·2	0·0

CHARTS AC 111, 1612, 5615; Imray C24

SHELTER Good S of The Heugh in 3-6m, but ⚓ is uncomfortable in fresh W/SW winds esp at sp flood, and foul in places; trip-line advised. Better shelter in The Ouse on sand/mud if able to dry out; but not in S/SE winds. For detailed information for trailer sailers see www.dca.uk.com.

NAVIGATION WPT 55°39′·76N 01°44′·88W, 260°/1·55M to Old Law E bn. From N identify Emmanuel Hd, conspic W △ bn, then appr via Goldstone Chan leaving Plough Seat PHM buoy to stbd. From S, clear Farne Is thence to WPT. Outer ldg bns lead 260° close past Ridge ECM and Triton SHM buoys. Possible overfalls in chan across bar (1·6m) with sp ebb up to 4kn. Inner ldg marks lead 310° to ⚓. Inshore route, round Castle Pt via Hole Mouth and The Yares, may be more sheltered, but is not for strangers.

Continued overleaf

HOLY ISLAND *continued*

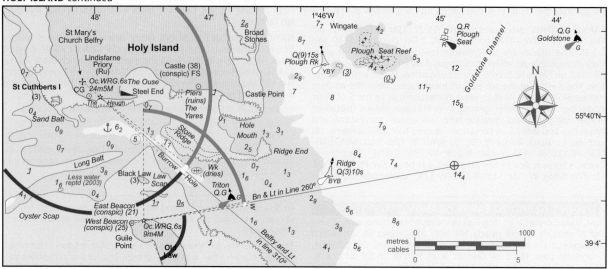

LIGHTS AND MARKS See 9.5.4 and chartlet. Outer leading marks/lts are Old Law beacons (conspic), 2 reddish obelisks 21/25m on 260°; E beacon has directional light, Oc WRG 6s. Inner leading marks/lights are The Heugh tower, B △, on with St Mary's church belfry 310°. The Heugh has directional light, Oc WRG 6s. Directional lights Oc WRG are aligned on 260° and 309·5° respectively.

COMMUNICATIONS (Code 01289) CGOC (01262) 672317; HM 389248.

FACILITIES Limited. ► (contact HM prior to launching) £5.00, ⚓ on village green, ✕ ☐ limited ▦ ☎ ☎ from Beal (5M); bus (occas) to Berwick. Note Lindisfarne is ancient name; Benedictine Abbey (ruins) and Castle (NT) are worth visiting. Causeway to mainland, covers at HW, is usable approx HW+3½ to HW-2.

9.5.22 BERWICK-UPON-TWEED

Northumberland 55°45'·87N 01°59'·05W ❀❀⚓⚓❀❀

CHARTS AC 111, 160, 1612, 5615; Imray C24

TIDES +0348 Dover; ML 2·5; Duration 0620

Standard Port NORTH SHIELDS (◄—)

Times				Height (metres)			
High Water		Low Water		MHWS	MHWN	MLWN	MLWS
0200	0800	0100	0800	5·0	3·9	1·8	0·7
1400	2000	1300	2000				
Differences BERWICK-UPON-TWEED							
−0053	−0053	−0109	−0109	−0·3	−0·1	−0·5	−0·1

SHELTER Good shelter or ⚓ except in strong E/SE winds. ⌒ may be available in Tweed Dock (no dock gates) which is primarily used by commercial shipping. If possible contact HM beforehand to avoid disappointment. Alternative temporary ⌒ possible at W end of Fish Jetty (1·2m).

NAVIGATION WPT 55°45'·62N 01°58'·0W, 294°/0·65M to breakwater lighthouse.

- On-shore winds and ebb tides cause very confused seas over the bar, minimum depth 0·6m. From HW−2 to HW+1 strong flood tide sets S across the entrance; keep well up to breakwater.
- The sands at the mouth of the Tweed shift so frequently that local knowledge is essential to enter the harbour.

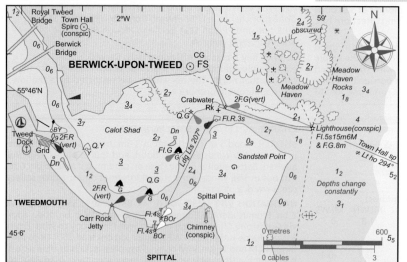

Town Hall Spire and Lt ho in line at 294°. Keep close to breakwater to avoid Sandstell Point. When past PHM buoy and S of Crabwater Rk, keep Spittal Bns (Or ▽ on B Or Bn) in line brg 207°; best water may be further W. Berwick Bridge (first/lowest) has about 3m clearance.

LIGHTS AND MARKS See 9.5.4 and chartlet.

COMMUNICATIONS (Code 01289) CGOC (01262) 672317; Police 101; Dr 307484. HM 307404.

VHF Ch 12 16 (0800-1700).

FACILITIES Tweed Dock ☎307404, ⌒ £8.00 ► ► ⚓ ▷ ⚒ ✕ ⊞ ☐ △ ☐(3t) Mobile.

Town ☎ ☎ ✉ Ⓑ ▦ ✕ ☐ ⇌ and ✈ (Newcastle or Edinburgh).

South East Scotland

Eyemouth to Buchan Ness

9.6.1	**Contents**		363
9.6.2	**Area map and Distance table**		364
9.6.3	**Tidal stream charts**		366
9.6.4	**Lights, buoys and waypoints**		368
9.6.5	**Passage information**		370

Berwick-upon-Tweed to Bass Rock – 370 • Firth of Forth south shore, River Forth to Kincardine, Firth of Forth north shore, Forth to Norway and Baltic – 372 • Fife Ness to Montrose – 380 • Montrose to Buchan Ness – 383

9.6.6	**Eyemouth**		370

Burnmouth • St Abbs

9.6.7	**Dunbar**		371

North Berwick

9.6.8	**Firth of Forth**		373

Port Edgar • Granton • Inchcolm • Aberdour • Burntisland • Kirkcaldy • Methil• Elie • St Monans • Fisherrow • Cramond

9.6.9	**Forth and Clyde Canal (Leith, tide tables/curves)**		375
9.6.10	**Anstruther**		379

Pittenweem • Crail • Isle of May

9.6.11	**River Tay**		381

Perth • St Andrews

9.6.12	**Arbroath**		382
9.6.13	**Montrose**		382

Johnshaven • Gourdon

9.6.14	**Stonehaven**		383
9.6.15	**Aberdeen, tide tables/curves**		384

DISTANCE TABLE

Approximate distances in nautical miles are by the most direct route while avoiding dangers and allowing for TSS. Places in *italics* are in adjoining areas; places in **bold** are in 9.0.7, Distances across the North Sea.

		1	2	3	4	5	6	7	8	9	10	11	12	13	14	15	16	17	18
1	*Berwick-upon-Tweed*	**1**																	
2	Eyemouth	10	**2**																
3	Dunbar	26	17	**3**															
4	North Berwick	35	25	9	**4**														
5	Granton	54	44	27	19	**5**													
6	**Port Edgar**	58	50	34	26	7	**6**												
7	Burntisland	53	43	26	18	5	8	**7**											
8	Anstruther	38	29	14	10	23	29	22	**8**										
9	Fife Ness	38	29	17	14	28	34	27	5	**9**									
10	Bell Rock	43	36	27	25	40	47	39	17	12	**10**								
11	**Dundee**	58	49	37	34	48	54	47	25	20	20	**11**							
12	Arbroath	51	44	34	31	45	51	44	22	17	10	15	**12**						
13	Montrose	59	51	43	41	55	61	54	32	27	17	27	12	**13**					
14	Stonehaven	72	66	58	58	72	78	71	49	44	32	45	30	20	**14**				
15	**Aberdeen**	82	78	73	70	84	90	83	61	56	44	57	42	32	13	**15**			
16	Peterhead	105	98	93	95	106	108	105	83	78	68	80	64	54	35	25	**16**		
17	*Fraserburgh*	121	114	109	108	122	128	121	99	94	83	96	79	68	51	39	16	**17**	
18	*Wick*	178	171	166	165	179	185	178	156	151	140	153	136	125	108	96	72	57	**18**

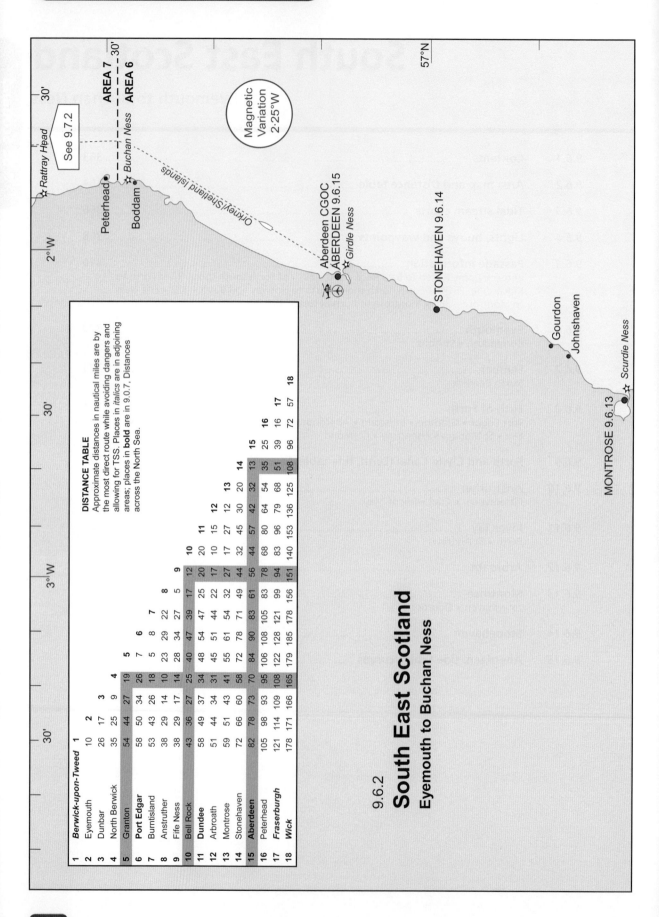

Magnetic Variation 2·25°W

See 9.7.2

Rattray Head
AREA 7
AREA 6
Buchan Ness
Peterhead
Boddam
Orkney/Shetland Islands

Aberdeen CGOC
ABERDEEN 9.6.15
Girdle Ness

STONEHAVEN 9.6.14

Gourdon
Johnshaven

Scurdie Ness
MONTROSE 9.6.13

30'
30'
57°N
2°W
30'
3°W
30'

9.6.2

South East Scotland
Eyemouth to Buchan Ness

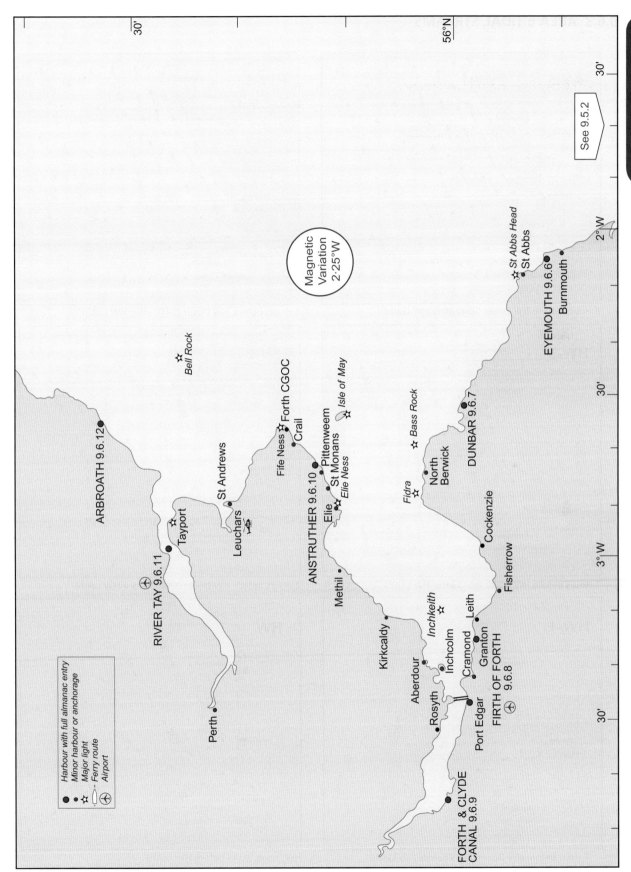

See 9.5.2

Magnetic
Variation
2·25°W

56°N

30'

30'

2°W

St Abbs Head ☆
St Abbs ●
EYEMOUTH 9.6.6 ●
Burnmouth ●

Bell Rock ☆

Forth CGOC ☆
Fife Ness ☆
Crail ●

Isle of May ⊕

Pittenweem ●
St Monans ●
Elie ●
Elie Ness ☆

Bass Rock ☆

St Andrews ●

Fidra ☆
North Berwick ●
DUNBAR 9.6.7 ●

ARBROATH 9.6.12 ●

Leuchars ✈

ANSTRUTHER 9.6.10

Cockenzie ●

Tayport ☆

Methil ●

Fisherrow ●
3°W

RIVER TAY 9.6.11 ⊕

Leith ●
Cramond ●
Granton ●

Kirkcaldy ●

Inchkeith

Inchcolm ●

Aberdour ●

FIRTH OF FORTH 9.6.8 ⊕

Rosyth ●

Port Edgar ●

Perth ●

30'

FORTH & CLYDE
CANAL 9.6.9 ●

Harbour with full almanac entry ●
Minor harbour or anchorage ● ☆
Major light ☆
Ferry route ⊢
Airport ⊕

9.6.3 AREA 6 TIDAL STREAMS

5 Hours before HW Dover (0340 after HW Leith)

4 Hours before HW Dover (0440 after HW Leith)

3 Hours before HW Dover (0540 after HW Leith)

2 Hours before HW Dover (0545 before HW Leith)

1 Hour before HW Dover (0445 before HW Leith)

HW Dover (0345 before HW Leith)

Northward 9.7.3   Southward 9.5.3

1 Hour after HW Dover (0245 before HW Leith)

2 Hours after HW Dover (0145 before HW Leith)

3 Hours after HW Dover (0045 before HW Leith)

4 Hours after HW Dover (0015 after HW Leith)

5 Hours after HW Dover (0115 after HW Leith)

6 Hours after HW Dover (0215 after HW Leith)

9.6.4 LIGHTS, BUOYS AND WAYPOINTS

Bold print = light with a nominal range of 15M or more. CAPITALS = place or feature. *CAPITAL ITALICS* = light-vessel, light float or Lanby. *Italics* = Fog signal. ***Bold italics*** = Racon. Some marks/buoys are fitted with AIS; see relevant charts.

BERWICK-UPON-TWEED TO BASS ROCK

BURNMOUTH
Ldg lts 241°. Front, FR 29m 4M; 55°50'·53N 02°04'·25W. Rear, 45m from front, FR 35m 4M. Both on W posts (unclear by day).

EYEMOUTH
Blind Buss ⚓ Q; 55°52'·80N 02°05'·25E.
Ldg Lts 174°. Front, W Bkwtr Hd ⚲, FG 9m 6M; 55°52'·47N 02°05'·29W. Rear, elbow 55m from front, FG 10m 6M.

ST ABBS to DUNBAR and BASS ROCK
St Abbs Hd ☆ 55°54'·96N 02°08'·29W; Fl 10s 68m **26M**; W twr; ***Racon (T) 18M***.
Barns Ness Tower (Lt ho disused 36m); 55°59'·22N 02°26'·76W.
Torness Pwr Stn Ldg Lts 215°. Front, Iso G 4s 6m 6M; W □ with B vert stripe on plinth; 55°58'·27N 02°24'·86W. Rear, Iso G 4s 22m 3M; W □ with B vert stripe on post; sync with front.
Dunbar Hbr Ldg Lts (Bayswell Hill) 198°. Front, Oc G 6s 15m 3M; 3 Or △ on col; 56°00'·26N 02°31'·21W. Rear, Oc G 6s 22m 3M; 3 Or ▽; sync with front; both lts 188°-G(intens)-208°.
Bass Rock, S side, ⚲ Fl (3) 20s 46m 10M; W twr; vis: 241°-107°; 56°04'·61N 02°38'·48W.

FIRTH OF FORTH AND SOUTH SHORE

(Direction of buoyage East to West)

NORTH BERWICK
Outfall ⚓ Fl Y; 56°04'·29N 02°40'·89W.

Fidra ☆ 56°04'·39N 02°47'·13W; Fl (4) 30s 34m **15M**; W twr; obsc by Bass Rock, Craig Leith and Lamb Island.

PORT SETON, COCKENZIE, FISHERROW and S CHANNEL
Port Seton, E Pier Hd ⚲ Iso WR 4s 10m W9M, R6M; vis: shore - R - 105°- W - 225°- R - shore; *Bell (occas)*; 55°58'·40N 02°57'·23W.
Fisherrow E Pier Hd ⚲ Oc 6s 5m 6M; 55°56'·79N 03°04'·11W.
Narrow Deep ⚲ Fl (2) R 10s; 56°01'·46N 03°04'·59W.
Herwit ⚲ Fl (3) G 10s; 56°01'·05N 03°06'·52W.
North Craig ⚓ Q (3) 10s 56°01'·02N 03°03'·52W.
Craigh Waugh ⚓ Fl (2) 10s;56°00'·26N 03°04'·47W.
Diffuser Hds (Outer) ⚓ 55°59'·81N 03°07'·84W.

LEITH and GRANTON
Leith Approach ⚲ Fl R 3s; 55°59'·95N 03°11'·51W.
East Bkwtr Hd ⚲ Iso R 4s 7m 9M; 55°59'·48N 03°10'·94W.
GrantonE Pier Head ⚲ Fl R 2s 5m 6M; 55°59'·28N 03°13'·27W.

NORTH CHANNEL and MIDDLE BANK
Inchkeith Fairway ⚲ Iso 2s; ***Racon (T) 5M***; 56°03'·49N 03°00'·10W.
No. 1 ⚲ Fl G 9s; 56°03'·22N 03°03'·71W.
No. 2 ⚲ Fl R 9s; 56°02'·90N 03°03'·72W.
No. 3 ⚲ Fl G 6s; 56°03'·22N 03°06'·10W.
No. 4 ⚲ Fl R 6s; 56°02'·89N 03°06'·11W.
No. 5 ⚲ Fl G 3s; 56°03'·18N 03°07'·88W.
No. 6 ⚲ Fl R 3s; 56°03'·05N 03°08'·44W.
No. 8 ⚲ Fl R 9s 56°02'·95N 03°09'·62W.
Inchkeith ⚲ 56°02'·01N 03°08'·17W; Fl 15s 67m 14M; stone twr.
Pallas Rock ⚓ VQ (9) 10s 56°01'·50N 03°09'·30W.
East Gunnet ⚓ Q (3) 10s; 56°01'·41N 03°10'·38W.
West Gunnet ⚓ Q (9) 15s 56°01'·34N 03°11'·06W.
No. 7 ⚲ QG; ***Racon (T) 5M***; 56°02'·80N 03°10'·97W.
No. 9 ⚲ Fl G 6s; 56°02'·32N 03°13'·48W.
No. 10 ⚲ Fl R 6s; 56°02'·07N 03°13'·32W.

No. 11 ⚲ Fl G 3s 56°02'·08N 03°15'·26W.
No. 12 ⚲ Fl R 3s; 56°01'·78N 03°15'·15W.
No. 13 ⚲ Fl G 9s; 56°01'·77N 03°16'·94W.
No. 14 ⚲ Q R; 56°01'·51N 03°16'·90W.
Oxcars ⚲ Fl (2) WR 7s 16m W13M, R12M; W twr; R band; vis: 072°-W-087°- R-196°-W-313°-R-072°; 56°01'·36N 03°16'·84W.
Inchcolm E Pt ⚲ Fl (3) 15s 20m 10M; Gy twr; part obsc by land 075°-145·5°; 56°01'·72N 03°17'·83W.
No. 15 ⚲ Fl G 6s; 56°01'·39N 03°18'·95W.

MORTIMER'S DEEP
Hawkcraig Point Ldg Lts 292°. Front, Iso 5s 12m 14M; W twr; vis: 282°-302°; 56°03'·03N 03°17'·07W. Rear, 96m from front, Iso 5s 16m 14M; W twr; vis: 282°-302°.
Inchcolm S Lts in line 066°. Front, 84m from rear, Q 7m 7M; W twr; vis: 062·5°-082·5°; 56°01'·78N 03°18'·28W. Common Rear, Iso 5s 11m 7M; W twr; vis: 062·5°-082·5°; 56°01'·80N 03°18'·13W. N Lts in line 076·7°. Front, 80m from rear, Q 7m 7M; W twr; vis: 062·5°-082·5°.

APPROACHES TO FORTH BRIDGES
No. 17 ⚲ Fl G 3s; 56°01'·23N 03°19'·84W.
No. 16 ⚲ Fl R 3s; 56°00'·87N 03°19'·60W.
No. 19 ⚲ Fl G 9s; 56°00'·71N 03°22'·47W.
Beamer Rk W ⚲ Fl 3s 5m 9M; W post; vis: 050°-230°; 56°00'·30N 03°24'·79W.
Beamer Rk E ⚲ Fl 3s 5m 9M; W post; vis: 230°-050°; 56°00'·27N 03°24'·72W.

PORT EDGAR
W Bkwtr Hd ⚲ Fl R 4s 4m 8M; 55°59'·86N 03°24'·78W. W blockhouse.

FIRTH OF FORTH – NORTH SHORE (INWARD)

BURNTISLAND
W Pier Outer Hd ⚲ Fl (2) R 6s 7m; W twr; 56°03'·22N 03°14'·26W.
E Pier Outer Hd ⚲ Fl (2) G 6s 7m 5M; 56°03'·24N 03°14'·17W.

ABERDOUR, BRAEFOOT BAY and INCHCOLM
Hawkcraig Pt ⚲ (see **MORTIMER'S DEEP** above).
Braefoot Bay Terminal, W Jetty. Ldg Lts 247·3°. **Front**, Fl 3s 6m **15M**; W △ on E dolphin; vis: 237·2°-257·2°; 56°02'·16N 03°18'·71W; 4 dolphins with 2 FG (vert). **Rear**, 88m from front, Fl 3s 12m **15M**; W ▽ on appr gangway; vis: 237·2°-257·2°; sync with front.

INVERKEITHING BAY
St David's ⚓ Fl G 5s 3m 7M; Or □, on pile; 56°01'·37N 03°22'·29W.
Channel ⚲ QG; 56°01'·43N 03°23'·02W.

ROSYTH
Main Chan Dir lt 323·5°. Bn 'A' Oc WRG 7m 4M; R □ on W post with R bands; vis: 318°-G-321°-321°-W-326°-R-328° (H24); 56°01'·19N 03°25'·61W.
Dir lt 115°, Bn 'C' Oc WRG 6s 7m 4M; W ▽ on W Bn; vis: 110°- R -113° W -116·5° - G -120°; 56°00'·61N 03°24'·25W.
S Arm Jetty Hd ⚲ L Fl (2) WR 12s 5m W9M; R6M; vis: 010°-W-280°-R-010°; 56°01'·09N 03°26'·58W.

RIVER FORTH

ROSYTH to GRANGEMOUTH
Dhu Craig ⚲ Fl G 5s; 56°00'·74N 03°27'·23W.
Blackness ⚲ QR; 56°01'·06N 03°30'·30W.
Tancred Bank ⚲ Fl (2) R 10s; 56°01'·58N 03°31'·91W.
Dods Bank ⚲ Fl R 3s; 56°02'·03N 03°34'·07W.
Bo'ness ⚲ Fl R 10s; 56°02'·23N 03°35'·38W.
Torry ⚲ Fl G 10s 5m 7M; G ○ structure; 56°02'·46N 03°35'·28W.
Bo'ness Bcns ⚲ 2 QR 3m 2M; 56°01'·85N 03°36'·22W.
Bo'ness Hbr ⚓ ; 56°01'·26N 03°36'·46W.

GRANGEMOUTH
Grangemouth App No. 1 ⚓ Fl (3) 10s 4m 6M;56°02'·12N 03°38'·10W.
Hen & Chickens ⬦ Fl (3) G 10s; 56°02'·35N 03°38'·08W.

FIRTH OF FORTH – NORTH SHORE (OUTWARD)
KIRKCALDY and METHIL
Kirkcaldy W Pier Hd ⚡ Fl R 5s 7m 3M; 56°06'·80N 03°08'·97W.
Kirkcaldy E Pier Hd ⚡ Fl G 5s 12m 3M; 56°06'·78N 03°08'·90W.
Methil Outer Pier Hd ⚡ Oc G 6s 8m 5M; W twr; vis: 280°-100°; 56°10'·76N 03°00'·48W.

ELIE and ST MONANS
Elie Ness ☆ 56°11'·04N 02°48'·77W; Fl 6s 15m **17M**; W twr.
St Monans Bkwtr Hd ⚡ Oc WRG 6s 5m W7M, R4M, G4M; vis: 282°-G-355°-W-026°-R-038°; 56°12'·20N 02°45'·94W.

PITTENWEEM and ANSTRUTHER EASTER
Pittenweem, Ldg Lts 037° Middle Pier Hd. Front, FR 4m 5M. Rear, FR 8m 5M. Both Gy Cols, Or stripes; 56°12'·69N 02°43'·69W.
Pittenweem, E Bkwtr Hd ⚡ Fl (2) RG 5s 9m R9M, G6M; vis: 265°-R-345°-G-055°; *Horn 90s (occas);* 56°12'·63N 02°43'·74W.
Anstruther, Ldg Lts 019°. Front FG 7m 4M; 56°13'·28N 02°41'·76W. Rear, 38m from front, FG 11m 4M, (both W masts).

MAY I, CRAIL, ST ANDREWS and FIFE NESS to MONTROSE
Isle of May ☆ 56°11'·12N 02°33'·46W(Summit); Fl (2) 15s 73m **22M**; ☐ twr on stone dwelling.
Crail, Ldg Lts 295°. Front, FR 24m 6M (not lit when hbr closed); 56°15'·46N 02°37'·84W. Rear, 30m from front, FR 30m 6M.
Fife Ness ☆ 56°16'·74N 02°35'·19W; Iso WR 10s 12m **W15M,** R12M; W bldg; vis: 143°-W-197°-R-217°-W-023°; 002320798.
N Carr ⚐ Q (3) 10s 3m 5M; 56°18'·05N 02°32'·94W.
St Andrews ⚡ 56°20'·38N 02°47'·07W; Dir lt 277·7; Iso WRG 2s 17m 9M; vis: 272°-G-276°-AlWG-277°-W-278°-AlWR-279°-R-283°.
St Andrews N Bkwtr Bn ⚡ Fl G 3M 56°20'·36N 02°46'·77W.

Bell Rk ☆ 56°26'·08N 02°23'·21W; Fl 5s 28m **18M**; *Racon (M) 18M*.
Met Mast ⚡ Mo(U) 15s 22m 10M; Aero FR (116m), *Horn Mo(U) 30s;* 56°26'·40N 02°14'·49W; 992351153.

RIVER TAY, TAYPORT, DUNDEE and PERTH
Tay Fairway ⊚ L Fl 10s; *Bell;* 56°28'·30N 02°36'·60W.
Middle ⬦ Fl G 3s; 56°28'·08N 02°38'·24W.
Middle ⬰ Fl (2) R 6s 56°27'·82N 02°37'·97W.
Abertay N ⚐ Q (3) 10s; *Racon (T) 8M;* 56°27'·39N 02°40'·36W.
Abertay S (Elbow) ⬰ Fl R 6s 56°27'·13N 02°39'·83W.

Tayport High Lt ⚡ 56°27'·17N 02°53'·95W; Dir lt 269°; Iso WRG 3s 24m W14M, R14M, G14M; W twr vis: 267°-G-268°-W-270°-R-271°.

ARBROATH
Ldg lts 299·2°. Front , FR 7m 5M; W col; 56°33'·29N 02°35'·16W. Rear, 50m from front, FR 13m 5M; W col.

MONTROSE
Scurdie Ness ☆ 56°42'·10N 02°26'·24W; Fl (3) 20s 38m **23M**; W twr; *Racon (T) 14-16M*.
Outer Ldg Lts 271·5°; Front, FR 11m 5M; W twin pillars, R bands; 56°42'·21N 02°27'·41W; Rear, 272m from front, FR 18m 5M; W twr, R cupola.
Inner Ldg Lts 265°; Front FG 21m 5M; Rear FG 33m 5M.

MONTROSE TO RATTRAY HEAD
JOHNSHAVEN and GOURDON HARBOUR
Johnshaven, Ldg Lts 316°. Front, FR 5m; 56°47'·62N 02°20'·26W. Rear, 85m from front, FG 20m; shows R when unsafe to ent hbr.
Gourdon Hbr, Ldg Lts 358°. Front, FR 5m 5M; W twr; shows G when unsafe to enter; *Siren (2) 60s* (occas); 56°49'·69N 02°17'·24W. Rear, 120m from front, FR 30m 5M; W twr.
Gourdon E B'water ⚡ Q 3m 3M; 49°57'·60N 02°17'·19W.
Todhead Lighthouse (disused), white tower, 13m.

STONEHAVEN to GIRDLE NESS
Outer Pier Hd ⚡ Iso WRG 4s 7m 5M; vis: 214°-G-246°-W-268°-R-280°; 56°57'·59N 02°12'·00W.
Girdle Ness ☆ Fl (2) 20s 56m **22M**; obsc by Greg Ness when brg more than about 020°; *Racon (G) 25M*; 57°08'·34N 02°02'·91W.

ABERDEEN
Fairway ⊚ Mo (A) 5s; *Racon (T) 7M*; 57°09'·31N 02°01'·95W.
Torry Ldg lts 237·2°, Front FR or G 14m 5M; R when ent safe, FG when dangerous to navigate; W twr; 57°08'·38N 02°04'·50W.
Rear, 220m from front, FR or G 19m 5M; R when ent safe, FG when dangerous to navigate; W twr; 57°08'·31N 02°04'·68W.
Aberdeen Harbour Ent ⚡ 57°08'·38N 02°04'·48W; Dir lt 237·2° Fl WRG 1s 7m 9M; metal post; vis: 235·2°-QG-236·2°-AlQGW-236·7°-QW-237·7°-AlQRW-238·2°-QR-239·2°.
S Bkwtr Hd ⚡ Fl (3) R 8s 23m 7M; 57°08'·69N 02°03'·34W.
N Pier Hd ⚡ Iso G 4s 11m 10M; W twr; 57°08'·74N 02°03'·69W.
In fog FY 10m (same twr) vis: 136°-336°; *Bell (3) 12s.*
S Skates Nose Jetty Hd ⚡ Q R 4m 4M; 57°08'·49N 02°04'·06W.
Abercrombie Jetty Hd ⚡ Oc G 4s 5m 4M; 57°08'·53N 02°04'·15W.

9.6.5 PASSAGE INFORMATION

More passage information is threaded between the harbours in this area. Admiralty Leisure Folios 5615 and 5617 cover the area from Eyemouth to Peterhead. **Bibliography:** *North and East Scotland* – Farne Islands to Cape Wrath – (Lawrence/Imray), the Royal Northumberland *YC's Sailing Directions Humber to Rattray Head* and *Forth Yacht Clubs Association Pilot Handbook* for the the Firth of Forth. The *North Sea (West) Pilot* (Admiralty NP54) covers the entire area.

A *'Rover Ticket'*, £50 from Aberdeenshire and Moray Councils, allows berthing for one week from the date of arrival at the first harbour. It includes, Johnshaven, Gourdon, Stonehaven, Rosehearty, Banff, Portsoy, Cullen, Portknockie, Findochty, Hopeman and Burghead.

BERWICK-UPON-TWEED TO BASS ROCK

(AC 160,175) From Berwick-upon-Tweed to the Firth of Forth there is no good harbour which can be approached with safety in strong onshore winds. So, if on passage with strong winds from N or E, plan accordingly and keep well to seaward. In late spring and early summer fog (haar) is likely in onshore winds.

The coast North from Berwick is rocky with cliffs rising in height to Burnmouth, then diminishing gradually to Eyemouth. Keep 5ca offshore to clear outlying rocks. Burnmouth, although small, has more alongside space than Eyemouth, which is a crowded fishing hbr. 2·1M NW is St Abb's Harbour, with temp ⚓ in offshore winds in Coldingham Bay close to the South. St Abb's Hd (Fl 10s 68m 26M) is a bold, steep headland, 92m high, with no offlying dangers.

9.6.6 EYEMOUTH

Borders 55°52'·52N 02°05'·29W ✲✲⬦⬦⬦✿✿✿

CHARTS AC 160, 1612, 5615; Imray C24

TIDES +0330 Dover; ML No data; Duration 0610

Standard Port LEITH (⟶)

Times				Height (metres)			
High Water		Low Water		MHWS	MHWN	MLWN	MLWS
0300	0900	0300	0900	5·6	4·4	2·0	0·8
1500	2100	1500	2100				
Differences EYEMOUTH							
−0005	+0007	+0012	+0008	−0·4	−0·3	0·0	+0·1

SHELTER Good in all weathers except uncomfortable surge in N over F5. Busy FV hbr which encourages yachtsman to visit. Berth as directed by HM. The options are:

• Middle Quay (100m pontoon), depth 0.9m bottom soft mud.

• On the W wall of the N-pointing jetty by the LB.

• On W wall of the centre jetty W of Gunsgreen House (conspic).

• ⚓ in bay only in offshore winds.

NAVIGATION WPT 55°52'·80N 02°05'·34W, 174°/3ca to E bkwtr lt.

• Entry should not be attempted in strong N to E winds. F.R (occas) lt ● or R flag = unsafe to enter indicates unsafe to enter bay or hbr.

• Caution, the entrance is 17m wide and is maintained to 2m depth with soundings taken monthly due to frequent sand movement and silting.

• Appr can be made N or S of Hurkars:
From the N, beware Blind Buss 1·2m.
From the S, approach on 250° midway between Hurkars and Hettle Scar (no ldg marks).

• Approaching vessels are advised to contact HM (mob 07885 742505) for latest information.

LIGHTS AND MARKS St Abbs Hd lt ho Fl 10s 68m 26M is 3M NW. Ldg lts 174° both FG 9/10m 6M, orange columns on W pier.

COMMUNICATIONS (Code 01890) CGOC (01224) 592334; Police 101; Dr 750599, ⊞ (0131) 536 1000. HM 750223, Mobile 07885 742505.

VHF Ch 16 12 (HO). Advisable to listen to VHF Ch 6 for FVs entering/ leaving port.

▶ *The stream runs strongly round the Hd, causing turbulence with wind against tide; this can be largely avoided by keeping close inshore. The ESE-going stream begins at HW Leith –0345, and the WNW-going at HW Leith +0240.* ◀

There is a good ⚓ in Pettico Wick, on NW side of St Abb'sHd, in S winds, but dangerous if the wind shifts onshore. There are no off-lying dangers between St Abbs Hd and Fast Castle Hd, 3M WNW. Between Fast Castle Hd and Barns Ness, about 8M NW, is the attractive little harbour of Cove; but it dries and should only be approached in ideal conditions.

Torness Power Station (conspic, lt on bkwtr) and 2 SPM Fl(5) Y 20s 5ca off Chapel Pt is 1·75M SE of Barns Ness (lt ho disused); this lies 2·5M ESE of Dunbar and is fringed with rocks. Tidal streams as for St Abbs Hd. Conspic chys (119m) are 7½ca WSW inland of Barns Ness. Between here and Dunbar keep at least 2½ca offshore to clear rocky patches. Sicar Rk (7·9m) lies about 1·25M ENE of Dunbar, and seas break on it in onshore gales.

The direct course from Dunbar to Bass Rk (Fl(3) 20s 46m 10M) is clear of all dangers; inshore of this line beware Wildfire Rks (dry) on NW side of Bellhaven Bay. In offshore winds there is ⚓ in Scoughall Road. 1M ESE of Gin Hd Carr bn (stone tr surmounted by cross) marks the N end of Great Carr, a ledge of rocks, nearly covering at HW. Drying ledges of rks extend 1M SE of Great Carr, up to 3ca offshore. Keep at least 5ca off Carr bn in strong onshore winds. Tantallon Castle (ruins) is on cliff edge 1M W of Great Car. Bass Rk (lt) lies 1·25M NNE of Gin Hd, and is a sheer, conspic rk (115m) with no offlying dangers; landing difficult due to swell.

EYEMOUTH

FACILITIES Hbr www.eyemouth-harbour.co.uk ⚓ ⌂. For 10m LOA £14.80 (quay), £18.90 (pontoon) discount >3 nts. ⚓ ⬦ ⓞ 🅿 🛢 (at harbour bldg). ⚒ 🔧 ⚖ Ⓔ 🚿 🔋(12t mob) wi-fi.

Town LB, ⬦ 🏪 🍴 ✗ 🗑 ✉ 🛢 🅰 Ⓑ. 🚆 (bus to Berwick-on-Tweed and Edinburgh), ✈ (Edinburgh).

ADJACENT HARBOURS

BURNMOUTH, Borders, **55°50´·61N 02°04´·10W.** AC 160. HW +0315 on Dover, –0025 on Leith; Duration 0615. Use Eyemouth tides 9.6.6. From S beware Quarry Shoal Rks; and E & W Carrs from N. 2 W posts (unclear by day, FR 29/35m 4M) 45m apart, lead 253° to close N of the hbr; as hbr mouth opens, enter on about 185° with outer hbr ent in line with 2FG (vert). Min depth at ent at LWS is 0·6m. Shelter is good especially in inner hbr (dries). With on-shore winds, swell makes outer hbr uncomfortable. HM (018907) 81283 (home). Facilities: ⌕ £10 (all LOA), ⚓ limited 🛒.

ST ABBS, Borders, **55°54´·10N 02°07´·74W.** AC 175. HW +0330 on Dover, –0017 on Leith; HW –0·6m on Leith; Duration 0605. Ldg line (about 228°) S face of Maw Carr on village hall (conspic R roof) leads SW until the hbr ent opens to port and the 2nd ldg line (about 176°) can be seen 2FR 2/4m 1M, or Y LB ho visible through ent. On E side of ent chan, beware Hog's Nose and on W side the Maw Carr. Shelter good. In strong on-shore winds outer hbr suffers from waves breaking over E pier. Inner hbr (dries) is best but often full of FVs. HM (0775 1136758) directs visitors. Facilities: ⌕ £15 (all LOA) ⌐ (launching £15), ⚓ ⊕ at quay, ✉ 🛒 ✗ more facilities & 🏠 at Coldingham.

9.6.7 DUNBAR

East Lothian **56°00´·39N 02°31´·09W** ❀❀◊◊☆☆☆

CHARTS AC 175, 734, 5615; Imray C24, C23, C27

TIDES +0330 Dover; ML 3·0; Duration 0600

Standard Port LEITH (→)

Times				Height (metres)			
High Water		Low Water		MHWS	MHWN	MLWN	MLWS
0300	0900	0300	0900	5·6	4·4	2·0	0·8
1500	2100	1500	2100				
Differences DUNBAR							
–0005	+0003	+0003	–0003	–0·3	–0·3	0·0	+0·1
FIDRA							
–0001	0000	–0002	+0001	–0·2	–0·2	0·0	0·0

SHELTER Outer (Victoria) Hbr is subject to surge in strong NW to NE winds. N side dries; ❶ berth signed by rubber ladder on W wall beneath castle, dries approx HW±2. Contact HM. Keep steps clear. Inner (Old or Cromwell) Hbr dries and is safe in strong onshore conditions; enter under bridge, lifted by HM on request.

NAVIGATION WPT 56°00´·60N 02°31´·49W, 132°/3ca to hbr ent, preferable entry in marginal conditions.

• Entry is dangerous in heavy on-shore swell with no access in strong winds from NW to E.
• Beware Wallace's Head Rock, Half Ebb Rock (2,m), 1½ca from ent. and Outer Buss 4ca E of ent.
• Min depth at ent 0·5m, but may be much less in abnormal circumstances.
• Keep to port on entry to avoid rockfall off castle.

LIGHTS AND MARKS Church and Castle ruin both conspic. From NE, ldg lts, Oc G 6s 15/22m 3M, synch, intens 188°-208°, 2 W △ on Or cols, lead 198° through the outer rks to the Roads; thence narrow ent opens with QR brg 132°. From NW, appr on brg 132° between bns on Wallaces Head and Half Ebb Rk.

COMMUNICATIONS (01368) CGOC (01224) 592334; Police 101; Dr 863704; ⊞ (0131) 536 1000. HM mob 07958 754858, Dep HM 07583 463991, harbourmaster@dunbarharbourtrust.co.uk.

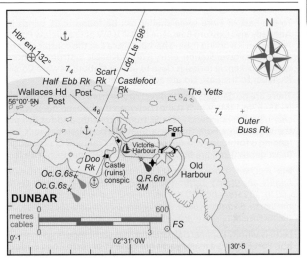

DUNBAR

FACILITIES
www.dunbarharbourtrust.co.uk **Quay** ⌐ (launch £10) ⌕ £16 (£20 >12m LOA), ⚓ ⚓.

N Wall ⌕; **Inner Hbr** ⚓ ⌐ ⌕.

Town ⚓ 🏠 ⛽ ⚒ ⚙ ⚛ Ⓑ 🛒 ✗ 🏠 ⇌ ✈ Edinburgh.

ADJACENT HARBOUR

NORTH BERWICK, East Lothian, **56°03´·74N 02°43´·04W.** AC 734. Fidra HW +0344 on Dover; ML 3·0m; Duration 0625. See 9.6.7 Fidra. Shelter good with winds from S to W but dangerous with on-shore winds. Very small drying harbour with entrance 8m wide. Max LOA normally 30'; LOA >40' only in emergency. From E or W, from position 0·25M S of Craigleith, steer S for Plattock Rks, thence SSW 40m off harbour wall (beware Maiden Rocks to the W) before turning 180° to port into hbr. Galloway Pier extends N from hbr wall.

Bkwtr light F R 7m 3M and FG 6m 3M (about 1ca S) in transit 188° (2 F R vert) when harbour closed). Facilities: ⚓ ⚓ on pier, ⌕ £15, HM www.nbharbour.org.uk ☎(01620) 893333, mob 0777 6467373.

East Lothian YC ☎(01620) 892698, ⚓ ✦⚙ 🏠.

Town 🏠 🏠 ⛽ ✉ Ⓑ 🛒 ⇌ and bus Edinburgh.

FIRTH OF FORTH – SOUTH SHORE

(AC 734, 735) Westward of Bass Rk, Craigleith (51m), Lamb Is (24m) and Fidra (31m) lie 5ca or more offshore, while the coast is generally foul. Craigleith is steep-to, and temporary ⚓ can be found on SE and SW sides; if passing inshore of it keep well to N side of chan. N Berwick hbr (dries) lies S of Craigleith, but is unsafe in onshore winds. Between Craigleith and Lamb Is, beware drying rocks up to 3ca from land. Lamb Is is 1·5M WNW of N Berwick and has a rocky ledge extending 2½ca SW. Fidra Is (Fl(4) 30s 34m 15M) is a bird reserve, nearly connected to the shore by rocky ledges, and should be passed to the N; passage and ⚓ on S side are tricky. ⚓ on E or W sides, depending on wind, in good weather.

To the East of Fidra some shelter can be found in SE winds in Aberlady and Gosford Bays. The best ⚓ is SW of Craigielaw Point. Port Seton is a drying fishing harbour 7½ca E of the conspic chys of Cockenzie Power Station; entry is not advisable in strong onshore winds. Cockenzie (dries) is close to power station; beware Corsik Rk 400m to the East. No attractions except boatyard.

There are no dangers on the direct course from Fidra to Inchkeith (Fl 15s 67m 14M), which stands between the buoyed deep water channels. Rocks extend 7½ca SE from Inchkeith, and for 5ca off the West side where there is a small harbour below the lt ho; landing is forbidden without permission. N Craig and Craig Waugh (least depth 0·2m) are buoyed shoals 2·5M SE from Inchkeith lt.

▶In N Chan, close to Inchkeith the W-going (flood) stream begins about HW Leith –0530, and the E-going at HW Leith +0030, Sp rates about 1kn. The streams gather strength towards the Forth Bridges, where they reach 2·25kn and there may be turbulence.◀

Leith is wholly commercial; Granton has yacht moorings in the E hbr; Port Edgar is a major yacht hbr close W of Forth Road Bridge. Hound Point oil terminal is an artificial 'island-jetty' almost in mid-stream, connected to the shore by underwater pipeline (☡). Yachts may pass the terminal on either side at least 30m off keeping well clear of manoeuvring ships.

CROSSING THE FIRTH OF FORTH

▶Tidal streams are quite weak in the outer part of the Firth, apart from the stream in the Tay, which attains 5kn in most places. Coastwise tidal streams between Fife Ness and Arbroath are weak.

Northbound. Leave before HW (Dover +0400) to be at N Carr at Dover +0600. Bound from Forth to Tay aim to arrive at Abertay By at LW slack (Dover –0200).

Southbound. Leave before LW (Dover –0200) to be at Bass Rk at HW Dover. If bound from Tay to Forth, leave late in ebb to pick up early flood off St Andrews to N Carr and into Forth.◀

RIVER FORTH TO KINCARDINE

(AC 736, 737, 738) The main shipping channel under the N span of the rail bridge N of Inchgarvie is busy with commercial traffic for Grangemouth and Rosyth. W of Beamer Rk, which supports the Queensferry Crossing the Firth widens as far as Bo'ness (small drying hbr) on the S shore where the channel narrows between drying mudbanks.

The Queensferry crossing construction continues to the W of the Forth Road Bridge between 56° 00′·76N, 3° 24′·44W (shore) and 55° 59′·61N, 3° 25′·16W (shore). There are now 3 pillars, one on Beamer Rk and two clear of the main navigable routes with the spans now under construction. The North span gives access to Rosyth while the South span allows traffic to pass upstream to Grangemouth. Works will be ongoing until the end of 2016 and numerous craft will be operating in the vicinity. Forth and Tay Navigation Service will broadcast information and vessels should maintain a listening watch on VHF Ch 71.

Charlestown (N bank) dries, but is a secure harbour. Grangemouth is industrially conspicuous. Beware of gas carriers, tankers and other cargo vessels; there are no facilities for yachts. Few yachts go beyond Kincardine swing bridge, clearance 6·5m, which is no longer opened. Clackmannanshire Bridge, clearance 6·5m, is about 4ca further up-river.

FIRTH OF FORTH – NORTH SHORE

(AC 734, 190) From Burntisland the N shore of Firth of Forth leads E to Kinghorn Ness. 1M SSW of Kinghorn Ness Blae Rk (SHM l) has least depth of 4·1m, and seas break on it in E gales.

▶ Rost Bank lies halfway between Kinghorn Ness and Inchkeith, with tide rips at Sp tides or in strong winds. ◀

From Kinghorn Ness to Kirkcaldy, drying rocks lie up to 3ca offshore. Kirkcaldy Hbr is effectively closed, but yachts can enter inner dock near HW by arrangement; entry is dangerous in strong E'lies, when seas break a long way out.

Between Kirkcaldy and Methil the only dangers more than 2ca offshore are The Rockheads, extending 4ca SE of Dysart, and marked by 2 SHM. Largo Bay is ⚓, well sheltered from N and E, but avoid gaspipes near the E side. Close SW of Elie, beware W Vows (dries) and E Vows (dries, bn). There is ⚓ close W of Elie Ness (9.6.8). Ox Rock (dries 2m) lies 5ca ENE of Elie Ness, and 2½ca offshore; otherwise there are no dangers more than 2ca offshore past St Monance, Pittenweem and Anstruther, but in bad weather the sea breaks on Shield Rk 4ca off Pittenweem. From Anstruther to Crail and on to Fife Ness keep 3ca offshore to clear Caiplie Rk and other dangers.

May Island (Fl(2) 15s 73m 22M) lies about 5M S of Fife Ness; its shores are bold except at NW end where rks extend 1ca off. ⚓ near N end at E or W Tarbert, on lee side according to winds; in good weather it is possible to land. Lt ho boats use Kirkhaven, close SE of lt ho.

FORTH TO NORWAY AND THE BALTIC

(AC 2182B, 2182C) Heading ENE'ly from the Firth of Forth the main hazards result from offshore industrial activities and their associated traffic. Particularly in summer months oil/gas exploration, movement of drilling rigs and pipe laying create potentially hazardous situations for small craft. Rig movements and many of the more intense activities are published in Notices to Mariners, but even so it is wise to avoid the gas and oil fields where possible and never to approach within 500m of installations. There are TSS to be avoided off the S and SW coast of Norway. Strong currents and steep seas may be experienced in the approaches to the Skagerrak.

9.6.8 FIRTH OF FORTH
E and W Lothian/City of Edinburgh/Fife

CHARTS AC 734, 735, 736, 737, 741, 5615; Imray C23, C27

TIDES +0350 (Granton) Dover; ML 3·3; Duration 0620

Standard Port LEITH (→)

Times				Height (metres)			
High Water		Low Water		MHWS	MHWN	MLWN	MLWS
0300	0900	0300	0900	5·6	4·4	2·0	0·8
1500	2100	1500	2100				
Differences COCKENZIE							
−0007	−0015	−0013	−0005	−0·2	0·0	ND	ND

GRANTON: Same as LEITH

SHELTER Port Edgar offers good shelter, but surge at LW esp in E winds. Silting in centre of entrance and close to wavebreak. Granton mostly dries but is open to violent swell in N'lies. Pontoons on E side of Middle Pier in about 2m and RFYC welcomes visitors. Pilot launches berth at seaward end. Leith is purely commercial. Rosyth should only be used in emergency. Forth Navigation Service controls Firth of Forth, Granton Hbr and all commercial docks.

NAVIGATION Port Edgar 56°N 03°24'·3W, 244°/3ca to W bkwtr. Do not enter E of wavebreak, beware of shallows at W end; 3kn sp limit. Granton WPT 56°00'·00N 03°13'·31W, 180°/0·72M to entry.

- Beware Hound Pt terminal; Forth Railway and Road Bridges; vessels bound to/from Rosyth and Grangemouth especially in local fog (haar). 12kn speed limit W of Forth Rly Bridge.
- On N shore, no vessel may enter Mortimer's Deep (Braefoot gas terminal) without approval from Forth Navigation Service.
- Protected chan runs from Nos 13 & 14 buoys (NNW of Oxcars) under the bridges (N of Inch Garvie and Beamer Rk), to Rosyth. When activated (occas) via Forth Ports plc, other vessels must clear the chan for Rosyth traffic.
- Caution: strong tidal streams.

LIGHTS AND MARKS See 9.6.4 and chartlets. Granton: R flag with W diagonal cross (or ● lt) on signal mast at middle pier hd = Entry prohibited.

COMMUNICATIONS (Code 0131) CGOC (01224) 592334; Police 101; Ⓗ Edinburgh Royal Infirmary 2292477; Forth and Tay Navigation Service (01324) 498584; QHM Rosyth (01383) 425050.

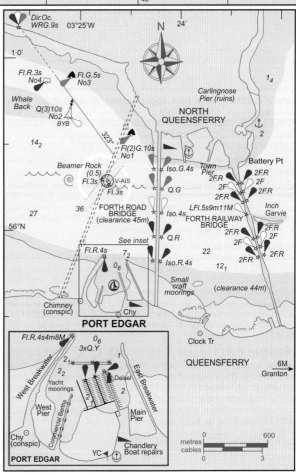

HMNB CLYDE (01436) 674321 ext 3206, for naval activity off Scotland; Forth Yacht Clubs Ass'n 5523452.

Call *Forth Navigation* (at Grangemouth) Ch **71** (calling and short messages, H24); **20** 12 will be requested if necessary. Traffic, nav and weather info available on request. Leith Hbr Radio Ch 12. Granton Marina, call *Boswell* Ch M. Port Edgar Marina Ch M **80** (Apr-Sept 0900-1930; Oct-Mar 0900-1630 LT); Rosyth Dockyard, call *QHM* Ch 74 13 73 (Mon-Fri: 0730-1700). Grangemouth Docks Ch 14 (H24).

FACILITIES
GRANTON Extensive development is taking place in NW of West Hrbr. HM via Leith ☎555 8866 ⚓ ⚓; **Royal Forth YC** ☎552 3006, ⚓ ⚓ ⚓ ⚓ ◻(5t) ⌂ **Forth Corinthian YC** ☎552 5939 ⚓ ⚓ ⚓ ✕ ⌂. **Town** ⛽ ⛽ 🛒 📧 Ⓑ 🍴 ✕ ⌂ ⇌ ✈ (buses to Edinburgh).

SOUTH QUEENSFERRY Port Edgar Marina ④ ☎3313330, www.portedgar.co.uk, (301 ⚓ +8⦿) £1.95, Access H24; ⚓ ⚓ ⚓ ◻ 🛢 🛢 ✕ ⚓ 🅱 Ⓔ ◬ ◻(5t) N end of main pier, ⌂ @; Port Edgar YC ✕ ⌂. **Town** ⛽ ⛽ 🍴 ✕ 📧 Ⓑ ⇌ (Dalmeny), ✈ Edinburgh.

ROSYTH Dockyard and ferry port; no facilities for yachts except in emergency. **Ferries:** Zeebrugge; 3/week; 17 hrs; Superfast (www.superfast.com).

GRANGEMOUTH HM ☎01324 498566 (H24); Port Office ☎498597 (HO). VHF Ch 14 16. Commercial port. An enclosed dock, ships in the appr to the locks are constrained by their draught. Yachts making for the R Carron should take care to remain clear. LPG carriers may be expected at the new gas storage terminal.

FIRTH OF FORTH SOUTH SHORE
FISHERROW, East Lothian, **55°56′·79N 03°04′·09W.** AC 734, 735. HW +0345 on Dover, −0005 on Leith; HW −0·1m on Leith; ML 3·0m; Duration 0620. Shelter good except in NW winds. Mainly a pleasure craft hbr, dries 5ca offshore. Appr dangerous in on-shore winds. High-rise block (38m) is conspic 9ca W of hbr. E pier lt, Oc 6s 5m 6M on metal framework tr. Berth on E pier. HM ☎(0131) 665 5900; **Fisherrow YC** ⚓. **Town** 🍴 ⛽ ⛽ from garage, ◬ 📧 Ⓑ ✕ ⌂.

CRAMOND, City of Edinburgh, **55°59′·80N 03°17′·49W.** AC 736. Tides as Leith (see 9.6.8). Cramond Island, approx 1M offshore, is connected to the S shore of the Firth by a drying causeway. A chan, virtually dries, marked by 7 SHM posts, leads W of Cramond Island to Cramond hbr at the mouth of R Almond, conspic white houses. ⚓ free or ⚓ off the Is. Seek local advice from: **Cramond Boat Club** ☎(0131) 336 1356 ⚓ ⚓ ⌂. **Village** 🍴 ✕ ⌂ Bus.

FIRTH OF FORTH NORTH SHORE
INCHCOLM, Fife, **56°01′·85N 03°17′·89W.** AC 736. Tides see 9.6.8. Best ⚓ in 4m, N of abbey (conspic); appr from NW or ESE, to land at pier close E (small fee). Meadulse Rks (dry) on N side. Ends of island foul. At SE end, lt Fl (3) 15s, obsc 075°-145°. No facilities. ☎0131 244 3101. Keep clear of large ships under way in Mortimer's Deep, which is well marked with ldg lts and lit lateral buoys.

ABERDOUR, Fife, **56°03′·00N 03°17′·49W.** AC 735, 736. HW +0345 on Dover; +0005 on Leith; HW 0·5m on Leith; ML 3·3m; Duration 0630. See 9.6.8. Good shelter except in SE winds when a swell occurs. The ⚓ between The Little Craigs and the disused pier is good but exposed to winds from E to SW. Temp berths, £7, are available in hbr (dries) alongside the quay wall. Beware Little Craigs (dries 2·2m). There are no lts/ marks. Facilities: ⚓(on pier). **Village** 🍴 ✕ ⌂. **Aberdour BC** ☎(01383) 860029.

BURNTISLAND, Fife, **56°03′·22N 03°14′·21W.** AC 735, 736. HW +0345 on Dover; +0005 on Leith; HW 0·5m on Leith; ML 3·3m; Duration 0630. Primarily a commercial harbour with two basins used by coasters oand offshore energy it is accessible at all states of tide and provides good shelter except in strong S'lies. When approaching from the E beware Black Rks, Heuchboy (dries 2·3m) and Kirkbush Rk (covers). Blae Rk offshore is adjacent to No 7 SHM. The small basin enclosed by the E'ly bwtr contains 25 F&A moorings of Burnt Island SC. The clubhouse is at the root of the bkwtr, contact through www.burntislandsailing.org.

KIRKCALDY, Fife, **56°06′·80N 03°08′·96W.** AC 741. HW +0340 on Dover, −0005 on Leith; HW −0·1m on Leith; ML 3·2m; Duration 0620. See 9.6.10. Hbr has outer and inner basin with good shelter; an emergency refuge except in strong E winds. Dredged in 2015 it serves the local flour mill and commercial aggregate tfc, with some local FVs. Charted depths are likely to have changed. Outer piers lit Fl G 5s /Fl R 5s respectively. W pier is high concrete wall and care needed when coming to the ⚓. Small craft should contact Forth Ports Authority ☎01383 421800, Forth Navigation (H24) Ch 71.

METHIL, Fife, **56°10′·75N 03°00′·55W.** AC 734,741. HW +0330 Dover; −0020 and −0·1 on Leith; ML 3m; Duration 0615. Commercial activity greatly reduced; infrastructure limited. Access at all states of the tide and provides shelter from E'lies. Ent is difficult in strong E and SE winds until within lee of bwtr. In bad weather seas can pile up in the shallow appr esp on a shallow bar which forms to seaward of the bwtr lt. The Fife Energy Park has a buoyed area and wind turbine on the foreshore W of the dock entrance. This is lit on each side by 2 F R (vert) and 2 F G (vert). Dock ☎(01333) 426725. Berthing by prior arrangement at Methil Boat Club ☎(01333) 421110. In emergency call Forth Navigation Ch 71 (H24) or Forth CG.

ELIE, Fife, **56°11′·20N 02°49′·29W.** AC 734. HW +0325 on Dover, −0015 on Leith; HW −0·1m on Leith; ML 3·0m; Duration 0620; Elie B provides good shelter from N winds for small craft but local knowledge is needed. Hbr dries; 3 short term waiting buoys available. Beware ledge off end of pier which dries. From E beware Ox Rk (dries 1m) 5M ENE of Elie Ness; from W beware rks off Chapel Ness, W Vows, E Vows (surmounted by cage bn) and Thill Rk, marked by PHM buoy. Lt: Elie Ness Fl 6s 15m 17M, W tr. HM (01333) 330219; ⚓ (3) drying £10/night, ⚓ ⚓ 🛢 ⚓ ⌂ SC. Police 08456 005702. Dr ☎330302; Services: 🛢(tanker) 🛢 ⌂. Elie & Earlsferry: 📧 Ⓑ 🍴 ✕ ⌂.

ST MONANS, Fife, **56°12′·25 N 02°45′·94W.** AC 734. HW +0335 on Dover, −0020 on Leith; HW −0·1m on Leith; ML 3·0m; Duration 0620. Shelter good except in strong SE to SW winds when scend occurs in the hbr (dries). Berth alongside E pier until contact with HM. From NE keep at least 2½ca from coast. Bkwtr hd Oc WRG 6s 5m 7/4M; E pier hd 2 FG (vert) 6m 4M. W pier hd 2 FR (vert)6m4M. Facilities: HM ☎07930 869538 (part time) if no reply, 01333 310836 (Anstruther HM assists); ⚓ (example for 10m LOA) £17.20 then £8.60/day thereafter, ⚓ ⌂, ⌂; Services: 🛢(tanker ☎730622) 🛢 ⌂. Police ☎08456 005702. **Village** 📧 Ⓑ 🍴 ✕ ⌂.

9.6.9 FORTH AND CLYDE CANAL

Lowland Carron River Ent. **51°02'·30N 03°41'·46W** 🌼🌺🌿🌸❀❀

These notes are for the convenience of those entering the canal at Grangemouth. Refer to 9.9.21 for the W entrance at Bowling Sea Lock, R Clyde. The eastern extension, by-passing low bridges on the R Carron, was completed autumn 2013, opened Mar 2014.

CHARTS AC 737, 741, 5615; Imray C27; *The Skipper's Guide* Forth & Clyde and Union Canal essential. Obtain from, Lowland Canals Office, Canal House, Applecross Street, Glasgow, G4 9SP. ☎(0141) 332 6936, www.scottishcanals.co.uk or www.waterscape.com.

TIDES Carron River +0325 Dover, 0030 Leith.

SHELTER Carron Sea Lock operates HW−4 to HW+1½, 0800-2000 and daylight hours. Temporary berthing at Grangemouth YC or at nearby Refuge Posts. Temporary ⚓s close WNW of Carron PHM and close SSW of Carron Bn SHM dependent on depth.

NAVIGATION Carron River entrance approached from Grangemouth Roads via Small Craft Recommended Tracks close N of Ship Manoeuvring Area.

- Canal passage has to be booked in advance, call *Carron Sea Lock* VHF Ch 74, ☎(01324)483034/07810 0794468.
- The canal is about 30M long with 39 locks. Allow a minimum of 21 hours underway for a passage to Bowling Sea Lock. Transit of the canal can be achieved in 2 days by reaching the Summit Pound at Lock 20 Wyndford on the first day.
- Canal can take vessels 20m LOA, 6m beam, 1.8m draught (add 0.1m/4inches to draught for freshwater), mast ht 3.0m. Masts should be unstepped before passing through Kerse Bridge

which is equipped with air draught gauges calibrated for the canal dimensions.
- Mast craneage at Bowling, Grangemouth YC and Port Edgar for larger yachts.
- Vessels should have a reliable engine and be capable of a minimum speed through the water of 4kn against adverse conditions. Ebb tide attains rates of up to 6kn in Carron River after heavy rainfall. 4mph speed limit throughout the canal.
- Access via Falkirk Wheel to Union Canal and Edinburgh, refer to *Scottish Canals Skipper's Guide.*

LOCKS Carron Sea Lock and 9.9.21 Bowling Sea Lock are operated by canal staff. At other locks staff available to assist. 5 day passage licence costs £5/m for boats with crew to assist at locks and £15/m without, including access to the Falkirk Wheel, Union Canal and Edinburgh.

LIGHTS AND MARKS Carron River channel is marked by lighted By PHM & Bn SHM and unlit By(s) PHM & SHM. Bkwtr/Training Bank to W & N of channel marked with Bn(s) & Bol(s).

BOAT SAFETY SCHEME *The Skipper's Guide* refers. At Carron Sea Lock transient/visiting craft staying no more than 28 days will be subject to a Dangerous Boat Check of gas and fuel systems. Also required to complete a boat condition declaration and provide evidence of insurance for £1M third party liability.

COMMUNICATIONS Carron Sea Lock (01324) 483034/07810794468. *Carron Sea Lock* VHF Ch **74**.

FACILITIES Carron Sea Lock ⚓ P 🚾 🛥 ⚓. **Falkirk Wheel (on Forth & Clyde)** ⚓ 🛒 🚿 🚾 ⛽. For details throughout the canal refer to *The Skipper's Guide* (using maps).

LEITH LAT 55°59'N LONG 3°11'W
TIMES AND HEIGHTS OF HIGH AND LOW WATERS

STANDARD TIME (UT)
For Summer Time add ONE hour in **non-shaded areas**

Dates in red are **SPRINGS**
Dates in blue are NEAPS

YEAR 2016

JANUARY

Time	m		Time	m
1 0009	1.6	**16** 0032	1.1
0708	4.7		0658	5.2
F 1211	2.0	SA 1245	1.5
1930	4.8	◐ 1912	5.3
2 0055	1.8	**17** 0122	1.3
0757	4.6		0758	5.0
SA 1308	2.3	SU 1346	1.8
◐ 2023	4.6	2017	5.0
3 0156	2.0	**18** 0229	1.6
0850	4.4		0907	4.8
SU 1427	2.4	M 1504	1.9
2121	4.5	2133	4.9
4 0317	2.1	**19** 0353	1.7
0948	4.4		1018	4.8
M 1557	2.4	TU 1625	1.8
2222	4.4	2247	4.9
5 0435	2.1	**20** 0511	1.7
1049	4.5		1127	4.9
TU 1705	2.2	W 1737	1.6
2325	4.6	2356	5.0
6 0532	1.9	**21** 0617	1.6
1150	4.7		1229	5.1
W 1758	1.9	TH 1843	1.4
7 0024	4.7	**22** 0057	5.2
0619	1.7		0711	1.4
TH 1245	4.9	F 1324	5.3
1844	1.7	1939	1.1
8 0115	5.0	**23** 0149	5.3
0701	1.5		0757	1.2
F 1331	5.2	SA 1411	5.5
1926	1.4	2027	0.9
9 0158	5.2	**24** 0235	5.5
0742	1.3		0838	1.1
SA 1412	5.4	SU 1454	5.6
2009	1.1	○ 2110	0.7
10 0238	5.4	**25** 0316	5.5
0824	1.1		0914	1.1
SU 1450	5.5	M 1535	5.6
● 2052	0.9	2147	0.7
11 0317	5.5	**26** 0356	5.5
0907	1.0		0945	1.1
M 1528	5.6	TU 1614	5.5
2136	0.7	2219	0.8
12 0356	5.6	**27** 0434	5.4
0950	0.9		1009	1.2
TU 1606	5.7	W 1651	5.4
2221	0.7	2243	0.9
13 0437	5.6	**28** 0510	5.2
1034	1.0		1031	1.3
W 1647	5.7	TH 1727	5.3
2305	0.7	2304	1.1
14 0520	5.6	**29** 0547	5.0
1116	1.1		1057	1.5
TH 1731	5.6	F 1804	5.1
2348	0.8	2332	1.3
15 0607	5.4	**30** 0625	4.9
1158	1.3		1129	1.7
F 1818	5.5	SA 1844	4.9
		31 0007	1.6
		0708	4.7
		SU 1209	1.9
		1929	4.8

FEBRUARY

Time	m		Time	m
1 0054	1.9	**16** 0153	1.7
0756	4.5		0840	4.7
M 1305	2.2	TU 1433	1.8
◑ 2024	4.4	2113	4.8
2 0156	2.1	**17** 0326	1.9
0853	4.4		0955	4.6
TU 1431	2.4	W 1609	1.9
2127	4.3	2231	4.7
3 0324	2.2	**18** 0458	1.9
0957	4.4		1110	4.7
W 1615	2.3	TH 1734	1.7
2236	4.4	2348	4.8
4 0453	2.1	**19** 0610	1.8
1104	4.5		1220	4.9
TH 1727	2.1	F 1843	1.4
2345	4.5		
5 0552	1.9	**20** 0053	5.0
1210	4.7		0703	1.6
F 1822	1.8	SA 1316	5.1
		1934	1.1
6 0046	4.8	**21** 0143	5.2
0641	1.6		0744	1.4
SA 1305	5.0	SU 1401	5.3
1909	1.4	2016	0.9
7 0135	5.1	**22** 0224	5.3
0725	1.3		0820	1.2
SU 1350	5.3	M 1441	5.5
1954	1.0	○ 2052	0.8
8 0218	5.4	**23** 0300	5.4
0809	1.0		0852	1.0
M 1430	5.6	TU 1517	5.5
● 2039	0.7	2124	0.7
9 0257	5.6	**24** 0334	5.4
0853	0.8		0921	1.0
TU 1509	5.8	W 1552	5.5
2125	0.4	2152	0.7
10 0337	5.8	**25** 0407	5.3
0938	0.6		0947	1.0
W 1548	5.9	TH 1625	5.4
2209	0.3	2215	0.8
11 0418	5.8	**26** 0440	5.2
1021	0.6		1010	1.1
TH 1629	5.9	F 1658	5.3
2252	0.3	2237	1.0
12 0501	5.7	**27** 0513	5.1
1103	0.7		1031	1.2
F 1713	5.9	SA 1731	5.2
2334	0.6	2300	1.2
13 0546	5.6	**28** 0549	5.0
1142	1.0		1054	1.4
SA 1800	5.7	SU 1807	5.0
		2326	1.4
14 0013	0.9	**29** 0627	4.8
0635	5.3		1123	1.7
SU 1222	1.3	M 1848	4.7
1852	5.4		
15 0055	1.3		
0732	5.0		
M 1313	1.6		
◑ 1956	5.0		

MARCH

Time	m		Time	m
1 0000	1.7	**16** 0127	1.8
0710	4.6		0817	4.7
TU 1203	1.9	W 1414	1.8
◑ 1936	4.5	2056	4.7
2 0052	2.0	**17** 0304	2.1
0803	4.4		0933	4.5
W 1311	2.2	TH 1600	1.8
2037	4.3	2214	4.6
3 0218	2.3	**18** 0442	2.1
0908	4.3		1050	4.6
TH 1504	2.3	F 1727	1.7
2151	4.3	2334	4.7
4 0409	2.3	**19** 0554	1.9
1021	4.3		1202	4.8
F 1652	2.1	SA 1832	1.4
2306	4.4		
5 0525	2.0	**20** 0040	4.9
1134	4.6		0645	1.7
SA 1757	1.7	SU 1300	5.0
		1918	1.2
6 0015	4.7	**21** 0128	5.1
0618	1.7		0723	1.4
SU 1236	4.9	M 1344	5.2
1848	1.3	1955	1.0
7 0109	5.1	**22** 0206	5.2
0704	1.3		0755	1.2
M 1324	5.3	TU 1421	5.3
1935	0.9	2026	0.8
8 0153	5.5	**23** 0239	5.3
0750	0.9		0825	1.0
TU 1406	5.6	W 1456	5.4
2022	0.5	○ 2054	0.8
9 0234	5.7	**24** 0310	5.3
0835	0.6		0855	0.9
W 1446	5.9	TH 1528	5.4
● 2107	0.2	2121	0.7
10 0315	5.9	**25** 0340	5.3
0920	0.4		0924	0.9
TH 1527	6.1	F 1559	5.4
2152	0.1	2147	0.8
11 0356	5.9	**26** 0411	5.3
1004	0.4		0950	1.0
F 1610	6.1	SA 1631	5.3
2235	0.2	2211	0.9
12 0440	5.8	**27** 0443	5.2
1046	0.5		1010	1.1
SA 1655	6.0	SU 1704	5.2
2316	0.4	2231	1.1
13 0525	5.6	**28** 0517	5.0
1126	0.7		1028	1.3
SU 1744	5.7	M 1739	5.0
2354	0.8	2252	1.4
14 0614	5.3	**29** 0554	4.9
1206	1.1		1053	1.5
M 1837	5.4	TU 1818	4.8
		2319	1.6
15 0033	1.3	**30** 0635	4.7
0709	5.0		1130	1.7
TU 1255	1.4	W 1903	4.6
◑ 1941	5.0		
		31 0003	1.9
		0722	4.5
		TH 1228	2.0
		◑ 2000	4.4

APRIL

Time	m		Time	m
1 0129	2.2	**16** 0410	2.2
0824	4.4		1021	4.5
F 1411	2.1	SA 1702	1.7
2111	4.3	2305	4.6
2 0324	2.3	**17** 0519	2.0
0940	4.4		1132	4.7
SA 1609	2.0	SU 1803	1.5
2229	4.5		
3 0451	2.0	**18** 0011	4.7
1056	4.5		0610	1.8
SU 1725	1.6	M 1231	4.9
2340	4.8	1848	1.3
4 0549	1.6	**19** 0101	4.9
1201	4.9		0648	1.6
M 1820	1.2	TU 1317	5.0
		1922	1.1
5 0038	5.1	**20** 0139	5.1
0637	1.2		0711	1.1
TU 1253	5.3	W 1355	5.2
1910	0.8	1952	1.0
6 0126	5.5	**21** 0212	5.2
0724	0.9		0754	1.1
W 1339	5.6	TH 1430	5.3
1958	0.4	2020	0.9
7 0209	5.8	**22** 0243	5.2
0812	0.5		0828	1.0
TH 1422	5.9	F 1502	5.3
● 2045	0.2	○ 2049	0.9
8 0251	5.9	**23** 0313	5.3
0859	0.3		0900	0.9
F 1506	6.1	SA 1534	5.3
2131	0.1	2118	0.9
9 0334	6.0	**24** 0344	5.3
0945	0.2		0930	1.0
SA 1551	6.1	SU 1606	5.2
2215	0.2	2146	1.0
10 0418	5.9	**25** 0417	5.2
1030	0.3		0956	1.1
SU 1639	6.0	M 1640	5.2
2257	0.5	2211	1.2
11 0505	5.6	**26** 0451	5.1
1113	0.6		1017	1.2
M 1729	5.7	TU 1716	5.0
2337	0.9	2232	1.4
12 0555	5.3	**27** 0527	5.0
1156	1.0		1041	1.3
TU 1824	5.3	W 1756	4.9
		2300	1.6
13 0016	1.4	**28** 0607	4.8
0651	5.0		1118	1.5
W 1247	1.3	TH 1841	4.7
1927	5.0	2343	1.9
14 0108	1.9	**29** 0654	4.7
0757	4.7		1217	1.7
TH 1402	1.7	F 1935	4.6
◐ 2037	4.7		
15 0236	2.2	**30** 0106	2.1
0909	4.6		0751	4.6
F 1542	1.8	SA 1346	1.9
2149	4.5	◐ 2040	4.5

Chart Datum: 2·90 metres below Ordnance Datum (Newlyn). HAT is 6·3 metres above Chart Datum.

》**FREE** monthly updates. Register at 《
www.reedsnauticalalmanac.co.uk

LEITH LAT 55°59'N LONG 3°11'W
TIMES AND HEIGHTS OF HIGH AND LOW WATERS

STANDARD TIME (UT)
For Summer Time add ONE hour in **non-shaded areas**

Dates in red are **SPRINGS**
Dates in blue are **NEAPS**

YEAR 2016

MAY

Time	m	Time	m
1 0247	2.1	**16** 0426	2.1
0902	4.5	1050	4.6
SU 1527	1.8	M 1714	1.6
2155	4.6	2326	4.6
2 0413	1.9	**17** 0520	1.9
1019	4.6	1150	4.7
M 1647	1.5	TU 1801	1.5
2306	4.8		
3 0514	1.6	**18** 0021	4.7
1125	4.9	0604	1.7
TU 1747	1.1	W 1242	4.8
		1837	1.4
4 0006	5.2	**19** 0104	4.9
0606	1.2	0645	1.5
W 1222	5.3	TH 1324	5.0
1840	0.8	1910	1.2
5 0057	5.5	**20** 0141	5.0
0656	0.9	0723	1.3
TH 1312	5.6	F 1401	5.1
1932	0.5	1944	1.1
6 0143	5.7	**21** 0214	5.2
0747	0.6	0800	1.1
F 1359	5.9	SA 1436	5.2
● 2022	0.3	○ 2017	1.0
7 0228	5.9	**22** 0246	5.2
0839	0.3	0836	1.0
SA 1446	6.0	SU 1510	5.2
2110	0.3	2051	1.0
8 0313	5.9	**23** 0320	5.3
0929	0.3	0910	1.0
SU 1534	6.0	M 1545	5.2
2156	0.4	2124	1.1
9 0359	5.8	**24** 0354	5.3
1016	0.3	0944	1.0
M 1624	5.9	TU 1620	5.2
2239	0.7	2158	1.2
10 0448	5.6	**25** 0429	5.2
1102	0.6	1017	1.1
TU 1716	5.6	W 1657	5.1
2320	1.1	2230	1.3
11 0538	5.4	**26** 0506	5.1
1147	0.9	1050	1.2
W 1810	5.3	TH 1737	5.0
2359	1.5	2305	1.5
12 0634	5.1	**27** 0547	5.0
1235	1.2	1130	1.3
TH 1909	5.0	F 1823	4.9
		2350	1.7
13 0043	1.8	**28** 0633	4.9
0735	4.8	1223	1.5
F 1338	1.5	SA 1914	4.8
◑ 2010	4.7		
14 0152	2.1	**29** 0057	1.9
0839	4.7	0726	4.8
SA 1502	1.7	SU 1330	1.5
2114	4.5	◑ 2015	4.7
15 0319	2.2	**30** 0215	1.9
0944	4.6	0831	4.7
SU 1615	1.7	M 1451	1.5
2221	4.5	2124	4.8
		31 0334	1.8
		0944	4.8
		TU 1610	1.4
		2233	4.9

JUNE

Time	m	Time	m
1 0440	1.6	**16** 0523	1.9
1053	5.0	1156	4.6
W 1715	1.2	TH 1750	1.6
2335	5.1		
2 0537	1.3	**17** 0019	4.7
1154	5.3	0610	1.7
TH 1812	0.9	F 1248	4.8
		1832	1.5
3 0030	5.4	**18** 0105	4.9
0631	1.0	0654	1.5
F 1249	5.5	SA 1331	4.9
1907	0.7	1911	1.3
4 0121	5.6	**19** 0145	5.1
0727	0.7	0734	1.3
SA 1341	5.7	SU 1411	5.1
2001	0.6	1949	1.2
5 0208	5.7	**20** 0222	5.2
0822	0.5	0813	1.1
SU 1431	5.8	M 1448	5.2
● 2051	0.5	○ 2027	1.1
6 0256	5.8	**21** 0258	5.3
0915	0.4	0852	1.0
M 1521	5.8	TU 1524	5.3
2138	0.6	2106	1.1
7 0343	5.7	**22** 0334	5.4
1003	0.4	0932	0.9
TU 1611	5.7	W 1601	5.3
2221	0.8	2146	1.1
8 0432	5.6	**23** 0410	5.4
1050	0.5	1012	0.9
W 1704	5.5	TH 1639	5.3
2301	1.1	2226	1.2
9 0522	5.4	**24** 0448	5.4
1133	0.8	1053	0.9
TH 1752	5.3	F 1720	5.3
2336	1.4	2308	1.3
10 0613	5.2	**25** 0529	5.3
1214	1.1	1136	1.0
F 1844	5.0	SA 1805	5.2
		2351	1.4
11 0009	1.7	**26** 0614	5.2
0706	5.0	1221	1.1
SA 1257	1.4	SU 1854	5.1
1936	4.8		
12 0056	2.0	**27** 0042	1.6
0803	4.8	0705	5.1
SU 1352	1.7	M 1314	1.3
◑ 2031	4.6	◑ 1950	4.9
13 0203	2.2	**28** 0144	1.7
0900	4.6	0804	5.0
M 1504	1.8	TU 1419	1.4
2127	4.5	2055	4.9
14 0323	2.2	**29** 0256	1.8
0958	4.5	0915	4.9
TU 1610	1.8	W 1535	1.4
2226	4.5	2204	4.9
15 0428	2.1	**30** 0408	1.7
1058	4.5	1026	5.0
W 1704	1.7	TH 1648	1.3
2325	4.6	2309	5.0

JULY

Time	m	Time	m
1 0514	1.4	**16** 0540	1.9
1133	5.1	1206	4.6
F 1751	1.2	SA 1801	1.7
2 0009	5.2	**17** 0026	4.8
0615	1.2	0628	1.7
SA 1234	5.3	SU 1301	4.8
1850	1.1	1845	1.5
3 0104	5.4	**18** 0116	5.0
0715	0.9	0712	1.4
SU 1329	5.5	M 1346	5.0
1945	0.9	1926	1.3
4 0155	5.6	**19** 0159	5.2
0812	0.6	0754	1.1
M 1420	5.6	TU 1426	5.2
● 2035	0.8	○ 2007	1.2
5 0243	5.7	**20** 0238	5.4
0903	0.5	0835	0.9
TU 1509	5.7	W 1504	5.4
2120	0.8	2049	1.0
6 0330	5.7	**21** 0315	5.5
0950	0.4	0918	0.7
W 1556	5.6	TH 1542	5.5
2202	0.9	2132	0.9
7 0416	5.6	**22** 0352	5.6
1033	0.5	1002	0.6
TH 1642	5.5	F 1620	5.5
2238	1.1	2215	0.9
8 0501	5.5	**23** 0430	5.6
1111	0.7	1045	0.6
F 1727	5.3	SA 1701	5.5
2306	1.3	2258	1.0
9 0547	5.3	**24** 0511	5.6
1143	1.0	1127	0.7
SA 1812	5.1	SU 1745	5.4
2331	1.5	2339	1.2
10 0632	5.1	**25** 0555	5.5
1210	1.3	1209	0.9
SU 1857	4.9	M 1833	5.3
11 0006	1.8	**26** 0022	1.4
0719	4.9	0644	5.3
M 1247	1.5	TU 1253	1.1
1943	4.7	◑ 1925	5.1
12 0056	2.0	**27** 0114	1.6
0810	4.7	0741	5.1
TU 1338	1.8	W 1349	1.4
◑ 2034	4.5	2028	4.9
13 0204	2.2	**28** 0223	1.7
0904	4.5	0851	5.0
W 1448	2.0	TH 1505	1.6
2128	4.4	2139	4.8
14 0329	2.2	**29** 0345	1.7
1002	4.4	1007	4.9
TH 1607	2.0	F 1629	1.6
2226	4.5	2249	4.9
15 0443	2.1	**30** 0501	1.6
1104	4.5	1118	4.9
F 1710	1.9	SA 1740	1.5
2327	4.6	2355	5.1
		31 0610	1.3
		1226	5.1
		SU 1841	1.4

AUGUST

Time	m	Time	m
1 0054	5.3	**16** 0046	4.9
0711	1.0	0652	1.4
M 1323	5.3	TU 1320	5.0
1933	1.2	1906	1.4
2 0146	5.5	**17** 0134	5.2
0805	0.8	0735	1.1
TU 1412	5.5	W 1402	5.3
● 2019	1.0	1948	1.1
3 0231	5.6	**18** 0214	5.5
0852	0.6	0818	0.8
W 1456	5.6	TH 1441	5.5
2101	0.9	○ 2031	0.9
4 0315	5.7	**19** 0252	5.7
0933	0.5	0902	0.5
TH 1539	5.5	F 1520	5.7
2138	0.9	2115	0.7
5 0356	5.6	**20** 0329	5.8
1011	0.6	0946	0.4
F 1619	5.5	SA 1559	5.8
2210	1.0	2158	0.7
6 0437	5.5	**21** 0409	5.9
1042	0.7	1029	0.3
SA 1658	5.3	SU 1640	5.8
2233	1.2	2241	0.7
7 0516	5.4	**22** 0451	5.9
1106	0.9	1110	0.5
SU 1736	5.1	M 1723	5.6
2255	1.4	2321	0.9
8 0555	5.2	**23** 0536	5.7
1128	1.2	1150	0.8
M 1815	5.0	TU 1810	5.4
2324	1.6		
9 0635	5.0	**24** 0002	1.2
1159	1.4	0624	5.5
TU 1857	4.8	W 1232	1.1
		1901	5.2
10 0002	1.8	**25** 0050	1.5
0720	4.7	0722	5.2
W 1242	1.7	TH 1323	1.5
◑ 1944	4.6	◑ 2004	4.9
11 0056	2.1	**26** 0200	1.7
0812	4.5	0834	4.9
TH 1339	2.0	F 1444	1.8
2037	4.4	2118	4.8
12 0214	2.3	**27** 0332	1.8
0912	4.4	0954	4.8
F 1501	2.2	SA 1618	1.9
2137	4.4	2233	4.8
13 0356	2.3	**28** 0459	1.7
1016	4.3	1112	4.8
SA 1632	2.1	SU 1735	1.8
2242	4.5	2345	5.1
14 0512	2.1	**29** 0611	1.4
1124	4.5	1222	5.0
SU 1734	1.9	M 1835	1.6
2348	4.7		
15 0607	1.8	**30** 0046	5.2
1228	4.7	0709	1.1
M 1823	1.7	TU 1317	5.2
		1921	1.4
		31 0136	5.4
		0755	0.8
		W 1402	5.4
		2001	1.2

Chart Datum: 2·90 metres below Ordnance Datum (Newlyn). HAT is 6·3 metres above Chart Datum.

》》 **FREE** monthly updates. Register at 《 www.reedsnauticalalmanac.co.uk

377

STANDARD TIME (UT)
For Summer Time add ONE hour in **non-shaded areas**

LEITH LAT 55°59'N LONG 3°11'W
TIMES AND HEIGHTS OF HIGH AND LOW WATERS

Dates in **red** are **SPRINGS**
Dates in blue are NEAPS

YEAR 2016

SEPTEMBER

Day	Time m		Day	Time m	
1 TH ●	0218 5.6 / 0834 0.7 / 1441 5.5 / 2037 1.0		**16** F ○	0147 5.6 / 0757 0.7 / 1415 5.7 / 2009 0.8	
2 F	0257 5.6 / 0910 0.6 / 1517 5.5 / 2110 1.0		**17** SA	0226 5.8 / 0841 0.4 / 1454 5.9 / 2053 0.6	
3 SA	0334 5.6 / 0941 0.6 / 1552 5.4 / 2139 1.0		**18** SU	0305 6.0 / 0925 0.2 / 1534 6.0 / 2138 0.5	
4 SU	0409 5.5 / 1007 0.7 / 1626 5.4 / 2202 1.1		**19** M	0346 6.1 / 1008 0.2 / 1616 5.9 / 2221 0.6	
5 M	0444 5.4 / 1028 0.9 / 1701 5.2 / 2223 1.2		**20** TU	0430 6.0 / 1050 0.4 / 1700 5.8 / 2304 0.8	
6 TU	0519 5.2 / 1049 1.1 / 1737 5.1 / 2246 1.4		**21** W	0517 5.8 / 1131 0.8 / 1747 5.5 / 2346 1.1	
7 W	0556 5.0 / 1115 1.4 / 1815 4.9 / 2315 1.7		**22** TH	0608 5.5 / 1211 1.2 / 1840 5.2	
8 TH	0638 4.8 / 1149 1.7 / 1859 4.7 / 2356 2.0		**23** F ◑	0035 1.4 / 0708 5.2 / 1304 1.7 / 1944 4.9	
9 F ◑	0726 4.6 / 1240 2.1 / 1950 4.5		**24** SA	0147 1.7 / 0822 4.9 / 1428 2.1 / 2101 4.7	
10 SA	0105 2.2 / 0825 4.4 / 1400 2.3 / 2052 4.4		**25** SU	0327 1.8 / 0942 4.7 / 1607 2.1 / 2217 4.7	
11 SU	0251 2.4 / 0933 4.3 / 1547 2.4 / 2200 4.4		**26** M	0456 1.7 / 1101 4.8 / 1723 2.0 / 2330 4.9	
12 M	0441 2.2 / 1045 4.4 / 1707 2.1 / 2310 4.6		**27** TU	0604 1.4 / 1211 5.0 / 1820 1.8	
13 TU	0543 1.8 / 1154 4.7 / 1800 1.8		**28** W	0031 5.2 / 0655 1.2 / 1304 5.2 / 1902 1.5	
14 W	0013 4.9 / 0630 1.4 / 1250 5.1 / 1844 1.4		**29** TH	0119 5.4 / 0736 1.0 / 1345 5.3 / 1937 1.3	
15 TH	0104 5.3 / 0714 1.0 / 1335 5.4 / 1926 1.1		**30** F	0159 5.5 / 0810 0.8 / 1420 5.4 / 2008 1.1	

OCTOBER

Day	Time m		Day	Time m	
1 SA ●	0235 5.5 / 0840 0.8 / 1453 5.5 / 2040 1.0		**16** SU ○	0159 5.9 / 0816 0.4 / 1428 6.0 / 2030 0.5	
2 SU	0309 5.5 / 0907 0.8 / 1524 5.4 / 2109 1.0		**17** M	0241 6.1 / 0902 0.2 / 1510 6.1 / 2117 0.4	
3 M	0342 5.5 / 0932 0.8 / 1556 5.4 / 2135 1.1		**18** TU	0325 6.2 / 0947 0.3 / 1553 6.0 / 2204 0.5	
4 TU	0415 5.4 / 0954 1.0 / 1628 5.3 / 2157 1.2		**19** W	0412 6.1 / 1031 0.5 / 1639 5.8 / 2249 0.7	
5 W	0449 5.2 / 1015 1.2 / 1702 5.2 / 2218 1.4		**20** TH	0501 5.9 / 1113 0.9 / 1728 5.6 / 2335 1.0	
6 TH	0525 5.1 / 1038 1.4 / 1739 5.0 / 2243 1.6		**21** F	0555 5.5 / 1155 1.4 / 1822 5.3	
7 F	0605 4.9 / 1106 1.7 / 1820 4.8 / 2318 1.8		**22** SA ◑	0027 1.3 / 0656 5.2 / 1247 1.9 / 1927 5.0	
8 SA	0650 4.6 / 1147 2.0 / 1908 4.6		**23** SU	0137 1.7 / 0808 4.9 / 1406 2.2 / 2041 4.8	
9 SU ◑	0015 2.1 / 0745 4.5 / 1307 2.3 / 2007 4.5		**24** M	0312 1.8 / 0923 4.7 / 1541 2.3 / 2154 4.8	
10 M	0155 2.3 / 0852 4.4 / 1457 2.4 / 2118 4.5		**25** TU	0435 1.7 / 1037 4.7 / 1654 2.1 / 2304 4.9	
11 TU	0352 2.2 / 1006 4.5 / 1631 2.2 / 2231 4.6		**26** W	0540 1.5 / 1146 4.9 / 1751 1.9	
12 W	0509 1.8 / 1117 4.8 / 1729 1.9 / 2337 4.9		**27** TH	0005 5.1 / 0629 1.3 / 1240 5.1 / 1832 1.7	
13 TH	0600 1.4 / 1216 5.1 / 1816 1.5		**28** F	0055 5.2 / 0707 1.2 / 1321 5.2 / 1906 1.5	
14 F	0031 5.3 / 0645 1.0 / 1305 5.5 / 1859 1.1		**29** SA	0135 5.3 / 0737 1.1 / 1355 5.3 / 1938 1.3	
15 SA	0117 5.6 / 0730 0.6 / 1348 5.8 / 1944 0.8		**30** SU ●	0211 5.4 / 0804 1.0 / 1427 5.4 / 2011 1.1	
			31 M	0245 5.4 / 0831 1.0 / 1457 5.4 / 2043 1.1	

NOVEMBER

Day	Time m		Day	Time m	
1 TU	0317 5.4 / 0859 1.0 / 1528 5.4 / 2113 1.1		**16** W	0308 6.1 / 0928 0.5 / 1534 6.0 / 2150 0.5	
2 W	0350 5.3 / 0926 1.1 / 1600 5.4 / 2139 1.2		**17** TH	0357 6.0 / 1013 0.7 / 1621 5.9 / 2238 0.6	
3 TH	0424 5.2 / 0951 1.3 / 1634 5.3 / 2203 1.3		**18** F	0448 5.8 / 1057 1.1 / 1711 5.6 / 2326 0.9	
4 F	0500 5.1 / 1015 1.5 / 1710 5.1 / 2229 1.5		**19** SA	0542 5.5 / 1140 1.5 / 1805 5.3	
5 SA	0539 5.0 / 1042 1.7 / 1749 5.0 / 2303 1.7		**20** SU	0016 1.2 / 0641 5.2 / 1226 1.9 / 1907 5.1	
6 SU	0623 4.8 / 1120 2.0 / 1835 4.8 / 2355 1.9		**21** M ◑	0117 1.5 / 0745 4.9 / 1328 2.2 / 2015 4.9	
7 M ●	0715 4.6 / 1226 2.3 / 1929 4.6		**22** TU	0235 1.7 / 0852 4.7 / 1452 2.3 / 2122 4.8	
8 TU	0116 2.1 / 0816 4.6 / 1410 2.4 / 2035 4.6		**23** W	0354 1.8 / 0959 4.7 / 1606 2.3 / 2227 4.8	
9 W	0254 2.0 / 0928 4.6 / 1544 2.2 / 2150 4.7		**24** TH	0458 1.7 / 1105 4.7 / 1705 2.1 / 2329 4.9	
10 TH	0423 1.8 / 1039 4.8 / 1651 1.9 / 2258 4.9		**25** F	0550 1.6 / 1203 4.9 / 1752 1.9	
11 F	0523 1.4 / 1141 5.1 / 1743 1.6 / 2356 5.3		**26** SA	0022 5.0 / 0628 1.5 / 1249 5.0 / 1832 1.7	
12 SA	0614 1.1 / 1233 5.5 / 1831 1.2		**27** SU	0108 5.1 / 0659 1.4 / 1327 5.2 / 1909 1.5	
13 SU	0047 5.6 / 0702 0.8 / 1320 5.8 / 1919 0.9		**28** M	0147 5.2 / 0729 1.3 / 1401 5.3 / 1945 1.3	
14 M ○	0134 5.9 / 0751 0.5 / 1404 6.0 / 2009 0.6		**29** TU ●	0222 5.3 / 0800 1.2 / 1433 5.4 / 2020 1.2	
15 TU	0221 6.1 / 0840 0.4 / 1448 6.0 / 2100 0.5		**30** W	0256 5.3 / 0832 1.2 / 1505 5.4 / 2053 1.2	

DECEMBER

Day	Time m		Day	Time m	
1 TH	0330 5.3 / 0904 1.2 / 1538 5.4 / 2126 1.2		**16** F	0345 5.9 / 0959 0.8 / 1606 5.8 / 2228 0.5	
2 F	0404 5.3 / 0936 1.3 / 1612 5.3 / 2157 1.2		**17** SA	0435 5.8 / 1042 1.1 / 1656 5.7 / 2314 0.7	
3 SA	0440 5.2 / 1007 1.4 / 1648 5.2 / 2229 1.3		**18** SU	0526 5.5 / 1121 1.4 / 1747 5.5 / 2358 1.0	
4 SU	0519 5.1 / 1038 1.6 / 1726 5.1 / 2304 1.5		**19** M	0619 5.2 / 1155 1.7 / 1842 5.2	
5 M	0602 5.0 / 1113 1.8 / 1809 5.0 / 2349 1.6		**20** TU	0042 1.4 / 0714 5.0 / 1233 2.0 / 1940 5.0	
6 TU	0650 4.9 / 1205 2.0 / 1858 4.9		**21** W ◑	0133 1.7 / 0811 4.7 / 1332 2.2 / 2040 4.8	
7 W ◑	0049 1.7 / 0745 4.8 / 1326 2.2 / 1957 4.8		**22** TH	0240 1.9 / 0909 4.6 / 1453 2.4 / 2140 4.7	
8 TH	0205 1.8 / 0851 4.7 / 1453 2.2 / 2107 4.8		**23** F	0352 2.0 / 1009 4.5 / 1608 2.3 / 2241 4.6	
9 F	0331 1.7 / 1001 4.9 / 1608 2.0 / 2220 4.9		**24** SA	0452 1.9 / 1110 4.6 / 1709 2.2 / 2341 4.7	
10 SA	0444 1.5 / 1106 5.1 / 1710 1.7 / 2325 5.2		**25** SU	0541 1.8 / 1208 4.7 / 1800 1.9	
11 SU	0544 1.2 / 1204 5.3 / 1805 1.3		**26** M	0035 4.8 / 0622 1.7 / 1256 4.9 / 1844 1.7	
12 M	0022 5.5 / 0638 1.0 / 1257 5.6 / 1859 1.0		**27** TU	0121 5.0 / 0700 1.5 / 1336 5.1 / 1924 1.5	
13 TU	0115 5.7 / 0731 0.8 / 1345 5.8 / 1954 0.7		**28** W	0201 5.1 / 0736 1.4 / 1412 5.3 / 2001 1.3	
14 W ○	0205 5.9 / 0823 0.7 / 1431 5.9 / 2048 0.5		**29** TH ●	0237 5.2 / 0812 1.3 / 1447 5.4 / 2037 1.2	
15 TH	0255 6.0 / 0912 0.7 / 1518 5.9 / 2139 0.5		**30** F	0312 5.3 / 0848 1.2 / 1521 5.4 / 2114 1.1	
			31 SA	0346 5.3 / 0925 1.2 / 1555 5.4 / 2151 1.0	

Chart Datum: 2·90 metres below Ordnance Datum (Newlyn). HAT is 6·3 metres above Chart Datum.

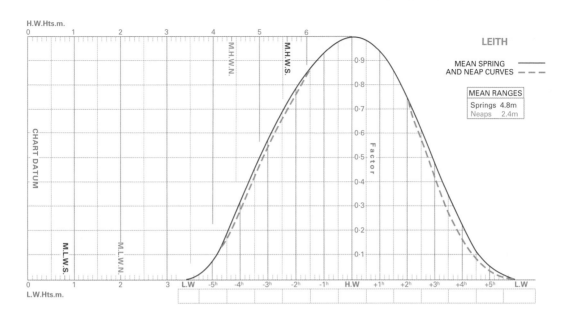

LEITH

MEAN SPRING ————
AND NEAP CURVES – – –

MEAN RANGES
Springs 4·8m
Neaps 2·4m

9.6.10 ANSTRUTHER

Fife **56°13'·15N 02°41'·82W** ✿✿✿✿✿✿✿✿

CHARTS AC 175, 734, 5615; Imray C24, C23, C27

TIDES +0315 Dover; ML 3·1; Duration 0620

Standard Port LEITH (◄——)

Times				Height (metres)			
High Water		Low Water		MHWS	MHWN	MLWN	MLWS
0300	0900	0300	0900	5·6	4·4	2·0	0·8
1500	2100	1500	2100				
Differences ANSTRUTHER EASTER							
–0018	–0012	–0006	–0008	–0·3	–0·2	0·0	0·0

SHELTER Good, but dangerous to enter in strong E to S winds. Hbr dries, soft mud. Pontoon berths for bilge keel craft only.

NAVIGATION WPT 56°12'·59N 02°42'·20W, 019° to ent, 0·60M. Beware FVs and creels in the area. Do not go N of W Pier Lt due to rks extending N & W across the mouth of Dreel Burn.

LIGHTS AND MARKS Conspic tr on W pier. Ldg lts 019°, both FG 7/11m 4M. Pier lts as chartlet. Horn (3) 60s in conspic tr.

COMMUNICATIONS (Code 01333) CGOC 450666; Police: 101, Dr 310352; Ⓗ St Andrews 01334 472327, Kirkaldy 01592 643355; ⊖ 0800 595000. HM 310836 (HO).

ADJACENT HARBOURS AND ANCHORAGE

PITTENWEEM, Fife, **56°12'·60N 02°43'·79W**. AC 734. HW +0325 Dover; –0015 and –0·1m on Leith; ML 3m; Duration 0620. Busy fishing hbr, dredged 1-2m, access all tides, but seek advice before entering at LW springs. Yachts not encouraged; contact HM for berth at W end of inner hbr, but only for emergency use. Outer hbr dries to rock; is only suitable for temp stop in calm weather. Appr 037° on ldg marks/lts, W cols, both FR 3/8m 5M. Rks to port marked by bn, Fl R 4s 3m 2M, and 3 unlit bns. VHF Ch 11 (0700-2100 Mon-Fri) or Forth CG Ch 16 (other times). HM ☎(01333) 312591. Facilities: ⚓ ⚒ ⛽(tanker: ☎730622) ⚓🛒⚒.

CRAIL, Fife, **56°15'·35N 02°37'·29W**. AC 175. HW +0320 on Dover, –0020 on Leith; HW –0·2m on Leith; ML 3·0m; Duration 0615. Good shelter for boats able to take the ground, but limited manoeuvring area. Appr between S pier and bn on rks to S following ldg line 295°, two W concrete pillars with FR lts, 24/30m 6M. Turn 150° to stbd for ent. Call Forth CG before entering. Beware FVs and creel ends. HM ☎07540 672809 (part-time, Mon-Fri). Facilities: ⚓, entry £16·00/craft then £8/day, ⚒⚓⚒ ⚓. **Village** ✉ Ⓑ🛒 ✕⚒.

Anstruther Hbr VHF Ch 11 16 (HO) or Forth CG 16 (OT).

FACILITIES Harbour: wall ⚓ 22+8♥ <10m LOA: entry £16·00 then £8/day; pontoons (no fin keels, 100+♥ check availability with HM) 1st night £23·90, then £13·70 per day; ⚓ ⚒ ⚓ ⚒ WC ⚒ 0800-2100; **Services** ⚒(tanker ☎730622) ⚓ ⚒ Ⓔ (☎311459) ACA.
Town ⚒ ⚒ ⚓ (☎01383 622444), SC (08456 0057023492) ✉ Ⓑ 🛒 ✕ ⚒ ⇌ (bus Cupar or Leuchars) ✈(Edinburgh/Dundee).

ISLE OF MAY, Fife, 56°11'·40N 02°33'·69W. AC 734. HW +0325 on Dover, –0025 on Leith. In settled weather only, and depending on the wind, ⚓ at E or W Tarbert in 4m; landing at Altarstanes. Near the SE tip there is a tiny hbr at Kirkhaven, with narrow, rky ent; yachts can moor fore-and-aft to rings in rks, in about 1-1·5m. SDs are needed. Beware Norman Rk to N of Island, and Maiden Hair Rk to S. At the summit, a 3 tr on stone ho, Fl (2) 15s 73m 22M. The island is a National Nature Reserve managed by Scottish National Heritage (☎01334 654038). It is a bird/seal colony sanctuary. **Sensitive seasons**: (a) breeding birds Apr to Sep; (b) breeding seals Oct to Jan; (c) moulting seals Jan to Mar. Land only at Alterstanes or (where restrictions may apply) Kirkhaven. Warden in residence Easter to end October when visitor centre open 1000-1730 BST. Contact on VHF Ch 16 or 6, or on first landing. No overnight accommodation or camping. Avoid marked out areas to minimise disturbance to wildlife and ongoing conservation experiments.

FIFE NESS TO MONTROSE

(AC 190) Fife Ness is fringed by rocky ledges, and a reef extends 1M NE to N Carr Rk (dries 1·4m, marked by bn). In strong onshore winds keep to seaward of N Carr ECM lit buoy. From here keep 5ca offshore to clear dangers entering St Andrews Bay ⚓; the little harbour dries, and should not be approached in onshore winds.

Northward from Firth of Forth to Rattray Head the coast is mostly rocky and steep-to, and there are no out-lying dangers within 2M of the coast except those off the River Tay and Bell Rock. In an on-shore blow there are few safe havens; both yachts and crews need to be prepared for offshore cruising rather than coast-crawling.

The **River Tay** (AC 1481) is approached from the NE via Fairway buoy; it is dangerous to cut corners from the S. The Bar, NE of Abertay lt buoy, is dangerous in heavy weather, particularly in strong onshore wind or swell. Abertay Sands extend nearly 4M East of Tentsmuir Point on the South side of the buoyed chan. Elbow is a shoal extension eastward. Gaa Sands, running 1¾M East from Buddon Ness, are marked by Abertay lit buoy (Racon) on the North side of the channel. Passage across Abertay and Gaa Sands is very dangerous. The estuary is shallow, with many shifting sandbanks.

Tayport is a good passage stop and although it dries, is the best yacht harbour in the Tay.

▶ *2M E of Bell Rk the S-going stream begins HW Aberdeen –0220, and the N-going at HW Aberdeen +0405, Sp rates 1kn. W of Bell Rock the streams begin earlier.* ◀

Bell Rock (Fl 5s 28m 18M, Racon(M)) lies 11½M E of Buddon Ness.

▶ *S of Buddon Ness the W-going (flood) stream begins about HW Aberdeen –0400, and the E-going at about HW Aberdeen +0230, Sp rates 2kn.* ◀

N from Buddon Ness the coast is sandy. 1·25M SW of Arbroath beware Elliot Horses, rocky patches, which extend about 5ca offshore with with min depth 1·9m. Between Whiting Ness and Scurdie Ness, 9·5M NNE, the coast is clear of out-lying dangers, but is mostly fringed with drying rocks up to 1ca off. In offshore winds there is temp ⚓ in SW of Lunan B, off Ethie Haven.

Scurdie Ness (Fl(3) 20s 38m 23M, Racon T) is conspicuous on S side of entrance to Montrose. Scurdie Rks (dry) extend 2ca E of the Ness. On N side of the channel Annat Bank dries up to about 5ca East of the shore, opposite Scurdie Ness (AC 1438).

▶ *The in-going stream begins at HW Aberdeen –0500, and the outgoing at HW Aberdeen +0115; **both streams are very strong, up to 7kn at sp**, and there is turbulence off the ent on the ebb. The ent is dangerous in strong onshore winds, with breaking seas extending to Scurdie Ness on the ebb. In marginal conditions the last quarter of the flood is best time to enter.* ◀

9.6.11 RIVER TAY

Fife/Angus Tayport (56°27'·10N 02°52'·87W) ❄️⚓🏴🏴🏵️🏵️

CHARTS AC 190, 1481, 5617; Imray C23

TIDES +0401 (Dundee) Dover; ML 3·1; Duration 0610

Standard Port ABERDEEN (→)

Times				Height (metres)			
High Water		Low Water		MHWS	MHWN	MLWN	MLWS
0000	0600	0100	0700	4·3	3·4	1·6	0·6
1200	1800	1300	1900				
Differences BAR							
+0100	+0100	+0050	+0110	+0·9	+0·8	+0·3	+0·1
DUNDEE							
+0140	+0120	+0055	+0145	+1·2	+1·0	+0·5	+0·4
NEWBURGH							
+0215	+0200	+0250	+0335	−0·2	−0·4	−1·1	−0·5
PERTH							
+0220	+0225	+0510	+0530	−0·9	−1·4	−1·2	−0·3

NOTE: At Perth LW time differences give the start of the rise, following a LW stand of about 4 hours.

SHELTER Good in Tay Estuary, but ent is dangerous in strong E/SE winds or on-shore swell. **Tayport** is best place for yachts on passage. Harbour partly dries except W side of NE pier; S side has many yacht moorings. **Dundee** Commercial harbour. Possible moorings off Royal Tay YC. ⚓s as chartlet: the ⚓ off the city is exposed/landing difficult. Off S bank good shelter at Woodhaven and ⚓s from Wormit BC. There are other ⚓s up-river at Balmerino, Newburgh and Inchyra.

NAVIGATION WPT Tay Fairway SWM buoy, 56°28'·30N 02°36'·60W, 239°/1M to Middle Bar buoys. Chan is well buoyed, least depth 5·2m. Keep N of Larick, a conspic disused lt bn.

> Beware strong tidal streams. No passage across Abertay or Gaa Sands; charted depths are unreliable.

LIGHTS AND MARKS See 9.6.4 and chartlet. 'Abertay' ECM buoy (Racon), at E end of Gaa Sands is a clear visual mark.

COMMUNICATIONS (Code 01382) CGOC (01224) 592334; Tayport Hbr Trust, Berthing Master 553799; Police 101; Dr 221976; ⊕ 660111. Forth & Tay Navigation Service (01324) 498584; HM (Dundee) 224121; HM (Perth) (01738) 624056.
Forth & Tay Navigation Service VHF Ch **71**; *Dundee Hbr Radio* VHF Ch **12** 16 (H24); local nav warnings, weather, vis and tides on request. Royal Tay YC, Ch M.

FACILITIES
N BANK: Royal Tay YC (Broughty Ferry) ☎477516, ⚓s free, ✗ 🛏; Services ⚓⚓🔧⚒🛢️🅴🛒(2t)🛢️ ACA.
Dundee City 🏨🏨🏤✉️🅱️🛢️🛒✗🛏⚡✈️

S BANK: Tayport Hbr Access HW±4, hbr dries 0.6m above CD with soft mud. Ent SWM at edge of chan 50m from hbr; ldg marks (185°) R △ on slipway and post. R/Wpost, F W on East wall. ⚓(88 + 7❶) £12.50 <35ft, £15.00 <45ft/day; (phil.tayportharbour@gmail.com) ⚓⚓⚓🗲✉️🅱️🛒✗🛏⚡ bus to St Andrews/Dundee.
Wormit Boating Club ☎541400 ⚓s free, ⚓⚓⚓🛒.

ADJACENT HARBOURS
PERTH, Perth & Kinross, 56°22'·89N 03°25'·74W. AC 1481; OS 53, 58. Tides, see 9.6.12. FYCA Pilot Handbook needed. Leave Tay Rly bridge about HW Dundee −2 to carry the tide for 16·5M to Perth. Keep clear of coasters which have to travel at speed and are constrained by their draught. Lit chan favours S bank for 9M to Newburgh. ⚓ to N of chan at W of town in 2m, piers are to be refurbished and only used with caution, ⚓ newly repaired. Care is required due to mudbanks mid-stream; keep S of Mugdrum Is. Up-river, power cables clearance 33m and Friarton Bridge 26m. Keep S of Willow Is, past gasworks to Hbr on W bank. Hbr has approx 1·5m. See HM, ☎(01738) 624056, for berth. VHF Ch 09 16. ⚓🏨🏨 all city amenities, ≈ ✈.

ST ANDREWS, Fife, 56°20'·32N 02°46'·79W. AC 190. HW −0015 Leith. Small drying hbr 7M S of Tay Estuary /8M NW of Fife Ness. In strong onshore winds breaking seas render appr/ent impossible. Appr near HW on 270°, N bkwtr bn in transit with cathedral twr; Dir Iso WRG 2s (277·5°) on lookout bldg. A recce by dinghy is useful. Best water is about 10m S of the bkwtr. 8m wide ent to inner hbr (drying 2·5m) has lock gates, usually open, and sliding footbridge; berth on W side. Facilities: ⚓ SC; all amenities of university town, inc golf course.

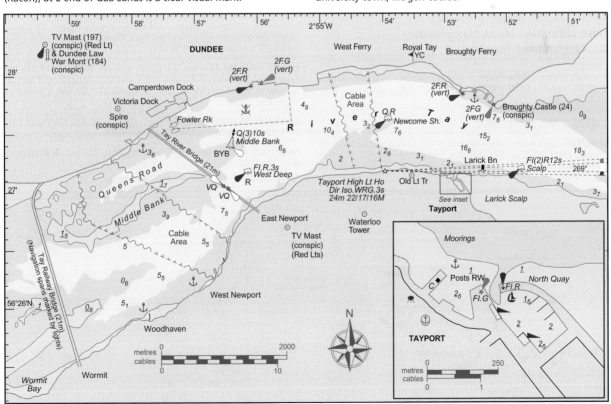

9.6.12 ARBROATH

Angus 56°33'·22N 02°34'·99W ✿✿✿✿✿✿✿

CHARTS AC 190, 1438, 5617; Imray C23

TIDES +0317 Dover; ML 2·9; Duration 0620

Standard Port ABERDEEN (→)

Times				Height (metres)			
High Water		Low Water		MHWS	MHWN	MLWN	MLWS
0000	0600	0100	0700	4·3	3·4	1·6	0·6
1200	1800	1300	1900				
Differences ARBROATH							
+0056	+0037	+0034	+0055	+1·0	+0·8	+0·4	+0·2

SHELTER Good, especially in Inner Basin with lock gates, afloat pontoon berths with 2.5m depth maintained. Ent can be dangerous in moderate SE swell. Inside the entrance, turn to starboard and then to starboard again. HM manned 0700 - 2000 (2200 on selected weekends), during which time lock opens approx HW±3.

NAVIGATION WPT 56°32'·98N 02°34'·21W, 299°/0·5M to ent.

> Entry should not be attempted LW±2½. Beware Knuckle rks to stbd and Cheek Bush rks to port on entering.

LIGHTS AND MARKS Ldg lts 299°, both FR 7/13m 5M, or twin trs of St Thomas' ✠ visible between N pier lt ho and W bkwtr bn. Hbr entry sigs: Fl G 3s on E pier = Entry safe. Same lt shows FR when hbr closed, entry dangerous. Siren (3) 60s at E pier lt is occas, for FVs. Inner Basin Lock Gates, FR = closed, FG = open >2.5m over sill.

COMMUNICATIONS (Code 01241) CGOC (01224) 592334; Police 101; Dr 876836. HM 872166; harbourmaster@angus,gov.uk.

Arbroath Harbour VHF Ch **11** 16.

FACILITIES Inner Basin 59⚓ inc Ⓥ £14·50/craft, 🚾 ▶ ⚒ 🔌 Ⓔ ♨.
Pier ⚓▬▬ ⚓ £13/craft, ⚓ 🛢 🛢(Oil Pier) ⚓(8t).
Town 🏨🏪🛒 ✉ Ⓑ ✕ 🖵 ⛟ ✈ (Dundee).

9.6.13 MONTROSE

Angus 56°42'·19N 02°26'·60W ✿✿✿✿✿✿

CHARTS AC 190, 1438, 5617; Imray C23

TIDES +0320 Dover; ML 2·9; Duration 0645

Standard Port ABERDEEN (→)

Times				Height (metres)			
High Water		Low Water		MHWS	MHWN	MLWN	MLWS
0000	0600	0100	0700	4·3	3·4	1·6	0·6
1200	1800	1300	1900				
Differences MONTROSE							
+0050	+0045	+0035	+0030	+0·6	+0·4	+0·3	+0·2

SHELTER Good; yachts are welcome in this busy commercial port. Contact HM for ⚓, usually available, but beware wash from other traffic. Double mooring lines advised due to strong tidal streams (up to 6kn).

NAVIGATION WPT 56°42'·18N 02°25'·11W, 271°/1·25M to front ldg lt. Beware Annat Bank to N and Scurdie Rks to S of ent chan. In quiet weather best access is LW to LW+1, but in strong onshore winds only safe access would be from HW –2 to HW.

> Ent is dangerous with strong onshore winds against any ebb tide when heavy overfalls develop.

LIGHTS AND MARKS See 9.6.4 and chartlet two sets of ldg lts: Outer 271·5°, both FR 11/18m 5M, front W twin pillars, R bands; rear W tr, R cupola. Inner 265°, both FG 21/33m 5M, Orange △ front and ▽ rear.

COMMUNICATIONS (Code 01674) CGOC (01224) 592334; Police 101; Dr 672554. HM 672302. VHF Ch **12** 16 (H24).

FACILITIES N Quay ☎672302, ⚓ £11/24hrs, £27/wk; ⚓ 🛢(tanker) 🛢 ⚒ 🔌 ⚓(40t) ♨.
Town 🏨✉ Ⓑ 🛒 ✕ 🖵 ⛟ ✈ (Aberdeen).

MONTROSE TO BUCHAN NESS

(AC 210) North from Montrose the coast is sandy for 5M to Milton Ness, where there is an ⚓ on the S side in N winds. Johnshaven, 2M NE, is a small harbour (dries) with a tight entrance, which should not be approached with onshore wind or swell. 5ca NE, off Brotherton Castle, drying rks extend 4ca offshore. Gourdon has a small harbour (mostly dries) approached on leading line between rocky ledges; with storm gates to inner basin. Outside the harbour rocks extend both sides of the entrance, and the sea breaks heavily in strong E winds. Keep a sharp lookout for lobster pot dan buoys between Montrose and Stonehaven.

North to Inverbervie the coast is fringed with rocky ledges up to 2ca offshore. Just N of Todhead Pt is Catterline, a small bay which forms a natural ⚓ in W winds, but open to E. Downie Pt, SE of Stonehaven, should be rounded 1ca off. The Bay is encumbered by rocky ledges up to 2ca from shore and exposed to the E; ⚓ 6ca E of Bay Hotel or berth afloat in outer harbour.

From Garron Pt to Girdle Ness (Fl(2) 20s 56m 22M, Racon G) the coast is mostly steep-to. Fishing nets may be met off headlands during fishing season. Craigmaroinn and Seal Craig (dry) are reefs 3ca offshore SE of Portlethen, a fishing village with landing sheltered by rocks. Cove Bay has a very small fishing harbour, off which there is ⚓ in good weather; Mutton Rk (dries 2·1m) lie 1½ca offshore. From Cove to Girdle Ness keep 5ca offshore, avoiding Hasman Rks (dries 3·4m) 1ca off Altens. Greg Ness and Girdle Ness, at SE corner of Aberdeen Bay, are fringed by rocks. Girdlestone is a rocky patch, depth less than 2m, 2ca ENE of the lt ho. A drying patch extends 2ca SE of lt ho.

▶ Off Girdle Ness the S-going stream begins at HW Aberdeen –0430, and the N-going at HW Aberdeen +0130, sp rates 2·5kn. A race forms on S-going stream. ◀

(AC 213) From Aberdeen there are few offshore dangers to Buchan Ness. Black Dog and Drums Links Firing Ranges lies 4½ and 8¾M N of Aberdeen; red flags and lights are shown when firing.

R Ythan, 1·75M SSW of Hackley Hd, is navigable by small craft, but channel shifts constantly. 3M North is the very small harbour of Collieston (mostly dries), only accessible in fine weather. 4¾M NNE of Hackley Head The Skares, rocks (marked by lit PHM) extend 3½ca from South arm of Cruden Bay, where there is ⚓ in offshore winds. On N side of Cruden B is Port Erroll (dries 2·5m).

HARBOURS SOUTH OF STONEHAVEN

JOHNSHAVEN, Aberdeenshire, 56°47'·60N 02°20'·07W. AC 210. HW +0245 on Dover; +0045 and +0·4m on Aberdeen; ML 2·7m; Duration 0626. Very small, attractive drying hbr 6·5M N of Montrose.

> Entry impossible in strong onshore winds; strictly a fair weather visit with great caution. Even in calm weather swell is a problem inside the hbr.

Appr from 5ca SE at near HW. Conspic W shed at N end of hbr. Ldg marks/lts on 316°: front, R structure with FR 5m; rear is G structure, 20m up the hill and 85m from front, with FG (FR when entry unsafe). Transit leads between rky ledges to very narrow (20m) ent. Turn 90° port into Inner Basin (dries 2·5m) and berth on outer wall or secure to mooring chains, rigged NE/SW. Stonehaven HM ☎(01569) 762741 (part-time). Facilities: ⚓ ☏ £20/night, ⛽ ✈ ⚒ ◻(5t) ⊠ 🛒 ✕ 🅿. Bus to Montrose/Aberdeen.

For details of Rover Ticket see 9.6.5.

GOURDON, Aberdeenshire, 56°49'·49N 02°17'·21W. AC 210. HW +0240 on Dover; +0035 on Aberdeen; HW +0·4m on Aberdeen; ML 2·7m; Duration 0620. Shelter good in inner W hbr (dries about 2m; protected by storm gates); access from about mid-flood. E (or Gutty) hbr is rky, with difficult access. Beware rky ledges marked by bn and extending 200m S from W pier end. **A dangerous rk dries on the ldg line** about 1½ca S of pier heads. Ldg marks/lts 358°, both FR 5/30m 5M, 2 W trs; front lt shows G when not safe to enter. W pier hd Fl WRG 3s 5m 9/7M, vis 180°-G-344°-W-354°-R-180°. Essential to keep in W sector until R Ldg Lts are aligned. E bkwtr hd Q 3m 7M. HM ☎(01569) 762741 (part-time, same as 9.6.14). Facilities: ⚓ ⚒ ✈ from standpipe, ⛽ ◻ ⚒ 🛒 ✕ 🅿.

9.6.14 STONEHAVEN

Aberdeenshire 56°57'·57N 02°12'·02W ✿✿✿❂❁❁

CHARTS AC 210, 1438, 5617; Imray C23

TIDES +0235 Dover; ML 2·6; Duration 0620

Standard Port ABERDEEN (→)

Times				Height (metres)			
High Water		Low Water		MHWS	MHWN	MLWN	MLWS
0000	0600	0100	0700	4·3	3·4	1·6	0·6
1200	1800	1300	1900				
Differences STONEHAVEN							
+0013	+0008	+0013	+0009	+0·2	+0·2	+0·1	0·0

SHELTER Good, especially from offshore winds. Berth in Outer harbour on W side of breakwater (3m to 2m) or N wall; sandbank forms with varying depths (not <0.6m to date) in middle to W side. Or ⚓ outside in fair weather. Hbr speed limit 3kn. Inner hbr dries 3·4m and in bad weather is closed, indicated by FG(occas) lt as shown. Do not go S of leading line, to clear rocks close E of inner harbour wall.

NAVIGATION WPT 56°57'·69N 02°11'·11W, 258°/0·5M to bkwtr lt. Give Downie Pt a wide berth. *Do not enter in strong on-shore winds.*

LIGHTS AND MARKS N pier Iso WRG 4s 7m 5M; appr in W sector, 246°-268°. Inner hbr ldg lts 273° only apply to inner hbr: front FW 6m 5M; rear FR 8m 5M. FG on SE pier is shown when inner hbr is closed by a boom in bad weather. Conspic monument on hill top to S of hbr.

COMMUNICATIONS (Code 01569) CGOC (01224) 592334; Police 101; Dr 762945; Maritime Rescue Institute 765768.

HM (part-time) 762741, Mobile 07741050210; VHF Ch 11.

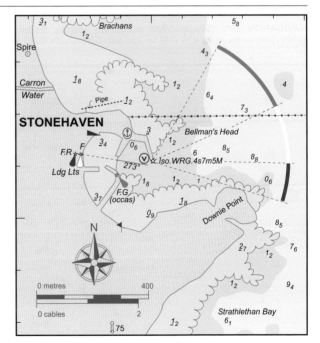

FACILITIES ⚓⚓ ⚓ (£15) ☏ £20/night ⚒ ⛽ 🚻/🚿 ⚓ ◻(1½t) ⛽(tanker, Fri early am). **Aberdeen & Stonehaven SC** ⚓ 🅿.
Town 🅿 ⛽ ⚒ ✕ 🅿 Ⓗ ⊠ Ⓑ ≷ ✈ (Aberdeen).
Berthing Fees: *For details of a Rover Ticket see 9.6.5.*

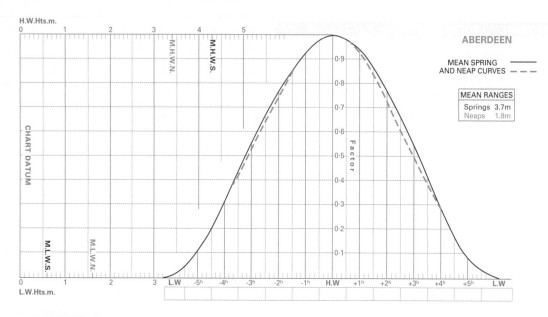

9.6.15 ABERDEEN

Aberdeenshire **57°08'·70N 02°03'·59W** ✿✿✿✿♨✿✿

CHARTS AC 210, 1446, 5617; Imray C23

TIDES +0231 Dover; ML 2·5; Duration 0620

SHELTER Good in hbr; open at all tides, but do not enter in strong NE/ESE winds. Call Aberdeen VTS when 3M off for permission to enter VTS area and berthing availability.

Yachts are not encouraged in this busy commercial port, but usually lie on N side of Albert Basin alongside floating linkspan.

⚓ in Aberdeen Bay gives some shelter from S and W winds. Peterhead is 25M to N; Stonehaven is 13M S.

NAVIGATION WPT Fairway SWM buoy, Mo (A) 5s, Racon, 57°09'·31N 02°01'·96W, 236°/1M to hbr ent. Give Girdle Ness a berth of at least ¼M (more in bad weather). Do not pass close round pier hds. Strong tidal streams and, with river in spate, overfalls. Chan dredged to 6m on ldg line.

LIGHTS AND MARKS Dir lt 237·2° Fl WRG 1s 7m 9M. Ldg lts FR 237·2° 5M (FG when unsafe). Traffic sigs at root of N pier:

🔴	=	Entry prohib
⚫	=	Dep prohib
⚫ & 🔴	=	Port closed

COMMUNICATIONS (Code 01224) CGOC 592334; ☎ 722334; Police 101.

Aberdeen VTS 597000; VHF Ch **12** 16 (H24).

FACILITIES Services ⛴ £18/craft for up to 5 days, ⚓ 🅱ℹ️ Ⓔ ACA. **City** all amenities, ⇌ ✈. **Ferries:** Kirkwall and Lerwick (www.northlinkferries.co.uk).

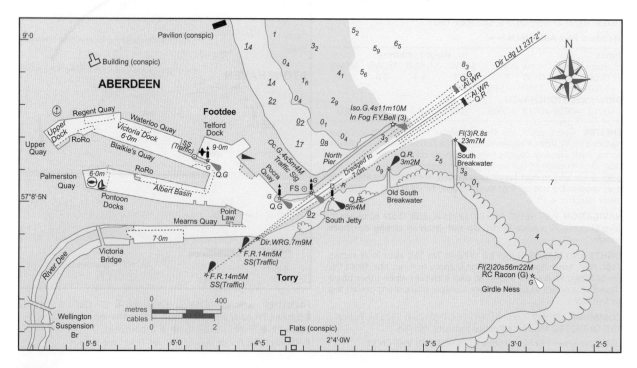

SE Scotland

STANDARD TIME (UT)
For Summer Time add ONE hour in **non-shaded areas**

ABERDEEN LAT 57°09'N LONG 2°04'W

TIMES AND HEIGHTS OF HIGH AND LOW WATERS

Dates in red are **SPRINGS**
Dates in blue are **NEAPS**

YEAR **2016**

JANUARY

Time	m	Time	m
1 0557	3.6	**16** 0549	3.9
1130	1.8	1136	1.3
F 1805	3.7	SA 1757	4.1
2 0016	1.5	**17** 0019	1.1
0648	3.5	0650	3.8
SA 1227	1.9	SU 1241	1.5
1900	3.6	1903	3.9
3 0114	1.6	**18** 0125	1.3
0745	3.4	0757	3.7
SU 1336	2.0	M 1355	1.6
2002	3.5	2015	3.8
4 0219	1.7	**19** 0239	1.4
0850	3.4	0909	3.7
M 1451	2.0	TU 1515	1.5
2112	3.5	2133	3.8
5 0327	1.7	**20** 0357	1.4
0955	3.5	1031	3.8
TU 1604	1.9	W 1631	1.4
2219	3.6	2246	3.9
6 0427	1.6	**21** 0502	1.3
1050	3.7	1120	4.0
W 1658	1.7	TH 1731	1.2
2313	3.7	2346	4.0
7 0515	1.5	**22** 0554	1.2
1135	3.9	1210	4.1
TH 1741	1.4	F 1821	1.0
2359	3.9		
8 0556	1.3	**23** 0037	4.1
1215	4.0	0639	1.1
F 1821	1.2	SA 1254	4.3
		1905	0.8
9 0039	4.0	**24** 0122	4.2
0634	1.1	0719	1.0
SA 1253	4.2	SU 1334	4.4
1859	1.0	○ 1946	0.7
10 0119	4.1	**25** 0202	4.2
0713	1.0	0755	1.0
SU 1329	4.3	M 1412	4.4
● 1938	0.8	2024	0.7
11 0158	4.2	**26** 0240	4.2
0752	0.9	0830	1.0
M 1406	4.4	TU 1448	4.4
2019	0.7	2100	0.7
12 0239	4.3	**27** 0317	4.1
0832	0.9	0903	1.1
TU 1445	4.5	W 1524	4.3
2100	0.6	2134	0.8
13 0321	4.3	**28** 0353	4.0
0913	0.9	0935	1.2
W 1526	4.5	TH 1559	4.2
2143	0.7	2208	1.0
14 0406	4.2	**29** 0429	3.9
0955	1.0	1008	1.3
TH 1610	4.4	F 1636	4.0
2229	0.7	2243	1.2
15 0454	4.1	**30** 0508	3.7
1042	1.2	1045	1.5
F 1700	4.3	SA 1717	3.8
2320	0.9	2323	1.4
		31 0553	3.6
		1128	1.7
		SU 1806	3.6

FEBRUARY

Time	m	Time	m
1 0012	1.6	**16** 0057	1.3
0645	3.4	0728	3.6
M 1225	1.8	TU 1329	1.5
◑ 1904	3.5	1957	3.7
2 0113	1.7	**17** 0214	1.5
0745	3.4	0844	3.6
TU 1340	2.0	W 1458	1.5
2012	3.4	2122	3.6
3 0225	1.8	**18** 0344	1.6
0854	3.4	1004	3.6
W 1503	1.9	TH 1622	1.4
2129	3.4	2242	3.7
4 0341	1.7	**19** 0453	1.5
1005	3.5	1109	3.8
TH 1620	1.7	F 1723	1.2
2240	3.5	2342	3.8
5 0444	1.6	**20** 0544	1.3
1103	3.7	1200	4.0
F 1715	1.5	SA 1811	1.0
2334	3.7		
6 0532	1.4	**21** 0029	4.0
1150	3.9	0626	1.2
SA 1759	1.2	SU 1242	4.1
		1852	0.8
7 0019	3.9	**22** 0110	4.0
0614	1.1	0703	1.0
SU 1231	4.1	M 1319	4.2
1840	0.9	○ 1928	0.7
8 0101	4.1	**23** 0145	4.1
0655	0.9	0737	1.0
M 1310	4.3	TU 1354	4.3
● 1921	0.6	2002	0.7
9 0141	4.3	**24** 0218	4.1
0736	0.8	0808	0.9
TU 1348	4.5	W 1426	4.3
2003	0.4	2034	0.7
10 0221	4.4	**25** 0249	4.1
0816	0.7	0838	0.9
W 1428	4.6	TH 1458	4.2
2044	0.3	2104	0.7
11 0303	4.4	**26** 0320	4.0
0857	0.7	0908	1.0
TH 1509	4.6	F 1530	4.2
2127	0.4	2135	0.9
12 0346	4.3	**27** 0352	3.9
0938	0.7	0938	1.1
F 1553	4.5	SA 1603	4.0
2211	0.5	2206	1.0
13 0432	4.2	**28** 0427	3.8
1023	0.9	1010	1.2
SA 1642	4.4	SU 1640	3.9
2258	0.7	2241	1.2
14 0523	4.0	**29** 0505	3.6
1113	1.1	1048	1.4
SU 1737	4.2	M 1724	3.7
2353	1.0	2322	1.4
15 0622	3.8		
1215	1.3		
M 1842	3.9		
◐			

MARCH

Time	m	Time	m
1 0552	3.5	**16** 0033	1.4
1134	1.6	0703	3.6
TU 1818	3.5	W 1311	1.4
◑		1943	3.6
2 0015	1.7	**17** 0153	1.6
0650	3.4	0819	3.5
W 1240	1.8	TH 1442	1.5
1925	3.3	2111	3.5
3 0128	1.8	**18** 0328	1.7
0759	3.3	0943	3.5
TH 1407	1.8	F 1608	1.3
2042	3.3	2231	3.6
4 0253	1.8	**19** 0438	1.6
0914	3.3	1051	3.7
F 1536	1.7	SA 1707	1.1
2203	3.4	2329	3.7
5 0411	1.6	**20** 0527	1.4
1025	3.5	1142	3.8
SA 1644	1.4	SU 1753	0.9
2306	3.6		
6 0507	1.4	**21** 0013	3.8
1120	3.8	0607	1.2
SU 1733	1.1	M 1223	4.0
2355	3.9	1830	0.8
7 0553	1.1	**22** 0049	3.9
1205	4.1	0642	1.1
M 1817	0.7	TU 1258	4.1
		1904	0.7
8 0038	4.2	**23** 0121	4.0
0634	0.8	0714	0.9
TU 1246	4.3	W 1331	4.2
1900	0.4	○ 1936	0.7
9 0119	4.4	**24** 0151	4.1
0716	0.6	0745	0.9
W 1327	4.5	TH 1402	4.2
● 1942	0.2	2006	0.7
10 0200	4.5	**25** 0221	4.1
0757	0.5	0814	0.8
TH 1408	4.7	F 1433	4.2
2024	0.2	2035	0.7
11 0241	4.5	**26** 0250	4.0
0838	0.6	0843	0.9
F 1451	4.7	SA 1504	4.1
2107	0.2	2104	0.8
12 0324	4.4	**27** 0320	4.0
0920	0.5	0912	1.0
SA 1536	4.6	SU 1537	4.0
2150	0.4	2134	1.0
13 0409	4.2	**28** 0353	3.9
1005	0.7	0944	1.1
SU 1625	4.4	M 1613	3.8
2237	0.7	2207	1.2
14 0459	4.0	**29** 0429	3.7
1056	0.9	1020	1.2
M 1722	4.1	TU 1654	3.7
2330	1.1	2245	1.4
15 0556	3.8	**30** 0511	3.6
1156	1.2	1104	1.4
TU 1828	3.8	W 1746	3.5
◐		2334	1.6
		31 0605	3.4
		1203	1.6
		TH 1852	3.3
		◑	

APRIL

Time	m	Time	m
1 0044	1.8	**16** 0258	1.8
0715	3.3	0911	3.5
F 1325	1.7	SA 1541	1.3
2006	3.3	2206	3.5
2 0212	1.8	**17** 0411	1.6
0829	3.3	1022	3.6
SA 1453	1.6	SU 1640	1.2
2125	3.4	2303	3.6
3 0334	1.6	**18** 0501	1.5
0942	3.5	1114	3.7
SU 1607	1.3	M 1725	1.0
2234	3.6	2346	3.7
4 0437	1.4	**19** 0541	1.3
1045	3.7	1156	3.8
M 1703	1.0	TU 1802	0.9
2327	3.9		
5 0527	1.1	**20** 0022	3.8
1135	4.0	0616	1.1
TU 1751	0.6	W 1232	3.9
		1835	0.8
6 0012	4.2	**21** 0054	3.9
0611	0.8	0649	1.0
W 1221	4.3	TH 1305	4.0
1835	0.3	1907	0.8
7 0055	4.4	**22** 0124	4.0
0654	0.5	0720	0.9
TH 1304	4.5	F 1337	4.1
● 1919	0.2	○ 1937	0.8
8 0137	4.5	**23** 0153	4.1
0736	0.3	0751	0.8
F 1348	4.7	SA 1409	4.1
2002	0.1	2007	0.8
9 0218	4.5	**24** 0223	4.0
0819	0.3	0821	0.9
SA 1433	4.7	SU 1442	4.0
2046	0.2	2037	0.9
10 0301	4.4	**25** 0254	4.0
0904	0.4	0851	1.0
SU 1520	4.5	M 1516	3.9
2130	0.5	2109	1.0
11 0347	4.3	**26** 0326	3.9
0950	0.5	0925	1.0
M 1612	4.3	TU 1552	3.8
2217	0.8	2142	1.1
12 0437	4.1	**27** 0402	3.8
1042	0.8	1002	1.1
TU 1710	4.0	W 1634	3.7
2309	1.2	2221	1.3
13 0534	3.8	**28** 0443	3.7
1142	1.0	1046	1.2
W 1816	3.7	TH 1725	3.5
		2310	1.5
14 0011	1.5	**29** 0535	3.5
0639	3.6	1143	1.4
TH 1253	1.3	F 1828	3.4
◑ 1927	3.5		
15 0127	1.7	**30** 0015	1.7
0751	3.5	0640	3.4
F 1417	1.4	SA 1256	1.4
2048	3.4	◑ 1937	3.4

Chart Datum: 2·25 metres below Ordnance Datum (Newlyn). HAT is 4·8 metres above Chart Datum.

ABERDEEN LAT 57°09'N LONG 2°04'W

TIMES AND HEIGHTS OF HIGH AND LOW WATERS

STANDARD TIME (UT)
For Summer Time add ONE hour in **non-shaded areas**

Dates in **red** are SPRINGS
Dates in **blue** are NEAPS

YEAR **2016**

MAY				JUNE				JULY				AUGUST			
Time	m	Time	m	Time	m	Time	m	Time	m	Time	m	Time	m	Time	m
1 0137 1.7 0752 3.4 SU 1416 1.3 2049 3.5		**16** 0326 1.7 0939 3.5 M 1600 1.3 2223 3.5		**1** 0327 1.4 0936 3.8 W 1559 0.9 2227 3.9		**16** 0431 1.6 1044 3.5 TH 1651 1.3 2312 3.6		**1** 0406 1.2 1019 3.9 F 1639 1.0 2301 4.0		**16** 0445 1.6 1059 3.5 SA 1701 1.4 2321 3.7		**1** 0557 0.9 1213 4.1 M 1817 1.0		**16** 0546 1.2 1206 3.8 TU 1800 1.2	
2 0256 1.6 0902 3.5 M 1529 1.1 2159 3.7		**17** 0425 1.6 1036 3.6 TU 1649 1.2 2310 3.6		**2** 0429 1.1 1039 4.0 TH 1658 0.7 2321 4.1		**17** 0517 1.4 1131 3.7 F 1733 1.2 2352 3.8		**2** 0509 1.0 1122 4.1 SA 1737 0.9 2353 4.1		**17** 0531 1.4 1147 3.7 SU 1744 1.3		**2** 0031 4.2 0645 0.7 TU 1302 4.2 ● 1901 0.9		**17** 0016 4.0 0625 0.9 W 1247 4.0 1839 1.0	
3 0403 1.3 1009 3.8 TU 1631 0.9 2257 3.9		**18** 0510 1.4 1123 3.7 W 1729 1.1 2349 3.8		**3** 0525 0.9 1135 4.2 F 1751 0.6		**18** 0557 1.2 1213 3.8 SA 1810 1.1		**3** 0604 0.8 1217 4.2 SU 1827 0.8		**18** 0003 3.9 0610 1.2 M 1230 3.8 1823 1.1		**3** 0115 4.3 0729 0.6 W 1345 4.2 1940 0.9		**18** 0054 4.2 0704 0.7 TH 1325 4.2 ○ 1918 0.8	
4 0458 1.1 1105 4.0 W 1723 0.6 2346 4.2		**19** 0548 1.2 1202 3.8 TH 1805 1.0		**4** 0009 4.2 0615 0.7 SA 1227 4.3 1839 0.5		**19** 0029 3.9 0633 1.1 SU 1252 3.9 1845 1.0		**4** 0041 4.3 0654 0.6 M 1308 4.3 ● 1913 0.8		**19** 0041 4.0 0648 1.0 TU 1309 4.0 ○ 1900 1.0		**4** 0155 4.4 0809 0.5 TH 1426 4.2 2018 0.9		**19** 0131 4.4 0744 0.5 F 1404 4.3 1957 0.7	
5 0547 0.8 1155 4.3 TH 1812 0.4		**20** 0023 3.9 0623 1.1 F 1239 3.9 1838 0.9		**5** 0055 4.4 0703 0.5 SU 1317 4.4 ● 1926 0.5		**20** 0104 4.0 0708 1.0 M 1329 3.9 ○ 1921 1.0		**5** 0127 4.4 0741 0.5 TU 1357 4.3 1956 0.8		**20** 0117 4.1 0726 0.8 W 1347 4.1 1938 0.9		**5** 0234 4.4 0847 0.5 F 1505 4.1 2053 1.0		**20** 0209 4.5 0824 0.4 SA 1443 4.4 2036 0.7	
6 0030 4.3 0632 0.5 F 1243 4.5 ● 1857 0.3		**21** 0056 4.0 0656 1.0 SA 1314 3.9 ○ 1910 0.9		**6** 0140 4.4 0751 0.4 M 1406 4.4 2011 0.6		**21** 0138 4.1 0744 0.9 TU 1405 4.0 1957 1.0		**6** 0211 4.4 0826 0.4 W 1443 4.2 2038 0.9		**21** 0153 4.3 0804 0.7 TH 1425 4.1 2017 0.9		**6** 0312 4.3 0923 0.6 SA 1542 4.0 2127 1.1		**21** 0249 4.6 0905 0.4 SU 1524 4.3 2117 0.7	
7 0114 4.5 0718 0.4 SA 1330 4.6 1942 0.3		**22** 0128 4.0 0729 0.9 SU 1348 4.0 1942 0.9		**7** 0225 4.4 0838 0.4 TU 1456 4.3 2055 0.8		**22** 0212 4.1 0821 0.8 W 1443 4.0 2033 1.0		**7** 0254 4.3 0909 0.5 TH 1529 4.1 2118 1.0		**22** 0230 4.3 0844 0.6 F 1505 4.2 2056 0.9		**7** 0349 4.2 0959 0.8 SU 1620 3.9 2201 1.2		**22** 0331 4.5 0947 0.5 M 1608 4.2 2159 0.8	
8 0157 4.5 0803 0.3 SU 1418 4.6 2027 0.4		**23** 0159 4.1 0801 0.9 M 1423 4.0 2015 0.9		**8** 0311 4.3 0925 0.5 W 1547 4.2 2140 0.9		**23** 0248 4.1 0859 0.8 TH 1522 4.0 2112 1.0		**8** 0337 4.2 0951 0.6 F 1614 4.0 2158 1.1		**23** 0309 4.3 0925 0.6 SA 1547 4.1 2136 0.9		**8** 0427 4.0 1034 1.0 M 1659 3.7 2238 1.4		**23** 0416 4.4 1032 0.7 TU 1656 4.1 2247 1.0	
9 0241 4.4 0850 0.4 M 1507 4.4 2112 0.6		**24** 0231 4.0 0835 0.9 TU 1458 3.9 2049 1.0		**9** 0359 4.2 1013 0.6 TH 1639 4.0 2225 1.2		**24** 0326 4.1 0939 0.8 F 1605 3.9 2152 1.1		**9** 0421 4.1 1033 0.8 SA 1659 3.8 2238 1.3		**24** 0350 4.3 1007 0.6 SU 1631 4.1 2219 1.0		**9** 0508 3.8 1113 1.2 TU 1742 3.6 2320 1.6		**24** 0508 4.2 1123 0.9 W 1751 3.9 2344 1.2	
10 0328 4.3 0938 0.5 TU 1600 4.2 2159 0.9		**25** 0305 4.0 0911 0.9 W 1537 3.9 2125 1.1		**10** 0449 4.0 1103 0.8 F 1733 3.8 2314 1.4		**25** 0408 4.1 1023 0.8 SA 1652 3.9 2238 1.2		**10** 0506 3.9 1116 1.0 SU 1745 3.6 2322 1.5		**25** 0436 4.2 1053 0.7 M 1721 3.9 2308 1.2		**10** 0556 3.6 1159 1.4 W 1832 3.5 ◐		**25** 0610 4.0 1223 1.2 TH 1855 3.7 ◑	
11 0418 4.1 1029 0.7 W 1658 4.0 2249 1.2		**26** 0342 3.9 0950 1.0 TH 1619 3.8 2206 1.2		**11** 0543 3.8 1156 1.1 SA 1827 3.6		**26** 0455 4.0 1113 0.9 SU 1745 3.8 2331 1.3		**11** 0554 3.7 1203 1.2 M 1834 3.5 ◑		**26** 0528 4.1 1146 0.9 TU 1817 3.8 ◑		**11** 0014 1.7 0652 3.5 TH 1256 1.6 1929 3.4		**26** 0055 1.4 0721 3.8 F 1336 1.4 2007 3.6	
12 0513 3.9 1126 0.9 TH 1758 3.7 2346 1.5		**27** 0424 3.8 1035 1.0 F 1709 3.7 2253 1.4		**12** 0009 1.6 0638 3.6 SU 1253 1.3 ◑ 1923 3.4		**27** 0550 3.9 1210 1.0 M 1844 3.7 ◐		**12** 0014 1.7 0646 3.6 TU 1256 1.4 ◐ 1927 3.4		**27** 0007 1.3 0628 4.0 W 1248 1.1 1920 3.7		**12** 0125 1.9 0757 3.4 F 1405 1.7 2034 3.4		**27** 0218 1.5 0842 3.7 SA 1502 1.6 2125 3.7	
13 0614 3.7 1228 1.1 F 1902 3.5 ◐		**28** 0513 3.7 1128 1.1 SA 1807 3.6 2353 1.5		**13** 0112 1.7 0738 3.5 M 1354 1.4 2025 3.4		**28** 0034 1.4 0652 3.8 TU 1315 1.1 1947 3.7		**13** 0116 1.8 0744 3.4 W 1356 1.5 2027 3.4		**28** 0116 1.4 0737 3.8 TH 1357 1.2 2028 3.7		**13** 0247 1.9 0912 3.3 SA 1522 1.7 2146 3.5		**28** 0347 1.4 1007 3.7 SU 1622 1.5 2237 3.8	
14 0052 1.7 0718 3.5 SA 1339 1.3 2010 3.4		**29** 0614 3.7 1232 1.2 SU 1911 3.5 ◑		**14** 0221 1.8 0843 3.4 TU 1501 1.4 2129 3.4		**29** 0144 1.4 0759 3.8 W 1423 1.1 2054 3.7		**14** 0227 1.8 0851 3.4 TH 1503 1.6 2132 3.4		**29** 0232 1.4 0851 3.8 F 1514 1.3 2141 3.7		**14** 0408 1.7 1025 3.4 SU 1629 1.6 2247 3.6		**29** 0455 1.2 1115 3.9 M 1719 1.3 2333 4.0	
15 0208 1.8 0828 3.5 SU 1455 1.3 2123 3.4		**30** 0104 1.6 0720 3.6 M 1344 1.2 2017 3.6		**15** 0333 1.7 0948 3.5 W 1602 1.4 2225 3.5		**30** 0256 1.4 0909 3.8 TH 1533 1.0 2201 3.8		**15** 0343 1.8 1000 3.4 F 1609 1.5 2232 3.5		**30** 0353 1.3 1009 3.8 SA 1628 1.2 2247 3.9		**15** 0503 1.5 1121 3.6 M 1719 1.4 2335 3.8		**30** 0548 0.9 1207 4.0 TU 1805 1.2	
												31 0218 1.5 0828 3.7 TU 1454 1.1 2125 3.7				**31** 0502 1.1 1117 3.9 SU 1728 1.1 2343 4.1		**31** 0019 4.2 0632 0.7 W 1250 4.1 1845 1.1	

Chart Datum: 2·25 metres below Ordnance Datum (Newlyn). HAT is 4·8 metres above Chart Datum.

FREE monthly updates. Register at
www.reedsnauticalalmanac.co.uk

STANDARD TIME (UT)
For Summer Time add ONE hour in **non-shaded areas**

ABERDEEN LAT 57°09′N LONG 2°04′W
TIMES AND HEIGHTS OF HIGH AND LOW WATERS

Dates in red are **SPRINGS**
Dates in blue are **NEAPS**

YEAR **2016**

SE Scotland

SEPTEMBER

Time	m	Time	m
1 0059	4.3	**16** 0027	4.3
0711	0.6	0639	0.6
TH 1328	4.2	F 1300	4.4
● 1921	1.0	○ 1855	0.8
2 0136	4.4	**17** 0106	4.5
0747	0.6	0720	0.6
F 1403	4.2	SA 1339	4.5
1954	0.9	1935	0.6
3 0210	4.4	**18** 0146	4.7
0820	0.6	0801	0.3
SA 1436	4.2	SU 1419	4.6
2026	0.9	2015	0.6
4 0244	4.3	**19** 0227	4.7
0852	0.7	0842	0.3
SU 1508	4.1	M 1500	4.5
2057	1.0	2056	0.6
5 0317	4.2	**20** 0311	4.7
0923	0.8	0924	0.4
M 1541	4.0	TU 1544	4.4
2128	1.1	2140	0.7
6 0351	4.1	**21** 0358	4.5
0954	1.0	1010	0.7
TU 1615	3.9	W 1631	4.2
2200	1.3	2229	0.9
7 0429	3.9	**22** 0452	4.3
1028	1.2	1101	1.0
W 1654	3.7	TH 1727	4.0
2238	1.5	2328	1.2
8 0513	3.7	**23** 0557	4.0
1109	1.5	1202	1.4
TH 1740	3.6	F 1833	3.8
2324	1.7	◗	
9 0607	3.5	**24** 0040	1.4
1200	1.7	0711	3.7
F 1837	3.5	SA 1319	1.7
◗		1946	3.7
10 0028	1.8	**25** 0207	1.5
0713	3.4	0835	3.6
SA 1311	1.9	SU 1450	1.8
1943	3.4	2108	3.7
11 0154	1.9	**26** 0337	1.4
0827	3.3	1001	3.7
SU 1435	1.9	M 1609	1.7
2056	3.4	2222	3.8
12 0323	1.8	**27** 0442	1.2
0947	3.4	1105	3.8
M 1554	1.8	TU 1704	1.5
2207	3.6	2317	4.0
13 0431	1.5	**28** 0532	1.0
1051	3.6	1152	4.0
TU 1651	1.5	W 1748	1.3
2303	3.8		
14 0518	1.2	**29** 0001	4.1
1139	3.9	0612	0.9
W 1735	1.3	TH 1232	4.1
2347	4.1	1824	1.2
15 0600	0.9	**30** 0039	4.3
1221	4.1	0648	0.8
TH 1816	1.0	F 1305	4.2
		1858	1.0

OCTOBER

Time	m	Time	m
1 0113	4.3	**16** 0040	4.6
0720	0.7	0654	0.4
SA 1336	4.2	SU 1314	4.6
● 1930	1.0	○ 1912	0.6
2 0145	4.3	**17** 0123	4.8
0751	0.7	0737	0.3
SU 1406	4.2	M 1355	4.7
2000	1.0	1955	0.5
3 0217	4.3	**18** 0207	4.8
0821	0.8	0820	0.4
M 1436	4.2	TU 1437	4.6
2030	1.0	2038	0.5
4 0249	4.2	**19** 0253	4.7
0850	0.9	0903	0.6
TU 1506	4.1	W 1521	4.5
2059	1.1	2125	0.7
5 0322	4.1	**20** 0344	4.5
0920	1.1	0950	0.9
W 1539	4.0	TH 1609	4.3
2131	1.2	2216	0.9
6 0358	3.9	**21** 0440	4.3
0952	1.3	1041	1.2
TH 1615	3.9	F 1705	4.1
2207	1.4	2315	1.1
7 0440	3.8	**22** 0546	4.0
1031	1.5	1143	1.6
F 1656	3.7	SA 1811	3.9
2250	1.6	◗	
8 0531	3.6	**23** 0026	1.3
1117	1.7	0658	3.7
SA 1749	3.6	SU 1257	1.8
2348	1.7	1923	3.7
9 0636	3.4	**24** 0148	1.4
1222	1.9	0818	3.6
SU 1857	3.5	M 1424	1.9
		2041	3.7
10 0106	1.8	**25** 0313	1.4
0740	3.4	0940	3.7
M 1349	2.0	TU 1544	1.8
2009	3.5	2155	3.8
11 0235	1.7	**26** 0418	1.3
0905	3.5	1042	3.8
TU 1512	1.9	W 1640	1.6
2121	3.6	2252	3.9
12 0349	1.5	**27** 0507	1.1
1015	3.7	1128	3.9
W 1616	1.6	TH 1724	1.5
2224	3.8	2337	4.0
13 0444	1.2	**28** 0546	1.0
1108	4.0	1206	4.0
TH 1705	1.3	F 1801	1.3
2314	4.1		
14 0530	0.9	**29** 0014	4.1
1152	4.2	0620	1.0
F 1749	1.0	SA 1238	4.1
2358	4.4	1834	1.2
15 0613	0.6	**30** 0049	4.2
1233	4.5	0652	0.9
SA 1831	0.8	SU 1309	4.2
		● 1906	1.1
		31 0121	4.2
		0722	0.9
		M 1338	4.2
		1937	1.0

NOVEMBER

Time	m	Time	m
1 0153	4.2	**16** 0151	4.8
0752	0.9	0801	0.5
TU 1408	4.2	W 1416	4.7
2007	1.0	2024	0.5
2 0226	4.2	**17** 0240	4.7
0822	1.0	0846	0.7
W 1438	4.2	TH 1502	4.6
2037	1.1	2113	0.6
3 0300	4.1	**18** 0332	4.5
0853	1.2	0933	1.0
TH 1510	4.1	F 1551	4.4
2110	1.2	2204	0.8
4 0336	4.0	**19** 0429	4.3
0926	1.3	1023	1.3
F 1545	4.0	SA 1645	4.2
2146	1.3	2301	1.0
5 0417	3.8	**20** 0531	4.0
1002	1.5	1119	1.6
SA 1625	3.9	SU 1747	4.0
2228	1.4		
6 0505	3.7	**21** 0004	1.2
1047	1.7	0635	3.8
SU 1713	3.7	M 1225	1.8
2321	1.6	◗ 1852	3.8
7 0605	3.5	**22** 0114	1.4
1146	1.9	0745	3.6
M 1815	3.6	TU 1339	1.9
●		2002	3.7
8 0029	1.7	**23** 0229	1.5
0713	3.5	0859	3.6
TU 1305	1.9	W 1459	1.9
1926	3.6	2115	3.7
9 0149	1.6	**24** 0340	1.4
0824	3.6	1004	3.7
W 1426	1.9	TH 1605	1.8
2035	3.7	2217	3.8
10 0303	1.4	**25** 0433	1.4
0934	3.7	1055	3.8
TH 1535	1.7	F 1654	1.6
2141	3.9	2306	3.9
11 0405	1.2	**26** 0515	1.3
1033	4.0	1135	3.9
F 1632	1.4	SA 1734	1.5
2239	4.1	2347	4.0
12 0458	0.9	**27** 0551	1.2
1122	4.2	1210	4.0
SA 1721	1.1	SU 1810	1.3
2330	4.4		
13 0546	0.7	**28** 0024	4.1
1207	4.4	0624	1.1
SU 1808	0.8	M 1243	4.1
		1844	1.2
14 0017	4.6	**29** 0100	4.1
0631	0.5	0657	1.1
M 1250	4.6	TU 1314	4.2
○ 1852	0.6	● 1916	1.1
15 0104	4.7	**30** 0134	4.1
0716	0.5	0728	1.1
TU 1333	4.7	W 1345	4.2
1938	0.5	1948	1.1

DECEMBER

Time	m	Time	m
1 0208	4.1	**16** 0230	4.6
0800	1.1	0832	0.8
TH 1417	4.2	F 1447	4.6
2021	1.1	2102	0.6
2 0242	4.1	**17** 0321	4.4
0833	1.2	0917	1.0
F 1449	4.2	SA 1534	4.5
2055	1.1	2151	0.7
3 0319	4.0	**18** 0414	4.2
0907	1.3	1003	1.2
SA 1524	4.1	SU 1624	4.3
2132	1.2	2241	0.9
4 0359	3.9	**19** 0508	4.0
0944	1.4	1051	1.5
SU 1603	4.0	M 1718	4.1
2213	1.3	2333	1.1
5 0444	3.8	**20** 0603	3.8
1026	1.6	1144	1.7
M 1647	3.9	TU 1814	3.9
2301	1.3		
6 0537	3.7	**21** 0030	1.3
1118	1.7	0700	3.6
TU 1741	3.8	W 1245	1.9
2359	1.4	◖ 1914	3.7
7 0639	3.6	**22** 0131	1.5
1224	1.8	0803	3.5
W 1846	3.8	TH 1353	2.0
◖		2020	3.6
8 0108	1.4	**23** 0238	1.6
0745	3.6	0910	3.5
TH 1340	1.8	F 1509	1.9
1954	3.8	2129	3.6
9 0219	1.4	**24** 0346	1.6
0852	3.7	1010	3.6
F 1452	1.7	SA 1616	1.8
2101	3.9	2229	3.7
10 0326	1.2	**25** 0439	1.5
0957	3.9	1100	3.7
SA 1558	1.5	SU 1706	1.6
2207	4.1	2319	3.8
11 0428	1.0	**26** 0522	1.4
1054	4.1	1142	3.9
SU 1656	1.2	M 1747	1.5
2306	4.3		
12 0523	0.9	**27** 0002	3.9
1144	4.3	0559	1.3
M 1749	1.0	TU 1219	4.0
		1823	1.3
13 0000	4.4	**28** 0041	4.0
0613	0.7	0635	1.2
TU 1231	4.5	W 1254	4.1
1838	0.7	1858	1.2
14 0051	4.6	**29** 0117	4.0
0701	0.7	0709	1.1
W 1316	4.6	TH 1327	4.2
○ 1926	0.6	● 1932	1.1
15 0140	4.6	**30** 0152	4.1
0747	0.7	0742	1.1
TH 1401	4.6	F 1359	4.2
2014	0.5	2007	1.0
		31 0228	4.1
		0817	1.1
		SA 1432	4.3
		2042	1.0

Chart Datum: 2·25 metres below Ordnance Datum (Newlyn). HAT is 4·8 metres above Chart Datum.

》》 **FREE** monthly updates. Register at 《
www.reedsnauticalalmanac.co.uk

387

Come what may!

PANTAENIUS
Sail & Motor Yacht Insurance

North East Scotland

**Buchan Ness to Cape Wrath
including Orkney and Shetland Islands**

9.7.1	Contents	389
9.7.2	Area map and Distance table	390
9.7.3	Tidal stream charts	392
9.7.4	Lights, buoys and waypoints	394
9.7.5	Passage information	396
	Buchan Ness to Rattray Head – 396 • Moray Firth: south coast – 397 • Moray Firth: north west coast – 407 • Pentland Firth – 411 • Pentland Firth to Cape Wrath, Orkney Islands – 416 • Shetland Islands – 419	
9.7.6	Peterhead	397
	Boddam	
9.7.7	Fraserburgh	398
	Rosehearty	
9.7.8	Macduff/Banff	399
	Portsoy	
9.7.9	Whitehills	399
	Cullen • Portknockie • Findochty	
9.7.10	Buckie	400
9.7.11	Lossiemouth	400
9.7.12	Hopeman	401
9.7.13	Burghead	401
9.7.14	Findhorn	402
9.7.15	Nairn	402
	Invergordon, tide tables/curves	403
9.7.16	Inverness	407
	Caledonian Canal • Fortrose, Avoch Cromarty, Cromarty Firth, Dornoch Firth, Golspie, Lybster	
9.7.17	Portmahomack	409
9.7.18	Helmsdale	410
9.7.19	Wick, tide tables/curves	410
9.7.20	Scrabster	415
	Kyle of Tongue • Loch Eriboll	
9.7.21	Orkney Islands	416
	Houton Bay • Shapinsay • Auskerry • Pierowall • Rapness	
9.7.22	Stromness	418
9.7.23	Kirkwall	418
9.7.24	Stronsay	419
9.7.25	Shetland Islands	420
	Scalloway • Foula • Out Skellies • Yell • Unst • Fair Isle	
9.7.26	Lerwick, tide tables/curves	422

NE Scotland

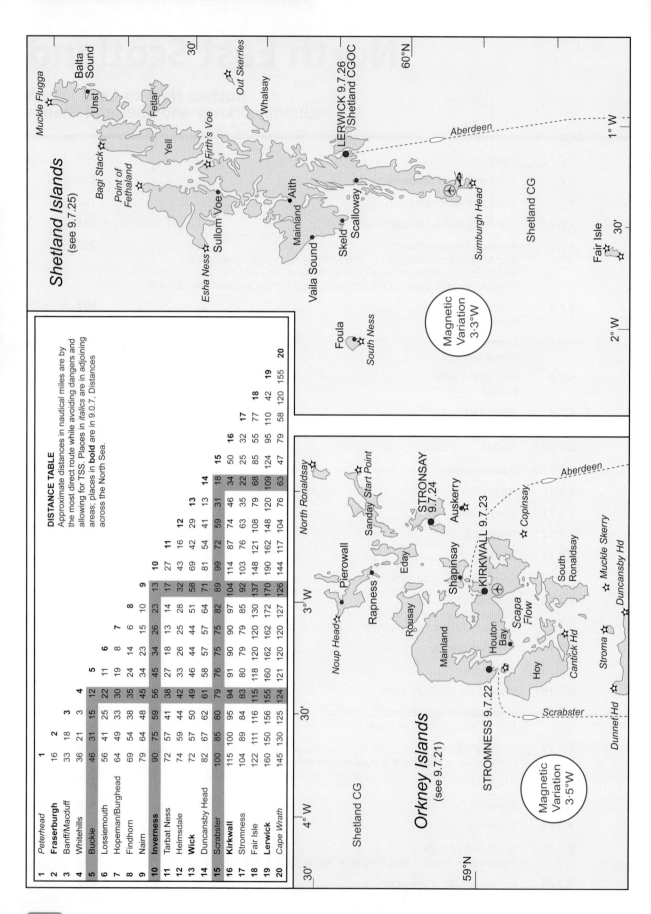

Shetland Islands (see 9.7.25)

Orkney Islands (see 9.7.21)

STROMNESS 9.7.22
KIRKWALL 9.7.23
STRONSAY 9.7.24
LERWICK 9.7.26 Shetland CGOC

Magnetic Variation 3·3°W

Magnetic Variation 3·5°W

DISTANCE TABLE

Approximate distances in nautical miles are by the most direct route while avoiding dangers and allowing for TSS. Places in *italics* are in adjoining areas; places in **bold** are in 9.0.7, Distances across the North Sea.

		1	2	3	4	5	6	7	8	9	10	11	12	13	14	15	16	17	18	19	20
1	*Peterhead*	**1**																			
2	**Fraserburgh**	16	**2**																		
3	Banff/Macduff	33	18	**3**																	
4	Whitehills	36	21	3	**4**																
5	**Buckie**	46	31	15	12	**5**															
6	Lossiemouth	56	41	25	22	11	**6**														
7	Hopeman/Burghead	64	49	33	30	19	8	**7**													
8	Findhorn	69	54	38	35	24	14	6	**8**												
9	Nairn	79	64	48	45	34	23	15	10	**9**											
10	**Inverness**	90	75	59	56	45	34	26	23	13	**10**										
11	Tarbat Ness	72	57	41	38	27	18	13	14	17	27	**11**									
12	Helmsdale	74	59	44	42	33	26	25	28	32	43	16	**12**								
13	**Wick**	72	57	50	49	46	44	44	51	58	69	42	29	**13**							
14	Duncansby Head	82	67	62	61	58	57	57	64	71	81	54	41	13	**14**						
15	**Scrabster**	100	85	80	79	76	75	75	82	89	99	72	59	31	18	**15**					
16	**Kirkwall**	115	100	95	94	91	90	90	97	104	114	87	74	46	34	50	**16**				
17	Stromness	104	89	84	83	80	79	79	85	92	103	76	63	35	22	25	32	**17**			
18	Fair Isle	122	111	116	115	118	120	120	130	137	148	121	108	79	68	85	55	77	**18**		
19	**Lerwick**	160	150	156	155	160	162	162	172	170	190	162	148	120	109	124	95	110	42	**19**	
20	*Cape Wrath*	145	130	125	124	121	120	120	127	126	144	117	104	76	63	47	79	58	120	155	**20**

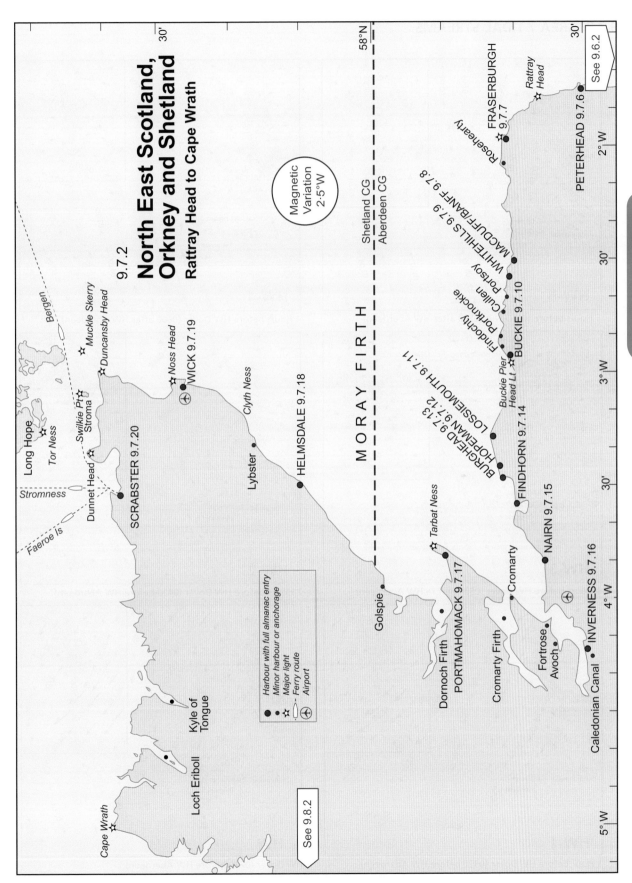

9.7.2

North East Scotland, Orkney and Shetland

Rattray Head to Cape Wrath

9.7.3 AREA 7 TIDAL STREAMS

HW-5
5 Hours before HW Dover (0500 after HW Aberdeen)

HW-4
4 Hours before HW Dover (0600 after HW Aberdeen)

HW-3
3 Hours before HW Dover (0520 before HW Aberdeen)

HW-2
2 Hours before HW Dover (0420 before HW Aberdeen)

HW-1
1 Hour before HW Dover (0320 before HW Aberdeen)

HW
HW Dover (0220 before HW Aberdeen)

Southward 9.6.3   Westward 9.8.3

HW+1
1 Hour after HW Dover (0120 before HW Aberdeen)

HW+2
2 Hours after HW Dover (0020 before HW Aberdeen)

HW+3
3 Hours after HW Dover (0040 after HW Aberdeen)

HW+4
4 Hours after HW Dover (0140 after HW Aberdeen)

HW+5
5 Hours after HW Dover (0240 after HW Aberdeen)

HW+6
6 Hours after HW Dover (0340 after HW Aberdeen)

NE Scotland

9.7.4 LIGHTS, BUOYS AND WAYPOINTS

Bold print = light with a nominal range of 15M or more. CAPITALS = place or feature. *CAPITAL ITALICS* = light-vessel, light float or Lanby. *Italics* = Fog signal. ***Bold italics*** = Racon. Some marks/buoys are fitted with AIS (<u>MMSI No</u>); see relevant charts.

BUCHAN NESS TO INVERNESS

PETERHEAD and RATTRAY HEAD

Buchan Ness ☆ Fl 5s 40m **18M**; W twr, R bands; ***Racon (O) 14-16M***; 57°28'·23N 01°46'·51W.

Kirktown Ldg lts 314°. Front, Oc R 6s 14m 10M; Or △ on lattice mast; 57°30'·22N 01°47'·21W. Rear, Oc R 6s 21m 10M (sync with front); Or ▽ on lattice mast.

S Bkwtr Hd ⚓ Fl (2) R 12s 24m 7M; 57°29'·79N 01°46'·54W.

N Bkwtr Hd ⚓ Iso RG 6s 19m 11M; W tripod; vis: 171°-R-236°-G-171°; *Horn 30s*; 57°29'·84N 01°46'·32W.

Rattray Hd ☆ 57°36'·61N 01°49'·03W Fl (3) 30s 28m **18M**; W twr; ***Racon (M) 15M***; *Horn (2) 45s*.

Smith Bank Met Mast ⬡ Mo(U) 15s 17m 10M Mast (101); Horn Mo(U) 30s; 58°10'·93N 02°49'·22W.

FRASERBURGH

Fraserburgh Ldg lt 291°; Iso R 2s 12m 9M; 57°41'·57N 02°00'·13W. Rear, 75m from front, Iso R 2s 17m 9M.

Fraserburgh, Balaclava Bkwtr Head ⚓ Fl (2) G 8s 26m 6M; dome on W twr; vis: 178°-326°; 57°41'·51N 01°59'·70W.

Kinnaird Hd ☆ 57°41'·87N 02°00'·26W Fl 5s 25m **22M**; vis: 092°-297°.

MACDUFF, BANFF and WHITEHILLS

Macduff Pier Hd ⚓ Fl (2) WRG 6s 12m W9M, R7M; W twr; vis: shore-G-115°-W-174°-R-210°; 57°40'·26N 02°30'·01W.

Macduff Ldg Lts 130°, Front FR 28m 3M; 57°40'·18N 02°29'·85W. Rear, 220m from front, FR 55m 3M; both Or △ on mast.

Banff Ldg Lts 295°, Front Fl R 4s; 57°40·30N 02°31·36W. Rear, QR; both vis: 210°-345°.

Whitehills Pier Hd ⚓ 57°40'·80N 02°34'·88W Fl WR 3s 7m W9M, R6M; W twr; vis: 132°-R-212°-W-245°.

PORTSOY and FINDOCHTY

Portsoy Pier Ldg Lts 173°, Front Fl G 4s 20m 3M; post; 57°41'·09N 02°41'·40W; Rear Q G 22m 3M; R △ on BW post.

Findochty Middle Pier Ldg Lts 166°, Front FR 6m 3M; 57°41'·90N 02°54'·20W. Rear FR 10m 3M.

BUCKIE

West Muck ⚓ QR 5m 7M; tripod; 57°41'·06N 02°58'·01W.

N Pier 60m from Hd ☆ 57°40'·9N 02°57'·5W Oc R 10s 15m **15M** W twr.

LOSSIEMOUTH, HOPEMAN and BURGHEAD

Lossiemouth S Pier Hd ⚓ Fl R 6s 11m 5M; 57°43'·42N 03°16'·69W.

Halliman Skerries ⚓ Q; ***Racon M***, 57°44'·33N 03°18'·57W.

Hopeman Ldg Lts 081°, Front,FR 3m; 57°42'·71N 03°26'·18W. Rear, 10m from front, FR 4m.

Burghead N Bkwtr Hd ⚓ Oc 8s 7m 5M; 57°42'·09N 03°30'·03W.

FINDHORN, NAIRN and INVERNESS FIRTH

Findhorn Landfall ⚓ LF 10s 57°40'·34N 03°38'·77W.

Nairn E Pier Hd ⚓ Oc WRG 4s 6m 5M; 8-sided twr; vis: shore-G-100°-W-207°-R-shore; 57°35'·62N 03°51'·65W.

Riff Bank E ⚓ Fl Y 10s 3m 5M 57°38'·38N 03°58'·18W; <u>AIS</u>.

SOUTH CHANNEL

Riff Bank S ⚓ Q (6) + L Fl 15s; 57°36'·73N 04°00'·97W.

Chanonry ⚓ 57°34'·45N 04°05'·56W Oc 6s 12m 12M; W twr; vis: 148°-073°.

Munlochy ⚓ L Fl 10s; 57°32'·91N 04°07'·65W .

Petty Bank ⚓ Fl R 5s 57°31'·58N 04°08'·98W.

Meikle Mee ⚓ Fl G 3s 57°30'·26N 04°12'·02W.

Longman Pt ⚓ Fl WR 2s 7m W5M, R4M; vis: 078°-W-258°-R-078°; 57°29'·99N 04°13'·31W.

Craigton Point ⚓ Fl WRG 4s 6m W11M, R7M, G7M; vis: 312° - W - 048° - R - 064° - W - 085° - G - shore; 57°30'·05N 04°14'·09W.

Bridge Centre , Or △; ***Racon (K) 6M***; 57°29'·97N 04°13'·79W.

INVERNESS and CALEDONIAN CANAL

R. Ness Outer ⚓ QR 3m 4M; 57°29'·83N 04°13'·93W.

Carnarc Pt ⚓ Fl G 2s 8m 4M; G lattice twr; 57°29'·72N 04°14'·25W.

Clachnaharry, S Tr'ng Wall Hd ⚓ Iso G 4s 5m 2M; tfc sigs; 57°29'·43N 04°15'·86W.

INVERNESS TO DUNCANSBY HEAD

CROMARTY FIRTH and INVERGORDON

Fairway ⚓ L Fl 10s; ***Racon (M) 5M***; 57°39'·96N 03°54'·19W.

Cromarty Bank ⚓ Fl (2) G 10s; 57°40'·66N 03°56'·78W.

Buss Bank ⚓ Fl R 3s 57°40'·97N 03°59'·54W.

Cromarty - The Ness (Lt ho disused W tr 13m); 57°40'·98N 04°02'·20W

Nigg Oil Terminal Pier Hd ⚓ Oc G 5s 31m 5M; Gy twr, floodlit; 57°41'·54N 04°02'·60W.

DORNOCH FIRTH to LYBSTER

Tarbat Ness ☆ 57°51'·88N 03°46'·76W Fl (4) 30s 53m **24M**; W twr, R bands; ***Racon (T) 14-16M***.

Lybster, S Pier Hd ⚓ Oc R 6s 10m 3M; 58°17'·79N 03°17'·41W.

Clyth Ness Lt Ho (unlit); W twr, R band; 58°18'·64N 03°12'·74W.

WICK

S Pier Hd ⚓ Fl WRG 3s 12m W12M, R9M, G9M; W 8-sided twr; vis: 253°-G-270°-W-286°-R-329°; *Bell (2) 10s* (occas); 58°26'·34N 03°04'·73W.

Dir lt 288·5° F WRG 9m W10M, R7M, G7M; col, N end of bridge; vis: 283·5°-G-287·2°-W-289·7°-R-293·5°; 58°26'·54N 03°05'·34W

Noss Hd ☆ 58°28'·71N 03°03'·09W Fl WR 20s 53m **W25M, R21M**; W twr; vis: shore-R-191°-W-shore.

DUNCANSBY HEAD TO CAPE WRATH

Duncansby Hd ☆ 58°38'·65N 03°01'·58W Fl 12s 67m **22M**; W twr; ***Racon (T)***.

Muckle Skerry ☆ 58°41'·41N 02°55'·49W Fl (3) 30s 52m **23M**; W twr; <u>992351086</u>.

Lother Rock ⚓ Fl 2s 13m 6M; ***Racon (M) 10M***; 58°43'·79N 02°58'·69W.

Swona ⚓ Fl 8s 17m 9M; vis: 261°-210°; 58°44'·25N 03°04'·24W.

Swona N Hd ⚓ Fl (3) 10s 16m 10M; 58°45'·11N 03°03'·10W.

Stroma ☆☆, Swilkie Point 58°41'·75N 03°07'·01W Fl (2) 20s 32m **20M**; W twr; <u>992351092</u>.

Dunnet Hd ☆ 58°40'·28N 03°22'·60W Fl (4) 30s 105m **23M**.

THURSO, SCRABSTER and CAPE WRATH

Thurso Ldg Lts 195°. Front, FG 5m 4M; Gy post; 58°35'·96N 03°30'·76W. Rear, FG 6m 4M; Gy mast.

Scrabster Q. E. Pier Hd ⚓ Fl (2) 4s 8m 8M 58°36'·66N 03°32'·31W.

Strathy Pt Lt Ho (disused) W twr on W dwelling; 58°36'·04N 04°01'·12W.

Sule Skerry ☆ 59°05'·09N 04°24'·38W Fl (2) 15s 34m **21M**; W twr; ***Racon (T)***.

North Rona ☆ 59°07'·27N 05°48'·91W Fl (3) 20s 114m **22M**.

Sula Sgeir ⚓ Fl 15s 74m 11M; □ structure; 59°05'·61N 06°09'·57W.

Loch Eriboll, White Hd ⚓ Fl WR10s 18m W13M, R12M; W twr and bldg; vis: 030°-W-172°-R-191°-W-212°; 58°31'·01N 04°38'·90W.

Cape Wrath ☆ 58°37'·54N 04°59'·99W Fl (4) 30s 122m **22M**; W twr.

ORKNEY ISLANDS

Tor Ness ☆ 58°46'·78N 03°17'·86W Fl 5s 21m **17M**; W twr.
Cantick Hd (S Walls, SE end) ⚡ 58°47'·23N 03°07'·88W Fl 20s 35m 13M; W twr.

SCAPA FLOW and APPROACHES

Long Hope, S Ness Pier Hd ⚡ Fl WRG 3s 6m W7M, R5M, G5M; vis: 082°-G- 242°-W- 252°-R-082°; 58°48'·05N 03°12'·35W.

Hoxa Head ⚡ Fl WR 3s 15m W9M, R6M; W twr; vis: 026°-W-163°-R-201°-W-215°; 58°49'·31N 03°02'·09W.
Nevi Skerry ⚓ Fl (2) 6s 7m 6M; 58°50'·67N 03°02'·70W.
Flotta Grinds ⚓ Fl (2) R; 58°50'·97N 03°00'·77W; 992351083.
Rose Ness ⚡ 58°52'·33N 02°49'·97W Fl 6s 24m 8M; W twr.
Barrel of Butter ⚡ Fl (2) 10s 6m 7M; 58 53'·40N 03°07'·62W.
Cava ⚡ Fl WR 3s 11m W10M, R8M; W ○ twr; vis: 351°-W-143°-196°-W-251°-R-271°-R-298°; 58°53'·21N 03°10'·70W .
Holm of Houton ⚡ 58°54'·62N 03°10'·97W; Fl R 3s 7m 5M.
Houton Bay Ldg Lts 316°. Front ⚓ Fl G 3s 8m. Rear ⚓, 200m from front, FG 16m; vis: 312°- 320°; 58°54'·97N 03°11'·56W.

CLESTRAN SOUND and HOY SOUND

Graemsay Is Hoy Sound ☆ Ldg Lts 104°. Front Low, Iso 3s 17m 12M; W twr; vis: 070°-255°, 58°56'·42N 03°18'·60W. **High Rear**, 1·2M from front, Oc WR 8s 35m **W20M, R16M**; W twr; vis: 097°-R-112°-W-163°-R-178°-W-332°; obsc on Ldg line within 0·5M.
Skerry of Ness ⚡ Fl WG 4s 7m W7M, G4M; vis: shore -W-090°-G-shore; 58°56'·95N 03°17'·83W.

STROMNESS

Ldg Lts 317°. Front FR 29m 11M; W twr; 58°57'·61N 03°18'·15W. Rear, 55m from front, FR 39m 11M; vis: 307°-327°; H24.

AUSKERRY

Copinsay ⚡ Fl (5) 30s 79m 14M; W twr; 58°53'·77N 02°40'·35W.

Auskerry ☆ 59°01'·51N 02°34'·34W Fl 20s 34m **20M**; W twr.
Helliar Holm, S end ⚡ Fl WRG 10s 18m W14M, R11M, G11M; W twr; vis: 256°-G-276°-W-292°-R-098°-W-116°-G-154°; 59°01'·13N 02°54'·09W.
Balfour Pier Shapinsay ⚡ Fl (2) WRG 5s 5m W3M, R2M, G2M; vis: 270°-G-010°-W-020°-R-090°; 59°01'·86N 02°54'·49W.

KIRKWALL

Thieves Holm, ⚡ Q R 8M; 59°01'·09N 02°56'·21W.

Pier N end ☆ 58°59'·29N 02°57'·72W Iso WRG 5s 8m **W15M**, R13M, G13M; W twr; vis: 153°-G-183°-W-192°-R-210°.

WIDE FIRTH

Linga Skerry ⚓ Q (3) 10s; 59°02'·39N 02°57'·56W.
Boray Skerries ⚓ Q (6) + L Fl 15s; 59°03'·65N 02°57'·66W.
Skertours ⚓ Q; 59°04'·11N 02°56'·72W.
Galt Skerry ⚓ Q; 59°05'·21N 02°54'·20W.
Seal Skerry ⚡ Fl R 3s 5m 3M; platform; 59°05'·21N 02°54'·20W.
Brough of Birsay ☆ 59°08'·19N 03°20'·41W Fl (3) 25s 52m **18M**.
Papa Stronsay NE end, The Ness ⚡ Fl(4)20s 8m 9M; W twr; 59°09'·34N 02°34'·93W.

SANDAY ISLAND and NORTH RONALDSAY

Quiabow ⚓ Fl (2) G 12s; 59°09'·82N 02°36'·30W.
Start Pt ☆ 59°16'·69N 02°22'·71W Fl (2) 20s 24m **18M**.
Kettletoft Pier Hd ⚡ Fl WRG 3s 7m W7M, R5M, G5M; vis: 351°-W- 011°-R-180°-G-351°; 59°13'·80N 02°35'·86W.
N Ronaldsay ☆ NE end, 59°23'·34N 02°22'·91W Fl 10s 43m **24M**; R twr, W bands; *Racon (T) 14-17M*.

EDAY and EGILSAY

Calf Sound ⚡ Fl (3) WRG 10s 6m W8M, R6M, G6M; W twr; vis: shore-R-215°-W-222°-G-301°-W-305°; 59°14'·21N 02°45'·82W.
Backaland Pier ⚡ 59°09'·43N 02°44'·88W Fl R 3s 5m 4M; vis: 192°-250°.
Egilsay Graand ⚓ Q (6) + L Fl 15s; 59°06'·86N 02°54'·42W.

WESTRAY and PIEROWALL

Noup Head ☆ 59°19'·86N 03°04'·23W Fl 30s 79m **20M**; W twr; vis: about 335°-282° but partially obsc 240°-275°.

Pierowall E Pier Head ⚡ Fl WRG 3s 7m W11M, R7M, G7M; vis: 254°-G-276°-W-291°-R-308°-G-215°; 59°19'·35N 02°58'·53W.
Papa Westray, Moclett Bay Pier Head ⚡ Fl WRG 5s 7m W5M, R3M, G3M; vis: 306°-G-341°-W-040°-R-074°; 59°19'·60N 02°53'·52W.

SHETLAND ISLES

FAIR ISLE

Skadan South ☆, 59°30'·84N 01°39'·16W Fl (4) 30s 32m **22M**; W twr; vis: 260°-146°, obsc inshore 260°-282°.

Skroo ☆ N end 59°33'·13N 01°36'·58W Fl (2) 30s 80m **22M**; W twr; vis: 086·7°-358°.

MAINLAND, SOUTH

Sumburgh Head ☆ 59°51'·21N 01°16'·58W Fl (3) 30s 91m **23M**. Lt Ho disused 59°51'·48N 01°16'·28W; W twr.
Mousa, Perie Bard ⚡ Fl 3s 20m 10M; 59°59'·84N 01°09'·51W.

BRESSAY and LERWICK

Bressay, Kirkabister Ness ⚡ 60°07'·20N 01°07'·31W; Fl (2) 10s 18m 10M. Adjacent Lt Ho disused.
Maryfield Ferry Terminal ⚡ Oc WRG 6s 5m 5M; vis: W008°-R013°-G-111°-008°; 60°09'·43N 01°07'·45W.

North Ness ⚡ Iso WG 4s 4m 5M; vis: shore-W-158°-G-216°-W-301°; 60°09'·57N 01°08'·77W
Loofa Baa ⚓ Q (6) + L Fl 15s 4m 5M; 60°09'·72N 01°08'·79W.
Soldian Rock ⚓ Q (6) + L Fl 15s 60°12'·51N 01°04'·73W.
N ent Dir lt 215°, Oc WRG 6s 27m 8M; Y △, Or stripe; vis: 211°-R-214°-W-216°-G-221°; 60°10'·47N 01°09'·53W.
Rova Hd ⚡ 60°11'·46N 01°08'·60W Fl (3) WRG 18s 12m W12M, R9M, G9M; W twr; vis: 090°-R-182°-W-191°-G-213°-R-241°-W-261·5°-G-009°-R-040°. Same structure and synchronised: Fl (3) WRG 18s 14m **W16M**, R13M, G13M; vis: 176·5°-R-182°-W-191°-G-196·5°.
Dales Voe ⚡ Fl (2) WRG 8s 5m W4M, R3M, G3M; vis: 220°-G-227°-W-233°-R-240°; 60°11'·79N 01°11'·23W.
Hoo Stack ⚡ Fl (4) WRG 12s 40m W7M, R5M, G5M; W pylon; vis: 169°-R-180°-W-184°-G-193°-W-169°. Same structure, Dir lt 182°. Fl (4) WRG 12s 33m W9M, R6M, G6M; vis: 177°- R-180°-W-184°-W-187°; synch with upper lt; 60°14'·96N 01°05'·38W.
Mull (Moul) of Eswick ⚡ Fl WRG 3s 50m W9M, R6M, G6M; W twr; vis: 028°-R-200°-W-207°-G-018°-W-028°; 60°15'·74N 01°05'·90W.
Inner Voder ⚓ Q (9) 15s; 60°16'·44N 01°05'·18W; AIS.

WHALSAY and SKERRIES

Symbister Ness ⚡ Fl (2) WG 12s 11m W8M, G6M; W twr; vis: shore-W-203°-G-shore; 60°20'·43N 01°02'·29W.
Suther Ness ⚡ Fl WRG 3s 10m W10M, R8M, G7M; vis: shore -W-038°-R-173°-W-206°-G-shore; 60°22'·12N 01°00'·20W.
Bound Skerry ☆ 60°25'·47N 00°43'·72W Fl 20s 44m **20M**; W twr.
South Mouth. Ldg Lts 014°. Front, FY3m 2M; 60°25'·33N 00°45'·01W. Rear, FY 12m 2M.
Muckle Skerry ⚡ Fl (2) WRG 10s 15m W7M, R5M, G5M; W twr; vis: 046°-W-192°-R-272°-G-348°-W-353°-R-046°; 60°26'·41N 00°51'·84W.

YELL SOUND

S ent, Lunna Holm ⚡ Fl (3) WRG 15s 19m W10M,R7M,G7M; W ○twr; vis: shore-R-090°-W-094°-G-209°-W-275°-R-shore; 60°27'·34N 01°02'·52W.

Firths Voe ☆, N shore 60°27'·21N 01°10'·63W Oc WRG 8s 9m **W15M**, R10M, G10M; W twr; vis: 189°-W-194°-G-257°-W-261°-R-339°-W-066°.
Linga Is. Dir lt 150° ⚡ Q (4) WRG 8s 10m W9M, R9M, G9M; vis: 145°-R-148°-W-152°-G-155°. Q (4) WRG 8s 10m W7M, R4M, G4M; same structure; vis: 052°-R-146°, 154°-G-196°-W-312°; synch; 60°26'·80N 01°09'·13W.
The Rumble Bn ⚡ R Bn; Fl 10s 8m 4M; *Racon (O)*; 60°28'·16N 01°07'·26W.

Yell, Ulsta Ferry Term. Bkwtr Hd ⚓ Oc RG 4s 7m R5M, G5M; vis: shore-G-354°, 044°-R-shore. Same structure; Oc WRG 4s 5m W8M, R5M, G5M; vis: shore-G-008°-W-036°-R-shore; 60°29'·74N 01°09'·52W.

Toft Ferry Terminal ☆,Dir lt 241° (H24); Dir Oc WRG 10s 8m **W16M,** R10M, G10M; vis: 236° -G-240°-W-242°-R-246°; by day W2M, R1M, G1M. 60°27'·96N 01°12'·34W.

Ness of Sound, W side ⚓ Fl (3) WRG 12s 18m W9M, R6M, G6M; vis: shore-G-345°-W-350°-R-160°-W-165°-G-shore; 60°31'·34N 01°11'·28W.

Brother Is. Dir lt 329°, Fl (4) WRG 8s 16m W10M, R7M, G7M; vis: 323·5°-G-328°-W-330°-R-333·5°; 60°30'·95N 01°14'·11W.

Mio Ness ⚓ Q (2) WR 10s 12m W7M, R4M; W ○ twr; vis: 282° - W - 238° - R - 282°; 60°29'·66N 01°13'·68W.

Tinga Skerry ⚓ Q(2)G 10s 9m 5M. W ○ twr; 60°30'·48N 01°14'·86W.

YELL SOUND, NORTH ENTRANCE

Bagi Stack ⚓ Fl (4) 20s 45m 10M; 60°43'·53N 01°07'·54W.

Gruney Is ⚓ Fl WR 5s 53m W8M, R6M; W twr; vis: 064°-R-180°-W-012°; *Racon (T) 14M;* 60°39'·15N 01°18'·17W.

Pt of Fethaland ☆ 60°38'·05N 01°18'·70W Fl (3) WR 15s 65m **W24M, R20M;** vis 080°-R-103°-W-160°-206°-W-340°.

Muckle Holm ⚓ Fl (4) 10s 32m 10M 60°34'·83N 01°16'·01W.

Little Holm ⚓ Iso 4s 12m 6M; W twr; 60°33'·42N 01°15'·88W.

Outer Skerry ⚓ Fl 6s 12m 8M; 60°33'·04N 01°18'·32W.

Quey Firth ⚓ Oc WRG 6s 22m W12M, R8M, G8M; W twr; vis: shore (through S & W)-W-290°-G-327°-W-334°-W-shore; 60°31'·43N 01°19'·58W.

Lamba, S side ⚓Fl WRG 3s 30m W8M, R5M, G5M; W twr; vis: shore-G-288°-W-293°-R-327°-W-044°-R-140°-W-shore. Dir lt 290·5° Fl WRG 3s 24m W10M, R7M, G7M; vis: 285·5°-G-288°-W-293°-W-295·5°; 60°30'·73N 01°17'·84W.

SULLOM VOE

Gluss Is ☆ Ldg Lts 194·7° (H24). **Front,** 60°29'·77N 01°19'·44W F 39m **19M;** □ on Gy twr; **Rear,** 0·75M from front, F 69m **19M;** □ on Gy twr; both Lts 9M by day.

Little Roe ⚓ Fl (3) WR 10s 16m W5M, R4M; W structure, Or band; vis: 036°-R-095·5°-W-036°; 60°29'·99N 01°16'·46W.

Skaw Taing ⚓ Ldg Lts 150·5°. Front, Oc WRG 5s 21m W8M, R5M, G5M; Or and W structure; vis: 049°-W-078°-G-147°-W-154°-R-169°-W-288°; 60°29'·10N 01°16'·86W. Rear, 195m from front, Oc 5s 35m 8M; vis: W145°-156°.

Ness of Bardister ⚓ Oc WRG 8s 20m W9M, R6M, G6M; Or &W structure; vis: 180·5°- W-240°- R-310·5°- W-314·5°- G-030·5°; 60°28'·19N 01°19'·63W.

Fugla Ness. Lts in line 212·3°. Rear, 60°27'·25N 01°19'·74W Iso 4s 45m 14M. Common front 60°27'·45N 01°19'·57W Iso 4s 27m 14M; synch with rear Lts. Lts in line 203°. Rear, 60°27'·26N 01°19'·81W Iso 4s 45m 14M.

Sella Ness ☆ Dir lt 133·5°; 60°26'·76N 01°16'·66W Oc WRG 10s 19m **W16M,** R3M, G3M; vis: 123·5° -G- 130·5°-Al WG (white phase increasing with brg)-132·5°-W-134·5°-Al WR(R phase inc

with brg)-136·5°-R-143·5°; H24. By day Oc WRG 10s 19m W2M, R1M,G1M as above.

EAST YELL, UNST and BALTA SOUND

Whitehill ⚓ Fl WR 3s 24m W9M, R6M; W clad metal frame twr; vis: 147°-W-163°-R-211°-W-352°-R-003°; 60°34'·80N 01°00'·22W.

Fetlar. Hamars Ness ⚓ Iso G 6s 5m 3M; 60°37'·78N 00°55'·73W.

Balta Sound ⚓Fl WR 10s 17m 10M, R7M; vis: 249°-W-008°-R-058°-W-154°; 60°44'·43N 00°47'·68W.

Holme of Skaw ⚓ Fl 5s 8m 8M; white metal framework twr; 60°49'·87N 00°46'·30W.

Muckle Flugga ☆ 60°51'·32N 00°53'·13W Fl (2) 20s 66m **22M.**

Yell. Cullivoe Bkwtr Hd ⚓ Fl (2) WRG 10s 3m 4M; vis: 080°-G-294°-W-355°-R-080°; 60°41'·86N 00°59'·70W.

Unst. Head of Mula ⚓ Fl WRG 5s 48m W10M, G7M, R7M; metal framework twr; vis: 292°-G-357°-W-002°-R-157°-W-161·5; 60°40'·76N 00°57'·58W.

MAINLAND, WEST

Esha Ness ☆ 60°29'·34N 01°37'·65W Fl 12s 61m **25M.**

Ness of Hillswick ⚓ Fl (4) WR 15s 34m W9M, R6M; vis: 217°-W-093°-R-114°; 60°27'·21N 01°29'·80W.

Muckle Roe, Swarbacks Minn ⚓ Fl WR 3s 30m W9M, R6M; vis: 314°-W-041°-R-075°-W-137°; 60°20'·98N 01°27'·07W.

W Burra Firth Outer ⚓ Oc WRG 8s 27m W9M, R7M, G7M; vis: 136°-G-142°-W-150°-R-156°. H24; 60°17'·79N 01°33'·56W.

W Burra Firth Inner ☆ 60°17'·78N 01°32'·17W F WRG 9m **W15M,** R9M, G9M; vis: 095°-G-098°-W-102°-105°; H24.

Ve Skerries ⚓ Fl (2) 20s 17m 11M; W twr; *Racon (T) 15M;* 60°22'·36N 01°48'·78W.

Papa Stour Housa Voe Dir lt 228° ⚓ F WRG 2m W9M, R7M, G7M; vis: 219°- G-226°- W-230°-R-239°; 60°19'·58N 01°40'·47W.

Rams Head ⚓ Fl WRG 8s 16m W9M, R6M; G6M; W house; vis: 265°-G-355°-W-012°-R-090°-W-136°, obsc by Vaila I when brg more than 030°; 60°11'·96N 01°33'·47W.

North Havra ⚓ Fl(3) WRG 12s 24m W11M, R8M, G8M; W twr; vis: 001°-G-053·5°-W-060·5°-R-144°, 274°- G-334°-W-337·5°-R -001°; 60°09'·85N 01°20'·29W.

SCALLOWAY

Bullia Skerry ⚓ Fl 5s 5m 5M; stainless steel pillar & platform 60°06'·55N 01°21'·57W.

Point of the Pund ⚓ Fl WRG 5s 20m W7M, R5M, G5M; W GRP twr; vis: 267°-W-350°-R-090°-G-111°-R-135°-W-140°-G-177°; 60°07'·99N 01°18'·31W.

Whaleback Skerry ⚓ Q; 60°07'·95N 01°18'·90W.

Blacks Ness Pier SW corner ⚓ Oc WRG 10s 10m W11M, G8M, R8M; vis: 052°-G-063·5°-W-065·5°-R-077°; 60°08'·02N 01°16'·59W.

Fugla Ness ⚓ Fl (2) WRG 10s 20m W10M, R7M, G7M; W twr; vis: 014°-G-032°-W-082°-R-134°-W-shore; 60°06'·38N 01°20'·85W.

FOULA

South Ness ☆ 60°06'·75N 02°03'·86W Fl WR (3) 15s 36m **W18M** R14M; W twr; vis: 221°-W-255°-R-277°-W-123°-obscured-221°.

9.7.5 PASSAGE INFORMATION

More passage information is threaded between the harbours in this area. Admiralty Leisure Folio: 5617 Fraserburgh to Inverness and the Caledonian Canal. **Bibliography:** *North and East Scotland* Farne Islands to Cape Wrath(Imray/Lawrence). *N&NE Scotland and Orkney Islands. Shetland Islands* (Both Clyde Cruising Club sailing directions and anchorages). *Shetland Islands Pilot* (Imray/ Buchanan). The *N Coast of Scotland Pilot* (Admiralty NP52).

A £50 'Rover ticket' (Aberdeenshire and Moray Councils) allows berthing for one week from arrival at the first harbour at other ports in the scheme. Orkney Marinas 'Weekly Rover' costs £11/m, to berth in their 3 marinas or on Port Authority Harbours, piers and moorings except St Margaret's Hope (9.7.21) which is independent.

BUCHAN NESS TO RATTRAY HEAD

Buchan Ness (lt, fog sig, Racon) is a rky peninsula. 2ca N is Meikle Mackie islet, close W of which is the small hbr of Boddam (dries). 3ca NE of Meikle Mackie is The Skerry, a rk 6m high on S side of Sandford B; rks on which the sea breaks extend 2ca NNE. The chan between The Skerry and the coast is foul with rks and not advised. Peterhead is easy to enter in almost all conditions and is an excellent passage port with marina at SW corner of the Bay.

Numerous submarine pipelines lead ashore to St Fergus Terminal in Rattray Bay. **Notes on offshore installations, see 3.14 and 9.5.5.** Rattray Hd (light on The Ron, rk 2ca E of Hd) has rocky foreshore, drying to 2ca off. Rattray Briggs is a detached reef, depth 0·2m, 2ca E of lt ho. Rattray Hard is a rky patch, depth 10·7m, 1·5M ENE of the lt ho, which raises a dangerous sea during onshore gales.

▶ Off Rattray Hd the S-going stream begins at HW Aberdeen –0420, and the N-going at HW Aberdeen +0110, sp rates 3kn. In normal conditions keep about 1M E of Rattray Hd, but pass 5M off in bad weather, preferably at slack water. ◀

Conspic radio masts with red aero lights lie 2·5M WNW and 2·2M W of lighthouse.

MORAY FIRTH: SOUTH COAST

(AC 115, 222, 223) Crossing the Moray Firth from Rattray Hd (lt, fog sig) to Duncansby Hd (lt, Racon) heavy seas may be met in strong W winds. Most hbrs in the Firth are exposed to NE-E winds. For oil installations, see 9.5.5; the Beatrice Field is 20M S of Wick.

▶ Tidal streams attain 3kn at Sp close off Rattray Hd. 5M NE of the Head the NE-going stream begins at HW Aberdeen +0140, and the SE-going stream at HW Aberdeen –0440, Sp rates 2kn. Except in the inner part of the Moray Firth streams are weak. ◀

In late spring/early summer fog (haar) is likely in onshore winds. In strong winds the sea breaks over Steratan Rk and Colonel Rk, respectively 3M E and 1M ENE of Fraserburgh. Banff B is shallow; N of Macduff beware Collie Rks. Banff hbr dries, and should not be approached in fresh NE-E winds, when seas break well offshore; Macduff is a possible alternative.

From Meavie Pt to Scar Nose dangers extend up to 3ca from shore in places. Beware Caple Rk (depth 0·2m) 7½ca W of Logie Hd.

Spey B is clear of dangers more than 7½ca from shore; ⚓ here, but only in offshore winds. Beware E Muck (dries) 5ca SW of

Craigenroan, an above-water rky patch 5ca SW of Craig Hd, and Middle Muck and W Muck in approach to Buckie; Findochty & Portknockie are 2 and 3.5M ENE. Halliman Skerries (dry; bn) lie 1·5M WNW of Lossiemouth. Covesea Skerries (dry) lie 5ca NW of their lt ho (disused).

To the SW of a line between Helmsdale and Lossiemouth is a EU Special Area of Conservation to protect a vulnerable population of bottlenose dolphins. Mariners are advised to proceed at a safe, constant speed through the area and avoid disturbing them.

Inverness Firth is approached between Nairn and S Sutor. In heavy weather there is a confused sea with overfalls on Guillam Bank, 9M S of Tarbat Ness. The sea also breaks on Riff Bank (S of S Sutor) which dries in places. Chans run both N and S of Riff Bank.

▶ Off Fort George, on E side of ent to Inverness Firth (AC1078), the SW-going stream begins HW Aberdeen +0605, sp rate 2·5kn; the NE-going stream begins at HW Aberdeen –0105, sp rate 3·5kn. There are eddies and turbulence between Fort George and Chanonry Pt when stream is running hard. Tidal streams in the Inverness Firth and approaches are not strong, except in the Cromarty Firth Narrows, the Fort George Narrows and the Kessock Road, including off the entrance to the Caledonian Canal. ◀

There is a firing range between Nairn and Inverness marked, when in operation, by flags at Fort George. Much of Inverness Firth is shallow, but a direct course from Chanonry Pt to Kessock Bridge, via Munlochy SWM and Meikle Mee SHM lt buoys, carries a least depth of 2·1m. Meikle Mee bank dries 0·2m.

9.7.6 PETERHEAD

Aberdeenshire **57°29'·81N 01°46'·42W** ❁❁❁❁♨♨✿✿

CHARTS AC 213, 1438, 5617; Imray C23

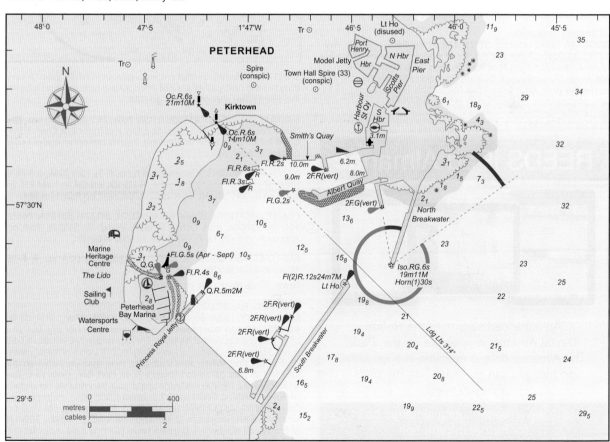

Continued overleaf

PETERHEAD *continued*

TIDES +0140 Dover; ML 2·3; Duration 0620

Standard Port ABERDEEN (←—)

Times				Height (metres)			
High Water		Low Water		MHWS	MHWN	MLWN	MLWS
0000	0600	0100	0700	4·3	3·4	1·6	0·6
1200	1800	1300	1900				
Differences PETERHEAD							
–0035	–0045	–0035	–0040	–0·3	–0·2	0·0	+0·1

SHELTER Good in marina (2·8m). A useful passage hbr, also a major fishing and oil/gas industry port. Access any weather/tide.

NAVIGATION WPT 57°29'·44N 01°45'·75W, 314°/0·5M to ent. No dangers. 5kn speed limit in Bay; 4kn in marina. Chan between marina bkwtr and SHM lt buoy is <30m wide.

LIGHTS AND MARKS Power stn chy (183m) is conspic 1·25M S of entrance. Ldg marks 314°, front Or △, rear Or ▽, on lattice masts; lights as chartlet. Marina E bkwtr ☆ Fl R 4s 6m 2M; W bkwtr hd, QG 5m 2M, vis 185°-300°. Buchan Ness ☆ Fl 5s 40m 28M is 1·6M S of entrance.

COMMUNICATIONS (Code 01779) CGOC (01224) 592334; Police 101; Dr 474841; ☒ 472316. Marina Manager (Port Authority: www.peterheadport.co.uk) 477868; Hr Control (H24) 483630.

All vessels, including yachts, **must** call *Peterhead Harbour Radio* VHF Ch **14** for clearance to enter/depart the Bay.

FACILITIES Peterhead Bay Marina marina@peterheadport.co.uk ☎477868, access all tides, 2·3m at ent, max LOA 20m. Pontoons are 'E' to 'A' from entrance.150⌒, £13 up to 6m LOA + £1/m (7 days for price of 5); ▣ ▟ ⛽; ⛟ from bowser at end of Princess Royal jetty (☆ Q R); ☷ ✕ at caravan site. **Peterhead SC** ☎(01358) 751340 (Sec); **Services** ⚓ ▟ ⚒ ⚒ ⚒ Ⓔ ⌂. **Town** ⛽ ⊠ ☷ 🛒 ✕ bus to Aberdeen for ⇌ & ✈.

ADJACENT HARBOUR
BODDAM, Aberdeenshire, **57°28'·47N 01°46'·56W**. AC 213. HW +0145 on Dover; Tides as 9.7.6. Good shelter in the lee of Meikle Mackie, the island just N of Buchan Ness, Fl 5s 40m 28M Horn (3) 60s. Inner hbr dries/unlit. Beware rks around Meikle Mackie and to the SW of it. Appr from 1½ca NW of The Skerry. Yachts on S side of outer hbr. All facilities at Peterhead, 2M N.

9.7.7 FRASERBURGH

Aberdeenshire **57°41'·50N 01°59'·79W** ❄❄🌊🌊🌸🌸

CHARTS AC 115, 222, 1462, 5617; Imray C23

TIDES +0120 Dover; ML 2·3; Duration 0615

Standard Port ABERDEEN (←—)

Times				Height (metres)			
High Water		Low Water		MHWS	MHWN	MLWN	MLWS
0000	0600	0100	0700	4·3	3·4	1·6	0·6
1200	1800	1300	1900				
Differences FRASERBURGH							
–0122	–0118	–0115	–0115	–0·5	–0·4	–0·1	+0·2

SHELTER A safe refuge, but ent is dangerous in NE/SE gales. A very busy fishing hbr; yachts are not encouraged but may find a berth in S Hbr (3.2m). FVs come and go H24.

NAVIGATION WPT 57°41'·30N 01°58'·80W, 291°/0·57M to ent. The Outer Ent Chan is maintained to 8·9m and Inner Ent Chan 5·9m. Good lookout on entering/leaving. Yachts can enter under radar control in poor vis.

LIGHTS AND MARKS Kinnairds Hd lt ho, Fl 5s 25m 22M, is 0·45M NNW of ent. Cairnbulg Briggs bn, Fl (2) 10s 9m 6M, is 1·8M ESE of ent. Ldg lts 291°: front Iso R 2s 12m 9M; rear Iso R 2s 17m 9M.

COMMUNICATIONS (Code 01346) CGOC (01224) 592334; Police 101; Dr 518088. Port Office 515858; Watch Tr 515926. Call on approach VHF Ch **12** 16 (H24) for directions/berth.

FACILITIES **Port** ☎515858, ⌒£10 (in S Hbr) any LOA, ⚓ ⚓ ▟ 🕰 ⚒ ⚒ Ⓔ △ ⌂(30t/70t mob) ⛽ 🛒 ✕ ⊡

Town ⊠ Ⓑ ⇌ ✈ (bus to Aberdeen).

ADJACENT HARBOURS
ROSEHEARTY, Aberdeen, **57°42'·08N 02°06'·87W**. ❄❄🌊🌊🌸🌸. AC 222, 213. HW Aberdeen –1. E pier and inner hbr dry, but end of W pier is accessible at all tides. Ent exposed in N/E winds; in E/SE winds hbr can be uncomfortable. Ldg lts, 219°, on BW metal poles; rks E of ldg line. When 30m from pier, steer midway between ldg line and W pier (W round tower), Fl 4s. Port Rae, close to E, has unmarked rks; local knowledge req'd. ⌒£10 any LOA. *For details of Rover berthing ticket see 9.7.5.* **Town** ⊠ ☷ ✕ ⌂. Firing range: for info ☎(01346) 571634; see also 9.7.5. **Pennan Bay**, 5M W: ⚓ on sand between Howdman (2·9m) and Tamhead (2·1m) rks, 300m N of hbr (small craft only). **Gardenstown** (Gamrie Bay). Appr from E of Craig Dagerty rk (4m, conspic). Harbour dries, or ⚓ off.

9.7.8 MACDUFF/BANFF

Aberdeenshire Macduff 57°40'·25N 02°30'·03W ⊛⊛✲☆☆
Banff 57°40'·22N 02°31'·27W ⊛⊛✲☆☆

CHARTS AC 115, 222, 1462, 5617; Imray C23

TIDES + 0055 Dover; ML 2·0; Duration 0615

Standard Port ABERDEEN (←—)

Times				Height (metres)			
High Water		Low Water		MHWS	MHWN	MLWN	MLWS
0200	0900	0400	0900	4·3	3·4	1·6	0·6
1400	2100	1600	2100				
Differences BANFF							
–0100	–0150	–0150	–0050	–0·4	–0·2	–0·1	+0·2

SHELTER Macduff: Reasonably good, but ent not advised in strong NW winds. Slight/moderate surge in outer hbr with N/NE gales. Hbr ent is 17m wide with 3 basins; approx 2·6m in outer hbr and 2m inner hbr. A busy cargo/fishing port with limited space for yachts. **Banff**: Popular hbr which dries. When Macduff ent is very rough in strong NW/N winds, Banff can be a safe refuge; pontoon berths in both basins. Best to contact HM beforehand. In strong E/ENE winds Banff is unusable.

NAVIGATION WPT 57°40'·48N 02°30'·59W, 127°/0·4M to **Macduff** ent. WPT 57°40'·11N 02°30'·85W, 115°/0·25M to **Banff** ent. Beware Feachie Craig, Collie Rks and rky coast N and S of hbr ent.

LIGHTS AND MARKS Macduff: Ldg lts/marks 127° both FR 28/55m 3M, orange △s. Pier hd lt, Fl (2) WRG 6s 12m 9/7M, W tr; shore-G-115°-W-174°-R-210°. **Banff**: Fl 4s end of New Quay and ldg lts 295° rear Fl R 2s, front Fl R 4s.

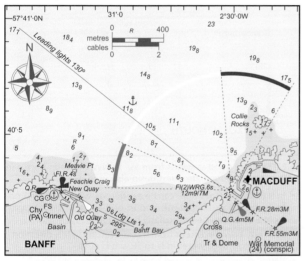

COMMUNICATIONS (Code 01261) CGOC (01224) 592334; Police 101; Dr (**Banff**) 812027. HM (**Macduff**) 832236, Watch tr 833962; HM (**Banff**) 815544.

Macduff Ch 12 16 (H24); **Banff** Ch 12.

FACILITIES Macduff: **Hbr** £10 any LOA; ⌐ ⏚ ✕ ⌁ ☎ ⌂. **Town** ☎ ☎ ⌧ 🛒 ✕ ⌂ ⇌ (bus to Keith).

Banff: Hbr www.banffmarina.com ❶ in Inner Hbr marina, £10 any LOA, *Rover berthing ticket see 9.7.5*; ⌐ ⏚ ⫶Ⅾ⫶ ◎ 🚾; **Banff SC**: ⌂. **Town**, ☎ ☎ 🛢 🛒 ✕ ⌂ ⇌ Ⓑ ✈ (Aberdeen).

ADJACENT HARBOURS

PORTSOY, Aberdeenshire, **57°41'·34N 02°41'·59W**. ⊛⊛✲☆☆. AC 222. HW +0047 on Dover; –0132 and Ht –0·3m on Aberdeen. Small drying hbr; ent exposed to NW/NE'lies. New Hbr to port of ent partially dries; Inner Hbr dries to clean sand. Ldg lts 173°, front Fl G 4s 20m 3M, metal post; rear Q G 22m 3M, R △ on BW post. HM ☎(01261) 815544. Facilities: few but hosts Scottish Traditional Boat Festival late June/early July. ⚓ £10, *for details of Rover berthing ticket see 9.7.5*, ⏚ ⌐ ⌧ 🛒 ✕ ⌂.

Sandend Bay, 1·7M W (57°41'N 02°44'·5W). ⚓ on sand E of hbr.

9.7.9 WHITEHILLS

Aberdeenshire **57°40'·80N 02°34'·87W** ⊛⊛✲☆☆☆

CHARTS AC 115, 222, 5617; Imray C23

TIDES +0050 Dover; ML 2·4; Duration 0610

Standard Port ABERDEEN (←—)

Times				Height (metres)			
High Water		Low Water		MHWS	MHWN	MLWN	MLWS
0200	0900	0400	0900	4·3	3·4	1·6	0·6
1400	2100	1600	2100				
Differences WHITEHILLS							
–0122	–0137	–0117	–0127	–0·4	–0·3	+0·1	+0·1

SHELTER Safe. In strong NW/N winds beware surge in the narrow ent and outer hbr, when ent is best not attempted. See HM for vacant pontoon berth.

NAVIGATION WPT 57°41'·98N 02°34'·89W, 180°/1·2M to bkwtr lt. Reefs on S side of chan marked by 2 rusty/white SHM bns. Beware fishing floats. Narrow entrance 1·7m below CD, Outer Hbr pontoon berths 2m and Inner Hbr 1.5m–1·8m.

LIGHTS AND MARKS Fl WR 3s on pier hd, vis 132°-R-212°-W-245°; approach in R sector.

COMMUNICATIONS (Code 01261) CGOC (01224) 592334; Police 101; Dr 812027. HM www.whitehillsharbour.co.uk 861291, mobile 07906 135786.

Whitehills Hbr Radio VHF Ch **14** 16.

FACILITIES Marina (47 + ❶), £15 (decreasing to £30/3nights), Quayside £12 ☎ ⌂ ⏚ ✕ ⌁ ☎ ⌂ @; **Town** ⌧ 🛒 ✕ ⌂ ⇌ (bus to Keith), ✈ (Aberdeen).

ADJACENT HARBOURS

CULLEN, Moray, **57°41'·63N 02°49'·29W**. ⊛✲✲☆☆. AC 222. HW +0045 on Dover, HW –0135 & –0·3m on Aberdeen; Duration 0555; ML 2·4m. Shelter good, but ent hazardous in strong W/N winds. Appr on 180° toward conspic viaduct and W bn on N pier. Caple Rk, 0·2m, is 5ca NE of hbr. Small drying hbr, best for shoal draught. Moor S of Inner jetty if < 1m draught with F G bn close S of the root of N pier. Beware moorings across inner basin ent. Pontoons in inner hbr (1·8m at HW Np). *For Rover berthing ticket see 9.7.5.* HM ☎(01542) 831700. **Town** 🛢 (hardware store), ⌧ 🛒✕ ⌂.

PORTKNOCKIE, Moray, **57°42'·28N 02°51'·79W**. ⊛⊛✲☆☆☆. AC 222. HW +0045 on Dover; –0135 and ht –0·3m Aberdeen; ML 2·3m; Duration 0555; access H24. Good shelter in one of the safest hbrs on S side of Moray Firth, but scend is often experienced; care needed in strong NW/N winds. FW ldg lts, on white-topped poles, lead approx 151°, to ent. Orange street lts surround the hbr. Limited ABs on N Quay; most of inner hbr dries. HM ☎(01542) 840833 (home, p/time); Facilities: ⌐ ⚓; *for details of Rover berthing ticket see 9.7.5*, ⏚ ✕; Dr ☎ 840272. **Town** Ⓑ ⌧ 🛒 ⌂.

NE Scotland

9.7.10 BUCKIE

Moray **57°40'·84N 02°57'·63W** ❀❀❀❀💧💧💧🌸🌸

CHARTS AC 223, 1462, 5617; Imray C23

TIDES +0040 Dover; ML 2·4; Duration 0550
Standard Port ABERDEEN (←—)

Times				Height (metres)			
High Water		Low Water		MHWS	MHWN	MLWN	MLWS
0200	0900	0400	0900	4·3	3·4	1·6	0·6
1400	2100	1600	2100				
Differences BUCKIE							
–0130	–0145	–0125	–0140	–0·2	–0·2	0·0	+0·1

SHELTER Good in all weathers.

In strong NNW to NE winds there is a dangerous swell over the bar at hbr ent, which is 24m wide.

Access H24. Berth in No 4 basin as directed by HM.

NAVIGATION Beware W, Middle and E Muck Rks, 3ca off shore. WPT 57°41'·0N 02°58'·6W, 110°/5ca to ent. The recommended track is 3.5 cables west of West Muck and approach Harbour at 115°. Approx 2 cables from entrance alter to 100° just as the 2FR (vert) and the Oc R 10s on the N breakwater come into line (096°). Traffic is controlled by VHF.

LIGHTS AND MARKS W Muck QR 5m 2M (tripod); N bkwtr, Oc R 10s 15m 15M (W twr); N bwtr hd 2FR(vert), S bwtr hd 2FG(vert). Harbour Lt Iso WG 2s 20m 16/12M, (W tr, R top inshore of hbr).

COMMUNICATIONS (Code 01542) CGOC (01224) 592334; Police 101; Dr 831555. HM 831700, mob 07842 532360. VHF Ch 12 16 (H24).

FACILITIES ⛴£17·50 or £8·50 <12hrs ⚓🔧⛽💧🚿(delivery)🛢🔨🔧 🛒🚿🏳(50t) 🛟(15t) 🏪. **Town** 🍴🏧ⓑ📮🗑 at Strathlene caravan site 1½M E, ⇌ (bus to Elgin), ✈ (Aberdeen or Inverness).

ADJACENT HARBOUR (2M ENE OF BUCKIE)
FINDOCHTY, Moray, **57°41'·94N 02°54'·29W.** AC 222. HW +0045 on Dover, HW –0135 & ht –0·2m on Aberdeen; ML 2·3m; Duration 0550. Ent is about 2ca W of conspic church belfry. 1ca N of ent, leave Beacon Rock (3m high) to stbd. Ldg lts, FR, lead 166° into Outer Basin which dries 0·2m and has many rky outcrops. Ent faces N and is 20m wide; unlit white bn at hd of W pier. Good shelter in inner basin for 100 small craft/yachts on 3 pontoons (the 2 W'ly pontoons dry); possible ⛴ on Sterlochy Pier in 2m; ⚓ (£10) *for details of Rover berthing ticket see 9.7.5.* HM mob 07900 920445. **Town** 📮ⓑ🍴✕🏪.

9.7.11 LOSSIEMOUTH

Moray **57°43'·41N 03°16'·63W** ❀❀💧💧💧🌸🌸🌸

CHARTS AC 223, 1462, 5617; Imray C23

TIDES +0040 Dover; ML 2·3; Duration 0605
Standard Port ABERDEEN (←—)

Times				Height (metres)			
High Water		Low Water		MHWS	MHWN	MLWN	MLWS
0200	0900	0400	0900	4·3	3·4	1·6	0·6
1400	2100	1600	2100				
Differences LOSSIEMOUTH							
–0125	–0200	–0130	–0130	–0·2	–0·2	0·0	0·0

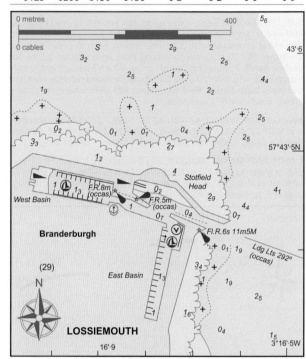

SHELTER Very good in winds from SSE to NW. Pontoon berths in both basins, dredged 2m. West Basin also used by FVs.

- In N to SE winds >F6 appr can be dangerous, with swell tending to break across hbr mouth and in outer hbr. Residual swell from E to SE can be particularly hazardous.
- Near ent, beware current from R Lossie setting in N'ly direction, causing confused water in N to SE winds at sp.
- Vessels drawing >2m would have little clearance at LWS ±2.

NAVIGATION WPT 57°43'·38N 03°16'·09W, 277°/0·3M to ent. Rks to N and S of hbr ent; appr from E. Approx 0·4m in entrance at LWS.

LIGHTS AND MARKS Halliman Skerries (drying) 1NM W of harbour entrance are marked by NCM, Q; *Racon M*, unlit bn and disused Lt Ho on adjacent mainland shore. Ldg lts 292°, both FR 5/8m; S pier hd Fl R 6s 11m 5M.

COMMUNICATIONS (Code 01343) CGOC (01224) 592334; Police 101; Dr 812277. VHF Ch 12 16 HO.

FACILITIES **Lossiemouth Marina** ☎813066 (95 inc ⓥ) ⛴£20 for first 24hrs and £15 thereafter/craft 🔧🛢🗑 wi-fi.

Hbr ⚓(£50 pa) 🔧🔧🛒🚿🏳(25t) 🛟(1t).
Hbr Service Stn ☎813001 📷📷🛢 Mon-Fri 0800-2030, Sat 0800-1930, Sun 0930-1900.
Town 📮ⓑⓒ🍴✕🏪⇌(bus to Elgin) ✈(Inverness).

9.7.12 HOPEMAN

Moray **57°42'·70N 03°26'·31W** ❀❀♠♦♦✿✿

CHARTS AC 223, 1462, 5617; Imray C23

TIDES +0050 Dover; ML 2·4; Duration 0610

Standard Port ABERDEEN (←)

Times				Height (metres)			
High Water		Low Water		MHWS	MHWN	MLWN	MLWS
0200	0900	0400	0900	4·3	3·4	1·6	0·6
1400	2100	1600	2100				
Differences HOPEMAN							
–0120	–0150	–0135	–0120	–0·2	–0·2	0·0	0·0

SHELTER Once in Inner Basin, shelter good from all winds; but hbr dries. Ent is difficult in winds from NE to SE. A popular yachting hbr with �.

NAVIGATION WPT 57°42'·66N, 03°26'·59W, 083° to ent, 0·17M.

Dangerous rks lie off hbr ent. Do not attempt entry in heavy weather. Beware lobster pot floats E and W of hbr (Mar-Aug).

LIGHTS AND MARKS See 9.7.4 and chartlet.

COMMUNICATIONS (Code 01343) CGOC (01224) 592334; Police 101; Dr 543141. HM 835337 (part-time).

Burghead Radio Ch 14 (HX).

FACILITIES Hbr ⌣£9, *for details of Rover Ticket see 9.7.5,* ▬ ⚓ ⚒ ✎ ⚙.

Town ✉ Ⓑ 🛒 ✕ 🗔 ⇌ (bus to Elgin), ✈ (Inverness).

9.7.13 BURGHEAD

Moray **57°42'·06N 03°30'·02W** ❀❀♠♦♦✿✿

CHARTS AC 223, 1462, 5617; Imray C23

TIDES +0035 Dover; ML 2·4; Duration 0610

Standard Port ABERDEEN (←)

Times				Height (metres)			
High Water		Low Water		MHWS	MHWN	MLWN	MLWS
0200	0900	0400	0900	4·3	3·4	1·6	0·6
1400	2100	1600	2100				
Differences BURGHEAD							
–0120	–0150	–0135	–0120	–0·2	–0·2	0·0	0·0

SHELTER One of the few Moray Firth hbrs accessible in strong E winds. Go alongside where available and contact HM. Can be very busy with FVs.

NAVIGATION WPT 57°42'·28N 03°30'·39W, 137°/0·28M to N pier lt QR. 0·6m in ent chan and 1·4m in hbr, but depths variable due to sand movement. Advisable to contact HM if entering near LW. 3kn speed limit.

LIGHTS AND MARKS No ldg lts but night ent is safe after identifying the N pier lts: QR 3m 5M and Oc 8s 7m 5M.

COMMUNICATIONS (Code 01343) CGOC (01224) 592334; Police 101; Dr 812277. HM 835337.

Burghead Radio VHF Ch **14** 12 (HO and when vessel due).

FACILITIES Hbr ⌣£9, *for details of Rover Ticket see 9.7.5* ⚓ ▬ ⚒ ✎ ⛽ ⚙(50t) mobile.

Town ✉ Ⓑ 🛒 🗔 ⇌(bus to Elgin), ✈(Inverness).

HOPEMAN

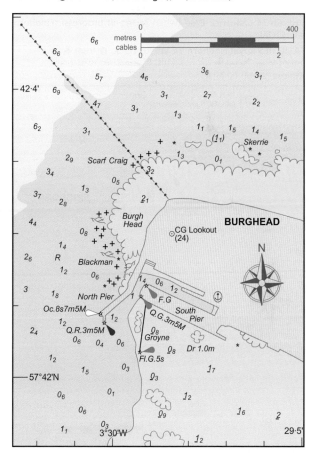

BURGHEAD

9.7.14 FINDHORN

Moray **57°39'·64N 03°37'·47W** ✿✿✿✿✿✿✿✿

CHARTS AC 223, 5617; Imray C23

TIDES +0110 Dover; ML 2·5; Duration 0615

Standard Port ABERDEEN (◄—)

Times				Height (metres)			
High Water		Low Water		MHWS	MHWN	MLWN	MLWS
0200	0900	0400	0900	4·3	3·4	1·6	0·6
1400	2100	1600	2100				
Differences FINDHORN							
–0120	–0150	–0135	–0130	0·0	–0·1	0·0	+0·1

SHELTER ⚓ in pool off boatyard or off N pier or dry out alongside, inside piers and ask at YC; or pick up Y ⚓ off N pier.

Do not attempt entry in strong NW/NE winds or with big swell running; expect breakers/surf either side of entrance.

NAVIGATION From the SWM Landfall Buoy the bar, which dries, is marked (Fl R marker); thence follow marked chan to entrance.

- Bar and sands shift frequently esp following bad weather.
- Buoys (inc SWM) often move to follow channel into the bay.
- Marks may be lifted Nov to early Apr.

Once past The Ee, turn port inside G buoys. The S part of Findhorn Bay dries extensively.

LIGHTS AND MARKS Unlit. There is a windsock on FS by The Ee. Boatyard building is conspic.

COMMUNICATIONS (Code 01309) CGOC (01224) 592334; Police 101; Dr 678866/678888; Findhorn Pilot (Derek Munro) 690802, mob 07747 840916; Findhorn BY 690099. Fairways Committee (via BY) 690099.

VHF Ch M *Chadwick Base* (when racing in progress). Findhorn BY: Ch 80.

FACILITIES Royal Findhorn YC ☎690247, ⚓ ⚓(free) ⚓ ☐.
Services ⚓ ⚓ ⚓ ⚓ ⚓(16t) ✗ ✎ ⚓ ⚓ ACA.
Kinloss (2½M) ⚓ ⚓ ⚓.

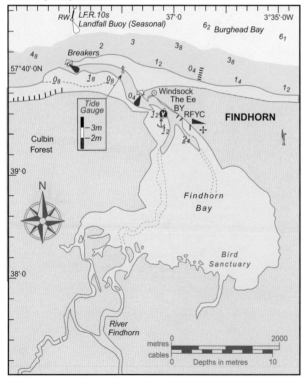

9.7.15 NAIRN

Highland **57°35'·61N 03°51'·65W** ✿✿✿✿✿✿✿

CHARTS AC 223, 1462, 5617; Imray C23

TIDES +0110 Dover; ML 2·2; Duration 0615

Standard Port INVERGORDON (—►)

Times				Height (metres)			
High Water		Low Water		MHWS	MHWN	MLWN	MLWS
0100	0700	0000	0700	4·3	3·3	1·6	0·7
1300	1900	1200	1900				
Differences NAIRN							
+0005	–0015	0000	–0015	0·0	0·0	0·0	0·0
McDERMOTT BASE							
+0015	–0005	+0015	0000	–0·1	0·0	+0·1	+0·2

SHELTER Good, but entry difficult in fresh NNE'ly. Pontoons in hbr with ⓥ berths. Best entry HW ± 1½. No commercial shipping.

NAVIGATION The beach dries to 100m off the pierheads. A mark is laid in the deepest water off the entrance to indicate line of approach. A gravel bar shows 200m NNW of ent at LAT. Inside, the best water is to the E side of the river chan which dries approx 1·9m.

- Buoy and sands shift frequently esp following bad weather.
- Mark may be lifted Nov to early Apr.

LIGHTS AND MARKS Lt ho on E pier hd, Oc WRG 4s 6m 5M, vis shore-G-100°-W-207°-R-shore. Keep in W sector.

COMMUNICATIONS (Code 01667) CGOC (01224) 592334; Police 101; Clinic 452096; Dr 453421. Hbr Office 456008, HM mob 07821 302944.

VHF Ch M (weekends only).

FACILITIES Nairn Basin ⚓ £15.28/48hrs ⚓ (launch £6) ⚓ (110 V) ⚓ ⚓; Nairn SC ☎453897, ☐.
Town ✉ ⑧ ⚓ ✗ ☐ ⚓ ✈ (Inverness).

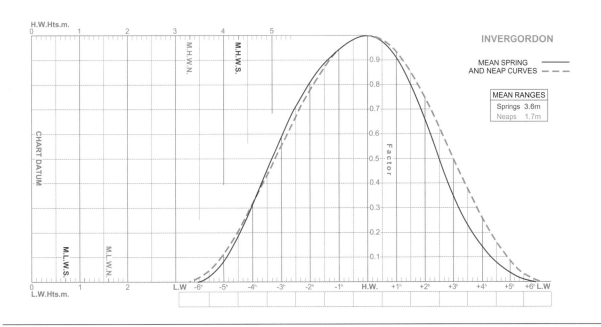

INVERGORDON

MEAN SPRING AND NEAP CURVES

MEAN RANGES	
Springs	3.6m
Neaps	1.7m

STANDARD TIME (UT)
For Summer Time add ONE hour in **non-shaded areas**

INVERGORDON LAT 57°41′N LONG 4°10′W
TIMES AND HEIGHTS OF HIGH AND LOW WATERS

Dates in **red** are **SPRINGS**
Dates in **blue** are **NEAPS**

YEAR 2016

JANUARY

Time	m		Time	m
1 0427	3.6		**16** 0422	4.0
0908	1.7		1009	1.3
F 1646	3.7		SA 1632	4.1
2233	1.6		☽ 2258	1.1
2 0513	3.5		**17** 0512	3.8
0959	1.9		1103	1.5
SA 1736	3.5		SU 1726	3.9
☾ 2349	1.7		2359	1.3
3 0605	3.4		**18** 0610	3.7
1107	2.0		1214	1.6
SU 1837	3.4		M 1830	3.8
4 0108	1.7		**19** 0136	1.4
0703	3.4		0717	3.6
M 1350	2.0		TU 1418	1.5
1938	3.4		1944	3.7
5 0206	1.7		**20** 0301	1.4
0802	3.5		0836	3.7
TU 1455	1.8		W 1532	1.3
2035	3.5		2108	3.8
6 0252	1.5		**21** 0405	1.3
0858	3.7		0952	3.9
W 1543	1.6		TH 1633	1.1
2128	3.7		2224	4.0
7 0337	1.4		**22** 0458	1.2
0950	3.9		1052	4.1
TH 1628	1.4		F 1724	0.9
2219	3.9		2322	4.1
8 0426	1.3		**23** 0541	1.1
1039	4.1		1142	4.3
F 1712	1.2		SA 1807	0.7
2307	4.0			
9 0516	1.1		**24** 0008	4.2
1125	4.2		0615	1.1
SA 1755	1.0		SU 1226	4.4
2352	4.2		○ 1845	0.6
10 0602	1.0		**25** 0050	4.2
1208	4.4		0643	1.1
SU 1837	0.8		M 1305	4.5
●			1920	0.6
11 0036	4.3		**26** 0127	4.2
0645	1.0		0710	1.1
M 1251	4.4		TU 1341	4.4
1918	0.7		1954	0.7
12 0120	4.3		**27** 0202	4.1
0727	0.9		0737	1.1
TU 1333	4.5		W 1417	4.3
2000	0.7		2026	0.8
13 0204	4.3		**28** 0236	4.0
0807	1.0		0759	1.2
W 1516	4.5		TH 1452	4.2
2042	0.7		2057	1.0
14 0249	4.2		**29** 0310	3.8
0847	1.0		0808	1.3
TH 1459	4.4		F 1527	4.0
2125	0.8		2114	1.2
15 0335	4.1		**30** 0343	3.7
0926	1.2		0830	1.4
F 1544	4.3		SA 1557	3.8
2209	0.9		2121	1.4
			31 0415	3.6
			0905	1.6
			SU 1626	3.6
			2205	1.6

FEBRUARY

Time	m		Time	m
1 0455	3.4		**16** 0543	3.6
0955	1.8		1153	1.5
M 1712	3.4		TU 1812	3.7
☾ 2321	1.8			
2 0553	3.3		**17** 0118	1.6
1120	2.0		0646	3.5
TU 1837	3.3		W 1403	1.5
			1929	3.5
3 0125	1.8		**18** 0253	1.6
0706	3.3		0802	3.6
W 1411	1.9		TH 1522	1.3
1954	3.3		2103	3.6
4 0224	1.7		**19** 0359	1.4
0812	3.5		0931	3.7
TH 1509	1.7		F 1624	1.1
2054	3.5		2221	3.8
5 0314	1.5		**20** 0452	1.3
0910	3.7		1038	4.0
F 1602	1.4		SA 1714	0.9
2150	3.7		2315	4.0
6 0405	1.3		**21** 0532	1.2
1005	3.9		1128	4.2
SA 1652	1.1		SU 1755	0.7
2244	3.9		2357	4.1
7 0458	1.1		**22** 0601	1.1
1058	4.2		1209	4.3
SU 1738	0.9		M 1827	0.6
2335	4.2		○	
8 0547	0.9		**23** 0033	4.1
1147	4.4		0619	1.0
M 1822	0.6		TU 1245	4.4
●			1855	0.6
9 0021	4.3		**24** 0105	4.1
0630	0.8		0643	1.0
TU 1234	4.6		W 1318	4.4
1903	0.4		1923	0.6
10 0106	4.4		**25** 0135	4.1
0712	0.7		0713	0.9
W 1318	4.7		TH 1350	4.3
1945	0.3		1953	0.7
11 0150	4.5		**26** 0204	4.0
0753	0.6		0743	1.0
TH 1402	4.7		F 1422	4.2
2026	0.4		2022	0.9
12 0234	4.4		**27** 0233	3.9
0833	0.7		0757	1.1
F 1446	4.6		SA 1451	4.0
2108	0.5		2039	1.1
13 0318	4.3		**28** 0300	3.8
0912	0.9		0757	1.2
SA 1531	4.4		SU 1516	3.8
2149	0.8		2040	1.3
14 0402	4.1		**29** 0330	3.6
0953	1.1		0821	1.4
SU 1617	4.2		M 1545	3.6
2233	1.1		2105	1.5
15 0449	3.9			
1043	1.3			
M 1709	3.9			
☽ 2329	1.4			

MARCH

Time	m		Time	m
1 0408	3.5		**16** 0521	3.6
0856	1.6		1146	1.5
TU 1627	3.4		W 1758	3.6
☾ 2156	1.7			
2 0458	3.4		**17** 0114	1.7
0945	1.8		0621	3.5
W 1725	3.2		TH 1351	1.5
			1914	3.4
3 0029	1.8		**18** 0242	1.7
0603	3.3		0732	3.5
TH 1329	1.8		F 1507	1.3
1856	3.2		2056	3.5
4 0158	1.7		**19** 0346	1.5
0720	3.3		0902	3.6
F 1435	1.6		SA 1608	1.1
2018	3.4		2209	3.7
5 0253	1.6		**20** 0437	1.4
0828	3.5		1017	3.8
SA 1532	1.3		SU 1657	0.9
2122	3.6		2259	3.9
6 0345	1.3		**21** 0517	1.3
0928	3.8		1107	4.0
SU 1626	1.0		M 1735	0.8
2221	3.9		2338	4.0
7 0439	1.1		**22** 0542	1.2
1027	4.1		1147	4.2
M 1716	0.7		TU 1803	0.7
2315	4.2			
8 0528	0.8		**23** 0011	4.1
1123	4.4		0553	1.1
TU 1801	0.4		W 1241	4.3
			○ 1825	0.7
9 0004	4.4		**24** 0039	4.1
0612	0.6		0616	0.9
W 1213	4.6		TH 1251	4.3
● 1843	0.2		1851	0.7
10 0049	4.6		**25** 0106	4.1
0654	0.4		0650	0.9
TH 1300	4.8		F 1323	4.2
1925	0.1		1921	0.7
11 0132	4.6		**26** 0134	4.1
0735	0.4		0723	0.9
F 1346	4.8		SA 1353	4.1
2006	0.2		1951	0.8
12 0215	4.6		**27** 0201	4.0
0816	0.5		0747	1.0
SA 1430	4.7		SU 1422	4.0
2047	0.4		2011	1.0
13 0258	4.4		**28** 0229	3.9
0857	0.7		0733	1.1
SU 1516	4.5		M 1449	3.8
2128	0.8		2010	1.2
14 0342	4.2		**29** 0300	3.7
0940	0.9		0752	1.2
M 1603	4.2		TU 1521	3.6
2211	1.2		2020	1.4
15 0429	3.9		**30** 0338	3.6
1031	1.2		0825	1.4
TU 1656	3.9		W 1603	3.4
☽ 2303	1.5		2052	1.6
			31 0426	3.4
			0909	1.6
			TH 1658	3.3
			☾ 2306	1.8

APRIL

Time	m		Time	m
1 0526	3.3		**16** 0219	1.8
1158	1.7		0704	3.5
F 1813	3.2		SA 1443	1.2
			2025	3.4
2 0123	1.8		**17** 0320	1.6
0638	3.4		0816	3.5
SA 1400	1.5		SU 1541	1.1
1941	3.3		2141	3.6
3 0228	1.6		**18** 0412	1.5
0749	3.5		0937	3.7
SU 1459	1.2		M 1630	1.0
2052	3.6		2231	3.8
4 0322	1.3		**19** 0453	1.4
0853	3.8		1035	3.9
M 1557	0.9		TU 1708	0.9
2156	3.9		2309	3.9
5 0416	1.1		**20** 0519	1.2
0955	4.1		1116	4.0
TU 1650	0.6		W 1733	0.9
2253	4.2		2341	4.0
6 0506	0.8		**21** 0527	1.1
1057	4.4		1150	4.1
W 1738	0.3		TH 1752	0.8
2343	4.5			
7 0552	0.5		**22** 0009	4.1
1151	4.7		0552	1.0
TH 1822	0.1		F 1223	4.2
●			○ 1819	0.8
8 0028	4.6		**23** 0037	4.1
0635	0.3		0628	0.9
F 1241	4.8		SA 1255	4.1
1904	0.1		1852	0.8
9 0113	4.7		**24** 0106	4.1
0718	0.3		0703	0.9
SA 1328	4.8		SU 1327	4.1
1946	0.2		1923	0.9
10 0156	4.6		**25** 0136	4.0
0800	0.4		0732	0.9
SU 1414	4.7		M 1357	3.9
2027	0.5		1949	1.0
11 0239	4.5		**26** 0206	3.9
0843	0.6		0731	1.1
M 1502	4.5		TU 1429	3.8
2108	0.9		1956	1.2
12 0324	4.2		**27** 0239	3.8
0930	0.8		0736	1.1
TU 1551	4.2		W 1505	3.7
2149	1.3		1958	1.4
13 0410	3.9		**28** 0318	3.7
1024	1.1		0809	1.2
W 1644	3.8		TH 1548	3.5
2237	1.6		2033	1.5
14 0502	3.7		**29** 0404	3.6
1152	1.3		0855	1.4
TH 1743	3.6		F 1641	3.4
☽			2144	1.7
15 0057	1.8		**30** 0501	3.5
0559	3.5		1126	1.5
F 1332	1.4		SA 1748	3.3
1853	3.4		☽	

Chart Datum: 2·10 metres below Ordnance Datum (Local). HAT is 5·0 metres above Chart Datum.

FREE monthly updates. Register at
www.reedsnauticalalmanac.co.uk

STANDARD TIME (UT)
For Summer Time add ONE hour in **non-shaded areas**

INVERGORDON LAT 57°41'N LONG 4°10'W
TIMES AND HEIGHTS OF HIGH AND LOW WATERS

Dates in red are **SPRINGS**
Dates in blue are **NEAPS**

YEAR 2016

NE Scotland

MAY		JUNE		JULY		AUGUST	
Time m	Time m	Time m	Time m	Time m	Time m	Time m	Time m
1 0013 1.8 0606 3.5 SU 1319 1.4 1906 3.4	**16** 0245 1.8 0735 3.5 M 1505 1.2 2041 3.5	**1** 0218 1.4 0750 3.8 W 1459 0.9 2057 3.8	**16** 0336 1.7 0852 3.6 TH 1542 1.4 2127 3.7	**1** 0303 1.2 0837 3.9 F 1541 1.0 2137 4.0	**16** 0321 1.6 0912 3.6 SA 1526 1.4 2135 3.7	**1** 0455 0.8 1055 4.1 M 1721 1.0 2318 4.3	**16** 0425 1.2 1030 3.9 TU 1637 1.2 2239 4.1
2 0151 1.6 0714 3.6 M 1427 1.1 2020 3.6	**17** 0337 1.6 0836 3.6 TU 1554 1.2 2141 3.6	**2** 0325 1.1 0858 4.0 TH 1600 0.8 2202 4.1	**17** 0407 1.5 0947 3.7 F 1605 1.3 2216 3.8	**2** 0407 1.0 0953 4.0 SA 1640 0.9 2239 4.2	**17** 0405 1.4 1006 3.7 SU 1610 1.3 2225 3.9	**2** 0544 0.6 1149 4.3 TU 1802 1.0 ●	**17** 0513 1.0 1120 4.1 W 1725 1.1 2327 4.3
3 0254 1.3 0820 3.8 TU 1527 0.9 2128 3.9	**18** 0420 1.5 0938 3.7 W 1632 1.2 2225 3.8	**3** 0424 0.9 1008 4.2 F 1655 0.6 2259 4.3	**18** 0434 1.3 1039 3.8 SA 1640 1.2 2301 4.0	**3** 0504 0.8 1101 4.2 SU 1731 0.8 2332 4.4	**18** 0450 1.2 1057 3.9 M 1657 1.2 2311 4.1	**3** 0006 4.5 0627 0.5 W 1236 4.3 1836 0.9	**18** 0557 0.7 1206 4.3 TH 1809 0.9 ○
4 0351 1.1 0925 4.1 W 1624 0.6 2228 4.2	**19** 0448 1.4 1031 3.8 TH 1656 1.1 2302 3.9	**4** 0516 0.6 1112 4.4 SA 1744 0.5 2349 4.5	**19** 0511 1.2 1124 3.9 SU 1721 1.1 2341 4.1	**4** 0554 0.6 1157 4.4 M 1815 0.8 ●	**19** 0534 1.0 1143 4.0 TU 1744 1.1 ○ 2354 4.2	**4** 0049 4.5 0705 0.5 TH 1317 4.3 1906 1.0	**19** 0012 4.5 0640 0.5 F 1249 4.4 1850 0.8
5 0444 0.8 1031 4.3 TH 1716 0.4 2321 4.4	**20** 0501 1.2 1114 4.0 F 1716 1.0 2336 4.0	**5** 0605 0.5 1208 4.5 SU 1829 0.5 ●	**20** 0552 1.0 1205 4.0 M 1803 1.0 ○	**5** 0020 4.5 0640 0.6 TU 1247 4.4 1855 0.8	**20** 0617 0.8 1226 4.1 W 1827 1.0	**5** 0129 4.5 0741 0.5 F 1356 4.2 1934 1.0	**20** 0056 4.6 0721 0.4 SA 1331 4.5 1931 0.7
6 0533 0.5 1130 4.6 F 1802 0.3 ●	**21** 0531 1.0 1153 4.0 SA 1749 0.9 ○	**6** 0036 4.6 0651 0.4 M 1258 4.6 1911 0.6	**21** 0019 4.1 0631 0.9 TU 1245 4.0 1842 1.0	**6** 0106 4.5 0724 0.4 W 1334 4.4 1932 0.9	**21** 0035 4.3 0658 0.7 TH 1307 4.2 1907 0.9	**6** 0207 4.4 0816 0.6 SA 1432 4.1 2000 1.1	**21** 0139 4.7 0803 0.4 SU 1413 4.4 2011 0.7
7 0008 4.6 0618 0.4 SA 1223 4.7 1845 0.3	**22** 0009 4.1 0608 0.9 SU 1230 4.0 1825 0.9	**7** 0121 4.6 0736 0.4 TU 1347 4.5 1952 0.8	**22** 0056 4.2 0711 0.9 W 1323 4.0 1920 1.0	**7** 0149 4.5 0806 0.5 TH 1418 4.3 2005 1.1	**22** 0116 4.4 0740 0.6 F 1349 4.2 1947 0.9	**7** 0244 4.3 0850 0.8 SU 1507 4.0 2022 1.2	**22** 0222 4.6 0844 0.5 M 1455 4.3 2050 0.8
8 0054 4.7 0703 0.3 SU 1312 4.7 1927 0.4	**23** 0042 4.1 0646 0.9 M 1304 4.0 1901 0.9	**8** 0206 4.5 0823 0.5 W 1435 4.3 2030 1.1	**23** 0133 4.2 0750 0.8 TH 1404 4.0 1956 1.1	**8** 0231 4.4 0848 0.6 F 1500 4.1 2032 1.2	**23** 0157 4.4 0821 0.6 SA 1431 4.2 2026 1.0	**8** 0321 4.1 0919 1.1 M 1542 3.8 2036 1.4	**23** 0306 4.5 0926 0.7 TU 1538 4.2 2132 1.0
9 0138 4.6 0747 0.4 M 1400 4.6 2009 0.7	**24** 0115 4.1 0720 0.9 TU 1338 4.0 1933 1.0	**9** 0251 4.3 0911 0.7 TH 1522 4.1 2103 1.3	**24** 0212 4.1 0829 0.9 F 1444 3.9 2030 1.2	**9** 0313 4.2 0930 0.9 SA 1542 3.9 2046 1.4	**24** 0239 4.4 0902 0.7 SU 1514 4.1 2104 1.0	**9** 0358 3.9 0920 1.3 TU 1616 3.6 2105 1.6	**24** 0351 4.3 1009 1.0 W 1624 4.0 2221 1.3
10 0222 4.5 0833 0.5 TU 1448 4.4 2049 1.0	**25** 0148 4.0 0749 1.0 W 1414 3.9 1957 1.2	**10** 0336 4.1 1003 0.9 F 1609 3.9 2116 1.5	**25** 0253 4.1 0910 0.9 SA 1528 3.9 2105 1.3	**10** 0355 4.0 1009 1.1 SU 1623 3.7 2110 1.5	**25** 0322 4.3 0944 0.8 M 1558 4.0 2145 1.2	**10** 0435 3.6 0943 1.5 W 1653 3.5 ● 2152 1.8	**25** 0441 4.0 1101 1.3 TH 1716 3.8 ◑ 2324 1.5
11 0307 4.3 0923 0.8 W 1538 4.1 2128 1.3	**26** 0224 3.9 0750 1.1 TH 1454 3.8 1958 1.3	**11** 0422 3.9 1059 1.1 SA 1656 3.7 2149 1.7	**26** 0336 4.0 0954 1.0 SU 1615 3.8 2155 1.4	**11** 0438 3.8 1049 1.3 M 1705 3.6 2151 1.7	**26** 0407 4.1 1029 1.0 TU 1645 3.8 ◑ 2235 1.3	**11** 0521 3.4 1042 1.7 TH 1746 3.4 2308 1.9	**26** 0540 3.8 1220 1.5 F 1818 3.6
12 0354 4.0 1020 1.0 TH 1629 3.8 2202 1.6	**27** 0304 3.8 0807 1.1 F 1538 3.7 2030 1.4	**12** 0510 3.7 1207 1.3 SU 1745 3.5 ◑ 2246 1.9	**27** 0424 3.9 1048 1.1 M 1706 3.7 ◑ 2256 1.5	**12** 0525 3.6 1141 1.5 TU 1753 3.4 ◑ 2250 1.8	**27** 0457 3.9 1125 1.2 W 1738 3.7 2339 1.5	**12** 0632 3.3 1311 1.8 F 1856 3.3	**27** 0108 1.5 0653 3.6 SA 1413 1.6 1935 3.6
13 0443 3.8 1137 1.2 F 1723 3.6 ◑ 2254 1.9	**28** 0349 3.7 0903 1.2 SA 1628 3.6 2202 1.6	**13** 0602 3.6 1318 1.4 M 1839 3.4	**28** 0517 3.8 1153 1.1 TU 1805 3.6	**13** 0619 3.5 1255 1.6 W 1848 3.4	**28** 0556 3.8 1242 1.3 TH 1843 3.6	**13** 0148 1.9 0740 3.3 SA 1414 1.7 2000 3.4	**28** 0242 1.4 0821 3.6 SU 1525 1.5 2059 3.8
14 0536 3.6 1300 1.3 SA 1822 3.4	**29** 0441 3.7 1105 1.3 SU 1727 3.5 ◑ 2327 1.6	**14** 0010 1.5 0658 3.5 TU 1418 1.4 1935 3.4	**29** 0004 1.5 0617 3.8 W 1315 1.2 1912 3.6	**14** 0054 1.9 0718 3.4 TH 1359 1.6 1946 3.4	**29** 0111 1.5 0706 3.7 F 1418 1.3 1958 3.7	**14** 0247 1.7 0840 3.4 SU 1502 1.6 2056 3.6	**29** 0351 1.1 0947 3.8 M 1624 1.3 2210 4.0
15 0143 1.9 0634 3.5 SU 1408 1.3 1927 3.4	**30** 0540 3.6 1230 1.2 M 1835 3.5	**15** 0251 1.8 0756 3.5 W 1506 1.4 2032 3.5	**30** 0134 1.4 0725 3.8 TH 1434 1.1 2026 3.8	**15** 0233 1.8 0817 3.4 F 1445 1.5 2042 3.5	**30** 0248 1.3 0825 3.7 SA 1530 1.2 2116 3.8	**15** 0337 1.5 0936 3.6 M 1548 1.4 2149 3.9	**30** 0447 0.9 1050 4.1 TU 1711 1.2 2305 4.3
	31 0046 1.6 0644 3.7 TU 1352 1.1 1947 3.6			**31** 0357 1.1 0947 3.9 SU 1631 1.1 2223 4.1			**31** 0532 0.7 1138 4.2 W 1748 1.1 2349 4.4

Chart Datum: 2·10 metres below Ordnance Datum (Local). HAT is 5·0 metres above Chart Datum.

›› FREE monthly updates. Register at ‹‹
www.reedsnauticalalmanac.co.uk

405

STANDARD TIME (UT)
For Summer Time add ONE hour in **non-shaded areas**

INVERGORDON   LAT 57°41'N   LONG 4°10'W
TIMES AND HEIGHTS OF HIGH AND LOW WATERS

Dates in red are **SPRINGS**
Dates in blue are NEAPS

YEAR 2016

SEPTEMBER

Time	m		Time	m
1 0609	0.6	**16** 0534	0.6
1219	4.3	1143	4.4
TH 1814	1.1	F 1749	0.8
●		○ 2347	4.6
2 0028	4.5	**17** 0618	0.4
0640	0.6	1227	4.5
F 1256	4.3	SA 1831	0.6
1835	1.0		
3 0104	4.5	**18** 0033	4.8
0710	0.6	0700	0.2
SA 1328	4.3	SU 1310	4.7
1902	1.0	1912	0.6
4 0137	4.4	**19** 0119	4.9
0739	0.7	0741	0.3
SU 1358	4.2	M 1352	4.7
1932	1.0	1953	0.6
5 0210	4.3	**20** 0203	4.8
0810	0.8	0823	0.4
M 1428	4.0	TU 1435	4.5
1958	1.2	2035	0.7
6 0242	4.1	**21** 0249	4.6
0834	1.1	0905	0.7
TU 1455	3.9	W 1518	4.3
2006	1.3	2118	0.9
7 0311	3.9	**22** 0336	4.4
0832	1.3	0948	1.1
W 1518	3.7	TH 1604	4.1
2024	1.5	2209	1.2
8 0337	3.7	**23** 0428	4.1
0853	1.5	1040	1.5
TH 1550	3.6	F 1656	3.8
2101	1.7	◑ 2316	1.5
9 0414	3.5	**24** 0528	3.8
0938	1.7	1206	1.8
F 1636	3.4	SA 1757	3.7
◑ 2209	1.9		
10 0508	3.3	**25** 0110	1.5
1122	1.9	0642	3.6
SA 1740	3.4	SU 1407	1.8
		1914	3.6
11 0110	1.9	**26** 0234	1.3
0649	3.2	0818	3.6
SU 1347	1.9	M 1515	1.6
1906	3.4	2042	3.8
12 0216	1.7	**27** 0338	1.1
0805	3.4	0940	3.8
M 1441	1.7	TU 1611	1.5
2014	3.6	2153	4.0
13 0307	1.5	**28** 0431	0.9
0906	3.6	1036	4.0
TU 1529	1.5	W 1657	1.3
2111	3.8	2246	4.2
14 0356	1.2	**29** 0514	0.8
1003	3.9	1120	4.2
W 1616	1.3	TH 1731	1.2
2204	4.1	2329	4.4
15 0446	0.9	**30** 0547	0.8
1056	4.2	1157	4.3
TH 1704	1.0	F 1751	1.2
2257	4.4		

OCTOBER

Time	m		Time	m
1 0004	4.4	**16** 0555	0.4
0610	0.7	1204	4.7
SA 1229	4.3	SU 1812	0.6
● 1805	1.1	○	
2 0036	4.4	**17** 0011	4.9
0634	0.7	0638	0.3
SU 1257	4.3	M 1248	4.8
1834	1.0	1854	0.5
3 0105	4.4	**18** 0059	4.9
0703	0.8	0721	0.3
M 1323	4.2	TU 1331	4.8
1907	1.0	1937	0.5
4 0136	4.3	**19** 0145	4.8
0733	0.9	0803	0.6
TU 1349	4.1	W 1415	4.6
1935	1.1	2021	0.7
5 0205	4.1	**20** 0233	4.6
0756	1.1	0845	0.9
W 1413	4.0	TH 1459	4.4
1942	1.3	2107	0.9
6 0233	3.9	**21** 0322	4.4
0800	1.3	0928	1.3
TH 1440	3.9	F 1546	4.2
1953	1.4	2201	1.2
7 0305	3.7	**22** 0416	4.1
0814	1.5	1017	1.6
F 1516	3.7	SA 1638	3.9
2025	1.6	◐ 2314	1.4
8 0345	3.5	**23** 0515	3.8
0846	1.7	1152	1.9
SA 1601	3.6	SU 1738	3.7
2119	1.8		
9 0437	3.4	**24** 0058	1.5
1011	2.0	0625	3.6
SU 1659	3.5	M 1348	1.9
◑ 2343	1.9	1849	3.7
10 0548	3.3	**25** 0215	1.3
1312	2.0	0756	3.6
M 1812	3.4	TU 1454	1.8
		2011	3.7
11 0140	1.7	**26** 0316	1.2
0722	3.4	0915	3.8
TU 1416	1.8	W 1549	1.6
1928	3.6	2124	3.9
12 0235	1.4	**27** 0407	1.1
0832	3.6	1009	4.0
W 1506	1.6	TH 1635	1.5
2032	3.8	2218	4.1
13 0325	1.1	**28** 0449	1.0
0932	3.9	1052	4.1
TH 1554	1.3	F 1709	1.4
2130	4.1	2301	4.2
14 0418	0.8	**29** 0519	1.0
1028	4.3	1128	4.2
F 1642	1.0	SA 1728	1.3
2227	4.4	2335	4.3
15 0508	0.6	**30** 0537	1.0
1118	4.5	1157	4.3
SA 1728	0.8	SU 1740	1.1
2321	4.7	●	
		31 0005	4.3
		0558	0.9
		M 1224	4.3
		1810	1.1

NOVEMBER

Time	m		Time	m
1 0035	4.3	**16** 0041	4.8
0630	0.9	0702	0.5
TU 1251	4.3	W 1312	4.8
1845	1.0	1925	0.5
2 0106	4.2	**17** 0130	4.8
0702	1.0	0745	0.7
W 1318	4.2	TH 1357	4.7
1916	1.1	2011	0.6
3 0137	4.1	**18** 0219	4.6
0728	1.1	0828	1.0
TH 1345	4.1	F 1443	4.5
1933	1.2	2100	0.8
4 0208	3.9	**19** 0310	4.3
0740	1.3	0910	1.3
F 1415	4.0	SA 1530	4.3
1937	1.3	2154	1.1
5 0244	3.8	**20** 0402	4.1
0749	1.5	0951	1.7
SA 1452	3.8	SU 1621	4.0
2006	1.5	2302	1.3
6 0326	3.6	**21** 0457	3.8
0818	1.7	1039	1.9
SU 1536	3.7	M 1716	3.8
2056	1.6	◐	
7 0416	3.5	**22** 0028	1.4
0910	1.9	0556	3.6
M 1630	3.6	TU 1311	2.0
◐ 2258	1.7	1816	3.7
8 0518	3.4	**23** 0143	1.4
1130	2.0	0703	3.6
TU 1735	3.5	W 1421	1.9
		1923	3.7
9 0048	1.6	**24** 0244	1.4
0634	3.5	0821	3.6
W 1334	1.9	TH 1518	1.8
1845	3.6	2032	3.8
10 0157	1.4	**25** 0335	1.3
0751	3.6	0925	3.8
TH 1435	1.6	F 1606	1.6
1954	3.8	2134	3.9
11 0253	1.1	**26** 0418	1.3
0858	3.9	1012	3.9
F 1528	1.4	SA 1644	1.5
2056	4.1	2222	4.0
12 0349	0.9	**27** 0447	1.3
0958	4.2	1049	4.1
SA 1619	1.1	SU 1707	1.4
2157	4.4	2300	4.1
13 0444	0.7	**28** 0456	1.2
1052	4.5	1121	4.2
SU 1709	0.8	M 1722	1.3
2257	4.6	2335	4.1
14 0533	0.5	**29** 0524	1.1
1141	4.7	1153	4.3
M 1755	0.6	TU 1753	1.1
○ 2351	4.8	●	
15 0619	0.5	**30** 0009	4.2
1227	4.8	0601	1.1
TU 1840	0.5	W 1225	4.3
		1829	1.1

DECEMBER

Time	m		Time	m
1 0043	4.1	**16** 0118	4.6
0637	1.1	0730	0.8
TH 1256	4.3	F 1342	4.7
1904	1.1	2002	0.5
2 0116	4.1	**17** 0207	4.5
0711	1.2	0811	1.0
F 1327	4.2	SA 1428	4.5
1935	1.2	2049	0.7
3 0150	4.0	**18** 0255	4.3
0735	1.3	0849	1.3
SA 1400	4.1	SU 1514	4.4
1956	1.2	2139	0.9
4 0229	3.9	**19** 0343	4.1
0740	1.4	0918	1.5
SU 1436	4.0	M 1600	4.2
2006	1.3	2231	1.2
5 0311	3.8	**20** 0431	3.9
0803	1.6	0938	1.7
M 1519	3.9	TU 1648	3.9
2057	1.4	2331	1.4
6 0400	3.7	**21** 0519	3.7
0849	1.7	1023	1.9
TU 1608	3.8	W 1740	3.8
2230	1.5	◐	
7 0454	3.6	**22** 0049	1.5
1040	1.8	0610	3.5
W 1705	3.7	TH 1128	2.0
◐ 2347	1.5	1837	3.6
8 0557	3.6	**23** 0159	1.6
1202	1.8	0707	3.5
TH 1808	3.7	F 1436	2.0
		1938	3.6
9 0107	1.4	**24** 0254	1.6
0707	3.6	0807	3.5
F 1340	1.7	SA 1530	1.8
1916	3.8	2038	3.6
10 0218	1.2	**25** 0337	1.6
0818	3.8	0904	3.7
SA 1457	1.4	SU 1614	1.7
2024	4.0	2133	3.7
11 0322	1.0	**26** 0359	1.5
0925	4.1	0957	3.9
SU 1557	1.2	M 1644	1.5
2131	4.2	2223	3.9
12 0422	0.9	**27** 0412	1.4
1026	4.3	1043	4.0
M 1652	0.9	TU 1708	1.3
2237	4.4	2307	4.0
13 0515	0.7	**28** 0454	1.3
1120	4.5	1124	4.2
TU 1742	0.7	W 1741	1.2
2335	4.6	2347	4.0
14 0603	0.7	**29** 0539	1.2
1209	4.7	1202	4.2
W 1830	0.5	TH 1818	1.1
○		●	
15 0028	4.7	**30** 0024	4.1
0647	0.7	0621	1.1
TH 1256	4.7	F 1238	4.3
1915	0.5	1856	1.0
		31 0101	4.1
		0700	1.1
		SA 1313	4.2
		1934	1.0

Chart Datum: 2·10 metres below Ordnance Datum (Local). HAT is 5·0 metres above Chart Datum.

》》 **FREE** monthly updates. Register at 《
www.reedsnauticalalmanac.co.uk 《

MORAY FIRTH: NORTH WEST COAST

(AC 115) Cromarty Firth (AC 1889, 1890) is entered between the North and South Sutors, both fringed by rks, some of which dry.

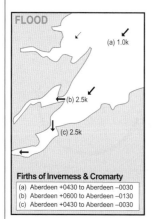

FLOOD

(a) 1.0k
(b) 2.5k
(c) 2.5k

Firths of Inverness & Cromarty

(a)	Aberdeen +0430 to Aberdeen −0030
(b)	Aberdeen +0600 to Aberdeen −0130
(c)	Aberdeen +0430 to Aberdeen −0030

EBB

(a) 1.0k
(b) 2.5k
(c) 3.5k

Firths of Inverness & Cromarty

(a)	Aberdeen −0030 to Aberdeen +0430
(b)	Aberdeen −0130 to Aberdeen +0600
(c)	Aberdeen −0030 to Aberdeen +0430

▶ *Off the entrance the in-going stream starts at HW Aberdeen +0600, and the out-going at HW Aberdeen −0130, sp rates 1·5 kn, stronger in the narrows between Nigg and Cromarty* ◀

Buss Bank buoy, Fl R 3s, marks the entrance, and there are good sheltered ⚓s within the firth. Keep 100m clear of the Nigg Oil Terminal on the north side. A buoyed deep-water channel runs to Invergordon; outside the channel there are shallows and drying banks. Beware large ships and tugs using Invergordon.

The coast NE to Tarbat Ness (lt) is fringed with rocks. Beware Three Kings (dries) about 3M NE of N Sutor.

▶ *Culloden Rk, a shoal with depth of 1·8m, extends 2½ca NE of Tarbat Ness, where stream is weak.* ◀

Beware salmon nets between Tarbat Ness and Portmahomack. Dornoch Firth is shallow, with shifting banks, and in strong E'lies the sea breaks heavily on the bar E of Dornoch Pt.

At Lothbeg Pt, 5M SW of Helmsdale, a rocky ledge extends 5ca offshore. Near Berriedale, 7M NE of Helmsdale, The Pinnacle, a detached rk 61m high, stands close offshore. The Beatrice oil field lies on Smith Bank, 28M NE of Tarbat Ness, and 11M off Caithness coast. Between Dunbeath and Lybster there are no dangers more than 2ca offshore. Clyth Ness is fringed by detached and drying rks. From here to Wick the only dangers are close inshore. There is ⚓ in Sinclair's B in good weather, but Freswick B further N is better to await the tide in Pentland Firth (beware wreck in centre of bay). Stacks of Duncansby and Baxter Rk (depth 2·7m) lie 1M and 4ca S of Duncansby Hd.

9.7.16   INVERNESS

Highland   57°29'·73N 04°14'·17W   ❄❄⚓⚓⚓⚓❀❀

CHARTS AC 223, 1077, 1078, 5617; Imray C23

TIDES +0100 Dover; ML 2·7; Duration 0620

Standard Port INVERGORDON (⟵)

Times				Height (metres)			
High Water		Low Water		MHWS	MHWN	MLWN	MLWS
0100	0700	0000	0700	4·3	3·3	1·6	0·7
1300	1900	1200	1900				
Differences INVERNESS							
+0010	+0015	+0015	+0010	+0·3	+0·2	+0·1	+0·1
FORTROSE							
0000	+0010	+0010	−0010	0·0	+0·1	ND	ND
CROMARTY							
−0005	0000	0000	−0005	0·0	0·0	0·0	0·0
DINGWALL							
+0020	+0015	ND	ND	0·0	+0·1	ND	ND

SHELTER Good in all weathers. Berth at Inverness Marina (2.5m at CD) or at one of the 2 marinas in the Caledonian Canal, entr to which can be difficult in strong tidal streams. The sea lock is normally available HW±4 in canal hours; the gates cannot be opened LW±2.

NAVIGATION WPT Meikle Mee SHM By Fl G 3s, 57°30'·25N 04°12'·03W, 250°/0·74M to Longman Pt bn. Inverness Firth is deep from Chanonry Pt to Munlochy SWM buoy, but shoal (2·1m) to Meikle Mee buoy. Meikle Mee partly dries. Beware bird rafts S of Avoch (off chartlet). Tidal streams are strong S of Craigton Pt (E-going stream at sp exceeds 5kn). Ent to R Ness is narrow, dredged to 3m CD. For the entrance to the Caledonian Canal, keep to N Kessock bank until clear of unmarked shoals on S bank. Care must be taken to avoid Carnarc Pt W of R mouth.

LIGHTS AND MARKS Longman Pt bn Fl WR 2s 7m 5/4M, vis 078°-W-258°-R-078°. Craigton Pt lt, Fl WRG 4s 6m 11/7M vis 312°-W-048°-R-064°-W-085°-G-shore. Caledonian Canal ent marked by QR and Iso G 4s on ends of training walls.

COMMUNICATIONS (Code 01463) CGOC (01224) 592334; Police 101; Dr 234151. HM 715715; Clachnaharry Sea Lock 713896; Canal Office 233140.
Inverness Hbr Office VHF Ch 12 (Mon-Fri: 0800 -1700 LT). *Inverness Marina* Ch 12. Caledonian Canal: Ch 74 is used by all stations. Call: *Clachnaharry Sea Lock*; or for office: *Caledonian Canal*.

FACILITIES **Inverness Marina** www.invernessmarina.com ☎220501 Dredged to 3m 150� inc 20🅥 £2.00, 🔥🚿⛽⚓🚻♿(45t) ⚒🛒🔧✕🚮. **City** All domestic facilities, ⇌ ✈.

MINOR HARBOURS IN INVERNESS FIRTH

FORTROSE, Highland, **57°34'·71N 04°08'·04W**. AC 1078. Tides 9.7.16. HW +0055 on Dover; ML 2·5m; Duration 0620. Small drying unlit hbr, soft silt bottom, well protected by Chanonry Ness to E, limited space. Follow ldg line 296°, Broomhill Ho (conspic on hill to NW) in line with school spire until abeam SPM buoy; then turn W to avoid Craig an Roan rks (1·8m) ESE of ent. Chanonry Pt lt, Oc 6s 12m 15M, obscd 073°-shore. HM ☎(01381) 620311; Dr ☎622000. Facilities: �⌐£5, 40⚓⚓🔧🛒 scrubbing grid ⚓⚓⛽✉Ⓑ🛒✕. **Chanonry SC** (near pier) ☎ 01381 621973.

AVOCH, Highland, **57°34'·03N 04°09'·94W**. AC 1078. Tides as Fortrose (1·25M to the ENE). Hbr dries, mostly on the N side, but is bigger than Fortrose. Small craft may stay afloat at neaps against the S pier (2FR (vert)). ⌐ may be available on drying pontoons; one ⚓ (max 35ft LOA) off hbr ent. HM ☎07779 833951. Facilities: ⌐£7 ⚓ 🔧. **Village** 🏨🏪🔧✉🛒✕🚮.

NORTH KESSOCK PIER, Highlands, **57°30'·06N 4°14'·87W**. AC 1078. Tides as Inverness (1 M to the ESE). Unlit tidal slipway, suitable for leisure craft up to 20 metres. 2m depth at MLWS. Mud and rounded stones bottom, sheltered from SW by Bkwtr. Ideal if awaiting sea lock at Caledonian Canal. Approach on N shore to avoid shallows at Inverness Harbour and South Kessock. Facilities: 🔧 ⌐ (tidal) ⚓ (by arrangement). **Village** LB (inshore), Ⓑ🛒✕🚮 Dolphin & Seal Centre, Bus. Jim Prentice ☎07802 915741.

INVERNESS *continued*

CALEDONIAN CANAL

Clachnaharry Sea Lock 57°29'·44N 04°15'·84W. These notes are for the convenience of those entering the Canal at Inverness. They should be used with the main information/chartlet in 9.8.17.

CHARTS AC 1791, 1078, 5617. *Scottish Canals Skipper's Guide.*

TIDES Differences: Clachnaharry +0116 on Dover.

SHELTER Clachnaharry sea lock operates HW±4 (sp) within canal hours. The road and rail swing bridges may cause delays up to 25 mins. Seaport and Caley marinas: see Facilities.

NAVIGATION The 60M Caledonian Canal consists of 38M through three lochs (Lochs Ness, Oich and Lochy), connected by 22M through canals. It can take vessels 45m LOA, 10m beam, 4m draught and max mast ht 27·4m. The passage normally takes two full days, possibly longer in the summer; 13 hrs is absolute minimum.

Speed limit is 5kn in canal sections. There are 10 swing bridges; road tfc has priority at peak hrs. Do not pass bridges without the keeper's instructions. From Clachnaharry sea lock to Loch Ness (Bona Ferry lt ho) is approx 7M, via Muirtown and Dochgarroch locks.

LOCKS All 29 locks are manned and operate early May to early Oct, 0800-1800LT daily. Dues: see 9.8.17. For regulations and *Scottish Canals Skipper's Guide* apply: Canal Manager, Muirtown Wharf, Inverness IV3 5LS, www. scottishcanals.co.uk ☎01463 725500.

BOAT SAFETY SCHEME Visiting vessels will be checked for apparent dangerous defects eg leaking gas or fuel, damaged electrical cables, taking in water, risk of capsize. £1M 3rd party insurance is required. For details see 9.8.17.

LIGHTS AND MARKS Chans are marked by posts, cairns and unlit buoys, PHM on the NW side of the chan and SHM on the SE side.

COMMUNICATIONS Clachnaharry sea lock ☎(01463) 713896; Canal Office, Inverness ☎(01463) 233140.

Sea locks and main lock flights operate VHF Ch **74** (HO).

FACILITIES Seaport Marina www.scottishcanals.co.uk ☎(01463) 725500, 80⌂+20♥, ⬛⚓⚓⛴⛽⬛⬦⬛✂✎⬚⬛(40t).

Caley Marina www.caleymarina.com ☎(01463) 236539, 25⌂+25♥ £1.50, ⬦✂✎⬚⬛(20t) ⬛ ACA.

MINOR HARBOURS FROM CROMARTY FIRTH TO WICK

CROMARTY FIRTH, Highland, **57°41'·18N 04°02'·09W**. AC 1889, 1890. HW +0100 on Dover –0135 on Aberdeen; HW height 0.0m on Aberdeen; ML 2·5m; Duration 0625. See 9.7.16. Excellent hbr extending 7·5M W, past Invergordon, then 9M SW. Good shelter always available, depending on wind direction. Beware rks and reefs round N and S Sutor at the ent; many unlit oil rig mooring buoys. *Cromarty Firth Port Control* VHF Ch 11 16 13 (H24) ☎(01349) 852308.

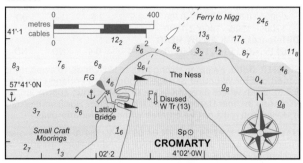

CROMARTY HARBOUR, Highland. **57°41'·00N 04°02'·35W**. From WPT 57°41'·10N 04°02'·50W, 142°/1.3ca to entrance. When inside harbour beware of set towards lattice bridge, tide runs at up to 4kts. Sheltered from N through E to S. Approx 20 ⌒ on 50m central pontoon, £20 o/n, depths range from 1·6m to 1m at inner end. 2 ⚓ 2ca W of S pier hd, or ⚓ in approx 6m with good holding but exposed to the West. Facility administered by Cromarty Harbour Trust, www.cromartyharbour.org. Barry Pickles ☎(07453 695648) (office not normally manned) or call Ch M/16. ⚓ ⊞ ⚲ available. **Cromarty Boat Club** www.cromartyboatclub.org 2⚓ in 0·5m–1·5m; ⛴/⚲ ⊡ (owned by Boat Club; key held at Royal Hotel). ⊠ ⚲. Local ferry to Nigg where an energy park dominates the foreshore.

INVERGORDON W HARBOUR, Highland. **57°41'·12N 04°10'·05W**. ⌒ on pontoons, ⚓ ⚒ ⚲ ⚑ ⚫ ⌐(3t) ⊠ ⚲ ✕ ⌂. Dinghy landing at Invergordon Boat Club (1·5M W), ☎(01349) 893772.

DORNOCH FIRTH, Highland. **57°51'·28N 03°59'·39W**. AC 115, 223. HW +0115 on Dover; ML 2·5m; Duration 0605; see 9.7.17. Excellent shelter but difficult entry. There are many shifting sandbanks, especially near the ent, from N edge of Whiteness Sands to S edge of Gizzen Briggs. ⚓s in 7m ¾M ESE of Dornoch Pt (sheltered from NE swell by Gizzen Briggs); in 7m 2ca SSE of Ard na Cailc; in 3·3m 1M below Bonar Bridge. Firth extends 15M inland, but AC coverage ceases ¼M E of Ferry Pt. The A9 road bridge, 3·3M W of Dornoch Pt, with 11m clearance, has 3 spans lit on both sides; span centres show Iso 4s, N bank pier Iso G 4s, S bank pier Iso R 4s and 2 midstream piers QY. Tarbat Ness lt ho Fl (4) 30s 53m 24M. Fl R 5s lt shown when Tain firing range active. Very limited facilities at Ferrytown and Bonar Bridge. CGOC ☎01224 592334. **Dornoch**: Dr ⚕ ⊠ Ⓑ ⚲ ✕ ⌂.

GOLSPIE, Highland, **57°58'·71N 03°56'·79W**. AC 223. HW +0045 on Dover; ML 2·3m; Duration 068. See 9.7.17. Golspie pier projects 60m SE across foreshore with arm projecting SW at the head, giving shelter during NE winds. Beware The Bridge, a bank (0·3m to 1·8m) running parallel to the shore ¼M to seaward of pier hd. Seas break heavily over The Bridge in NE winds. There are no lts. To enter, keep Duke of Sutherland's Memorial in line 316° with boathouse SW of pier, until church spire in village is in line 006° with hd of pier, then keep on those marks. Harbour gets very congested; good ⚓ off pier. HM ☎01431 821692 ⚓ ⚒. **Town** Ⓗ Dr ⚕ ⚕ ⚫ ⊠ Ⓑ ⚲ ✕ ⌂ ≈.

LYBSTER, Highland, **58°17'·72N 03°17'·39W**. AC 115. HW +0020 on Dover; HW -0150 sp, -0215 np; HW ht -0·6m on Aberdeen; ML 2·1m; Duration 0620. Excellent shelter in SW corner of inner hbr; ⌒ on W side of pier in about 1·2m, £7.00/wk. Most of hbr dries to sand/mud and is much used by FVs; no bollards on N wall. Appr on about 350°. Beware rks close on E side of ent; narrow (10m) ent is difficult in strong E to S winds. Min depth 2·5m in ent. S pier hd, Oc R 6s 10m 3M, occas in fishing season. ⚓ on W quay. ⛴ ⊡ at Waterlines Visitor Centre, 01593 721520. **Town** ⚲ ✕ ⌂.

9.7.17 PORTMAHOMACK

Highland **57°50'·25N 03°50'·00W** ✿✿◈✿✿

CHARTS AC 115, 223; Imray C23

TIDES +0035 Dover; ML 2·5; Duration 0600

Standard Port ABERDEEN (◀——)

Times				Height (metres)			
High Water		Low Water		MHWS	MHWN	MLWN	MLWS
0300	0800	0200	0800	4·3	3·4	1·6	0·6
1500	2000	1400	2000				
Differences PORTMAHOMACK							
–0120	–0210	–0140	–0110	–0·2	–0·1	+0·1	+0·1
MEIKLE FERRY (Dornoch Firth)							
–0100	–0140	–0120	–0055	+0·1	0·0	–0·1	0·0
GOLSPIE							
–0130	–0215	–0155	–0130	–0·3	–0·3	–0·1	0·0

SHELTER Good, but uncomfortable in SW/NW winds. Hbr dries, access only at HW, but good ⚓ close SW of pier.

NAVIGATION WPT 57°53'·00N 03°50'·00W, 180°/2.7M to 1ca W of hbr ent. Beware Curach Rks which lie from 2ca SW of pier to the shore. Rks extend N and W of the pier. Beware lobster pot floats and salmon nets N of hbr. Tain firing & bombing range is about 3M to the W, S of mouth of Dornoch Firth; R flags, R lts, shown when active.

LIGHTS AND MARKS Tarbert Ness lt ho Fl (4) 30s 53m 24M, W twr R bands, is 2·6M to NE of hbr. Pier hd 2 FR (vert) 7m 5M.

COMMUNICATIONS (Code 01862) CGOC (01224) 592334; Dr 892759. No HM; enquiries: harbours@highland.gov.uk (01571) 844807.

FACILITIES **Hbr**, ⌒ <5m £9, 5-7m £12, 7-10m £16, some drying pontoon berths, ⚓ ⚒ ⚓ ⊞. **Town** ⊠ ⚲ ⌂ ✕ ≈ (bus to Tain), ✈ (Inverness).

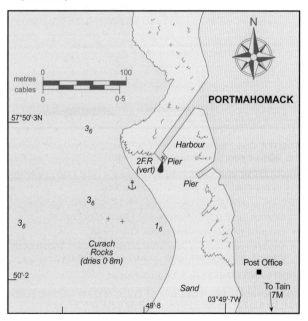

9.7.18 HELMSDALE

Highland 58°06'·83N 03°38'·89W ✿✿✿✿✿✿

CHARTS AC 115, 1462; Imray C23

TIDES +0035 Dover; ML 2·2; Duration 0615

Standard Port WICK (→)

Times				Height (metres)			
High Water		Low Water		MHWS	MHWN	MLWN	MLWS
0000	0700	0200	0700	3·5	2·8	1·4	0·7
1200	1900	1400	1900				
Differences HELMSDALE							
+0025	+0015	+0035	+0030	+0·5	+0·3	+0·1	−0·1

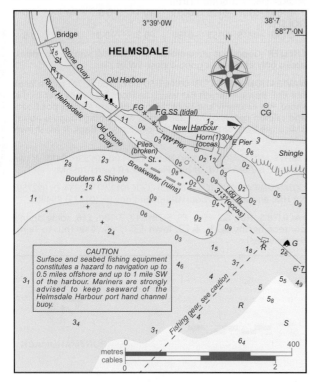

CAUTION
Surface and seabed fishing equipment constitutes a hazard to navigation up to 0.5 miles offshore and up to 1 mile SW of the harbour. Mariners are strongly advised to keep seaward of the Helmsdale Harbour port hand channel buoy.

SHELTER Good, except in strong E/SE'lies. ⌒ on NW pier, approx 1m. 80m of pontoon berths through HM.

NAVIGATION WPT 58°06'·59N 03°38'·39W, 313°/0·35M to ent; ldg marks are black masts with Or □ topmarks.

• Surface and seabed fishing gear, hazardous to navigation, extends up to 0.5M offshore and 1.0M SW of hbr.

• Essential to pass seaward of Hbr Chan PHM.

• Beware spate coming downriver after heavy rain. Shoals both sides of chan and bar build up when river in spate.

LIGHTS AND MARKS See 9.7.4 and chartlet.

COMMUNICATIONS (Code 01431) CGOC (01224) 592334; Police 101; Dr 821221, or 821225 (Home). HM 821692 (Office).

VHF Ch 13 16.

FACILITIES ⌒ £9<5m>£12<7m>£16<10m, ⚓ (See HM), ⎯ ⛽ ⚓.
Town ⛽(hardware store), ✉ ⛏ ✗ ⎅ Ⓑ(Brora) ⇌.

9.7.19 WICK

Highland 58°26'·38N 03°04'·72W ✿✿✿✿✿✿

CHARTS AC 115, 1462; Imray C23, C68

TIDES +0010 Dover; ML 2·0; Duration 0625. Wick is a Standard Port. Daily tidal predictions are given below.

Standard Port WICK (→)

Times				Height (metres)			
High Water		Low Water		MHWS	MHWN	MLWN	MLWS
0000	0700	0200	0700	3·5	2·8	1·4	0·7
1200	1900	1400	1900				
Differences DUNCANSBY HEAD							
−0115	−0115	−0110	−0110	−0·4	−0·4	ND	ND

SHELTER Good, except in strong NNE to SSE winds, to await right conditions for W-bound passage through the Pentland Firth (see 9.7.5). Berth where directed in the Outer Harbour, approx 2m, or in marina in Inner Harbour, 2·0m. NB: The River Hbr (commercial) is leased and must not be entered without prior approval.

NAVIGATION WPT 58°26'·18N 03°03'·39W, 284°/0·72M to S pier. From the N, open up hbr ent before rounding North Head so as to clear drying Proudfoot Rks.

Hbr ent is dangerous in strong E'lies as craft have to turn 90° to port at the end of S pier.

On S side of bay, unlit NCM, 300m ENE of LB slip, marks end of ruined bkwtr.

LIGHTS AND MARKS S pier lt, Fl WRG 3s 12m 12/9M, 253°-G-270°-W-286°-R-329°, Bell (2) 10s (fishing). Ldg lts, both FR 5/8m, lead 234° into outer hbr. Traffic signals:

B ● (●) at CG stn on S Head = hbr closed by weather.
B ● (●) at S pier head = caution; hbr temp obstructed.

COMMUNICATIONS (Code 01955) HM 602030; CGOC (01224) 592334; Police 101; Ⓗ 602434, 602261.

VHF Ch 14 16 (Mon-Fri 0800-1700).

FACILITIES Poss temporary ⌒ on pontoons in Outer Harbour. **Wick Marina** (Inner Harbour) dredged to 2·0m 70⌒ inc 15 ⓥ; £20 < 10m; max 25m LOA. ⛽ 50m from marina ⛽. **Fish Jetty** ⚓ ⛽ ⚓. **Services** ⎯ ⚒ ⚓ ⛽ ⛽(15/100t).
Town 🏪 ⛽(at Heat Centre) ✉ ⒷⒷ ✗ ⎅ ⇌ ✈.

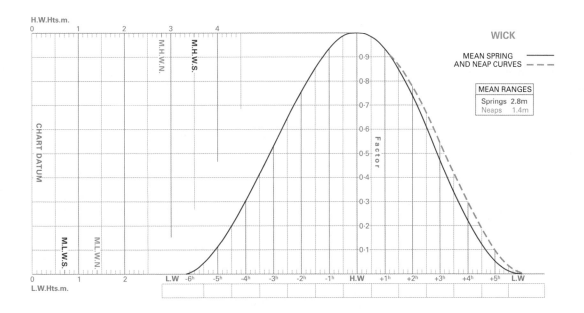

WICK

MEAN SPRING ——————
AND NEAP CURVES – – – –

MEAN RANGES	
Springs	2.8m
Neaps	1.4m

PENTLAND FIRTH

(AC 2162, 2581) ▶ *This potentially dangerous chan should only be attempted with moderate winds (less than F4), good vis, no swell and a fair np tide. In such conditions it presents few problems. A safe passage depends on a clear understanding of tidal streams and correct timing. The Admiralty Tidal Stream Atlas for Orkney and Shetland (NP 209) gives large scale vectors and is essential. Even in ideal conditions the races off Duncansby Hd, Swilkie Pt (N end of Stroma), and Rks of Mey (Merry Men of Mey) must be avoided as they are always dangerous to small craft. Also avoid the Pentland Skerries, Muckle Skerry, Old Head, Lother Rock (S Ronaldsay), and Dunnet Hd on E-going flood. For passages across the Firth see CCC SDs for Orkney.* ◀

At E end the Firth is entered between Duncansby Hd and Old Hd (S Ronaldsay), between which lie Muckle Skerry and the Pentland Skerries. Near the centre of Firth are the Islands of Swona (N side) and Stroma (S side). Outer Sound (main chan, 2·5M wide) runs between Swona and Stroma; Inner Sound (1·5M wide) between Stroma and the mainland. Rks of Mey extend about 2ca N of St John's Pt. The W end of the Firth is between Dunnet Hd and Tor Ness (Hoy).

▶*Tide flows strongly around and through the Orkney Islands. The Pentland Firth is a dangerous area for all craft, tidal flows reach 12 knots between Duncansby Head and S Ronaldsay. W of Dunnet Hd and Hoy is less violent. There is little tide within Scapa Flow. Tidal streams reach 8-9kn at sp in the Outer Sound, and 9-12kn between Pentland Skerries and Duncansby Hd. The resultant dangerous seas, very strong eddies and violent races should be avoided by yachts at all costs.*

The E-going stream begins at HW Aberdeen +0500, and the W-going at HW Aberdeen –0105. **Duncansby Race** *extends ENE towards Muckle Skerry on the SE-going stream, but by HW Aberdeen –0440 it extends NW from Duncansby Hd. Note: HW at Muckle Skerry is the same time as HW Dover. A persistent race off* **Swilkie Pt,** *at N end of Stroma,* **is very dangerous with a strong W'ly wind over a W-going stream. The most dangerous and extensive race in the Firth is Merry Men of Mey.** *It forms off St John's Pt on W-going stream at HW Aberdeen –0150 and for a while extends right across to Tor Ness with heavy breaking seas even in fine weather.* ◀

FLOOD	EBB
Pentland Firth & Orkneys	**Pentland Firth & Orkneys**
(a) Aberdeen +0500 to Aberdeen –0105	(a) Aberdeen –0105 to Aberdeen +0500
(b) Aberdeen +0310 to Aberdeen –0330	(b) Aberdeen –0330 to Aberdeen +0310
(c) Aberdeen +0400 to Aberdeen –0150	(c) Aberdeen –0150 to Aberdeen +0400

(c) 3.0k

(b) 3.0k

(a) 12.0k

(c) 3.0k

(b) 3.0k

(a) 8.0k

Passage Westward: This is the more difficult direction due to prevailing W winds. ▶ *Freswick B, 3·5M S of Duncansby Hd, is a good waiting anch; here an eddy runs N for 9 hrs. Round Duncansby Hd close in at HW Aberdeen –0220, as the ebb starts to run W. Take a mid-course through the Inner Sound to appr the Rks of Mey from close inshore. Gills Bay is a temp anch if early; do not pass Rks of Mey until ebb has run for at least 2 hrs. Pass 100m N of the Rks (awash).* ◀

Passage Eastward: ▶ *With a fair wind and tide, no race forms and the passage is easier. Leave Scrabster at local LW+1 so as to be close off Dunnet Hd not before HW Aberdeen +0340 when the E-going flood starts to make. If late, give the Hd a wide berth. Having rounded the Rks of Mey, steer S initially to avoid being set onto the rky S tip of Stroma, marked by unlit SCM bn.* ◀ Then keep mid-chan through the Inner Sound and maintain this offing to give Duncansby Hd a wide berth.

STANDARD TIME (UT)
For Summer Time add ONE hour in **non-shaded areas**

WICK   LAT 58°26'N   LONG 3°05'W
TIMES AND HEIGHTS OF HIGH AND LOW WATERS

Dates in red are SPRINGS
Dates in blue are NEAPS

YEAR 2016

JANUARY

#	Time	m	#	Time	m
1 F	0341 0904 1552 2204	2.9 1.5 3.1 1.3	**16** SA ☽	0338 0916 1548 2207	3.2 1.2 3.4 1.0
2 SA ☽	0434 1001 1647 2306	2.8 1.6 2.9 1.4	**17** SU	0437 1024 1651 2319	3.1 1.3 3.2 1.1
3 SU	0533 1117 1750	2.8 1.7 2.9	**18** M	0544 1144 1802	3.0 1.4 3.1
4 M	0013 0635 1239 1855	1.4 2.8 1.7 2.8	**19** TU	0035 0653 1309 1917	1.2 3.0 1.4 3.1
5 TU	0121 0737 1354 1959	1.4 2.9 1.6 2.9	**20** W	0151 0806 1423 2033	1.2 3.1 1.2 3.2
6 W	0218 0833 1448 2056	1.3 3.0 1.4 3.0	**21** TH	0252 0909 1521 2137	1.1 3.2 1.0 3.3
7 TH	0304 0921 1531 2145	1.2 3.2 1.3 3.1	**22** F	0341 1001 1609 2229	1.1 3.4 0.9 3.4
8 F	0343 1003 1610 2229	1.1 3.3 1.1 3.3	**23** SA	0423 1047 1652 2314	1.0 3.5 0.7 3.4
9 SA	0420 1043 1647 2310	1.0 3.5 0.9 3.4	**24** SU ○	0501 1129 1731 2355	1.0 3.6 0.7 3.4
10 SU ●	0457 1122 1724 2351	0.9 3.6 0.8 3.5	**25** M	0537 1208 1809	0.9 3.6 0.6
11 M	0535 1202 1804	0.9 3.7 0.7	**26** TU	0033 0611 1245 1844	3.4 0.9 3.6 0.7
12 TU	0033 0615 1243 1845	3.5 0.8 3.7 0.6	**27** W	0110 0644 1320 1919	3.3 1.0 3.5 0.7
13 W	0116 0656 1325 1928	3.5 0.9 3.7 0.6	**28** TH	0145 0716 1355 1953	3.2 1.0 3.4 0.9
14 TH	0200 0738 1408 2014	3.4 0.9 3.6 0.7	**29** F	0219 0748 1430 2028	3.1 1.1 3.3 1.0
15 F	0247 0824 1455 2106	3.3 1.1 3.5 0.8	**30** SA	0256 0823 1508 2107	3.0 1.3 3.1 1.2
			31 SU	0337 0903 1551 2157	2.9 1.4 3.0 1.3

FEBRUARY

#	Time	m	#	Time	m
1 M ☽	0427 0959 1646 2304	2.8 1.6 2.8 1.5	**16** TU	0513 1119 1743	2.9 1.3 3.0
2 TU	0528 1127 1756	2.7 1.7 2.7	**17** W	0010 0626 1255 1904	1.3 2.9 1.3 2.9
3 W	0019 0637 1258 1909	1.5 2.7 1.6 2.7	**18** TH	0139 0746 1417 2028	1.3 2.9 1.2 3.0
4 TH	0134 0744 1415 2020	1.4 2.8 1.5 2.8	**19** F	0245 0855 1515 2130	1.2 3.1 1.0 3.1
5 F	0235 0844 1507 2119	1.3 3.0 1.3 3.0	**20** SA	0332 0948 1600 2219	1.2 3.3 0.8 3.2
6 SA	0321 0935 1550 2207	1.2 3.2 1.0 3.2	**21** SU	0411 1033 1638 2300	1.1 3.4 0.7 3.3
7 SU	0401 1020 1629 2252	1.0 3.4 0.8 3.4	**22** M	0445 1112 1713 2337	0.9 3.5 0.6 3.3
8 M ●	0440 1102 1708 2334	0.9 3.6 0.6 3.5	**23** TU	0517 1149 1746	0.9 3.5 0.6
9 TU	0519 1144 1748	0.7 3.7 0.4	**24** W	0011 0549 1223 1818	3.3 0.8 3.5 0.6
10 W	0017 0559 1227 1829	3.6 0.6 3.8 0.3	**25** TH	0044 0620 1255 1849	3.3 0.8 3.5 0.7
11 TH	0100 0640 1309 1912	3.6 0.6 3.8 0.4	**26** F	0115 0650 1327 1919	3.2 0.9 3.4 0.8
12 F	0143 0721 1352 1955	3.5 0.7 3.7 0.5	**27** SA	0145 0720 1358 1950	3.2 0.9 3.3 0.9
13 SA	0227 0805 1438 2042	3.4 0.8 3.6 0.7	**28** SU	0218 0751 1432 2023	3.1 1.1 3.1 1.0
14 SU	0314 0853 1529 2137	3.2 1.0 3.4 0.9	**29** M	0253 0825 1511 2101	2.9 1.2 3.0 1.2
15 M ☽	0408 0955 1629 2247	3.0 1.2 3.2 1.2			

MARCH

#	Time	m	#	Time	m
1 TU ☽	0335 0908 1559 2154	2.8 1.4 2.8 1.4	**16** W	0445 1103 1728 2346	2.9 1.2 2.8 1.4
2 W	0428 1016 1702 2320	2.7 1.5 2.7 1.5	**17** TH	0601 1242 1853	2.8 1.2 2.8
3 TH	0538 1202 1824	2.7 1.5 2.6	**18** F	0124 0722 1404 2016	1.4 2.8 1.1 2.8
4 F	0047 0654 1334 1943	1.5 2.7 1.4 2.7	**19** SA	0231 0834 1459 2115	1.3 3.0 0.9 3.0
5 SA	0202 0804 1438 2050	1.4 2.8 1.2 2.9	**20** SU	0317 0928 1542 2200	1.2 3.1 0.8 3.1
6 SU	0256 0903 1524 2144	1.2 3.1 0.9 3.1	**21** M	0353 1012 1616 2239	1.1 3.2 0.7 3.2
7 M	0340 0953 1606 2230	1.0 3.3 0.6 3.3	**22** TU	0424 1050 1648 2313	0.9 3.3 0.6 3.2
8 TU	0420 1039 1647 2314	0.8 3.5 0.4 3.5	**23** W ○	0455 1125 1719 2345	0.8 3.4 0.6 3.3
9 W ●	0500 1123 1728 2356	0.7 3.7 0.2 3.6	**24** TH	0526 1158 1749	0.8 3.4 0.6
10 TH	0541 1207 1809	0.4 3.8 0.2	**25** F	0015 0556 1230 1819	3.3 0.7 3.4 0.6
11 F	0039 0621 1251 1851	3.6 0.4 3.8 0.2	**26** SA	0045 0625 1300 1848	3.2 0.8 3.3 0.7
12 SA	0122 0703 1335 1934	3.6 0.5 3.8 0.4	**27** SU	0115 0655 1331 1918	3.2 0.8 3.2 0.8
13 SU	0205 0747 1422 2019	3.4 0.6 3.6 0.7	**28** M	0145 0726 1405 1949	3.1 0.9 3.1 1.0
14 M	0250 0835 1513 2110	3.2 0.8 3.3 1.0	**29** TU	0219 0759 1443 2024	3.0 1.0 2.9 1.1
15 TU ☽	0342 0937 1613 2218	3.0 1.0 3.0 1.2	**30** W	0259 0840 1528 2110	2.9 1.2 2.8 1.3
			31 TH ☽	0348 0936 1628 2226	2.8 1.3 2.6 1.5

APRIL

#	Time	m	#	Time	m
1 F	0451 1119 1747	2.7 1.4 2.6	**16** SA	0053 0649 1337 1949	1.5 2.8 1.0 2.7
2 SA	0005 0609 1250 1908	1.5 2.7 1.3 2.7	**17** SU	0204 0802 1432 2048	1.4 2.9 0.9 2.9
3 SU	0125 0723 1402 2018	1.4 2.8 1.0 2.9	**18** M	0251 0859 1513 2133	1.2 3.0 0.8 3.0
4 M	0226 0827 1454 2116	1.1 3.0 0.8 3.1	**19** TU	0327 0944 1547 2211	1.1 3.1 0.8 3.1
5 TU	0313 0923 1539 2205	0.9 3.2 0.5 3.3	**20** W	0359 1023 1619 2245	1.0 3.2 0.7 3.2
6 W	0356 1013 1622 2250	0.7 3.5 0.3 3.5	**21** TH	0431 1058 1650 2317	0.8 3.2 0.6 3.2
7 TH ●	0438 1100 1704 2334	0.5 3.7 0.2 3.6	**22** F ○	0502 1132 1721 2347	0.8 3.3 0.6 3.3
8 F	0520 1146 1747	0.3 3.8 0.1	**23** SA	0533 1204 1751	0.7 3.3 0.7
9 SA	0017 0603 1233 1829	3.6 0.3 3.8 0.2	**24** SU	0017 0603 1236 1821	3.3 0.7 3.2 0.7
10 SU	0100 0647 1319 1913	3.6 0.4 3.7 0.4	**25** M	0048 0634 1309 1851	3.2 0.8 3.1 0.8
11 M	0143 0732 1407 1957	3.4 0.5 3.5 0.7	**26** TU	0120 0707 1344 1924	3.2 0.8 3.0 1.0
12 TU	0228 0823 1459 2047	3.3 0.7 3.2 1.0	**27** W	0154 0742 1423 2000	3.1 0.9 2.9 1.1
13 W	0319 0926 1559 2152	3.1 0.9 3.0 1.3	**28** TH	0234 0824 1509 2046	3.0 1.0 2.8 1.3
14 TH ☽	0420 1047 1711 2316	2.9 1.1 2.8 1.5	**29** F	0321 0920 1606 2153	2.9 1.1 2.7 1.4
15 F	0533 1217 1830	2.8 1.1 2.7	**30** SA ☽	0420 1048 1719 2328	2.8 1.2 2.7 1.4

Chart Datum: 1·71 metres below Ordnance Datum (Newlyn). HAT is 4·0 metres above Chart Datum.

FREE monthly updates. Register at
www.reedsnauticalalmanac.co.uk

STANDARD TIME (UT)
For Summer Time add ONE
hour in **non-shaded areas**

WICK LAT 58°26'N LONG 3°05'W
TIMES AND HEIGHTS OF HIGH AND LOW WATERS

Dates in red are **SPRINGS**
Dates in blue are **NEAPS**

YEAR 2016

NE Scotland

MAY			JUNE			JULY			AUGUST		
Time m		Time m	Time m		Time m	Time m		Time m	Time m		Time m
1 0533 2.8 / 1212 1.1 / SU 1836 2.7		**16** 0117 1.4 / 0716 2.8 / M 1351 1.0 / 2007 2.8	**1** 0117 1.1 / 0721 3.0 / W 1352 0.7 / 2014 3.1		**16** 0218 1.3 / 0827 2.8 / TH 1438 1.0 / 2059 2.9	**1** 0157 1.1 / 0804 3.1 / F 1429 0.8 / 2050 3.2		**16** 0234 1.3 / 0842 2.8 / SA 1449 1.2 / 2106 3.0	**1** 0346 0.8 / 1005 3.2 / M 1602 0.9 / 2224 3.4		**16** 0338 1.0 / 0954 3.0 / TU 1547 1.1 / 2203 3.3
2 0046 1.3 / 0646 2.8 / M 1324 0.9 / 1944 2.9		**17** 0213 1.3 / 0818 2.9 / TU 1436 0.9 / 2057 2.9	**2** 0217 0.1 / 0825 3.2 / TH 1447 0.6 / 2111 3.2		**17** 0303 1.2 / 0918 2.9 / F 1518 1.0 / 2141 3.0	**2** 0257 0.9 / 0911 3.2 / SA 1523 0.8 / 2146 3.3		**17** 0320 1.2 / 0934 2.9 / SU 1531 1.1 / 2150 3.1	**2** 0432 0.6 / 1054 3.3 / TU 1644 0.9 / ● 2309 3.5		**17** 0415 0.8 / 1037 3.2 / W 1625 0.9 / 2245 3.4
3 0152 1.1 / 0752 3.0 / TU 1422 0.7 / 2045 3.1		**18** 0255 1.2 / 0909 2.9 / W 1514 0.9 / 2138 3.0	**3** 0311 0.8 / 0925 3.3 / F 1537 0.5 / 2203 3.4		**18** 0343 1.1 / 1001 3.0 / SA 1555 0.9 / 2219 3.1	**3** 0351 0.7 / 1011 3.3 / SU 1611 0.7 / 2236 3.4		**18** 0400 1.0 / 1018 3.0 / M 1609 1.0 / 2230 3.2	**3** 0514 0.5 / 1139 3.4 / W 1722 0.8 / 2351 3.6		**18** 0452 0.6 / 1118 3.4 / TH 1702 0.8 / ○ 2326 3.6
4 0245 0.9 / 0852 3.2 / W 1512 0.5 / 2137 3.3		**19** 0332 1.1 / 0952 3.0 / TH 1548 0.8 / 2214 3.1	**4** 0400 0.6 / 1021 3.5 / SA 1624 0.5 / 2251 3.5		**19** 0419 0.9 / 1041 3.1 / SU 1630 0.9 / 2255 3.2	**4** 0440 0.5 / 1103 3.4 / M 1656 0.7 / ● 2323 3.5		**19** 0437 0.9 / 1059 3.1 / TU 1645 0.9 / ○ 2309 3.4	**4** 0554 0.5 / 1219 3.4 / TH 1759 0.8		**19** 0530 0.5 / 1158 3.5 / F 1740 0.7
5 0332 0.7 / 0947 3.4 / TH 1558 0.3 / 2225 3.5		**20** 0406 0.9 / 0653 0.8 / F 1622 0.8 / 2248 3.2	**5** 0448 0.4 / 1113 3.5 / SU 1709 0.5 / ● 2337 3.6		**20** 0455 0.8 / 1119 3.1 / M 1704 0.9 / ○ 2330 3.3	**5** 0526 0.4 / 1151 3.4 / TU 1739 0.7		**20** 0513 0.7 / 1139 3.2 / W 1722 0.8 / 2347 3.5	**5** 0030 3.6 / 0631 0.5 / F 1258 3.3 / 1833 0.8		**20** 0006 3.7 / 0609 0.4 / SA 1240 3.5 / 1819 0.6
6 0417 0.5 / 1038 3.6 / F 1642 0.3 / ● 2311 3.6		**21** 0440 0.8 / 1106 3.1 / SA 1654 0.7 / ○ 2320 3.2	**6** 0536 0.4 / 1203 3.5 / M 1754 0.6		**21** 0529 0.8 / 1157 3.2 / TU 1739 0.8	**6** 0007 3.6 / 0611 0.4 / W 1238 3.4 / 1819 0.8		**21** 0550 0.6 / 1218 3.3 / TH 1759 0.8	**6** 0108 3.5 / 0707 0.6 / SA 1335 3.2 / 1907 0.9		**21** 0048 3.7 / 0649 0.3 / SU 1321 3.5 / 1900 0.6
7 0502 0.4 / 1128 3.7 / SA 1726 0.3 / 2355 3.6		**22** 0513 0.8 / 1141 3.2 / SU 1726 0.8 / 2353 3.3	**7** 0022 3.6 / 0623 0.4 / TU 1251 3.5 / 1837 0.7		**22** 0006 3.3 / 0605 0.7 / W 1235 3.2 / 1815 0.8	**7** 0050 3.6 / 0654 0.5 / TH 1321 3.3 / 1858 0.9		**22** 0026 3.5 / 0628 0.5 / F 1259 3.3 / 1838 0.8	**7** 0144 3.4 / 0742 0.7 / SU 1410 3.1 / 1940 1.0		**22** 0130 3.7 / 0731 0.3 / M 1403 3.4 / 1942 0.7
8 0547 0.3 / 1216 3.7 / SU 1810 0.4		**23** 0545 0.7 / 1215 3.2 / M 1758 0.8	**8** 0107 3.5 / 0710 0.4 / W 1339 3.3 / 1920 0.9		**23** 0043 3.4 / 0643 0.7 / TH 1315 3.2 / 1853 0.9	**8** 0132 3.5 / 0736 0.6 / F 1403 3.2 / 1936 1.0		**23** 0107 3.5 / 0709 0.5 / SA 1341 3.3 / 1919 0.8	**8** 0221 3.3 / 0817 0.9 / M 1447 3.0 / 2015 1.2		**23** 0214 3.6 / 0815 0.6 / TU 1448 3.3 / 2028 0.9
9 0039 3.6 / 0633 0.3 / M 1304 3.6 / 1854 0.6		**24** 0025 3.3 / 0618 0.7 / TU 1251 3.1 / 1831 0.9	**9** 0151 3.4 / 0759 0.6 / TH 1427 3.1 / 2004 1.1		**24** 0122 3.3 / 0723 0.7 / F 1356 3.1 / 1934 0.9	**9** 0213 3.3 / 0818 0.7 / SA 1446 3.0 / 2015 1.1		**24** 0148 3.5 / 0752 0.5 / SU 1424 3.2 / 2002 0.9	**9** 0259 3.1 / 0855 1.0 / TU 1527 2.9 / 2054 1.3		**24** 0302 3.4 / 0905 0.8 / W 1539 3.1 / 2124 1.1
10 0124 3.5 / 0721 0.4 / TU 1353 3.4 / 1939 0.8		**25** 0100 3.2 / 0653 0.8 / W 1328 3.1 / 1906 0.9	**10** 0237 3.2 / 0850 0.7 / F 1517 3.0 / 2050 1.2		**25** 0203 3.3 / 0807 0.6 / SA 1441 3.0 / 2019 1.0	**10** 0255 3.2 / 0901 0.9 / SU 1530 2.9 / 2056 1.3		**25** 0231 3.4 / 0838 0.6 / M 1511 3.1 / 2049 1.0	**10** 0341 2.9 / 0941 1.2 / W 1614 2.8 / ◐ 2147 1.5		**25** 0358 3.2 / 1010 1.1 / TH 1639 3.0 / ◐ 2243 1.2
11 0209 3.3 / 0813 0.6 / W 1445 3.1 / 2027 1.1		**26** 0136 3.2 / 0732 0.8 / TH 1409 3.0 / 1946 1.0	**11** 0327 3.1 / 0944 0.9 / SA 1611 2.8 / 2145 1.4		**26** 0248 3.2 / 0857 0.8 / SU 1532 3.0 / 2112 1.1	**11** 0341 3.0 / 0949 1.0 / M 1618 2.8 / 2147 1.4		**26** 0320 3.3 / 0932 0.8 / TU 1604 3.0 / ◐ 2148 1.1	**11** 0434 2.8 / 1043 1.4 / TH 1711 2.7 / 2310 1.6		**26** 0507 3.0 / 1130 1.2 / F 1750 2.9
12 0259 3.1 / 0913 0.8 / TH 1542 2.9 / 2123 1.3		**27** 0217 3.1 / 0816 0.9 / F 1455 2.9 / 2033 1.2	**12** 0423 2.9 / 1043 1.0 / SU 1709 2.7 / ◐ 2250 1.5		**27** 0339 1.1 / 0958 0.8 / M 1630 2.9 / ◐ 2218 1.2	**12** 0432 2.9 / 1044 1.2 / TU 1713 2.7 / ◐ 2255 1.5		**27** 0416 3.1 / 1038 0.9 / W 1706 2.9 / 2304 1.2	**12** 0541 2.7 / 1157 1.4 / F 1818 2.7		**27** 0015 1.3 / 0627 2.9 / SA 1259 1.3 / 1907 2.9
13 0355 3.0 / 1021 1.0 / F 1646 2.8 / ◐ 2234 1.4		**28** 0303 3.0 / 0911 0.9 / SA 1549 2.8 / 2132 1.3	**13** 0523 2.8 / 1147 1.1 / M 1809 2.7		**28** 0438 3.1 / 1108 0.9 / TU 1735 2.9 / 2333 1.2	**13** 0531 2.8 / 1147 1.3 / W 1812 2.7		**28** 0523 3.1 / 1151 0.9 / TH 1814 2.9	**13** 0041 1.5 / 0654 2.6 / SA 1314 1.4 / 1925 2.8		**28** 0144 1.2 / 0752 3.0 / SU 1415 1.3 / 2022 3.1
14 0501 2.8 / 1135 1.1 / SA 1754 2.7 / 2356 1.5		**29** 0358 2.9 / 1023 1.0 / SU 1654 2.8 / ◐ 2253 1.3	**14** 0004 1.5 / 0625 2.8 / TU 1254 1.1 / 1911 2.7		**29** 0546 3.0 / 1217 0.9 / W 1841 2.9	**14** 0012 1.5 / 0634 2.7 / TH 1255 1.3 / 1914 2.7		**29** 0025 1.2 / 0637 3.0 / F 1309 1.1 / 1925 3.0	**14** 0202 1.4 / 0806 2.7 / SU 1419 1.3 / 2027 2.9		**29** 0249 1.0 / 0903 3.1 / M 1509 1.2 / 2121 3.3
15 0609 2.8 / 1251 1.1 / SU 1904 2.7		**30** 0503 2.9 / 1139 0.9 / M 1805 2.8	**15** 0120 1.4 / 0728 2.8 / W 1351 1.1 / 2009 2.8		**30** 0046 1.2 / 0655 3.0 / TH 1326 0.9 / 1947 3.0	**15** 0133 1.4 / 0740 2.7 / F 1358 1.2 / 2014 2.8		**30** 0147 1.1 / 0754 3.0 / SA 1419 1.1 / 2034 3.1	**15** 0256 1.2 / 0906 2.9 / M 1507 1.2 / 2119 3.1		**30** 0338 0.8 / 0957 3.2 / TU 1552 1.1 / 2210 3.4
		31 0008 1.3 / 0613 2.9 / TU 1249 0.9 / 1911 2.9						**31** 0253 1.0 / 0906 3.1 / SU 1516 1.0 / 2134 3.3			**31** 0420 0.7 / 1041 3.3 / W 1628 1.0 / 2252 3.5

Chart Datum: 1·71 metres below Ordnance Datum (Newlyn). HAT is 4·0 metres above Chart Datum.

》》 FREE monthly updates. Register at 《
www.reedsnauticalalmanac.co.uk

413

STANDARD TIME (UT)
For Summer Time add ONE hour in **non-shaded areas**

WICK   LAT 58°26'N   LONG 3°05'W
TIMES AND HEIGHTS OF HIGH AND LOW WATERS

Dates in red are SPRINGS
Dates in blue are NEAPS

YEAR 2016

SEPTEMBER

Time	m		Time	m
1 0456	0.6	**16**	0427	0.5
1120	3.4		1053	3.5
TH 1702	0.9	F	1640	0.7
● 2331	3.6	○	2301	3.7
2 0530	0.5	**17**	0506	0.4
1157	3.4		1135	3.6
F 1735	0.8	SA	1719	0.6
			2344	3.8
3 0007	3.6	**18**	0546	0.3
0603	0.5		1216	3.7
SA 1231	3.4	SU	1759	0.5
1807	0.8			
4 0042	3.5	**19**	0026	3.9
0635	0.6		0626	0.3
SU 1303	3.3	M	1258	3.7
1838	0.9		1840	0.6
5 0115	3.4	**20**	0111	3.8
0707	0.7		0708	0.4
M 1335	3.2	TU	1341	3.6
1909	1.0		1923	0.7
6 0147	3.3	**21**	0156	3.7
0738	0.9		0752	0.7
TU 1407	3.1	W	1425	3.4
1941	1.1		2010	0.8
7 0221	3.2	**22**	0246	3.5
0810	1.0		0841	0.9
W 1442	3.0	TH	1515	3.2
2015	1.2		2108	1.0
8 0300	3.0	**23**	0344	3.2
0847	1.2		0945	1.2
TH 1523	2.9	F	1615	3.0
2058	1.4	◗	2232	1.2
9 0347	2.8	**24**	0456	3.0
0938	1.4		1111	1.5
F 1614	2.8	SA	1729	2.9
◗ 2206	1.5			
10 0449	2.7	**25**	0008	1.3
1101	1.6		0619	2.9
SA 1721	2.7	SU	1248	1.5
2351	1.6		1847	3.0
11 0610	2.6	**26**	0135	1.1
1228	1.6		0745	2.9
SU 1836	2.8	M	1404	1.4
			2004	3.1
12 0122	1.5	**27**	0237	0.9
0728	2.7		0851	3.1
M 1345	1.5	TU	1455	1.3
1945	2.9		2103	3.2
13 0225	1.2	**28**	0322	0.9
0835	2.9		0940	3.2
TU 1440	1.3	W	1534	1.2
2044	3.1		2150	3.4
14 0310	1.0	**29**	0359	0.8
0927	3.1		1021	3.3
W 1522	1.1	TH	1607	1.0
2133	3.3		2231	3.5
15 0349	0.8	**30**	0432	0.7
1011	3.3		1057	3.4
TH 1601	0.9	F	1639	0.9
2218	3.5		2307	3.5

OCTOBER

Time	m		Time	m
1 0503	0.6	**16**	0441	0.4
1130	3.4		1110	3.7
SA 1710	0.9	SU	1657	0.6
● 2341	3.5	○	2321	3.9
2 0534	0.6	**17**	0522	0.3
1201	3.4		1152	3.8
SU 1741	0.8	M	1739	0.5
3 0014	3.5	**18**	0006	3.9
0604	0.7		0604	0.4
M 1232	3.4	TU	1236	3.8
1812	0.9		1822	0.5
4 0046	3.4	**19**	0053	3.9
0634	0.8		0647	0.5
TU 1302	3.3	W	1319	3.7
1842	0.9		1908	0.6
5 0118	3.3	**20**	0141	3.7
0704	0.9		0732	0.8
W 1332	3.3	TH	1404	3.5
1913	1.0		1957	0.8
6 0151	3.2	**21**	0232	3.4
0734	1.1		0821	1.1
TH 1406	3.2	F	1454	3.3
1947	1.2		2058	1.0
7 0229	3.0	**22**	0332	3.2
0809	1.3		0922	1.4
F 1445	3.0	SA	1553	3.1
2027	1.3	◗	2219	1.2
8 0314	2.9	**23**	0443	3.0
0852	1.5		1045	1.6
SA 1532	2.9	SU	1705	3.0
2122	1.5		2348	1.2
9 0411	2.7	**24**	0602	2.9
1002	1.6		1220	1.6
SU 1633	2.8	M	1820	3.0
◗ 2305	1.5			
10 0529	2.7	**25**	0111	1.2
1143	1.7		0721	2.9
M 1749	2.8	TU	1339	1.6
			1934	3.1
11 0035	1.4	**26**	0212	1.1
0649	2.8		0826	3.0
TU 1304	1.6	W	1432	1.4
1902	2.9		2035	3.2
12 0145	1.2	**27**	0257	1.0
0758	2.9		0914	3.2
W 1406	1.4	TH	1511	1.3
2005	3.1		2124	3.3
13 0236	1.0	**28**	0333	0.9
0855	3.2		0954	3.3
TH 1453	1.2	F	1544	1.2
2059	3.3		2205	3.4
14 0319	0.7	**29**	0404	0.8
0942	3.4		1030	3.4
F 1535	0.9	SA	1615	1.0
2148	3.6		2241	3.4
15 0400	0.5	**30**	0435	0.8
1027	3.6		1102	3.4
SA 1616	0.7	SU	1647	1.0
2235	3.8	●	2316	3.5
		31	0505	0.8
			1133	3.5
		M	1718	0.9
			2348	3.5

NOVEMBER

Time	m		Time	m
1 0536	0.8	**16**	0545	0.5
1203	3.5		1215	3.8
TU 1750	0.9	W	1809	0.5
2 0021	3.4	**17**	0038	3.8
0606	0.9		0629	0.7
W 1234	3.4	TH	1300	3.7
1821	1.0		1857	0.6
3 0054	3.3	**18**	0128	3.7
0636	1.0		0714	0.9
TH 1305	3.4	F	1346	3.6
1853	1.0		1948	0.8
4 0128	3.2	**19**	0220	3.4
0707	1.2		0802	1.2
F 1339	3.3	SA	1435	3.4
1927	1.1		2047	0.9
5 0206	3.1	**20**	0317	3.2
0742	1.3		0857	1.4
SA 1417	3.2	SU	1531	3.3
2008	1.2		2156	1.1
6 0250	3.0	**21**	0421	3.0
0824	1.5		1006	1.6
SU 1502	3.1	M	1635	3.1
2059	1.4	◖	2310	1.2
7 0344	2.9	**22**	0529	2.9
0922	1.6		1126	1.7
M 1557	3.0	TU	1744	3.1
◖ 2220	1.4			
8 0453	2.8	**23**	0028	1.2
1056	1.7		0638	2.9
TU 1706	2.9	W	1252	1.7
2348	1.4		1852	3.0
9 0610	2.8	**24**	0134	1.2
1218	1.6		0745	3.0
W 1819	3.0	TH	1355	1.6
			1957	3.1
10 0100	1.2	**25**	0223	1.1
0718	3.0		0839	3.1
TH 1326	1.4	F	1441	1.4
1925	3.1		2051	3.2
11 0159	1.0	**26**	0301	1.1
0819	3.2		0923	3.2
F 1421	1.2	SA	1519	1.3
2024	3.4		2136	3.3
12 0249	0.8	**27**	0336	1.0
0912	3.4		1000	3.3
SA 1508	1.0	SU	1553	1.2
2119	3.6		2215	3.3
13 0334	0.6	**28**	0408	1.0
1000	3.6		1035	3.4
SU 1553	0.8	M	1627	1.1
2211	3.7		2252	3.4
14 0418	0.5	**29**	0440	1.0
1046	3.7		1107	3.5
M 1637	0.6	TU	1700	1.0
○ 2300	3.9	●	2326	3.4
15 0501	0.5	**30**	0512	1.0
1130	3.8		1139	3.5
TU 1723	0.5	W	1733	1.0
2349	3.9			

DECEMBER

Time	m		Time	m
1 0001	3.4	**16**	0026	3.7
0543	1.0		0614	0.8
TH 1211	3.5	F	1244	3.8
1805	1.0		1847	0.6
2 0036	3.3	**17**	0116	3.6
0615	1.1		0658	1.0
F 1244	3.5	SA	1330	3.7
1839	1.0		1936	0.7
3 0111	3.3	**18**	0204	3.4
0649	1.1		0742	1.1
SA 1319	3.4	SU	1416	3.6
1915	1.0		2027	0.9
4 0150	3.2	**19**	0255	3.2
0725	1.2		0827	1.3
SU 1357	3.3	M	1505	3.4
1955	1.1		2121	1.0
5 0232	3.1	**20**	0348	3.1
0807	1.4		0918	1.5
M 1440	3.2	TU	1600	3.2
2043	1.2		2220	1.2
6 0322	3.0	**21**	0445	2.9
0857	1.5		1021	1.6
TU 1530	3.1	W	1659	3.1
2145	1.4	◖	2325	1.3
7 0421	2.9	**22**	0546	2.8
1007	1.6		1135	1.7
W 1630	3.1	TH	1802	3.0
◖ 2304	1.2			
8 0531	2.9	**23**	0034	1.3
1131	1.6		0649	2.8
TH 1739	3.1	F	1258	1.7
			1907	3.0
9 0016	1.3	**24**	0137	1.3
0639	3.0		0752	2.9
F 1243	1.5	SA	1404	1.6
1847	3.2		2011	3.1
10 0122	1.0	**25**	0227	1.3
0743	3.2		0846	3.0
SA 1348	1.3	SU	1453	1.4
1952	3.3		2105	3.1
11 0220	0.9	**26**	0308	1.2
0842	3.3		0930	3.2
SU 1444	1.1	M	1533	1.3
2054	3.5		2150	3.1
12 0312	0.8	**27**	0345	1.1
0936	3.5		1009	3.3
M 1535	0.8	TU	1610	1.2
2152	3.6		2230	3.2
13 0359	0.7	**28**	0420	1.1
1026	3.7		1044	3.4
TU 1624	0.7	W	1645	1.1
2246	3.7		2307	3.3
14 0445	0.7	**29**	0453	1.1
1113	3.8		1119	3.5
W 1712	0.6	TH	1719	1.0
○ 2337	3.8	●	2344	3.3
15 0530	0.7	**30**	0526	1.0
1159	3.8		1153	3.5
TH 1800	0.5	F	1752	0.9
		31	0019	3.3
			0559	1.0
		SA	1228	3.5
			1827	0.9

Chart Datum: 1·71 metres below Ordnance Datum (Newlyn). HAT is 4·0 metres above Chart Datum.

⟩⟩ **FREE** monthly updates. Register at ⟨
www.reedsnauticalalmanac.co.uk ⟨

9.7.20 SCRABSTER

Highland **58°36'·61N 03°32'·61W** ❀❀❀♦♦♦♦☘☘☘

CHARTS AC 1954, 2162, 1462; Imray C68

TIDES -0240 Dover; ML 3·2; Duration 0615

Standard Port WICK (←—)

Times				Height (metres)			
High Water		Low Water		MHWS	MHWN	MLWN	MLWS
0200	0700	0100	0700	3·5	2·8	1·4	0·7
1400	1900	1300	1900				
Differences SCRABSTER							
–0255	–0225	–0240	–0230	+1·5	+1·2	+0·8	+0·3
GILLS BAY							
–0150	–0150	–0202	–0202	+0·7	+0·7	+0·6	+0·3
STROMA							
–0115	–0115	–0110	–0110	–0·4	–0·5	–0·1	–0·2
LOCH ERIBOLL (Portnancon)							
–0340	–0255	–0315	–0255	+1·6	+1·3	+0·8	+0·4
KYLE OF DURNESS							
–0350	–0350	–0315	–0315	+1·1	+0·7	+0·4	–0·1
SULE SKERRY (59°05'N 04°24'W)							
–0320	–0255	–0315	–0250	+0·4	+0·3	+0·2	+0·1
RONA (59°08'N 05°49'W)							
–0350	–0350	–0350	–0340	–0·3	–0·4	–0·2	–0·4

SHELTER Very good except for swell in NW and N winds. Yachts usually lie in the Inner (0·9–1·2m) or Centre Basins (0·9–2·7m). Pontoons in inner hbr. ⚓ is not advised.

• A good hbr to await the right conditions for E-bound passage through Pentland Firth (see 9.7.5).

• Beware floating creel lines in E of hbr.

NAVIGATION WPT 58°36'·58N 03°32'·09W, 278°/0·25M to E pier lt. Can be entered H24 in all weathers. Beware FVs and the Orkney ferries.

LIGHTS AND MARKS No ldg marks/lts but entry is simple once the conspic ice plant tr and/or pier lts have been located. Do not confuse hbr lts with the shore lts of Thurso.

COMMUNICATIONS (Code 01847) CGOC (01224) 592334; Police 101; Dr 893154. HM 892779 mobile 07803 290366.

Call HM VHF Ch 12 16 (H24) for berthing directions, before entering hbr. From the W reception is very poor due to masking by Holborn Head.

FACILITIES www.scrabster.co.uk ⚓£13·00/night on quay, pontoon by request ⬡ ⬟ ⚓ 🔧 🛢 ⬡ ⬡ ⬡ ⬟(15/30/99t). **Pentland Firth YC** ⚓ ⬡ ✕ ⬟ (keys held by Duty HM).**Thurso** 🛢 ⬟ ✕ ⬟ ✉ Ⓑ ⬟ ✈ (Wick). **Ferries:** Stromness 3/day 1½ Hrs; Northlink (www.northlinkferries.co.uk). Faroe Islands (Torshavn) weekly (seasonal) 12 Hrs; Bergen weekly (seasonal) 16 Hrs; Smyril (www.smyril.co.uk).

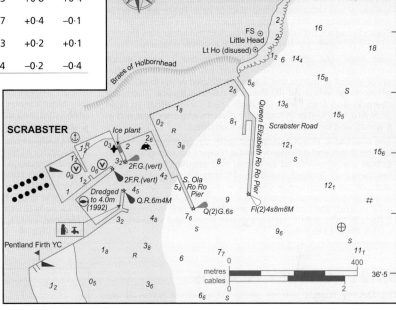

ANCHORAGES BETWEEN SCRABSTER AND CAPE WRATH

KYLE OF TONGUE, Highland, **58°31'·97N 04°22'·67W** (ent). AC 2720, 1954. HW +0050 on Ullapool; HW ht –0·4m; see 9.7.20. The Kyle runs about 7M inland.

Entry (see 9.7.5) should not be attempted in strong N winds.

⚓ at Talmine (W of Rabbit Is) protected from all but NE winds; at Skullomie Hr, protected from E'lies; off Mol na Coinnle, a small bay on SE side of Eilean nan Ron, protected from W and N winds; off S of Rabbit Is, protected from W to N winds. No ldg lts/marks. Facilities: Limited 🍴 at Talmine (½M from slip) or at Coldbachie (1½M from Skullomie).

LOCH ERIBOLL, Highland, **58°32'·58N 04°37'·48W**. AC 2076. HW –0345 on Dover; ML 2·7m. See 9.7.20. Enter between Whiten Hd and Klourig Is in W sector of White Hd lt. See 9.7.4.

In SW winds fierce squalls funnel down the loch.

Yachts can enter drying Rispond Hbr and dry out alongside; no lts/marks and very limited facilities. Good ⚓s at: Rispond Bay on W side of loch, ent good in all but E winds, in approx 5m; off Portnancon in 5·5m; at the head of the loch; Camus an Duin and in bays to N and S of peninsula at Heilam on E side of loch. Beware numerous marine farms in the S half of the loch.

PENTLAND FIRTH TO CAPE WRATH

(AC 1954) Dunnet B, S of Dunnet Hd (lt) gives temp ‡ in E or S winds, but dangerous seas enter in NW'lies. On W side of Thurso B is Scrabster sheltered from S and W. ▶ *Between Holborn Hd and Strathy Pt the E-going stream begins at HW Ullapool –0150, and the W-going at HW Ullapool +0420, sp rates 1·8kn. Close to Brims Ness off Ushat Hd the sp rate is 3kn, and there is often turbulence.* ◀

SW of Ushat Hd the Dounreay power stn is conspic, near shore. Dangers extend 2½ca seaward off this coast.▶ *Along E side of Strathy Pt (lt) an eddy gives almost continuous N-going stream, but there is usually turbulence off the Pt where this eddy meets the main E or W stream.*◀ Several small bays along this coast give temp ‡ in offshore winds, but should not be used or approached with wind in a N quarter.

Kyle of Tongue is entered from E through Caol Raineach, S of Eilean nan Ron, or from N between Eilean Iosal and Cnoc Glass. There is no chan into the kyle W of Rabbit Is, to which a drying spit extends 0·5M NNE from the mainland shore. Further S there is a bar across entrance to inner part of kyle. There are ‡s on SE side of Eilean nan Ron, SE side of Rabbit Is, off Skullomie, or S of Eilean Creagach off Talmine. Approach to the latter runs close W of Rabbit Islands, but beware rks to N and NW of them.

Loch Eriboll (AC 2076) provides secure ‡s, but in strong winds violent squalls blow down from mountains. Eilean Cluimhrig lies on W side of entrance; the E shore is fringed with rks up to 2ca offshore. At White Hd (lt) the loch narrows to 6ca. There are chans W and E of Eilean Choraidh. Best ‡s in Camas an Duin (S of Ard Neackie) or in Rispond B close to entrance (but not in E winds, and beware Rispond Rk which dries).

The coast to C Wrath is indented, with dangers extending 3ca off the shore and offlying rks and Is. Once a yacht has left Loch Eriboll she is committed to a long and exposed passage until reaching Loch Inchard. The Kyle of Durness is dangerous if the wind or sea is onshore. ▶ *Give Cape Wrath a wide berth when wind-against-tide which raises a severe sea.*◀ A firing exercise area extends 8M E of C. Wrath, and 4M offshore. When in use, R flags or pairs of R lts (vert) are shown from E and W limits, and yachts should keep clear.

ORKNEY ISLANDS

(AC 2249, 2250) The Islands are mostly indented and rocky, but with sandy beaches especially on NE sides. ▶*Pilotage is easy in good vis, but in other conditions great care is needed since tides run strongly. For details refer to* Clyde Cruising Club's Orkney Sailing Directions *and the* Admiralty Tidal Atlas NP 209. *When cruising in Orkney it is essential to understand and use the tidal streams to the best advantage, avoiding the various tide races and overfalls, particularly near sp.*◀

A good engine is needed since, for example, there are many places where it is dangerous to get becalmed. Swell from the Atlantic or North Sea can contribute to dangerous sea conditions, or penetrate to some of the ‡s. During summer months winds are not normally unduly strong, and can be expected to be Force 7 or more on about two days a month. But in winter the wind reaches this strength for 10-15 days per month, and gales can be very severe in late winter and early spring. Cruising conditions are best near midsummer, when of course the hours of daylight are much extended.

Stronsay Firth and Westray Firth run SE/NW through the group. The many good ‡s include: Deer Sound (W of Deer Ness); Bays of Firth and Isbister, and off Balfour in Elwick B (all leading fm Wide Firth); Rysa Snd, B of Houton, Hunda Snd (Scapa Flow); Rousay Sound; and Pierowall Road (Westray). Plans for some of these are on AC 2622. The European Marine Energy test site with underwater turbines lies between Eday and Muckle Green Holm, and an offshore wave test site marked by cardinal buoys NNW of Hoy Mouth (emec.org.uk). A major oil terminal and associated prohibited area is at Flotta, on the S side of Scapa Flow.

▶ *Tide races or dangerous seas occur at the entrances to most of the firths or sounds when the stream is against strong winds. This applies particularly to Hoy Sound, Eynhallow Sound, Papa Sound (Westray), Lashy Sound, and North Ronaldsay Firth. Also off Mull Head, over Dowie Sand, between Muckle Green Holm and War Ness (where violent turbulence may extend right across the firth), between Faraclett Head and Wart Holm, and off Sacquoy Hd. Off War Ness the SE-going stream begins at HW Aberdeen +0435, and the NW-going at HW Aberdeen –0200, sp rates 7kn.* ◀

9.7.21 ORKNEY ISLANDS

The Orkney Islands number about 70, of which some 24 are inhabited. They extend 5 to 50M NNE from Duncansby Hd, are mostly low-lying, but Hoy in the SW of the group reaches 475m (1560ft). Coasts are rocky and much indented, but there are many sandy beaches. A passage with least width of about 3M runs NW/SE through the group. The islands are separated from Scotland by the Pentland Firth, a very dangerous stretch of water. The principal island is Mainland (or Pomona) on which stands Kirkwall, the capital. For inter island ferries see www.orkneyferries.co.uk.

Severe gales blow in winter and early spring. The climate is mild but windy, and very few trees grow. There are LBs at Longhope, Stromness and Kirkwall.

CHARTS AC 2162, 2250 and 2249, at medium scale. For larger scale charts, see under individual hbrs; Imray C68

TIDES Wick (9.7.19) is the Standard Port. Tidal streams are strong, particularly in Pentland Firth and in the islands' firths and sounds.

SHELTER/MARINA CHARGES There are piers (fender board advised) at all main islands www.sailnorthscotland.com. For details of charges at marinas or piers throughout the Orkney Islands, except St Margaret's Hope, www.orkneymarinas.co.uk. Some of the many ‡s are listed:

MAINLAND **Scapa Bay**: good except in S winds. No yacht berths alongside pier due to heavy hbr traffic. Only ents to Scapa Flow are via Hoy Snd, Hoxa Snd or W of Flotta.
St Marys (Holm/Ham): 58°53'·80N 02°54'·50W; N side of Kirk Sound; ‡ in B of Ayre or berth E side of Pier, HW±4. ⌂ ⌂ ⊠ ✕ ⌂ Bus to Kirkwall/Burwick Ferry.

Kirk Sound (E ent): ⚓ N of Lamb Holm; beware fish cages.

Deer Sound: ⚓ in Pool of Mirkady or off pier on NW side of sound; very good shelter, no facilities.

Hunda Sound: good ⚓ in all winds.

SOUTH RONALDSAY/BURRAY St Margaret's Hope/Water Sound: ❀❀✿⚓⚓✿✿. Ldg Lts, F.G, 174°. ⚓ in centre of bay; or ⌑ £7/day at pier (but limited space and keep clear of ferry berth on S side of pier), beware salmon farm. HM ☎01856 831440, mob 07879 688040; Dr 831206. ⚓ ⌂ ⌂ ⌂ ✉ 🛒 ✕ 🛏 Bus to Kirkwall, vehicle ferry to Gill's Bay/Caithness.

Widewall Bay: ⚓ sheltered except SW'lies.

FLOTTA 58°50'·20N 03°07'·90W; berth on Sutherland Pier, SW of oil terminal. HM ☎01856 701411. ⌂ ⌂ 🛒 ✉.

HOY Long Hope: 58°48'·07N 03°12'·25W; ⚓ E of Pier, used by ferry, or berth on pier (safest at slack water). 2 ⚓'s < 18 tonnes close to Pier. ☎01856 701263; Dr 701209. Facilities: ⚓ ⌂ ⌂ ⌂ ✉ 🛒 🛏.

Lyness: 58°50'·20N 03°11'·40W; berth on Pier; not piles(disused); ⚓ in Ore Bay. HM ☎01856 791387. ⚓ ⌂ ⌂ ✉ 🛏. Avoid fish cages.

Pegal Bay: good ⚓ except in strong W winds.

ROUSAY Wyre Sound: 59°09'·40N 02°44'·70W; ⚓ E of Rousay Pier or berth on it ✉ 🛒 ✕. Piermaster ☎01856 821261.

EDAY Fersness Bay: good holding, sheltered from S winds. Dr ☎ 01857 622243.

Backaland Bay: 59°09'·40N 02°44'·70W: berth on Pier clear of ferry; Piermaster ☎01856 622282; or ⚓ to NW. Beware cross tides. ⚓ ⚑ ⚑ ✉ 🛒.

Calf Sound: ⚓ in Carrick B; good shelter from SW-NW'lies.

PAPA WESTRAY Bay of Moclett 58°21'·40N 02°26'·30W: Good ⚓ but open to S. Pier. Piermaster ☎01857 644259;

South Wick: ⚓ off the old pier or ESE of pier off Holm of Papa. Backaskaill: ⌂ ⌂ ✉ 🛒.

SANDAY Loth Bay: 59°11'·50N 02°41'·80W; berth on Pier clear of ferry. Piermaster ☎01857 600227; Beware strong tides.

Kettletoft Bay: 59°13'·90N 02°35'·80W; ⚓ in bay or berth on Pier, very exposed to SE'lies. HM ☎01857 600227, Dr 600221; ⚓ ⌂ ⌂ ⚑ ✉ Ⓑ 🛒 hotel.

North Bay: on NW side of island, exposed to NW.

Otterswick: good ⚓ except in N or E winds.

NORTH RONALDSAY South Bay: 58°21'·40N 02°26'·30W; ⚓ in middle of bay or berth on Pier. Piermaster ☎01857 633239; open to S & W and swell. ✉ 🛒.

Linklet Bay: ⚓ off jetty at N of bay, open to E.

NAVIGATION From the mainland, appr from Scrabster to Stromness and Scapa Flow via Hoy Mouth and Hoy Sd. From the Moray Firth keep well E of the Pentland Skerries if going N to Kirkwall. If bound for Scapa Flow via Hoxa Sd, keep close to Duncansby Hd, passing W of the Pentland Skerries and between Swona and S Ronaldsay. Keep clear of Lother Rk (dries 1·8m) off SW tip of S Ronaldsay. Time this entry for slack water in the Pentland Firth (about HW Aberdeen –1¾ and +4). Beware of tankers off Flotta oil terminal and in S part of Scapa Flow, where are the remains of the German WW1 Battle Fleet; classified as Historic Wrecks (see 9.0.3) protected, but authorised diving allowed.

Elsewhere in Orkney navigation is easy in clear weather, apart from the strong tidal streams in all the firths and sounds. Beware races and overfalls off Brough of Birsay (Mainland), Noup Head (Westray) and Dennis Head (N Ronaldsay). Keep a good lookout for the many lobster pots (creels).

LIGHTS AND MARKS The main hbrs and sounds are well lit; for details see 9.7.4. Powerful lts are shown offshore from Cantick Hd, Graemsay Island, Copinsay, Auskerry, Kirkwall, Brough of Birsay, Sanday Island, N Ronaldsay and Noup Hd.

Orkney Hbrs Navigation Service (call: *Orkney Hbr Radio*, Ch 09 11 20 16 (H24)) covers Scapa Flow and apprs, Wide Firth, Shapinsay Sound and Kirkwall Bay.

COMMUNICATIONS Area Code for islands SW of Stronsay and Westray Firths is 01856; islands to the NE are 01857.

MEDICAL SERVICES Doctors are available at Kirkwall, Stromness, Rousay, Hoy, Shapinsay, Eday, St and N Ronaldsay, Stronsay, Sanday and Westray (Pierowall); Papa Westray is looked after by Westray. The only hospital (and dentist) are at Kirkwall. Serious cases are flown to Aberdeen (1 hour).

MARINE FARMS Fish cages/farms approx 30m x 50m may be found anywhere in sheltered waters within anchoring depths. Some are well buoyed, others are marked only by poles. Too many to list but the following will show the scale of the operations:

Beware **salmon cages** (may be marked by Y buoys/lts) at:

Kirkwall Bay	Toy Ness (Scapa Flow)
Rysa Sound	St Margaret's Hope
Bring Deeps	Backaland Bay (Eday)
Pegal Bay (Hoy)	Hunda Sound
Lyrawa Bay	Kirk Sound
Ore Bay (Hoy)	Carness Bay
Widewall Bay (S Ronaldsay)	Bay of Ham
	Bay of London (Eday)

Beware **oysters and longlines** at:

Widewall Bay	Bay of Firth
Swanbister Bay	Damsay Sound
Water Sound	Millburn Bay (Gairsay)
Hunda Sound	Pierowall
Deer Sound	Bay of Skaill (Westray)
Inganess Bay	Longhope

MINOR HARBOURS IN THE ORKNEY ISLANDS

HOUTON BAY, Mainland, **58°54'·85N 03°11'·33W**. AC 35, 2568. HW –0140 on Dover, –0400 on Aberdeen; HW ht +0·3m on Kirkwall; ML 1·8m; Duration 0615. ⚓ in the bay in approx 5·5m at centre, sheltered from all winds. Ent is to the E of Holm of Houton marked by bn Fl R 3s 7m 5M; chan dredged 3·5m for 15m each side of ldg line. Keep clear of shipping/ferries plying to Flotta. Ldg lts 316°: front Fl G 3s 8m, rear FG 16m; both R △ on W pole, B bands. Ro Ro terminal in NE corner marked by Iso R 4s with SHM Fl G on edge of ldg line. Bus to Kirkwall; 〰 close E of piers. Yachtsmen may contact M.Grainger ☎01856 811397 for help.

SHAPINSAY, Orkney Islands, **59°01'·97N 02°54'·10W**. AC 2249, 2584. HW –0015 on Dover, –0330 on Aberdeen; HW ht –1·0m on Aberdeen. Good shelter in Elwick Bay off Balfour on SW end of island in 2·5-3m. Enter bay passing W of Helliar Holm which has lt Fl WRG 10s on S end. Keep mid-chan. Balfour Pier lt Q WRG 5m 3/2M; vis 270°-G-010°-W-020°-R-090°. Piermaster ☎01856 711358; Tides in The String reach 5kn at sp. Facilities: ⚓ ⌂ ⌂ ✉ shop, 🛏.

AUSKERRY, Orkney Islands, **59°02'·02N 02°34'·65W**. AC 2250. HW –0010 on Dover, -0315 on Aberdeen; HW ht -1m on Aberdeen. Small island at ent to Stronsay Firth with small hbr on W side. Safe ent and good shelter except in SW winds. Ent has 3·5m; 1·2m alongside pier. Yachts can lie secured between ringbolts at ent and the pier. Auskerry Sound and Stronsay Firth are dangerous with wind over tide. Auskerry lt at S end, Fl 20s 34m 18M, W tr. No facilities.

PIEROWALL, Westray, **59°19'·32N 02°58'·51W**. AC 2250, 2622. HW –0135 on Dover; ML 2·2m; Duration 0620. See 9.7.24. The bay is a good ⚓ in 2-7m and well protected. Marina pontoons in deep water alongside pier at Gill Pt. From S, beware Skelwick Skerry rks, and from the N the rks extending approx 1ca off Vest Ness. The N ent via Papa Sound needs local knowledge; tide race on the ebb. A dangerous tide race runs off Mull Hd at the N of Papa Westray. Lights: E Pier Hd Fl WRG 3s 7m 11/7M. W Pier Hd 2 FR (vert) 4/6m 3M. VHF Ch 16. HM ☎01857 677216. **Marina** ☎07810 465784 www. orkneymarinas.co.uk, £2.00m (see 9.7.5) ⚓ ⚓. Services: ⌂ ⌂ ⚓ ✉ Ⓑ 🛒 🛏 ✕. Dr 01857 677209.

RAPNESS: 58°14'·90N 02°51'·50W, berth on Pier clear of Ro-Ro. Piermaster ☎01857 677212; Open to SSW.

9.7.22 STROMNESS

Orkney Islands, Mainland **58°57'·78N 03°17'·72W**
⊕⊕⊚♦♦♦♧♧♧

CHARTS AC 2249, 2568; Imray C68

TIDES −0145 Dover; ML 2·0; Duration 0620

Standard Port WICK (←—)

Times				Height (metres)			
High Water		Low Water		MHWS	MHWN	MLWN	MLWS
0000	0700	0200	0700	3·5	2·8	1·4	0·7
1200	1900	1400	1900				
Differences STROMNESS							
−0225	−0135	−0205	−0205	+0·1	−0·1	0·0	0·0
ST MARY'S (Scapa Flow)							
−0140	−0140	−0140	−0140	−0·2	−0·2	0·0	−0·1
BURRAY NESS (Burray)							
+0005	+0005	+0015	+0015	−0·2	−0·3	−0·1	−0·1
WIDEWALL BAY (S Ronaldsay)							
−0155	−0155	−0150	−0150	+0·1	−0·1	−0·1	−0·3
BUR WICK (S Ronaldsay)							
−0100	−0100	−0150	−0150	−0·1	−0·1	+0·2	+0·1
MUCKLE SKERRY (Pentland Firth)							
−0025	−0025	−0020	−0020	−0·9	−0·8	−0·4	−0·3

SHELTER Very good. Northern Lights Board have sole use of pier near to ldg lts. Marina in N of hbr; or ⚓ where shown.

NAVIGATION WPT 58°56'·93N 03°17'·00W, 317°/0·9M to front ldg lt. Turn N to pass between N Pier and Coplands Pier to marina.

- Entry from the W should not be attempted with strong wind against tide due to heavy overfalls.
- If entering against the ebb, stand on to avoid being swept onto Skerry of Ness.
- Tides in Hoy Sound are >7kn at springs. No tidal stream in hbr.

LIGHTS AND MARKS For Hoy Sound, ldg lts 104° on Graemsay Is: front lso 3s 17m 15M, W tr; rear Oc WR 8s 35m 20/16M, ldg sector is 097°-R-112°. Skerry of Ness, Fl WG 4s 7m 7/4M; shore-W-090°-G-shore. Hbr ldg lts 317°, both FR 29/39m 11M (H24), W trs, vis 307°-327°. Coplands Pier 2 F G. Stromness Marina ent. Fl(2)R 5s.

COMMUNICATIONS (Code 01856) CGOC 01595 692976; Police 850222; Dr 850205; Dentist 850658. HM 07810 465825; Fuel 851286. For advice Yachtsmen may contact: **Orkney Marinas** ☎871313.

VHF Ch 14 16 (0900-1700 LT). (See also 9.7.23.)

FACILITIES **Stromness Marina:** ☎07810 465825, 10♥ £2.00m (see 9.7.5), ▬ 🚿 ♦ ▮⛽ 🚽 📞(0800-2100 ☎8851286) ⚒ ⚓ 🏪 ⓔ 🛟(mob 30t). **Town** ⚓ 🏠🏪⛽✉🏦®🛒 ✕🍴⇌ (Ferries to Scrabster, bus to Thurso), ✈ (Kirkwall).

9.7.23 KIRKWALL

Orkney Islands, Mainland **58°59'·30N 02°57'·70W**
⊕⊕⊚♦♦♧♧♧

CHARTS AC 2250, 2249, 2584, 1553; Imray C68

TIDES −0045 Dover; ML 1·8; Duration 0620

Standard Port WICK (←—)

Times				Height (metres)			
High Water		Low Water		MHWS	MHWN	MLWN	MLWS
0000	0700	0200	0700	3·5	2·8	1·4	0·7
1200	1900	1400	1900				
Differences KIRKWALL							
−0042	−0042	−0041	−0041	−0·5	−0·4	−0·1	−0·1
DEER SOUND							
−0040	−0040	−0035	−0035	−0·3	−0·3	−0·1	−0·1
TINGWALL							
−0200	−0125	−0145	−0125	−0·4	−0·4	−0·1	−0·1

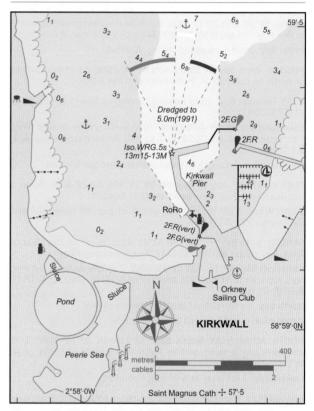

SHELTER Good except in N winds or W gales when there is a surge at the ent. Berth in marina on E of main pier, or SW end of main pier in inner hbr (very full in Jun/Jul). Safe ⚓ between pier and Crow Ness Pt. Hatston RoRo terminal in NW of bay (2 FG (vert)).

NAVIGATION WPT 59°01'·37N 02°57'·10W, 188°/2·2M to pier hd lt. Appr in W sector of pier hd lt, leaving newly extended Hatston Pier (2FG) / commercial terminal to W. Bay is shoal to SW beyond hbr.

LIGHTS AND MARKS See 9.7.4 and chartlet. Appr with St Magnus Cathedral (conspic) brg about 185°.

COMMUNICATIONS (Code 01856) CGOC (01595) 692976; ⚕ 873802; Police 872241; Dr 888000 (Ⓗ). HM 872292; Port Office 873636; Fuel 873105.

Kirkwall Hbr Radio VHF Ch 14 16 (0800-1700 LT). Orkney Hbrs Navigation Service, call: *VTS Orkney* Ch 09 11 20 16 (0915–1715).

FACILITIES **Kirkwall Marina** ☎879600 (berths 07810 465835), 71 inc ♥ £2.00m (see 9.7.5). **Pier** 🚿 ▮(HO ☎873105) ♦ 🛟(mob 25 t) ⚓; **N and E Quays** M; **Orkney SC** ☎872331, ⚓ ▬ 🛟⚡(£15/4days). **Town** 🚿♦▮⛽🏪⚒✂®🏦⛽✉Ⓑ🛒✕🍴 @ Library(HO). Ferries to Scrabster, Aberdeen and Shetland, ✈.

9.7.24 STRONSAY

Orkney Islands, Stronsay 59°08'·57N 02°36'·01W ✿✿✿✿✿✿✿✿✿

CHARTS AC 2250, 2622; Imray C68

TIDES As Dover; ML 1·7; Duration 0620

Standard Port WICK (←—)

Times				Height (metres)			
High Water		Low Water		MHWS	MHWN	MLWN	MLWS
0000	0700	0200	0700	3·5	2·8	1·4	0·7
1200	1900	1400	1900				
Differences WHITEHALL (Stronsay)							
–0030	–0030	–0025	–0030	–0·1	0·0	+0·2	+0·2
LOTH (Sanday)							
–0045	–0045	–0055	–0105	–0·4	–0·3	+0·1	+0·2
EGILSAY (Rousay Sound)							
–0125	–0125	–0125	–0125	–0·1	0·0	+0·2	+0·1
KETTLETOFT PIER (Sanday)							
–0030	–0025	–0025	–0025	0·0	0·0	+0·2	+0·2
RAPNESS (Westray)							
–0205	–0205	–0205	–0200	+0·1	+0·1	+0·2	0·0
PIEROWALL (Westray)							
–0150	–0150	–0145	–0145	+0·2	0·0	0·0	–0·1

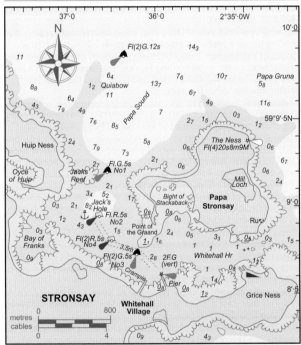

SHELTER Good from all winds. Good ⚓ between seaward end of piers, or berth on outer end of W pier and contact HM. The extended E pier head is berth for Ro-Ro ferry. There are many other sheltered ⚓s around the bay.

NAVIGATION WPT 59°09'·82N 02°36'·31W, Quiabow SHM lt buoy, 189°/6·5ca to No 1 lt buoy. 800m NE of Huip Ness is Quiabow, a submerged rk. Jack's Reef extends 400m E from Huip Ness, and is marked by No 1 SHM lt buoy. A bank extends 350m SW from Papa Stronsay. Crampie Shoal is in mid-chan, marked by No 3 buoy. The buoyed chan to Whitehall pier is dredged 3·5m. Spit to E of Whitehall pier extends 400m N. The E ent is narrow and shallow and should not be attempted.

LIGHTS AND MARKS See chartlet.

COMMUNICATIONS (Code 01857) CGOC (01595) 692976; Police (01856) 872241; Dr 616321. Piermaster 07901 984328. See Kirkwall for VHF comms.

FACILITIES W Pier ⚓✕🛒; Main (E) Pier ⚓✕⛽🛒 clear of ferry. **Village (Whitehall)** ⛽🍴✉🅱🚿🛒🚈 (Ferry to Scrabster, bus to Thurso) ✈.

(AC 3281, 3282, 3283) ▶ *These Islands mostly have bold cliffs and are relatively high, separated by narrow sounds through which the tide runs strongly, so that in poor vis great care is needed. Avoid sp tides, swell and wind against tide conditions. The tidal flow around the Shetland Islands rotates as the cycle progresses. When the flood begins, at –0400 HW Dover, the tidal flow is to the E, at HW Dover it is S, at Dover +0300 it is W, and at –0600 it is N.* ◀

Between Mainland and Whalsay passage can be made E or W of W Linga with spring rates of about 2kts with weaker streams to be found to the W in Lunning Sound.

▶ *Slack water at Lunning Sound occurs simultaneously at Lerwick, either 4 hours before or 2 hours after HW Lerwick.* ◀

Passage further Northward can be made through Yell Sound or Colgrave/Bluemell Sounds with shelter at Cullivoe.

▶ *Tides in Yell Sound are slack at the same time as high and low water in Lerwick. The stream normally runs at about 4 kts between the islands at the narrowest part between Yell and the Mainland but may reach 8 kts at springs. Colgrave Sound to the W of Yell experiences somewhat less tidal stream.* ◀

Bluemull Sound between Yell and Unst is a longer narrower passage. Well marked, passage is best attempted in daylight keeping to the middle of the channel clear of the fishing vessels en-route to Cullivoe.

▶ *The stream runs at up to 6 knts at springs for longer duration and may be carried throughout a transit.* ◀

Although there are many secluded and attractive ⚓s, remember that the weather can change very quickly, with sudden shifts of wind. Also beware salmon fisheries and mussel rafts (unlit) in many Voes, Sounds and hbrs. Lerwick is the busy main port and capital; other hbrs are Scalloway, Walls (Vaila Sound), Uyesound (S Unst), Mid Yell (E Yell), Cullivoe (N Yell), Burravoe (S Yell), Symbister (Whalsay, Out Skerries (Skerries Isle) and Balta Sound. Refer to the CCC's *Shetland Sailing Directions*.

▶ *Coming from the S, beware a most violent and dangerous race (roost) off Sumburgh Hd (at S end of Mainland) on both streams. Other dangerous areas include between Ve Skerries and Papa Stour; the mouth of Yell Sound with strong wind against N-going stream; and off Holm of Skaw (N end of Unst). Tidal streams run mainly NW/SE and are not strong except off headlands and in the major sounds; the Admiralty Tidal Atlas NP 209 gives detail. The sp range is about 2m.* ◀

The 50M passage from Orkney can conveniently be broken by a stop at Fair Isle (North Haven). ▶ *Note that races form off both ends of the Is, especially S (Roost of Keels).* ◀

Recommended Traffic Routes: NW-bound ships pass to the NE (no closer than 10M to Sumburgh Hd) or SW of Fair Isle; SE-bound ships pass no closer than 5M off N Ronaldsay (Orkney). Lerwick to Bergen, Norway, is about 210M.

FLOOD
Shetland Islands
(a) Dover –0410 to Dover +0020
(b) Dover –0400 to Dover +0030
(c) Dover –0530 to Dover +01 00
(d) Dover –0400 to Dover –0200

(b) 1.5k
(c) 6.0k
(d)
(a) Violent Race

EBB
Shetland Islands
(a) Dover +0050 to Dover –0410
(b) Dover +0130 to Dover –0500
(c) Dover +01 00 to Dover –0530
(d) Dover +0200 to Dover +0500

(b) 1.5k
(c) 6.0k
(d)
(a) Violent Race

9.7.25 SHETLAND ISLANDS

The Shetland Islands number about 100 islands, holms and rks with fewer than 20 inhabited. Lying 90 to 150M NNE of the Scottish mainland, the biggest island is Mainland with Lerwick (9.7.26), the capital, on the E side. Scalloway, the only other town and old capital, is on the W side. At the very S is Sumburgh airport, with other airstrips at Baltasound, Scalsta and Tingwall. Two offlying islands of the Shetland group are Fair Isle, 20M SSW of Sumburgh Hd and owned by the NT for Scotland, and Foula, 12M WSW of Mainland. Mariners in the vicinity of inter island ferry ports should check carefully for ferries entering or leaving the following ports and allow them free passage. (Fair Isle North Haven, Grutness, Dury Voe, Vidlin Voe, Symbister Out Skerries, Mossbank, Ulsta, Gutcher, Belmont , NW Fetlar and West Burrafirth.)

There are LBs at Lerwick and Aith. The CGOC Lerwick ☎(01595) 692976, with an Auxillary Stn at Fair Is.

CHARTS AC: medium scale 3281, 3282, 3283; larger scale 3271, 3272, 3292, 3293, 3294, 3295, 3297, 3298.

TIDES Standard Port Lerwick. Tidal streams run mostly N-S or NW-SE; in open waters to the E and W they are mostly weak. Rates >6kn can cause dangerous disturbances at the N and S extremities of the islands and in the two main sounds (Yell Sound and BlueMull/Colgrave Sounds). Keep 3M off Sumburgh Head to clear a dangerous race (Röst) or pass close inshore.

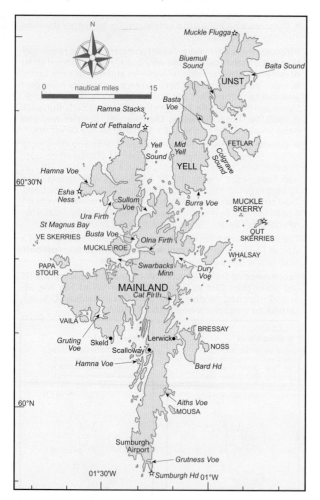

SHELTER Weather conditions are bad in winter; yachts should only visit Apr-Sept. Around mid-summer it is daylight H24. Some small marinas are asterisked* in the column on the right (see www. visitshetland.com); they may have room for visitors. Of the many ⚓s, the following are safe to enter in most conditions.

NAVIGATION A careful lookout must be kept for marine farms, mostly marked by Y buoys and combinations of Y lts. Clyde CC's *Shetland Sailing Directions and Anchorages* are essential for visitors. Local magnetic anomalies may be experienced.

LIGHTS AND MARKS See 9.7.4. Powerful lighthouses at Fair I, Sumburgh Hd, Kirkabister Ness, Bound Skerry, Muckle Flugga, Pt of Fethaland, Esha Ness and Foula. All inter-island ferry ports have leading sectored lights.

COMMUNICATIONS (Code 01595; 01806 for Sullom Voe) CGOC 692976; Sullom Voe Port Control (01806) 242551, 🖷 242237; ⌨ 692239; Forecaster (01806) 242069; Sumburgh Airport (01950) 460654. Port Radio services see 9.7.26. No Coast Radio Station.

FACILITIES All stores obtainable in Lerwick and Scalloway. Elsewhere yachts should be prepared for extended offshore cruising and obtain water and diesel as and when they can:

MAINLAND (anti-clockwise from Sumburgh Head)

Grutness Voe: 1·5M N of Sumburgh Hd, a convenient passage ⚓, open to NE. Beware 2 rocks awash in mid-ent.

Pool of Virkie*: 59°53'N 01°17W; Close N of Grutness Voe. Ness Boating Club Marina ☎01950 477260, ⌂ (pontoon 1🅥< 1m £10/craft/week), ⚓ ⚓ 🔱 WC. Shallow and narrow/entrance.

Cat Firth: excellent shelter, ⚓ in approx 6m. Facilities: ✉ (Skellister), ⚓ ✉ 🍴 (Lax Firth).

Grunna Voe: off S side of Dury Voe, good shelter and holding, ⚓ in 5-10m; beware prohib ⚓ areas. Facilities: ⚓ ✉ 🍴 (Lax Firth).

South of Yell Sound*: Tides –0025 on Lerwick. W of Lunna Ness, well protected ⚓s with good holding include: Boatsroom Voe, W Lunna Voe (small hotel, ⚓), Colla Firth* and Dales Voe. No facilities.

Vidlin: marina shop, bus service.

Sullom Voe: Tides –0130 on Lerwick. 6·5M long deep water voe, Commercial harbour. Contact Sullom Voe VTS (Ch14) prior to entry. ⚓ S of the narrows. Facilities at Brae: ⚓ 🛢 🗡 ⚒ 🛒 ✉ 🍴.

Hamna Voe: Tides –0200 on Lerwick; very good shelter. Ldg line 153° old house on S shore with prominent rock on pt of W shore 3ca within ent. Almost land-locked; ⚓ in 6m approx, bottom foul with old moorings. Facilities: 🅥s, ⚓ (at pier lit Fl R 3s), 🛢 (1·5M), ✉ (0·5M), cafe, WC/🌡 🔌 at nearby campsite.

Ura Firth: NE off St Magnus Bay, ⚓ off Hills Wick on W side or in Hamar Voe (no facilities) on E side, which has excellent shelter/good holding in all weathers. Facilities: **Hills Wick** ⚓ 🛢 🗡 🔧 ⚒ ✉ 🍴 ✕ 🛒.

Busta Voe: N of Swarbacks Minn. 4🅥 (< 12m LOA) at Delting Marina in 2m (☎01806 522479); Brae (½M): 🛢 🗡 🔧 ⚒ ✉ 🍴 ✕.

Olna Firth: NE of Swarbacks Minn, beware rk 1ca off S shore which dries. ⚓ in firth, 4-8m or in Gon Firth or go alongside marina or pier at Voe. Facilities: (Voe) ⚓ 🛢 ✉ 🍴 🛒.

Swarbacks Minn*: a large complex of voes and isles SE of St Magnus Bay. Best ⚓ Uyea Sound or Aith Voe, both well sheltered and good holding. No facilities.

Aith Voe*: 60°17·3'N 01°22·3W; Good shelter from all directions ⌂ depth LW 1.2m (marina pontoon 2 🅥s) otherwise on pier access H24, ⚓ ⚓ ⚓ 🔱 🗡 WC/🌡 (leisure centre) ✉ 🍴 ✕ 🛒.

Vaila Sound (Walls), 60°13'·65N 01°33'·87W. AC 3295. Tides approx as Scalloway (above); see 9.7.26. Very good shelter. Appr to E of Vaila island (do not attempt Wester Sound) in the W sector (355°-012°) of Rams Head lt, Fl WRG 8s 16m 9/6M. Gruting Voe lies to the NE. Enter Easter Sound and go N for 1·5M, passing E of Linga islet, to Walls at the head of Vaila Voe. Navigate by echo sounder. Beware fish farms. ⚓ N of Salt Ness in 4-5m in mud. Temporary ⌂ on Bayhaa pier (covers). Close E, ⌂£1 Peter Georgeson's pontoon. **Marina**: Sec ☎(01595) 809273, ⚓. **Walls Regatta Club** welcomes visitors; ⚓ 🌡 🛒. **Village**: 🛢 🛢 ✉ 🍴.

Gruting Voe*: HW –0150 on Lerwick, ⚓ in main voe or in Seli, Scutta or Browland* voes. Facilities: ✉ 🍴 at Bridge of Walls (head of Browland Voe).

Skeld, 60°09'·65N 01°27'W. AC 3283, 3294. Tides approx as Scalloway. Appr from Skelda Voe which is subject to heavy seas and swell in strong S'lies. Beware Braga Rk (dries 2m) and drying rk 1ca S of it. At E of ent also Snap Rk (dries 3.5m) with isolated rks inshore. **Marina** offers good shelter as virtually landlocked, welcomes visitors; 8🅥 £10/craft/week; James Scott ☎(01595) 860287 ⚓ 🔱 WC/🌡 🛢 🛢 bus to Lerwick 2/daily.

UNST Balta Sound: 60°44'·32N 00°48'·12W. AC 3293. HW –0105 on Dover; ML 1·3; Duration 0640. See 9.7.25. Balta Sound is a large almost landlocked inlet with good shelter from all winds. Beware fish farms and bad holding on kelp. Safest and main entry is via S Chan between Huney Is and Balta Is; inner chan marked by two PHM lt buoys. N Chan is deep but narrow; keep to Unst shore. ⚓ off Sandisons Wharf (2FG vert) in approx 6m pontoon on W side of pier is derelict and reported unsafe (Oc WRG 10s 5m 2M, 272°-G-282°-W-287°-R-297°). Small boat marina has single shallow Ⓥ berth. VHF Ch 16; 20 (HO or as required). Facilities: ⚓ ⓘ ✕ ⚒ 🏪 🏠 hotel. **Baltasound village,** ✉ 🍴 ✕ 🏠.

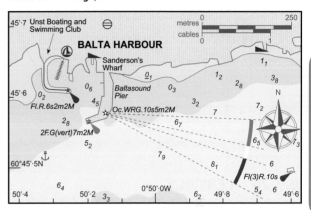

FAIR ISLE: 59°32'·37N 01°36'·21W. AC 3299. HW -0030 on Dover; ML 1·4m; Duration 0620. See 9.7.26. Good shelter in North Haven, except in NE winds. ⚓ on pier or ⚓ in approx 2m. Beware strong cross-tides in the apprs; rocks all round Fair Isle, esp in S Haven and S Harbour which are hazardous. Ldg marks 199° into N Haven: front, Stack of N Haven (a dark 'tooth' sticking up from the jumble of blocks which form the bkwtr) in transit with conspic summit of Sheep Craig (rear). Dir lt into N Haven, Oc WRG 8s 10m 6M; vis 204°-G-208°-W-211°-R-221°. Lights: N tip, Skroo Fl(2) 30s 80m **22M**, S tip, Skadan Fl(4) 30s 31m **22M**. Facilities: ✉ 🍴 at N Shriva, ✈, bi-weekly mail boat to Grutness. No fuel or water. Bird observatory offers 🚾 🍴/meals. Use large fenders supplied against pier to protect from scend/swell.

Scalloway, 60°08'·02N 01°16'·59W. AC 3294. HW –0200 on Dover; ML 0·9m; Duration 0620. See 9.7.26. A busy fishing port; good shelter and ⚓ in all weathers. Care is needed negotiating the islands in strong SW'lies. The N Chan is easier and safer than the S Chan, both are well lit and marked. Castle and warehouse (both conspic) lead 054° through S Chan. Dir Oc WRG 10s on hbr quay leads 064·5° into hbr (dredged 8.5m). Lts as chartlet. Ⓥ pontoon in 3m off SBC is best option; marina close N or new marina in E Voe (though mostly full of local boats). ⚓s in hbr 6–10m or in Hamna Voe (W Burra). Info at www.shetland.gov.uk/ports/scalloway or e-mail scalloway.harbour@shetland.gov.uk (cc to sullomvoevts@shetland.gov.uk). Call *Scalloway Hbr Radio* VHF Ch **12** 16 (Mon-Fri 0600-1800; Sat 0600-1230LT). Piermaster ☎(01595) 744221. Facilities: **Scalloway Boat Club** (SBC) ☎880409; ⚓, 🏠. **Town** ⚓ ⚒ 🏪 🏪 ✕ ⚒ ⚓ 🏪 🏪 🏪 🏪 ✉ 🍴 ✕ 🏠 bus to Lerwick.

Hamna Voe*: 60°06·3'N 01°20·3W (West Burra) Entered between Fugla Ness and Alta Ness. Open to the NW; good shelter in small marina. Small village, bus to Lerwick and all other routes.

WHALSAY*: 60°20'·60N 01°01'·60W; Fishing hbr with small marina at Symbister welcomes visitors. ⚓ ⓘ ⚓ (pontoon or pier 20Ⓥs) £7.17/craft/4 days; ⚒, 🏪 ⚓ ⓘ 🏪 🍴 ✕.

FOULA: 60°08'·02N 02°02'·92W. AC 3283. HW -0150 on Dover; ML 1·3m. See 9.7.26. Highest ground is 416m. S Ness sectored Lt ho, Fl WR(3) 15s, is at the S tip. Beware both Foula Shoal (7·6m) 4·3M E of the island and Hœvdi Grund (1·4m), 2M to the SE of Ham Voe. **Ham Voe:** is a narrow inlet on the E coast with a quay; rks on both sides. Two R ▲ ldg marks, approx 270°, are hard to see. ☆ 2 FG (vert) on pierhead. Berthing or landing is only possible in settled weather with no swell. Take advice from mail boat skipper out of Walls and call Foula, ☎(01595) 753222. Small ✈. No other facilities.

OUT SKERRIES: 60°25'·3N 00°45'·1W AC 3282. Good shelter at pier. The NE entrance is buoyed and is the easier and has a leading light (Local ferry port). There are two Historic Wrecks (*Kennemerland* and *Wrangels Palais*) on Out Skerries at 60°25'·2N 00°45'·0W and 60°25'·5N 00°43'·3W (see 9.0.3). Shops, 🚾 🍴 at pier. Beware strong tidal conditions off the NE entrance ferries approaching or leaving. Pier (£2/craft/night); ⚓ ⓘ 🏪 (by arrangement), toilet/🍴 wi-fi; Village ✉ shop. Good ⚓ at Bruray.

YELL: Mid Yell Voe*: 60°36'N 01°03·4W; tides –0040 on Lerwick, enter via S Sd or Hascosay Sd, good ⚓ in wide part of voe 2·5-10m. Berth at marina £5/week or Pier £7.17/4 days/craft. Facilities: ✉ ⚓ at pier on S side, 🏪 ✕ ⚒ 🏪 🍴.
Basta Voe: good ⚓ above shingle bank in 5-15m; good holding in places. Facilities: ⚓ 🍴 Hotel, ✉.
Blue Mull Sound/Cullivoe*: 60°42'N 00°59'·7W; Hbr with small craft marina depth 1.2m ⚓£5/craft/day, ⚓ ⚓ 🏪. Pier ⚓ 20Ⓥs £7.17/craft/4 days; ⚓ ⓘ 🏪 🚾 🍴 ✕. **Town** ✉ 🍴 ✕ 🏠. ⚓ off pier.
Burra Voe*: 60°29'·8N 01°02'·4W; Entrance to voe has 2.5m bar. Good ⚓ at head of voe, but buoy ⚓ as precaution against fouling old moorings. ⚓ at small marina £5/visit or on Pier £12/visit , ⚓ ⓘ 🚾 ✉ 🍴 🏠.

FETLAR: Hamars Ness 60°37'·78N 00°55'·76W; small hbr with new b'water (Iso G 6s 5m 3M) at W end of rocky bay. Appr W of WCM.

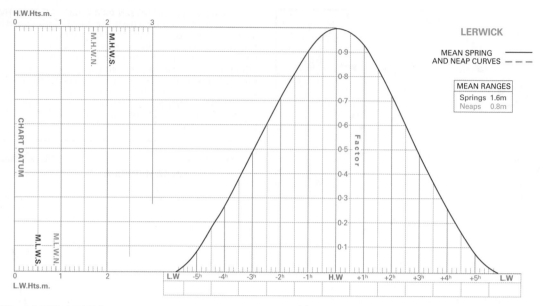

LERWICK

MEAN SPRING ———
AND NEAP CURVES - - -

MEAN RANGES
Springs 1.6m
Neaps 0.8m

9.7.26 LERWICK

Shetland Is, Mainland **60°09'·26N 01°08'·42W** ✿✿✿✿✿✿

CHARTS AC 3283, 3272, 3271

TIDES –0001 Dover; ML 1·4; Duration 0620

Standard Port LERWICK (←)

Times				Height (metres)			
High Water		Low Water		MHWS	MHWN	MLWN	MLWS
0000	0600	0100	0800	2·1	1·7	0·9	0·5
1200	1800	1300	2000				
Differences FAIR ISLE							
–0006	–0015	–0031	–0037	+0·1	0·0	+0·1	+0·1
SUMBURGH (Grutness Voe)							
+0006	+0008	+0004	–0002	–0·3	–0·3	–0·2	–0·1
DURY VOE							
–0015	–0015	–0010	–0010	0·0	–0·1	0·0	–0·2
BURRA VOE (YELL SOUND)							
–0025	–0025	–0025	–0025	+0·2	+0·1	0·0	–0·1
BALTA SOUND							
–0040	–0045	–0040	–0045	+0·3	+0·2	+0·1	0·0
BLUEMULL SOUND							
–0135	–0135	–0155	–0155	+0·5	+0·2	+0·1	0·0
SULLOM VOE							
–0135	–0125	–0135	–0120	0·0	0·0	–0·2	–0·2
HILLSWICK (URA FIRTH)							
–0220	–0220	–0200	–0200	–0·1	–0·1	–0·1	–0·1
SCALLOWAY							
–0150	–0150	–0150	–0150	–0·5	–0·4	–0·3	0·0
FOULA (23M West of Scalloway)							
–0140	–0140	–0130	–0130	–0·1	–0·1	0·0	0·0

SHELTER Good. HM allocates berths in Small Dock or Albert Dock. FVs occupy most alongside space. ⚓ prohib for about 2ca off the waterfront. Gremista marina in N hbr, is mainly for local boats, and is about 1M from the town.

NAVIGATION WPT 60°05'·97N 01°08'·62W, 010°/3·5M to Maryfield lt, in W sector. From S, Bressay Sound is clear of dangers. From N, WPT 60°11'·60N 01°07'·88W, 215°/1·34M to N ent Dir lt (Oc WRG 6s, 214°-W-216°). Beware Soldian Rk (dries), Nive Baa (0·6m), Green Holm (10m) and The Brethren (two rocks 2m and 1·5m).

LIGHTS AND MARKS Kirkabister Ness, Fl (2) 20s 18m 10M; Cro of Ham, Fl 3s 3M; Maryfield, Oc WRG 6s, 008°-W-013°; all on Bressay. Twageos Pt, L Fl 6s 8m 6M. Loofa Baa SCM lt bn, as on chartlet. 2 SHM lt buoys mark Middle Ground in N Hbr.

COMMUNICATIONS (Code 01595) CGOC 692976; ⚓692239; Police 692110; Dr 693201. HM 692991.

Lerwick Harbour VHF Ch **12** 16 (H24) for Vessel Information Service. Other stns: *Sullom Voe Hbr Radio* broadcasts traffic info and local forecasts on request Ch **14** 12 20 16 (H24).

FACILITIES Hbr ⚓ ↖ ↘ ⚓ 🏺 ♦ ⚒ ✕ ⚒.
Lerwick Port Authority ☎692991, Ⓥpontoons £0.60, ⚓ ⚓ ⊃ ♦ ↖
🚻 ramps/**WC**.
Lerwick Boating Club ☎692407, ↖ ◊ ◻ ◻.
Services ⚒ Ⓑ Ⓔ ✕ ↘ ♖ ◭ ◬ ♦ ACA.
Town ✉ Ⓑ ♨ ✕ ◻ @ **Ferries:** Aberdeen via Kirkwall; daily; overnight; North Link (www.northlinkferries.co.uk), ✈. See www. shetlandtimes.co.uk and www.lerwickboating.co.uk.

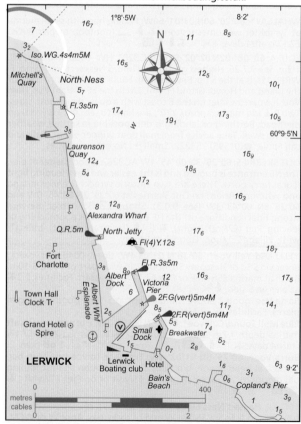

STANDARD TIME (UT)
For Summer Time add ONE hour in **non-shaded areas**

LERWICK LAT 60°09'N LONG 1°08'W
TIMES AND HEIGHTS OF HIGH AND LOW WATERS

Dates in red are **SPRINGS**
Dates in blue are NEAPS

YEAR 2016

JANUARY

Day	Time m	Time m	Time m	Time m
1 F	0317 1.8	0904 1.1	1526 1.9	2152 0.9
16 SA	0321 2.0	0905 0.8	1527 2.1	2147 0.7 ◗
2 SA	0405 1.8	1001 1.1	1616 1.8	2252 1.0 ◑
17 SU	0418 1.9	1007 0.9	1630 2.0	2257 0.8
3 SU	0501 1.7	1119 1.2	1719 1.8	2358 1.0
18 M	0524 1.8	1129 1.0	1743 1.9	
4 M	0615 1.7	1236 1.1	1842 1.8	
19 TU	0021 0.8	0637 1.9	1257 0.9	1903 1.9
5 TU	0100 1.0	0722 1.8	1340 1.1	1946 1.8
20 W	0134 0.8	0747 1.9	1406 0.9	2015 1.9
6 W	0154 0.9	0813 1.9	1429 1.0	2037 1.9
21 TH	0233 0.8	0847 2.0	1503 0.7	2117 2.1
7 TH	0240 0.9	0856 2.0	1512 0.9	2122 2.0
22 F	0323 0.8	0939 2.1	1551 0.6	2211 2.1
8 F	0322 0.8	0937 2.1	1551 0.7	2204 2.0
23 SA	0407 0.7	1026 2.2	1635 0.5	2257 2.1
9 SA	0402 0.8	1016 2.2	1631 0.6	2247 2.1
24 SU	0447 0.7	1108 2.3	1715 0.5	2338 2.1 ○
10 SU	0442 0.7	1056 2.3	1710 0.5	2329 2.2 ●
25 M	0526 0.7	1147 2.3	1754 0.4	
11 M	0522 0.7	1136 2.3	1751 0.5	
26 TU	0016 2.1	0602 0.7	1224 2.3	1830 0.5
12 TU	0012 2.2	0602 0.6	1217 2.3	1832 0.4
27 W	0051 2.1	0637 0.7	1259 2.2	1906 0.5
13 W	0056 2.2	0643 0.6	1259 2.3	1915 0.4
28 TH	0125 2.0	0710 0.8	1333 2.2	1942 0.6
14 TH	0142 2.1	0727 0.7	1344 2.3	2001 0.5
29 F	0159 1.9	0743 0.8	1408 2.1	2017 0.7
15 F	0230 2.0	0814 0.8	1432 2.2	2051 0.6
30 SA	0235 1.9	0818 0.9	1445 1.9	2057 0.8
31 SU	0314 1.8	0859 1.0	1528 1.8	2144 0.9

FEBRUARY

Day	Time m	Time m	Time m	Time m
1 M	0400 1.7	0956 1.1	1619 1.7	2249 1.0 ◗
16 TU	0452 1.8	1104 0.9	1723 1.8	2359 0.9
2 TU	0455 1.7	1129 1.1	1723 1.7	
17 W	0610 1.8	1243 0.9	1852 1.7	
3 W	0007 1.0	0607 1.7	1252 1.1	1854 1.7
18 TH	0121 0.9	0729 1.8	1358 0.8	2010 1.8
4 TH	0114 1.0	0728 1.8	1355 1.0	2005 1.7
19 F	0225 0.9	0834 1.9	1455 0.7	2111 1.9
5 F	0211 0.9	0824 1.9	1446 0.9	2057 1.9
20 SA	0314 0.8	0927 2.0	1541 0.6	2200 2.0
6 SA	0259 0.8	0911 2.0	1531 0.7	2143 2.0
21 SU	0355 0.7	1012 2.1	1621 0.5	2241 2.0
7 SU	0343 0.7	0955 2.1	1612 0.5	2228 2.1
22 M	0432 0.7	1052 2.2	1657 0.4	2318 2.1 ○
8 M	0425 0.6	1037 2.2	1653 0.4	2312 2.2 ●
23 TU	0507 0.6	1128 2.2	1731 0.4	2351 2.1
9 TU	0505 0.5	1120 2.3	1734 0.3	2356 2.2
24 W	0540 0.6	1201 2.2	1804 0.4	
10 W	0546 0.5	1202 2.4	1815 0.2	
25 TH	0023 2.0	0611 0.6	1233 2.2	1835 0.5
11 TH	0039 2.2	0627 0.5	1245 2.4	1857 0.3
26 F	0053 2.0	0641 0.6	1304 2.1	1907 0.5
12 F	0123 2.2	0709 0.5	1329 2.3	1941 0.3
27 SA	0123 1.9	0712 0.7	1335 2.0	1939 0.6
13 SA	0208 2.1	0754 0.6	1416 2.2	2028 0.5
28 SU	0154 1.9	0744 0.8	1409 1.9	2013 0.7
14 SU	0256 2.0	0843 0.7	1509 2.1	2120 0.7
29 M	0229 1.8	0822 0.8	1449 1.8	2054 0.8
15 M	0349 1.9	0942 0.8	1609 1.9	2226 0.8 ◗

MARCH

Day	Time m	Time m	Time m	Time m
1 TU	0312 1.7	0908 0.9	1538 1.7	2145 1.0 ◗
16 W	0422 1.7	1049 0.8	1707 1.7	2338 1.0
2 W	0404 1.7	1015 1.0	1638 1.6	2304 1.0
17 TH	0541 1.7	1228 0.8	1841 1.6	
3 TH	0507 1.6	1203 1.0	1756 1.6	
18 F	0105 1.0	0707 1.7	1342 0.7	1958 1.7
4 F	0036 1.0	0627 1.6	1319 0.9	1929 1.6
19 SA	0209 0.9	0814 1.8	1439 0.6	2055 1.8
5 SA	0142 0.9	0747 1.7	1417 0.8	2030 1.8
20 SU	0257 0.8	0906 1.9	1526 0.5	2139 1.8
6 SU	0236 0.8	0842 1.9	1505 0.6	2120 1.9
21 M	0337 0.7	0950 2.0	1600 0.5	2217 1.9
7 M	0322 0.7	0930 2.0	1549 0.4	2206 2.0
22 TU	0412 0.6	1028 2.0	1633 0.4	2251 2.0
8 TU	0404 0.5	1015 2.2	1631 0.3	2250 2.2
23 W	0445 0.5	1103 2.1	1704 0.4	2323 2.0 ○
9 W	0445 0.4	1059 2.3	1712 0.1	2334 2.2 ●
24 TH	0515 0.5	1135 2.1	1735 0.4	2352 2.0
10 TH	0526 0.3	1143 2.4	1754 0.1	
25 F	0545 0.5	1206 2.1	1804 0.4	
11 F	0017 2.2	0608 0.3	1227 2.4	1836 0.1
26 SA	0021 2.0	0614 0.5	1236 2.0	1834 0.5
12 SA	0101 2.2	0651 0.3	1313 2.3	1919 0.3
27 SU	0049 1.9	0645 0.6	1306 2.0	1905 0.6
13 SU	0144 2.1	0736 0.4	1401 2.2	2005 0.5
28 M	0119 1.9	0718 0.6	1340 1.9	1939 0.7
14 M	0230 2.0	0825 0.5	1453 2.0	2056 0.7
29 TU	0152 1.8	0756 0.7	1419 1.8	2017 0.8
15 TU	0322 1.8	0942 0.7	1553 1.8	2159 0.8 ◗
30 W	0233 1.8	0840 0.8	1508 1.7	2105 0.9
31 TH	0323 1.7	0939 0.9	1608 1.6	2210 1.0 ◗

APRIL

Day	Time m	Time m	Time m	Time m
1 F	0425 1.6	1107 0.9	1721 1.5	2354 1.0
16 SA	0036 1.0	0634 1.6	1314 0.7	1929 1.6
2 SA	0539 1.6	1239 0.8	1851 1.6	
17 SU	0142 0.9	0742 1.7	1410 0.6	2024 1.7
3 SU	0110 0.9	0705 1.7	1343 0.6	2000 1.7
18 M	0232 0.8	0835 1.8	1454 0.5	2107 1.8
4 M	0208 0.8	0810 1.8	1435 0.5	2052 1.9
19 TU	0313 0.7	0920 1.8	1531 0.5	2145 1.8
5 TU	0256 0.6	0902 2.0	1522 0.3	2140 2.0
20 W	0348 0.6	0959 1.9	1604 0.4	2219 1.9
6 W	0340 0.4	0949 2.1	1606 0.2	2225 2.1
21 TH	0420 0.5	1034 1.9	1634 0.4	2251 1.9
7 TH	0423 0.3	1036 2.2	1649 0.1	2309 2.2 ●
22 F	0450 0.5	1108 2.0	1704 0.4	2321 2.0 ○
8 F	0506 0.2	1122 2.3	1731 0.1	2353 2.2
23 SA	0520 0.5	1140 2.0	1734 0.4	2350 2.0
9 SA	0549 0.2	1209 2.3	1814 0.2	
24 SU	0551 0.5	1211 1.9	1806 0.5	
10 SU	0036 2.2	0633 0.3	1257 2.2	1858 0.3
25 M	0019 2.0	0623 0.5	1243 1.9	1838 0.6
11 M	0121 2.1	0720 0.3	1347 2.1	1944 0.5
26 TU	0050 1.9	0658 0.5	1318 1.8	1913 0.6
12 TU	0207 2.0	0811 0.4	1441 1.9	2035 0.7
27 W	0125 1.9	0737 0.6	1359 1.7	1952 0.7
13 W	0258 1.9	0911 0.6	1539 1.7	2137 0.6
28 TH	0205 1.8	0823 0.7	1448 1.7	2040 0.7
14 TH	0356 1.7	1031 0.7	1649 1.6	2308 1.0 ◗
29 F	0254 1.7	0918 0.7	1548 1.6	2142 0.9
15 F	0508 1.7	1201 0.7	1816 1.6	
30 SA	0355 1.6	1030 0.7	1656 1.6	2306 0.9 ◗

Chart Datum: 1·22 metres below Ordnance Datum (Local). HAT is 2·5 metres above Chart Datum.

NE Scotland

STANDARD TIME (UT)
For Summer Time add ONE hour in **non-shaded areas**

LERWICK LAT 60°09'N LONG 1°08'W
TIMES AND HEIGHTS OF HIGH AND LOW WATERS

Dates in red are SPRINGS
Dates in blue are NEAPS

YEAR 2016

MAY

Day	Time m	Time m	Time m	Time m
1	0505 1.6	1157 0.7	SU 1815 1.6	
2	0033 0.8	0624 1.7	M 1307 0.6	1927 1.7
3	0136 0.7	0736 1.8	TU 1404 0.4	2023 1.8
4	0228 0.6	0833 1.9	W 1454 0.3	2112 2.0
5	0316 0.4	0924 2.1	TH 1541 0.2	2159 2.1
6	0401 0.3	1014 2.2	F 1626 0.2 ●	2245 2.2
7	0446 0.2	1103 2.2	SA 1710 0.2	2330 2.2
8	0532 0.2	1153 2.2	SU 1754 0.2	
9	0015 2.2	0618 0.2	M 1244 2.1	1839 0.4
10	0101 2.1	0706 0.2	TU 1335 2.0	1926 0.5
11	0148 2.0	0757 0.4	W 1427 1.9	2016 0.7
12	0237 1.9	0855 0.5	TH 1521 1.7	2113 0.8
13	0330 1.8	1003 0.6	F 1621 1.6 ☽	2226 0.9
14	0433 1.7	1119 0.7	SA 1733 1.5	2349 0.9
15	0549 1.6	1230 0.7	SU 1844 1.6	
16	0101 0.9	0700 1.6	M 1330 0.7	1941 1.6
17	0158 0.8	0756 1.7	TU 1418 0.6	2028 1.7
18	0242 0.7	0844 1.7	W 1457 0.6	2108 1.8
19	0320 0.7	0926 1.8	TH 1531 0.5	2145 1.9
20	0353 0.6	1005 1.8	F 1603 0.5	2220 1.9
21	0425 0.5	1041 1.9	SA 1635 0.5 ○	2252 2.0
22	0457 0.5	1116 1.9	SU 1708 0.5	2324 2.0
23	0530 0.5	1150 1.9	M 1742 0.5	2356 2.0
24	0605 0.5	1225 1.9	TU 1817 0.6	
25	0029 2.0	0642 0.5	W 1304 1.8	1854 0.6
26	0106 1.9	0723 0.5	TH 1346 1.8	1936 0.7
27	0147 1.9	0809 0.5	F 1435 1.7	2024 0.8
28	0236 1.8	0902 0.6	SA 1531 1.7	2120 0.8
29	0333 1.7	1004 0.6	SU 1633 1.6 ☽	2229 0.8
30	0439 1.7	1117 0.6	M 1743 1.6	2350 0.8
31	0551 1.7	1231 0.5	TU 1853 1.7	

JUNE

Day	Time m	Time m	Time m	Time m
1	0101 0.7	0704 1.8	W 1334 0.5	1953 1.8
2	0201 0.6	0807 1.9	TH 1429 0.4	2046 1.9
3	0254 0.5	0903 2.0	F 1519 0.3	2136 2.0
4	0343 0.4	0957 2.1	SA 1607 0.3	2224 2.1
5	0431 0.3	1050 2.1	SU 1653 0.3 ●	2311 2.2
6	0518 0.2	1142 2.1	M 1738 0.4	2358 2.2
7	0605 0.2	1233 2.1	TU 1823 0.5	
8	0044 2.1	0652 0.2	W 1322 2.0	1908 0.6
9	0130 2.1	0741 0.3	TH 1409 1.9	1954 0.7
10	0216 2.0	0832 0.5	F 1456 1.8	2044 0.8
11	0303 1.9	0927 0.6	SA 1545 1.6	2140 0.9
12	0354 1.7	1027 0.7	SU 1641 1.6 ☽	2247 0.9
13	0454 1.7	1131 0.7	M 1746 1.6	
14	0002 0.9	0606 1.6	TU 1235 0.7	1851 1.6
15	0111 0.9	0712 1.6	W 1331 0.7	1945 1.7
16	0205 0.8	0806 1.7	TH 1417 0.7	2031 1.7
17	0248 0.8	0853 1.7	F 1457 0.7	2112 1.8
18	0326 0.7	0936 1.8	SA 1534 0.6	2150 1.8
19	0401 0.7	1015 1.9	SU 1610 0.6	2226 2.0
20	0437 0.5	1054 1.9	M 1646 0.6 ○	2302 2.0
21	0513 0.5	1132 1.9	TU 1723 0.6	2337 2.0
22	0550 0.4	1211 1.9	W 1801 0.6	
23	0014 2.0	0629 0.4	TH 1252 1.9	1840 0.6
24	0053 2.0	0710 0.4	F 1335 1.9	1922 0.6
25	0135 2.0	0755 0.4	SA 1421 1.8	2008 0.7
26	0221 1.9	0844 0.5	SU 1513 1.8	2059 0.7
27	0315 1.9	0939 0.5	M 1610 1.7 ☽	2159 0.8
28	0416 1.8	1044 0.6	TU 1713 1.7	2311 0.8
29	0524 1.6	1158 0.6	W 1822 1.7	
30	0031 0.8	0637 1.8	TH 1309 0.6	1927 1.8

JULY

Day	Time m	Time m	Time m	Time m
1	0140 0.7	0747 1.9	F 1410 0.6	2025 1.9
2	0238 0.6	0849 1.9	SA 1504 0.5	2119 2.0
3	0331 0.4	0947 2.0	SU 1553 0.5	2210 2.1
4	0420 0.3	1042 2.1	M 1639 0.5 ●	2258 2.2
5	0507 0.3	1133 2.1	TU 1723 0.5	2344 2.2
6	0552 0.3	1219 2.1	W 1806 0.5	
7	0028 2.2	0636 0.3	TH 1303 2.0	1847 0.6
8	0110 2.1	0719 0.4	F 1344 1.9	1929 0.7
9	0151 2.0	0802 0.5	SA 1425 1.8	2010 0.8
10	0232 1.9	0847 0.6	SU 1506 1.7	2053 0.8
11	0315 1.8	0934 0.7	M 1550 1.7	2144 0.9
12	0402 1.7	1014 0.8	TU 1640 1.6 ☽	2252 1.0
13	0459 1.6	1131 0.8	W 1744 1.6	
14	0010 1.0	0615 1.6	TH 1235 0.9	1856 1.6
15	0120 0.9	0727 1.6	F 1333 0.8	1953 1.7
16	0215 0.9	0822 1.7	SA 1423 0.8	2040 1.8
17	0259 0.8	0909 1.8	SU 1507 0.8	2122 1.9
18	0339 0.7	0952 1.9	M 1548 0.7	2202 2.0
19	0417 0.6	1033 1.9	TU 1627 0.6 ○	2241 2.1
20	0455 0.5	1114 2.0	W 1706 0.6	2320 2.1
21	0534 0.4	1155 2.0	TH 1745 0.5	2359 2.2
22	0614 0.3	1236 2.0	F 1825 0.6	
23	0039 2.2	0654 0.3	SA 1319 2.0	1906 0.6
24	0121 2.1	0737 0.4	SU 1404 2.0	1950 0.6
25	0206 2.1	0823 0.4	M 1452 1.9	2038 0.7
26	0257 2.0	0915 0.5	TU 1545 1.8 ☽	2133 0.7
27	0355 1.9	1014 0.6	W 1644 1.8	2242 0.8
28	0501 1.8	1131 0.7	TH 1753 1.8	
29	0010 0.8	0619 1.8	F 1252 0.8	1906 1.8
30	0128 0.7	0737 1.8	SA 1359 0.7	2012 1.9
31	0232 0.6	0845 1.9	SU 1456 0.7	2109 2.0

AUGUST

Day	Time m	Time m	Time m	Time m
1	0325 0.5	0944 2.0	M 1544 0.6	2200 2.1
2	0412 0.4	1035 2.0	TU 1627 0.6	● 2247 2.2
3	0455 0.3	1120 2.1	W 1708 0.6	2329 2.2
4	0535 0.3	1200 2.1	TH 1746 0.6	
5	0009 2.2	0614 0.3	F 1238 2.0	1823 0.6
6	0046 2.2	0652 0.4	SA 1314 2.0	1859 0.6
7	0122 2.1	0729 0.5	SU 1349 1.9	1934 0.7
8	0158 2.0	0805 0.6	M 1424 1.8	2010 0.8
9	0236 1.9	0844 0.7	TU 1502 1.8	2051 0.9
10	0317 1.8	0928 0.8	W 1546 1.7	☽ 2145 1.0
11	0406 1.7	1028 0.9	TH 1637 1.6	2310 1.0
12	0507 1.6	1144 1.0	F 1743 1.6	
13	0034 1.0	0636 1.6	SA 1254 1.0	1908 1.7
14	0140 0.9	0751 1.7	SU 1353 0.9	2007 1.8
15	0232 0.8	0843 1.8	M 1443 0.8	2054 1.9
16	0315 0.7	0928 1.9	TU 1527 0.7	2137 2.0
17	0356 0.6	1011 2.0	W 1607 0.6	2219 2.1
18	0435 0.4	1053 2.1	TH 1647 0.5	○ 2259 2.2
19	0514 0.3	1134 2.1	F 1726 0.5	2340 2.3
20	0553 0.3	1216 2.2	SA 1806 0.5	
21	0021 2.3	0634 0.2	SU 1259 2.2	1846 0.5
22	0104 2.3	0716 0.3	M 1342 2.1	1930 0.5
23	0149 2.2	0801 0.4	TU 1428 2.0	2017 0.6
24	0239 2.1	0850 0.6	W 1519 1.9	2112 0.7
25	0338 2.0	0949 0.7	TH 1618 1.8	☽ 2222 0.8
26	0446 1.8	1110 0.9	F 1729 1.8	
27	0002 0.8	0610 1.8	SA 1244 0.9	1850 1.8
28	0124 0.8	0736 1.8	SU 1353 0.9	2001 1.9
29	0226 0.7	0842 1.9	M 1448 0.8	2059 2.0
30	0317 0.6	0935 2.0	TU 1533 0.7	2147 2.1
31	0359 0.5	1020 2.0	W 1612 0.7	2230 2.2

Chart Datum: 1·22 metres below Ordnance Datum (Local). HAT is 2·5 metres above Chart Datum.

》》 **FREE** monthly updates. Register at 《
www.reedsnauticalalmanac.co.uk

STANDARD TIME (UT)
For Summer Time add ONE hour in **non-shaded areas**

LERWICK   LAT 60°09'N   LONG 1°08'W
TIMES AND HEIGHTS OF HIGH AND LOW WATERS

Dates in red are **SPRINGS**
Dates in blue are NEAPS

YEAR 2016

NE Scotland

SEPTEMBER

Time m	Time m
1 0438 0.4 / 1100 2.1 / TH 1649 0.6 / ● 2309 2.2	**16** 0410 0.4 / 1028 2.2 / F 1624 0.5 / ○ 2235 2.3
2 0514 0.4 / 1136 2.1 / F 1724 0.6 / 2345 2.2	**17** 0450 0.3 / 1110 2.2 / SA 1704 0.4 / 2318 2.4
3 0548 0.4 / 1209 2.1 / SA 1757 0.6	**18** 0531 0.2 / 1152 2.3 / SU 1745 0.4
4 0018 2.2 / 0621 0.4 / SU 1241 2.0 / 1829 0.6	**19** 0001 2.4 / 0611 0.2 / M 1235 2.3 / 1827 0.4
5 0051 2.1 / 0653 0.5 / M 1311 2.0 / 1901 0.7	**20** 0045 2.4 / 0654 0.3 / TU 1318 2.2 / 1911 0.5
6 0124 2.1 / 0726 0.6 / TU 1343 1.9 / 1934 0.8	**21** 0133 2.3 / 0739 0.5 / W 1404 2.1 / 1959 0.6
7 0158 2.0 / 0800 0.8 / W 1418 1.9 / 2011 0.9	**22** 0225 2.1 / 0828 0.7 / TH 1455 2.0 / 2056 0.7
8 0237 1.8 / 0839 0.9 / TH 1459 1.8 / 2057 1.0	**23** 0325 2.0 / 0927 0.9 / F 1554 1.9 / ◑ 2211 0.8
9 0325 1.7 / 0929 1.0 / F 1548 1.7 / ◐ 2205 1.0	**24** 0435 1.8 / 1055 1.0 / SA 1706 1.8 / 2354 0.8
10 0423 1.6 / 1046 1.1 / SA 1648 1.7 / 2351 1.0	**25** 0604 1.8 / 1232 1.0 / SU 1833 1.8
11 0538 1.6 / 1218 1.1 / SU 1804 1.7	**26** 0112 0.8 / 0728 1.8 / M 1341 1.0 / 1945 1.9
12 0104 1.0 / 0716 1.7 / M 1324 1.0 / 1929 1.8	**27** 0213 0.7 / 0830 1.9 / TU 1433 0.9 / 2041 2.0
13 0201 0.8 / 0815 1.8 / TU 1417 0.9 / 2024 1.9	**28** 0300 0.6 / 0917 2.0 / W 1516 0.8 / 2127 2.1
14 0248 0.7 / 0902 1.9 / W 1502 0.8 / 2110 2.1	**29** 0340 0.5 / 0958 2.0 / TH 1553 0.7 / 2208 2.2
15 0330 0.5 / 0945 2.1 / TH 1544 0.6 / 2153 2.2	**30** 0415 0.5 / 1034 2.1 / F 1627 0.6 / 2245 2.2

OCTOBER

Time m	Time m
1 0448 0.5 / 1107 2.1 / SA 1659 0.6 / ● 2318 2.2	**16** 0425 0.3 / 1044 2.3 / SU 1642 0.4 / ○ 2255 2.4
2 0519 0.5 / 1137 2.1 / SU 1731 0.6 / 2350 2.2	**17** 0507 0.2 / 1127 2.4 / M 1725 0.4 / 2341 2.5
3 0550 0.5 / 1207 2.1 / M 1801 0.6	**18** 0550 0.3 / 1210 2.3 / TU 1809 0.4
4 0021 2.1 / 0620 0.6 / TU 1236 2.1 / 1832 0.7	**19** 0029 2.4 / 0633 0.2 / W 1255 2.3 / 1855 0.4
5 0053 2.1 / 0651 0.7 / W 1306 2.0 / 1905 0.8	**20** 0120 2.3 / 0719 0.6 / TH 1342 2.2 / 1945 0.6
6 0127 2.1 / 0724 0.8 / TH 1339 2.0 / 1942 0.8	**21** 0215 2.1 / 0809 0.8 / F 1434 2.1 / 2043 0.7
7 0205 1.9 / 0801 0.9 / F 1418 1.9 / 2026 0.9	**22** 0314 2.0 / 0908 1.0 / SA 1532 2.0 / ◑ 2158 0.8
8 0253 1.8 / 0847 1.0 / SA 1507 1.8 / 2124 1.0	**23** 0421 1.8 / 1032 1.1 / SU 1641 1.9 / 2333 0.8
9 0351 1.7 / 0951 1.1 / SU 1607 1.8 / ◐ 2257 1.0	**24** 0545 1.8 / 1207 1.1 / M 1805 1.9
10 0500 1.7 / 1136 1.1 / M 1717 1.7	**25** 0049 0.8 / 0704 1.8 / TU 1317 1.1 / 1917 1.9
11 0025 1.0 / 0629 1.7 / TU 1251 1.1 / 1841 1.8	**26** 0149 0.6 / 0803 1.9 / W 1410 1.0 / 2014 2.0
12 0126 0.8 / 0742 1.8 / W 1347 0.9 / 1948 1.9	**27** 0236 0.7 / 0849 1.9 / TH 1453 0.9 / 2100 2.1
13 0216 0.7 / 0832 2.0 / TH 1435 0.8 / 2039 2.1	**28** 0315 0.6 / 0928 2.0 / F 1530 0.8 / 2141 2.1
14 0301 0.5 / 0917 2.1 / F 1518 0.7 / 2126 2.2	**29** 0349 0.6 / 1004 2.1 / SA 1604 0.7 / 2218 2.2
15 0343 0.4 / 1001 2.2 / SA 1600 0.5 / 2210 2.4	**30** 0420 0.6 / 1036 2.1 / SU 1636 0.7 / ● 2252 2.2
	31 0450 0.6 / 1107 2.2 / M 1706 0.7 / 2325 2.2

NOVEMBER

Time m	Time m
1 0520 0.6 / 1137 2.2 / TU 1737 0.7 / 2357 2.1	**16** 0531 0.4 / 1150 2.4 / W 1754 0.4
2 0551 0.7 / 1206 2.2 / W 1809 0.7	**17** 0017 2.4 / 0615 0.5 / TH 1236 2.4 / 1842 0.4
3 0029 2.1 / 0622 0.8 / TH 1237 2.1 / 1843 0.7	**18** 0110 2.3 / 0702 0.7 / F 1324 2.3 / 1933 0.5
4 0104 2.0 / 0656 0.8 / F 1310 2.1 / 1921 0.8	**19** 0204 2.1 / 0752 0.9 / SA 1415 2.2 / 2029 0.6
5 0143 1.9 / 0734 0.9 / SA 1348 2.0 / 2005 0.9	**20** 0259 2.0 / 0847 1.0 / SU 1510 2.1 / 2136 0.8
6 0229 1.8 / 0818 1.0 / SU 1435 1.9 / 2058 0.9	**21** 0358 1.9 / 0955 1.1 / M 1611 2.0 / ◑ 2254 0.8
7 0326 1.8 / 0916 1.1 / M 1533 1.8 / ◑ 2208 1.0	**22** 0507 1.8 / 1121 1.2 / TU 1723 1.9
8 0431 1.7 / 1036 1.1 / TU 1640 1.8 / 2336 0.9	**23** 0009 0.9 / 0621 1.8 / W 1238 1.1 / 1837 1.9
9 0546 1.7 / 1208 1.1 / W 1755 1.8	**24** 0113 0.8 / 0723 1.8 / TH 1339 1.0 / 1937 1.9
10 0045 0.8 / 0701 1.8 / TH 1312 1.0 / 1909 1.9	**25** 0204 0.8 / 0812 1.9 / F 1427 1.0 / 2028 2.0
11 0142 0.7 / 0759 2.0 / F 1405 0.8 / 2008 2.1	**26** 0245 0.8 / 0855 2.0 / SA 1507 0.9 / 2112 2.0
12 0231 0.6 / 0848 2.1 / SA 1452 0.7 / 2059 2.2	**27** 0321 0.8 / 0932 2.1 / SU 1542 0.8 / 2151 2.1
13 0317 0.5 / 0934 2.3 / SU 1538 0.6 / 2147 2.4	**28** 0353 0.7 / 1007 2.1 / M 1614 0.8 / 2228 2.1
14 0402 0.4 / 1019 2.3 / M 1622 0.5 / ○ 2236 2.4	**29** 0424 0.7 / 1040 2.2 / TU 1646 0.7 / ● 2303 2.1
15 0446 0.4 / 1104 2.4 / TU 1708 0.4 / 2326 2.4	**30** 0456 0.7 / 1112 2.2 / W 1718 0.7 / 2337 2.1

DECEMBER

Time m	Time m
1 0528 0.8 / 1144 2.2 / TH 1752 0.7	**16** 0008 2.3 / 0601 0.6 / F 1222 2.4 / 1830 0.4
2 0012 2.1 / 0601 0.8 / F 1216 2.2 / 1827 0.7	**17** 0059 2.3 / 0646 0.7 / SA 1309 2.3 / 1919 0.5
3 0048 2.0 / 0636 0.9 / SA 1250 2.2 / 1906 0.7	**18** 0148 2.1 / 0733 0.8 / SU 1357 2.3 / 2010 0.6
4 0127 2.0 / 0715 0.9 / SU 1328 2.1 / 1949 0.8	**19** 0236 2.0 / 0821 0.9 / M 1445 2.1 / 2104 0.7
5 0212 1.9 / 0759 1.0 / M 1413 2.0 / 2038 0.8	**20** 0326 1.9 / 0914 1.0 / TU 1535 2.0 / 2203 0.8
6 0304 1.9 / 0851 1.0 / TU 1506 2.0 / 2135 0.8	**21** 0420 1.8 / 1017 1.1 / W 1633 1.9 / ◑ 2311 0.9
7 0403 1.8 / 0953 1.1 / W 1608 1.9 / ◑ 2244 0.9	**22** 0523 1.7 / 1137 1.1 / TH 1742 1.8
8 0509 1.8 / 1112 1.1 / TH 1717 1.9	**23** 0020 0.9 / 0631 1.8 / F 1255 1.1 / 1853 1.8
9 0001 0.9 / 0620 1.9 / F 1232 1.0 / 1830 2.0	**24** 0121 0.9 / 0730 1.8 / SA 1355 1.1 / 1952 1.9
10 0107 0.7 / 0726 2.0 / SA 1334 0.9 / 1937 2.1	**25** 0210 0.9 / 0820 1.9 / SU 1442 1.0 / 2042 1.9
11 0204 0.7 / 0821 2.1 / SU 1429 0.8 / 2035 2.2	**26** 0251 0.9 / 0902 2.0 / M 1521 0.9 / 2126 2.0
12 0255 0.6 / 0911 2.2 / M 1519 0.6 / 2130 2.3	**27** 0327 0.8 / 0941 2.1 / TU 1556 0.9 / 2206 2.0
13 0343 0.5 / 1000 2.3 / TU 1608 0.5 / 2223 2.3	**28** 0402 0.8 / 1017 2.1 / W 1630 0.8 / 2244 2.1
14 0430 0.5 / 1047 2.4 / W 1655 0.4 / ○ 2316 2.4	**29** 0436 0.8 / 1052 2.2 / TH 1703 0.7 / ● 2321 2.1
15 0516 0.5 / 1135 2.4 / TH 1743 0.4	**30** 0511 0.8 / 1126 2.2 / F 1738 0.6 / 2358 2.1
	31 0546 0.8 / 1200 2.2 / SA 1815 0.6

Chart Datum: 1·22 metres below Ordnance Datum (Local). HAT is 2·5 metres above Chart Datum.

》FREE monthly updates. Register at 《
www.reedsnauticalalmanac.co.uk

425

North West Scotland

Cape Wrath to Crinan Canal

9.8.1	Contents	427
9.8.2	Area map and Distance table	428
9.8.3	Tidal stream charts	430
9.8.4	Lights, buoys and waypoints	432
9.8.5	Passage information	435

Outer Hebrides • Cape Wrath to Ullapool • Ullapool to Loch Torridon – 435 • Skye and the Inner Sound – 447 • Kyle Akin to Mallaig, Ardnamurchan to Crinan, Firth of Lorne and Sound of Luing – 451 • The Small Isles – 452

9.8.6	Stornoway, tide tables/curves	436

Loch Shell (Lewis) • Shiant Islands • E Loch Tarbert (Harris) • Sound of Harris • **N Uist**: Loch Maddy, Loch Eport • **S Uist**: Loch Carnan, Loch Skipport, Loch Eynort, Loch Boisdale • Acairseid Mhór (Eriskay) • North Bay, Castlebay (Barra) • Loch Rodel • Kallin • Flannan, St Kilda, Rockall and Monach Islands

9.8.7	Kinlochbervie	442

Loch Laxford • Loch Inver

9.8.8	Ullapool, tide tables/curves	443

Loch Kanaird • Summer Isles • Little Loch Broom • Loch Ewe • Loch Gairloch

9.8.9	Portree	448

Loch Torridon • Loch A'Bhraige (Rona) • Acairseid Mhór • **Around Skye**: Loch Dunvegan • Loch Harport • Soay • Armadale Bay • Crowlin Islands

	Submarine exercise areas	449
9.8.10	Plockton	449

Loch Carron

9.8.11	Loch Alsh	450

⚓s in The Sound of Sleat • Loch Hourn • Loch Nevis

9.8.12	Mallaig	452

⚓s in The Small Islands: Canna, Rum, Eigg, Muck

9.8.13	Loch Sunart	453

Arinagour (Coll) • Gott Bay (Tiree)

9.8.14	Tobermory	454

Sound of Mull • Craignure • **Around Mull**: Treshnish Isles • Staffa • Gometra • Loch na Keal • Loch Lathaich • Sound of Iona • Tinker's Hole • Carsaig Bay • Loch Spelve • Lynn of Lorn • Loch Creran • Loch Leven • Corran Narrows

9.8.15	Loch Aline	455
9.8.16	Fort William/Corpach	455
9.8.17	Caledonian Canal (Oban, tide tables/curves)	456
9.8.18	Oban	460

Dunstaffnage • Loch Etive • Loch Feochan • Puilladobhrain • Cuan Sound • Ardinamar

9.8.19	Loch Melfort	462
9.8.20	Craobh Marina (Loch Shuna)	463
9.8.21	Loch Craignish	463

Ardfern

9.8.22	Ferries on the west coast of Scotland	464

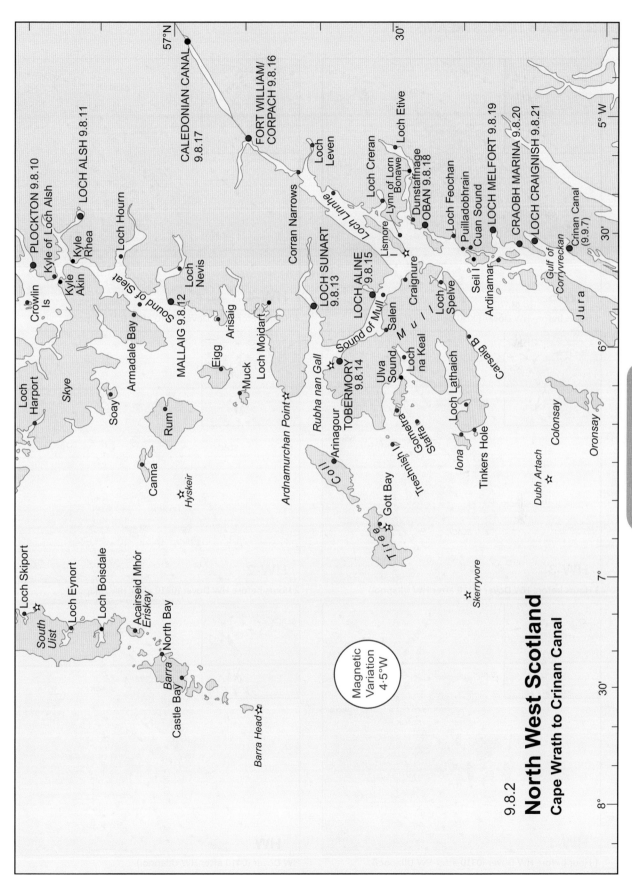

57°N

CALEDONIAN CANAL 9.8.17

FORT WILLIAM/ CORPACH 9.8.16

PLOCKTON 9.8.10

LOCH ALSH 9.8.11

Kyle of Loch Alsh

Kyle Rhea

Loch Hourn

Loch Leven

Loch Creran

Loch Etive

Dunstaffnage

OBAN 9.8.18

Loch Feochan

LOCH MELFORT 9.8.19

CRAOBH MARINA 9.8.20

LOCH CRAIGNISH 9.8.21

Crinan Canal (9.9.7)

Corran Narrows

Loch Linnhe

LOCH SUNART 9.8.13

LOCH ALINE 9.8.15

Lismore I

Lynn of Lorn

Bonawe

Puilladobhrain

Cuan Sound

Seil I

Craignure

Ardinamar

30'

Crowlin Is

Kyle Akin

Sound of Sleat

MALLAIG 9.8.12

Loch Nevis

Arisaig

Loch Moidart

Rubha nan Gall

TOBERMORY 9.8.14

Sound of Mull

Salen

Mull

Loch na Keal

Loch Spelve

Carsaig B

Jura

Loch Harport

Skye

Soay

Eigg

Muck

Ardnamurchan Point

Arinagour

Coll

Ulva Sound

Gometra

Staffa

Loch Lathaich

Iona

Gulf of Corryvreckan

Canna

Rum

Hyskeir

Treshnish I

Tinkers Hole

Colonsay

Oronsay

Dubh Artach

Loch Skiport

Loch Eynort

Loch Boisdale

Acairseid Mhór

Eriskay

South Uist

North Bay

Barra

Tiree

Gott Bay

Skerryvore

Magnetic Variation 4·5°W

Castle Bay

Barra Head

9.8.2

North West Scotland
Cape Wrath to Crinan Canal

5° W

6°

7°

30'

30'

8°

NW Scotland

9.8.3 AREA 8 TIDAL STREAMS

HW−5
5 Hours before HW Dover (0050 before HW Ullapool)

HW−4
4 Hours before HW Dover (0010 after HW Ullapool)

HW−3
3 Hours before HW Dover (0110 after HW Ullapool)

HW−2
2 Hours before HW Dover (0210 after HW Ullapool)

HW−1
1 Hour before HW Dover (0310 after HW Ullapool)

HW
HW Dover (0410 after HW Ullapool)

Eastward 9.7.3   Southward 9.9.3   Mull of Kintyre 9.9.11

1 Hour after HW Dover (0510 after HW Ullapool)

2 Hours after HW Dover (0615 before HW Ullapool)

3 Hours after HW Dover (0515 before HW Ullapool)

4 Hours after HW Dover (0415 before HW Ullapool)

5 Hours after HW Dover (0315 before HW Ullapool)

6 Hours after HW Dover (0215 before HW Ullapool)

NW Scotland

9.8.4 LIGHTS, BUOYS AND WAYPOINTS

Bold print = light with a nominal range of 15M or more. CAPITALS = place or feature. *CAPITAL ITALICS* = light-vessel, light float or Lanby. *Italics* = Fog signal. ***Bold italics*** = Racon. Some marks/buoys are fitted with AIS (<u>MMSI No</u>); see relevant charts.

CAPE WRATH TO LOCH TORRIDON

Cape Wrath ☆ 58°37'·54N 04°59'·99W Fl (4) 30s 122m **22M**; W twr.

LOCH INCHARD and LOCH LAXFORD

Kinlochbervie Dir lt 327° ☆ 58°27'·49N 05°03'·08W WRG 15m **16M** vis: 326°-FG-326·5°-AltGW-326·8°-FW-327·3°-AltRW-327·5°-FR-328°.
Creag Mhòr Dir lt 147°; Iso WRG 2s 16m 4M; vis: 136·5°-R -146·5°-W-147·5°-G-157·5°; 58°26'·99N 05°02'·45W.
Stoer Head ☆ 58°14'·43N 05°24'·07W Fl 15s 59m **24M**; W twr.

LOCH INVER, SUMMER ISLES and ULLAPOOL

Soyea I ⚲ Fl (2) 10s 34m 6M; 58°08'·56N 05°19'·67W.
Glas Leac ⚲ Fl WRG 3s 7m 5M; vis: 071°- W-078°-R-090°-G-103°-W-111°, 243°-W-247°-G-071°; 58°08'·68N 05°16'·36W.
Rubha Cadail ⚲ Fl WRG 6s 11m W9M, R6M, G6M; W twr; vis: 311°-G-320°-W-325°-R-103°-W-111°-G-118°-W-127°-R-157°-W-199°; 57°55'·51N 05°13'·40W.
Ullapool Pt ⚲ Iso R 4s 8m 6M; W twr; vis: 258°-108°; 57°53'·59N 05°09'·93W.
Cailleach Head ⚲ Fl (2) 12s 60m 9M; W twr; vis: 015°-236°; 57°55'·81N 05°24'·23W.

LOCH EWE and LOCH GAIRLOCH

Fairway ⚲ L Fl 10s; 57°51'·98N 05°40'·09W.
Rubha Reidh ☆ 57°51'·52N 05°48'·72W Fl (4) 15s 37m **24M**.
Glas Eilean ⚲ Fl WRG 6s 9m W6M, R4M; vis: 080°-W-102°-R-296°-W-333°-G-080°; 57°42'·79N 05°42'·42W.
Sgeir Dhubh Mhór ⚲; 57°42'·02N 05°42'·72W

OUTER HEBRIDES – EAST SIDE

LEWIS

Butt of Lewis ☆ 58°30'·89N 06°15'·84W Fl 5s 52m **25M**; R twr; vis: 056°-320°.
Tiumpan Head ☆ 58°15'·66N 06°08'·29W Fl (2) 15s 55m **25M**; W twr.
Broad Bay Tong ⚓ Ldg Lts 320°, Oc R 8s 8m 4M; 58°14'·48N 06°19'·98W. Rear, 70m from front, Oc R 8s 9m 4M.

STORNOWAY

Arnish Point ⚲ Fl WR 10s 17m W9M, R7M; W ○ twr; vis: 088°-W-198°-R-302°-W-013°; 58°11'·50N 06°22'·16W.
Sandwick Bay, NW side ⚲ Oc WRG 6s 10m 9M; vis: 334°-G-341°-W-347°-R-354°; 58°12'·20N 06°22'·11W.
No. 1 Pier SW corner ⚲ Q WRG 5m 11M; vis: shore-G-335°-W-352°-R-shore; 58°12'·36N 06°23'·43W.
Creed Estuary ⚲ Iso WRG 10s 24m 5M; vis: 277°-G-282°-W-290°-R-295°; 58°12'·03N 06°23'·47W.
No. 3 Pier ⚲ Q (2) G 10s 7m 2M 58°12'·31N 06°23'·28W.
Glumaig Hbr ⚲ Iso WRG 3s 8m 3M; 58°11'·27N 06°22'·9W; grey framework twr; vis: 150°-G-174°-W-180°-R-205°.

LOCH ERISORT, LOCH SHELL and EAST LOCH TARBERT

Shiants ▲ QG; 57°54'·57N 06°25'·70W.
Sgeir Inoe ▲ Fl G 6s; ***Racon (M) 5M***; 57°50'·93N 06°33'·93W.
Eilean Glas (Scalpay) ☆ 57°51'·41N 06°38'·55W Fl (3) 20s 43m **23M**; W twr, R bands; ***Racon (T) 16-18M***; <u>992356014</u>.
Sgeir Graidach ⚑ Q (6) + L Fl 15s; 57°50'·36N 06°41'·37W.
Sgeir Ghlas ⚲ Iso WRG 4s 9m W9M, R6M, G6M; W ○ twr; vis: 282°-G-319°-W-323°-R-153°-W-164°-G-171°; 57°52'·36N 06°45'·24W.
Tarbert ⚲ Dir Iso WRG 4s 9m 4M; vis : 290°-Iso G-297°-Al WG-300°-Iso W-303°-Al WR-306°-Iso R306°-313°; 57°53'·82N 06°47'·93W.

SOUND OF HARRIS, LEVERBURGH and BERNERAY

No.1 ▲ QG; 57°41'·20N 07°02'·67W.
No. 3 ▲ Fl G 5s; 57°41'·86N 07°03'·44W.

No. 4 ⚑ Fl R 5s; 57°41'·76N 07°03'·63W.
Suilven ⚑ Fl (3)R 10s; 57°41'·68N 07°04'·36W.
Cabbage ⚑ Fl (2) R 6s; ***Racon (T) 5M (3cm)***; 57°42'·13N 07°03'·96W.
Leverburgh Ldg Lts 014·7°. Front, Q 10m 4M 57°46'·23N 07°02'·04W. Rear, Oc 3s 12m 4M.
Jane's Tower ⚑ Q (2) G 5s 6m 4M; vis: obscured 273°-318°; 57°45'·76N 07°02'·12W.
Leverburgh Reef ⚑ Fl R 2s 4m 57°45'·97N 07°01'·86W.
Laimhrig Mhor ⚲ Q G; 57°45'·92N 07°01'·65W.
Leverburgh Pier Ldg Lts 063°. Front, Iso Bu 4s 8s 6m 4M; 57°46'·00N 07°01'·61W, metal col. Rear, Iso Bu 4s 8s 7m 4M.

NORTH UIST

Fairway ⚲ L Fl 10s; 57°40'·23N 07°01'·39W.
Vallay Island ⚲ Fl WRG 3s 4m 8M; vis: 206°-W-085°-G-140°-W-145°-R-206°; 57°39'·69N 07°26'·42W.
Griminish Hbr Ldg Lts 183°. Front, QG 6m 4M; 57°39'·38N 07°26'·75W. Rear, 110m from front, QG 7m 4M.

LOCH MADDY and GRIMSAY

Weaver's Pt ⚲ 57°36'·49N 07°06'·00W Fl 3s 24m 7M; W hut.
Glas Eilean Mòr ⚲ 57°35'·95N 07°06'·70W Fl (2) G 4s 8m 5M.
Vallaquie Is Dir lt 207·5°/ 255·5° ⚲ 57°35'·50N 07°09'·40W; W pillar; vis:Fl (3) WRG 8s 11m W5M, R5M, G5M; shore - G -205°-W-210°-R-240°; Fl (3) WRG 8s 11m W8M, R8M, G8M; 249°-G-254°-W-257°-R-262°.
Lochmaddy Ldg Lts 298°. Front, Ro-Ro Pier 2 FG (vert) 8m 4M. Rear, 110m from front, Oc G 8s 10m 4M; vis: 284°-304°; 57°35'·76N 07°09'·36W.
Grimsay No. 1 ⚑ Fl (2) R 8s; 57°28'·26N 07°11'·82W.

SOUTH UIST and LOCH CARNAN

Landfall ⚑ L Fl 10s; 57°22'·27N 07°11'·52W.
Ldg Lts 222°. Front Fl R 2s 7m 5M; W ◇ on post; 57°22'·00N 07°16'·34W. Rear, 58m from front, Iso R 10s 11m 5M; W ◇ on post.
Ushenish ☆ (S Uist) 57°17'·90N 07°11'·57W Fl W 20s 54m **W19M**; W twr; vis: 193°- 018°.

LOCH BOISDALE

MacKenzie Rk ⚑ Fl (3) R 15s 3m 4M; 57°08'·24N 07°13'·71W.
Calvay E End ⚲ Fl (2) WRG 10s 16m W7M, R7M, G7M; W twr; vis: 111°-W-190°-G-202°-W-286°-R-111°; 57°08'·53N 07°15'·38W; <u>992351139</u>.
Gasay I ⚲ Fl WR 5s 10m W7M, R7M; W twr; vis: 120°-W-284°-R-120°; 57°08'·93N 07°17'·39W.
Loch Boisdale Dir lt ⚲, 292·5°; WRG 4s 4m 5M; vis: 287·5°-IsoG-290·5°-AltWG-292°-IsoW-293°-AltW-294·5°-IsoR-298·5°; 57°09'·19N 07°18'·24W.
2 FG 8m 3M on dn; 57°09'·13N 07°18'·18W.

LUDAIG and ERISKAY

Ludaig Bwtr ⚲ 2 FR (vert) 6m 3M; 57°06'·17N 07°19'·49W.
Acairseid Mhor Ldg Lts 285°. Front, Oc R 6s 9m 4M; 57°03'·89N 07°17'·25W. Rear, 24m from front, Oc R 6s 10m 4M.

BARRA, CASTLEBAY and VATERSAY SOUND

Drover Rocks ⚑ Q (6) + L Fl 15s; 57°04'·08N 07°23'·54W.
Binch Rock ⚑ Q (6) + L Fl 15s; 57°01'·60N 07°17'·12W.
Curachan ⚑ Q (3) 10s; 56°58'·56N 07°20'·51W.
Ardveenish ⚲ Oc WRG 6m 9/6M; vis: 300°-G-304°-W-306°-R-310°; 57°00'·21N 07°24'·43W.
Bo Vich Chuan ⚑ Q(6)+L Fl 15s; ***Racon (M) 5M***; 56°56'·15N 07°23'·31W.
Channel Rk ⚲ Fl WR 6s 4m W6M, R4M; vis: 121·5°-W-277°-R-121·5°; 56°56'·24N 07°28'·94W.
Sgeir a Scape ▲ Fl (2) G 8s; 56°56'·25N 07°27'·21W.
Rubha Glas. Ldg Lts 295°. Front ⚲ FBu 9m 6M; Or △ on W twr; 56°56'·77N 07°30'·64W. Rear, 457m from front, FBu 15m 6M; Or ▽ on W twr; vis: 15° and 8° respectively either side of ldg line.
Barra Hd ☆ 56°47'·12N 07°39'·21W Fl 15s 208m **18M**; W twr; obsc by islands to NE; <u>992351095</u>.

OUTER HEBRIDES – WEST SIDE

Flannan I ☆, Eilean Mór Fl (2) 30s 101m **20M**; W twr; 58°17'·29N 07°35'·29W, obsc in places by Is to W of Eilean Mór.
Rockall ⚡ Fl 15s 19m 8M (unreliable); 57°35'·76N 13°41'·27W.
Gasker Lt ⚡ Fl (3) 10s 38m 10M; 57°59'·05N 07°17'·20W.
Whale Rock ⚓ Q (3) 10s 5m 5M; *Racon (T)*; 57°54'·40N 07°59·91W.
Haskeir I ☆ 57°41'·96N 07°41·31W Fl 20s 44m **24M**; W twr; *Racon (M) 17–15M*; 992351140.
Monach Isles ☆ Fl (2) 15s 47m **18M**; R brick twr; 57°31'·55N 07°41'·68W.

EAST LOCH ROAG

Aird Laimishader Carloway ⚡ Fl 6s 63m 8M; W hut; obsc on some brgs; 58°17'·06N 06°49'·50W.
Ardvanich Pt ⚡ Fl G 3s 4m 2M; 58°13'·48N 06°47'·68W.
Tidal Rk ⚡ Fl R 3s 2m 2M (sync with Ardvanich Pt above); 58°13'·45N 06°47'·57W.
Grèinam ⚡ Fl WR 6s 8m W8M, R7M; W bn; vis: R143°-169°, W169°-143°; 58°13'·30N 06°46'·16W.

LOCH ROAG

Rubha Domhain ⚡ Fl G 5s 9m 1M; post; 58°12'·96N 06°45'·40W.
Bogha na Muilne ⚓ Fl (2) 10s; 58°12'·70N 06°54'·23W.
Mia Vaig Bay ⚓ Fl G 2s 2m 1M; bn; 58°11'·83N 06°56'·60W.

NORTH UIST and SOUTH UIST

Vallay I ⚡ Fl WRG 3s 8M; vis: 206°-W-085°-G-140°-W-145°-R-206°; 57°39'·70N 07°26'·34W.
Falconet twr ⚡ FR 25m 8M (3M by day); shown 1hr before firing, changes to Iso R 2s 15 min before firing until completion; 57°22'·04N 07°23'·58W.

ST KILDA

Ldg Lts 270°. Front, Oc 5s 26m 3M; 57°48'·32N 08°34'·31W. Rear, 100m from front, Oc 5s 38m 3M; sync.

LOCH TORRIDON TO MALLAIG

LITTLE MINCH and W SKYE

Eugenie Rock ⚓ Q 6 + LF 15s; 57°46'·47N 06°27'·28W.
Eilean Troddday ⚡ Fl (2) WRG 10s 52m W12M, R9M, G9M; W Bn; vis: W062°-R088°-130°-W-322°-G-062°; 57°43'·63N 06°17'·93W; 992351084.
Uig, Edward Pier Hd ⚡ 57°35'·09N 06°22'·29W Iso WRG 4s 9m W7M, R4M, G4M; vis: 180-W-006°-G-050°-W-073°-R-180°.
Waternish Pt ⚡ Fl 20s 21m 8M; W twr; 57°36'·48N 06°37'·99W.
Loch Dunvegan, Uiginish Pt ⚡ Fl WRG 3s 16m W7M,R5M,G5M; W lattice twr; vis: 041°-G-132°-W-145°-R-148°-W-253°-R-263°-W-273°-G-306°, (>148° obsc by Fiadhairt Pt); 57°26'·84N 06°36'·53W.
Neist Point ☆ 57°25'·41N 06°47'·30W Fl 5s 43m **16M**; W twr.
Loch Harport, Ardtreck Pt ⚡ 57°20'·38N 06°25'·80W Fl 6s 18m 9M; small W twr.

RONA, LOCH A'BHRAIGE and INNER SOUND

Na Gamhnachain ⚓ Q; 57°35'·89N 05°57'·71W.
Rona NE Point ☆ 57°34'·69N 05°57'·54W Fl 12s 69m **19M**; W twr; vis: 050°-358°; 992351045.
Loch A'Bhraige, Sgeir Shuas ⚡ Fl R 2s 6m 3M; vis: 070°-199°; 57°35'·02N 05°58'·61W.
Ldg Lts 136·5°. Front, ⚓ Q WRG 3m W4M, R3M; vis: 135°-W- 138°-R-318°-G-135°; 57°34'·41N 05°58'·09W. Rear, ⚓ Iso 6s 28m 5M.
Ru Na Lachan ⚡ Oc WR 8s 21m 10M; twr; vis: 337°-W-022°-R-117°-W-162°; 57°29'·02N 05°52'·15W.

SOUND OF RAASAY, PORTREE and CROWLIN ISLANDS

Sgeir Mhór ⚓Fl G 5s; 57°24'·57N 06°10'·53W.
Eilean Beag ⚡ Fl 6s 32m 6M; W Bn; 57°21'·21N 05°51'·42W.
Eyre Pt ⚡ Fl WR 3s 6m W9M R6M; W clad framework twr; vis: 215°-W-267°-R-288°-W-063°; 57°20'·01N 06°01'·28W.
Sgeir Thraid ⚡ Q 12m 3M; metal twr; 57°19'·82N 05°56'·51W.
Sgeir Ghobhlach ⚡ Fl(3)10s 10m 3M; metal twr; 57°15'·69N 05°52'·25W.

LOCH CARRON

Sgeir Golach ⚡ Fl 10s 5m 3M; 57°21'·20N 05°39'·01W.
Eilean a Chait Lt ho (dis) 13M; 57°20'·96N 05°38'·87W.
Bogha Dubh Sgeir ⚓ Fl (2)R 6s 4m 2M; 57°20'·92N 05°37'·85W.

RAASAY and LOCH SLIGACHAN

Suisnish ⚡ 2 FG (vert) 8m 2M; 57°19'·87N 06°03'·91W.
Eyre Point ⚡ Fl WR 3s 6m W9M, R6M; W twr; vis: 215°-W-266°-R- 288°-W-063°; 57°20'·01N 06°01'·29W.

KYLEAKIN and KYLE OF LOCH ALSH

Carragh Rk ⚓ Fl (2) G 12s; *Racon (T) 5M*; 57°17'·18N 05°45'·36W.
Bow Rk ⚓ Fl (2) R 12s; 57°16'·71N 05°45'·85W.
Fork Rks ⚓ Fl G 6s; 57°16'·85N 05°44'·93W.
Black Eye Rk ⚓ Fl R 6s; 57°16'·72N 05°45'·31W.
Eileanan Dubha East ⚡ Fl (2) 10s 9m 8M; vis: obscured 104°-146°; 57°16'·56N 05°42'·32W.
8 Metre Rock ⚡ Fl G 6s 5m 4M; 57°16'·60N 05°42'·69W.
String Rock ⚓ Fl R 6s; 57°16'·50N 05°42'·89W.
Sgeir-na-Caillich ⚡ Fl (2) R 6s 3m 4M; 57°15'·59N 05°38'·90W.

SOUND OF SLEAT

Kyle Rhea ⚡ Fl WRG 3s 7m W8M, R5M, G5M; W Bn; vis: shore-R-219°-W-228°-G-338°-W-346°-R-shore; 57°14'·22N 05°39'·93W.
Sandaig I, NW point ⚡ Fl 6s 13m 8M; W twr; 57°10'·05N 05°42'·29W.
Ornsay, SE end ⚡ Oc 8s 18m 12M; W twr; vis: 157°-030°; 57°08'·60N 05°46'·85W.
Pt. of Sleat ⚡ Fl 3s 20m9M; W twr; 57°01'·08N 06°01'·08W.

MALLAIG and LOCH NEVIS ENTRANCE

Sgeir Dhearg ⚓ QG; 57°00'·74N 05°49'·50W.
Northern Pier E end ⚡ Iso WRG 4s 6m W9M, R6M, G6M; Gy twr; vis: 181°-G-185°-W-197°-R-201°. Fl G 3s 14m 6M; same structure; 57°00'·47N 05°49'·50W.
Sgeir Dhearg ⚡ 57°00'·63N 05°49'·61W Fl (2) WG 8s 6m 5M; Gy Bn; vis: 190°-G-055°-W-190°.

SMALL ISLES AND WEST OF MULL

CANNA and RUM

Canna, E end Sanday Is ⚡ Fl 10s 32m 9M; W twr; vis: 152°-061°; 57°02'·82N 06°28'·02W.
Loch Scresort ⚓ Q; 57°00'·79N 06°14'·61W.

HYSKEIR, EIGG, MUCK and ARISAIG

Humla ⚓ Fl G 6s 3m 4M 57°00'·46N 06°37'·39W.
Hyskeir ☆ 56°58'·16N 06°40'·83W Fl (3) 30s 41m **24M**; W twr. *Racon (T) 14–17M*; 992351094.
SE point Eigg (Eilean Chathastail) ⚡ Fl 6s 24m 8M; W twr; vis: 181°-shore; 56°52'·25N 06°07'·28W.
Eigg, Sgeir nam Bagh (Ferry Terminal) ⚡ Dir 245°; Fl WRG 3s 9m W14, R11, G11; H24; steel pole; vis: 242·5°-G-244°-W-246°-R-247·5°. 2FR(vert) on same structure; 56°52'·79N 06°07'·56W.
Isle of Muck (Port Mor) ⚡ Dir Fl WRG 3s 7m W14, R11, G11, by day W1, R1, G1; steel twr; vis: 319·5°-G-321°-W-323°-R-324·5°; 56°49'·96N 06°13'·64W.
Bogha Ruadh⚡ Fl G 5s 4m 3M; 56°49'·56N 06°13'·05W.
Bo Faskadale ⚓ Fl (3) G 18s; 56°48'·18N 06°06'·37W.
Ardnamurchan ☆ 56°43'·63N 06°13'·58W Fl (2) 20s 55m **22M**; Gy twr; vis: 002°-217°.
Cairns of Coll, Suil Ghorm ⚡ Fl 12s 23m 10M; W twr; 56°42'·26N 06°26'·75W.

TIREE, COLL and ARINAGOUR

Loch Eatharna, Pier Head ⚡ Dir 325°; Oc WRG 7s 6m 2M; vis: 316°-G-322°-W-328°-R-334°; 56°36'·86N 06°31'·29W.
Loch Eatharna, Bogha Mór ⚓ Fl G 6s; 56°36'·65N 06°30'·90W.
Roan Bogha ⚓ Q (6) + L Fl 15s 3m 5M; 56°32'·23N 06°40'·18W.
Placaid Bogha ⚓ Fl G 4s; 56°33'·22N 06°44'·06W.
Scarinish ⚡ Fl 3s 11m 12M; W □ twr; vis: 210°-030°; S side of ent, 56°30'·02N 06°48'·26W.
Cairn na Burgh More (Treshnish Is), Fl (3) 15s 36m 8M; solar panels on framework tr; 56°31'·05N 06°22'·95W.

Gott Bay Ldg Lts 286·5°. Front FR 8m; 56°30'·61N 06°47'·82W. Rear 30m from front FR 11m.

Skerryvore ☆ Fl 10s 46m **23M**; Gy twr; *Racon (M) 18M*. 56°19'·36N 07°06'·88W; 992351091.

LOCH NA KEAL and LOCH NA LÀTHAICH

Sgeir a Charraigein ⚓ Q(6) + LFl 15s; 56°28'·29N 06°07'·71W.
Eileanan na Liathanaich, SE end ⚓ Fl WR 6s 12m W8M, R6M; vis: R088°- W108°-088°; 56°20'·56N 06°16'·38W.

DUBH ARTACH

Dubh Artach ☆ 56°07'·94N 06°38'·08W Fl (2) 30s 44m **20M**; Gy twr, R band; 992351088.

SOUND OF MULL

LOCH SUNART, TOBERMORY and LOCH ALINE

Ardmore Pt ⚓ Fl (2) 10s 18m 13M; 56°39'·37N 06°07'·70W.
New Rks ▲ Fl G 6s 56°39'·05N 06°03'·30W.
Rubha nan Gall ☆ 56°38'·33N 06°04'·00W Fl 3s 17m **15M**; W twr.
Tobermory Slip ⚓ Iso WRG 3s 5m 4M; vis: 220°-G-232°-W-244°-R-256°; metal column; 56°37'·41N 06°03'·77W.
Eileanan Glasa (Dearg Sgeir) ⚓ 56°32'·25N 05°54'·80W Fl 6s 7m 8M; W ○ twr.
Fiunary Spit ▲ Fl G 6s; 56°32'·66N 05°53'·17W.
Fishnish Ferry Slip ⚓ Iso WRG 3s 5m 4M; vis: 158°-G-170°-W-195°-R-230°; metal column; 56°30'·89N 05°48'·61W.
Loch Aline, ⚓ Dir 357°; Oc WRG 6s 2m 4M; vis: 353°-G-356°-W-358°-R-002°; Or concrete plinth; 56°32'·38N 05°46'·47W.
Ardtornish Pt ⚓ Fl (2) WRG 10s 8m W8M, R6M, G6M; W twr; vis: G shore- 301°-W-308°-R-342°-W-057°-R-095°-W-108°-G-shore; 56°31'·09N 05°45'·21W.
Craignure Ldg Lts 240·9°. Front, FR 10m; 56°28'·26N 05°42'·28W. Rear, 150m from front, FR 12m; vis: 225·8°-255·8°.

MULL TO CALEDONIAN CANAL AND OBAN

Lismore ☆, SW end 56°27'·34N 05°36'·45W Fl 10s 31m **17M**; W twr; vis: 237°-208°.
Lady's Rk ⚓ Fl 6s 12m 5M; Red clad lattice tower on W base 56°26'·92N 05°37'·05W; 992351093.
Duart Pt ⚓ Fl (3) WR 18s 14m W5M, R3M; vis: 162°-W-261°-W-275°-W-353°-R-shore; 56°26'·84N 05°38'·77W.

LOCH LINNHE

Corran Shoal ⚓ QR 56°43'·69N 05°14'·39W.
Corran Pt ⚓ Iso WRG 4s 12m W10M, R7M, G7M; W twr; vis: shore-R-195°-W-215°-G-020-W-030°-R-shore; 56°43'·25N 05°14'·54W.
Corran Narrows NE ⚓ Fl 5s 4m 4M; W twr; vis: S shore-214°; 56°43'·62N 05°13'·90W.
Clovullin Spit ⚓ Fl (2) R 15s; 56°42'·29N 05°15'·56W.
Cuil-cheanna Spit ▲ Fl G 6s; 56°41'·17N 05°15'·72W.
Mc Lean Rk ⚓ QR 56°49'·80N 05°07'·04W.
Eilean na Creiche ⚓ Fl (2) 5s; 56°50'·25N 05°07'·02W.
Eilean na Craobh ⚓ Fl (2) R 6s 4m 2M; 56°50'·43N 05°07'·95W.

FORT WILLIAM and CALEDONIAN CANAL

Corpach, Caledonian Canal Lock ent ⚓ Iso WRG 4s 6m 5M; W twr; vis: G287°- W310°- R335°-030°; 56°50'·52N 05°07'·44W.

LYNN OF LORN

Sgeir Bhuidhe Appin ⚓ Fl (2) WR 7s 8m W9M R6M; W Bn; vis: W013·5°- R184°-220°; 56°33'·63N 05°24'·65W.
Appin Point ▲ Fl G 6s; 56°32'·69N 05°25'·97W.
Dearg Sgeir, off Aird's Point ⚓ Fl WRG 2s 2m W3M, R1M, G1M; vis: 196°-R-246°-W-258°-G-041°-W-058°-R-093°-W-139°; 56°32'·20N 05°25'·22W.
Rubha nam Faoileann (Eriska) ⚓ QG 2m 2M; G col; vis: 128°-329°; 56°32'·20N 05°24'·11W.
Branra Rk ⚓ Fl(2) 10s 4m 5M; metal frame on W twr; 56°32'·02N 05°26'·60W.

OBAN

N spit of Kerrera ⚓ Fl R 3m 4M; W post; 56°25'·32N 05°29'·29W.
Dunollie ⚓ Fl (2) WRG 6s 6m W8M, R6M, G6M; vis: 351°-G- 020°-W-047°-R-120°-W-138°-G-143°; 56°25'·37N 05°29'·05W.
Rubh'a'Chruidh ⚓ Q R 3m 2M; W col, R bands; 56°25'·49N 05°29'·56W.
Kerrera ⚓ QR; 56°24'·15N 05°30'·81W.
Corran Ledge ⚓ VQ (9) 10s; 56°25'·19N 05°29'·11W.
Oban Bay ⚓ Q; 56°24'·92N 05°29'·23W.
Sgeir Rathaid ⚓ Q(6)+L Fl 15s; 56°24'·75N 05°29'·37W.
North Pier ⚓ 2 F G (vert); metal post; 56°24'·89N 05°28'·49W.
Ro-Ro Jetty ⚓ 2 F G (vert); post on dn; 56°24'·69N 05°28'·65W.
Northern Lt Pier ⚓ 2 F G (vert); metal post; 56°24'·71N 05°28'·91W.

OBAN TO LOCH CRAIGNISH

Heather Is ⚓ Fl R 2·5s 11m 2M; 56°24'·41N 05°30'·24W.
Ferry Rks SE ⚓ Q (3) 10s; 56°23'·99N 05°30'·53W.
Sgeirean Dubha ⚓ Fl (2) 12s 7m 5M; W ○ twr; 56°22'·81N 05°32'·27W.
Sgeir an Fheuran ▲ Fl G 3s; 56°22'·80N 05°31'·94W.
Bogha Nuadh ⚓ Q (6) + LFl 15s; 56°21'·69N 05°37'·87W; 992351090.
Bogha Ghair ⚓ Q (3) 10s; 56°16'·50N 05°40'·44W.
Bono Rock ⚓ Q (9) 15s; 56°16'·21N 05°41'·22W.
Fladda ⚓ Fl (2) WRG 9s 13m W11M, R9M, G9M; W twr; vis: 169°-R-186°-W-337°-G-344°-W-356°-R-026°; 56°14'·89N 05°40'·83W.
Dubh Sgeir (Luing) ⚓ Fl WRG 6s 9m W6M, R4M. G4M; W twr; vis: W000°- R010°- W025°- G199°-000°; *Racon (M) 5M*; 56°14'·76N 05°40'·20W.
The Garvellachs, Eileach an Naoimh, SW end ⚓ Fl 6s 21m 9M; W Bn; vis: 240°-215°; 56°13'·04N 05°49'·06W.

LOCH MELFORT and CRAOBH HAVEN

Melfort Pier ⚓ Dir FR 6m 3M; (Apr -Nov); 56°16'·14N 05°30'·19W.
⚓ 56°12'·88N 05°33'·59W.
Craobh Marina Bkwtr Hd ⚓ Iso WRG 5s 10m, W5M, R3M, G3M; vis:114°-G-162°-W-183°-R-200°; 56°12'·78N 05°33'·52W.

9.8.5 PASSAGE INFORMATION

More passage information is threaded between the harbours in this area. It is essential to carry large scale charts and current pilot books for this area which is only partially covered between Ardnamurchan Point and the Shiant Isles by Admiralty Leisure Folio 5616. **Bibliography:** *Clyde Cruising Club Sailing Directions-Kintyre to Arnamurchan, CCC Sailing Directions-Ardnamurchan to Cape Wrath, CCC Sailing Directions-Outer Hebrides*(all Imray/Clyde Cruising Club, Lawrence). *The Yachtsman's Pilot to Skye and Northwest Scotland* (Imray/Lawrence). The *North West Coast of Scotland Pilot* (Admiralty NP66B) covers the whole area.

Scotland's west coast provides splendid, if boisterous, sailing and matchless scenery. In summer the long daylight hours and warmth of the Gulf Stream compensate for the lower air temperatures and higher wind speeds experienced when depressions run typically north of Scotland. Inshore winds are often unpredictable, due to geographical effects of lochs, mountains and islands offshore; calms and squalls can alternate rapidly.

Good anchors, especially on kelp/weed, are essential. ⚓s are listed but it should not be assumed that these will always be available. Particularly in N of area, facilities are very dispersed. A 'Rover Ticket' from Highland Council allows berthing for 15 days; the scheme includes: Kinlochbervie, Lochinver, Gairloch, Kyle of Lochalsh, Kyleakin, Portree and Uig. VHF communications with shore stations may be limited by high ground. Beware ever more fish farms in many inlets. Local magnetic anomalies occur in Kilbrannan Sound, Passage of Tiree, Sound of Mull, Canna, and East Loch Roag. Submarines exercise throughout these waters.

OUTER HEBRIDES

(AC 1785, 1794, 1795) The E sides of these Is have many good, sheltered ⚓s, but W coasts give little shelter. The CCC's *Outer Hebrides SDs* or *The Western Isles* (Imray) are advised.

▶ *The Minches and Sea of the Hebrides can be very rough, particularly in the Little Minch between Skye and Harris, and around Shiant Is where tide runs locally 4kn at sp, and heavy overfalls can occur. The NE going stream on both shores begins at HW Ullapool -0340 (HW Dover +0430), with the strongest flow from mid channel to the Skye coast. There is a W going counter tide E of Vaternish Pt.*

The SW going stream on both shores begins at HW Ullapool +0240 (HW Dover –0130), with the strongest flow from mid channel to the Skye coast, sp rates 2·5kn. The E going stream in Sound of Scalpay runs at up to 2k.The E going flood and W going ebb in Sound of Scalpay run at up to 2k.◀

▶ **Sound of Harris.** *The behaviour of tidal streams is complicated and varies from day to night, springs to neaps, and winter to summer. In the NE part of the Sound, at springs, the stream sets SE from Ullapool +0535 to –0035. At neaps, in the summer, the stream sets SE during the day and NW during the night. To the SW of the Sound, the NW stream starts at Ullapool –0545; the SE stream starts at Ullapool +0025. The approximate maximum rates are as shown. For more detailed information refer to the Admiralty West Coast of Scotland Pilot.◀*

From N to S, the better harbours in the Outer Hebrides include:

Lewis. Stornoway; Loch Grimshader (beware Sgeir a'Chaolais, dries in entrance); Loch Erisort; Loch Odhairn; Loch Shell. Proceeding S from here, or to E Loch Tarbert beware Sgeir Inoe (dries 2·3m) 3M ESE of Eilean Glas lt ho at SE end of Scalpay.

Harris. E Loch Tarbert; Loch Scadaby; Loch Stockinish; Loch Finsby; W Loch Tarbert; Loch Rodel (⚓). A well marked ferry channel connects Leverburgh (South Harris) to Berneray.

N Uist. Loch Maddy (⚓); Loch Eport, Kallin Hbr (⚓).

S Uist. Loch Carnan (⚓); Loch Skiport; Loch Eynort; Loch Boisdale (⚓).

Barra. Castlebay (⚓), and Berneray, on N side, E of Shelter Rk.

There are numerous ⚓s. Bratanish Mor in L Roag is very sheltered and provides respite on W coast

Activity at the Hebrides Range, S. Uist ☎(01870) 604441, is broadcast daily at 0950LT and Mon-Fri 1100-1700LT on VHF Ch 12 (Ch 73 in emergency) and on MF 2660 kHz.

CAPE WRATH TO ULLAPOOL

(AC 1785, 1794) C Wrath (lt fog sig) is a steep headland (110m).

▶ *To N of it the E-going stream begins at HW Ullapool – 0350, and W-going at HW Ullapool + 0235, sp rates 3kn. Eddies close inshore cause almost continuous W-going stream E of Cape, and N-going stream SW of it. Where they meet is turbulence, with dangerous seas in bad weather.* ◀

Duslic Rk, 7ca NE of lt ho, dries 3·4m. 6M SW of C Wrath, islet of Am Balg (45m)is foul for 2ca around.

There are ⚓ in Loch Inchard (AC 2503), the best shelter being in Kinlochbervie on N shore; also good ⚓s among Is along S shore of Loch Laxford, entered between Ardmore Pt and Rubha Ruadh. Handa Is to WSW is a bird sanctuary. Handa Sound is navigable with care, but beware Bogha Morair in mid-chan and associated overfalls.

▶ *Tide turns 2hrs earlier in the Sound than offshore. Strong winds against tide raise a bad sea off Pt of Stoer.*◀

The best shelter is 8M S at Loch Inver (AC 2504), with good ⚓ off hotel near head of loch. ⚓s to the N of Loch Inver may be found at Kylesku, Culkein Drumbeg, Drumbeg, Loch Nedd and Loch Roe. To the S lies Enard Bay.

ULLAPOOL TO LOCH TORRIDON

(AC 1794, 2210, 5616) The Summer Isles (AC 2501), 12M NW of the major fishing port of Ullapool, offer some sheltered ⚓s and tight approaches. The best include the bay on E side of Tanera Mor; off NE of Tanera Beg (W of Eilean Fada Mor); and in Caolas Eilean Ristol, between the island and mainland.

Loch Ewe (AC 3146) provides good shelter and easy access. Best ⚓s are in Poolewe Bay (beware Boor Rks off W shore) and in SW corner of Loch Thuirnaig (entering, keep close to S shore to avoid rocks extending from N side). Off Rubha Reidh (lt) seas can be dangerous.

▶ *The NE-going stream begins at HW Ullapool –0335; the SW-going at HW Ullapool +0305. Sp rates 3kn, but slacker to SW of point.* ◀

Longa Is lies N of ent to Loch Gairloch (AC 2528). The channel N of it is navigable but narrow at E end. Outer loch is free of dangers, but exposed to swell. Best ⚓ is on S side of loch in Caolas Bad a' Chrotha, W of Eilean Horrisdale.

Entering L Torridon (AC 2210) from S or W beware Murchadh Breac (dries 1·5m) 3ca NNW of Rubha na Fearna. Best ⚓s are SW of Eilean Mor (to W of Ardheslaig); in Loch a 'Chracaich, 7ca further SE; E of Shieldaig Is; and near head of Upper L Torridon.

▶ *Streams are weak except where they run 2-3kn in narrows between L Shieldaig and Upper L Torridon.*◀

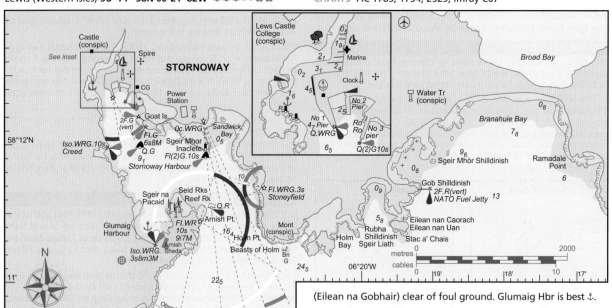

9.8.6 STORNOWAY

Lewis (Western Isles) 58°11'·58N 06°21'·82W ✳✳✳◊◊✿✿   CHARTS AC 1785, 1794, 2529; Imray C67

TIDES

Standard Port STORNOWAY (⟶)

Times				Height (metres)			
High Water		Low Water		MHWS	MHWN	MLWN	MLWS
0100	0700	0300	0900	4·8	3·7	2·0	0·7
1300	1900	1500	2100				

Differences CARLOWAY (W Lewis)

−0040	+0020	−0035	−0015	−0.7	−0.5	−0.3	0.0

LITTLE BERNERA (W Lewis)

−0021	−0011	−0017	−0027	−0.5	−0.6	−0.4	−0.2

WEST LOCH TARBERT (W Harris)

−0015	−0015	−0046	−0046	−1.1	−0.9	−0.5	0.0

SHELTER Good. New marina (2014), max LOA 24m, in Inner Hbr, has depths to 3m. Visitors should report to Stornoway Port Authority; large boats may request berth on 80m pontoon on Esplanade Quay. Ferries use W side No 3 pier, and commercial vessels on Nos 1 and 2 Piers. South'ly swells can make anchoring uncomfortable. ‡ on chartlet at: Poll nam Portan to W of inner chan in bay NW of Goat Is

(Eilean na Gobhair) clear of foul ground. Glumaig Hbr is best ‡.

NAVIGATION WPT 58°09'·98N 06°20'·87W, 343°/2·3M to Oc WRG lt. Access to the harbour is straightforward at all states of the tide. Reef Rk, N of Arnish Pt to W of entrance, indicated by PHM, QR. To E side of entrance an unlit G bn marks the Beasts of Holm, a rky patch. This lies off Holm Pt, on which is a conspic memorial. A local magnetic anomaly exists over a small area in mid-hbr, 1·75ca N of Seid Rks PHM bn.

LIGHTS AND MARKS Arnish sheds are conspic 3ca SW of Arnish Pt lt, W sector 302°-013° covers ent. Then in turn follow W sectors of: Sandwick Bay lt (close E of water tr, 3 power stn chys; all conspic), then Creed Estuary, Iso WRG 10s, across the harbour, and No 1 Pier, Q WRG. Lit bns in inner hbr mark appr chan to marina.

COMMUNICATIONS (Code 01851) CGOC 702013; Police 101; Dr 703145; ⊖ 703626. HM 702688. HM VHF Ch 12 16 (H24).

FACILITIES Marina 80⌷ inc 8♥, £2/m/day, £10/m/wk; ⚓ ⌷ wi-fi. **Piers 1 & 2** ⚓ ⌷(10t) ➤. **Pier 3** ⚓ ⌷(mob) ⌷(road tanker). **Services** ⚒ ⚒ ⌷⌷⌷ ACA. **Town** ⌷⌷⌷⌷⌷ ⌷ ⌷⌷⌷⌷⌷ ⌷ (Ferry to Ullapool, bus to Garve), ✈.

STANDARD TIME (UT)
For Summer Time add ONE hour in **non-shaded areas**

STORNOWAY LAT 58°12′N LONG 6°23′W
TIMES AND HEIGHTS OF HIGH AND LOW WATERS

Dates in red are **SPRINGS**
Dates in blue are **NEAPS**

YEAR 2016

JANUARY

Day	Time m	Day	Time m
1 F	0506 1.9 / 1113 4.1 / 1754 1.8	**16** SA	0510 1.4 / 1128 4.5 / 1746 1.3
2 SA	0010 3.7 / 0556 2.1 / 1223 3.9 / 1850 2.0	**17** SU	0013 4.0 / 0610 1.7 / 1239 4.3 / 1849 1.6
3 SU	0123 3.7 / 0658 2.3 / 1342 3.8 / 2000 2.1	**18** M	0131 3.9 / 0725 1.9 / 1355 4.2 / 2005 1.7
4 M	0233 3.7 / 0816 2.3 / 1453 3.8 / 2113 2.1	**19** TU	0248 3.9 / 0849 1.9 / 1510 4.2 / 2125 1.7
5 TU	0334 3.8 / 0931 2.2 / 1553 3.9 / 2213 1.9	**20** W	0358 4.1 / 1008 1.7 / 1617 4.3 / 2234 1.6
6 W	0425 4.0 / 1031 2.0 / 1642 4.0 / 2300 1.7	**21** TH	0455 4.4 / 1111 1.5 / 1714 4.4 / 2329 1.4
7 TH	0506 4.3 / 1118 1.8 / 1723 4.2 / 2341 1.5	**22** F	0541 4.6 / 1203 1.2 / 1801 4.6
8 F	0543 4.5 / 1200 1.5 / 1800 4.4	**23** SA	0015 1.2 / 0621 4.8 / 1249 1.0 / 1841 4.7
9 SA	0020 1.3 / 0617 4.7 / 1240 1.3 / 1835 4.5	**24** SU	0057 1.0 / 0658 4.9 / 1330 0.9 / 1919 4.7
10 SU	0058 1.1 / 0652 4.9 / 1319 1.0 / 1912 4.7	**25** M	0135 1.0 / 0732 5.0 / 1408 0.8 / 1954 4.7
11 M	0135 1.0 / 0728 5.1 / 1358 0.9 / 1951 4.8	**26** TU	0211 1.0 / 0805 4.9 / 1443 0.9 / 2027 4.6
12 TU	0213 0.9 / 0807 5.1 / 1439 0.8 / 2033 4.7	**27** W	0245 1.0 / 0836 4.8 / 1518 1.0 / 2101 4.5
13 W	0253 0.9 / 0849 5.1 / 1520 0.8 / 2117 4.6	**28** TH	0319 1.1 / 0908 4.7 / 1552 1.1 / 2135 4.3
14 TH	0334 1.0 / 0935 5.0 / 1604 0.9 / 2206 4.4	**29** F	0354 1.3 / 0941 4.4 / 1628 1.4 / 2213 4.1
15 F	0419 1.2 / 1027 4.8 / 1652 1.1 / 2304 4.2	**30** SA	0430 1.6 / 1018 4.2 / 1706 1.6 / 2300 3.9
		31 SU	0509 1.8 / 1105 4.0 / 1749 1.9

FEBRUARY

Day	Time m	Day	Time m
1 M	0004 3.7 / 0556 2.1 / 1214 3.7 / 1844 2.1	**16** TU	0102 3.9 / 0656 1.8 / 1337 4.0 / 1932 1.9
2 TU	0126 3.6 / 0700 2.3 / 1347 3.6 / 2000 2.2	**17** W	0227 3.8 / 0829 1.9 / 1459 3.9 / 2105 1.9
3 W	0241 3.7 / 0828 2.3 / 1507 3.6 / 2126 2.1	**18** TH	0345 4.0 / 1001 1.8 / 1612 4.0 / 2224 1.8
4 TH	0346 3.8 / 0952 2.2 / 1612 3.8 / 2229 1.9	**19** F	0445 4.2 / 1105 1.5 / 1708 4.2 / 2319 1.5
5 F	0437 4.1 / 1052 1.9 / 1701 4.0 / 2317 1.6	**20** SA	0532 4.4 / 1154 1.2 / 1751 4.4
6 SA	0519 4.4 / 1139 1.5 / 1742 4.3 / 2359 1.3	**21** SU	0003 1.3 / 0609 4.6 / 1237 1.0 / 1827 4.5
7 SU	0557 4.7 / 1221 1.1 / 1820 4.5	**22** M	0042 1.1 / 0642 4.8 / 1313 0.8 / 1900 4.6
8 M	0039 1.0 / 0633 5.0 / 1302 0.8 / 1856 4.8	**23** TU	0118 0.9 / 0712 4.9 / 1347 0.8 / 1930 4.7
9 TU	0119 0.8 / 0710 5.2 / 1342 0.5 / 1935 4.9	**24** W	0151 0.9 / 0741 4.9 / 1419 0.8 / 2000 4.6
10 W	0158 0.6 / 0750 5.3 / 1423 0.4 / 2015 4.9	**25** TH	0222 0.9 / 0809 4.8 / 1450 0.9 / 2029 4.5
11 TH	0238 0.5 / 0831 5.3 / 1503 0.4 / 2057 4.8	**26** F	0253 1.0 / 0837 4.6 / 1521 1.0 / 2100 4.4
12 F	0318 0.6 / 0911 5.2 / 1545 0.5 / 2142 4.6	**27** SA	0325 1.1 / 0907 4.5 / 1552 1.2 / 2133 4.2
13 SA	0402 0.8 / 1005 4.9 / 1630 0.8 / 2235 4.4	**28** SU	0357 1.3 / 0940 4.2 / 1625 1.4 / 2211 4.0
14 SU	0449 1.2 / 1103 4.6 / 1719 1.2 / 2340 4.1	**29** M	0433 1.6 / 1019 4.0 / 1702 1.7 / 2302 3.8
15 M	0545 1.5 / 1215 4.2 / 1817 1.6		

MARCH

Day	Time m	Day	Time m
1 TU	0514 1.8 / 1113 3.7 / 1746 1.9	**16** W	0038 3.8 / 0636 1.7 / 1324 3.8 / 1902 1.9
2 W	0016 3.6 / 0606 2.1 / 1241 3.5 / 1849 2.2	**17** TH	0206 3.8 / 0814 1.9 / 1447 3.7 / 2043 2.0
3 TH	0145 3.6 / 0725 2.2 / 1420 3.5 / 2028 2.2	**18** F	0326 3.9 / 0948 1.7 / 1600 3.8 / 2207 1.9
4 F	0300 3.7 / 0908 2.1 / 1537 3.6 / 2155 2.0	**19** SA	0428 4.2 / 1050 1.5 / 1654 4.0 / 2301 1.6
5 SA	0401 3.9 / 1022 1.8 / 1635 3.9 / 2251 1.7	**20** SU	0513 4.3 / 1136 1.2 / 1734 4.2 / 2344 1.4
6 SU	0450 4.2 / 1114 1.4 / 1720 4.2 / 2336 1.3	**21** M	0549 4.4 / 1215 1.0 / 1808 4.4
7 M	0531 4.6 / 1158 1.0 / 1759 4.5	**22** TU	0021 1.1 / 0620 4.6 / 1250 0.8 / 1837 4.5
8 TU	0018 0.9 / 0609 5.0 / 1241 0.6 / 1836 4.8	**23** W	0056 1.0 / 0648 4.7 / 1322 0.7 / 1905 4.6
9 W	0058 0.6 / 0648 5.2 / 1321 0.3 / 1914 5.0	**24** TH	0127 0.9 / 0715 4.7 / 1352 0.7 / 1933 4.6
10 TH	0138 0.4 / 0728 5.4 / 1402 0.1 / 1953 5.1	**25** F	0158 0.8 / 0742 4.6 / 1421 0.8 / 2001 4.6
11 F	0219 0.3 / 0810 5.4 / 1442 0.1 / 2035 5.0	**26** SA	0228 0.9 / 0809 4.5 / 1450 0.9 / 2030 4.5
12 SA	0300 0.4 / 0855 5.2 / 1524 0.3 / 2119 4.7	**27** SU	0258 1.0 / 0838 4.4 / 1520 1.0 / 2101 4.3
13 SU	0344 0.6 / 0945 4.8 / 1607 0.7 / 2210 4.4	**28** M	0330 1.2 / 0911 4.2 / 1551 1.3 / 2137 4.1
14 M	0431 1.0 / 1045 4.4 / 1653 1.1 / 2314 4.1	**29** TU	0405 1.4 / 0951 3.9 / 1625 1.5 / 2223 3.9
15 TU	0525 1.4 / 1200 4.1 / 1749 1.6	**30** W	0444 1.6 / 1042 3.7 / 1706 1.8 / 2327 3.7
		31 TH	0533 1.9 / 1202 3.5 / 1802 2.0

APRIL

Day	Time m	Day	Time m
1 F	0054 3.6 / 0644 2.0 / 1341 3.4 / 1933 2.2	**16** SA	0255 3.8 / 0919 1.7 / 1535 3.7 / 2134 2.0
2 SA	0215 3.7 / 0824 2.0 / 1501 3.6 / 2114 2.0	**17** SU	0358 3.9 / 1021 1.5 / 1629 3.8 / 2232 1.7
3 SU	0321 3.9 / 0946 1.7 / 1604 3.8 / 2219 1.7	**18** M	0446 4.1 / 1107 1.3 / 1709 4.0 / 2316 1.5
4 M	0415 4.2 / 1044 1.3 / 1652 4.2 / 2308 1.3	**19** TU	0522 4.2 / 1146 1.1 / 1742 4.2 / 2355 1.3
5 TU	0501 4.6 / 1131 0.8 / 1734 4.5 / 2352 0.8	**20** W	0554 4.4 / 1220 0.9 / 1812 4.4
6 W	0544 4.9 / 1215 0.4 / 1813 4.8	**21** TH	0029 1.1 / 0622 4.4 / 1253 0.8 / 1840 4.5
7 TH	0035 0.5 / 0624 5.2 / 1257 0.2 / 1852 5.1	**22** F	0102 1.0 / 0650 4.5 / 1323 0.8 / 1908 4.6
8 F	0117 0.3 / 0707 5.3 / 1339 0.0 / 1932 5.1	**23** SA	0133 0.9 / 0717 4.5 / 1353 0.8 / 1936 4.6
9 SA	0159 0.2 / 0751 5.3 / 1420 0.1 / 2014 5.0	**24** SU	0204 0.9 / 0746 4.4 / 1422 0.9 / 2005 4.5
10 SU	0243 0.3 / 0838 5.1 / 1502 0.3 / 2059 4.8	**25** M	0235 1.0 / 0817 4.3 / 1452 1.0 / 2037 4.4
11 M	0327 0.5 / 0930 4.7 / 1545 0.7 / 2150 4.5	**26** TU	0308 1.1 / 0853 4.1 / 1524 1.2 / 2115 4.2
12 TU	0416 0.9 / 1033 4.3 / 1632 1.2 / 2254 4.2	**27** W	0344 1.2 / 0935 3.9 / 1559 1.4 / 2200 4.0
13 W	0511 1.3 / 1147 3.9 / 1725 1.6	**28** TH	0424 1.4 / 1029 3.7 / 1641 1.7 / 2300 3.9
14 TH	0015 3.9 / 0620 1.6 / 1306 3.7 / 1835 2.0	**29** F	0514 1.7 / 1143 3.6 / 1735 1.9
15 F	0138 3.8 / 0750 1.8 / 1424 3.6 / 2008 2.1	**30** SA	0016 3.8 / 0619 1.8 / 1309 3.5 / 1856 2.0

Chart Datum: 2·71 metres below Ordnance Datum (Newlyn). HAT is 5·5 metres above Chart Datum.

NW Scotland

FREE monthly updates. Register at www.reedsnauticalalmanac.co.uk

437

STORNOWAY LAT 58°12'N LONG 6°23'W

TIMES AND HEIGHTS OF HIGH AND LOW WATERS

STANDARD TIME (UT)
For Summer Time add ONE hour in **non-shaded areas**

Dates in red are SPRINGS
Dates in blue are NEAPS

YEAR **2016**

MAY

	Time	m		Time	m
1 SU	0134 0746 1425 2030	3.8 1.8 3.6 1.9	**16** M	0317 0938 1552 2150	3.8 1.6 3.7 1.9
2 M	0242 0907 1530 2142	3.9 1.5 3.9 1.6	**17** TU	0409 1029 1636 2241	3.9 1.5 3.9 1.7
3 TU	0341 1009 1623 2237	4.2 1.2 4.2 1.3	**18** W	0450 1111 1713 2323	4.0 1.3 4.1 1.5
4 W	0432 1101 1708 2326	4.5 0.8 4.5 0.9	**19** TH	0526 1149 1746	4.1 1.1 4.3
5 TH	0519 1148 1750	4.8 0.5 4.8	**20** F	0001 0557 1223 1816	1.3 4.2 1.0 4.4
6 F	0012 0603 1234 1831	0.6 5.1 0.3 5.0 ●	**21** SA	0036 0628 1256 1845	1.1 4.3 1.0 4.5 ○
7 SA	0057 0648 1317 1913	0.4 5.1 0.2 5.1	**22** SU	0110 0658 1327 1915	1.0 4.3 0.9 4.5
8 SU	0142 0735 1400 1956	0.3 5.1 0.3 5.0	**23** M	0143 0729 1358 1946	1.0 4.3 1.0 4.5
9 M	0228 0824 1443 2043	0.3 4.9 0.5 4.8	**24** TU	0216 0803 1431 2021	1.0 4.2 1.0 4.5
10 TU	0314 0918 1527 2134	0.5 4.6 0.8 4.5	**25** W	0251 0841 1505 2100	1.0 4.1 1.2 4.4
11 W	0403 1019 1613 2234	0.8 4.2 1.2 4.3	**26** TH	0329 0926 1542 2146	1.1 4.0 1.3 4.2
12 TH	0457 1126 1704 2345	1.2 3.9 1.6 4.0	**27** F	0411 1019 1626 2242	1.3 3.9 1.5 4.1
13 F	0559 1236 1806 ◑	1.5 3.7 1.9	**28** SA	0500 1124 1719 2348	1.4 3.7 1.7 4.0
14 SA	0100 0712 1347 1921	3.8 1.7 3.6 2.0	**29** SU	0601 1238 1828 ◐	1.5 3.7 1.8
15 SU	0212 0831 1455 2043	3.8 1.7 3.6 2.0	**30** M	0058 0713 1350 1950	4.0 1.5 3.7 1.8
			31 TU	0207 0828 1457 2104	4.0 1.4 3.9 1.6

JUNE

	Time	m		Time	m
1 W	0309 0935 1555 2207	4.2 1.2 4.1 1.4	**16** TH	0414 1033 1641 2249	3.8 1.6 4.0 1.7
2 TH	0406 1033 1646 2302	4.4 0.9 4.4 1.1	**17** F	0457 1116 1719 2332	3.9 1.4 4.1 1.5
3 F	0459 1125 1732 2353	4.7 0.7 4.7 0.8	**18** SA	0535 1154 1753	4.0 1.3 4.3
4 SA	0548 1213 1816	4.8 0.6 4.9	**19** SU	0011 0609 1230 1825	1.3 4.2 1.2 4.5
5 SU	0042 0636 1259 1859	0.6 4.9 0.5 5.0 ●	**20** M	0048 0642 1305 1857	1.2 4.2 1.1 4.6 ○
6 M	0130 0724 1344 1943	0.5 4.9 0.5 4.9	**21** TU	0124 0716 1339 1930	1.1 4.3 1.0 4.6
7 TU	0216 0813 1427 2028	0.5 4.7 0.7 4.8	**22** W	0200 0752 1414 2007	1.0 4.3 1.0 4.6
8 W	0303 0904 1511 2116	0.6 4.5 0.9 4.6	**23** TH	0238 0831 1451 2046	0.9 4.3 1.1 4.6
9 TH	0350 0956 1555 2207	0.8 4.2 1.2 4.4	**24** F	0317 0914 1530 2131	0.9 4.2 1.1 4.5
10 F	0438 1052 1641 2304	1.1 4.0 1.5 4.1	**25** SA	0359 1004 1613 2222	1.0 4.1 1.3 4.4
11 SA	0529 1153 1732	1.3 3.7 1.7	**26** SU	0446 1101 1703 2321	1.1 4.0 1.5 4.3
12 SU	0009 0626 1257 1831 ◐	3.9 1.6 3.6 1.9	**27** M	0540 1207 1803 ◐	1.3 3.9 1.6
13 M	0118 0730 1402 1941	3.8 1.7 3.6 2.1	**28** TU	0027 0642 1319 1914	4.2 1.4 3.8 1.7
14 TU	0224 0839 1504 2055	3.7 1.8 3.6 2.0	**29** W	0137 0753 1428 2031	4.1 1.4 3.9 1.7
15 W	0324 0942 1557 2158	3.7 1.7 3.8 1.9	**30** TH	0245 0904 1533 2142	4.2 1.3 4.1 1.5

JULY

	Time	m		Time	m
1 F	0349 1011 1630 2246	4.3 1.2 4.3 1.3	**16** SA	0429 1044 1652 2305	3.8 1.7 4.0 1.7
2 SA	0448 1108 1720 2342	4.4 1.1 4.5 1.0	**17** SU	0513 1128 1730 2349	3.9 1.5 4.2 1.5
3 SU	0541 1200 1806	4.6 0.9 4.7	**18** M	0551 1207 1805	4.1 1.3 4.5
4 M	0033 0629 1247 1849 ●	0.8 4.7 0.8 4.9	**19** TU	0028 0626 1245 1838	1.3 4.2 1.1 4.6 ○
5 TU	0121 0715 1331 1931	0.6 4.7 0.8 4.9	**20** W	0107 0700 1322 1913	1.0 4.4 1.0 4.8
6 W	0206 0759 1413 2012	0.6 4.7 0.8 4.9	**21** TH	0145 0736 1358 1950	0.8 4.5 0.9 4.9
7 TH	0249 0842 1453 2052	0.6 4.5 0.9 4.7	**22** F	0223 0815 1436 2029	0.7 4.5 0.9 4.9
8 F	0330 0925 1533 2133	0.8 4.3 1.1 4.5	**23** SA	0302 0856 1515 2112	0.7 4.5 0.9 4.8
9 SA	0412 1009 1614 2216	1.0 4.1 1.3 4.3	**24** SU	0343 0942 1557 2159	0.7 4.4 1.0 4.7
10 SU	0454 1058 1656 2305	1.2 3.9 1.6 4.0	**25** M	0427 1034 1644 2255	0.9 4.2 1.2 4.5
11 M	0539 1155 1743	1.5 3.7 1.8	**26** TU	0516 1136 1738 ◐	1.1 4.0 1.5
12 TU	0006 0630 1301 1839 ◐	3.8 1.7 3.6 2.0	**27** W	0001 0612 1250 1844	4.3 1.3 3.9 1.7
13 W	0121 0731 1409 1949	3.7 1.9 3.6 2.1	**28** TH	0115 0720 1406 2005	4.1 1.5 3.9 1.8
14 TH	0232 0843 1512 2107	3.6 1.9 3.7 2.1	**29** F	0230 0839 1518 2128	4.0 1.6 4.0 1.7
15 F	0335 0951 1606 2214	3.7 1.8 3.8 2.0	**30** SA	0342 0956 1622 2240	4.1 1.5 4.2 1.4
			31 SU	0445 1059 1714 2337	4.2 1.3 4.5 1.2

AUGUST

	Time	m		Time	m
1 M	0537 1151 1758	4.4 1.2 4.7	**16** TU	0531 1144 1742	4.1 1.4 4.5
2 TU	0026 0622 1236 1838 ●	0.9 4.6 1.0 4.9	**17** W	0007 0606 1224 1816	1.2 4.3 1.1 4.8
3 W	0111 0702 1317 1915	0.7 4.6 0.9 4.9	**18** TH	0046 0641 1302 1851 ○	0.9 4.6 0.9 5.0
4 TH	0151 0739 1356 1950	0.6 4.7 0.8 4.9	**19** F	0125 0716 1339 1928	0.6 4.7 0.7 5.1
5 F	0229 0815 1432 2024	0.6 4.6 0.9 4.8	**20** SA	0203 0754 1418 2007	0.5 4.8 0.6 5.2
6 SA	0305 0850 1508 2057	0.7 4.4 1.0 4.6	**21** SU	0242 0834 1457 2050	0.4 4.8 0.6 5.1
7 SU	0340 0926 1544 2131	0.9 4.3 1.2 4.4	**22** M	0322 0917 1539 2136	0.5 4.6 0.8 4.9
8 M	0416 1004 1620 2208	1.2 4.1 1.4 4.1	**23** TU	0405 1006 1624 2231	0.7 4.4 1.1 4.6
9 TU	0454 1050 1700 2254	1.4 3.8 1.7 3.9	**24** W	0451 1107 1716 2340	1.0 4.2 1.4 4.3
10 W	0535 1152 1746 ◑	1.7 3.7 2.0	**25** TH	0545 1226 1822 ◑	1.4 4.0 1.7
11 TH	0002 0626 1311 1845	3.7 1.9 3.6 2.2	**26** F	0102 0653 1350 1948	4.0 1.7 3.9 1.9
12 F	0136 0736 1424 2009	3.5 2.1 3.6 2.3	**27** SA	0224 0821 1508 2124	3.9 1.8 4.0 1.8
13 SA	0255 0903 1529 2137	3.5 2.1 3.8 2.1	**28** SU	0340 0949 1615 2237	4.0 1.8 4.2 1.5
14 SU	0400 1012 1622 2239	3.6 1.9 4.0 1.9	**29** M	0442 1052 1706 2331	4.1 1.6 4.4 1.2
15 M	0451 1102 1704 2326	3.8 1.7 4.2 1.6	**30** TU	0530 1140 1747	4.3 1.3 4.6
			31 W	0015 0609 1222 1822	1.0 4.5 1.1 4.8

Chart Datum: 2·71 metres below Ordnance Datum (Newlyn). HAT is 5·5 metres above Chart Datum.

STANDARD TIME (UT)
For Summer Time add ONE hour in **non-shaded areas**

STORNOWAY   LAT 58°12'N   LONG 6°23'W
TIMES AND HEIGHTS OF HIGH AND LOW WATERS

Dates in red are **SPRINGS**
Dates in blue are **NEAPS**

YEAR 2016

NW Scotland

SEPTEMBER

Day	Time	m	Time	m	Day	Time	m	Time	m
1	0055 / 0643	0.8 / 4.6	TH 1300 / ● 1854	1.0 / 4.9	16	0021 / 0617	0.8 / 4.7	F 1238 / ○ 1827	0.8 / 5.2
2	0131 / 0715	0.7 / 4.7	F 1335 / 1925	0.9 / 4.9	17	0100 / 0653	0.5 / 5.0	SA 1317 / 1905	0.6 / 5.4
3	0204 / 0746	0.7 / 4.7	SA 1408 / 1954	0.9 / 4.8	18	0139 / 0730	0.3 / 5.1	SU 1357 / 1945	0.5 / 5.4
4	0236 / 0816	0.8 / 4.6	SU 1441 / 2023	1.0 / 4.7	19	0219 / 0810	0.2 / 5.0	M 1437 / 2028	0.5 / 5.3
5	0308 / 0847	0.9 / 4.4	M 1513 / 2053	1.1 / 4.5	20	0259 / 0853	0.4 / 4.9	TU 1520 / 2116	0.7 / 5.0
6	0339 / 0920	1.1 / 4.3	TU 1547 / 2126	1.3 / 4.2	21	0342 / 0941	0.7 / 4.6	W 1606 / 2213	1.0 / 4.6
7	0413 / 0958	1.4 / 4.0	W 1622 / 2205	1.6 / 4.0	22	0428 / 1043	1.1 / 4.3	TH 1659 / 2328	1.4 / 4.2
8	0449 / 1049	1.7 / 3.8	TH 1703 / 2300	1.9 / 3.7	23	0520 / 1207	1.5 / 4.0	F 1806 / ◑	1.7 /
9	0532 / 1206	1.9 / 3.7	F 1755 / ◑	2.1 /	24	0054 / 0629	3.9 / 1.9	SA 1335 / 1938	3.9 / 1.9
10	0037 / 0631	3.5 / 2.2	SA 1334 / 1912	3.6 / 2.3	25	0217 / 0805	3.8 / 2.1	SU 1455 / 2118	4.0 / 1.8
11	0214 / 0807	3.5 / 2.3	SU 1447 / 2055	3.7 / 2.2	26	0333 / 0937	3.9 / 2.0	M 1601 / 2226	4.2 / 1.5
12	0328 / 0938	3.6 / 2.1	M 1547 / 2209	3.9 / 1.9	27	0432 / 1037	4.1 / 1.7	TU 1651 / 2315	4.4 / 1.3
13	0423 / 1035	3.8 / 1.8	TU 1634 / 2259	4.2 / 1.6	28	0516 / 1123	4.3 / 1.5	W 1729 / 2355	4.6 / 1.1
14	0506 / 1119	4.1 / 1.5	W 1714 / 2341	4.5 / 1.2	29	0551 / 1202	4.5 / 1.2	TH 1802	4.7
15	0542 / 1159	4.4 / 1.1	TH 1750	4.9	30	0031 / 0621	0.9 / 4.6	F 1238 / 1831	1.1 / 4.8

OCTOBER

Day	Time	m	Time	m	Day	Time	m	Time	m
1	0105 / 0650	0.8 / 4.7	SA 1311 / ● 1859	1.0 / 4.8	16	0034 / 0629	0.4 / 5.1	SU 1254 / ○ 1842	0.6 / 5.4
2	0136 / 0718	0.8 / 4.7	SU 1343 / 1926	1.0 / 4.8	17	0115 / 0708	0.3 / 5.2	M 1336 / 1925	0.4 / 5.4
3	0206 / 0746	0.8 / 4.7	M 1414 / 1954	1.0 / 4.7	18	0156 / 0749	0.3 / 5.2	TU 1418 / 2010	0.5 / 5.3
4	0236 / 0815	1.0 / 4.6	TU 1445 / 2023	1.1 / 4.5	19	0238 / 0832	0.5 / 5.0	W 1503 / 2101	0.6 / 5.0
5	0306 / 0846	1.2 / 4.4	W 1517 / 2055	1.3 / 4.3	20	0321 / 0922	0.8 / 4.7	TH 1551 / 2201	0.9 / 4.6
6	0337 / 0921	1.4 / 4.3	TH 1552 / 2133	1.5 / 4.0	21	0407 / 1023	1.2 / 4.4	F 1645 / 2316	1.3 / 4.2
7	0412 / 1006	1.7 / 4.0	F 1631 / 2224	1.8 / 3.8	22	0500 / 1146	1.6 / 4.2	SA 1753 / ◑	1.7 /
8	0451 / 1110	1.9 / 3.8	SA 1720 / 2349	2.0 / 3.6	23	0039 / 0607	3.9 / 2.0	SU 1312 / 1920	4.0 / 1.9
9	0544 / 1241	2.2 / 3.7	SU 1828 / ◑	2.2 /	24	0158 / 0737	3.8 / 2.2	M 1430 / 2052	4.0 / 1.8
10	0130 / 0710	3.5 / 2.4	M 1401 / 2006	3.8 / 2.2	25	0312 / 0907	3.9 / 2.0	TU 1537 / 2200	4.2 / 1.6
11	0249 / 0854	3.6 / 2.2	TU 1506 / 2130	4.0 / 1.9	26	0411 / 1010	4.0 / 1.9	W 1628 / 2248	4.3 / 1.4
12	0349 / 1000	3.9 / 1.9	W 1558 / 2225	4.3 / 1.5	27	0454 / 1057	4.2 / 1.7	TH 1707 / 2328	4.5 / 1.2
13	0436 / 1048	4.2 / 1.5	TH 1642 / 2311	4.6 / 1.1	28	0528 / 1137	4.4 / 1.4	F 1739	4.6
14	0515 / 1131	4.6 / 1.2	F 1723 / 2353	4.9 / 0.7	29	0004 / 0558	1.1 / 4.6	SA 1213 / 1808	1.3 / 4.7
15	0552 / 1212	4.9 / 0.8	SA 1802	5.2	30	0037 / 0626	1.0 / 4.7	SU 1247 / ● 1836	1.2 / 4.7
					31	0108 / 0654	1.0 / 4.8	M 1319 / 1904	1.1 / 4.7

NOVEMBER

Day	Time	m	Time	m	Day	Time	m	Time	m
1	0138 / 0722	1.0 / 4.8	TU 1350 / 1932	1.1 / 4.6	16	0136 / 0732	0.4 / 5.3	W 1403 / 1958	0.5 / 5.2
2	0208 / 0751	1.1 / 4.7	W 1421 / 2002	1.2 / 4.5	17	0220 / 0818	0.6 / 5.1	TH 1450 / 2050	0.7 / 4.9
3	0238 / 0822	1.2 / 4.6	TH 1454 / 2036	1.3 / 4.3	18	0304 / 0907	0.9 / 4.9	F 1540 / 2149	0.9 / 4.6
4	0310 / 0857	1.4 / 4.4	F 1529 / 2115	1.5 / 4.1	19	0351 / 1005	1.2 / 4.6	SA 1633 / 2256	1.2 / 4.2
5	0344 / 0940	1.6 / 4.3	SA 1609 / 2206	1.7 / 3.9	20	0442 / 1116	1.6 / 4.3	SU 1734	1.6
6	0423 / 1036	1.9 / 4.1	SU 1656 / 2316	1.9 / 3.7	21	0009 / 0541	4.0 / 2.0	M 1235 / ◑ 1845	4.1 / 1.8
7	0513 / 1152	2.1 / 3.9	M 1757 / ●	2.0 /	22	0123 / 0654	3.8 / 2.2	TU 1351 / 2005	4.1 / 1.9
8	0045 / 0626	3.6 / 2.3	TU 1313 / 1919	3.9 / 2.1	23	0234 / 0816	3.8 / 2.2	W 1459 / 2117	4.1 / 1.8
9	0204 / 0801	3.7 / 2.2	W 1421 / 2042	4.0 / 1.9	24	0336 / 0929	3.9 / 2.1	TH 1555 / 2212	4.1 / 1.7
10	0310 / 0917	3.9 / 2.0	TH 1520 / 2146	4.3 / 1.5	25	0424 / 1024	4.1 / 1.9	F 1639 / 2257	4.2 / 1.5
11	0403 / 1014	4.2 / 1.6	F 1610 / 2238	4.6 / 1.2	26	0502 / 1108	4.3 / 1.7	SA 1715 / 2335	4.4 / 1.4
12	0448 / 1102	4.6 / 1.3	SA 1656 / 2325	4.9 / 0.8	27	0535 / 1147	4.5 / 1.5	SU 1748	4.5
13	0529 / 1148	4.9 / 0.9	SU 1740	5.2	28	0010 / 0605	1.3 / 4.6	M 1224 / 1818	1.4 / 4.5
14	0009 / 0609	0.6 / 5.1	M 1233 / ○ 1824	0.7 / 5.3	29	0043 / 0635	1.2 / 4.7	TU 1257 / ● 1847	1.3 / 4.6
15	0053 / 0650	0.4 / 5.3	TU 1318 / 1910	0.5 / 5.3	30	0115 / 0704	1.2 / 4.8	W 1330 / 1917	1.2 / 4.5

DECEMBER

Day	Time	m	Time	m	Day	Time	m	Time	m
1	0146 / 0734	1.2 / 4.8	TH 1403 / 1949	1.2 / 4.5	16	0206 / 0806	0.7 / 5.2	F 1440 / 2039	0.7 / 4.9
2	0217 / 0806	1.3 / 4.7	F 1437 / 2024	1.3 / 4.4	17	0250 / 0853	0.9 / 5.0	SA 1528 / 2130	0.8 / 4.6
3	0250 / 0842	1.4 / 4.6	SA 1513 / 2104	1.4 / 4.2	18	0334 / 0942	1.2 / 4.8	SU 1616 / 2224	1.1 / 4.3
4	0325 / 0923	1.5 / 4.5	SU 1553 / 2151	1.5 / 4.1	19	0420 / 1037	1.5 / 4.5	M 1707 / 2323	1.4 / 4.0
5	0405 / 1013	1.7 / 4.3	M 1638 / 2249	1.6 / 3.9	20	0510 / 1141	1.8 / 4.2	TU 1802	1.7
6	0452 / 1114	1.9 / 4.2	TU 1732	1.8	21	0029 / 0606	3.8 / 2.0	W 1253 / ◑ 1905	4.0 / 1.9
7	0001 / 0552	3.8 / 2.1	W 1226 / ◑ 1839	4.1 / 1.8	22	0139 / 0713	3.7 / 2.2	TH 1405 / 2015	3.9 / 2.0
8	0118 / 0711	3.8 / 2.1	TH 1337 / 1954	4.2 / 1.8	23	0247 / 0829	3.8 / 2.3	F 1510 / 2124	3.9 / 1.9
9	0228 / 0830	4.0 / 2.0	F 1442 / 2105	4.3 / 1.6	24	0345 / 0941	3.9 / 2.2	SA 1605 / 2220	4.0 / 1.8
10	0330 / 0938	4.2 / 1.8	SA 1541 / 2207	4.5 / 1.3	25	0433 / 1037	4.1 / 2.0	SU 1651 / 2306	4.1 / 1.7
11	0423 / 1036	4.5 / 1.4	SU 1634 / 2300	4.8 / 1.0	26	0512 / 1123	4.3 / 1.8	M 1729 / 2345	4.2 / 1.5
12	0510 / 1128	4.8 / 1.1	M 1725 / 2349	5.0 / 0.8	27	0547 / 1202	4.5 / 1.6	TU 1803	4.3
13	0554 / 1217	5.0 / 0.9	TU 1813	5.1	28	0021 / 0618	1.4 / 4.6	W 1239 / 1835	1.4 / 4.4
14	0036 / 0637	0.7 / 5.2	W 1306 / ○ 1901	0.7 / 5.2	29	0055 / 0649	1.3 / 4.7	TH 1314 / ● 1906	1.3 / 4.5
15	0122 / 0721	0.7 / 5.3	TH 1353 / 1949	0.6 / 5.1	30	0128 / 0720	1.2 / 4.8	F 1349 / 1939	1.2 / 4.5
					31	0201 / 0753	1.2 / 4.8	SA 1424 / 2013	1.1 / 4.5

Chart Datum: 2·71 metres below Ordnance Datum (Newlyn). HAT is 5·5 metres above Chart Datum.

》》 **FREE** monthly updates. Register at 《
www.reedsnauticalalmanac.co.uk

439

HARBOURS/ANCHORAGES ON THE EAST OF THE OUTER HEBRIDES

Visitor moorings mentioned in this section are inspected annually, have pick up buoys, can take yachts <15tons, and there is no charge <7days. Refer to www.w-isles.gov.uk/harbour master.

TIDES

Standard Port STORNOWAY (←)

Times				Height (metres)			
High Water		Low Water		MHWS	MHWN	MLWN	MLWS
0100	0700	0300	0900	4·8	3·7	2·0	0·7
1300	1900	1500	2100				

Differences East side of Outer Hebrides, N to S

LOCH SHELL (Harris)							
−0013	0000	0000	−0017	0·0	−0·1	−0·1	0·0
EAST LOCH TARBERT (Harris)							
−0025	−0010	−0010	−0020	+0·2	0·0	+0·1	+0·1
LOCH MADDY (N Uist)							
−0044	−0014	−0016	−0030	0·0	−0·1	−0·1	0·0
LOCH CARNAN (S Uist)							
−0050	−0010	−0020	−0040	−0·3	−0·5	−0·1	−0·1
LOCH SKIPORT (S Uist)							
−0100	−0025	−0024	−0024	−0·2	−0·4	−0·3	−0·2
LOCH BOISDALE (S Uist)							
−0055	−0030	−0020	−0040	−0·7	−0·7	−0·3	−0·2
BARRA (North Bay)							
−0103	−0031	−0034	−0048	−0·6	−0·5	−0·2	−0·1
CASTLE BAY (Barra)							
−0115	−0040	−0045	−0100	−0·5	−0·6	−0·3	−0·1
BARRA HEAD (Berneray)							
−0115	−0040	−0045	−0055	−0·8	−0·7	−0·2	+0·1

SHELTER

LOCH SHELL, Lewis, 57°59′·00N 06°25′·50W. AC 1794. HW −0437 on Dover; ML 2·7m. See 9.8.6. Pass S of Eilean Iuvard; beware rks to W of Is. ‡ in Tob Eishken, 2½M up loch on N shore (beware rk awash on E side of ent), or at head of loch (exposed to E winds; dries some distance). **Facilities:** ⊠/Stores at Lemreway.

SHIANT ISLANDS, Lewis, 57°53′·68N 06°21′·37W. AC 1794, 1795. Tides as Loch Shell 9.8.6. Beware strong tidal streams and overfalls in Sound of Shiant. Strictly a fair weather ‡; in W winds ‡ E of Mol Mor, isthmus between Garbh Eileen (159m) and Eileen an Tighe. In E winds ‡ W of Mol Mor. No lights or facilities.

EAST LOCH TARBERT, Harris, 57°49′·98N 06°41′·07W. AC 2905. HW −0446 on Dover; ML 3·0m; Duration 0605. See 9.8.6. Appr via Sound of Scalpay; beware Elliot Rk (2m) 2½ca SSW of Rubha Crago. A bridge (57°52′·80N 06°41′·73W) joins Scalpay to Harris (20m ht). Bridge lts: Centre Oc 6s; N side Iso G 4s 35m; S side Iso R 4s 35m. Eilean Glas lt ho at E end of Scalpay, Fl (3) 20s 43m 23M; W tr, R bands. In Sound of Scalpay, stream sets W from HW +3, and E from HW −3. ‡ off Tarbert WSW of Steamer pier in 2·5m. ☎(01589) 502444. **Facilities:** ⚓ 🗑 🏠 Dr, ⊠ 🛒 ✕ 🖵 ferry to Uig. Alternatively Scalpay N Hbr gives good shelter. Beware rk 5ca off Aird an Aiseig, E side of ent. SHM buoy marks wk off Coddem; 5ca E of the buoy is a rk, depth 1·1m. Ldg Lt Dir Iso WRG 4s 9m 4M. Both piers have 2FG (vert) lts; ‡ 7ca N, in 3m. **Facilities:** ⚓ ⊠ 🛒 Ferry to Harris.

LOCH STOCKINISH, Harris. AC 1757. Entr via Loch Chlual marked by metal Bn Fl R 2s at 57°48′·49N 06°48′·79W and SCM Q(6)+L Fl 15s atop B/Y perch at 57°48′·56N 06°48′·81W; pier landing and fish farms in loch.

LOCH RODEL, Harris. AC 2802. 3🛟s at 57°44′·2N 06°57′·4W in Poll an Tigh-mhàil; enter from SW past jetties. No lts.

SOUND OF HARRIS, 57°43′N 06°58′W. Passages through this difficult Sound are detailed in the *W Coast of Scotland Pilot*. The Stanton and Outer Stromay Chans off the Harris shore are the most feasible for yachts. AC 2802 shows the ferry routes from Leverburgh to Berneray.

LOCH MADDY, North Uist, 57°35′·98N 07°06′·07W. AC 2825. HW −0500 on Dover. See 9.8.6. With strong wind against tide there can be bad seas off ent. Appr clear, but from S beware submerged rk ½ca N of Leacnam Madadh. Lts: Weaver's Pt Fl 3s 21m 7M; Glas Eilean Mor Fl (2) G 4s 8m 5M; Rubna Nam Pleac Fl R 4s 7m 5M. Inside loch: Ruigh Liath QG 6m 5M; Vallaquie Is Dir Fl (3) WRG 8s. Ferry pier ldg lts 298°: front 2FG(vert) 4M; rear Oc G 8s 10m 4M, vis 284°-304°. Pontoon lit Fl R 3s. 2🛟s Bagh Aird nam Madadh; 2 🛟s W of and 4 🛟s SW of ferry pier ☎(01870) 602425; 2🛟s E of Oronsay. ‡s: clear S of ferry pier; NE of Vallaquie Is; Charles Hbr; Oronsay (‡ not advised due to moorings), tidal berth on private pier; Sponish Hbr; Loch Portain. VHF Ch 12 16. Port Manager ☎(01876) 5003337(day), 5003226(night). **Facilities:** Lochmaddy, 26🛟 2🛟 on lit pontoon (Fl R 3s 2M) £2/night inc ⚡, ⚓ 🗑 🏠 🛢 ⊠ Ⓑ 🛒; Loch Portain ⊠ Shop, Bus to Benbecula.

LOCH EPORT, North Uist, 57°33′·45N 07°08′·12W. AC 2825, but not the head of loch. Tides, approx as L Maddy; 3kn sp stream. On the S side of ent are rks, some drying. The ent proper is clean but very narrow (about 100m) for 5ca, then widens. Follow the charted clearing line 082°. Best ‡s are: Bàgh a' Bhiorain (S of chan; line up cairn and Bu boulder on 129°); and Acairseid Lee (N bank) E or W of Deer Is. ⊠ 🛒 ✕ 🖵 at Clachan, head of loch.

KALLIN, Grimsay. 57°28′·9N 07°12′·2W. AC 2904. 1🛟 in NE of hbr. 3 chan lt buoys and 2 FR (vert) on hbr bkwtr. ☎(01870) 602425.

LOCH CARNAN, South Uist, 57°22′·03N 07°16′·39W. AC 2825. Tides, see 9.8.6. SWM buoy, L Fl 10s, at 57°22′·30N 07°11′·57W is almost 2M E of app chan proper, marked by Nos 1 and 2 buoys, Fl G 2·5s and Fl R 2s, at 57°22′·45N 07°14′·90W. Round No 3 PHM buoy, Fl R 5s, between Gasay and Taigh Iamain, then pick up ldg lts 222° to Sandwick Quay; front Fl R 2s, rear Iso R 10s, both 5M, W ◊s on posts. Power stn and 2 chys are conspic close to SE of quay. ☎(01870) 602425 for permission to berth on the quay (MoD property). ‡ or 2🛟s about 2ca WNW of the quay in deep water. ☎(01870) 610238. The passage S of Gasay is unmarked and needs careful pilotage.

LOCH SKIPPORT, South Uist, 57°19′·98N 07°13′·67W. AC 2904, 2825. HW −0602 on Dover; see 9.8.6. Easy ent 3M NNE of Hecla (604m). No lights, but 2¼M SSE is Usinish lt ho Fl WR 20s 54m 19/15M. ‡s at: Wizard Pool in 7m; beware Float Rk, dries 2·3m; on N side of Caolas Mor in 7m; Bagh Charmaig in 5m. Linne Arm has narrow ent, many fish farms and poor holding. No facilities.

LOCH EYNORT, South Uist, 57°13′·13N 07°16′·87W. AC 2825. Tides: interpolate between Lochs Skipport and Boisdale, see 9.8.6. ‡s in the outer loch at Cearcdal Bay and on the N side just before the narrows are exposed to the E. The passage to Upper L Eynort is very narrow and streams reach 5-7kn; best not attempted unless local fishermen offer guidance. Good ‡ inside at Bàgh Lathach.

LOCH BOISDALE, South Uist, 57°08′·78N 07°16′·07W. AC 2770. HW −0455 on Dover; ML 2·4m; Duration 0600. See 9.8.6. Good shelter except in SE gales when swell runs right up the 2M loch. From N, approach between Rubha na Cruibe and Calvay Is leading line 245°: Hollisgeir (0·3m) on with pier (ru). From S beware Clan Ewan Rk, which dries 1·2m, and McKenzie Rk (2·4m), marked by PHM lt buoy Fl (3) R 15s. A new harbour with marina is formed by a causeway from Rubha Bhuailt to Gasay, with two b'waters (G 6s and Fl R 6s with traffic sigs) and an entrance to the N with access from the channel to Boisdale Hbr which lies N of Gasay Is. Off E end beware rock and further development. ‡ off pier in approx 4m, or SW of Gasay Is in approx 9m. 4🛟s NE of pier; ☎(01870) 602425. There are fish cages W of Rubha Bhuailt. Lts: Dir Lt 292·5° WRG 4s 4m 5M, Ro-Ro jetty hd Iso RG 4s 8m 2M; close SE, Eilean Dubh Fl (2) R 5s. E Calvay Is Fl (2) WRG 10s 16m 7M. Gasay Is Fl WR 5s 10m 7M, opposite shore, Fl G 6s 3m 3M; See 9.8.4. ☎(0187) 700288. **Facilities:** ⚓ (on pier), 🏠 ⊠ ✕ 🖵. Ferry to mainland.

ACAIRSEID MHÓR, Eriskay, ⊕ 57°03′·78N 07°16′·35W. AC 2770. Tides approx as for North Bay (Barra), see 9.8.6. Ben Scrien (183m) is a conspicuous, pointed peak N of the harbour. Leading lts 285°, both Oc R 6s 9/10m 4M, W △ ▽ on orange posts, lead for 0·5M from the waypoint between two drying rocks into the outer loch. A SHM buoy, Fl G 6s, marks a rock drying 3m. Possible 🛟 on pontoon; 2🛟s (perhaps submerged at HW) are at 57°03′·95N 07°17′·40W on S side of inner loch, opp pier, 2 FG (vert). ☎(01870) 602425. ⊠ 🛒 ✕ 🖵 at Haun, 1·5M at N end of island.

NORTH BAY, Barra, 57°00′·11N 07°24′·67W. AC 2770. Tides see 9.8.6. Well marked approach to inlet sheltered from both S and W winds. There are 🛟s in N part of Bay Hirivagh, ‡ 1ca WNW of Black Island or temporarily 🛟 on the quay in 4·5m. WPT 56°58′·68N 07°20′·31W is about 200m NE of Curachan ECM buoy, Q (3) 10s, and in the white sector (304°-306°) of Ardveenish Dir ☆ 305°, Oc WRG 3s, 2·5M to the WNW. ⚓ 🛒 🖵 bus to Castlebay.

CASTLE BAY, Barra, **56°56´·78N 07°29´·67W**. AC 2769. HW –0525 on Dover; ML 2·3m; Duration 0600. See 9.8.6. Very good shelter & holding. Best ⚓ is in approx 8m NW of Kiessimul Castle; NE of castle are rocks. About 20⚓s lie to W of the pier; they may be encroaching on the recommended ⚓ ☎(01870) 602425. Alternatively ⚓ in Vatersay Bay in approx 9m. W end of Vatersay Sound is closed by a causeway. Beware rocks NNW of Sgeir Dubh a conspic W/G tower, Q(3)G 6s 6m 5M, which leads 283° in transit with Sgeir Liath bn. Channel Rock, 2ca to the South, is marked by Fl WR 6s 4m 6/4M. Close-in leading lts 295°, both FBu 6M on W framework towers: front 9m Or △ on Rubha Glas; rear 457m from front, 15m Or ▽, vis: 15° and 8° respectively either side of ldg line. ☎(01871) 810306.
Facilities: ⚓ 🚢 🏧 ✉ 🍺 ✕ 🍴. Ferry to mainland.

ISLANDS WEST OF THE OUTER HEBRIDES (N TO S)

TIDES
Standard Port STORNOWAY (⟵)

Times				Height (metres)			
High Water		Low Water		MHWS	MHWN	MLWN	MLWS
0100	0700	0300	0900	4·8	3·7	2·0	0·7
1300	1900	1500	2100				
Differences West side of Outer Hebrides, N to S							
SCOLPAIG (W North Uist)							
–0033	–0033	–0040	–0040	–1·0	–0·9	–0·5	0·0
BALIVANICH (W Benbecula)							
–0103	–0017	–0031	–0045	–0·7	–0·6	–0·5	–0·2
FLANNAN ISLES							
–0026	–0016	–0016	–0026	–0·9	–0·7	–0·6	–0·2
VILLAGE BAY (St Kilda)							
–0040	–0040	–0045	–0045	–1·4	–1·1	–0·8	–0·3
ROCKALL							
–0055	–0055	–0105	–0105	–1·8	–1·5	–0·9	–0·2
SHILLAY (Monach Islands)							
–0103	–0043	–0047	–0107	–0·6	–0·7	–0·7	–0·3

SHELTER
FLANNAN ISLES, Western Isles, centred on **58°17´·28N 07°35´·27W** (Eilean Mór). AC 2524, 2721. Tides, as above. Uninhabited group of several rky islets, 18M WNW of Gallan Head (Lewis). The main islet is Eilean Mór where landing can be made on SW side in suitable conditions. Lt ho, Fl (2) 30s 101m 20M, is a 23m high W tr on NE tip of Eilean Mór; the lt is obscured by islets to the W which are up to 57m high. No recommended ⚓s and the few charted depths are by lead-line surveys.

ST KILDA, Western Isles, **57°48´·28N 08°33´·07W**. AC 2721, 2524. Tides at Village Bay, Hirta: HW –0510 on Dover; ML 1·9m; Duration 0615; see above. A group of four isles and three stacks, the main island is Hirta. The facility is now manned by a civilian company, Qinetiq ☎(01870) 604443, based at South Uist. Hirta is owned by National Trust for Scotland, who employ a Seasonal Warden, ☎01870 604628. ⚓ in Village Bay, SE-facing, in approx 5m about 1·5ca off the pier. Ldg lts 270°, both Oc 5s 26/38m 3M. If wind is between NE and SSW big swells enter the bay; good holding, but untenable if winds strong. Levenish Is (55m) is 1·5M E of Hirta with offlying rks. Courtesy call to *Kilda Radio* VHF Ch 16 **12** 73 (HJ) before landing; or Qinetic ☎(01870) 604406 (HO), 604612 (OT). Alternative ⚓ at Glen Bay on N side is only safe in S & E winds. Facilities: ⚓ from wells near landings.

ROCKALL, 57°35´·7N 13°41´·2W. AC 1128, 2524. Tides, as above. A 19m high granite rock, 200M W of N Uist. Best access by helicopter. Lt, Fl 15s 13M, is often extinguished for long periods due to weather damage. Helen's Reef, 1·4m, on which the sea breaks is 2M ENE.

MONACH ISLANDS (Heisker Is), centred on **57°31´·28N 07°38´·07W**. AC 2721, 2722. Tides, see 9.8.6 Shillay. The group lies 5M SW of N Uist and 8M WNW of Benbecula. The 5 main islands (W-E) are Shillay, Ceann Iar, Shivinish, Ceann Ear and Stockay; all uninhabited. Many rky offliers from NW through N to SE of the group. On Shillay there is a conspic red brick lt ho, Fl (2) 15s. ⚓s at: E of lt ho; Croic Hbr, bay N of Shivinish; and S Hbr on W side of Shivinish.

9.8.7 KINLOCHBERVIE

Highland 58°27′·26N 05°02′·78W ✵✵✵✵⚓⚓❀❀

CHARTS AC 1954, 1785, 2503; Imray C67

TIDES −0400 Dover; ML 2·7; Duration 0610

Standard Port ULLAPOOL (→)

Times				Height (metres)			
High Water		Low Water		MHWS	MHWN	MLWN	MLWS
0000	0600	0300	0900	5·2	3·9	2·1	0·7
1200	1800	1500	2100				
Differences LOCH BERVIE							
+0017	+0020	+0015	+0015	−0·4	−0·2	−0·1	+0·1

SHELTER Very good in Kinlochbervie Hbr off the N shore of Loch Inchard. A useful passage port, only 14.5 track miles S of Cape Wrath. It is an active FV port, but yachts ⚓ on pontoon (contact HM for details); NE side is shoal/foul. If full, ⚓ at Loch Clash, open to W, but sheltered in winds n through SE; landing jetty in 2·7m. Other ⚓s at: Camus Blair on S shore, 5ca SW of hbr ent, and up the loch at L Sheigra, Achriesgill Bay and 5ca short of the head of the loch.

NAVIGATION WPT 58°27′·34N 05°05′·08W (at mouth of Loch Inchard), 100°/1·3M to hbr ent. The sides of the loch are clean, but keep to N side of Loch Inchard to clear Bodha Ceann na Saile NCM and rk (3m depth) almost in mid-chan.

LIGHTS AND MARKS From offshore in good vis Ceann Garbh, a conspic mountain 900m (6M inland), leads 110° toward ent of Loch Inchard. Rubha na Leacaig, Fl (2) 10s 30m 8M, marks N side of loch ent. Dir ✦ WRG (H24) 15m 16M, Y framework tr (floodlit) leads 327° into hbr; see 9.8.4 for vis sectors. The 25m wide ent chan (and hbr) is dredged 4m and marked by 2 PHM poles, Fl R 4s and QR, and by a SHM pole, Fl G 4s. On S shore of loch Creag Mhòr, Dir Oc lt WRG 2.8s 16m 9M, is aligned 147°/327° with hbr ent chan; see 9.8.4.

COMMUNICATIONS (Code 01971) CGOC (01851) 702013; ⊖ (0141) 887 9369 (H24); Police 101; Dr 502002. HM 521235, mob 07901 514350.

VHF Ch **14** 16 HX. Ch 06 is used by FVs in the Minches.

FACILITIES harbours@highland.gov.uk ⚓ (pontoon) £1.50<10m for 48hrs to E side of hbr; 8 ♥ in new development with ⛟ and ⚓ to W side. ⚓ 🛢 at FV quay, 🛒🔋🛢⚒✉🛒🍴✕ℙ (Mission & Hbr Office), ♿. In summer, bus to Inverness.

ANCHORAGES & HBRS BETWEEN KINLOCHBERVIE & ULLAPOOL

LOCH LAXFORD Highland, 58°24′·78N 05°07′·18W. AC 2503. HW −0410 on Dover. ML 2·7m. See 9.8.7. Ent between Rubha Ruadh and Ardmore Pt, 1M ENE, clearly identified by 3 isolated mountains (N-S) Ceann Garbh, Ben Arkle and Ben Stack. The many ⚓s in the loch include: Loch a'Chadh-fi, on N/NE sides of islet (John Ridgeway's Adventure School on Pt on W side of narrows has moorings); Bagh nah-Airde Beag, next bay to E, (beware rk 5ca off SE shore which covers at MHWS); Weaver's Bay on SW shore, 3M from ent (beware drying rk off NW Pt of ent); Bagh na Fionndalach Mor on SW shore (4-6m); Fanagmore Bay on SW shore (beware head of bay foul with old moorings). Beware many fish farming cages. Facilities: none, nearest stores at Scourie (5M).

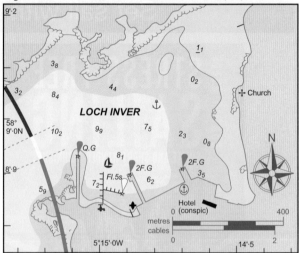

LOCH INVER Highland, 58°08′·98N 05°15′·08W. AC 2504. HW −0433 on Dover; ML 3·0m. See 9.8.7. Good shelter in all weathers at head of loch in busy fishing hbr on S side. Appr N or S of Soyea Is, Fl (2) 10s 34m 6M; beware rock drying 1·7m about 50m off Kirkaig Point (S side of ent). Glas Leac, a small islet 7ca WSW of hbr, may be passed on either side. ✦, Fl WRG 3s, has 3 different approaches (see 9.8.4) covering the chans N and S of Soyea Is and into the hbr. The church, hotel (S side) and white ho (N side) are all conspic. A 22-berth marina is between the breakwater (QG) and the first FV pier. Or, in W'ly gales, ⚓ in the lee of bkwtr in about 8m; or where HM directs. Other ⚓s on S shore of Loch Inver. VHF Ch 12 16. HM ☎(01571) 844247; mob 07958 734610. Health Centre 01571 844452. Facilities: ⚓ 🛒🛢🔋🛒ℙ (Leisure Centre), 🛢 🍴✉🛢.

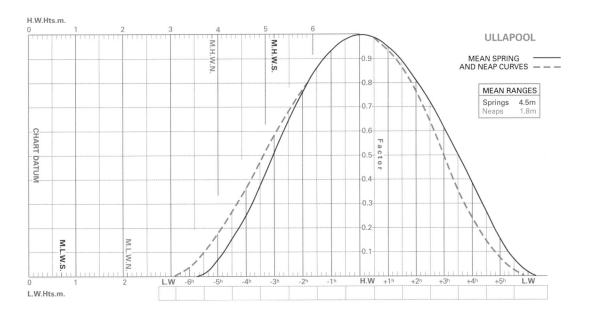

ULLAPOOL

MEAN SPRING
AND NEAP CURVES

MEAN RANGES	
Springs	4.5m
Neaps	1.8m

9.8.8 ULLAPOOL

Highland 57°53'·70N 05°09'·38W ✿✿✿✿⚓⚓✿✿

CHARTS AC 1794, 2500, 2501, 2509; Imray C67

TIDES –0415 Dover; ML 3·0; Duration 0610

Standard Port ULLAPOOL (→)

Times				Height (metres)			
High Water		Low Water		MHWS	MHWN	MLWN	MLWS
0000	0600	0300	0900	5·2	3·9	2·1	0·7
1200	1800	1500	2100				
Differences LOCH LAXFORD							
+0015	+0015	+0005	+0005	–0·3	–0·4	–0·2	0·0
BADCALL BAY							
+0005	+0005	+0005	+0005	–0·7	–0·5	–0·5	+0·2
LOCH NEDD							
0000	0000	0000	0000	–0·3	–0·2	–0·2	0·0
LOCH INVER							
–0005	–0005	–0005	–0005	–0·2	0·0	0·0	+0·1
SUMMER ISLES (Tanera Mor)							
–0005	–0005	–0010	–0010	–0·1	+0·1	0·0	+0·1
LOCH EWE (Mellon Charles, 57°51'N 05°38'W)							
–0005	–0010	–0005	–0005	–0·1	–0·1	–0·1	+0·1
LOCH GAIRLOCH							
–0012	–0011	–0011	–0011	–0·2	+0·1	–0·2	+0·2

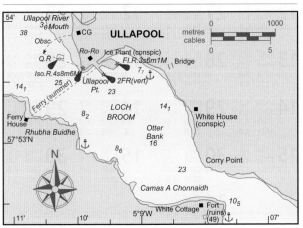

SHELTER A commercial port but yachts welcome. Good ⚓ E of pier, 8⚓ or possible ⬭ on pier (consult HM). Convenient landing inside pier, which may be busy with tourist vessels. Loch Kanaird (N of ent to Loch Broom) has good ⚓ E of Isle Martin. Possible ⚓s 6ca S of Ullapool Pt, and beyond the narrows 3ca ESE of W cottage. The upper loch is squally in strong winds.

NAVIGATION WPT L Broom ent 57°55'·78N 05°15'·08W, 129°/3.5M to Ullapool Pt lt. N of Ullapool Pt extensive drying flats off the mouth of Ullapool R are marked by QR buoy. Beware fish pens and unlit buoys SE of narrows off W shore.

LIGHTS AND MARKS Rhubha Cadail, N of L. Broom ent, Fl WRG 6s 11m 9/6M. Cailleach Hd, W of ent, Fl (2) 12s 60m 9M. Ullapool Pt Iso R 4s 8m 6M; grey mast, vis 258°-108°.

COMMUNICATIONS (Code 01854) CG (non-emergency) 613076; Police 101; Dr Dr 08454 242424; Dentist 613289 or 612660; ⊜ (0141) 887 9369. HM 612724, mob 07734 004843.

Ullapool Harbour Ch 14 16 (H24).

FACILITIES Pier ⚓ ♣ ⚓; **Loch Broom SC** Services: ♠ ⚒ & GRP repairs Mobile Marine Services ☎633719, mob 07866 516067, ⑤ Ⓔ ✕ A.Morgan ☎666383; **Town** ⏚ ⏚ (0700-2100), @ at Captains Cabin, Library and Ceilidh Place Hotel, all domestic facilities, ⏚ at swimming pool, ☒ ✕ ▭ ⇌ (bus to Garve). Daily buses to Inverness (✈)(no Sunday bus service. **Ferries:** Stornoway; 2-3/day; 2¾ Hrs; Calmac (www.calmac.co.uk).

ADJACENT ANCHORAGES

LOCH KANAIRD, 57°56'·60N 05°12'·20W. AC 2500. HW –0425 Dover; See 9.8.8; streams are weak and irreg. Situated E of the entrance to Loch Broom, S of the Summer Is and protected by Isle Martin. Good ⚓ (tripping line recommended) ENE of Sgeir Mhor or E of Isle Martin, N of Rubha Beag (also ⬭ on pontoon). Beware cables and marine farms. Ent from S (1ca wide) between two rocky spits, leading line E extremity of Isle Martin in transit with dark streak on cliff behind brg 000°. Ent from N, deep and clear of dangers. Facilities: Ardmair Boat Centre P. Fraser ☎(01854) 612054, ♠ ⚓ ⏚ some repairs. Isle Martin, Landing permitted (voluntary donation to Trust ☎(01854) 612531).

NW Scotland

ULLAPOOL   LAT 57°54'N   LONG 5°09'W
TIMES AND HEIGHTS OF HIGH AND LOW WATERS

STANDARD TIME (UT)
For Summer Time add ONE hour in **non-shaded areas**

Dates in red are **SPRINGS**
Dates in blue are NEAPS

YEAR 2016

JANUARY

Time	m	Time	m
1 0516 2.0 / 1119 4.3 / F 1759 2.0		**16** 0519 1.5 / 1134 4.8 / SA 1752 1.4 ☽	
2 0014 4.1 / 0606 2.2 / SA 1227 4.1 / 1855 2.2 ◑		**17** 0018 4.4 / 0619 1.8 / SU 1246 4.6 / 1855 1.7	
3 0125 4.0 / 0709 2.4 / SU 1346 4.0 / 2005 2.3		**18** 0133 4.3 / 0733 2.0 / M 1400 4.5 / 2010 1.8	
4 0236 4.0 / 0826 2.5 / M 1459 4.0 / 2118 2.2		**19** 0250 4.3 / 0856 2.0 / TU 1516 4.5 / 2130 1.8	
5 0339 4.1 / 0940 2.4 / TU 1600 4.2 / 2218 2.1		**20** 0402 4.5 / 1014 1.8 / W 1625 4.6 / 2240 1.7	
6 0430 4.4 / 1037 2.1 / W 1649 4.3 / 2306 1.8		**21** 0500 4.7 / 1117 1.6 / TH 1721 4.8 / 2335 1.5	
7 0511 4.6 / 1124 1.9 / TH 1730 4.5 / 2347 1.6		**22** 0547 5.0 / 1209 1.3 / F 1808 4.9	
8 0548 4.8 / 1206 1.6 / F 1807 4.7		**23** 0022 1.3 / 0627 5.2 / SA 1255 1.1 / 1849 5.1	
9 0026 1.4 / 0623 5.1 / SA 1246 1.3 / 1842 4.9		**24** 0105 1.1 / 0704 5.3 / SU 1336 0.9 / ○ 1926 5.1	
10 0105 1.2 / 0658 5.3 / SU 1325 1.1 / ● 1919 5.1		**25** 0143 1.0 / 0738 5.3 / M 1413 0.9 / 2001 5.1	
11 0143 1.0 / 0734 5.4 / M 1405 0.9 / 1958 5.1		**26** 0219 1.0 / 0811 5.3 / TU 1449 0.9 / 2035 5.0	
12 0221 1.0 / 0813 5.5 / TU 1445 0.8 / 2039 5.1		**27** 0254 1.1 / 0842 5.1 / W 1523 1.0 / 2108 4.8	
13 0301 1.0 / 0855 5.4 / W 1526 0.8 / 2123 5.0		**28** 0328 1.2 / 0914 5.0 / TH 1558 1.2 / 2143 4.7	
14 0342 1.1 / 0941 5.3 / TH 1610 1.0 / 2213 4.8		**29** 0402 1.4 / 0947 4.7 / F 1633 1.5 / 2220 4.4	
15 0428 1.3 / 1033 5.1 / F 1658 1.2 / 2310 4.6		**30** 0438 1.7 / 1024 4.5 / SA 1711 1.7 / 2306 4.2	
		31 0518 2.0 / 1110 4.2 / SU 1754 2.0	

FEBRUARY

Time	m	Time	m
1 0009 4.0 / 0605 2.2 / M 1220 4.0 / ◑ 1848 2.2		**16** 0104 4.2 / 0704 1.9 / TU 1341 4.3 / 1939 2.0	
2 0128 3.9 / 0710 2.4 / TU 1350 3.9 / 2005 2.3		**17** 0229 4.2 / 0836 2.1 / W 1505 4.2 / 2111 2.0	
3 0243 3.9 / 0837 2.5 / W 1511 3.9 / 2131 2.3		**18** 0350 4.3 / 1005 1.9 / TH 1619 4.3 / 2230 1.9	
4 0350 4.1 / 0958 2.3 / TH 1618 4.1 / 2235 2.0		**19** 0451 4.5 / 1110 1.6 / F 1715 4.5 / 2326 1.6	
5 0442 4.4 / 1057 2.0 / F 1707 4.3 / 2324 1.7		**20** 0538 4.7 / 1200 1.3 / SA 1759 4.7	
6 0525 4.7 / 1144 1.6 / SA 1748 4.6		**21** 0011 1.4 / 0615 5.0 / SU 1242 1.1 / 1835 4.9	
7 0006 1.4 / 0602 5.0 / SU 1227 1.2 / 1826 4.9		**22** 0050 1.1 / 0648 5.1 / M 1319 0.9 / ○ 1907 5.0	
8 0047 1.1 / 0639 5.3 / M 1308 0.9 / ● 1903 5.2		**23** 0126 1.0 / 0718 5.2 / TU 1353 0.8 / 1938 5.1	
9 0126 0.8 / 0717 5.5 / TU 1349 0.6 / 1941 5.3		**24** 0159 0.9 / 0747 5.2 / W 1425 0.8 / 2007 5.0	
10 0206 0.6 / 0756 5.7 / W 1429 0.4 / 2021 5.4		**25** 0230 0.9 / 0815 5.1 / TH 1456 0.9 / 2037 4.9	
11 0245 0.6 / 0837 5.7 / TH 1509 0.4 / 2103 5.3		**26** 0301 1.0 / 0843 5.0 / F 1526 1.0 / 2107 4.8	
12 0326 0.7 / 0922 5.5 / F 1551 0.6 / 2149 5.0		**27** 0332 1.2 / 0913 4.8 / SA 1557 1.2 / 2140 4.6	
13 0410 0.9 / 1011 5.2 / SA 1636 0.9 / 2241 4.7		**28** 0405 1.4 / 0947 4.6 / SU 1631 1.5 / 2218 4.3	
14 0457 1.2 / 1110 4.9 / SU 1725 1.3 / 2346 4.4		**29** 0441 1.7 / 1026 4.3 / M 1708 1.8 / 2307 4.1	
15 0553 1.6 / 1221 4.5 / M 1824 1.7 ◐			

MARCH

Time	m	Time	m
1 0522 2.0 / 1118 4.0 / TU 1753 2.1 ◐		**16** 0041 4.2 / 0643 1.8 / W 1327 4.1 / 1912 2.1	
2 0019 3.9 / 0615 2.2 / W 1246 3.8 / 1857 2.3		**17** 0209 4.0 / 0819 2.0 / TH 1453 4.0 / 2051 2.2	
3 0147 3.8 / 0734 2.4 / TH 1423 3.8 / 2035 2.4		**18** 0333 4.1 / 0952 1.9 / F 1607 4.1 / 2214 2.0	
4 0304 3.9 / 0914 2.3 / F 1542 3.9 / 2201 2.1		**19** 0435 4.3 / 1055 1.6 / SA 1700 4.3 / 2309 1.7	
5 0407 4.2 / 1027 1.9 / SA 1641 4.2 / 2258 1.8		**20** 0520 4.5 / 1141 1.3 / SU 1741 4.6 / 2352 1.4	
6 0456 4.5 / 1119 1.5 / SU 1725 4.6 / 2343 1.4		**21** 0556 4.7 / 1221 1.1 / M 1815 4.7	
7 0537 4.9 / 1204 1.0 / M 1804 4.9		**22** 0029 1.2 / 0627 4.9 / TU 1256 0.9 / 1844 4.9	
8 0025 1.0 / 0616 5.3 / TU 1247 0.6 / 1842 5.2		**23** 0104 1.0 / 0655 5.0 / W 1328 0.8 / ○ 1913 5.0	
9 0106 0.6 / 0655 5.6 / W 1327 0.3 / ● 1920 5.4		**24** 0135 0.9 / 0722 5.0 / TH 1358 0.8 / 1940 5.0	
10 0146 0.4 / 0735 5.8 / TH 1408 0.1 / 2000 5.5		**25** 0205 0.9 / 0749 5.0 / F 1427 0.8 / 2008 4.9	
11 0226 0.3 / 0817 5.7 / F 1449 0.2 / 2041 5.4		**26** 0235 0.9 / 0816 4.9 / SA 1456 0.9 / 2037 4.8	
12 0308 0.4 / 0902 5.5 / SA 1530 0.4 / 2126 5.1		**27** 0305 1.1 / 0846 4.7 / SU 1526 1.1 / 2108 4.7	
13 0351 0.7 / 0952 5.2 / SU 1614 0.8 / 2217 4.8		**28** 0337 1.2 / 0919 4.5 / M 1558 1.3 / 2144 4.4	
14 0438 1.0 / 1052 4.8 / M 1701 1.2 / 2321 4.4		**29** 0412 1.5 / 0958 4.3 / TU 1633 1.6 / 2228 4.2	
15 0533 1.5 / 1206 4.4 / TU 1757 1.7 ◐		**30** 0452 1.7 / 1048 4.0 / W 1715 1.9 / 2330 4.0	
		31 0542 2.0 / 1206 3.8 / TH 1812 2.2 ◐	

APRIL

Time	m	Time	m
1 0058 3.9 / 0652 2.2 / F 1344 3.7 / 1943 2.3		**16** 0304 4.0 / 0922 1.8 / SA 1542 4.0 / 2143 2.0	
2 0219 3.9 / 0830 2.1 / SA 1504 3.9 / 2122 2.1		**17** 0407 4.2 / 1025 1.6 / SU 1635 4.3 / 2240 1.8	
3 0327 4.1 / 0952 1.8 / SU 1608 4.2 / 2227 1.7		**18** 0453 4.3 / 1112 1.4 / M 1715 4.4 / 2325 1.5	
4 0422 4.5 / 1049 1.4 / M 1657 4.5 / 2316 1.3		**19** 0530 4.5 / 1152 1.2 / TU 1749 4.6	
5 0508 4.9 / 1137 0.9 / TU 1739 4.9		**20** 0003 1.3 / 0601 4.7 / W 1227 1.0 / 1819 4.7	
6 0000 0.9 / 0550 5.3 / W 1221 0.5 / 1818 5.2		**21** 0037 1.1 / 0630 4.8 / TH 1259 0.9 / 1847 4.9	
7 0043 0.5 / 0631 5.6 / TH 1304 0.2 / ● 1858 5.5		**22** 0109 1.0 / 0658 4.9 / F 1330 0.9 / ○ 1915 4.9	
8 0125 0.3 / 0714 5.7 / F 1345 0.1 / 1938 5.5		**23** 0140 0.9 / 0725 4.8 / SA 1359 0.9 / 1943 4.9	
9 0207 0.2 / 0758 5.6 / SA 1427 0.1 / 2020 5.4		**24** 0211 1.0 / 0754 4.7 / SU 1429 1.0 / 2012 4.8	
10 0250 0.3 / 0845 5.4 / SU 1509 0.4 / 2106 5.2		**25** 0242 1.0 / 0825 4.6 / M 1459 1.1 / 2044 4.7	
11 0335 0.6 / 0938 5.1 / M 1553 0.8 / 2157 4.8		**26** 0315 1.1 / 0900 4.5 / TU 1532 1.3 / 2121 4.5	
12 0423 0.9 / 1040 4.6 / TU 1641 1.2 / 2301 4.5		**27** 0350 1.3 / 0942 4.3 / W 1608 1.5 / 2206 4.3	
13 0518 1.4 / 1152 4.3 / W 1735 1.7		**28** 0431 1.5 / 1035 4.0 / TH 1650 1.8 / 2305 4.1	
14 0018 4.2 / 0626 1.7 / TH 1309 4.0 / ◑ 1846 2.1		**29** 0521 1.7 / 1147 3.9 / F 1746 2.0	
15 0142 4.0 / 0754 1.9 / F 1430 3.9 / 2019 2.2		**30** 0022 4.0 / 0626 1.9 / SA 1312 3.8 / ◐ 1906 2.1	

Chart Datum: 2·75 metres below Ordnance Datum (Newlyn). HAT is 5·9 metres above Chart Datum.

》 **FREE** monthly updates. Register at 《
www.reedsnauticalalmanac.co.uk

STANDARD TIME (UT)
For Summer Time add ONE hour in **non-shaded areas**

ULLAPOOL   LAT 57°54′N   LONG 5°09′W
TIMES AND HEIGHTS OF HIGH AND LOW WATERS

Dates in red are **SPRINGS**
Dates in blue are **NEAPS**

YEAR 2016

MAY

#	Time	m	#	Time	m
1 SU	0140 / 0751 / 1428 / 2040	4.0 / 1.9 / 3.9 / 2.0	16 M	0326 / 0942 / 1558 / 2159	4.0 / 1.7 / 4.0 / 1.9
2 M	0248 / 0913 / 1534 / 2150	4.2 / 1.6 / 4.2 / 1.7	17 TU	0417 / 1034 / 1642 / 2249	4.1 / 1.6 / 4.2 / 1.7
3 TU	0348 / 1015 / 1627 / 2245	4.5 / 1.3 / 4.5 / 1.3	18 W	0458 / 1117 / 1719 / 2331	4.3 / 1.4 / 4.4 / 1.5
4 W	0439 / 1107 / 1713 / 2334	4.8 / 0.9 / 4.9 / 0.9	19 TH	0534 / 1155 / 1752	4.4 / 1.2 / 4.6
5 TH	0526 / 1155 / 1756	5.1 / 0.5 / 5.2	20 F	0008 / 0605 / 1230 / 1822	1.3 / 4.5 / 1.1 / 4.7
6 F ●	0019 / 0610 / 1240 / 1837	0.6 / 5.4 / 0.3 / 5.4	21 SA ○	0043 / 0635 / 1302 / 1852	1.2 / 4.6 / 1.0 / 4.8
7 SA	0105 / 0656 / 1324 / 1919	0.4 / 5.5 / 0.2 / 5.4	22 SU	0116 / 0737 / 1334 / 1922	1.1 / 4.6 / 1.0 / 4.9
8 SU	0150 / 0742 / 1407 / 2003	0.3 / 5.4 / 0.3 / 5.3	23 M	0149 / 0737 / 1406 / 1953	1.0 / 4.6 / 1.0 / 4.8
9 M	0235 / 0832 / 1451 / 2050	0.4 / 5.2 / 0.5 / 5.1	24 TU	0223 / 0811 / 1438 / 2027	1.0 / 4.6 / 1.1 / 4.8
10 TU	0321 / 0926 / 1536 / 2141	0.6 / 4.9 / 0.9 / 4.8	25 W	0258 / 0849 / 1513 / 2106	1.1 / 4.5 / 1.2 / 4.7
11 W	0410 / 1026 / 1623 / 2241	0.9 / 4.6 / 1.2 / 4.5	26 TH	0335 / 0933 / 1551 / 2152	1.2 / 4.3 / 1.4 / 4.5
12 TH	0504 / 1131 / 1715 / 2350	1.2 / 4.2 / 1.6 / 4.2	27 F	0418 / 1025 / 1635 / 2247	1.3 / 4.2 / 1.6 / 4.4
13 F ◑	0605 / 1239 / 1818	1.6 / 4.0 / 2.0	28 SA	0507 / 1129 / 1729 / 2354	1.5 / 4.1 / 1.8 / 4.2
14 SA	0104 / 0717 / 1352 / 1934	4.0 / 1.8 / 3.9 / 2.1	29 SU ◐	0606 / 1242 / 1839	1.6 / 4.0 / 1.9
15 SU	0220 / 0835 / 1501 / 2055	4.0 / 1.8 / 3.9 / 2.1	30 M	0106 / 0718 / 1354 / 1959	4.2 / 1.6 / 4.0 / 1.9
			31 TU	0214 / 0834 / 1500 / 2113	4.3 / 1.5 / 4.2 / 1.7

JUNE

#	Time	m	#	Time	m
1 W	0316 / 0941 / 1559 / 2215	4.5 / 1.3 / 4.5 / 1.4	16 TH	0423 / 1038 / 1646 / 2256	4.1 / 1.7 / 4.3 / 1.8
2 TH	0413 / 1039 / 1650 / 2310	4.7 / 1.0 / 4.7 / 1.1	17 F	0505 / 1122 / 1724 / 2339	4.2 / 1.5 / 4.4 / 1.6
3 F	0506 / 1132 / 1737	5.0 / 0.8 / 5.0	18 SA	0542 / 1201 / 1759	4.3 / 1.4 / 4.6
4 SA	0000 / 0555 / 1221 / 1821	0.8 / 5.2 / 0.6 / 5.2	19 SU	0017 / 0616 / 1237 / 1831	1.4 / 4.5 / 1.2 / 4.8
5 SU ●	0049 / 0644 / 1312 / 1905	0.6 / 5.2 / 0.5 / 5.3	20 M ○	0055 / 0649 / 1312 / 1903	1.2 / 4.6 / 1.1 / 4.9
6 M	0136 / 0732 / 1352 / 1950	0.5 / 5.2 / 0.6 / 5.3	21 TU	0131 / 0723 / 1347 / 1937	1.1 / 4.6 / 1.1 / 4.9
7 TU	0223 / 0821 / 1436 / 2035	0.5 / 5.1 / 0.7 / 5.1	22 W	0207 / 0759 / 1422 / 2013	1.0 / 4.7 / 1.1 / 4.9
8 W	0309 / 0911 / 1520 / 2123	0.6 / 4.9 / 0.9 / 4.9	23 TH	0244 / 0838 / 1459 / 2053	1.0 / 4.6 / 1.1 / 4.9
9 TH	0356 / 1003 / 1605 / 2214	0.9 / 4.6 / 1.2 / 4.6	24 F	0323 / 0921 / 1538 / 2137	1.0 / 4.6 / 1.2 / 4.8
10 F	0444 / 1058 / 1652 / 2310	1.1 / 4.3 / 1.5 / 4.4	25 SA	0405 / 1010 / 1622 / 2228	1.1 / 4.4 / 1.4 / 4.7
11 SA	0535 / 1157 / 1743	1.4 / 4.1 / 1.8	26 SU	0452 / 1107 / 1712 / 2328	1.2 / 4.3 / 1.5 / 4.5
12 SU ◐	0013 / 0631 / 1300 / 1843	4.1 / 1.7 / 3.9 / 2.0	27 M ◑	0545 / 1212 / 1812	1.3 / 4.2 / 1.7
13 M	0122 / 0736 / 1406 / 1953	4.0 / 1.8 / 3.9 / 2.1	28 TU	0035 / 0647 / 1322 / 1923	4.4 / 1.5 / 4.2 / 1.8
14 TU	0231 / 0844 / 1509 / 2105	3.9 / 1.9 / 3.9 / 2.1	29 W	0144 / 0758 / 1430 / 2039	4.2 / 1.5 / 4.2 / 1.7
15 W	0333 / 0947 / 1602 / 2206	4.0 / 1.8 / 4.1 / 2.0	30 TH	0251 / 0910 / 1535 / 2150	4.4 / 1.5 / 4.4 / 1.6

JULY

#	Time	m	#	Time	m
1 F	0356 / 1017 / 1634 / 2253	4.6 / 1.3 / 4.6 / 1.3	16 SA	0436 / 1049 / 1656 / 2310	4.0 / 1.8 / 4.3 / 1.8
2 SA	0455 / 1115 / 1725 / 2348	4.7 / 1.1 / 4.9 / 1.0	17 SU	0520 / 1134 / 1735 / 2354	4.2 / 1.6 / 4.5 / 1.5
3 SU	0548 / 1207 / 1812	4.9 / 1.0 / 5.1	18 M	0558 / 1214 / 1811	4.4 / 1.4 / 4.7
4 M ●	0039 / 0637 / 1255 / 1855	0.8 / 5.0 / 0.8 / 5.2	19 TU ○	0034 / 0633 / 1252 / 1845	1.3 / 4.6 / 1.2 / 4.9
5 TU	0127 / 0723 / 1339 / 1937	0.6 / 5.1 / 0.8 / 5.2	20 W	0113 / 0708 / 1329 / 1920	1.1 / 4.7 / 1.0 / 5.1
6 W	0212 / 0807 / 1422 / 2018	0.6 / 5.0 / 0.8 / 5.2	21 TH	0151 / 0744 / 1407 / 1956	0.9 / 4.9 / 0.9 / 5.2
7 TH	0255 / 0850 / 1502 / 2059	0.7 / 4.9 / 0.9 / 5.0	22 F	0229 / 0822 / 1444 / 2035	0.7 / 4.9 / 0.9 / 5.2
8 F	0336 / 0933 / 1543 / 2140	0.8 / 4.7 / 1.1 / 4.8	23 SA	0308 / 0903 / 1524 / 2118	0.7 / 4.9 / 0.9 / 5.1
9 SA	0417 / 1016 / 1623 / 2223	1.0 / 4.4 / 1.4 / 4.5	24 SU	0349 / 0948 / 1606 / 2206	0.8 / 4.7 / 1.1 / 5.0
10 SU	0459 / 1104 / 1706 / 2311	1.3 / 4.2 / 1.6 / 4.3	25 M	0433 / 1040 / 1653 / 2302	0.9 / 4.6 / 1.3 / 4.7
11 M	0544 / 1200 / 1753	1.6 / 4.0 / 1.9	26 TU ◐	0522 / 1142 / 1747	1.1 / 4.4 / 1.5
12 TU ◑	0012 / 0635 / 1304 / 1849	4.0 / 1.8 / 3.9 / 2.1	27 W	0008 / 0618 / 1254 / 1853	4.5 / 1.4 / 4.2 / 1.7
13 W	0124 / 0737 / 1411 / 2000	3.9 / 2.0 / 3.9 / 2.2	28 TH	0121 / 0726 / 1407 / 2012	4.4 / 1.6 / 4.2 / 1.8
14 TH	0237 / 0849 / 1516 / 2116	3.8 / 2.1 / 3.9 / 2.2	29 F	0236 / 0845 / 1521 / 2134	4.3 / 1.7 / 4.3 / 1.7
15 F	0343 / 0956 / 1611 / 2220	3.9 / 2.0 / 4.1 / 2.0	30 SA	0349 / 1002 / 1626 / 2245	4.4 / 1.6 / 4.5 / 1.5
			31 SU	0452 / 1106 / 1720 / 2343	4.6 / 1.4 / 4.7 / 1.2

AUGUST

#	Time	m	#	Time	m
1 M	0545 / 1158 / 1805	4.8 / 1.2 / 5.0	16 TU	0537 / 1151 / 1747	4.4 / 1.5 / 4.8
2 TU ●	0032 / 0629 / 1244 / 1844	1.0 / 4.9 / 1.0 / 5.1	17 W	0012 / 0612 / 1231 / 1822	1.3 / 4.7 / 1.2 / 5.1
3 W	0117 / 0709 / 1326 / 1921	0.8 / 5.0 / 0.9 / 5.2	18 TH ○	0052 / 0647 / 1309 / 1858	0.9 / 4.9 / 0.9 / 5.3
4 TH	0157 / 0747 / 1404 / 1956	0.7 / 5.0 / 0.9 / 5.2	19 F	0130 / 0723 / 1347 / 1935	0.7 / 5.1 / 0.7 / 5.5
5 F	0235 / 0823 / 1441 / 2030	0.7 / 5.0 / 0.9 / 5.1	20 SA	0209 / 0800 / 1425 / 2014	0.5 / 5.2 / 0.6 / 5.5
6 SA	0311 / 0858 / 1517 / 2103	0.8 / 4.8 / 1.0 / 4.9	21 SU	0248 / 0840 / 1505 / 2056	0.5 / 5.2 / 0.7 / 5.4
7 SU	0346 / 0933 / 1552 / 2137	1.0 / 4.6 / 1.2 / 4.7	22 M	0328 / 0923 / 1546 / 2143	0.5 / 5.0 / 0.8 / 5.2
8 M	0422 / 1011 / 1629 / 2215	1.2 / 4.4 / 1.5 / 4.4	23 TU	0411 / 1012 / 1632 / 2238	0.7 / 4.8 / 1.1 / 4.9
9 TU	0459 / 1056 / 1708 / 2301	1.5 / 4.2 / 1.8 / 4.1	24 W	0457 / 1113 / 1724 / 2348	1.1 / 4.5 / 1.4 / 4.6
10 W ◐	0541 / 1156 / 1755	1.8 / 4.0 / 2.1	25 TH ◑	0552 / 1229 / 1830	1.5 / 4.3 / 1.8
11 TH	0009 / 0631 / 1311 / 1854	3.9 / 2.1 / 3.9 / 2.3	26 F	0107 / 0700 / 1351 / 1955	4.3 / 1.8 / 4.2 / 2.0
12 F	0138 / 0742 / 1425 / 2017	3.8 / 2.2 / 3.9 / 2.4	27 SA	0229 / 0828 / 1513 / 2128	4.2 / 2.0 / 4.2 / 1.9
13 SA	0259 / 0908 / 1532 / 2141	3.8 / 2.2 / 4.2 / 2.2	28 SU	0347 / 0955 / 1621 / 2241	4.3 / 1.9 / 4.4 / 1.6
14 SU	0406 / 1018 / 1627 / 2243	3.9 / 2.1 / 4.2 / 2.0	29 M	0449 / 1059 / 1712 / 2336	4.5 / 1.6 / 4.7 / 1.3
15 M	0457 / 1109 / 1710 / 2330	4.1 / 1.8 / 4.5 / 1.6	30 TU	0537 / 1148 / 1753	4.7 / 1.4 / 4.9
			31 W	0020 / 0616 / 1230 / 1829	1.0 / 4.9 / 1.1 / 5.1

Chart Datum: 2·75 metres below Ordnance Datum (Newlyn). HAT is 5·9 metres above Chart Datum.

NW Scotland

》》 **FREE** monthly updates. Register at 《 www.reedsnauticalalmanac.co.uk 《

445

STANDARD TIME (UT)
For Summer Time add ONE hour in **non-shaded areas**

ULLAPOOL   LAT 57°54'N   LONG 5°09'W
TIMES AND HEIGHTS OF HIGH AND LOW WATERS

Dates in red are **SPRINGS**
Dates in blue are **NEAPS**

YEAR **2016**

SEPTEMBER

Time m	Time m
1 0100 0.8 / 0650 5.0 / TH 1308 1.0 / ● 1901 5.2	**16** 0026 0.8 / 0623 5.1 / F 1246 0.8 / ○ 1833 5.5
2 0136 0.7 / 0723 5.1 / F 1343 0.9 / 1931 5.2	**17** 0106 0.5 / 0659 5.4 / SA 1324 0.6 / 1911 5.7
3 0210 0.7 / 0753 5.0 / SA 1416 0.9 / 2000 5.1	**18** 0145 0.3 / 0736 5.5 / SU 1404 0.5 / 1951 5.7
4 0242 0.8 / 0823 4.9 / SU 1448 1.0 / 2030 5.0	**19** 0225 0.3 / 0816 5.4 / M 1444 0.5 / 2035 5.6
5 0313 1.0 / 0854 4.8 / M 1521 1.1 / 2100 4.8	**20** 0306 0.4 / 0859 5.2 / TU 1527 0.7 / 2123 5.3
6 0345 1.2 / 0927 4.6 / TU 1554 1.4 / 2133 4.5	**21** 0348 0.7 / 0948 4.9 / W 1613 1.0 / 2220 4.9
7 0418 1.5 / 1004 4.4 / W 1630 1.7 / 2212 4.3	**22** 0435 1.1 / 1049 4.6 / TH 1706 1.4 / 2334 4.5
8 0455 1.8 / 1053 4.1 / TH 1711 2.0 / 2306 4.0	**23** 0529 1.6 / 1210 4.3 / F 1813 1.8 ◗
9 0539 2.1 / 1208 3.9 / F 1803 2.2 ◗	**24** 0057 4.2 / 0638 2.0 / SA 1337 4.2 / 1943 2.0
10 0039 3.8 / 0639 2.3 / SA 1335 3.9 / 1918 2.4	**25** 0222 4.1 / 0813 2.2 / SU 1501 4.2 / 2120 1.9
11 0214 3.7 / 0815 2.4 / SU 1450 4.0 / 2059 2.3	**26** 0340 4.2 / 0944 2.1 / M 1608 4.4 / 2229 1.7
12 0332 3.9 / 0944 2.2 / M 1552 4.2 / 2212 2.0	**27** 0438 4.4 / 1045 1.8 / TU 1658 4.7 / 2319 1.4
13 0428 4.1 / 1041 1.9 / TU 1640 4.5 / 2303 1.6	**28** 0522 4.7 / 1131 1.5 / W 1736 4.9
14 0511 4.5 / 1126 1.5 / W 1720 4.8 / 2346 1.2	**29** 0001 1.1 / 0557 4.9 / TH 1210 1.3 / 1809 5.0
15 0548 4.8 / 1206 1.2 / TH 1757 5.2	**30** 0038 1.0 / 0628 5.0 / F 1246 1.1 / 1838 5.1

OCTOBER

Time m	Time m
1 0111 0.9 / 0657 5.1 / SA 1319 1.0 / ● 1906 5.2	**16** 0040 0.5 / 0635 5.5 / SU 1301 0.6 / ○ 1849 5.8
2 0142 0.8 / 0725 5.1 / SU 1350 1.0 / 1933 5.1	**17** 0121 0.3 / 0713 5.6 / M 1343 0.4 / 1931 5.8
3 0212 0.9 / 0753 5.1 / M 1421 1.1 / 2001 5.0	**18** 0203 0.3 / 0754 5.6 / TU 1425 0.5 / 2017 5.6
4 0242 1.0 / 0822 4.9 / TU 1452 1.2 / 2030 4.8	**19** 0245 0.6 / 0839 5.4 / W 1510 0.7 / 2108 5.3
5 0312 1.2 / 0852 4.8 / W 1524 1.4 / 2103 4.6	**20** 0328 0.8 / 0928 5.1 / TH 1558 1.0 / 2208 4.9
6 0344 1.5 / 0928 4.6 / TH 1559 1.6 / 2141 4.3	**21** 0416 1.3 / 1030 4.7 / F 1652 1.4 / 2321 4.5
7 0419 1.7 / 1011 4.3 / F 1638 1.9 / 2230 4.1	**22** 0510 1.7 / 1149 4.4 / SA 1759 1.8 ◗
8 0500 2.0 / 1112 4.1 / SA 1727 2.1 / 2350 3.9	**23** 0041 4.2 / 0618 2.1 / SU 1315 4.3 / 1924 2.0
9 0554 2.3 / 1243 4.0 / SU 1834 2.3 ◗	**24** 0203 4.1 / 0747 2.3 / M 1438 4.3 / 2055 2.0
10 0130 3.8 / 0720 2.5 / M 1404 4.0 / 2011 2.3	**25** 0319 4.2 / 0916 2.2 / TU 1545 4.4 / 2203 1.8
11 0251 3.9 / 0902 2.3 / TU 1511 4.2 / 2134 2.0	**26** 0416 4.4 / 1018 2.0 / W 1635 4.6 / 2253 1.5
12 0353 4.2 / 1007 2.0 / W 1605 4.5 / 2230 1.6	**27** 0459 4.6 / 1105 1.7 / TH 1714 4.7 / 2334 1.3
13 0440 4.5 / 1056 1.6 / TH 1649 4.9 / 2316 1.2	**28** 0534 4.8 / 1145 1.5 / F 1746 4.9
14 0520 4.9 / 1138 1.2 / F 1729 5.3 / 2358 0.8	**29** 0010 1.2 / 0604 4.9 / SA 1221 1.3 / 1815 5.0
15 0557 5.3 / 1220 0.8 / SA 1809 5.6	**30** 0044 1.1 / 0633 5.1 / SU 1254 1.2 / ● 1843 5.0
	31 0115 1.0 / 0701 5.1 / M 1326 1.2 / 1911 5.0

NOVEMBER

Time m	Time m
1 0145 1.1 / 0729 5.1 / TU 1356 1.2 / 1939 4.9	**16** 0143 0.5 / 0738 5.6 / W 1410 0.5 / 2005 5.6
2 0215 1.2 / 0757 5.0 / W 1428 1.2 / 2009 4.8	**17** 0227 0.6 / 0824 5.5 / TH 1457 0.7 / 2057 5.3
3 0245 1.3 / 0828 4.9 / TH 1500 1.4 / 2043 4.7	**18** 0312 0.9 / 0914 5.2 / F 1546 1.0 / 2155 4.9
4 0317 1.5 / 0903 4.7 / F 1535 1.6 / 2122 4.4	**19** 0359 1.3 / 1011 4.9 / SA 1639 1.3 / 2301 4.6
5 0352 1.7 / 0945 4.5 / SA 1615 1.8 / 2210 4.2	**20** 0451 1.7 / 1120 4.6 / SU 1740 1.7
6 0432 2.0 / 1039 4.3 / SU 1702 2.0 / 2318 4.0	**21** 0011 4.3 / 0551 2.1 / M 1238 4.4 / ◗ 1850 1.9
7 0523 2.2 / 1155 4.2 / M 1802 2.2 ●	**22** 0126 4.2 / 0706 2.3 / TU 1357 4.3 / 2008 2.0
8 0047 3.9 / 0636 2.4 / TU 1317 4.2 / 1923 2.2	**23** 0241 4.2 / 0827 2.3 / W 1508 4.3 / 2121 1.9
9 0206 4.0 / 0810 2.4 / W 1427 4.3 / 2047 2.0	**24** 0342 4.3 / 0938 2.2 / TH 1603 4.4 / 2217 1.8
10 0312 4.2 / 0926 2.1 / TH 1526 4.6 / 2152 1.6	**25** 0429 4.4 / 1032 2.0 / F 1646 4.5 / 2302 1.6
11 0407 4.6 / 1021 1.7 / F 1617 4.9 / 2244 1.3	**26** 0508 4.6 / 1116 1.8 / SA 1723 4.7 / 2341 1.5
12 0452 4.9 / 1110 1.3 / SA 1703 5.2 / 2331 0.9	**27** 0541 4.8 / 1155 1.6 / SU 1755 4.8
13 0533 5.3 / 1155 1.0 / SU 1747 5.5	**28** 0016 1.4 / 0611 5.0 / M 1230 1.5 / 1825 4.9
14 0015 0.6 / 0614 5.5 / M 1240 0.7 / ○ 1831 5.7	**29** 0050 1.3 / 0641 5.1 / TU 1304 1.3 / ● 1855 4.9
15 0100 0.5 / 0655 5.7 / TU 1325 0.5 / 1917 5.7	**30** 0121 1.2 / 0710 5.1 / W 1337 1.3 / 1925 4.9

DECEMBER

Time m	Time m
1 0153 1.3 / 0740 5.1 / TH 1409 1.3 / 1957 4.8	**16** 0214 0.8 / 0813 5.5 / F 1447 0.7 / 2045 5.3
2 0225 1.3 / 0812 5.0 / F 1443 1.3 / 2031 4.7	**17** 0258 1.0 / 0859 5.4 / SA 1534 0.9 / 2136 5.0
3 0258 1.5 / 0848 4.9 / SA 1519 1.4 / 2110 4.6	**18** 0343 1.2 / 0949 5.1 / SU 1622 1.2 / 2230 4.7
4 0333 1.6 / 0928 4.8 / SU 1558 1.6 / 2156 4.4	**19** 0430 1.5 / 1043 4.8 / M 1712 1.5 / 2328 4.4
5 0413 1.8 / 1017 4.6 / M 1643 1.7 / 2252 4.3	**20** 0520 1.9 / 1145 4.5 / TU 1808 1.8
6 0501 2.0 / 1117 4.5 / TU 1737 1.9	**21** 0032 4.2 / 0617 2.2 / W 1256 4.3 / ◗ 1910 2.0
7 0004 4.2 / 0601 2.2 / W 1231 4.4 / ◗ 1843 1.9	**22** 0143 4.1 / 0725 2.3 / TH 1411 4.2 / 2020 2.1
8 0121 4.2 / 0719 2.2 / TH 1343 4.4 / 1959 1.9	**23** 0253 4.1 / 0840 2.4 / F 1519 4.2 / 2129 2.1
9 0230 4.3 / 0839 2.1 / F 1448 4.6 / 2110 1.7	**24** 0351 4.3 / 0949 2.3 / SA 1613 4.3 / 2225 2.0
10 0332 4.5 / 0946 1.8 / SA 1547 4.8 / 2212 1.4	**25** 0438 4.4 / 1044 2.1 / SU 1658 4.4 / 2311 1.8
11 0426 4.8 / 1043 1.5 / SU 1641 5.1 / 2306 1.1	**26** 0518 4.6 / 1129 1.9 / M 1736 4.5 / 2351 1.6
12 0514 5.1 / 1135 1.2 / M 1731 5.3 / 2355 0.9	**27** 0553 4.8 / 1208 1.7 / TU 1811 4.7
13 0559 5.4 / 1224 0.9 / TU 1820 5.5	**28** 0027 1.5 / 0624 5.0 / W 1245 1.5 / 1842 4.8
14 0043 0.7 / 0643 5.6 / W 1312 0.7 / ○ 1908 5.5	**29** 0102 1.4 / 0656 5.1 / TH 1320 1.4 / ● 1914 4.8
15 0129 0.7 / 0727 5.6 / TH 1400 0.6 / 1956 5.5	**30** 0136 1.3 / 0727 5.1 / F 1355 1.3 / 1946 4.9
	31 0209 1.3 / 0759 5.2 / SA 1430 1.2 / 2020 4.9

Chart Datum: 2·75 metres below Ordnance Datum (Newlyn). HAT is 5·9 metres above Chart Datum.

ADJACENT ISLANDS

SUMMER ISLES, 58°01′N 05°25′W. AC 2509, 2501. HW –0425 Dover; See 9.8.8; streams are weak and irregular. In the N apps to Loch Broom some 30 islands and rks, the main ones being Eilean Mullagrach, Isle Ristol, Glas-leac Mor, Tanera Beg, Eilean a' Char, Eilean Fada Mor. Beware rocks at S end of Horse Sound. ⚓s:

Isle Ristol, ⚓ to S of drying causeway and clear of moorings; beware landings frequented by local FVs. Close to slip is lt Fl G 3s. Facilities: 🖥 (at Polbain 2M).

Tanera Beg, ⚓ in the chan to the E inside Eilean Fada Mor.

Tanera Mor, on E side, ⚓ in bay to N off pier or in W corner off stone jetty; beware many moorings and fish pens. Also poss ⚓ close NW of Is and close E of Eilean na Saille, but N of drying rock. Facilities: ☎(01854) 622272 B Wilder for ⚓ 🔧 📶 Tea room.

Badentarbat Bay, Temp'y ⚓ NW part of bay ESE of pier which is busy with marina farm activity. Beware N part of bay is shoal. Suitable for Achiltibuie or Polbain on mainland. Facilities: 🛢🔧🛥 (emerg) ☎(01854) 622261 🖥 ✕.

LITTLE LOCH BROOM, 57°54′·98N 05°24′·68W. AC 2500. HW –0415 Dover as per Ullapool, see 9.8.8; streams are weak (approx 1kn). Between Loch Broom and Gruinard Bay, entered S of Cailleach Hd it is deep (115m in centre) and shoreline steep to. In approaches beware of Ardross Rk (depth 0.6m) 0.38M ENE of Stattic Pt.

Camusnagaul, on S side of loch 1.6M from head and Dundonnel Hotel, ⚓ to E of hamlet.

Scoraig, inside narrows on N shore in N'lies, ⚓ off jetty clear of submarine cables and marine farm. ⚒ C.Dawson ☎01845 613380.

ANCHORAGES BETWEEN LITTLE LOCH BROOM & LOCH GAIRLOCH

LOCH EWE, 57°52′·0N 05°40′·0W (SWM buoy, L Fl 10s). AC 2509, 3146. Tides: See 9.8.8; HW –0415 on Dover; ML 2·6m; Duration 0610. Shelter in all winds. Easy ent with no dangers in loch. Rhubha Reidh lt, Fl (4) 15s 37m 24M, W tr, is 4·5M W of ent. No 1 buoy Fl (3) G 10s. Loch approx 7M long with Isle Ewe and 2 small islets about 2M from ent in centre; can be passed on either side. Temp ⚓ in bay of Isle of Ewe. Beware unlit buoy A2 between E side Isle Ewe & Aultbea Pier, 2 FG (vert). NATO fuelling jetty and dolphins, all Fl G 4s.

Loch Thurnaig, 1.7M SSE of Isle of Ewe, sheltered ⚓ off Ob na Bà Rùaidhe, SE of pier to W of drying reef. Beware of numerous marine farms in approaches.

Aultbea, ⚓ off Aultbea Hotel on E shore or in NW'lies close E of pier. N part of bay is shoal. Facilities: Dr, 🖂🖥✕🖵.

Poolewe Bay, ⚓ in SW part of bay in 3·5m, at head of loch: Boor Rks(3m) with drying rocks extending 1ca NW lie off W shore about 9ca NW of Poolewe. Head of Loch dries for a considerable distance. Facilities: 🔧🔨🛥 on pier, 🖂 (at garage) 🛢🖂🖥✕🖵.

Inverewe Gdns on NE side of bay, ⚓ to SW of Jetty at Port na Cloiche Gile.

LOCH GAIRLOCH, 57°43′N 05°45′W. AC 228, 2528. HW –0440 on Dover. See 9.8.8. A wide loch facing W. Ent clear of dangers but Submarine Exercise Area (see 9.8.22) to S of Longa Is. Quite heavy seas enter in bad weather. Lts: Glas Eilean Fl WRG 6s 9m 6/4M, Gairloch Pier Hd, QR 9m. HM ☎(01445) 712140.

Badachro, good shelter SW of Eilean Horrisdale on SW side of loch but busy with FVs and local craft. Unlit PHM/SHM mark rock ledges in approach. Passage to S of Eilean Horrisdale is possible with caution. ⚓ on E side NNW of Sgeir Dhubh Bhea. ✕ 🖵(Bad Inn). HM ☎(01445) 741255 for M availability.

Loch Shieldaig, 6 Blue ⚓ (with payment instructions) at SE end of the loch with many moorings for local craft. If congested ⚓ to SSW of Eilean an t-Sabhail. Beware marine farms.

Flowerdale Bay, ⚓ in approx 6m near Gairloch pier or berth on pontoons (2m depth N side, 3m S side) at pier. HM ☎(01445) 712140. VHF Ch 16 (occas). Gairloch Pier: 🛢 fees charged. 🔧📶🛥 🖂🛢🖂🖥🖵 SC, Hotel.

9.8.9 PORTREE

Skye (Highland) **57°24'·73N 06°11'·07W** ✳✳✳◊◊⚓⚓⚓

CHARTS AC 2209, 2534, 5616; Imray C66

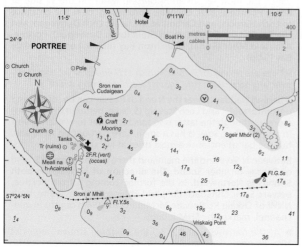

TIDES −0445 Dover; ML no data; Duration 0610

Standard Port ULLAPOOL (→)

Times				Height (metres)			
High Water		Low Water		MHWS	MHWN	MLWN	MLWS
0000	0600	0300	0900	5·2	3·9	2·1	0·7
1200	1800	1500	2100				
Differences PORTREE (Skye)							
−0025	−0025	−0025	−0025	+0·1	−0·2	−0·2	0·0
SHIELDAIG (Loch Torridon)							
−0020	−0020	−0015	−0015	+0·4	+0·3	+0·1	0·0
LOCH A'BHRAIGE (Rona)							
−0020	0000	−0010	0000	−0·1	−0·1	−0·1	−0·2
LOCH SNIZORT (Uig Bay, Skye)							
−0045	−0020	−0005	−0025	+0·1	−0·4	−0·2	0·0
LOCH DUNVEGAN (Skye)							
−0105	−0030	−0020	−0040	0·0	−0·1	0·0	0·0
LOCH HARPORT (Skye)							
−0115	−0035	−0020	−0100	−0·1	−0·1	0·0	+0·1
SOAY (Camus nan Gall)							
−0055	−0025	−0025	−0045	−0·4	−0·2	ND	ND

SHELTER Secure in all but strong S to SW'lies, when Camas Bàn is more sheltered. In the NW of the bay there are 8 ⚓s for <15 tons. Short stay pontoon on pier.

NAVIGATION WPT 57°24'·58N 06°10'·07W, 275°/0·72M to pier. From the S, avoid rks off An Tom Pt (1·5M to E, off chartlet).

LIGHTS AND MARKS Only lts are a SHM buoy Fl G 5s marking Sgeir Mhór, 2 FR (vert) 6m 4M (occas) on the pier and a SPM buoy Fl Y 5s.

COMMUNICATIONS (Code 01478) CGOC (01851) 702013; ⊖ (0141) 887 9369; Police 101; Dr 612013; ⊞ 612704. HM 612926; Moorings 612341.

VHF Ch 16 12 (occas).

FACILITIES Pier ⤢ ⌖£14 ⚓ ⛽(Mon-Fri).
Town ⌖ 🛢 ⊠ ⑧ ⚒ ⊡ ✕ ⌷ bus to Kyle of Lochalsh, ⇌.

ANCHORAGES AROUND OR NEAR SKYE

LOCH TORRIDON Highland, **57°36'N 05°49'W**. AC 2210. Tides, see 9.8.9. Three large lochs: ent to outer loch (Torridon) is 3M wide, with isolated Sgeir na Trian (2m) almost in mid-chan; ⚓s on SW side behind Eilean Mór and in L Beag. L Sheildaig is middle loch with good ⚓ and 2 ⚓s (up to 10 tons) between the Is and village. 1M to the N, a 2ca wide chan leads into Upper L Torridon; many fish cages and prone to squalls. Few facilities, except Shieldaig: ⚓ ⊠ ⚒ ⌷ Garage; Diabeg: ✕.

LOCH A'BHRAIGE Rona (Highland), **57°34'·6N 05°57'·9W**. AC 2479, 2534. HW −0438 on Dover; ML 2·8m; Duration 0605. See 9.8.9. A good ⚓ in NW of the island, safe except in NNW winds. Beware rks on NE side up to 1ca off shore. Hbr in NE corner of loch head. Ldg lts 137°, see 9.8.4. Facilities: jetty, ⚓ and a helipad, all owned by MOD (DRA). Before ent, call *Rona Range Control* VHF Ch 13.

Acarseid Mhór is ⚓ on W of Rona. One ⚓ £12, ☎07831 293963. App S of Eilean Garbh marked by W arrow. SD sketch of rks at ent is necessary. ⚓ ⌖ ⌾. At **Churchton Bay**, SW tip of Raasay, there are 4 HIE ⚓s; ☎(01478) 612341; ⤢ ⊡ ⌾ ✕.

LOCH DUNVEGAN 4 ⚓s off Stein, **57°30'·9N 06°34'·5W**. 2 ⚓s off Dunvegan, 57°26'·3N 06°35'·2W. Fuel, ⚓ ✕. ☎(01478) 612341.

LOCH HARPORT Skye (Highland), **57°20'·6N 06°25'·8W**. AC 1795. HW −0447 (sp), −0527 (np) on Dover. See 9.8.9. On E side of Loch Bracadale, entered between Oronsay Is and Ardtreck Pt (W lt ho, Fl 6s 18m 9M). SW end of Oronsay has conspic rk pillar, called The Castle; keep ¼M off-shore here and off E coast of Oronsay which is joined to Ullinish Pt by drying reef. ⚓ Oronsay Is, N side of drying reef (4m), or on E side, but beware rk (dries) 0·5ca off N shore of Oronsay. Fiskavaig Bay 1M S of Ardtreck (7m); Loch Beag on N side of loch, exposed to W winds (beware fish farm); 1 ⚓ off the distillery and 2 ⚓ off *The Old Inn* at Carbost on SW shore. Facilities: Carbost ⚓ ⊠ ⚒ ✕ ⌷ ⌖(garage). Port na Long ⚓ ⚒ ⌷.

LOCH SLIGACHAN Skye (Highland), **57°19'·0N 06°06'·0W**. AC 2534. Tides see 9.8.9. Differences as for Portree. Steep sided sheltered Loch S of Narrows of Raasay. In sheltered weather ⚓ outside in Balmeanach Bay. Entr marked by ECM and SHM. Sconser harbour with ⤢ pier and pontoon (lit Fl R 3s 2m 3M) inside narrows used to service fish farms and shellfish beds, which should be avoided.

SOAY HARBOUR Skye (Highland), **57°09'·5N 06°13'·4W**. AC 2208. Tides see 9.8.9. Narrow inlet on NW side of Soay; enter from Soay Sound above half flood to clear bar, dries 1·0m. Appr on 135° from 5ca out to avoid reefs close each side of ent. Cross bar slightly E of mid-chan, altering 20° stbd for best water; ent is 15m wide between boulder spits marked by W poles with Or tops. ⚓ in 3m mid-pool or shoal draught boats can enter inner pool. Good shelter and holding. Camas nan Gall (poor holding); no other facilities.

ARMADALE BAY Skye (Highland), 4M NW of Mallaig. AC 2208. 6 ⚓s £12 at **57°04'·0N 05°53'·6W**, ☎(01471) 844216. Bay is sheltered from SE to N winds but subject to swell in winds from N to SE. From S, beware the Eilean Maol & Sgorach rocks. Ferry pier, Oc R 6s 6m 6M, with conspic W shed. Facilities: Skye Yachts ☎01471 844216, M (£12·50), ⚓ at pontoon ⛽ ⚒ (not Sat pm or Sun), Free pontoon for tenders, ferry to Mallaig for ⇌. Ardvasar ¾ mile: ⤢ HW±3 at ⌂(launching/parking £5), ⌖(not Sun) 🛢 🚻 ⌾ ✕ ⌷.

CROWLIN ISLANDS Highland, **57°21'·1N 05°50'·6W**. AC 2209, 2498. HW −0435 on Dover, −0020 on Ullapool; HW +0·3m on Ullapool. See 9.8.8. ⚓ between Eilean Meadhonach and Eilean Mor, appr from N, keep E of Eilean Beg. Excellent shelter except in strong N winds. There is an inner ⚓ with 3½m but ent chan dries. Eilean Beg lt ho Fl 6s 32m 6M, W tr. No facilities.

SUBMARINE EXERCISE INFORMATION

Details of submarine activity in the Exercise Areas north of Mull are broadcast at 0710 and 1910 UT by Stornoway CG on a specified VHF Ch after an initial announcement on Ch 16. Areas are referred to by the names given below, rather than by the numbers indicated on the chartlet. For Areas 22–81 (South of Mull), see page 482.

SUBFACTS

General information on SUBFACTS is also broadcast twice daily at 0620 & 1820 UT on Navtex. Stornoway CGOC will provide SUBFACTS on request; call on Ch 16.

A 'Fisherman's hotline' (☎(01436) 677201) is available on a 24 hour basis and may be used for any queries relating to SUBFACTS from any mariner.

Submarines on the surface and at periscope depth always listen on Ch 16.

1	Tiumpan	14	Raasay
2	Minch North	15	Neist
3	Stoer	16	Bracadale
4	Shiant	17	Ushenish
5	Minch South	18	Hebrides North
6	Ewe	19	Canna
7	Trodday	20	Rhum
8	Rona West	21	Sleat
9	Rona North	22	Barra
10	Lochmaddy	23	Hebrides Central
11	Dunvegan	24	Hawes
12	Portree	25	Eigg
13	Rona South	26	Hebrides South

9.8.10 PLOCKTON

Highland **57°20'·52N 05°38'·47W** ❀❀❀⚓⚓⚓❀❀❀

CHARTS AC 2209, 2528, 5616; Imray C66

TIDES –0435 Dover; ML 3·5m; Duration 0600; See 9.8.9

Standard Port ULLAPOOL (→)

Times				Height (metres)			
High Water		Low Water		MHWS	MHWN	MLWN	MLWS
0000	0600	0300	0900	5·2	3·9	2·1	0·7
1200	1800	1500	2100				
Differences PLOCKTON							
+0005	–0025	–0005	–0010	+0·5	+0·5	+0·5	+0·2

SHELTER Good, exposed only to N/NE'lies. 9 Y ⚓s or ⚓ in centre of bay in approx 3·5m. Inner part of bay shoals and dries to the SW.

NAVIGATION WPT 57°21'·16N 05°39'·44W; thence towards Bogha Dubh Sgeir Bn, between Eilean a Chait disused Lt ho and Sgeir Golach Bn / High Stone (1m) to the North. Hawk Rock (0.1m) will be cleared when Duncraig Castle bears 158°. Alter S to the ⚓. Beware Plockton Rks (3·1m) on E side of bay.

LIGHTS AND MARKS See 9.8.4 and chartlet. Old lt ho (13m) on Eilean a Chait is conspic, as is Duncraig Castle. Sgeir Golach Bn a twr with W panels Fl W 10s, Dubh Sgeir Bn a tower with R panels Fl(2) R 6s and Plockton Rks a PHM Bn Fl R 3s.

COMMUNICATIONS HM (01599) 534589, 🖷 534167, Mobile 07802 367253 (at Kyle of Lochalsh); CGOC (01851) 702013; ⊜ (0141) 887 9369; Police101; Dr via HM.

FACILITIES Village ⚓£5, ⚓⚓ at 50m dinghy pontoon (H24 except MLWS), Ⓔ ⚓ 🖳 ✕ 🗄 🛒 airstrip, (bus to Kyle of Lochalsh).

ADJACENT MOORINGS

LOCH CARRON Strome Narrows a re not buoyed or lit. Appr on 328° between Sgeir Chreagach and Sgeir Fhada. Fish farms on W shore of Loch Kishorn.

9.8.11 LOCH ALSH
Highland 57°16′·68N 05°42′·87W ✿✿✿◊◊✿✿✿

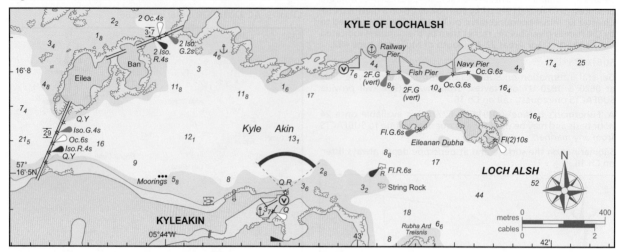

CHARTS AC 2540, 5616; Imray C66

TIDES –0450 Dover; ML 3·0m; Duration 0555; See 9.8.9

Standard Port ULLAPOOL (→)

Times				Height (metres)			
High Water		Low Water		MHWS	MHWN	MLWN	MLWS
0000	0600	0300	0900	5·2	3·9	2·1	0·7
1200	1800	1500	2100				
Differences KYLE OF LOCHALSH							
–0040	–0020	–0005	–0025	+0·1	0·0	0·0	–0·1
DORNIE BRIDGE (Loch Alsh)							
–0040	–0010	–0005	–0020	+0·1	–0·1	0·0	0·0
GLENELG BAY (Kyle Rhea)							
–0105	–0035	–0035	–0055	–0·4	–0·4	–0·9	–0·1
LOCH HOURN							
–0125	–0050	–0040	–0110	–0·2	–0·1	–0·1	+0·1

SHELTER **Kyle of Lochalsh:** ⌂ on Railway Pier, with FVs, or on 40m L-shaped pontoon (seasonal), close W of Railway Pier; or ⚓ off the hotel in 11m. **Kyle Akin:** 3 free ⚓s are subject to tidal stream. Lit Ⓥ pontoon on NW side of hbr for yachts, shoal on N side.

NAVIGATION WPT (from Inner Sound) 57°16′·98N 05°45′·77W, 123°/0·75M to bridge. Chan to bridge is marked by 2 PHM and 2 SHM lt buoys. Bridge to Skye, 29m clearance, is lit Oc 6s in centre of main span, Iso R 4s on S pier and Iso G4s on N pier. The secondary NE span is lit, but has only 3.7m clearance.

LIGHTS AND MARKS See chartlet. Direction of buoyage is N in Kyle Rhea, thence E up Loch Alsh; but W through Kyle Akin and the bridge, ie SHMs are on N side of channel.

COMMUNICATIONS (Code 01599) CGOC (01851) 702013; ⊜ (0141) 887 9369; Police 101; Dr, Ⓗ: via HM. Skye bridge 534844. HM 534306, Mobile 07990 971161, Deputy 07771 816900. VHF Ch **11** 16.

FACILITIES **Kyle of Lochalsh:** www.kyleandlochalsh.org.uk/index.asp ⌂pontoon, £15 1st night then £10, ⚓ ⚑ (Fish pier via HM, Mon-Fri) 🛠️⛽🔌Ⓔ🔋Ⓑ⊠🛒🗑️✕🗐🏴 £2 (honesty box). 🚃 (useful railhead), Bus to Glasgow & Inverness; buses every ½hr to/from Kyleakin.
Kyleakin: ☎534167 or VHF Ch 11, 120m ⌂ (£14/craft) depth varies from 3m to drying. ⚓⚓✕🗐.

ANCHORAGES IN LOCH ALSH (see also 9.8.9)

⚓s, safe depending on winds, are (clockwise from Kyle): Avernish B, (2ca N of Racoon Rk) in 3m clear of power cables, open to SW; NW of Eilean Donnan Cas (conspic); in Ratagan B at head of L Duich in 7m; in Totaig B facing Loch Long ent in 3·5m; on S shore in Ardintoul B in 5·5m; at head of Loch na Béiste in 7m close inshore and W of fish cages.

SOUND OF SLEAT
There are ⚓s at: **Glenelg Bay (57°12′·57N 05°37′·95W)** SW of pier out of the tide, but only moderate holding. Usual facilities in village. At **Sandaig Bay (57°10′·0N 05°41′·4W)**, exposed to SW. Sandaig Is are to NW of the bay; beware rks off Sgeir nan Eun. Eilean Mór lit Fl 6s. At **Isleornsay Hbr (57°09′N 05°48′W)** 5⚓s at Duisdale (NW corner) 2ca N of pier, 2FR (vert) and floodlit. Give drying N end of Ornsay Is a wide berth. Lts: SE tip of Ornsay, Oc 8s 18m 15M, W tr; N end, Fl R 6s 8m 4M. Facilities: ⚓ ⊠ Hotels.

LOCH HOURN Highland, **57°08′N 05°42′W.** AC 2208, 2541. Tides see 9.8.9. Ent is S of Sandaig Is and opposite Isle Ornsay, Skye. Loch extends 11M inland via 4 narrows to Loch Beag; it is scenically magnificent, but violent squalls occur in strong winds. Sgeir Ulibhe, drying 2·1m, bn, lies almost in mid-ent; best to pass S of it to clear Clansman Rk, 2·1m, to the N. ⚓s on N shore at Eilean Ràrsaidh and Camas Bàn, within first 4M. For pilotage further E, consult SDs. Facilities at Arnisdale (Camas Bàn): ⚓ 🛒 ✕ 2⚓s for patrons, 🗐⊠. Doune Marine (01687 462667) ⚓ 🗐.

LOCH NEVIS Highland, **57°02′·2N 05°43′·3W.** AC 2208, 2541. HW –0515 on Dover. See 9.8.9. Beware rocks Bogha cas Sruth (1·8m), Bogha Don and Sgeirean Glasa both marked by bns. Good ⚓ NE of Eilean na Glaschoille, except in S winds; 9 Or ⚓s off Inverie, £5 (free if dining), 10m max LOA. Call *Old Forge* VHF Ch 16, 12. ⚓ ✕ 🗐🏴🗑️. In strong winds expect violent unpredictable squalls. Enter the inner loch with caution, and ⚓ N or SE of Eilean Maol.

ARISAIG, (Loch nan Ceall), Highland, **56°53′·64N 05°55′·77W** (ent). AC 2207. HW –0515 on Dover; +0030 and +0·9m on Oban. Exposed to strong winds, especially from SE to NW. SDs essential. W mark on Rubh' Arisaig on S side of ent. S Chan is winding, marked by 8 perches. Appr HW±4 to avoid strongest streams LW±1½. Caution: many unmarked rks; no lts. Sheltered ⚓ at head of loch, clear of moorings. Call **Arisaig Marine** ☎(01687) 450224, VHF Ch 16, M; few ⚓s, pontoon (loading, ⚓ & fuel) ⚓ 🛠️🔧🔌🔋⚓(10t)🔋. **Village** ⚓ at hotel, ⛽⊠🛒✕🗐🚃.

KYLE AKIN TO MALLAIG

Approaching Kyle Akin (AC 2540) from W, beware dangerous rks to N, off Bleat Is (at S side of entrance to Loch Carron); on S side of chan, Bogha Beag (dries 1·2m) and Black Eye Rk (depth 3·2m), respectively 6ca and 4ca W of bridge. Pass at least 100m N or S of Eileanan Dubha in Kyle Akin. On S side of chan String Rk (dries) is marked by PHM lt buoy. Kyle Rhea connects Loch Alsh with NE end of Sound of Sleat.

▶ The tidal streams are very strong: N-going stream begins HW Ullapool + 0600, sp rate 6-7kn; S-going stream begins at HW Ullapool, sp rate 8kn. N-going stream in Kyle Rhea begins HW Dover +0140 (HW Ullapool +0555) and runs for 6 hours. The E-going stream in Kyle Akin begins (Sp) HW Dover +0350 (HW Ullapool –0415). (Nps) HW Dover –0415 (HW Ullapool).
S-going stream in Kyle Rhea begins HW Dover –0415 (HW Ullapool) and runs for 6 hours. The W-going stream in Kyle Akin begins (Sp) HW Dover –0015 (HW Ullapool +0400). (Nps) HW Dover +0140 (HW Ullapool +0555). Eddies form both sides of the Kyle with dangerous overfalls off S end in fresh S'ly winds on S-going stream. ◀
Temp ⚓ in Sandaig Bay, 3M to SW.

The Sound of Sleat widens to 4M off Point of Sleat and is exposed to SW winds unless Eigg and Muck give a lee. Mallaig is a busy fishing and ferry harbour, convenient for supplies. Further S the lochs require intricate pilotage. 6M NE of Ardnamurchan Pt (lt, fog sig) are Bo Faskadale rks, drying 0·5m and marked by SHM lt buoy, and Elizabeth Rock with depth of 0·7m.

▶ Ardnamurchan Pt is an exposed headland onto which the ebb sets. With onshore winds, very heavy seas extend 2M offshore and it should be given a wide berth. Here the N-going stream begins at HW Oban – 0525, and the S-going at HW Oban + 0100, sp rates 1·5kn. ◀

ARDNAMURCHAN TO CRINAN

(AC 2171, 2169, 5616) S of Ardnamurchan the route lies either W of Mull via Passage of Tiree (where headlands need to be treated with respect in bad weather); or via the more sheltered Sound of Mull and Firth of Lorne. The former permits a visit to Coll and Tiree, where best ⚓s are at Arinagour (🅰s) and Gott Bay respectively. Beware Cairns of Coll, off the N tip.

The W coast of Mull is rewarding in settled weather, but careful pilotage is needed. Beware tide rip off Caliach Pt (NW corner) and Torran Rks off SW end of Mull (large scale AC 2617 required). Apart from the attractions of Iona and of Staffa (Fingal's Cave), the remote Treshnish Is are worth visiting. The best ⚓s in this area are at Ulva, Gometra, Bull Hole and Tinker's Hole in Iona Sound. The usual passage through Iona Sound avoids overfalls W of Iona, but heed shoal patches. Loch Lathaich on the N side of Ross of Mull is 5M to the E; a good base with ⚓ at Bunessan.

The Sound of Mull gives access to Tobermory, Dunstaffnage Bay, Oban, and up Loch Linnhe through Corran Narrows (where tide runs strongly) to Fort William and to Corpach for the Caledonian Canal. Apart from these places, there are dozens of ⚓s in the sheltered lochs inside Mull, as for example in Loch Sunart with 🅰s at Kilchoan.

FIRTH OF LORNE AND SOUND OF LUING

▶ **Firth of Lorne:** the N-going stream begins at HW Oban +0430 (HW Dover –0100). The W-going stream in the Sound of Mull begins at HW Oban –0550 (HW Dover +0105). The ingoing tides at Lochs Feochan, Etive and Creran begin at HW Oban –0350, +0430 and +0600. The S-going stream in the Firth of Lorne begins at HW Oban –0155 (HW Dover +0500). The E-going stream in the Sound of Mull begins at HW Oban –0025 (HW Dover +0555).◀

On the mainland shore Puilladobhrain is a sheltered ⚓. Cuan Sound is a useful short cut to Loch Melfort, and Craobh Marina. Good shelter, draught permitting, in Ardinamar B, SW of Torsa.

Sound of Luing (AC 2326) between Fladda (lt), Lunga and Scarba on the W side, and Luing and Dubh Sgeir (lt) on the E side, is the normal chan to or from Sound of Jura, despite dangers at the N end and strong tidal streams.

▶ The N- and W-going flood begins at HW Oban + 0430; the S- and E-going ebb at HW Oban –0155. Sp rates are 2·5kn at S end of Sound, increasing to 6kn or more in Islands off N entrance, where there are eddies, races and overfalls. The N- or W-going stream begins in the following sequence: ◀

Dorus Mor: HW Oban +0330. Sp: 8kn.
Corryvreckan: HW Oban +0410. Sp: 8.5kn.
Cuan Sound: HW Oban +0420. Sp: 6kn.
Sound of Jura: HW Oban +0400. Sp: 4kn.
Sound of Luing: HW Oban +0430. Sp: 7kn.

The S- or E-going stream begins as follows:
Dorus Mor: HW Oban –0215. Sp: 8kn.
Corryvreckan: HW Oban –0210. Sp: 8.5kn.
Cuan Sound: HW Oban –0200. Sp: 6kn.
Sound of Jura: HW Oban –0205. Sp: 4kn.
Sound of Luing: HW Oban –0155. Sp: 7kn.

▶ From the N, beware very strong streams, eddies and whirlpools in Dorus Mór, off Craignish Pt. Streams begin to set W and N away from Dorus Mór at HW Oban +0345, and E and S towards Dorus Mór at HW Oban –0215, sp rates 7kn.◀

At N end of Sound of Jura (AC 2326) is Loch Craignish. For Gulf of Corryvrecken, Colonsay, Islay, Loch Crinan and passage south through the Sound of Jura, see 9.9.5.

9.8.12 MALLAIG

Highland **57°00´·47N 05°49´·47W** 🌊🌊🌊⚓⚓⚓☆☆

CHARTS AC 2208, 2541, 5616; Imray C65, C66

TIDES –0515 Dover; ML 2·9; Duration 0605

Standard Port OBAN (→)

Times				Height (metres)			
High Water		Low Water		MHWS	MHWN	MLWN	MLWS
0000	0600	0100	0700	4·0	2·9	1·8	0·7
1200	1800	1300	1900				
Differences MALLAIG							
+0017	+0017	+0030	+0024	+1·0	+0·7	+0·3	+0·1
INVERIE BAY (Loch Nevis)							
+0030	+0020	+0035	+0020	+1·0	+0·9	+0·2	0·0
BAY OF LAIG (Eigg)							
+0015	+0030	+0040	+0005	+0·7	+0·6	–0·2	–0·2
LOCH MOIDART							
+0015	+0015	+0040	+0020	+0·8	+0·6	–0·2	–0·2
LOCH EATHARNA (Coll)							
+0025	+0010	+0015	+0025	+0·4	+0·3	ND	ND
GOTT BAY (Tiree)							
0000	+0010	+0005	+0010	0·0	+0·1	0·0	0·0

SHELTER Good in SW'lies but open to N. Access H24. Small marina or poss berth on Fish Pier. 12 ⚓s or ⚓ in SE part of hbr clear of moorings. Despite downturn in fishing, Mallaig is a reasonably busy FV harbour, also Skye and Small Isles ferries.

NAVIGATION WPT 57°00´·75N 05°49´·40W, 191°/0·3M to Steamer Pier lt, passing E of Sgeir Dhearg lt bn. The former W chan is permanently closed to navigation.

LIGHTS AND MARKS As chartlet & 9.8.4. Town lts may obscure Sgeir Dhearg lt. Outer Hbr Pier Hds floodlit. IPTS (3 FR vert at pier hd) when ferries manoeuvring, no other traffic allowed except with HM's permission.

COMMUNICATIONS (Code 01687) CGOC (01851) 702013; ⊖ (0141) 887 9369; Police 101. HM 462154, outside office hrs 462411; Dr 462202.
Mallaig Hbr Radio VHF Ch 09 16 (HO).

FACILITIES ⌇ in marina, £2.40/m/night, short stay <4hrs £12. ⫶ £3.60 /night. If marina full berth on Fish Pier, £6 for ⚓/fuel/stores. M £15/night ⚓ free but limited space; seek HM permission. ⚓ ◣ ⛵ ⚓ WC/🚿 📦 ⚒ ⚓ ☎ Ⓔ ⚓(mob 10t) ⚓ ACA, wi-fi.
Town 🏧 🛒 ✕ 🏪 ✉ Ⓑ ⇌.

THE SMALL ISLES

(AC 2207, 2208, 5616) These consist of Canna, Rum, Eigg and Muck. The hbrs at Eigg (SE end), Rhum (Loch Scresort) and Canna (between Canna and Sanday) are all exposed to E'lies; Canna has best shelter and is useful for the Outer Hebrides. Dangers extend SSW from Canna: at 1M Jemina Rk (depth 1·5m) and Belle Rk (depth 3·6m); at 2M Humla Rk (5m high), marked by buoy and with offlying shoals close W of it; at 5M Hyskeir (lt, fog sig), the largest of a group of small islands; and at 7M Mill Rks (with depths of 1·8m).

▶ *The tide runs hard here, and in bad weather the sea breaks heavily up to 15M SW of Canna. Between Skerryvore and Neist Pt the stream runs generally N and S, starting N-going at HW Ullapool + 0550, and S-going at HW Ullapool –0010. It rarely exceeds 1kn, except near Skerryvore, around headlands of The Small Isles, and over rks and shoals.* ◀

1M off the N side of Muck are Godag Rks, some above water but with submerged dangers extending 2ca further N. Most other dangers around the Small Isles are closer inshore, but there are banks on which the sea breaks heavily in bad weather. A local magnetic anomaly exists about 2M E of Muck.

ANCHORAGES IN THE SMALL ISLES

CANNA The Small Isles, **57°03´·3N 06°29´·4W**. AC 1796, 2208. HW –0457 (Sp), –0550 (Np) on Dover; HW –0035 and –0·4m on Ullapool; Duration 0605. Good shelter, except in strong E'lies, in hbr between Canna and Sanday Is. 10⚓s (£10) around edge of the bay and middle is clear to ⚓ but holding is poor due to kelp. Appr along Sanday shore, keeping N of Sgeir a' Phuirt, dries 4·6m. Ldg marks as in SDs. ⚓ in 3–4m W of Canna pier, off which beware drying rky patch. ⚓ Lt advised due to FVs. Conspic bn, Fl W10s 32m 9M, vis 152°-061°, at E end of Sanday Is. Magnetic anomaly off NE Canna. Facilities: ⚓ ✕ (☎01687 460164), shower at farm (01687 462963) with limited ✉ 🛒 Note: NTS manage island.

RUM The Small Isles, **57°00´·1N 06°15´·7W**. AC 2207, 2208. HW –0500 on Dover; –0035 and –0·3m on Ullapool; ML 2·4m; Duration 0600. SNH owns Is. The mountains (809m) are unmistakable. Landing is only allowed at L Scresort on E side; no dogs beyond village limit. Beware drying rks 1ca off N point of ent and almost 3ca off S point. The head of the loch dries 2ca out; ⚓ off slip on S side or further in, to NE of jetty. Hbr is open to E winds/swell. Facilities: ⚓ WC ⚓ (eco) by landing pier, ✉ 🛒 (limited) ✕ 🏪 ferry to Mallaig.

EIGG HARBOUR The Small Isles, **56°52´·6N 06°07´·6W**. AC 2207. HW –0523 on Dover. See 9.8.12. Entering from N or E use Sgeir nam Bagh (Ferry Terminal) Dir lt 245° Fl WRG 3s, leads between Garbh Sgeir Bn Fl (2) G 10s & Flod Sgeir Bn Fl R 5s. An Sgùrr is a conspic 391m high peak/ridge brg 281°/1·3M from Sgeir nam Bagh. Most of hbr dries, but good ⚓ 1ca NE of Galmisdale Pt pier, except in NE winds when yachts should go through the narrows and ⚓ in South Bay in 6–8m; tide runs hard in the narrows. Also ⚓ in 2·5m at Poll nam Partan, about 2ca N of Flod Sgeir. SE point of Eilean Chathastail Fl 6s 24m 8M, vis 181°-shore, W tr. VHF Ch 08 *Eigg Hbr*. HM ☎ via (01687) 482428. ⚓ repairs. **Pierhead** 📍 ◣ ✉ 🛒 ✕. **Cleadale village** (2M to N) 🏪.

MUCK The Small Isles, **56°49´·8N 06°13´·3W**. AC 2207. Tides approx as Eigg, 9.8.12. Port Mór at SE end is the main hbr, with a deep pool inside offlying rks, but open to S'lies. Approach: Dir lt Fl WRG 3s 7m 11M (vis 1M by day) leads between Dubh Sgeir (Fl(2) R 10s) and Bogha Ruadh (Fl(2) G 10s) rks; see 9.8.4. ⚓ towards the NW side of inlet; NE side has drying rks. To N of Is, Bagh a' Ghallanaich is ⚓ protected from S. Few facilities.

9.8.13 LOCH SUNART

Highland 56°39'·49N 06°00'·07W ✿✿✿◊◊✿✿

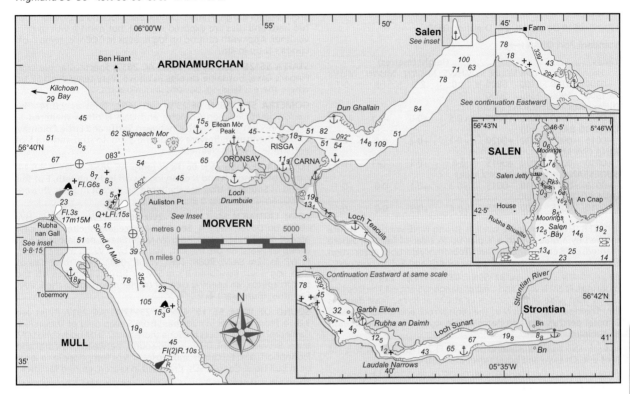

CHARTS AC 5611, 2171, 2392, 2394; Imray C65, 2800

TIDES Salen –0500 Dover; ML 2·0

Standard Port OBAN (→)

Times				Height (metres)			
High Water		Low Water		MHWS	MHWN	MLWN	MLWS
0100	0700	0100	0800	4·0	2·9	1·8	0·7
1300	1900	1300	2000				
Differences SALEN (Loch Sunart)							
–0015	+0015	+0010	+0005	+0·6	+0·5	–0·1	–0·1

SHELTER ⌀s at Kilchoan Bay (2M N of Ardmore Pt). ⌀s in Loch Drumbuie (S of Oronsay) sheltered in all winds; in Sailean Mór (N of Oronsay) convenient and easy ent; between Oronsay and Carna; in Loch Teacuis (very tricky ent); E of Carna; Salen Bay, with ⌀s and pontoon, open only to SSE (see facilities); Garbh Eilean (NW of Rubha an Daimh), and E of sand spit by Strontian R.

NAVIGATION West WPT, 56°39'·69N 06°03'·07W, 083°/3·7M to Creag nan Sgarbh (NW tip of Orinsay). South WPT, 56°38'·00N 06°00'·65W, 1M S of Auliston Pt; there are extensive rky reefs, The Stirks W of this pt. AC 2394 and detailed directions are needed to navigate the 17M long loch, particularly in its upper reaches. Beware Ross Rk, S of Risga; Broad Rk, E of Risga; Dun Ghallain Rk; shoals extending 3ca NNW from Eilean mo Shlinneag off S shore; drying rock 1ca W of Garbh Eilean and strong streams at sp in Laudale Narrows. Fish farms on both sides of the loch.

LIGHTS AND MARKS Unlit. Transits as on the chart: from W WPT, Risga on with N tip of Oronsay at 083°; N pt of Carna on with top of Risga 092°. From S WPT, Ben Hiant bearing 354°, thence Eilean Mor Peak at 052°. Further up the loch, 339° and 294°, as on chartlet, are useful. Other transits are shown on AC 2394.

COMMUNICATIONS (Code 01967) CGOC (01851) 702013; ⊖ (0141) 887 9369; Dr 431231. No VHF, except at Salen Bay (below).

FACILITIES

Kilchoan Bay (56°41'·5N 06°07'·3W); ☎(01972) 510305; 4 ⌀s £13/night, 🚻 ⌀ 🏠 🏠 🍺 hotel (1M).

Salen Bay (56°42'·39N 05°46'·17W); www.salenjetty.co.uk ☎(01967) 431510, mob 07909944494. VHF Ch 80 (occas). Keep on E side of entrance to avoid a drying reef on W side. Pontoon 8 ◊, O/night £10< 6m <£2·40 /m, £5 <4hrs; 4 R ⌀s(15t) (other moorings private); buoy advised, as bottom is generally foul. ⚓ ⚓ 🛢 ⌁ 🍺 ✕ 🗐.

Acharacle (2½M), ⊠ 🍺 ⇌ (bus to L. Ailort/Ft William), ✈ Oban.

Strontian ⚓ 🏠 ⊠ 🍺 🛢 🗐 hotel, bus to Ft William.

ANCHORAGES IN COLL AND TIREE (Argyll and Bute)

ARINAGOUR, Loch Eatharna, Coll, 56°37'·0N 06°31'·2W. AC 5611, 2171, 2474. HW –0530 on Dover; ML 1·4m; Duration 0600; see 9.8.12. Good shelter except with SE swell or strong winds from ENE to SSW. Enter at SHM buoy, Fl G 6s, marking Bogha Mòr. Thence NW to Arinagour ferry pier, 2 FR(vert) 10m and Dir Oc WRG 7s. Beware McQuarrie's Rk (dries 2·9m) 1ca E of pier hd and unmarked drying rks further N on E side of fairway. ✠ and hotel are conspic ahead. Continue N towards old stone pier; ⚓ S of it or use 12⌀s S of Dgeir Dubh for boats <10T at £10/night, pay at ferry terminal. Also ⚓ E of Eilean Eatharna. Piermaster ☎(01879) 230347; VHF Ch 31. Facilities: HIE Trading Post ☎230349, ⚓ ⚓ 🛢 ⌁ 🍺 ✕. **Village** 🗐 ⚓ 🏠 ⊠ Ferry to Oban (⇌).

GOTT BAY, Tiree, 56°30'·74N 06°48'·07W. AC 5611, 2474. Tides as 9.8.12; HW –0540 on Dover. HM ☎(01879) 230337, VHF Ch 31. Adequate shelter in calm weather, but exposed in winds ENE to S. (in which case use Wilson hbr in Balephetrish Bay IM W, Ldg lts F.G/Oc.G). The bay, at NE end of island, can be identified by conspic latticed tr at Scarinish about 8ca SW of ent, with lt Fl 3s 11m 16M close by (obscd over hbr). Appr on NW track, keeping to SW side of bay which is obstructed on NE side by Soa Is and drying rks. The ferry pier at S side of ent has FR ldg lts 286½°. ⚓ 1ca NW of pier head in 3m on sand; ⚓ at pier. Facilities: 🏠 🏠 🛢 🍺 ✕ 🗐 ⊠ at Scarinish (½M), Ferry to Oban (⇌).

9.8.14 TOBERMORY

Mull (Argyll and Bute) 56°37'·19N 06°03'·87W ✱✱✱✱🚢🚢🚢🏵🏵🏵

CHARTS AC 5611, 2171, 2390, 2474; Imray C65, 2800

TIDES −0519 Dover; ML 2·4; Duration 0610

Standard Port OBAN (→)

Times				Height (metres)			
High Water		Low Water		MHWS	MHWN	MLWN	MLWS
0100	0700	0100	0800	4·0	2·9	1·8	0·7
1300	1900	1300	2000				
Differences TOBERMORY (Mull)							
+0025	+0010	+0015	+0025	+0·5	+0·6	+0·1	+0·2
CARSAIG BAY (S Mull)							
−0015	−0005	−0030	+0020	+0·1	+0·2	0·0	−0·1
IONA (SW Mull)							
−0010	−0010	−0020	+0015	0·0	+0·1	−0·3	−0·2
BUNESSAN (Loch Lathaich, SW Mull)							
−0015	−0015	−0010	−0015	+0·3	+0·1	0·0	−0·1
ULVA SOUND (W Mull)							
−0010	−0015	0000	−0005	+0·4	+0·3	0·0	−0·1

SHELTER Good, but some swell in strong N/NE winds. 26 🛟s are marked by a blue ⬤ with a Y pick-up. ⚓ clear of fairway, where marked; at SE end of The Doirlinn. Tobermory Harbour Association (THA) manages local and visitors moorings and ⬤ on pontoon.

NAVIGATION WPT 56°37'·59N 06°03'·17W, 224°/0·5M to moorings. N ent is wide and clear of dangers. S ent via The Doirlinn is only 80m wide at HW, and dries at LW; at HW±2 least depth is 2m. The clear channel is N of the central rock and is marked with 2 metal perches. No ⚓ in harbour fairway (buoyed). Keep wash to a minimum.

LIGHTS AND MARKS Rhubha nan Gall, Fl 3s 17m 15M, W tr is 1M N of ent. Ch spire and hotel turret are both conspic.

COMMUNICATIONS (Code 01688) CGOC (01851) 702013; Local CG 302200; Police 101; Dr 302013. Cal-Mac Piermaster 302017; THA Moorings Officer ☎302876, mob 07917 832497.

VHF Ch 16 12 (HO), M.

FACILITIES 25🛟s £15/night ⚓<25'; 50⬤ on pontoon £2.25 inc ⚡; 🚿⚓; Taigh Solais (THA Harbour Building) 🚻📶📷 wi-fi.
Western Isles YC ☎302371. Town 🏨🛒⚓⛽🏧 ACA ✉ⓑ🍴✗ ⬤ Divers, ⛴ (ferry to Oban), ✈ (seaplane to Glasgow; helipad on golf course).

SOUND OF MULL Dearg Sgeir Fl 6s 7m 8M marks Eileann Glasa, 2·3M to SE Rubha na Leitrach, 2 F R (vert), marks N ent to Fishnish Bay.

SALEN BAY (Mull, 56°31'·44N 05°56'·82W). Beware Bogha, drying rks 6ca E of the bay; 🛟s £10; ⚓ at Salen Jetty, fee £1/person. Facilities: ⚓🚿. Village: Dr, 🛒ⓔ⛽✉🍴✗.

CRAIGNURE, (Mull, 56°28'·37N 05°42'·25W) ldg lts 241°, both FR 10/12m. Facilities: ⚓✉🍴⬤ Ferry to Oban. Tides, see 9.8.15.

ANCHORAGES on WEST and SOUTH COASTS OF MULL
(Anti-clockwise from the north. SDs essential) (See AC 5611)

TRESHNISH ISLES 56°29'N 06°31'W. AC 2652. The main Is (N to S) are: Cairn na Burgh, Fladda, Lunga, Bac Mòr and Bac Beag. Tides run hard and isles are exposed to swell, but merit a visit in calm weather. Appr with caution on ldg lines as in CCC SDs; temp ⚓ off Lunga's N tip in 4m.

STAFFA 56°25'·96N 06°20'·34W. AC 2652. Spectacular isle with Fingal's Cave, but same caveats as above. Very temp ⚓ off SE tip where there is landing; beware unmarked rks.

GOMETRA 56°28'·85N 06°23'W. AC 2652. Tides as 9.8.14. The narrow inlet between Gometra and Ulva Is offers sheltered ⚓, except in S'lies. Appr on 020° between Staffa and Little Colonsay, or N of the former. Beware rks drying 3·2m, to stbd and 5ca S of ent. Inside, E side is cleaner.

LOCH NA KEAL 56°26'N 06°18'W (ent). AC 2652. Tides in 9.8.14. Appr S of Geasgill Is and N of drying rks off Inch Kenneth; E of this Is and in Sound of Ulva are sheltered ⚓s, except in S'lies. Beware MacQuarrie's Rk, dries 0·8m.

LOCH LATHAICH Mull, 56°19'·29N 06°15'·47W. AC 2617. HW −0545 on Dover; ML 2·4. See 9.8.14. Excellent shelter with easy access; good base for cruising W Mull. Eilean na Liathanaich (a group of islets) lie off the ent, marked by a W bn at the E end, Fl WR 6s 12m 8/6M, R088°-108°, W108°-088°. Keep to W side of loch and ⚓ off Bendoran BY in SW, or SE of Eilean Ban off the pier in approx 5m keeping clear of creeler moorings to the E side. Facilities: (Bunessan) ⚓✉🛒✗⬤.

SOUND OF IONA 56°19'·45N 06°23'·12W. AC 2617. Tides see 9.8.14. From N, enter in mid-chan; from S keep clear of Torran Rks. Cathedral brg 012° closes the Iona shore past 2 SHM buoys and SCM buoy. Beware a bank 0·1m in mid-sound, between cathedral and Fionnphort; also tel cables and ferries. ⚓ S of ferry close in to Iona, or in Bull Hole. Consult SDs. Crowded in season; limited facilities.

TINKER'S HOLE Ross of Mull, 56°17'·49N 06°23'·07W. AC 2617. Beware Torran Rks, reefs extending 5M S and SW of Erraid. Usual app from S, avoiding Rankin's Rks, drying 0·8m, and rk, dries 2·3m, between Eilean nam Muc and Erraid. Popular ⚓ in mid-pool between Eilean Dubh and Erraid.

CARSAIG BAY Ross of Mull, 56°19'·19N 05°59'·07W. Tides see 9.8.14. AC 2386. Temp, fair weather ⚓s to N of Gamhnach Mhòr, reef 2m high, or close into NW corner of bay. Landing at stone quay on NE side. No facilities.

LOCH SPELVE Mull, 56°23'N 05°41'W. AC 2387. Tides as Oban. Landlocked water, prone to squalls off surrounding hills. Ent narrows to ½ca due to shoal S side and drying rk N side marked by G pole, Fl G 5s 3m 2M, 56°23'·24N 05°42'·04W. CCC SDs give local ldg lines. ⚓s in SW and NW arms, clear of fish farms. Pier at Croggan; no facilities.

ANCHORAGES ALONG LOCH LINNHE
(AC 5611, 2378, 2379, 2380)

LYNN OF LORN 56°33'N 05°25'·2W. At NE end are ⚓s off Port Appin, clear of ferry and cables (beware Appin Rks); and in Airds Bay, open to SW. At NW tip of Lismore, Port Ramsey offers good ⚓s between the 3 main islets.
Linnhe Marina ☎(01631) 730401 offers pontoon and moorings.

LOCH CRERAN 56°32'·14N 05°25'·22W. Tides 9.8.16. Enter at Airds Pt, Dir lt 050°, Fl WRG 2s, W vis 041-058°. Chan turns 90° stbd with streams of 4kn. Sgeir Callich, rky ridge extends NE to SHM By, Fl G 3s; ⚓ W of it. 🛟s (max LOA 7m) off Barcaldine; also 3🛟s (max LOA 9m) off Creagan Inn, ☎(01631) 573250. Bridge has 12m clearance. Beware of extensive marine farms in loch.

LOCH LEVEN 56°42'N 05°12'W. Tides 9.8.16. Ballachulish Bay ⚓s: N shore Onich and off St Brides, S shore Kentallen Bay (deep). Approach Ballachullish Bridge (16m clnce) on 114°. Moorings are 4ca to ENE with sheltered ⚓ at Poll an Dùnan, ent W of perch. Loch is navigable 7M to Kinlochleven.
Ballachulish: Hotels, ✉ⓑ🛒✗.

CORRAN NARROWS 56°43'·27N 05°14'·34W. AC 2372. Sp rate 6kn. Well buoyed; Corran Pt lt ho, Iso WRG 4s, and lt bn 5ca NE. ⚓ 5ca NW of Pt, off Camas Aiseig pier/slip.

9.8.15 LOCH ALINE

Highland **56°32'·09N 05°46'·47W** ✿✿✿✿✿✿✿✿✿✿

CHARTS AC 5611, 2390; Imray C65, 2800

TIDES –0523 Dover; Duration 0610

Standard Port OBAN (→)

Times				Height (metres)			
High Water		Low Water		MHWS	MHWN	MLWN	MLWS
0100	0700	0100	0800	4·0	2·9	1·8	0·7
1300	1900	1300	2000				
Differences LOCH ALINE							
+0012	+0012	ND	ND	+0·5	+0·3	ND	ND
SALEN (Sound of Mull)							
+0045	+0015	+0020	+0030	+0·2	+0·2	–0·1	0·0
CRAIGNURE (Sound of Mull)							
+0030	+0005	+0010	+0015	0·0	+0·1	–0·1	–0·1

SHELTER Very good. ‡s in SE end of loch and in N and E part of loch. Temp berth on old stone slip on W side of ent, depth and ferries permitting. Walk ashore pontoons N of sand mine pier.

NAVIGATION WPT 56°31'·49N 05°46'·37W, 176°/0·9M toldg bn. Bn (bright orange) leads 357°, 100m W of Bogha Lurcain, drying rk off Bolorkle Pt on E side of ent. Buoyed ent is easy, but narrow with a bar (min depth 2·1m); stream runs 2½kn at springs. Beware coasters (which completely fill the ent) from the sand mine going to/from the jetty and ferries to/from Mull. Last 5ca of loch dries.

LIGHTS AND MARKS Ardtornish Pt lt ho, 1M SSE of ent, Fl (2) WRG 10s 7m 8/5M. Lts and buoys as chartlet. War memorial, 9m high, stands on W side of ent. Dir Ldg lt, Oc WRG 6s, 357°, co-located with front ldg bn (Or). 1M up the loch on E side a Y bn with Y ● topmark marks a reef, and ½M further up similar bn marks a larger reef on W side. Clock tr is very conspic at head of loch.

COMMUNICATIONS (Code 01967) CGOC (01851) 702013; ⊖ (0141) 887 9369; Ⓗ (01631) 563727; Dr 421252. No VHF.

FACILITIES Pontoons, www.lochalineharbour.co.uk ☎ 07583 800500, 26⌑ inc Ⓥ £2·10, ⓌⒸ 🏁 Ⓞ. Village: ⚓ at pier, 🔧 🏪 🍴 🖂 ✕ 🏧 (Ⓑ ⇌ ✈ at Oban), Ferry to Fishnish Bay (Mull).

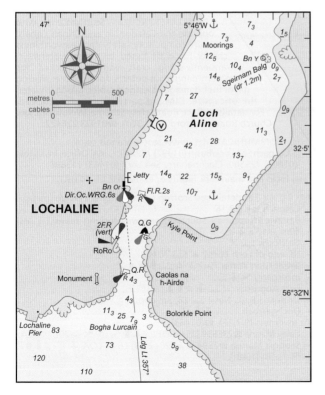

9.8.16 FORT WILLIAM/CORPACH

Highland **56°48'·99N 05°07'·07W** (off Fort William)

CHARTS AC 5611, 2372, 2380, 5617; Imray C65, C63, 2800

TIDES –0535 Dover; ML 2·3; Duration 0610

Standard Port OBAN (→)

Times				Height (metres)			
High Water		Low Water		MHWS	MHWN	MLWN	MLWS
0100	0700	0100	0800	4·0	2·9	1·8	0·7
1300	1900	1300	2000				
Differences CORPACH							
0000	+0020	+0040	0000	0·0	0·0	–0·2	–0·2
LOCH EIL (Head)							
+0025	+0045	+0105	+0025	ND	ND	ND	ND
CORRAN NARROWS							
+0007	+0007	+0004	+0004	+0·4	+0·4	–0·1	0·0
LOCH LEVEN (Head)							
+0045	+0045	+0045	+0045	ND	ND	ND	ND
LOCH LINNHE (Port Appin)							
–0005	–0005	–0030	0000	+0·2	+0·2	+0·1	+0·1
LOCH CRERAN (Barcaldine Pier)							
+0010	+0020	+0040	+0015	+0·1	+0·1	0·0	+0·1
LOCH CRERAN (Head)							
+0015	+0025	+0120	+0020	–0·3	–0·3	–0·4	–0·3

SHELTER Exposed to winds SW thro' N to NE. ‡s off Fort William pier; in Camus na Gall; SSW of Eilean A Bhealaidh; and off Corpach Basin, where there is also a waiting pontoon. The sea lock is normally available HW±4 during canal hrs. For Caledonian Canal see 9.8.17.

NAVIGATION Corpach WPT 56°50'·29N 05°07'·07W, 315°/0·32M to lock ent.

> Beware McLean Rk, dries 0·3m, buoyed, 8ca N of Fort William. Lochy Flats dry 3ca off the E bank. In Annat Narrows at ent to Loch Eil streams reach 5kn.

LIGHTS AND MARKS Iso WRG 4s lt is at N jetty of sea-lock ent, W310°-335°. A long pier/viaduct off Ft William is unlit.

COMMUNICATIONS (Code 01397) CGOC (01851) 702013; ⊖ 702948; Police 101; Dr 703136. HM 772249.

Corpach Lock VHF Ch **74** 16 (during canal hours).

FACILITIES **Fort William: Pier** ☎703881 ⚓ 🔧 ⛴. **Town:** 🏪 ✕ 🔨 🏧 🛢 ACA 🖂 Ⓑ 🍴 ✕ 🏧 YC Ⓗ ⇌. **Corpach: Corpach Basin** ☎772249 🔧 ⛴ ⚓ 🔧. **Lochaber YC** ☎703576 ⚓ 🔧 ⚓. **Services** ✕ 🔧 🏧 🔨 Divers. **Village:** 🏪 🏪 🖂 Ⓑ 🍴 ✕ 🏧 ⇌.

9.8.17 CALEDONIAN CANAL
Highland ✦✦✦✦✦ ✦✦✦

CHARTS AC 1791, 5617; Imray C63; *Scottish Canals Skippers Guide*

TIDES Tidal differences: Corpach –0455 on Dover; See 9.8.16. Clachnaharry: +0116 on Dover; See 9.7.16.

SHELTER Corpach is at the SW ent of the Caledonian Canal, see 9.8.16; the sea locks at both ends do not open LW±2 at springs. For best shelter transit the sea lock and lie above the double lock. Numerous pontoons along the canal; cost is included in the canal dues. For Inverness see 9.7.16.

NAVIGATION The 60M canal consists of 38M through 3 lochs, (Lochs Lochy, Oich and Ness), connected by 22M of canal.

Loch Oich is part of a hydro-electric scheme which may vary the water level. The passage normally takes two full days, possibly longer in the summer; absolute minimum is 14 hrs. Speed limit is 5kn in the canal sections. There are 10 swing bridges; road traffic has priority at peak hrs. Do not pass bridges without the keeper's instructions.

LOCKS There are 29 locks: 14 between Loch Linnhe (Corpach), via Lochs Lochy and Oich up to the summit (106ft above sea level); and 15 locks from the summit down via Loch Ness to Inverness.

2012 information

Hours: Winter Mon-Fri 0900-1600
         Spring Mon-Sun 0830-1730
         Summer Mon-Sun 0800-1730
         Autumn Mon-Sun 0830-1730

Dues payable at Corpach: transit/lock fee outward £16/m for 8 day passage or less; return £12/m. £20 per metre is charged for a 2 week sojourn. For regulations and useful booklet *Scottish Canals Skipper's Guide* apply: Canal Manager, Canal Office, Seaport Marina, Muirtown Wharf, Inverness IV3 5LE, ☎(01463) 233140, or download from: www. waterscape.com

LIGHTS AND MARKS See 9.8.16 and 9.7.16 for ent lts. Channel is marked by posts, cairns and unlit buoys, PHM on the NW side of the chan and SHM on the SE side.

BOAT SAFETY SCHEME Scottish Canals, who operate the Caledonian and Crinan Canals, require compulsory safety and seaworthiness checks for craft based on these waterways. Transient/visiting craft will be checked for apparent dangerous defects, for example: leaking gas or fuel, damaged electrical cables, taking in water, risk of capsize. £1M 3rd party insurance is required. For details contact: Boat Safety Scheme, Willow Grange, Church Road, Watford WD1 3QA. ☎01923 226422.

COMMUNICATIONS www.scottishcanals.co.uk. Corpach Sea Lock/Basin (01397) 772249; Canal Office, Inverness (01463) 725500; Clachnaharry Sea Lock (01463) 713896.

Sea locks and main lock flights operate VHF Ch **74** (HO).

FACILITIES For details see *Scottish Canals Skipper's Guide* (maps).
– Corpach see 9.8.16.
– Banavie (Neptune's Staircase; one-way locking takes 1½ hrs) 60m jetty ⬡ ⚓ ◫ ⛽ 🛒 ⊠.
– Gairlochy ⌂ ✕.
– NE end of Loch Lochy ⚓ ⌂ 🛒 ✕.
– Great Glen Water Park ⬡ ⚓ ⌂ 🛒 ⬛ ✕.
– Invergarry ⚓ ⚓ ⌂.
– Fort Augustus ⬡ ⌂ ⚓ ⛽ 🔥 🔥 ⚲ 🔌 ⊠ 🛒 ⊡.
– Urquhart B. (L. Ness) ⌂ £6, 3m depth, ⛽ ⚓ 🛒 ⊡.
– Dochgarroch ⬡ ⛽ ⬛ 🛒.
At Inverness (9.7.16), ☎(01463):
– Caley Marina ☎236539 (25+25Ⓥ) ⛽ ⚓ ⬛ ⚲ 🔌 ✕ ⊠(20t) ⚓ ACA.
– Seaport Marina ☎239475 ⬡ (20+20Ⓥ) £6 all LOA, ⚓ ⛽ ◫ ⬛ ⬛ ⊡ ✕ ⚲ 🔌 ⊠(40t).

STANDARD TIME (UT)
For Summer Time add ONE hour in **non-shaded areas**

OBAN LAT 56°25'N LONG 5°29'W

TIMES AND HEIGHTS OF HIGH AND LOW WATERS

Dates in red are **SPRINGS**
Dates in blue are **NEAPS**

YEAR 2016

NW Scotland

JANUARY

Time	m	Time	m
1 0401	1.5	**16** 0358	1.1
0943	3.4	1000	3.6
F 1647	1.9	SA 1636	1.3
2215	3.2	◑ 2214	3.2
2 0449	1.7	**17** 0459	1.3
1032	3.2	1107	3.3
SA 1743	2.0	SU 1741	1.5
◑ 2317	3.0	2324	3.1
3 0543	1.8	**18** 0610	1.4
1139	3.1	1246	3.2
SU 1848	2.1	M 1852	1.5
4 0051	3.0	**19** 0100	3.0
0641	1.9	0729	1.5
M 1313	3.1	TU 1419	3.3
1959	2.0	2004	1.5
5 0214	3.1	**20** 0232	3.2
0744	1.9	0851	1.4
TU 1431	3.2	W 1527	3.4
2104	1.9	2112	1.4
6 0309	3.2	**21** 0336	3.4
0849	1.8	1001	1.3
W 1525	3.4	TH 1619	3.6
2152	1.7	2209	1.2
7 0355	3.4	**22** 0422	3.6
0947	1.6	1056	1.1
TH 1610	3.5	F 1700	3.7
2231	1.4	2257	1.0
8 0436	3.6	**23** 0503	3.8
1036	1.4	1141	1.0
F 1651	3.7	SA 1737	3.8
2307	1.2	2340	1.0
9 0515	3.8	**24** 0541	4.0
1120	1.2	1223	1.0
SA 1729	3.8	SU 1811	3.9
2343	1.0	○	
10 0552	4.0	**25** 0020	0.8
1202	1.0	0617	4.1
SU 1804	3.9	M 1302	1.0
●		1844	3.9
11 0019	0.9	**26** 0059	0.8
0628	4.0	0653	4.1
M 1243	0.9	TU 1339	1.1
1837	3.9	1917	3.9
12 0058	0.8	**27** 0136	0.8
0703	4.1	0727	4.0
TU 1324	0.8	W 1414	1.2
1912	3.9	1949	3.8
13 0137	0.7	**28** 0212	1.0
0741	4.0	0800	3.9
W 1406	0.9	TH 1448	1.4
1951	3.8	2021	3.7
14 0220	0.8	**29** 0248	1.1
0822	3.9	0832	3.7
TH 1451	1.0	F 1522	1.6
2033	3.6	2054	3.5
15 0306	0.9	**30** 0325	1.4
0908	3.8	0905	3.6
F 1540	1.1	SA 1600	1.7
2120	3.4	2131	3.4
		31 0405	1.6
		0941	3.4
		SU 1645	1.9
		2213	3.2

FEBRUARY

Time	m	Time	m
1 0452	1.8	**16** 0546	1.4
1024	3.2	1211	3.0
M 1742	2.0	TU 1819	1.5
◑ 2311	3.0		
2 0549	1.9	**17** 0025	2.9
1126	3.0	0706	1.5
TU 1849	2.1	W 1405	3.0
		1935	1.6
3 0101	2.9	**18** 0215	3.0
0654	2.0	0846	1.5
W 1342	3.0	TH 1527	3.1
2003	2.0	2053	1.5
4 0238	3.0	**19** 0332	3.2
0805	1.9	1002	1.4
TH 1505	3.1	F 1620	3.3
2113	1.8	2156	1.3
5 0334	3.2	**20** 0415	3.4
0917	1.7	1052	1.2
F 1555	3.3	SA 1654	3.5
2204	1.5	2244	1.1
6 0419	3.5	**21** 0451	3.7
1017	1.4	1133	1.1
SA 1637	3.5	SU 1724	3.7
2246	1.2	2325	0.9
7 0459	3.7	**22** 0526	3.8
1105	1.1	1209	1.0
SU 1715	3.7	M 1755	3.8
2324	0.9	○	
8 0538	4.0	**23** 0003	0.8
1148	0.8	0600	4.0
M 1750	3.9	TU 1243	0.9
●		1826	3.9
9 0002	0.6	**24** 0040	0.7
0614	4.1	0633	4.0
TU 1229	0.6	W 1316	1.0
1823	4.0	1856	3.9
10 0042	0.4	**25** 0114	0.7
0650	4.2	0704	4.0
W 1309	0.5	TH 1347	1.0
1858	4.0	1925	3.9
11 0123	0.4	**26** 0147	0.9
0727	4.2	0734	3.9
TH 1350	0.5	F 1417	1.2
1935	3.9	1954	3.8
12 0205	0.4	**27** 0218	1.0
0806	4.0	0803	3.8
F 1432	0.6	SA 1446	1.4
2015	3.8	2024	3.7
13 0251	0.6	**28** 0249	1.2
0849	3.8	0831	3.6
SA 1518	0.8	SU 1515	1.5
2058	3.6	2055	3.5
14 0340	0.8	**29** 0323	1.5
0936	3.5	0901	3.4
SU 1610	1.1	M 1549	1.7
2147	3.3	2132	3.3
15 0438	1.1		
1035	3.2		
M 1710	1.3		
◑ 2249	3.0		

MARCH

Time	m	Time	m
1 0403	1.7	**16** 0527	1.3
0937	3.2	1145	2.8
TU 1640	1.9	W 1750	1.4
◑ 2217	3.0	2356	2.8
2 0459	1.9	**17** 0650	1.6
1026	3.0	1306	2.8
W 1752	2.0	TH 1907	1.5
2326	2.9		
3 0613	2.0	**18** 0156	2.9
1151	2.8	0843	1.5
TH 1910	2.0	F 1522	2.9
		2031	1.5
4 0200	2.8	**19** 0327	3.1
0732	1.9	0953	1.4
F 1444	2.9	SA 1609	3.1
2028	1.8	2137	1.3
5 0311	3.1	**20** 0358	3.3
0853	1.6	1038	1.2
SA 1538	3.1	SU 1634	3.3
2132	1.5	2225	1.1
6 0358	3.4	**21** 0430	3.5
0957	1.3	1115	1.1
SU 1619	3.4	M 1701	3.5
2220	1.1	2305	0.9
7 0439	3.7	**22** 0503	3.7
1046	0.9	1147	1.0
M 1656	3.7	TU 1731	3.7
2303	0.7	2341	0.8
8 0517	4.0	**23** 0537	3.9
1129	0.6	1218	0.9
TU 1730	3.9	W 1802	3.9
2343	0.4	○	
9 0555	4.2	**24** 0015	0.7
1210	0.4	0608	4.0
W 1804	4.0	TH 1248	0.9
●		1831	3.9
10 0024	0.2	**25** 0048	0.8
0632	4.3	0639	4.0
TH 1250	0.3	F 1317	1.0
1839	4.1	1859	3.9
11 0107	0.1	**26** 0120	0.9
0709	4.2	0708	3.9
F 1330	0.3	SA 1346	1.1
1917	4.0	1927	3.8
12 0150	0.2	**27** 0150	1.0
0748	4.1	0736	3.8
SA 1412	0.4	SU 1413	1.2
1956	3.8	1956	3.7
13 0236	0.4	**28** 0218	1.2
0829	3.8	0802	3.6
SU 1457	0.6	M 1435	1.4
2038	3.6	2027	3.5
14 0325	0.7	**29** 0246	1.4
0915	3.5	0831	3.4
M 1547	0.9	TU 1502	1.6
2125	3.3	2101	3.3
15 0421	1.0	**30** 0322	1.6
1010	3.1	0906	3.2
TU 1644	1.2	W 1544	1.7
◑ 2223	3.0	2144	3.1
		31 0415	1.8
		0953	3.0
		TH 1656	1.9
		◑ 2246	2.9

APRIL

Time	m	Time	m
1 0538	1.9	**16** 0123	2.8
1108	2.8	0821	1.6
F 1823	1.9	SA 1449	2.8
		1956	1.5
2 0058	2.9	**17** 0245	3.0
0705	1.8	0928	1.5
SA 1408	2.8	SU 1535	3.0
1943	1.7	2105	1.4
3 0237	3.1	**18** 0326	3.2
0827	1.6	1012	1.3
SU 1511	3.1	M 1601	3.2
2054	1.4	2156	1.2
4 0329	3.4	**19** 0400	3.4
0931	1.2	1048	1.1
M 1553	3.3	TU 1630	3.4
2150	1.0	2237	1.0
5 0412	3.8	**20** 0435	3.6
1022	0.8	1119	1.1
TU 1630	3.6	W 1702	3.6
2237	0.7	2313	0.9
6 0453	4.0	**21** 0509	3.7
1106	0.5	1148	1.0
W 1706	3.9	TH 1734	3.8
2322	0.4	2347	0.9
7 0532	4.2	**22** 0542	3.8
1147	0.3	1217	0.9
TH 1742	4.0	F 1805	3.9
●		○	
8 0005	0.2	**23** 0020	0.9
0610	4.3	0614	3.9
F 1228	0.2	SA 1247	1.0
1819	4.1	1834	3.9
9 0050	0.1	**24** 0052	0.9
0650	4.2	0644	3.8
SA 1310	0.2	SU 1317	1.0
1858	4.0	1904	3.8
10 0135	0.2	**25** 0124	1.0
0730	4.0	0713	3.7
SU 1352	0.3	M 1344	1.1
1939	3.9	1934	3.7
11 0221	0.4	**26** 0153	1.2
0812	3.8	0741	3.6
M 1437	0.6	TU 1408	1.3
2021	3.7	2005	3.6
12 0311	0.7	**27** 0223	1.3
0857	3.4	0811	3.4
TU 1526	0.8	W 1438	1.4
2107	3.4	2041	3.4
13 0406	1.1	**28** 0301	1.5
0951	3.0	0849	3.2
W 1620	1.1	TH 1520	1.6
2203	3.1	2125	3.2
14 0510	1.4	**29** 0355	1.6
1119	2.8	0939	3.0
TH 1722	1.4	F 1621	1.7
◑ 2325	2.9	2225	3.1
15 0633	1.6	**30** 0512	1.7
1325	2.7	1051	2.8
F 1834	1.5	SA 1740	1.7
		◑ 2356	3.0

Chart Datum: 2·10 metres below Ordnance Datum (Newlyn). HAT is 4·5 metres above Chart Datum.

》》 **FREE** monthly updates. Register at 《
www.reedsnauticalalmanac.co.uk

STANDARD TIME (UT)
For Summer Time add ONE hour in **non-shaded areas**

OBAN LAT 56°25′N LONG 5°29′W
TIMES AND HEIGHTS OF HIGH AND LOW WATERS

Dates in red are **SPRINGS**
Dates in blue are **NEAPS**

YEAR 2016

MAY

Day	Time	m		Day	Time	m
1 SU	0637	1.7		**16** M	0150	2.9
	1247	2.8			0848	1.6
	1900	1.6			1444	2.9
					2016	1.5
2 M	0150	3.2		**17** TU	0244	3.1
	0755	1.5			0937	1.5
	1429	3.0			1522	3.1
	2014	1.4			2115	1.4
3 TU	0254	3.5		**18** W	0325	3.2
	0900	1.2			1015	1.4
	1520	3.3			1556	3.3
	2117	1.0			2201	1.3
4 W	0343	3.7		**19** TH	0403	3.4
	0953	0.9			1048	1.2
	1601	3.6			1631	3.5
	2211	0.7			2239	1.2
5 TH	0427	4.0		**20** F	0440	3.6
	1040	0.6			1117	1.1
	1640	3.8			1706	3.7
	2301	0.5			2315	1.1
6 F	0509	4.1		**21** SA	0516	3.7
	1124	0.4			1147	1.1
	1720	4.0			1740	3.8
	● 2348	0.3			○ 2351	1.0
7 SA	0551	4.2		**22** SU	0551	3.7
	1207	0.3			1218	1.0
	1801	4.1			1813	3.8
8 SU	0034	0.3		**23** M	0026	1.0
	0632	4.1			0624	3.7
	1250	0.3			1251	1.0
	1842	4.0			1844	3.8
9 M	0121	0.3		**24** TU	0102	1.1
	0714	4.0			0655	3.7
	1334	0.4			1321	1.1
	1924	3.9			1917	3.7
10 TU	0209	0.5		**25** W	0137	1.1
	0757	3.7			0727	3.6
	1418	0.6			1351	1.2
	2007	3.7			1951	3.6
11 W	0258	0.8		**26** TH	0212	1.2
	0843	3.4			0801	3.4
	1506	0.8			1425	1.3
	2052	3.5			2029	3.5
12 TH	0351	1.1		**27** F	0253	1.4
	0934	2.8			0842	3.3
	1557	1.1			1508	1.4
	2143	3.2			2114	3.4
13 F	0451	1.4		**28** SA	0345	1.5
	1043	2.8			0932	3.1
	1653	1.3			1601	1.4
	◐ 2247	3.0			2210	3.3
14 SA	0604	1.6		**29** SU	0451	1.5
	1237	2.7			1035	3.0
	1755	1.5			1708	1.5
					◑ 2322	3.2
15 SU	0021	2.9		**30** M	0607	1.5
	0735	1.7			1156	2.9
	1353	2.8			1823	1.5
	1905	1.5				
				31 TU	0056	3.3
					0721	1.4
					1332	3.0
					1937	1.3

JUNE

Day	Time	m		Day	Time	m
1 W	0217	3.4		**16** TH	0243	3.1
	0827	1.2			0936	1.6
	1442	3.2			1522	3.2
	2047	1.1			2111	1.5
2 TH	0316	3.6		**17** F	0330	3.3
	0925	1.0			1014	1.4
	1534	3.5			1602	3.4
	2148	0.9			2200	1.4
3 F	0406	3.8		**18** SA	0413	3.4
	1016	0.8			1048	1.3
	1620	3.7			1641	3.5
	2243	0.7			2243	1.3
4 SA	0452	4.0		**19** SU	0454	3.6
	1104	0.6			1121	1.2
	1703	3.9			1719	3.7
	2333	0.5			2325	1.2
5 SU	0536	4.0		**20** M	0533	3.7
	1149	0.5			1154	1.1
	1747	4.0			1756	3.8
	●				○	
6 M	0022	0.5		**21** TU	0005	1.1
	0619	4.0			0609	3.7
	1234	0.5			1229	1.0
	1829	4.0			1830	3.8
7 TU	0110	0.6		**22** W	0045	1.0
	0702	3.9			0643	3.7
	1317	0.5			1303	1.0
	1911	4.0			1905	3.8
8 W	0157	0.7		**23** TH	0124	1.0
	0745	3.7			0716	3.6
	1401	0.7			1337	1.0
	1953	3.8			1940	3.8
9 TH	0244	0.9		**24** F	0203	1.0
	0828	3.5			0751	3.5
	1446	0.8			1414	1.0
	2036	3.6			2019	3.7
10 F	0333	1.2		**25** SA	0245	1.1
	0912	3.2			0832	3.4
	1532	1.1			1456	1.1
	2120	3.4			2102	3.6
11 SA	0424	1.4		**26** SU	0333	1.2
	1003	3.0			0919	3.3
	1620	1.3			1545	1.2
	2209	3.2			2153	3.5
12 SU	0521	1.6		**27** M	0429	1.3
	1108	2.9			1013	3.1
	1713	1.5			1643	1.3
	◐ 2308	3.0			◑ 2255	3.3
13 M	0627	1.7		**28** TU	0536	1.4
	1239	2.8			1121	3.0
	1809	1.6			1752	1.4
14 TU	0024	2.9		**29** W	0015	3.3
	0743	1.8			0647	1.4
	1348	2.9			1247	3.0
	1910	1.6			1907	1.4
15 W	0144	3.0		**30** TH	0147	3.3
	0848	1.7			0757	1.3
	1439	3.0			1413	3.2
	2013	1.6			2022	1.3

JULY

Day	Time	m		Day	Time	m
1 F	0258	3.5		**16** SA	0301	3.1
	0901	1.1			0941	1.6
	1517	3.4			1538	3.2
	2131	1.1			2122	1.6
2 SA	0355	3.6		**17** SU	0352	3.3
	0957	1.0			1023	1.4
	1608	3.6			1622	3.4
	2231	0.9			2217	1.4
3 SU	0445	3.8		**18** M	0437	3.5
	1048	0.8			1100	1.2
	1654	3.8			1703	3.6
	2324	0.8			2304	1.2
4 M	0530	3.8		**19** TU	0519	3.6
	1135	0.7			1135	1.0
	1737	3.9			1741	3.8
	●				○ 2348	1.0
5 TU	0013	0.7		**20** W	0557	3.7
	0612	3.9			1209	0.9
	1220	0.6			1818	3.9
	1818	4.0				
6 W	0059	0.7		**21** TH	0029	0.9
	0652	3.8			0630	3.8
	1302	0.6			1245	0.8
	1858	4.0			1852	4.0
7 TH	0144	0.8		**22** F	0109	0.8
	0730	3.7			0702	3.8
	1344	0.7			1321	0.7
	1937	3.9			1927	4.0
8 F	0226	1.0		**23** SA	0149	0.8
	0808	3.6			0737	3.7
	1424	0.8			1400	0.7
	2015	3.8			2004	3.9
9 SA	0308	1.2		**24** SU	0230	0.8
	0845	3.4			0815	3.6
	1505	1.0			1442	0.8
	2052	3.6			2046	3.8
10 SU	0350	1.4		**25** M	0315	0.9
	0923	3.3			0859	3.4
	1546	1.2			1529	0.9
	2131	3.4			2133	3.6
11 M	0434	1.6		**26** TU	0407	1.1
	1007	3.1			0949	3.3
	1631	1.4			1623	1.1
	2215	3.2			◑ 2229	3.4
12 TU	0524	1.8		**27** W	0508	1.3
	1102	3.0			1050	3.1
	1720	1.6			1730	1.3
	◐ 2308	3.0			2344	3.2
13 W	0623	1.9		**28** TH	0617	1.4
	1222	2.9			1215	3.0
	1815	1.7			1845	1.4
14 TH	0024	2.9		**29** F	0129	3.1
	0733	1.9			0730	1.4
	1349	2.9			1359	3.1
	1915	1.8			2005	1.4
15 F	0153	3.0		**30** SA	0252	3.2
	0845	1.8			0842	1.3
	1450	3.1			1515	3.3
	2019	1.7			2125	1.3
				31 SU	0356	3.4
					0944	1.1
					1608	3.5
					2229	1.1

AUGUST

Day	Time	m		Day	Time	m
1 M	0445	3.6		**16** TU	0423	3.4
	1037	0.9			1039	1.2
	1650	3.7			1646	3.7
	2320	1.0			2248	1.2
2 TU	0526	3.7		**17** W	0503	3.6
	1123	0.8			1115	1.0
	1728	3.9			1724	3.9
	●				2331	0.9
3 W	0004	0.9		**18** TH	0539	3.8
	0602	3.8			1149	0.7
	1206	0.7			1801	4.0
	1805	4.0			○	
4 TH	0045	0.8		**19** F	0011	0.7
	0637	3.9			0612	3.9
	1246	0.6			1225	0.5
	1841	4.0			1835	4.1
5 F	0124	0.9		**20** SA	0051	0.6
	0710	3.8			0643	3.9
	1324	0.7			1303	0.4
	1916	4.0			1909	4.1
6 SA	0202	1.0		**21** SU	0130	0.5
	0743	3.8			0717	3.9
	1400	0.8			1343	0.4
	1949	3.9			1946	4.1
7 SU	0237	1.1		**22** M	0211	0.6
	0814	3.6			0755	3.8
	1435	1.0			1426	0.6
	2021	3.7			2026	3.9
8 M	0313	1.3		**23** TU	0255	0.7
	0846	3.5			0836	3.6
	1512	1.2			1513	0.8
	2054	3.5			2110	3.6
9 TU	0350	1.5		**24** W	0344	0.9
	0922	3.3			0924	3.3
	1552	1.4			1607	1.0
	2129	3.3			2204	3.3
10 W	0434	1.7		**25** TH	0442	1.2
	1004	3.1			1023	3.1
	1637	1.7			1713	1.3
	◐ 2209	3.1			◑ 2320	3.1
11 TH	0528	1.9		**26** F	0551	1.4
	1102	3.0			1152	2.9
	1731	1.8			1829	1.5
	2303	3.0				
12 F	0634	2.0		**27** SA	0123	3.0
	1249	2.9			0707	1.5
	1833	1.9			1358	3.0
					2000	1.5
13 SA	0102	2.9		**28** SU	0253	3.1
	0750	1.9			0826	1.4
	1426	3.0			1519	3.2
	1942	1.9			2132	1.4
14 SU	0242	3.2		**29** M	0356	3.3
	0906	1.7			0933	1.2
	1522	3.2			1606	3.4
	2055	1.7			2229	1.2
15 M	0339	3.2		**30** TU	0439	3.5
	0959	1.5			1024	1.0
	1606	3.4			1634	3.7
	2159	1.5			2312	1.1
				31 W	0511	3.6
					1108	0.8
					1712	3.9
					2349	0.9

Chart Datum: 2·10 metres below Ordnance Datum (Newlyn). HAT is 4·5 metres above Chart Datum.

STANDARD TIME (UT)
For Summer Time add ONE hour in **non-shaded areas**

OBAN   LAT 56°25'N   LONG 5°29'W
TIMES AND HEIGHTS OF HIGH AND LOW WATERS

Dates in red are **SPRINGS**
Dates in blue are **NEAPS**

YEAR 2016

NW Scotland

SEPTEMBER

Time	m	Time	m
1 0543	3.8	**16** 0516	3.8
1148	0.7	1126	0.6
TH 1746	4.0	F 1737	4.2
●		○ 2349	0.6
2 0025	0.9	**17** 0548	4.0
0614	3.9	1204	0.4
F 1225	0.7	SA 1812	4.3
1818	4.1		
3 0059	0.9	**18** 0028	0.4
0645	3.9	0620	4.0
SA 1300	0.7	SU 1244	0.3
1849	4.1	1847	4.3
4 0132	1.0	**19** 0107	0.4
0714	3.9	0655	4.0
SU 1333	0.8	M 1325	0.3
1920	4.0	1924	4.2
5 0204	1.1	**20** 0149	0.4
0743	3.8	0733	3.9
M 1406	1.0	TU 1410	0.5
1950	3.9	2004	4.0
6 0236	1.3	**21** 0233	0.6
0812	3.7	0815	3.7
TU 1439	1.2	W 1458	0.7
2019	3.7	2049	3.6
7 0310	1.5	**22** 0322	0.9
0845	3.5	0902	3.5
W 1514	1.5	TH 1553	1.1
2049	3.5	2141	3.3
8 0348	1.7	**23** 0419	1.2
0922	3.3	1000	3.2
TH 1557	1.7	F 1659	1.4
2123	3.2	◐ 2259	3.0
9 0439	1.9	**24** 0526	1.4
1010	3.1	1134	3.0
F 1651	1.9	SA 1818	1.6
◐ 2207	3.0		
10 0546	2.0	**25** 0117	2.9
1131	2.9	0642	1.5
SA 1759	2.1	SU 1354	3.0
2324	2.8	2005	1.6
11 0704	2.0	**26** 0250	3.0
1405	3.0	0806	1.5
SU 1916	2.0	M 1518	3.2
		2128	1.5
12 0228	2.9	**27** 0348	3.2
0825	1.8	0915	1.3
M 1504	3.2	TU 1553	3.4
2035	1.8	2217	1.3
13 0324	3.1	**28** 0421	3.4
0927	1.6	1005	1.1
TU 1547	3.5	W 1618	3.7
2140	1.5	2255	1.1
14 0405	3.4	**29** 0447	3.6
1011	1.2	1047	0.9
W 1625	3.7	TH 1648	3.8
2228	1.1	2328	1.0
15 0442	3.6	**30** 0516	3.8
1049	0.9	1125	0.8
TH 1701	4.0	F 1720	4.0
2310	0.8	2359	1.0

OCTOBER

Time	m	Time	m
1 0547	3.9	**16** 0521	4.1
1200	0.8	1142	0.4
SA 1751	4.1	SU 1747	4.4
●		○	
2 0029	0.9	**17** 0003	0.4
0616	4.0	0557	4.2
SU 1234	0.8	M 1225	0.3
1821	4.1	1825	4.4
3 0100	1.0	**18** 0045	0.3
0645	4.0	0634	4.2
M 1306	0.9	TU 1309	0.4
1851	4.0	1904	4.2
4 0132	1.1	**19** 0127	0.4
0714	4.0	0715	4.0
TU 1337	1.1	W 1355	0.5
1921	3.9	1945	4.0
5 0203	1.3	**20** 0212	0.6
0743	3.8	0757	3.8
W 1408	1.3	TH 1445	0.8
1949	3.7	2030	3.6
6 0233	1.5	**21** 0301	0.9
0815	3.6	0845	3.6
TH 1441	1.6	F 1540	1.1
2017	3.5	2122	3.3
7 0304	1.7	**22** 0356	1.1
0851	3.4	0941	3.3
F 1519	1.8	SA 1644	1.5
2049	3.3	◐ 2236	3.0
8 0346	1.9	**23** 0459	1.4
0934	3.2	1106	3.1
SA 1613	2.0	SU 1803	1.7
2131	3.1		
9 0454	2.0	**24** 0055	2.8
1038	3.0	0611	1.6
SU 1728	2.1	M 1328	3.1
2239	2.9	1950	1.7
10 0615	2.0	**25** 0226	2.9
1327	3.0	0733	1.6
M 1850	2.1	TU 1454	3.0
		2106	1.6
11 0159	2.9	**26** 0322	3.1
0735	1.9	0845	1.4
TU 1434	3.2	W 1526	3.4
2010	1.8	2153	1.4
12 0258	3.1	**27** 0351	3.3
0844	1.6	0938	1.2
W 1519	3.5	TH 1550	3.6
2113	1.5	2229	1.3
13 0339	3.4	**28** 0416	3.5
0936	1.3	1021	1.1
TH 1557	3.8	F 1620	3.8
2201	1.1	2301	1.2
14 0414	3.6	**29** 0446	3.7
1020	0.9	1058	1.0
F 1634	4.1	SA 1652	3.9
2244	0.8	2330	1.1
15 0448	3.9	**30** 0518	3.9
1101	0.6	1133	1.0
SA 1710	4.3	SU 1724	4.0
2324	0.5	● 2359	1.1
		31 0549	4.0
		1206	1.0
		M 1755	4.1

NOVEMBER

Time	m	Time	m
1 0030	1.1	**16** 0024	0.4
0619	4.1	0618	4.2
TU 1239	1.1	W 1256	0.5
1826	4.0	1847	4.2
2 0102	1.1	**17** 0108	0.5
0650	4.0	0700	4.1
W 1312	1.2	TH 1343	0.7
1857	3.9	1930	4.0
3 0134	1.3	**18** 0154	0.6
0721	3.9	0744	4.0
TH 1344	1.4	F 1433	0.9
1926	3.8	2015	3.7
4 0203	1.4	**19** 0242	0.8
0753	3.8	0831	3.7
F 1416	1.6	SA 1526	1.2
1955	3.6	2104	3.4
5 0232	1.6	**20** 0334	1.1
0828	3.6	0923	3.5
SA 1452	1.8	SU 1625	1.5
2029	3.4	2205	3.1
6 0308	1.8	**21** 0430	1.4
0910	3.4	1028	3.2
SU 1542	1.9	M 1736	1.7
2113	3.2	◑ 2355	2.9
7 0402	1.9	**22** 0534	1.5
1007	3.2	1215	3.1
M 1654	2.0	TU 1904	1.8
◑ 2216	3.0		
8 0519	2.0	**23** 0130	2.9
1137	3.2	0645	1.6
TU 1816	2.0	W 1353	3.2
2359	2.9	2025	1.8
9 0638	1.9	**24** 0231	3.0
1346	3.3	0758	1.6
W 1933	1.8	TH 1444	3.3
		2118	1.7
10 0214	3.1	**25** 0310	3.2
0752	1.7	0900	1.5
TH 1442	3.6	F 1518	3.4
2037	1.5	2158	1.5
11 0303	3.3	**26** 0343	3.4
0856	1.4	0949	1.4
F 1526	3.8	SA 1551	3.6
2129	1.2	2232	1.3
12 0342	3.6	**27** 0416	3.6
0948	1.1	1029	1.3
SA 1606	4.1	SU 1624	3.8
2215	0.9	2302	1.3
13 0419	3.9	**28** 0451	3.8
1036	0.8	1105	1.3
SU 1646	4.3	M 1659	3.9
2259	0.6	2332	1.2
14 0457	4.1	**29** 0525	3.9
1123	0.6	1140	1.2
M 1726	4.4	TU 1734	4.0
○ 2341	0.5	●	
15 0537	4.2	**30** 0004	1.2
1209	0.5	0558	4.0
TU 1806	4.4	W 1215	1.2
		1808	4.0

DECEMBER

Time	m	Time	m
1 0038	1.2	**16** 0054	0.6
0632	4.0	0649	4.2
TH 1251	1.3	F 1334	0.8
1840	3.9	1919	4.0
2 0111	1.2	**17** 0139	0.6
0704	3.9	0733	4.1
F 1326	1.4	SA 1421	1.0
1912	3.8	2002	3.8
3 0142	1.3	**18** 0224	0.8
0738	3.8	0816	3.9
SA 1400	1.5	SU 1509	1.2
1943	3.7	2045	3.5
4 0212	1.4	**19** 0311	1.0
0813	3.7	0901	3.6
SU 1436	1.6	M 1600	1.5
2019	3.5	2132	3.3
5 0247	1.6	**20** 0400	1.3
0853	3.6	0949	3.4
M 1521	1.7	TU 1655	1.7
2101	3.3	2228	3.1
6 0333	1.7	**21** 0452	1.5
0943	3.4	1047	3.2
TU 1619	1.8	W 1758	1.9
2155	3.2	◑ 2351	3.0
7 0433	1.7	**22** 0549	1.6
1048	3.3	1206	3.1
W 1732	1.9	TH 1912	1.9
◑ 2305	3.1		
8 0546	1.8	**23** 0121	3.0
1218	3.3	0652	1.7
TH 1847	1.8	F 1340	3.1
		2026	1.9
9 0037	3.1	**24** 0221	3.1
0702	1.7	0801	1.8
F 1354	3.5	SA 1439	3.2
1955	1.6	2121	1.8
10 0210	3.2	**25** 0308	3.2
0815	1.5	0906	1.7
SA 1454	3.7	SU 1522	3.4
2056	1.3	2202	1.6
11 0308	3.5	**26** 0349	3.4
0920	1.2	0957	1.6
SU 1543	3.9	M 1601	3.5
2148	1.1	2237	1.5
12 0355	3.8	**27** 0428	3.6
1017	1.0	1039	1.5
M 1628	4.1	TU 1640	3.7
2237	0.8	2310	1.3
13 0440	4.0	**28** 0506	3.8
1109	0.8	1118	1.4
TU 1711	4.2	W 1719	3.8
2323	0.7	2344	1.2
14 0523	4.1	**29** 0543	3.9
1158	0.7	1156	1.3
W 1754	4.2	TH 1756	3.9
○		●	
15 0008	0.6	**30** 0018	1.2
0606	4.2	0619	4.0
TH 1246	0.7	F 1234	1.2
1836	4.1	1829	3.9
		31 0052	1.1
		0652	4.0
		SA 1311	1.2
		1900	3.8

Chart Datum: 2·10 metres below Ordnance Datum (Newlyn). HAT is 4·5 metres above Chart Datum.

》》 FREE monthly updates. Register at 《
www.reedsnauticalalmanac.co.uk

459

9.8.18 OBAN

Argyll and Bute **56°24'·99N 05°29'·07W** ❄❄❄◊◊◊❀❀

CHARTS AC 2171, 2387, 2388, 1790; Imray C65, 2800

TIDES –0530 Dover; ML 2·4; Duration 0610

Standard Port OBAN (←—)

Times				Height (metres)			
High Water		Low Water		MHWS	MHWN	MLWN	MLWS
0100	0700	0100	0800	4·0	2·9	1·8	0·7
1300	1900	1300	2000				
Differences DUNSTAFFNAGE BAY							
+0005	0000	0000	+0005	+0·1	+0·1	+0·1	+0·1
CONNEL							
+0020	+0005	+0010	+0015	–0·3	–0·2	–0·1	+0·1
BONAWE							
+0150	+0205	+0240	+0210	–2·0	–1·7	–1·3	–0·5

SHELTER Good shelter in bay except in strong SW/NW winds, but Oban Marina in Ardantrive Bay (Kerrera) is well sheltered from all directions, but as yet has no water on pontoons. The only ⚓s convenient for town are in deep water off the Esplanade and NW of Sailing Club, beware of moorings. Alternatively ⚓ at N end of Kerrera to S of Rubh'a Chruidh Is or Horseshoe Bay and Little Horseshoe Bay, Kerrera Sound. Beware fish farms S of Ardantrive Bay.

NAVIGATION **WPT (N)** 56°25'·85N 05°30'·0W, 132°/0·7M to Dunollie lt; **WPT (S)** 56°22'N 05°33'W, 035°/1M to 1·5ca E of Sgeirean Dubha, thence through Kerrera Sound to Oban Bay. Ferry Rks can be passed on either side. Direction of the buoyage is **NE**, thus pass to the E'ward of PHM unlit buoy or to W'ward of the SHM QG buoy. The N'ly SHM buoy must be passed to the W'ward. In Oban Bay beware Sgeir Rathaid, buoyed, in middle of the bay and ferries running to/from Railway Pier.

LIGHTS AND MARKS See 9.8.4 and chartlet.

COMMUNICATIONS (Code 01631) CGOC (01851) 702013; ⊜ 08457 231110; Police 101; Dr 563175; N Pier ☎562892.
North Pier Ch 12 16 (0900-1700); Railway Pier, *CalMac* Ch 06 12 16; *Oban Marina* Ch 80.

FACILITIES **North Pier** ⚓ ⚓ ⚓ ⊠(15t mob) via Piermaster.
Railway Pier ⚓ ⚓ ⚓(H24, ☎562849) ▲ ⚓ ACA ⚲ 🖾 divers.
Oban Sailing Club 16 ⚓s £15/night, pay at Y honesty box £5 until 1600; short stay ⌿ (with ⚓), £2/half hour; free for 30 mins.
Oban Marina (Ardantrive Bay) www.obanmarina.com ☎565333; 30 ⚓s £18/nt (20t); £30/nt (50t); 100⌿ £2.35+⟐▷ ▲ ⚓ 🖾 ⚓ Ⓔ 🖾 ✕ ⟐ ⚓(limited) ⊠ ▣(50t). Water taxi to/from Oban (hrly, 0830-2100 Apr, 0730-2300, Oct-Apr 2hrly), free for boat crews.
Town ⌂ (½M), all facilities, Ⓗ, ⇌ Connel (airstrip) 3M, Glasgow (✈) 3hrs, ferries to Inner & Outer Hebrides.
Heather Is ⚓s 3Y 15t, 1 Gy 20t, 1 Or 50t; £10/nt (pay Ardantrive Fm).

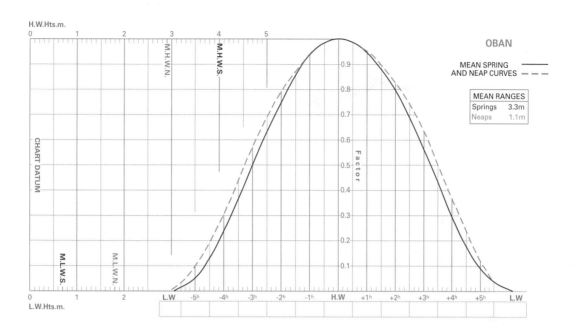

ADJACENT HARBOUR ON MAINLAND TO N OF OBAN

DUNSTAFFNAGE Marina, Argyll and Bute, **56°27'·04N 05°25'·97W**. AC 2388, 5611. Within the bay the tidal stream is rotary setting mainly E through the marina and running strongly through the moorings. Enter between Rubha Garbh and Eilean Mór. No navigational hazards, W and SW sides of bay dry. Use buoyed fairway, keeping clear of moorings, spd limit 4kn.

☎(01631) 566555, VHF Ch37/M1 ⚓ (launch H24 except LWS ±2hrs) 200⬤ inc ❶ £2·45+£3 ⬙ ▣ ⚓(0830-2000 summer) ⬛ ⚒ ⬠ ⬡ ⌂ ◪(40t) ⬙(masting) ✕ 🛒 (½M) ⬙ (¾M), bus, ⇌ Oban (3M).

ANCHORAGES ON MAINLAND SHORE OF FIRTH OF LORN

LOCH FEOCHAN, Argyll and Bute, **56°21'·39N 05°29'·77W**. AC 5611, 2387. HW = HW Oban; flood runs 4 hrs, ebb for 8 hrs. Caution: strong streams off Ardentallan Pt. Good shelter, 5M S of Oban and 1·5M SE of Kerrera. Best appr at local slack LW = LW Oban +0200. Narrow buoyed channel, ⚓ off pier, or moor off.

Ardoran Marine ☎(01631) 566123, 📠 566611; 4 ⚓s £10·00 <11m, ⚓ ⚒ ⬠ ⬡ ⌂ ◫.

PUILLADOBHRAIN, Argyll and Bute, **56°19'·47N 05°35'·22W**. AC 5611, 2386/2387. Tides as Oban. Popular ⚓ on the SE shore of the Firth of Lorne, approx 7M S of Oban, sheltered by the islets to the W of it. At N end of Ardencaple Bay identify Eilean Dùin (18m) and steer SE keeping 1½ca off to clear a rk awash at its NE tip. Continue for 4ca between Eilean nam Beathach, with Orange drum on N tip, and Dun Horses rks drying 2·7m. Two W cairns on E side of Eilean nam Freumha lead approx 215° into the inner ⚓ in about 4m. Landing at head of inlet. Nearest facilities: ⬜ ☎ at Clachan Br (½M); ☎ ✉ at Clachan Seil.

CUAN SOUND, Argyll and Bute, **56°15'·84N 05°37'·47W**. AC 2386, 2326. Tides see 9.8.18 SEIL SOUND. Streams reach 6kn at sp; N-going makes at HW Oban +0420, S-going at HW Oban −2. The Sound is a useful doglegged short cut from Firth of Lorne to Lochs Melfort and Shuna, but needs care due to rks fish farms and tides. There are ⚓s at either end to await the tide. At the 90° dogleg, pass close N of Cleit Rk onto which the tide sets; it is marked by a Y △ perch. The chan is only ¾ca wide here due to rks off Seil. Overhead cables (35m) cross from Seil to Luing. There are ⚓s out of the tide to the S of Cleit Rk. No lts/facilities. See CCC SDs.

ARDINAMAR, Luing/Torsa, **56°14'·92N 05°37'·04W**. AC 2326. HW −0555 on Dover; ML 1·7m; see 9.8.18 SEIL SOUND. A small cove and popular ⚓ between Luing and Torsa, close W of ent to L. Melfort. Appr on brg 290°. Narrow, shallow (about 1m CD) ent has drying rks either side, those to N marked by 2 SHM perches. Keep about 15m S of perches to ⚓ in 2m in centre of cove; S part dries. Few facilities: ✉ ⬛ ☎ at Cullipool 1·5M WNW. ⬛ at Cuan Sound ferry 2M NNW.

9.8.19 LOCH MELFORT

Argyll and Bute 56°14'·59N 05°34'·07W ✿✿✿✿⬡⬡⬡✿✿✿

CHARTS AC 5611, 2169, 2326; Imray C65, 2800

TIDES Loch Shuna –0615 Dover; ML Loch Melfort 1·7; Duration Seil Sound 0615

Standard Port OBAN (←—)

Times				Height (metres)			
High Water		Low Water		MHWS	MHWN	MLWN	MLWS
0100	0700	0100	0800	4·0	2·9	1·8	0·7
1300	1900	1300	2000				
Differences LOCH MELFORT							
–0055	–0025	–0040	–0035	–1·2	–0·8	–0·5	–0·1
SEIL SOUND							
–0035	–0015	–0040	–0015	–1·3	–0·9	–0·7	–0·3

SHELTER Good at Kilmelford Yacht Haven in Loch na Cille; access at all tides for 3m draught, W lts at end of pier and along its length. Or lt at Melfort Pier (Fearnach Bay at N end of loch): pier/pontoon in 2m, but chan to inner hbr dries; good ⚓ in N winds. ⚓'ge sheltered from S to W at Kames Bay clear of moorings and rocks. Beware extensive marine farm activity inshore NE of Arduaine Point.

NAVIGATION WPT 56°13'·90N 05°34'·80W, 000°/3·5ca to Eilean Gamhna. Pass either side of Eilean Gamhna. 8ca NE lies Campbell Rk (1·8m). A rk drying 1m lies 1½ca ESE of the FS on Eilean Coltair. The S side of L Melfort is mostly steep-to, except in Kames Bay. At Loch na Cille, beware drying reef ¾ca off NE shore (PHM perch), and rk near S shore (SHM perch); boats may obscure perches.

LIGHTS AND MARKS A Dir FR ☆ 6m 3M on Melfort pier (also depth gauge) and a Dir FG ☆ close NE on the shore are not ldg lts, nor do they form a safe transit. Approach on a N'ly track keeping them an equal angle off each bow.

COMMUNICATIONS (Code 01852) Police 101; Ⓗ (01631) 567500; CGOC (01851) 702013; ⊜ (0141) 887 9369.

Kilmelford VHF Ch **80** M (HO).

FACILITIES

Melfort Pier melharbour@aol.com ☎200333, ⌂ < 6m or ⚓ £12/day, ▬ ▫ @ wi-fi, hotel accommodation.

Kilmelford Yacht Haven ☎ 200248, 🗺 200343, ▬ HW±4; ⌂£16/craft (any LOA) 🅽(Mon-Sat 0800-2000) 🔋 ▫ ▣(20t) ⚒ ⚙ ▣.

Kilmelford Village (1¼M): ✉ 🍺 ✗ 🍴.

9.8.20 CRAOBH MARINA (L SHUNA)

Argyll & Bute **56°12′·80N 05°33′·54W** ✱✱✱⚓⚓⚓⚓❀❀❀

CHARTS AC 5611, 2169, 2326; Imray C65, 2800

TIDES HW Loch Shuna –0100 Oban; –0615 Dover; Seil Sound Duration 0615, ML 1·4. For tidal figures see 9.8.19.

SHELTER Very good. Craobh (pronounced Croove) Marina (access H24) on SE shore of Loch Shuna is enclosed by N and S causeways between islets. The ent is between 2 bkwtrs on the N side. In the marina, a shoal area S of the E bkwtr is marked by 9 PHM and 2 SHM buoys. A Y perch in W corner of hbr marks a spit; elsewhere ample depth. There are ⚓s in Asknish Bay 1M to the N, and in the bays E of Eilean Arsa and at Bàgh an Tigh-Stòir, S of Craobh.

NAVIGATION WPT 56°13′·01N 05°33′·57W, 173°/2ca to ent. Tidal streams in Loch Shuna are weak. Beware fish farm 2ca N of Shuna, lobster pots in appr's and unmarked rks (dr 1·5m) 4ca NNE of ent. An unlit SHM buoy marks a rk (1m) 150m NNW of the W bkwtr. 1M N of marina, Eich Donna, an unmarked reef (1·5m), lies between Eilean Creagach and Arduaine Pt.

LIGHTS AND MARKS The W sector, 162°-183°, of Dir Lt, Iso WRG 5s 10m 5/3M, on E bkwtr hd leads 172° between the close-in rks above. Multi coloured marina buildings are conspic.

COMMUNICATIONS (Code 01852) CGOC (01851) 702013; ⊖ (0141) 887 9369; Police 101; Ⓗ (01546) 602323.

VHF Ch M 80 (summer 0830-2000; winter 0830-1800).

FACILITIES Craobh Marina ☎500222 mob 07917 805386; ⚓ (launch/recovery £3.70/m); (200+50 Ⓥ) £2·30; ⛽🚿⚓⚑🚻⊡(30t)

🅰(12t) ⚒ 🔧 Ⓑ Ⓔ ⊙ ✕ SC, Divers.

Village: ✉(Kilmelford) Ⓑ(Fri) 🛒 🗑 ⇌ (Oban by bus) ✈ (Glasgow).

9.8.21 LOCH CRAIGNISH

Argyll and Bute **56°07′·99N 05°35′·07W** ✱✱✱⚓⚓⚓❀❀❀

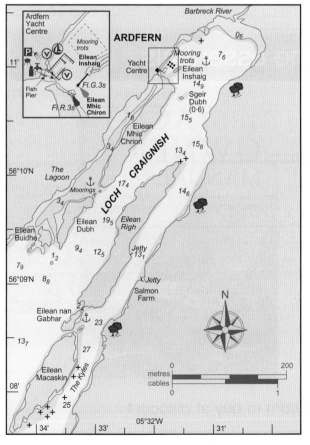

CHARTS AC 5611, 2169, 2326; Imray C65, C63, 2800

TIDES +0600 Dover; ML (Loch Beag)1·2; Duration (Seil Sound) 0615

Standard Port OBAN (←—)

Times				Height (metres)			
High Water		Low Water		MHWS	MHWN	MLWN	MLWS
0100	0700	0100	0800	4·0	2·9	1·8	0·7
1300	1900	1300	2000				
Differences LOCH BEAG (3M SSW of Ardfern)							
–0110	–0045	–0035	–0045	–1·6	–1·2	–0·8	–0·4

NOTE: HW Ardfern is approx HW Oban –0045; times/heights much affected by local winds and barometric pressure

SHELTER Beware squalls in E'lies, especially on E side of loch. Good at Ardfern, 56°11′·0N 05°31′·8W, access H24 appr chan unlit; ⚓s at:
- Eilean nan Gabhar; appr from E chan & ⚓ E of island.
- Eilean Righ; midway up the E side of the island.
- Eilean Dubh in the 'lagoon' between the island and mainland bounded by moorings to the NEand SW of the entrance.

NAVIGATION WPT 56°07′·59N 05°35′·37W (off chartlet) between Dorus Mór and Liath-sgier Mhòr.

Beware: strong tidal streams (up to 8kn) in Dorus Mór; a reef extending 1ca SSW of Eilean Macaskin; rk 1½ca SSW of Eilean Dubh; fish cages especially on E side of loch; a drying rock at N end of Ardfern ⚓ with a rock awash ¼ca E of it. (These 2 rks are ½ca S of the more S'ly of little islets close to mainland.) The main fairway is free from hazards, except for Sgeir Dhuhb (0·6m), an unmarked rk 3½ca SSE of Ardfern, with a reef extending about ½ca all round. Ardfern is 1ca W of Eilean Inshaig.

LIGHTS AND MARKS Ent to Ardfern Yacht Centre marked by PHM buoy, Fl R 3s, and end of floating breakwater, Fl G 3s, which extends 90m SW from Eilean Inshaig.

COMMUNICATIONS CGOC (01851) 702013; ⊖ (0141) 887 9369; Dr (01546) 602921; Ⓗ (01546) 602449.

Ardfern Yacht Centre VHF Ch 80 M (office hrs).

FACILITIES Ardfern Yacht Centre www.ardfernyacht.co.uk ☎ 01852 500247; 12 Ⓥ £1·40) 87+20 Ⓥ ⊂£2·20; ⚓ (HW±3, £1.75/m) ⚡⚓(HO) ⛽⚒🔧⚓⚑🚻(40t) 🅰(25t) ⚓ ACA.

Village ✉ 🛒 ✕ 🗑 Ⓑ(Fri) ⇌(Oban) ✈(Glasgow).

NW Scotland

9.8.22   FERRIES

Many ferries ply between mainland and island harbours. This summary may prove useful when plans or crews change in remote places. It covers Areas 8 and 9. The major operator is Caledonian MacBrayne: Head Office, The Ferry Terminal, Gourock PA19 1QP. For reservations: ☎08000 665000; www.calmac.co.uk. Many routes are very short and may not be pre-bookable; seasonal routes are marked *.

CalMac

From	To	Time	Remarks
Areas 8			
Berneray	Leverburgh	1¼	
Ullapool	Stornoway	2¾ hrs	
Uig (Skye)	Tarbert (Harris)	1¾ hrs	Not Sun
Uig	Lochmaddy (N Uist)	1¾ hrs	
Oban	Castlebay/Lochboisdale	5-7 hrs	
Sconser (Skye)	Raasay	15 mins	Not Sun
Mallaig*	Armadale (Skye)	20 mins	
Mallaig	Eigg-Muck-Rum-Canna	Varies	Not Sun
Oban	Coll-Tiree	Varies	Not Thurs
Tobermory	Kilchoan	35 mins	
Fionnphort	Iona	5 mins	
Lochaline	Fishnish (Mull)	15 mins	
Oban	Craignure (Mull)	45 mins	
Oban	Lismore	50 mins	Not Sun
Areas 8/9			
Oban	Colonsay	2¼ hrs	Sun/W/Fri
Area 9			
Kennacraig	Port Askaig/Colonsay	Varies	Wed
Kennacraig	Port Ellen	2h 10m	
Kennacraig	Port Askaig	2 hrs	
Tayinloan	Gigha	20 mins	
Ardrossan	Brodick	55 mins	
Claonaig	Lochranza (Arran)	30 mins	
Largs	Cumbrae Slip	10 mins	
Tarbert (L Fyne)	Portavadie*	25 mins	
Colintraive	Rhubodach (Bute)	5 mins	
Wemyss Bay	Rothesay (Bute)	35 mins	
Gourock	Dunoon	20 mins	

Other Island Ferry Operators

From	To	Operator	Telephone
Areas 8			
Corran - V	Ardgour	Highland Council	01855 841243
Easdale - P	Seil	Area Manager	01631 562125
Firth of Lorn - P	Colonsay	K & ✉ Byrne	01951 200320
	Uisken (Ross of Mull)		
	Scalasaig (Colonsay)		
	Tarbert (Jura)		
	Port Askaig (Islay)		
Jura - V	Port Askaig (Islay)	Serco Denholm	01496 840681
Kerrara - P	Oban	Oban Yachts	01631 565333
Kilgregan - P	Gourock	Clyde Marine Motoring	01475 721281
Lismore - P	Port Appin	Area Manager	01631 562125
Loch Nevis - P	Mallaig	Bruce Watt	01687 462233
Luing - P/V	Seil	Area Manager	01631 562125
Morvern	Sunart	Pre-book via	01688 302851
Mull	Drimnin	Sound of Mull	or mobile
Ardnamurchan	Tobermoray	Transport	07799 608199
	Kilchoan		
Skye - V	Glenelg	IOSFCIC	01599 522236
Staffa - P	Iona	D Kirkpatrick	01681 700373
Staffa - P	Mull	Gordon Grant	01681 700338
Staffa - P	Mull	Turus Mara	01688 400242

P = Passenger only   V = Cars and Passengers

South West Scotland

Crinal Canal to Mull of Galloway

9.9.1	Contents	465
9.9.2	Area map and Distance table	466
9.9.3	Tidal stream charts	468
9.9.4	Lights, buoys and waypoints	470
9.9.5	Passage information	472
	Corryvreckan to Crinan • West of Jura to Islay, Sound of Jura to Gigha – 472 • Mull of Kintyre to Upper Loch Fyne – 475 • Arran, Bute and Firth of Clyde – 481 • Firth of Clyde to Mull of Galloway, Scotland SW coast – 486	
9.9.6	Port Ellen	473
	Scalasaig (Colonsay) • Loch Tarbert (Jura) • Port Askaig (Islay) • Craighouse (Jura)• Loch Sween • West Loch Tarbert (Kintyre) • Gigha Island	
9.9.7	Crinan Canal	474
9.9.8	Ardrishaig	475
9.9.9	Tarbert	476
9.9.10	Portavadie	476
9.9.11	Tidal stream charts for Mull of Kintyre	477
9.9.12	Campbeltown	478
	Carradale Bay	
9.9.13	Lamlash	479
	Brodick (Arran) • Loch Ranza (Arran) • St Ninian's Bay (Bute) • West Kyle • Caladh Harbour • Loch Riddon • Burnt Islands • Kilchattan Bay (Bute)	
9.9.14	Rothesay	480
	Millport (Cumbraes)	
9.9.15	Port Bannatyne	480
9.9.16	Fairlie Quay	481
9.9.17	Largs	481
9.9.18	Submarine Exercise Areas	482
9.9.19	Inverkip (Kip Marina)	483
	Dunoon • Gourock	
9.9.20	Holy Loch	483
9.9.21	Gareloch/Rhu	484
	Loch Long/Loch Goil	
9.9.22	River Clyde, Greenock tide tables/curves	485
	Greenock • Dumbarton (Sandpoint) • Bowling Harbour	
9.9.23	Ardrossan	490
	Irvine • Ayr	
9.9.24	Troon	491
	Girvan	
9.9.25	Loch Ryan/Stranraer	492
	Drummore • Isle of Whithorn • Garlieston • Pt William	
9.9.26	Portpatrick	493
9.9.27	Kirkcudbright	494
	Kippford	

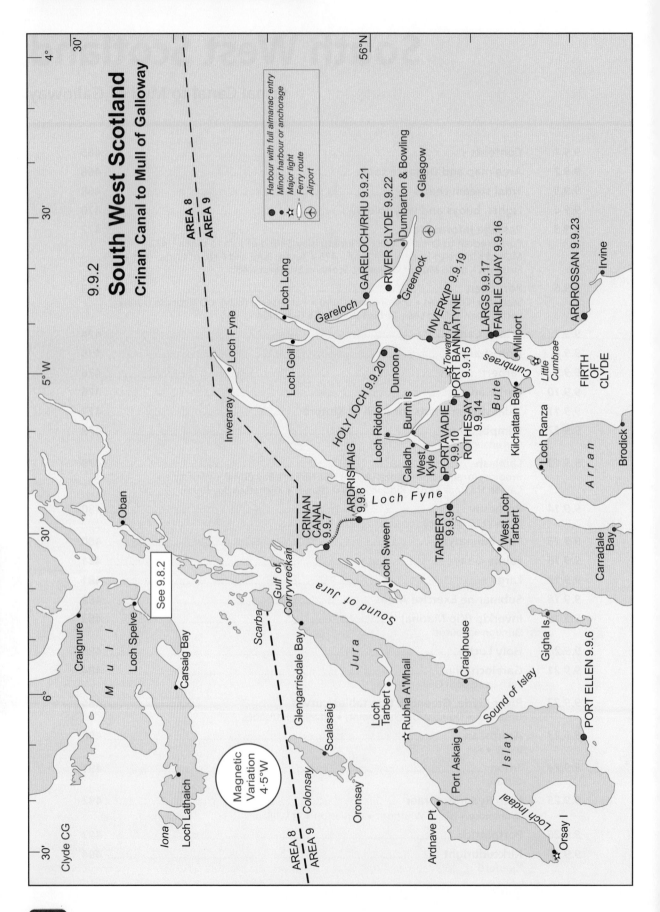

9.9.2 South West Scotland
Crinan Canal to Mull of Galloway

Harbour with full almanac entry
Minor harbour or anchorage
Major light
Ferry route
Airport

AREA 8
AREA 9

Magnetic Variation 4·5°W

See 9.8.2

Clyde CG

Mull

Iona
Loch Lathaich

Craignure

Loch Spelve

Carsaig Bay

Oban

Scarba

Gulf of Corryvreckan

CRINAN CANAL 9.9.7

ARDRISHAIG 9.9.8

Glengarrisdale Bay

Jura

Scalasaig

Colonsay

Oronsay

Loch Tarbert

Rubha A'Mhail

Sound of Jura

Loch Sween

Loch Fyne

Inveraray

Loch Fyne

Loch Goil

Loch Long

Gareloch

GARELOCH/RHU 9.9.21

RIVER CLYDE 9.9.22

Dumbarton & Bowling

Glasgow

Greenock

INVERKIP 9.9.19

Toward Pt

PORT BANNATYNE 9.9.15

Dunoon

Loch Riddon

Caladh
Burnt Is
West Kyle

PORTAVADIE 9.9.10

ROTHESAY 9.9.14

Bute

Kilchattan Bay

HOLY LOCH 9.9.20

LARGS 9.9.17

FAIRLIE QUAY 9.9.16

Millport

Cumbraes

Little Cumbrae

ARDROSSAN 9.9.23

Irvine

FIRTH OF CLYDE

TARBERT 9.9.9

West Loch Tarbert

Loch Ranza

Arran

Carradale Bay

Brodick

Craighouse

Gigha Is

Sound of Islay

Port Askaig

Islay

Ardnave Pt

Loch Indaal

PORT ELLEN 9.9.6

Orsay I

AREA 8
AREA 9

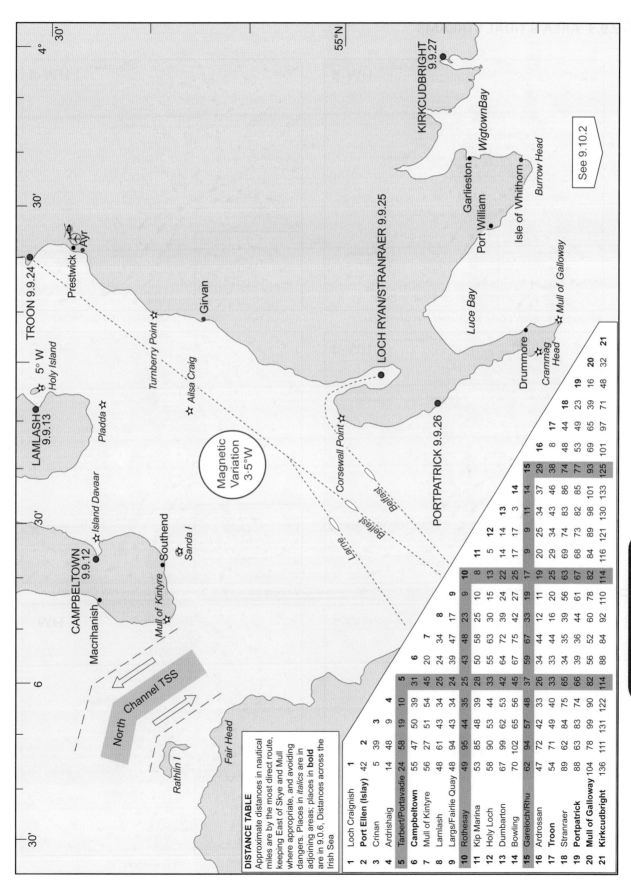

KIRKCUDBRIGHT 9.9.27

WigtownBay

Garlieston

Burrow Head

See 9.10.2

LOCH RYAN/STRANRAER 9.9.25

Port William

Isle of Whithorn

Luce Bay

PORTPATRICK 9.9.26

Drummore

Crammag Head

Mull of Galloway

Corsewall Point

Belfast

Belfast

Larne

TROON 9.9.24

Prestwick

Ayr

5° W

Holy Island

Turnberry Point

Girvan

☆ Ailsa Craig

Magnetic Variation 3·5°W

55°N

LAMLASH 9.9.13

Pladda ☆

North Channel TSS

CAMPBELTOWN 9.9.12

Macrihanish

☆ Island Davaar

Southend

Mull of Kintyre

Sanda I

Fair Head

Rathlin I

SW Scotland

DISTANCE TABLE
Approximate distances in nautical miles are by the most direct route, keeping East of Skye and Mull where appropriate, and avoiding dangers. Places in *italics* are in adjoining areas; places in **bold** are in 9.0.6, Distances across the Irish Sea

		1	2	3	4	5	6	7	8	9	10	11	12	13	14	15	16	17	18	19	20	21
1	Loch Craignish	1																				
2	**Port Ellen (Islay)**	42	**2**																			
3	Crinan	5	39	**3**																		
4	Ardrishaig	14	48	9	**4**																	
5	Tarbert/Portavadie	24	58	19	10	**5**																
6	**Campbeltown**	55	47	50	39	31	**6**															
7	Mull of Kintyre	56	27	51	54	45	20	**7**														
8	Lamlash	48	61	43	34	25	24	34	**8**													
9	Largs/Fairlie Quay	48	94	43	34	24	39	47	17	**9**												
10	Rothesay	49	95	44	35	25	43	48	23	9	**10**											
11	Kip Marina	53	85	48	39	28	50	58	25	10	8	**11**										
12	Holy Loch	58	90	53	44	33	55	63	30	15	13	5	**12**									
13	Dumbarton	67	99	62	53	44	64	72	39	24	22	14	14	**13**								
14	Bowling	70	102	65	56	45	67	75	42	27	25	17	17	3	**14**							
15	**Gareloch/Rhu**	62	94	57	48	37	59	67	33	19	17	9	9	11	14	**15**						
16	Ardrossan	47	72	42	33	26	34	44	12	11	19	20	25	34	37	29	**16**					
17	**Troon**	54	71	49	40	33	33	44	16	20	25	29	34	43	46	38	8	**17**				
18	Stranraer	89	62	84	75	65	34	35	39	56	63	69	74	83	86	74	48	44	**18**			
19	**Portpatrick**	88	63	83	74	66	39	36	44	61	67	68	73	82	85	77	53	49	23	**19**		
20	**Mull of Galloway**	104	78	99	90	82	56	52	60	78	82	84	89	98	101	93	69	65	39	16	**20**	
21	**Kirkcudbright**	136	111	131	122	114	88	84	92	110	114	116	121	130	133	125	101	97	71	48	32	**21**

467

9.9.3 AREA 9 TIDAL STREAMS

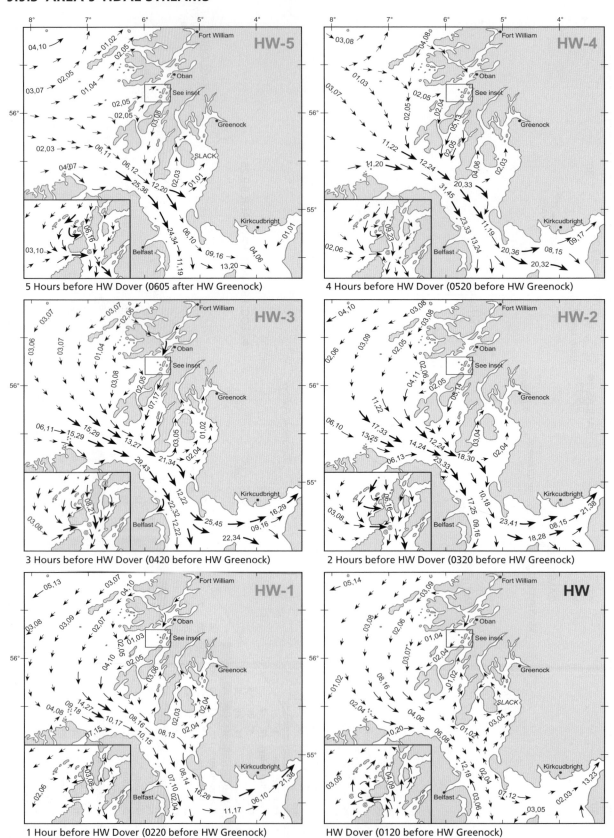

5 Hours before HW Dover (0605 after HW Greenock)

4 Hours before HW Dover (0520 before HW Greenock)

3 Hours before HW Dover (0420 before HW Greenock)

2 Hours before HW Dover (0320 before HW Greenock)

1 Hour before HW Dover (0220 before HW Greenock)

HW Dover (0120 before HW Greenock)

Northward 9.8.3   Mull of Kintyre 9.9.11   Irish Sea 9.10.3   North Ireland 9.13.3

1 Hour after HW Dover (0020 before HW Greenock)

2 Hours after HW Dover (0040 after HW Greenock)

3 Hours after HW Dover (0140 after HW Greenock)

4 Hours after HW Dover (0240 after HW Greenock)

5 Hours after HW Dover (0340 after HW Greenock)

6 Hours after HW Dover (0440 after HW Greenock)

SW Scotland

9.9.4 LIGHTS, BUOYS AND WAYPOINTS

Bold print = light with a nominal range of 15M or more. CAPITALS = place or feature. *CAPITAL ITALICS* = light-vessel, light float or Lanby. *Italics* = Fog signal. ***Bold italics*** = Racon. Some marks/buoys are fitted with AIS (<u>MMSI No</u>); see relevant charts.

COLONSAY TO ISLAY

COLONSAY
Scalasaig, Rubha Dubh ⚓ Fl (2) WR 10s 8m W8M, R6M; W bldg; vis: shore-R- 230°-W-337°-R-354°; 56°04'·01N 06°10'·90W.

SOUND OF ISLAY
Rhubh' a Mháil (Ruvaal) ☆ 55°56'·18N 06°07'·46W Fl (3) 15s 45m **19M**; W twr.
Carragh an t'Struith ⚓ Fl 3s 8m 9M; W twr; vis: 354°-180°; 55°52'·30N 06°05'·78W.
Carraig Mòr ⚓ Fl (2) WR 6s 7m W8M, R6M; W twr; vis: shore-R-175°-W-347°-R-shore; 55°50'·42N 06°06'·13W.
McArthur's Hd ⚓ Fl (2) WR 10s 39m W13M, R10M; W twr; W in Sound of Islay from NE coast,159°-R-244°-W-E coast of Islay; 55°45'·84N 06°02'·90W.
Orsay Is, **Rhinns of Islay** ☆ 55°40'·40N 06°30'·84W Fl 5s 46m **24M**; W twr; vis: 256°-184°.

PORT ELLEN and LOCH INDAAL
Carraig Fhada ⚓ Fl WRG 3s 20m W8M, R6M, G6M; W☐twr; vis: W shore- 248°-G-311°-W-340°-R-shore; 55°37'·22N 06°12'·71W.
Otter Gander ⚓ VQ (3) 5s; 55°36'·60N 06°12'·34W.
Fhada ⚓ Q (3) 10s; 55°37'·21N 06°12'·41W.
Sgeir nan Ron ⚓ Q R; 55°45'·53N 06°17'·32W.
Hbr entrance ⚓ Q G; 55°37'·55N 06°11'·45W.
Rubh'an Dùin ⚓ Fl (2) WR 7s 15m W11M, R8M; W twr; vis: shore-R-218°-W-249°-R-350°-W-shore; 55°44'·70N 06°22'·28 W.

JURA TO MULL OF KINTYRE

SOUND OF JURA, CRAIGHOUSE, L SWEEN and GIGHA
Skervuile ⚓ Fl 15s 22m 9M; 55°36'·60N 06°12'·34W; W twr.
Nine Foot Rk ⚓ Q (3) 10s; 55°52'·44N 05°52'·91W.
Goat Rock ⚓ VQ (3) 5s; 55°50'·12N 05°55'·67W.
Eilean nan Gabhar ⚓ Fl 5s 9m 8M; 55°50'·01N 05°56'·25W.
Craighouse ⚓ Fl R 6s 3m 5M; 55°49'·98N 05°56'·43W.
Na Cùiltean ⚓ Fl 10s 9m 9M; 55°48'·64N 05°54'·90W.
Gamhna Gigha ⚓ Fl (2) 6s 7m 5M; 55°43'·78N 05°41'·08W.

WEST LOCH TARBERT
Dunskeig Bay ⚓ Q (2) 10s 11m 8M; 55°45'·22N 05°35'·00W.
Eileen Tráighe (off S side) ⚓ Fl (2) R 5s 5m 3M; R post; 55°45'·37N 05°35'·75W.

MULL OF KINTYRE
Mull of Kintyre ☆ 55°18'·64N 05°48'·25W Fl (2) 20s 91m **24M**; W twr on W bldg; vis: 347°-178°.

CRINAN CANAL and ARDRISHAIG
Crinan, E of lock ent ⚓ Fl WG 3s 8m 4M; W twr, R band; vis: shore-W-146°-G-shore; 56°05'·48N 05°33'·37W.
Ardrishaig Bkwtr Hd ⚓ L Fl WRG 6s 9m 4M; vis: 287°-G-339°-W-350°-R-035°; 56°00'·76N 05°26'·59W.

LOCH FYNE TO SANDA ISLAND

EAST LOCH TARBERT
Eilean na Beithe ⚓ Fl WRG 3s 7m 5M; vis: G036°- W065°- R078°-106°; 55°52'·68N 05°19'·62W.

KILBRANNAN SOUND, CRANNAICH and CARRADALE BAY
Port Crannaich Bkwtr Hd ⚓ Fl R 10s 5m 6M; vis: 099°-279°; 55°35'·60N 05°27'·84W.

CAMPBELTOWN LOCH to SANDA ISLAND
Davaar N Pt ☆ 55°25'·69N 05°32'·42W Fl (2) 10s 37m **23M**; W twr; vis: 073°-330°.
Sanda Island ☆ 55°16'·50N 05°35'·01W Fl 10s 50m **15M**; W twr; vis: 242°-121°; <u>992351116</u>.
Patersons Rock ⚓ Fl (3) R 18s; 55°16'·90N 05°32'·48W.

KYLES OF BUTE TO RIVER CLYDE

WEST KYLE
Ardlamont Point No. 47 ⚓ Fl R 4s; 55°49'·59N 05°11'·76.
Carry Rk No 46 ⚓ Fl R 4s; 55°51'·40N 05°12'·14W.

BURNT ISLAND CHANNEL
⚓ Fl R 4s; 55°55'·80N 05°10'·46W.
⚓ Fl G 3s; 55°55'·79N 05°10'·48W.
⚓ Fl R 2s; 55°55'·76N 05°10'·39W.
⚓ Fl G 5s; 55°55'·74N 05°10'·38W.
Creyke Rk No 45 ⚓ ; 55°55'·68N 05°10'·89W.
Beere Rk No 44 ⚓ ; 55°55'·10N 05°10'·63W.
Wood Fm Rk ⚓ ; 55°55'·42N 05°10'·31W.

EAST KYLE
Colintrave Pt ⚓ Fl R 3s 7m 2M; metal post; 55°55'·39N 05°09'·22W.
Ardmaleish Point No. 41 ⚓ Q; 55°53'·02N 05°04'·70W.

ROTHESAY SOUND
Ardyne ⚓ Fl G 3s; 55°52'·10N 05°03'·20W.
Bogany Point No. 36 ⚓ Fl R 4s; 55°50'·78N 05°01'·41W.

FIRTH OF CLYDE
Toward Pt ☆ 55°51'·73N 04°58'·79W Fl 10s 21m **22M**; W twr.
No. 34 ⚓ Q (3) 10s; 55°51'·44N 04°59'·11W.
Skelmorlie ⚓ Iso 5s; 55°51'·65N 04°56'·34W.

WEMYSS and INVERKIP
Cowal ⚓ L Fl 10s; 55°56'·00N 04°54'·83W.
The Gantocks ⚓ Fl R 6s 12m 6M; ○ twr; 55°56'·45N 04°55'·08W.
The Gantocks No. 31 ⚓ Q; 55°56'·56N 04°55'·13W.

DUNOON
Cloch Point ⚓ Fl 3s 24m 8M; W ○ twr, B band, W dwellings; 55°56'·55N 04°52'·74W.

LOCH LONG and LOCH GOIL
Loch Long ⚓ Oc 6s; 55°59'·15N 04°52'·42W.
Baron's Pt No. 3 ⚓ Oc (2) Y 10s 5m 3M; 55°59'·18N 04°51'·12W.
Ravenrock Pt ⚓ Fl 4s 12m 10M; W twr on W col. Dir lt 204°, WRG 9m (same twr); vis: 201·5°-F R-203°-Al WR(W phase incr with brg)-203·5°-FW-204·5°-Al WG(G phase incr with brg)-205°-FG-206·5°; 56°02'·14N 04°54'·39W.
Port Dornaige ⚓ Fl 6s 8m 11M; W col; vis: 026°-206°; 56°03·75N 04°53'·65W.
Rubha Ardnahein ⚓ Fl R 5s 3m 3M; vis: 132°-312°; 56°06'·15N 04°53'·60W.
The Perch, Ldg Lts 318° Front, F WRG 3m 5M; vis: 311°-G-317°-W-320°-R-322°. Same structure, Fl R 3s 3m 3M; vis: 187°-322°; 56°06'·90N 04°54'·31W. Rear, 700m from front, F 7m 5M; vis: 312°-322·5°.
Cnap Pt ⚓ Ldg Lts 031°. Front, Q 8m 10M; W col; 56°07'·40N 04°49'·96W. Rear, 87m from front F 13m; R line on W twr.

GOUROCK
Ashton ⚓ Iso 5s; 55°58'·10N 04°50'·65W.
Rosneath Patch ⚓ Fl (2) 10s 5m 10M; 55°58'·52N 04°47'·45W.

ROSNEATH, RHU NARROWS and GARELOCH
Ldg Lts 356°. **Front, No. 7N** ⚓ 56°00'·05N 04°45'·36W Dir lt 356°. WRG 5m **W16M**, R13M, G13M; vis: 353°-Al WG-355°- FW-357°-Al WR-000°-FR-002°.
Dir lt 115° WRG 5m **W16M**, R13M, G13M; vis: 111°-Al WG-114°-FW- 116°-Al WR-119°-FR-121°. Passing lt Oc G 6s 6m 3M;

G △ on G pile . Rear, Ardencaple Castle Centre ⚡ 56°00'·54N 04°45'·43W 2 FG (vert) 26m 12M; twr on Castle NW corner; vis: 335°-020°.

No. 8N Lt Bn ⚓ 55°59'·09N 04°44'·21W Dir lt 080° WRG 4m; **W16M**, R13M,G13M;vis: 075°-FG-077·5°-Al WG-079·5°-FW-080·5°-AltWR-082·5°-FR-085°. **Dir lt 138°** WRG 4m **W16M**, R13M, G13M; vis: 132°-FG-134°-Al WG- FW137°-139°-Al WR-142°. Passing lt Fl Y 3s 6m 3M.

Gareloch No. 1 Lt Bn ⚡ VQ (4) Y 5s 9m; Y 'X' on Y structure; 55°59'·12N 04°43'·89W.

No. 3N Lt Bn ⚓ 56°00'·07N 04°46'·72W Dir lt 149° WRG 9m **W16M**, R13M, G13M F & Al; vis: 144°-FG-145°-Al WG-148°-FW-150°-Al WR-153°-FR-154°. Passing lt Oc R 8s 9m 3M.

Rosneath DG Jetty ⚡ 2 FR (vert) 5M; W col; vis: 150°-330°; 56°00'·39N 04°47'·51W.

Rhu Pt ⚓ Dir lt 318° WRG **W16M**, R13M, G13M; vis: 315°-Al WG-317°-F-319°-Al WR-321°-FR-325°; Dir W; vis: 325°-QW-340°-FW-345°-OcW-350°; 56°00'·95N 04°47'·20W.

Limekiln No. 2N Lt Bn ⚓ 56°00'·67N 04°47'·64W Dir lt 295° WRG 5m **W16M**, R13M, G13M F & Al; R □ on R Bn; vis: 291°-Al WG- 294°-FW- 296°-Al WR-299°-FR-301°.

Mambeg Dir lt 331°, Q (4) WRG 8s 10m 14M; vis: 328·5°-G-330°-W- 332°-R-333°; H24; 56°03'·74N 04°50'·47W.

GREENOCK and PORT GLASGOW
Anchorage Lts in line 196°. Front, FG 7m 12M; Y col; 55°57'·62N 04°46'·58W. Rear, 32m from front, FG 9m 12M. Y col.

Lts in line 194·5°. Front, FG 18m; 55°57'·45N 04°45'·91W. Rear, 360m from front, FG 33m.

Steamboat Quay, W end ⚡ FG 12m 12M; B&W chequered col; vis 210°-290°; 55°56'·25N 04°41'·44W. From here to Glasgow Lts on S bank are Fl G and Lts on N bank are Fl R.

CLYDE TO MULL OF GALLOWAY

HUNTERSTON CHANNEL
'C' ▲ Fl G 5s; 55°48'·10N 05°55'·30W.
Hun 1 ⚏ (Y) Fl (4) Y 15s; 55°48'·10N 04°54'·21W.
Hun 3 ⚏ Fl R 2s; 55°47'·60N 04°53'·52W.

LARGS and FAIRLIE
Approach ⚬ L Fl 10s; 55°46'·40N 04°51'·85W.
Fairlie Patch ▲ Fl G 1·5s; 55°45'·38N 04°52'·34W.

MILLPORT and GREAT CUMBRAE
Tomont End Daymark 55°47'·56N 04°54'·10W.
Tattie Pier ⚡ WRG 3s 5m 4M; vis: 222°-G-234°-W-246°-R-258°; metal column; 55°47'·20N 04°53'·89W.
Ldg Lts 333°. Pier Head front, 55°45'·04N 04°55'·85W FR 7m 5M. Rear, 137m from front, FR 9m 5M.
Mountstuart ⚓ L Fl 10s; 55°48'·00N 04°57'·57W.
Runnaneun Pt (Rubha'n Eun) ⚡ Fl R 6s 8m 12M; W twr; 55°43'·79N 05°00'·23W.
Little Cumbrae Is, Cumbrae Elbow ⚡ Fl 6s 28m 14M; W twr; vis: 334°-193°; 55°43'·22N 04°58'·06W.

ARDROSSAN
Approach Dir lt 055°, WRG 15m W14M, R11M, G11M; vis: 050°-F G-051·2°-Alt WG(W phase inc with Brg)- 053·8°-FW-056·2°-Alt WR(R phase inc with brg)-058·8°-FR-060°; 55°38'·66N 04°49'·22W. Same structure FR 13m 6M; vis: 325°-145°.
Lt ho Pier Hd ⚡ Iso WG 4s 11m 9M; W twr; vis: 035°-W-317°-G-035°; 55°38'·47N 04°49'·57W.

IRVINE
Ldg Lts 051°. Front, FG 10m 5M; 55°36'·40N 04°41'·57W. Rear, 101m from front, FR 15m 5M; G masts, both vis: 019°-120°.

TROON
Troon Approach ▲ Fl G 4s; 55°33'·06N 04°41'·34W.
W Pier Hd ⚡ Fl (2) WG 5s 11m 9M; W twr; vis: 036°-G-090°-W-036°; 55°33'·07N 04°41'·02W.
Lady I ⚡ Fl 2s 19m 11M; W Tr R vert stripes; *Racon (T) 13-11M*; 55°31'·63N 04°44'·05W.

ARRAN, RANZA, LAMLASH and BRODICK
Pillar Rk Pt ☆ (Holy Island), 55°31'·04N 05°03'·65W Fl (2) 20s 38m **18M**; W □ twr.
Holy Is SW end ⚡ Fl G 3s 14m 6M; W twr; vis: 282°-147°; 55°30'·73N 05°04'·21W.
Pladda ☆ 55°25'·50N 05°07'·12W Fl (3) 30s 40m **17M**; W twr.

AYR and AILSA CRAIG
S Pier Hd ⚡ Q 7m 7M; R twr; vis: 012°-161°. Also FG 5m 5M; vis: 012°-082°; 55°28'·17N 04°38'·74W.
Ldg Lts 098°. Front, FR 10m 5M; Tfc sigs; 55°28'·15N 04°38'·38W. Rear, 130m from front Oc R 10s 18m 9M.
Maidens Hbr, E side ⚡ Fl G 5s 4m 3M; 55°20'·24N 04°49'·20W.
Maidens Hbr, W side ⚡ Fl R 3s 4m 2M; 55°20'·23N 04°49'·19W.
Turnberry Point ⚡, near castle ruins 55°19'·57N 04°50'·69W Fl 15s 29m 12M; W twr.
Ailsa Craig ☆ 55°15'·12N 05°06'·52W Fl 4s 18m **17M**; W twr; vis: 145°-028°.

GIRVAN
S Pier Hd ⚡ 2 FG (vert) 8m 4M; W twr; 55°14'·72N 04°51'·90W.

LOCH RYAN and STRANRAER
Milleur Point ⚓ Q; 55°01'·28N 05°05'·66W.
Fairway ⚬ Iso 4s; 54°59'·77N 05°03'·82W.
Forbes Shoal ⚏ QR; 54°59'·47N 05°02'·96W.
Loch Ryan W ▲ QG;54°59'·23N 05°03'·24W .
Cairn Pt ⚡ Fl (2) R 10s 14m 12M; W twr; 54°58'·46N 05°01'·85W.
Cairnryan ⚡ Fl R 5s 5m 5M; 54°57'·77N 05°00'·99W.
Stranraer No.1 ⚓ Oc G 6s; 54°56'·67N 05°01'·32W.
No. 3 ⚓ QG; 54°55'·87N 05°01'·60W.
No. 5 ⚓ Fl G 3s; 54°55'·08N 05°01'·86W.
E Pier Hd ⚡ 2 FR (vert) 9m; 54°54'·61N 05°01'·60W.
Corsewall Point ☆ 55°00'·41N 05°09'·58W Fl (5) 30s 34m **22M**; W twr; vis: 027°-257°.
Black Head Old Lighthouse (disused); W tower, 22m; 54°51'·70N 05°08'·80W.

PORTPATRICK
Ldg Lts 050·5°. Front, FG (occas); 54°50'·50N 05°07'·02W. Rear, 68m from front, FG 8m (occas).
Crammag Hd ☆ 54°39'·90N 04°57'·92W Fl 10s 35m **18M**; W twr.
Mull of Galloway ☆, SE end 54°38'·08N 04°51'·45W Fl 20s 99m **28M**; W twr; vis: 182°-105°.

MULL OF GALLOWAY and WIGTOWN BAY
Port William Ldg Lts 105°. Front, Pier Hd Fl G 3s 7m 3M; 54°45'·66N 04°35'·28W. Rear, 130m from front, FG 10m 2M.
Isle of Whithorn ⚡ Fl WR 3s 20m 6/4M; vis: 310°-W-005°- R-040°; 54°41'·79N 04°21'·54W.
Whithorn Ldg Lts 335°. Front, Oc R 8s 7m 7M; Or ♦; 54°42'·01N 04°22'·05W. Rear, 35m from front, Oc R 8s 9m 7M; Or ♦, synch.
Little Ross ⚡ Fl 5s 50m 12M; W twr; obsc in Wigtown B when brg more than 103°; 54°45'·93N 04°05'·10W.

KIRKCUDBRIGHT BAY and KIPPFORD
Little Ross NNE end of Is ⚡ Fl (2) 5s 21m 5M; Stone bcn; 54°46'·06N 04°05'·02W.
Hestan I, E end ⚡ Fl (2) 10s 42m 9M; 54°49'·95N 03°48'·53W.

9.9.5 PASSAGE INFORMATION

More passage information is threaded between the harbours in this area. Conditions in the SW of Scotland are generally less rugged south of Mull but good charts and up to date Pilots are still essential. Admiralty Leisure Folios 5611 covers the Mull of Kintyre to Point of Ardnamurchan, 5610, the Firth of Clyde. 5613, Irish Sea eastern part including Isle of Man, covers the Scottish coast south of the Clyde to Kirkcudbright. **Bibliography:** *The Yachtsman's Pilot, Clyde to Colonsay* (Imray/Lawrence). *Clyde Cruising Club Sailing Directions-Firth of Clyde including Solway Firth and North Channel* (Imray/Clyde Cruising Club, Lawrence). The *South West Coast of Scotland Pilot* (Admiralty NP66A) covers the whole area.

Submarines regularly exercise in this area, see **9.9.18, SUBMARINE EXERCISE AREAS** and SUBFACTS for information on this activity.

Some of the following more common *Gaelic* terms may help with navigation: *Acairseid*: anchorage. *Ailean*: meadow. *Aird, ard*: promontory. *Aisir, aisridh*: passage between rocks. *Beag*: little. *Beinn*: mountain. *Bo, boghar, bodha*: rock. *Cala*: harbour. *Camas*: channel, bay. *Caol*: strait. *Cladach*: shore, beach. *Creag*: cliff. *Cumhamn*: narrows. *Dubh, dhubh*: black. *Dun*: castle. *Eilean, eileanan*: island. *Garbh*: rough. *Geal, gheal*: white. *Glas, ghlas*: grey, green. *Inis*: island. *Kyle*: narrow strait. *Linn, Linne*: pool. *Mor, mhor*: large. *Mull*: promontory. *Rinn, roinn*: point. *Ruadh*: red, brown. *Rubha, rhu*: cape. *Sgeir*: rock. *Sruth*: current. *Strath*: river valley. *Tarbert*: isthmus. *Traigh*: beach. *Uig*: bay.

CORRYVRECKAN TO CRINAN

(AC 2326, 2343) Between Scarba and Jura is the Gulf of Corryvreckan (AC 2343) which is best avoided, and should never be attempted by small craft except at slack water and in calm conditions. (In any event the Sound of Luing is always a safer and not much longer alternative.) The Gulf has a least width of 6ca and is free of dangers apart from its very strong tides which, in conjunction with a very uneven bottom, cause extreme turbulence. This is particularly dangerous with strong W winds over a W-going (flood) tide which spews out several miles to seaward of the gulf, with overfalls extending 5M from the W of ent (The Great Race). Keep to the S side of the gulf to avoid the worst turbulence and the whirlpool known as The Hag, in depths of only 29m, as opposed to more than 100m in the fairway.

▶ *The W-going stream in the gulf begins at HW Oban + 0410, and the E-going at HW Oban − 0210. Sp rate W-going is 8·5kn, and E-going about 6kn. The range of tide at sp can vary nearly 2m between the E end of the gulf (1·5m) and the W end (3·4m), with HW 30mins earlier at the E end. Slack water occurs at HW Oban +0400 and −0230 and lasts almost 1 hr at Nps, but only 15 mins at sps. On the W-going (flood) stream eddies form both sides of the gulf, but the one on the N (Scarba) shore is more important. Where this eddy meets the main stream off Camas nam Bairneach there is violent turbulence, with heavy overfalls extending W at the division of the eddy and the main stream.* ◀

There are temp ⚓s with the wind in the right quarter in Bàgh Gleann a' Mhaoil in the SE corner of Scarba, and in Bàgh Gleann nam Muc at N end of Jura but the latter has rks in approaches E and SW of Eilean Beag.

SE of Corryvreckan is Loch Crinan, which leads to the Crinan Canal. Beware Black Rk, 2m high and 2ca N of the canal sea lock, and dangers extending 100m from the rock.

WEST OF JURA TO ISLAY

(AC 2481, 2168) The W coasts of Colonsay and Oronsay (AC 2169) are fringed with dangers up to 2M offshore. A narrow drying channel crossed by an overhead cable (10m) separates them.

The Sound of Islay presents no difficulty; hold to the Islay shore, where all dangers are close in. There are ⚓s in the Sound, but holding is mostly poor. The best places are alongside at Port Askaig, or at ⚓ off the distillery in Bunnahabhain B, 2·5M to the N.

▶ *The N-going stream begins at HW Oban + 0440, and the S-going at HW Oban − 0140. Main flood begins HW Oban +0545. Streams turn approx 1 hr earlier in Gigha Sd and at Kintyre and Jura shores. S going stream for 9hrs close inshore between Gigha and Machrihanish starting HW Oban −0530.*

Main ebb begins HW Oban −0015. Streams turn 1 hr earlier in Gigha Sd, Kintyre and Jura shores. Overfalls off McArthur's Hd. The Sp rates are 2·5kn at N entrance and 1·5kn at S entrance, but reaching 5kn in the narrows off Port Askaig. There are overfalls off McArthur's Hd (Islay side of S entrance) during the S-going stream. ◀

The N coast of Islay and Rhinns of Islay are very exposed. In the N there is an ⚓ SE of Nave Island at entrance to Loch Gruinart; beware Balach Rks which dry, just to N.

▶ *To the SW off Orsay (lt), Frenchman's Rks and W Bank there is a race and overfalls which should be cleared by 3M. Here the NW-going stream begins at HW Oban + 0530, and the SE-going at HW Oban − 0040; Sp rates are 6-8kn inshore, but decrease to 3kn 5M offshore.* ◀

Loch Indaal gives some shelter; beware rks extending from Laggan Pt on E side of ent. Off the Mull of Oa there are further overfalls. Port Ellen, the main hbr on Islay, has HIE ⚓s; there are some dangers in approach, and it is exposed to S; see AC 2474.

SOUND OF JURA TO GIGHA

(AC 2397, 2396, 2168) From Crinan to Gigha the Sound of Jura is safe if a mid-chan course is held. Ruadh Sgeir (lt) are rocky ledges in mid-fairway, about 3M W of Crinan. Loch Sween (AC 2397) can be approached N or SE of MacCormaig Islands, where there is an attractive ⚓ on NE side of Eilean Mor, but exposed to NE. Coming from N beware Keills Rk and Danna Rk. Sgeirea a Mhain is a rk in fairway 1·5M NE of Castle Sween (conspic on SE shore). ⚓ at Tayvallich, near head of loch on W side.

W Loch Tarbert (AC 2477) is long and narrow, with good ⚓s and lts near ent, but unmarked shoals. On entry give a berth of at least 2½ca to Eilean Traighe off N shore, E of Ardpatrick Pt. Dun Skeig, an isolated hill, is conspic on S shore. Good ⚓ near head of loch, 1M by road from E Loch Tarbert, Loch Fyne.

On W side of Sound, near S end of Jura, are The Small Is (AC 2396) across the mouth of Loch na Mile. Beware Goat Rk (dries 0·3m) 1·5ca off S'most Is, Eilean nan Gabhar, behind which is good ⚓. Also possible to go alongside Craighouse Pier (HIE ⚓). Another ⚓ is in Lowlandman's B, about 3M to N, but exposed to S winds; Ninefoot Rks with depth of 2·4m and ECM lt buoy lie off ent. Skervuile (lt) is a reef to the E, in middle of the Sound.

S of W Loch Tarbert, and about 2M off the Kintyre shore, is Gigha Is (AC 2475). Good ⚓s on E side in Druimyeon B and Ardminish B (HIE ⚓s), respectively N and S of Ardminish Pt. Outer and Inner Red Rks (least depth 2m) lie 2M SW of N end of Gigha Is. Dangers extend 1M W off S end of Gigha Is. Gigalum Is and Cara Is are off the S end. Gigha Sound needs very careful pilotage, since there are several dangerous rks, some buoyed/lit, others not.

▶ *The N-going stream begins at HW Oban + 0430, and S-going at HW Oban − 0155, sp rates 1·3kn.* ◀

HARBOURS AND ANCHORAGES IN COLONSAY, JURA, ISLAY AND THE SOUND OF JURA

SCALASAIG, Colonsay, 56°04′·14N 06°10′·86W. AC 2169. HW +0542 on Dover; ML 2·2m. See 9.9.6. Conspic monument ½M SW of hbr. Beware group of rks N of pier hd marked by bn. 2❷ berths run by CMAL on N side of pier, inner end approx 2·5m. Inner hbr to SW of pier is safe, but dries. Ldg lts 262°, both FR 8/10m on pier. Also ⬇ clear of cable in **Loch Staosnaig**; SW of Rubha Dubh lt, Fl (2) WR 10s 8m 8/6M; shore-R-230°-W-337°-R-354°. Facilities: 🏠 🅿 🛒 (all at ✉), ⛽, Hotel ☎(01951) 200316, Dr (0951) 200328.

LOCH TARBERT, W Jura, 55°57′·69N 06°00′·06W. AC 2169, 2481. Tides as Rubha A'Mhàil (N tip of Islay). See 9.9.6. HW –0540 on Dover; ML 2·1m; Duration 0600. Excellent shelter inside the loch, but subject to squalls in strong winds; ⬇ outside in Glenbatrick Bay in approx 6m in S winds, or at Bagh Gleann Righ Mor in approx 2m in N winds. To enter inner loch via Cumhann Beag, there are four pairs of ldg marks (W stones) at approx 120°, 150°, 077°, and 188°, the latter astern, to be used in sequence; pilot book required. There are no facilities.

PORT ASKAIG, Islay, 55°50′·87N 06°06′·26W. AC 2168, 2481. HW +0610 on Dover; ML 1·2m. See 9.9.6. Hbr on W side of Sound of Islay. ⬇ close inshore in 4m or secure to ferry pier. Beware strong tide/eddies. ☆ FR at LB. Facilities: ⛽(on pier) 🏠 🅿 ✉ 🛒 ✗ Hotel ferries to Jura and Kintyre. Other ⬇s in the Sound at: Bunnahabhain (2M N); Whitefarland Bay, Jura, opp Caol Ila distillery; NW of Am Fraoch Eilean (S tip of Jura); Aros Bay, N of Ardmore Pt.

CRAIGHOUSE, SE Jura, 55°49′·99N 05°56′·31W. AC 2168, 2481, 2396. HW +0600 on Dover; ML 0·5m; Duration 0640 np, 0530 sp. See 9.9.6. Good shelter, but squally in W winds. Enter between lit bn on SW end of Eilean nan Gabhar, Fl 5s 7m 8M vis 225°-010°, and unlit bn close SW. There are 16 Jura Dev Trust ⚓s (£10/night) N of pier (☎(01496) 810332), where yachts may berth alongside; or ⬇ in 5m at the N end of Loch na Mile (poor holding in heavy weed). Facilities: very limited, short stay pontoon ⛽ 🏠 🅿 🅿 ✉ 🛒 ✗ 🏠. **Lowlandman's Bay** is 1M further N, with ECM buoy, Q (3) 10s, marking Nine Foot Rk (2·4m) off the ent. ⬇ to SW of conspic houses, off stone jetty.

LOCH SWEEN, Argyll and Bute, 55°55′·69N 05°41′·26W. AC 2397. HW +0550 on Dover; ML 1·5m; Duration = 0610. See 9.9.7 Carsaig Bay. Off the ent to loch, **Eilean Mòr** (most SW'ly of MacCormaig Isles) has tiny ⬇ on N side in 3m; local transit marks keep clear of two rks, 0·6m and 0·9m. Inside the loch, beware Sgeirean a'Mhain, a rk in mid-chan to S of Taynish Is, 3M from ent. Good shelter in Loch a Bhealaich (⬇ outside **Tayvallich** in approx 7m on boulders) or enter inner hbr to ⬇ W of central reef. There are no lights. Facilities: 🅿 🏠 ✉ with ⛽ 🛒 ✗. Close to NE are ⬇s at **Caol Scotnish** and **Fairy Is**, the former obstructed by rks 3ca from ent.

WEST LOCH TARBERT, Argyll and Bute, (Kintyre), 55°45′N 05°36′W. AC 2476. Tides as Gigha Sound, 9.9.6. Good shelter. En S of Eilean Traighe, Fl (2) R 5s, and NW of Dun Skeig, Q (2) 10s, where there is conspic conical hill (142m). Avoid shallows inshore to each side. Loch is lit for 5M by 3 bns, QG, QR and QG in sequence, up to Kennacraig ferry pier, 2FG (vert). PHM buoy, QR, is 2½ca NW of pier. Caution: many drying rks and fish farms outside the fairway and near head of loch. ⬇s are NE of Eilean Traighe (beware weed and ferry wash); near Rhu Pt, possible ⚓s; NE of Eilean dà Gallagain, and at loch hd by pier (ru). Tarbert (9.9.9) is 1·5M walk/bus.

GIGHA ISLAND, Argyll and Bute, 55°40′·6N 05°44′·0W. AC 2168, 2475. HW +0600 on Dover; ML 0·9m; Duration 0530. See 9.9.6. **Ardminish Bay**: 23 Gigha Heritage Trust ⚓s in the centre (£10) 17 rated at 15t and 6 at 25t or ⬇ NE of Ferry Jetty. Refurbished pontoon and walkway for landing. Reefs extend off both points, the S'ly reef marked by an unlit PHM buoy. Kiln Rk (dries 1·5m) is close NE of the old ferry jetty. **Druimyeon Bay** is more sheltered in E'lies, but care needed entering from S. ⬇s sheltered from winds in Kilnaughton Bay, Port Mór (S-W), Bàgh na Dòirlinne (SE-S), W Tarbert Bay (NE). Caolas Gigalum (⬇ 50m SE of pier) is safe in all but NE-E winds. Beware many rks in Gigha Sound. Lts: Fl (2) 6s, on Gamhna Gigha (off NE tip); WCM buoy Fl (9) 15s marks Gigalum Rks, at S end of Gigha. **Ardminish** ☎(01583) 505254: ⛽ 🏠 🅿 🅿 ✉ 🛒 ✗ 🏠.

9.9.6 PORT ELLEN

Islay (Argyll and Bute) 55°37′·29N 06°12′·26W ❋❋◊◊✿✿

CHARTS AC 2168, 2476; Imray C64

TIDES HW +0620 np, +0130 sp on Dover; ML 0·6. Sea level is much affected by the weather, rising by 1m in S/E gales; at nps the tide is sometimes diurnal and range negligible.

Standard Port OBAN (◀—)

Times				Height (metres)			
High Water		Low Water		MHWS	MHWN	MLWN	MLWS
0100	0700	0100	0800	4·0	2·9	1·8	0·7
1300	1900	1300	2000				
Differences PORT ELLEN (S Islay)							
–0530	–0050	–0045	–0530	–3·1	–2·1	–1·3	–0·4
SCALASAIG (E Colonsay)							
–0020	–0005	–0015	+0005	–0·3	–0·2	–0·3	0·0
GLENGARRISDALE BAY (N Jura)							
–0020	0000	–0010	0000	–0·4	–0·2	0·0	–0·2
CRAIGHOUSE (SE Jura)							
–0230	–0250	–0150	–0230	–3·0	–2·4	–1·3	–0·6
RUBHA A'MHÀIL (N Islay)							
–0020	0000	+0005	–0015	–0·3	–0·1	–0·3	–0·1
ARDNAVE POINT (NW Islay)							
–0035	+0010	0000	–0025	–0·4	–0·2	–0·3	–0·1
ORSAY ISLAND (SW Islay)							
–0110	–0110	–0040	–0040	–1·4	–0·6	–0·5	–0·2
BRUICHLADDICH (Islay, Loch Indaal)							
–0105	–0035	–0110	–0110	–1·8	–1·3	–0·4	+0·3
PORT ASKAIG (Sound of Islay)							
–0030	–0035	–0015	–0025	–1·8	–1·3	–0·7	–0·2
GIGHA SOUND (Sound of Jura)							
–0450	–0210	–0130	–0410	–2·5	–1·6	–1·0	–0·1
MACHRIHANISH							
–0520	–0350	–0340	–0540	Mean range 0·5 metres.			

SHELTER Good shelter on pontoons (3m at MLWS), but in S winds swell sets into the bay. W of Rubha Glas rocks marked by 3 bns with reflective topmarks. In W'lies ⬇ in Kilnaughton Bay, N of Carraig Fhada lt ho; or 4M ENE at Loch-an-t-Sàilein.

NAVIGATION WPT 55°36′·69N 06°12′·06W, 326°/0·63M to Carraig Fhada lt ho. Beware Otter Rk 4M SE of hbr marked by ECM, rks on both sides of ent and in NE corner of bay. Keep close to pier.

LIGHTS AND MARKS On W side 10 radio masts (103m) and Carraig Fhada lt ho (conspic), Fl WRG 3s 20m 8/6M; keep in W sector until past the SHM lt buoy. Sgeir nan Ròn Bn Q R, dolphin off Ro-Ro Pier 2 FG (vert) and marina pontoon Fl G 4s. Limit of dredged area around pier and pontoons area marked by buoys.

COMMUNICATIONS CGOC (02891) 463933. Moorings are run by local community.

FACILITIES Marina (30⬇ inc ❷, ⚓ in bay), portellenmarina.co.uk ☎300301 for fuel/assistance, dredged to 3m, ⚓ £12<10m>£14, 🚻/🚿 🅿 🏠 🅿. **Village:** ⛽ 🅿 ✉ 🛒 ✗ 🏠 @.

9.9.7 CRINAN CANAL

Argyll and Bute 56°05'·50N 05°33'·38W Crinan ✿✿✿✿✿✿✿✿✿

CHARTS AC 2326, 2476; Imray C65, C63.

TIDES –0608 Dover; ML 2·1; Duration 0605; HW Crinan is at HW Oban –0045

Standard Port OBAN (←)

Times				Height (metres)			
High Water		Low Water		MHWS	MHWN	MLWN	MLWS
0100	0700	0100	0800	4·0	2·9	1·8	0·7
1300	1900	1300	2000				
Differences CARSAIG BAY (56°02'N 05°38'W)							
–0105	–0040	–0050	–0050	–2·1	–1·6	–1·0	–0·4

NOTE: In the Sound of Jura, S of Loch Crinan, the rise of tide occurs mainly during the 3½ hrs after LW; the fall during the 3½ hrs after HW. At other times the changes in level are usually small and irregular.

SHELTER Complete shelter in canal basin; yachts are welcome. Good shelter in Crinan Hbr (E of Eilean da Mheinn) but full of moorings, contact Crinan Boatyard. Except in strong W/N winds, ⊥ E of the canal ent, clear of fairway or SW of Eilean da Mheinn. Good holding in Gallanach Bay on N side of L Crinan in about 3m.

NAVIGATION WPT 56°05'·70N 05°33'·64W, 146°/0·27M to Fl WG 3s lt. Beware Black Rock (2m high) in appr NE of ldg line 146° to dir Fl WG 3s lt. Off NW corner of chartlet, no ⊥ in a nearly rectangular shellfish bed, 6ca by 6ca. SPM lt buoys mark each corner: Fl (4) Y 12s at the NE and NW corners, Fl Y 6s at the SW and SE corners; the latter being about 100m NE of Black Rock.

CANAL Canal is 9M (14.5km) long with 15 locks and 7 opening bridges. Transit time is at least 6 hrs, observing the 4kn speed limit. Entry at all tides. Max LOA: 26·82m, 6.09m beam, 2·7m draught (add 0.10m to salt water draught. If >2.2m contact Canal Office or Sea Locks 24 hrs in advance to confirm canal level will allow passage – there may be some delay to raise water levels) mast 28.95m. The canal operates Mon-Fri in winter months increasing to 6/7 days in peak season. Contact Crinan Canal Office for timings or consult www.scottishcanals.co.uk. Periods of closure may be expected during winter months. It is closed between Christmas and New Year. Long warps and fenders are essential; sea locks can be turbulent at low tide and care is to be taken handling warps – always take a turn round a suitable cleat. Do not pump out bilges or heads in the canal. Water points are supplied but not hoses. Bellanoch pontoons / moorings are reserved for long term berth holders and should not be used by those in transit.

BOAT SAFETY SCHEME Transit craft may be subject to random safety checks on gas systems. £1M 3rd party insurance is required. See also 9.8.17.

LIGHTS AND MARKS Crinan Hotel is conspic W bldg. A conspic chy, 3ca SW of hotel, leads 187° into Crinan Hbr. E of sea-lock: Dir Fl WG 3s 8m 4M, vis 114°-W-146°-G-280°. Ent: 2 FG (vert) and 2FR (vert).

COMMUNICATIONS (Code 01546) CGOC (02891) 463933; Police 101; Dr 462001. Canal Office 603210 (HO). Sea lock 830285 (for advice at w/e and if no response to VHF – staff carry phone). VHF Ch 74 16 only at sea locks.

FACILITIES Canal Office (Mon-Fri), www.scottishcanals.co.uk. All fees are under review; contact office for details. ⚓⚓⚓ *Skippers Guide* is essential reading; **Sea Basin** ⊇ (overnight rate available). Use shore toilets, not yacht heads, whilst in canal.

Services: Crinan Boatyard www.crinanboatyard.co.uk ☎830232; ➤ ⚓⚓⚓⚓ ⚒ ⚓⚓⚓⚓ C(5t), ✗.

Village ✉ Ⓑ ⚓⚓ ✗ (Ardrishaig), ⇌ (Oban), ✈ (Glasgow).

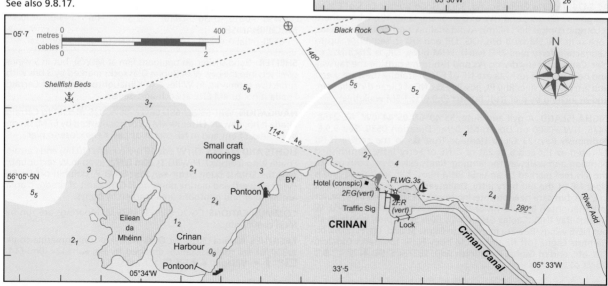

MULL OF KINTYRE

(AC 2126, 2199, 2798) From Crinan to Mull of Kintyre is about 50M. Passage round the Mull of Kintyre within 2 or 3 miles of the coast requires great care. Races exist S and SW of the Mull, and fresh to strong S winds cause dangerous breaking seas especially during E-going (flood) tidal streams. Best to keep at least 3 miles offshore and S of Sanda Island.

▶ *In Sanda Sound the W-going stream starts HW Dover –0110, and the E-going stream starts HW Dover +0500. Sp rates 5kn. Close W of the Mull the N-going stream starts at HW Dover –0130, and the S-going stream at HW Dover +0430. Sp rates 5kn. Careful timing is needed, especially W-bound.* ◀

The Traffic Separation Scheme in the North Channel, is only 2M W of the Mull and limits sea-room in the ITZ.

Sanda Sound separates Sanda Island and its rocks and islets from Kintyre. On the mainland shore beware Macosh Rks (dry, PHM lt buoy), forming part of Barley Ridges, 2ca offshore; and Arranman's Barrels, drying and submerged, marked by PHM lt buoy. A tidal turbine is sited 4ca SE off Rat Stane just outside a line joining the 2 buoys, with a lattice twr SPM, X topmark, Fl Y 5s. Sanda Is provides useful stopping point while waiting for the tide. Beware Paterson's Rk (9ca E, dries). There is good ⚓ on N side in Sanda Roads. Approach with Boat Ho bearing 155°; beware strong tidal stream across entrance, and drying rocks in middle of bay. See Plan on AC 2126.

▶ *Once E of Mull of Kintyre, tidal conditions and pilotage much improve.* ◀

MULL OF KINTYRE TO UPPER LOCH FYNE

(AC 2126, 2383, 2381, 2382). Campbeltown is entered N of Island Davaar (lt). 1·5M N off lt ho is Otterard Rk (depth 3·8m), with Long Rk (dries 1·1m) 5ca W of it; only Otterard Rock is buoyed.

▶ *E of Island Davaar tide runs 3kn at Sp, and there are overfalls.* ◀

Kilbrannan Sound runs 21M from Island Davaar to Skipness Pt, where it joins Inchmarnock Water, Lower Loch Fyne and Bute Sound.

▶ *There are few dangers apart from overfalls off Carradale Pt and on Erins Bank, 10M S of Skipness, on S-going stream.* ◀

Good ⚓ in Carradale Bay, off Torrisdale Castle.

Lower L Fyne (AC 2381) is mainly clear of dangers to East L Tarbert. On E shore beware rks off Ardlamont Pt; 4M to NW is Skate Is which is best passed to W. 3M S of Ardrishaig beware Big Rk (depth 2·1m). Further N, at entrance to Loch Gilp (mostly dries), note shoals (least depth 1·5m) round Gulnare Rk, PHM lt buoy; also Duncuan Is with dangers extending SW to Sgeir Sgalag (depth 0·6m), buoyed.

Where Upper L Fyne turns NE (The Narrows) it is partly obstructed by Otter Spit (dries 0·9m), extending 8ca WNW from E shore and marked by lt bn. The stream runs up to 2kn here. A buoyed/lit rk, depth less than 2m, lies about 7ca SW of Otter Spit bn. In Upper L Fyne (AC 2382) off Minard Pt, the chan between rks and islands in the fairway is buoyed/lit.

9.9.8 ARDRISHAIG

Argyll and Bute 56°00'·78N 05°26'·62W ✻✻♦♦♧♧

CHARTS AC 5610, 2131, 2381; Imray C63, 2900

TIDES +0120 Dover; ML 1·9; Duration 0640

Standard Port GREENOCK (→)

Times				Height (metres)			
High Water		Low Water		MHWS	MHWN	MLWN	MLWS
0000	0600	0000	0600	3·4	2·8	1·0	0·3
1200	1800	1200	1800				
Differences ARDRISHAIG							
+0006	+0006	-0015	+0020	0·0	0·0	+0·1	−0·1
INVERARAY							
+0011	+0011	+0034	+0034	−0·1	+0·1	−0·5	−0·2

SHELTER Hbr is sheltered except from strong E'lies; do not berth on pier or ⚓ due to commercial vessels H24. Sea lock into the Crinan Canal is usually left open, however, a waiting pontoon has been installed outside the lock, with restricted depth, for use in fair weather only. Access at all tides. Complete shelter in the canal basin, or beyond lock No 2. Also ⚓ 2ca N of hbr, off the W shore of L Gilp. Head of Loch dries for nearly 1M.

NAVIGATION WPT No 48 PHM buoy, Fl R 4s, 56°00'·18N 05°26'·31W, 345°/0·6M to bkwtr lt. Dangerous drying rocks to E of appr chan are marked by No 49 Fl.G.5s SHM buoy.

LIGHTS AND MARKS Conspic W Ho on with block of flats leads 315°between Nos 48 and 49 buoys. Bkwtr lt, L Fl WRG 6s, W339°-350°. Pier is floodlit. Other lights as plan.

COMMUNICATIONS (Code 01546) CGOC (02891) 463933; Police 101; Dr 462001. Canal Office 603210 (Mon-Fri); Sea lock 602458.

VHF Ch **74** 16.

FACILITIES Pier/Hbr ☎603210, ☺ (fees under review) ⚓ ⚓.

Crinan Canal www.scottishcanals.co.uk **Sea Lock** ☎602458, ☺ ⚓⚓⚓ ✗ 🖵; dues, see 9.9.7. **Services:** ⛽ ✗ ✎ 🖳 🚿 🛢.

Village: ✉ Ⓑ 🛒 ✗ 🖵 ≈ (Oban), ✈ (Glasgow or Campbeltown).

9.9.9 TARBERT

Tarbert (entrance) **55°52'·05N 05°24'·22W** ❀❀❀❀♨♨♨♨♤♤

CHARTS AC 5610, 2131, 2381; Imray C63, 2900

TIDES +0120 Dover; ML 1·9; Duration 0640

Standard Port GREENOCK (→)

Times				Height (metres)			
High Water		Low Water		MHWS	MHWN	MLWN	MLWS
0000	0600	0000	0600	3·4	2·8	1·0	0·3
1200	1800	1200	1800				
Differences EAST LOCH TARBERT							
–0005	–0005	0000	–0005	+0·2	+0·1	0·0	0·0

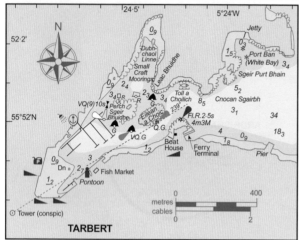

TARBERT

SHELTER Very good in all weathers on new pontoons. Access H24. 100 Ⓥ ◖ on pontoons (call on VHF Ch 14 for berth); only use Fish Quay if directed by HM.

NAVIGATION WPT 55°52'·02N 05°23'·03W, 270°/0·7M to Fl R 2·5s lt. Ent is 60m wide and well marked. Cock Isle divides the ent in half: Main hbr to the S, Buteman's Hole to the N, where ⚓s are fouled by heavy moorings and lost chains. Speed limit 3kn.

LIGHTS AND MARKS Outer ldg lts 252° to S ent: Fl R 2·5s on with Cock Is lt QG. Inner ldg line 239°: same QG, G column, on with conspic ⊞ tr. Note: The W sector, 065°-078°, of Eilean na Beithe ☀, Fl WRG 3s 7m 5M (on E shore of Lower Loch Fyne), could be used to position for the initial appr to Tarbert.

COMMUNICATIONS (Code 01880) CGOC (02891) 463933; Police 101; Ⓗ (01546) 602323. HM 820344, www.tarbertharbour.co.uk. VHF Ch **14** (0900-1700LT).

FACILITIES Yacht Berthing Facility, www.tarbertharbour.co.uk, ◖ £2.54, ⚓ ⛽.

Fish Quay ⚓ ◢(Mon-Fri 0800-1800); **Tarbert YC** ◣.

Town 🛏 🛏(0800-1800) ▲ ⚓ ✕ ⚒ ⊞ ⚠ ⛽ ⛴ ⚒ ACA, 🛒 ✉ Ⓑ ✕ 🖵 ⇌ (bus to Glasgow), ✈ (Glasgow/Campbeltown).

ANCHORAGES IN LOCH FYNE
• *Beware discarded wires on seabed throughout the area.*
In Lower L Fyne there is a ⚓ at Kilfinan Bay, ☎(01700) 821201. In Upper Loch Fyne: 10 ⚓ for patrons of the Oyster Bar at Cairndow 6M N of Inverary; ⚓ SSW of pier in 4m or dry out NW of the pier; ⚓ (on pier), ▲ ✉ 🛒 ✕ 🖵 bus to Glasgow. On NW bank at Port Ann, Loch Gair and Minard Bay; and on SE bank at Otter Ferry, Sailors' Bothy run by Otter Ferry Pier Association with dinghy pontoon, 🚿 and toilets (15⚓s at Oyster Catcher Inn), Strachur Bay (pontoon and 5⚓s off Creggans Inn, ☎(01369) 860279) and St Catherine's.

9.9.10 PORTAVADIE

Tarbert (entrance) **55°52'·23N 05°18'·98W** ❀❀❀❀♨♨♨♨♤♤

CHARTS AC 5610, 2131, 2381; Imray C63, 2900

TIDES +0120 Dover; ML 1·9; Duration 0640

Standard Port GREENOCK (→)

Times				Height (metres)			
High Water		Low Water		MHWS	MHWN	MLWN	MLWS
0000	0600	0000	0600	3·4	2·8	1·0	0·3
1200	1800	1200	1800				
Differences PORTAVADIE							
–0005	–0005	0000	–0005	+0·2	+0·1	0·0	0·0

SHELTER Very good in a well protected deep water facility. Access H24. 230 Ⓥ ◖ on pontoons with assistance available (call on VHF 80 for berth). If arriving after hours use S, A, B or ◖ pontoons.

NAVIGATION WPT 55°52'·00N 05°19'·20W, 027°/0·27M to entrance. When approaching from north pass SW of Eilean Buidhe which has fish farm to the NE. Inshore passage may be used with local knowledge. From the S Clear Sgat Mor and round Rubha Stillaig avoiding fish farm inshore. Portavadie has a narrow entrance between lit breakwaters. Speed limit 3kn.

LIGHTS AND MARKS Eilean na Beithe to North Fl WRG 3s 7m 5M 036°-G-065°-W-078°-R-106° marks access to Glenan Bay and inshore passage NE of Eilean Buidhe. Sgat Mor Fl 3s 9m 12M. Entrance marked by beacons PH Fl R 2s 6m 2M; SH Fl (2) G 8s 6m 4M.

COMMUNICATIONS (Code 01880) CGOC (02891) 463933; Police 101. VHF Ch **80** (0900-1700LT).

FACILITIES Marina, www.portavadiemarina.com ☎811075, 206 ◣ ◖ inc Ⓥ £2.15, ⚒ ▲ ✕ ⊞ 🍴 ⛽(12t) ⚠ ✕ 🖵 wi-fi, small shop, ✈ (Glasgow). Other local shops at Kames (2.5M).

ADJACENT ANCHORAGES
Glenan Bay to NW ⚓ in 15m steep to bay open to SW.

Asgog B1.5M to S entered between Sgat Mor and Sgat Beag. ⚓ in 15M to E of Eilean Aoidhe.

9.9.11 TIDAL STREAMS AROUND THE MULL OF KINTYRE

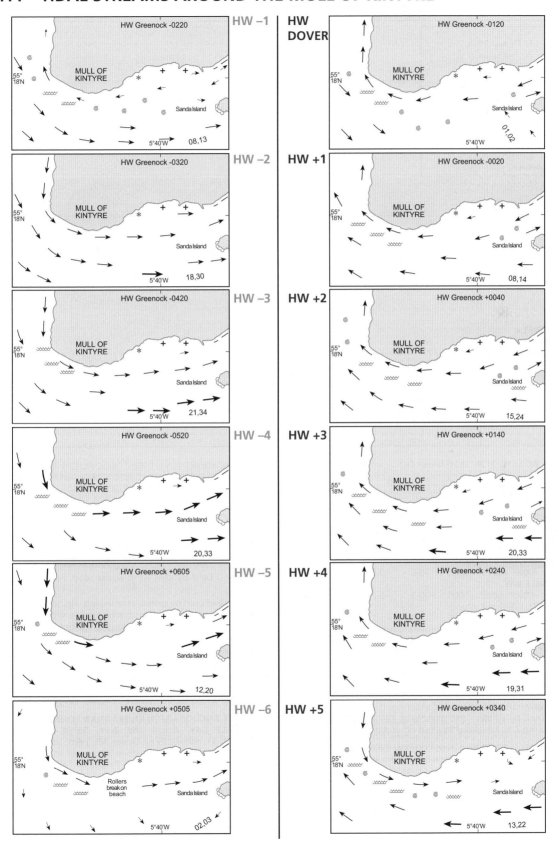

9.9.12 CAMPBELTOWN

Argyll & Bute 55°25'·90N 05°32'·56W Hbr ent ⚓⚓⚓♨♨♧♧♧

CHARTS AC 5610, 2126, 1864; Imray C63

TIDES +0125 Dover; ML 1·8; Duration 0630

Standard Port GREENOCK (→)

Times				Height (metres)			
High Water		Low Water		MHWS	MHWN	MLWN	MLWS
0000	0600	0000	0600	3·4	2·8	1·0	0·3
1200	1800	1200	1800				
Differences CAMPBELTOWN							
−0025	−0005	0015	+0005	−0·5	−0·3	+0·1	+0·2
CARRADALE BAY							
−0015	−0005	−0005	+0005	−0·3	−0·2	+0·1	+0·1
SOUTHEND, (Mull of Kintyre)							
−0030	−0010	+0005	+0035	−1·3	−1·2	−0·5	−0·2

SHELTER Good, but gusts off the hills in strong SW'lies. Main harbour dredged to 9m for commercial vessels. Yacht pontoon dredged 3·0m is close NW of Old Quay and gives excellent sheltered berthing. Yachts >12m LOA should notify ETA to the berthing master by ☎. Good ⚓ on S side of loch, E of front leading light. ⚓ possible in N of harbour, but is exposed to SE winds. Holding is poor and great care must be taken not to foul local SC moorings (large Y buoys). Some may be used by visitors - £10 payable at Royal Hotel. S of Island Davaar temp ⚓ in Kildalloig Bay, but no access to the loch.

NAVIGATION WPT 55°26'·24N 05°31'·61W, 240°/1·4M to first chan buoys. The ent is easily identified by radio masts N and NE of Trench Pt and conspic lt ho on N tip of Island Davaar off which streams are strong (4kn sp). Caution: The Doirlinn, a bank drying 2·5m which covers at HW, is close S of the ldg line.

LIGHTS AND MARKS See 9.9.5 and chartlet. Otterard Rk (3·8m depth, off chartlet, 1·5M NNE of Island Davaar), is marked by ECM lt buoy.

COMMUNICATIONS (Code 01586) CGOC (02891) 463933; Police 101; Dr 552105. HM 552552 mobile 07825 732862, Asst HM 551242 mobile 07876 580950. Berthing mobile 07798 524821.

VHF Ch 13 16 (Mon-Fri 0900-1700).

FACILITIES Yacht pontoon (10+20Ⓥ) £1.70, ⚓ 🛢.

Aqualibrium: dedicated facilities for yacht crews, opposite pontoon: @ 🚾/♿ ▢ ✕ swimming pool, gym, sauna, creche and library.

Royal Hotel – payment of berthing fees: @ 🚾/♿ ▢ ✕.

Old Quay D (and by cans from across road, opposite pontoon), ⚓ 🍺 LB.

Ferry Terminal used by commercial vessels. No berthing alongside without HM's permission. The attached public slipway has a drying out berth (with timber keel blocks and upright leaning posts) for FVs and yachts, ⚓ 🛢.

Campbeltown SC ⛵ (dinghies), 🏁 regular racing.

Town 🏬 🏨 🛢 ▢ ⚒ 📷 Ⓔ ♿(10t) ⚙ ACA, 🍽(2) ✕ 🛏 ✉ Ⓑ; 🛁🚾 about 100m from yacht pontoon (keypad, number displayed on pontoon). Bus 3/day to Glasgow, ✈ twice daily to Glasgow, (nearest ⇌ Arrochar, 90 miles N by road). Car hire/taxi.

ADJACENT ANCHORAGE IN KILBRANNAN SOUND

CARRADALE BAY, Argyll and Bute, **55°34'·40N 05°28'·66W**. AC 5610, 2131, HW+0115 on Dover. ML 1·8m. See 9.9.12. Good ⚓ 1 M S in 7m off Torrisdale Castle in SW corner. In N & E winds ⚓ in NE corner of bay, W of Carradale Pt. 3ca E of this Pt, a PHM buoy Fl (2) R 12s marks Crubon Rk.

In S & E winds a swell sets into bay, when good shelter can be found 1M N in **Carradale Harbour** (Port Crannaich); if full of FVs, use 4⚓ immediately to N or ⚓ 100m N of Hbr. Bkwtr lt, Fl R 10s 5m 6M. Call HM Campbeltown for overnight berth, £10 pay in honesty box. Facilities: ⚓ on pier, ✉ 🛒 ✕ 🍺.

9.9.13 LAMLASH

Isle of Arran, N Ayrshire 55°32'·00N 05°07'·06W ✵✵✵✵♨♨♧♧♧

CHARTS AC 5610, 2131, 2220, 1864; Imray C63

TIDES +0115 Dover; ML no data; Duration 0635

Standard Port GREENOCK (→)

Times				Height (metres)			
High Water		Low Water		MHWS	MHWN	MLWN	MLWS
0000	0600	0000	0600	3·4	2·8	1·0	0·3
1200	1800	1200	1800				
Differences LAMLASH							
–0016	–0036	–0024	–0004	–0·2	–0·2	ND	ND
BRODICK BAY							
–0013	–0013	–0008	–0008	–0·2	–0·1	0·0	+0·1
LOCH RANZA							
–0015	–0005	–0010	–0005	–0·4	–0·3	–0·1	0·0

SHELTER Very good in all weathers. Lamlash is a natural hbr with sheltered ⚓s as follows: off Lamlash except in E'lies – depth may be 20m but shoals rapidly to the W; off Kingscross Point, good except in strong N/NW winds; off the NW of Holy Island in E'lies. ♨ off Lamlash Pier. Dry out against pier if in need of repairs. See also Brodick 5M N, and Loch Ranza 14M N (RH col).

NAVIGATION WPT 55°32'·63N 05°03'·06W, 270°/1M to N Chan buoy (Fl R 6s).

Beware: submarines exercise frequently in this area (see 9.9.18), and also wreck of landing craft (charted) off farmhouse on Holy Is.

LIGHTS AND MARKS See 9.9.5 and chartlet.

COMMUNICATIONS (Code 01770) CGOC (02891) 463933; Police 101; 🏥 600777. No VHF.

FACILITIES Lamlash Old Pier ⚓ (£3–£5),🔌 ⚓ ✕ ⚓; **Arran YC** ☎01770 600333 25♨ £10. **Village** (Lamlash/Brodick) 🏪 🏧 ✕ 🍴 ✉ 🛒 ⇌ (bus to Brodick, Ferry to Ardrossan), ✈ (Glasgow or Prestwick).

OTHER HARBOURS ON ARRAN (AC 2724, 2131, 2126)

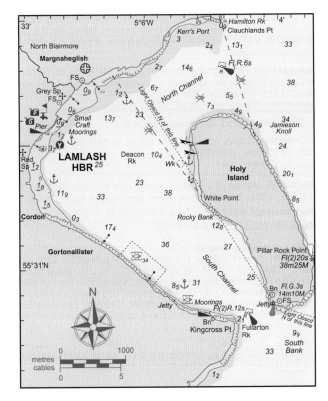

BRODICK, Arran, 55°35'·50N 05°08'·66W. AC 2131, 2220, 1864. HW +0115 on Dover; ML 1·8m; Duration 0635. See 9.9.13. Shelter good except in E winds. ⚓ W of ferry pier in 3m; on NW side just below Castle in 4·5m, or further N off Merkland Pt in 3-4m. Also 10♨s. There are no navigational dangers but the bay is in a submarine exercise area; see 9.9.18. Lts: Ro Ro pier dolphin 2 F R (vert) 9m 4M and 2 F G (vert) 9m 4M on pier hd. Facilities: 🏪 🏧 ⚓ (at pier hd) ✕ ✉ Ⓑ 🛒 ✕ 🍴. Ferry to Ardrossan.

LOCH RANZA, Arran, 55°42'·60N 05°17'·96W. AC 2131, 2383, 2221. HW +0120 on Dover; ML 1·7m; Duration 0635. See 9.9.13. Good shelter, but swell enters loch in N'lies. The 850m mountain 4M to S causes fierce squalls in the loch in S winds. Beware Screda Reef extending SW off Newton Pt. 12♨s. ⚓ in 5m off castle (conspic); holding is suspect in soft mud. Dinghy landing and temp ⚓ on pontoon at Lochranza. 2F G lts on RoRo pier at Coillemore. S shore dries. Facilities: 🏪 ⚓ at ferry slip, ✉ 🛒 ✕. Ferry to Claonaig.

HARBOURS AND ANCHORAGES AROUND BUTE (Clockwise from Garroch Head, S tip of Bute) (AC 2131, 1906/7)

ST NINIAN'S BAY, Bute, 55°48'·15N 05°07'·86W. AC 2221, 2383. Inchmarnock Is gives some shelter from the W, but Sound is exposed to S'lies. At S end, beware Shearwater Rk, 0·9m, almost in mid-sound. ⚓ in about 7m, 2ca E of St Ninian's Pt; beware drying spit to S of this Pt. Or ⚓ off E side of Inchmarnock, close abeam Midpark Farm.

WEST KYLE, Bute, 55°54'N 05°12'·7W. AC1906. Tides, see 9.9.14 (Tighnabruaich). On W bank PHM buoys, each Fl R 4s, mark Ardlamont Pt, Carry Pt and Rubha Ban; N of which are two Fl Y buoys (fish farms). ⚓ close off Kames or Tighnabruaich, where space allows; or in Black Farland Bay (N of Rubha Dubh). Some ♨s off Kames and Royal Hotels, maintained by them on a 'use at own risk basis': Kames Hotel ☎(01700) 811489; Kyles of Bute Hotel 811350; Royal Hotel 811239; and Tighnabruaich Hotel 811615. Facilities: ⚓ 🏪 🍴 🛒.

CALADH HARBOUR, Argyll and Bute, 55°56'·00N, 05°11'·73W. AC 1906. HW (Tighnabruaich) +0015 on Dover; ML 2·1m. See 9.9.14. Perfectly sheltered natural hbr on W side of ent to Loch Riddon. Enter Caladh Hbr to N or S of Eilean Dubh; keep to the middle of the S passage. When using the N ent, keep between R and G bns to clear a drying rk marked by perch. ⚓ in the middle of hbr clear of moorings. No facilities/stores; see West Kyle above.

LOCH RIDDON, Argyll and Bute, 55°57'N 05°11'·6W. AC 1906. Tides, see 9.9.14. Water is deep for 1·3M N of Caladh and shore is steep-to; upper 1·5M of loch dries. ⚓ on W side close N of Ormidale pier; on E side at Salthouse; off Eilean Dearg (One Tree Is); and at NW corner of Fearnoch Bay.

BURNT ISLANDS, Bute, 55°55'·76N 05°10'·39W. AC 1906. Tides, see 9.9.17. The three islands (Eilean Mor, Fraoich and Buidhe) straddle the East Kyle. There are 2 channels: North, between Buidhe and the other 2 islets, is narrow, short and marked by 2 SHM buoys (the NW'ly one is Fl G 3s), and one PHM buoy, Fl R 2s. South chan lies between Bute and Fraoich/Mor; it is unlit, but marked by one PHM and two SHM buoys. Depths may be less than charted. A SHM buoy, Fl G 3s, is off Rubha a' Bhodaich, 4ca ESE. Direction of buoyage is to SE. Sp streams reach 5kn in N Chan and 3kn in S Chan. ⚓ in Wreck Bay, Balnakailly Bay or in the lee of Buidhe and Mor in W'lies; also W of Colintraive Pt, clear of ferry and cables. There are 6♨s off the hotel, ☎(01700) 841207.

KILCHATTAN BAY, Bute, 55°45'N 05°01'·1W. AC 1907. Bay is deep, but dries 3ca off the W shore. Temp ⚓s only in offshore winds: off the village on SW side, or on N side near Kerrytonlia Pt. Rubh' an Eun Lt, Fl R 6s, is 1·1M to SSE. Facilities: ⚓ ✉ 🛒 bus to Rothesay.

9.9.14 ROTHESAY

Isle of Bute, Argyll and Bute **55°50´·32N 05°03´·08W**
❋❋❋♨♨✿✿✿

CHARTS AC 5610, 2131, 1907, 1906, 1867; Imray C63, 2900

TIDES +0100 Dover; ML 1·9; Duration 0640

Standard Port GREENOCK (⟶)

Times				Height (metres)			
High Water		Low Water		MHWS	MHWN	MLWN	MLWS
0000	0600	0000	0600	3·4	2·8	1·0	0·3
1200	1800	1200	1800				
Differences ROTHESAY BAY							
−0020	−0015	−0010	−0002	+0·2	+0·2	+0·2	+0·2
RUBHA A'BHODAICH (Burnt Is)							
−0020	−0010	−0007	−0007	−0·2	−0·1	+0·2	+0·2
TIGHNABRUAICH							
+0007	−0010	−0002	−0015	0·0	+0·2	+0·4	+0·5

SHELTER Good in Outer and Inner Harbours (max 2m draught and very limited space for LOA > 40ft), and on pontoons S of West Arm, but beware of very strong easterly flow towards the rocks on the E side of the harbour due to wash during ferry manoeuvres (and whilst berthed). No berthing on north and east faces of Main Pier. Good ⚓ off Isle of Bute SC in west side of bay. Exposed to N/NE; **Port Bannatyne Marina** (Kames Bay) offers better shelter.

NAVIGATION WPT 55°51´·00N 05°02´·76W, 194°/0·69M to Outer hbr ent. From E keep 1ca off Bogany Pt PHM buoy, Fl R 4s. Call HM for bridge openings to Inner Harbour. Entry/exit to/from hbr controlled by traffic lights. Extreme care needed; very limited room to manoeuvre.

LIGHTS AND MARKS Lts as chartlet, hard to see against shore lts. Conspicuous church spire leads 190° to outer harbour. Port Sigs:

● (3 vert) = Harbour closed to all traffic movements.

● (2 vert) + ○ = Harbour open.

COMMUNICATIONS (Code 01700) CGOC (02891) 463933; Police 101; Dr 503985; Ⓗ 503938. HM 503842; Bute Berthing Co mob 07799 724225. *Bute Berthing* M; *Rothesay Harbour* Ch **12** 16.

FACILITIES 68 ⬭ Ⓥ (Bute Berthing Co) call ahead for availability. £16 for 8m LOA, £21 for 11m LOA, short stay £6, ⬕ ⚓ ⚓ ⚒ ⛽ on Pier; 🛢 by arrangement weekdays (min 200 galls).

Town (Wed - early closing) ☎ ☎ (closes 1800) @(0900-1700) ✉ Ⓑ 🛒 ✕ 🏧 ⇌ (ferry to Wemyss Bay) ✈(Glasgow).

9.9.15 PORT BANNATYNE

Isle of Bute, Argyll and Bute **55°51´·75N 05°05´·17W**
❋❋♨♨✿✿✿

CHARTS AC 5610, 2131, 1907, 1906, 1867; Imray C63, 2900

TIDES As for Rothesay

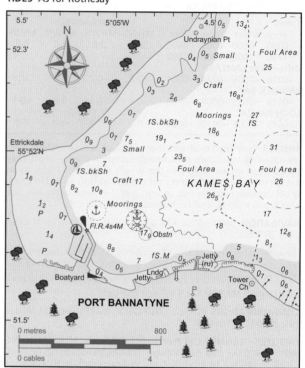

SHELTER Well sheltered from westerly winds and protected by stone breakwater. Marina situated in SW corner of Kames Bay provides secure alongside berth in 2.4m. Alternatively ⚓ in 11m to NE of b'water clear of small craft along N shore. moorings NE of marina – beware sunken fishing boat. Beware drying rks 1ca off Ardbeg Pt, and shoal water, marked by a row of piles, close W of marina breakwater.

NAVIGATION WPT 55°52´·00N 05°04´·00W, 252°/0·7M to marina entrance, which is accessible at all states of the tide. Approach marina keeping at least 20m clear of the end of the breakwater. Avoid no ⚓/fishing areas to ENE of breakwater.

LIGHTS AND MARKS Breakwater lt Fl R 4s 2M.

COMMUNICATIONS (Code 01700) CGOC (02891) 463933; Police 101; Dr 503985; Ⓗ 503938. Marina VHF Ch 37 (M1).

FACILITIES Marina, www.portbannatynemarina.co.uk, ☎503116, 105⬭ inc Ⓥ in 2·5m, £2·00/m, £6 for short stay (4 hrs) ⊡ wi-fi.

Port Boatyard ☎502719 ⚓ ⚓ ⚒ 🛢 ⚒ ⚒ ⛽ Ⓔ 🏠 🚿 ⛴(15t) ⚓.

Town 🍴 ✉ 🏧 ✕ bus to Rothesay.

GREAT CUMBRAE ISLAND

MILLPORT, Great Cumbrae, N Ayrshire, **55°45´·00N 04°55´·82W**. AC 5610, 1867, 1907. HW +0100 on Dover; ML 1·9m; Duration 0640. See 9.9.17. Good shelter, except in S'lies. ⚓ in approx 3m S of pier or E of the Eileans. 12 ⚓s 1½ca SSE of pier. Ldg marks: pier hd on with ✠ twr 333°; or ldg lts 333°, both FR 7/9m 5M, between the Spoig and the Eileans, QG. Unmarked, drying rk is close E of ldg line. HM ☎(01475) 530826.

Town ⚓ ☎ ☎ 🛢 ⚓ ⚓ ✉ 🍴 ✕ 🏧. Possible temporary ⬭ on pontoon belonging to SportScotland off NE of island (approx 55°47´·1N 04°53´·8W) in 3.5m–4.5m; no services.

9.9.16 FAIRLIE QUAY

N Ayrshire **55°46′·07N 04°51′·78W** ❀◊❀❀

CHARTS AC 5610, 2131, 1907, 1867; Imray C63, 2900

TIDES +0105 Dover; ML 1·9; Duration 0640

Standard Port GREENOCK (→)

Times				Height (metres)			
High Water		Low Water		MHWS	MHWN	MLWN	MLWS
0000	0600	0000	0600	3·4	2·8	1·0	0·3
1200	1800	1200	1800				
Differences MILLPORT							
–0005	–0025	–0025	–0005	0·0	–0·1	0·0	+0·1

SHELTER Access H24. Poor shelter; pontoon subject to swell in fresh conditions and wash from passing vessels. Craft should not be left unsupervised unless calm settled weather is forecast. Cumbrae Island gives shelter from westerlies; Largs Channel is open to S or N winds.

NAVIGATION WPT 55°46′·08N 04°51′·84W. From S beware Hunterston Sands, Southannan Sands and outfalls from Hunterston Power Stn (conspic).

LIGHTS AND MARKS Lts as on chartlet. Fairlie Quay, 2 FG (vert) N & S ends of Pier.

COMMUNICATIONS (Code 01475) CGOC (02891) 463933; Dr 673380; Ⓗ 733777; Police 101. Fairlie Quay 568267.

Fairlie Quay Marina Ch **80** M (H24).

FACILITIES Fairlie Quay Fuel H24: 🛢 ⛽ ⚒ ⚓ ⌕ ⊿ 🗑 🏁 (80t) ⛴. Fairlie YC; **Town** ✉ Ⓑ 🛒 ✕ 🗋 ⇌ (Fairlie & Largs; 50 mins to Glasgow), ✈.

9.9.17 LARGS

N Ayrshire **55°46′·40N 04°51′·84W** ❀❀❀◊◊◊❀❀

CHARTS AC 5610, 2131, 1907, 1867; Imray C63, 2900

TIDES +0105 Dover; ML 1·9; Duration 0640

Standard Port GREENOCK (→)

Times				Heights (metres)			
High Water		Low Water		MHWS	MHWN	MLWN	MLWS
0000	0600	0000	0600	3·4	2·8	1·0	0·3
1200	1800	1200	1800			.	
Differences MILLPORT							
–0005	–0025	–0025	–0005	0·0	–0·1	0·0	+0·1

SHELTER Excellent in Largs Yacht Haven, access all tides (2·5m in ent; 3m deep berths). 7 pontoons; Ⓥ on 'C/D'. Cumbrae Is gives shelter from W'lies; Largs Chan is open to S or N winds.

NAVIGATION WPT 55°46′·40N 04°51′·84W, SWM lt buoy, off ent. From S beware Hunterston Sands, Southannan Sands and outfalls from Hunterston Power Stn (conspic). From the S bkwtr to Fairlie Quay is a restricted, no ⚓ area.

LIGHTS AND MARKS 'Pencil' monument (12m) conspic 4ca N of ent. Lts as on chartlet. Largs Pier, 2 FG (vert) when vessel expected.

COMMUNICATIONS (Code 01475) CGOC (02891) 463933; Dr 673380; Ⓗ 733777; Police 101; Largs SC 670000.

Largs Yacht Haven Ch **80** M (H24).

FACILITIES Largs Yacht Haven www.yachthavens.com ☎ 675333, (730⊆ inc Ⓥ), £2·90; ⚓ ⮤(H24) 🗑 🛢 ⮤(H24) 🛢 ⚒ ⚓ ⌕ 🗑 Ⓔ ⊿ 🗑 🏁(70t) ⚓(17t) Divers, ice, ✕ 🗋. **Town** (1M) ✉ Ⓑ 🛒 ✕ 🗋 ⇌ (50 mins to Glasgow), ✈.

ARRAN, BUTE AND FIRTH OF CLYDE

(AC1906, 1907) Arran's mountains tend to cause squalls or calms, but there are good ⚓s at Lamlash, Brodick and Loch Ranza. Sannox Rock (depth 1·5m) is 2½ca off Arran coast 8M N of Lamlash. 1ca off W side of Inchmarnock is Tra na-h-uil, a rock drying 1·5m. In Inchmarnock Sound, Shearwater Rk (depth 0·9m) lies in centre of S entrance.

The Kyles of Bute are attractive channels N of Bute, and straight-forward apart from the Burnt Islands where it is best to take the north channel, narrow but well buoyed, passing between Eilean Buidhe and Eilean Fraoich/Eilean Mor. ►*Spring stream may reach 5kn.*◄ Caladh Hbr is a beautiful ⚓ 7ca NW of Burnt Is.

The N lochs in Firth of Clyde are less attractive. Loch Goil is worth a visit; Loch Long is squally with few ⚓s but no hidden dangers; Gareloch has Rhu Marina, Faslane submarine base and small craft moorings at the northern end. ► *Navigation in Firth of Clyde is straightforward since tidal streams are weak, seldom exceeding 1kn. It is however a busy area with many large commercial vessel movements and extensive naval training which takes place involving submarines, ships and aircraft. All users are recommended to read The Clydeport CLYDE LEISURE NAVIGATION GUIDE – www.clydeport.co.uk.* ◄ Channels are well marked; but beware unlit moorings. There are marinas on the mainland at Largs, Holy Loch, Portavadie and Inverkip. Rothesay Hbr on E Bute, and Kilchattan B (⚓ 6M to S) are both sheltered from SSE to WNW.

9.9.18 SUBMARINE EXERCISE AREAS

Submarines keep a constant listening watch on Ch 16. Submarines on the surface comply strictly with IRPCS; submarines at periscope depth will not close to within 1500 yds of a FV without her express agreement. See 3.13.1 for more guidance.

A 'Fisherman's hotline' (☎(01436) 677201) is available on a 24 hour basis and may be used for any queries relating to SUBFACTS from any mariner.

SUBFACTS broadcasts detail submarine activity in the exercise areas between Barra and the Isle of Man. They are transmitted by CGOC Belfast at 0710 and 1910 on a specified VHF Ch after an initial announcement on Ch 16.

The areas are referred to by the names given in the table for Areas 1–26 (North of Mull), see 9.8.22.

General information on SUBFACTS is also broadcast twice daily at 0620 & 1820 UT on Navtex.

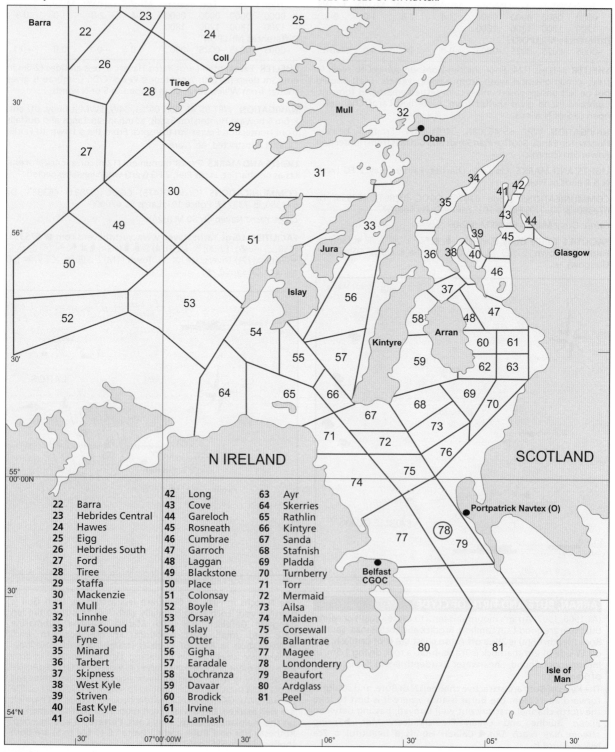

22	Barra	42	Long	63	Ayr
23	Hebrides Central	43	Cove	64	Skerries
24	Hawes	44	Gareloch	65	Rathlin
25	Eigg	45	Rosneath	66	Kintyre
26	Hebrides South	46	Cumbrae	67	Sanda
27	Ford	47	Garroch	68	Stafnish
28	Tiree	48	Laggan	69	Pladda
29	Staffa	49	Blackstone	70	Turnberry
30	Mackenzie	50	Place	71	Torr
31	Mull	51	Colonsay	72	Mermaid
32	Linnhe	52	Boyle	73	Ailsa
33	Jura Sound	53	Orsay	74	Maiden
34	Fyne	54	Islay	75	Corsewall
35	Minard	55	Otter	76	Ballantrae
36	Tarbert	56	Gigha	77	Magee
37	Skipness	57	Earadale	78	Londonderry
38	West Kyle	58	Lochranza	79	Beaufort
39	Striven	59	Davaar	80	Ardglass
40	East Kyle	60	Brodick	81	Peel
41	Goil	61	Irvine		
		62	Lamlash		

9.9.19 INVERKIP (KIP MARINA)

Inverclyde **55°54'·50N 04°53'·00W** ❀❀❀♨♨♨♧♧

CHARTS AC 5610, 2131, 1907; Imray C63, 2900

TIDES +0110 Dover; ML 1·8; Duration 0640

Standard Port GREENOCK (→)

Times				Height (metres)			
High Water		Low Water		MHWS	MHWN	MLWN	MLWS
0000	0600	0000	0600	3·4	2·8	1·0	0·3
1200	1800	1200	1800				
Differences WEMYSS BAY							
–0005	–0005	–0005	–0005	0·0	0·0	+0·1	+0·1

SHELTER Excellent inside marina. Chan and marina are dredged 3·5m; accessible H24. Inverkip Bay is exposed to SW/NW winds.

NAVIGATION WPT 55°54'·49N 04°52'·95W, Kip SHM buoy, Fl G 5s, at ent to buoyed chan; beware shifting bank to the N.

LIGHTS AND MARKS SHM 55°54'·55N 04°54'·47W Fl G 1·06M 093° to entr which is ½M N of conspic chmy (238m). SPM buoy marks sewer outfall off Ardgowan Pt. From Kip SHM buoy, 3 SHM and 3 PHM buoys mark 365m long appr chan.

COMMUNICATIONS (Code 01475) CGOC (02891) 463933; Police 101; Dr 520248; Ⓗ 33777.

Kip Marina Ch **80** M (H24). *Clydeport Estuary Radio* VHF Ch **12** 16 (H24). Info on weather and traffic available on request.

FACILITIES Kip Marina www.scottishmarinas.co.uk ☎521485; 625⊆ inc Ⓥ, £2.25, £12 <5hrs, 🏮🛢️▣🔧🔨🈱Ⓔ♧🏗️(50t) 🛥️🛢️. Diver, 🛒✕ YC. **Town** ✉Ⓑ (Gourock), ⇌ ✈ (Glasgow).

ADJACENT ANCHORAGE

DUNOON, Argyll and Bute, **55°56'·70N 04°55'·20W**. AC 5610, 2131, 1907, 1994. Use Greenock tides. Temp ⚓ in West or East Bays (S and N of Dunoon Pt). The former is open to the S; the latter more shoal. The Gantocks, drying rks 3ca SE of Dunoon Pt, have W ○ bn tr, Fl R 6s 12m 6M. 2FR (vert) on ferry pier. Facilities: 🈱🈱🏮 ✉🛒✕⊑ ferry to Gourock. 3⚓s off Innellan, 3·7M S of Dunoon, ☎(01369) 830445.

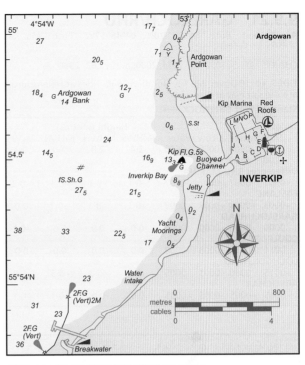

ANCHORAGE IN THE UPPER FIRTH OF CLYDE

GOUROCK, Inverclyde, **55°58'·00N 04°49'·00W**. AC 5610, 2131, 1994, Imray C63, 2900 Series; OS 63. Use Greenock tides. ⚓ in West Bay, beware of moorings. It is exposed in NW/NE'lies. Beware foul ground in Gourock Bay. No navigational dangers, but much shipping and ferries in the Clyde. The S edge of the Firth of Clyde recommended channel lies 2ca N of Kempock Pt. The pier at the S end of Gourock Bay is disused and unsafe. Facilities: Royal Gourock YC ☎632983 ⚓🛒🛢️⚓🔨♧🛒✕⊑. **Town:** 🈱🈱✉Ⓑ🛒✕ ⊑ ⇌ Ferry (to Dunoon) ✈ (Glasgow).

9.9.20 HOLY LOCH

Argyll **55°59'·03N 04°56'·82W** ❀❀❀♨♨♨♧♧♧

CHARTS AC 5610, 2131, 1994; Imray C63, 2900

TIDES +0122 Dover; ML 2·0; Duration 0640

Standard Port GREENOCK (→)

SHELTER Good, but exposed to to E/NE.

NAVIGATION WPT 55°59'·03N 04°55'·75W 270° to marina ent 0.60M. No offshore dangers. Holy Loch is controlled by QHM Clyde but navigation is not restricted; keep well clear of ferries when entering or leaving the loch.

LIGHTS AND MARKS No 30 SCM 55°58'·75N 04°53'·82W off Strone Pt. 2FR at SE end of pier. Marina floodlit.

COMMUNICATIONS (Code 01369) CGOC (02891) 463933; Police 101; Dr 703279; Ⓗ 704341.

Holy Loch Marina Ch **80** M.

FACILITIES Marina www.holylochmarina.co.uk ☎701800, 270⊆+ Ⓥ in 3m £2.64, 2nd night ½ price; short stay £5.50. 🏮🏮▣🛢️☔🏗️(25t) 🛥️🛢️✕. Possible ⊆ on breakwater pontoon for larger yachts in depth >5m, but open to the E. Commercial timber yard operating alongside marina facilities. **Holy Loch SC** ☎702707.

Sandbank (town) PO, 🛒⊑✕ Ferries from Hunter's Quay to McInroy's Pt, Gourock, connecting with trains and coaches to Glasgow Airport. Good local area bus service.

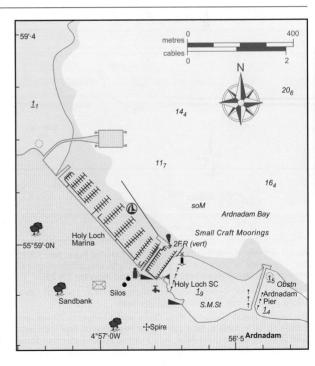

9.9.21 GARELOCH/RHU

Argyll and Bute **56°00'·70N 04°46'·57W** (Rhu Marina)
❋❋❋◊◊◊☆☆☆

CHARTS AC 5610, 2131, 1994, 2000; Imray C63, 2900

TIDES +0110 Dover; ML 1·9; Duration 0640. Tides at Helensburgh are the same as at Greenock.

Standard Port GREENOCK (→)

Times				Height (metres)			
High Water		Low Water		MHWS	MHWN	MLWN	MLWS
0000	0600	0000	0600	3·4	2·8	1·0	0·3
1200	1800	1200	1800				
Differences RHU MARINA							
–0007	–0007	–0007	–0007	–0·1	–0·1	–0·1	–0·2
FASLANE							
+0003	+0003	+0003	+0003	+0·1	+0·1	+0·1	0·0
GARELOCHHEAD							
0000	0000	0000	0000	0·0	0·0	0·0	–0·1
COULPORT							
–0011	–0011	–0008	–0008	0·0	0·0	0·0	0·0
LOCHGOILHEAD							
+0015	0000	–0005	–0005	–0·2	–0·3	–0·3	–0·3
ARROCHAR							
–0005	–0005	–0005	–0005	0·0	0·0	–0·1	–0·1

BYELAWS Loch Long and Gareloch are Dockyard Ports under the jurisdiction of the Queen's Harbour Master. All submarines and other warships at anchor or underway have an Exclusion Zone around them enforced by MOD Police.

SHELTER Rhu Marina rock bkwtr 1m above MHWS, now extended with submerged section (end marked by SHM Fl G 5s) replacing floating wavebreak, protects the berths from strong winds from E to S. Helensburgh Pier is a temp'y drying berth, rather exposed, used by occas steamers. ♣ E of marina or in Rosneath Bay. Moorings N of the narrows at Stroul B & Clynder; and at the head of the loch. Clyde Naval Base at Faslane must be avoided by yachts.

NAVIGATION WPT 55°59'·29N 04°45'·26W, 356°/1·3M to bn No 7. Beaches between Cairndhu Pt and Helensburgh Pier are strewn with large boulders above/below MLWS. Gareloch ent is via Rhu Narrows. Marina entrance not easy to find at night between pontoon breakwater and Fl G Bn of marina entrance; beware cross-tides. There are large unlit MoD buoys and barges off W shore of Gareloch; for Garelochhead keep to W shore until well clear of Faslane Naval Base.

LIGHTS AND MARKS Ldg/dir lts into Gareloch 356°, 318°, 295°, 329° and 331°. Conspic ✠ tr at Rhu Point.

COMMUNICATIONS (Code 01436) CGOC (02891) 463933; Police 101; Ⓗ (01389) 754121; Dr 672277. QHM 674321.
MoD Police/QHM Ch 16/73. Rhu Marina Ch **80** M (H24 in season).

FACILITIES Rhu Marina, sbell@quaymarinas.com ☎820238; ⚓ (235) ⚓ £2.40 (short stay £12) pontoons A-J, ⚓ ⚓ ⚓ ⚓ ⚓(35t) ⚓ ⚓ ⚓ ⚓ ⚓ ✗ wi-fi.
Royal Northern and Clyde YC ☎820322; club ⚓s, ⚓ ✗ ⚓; **Helensburgh SC** ☎672778 ⚓ (dinghies), ⚓ ♣. Town all services, ⇌ ✈ (Glasgow). **DRB Marine** (Rosneath) ☎831231, use jetty only to load, some ⚓s, ⚓(40t). **Silvers Yard** private moorings.

Naval activity: Beware submarines from Faslane Base. See 9.9.18 for submarine activity (Subfacts) in the Clyde and offshore or call FOSNNI Ops ☎(01436) 674321 Ext 3206 and monitor Ch 73.

Protected Areas: Vessels are never allowed within 150m of naval shore installations at Faslane and Coulport.

Restricted Areas (Faslane, Rhu Chan and Coulport): These are closed to all vessels during submarine movements (see opposite and W Coast of Scotland Pilot, App 2). MoD Police patrols enforce areas which are shown on charts. The S limit of Faslane Restricted area is marked by two Or posts with X topmarks on Shandon foreshore. The W limit is marked by Iso WRG 4s, vis W356°-006°, at Gareloch Oil fuel depot N jetty. The following signals are shown when restrictions are in force:

Entrance to Gareloch

Day & Night: ● ● ● (vert), supplemented by R flag with W diagonal bar.

Faslane and Coulport
Day & Night: ● ● ● (vert), supplemented by International Code pendant over pendant Nine.

LOCH LONG and LOCH GOIL, Argyll and Bute, approx 56°00′N 04°52′·5W to 56°12′·00N 04°45′·00W. AC *5610*, 3746. Tides: See 9.9.21 for differences. ML 1·7m; Duration 0645.

SHELTER Loch Long is about 15M long. Temp ⚓s (south to north) at: Cove, Blairmore (not in S'lies), Ardentinny, Portincaple, Coilessan (about 1M S of Ardgartan Pt), and near head of loch (Arrochar) on either shore. In Loch Goil ⚓ at Swines Hole and off Carrick Castle (S of the pier, in N'lies a swell builds). Avoid ⚓ near Douglas Pier. The head of the loch is crowded with private/dinghy moorings, and is either too steep-to or too shallow to ⚓. The loch is frequently closed to navigation due to the trial range half way down.

LIGHTS Coulport Jetty, 2FG (vert) each end; Covered Berth L Fl G (SW corner) with 3F G(vert) and F G to NE; Port Dornaige Fl 6s 8m 11M, vis 026°-206°; Dog Rock (Carraig nan Ron) Fl 2s 11M; Finnart Oil Terminal has FG lts and ldg lts 031° QW/FW on Cnap Pt. Upper Loch Long is unlit. Loch Goil ent is marked by 2 PHM buoys (Fl R 3s and QR), a SHM buoy (QG) and ldg lts 318°: front (The Perch) Dir FWRG and Fl R 3s; rear FW. Rubha Ardnahein Fl R5s.

FACILITIES Loch Long (Cove), Cove SC, ⚓ 🛒 🏠; ⚓(pier); (Portincaple) ✉ 🛒 hotel, ⚓; (Ardentinny) shop, 🛒 ✕ hotel, ⛽; (Blairmore) ⚓ ✉ 🛒; (Arrochar) shops, hotel, ⚓ ⛽ ✉. Loch Goil (Carrick Castle) has 2 blue ⚓s (£10/night, max 15t, free wi-fi), 🚾. Lochgoilhead: ⚓ ⛽ ✉ 🛒.

9.9.22   RIVER CLYDE

Dunbartonshire **55°57′·50N 04°45′·70W** (Tail of the Bank)
✿✿✿✿⚓⚓✿✿

CHARTS AC 5610, 2131, 1994, 2007: *BW Skipper's Guide for Forth and Clyde Canal, BW Skipper's Handbook for Lowland Canals*; Imray C63, 2900

TIDES +0122 Dover; ML 2·0; Duration 0640

Standard Port GREENOCK (→)

Times				Height (metres)			
High Water		Low Water		MHWS	MHWN	MLWN	MLWS
0000	0600	0000	0600	3·4	2·8	1·0	0·3
1200	1800	1200	1800				
Differences PORT GLASGOW (55°56′·10N 04°40′·50W)							
+0010	+0005	+0010	+0020	+0·2	+0·1	0·0	0·0
DUMBARTON							
+0015	+0010	+0020	+0040	+0·4	+0·3	+0·1	0·0
BOWLING							
+0020	+0010	+0030	+0055	+0·6	+0·5	+0·3	+0·1
RENFREW							
+0025	+0015	+0035	+0100	+0·9	+0·8	+0·5	+0·2
GLASGOW							
+0025	+0015	+0035	+0105	+1·4	+1·1	+0·8	+0·4

SHELTER Good at Greenock (James Watt Dock Marina), Dumbarton (Sandpoint Marina) and Bowling Harbour.

NAVIGATION WPT 55°57′·50N 04°45′·70W, 1·2M to James Watt Dock (JWD) entrance, which experiences strong tidal streams flowing in the dredged channel across the entrance. Sandpoint Marina and Bowling Harbour are, respectively, about 6M and 9M further up-river. The density of shipping and numerous ferries in the Clyde warrant particular attention. Large vessels manoeuvre at the Greenock Container / Ocean Terminal and masters are to maintain communication with Harbour Control.

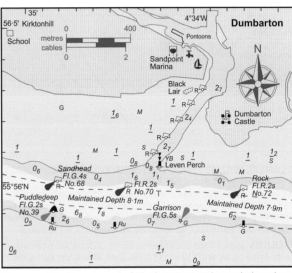

The speed limit upstream of Cloch point is 12kn and 5kn where boards mark Dead Slow. Vessels under sail are not relieved of their obligations under Rules for Narrow Channels and are reminded that E of Bowling vessels may proceed under power alone.
All vessels planning an up-river passage are to contact their berth or destination before proceeding. Keep just inboard of chan lateral marks as depths shoal rapidly outboard. For Dumbarton cross the river at 90° at No 39 SHM buoy. Depths in R Leven may be less than charted. For Bowling Hbr cross at No 45 SHM buoy.

LIGHTS AND MARKS The River Clyde is well buoyed/lit.

COMMUNICATIONS (Codes: Greenock 01475; Glasgow 0141, followed by 7 digit Tel No). CGOC (02891) 463933; Police 101;

Greenock

MEAN SPRING
AND NEAP CURVES

MEAN RANGES
Springs 3.1m
Neaps 1.8m

Dr 634617; ⊞ 01389 754121. HM 725775; Estuary Control 726221; British Waterways Board 332 6936.

All vessels seeking passage upstream beyond No 1 buoy are to seek permission from *Clyde Harbour Control* Ch **12** 16 (H24) at charted reporting points. Weather/traffic info on request. James Watt Dock Ch 80; Marinas Ch M. Bowling lockkeeper Ch 74, HW±2.

FACILITIES James Watt Dock Marina 55°56'·72N 04°44'·01W. info@jwdmarina.co.uk ☎07710 611117; 65⛵ inc Ⓥ in 4.5m, £2.15, short stay £11.50/craft. ⚓🚻 🛒▲⚓✕🏪 at marina entrance.
Greenock ✕🔧🔭⚓🏪🔌✕🏪.

Sandpoint Marina 55°56'·48N 04°34'·07W www.sandpoint-marina.co.uk ☎01389 762396/731500. Limited visitors' berths, £10 per night. VHF Ch **M**. ⚓ 🔭🛒🖥(40t) ⚓ ☎01389 742438. **Town** ✉ Ⓑ 🛒✕🏪🚉 30 mins to Glasgow.

Bowling Basin 55°55'·8N 04°29'·0W, W end of Forth & Clyde Canal, see 9.6.9. ☎01389 877969. Basin Lock access dries 0·7m. Call VHF Ch 74 for t lock schedule. *Skipper's Brief* www.scottishcanals.co.uk.
Facilities ⚓ £3.75, 🔧🖥/🚻 ⚓ by prior notice, 🛠(mast). **Town** 🔭 🖂 Ⓑ 🛒✕🏪🚉.

There are no yachting facilities on R Clyde E of Bowling Basin.

GLASGOW 55°55'·94N 04°34'·25W
Clyde Yacht Clubs Association is at 8 St James St, Paisley. ☎8878296. Clyde CC Suite 408, Pentagon Centre, 36 Washington St, Glasgow, G3 8AZ, ☎221 2774.

FIRTH OF CLYDE TO MULL OF GALLOWAY

(AC 2131, 2126, 2199, 2198) Further S the coast is less inviting, with mostly commercial hbrs until Ardrossan and Troon, NW of which there are various dangers: Troon Rk (depth 5·6m, but sea can break), Lappock Rk (dries 0·6m, marked by bn), and Mill Rk (dries 0·4m, buoyed). Lady Isle (lt, racon) is 2M WSW of Troon.

▶ *There is a severe race off Bennane Hd (8M SSE of Ailsa Craig, conspic) when tide is running strongly.*◀

Loch Ryan offers little for yachtsmen but there is a small marina at Stranraer and ⚓ S of Kirkcolm Pt, inside the drying spit which runs in SE direction 1·5M from the point. There is also a useful ⚓ in Lady Bay, sheltered except from NE. Portpatrick is a useful passage harbour, but not in onshore winds.

▶ **North Channel South:** *Irish coast – the S-going flood begins HW Belfast –0600 (HW Dover +0610). Counter tide off Donaghadee and Island Magee for last 3 hrs of flood. Scottish coast between Corsewall Pt and Mull of Galloway – HW Greenock +0310 (HW Dover +0430). Sp rate off Corsewall Pt is 2–3kn, increasing to*

5kn off and S of Black Hd. Races occur off Copeland Is. Morroch B, Money Hd Mull of Logan and Mull of Galloway.

*Irish coast – N-going ebb begins at HW Belfast (HW Dover –0015). Scottish coast – HW Greenock –0250 (HW Dover –0130). Races off Copeland Is and Mull of Galloway. Flood begins 2 hrs early close inshore N of Mull of Galloway. A race SSE of Crammag Hd is bad if wind against tide. Mull of Galloway (lt) is a high (82m), steep-to headland with a **dangerous race extending nearly 3M to S**. On E-going stream it extends NNE into Luce B; on W-going stream it extends SW and W. Give the race a wide berth, or pass close inshore at slack water nps and calm weather. SW wind >F4 against W-going stream, do not attempt inshore route.*◀

SCOTLAND – SW COAST

The Scares, two groups of rocks, lie at the mouth of Luce Bay which elsewhere is clear more than 3ca offshore; but the whole bay is occupied by a practice bombing range, marked by 12 DZ SPM lt buoys. Good ⚓ at E Tarbert B awaiting the tide around Mull of Galloway, or good shelter in drying hbr of Drummore.

▶ *Off Burrow Hd there is a bad race in strong W winds with a W-going tide.*◀

Luce Bay Firing Range (D402/403) lies at the NW end. For info on activity ☎(01776) 888792. In Wigtown B the best ⚓ is in Isle of Whithorn B, but exposed to S. It is also possible to dry out in Garlieston.

A tank firing range, between the E side of ent to Kirkcudbright Bay and Abbey Hd, 4M to E, extends 14M offshore. If unable to avoid the area, cross it at N end close inshore. For information contact the Range safety boat 'Gallovidian' on VHF Ch 16, 73. The range operates 0900-1600LT Mon-Fri, but weekend and night firing may also occur.

STANDARD TIME (UT)
For Summer Time add ONE hour in **non-shaded areas**

GREENOCK   LAT 55°57'N   LONG 4°46'W
TIMES AND HEIGHTS OF HIGH AND LOW WATERS

Dates in red are **SPRINGS**
Dates in blue are NEAPS

YEAR 2016

JANUARY

Time	m	Time	m
1 0451	3.1	**16** 0429	3.4
1016	0.9	1008	0.6
F 1645	3.4	SA 1648	3.6
2238	0.8	◗ 2248	0.4
2 0538	3.0	**17** 0516	3.3
1110	1.1	1106	0.7
SA 1730	3.3	SU 1742	3.4
◗ 2336	1.0	2352	0.6
3 0630	2.9	**18** 0611	3.2
1213	1.2	1213	0.9
SU 1822	3.1	M 1850	3.2
4 0042	1.0	**19** 0104	0.7
0732	2.9	0724	3.0
M 1324	1.2	TU 1331	0.9
1922	3.0	2019	3.1
5 0150	1.0	**20** 0216	0.7
0842	3.0	0857	3.1
TU 1432	1.1	W 1446	0.8
2031	3.0	2144	3.1
6 0249	1.0	**21** 0319	0.6
0948	3.1	1009	3.2
W 1526	1.0	TH 1548	0.6
2139	3.1	2249	3.3
7 0339	0.9	**22** 0414	0.5
1041	3.3	1105	3.4
TH 1612	0.8	F 1640	0.5
2236	3.2	2343	3.4
8 0421	0.8	**23** 0502	0.4
1125	3.4	1152	3.5
F 1652	0.6	SA 1725	0.3
2324	3.3		
9 0500	0.7	**24** 0032	3.4
1204	3.5	0546	0.4
SA 1729	0.5	SU 1235	3.6
		○ 1804	0.3
10 0008	3.4	**25** 0116	3.4
0536	0.6	0626	0.4
SU 1240	3.6	M 1314	3.7
● 1805	0.4	1841	0.3
11 0052	3.4	**26** 0155	3.3
0614	0.5	0704	0.4
M 1316	3.7	TU 1351	3.7
1843	0.3	1916	0.3
12 0135	3.5	**27** 0231	3.3
0656	0.5	0740	0.4
TU 1354	3.8	W 1426	3.7
1925	0.3	1951	0.4
13 0218	3.5	**28** 0304	3.3
0740	0.4	0816	0.5
W 1434	3.8	TH 1501	3.7
2010	0.2	2028	0.4
14 0301	3.5	**29** 0340	3.3
0827	0.4	0854	0.6
TH 1517	3.8	F 1537	3.6
2058	0.2	2108	0.5
15 0344	3.5	**30** 0417	3.2
0916	0.5	0935	0.7
F 1601	3.7	SA 1614	3.5
2150	0.3	2151	0.7
		31 0457	3.1
		1019	0.8
		SU 1654	3.3
		2237	0.8

FEBRUARY

Time	m	Time	m
1 0540	3.0	**16** 0539	3.2
1111	1.0	1147	0.7
M 1739	3.1	TU 1819	3.1
◗ 2332	1.0		
2 0631	2.9	**17** 0041	0.7
1214	1.1	0638	3.0
TU 1833	3.0	W 1311	0.8
		1955	2.9
3 0040	1.1	**18** 0159	0.8
0736	2.8	0825	2.9
W 1331	1.2	TH 1433	0.8
1937	2.9	2140	2.9
4 0156	1.1	**19** 0306	0.7
0858	2.9	0954	3.0
TH 1446	1.0	F 1537	0.6
2053	2.9	2244	3.1
5 0303	1.1	**20** 0402	0.5
1008	3.0	1051	3.2
F 1543	0.8	SA 1628	0.4
2207	3.0	2335	3.2
6 0354	0.8	**21** 0450	0.4
1059	3.2	1138	3.4
SA 1628	0.6	SU 1711	0.3
2305	3.1		
7 0437	0.6	**22** 0020	3.3
1141	3.4	0531	0.3
SU 1707	0.4	M 1221	3.5
2353	3.3	○ 1749	0.3
8 0517	0.5	**23** 0100	3.3
1220	3.5	0609	0.3
M 1746	0.2	TU 1258	3.6
●		1822	0.3
9 0037	3.3	**24** 0136	3.3
0557	0.3	0643	0.3
TU 1300	3.6	W 1332	3.6
1825	0.1	1852	0.3
10 0121	3.4	**25** 0207	3.3
0638	0.2	0714	0.3
W 1340	3.7	TH 1404	3.6
1906	0.0	1923	0.3
11 0204	3.5	**26** 0237	3.3
0722	0.2	0746	0.4
TH 1422	3.8	F 1436	3.6
1951	0.0	1957	0.4
12 0245	3.5	**27** 0309	3.3
0808	0.2	0820	0.4
F 1504	3.8	SA 1510	3.5
2038	0.0	2032	0.4
13 0327	3.5	**28** 0343	3.3
0855	0.2	0857	0.5
SA 1547	3.7	SU 1545	3.5
2128	0.1	2109	0.5
14 0408	3.5	**29** 0418	3.2
0946	0.3	0938	0.6
SU 1632	3.6	M 1623	3.3
2223	0.3	2151	0.7
15 0451	3.3		
1031	0.5		
M 1721	3.4		
◗ 2326	0.5		

MARCH

Time	m	Time	m
1 0456	3.1	**16** 0513	3.2
1024	0.8	1127	0.6
TU 1705	3.1	W 1800	2.9
◗ 2239	0.9		
2 0539	2.9	**17** 0021	0.8
1121	0.9	0607	2.9
W 1755	2.9	TH 1256	0.7
2338	1.0	1941	2.7
3 0637	2.7	**18** 0140	0.8
1231	1.0	0747	2.8
TH 1856	2.8	F 1415	0.7
		2130	2.8
4 0053	1.1	**19** 0247	0.8
0756	2.7	0932	2.9
F 1357	1.0	SA 1518	0.5
2012	2.8	2229	3.0
5 0218	1.0	**20** 0343	0.6
0927	2.8	1030	3.1
SA 1508	0.7	SU 1608	0.4
2140	2.9	2316	3.1
6 0323	0.8	**21** 0430	0.4
1028	3.1	1117	3.3
SU 1559	0.4	M 1650	0.3
2245	3.1	2358	3.2
7 0412	0.6	**22** 0512	0.3
1115	3.3	1158	3.4
M 1642	0.2	TU 1727	0.2
2334	3.2		
8 0455	0.3	**23** 0036	3.3
1157	3.5	0548	0.3
TU 1722	0.0	W 1236	3.4
		○ 1758	0.0
9 0019	3.3	**24** 0109	3.3
0536	0.2	0620	0.3
W 1241	3.6	TH 1308	3.4
● 1802	-0.2	1826	0.0
10 0103	3.4	**25** 0139	3.3
0618	0.0	0648	0.3
TH 1324	3.7	F 1339	3.4
1845	-0.2	1855	0.3
11 0145	3.5	**26** 0208	3.3
0702	0.0	0717	0.3
F 1407	3.8	SA 1410	3.4
1929	-0.2	1926	0.4
12 0226	3.5	**27** 0239	3.4
0747	0.0	0749	0.3
SA 1450	3.8	SU 1443	3.4
2016	-0.1	1959	0.4
13 0306	3.6	**28** 0310	3.3
0834	0.0	0824	0.4
SU 1533	3.7	M 1548	3.4
2106	0.1	2037	0.4
14 0346	3.5	**29** 0342	3.3
0924	0.2	0905	0.5
M 1617	3.5	TU 1555	3.3
2200	0.3	2119	0.6
15 0427	3.4	**30** 0417	3.1
0951	0.4	0951	0.6
TU 1704	3.3	W 1637	3.1
◗ 2303	0.6	2207	0.7
		31 0456	3.0
		1047	0.8
		TH 1725	2.9
		◗ 2304	0.9

APRIL

Time	m	Time	m
1 0547	2.8	**16** 0111	0.9
1153	0.8	0708	2.8
F 1824	2.8	SA 1348	0.6
		2102	2.7
2 0012	1.1	**17** 0218	0.8
0701	2.7	0856	2.9
SA 1312	0.8	SU 1448	0.5
1938	2.7	2201	2.9
3 0133	1.0	**18** 0315	0.7
0838	2.8	0958	3.1
SU 1427	0.6	M 1539	0.4
2108	2.8	2247	3.1
4 0247	0.8	**19** 0404	0.5
0952	3.0	1047	3.2
M 1525	0.3	TU 1622	0.3
2218	3.0	2328	3.2
5 0343	0.5	**20** 0446	0.4
1044	3.2	1128	3.3
TU 1612	0.0	W 1658	0.3
2310	3.2		
6 0431	0.3	**21** 0005	3.2
1131	3.4	0522	0.3
W 1656	-0.2	TH 1206	3.3
2356	3.3	1730	0.3
7 0515	0.1	**22** 0039	3.3
1217	3.6	0554	0.3
TH 1739	-0.3	F 1239	3.3
●		○ 1759	0.4
8 0040	3.4	**23** 0110	3.3
0558	-0.1	0622	0.3
F 1304	3.6	SA 1310	3.3
1823	-0.3	1827	0.4
9 0123	3.5	**24** 0140	3.4
0642	-0.1	0651	0.3
SA 1350	3.7	SU 1342	3.3
1908	-0.2	1858	0.4
10 0205	3.6	**25** 0209	3.4
0727	-0.1	0722	0.3
SU 1435	3.7	M 1416	3.4
1956	-0.1	1933	0.4
11 0245	3.6	**26** 0240	3.4
0814	-0.1	0759	0.3
M 1520	3.6	TU 1453	3.3
2046	0.1	2012	0.5
12 0325	3.6	**27** 0312	3.4
0904	0.1	0840	0.4
TU 1604	3.4	W 1532	3.3
2140	0.3	2057	0.5
13 0406	3.4	**28** 0346	3.3
1000	0.3	0928	0.5
W 1653	3.2	TH 1614	3.2
2242	0.6	2146	0.7
14 0451	3.2	**29** 0425	3.1
1109	0.5	1023	0.6
TH 1750	2.9	F 1702	3.0
◗ 2356	0.8	2243	0.8
15 0545	3.0	**30** 0512	3.0
1233	0.6	1127	0.6
F 1922	2.7	SA 1758	2.9
		◗ 2347	0.9

Chart Datum: 1·62 metres below Ordnance Datum (Newlyn). HAT is 3·9 metres above Chart Datum.

STANDARD TIME (UT)
For Summer Time add ONE hour in **non-shaded areas**

GREENOCK — LAT 55°57′N LONG 4°46′W
TIMES AND HEIGHTS OF HIGH AND LOW WATERS

Dates in red are **SPRINGS**
Dates in blue are NEAPS

YEAR 2016

MAY

Day	Time	m	Time	m
1 SU	0620 / 1239 / 1908	2.8 / 0.6 / 2.8	**16** M 0139 / 0758 / 1410 / 2116	0.9 / 2.9 / 0.6 / 2.8
2 M	0058 / 0752 / 1350 / 2031	0.9 / 2.8 / 0.4 / 2.9	**17** TU 0239 / 0913 / 1502 / 2207	0.8 / 3.0 / 0.5 / 3.0
3 TU	0211 / 0913 / 1451 / 2145	0.8 / 3.0 / 0.2 / 3.1	**18** W 0331 / 1007 / 1546 / 2251	0.6 / 3.1 / 0.4 / 3.1
4 W	0314 / 1013 / 1543 / 2241	0.5 / 3.2 / 0.0 / 3.2	**19** TH 0416 / 1052 / 1625 / 2330	0.5 / 3.2 / 0.4 / 3.2
5 TH	0406 / 1105 / 1631 / 2330	0.3 / 3.4 / -0.2 / 3.4	**20** F 0454 / 1131 / 1700	0.4 / 3.2 / 0.4
6 F	0453 / 1154 / 1717	0.1 / 3.5 / -0.2	**21** SA 0007 / 0528 / 1207 / 1732	3.3 / 0.4 / 3.2 / 0.5
7 SA	0016 / 0538 / 1244 / 1802	3.5 / -0.1 / 3.6 / -0.2	**22** SU 0041 / 0559 / 1240 / 1803	3.4 / 0.4 / 3.2 / 0.5
8 SU	0101 / 0623 / 1333 / 1849	3.6 / -0.1 / 3.6 / -0.1	**23** M 0113 / 0629 / 1314 / 1836	3.4 / 0.4 / 3.3 / 0.5
9 M	0144 / 0709 / 1420 / 1938	3.6 / -0.1 / 3.6 / 0.0	**24** TU 0143 / 0702 / 1351 / 1913	3.5 / 0.3 / 3.3 / 0.5
10 TU	0225 / 0756 / 1507 / 2028	3.6 / -0.1 / 3.5 / 0.2	**25** W 0214 / 0739 / 1430 / 1954	3.5 / 0.3 / 3.3 / 0.5
11 W	0306 / 0846 / 1553 / 2121	3.6 / 0.1 / 3.3 / 0.4	**26** TH 0248 / 0822 / 1512 / 2040	3.5 / 0.3 / 3.3 / 0.5
12 TH	0348 / 0941 / 1642 / 2219	3.5 / 0.3 / 3.1 / 0.6	**27** F 0325 / 0910 / 1555 / 2130	3.4 / 0.4 / 3.2 / 0.6
13 F	0433 / 1044 / 1738 / 2323	3.3 / 0.4 / 2.9 / 0.8	**28** SA 0404 / 1004 / 1643 / 2225	3.3 / 0.4 / 3.1 / 0.7
14 SA	0524 / 1157 / 1847	3.1 / 0.6 / 2.7	**29** SU 0451 / 1105 / 1736 / 2324	3.2 / 0.5 / 3.0 / 0.8
15 SU	0032 / 0630 / 1309 / 2009	0.9 / 2.9 / 0.6 / 2.7	**30** M 0551 / 1211 / 1839	3.0 / 0.4 / 3.0
			31 TU 0029 / 0713 / 1318 / 1953	0.8 / 3.0 / 0.4 / 3.0

JUNE

Day	Time	m	Time	m
1 W	0139 / 0836 / 1420 / 2109	0.7 / 3.1 / 0.2 / 3.1	**16** TH 0252 / 0912 / 1507 / 2206	0.8 / 2.9 / 0.6 / 3.0
2 TH	0245 / 0944 / 1517 / 2212	0.6 / 3.2 / 0.1 / 3.2	**17** F 0342 / 1008 / 1552 / 2254	0.7 / 3.0 / 0.6 / 3.2
3 F	0343 / 1041 / 1609 / 2306	0.4 / 3.4 / 0.0 / 3.3	**18** SA 0425 / 1053 / 1631 / 2336	0.5 / 3.1 / 0.5 / 3.3
4 SA	0435 / 1135 / 1658 / 2355	0.2 / 3.4 / -0.1 / 3.5	**19** SU 0504 / 1133 / 1708	0.5 / 3.1 / 0.5
5 SU	0523 / 1226 / 1746	0.0 / 3.5 / 0.0	**20** M 0014 / 0539 / 1211 / 1742	3.4 / 0.4 / 3.2 / 0.5
6 M	0041 / 0608 / 1318 / 1833	3.5 / 0.0 / 3.5 / 0.0	**21** TU 0048 / 0612 / 1249 / 1817	3.4 / 0.4 / 3.2 / 0.5
7 TU	0125 / 0654 / 1407 / 1921	3.6 / -0.1 / 3.4 / 0.1	**22** W 0120 / 0646 / 1329 / 1855	3.5 / 0.3 / 3.3 / 0.5
8 W	0208 / 0740 / 1454 / 2010	3.6 / 0.0 / 3.3 / 0.3	**23** TH 0153 / 0724 / 1411 / 1938	3.5 / 0.3 / 3.3 / 0.5
9 TH	0249 / 0826 / 1540 / 2059	3.6 / 0.1 / 3.2 / 0.4	**24** F 0229 / 0806 / 1454 / 2024	3.5 / 0.3 / 3.3 / 0.4
10 F	0330 / 0916 / 1627 / 2150	3.5 / 0.2 / 3.1 / 0.5	**25** SA 0308 / 0853 / 1538 / 2113	3.5 / 0.3 / 3.3 / 0.5
11 SA	0413 / 1010 / 1715 / 2244	3.4 / 0.4 / 3.0 / 0.7	**26** SU 0349 / 0945 / 1624 / 2205	3.5 / 0.3 / 3.2 / 0.5
12 SU	0459 / 1110 / 1806 / 2343	3.2 / 0.5 / 2.9 / 0.8	**27** M 0434 / 1042 / 1714 / 2301	3.4 / 0.3 / 3.2 / 0.6
13 M	0551 / 1216 / 1904	3.0 / 0.6 / 2.8	**28** TU 0528 / 1144 / 1810	3.2 / 0.3 / 3.1
14 TU	0047 / 0653 / 1320 / 2007	0.9 / 2.9 / 0.7 / 2.8	**29** W 0002 / 0637 / 1249 / 1915	0.7 / 3.1 / 0.4 / 3.0
15 W	0153 / 0804 / 1418 / 2111	0.9 / 2.9 / 0.7 / 2.9	**30** TH 0109 / 0801 / 1354 / 2033	0.7 / 3.1 / 0.3 / 3.0

JULY

Day	Time	m	Time	m
1 F	0220 / 0919 / 1456 / 2146	0.7 / 3.1 / 0.3 / 3.1	**16** SA 0306 / 0915 / 1519 / 2217	0.8 / 2.9 / 0.7 / 3.1
2 SA	0325 / 1025 / 1553 / 2246	0.5 / 3.2 / 0.2 / 3.3	**17** SU 0357 / 1016 / 1605 / 2307	0.7 / 3.0 / 0.7 / 3.2
3 SU	0421 / 1122 / 1645 / 2339	0.3 / 3.3 / 0.1 / 3.4	**18** M 0441 / 1105 / 1645 / 2348	0.5 / 3.1 / 0.6 / 3.3
4 M	0511 / 1216 / 1733	0.1 / 3.4 / 0.1	**19** TU 0519 / 1147 / 1722	0.4 / 3.1 / 0.5
5 TU	0026 / 0557 / 1307 / 1820	3.5 / 0.1 / 3.3 / 0.2	**20** W 0025 / 0554 / 1229 / 1758	3.4 / 0.3 / 3.2 / 0.5
6 W	0111 / 0640 / 1355 / 1905	3.6 / 0.0 / 3.3 / 0.2	**21** TH 0059 / 0629 / 1310 / 1837	3.5 / 0.2 / 3.2 / 0.4
7 TH	0152 / 0722 / 1440 / 1949	3.6 / 0.1 / 3.2 / 0.3	**22** F 0135 / 0706 / 1353 / 1920	3.6 / 0.2 / 3.3 / 0.4
8 F	0231 / 0804 / 1522 / 2033	3.6 / 0.1 / 3.2 / 0.4	**23** SA 0213 / 0748 / 1436 / 2005	3.6 / 0.1 / 3.3 / 0.3
9 SA	0310 / 0846 / 1601 / 2117	3.6 / 0.2 / 3.1 / 0.5	**24** SU 0253 / 0833 / 1520 / 2052	3.6 / 0.1 / 3.3 / 0.3
10 SU	0349 / 0932 / 1642 / 2203	3.5 / 0.4 / 3.0 / 0.6	**25** M 0334 / 0922 / 1604 / 2142	3.6 / 0.1 / 3.3 / 0.4
11 M	0429 / 1022 / 1725 / 2253	3.3 / 0.5 / 3.0 / 0.7	**26** TU 0418 / 1017 / 1650 / 2236	3.5 / 0.2 / 3.3 / 0.5
12 TU	0512 / 1118 / 1811 / 2348	3.2 / 0.7 / 2.9 / 0.9	**27** W 0507 / 1117 / 1739 / 2335	3.3 / 0.4 / 3.2 / 0.6
13 W	0601 / 1220 / 1904	3.0 / 0.8 / 2.8	**28** TH 0606 / 1224 / 1838	3.1 / 0.5 / 3.1
14 TH	0053 / 0657 / 1325 / 2006	1.0 / 2.9 / 0.8 / 2.8	**29** F 0044 / 0728 / 1335 / 1957	0.7 / 3.0 / 0.5 / 3.0
15 F	0204 / 0804 / 1426 / 2115	1.0 / 2.8 / 0.9 / 2.9	**30** SA 0202 / 0904 / 1443 / 2126	0.8 / 3.0 / 0.5 / 3.0
			31 SU 0314 / 1019 / 1543 / 2234	0.6 / 3.1 / 0.4 / 3.2

AUGUST

Day	Time	m	Time	m
1 M	0413 / 1117 / 1635 / 2327	0.4 / 3.2 / 0.3 / 3.4	**16** TU 0416 / 1043 / 1622 / 2322	0.6 / 3.0 / 0.6 / 3.3
2 TU	0503 / 1209 / 1723	0.2 / 3.3 / 0.3	**17** W 0456 / 1129 / 1700	0.4 / 3.1 / 0.5
3 W	0014 / 0546 / 1257 / 1806	3.5 / 0.1 / 3.3 / 0.3	**18** TH 0001 / 0532 / 1211 / 1737	3.4 / 0.2 / 3.2 / 0.4
4 TH	0057 / 0625 / 1341 / 1847	3.6 / 0.1 / 3.2 / 0.3	**19** F 0038 / 0608 / 1253 / 1816	3.5 / 0.1 / 3.3 / 0.3
5 F	0135 / 0702 / 1420 / 1925	3.6 / 0.1 / 3.2 / 0.3	**20** SA 0117 / 0645 / 1335 / 1858	3.6 / 0.1 / 3.3 / 0.3
6 SA	0211 / 0737 / 1454 / 2003	3.6 / 0.2 / 3.2 / 0.4	**21** SU 0157 / 0726 / 1417 / 1943	3.7 / 0.0 / 3.4 / 0.3
7 SU	0246 / 0814 / 1528 / 2041	3.6 / 0.3 / 3.1 / 0.4	**22** M 0238 / 0811 / 1459 / 2030	3.7 / 0.0 / 3.4 / 0.2
8 M	0321 / 0853 / 1604 / 2122	3.5 / 0.4 / 3.1 / 0.5	**23** TU 0319 / 0859 / 1541 / 2119	3.7 / 0.1 / 3.4 / 0.3
9 TU	0358 / 0936 / 1643 / 2206	3.4 / 0.5 / 3.1 / 0.7	**24** W 0402 / 0951 / 1624 / 2212	3.6 / 0.3 / 3.4 / 0.4
10 W	0436 / 1023 / 1725 / 2255	3.3 / 0.7 / 3.0 / 0.8	**25** TH 0448 / 1052 / 1711 / 2312	3.4 / 0.5 / 3.3 / 0.6
11 TH	0519 / 1118 / 1813 / 2355	3.1 / 0.9 / 2.9 / 1.0	**26** F 0542 / 1204 / 1806	3.1 / 0.6 / 3.1
12 F	0611 / 1225 / 1910	2.9 / 1.0 / 2.8	**27** SA 0026 / 0702 / 1323 / 1923	0.8 / 2.9 / 0.7 / 3.0
13 SA	0108 / 0713 / 1340 / 2021	1.1 / 2.8 / 1.0 / 2.8	**28** SU 0153 / 0904 / 1434 / 2113	0.8 / 2.9 / 0.7 / 3.0
14 SU	0228 / 0826 / 1446 / 2138	1.0 / 2.8 / 0.9 / 3.0	**29** M 0306 / 1017 / 1533 / 2222	0.7 / 3.0 / 0.6 / 3.2
15 M	0329 / 0944 / 1539 / 2237	0.8 / 2.9 / 0.8 / 3.1	**30** TU 0404 / 1111 / 1624 / 2314	0.5 / 3.2 / 0.4 / 3.4
			31 W 0451 / 1158 / 1709 / 2358	0.3 / 3.3 / 0.3 / 3.5

Chart Datum: 1·62 metres below Ordnance Datum (Newlyn). HAT is 3·9 metres above Chart Datum.

FREE monthly updates. Register at
www.reedsnauticalalmanac.co.uk

STANDARD TIME (UT)
For Summer Time add ONE hour in **non-shaded areas**

GREENOCK LAT 55°57'N LONG 4°46'W
TIMES AND HEIGHTS OF HIGH AND LOW WATERS

Dates in red are **SPRINGS**
Dates in blue are **NEAPS**

YEAR **2016**

SEPTEMBER

Day	Time m	Time m	Time m	Time m
1 TH	0531 0.2	1240 3.3	1750 0.3 ●	
2 F	0038 3.6	0607 0.2	1319 3.3	1826 0.3
3 SA	0115 3.6	0638 0.2	1352 3.2	1859 0.4
4 SU	0148 3.6	0709 0.3	1422 3.2	1931 0.4
5 M	0220 3.6	0741 0.4	1453 3.3	2006 0.5
6 TU	0253 3.6	0816 0.4	1527 3.3	2043 0.5
7 W	0327 3.5	0853 0.5	1603 3.2	2123 0.6
8 TH	0404 3.4	0934 0.7	1642 3.1	2209 0.8
9 F	0445 3.2	1021 0.9	1727 3.0	◑2303 1.0
10 SA	0535 3.0	1122 1.1	1821 2.9	
11 SU	0014 1.1	0635 2.8	1243 1.2	1929 2.8
12 M	0144 1.1	0749 2.8	1407 1.1	2054 2.9
13 TU	0256 0.9	0916 2.9	1509 0.9	2203 3.1
14 W	0347 0.6	1022 3.1	1555 0.7	2251 3.3
15 TH	0428 0.3	1110 3.2	1635 0.5	2333 3.5
16 F	0506 0.1	1152 3.3	1714 0.3 ○	
17 SA	0014 3.6	0543 0.0	1233 3.4	1754 0.2
18 SU	0056 3.7	0622 -0.1	1314 3.5	1836 0.2
19 M	0139 3.8	0704 -0.1	1356 3.5	1921 0.1
20 TU	0222 3.8	0748 0.0	1437 3.6	2007 0.2
21 W	0305 3.8	0836 0.1	1519 3.6	2057 0.3
22 TH	0348 3.6	0929 0.4	1601 3.5	2150 0.5
23 F	0434 3.4	1031 0.6	1647 3.4	◑2254 0.7
24 SA	0527 3.1	1148 0.8	1741 3.2	
25 SU	0017 0.8	0653 2.8	1310 0.9	1857 3.0
26 M	0143 0.8	0901 2.9	1419 0.9	2055 3.0
27 TU	0251 0.7	1006 3.1	1517 0.7	2202 3.2
28 W	0346 0.5	1055 3.3	1607 0.5	2253 3.4
29 TH	0431 0.4	1138 3.4	1650 0.4	2336 3.5
30 F	0510 0.3	1216 3.4	1729 0.4	

OCTOBER

Day	Time m	Time m	Time m	Time m
1 SA	0015 3.6	0543 0.3	1251 3.4	●1803 0.4
2 SU	0050 3.6	0613 0.4	1321 3.4	1833 0.4
3 M	0121 3.6	0641 0.4	1349 3.4	1902 0.5
4 TU	0153 3.6	0711 0.6	1420 3.4	1934 0.5
5 W	0225 3.6	0744 0.5	1453 3.5	2009 0.6
6 TH	0300 3.5	0819 0.6	1527 3.4	2048 0.7
7 F	0337 3.4	0859 0.8	1604 3.3	2133 0.8
8 SA	0418 3.3	0944 1.0	1645 3.2	2225 1.0
9 SU	0505 3.1	1040 1.2	1736 3.0	◑2331 1.1
10 M	0603 2.9	1151 1.3	1843 2.9	
11 TU	0054 1.1	0716 2.8	1318 1.3	2005 2.9
12 W	0214 0.9	0844 2.9	1431 1.1	2122 3.1
13 TH	0311 0.6	0955 3.1	1524 0.8	2217 3.4
14 F	0356 0.3	1045 3.3	1609 0.5	2304 3.5
15 SA	0437 0.1	1128 3.5	1651 0.3	2349 3.7
16 SU	0518 0.0	1210 3.6	1733 0.2 ○	
17 M	0035 3.8	0559 -0.1	1253 3.6	1816 0.1
18 TU	0121 3.8	0643 0.0	1335 3.7	1901 0.1
19 W	0206 3.8	0729 0.1	1417 3.7	1948 0.2
20 TH	0251 3.8	0817 0.3	1459 3.7	2038 0.3
21 F	0336 3.6	0911 0.5	1542 3.6	2132 0.5
22 SA	0424 3.4	1012 0.8	1628 3.5	◑2238 0.7
23 SU	0521 3.1	1127 1.0	1721 3.3	
24 M	0002 0.8	0648 2.9	1246 1.1	1834 3.1
25 TU	0121 0.8	0837 2.9	1354 1.2	2019 3.1
26 W	0226 0.7	0940 3.1	1452 0.8	2131 3.2
27 TH	0320 0.6	1028 3.3	1543 0.7	2223 3.4
28 F	0405 0.5	1110 3.4	1627 0.5	2307 3.5
29 SA	0444 0.4	1147 3.5	1705 0.5	2346 3.5
30 SU	0518 0.5	1220 3.5	●1739 0.5	
31 M	0021 3.5	0547 0.5	1251 3.5	1809 0.5

NOVEMBER

Day	Time m	Time m	Time m	Time m
1 TU	0054 3.5	0616 0.6	1321 3.6	1837 0.6
2 W	0126 3.5	0645 0.6	1352 3.6	1908 0.6
3 TH	0200 3.5	0718 0.6	1424 3.6	1943 0.6
4 F	0236 3.5	0754 0.7	1458 3.6	2022 0.7
5 SA	0314 3.3	0835 0.8	1533 3.5	2107 0.8
6 SU	0355 3.3	0921 1.0	1612 3.3	2159 0.9
7 M	0441 3.2	1015 1.1	1659 3.2	●2300 1.0
8 TU	0535 3.0	1118 1.3	1800 3.1	
9 W	0011 1.0	0643 3.0	1232 1.3	1917 3.1
10 TH	0127 0.8	0804 3.0	1347 1.1	2037 3.2
11 F	0230 0.6	0919 3.2	1449 0.9	2142 3.4
12 SA	0323 0.4	1015 3.4	1541 0.6	2236 3.6
13 SU	0410 0.2	1103 3.5	1628 0.4	2326 3.7
14 M	0455 0.0	1148 3.6	1714 0.2 ○	
15 TU	0015 3.8	0539 0.0	1233 3.7	1758 0.2
16 W	0104 3.8	0625 0.1	1317 3.8	1844 0.1
17 TH	0153 3.8	0712 0.2	1400 3.9	1932 0.2
18 F	0240 3.7	0801 0.4	1443 3.8	2021 0.3
19 SA	0327 3.5	0854 0.6	1526 3.7	2115 0.5
20 SU	0417 3.4	0951 0.8	1612 3.6	2217 0.6
21 M	0512 3.1	1057 1.0	1704 3.4	◑2330 0.8
22 TU	0622 3.0	1209 1.1	1806 3.2	
23 W	0046 0.9	0746 2.9	1318 1.1	1926 3.1
24 TH	0151 0.8	0857 3.0	1420 1.0	2044 3.2
25 F	0246 0.7	0951 3.2	1513 0.8	2145 3.3
26 SA	0334 0.7	1035 3.3	1559 0.7	2233 3.4
27 SU	0415 0.6	1115 3.5	1640 0.6	2315 3.4
28 M	0451 0.6	1151 3.5	1716 0.6	2353 3.4
29 TU	0524 0.6	1225 3.6	○1749 0.6	
30 W	0027 3.5	0555 0.7	1258 3.6	1819 0.6

DECEMBER

Day	Time m	Time m	Time m	Time m
1 TH	0102 3.5	0625 0.7	1329 3.7	1850 0.6
2 F	0138 3.5	0658 0.7	1401 3.7	1924 0.6
3 SA	0215 3.5	0735 0.7	1435 3.7	2003 0.7
4 SU	0255 3.5	0817 0.8	1511 3.6	2047 0.7
5 M	0336 3.4	0903 0.9	1550 3.5	2137 0.8
6 TU	0421 3.3	0954 1.0	1634 3.4	2233 0.8
7 W	0510 3.2	1051 1.1	1726 3.3	◑2336 0.8
8 TH	0608 3.1	1156 1.1	1833 3.2	
9 F	0044 0.8	0719 3.1	1306 1.1	1953 3.2
10 SA	0151 0.6	0837 3.2	1415 0.9	2108 3.3
11 SU	0252 0.5	0944 3.3	1515 0.7	2211 3.5
12 M	0346 0.3	1039 3.5	1609 0.5	2307 3.6
13 TU	0435 0.2	1129 3.6	1658 0.3	
14 W	0000 3.7	0523 0.2	1217 3.7	○1744 0.2
15 TH	0052 3.7	0610 0.2	1302 3.8	1831 0.2
16 F	0142 3.7	0658 0.3	1346 3.9	1918 0.2
17 SA	0230 3.6	0746 0.4	1429 3.9	2005 0.3
18 SU	0317 3.5	0835 0.5	1512 3.8	2054 0.4
19 M	0404 3.3	0926 0.7	1556 3.7	2147 0.5
20 TU	0452 3.2	1020 0.8	1643 3.5	2245 0.7
21 W	0543 3.1	1121 1.0	1733 3.3	◑2352 0.8
22 TH	0641 3.0	1229 1.1	1830 3.2	
23 F	0102 0.9	0747 2.9	1337 1.1	1936 3.1
24 SA	0205 0.9	0855 3.0	1438 1.0	2048 3.1
25 SU	0258 0.9	0954 3.2	1529 0.9	2152 3.2
26 M	0345 0.8	1042 3.3	1614 0.7	2243 3.2
27 TU	0426 0.7	1124 3.5	1654 0.7	2326 3.3
28 W	0503 0.7	1203 3.6	1731 0.6	
29 TH	0005 3.3	0537 0.7	1238 3.6	●1804 0.6
30 F	0043 3.4	0609 0.7	1310 3.7	1836 0.6
31 SA	0120 3.4	0642 0.7	1342 3.7	1909 0.6

Chart Datum: 1·62 metres below Ordnance Datum (Newlyn). HAT is 3·9 metres above Chart Datum.

SW Scotland

》》 **FREE** monthly updates. Register at 《
www.reedsnauticalalmanac.co.uk

489

9.9.23 ARDROSSAN

N Ayrshire 55°38'·50N 04°49'·61W ❄❄⚓⚓⚓❁❁

CHARTS AC 5610, 2126, 2221, 2491, 1866; Imray C63

TIDES +0055 Dover; ML 1·9; Duration 0630

Standard Port GREENOCK (←—)

Times				Height (metres)			
High Water		Low Water		MHWS	MHWN	MLWN	MLWS
0000	0600	0000	0600	3·4	2·8	1·0	0·3
1200	1800	1200	1800				
Differences ARDROSSAN							
–0020	–0010	–0010	–0010	–0·2	–0·2	+0·1	+0·1
IRVINE							
–0020	–0020	–0030	–0010	–0·3	–0·3	–0·1	0·0

SHELTER Good in marina (formerly Eglinton Dock), access at all tides over sill, 5·2m least depth. A storm gate is fitted; max acceptable beam is 8·6m (28ft). Strong SW/NW winds cause heavy seas in the apprs and the hbr may be closed in SW gales. Ferries berth on both sides of Winton Pier.

NAVIGATION WPT 55°38'·13N 04°50'·55W, 055°/0·65M to hbr ent. From the W/NW keep clear of low-lying Horse Isle (conspic W tower on its S end) ringed by drying ledges. The passage between Horse Isle and the mainland is obstructed by unmarked drying rks and should not be attempted. Be aware of following dangers: From the S/SE, Eagle Rk 3ca S of hbr ent, marked by SHM buoy, Fl G 5s. 3ca SE of Eagle Rk lies unmarked Campbell Rk (0·2m). W Crinan Rk (1·1m) is 300m W of hbr ent, marked by PHM buoy, Fl R 4s.

LIGHTS AND MARKS Dir lt WRG 15m W14M, R/G11M (see 9.9.4); W sector leads 055° to hbr ent between lt ho and detached bkwtr. Lt ho Iso WG 4s 11m 9M, 317°-G-035°, W elsewhere. On S end of detached bkwtr, Fl R 5s 7m 5M.

Traffic Signals, shown H24 from control twr at ent to marina:

3 F ● lts (vert) = hbr and marina closed; no entry/exit for commercial and pleasure vessels.

3 F ● lts (vert) = marina open, hbr closed; pleasure craft may enter/exit the marina, no commercial movements.

2 F ● lts over 1 F ● = hbr open, marina closed; in severe weather marina storm gate is closed. Commercial vessels may enter/exit hbr, subject to approval by Hbr Control on VHF. Pleasure craft must clear the approach channel, ferry turning area (between the detached bkwtr and Winton Pier) and the outer basin. Yachts may not manoeuvre under sail alone until seaward of the outer breakwater.

COMMUNICATIONS (Code 01294) CGOC (02891) 463933; Police 101; Dr 463011. Marina www. clyde marina.com ☎607077.

Clyde Marina VHF Ch **80** M. *Hbr Control* Ch 12 14 16 (H24).

FACILITIES Clyde Marina (252⌑ inc 50🆅) £2·60, short stay £9<4hrs, access H24, office hrs 0900-1800 🛢 ⚓ 🖭 Ⓔ 🏴(50t) 🅿 ✕. Town 🛢 ⊠ Ⓑ 🛒 🗓 ✕ 🗔. ⇌ ✈ (Prestwick/Glasgow). Ferry to Brodick (Arran).

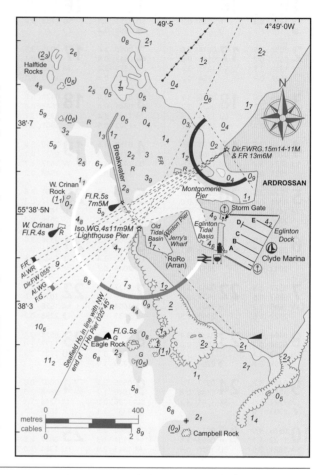

ADJACENT HARBOURS

IRVINE, N Ayrshire, **55°36'·17N 04°42'·07W**. AC 2126, 2220, 1866. HW +0055 on Dover. Tides: see 9.9.23.

Do not attempt entrance in heavy onshore weather. Good shelter once across the bar which carries approx 0·5m. The IB-B SPM buoy, Fl Y 3s, is 1·15M from hbr ent, close NW of ldg line. 5 blocks of flats and chimneys are conspic ENE of entrance. Ldg lts 051°: front, FG 10m 5M; rear, FR 15m 5M. The S ent groyne is lit, Fl G 3s; groynes inside the ent have unlit perches. White Pilot tower with mast is conspic 3ca inside entrance.

HM ☎(01294) 487286, 🖷 487111, VHF Ch 12 (0800-1600 Tues and Thurs; 0800-1300 Wed).

⚓ prohib. Visitors' pontoons on N side or on S side at visitors' quay (2·2m). Berths available above opening footbridge which opens at 5 mins notice ☎08708 403123 or VHF Ch 12 (call *Irvine Bridge*). **Quay** ⌑, 5m<£4<10m<£6<15m £9, ⚓ ⚓ 🏴⚓ 🅰 🛢(3t).
Town 🏧 🏧 (1·5km), 🛢 🛒 ✕ ⇌ ✈ (Prestwick).

AYR, S Ayrshire, **55°28'·22N 04°38'·78W**. AC 5610, 2126, 2220, 1866. HW +0050 on Dover; ML 1·8m. Duration 0630. See 9.9.24.

Good shelter in S dock although berth space limited. After heavy rains large amounts of debris may be washed down the R Ayr. From the W, harbour entrance lies between conspic gasholder to the N and townhall spire to the S. St Nicholas SHM buoy, Fl G 2s, warns of shoals and a separate drying rock (0·8m) 150m S of entrance.

Ldg lts 098°: front, by Pilot Stn, FR 10m 5M R tower, rear (130m from front), Oc R 10s 18m 9M. Traffic signals (near front ldg lt):
        2 ● (vert) = hbr closed to incoming traffic.

N Bkwtr Hd, QR 9m 5M. S Pier Hd, Q 7m 7M, vis 012°-161°, and FG 5m 5M, same structure, vis 012°-082°, over St Nicholas Rk.

HM ☎(01292) 281687; VHF Ch 14 16.

Ayr Y & CC at S dock, ⚓ limited ⌑ on pontoon on N side of harbour (access key, £10 deposit, from Ship Inn).
Services ✕ ⚓ 🖭 🛢 🅿.
Town ⚓ 🏧 🏧 🛢 ⊠ Ⓑ 🛒 ✕ 🗔 ⇌.

9.9.24 TROON

S Ayrshire 55°33'·10N 04°40'·97W ❄❄⚓⚓🌼🌼

CHARTS AC 5610, 2126, 2220, 1866; Imray C63

TIDES +0050 Dover; ML 1·9; Duration 0630

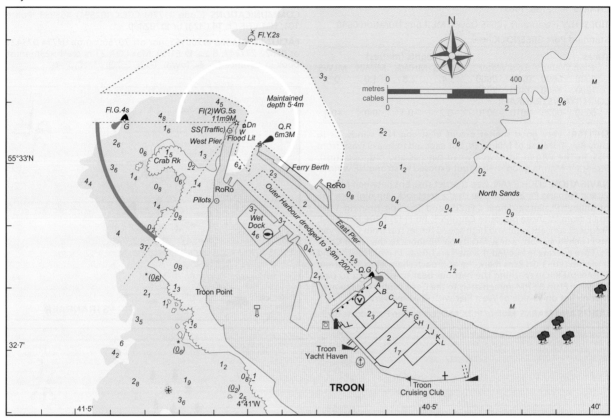

HARBOUR ON THE FIRTH OF CLYDE SOUTH OF TROON

GIRVAN, S Ayrshire, **55°14'·77N 04°51'·87W**. AC 2199, 1866. HW +0043 on Dover; ML 1·8m; Duration 0630. See 9.9.24.

Good shelter. Brest Rks, 3·5M N of the harbour extend 6ca offshore. Bar carries approx 1·5m. Beware Girvan Patch, 1·7m, 4ca SW of ent. Ch spire (conspic) brg 104° leads between N Bkwtr, Fl R 4s 5m 2M, and S pier, 2 FG (vert) 8m 4M. Inner N groyne, Iso 4s 3m 4M. Traffic sig at root of S pier: 2 B discs (hor)/ 2 ● (hor) = hbr shut.

HM ☎(01465) 713648; VHF Ch 12 16 (HO)

35 ⌣ (£2·00) on fingers around perimeter of inner harbour (1·7m). Coasters and FVs berth on adjacent quay: ⚓ ▬.

Town 🅿 🚲 Ⓑ ⊠ 🍴 ✕ ⇌.

Standard Port GREENOCK (⟵)

Times				Height (metres)			
High Water		Low Water		MHWS	MHWN	MLWN	MLWS
0000	0600	0000	0600	3·4	2·8	1·0	0·3
1200	1800	1200	1800				
Differences TROON							
−0025	−0025	−0020	−0020	−0·2	−0·2	0·0	0·0
AYR							
−0025	−0025	−0030	−0015	−0·4	−0·3	+0·1	+0·1
GIRVAN							
−0025	−0040	−0035	−0010	−0·3	−0·3	−0·1	0·0

SHELTER Complete in marina (2·4m at ent, 1·6m at SE end); speed limit 5kn. *Strong W winds cause heavy seas in the hbr approach.*

NAVIGATION WPT 55°33'·20N 04°42'·00W, 103°/0·6M to W pier lt. Appr in sector SW to NW. Beware Lady Isle, Fl (4) 30s 19m 8M, W bn, 2·2M SW; Troon Rock (5·6m, occas breaks) 1·1M W; Lappock Rock (0·6m, bn with G barrel topmark) 1·6M NNW; Mill Rock (0·4m) ½M NNE of hbr ent, marked by unlit PHM buoy. Beware wash from high speed ferries in approaches.

LIGHTS AND MARKS No ldg lts. Traffic signals when Fast Ferries arriving/departing. 14m SE of W pier hd lt, Fl (2) WG 5s, there is a floodlit dolphin, W with dayglow patches. A SHM lt buoy, FG, marks the chan in the ent to marina.

COMMUNICATIONS (Code 01292) CGOC (02891) 463933; Police 101; Dr 313593; Ⓗ 610555 (Ayr). HM 281687; Troon CC 311865; Troon YC 316770. HM Ch 14. Marina Ch **80** M (H24).

FACILITIES Troon Yacht Haven ⚓ www.yachthavens.com ☎315553 access all tides 300⌣+50Ⓥ, £2·70 (short stay £10·20). Ⓥ berth on pontoon **A**, first to stbd. LOA 36m x 3m draught ⌣ at hammerheads. ▬ 🔧 🔩(H24) ⚓ ⚒ 🔋 🛢 Ⓔ ⚙ ⛽(daily pick-up) 🚤 (50t) 🏗(25t) 🛒 🍴 ✕ ☎311523.

Town ⊠ Ⓑ ⇌ ✈ (Prestwick/Glasgow). Fast RoRo ferry to Larne.

9.9.25 LOCH RYAN/STRANRAER

Dumfries and Galloway 55°01′N 05°05′W ✿✿🚢🚣🏴🏴

CHARTS AC 2198, 1403

TIDES HW (Stranraer) +0055 Dover; ML 1.6m; Duration 0640

Standard Port GREENOCK (←—)

Times				Heights (metres)			
High Water		Low Water		MHWS	MHWN	MLWN	MLWS
0000	0600	0200	0800	3.4	2.8	1.0	0·3
1200	1800	1400	2000				
LOCH RYAN/STRANRAER							
–0030	–0025	–0010	–0010	–0·2	–0·1	0·0	+0·1

SHELTER Very good shelter except in strong NW winds. ⚓s in Lady Bay, 1·3M SSE of Milleur Pt, but exposed to heavy wash from HSS; in The Wig in 3m (avoid weed patches and old moorings); Stranraer Marina good shelter, but exposed to strong NE winds.

NAVIGATION LOCH RYAN: (see chartlet rhs). Ent between Milleur Pt and Finnarts Pt. Beware The Beef Barrel rock, 1m high 6ca SSE of Milleur Pt, Forbes Shoal ⚓ Fl(2)5s and The Spit running 1·5M to SE from W shore opposite Cairn Pt lt ho Fl (2) R10s 14m 12M. Frequent conventional and high-speed ferries run from Cairnryan and Loch Ryan Port, a new facility on NE shore to the NE of Cairn Pt. They operate in buoyed channels and areas, in which anchoring is prohibited. 3 Gn Bcns mark the approach channel to Stranraer. STRANRAER: On reaching the harbour bear to starboard to pass between PHM & SHM immediately to the E of a rubble breakwater. Marina lies on SE side of West Pier with ❶ at seaward end.

LIGHTS AND MARKS Milleur Pt NCM Q; Loch Ryan Port ldg lts Iso 2s

9M; Cairn Pt Fl(2) R 10s 14m; Cairnryan ferry terminal Fl R 5s 5m 5M. Stranraer, centre pier hd 2FBu (vert), E pier hd 2FR (vert); PHM Fl R 5s, SHM Fl G 5s.

COMMUNICATIONS (Code 01776) CGOC (02891) 463933; Police 101; HM VHF Ch **14** (H24) or ☎702460.

FACILITIES Marina ☎706565, hbr off 707500; mob 07734 073421 or 07827 277247. 8 ⛴ ❶ £1.50. Max LOA 27m; craft >23m must book in advance 🔌 ⚓. **Town:** 🏧🏨🏬✉🅑🍺✕🛒🚉.

STRANRAER

HARBOURS FROM MULL OF GALLOWAY TO WIGTOWN BAY

DRUMMORE, Dumfries and Galloway, **54°41'·57N 04°53'·49W.** AC 2094, 5613. HW +0040m on Dover; ML 3.3m; Duration 0610. Owned by Drummore Hbr Trust the small drying harbour has now silted almost entirely rendering of little interest to passing mariners. Temp'y ⚓ 5.5ca NNE of hbr ent in 4m. Facilities in village: 🏠 🏠 🛢 ✉ 🛒 ✕ 🛏 ⇌ (bus to Stranraer), ✈ (Carlisle).

PORT WILLIAM, Dumfries and Galloway, **54°45'·70N 04°35'·10W.** AC 1826, 2094, 5613. HW +0035 on Dover; ML 3.7m; Duration 0545. See 9.9.27. Sheltered harbour dries having approx 1m at HW±2. ⚓ in 3m approx 0.4M off hbr ent. Ldg lts F G 2M on shore Front lt on pier Fl G 3s 3M. HM ☎07734 073419.

ISLE OF WHITHORN, Dumfries and Galloway, **54°41'·91N 04°21'·88W.** AC 1826, 2094, 5613. HW +0035 on Dover; ML 3.7m; Duration 0545. See 9.9.27. Shelter good but hbr dries, having approx 2.5m at HW±3. On W side of ent beware the Skerries ledge. St Ninian's Tr (Fl WR 3s 20m 6/4M viz: 310°-W-005°-R-040°; W □ tr) is conspic at E side of ent. E pier hd has QG 4m 5M; ldg lts 335°, both Oc R 8s 7/9m 7M, synch, Or masts and ◇. HM ☎07734 073420, www.isleofwhithorn.com; Facilities: ⌒ on quay £4.83/yacht, 2 ━ (launching £1.00), 🛢 🛢(by arrangement with HM), ⚓ ✕ ✎ 🛢 ✉ 🛒 🛏 VHF Ch 08 (occas).

GARLIESTON, Dumfries and Galloway, **54°47'·36N 04°21'·83W.** AC 1826, 2094, 5613. HW +0035 on Dover; ML no data; Duration 0545. See 9.9.27. Hbr affords complete shelter but dries. Pier hd lt 2FR (vert) 5m 3M. Beware rky outcrops in W side of bay marked by a perch. HM ☎(01988) 600295, Mobile 07734 073422. Facilities: ⚓ £3, ⚓ 🛏 on quay, ━. **Town:** 🏠 🏠 🛒 ✎.

9.9.26 PORTPATRICK

Dumfries and Galloway 54°50'·42N 05°07'·18W ❀❀◊◊✿✿

CHARTS AC 2724, 2198; Imray C62

TIDES +0032 Dover; ML 2·1; Duration 0615

Standard Port LIVERPOOL (⟶)

Times				Heights (metres)			
High Water		Low Water		MHWS	MHWN	MLWN	MLWS
0000	0600	0200	0800	9·4	7·5	3·2	1·1
1200	1800	1400	2000				
Differences PORTPATRICK							
+0038	+0032	0009	−0008	−5·5	−4·4	−2·0	−0·6

SHELTER Good in tiny Inner harbour (on rough stone wall), but entrance is difficult in strong SW/NW winds.

NAVIGATION WPT 54°50'·00N 05°08'·07W, 050°/0·7M to ent. Ent to outer hbr by short narrow chan with hazards either side, including rky shelf covered at HW. Beware cross tides off ent, up to 3kn springs. Barrel buoy (a mooring buoy) marks end of Half Tide Rk; do not cut inside.

LIGHTS AND MARKS Old Lt Ho (W tower, 22m) on Black Head is conspic 1·6M north of entrance. Ldg lts 050·5°, FG (H24) 6/8m: Front on sea wall; rear on bldg; 2 vert orange stripes by day. Conspic features include: TV mast 1M NE, almost on ldg line; large hotel on cliffs about 1½ca NNW of hbr ent; Dunskey Castle (ru) 4ca SE of hbr.

COMMUNICATIONS (Code 01776) CGOC (02891) 463933; Police 101; ℍ 702323; RNLI Coxswain (for local advice) 07771 741717. HM 810355.

FACILITIES Hbr ⚓ ━ ━ (small craft), ⌒£16, ⚓. **Village** 🏠 🛢 (bulk tanker) 🛢 ✉ 🛒 ✕ 🛏 ⇌ (bus to Stranraer Ⓑ), ✈ (Prestwick).

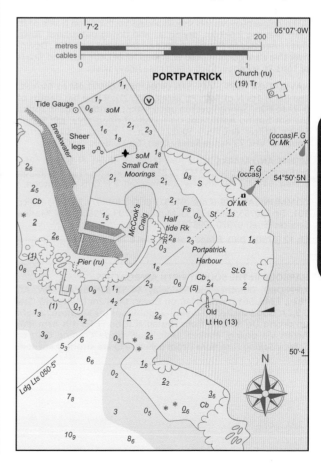

9.9.27 KIRKCUDBRIGHT

Dumfries and Galloway **54°50'·33N 04°03'·47W** ❀❀◊◊❁❁❁

CHARTS AC 1826, 2094, 1346, 1344, 5613; Imray C62

TIDES +0030 Dover; ML 4·1; Duration 0545

Standard Port LIVERPOOL (9.10.17) (⟶)

Times				Height (metres)			
High Water		Low Water		MHWS	MHWN	MLWN	MLWS
0000	0600	0200	0800	9·4	7·5	3·2	1·1
1200	1800	1400	2000				
Differences KIRKCUDBRIGHT BAY							
+0020	+0020	+0005	−0005	−1·9	−1·6	−0·8	−0·3
DRUMMORE							
+0035	+0045	+0010	+0015	−3·5	−2·6	−1·2	−0·5
PORT WILLIAM							
+0035	+0035	+0020	−0005	−3·0	−2·3	−1·1	ND
GARLIESTON							
+0030	+0040	+0025	0000	−2·4	−1·8	−0·8	ND
ISLE OF WHITHORN							
+0025	+0030	+0020	0000	−2·5	−2·1	−1·1	−0·4
HESTAN ISLET (Kippford)							
+0030	+0030	+0015	+0020	−1·1	−1·2	−0·8	−0·2
SOUTHERNESS POINT							
+0035	+0035	+0025	+0005	−0·8	−0·8	ND	ND
ANNAN WATERFOOT							
+0055	+0110	+0215	+0305	−2·3	−2·7	−3·0	CD+
TORDUFF POINT							
+0110	+0145	+0515	+0405	−4·2	−5·0	CD+	CD+
REDKIRK							
+0115	+0220	+0710	+0440	−5·6	−6·3	CD+	CD+

NOTES: At Annan Waterfoot, Torduff Pt and Redkirk the LW time differences are for the start of the rise, which at sp is very sudden. CD+ = At LW the tide does not usually fall below CD; ND = No Data.

SHELTER Very good. Depths at LW: 1·0–2·0m at the floating pontoon 300m below quay, visitors should always berth/raft on N side of it. Craft drawing 2·0–3·0m, drying ♦, and drying out against Town Quay (used H24 by FVs) by advance arrangement with HM. Downriver there are good ⌇s close W, N, and NE of Little Ross and ½ca N of Torrs Pt, except in S'lies which raise heavy swell.

NAVIGATION WPT 54°45'·51N 04°04'·08W, 005°/1·4M to Torrs Pt. The Bar is 1ca N of Torrs Pt; access HW±2½. R Dee has depths of 0·1m to 4·3m. Spring tides run up to 3–4kn. A firing range straddles the ent but no restrictions on transit of firing area; call Range Safety Officer ☎(01557) 500271 (out of hours), 830236 office hours) or VHF 73 Range Safety Craft, Gallavidian, on Ch 16.

LIGHTS AND MARKS Little Ross Lt ho, W of ent, Fl 5s 50m 12M, NNE end of Is, Stone bn Fl (2) 5s 21m 5M. No 1 bn, Fl 3s 7m 3M, on ♦ shed (54°47'·70N). River well lit/marked, 2 Fl G lts at pontoon.

COMMUNICATIONS (Code 01557) CGOC (01407) 762051; Police 101; Dr 330755; HM 331135 Mobile 07709 479663; GM Marine Services (Fuel) 07970 109814.

Ch 12 (0730-1700). *Range Control* Ch 16 73.

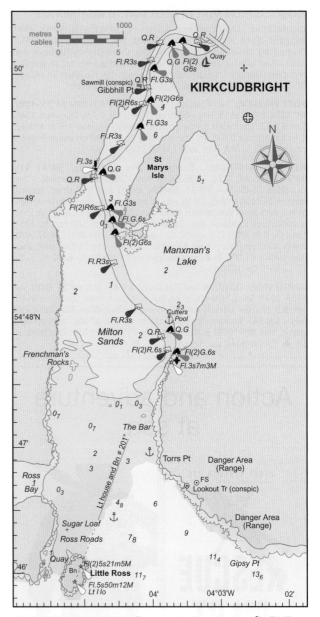

FACILITIES Pontoon/jetty ⌇ max LOA 60m £1.40 inc ⌾ ⊕ ⚓
Town Quay ⚓ ♠ ▮(by arrangement with HM) ⚒ ⊞ ⚙ ▷(15t).
KYC ☎330963; **SC** ☎330032 ▲ ▬ ⚓
Town ✉ ⑧ ⊟ ✕ ⌷ ⇌ (Dumfries 30M), ✈ (Glasgow 100M).

HARBOUR IN THE SOLWAY FIRTH APPROACHES

KIPPFORD, Dumfries and Galloway, **54°52'·36N 03°48'·93W**. AC 1826, 1346, 5613. HW +0040 on Dover; ML 4·2m (Hestan Is). See 9.9.27. Good shelter on drying ♦/pontoons off Kippford, 2·75M up drying Urr Estuary from Hestan Is lt ho Fl (2) 10s 42m 9M. Access via drying, marked, unlit chan. *Clyde Cruising Club* or *Solway Sailing Directions* (from Solway YC) are strongly advised. Beware Craig Roan on E side of ent. Temp ⌇s NE or W of Hestan Is to await tide. VHF: Ch M call *Kippford Startline* (YC) HW±2 in season. Ch 16 *Kippford Slipway* (Pilotage). Facilities: **Solway YC** www.thesyc.com ☎(01556) 600221; ⌇, ⊕, ⚓ M; **Services:** ⌇ £6, ▲ ▬ ⌾ ♠ ⚙ ⚒ ⌂ *Kippford Slipway Ltd* (01556) 620249. **Town** ▬ ⌾△✉⌷⌷.

NW England, IoM & N Wales

Mull of Galloway to Bardsey Island

9.10.1	**Contents**	495
9.10.2	**Area map and Distance table**	496
9.10.3	**Tidal stream charts**	498
9.10.4	**Lights, buoys and waypoints**	500
9.10.5	**Passage information**	501
	Solway Firth – 501 • Isle of Man, Barrow to Conwy – 504 • Menai Strait, Anglesey to Bardsey Island – 516	
9.10.6	**Maryport**	502
9.10.7	**Workington**	502
	Silloth • Harrington	
9.10.8	**Whitehaven**	503
	Ravenglass	
9.10.9	**Isle of Man**	503
	Port Erin • Castletown • Laxey	
9.10.10	**Peel**	505
9.10.11	**Port St Mary**	505
	Derby Haven	
9.10.12	**Douglas**	506
9.10.13	**Ramsey**	506
9.10.14	**Glasson Dock**	507
	Barrow-in-Furness • Heysham	
9.10.15	**Fleetwood**	508
9.10.16	**River Ribble/Preston**	509
9.10.17	**Liverpool, tide tables/curves**	510
	River Alt • Rhyl	
9.10.18	**Conwy and Deganwy**	515
9.10.19	**Menai Strait**	518
	Beaumaris • Bangor • The Swellies • Port Dinorwic • Caernarfon	
9.10.20	**Holyhead, tide tables/curves**	520
	Porth Dinllaen • Aberdaron • Bardsey Island	

NW England

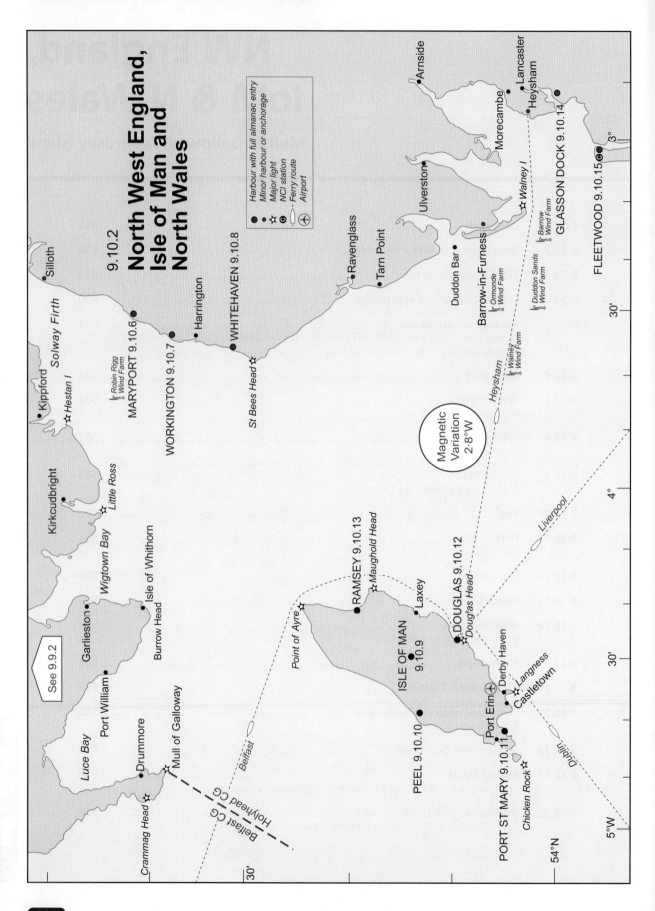

9.10.2
North West England, Isle of Man and North Wales

Magnetic Variation 2·8°W

See 9.9.2

Harbour with full almanac entry
Minor harbour or anchorage
Major light
NCI station
Ferry route
Airport

Silloth
Solway Firth
Kippford
Hestan I
Robin Rigg Wind Farm
MARYPORT 9.10.6
Harrington
WORKINGTON 9.10.7
WHITEHAVEN 9.10.8
St Bees Head
Ravenglass
Tarn Point
Ulverston
Duddon Bar
Barrow-in-Furness
Ormonde Wind Farm
Duddon Sands Wind Farm
Walney I
Barrow Wind Farm
Walney Wind Farm
GLASSON DOCK 9.10.14
FLEETWOOD 9.10.15
Arnside
Morecambe
Lancaster
Heysham
Heysham

Kirkcudbright
Little Ross
Wigtown Bay
Isle of Whithorn
Burrow Head
Garlieston
Port William
Luce Bay
Drummore
Mull of Galloway
Crammag Head
Belfast CG Holyhead CG
Belfast

Point of Ayre
Maughold Head
RAMSEY 9.10.13
Laxey
DOUGLAS 9.10.12
Douglas Head
ISLE OF MAN 9.10.9
Derby Haven
Langness
Castletown
Port Erin
PEEL 9.10.10
PORT ST MARY 9.10.11
Chicken Rock
Liverpool
Dublin

5°W
4°
30'
54°N
30'
3°
30'

496

DISTANCE TABLE
Approximate distances in nautical miles are by the most direct route, whilst avoiding dangers and allowing for Traffic Separation Schemes. Places in *italics* are in adjoining areas; places in **bold** are in 9.0.6. Distances across the Irish Sea.

1. *Fishguard*
2. *Bardsey Island*
3. **Holyhead**
4. *Menai (Caernarfon)*
5. **Menai (Beaumaris)**
6. **Conwy/Deganwy**
7. **Liverpool**
8. *River Ribble Ent*
9. **Fleetwood**
10. **Glasson Dock**
11. Ramsey
12. Douglas
13. **Port St Mary**
14. *Peel*
15. **Point of Ayre**
16. Whitehaven
17. Workington
18. Maryport
19. Kirkcudbright
20. *Mull of Galloway*

Magnetic Variation 2·4°W

RIVER RIBBLE 9.10.16
LIVERPOOL 9.10.17
Formby
R Alt
Belfast
Douglas, IOM
Dublin
Hilbre I
Burbo Bank Wind Farm
North Hoyle Wind Farm
Mostyn Quay
Gwynt-y-Môr Wind Farm
Rhyl Flats Wind Farm
Great Ormes Head
Llandudno
Colwyn Bay
CONWY AND DEGANWY 9.10.18
Bangor
MENAI STRAIT 9.10.19
Port Dinorwic
Caernarfon
Trefor
Porth Dinllaen
Porth Ysgaden

Anglesey
Moelfre
Beaumaris
The Swellies
Point Lynas
Amlwch
Cemaes Bay
Skerries
Skerries TSS
South Stack
HOLYHEAD 9.10.20
Holyhead CGOC
Treaddur Bay
Valley
Port Trecastell
Dublin / Dun Laoghaire
Dublin

See 9.11.2

MENAI STRAIT (inset)
BEAUMARIS
RAYC
BANGOR
Menai Suspension Bridge
Road and Railway Bridge
The Swellies
PORT DINORWIC
PLAS MENAI
CAERNARFON
Abermenai Pt
Fort Belan
The Bar
Llanddwyn Is.
53°15'N — 53°06'N
04°07'W — 04°21'W

53°30'N · 53°N · 5°W · 4° · 30'

9.10.3 AREA 10 TIDAL STREAMS

5 Hours before HW Dover (0515 before HW Liverpool)

4 Hours before HW Dover (0415 before HW Liverpool)

3 Hours before HW Dover (0315 before HW Liverpool)

2 Hours before HW Dover (0215 before HW Liverpool)

1 Hour before HW Dover (0115 before HW Liverpool)

HW Dover (0015 before HW Liverpool)

Northward 9.9.3   Southward 9.11.3   North Ireland 9.13.3   Mull of Kintyre 9.9.11   South Ireland 9.12.3

1 Hour after HW Dover (0045 after HW Liverpool)

2 Hours after HW Dover (0145 after HW Liverpool)

3 Hours after HW Dover (0245 after HW Liverpool)

4 Hours after HW Dover (0345 after HW Liverpool)

5 Hours after HW Dover (0445 after HW Liverpool)

6 Hours after HW Dover (0545 after HW Liverpool)

9.10.4 LIGHTS, BUOYS AND WAYPOINTS

Bold print = light with a nominal range of 15M or more. CAPITALS = place or feature. *CAPITAL ITALICS* = light-vessel, light float or Lanby. *Italics* = Fog signal. ***Bold italics*** = Racon. Some marks/buoys are fitted with AIS (<u>MMSI No</u>); see relevant charts.

SOLWAY FIRTH TO BARROW-IN-FURNESS
SILLOTH and MARYPORT
Lees Scar ⚓ Fl G 5s 11m 4M; W piles; 54°51'·78N 03°24'·79W.
Groyne Hd ⚓ 2 FG (vert) 4m 4M; Fl Bu tfc signals close by; 54°52'·14N 03°23'·93W.
Maryport S Pier Hd ⚓.Fl 1·5s 10m 6M; 54°43'·07N 03°30'·64W.

WORKINGTON and HARRINGTON
N Workington ⚓ Q; 54°40'·10N 03°38'·18W.
S Workington ⚓ VQ (6) + L Fl 10s; 54°37'·01N 03°38'·58W.
South Pier ⚓ Fl 5s 11m 5M; R bldg; 54°39'·12N 03°34'·67W.
Ldg Lts 131·8°. Front, FR 10m 3M; 54°38'·92N 03°34'·19W. Rear, 134m from front, FR 12m 3M.

WHITEHAVEN
W Pier Hd ⚓ Fl G 5s 16m 8M; W ○ twr; 54°33'·17N 03°35'·92W.
N Pier Hd ⚓ Fl R 5s 8m 10M; W ○ twr; 54°33'·17N 03°35'·75W.
Saint Bees Hd ☆ 54°30'·81N 03°38'·23W Fl (2) 20s 102m **18M**; W○ twr; obsc shore-340°.

RAVENGLASS
Blockhouse ⚓ FG; (Eskdale Range); 54°20'·16N 03°25'·34W.
Selker ⚓ Fl (3) G 10s; *Bell;* 54°16'·14N 03°29'·58W.

BARROW-IN-FURNESS
Lightning Knoll ⚓ L Fl 10s; 53°59'·83N 03°14'·28W.
Halfway Shoal ⚓ QR 19m 10s; R&W chequer Bn; 54°01'·46N 03°11'·88W.
Isle of Walney ☆ 54°02'·92N 03°10'·64W Fl 15s 21m **23M**; stone twr; obsc 122°-127° within 3M of shore.
Walney Chan Ldg Lts 040·7°. No.1 Front ⚓, Q 7m 10M; B Pile; 54°03'·19N 03°09'·22W. No. 2 Rear ⚓, 0·61M from front, Iso 2s 13m 10M; Pile.
Rampside Sands Ldg Lts 005·1°. No. 3 Front ⚓, Q 9m10M; W ○ twr; 54°04'·41N 03°09'·79W. No. 4 Rear ⚓, 0·77M from front, Iso 2s 14m 6M; R col, W face.

ISLE OF MAN
Whitestone Bank ⚓ Q (9) 15s; 54°24'·58N 04°20'·41W.
Point of Ayre ☆ 54°24'·94N 04°22'·13W Fl (4) 20s 32m **19M**; W twr, two R bands, ***Racon (M) 13-15M***.
Low Lt Ho (unlit), RW twr, B base, 54°25'·03N 04°21'·86W.

PEEL
Peel Bkwtr Hd ⚓ Oc 7s 11m 6M; W twr; 54°13'·67N 04°41'·69W.

PORT ERIN and PORT ST MARY
Ldg Lts 099·1°. Front, 54°05'·23N 04°45'·57W FR 10m 5M; W twr, R band. Rear, 39m from front, FR 19m 5M; W col, R band.
Calf of Man Lighthouse (disused), white 8-sided tower.
Chicken Rk ☆ Fl 5s 38m **20M**; ***Racon (C)*** 54°02'·27N 04°50'·32W
Alfred Pier Hd ⚓ Oc R 10s 8m 6M; 54°04'·33N 04°43'·82W.

CASTLETOWN and DERBY HAVEN
Dreswick Pt ⚓ Fl (2) 30s 23m 12M; W twr; 54°03'·29N 04°37'·45W.
New Pier Hd ⚓ Oc R 15s 8m 5M; 54°04'·33N 04°38'·97W.
Derby Haven, Bkwtr SW end ⚓Iso G 2s 5m 5M; W twr, G band; 54°04'·58N 04°37'·06W.

DOUGLAS
Douglas Head ☆ 54°08'·60N 04°27'·95W Fl 10s 32m **24M**; W twr; obsc brg more than 037°. FR Lts on radio masts 1 and 3M West.
No. 1 ⚓ Q (3) G 5s; 54°09'·04N 04°27'·68W.
Princess Alexandra Pier Hd ⚓ Fl R 5s 16m 8M; R mast; *Whis (2) 40s;* 54°08'·84N 04°27'·85W.

Ldg Lts 229·3°, Front ⚓, Oc Bu 10s 9m 5M; W △ R border on mast; 54°08'·72N 04°28'·25W. Rear ⚓, 62m from front, Oc Bu 10s 12m 5M; W ▽ on R border; synch with front.
Victoria Pier Hd ⚓ Iso G 10s 10m 3M; W col; vis: 225°-327°; Intnl Port Tfc Signals; 54°08'·84N 04°28'·08W.
Conister Rk Refuge ⚓ Q 3M; vis: 234°-312°; 54°09'·03N 04°28'·12W.

LAXEY to RAMSEY
Laxey Pier Hd ⚓ Oc R 3s 7m 5M; W twr, R band; obsc when brg less than 318°; 54°13'·50N 04°23'·43W.
Maughold Head ☆ 54°17'·72N 04°18'·58W Fl (3) 30s 65m **21M**.
Bahama ⚓ Q (6) + L Fl 10s; 54°20'·01N 04°08'·57W.
Queens Pier Dn ⚓ Fl R 5s; 54°19'·28N 04°21'·95W.
King William Bank ⚓ Q (3) 10s; 54°26'·01N 04°00'·08W.

BARROW TO RIVERS MERSEY AND DEE
MORECAMBE
Met Mast ⚓ Fl Y 5s 20m 5M; 54°00'·13N 03°33'·38W; Aero FR (80m), *Horn 30s.*
Lightning Knoll ⚓ L Fl 10s; 53°59'·84N 03°14'·28W.
Morecambe ⚓ Q (9) 15s; 53°52'·00N 03°22'·00W.
Lune Deep ⚓ Q (6) + L Fl 15s; ***Racon (T);*** 53°56'·07N 03°12·90W.
Lts in line about 090°. Front, FR 10m 2M; G mast; 54°04'·41N 02°52'·63W. Rear, 140m from front, FR 14m 2M; G mast.

HEYSHAM
SW Quay Ldg Lts 102·2°. Front ⚓, both F Bu 11/14m 2M; Or & B ♦ on masts; 54°01'·91N 02°55'·22W.

RIVER LUNE, GLASSON DOCK and FLEETWOOD
R Lune ⚓ Q (9) 15s; 53°58'·63N 03°00'·03W.
Fairway No. 1(Fleetwood) ⚓ 53°57'·67N 03°02'·03W Q; *Bell.*
Fleetwood Esplanade Ldg Lts 156°. Front, Iso G 2s 14m 9M; 53°55'·71N 03°00'·56W. Rear, 320m from front, Iso G 4s 28m 9M. Both vis on Ldg line only. (H24) (chan liable to change).

RIVER RIBBLE
Gut ⚓ L Fl 10s; 53°41'·74N 03°08'·98W.
Perches show Fl R on N side, and Fl G on S side of chan.
S side, 14¼M Perch ⚓ Fl G 5s 6m 3M; 53°42'·75N 03°04'·90W.
Jordan's Spit ⚓ Q (9) 15s; 53°35'·76N 03°19'·28W.
FT ⚓ Q; 53°34'·56N 03°13'·20W.

RIVER MERSEY and LIVERPOOL
Bar ⚓ Fl.5s 12M; ***Racon (T) 10M;*** 53°32'·01N 03°20'·98W.
Q1 ⚓ VQ; 53°31'·00N 03°16'·72W.
Q2 ⚓ VQ R; 53°31'·47N 03°14'·95W.
Q3 ⚓ Fl G 3s; 53°30'·95N 03°15'·10W.
Formby ⚓ Iso 4s; 53°31'·13N 03°13'·48W.
C4 ⚓ Fl R 3s; 53°31'·82N 03°08'·51W.
Crosby ⚓ Oc 5s; 53°30'·72N 03°06'·29W.
C14 ⚓ Fl R 3s; R hull; 53°29'·91N 03°05'·34W.
Brazil ⚓ QG; 53°26'·85N 03°02'·23W.

RIVER DEE, MOSTYN and CONNAH'S QUAY
HE1 ⚓ Q (9) 15s; 53°26'·33N 03°18'·08W.
HE2 ⚓ Fl G 2·5s; 53°24'·90N 03°12'·88W.
HE3 ⚓ Q G; 53°24'·55N 03°12'·74W.
West Hoyle Spit (Earwig) Dir Lt 090. ⚓ 53°21'·21N 03°24'·07W; Iso WRG 2s W10M, R7M, G7M (day W3M, R2M, G2M). vis: 087·5°-G-089·5°-W-090·5°-R-092·5°; occasional.
Mostyn Dir Lt ⚓ 177·5°. 53°19'·23N 03°16'·38W Iso WRG 2s 11m W 10M R,G 7M, by day W 3M R,G 2M; vis: 175°-G-177°-W-178°-R-180°; H24.

WALES – NORTH COAST AND INNER PASSAGE
South Hoyle Outer ⚓ Fl R 2·5s; 53°21'·47N 03°24'·70W.
Prestatyn ⚓ QG; 53°21'·51N 03°28'·51W.
Inner Passage ⚓ Fl R 5s; 53°21'·91N 03°31'·95W.
Mid Patch Spit ⚓ QR; 53°22'·25N 03°32'·67W.
Mast ⚓ Mo (U) 15s 12m 10M & 2FR (vert); Horn Mo (U) 30s;

Mast (80); 53°28'·84N 03°30'·50W.
W Constable ⚓ Q (9) 15s; *Racon (M) 10M*; 53°23'·14N 03°49'·26W.
N Constable ⚓ VQ; 53°23'·76N 03°41'·42W.

RHYL, LLANDUDNO and CONWY
River Clwyd Outfall ⚓ Q R 7m 2M; 53°19'·45N 03°30'·34W.
Rhyll SWM ⚓ Mo(A) 10s; 53°19'·60N 03°32'·00W.
Llandudno Pier Hd ⚡ 2 FG (vert) 8m 4M; 53°19'·90N 03°49'·51W.
Great Ormes Hd Lt Ho, (unlit); 53°20'·56N 03°52'·17W.
Conwy Fairway ⚓ L Fl 10s; 53°17'·95N 03°55'·58W.
C1 ⚓ Fl G 10s; 53°17'·83N 03°54'·58W.
C2 ⚓ Fl (2) R 10s; 53°17'·94N 03°54'·52W.
River Conwy ent, ⚡ L Fl G 15s 5m 2M; 53°18'·03N 03°50'·86W.

ANGLESEY
Point Lynas ☆ 53°24'·98N 04°17'·35W Oc 10s 39m **18M**;
W castellated twr; vis: 109°-315°; H24.

AMLWCH to HOLYHEAD BAY
Main Bkwtr ⚡ Fl G 15s 11m 3M; W mast; vis: 141°-271°;
53°25'·02N 04°19'·91W.
Furlong ⚓ Fl G 2·5s; 53°25'·41N 04°30'·47W.
Archdeacon Rock ⚓ Q; 53°26'·71N 04°30'·87W.
Victoria Bank ⚓ VQ; 53°25'·61N 04°31'·37W.
Coal Rk ⚓ Q (6) + L Fl 15s; 53°25'·91N 04°32'·79W.
Ethel Rk ⚓ VQ; 53°26'·64N 04°33'·67W.
The Skerries ☆ 53°25'·27N 04°36'·50W Fl (2) 15s 36m **20M**; W ○
twr, R band; *Racon (T) 25M*. Iso R 4s 26m 10M; same twr; vis:
233°-252°, sector unreliable > 4M; *Horn (2) 60s*. H24 in periods
of reduced visibility; 992351084.
Langdon ⚓ Q (9) 15s; 53°22'·74N 04°38'·74W.
Bolivar ⚓ Fl G 2·5s; 53°21'·51N 04°35'·33W.
Wk ⚓ Fl (2) R 10s; 53°20'·43N 04°36'·60W.

HOLYHEAD to SOUTH STACK
Bkwtr Head ⚡ Fl (3) G 10s 21m 14M; W □ twr, B band; Fl Y vis:
174°-226°; *Siren 20s;* 53°19'·86N 04°37'·16W.

Spit ⚓ Fl G 3s; 53°19'·79N 04°37'·15W.
South Stack ☆ 53°18'·41N 04°41'·98W Fl 10s 60m **24M**; (H24);
W ○ twr; obsc to N by N Stack and part obsc in Penrhos
bay; *Horn 30s*. Fog Det lt vis: 145°-325°.

MENAI STRAIT TO BARDSEY ISLAND
Ten Feet Bank ⚓ QR; 53°19'·47N 04°02'·82W.
Dinmor ⚓ QG; 53°19'·34N 04°03'·32W.
Trwyn-Du ⚡ Fl 5s 19m 12M; W ○ castellated twr, B bands;
vis: 101°-023°; *Bell (1) 30s*, sounded continuously; 53°18'·77N
04°02'·44W. FR on radio mast 2M SW.

APPROACHES to BEAUMARIS and BANGOR
(Direction of buoyage ⌂ NE to SW)
Perch Rock ⚓ Fl R 5s; 53°18'·73N 04°02'·09W.

PORT DINORWIC and CAERNARFON
Port Dinorwic Pier Hd ⚡ 53°11'·21N 04°12'·60W F WR 5m 2M;
vis: 225°-R- 357°-W-225°.

(Direction of buoyage ⌂ SW to NE)
Caernarfon N Pier Hd ⚡ 2 FG (vert) 5m 2M; 53°08'·72N
04°16'·56W.
Abermenai Point ⚡ Fl WR 3·5s 6m 3M; W mast; vis: 065°-R-245°-
W-065°; 53°07'·62N 04°19'·72W.
C1 ⚓ Fl G 5s; 53°06'·85N 04°24'·35W.
C2 ⚓ Fl R 10s; 53°06'·95N 04°24'·46W.
Caernarfon Bar ⚓ L Fl 10s; 53°06'·45N 04°25'·00W.
Llanddwyn I ⚡Fl WR 2·5s 12m W7M, R4M; W twr; vis: 280°-R-
015°-W-120°; 53°08'·05N 04°24'·79W.

PORTH DINLLÄEN
CG Stn ⚡ FR when firing 10M N; 52°56'·82N 04°33'·89W.
Careg y Chwislen ⚓ Fl(2) 10s 2M 52°56'·99N 04°33'·51W.

Bardsey I ☆ 52°44'·97N 04°48'·02W Fl (5) 15s 39m **26M**; W□twr,
R bands; obsc by Bardsey Is 198°-250° and in Tremadoc B when
brg < 260°.

9.10.5 PASSAGE INFORMATION

More passage information is threaded between the harbours
in this area. Admiralty Leisure Folio 5613, Irish Sea eastern
part including Isle of Man, covers the Mull of Galloway to
Great Ormes Head including the Isle of Man. Folio 5609, North
West Wales including Menai Strait, covers Colwyn Bay to
Aberystwyth, including Anglesey. **Bibliography:** *Irish Sea Pilot*
(Imray/Rainsbury). *Cruising Anglesey and Adjoining Waters*
(Imray/Morris).The *West Coasts of England and Wales Pilot*
(Admiralty NP37) covers the whole area.

SOLWAY FIRTH

(AC 1346) Between Abbey Hd and St Bees Hd lies the Solway
Firth, most of which is encumbered by shifting sandbanks. *Solway
SDs* are essential. Off the entrances to the Firth, and in the
approaches to Workington, beware shoals over which strong W
winds raise a heavy sea. The area N of Silloth is unsurveyed but
buoyed channels are navigable as far as Dumfries and Annan on
the N shore. Buoys are laid primarily for the aid of pilots.

▶ *Local knowledge is required, particularly in the upper Firth,
where streams run very strongly in the channels when the
banks are dry, and less strongly over the banks when covered.
In Powfoot Channel the flood stream begins at HW L'pool –
0300, and the ebb at HW L'pool + 0100,* **Sp rates up to 6kn.** ◀

The River Nith, approached between Borron Point and
Blackshaw Spit, leads to Carsethorn, Glencaple, Kingholm Quay
and Dumfries. There are 2 ⚓s at Caresthorn; Kingholm Quay
offers drying berths, ⚋ and on pontoons. Local knowledge is
essential, but pilotage for visiting yachts is available. Contact

cargo@nith-navigation.co.uk, ☎07801 321457 or 07860 522598,
or visit: www.nith-navigation.co.uk.

On the south side of the Solway Firth are Maryport, Workington,
Harrington and Whitehaven. South of St Bees Head lies the
Nuclear processing plant at Sellafield. The large shallow drying
harbour at Ravenglass lies south again with Eskmeals Firing
Range. Selker rocks and Black Leg Rk are notable shoals
between Ravenglass and the Duddon Estuary. This dries at LW
and is only navigable by small shoal draught craft. The Isle of
Walney has an airfield and shelters Barrow -in-Furness which
is reached by Walney Chan on rounding South East Point.

Morecambe Bay is shallow and mostly dries. Heysham harbour
is exclusively for commercial traffic. The R Lune gives access to
Port Glasson and the R Wyre to Fleetwood. The Channels which
traverse the extensive offshore sandbanks are buoyed and shift
frequently. Shell Flat extends 11M off the beach at Blackpool.

WIND FARMS

Robin Rigg wind farm is approximately 7M WNW of Maryport.
Be aware of Walney, Ormonde Barrow and West of Duddon
Sands wind farms between 6M and 12M off Walney Island.
Burbo Bank wind farm is directly S of the approaches to
Liverpool with an extension planned to the W. North Hoyle,
Gwynt y Mor and Rhyl Flats wind farms WSW lie between 6M
and 12M off the coast of N Wales.

An extensive area has been earmarked for development
between Anglesey and the Isle of Man. Known collectively as
the Celtic Array this will consist of 3 large wind farms. the first
of these, Rhiannon, is undergoing public consultation.

9.10.6 MARYPORT

Cumbria 54°43′·03N 03°30′·38W ✿✿♨♨♨✿✿✿

CHARTS AC 1826, 1346, 2013, 5613; Imray C62

TIDES +0038 Dover; ML no data; Duration 0550
Standard Port LIVERPOOL (→)

Times				Height (metres)			
High Water		Low Water		MHWS	MHWN	MLWN	MLWS
0000	0600	0200	0800	9·4	7·5	3·2	1·1
1200	1800	1400	2000				
Differences MARYPORT							
+0021	+0036	+0017	+0002	−0.8	−0.9	−0.7	−0.2
SILLOTH							
+0035	+0045	+0040	+0050	−0.2	−0.4	−0.9	−0.3

SHELTER Good in marina, sill 3·1m above CD, 2.5m above sill for entry; at other times Workington is a refuge. Elizabeth Dock gates u/s. Elizabeth Basin dries 2m; commercial, not used by yachts.

NAVIGATION WPT 54°43′·09N 03°32′·47W, 090°/1M to S pier. Overfalls at ent in W/SW winds over ebb. At HW−3 1·8m over bar at ent and in river chan; mud banks cover HW −2. Max speed 4kn.

LIGHTS AND MARKS SHM bn Fl G 5s marks outfall 6ca SW of S pier. SSS displays IPTS signals 2 and 5 (see IPTS 5.20)

COMMUNICATIONS (Code 01900) CGOC (01407) 762051; Police 101; Dr 815544; ⊞ 812634. HM 817440; Hbr Authority 604351.
Port VHF Ch **12** 16. Marina (H24) Ch **12** 16.

FACILITIES Maryport Marina www.maryportmarina.com ☎814431 190⌂ inc ✔ £20.80/craft inc ⌁ ⚓ (±3HW) 🚽 🛢 🛠 ⛽ 🚿 🛒(25t) 🅿 **Maryport Yachting Association** www.maryportyachting.co.uk. **Town** 🏬 🏧 ✉ Ⓑ 🍴 ✕ 🗐 ⇌ ✈ (Manchester/Newcastle).

MARYPORT HARBOUR

9.10.7 WORKINGTON

Cumbria 54°39′·03N 03°34′·38W ✿♨✿

CHARTS AC 1826, 1346, 2013, 5613; Imray C62

TIDES +0025 Dover; ML 4·5; Duration 0545
Standard Port LIVERPOOL (→)

Times				Height (metres)			
High Water		Low Water		MHWS	MHWN	MLWN	MLWS
0000	0600	0200	0800	9·4	7·5	3·2	1·1
1200	1800	1400	2000				
Differences WORKINGTON							
+0029	+0027	+0014	+0004	−1.1	−1.1	−0.5	−0.1

SHELTER Good. Ent and chan to Prince of Wales Dock are maintained at 1·2m. Berth/moor by prior arrangement with Vanguard SC (☎07707 785328) in drying Tidal Dock; no ⚓ in Turning Basin. Prince of Wales Dock for commercial vessles only. Low (1·8m) fixed railway bridge across ent to inner Tidal Hbr.

NAVIGATION WPT 54°39′·59N 03°35′·38W, 131°/1M to front ldg lt.

> • Tide sets strongly across ent. In periods of heavy rain a strong freshet from R Derwent may be encountered in the hbr ent.

LIGHTS AND MARKS Workington Bank, least depth 5·5m, is 2M W of hbr ent; and is marked by NCM and SCM lt buoys (see 9.10.4). Ldg lts 132°, both FR 10/12m 3M, on W pyramidal trs with R bands. Two sets of F Bu lts in line mark NE and SW edges of chan. There are 16 wind-turbines on shore between ¾M and 2M NE of hbr ent.

COMMUNICATIONS (Code 01900) CGOC (01407) 762051; Police 101; Dr 64866; ⊞ 602244. HM 602301.
VHF Ch 14 16 (HW−2½ to HW+2 approx).

FACILITIES Dock ⚓ 🛢 🛠 ⌁.
Vanguard SC www.vanguard-workington.co.uk ☎01228 674238 ⚓ ⚓.
Town 🏬 ✉ Ⓑ 🍴 ✕ 🗐 ⇌ ✈ (Carlisle).

WORKINGTON

MINOR HARBOUR 10M NNE OF MARYPORT

SILLOTH, Cumbria, 54°52′·16N 03°23′·86W. AC 1826, 1346, 2013, 5613. HW −0050 on Dover; ML no data; Duration 0520. See 9.10.6. Commercial harbour which welcomes occasional visiting yachts. Appr via English or Middle Chans; requires local knowledge, Beware constantly shifting chans and banks, and unsurveyed areas. East Cote Dir lt 052° FG 15m 12M; vis 046°-058°, intens 052°. Ldg lts, both F, 115°. Groyne 2 FG (vert). Outer hbr dries 2·8m; lock into New Dock (4m) HW−1½ to HW+½. Ent marked by Oc R/G lts. Subtract 1m from height of tide to calculate draught over sill. £24 covers entry/exit and 1 week stay. Traffic sigs on mast at New Dock; must receive permission before proceeding. VHF Ch 16 12. HM ☎(01697) 331358. ⚓ WSW of ent in about 4m off Lees Scar, Fl G 5s 11m 4M; exposed to SW winds. Facilities: ⚓ ✉ Ⓑ ✕ 🗐 🍴.

MINOR HARBOUR 2M S OF WORKINGTON

HARRINGTON, Cumbria, 54°36′·77N 03°34′·29W. AC 1826, 1346, 2013, 5613. HW +0025 on Dover; Duration 0540; Use Diff's Workington 9.10.7. Good shelter in small hbr only used by local FVs and yachts; dries 3ca offshore. Ent, marked by PHM perches on N side, difficult in strong W winds. Berth on N wall of inner hbr (free). Call ☎(01946) 823741 Ext 148 for moorings. Limited facilities. SC ☎(01946) 830600.

9.10.8   WHITEHAVEN

Cumbria **54°33'·18N 03°35'·82W** ✿✿✿⚓⚓✿✿

CHARTS AC 1826, 1346, 2013, 5613; Imray C62

TIDES +0015 Dover; ML 4·5; Duration 0550

Standard Port LIVERPOOL (→)

Times				Height (metres)			
High Water		Low Water		MHWS	MHWN	MLWN	MLWS
0000	0600	0200	0800	9·4	7·5	3·2	1·1
1200	1800	1400	2000				
Differences WHITEHAVEN							
+0010	+0020	+0005	0000	−1·4	−1·2	−0·8	−0·1
TARN POINT (54°17'N 03°25'W)							
+0010	+0010	+0005	−0005	−1·1	−1·1	−0·7	−0·2
DUDDON BAR (54°09'N 03°20'W)							
+0007	+0007	+0005	−0001	−0·9	−0·9	−0·6	−0·2

SHELTER Very good, entry safe in most weathers. One of the more accessible ports of refuge in NW England, with a marina in the Inner Hbr. Appr chan across the outer hbr is dredged to 1·0m above CD. Sea lock (30m x 13.7m), with sill at CD, maintains 7m within inner hbr. N Hbr is for commercial and FV use.

NAVIGATION WPT 54°33'·34N 03°36'·21W, 133°/4½ca to W pier hd. There are no hazards in the offing. Hold closer to the North Pier, and beware bar at end of W Pier.

LIGHTS AND MARKS Several tall chimneys are charted within 1·5M S of hbr. St Bees Head Lt ho is 2·7M SSW of hbr ent. SHM bn, Fl G 2·5s, 4½ca S of W pierhead marks sewer outfall. IPTS sigs 1-3 shown from N side of lock ent to seaward only at present: priority to inward bound vessels.

COMMUNICATIONS (Code 01946) CGOC (01407) 762051; Police 101; HM 692435; Sealock 694672.

HM VHF Ch 12 16.

FACILITIES Marina ⊕ www.whitehavenmarina.co.uk ☎692435 285 inc ♥ in Inner Harbour (Lowther Marina) and Queen's Dock (Queen's Marina) £2.20, ▅ ⬚ ▮♠✗ ⚒ ⊞ Ⓔ ℞ ▤(45t) ⚓ SC.
Town Market days Thurs, Sat; ⌂ ⊡ ◻ ⊠ Ⓑ �ℳ ✗ ⌂ ⇌.

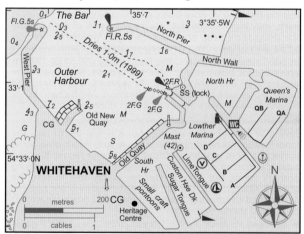

MINOR HARBOUR BETWEEN ST BEES HD AND MORECAMBE BAY

RAVENGLASS, Cumbria, **54°20'·00N 03°26'·80W** (drying line). AC 1346, 1826, 5613. HW +0020 on Dover; ML no data; Duration 0545. See 9.10.8 (Tarn Pt). Large drying hbr, into which R's Mite, Irt and Esk flow; approx 2·5m in ent at HW–2. Sellafield power stn with WCM lt buoy and outfall buoys are 5M NNW. FG ☆ (occas) on blockhouse at S side of ent. *Solway Sailing Directions* with pilotage notes by Ravenglass Boating Ass'n, local knowledge advisable. From N beware Drigg Rk and from S Selker Rks, SHM lt buoy, 5M SSW of ent. Firing range D406 close S at Eskmeals; Mon-Thur 0800-1600LT (1500 Fri). When in use R flags flown, R lts at night; *Eskmeals Gun Range* VHF Ch 16 11, ☎(01229) 712245/5.
Village: ▅ ⚓ ⊠ ◻ ℳ.

9.10.9   ISLE OF MAN

Charts AC Irish Sea 1826, 1411; 2094 (small scale); 2696 (ports), 5613; Imray C62, Y70

The Isle of Man is one of the British Islands, set in the Irish Sea roughly equidistant from England, Scotland, Wales and Ireland but it is not part of the UK. It has a large degree of self-government. Lights are maintained by the Northern Lighthouse Board. Manx harbours are administered by the IoM Government.

Distances See 9.10.2 for distances between ports in Area 10 and 9.0.6 for distances across the Irish Sea, North Channel and St George's Channel.

Harbours and anchorages Most of the harbours are on the E and S sides, but a visit to the W coast with its characteristic cliffs is worth while. The four main harbours are described in an anti-clockwise direction from Peel on the W coast to Ramsey in the NE. There are good ⚓s at: Port Erin in the SW, Castletown and Derby Haven in the SE, and Laxey Bay in the E. All IOM harbours charge the same overnight berthing rate, £1.18/m (multi-hulls £1.76/m).

R/T If contact with local Harbour Offices cannot be established on VHF, vessels should call *Douglas Hbr Control* Ch **12** 16 for urgent messages or other info.

Coastguard Call Holyhead CGOC Ch 16 67 86; the Snaefell (IoM) aerial is linked to the CG network.

Customs The IOM is under the same customs umbrella as the rest of the UK, and there are no formalities on arriving from or returning to the UK.

Weather Forecasts can be obtained direct from the forecaster at Ronaldsway Met Office ☎0900 6243 3200, H24. **Douglas Hbr Control** (9.10.12) can supply visibility and wind info on request.

MINOR HARBOURS IN THE ISLE OF MAN

PORT ERIN, Isle of Man, **54°05'·31N 04°46'·34W**. AC 2094, 2696, 5613. HW –0020 on Dover; ML 2·9m; Duration 0555. See 9.10.15. From S and W, Milner's Twr on Bradda Hd is conspic. Ldg lts (FR 10/19m 5M, lead 099° into the bay. Beware the ruined bkwtr (dries 2·9m) extending N from the SW corner, marked by an unlit SHM buoy. A small hbr on the S side dries 0·8m. Raglan Pier (E arm of hbr), Oc G 5s 8m 5M. 2⚓s W of Raglan Pier. Good ⚓ in 3-8m N of Raglan Pier, but exposed to W'lies. Call Port St. Mary Harbour (VHF Ch 12) ☎833205.

CASTLETOWN BAY, Isle of Man, **54°03'·51N 04°38'·57W**. AC 2094, 2696, 5613. HW +0025 on Dover; ML 3·4m; Duration 0555. The bay gives good shelter except in SE to SW winds. From the E, keep inside the race off Dreswick Pt, or give it a wide berth. Beware Lheeah-rio Rks in W of bay, marked by PHM buoy, Fl R 3s, Bell. Hbr dries 3·1m to level sand. Berth in outer hbr or go via swing footbridge (manually opened) into inner hbr below fixed bridge. ⚓ between Lheeah-rio Rks and pier in 3m; or NW of Langness Pt. Lts: Langness lt, on Dreswick Pt, Fl (2) 30s 23m 12M. Hbr S (New) Pier, Oc R 15s 8m 5M; Inner S pier (Irish Quay), Oc R 4s 5m 5M, vis 142°-322°. 150m NW is swing bridge marked by 2 FR (hor). N pier, Oc G 4s 3m (W metal post on concrete column). VHF Ch 12 16 (when vessel due). HM ☎823549; Dr 823597. Facilities: **Outer hbr** ▅ ▅ ◻ ▤(mob); **Irish Quay** ⚓ ◻ ▤; **Inner hbr** ⚓ ◻ ▤; **Town** ⌂ ⊡ ▮✗.

LAXEY, Isle of Man, **54°13'·46N 04°23'·32W**. AC 2094, 5613. HW +0025 on Dover; +0010 and –2·0m on Liverpool; ML 4m; Duration 0550. The bay gives good shelter in SW to N winds. 2 yellow ⚓s (seasonal) are close E of hbr ent; 2 more are 1M S in Garwick Bay. ⚓ about 2ca S of pier hds or in Garwick Bay. The hbr dries 3·0m to rk and is only suitable for small yachts. Beware rks on N side of the narrow ent. Keep close to pier after entering to avoid training wall on NE side. ◻ on inside of pier; inner basin is full of local boats. Pier hd lt Oc R 3s 7m 5M, obsc when brg <318°. Bkwtr hd lt Oc G 3s 7m 5M. Harbour Office ☎861663. Facilities: ⚓ ⊠ Ⓑ ✗ ◻.

ISLE OF MAN

(Charts 2094, 2696, 5613) For pilotage information and harbours of IoM, see *IOM Sailing Directions, Tidal Streams and Anchorages*, published by Hunter Publications. For notes on crossing the Irish Sea, see 9.13.5. There are four choices rounding S of IoM:

- In bad weather, or at night, keep S of Chicken Rk.
- In good conditions, pass between Chicken Rk and Calf of Man.
- By day, with winds <F 3 and a reliable engine giving >5kn, Calf Snd between Calf of Man and IoM is navigable. Pass between Kitterland Is and Thousla Rk, marked by bn Fl R 3s 9m 4M.
- Little Sound, a minor unmarked channel, runs E of Kitterland Is.

▶ *The stream runs strongly through Calf Sound, N-going at HW Liverpool –0145, and S-going at HW Liverpool +0345, Sp rate 3·5kn. W of Calf of Man the stream runs N and S, but changes direction off Chicken Rk running E/W between Calf of Man and Langness Pt. Overfalls extend E from Chicken Rk on E-going stream, beginning at HW L'pool + 0610, and N from the rock on W-going stream, which begins at HW L'pool. The Skerranes (drying) extend 1ca SW Of Langness Pt (lt). Tidal stream run strongly, with eddies and a race. A dangerous sea builds to the E of Langness peninsula in strong winds. Here the NE-going stream begins at HW L'pool +0545, and the SW-going at HW L'pool –0415, sp rates 2·25kn.* ◀

⚓ (exposed to the E) in Derby Haven, N of St Michael's Is. From here to Douglas and on to Maughold Hd (lt), there are no dangers more than 4ca offshore.

▶ *Inshore the SW-going stream runs for 9 hrs and the NE-going for 3 hrs, since an eddy forms during the second half of the main NE-going stream. Off Maughold Hd the NE-going stream begins at HW L'pool +0500, and the SW-going at HW L'pool –0415.* ◀

SE, E and NW of Pt of Ayre are dangerous banks, on which seas break in bad weather. These are Whitestone Bank (least depth 2·0m), Bahama Bank (1·5m), Ballacash Bank (2·7 m), King William Banks (3·3m), and Strunakill Bank (6·7m).

▶ *At Point of Ayre E-going stream begins HW Liverpool –0600 (HW Dover –0545). W-going stream at Point of Ayre begins HW L'pool*

(HW Dover +0015). A counter tide runs inside the banks to the East. In Ramsey Bay the S-going tide runs for 3 hours from L'pool +0515 (Dover +0530). In Ramsey Bay the N-going tide runs for 9 hours from L'pool –0345 (Dover –0330). ◀

The W coast of IOM has few pilotage hazards. A spit with depth of 1·4m runs 2ca offshore from Rue Pt. Jurby Rk (depth 2·7m) lies 3ca off Jurby Hd. Craig Rk (depth 4m) and shoals lie 2·5M NNE of Peel.

BARROW TO CONWY

(AC 2010, 1981, 1978, 5613) Ent to Barrow-in-Furness (AC 3164) is about 1M S of Hilpsford Pt at S end of Walney Island where the lt ho is prominent. W winds cause rough sea in the entrance. Moorings and ⚓ off Piel and Roa Islands, but space is limited.

▶ *The stream sets across the narrow channel, which is well marked, but shallow in patches. Stream runs hard on ebb. Coming from the S it is possible with sufficient rise of tide to cross the sands between Fleetwood and Barrow.* ◀

Barrow Wind Farm (30 turbines) marked by FR and Fl Y lights and fog signals on some structures is centred approx 7·6M SSW of Barrow-in-Furness. Further W lies the W Duddon Sands Wind Farm (108 turbines). Its W'rn extremity is marked by a tall Met Mast. To the North Walney and Ormonde (30 turbines) wind farms straddle the northern approaches to Barrow and the Lune.

Lune Deep, 2M NW of Rossall Pt, is the entrance to Morecambe Bay (chart 2010), and gives access to the ferry/commercial port of Heysham, Glasson Dock, and Fleetwood; it is well buoyed.

▶ *Streams run 3·5kn at sp.* ◀

Most of the Bay is encumbered with drying sands, intersected by chans which are subject to change. S of Morecambe B, beware shoals and drying outfall (2·0m) extending 3M W of Rossall Pt. Further S, R. Ribble gives access via a long drying chan to the marina at Preston.

Queen's Channnel and Crosby Chan (AC 1951 & 1978) are entered E of the Bar light float. They are well buoyed, dredged and preserved by training banks, and give main access to R Mersey and Liverpool. Keep clear of commercial shipping. From the N the old Formby chan is abandoned, but possible near HW. Towards HW in moderate winds a yacht can cross the training bank (level varies between 2m and 3m above CD) E of Great Burbo Bank, if coming from the W; Rock Chan, which is unmarked and dries in parts, may also be used but beware wrecks. W of Bar Lt Vessel the Douglas oil field, an IMO Area to be Avoided centred on 53°32'·23N 03°34'·70W, effectively separates the traffic flow.

In good weather at neaps, the Dee Estuary (AC 1953, 1978) is accessible for boats able to take the ground. But most of the estuary dries and banks extend 6M seaward. Chans shift, and buoys are moved as required. Main ent is Welsh Chan, but if coming from N, Hilbre Swash runs W of Hilbre Is.

▶ *Stream runs hard in chans when banks are dry.* ◀

Sailing W from the Dee on the ebb, it is feasible to take the Inner Passage (buoyed) S of West Hoyle Spit, which gives some protection from onshore winds at half tide or below. Rhyl is a tidal hbr marked offshore by a SWM, not accessible in strong onshore winds, but gives shelter for yachts able to take the ground. Abergele Rd, Colwyn B and Llandudno B are possible ⚓s in settled weather and S winds. Conwy and Deganwy harbours offer good shelter in marinas.

▶ *Between Point of Ayr and Great Ormes Head the E-going stream begins at HW Liverpool +0600, and the W-going at HW Liverpool –0015, sp rates 3kn.* ◀

North Hoyle Wind Farm consists of 30 turbines and is centred on 53°25'·00N 03°27'·00W. Each turbine is 58m high, with 40m diameter blades and clearance of 18m. Many of them are lit. Gwynt-y-Mor Wind Farm centred on 53°28'·20N 03°37'·70W. The area is delineated by cardinal and yellow buoys. Underwater obstructions will exist throughout. Vessels are advised to keep well clear and not enter the area.

9.10.10 PEEL

Isle of Man 54°13'·61N 04°41'·68W ❁❁⚓⚓♻♻♻

CHARTS AC 2094, 2696, 5613; Imray C62; Y70

TIDES +0005 Dover; ML 2·9; Duration 0545

Standard Port LIVERPOOL (→)

Times				Height (metres)			
High Water		Low Water		MHWS	MHWN	MLWN	MLWS
0000	0600	0200	0700	9·4	7·5	3·2	1·1
1200	1800	1400	1900				
Differences PEEL							
+0010	+0010	-0020	-0030	-4·2	-3·2	-1·7	-0·7

SHELTER Good, except in strong NW to NE winds when ent should not be attempted. 4 Y ⚓s (seasonal) off S groyne in about 2m. Fin keelers may be able to berth on N bkwtr in 5m. The marina is in the Inner Harbour with berthing on pontoons and alongside harbour wall for boats up to 15m LOA. Pass dimensions on Ch 12. Approach dries 1·4m; access approx HW±2 when flapgate lowered and through swing bridge. Maintained depth 2·5m.

NAVIGATION WPT 54°14'·00N 04°41'·78W, 207°/5ca to groyne lt. Beware groyne on S side of harbour ent, submerged at half tide.

LIGHTS AND MARKS Power stn chy (83m, grey with B top) at S end of inner hbr is conspic from W and N; chy brg 203° leads to hbr ent. Groyne lt and Clock Tr (conspic) in transit 200° are almost on same line. Peel Castle and 2 twrs are conspic on St Patrick's Isle to NW of hbr. 207° ldg lts, Fl R 5s. The E and W abutments to the Inner Hbr ent are marked by lts Fl R 3s and Fl G 3s respectively. Traffic signals at marina entrance. Swing bridge has FR lts at centre span. Fl Y lts = flap gate being raised lowered; bridge opening/closing.

COMMUNICATIONS (Code 01624) CGOC (01407) 762051; ♻ 0900 6243 322; Police 697327; Dr 843636. Harbour Office 842338, Mobile 07624 495036.
Peel Hbr VHF Ch **12** 16 (HW±2), HO for Swing Bridge – other times call *Douglas Hbr Control* Ch 12 for remote bridge opening.

FACILITIES **Outer Hbr,** ⌇ see 9.10.9; **Marina** 120⌇ inc ♥, £1.72; access via flapgate; ⚓ ⛴ ⚓ ⚓ ✕ ⚓ 🅿 🛒 ⛽(mob); **Peel Sailing & Cruising Club** ☎842390 📶 (key from Harbour Office), ✕ 🅿 🖬. **Town** ⚓ 🏪 🏧 ⚓ 🅿 ACA ✉ ⑧ 🛒 ✕ 🖂 ⇄ (bus to Douglas & ferry), ✈ Ronaldsway.

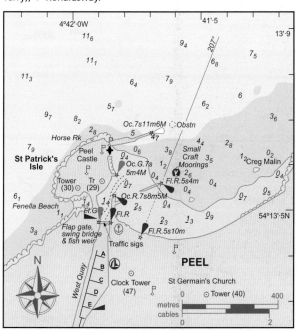

PEEL

9.10.11 PORT ST MARY

Isle of Man **54°04'·43N 04°43'·73W** ❁❁⚓⚓♻♻♻

CHARTS AC 2094, 2696, 5613; Imray C62; Y70

TIDES +0020 Dover; ML 3·2; Duration 0605

Standard Port LIVERPOOL (→)

Times				Height (metres)			
High Water		Low Water		MHWS	MHWN	MLWN	MLWS
0000	0600	0200	0700	9·4	7·5	3·2	1·1
1200	1800	1400	1900				
Differences PORT ST MARY							
+0010	+0020	-0015	-0035	-3·5	-2·7	-1·6	-0·6
CALF SOUND							
+0010	+0010	-0020	-0030	-3·3	-2·7	-1·2	-0·5
PORT ERIN							
+0018	+0010	-0013	-0028	-4·1	-3·3	-1·6	-0·6

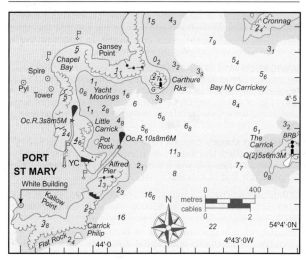

SHELTER Very good except in E or SE winds. ⚓ S of Gansey Pt, but poor holding. 5 yellow ⚓s (seasonal) between Alfred Pier and Little Carrick.

NAVIGATION WPT 54°04'·21N 04°43'·37W, 295°/0·3M to Alfred Pier lt. Rocky outcrops to SE of pier to 2ca offshore. Beware lobster/crab pots, especially between Calf of Man and Langness Pt.

LIGHTS AND MARKS Alfred Pier, Oc R 10s, in transit 295° with Inner Pier, Oc R 3s, W twr + R band, leads clear S of The Carrick Rk, in centre of bay, which is marked by IDM bn, Q (2) 5s 6m 3M. A conspic TV mast (133m), 5ca WNW of hbr, in transit with Alfred Pier lt leads 290° towards the hbr and also clears The Carrick rock.

COMMUNICATIONS (Code 01624) CGOC (01407) 762051; ♻ 0900 6243 322, Police 697327; Dr 832226. HM 833205.
Port St Mary Hbr VHF Ch 12 16 (when vessel due or through Douglas Hbr Control Ch 12).

FACILITIES **Alfred Pier** ⌇ see 9.10.9; ⚓ ⚓ ⛴ ⚓ ⛽(mob); **Inner Hbr** dries 2·4m on sand; ⌇ ⚓ ⚓ ⛴ ✕ 🅿 △; **Isle of Man YC** ☎832088, ⛴ 📶 🖬. **Town** 🏧 ✉ ⑧ 🛒 ✕ 🖂 ⇄ (bus to Douglas, qv for ferries), ✈ Ronaldsway.

MINOR HARBOUR EAST OF CASTLETOWN

DERBY HAVEN, Isle of Man, 54°04'·65N 04°36'·45W. Tides & charts as above. Rather remote bay exposed only to NE/E winds. ⚓ in centre of bay, NW of St Michael's Island, in 3-5m. A detached bkwtr on NW side of the bay gives shelter to craft able to dry out behind it. Lts: Iso G 2s on S end of bkwtr. Aero FR (occas) at Ronaldsway airport, NW of bay. Facilities: at Castletown (1½M) or Port St Mary.

9.10.12 DOUGLAS

Isle of Man 54°08'·87N 04°27'·96W ✿✿✿❀⚓⚓✿✿✿

CHARTS AC 2094, 2696, 5613; Imray C62; Y70

TIDES +0009 Dover; ML 3·8; Duration 0600

Standard Port LIVERPOOL (→)

Times				Height (metres)			
High Water		Low Water		MHWS	MHWN	MLWN	MLWS
0000	0600	0200	0700	9·4	7·5	3·2	1·1
1200	1800	1400	1900				
Differences DOUGLAS							
+0010	+0020	−0020	−0030	−2·5	−2·1	−0·8	−0·3

SHELTER Good except in NE winds. Very heavy seas run in during NE gales. Outer hbr: Victoria and King Edward VIII piers are for commercial vessels/ferries. At inner end of Battery Pier in summer about 18 boats can raft up on pontoon; untenable in NE/E winds. Complete shelter in inner hbr, with possible pontoon berths, flapgate lowers on the flood and rises on the ebb at 4·4m above CD, lifting bridge opens every ½H subject to tide/road traffic conditions (request opening on VHF Ch12).

NAVIGATION WPT 54°09'·01N 04°27'·67W (abeam No 1 SHM buoy, Q (3) G 5s), 229°/0·47M to front ldg lt. Appr from NE of No 1 buoy (to avoid overfalls E of Princess Alexandra Pier) and await port entry sig, or call on VHF Ch 12. There is no bar. Keep clear of large vessels and ferries. Beware concrete step at end of dredged area (◊ mark on King Edward VIII Pier) and cill at ent to inner hbr.

LIGHTS AND MARKS Douglas Head Fl 10s 32m 24M. Ldg lts 229°, both Oc Bu 10s 9/12m 5M, synch; front W △; rear W ▽, both on R border. IPTS Nos 2, 3 and 5 shown from mast on Victoria Pier. Dolphin at N end of Alexandra Pier 2FR (vert).

COMMUNICATIONS (Code 01624) CGOC (01407) 762051; ☎ 0900 6243 322, Police 697327; Ⓗ 650000. Harbour Control 686628 (H24).

Douglas Hbr Control VHF Ch **12** 16 (H24); also broadcasts nav warnings for IoM ports and coastal waters on Ch 12 including weather and tidal info on request.

FACILITIES Outer Hbr ⬭ see 9.10.9, ⚓ ⬟ ⚓ at pontoon, 🐟 ⬚(mob); **Inner Hbr** (Pontoons, N and S Quays) 110⬭; Ⓥ according to space, ⬟ ⚓ ⬭ ⬚; **Douglas Bay YC** ☎673965, 🔧 ⬟ ⬚ 0930-2300 ⬭. **Services** ⬚ 🔧 🔨 ⬚ ⬚ ACA, Divers, Kos. **Town** www.gov.im 🛢 ✉ Ⓑ ⬚ ⬚ ✕ ⬚ ✈ Ronaldsway. **Ferries** Liverpool; 16/week; 2½ hrs; IoM. Heysham; 16/week; 3½ hrs; Steam Packet (www.steam-packet.com). Also to Dublin and Belfast.

9.10.13 RAMSEY

Isle of Man 54°19'·44N 04°22'·49W ✿✿✿⚓⚓✿✿

CHARTS AC 2094, 2696, 5613; Imray C62; Y70

TIDES +0020 Dover; ML 4·2; Duration 0545

Standard Port LIVERPOOL (→)

Times				Height (metres)			
High Water		Low Water		MHWS	MHWN	MLWN	MLWS
0000	0600	0200	0700	9·4	7·5	3·2	1·1
1200	1800	1400	1900				
Differences RAMSEY							
+0010	+0020	−0010	−0020	−2·0	−1·6	−0·9	−0·2

SHELTER Very good in inner harbour but entrance difficult in strong NE-E winds. Hbr dries 1·8m–6m. Access approx HW −2½ to HW +2. Berth on Town quay (S side) or as directed by HM on entry. 5 yellow ⚓s are close to Queen's Pier hd (seasonal). Note: Landing on Queen's Pier is prohibited.

NAVIGATION WPT 54°19'·44N 04°21'·89W, 270°/0·37M to ent. The foreshore dries out 1ca to seaward of the pier hds.

LIGHTS AND MARKS No ldg lts/marks. Relative to hbr ent, Pt of Ayre, Fl (4) 20s 32m 19M, is 5·5M N; Maughold Hd, Fl (3) 30s 65m 21M, is 3M SE; Albert Tr (☐ stone tr 14m, on hill 130m) is conspic 7ca S; and Snaefell (617m) bears 220°/5M. Inside the hbr an Iso G 4s, G SHM post, marks the S tip of Mooragh Bank; it is not visible from seaward. 2FR (hor) on each side mark the centre of swing bridge.

COMMUNICATIONS (Code 01624) CGOC (01407) 762051; ☎ 0900 6243 322, Police 812234; Dr 813881; Ⓗ 811811. Harbour Office 812245, mob 07624 460304.

Ramsey Hbr VHF Ch **12** 16 (0800-1600LT and when a vessel is due); Other times call *Douglas Hbr Control* Ch **12**.

FACILITIES Outer Hbr: E Quay ☎812245, only for commercial vessels (frequent movements H24); **Town Quay** (S side) ⬭ see 9.10.9, ⚓ ⬚. **Inner Hbr, W Quay** ⬭ ⬚ ⚓ ⬟ (Grid); **N Quay** ⬭ 🛢⚓; **Shipyard Quay** ⬟; **Old Hbr** ⚓ ⬟ ⬭; **Manx S&CC** ☎813494, ⬭; **Services** ⬚ 🔧 🔨 ⬚ Ⓔ. **Town** 🛢 Kos, ✉ Ⓑ ⬚ ⬚ ✕ ⬚ @ in Library, ⇌ (bus to Douglas for ferries), ✈ Ronaldsway.

9.10.14 GLASSON DOCK

Lancashire 53°59'·98N 02°50'·93W ✿ ✿ ⚓⚓⚓ ✿✿

CHARTS AC 1826, 1552, 2010, 5613; Imray C62

TIDES +0020 Dover; ML No data; Duration 0535

Standard Port LIVERPOOL (→)

Times				Height (metres)			
High Water		Low Water		MHWS	MHWN	MLWN	MLWS
0000	0600	0200	0700	9·4	7·5	3·2	1·1
1200	1800	1400	1900				
Differences BARROW-IN-FURNESS (Ramsden Dock)							
+0020	+0020	+0010	+0010	−0·1	−0·4	0·0	+0·1
ULVERSTON							
+0025	+0045	ND	ND	−0·1	−0·2	ND	ND
ARNSIDE							
+0105	+0140	ND	ND	+0·4	+0·1	ND	ND
MORECAMBE							
+0010	+0015	+0025	+0010	+0·1	−0·1	−0·3	0·0
HEYSHAM							
+0014	+0012	+0002	−0003	+0·2	−0·1	−0·1	+0·1
GLASSON DOCK							
+0025	+0035	+0215	+0235	−2·8	−3·1	ND	ND
LANCASTER							
+0115	+0035	DR	DR	−5·1	−5·0	DR	DR

NOTE: At Glasson Dock LW time differences give the end of a LW stand which lasts up to 2 hrs at sp.

SHELTER Very good in marina, sea lock (9m wide) opens into Glasson Dock HW−¾ to HW (Liverpool) (daylight hrs only Nov-Mar) Inner lock/swing bridge lead into BWML basin. There are no safe LW berths on the outer quay walls.

NAVIGATION WPT 53°58'·63N 03°00'·03W (Lune WCM), 4·2M to Plover Scar. Enter buoyed chan at HW−1½ (−1¼ Nps). It changes frequently. For latest info call Port Commission (☎01524 751724).

The flood sets strongly SE onto Plover Scar (5kts Sp). ⚓ is prohibited between Sunderland Pt and No 10 PHM. Beyond this a training wall marked by PHM lt buoys/bn. Yachts may have to cross this to seek out the deepest water. It extends to lock ent where there is a W'ly eddy and off which the river virtually dries – do not ⚓. R Lune is navigable to Lancaster.

• **It is dangerous to proceed downriver later than HW +1.**

LIGHTS AND MARKS Plover Scar, Fl 2s, on with Cockersand, FW, leads 084° up to Nos 2/3 buoys; then follow buoyed chan which shifts.

Tfc Sigs at Glasson DockheadW side
- ● = lock manned, but shut – No entry
- ◗ = lock open, clear to enter
- ◔ Fl = gate opening or closing

Tfc Sigs inside lock: ○ = gate open ● = gate shut.

COMMUNICATIONS (Code 01524) CGOC (01407) 762051; Police 101 Ⓗ 765944; HM 07910 315606.

VHF Ch 16 69 (HW−1½ to HW Liverpool).

FACILITIES Glasson Basin Marina www.bwml.co.uk ☎751491; 240 +20Ⓥ (£12·50/craft); ⚓ ▮ 🅰 ⚒ ▣(50t) ⚓.

Glasson SC ☎751089 ⚓ ▬ 🖵; **Lune CC** Access only near HW.

Town 🏧 ✉ 🛒 ✕ 🖵 🚌 (bus to Lancaster 4M), ✈ (Blackpool).

ADJACENT HARBOURS IN MORECAMBE BAY

BARROW-IN-FURNESS, Cumbria, 54°05'·64N 03°13'·44W. AC 1826, 2010, 3164, 5613. HW +0030 on Dover; ML 5·0m; Duration 0530 (→). Good shelter but open to SE winds. Drying moorings off Piel and Roa Islands or ⚓ clear of fairway. Marks/lts: Walney Island Lt ho (conspic stone tr), Fl 15s 21m 23M (obsc 122°-127° within 3M of shore). Directions: From Lightning Knoll SWM lt buoy, ldg lts, front Q 7m 14M; rear (6ca in front), Iso 2s 13m 14M (lattice structures) lead 041°/3·7M past Halfway Shoal bn, QR 16m 10M (RY chequers), to Bar lt buoy abeam Walney Is lt ho in dredged chan (call HM for latest depth). Inner Chan ldg lts, front Q 9m 14M, rear Iso 2s 14m 14M, lead 005° past Piel Is. 3M NW at Barrow, commercial docks reached via buoyed/lit Walney Chan, dredged to 2·5m, which must be kept clear. HM ☎(01229) 822911; VHF *Barrow Port Radio* Ch 12 16 (H24).

Piel Is, ruined castle, jetty and moorings on E side. Facilities: Inn.

Roa Is, 5ca N, has jetty at S end and moorings on E & W sides. ▬ (Roa Is Boat Club ☎(01229) 825291). Causeway joins to mainland. Facilities: Hotel at Rampside, 🛒.

HEYSHAM, Lancashire, 54°02'·01N 02°55'·96W. AC 1826, 2010, 1552, 5613. HW +0015 on Dover; ML 5·1m; Duration 0545. See 9.10.14. Good shelter, but yachts not normally accepted without special reason. Beware high speed ferries and oil rig supply ships. Ldg lts 102°, both F Bu 11/14m 2M, Y+B ◇ on masts. S jetty lt 2 FG (vert), Siren 30s. S pier hd ,Oc G 7·5s 9m 6M. N pier hd, 2FR (vert) 11m, obsc from seaward. Ent sigs: R flag or ● = no entry; no sig = no dep; 2 R flags or 2 ● = no ent or dep. VHF Ch 14 74 16 (H24). HM ☎(01524) 852373. Facilities: ⚓ 🛒 ✕ 🖵 at Morecambe (2M).

9.10.15 FLEETWOOD

Lancashire 53°55′·49N 03°00′·15W ❀❀❀❀♦♦♦❁❁

CHARTS AC 1826, 2010, 1552, 5613; Imray C62

TIDES +0015 Dover; ML 5.2; Duration 0530

Standard Port LIVERPOOL (→)

Times				Height (metres)			
High Water		Low Water		MHWS	MHWN	MLWN	MLWS
0000	0600	0200	0700	9·4	7·5	3·2	1·1
1200	1800	1400	1900				
Differences WYRE LIGHTHOUSE							
–0005	–0005	0000	–0005	–0·2	–0·2	ND	ND
FLEETWOOD							
–0004	–0004	–0006	–0006	0·0	–0·2	–0·1	+0·1
BLACKPOOL							
–0010	0000	–0010	–0020	–0·5	–0·5	–0·4	–0·1

SHELTER Excellent in Fleetwood Haven Marina. Sheltered ⚓, clear of turning circle, off Knott End pier on E bank to await tide. Passage up-river to Skippool (5M) needs local knowledge and is only possible with shoal draught.

NAVIGATION WPT 53°57′·67N 03°02′·05W, Fairway NCM buoy, Q, 134°/3ca to No 3 SHM buoy, VQ.G. Regular dredging is no longer undertaken to maintain Fleetwood channel. Depths are liable to frequent change owing to siltation. Call HM for latest info.

Ideally arrive at WPT about HW-1½ if bound for Wyre Dock marina, 3M up-river, which can be entered during freeflow on all tides. Call *Fleetwood Dock Radio* before entering buoyed Dock Channel. Waiting pontoon outside lock on East Jetty.

LIGHTS AND MARKS Chan is well buoyed/lit. Ldg lts, front Iso G 2s; rear Iso G 4s, 156° (only to be used between Nos 8 buoy and 11 perch bn). Lock sigs: ● lights = 'Stop'; ● lights = 'Go'.

COMMUNICATIONS (Code 01253) CGOC (01407) 762051; Police 101; Dr 873312. HM (ABP) 872323.

Fleetwood Dock Radio, Ch 12 (HW±2), before entering channel.

FACILITIES

Fleetwood Haven Marina ☎879062 ④ 340 inc ❻ (max 20m LOA) in two basins, £2.35, ⛽ ⚓ ☐ 🛢(0800–1700) ⚓ ✕ ⚓ 🗙 ⚓ ▣(75t) 🛒 ✕.

River Wyre YC ☎811948, 1❻.

Blackpool & Fleetwood YC (Skippool) ☎884205, ⚓ ⚓ ⚓ ☐.

Services ✕ ✎ 🆑 Ⓔ(☎823535) ⚓ ⚓.

Town 🏠 🏠 ✉ Ⓑ 🍴 ✕ ☐ ⇌ (Poulton-le-Fylde or Blackpool), ✈ (Blackpool).

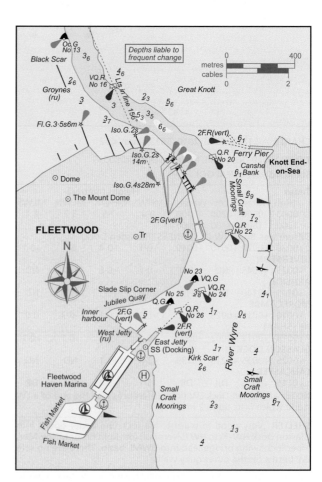

9.10.16 RIVER RIBBLE/PRESTON

Lancashire 53°43´·51N 03°00´·08W ❀❀♦♦♦♦❀❀

CHARTS AC 1826, 1981, 5613; Imray C62

TIDES +0013 Dover; ML No data; Duration 0520

Standard Port LIVERPOOL (→)

Times				Height (metres)			
High Water		Low Water		MHWS	MHWN	MLWN	MLWS
0000	0600	0200	0700	9·4	7·5	3·2	1·1
1200	1800	1400	1900				
Differences PRESTON							
+0015	+0015	+0330	+0305	−4·1	−4·2	−3·1	−1·0

LW time differences give the end of a LW stand lasting 3½ hrs.

SOUTHPORT							
−0015	−0005	ND	ND	−0·4	−0·4	ND	ND
FORMBY							
−0010	−0005	−0025	−0025	−0·4	−0·2	−0·3	−0·1

SHELTER Good in Preston Marina (5m depth) 15M upriver. Possible drying berths on the N bank at Lytham or Freckleton, or 2M up R Douglas access only near HW.

NAVIGATION WPT **A** 53°41´·75N 03°08´·90W, Gut SWM lt buoy. The chartlet shows only the outer 6M of the estuary. Due to changes to the Gut Channel and South Gut, best water is now in the Gut Channel which almost dries but not to the extent shown on AC 1981. It is navigable approx HW±2. South Gut is no longer discernible and should not be used without local knowledge.

Leave WPT **A** at HW Liverpool −2 then track 070°/2·6M to WPT **B** (53°42´·90N 03°05´·00W). Continue up Gut Channel leaving the first 3 perches about 100m to starboard. Leave 11½M perch 100m to port. The river trends 080° between training walls (dry 3m), marked by lit perches. The remaining channel is straightforward but night entry is not advised.

LIGHTS AND MARKS 14½M perch is the most seaward channel mark. Up-river of 11½M perch there is a PHM buoy and perch (off chartlet), both Fl R 5s. 4 SHM perches, Fl G 5s, lead to 5M perch, Fl

(2) G 10s, marking mouth of unlit R Douglas. 3M and 2M perches are Fl G 5s; 1M perch is Fl G 10s. Warton airfield beacon, Mo (WQ) G 9s occas, is N abeam 6M perch, Fl G 5s.

COMMUNICATIONS (Code 01772) CGOC (01407) 762051; Police 101; Ⓗ 710408. Preston locks 726871. At Preston, for locks call *Riversway* Ch **16** 14 (HW −1½ to +2); Marina Ch 80 (HO); Douglas BY Ch 16 when vessel due.

FACILITIES Waiting berth (2m draught) outside storm gates (usually open). Traffic lts at marina lock; plan below. Lockmaster on duty HW±2, on daylight tides (0700-2000 Apr-Sep by Lavers Liverpool Tide Tables). This and tidal height should permit locking from -1 HW +2 on demand. Swing bridge opens in unison with locks. From 1 Oct-31 Mar contact marina to book locking 24 hrs in advance.

Preston Marina www.prestonmarina. co.uk ☎733595, mob 07770 505094, £1·60, ♿ ⛽ 🛢 ⚒ 🔧 🛠(25t, 45t by arrangement) 🅿 ACA 🛒 ✗.

Douglas BY, ☎812462 mob 07740 780899, ⚓ ⚓£0·70 ⛽ 🛢 ⚒ 🔧 🛠(20t) 🅿.

Freckleton BY ☎632439 mob 07957 820881, ⚓ ⚓£5/yacht ⛽ 🔧 🅿.

Ribble Cruising Club ☎(01253) 739983.

White Tr

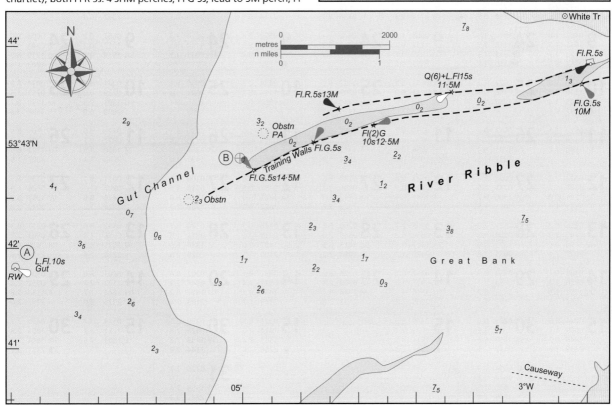

LIVERPOOL (GLADSTONE DOCK) LAT 53°27'N LONG 3°01'W
TIMES AND HEIGHTS OF HIGH AND LOW WATERS

STANDARD TIME (UT)
For Summer Time add ONE hour in **non-shaded** areas

Dates in red are SPRINGS
Dates in blue are NEAPS

YEAR **2016**

JANUARY

Day	Time m	Time m	Time m	Time m		Day	Time m	Time m	Time m	Time m
1	0314 8.0	0943 2.9	F 1534 8.1	2226 2.9		**16**	0314 8.7	0956 2.1	SA 1539 8.9	☽ 2229 2.0
2	0404 7.6	1030 3.3	SA 1629 7.7	☽ 2319 3.2		**17**	0411 8.3	1052 2.4	SU 1640 8.5	2329 2.3
3	0504 7.3	1131 3.5	SU 1734 7.4			**18**	0518 8.0	1200 2.6	M 1751 8.3	
4	0026 3.3	0612 7.3	M 1245 3.6	1843 7.4		**19**	0042 2.5	0634 7.9	TU 1320 2.7	1906 8.2
5	0135 3.2	0720 7.4	TU 1357 3.3	1948 7.6		**20**	0200 2.5	0748 8.1	W 1438 2.4	2018 8.4
6	0235 2.9	0819 7.8	W 1457 2.9	2044 7.9		**21**	0310 2.2	0853 8.5	TH 1546 2.0	2120 8.6
7	0327 2.5	0908 8.2	TH 1548 2.5	2131 8.3		**22**	0410 1.9	0948 8.8	F 1644 1.6	2213 8.9
8	0413 2.1	0951 8.6	F 1636 2.0	2212 8.6		**23**	0502 1.7	1036 9.1	SA 1734 1.3	2259 9.1
9	0457 1.8	1031 9.0	SA 1721 1.7	2252 8.9		**24**	0547 1.5	1118 9.3	SU 1818 1.2	○ 2340 9.2
10	0540 1.5	1114 9.3	SU 1805 1.4	● 2332 9.1		**25**	0627 1.4	1156 9.4	M 1858 1.1	
11	0622 1.3	1149 9.5	M 1848 1.2			**26**	0017 9.2	0702 1.4	TU 1233 9.4	1933 1.3
12	0012 9.3	0704 1.2	TU 1231 9.6	1931 1.1		**27**	0052 9.1	0734 1.6	W 1306 9.2	2005 1.5
13	0055 9.3	0745 1.3	W 1314 9.6	2012 1.1		**28**	0125 8.9	0804 1.8	TH 1339 9.0	2035 1.8
14	0138 9.2	0826 1.4	TH 1358 9.5	2054 1.3		**29**	0159 8.6	0833 2.1	F 1413 8.7	2103 2.1
15	0224 9.0	0909 1.7	F 1446 9.2	2139 1.6		**30**	0235 8.3	0905 2.5	SA 1450 8.3	2135 2.5
						31	0314 7.9	0942 2.8	SU 1532 7.9	2215 2.9

FEBRUARY

Day	Time m	Time m	Time m	Time m		Day	Time m	Time m	Time m	Time m
1	0402 7.5	1030 3.2	M 1624 7.5	☽ 2310 3.3		**16**	0449 8.0	1135 2.6	TU 1726 8.0	
2	0503 7.2	1135 3.5	TU 1734 7.2			**17**	0013 2.7	0608 7.7	W 1300 2.8	1847 7.8
3	0026 3.4	0618 7.2	W 1259 3.5	1854 7.2		**18**	0139 2.8	0730 7.8	TH 1425 2.6	2007 7.9
4	0147 3.2	0732 7.4	TH 1416 3.2	2006 7.5		**19**	0257 2.6	0841 8.2	F 1537 2.1	2112 8.3
5	0252 2.8	0834 7.9	F 1519 2.6	2103 8.0		**20**	0400 2.2	0937 8.6	SA 1635 1.7	2204 8.6
6	0347 2.3	0924 8.4	SA 1613 2.1	2151 8.5		**21**	0451 1.8	1023 8.9	SU 1722 1.4	2246 8.9
7	0436 1.8	1009 8.9	SU 1703 1.5	2234 8.9		**22**	0533 1.6	1102 9.2	M 1801 1.2	○ 2323 9.1
8	0523 1.4	1051 9.4	M 1750 1.1	● 2315 9.3		**23**	0608 1.4	1137 9.3	TU 1836 1.1	2356 9.1
9	0608 1.0	1132 9.7	TU 1835 0.7	2357 9.6		**24**	0640 1.4	1210 9.3	W 1907 1.2	
10	0651 0.8	1214 9.9	W 1918 0.6			**25**	0027 9.1	0709 1.4	TH 1241 9.2	1935 1.3
11	0039 9.7	0733 0.8	TH 1258 10.0	1959 0.6		**26**	0058 9.0	0737 1.5	F 1312 9.1	2001 1.5
12	0122 9.6	0813 0.9	F 1341 9.9	2039 0.8		**27**	0128 8.8	0805 1.7	SA 1342 8.8	2028 1.8
13	0206 9.3	0854 1.2	SA 1427 9.5	2120 1.2		**28**	0200 8.6	0835 2.0	SU 1413 8.5	2058 2.2
14	0252 8.9	0937 1.6	SU 1516 9.1	2205 1.8		**29**	0234 8.2	0909 2.4	M 1448 8.1	2133 2.6
15	0345 8.5	1029 2.1	M 1614 8.5	☽ 2301 2.3						

MARCH

Day	Time m	Time m	Time m	Time m		Day	Time m	Time m	Time m	Time m
1	0313 7.8	0950 2.9	TU 1530 7.6	☽ 2218 3.1		**16**	0424 8.0	1117 2.5	W 1705 7.8	2348 2.9
2	0404 7.4	1044 3.2	W 1630 7.2	2324 3.4		**17**	0544 7.6	1243 2.8	TH 1830 7.5	
3	0517 7.1	1203 3.4	TH 1756 7.0			**18**	0118 3.0	0709 7.6	F 1409 2.6	1952 7.7
4	0054 3.4	0642 7.2	F 1334 3.2	1924 7.3		**19**	0240 2.8	0822 8.0	SA 1520 2.2	2057 8.1
5	0215 3.0	0756 7.7	SA 1447 2.6	2032 7.8		**20**	0343 2.3	0918 8.4	SU 1615 1.8	2146 8.5
6	0318 2.4	0854 8.3	SU 1548 2.0	2125 8.4		**21**	0432 2.0	1002 8.7	M 1700 1.5	2225 8.8
7	0413 1.8	0943 8.9	M 1641 1.4	2211 9.0		**22**	0511 1.7	1040 9.0	TU 1736 1.3	2259 9.0
8	0503 1.2	1028 9.5	TU 1730 0.8	2254 9.5		**23**	0544 1.5	1113 9.1	W 1808 1.2	2330 9.1
9	0550 0.8	1111 9.9	W 1816 0.4	● 2337 9.8		**24**	0614 1.4	1144 9.2	TH 1837 1.2	
10	0634 0.5	1154 10.1	TH 1859 0.2			**25**	0000 9.1	0641 1.3	F 1215 9.1	1903 1.3
11	0019 9.9	0716 0.4	F 1238 10.2	1940 0.2		**26**	0030 9.1	0710 1.4	SA 1245 9.0	1931 1.4
12	0102 9.8	0757 0.5	SA 1322 10.0	2020 0.5		**27**	0100 8.9	0739 1.5	SU 1314 8.8	1959 1.7
13	0145 9.5	0838 0.8	SU 1407 9.6	2100 1.1		**28**	0130 8.7	0811 1.8	M 1344 8.5	2029 2.0
14	0231 9.1	0921 1.4	M 1456 9.1	2143 1.7		**29**	0202 8.4	0844 2.1	TU 1417 8.2	2103 2.4
15	0321 8.5	1011 2.0	TU 1553 8.4	☽ 2236 2.4		**30**	0239 8.1	0922 2.5	W 1458 7.8	2144 2.9
						31	0326 7.7	1011 2.9	TH 1553 7.4	☽ 2242 3.3

APRIL

Day	Time m	Time m	Time m	Time m		Day	Time m	Time m	Time m	Time m
1	0433 7.4	1122 3.2	F 1712 7.1			**16**	0050 3.1	0637 7.6	SA 1341 2.6	1923 7.6
2	0006 3.4	0557 7.3	SA 1252 3.0	1842 7.3		**17**	0209 2.9	0750 7.8	SU 1449 2.3	2029 7.9
3	0135 3.0	0716 7.7	SU 1412 2.5	1957 7.8		**18**	0312 2.6	0848 8.2	M 1543 2.0	2117 8.3
4	0245 2.4	0821 8.3	M 1517 1.9	2056 8.5		**19**	0401 2.4	0933 8.5	TU 1627 1.7	2156 8.6
5	0345 1.8	0914 9.0	TU 1614 1.2	2145 9.1		**20**	0440 1.9	1011 8.7	W 1703 1.5	2230 8.8
6	0438 1.2	1002 9.5	W 1706 0.7	2230 9.5		**21**	0513 1.7	1044 8.9	TH 1735 1.4	2301 8.9
7	0527 0.7	1047 9.9	TH 1753 0.3	● 2314 9.9		**22**	0544 1.5	1117 8.9	F 1805 1.3	○ 2332 9.0
8	0613 0.4	1132 10.2	F 1837 0.1	2357 10.0		**23**	0614 1.4	1148 9.0	SA 1834 1.3	
9	0657 0.2	1217 10.2	SA 1919 0.2			**24**	0003 9.0	0644 1.4	SU 1219 8.9	1903 1.4
10	0041 9.9	0739 0.3	SU 1303 9.9	2000 0.6		**25**	0034 8.9	0717 1.5	M 1250 8.7	1934 1.6
11	0125 9.6	0822 0.7	M 1350 9.5	2040 1.1		**26**	0105 8.8	0750 1.7	TU 1322 8.5	2007 1.9
12	0211 9.1	0907 1.2	TU 1439 8.9	2124 1.8		**27**	0139 8.6	0826 1.9	W 1357 8.3	2042 2.3
13	0302 8.6	0958 1.8	W 1535 8.3	2216 2.4		**28**	0217 8.3	0904 2.3	TH 1439 8.0	2123 2.6
14	0403 8.0	1102 2.4	TH 1644 7.7	☽ 2325 3.0		**29**	0304 8.0	0952 2.6	F 1533 7.6	2217 3.0
15	0518 7.7	1221 2.7	F 1803 7.5			**30**	0406 7.7	1055 2.8	SA 1644 7.4	☽ 2331 3.1

Chart Datum: 4·93 metres below Ordnance Datum (Newlyn). HAT is 10·3 metres above Chart Datum.

STANDARD TIME (UT)
For Summer Time add ONE hour in **non-shaded areas**

LIVERPOOL (GLADSTONE DOCK) LAT 53°27'N LONG 3°01'W
TIMES AND HEIGHTS OF HIGH AND LOW WATERS

Dates in red are **SPRINGS**
Dates in blue are **NEAPS**

YEAR 2016

MAY

Time	m	Time	m
1 0523 1215 SU 1806	7.7 2.7 7.5	**16** 0122 0705 M 1405 1945	3.1 7.7 2.5 7.7
2 0054 0639 M 1334 1921	2.9 7.9 2.3 8.0	**17** 0226 0806 TU 1500 2038	2.9 7.9 2.3 8.0
3 0209 0746 TU 1443 2024	2.4 8.4 1.8 8.5	**18** 0318 0855 W 1546 2121	2.6 8.1 2.1 8.3
4 0313 0843 W 1544 2117	1.8 9.0 1.3 9.0	**19** 0401 0937 TH 1625 2158	2.3 8.4 1.9 8.6
5 0410 0935 TH 1638 2205	1.3 9.4 0.8 9.5	**20** 0438 1014 F 1700 2232	2.0 8.6 1.7 8.8
6 0502 1024 F 1728 ● 2251	0.8 9.8 0.5 9.7	**21** 0513 1049 SA 1733 ○ 2305	1.7 8.7 1.6 8.9
7 0551 1112 SA 1814 2337	0.5 10.0 0.4 9.9	**22** 0548 1123 SU 1806 2338	1.6 8.8 1.5 9.0
8 0638 1159 SU 1858	0.3 9.9 0.4	**23** 0622 1156 M 1840	1.5 8.8 1.5
9 0022 0723 M 1247 1940	9.8 0.4 9.7 0.7	**24** 0011 0658 TU 1230 1914	9.0 1.5 8.7 1.6
10 0108 0808 TU 1334 2023	9.5 0.7 9.3 1.2	**25** 0046 0735 W 1305 1950	8.9 1.6 8.6 1.8
11 0154 0854 W 1423 2107	9.1 1.2 8.8 1.8	**26** 0122 0813 TH 1344 2028	8.8 1.7 8.4 2.1
12 0243 0945 TH 1516 2156	8.7 1.7 8.3 2.4	**27** 0203 0853 F 1427 2110	8.6 2.0 8.2 2.3
13 0340 1042 F 1617 ◗ 2256	8.2 2.2 7.8 2.9	**28** 0250 0940 SA 1519 2201	8.3 2.2 8.0 2.6
14 0445 1150 SA 1726	7.8 2.6 7.5	**29** 0348 1036 SU 1623 ◗ 2304	8.1 2.3 7.8 2.7
15 0008 0555 SU 1300 1838	3.1 7.6 2.6 7.5	**30** 0455 1144 M 1735	8.1 2.3 7.8
		31 0017 0606 TU 1258 1847	2.7 8.2 2.2 8.1

JUNE

Time	m	Time	m
1 0131 0714 W 1409 1953	2.4 8.5 1.8 8.5	**16** 0227 0812 TH 1459 2041	2.9 7.8 2.5 8.0
2 0240 0815 TH 1513 2051	2.0 8.9 1.5 8.9	**17** 0318 0901 F 1545 2124	2.6 8.0 2.2 8.3
3 0342 0912 F 1611 2143	1.5 9.2 1.1 9.2	**18** 0402 0944 SA 1625 2203	2.3 8.3 2.0 8.6
4 0439 1005 SA 1704 2232	1.1 9.5 0.9 9.5	**19** 0444 1023 SU 1704 2240	2.0 8.5 1.8 8.8
5 0532 1055 SU 1753 ● 2320	0.8 9.6 0.7 9.6	**20** 0524 1100 M 1742 ○ 2315	1.7 8.6 1.6 8.9
6 0622 1144 M 1840	0.6 9.6 0.8	**21** 0604 1136 TU 1820 2352	1.5 8.7 1.5 9.1
7 0006 0709 TU 1232 1923	9.7 0.6 9.5 1.0	**22** 0643 1212 W 1859	1.4 8.8 1.5
8 0052 0755 W 1319 2006	9.5 0.8 9.2 1.4	**23** 0029 0723 TH 1251 1937	9.1 1.4 8.8 1.6
9 0137 0841 TH 1405 2048	9.2 1.2 8.8 1.8	**24** 0109 0803 F 1332 2017	9.0 1.5 8.7 1.7
10 0223 0926 F 1452 2132	8.8 1.6 8.4 2.3	**25** 0151 0844 SA 1416 2059	8.9 1.6 8.6 2.0
11 0311 1014 SA 1542 2220	8.4 2.1 8.0 2.7	**26** 0237 0928 SU 1505 2146	8.8 1.8 8.4 2.2
12 0405 1107 SU 1639 ◗ 2316	8.0 2.5 7.6 3.1	**27** 0329 1018 M 1601 ◗ 2241	8.6 1.9 8.2 2.4
13 0506 1207 M 1742	7.7 2.7 7.4	**28** 0429 1117 TU 1706 2346	8.4 2.1 8.1 2.5
14 0021 0610 TU 1309 1848	3.2 7.6 2.8 7.4	**29** 0536 1225 W 1817	8.3 2.1 8.1
15 0127 0714 W 1408 1949	3.2 7.6 2.7 7.7	**30** 0058 0646 TH 1338 1926	2.4 8.4 2.0 8.3

JULY

Time	m	Time	m
1 0212 0753 F 1447 2030	2.2 8.6 1.8 8.6	**16** 0235 0826 SA 1506 2051	3.0 7.7 2.6 8.0
2 0321 0855 SA 1550 2127	1.8 8.9 1.6 8.9	**17** 0329 0916 SU 1554 2136	2.6 8.0 2.3 8.4
3 0423 0952 SU 1646 2219	1.4 9.1 1.3 9.2	**18** 0418 0959 M 1639 2216	2.2 8.3 1.9 8.7
4 0519 1044 M 1737 ● 2307	1.1 9.3 1.2 9.4	**19** 0503 1039 TU 1722 ○ 2255	1.8 8.6 1.7 9.0
5 0610 1133 TU 1824 2353	0.9 9.3 1.1 9.5	**20** 0547 1117 W 1804 2333	1.5 8.8 1.5 9.2
6 0657 1218 W 1908	0.8 9.3 1.2	**21** 0630 1156 TH 1845	1.2 9.0 1.3
7 0036 0740 TH 1301 1948	9.4 0.9 9.1 1.4	**22** 0012 0711 F 1236 1926	9.4 1.1 9.1 1.3
8 0118 0822 F 1342 2026	9.2 1.2 8.8 1.7	**23** 0053 0752 SA 1317 2006	9.4 1.1 9.1 1.3
9 0157 0901 SA 1421 2102	8.9 1.6 8.5 2.1	**24** 0135 0832 SU 1400 2046	9.4 1.2 9.0 1.5
10 0238 0938 SU 1503 2139	8.6 2.0 8.2 2.5	**25** 0220 0913 M 1446 2130	9.2 1.4 8.7 1.8
11 0322 1017 M 1549 2220	8.2 2.4 7.8 2.9	**26** 0308 0958 TU 1538 ◗ 2219	9.0 1.7 8.5 2.1
12 0412 1103 TU 1644 ◗ 2312	7.8 2.8 7.5 3.2	**27** 0404 1051 W 1638 2320	8.6 2.0 8.2 2.4
13 0511 1201 W 1747	7.5 3.0 7.3	**28** 0510 1157 TH 1751	8.3 2.3 8.0
14 0019 0618 TH 1308 1855	3.4 7.3 3.1 7.4	**29** 0034 0624 F 1314 1906	2.5 8.2 2.4 8.0
15 0131 0726 F 1411 1958	3.3 7.4 2.9 7.6	**30** 0155 0739 SA 1429 2017	2.4 8.2 2.2 8.3
		31 0309 0847 SU 1536 2118	2.1 8.5 2.0 8.7

AUGUST

Time	m	Time	m
1 0414 0946 M 1635 2210	1.7 8.8 1.7 9.1	**16** 0353 0936 TU 1615 2152	2.2 8.2 2.0 8.8
2 0510 1036 TU 1725 ● 2256	1.3 9.0 1.4 9.3	**17** 0442 1017 W 1702 2232	1.7 8.7 1.6 9.2
3 0559 1121 W 1810 2338	1.0 9.2 1.3 9.4	**18** 0529 1057 TH 1746 ○ 2312	1.3 9.1 1.3 9.5
4 0642 1201 TH 1850	0.9 9.2 1.3	**19** 0613 1136 F 1829 2352	1.0 9.3 1.0 9.7
5 0017 0720 F 1239 1925	9.4 1.0 9.1 1.4	**20** 0656 1216 SA 1911	0.7 9.5 0.9
6 0053 0756 SA 1314 1958	9.3 1.2 8.9 1.6	**21** 0033 0736 SU 1258 1951	9.8 0.7 9.5 1.0
7 0128 0828 SU 1348 2029	9.1 1.5 8.7 1.9	**22** 0116 0816 M 1340 2031	9.8 0.8 9.3 1.2
8 0202 0858 M 1423 2059	8.8 1.9 8.4 2.3	**23** 0200 0855 TU 1425 2113	9.5 1.1 9.0 1.6
9 0239 0928 TU 1503 2133	8.4 2.3 8.1 2.7	**24** 0247 0938 W 1515 2200	9.2 1.6 8.6 2.0
10 0320 1003 W 1549 ◗ 2217	8.0 2.7 7.7 3.1	**25** 0342 1028 TH 1614 ◗ 2300	8.7 2.1 8.2 2.4
11 0411 1052 TH 1646 2316	7.5 3.1 7.3 3.4	**26** 0449 1133 F 1730	8.2 2.6 7.8
12 0517 1201 F 1758	7.2 3.4 7.2	**27** 0019 0609 SA 1257 1852	2.7 7.9 2.8 7.8
13 0034 0635 SA 1321 1913	3.5 7.1 3.3 7.4	**28** 0146 0731 SU 1419 2008	2.6 7.9 2.6 8.2
14 0153 0749 SU 1429 2017	3.2 7.3 3.0 7.8	**29** 0302 0842 M 1528 2109	2.2 8.3 2.3 8.6
15 0258 0848 M 1525 2108	2.8 7.8 2.5 8.3	**30** 0406 0939 TU 1625 2159	1.7 8.6 1.9 9.0
		31 0459 1025 W 1712 2241	1.4 8.9 1.6 9.2

Chart Datum: 4·93 metres below Ordnance Datum (Newlyn). HAT is 10·3 metres above Chart Datum.

》FREE monthly updates. Register at 《
www.reedsnauticalalmanac.co.uk

511

NW England

LIVERPOOL (GLADSTONE DOCK) LAT 53°27'N LONG 3°01'W

STANDARD TIME (UT)
For Summer Time add ONE hour in **non-shaded areas**

TIMES AND HEIGHTS OF HIGH AND LOW WATERS

Dates in red are **SPRINGS**
Dates in blue are **NEAPS**

YEAR **2016**

SEPTEMBER

Time	m		Time	m
1 0543	1.1	**16** 0506	1.1
1104	9.1	1033	9.3
TH 1752	1.4	F 1725	1.1
● 2319	9.4	○ 2248	9.8
2 0621	1.1	**17** 0552	0.7
1140	9.2	1113	9.6
F 1827	1.4	SA 1810	0.8
2353	9.4	2329	10.0
3 0655	1.1	**18** 0635	0.5
1212	9.2	1154	9.8
SA 1859	1.4	SU 1852	0.7
4 0026	9.3	**19** 0011	10.1
0724	1.3	0716	0.5
SU 1244	9.1	M 1236	9.8
1927	1.6	1933	0.7
5 0057	9.1	**20** 0055	10.0
0752	1.5	0756	0.7
M 1315	8.9	TU 1319	9.6
1955	1.8	2014	1.0
6 0128	8.9	**21** 0140	9.7
0819	1.8	0836	1.1
TU 1347	8.6	W 1404	9.2
2024	2.1	2057	1.4
7 0201	8.5	**22** 0228	9.2
0846	2.2	0918	1.7
W 1422	8.3	TH 1454	8.7
2057	2.5	2146	2.0
8 0236	8.1	**23** 0323	8.6
0919	2.6	1008	2.3
TH 1504	7.9	F 1554	8.2
2136	2.9	◑ 2248	2.5
9 0318	7.6	**24** 0433	8.0
1002	3.1	1116	2.9
F 1550	7.5	SA 1712	7.8
◐ 2228	3.3		
10 0416	7.2	**25** 0010	2.8
1103	3.5	0556	7.7
SA 1701	7.2	SU 1243	3.1
2343	3.5	1836	7.8
11 0540	6.9	**26** 0136	2.6
1229	3.6	0721	7.8
SU 1825	7.2	M 1407	2.9
		1953	8.1
12 0111	3.4	**27** 0250	2.2
0708	7.1	0831	8.2
M 1351	3.2	TU 1515	2.4
1939	7.6	2054	8.5
13 0225	2.8	**28** 0350	1.8
0817	7.7	0924	8.6
TU 1455	2.7	W 1609	2.1
2037	8.2	2141	8.9
14 0325	2.2	**29** 0439	1.5
0909	8.3	1006	8.9
W 1550	2.1	TH 1652	1.8
2124	8.8	2221	9.1
15 0418	1.6	**30** 0519	1.3
0952	8.8	1042	9.1
TH 1639	1.6	F 1728	1.6
2207	9.4	2255	9.3

OCTOBER

Time	m		Time	m
1 0553	1.3	**16** 0527	0.6
1114	9.2	1049	9.8
SA 1800	1.5	SU 1747	0.8
● 2327	9.3	○ 2306	10.2
2 0623	1.3	**17** 0612	0.4
1145	9.2	1132	10.0
SU 1829	1.5	M 1832	0.6
2358	9.2	2350	10.2
3 0651	1.4	**18** 0654	0.4
1214	9.2	1215	10.0
M 1857	1.6	TU 1915	0.6
4 0028	9.1	**19** 0035	10.1
0718	1.6	0736	0.7
TU 1245	9.0	W 1259	9.7
1926	1.8	1959	0.9
5 0059	8.9	**20** 0123	9.7
0745	1.8	0817	1.2
W 1316	8.8	TH 1345	9.4
1956	2.0	2044	1.3
6 0130	8.6	**21** 0212	9.2
0814	2.2	0901	1.8
TH 1348	8.5	F 1436	8.8
2029	2.4	2135	1.9
7 0203	8.2	**22** 0308	8.5
0847	2.6	0952	2.4
F 1424	8.1	SA 1537	8.3
2107	2.8	◑ 2238	2.4
8 0242	7.8	**23** 0416	8.0
0927	3.0	1058	3.0
SA 1509	7.7	SU 1651	7.9
2155	3.1	2355	2.7
9 0334	7.4	**24** 0536	7.6
1021	3.5	1221	3.2
SU 1612	7.4	M 1811	7.8
◐ 2302	3.4		
10 0450	7.1	**25** 0114	2.6
1141	3.6	0657	7.7
M 1736	7.3	TU 1342	3.0
		1926	8.0
11 0029	3.3	**26** 0225	2.3
0621	7.2	0807	8.0
TU 1310	3.4	W 1448	2.7
1856	7.7	2027	8.4
12 0149	2.8	**27** 0322	2.0
0738	7.7	0859	8.4
W 1421	2.8	TH 1541	2.3
2000	8.2	2115	8.7
13 0253	2.2	**28** 0409	1.8
0836	8.3	0941	8.7
TH 1519	2.2	F 1623	2.1
2052	8.9	2154	8.9
14 0349	1.6	**29** 0448	1.6
0923	9.0	1016	8.9
F 1612	1.6	SA 1659	1.9
2138	9.5	2229	9.1
15 0439	1.0	**30** 0521	1.6
1007	9.5	1047	9.1
SA 1701	1.1	SU 1730	1.8
2222	9.9	● 2301	9.1
		31 0551	1.5
		1118	9.2
		M 1800	1.7
		2332	9.1

NOVEMBER

Time	m		Time	m
1 0619	1.6	**16** 0634	0.6
1148	9.2	1157	10.0
TU 1830	1.7	W 1900	0.7
2 0003	9.0	**17** 0020	10.0
0648	1.7	0717	0.8
W 1219	9.1	TH 1243	9.8
1901	1.8	1946	0.9
3 0035	8.9	**18** 0109	9.6
0718	1.8	0801	1.3
TH 1250	8.9	F 1330	9.5
1934	1.9	2033	1.2
4 0107	8.6	**19** 0159	9.1
0750	2.1	0846	1.8
F 1323	8.7	SA 1421	9.0
2009	2.2	2125	1.7
5 0140	8.3	**20** 0253	8.6
0824	2.5	0936	2.4
SA 1359	8.4	SU 1517	8.5
2047	2.6	2222	2.2
6 0220	8.0	**21** 0353	8.1
0904	2.9	1035	2.9
SU 1443	8.1	M 1621	8.1
2133	2.9	◑ 2328	2.6
7 0309	7.7	**22** 0501	7.7
0954	3.3	1146	3.2
M 1540	7.8	TU 1732	7.9
◐ 2232	3.1		
8 0414	7.4	**23** 0038	2.7
1102	3.5	0615	7.6
TU 1654	7.6	W 1300	3.2
2348	3.1	1843	7.9
9 0536	7.4	**24** 0145	2.6
1225	3.3	0726	7.8
W 1812	7.8	TH 1407	3.0
		1948	8.0
10 0107	2.8	**25** 0243	2.4
0654	7.8	0823	8.1
TH 1341	2.9	F 1502	2.8
1920	8.3	2040	8.3
11 0216	2.2	**26** 0332	2.2
0759	8.3	0909	8.4
F 1445	2.3	SA 1548	2.5
2018	8.4	2124	8.5
12 0316	1.7	**27** 0412	2.0
0853	8.9	0946	8.7
SA 1542	1.7	SU 1626	2.2
2110	9.4	2202	8.7
13 0411	1.2	**28** 0448	1.9
0941	9.4	1021	8.9
SU 1635	1.2	M 1701	2.0
2158	9.8	2237	8.9
14 0501	0.8	**29** 0520	1.8
1026	9.8	1053	9.0
M 1725	0.9	TU 1735	1.9
○ 2245	10.1	● 2310	8.9
15 0548	0.6	**30** 0552	1.7
1112	10.0	1125	9.1
TU 1813	0.7	W 1808	1.8
2333	10.1	2343	8.9

DECEMBER

Time	m		Time	m
1 0624	1.7	**16** 0009	9.8
1158	9.1	0702	1.0
TH 1842	1.8	F 1231	9.8
		1935	0.8
2 0016	8.8	**17** 0057	9.5
0657	1.8	0747	1.3
F 1231	9.0	SA 1317	9.5
1918	1.8	2022	1.1
3 0050	8.7	**18** 0144	9.2
0732	2.0	0831	1.7
SA 1306	8.9	SU 1404	9.2
1955	2.0	2109	1.5
4 0125	8.5	**19** 0232	8.7
0808	2.3	0916	2.2
SU 1343	8.7	M 1452	8.8
2034	2.2	2157	2.0
5 0205	8.3	**20** 0321	8.3
0848	2.6	1003	2.9
M 1425	8.5	TU 1544	8.4
2117	2.5	2250	2.5
6 0251	8.0	**21** 0416	7.8
0934	2.9	1057	3.1
TU 1516	8.2	W 1643	8.0
2209	2.7	◑ 2348	2.8
7 0348	7.8	**22** 0519	7.6
1032	3.1	1201	3.3
W 1619	8.1	TH 1748	7.7
◐ 2312	2.7		
8 0457	7.7	**23** 0051	2.9
1142	3.1	0627	7.5
TH 1730	8.1	F 1310	3.4
		1855	7.7
9 0024	2.6	**24** 0153	2.9
0611	7.9	0734	7.6
F 1258	2.9	SA 1413	3.2
1841	8.3	1958	7.8
10 0137	2.3	**25** 0248	2.7
0722	8.3	0830	7.9
SA 1409	2.5	SU 1508	2.9
1946	8.7	2051	8.1
11 0244	1.9	**26** 0335	2.5
0823	8.7	0916	8.3
SU 1513	2.0	M 1554	2.6
2045	9.1	2136	8.3
12 0343	1.5	**27** 0416	2.2
0918	9.2	0955	8.6
M 1612	1.5	TU 1635	2.3
2139	9.5	2215	8.5
13 0438	1.1	**28** 0454	2.0
1008	9.5	1032	8.8
TU 1706	1.1	W 1713	2.0
2230	9.7	2251	8.7
14 0529	0.8	**29** 0530	1.8
1056	9.8	1106	9.0
W 1758	0.8	TH 1751	1.8
○ 2320	9.8	● 2326	8.8
15 0617	0.9	**30** 0606	1.7
1144	9.9	1141	9.1
TH 1847	0.7	F 1828	1.7
		31 0000	8.9
		0642	1.7
		SA 1216	9.2
		1906	1.6

Chart Datum: 4·93 metres below Ordnance Datum (Newlyn). HAT is 10·3 metres above Chart Datum.

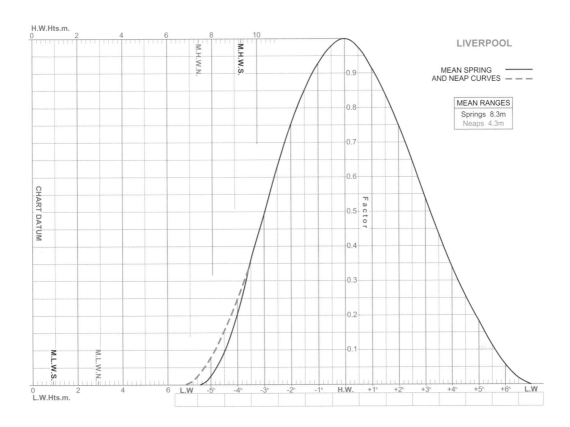

LIVERPOOL

MEAN SPRING AND NEAP CURVES

MEAN RANGES
Springs 8.3m
Neaps 4.3m

9.10.17 LIVERPOOL

Merseyside **53°24'·22N 03°00'·28W** (Liver Bldg) ✿✿✿♨♨♨✿✿

CHARTS AC 1826, 1978, 1951, 3490, 5613; Imray C62, C52

TIDES +0015 Dover; ML 5·2; Duration 0535

Standard Port LIVERPOOL (GLADSTONE DOCK) (←)

Times				Height (metres)			
High Water		Low Water		MHWS	MHWN	MLWN	MLWS
0000	0600	0200	0700	9·4	7·5	3·2	1·1
1200	1800	1400	1900				
Differences EASTHAM (River Mersey)							
+0014	+0014	+0006	+0006	+0·2	0·0	−0·4	−0·5
HALE HEAD (River Mersey)							
+0035	+0030	ND	ND	−2·5	−2·6	ND	ND
WIDNES (River Mersey)							
+0045	+0050	+0355	+0340	−4·3	−4·5	−2·8	−0·6
FIDDLER'S FERRY (River Mersey)							
+0105	+0120	+0535	+0445	−6·0	−6·4	−2·7	−0·6
HILBRE ISLAND (River Dee)							
−0011	−0008	−0013	−0018	−0·4	−0·3	+0·1	+0·2
MOSTYN DOCKS (River Dee)							
−0015	−0010	−0025	−0025	−0·9	−0·8	ND	ND
CONNAH'S QUAY (River Dee)							
+0005	+0020	+0350	+0335	−4·7	−4·5	DR	DR
CHESTER (River Dee)							
+0110	+0110	+0455	+0455	−5·4	−5·5	DR	DR
COLWYN BAY							
−0015	−0015	ND	ND	−1·6	−1·4	ND	ND
LLANDUDNO							
−0009	−0021	−0031	−0038	−1·7	−1·6	−0·9	−0·6

NOTE: LW time differences at Connah's Quay give the end of a LW stand lasting about 3¾hrs at sp and 5hrs at nps. A bore occurs in the R Dee at Chester.

SHELTER Good at marinas in Brunswick/Coburg docks and in Canning/Albert Docks. Fair weather ⚓ on the SW side of river. Strong winds from any direction raise a swell on the bar. Wind against tide causes steep, breaking seas in outer reaches of River Mersey with disturbed waters into the Crosby Channel.

NAVIGATION WPT 53°32'·02N 03°20'·98W, Bar light float, 111°/2·8M to Q1 NCM lt F. From Q1 to marina is 15·5M via Queens and Crosby Channels. Both have training banks which cover and it is unsafe to navigate between these and the floats/buoys. Sp tidal streams exceed 5kn within the river. Elsewhere great caution is needed as the whole area (R Dee, R Mersey to R Alt and N to Morecambe Bay) is littered with sandbanks and damaged training walls. Local knowledge is advisable inshore outside marked passages.

Leeds and Liverpool Canal (BWB) gives access to E Coast, ent at Stanley Dock. Max draught 1·0m, air draught 2·2m, beam 4·3m, LOA 18·3m. Liverpool to Goole 161M, 103 locks.

Manchester Ship Canal, ent at Eastham Locks (outer wall bn Iso YR 4s), leads 31M to Salford Quays or R Weaver for boatyards at Northwich. Obtain licence from MSC Co ☎0151 327 1461.

R Dee: WPT 53°25'·13N 03°13'·17W, Hilbre Swash HE2 ECM buoy, Q (3) 10s, (chan shifts). From the W use Welsh Chan. The Dee estuary mostly dries.

LIGHTS AND MARKS Bar SWM lt float, Fl 5s Racon. Keep 5ca clear of SPM, 5ca W of Bar lt float. Formby SWM buoy, Iso 4s, is at the ent to Queen's Chan which is marked by light floats and buoys, and 3 NCM buoys, numbered Q1-Q12. Crosby Chan is similarly marked, C1-C23. From Crosby SWM buoy, Oc 5s, the track up-river is approx 145°. The Liver Bldg (twin spires) and Port of Liverpool Bldg (dome) are conspic on the E bank opposite Birkenhead Docks.

COMMUNICATIONS (Code 0151) CGOC (01407) 762051; Police 101; Ⓗ 709 0141. Port Ops 949 6134/5; 🚒949 6095.

Contact *Mersey VTS* Ch 12 16 (H24) before Bar lighted Buoy. Nav /gale warnings are broadcast on receipt on Ch 12. Movements, nav warnings and weather are b/cast on Ch 09 at HW−3 and −2. *Canning Dock/Liverpool Marina* Ch **M**. Eastham Locks Ch 07 (H24). Manchester Ship Canal Ch 14. Liverpool NTM for further detail.

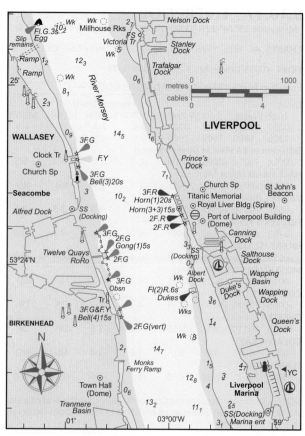

FACILITIES Liverpool Marina entrance is 1M S of the Liver Bldg, abeam Pluckington Bank WCM buoy. **IPTS** (sigs 2, 3 & 5) from conspic black control bldg at lock ent. 2m depth over sill at Brunswick Dock lock (width 8m) approx HW ±2¼ sp, ±1½ nps, 0600-2200 Mar-Oct. Pontoons inside, min depth 3·5m; night locking by arrangement. www.liverpoolmarina.co.uk ☎7076777 (0600-2200 Mar–Oct). 350 ⬦, £2·20; ⬦ ⬦ ⬦ ⬦ ⬦ (½M) ⬦ ⬦ ⬦ ⬦ ⬦ ⬦ (60t), ⬦ @ wi-fi, ⬦ ✕ (☎7076888).

Albert Dock ☎7096558; access HW–2 to HW via Canning Dock, but not on every tide. VHF Ch M when manned, ⬦ ⬦ ⬦.

Royal Mersey YC ☎6453204, ⬦ ⬦ ⬦ ⬦ ⬦ ⬦ ✕ ⬦.

W Kirby SC ☎6255579, M (in Dee Est) ⬦ ⬦ (at HW), ⬦ ⬦ (30 ton) ⬦.

Hoylake SC ☎6322616, ⬦ ⬦ ⬦ ⬦.

City all facilities, ⬦ ✈.

ADJACENT ANCHORAGE

RIVER ALT, Merseyside, **53°31′·42N 03°03′·80W**. AC 1978, 1951, 5613. HW –0008 on Dover. Good shelter but only for LOA <8·5m x 1·2m draught on a HW of at least 8m. Mersey E training wall can be crossed HW±2. Entrance to channel (shifts frequently) is E of C14 PHM lt float. The channel runs SE of, and parallel to the large fixed marks on the training wall (outer is a R basket; inner a Y cross), then between R and G perches. Inadvisable to ⬦ in R Alt without local knowledge; pick up a free mooring off the SC and contact club. Facilities are very limited. **Blundell Sands SC** ☎(0151) 9292101 (occas), ⬦ ⬦ (at HW) ⬦ ⬦.

ADJACENT HARBOUR

RHYL, Denbigh, **53°19′·0N 03°30′·4W**. AC 1978, 1951, 5613. Good shelter for vessels able to take the ground on pontoons in outer hbr. Inner hbr dries. Strong tides rip across ent between SWM and Hope's Nose, but training and harbour walls now renovated to make entry achievable with care. Lifting bridge divides hbrs and opens on req Ch 14. Contact HM prior to visit: ☎01745 360625.

9.10.18 CONWY AND DEGANWY

Conwy 53°17'·48N 03°50'·23W (marina) 🚩🚩♦♦♦♦🌸🌸🌸
Deganwy 53°17'·41N 03°49'·66W (marina) 🚩🚩♦♦♦♦🌸🌸🌸

CHARTS AC 5609, 1826, 1977, 1978, 1463; Imray C52

TIDES –0015 Dover; ML 4·3; Duration 0545

Standard Port HOLYHEAD (→)

Times				Height (metres)			
High Water		Low Water		MHWS	MHWN	MLWN	MLWS
0000	0600	0500	1100	5·6	4·4	2·0	0·7
1200	1800	1700	2300				
Differences CONWY							
+0025	+0035	+0120	+0105	+2·3	+1·8	+0·6	+0·4

NOTE: HW Conwy is approx HW Liverpool –0040 sp and –0020 nps.

SHELTER Good in river, except in NW gales. Good in marinas.

NAVIGATION Hbr guide www.conwy.gov.uk/harbourandseaboard WPT 53°17'·95N 03°55'·58W, Fairway buoy LFl 10s; 095°/6·2ca to channel between C1/C2 buoys.

After C1/C2 buoys the channel up to Deganwy is marked by lit lateral marks in sequence: C2A, C4, then paired SHM/PHM: C3/C6, C5/C8, C7/C10, C9/C12, C11/C14. C16 (Fl R 6s) at narrows into R Conwy. A bar, dries 1·0m, forms at The Scabs adjacent to No 6 PHM. Perch Lt bcn Fl G 15s 5M marks turn to river ent. 10kn speed limit above.

Beware unlit moorings throughout the harbour, Beacons Jetty in the narrows and unlit pontoons upstream from Bodlondeb Pt. Spring ebb runs at 5kn and vacant moorings may be submerged.

North Deep, a buoyed inshore passage (close SW of Gt Orme's Hd to Deganwy Point), is only advised with local knowledge.

LIGHTS AND MARKS Penmaenmawr SPM outfall buoy (SW of C1 buoy) is not a channel mark. The 4 towers of Conwy castle and the 2 adjacent bridge towers are conspic once in the river.

COMMUNICATIONS (Code 01492) CGOC (01407) 762051; Police 101; Dr 592424; HM 596253.

HM Ch **14**. Summer 0900-1700LT daily; winter, same times Mon-Fri). Marinas, pre-call Ch 80 (H24) for a berth. N Wales CC launch Ch **M** water taxi.

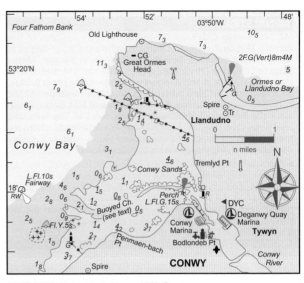

FACILITIES Beacons Jetty ⚓ HW±3.

Conwy Marina ⚓ www.quaymarinas.com ☎593000. Access when height of tide above CD exceeds 3·5m; waiting pontoon outside. Gate times are on an electronic noticeboard. Obey R/G entry lts. 505⚓ inc 🅥, £2·60m (short stay £10·95) ⚓ ⚓ ⚓ (0700-2200) ⚓ 🔧 ⚒ ⚓ Ⓔ ⚓ ♨ 🅿 (30t) ⚓ ✕ 🖥 wi-fi.

Deganwy Quay Marina ⚓ www.quaymarinas.com ☎576888. Access when height of tide above CD exceeds 4m, approx HW ±3¼. Keep close to 3 lit PHM lt marking the appr channel (see below). Min depth in chan is 2·5m when gate open. Obey R/G entry lights. 150⚓ inc 🅥, £2·60 (short stay £10·95) ⚓ ⚓ ⚓ (0700-2200) ⚓ 🔧 🅿 ✕ ⚒ Ⓔ 🅿(20t) ✕ 🖥 wi-fi.

Harbour Above Bodlondeb Pt contact HM for: ⚓'s, pontoon ⚓£13; **Town Quay** 3 drying ⚓ £11<8m>£12·50<12·3m; ⚓ ⚓ 🔧 (Town Quay HW±2, call Ch 14 to confirm).

N Wales Cruising Club (Conwy) ☎593481, ⚓ ⚓ ⚓ 🖥.
Conwy YC (Deganwy) ☎583690, ⚓ ⚓ ⚓ ⚓ ✕ 🖥.

Town 🏧 🏧 ✉ Ⓑ 🛒 ✕ 🖥 ⚞ ✈ (Liverpool).

MENAI STRAIT

(AC 1464, Imray C52) From NE to SW the main features of this narrow, 16 mile channel are: Puffin Is, where the Strait begins at its North-East entrance; Beaumaris; Garth Pt at Bangor; Menai Suspension Bridge (29m); The Swellies, a narrow 1M stretch with strong tidal streams and dangers mid-stream; Britannia Bridge, both road and rail (26m), with cables close W at elevation of 21m; Port Dinorwic; Caernarfon; Abermenai Pt and Fort Belan, which mark the SW end of Strait; and Caernarfon Bar. The following brief notes cover only very basic pilotage. For detailed directions see the UKHO *West Coasts of England and Wales Pilot*, or *Cruising Anglesey and Adjoining Waters* by Ralph Morris.

▶*The Swellies should be taken near local HW slack, and an understanding of tidal streams is essential. The tide is about 1 hour later, and sp range about 2·7m more, at the NE end of the Strait than at the SW end. Levels differ most at 1hr after HW at the NE end where sea level is > 1·8m above the level at the SW end and at about 4¾hrs before HW at the NE end where sea level is > 1·8m below the level at the SW end.*

Normally the stream runs as follows (referred to HW Dover).

Interval fm HW	Direction
–0110 to +0340	SW between Garth Pt and Abermenai Pt.
+0340 to +0510	Outwards from The Swellies, (NE towards Garth Pt / SW towards Abermenai Pt)
+0510 to –0110	NE between Abermenai Pt and Garth Pt.

See diagram on adjacent page.

*Sp rates are generally about 3kn, **but more in narrows, eg 5kn off Abermenai Pt, 6kn between the bridges, and 8kn through The Swellies.** Timings and rates of streams may be affected by strong winds in either direction and storm surges in the Irish Sea.*◀

Note: Direction of buoyage becomes NE off Caernarfon.

From SW to NE, aim to arrive at the Swellies early HW Dover –0300 to take the last of the E-going tidal stream to pass Menai bridge between HW Dover –0230 to –0200. Better to arrive too early, and wait, than too late and risk not making it through the narrows before the stream turns to the W. Approach the S arch of Britannia Bridge keeping the white pyramid in line with the centre of the arch. Once through the bridge run parallel to the shore until abeam the pyramid. At this point, look astern and keep the white leading marks (beneath the S arch of the bridge) in line. When Price's Point is abeam, turn to starboard to pass 20-30m S of Swelly Perch rock, then head just S of the Anglesey (ie N) pillar of the Menai Suspension Bridge until clear of the Platters (cottage abeam to starboard). Then head for the centre of the bridge.

From the NE, pilotage is the reverse of the above, passing Beaumaris in time to arrive at the Swellies at about HW Dover –0200, though the timing is not quite so critical. If too early, passage will be slow until the stream turns, but if too late you may have difficulty with the strong, but favourable, stream.

▶ *Caernarfon Bar is impassable even in moderately strong winds against the ebb, and the narrows at Abermenai Pt demand a fair tide, or slackish water, since the tide runs strongly here. Going seaward on first of the ebb, when there is water over the banks, it may not be practicable to return to the Strait if conditions on the bar are bad.*◀

Leaving Abermenai Pt on the last of the ebb means banks to seaward are exposed and there is little water in the channel or over the bar.

ANGLESEY TO BARDSEY ISLAND

(AC 1977, 1970, 1971 Imray C52) On N coast of Anglesey a race extends 5ca off Pt Lynas (lt, Oc 10s) on the E-going stream. Amlwch is a small harbour (partly dries) 1·5M W of Pt Lynas. A drying rock lies 100m offshore on W side of approach, which should not be attempted in strong onshore winds. From here to Carmel Head beware E Mouse (and shoals to SE), Middle Mouse, Harry Furlong's Rocks (dry) marked by SHM Fl G 2½s, Victoria Bank (least depth 1·8m) and W Mouse (with dangers to W & SW). Further offshore are the shoals of Archdeacon Rk, Coal Rk and Ethel Rk which lie in the red sector of Skerries Lt Ho.

Cardinal buoys mark their position. The outermost of these dangers is 2M offshore. Overfalls and races occur at headlands and over the many rocks and shoals along this coast.

In good conditions by day and at slack water, Carmel Hd can be rounded close inshore following a track outside East Mouse, inshore of Middle & West Mouse, staying N of Furlong SHM and passing S of Victoria Bank. Short, steep seas break here in even moderate winds and can be dangerous. The best passage at night or in bad weather, is to pass 1M off Skerries, in the TSS ITZ. Holyhead (port of refuge), is accessible H24 in all weathers. The marina is within New Harbour; beware fast ferries across the entrance and note the TSS.

▶*Between Carmel Hd and The Skerries the NE-going stream begins at HW Dover +0530 (HW Holyhead +0615), and the SW-going at HW Dover –0045 (HW Holyhead +0005), **Sp rates 5kn.** 1M NW of Skerries the stream turns 0130 later, running less strongly. **Flood and Ebb tides close to the coast runs at over 5k springs,** and at about 2.5kn 7 miles offshore. The brief period of slack water offshore is HW Dover –0100 and +0500. There is no significant counter tide in Holyhead Bay, where slack water lasts longer.*◀

▶*Races occur off N Stack and S Stack. There is a severe and dangerous race off South Stack starting at HW -0600 on the flood (Sp rate 5kn) extending 8nm to the Skerries and another dangerous race off Carmel Hd (Sp rate 6kn). Races are less extensive on SSW-going stream, which begins at HW Holyhead +0020, (Sp rate 5kn).*◀

The W coast of Anglesey is rugged with rks, some drying, up to 1·5M offshore. Races run off Penrhyn Mawr and Rhoscolyn Hd, between which is Maen Piscar a dangerous, drying rock (1·2m).

▶*A W-going eddy occurs E of Penrhyn Mawr during the SE-going stream and eddies form in Abraham's Bosom and Gogarth Bay.*◀

E of Llanddwyn Is, Pilot's Cove is a good ⚓ to await suitable conditions.

On the Lleyn Peninsula Porth Dinllaen is a good ⚓, but exposed to NW through N to SE. Braich y Pwll is the steep, rky point at end of Lleyn Peninsula (AC 1971). About 1M N of it and up to 1M offshore lie The Tripods, a bank on which there are overfalls and steep seas with wind against tide.

▶*Bardsey Sound, 1·5M wide, can be used by day in moderate winds. **Stream reaches 8kn at Sp, and passage should be made at slack water,** which occurs about 2¼ hrs after local high and low water: HW Dover –0100 (HW Holyhead –0010) going E, or westbound LW Dover –0100 (LW Holyhead +0030). An eddy ('Young flood') flows W from N of Carreg Ddu 2hrs before the tide turns allowing approach to the Snd at HW Dover +0500.*◀

Avoid Carreg Ddu on N side and Maen Bugail Rock (dries 4·1m) on S side of Sound, where there are dangerous races. If passing outside Bardsey Is make a good offing to avoid overfalls which extend 1·5M W and 2·5M S of the island. Turbulence occurs over Bastram Shoal, Devil's Tail and Devil's Ridge, which lie SSE and E of Bardsey Island respectively.

FLOOD Bardsey Sound

The tide turns to the NW or NE (flood) as follows:
(a): Dover +0300;
(b): Dover +0500;
(c): Dover +0600;
These times are approximate.

EBB Bardsey Sound

The tide turns to the SW or SE (ebb) as follows:
(a): Dover –0300;
(b): Dover –0100;
(c): Dover +0030;
These times are approximate.

NOTE: There is a strong eddy down tide off Bardsey Island and overfalls throughout the area.

FLOOD (T) : turning → : < 2k ➡ : 2-4k ⫸ : 4k +

EBB (T) : turning → : < 2k ➡ : 2-4k ⫸ : 4k +

LOCAL LW: Caernarfon HW Dover –0555. Port Dinorwic –0620. Menai –0540. Beaumaris –0605.

HW DOVER –0500

HW DOVER –0400

HW DOVER –0300

HW DOVER –0200

SLACK WATER IN THE SWELLIES: HW Dover –0200 to –0230.

HW DOVER –0100

LOCAL HW: Belan: HW Dover –0115. Caernarfon: –0105. Port Dinorwic: –0050.

LOCAL HW: Menai –0005. Beaumaris –0010.

HW DOVER +0100

HW DOVER +0200

HW DOVER +0300

HW DOVER +0400

HW DOVER +0500

LOCAL LW: Belan: HW Dover +0520.

9.10.19   MENAI STRAIT
Gwynedd/Isle of Anglesey ❀⚓⚓⚓✿✿

CHARTS AC 5609, 1464, Imray C52. NOTE: The definitive Pilot book is *Cruising Anglesey and Adjoining Waters* by Ralph Morris.

TIDES –0035 Dover (Sp), 0020 (Np)
Standard Port HOLYHEAD (⟶)

Times				Height (metres)			
High Water		Low Water		MHWS	MHWN	MLWN	MLWS
0000	0600	0500	1100	5·6	4·4	2·0	0·7
1200	1800	1700	2300				
Differences BEAUMARIS							
+0025	+0010	+0055	+0035	+2·0	+1·6	+0·5	+0·1
MENAI BRIDGE							
+0030	+0010	+0100	+0035	+1·7	+1·4	+0·3	0·0
PORT DINORWIC							
–0015	–0025	+0030		0·0	0·0	0·0	+0·1
CAERNARFON							
–0030	–0030	+0015	–0005	–0·4	–0·4	–0·1	–0·1
FORT BELAN							
–0040	–0015	–0025	–0005	–1·0	–0·9	–0·2	–0·1
LLANDDWYN ISLAND							
–0115	–0055	–0030	–0020	–0·7	–0·5	–0·1	0·0

NORTH EAST ENTRANCE

SHELTER Reasonable ⚓ N of Trwyn Du in Southerly winds or in outer road in settled conditions to await a favourable tide.

NAVIGATION WPT 53°19'·48N 04°03'·28W, 137°/1M to Perch Rk.
- In N'ly gales seas break on Ten Feet Bank and there are high, steep breaking seas in Puffin Sound.
- In N Strait keep to buoyed chan, nearer Anglesey taking care to avoid the SHM marking a wreck W of B1 SHM.

LIGHTS AND MARKS See 9.10.4 and chartlet. At NE end of Strait, Trwyn-Du Lt ho, Conspic tr on Puffin Is and Perch Rk PHM bn. The channel is laterally buoyed (direction SW) as far as Caernarfon.

BEAUMARIS   53°15'·66N 04°05'·38W

TIDES Beaumaris –0025 Dover; ML Beaumaris 4·2; Duration 0540

SHELTER Reasonable off Beaumaris except from NE winds. ⚓ S of B10 PHM buoy.

LIGHTS AND MARKS Beaumaris pier has light 2Fl (3) G 10s 6M.

COMMUNICATIONS (Code 01248); Dr (emergency) 384001; Police 101; CGOC (01407) 762051.

FACILITIES Royal Anglesey YC www.royalangleseyyc.org.uk
☎810295, ⚓ ⚓ ✕ ⬜.
North West Venturers' YC, at Gallows Pt www.nwyc.org.uk ⚓.
Power Marine ⚓ ✕ ⚒ Ⓗ Ⓔ ▣(20t) ⬚(2t).
Town ⓅⒾ⬚✉Ⓑ ⇌ (bus to Bangor), ✈ (Anglesey/Liverpool).

BANGOR   53°14'·46N 04°07'·58W

SHELTER Good, but open to winds from N to NE, at Dickies BY or Port Penrhyn dock. Dock and approaches dry approx 3m.

COMMUNICATIONS (Code 01248) Penrhyn HM ☎352525.

FACILITIES Dickies, Port Penrhyn, www.dickies.co.uk ☎363400, 5⬚ £15/night, ⚓ Ⓓ ✕ ⚒ Ⓗ Ⓔ ⚓ ⊼ ▣(30t) ⬚ Ⓘ.
Town (1M) ⓅⒾ ⬚ ✉ Ⓑ ⇌ ✈ (Anglesey/Liverpool).

MENAI BRIDGE   53°13'·20N 04°09'·79W

COMMUNICATIONS (Code 01248) Piermaster Menai Bridge 712312, mobile 07990 531595. At Menai Bridge, all moorings are private. Pier is owned and managed by Bangor University.

FACILITIES St George's Pier (Fl G 10s)⚓ at all tides.

The Marine Club, Porth Daniel, ☎717484, ⚓ Ⓓ ▲ ⚓ ⚓ Ⓘ ⊼

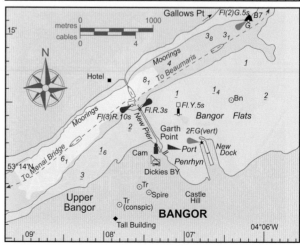

THE SWELLIES

NAVIGATION

- The passage should only be attempted at slack HW, which is –0145 HW Dover.
- The shallow rky narrows between the bridges may be dangerous for yachts at other times, when the stream can reach 8kn.
- At slack HW Sp there is 3m over The Platters and the outcrop off Price Pt, which can be ignored.
- For shoal-draught boats passage is also possible at slack LW Nps, when there are depths of 0·5m close E of Britannia Bridge.
- The bridges and power cables have a least clearance of 21m at MHWS. Night passage is not recommended.

LIGHTS AND MARKS See 9.10.4. and chartlet.
Britannia Bridge, E side ldg lts 231°, both FW. Br lts, both sides: Centre span Iso 5s 27m 3M; S end, FR 21m 3M; N end, FG 21m 3M.

PORT DINORWIC/Y Felinheli 53°11'·20N 04°13'·70W

LIGHTS AND MARKS Pier Hd F WR 5m 2M, vis R225°- W357°-225°.

COMMUNICATIONS(Code 01248); Dr 670423. Marina 671500.
Dinorwic Marina VHF Ch, Plas Menai Centre *Menai Base* Ch **80** M.

FACILITIES Port Dinorwic Marina ⌖ portdinorwic.co.uk ☎ 671500; **Tidal basin** dries at Sp; lock (width 9·7m) opens HW±2. ⚓ 180⌒ ⛽⚓🅿️🅰️⚓⚒⚓🅱️🅴⚓🅿️(15t).

Dinas Boatyard ☎671642, ⚓⚒⚓🅱️⚓ all 🏠 facilities 🅰️.

Town 🅴 ✉ (Bangor or Caernarfon), 🅱️ ⇌ (Bangor), ✈ (Liverpool).

Plas Menai (between Port Dinorwic and Caernarfon), Sailing and Sports Centre ☎670964, 🅰️, day moorings only.

CAERNARFON 53°08'·51N 04°16'·83W ⚓🌸🅰️🅰️🅰️🌸🌸🌸

SHELTER Good in Victoria Dock Marina, access HW±2 via lifting cill, traffic lts (HO); pontoons at SW end in 2m, 🅥 to starboard of ent. In river hbr (S of conspic castle) berth against wall, dries to mud/gravel, access HW±3 via swing bridge; for opening sound B (—···). ⚓ off Foel Ferry only with local knowledge; temp ⚓ off Abermenai Pt, sheltered from W, but strong streams. 🅰 waiting 1½ca SW of C9.

COMMUNICATIONS (Code 01286) Police 101; Ⓗ 01248 384384; Dr (emergency) 01248 384001. HM 672118, Mobile 07786 730865.
Victoria Dock Marina Ch **80**. *Caernarfon Hbr* Ch **14** 16 (HJ).

FACILITIES River £9·60<20'<£13·20<30'<£18 /craft, ⚓⚓🅰️⚓.
Victoria Dock ☎672346, 46 ⌒ £2·35, ⚓🅰️⚓ ⚓.
Caernarfon SC ☎(01248) 672861,⚓ 🅰️.
Royal Welsh YC ☎(01248) 672599, 🅰️.

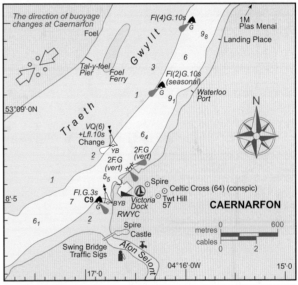

SOUTH WEST ENTRANCE – CAERNARFON BAR

NAVIGATION A dangerous sea can build in even a moderate breeze against tide, especially if a swell is running. Bar Chan should not be used at any time other than local HW-3 to HW+3. Caernarfon Bar shifts often and unpredictably see website at www.caernarfon-hbr.demon.co.uk/buoy_e.htm. **Direction of buoyage changes to NE at Caernarfon.**

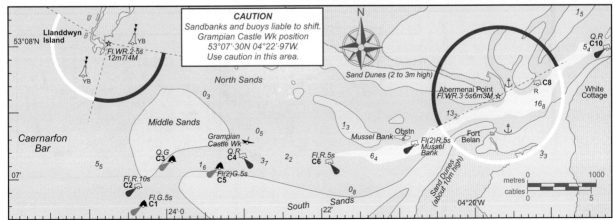

9.10.20 HOLYHEAD
Isle of Anglesey 53°19'·72N 04°37'·07W ❀❀❀❀♨♨❀❀❀

HOLYHEAD

CHARTS AC 5609, 1826, 1970, 1977, 1413, 2011; Imray C61, C52

TIDES –0050 Dover; ML 3·2; Duration 0615

Standard Port HOLYHEAD (→)

Times				Height (metres)			
High Water		Low Water		MHWS	MHWN	MLWN	MLWS
0000	0600	0500	1100	5·6	4·4	2·0	0·7
1200	1800	1700	2300				
Differences TRWYN DINMOR (W of Puffin Is)							
+0025	+0015	+0050	+0035	+1·9	+1·5	+0·5	+0·2
MOELFRE (NE Anglesey)							
+0025	+0020	+0050	+0035	+1·9	+1·4	+0·5	+0·2
AMLWCH (N Anglesey)							
+0020	+0010	+0035	+0025	+1·6	+1·3	+0·5	+0·2
CEMAES BAY (N Anglesey)							
+0020	+0025	+0040	+0035	+1·0	+0·7	+0·3	+0·1
TREARDDUR BAY (W Anglesey)							
–0045	–0025	–0015	–0015	–0·4	–0·4	0·0	+0·1
PORTH TRECASTELL (SW Anglesey)							
–0045	–0025	–0005	–0015	–0·6	–0·6	0·0	0·0
TREFOR (Lleyn peninsula)							
–0115	–0100	–0030	–0020	–0·8	–0·9	–0·2	–0·1
PORTH DINLLAEN (Lleyn peninsula)							
–0120	–0105	–0035	–0025	–1·0	–1·0	–0·2	–0·2
PORTH YSGADEN (Lleyn peninsula)							
–0125	–0110	–0040	–0035	–1·1	–1·0	–0·1	–0·1
BARDSEY ISLAND							
–0220	–0240	–0145	–0140	–1·2	–1·2	–0·5	–0·1

SHELTER Accessible at all times. Good in marina, least depth 2m, and on YC moorings which are uncomfortable in strong NE-E winds. Temporary visitors' berths on E side of marina floating breakwater are exposed to the east. ↓ not recommended – foul ground. Strong NW-NE winds raise an uncomfortable sea.

NAVIGATION WPT 53°20′·12N 04°37′·27W, 165°/0·28M to Bkwtr Lt ho. Small craft should leave Spit SHM buoy to starboard and maintain a safe distance from bkwtr to allow for obstns before heading for marina/moorings. Yachts should keep well clear of commercial shipping and the Outer Platters. Beware numerous fishing markers close to the breakwater. No ↓ in the fairways.

Frequent HSS and Ro-Ro ferries to and from Ireland use inner hbr south of aluminium jetty and comply with a mini-TSS 100-500m off bkwtr hd.

LIGHTS AND MARKS Ldg marks 165°: bkwtr lt ho on with chy (127m, grey + B top; R lts); chy and Holyhead Mountain (218m) are conspic from afar. Cranes on aluminium jetty conspic closer to.

COMMUNICATIONS (Code 01407) CGOC 762051; Police 101; Dr via CGOC. Marina 764242; Port Control 763071.

Holyhead Marina and *Holyhead SC*: Ch **M**. Monitor *Holyhead* Ch 14 16 (H24) for ferry traffic. Broadcast of local nav info/warnings 1200 (UT) daily on Ch 14.

FACILITIES Marina www.holyheadmarina.co.uk, ☎764242; 500⌓ + ✔, £2.60, short stay £7/craft, ⊾ ⊙ ⌁ (0800-1800) ⌁ ⚒ ⊞ Ⓔ ⌇ ☗(15t) ⌾ ✕ ⌸ wi-fi.

Inner Hbr ☎762304, used by Stena HSS; not advised for yachts.

Holyhead SC (HSC) ☎762526; M £8.00, ⊾ ⌁ ⏚ launch (call Ch M: 0900-2100, Fri/Sat to 2330) ✕ ⌸.

Town ⌚ ⊠ Ⓑ ⎈ ⊙ ✕ ⌸ ⇌ ✈ (Anglesey/Liverpool). Ferry/HSS to Dun Laoghaire and Dublin.

MINOR HARBOURS (HW –0240 on Dover, see table (←)).
May be useful, in settled conditions only, on passage N, see Pl.

Trearddur Bay (3M south) 53°16′·66N 04°37′·50W. AC 1970, 5609,Imray C52. ⚓ (private, small craft only) or poss ↓; open to prevailing SW'lies.

PORTH DINLLAEN, Gwynedd, 52°56′·68N 04°33′·66W. AC 1971, 1512. Shelter good in S to W winds but strong NNW to NNE winds cause heavy seas in the bay. Beware Carreg-y-Chad (1·8m) 0·75M SW of the point, and Carreg-y-Chwislen (dries, with unlit IDM Bn) 2ca ENE of the point. From N, Garn Fadryn (369m) brg 182° leads into the bay. Best ↓ 1ca S of LB ho in approx 2m.
HM ☎(01758) 720276. Facilities: ⎈ ⌸ by landing stage.

Morfa Nefyn (1M) ⌚ ⎈ ✕ ⌸.

ABERDARON BAY, Gwynedd, 52°47′·75N 04°42′·82W. AC 5609, 1410, 1411, 1971, 5609. In SE appr to bay are 2 islets Ynys Gwylan-fawr and Ynys Gwylan-bâch with a deep water passage between them, also passage between Ynys Gwylan-fawr and Trwyn Gwningaer (shore). ↓ 4ca SSW of church in 5-10m, limited protection and subject to swell at times but sheltered in winds from W through N to NE. **Do not leave craft unattended due to poor holding ground**. Dinghy landing with care on shingle beach at Porth Simdddy or Meudwy, W side of bay.

Village limited facilities: ⏚ ⊠ ✕ ⌸ ⇌ (Pwllheli).

BARDSEY ISLAND – YNYS ENNLI, Gwynedd, 52°45′·00N 04°47′·40W. AC 5609, 1410, 1411, 1971. Island of historical interest with ruined abbey and religious settlement since 6th century (now a managed farm and nature reserve, see 9.0.3). ↓ W of Pen Cristin in appr to Henllwyn Cove in 4m is sheltered in winds from WSW through N to NE but is subject to swell at times. Small craft may ↓ in the cove with caution, but beware isolated rocks and shoals. Landing at jetty in rocks marked by small boathouse at top of slip.

N Wales

STANDARD TIME (UT)
For Summer Time add ONE hour in **non-shaded areas**

HOLYHEAD LAT 53°19′N LONG 4°37′W
TIMES AND HEIGHTS OF HIGH AND LOW WATERS

Dates in red are **SPRINGS**
Dates in blue are **NEAPS**

YEAR 2016

JANUARY

#	Time	m	#	Time	m
1 F	0235 / 0835 / 1450 / 2117	4.8 / 1.9 / 4.9 / 1.8	**16** SA	0230 / 0836 / 1452 / 2112	5.2 / 1.3 / 5.4 / 1.2
2 SA	0327 / 0927 / 1544 / 2215	4.6 / 2.2 / 4.7 / 2.0	**17** SU	0330 / 0938 / 1555 / 2218	4.9 / 1.6 / 5.2 / 1.4
3 SU	0430 / 1031 / 1650 / 2321	4.4 / 2.4 / 4.6 / 2.1	**18** M	0442 / 1051 / 1709 / 2331	4.8 / 1.7 / 5.0 / 1.6
4 M	0540 / 1143 / 1802	4.4 / 2.4 / 4.6	**19** TU	0600 / 1208 / 1827	4.8 / 1.7 / 5.0
5 TU	0026 / 0647 / 1250 / 1908	2.1 / 4.6 / 2.2 / 4.7	**20** W	0043 / 0712 / 1319 / 1937	1.5 / 4.9 / 1.6 / 5.1
6 W	0123 / 0743 / 1346 / 2003	1.9 / 4.8 / 2.0 / 4.8	**21** TH	0149 / 0813 / 1422 / 2037	1.4 / 5.2 / 1.3 / 5.4
7 TH	0211 / 0828 / 1433 / 2048	1.7 / 5.0 / 1.7 / 5.0	**22** F	0245 / 0903 / 1515 / 2127	1.2 / 5.4 / 1.1 / 5.4
8 F	0253 / 0908 / 1514 / 2128	1.5 / 5.3 / 1.4 / 5.4	**23** SA	0332 / 0948 / 1601 / 2211	1.1 / 5.6 / 0.9 / 5.5
9 SA	0332 / 0946 / 1554 / 2207	1.2 / 5.5 / 1.1 / 5.4	**24** SU	0414 / 1029 / 1643 / 2252	1.0 / 5.7 / 0.8 / 5.5
10 SU	0410 / 1023 / 1633 / 2245	1.0 / 5.7 / 0.9 / 5.5	**25** M	0452 / 1107 / 1721 / 2329	0.9 / 5.8 / 0.8 / 5.5
11 M	0449 / 1102 / 1713 / 2325	0.9 / 5.8 / 0.8 / 5.6	**26** TU	0529 / 1144 / 1758	0.9 / 5.7 / 0.8
12 TU	0529 / 1142 / 1754	0.8 / 5.9 / 0.7	**27** W	0005 / 0603 / 1220 / 1833	5.4 / 1.0 / 5.6 / 1.0
13 W	0007 / 0610 / 1225 / 1837	5.6 / 0.8 / 5.9 / 0.7	**28** TH	0040 / 0638 / 1254 / 1908	5.3 / 1.2 / 5.5 / 1.1
14 TH	0052 / 0654 / 1310 / 1924	5.5 / 0.9 / 5.8 / 0.8	**29** F	0115 / 0714 / 1328 / 1945	5.1 / 1.4 / 5.3 / 1.4
15 F	0139 / 0742 / 1358 / 2014	5.4 / 1.1 / 5.6 / 1.0	**30** SA	0151 / 0752 / 1406 / 2025	4.9 / 1.6 / 5.1 / 1.6
			31 SU	0232 / 0836 / 1449 / 2112	4.7 / 1.9 / 4.8 / 1.9

FEBRUARY

#	Time	m	#	Time	m
1 M	0322 / 0928 / 1543 / 2210	4.5 / 2.1 / 4.6 / 2.1	**16** TU	0411 / 1027 / 1647 / 2307	4.8 / 1.6 / 4.8 / 1.7
2 TU	0427 / 1035 / 1657 / 2322	4.4 / 2.3 / 4.4 / 2.2	**17** W	0535 / 1150 / 1814	4.6 / 1.7 / 4.7
3 W	0546 / 1154 / 1819	4.4 / 2.3 / 4.4	**18** TH	0027 / 0657 / 1309 / 1933	1.8 / 4.7 / 1.6 / 4.8
4 TH	0036 / 0659 / 1306 / 1930	2.1 / 4.5 / 2.1 / 4.6	**19** F	0140 / 0803 / 1416 / 2034	1.6 / 5.0 / 1.4 / 5.0
5 F	0138 / 0757 / 1404 / 2024	1.9 / 4.8 / 1.8 / 4.8	**20** SA	0237 / 0855 / 1508 / 2120	1.4 / 5.2 / 1.1 / 5.2
6 SA	0228 / 0844 / 1452 / 2109	1.6 / 5.1 / 1.4 / 5.1	**21** SU	0322 / 0937 / 1550 / 2159	1.2 / 5.4 / 0.9 / 5.3
7 SU	0311 / 0924 / 1534 / 2149	1.2 / 5.4 / 1.0 / 5.4	**22** M	0401 / 1014 / 1627 / 2234	1.0 / 5.6 / 0.8 / 5.4
8 M	0351 / 1004 / 1614 / 2228	0.9 / 5.7 / 0.7 / 5.6	**23** TU	0435 / 1048 / 1701 / 2307	0.9 / 5.6 / 0.7 / 5.4
9 TU	0431 / 1043 / 1655 / 2308	0.7 / 5.9 / 0.5 / 5.7	**24** W	0507 / 1121 / 1732 / 2339	0.9 / 5.6 / 0.8 / 5.4
10 W	0511 / 1124 / 1736 / 2350	0.5 / 6.1 / 0.3 / 5.8	**25** TH	0539 / 1153 / 1804	0.9 / 5.6 / 0.9
11 TH	0553 / 1207 / 1819	0.5 / 6.1 / 0.3	**26** F	0010 / 0610 / 1225 / 1835	5.4 / 1.0 / 5.5 / 1.0
12 F	0032 / 0636 / 1252 / 1903	5.7 / 0.5 / 6.0 / 0.5	**27** SA	0042 / 0643 / 1256 / 1908	5.3 / 1.1 / 5.3 / 1.2
13 SA	0118 / 0722 / 1338 / 1952	5.5 / 0.7 / 5.8 / 0.8	**28** SU	0115 / 0717 / 1330 / 1943	5.1 / 1.3 / 5.1 / 1.4
14 SU	0206 / 0813 / 1430 / 2046	5.3 / 1.0 / 5.5 / 1.1	**29** M	0151 / 0756 / 1408 / 2024	4.9 / 1.6 / 4.9 / 1.7
15 M	0302 / 0914 / 1530 / 2150	5.0 / 1.4 / 5.1 / 1.5			

MARCH

#	Time	m	#	Time	m
1 TU	0233 / 0842 / 1454 / 2114	4.7 / 1.9 / 4.6 / 2.0	**16** W	0345 / 1009 / 1631 / 2245	4.8 / 1.6 / 4.7 / 1.9
2 W	0327 / 0942 / 1558 / 2222	4.5 / 2.1 / 4.3 / 2.2	**17** TH	0510 / 1133 / 1803	4.6 / 1.7 / 4.5
3 TH	0444 / 1100 / 1730 / 2346	4.3 / 2.2 / 4.3 / 2.2	**18** F	0009 / 0638 / 1255 / 1923	1.9 / 4.6 / 1.6 / 4.6
4 F	0611 / 1224 / 1855	4.4 / 2.1 / 4.4	**19** SA	0124 / 0747 / 1400 / 2021	1.8 / 4.8 / 1.4 / 4.8
5 SA	0102 / 0721 / 1332 / 1957	2.0 / 4.7 / 1.7 / 4.7	**20** SU	0221 / 0838 / 1450 / 2105	1.5 / 5.1 / 1.2 / 5.0
6 SU	0200 / 0815 / 1425 / 2045	1.6 / 5.0 / 1.3 / 5.1	**21** M	0305 / 0918 / 1530 / 2140	1.3 / 5.3 / 1.0 / 5.2
7 M	0247 / 0859 / 1509 / 2127	1.2 / 5.4 / 0.8 / 5.4	**22** TU	0341 / 0952 / 1604 / 2212	1.1 / 5.4 / 0.8 / 5.3
8 TU	0329 / 0940 / 1551 / 2207	0.8 / 5.7 / 0.5 / 5.7	**23** W	0413 / 1024 / 1635 / 2242	0.9 / 5.5 / 0.8 / 5.4
9 W	0409 / 1021 / 1632 / 2247	0.5 / 6.0 / 0.2 / 5.8	**24** TH	0443 / 1055 / 1705 / 2311	0.9 / 5.5 / 0.8 / 5.4
10 TH	0450 / 1103 / 1714 / 2329	0.3 / 6.2 / 0.1 / 5.9	**25** F	0513 / 1126 / 1734 / 2341	0.8 / 5.5 / 0.8 / 5.4
11 F	0533 / 1147 / 1757	0.2 / 6.2 / 0.1	**26** SA	0544 / 1157 / 1804	0.8 / 5.4 / 1.0
12 SA	0012 / 0616 / 1232 / 1842	5.9 / 0.3 / 6.1 / 0.3	**27** SU	0012 / 0615 / 1229 / 1836	5.3 / 1.0 / 5.3 / 1.1
13 SU	0057 / 0703 / 1320 / 1930	5.7 / 0.5 / 5.8 / 0.7	**28** M	0045 / 0649 / 1302 / 1909	5.2 / 1.2 / 5.1 / 1.3
14 M	0144 / 0755 / 1411 / 2023	5.4 / 0.8 / 5.4 / 1.1	**29** TU	0120 / 0726 / 1339 / 1948	5.0 / 1.4 / 4.9 / 1.6
15 TU	0238 / 0855 / 1512 / 2126	5.1 / 1.2 / 5.0 / 1.5	**30** W	0159 / 0810 / 1422 / 2035	4.8 / 1.6 / 4.6 / 1.9
			31 TH	0249 / 0906 / 1521 / 2139	4.6 / 1.9 / 4.4 / 2.1

APRIL

#	Time	m	#	Time	m
1 F	0356 / 1019 / 1647 / 2302	4.4 / 2.0 / 4.3 / 2.2	**16** SA	0607 / 1227 / 1858	4.6 / 1.6 / 4.5
2 SA	0524 / 1144 / 1818	4.4 / 1.9 / 4.4	**17** SU	0055 / 0717 / 1332 / 1956	1.9 / 4.7 / 1.5 / 4.7
3 SU	0023 / 0641 / 1256 / 1926	2.0 / 4.6 / 1.6 / 4.7	**18** M	0154 / 0809 / 1422 / 2039	1.7 / 4.9 / 1.3 / 4.9
4 M	0127 / 0740 / 1353 / 2017	1.6 / 5.0 / 1.1 / 5.1	**19** TU	0238 / 0850 / 1502 / 2114	1.5 / 5.1 / 1.1 / 5.1
5 TU	0218 / 0829 / 1441 / 2101	1.2 / 5.4 / 0.7 / 5.4	**20** W	0315 / 0925 / 1536 / 2145	1.3 / 5.2 / 1.0 / 5.2
6 W	0303 / 0914 / 1525 / 2143	0.7 / 5.7 / 0.4 / 5.7	**21** TH	0347 / 0957 / 1606 / 2215	1.1 / 5.3 / 0.9 / 5.3
7 TH	0346 / 0957 / 1608 / 2224	0.4 / 6.0 / 0.1 / 5.9	**22** F	0417 / 1029 / 1636 / 2244	1.0 / 5.4 / 0.9 / 5.4
8 F	0429 / 1041 / 1652 / 2307	0.2 / 6.2 / 0.0 / 6.0	**23** SA	0448 / 1100 / 1706 / 2315	0.9 / 5.4 / 0.9 / 5.4
9 SA	0513 / 1127 / 1736 / 2351	0.1 / 6.2 / 0.1 / 5.9	**24** SU	0519 / 1132 / 1737 / 2347	0.9 / 5.3 / 1.0 / 5.4
10 SU	0559 / 1214 / 1822	0.2 / 6.0 / 0.4	**25** M	0552 / 1205 / 1809	1.0 / 5.2 / 1.1
11 M	0037 / 0647 / 1304 / 1910	5.7 / 0.4 / 5.7 / 0.7	**26** TU	0020 / 0627 / 1240 / 1844	5.4 / 1.1 / 5.1 / 1.3
12 TU	0126 / 0740 / 1357 / 2003	5.5 / 0.7 / 5.3 / 1.2	**27** W	0057 / 0705 / 1318 / 1923	5.2 / 1.3 / 4.9 / 1.5
13 W	0219 / 0840 / 1458 / 2105	5.1 / 1.1 / 4.9 / 1.6	**28** TH	0137 / 0749 / 1403 / 2010	5.0 / 1.5 / 4.7 / 1.7
14 TH	0323 / 0951 / 1614 / 2219	4.8 / 1.5 / 4.6 / 1.9	**29** F	0225 / 0843 / 1459 / 2110	4.8 / 1.6 / 4.5 / 1.9
15 F	0442 / 1110 / 1740 / 2341	4.6 / 1.6 / 4.5 / 2.0	**30** SA	0327 / 0950 / 1615 / 2226	4.6 / 1.7 / 4.4 / 2.0

Chart Datum: 3·05 metres below Ordnance Datum (Newlyn). HAT is 6·3 metres above Chart Datum.

STANDARD TIME (UT)
For Summer Time add ONE hour in **non-shaded areas**

HOLYHEAD LAT 53°19'N LONG 4°37'W
TIMES AND HEIGHTS OF HIGH AND LOW WATERS

Dates in red are SPRINGS
Dates in blue are NEAPS

YEAR 2016

MAY

Time	m	Time	m
1 0444	4.6	**16** 0012	2.0
1108	1.7	0632	4.6
SU 1741	4.5	M 1251	1.6
2345	1.9	1916	4.6
2 0601	4.8	**17** 0113	1.9
1219	1.4	0729	4.7
M 1851	4.8	TU 1344	1.5
		2003	4.8
3 0052	1.6	**18** 0203	1.7
0704	5.0	0815	4.9
TU 1320	1.1	W 1427	1.3
1946	5.1	2041	5.0
4 0147	1.2	**19** 0243	1.5
0758	5.4	0854	5.0
W 1412	0.7	TH 1504	1.2
2034	5.4	2115	5.1
5 0236	0.8	**20** 0319	1.3
0847	5.7	0929	5.1
TH 1459	0.4	F 1537	1.1
2118	5.7	2147	5.2
6 0323	0.5	**21** 0351	1.2
0934	5.9	1003	5.2
F 1545	0.2	SA 1609	1.0
● 2202	5.8	○ 2219	5.3
7 0409	0.3	**22** 0424	1.1
1022	6.0	1036	5.2
SA 1631	0.2	SU 1640	1.0
2247	5.9	2251	5.4
8 0456	0.2	**23** 0458	1.0
1110	6.0	1110	5.2
SU 1717	0.3	M 1713	1.1
2333	5.9	2325	5.4
9 0544	0.3	**24** 0533	1.0
1159	5.8	1145	5.2
M 1804	0.5	TU 1748	1.1
10 0020	5.8	**25** 0001	5.4
0634	0.5	0609	1.1
TU 1250	5.6	W 1223	5.1
1853	0.8	1825	1.2
11 0109	5.5	**26** 0039	5.3
0726	0.7	0649	1.2
W 1342	5.2	TH 1303	5.0
1944	1.4	1906	1.4
12 0201	5.2	**27** 0121	5.2
0823	1.1	0734	1.3
TH 1440	4.9	F 1349	4.8
2041	1.6	1953	1.6
13 0300	5.0	**28** 0209	5.0
0927	1.4	0826	1.4
F 1547	4.6	SA 1443	4.7
◐ 2147	1.9	2049	1.7
14 0407	4.7	**29** 0305	4.9
1037	1.6	0927	1.5
SA 1701	4.5	SU 1549	4.6
2301	2.0	◑ 2157	1.8
15 0521	4.6	**30** 0412	4.9
1147	1.7	1036	1.6
SU 1815	4.5	M 1704	4.6
		2310	1.7
		31 0524	4.9
		1145	1.3
		TU 1815	4.8

JUNE

Time	m	Time	m
1 0018	1.5	**16** 0119	1.9
0631	5.1	0733	4.7
W 1249	1.1	TH 1347	1.6
1916	5.0	2004	4.8
2 0118	1.2	**17** 0207	1.7
0731	5.3	0821	4.8
TH 1346	0.8	F 1430	1.5
2009	5.3	2045	5.0
3 0213	0.9	**18** 0249	1.5
0825	5.5	0902	4.9
F 1437	0.6	SA 1508	1.3
2058	5.5	2121	5.1
4 0304	0.7	**19** 0327	1.3
0917	5.7	0939	5.1
SA 1527	0.5	SU 1543	1.2
2145	5.7	2155	5.3
5 0353	0.5	**20** 0403	1.2
1007	5.8	1015	5.2
SU 1614	0.5	M 1618	1.1
● 2231	5.8	○ 2230	5.4
6 0443	0.4	**21** 0439	1.0
1056	5.8	1050	5.2
M 1702	0.5	TU 1653	1.1
2318	5.8	2306	5.5
7 0532	0.4	**22** 0516	1.0
1146	5.7	1128	5.2
TU 1749	0.7	W 1730	1.0
		2344	5.5
8 0005	5.7	**23** 0555	0.9
0620	0.5	1207	5.2
W 1235	5.5	TH 1809	1.1
1835	0.9		
9 0052	5.6	**24** 0024	5.5
0710	0.8	0635	0.9
TH 1324	5.2	F 1249	5.1
1923	1.2	1851	1.2
10 0140	5.3	**25** 0106	5.4
0801	1.0	0719	1.0
F 1415	4.9	SA 1334	5.0
2013	1.5	1937	1.3
11 0230	5.1	**26** 0153	5.3
0855	1.3	0808	1.1
SA 1509	4.7	SU 1424	4.9
2107	1.8	2029	1.4
12 0325	4.8	**27** 0244	5.2
0953	1.6	0903	1.2
SU 1610	4.5	M 1522	4.8
◐ 2209	2.0	◑ 2129	1.6
13 0426	4.6	**28** 0344	5.1
1056	1.7	1006	1.3
M 1715	4.4	TU 1631	4.8
2317	2.1	2238	1.6
14 0532	4.5	**29** 0452	5.0
1159	1.8	1115	1.3
TU 1821	4.5	W 1743	4.8
		2348	1.5
15 0021	2.1	**30** 0603	5.1
0637	4.6	1222	1.2
W 1257	1.7	TH 1850	5.0
1917	4.6		

JULY

Time	m	Time	m
1 0055	1.4	**16** 0129	2.0
0710	5.2	0747	4.6
F 1325	1.1	SA 1356	1.7
1950	5.2	2013	4.8
2 0156	1.1	**17** 0220	1.7
0811	5.3	0836	4.8
SA 1422	0.9	SU 1440	1.5
2043	5.4	2055	5.1
3 0252	0.9	**18** 0303	1.4
0906	5.5	0917	5.0
SU 1514	0.8	M 1520	1.3
2132	5.6	2133	5.3
4 0344	0.7	**19** 0342	1.2
0957	5.6	0955	5.1
M 1602	0.7	TU 1557	1.1
● 2219	5.7	○ 2209	5.4
5 0433	0.5	**20** 0420	1.0
1045	5.6	1032	5.3
TU 1648	0.7	W 1634	1.0
2304	5.8	2246	5.6
6 0520	0.5	**21** 0458	0.8
1131	5.5	1110	5.4
W 1732	0.8	TH 1712	0.9
2348	5.7	2325	5.7
7 0605	0.6	**22** 0537	0.7
1216	5.4	1150	5.4
TH 1815	0.9	F 1752	0.8
8 0031	5.6	**23** 0006	5.7
0648	0.8	0617	0.7
F 1300	5.2	SA 1232	5.4
1856	1.1	1833	0.9
9 0114	5.4	**24** 0048	5.7
0731	1.0	0700	0.7
SA 1342	5.0	SU 1315	5.3
1939	1.4	1918	1.0
10 0156	5.2	**25** 0133	5.6
0815	1.3	0747	0.9
SU 1425	4.8	M 1403	5.2
2024	1.6	2007	1.2
11 0240	5.0	**26** 0222	5.4
0903	1.5	0839	1.0
M 1513	4.6	TU 1456	5.0
2113	1.9	◑ 2104	1.4
12 0329	4.7	**27** 0319	5.2
0956	1.8	0939	1.3
TU 1609	4.4	W 1600	4.8
◐ 2212	2.1	2211	1.6
13 0429	4.5	**28** 0427	5.0
1058	1.9	1049	1.4
W 1714	4.4	TH 1715	4.8
2320	2.2	2326	1.6
14 0537	4.4	**29** 0544	4.9
1202	2.0	1202	1.5
TH 1822	4.4	F 1831	4.8
15 0029	2.1	**30** 0040	1.5
0646	4.5	0659	5.0
F 1303	1.9	SA 1312	1.4
1923	4.6	1938	5.0
		31 0149	1.3
		0806	5.1
		SU 1414	1.2
		2035	5.2

AUGUST

Time	m	Time	m
1 0248	1.0	**16** 0238	1.5
0902	5.3	0855	4.9
M 1507	1.1	TU 1456	1.4
2123	5.5	2109	5.3
2 0338	0.8	**17** 0319	1.1
0950	5.4	0934	5.2
TU 1552	0.9	W 1535	1.1
● 2207	5.7	2146	5.5
3 0423	0.7	**18** 0358	0.8
1033	5.5	1011	5.4
W 1634	0.8	TH 1613	0.8
2249	5.7	○ 2224	5.8
4 0504	0.6	**19** 0436	0.6
1113	5.5	1049	5.6
TH 1713	0.8	F 1651	0.7
2328	5.7	2303	5.9
5 0543	0.7	**20** 0515	0.4
1152	5.4	1128	5.7
F 1751	0.9	SA 1731	0.6
		2344	6.0
6 0006	5.6	**21** 0556	0.4
0620	0.8	1210	5.6
SA 1229	5.3	SU 1812	0.6
1827	1.0		
7 0043	5.5	**22** 0027	5.9
0657	1.0	0638	0.5
SU 1305	5.1	M 1253	5.5
1903	1.3	1856	0.8
8 0119	5.3	**23** 0112	5.8
0734	1.2	0724	0.7
M 1342	5.0	TU 1340	5.4
1942	1.5	1945	1.1
9 0156	5.1	**24** 0201	5.5
0814	1.5	0815	1.0
TU 1422	4.8	W 1431	5.1
2025	1.8	2042	1.3
10 0238	4.8	**25** 0258	5.2
0859	1.8	0914	1.3
W 1509	4.6	TH 1535	4.9
◐ 2115	2.0	◑ 2150	1.6
11 0329	4.6	**26** 0408	4.9
0954	2.0	1027	1.6
TH 1609	4.4	F 1654	4.7
2218	2.2	2312	1.7
12 0438	4.4	**27** 0534	4.8
1102	2.2	1148	1.8
F 1724	4.4	SA 1819	4.8
2335	2.3		
13 0558	4.3	**28** 0033	1.6
1216	2.1	0657	4.8
SA 1838	4.5	SU 1304	1.7
		1930	5.0
14 0050	2.1	**29** 0145	1.4
0713	4.5	0805	5.0
SU 1321	2.0	M 1408	1.5
1940	4.7	2027	5.2
15 0150	1.8	**30** 0242	1.1
0810	4.7	0857	5.2
M 1412	1.7	TU 1458	1.2
2028	5.0	2113	5.5
		31 0328	0.9
		0939	5.3
		W 1539	1.1
		2153	5.6

Chart Datum: 3·05 metres below Ordnance Datum (Newlyn). HAT is 6·3 metres above Chart Datum.

N Wales

》》 FREE monthly updates. Register at 《
www.reedsnauticalalmanac.co.uk

523

STANDARD TIME (UT)
For Summer Time add ONE hour in **non-shaded areas**

HOLYHEAD LAT 53°19'N LONG 4°37'W
TIMES AND HEIGHTS OF HIGH AND LOW WATERS

Dates in red are **SPRINGS**
Dates in blue are **NEAPS**

YEAR 2016

SEPTEMBER

Time	m		Time	m
1 0407	0.8	**16** 0331	0.7	
1016	5.4	0947	5.6	
TH 1616	0.9	F 1548	0.7	
● 2229	5.7	○ 2159	5.9	
2 0443	0.7	**17** 0410	0.4	
1051	5.5	1025	5.8	
F 1651	0.9	SA 1627	0.5	
2304	5.7	2239	6.1	
3 0517	0.7	**18** 0450	0.3	
1124	5.5	1105	5.9	
SA 1724	0.9	SU 1708	0.4	
2338	5.6	2321	6.2	
4 0550	0.8	**19** 0532	0.3	
1157	5.4	1147	5.9	
SU 1757	1.0	M 1751	0.4	
5 0011	5.5	**20** 0005	6.1	
0622	1.0	0615	0.4	
M 1230	5.3	TU 1231	5.8	
1830	1.2	1836	0.6	
6 0044	5.3	**21** 0052	5.9	
0655	1.2	0702	0.7	
TU 1303	5.1	W 1318	5.5	
1905	1.4	1926	0.9	
7 0118	5.1	**22** 0143	5.6	
0731	1.5	0753	1.1	
W 1339	5.0	TH 1410	5.3	
1944	1.7	2025	1.3	
8 0156	4.9	**23** 0242	5.2	
0811	1.8	0854	1.5	
TH 1421	4.7	F 1514	5.0	
2030	2.0	◑ 2136	1.6	
9 0241	4.6	**24** 0357	4.8	
0900	2.0	1010	1.9	
F 1514	4.5	SA 1636	4.8	
◑ 2128	2.2	2300	1.7	
10 0343	4.4	**25** 0528	4.7	
1005	2.3	1134	2.0	
SA 1627	4.4	SU 1804	4.8	
2244	2.3			
11 0512	4.3	**26** 0023	1.7	
1126	2.3	0652	4.8	
SU 1752	4.5	M 1252	1.9	
		1917	5.0	
12 0008	2.2	**27** 0133	1.5	
0638	4.4	0756	5.0	
M 1244	2.1	TU 1355	1.6	
1903	4.7	2013	5.2	
13 0116	1.9	**28** 0227	1.2	
0742	4.7	0844	5.2	
TU 1342	1.8	W 1442	1.4	
1957	5.0	2056	5.4	
14 0208	1.5	**29** 0310	1.0	
0829	5.0	0922	5.3	
W 1428	1.4	TH 1521	1.2	
2040	5.4	2133	5.6	
15 0252	1.1	**30** 0346	0.9	
0909	5.3	0955	5.4	
TH 1509	1.1	F 1555	1.1	
2120	5.7	2206	5.6	

OCTOBER

Time	m		Time	m
1 0419	0.9	**16** 0344	0.4	
1026	5.5	1000	5.9	
SA 1626	1.0	SU 1604	0.5	
● 2238	5.6	○ 2215	6.2	
2 0449	0.9	**17** 0426	0.2	
1056	5.5	1042	6.0	
SU 1657	1.0	M 1647	0.4	
2310	5.6	2300	6.3	
3 0519	0.9	**18** 0509	0.3	
1127	5.5	1125	6.0	
M 1728	1.1	TU 1732	0.4	
2341	5.5	2346	6.2	
4 0550	1.1	**19** 0555	0.5	
1158	5.4	1211	5.9	
TU 1801	1.2	W 1820	0.6	
5 0013	5.4	**20** 0035	5.9	
0621	1.3	0642	0.8	
W 1231	5.3	TH 1259	5.7	
1835	1.4	1912	0.9	
6 0047	5.2	**21** 0129	5.5	
0655	1.5	0735	1.2	
TH 1306	5.1	F 1353	5.4	
1912	1.6	2012	1.2	
7 0123	5.0	**22** 0229	5.1	
0733	1.8	0836	1.6	
F 1345	4.9	SA 1456	5.1	
1955	1.9	◑ 2122	1.5	
8 0206	4.7	**23** 0344	4.8	
0818	2.0	0949	2.0	
SA 1434	4.7	SU 1614	4.9	
2050	2.1	2242	1.7	
9 0303	4.4	**24** 0511	4.7	
0919	2.3	1111	2.1	
SU 1538	4.5	M 1739	4.9	
◑ 2201	2.2			
10 0426	4.3	**25** 0000	1.7	
1039	2.4	0631	4.7	
M 1704	4.5	TU 1228	2.0	
2325	2.1	1851	5.0	
11 0559	4.4	**26** 0109	1.6	
1201	2.2	0734	4.9	
TU 1821	4.7	W 1330	1.8	
		1948	5.1	
12 0037	1.8	**27** 0202	1.4	
0707	4.7	0821	5.1	
W 1306	1.9	TH 1419	1.6	
1920	5.1	2032	5.3	
13 0134	1.4	**28** 0245	1.2	
0758	5.1	0858	5.3	
TH 1357	1.5	F 1458	1.4	
2008	5.4	2108	5.4	
14 0221	1.0	**29** 0320	1.1	
0841	5.4	0930	5.4	
F 1441	1.1	SA 1531	1.3	
2051	5.8	2141	5.5	
15 0303	0.7	**30** 0352	1.0	
0921	5.7	1000	5.5	
SA 1522	0.7	SU 1602	1.2	
2133	6.0	● 2213	5.5	
		31 0422	1.0	
		1030	5.5	
		M 1633	1.1	
		2244	5.5	

NOVEMBER

Time	m		Time	m
1 0451	1.1	**16** 0451	0.4	
1100	5.6	1108	6.1	
TU 1704	1.2	W 1717	0.4	
2316	5.5	2332	6.1	
2 0522	1.2	**17** 0538	0.6	
1131	5.5	1155	6.0	
W 1737	1.2	TH 1808	0.6	
2348	5.4			
3 0554	1.3	**18** 0022	5.8	
1205	5.4	0627	0.9	
TH 1812	1.4	F 1244	5.8	
		1900	0.8	
4 0023	5.2	**19** 0116	5.5	
0627	1.5	0718	1.2	
F 1240	5.3	SA 1337	5.5	
1849	1.5	1958	1.1	
5 0100	5.0	**20** 0214	5.2	
0705	1.7	0815	1.6	
SA 1320	5.1	SU 1435	5.3	
1931	1.7	2101	1.4	
6 0143	4.8	**21** 0321	4.9	
0749	2.0	0920	1.9	
SU 1406	4.9	M 1543	5.0	
2022	1.9	◑ 2212	1.7	
7 0236	4.6	**22** 0436	4.7	
0845	2.2	1034	2.1	
M 1504	4.8	TU 1658	4.9	
◑ 2126	2.0	2324	1.8	
8 0347	4.5	**23** 0552	4.7	
0957	2.3	1147	2.2	
TU 1617	4.7	W 1810	4.9	
2242	2.0			
9 0514	4.5	**24** 0030	1.7	
1117	2.2	0657	4.8	
W 1735	4.8	TH 1253	2.1	
2355	1.8	1911	5.0	
10 0627	4.8	**25** 0127	1.6	
1226	1.9	0748	4.9	
TH 1841	5.1	F 1346	1.9	
		2000	5.1	
11 0056	1.4	**26** 0213	1.5	
0724	5.1	0829	5.1	
F 1323	1.5	SA 1430	1.7	
1935	5.4	2041	5.2	
12 0149	1.0	**27** 0251	1.4	
0812	5.4	0904	5.3	
SA 1412	1.2	SU 1506	1.5	
2023	5.7	2117	5.3	
13 0235	0.7	**28** 0325	1.3	
0855	5.7	0936	5.4	
SU 1458	0.8	M 1540	1.4	
2109	6.0	2150	5.4	
14 0320	0.5	**29** 0357	1.2	
0938	5.9	1007	5.5	
M 1543	0.5	TU 1612	1.3	
○ 2155	6.2	● 2223	5.4	
15 0405	0.4	**30** 0428	1.2	
1022	6.1	1038	5.6	
TU 1629	0.4	W 1645	1.2	
2243	6.2	2255	5.4	

DECEMBER

Time	m		Time	m
1 0500	1.2	**16** 0525	0.7	
1111	5.6	1141	6.0	
TH 1719	1.2	F 1757	0.6	
2329	5.3			
2 0533	1.3	**17** 0010	5.8	
1145	5.5	0612	0.9	
F 1754	1.3	SA 1230	5.9	
		1846	0.7	
3 0005	5.2	**18** 0101	5.5	
0608	1.4	0700	1.1	
SA 1222	5.5	SU 1318	5.7	
1832	1.4	1938	1.0	
4 0044	5.1	**19** 0152	5.2	
0646	1.6	0750	1.5	
SU 1301	5.3	M 1409	5.4	
1913	1.5	2032	1.3	
5 0126	5.0	**20** 0246	4.9	
0729	1.7	0844	1.8	
M 1345	5.2	TU 1503	5.1	
2001	1.6	2130	1.6	
6 0215	4.8	**21** 0346	4.7	
0820	1.9	0944	2.0	
TU 1437	5.1	W 1604	4.9	
2057	1.7	◑ 2234	1.8	
7 0315	4.7	**22** 0453	4.6	
0922	2.0	1052	2.2	
W 1539	5.0	TH 1712	4.7	
◑ 2202	1.7	2339	1.9	
8 0428	4.7	**23** 0602	4.5	
1034	2.0	1201	2.2	
TH 1650	5.0	F 1820	4.7	
2313	1.6			
9 0544	4.8	**24** 0040	1.9	
1145	1.9	0704	4.7	
F 1800	5.1	SA 1304	2.1	
		1921	4.8	
10 0019	1.4	**25** 0135	1.8	
0648	5.0	0754	4.9	
SA 1250	1.6	SU 1357	2.0	
1903	5.3	2011	4.9	
11 0119	1.2	**26** 0221	1.7	
0744	5.3	0836	5.1	
SU 1346	1.3	M 1441	1.7	
1959	5.6	2053	5.0	
12 0212	0.9	**27** 0300	1.5	
0834	5.6	0913	5.2	
M 1438	1.0	TU 1519	1.5	
2051	5.8	2131	5.2	
13 0302	0.7	**28** 0335	1.4	
0921	5.8	0947	5.3	
TU 1528	0.7	W 1554	1.4	
2141	6.0	2205	5.3	
14 0350	0.6	**29** 0408	1.3	
1007	6.0	1020	5.5	
W 1618	0.5	TH 1628	1.2	
○ 2231	6.0	● 2239	5.3	
15 0437	0.6	**30** 0442	1.2	
1054	6.1	1053	5.6	
TH 1707	0.5	F 1703	1.1	
2321	5.9	2314	5.3	
		31 0516	1.2	
		1129	5.6	
		SA 1739	1.1	
		2350	5.3	

Chart Datum: 3·05 metres below Ordnance Datum (Newlyn). HAT is 6·3 metres above Chart Datum.

South Wales & Bristol Channel

Abersoch to Land's End

9.11.1	**Contents**	525
9.11.2	**Area map and Distance table**	526
9.11.3	**Tidal stream charts**	528
9.11.4	**Lights, buoys and waypoints**	530
9.11.5	**Passage information**	532
	Cardigan Bay, Ramsey Sound, The Bishops, The Smalls to Milford Haven – 532 • Milford Haven to Mumbles Head, Firing Ranges – Linney Head to Burry Inlet – Mumbles Head to Cardiff – Severn Estuary – 544 • Avonmouth to Hartland Point – 556 • Hartland Point to Land's End – 559	
9.11.6	**Abersoch**	533
9.11.7	**Pwllheli**	534
	Llanbedrog • Mochras	
9.11.8	**Porthmadog**	535
9.11.9	**Barmouth**	535
9.11.10	**Aberdovey**	536
	Aberaeron • New Quay	
9.11.11	**Aberystwyth**	536
	Port Cardigan	
9.11.12	**Fishguard**	537
	Solva • St Brides Bay • Skomer Island	
9.11.13	**Milford Haven, tide tables/curves**	538
9.11.14	**Tenby**	545
	Saundersfoot • Carmarthen	
9.11.15	**Burry Port**	546
9.11.16	**Swansea**	547
	Mumbles • River Neath • Porthcawl	
9.11.17	**Barry**	548
9.11.18	**Cardiff/Penarth**	548
	Newport	
9.11.19	**Sharpness**	550
	The Severn Bridges	
9.11.20	**Bristol (City Docks) (Avonmouth, tide tables/curves)**	554
9.11.21	**Portishead**	556
	Weston-super-Mare	
9.11.22	**Burnham-on-Sea**	557
9.11.23	**Watchet**	557
9.11.24	**Ilfracombe**	558
	Minehead • Porlock Weir • Lynmouth • Watermouth	
9.11.25	**Rivers Taw and Torridge**	559
	Clovelly • Lundy Island • Bude • Boscastle	
9.11.26	**Padstow**	561
	Port Isaac • Newquay • Portreath • Hayle • St Ives	

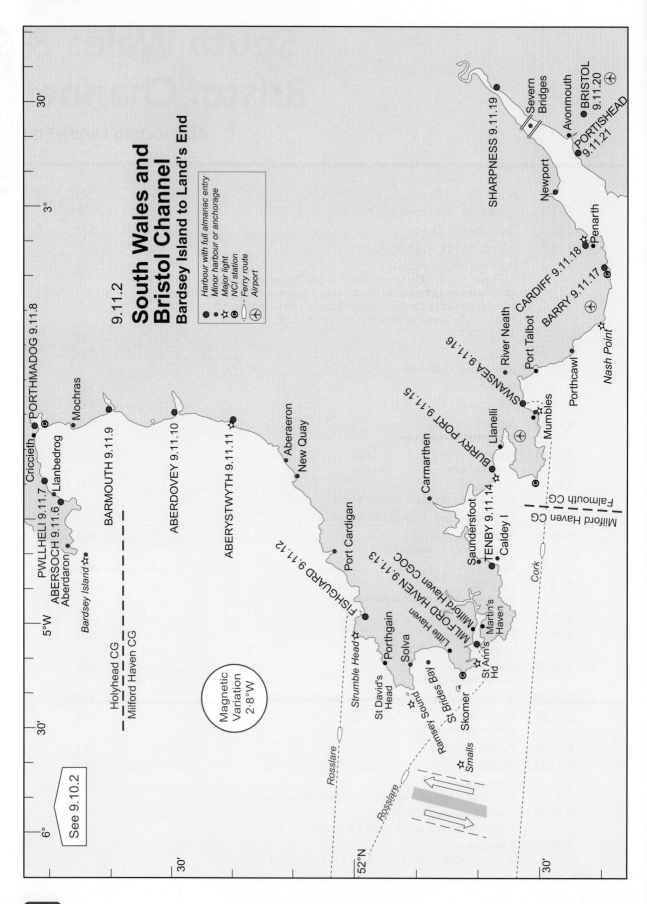

9.11.2

South Wales and Bristol Channel
Bardsey Island to Land's End

Legend:
- ● Harbour with full almanac entry
- • ☆ Minor harbour or anchorage
- ☆ Major light
- Ⓒ NCI station
- ⬭ Ferry route
- ✈ Airport

See 9.10.2

Magnetic Variation 2·8°W

Holyhead CG
Milford Haven CG

Bardsey Island

5°W

PWLLHELI 9.11.7
ABERSOCH 9.11.6
Aberdaron
Criccieth
PORTHMADOG 9.11.8
Llanbedrog
Mochras

BARMOUTH 9.11.9

ABERDOVEY 9.11.10

ABERYSTWYTH 9.11.11

Aberaeron
New Quay

Port Cardigan

FISHGUARD 9.11.12

Strumble Head
St David's Head
Porthgain
Solva
Ramsey Sound
St Brides Bay
Skomer
Smalls
Little Haven
Martin's Haven
St Ann's Hd
MILFORD HAVEN 9.11.13
Milford Haven CGOC

Rosslare
Rosslare

Cork

Milford Haven CG
Falmouth CG:

Caldey I
TENBY 9.11.14
Saundersfoot
Carmarthen
Llanelli
BURRY PORT 9.11.15
Mumbles
SWANSEA 9.11.16
River Neath
Port Talbot
Porthcawl
Nash Point
BARRY 9.11.17
CARDIFF 9.11.18
Penarth

Newport
SHARPNESS 9.11.19

Severn Bridges
Avonmouth
BRISTOL 9.11.20
PORTISHEAD 9.11.21

52°N

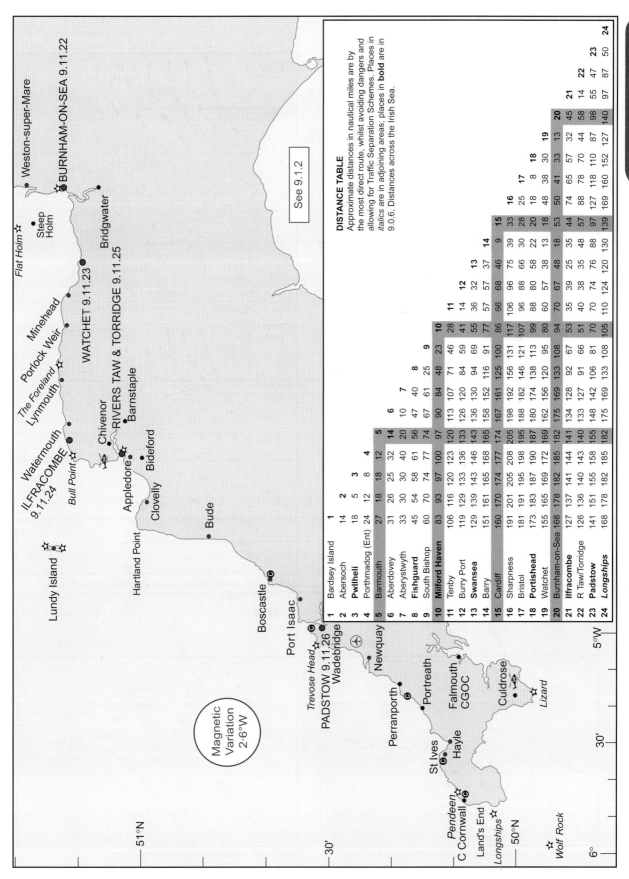

Map labels: Weston-super-Mare · BURNHAM-ON-SEA 9.11.22 · Flat Holm · Steep Holm · Bridgwater · Minehead · WATCHET 9.11.23 · Porlock Weir · The Foreland · Lynmouth · RIVERS TAW & TORRIDGE 9.11.25 · Barnstaple · Chivenor · Watermouth · ILFRACOMBE 9.11.24 · Bull Point · Appledore · Bideford · Clovelly · Bude · Hartland Point · Lundy Island · Boscastle · Port Isaac · Trevose Head · PADSTOW 9.11.26 · Wadebridge · Newquay · Perranporth · Portreath · Falmouth CGOC · Culdrose · Lizard · St Ives · Hayle · Pendeen · C Cornwall · Land's End · Longships · Wolf Rock · Magnetic Variation 2·6'W · See 9.1.2 · 51°N · 30' · 50°N · 5°W · 30' · 6°

DISTANCE TABLE

Approximate distances in nautical miles are by the most direct route, whilst avoiding dangers and allowing for Traffic Separation Schemes. Places in *italics* are in adjoining areas; places in **bold** are in 9.0.6. Distances across the Irish Sea.

No.	Place	1	2	3	4	5	6	7	8	9	10	11	12	13	14	15	16	17	18	19	20	21	22	23	24
1	Bardsey Island	**1**																							
2	Abersoch	14	**2**																						
3	**Pwllheli**	18	5	**3**																					
4	Porthmadog (Ent)	24	12	8	**4**																				
5	Barmouth	27	18	18	12	**5**																			
6	Aberdovey	31	26	25	32	14	**6**																		
7	Aberystwyth	33	30	30	40	20	10	**7**																	
8	**Fishguard**	45	54	58	61	56	47	40	**8**																
9	South Bishop	60	70	74	77	74	67	61	25	**9**															
10	**Milford Haven**	83	93	97	100	97	90	84	48	23	**10**														
11	Tenby	106	116	120	123	120	113	107	71	46	28	**11**													
12	Burry Port	119	129	133	136	133	126	120	84	59	41	14	**12**												
13	**Swansea**	129	139	143	146	143	136	130	94	69	55	36	32	**13**											
14	Barry	151	161	165	168	165	158	152	116	91	77	57	57	37	**14**										
15	**Cardiff**	160	170	174	177	174	167	161	125	100	86	66	68	46	9	**15**									
16	Sharpness	191	201	205	208	205	198	192	156	131	117	106	96	75	39	33	**16**								
17	Bristol	181	191	195	198	195	188	182	146	121	107	96	88	66	30	28	25	**17**							
18	**Portishead**	173	183	187	190	187	180	174	138	113	99	88	80	58	22	20	18	8	**18**						
19	Watchet	155	165	169	172	169	162	156	120	95	80	60	57	38	13	18	48	38	30	**19**					
20	**Burnham-on-Sea**	168	178	182	185	182	175	169	133	108	94	75	67	48	18	18	53	50	41	13	**20**				
21	Ilfracombe	127	137	141	144	141	134	128	92	67	53	35	39	25	35	44	57	74	65	57	45	**21**			
22	R Taw/Torridge	126	136	140	143	140	133	127	91	66	51	40	38	35	48	57	88	78	70	44	58	14	**22**		
23	**Padstow**	141	151	155	158	155	148	142	106	81	70	70	74	76	88	97	127	110	87	110	98	55	47	**23**	
24	*Longships*	168	178	182	185	182	175	169	133	108	105	110	124	130	139	140	169	152	160	127	140	97	87	50	**24**

9.11.3 AREA 11 TIDAL STREAMS

5 Hours before HW Dover (HW Milford Haven)

4 Hours before HW Dover (0100 after HW Milford Haven)

3 Hours before HW Dover (0200 after HW Milford Haven)

2 Hours before HW Dover (0300 after HW Milford Haven)

1 Hour before HW Dover (0400 after HW Milford Haven)

HW Dover (0500 after HW Milford Haven)

Southward 9.1.3   Northward 9.10.3   South Ireland 9.12.3

1 Hour after HW Dover (0600 after HW Milford Haven)

2 Hours after HW Dover (0525 before HW Milford Haven)

3 Hours after HW Dover (0425 before HW Milford Haven)

4 Hours after HW Dover (0325 before HW Milford Haven)

5 Hours after HW Dover (0225 before HW Milford Haven)

6 Hours after HW Dover (0125 before HW Milford Haven)

9.11.4 LIGHTS, BUOYS AND WAYPOINTS

Bold print = light with a nominal range of 15M or more. CAPITALS = place or feature. *CAPITAL ITALICS* = light-vessel, light float or Lanby. *Italics* = Fog signal. ***Bold italics*** = Racon. Many marks/buoys are fitted with with AIS (MMSI No); see relevant charts.

CARDIGAN BAY (SEE ALSO 9.10.4)

Bardsey I ☆ 52°44'·98N 04°48'·01W Fl R 10s 39m **18M**; W ☐ twr, R bands; obsc by Bardsey I 198°-250° and in Tremadoc B when brg less than 260°.
St Tudwal's ⚡ Fl WR 15s 46m W14, R10M; vis: 349°-W-169°-R-221°-W-243°-R-259°-W-293°-R-349°; obsc by East I 211°-231°; 52°47'·92N 04°28'·30W.

PWLLHELI and PORTHMADOG
Pwllheli App ⚓ Iso 2s; 52°53'·03N 04°22'·90W.
Porthmadog Fairway ⚓ L Fl 10s; 52°53'·01N 04°11'·10W.

BARMOUTH and ABERDOVEY
Diffuser ⚓ Fl Y 5s; 52°43'·19N 04°05'·38W.
Barmouth Outer ⚓ L Fl 10s; 52°42'·62N 04°04'·83W.
Aberdovey Outer ⚓ Iso 4s; 52°32'·00N 04°05'·56W.
Cynfelyn Patches, Patches ⚓ Q (9) 15s; 52°25'·83N 04°16'·41W.

ABERYSTWYTH, ABERAERON and NEW QUAY
Aberystwyth S Bkwtr Hd ⚡ Fl (2) WG 10s 12m 10M; vis: 030°-G-053°-W-210°; 52°24'·40N 04°05'·52W.
Ldg Lts 133°. Front, FR 4m 5M; 52°24'·37N 04°05'·39W. Rear, 52m from front, FR 7m 6M.
Aberaeron N Pier ⚡ Fl (4) WRG 15s 10m 6M; vis: 050°-G-104°-W-178°-R-232°; 52°14'·61N 04°15'·87W.
Carreg Ina ⚓ Q; 52°13'·25N 04°20'·75W.
New Quay Pier Hd ⚡ Fl WG 3s 12m W8M, G5M; G △; vis: 135°-W- 252°-G-295; 52°12'·95N 04°21'·35W.

CARDIGAN and FISHGUARD
Cardigan Channel ⚓ Fl (2) 5s; 52°06'·44N 04°41'·43W.
Cardigan Cliff Base Bn 1 ⚡ Fl R 4s 3M; 52°07'·05N 04°41'·39W.
Cardigan Cliff Base Bn 2 ⚡ Fl R 2s 3M; 52°06'·91N 04°41'·34W.
Fishguard N Bkwtr Hd ⚡ Fl G 4·5s 18m 13M; *Bell (1) 8s*; 52°00'·76N 04°58'·23W.
Strumble Head ☆ 52°01'·79N 05°04'·43W Fl (4) 15s 45m **26M**; vis: 038°-257°; (H24).

BISHOPS and SMALLS
South Bishop ☆ 51°51'·14N 05°24'·74W Fl 5s 44m **16M**; W ○ twr; *Horn (3) 45s*; ***Racon (O)10M***; (H24).
The Smalls ☆ 51°43'·27N 05°40'·19W Fl (3) 15s 36m **18M**; ***Racon (T)***; *Horn (2) 60s*. Same twr, Iso R 4s 33m 13M; vis: 253°-285° over Hats & Barrels Rk; both Lts shown H24 in periods of reduced visibility; 992351123.
Skokholm I ⚡, 51°41'·64N 05°17'·22W Fl WR 10s 54m W8M, R8M; vis: 301°-W-154°-R-301°; partially obsc 226°-258°.

WALES – SOUTH COAST – BRISTOL CHANNEL
MILFORD HAVEN
St Ann's Head ☆ 51°40'·87N 05°10'·42W Fl WR 5s 48m **W18M, R17M**, R14M; W 8-sided twr; vis: 233°-W-247°-R-332°-W131°, partially obscured 124°-129°; *Horn (2) 60s*.
W Blockhouse Point ⚓ Ldg Lts 022·5°. Front, F 54m 13M (day 10M); B stripe on W twr; vis: 004·5°-040·5°; intens on lead line. vis: 004·5°-040·5°; ***Racon (Q)***; 51°41'·31N 05°09'·56W.
Watwick Point Common Rear ☆, 0·5M from front, F 80m **15M** (day 10M); vis: 013·5°-031·5°; ***Racon (Y)***.
W Blockhouse Point ⚓ Q WR 21m W9M, R7M; R lantern on W base: vis: 220°-W-250°-R-020°-W-036°-R-049°; 51°41'·31N 05°09'·56W.
Dale Fort ⚡ Fl (2) WR 5s 20m W5M, R3M; vis: 222°-R-276°-W-019°; 51°42'·16N 05°09'·01W.
Gt Castle Hd ⚓ F WRG 27m W5M, R3M, G3M; vis: 243°-R-281°-G-299°-W-029°; 51°42'·68N 05°07'·06W. Co-located Dir WRG (040°); vis: 038·25°-G-039°-AlWG-039·5°-W-040·5°-AlWR-041°-R-041·75°.

(**not** used in conjunction with the front ldg light below)
Ldg Lts ≠ 039·7° **Front**, Oc 4s 27m **15M**; vis: 031·2°-048·2°. **Rear**, 890m from front. Oc 8s 53m **15M** (day 10M); vis: 031·2°-048·2°.
St Ann's ⚓ Fl R 2·5s; 51°40'·25N 05°10'·51W.
Mid Channel Rks ⚓ Q (9) 15s; 51°40'·18N 05°10'·14W.
Mid Channel Rk ⚡ Fl (3) G 7s 18m 8M; 51°40'·31N 05°09'·83W.
Rows Rks ⚓ Q R; 51°40'·22N 05°09'·02W.
Sheep ⚓ QG; 51°40'·06N 05°08'·31W.
Millbay ⚓ Fl (2) R 5s; 51°41'·05N 05°09'·45W.
W Chapel ⚓ Fl G 10s; 51°40'·98N 05°08'·67W.
E Chapel ⚓ Fl R 5s; 51°40'·87N 05°08'·15W.
Rat ⚓ Fl G 5s; 51°40'·80N 05°07'·86W.
Angle ⚓ VQ; 51°41'·63N 05°08'·27W.
Thorn Rock ⚓ Q (9) 15s; 51°41'·53N 05°07'·76W.
Turbot Bank ⚓ VQ (9) 10s; 51°37'·41N 05°10'·08W.
ODAS Fl(5) Y 20s; 51°36'·70N 05°08'·70W.
St Gowan ⚓ Q (6) + L Fl 15s, ***Racon (T) 10M***; 51°31'·93N 04°59'·77W; 992351126.

TENBY to SWANSEA BAY
Caldey I ⚡ Fl (3) WR 20s 65m W13M, R9M; vis: R173°- W212°-R088°-102°; 51°37'·90N 04°41'·08W.
Woolhouse ⚓ Q (6) + L Fl 15s; 51°39'·35N 04°39'·69W.
Burry Port Inlet ⚡ Fl 5s 7m 6½M; 51°40'·62N 04°15'·06W.
W. Helwick ⚓ (9) 15s; ***Racon (T) 10M***; 51°31'·40N 04°23'·65W.
E. Helwick ⚓ VQ (3) 5s; *Bell*; 51°31'·80N 04°12'·68W.

SWANSEA BAY and SWANSEA
Ledge ⚓ VQ (6) + L Fl 10s; 51°29'·93N 03°58'·77W.
Mixon ⚓ Fl (2) R 5s; *Bell*; 51°33'·12N 03°58'·78W.
Outer Spoil Gnd ⚓ Fl Y 2·5s; 51°32'·11N 03°55'·73W.
Grounds ⚓ VQ (3) 5s; 51°32'·81N 03°53'·47W.
Mumbles ☆ 51°34'·01N 03°58'·27W Fl (4) 20s 35m **15M**; W twr; *Horn (3) 60s*.
SW Inner Green Grounds ⚓ Q(6)+L Fl 15s; *Bell*; 51°34'·06N 03°57'·03W.
Swansea Lts in line 020°. Front: E Breakwater head, Oc G 4s 5m 2M & 2FG(vert) 10m 6M; 51°36'·38N 03°55'·62W. Rear, 3·1ca from front: FG 6M.

SWANSEA BAY, RIVER NEATH and PORT TALBOT
Neath App Chan ⚓ Fl G 5s; 51°35'·71N 03°52'·83W.
Cabenda ⚓ VQ (6) + L Fl 10s; ***Racon (Q)***; 51°33'·36N 03°52'·23W.
Ldg Lts 059·8° (occas). Front, Oc R 3s 12m 6M; 51°34'·92N 03°48'·10W. Rear, 400m from front, Oc R 6s 32m 6M.

BRISTOL CHANNEL – NORTH SHORE (EASTERN PART)
W Scar ⚓ Q (9) 15s, *Bell*, ***Racon (T) 10M***; 51°28'·31N 03°55'·57W.
South Scar (S SCAR) ⚓ Q (6) + L Fl 15s; 51°27'·61N 03°51'·58W.
Hugo ⚓ Q R; 51°28'·55N 03°48'·03W.
E. Scarweather ⚓ Q (3) 10s; *Bell*; 51°27'·98N 03°46'·76W.

PORTHCAWL
Fairy ⚓ Q (9) 15s; *Bell*; 51°27'·86N 03°42'·07W.
Porthcawl Bkwtr Hd ⚡ 51°28'·39N 03°41'·98W F WRG 10m W6M, R4M, G4M; vis: 302°-G-036°-W-082°-R-122°.
W Nash ⚓ VQ (9) 10s ; *Bell*; 51°25'·99N 03°45'·95W.
Nash ☆ 51°24'·03N 03°33'·06W Fl (2) WR 15s 56m **W21M, R16M**; vis: 280°-R-290°-W-100°-R-120°-W-128°.
Breaksea ⚓ L Fl 10s; ***Racon (T) 10M***; 51°19'·88N 03°19'·08W; 992351124.

BARRY
W Bkwtr Hd ⚡ Fl 2·5s 12m 10M; 51°23'·46N 03°15'·52W.
N. One Fathom ⚓ Q; 51°20'·94N 03°12'·17W.
Mackenzie ⚓ QR; 51°21'·75N 03°08'·24W.

CARDIFF and PENARTH ROADS
Lavernock Outfall ⚓ Fl Y 5s; 51°23'·95N 03°09'·50W.
Ranie ⚓ Fl (2) R 5s; 51°24'·23N 03°09'·39W.
S Cardiff ⚓ Q (6) + L Fl 15s; *Bell*; 51°24'·18N 03°08'·57W.
Mid Cardiff ⚓ Fl (3) G 10s; 51°25'·60N 03°08'·09W.

Cardiff Spit ⌀ QR; 51°24'·57N 03°07'·12W.
N Cardiff ▲ QG; 51°26'·52N 03°07'·19W.
Wrach Chan Dir lt 348·5°. Oc WRG 10s 5m; W3M, R3M, G3M;
vis: 344·5°-G-347°-W-350°-R-352°; H24; 51°27'·16N 03°09'·75W.
Outer Wrach ↯ Q (9) 15s; 51°26'·20N 03°09'·46W.

FLATHOLM to THE BRIDGE
Flat Holm ☆, SE Pt 51°22'·54N 03°07'·14W Fl (3) WR 10s 50m **W15M**,
R12M; W ○ twr; vis: 106°-R-140°-W-151°-R-203°-W-106°; (H24).
Monkstone ⌀ Fl 5s 13m 12M; R column on W ○ twr; 51°24'·89N
03°06'·01W.
Tail Patch ▲ QG; 51°23'·53N 03°03'·65W.
Hope ↯ Q (3) 10s; 51°24'·84N 03°02'·68W.
NW Elbow ↯ VQ (9) 10s; *Bell;* 51°26'·28N 02°59'·93W.
EW Grounds ↯ L Fl 10s 7M; *Whis; Racon (T) 7M;* 51°27'·12N
02°59'·95W.

NEWPORT DEEP, RIVER USK and NEWPORT
Newport Deep ▲ Fl (3) G 10s; *Bell;*51°29'·36N 02°59'·12W.
East Usk ☆ 51°32'·40N 02°58'·01W; Fl (2) WRG 10s 11m
W11M,R10M, G10M; vis: 284°-W-290° -obscured shore-324°-R-
017°-W-037°-G-115°-W-120°. Also Oc WRG 10s 10m W11M,
R9M, G9M; vis: 013°-G-017°-W- 019°-R-023°.
Julians Pill Ldg Lts 062°. Front, FG 5m 4M; 51°33'·30N
02°57'·94W. Rear, 61m from front, FG 8m 4M.

BRISTOL CHANNEL – SOUTH SHORE (EASTERN PART)
BRISTOL DEEP
N Elbow ▲ QG; *Bell;* 51°26'·97N 02°58'·65W.
S Mid Grounds ↯ VQ (6) + L Fl 10s; 51°27'·62N 02°58'·68W.
E Mid Grounds ⌀ Fl R 5s; 51°27'·75N 02°54'·98W.
Clevedon ↯ VQ; 51°27'·39N 02°54'·93W.
Welsh Hook ↯ Q (6) + L Fl 15s; 51°28'·53N 02°51'·86W.
Avon ▲ Fl G 2·5s; 51°27'·92N 02°51'·73W.
Black Nore Pt Lt Ho; W ○ twr on frame; 51°29'·09N 02°48'·05W.
Newcome ⌀ 51°30'·01N 02°46'·71W Fl (3) R 10s.
Denny Shoal ↯ VQ (6) + L Fl 10s; 51°30'·15N 02°45'·45W.
Cockburn ⌀ 51°30'·45N 02°44'·10W Fl R 2·5s.
W Dock Middle ▲ 51°29.95'·01N 02°44'·20W Q G.
W Dock Outer ▲ 51°30'·00N 02°44'·85W Fl G 5s.
Firefly ▲ Fl (2) G 5s; 51°29'·96N 02°45'·35W.
Portishead Point ☆ 51°29'·68N 02°46'·42W Q (3) 10s 9m **16M**;
B twr, W base; vis: 060°-262°; *Horn 20s.*

PORTISHEAD
Pier Hd ⌀ Iso G 2s 5m 3M; 51°29'·69N 02°45'·27W.
Seabank. Dir WG 6m 5M; vis: 085·8°-FG-086·7°-AltWG-086·9°-FW-
089·7°. 5M; vis: 070·3°-103·3°; by day 1M, vis: 076·8°-096·8°.
Lts in line 086·8°. Q(8) 5M 10/16m; vis: 070·3°-103·3°; day 1M vis:
076·8°-096·8° Front 51°30'·07N 02°43'·80W; Rear, 500m behind.
Knuckle Lts in line 099·6°, Oc G 5s 6m 6M; 51°29'·94N 02°43'·67W.
Rear, 165m from front, FG 13m 6M; vis: 044°-134°.

AVONMOUTH
Royal Edward Dock N Pier Hd ⌀ Fl 4s 15m 10M; vis: 060°-228·5°;
51°30'·49N 02°43'·09W.
King Road Ldg Lts 072·4°. N Pier Hd ⌀ Front, Oc R 5s 5m 9M; W
obelisk, R bands; vis: 062°-082°; 51°30'·49N 02°43'·09W. Rear
⌀, 546m from front, QR 15m 10M; vis: 066°- 078°.

RIVER AVON, CUMBERLAND BASIN and AVON BRIDGE
S Pier Hd ⌀ Oc RG 30s 9m 10M and FBu 4m 1M; vis: 294°-R-
036°-G-194°; 51°30'·37N 02°43'·10W. *Bell(1) 10s.*
Ldg Lts 127·2°. Front ↯, Iso R 2s 6m 3M, vis: 010°-160°; 51°30'·10N
02°42'·59W. Rear↯ , Iso R 2s10m 3M; vis: 048°-138°.

BRISTOL CHANNEL (SOUTH SHORE)
WESTON-SUPER-MARE
Pier Hd ⌀ 2 FG (vert) 6m; 51°20'·88N 02°59'·26W.
E Culver ↯ Q (3) 10s; 51°18'·00N 03°15'·44W.
W Culver ↯ VQ (9) 10s; 51°17'·47N 03°19'·00W.
Gore ⌀ Iso 5s; *Bell;* 51°13'·94N 03°09'·79W.

BURNHAM-ON-SEA and RIVER PARRETT
Ent ⌀ Fl 7·5s 7m 12M; vis: 074°-164°; 51°14'·89N 03°00'·36W;
Dir lt 076°. F WRG 4m W12M, R10M, G10M; vis: 071°-G- 075°-
W-077°- R-081°.
DZ No. 1 ⌀ Fl Y 2·5s; 51°15'·28N 03°09'·49W.
DZ No. 2 ⌀ Fl Y 10s; 51°13'·77N 03°19'·86W.
DZ No. 3 ⌀ Fl Y 5s; 51°16'·34N 03°14'·98W.

WATCHET, MINEHEAD and PORLOCK WEIR
Watchet W Bkwtr Hd ⌀ Oc G 3s 9m 9M; 51°11'·03N 03°19'·74W.
Watchet E Pier ⌀ 2 FR (vert) 3M; 51°11'·01N 03°19'·72W.
Minehead Bkwtr Hd ⌀ Fl (2) G 5s 4M; vis: 127°-262°; 51°12'·81N
03°28'·36W.
Lynmouth Foreland ☆ 51°14'·73N 03°47'·21W Fl (4) 15s 67m
18M; W ○ twr; vis: 083°-275°; (H24).

LYNMOUTH and WATERMOUTH
River Training Arm ⌀ 2 FR (vert) 6m 5M; 51°13'·90N 03°49'·83W.
Harbour Arm ⌀ 2 FG (vert) 6m 5M; 51°13'·92N 03°49'·84W.
Sand Ridge ▲ Q G; 51°15'·01N 03°49'·77W.
Copperas Rock ▲ Fl G 2·5s; 51°13'·78N 04°00'·60W.
Watermouth ⌀ Oc WRG 5s 1m 3M; W △; vis: 149·5°-G-151·5°-W-
154·5°-R-156·5°; 51°12'·93N 04°04'·60W.

ILFRACOMBE to BAGGY POINT
Ldg Lts 188°. Front, Oc 10s 8m 3M; 51°12'·53N 04°06'·65W.
Rear, Oc 10s 6m 3M.
Horseshoe ↯ Q; 51°15'·02N 04°12'·96W.
Bull Point ☆ 51°11'·94N 04°12'·09W Fl (3) 10s 54m **20M**; W ○ twr,
obscd shore-056°. Same twr; FR 48m 12M; vis: 058°-096°.
Morte Stone ▲ Fl G 5s; 51°11'·30N 04°14'·95W.
Baggy Leap ▲ Fl (2) G 10s; 51°08'·92N 04°16'·97W.

BIDEFORD, RIVERS TAW and TORRIDGE
Bideford Fairway ↯ L Fl 10s; *Bell;* 51°05'·25N 04°16'·25W.
Bideford Bar ▲ Q G; 51°04'·89N 04°14'·62W.
Instow ☆ Ldg Lts 118°. **Front,** 51°03'·62N 04°10'·66W Oc 6s
22m **15M**; vis: 103·5°-132°. **Rear,** 427m from front, Oc 10s 38m
15M; vis: 103°-132·5°; (H24).
Crow Pt ⌀ Fl WR 2. 5s 8m W6M R5M; vis: 225°-R-232°-W-237°-
R-358°-W- 015°-R-045°; 51°03'·96N 04°11'·39W.

LUNDY
Near North Point ☆ 51°12'·10N 04°40'·65W Fl 15s 48m **17M**;
vis: 009°-285°.

South East Point ☆ 51°09'·72N 04°39'·37W Fl 5s 53m **15M**;
vis: 170°-073°.
Hartland Point ⌀ 51°01'·3N 04°31'·6W Fl (6) 15s 30m 8M;
adjacent to Hartland Point Light House (disused).

NORTH CORNWALL
BUDE, PADSTOW and NEWQUAY
Compass Point twr 50°49'·71N 04°33'·42W.
Stepper Point (Padstow) ⌀ L Fl 10s 12m 4M; 50°34'·12N
04°56'·72W.
Trevose Head ☆ 50°32'·94N 05°02'·13W Fl 7·5s 62m **21M**.
North Pier Hd (Newquay) ⌀ 2 FG (vert) 5m 2M; 50°25'·07N
05°05'·19W.

HAYLE and ST IVES
The Stones ↯ Q; 50°15'·64N 05°25'·51W.
Godrevy I ⌀ Fl WR 10s 28m 8M; metal post; vis: 022°-W-101°-
R-145°-W-272°; 50°14'·56N 05°24'·01W;
adjacent to Godrevy Light House (disused).
Hayle App ⌀ QR; 50°12'·26N 05°26'·30W.
St Ives App ▲ 50°12'·85N 05°28'·42W.
Bann Shoal ↯ Fl G 2·5s; *Racon (B) 10M;* 50°20'·03N 05°51'·11W;
992351053.
Pendeen ☆ 50°09'·90N 05°40'·30W Fl (4) 15s 59m **16M**; vis: 042°-
240°; between Gurnard Hd and Pendeen it shows to coast.

9.11.5 PASSAGE INFORMATION

More passage information is threaded between the harbours in this area. Admiralty Leisure Folio 5620 covers South West Wales, 5608 the Bristol Channel and 5603 the approach to Bristol Channel from Land's End. For additional tidal information east of Ilfracombe/Swansea see *Arrowsmith's Bristol Channel Tide Tables* from J.W. Arrowsmith Ltd ☎(0117) 9667545. **Bibliography:** *Bristol Channel and River Severn Cruising Guide*, Milford Haven to St Ives (Imray/Cumberlidge). The *West Coasts of England and Wales Pilot* (Admiralty NP37) covers the whole area.

It is useful to know some Welsh words with navigational significance.

Aber: estuary.	Afon: river.
Bach, bychan, fach: little.	Borth: cove.
Bryn: hill.	Careg, craig: rock.
Coch, goch: red.	Dinas: fort.
Ddu: black.	Fawr, Mawr: big.
Ffrydiau: tide rip.	Llwyd: grey.
Moel: bare conical hill.	Mor: sea.
Morfa: sandy shore.	Mynydd: mountain.
Penrhyn: headland.	Porth: cove.
Ynys, Ynysoedd: island(s).	

CARDIGAN BAY

(AC 1971, 1972, 1973) Harbours are mostly on a lee shore, and/or have bars which make them dangerous to approach in bad weather. Abersoch and Pwllheli offer best shelter from the prevailing westerlies. There may be overfalls off Trwyn Cilan, SW of St Tudwal's Is (lit). In N part of the bay there are three major dangers to coasting yachts: St Patrick's Causeway (Sarn Badrig) runs 12M SW from Mochras Pt. It is mostly large loose stones, and dries (up to 1·5m) for much of its length. In strong winds the sea breaks heavily at all states of tide. The outer end is marked by a WCM light buoy. At the inner end there is a channel about 5ca offshore, which can be taken with care at half tide.

Sarn-y-Bwch runs 4M WSW from Pen Bwch Pt. It is consists of rocky boulders, drying in places close inshore and with least depth 0·3m over 1M offshore. There is a WCM buoy off the W end. NW of Aberystwyth, Sarn Cynfelyn and Cynfelyn Patches extend 6·5M offshore, with depths of 1·5m in places. A WCM buoy is at the outer end. Almost halfway along the bank is Main Channel, 3ca wide, running roughly N/S but not marked.

Aberporth MOD Range occupies much of Cardigan Bay. It is usually active Mon-Fri 0900-1630LT, and some weekends, but times are subject to change. Active danger areas vary in size and location and may include the area off the Range head to the west of Aberporth. Beware of targets and buoys within the danger area, some unlit. Range activity is broadcast on VHF Ch 16 one hour before live firings. Flags are flown either side of the Range head to signify activity in the inner (yellow flags) or outer (red flags) danger areas. If passing through the area during operational hours contact *Aberporth Marine Control* on Ch 16 or Ch 13. Further information may be obtained from Aberporth Marine Control, ☎ (01239) 813760, or Aberporth Range Control ☎(01239) 813480.

If on passage N/S through St George's Chan (ie not bound for Cardigan B or Milford Haven) the easiest route, and always by night, is W of the Bishops and the Smalls, noting the TSS. If bound to/from Milford Haven or Cardigan Bay, passage inside both the Smalls and Grassholm is possible.

RAMSEY SOUND, THE BISHOPS, THE SMALLS TO MILFORD HAVEN

(AC 1478, 1482) The Bishops and the Clerks are islets and rocks 2·5M W and NW of Ramsey Island, a bird sanctuary SSW of St David's Head. N Bishop is the N'ly of the group, 3ca ENE of which is Bell Rock (depth 1·9m). S Bishop (Lt, fog sig) is 3M to the SSW.

Between S Bishop and Ramsey Is the dangers include Daufraich with offlying dangers to the E and heavy overfalls; Llech Isaf and Llech Uchaf drying rocks are further ENE. Carreg Rhoson and other dangers are between Daufraich and N Bishop. The navigable routes between most of these islets and rocks trend NE/SW. Use only by day, in good visibility and with local knowledge. The N/S route close W of Ramsey Is is considered easier than Ramsey Sound.

▶ *2M W of The Bishops the S-going stream begins at Milford Haven +0400, and the N-going at Milford Haven –0230, sp rates 2kn. Between The Bishops and Ramsey Is the SW-going stream begins at Milford Haven +0330, and the NE-going at Milford Haven –0300, sp rates 5kn. Ramsey Sound should be taken at slack water. S-going stream begins at Milford Haven +0300, and the N-going at Milford Haven –0330, sp rates 6kn at The Bitches, where the channel is narrow (2ca), decreasing N & S.* ◀

FLOOD SW Wales

> The tide turns to the S or SE (Bristol Channel flood) as follows:
> (a): Milford Haven +0330;
> (b) & (c): Milford Haven +0400;
> (d): Milford Haven +0200.

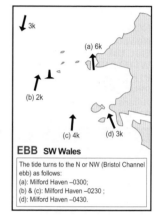

EBB SW Wales

> The tide turns to the N or NW (Bristol Channel ebb) as follows:
> (a): Milford Haven –0300;
> (b) & (c): Milford Haven –0230 ;
> (d): Milford Haven –0430.

The Bitches are rocks up to 4m high extending 2ca from E side of Ramsey Island. Other dangers are: Gwahan and Carreg-gafeiliog, both 3m high at N end of Sound, to W and E; Horse Rock (dries 0·9m) almost in mid-channel about 5ca NNE of The Bitches, with associated overfalls; Shoe Rock (dries 3m) at SE end of channel; and rocks extending 5ca SSE from S end of Ramsey Island.

St Brides Bay is a regular anchorage for tankers, but ⚓ only in settled weather/offshore winds as it is a trap in westerlies. Solva is a little hbr with shelter for boats able to take the ground, or ⚓ behind Black Rk (dries 3·6m) in the middle of the entrance.

The Smalls Lt, where there is a Historic Wreck (see 9.0.3) is 13M W of the Welsh mainland (Wooltack Pt). 2M and 4M E of The Smalls are the Hats and Barrels, rocky patches on which the sea breaks.

▶ *7M E of The Smalls is Grassholm Island with a race either end and strong tidal eddies so that it is advisable to pass about 1M off. The chan between Barrels and Grassholm is 2·5M wide, and here the S-going stream begins at Milford Haven +0440, and the N-going at Milford Haven –0135, sp rates 5kn.* ◀

Five miles of clear water lie between Grassholm and Skomer Is/ Skokholm Is to the E. Wildgoose Race, which forms W of Skomer and Skokholm is very dangerous, so keep 2M W of these two Is. To E of Skomer Island Midland Island, and between here and Wooltack Point is Jack Sound, least width about 1ca. Do not attempt it without AC 1482, detailed pilotage directions, and only at slack water neaps.

▶ *Correct timing is important. The S-going stream begins at Milford Haven +0200, and the N-going at Milford Haven –0430, sp rates 6-7kn.* ◀

Rocks which must be identified include, from N to S: On E side of chan off Wooltack Pt, Tusker Rock (2m), steep-to on its W side; and off Anvil Point, The Cable (2·4m), The Anvil and Limpet Rocks (3·7m). On the W side lie the Crabstones (3·7m) and the Blackstones (1·5m).

9.11.6 ABERSOCH
Gwynedd 52°49'·29N 04°29'·20W (Ⓐ) ❀❀♢♢✿✿

CHARTS AC 5609, 1410, 1411, 1971, 1512; Imray C61, C52, C51

TIDES –0315 Dover; ML 2·5; Duration 0520; Zone 0 (UT)

Standard Port MILFORD HAVEN (→)

Times				Height (metres)			
High Water		Low Water		MHWS	MHWN	MLWN	MLWS
0100	0800	0100	0700	7·0	5·2	2·5	0·7
1300	2000	1300	1900				
Differences ST TUDWAL'S ROADS							
+0155	+0145	+0240	+0310	–2·2	–1·9	–0·7	–0·2
ABERDARON							
+0210	+0200	+0240	+0310	–2·4	–1·9	–0·6	–0·2

SHELTER There are few moorings for visitors. Apply to HM or SC. ⚓ in St Tudwal's Roads clear of moored yachts; sheltered from SSE through S to NE.

NAVIGATION WPT 52°48'·52N 04°26'·13W, 293°/2·4M to YC jetty. There are no navigational dangers, but steer well clear of the drying rks to the E of East Island; Carred y Trai buoy Fl R 2·5s is 2ca E of these rks (just off chartlet). St Tudwal's islands themselves are fairly steep-to, except at N ends. St Tudwal's Sound is clear of dangers.

LIGHTS AND MARKS The only lt is on St Tudwal's West Island, Fl WR 15s 46m 14/10M (see chartlet and 9.11.4).

COMMUNICATIONS (Code 01758) CGOC (01407) 762051; Police 101; Dr 612535. HM 712203.

S Caernarfon YC Ch **80** M.

FACILITIES **South Caernarvonshire YC** ☎712338, ⚓ ⚓ ⚓ ⚓ ⚓ ✕ ⌂ (May-Sept).

Abersoch Power Boat Club ☎812027.

Services ⚓ ⚓ ✕ ⚓ ⚓ ⚓ ⚓(12t).

Town ⚓ ACA, ✉ Ⓑ ⚓ ✕ ⌂ ≢ (Pwllheli), ✈ (Chester).

9.11.7 PWLLHELI

Gwynedd 52°53′·23N 04°23′·75W

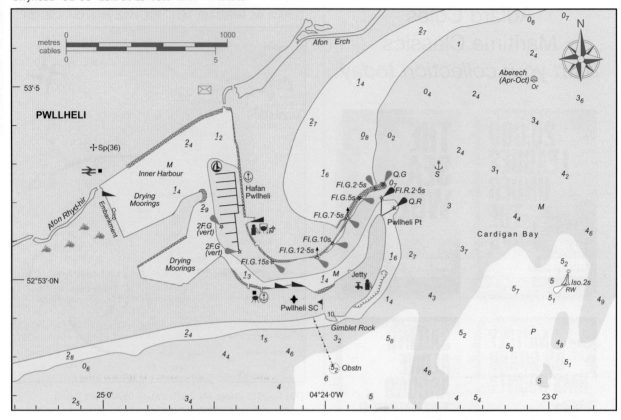

CHARTS AC 1410, 1971, 1512; Imray C61, C52, C51

TIDES –0315 Dover; ML 2·6; Duration 0510

Standard Port MILFORD HAVEN (→)

Times				Height (metres)			
High Water		Low Water		MHWS	MHWN	MLWN	MLWS
0100	0800	0100	0700	7·0	5·2	2·5	0·7
1300	2000	1300	1900				
Differences PWLLHELI							
+0210	+0150	+0245	+0320	−1·9	−1·6	−0·6	−0·1
CRICCIETH							
+0210	+0155	+0255	+0320	−2·0	−1·8	−0·7	−0·3

SHELTER Good in hbr and marina. Drying moorings in inner hbr (SW and NW bights).

NAVIGATION WPT 52°53′·02N 04°23′·07W, SWM lt buoy 299°/0·47M to QG lt at head of Training Arm.

- Ent is safe in most winds, but in strong E to SW winds sea breaks on offshore shoals.
- Ent subject to silting; 3 tide gauges. Average tidal stream 2kn.
- No ⚓ in hbr; 4kn speed limit.

LIGHTS AND MARKS See 9.11.4 and chartlet. Gimblet Rock (30m) is conspic conical rock 3ca SW of ent.

COMMUNICATIONS (Code 01758) CGOC (01407) 762051; Ⓗ Bangor (A&E) 01248 384384; Police 101; Dr 701457. HM ☎704081. Marina: Ch 80 M H24. HM: VHF Ch 16 (0900-1715).

FACILITIES Marina www.hafanpwllheli.co.uk ☎701219; 420⌑, £2·63+ ⌁£3·22, £13·30 <4hrs. Pontoons are numbered 4–12 from

S–N; Pile berths on S side of appr channel.

Services 🅿🛥⚓🔧🛢⛽♨⚒🗜 Ⓔ ☒ ▣(50t) ⛴(14t) ⛽ wi-fi,.

Marina Boat Club ☎612271, ⚓⛽⌑.

Harbour 6 ⚓, £8/craft/day, ⚓⛽🛥🚿⚓⛴⚒⛵⛺.

Town ☒ Ⓑ ⛽ ✕ ⌂ ⇌ ✈ (Chester).

YACHT CLUBS South Caernarvonshire YC ☎712338.
Pwllheli SC pwllhelisailingclub@btinternet.com ☎614442.
Marina Boat Club ☎612271, ⛽⚓ ⌂.

ADJACENT HARBOURS AND ANCHORAGES

LLANBEDROG, Gwynedd, 52°51′·14N 04°28′·09W. AC 5609, 1410, 1411, 1971, 1512. HW –0245 on Dover, see 9.11.6 (St Tudwal's Roads). Good ⚓ N of Trwyn Llanbedrog (steep sided headland 131m) 1.0ca ENE of Bcn PHM in 2-3m. Beach has numerous conspicuous bathing huts. Sheltered in winds from N through W to SW. ⚓ well clear of local craft moorings as they have a long scope. Dinghy landing near boathouse.

Village limited facilities: ⛽ ▣ ▣ (½M) ✕ ⌂ ⇌ (Pwllheli).

MOCHRAS, Gwynedd, 52°49′·57N 04°07′·77W. AC 1971, 1512. HW –0245 on Dover. Small yacht hbr in drying Mochras lagoon on SE side of Shell Is. Bar, about 2ca seaward. Appr advised HW±2, but HW±1 at sp. 3 grey posts, R topmarks, mark NE side of chan. Ebb tide runs strongly in the narrow ent between the sea wall and Shell Is (lt Fl WRG 4s, 079°-G-124°-W-134°-R-179°; Mar-Nov); at sp beware severe eddies inside. Inside ent, buoyed chan runs WSW to Shell Island Yacht Hbr ☎01341 241453.

Facilities: ⛽⚓ ⛽ 🏴 ✕ ⌂. Shifting chan, marked by posts and buoys, runs NE to Pensarn, where permanent moorings limit space. Drying ⌑ ⇌.

9.11.8 PORTHMADOG

Gwynedd **52°55'·32N 04°07'·77W** ✿♙♙♙✿✿✿

CHARTS AC 1410, 1971, 1512; Imray C61, C51, C52

TIDES −0247 Dover; ML no data; Duration 0455

Standard Port MILFORD HAVEN (⟶)

Times				Height (metres)			
High Water		Low Water		MHWS	MHWN	MLWN	MLWS
0100	0800	0100	0700	7·0	5·2	2·5	0·7
1300	2000	1300	1900				
Differences PORTHMADOG							
+0235	+0210	ND	ND	−1·9	−1·8	ND	ND

SHELTER Inner hbr (N of Cei Ballast): Good all year round; visitors' drying ⚓ adjacent Madoc YC or afloat rafted on moored yachts. Outer hbr: Summer only, exposed to S winds. Speed limit 6kn in hbr upstream of No 8 buoy.

NAVIGATION WPT, 52°53'·01N 04°11'·10W, (SWM) Head ESE to enter laterally buoyed chan, which turns N at No 4 buoy to the E of N bank before turning NE off Ynys Gyngar (W Ho conspic).

- Channel shifts frequently and may divide but is well marked.
- Channel varies from that charted.
- In SW'lies, short steep-sided seas, especially on the ebb.
- Latest info from HM on request. Advise entering HW±1½.

LIGHTS AND MARKS Fairway buoy RW, L Fl 10s. SPM Fl Y 5s marks outfall 6 Ca to NW. Chan marker buoys Nos 1–9, 12,14, 15, 17 lit (May-Oct), have R/G reflective top marks and reflective numbers. Moel-y-Gest is conspic hill (259m) approx 1M WNW of harbour. Harlech Castle (ru) is about 3M SE of approach channel.

COMMUNICATIONS (Code 01766) CGOC (01407) 762051; Police 101; Dr 512284. HM 512927, mobile (07879) 433147; Pilot 530684. HM Ch 16 **12** (HO and when vessel due). Madoc YC Ch M.

FACILITIES Harbour £8·00 all craft, pilot, ⚓ access HW±4 (£10), ⚓(quay), ⌧ ⌕ ⚓(8t) ⚓ ACA.
Porthmadog SC ☎513546, ⚓⚓⚓.
Madoc YC ☎512976, ♿ ⚓⚓⚓⚓⚓⚓.
Town ⚓⚓⌧⚓⚓⚓✈ (Chester).

9.11.9 BARMOUTH

Gwynedd **52°42'·97N 04°03'·07W** ✿✿♙♙✿✿✿

CHARTS AC 5609, 1410, 1971, 1484; Imray C61, C51

TIDES −0305 Dover; ML 2·6; Duration 0515

Standard Port MILFORD HAVEN (⟶)

Times				Height (metres)			
High Water		Low Water		MHWS	MHWN	MLWN	MLWS
0100	0800	0100	0700	7·0	5·2	2·5	0·7
1300	2000	1300	1900				
Differences BARMOUTH							
+0207	+0200	+0300	+0233	−2·0	−1·5	−0·6	0·0

SHELTER Good. Entry safe HW±2½, but impossible in strong SW'lies. Exposed ⚓ W of Barmouth Outer buoy in 6 to 10m. In hbr there are 5 ⚓s, a pontoon for up to 10 boats <10m able to take the ground to the N of the quay, at the W end of which is a ⚓ berth that dries at half-tide. Secure as directed by HM because of submarine cables and strong tidal streams. Drying ⚓ inside Penrhyn Pt.

NAVIGATION WPT, Barmouth Outer SWM buoy, 52°42'·72N 04°05'·02W, 082°/0·8M to Y Perch lt, QR. Appr from SW between St Patrick's Causeway (Sarn Badrig) and Sarn-y-Bwch. Barmouth can be identified by Cader Idris, a mountain 890m high, 5M ESE. Fegla Fawr, a rounded hill, lies on S side of hbr.

- The Bar, 0·75M W of Penrhyn Pt, min depth 0·3m, is constantly changing. Chan is marked by buoys fitted with radar reflectors and reflective tape and moved as required.
- Spring ebb runs 3–5kn.
- The estuary and river (Afon Mawddach) are tidal and can be navigated for about 7M above railway br (clearance approx 5·5m); but chan is not buoyed, and sandbanks move constantly – local knowledge essential.

LIGHTS AND MARKS Outer SWM L Fl 10s; Bar SHM Fl G; No2 PHM Fl R. Y Perch lt, QR. Y Perch Sth Cardinal bn Q6 +LFl 15s marks S end of stony ledge extending 3ca SW from Ynys y Brawd across N Bank, 50m to SE PHM, Q R, marks channel. Ynys y Brawd groyne, with bn, Fl R 5s 5M at SE end. NW end of railway bridge is lit, 2 FR (hor).

COMMUNICATIONS (Code 01341) CGOC (01407) 762051; Police 101; Dr 280521. HM 280671 mob 07795 012747; Local information www.barmouthwebcam.co.uk

Call *Barmouth Hbr* VHF Ch **12** 16 (Apr-Sep 0900-1700 later for HW; Oct-Mar 0900-1600); wind and sea state are available.

FACILITIES 3 drying ⚓ M (call HM in advance for deep water ⚓), 8 ⚓ inc 6⚓ on pontoon/quay £10/£6 yacht (inc ⚓), ♿ ⚓⚓ 2 ⚓ (HW±3, £10/day). ⚓⚓ ⌧ Ⓔ at ⌕. **Merioneth YC** ☎280000.
Town ⚓⚓⚓⌧ⒷⓄ⚓ ✕ ⚓ ✈ (Manchester/ L'pool); Ferry to Penrhyn Pt for **Fairbourne** ⌧⚓ ⚓.

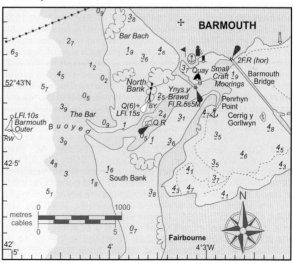

9.11.10 ABERDOVEY

Gwynedd 52°32'·57N 04°02'·72W (Jetty) ❀❀⚓⚓⚓✿✿✿

CHARTS AC 5609, 1410, 1972, 1484; Imray C61, C51

TIDES –0320 Dover; ML 2·6; Duration 0535. For differences see 9.11.11.

SHELTER Good except in strong W/SW winds. Berth on jetty in 3m; to the E there is heavy silting.

NAVIGATION WPT Aberdovey Outer SWM buoy, 52°32'·00N 04°05'·56W, 093°/0.43M to Bar buoy.

- Bar and channel constantly shift and are hazardous below half-tide; buoys moved accordingly.
- Enter channel at gateway between Bar SHM and PHM Fl R 5s. Visitors should call HM on VHF before entering.

LIGHTS AND MARKS See 9.11.4 and chartlet. No daymarks. Lts may be unreliable. Submarine cables (prohib ⚓s) marked by bns with R ◇ topmarks.

COMMUNICATIONS (Code 01654) CGOC (01646) 690909; Police 101; Dr 710414; ⊞ 710411. HM 767626, mob 07879 433148.

Call *Aberdovey Hbr* VHF Ch 12 16.

FACILITIES Jetty/Wharf ⚓ ⚓ 5◻ £8·00, 🔌 ⚓ 🔧.
Dovey YC ☎767607, ⚓ (£4.50 if engine <10hp) ⚓ 🔧 ▢.
Ynyslas BY (S shore) 🚩(16t) ⚒.
Town 🏪 🏦 🏧 ✉ Ⓑ 🍺 ✕ ▢ ⇌ ✈ (Birmingham).

OTHER HARBOURS IN SOUTH PART OF CARDIGAN BAY

ABERAERON, Ceredigion, **52°14'·62N 04°15'·94W**. AC 1410, 1972, 1484. HW –0325 on Dover; +0140 and –1·9m on Milford Haven; ML 2·7m; Duration 0540. A small, popular drying hbr at the mouth of the R Aeron. Short drying piers extend each side of river ent. In strong NW'lies there is little shelter. ◻£8·50/craft on NW wall, only in calm conditions. Foul ground with depths of 1·5m extend 3ca offshore to SW of Aberaeron. Beware Carreg Gloyn (0·3m) 4ca WSW of hbr, and Sarn Cadwgan (1·8m) shoals 5ca N of the hbr ent. Lts (see 9.11.4 for sectors): N pier Fl (4) WRG 15s 10m 6M. S pier Fl (3) G 10s 11m 6M. VHF 14 16. HM ☎(01545) 571645 ⚓ 🔧(from fishermen). YC ☎570077. Town 🏪.

NEW QUAY, Ceredigion, **52°12'·92N 04°21'·22W**. AC 1410, 1972, 1484. HW –0335 on Dover; Duration 0540; see 9.11.11. Good shelter in offshore winds, but untenable in North Westerlies. Carreg Ina, rks drying 1·3m, are marked by NCM buoy, Q, on E side of bay. 2 Y bns mark sewer outfall running 7ca NNW from Ina Pt. The hbr (dries 1·6m) is protected by a pier; Lt, Fl WG 3s 12m 8/5M; 135°-W-252°-G-295°. Groyne extends 80m SSE of pier head to a starboard bn; close ENE of which is a ECM bn, Q (3) 10s. HM ☎(01545) 560368; Dr ☎560203; YC ☎560516. VHF Ch 14 16. Facilities: ⚓ 🔧(from fishermen) ⚓(£8·50). Town 🏪(3M) ✉ ✕ ▢.

9.11.11 ABERYSTWYTH

Ceredigion **52°24'·42N 04°05'·47W** ❀❀⚓⚓⚓✿✿✿

CHARTS AC 5609, 5620, 1410, 1972, 1484; Imray C61, C51

TIDES – 0330 Dover; ML 2·7; Duration 0540
Standard Port MILFORD HAVEN (→)

Times				Height (metres)			
High Water		Low Water		MHWS	MHWN	MLWN	MLWS
0100	0800	0100	0700	7·0	5·2	2·5	0·7
1300	2000	1300	1900				
Differences ABERDOVEY							
+0215	+0200	+0230	+0305	–2·0	–1·7	–0·5	0·0
ABERYSTWYTH							
+0145	+0130	+0210	+0245	–2·0	–1·7	–0·7	0·0
NEW QUAY							
+0150	+0125	+0155	+0230	–2·1	–1·8	–0·6	–0·1
ABERPORTH							
+0135	+0120	+0150	+0220	–2·1	–1·8	–0·6	–0·1

SHELTER Good, in marina (1·7m) on E side of chan; or dry against Town Quay.

NAVIGATION WPT 52°24'·83N 04°06'·22W,133/0·6M° to ent. Appr dangerous in strong on-shore winds. The Bar, close off S pier hd, has 0·7m least depth. From N, beware Castle Rks, in R sector of N bkwtr lt; also rks drying 0·5m W of N bkwtr and boulders below S pier hd. Turn 90° port inside narrow ent. E edge of inner hbr chan 0·3m is defined by WCM bn which in line with Y daymark gives a daytime lead into the harbour.

LIGHTS AND MARKS N Bkwtr Hd ≠ 140° Wellington Mon't (on top Pendinas, conspic hill 120m high) clears to S of Castle Rks. Ldg lts 133°, both FR on Ystwyth Bridge, white daymarks. WCM bn on with Y daymark leads 100° across bar into harbour entrance.

COMMUNICATIONS (Code 01970) CGOC (01646) 690909; Police 101; Dr 624855. HM 611433, Mobile 07974 023965; Marina 611422.

HM Ch 14 16. Marina Ch 80.

FACILITIES Marina (Y Lanfa), ☎611422 ⚓ (launch £10), ◻88+15❷ £2·80+£3🔌 ⛽ ⚓ 🔧 🚩(10t) ⚒(15t by arrangement).
Town Quay ⚓ ◻ £8·50/craft, ⚓ 🔋 ⚒ ⚒ 🔲 Ⓔ ⚒(25t).
YC ☎612907, ⚓ ⚓ ▢.
Town 🏪 🏦 🏧 ✉ Ⓑ 🍺 ✕ ▢ ⇌ ✈ (Swansea).

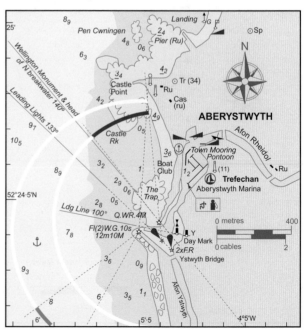

MINOR HARBOUR, 12M NE of Fishguard

PORT CARDIGAN, Ceredigion, **52°07'·02N 04°42'·07W**. AC 1973, 1484. HW −0405 on Dover; ML 2·4m; Duration 0550. Good shelter, but ent dangerous in strong N/NW winds. **Large scale charts are essential;** contact Afon Teifi Fairways Ltd, www.afonteififairways.com or Mooring Master (see below) in advance. The bar has 0·5m or less. ⚓ near Hotel (conspic) on E side of ent (2 bns Fl.R.4s, Fl.R.2s). Chan is usually close to E side marked by W posts; IDM (Fl (2) 5s) should be left to stbd when clear of the bar. ⚓ off Pen-yr-Ergyd. Inside spit pontoon or ⚓s off Teifi Boating Club, Mooring Master ☎(01239) 613966, mob 07799 284206. Upstream to Bryn-Du chan is marked by SWM but shifts constantly. St Dogmaels chan lies between ⚓s and pontoon, Mooring Master ☎07909 830752. Above the Ferry Inn chan runs up to the old bridge (3m clearance) through Netpool with deeper water moorings, Mooring Master ☎07814 035438.

Teifi Boating Club ☎613846, ⛽ 🚻; Services: ⚓ ✕.

Town Ⓑ 🛒 ⛽ 🏪 🔁 ⚓ ⚓ ✕ 🚻.

9.11.12 FISHGUARD

Pembrokeshire **52°00'·12N 04°58'·40W**
Commercial Hbr ✿✿✿◊◊◊; Lower Hbr ✿✿✿◊✿✿✿

CHARTS AC 5620, 1178, 1410, 1973, 1484; Imray C61, C51, C60

TIDES −0400 Dover; ML 2·6; Duration 0550

Standard Port MILFORD HAVEN (→)

Times				Height (metres)			
High Water		Low Water		MHWS	MHWN	MLWN	MLWS
0100	0800	0100	0700	7·0	5·2	2·5	0·7
1300	2000	1300	1900				
Differences FISHGUARD							
+0115	+0100	+0110	+0135	−2·2	−1·8	−0·5	+0·1
PORT CARDIGAN							
+0140	+0120	+0220	+0130	−2·3	−1·8	−0·5	0·0
CARDIGAN (Town)							
+0220	+0150	ND	ND	−2·2	−1·6	ND	ND
PORTHGAIN							
+0055	+0045	+0045	+0100	−2·5	−1·8	−0·6	0·0
RAMSEY SOUND							
+0030	+0030	+0030	+0030	−1·9	−1·3	−0·3	0·0
SOLVA							
+0015	+0010	+0035	+0015	−1·5	−1·0	−0·2	0·0
LITTLE HAVEN							
+0010	+0010	+0025	+0015	−1·1	−0·8	−0·2	0·0
MARTIN'S HAVEN							
+0010	+0010	+0015	+0015	−0·8	−0·5	+0·1	+0·1
SKOMER IS							
−0005	−0005	+0005	+0005	−0·4	−0·1	0·0	0·0

SHELTER Good, except in strong NW/NE winds. Beware large swell, especially In N winds. No ⚓, except SW of ferry quay. Good holding in most of the bay. At Lower town ⚓ off Saddle or Castle Points in 1·4m. Strong S'lies funnel down the hbr.

NAVIGATION WPT 52°01'·02N 04°57'·57W, 237°/0·48M to N bkwtr lt. *Keep at least 1ca clear of N bkwtr head; beware of ferries and High-Speed SeaCats manoeuvring.*

LIGHTS AND MARKS See 9.11.4 and chartlet. Strumble Hd Lt ho, Fl (4) 15s 45m 26M, is approx 4M WNW of hbr. The SHM (bn) at Aber Gwaun is very small.

COMMUNICATIONS (Code 01348) CGOC (01646) 690909; Police 101; Dr 872802. Commercial Hbr Supervisor 404425.

HM Ch **14** 16. Goodwick Marine Ch M (occas).

FACILITIES Fishguard (Lower town) dries 3·2m; limited ⚓, £5/night, 15⚓ (drying, max 5 tons), ⚓ ⚓ ⚓ ⚓ (from Goodwick Marine) ✕ ⚓ Ⓔ 🚿 ⚓ ACA.

Fishguard Bay YC ⚓ ⚓ 🚻.

Town 🏛 🏨 🍴 📮 Ⓑ ⚓ 🛒 ✕ 🚻 @ ⚓ ✈ (Cardiff), Ferry to Rosslare; 2/day; 1¾hrs; Stena Line (www.stenaline.ie).

HARBOURS IN ST BRIDES BAY (See AC 5620)

SOLVA, Pembrokeshire, **51°52'·02N 05°11'·67W**. AC 1478. HW −0450 on Dover; ML 3·2m; Duration 0555. See 9.11.12. Good shelter for small boats that can take the ground. Avoid in strong S winds. Black Scar, Green Scar and The Mare are rks 5ca S. Ent via SSE side; best water near Black Rk in centre of ent. Beware stone spit at Trwyn Caws on W just inside ent. There are 9 Or ⚓s drying on hard sand (£6.00), drying/rafting ABs for <9·5m LOA; or ⚓ behind the rk in approx 3m. Yachts can go up to the quay (£7.50) Facilities: 🍴 🛒 ✕ 🚻 in village. ⚓ on quay, ⚓ (launching £5). HM (01437) 721703 mob 07974 020139, VHF Ch 16 8, ⚓ ⚓. **Solva Boat Owners Assn** ☎721489 mob 07974 020139.

ST BRIDES BAY, Pembrokeshire, **51°49'·02N 05°10'·07W**. AC 1478. HW (Little Haven) −0450 on Dover; ML 3·2m; Duration 0555. See 9.11.12. A SPM buoy, Fl (5) Y 20s, is midway between Ramsey and Skomer islands at 51°48'·2N 05°20'·0W. Keep at least 100m offshore 1/9-28/2 to avoid disturbing seals, and ditto nesting sea birds 1/3-31/7. Many good ⚓s, especially between Little Haven and Borough Head in S or E winds or between Solva and Dinas Fawr in N or E winds. In W'lies boats should shelter in Solva (above), Skomer (below) or Pendinas Bach. Tankers ⚓ in mouth of bay. For apprs from the N or S see 9.11.5. Facilities: (Little Haven) ⚓ (cans) ⚓ 🛒 ✕ 🚻.

SKOMER, Pembrokeshire, **51°44'·42N 05°16'·77W**. AC 1478, 2878. HW −0455 Dover. See 9.11.12. The island is a National Nature Reserve (fee payable to Warden on landing) and also a Marine Nature Reserve, extending to Marloes Peninsula. Keep at least 100m offshore 1/9-28/2 to avoid disturbing seals and ditto nesting sea birds 1/3-31/7. There is a 5kn speed limit within 100m of the island. ⚓ in S Haven (N Haven ⚓ due to eel grass). Enter N Haven close to W shore, and land on W side of bay on beach or at steps. In N Haven pick up ⚓s provided. No access to the island from S Haven. For Jack Sound see 9.11.5. There are no lts, marks or facilities. For info, Marine Conservation Officer ☎(01646) 636736.

9.11.13 MILFORD HAVEN

Pembrokeshire 51°40'·13N 05°08'·16W ✿✿✿✿✿✿✿✿✿

CHARTS AC 5620, 1410, 1178, 1478, 2878, 3273/4/5; Imray C60, C13, 2600

TIDES –0500 Dover; ML 3·8; Duration 0605

Standard Port MILFORD HAVEN (→)

Times				Height (metres)			
High Water		Low Water		MHWS	MHWN	MLWN	MLWS
0100	0800	0100	0700	7·0	5·2	2·5	0·7
1300	2000	1300	1900				
Differences DALE ROADS							
–0005	–0005	–0008	–0008	0·0	0·0	0·0	–0·1
NEYLAND							
+0002	+0010	0000	0000	0·0	0·0	0·0	0·0
HAVERFORDWEST							
+0010	+0025	DR	DR	–4·8	–4·9	DR	DR
LLANGWM (Black Tar)							
+0010	+0020	+0005	0000	+0·1	+0·1	0·0	–0·1

SHELTER Very good in various places round the hbr, especially in Milford Marina and Neyland Yacht Haven. ⚓s in Dale Bay; off Chapel Bay and Angle Pt on S shore; off Scotch Bay, E of Milford marina; and others beyond Pembroke Dock. Free pontoons (May–Oct) include: Dale Bay, Gelliswick Bay; waiting pontoons off Milford Dock and Hobbs Pt (for Pembroke Dock); and drying pontoons at Dale Beach, Hazelbeach, Neyland and Burton; mainly intended for tenders. It is possible to dry out safely at inshore areas of Dale, Sandy Haven and Angle Bay, depending on weather.

NAVIGATION WPT (W) 51°40'·25N 05°10'·65W. WPT (E) 51°40'·05N 05°08'·06W.

- The tide sets strongly across the ent to the Haven particularly at sp. In bad weather avoid passing over Mid Chan Rks and St Ann's Hd shoal, where a confused sea and swell will be found. Give St Ann's Head a wide berth especially on the ebb, when East Chan by Sheep Island is better.
- Beware large tankers entering/departing the Haven and ferries moving at high speed in the lower Haven. **Keep outside the main shipping channels if possible.**
- Milford Haven Port Authority has a jetty, Port Control and offices near Hubberston Pt. Their launches have G hulls and W upperworks with 'PILOT' in black letters and fly a Pilot flag while on patrol; Fl Bu lt at night. Their instructions must be obeyed. No vessel may pass within 100m of any terminal or any tanker, whether at ⚓ or under way.

River Cleddau is navigable 6M to Picton Pt, at junction of West and East arms, at all tides for boats of moderate draught. Clearance under Cleddau Bridge above Neyland is 36m; and 25m under power cable 1M upstream. Chan to Haverfordwest has 2m at HW and clearances of only 6m below cables and bridge; only feasible for shoal draught/lifting keel and unmasted craft.

Firing Ranges to the S and SE, see 9.11.14 and AC 1076.

LIGHTS AND MARKS See 9.11.4 and chartlets. Y SPM have been established in W Angle, Gelliswick, Dale Bays and off Jenkins Pt to prohibit high speed craft and water skiing.

Milford Haven leading lights are specifically for VLCCs entering/leaving hbr and proceeding to/from the oil refineries. The Haven is very well buoyed and lit as far as Cleddau bridge.

Milford Dock ldg lts 348°, both F.Bu, with W ○ daymarks. Dock entry sigs on E side of lock: 2 FG (vert) = gates open, vessels may enter. VHF Ch 18 is normally used for ent/exit by day and night.

COMMUNICATIONS (Code 01646) Lock 696310; CGOC 690909; Police 101; ⊞ Haverfordwest (01437) 764545; Dr 690674. Port Control 696136/7.

Monitor *Milford Haven Port Control* Ch 12 (H24) while underway in the Haven. *Milford Haven Patrol* launches Ch11/12 (H24). To lock in to Milford Dock marina call *Pierhead,* Ch 14, for clearance, then call *Milford Marina,* Ch ⚓ for a berth. Neyland Yacht Haven Ch 80/M.

Broadcasts: Nav warnings on Ch 12 14 at 0300 and 1500 (both UT). Gale warnings on receipt. Tide hts and winds on request. Bcsts on Ch 16 67 at 0335 then every 4 hrs to 2335.

FACILITIES Marinas/Berthing (from seaward):

Dale Bay YC Pontoon ⚓ ⚓ ✕ 🛒.

Milford Haven Port Authority jetty ☎696133 (occas use by visitors with approval from HM), ⚓ ⚓.

Milford Marina ☎696312, Pierhead ☎696310. VHF Ch 14, 80/M. Lock hrs summer (Apr–Sep): ent HW–5, exit HW–4, free flow HW–2¼ to HW–¼, ent+3½, exit+4½; winter (Oct–Mar): ent HW–3¼, free flow HW–2¼ to HW–¼, ent+2½, exit+3; waiting pontoon or shelter in lock, 3·5m water at MLWS. 280⚓ inc ⚓, £2·85 ⚓ ⚓. ⚓ ⚓(H24) ⚓ ✕ ⚓ ⚓ ⚓ ⚓ ⚓ ⚓ ⚓(14t) ⚓ ⚓ ⚓ ⚓ ⚓ ✕ ⚓ ice, wi-fi.

Lawrenny Yacht Station ☎651212/651065, ⚓ £5, 100⚓; ⚓ ⚓ ⚓ (launch £5) ⚓ ⚓ ⚓ ⚓ ⚓ ✕ ⚓ ⚓ ⚓ ⚓(15t) ⚓ 🛒 ✕.

Neyland Yacht Haven ⚓, www.yachthavens.com ☎601601, 420⚓ inc ⚓ £2·50, ⚓ ⚓ ✕ ⚓ ⚓ ⚓ ⚓ ⚓ ⚓(20t) ⚓ ⚓ ✕; Access lower basin H24, but upper basin depth limited (sill + depth gauges and R/G lit perches); marina and approaches dredged annually Oct/March – proceed with extreme caution.

Dale Sailing Co (at Neyland) ⚓ ⚓ LPG ✕ ⚓ ⚓ ⚓(35t) ⚓.

East Llannion Marine: access HW±3, ⚓ scrubbing piles ⚓ ⚓(30t).

Rudder's BY, small but useful, is just upstream of Burton Pt.

Yacht Clubs: Dale YC ☎636362; Pembrokeshire YC ☎692799; Neyland YC ☎600267; Pembroke Haven YC ☎684403; Lawrenny YC ☎651212.

Towns: Milford Haven, ⚓ ⑧ ⚓. Pembroke Dock; ⚓ ⑧ ⚓ ⊞. Neyland, ⚓ ⑧. Haverfordwest, ⊞ ⚓ ⑧ ⚓ ✈ (Swansea or Cardiff). Ferry: Pembroke Dock–Rosslare; 2/day; Irish Ferries.

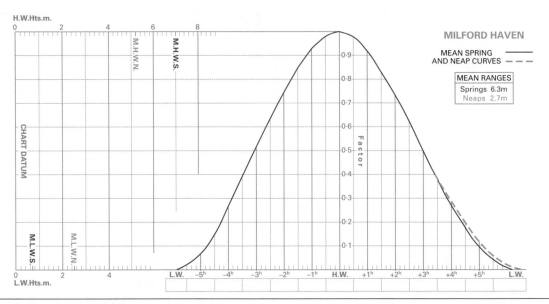

H.W.Hts.m.

MILFORD HAVEN

MEAN SPRING
AND NEAP CURVES

MEAN RANGES
Springs 6.3m
Neaps 2.7m

CHART DATUM

M.H.W.N.
M.H.W.S.

Factor

0·9
0·8
0·7
0·6
0·5
0·4
0·3
0·2
0·1

M.L.W.S.
M.L.W.N.

L.W. −5ʰ −4ʰ −3ʰ −2ʰ −1ʰ H.W. +1ʰ +2ʰ +3ʰ +4ʰ +5ʰ L.W.

LW.Hts.m.

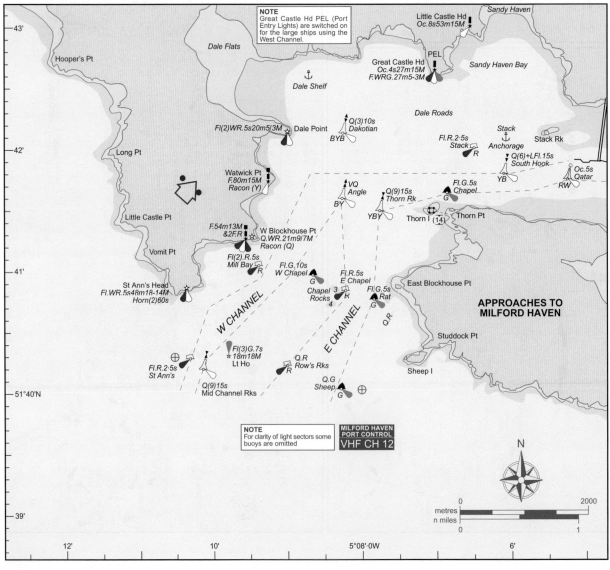

NOTE
Great Castle Hd PEL (Port
Entry Lights) are switched on
for the large ships using the
West Channel.

Little Castle Hd
Oc.8s53m15M

Sandy Haven

Hooper's Pt

Dale Flats

Great Castle Hd
Oc.4s27m15M
F.WRG.27m5-3M

PEL

Sandy Haven Bay

Dale Shelf

Dale Roads

Fl(2)WR.5s20m5/3M
Dale Point

Q(3)10s
Dakotian
BYB

Stack
Anchorage

Stack Rk

Long Pt

Fl.R.2·5s
Stack
R

Stack

Q(6)+LFl.15s
South Hook
YB

Oc.5s
Qatar
RW

Little Castle Pt

Watwick Pt
F.80m15M
Racon (Y)

VQ
Angle
BY

Fl.G.5s
Chapel
G

Q(9)15s
Thorn Rk

Vomit Pt

F.54m13M
&2F.R

W Blockhouse Pt
Q.WR.21m9/7M
Racon (Q)

YBY

Thorn I (14)

Thorn Pt

Fl(2).R.5s
Mill Bay
R

Fl.G.10s
W Chapel
G

Fl.R.5s
E Chapel

East Blockhouse Pt

St Ann's Head
Fl.WR.5s48m18-14M
Horn(2)60s

Chapel
Rocks
3
4

Fl.G.5s
Rat
G

**APPROACHES TO
MILFORD HAVEN**

Q.R

Studdock Pt

W CHANNEL

E CHANNEL

Fl(3)G.7s
★18m18M
Lt Ho

Q.R
Row's Rks
R

Sheep I

Fl.R.2·5s
St Ann's

Q.G
Sheep
G

Q(9)15s
Mid Channel Rks

51°40'N

NOTE
For clarity of light sectors some
buoys are omitted

MILFORD HAVEN
PORT CONTROL
VHF CH 12

N

39'

metres   0   2000
n miles   0   1

12'   10'   5°08'·0W   6'

STANDARD TIME (UT)
For Summer Time add ONE hour in **non-shaded areas**

MILFORD HAVEN LAT 51°42'N LONG 5°03'W
TIMES AND HEIGHTS OF HIGH AND LOW WATERS

Dates in red are SPRINGS
Dates in blue are NEAPS

YEAR 2016

JANUARY

Day	Time m	Time m	Time m	Time m
1 F	0419 2.1	1030 5.9	1650 2.2	2256 5.5
16 SA ☽	0426 1.4	1039 6.5	1658 1.5	2305 6.1
2 SA	0506 2.4	1117 5.6	1742 2.5	2350 5.3
17 SU	0522 1.8	1136 6.2	1759 1.8	
3 SU	0606 2.6	1217 5.4	1850 2.6	
18 M	0008 5.8	0630 2.0	1245 5.9	1912 2.0
4 M	0100 5.2	0719 2.7	1332 5.3	2003 2.6
19 TU	0122 5.7	0753 2.1	1400 5.8	2031 2.0
5 TU	0215 5.3	0833 2.6	1442 5.4	2110 2.4
20 W	0239 5.8	0911 1.9	1514 6.0	2142 1.8
6 W	0317 5.6	0935 2.3	1540 5.7	2204 2.1
21 TH	0349 6.1	1018 1.6	1620 6.2	2241 1.5
7 TH	0408 5.9	1026 1.9	1628 6.0	2249 1.7
22 F	0449 6.4	1113 1.3	1715 6.5	2331 1.2
8 F	0452 6.3	1111 1.6	1712 6.4	2331 1.4
23 SA	0538 6.8	1201 1.1	1802 6.8	
9 SA	0534 6.7	1152 1.3	1753 6.6	
24 SU ○	0016 1.0	0621 7.0	1244 0.9	1843 6.9
10 M ●	0011 1.2	0614 6.9	1234 1.0	1834 6.9
25 M	0056 0.9	0700 7.0	1322 0.9	1921 6.9
11 M	0052 1.0	0654 7.1	1315 0.8	1915 7.0
26 TU	0133 0.9	0737 7.0	1357 0.9	1956 6.8
12 TU	0132 0.8	0736 7.2	1357 0.7	1957 7.0
27 W	0207 1.0	0811 6.9	1430 1.1	2029 6.6
13 W	0214 0.8	0818 7.2	1439 0.8	2039 6.9
28 TH	0240 1.2	0844 6.7	1502 1.3	2102 6.4
14 TH	0255 0.9	0902 7.1	1522 0.9	2124 6.7
29 F	0311 1.4	0916 6.5	1532 1.6	2135 6.1
15 F	0339 1.1	0948 6.8	1607 1.2	2211 6.4
30 SA	0343 1.7	0950 6.1	1605 1.9	2210 5.8
31 SU	0418 2.0	1028 5.8	1643 2.2	2252 5.5

FEBRUARY

Day	Time m	Time m	Time m	Time m
1 M ☽	0502 2.3	1113 5.5	1733 2.5	2346 5.2
16 TU	0601 2.0	1216 5.7	1840 2.1	
2 TU	0604 2.6	1214 5.2	1847 2.7	
17 W	0052 5.5	0728 2.2	1336 5.5	2008 2.2
3 W	0100 5.1	0729 2.7	1337 5.1	2014 2.6
18 TH	0219 5.5	0857 2.1	1500 5.6	2128 2.0
4 TH	0226 5.2	0850 2.5	1457 5.3	2126 2.3
19 F	0338 5.8	1008 1.8	1610 5.9	2230 1.7
5 F	0333 5.6	0954 2.1	1558 5.7	2221 1.9
20 SA	0438 6.2	1103 1.4	1704 6.3	2319 1.3
6 SA	0425 6.1	1046 1.6	1649 6.2	2309 1.4
21 SU	0525 6.6	1148 1.1	1747 6.6	
7 SU	0512 6.6	1132 1.2	1734 6.6	2353 1.0
22 M ○	0001 1.1	0605 6.8	1227 0.9	1825 6.8
8 M ●	0556 7.0	1216 0.8	1818 7.0	
23 TU	0038 0.9	0641 6.9	1301 0.8	1859 6.8
9 TU	0035 0.7	0638 7.3	1300 0.5	1900 7.2
24 W	0111 0.8	0715 7.0	1333 0.8	1931 6.8
10 W	0118 0.5	0721 7.5	1342 0.3	1942 7.3
25 TH	0143 0.9	0746 6.9	1403 0.9	2002 6.7
11 TH	0200 0.4	0803 7.5	1424 0.3	2024 7.3
26 F	0213 1.0	0815 6.8	1431 1.1	2031 6.6
12 F	0241 0.5	0846 7.4	1506 0.5	2107 7.0
27 SA	0242 1.1	0845 6.6	1459 1.3	2101 6.4
13 SA	0323 0.7	0930 7.1	1545 0.8	2151 6.7
28 SU	0311 1.4	0915 6.3	1528 1.6	2132 6.1
14 SU	0407 1.1	1017 6.7	1634 1.3	2240 6.3
29 M	0342 1.7	0947 6.0	1600 1.9	2207 5.7
15 M ☽	0457 1.5	1110 6.2	1728 1.8	2338 5.8

MARCH

Day	Time m	Time m	Time m	Time m
1 TU ☽	0419 2.1	1026 5.6	1640 2.3	2252 5.4
16 W	0538 2.0	1151 5.6	1812 2.2	
2 W	0508 2.4	1118 5.2	1737 2.6	2357 5.1
17 TH	0027 5.4	0706 2.3	1314 5.3	1945 2.4
3 TH	0624 2.7	1234 5.0	1913 2.7	
18 F	0157 5.4	0839 2.2	1442 5.4	2109 2.2
4 F	0127 5.1	0804 2.6	1410 5.1	2045 2.5
19 SA	0319 5.7	0951 1.9	1553 5.8	2211 1.8
5 SA	0253 5.4	0920 2.2	1526 5.6	2151 2.0
20 SU	0418 6.1	1044 1.5	1644 6.1	2259 1.4
6 SU	0355 6.0	1019 1.6	1623 6.1	2243 1.4
21 M	0504 6.4	1126 1.2	1725 6.5	2338 1.2
7 M	0446 6.5	1108 1.1	1711 6.6	2330 0.9
22 TU	0543 6.7	1202 1.0	1801 6.7	
8 TU	0533 7.0	1155 0.6	1756 7.1	
23 W ○	0013 1.0	0616 6.8	1235 0.9	1833 6.8
9 W ●	0015 0.5	0617 7.4	1239 0.2	1839 7.4
24 TH	0046 0.9	0648 6.9	1305 0.9	1904 6.8
10 TH	0059 0.2	0701 7.7	1323 0.1	1922 7.5
25 F	0116 0.9	0718 6.8	1334 0.9	1933 6.8
11 F	0142 0.1	0744 7.7	1405 0.1	2004 7.5
26 SA	0146 0.9	0747 6.8	1402 1.0	2002 6.7
12 SA	0224 0.2	0827 7.6	1446 0.3	2047 7.2
27 SU	0214 1.1	0815 6.6	1430 1.2	2031 6.5
13 SU	0306 0.5	0911 7.2	1528 0.7	2130 6.9
28 M	0244 1.3	0845 6.4	1458 1.5	2101 6.2
14 M	0349 0.9	0956 6.7	1612 1.2	2217 6.4
29 TU	0315 1.6	0916 6.1	1530 1.8	2135 5.9
15 TU ☽	0437 1.5	1048 6.1	1703 1.8	2313 5.8
30 W	0351 1.9	0953 5.7	1607 2.1	2217 5.6
31 TH ☽	0436 2.2	1042 5.4	1658 2.4	2316 5.3

APRIL

Day	Time m	Time m	Time m	Time m
1 F	0541 2.5	1153 5.1	1819 2.6	
16 SA	0126 5.4	0810 2.3	1411 5.3	2038 2.3
2 SA	0041 5.2	0719 2.5	1326 5.1	2002 2.5
17 SU	0245 5.6	0920 2.0	1521 5.6	2140 2.0
3 SU	0211 5.4	0843 2.1	1450 5.5	2116 2.0
18 M	0346 5.9	1013 1.7	1613 6.0	2228 1.6
4 M	0321 6.0	0947 1.5	1552 6.1	2214 1.4
19 TU	0433 6.2	1055 1.4	1655 6.3	2308 1.4
5 TU	0417 6.5	1040 1.0	1644 6.7	2304 0.9
20 W	0512 6.5	1131 1.2	1732 6.5	2344 1.2
6 W	0506 7.1	1129 0.5	1731 7.1	2351 0.4
21 TH	0547 6.6	1204 1.1	1805 6.7	
7 TH ●	0553 7.5	1215 0.2	1816 7.5	
22 F ○	0017 1.1	0619 6.7	1236 1.0	1835 6.7
8 F	0037 0.2	0638 7.7	1301 0.0	1900 7.6
23 SA	0049 1.0	0650 6.7	1306 1.0	1905 6.8
9 SA	0122 0.1	0723 7.7	1344 0.1	1944 7.6
24 SU	0119 1.0	0720 6.7	1335 1.1	1935 6.7
10 SU	0205 0.2	0807 7.5	1426 0.3	2027 7.3
25 M	0150 1.1	0749 6.6	1405 1.2	2006 6.6
11 M	0248 0.5	0852 7.1	1508 0.7	2111 6.9
26 TU	0222 1.2	0821 6.4	1436 1.4	2038 6.4
12 TU	0333 0.9	0938 6.6	1552 1.3	2158 6.4
27 W	0255 1.5	0854 6.2	1509 1.7	2113 6.1
13 W	0421 1.5	1028 6.1	1642 1.8	2252 5.9
28 TH	0333 1.7	0933 5.9	1548 2.0	2157 5.8
14 TH ☽	0520 2.0	1129 5.5	1747 2.3	
29 F	0419 2.0	1022 5.6	1639 2.2	2254 5.6
15 F	0001 5.5	0641 2.3	1247 5.2	1914 2.4
30 SA ☽	0520 2.2	1129 5.3	1749 2.4	

Chart Datum: 3·71 metres below Ordnance Datum (Newlyn). HAT is 7·9 metres above Chart Datum.

STANDARD TIME (UT)
For Summer Time add ONE hour in **non-shaded areas**

MILFORD HAVEN LAT 51°42′N LONG 5°03′W
TIMES AND HEIGHTS OF HIGH AND LOW WATERS

Dates in red are **SPRINGS**
Dates in blue are NEAPS

YEAR **2016**

MAY

Time	m	Time	m
1 0010	5.5	**16** 0158	5.5
0642	2.2	0833	2.2
SU 1252	5.3	M 1436	5.5
1921	2.2	2056	2.2
2 0133	5.6	**17** 0302	5.7
0805	2.0	0930	1.9
M 1413	5.6	TU 1532	5.7
2039	1.9	2149	1.9
3 0245	6.0	**18** 0353	6.0
0913	1.5	1016	1.7
TU 1519	6.1	W 1618	6.0
2142	1.5	2233	1.7
4 0345	6.5	**19** 0437	6.2
1010	1.0	1056	1.5
W 1614	6.6	TH 1658	6.3
2236	1.0	2312	1.4
5 0439	7.0	**20** 0515	6.4
1102	0.6	1132	1.3
TH 1705	7.1	F 1734	6.5
2327	0.6	2348	1.3
6 0529	7.3	**21** 0550	6.5
1151	0.3	1206	1.2
F 1753	7.4	SA 1808	6.6
●		○	
7 0015	0.3	**22** 0022	1.2
0616	7.5	0623	6.6
SA 1238	0.2	SU 1239	1.2
1839	7.5	1840	6.7
8 0102	0.2	**23** 0056	1.1
0703	7.5	0656	6.6
SU 1323	0.3	M 1312	1.2
1924	7.5	1913	6.7
9 0148	0.3	**24** 0130	1.1
0749	7.3	0729	6.5
M 1407	0.5	TU 1345	1.2
2009	7.3	1946	6.6
10 0233	0.6	**25** 0205	1.2
0835	7.0	0803	6.4
TU 1451	0.9	W 1419	1.4
2054	6.9	2022	6.5
11 0318	1.0	**26** 0242	1.4
0921	6.6	0840	6.3
W 1535	1.3	TH 1456	1.5
2141	6.5	2101	6.3
12 0406	1.4	**27** 0322	1.5
1010	6.1	0922	6.1
TH 1623	1.8	F 1538	1.7
2232	6.0	2146	6.1
13 0500	1.9	**28** 0408	1.7
1105	5.6	1011	5.8
F 1720	2.2	SA 1627	2.0
◑ 2332	5.7	2241	5.9
14 0607	2.2	**29** 0504	1.9
1211	5.3	1111	5.6
SA 1832	2.4	SU 1729	2.1
		◑ 2347	5.8
15 0044	5.5	**30** 0613	1.9
0723	2.3	1223	5.6
SU 1327	5.3	M 1846	2.1
1949	2.4		
		31 0059	5.9
		0729	1.8
		TU 1338	5.8
		2003	1.9

JUNE

Time	m	Time	m
1 0211	6.1	**16** 0307	5.6
0839	1.5	0932	2.0
W 1446	6.1	TH 1537	5.7
2110	1.5	2153	2.0
2 0315	6.4	**17** 0358	5.8
0941	1.2	1019	1.8
TH 1546	6.5	F 1623	6.0
2210	1.2	2238	1.7
3 0413	6.8	**18** 0442	6.1
1037	0.9	1100	1.6
F 1641	6.8	SA 1704	6.3
2305	0.8	2319	1.5
4 0507	7.0	**19** 0522	6.3
1129	0.7	1139	1.4
SA 1733	7.1	SU 1742	6.5
2356	0.6	2357	1.3
5 0559	7.2	**20** 0559	6.4
1218	0.5	1215	1.3
SU 1822	7.3	M 1818	6.6
●		○	
6 0045	0.5	**21** 0035	1.2
0647	7.2	0635	6.5
M 1306	0.5	TU 1252	1.2
1908	7.3	1854	6.7
7 0133	0.5	**22** 0113	1.1
0734	7.1	0712	6.6
TU 1351	0.7	W 1328	1.1
1954	7.2	1931	6.8
8 0218	0.7	**23** 0151	1.1
0819	6.9	0750	6.6
W 1434	0.9	TH 1406	1.2
2038	6.9	2010	6.7
9 0303	1.0	**24** 0230	1.1
0904	6.5	0830	6.5
TH 1517	1.3	F 1446	1.3
2123	6.6	2051	6.6
10 0347	1.4	**25** 0312	1.2
0948	6.2	0912	6.4
F 1601	1.6	SA 1528	1.4
2208	6.2	2136	6.5
11 0433	1.7	**26** 0357	1.4
1035	5.8	0959	6.2
SA 1648	2.0	SU 1615	1.6
2257	5.9	2226	6.3
12 0525	2.1	**27** 0447	1.5
1128	5.5	1053	6.0
SU 1743	2.3	M 1709	1.8
◑ 2354	5.6	◑ 2324	6.1
13 0625	2.3	**28** 0546	1.7
1231	5.3	1155	5.8
M 1848	2.4	TU 1815	1.9
14 0100	5.4	**29** 0030	6.0
0731	2.3	0655	1.8
TU 1340	5.3	W 1306	5.8
1957	2.4	1930	1.9
15 0208	5.5	**30** 0141	6.0
0836	2.2	0808	1.7
W 1443	5.5	TH 1417	5.9
2100	2.2	2044	1.7

JULY

Time	m	Time	m
1 0250	6.2	**16** 0318	5.5
0917	1.5	0941	2.1
F 1523	6.2	SA 1548	5.7
2150	1.4	2206	2.0
2 0354	6.4	**17** 0411	5.8
1018	1.2	1030	1.8
SA 1624	6.5	SU 1636	6.1
2250	1.1	2253	1.7
3 0453	6.7	**18** 0456	6.1
1113	1.0	1113	1.5
SU 1719	6.8	M 1718	6.4
2343	0.9	2335	1.4
4 0546	6.9	**19** 0537	6.4
1203	0.8	1153	1.3
M 1809	7.0	TU 1758	6.7
●		○	
5 0033	0.7	**20** 0016	1.2
0635	7.0	0617	6.6
TU 1251	0.7	W 1233	1.1
1855	7.1	1836	6.9
6 0119	0.7	**21** 0056	1.0
0720	7.0	0656	6.8
W 1334	0.8	TH 1313	0.9
1939	7.1	1916	7.0
7 0202	0.8	**22** 0136	0.8
0802	6.8	0736	6.8
TH 1416	0.9	F 1353	0.9
2020	6.9	1957	7.1
8 0243	1.0	**23** 0217	0.8
0842	6.6	0817	6.8
F 1455	1.2	SA 1433	0.9
2100	6.7	2038	7.0
9 0322	1.3	**24** 0258	0.9
0921	6.3	0859	6.7
SA 1533	1.5	SU 1515	1.0
2138	6.4	2122	6.8
10 0400	1.6	**25** 0341	1.0
1001	6.0	0943	6.5
SU 1611	1.8	M 1559	1.3
2218	6.1	2209	6.6
11 0439	1.9	**26** 0427	1.3
1043	5.7	1032	6.2
M 1653	2.1	TU 1648	1.6
2303	5.7	◑ 2301	6.3
12 0525	2.2	**27** 0521	1.6
1132	5.4	1129	5.9
TU 1745	2.4	W 1748	1.8
◑ 2356	5.4		
13 0622	2.4	**28** 0003	6.0
1233	5.2	0626	1.9
W 1851	2.5	TH 1238	5.7
		1903	2.0
14 0102	5.3	**29** 0116	5.8
0731	2.5	0744	1.9
TH 1346	5.2	F 1354	5.7
2003	2.5	2026	2.0
15 0215	5.3	**30** 0232	5.9
0841	2.4	0901	1.8
F 1453	5.4	SA 1509	5.9
2110	2.3	2140	1.7
		31 0343	6.1
		1007	1.6
		SU 1615	6.3
		2242	1.4

AUGUST

Time	m	Time	m
1 0445	6.4	**16** 0431	6.0
1103	1.3	1049	1.6
M 1711	6.6	TU 1654	6.4
2335	1.1	2313	1.4
2 0537	6.7	**17** 0515	6.4
1152	1.0	1132	1.2
TU 1758	6.9	W 1736	6.8
●		2355	1.0
3 0021	0.9	**18** 0557	6.8
0622	6.8	1213	0.9
W 1236	0.9	TH 1817	7.1
1841	7.1	○	
4 0104	0.8	**19** 0036	0.7
0703	6.9	0637	7.0
TH 1316	0.8	F 1254	0.7
1920	7.1	1857	7.3
5 0142	0.8	**20** 0118	0.5
0741	6.8	0718	7.2
F 1353	0.9	SA 1335	0.6
1957	7.0	1939	7.4
6 0218	0.9	**21** 0159	0.5
0816	6.7	0759	7.2
SA 1428	1.1	SU 1416	0.6
2032	6.8	2020	7.4
7 0251	1.2	**22** 0241	0.6
0850	6.5	0841	7.0
SU 1501	1.3	M 1458	0.8
2105	6.5	2103	7.1
8 0323	1.4	**23** 0322	0.8
0924	6.2	0924	6.8
M 1533	1.6	TU 1540	1.1
2139	6.2	2149	6.8
9 0355	1.7	**24** 0406	1.2
0959	5.9	1010	6.4
TU 1607	1.9	W 1628	1.5
2215	5.9	2239	6.3
10 0430	2.1	**25** 0457	1.6
1039	5.6	1105	6.0
W 1648	2.3	TH 1726	1.9
◑ 2259	5.5	◑ 2340	5.9
11 0516	2.4	**26** 0601	2.0
1129	5.3	1214	5.7
TH 1745	2.6	F 1846	2.2
2355	5.2		
12 0623	2.6	**27** 0056	5.6
1237	5.1	0726	2.2
F 1905	2.7	SA 1338	5.6
		2017	2.2
13 0112	5.1	**28** 0221	5.6
0748	2.7	0851	2.1
SA 1404	5.2	SU 1501	5.8
2027	2.6	2135	1.8
14 0237	5.2	**29** 0338	5.9
0903	2.4	0959	1.8
SU 1514	5.5	M 1608	6.2
2134	2.2	2235	1.5
15 0341	5.6	**30** 0437	6.3
1001	2.0	1053	1.4
M 1608	5.9	TU 1700	6.6
2227	1.8	2324	1.2
		31 0524	6.6
		1138	1.1
		W 1744	6.9

Chart Datum: 3·71 metres below Ordnance Datum (Newlyn). HAT is 7·9 metres above Chart Datum.

STANDARD TIME (UT)
For Summer Time add ONE hour in **non-shaded areas**

MILFORD HAVEN   LAT 51°42′N   LONG 5°03′W
TIMES AND HEIGHTS OF HIGH AND LOW WATERS

Dates in red are **SPRINGS**
Dates in blue are **NEAPS**

YEAR **2016**

S Wales

SEPTEMBER

Day	Time m	Day	Time m
1 TH ●	0006 1.0 / 0605 6.8 / 1218 0.9 / 1822 7.0	16 F ○	0533 7.0 / 1150 0.8 / 1753 7.3
2 F	0043 0.9 / 0642 6.9 / 1254 0.9 / 1858 7.1	17 SA	0014 0.5 / 0615 7.3 / 1233 0.5 / 1835 7.6
3 SA	0117 0.9 / 0716 6.9 / 1327 0.9 / 1930 7.0	18 SU	0057 0.3 / 0657 7.5 / 1315 0.4 / 1918 7.7
4 SU	0149 0.9 / 0747 6.8 / 1359 1.0 / 2001 6.9	19 M	0139 0.3 / 0738 7.5 / 1357 0.4 / 2000 7.6
5 M	0218 1.1 / 0818 6.6 / 1429 1.2 / 2031 6.6	20 TU	0221 0.4 / 0820 7.3 / 1439 0.6 / 2044 7.3
6 TU	0247 1.4 / 0848 6.4 / 1523 1.5 / 2102 6.4	21 W	0302 0.8 / 0904 7.0 / 1523 1.0 / 2129 6.9
7 W	0315 1.7 / 0919 6.1 / 1529 1.8 / 2134 6.0	22 TH	0346 1.2 / 0950 6.5 / 1611 1.5 / 2220 6.3
8 TH	0347 2.0 / 0954 5.8 / 1605 2.2 / 2212 5.6	23 F	0436 1.8 / 1045 6.0 / 1710 2.0 / ◑ 2321 5.8
9 F	0425 2.4 / 1037 5.4 / 1652 2.5 / ◔ 2301 5.4	24 SA	0541 2.2 / 1155 5.6 / 1834 2.3
10 SA	0519 2.7 / 1139 5.1 / 1805 2.8	25 SU	0040 5.4 / 0712 2.4 / 1324 5.5 / 2009 2.3
11 SU	0013 5.0 / 0650 2.9 / 1307 5.1 / 1945 2.8	26 M	0210 5.5 / 0840 2.3 / 1449 5.8 / 2125 2.0
12 M	0150 5.0 / 0825 2.6 / 1436 5.4 / 2102 2.4	27 TU	0326 5.8 / 0946 1.9 / 1554 6.2 / 2222 1.6
13 TU	0309 5.5 / 0931 2.2 / 1538 5.9 / 2159 1.9	28 W	0421 6.2 / 1037 1.5 / 1642 6.6 / 2306 1.3
14 W	0404 6.0 / 1022 1.7 / 1627 6.4 / 2247 1.3	29 TH	0505 6.6 / 1119 1.3 / 1723 6.8 / 2344 1.1
15 TH	0450 6.5 / 1107 1.2 / 1711 6.9 / 2331 0.9	30 F	0543 6.8 / 1155 1.1 / 1759 7.0

OCTOBER

Day	Time m	Day	Time m
1 SA ●	0017 1.0 / 0617 6.9 / 1229 1.0 / 1832 7.0	16 SU ○	0551 7.4 / 1210 0.5 / 1813 7.7
2 SU	0049 1.0 / 0648 6.9 / 1300 1.0 / 1902 7.0	17 M	0034 0.3 / 0634 7.6 / 1255 0.3 / 1856 7.8
3 M	0119 1.0 / 0718 6.9 / 1330 1.1 / 1932 6.9	18 TU	0118 0.3 / 0717 7.6 / 1339 0.4 / 1941 7.7
4 TU	0147 1.2 / 0747 6.8 / 1359 1.2 / 2001 6.7	19 W	0201 0.4 / 0801 7.5 / 1423 0.6 / 2026 7.3
5 W	0215 1.4 / 0816 6.6 / 1428 1.5 / 2030 6.5	20 TH	0244 0.8 / 0846 7.1 / 1508 1.0 / 2113 6.9
6 TH	0243 1.6 / 0847 6.3 / 1459 1.8 / 2101 6.1	21 F	0329 1.3 / 0934 6.6 / 1557 1.5 / 2204 6.3
7 F	0314 1.9 / 0920 6.0 / 1535 2.1 / 2136 5.8	22 SA	0419 1.8 / 1029 6.2 / 1657 2.0 / ◑ 2304 5.8
8 SA	0350 2.3 / 1000 5.7 / 1619 2.5 / 2222 5.4	23 SU	0523 2.3 / 1136 5.7 / 1817 2.4
9 SU	0438 2.6 / 1056 5.3 / 1721 2.7 / ◔ 2329 5.1	24 M	0020 5.4 / 0649 2.5 / 1300 5.6 / 1947 2.4
10 M	0555 2.9 / 1217 5.2 / 1900 2.8	25 TU	0145 5.4 / 0815 2.4 / 1421 5.7 / 2100 2.1
11 TU	0102 5.1 / 0742 2.7 / 1354 5.4 / 2024 2.4	26 W	0259 5.7 / 0921 2.1 / 1526 6.1 / 2155 1.8
12 W	0230 5.4 / 0856 2.3 / 1501 5.9 / 2127 1.9	27 TH	0355 6.1 / 1011 1.8 / 1615 6.4 / 2239 1.5
13 TH	0331 6.0 / 0952 1.7 / 1555 6.5 / 2218 1.3	28 F	0439 6.4 / 1052 1.5 / 1656 6.6 / 2316 1.3
14 F	0421 6.6 / 1040 1.2 / 1643 7.0 / 2305 0.8	29 SA	0516 6.7 / 1128 1.3 / 1732 6.8 / 2349 1.2
15 SA	0507 7.1 / 1126 0.8 / 1728 7.4 / 2350 0.5	30 SU ●	0550 6.8 / 1202 1.2 / 1805 6.9
		31 M	0021 1.2 / 0621 6.9 / 1234 1.2 / 1835 6.9

NOVEMBER

Day	Time m	Day	Time m
1 TU	0051 1.2 / 0651 6.9 / 1304 1.2 / 1905 6.8	16 W	0059 0.4 / 0700 7.6 / 1323 0.5 / 1925 7.6
2 W	0120 1.2 / 0721 6.8 / 1335 1.3 / 1935 6.7	17 TH	0144 0.6 / 0746 7.5 / 1410 0.6 / 2012 7.3
3 TH	0149 1.4 / 0751 6.7 / 1406 1.5 / 2006 6.5	18 F	0229 0.9 / 0832 7.2 / 1456 1.0 / 2059 6.8
4 F	0220 1.6 / 0823 6.5 / 1439 1.7 / 2038 6.3	19 SA	0315 1.3 / 0920 6.8 / 1546 1.4 / 2149 6.4
5 SA	0252 1.9 / 0857 6.2 / 1515 2.0 / 2114 6.0	20 SU	0404 1.8 / 1012 6.3 / 1641 1.9 / 2244 5.9
6 SU	0329 2.2 / 0937 5.9 / 1559 2.3 / 2159 5.6	21 M	0501 2.2 / 1111 5.9 / 1747 2.2 / ◑ 2348 5.5
7 M	0415 2.5 / 1030 5.6 / 1655 2.5 / ◔ 2301 5.4	22 TU	0612 2.5 / 1221 5.7 / 1905 2.4
8 TU	0520 2.7 / 1141 5.5 / 1815 2.6	23 W	0103 5.4 / 0730 2.5 / 1337 5.7 / 2017 2.3
9 W	0021 5.3 / 0653 2.7 / 1305 5.6 / 1941 2.3	24 TH	0216 5.5 / 0840 2.3 / 1444 5.8 / 2117 2.1
10 TH	0146 5.5 / 0815 2.3 / 1420 5.9 / 2049 1.9	25 F	0317 5.8 / 0935 2.1 / 1539 6.1 / 2204 1.8
11 F	0254 5.8 / 0917 1.8 / 1521 6.5 / 2146 1.4	26 SA	0405 6.1 / 1021 1.8 / 1624 6.3 / 2245 1.6
12 SA	0350 6.5 / 1012 1.3 / 1614 6.9 / 2238 1.0	27 SU	0446 6.4 / 1100 1.6 / 1703 6.5 / 2321 1.5
13 SU	0440 7.0 / 1101 0.9 / 1703 7.3 / 2326 0.6	28 M	0523 6.6 / 1136 1.5 / 1739 6.6 / 2355 1.4
14 M ○	0528 7.4 / 1149 0.6 / 1751 7.6	29 TU ●	0557 6.7 / 1210 1.4 / 1812 6.7
15 TU	0013 0.4 / 0614 7.6 / 1237 0.4 / 1838 7.7	30 W	0027 1.3 / 0629 6.8 / 1244 1.3 / 1844 6.7

DECEMBER

Day	Time m	Day	Time m
1 TH	0059 1.3 / 0701 6.8 / 1316 1.3 / 1916 6.6	16 F	0130 0.6 / 0734 7.4 / 1359 0.7 / 2000 7.2
2 F	0131 1.4 / 0733 6.8 / 1350 1.4 / 1949 6.5	17 SA	0215 0.8 / 0820 7.2 / 1445 0.9 / 2045 6.9
3 SA	0203 1.5 / 0807 6.6 / 1425 1.5 / 2024 6.4	18 SU	0300 1.1 / 0905 6.9 / 1530 1.2 / 2131 6.5
4 SU	0238 1.7 / 0843 6.5 / 1503 1.7 / 2102 6.2	19 M	0344 1.5 / 0951 6.5 / 1617 1.6 / 2217 6.1
5 M	0317 1.9 / 0924 6.2 / 1546 1.9 / 2146 5.9	20 TU	0431 1.9 / 1039 6.2 / 1707 2.0 / 2309 5.7
6 TU	0401 2.1 / 1013 6.0 / 1636 2.1 / 2240 5.7	21 W ◑	0524 2.3 / 1134 5.8 / 1806 2.3
7 W	0457 2.3 / 1114 5.8 / 1739 2.2 / ◑ 2347 5.6	22 TH	0009 5.4 / 0627 2.5 / 1239 5.6 / 1913 2.5
8 TH	0608 2.4 / 1224 5.8 / 1856 2.2	23 F	0119 5.3 / 0739 2.5 / 1349 5.5 / 2023 2.4
9 F	0102 5.6 / 0729 2.3 / 1339 6.0 / 2009 1.9	24 SA	0227 5.5 / 0847 2.4 / 1453 5.6 / 2122 2.2
10 SA	0215 5.9 / 0841 1.9 / 1446 6.3 / 2114 1.6	25 SU	0326 5.7 / 0944 2.3 / 1548 5.8 / 2211 2.0
11 SU	0319 6.3 / 0944 1.5 / 1546 6.7 / 2213 1.2	26 M	0415 6.0 / 1031 1.9 / 1634 6.1 / 2254 1.8
12 M	0416 6.8 / 1040 1.1 / 1642 7.0 / 2306 0.9	27 TU	0457 6.3 / 1112 1.7 / 1715 6.3 / 2332 1.5
13 TU	0508 7.1 / 1133 0.8 / 1735 7.3 / 2356 0.7	28 W	0535 6.5 / 1150 1.5 / 1752 6.6
14 W ○	0559 7.4 / 1223 0.6 / 1825 7.4	29 TH ●	0007 1.4 / 0610 6.7 / 1226 1.3 / 1827 6.6
15 TH	0044 0.6 / 0647 7.5 / 1312 0.5 / 1913 7.4	30 F	0042 1.5 / 0645 6.8 / 1302 1.2 / 1901 6.6
		31 SA	0116 1.2 / 0719 6.8 / 1337 1.2 / 1936 6.6

Chart Datum: 3·71 metres below Ordnance Datum (Newlyn). HAT is 7·9 metres above Chart Datum.

》》 **FREE** monthly updates. Register at 《
www.reedsnauticalalmanac.co.uk

MILFORD HAVEN TO MUMBLES HEAD

(AC 1179, 1076) Milford Haven is a long natural, all-weather harbour with marinas beyond the oil terminals. Beware Turbot Bank (WCM lt buoy) 3M S of the entrance. Crow Rk (dries 5·5m) is 5ca SSE of Linney Hd, and The Toes are dangerous submerged rocks close W and SE of Crow Rock. There is a passage inshore of these dangers. There are overfalls on St Gowan Shoals which extend 4M SW of St Govan's Hd, and the sea breaks on the shallow patches in bad weather. There are firing areas from Linney Hd to Carmarthen Bay.

Caldey Is (lit) lies S of Tenby. Off its NW pt is St Margaret's Is connected by a rocky reef. Caldey Sound, between St Margaret's Is and Giltar Pt (AC 1482), is buoyed, but beware Eel Spit near W end of Caldey Is where there can be a nasty sea with wind against tide, and Woolhouse Rocks (dry 3·6m) 1ca NE of Caldey Is. Saundersfoot harbour (dries) is 2M N of Tenby, with an ⚓ well sheltered from N and W but subject to swell. Streams are weak here. Carmarthen Bay has no offshore dangers for yachts, other than the extensive drying sands at head of the bay and on its E side off Burry Inlet.

S of Worms Head, Helwick Sands (buoyed at each end) extend 7M W from Port Eynon Pt; least depth of 1·3m is near their W end. Stream sets NE/SW across the sands. There is a narrow channel inshore, close to Port Eynon Pt.

▶ *Between here and Mumbles Hd the stream runs roughly along coast, sp rates 3kn off headlands, but there are eddies in Port Eynon Bay and Oxwich Bay (both yacht anchorages), and over-falls SSE of Oxwich Pt.* ◀

FIRING RANGES – LINNEY HEAD TO BURRY INLET

For daily info on all range firing times call *Milford Haven CG* Ch 16/67 or ☎01646 690909.

Castlemartin Range Danger Area extends 12M WNW from Linney Hd, thence in an anti-clockwise arc to a point 12M S of St Govan's Hd. The exact Danger Area operative on any one day depends on the ranges/ammunition used; it is primarily a tank range. When firing is in progress R flags are flown (Fl R lts at night) along the coast from Freshwater West to Linney Head to St Govan's Head. Yachts are requested to keep clear of ranges when active.

Firing takes place on weekdays 0900–1630, exceptionally to 1700. Night firing takes place on Mon to Thurs, up to 2359, depending on the hours of darkness. In Jan only small arms are usually fired and the danger area is reduced.

Days/times of firing are published locally and can be obtained by VHF from *Castlemartin Range* Ch 16 or ☎01646 662367 (H24 answering service); Range safety launches Ch 16 or 12; and Milford Haven CG Ch 16. Also from the Range Office ☎(01646) 662287 or Warren Tower 01646 662336.

Manorbier Range (further E) covers a sector arc radius 12M centred on Old Castle Head; E/W extent is approximately between St Govan's Hd and Caldey Is (see AC Q6402). It is usually active Mon-Fri 0900-1700LT, occasionally Sat and Sun, and is primarily a surface to air missile range, but active areas depend on the weapons in use on any given day. On firing days warnings are broadcast on Ch 16, 73 at 0830, 1430 and on request; red flags are flown either side of Old Castle Hd. Yachts on passage should either keep 12M offshore or transit close inshore via Stackpole Hd, Trewent Pt, Priest's Nose and Old Castle Hd. Firing days/times are available from local HMs and YCs. For further info call: *Manorbier Range Control* Ch 16, 73 (also manned by Range safety launches); *Milford Haven CG* Ch 16; or Range Control ☎(01834) 871282 ext 209, 🖷 871283.

Penally Range (further E at Giltar Pt) is for small arms only and seldom interferes with passage through Caldey Sound. Info ☎(01834) 843522.

Pendine Range (between Tenby and Burry Inlet) is a MOD range for testing explosive devices. It is usually possible to steer the rhumb line course from Tenby to Worms Hd without interference. Info ☎(01994) 453243. Broadcasts on VHF Ch 16, 73 at 0900 and 1400LT. Range active 0800-1615.

Pembrey Range (approx 5M NW of Burry Inlet) is used for bombing practice by the RAF. Info ☎(01554) 891224.

MUMBLES HEAD TO CARDIFF

(AC 1165, 1182) Off Mumbles Hd (Lt, fog sig) beware Mixon Shoal (dries 0·3m), marked by PHM buoy. In good conditions pass N of shoal, 1ca off Mumbles Hd. ⚓ N of Mumbles Hd, good holding but exposed to swell. At W side of Swansea Bay, Green Grounds, rky shoals, lie in appr's to Swansea.

Scarweather Sands, much of which dry (up to 3·3m) with seas breaking heavily, extend 7M W from Porthcawl (9.11.16) and are well buoyed (AC 1161). There is a chan between the sands and coast to E, but beware Hugo Bank (dries 2·6m) and Kenfig Patches (0·5m) with o'falls up to 7ca offshore between Sker Pt and Porthcawl.

Nash Sands extend 7·5M WNW from Nash Pt. Depths vary and are least at inshore end (dries 3m), but Nash Passage, 1ca wide, runs close inshore between E Nash ECM buoy and ledge off Nash Pt. ▶ *On E-going stream there are heavy overfalls off Nash Pt and at W end of Nash Sands. Between Nash Pt and Breaksea Pt the E-going stream begins at HW Avonmouth +0535, and the W-going at HW Avonmouth –0035, sp rates 3kn. Off Breaksea Pt there may be overfalls.* ◀

From Rhoose Pt to Lavernock Pt the coast is fringed with foul ground. Lavernock Spit extends 1·75M S of Lavernock Pt, and E of the spit is main chan to Cardiff; the other side of the chan being Cardiff Grounds, a bank drying 5·4m which lies parallel with the shore and about 1·5M from it.

SEVERN ESTUARY

(AC 1176, 1166) Near the centre of the Bristol Channel, either side of the buoyed fairway, are the islands of Flat Holm (Lt, fog sig) and Steep Holm. 7M SW of Flat Holm lies Culver Sand (0·9m), 3M in length, buoyed with W and ECM. Monkstone Rk (Lt, dries) is 2M NW of the buoyed channel to Avonmouth and Bristol. Extensive drying banks cover the North shore of the estuary, beyond Newport and the Severn bridges (AC 1176). Call Bristol VTS VHF Ch 12 when passing reporting points inwards/outwards.

▶ *The tidal range in the Bristol Channel is exceptional, 12·2m at sp and 6·0m np. Care must be taken to understand the changes in chart datum in the upper reaches of the estuary. Tidal streams are very powerful, particularly above Avonmouth. Between Flat and Steep Holm the E-going stream begins at HW Avonmouth –0610, sp 3kn, and the W-going at HW Avonmouth +0015, sp 4kn.* ◀

The entrance to the R Avon is just to the S of Avonmouth S Pier Hd. Bristol City Docks lie some 6M up-river. Approach the entrance via King Road and the Newcombe and Cockburn lt buoys and thence via the Swash channel into the Avon. The entrance dries at LW but the river is navigable at about half tide.

▶ *Tidal streams are strong in the approaches to Avonmouth, up to 5kn at sp. The tide is also strong in the R. Avon which is best entered no earlier than HW Avonmouth – 0200.* ◀

From Avonmouth it is 16M to Sharpness, which you should aim to reach at about HW Avonmouth. The passage between Avonmouth and Sharpness requires considerable attention to timing and tidal conditions if it is to be undertaken safely. Plan your passage to arrive off Sharpness no earlier than HW Sharpness –1 and no later than HW Sharpness. Information to assist in passage planning is available from Gloucester Harbour Trustees (see entry for Sharpness). Pilots are available (see entry for Sharpness). To ensure the safety of navigation of all vessels, the fairway between Avonmouth approaches and Sharpness old Dock entrance must be recognised as a 'narrow channel' and attention is drawn to the provisions of rule 9 of the 'collision regulations'.

▶*Spring streams can run 8kn at the Shoots, and 6kn at the Severn bridges. At the Shoots the flood begins at HW Avonmouth –0430 and the ebb at HW Avonmouth +0045. The Severn Bore can usually be seen if Avonmouth range is 13·5m or more. For tidal streams 'Arrowsmiths' is recommended, from Arrowsmiths, Winterstoke Rd, Bristol, BS3 2NT.* ◀

9.11.14 TENBY

Pembrokeshire 51°40'·42N 04°41'·93W ✿✿♤♤✿✿✿

CHARTS AC 5620, 1179, 1076, 1482; Imray C60, 2600

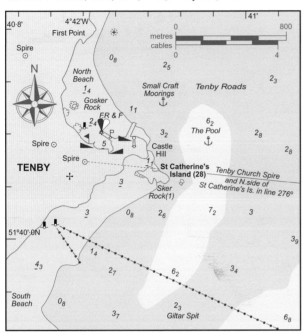

TIDES −0510 Dover; ML 4·5; Duration 0610

Standard Port MILFORD HAVEN (←)

Times				Height (metres)			
High Water		Low Water		MHWS	MHWN	MLWN	MLWS
0100	0800	0100	0700	7·0	5·2	2·5	0·7
1300	2000	1300	1900				
Differences TENBY							
−0015	−0010	−0015	−0020	+1·4	+1·1	+0·5	+0·2
STACKPOLE QUAY (7M W of Caldey Island)							
−0005	+0025	−0010	−0010	+0·9	+0·7	+0·2	+0·3

SHELTER Good, but hbr dries up to 5m. ⚓ in deep water off North Beach. Sheltered ⚓s, depending on wind direction, to NE in Tenby Roads, in Lydstep Haven (2·5M SW), and around Caldey Is as follows: Priory Bay to the N but shallow, Jones Bay (NE), Drinkim Bay (E) or Sandtop Bay (W).

NAVIGATION WPT 51°40'·02N 04°38'·08W, 279°/2·2M to monument on Castle Hill. The ⊕ WPT (off chartlet) is 2ca W of DZ2 SPM lit buoy. Beware Woolhouse Rks (3·6m) 1·5M SExE of the Hbr, marked by lit SCM buoy; and Sker Rk (1m high) closer in off St Catherine's Island (28m). From the W, Caldey Sound is navigable with care between Eel Pt SHM and Giltar Spit PHM lit buoys. Approaching Tenby Roads, keep outside the line of mooring buoys. For adjacent Firing ranges, see overleaf. Caldey Island is private and no landing without permission from the Abbot.

LIGHTS AND MARKS See 9.11.4 and chartlet. Church spire transit N side of St Catherine's Is 276°. FR 7m 7M on pier hd. Inside hbr, FW 6m 1M marks landing steps PHM beacon (unlit) marks outcrop from Gosker Rk on beach close N of hbr ent. Hbr is floodlit.

COMMUNICATIONS (Code 01834) CGOC (01646) 690909; Police 101; Dr 844161; 🏥 842040. HM 842717 (end May-end Sept), Mobile 07977 609947.

Ch 16, 80 (listening during HO).

FACILITIES Hbr ⚓ (<4·2m) ⚐. Town 🏴 🏴 ⚓ ⊠ Ⓑ 🗑 🛒 ✕ 🖵 ⇌ ✈ (Swansea; and a small airfield at Haverfordwest).

OTHER ADJACENT HARBOURS (See AC 5620)

SAUNDERSFOOT, Pembrokeshire, 51°42'·60N 04°41'·76W. AC 1179, 1076, 1482. HW −0510 on Dover; ML 4·4m; Duration 0605. See 9.11.14. A half-tide hbr with good shelter, but there may be a surge in prolonged E winds. On appr, beware buoys marking restricted area (power boats, etc) between Coppett Hall Pt and Perry's Pt. ⚐ may be available (see HM), or moorings in the middle. Pier hd lt Fl R 5s 6m 7M on stone cupola. VHF: HM 11 16. HM ☎(01834) 812094/(Home 831389). Facilities: ⚓ 6⚐⚓ (on SW wall) 🏴 🏴 🔧 🖵 ⚑. Town ⊠ Ⓑ 🛒 ✕ 🖵 ⇌ (Tenby/Saundersfoot).

CARMARTHEN, Carmarthenshire, 51°46'·27N 04°22'·53W. AC 1179, 1076. HW −0455 on Dover. See 9.11.15. R Towy & Taf dry. Beware Carmarthen Bar in S winds F4 and over with strong sp streams. Nav info is available from Carmarthen Bar Navigation Committee ☎ (01267) 231250 or YC's. Appr on N'ly hdg toward Wharley Pt, leaving DZ8 & 9 buoys 5ca to stbd. Chan shifts frequently and is unmarked so local knowledge reqd unless conditions ideal. ⚓ in mid-stream or ⚓s at R Towy YC off Ferryside (7M below Carmarthen) or R Towey Boat Club 1M N on W bank. 4 power lines cross in last 2·5M before Carmarthen, clearance 7·4m. **R Towy YC** ☎(01267) 238356, ⚓ ⚓ 🖵. **R Towy BC** ☎(01267) 238316. Town ⚓ ⊠ Ⓑ 🖵 🛒 ⇌ ✈ (Cardiff).

9.11.15 BURRY PORT

Carmarthenshire **51°40'·52N 04°14'·93W** (Burry Port) 🌸🌀💧💧🌸

CHARTS AC 5620, 5608 1179, 1076; Imray C59, C60, 2600

TIDES −0500 Dover; ML 4·7; Duration 0555

Standard Port MILFORD HAVEN (←—)

Times				Height (metres)			
High Water		Low Water		MHWS	MHWN	MLWN	MLWS
0100	0800	0100	0700	7·0	5·2	2·5	0·7
1300	2000	1300	1900				
Differences BURRY PORT							
+0003	+0003	+0007	+0007	+1·6	+1·4	+0·5	+0·4
LLANELLI							
−0003	−0003	+0150	+0020	+0·8	+0·6	ND	ND
FERRYSIDE							
0000	−0010	+0220	0000	−0·3	−0·7	−1·7	−0·6
CARMARTHEN							
+0010	0000	DR	DR	−4·4	−4·8	DR	DR

SHELTER Good in Burry Port Marina via lock with flapgate controlled by R/G lts. Access HW±2 (max) over sill depth 2·5m. Sp tides run hard. Note: If bad weather precludes access, especially in W'lies, see 9.11.14 for ⚓s around Caldey Island.

NAVIGATION WPT 51°36'·50N 04°23'·00W, 090°/2·5M to Burry Holms, then about 5M to Burry Port over Lynch Sands which dry approx 1·5m. Carmarthen Bar and the approaches to Burry Port should not be attempted in W winds >F5, nor at night. Depths are changeable. Pendine Range may be active (look for red flags) but clear range procedures are in force and there are no restrictions on navigation.

From close NW of Burry Holms, with sufficient rise of tide, track 015° for 2·5M with the W edge of Burry Holms in transit astern with Worms Head. When Whiteford lighthouse (disused) bears 086° alter to approx 065° until Barrel Post bears 015°.

Near approaches to the harbour are marked by a buoyed channel, dries 3·2m. Barrel Post should not be rounded too close.

LIGHTS AND MARKS Whiteford lt ho is conspic, but no longer lit. On head of W bkwtr is Barrel post; 1½ca N is conspic old lt ho (W tr, R top) Fl 5s 7m 6½M, and flagstaff.

COMMUNICATIONS (Code 01554); CGOC (01646) 690909; Pendine Range (01994) 453243 Ext 240; Police 101; Dr 832240. HM 835691, mob (emergencies only) 07817 395710. Marina Ch M (HO).

FACILITIES **Burry Port Marina** 450 inc ♥, £1·65, max LOA 11m, max draught 1·5m; **Burry Port YC** ⌂; **Services** ⛟⚓🔧🔨⛽🛒. **Town** 🍴📞 (approx 1M), 🏦✉️Ⓑ🍺✕🅿️🚆✈ (Cardiff).

9.11.16 SWANSEA

Swansea 51°36'·43N 03°55'·67W ✿✿✿✿◊◊◊◊❀❀❀

CHARTS AC 5608, 1179, 1165, 1161; Imray C59, 2600

TIDES –0500 Dover; ML 5·2; Duration 0620

Standard Port MILFORD HAVEN (←)

Times				Height (metres)			
High Water		Low Water		MHWS	MHWN	MLWN	MLWS
0100	0800	0100	0700	7·0	5·2	2·5	0·7
1300	2000	1300	1900				
Differences SWANSEA							
+0004	+0006	–0006	–0003	+2·6	+2·1	+0·7	+0·3
MUMBLES							
+0001	+0003	–0012	–0005	+2·5	+2·0	+0·8	+0·4
PORT TALBOT							
0000	+0004	–0010	–0005	+2·7	+2·1	+1·0	+0·4
PORTHCAWL							
+0005	+0010	–0010	–0005	+2·9	+2·3	+0·8	+0·3

SHELTER Very good in marina.

NAVIGATION WPT 51°35'·53N 03°56'·08W (SHM buoy) 020°/0·92M to E bkwtr lt. Keep seaward of Mixon Shoal. When N of SW Inner Green Grounds (SWIGG) SCM lt buoy keep to W of dredged chan and clear of commercial ships. *Yachts must motor in approaches and in harbour*, max speed 4kn. In Swansea Bay tidal streams flow anti-clockwise for 9½ hrs (Swansea HW –3½ to +6), with at times a race off Mumbles Hd. From HW–6 to –3 the stream reverses, setting N past Mumbles Hd towards Swansea.

LOCKS Enter via R Tawe barrage locks, which operate on request HW±4½ (co-ordinated with the marina lock), 0700-2200BST; out of season, 0700-1900UT, but to 2200 at w/ends. There are pontoons in both locks.

Vessels usually exit Tawe barrage locks at H+00, and enter at H+30. Locks are closed when ht of tide falls to 1·5m above CD, usually at MLWS. At sp, do not enter river until LW+1½. Two large Or holding buoys below barrage in mid-stream; also, at W side of barrage lock, a landing pontoon (dries, foul ground).

LIGHTS AND MARKS See 9.11.4 and chartlet. Mumbles Hd, Fl (4) 20s35m16M, is 3M SSW of hbr ent. A conspic TV mast (R lts) NNE of hbr is almost aligned with the fairway. Ldg lts 020° mark E side of chan dredged 3m. When N of Swansea Middle West (QR) and Swansea Middle East (QG) chan buoys stay inside dredged chan. Barrage lock lit by 2FR/FG (vert) to seaward.

Port Traffic sigs are conspic at W side of ent to King's Dock; there are 9 lts, ● or ◐, arranged in a 3 x 3 frame. Yachts arriving must obey the middle lt in left column:

- ● = Do not enter the river; hold SW of W Pier.
- ◐ = Yachts may enter the river, keeping to mid-chan, then to W of holding buoys.

Lock Master will advise on tfc movements Ch 18.

Lock sigs for barrage and marina locks alike are:

- ●] = Lock closed. Do not proceed
- ● = Wait
- ◐ = Enter with caution
- ●] = Free flow operating; proceed with caution

COMMUNICATIONS (Code 01792) CGOC 366534; Police 101; ⊞ 205666; Dr 653452; DVLA (for SSR) 783355. HM 653787; Barrage 456014.

Barrage, *Tawe Lock* Ch 18. *Swansea Marina* Ch 80. For commercial docks call *Swansea Docks Radio* VHF Ch14 (H24).

FACILITIES Swansea Marina (340+50Ⓥ) ☎470310, £1·85, lock 9m wide 🅿️🅘⛽🛁♨️⚓🔧🔦🅑 Ⓔ △ ▣(25t) ▸(1t) ⚓ ACA, ice, ✕ 🛒 wi-fi.

Swansea Yacht & Sub Aqua Club (SY & SAC) ☎654863, No visitors, ⚓🔧🛥 ▣(5t static) ✕ 🛒.

City ✉ Ⓑ 🖳 ✕ 🛒 ⇌ ✈.

ADJACENT HARBOURS AND ANCHORAGES

Mumbles, 51°34'·2N 03°58'·2W. Good ⚓ in W'lies 5ca N of Mumbles **Bristol Chan YC** ☎(01792) 366000, ⚓ 🛥; **Mumbles YC** ☎369321, ⚓🔧🛥🅿️🛒 (hire).

R Neath, 51°37'·88N 03°49'·97W. Ent over bar HW±2½ via 1·5M chan, marked/lit training wall to stbd. Tfc info from *Neath Pilot* VHF Ch 77, if on stn. **Monkstone Marina**, W bank just S of bridge, dries 4m: ⚓, 2 Y ❁s, 🛥⚓🅿️▣(15t) ✕ 🛒. Visitors welcome. **Monkstone C & SC**, ☎(01792) 812229; VHF Ch **M** (occas).

PORTHCAWL, Bridgend, 51°28'·48N 03°42'·02W. AC 1165, 1169. HW –0500 on Dover; ML 5·3m. See 9.11.16. Hbr protected by bkwtr running SE from Porthcawl Pt. Beware rk ledges and appr drying 5m above CD. Porthcawl lt ho, F WRG (see 9.11.4). Tidal streams off bkwtr can reach 6kn at Sp. 3 ❁s or ⚓ approx 3ca SSE of lt ho. Look out for swimmers. Marina access gate opens HW±3, R/G tfc lts, sill 3·5m above CD; Ⓥ by arrangement, HM ☎(01656) 815715 VHF Ch 80. Facilities: 🛥 70⚓🅿️🗑. **Porthcawl Hbr B C** ☎782342. **Town** 🏪🅿️⚓ ✉ Ⓑ 🖳 ✕ 🛒 ⇌(Bridgend) ✈(Cardiff).

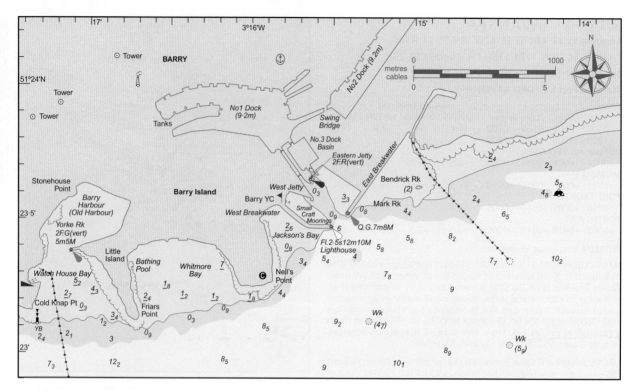

9.11.17 BARRY

Vale of Glamorgan **51°23'·48N 03°15'·45W** ❀❀❀♦♦♦❀❀

CHARTS AC 5608, 1179, 1152, 1182; Imray C59, 2600

TIDES –0423 Dover; ML 6·1; Duration 0630

Standard Port BRISTOL (AVONMOUTH) (→)

Times				Height (metres)			
High Water		Low Water		MHWS	MHWN	MLWN	MLWS
0600	1100	0300	0800	13·2	9·8	3·8	1·0
1800	2300	1500	2000				
Differences BARRY							
–0025	–0025	–0130	–0045	–1·5	–1·1	0·0	+0·2
FLAT HOLM							
–0015	–0015	–0035	–0035	–1·4	–1·0	–0·5	0·0
STEEP HOLM							
–0020	–0020	–0040	–0040	–1·7	–1·1	–0·5	–0·4

SHELTER Good, but in strong E/SE winds avoid Barry; No 1 Dock is not available for pleasure craft. Access H24 to the Outer hbr. No ⌐ or ⚓, but Barry YC administers all moorings and welcomes visitors. Temporary mooring may be available from Bristol Pilots on request.

NAVIGATION WPT 51°23'·03N 03°15'·08W, 332°/0·53M to entrance. Beware heavy commercial traffic. Approaching from E keep well out from the shore. Strong tidal stream across entrance.

LIGHTS AND MARKS Welsh Water Barry West PHM buoy, Fl R 5s, and Merkur PHM buoy, Fl R 2·5s, lie respectively 217°/1·5M and 191°/1·65M from harbour entrance. W bkwtr Fl 2·5s 10M. E bkwtr QG 8M.

COMMUNICATIONS (Code 01446) CGOC (01646) 690909; Police 101; Dr 733355. HM 0870 609 6699. BY 678186.

Barry Radio VHF Ch **11** 10 16 (HW–4 to HW+3); tidal info on request. *Bristol Pilot* via Ch 16 may advise on vacant moorings.

FACILITIES **Barry YC** (130) ☎735511, approaches dry approx 2m; ♠▬⚓⚒✖🔧🏪⛽🛒🚾⚓(50t) ⌐. **Town** 📞📮(2M) 🛢(2M) 🏧🛒🚂✉ Ⓑ✖🗓⇌ ✈ (Cardiff).

9.11.18 CARDIFF/PENARTH

Vale of Glamorgan **51°26'·74N 03°09'·92W** (marina) ❀❀♦♦♦❀❀❀

CHARTS AC 5608, 1179, 1176, 1182; Imray C59, 2600

TIDES –0425 Dover; ML 6·4; Duration 0610

Standard Port BRISTOL (AVONMOUTH) (→)

Times				Height (metres)			
High Water		Low Water		MHWS	MHWN	MLWN	MLWS
0600	1100	0300	0800	13·2	9·8	3·8	1·0
1800	2300	1500	2000				
Differences CARDIFF							
–0015	–0015	–0035	–0030	–0·9	–0·7	+0·2	+0·2
NEWPORT							
–0005	–0010	–0015	–0015	–0·9	–0·9	–0·2	–0·2
CHEPSTOW (River Wye)							
+0020	+0020	ND	ND	ND	ND	ND	ND

NOTE: At Newport the ht of LW does not normally fall below MLWS. Tidal hts are based on a minimum river flow; max flow may raise ht of LW by as much as 0·3m.

SHELTER Very good in marinas. See above for barrage locks. Waiting trot berths in outer hbr; or ⚓ off Penarth seafront in W'lies; in E'lies cramped ⚓ off Alexandra Dock ent in 2m.

NAVIGATION WPT 51°24'·03N 03°08'·81W (2½ca SW of S Cardiff SCM lt buoy), 349°/2·9M to barrage locks.

The outer approaches from W or SW are via Breaksea lt float and N of One Fathom Bk. Keep S of Lavernock Spit SCM lt buoy and NW of Flat Holm and Wolves NCM lt buoy drying rk. From NE, drying ledges and shoals extend >1M offshore. From E, appr via Monkstone Lt ho and S Cardiff SCM buoy. Ranny Spit (dries 0·4m) is 3½ca to the W, and Cardiff Grounds (dries 5·4m) 3½ca to the E.

The Wrach Chan is buoyed/lit and dredged 1·2m; it passes 1½ca E of Penarth Hd. The locks appr chan, buoyed, is dredged 0·7m below CD.

Do not impede merchant ships, especially those entering/leaving Alexandra Dock.

Cardiff Bay Barrage and Marina lock

- Call *Barrage Control* VHF Ch 18 or ☎02920 700234 to request lock-in or lock-out. Waiting berth on a barge in outer hbr.
- If entering near LW ask Barrage Control for up-to-date depths.
- Subject to VHF instructions, enter the outer hbr (Wpt 51°26'·71N 03°09'·84W) and lock in.
- IPTS (sigs 1, 2, 3, 5) are shown at lock ent.
- Departures on H and H +30. Arrivals at H +15 and H +45.
- Lock into marina operates H24 on free-flow from the bay.

LIGHTS AND MARKS Directional light, Oc WRG 10s 5m3M, on a W metal post marks Wrach Channel. SPM Fl Y 3s mark the 5kt limit.

COMMUNICATIONS (Code 02920) CGOC (01646) 690909; ⚓ 397020; Police101; Dr 415258; HM 400500; Barrage control 700234; Cardiff Bay Authority 877900.

Port VHF Ch **14** 16 (HW–4 to HW+3). *Barrage Control* Ch 18 H24. Penarth Quays Marina Ch 80 H24. Cardiff Marina Ch M.

FACILITIES Penarth Quays Marina ⓥ www.quaymarinas.com ☎705021, H24, 350◯+ⓥ welcome, £1.80 (short stay £11), max draught 3m, lock width 7·3m; 🛢 🛒 🏗(0930-1630, E pontoon) 🛢 ✕ ⚓ 🏧 Ⓔ 🏴 🔧(20t) 🛒(20t) 🅟 ACA ✕.

Penarth YC ☎ 708196, 🔧 ⚓ 🏠.

Mermaid Quay ⓥ on pontoons (£1/craft/hr) near city centre; many Rs and Bars.

Cardiff Marina www.themarinegroup.co.uk ☎396078, 320◯ inc 50ⓥ £2·20, 🖥 🛢 ⛽ 🛒(4½t) 🚂(short walk).

Cardiff YC ☎463697, ⚓ 🔧(floating pontoon) ⚓ 🔧 🏠.

Cardiff Bay YC ☎226575, ⚓ 🔧 🔧 ⚓ 🛒 🏠.

City ✉ Ⓑ 🛒 ✕ 🏠 ⇄ ✈ (15 mins).

ADJACENT HARBOUR

NEWPORT, Newport, **51°32'·95N 02°59'·13W**. AC 5608, 1179, 1152, 1176. HW –0425 on Dover; ML 6·0m; Duration 0620. See 9.11.18. A commercial port controlled by ABP, but a safe shelter for yachts. Enter R Usk over bar (approx 0·5m) E of West Usk buoy, QR, and follow buoyed and lit chan to S Lock ent; turn NE (ldg lts 062°) for yacht moorings on S side between power stn pier and YC. Beware overhead cables in Julian's Pill, clearance 3·8m.

Lights, see 9.11.4. East Usk lt ho Fl (2) WRG 10s. Ldg lts 062°, both FG. Alexandra Dock, S lock W pier head 2 FR (vert) 9/7m 6M. E pier head 2 FG (vert) 9/7m 6M.

Port VHF, *Newport Radio*, and VTS: Ch 09, **71**, 74 (HW ±4). HM (ABP) ☎0870 609 6699. Facilities: 🅟 Ⓔ 🔧 ✕ Ⓔ. **Newport and Uskmouth SC** 🏠 M. **Town** All facilities.

THE SEVERN BRIDGES (AC 1166) The Second Severn Crossing (37m clearance), from 51°34'·88N 02°43'·80W to 51°34'·14N 02°39'·82W, is 4M upriver from Avonmouth and 3M below the Severn Bridge. The following brief directions, coupled with strong tidal streams and shifting banks, emphasize the need for local knowledge.

Going upriver, pass both bridges at about HW Avonmouth –1¼ (see also 9.11.19); max sp stream is 8kn at The Shoots and 6kn at the Severn Bridge (set across the chan when the banks are covered).

Redcliffe F Bu ldg lts in transit with Charston Rock lt lead 013° through The Shoots, a narrow passage between English Stones (6·2m) and the rocky ledges (5·1m) off the Welsh shore. 5ca S of the Second Crossing, the chan is marked by Lower Shoots WCM bn and Mixoms PHM bn.

No vessel may navigate between the shore and the nearer Tower of the 2nd Crossing, except in emergency.

4ca N of the 2nd Crossing, leave the 013° transit before passing Old Man's Hd WCM bcn and Lady Bench PHM bcn. From abeam Charston Rk, keep Chapel Rk, Fl 2·6s, brg 050° until the E twr of Severn Bridge bears 068°; maintain this brg until Lyde Rk, QWR, bears about 355°, when alter 010° to transit the bridge (36m clearance) close to rks drying 1m.

Radar Warning. In certain conditions and tidal states, radar displays may show misleading echoes in the vicinity of the 2nd Severn Crossing. *Racon (O)* is at centre span of crossing.

9.11.19 SHARPNESS

Gloucestershire 51°43'·03N 02°29'·08W ❀❀♦♦❀❀

CHARTS AC 5608, 1166, Imray C59, 2600

TIDES –0315 Dover; Duration 0415. Note: The tidal regime is irregular and deviates from Avonmouth curve.

Standard Port BRISTOL (AVONMOUTH) (→)

Times				Height (metres)			
High Water		Low Water		MHWS	MHWN	MLWN	MLWS
0000	0600	0000	0700	13·2	9·8	3·8	1·0
1200	1800	1200	1900				
Differences SUDBROOK (Second Severn Crossing)							
+0010	+0010	+0025	+0015	+0·2	+0·1	–0·1	+0·1
BEACHLEY/AUST (Severn Bridge)							
+0010	+0015	+0040	+0025	–0·2	–0·2	–0·5	–0·3
INWARD ROCKS (River Severn)							
+0020	+0020	+0105	+0045	–1·0	–1·1	–1·4	–0·6
NARLWOOD ROCKS							
+0025	+0025	+0120	+0100	–1·9	–2·0	–2·3	–0·8
WHITE HOUSE							
+0025	+0025	+0145	+0120	–3·0	–3·1	–3·6	–1·0
BERKELEY							
+0030	+0045	+0245	+0220	–3·8	–3·9	–3·4	–0·5
SHARPNESS DOCK							
+0035	+0050	+0305	+0245	–3·9	–4·2	–3·3	–0·4
WELLHOUSE ROCK							
+0040	+0055	+0320	+0305	–4·1	–4·4	–3·1	–0·2

SHELTER The gates to the basin and the lock into the commercial dock are generally closed and are opened only for commercial shipping movements and other craft which have made the required booking and given appropriate notice (min. 24h). It is essential for arrangements to be made by contacting the Pierhead office which is generally manned during daylight tides, HW–4 to HW+1 (if unmanned leave a message). Plan your passage to arrive no sooner than HW–1 and no later than HW. Detailed information on lock fees and canal licences is available from the Pierhead office. Comprehensive information on the passage from Avonmouth is available from Gloucester Harbour Trustees. Fog signals are available from Sharpness Radio on VHF Ch 13 or Pierhead office.

NAVIGATION Use transits shown on AC 1166. Study guidance notes from Gloucester Harbour Trustees in good time.

- Leave King Road, Avonmouth (17M downriver) not before HW Sharpness –3, to be off Hbr ent about HW –½. Stem strong flood S of F Bu lt; beware tidal eddy.
- Do not proceed above the Severn Bridge except HW–1¼/+2. There is a 5H flood and a 7H ebb.

LIGHTS AND MARKS Lights as chartlet, but night passage not advised without local knowledge/pilot (07774 226143).

COMMUNICATIONS (Code 01453) Police 101; ⊞ 810777. Pierhead 511968 (HW–4 to HW+1). Gloucester Harbour Trustees www.gloucesterharbourtrustees.org ☎811913 (HO).

Call Gloucester and Sharpness Canal Ch 74 for bridges (no locks). Call *Bristol VTS* VHF Ch 12 when passing reporting points inwards/outwards E/W Grounds PHM, Welsh Hook SCM and at Lower Shoots Bcn.

FACILITIES Sharpness Marina sharpness@f2s.com ☎811476. 170 inc 2 **Ⓥ**, £8 all LOA. ⚓ ⛽ ⚒ ⚓ ⚓ ⊞ ⚓ ⚓. **Town** ⛽ (above Fretherne Bridge), ✉ Ⓑ(Berkeley) 🛒 ✕ 🗋, ⇌ (Stonehouse), ✈ (Bristol). Gloucester: ⛽ ⚓ ACA.

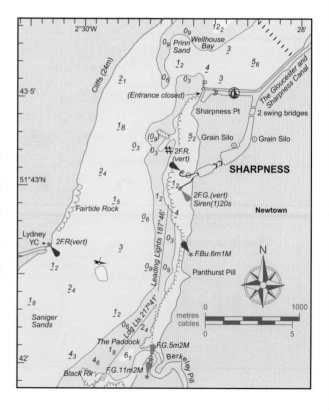

BRISTOL (AVONMOUTH) LAT 51°30'N LONG 2°44'W

TIMES AND HEIGHTS OF HIGH AND LOW WATERS

STANDARD TIME (UT)
For Summer Time add ONE hour in **non-shaded areas**

Dates in red are SPRINGS
Dates in blue are NEAPS

YEAR **2016**

Bristol Channel

JANUARY
Time	m		Time	m
1 0513	2.6	**16**	0528	2.0
1113	11.1		1129	12.3
F 1735	2.9	SA 1757	2.0	
2335	10.6	☽ 2352	11.8	
2 0553	3.0	**17**	0611	2.4
1156	10.5		1222	11.6
SA 1819	3.2	SU 1845	2.5	
☽				
3 0022	10.1	**18**	0049	11.1
0643	3.5		0705	3.0
SU 1253	10.1	M 1332	11.0	
1915	3.6		1949	3.1
4 0129	9.8	**19**	0207	10.7
0749	3.8		0833	3.5
M 1408	9.9	TU 1456	10.8	
2025	3.7		2127	3.3
5 0248	9.9	**20**	0330	10.8
0904	3.7		1013	3.3
TU 1522	10.2	W 1609	11.1	
2141	3.5		2246	2.8
6 0357	10.4	**21**	0439	11.3
1017	3.2		1122	2.6
W 1624	10.8	TH 1712	11.7	
2252	2.9		2349	2.1
7 0454	11.2	**22**	0537	12.1
1121	2.5		1220	1.9
TH 1718	11.5	F 1807	12.4	
2352	2.3			
8 0544	12.0	**23**	0044	1.5
1217	2.0		0628	12.8
F 1806	12.1	SA 1314	1.4	
			1856	12.9
9 0045	1.8	**24**	0136	1.1
0629	12.6		0714	13.2
SA 1310	1.6	SU 1404	1.1	
1851	12.6	○ 1941	13.2	
10 0136	1.6	**25**	0223	0.9
0712	13.0		0757	13.4
SU 1400	1.5	M 1450	1.1	
● 1935	12.9		2022	13.2
11 0223	1.5	**26**	0306	1.0
0755	13.2		0836	13.3
M 1447	1.4	TU 1529	1.3	
2019	13.1		2100	13.0
12 0306	1.4	**27**	0341	1.3
0837	13.4		0911	13.0
TU 1529	1.3	W 1600	1.6	
2101	13.2		2132	12.6
13 0344	1.5	**28**	0406	1.7
0919	13.4		0942	12.6
W 1607	1.3	TH 1620	2.0	
2142	13.1		2200	12.2
14 0418	1.5	**29**	0424	1.9
1001	13.2		1011	12.2
TH 1642	1.4	F 1637	2.2	
2223	12.9		2227	11.7
15 0452	1.7	**30**	0446	2.1
1043	12.9		1040	11.7
F 1717	1.6	SA 1701	2.3	
2305	12.5		2257	11.2
		31	0516	2.4
			1112	11.1
		SU 1734	2.6	
			2332	10.7

FEBRUARY
Time	m		Time	m
1 0554	2.9	**16**	0019	11.2
1152	10.5		0633	2.9
M 1817	3.1	TU 1259	10.8	
☽			1908	3.2
2 0019	10.1	**17**	0132	10.4
0645	3.4		0740	3.7
TU 1250	9.9	W 1430	10.2	
1917	3.7		2043	3.7
3 0128	9.7	**18**	0306	10.2
0803	3.8		0948	3.7
W 1411	9.7	TH 1550	10.4	
2043	3.9		2223	3.3
4 0259	9.8	**19**	0420	10.7
0929	3.6		1103	3.0
TH 1538	10.1	F 1656	11.1	
2207	3.4		2328	2.5
5 0417	10.6	**20**	0521	11.6
1044	3.0		1201	2.1
F 1647	11.0	SA 1752	12.0	
2319	2.7			
6 0516	11.5	**21**	0023	1.7
1149	2.2		0611	12.4
SA 1743	11.9	SU 1254	1.4	
			1839	12.7
7 0020	2.0	**22**	0114	1.1
0607	12.4		0657	13.0
SU 1248	1.7	M 1343	1.0	
1832	12.6	○ 1922	13.1	
8 0117	1.5	**23**	0202	0.8
0654	13.1		0738	13.3
M 1343	1.3	TU 1428	0.9	
● 1919	13.2		2002	13.2
9 0209	1.2	**24**	0245	0.6
0739	13.6		0815	13.3
TU 1434	1.0	W 1508	0.9	
2004	13.6		2036	13.0
10 0257	1.0	**25**	0321	1.0
0823	13.9		0848	13.1
W 1520	0.8	TH 1540	1.4	
2047	13.8		2106	12.7
11 0338	0.8	**26**	0348	1.4
0906	14.0		0917	12.7
TH 1600	0.7	F 1600	1.8	
2128	13.8		2133	12.4
12 0414	0.9	**27**	0404	1.7
0947	13.9		0943	12.4
F 1634	0.9	SA 1611	1.9	
2208	13.5		2158	12.0
13 0445	1.1	**28**	0419	1.8
1028	13.5		1009	12.0
SA 1704	1.2	SU 1629	2.0	
2247	13.0		2223	11.6
14 0515	1.5	**29**	0444	2.0
1110	12.7		1037	11.5
SU 1736	1.7	M 1658	2.2	
2329	12.2		2254	11.1
15 0549	2.1			
1157	11.8			
M 1815	2.4			
☽				

MARCH
Time	m		Time	m
1 0517	2.4	**16**	0608	2.8
1112	10.9		1232	10.5
TU 1734	2.7	W 1838	3.3	
☽ 2335	10.5			
2 0559	3.0	**17**	0102	10.1
1201	10.2		0708	3.7
W 1822	3.4	TH 1407	9.8	
			1959	4.0
3 0032	9.9	**18**	0243	9.9
0701	3.7		0920	4.0
TH 1310	9.7	F 1529	10.0	
1936	4.0		2158	3.7
4 0155	9.6	**19**	0357	10.4
0839	3.9		1040	3.2
F 1446	9.8	SA 1634	10.8	
2124	3.8		2304	2.7
5 0336	10.1	**20**	0458	11.3
1009	3.3		1137	2.2
SA 1616	10.6	SU 1729	11.7	
2249	3.0		2358	1.8
6 0448	11.2	**21**	0549	12.1
1124	2.4		1227	1.5
SU 1719	11.7	M 1816	12.4	
2357	2.2			
7 0544	12.3	**22**	0047	1.1
1227	1.6		0633	12.8
M 1812	12.7	TU 1315	1.0	
			1857	12.9
8 0056	1.5	**23**	0134	0.8
0633	13.2		0712	13.1
TU 1324	1.0	W 1400	0.8	
1859	13.5	○ 1934	13.1	
9 0151	0.9	**24**	0218	0.7
0720	13.9		0748	13.1
W 1416	0.6	TH 1440	1.0	
● 1944	14.0		2008	13.0
10 0240	0.6	**25**	0255	0.9
0805	14.3		0821	12.9
TH 1503	0.3	F 1514	1.3	
2028	14.2		2038	12.7
11 0323	0.4	**26**	0325	1.3
0848	14.4		0850	12.6
F 1543	0.3	SA 1536	1.7	
2109	14.2		2105	12.4
12 0400	0.4	**27**	0342	1.6
0930	14.3		0917	12.3
SA 1617	0.5	SU 1546	1.9	
2149	13.9		2130	12.1
13 0431	0.7	**28**	0355	1.7
1010	13.7		0941	12.0
SU 1646	1.0	M 1602	1.9	
2227	13.3		2154	11.8
14 0458	1.3	**29**	0419	1.8
1050	12.8		1009	11.7
M 1714	1.6	TU 1630	2.0	
2307	12.3		2225	11.4
15 0528	2.0	**30**	0450	2.1
1134	11.7		1045	11.2
TU 1749	2.4	W 1704	2.4	
☽ 2353	11.2		2306	10.9
		31	0530	2.6
			1132	10.6
		TH 1748	3.1	
		☽ 2359	10.2	

APRIL
Time	m		Time	m
1 0624	3.3	**16**	0216	9.8
1235	10.0		0823	3.9
F 1851	3.7	SA 1500	9.9	
			2117	3.8
2 0113	9.8	**17**	0327	10.3
0749	3.7		1005	3.3
SA 1401	9.8	SU 1603	10.5	
2035	3.9		2230	3.0
3 0252	10.1	**18**	0426	11.0
0933	3.3		1104	2.5
SU 1541	10.5	M 1657	11.3	
2217	3.2		2325	2.1
4 0417	11.1	**19**	0517	11.8
1056	2.4		1154	1.8
M 1652	11.6	TU 1744	12.1	
2331	2.2			
5 0518	12.3	**20**	0014	1.4
1202	1.6		0601	12.3
TU 1747	12.7	W 1242	1.3	
			1825	12.6
6 0031	1.4	**21**	0101	1.0
0609	13.3		0641	12.7
W 1300	0.9	TH 1326	1.1	
1836	13.6		1902	12.8
7 0126	0.8	**22**	0145	1.0
0657	14.0		0717	12.7
TH 1352	0.4	F 1408	1.1	
● 1922	14.2	○ 1937	12.8	
8 0216	0.4	**23**	0224	1.1
0743	14.4		0752	12.6
F 1440	0.1	SA 1443	1.4	
2006	14.4		2009	12.6
9 0301	0.2	**24**	0257	1.2
0828	14.5		0823	12.5
SA 1521	0.1	SU 1510	1.7	
2048	14.4		2038	12.4
10 0340	0.3	**25**	0320	1.7
0910	14.3		0852	12.2
SU 1557	0.4	M 1525	1.9	
2129	14.0		2106	12.2
11 0414	0.7	**26**	0337	1.8
0952	13.7		0920	12.0
M 1627	1.0	TU 1543	1.9	
2208	13.3		2133	11.9
12 0443	1.3	**27**	0401	1.8
1033	12.7		0951	11.8
TU 1656	1.7	W 1611	2.0	
2248	12.3		2207	11.6
13 0512	2.0	**28**	0434	2.0
1116	11.6		1029	11.4
W 1729	2.5	TH 1646	2.3	
2333	11.2		2248	11.2
14 0551	2.7	**29**	0514	2.4
1211	10.5		1115	10.9
TH 1815	3.3	F 1730	2.8	
☽			2341	10.7
15 0039	10.2	**30**	0606	2.9
0647	3.6		1215	10.4
F 1340	9.8	SA 1828	3.4	
1926	4.0	☽		

Chart Datum: 6·50 metres below Ordnance Datum (Newlyn). HAT is 14·7 metres above Chart Datum.

》》 FREE monthly updates. Register at 《
www.reedsnauticalalmanac.co.uk

551

STANDARD TIME (UT)
For Summer Time add ONE hour in **non-shaded areas**

BRISTOL (AVONMOUTH) LAT 51°30'N LONG 2°44'W
TIMES AND HEIGHTS OF HIGH AND LOW WATERS

Dates in red are SPRINGS
Dates in blue are NEAPS

YEAR 2016

MAY

Time	m	Time	m
1 0049	10.3	**16** 0247	10.2
0718	3.3	0853	3.5
SU 1331	10.2	M 1522	10.2
1954	3.6	2133	3.3
2 0215	10.4	**17** 0346	10.7
0855	3.1	1013	3.0
M 1501	10.6	TU 1617	10.9
2141	3.2	2240	2.6
3 0341	11.2	**18** 0437	11.2
1022	2.4	1111	2.3
TU 1619	11.5	W 1705	11.5
2259	2.3	2334	2.0
4 0448	12.2	**19** 0524	11.8
1131	1.7	1201	1.8
W 1719	12.6	TH 1749	12.1
5 0002	1.5	**20** 0022	1.5
0543	13.1	0605	12.2
TH 1231	1.1	F 1248	1.5
1810	13.4	1828	12.4
6 0059	0.9	**21** 0108	1.3
0633	13.8	0645	12.4
F 1325	0.5	SA 1332	1.4
● 1858	14.0	○ 1905	12.5
7 0150	0.5	**22** 0150	1.3
0721	14.2	0722	12.4
SA 1414	0.3	SU 1411	1.5
1943	14.3	1941	12.5
8 0238	0.3	**23** 0229	1.5
0808	14.2	0758	12.3
SU 1458	0.3	M 1445	1.7
2027	14.2	2015	12.4
9 0321	0.4	**24** 0301	1.7
0852	14.0	0833	12.2
M 1537	0.6	TU 1510	1.9
2110	13.8	2048	12.3
10 0358	0.8	**25** 0326	1.8
0936	13.4	0906	12.1
TU 1611	1.1	W 1533	2.0
2152	13.1	2120	12.1
11 0430	1.4	**26** 0353	1.9
1018	12.6	0941	12.0
W 1641	1.7	TH 1602	2.0
2232	12.3	2157	11.9
12 0501	2.0	**27** 0427	1.9
1101	11.4	1020	11.7
TH 1714	2.4	F 1639	2.2
2317	11.3	2239	11.6
13 0538	2.7	**28** 0508	2.2
1150	10.7	1106	11.4
F 1756	3.1	SA 1722	2.5
◐		2330	11.2
14 0013	10.4	**29** 0557	2.5
0626	3.3	1200	11.0
SA 1300	10.0	SU 1816	2.9
1854	3.6	◐	
15 0136	10.0	**30** 0031	10.9
0732	3.6	0659	2.8
SU 1419	9.9	M 1306	10.8
2009	3.7	1927	3.0
		31 0146	10.9
		0818	2.8
		TU 1425	10.9
		2102	3.0

JUNE

Time	m	Time	m
1 0307	11.3	**16** 0350	10.6
0946	2.5	1006	2.9
W 1544	11.4	TH 1621	10.9
2227	2.5	2241	2.6
2 0417	11.9	**17** 0442	11.1
1100	1.9	1112	2.4
TH 1650	12.2	F 1710	11.5
2333	1.8	2339	2.1
3 0518	12.7	**18** 0530	11.6
1202	1.3	1206	2.0
F 1746	13.0	SA 1755	12.0
4 0032	1.3	**19** 0030	1.7
0611	13.3	0613	12.0
SA 1259	0.9	SU 1256	1.7
1836	13.5	1836	12.3
5 0126	0.8	**20** 0118	1.5
0701	13.6	0656	12.2
SU 1351	0.6	M 1341	1.6
● 1923	13.8	○ 1917	12.5
6 0217	0.6	**21** 0202	1.5
0750	13.7	0737	12.3
M 1439	0.6	TU 1423	1.7
2009	13.8	1956	12.6
7 0304	0.7	**22** 0244	1.6
0836	13.6	0816	12.4
TU 1522	0.8	W 1500	1.8
2054	13.6	2034	12.5
8 0345	1.0	**23** 0321	1.7
0921	13.2	0855	12.3
W 1559	1.2	TH 1532	1.9
2136	13.1	2112	12.5
9 0421	1.5	**24** 0354	1.8
1003	12.6	0934	12.3
TH 1630	1.7	F 1602	1.9
2217	12.4	2150	12.4
10 0451	2.0	**25** 0427	1.8
1044	11.8	1014	12.2
F 1700	2.2	SA 1637	2.0
2257	11.6	2232	12.2
11 0522	2.4	**26** 0505	1.9
1124	11.1	1057	12.0
SA 1736	2.7	SU 1717	2.1
2341	10.9	2319	11.9
12 0601	2.9	**27** 0549	2.1
1212	10.4	1145	11.6
SU 1821	3.1	M 1804	2.5
◑		◑	
13 0037	10.3	**28** 0013	11.5
0650	3.2	0640	2.4
M 1314	10.0	TU 1243	11.2
1917	3.4	1902	2.8
14 0148	10.1	**29** 0119	11.2
0749	3.4	0745	2.6
TU 1424	10.0	W 1353	11.0
2023	3.5	2021	3.1
15 0253	10.2	**30** 0236	11.2
0855	3.3	0910	2.7
W 1526	10.3	TH 1513	11.1
2133	3.2	2157	2.8

JULY

Time	m	Time	m
1 0351	11.5	**16** 0359	10.5
1031	2.4	1021	3.0
F 1624	11.7	SA 1632	10.9
2309	2.3	2257	2.6
2 0456	12.0	**17** 0456	11.1
1138	1.9	1127	2.4
SA 1724	12.3	SU 1724	11.6
		2356	2.0
3 0010	1.7	**18** 0546	11.7
0553	12.6	1224	2.0
SU 1237	1.4	M 1811	12.2
1818	13.0		
4 0107	1.2	**19** 0049	1.7
0645	13.1	0632	12.2
M 1332	1.0	TU 1316	1.7
● 1907	13.4	○ 1855	12.6
5 0200	0.9	**20** 0141	1.5
0735	13.3	0717	12.5
TU 1422	0.8	W 1405	1.6
1954	13.5	1938	12.9
6 0249	0.9	**21** 0230	1.4
0822	13.3	0801	12.7
W 1508	0.9	TH 1451	1.6
2038	13.4	2020	13.0
7 0334	1.0	**22** 0314	1.4
0906	13.1	0843	12.8
TH 1548	1.2	F 1531	1.6
2120	13.1	2101	13.0
8 0411	1.4	**23** 0353	1.4
0946	12.7	0923	12.8
F 1619	1.6	SA 1605	1.6
2158	12.6	2141	13.0
9 0439	1.9	**24** 0427	1.5
1021	12.1	1003	12.6
SA 1644	2.0	SU 1636	1.7
2232	12.0	2221	12.8
10 0502	2.2	**25** 0459	1.5
1054	11.5	1043	12.5
SU 1711	2.3	M 1709	1.8
2306	11.4	2304	12.4
11 0531	2.5	**26** 0535	1.8
1129	10.9	1127	12.1
M 1745	2.7	TU 1748	2.2
2345	10.8	◑ 2352	11.9
12 0609	2.8	**27** 0618	2.2
1211	10.4	1218	11.5
TU 1830	3.1	W 1836	2.7
◑			
13 0036	10.2	**28** 0052	11.2
0657	3.2	0712	2.7
W 1308	10.0	TH 1324	10.9
1927	3.4	1942	3.3
14 0143	9.9	**29** 0211	10.8
0759	3.4	0834	3.1
TH 1421	9.9	F 1447	10.7
2036	3.5	2132	3.4
15 0255	10.0	**30** 0331	10.9
0909	3.4	1009	3.0
F 1531	10.2	SA 1604	11.1
2149	3.2	2251	2.8
		31 0440	11.4
		1119	2.4
		SU 1708	11.8
		2354	2.1

AUGUST

Time	m	Time	m
1 0540	12.1	**16** 0522	11.5
1219	1.7	1158	2.2
M 1804	12.5	TU 1748	12.1
2 0051	1.4	**17** 0027	1.8
0632	12.7	0611	12.3
TU 1314	1.2	W 1255	1.8
● 1853	13.1	1834	12.8
3 0144	1.0	**18** 0122	1.4
0720	13.1	0658	12.8
W 1405	0.9	TH 1348	1.5
1939	13.4	○ 1919	13.3
4 0233	0.8	**19** 0214	1.2
0805	13.2	0742	13.2
TH 1452	0.8	F 1438	1.3
2021	13.4	2003	13.6
5 0317	0.9	**20** 0302	1.0
0847	13.1	0826	13.4
F 1532	1.0	SA 1521	1.2
2100	13.2	2044	13.7
6 0354	1.3	**21** 0343	1.0
0923	12.8	0907	13.4
SA 1604	1.4	SU 1558	1.2
2134	12.8	2125	13.6
7 0421	1.7	**22** 0418	1.1
0954	12.3	0946	13.3
SU 1625	1.8	M 1629	1.3
2203	12.3	2205	13.4
8 0438	2.1	**23** 0448	1.3
1022	11.8	1026	13.0
M 1644	2.1	TU 1657	1.6
2232	11.8	2246	13.0
9 0458	2.3	**24** 0517	1.7
1050	11.3	1107	12.3
TU 1709	2.4	W 1729	2.1
2302	11.1	2331	12.0
10 0526	2.6	**25** 0554	2.3
1123	10.7	1154	11.5
W 1743	2.8	TH 1810	2.8
◑ 2339	10.7	◑	
11 0604	3.0	**26** 0027	11.0
1206	10.1	0641	3.0
TH 1830	3.4	F 1258	10.6
		1909	3.6
12 0032	9.9	**27** 0151	10.3
0658	3.6	0801	3.7
F 1310	9.7	SA 1430	10.3
1941	3.9	2116	3.9
13 0151	9.6	**28** 0318	10.4
0817	3.9	0953	3.5
SA 1440	9.7	SU 1550	10.7
2105	3.8	2238	3.2
14 0318	9.9	**29** 0428	11.0
0940	3.6	1104	2.7
SU 1557	10.3	M 1655	11.5
2222	3.2	2339	2.2
15 0426	10.7	**30** 0526	11.8
1054	2.9	1201	1.9
M 1657	11.2	TU 1749	12.4
2329	2.4		
		31 0033	1.4
		0617	12.6
		W 1254	1.2
		1836	13.1

Chart Datum: 6·50 metres below Ordnance Datum (Newlyn). HAT is 14·7 metres above Chart Datum.

›› FREE monthly updates. Register at ‹‹
www.reedsnauticalalmanac.co.uk

STANDARD TIME (UT)
For Summer Time add ONE hour in **non-shaded areas**

BRISTOL (AVONMOUTH)   LAT 51°30'N   LONG 2°44'W
TIMES AND HEIGHTS OF HIGH AND LOW WATERS

Dates in red are SPRINGS
Dates in blue are NEAPS

YEAR 2016

Bristol Channel

SEPTEMBER

Time	m		Time	m
1 0123	0.9	**16** 0101	1.3
0702	13.1	0636	13.1
TH 1344	0.8	F 1327	1.3
● 1920	13.4	○ 1857	13.7
2 0211	0.7	**17** 0154	0.9
0744	13.3	0721	13.6
F 1429	0.7	SA 1417	1.0
1959	13.5	1942	14.1
3 0254	0.8	**18** 0241	0.7
0822	13.2	0804	13.9
SA 1510	0.9	SU 1502	0.8
2035	13.3	2024	14.2
4 0331	1.2	**19** 0324	0.6
0855	12.9	0846	14.0
SU 1542	1.3	M 1541	0.9
2106	12.9	2106	14.1
5 0358	1.7	**20** 0400	0.8
0924	12.5	0926	13.7
M 1602	1.8	TU 1614	1.1
2133	12.4	2147	13.7
6 0411	2.1	**21** 0430	1.2
0949	12.0	1006	13.3
TU 1615	2.1	W 1642	1.5
2158	11.9	2227	13.0
7 0423	2.3	**22** 0458	1.7
1014	11.6	1047	12.5
W 1635	2.4	TH 1711	2.2
2224	11.4	2311	11.9
8 0446	2.5	**23** 0531	2.5
1042	11.0	1132	11.4
TH 1704	2.6	F 1749	3.0
2256	10.8	◑
9 0518	2.9	**24** 0006	10.8
1118	10.4	0615	3.3
F 1742	3.2	SA 1237	10.4
◑ 2339	10.1	1845	3.9
10 0602	3.5	**25** 0136	10.0
1211	9.8	0733	4.1
SA 1838	3.9	SU 1417	10.0
		2101	4.1
11 0045	9.5	**26** 0303	10.0
0711	4.1	0936	3.8
SU 1336	9.4	M 1533	10.5
2018	4.2	2220	3.3
12 0230	9.5	**27** 0410	10.8
0858	4.1	1043	2.9
M 1522	9.9	TU 1635	11.4
2150	3.6	2318	2.3
13 0357	10.3	**28** 0506	11.7
1025	3.3	1138	2.0
TU 1630	11.0	W 1728	12.3
2304	2.7	
14 0458	11.4	**29** 0008	1.5
1134	2.5	0554	12.5
W 1724	12.1	TH 1229	1.2
		1813	13.0
15 0005	1.8	**30** 0057	1.0
0549	12.4	0638	13.0
TH 1233	1.8	F 1316	0.8
1812	13.0	1855	13.3

OCTOBER

Time	m		Time	m
1 0142	0.7	**16** 0127	0.8
0717	13.2	0657	13.9
SA 1401	0.8	SU 1351	0.9
● 1933	13.4	○ 1919	14.3
2 0225	0.8	**17** 0216	0.5
0753	13.2	0741	14.2
SU 1441	1.0	M 1438	0.7
2007	13.2	2003	14.5
3 0302	1.2	**18** 0259	0.5
0825	12.9	0824	14.3
M 1515	1.4	TU 1519	0.7
2037	12.8	2046	14.3
4 0329	1.7	**19** 0338	0.7
0853	12.5	0906	14.0
TU 1537	1.9	W 1556	1.0
2104	12.4	2129	13.8
5 0343	2.1	**20** 0411	1.2
0918	12.1	0947	13.4
W 1548	2.1	TH 1627	1.5
2129	12.0	2211	13.0
6 0352	2.3	**21** 0441	1.8
0942	11.7	1029	12.6
TH 1607	2.3	F 1658	2.2
2154	11.6	2255	11.9
7 0415	2.4	**22** 0514	2.6
1010	11.3	1115	11.5
F 1635	2.5	SA 1735	3.0
2226	11.1	◑ 2349	10.8
8 0446	2.7	**23** 0557	3.4
1046	10.7	1218	10.5
SA 1710	3.0	SU 1830	3.8
2308	10.4	
9 0526	3.3	**24** 0114	10.0
1135	10.1	0705	4.1
SU 1758	3.7	M 1354	10.1
◑		2019	4.2
10 0007	9.8	**25** 0238	10.0
0622	4.0	0903	4.0
M 1247	9.6	TU 1507	10.5
1919	4.2	2150	3.6
11 0134	9.5	**26** 0342	10.6
0802	4.3	1013	3.2
TU 1434	9.8	W 1606	11.2
2111	3.9	2247	2.7
12 0320	10.2	**27** 0437	11.4
0950	3.7	1107	2.3
W 1557	10.9	TH 1658	12.0
2233	2.9	2337	1.9
13 0429	11.3	**28** 0525	12.2
1105	2.7	1157	1.6
TH 1656	12.0	F 1744	12.6
2338	2.0	
14 0523	12.4	**29** 0024	1.3
1206	1.9	0608	12.8
F 1747	13.1	SA 1244	1.2
		1825	13.0
15 0035	1.3	**30** 0109	1.0
0611	13.3	0646	13.0
SA 1301	1.3	SU 1328	1.0
1833	13.8	● 1903	13.1
		31 0152	1.1
		0722	13.0
		M 1409	1.2
		1938	13.0

NOVEMBER

Time	m		Time	m
1 0229	1.3	**16** 0235	0.5
0755	12.8	0804	14.3
TU 1444	1.5	W 1459	0.7
2010	12.7	2029	14.2
2 0300	1.7	**17** 0318	0.7
0825	12.6	0848	14.1
W 1511	1.9	TH 1539	1.0
2040	12.4	2114	13.8
3 0319	2.1	**18** 0355	1.1
0853	12.2	0931	13.5
TH 1528	2.2	F 1616	1.5
2107	12.1	2158	13.0
4 0331	2.3	**19** 0429	1.7
0920	11.9	1015	12.7
F 1548	2.3	SA 1650	2.1
2134	11.7	2242	12.1
5 0355	2.4	**20** 0502	2.4
0949	11.6	1100	11.8
SA 1617	2.4	SU 1725	2.8
2208	11.4	2331	11.1
6 0427	2.6	**21** 0542	3.1
1027	11.1	1155	10.9
SU 1653	2.8	M 1811	3.4
2250	10.9	◐
7 0507	3.0	**22** 0037	10.3
1115	10.6	0634	3.7
M 1739	3.3	TU 1314	10.3
◐ 2344	10.3	1914	3.9
8 0558	3.6	**23** 0158	10.0
1218	10.2	0748	4.0
TU 1844	3.8	W 1429	10.4
		2044	3.8
9 0056	10.0	**24** 0304	10.3
0713	4.0	0919	3.7
W 1342	10.2	TH 1529	10.8
2020	3.8	2201	3.3
10 0228	10.2	**25** 0400	10.9
0902	3.8	1025	3.0
TH 1514	10.8	F 1622	11.4
2153	3.1	2257	2.4
11 0351	11.1	**26** 0450	11.6
1028	3.0	1119	2.3
F 1622	11.8	SA 1710	11.9
2305	2.2	2347	2.0
12 0452	12.2	**27** 0535	12.1
1134	2.1	1207	1.8
SA 1718	12.8	SU 1753	12.4
13 0004	1.5	**28** 0033	1.6
0544	13.1	0615	12.6
SU 1231	1.4	M 1252	1.5
1809	13.6	1833	12.6
14 0059	0.9	**29** 0117	1.4
0632	13.9	0652	12.8
M 1324	1.0	TU 1335	1.4
○ 1856	14.2	● 1910	12.7
15 0149	0.6	**30** 0157	1.5
0719	14.2	0728	12.7
TU 1413	0.7	W 1414	1.6
1943	14.3	1946	12.6

DECEMBER

Time	m		Time	m
1 0232	1.7	**16** 0302	0.7
0802	12.6	0833	14.0
TH 1448	1.8	F 1527	0.9
2020	12.4	2101	13.7
2 0300	2.0	**17** 0343	1.0
0835	12.4	0918	13.6
F 1515	2.1	SA 1607	1.3
2052	12.2	2145	13.2
3 0321	2.2	**18** 0419	1.5
0906	12.1	1001	13.0
SA 1539	2.2	SU 1641	1.8
2123	12.0	2227	12.4
4 0345	2.3	**19** 0451	2.1
0939	11.9	1042	12.3
SU 1608	2.3	M 1712	2.4
2159	11.7	2308	11.6
5 0418	2.4	**20** 0523	2.6
1017	11.6	1126	11.5
M 1645	2.5	TU 1747	2.9
2240	11.4	2353	10.8
6 0457	2.7	**21** 0603	3.1
1102	11.3	1217	10.8
TU 1729	2.8	W 1830	3.3
2329	11.0	◐
7 0545	3.1	**22** 0051	10.2
1157	10.9	0653	3.5
W 1823	3.1	TH 1326	10.3
◐		1925	3.6
8 0028	10.7	**23** 0205	10.0
0645	3.4	0755	3.7
TH 1305	10.7	F 1436	10.2
1933	3.3	2031	3.7
9 0142	10.6	**24** 0310	10.2
0805	3.6	0907	3.6
F 1427	10.9	SA 1537	10.5
2104	3.2	2148	3.4
10 0307	11.0	**25** 0407	10.7
0945	3.2	1023	3.1
SA 1545	11.5	SU 1631	11.0
2228	2.6	2258	2.8
11 0419	11.8	**26** 0458	11.3
1102	2.5	1124	2.5
SU 1650	12.3	M 1719	11.6
2334	1.9	2353	2.2
12 0519	12.6	**27** 0543	11.9
1204	1.8	1215	2.0
M 1746	13.1	TU 1804	12.0
13 0032	1.3	**28** 0041	1.8
0611	13.4	0625	12.4
TU 1300	1.2	W 1302	1.7
1837	13.7	1845	12.3
14 0126	0.8	**29** 0127	1.6
0700	13.9	0705	12.6
W 1353	0.9	TH 1347	1.6
○ 1927	14.0	● 1925	12.5
15 0216	0.7	**30** 0209	1.7
0747	14.1	0743	12.7
TH 1442	0.8	F 1429	1.7
2014	14.0	2003	12.5
		31 0247	1.6
		0820	12.6
		SA 1506	1.9
		2040	12.4

Chart Datum: 6·50 metres below Ordnance Datum (Newlyn). HAT is 14·7 metres above Chart Datum.

» FREE monthly updates. Register at «
www.reedsnauticalalmanac.co.uk

553

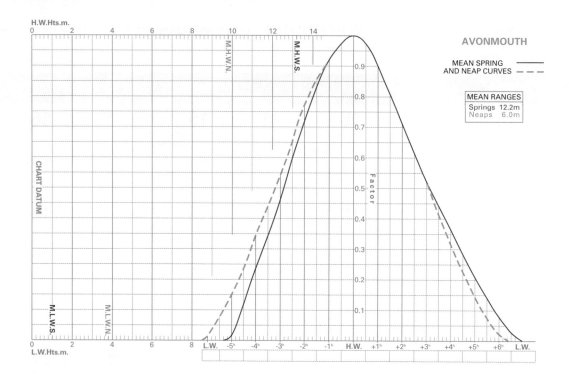

AVONMOUTH

MEAN SPRING
AND NEAP CURVES

MEAN RANGES
Springs 12.2m
Neaps 6.0m

9.11.20 BRISTOL (CITY DOCKS)

Ent to R Avon 51°30'·44N 02°43'·31W ❄❄◊◊◊❀❀❀

CHARTS AC 5608, 1179, 1176, 1859; Imray C59, 2600

TIDES –0410 on Dover; ML 7·0; Duration 0620. NB very large range

Standard Port BRISTOL (AVONMOUTH) (←)

Times				Height (metres)			
High Water		Low Water		MHWS	MHWN	MLWN	MLWS
0200	0800	0300	0800	13·2	9·8	3·8	1·0
1400	2000	1500	2000				
Differences SHIREHAMPTON (R Avon, 51°29'N 02°41'W)							
0000	0000	+0035	+0010	–0·7	–0·7	–0·8	0·0
SEA MILLS (R Avon, 51°29'N 02°39'W)							
+0005	+0005	+0105	+0030	–1·4	–1·5	–1·7	–0·1
CUMBERLAND BASIN (Ent)							
+0010	+0010	DR	DR	–2·9	–3·0	DR	DR

SHELTER Excellent in Bristol Floating Harbour. Avonmouth and Royal Portbury Docks are prohib to yachts, except in emergency. If early/late on the tide, ⚓ as close inshore as depth allows in Portishead Pool between Portishead Dock and Royal Portbury Dock, but avoid in onshore winds.

NAVIGATION From the south and west: from Avon SHM buoy (51°27'·92N 02°51'·73W) either follow the Bristol Deep shipping channel taking care not to impede the safe passage of commercial traffic or stay well clear to the S passing close to Portishead Pt. If to the north, cross the shipping channel by the most direct route keeping clear of inbound/outbound traffic.

From the north: Conform to the channel taking care not to impede the safe passage of commercial traffic.

- Beware large ships transiting/turning in the main channel.
- Beware ships and tugs leaving the docks at any time.
- When on passage up/down the Severn estuary without stopping at Bristol or Portishead, use the *Offshore Route*.

Entering R Avon keep N of shallows (Swash Bank) between Royal Portbury Dock and Avonmouth Docks. Speed limit in river: 6kn if draught > 2m; 9kn if draught < 2m.

LIGHTS AND MARKS Ldg lts 127°, both Iso R 2s, to R Avon ent abeam S pier lt, Oc RG 30s. St George ldg lts 173°, both Oc G 5s, front/rear W/Or posts, clear Swash Bank (dries). Upriver, G or R ☆s mark the outside of bends.

Entry signals to Bristol Hbr may be shown from E bank, 1½ and 2½ca S of Clifton Suspension Bridge: ● = continue with caution; ● = stop and await orders.

LOCKS Floating Harbour (approx 7M upriver) is accessed via Entrance Lock into Cumberland Basin and Junction Lock into Floating Hbr. Aim to reach the Ent lock no later than HW –0015. Ent lock opens approx HW –2½, –1½ and –¼hr for arrivals; departures approx 15 mins after these times. Plimsoll swing bridge opens in unison with lock, except weekday rush hours 0800–0900 and 1700–1800.

Junction lock is always open, unless ht of HW is >9·6m ('stopgate' tide) when it closes; read special instructions. If you miss the last lock-in, call Ch 12 for advice.

BRIDGES Prince St Bridge opens (by arrangement only) summer 0915–2130, winter 0915–1830. Call Duty Office ☎9031484 or VHF Ch 73. Same for Redcliffe Bridge, but only opened in exceptional circumstances. Clearances above HAT: St Augustine's Reach 1·8m & 0·7m; Prince St 0·7m; Redcliffe 2·1m; Bristol 2·6m. The R Avon leads to Netham lock ☎9776590, Keynsham, Bath and the Kennett & Avon canal.

COMMUNICATIONS (Code 0117) CGOC (01646) 690909; Police 101; Cumberland Basin Dock Master 9273633 Ⓗ 923 0000.

Bristol VTS, Ch 12, ☎0117 980 2638.

- All vessels are to maintain a listening watch on Ch 12 in Bristol VTS area and monitor commercial ship movements.
- At Black Rks (1M to run) call City Docks Radio Ch 14, 11 low power, (HW–3 to HW+1) for locking instructions.
- For berth: Bristol Floating Hbr Ch 73; Bristol Marina Ch 80.

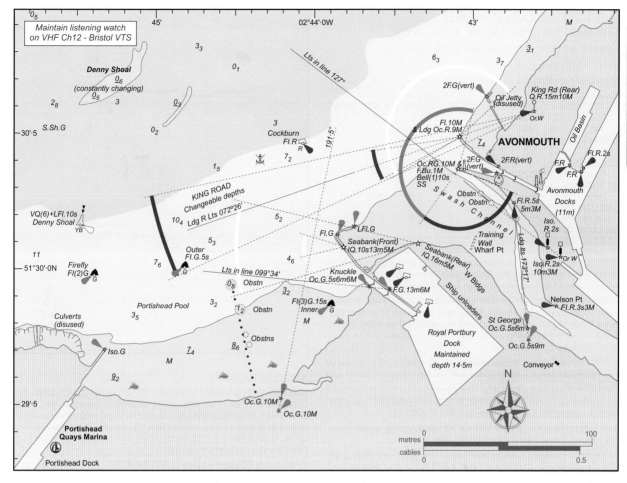

FACILITIES (from seaward) **Portishead Cruising Club** at Crockerne Pill, S bank, 5ca above M5 bridge; www.portisheadcruisingclub.org.uk ☎01275 373988, Drying ⚓ (soft mud) ⚓ grid (£5/day).

Bristol Hbr HM, Underfall Yard, Cumberland Rd, Bristol BS1 6XG. www.bristol-city.gov.uk ☎9031484; approx 40 Ⓥ consult website and HM in advance for detailed info on lock and bridge opening times ⚓ near ent to St Augustine's Reach, £2.67 inc licence fee. ⚓ ⬇ ☝ ⚓ ⚓ (launching £9.22) ⚓ ☝ ☕ ✕ ⚓; ferries ply between various landing stages around the hbr.

Baltic Wharf Leisure Centre contact HM ☎9031484, ⚓ ⚓ ⚓.

Bristol Marina Access–3 HW+1. www.bristolmarina.co.uk ☎9213198, 170⚓, inc about 5 Ⓥ (21m max LOA) £2.00. ⚓ ⚓ (£2.25/m) ⚓ ⚓ ☝ ⚓ ⚓ ☕ Ⓔ ✕ ⚓ ⚓ (6/12t) ⚓ (50t) ⚓.

Bathurst Basin ⚓. **Cabot CC** ☎9268318, ⚓ ⚓ ⚓ ⚓.

City All domestic facilities, ACA, ⚓ ✈.

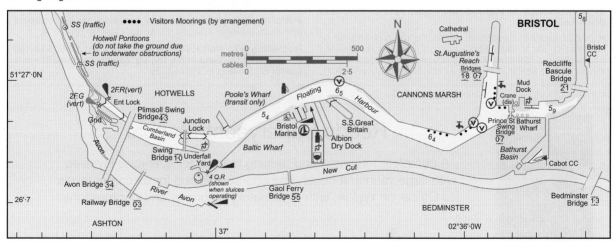

9.11.21 PORTISHEAD

Somerset **51°29'·56N 02°45'·41W** ✿✿✿◊◊◊◊✿✿✿

CHARTS AC 5608, 1176, 1859; Imray C59, 2600

TIDES –0405 Dover; ML 6·8

Standard Port BRISTOL (AVONMOUTH) (←—)

Times				Height (metres)			
High Water		Low Water		MHWS	MHWN	MLWN	MLWS
0200	0800	0300	0800	13·2	9·8	3·8	1·0
1400	2000	1500	2000				
Differences PORTISHEAD							
–0002	0000	ND	ND	–0·1	–0·1	ND	ND
CLEVEDON							
–0010	–0020	–0025	–0015	–0·4	–0·2	+0·2	0·0
ST THOMAS HEAD							
0000	0000	–0030	–0030	–0·4	–0·2	+0·1	+0·1
ENGLISH AND WELSH GROUNDS							
–0008	–0008	–0030	–0030	–0·5	–0·8	–0·3	0·0
WESTON-SUPER-MARE							
–0020	–0030	–0130	–0030	–1·2	–1·0	–0·8	–0·2

SHELTER Good in marina. Access by lock 9m x 40m x 5m min depth over cill HW ±3½ (minimum). ♥s available in marina. ⚓ 1ca NE of pier hd (sheltered from SE - W) to await tide for marina or Sharpness.

NAVIGATION WPT 51°29'·96N 02°45'·35W, Firefly SHM lt buoy, 168°/500m to pier hd. Firefly Rks (0·9m) are close W of the 168° appr track. Appr's dry to mud and are exposed to N/NE winds. Close inshore a W-going eddy begins at HW –3 whilst the flood is still making E.

LIGHTS AND MARKS Portishead Pt, Q (3) 10s9m 16M, is 7ca W of Portishead pierhd, Iso G 2s 5m 3M. Lock ent has 2FG(vert) and 2FR(vert) lts.

COMMUNICATIONS (Code 01275) CGOC (01646) 690909; Police 101; Health centre 841630; ⊞ (Clevedon) 01275 872212.

See Bristol. Monitor *Bristol VTS* Ch **12** for shipping movements. Portishead Quays Marina VHF Ch 80 24H.

FACILITIES Marina ⊕www.quaymarinas.com ☎841941 (200⌂+♥) £2·60 (min charge £21); ⟡ ⌿ ▲ ▲ ♠ ⚓ ⊠ ⚒ ⊡ ◬ ♨ ◪(35t) ◐ ⌑ wi-fi.

Portishead Cruising Club ☎373988, ⌑.

Town ✉ Ⓑ ⇌ ✈ (Bristol). 3M to Junction 19 of M5.

ADJACENT HARBOUR

WESTON-SUPER-MARE, Somerset, **51°21'·03N 02°59'·28W**. AC 5608, 1179, 1176,1152. HW –0435 on Dover; Duration 0655; ML 6·1m. See 9.11.21. Good shelter, except in S'lies, in Knightstone Hbr (dries) at N end of bay; access HW±1½. Causeway at ent marked by bn. Grand Pier hd 2 FG (vert) 6/5m. Or ⚓ in good weather in R Axe (dries), enter only near HW.

Weston Bay YC (sited on beach) ☎01934 413366, VHF Ch 80, ⚓ ⌑.

Uphill Boat Centre ☎01934 418617. ⚓(lower R Axe-drying) ⌿ ⌂ ⚓ ▲ ▲ ✕ ⚒ ◐ ⊡ ◪(10t).

Town ⚓ ⟐ ✉ Ⓑ ⊞✕ ⌑ ⇌ .

AVONMOUTH TO HARTLAND POINT

(AC 1152, 1165) From Avonmouth to Sand Pt, the part-drying English Grounds extend 3M off the S shore. Portishead Quays Marina offers good shelter. Extensive mud flats fill the bays S to Burnham-on-Sea. Westward, the S shore of Bristol Chan is cleaner than N shore. But there is less shelter since the approaches to harbours such as Watchet, Minehead, Porlock Weir and Watermouth dry out.

▶ *In bad weather dangerous overfalls occur NW and NE of Foreland Pt. 5M to W there is a race off Highveer Pt. Between Ilfracombe and Bull Pt the E-going stream begins at HW Milford Haven +0540, and the W-going at HW Milford Haven –0025,* *sp rates 3kn. Overfalls occur up to 1·5M N of Bull Pt and over Horseshoe Rks, which lie 3M N. There is a dangerous race off Morte Pt, 1·5M to W of Bull Pt.* ◀

Shelter is available under the lee of Lundy Island; but avoid bad races to NE (White Horses), NW (Hen and Chickens), and SE; also overfalls over NW Bank.

▶ *W of Lundy streams are moderate, but strong around the island and much stronger towards the Bristol Channel proper.* ◀

Proceeding WSW from Rivers Taw/Torridge, keep 3M off to avoid the race N of Hartland Pt (Lt, fog sig, conspic radome). There is shelter off Clovelly in S/SW winds.

9.11.22 BURNHAM-ON-SEA

Somerset **51°14'·23N 03°00'·33W** ❀⚓♧♧

CHARTS AC 5608, 1179, 1152; Imray C59, 2600

TIDES −0435 Dover; ML 5·4; Duration 0620
Standard Port BRISTOL (AVONMOUTH) (←)

Times				Height (metres)			
High Water		Low Water		MHWS	MHWN	MLWN	MLWS
0200	0800	0300	0800	13·2	9·8	3·8	1·0
1400	2000	1500	2000				
Differences BURNHAM-ON-SEA							
−0020	−0025	−0030	0000	−2·3	−1·9	−1·4	−1·1
BRIDGWATER							
−0015	−0030	+0305	+0455	−8·6	−8·1	DR	DR

SHELTER Ent is very choppy in strong winds, especially from SW to W and from N to NE. ⚓ S of town jetty or, for best shelter, ▼/⚓ in R Brue (dries). Contact HM by ☎ in advance.

NAVIGATION WPT 51°13'·46N 03°09'·80W, 0·5M S of Gore SWM buoy. Enter HW−2 to HW; not advised at night. From WPT track 090°/1·5M until abeam LH edge of Hinkley Pt C Power Station; then 070°/2M until close abeam No1 PHM buoy; 076°/2M in W sector of Low Lt; 112° on ldg lts; then 180° into river channel.

Banks and depths change frequently. Beware unmarked fishing stakes outside approach channels.

LIGHTS AND MARKS Lower lt ho Dir 076° as chartlet. Ldg lts/ marks 112° (moved as chan shifts): front FR 6m 3M, R stripe on ▢W background on sea wall; rear FR 12m 3M, RH edge of church tr.

COMMUNICATIONS (Code 01278) Police 101; Ⓗ 773100. HM 01934 822666 H24; harbour.master@sedgemoor.gov.uk.

FACILITIES Burnham-on-Sea MB&SC ☎792911 (Wed eve and Sun am) ⚓ two drying pontoon berths in River Brue (free) ⟍ ⚓ ⊕ ⛟ ⌂. **Services** ⚒ ⚓ ⊞ ACA (Bristol).

Town 🛢 ✉ Ⓑ ⇌ (Highbridge), ✈(Bristol).

ADJACENT FACILITY
Combwich A cargo wharf has been constructed to facilitate the inload of large indivisible loads to Hinkley Pt C Power Station. The approaches can be busy on appropriate tides. Note: There is no access to Bridgwater Marina from sea or the R Parrett.

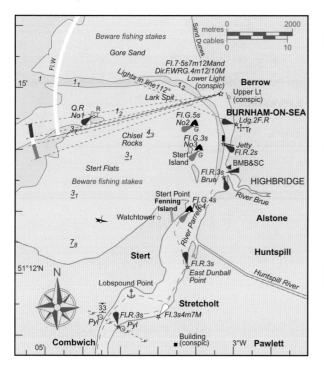

9.11.23 WATCHET

Somerset **51°11'·03N 03°19'·72W** ❀⚓♧♧♧

CHARTS AC 5608, 1179, 1152, 1160; Imray C59, 2600

TIDES −0450 Dover; ML 5·9; Duration 0655
Standard Port BRISTOL (AVONMOUTH) (←)

Times				Height (metres)			
High Water		Low Water		MHWS	MHWN	MLWN	MLWS
0200	0800	0300	0800	13·2	9·8	3·8	1·0
1400	2000	1500	2000				
Differences HINKLEY POINT							
−0032	−0028	−0055	−0049	−1·4	−1·1	−0·1	0·0
WATCHET							
−0035	−0050	−0145	−0040	−1·9	−1·5	+0·1	+0·1
MINEHEAD							
−0037	−0052	−0155	−0045	−2·6	−1·9	−0·2	0·0
PORLOCK BAY							
−0045	−0055	−0205	−0050	−3·0	−2·2	−0·1	−0·1
LYNMOUTH							
−0055	−0115	ND	ND	−3·6	−2·7	ND	ND

SHELTER Good, but open to N and E winds. Outer hbr (controlled by HM on VHF) dries 6·5m, but has about 6m depth at MHWS. Marina entered through a dropping sill gate (width 7m) which is open when the tide level is at or above CD +6.92m (retained water level). Min clearance over the gate is 2.5m. Approx gate opening times HW ±2½H sp, HW ±1½H np. **Outer harbour** 6 drying ⚓ on W wall (free), also used by commercial shipping.

NAVIGATION WPT 51°12'·03N 03°18'·88W, 208°/1·1M to hbr ent. Rks/mud dry 5ca to seaward. Beware tidal streams 4-5kn at sp offshore and around W pier hd. The marina entrance is controlled by automatic stop/go R/G lts. Entrance max width 7m. Min retained water depths vary 1.5–3.0m. Marina subject to constant silting and some berths have limited accessibility. 3kn speed limit in marina.

LIGHTS AND MARKS Two unlit radio masts (206m) bearing 208° Station is conspic 7·5M to the E. W pier hd Oc G 3s 9m 9M on Red (R) tr. E pier hd 2 FR (vert) 3M.

COMMUNICATIONS (Code 01984) CGOC (01646) 690909; Police 101. HM 01643 703704 and ask for HM (WSDC at Minehead) see **Marina** ☎ below; Watchet Boat Owners Association (01643) 702569. Outer Harbour Ch 16 (only when vessel expected); Marina Ch 80.

FACILITIES Watchet Harbour Marina www.watchetharbour.co.uk ☎631264, mob 07969 138938, 250⚓ inc ▼ £2·25 inc ⊕, 🛢 🛢 ⊞ ⚓ ⛾(20t) ⚓ ACA. **Town** ✉ ⛟ ⚒ ⌂. At Williton (2M): 🏧 🏦 🛢; Ⓗ (Minehead 8M), ⇌ (Taunton 18M), ✈ (Bristol).

ADJACENT JETTY
Hinkley Pt C Power Station. The power station redevelopment includes the construction of a jetty and dolphins on Hinkley Pt. Craft are not allowed within 100m except in emergency.

HARBOURS ON THE SOUTH SHORE OF THE BRISTOL CHANNEL

MINEHEAD, Somerset, **51°12´·79N 03°28´·37W**. AC 1179, 1165, 1160. HW –0450 on Dover. ML 5·7m. See 9.11.23. Small hbr, dries 7·5m. Good shelter within pier curving E and then SE, over which seas may break in gales at MHWS; exposed to E'lies. Best appr from N or NW; beware The Gables, shingle bank (dries 3·7m) about 5ca ENE of pier. Keep E of a sewer outfall which passes ½ca E of pier hd and extends 1¾ca NNE of it; outfall is protected by rk covering, drying 2·8m and N end marked by SHM Bn QG 6m 7M. There are 2 R ⚓s at hbr ent just seaward of 3 posts or Med moor on quay, £1·80/m/day, at inner end of pier. Hbr gets very crowded. Holiday camp is conspic 6ca SE. Pierhd lt Fl (2) G 5s 4M, vis 127°-262°. VHF Ch 16 12 14 (occas). HM ☎(01643) 708101; Facilities: **Hbr** ⚓ (launching £7.00). **Town** ⛽🅿️⛴🛒🔧🔍🏪✉️Ⓑ🍴✕🏬⇌ (Taunton).

PORLOCK WEIR, Somerset, **51°13´·17N 03°37´·64W**. AC 1179, 1165, 1160. HW –0500 on Dover; ML 5·6m. See 9.11.23. Access only near HW. Ent chan (250°), about 15m wide marked by withies (3 PHM and 1 SHM), between shingle bank/wood pilings to stbd and sunken wooden wall to port is difficult in any seas. A small pool (1m) just inside ent is for shoal draught boats; others dry out on pebble banks.

Or turn 90° stbd, via gates (but opening bridge usually closed), into inner drying dock with good shelter. No lts. HM ☎01643 863187 (not local). **Porlock Weir SC**. Facilities: ⚓ and limited 🛒.

LYNMOUTH, Devon, **51°14´·16N 03°49´·79W**. AC 1160, 1165. HW–0515 on Dover. See 9.11.23. Tiny hbr, dries approx 5m, only suitable in settled offshore weather. Appr from Sand Ridge SHM lt buoy, 1·6M W of Foreland Pt and 9ca N of hbr ent. The narrow appr channel between drying boulder ledges is marked by 7 unlit posts. After first 2 posts keep 10m away from next SH post then next 2PH posts keep to middle of ent. Hbr ent is between piers, 2FR/FG lts, on W side of river course. Berth on E pier, which covers (beware) at MHWS. Resort facilities. Admin by Council ☎01598 752384.

WATERMOUTH, Devon, **51°13´·03N 04°04´·6W**. AC 1179, 1165. HW –0525 on Dover; ML 4·9m; Duration 0625. Use 9.11.24. Good shelter in drying hbr, but heavy surge runs in strong NW winds. Dir lt 153° Oc WRG 5s 1m, W sector 151·5°-154·5°, W △ on structure, 1½ca inside ent on S shore. Bkwtr, covered at high tide, has Y poles along its length and a G pole with conical topmark at the end. 9 Y ⚓s with B handles £6.50. HM ☎(01271) 865422. Facilities: **Hbr** ⚓ (cans) ⛽🔧(12t) 🔍; **YC**, 🏬. **Combe Martin** 1½M all facilities.

9.11.24 ILFRACOMBE
Devon **51°12´·65N 04°06´·65W** ✿✿⚓⚓✿✿

CHARTS AC 5608, *1179, 1165,* 1160; Imray C59, 2600

TIDES –0525 Dover; ML 5·0; Duration 0625

Standard Port MILFORD HAVEN (←)

Times				Height (metres)			
High Water		Low Water		MHWS	MHWN	MLWN	MLWS
0100	0700	0100	0700	7·0	5·2	2·5	0·7
1300	1900	1300	1900				
Differences ILFRACOMBE							
–0016	–0016	–0041	–0031	+2·2	+1·7	+0·6	+0·3
LUNDY ISLAND							
–0025	–0025	–0035	–0030	+1·0	+0·8	+0·2	+0·1

SHELTER Good except in NE/E winds. SW gales can cause surge in hbrs, which dry. 8 ⚓s in outer hbr, dries 2·7m. Or ⚓ clear of pier. Visitors ⚓ on N snd S walls of Inner Hbr, dries approx 4m, or dry out on chains off foot of N Pier. Call beforehand to confirm berth availability.

NAVIGATION WPT 51°13´·23N 04°06´·67W, 180°/0·55M to pier hd. From E, beware Copperas Rks (4M to E), and tide rips on Buggy Pit, 7ca NE of ent. On entry keep toward Pier to clear drying ledges and lobster keep-pots obstructing Hbr ent on SE side.

LIGHTS AND MARKS See 9.11.4 and chartlet.

COMMUNICATIONS (Code 01271) CGOC (01646) 690909; Police 101; Dr 863119. HM 862108 mob 07775 532606.

Ilfracombe Hbr VHF Ch 12 16 (not H24).

FACILITIES Hbr ⚓ (launching £5.84/day, £1·45/week) ⚓ and ⚓s £1·32 (inc use of 🔍) ⚓ 🔌(S Quay) ⛽🔧🔍📷🔍. **Ilfracombe YC** ☎863969, ⚓🔋🔧(35t as arranged) 🏬 **Town** ✉️Ⓑ🛒✕🏬 bus to Barnstaple (⇌), ✈ (Exeter).

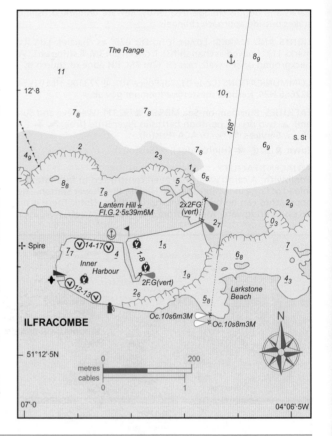

HARTLAND POINT TO LAND'S END

(AC 1156, 1149) The N coast of Cornwall and SW approaches to Bristol Chan are very exposed. Yachts need to be sturdy and well equipped since, if bad weather develops, no shelter may be at hand.

Bude dries, and is not approachable in W winds; only accessible in calm weather or offshore winds. Boscastle is a tiny hbr (dries) 3M NE of Tintagel Hd. Only approach in good weather or offshore winds; ⚓ off or dry out alongside.

Padstow can be a refuge, but in strong NW winds the sea breaks on bar and prevents entry. Off Trevose Hd beware Quies Rks which extend 1M to W. From here S the coast is relatively clear to Godrevy Is, apart from Bawden Rks 1M N of St Agnes Hd. Newquay Bay is good ⚓ in offshore winds, and the hbr (dries) is sheltered but uncomfortable in N winds. Off Godrevy Is are The Stones, drying rky shoals extending 1·5M offshore, marked by NCM lt buoy.

In St Ives Bay (AC 1168), Hayle is a commercial port (dries); seas break heavily on bar at times, especially with a ground swell. ▶ *Stream is strong, so enter just before HW.* ◀ The bottom is mostly sand. St Ives (dries) gives shelter from winds E to SW, but is very exposed to N; there is sometimes a heavy breaking sea if there is ground swell. Keep clear of the renewable energy development area centred approx 325°/10M from St Ives Bay (see AC 2565).

From St Ives to Land's End coast is rugged and exposed. There are o'falls SW of Pendeen Pt. Vyneck Rks lie awash about 3ca NW of C Cornwall. The Brisons are two high rky islets 5ca SW of C Cornwall, with rky ledges inshore and to the S. The Longships group of rks is about 1M W of Land's End. The inshore passage (001° on Brisons) is about 4ca wide with unmarked drying rks on the W side; only to be considered in calm weather and beware of fishing gear marker buoys.

For Isles of Scilly, South Cornwall and Tides, see 9.1.5.

9.11.25 RIVERS TAW AND TORRIDGE

Devon 51°04'·37N 04°12'·88W ✦✦🌊🌊🌊🏵🏵🏵

CHARTS AC 5608, 1179, 1164, 1160; Imray C58, 2600

TIDES –0525 (Appledore) Dover; ML 3·6; Duration 0600

Standard Port MILFORD HAVEN (⟵)

Times				Height (metres)			
High Water		Low Water		MHWS	MHWN	MLWN	MLWS
0100	0700	0100	0700	7·0	5·2	2·5	0·7
1300	1900	1300	1900				
Differences APPLEDORE							
–0020	–0025	+0015	–0045	+0·5	0·0	–0·9	–0·5
YELLAND MARSH (R Taw)							
–0010	–0015	+0100	–0015	+0·1	–0·4	–1·2	–0·6
FREMINGTON (R Taw)							
–0010	–0015	+0030	–0030	–1·1	–1·8	–2·2	–0·5
BARNSTAPLE (R Taw)							
0000	–0015	–0155	–0245	–2·9	–3·8	–2·2	–0·4
BIDEFORD (R Torridge)							
–0020	–0025	0000	0000	–1·1	–1·6	–2·5	–0·7
CLOVELLY							
–0030	–0030	–0020	–0040	+1·3	+1·1	+0·2	+0·2

SHELTER Very well protected, but ent is dangerous in strong on-shore winds and/or swell. Yachts can ⚓ or pick up RNLI buoy (please donate to RNLI, Boathouse ☎473969) in Appledore Pool N of Skern Pt where spring stream can reach 5kn. Bideford quay dries to hard sand; used by commercial shipping.

NAVIGATION WPT 51°05'·43N 04°16'·11W, 118°/0·9M to Bar.

Bar and sands constantly shift; buoys are moved occasionally to comply. For advice on bar contact Bideford HM. Least depths over bar vary from 0·1m and 0·4m. Following 2014 storms a bank has formed which just dries at LW on the leading line. Estuary dries and access is only feasible from HW–2 to HW. Night entry not advised for strangers. Once tide is ebbing, breakers quickly form between Bideford Bar SHM lt buoy and Middle Ridge SHM lt buoy. Hold the ldg line 118° only up to Outer Pulley where chan deviates stbd toward Pulley buoy and Grey Sand Hill, thence to Appledore Pool. 2M passage to Bideford is not difficult. Barnstaple (7M): seek local advice or take pilot.

LIGHTS AND MARKS See 9.11.4 and chartlet. Ldg marks are W trs, lit H24. R Torridge: Lt QY at E end of Bideford bridge, 2FR vert & 2FG vert indicate preferred chan; then SHM bn, QG, on W bank.

COMMUNICATIONS (Appledore/Bideford: Code 01237) CGOC (01646) 690909; Police 101; Dr 474994. (Instow/Barnstaple: Code 01271) Dr 372672. Appledore HM 428700; Bideford HM/Pilot 475834, mob 07967 333725.

2 Rivers Port/Pilots VHF Ch 12 16 (From HW–2 occasional).

FACILITIES
APPLEDORE: no ⛽; slips at town quay. **Services** ⚓ ⛽ ⚒ ⚙ 🖥 Ⓔ 🛠 ⛴(70t) ⚓.

BIDEFORD: some ⛽(contact HM) ⚓ ⛽ ⚒ ✕ 🗑. Ferry to Lundy Is.

INSTOW: North Devon YC, ☎861390, ⚓ ⚓ ✕ 🗑; Services ⚓ ⛽ via Instow Marine ☎861081, ⚓ ⛽ ⚙ Ⓔ ⛴(4t) ⚓. **Town** ⚓ ✕ 🗑.

BARNSTAPLE: ⛽ (free for short stay): limited facilities; ⚓ 🏦 🏧 ⚓ 🗑 🗑.

Towns ✉ (all four), Ⓑ (Barnstaple, Bideford) 🚂 (Barnstaple) ✈ (Exeter).

CLOVELLY, Devon, 51°00'·18N 04°23'·77W ✦🌊🌊🏵🏵🏵. AC 1164. Tides see above. HW –0524 on Dover. Tiny drying hbr, 5M E of Hartland Pt, is sheltered from S/SW winds; useful to await the tide into Bideford or around Hartland Pt. Some ⛽ (max LOA 12m) £7 on pier, access only near HW; or ⚓ off in 5m. Lt Fl G 5s 5m 5M on hbr wall. HM mob: 07975 501830. Facilities: ⚓ ⚓ ✉ limited ⚒ 🏦 🏧.

RIVERS TAW & TORRIDGE *continued*

ISLAND IN BRISTOL CHANNEL, 10M NNW of Hartland Pt

LUNDY ISLAND, Devon, **51°09′·83N 04°39′·27W**. AC 5608, *1179, 1164*. HW –0530 on Dover; ML 4·3m; Duration 0605. See 9.11.24. Shore Office ☎01237 470074, Island ☎01237 431831. Beware bad tide races, esp on E-going flood, off the N and SE tips of the island; and to the SW on the W-going ebb. A violent race forms over Stanley Bank 3M NE of the N tip. Shelter good in lee of island's high ground (145m). In SSW to NW winds, usual ⚓ is close inshore to NW of SE Pt and Rat Island, clear of ferry. In N'lies ⚓ in The Rattles, small bay on S side. In E'lies Jenny's Cove is safe if no W'ly swell. Lts: NW Pt, Fl 15s 48m 17M, vis 009°-285°, W ○ tr. On SE Pt, Fl 5s 53m 15M, vis 170°-073°, W ○ tr. Two Historic Wrecks (see 9.0.3) lie on the E side of island, at 51°11′N 04°39′·4W, and 4ca further E. Facilities: Landing by the ⚓ off SE end of island or using N side of jetty to disembark passengers only, boats may not remain alongside; £5.00/person landing fee. The waters around Lundy are a Marine Nature Reserve. **Lundy Co**: ▮ ✉ ▱ limited stores.

MINOR HARBOURS ON THE NW COAST OF CORNWALL

BUDE, Cornwall, **50°49′·93N 04°33′·37W**. AC 5608, 1156. HW –0540 on Dover. Duration 0605. See 9.11.26. Limited shelter in drying hbr, access near HW in daylight, quiet weather, no swell conditions but sea-lock gives access to canal basin with 2m. Conspic W radar dish aerials 3·3M N of hbr. Outer ldg marks 075°, front W spar with Y ◇ topmark, rear W flagstaff; hold this line until inner ldg marks in line at 131°, front W pile, rear W spar, both with Y △ topmarks. There are no lts. VHF Ch 16 12 (when vessel expected). Advise HM of ETA with 24H notice ☎(01288) 353111; ⚓ on quay; **Town** (½M); ▮ ✉ Ⓑ ☕ ✕ ▱.

BOSCASTLE, Cornwall, **50°41′·48N 04°42·17W**. AC 5608, 1156. HW –0543 on Dover; see 9.11.26. A tiny, picturesque hbr, almost a land-locked cleft in the cliffs. Access near HW, but not in onshore winds when swell causes surge inside. An E'ly appr, S of Meachard Rk (37m high, 2ca NW of hbr), is best. 2 short bkwtrs at ent; moor as directed on drying S quay. HM ☎01840 250200.

9.11.26 PADSTOW

Cornwall 50°32'·51N 04°56'·17W ❀❀❀❀❀❀❀❀

CHARTS AC 5608, 1156, 1168; Imray C58, 2400

TIDES –0550 Dover; ML 4·0; Duration 0600

Standard Port MILFORD HAVEN (←)

Times				Height (metres)			
High Water		Low Water		MHWS	MHWN	MLWN	MLWS
0100	0700	0100	0700	7·0	5·2	2·5	0·7
1300	1900	1300	1900				
Differences BUDE							
–0040	–0040	–0035	–0045	+0·7	+0·6	ND	ND
BOSCASTLE							
–0045	–0010	–0110	–0100	+0·3	+0·4	+0·2	+0·2
PADSTOW							
–0055	–0050	–0040	–0050	+0·3	+0·4	+0·1	+0·1
WADEBRIDGE (R Camel)							
–0052	–0052	+0235	+0245	–3·8	–3·8	–2·5	–0·4
NEWQUAY							
–0100	–0110	–0105	–0050	0·0	+0·1	0·0	–0·1
PERRANPORTH							
–0100	–0110	–0110	–0050	–0·1	0·0	0·0	+0·1
ST IVES							
–0050	–0115	–0105	–0040	–0·4	–0·3	–0·1	+0·1
CAPE CORNWALL							
–0130	–0145	–0120	–0120	–1·0	–0·9	–0·5	–0·1

NOTE: At Wadebridge LW time differences give the start of the rise, following a LW stand of about 5 hours.

SHELTER Good in inner hbr 3m+, access approx HW±2 sp, ±1½ nps via tidal gate. If too late for gate, moor in the Pool or ⚓ close N in 1·5m LWS. Drying moorings available for smaller craft. Good ⌓ at Wadebridge 4.5M up R Camel.

NAVIGATION WPT 50°34'·56N 04°56'·07W, 044° Stepper Pt 6ca. From SW, beware Quies Rks, Gulland Rk, The Hen, Gurley Rk, Chimney

Rks and a wreck 5ca W of Stepper Pt (all off the chartlet). From N, keep well off Newland Island and its offlying reef.

- Best appr HW–2½, do not try LW±1½; least depth on the bar is 0·5m at MLWS. Waiting ⚓s in Port Quin Bay in lee of Rumps Pt and Mother Ivey's Bay 3M WSW Stepper Pt.
- Shifting banks in estuary require care and a rising tide (ditto the drying R Camel to Wadebridge, 4M). In doubt, consult HM.
- Identify the first 2 chan buoys before entry. In strong onshore winds or heavy ground swell, seas can break on Doom Bar and in the adjacent chan.
- S of St Saviour's Pt the chan lies very close to W shore.

LIGHTS AND MARKS See 9.11.4 and chartlet.

COMMUNICATIONS (Codes 01841) CGOC (01326) 317575; Police 101; Dr 532346. HM padstowharbour@btconnect.com , ☎532239; Ferry to Rock, (01326) 317575 it also acts as water taxi.

HM VHF Ch 12 16 (Mon-Fri 0800–1700 and HW±2). Ferry Ch 12.

FACILITIES Hbr ⌓ £2·00, M £1·01, ⌁ (£4·00 <10 hp engine<£6·00) ⚓ ⚓ ♠ ♠ ⊡(♦) ⚓ ⊡. **Services** ✕ ✎ ⊡ ⌂ ⌂(60t) ⚓.
Town ⊡ ⌁ ⊠ Ⓑ ⊞ ✕ ⌂ Bus (to Bodmin ⇌/✈ Newquay). **Rock SC** ☎(01208) 862431, ⌁.
Wadebridge HM as Padstow; ⌓ only; **Town** ⊠ Ⓑ ⊞ ✕ ⌂.

MINOR HBRS BETWEEN BOSCASTLE AND LAND'S END

PORT ISAAC, Cornwall, **50°35'·75N 04°49'·57W**. AC 1156, 1168. HW –0548 on Dover; ML 4·0m. Small drying hbr. Conspic ✠ tr bears 171°/1·3M. Rks close E of 50m wide ent between short bkwtrs. HM ☎01208 880321, mob 07855 429422; ⊠ ⊞ ✕ ⌂ LB.

NEWQUAY, Cornwall, **50°25'·06N 05°05'·19W**. AC 1149, 1168. HW –0604 on Dover; ML 3·7m; see 9.11.26. Ent to drying hbr ('The Gap') between two walls, is 23m wide. Beware Old Dane Rk and Listrey Rk outside hbr towards Towan Hd. Swell causes a surge in the hbr. Enter HW±2 but not in strong onshore winds. Lts: N pier 2 FG (vert) 2M; S pier 2 FR (vert) 2M. VHF Ch 08 16 14. Facilities: ♠ ⚓. **Town**; ⌁ ⚓ ⊡ ⊞ ✕ ⌂. Note: Shoal draught boats can dry out in Gannel Creek, close S of Newquay, but only in settled weather. Beware causeway bridge halfway up creek.

PORTREATH, Cornwall, **50°15'·88N 05°17'·57W**. AC 1149. HW –0600 on Dover. Conspic W daymark (38m) at E side of ent to small drying hbr. Gull Rk (23m) is 3ca W of ent and Horse Rk is close N. Keep close to pier on W side of chan. ⌓ in either of 2 basins, both dry. ⊠ ⊞ ✕ ⌂.

HAYLE, Cornwall, **50°12'·14N 05°26'·13W** (NCM) ❀❀❀. AC 1149, 1168. HW –0605 on Dover; ML 3·6m; Duration 0555. See 9.11.26. Drying hbr gives very good shelter, but is not advised for yachts. In ground swell dangerous seas break on the bar, drying 2·7m; approx 4m at ent @ MHWS. Cross in good weather HW±1. Marks do not necessarily indicate best water. Ldg marks/lts 180°: both W □ R horiz band, ☆ FW 17/23m 4M. NCM buoy, Q is about 7ca N of the front ldg lt. Training wall on W side of ent chan is marked by 5 lit perches, all Oc G 4s. The hbr is divided by long central island (about 700m long, with lt bn QG at NW end) which should be left to stbd. Follow the SE arm of hbr to Hayle; the S arm leads to Lelant Quay. Visitors must contact HM ☎(01736) 754043, Mob 07500993867, VHF Ch 16 prior to arrival. ⌓£5. Facilities: ⚓ (can) ⊡ ⊡ ♠ Ⓑ ⊞ ✕ ⌂ ⇌.

ST IVES, Cornwall, **50°12'·79N 05°28'·67W** ❀❀❀❀❀❀❀❀. AC 1149, 1168. HW –0610 on Dover; ML 3·6m; Duration 0555. See 9.11.26. Drying hbr with about 4·5m @ MHWS. Good shelter except in onshore winds when heavy swell works in. ⚓ in 3m between the hbr and Porthminster Pt to S and drying Or ⚓s in hbr. From the NW beware Hoe Rk off St Ives Hd, and obstn drying 0·9m at collapsed b'water E of Smeaton pier inshore of unlit SHM. From SE off Porthminster Pt avoid The Carracks and obstn drying 1·0m in Carbis Bay. Keep E of SHM buoy about 1½ ca ENE of E pier. Lts: E pier hd 2 FG (vert) 8m 5M. W pier hd 2 FR (vert) 5m 3M. VHF Ch 12 16 (occas). HM ☎ (01736) 795018. Facilities: ⚓s £12.13; **E Pier** ⚓. **Town** ♠ Ⓑ ⊠ ⊞ ⊡ ✕ ⌂ ⇌.

South Ireland

Malahide clockwise to Liscannor Bay

9.12.1	Contents	563
9.12.2	Area map	564
9.12.3	Tidal stream charts	566
9.12.4	Lights, buoys and waypoints	568
9.12.5	Passage information including Area 12 Distance table	570
	Malahide to Tuskar Rock – 572 • Tuskar Rock to Cobh – 581 • Cobh to Mizen Head – 590 • West coast, Mizen Head to Dingle Bay – 592 • Dingle Bay to Liscannor Bay – 598 • Shannon Estuary – 599	
9.12.6	Special notes for Ireland (including useful Gaelic words)	570
9.12.7	Malahide	572
9.12.8	Howth	573
9.12.9	Dublin (tide tables/curves)	574
9.12.10	Dun Laoghaire	578
	Greystones	
9.12.11	Wicklow	579
9.12.12	Arklow	579
	Courtown • Polduff Pier	
9.12.13	Wexford	580
9.12.14	Rosslare	581
9.12.15	Kilmore Quay	582
9.12.16	Waterford	583
9.12.17	Dunmore East	584
	Dungarvan Bay	
9.12.18	Youghal	584
	Ballycotton	
9.12.19	Cork Harbour (Cobh, tide tables/curves)	585
9.12.20	Kinsale	590
	Oysterhaven	
9.12.21	Courtmacsherry	591
9.12.22	Glandore/Unionhall	591
9.12.23	Castletownshend	591
	Barloge Creek	
9.12.24	Baltimore	593
	Horseshoe Harbour • Clear Island, North Harbour • Roaringwater Bay	
9.12.25	Schull	594
9.12.26	Crookhaven	594
	Goleen • Dunmanus Bay	
9.12.27	Bantry Bay	595
	Castletown • Lawrence Cove • Lonehort Harbour • Adrigole • Glengarriff • Bantry	
9.12.28	Kenmare River	596
	Garnish Bay • Ballycrovane • Cleanderry • Ardgroom • Kilmakilloge • Ormond's Harbour • Kenmare • Dunkerron Harbour • Sneem • Darrynane	
9.12.29	Dingle Bay	597
9.12.30	Valentia Harbour	597
9.12.31	Dingle	598
	Portmagee • Ventry • Smerwick Harbour • Brandon Bay • Tralee Bay	
9.12.32	Fenit Harbour	599
9.12.33	Shannon Estuary	600
9.12.34	Kilrush (Tarbert Island, tide tables/curves)	601

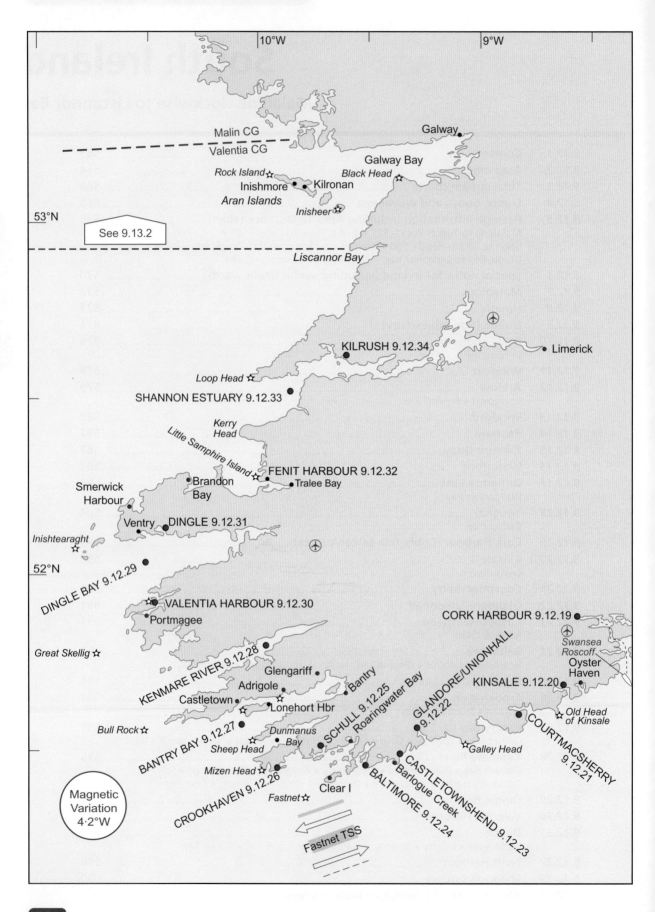

Malin CG
Valentia CG
Galway

Galway Bay

Rock Island ☆
Inishmore ● Kilronan ● *Black Head* ☆
Aran Islands
Inisheer ☆

53°N

See 9.13.2

Liscannor Bay

KILRUSH 9.12.34 ● ● Limerick

Loop Head ☆
SHANNON ESTUARY 9.12.33 ●

Kerry Head

Little Samphire Island ☆
FENIT HARBOUR 9.12.32
● Brandon ● Tralee Bay
Bay

Smerwick Harbour
● Ventry ● DINGLE 9.12.31

Inishtearaght ☆

52°N

DINGLE BAY 9.12.29 ●

☆● VALENTIA HARBOUR 9.12.30
● Portmagee

CORK HARBOUR 9.12.19 ●

Great Skellig ☆

Swansea Roscoff
Oyster Haven

KENMARE RIVER 9.12.28 ●
● Glengariff
● Bantry
GLANDORE/UNIONHALL
KINSALE 9.12.20 ●

● Adrigole
☆ Lonehort Hbr
SCHULL 9.12.25
Roaringwater Bay
9.12.22
COURTMACSHERRY 9.12.21

● Castletown
☆ *Old Head of Kinsale*

Bull Rock ☆
BANTRY BAY 9.12.27
Dunmanus Bay
☆ *Galley Head*

Sheep Head
CASTLETOWNSHEND 9.12.23

Mizen Head ☆
Barlogue Creek

CROOKHAVEN 9.12.26
Clear I
BALTIMORE 9.12.24

Fastnet ☆

Magnetic Variation 4·2°W

Fastnet TSS

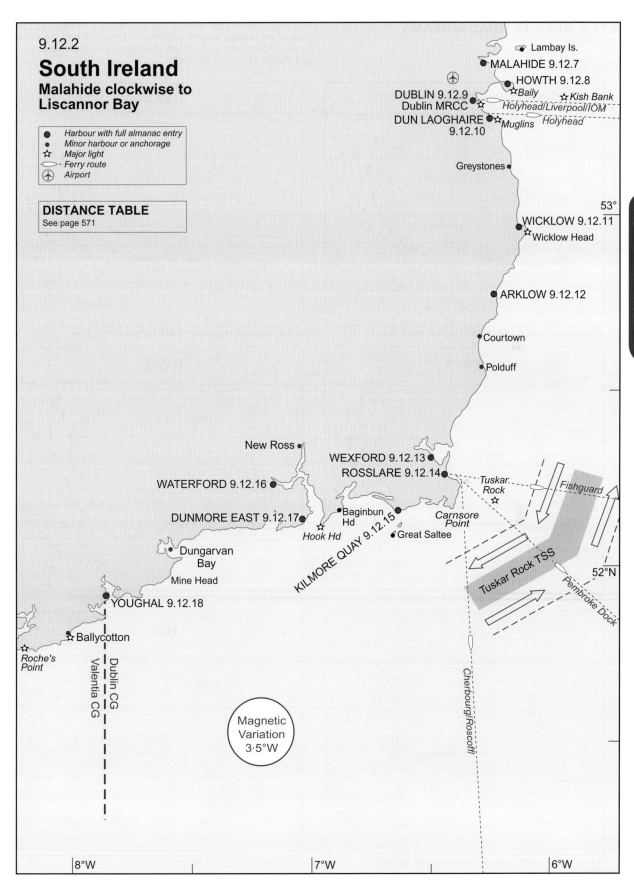

9.12.2

South Ireland
Malahide clockwise to Liscannor Bay

● Harbour with full almanac entry
● Minor harbour or anchorage
☆ Major light
⬭ Ferry route
✈ Airport

DISTANCE TABLE
See page 571

S Ireland

Lambay Is.
MALAHIDE 9.12.7
HOWTH 9.12.8
☆ Baily ☆ Kish Bank
DUBLIN 9.12.9
Dublin MRCC Holyhead/Liverpool/IOM
DUN LAOGHAIRE ☆ Muglins Holyhead
9.12.10

Greystones

53°

WICKLOW 9.12.11
☆ Wicklow Head

ARKLOW 9.12.12

Courtown

Polduff

New Ross
WEXFORD 9.12.13
ROSSLARE 9.12.14
Tuskar Rock Fishguard
WATERFORD 9.12.16
DUNMORE EAST 9.12.17
Baginbun Hd
Hook Hd Carnsore Point
Great Saltee
KILMORE QUAY 9.12.15 Tuskar Rock ☆
Dungarvan Bay Tuskar Rock TSS
Mine Head 52°N
Pembroke Dock
YOUGHAL 9.12.18
☆ Ballycotton
☆ Roche's Point
Dublin CG
Valentia CG
Cherbourg/Roscoff

Magnetic Variation 3·5°W

8°W 7°W 6°W

9.12.3 AREA 12 TIDAL STREAMS

The tidal arrows (with no rates shown) off the S and W coasts of Ireland are printed by kind permission of the Irish Cruising Club, to whom the Editor is indebted. They have been found accurate, but should be used with caution.

5 Hours before HW Dover (0530 before HW Dublin)

4 Hours before HW Dover (0430 before HW Dublin)

3 Hours before HW Dover (0330 before HW Dublin)

2 Hours before HW Dover (0230 before HW Dublin)

1 Hour before HW Dover (0130 before HW Dublin)

HW Dover (0030 before HW Dublin)

Northward 9.13.3   South Irish Sea 9.11.3

The tidal arrows (with no rates shown) off the S and W coasts of Ireland are printed by kind permission of the Irish Cruising Club, to whom the Editor is indebted. They have been found accurate, but should be used with caution.

1 Hour after HW Dover (0030 after HW Dublin)

2 Hours after HW Dover (0130 after HW Dublin)

3 Hours after HW Dover (0230 after HW Dublin)

4 Hours after HW Dover (0330 after HW Dublin)

5 Hours after HW Dover (0430 after HW Dublin)

6 Hours after HW Dover (0530 after HW Dublin)

S Ireland

9.12.4 LIGHTS, BUOYS AND WAYPOINTS

Bold print = light with a nominal range of 15M or more. CAPITALS = place or feature. *CAPITAL ITALICS* = light-vessel, light float or Lanby. *Italics* = Fog signal. *Bold italics* = Racon. Many marks/buoys are fitted with AIS (MMSI No); see relevant charts.

LAMBAY ISLAND TO TUSKAR ROCK

MALAHIDE and LAMBAY ISLAND
Taylor Rks ⚓ Q; 53°30'·21N 06°01'·87W.
Burren Rocks ⚓ Fl G 5s; 53°29'·35N 06°02'·35W.
Malahide approach ⚓ L Fl 10s; 53°27'·12N 06°06'·87W.

HOWTH
Rowan Rocks ⚓ Q (3) 10s; 53°23'·88N 06°03'·27W.
Howth ⚓ 53°23.727'N 006°03.593'W; 992501250.
E Pier Hd ⚓ Fl WR 7·5s 13m W12M, R9M; W twr; vis: W256°-R295°-256°; 53°23'·64N 06°04'·02W.
Baily ☆ 53°21'·69N 06°03'·16W Fl 15s 41m **20M**; twr. Fog Det Lt VQ; 992501010.
Rosbeg E ⚓ Q (3) 10s; 53°21'·02N 06°03'·45W.
Rosbeg S ⚓ Q (6) + L Fl 15s; 53°20'·22N 06°04'·17W.

PORT OF DUBLIN
Dublin Bay ⚓ Mo (A) 10s; *Racon (M)*; 53°19'·92N 06°04'·65W; 992351301.
No. 1 ⚓ Fl (3) G 5s; 53°20'·30N 06°05'·56W.
No. 3 ⚓ IQ G; 53°20'·57N 06°06'·76W.
No. 4 ⚓ IQ R; 53°20'·48N 06°06'·93W.
No. 5 ⚓ Fl G 2s; 53°20'·64N 06°08'·60W.
No. 6 ⚓ Fl R 2s; 53°20'·56N 06°08'·75W.
Great S Wall Hd Poolbeg ⚓ Fl R 4s 20m 10M *(sync with N.Bull)*; R ○ twr; 53°20'·52N 06°09'·08W.
N Bull ⚓ Fl G 4s 15m 10M; G ○ twr; 53°20'·70N 06°08'·98W.
N Bank ☆ 53°20'·69N 06°10'·59W Oc G 8s 10m **16M**; G □ twr.

DUN LAOGHAIRE
E Bkwtr Hd ⚓ Fl (2) R 8s 16m **17M**; granite twr, R lantern 7M; 53°18'·14N 06°07'·62W. Fog Det Lt VQ.
Outfall ⚓ Fl Y 5s; 53°18'·41N 06°08'·35W.
Muglins ⚓ Fl R 5s 14m 11M; 53°16'·52N 06°04'·58W.

OFFSHORE MARKS
Bennett Bank ⚓ Q(6)+ L Fl 15s; 53°20'·17N 05°55'·11W; 992501012.
Kish Bank ☆ 53°18'·65N 05°55'·54W Fl (2) 20s 29m **21M** (H24); W twr, R band; *Racon (T) 15M*; 992501017.
N Kish ⚓ VQ; 53°18'·56N 05°56'·44W.
E Kish ⚓ Fl (2) R 10s; 53°14'·35N 05°53'·56W; 992501020.
N Burford ⚓ Q 53°20.51'N 006° 01.49'W; 992501016.
S Burford ⚓ VQ(6)+L Fl 10s; 53°18.06'N 006° 01.30'W; 992501018.
E Codling ⚓ Fl (4) R 10s; 53°08'·54N 05°46'·07W; 992501024.
W Codling ⚓ Fl G 10s; 53°06'·97N 05°54'·51W.
Moulditch ⚓ Fl R 10s; 53°08'·43N 06°01'·23W; 992501022.
S Codling ⚓ VQ (6) + L Fl 10s; 53°04'·74N 05°49'·76W.
Codling ⚓ Q(3) 10s; 53°03.020'N 005°40.815'W, *Racon (G) 10M*, 992501028.
Breaches Shoal ⚓ Fl (2) R 6s; 53°05'·67N 05°59'·81W.
North India ⚓ Q; 53°03'·12N 05°53'·46W.
South India ⚓ Q (6) + L Fl 15s; 53°00'·36N 05°53'·31W; 992501030.

WICKLOW to ARKLOW
Wicklow ⚓ Fl (4) Y 10s; 52°59'·54N 06°01'·29W.
E Pier Hd ⚓ Fl WR 5s 11m 6M; W twr, R base and cupola; vis: 136°-R-293°-W-136°; 52°58'·99N 06°02'·07W.
W Packet Quay Hd ⚓ Fl WG 10s 5m 6M; vis: 076°-G-256°-W-076°; 52°58'·88N 06°02'·08W.
Wicklow Hd ☆ 52°57'·95N 05°59'·89W Fl (3) 15s 37m **23M**; W twr; 992501031.

Horseshoe ⚓ Fl R 3s; 52°56'·84N 05°58'·47W; 992501032.
N Arklow ⚓ Q; 52°53'·86N 05°55'·21W; 992501034.
Arklow Bank Wind Farm from 52°48'·47N 05°56'·57W to 52°46'·47N 05°57'·11W, N and S Turbines Fl Y 5s14m 10M. No 7 Turbine 992501302. Other turbines Fl Y 5s. See 9.12.5.

ARKLOW to WEXFORD
S Pier Hd ⚓ Fl WR 6s 11m 13M; twr; vis: R shore- W223°- R350°-shore; 52°47'·61N 06°08'·22W.
Roadstone Bkwtr Hd ⚓ QY; 52°46'·65N 06°08'·23W.
S Arklow ⚓ Q (6) + LFl 15s; *Racon (O) 10M*; 52°40'·20N 05°58'·89W; 992501040.
No. 2 Glassgorman ⚓ Fl (4) R 10s; 52°45'·35N 06°05'·34W; 992501038.
No. 1 Glassgorman ⚓ Fl (2) R 6s; 52°37'·69N 06°07'·34W.
N Blackwater ⚓ Q; 52°32'·22N 06°09'·51W; 992501046.
No. 6 Rusk ⚓ Fl R 3s; 52°32'·65N 06°10'·41W.
No. 4 Rusk ⚓ Fl (2) R 5s; 52°31'·07N 06°10'·86W.
No. 2 Rusk ⚓ Fl (2) R 5s; 52°28'·64N 06°12'·61W.
No. 1 Rusk ⚓ Fl (2) G 5s; 52°28'·54N 06°11'·80W; 992501048.
W Blackwater ⚓ Fl G 6s; 52°25'·87N 06°13'·56W.
SE Blackwater ⚓ Q (3) 10s, 52°25'·62N 06°08'·42W; *Racon (M) 10M*; 992501050.
S Blackwater ⚓ Q (6) + L Fl 15s; 52°22'·76N 06°12'·87W.
North Long ⚓ Q; 52°21'·44N 06°17'·04W.
West Long ⚓ QG; 52°18'·18N 06°17'·96W.
Lucifer ⚓ VQ (3) 5s; 52°17'·02N 06°12'·67W; 992501054.

ROSSLARE
S Long ⚓ Fl (2) G 6s (sync with Splaugh); 52°14'·74N 06°15'·80W.
Splaugh ⚓ Fl (2) R 6s (sync with S Long); 52°14'·37N 06°16'·76W; 992501062.
South Holdens ⚓ Fl G 3s; 52°15'·14N 06°17'·24W. (sync Calmines).
Calmines ⚓ Fl R 3s; 52°15'·01N 06°17'·77W. (sync S Holdens).
W Holdens ⚓ Fl (3) G 10s; 52°15'·77N 06°18'·74W.

Rosslare Pier Hd ⚓ Oc WRG 5s 15m W13M, R10M, G10M; R twr; vis:098°-G-188°-W-208°-G-246°-G-283°-W-286°-R-320°; 52°15'·43N 06°20'·29W.
Ballygeary ⚓ Oc WR 1·7s 7m 4M vis: shore-R-152°-W-200°-W(unintens)-205°; 52°15'·25N 06°20'·48W.

TUSKAR ROCK TO OLD HEAD OF KINSALE
Tuskar ☆ 52°12'·17N 06°12'·44W Q (2) 7·5s 33m **24M**; W twr; *Racon (T) 18M*; 992501065.
S Rock ⚓ Q (6)+L Fl 15s; 52°10'·80N 06°12'·84W; 992501068.
Fundale ⚓ Fl (2) R 10s; 52°11'·04N 06°19'·78W.
Barrels ⚓ Q (3) 10s; 52°08'·32N 06°22'·05W; 992501070.

KILMORE
Kilmore Quay SWM ⚓ Iso 10s; (Apr-Sep); 52°09'·20N 06°35'·30W.
Kilmore Bkwtr Hd ⚓ Q RG 7m 5M; vis: 269°-R-354°-G-003°-R-077°; 52°10'·20N 06°35'·15W.
Ldg lts 007·9°, Front 52°10'·37N 06°35'·08W Oc 4s 3m 6M. Rear, 100m from front, Oc 4s 6m 6M; sync with front.
Bore Rks ⚓ Q (3) 10s; 52°06.07'N 006°31.87'W; 992501072.
Coningbeg ⚓ Q(6)+LFl 15s; 52°03'·20N 06°38'·57W; *Racon (T) 7M*; 992501074.
Red Bank W ⚓ 52°04.499'N 006°41.652'W; 992501076.
'M5' ODAS 35 ⚓ 51°41'·40N 06°42'·24W; Fl (5) Y 20s.

WATERFORD
Hook Hd ☆ 52°07'·42N 06°55'·77W Fl 3s 46m **23M**; W twr, two B bands; *Racon (K) 10M*. Fog Det Lt VQ. 992501079.
Waterford ⚓ Fl R 3s. Fl (3) R 10s; 52°08'·95N 06°57'·00W.
Duncannon Dir lt ⚓ F WRG 13m 10M, white tower on fort, 359·5°-FG-001·2°-Alt GW-001·7°-FW-002·4°-Alt WR-002·9°-FR-004·5°; 52°13'·23N 06°56'·25W; Oc WR 4s 13m W 9M R 7M on same tower, 119°-R-149°-W-172°.

Passage Pt ⚲ Fl WR 5s 7m W6M, R5M; R pile structure; vis: W shore- R127°-302°; 52°14'·26N 06°57'·77W.

Cheek Pt ⚲ Q WR 6m 5M; W mast; vis: W007°-R289°-007°; 52°16'·12N 06°59'·38W.

Sheagh ⚲ Fl R 3s 29m 3M; Gy twr; vis: 090°-318°; 52°16'·29N 06°59'·34W.

Snowhill Point Ldg lts 255°. Front, Fl WR 2·5s 5m 3M; vis: W222°- R020°- W057°-107°; 52°16'·39N 07°00'·91W. Rear, Flour Mill, 750m from front, Q 12m 5M.

Queen's Chan Ldg lts 098°. Front, QR 8m 5M; B twr, W band; vis: 030°-210°;52°15'·32N 07°02'·38W. Rear, 550m from front, Q 15m 5M; W mast.

Beacon Quay ⚲ Fl G 3s 9m; vis: 255°-086°; 52°15'·50N 07°04'·21W.

Cove ⚲ Fl WRG 6s 6m 2M; W twr; vis: R111°- G161°- W234°-111°; 52°15'·05N 07°05'·16W.

Smelting Ho Pt ⚲ Q 8m 3M; W mast; 52°15'·15N 07°05'·27W.

Ballycar ⚲ Fl RG 3s 5m; vis: G127°- R212°-284°; 52°15'·06N 07°05'·51W.

DUNMORE EAST

East Pier Head ☆ 52°08'·93N 06°59'·34W Fl WR 8s 13m **W17M**, R13M; Gy twr, vis: W225°- R310°-004°.

Dunmore East 1 ▲ Q G; 52°09'·06N 06°59'·38W.

Dunmore East 2 ▲ Fl G 2s; 52°09'·00N 06°59'·47W.

W Wharf ⚲ Fl G 2s 6m 4M; vis: 165°-246°; 52°08'·97N 06°59'·45W.

DUNGARVAN

Ballinacourty Pt ⚲ Fl (2) WRG 10s 16m W10M, R8M, G8M; W twr; vis: G245°- W274°- R302°- W325°-117°; 52°04'·69N 07°33'·18W.

Helvick ⚰ Q (3) 10s; 52°03'·61N 07°32'·25W.

Mine Head ⚲ 51°59'·56N 07°35'·23W Fl (4) 30s 87m 12M; W twr, B band; vis: 228°-052°; 992501085.

YOUGHAL

Bar Rocks ⚰ Q (6) + L Fl 15s; 51°54'·85N 07°50'·05W.

Blackball Ledge ⚰ 51°55'·34N 07°48'·53W Q (3) 10s.

W side of ent ☆ 51°56'·57N 07°50'·53W Fl WR 2·5s 24m **W17M**, R13M; W twr; vis: W183°- R273°- W295°- R307°- W351°-003°.

BALLYCOTTON

Ballycotton ☆ 51°49'·50N 07° 59'·13W Fl WR 10s 59m **W18M, R14M**; B twr, within W walls, B lantern; vis: 238°-W-048°-R-238°; 992501093.

The Smiths ⚰ Fl (3) R 10s; 51°48'·62N 08°00'·71W.

Power ⚰ Q (6) + L Fl 15s; 51°45'·59N 08°06'·67W; 992501098.

CORK

Cork ⚰ L Fl 10s; 51°42'·92N 08°15'·60W; *Racon (T) 7M*; 992501100.

Daunt Rock ⚰ Fl (2) R 6s; 51°43'·52N 08°17'·66W; 992501102.

Fort Davis Ldg lts 354·1°. Front, 51°48'·82N 08°15'·80W Dir WRG 29m **17M**; vis: FG351·5°-AlWG352·25°-FW353°-AlWR355°-FR355·75°-356·5°. Rear, Dognose Quay, 203m from front, Oc 5s 37m 10M; Or 3, synch with front.

Roche's Pt ☆ 51°47'·59N 08°15'·29W Fl WR 3s 30m **W20M, R16M**; vis: shore-R-292°-W-016°-R-033°, 033°- W (unintens)-159°-R-shore; 992501099.

Outer Hbr Rk E2 ⚰ Fl R 2·5s; 51°47'·52N 08°15'·67W.

Chicago Knoll E1 ▲ Fl G 5s; 51°47'·66N 08°15'·54W.

W1 ▲ Fl G 10s; 51°47'·69N 08°16'·05W.

W2 ⚰ Fl R 10s; 51°47'·69N 08°16'·34W.

White Bay Ldg lts 034·6°. Front, Oc R 5s 11m 5M; W hut; 51°48'·53N 08°15'·22W. Rear, 113m from front, Oc R 5s 21m 5M; W hut; synch with front.

Spit Bank Pile ⚲ Iso WR 4s 10m W10M, R7M; W house on R piles; vis: R087°- W196°- R221°- 358°; 51°50'·72N 08°16'·45W.

KINSALE and OYSTER HAVEN

Bulman ⚰ Q (6) + L Fl 15s; 51°40'·14N 08°29'·74W; 992501104.

Charlesfort ⚲ Fl WRG 5s 18m W9M, R6M, G7M; vis: G348°-W358°- R004°-168°; H24; 51°41'·75N 08°29'·84W.

OLD HEAD OF KINSALE TO MIZEN HEAD

Old Head of Kinsale ☆, S point 51°36'·29N 08°32'·02W Fl (2) 10s 72m **20M**; B twr, two W bands; 992501107.

COURTMACSHERRY

Barrel Rock ⚐ 51°37'·01N 08°37'·30W.

Black Tom ▲ Fl G 5s; 51°36'·41N 08°37'·95W; 992501110.

Wood Pt (Land Pt) ⚲ Fl (2) WR 5s 15m 5M; vis: W315°- R332°-315°; 51°38'·16N 08°41'·00W.

Galley Head ☆ summit 51°31'·80N 08°57'·21W Fl (5) 20s 53m **23M**; W twr; vis: 256°-065°.

GLANDORE and CASTLETOWNSHEND

Reen Point ⚲ Fl WRG 10s 9m W5M, R3M, G3M; W twr; vis: shore-G-338°-W-001°-R-shore; 51°30'·98N 09°10'·50W.

Kowloon Bridge ⚰ Q (6) + L Fl 15s; 51°27'·58N 09°13'·75W; 992501118.

BALTIMORE and FASTNET

Barrack Pt ⚲ Fl (2) WR 6s 40m W6M, R3M; vis: R168°- W294°-038°; 51°28'·33N 09°23'·65W.

Loo Rock ▲ Fl G 3s; 51°28'·44N 09°23'·46W; 992501119.

Lousy Rks ⚐ 51°28'·93N 09°23'·39W.

Wallis Rk ⚰ Fl (3) R 10s; 51°28'·93N 09°22'·99W.

Lettuce Pt ⚰ VQ (3) 5s; 51°29'·32N 09°23'·81W.

Mealbeg ⚰ Q (6) + L Fl 15s; 51°29'·68N 09°24'·69W.

Inane Pt ⚰ Q (6) + L Fl 15s; 51°30'·02N 09°23'·66W.

Hare Is ⚰ Q; 51°30'·17N 09°25'·38W.

Fastnet ☆, W end 51°23'·35N 09°36'·19W Fl 5s 49m **27M**; Gy twr, *Racon (G) 18M*. Fog Det Lt VQ; 992501123.

SCHULL and LONG ISLAND CHANNEL

Amelia ⚑; 51°29.979'N 009°31.461'W; 992501120.

Copper Point ⚲ 51°30'·25N 09°32'·06W Q (3) 10s 16m 8M.

Ldg lts 346° Front, Oc 5s 5m 11M, W mast; 51°31'·68N 09°32'·43W. Rear, 91m from front, Oc 5s 8m11M; W mast.

CROOKHAVEN

Rock Is Pt ⚲ L Fl WR 8s 20m W13M, R11M; W twr; vis: W over Long Is B to 281°-R-340°; inside harbour 281°-R-348°- towards N shore; 51°28'·59N 09°42'·27W; 992501124.

MIZEN HEAD TO DINGLE BAY

Mizen Head ⚲ Iso 4s 55m 12M; vis: 313°-133°; 51°26'·99N 09°49'·23W; 992501127.

Sheep's Hd ☆ 51°32'·59N 09°50'·92W Fl (3) WR 15s 83m **W15M** R9M; W bldg; vis: 011·3°-R-016·6°-W-212·5°-(partially obs) 233° (shore); 992501129.

BANTRY BAY, CASTLETOWN BEARHAVEN, WHIDDY ISLE, BANTRY and GLENGARIFF

Roancarrigmore ⚲ Fl WR 5s 13m W11M, R9M, R(unintens)5M; s/steel twr; vis: 312°-W-050°-R-122°-R(unintens)-207°-obsc-246°-R-312°; 51°39'·18N 09°44'·82W. Old Lt ho W twr, B band; 992501130.

Ardnakinna Pt ⚲ 51°37'·10N 09°55'·09W Fl (2) WR 10s 62m W14M, R9M; W ○ twr; vis: 319°-R- 348°-W- 066°-R-shore; 992501131.

Walter Scott Rk ⚰ Q (6) + L Fl 15s; 51°38'·52N 09°54'·20W; 992501128.

Castletown (Dinish Is) Dir lt 023.25° ⚲ Oc WRG 5s 7m W15M, R12M, G12M; W hut, R stripe; vis: 019·5°-G-023°-W-023·5°-R-027°; 51°38'·78N 09°54'·32W; 992501125.

Castletown Ldg lts 008°. Front, Oc Bu 6s 4m 6M; W col, R stripe; vis: 005°-013°; 51°39'·16N 09°54'·40W. Rear, 80m from front, Oc Bu 6s 7m 6M; W with R stripe; vis: 005°-013°.

Bull Rock ☆ 51°35'·52N 10°18'·07W Fl 15s 91m **18M**; W twr; 992501131.

KENMARE RIVER, DARRYNANE and BALLYCROVANE
Illaunnameanla (Ballycrovane) ⚡ Fl R 3s; 51°42'·60N 09°57'·50W.
Book Rks ⚓ Fl(2) R 10s; 51°46'·54N 09°49'·74W.
Cuskeal ⚓ Fl R 5s; 51°46'·45N 09°48'·85W.
Bunaw Ldg Lts 041°. Front, Oc R 3s 9m 2M; B col, Y bands; 51°46'·80N 09°48'·40W. Rear, 200m from front Iso R 2s 11m 2M.
Maiden Rk ⚑ Fl G 5s; 51°48'·98N 09°47'·98W; 992501134.
Bat Rk ⚑ Fl G 5s; 51°50'·89N 09°40'·89W.
Carrignaronebeg ⚓ Fl R 5s; 51°50'·99N 09°41'·39W.
No1 ⚑ Fl G 5s; 51°52'·09N 09°36'·35W.
Carrignarone (Seal Rk) Fl 5s 2m 2M; 51°48'·58N 09°52'·65W.
Castlecove Ldg Lts 045°. Front, Oc 2s 5m 6M; concrete post W front; 51°46'·29N 10°02'·04W. Rear, 400m from front Oc 2s 18m 6M.
West Cove Hbr ⚡ Fl R 3s; post W front; 51°46'·00N 10°02'·93W.
West Cove Hbr Ldg Lts 312°. Front, Fl 2s 2m 6M; concrete post; 51°46'·11N 10°03'·08W. Rear, 200m from front Fl 2s 4m 6M.
Darrynane Ldg lts 034°. Front, Oc 3s 10m 4M; 51°45'·90N 10°09'·20W. Rear, Oc 3s 16m 4M.

SKELLIG ISLANDS
Skelligs Rock ⚡ Fl (3) 15s 53m 12M; W twr; vis: 262°-115°; part obsc within 6M 110°-115°; 51°46'·11N 10°32'·52W; 992501137.

VALENTIA and PORTMAGEE
Fort (Cromwell) Point ☆ 51°56'·02N 10°19'·28W Fl WR 2s 16m **W17M, R15M**; W twr; vis: 304°-R-351°,102°-W-304°; obsc from seaward by Doulus Head when brg more than 180°; 992501141.
Dir lt 141° ⚡ Oc WRG 4s 25m W11M, R8M, G8M (by day: W3M, R2M, G2M); W twr, R stripe; vis:136°-G-140°-W-142°-R-146°; 51°55'·51N 10°18'·42W.
The Foot ⚓ VQ (3) 5s; 51°55'·72N 10°17'·07W; 992501140.

DINGLE BAY TO LOOP HEAD

DINGLE BAY, VENTRY and DINGLE
Ldg lts 182° Oc 3s. Front 52°07'·41N 10°16'·59W, rear 100m behind.
Inishtearaght ☆, W end Blasket Islands 52°04'·54N 10°39'·67W Fl (2) 20s 84m **19M**; W twr; vis: 318°-221°; *Racon (O)*; 992501143.

BRANDON BAY, TRALEE BAY and FENIT HARBOUR
Little Samphire Is ☆ 52°16'·26N 09°52'·91W Fl WRG 5s 17m **W16M**, R13M; G13M; Bu ○ twr; vis: 262°-R-275°, 280-R-090°-G-140°-W-152°-R-172°.
Great Samphire I ⚡ QR 15m 3M; vis: 242°-097°; 52°16'·15N 09°51'·81W.
Fenit Hbr Pier Hd ⚡ 2 FR (vert) 12m 3M; vis: 148°-058°; 52°16'·24N 09°51'·55W.

SHANNON ESTUARY
Ballybunnion ⚓ VQ; *Racon (M) 6M*; 52°32'·52N 09°46'·93W; 992501146.
Kilstiffin ⚓ Fl R 3s; 52°33'·80N 09°43'·83W; 992501148.

Kilcredaun ⚓ Q R (sync); 52°34'·42N 09°41'·16W.
Tail of Beal ⚑ Q G (sync); 52°34'·37N 09°40'·71W; 992501150.
Kilcredaun Head ▯, W twr; 52°34'·78N 09°42'·58W.
Carrigaholt ⚓ Fl (2) R 6s (sync); 52°34'·90N 09°40'·47W.
Beal Spit ⚑ Fl (2) G 6s (sync); 52°34'·80N 09°39'·94W.
Beal Bar ⚑ Fl G 3s (sync); 52°35'·18N 09°39'·05W.
Doonaha ⚓ Fl R 3s (sync); 52°35'·54N 09°39'·01W; 992501154.
Letter Point ⚓ Fl R 7s; 52°35'·44N 09°35'·89W.
Asdee ⚓ Fl R 5s; 52°35'·09N 09°34'·55W.
Rineanna ⚓ QR; 52°35'·59N 09°31'·24W.
North Carraig ⚓ Q; 52°35'·60N 09°29'·76W.
Scattery Is, Rineanna Pt ⚡ Fl (2) 8s 15m 10M; W twr; vis: 208°-092°; 52°36'·35N 09°31'·07W.

KILRUSH
Marina Ent Chan Ldg lts 355°. Front, Oc 3s; 52°37'·99N 09°30'·27W. Rear, 75m from front, Oc 3s.
Tarbert Is N Point ⚡ Q WR 4s 18m W14M, R10M; W ○ twr; vis: W069°- R277°- W287°-339°; 52°35'·52N 09°21'·83W.
Tarbert (Ballyhoolahan Pt) Ldg lts 128·2° ⚓. Front,Iso 3s 13m 3M; △ on W twr; vis: 123·2°-133·2°; 52°34'·35N 09°18'·80W. Rear, 400m from front, Iso 5s 18m 3M; G stripe on W Bn.
Garraunbaun Pt ⚡ Fl (3) WR 10s 16m W8M, R5M; W □ col, vis: R shore - W072°- R242°- shore; 52°35'·62N 09°13'·94W.
Rinealon Pt ⚡ Fl 2·5s 4m 7M; B col, W bands; vis: 234°-088°; 52°37'·12N 09°09'·82W.

FOYNES
W Chan Ldg lts 107·9° (may be moved for changes in chan). Front, Oc 4s 34m 12M; 52°36'·91N 09°06'·59W . Rear, Oc 4s 39m 12M.

RIVER SHANNON
Beeves Rock ⚡ Fl WR 5s 12m W12M, R9M; vis: 064·5°-W-091°-R-238°-W-265°-W(unintens)-064·5°; 52°39'·01N 09°01'·35W .
North Channel Ldg lts 093°. Front, Tradree Rock Fl R 2s 6m 5M; W Trs; vis: 246°-110°; 52°41'·00N 08°49'·87W. Rear 0·65M from front, Iso 6s 14m 5M; W twr, R bands; vis: 327°-190°.
N side, Ldg lts 061°. Front, 52°40'·72N 08°45'·27W, Crawford Rock 490m from rear, Fl R 3s 6m 5M. Crawford No. 2, Common Rear, Iso 6s 10m 5M; 52°40'·85N 08°44'·88W.
Ldg lts 302·1°, Flagstaff Rock, 670m from rear, Fl R 7s 7m 5M; 52°40'·66N 08°44'·40W.
Ldg lts 106·5°. Meelick Rk, Front Iso R 4s 6m 5M; 52°40'·24N 08°42'·32W. Meelick No. 2, rear 275m from front Iso R 4s 9m 5M; both W beacons.

LOOP HEAD
Loop Head ☆ 52°33'·67N 09°55'·94W Fl (4) 20s 84m **23M**; vis: 280°-218°; 992501161.
Mal Bay ⚓ Fl(5) Y 20s 52°46.42'N 009°35.15'W; *AIS*.

9.12.5 PASSAGE INFORMATION
More passage information is threaded between the harbours in this area.The latest editions of the Irish Cruising Club Sailing Directions are strongly recommended for all Irish waters (www. irishcruisingclub.com). Particularly useful on the N and W coast where other information is scarce they are published in two volumes, *E and N coasts of Ireland* and *S and W coasts of Ireland*. (Both available from Imray in UK or Todd Chart Services, Bangor, Co Down, ☎028 9146 6640 (www.toddchart.com). See also *Cruising Cork and Kerry* (Imray/Swanson).

For notes on crossing the Irish Sea, see 9.13.5; and for Distances across it see 9.0.6. More Passage Information is threaded between the harbours of this Area.

9.12.6 SPECIAL NOTES FOR IRELAND
Céad Míle Fáilte! One hundred thousand Welcomes!

Lifejackets: It is compulsory to wear lifejackets in Irish waters on all craft of <7m LOA. Children under 16 years must wear a lifejacket or personal flotation device at all times on deck when underway. Every vessel, irrespective of size, must carry a lifejacket or personal flotation device for each person on board. These regulations are mandatory.

Irish Customs: First port of call should preferably be at Customs posts in one of the following hbrs: Dublin, Dun Laoghaire, Waterford, New Ross, Cork, Ringaskiddy, Bantry, Foynes, Limerick, Galway, Sligo and Killybegs. Yachts may, in fact, make their first call anywhere and if no Customs officer arrives within a reasonable time, the skipper should inform the nearest Garda (Police) station of the yacht's arrival. Only yachts from non-EU member states should fly flag Q or show ● over ○ lts on arrival. Passports are not required by UK citizens. All current Northern Ireland 5 and 6 digit telephone numbers have become 8 digits **028 9012 3456**.

Telephone: To call the **Republic of Ireland (RoI)** from the UK, dial **00 - 353**, then the area code minus the initial 0, followed by the number. To call UK from the RoI: dial 00-44, followed by the area code minus the initial 0, then the number.

Salmon drift nets are now illegal and should not be encountered. Draft netting (across a river or estuary) is still permitted, but is rare.

Liquefied petroleum gas: In RoI LPG is supplied by Kosan, a sister company of Calorgas Ltd, but the smallest bottle is taller than the Calor equivalent fitted in most yachts with different connections. Calor bottles can be filled in larger towns. Camping Gaz is widely available.

Useful Irish websites:

Commissioners of Irish Lights	www.cil.ie
Irish Sailing Association	www.sailing.ie
Irish Tourist Board	www.discoverireland.com
Sea Area Forecast	www.met.ie/forecasts/sea-area.asp

Currency is the Euro (€), readily available from ATMs. VAT is 23%.

Access by air: Intl airports in RoI are at Dublin, Cork & Shannon. Regional airports in Waterford, Farranfore (Kerry), Knock (Mayo) and Carrickfinn (Donegal) provide some scheduled flights between Ireland and Britain. For details, see http://www.irishairports.com/.

Northern Ireland: Belfast CG (MRCC) is at Bangor, Co Down, ☎(028 91) 463933. HM Customs (⊖) should be contacted H24 on ☎ (028 90) 358250 at the following ports, if a local Customs Officer is not available: Belfast, Warrenpoint, Kilkeel, Ardglass, Portavogie, Larne, Londonderry, Coleraine. Northern Ireland's main airport is Belfast (Aldergrove).

AIS stations: 73 AIS stations have been established on navigational marks around the coast of Ireland under the control of the Commissioners of Irish Lights. These are located at lighthouses, and in more remote areas fitted to dedicated lit Y buoys offshore.

S Ireland

GAELIC: It helps to understand some of the more common words for navigational features (courtesy of the Irish Cruising Club):

Ail, alt	cliff, height	Cladach	shore	Inish, illaun	island	Mara	of the sea
Aird, ard	height, high	Cuan, coon	harbour	Inver	river mouth	More, mor	big
Anna, annagh	marsh	Derg, dearg	red	Keal, keel	narrow place,	Rannagh	point
Ath	ford	Drum	hill, ridge		sound	Ron, roan	seal
Bal, Bally	town	Duff, dubh	black	Kill	church, cell	Roe, ruadh	red
Barra	sandbank	Dun, doon	fort	Kin, ken	promontory,	Scolt	split, rky gut
Bel, beal	mouth, strait	Ennis	island		head	Scrow	boggy,
Beg	little	Fad, fadda	long	Knock	hill		grassy sward
Ben, binna	hill	Fan	slope	Lag	hollow	Slieve	mountain
Bo	sunken rock	Fin	white	Lahan	broad	Slig	shells
Boy, bwee	yellow	Freagh, free	heather	Lea	grey	Stag, stac	high rock
Bullig	shoal, round	Gall	stranger	Lenan	weed-	Tawney	low hill
	rock, breaker	Glas, glass	green		covered rock	Tigh, ti	house
Bun	end, river mouth	Glinsk	clear water	Lis	ancient fort	Togher	causeway
Caher	fort	Gorm	blue	Long, luing	ship	Tra, traw	strand
Camus	bay, river bend	Gub	point of land	Maan	middle	Turlin	boulder, beach
Carrick	rock	Hassans	swift current	Maol, mwee	bare	Vad, bad	boat

---

DISTANCE TABLE
Approximate distances in nautical miles are by the most direct route, whilst avoiding dangers and allowing for Traffic Separation Schemes. Places in *italics* are in adjoining areas; places in **bold** are in 9.0.6, Distances across the Irish Sea.

1	*Carlingford Lough*	1																										
2	Malahide	36	2																									
3	Howth	39	4	3																								
4	**Dublin/Dun Laoghaire**	48	12	8	4																						
5	Wicklow	63	29	25	21	5																						
6	**Arklow**	75	43	37	36	15	6																					
7	Wexford (Bar Buoy)	104	71	66	62	43	30	7																				
8	**Rosslare**	108	75	70	66	47	34	4	8																			
9	Kilmore Quay	128	95	90	86	67	54	25	21	9																		
10	Waterford	154	127	116	117	99	84	51	47	32	10																	
11	**Dunmore East**	139	112	101	102	84	69	36	32	17	15	11																
12	Youghal	172	143	134	133	115	100	69	65	51	52	37	12															
13	Crosshaven	192	165	154	155	137	122	89	85	69	74	59	25	13														
14	Kinsale	202	178	164	168	150	135	99	95	79	84	69	35	17	14													
15	Courtmacsherry	201	177	163	167	149	136	109	105	85	90	75	44	26	13	15												
16	Glandore	213	189	175	179	161	148	121	117	97	102	87	56	38	25	23	16											
17	Castletownshend	224	200	186	190	172	159	132	128	108	113	98	67	49	36	26	6	17										
18	**Baltimore**	239	205	201	196	177	164	136	132	112	117	102	70	54	42	34	16	12	18									
19	**Fastnet Rock**	250	216	212	207	189	174	148	144	122	127	112	78	60	49	42	24	20	10	19								
20	Schull	256	222	218	213	195	180	154	150	128	133	118	84	66	55	48	30	26	14	9	20							
21	Crookhaven	256	222	218	213	195	180	154	150	128	133	118	84	66	55	47	29	25	16	8	10	21						
22	Bantry	286	252	237	243	225	210	184	180	158	163	148	114	96	85	78	60	56	45	36	41	35	22					
23	Kenmare R (Scariff Is)	285	251	236	242	224	209	183	179	157	162	147	113	95	84	77	59	55	44	35	40	34	42	23				
24	Valentia Hbr	295	261	257	252	242	227	192	188	175	180	165	131	113	102	96	78	74	56	54	60	53	55	20	24			
25	Dingle	308	274	270	265	246	233	205	201	181	186	171	139	123	111	104	86	82	69	62	67	61	63	28	13	25		
26	Fenit Hbr	344	310	306	301	283	268	242	238	216	221	206	172	154	143	136	118	114	105	95	99	93	100	60	47	45	26	
27	Kilrush	361	333	323	318	299	286	258	254	234	239	224	192	176	164	155	137	133	122	114	116	109	116	78	66	64	32	27
28	*Slyne Head*	317	355	351	346	328	313	287	283	261	266	251	217	199	188	180	162	158	153	139	141	133	144	103	97	95	70	75 28

9.12.7 MALAHIDE

Dublin 53°27'·20N 06°08'·90W ❄ ⊛ ♤♤♤♤ ♧♧♧

CHARTS AC 1468, 633, 5621; Imray C61, C62

TIDES +0030 Dover; ML 2·4; Duration 0615

Standard Port DUBLIN (NORTH WALL) (→)

Times				Height (metres)			
High Water		Low Water		MHWS	MHWN	MLWN	MLWS
0000	0700	0000	0500	4·1	3·4	1·5	0·7
1200	1900	1200	1700				
Differences MALAHIDE							
+0002	+0003	+0009	+0009	+0·1	−0·2	−0·4	−0·2

SHELTER Excellent in marina, dredged approx 2·3m. Visitors berth on pontoon **A**. Moorings preclude ⚓ in the inlet.

NAVIGATION WPT: SWM LFl 10s 53°27'·12N 06°06'·87W. The approach channel, which is well marked by 12 lit PHM and SHM, lies between drying sandbanks. The entrance is straightforward, but there is less than 1m of water below CD at some points of the chan, so adequate height of tide is required for safe entry. The flood reaches 3kn sp, the ebb 3½kn and a strong flow may be experienced through the marina at some states of the tide. Steep seas are formed in strong onshore winds against the full ebb, during which extreme caution needs to be exercised. The speed limit is 5kn in the fairway and 4kn in the marina.

LIGHTS AND MARKS The marina and apartment blocks close S of it are visible from the WPT. The marina enrance is marked by occulting red and green lights to each side.

COMMUNICATIONS (Code 01) MRCC 6620922/3; ⊜ 8746571; Police 666 4600; Dr 845 5994; Ⓗ 837 7755.

Malahide Marina, Ch **M** 80 (H24). MYC, call *Yacht Base* Ch **M** (occas).

FACILITIES Marina ⚓ (349) ☎ 8454129, 20♥ €3·90, ⚡(charge varies) ⊚ ⌂/WC ⛽ ⚓(H24) ⚓ Kos ⚒ Ⓔ ⌂ ▣(30t) ✕ ⌂ ice.
Malahide YC ☎ 8453372, ⚓ (for craft <5.5m, launching HW ±2).
Town ⊚ ✉ Ⓑ ⇌ ✈ (Dublin).

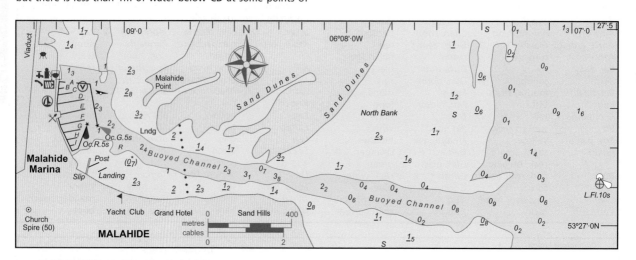

Church Spire (50)

MALAHIDE

MALAHIDE TO TUSKAR ROCK

(AC 1468, 1787) Malahide, 4M from both Lambay Is and Howth, can be entered in most weather via a channel between drying sandbanks marked by R & G fixed perches. Ireland's Eye, a rky island which rises steeply to a height of 99m, lies about 7½ca N of Howth with reefs running SE and SW from Thulla Rk at its SE end. Ben of Howth, on N side of Dublin Bay, is steep-to, with no dangers more than 1ca offshore.

Rosbeg Bank lies on the N side of Dublin Bay. Burford Bank, on which the sea breaks in E gales, and Kish Bank lie offshore in the approaches. ▶ *The N-going stream begins at HW Dublin –0600, and the S-going at HW Dublin, sp rates 3kn.* ◀

From Dublin to Carnsore Pt as a cruising area, hbr facilities are being improved. The shallow offshore banks cause dangerous overfalls and dictate the route which is sheltered from the W winds. ▶ *Tidal streams run mainly N and S, but the N-going flood sets across the banks on the inside, and the S-going ebb sets across them on the outside.* ◀

Leaving Dublin Bay, yachts normally use Dalkey Sound, but with a foul tide or light wind it is better to use Muglins Sound. Muglins (lt) is steep-to except for a rk about 1ca WSW of the lt. Beware Leac Buidhe (dries) 1ca E of Clare Rk. The inshore passage is best as far as Wicklow. Thereafter yachts may either route offshore, passing east of Arklow Bank and its Lanby to fetch Tuskar Rock or Greenore Pt, or keep inshore of Arklow Bank, avoiding Glassgorman Banks. Stay E of the Blackwater Bank to Lucifer ECM buoy then pass W of Tuskar Rock to round Carnsore Pt to the SW. Avoid the Rusk Channel except in settled weather. Arklow is safe in offshore winds; Wexford has a difficult entrance. Rosslare lacks yacht facilities, but provides good shelter to wait out a SW'ly blow.

Arklow Bank Wind Farm, 7 miles east of Arklow, has 7 wind turbines, each 72.8m high with 104m diameter blades. They are lit, No 7 is fitted with an AIS transmitter.

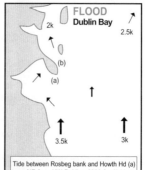

Tide between Rosbeg bank and Howth Hd (a) runs NE from HW Dublin +0300 for 9h30. In Howth Sd (b) the stream is NW going from +0430 to –0130. New flood and ebb tides begin close to the S shore and N of Baily up to 1h before HW Dublin.

The tide between Rosbeg bank and Howth Hd (a) runs SW from HW Dublin for 3h. In Howth Sd (b) the stream is SE going from –0130 to +0430. Strengths of streams increase S of Dublin Bay, and decrease N of it.

9.12.8 HOWTH

Dublin **53°23′·60N 06°04′·00W** ✿✿✿✿≈≈≈✿✿✿

CHARTS AC 1468, 1415, 5621; Imray C61, C62

TIDES +0025 Dover; ML 2·4; Duration 0625

Standard Port DUBLIN (NORTH WALL) (→)

Times				Height (metres)			
High Water		Low Water		MHWS	MHWN	MLWN	MLWS
0000	0700	0000	0500	4·1	3·4	1·5	0·7
1200	1900	1200	1700				
Differences HOWTH							
–0007	–0005	+0001	+0005	0·0	–0·1	–0·2	–0·2

SHELTER Excellent in marina, at all states of tide. After a severe ENE'ly storm, expect a dangerous scend in the app chan which has 11🛟, 6G & 5R. Caution: many moorings in E part of outer hbr. No ent to FV basin for yachts. Inside the inner hbr keep strictly to chan to avoid drying shoals either side and a substantial wavebreak (gabion). Marina depth 1·8m. R posts mark the outer limit of dredging around the marina. 4kn speed limit. There is a fair weather ⚓ in 2-3m at Carrigeen Bay, SW side of Ireland's Eye.

NAVIGATION WPT Howth SHM buoy, Fl G 5s, 53°23′·72N 06°03′·53W, 251°/0·27M to E pier lt. From the S beware Casana Rk 4ca S of the Nose of Howth and Puck's Rks extending about 50m off the Nose. Ireland's Eye is 0·6M N of hbr, with the Rowan Rks SE and SW from Thulla, marked by Rowan Rks ECM, and S Rowan SHM lt buoys. The usual approach passes S of Ireland's Eye, but beware many lobster pots between the Nose of Howth and the hbr, and rocks off both pier hds. Howth Sound has 2·4m min depth; give way to FVs (constrained by draught) in the Sound and hbr entrance. 2·5m in marina approach channel.

LIGHTS AND MARKS Baily Lt Ho, Fl 15s 41m 26M, is 1·5M S of Nose of Howth. Disused lt ho on E pier is conspic. E pier lt, Fl (2) WR 7·5s 13m 12/9M; 256°-W-295°, R elsewhere. W sector leads safely to NE pierhead which should be rounded 50m off. Trawler Dock moles are lit QR and Fl G 3s. Marina channel has 8 lateral marks (some lit) and perches with W reflective tape (2 bands = port, 1 = stbd).

COMMUNICATIONS (Code 01) MRCC 6620922/3; ⊜8746571; Police 666 4900; Dr 832 3191; ⊞ 837 7755. HM 8322252.

Marina Ch **M** 80. *Howth Harbour* VHF Ch 16 (Mon-Fri 0700–2300).

FACILITIES Howth Marina ☎8392777 (HO 0800-2130 Apr-Oct) 350 inc
Ⓥ €3·65 (inc ⫣) 🖪 ⏤ ⛽(H24) ⬛ 🔧(H24) ⛽ ⚓ ⛴(15t).

Howth YC www.hyc.ie ☎8322141 🖪 ⚿ Scrubbing posts <20m LOA, ✕(☎8392100) ⛙(☎8320606).

Howth Boat Club (no facilities). **Services** LB, ⚓ Kos, 🔧 ⚓ ⛽ Ⓔ ⛽.
Town ⊠ Ⓑ 🖥 ⇌ ✈ (Dublin).

HOWTH

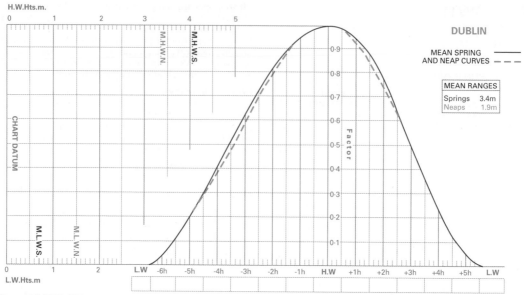

DUBLIN

MEAN SPRING ——————
AND NEAP CURVES – – – –

MEAN RANGES	
Springs	3.4m
Neaps	1.9m

9.12.9 DUBLIN

Dublin 53°20'·85N 06°14'·80W ✹✹✹✹♤♤✿✿✿

CHARTS AC 1468, 1415, 1447, 5621; Imray C61, C62

TIDES +0042 Dover; ML 2·4; Duration 0640

Standard Port DUBLIN (NORTH WALL) (→)

Times				Height (metres)			
High Water		Low Water		MHWS	MHWN	MLWN	MLWS
0000	0700	0000	0500	4·1	3·4	1·5	0·7
1200	1900	1200	1700				
Differences DUBLIN BAR and DUN LAOGHAIRE							
−0006	−0001	−0002	−0003	0·0	0·0	0·0	+0·1
GREYSTONES (53°09'N 06°04'W)							
−0008	−0008	−0008	−0008	−0·5	−0·4	ND	ND

SHELTER Dublin Port, excellent on Poolbeg YBC Marina, S of Alexandra Basin W.

NAVIGATION Dublin Port WPT 53°20'·50N 06°06'·71W, 275°/1·35M to hbr ent. Accessible H24, but the ent to R Liffey is rough in E'ly >F5 against ebb tide. Note Bull Wall covers 0.6 to 2.7m at HW.

For clearance to enter call *Dublin VTS* Ch 12. Conform to Dublin Port Company Small Craft Regulations (www.dublinport.ie).

Beware of shipping; small craft must keep outside fairway buoys. Yachtsmen wishing to visit Dublin City Moorings, N side between Samuel Beckett and Sean O'Casey Bridges, should check well in advance (+353 1 8183300), as restrictive bridge openings for East Link and especially Samuel Beckett Brs, may preclude this.

LIGHTS AND MARKS Entrance, N. Bull Lt Fl G 4s 15m 10M synchro with Gt. S. Wall Hd Poolbeg Lt Fl R 4s 20m 10M. Poolbeg power stn 2 R/W chimneys (VQ R) are conspic from afar.

COMMUNICATIONS (Code 00 353 1) MRCC 6620922/3; Coast/Cliff Rescue Service 2803900; ⊖2803992; ♨1550 123855; Police (Dublin) 6668000; Dr 2859244; Ⓗ2806901. HM Dublin 8748771.
Dublin VTS Ch **12** 13 16 (H24). Lifting bridge (call *Eastlink*) Ch 12. *Poolbeg Marina* Ch **37** 12 16. Dublin Coast Radio Stn 16 67 83.

FACILITIES Poolbeg YBC Marina www.poolbegmarina.ie ☎6689983 35🅥 (<20m LOA in 2.4m),€3·80, ⬛ ⚓⛽⏨ 🛠 📶 🔘 ✕ 🗑.
Dublin City, all needs, @ at library, ⇌ ✈.
Ferries Holyhead; 4/day; 3¼ hrs; Irish Ferries (www.irishferries.com) & Stena Line (www.stenaline.co.uk). Liverpool; 18/wk (seasonal); 7 hrs; P&O (www.poferries.com).

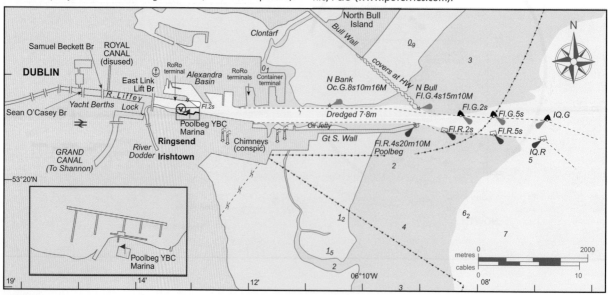

STANDARD TIME (UT)
For Summer Time add ONE hour in **non-shaded areas**

DUBLIN (NORTH WALL) LAT 53°21'N LONG 6°13'W
TIMES AND HEIGHTS OF HIGH AND LOW WATERS

Dates in **red** are **SPRINGS**
Dates in blue are NEAPS

YEAR 2016

JANUARY

Time	m	Time	m
1 0408	3.5	**16** 0357	3.8
0945	1.4	0933	1.0
F 1627	3.7	SA 1613	4.0
2223	1.2	◑ 2209	0.8
2 0508	3.4	**17** 0458	3.7
1044	1.6	1037	1.2
SA 1727	3.5	SU 1715	3.9
◑ 2319	1.3	2315	1.0
3 0612	3.3	**18** 0608	3.7
1146	1.6	1148	1.3
SU 1832	3.4	M 1826	3.8
4 0018	1.4	**19** 0028	1.1
0713	3.4	0721	3.7
M 1251	1.7	TU 1303	1.3
1935	3.4	1941	3.7
5 0121	1.4	**20** 0142	1.2
0810	3.5	0829	3.8
TU 1353	1.6	W 1414	1.2
2032	3.5	2053	3.7
6 0218	1.4	**21** 0249	1.1
0900	3.6	0932	3.9
W 1447	1.4	TH 1517	1.0
2122	3.6	2157	3.8
7 0305	1.2	**22** 0345	1.0
0944	3.8	1026	4.0
TH 1530	1.3	F 1609	0.8
2206	3.7	2251	3.9
8 0344	1.1	**23** 0431	0.9
1022	3.9	1113	4.1
F 1606	1.1	SA 1655	0.7
2245	3.8	2335	3.9
9 0419	0.9	**24** 0511	0.8
1057	4.0	1153	4.1
SA 1641	0.9	SU 1736	0.6
2322	3.9	○	
10 0454	0.8	**25** 0012	3.9
1132	4.1	0547	0.8
SU 1717	0.7	M 1227	4.1
● 2359	4.0	1814	0.6
11 0531	0.7	**26** 0044	3.8
1210	4.2	0621	0.8
M 1756	0.6	TU 1259	4.1
		1851	0.6
12 0039	4.0	**27** 0117	3.8
0610	0.7	0656	0.8
TU 1253	4.3	W 1335	4.0
1838	0.5	1929	0.7
13 0123	4.0	**28** 0154	3.7
0654	0.7	0733	0.9
W 1338	4.3	TH 1413	4.0
1925	0.5	2008	0.8
14 0211	4.0	**29** 0233	3.7
0742	0.8	0813	1.0
TH 1426	4.2	F 1454	3.8
2015	0.6	2049	0.9
15 0302	3.9	**30** 0315	3.6
0835	0.9	0856	1.2
F 1518	4.2	SA 1537	3.7
2110	0.7	2133	1.1
		31 0402	3.4
		0946	1.3
		SU 1625	3.5
		2222	1.3

FEBRUARY

Time	m	Time	m
1 0457	3.3	**16** 0540	3.6
1046	1.5	1123	1.2
M 1723	3.3	TU 1809	3.6
◑ 2320	1.4		
2 0607	3.2	**17** 0000	1.3
1153	1.6	0657	3.5
TU 1839	3.2	W 1242	1.3
		1929	3.5
3 0025	1.5	**18** 0123	1.3
0719	3.3	0811	3.6
W 1301	1.6	TH 1401	1.2
1950	3.3	2044	3.6
4 0131	1.5	**19** 0238	1.2
0820	3.4	0918	3.7
TH 1405	1.5	F 1507	1.0
2050	3.4	2150	3.6
5 0230	1.3	**20** 0334	1.1
0911	3.6	1015	3.9
F 1458	1.2	SA 1558	0.8
2140	3.5	2244	3.7
6 0317	1.1	**21** 0418	0.9
0955	3.8	1102	4.0
SA 1541	1.0	SU 1641	0.7
2222	3.7	2326	3.8
7 0357	0.9	**22** 0457	0.8
1034	4.0	1141	4.0
SU 1619	0.7	M 1719	0.6
2301	3.9	○ 2358	3.8
8 0435	0.7	**23** 0531	0.7
1112	4.2	1210	4.0
M 1657	0.5	TU 1754	0.6
● 2339	4.0		
9 0513	0.5	**24** 0021	3.8
1150	4.3	0602	0.7
TU 1737	0.3	W 1237	4.0
		1827	0.6
10 0018	4.1	**25** 0049	3.8
0552	0.4	0633	0.7
W 1232	4.4	TH 1308	4.0
1819	0.2	1859	0.6
11 0101	4.1	**26** 0121	3.7
0634	0.4	0704	0.7
TH 1317	4.4	F 1342	3.9
1904	0.2	1932	0.7
12 0146	4.1	**27** 0155	3.7
0721	0.5	0736	0.8
F 1404	4.3	SA 1419	3.8
1953	0.4	2005	0.8
13 0235	4.0	**28** 0233	3.6
0812	0.6	0812	0.9
SA 1455	4.2	SU 1459	3.7
2046	0.5	2042	1.0
14 0328	3.9	**29** 0315	3.5
0909	0.8	0853	1.1
SU 1550	4.0	M 1544	3.5
2143	0.8	2124	1.2
15 0428	3.7		
1012	1.0		
M 1653	3.8		
◑ 2246	1.0		

MARCH

Time	m	Time	m
1 0403	3.4	**16** 0514	3.6
0941	1.3	1102	1.1
TU 1635	3.3	W 1756	3.5
◑ 2217	1.4	2333	1.3
2 0500	3.2	**17** 0634	3.5
1047	1.5	1220	1.2
W 1739	3.2	TH 1916	3.4
2329	1.5		
3 0614	3.2	**18** 0059	1.4
1210	1.5	0749	3.5
TH 1904	3.1	F 1343	1.1
		2031	3.5
4 0046	1.5	**19** 0219	1.3
0734	3.2	0858	3.6
F 1323	1.4	SA 1449	1.0
2016	3.3	2136	3.6
5 0154	1.4	**20** 0315	1.2
0835	3.4	0956	3.8
SA 1425	1.1	SU 1539	0.8
2112	3.5	2227	3.6
6 0249	1.1	**21** 0359	1.0
0925	3.7	1043	3.9
SU 1514	0.8	M 1620	0.7
2158	3.7	2307	3.7
7 0334	0.8	**22** 0436	0.8
1008	3.9	1120	3.9
M 1556	0.5	TU 1656	0.6
2238	3.9	2336	3.7
8 0414	0.5	**23** 0510	0.7
1049	4.2	1149	3.9
TU 1636	0.2	W 1729	0.6
2317	4.1	○ 2357	3.7
9 0453	0.3	**24** 0541	0.7
1129	4.3	1214	3.9
W 1717	0.1	TH 1800	0.6
● 2356	4.2		
10 0533	0.2	**25** 0021	3.7
1211	4.4	0609	0.7
TH 1759	0.0	F 1243	3.9
		1829	0.6
11 0037	4.2	**26** 0050	3.8
0615	0.2	0636	0.7
F 1256	4.4	SA 1315	3.8
1843	0.1	1857	0.7
12 0122	4.1	**27** 0123	3.8
0701	0.2	0706	0.7
SA 1344	4.3	SU 1351	3.7
1930	0.3	1928	0.8
13 0210	4.0	**28** 0201	3.7
0752	0.4	0740	0.6
SU 1436	4.1	M 1431	3.6
2022	0.5	2005	0.9
14 0302	3.9	**29** 0243	3.6
0850	0.6	0821	0.9
M 1532	3.9	TU 1515	3.5
2119	0.8	2048	1.1
15 0401	3.7	**30** 0329	3.5
0953	0.8	0908	1.1
TU 1638	3.7	W 1604	3.3
◑ 2221	1.1	2138	1.3
		31 0422	3.3
		1007	1.2
		TH 1704	3.2
		◑ 2243	1.4

APRIL

Time	m	Time	m
1 0526	3.2	**16** 0025	1.5
1125	1.3	0720	3.5
F 1820	3.1	SA 1314	1.1
		2006	3.4
2 0004	1.5	**17** 0146	1.4
0643	3.2	0829	3.6
SA 1245	1.2	SU 1420	1.0
1938	3.3	2110	3.5
3 0118	1.3	**18** 0246	1.3
0755	3.4	0927	3.7
SU 1351	1.0	M 1511	0.9
2039	3.5	2159	3.6
4 0218	1.1	**19** 0332	1.1
0852	3.7	1014	3.8
M 1445	0.7	TU 1553	0.8
2129	3.7	2237	3.6
5 0307	0.8	**20** 0411	0.9
0940	3.9	1052	3.8
TU 1532	0.4	W 1629	0.7
2213	3.9	2306	3.7
6 0350	0.5	**21** 0445	0.8
1025	4.1	1122	3.8
W 1614	0.1	TH 1702	0.7
2254	4.1	2331	3.7
7 0432	0.2	**22** 0517	0.8
1108	4.3	1151	3.8
TH 1656	0.0	F 1732	0.7
● 2334	4.2	○ 2355	3.8
8 0514	0.1	**23** 0545	0.7
1152	4.4	1219	3.8
F 1739	0.0	SA 1759	0.7
9 0016	4.2	**24** 0023	3.8
0558	0.1	0611	0.7
SA 1239	4.3	SU 1251	3.8
1823	0.1	1826	0.8
10 0101	4.2	**25** 0057	3.8
0645	0.2	0640	0.8
SU 1328	4.2	M 1327	3.7
1910	0.3	1859	0.8
11 0149	4.1	**26** 0135	3.8
0737	0.3	0716	0.8
M 1421	4.1	TU 1408	3.6
2002	0.6	1937	0.9
12 0241	3.9	**27** 0218	3.7
0834	0.5	0759	0.9
TU 1519	3.9	W 1453	3.6
2058	0.9	2022	1.0
13 0341	3.8	**28** 0305	3.6
0936	0.7	0848	1.0
W 1625	3.6	TH 1542	3.4
2158	1.1	2114	1.2
14 0451	3.6	**29** 0356	3.5
1042	0.9	0946	1.1
TH 1739	3.5	F 1639	3.3
◑ 2306	1.4	2216	1.3
15 0607	3.5	**30** 0456	3.4
1155	1.1	1056	1.1
F 1854	3.4	SA 1743	3.3
		◑ 2328	1.4

Chart Datum: 0·20 metres above Ordnance Datum (Dublin). HAT is 4·5 metres above Chart Datum.

》》 FREE monthly updates. Register at 《
www.reedsnauticalalmanac.co.uk

575

STANDARD TIME (UT)
For Summer Time add ONE hour in **non-shaded areas**

DUBLIN (NORTH WALL) LAT 53°21'N LONG 6°13'W
TIMES AND HEIGHTS OF HIGH AND LOW WATERS

Dates in red are SPRINGS
Dates in blue are NEAPS

YEAR 2016

MAY

Day	Time	m	Day	Time	m
1 SU	0605	3.4	16 M	0058	1.5
	1210	1.1		0749	3.6
	1900	3.4		1340	1.1
				2028	3.4
2 M	0040	1.3	17 TU	0205	1.4
	0715	3.5		0847	3.6
	1318	0.9		1435	1.0
	2004	3.5		2118	3.5
3 TU	0144	1.1	18 W	0258	1.2
	0817	3.7		0936	3.7
	1416	0.6		1520	0.9
	2059	3.7		2158	3.6
4 W	0238	0.8	19 TH	0340	1.1
	0912	3.9		1017	3.7
	1507	0.4		1558	0.9
	2147	3.9		2232	3.7
5 TH	0327	0.5	20 F	0417	1.0
	1002	4.1		1053	3.7
	1553	0.2		1632	0.8
	2232	4.1		2303	3.8
6 F	0412	0.3	21 SA	0450	0.9
	1050	4.3		1125	3.7
	1638	0.1		1703	0.8
●	2315	4.2	○	2330	3.8
7 SA	0457	0.2	22 SU	0520	0.9
	1137	4.3		1156	3.7
	1722	0.1		1731	0.8
	2359	4.2		2359	3.8
8 SU	0544	0.2	23 M	0547	0.8
	1225	4.3		1229	3.7
	1806	0.2		1800	0.8
9 M	0044	4.2	24 TU	0033	3.9
	0632	0.2		0618	0.8
	1315	4.2		1306	3.7
	1853	0.4		1835	0.9
10 TU	0132	4.1	25 W	0114	3.9
	0724	0.3		0656	0.8
	1408	4.0		1348	3.7
	1943	0.7		1915	0.9
11 W	0224	4.0	26 TH	0157	3.9
	0819	0.5		0741	0.8
	1505	3.8		1433	3.7
	2037	0.9		2001	1.0
12 TH	0322	3.8	27 F	0245	3.8
	0918	0.7		0832	0.9
	1607	3.6		1523	3.6
	2134	1.2		2054	1.1
13 F	0427	3.7	28 SA	0336	3.7
	1019	0.9		0929	0.9
	1713	3.5		1618	3.5
◐	2236	1.4		2152	1.2
14 SA	0536	3.6	29 SU	0432	3.7
	1124	1.0		1032	1.0
	1821	3.4		1719	3.5
	2343	1.5	◑	2257	1.2
15 SU	0645	3.5	30 M	0534	3.6
	1233	1.1		1139	0.9
	1928	3.4		1825	3.5
			31 TU	0005	1.2
				0640	3.7
				1246	0.8
				1930	3.6

JUNE

Day	Time	m	Day	Time	m
1 W	0110	1.1	16 TH	0214	1.4
	0745	3.8		0854	3.6
	1348	0.7		1441	1.1
	2030	3.8		2117	3.6
2 TH	0210	0.9	17 F	0305	1.3
	0846	3.9		0941	3.6
	1444	0.6		1524	1.1
	2124	3.9		2158	3.7
3 F	0305	0.7	18 SA	0346	1.2
	0942	4.1		1022	3.7
	1535	0.4		1601	1.0
	2214	4.0		2234	3.8
4 SA	0356	0.5	19 SU	0422	1.1
	1035	4.2		1100	3.7
	1623	0.4		1634	0.9
	2301	4.1		2306	3.9
5 SU	0445	0.4	20 M	0453	1.0
	1125	4.2		1134	3.7
	1708	0.4		1705	0.9
●	2345	4.2	○	2337	3.9
6 M	0533	0.3	21 TU	0524	0.9
	1214	4.2		1208	3.8
	1753	0.5		1738	0.8
7 TU	0030	4.2	22 W	0012	4.0
	0620	0.4		0558	0.8
	1303	4.1		1246	3.8
	1837	0.6		1814	0.8
8 W	0117	4.1	23 TH	0053	4.0
	0710	0.4		0638	0.8
	1352	3.9		1328	3.8
	1924	0.8		1855	0.8
9 TH	0206	4.0	24 F	0137	4.0
	0802	0.5		0723	0.7
	1444	3.8		1413	3.8
	2013	0.9		1941	0.9
10 F	0258	3.9	25 SA	0224	4.0
	0856	0.7		0813	0.7
	1538	3.6		1502	3.8
	2106	1.1		2032	1.0
11 SA	0355	3.8	26 SU	0314	4.0
	0951	0.9		0908	0.8
	1636	3.5		1554	3.7
	2202	1.3		2127	1.0
12 SU	0457	3.7	27 M	0408	3.9
	1048	1.0		1007	0.8
	1738	3.4		1651	3.6
◑	2302	1.4	◐	2228	1.1
13 M	0601	3.6	28 TU	0507	3.8
	1147	1.1		1111	0.9
	1839	3.3		1754	3.6
				2333	1.2
14 TU	0005	1.5	29 W	0612	3.8
	0703	3.5		1217	0.9
	1250	1.2		1901	3.6
	1937	3.4			
15 W	0112	1.5	30 TH	0041	1.1
	0801	3.5		0720	3.8
	1349	1.2		1323	0.9
	2030	3.5		2006	3.7

JULY

Day	Time	m	Day	Time	m
1 F	0147	1.1	16 SA	0225	1.4
	0827	3.9		0907	3.5
	1425	0.8		1448	1.3
	2106	3.9		2125	3.6
2 SA	0250	0.9	17 SU	0315	1.3
	0930	3.9		0954	3.6
	1522	0.7		1531	1.1
	2201	4.0		2206	3.8
3 SU	0346	0.7	18 M	0354	1.1
	1027	4.0		1035	3.7
	1613	0.7		1608	1.0
	2250	4.1		2242	3.9
4 M	0437	0.6	19 TU	0429	1.0
	1118	4.1		1112	3.8
	1658	0.6		1642	0.9
●	2335	4.2	○	2316	4.0
5 TU	0524	0.5	20 W	0502	0.8
	1204	4.0		1147	3.8
	1740	0.6		1716	0.8
				2351	4.1
6 W	0017	4.2	21 TH	0538	0.7
	0609	0.5		1224	3.9
	1248	4.0		1753	0.7
	1821	0.7			
7 TH	0059	4.1	22 F	0030	4.2
	0654	0.5		0618	0.6
	1331	3.9		1306	3.9
	1902	0.8		1833	0.7
8 F	0142	4.1	23 SA	0114	4.2
	0740	0.6		0702	0.5
	1415	3.8		1350	3.9
	1946	0.9		1918	0.7
9 SA	0227	4.0	24 SU	0200	4.2
	0828	0.7		0750	0.6
	1501	3.6		1437	3.9
	2033	1.0		2007	0.8
10 SU	0315	3.8	25 M	0250	4.1
	0917	0.9		0844	0.6
	1550	3.5		1528	3.8
	2124	1.2		2101	0.9
11 M	0408	3.7	26 TU	0343	4.0
	1008	1.0		0941	0.8
	1644	3.4		1623	3.8
	2219	1.4	◐	2201	1.0
12 TU	0507	3.5	27 W	0441	3.9
	1101	1.2		1043	0.9
	1745	3.3		1726	3.7
◐	2317	1.5		2306	1.1
13 W	0612	3.4	28 TH	0548	3.8
	1158	1.3		1151	1.0
	1847	3.3		1837	3.6
14 TH	0020	1.6	29 F	0018	1.2
	0716	3.4		0703	3.7
	1257	1.4		1302	1.1
	1945	3.4		1948	3.7
15 F	0124	1.5	30 SA	0132	1.2
	0814	3.4		0818	3.7
	1356	1.3		1412	1.1
	2038	3.5		2054	3.8
			31 SU	0241	1.0
				0925	3.8
				1513	1.0
				2153	3.9

AUGUST

Day	Time	m	Day	Time	m
1 M	0340	0.9	16 TU	0328	1.1
	1023	3.9		1012	3.7
	1604	0.9		1543	1.0
	2243	4.1		2217	3.9
2 TU	0430	0.7	17 W	0404	0.8
	1113	3.9		1049	3.8
	1648	0.8		1619	0.8
●	2326	4.1		2253	4.1
3 W	0515	0.6	18 TH	0440	0.6
	1155	3.9		1125	3.9
	1727	0.7		1654	0.6
			○	2329	4.2
4 TH	0003	4.1	19 F	0516	0.4
	0555	0.5		1202	4.0
	1231	3.9		1732	0.5
	1803	0.7			
5 F	0037	4.1	20 SA	0007	4.3
	0634	0.5		0555	0.3
	1305	3.8		1242	4.1
	1839	0.8		1811	0.5
6 SA	0113	4.1	21 SU	0049	4.4
	0714	0.6		0638	0.3
	1341	3.8		1324	4.1
	1917	0.8		1855	0.5
7 SU	0153	4.0	22 M	0135	4.3
	0754	0.7		0726	0.4
	1421	3.7		1411	4.0
	1957	1.0		1943	0.6
8 M	0234	3.9	23 TU	0224	4.2
	0838	0.9		0817	0.6
	1503	3.6		1501	3.9
	2042	1.1		2037	0.8
9 TU	0319	3.7	24 W	0318	4.1
	0924	1.0		0915	0.8
	1548	3.5		1556	3.8
	2132	1.3		2138	1.0
10 W	0409	3.5	25 TH	0418	3.9
	1014	1.2		1018	1.0
	1641	3.4		1700	3.7
◐	2228	1.4	◑	2245	1.1
11 TH	0511	3.4	26 F	0531	3.7
	1109	1.4		1128	1.2
	1746	3.3		1817	3.6
	2332	1.6			
12 F	0627	3.3	27 SA	0001	1.3
	1210	1.5		0655	3.6
	1858	3.3		1245	1.3
				1933	3.6
13 SA	0039	1.6	28 SU	0122	1.2
	0737	3.3		0813	3.6
	1313	1.5		1400	1.3
	2001	3.4		2043	3.8
14 SU	0146	1.5	29 M	0235	1.1
	0837	3.4		0921	3.7
	1413	1.4		1503	1.1
	2054	3.6		2144	3.9
15 M	0244	1.3	30 TU	0332	0.9
	0928	3.5		1019	3.8
	1503	1.2		1552	1.0
	2139	3.7		2234	4.0
			31 W	0419	0.7
				1105	3.9
				1634	0.9
				2315	4.1

Chart Datum: 0·20 metres above Ordnance Datum (Dublin). HAT is 4·5 metres above Chart Datum.

>> **FREE** monthly updates. Register at <<
www.reedsnauticalalmanac.co.uk

STANDARD TIME (UT)
For Summer Time add ONE hour in **non-shaded areas**

DUBLIN (NORTH WALL) LAT 53°21'N LONG 6°13'W
TIMES AND HEIGHTS OF HIGH AND LOW WATERS

Dates in red are **SPRINGS**
Dates in blue are **NEAPS**

YEAR **2016**

S Ireland

SEPTEMBER

#	Time m	#	Time m
1	0459 0.6 / 1143 3.9 / TH 1710 0.8 / ● 2347 4.1	16	0416 0.4 / 1101 4.1 / F 1632 0.5 / ○ 2305 4.3
2	0536 0.5 / 1211 3.8 / F 1744 0.7	17	0454 0.2 / 1138 4.2 / SA 1710 0.4 / 2344 4.4
3	0014 4.1 / 0610 0.6 / SA 1238 3.8 / 1815 0.7	18	0534 0.1 / 1217 4.2 / SU 1751 0.3
4	0046 4.0 / 0644 0.6 / SU 1309 3.8 / 1848 0.8	19	0026 4.5 / 0615 0.2 / M 1259 4.2 / 1834 0.3
5	0121 4.0 / 0719 0.7 / M 1344 3.8 / 1923 0.9	20	0112 4.4 / 0701 0.3 / TU 1346 4.1 / 1922 0.5
6	0159 3.9 / 0755 0.9 / TU 1423 3.7 / 2001 1.0	21	0202 4.2 / 0753 0.6 / W 1436 4.0 / 2017 0.7
7	0240 3.7 / 0835 1.0 / W 1504 3.6 / 2043 1.2	22	0258 4.0 / 0850 0.8 / TH 1533 3.9 / 2119 0.9
8	0326 3.6 / 0921 1.2 / TH 1546 3.5 / 2133 1.4	23	0402 3.8 / 0955 1.1 / F 1639 3.7 / ◑ 2229 1.1
9	0419 3.4 / 1017 1.4 / F 1645 3.3 / ◑ 2240 1.5	24	0522 3.6 / 1106 1.3 / SA 1758 3.6 / 2346 1.2
10	0530 3.2 / 1124 1.6 / SA 1759 3.3 / 2357 1.6	25	0646 3.5 / 1226 1.5 / SU 1916 3.7
11	0658 3.2 / 1234 1.6 / SU 1919 3.3	26	0109 1.2 / 0803 3.6 / M 1345 1.4 / 2026 3.8
12	0108 1.5 / 0807 3.3 / M 1340 1.5 / 2020 3.5	27	0221 1.1 / 0911 3.7 / TU 1446 1.3 / 2127 3.9
13	0212 1.3 / 0902 3.5 / TU 1434 1.3 / 2109 3.7	28	0316 0.9 / 1005 3.8 / W 1534 1.1 / 2217 4.0
14	0300 1.0 / 0947 3.7 / W 1517 1.0 / 2150 3.9	29	0400 0.7 / 1049 3.9 / TH 1614 0.9 / 2258 4.1
15	0339 0.7 / 1025 3.9 / TH 1555 0.7 / 2228 4.2	30	0439 0.6 / 1124 3.9 / F 1650 0.8 / 2328 4.1

OCTOBER

#	Time m	#	Time m
1	0513 0.6 / 1149 3.9 / SA 1723 0.8 / ● 2354 4.0	16	0432 0.2 / 1115 4.3 / SU 1650 0.3 / ○ 2324 4.5
2	0545 0.6 / 1212 3.8 / SU 1754 0.8	17	0513 0.1 / 1155 4.3 / M 1732 0.2
3	0022 4.0 / 0615 0.7 / M 1241 3.9 / 1823 0.8	18	0007 4.5 / 0555 0.2 / TU 1238 4.3 / 1817 0.3
4	0055 3.9 / 0645 0.8 / TU 1314 3.8 / 1854 0.9	19	0055 4.4 / 0641 0.4 / W 1325 4.2 / 1907 0.4
5	0131 3.9 / 0716 0.9 / W 1350 3.8 / 1928 1.0	20	0147 4.2 / 0732 0.6 / TH 1417 4.1 / 2002 0.6
6	0211 3.7 / 0753 1.1 / TH 1431 3.7 / 2008 1.1	21	0245 4.0 / 0829 0.9 / F 1515 4.0 / 2104 0.9
7	0255 3.6 / 0835 1.2 / F 1516 3.6 / 2054 1.3	22	0352 3.8 / 0932 1.2 / SA 1621 3.8 / ◑ 2211 1.1
8	0346 3.4 / 0928 1.4 / SA 1607 3.5 / 2153 1.4	23	0509 3.6 / 1042 1.4 / SU 1736 3.7 / 2325 1.2
9	0448 3.2 / 1037 1.6 / SU 1709 3.3 / ◑ 2311 1.5	24	0627 3.5 / 1159 1.5 / M 1850 3.7
10	0611 3.2 / 1154 1.7 / M 1826 3.3	25	0045 1.2 / 0742 3.6 / TU 1318 1.5 / 2000 3.8
11	0029 1.5 / 0729 3.3 / TU 1303 1.5 / 1936 3.5	26	0156 1.1 / 0848 3.7 / W 1421 1.4 / 2101 3.9
12	0135 1.2 / 0829 3.5 / W 1401 1.3 / 2032 3.7	27	0251 1.0 / 0941 3.8 / TH 1510 1.2 / 2152 4.0
13	0228 0.9 / 0917 3.7 / TH 1448 1.0 / 2119 4.0	28	0336 0.8 / 1024 3.9 / F 1552 1.1 / 2233 4.0
14	0312 0.6 / 0958 4.0 / F 1530 0.7 / 2201 4.2	29	0414 0.8 / 1058 3.9 / SA 1629 1.0 / 2305 4.0
15	0353 0.3 / 1036 4.1 / SA 1610 0.5 / 2242 4.4	30	0448 0.7 / 1124 3.9 / SU 1702 0.9 / ● 2334 4.0
		31	0520 0.8 / 1149 3.9 / M 1734 0.9

NOVEMBER

#	Time m	#	Time m
1	0002 3.9 / 0548 0.8 / TU 1217 3.9 / 1803 0.9	16	0540 0.3 / 1222 4.4 / W 1805 0.3
2	0034 3.9 / 0616 0.9 / W 1249 3.9 / 1832 0.9	17	0043 4.3 / 0625 0.5 / TH 1310 4.3 / 1855 0.4
3	0109 3.8 / 0646 1.0 / TH 1325 3.9 / 1904 0.9	18	0136 4.2 / 0715 0.7 / F 1402 4.2 / 1949 0.6
4	0148 3.7 / 0722 1.1 / F 1405 3.8 / 1943 1.1	19	0234 4.0 / 0809 1.0 / SA 1458 4.1 / 2048 0.8
5	0232 3.6 / 0804 1.2 / SA 1450 3.7 / 2029 1.2	20	0337 3.8 / 0909 1.2 / SU 1600 3.9 / 2150 1.0
6	0322 3.5 / 0855 1.4 / SU 1539 3.6 / 2123 1.3	21	0446 3.6 / 1012 1.4 / M 1708 3.8 / ◔ 2255 1.1
7	0419 3.4 / 0956 1.5 / M 1635 3.5 / ◑ 2230 1.4	22	0557 3.5 / 1122 1.6 / TU 1816 3.7
8	0528 3.3 / 1110 1.6 / TU 1739 3.5 / 2345 1.3	23	0007 1.2 / 0706 3.5 / W 1237 1.6 / 1922 3.7
9	0643 3.4 / 1222 1.5 / W 1848 3.6	24	0118 1.2 / 0810 3.6 / TH 1345 1.5 / 2024 3.8
10	0054 1.2 / 0748 3.6 / TH 1324 1.4 / 1950 3.7	25	0218 1.1 / 0905 3.7 / F 1441 1.4 / 2117 3.8
11	0154 0.9 / 0841 3.8 / F 1417 1.1 / 2045 4.0	26	0307 1.0 / 0949 3.8 / SA 1527 1.2 / 2202 3.9
12	0244 0.7 / 0928 4.0 / SA 1504 0.8 / 2135 4.2	27	0348 1.0 / 1025 3.9 / SU 1606 1.1 / 2239 3.9
13	0330 0.4 / 1012 4.2 / SU 1549 0.6 / 2222 4.3	28	0423 0.9 / 1057 3.9 / M 1642 1.0 / 2313 3.9
14	0413 0.3 / 1055 4.3 / M 1633 0.4 / ○ 2308 4.4	29	0455 0.9 / 1127 4.0 / TU 1714 1.0 / ● 2344 3.9
15	0456 0.2 / 1138 4.4 / TU 1718 0.3 / 2354 4.4	30	0525 0.9 / 1156 4.0 / W 1744 1.0

DECEMBER

#	Time m	#	Time m
1	0015 3.8 / 0552 1.0 / TH 1227 4.0 / 1812 1.0	16	0035 4.2 / 0612 0.6 / F 1257 4.4 / 1844 0.4
2	0050 3.8 / 0622 1.0 / F 1302 4.0 / 1845 1.0	17	0125 4.1 / 0659 0.7 / SA 1346 4.3 / 1935 0.5
3	0128 3.8 / 0658 1.1 / SA 1343 4.0 / 1924 1.0	18	0218 4.0 / 0748 0.9 / SU 1438 4.2 / 2027 0.7
4	0212 3.7 / 0740 1.2 / SU 1427 3.9 / 2008 1.1	19	0313 3.8 / 0842 1.1 / M 1532 4.0 / 2122 0.8
5	0259 3.6 / 0828 1.3 / M 1515 3.8 / 2059 1.1	20	0412 3.6 / 0939 1.3 / TU 1631 3.9 / 2219 1.0
6	0352 3.6 / 0924 1.4 / TU 1607 3.8 / 2157 1.1	21	0515 3.5 / 1040 1.5 / W 1734 3.7 / ◔ 2320 1.2
7	0451 3.5 / 1027 1.5 / W 1704 3.7 / ◑ 2302 1.2	22	0619 3.5 / 1145 1.6 / TH 1837 3.6
8	0557 3.5 / 1136 1.5 / TH 1806 3.7	23	0026 1.3 / 0720 3.5 / F 1257 1.6 / 1938 3.6
9	0012 1.1 / 0704 3.6 / F 1244 1.4 / 1911 3.8	24	0134 1.3 / 0817 3.6 / SA 1404 1.5 / 2036 3.6
10	0118 1.0 / 0805 3.8 / SA 1345 1.2 / 2013 3.9	25	0232 1.3 / 0907 3.7 / SU 1459 1.4 / 2127 3.7
11	0217 0.8 / 0900 3.9 / SU 1441 1.0 / 2112 4.1	26	0319 1.2 / 0951 3.8 / M 1543 1.3 / 2211 3.7
12	0310 0.6 / 0951 4.1 / M 1532 0.8 / 2206 4.2	27	0358 1.1 / 1029 3.9 / TU 1621 1.2 / 2250 3.7
13	0358 0.5 / 1039 4.3 / TU 1621 0.5 / 2257 4.3	28	0431 1.1 / 1104 3.9 / W 1654 1.1 / 2325 3.8
14	0444 0.5 / 1125 4.4 / W 1709 0.4 / ○ 2346 4.3	29	0502 1.0 / 1136 4.0 / TH 1724 1.0 / ● 2357 3.8
15	0528 0.5 / 1210 4.4 / TH 1756 0.4	30	0531 1.0 / 1206 4.0 / F 1753 0.9
		31	0030 3.8 / 0601 0.9 / SA 1241 4.1 / 1825 0.9

Chart Datum: 0·20 metres above Ordnance Datum (Dublin). HAT is 4·5 metres above Chart Datum.

》》 FREE monthly updates. Register at 《
www.reedsnauticalalmanac.co.uk

577

9.12.10 DUN LAOGHAIRE

Dublin (Port ent) 53°18'·16N 06°07'·68W ✿✿✿ⓘⓘⓘ✿✿✿

CHARTS AC 1468, 1415, 1447, 5621; Imray C61, C62

TIDES +0042 Dover; ML 2·4; Duration 0640
Standard Port DUBLIN (NORTH WALL) (←)

Times				Height (metres)			
High Water		Low Water		MHWS	MHWN	MLWN	MLWS
0000	0700	0000	0500	4·1	3·4	1·5	0·7
1200	1900	1200	1700				
Differences DUBLIN BAR and DUN LAOGHAIRE							
0000	0000	0002	+0003	0·0	+0·1	0·0	0·0
GREYSTONES (53°09'N 06°04'W)							
–0008	–0008	–0008	–0008	–0·5	–0·4	ND	ND

SHELTER Excellent. Breakwaters within main hbr protect 800+ berth marina from ferry wash. Ⓥ berths on end of pontoons or as directed. Additional berths inside Western Marina Breakwater. All YCs advertise ⚓s and/or pontoon berths.

NAVIGATION Dun Laoghaire WPT 53°18'·40N 06°07'·00W, 240°/0·47M to ent. Accessible H24 Keep clear of commercial traffic turning off St Michael's Pier. Beware drying rks around E Pier Hd. **TSS**: Two TSS ½M long, one at either end of the Burford Bank lead to/from an anti-clockwise circular TSS radius 2·3ca centred on Dublin Bay SWM lt buoy (53°19'·90N 06°04'·58W, 2·5M NE of Dun Laoghaire/ 2·75M ESE of Dublin Port). Do not impede large vessels and keep clear of the TSS and fairway.

LIGHTS AND MARKS Ro Ro Berth 2 x Fl.W = 'Large vessel under way; small craft keep clear of No 1 Fairway (out to 600m seaward of ent). Traffic signals (Fl W) repeated at head of E Marina Breakwater. Marina Breakwater Hds marked by sync lts Fl G 2s and Fl R 2s.

COMMUNICATIONS (Code 00 353 1) MRCC 6620922/3; Coast/Cliff Rescue Service 2803900; ⊖ 2803992; ⚓ 1550 123855; Police 6665000; Dr 2859244; Ⓗ 2806901. HM Dun Laoghaire 2801130.
Dun Laoghaire Hbr VHF Ch **14** 16 (H24); Marina Ch **M**; YCs Ch M. Dublin Coast Radio Stn 16 67 83.

FACILITIES Dun Laoghaire Harbour Marina, www.dlmarina.com ☎2020040; 800 ⌑ inc 20Ⓥ €3.75, €5.00<6hrs, ♿⊡♠▬. **Town** ⛽♨⚓⚒Ⓔ Ⓖⓘ🚂 ACA.

YACHT CLUBS (E→W):
Irish National SC ☎28444195.
National YC ☎2805725, ▬▲🔧⚓♠ ⌑(12t) ✕⬜.
Royal St George YC ☎2801811, Boatman ☎2801208; ⌑ by prior agreement, €3.60, ▬▲🔧⌑(5t) ✕⬜.
Royal Irish YC ☎2809452, ▬▲🔧⚓♠⌑(5t) ✕⬜.
Dun Laoghaire Motor YC ☎2801371, ▬⌑🔧✕⬜.

ADJACENT HARBOUR

GREYSTONES 53°09'·05N 06°03'·80W midway between Dun Laoghaire and Wicklow. Harbour ent open to large seas in NE winds, with scend in outer basin. The redeveloped harbour encloses **Greystones Marina**, info@greystonesharbourmarina.ie ☎353 (086) 2718161; 100 ⌑, Ⓥ facilities tbc in N Basin.

9.12.11   WICKLOW

Wicklow **52°58'·98N 06°02'·70W** ✿✿⊛♤♤✿✿

CHARTS AC 1468, 633, 5621; Imray C61

TIDES −0010 Dover; ML 1·7; Duration 0640

Standard Port DUBLIN (NORTH WALL) (←)

Times				Height (metres)			
High Water		Low Water		MHWS	MHWN	MLWN	MLWS
0000	0700	0000	0500	4·1	3·4	1·5	0·7
1200	1900	1200	1700				
Differences WICKLOW							
−0019	−0019	−0024	−0026	−1·4	−1·1	−0·4	0·0

SHELTER Very safe, access H24. Outer hbr is open to NE winds which cause a swell. Moorings in NW of hbr belong to SC and may not be used without permission. 4 berths on E Pier (2·5m) are convenient except in strong winds NW to NE, with fender boards/ladders provided. W pier is not recommended. ⚓ in hbr is restricted by ships' turning circle. Inner hbr (river) gives excellent shelter in 2·5m on N and S Quays, which are used by FVs. Packet Quay is for ships (2·5m), but may be used if none due; fender board needed. Yachts should berth on N or S quays as directed and/or space available.

NAVIGATION WPT 52°59'·20N 06°01'·80W, 220°/0·27M to ent. Appr presents no difficulty; keep in the R sector of the E pier lt to avoid Planet Rk and Pogeen Rk.

LIGHTS AND MARKS W pier head lt, Iso G 4s, is shown H24. ☆ Fl WG 10s on Packet Quay hd is vis 076°-G-256°-W-076°.

COMMUNICATIONS (Code 0404) MRCC (01) 6620922; Cliff Rescue

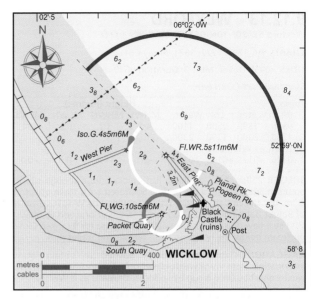

Service 086 8686009; ⊖ 67222; Police 60140; Dr 67381. HM 67455. Ch 12, **14**, 16. Wicklow SC Ch M 16 (occas).

FACILITIES **East Pier, S and N Quays,** ⚓ ⚲ €14.00/craft ⚓ 🗑 🚽 (bulk: see HM). **Wicklow SC** ☎67526, ⚓ (except LWS) ⚓ ⚓ ⚓ 🅿.

Service 🔋 Kos ⚓ ⊞ 🅱 ⚓. **Town** ✉ 🅱 🏦 🛒 🍴 ✕ 🖵 ⚌ ✈(Dublin).

9.12.12   ARKLOW

Wicklow **52°47'·60N 06°08'·20W** ✿✿⊛♤♤✿✿

CHARTS AC 1468, 633, 5621; Imray C61

TIDES −0150 Dover; ML 1·0; Duration 0640

Standard Port DUBLIN (NORTH WALL) (←)

Times				Height (metres)			
High Water		Low Water		MHWS	MHWN	MLWN	MLWS
0000	0700	0000	0500	4·1	3·4	1·5	0·7
1200	1900	1200	1700				
Differences ARKLOW (Note small Range)							
−0315	−0201	−0140	−0134	−2·7	−2·2	−0·6	−0·1
COURTOWN							
−0328	−0242	−0158	−0138	−2·8	−2·4	−0·5	0·0

SHELTER Good, access H24. **Arklow Roadstone Hbr**, 1M S of Arklow, is not for yachts.

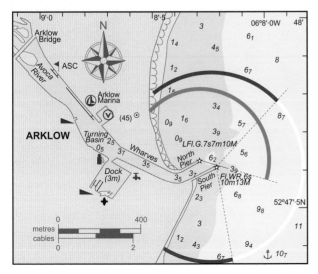

- Strong E'ly winds make the entrance unsafe and cause significant turbulence in the river.

A 60 berth marina lies on NE side of river. Good ⚓ in bay; avoid whelk pots. One ⚲ in 1.2m at ASC quay. ⚲ on SE wall of Dock (ent is 13·5m wide; 3m depth) in perfect shelter, but amidst FVs.

NAVIGATION WPT 52°47'·60N 06°07'·50W, 270°/0·40M to ent. No navigational dangers, but beware ebb setting SE across hbr ent; give S pierhd a wide berth. Ent is difficult without power, due to blanking by piers. 3kn speed limit. Night entry, see below. Caution: Upriver of Dock ent keep to NE side of river.

LIGHTS AND MARKS No ldg lts/marks, see 9.12.4 and chartlet Conspic factory chy 2·5ca NW of piers.

- Beware the Pier Hd lts are very difficult to see due to powerful orange flood lts near the root of both piers shining E/ENE onto them (to assist pilotage of departing shipping).
- Best advice to visitors is to approach from the NE with R sector just showing, or enter by day.

COMMUNICATIONS (Code 0402) MRCC (01) 6620922/3; ⊖ 67222; Cliff Rescue Service 32430; RNLI 32901; Police 32304; Dr 32421. HM 32466.

Marina Ch 16 12 (HJ); Arklow SC Ch 10.

FACILITIES **Dock** €15.00/craft (2nd day free), use 🚽/🛁 in LB Hse, 🔋 (Dock ent, HO); **Arklow SC** (NE bank, 500m upriver) 🛁; **Marina** ☎087 258 8078, 60⚲ inc🅥 €3·35. **Services** ⚓ ✕ ⚓ ⊞ ⚓(20t) Kos. **Town** 🏨 🚽 limited 🏦 ✉ 🅱 🛒 ✕ 🖵 @ ⚌ ✈(Dublin).

MINOR HARBOURS S OF ARKLOW
COURTOWN, Wexford, **52°38'·55N 06°13'·50W.** AC 1787. 9.6M S of Arklow is feasible in settled weather and offshore winds. Caution: 10m wide ent; only 1m (maintained by local YC) at MLWS due to silting. ⚲ on E wall or pick up vacant mooring.

POLDUFF PIER, Wexford, **52°34'·15N 06°11'·97W.** AC 1787. 14M S of Arklow. Pier, extends 100m NE from shore. NW side of pier has slipway from shore and 1m depth alongside. Local moorings W of pier. Good shelter for small boats in S to W winds. Swell in winds E of S. Appr in daylight only. Appr is clear from NE to E. Rks extend from shore E'ward 200m N of pier but steering along line of pier will clear them. Pub near pier. Village (Ballygarrett) 1·25M W.

9.12.13 WEXFORD

Wexford 52°20'·10N 06°27'·00W ✿⚓⚓✿✿

CHARTS AC 1787, 1772, 5621; Imray C61

TIDES –0450 Dover; ML 1·3; Duration 0630

Standard Port COBH (→)

Times				Height (metres)			
High Water		Low Water		MHWS	MHWN	MLWN	MLWS
0500	1100	0500	1100	4·1	3·2	1·3	0·4
1700	2300	1700	2300				
Differences WEXFORD							
+0126	+0126	+0118	+0108	–2·1	–1·7	–0·3	+0·1

SHELTER 250m of ⌒ available just below bridge though silting occurs in 50m closest to the bridge. Sheltered ⚓ off town quays in 2·3m, but streams are strong. ⚓s are provided by WHBTC, close N of Ballast Bank (☎053 9122039 or Ch 16/69) or proceed to quays. FVs use a new quay on E bank with ⚓ and space for ⛴tanker. Close upstream below the bridge is an all-tide ⬳ and trailer park.

NAVIGATION Visitors are strongly advised to obtain latest information from www.wexfordharbour.com another site www. wexfordharbour.info is run by sailing enthusiasts as is WHBTC.

- Entry is hazardous in strong winds between SE and NE when the sea breaks on the bar.
- For strangers best entry is HW±2. It is difficult at any time for vessels with a draught >1.3m.

- Maintain a course on the rhumb line between marks as channel is narrow in places. Beware the S training wall (covers) and pass well S of Black Man SHM Bn at end of N training wall.
- To Wexford Bar SWM (52°19'·14N 06°19·39W): make good 214°/2·7M from North Long Buoy; from S make good 353°/3·4M from West Holdens Buoy. SWM is on 10m contour and marks the start of the buoyed channel to Wexford.

LIGHTS AND MARKS The narrow channel frequently shifts and marks are moved accordingly; positions on the chartlet are not reliable. They are sequentially numbered from 2–30. Outer channel buoys Nos 2–9 in position from Apr–Oct are removed seasonally; see www.wexfordharbour.com The 2 church spires / Opera Ho (conspic) at Wexford give useful general orientation. There are no ldg lts.

COMMUNICATIONS (Code 053) MRCC (01) 6620922/3; ⊜ 9133116; Police 9122333; Dr 9122524; Ⓗ 9142233. Wexford Hbr BC 9122039. HM 9122300, email: harbourmaster@wexfordcoco.ie.
Wexford Hbr listening watch VHF 16, 09 (HO).

FACILITIES HM Capt Paul Murphy, based in Ballast Office on Crescent Quay, is the Marine Officer responsible for all Co. Wexford harbours except Rosslare Europort.
Wexford Quays ⬳ ⌒(free) ⚓ ⛴ delivery by road tanker ⚒ Ⓔ🅿.
Wexford Harbour Boat & Tennis Club, www.whbtc.ie ☎086 8131783, ⬳ 🏗(5t) 🚿.
Town ✉ Ⓑ 🛒 ⇌ ✈ (Waterford).

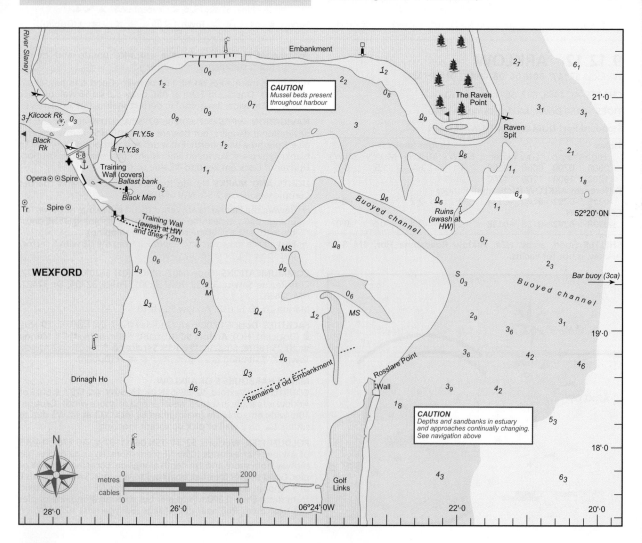

9.12.14   ROSSLARE

Dublin 52°15'·30N 06°20'·90W ❀❀❀❀◊✿✿

CHARTS AC 1787, 1772, 5621; Imray C61, C57

TIDES –0510 Dover; ML 1·1; Duration 0640

Standard Port COBH (→)

Times				Height (metres)			
High Water		Low Water		MHWS	MHWN	MLWN	MLWS
0500	1100	0500	1100	4·1	3·2	1·3	0·4
1700	2300	1700	2300				
Differences ROSSLARE EUROPORT							
+0045	+0035	+0015	–0005	–2·2	–1·8	–0·5	–0·1

SHELTER Useful passage shelter from SW'lies, but few facilities for yachts which may berth on E wall of marshalling area (⚓ on the chartlet, 3·7m), or ⚓ clear of harbour to the west. *Small craft harbour (5ca W) is not recommended in onshore winds and can become dangerous if winds freshen.* Rosslare Europort has 80 ferry movements per week.

NAVIGATION WPT 52°14'·70N 06°15'·80W (abeam S Long SHM buoy, Fl(2) G 6s), 287°/2·9M to 1ca N of breakwater light. Main approach from E, S and W is via S Shear, buoyed/lit chan to S of Holden's Bed, a shoal of varying depth; the tide sets across the chan. From S, beware rks off Greenore Pt, and overfalls here and over The Baillies. From the N, appr via N Shear. Tuskar TSS is approx 8M ESE of hbr. Yachts, bound N/S, will usually pass to the W of Tuskar Rk where the 3·5M wide chan lies to seaward of The Bailies. A passage inshore of The Bailies requires local knowledge and is not recommended at night. In heavy weather or poor vis, passage E of Tuskar Rk is advised.

LIGHTS AND MARKS Tuskar Rk, Q (2) 7·5s 33m 28M, is 5·8M SE of hbr. Water tr (R lt, 35m) is conspic 0·8M SSE of hbr ent. Bkwtr lt, Oc WRG 5s 15m 13/10M, see 9.12.4. Its two W sectors (188°-208° and 283°-286°) cover N and S Shear respectively. Note: Powerful floodlights in the hbr make identification of navigational lights difficult.

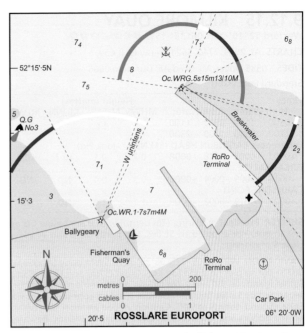

ROSSLARE EUROPORT

COMMUNICATIONS (Code 053) MRCC (01) 6620922/3; LB Lookout Stn 9133205; ⊖ 9133116; Police 053 9165200; Dr 9131154; Ⓗ 9142233. HM 9157921, mobile 087 2320251. Call on VHF Ch **12** (H24) before entering hbr.

FACILITIES Hbr Ops ☎9157929, No dues, ⚓ ⛽ (by hose on berths 2 and 3) 🚻 🚿 Kos ✎ 🖃 🛒 Divers. **Village** 🍴 ✕ 🏪 ✉ Ⓑ ≋ ✈ (Dublin). **Ferries** Irish Ferries (www.irish ferries.com): Pembroke Dock, 2/day, 4 hrs; Cherbourg, 2 or 3/week, 18½ hrs; Roscoff, 1 or 2/week, 17½ hrs (seasonal). Stena (www.stenaline.co.uk): Fishguard, 2/day, 1¾ hrs.

TUSKAR ROCK TO COBH

(AC 2049) Dangerous rks lie up to 2ca NW and 6½ca SSW of Tuskar Rk and there can be a dangerous race off Carnsore Pt. In bad weather or poor visibility, use the Inshore Traffic Zone of the Tuskar Rock TSS, passing to seaward of Tuskar Rk (lt), the Barrels ECM lt buoy and Coningbeg buoy.

FLOOD
Carnsore Point

(d) 1k
(e)
(c) 3.5k   (a) 2.5k
(b) 2.5k
2.5k

The tide turns to the NE or N (Irish Sea flood) as follows:
(a): HW Dover +0500;
(b): HW D +0520;
(c): HW D +0600;
(d): HW D –0600. NE going streams are shorter in duration and weaker than SE going - careful passage planning is essential.

EBB
Carnsore Point

(d) 2k
(e)
(c) 3.5k   (a) 3.5k
(b) 2.5k   2.5k

The tide turns to the SW or S as follows:
(a): HW Dover –0200;
(b): HW D –0020;
(c): HW D –0015;
(d): HW D –0300. Leaving Rosslare at HW D –0300 a yacht can carry a fair tide for about 8h until HW D +0515 off Hook Head.

NOTE: The tide turns on St Patrick's Bridge (e) up to 2 hours earlier than in Saltee Sound

If taking the inshore passage from Greenore Pt, stay inside The Bailies to pass 2ca off Carnsore Pt. Beware of lobster pots. Steer WSW to pass N of Black Rk and the Bohurs, S of which are extensive overfalls. The small hbr of Kilmore Quay has been rebuilt with a marina, but beware rks and shoals in the approaches.

Saltee Sound (AC 2740) is a safe passage, least width 3ca, between Great and Little Saltee, conspic islands to S and N. ▶ *Sebber Bridge extends 7½ca N from the NE point of Great Saltee and Jackeen Rk is 1M NW of the S end of Little Saltee, so care is needed through the sound, where the stream runs 3·5kn at sp.* ◀

There are several rks S of the Saltees, but clear passage between many of them. There are no obstructions on a direct course for a point 1M S of Hook Head, clear of Tower Race which forms when the ebb (from Waterford Hbr) is stronger than the W going stream. In strong W winds Tower Race can be violent. Overfalls occur off Waterford Hbr entrance (AC2046) when the W'ly stream begins off Hook Hd and meets the strong outgoing stream from Waterford Hbr.

Dunmore East (AC 2046) is a useful passage port at the mouth of Waterford Hbr. To the W, beware Falskirt, a dangerous rk off Swines Pt. There are few offlying rocks from Tramore Bay to Ballinacourty Pt on the N side of Dungarvan Bay (AC 2017). Helvick is a small sheltered hbr approached along the S shore of the bay, keeping S of Helvick Rk (ECM lt buoy) and other dangers to the N.

Mine Hd (lt) has two dangerous rks, The Rogue about 2½ca E and The Longship 1M SW. To the W is a subm'gd rk 100m SE of Ram Hd. ▶ *Here the W-going stream starts at HW Cobh +0230, and the E-going at HW Cobh –0215, sp rates 1·5kn.* ◀ Pass 2ca off Capel Island; the sound between it and Knockadoon Hd is not recommended.

The N side of Ballycotton B is foul up to 5ca offshore. Ballycotton Hbr is small and crowded, but usually there is sheltered ⚓ outside. Sound Rk and Small Is lie between the mainland and Ballycotton Is (lit). From Ballycotton to Cork keep at least 5ca off for dangers including The Smiths (PHM lt buoy) 1·5M WSW of Ballycotton Island. Pass between Hawk Rk, close off Power Hd, and Pollock Rk (PHM lt buoy) 1.25M SE.

9.12.15 KILMORE QUAY

Wexford 52°10'·25N 06°35'·15W ✿✿✿✿⚓⚓✿✿✿

CHARTS AC 2049, 2740, 5621; Imray C61, C57

TIDES –0535 Dover; ML No data; Duration 0605

Standard Port COBH (→)

Times				Height (metres)			
High Water		Low Water		MHWS	MHWN	MLWN	MLWS
0500	1100	0500	1100	4·1	3·2	1·3	0·4
1700	2300	1700	2300				
Differences BAGINBUN HEAD (5M NE of Hook Hd)							
+0003	+0003	–0008	–0008	–0·2	–0·1	+0·2	+0·2
GREAT SALTEE							
+0019	+0009	–0004	+0006	–0·3	–0·4	ND	ND
CARNSORE POINT							
+0029	+0019	–0002	+0008	–1·1	–1·0	ND	ND

SHELTER Excellent in marina (3m depth) with 0.8m below CD at LAT in ent. Hbr exposed to SE'lies. FVs berth on E and W piers close S of marina.

NAVIGATION WPT 52°09'·22N 06°35'·32W, SWM buoy (Apr-Sep), 007°/1·0M to pier hd lt, on ldg line. Great (57m) and Little (35m) Saltee Islands lie 3M and 1·7M to SSW and S, respectively, of hbr, separated by Saltee Sound. From the E, safest appr initially is via Saltee Sound, then N to WPT. Caution: In bad weather seas break on the Bohurs and The Bore, rks 2M E of Saltee Islands. Beware, particularly on E-going stream, Goose Rk (2·6m) close W of Little Saltee and Murroch's Rk (2·1m) 6ca NW of Little Saltee. St Patrick's Bridge, 650m E of the WPT, is a 300m wide E/W chan used by FVs and yachts, but carrying < 2m; care needed in strong SW winds when a standing wave can form. It is marked by a PHM buoy, Fl R 6s, and a SHM buoy, Fl G 6s, (laid Apr to mid-Sep); general direction of buoyage is E. From the W, appr is clear but keep at least 5ca off Forlorn Pt to avoid Forlorn Rk (1·5m).

LIGHTS AND MARKS See 9.12.4 and chartlet. Ldg lts/marks lead 008° to the hbr.

- From W, ldg lts are obsc'd by piers until S of hbr ent. Turn 90° port into hbr ent, just past ✿ QRG at W Quay Hd; do not overshoot into shoal water ahead. The R sectors of this ✿ QRG warn of Forlorn Rock and The Lings to W of the ldg line and shingle banks drying 0·6m close E of the ldg line.

A white-gabled church is conspic from afar, as are two 20m high flood lt pylons on the E quay.

COMMUNICATIONS (Code 053) MRCC (01) 6620922/3; Emergency/ Dr/Police 999; ⊖ 9133741. HM 9129955, mob 087 9001037, assistant. marineofficer@wexfordcoco.ie. VHF Ch 16, 09 (occas).

FACILITIES www.kilmorequaymarina.com or HM ☎(0800-2000); ⚓ ⏛ 20Ⓥ €2.50 (€25 min) Ⓓ ⚓ ♠(24/7) ⛽ ⚒ ⛽ cycle hire.
Village ⛽ ⚒ Ⓔ Ⓛ 🏧 📮 Ⓣ Ⓣ 3M away, ✉ 🚆 ✕ ⌹ Ⓗ(Wexford 15M).

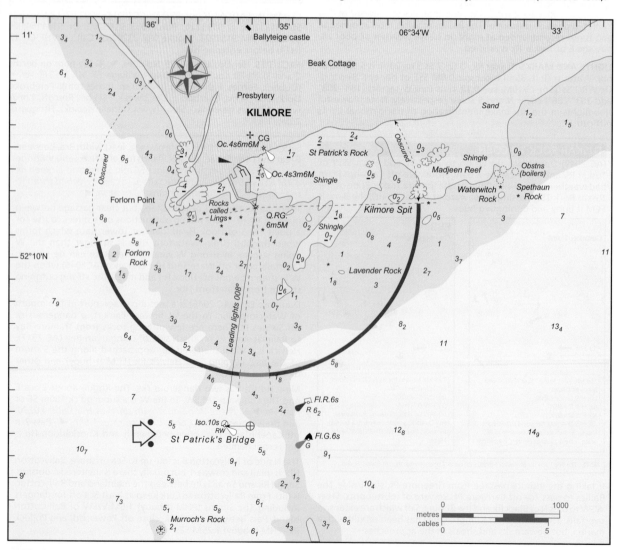

9.12.16 WATERFORD

Waterford 52°15'·50N 07°06'·00W ✳✳✳✳❀❀

CHARTS AC 2049, 2046, 5621, 5622; Imray C61, C57

TIDES –0520 Dover; ML 2·4; Duration 0605

Standard Port COBH (→)

Times				Height (metres)			
High Water		Low Water		MHWS	MHWN	MLWN	MLWS
0500	1100	0500	1100	4·1	3·2	1·3	0·4
1700	2300	1700	2300				
Differences WATERFORD							
+0053	+0032	+0015	+0100	+0·6	+0·6	+0·4	+0·2
CHEEKPOINT							
+0026	+0021	+0019	+0022	+0·5	+0·4	+0·3	+0·2
KILMOKEA POINT							
+0026	+0022	+0020	+0020	+0·2	+0·1	+0·1	+0·1
NEW ROSS							
+0100	+0030	+0055	+0130	+0·3	+0·4	+0·3	+0·4

SHELTER Very good on 4 long marina pontoons after Nanny pontoon (first one/private) on S bank, abeam cathedral spire. Beware strong tidal stream. Up the estuary are many excellent ⚓s: off S side of R Suir in King's Chan (only to be entered W of Little Is); and up the R Barrow near Marsh Pt (about 2M S of New Ross) and 0·5M S of New Ross fixed bridge.

NAVIGATION WPT 52°06'·50N 06°58'·50W, 002°/6·7M to Dir lt at Duncannon. From the E, keep clear of Brecaun reef (2M NE of Hook Hd). Keep about 1·5M S of Hook Hd to clear Tower Race and overfalls, especially HW Dover ±2. From the W beware Falskirt Rk (3m), 2ca off Swine Head and 2M WSW of Dunmore East (9.12.17). Marine Farms off Creadan Hd and Broomhill Pt; large ships go up to Waterford; beware ferry between Passage East and Ballyhack.

LIGHTS AND MARKS The estuary and R Suir are very well buoyed/lit all the way to Waterford. The estuary ent is between Dunmore East and Hook Hd. Duncannon dir lt, F WRG, leads 002° into the river. R Barrow is also well lit/marked up to New Ross. The railway swing bridge (now disused) at the river ent still opens at its W end. Call Bridge-keeper ☎086 816 7826 (no VHF).

COMMUNICATIONS (Code 051) MRCC (01) 6620922; ⊜ 832090; Police 305300; Dr 855411; ℍ848000.; HMs Waterford 301400, Mob 08725 98297 or 08722 24961; New Ross 421303.

Waterford and New Ross VHF Ch 14 16. No VHF for marina.

FACILITIES Marina ☎309900 all tide access for 1·5m draught. Access/berth, call mob 087 2384944; 100⌷ €16 (€80/wk)<7.6m>€20 (€95/wk)<10.7m>€30 (€130/wk)<13.7m. ⚓ ⌁ €2·50/10Kw, ⛽ ▭ at HM's office, ⛽(from Dunmore East by Coast2Coast ☎382797, mob 0872512170) ✕ ⚒ ⊞ ⊠ ℝ(Ballyhack ☎389164).

New Ross Marina newrossmarina@wexfordcoco.ie mob 086 388 9652; ⚓ ⌁ ⚓ 66⌷ €10 (€50/wk)<9.5m>€20 (€100/wk) < 10.5m>€30 (€120/wk)<13.5m; ⛽ (as above) ▮▭.

Waterford Boatyard (R Suir), ☎857755 or 086 1727502 for bridge lift on approach. Waterford CC ☎076 1102020 for Rice Bridge lift.

City all facilities, bus to Rosslare for UK ferries, ✈ (Waterford) for UK/Continent.

9.12.17 DUNMORE EAST

Waterford 52°08'·95N 06°59'·37W ✿✿✿✿◊◊✿✿

CHARTS AC 2049, 2046, 5621, 5622; Imray C61, C57

TIDES –0535 Dover; ML 2·4; Duration 0605

Standard Port COBH (⟶)

Times				Height (metres)			
High Water		Low Water		MHWS	MHWN	MLWN	MLWS
0500	1100	0500	1100	4·1	3·2	1·3	0·4
1700	2300	1700	2300				
Differences DUNMORE EAST							
+0008	+0003	0000	0000	+0·1	0·0	+0·1	+0·2
DUNGARVAN HARBOUR							
+0004	+0012	+0007	–0001	0·0	+0·1	–0·2	0·0

SHELTER Very good in harbour, but moorings open to the E. ⚓ N of the hbr. A busy FV hbr which may be a port of refuge for leisure craft. Large cruise liners ⚓ off. Waterford Harbour SC administer moorings. In bad weather berth alongside FVs at W Wharf, clear of ice plant, in at least 2m or go up R Suir to Waterford (9.12.16).

NAVIGATION WPT (see also 9.12.16) 52°08'·00N, 06°58'·00W, 317°/1·2M to bkwtr lt. Enter under power. From E, stay 1·5M off Hook Hd to clear Tower Race; then steer for hbr in R sector of E pier lt ho. In calm weather Hook Hd can be rounded 1ca off. From W, beware Falskirt Rk (off Swines Hd, 2M WSW) dries 3·0m. By night track E for Hook Hd until in R sector of E pier lt, then alter to N.

LIGHTS AND MARKS Lts as chartlet 2 lit SHM in appr see 9.12.4.

COMMUNICATIONS (Code 051) MRCC (01) 6620922/3; Duty CG (may be used for emergencies) 0868 501764; Police 383112; ⊖ 832090; Dr 855411. HM 383166, mobile 0877 931705; Waterford Hbr SC 383230. Tugs: Fastnet shipping 051 832946, South East Tug Services 051 852819.

VHF Ch **14** 16 (Pilot Station ☎051 383119).

FACILITIES Hbr ⚓ (piers) ⛴(Coast2Coast ☎051 382797, mob 0872512170) ⚓ scrubbing grid, ⛽ ⚓(220t).
Waterford Hbr SC ⚓ if available, ⚑ ✕ ⚓; visitors welcome.
Village ⛽ ⛽ Kos ✉ Ⓑ ⚓ ✕ ⚓ ✈ (Waterford).

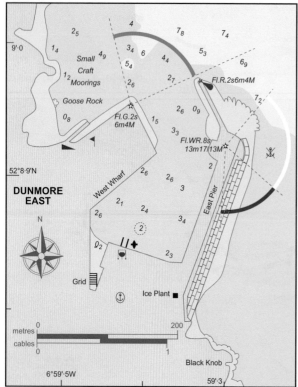

9.12.18 YOUGHAL

Cork 51°56'·54N 07°50'·20W ✿✿✿✿◊◊✿✿

CHARTS AC 2049, 2071, 5622; Imray C57

TIDES –0556 Dover; ML 2·1; Duration 0555

Standard Port COBH (⟶)

Times				Height (metres)			
High Water		Low Water		MHWS	MHWN	MLWN	MLWS
0500	1100	0500	1100	4·1	3·2	1·3	0·4
1700	2300	1700	2300				
Differences YOUGHAL							
0000	+0010	+0010	0000	–0·2	–0·1	–0·1	–0·1

SHELTER Good, but strong S'lies cause swell inside the hbr. Possible ⚓ in 1.3m (drying patches just off quay); seasonal short-stay pontoon, with ⚓ off landing jetty; 2 ⚓s (see chartlet), €20/night, owned by Aquatrek (call on Ch 69 or ☎086 8593482). ⚓ as chartlet; no dues. Strong tidal streams run throughout ⚓orages.

NAVIGATION WPT, East Bar, 51°55'·62N 07°48'·00W, 302°/1·8M to Fl WR 2·5s lt. Beware Blackball Ledge (ECM lt buoy) and Bar Rks (SCM lt buoy), both outside hbr ent in R sector of lt ho. From W, appr via West Bar (1·7m) is shorter route and E Bar has 2·0m.

- Both Bars in E to SSW'lies >F6 are likely to have dangerous seas.
- Beware salmon nets set during June-July, Mon-Thurs 0400-2100.
- Red Bank is continually changing, obtain local knowledge.

LIGHTS AND MARKS See 9.12.4 and chartlet. Water tr is conspic from seaward; clock tr and ✠ tr within hbr. Upriver, 175° transit of convent belfry tr/town hall clears W of Red Bank.

COMMUNICATIONS (Code 024) MRCC (066) 9476109; Coast Guard 93252 or mob 0868 501769; ⊖ (021) 4325000; Police 92200; Dr 92702. HM Mobile 0872511143; Youghal Shipping 92577 (for poss ⚓).
VHF Ch 14 16 0900-1700 and when ships expected.

FACILITIES Services ⚓ ⚓ ⚓ ⚓ (☎Youghal Shipping).
Town ⛽ ⛽ ⛽ ✉ Ⓑ ⚓ ✕ ⚓ Bus (Cork/W'ford) ✈(Cork). Wkly ferry to Roscoff www.brittanyferries.ie.

STANDARD TIME (UT)
For Summer Time add ONE hour in **non-shaded areas**

COBH   LAT 51°51'N   LONG 8°18'W
TIMES AND HEIGHTS OF HIGH AND LOW WATERS

Dates in red are **SPRINGS**
Dates in blue are NEAPS

YEAR 2016

S Ireland

JANUARY

Time	m		Time	m
1 0357	1.1	**16**	0355	0.8
0948	3.6		0954	3.9
F 1625	1.2	SA 1619		0.9
2203	3.5	◑ 2213		3.8
2 0447	1.2	**17**	0450	0.9
1036	3.5		1049	3.8
SA 1717	1.4	SU 1717		1.0
◔ 2255	3.4	2312		3.6
3 0543	1.3	**18**	0554	1.0
1132	3.4		1152	3.6
SU 1817	1.4	M 1825		1.1
2358	3.4			
4 0646	1.4	**19**	0020	3.5
1238	3.4		0707	1.1
M 1920	1.4	TU 1304		3.6
		1941		1.1
5 0108	3.4	**20**	0137	3.5
0747	1.3		0824	1.0
TU 1346	3.5	W 1417		3.6
2019	1.3	2056		1.0
6 0215	3.5	**21**	0252	3.7
0846	1.2		0936	0.9
W 1446	3.6	TH 1524		3.7
2115	1.1	2201		0.8
7 0312	3.7	**22**	0356	3.9
0940	1.1		1035	0.7
TH 1539	3.7	F 1621		3.9
2206	1.0	2255		0.6
8 0403	3.9	**23**	0448	4.0
1029	0.9		1124	0.5
F 1625	3.9	SA 1709		4.0
2251	0.8	2340		0.5
9 0448	4.0	**24**	0533	4.2
1113	0.7		1207	0.5
SA 1707	4.0	SU 1751		4.1
2332	0.6	○		
10 0530	4.2	**25**	0020	0.4
1154	0.6		0613	4.2
SU 1746	4.1	M 1245		0.5
●		1828		4.1
11 0012	0.5	**26**	0056	0.5
0610	4.2		0649	4.2
M 1234	0.6	TU 1320		0.6
1825	4.1	1903		4.1
12 0052	0.5	**27**	0130	0.5
0650	4.2		0724	4.1
TU 1314	0.5	W 1353		0.7
1905	4.1	1935		4.0
13 0134	0.5	**28**	0204	0.7
0732	4.2		0758	4.0
W 1357	0.6	TH 1426		0.8
1948	4.1	2009		3.9
14 0218	0.5	**29**	0238	0.8
0817	4.2		0832	3.9
TH 1441	0.6	F 1501		0.9
2033	4.0	2044		3.8
15 0305	0.6	**30**	0314	0.9
0904	4.1		0909	3.8
F 1528	0.7	SA 1538		1.1
2121	3.9	2123		3.7
		31	0355	1.1
			0949	3.6
		SU 1621		1.2
		2207		3.5

FEBRUARY

Time	m		Time	m
1 0444	1.3	**16**	0525	0.9
1035	3.5		1123	3.5
M 1715	1.4	TU 1754		1.1
◑ 2300	3.4	2351		3.4
2 0545	1.4	**17**	0638	1.1
1131	3.4		1237	3.3
TU 1821	1.4	W 1913		1.2
3 0005	3.3	**18**	0114	3.3
0654	1.4		0804	1.1
W 1242	3.3	TH 1359		3.4
1932	1.4	2039		1.1
4 0121	3.3	**19**	0237	3.5
0803	1.3		0924	0.9
TH 1359	3.4	F 1510		3.5
2038	1.3	2151		0.9
5 0234	3.5	**20**	0342	3.7
0907	1.1		1024	0.7
F 1506	3.5	SA 1607		3.7
2137	1.0	2245		0.6
6 0334	3.7	**21**	0434	3.9
1004	0.9		1112	0.5
SA 1601	3.8	SU 1654		3.9
2228	0.8	2328		0.5
7 0425	3.9	**22**	0516	4.1
1053	0.7		1152	0.4
SU 1647	3.9	M 1735		4.0
2313	0.5	○		
8 0510	4.1	**23**	0004	0.4
1137	0.5		0554	4.1
M 1730	4.1	TU 1226		0.4
● 2355	0.4	1809		4.1
9 0552	4.3	**24**	0036	0.4
1218	0.3		0627	4.1
TU 1810	4.2	W 1257		0.5
		1841		4.0
10 0035	0.3	**25**	0105	0.5
0634	4.3		0659	4.1
W 1259	0.3	TH 1325		0.6
1850	4.2	1910		4.0
11 0117	0.2	**26**	0133	0.6
0716	4.3		0729	4.0
TH 1340	0.3	F 1354		0.7
1932	4.2	1940		3.9
12 0200	0.3	**27**	0203	0.7
0759	4.2		0759	3.9
F 1423	0.4	SA 1424		0.8
2015	4.1	2013		3.9
13 0246	0.4	**28**	0236	0.8
0844	4.1		0832	3.8
SA 1508	0.5	SU 1458		0.9
2100	4.0	2048		3.8
14 0333	0.5	**29**	0313	1.0
0931	3.9		0908	3.7
SU 1555	0.7	M 1536		1.1
2149	3.8	2128		3.6
15 0425	0.7			
1022	3.7			
M 1649	0.9			
◑ 2243	3.6			

MARCH

Time	m		Time	m
1 0357	1.1	**16**	0503	0.9
0951	3.5		1058	3.4
TU 1623	1.3	W 1730		1.0
◑ 2216	3.5	2327		3.3
2 0453	1.3	**17**	0615	1.1
1043	3.4		1214	3.2
W 1726	1.4	TH 1850		1.1
2316	3.3			
3 0605	1.4	**18**	0055	3.2
1149	3.2		0743	1.1
TH 1844	1.4	F 1339		3.2
		2021		1.1
4 0031	3.3	**19**	0219	3.3
0721	1.3		0905	0.9
F 1310	3.2	SA 1451		3.4
1959	1.3	2134		0.8
5 0153	3.4	**20**	0322	3.6
0832	1.1		1005	0.7
SA 1429	3.4	SU 1547		3.6
2104	1.0	2227		0.6
6 0302	3.6	**21**	0412	3.8
0934	0.9		1051	0.5
SU 1532	3.6	M 1633		3.8
2201	0.7	2309		0.5
7 0357	3.9	**22**	0453	4.0
1028	0.6		1130	0.4
M 1623	3.9	TU 1713		4.0
2249	0.4	2343		0.4
8 0445	4.1	**23**	0530	4.0
1114	0.3		1202	0.4
TU 1708	4.1	W 1747		4.0
2334	0.2	○		
9 0530	4.3	**24**	0012	0.4
1158	0.1		0603	4.0
W 1750	4.2	TH 1231		0.5
●		1817		4.0
10 0016	0.1	**25**	0038	0.5
0612	4.3		0632	4.0
TH 1240	0.1	F 1256		0.5
1831	4.3	1845		4.0
11 0059	0.0	**26**	0103	0.6
0655	4.3		0700	3.9
F 1322	0.1	SA 1323		0.6
1913	4.3	1913		3.9
12 0142	0.1	**27**	0132	0.6
0738	4.3		0728	3.9
SA 1404	0.1	SU 1353		0.7
1956	4.2	1944		3.9
13 0227	0.2	**28**	0205	0.8
0823	4.1		0800	3.8
SU 1449	0.3	M 1426		0.8
2040	4.0	2018		3.8
14 0314	0.4	**29**	0242	0.9
0909	3.9		0835	3.7
M 1536	0.5	TU 1503		1.0
2127	3.8	2057		3.7
15 0405	0.6	**30**	0325	1.0
0959	3.6		0917	3.6
TU 1628	0.8	W 1548		1.2
◑ 2220	3.6	2143		3.5
		31	0417	1.2
			1008	3.4
		TH 1647		1.3
		◔ 2240		3.4

APRIL

Time	m		Time	m
1 0525	1.3	**16**	0031	3.2
1113	3.2		0716	1.0
F 1803	1.4	SA 1312		3.1
2352	3.3	1951		1.0
2 0643	1.3	**17**	0150	3.3
1230	3.2		0833	0.9
SA 1921	1.2	SU 1422		3.3
		2102		0.9
3 0113	3.4	**18**	0251	3.5
0756	1.1		0932	0.8
SU 1350	3.3	M 1517		3.5
2029	1.0	2156		0.7
4 0226	3.6	**19**	0341	3.7
0901	0.8		1020	0.6
M 1457	3.6	TU 1604		3.7
2129	0.7	2239		0.6
5 0325	3.8	**20**	0424	3.8
0958	0.5		1100	0.5
TU 1553	3.9	W 1644		3.8
2222	0.4	2314		0.5
6 0417	4.1	**21**	0502	3.9
1048	0.2		1133	0.5
W 1642	4.1	TH 1719		3.9
2310	0.1	2343		0.5
7 0504	4.2	**22**	0535	3.9
1135	0.1		1201	0.5
TH 1727	4.2	F 1751		3.9
● 2355	0.0	○		
8 0550	4.3	**23**	0009	0.5
1219	0.0		0605	3.9
F 1811	4.3	SA 1228		0.6
		1819		3.9
9 0040	-0.1	**24**	0036	0.6
0634	4.3		0633	3.9
SA 1303	0.0	SU 1256		0.6
1854	4.3	1849		3.9
10 0125	0.0	**25**	0106	0.7
0719	4.2		0702	3.9
SU 1348	0.1	M 1328		0.7
1938	4.2	1920		3.9
11 0211	0.1	**26**	0141	0.7
0804	4.1		0734	3.8
M 1433	0.2	TU 1403		0.8
2023	4.0	1955		3.8
12 0259	0.3	**27**	0220	0.8
0850	3.9		0812	3.7
TU 1521	0.4	W 1442		0.9
2110	3.8	2035		3.7
13 0349	0.5	**28**	0304	1.0
0940	3.6		0855	3.6
W 1613	0.7	TH 1528		1.0
2203	3.5	2121		3.6
14 0446	0.8	**29**	0355	1.1
1038	3.3		0946	3.5
TH 1713	0.9	F 1624		1.2
◑ 2308	3.3	2217		3.5
15 0555	1.0	**30**	0457	1.2
1150	3.1		1047	3.4
F 1828	1.1	SA 1732		1.2
		◑ 2324		3.4

Chart Datum: 0·13 metres above Ordnance Datum (Dublin). HAT is 4·5 metres above Chart Datum.

》》 FREE monthly updates. Register at 《
www.reedsnauticalalmanac.co.uk

585

COBH LAT 51°51′N LONG 8°18′W
TIMES AND HEIGHTS OF HIGH AND LOW WATERS

STANDARD TIME (UT)
For Summer Time add ONE hour in **non-shaded areas**

Dates in red are SPRINGS
Dates in blue are NEAPS

YEAR 2016

MAY

Time	m		Time	m
1 SU 0609	1.2	**16** M 0108	3.3
1159	3.3	0747	1.0
1846	1.1	1340	3.3
		2015	1.0
2 M 0039	3.5	**17** TU 0210	3.4
0721	1.0	0846	0.9
1314	3.4	1437	3.4
1955	0.9	2111	0.9
3 TU 0151	3.6	**18** W 0302	3.5
0827	0.8	0937	0.8
1422	3.6	1526	3.6
2057	0.7	2158	0.8
4 W 0253	3.8	**19** TH 0347	3.7
0927	0.5	1021	0.7
1521	3.9	1609	3.7
2154	0.4	2237	0.7
5 TH 0348	4.0	**20** F 0429	3.8
1022	0.3	1058	0.7
1615	4.1	1648	3.8
2247	0.2	2310	0.6
6 F 0440	4.2	**21** SA 0506	3.8
1112	0.1	1131	0.6
1704	4.2	1723	3.9
● 2336	0.0	○ 2341	0.6
7 SA 0529	4.3	**22** SU 0539	3.9
1200	0.0	1202	0.6
1752	4.3	1756	3.9
8 SU 0023	0.0	**23** M 0012	0.6
0615	4.3	0609	3.9
1247	0.0	1234	0.6
1837	4.3	1828	3.9
9 M 0110	0.0	**24** TU 0046	0.7
0701	4.2	0641	3.9
1333	0.1	1309	0.7
1922	4.2	1902	3.9
10 TU 0157	0.1	**25** W 0123	0.7
0747	4.0	0716	3.8
1419	0.2	1347	0.7
2007	4.0	1939	3.9
11 W 0245	0.3	**26** TH 0204	0.8
0834	3.8	0756	3.8
1507	0.4	1428	0.8
2055	3.8	2020	3.8
12 TH 0335	0.5	**27** F 0249	0.9
0922	3.6	0840	3.7
1558	0.7	1515	0.9
2146	3.6	2107	3.8
13 F 0429	0.8	**28** SA 0339	0.9
1016	3.4	0931	3.6
1655	0.9	1607	1.0
◐ 2245	3.4	2200	3.7
14 SA 0530	1.0	**29** SU 0435	1.0
1119	3.2	1028	3.5
1759	1.0	1708	1.0
2355	3.3	◐ 2302	3.6
15 SU 0639	1.0	**30** M 0540	1.0
1231	3.2	1132	3.5
1910	1.0	1815	1.0
		31 TU 0009	3.6
		0648	1.0
		1241	3.5
		1923	0.9

JUNE

Time	m		Time	m
1 W 0119	3.7	**16** TH 0216	3.4
0755	0.8	0846	1.0
1349	3.7	1441	3.5
2028	0.7	2109	0.9
2 TH 0223	3.8	**17** F 0307	3.5
0858	0.6	0936	0.9
1451	3.8	1530	3.6
2129	0.5	2155	0.9
3 F 0322	4.0	**18** SA 0354	3.7
0958	0.4	1021	0.8
1550	4.0	1615	3.7
2226	0.3	2237	0.8
4 SA 0419	4.1	**19** SU 0436	3.8
1053	0.3	1102	0.7
1645	4.1	1656	3.9
2319	0.2	2315	0.7
5 SU 0511	4.2	**20** M 0515	3.8
1144	0.2	1139	0.7
1735	4.2	1734	3.9
●		○ 2352	0.6
6 M 0009	0.1	**21** TU 0550	3.9
0559	4.2	1215	0.6
1232	0.1	1811	4.0
1822	4.2		
7 TU 0056	0.1	**22** W 0029	0.6
0646	4.1	0625	3.9
1319	0.2	1253	0.6
1907	4.1	1847	4.0
8 W 0143	0.2	**23** TH 0108	0.6
0731	4.0	0703	3.9
1404	0.3	1332	0.6
1952	4.0	1926	4.0
9 TH 0229	0.4	**24** F 0150	0.7
0816	3.9	0743	3.9
1451	0.5	1415	0.7
2037	3.9	2008	3.9
10 F 0316	0.6	**25** SA 0234	0.7
0901	3.7	0828	3.8
1538	0.6	1500	0.7
2123	3.7	2054	3.9
11 SA 0405	0.8	**26** SU 0322	0.8
0948	3.5	0916	3.7
1628	0.8	1549	0.8
2213	3.5	2144	3.8
12 SU 0457	0.9	**27** M 0413	0.8
1039	3.4	1009	3.6
1722	1.0	1643	0.8
◐ 2310	3.4	◐ 2239	3.7
13 M 0554	1.1	**28** TU 0511	0.9
1138	3.3	1107	3.6
1821	1.1	1745	0.9
		2341	3.7
14 TU 0014	3.3	**29** W 0616	0.9
0654	1.1	1211	3.6
1244	3.3	1852	0.9
1921	1.1		
15 W 0118	3.3	**30** TH 0048	3.7
0752	1.1	0725	0.9
1346	3.3	1319	3.6
2017	1.0	2001	0.8

JULY

Time	m		Time	m
1 F 0156	3.7	**16** SA 0224	3.4
0833	0.8	0854	1.1
1427	3.7	1452	3.5
2108	0.7	2116	1.0
2 SA 0301	3.8	**17** SU 0319	3.5
0938	0.6	0947	0.9
1531	3.9	1544	3.7
2210	0.5	2207	0.9
3 SU 0402	3.9	**18** M 0408	3.7
1037	0.5	1035	0.8
1630	4.0	1631	3.8
2306	0.4	2252	0.7
4 M 0456	4.0	**19** TU 0452	3.8
1130	0.3	1117	0.7
1722	4.1	1713	3.9
● 2356	0.3	○ 2333	0.6
5 TU 0546	4.1	**20** W 0532	3.9
1218	0.3	1156	0.5
1809	4.2	1752	4.0
6 W 0042	0.3	**21** TH 0012	0.5
0630	4.1	0609	4.0
1303	0.3	1235	0.5
1851	4.1	1831	4.1
7 TH 0125	0.3	**22** F 0052	0.5
0713	4.0	0648	4.0
1345	0.3	1315	0.5
1933	4.0	1910	4.1
8 F 0208	0.5	**23** SA 0133	0.5
0753	3.9	0728	4.0
1427	0.5	1357	0.5
2013	3.9	1952	4.1
9 SA 0250	0.6	**24** SU 0216	0.5
0833	3.8	0812	3.9
1509	0.6	1441	0.5
2054	3.8	2036	4.0
10 SU 0332	0.8	**25** M 0302	0.6
0914	3.6	0858	3.9
1551	0.8	1528	0.6
2136	3.6	2124	3.9
11 M 0416	0.9	**26** TU 0350	0.7
0956	3.5	0947	3.8
1636	1.0	1618	0.7
2221	3.5	◐ 2216	3.8
12 TU 0504	1.1	**27** W 0444	0.8
1044	3.4	1041	3.7
1726	1.1	1716	0.8
◐ 2312	3.4	2314	3.7
13 W 0558	1.2	**28** TH 0545	0.9
1140	3.3	1143	3.6
1823	1.2	1823	0.9
14 TH 0013	3.3	**29** F 0021	3.5
0657	1.2	0657	1.0
1246	3.3	1255	3.5
1922	1.2	1937	0.9
15 F 0121	3.3	**30** SA 0135	3.5
0757	1.2	0813	0.9
1352	3.4	1410	3.6
2021	1.1	2052	0.8
		31 SU 0246	3.6
		0925	0.8
		1520	3.7
		2200	0.7

AUGUST

Time	m		Time	m
1 M 0349	3.8	**16** TU 0341	3.6
1027	0.6	1007	0.8
1620	3.9	1605	3.8
2256	0.5	2228	0.7
2 TU 0444	3.9	**17** W 0428	3.8
1119	0.4	1053	0.6
1710	4.1	1650	4.0
● 2343	0.4	2311	0.5
3 W 0531	4.0	**18** TH 0510	4.0
1203	0.3	1134	0.4
1753	4.1	1731	4.1
		○ 2352	0.4
4 TH 0025	0.3	**19** F 0549	4.0
0612	4.1	1213	0.3
1244	0.3	1811	4.2
1832	4.1		
5 F 0104	0.3	**20** SA 0032	0.3
0651	4.1	0628	4.1
1321	0.4	1254	0.3
1909	4.1	1850	4.2
6 SA 0141	0.5	**21** SU 0113	0.3
0727	3.9	0709	4.1
1357	0.5	1336	0.3
1944	4.0	1932	4.2
7 SU 0216	0.6	**22** M 0155	0.4
0802	3.8	0752	4.0
1432	0.6	1419	0.3
2020	3.8	2015	4.1
8 M 0252	0.8	**23** TU 0240	0.4
0837	3.7	0836	4.0
1508	0.8	1505	0.5
2056	3.7	2102	4.0
9 TU 0329	0.9	**24** W 0327	0.6
0915	3.6	0924	3.8
1546	1.0	1555	0.6
2135	3.6	2152	3.8
10 W 0410	1.1	**25** TH 0419	0.8
0957	3.5	1017	3.7
1629	1.1	1651	0.8
◐ 2218	3.5	◐ 2249	3.6
11 TH 0459	1.2	**26** F 0520	1.0
1046	3.4	1119	3.5
1723	1.3	1758	1.0
2310	3.3	2358	3.4
12 F 0601	1.3	**27** SA 0635	1.1
1146	3.3	1237	3.4
1829	1.3	1919	1.1
13 SA 0017	3.2	**28** SU 0119	3.4
0709	1.3	0759	1.0
1300	3.2	1401	3.4
1937	1.3	2044	1.0
14 SU 0135	3.2	**29** M 0236	3.5
0815	1.2	0917	0.8
1414	3.4	1512	3.6
2041	1.1	2152	0.7
15 M 0245	3.4	**30** TU 0338	3.7
0915	1.0	1018	0.6
1515	3.6	1608	3.9
2138	0.9	2245	0.5
		31 W 0430	3.9
		1106	0.4
		1655	4.0
		2329	0.4

Chart Datum: 0·13 metres above Ordnance Datum (Dublin). HAT is 4·5 metres above Chart Datum.

STANDARD TIME (UT)
For Summer Time add ONE hour in **non-shaded areas**

COBH LAT 51°51'N LONG 8°18'W
TIMES AND HEIGHTS OF HIGH AND LOW WATERS

Dates in **red** are **SPRINGS**
Dates in blue are NEAPS

YEAR 2016

S Ireland

SEPTEMBER		OCTOBER		NOVEMBER		DECEMBER	
Time m	Time m	Time m	Time m	Time m	Time m	Time m	Time m
1 0514 4.0 1147 0.4 TH 1735 4.1 ●	**16** 0443 4.0 1109 0.3 F 1705 4.2 ○ 2329 0.3	**1** 0529 4.1 1158 0.5 SA 1746 4.1 ●	**16** 0500 4.3 1129 0.2 SU 1723 4.4 ○ 2349 0.2	**1** 0010 0.7 0605 4.0 TU 1223 0.7 1817 4.0	**16** 0018 0.2 0612 4.4 W 1245 0.2 1835 4.3	**1** 0016 0.8 0615 4.1 TH 1232 0.8 1825 4.0	**16** 0053 0.3 0647 4.4 F 1321 0.3 1908 4.2
2 0006 0.4 0552 4.1 F 1222 0.4 1810 4.1	**17** 0525 4.2 1151 0.2 SA 1747 4.3	**2** 0012 0.5 0601 4.1 SU 1225 0.5 1816 4.1	**17** 0544 4.3 1213 0.1 M 1807 4.4	**2** 0037 0.8 0634 4.0 W 1251 0.8 1845 4.0	**17** 0104 0.2 0659 4.3 TH 1332 0.3 1921 4.2	**2** 0048 0.8 0648 4.0 F 1307 0.9 1857 4.0	**17** 0139 0.4 0733 4.2 SA 1408 0.5 1953 4.1
3 0040 0.4 0627 4.0 SA 1254 0.4 1843 4.1	**18** 0010 0.2 0606 4.2 SU 1232 0.1 1828 4.3	**3** 0039 0.6 0630 4.0 M 1251 0.6 1844 4.0	**18** 0033 0.1 0628 4.3 TU 1258 0.1 1851 4.3	**3** 0107 0.8 0705 4.0 TH 1323 0.9 1915 3.9	**18** 0152 0.4 0746 4.2 F 1420 0.5 2008 4.0	**3** 0124 0.9 0723 4.0 SA 1345 1.0 1933 3.9	**18** 0226 0.5 0819 4.1 SU 1455 0.6 2038 3.9
4 0110 0.5 0658 4.0 SU 1323 0.5 1914 4.0	**19** 0052 0.2 0648 4.2 M 1315 0.2 1910 4.3	**4** 0105 0.7 0659 3.9 TU 1318 0.7 1912 3.9	**19** 0118 0.2 0713 4.3 W 1344 0.2 1936 4.2	**4** 0141 0.9 0739 3.9 F 1400 1.0 1950 3.9	**19** 0241 0.5 0835 4.0 SA 1511 0.7 2057 3.8	**4** 0204 0.9 0802 3.9 SU 1426 1.1 2014 3.8	**19** 0314 0.7 0906 3.9 M 1543 0.8 2124 3.7
5 0140 0.7 0729 3.9 M 1353 0.7 1945 3.9	**20** 0135 0.2 0731 4.2 TU 1359 0.2 1954 4.2	**5** 0134 0.8 0730 3.9 W 1348 0.6 1942 3.9	**20** 0204 0.3 0759 4.1 TH 1432 0.4 2023 4.0	**5** 0219 1.1 0818 3.8 SA 1441 1.2 2031 3.7	**20** 0333 0.8 0927 3.8 SU 1604 0.9 2150 3.6	**5** 0248 1.1 0846 3.9 M 1513 1.2 2101 3.7	**20** 0404 0.9 0956 3.7 TU 1634 1.0 2214 3.5
6 0210 0.8 0801 3.8 TU 1424 0.8 2017 3.8	**21** 0220 0.4 0816 4.0 W 1446 0.4 2040 4.0	**6** 0206 0.9 0804 3.8 TH 1423 1.0 2017 3.8	**21** 0253 0.5 0848 3.9 F 1522 0.7 2113 3.8	**6** 0305 1.2 0903 3.7 SU 1530 1.3 2119 3.6	**21** 0430 1.0 1026 3.6 M 1705 1.1 ◑ 2252 3.4	**6** 0339 1.2 0936 3.8 TU 1606 1.3 2155 3.6	**21** 0458 1.1 1050 3.5 W 1731 1.2 ◑ 2311 3.4
7 0244 0.9 0837 3.7 W 1459 1.0 2052 3.7	**22** 0308 0.5 0904 3.9 TH 1536 0.6 2130 3.8	**7** 0244 1.1 0842 3.7 F 1504 1.2 2057 3.6	**22** 0347 0.8 0942 3.7 SA 1619 0.9 ◑ 2210 3.5	**7** 0359 1.3 0956 3.6 M 1629 1.4 ◑ 2218 3.5	**22** 0536 1.1 1133 3.4 TU 1815 1.2	**7** 0437 1.2 1034 3.7 W 1707 1.3 ◑ 2257 3.6	**22** 0558 1.2 1151 3.4 TH 1833 1.3
8 0321 1.1 0916 3.6 TH 1540 1.1 2133 3.5	**23** 0401 0.8 0957 3.6 F 1632 0.9 ◑ 2228 3.5	**8** 0329 1.2 0927 3.5 SA 1554 1.3 2145 3.5	**23** 0448 1.0 1046 3.5 SU 1726 1.1 2320 3.3	**8** 0506 1.4 1100 3.5 TU 1739 1.4 2327 3.4	**23** 0004 3.3 0649 1.2 W 1247 3.4 1927 1.2	**8** 0543 1.3 1138 3.7 TH 1815 1.3	**23** 0017 3.3 0702 1.2 F 1258 3.4 1935 1.3
9 0407 1.2 1002 3.4 F 1631 1.3 ◑ 2222 3.4	**24** 0502 1.0 1102 3.4 SA 1740 1.1 2340 3.3	**9** 0426 1.4 1022 3.4 SU 1658 1.4 ◑ 2246 3.3	**24** 0602 1.1 1206 3.3 M 1848 1.2	**9** 0619 1.4 1212 3.5 W 1853 1.3	**24** 0117 3.4 0759 1.1 TH 1352 3.5 2030 1.1	**9** 0005 3.6 0652 1.2 F 1247 3.7 1924 1.1	**24** 0126 3.4 0803 1.2 SA 1359 3.5 2033 1.2
10 0507 1.4 1059 3.3 SA 1739 1.4 2324 3.2	**25** 0619 1.1 1225 3.3 SU 1906 1.1	**10** 0540 1.5 1131 3.3 M 1815 1.5	**25** 0042 3.3 0728 1.1 TU 1327 3.4 2009 1.1	**10** 0042 3.5 0729 1.2 TH 1325 3.7 2000 1.1	**25** 0218 3.5 0857 1.0 F 1446 3.7 2122 1.0	**10** 0115 3.7 0759 1.0 SA 1353 3.9 2029 0.9	**25** 0226 3.5 0859 1.1 SU 1453 3.6 2125 1.1
11 0623 1.4 1211 3.2 SU 1855 1.4	**26** 0105 3.3 0749 1.1 M 1351 3.4 2033 1.0	**11** 0001 3.3 0657 1.4 TU 1251 3.4 1930 1.3	**26** 0158 3.4 0841 1.0 W 1432 3.6 2111 0.9	**11** 0153 3.7 0832 0.9 F 1427 3.9 2059 0.8	**26** 0309 3.7 0945 0.9 SA 1532 3.8 2206 0.9	**11** 0220 3.9 0902 0.8 SU 1454 4.0 2130 0.7	**26** 0318 3.7 0949 1.0 M 1542 3.8 2211 0.9
12 0043 3.2 0737 1.3 M 1334 3.3 2006 1.2	**27** 0222 3.4 0906 0.9 TU 1458 3.6 2138 0.8	**12** 0123 3.4 0806 1.2 W 1405 3.6 2034 1.0	**27** 0256 3.6 0936 0.8 TH 1523 3.8 2200 0.8	**12** 0253 3.9 0928 0.7 SA 1522 4.1 2154 0.6	**27** 0354 3.9 1026 0.8 SU 1614 3.9 2244 0.8	**12** 0321 4.1 1001 0.6 M 1551 4.2 2225 0.5	**27** 0405 3.8 1032 0.9 TU 1626 3.9 2251 0.9
13 0206 3.4 0842 1.1 TU 1443 3.5 2107 1.0	**28** 0322 3.7 1003 0.7 W 1550 3.8 2227 0.6	**13** 0232 3.6 0905 0.9 TH 1503 3.9 2130 0.7	**28** 0344 3.8 1021 0.7 F 1606 3.9 2241 0.7	**13** 0347 4.1 1021 0.4 SU 1613 4.3 2244 0.4	**28** 0434 4.0 1101 0.8 M 1652 4.0 2317 0.8	**13** 0418 4.2 1056 0.4 TU 1645 4.3 2317 0.4	**28** 0447 4.0 1110 0.9 W 1705 4.0 2327 0.8
14 0309 3.6 0937 0.8 W 1536 3.8 2200 0.7	**29** 0410 3.9 1048 0.5 TH 1634 4.0 2308 0.5	**14** 0327 3.9 0957 0.6 F 1553 4.1 2220 0.5	**29** 0426 4.0 1059 0.6 SA 1645 4.0 2315 0.6	**14** 0437 4.3 1111 0.3 M 1702 4.4 ○ 2332 0.2	**29** 0510 4.0 1132 0.8 TU 1726 4.0 ● 2346 0.8	**14** 0511 4.3 1146 0.4 W 1735 4.3 ○	**29** 0525 4.1 1144 0.8 TH 1740 4.0 ●
15 0359 3.8 1026 0.6 TH 1623 4.0 2246 0.5	**30** 0452 4.0 1126 0.4 F 1712 4.1 2343 0.5	**15** 0415 4.1 1044 0.3 SA 1639 4.3 2305 0.3	**30** 0503 4.0 1131 0.6 SU 1719 4.1 ● 2344 0.6	**15** 0526 4.4 1158 0.2 TU 1749 4.4	**30** 0544 4.1 1201 0.7 W 1756 4.0	**15** 0006 0.3 0600 4.4 TH 1235 0.3 1822 4.3	**30** 0000 0.7 0600 4.1 F 1219 0.8 1812 4.0
			31 0536 4.1 1157 0.7 M 1750 4.1				**31** 0034 0.7 0634 4.1 SA 1254 0.8 1845 4.0

Chart Datum: 0·13 metres above Ordnance Datum (Dublin). HAT is 4·5 metres above Chart Datum.

» **FREE** monthly updates. Register at »
www.reedsnauticalalmanac.co.uk

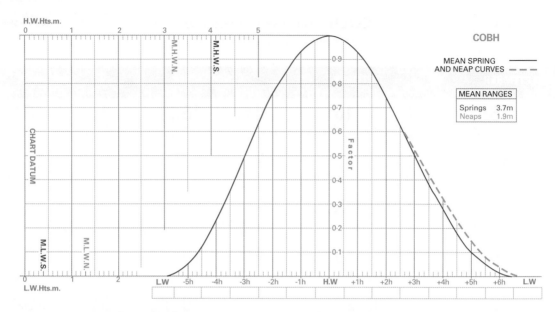

COBH

MEAN SPRING AND NEAP CURVES

MEAN RANGES	
Springs	3.7m
Neaps	1.9m

9.12.19 CORK HARBOUR

Cork 51°47'·50N 08°15'·54W ✿✿✿✿⚓⚓⚓✿✿✿

CHARTS AC 1765, 1777, 1773, 5622; Imray C57, C56

TIDES –0523 Dover; ML 2·3; Duration 0555

Standard Port COBH (←—)

Times				Height (metres)			
High Water		Low Water		MHWS	MHWN	MLWN	MLWS
0500	1100	0500	1100	4·1	3·2	1·3	0·4
1700	2300	1700	2300				
Differences BALLYCOTTON (15M ENE of Roche's Point)							
–0011	+0001	+0003	–0009	0·0	0·0	–0·1	0·0
RINGASKIDDY							
+0005	+0020	+0007	+0013	+0·1	+0·1	+0·1	+0·1
MARINO POINT							
0000	+0010	0000	+0010	+0·1	+0·1	0·0	0·0
CORK CITY							
+0005	+0010	+0020	+0010	+0·4	+0·4	+0·3	+0·2
ROBERTS COVE (approx 4M SW of Roche's Point)							
–0005	–0005	–0005	–0005	–0·1	0·0	0·0	+0·1

SHELTER Very good in all conditions, esp in Crosshaven and East Passage. There are 3 marinas at Crosshaven, plus a small marina and several ⚓s up the Owenboy River, notably at Drake's Pool. There are marinas at East Ferry on Great Island and at Monkstown. Cobh, Ringaskiddy and Cork City are commercial and ferry ports; contact HM/Port berthing Master for advice on berths. 100m yacht pontoon in 4·5m at Custom House Quay, Cork City.

NAVIGATION WPT 51°46'·57N, 08°15'·39W, 005°/1M to Roche's Pt lt; also on 354°ldg line. Safe hbr with no dangers for yachts; ent is deep and well marked. Sp rate is about 1½kn in ent. Main chan up to Cork and the chan to E Ferry are marked, but shoal outside navigable chan. Ent to Crosshaven/Owenboy River carries at least 3m at LWS, and the chan is buoyed.

LIGHTS AND MARKS See 9.12.4. The 24·5m high hammerhead water tr S of Crosshaven and the R/W power stn chy NE of Corkbeg Is are conspic from seaward. Two set Ldg lts/marks lead through The Sound, either side of Hbr Rk (5·2m), not a hazard for yachts; but do not impede merchant ships. The chan to Crosshaven is marked by C1 SHM lt buoy, C1A SHM lt buoy; C2A PHM lt buoy, C2 PHM lt buoy; and C4 PHM lt buoy.

COMMUNICATIONS (Code 021) MRCC (066) 9476109; IMES 4831448; ⊜ 4311024; Police 4831222; Dr 4831716; Ⓗ 4546400. HM/Port Berthing Master 4273125, info@portofcork.ie; Port Ops 4811380. *Cork Hbr Radio* (Port Ops) VHF Ch 12 14 16 (H24); *Crosshaven BY* Ch **M** (Mon–Fri: 0830–1730LT). *Royal Cork YC Marina* Ch **M** (0900–2359LT) and RCYC water taxi, *Salve Marine* Ch **M** 0830–1730.

FACILITIES Crosshaven BY Marina www.crosshavenboatyard.com ☎4831161, ⚓ 100⚲+20⍟ €2.50, ▯ ▮ 🛢 🛠 🔧 ⛽ Ⓔ ⚓ 🏗(40t) 🛳(1½t).

Salve Marine ☎4831145, ◣ ⚓ 45⚲+12⍟ €2.50, ▮ 🛢 Koz ⚒ 🛠 🔧 Ⓔ 🛋 🪑 🛒 🏦. **Village:** ⚓ 🏤 🏧 🍺 ✕.

Royal Cork YC Marina www.royalcork.com ☎4831023, 🛆 ◣ ▯ 170⚲+30⍟ €3.00, ▮ 🛢 🛠 Ⓔ 🅿 ✕ ▯ wi-fi.

Crosshaven Pier/Pontoon ⚲€21.00 any size.

East Ferry Marina ☎(086) 7357785, access all tides, max draught 5·5m ◣ 85⚲+15⍟ €2.50, ▮.

Crosshaven Village FV pier in 3·5m at Town quay, HW±4(for RIBs small power craft) Dr ◥ ◣ grid ⚓ ✉ Ⓒ 🍺 ✕ ▯.

Cork Harbour Marina, www.monkstownmarina.com ☎(087) 3669009, access all tides, max length 17m, 80⚲ inc⍟ facilities under development.

Cork City Marina, 150m ⍟ pontoon ≤12m €20.

Cork City All facilities. ⇌ ✈, Ferries: Roscoff; weekly; 13 hrs; Brittany (www.brittany-ferries.ie).

Aghada Pontoon with 0.4m at LAT. ⚓ (enquire at petrol stn). Appr by 2 SHM 'A1' FL G 5s and 'A2' FL(2)G 10s which should be passed close to remining on a steady straight track; Facilities: 🍺 ✕ ▯.

MINOR HARBOUR 16M ENE of YOUGHAL
DUNGARVAN BAY, Waterford, 52°05'·15N 07°36'·70W. ✿✿✿⚓⚓✿✿✿. AC 2017, 5622. HW –0542 on Dover; Duration 0600. See 9.12.17. A large bay, drying to the W, entered between Helvick Hd to the S and Ballynacourty Pt to the N. Approach in W sector (274°-302°) of this lt to clear Carricknamoan islet to the N, and Carrickapane Rk and Helvick Rk (ECM buoy Q (3) 10s) to the S. 5ca W of this buoy are The Gainers, a large unmarked rocky patch (dries 0·8m). Beware salmon nets. Off Helvick harbour are 8 Y ⚓s or ⚓ in approx 4m. Dungarvan town harbour is accessible via buoyed chan which almost dries and shifts, the buoys being moved to suit. Approach is difficult in SE'lies >F6. ⚓ in the pool below the town or ⚲ on pontoon (dries to soft mud), S bank below bridge; craft can stay overnight beyond double Y lines. Facilities: 🏤 🏧 Kos ✉ Ⓑ 🍺 ▯ ✕ ▯.

MINOR HARBOUR 15M ENE of ROCHE'S POINT
BALLYCOTTON, Cork, 51°49'·70N 08°00·19W. AC 2424, 5622. HW –0555 on Dover; Duration 0550. See above. Small, NE-facing harbour at W end of bay suffers from scend in strong SE winds; 3m in entrance and about 1·5m against piers. Many FVs alongside piers, on which yachts should berth, rather than ⚓ in hbr, which is foul with old ground tackle. 6 Y⚓s are outside harbour, or good ⚓ in offshore winds in 6m NE of pier, protected by Ballycotton Is. Lt ho Fl WR 10s 59m 21/17M, B tr in W walls; 238°-W-048°-R-238°. Facilities: ⚓ on pier. Village LB, Kos ✉ 🍺 Hotel ✕.

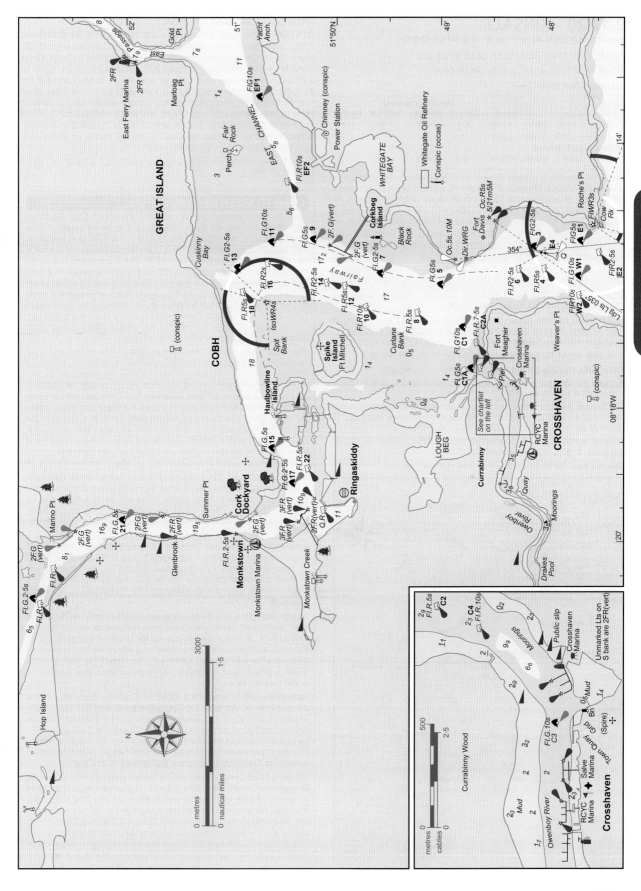

9.12.20 KINSALE

Cork 51°40'·80N 08°30'·00W ✦✦✦♒♒♒♒♒ ✿✿✿

CHARTS AC 1765, 2053, 5622, 5623; Imray C56

TIDES –0600 Dover; ML 2·2; Duration 0600

Standard Port COBH (←—)

Times				Height (metres)			
High Water		Low Water		MHWS	MHWN	MLWN	MLWS
0500	1100	0500	1100	4·1	3·2	1·3	0·4
1700	2300	1700	2300				
Differences KINSALE							
–0019	–0005	–0009	–0023	–0·2	0·0	+0·1	+0·2

SHELTER Excellent, except in very strong SE winds. Access H24 in all weathers/tides. Marinas at Kinsale YC and Castlepark; ♥'s berth on outside of pontoons in 10m in both cases; NNW of latter is FV pontoon and no ⚓ area. Possible ⌓ (Sun-Thurs) at Trident Hotel. 2 large ⚓s NE of Kinsale Bridge (contact HM). No ⚓ allowed in main channel or within 700m of Town Pier. Contact HM prior to ⚓. All craft have to pay harbour dues as well as berthing/launching fees.

NAVIGATION WPT 51°40'·00N 08°30'·00W, 001°/1·7M to Charles's Fort lt. Beware: Bulman Rk (0·9m; SCM lt buoy) 4ca S of Preghane Pt and Farmer Rk (0·6m) ¾ca off W bank. Harbour speed limit 6kn.

LIGHTS AND MARKS See 9.12.4 and chartlet. Chan is marked by PHM lt buoys. Marina lts are 2 FG or FR.

COMMUNICATIONS (Code 021) MRCC (066) 9476109; Coast/Cliff Rescue 0868 501804; ⊜ 6027700; Police 4779250; Dr 4772253; Ⓗ 4546400. HM kharbour@iol.ie, 4772503 (HO), 4773047 (OT).
KYC VHF Ch M 16. Castlepark Marina 06 16 M. HM **14** 16.

FACILITIES Kinsale BY ☎4774774, ♥ ⬧ ✕ ✎ ⊡ ⚑(30t).
Kinsale YC Marina www.kyc.ie ☎4772196, mob 087 6787377 170+50♥ €3·20, ⧄ ▬ ⬧ ✕ ⌂.
Castlepark Marina www.castleparkmarina.com ☎4774959, mob 0877502737, 130⌓ inc 20♥ €3·50, deals for longer stays, ▬ ⊡ ✕ ⌂ @.
Trident Hotel ⌓ ⊥ ⬧ Sovereign Sailing 4774145/087 6172555.
Services ⬧⬧⚓✎⊡ Ⓔ ⚠ Divers ⚓(30t).
Town LB ▬ 🕾🕾✉Ⓑ🛒✕⎕ (bus to Cork, ⇌, ✈).

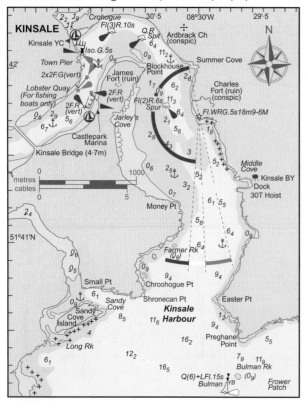

ANCHORAGE 2M EAST OF KINSALE

OYSTERHAVEN, Cork, 51°41'·20N 08°26'·90W. ✦✦◌ ✿✿✿ ✿✿. AC 1765, 2053, 5622. HW –0600 on Dover; ML 2·2m; Duration 0600. Use 9.12.19. Good shelter but ⚓'s subject to swell in S winds. Enter 0·5M N of Big Sovereign, a steep islet divided into two. Keep to S of Little Sovereign on E side of ent. There is foul ground off Ballymacus Pt on W side, and off Kinure Pt on E side. Pass W of Hbr Rk (0·9m) off Ferry Pt, the only danger within hbr. ⚓ NNW of Ferry Pt in 4–6m on soft mud/weed. NW arm shoals suddenly about 5ca NW of Ferry Pt. Also ⚓ up N arm of hbr in 3m off the W shore. Weed in higher reaches may foul ⚓. No lts, marks or VHF radio. Coast/Cliff Rescue Service ☎(021) 4770711.

Oysterhaven Yacht Harbour (strictly private) 51°42'·10N 08°26'·58W, ☎(021) 4770878.

COBH TO MIZEN HEAD

Near the easy entrance and excellent shelter of Cork Harbour (AC 1777), Ringabella Bay offers temp ⚓ in good weather. 7ca SE of Robert's Hd is Daunt Rk (3·5m) on which seas break in bad weather; marked by PHM lt buoy. Little Sovereign on with Reanies Hd 241° leads inshore of it. The Sovereigns are large rks off Oyster Haven, a good hbr but prone to swell in S'lies. The ent is clear except for Harbour Rk which must be passed on its W side. Bulman Rk (SCM lt buoy) is 4ca S of Preghane Pt at the ent to Kinsale's fine harbour. ▶ *The tide on the Cork coast turns to the NE at HW Dover +0045. There is an eddy 5M ESE of Old Head of Kinsale at HW Dover +0400. The ingoing Cork Harbour tide begins at HW Dover +0055.*

The tide turns SW at HW Dover +0500. The outgoing Cork Harbour tide begins at HW Dover –0540. Old Head of Kinsale (lt, fog sig) is quite steep-to, but a race extends 1M to SW on W-going stream, and to SE on E-going stream. ◀ There is an inshore passage in light weather, but in strong winds keep 2M off. (AC 2424) From Cork to Mizen Hd there are many natural hbrs. Only the best are mentioned here. ▶ *Offshore the stream seldom exceeds 1·5kn, but it is stronger off headlands causing races and overfalls with wind against tide. Prolonged W winds increase the rate/duration of the E-going stream, and strong E winds have a similar effect on the W-going stream.* ◀

In the middle of Courtmacsherry Bay are several dangers, from E to W: Blueboy Rk, Barrel Rk (with Inner Barrels closer inshore), and Black Tom; Horse Rk is off Barry's Pt at the W side of the bay. These must be avoided going to or from Courtmacsherry, where the bar breaks in strong S/SE winds, but the river carries 2·3m.

Beware Cotton Rk and Shoonta Rk close E of Seven Heads, off which rks extend 50m. Clonakilty B has little to offer. Keep at least 5ca off Galley Hd to clear Dhulic Rk, and further off in fresh winds. ▶ *Offshore the W-going stream makes at HW Cobh +0200, and the E-going at HW Cobh –0420, sp rates 1·5kn.* ◀

Across Glandore Bay there are good ⚓s off Glandore, or Union Hall. Sailing W from Glandore, pass outside or inside High Is and Low Is; if inside beware Belly Rk (awash) about 3ca S of Rabbit Is. Castle Haven, a sheltered and attractive hbr, is entered between Reen Pt (lt) and Battery Pt. Toe Head has foul ground 100m S, and 7½ca S is a group of rks called the Stags. Baltimore is 7M further W.

Fastnet Rk (lit) is nearly 4M WSW of C Clear; 2½ca NE of it is an outlying rk. An E/W TSS lies between 2 and 8M SSE of the Fastnet. Long Island Bay can be entered from C Clear or through Gascanane Sound, between Clear Is and Sherkin Is. Carrigmore Rks lie in the middle of this chan, with Gascanane Rk 1ca W of them. The chan between Carrigmore Rks and Badger Island is best. If bound for Crookhaven, beware Bullig Reef, N of Clear Is.

Schull is N of Long Island, inside which passage can be made W'ward to Crookhaven. This is a well sheltered hbr, accessible at all states of tide, entered between Rock Is lt Ho and Alderman Rks, ENE of Streek Hd. ⚓ off the village. ▶ *Off Mizen Hd (lt ho) the W-going stream starts at HW Cobh +0120, and the E-going at HW Cobh –0500. The sp rate is 4kn, which with wind against tide forms a dangerous race, sometimes reaching to Brow Hd or Three Castle Hd, with broken water right to the shore.* ◀

9.12.21 COURTMACSHERRY

Cork **51°38'·22N 08°40'·90W** ※※◈◈♢♢

CHARTS AC 2092, 2081, 5622, 5623; Imray C56

TIDES HW −0610 on Dover; Duration 0545

Standard Port COBH (←—)

Times				Height (metres)			
High Water		Low Water		MHWS	MHWN	MLWN	MLWS
0500	1100	0500	1100	4·1	3·2	1·3	0·4
1700	2300	1700	2300				
Differences COURTMACSHERRY							
−0025	−0008	−0008	−0015	−0·1	−0·1	0·0	+0·1

SHELTER Good shelter upriver, but in strong S/SE winds seas break on the bar (<2m), when ent must not be attempted. Dry out in small inner hbr or ⌓ afloat on jetty (FVs) or on yacht pontoon (37m). ⚓ NE of Ferry Pt in about 2·5m or N of pontoon. Weed may foul ⚓; best to moor using two ⚓s.

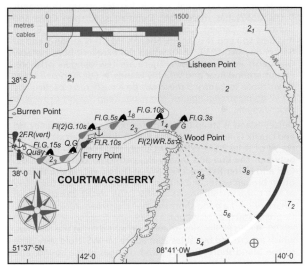

NAVIGATION WPT, 51°37'·50N 08°40'·17W, 324°/0·8M to Wood Pt. Appr in the W sector of Wood Pt lt, between Black Tom and Horse Rk (dries 3·6m); the latter is 3-4½ca E of Barry Pt on the W shore. Black Tom (2·3m), with SHM buoy Fl.G.5s 5ca SSE, is close NE of the appr. Further to NE, in centre of bay, Barrel Rk (dries 2·6m), has unlit perch (ruin). To NNW and E of it are Inner Barrels (0·5m) and Blueboy Rk. Hbr ent is between Wood Pt and SHM buoy 2ca NE, Fl G 3s. Chan (2m but 1·4m close inside ent) is marked by 5 SHM buoys and 1 PHM buoy (off Ferry Point). Upstream of Ferry Point there are moorings on both sides of the channel. The pier head is marked by 2 FR (vert) lights.

LIGHTS AND MARKS Wood Pt, Fl (2) WR 5s 15m 5M.

COMMUNICATIONS (023) Coast Rescue Service (Mob) 08685 01806; Dr 8846186 (HN 1850 335999); Police 8846122. HM (Mob) 0867394299; RNLI boathouse 8846600. No VHF.

FACILITIES Quay ♿ LB ⚓ ⌓ (min €15/craft) ⚓ ⚡ ♿ 🅿 (Pier 0930-1800 on request). **Village** ✕ 🏠 🛒 ⇌ ✈ (bus to Cork).

9.12.22 GLANDORE/UNIONHALL

Cork, **51°33'·70N 09°07'·20W** ※※◈◈♢♢

CHARTS AC 2092, 5622, 5623; Imray C56

TIDES Approx as for 9.12.23 Castletownshend

SHELTER Excellent. 12 Y⚓s or ⚓ 1½ca SW of Glandore Pier in 2m or 1ca NE of the New pier at Unionhall in 3m.

NAVIGATION WPT 51°32'·35N 09°05'·10W, 309°/1·2M to Outer Dangers. Approach between Adam Is and Goat's Hd, thence keep

---

E of Eve Is and W of the chain of rocks: Outer, Middle and Inner Dangers and Sunk Rk. Before altering W for Unionhall, stand on to clear mudbank 1ca off S shore. Speed limit 3kn.

LIGHTS AND MARKS See 9.12.4. Galley Hd, Fl (5) 20s, is 5M E of ent. Middle and Inner Dangers are marked by 2 SHM bns; and Sunk Rk by a SHM lt buoy, Fl G 5s.

COMMUNICATIONS (Code 028) Coast Rescue 33115; Police (023) 23088; Dr 23456; HM Glandore/Unionhall 34737, mob 0866081944.

HM Ch 06.

FACILITIES Glandore GHYC, ⚓s €14.00/craft (contact Glandore Inn ☎33468/33518), ⚓ ⌂ ✕ 🅿 Kos. Unionhall ⌓ (outside of FVs/drying ⌓ Old Quay) €15.00/craft, ⚓ D(occ on quayside) or 🅿 🅿 (Leap 2M) ⚒ ⚓ ♿ 🅆🅒 ⌂ ✕ 🅿 🛒.

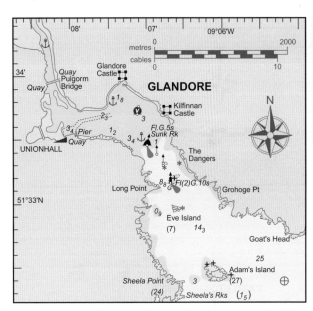

9.12.23 CASTLETOWNSHEND

Cork **51°30'·90N 09°10'·70W** ※※◈◈♢♢♢

CHARTS AC 2092, 2129, 5623; Imray C56

TIDES +0605 Dover; ML 2·2; Duration 0605

Standard Port COBH (←—)

Times				Height (metres)			
High Water		Low Water		MHWS	MHWN	MLWN	MLWS
0500	1100	0500	1100	4·1	3·2	1·3	0·4
1700	2300	1700	2300				
Differences CASTLETOWNSHEND							
−0020	−0030	−0020	−0050	−0·4	−0·2	+0·1	+0·3
CLONAKILTY BAY (5M NE of Galley Head)							
−0033	−0011	−0019	−0041	−0·3	−0·2	ND	ND

SHELTER Excellent ⚓ in midstream SE of Castletownshend slip, protected from all weathers and available at all tides; but the outer part of hbr is subject to swell in S winds. Or ⚓ N of Cat Island, or upstream as depth permits. Caution: An underwater cable runs E/W across the hbr from the slip close N of Reen Pier to the slip at Castletownshend.

NAVIGATION WPT 51°30'·28N, 09°10'·26W, 349°/7ca to Reen Pt lt. Enter between Horse Is (35m) and Skiddy Is (9m) both of which have foul ground all round. Black Rk lies off the SE side of Horse Is and is steep-to along its S side. Flea Sound is a narrow boat chan, obstructed by rks. Colonel's Rk (0·5m) lies close to the E shore, 2ca N of Reen Pt. Beware salmon nets.

Continued overleaf

S Ireland

CASTLETOWNSHEND *continued*

LIGHTS AND MARKS Reen Pt, Fl WRG 10s; a small slender W bn; vis shore-G-338°-W-001°-R-shore. A ruined tr is on Horse Is.

COMMUNICATIONS (Code 028) MRCC (066) 9746109; Coast/Cliff Rescue Service 21039; ⊖ Bantry (027) 50061; Police 36144; Dr 23456; Ⓗ 21677.

No VHF

FACILITIES Reen Pier ⤵ ⚓; **Sailing Club** ☎36100; **Castletownshend Village** ⤵ ⚓ ✉ Ⓑ(Skibbereen) 🛒 ✕ ▱ ⇌ ✈ (Cork).

BARLOGE CREEK, Cork, **51°29′·57N 09°17′·58W**. AC 2129. Tides approx as Castletownshend. A narrow creek, well sheltered except from S/SE winds. Appr with Gokane Pt brg 120°. Enter W of Bullock Is, keeping to the W side to clear rks S of the island. ⚓ W of the Is in 3m but weedy. No facilities.

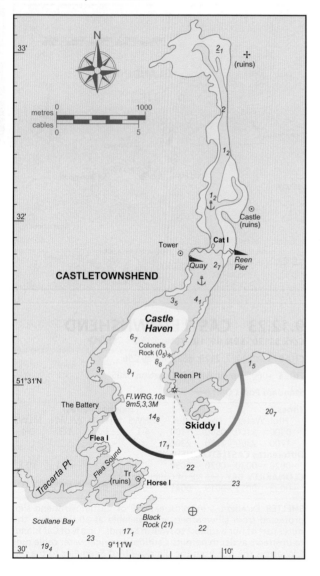

THE WEST COAST

This coast offers wonderful cruising, although exposed to the Atlantic and any swell offshore; but this diminishes mid-summer. In bad weather however the sea breaks dangerously on shoals with quite substantial depths. There is usually a refuge close by, but if caught out in deteriorating weather and poor vis, a stranger may need to make an offing until conditions improve, so a sound yacht and good crew are required. Even in mid-summer at least one gale may be met in a two-week cruise. Fog is less frequent than in the Irish Sea.

▶ *Tidal streams are weak, except round headlands.* ◀ There are few lights, so inshore navigation is unwise after dark, however coastal navigation is feasible at night in good visibility. Keep a good watch for lobster pots in inshore waters. Stores, fuel and water are not readily available.

MIZEN HEAD TO DINGLE BAY

(AC 2423) At S end of Dunmanus Bay Three Castle Hd has rks 1ca W, and sea can break on S Bullig 4ca off Hd. Dunmanus B (AC 2552) has three hbrs: Dunmanus, Kitchen Cove and Dunbeacon. Carbery, Cold and Furze Is lie in middle of B, and it is best to keep N of them. Sheep's Hd (lt) is at the S end of Bantry Bay (AC 1838, 1840) which has excellent hbrs, notably Glengariff and Castletown. There are few dangers offshore, except around Bear and Whiddy Islands. ▶ *Off Blackball Hd at W entrance to Bantry B there can be a nasty race, particularly on W-going stream against the wind. Keep 3ca off Crow Is to clear dangers.* ◀

Dursey Island is steep-to except for rk 7½ca NE of Dursey Hd and Lea Rk (1·4m)1½ca SW. The Bull (lt, fog sig, Racon) and two rks W of it lie 2·5M WNW of Dursey Hd. The Cow is midway between The Bull and Dursey Hd, with clear water each side. Calf and Heifer Rks are 7½ca SW of Dursey Hd, where there is often broken water. ▶ *2M W of The Bull the stream turns NW at HW Cobh +0150, and SE at HW Cobh –0420. Dursey Sound (chart 2495) is a good short cut, but the stream runs 4kn at sp; W-going starts at HW Cobh +0135, and E-going at HW Cobh –0450.* ◀ Flag Rk lies almost awash in mid-chan at the narrows, which are crossed by cables 25m above MHWS. Hold very close to the Island shore. Beware wind changes in the sound, and broken water at N entrance.

Kenmare R. (AC 2495) has attractive hbrs and ⚓s, but its shores are rky, with no lights. The best places are Sneem, Kilmakilloge and Ardgroom. Off Lamb's Head, Two Headed Island is steep-to; further W is Moylaun Is with a rk 300m SW of it. Little Hog (or Deenish) Island is rky 1·5M to W, followed by Great Hog (or Scariff) Is which has a rk close N, and a reef extending 2ca W.

Darrynane is an attractive, sheltered hbr NNW of Lamb Hd. The entrance has ldg lts and marks, but is narrow and dangerous in bad weather. Ballinskelligs Bay has an ⚓ N of Horse Is, which has two rks close off E end. Centre of bay is a prohib ⚓ (cables reported).

Rough water is met between Bolus Hd and Bray Hd with fresh onshore winds or swell. The SW end of Puffin Island is steep-to, but the sound to the E is rky and not advised. Great Skellig (lit) is 6M, and Little Skellig 5M WSW of Puffin Is. Lemon Rk lies between Puffin Is and Little Skellig. ▶ *Here the stream turns N at HW Cobh +0500, and S at HW Cobh –0110.* ◀ There is a rk 3ca SW of Great Skellig. When very calm it is possible to go alongside at Blind Man's Cove on NE side of Great Skellig, where there are interesting ruins.

9.12.24 BALTIMORE

Cork 51°28'·30N 09°23'·40W ✳✳✳✳◊◊❀❀❀

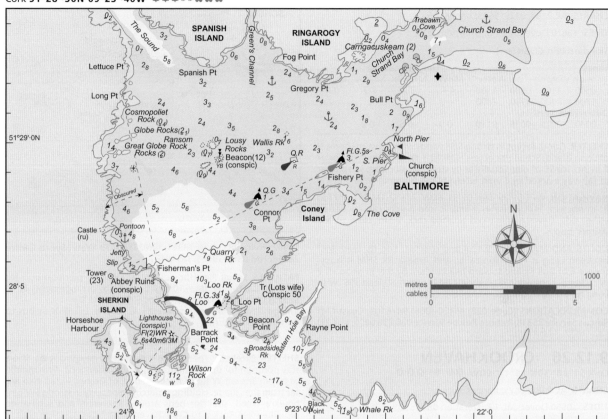

CHARTS AC 2129, 3725, 5623; Imray C56

TIDES −0605 Dover; ML 2·1; Duration 0610

Standard Port COBH (←)

Times				Height (metres)			
High Water		Low Water		MHWS	MHWN	MLWN	MLWS
0500	1100	0500	1100	4·1	3·2	1·3	0·4
1700	2300	1700	2300				
Differences BALTIMORE							
−0025	−0005	−0010	−0050	−0.6	−0.3	+0.1	+0.2

SHELTER Excellent. Access H24 from the S. At Baltimore, pontoon (Apr-Sep) on S Pier, ⌇ for up to 20 yachts in 2·4m. Inner Hbr, partly dries between N and S piers, mostly used by ferries, local boats and FVs; the latter also berth on N pier (addition to pier in progress). Ro Ro berth SE of Bull Pt. ⚓ about 200m W of S pier, or about 200m N of N Pier. Beware extensive moorings. In strong NW'lies ⚓ in Church Strand Bay. Its are required. Do not ⚓ in dredged chan between Wallis Rk buoy and N Pier, or in chan to Church Strand Bay. In strong W winds ⚓ in lee of Sherkin Is off Castle ruins or berth on pontoon (see Adjacent Anchorages, Sherkin Island).

NAVIGATION WPT 51°27'·93N 09°23'·46W, 000°/5ca to Loo Rk SHM lt buoy. No passage between buoy and mainland. Beware Lousy Rks (SCM bn) and Wallis Rk (PHM lt buoy) in middle of the bay. Cardinal buoys mark rocks in The Sound, which demands care in pilotage; AC 3725 and ICC SDs essential. SCM marks isolated rk off Inane Pt in the R Ilen, which is navigable on the flood for at least 4M above The Sound. Speed limit 6kn in R Ilen and N/NE of Wallace Rock.

LIGHTS AND MARKS See 9.12.4 and chartlet. Ent easily identified by Barrack Pt lt ho and conspic W tr (Lot's Wife) on Beacon Pt.

COMMUNICATIONS (Code 028) MRCC and Coast/Cliff Rescue Service (066) 9476109; ⊕ (027) 53210; Police 20102; Dr 23456/after hrs 1850 335999, Ⓗ 21677. HM 087 2351485.

VHF (HM) Ch 09, 16.

FACILITIES Berthing: May-Sept from €20/craft, ☎22145, 0872 351485 or www.atlanticboat.ie; ⌇ ⌇ ▮(☎20106) ⚓ on pontoon; Baltimore SC ☎20426, visitors welcome ▶ ⌇; Glenans Irish Sailing School ☎(01) 6611481, 028 20154; Sherkin Pontoon ⚓ access to shore in season. Services ▮ ⚓ Kos ⚒ ⊞ ⌖ (on R Ilen at Old Court), ⛽ ACA ⚑. Village ✉ ⌂ bus to Cork for ⇌ ✈.

ADJACENT ANCHORAGES

SHERKIN ISLAND on W side of harbour entrance. ⚓s: Horseshoe Harbour (keep well to the W in narrow entrance; unlit); and off Castle ruins, 5ca N, where there are also some ⌇ on pontoons (⚓ and electricity) belonging to nearby Islander's Rest Hotel.

CLEAR ISLAND, N HARBOUR, 51°26'·60N 09°30'·20W. AC 2129, 5623. Tides approx as Schull, 9.12.25. A tiny, partly drying inlet on N coast of Clear Is, exposed to N'ly swell. Rks either side of outer appr 196°. Inside the narrow (30m), rky ent keep to the E. Lie to 2 ⚓s on E side in about 1·5m or drying berth in Inner Hbr clear of ferry bad weather berth. Outer (N) bkwtr rebuilt with tidal gate (2015) which will permit vessels to remain afloat. Facilities ⌇ ⚓ ⊞ ⊞ ✉ ⚑ ✗ ⌂.

ROARINGWATER BAY, Long Island Bay AC 2129, 5623. Enter between Cape Clear and Mizen Hd, extends NE into Roaring- water Bay. Beware extensive, low-lying unlit fish farms. Safest appr, S of Schull, is via Carthy's Sound (51°30'N 09°30'W). From the SE appr via Gascanane Sound, but beware Toorane Rks, Anima Rk and outlying rks off many of the islands. Shelter in various winds at ⚓s clockwise from Horse Island: 3ca E and 7ca NE of E tip of Horse Is; in Ballydehob B 2m (Ballydehob has a good quay for tenders and most facilities); Poulgorm B 2m; 3ca SSW of Carrigvalish Rks in 6m. Rincolisky Cas (ru) is conspic on S side of bay. The narrow chan E of Hare Is and N of Sherkin Is has two ⚓s; also leads via The Sound into Baltimore hbr. Local advice useful. Temp'y fair weather ⚓s in Carthy's Islands. Rossbrin Cove safe ⚓ 2·5M E of Schull, but many local moorings; no access from E of Horse Is due to drying Horse Ridge.

S Ireland

9.12.25 SCHULL

Cork 51°30'·80N 09°32'·00W ✿✿✿✿◊◊✿✿✿

CHARTS AC 2184, 2129, 5623; Imray C56

TIDES +0610 Dover; ML 1·8; Duration 0610

Standard Port COBH (←)

Times				Height (metres)			
High Water		Low Water		MHWS	MHWN	MLWN	MLWS
0500	1100	0500	1100	4·1	3·2	1·3	0·4
1700	2300	1700	2300				
Differences SCHULL							
−0040	−0015	−0015	−0110	−0·9	−0·6	−0·2	0·0

SHELTER Good, except in strong S/SE winds when best shelter is N of Long Island. Schull Hbr access H24. 12 Y ⚓s in NE part of hbr or ⚓ in 3m 1ca SE of pier, usually lit by street lts all night; keep clear of fairway marked by 8 unlit lateral buoys (summer time only).

NAVIGATION WPT 51°29'·60N 09°31'·60W, 346°/2·1M to front ldg lt. In hbr ent, Bull Rk (dries 1·8m), R iron perch, can be passed either side. Beware unmarked isolated rock, 0·5m, close to the visitors' buoys.

LIGHTS AND MARKS See 9.12.4 and chartlet. Ldg lts, lead 346° between Long Is Pt, W conical tr, and Amelia Rk SHM lt buoy; thence E of Bull Rk and toward head of bay. By day in good vis 2 W radomes conspic on Mt Gabriel (2M N of Schull) lead 355° with Long Is Pt lt ho in transit.

COMMUNICATIONS (Code 028) MRCC (066) 9476109; Coast/Cliff Rescue Service 35318; Inshore Rescue Service 086 236 0206: ⊖ (027)

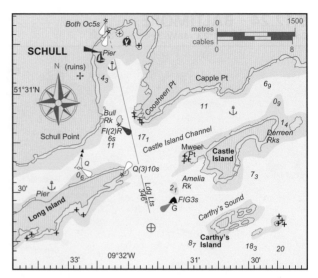

51562; Police 28111; Dr 28311; Ⓗ (027) 50133. Water Sports Centre 28554; HM 086 039105. No VHF.

FACILITIES Schull Pier/Hbr ⚓ (not at LWS) FV pier/pontoon for tenders, ⚏ tempy, ⚓s €10/night, ⚒ ♠ (by arrangement) ⚓ ⚴. Sailing Club ☎37352; Services Kos ⚱ ✕ ⚒ Ⓔ Ⓛ. Village 🏤 📮 (Ballydehob 5M), 🛒 @, Charts, ✉ Ⓑ ✕ ⚏ bus to Cork for ⇌ ✈ Ferries.

9.12.26 CROOKHAVEN

Cork 51°28'·50N 09°42'·00W ✿✿ ◊✿✿✿✿

CHARTS AC 2184, 5623; Imray C56

TIDES +0550 Dover; ML 1·8; Duration 0610

Standard Port COBH (←)

Times				Height (metres)			
High Water		Low Water		MHWS	MHWN	MLWN	MLWS
0500	1100	0500	1100	4·1	3·2	1·3	0·4
1700	2300	1700	2300				
Differences CROOKHAVEN							
−0057	−0033	−0048	−0112	−0·8	−0·6	−0·4	−0·1
DUNMANUS HARBOUR							
−0107	−0031	−0044	−0120	−0·7	−0·6	−0·2	0·0
DUNBEACON HARBOUR							
−0057	−0025	−0032	−0104	−0·8	−0·7	−0·3	−0·1

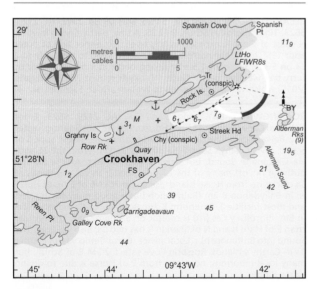

SHELTER Excellent. There are 8 Y ⚓s and 10 dayglow R ⚓s. Short stay pontoon for up to 4 boats (⚓/stores/passengers). ⚓s in middle of bay in 3m; off W tip of Rock Is; and E of Granny Is; last two are far from the village. Holding is patchy, especially in strong SW'lies; beware weed, shellfish beds around shoreline and submarine pipeline from Rock Is to Crookhaven.

NAVIGATION WPT 51°28'·50N 09°40'·50W, 274°/1M to Rock Is lt ho. Ent between this lt and NCM bn on Black Horse Rks (3½ca ESE). From S, keep 1ca E of Alderman Rks and ½ca off Black Horse Rks bn. Passage between Streek Hd and Alderman Rks is not advised for strangers. Inside the bay the shores are steep to. 1ca WSW of West end of Rock Is beware isolated rock, which is awash at LAT.

LIGHTS AND MARKS Lt ho on Rock Is (conspic W tr) L Fl WR 8s 20m 13/11M; vis outside hbr: W over Long Is Bay-281°-R-340°; vis inside hbr: 281°-R-348°-W-shore(N).

COMMUNICATIONS (Code 028) MRCC (066) 9476109; Coast/Cliff Rescue at Goleen 35318; ⊖ (027) 50061; Dr 35148; ✉ 35200. HM (O'Sullivan's Bar) 35319. No VHF.

FACILITIES Village ⚓s (☎086 356 7771) €8.00, pontoon for small craft ⚓ 🏤 ✂ ✉ Ⓑ(Schull) 🛒 ✕ ⚏, taxi to Goleen then bus to Cork, ✈ ⇌.

ADJACENT HARBOURS

GOLEEN (Kireal-coegea), Cork, 51°29'·65N 09°42'·21W. AC 2184, 5623. Tides as Crookhaven. A narrow inlet 6ca N of Spanish Pt; good shelter in fair weather, except from SE. 2 churches are easily seen, but ent not visible until close. Keep to S side of ent and ⚓ fore-and-aft just below quay, where ⚏ also possible. Facilities: 🏤🛒⚏.

DUNMANUS BAY, Cork. AC 2552, 5623. Tides see 9.12.26. Appr between Three Castle Hd and Sheep's Hd, Fl (3) WR 15s 83m 18/15M; no other lts. Ent to Dunmanus Hbr, 51°32'·70N 09°39'·86W, is 1ca wide; breakers both sides. ⚓ in 4m centre of B. Kitchen Cove, 51°35'·50N 09°38'·05W, best of the 3 hbrs; enter W of Owens Is and ⚓ 1ca NNW of it or 2ca further N in 3m. Exposed to S, but good holding. Quay at Ahakista village: ⚏(drying) ⚓ ⚰ 🛒 ✕ ⚏. Dunbeacon Hbr, 51°36'·35N 09°33'·60W, is shallow and rock-girt. Quay possible ⚏ in 2m (used by FVs), ⚓; ⚓ E or SE of Mannion Is. Durrus (¾M): 🏤🏤✕⚏.

9.12.27 BANTRY BAY

Cork 51°34'N 09°57'W ✵✵✵⚓⚓⚓✿✿✿

CHARTS AC 2552, 1840, 1838, 5623; Imray C56

TIDES +0600 Dover; ML 1·8; Duration 0610

Standard Port COBH (←)

Times				Height (metres)			
High Water		Low Water		MHWS	MHWN	MLWN	MLWS
0500	1100	0500	1100	4·1	3·2	1·3	0·4
1700	2300	1700	2300				
Differences BANTRY							
−0045	−0025	−0040	−0105	−0·7	−0·6	−0·2	+0·1
CASTLETOWN (Bearhaven)							
−0048	−0012	−0025	−0101	−0·8	−0·6	−0·2	0·0
BLACK BALL HARBOUR (51°36N 10°02W)							
−0115	−0035	−0047	−0127	−0·7	−0·6	−0·1	+0·1

SHELTER/NAVIGATION Bantry Bay extends 20M ENE from Sheep's Hd, Fl (3) WR 15s 83m 18/15M. Access is easy, but the Bay is exposed to W'lies. The shore is clear everywhere except off Bear Is and Whiddy Is. Some of the many well sheltered ⚓s on the N shore are detailed on this page. The S shore has few ⚓s.

CASTLETOWN BEARHAVEN 51°38'·80N 09°54'·45W. AC 1840, 5623. Castletownbere Fishery Harbour lies between Dinish Is and the mainland; also ⚓ at **Dunboy Bay,** W of Piper Snd (open to E). 4 Y 🅐s are laid Apr-Sep 3ca E of Dinish Is. Lts: At W ent, Ardnakinna Pt, Fl (2) WR 10s 62m 17/14M, H24. At E ent to Bearhaven: Roancarrigmore, Fl WR 5s 13m 11/9M. Appr W of Bear Is on 023·25° Dir lt, Oc WRG 5s 7m 15/12M (023°-W-023·5°); then inner ldg lts 008°, both Oc Bu 6s 4/7m 6M, vis 005°-013°, via ent chan which narrows to 50m where it is marked by lit bcns. Beware Walter Scott Rk (2·7m), SCM lt buoy, and Carrigaglos (0·6m high) S of Dinish Is. VHF Ch 14 16. HM ☎(027) 70220. Facilities: ⚓ 🅑 on quay; 🅓 on Dinish Is. **Town** 🅣 🅣 Kos ✕ ⚒ 🛒 ⌂ ⊠ Ⓑ 🛒 ✕ ⌂ @.

LAWRENCE COVE MARINA 51°38'·28N 09°49'·28W; AC 1840, 5623. *See inset on chartlet below.* Good shelter on N side of Bear Island.

Marina on S side of cove has NE/SW pontoon 90m long (40🛥 in 3-3·5m). From E keep clear of Palmer Rock and a shoal patch, both 1·8m. ☎027 75044; VHF Ch 16 M. 🛥(€1.52), ⚓ 🅓 🅑 🅑 🅓. Friendly welcome at the only marina between Kinsale and Cahersiveen/Dingle. There are 4 yellow 🅐s in 3m close S of Ardagh Point, or ⚓ in 4m to W of Turk Is. At Rerrin ⚓ ⬤ 🅡(storage) ⊠ 🛒 ✕ ⌂ bus to Cork.

LONEHORT HARBOUR 51°38'·12N 09°47'·80W. AC 1840, 5623. At E tip of Bear Is, good shelter but keep S at ent to clear unmarked rks; then turn ENE to ⚓ in 2·7m at E end of cove. Be aware of Firing Practice Area to east of harbour entrance (full clear range procedure in force).

ADRIGOLE 51°40'·51N 09°43'·22W. AC 1840, 5623. Good shelter, but squally in W/N gales. Beware Doucallia Rk, dries 1·2m, 1M SSW of ent. Beyond the 2ca wide ent, keep E of Orthons Is (rks on W side). 7 Y 🅐s NE of Orthons Is. ⚓s to suit wind direction: off pier on E shore 4m; N or NW of Orthons Is. Drumlave (½M E): ⚓ ⚒ (sailing school) 🅣 ⬤ ⊠ 🛒.

Trafrask Bay, 2M east, has Y🅐 at **51°40'·8N 09°40'·1W;** 🛒 ✕ ⌂.

GLENGARRIFF 51°44'·20N 09°31'·90W. AC 1838, 5623. Tides as Bantry. Beautiful ⚓ S of Bark Is in 7-10m; or to NE in 3m, where there are 6 Y 🅐s. Better for yachts than Bantry hbr. Ent between Big Pt and Gun Pt. Keep 1ca E of rks off Garinish Island (Illnacullen) and Ship Is; chan marked to either side by lateral marks LFl R/G 4s. Beware marine farms. HT cables removed allowing access N of Garinish with local knowledge; but rocky chan W of should not be attempted. Facilities: Eccles hotel, 🅟. **Village** ⚓ (ferry berth 1800-1900) 🅣 Kos ⊠ 🛒 ✕ ⌂ @.

BANTRY 51°40'·85N 09°27'·85W. AC 1838, 5623. Beware Gerane Rks ½M W of Whiddy Is. Appr via the buoyed/lit N chan to E of Horse and Chapel Is; keep 2ca off all islands to clear unlit marine farms. The S chan, fair weather only. Appr Relane Pt hdg 063° with HW mark S. Beach (seaward edge of airfield) ≠ Reenbeg Cliff (distant rounded hill) to clear Cracker Rk, then leave Blue Hill and S Beach to stbd. 🅐s call Bantry Hbr Ch 14 or ☎027 51253; VHF Ch 14 11 16 (H24). HM (027) 53277; ⊖ 50061; Police 50045; Dr 50405; Ⓗ 50133. MRCC (066) 9476109. Facilities: ⚓ **Pier** 🛥(drying) ⚓ 🅑; **Bantry Bay SC** ☎51724, Kos ⚒ ⊠ Ⓑ 🛒 ✕ ⌂ bus to Cork.

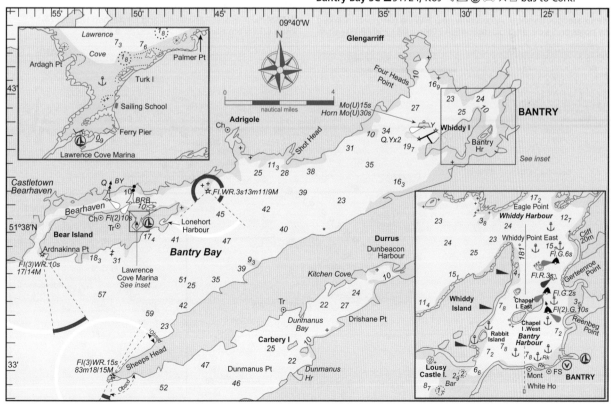

9.12.28 KENMARE RIVER

Kerry 51°45'·00N 10°00'·00W ✴✴✴✴⚓✿✿✿

CHARTS AC 2495; Imray C56

TIDES +0515 Dover; Duration Dunkerron 0620

Standard Port COBH (←—)

Times				Height (metres)			
High Water		Low Water		MHWS	MHWN	MLWN	MLWS
0500	1100	0500	1100	4·1	3·2	1·3	0·4
1700	2300	1700	2300				
Differences BALLYCROVANE HARBOUR (Coulagh Bay)							
−0116	−0036	−0053	−0133	−0·6	−0·5	−0·1	0·0
DUNKERRON HARBOUR							
−0117	−0027	−0050	−0140	−0·2	−0·3	+0·1	0·0
WEST COVE (51°46'N 10°03'W)							
−0113	−0033	−0049	−0129	−0·6	−0·5	−0·1	0·0
BALLINSKELLIGS BAY							
−0119	−0039	−0054	−0134	−0·5	−0·5	−0·1	0·0

SHELTER Garnish Bay (S of Long I): is only good in settled weather and W'ly winds. ⚓ either W or 1ca S of the Carrigduff concrete bn.

Ballycrovane: in NE of Coulagh B is a good ⚓, but open to W'ly swell which breaks on submerged rks in SE. N and E shores are foul 5ca NE of Bird Is.

Cleanderry: ent NE of Illaunbweeheen (Yellow Is) is only 7m wide and rky. ⚓ ENE of inner hbr. Beware marine farms.

Ardgroom: excellent shelter, but intricate ent over rky bar. Appr with B bn brg 135°; then 2 W bns (front on Black Rk, rear ashore) lead 099° over bar. Alter 206° as two bns astern come in transit. When clear, steer WNW to ⚓ 0·5 ca E of Reenavade pier. Beware marine farms off Ardgroom & Kilmakilloge.

Kilmakilloge: is a safe ⚓ in all winds. Beware mussel beds and rky shoals. On appr keep S side of ent, heading W of Spanish Is until past PHM buoy, Fl(2) R 10s, then steer approx 105°. Bunaw Hbr ldg lts 041°, front Oc R 3s, rear Iso R 2s (access only near HW; ⌚ for shoal draught). Keep S side of ent; ⚓ 2ca W of Carrigwee bn; S of Eskadawer Pt; or Collorus Hbr W side only.

Ormond's Harbour: good shelter except in SW or W winds, but beware rk 2½ca ENE of Hog Is. ⚓ in S half of bay.

Kenmare: good shelter. Access only near HW via narrow ch marked by poles on S side. Poss ⌚ (☎064 664 2059 or 087 250 8803) rafted

at end of pier (⚓ available) at N side of river, just below town. Beware very strong ebbs at springs and after heavy rains.

Dunkerron Harbour: ent between Cod Rks and The Boar to ⚓ 1ca NW of Fox Is in 3·2m; land at Templenoe pier. 4ca E of Reen Pt behind pier ⚓ available.

Sneem: enter between Sherky Is and Rossdohan Is. Hotel conspic NE of hbr. 3 Y ⚓s and ⚓ NE of Garinish Is, but uncomfortable if swell enters either side of Sherky Is. Beware of marine farms.

Darrynane: 1½M NW of Lamb's Hd, is appr'd from the S between Deenish and Moylaun Islands, but not with high SW swell. Enter with care on the ldg marks/lts 034°, 2 W bns, both Oc 3s 10/16m 4M. Safe ⚓ (3m) NE of Lamb's Is. Also ⚓s in Lehid Hbr (S shore), R Blackwater, Coongar Hbr & W Cove.

NAVIGATION WPT 51°40'·00N 10°17'·20W, 065°/15·6M to 0.5M S of Sherky Island. From SW, keep NW of The Bull and Dursey Is. From SE, Dursey Sound is possible in fair weather but narrow (beware Flag Rk 0·3m) and with cable car, 21m clearance. To clear dangerous rks off Coulagh Bay, keep twr on Dursey Is well open of Cod's Head 220°. From NW, there are 3 deep chans off Lamb's Head: between Scarriff Is and Deenish Is which is clear; between Deenish and Moylaun Is which has rky shoals; and between Moylaun and Two Headed Is which is clear and 4½ca wide. A night appr into the river is possible, but close appr to hbrs or ⚓s is not advised. Up-river from Sneem Hbr, keep clear of Maiden Rk(dries 0·5m); Church Rks and Lackeen Rks are unmarked; a buoyed chan lies between Carrignaronabeg and Bat Rock. Beware marine farms.

LIGHTS AND MARKS On Dursey Is: Old Watch Twr (conspic) 250m. Eagle Hill (Cod's Hd) 216m. Lights are as on chartlet.

COMMUNICATIONS (Code 064) MRCC (066) 9476109; ⊖ Bantry (027) 50061; Ⓗ 4108; Police 41177. No VHF.

FACILITIES
ARDGROOM Pallas Hbr 🛥 🛥 🛒 ✕ 🛥 Kos @ ✉ at Ardgoom village (2M SSW of Reenavade pier).

KILMAKILLOGE Bunaw Pier ⌚ 🛥 🗓; 2M to 🛥 Kos, ✉.

KENMARE ⌚. Town 🛥 🛥 Kos ⬛ ✕ 🛒 🛥 Ⓗ ✉ inc @ Ⓑ ⇌ (bus to Killarney) ✈ (Cork or Killarney).

SNEEM ⚓ at Hotel Parknasilla and Oysterbed Ho pier (⚓).
Town (2M from hbr), 🛥 🛥 ✕ 🛥 ✉ 🛒 ⬛ Kos.

9.12.29 DINGLE BAY

9.12.30 VALENTIA HARBOUR

Kerry 51°55′·7N 10°17′·1W ⚓⚓⚓☆☆ 🌢🌢🌢

CHARTS AC 2125; Imray C56

TIDES +0515 Dover

Standard Port COBH (←—)

Times				Height (metres)			
High Water		Low Water		MHWS	MHWN	MLWN	MLWS
0500	1100	0500	1100	4·1	3·2	1·3	0·4
1700	2300	1700	2300				
Differences VALENTIA HARBOUR (Knight's Town)							
–0118	–0038	–0056	–0136	–0·6	–0·4	–0·1	0·0

SHELTER Good ⚓s at: Glanleam B, 6ca S of Fort Pt in 4m; 1ca NW of LB slip in 2·5m (beware The Foot, spit drying 1·2m, marked by

ECM lt buoy) SE of Ferry Pier at Knight's Town (E end of the island) in 4m; in bay on the S side of Beginish Is in 3m.

NAVIGATION WPT 51°56′·84N 10°20′·15W 147°/1M to Fort Pt lt. Appr the NE end of Valentia Is, either via:

- Beware rock outcrop 2.6m 1 Ca ENE Reenadrolaun Pt and other shoal patches within 10m contour.
- Doulus Bay, N of Beginish Is (avoid if swell is running). Clear Black Rks by ½ca then head for E end of Beginish Is as soon as Reenard Pt opens to port. When Lamb Is and the N point of Beginish are about to open astern, steer to keep them so till Reenard Point opens to the E of Church Is. Then steer to pass between Church Is and the WCM buoy marking Passage Rk.
- Better ent, between Fort Pt and Beginish Is (easy access except in strong NW'lies). Beware Hbr Rk, 2·6m, 3ca SE of Fort Pt marked by ECM lt Bn.

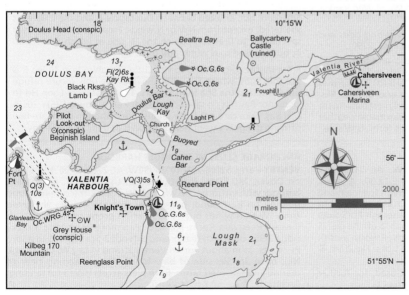

For **Cahersiveen Marina** from Valentia Is start as early on the tide possible to cross the Caher Bar, min depth 1·4m (buoyed), using the ldg lts/lines 019-199°, then keep to the middle of the river.

For appr S of Island via Portmagee chan see **Portmagee** overleaf for info on availability of the swing bridge across the chan.

LIGHTS AND MARKS See 9.12.4 and chartlet.

COMMUNICATIONS CG 9476109; ⊕ 7128540; Dr 9472121; Police 9472111; Ⓗ 9472100; Knightstown HM (066) 9476124;

Cahersiveen Marina ☎947 2777, VHF Ch **M** www.cahersiveenmarina.ie.

FACILITIES   Knight's Town Kos ⬚ ⚒ ⚓ 🏠 **Marina** ⚲ are readily available in expanding facility, some with 🚿. Ferry/bus to Cahersiveen (2½M) for usual shops; Bus to ⚆ (Tralee and Killarney) ✈ (Kerry, Shannon, Cork). **Cahersiveen Marina** 2M upriver, min depth 2.5m in basin, 93 inc Ⓥ €2.50, ⬚ 🛢 🛠 ⚒ ⚓ Ⓔ 🔌(14t) ✉ Ⓑ 🛒 ✕ 🛟.

9.12.31 DINGLE

Kerry 52°07'·14N 10°15'·48W ❀❀❀☆☆☆❀❀❀

CHARTS AC 2789, 2790; Imray C55, C56

TIDES +0540 Dover; ML 2·1m; Duration 0605

Standard Port COBH (←—)

Times				Height (metres)			
High Water		Low Water		MHWS	MHWN	MLWN	MLWS
0500	1100	0500	1100	4·1	3·2	1·3	0·4
1700	2300	1700	2300				
Differences DINGLE							
–0111	–0041	–0049	–0119	–0·1	0·0	+0·3	+0·4
SMERWICK HARBOUR							
–0107	–0027	–0041	–0121	–0·3	–0·4	ND	ND

SHELTER Excellent at marina (3·3m depth) in landlocked hbr. A busy fishing port.

NAVIGATION WPT 52°06'·20N 10° 15'·48W, 360°/1·06M to lt Fl G 3s. Straightforward entrance H24. Beware Crow Rk (dries 3·7m) and shoals lying between it and Reenbeg Pt (0·8M NE). Keep clear of rocky ledge SW of Black Pt indicated by lit SHM.

- **Castlemaine Hbr**, approx 15M E at the head of Dingle Bay, largely dries and should not be attempted without local knowledge or inspection at LW.

LIGHTS AND MARKS Eask Twr (195m, with fingerpost pointing E) is conspic 0·85M WSW of ent. Lt trs, Fl G 3s 20m 6M on NE side, and Fl 5s 5m 3M SW of ent. Ent chan, dredged 2·6m, is marked by 5 SHM lt buoys and 3 PHM lt buoys, as chartlet. Ldg lts, both Oc 3s, (W ◊s on B poles) lead from astern 182° to hbr bkwtrs. Sectored Dir lt 002° Oc RWG 4s on W side of Main Pier.

COMMUNICATIONS (Code 066) Coastguard (066) 9476109; ⊖ 7121480; Dr 9152225; ℍ9151455; Police 9151522. All emergencies: 999 or 112.

Ch 14 16 ⚓ but no calls required. Valentia Radio (Ch 24 28) will relay urgent messages to Dingle HM.

FACILITIES Marina dinglemarina.ie ☎087 9254115 ◣ (launch €2) 60◯+20🅥 €2.50, ⅃ ⊡ ♠ ◻(70t).

Town ⏃ ⏃ ♠ Kos ⚒ ⊞ Ⓔ ⚠ ⚓ ✉ Ⓑ 🛒 ✕ ◻ ⇌ Tralee (by bus) ✈ (Kerry 30M).

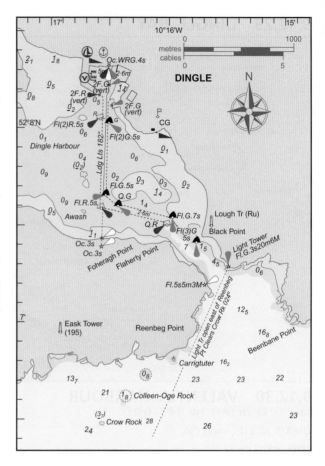

DINGLE BAY TO LISCANNOR BAY

(AC 2254) Dingle Bay (AC 2789, 2790) is wide and deep, with few dangers around its shores. Cahersiveen and Dingle have small marinas. The best ⚓s are at Portmagee and Ventry. At the NW ent to the bay, 2·5M SSW of Slea Hd, is Wild Bank (or Three Fathom Pinnacle), a shallow patch with overfalls. 3M SW of Wild Bank is Barrack Rk, which breaks in strong winds.

The Blasket Islands are very exposed, ► with strong tides and overfalls ◄ but worth a visit in settled weather (AC 2790).

Great Blasket and Inishvickillane each have ⚓ and landing on their NE side. Between lies Inishabro with a beach and impressive rock formations. Inishtearaght is the most W'ly Is (lt), but further W lie Tearaght Rks, and 3M S are Little Foze and Gt Foze Rks. Inishtooskert is the most N'ly. Blasket Sound is the most convenient N-S route, 1M wide, and easy in daylight and reasonable weather with fair wind or tide; extensive rks and shoals form its W side.

► The N-going stream starts at HW Galway +0430, and the S-going at HW Galway –0155, with sp rate 3kn. ◄

Between Blasket Sound and Sybil Pt there is a race in W or NW winds with N-going tide, and often a nasty sea. Sybil Pt has steep cliffs, and offlying rks extend 3½ca.

Smerwick hbr, entered between Duncapple Is and the E Sister, is sheltered, except from NW or N winds. From here the scenery is spectacular to Brandon Bay on the W side of which there is an ⚓, but exposed to N winds and to swell.

From Inishtearaght to Loop Hd, Little Samphire Is in Tralee B is the sole light, where Fenit hbr provides the only secure refuge until entering the Shannon Estuary. The coast from Loop Hd to Liscannor Bay has no safe ⚓, no lts, save an AIS buoy Fl(5)Y 20s in Mal Bay. Take care not to be set inshore, although there are few offlying dangers except near Mutton Is and in Liscannor Bay.

ANCHORAGE SW VALENTIA ISLAND

PORTMAGEE, Kerry, **51°53'·20N 10°22'·29W**. AC 2125. HW +0550 on Dover; ML 2·0m; Duration 0610; See 9.12.30. A safe ⚓ 2·5M E of Bray Head in Portmagee Sound between the mainland and Valentia Is. The ent to the Sound often has bad seas, but dangers are visible. Care required E of Reencaheragh Pt (S side) due rks either side. Deepest water is N of mid-chan. ⚓ off the pier (S side) in 5m, opposite Skelling Heritage Centre (well worth a visit). ◯ on pier is not recommended due to strong tides. Facilities: ⚓ ♨ ▷ (at hotel), ✕ ◻ Kos.

The swing bridge 1ca E of pier is currently closed; for further information check with Valentia RNLI or Valentia CG.

ANCHORAGE CLOSE W OF DINGLE HARBOUR

VENTRY Kerry, **52°06·70N 10°20'·30W**. AC 2789, 2790. HW +0540 on Dover; ML 2·1m; Duration 0605; Use 9.12.31. A pleasant harbour with entrance 1M wide 2M W of Eask Tr (conspic). Good holding on hard sand, open to swell from the SE; Sheltered from SW to N winds but in fresh w'lies prone to sharp squalls. Beware Reenvare Rocks 1ca SE of Parkmore Point; also a rocky ridge 2·9m, on which seas break, extends 2·5ca SSE of Ballymore Pt. No lts. ⚓s in about 4m off Ventry Strand (❊ brg W, the village NE); or in 3m SW side of bay, 1ca N of pier keeping clear of marine and seaweed farms. Both piers access HW±3 for landing. Facilities: ⏃ ◣ 🛒 Kos ✕ ◻ ✉.

ANCHORAGES BETWEEN THE BLASKETS AND KERRY HD

SMERWICK HARBOUR, Kerry, **52°13′·00N 10°24′·00W**. AC 2789. Tides 9.12.31. Adequate shelter in 1M wide bay, except from NW'ly when considerable swell runs in. Ent between The Three Sisters (150m hill) and Dunacapple Is to the NE. Beacon Fl.R.3s at pier end at Ballynagall Pt. ⚓s at: the W side close N or S of the Boat Hr in 3-10m; to the S, off Carrigveen Pt in 2·5m; or in N'lies at the bay in NE corner inside 10m line. Facilities at Ballynagall village on SE side: pier (0·5m), limited 🛒, 🚌 bus.

BRANDON BAY, Kerry, **52°16′·10N 10°09′·92W**. AC 2739. Tides as Fenit 9.12.32. A 4M wide bay, very exposed to the N, but in moderate SW-W winds there is safe ⚓ in 6m close E of drying Brandon Pier, 2FG (vert). Cloghane Inlet in SW of Bay is not advised. Facilities at Brandon: limited 🕿(1M) ⊠ 🛒 ✕ 🍺 bus.

TRALEE BAY, Kerry, **52°18′·00N 09°56′·00W**. AC 2739. HW –0612 on Dover; ML 2·6m; Duration 0605. See 9.12.32. Enter the bay passing 3M N of Magharee Islands. Pick up the W sector (W140°–152°) of Little Samphire Is lt, Fl WRG 5s 17m 16/13M; see 9.12.4 for sectors. Approach between Magharee Is and Mucklaghmore (30m high).

9.12.32 FENIT HARBOUR

Kerry 52°16′·20N 09°51′·61W ❀❀❀♨♦✿✿✿

CHARTS AC 2254, 2739; Imray C55

TIDES –0612 Dover; ML 2·6m; Duration 0605

Standard Port COBH (←—)

Times				Height (metres)			
High Water		Low Water		MHWS	MHWN	MLWN	MLWS
0500	1100	0500	1100	4·1	3·2	1·3	0·4
1700	2300	1700	2300				
Differences FENIT PIER (Tralee Bay)							
–0057	–0017	–0029	–0109	+0·5	+0·2	+0·3	+0·1

SHELTER Good shelter in marina (2.7m depth).

NAVIGATION Appr on 146° to WPT 52°16′·00N 09°53′·00W, Little Samphire Is, conspic lt ho, Fl WRG 5s 17m 16/13M, thence 7ca E to Samphire Is; lt QR 15m 3M vis 242°–097°, leaving S breakwater bn FlR 6m 3M to N. Fenit Pier Hd 2 FR (vert) 12m 3M, vis 148°–058°. Easy ent H24.

LIGHTS AND MARKS See 9.12.4 and chartlet.

COMMUNICATIONS (Code 066) ⊜7121480. Port Manager ☎7136231; VHF Ch 16 **14 M** (0900-2100UT). Tralee SC ☎7136119. *Neptune* (Tralee SC) VHF Ch 14 16.

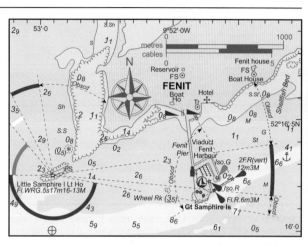

FACILITIES Marina www.fenitharbour.com, ⚓ 130 inc 15Ⓥ; max LOA 25m, €1.60 (min €16.00/craft) 🅿 🚿 🔌 ⟒(at LB stn) ▲ 🔧 🅿 ♨ 🅑 🛒.

Village 🕿 🛢 Ⓑ 🛒 ✕ 🍺.

Tralee (by bus, Fridays only) 🔧 ♨ 🅿 ✕ 🍺 ⇌ ✈ (Kerry 30M).

THE SHANNON ESTUARY

(AC 1819, 1547, 1548, 1549) The lower reaches of the Shannon are tidal for 50M, from its mouth between Loop Hd and Kerry Hd up to Limerick Dock, some 15M beyond the junction with R Fergus.
▶ *The tides and streams have roughly equal duration and similar flood and ebb streams. In the entrance the flood stream begins at HW Tarbert Is –0555, and the ebb at HW Tarbert Is +0015.* ◀

There are several ⚓s available for yachts on passage up or down the coast. Kilbaha Bay (AC 1819) is about 3M E of Loop Hd, and is convenient in good weather or N winds, but is open to SE and any swell. Carrigaholt B (AC 1547), entered about 1M N of Kilcredaun Pt, is well sheltered from W winds and has little tidal stream.

▶ *Off Kilcredaun Pt the ebb reaches 4kn at Sp, and in strong winds between S and NW a race forms. This can be skirted by keeping near the N shore, which is free from offlying dangers,* cheating the worst of the tide. When leaving the Shannon in strong W winds, pass Kilcredaun Pt at slack water, and again keep near the N shore. Loop Hd (lt) marks the N side of Shannon Est, and should be passed 3ca off. Here the stream runs SW from HW Tarbert Is +0300, and NE from HW Tarbert Is –0300. ◀

In N winds ⚓ SE of Querrin Pt (AC 1547), 4½M above Kilcredaun on N shore. At Kilrush there is a marina and ⚓s E of Scattery Is and N of Hog Is. Note that there are overfalls 0·75M S of Scattery Is with W winds and ebb tide.

▶ *Above the junction with R Fergus (AC 1540) the characteristics become more riverine, ie the flood stream is stronger than the ebb, but it runs for a shorter time. In the Shannon the stream is much affected by the wind. S and W winds increase the rate and duration of the flood stream, and reduce the ebb. Strong N or E winds have the opposite effect.* ◀

9.12.33 SHANNON ESTUARY

Clare (N); Kerry and Limerick (S) **52°35'·00N 09°40'·00W** ✵✵✵✵♤♤♧♧♧

CHARTS AC 1819, 1547, 1548, 1549, 1540; L. Derg 5080; L. Ree 5078; Imray C55. The ICC's *Sailing Directions for S & W Ireland* and/or Admiralty *Irish Coast Pilot* are essential.

TIDES

HW at	HW Tarbert Is	HW Dover
Kilbaha & Carrigaholt	−0030	+0605
Coney Island	+0045	−0500
Foynes	+0030	−0515
Limerick	+0110	−0435

At Limerick strong S-W winds increase the height and delay the times of HW; strong N-E winds do the opposite.

SHELTER The Shannon Estuary is 50M long (Loop Hd to Limerick). For boats on passage N/S the nearest ⚓ is Kilbaha Bay, 3M inside Loop Hd; it has ⚓s sheltered in winds from W to NE, but is exposed to swell and holding is poor. 6M further E, Carrigaholt Bay has ⚓s and good shelter from W'lies; ⚓ just N of the new quay, out of the tide. Kilrush Marina (9.12.34) with all facilities is 7M further E. From Kilcredaun Head in the W to the R Fergus ent (about 25M) there are ⚓s or ⚓s, protected from all but E winds, at Tarbert Is, Glin, Labasheeda, Killadysert (pontoon) and among the islands in the Fergus mouth. There is a pontoon off the YC S of Foynes Is and drying quays at Ballylongford Creek (Saleen), Knock and Clarecastle (S of Ennis, off chartlet). Yachts may enter Limerick Dock, but this is a primarily a commercial port.

NAVIGATION WPT 52°32'·52N 09°46'·93W, Ballybunnion NCM Q 062°/4·0M to intercept Ldg lts, Oc. 5s, 046·4° on Corlis Pt, passing between Kilcredaun PHM Q.R(sync) and Tail of Beal buoy SHM, Q.G(sync). For notes on entrance and tidal streams see 9.12.5. The ebb can reach 4kn. The lower estuary between Kerry and Loop Heads is 9M wide, narrowing to 2M off Kilcredaun Pt.

Here the chan is well buoyed in mid-stream and then follows the Kerry shore, S of Scattery Is. From Tarbert Is to Foynes Is the river narrows to less than 1M in places, before widening where the R Fergus joins from the N, abeam Shannon airport. Above this point the buoyed chan narrows and becomes shallower although there is a minimum of 2m at LWS. AC 1540 is essential for the final 15M stretch to Limerick.

RIVER SHANNON The Shannon, the longest navigable river in the UK or Ireland, is managed by Shannon Foynes Port Company up to Limerick. Upstream it effectively becomes an inland waterway; progress is restricted by locks and bridges. Info on navigation and facilities can be obtained from the Waterways Service, Dept of Art, Culture & the Gaeltacht, 51 St Stephen's Green, Dublin 2, ☎01-6613111; or from the Inland Waterways Association of Ireland, Kingston House, Ballinteer, Dublin 4, ☎01-983392; also from Tourist Offices and inland marinas.

LIGHTS AND MARKS See 9.12.4. There are QW (vert) aero hazard lts on tall chimneys at Money Pt power station 3M ESE of Kilrush. 2 chys at Tarbert Is are conspic (R lts).

COMMUNICATIONS Foynes Ch 12 13 16 (occas). *Shannon Estuary Radio* Ch 12 13 16 (HO).

FACILITIES Marine facilities are available at several communities on the Shannon; ⚓ ⚓ 🏠 🏠 🍺 can be found at many villages. ⚓s are at Foynes and Glin Pier (pontoon). E of Aughanish Is on S shore, R Deal is navigable 3M to Askeaton. Deal BC may offer berths (tidal constraints). Ent, marked by RW bn, is 8ca SE of Beeves Rk with 1m in buoyed chan. ⚓ at Massey's Pier ☎069 73100, ⚓s ⚓ 🍺 ⚓ ⚓ ⚓ ⚓.

Facilities at Foynes and Limerick inc:

Foynes ⚓ ⚓ ⚓ ⚓ 🏠 🏠 ✉ Ⓑ 🍺 ✗ 🍴.

Foynes YC ☎(069) 65261, ⚓(pontoon) ✗ 🍴.

Limerick: Hbr Commission ☎(069) 73100; ⊜ ☎415366. **City** all usual city amenities, ⇌ ✈ (Shannon).

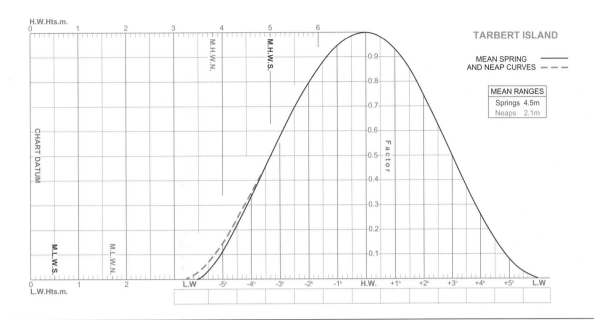

TARBERT ISLAND

MEAN SPRING ————
AND NEAP CURVES – – –

MEAN RANGES	
Springs	4.5m
Neaps	2.1m

9.12.34 KILRUSH

Clare 52°37'·90N 09°29'·70W ✹✹✹◊◊✿✿

CHARTS AC 1819, 1547; Imray C55

TIDES –0555 Dover; ML 2·6; Duration 0610

Standard Port TARBERT ISLAND (→)

Times				Height (metres)			
High Water		Low Water		MHWS	MHWN	MLWN	MLWS
0500	1000	0000	0600	5·0	3·8	1·7	0·5
1700	2200	1200	1800				
Differences KILRUSH							
–0010	–0010	–0005	–0005	0·0	–0·1	0·0	0·0

SHELTER Excellent in Kilrush Marina (2·7m), access via lock (width 9m) H24. Day ‡ in lee of Scattery Is.

NAVIGATION WPT 52°37'·18N 09°31'·64W, 063°/1M to SWM buoy. From seaward usual appr is N of Scattery Is (prominent Round Twr, 26m); beware Baurnahard Spit and Carrigillaun to the N. Follow ldg lts (355°) exactly in buoyed channel where the depth is reported to be 0·5m below CD. Coming downriver between Hog Is and mainland, beware unmarked Wolf Rk.

LIGHTS AND MARKS SWM buoy, L Fl 10s, at ent to buoyed chan, with ldg lts 355°, both Oc 3s, to lock. Fl G 3s on S side of lock.

COMMUNICATIONS (Code 06590) MRCC (066) 9476109; Coast/Cliff Rescue Service 51004; ⊖ (061) 415366; ⚓ (061) 62677; Police 51017; Dr (065) 51581 also 51470. Marina 52072 Mobile 086 2313870; Lock 52155. Marina Ch 80. Kilrush Ch 16, 12.

FACILITIES Marina ⊕ kilrushmarina.ie ☎ 52072, lock 52155; ⛵ 120◯ inc ⓥ €1.90 (min €16.00/craft) ⬛🛢🛢 Kos ⌂ ▣(45t) ⤵(26t) **Town** ⛽✕⚒⚓🍴🏧⊠ⓑ🛒✕🛏@⚡ (bus to Limerick) ✈ (Shannon).

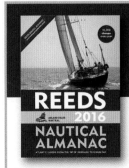

TARBERT ISLAND   LAT 52°35′N   LONG 9°21′W
TIMES AND HEIGHTS OF HIGH AND LOW WATERS

STANDARD TIME (UT)
For Summer Time add ONE hour in **non-shaded areas**

Dates in red are **SPRINGS**
Dates in blue are **NEAPS**

YEAR 2016

JANUARY

Time	m	Time	m
1 0331	1.5	**16** 0340	1.1
1004	4.2	1002	4.6
F 1611	1.5	SA 1615	1.0
2241	4.0	◑ 2241	4.4
2 0423	1.7	**17** 0440	1.3
1059	4.0	1105	4.3
SA 1710	1.6	SU 1719	1.2
◑ 2340	3.9	2349	4.2
3 0527	1.9	**18** 0553	1.4
1206	3.9	1223	4.2
SU 1818	1.7	M 1833	1.3
4 0045	3.9	**19** 0106	4.2
0642	1.9	0713	1.4
M 1317	3.9	TU 1345	4.2
1924	1.7	1947	1.3
5 0147	4.1	**20** 0219	4.3
0751	1.8	0829	1.3
TU 1420	4.0	W 1456	4.4
2022	1.6	2054	1.2
6 0242	4.2	**21** 0322	4.5
0849	1.5	0933	1.0
W 1512	4.2	TH 1555	4.6
2113	1.4	2152	1.0
7 0330	4.4	**22** 0416	4.7
0940	1.3	1026	0.8
TH 1559	4.4	F 1646	4.7
2159	1.2	2242	0.9
8 0414	4.6	**23** 0503	4.9
1026	1.0	1113	0.6
F 1642	4.6	SA 1731	4.9
2242	1.0	2326	0.7
9 0455	4.8	**24** 0546	5.0
1110	0.8	1156	0.5
SA 1723	4.7	SU 1813	4.9
2324	0.8	○	
10 0536	5.0	**25** 0006	0.7
1151	0.6	0624	5.1
SU 1804	4.8	M 1235	0.5
●		1851	4.9
11 0004	0.7	**26** 0043	0.7
0616	5.1	0701	5.1
M 1231	0.5	TU 1311	0.5
1844	4.9	1927	4.8
12 0044	0.6	**27** 0115	0.8
0656	5.1	0736	5.0
TU 1311	0.4	W 1344	0.7
1926	4.9	2001	4.8
13 0124	0.6	**28** 0147	0.9
0738	5.1	0811	4.9
W 1351	0.4	TH 1416	0.9
2009	4.8	2036	4.6
14 0205	0.7	**29** 0220	1.1
0822	5.0	0847	4.7
TH 1434	0.5	F 1449	1.2
2055	4.7	2113	4.5
15 0250	0.8	**30** 0256	1.4
0909	4.8	0925	4.4
F 1521	0.7	SA 1526	1.4
2144	4.5	2155	4.3
		31 0337	1.6
		1009	4.2
		SU 1612	1.7
		2244	4.1

FEBRUARY

Time	m	Time	m
1 0427	1.8	**16** 0521	1.5
1106	4.0	1156	4.1
M 1709	1.9	TU 1804	1.6
◑ 2344	4.0		
2 0533	2.0	**17** 0035	4.1
1216	3.9	0647	1.6
TU 1821	2.0	W 1327	4.1
		1928	1.6
3 0052	4.0	**18** 0159	4.2
0653	2.0	0813	1.4
W 1332	3.9	TH 1445	4.2
1936	1.9	2042	1.4
4 0159	4.1	**19** 0309	4.4
0810	1.8	0921	1.2
TH 1438	4.0	F 1546	4.4
2040	1.7	2141	1.2
5 0257	4.3	**20** 0405	4.6
0912	1.5	1014	0.9
F 1533	4.3	SA 1636	4.7
2134	1.4	2229	1.0
6 0348	4.6	**21** 0451	4.8
1004	1.1	1058	0.7
SA 1621	4.5	SU 1718	4.8
2221	1.1	2311	0.8
7 0434	4.8	**22** 0530	5.0
1050	0.8	1138	0.5
SU 1705	4.8	M 1755	4.9
2305	0.8	○ 2348	0.7
8 0517	5.0	**23** 0605	5.1
1132	0.5	1215	0.5
M 1747	5.0	TU 1829	5.0
● 2347	0.5		
9 0559	5.2	**24** 0022	0.7
1213	0.3	0638	5.1
TU 1828	5.1	W 1248	0.6
		1901	5.0
10 0028	0.4	**25** 0052	0.8
0640	5.3	0710	5.1
W 1254	0.2	TH 1318	0.7
1909	5.2	1932	4.9
11 0108	0.3	**26** 0121	0.9
0721	5.3	0743	4.9
TH 1334	0.2	F 1346	0.9
1951	5.1	2004	4.8
12 0149	0.4	**27** 0151	1.0
0804	5.2	0815	4.8
F 1415	0.3	SA 1416	1.1
2034	5.0	2038	4.6
13 0231	0.6	**28** 0223	1.2
0849	5.0	0848	4.6
SA 1459	0.6	SU 1449	1.4
2119	4.8	2114	4.4
14 0317	0.9	**29** 0259	1.5
0939	4.7	0926	4.3
SU 1549	1.0	M 1528	1.6
2212	4.5	2157	4.2
15 0412	1.2		
1039	4.4		
M 1649	1.3		
◑ 2315	4.2		

MARCH

Time	m	Time	m
1 0342	1.7	**16** 0455	1.4
1014	4.1	1135	4.0
TU 1617	1.9	W 1739	1.7
◑ 2251	4.0		
2 0439	1.9	**17** 0006	4.0
1120	3.8	0622	1.6
W 1724	2.1	TH 1310	3.9
2359	3.9	1909	1.7
3 0555	2.0	**18** 0137	4.0
1242	3.8	0754	1.5
TH 1847	2.0	F 1432	4.1
		2028	1.5
4 0114	4.0	**19** 0252	4.2
0724	1.9	0903	1.2
F 1404	3.9	SA 1531	4.3
2007	1.8	2125	1.3
5 0223	4.2	**20** 0347	4.5
0841	1.5	0953	0.9
SA 1507	4.2	SU 1618	4.6
2109	1.4	2210	1.0
6 0321	4.5	**21** 0431	4.7
0938	1.1	1036	0.7
SU 1558	4.5	M 1657	4.8
2159	1.0	2250	0.8
7 0410	4.8	**22** 0508	4.9
1026	0.7	1114	0.6
M 1644	4.8	TU 1732	4.9
2245	0.7	2325	0.7
8 0456	5.1	**23** 0542	4.9
1110	0.3	1149	0.6
TU 1727	5.1	W 1803	4.9
2328	0.3	○ 2357	0.7
9 0539	5.3	**24** 0613	4.9
1152	0.1	1220	0.6
W 1809	5.3	TH 1833	5.0
●			
10 0009	0.2	**25** 0027	0.7
0621	5.5	0643	5.0
TH 1233	0.0	F 1249	0.7
1849	5.4	1903	4.9
11 0050	0.1	**26** 0056	0.8
0703	5.5	0714	4.9
F 1314	0.0	SA 1317	0.9
1930	5.3	1934	4.9
12 0130	0.2	**27** 0125	0.9
0746	5.4	0745	4.8
SA 1355	0.2	SU 1346	1.1
2012	5.2	2006	4.7
13 0212	0.4	**28** 0155	1.1
0831	5.1	0816	4.6
SU 1437	0.6	M 1417	1.3
2056	4.9	2040	4.5
14 0257	0.7	**29** 0229	1.3
0920	4.7	0852	4.4
M 1525	1.0	TU 1453	1.5
2146	4.5	2118	4.3
15 0348	1.1	**30** 0309	1.5
1018	4.3	0935	4.1
TU 1623	1.4	W 1537	1.8
◑ 2246	4.2	2207	4.1
		31 0400	1.7
		1036	3.9
		TH 1639	2.0
		◑ 2311	3.9

APRIL

Time	m	Time	m
1 0510	1.8	**16** 0103	3.9
1157	3.8	0719	1.4
F 1803	2.0	SA 1404	4.0
		1959	1.6
2 0028	3.9	**17** 0220	4.0
0638	1.7	0829	1.3
SA 1328	3.9	SU 1503	4.2
1930	1.8	2056	1.3
3 0145	4.1	**18** 0316	4.2
0801	1.4	0921	1.0
SU 1438	4.2	M 1549	4.4
2040	1.4	2142	1.1
4 0250	4.4	**19** 0401	4.4
0905	1.0	1004	0.8
M 1532	4.5	TU 1629	4.6
2134	0.9	2221	0.9
5 0344	4.8	**20** 0439	4.6
0957	0.5	1042	0.7
TU 1620	4.9	W 1703	4.7
2221	0.5	2257	0.8
6 0432	5.1	**21** 0512	4.7
1044	0.2	1117	0.7
W 1704	5.2	TH 1734	4.8
2306	0.2	2330	0.7
7 0517	5.3	**22** 0544	4.8
1128	0.0	1150	0.7
TH 1747	5.4	F 1804	4.9
● 2349	0.0	○	
8 0601	5.4	**23** 0001	0.7
1210	-0.1	0614	4.8
F 1829	5.4	SA 1220	0.8
		1834	4.9
9 0030	-0.1	**24** 0032	0.8
0645	5.4	0646	4.8
SA 1253	0.0	SU 1250	0.9
1911	5.4	1906	4.8
10 0112	0.0	**25** 0102	0.8
0729	5.3	0718	4.7
SU 1334	0.2	M 1320	1.0
1953	5.2	1939	4.7
11 0154	0.3	**26** 0133	0.9
0815	5.0	0751	4.6
M 1417	0.5	TU 1352	1.2
2037	4.9	2014	4.6
12 0239	0.6	**27** 0207	1.0
0904	4.7	0828	4.4
TU 1504	1.0	W 1428	1.3
2125	4.5	2052	4.4
13 0329	1.0	**28** 0247	1.2
1001	4.3	0912	4.2
W 1600	1.4	TH 1511	1.5
2222	4.2	2138	4.2
14 0432	1.3	**29** 0336	1.3
1114	4.0	1008	4.0
TH 1714	1.7	F 1610	1.7
◑ 2336	3.9	2236	4.0
15 0553	1.5	**30** 0440	1.5
1244	3.8	1122	3.8
F 1841	1.7	SA 1727	1.8
		◑ 2348	4.0

Chart Datum: 0·30 metres below Ordnance Datum (Dublin). HAT is 5·5 metres above Chart Datum.

FREE monthly updates. Register at
www.reedsnauticalalmanac.co.uk

STANDARD TIME (UT)
For Summer Time add ONE hour in **non-shaded** areas

TARBERT ISLAND LAT 52°35'N LONG 9°21'W
TIMES AND HEIGHTS OF HIGH AND LOW WATERS

Dates in red are **SPRINGS**
Dates in blue are **NEAPS**

YEAR 2016

S Ireland

MAY

	Time	m		Time	m
1 SU	0600 1249 1852	1.4 3.9 1.6	**16** M	0132 0738 1422 2013	3.9 1.3 4.0 1.4
2 M	0105 0720 1405 2006	4.1 1.2 4.2 1.3	**17** TU	0232 0835 1511 2103	4.0 1.2 4.2 1.3
3 TU	0216 0828 1503 2105	4.3 0.8 4.5 0.9	**18** W	0321 0922 1553 2146	4.2 1.0 4.4 1.1
4 W	0314 0925 1554 2156	4.7 0.5 4.8 0.5	**19** TH	0403 1004 1630 2225	4.3 0.9 4.5 0.9
5 TH	0406 1016 1641 2243	5.0 0.2 5.1 0.2	**20** F	0439 1043 1703 2301	4.5 0.8 4.7 0.8
6 F	0455 1103 1726 2329	5.2 0.0 5.3 0.0	**21** SA	0513 1118 1735 2337	4.6 0.8 4.7 0.8
7 SA	0543 1148 1810	5.3 0.0 5.3	**22** SU	0547 1152 1808	4.6 0.8 4.8
8 SU	0013 0629 1233 1853	0.0 5.2 0.1 5.3	**23** M	0010 0620 1226 1842	0.7 4.6 0.9 4.8
9 M	0057 0715 1316 1937	0.0 5.1 0.3 5.1	**24** TU	0043 0655 1259 1917	0.7 4.6 0.9 4.7
10 TU	0140 0802 1400 2021	0.2 4.9 0.6 4.8	**25** W	0117 0733 1333 1954	0.8 4.5 1.0 4.6
11 W	0224 0851 1446 2108	0.5 4.6 0.9 4.5	**26** TH	0152 0813 1411 2034	0.8 4.4 1.1 4.5
12 TH	0312 0944 1538 2159	0.8 4.3 1.3 4.2	**27** F	0233 0857 1456 2119	0.9 4.3 1.3 4.4
13 F	0408 1048 1643 2302	1.1 4.0 1.5 4.0	**28** SA	0321 0950 1550 2213	1.0 4.1 1.4 4.2
14 SA	0516 1205 1759	1.3 3.8 1.7	**29** SU	0419 1055 1658 2317	1.1 4.0 1.5 4.1
15 SU	0017 0630 1320 1912	3.8 1.4 3.9 1.6	**30** M	0528 1212 1817	1.1 4.0 1.4
			31 TU	0030 0642 1330 1931	4.1 1.0 4.2 1.2

JUNE

	Time	m		Time	m
1 W	0142 0752 1433 2035	4.3 0.8 4.5 0.9	**16** TH	0234 0837 1513 2108	4.0 1.3 4.2 1.3
2 TH	0246 0854 1528 2132	4.5 0.6 4.7 0.6	**17** F	0323 0925 1555 2153	4.1 1.2 4.4 1.1
3 F	0343 0950 1619 2224	4.8 0.4 5.0 0.3	**18** SA	0405 1008 1633 2235	4.2 1.1 4.6 1.0
4 SA	0436 1041 1707 2313	4.9 0.3 5.1 0.1	**19** SU	0444 1049 1709 2314	4.4 1.0 4.6 0.8
5 SU	0527 1130 1754 2359	5.0 0.2 5.2 0.1	**20** M	0521 1128 1745 2351	4.5 0.9 4.7 0.7
6 M	0615 1216 1839	5.0 0.3 5.1	**21** TU	0559 1205 1822	4.5 0.8 4.8
7 TU	0043 0702 1301 1923	0.1 4.9 0.4 5.0	**22** W	0027 0637 1242 1859	0.7 4.6 0.8 4.8
8 W	0127 0749 1344 2006	0.2 4.8 0.6 4.9	**23** TH	0104 0717 1319 1938	0.6 4.6 0.8 4.8
9 TH	0210 0835 1428 2050	0.4 4.6 0.9 4.6	**24** F	0141 0759 1359 2019	0.6 4.6 0.9 4.7
10 F	0254 0923 1514 2136	0.7 4.3 1.1 4.4	**25** SA	0221 0843 1442 2103	0.6 4.5 1.0 4.6
11 SA	0341 1015 1606 2226	1.0 4.1 1.4 4.1	**26** SU	0307 0933 1533 2153	0.7 4.4 1.1 4.5
12 SU	0435 1116 1709 2325	1.2 3.9 1.6 3.9	**27** M	0359 1031 1633 2252	0.8 4.2 1.2 4.3
13 M	0538 1224 1818	1.4 3.8 1.7	**28** TU	0500 1140 1744	1.0 4.2 1.3
14 TU	0032 0643 1329 1922	3.8 1.4 3.9 1.6	**29** W	0000 0609 1256 1900	4.2 1.0 4.2 1.2
15 W	0138 0743 1425 2018	3.9 1.4 4.0 1.5	**30** TH	0113 0721 1406 2011	4.2 1.0 4.4 1.0

JULY

	Time	m		Time	m
1 F	0223 0828 1507 2114	4.4 0.8 4.6 0.8	**16** SA	0243 0847 1520 2124	3.9 1.5 4.2 1.4
2 SA	0325 0931 1602 2210	4.5 0.7 4.8 0.5	**17** SU	0333 0938 1604 2210	4.1 1.3 4.4 1.1
3 SU	0422 1027 1653 2301	4.7 0.6 5.0 0.3	**18** M	0418 1024 1645 2253	4.3 1.1 4.6 0.9
4 M	0515 1118 1741 2348	4.8 0.5 5.1 0.2	**19** TU	0500 1106 1725 2333	4.4 0.9 4.8 0.7
5 TU	0604 1204 1826	4.9 0.4 5.1	**20** W	0541 1147 1804	4.6 0.8 4.9
6 W	0032 0649 1248 1908	0.2 4.9 0.5 5.0	**21** TH	0012 0621 1227 1842	0.5 4.7 0.7 5.0
7 TH	0114 0733 1328 1949	0.3 4.8 0.6 4.9	**22** F	0050 0702 1305 1922	0.4 4.8 0.6 5.0
8 F	0153 0815 1407 2029	0.4 4.6 0.8 4.7	**23** SA	0128 0743 1345 2003	0.3 4.8 0.6 5.0
9 SA	0231 0856 1445 2109	0.7 4.5 1.0 4.5	**24** SU	0208 0827 1427 2046	0.4 4.7 0.7 4.8
10 SU	0310 0938 1526 2151	0.9 4.3 1.3 4.3	**25** M	0251 0914 1514 2135	0.5 4.6 0.9 4.7
11 M	0351 1025 1614 2239	1.2 4.1 1.5 4.1	**26** TU	0339 1007 1609 2230	0.7 4.4 1.1 4.4
12 TU	0441 1120 1714 2335	1.4 3.9 1.7 3.9	**27** W	0435 1111 1716 2336	0.9 4.3 1.3 4.2
13 W	0541 1225 1826	1.6 3.9 1.8	**28** TH	0542 1226 1836	1.1 4.2 1.3
14 TH	0040 0647 1332 1933	3.8 1.7 3.9 1.7	**29** F	0052 0658 1344 1954	4.1 1.2 4.2 1.2
15 F	0145 0751 1430 2032	3.8 1.6 4.1 1.6	**30** SA	0209 0814 1453 2103	4.1 1.2 4.4 1.0
			31 SU	0317 0921 1552 2201	4.4 1.0 4.6 1.0

AUGUST

	Time	m		Time	m
1 M	0415 1019 1644 2252	4.5 0.8 4.8 0.5	**16** TU	0356 1001 1623 2232	4.3 1.2 4.6 0.9
2 TU	0507 1108 1730 2337	4.7 0.6 5.0 0.3	**17** W	0441 1046 1705 2313	4.5 0.9 4.8 0.6
3 W	0552 1152 1812	4.8 0.6 5.1	**18** TH	0523 1128 1745 2353	4.7 0.7 5.0 0.3
4 TH	0018 0633 1233 1851	0.3 4.8 0.6 5.1	**19** F	0604 1209 1824	4.9 0.4 5.2
5 F	0057 0712 1309 1928	0.3 4.8 0.6 5.0	**20** SA	0032 0644 1249 1904	0.2 5.0 0.3 5.2
6 SA	0132 0748 1342 2003	0.5 4.7 0.8 4.8	**21** SU	0111 0725 1329 1945	0.1 5.0 0.4 5.2
7 SU	0204 0824 1414 2039	0.7 4.6 1.0 4.7	**22** M	0151 0808 1410 2029	0.2 5.0 0.5 5.0
8 M	0236 0900 1449 2116	0.9 4.4 1.2 4.4	**23** TU	0232 0853 1456 2116	0.4 4.8 0.7 4.8
9 TU	0310 0939 1528 2157	1.2 4.2 1.5 4.2	**24** W	0318 0943 1548 2211	0.7 4.5 1.0 4.5
10 W	0351 1025 1616 2246	1.5 4.0 1.7 3.9	**25** TH	0412 1044 1653 2317	1.1 4.3 1.3 4.2
11 TH	0441 1124 1718 2347	1.7 3.9 1.9 3.8	**26** F	0519 1202 1818	1.4 4.1 1.4
12 F	0545 1234 1839	1.9 3.8 1.9	**27** SA	0039 0643 1330 1945	4.0 1.5 4.1 1.3
13 SA	0057 0701 1346 1956	3.7 1.9 3.9 1.8	**28** SU	0204 0808 1446 2056	4.1 1.4 4.3 1.1
14 SU	0207 0812 1447 2057	3.8 1.7 4.1 1.5	**29** M	0314 0917 1545 2152	4.3 1.2 4.5 0.8
15 M	0306 0911 1538 2147	4.0 1.5 4.4 1.2	**30** TU	0410 1010 1635 2239	4.5 0.9 4.8 0.5
			31 W	0457 1055 1718 2321	4.7 0.7 4.9 0.4

Chart Datum: 0·30 metres below Ordnance Datum (Dublin). HAT is 5·5 metres above Chart Datum.

》》 FREE monthly updates. Register at 《
www.reedsnauticalalmanac.co.uk

603

TARBERT ISLAND   LAT 52°35'N   LONG 9°21'W
TIMES AND HEIGHTS OF HIGH AND LOW WATERS

STANDARD TIME (UT)
For Summer Time add ONE hour in **non-shaded areas**

Dates in red are SPRINGS
Dates in blue are NEAPS

YEAR 2016

SEPTEMBER

#	Time	m	#	Time	m
1 TH ●	0537 1136 1755 2359	4.8 0.6 5.0 0.4	**16** F ○	0502 1108 1724 2331	4.9 0.4 5.2 0.1
2 F	0613 1212 1830	4.9 0.6 5.0	**17** SA	0544 1149 1805	5.1 0.2 5.3
3 SA	0034 0646 1245 1903	0.4 4.9 0.7 4.9	**18** SU	0011 0624 1231 1846	0.0 5.2 0.1 5.4
4 SU	0105 0718 1314 1935	0.6 4.8 0.8 4.9	**19** M	0051 0706 1311 1928	0.0 5.2 0.2 5.3
5 M	0134 0750 1343 2008	0.8 4.7 1.0 4.7	**20** TU	0132 0748 1353 2012	0.1 5.1 0.3 5.1
6 TU	0203 0822 1415 2042	1.0 4.6 1.2 4.5	**21** W	0213 0832 1438 2100	0.4 4.9 0.6 4.8
7 W	0234 0858 1450 2118	1.2 4.4 1.4 4.3	**22** TH	0258 0922 1529 2155	0.8 4.6 0.9 4.4
8 TH	0310 0938 1532 2202	1.5 4.2 1.7 4.0	**23** F ◑	0351 1021 1635 2303	1.2 4.2 1.3 4.1
9 F ◑	0354 1030 1626 2300	1.7 4.0 1.9 3.8	**24** SA	0501 1141 1802	1.5 4.0 1.5
10 SA	0454 1139 1741	1.9 3.8 2.0	**25** SU	0030 0631 1316 1934	3.9 1.7 4.0 1.4
11 SU	0013 0611 1301 1912	3.7 2.0 3.8 1.9	**26** M	0158 0800 1434 2043	4.0 1.5 4.2 1.1
12 M	0134 0736 1413 2027	3.8 1.8 4.0 1.6	**27** TU	0304 0904 1531 2134	4.2 1.3 4.4 0.8
13 TU	0240 0845 1510 2121	4.0 1.5 4.3 1.2	**28** W	0356 0953 1617 2218	4.4 1.0 4.7 0.6
14 W	0333 0938 1558 2207	4.3 1.1 4.6 0.7	**29** TH	0438 1035 1658 2257	4.6 0.8 4.8 0.5
15 TH	0419 1024 1642 2250	4.6 0.8 4.9 0.4	**30** F	0515 1113 1733 2333	4.8 0.7 4.9 0.5

OCTOBER

#	Time	m	#	Time	m
1 SA ●	0548 1147 1805	4.8 0.6 4.9	**16** SU ○	0522 1128 1744 2348	5.2 0.1 5.3 -0.1
2 SU	0006 0617 1218 1835	0.5 4.8 0.7 4.9	**17** M	0605 1211 1828	5.3 0.0 5.4
3 M	0036 0647 1247 1906	0.6 4.8 0.8 4.8	**18** TU	0030 0646 1254 1911	0.0 5.3 0.1 5.3
4 TU	0104 0717 1316 1938	0.8 4.8 0.9 4.7	**19** W	0112 0729 1337 1957	0.1 5.2 0.2 5.0
5 W	0132 0749 1346 2010	1.0 4.7 1.1 4.5	**20** TH	0155 0814 1422 2046	0.4 4.9 0.5 4.7
6 TH	0203 0823 1420 2046	1.2 4.5 1.3 4.3	**21** F	0240 0903 1514 2141	0.8 4.6 0.9 4.3
7 F	0237 0901 1459 2128	1.4 4.3 1.5 4.1	**22** SA ◑	0333 1000 1617 2247	1.2 4.2 1.2 4.0
8 SA	0318 0948 1549 2222	1.7 4.1 1.7 3.8	**23** SU	0440 1116 1737	1.5 4.0 1.4
9 SU	0414 1051 1657 2333	1.9 3.9 1.8 3.7	**24** M ◑	0011 0606 1249 1905	3.8 1.7 3.9 1.4
10 M	0530 1211 1824	2.0 3.8 1.8	**25** TU	0135 0733 1407 2014	3.9 1.6 4.1 1.2
11 TU	0057 0657 1335 1948	3.8 1.8 4.0 1.5	**26** W	0239 0837 1504 2105	4.1 1.4 4.3 1.0
12 W	0211 0814 1439 2048	4.0 1.5 4.3 1.1	**27** TH	0329 0926 1550 2148	4.3 1.1 4.5 0.8
13 TH	0307 0911 1530 2138	4.4 1.1 4.6 0.7	**28** F	0411 1007 1630 2227	4.5 0.9 4.6 0.7
14 F	0355 1000 1617 2223	4.7 0.7 4.9 0.3	**29** SA	0447 1044 1706 2303	4.7 0.8 4.7 0.6
15 SA	0439 1045 1701 2306	5.0 0.7 5.2 0.1	**30** SU ●	0519 1119 1738 2336	4.8 0.7 4.8 0.7
			31 M	0548 1151 1809	4.8 0.7 4.8

NOVEMBER

#	Time	m	#	Time	m
1 TU	0006 0617 1222 1839	0.7 4.8 0.8 4.7	**16** W	0011 0629 1238 1857	0.1 5.3 0.1 5.2
2 W	0036 0649 1253 1911	0.9 4.8 0.9 4.7	**17** TH	0055 0714 1323 1944	0.2 5.1 0.2 5.0
3 TH	0106 0722 1324 1945	1.0 4.7 1.0 4.5	**18** F	0138 0759 1409 2033	0.5 4.9 0.4 4.7
4 F	0138 0757 1358 2022	1.2 4.6 1.1 4.3	**19** SA	0224 0846 1458 2125	0.8 4.6 0.7 4.4
5 SA	0213 0834 1437 2104	1.3 4.4 1.3 4.1	**20** SU	0313 0939 1555 2224	1.1 4.3 1.0 4.1
6 SU	0254 0919 1524 2155	1.5 4.2 1.5 3.9	**21** M ◑	0413 1042 1703 2336	1.5 4.1 1.3 3.9
7 M ◑	0345 1015 1625 2300	1.7 4.0 1.6 3.8	**22** TU	0526 1202 1817	1.6 3.9 1.4
8 TU	0454 1127 1742	1.8 3.9 1.6	**23** W	0052 0643 1322 1926	3.9 1.7 3.9 1.3
9 W	0018 0616 1250 1903	3.8 1.7 4.0 1.4	**24** TH	0158 0751 1425 2022	4.0 1.5 4.1 1.2
10 TH	0136 0735 1402 2010	4.1 1.5 4.2 1.0	**25** F	0251 0846 1515 2110	4.2 1.3 4.3 1.1
11 F	0236 0839 1500 2105	4.4 1.1 4.6 0.7	**26** SA	0336 0931 1558 2152	4.4 1.0 4.4 0.9
12 SA	0327 0932 1550 2154	4.7 0.7 4.9 0.4	**27** SU	0414 1012 1636 2230	4.5 1.0 4.6 0.9
13 SU	0415 1021 1638 2241	5.0 0.4 5.1 0.2	**28** M	0449 1050 1711 2306	4.7 0.9 4.6 0.9
14 M ○	0500 1108 1725 2326	4.8 0.1 5.2 0.1	**29** TU ●	0520 1126 1744 2340	4.8 0.9 4.7 0.9
15 TU	0545 1153 1811	5.3 0.0 5.3	**30** W	0553 1201 1817	4.8 0.8 4.7

DECEMBER

#	Time	m	#	Time	m
1 TH	0013 0626 1235 1850	0.9 4.8 0.8 4.6	**16** F	0040 0700 1311 1931	0.4 5.1 0.2 5.0
2 F	0046 0701 1308 1926	1.0 4.8 0.9 4.6	**17** SA	0123 0744 1355 2017	0.5 5.0 0.4 4.8
3 SA	0120 0737 1343 2004	1.1 4.7 1.0 4.4	**18** SU	0206 0829 1440 2104	0.8 4.8 0.6 4.5
4 SU	0156 0815 1421 2046	1.2 4.6 1.1 4.3	**19** M	0251 0915 1528 2153	1.0 4.5 0.9 4.3
5 M	0236 0858 1506 2134	1.3 4.4 1.2 4.2	**20** TU	0339 1006 1622 2249	1.3 4.3 1.1 4.1
6 TU	0324 0948 1559 2231	1.4 4.3 1.3 4.0	**21** W ◑	0436 1106 1724 2353	1.5 4.1 1.4 3.9
7 W ◑	0423 1051 1705 2339	1.6 4.1 1.4 4.0	**22** TH	0544 1219 1830	1.7 3.9 1.5
8 TH	0535 1205 1819	1.6 4.1 1.3	**23** F	0101 0654 1332 1932	3.9 1.7 4.0 1.5
9 F	0054 0653 1322 1929	4.1 1.4 4.2 1.1	**24** SA	0203 0757 1433 2027	4.1 1.6 4.1 1.4
10 SA	0202 0804 1428 2032	4.4 1.2 4.5 0.8	**25** SU	0255 0853 1523 2116	4.2 1.4 4.2 1.3
11 SU	0259 0906 1525 2128	4.6 0.9 4.7 0.6	**26** M	0340 0941 1607 2200	4.4 1.3 4.4 1.2
12 M	0351 1000 1618 2219	4.9 0.6 4.9 0.4	**27** TU	0420 1025 1646 2240	4.6 1.1 4.5 1.1
13 TU	0440 1051 1708 2308	5.1 0.3 5.1 0.3	**28** W	0456 1105 1723 2318	4.7 1.0 4.6 1.0
14 W ○	0528 1139 1757 2355	5.2 0.2 5.1 0.3	**29** TH ●	0532 1143 1758 2354	4.8 0.9 4.6 0.9
15 TH	0615 1226 1845	5.2 0.2 5.1	**30** F	0607 1219 1833	4.9 0.8 4.7
			31 SA	0029 0643 1254 1909	0.9 4.9 0.8 4.7

Chart Datum: 0·30 metres below Ordnance Datum (Dublin). HAT is 5·5 metres above Chart Datum.

604

FREE monthly updates. Register at
www.reedsnauticalalmanac.co.uk

North Ireland

Liscannor Bay clockwise to Lambay Island

9.13.1	Contents	605
9.13.2	Area map and Distance table	606
9.13.3	Tidal stream charts	608
9.13.4	Lights, buoys and waypoints	610
9.13.5	Passage information	612
	Crossing the Irish Sea, crossing to/from Scotland – 612 • Liscannor Bay to Slyne Head – 612 • Slyne Head to Eagle Island, Eagle Island to Bloody Foreland – 615 • Bloody Foreland to Fair Head – 622 • Fair Head to Lambay Island – 629	
9.13.6	Special notes for Ireland See 9.12.6	613
	Useful Gaelic words – 571	
9.13.7	Galway Bay, tide tables/curves	613
	Aran Islands • Galway Harbour	
9.13.8	Westport (Clew Bay)	619
	Clifden Bay • Ballynakill • Inishbofin • Killary Harbour • Inishturk • Clare Island • Blacksod Bay • Portnafrankagh • Broad Haven	
9.13.9	Sligo	620
	Killala Bay • Mullaghmore	
9.13.10	Killybegs	621
	Teelin Harbour • Inishkeel • Burtonport • Gweedore/Bunbeg • Sheep Haven • Mulroy Bay	
9.13.11	Lough Swilly	622
9.13.12	Lough Foyle (River Foyle, tide tables/curves)	623
9.13.13	River Bann and Coleraine	628
	Portstewart	
9.13.14	Portrush	625
9.13.15	Tidal streams around Rathlin Island	630
	Rathlin Island • Ballycastle • Red Bay • Carnlough • Glenarm	
9.13.16	Larne	632
9.13.17	Belfast Lough, tide tables/curves	633
	Carrickfergus • Belfast • Cultra • Ballyholme Bay • Groomsport Bay • Bangor • Donaghadee • Portavogie	
9.13.18	Strangford Lough	639
	Strangford • Portaferry • Quoile • Killyleagh • Sketrick • Kircubbin	
9.13.19	Ardglass	639
	Dundrum Bay • Annalong • Kilkeel	
9.13.20	Carlingford Lough	640
	Port Oriel • Balbriggan • Skerries	

N Ireland

605

1	*Kilrush*	318	361	364	335	332	323	309	291	276	274	270	251	220	203	191	183	142	119	75	76	**1**
2	Galway	348	366	338	309	307	297	284	266	253	251	245	227	192	178	166	163	104	94	49	**2**
3	Slyne Head	352	317	289	260	257	248	234	217	201	199	195	178	143	128	117	114	55	44	**3**
4	Westport	337	323	295	266	249	240	226	207	193	191	187	168	137	120	108	100	57	**4**
5	Eagle Island	297	262	234	205	198	193	175	162	147	145	136	123	88	72	62	59	**5**
6	Sligo	281	246	218	189	179	177	156	146	123	121	117	107	72	51	30	**6**
7	Killybegs	267	232	204	175	171	163	148	132	115	113	109	93	58	43	**7**
8	Burtonport	218	182	153	130	130	116	108	90	74	72	68	49	18	**8**
9	**Tory Island**	209	174	146	117	113	105	90	74	57	55	51	35	**9**
10	L Swilly (Fahan)	200	166	138	109	104	96	81	65	48	46	42	**10**
11	**Lough Foyle**	157	121	92	72	73	55	47	30	11	9	**11**
12	River Bann	155	120	92	63	65	53	40	24	5	**12**
13	Portrush	150	115	87	58	60	48	35	19	**13**
14	Altacarry Head	135	102	74	45	45	31	21	**14**
15	Carnlough	118	78	50	25	26	11	**15**
16	Larne	108	73	45	16	16	**16**
17	Carrickfergus	101	66	39	6	**17**
18	**Bangor**	96	61	34	**18**
19	Strangford L	71	36	**19**
20	Carlingford L	50	**20**
21	*Dun Laoghaire*	**21**

DISTANCE TABLE
Approximate distances in nautical miles are by the most direct route, whilst avoiding dangers and allowing for Traffic Separation Schemes. Places in *italics* are in adjoining areas; places in **bold** are in 9.0.6, Distances across the Irish Sea.

Magnetic Variation 4·8°W

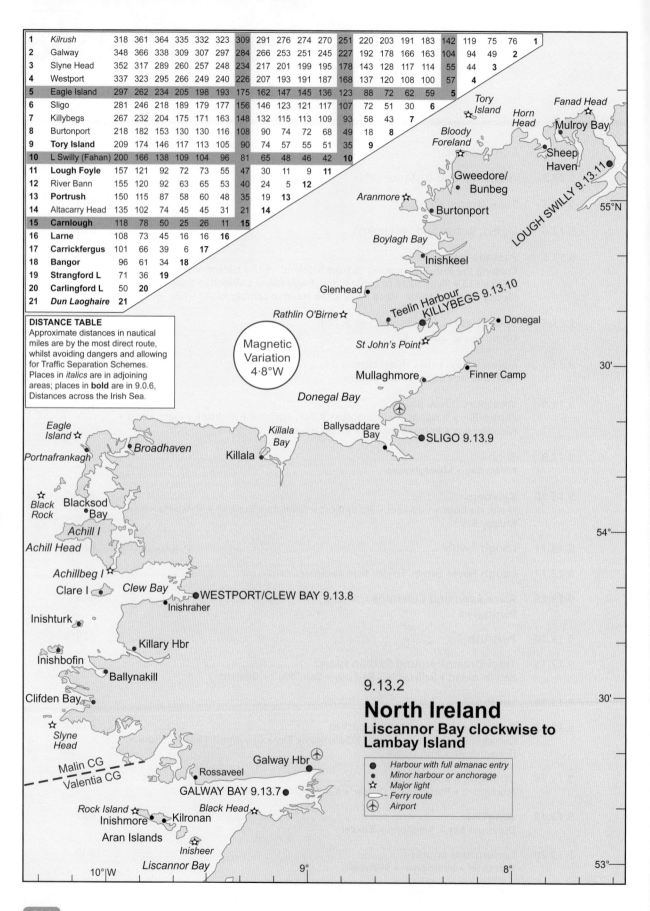

9.13.2
North Ireland
Liscannor Bay clockwise to Lambay Island

●	Harbour with full almanac entry
•	Minor harbour or anchorage
☆	Major light
⌐	Ferry route
✈	Airport

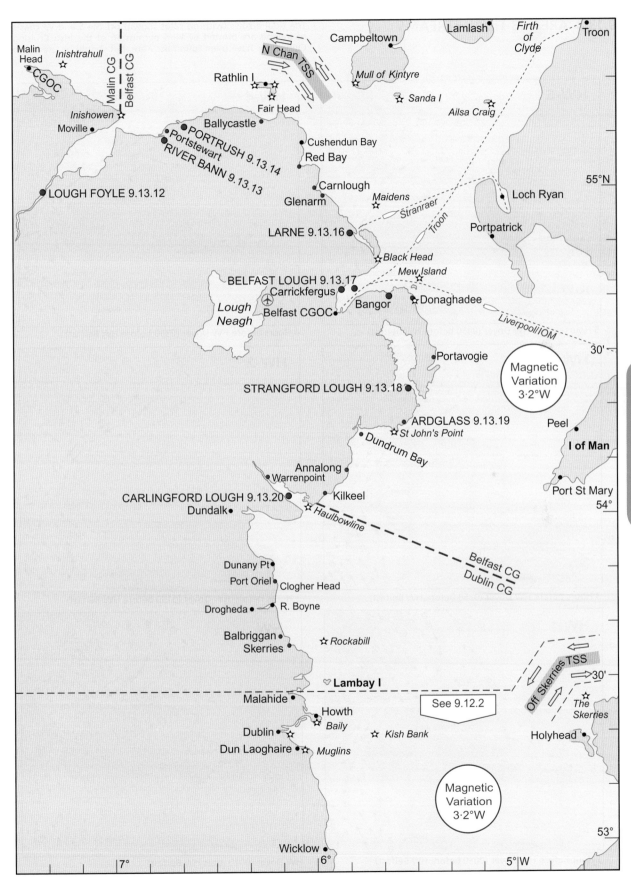

9.13.3 AREA 13 TIDAL STREAMS

The tidal arrows (with no rates shown) off the S and W coasts of Ireland are printed by kind permission of the Irish Cruising Club. They have been found accurate, but should be used with caution.

5 Hours before HW Dover (0450 before HW Belfast)

4 Hours before HW Dover (0350 before HW Belfast)

3 Hours before HW Dover (0250 before HW Belfast)

2 Hours before HW Dover (0150 before HW Belfast)

1 Hour before HW Dover (0050 before HW Belfast)

HW Dover (0010 after HW Belfast)

Rathlin Island 9.13.15   SW Scotland 9.9.3
Mull of Kintyre 9.9.11   South Ireland 9.12.3
North Irish Sea 9.10.3

The tidal arrows (with no rates shown) off the S and W coasts of Ireland are printed by kind permission of the Irish Cruising Club. They have been found accurate, but should be used with caution.

1 Hour after HW Dover (0110 after HW Belfast)

2 Hours after HW Dover (0210 after HW Belfast)

3 Hours after HW Dover (0310 after HW Belfast)

4 Hours after HW Dover (0410 after HW Belfast)

5 Hours after HW Dover (0510 after HW Belfast)

6 Hours after HW Dover (0610 after HW Belfast)

N Ireland

9.13.4 LIGHTS, BUOYS AND WAYPOINTS

Bold print = light with a nominal range of 15M or more. CAPITALS = place or feature. *CAPITAL ITALICS* = light-vessel, light float or Lanby. *Italics* = Fog signal. ***Bold italics*** = Racon. Many marks/buoys are fitted with AIS, <u>MMSI No</u>; see relevant charts.

LISCANNOR BAY TO SLYNE HEAD

GALWAY BAY and INISHMORE

Eeragh, Rock Is ☆ 53°08'·91N 09°51'·40W Fl 15s 35m **18M**; W twr, two B bands; vis: 297°-262°; <u>992501172</u>.

Straw ⚓ Fl (2) 5s 11m 12M; W twr; 53°07'·08N 09°37'·82W.

Killeany Ldg lts 192°. Front, Oc 5s 6m 3M; W col on W ☐ base; vis: 142°-197°, 53°06'·25N 09°39'·74W. Rear, 43m from front, Oc 5s 8m 2M; W col on W ☐ base; vis: 142°-197°.

Inishmaan, Ldg lts 199°, Oc 6s 8M; 53°06'·08N 09°34'·77W.

Killeaney ⬆ Fl G 3s; 53°07.33'N 009°38.37'W; <u>992501166</u>.

Inisheer ☆ 53°02'·76N 09°31'·61W Iso WR 12s 34m **W18M**, R11M; vis: 225°-W(partially vis >7M)-231°, 231°-W-245°-R-269°-W-115°; ***Racon (K) 13M***; <u>992501165</u>.

Finnis ⚓ Q (3) 10s; 53°02'·82N 09°29'·14W, <u>992501164</u>.

Black Hd ⚓ Fl WR 5s 20m W11M, R8M, W ☐ twr; vis: 045°-R268°-276°; 53°09'·25N 09°15'·83W.

GALWAY

Margaretta Shoal ⬆ Fl G 3s; *Whis;* 53°13'·68N 09°05'·99W, <u>992501170</u>.

Leverets ⚓ Q WRG 9m 10M; B ☐ twr, W bands; vis: 015°-G-058°-W-065°-R-103°-G-143·5°-W-146·5°-R-015°; 53°15'·33N 09°01'·90W.

Rinmore ⚓ Iso WRG 4s 7m 5M; W ☐ twr; vis: 359°-G-008°-W-018°-R-027°; 53°16'·12N 09°01'·97W.

Appr Chan Dir lt 325°, WRG 7m 3M; vis: 322·25°-FG-323·75°-AlGW-324·75°-FW-325·25°-AlRW-326·25°-FR-331·25°-FlR-332·25°; 53°16'·12N 09°02'·83W.

GALWAY to SLYNE HEAD

Barna Quay Hd ⚓ Fl 2 WRG 5s 6m W8M, R5M, G5M; vis:250°-G-344·5°-W-355·5°-R-090°; 53°14'·93N 09°08'·90W.

Spiddle Pier Hd ⚓ Fl WRG 3·5s 11m W8M, R8M, G8M; Y col; vis: 265°-G-308°-W-024°-R-066°; 53°14'·42N 09°18'·54W.

Cashla Bay Ent, W side ⚓ Fl (3) WR 10s 8m W6M, R3M; W col on concrete twr; vis: 216°-W-000°-R-069°; 53°14'·23N 09°35'·20W.

Cannon Rk ⬆ Fl G 5s; 53°14'·08N 09°34'·35W, <u>992501173</u>.

Lion Pt Dir lt 010°, Iso WRG 5s 6m (H24), G6M, W8M, R6M,(night), G2M, W3M, R2M(day); vis: 005°-G- 008·5°-W-011·5°-R-015°; 53°15'·83N 09°33'·98W; <u>992501171</u>.

Rossaveel Pier Ldg lts 116°. Front, 53°16'·02N 09°33'·38W Oc 3s 7m 3M; W mast. Rear, 90m from front, Oc 3s 8m 3M.

Kiggaul Bay ⚓ Fl WRG 3s 5m W5M, R3M, G3M; vis: 310°-G-329°-W-349°-R-059°; 53°14'·03N 09°43'·02W.

Croaghnakeela Is ⚓ Fl 3·7s 7m 5M; W col; vis: 034°-045°, 218°-286°, 311°-325°; 53°19'·40N 09°58'·21W.

Inishnee ⚓ Fl (2) WRG 10s 9m W5M, R3M, G3M; W col on W ☐ base; vis: 314°-G-017°-W-030°-R-080°-W-194°; 53°22'·75N 09°54'·53W.

Slyne Head, North twr, Illaunamid ☆ 53°23'·99N 10°14'·05W Fl (2) 15s 35m **19M**; B twr; <u>992501175</u>.

SLYNE HEAD TO EAGLE ISLAND

CLIFDEN BAY and INISHBOFIN

Carrickrana Rocks Bn, large W Bn; 53°29'·24N 10°09'·48W.

Cleggan Point ⚓ Fl (3) WRG 15s 20m W6M, R3M, G3M; W col on W hut; vis: shore-G-091°-W-124°-R-221°; 53°34'·48N 10°07'·69W.

Inishlyon Lyon Head ⚓ Fl WR 7·5s 13m W7M, R4M; W post; vis: 036°-W-058°-R-184°-W-325°-R-036°; 53°36'·74N 10°09'·56W.

Gun Rock ⚓ Fl (2) 6s 8m 4M; W col; vis: 296°-253°; 53°36'·59N 10°13'·23W.

Inishbofin Dir lt 021°, Oc WRG 6s 10m 11M; W pole; vis: 015·75°-G-020·75°-W-021·25°-R-026·25°; 53°36'·78N 10°13'·16W.

CLEW BAY and WESTPORT

Roonagh Quay Ldg lts 144°. Front 53°45'·75N 09°54'·23W. Rear, 54m from front, both Iso Bu 10s 9/15m.

Inishgort ⚓ L Fl 10s 10M; 53°49.59'N 009°40.25'W; <u>992501179</u>.

Cloughcormick ⚓ Q (9) 15s; 53°50'·56N 09°43'·20W.

Achillbeg I, S Point ☆ 53°51'·51N 09°56'·84W Fl WR 5s 56m **W16M, R18M**, R11M; W ○ twr on ☐ building; vis: 262°-R-281°-W-342°-R-060°-W-092°-R(intens)-099°-W-118°; <u>992501181</u>.

ACHILL SOUND

Ldg lts 330°, 53°52'·50N 09°56'·91W Whitestone Point. Front and rear both Oc 4s 5/6m; W ◇, B stripe.

Achill I Ldg lts 310°, Purteen, Oc 8s 5m; 53°57'·83N 10°05'·93W (PA). Rear, 46m from front Oc 8s 6m.

BLACKSOD BAY

Blacksod ⚓ Q (3) 10s; 54°05'·89N 10°03'·01W.

Blacksod Pier Root ⚓ Fl (2) WR 7·5s 13m W12M, R9M; W twr on bldg; vis: 189°-R-210°-W-018°; 54°05'·92N 10°03'·63W; <u>992501186</u>.

Carrigeenmore ⚓ VQ(3) 5s 3M; 54°06'·56N 10°03'·4W.

Black Rock ☆ 54°04'·06N 10°19'·23W Fl WR 12s 86m **W20M, R16M**; W twr; vis: 276°-W-212°-R-276°; <u>992501183</u>.

Eagle Is, W end ☆ 54°17'·02N 10°05'·56W Fl (3) 20s 67m **18M**; W twr; <u>992501189</u>.

EAGLE ISLAND TO RATHLIN O'BIRNE

BROAD HAVEN BAY

Rinroe Pt ⚓ Fl (2) 10s 5m 3M; 54°17'·83N 09°50'·59W.

Gubacashel Point ⚓ Iso WR 4s 27m **W17M**, R12M; 110°-R-133°-W-355°-R-021° W twr; 54°16'·06N 09°53'·33W.

KILLALA

Carrickpatrick ⚓ Q (3) 10s; 54°15'·56N 09°09'·14W; <u>992501192</u>.

Inishcrone Pier Root ⚓ Fl WRG 1·5s 8m 2M; vis: 098°-W-116°-G-136°-R-187°; 54°13'·21N 09°05'·79W.

Ldg lts 230°. Rinnaun Point. Front No. 1, Oc 10s 7m 5M; ☐ twr; 54°13'·21N 09°05'·79W. Rear, 150m from front, No. 2 Oc 10s 12m 5M; ☐ twr.

Inch Is Dir lt 215°, Fl WRG 2s 6m 3M; ☐ twr; vis: 205°-G-213°-W-217°-R-225°; 54°13'·29N 09°12'·30W.

Ldg lts 196°. Kilroe. Front, 54°12'·63N 09°12'·33W Oc 4s 5m 2M; ☐ twr. Rear,120m from front, Oc 4s 10m 2M; ☐ twr.

Ldg lts 236°. Pier. Front, Iso 2s 5m 2M; W ◇ on twr; 54°13'·02N 09°12'·84W. Rear, 200m from front, Iso 2s 7m 2M; W ◇ on pole.

SLIGO

Black Rock ⚓ Fl WR 5s 24m 10/8M; vis: 130°-W-107°-R-130° (R sector covers Wheat and Seal rks); W twr, B band; 54°18'·46N 08°37'·06W ⬆ Fl G 3s; *Whis;* 53°13'·68N 09°05'·99W; <u>992501195</u>.

Lower Rosses, (N of Cullaun Bwee) ⚓ Fl (2) WRG 10s 8m 10M; W hut on piles; vis: 061°-G-066°-W-070°-R-075°; shown H24; 54°19'·73N 08°34'·41W.

Ldg lts 125° Fl (3) 6s 7M (sync). Front, Metal Man; 54°18'·24N 08°34'·55W, 3m. Rear, Oyster Is, 365m from front, 13m; H24.

Wheat Rock ⚓ Q (6) + LFl 15s; 54°18'·84N 08°39'·10W; <u>992501196</u>.

DONEGAL BAY and KILLYBEGS

St John's Pt ⚓ Fl 6s 30m 14M; W twr; 54°34'·16N 08°27'·66W; <u>992501197</u>.

Bullockmore ⚓ Q (9) 15s; 54°33'·98N 08°30'·14W; <u>992501198</u>.

Rotten I ☆ 54°36'·97N 08°26'·41W Fl WR 4s 20m **W15M**, R11M; W twr; vis: W255°- R008°- W039°-208°.

New Landing Dir lt 338°, Oc WRG 8s 17m; vis: 328°-G-334°-Al WG-336°-W-340°-Al WR-342°-R-348°; 54°38'·14N 08°26'·38W.

Killybegs Outer ⚓ VQ (6) + L Fl 10s; 54°37'·92N 08°29'·15W.

RATHLIN O'BIRNE TO BLOODY FORELAND

S of ARAN and RUTLAND S CHANNEL

Rathlin O'Birne ⚓ Fl WR 15s 35m W12M, R10M; W twr; vis: 195°-R-307°-W-195°; ***Racon (O) 13M***, vis 284°-203°. 54°39'·82N 08°49'·95W; <u>992501203</u>.

Dawros Head ☆ L Fl 10s 39m 4M; 54°49'·63N 08°33'·64W.

Wyon Point ⚓ Fl (2) WRG 10s 8m W6M, R3M; W ☐ twr; vis: shore-G-021°-W-042°-R-121°-W-150°-R-shore; 54°56'·51N 08°27'·54W.

BURTONPORT
Ldg lts 068·1°. Front , FG 17m 1M; Gy Bn, W band; 54°58'·95N 08°26'·40W. Rear, 355m from front, FG 23m 1M; Gy Bn, Y band.

N SOUND OF ARAN and RUTLAND N CHANNEL
Rutland I Ldg lts 137·6°. Front, 54°58'·97N 08°27'·68W Oc 6s 8m 1M; W Bn, B band. Rear, 330m from front, Oc 6s 14m 1M.
Inishcoo Ldg lts 119·3°. Front, Iso 6s 6m 1M; W Bn, B band; 54°59'·43N 08°29'·63W. Rear, 248m from front, Iso 6s 11m 1M.
Ldg lts 186°. Front, Oc 8s 8m 3M; B Bn, W band; 54°58'·94N 08°29'·27W. Rear, 395m from front, Oc 8s 17m 3M; B Bn.
Aranmore, Rinrawros Pt ☆ 55°00'·90N 08°33'·67W Fl (2) 20s 71m **27M**; W twr; obsc by land about 234°-007° and about 013°. Auxiliary lt Fl R 3s 61m 13M, same twr; vis: 203°-234°; 992501205.

OWEY SOUND to INISHSIRRER
Cruit Is. Owey Sound Ldg lts 068·3°. Oc 10s. Front, 55°03'·06N 08°25'·85W. Rear, 107m from front.
Rinnalea Point ⚡ 55°02'·59N 08°23'·72W Fl 7·5s 19m 9M; ☐ twr; vis: 132°-167°.
Gola Is Ldg lts 171·2°. Oc 3s 2M (sync). Front, 9m; W Bn, B band; 55°05'·11N 08°21'·07W. Rear, 86m from front, 13m; B Bn, W band.
Glassagh. Ldg lts 137·4°. Front, Oc 8s 12m 3M; 55°06'·83N 08°18'·97W. Rear, 46m from front, Oc 8s 17m 3M; synch.
Inishsirrer, NW end ⚡ Fl 3·7s 20m 4M; W ☐ twr vis: 083°-263°; 55°07'·40N 08°20'·93W.

BLOODY FORELAND TO INISHTRAHULL
BLOODY FORELAND to SHEEPHAVEN
Bloody Foreland ⚡ Fl WG 7·5s 14m W6M, G4M; vis: 062°-W-232°-G-062°; 55°09'·51N 08°17'·03W.
Tory Island ☆ 55°16'·36N 08°14'·96W Fl (4) 30s 40m **18M**; B twr, W band; vis: 302°-277°; *Racon (M) 12-23M;* 992501211.
West Town Ldg lts 001° ⚡ Iso 2s 9m 7M △ on Y structure R stripe; 55°15'·79N 08°13'·50W. Rear Iso 2s 11m 7M ▽ ditto (sync).
Inishbofin Pier ⚡ Fl 8s 3m 3M; 55°10'·14N 08°10'·01W.
Ballyness Hbr. Ldg lts 119·5°. Front, Iso 4s 25m 1M; 55°09'·06N 08°06'·98W. Rear, 61m from front, Iso 4s 26m 1M.
Portnablahy Ldg lts 125·3°. Front Oc 6s 7m 2M; B col, W bands; 55°10'·79N 07°55'·65W. Rear, 81m from front, Oc 6s 12m 2M; B col, W bands.

MULROY BAY
Limeburner ⚓ Q Fl; 55°18'·54N 07°48'·40W; 992501214.
Ravedy Is ⚡ Fl 3s 9m 3M; vis 177°-357°; 55°15'·14N 07°46'·90W

LOUGH SWILLY, BUNCRANA and RATHMULLAN
Fanad Head ☆ 55°16'·57N 07°37'·92W Fl (5) WR 20s 39m **W18M**, R14M; W twr; vis 100°-R-110°-W-313°-R-345°-W-100°; 992501215.
Swillymore ▲ Fl G 3s; 55°15'·12N 07°35'·79W; 992501216.
Dunree ⚡ Fl (2) WR 5s 49m W12M, R9M; vis: 320°-R-328°-W-183°-R-196°; 55°11'·89N 07°33'·25W.
Buncrana Pier near Hd ⚡ Iso WR 4s 11m W13M, R10M; R twr, W band; vis: R shore- over Inch spit, 052°-W-139°-R-shore over White Strand Rock; 55°07'·60N 07°27'·88W.
Rathmullan Pier Hd ⚡ Fl G 3s 5M; vis: 206°-345°; 55°05'·70N 07°31'·66W.
Inishtrahull ☆ 55°25'·89N 07°14'·63W Fl (3) 15s 59m **19M**; W twr; obscd 256°-261° within 3M; *Racon (T) 24M 060°-310°.* Fog Det Lt VQ; 992501229.

INISHTRAHULL TO RATHLIN ISLAND
LOUGH FOYLE
Foyle ⚓ L Fl 10s; 55°15'·32N 06°52'·62W; *Racon (M)*; 992501230.
Tuns ⚓ Fl R 3s; 55°14'·00N 06°53'·46W; 992501232.
Inishowen Dunagree ☆ 55°13'·60N 06°55'·70W Fl (2) WRG 10s 28m **W18M**, R14M, G14M; W twr, 2 B bands; vis: 197°-G-211°-W-249°-R-000°. Fog Det lt VQ 16m vis: 270°.

Greencastle S Bkwtr Dir lt 042·5°. Fl (2) WRG 3s 4m W11M, R9M, G9M; vis 307°-G- 040°-W-045°-R-055°; 55°12'·17N 06°59'·13W.
McKinney's ⚓ Fl R 5s; 55°10'·9N 07°00'·5W.
Moville ⚡ Fl WR 2·5s 11m 4M; W house on G piles vis: 240°-W-064°-R-240°; 55°10'·98N 07°02'·13W.

RIVER BANN, COLERAINE and PORTRUSH
River Bann Ldg lts 165°. Front, Oc 5s 6m 2M; W twr; 55°09'·96N 06°46'·23W. Rear, 245m from front, Oc 5s 14m 2M; W ☐ twr.
W Mole ⚡ Fl G 5s 4m 2M; vis: 170°-000°; 55°10'·25N 06°46'·45W.
Portstewart Point ⚡ Oc R 10s 21m 5M; R ☐ hut; vis: 040°-220°; 55°11'·33N 06°43'·26W.
N Pier Hd ⚡ Fl R 3s 6m 3M; vis: 220°-160°; 55°12'·34N 06°39'·58W.
Skerries ⚓ Fl R 5s; 53°13'·90N 06°36'·90W; 992351096.

RATHLIN ISLAND
Rathlin W 0·5M NE of Bull Pt ⚡ Fl R 5s 62m **22M**; W twr, lantern at base; vis: 015°-225°; H24. Fog Det Lt VQ; 992351000.
Drake Wreck ⚓ 55°17'·00N 06°12'·48W Q (6) + L Fl 15s.
Manor House ⚡ Oc WRG 4s 5M; vis: 020°-G-023°-W-026°-R-029°; 55°17'·52N 06°11'·73W.
Rue Pt ⚡ Fl(2) 5s 16m 14M; W twr B bands; 55°15'·53N 06°11'·47W; 992351131.
Altacarry Head ☆ 55°18'·11N 06°10'·31W Fl (4) 20s 74m **26M**; W twr, B band; vis: 110°-006° and 036°-058°; *Racon (G) 15-27M;* 992320707.

FAIR HEAD TO LAMBAY ISLAND
RED BAY, CARNLOUGH and LARNE
Red Bay Pier ⚡ Fl 3s 10m 5M; 55°03'·93N 06°03'·21W.
Carnlough Hbr N Pier ⚡ Fl G 3s 4m 5M; 54°59'·59N 05°59'·29W.
East Maiden ☆ Fl (3) 15s 29m **23M**; W twr, B band; *Racon (M) 11-21M.* Auxiliary lt Fl R 5s 15m 8M; 54°55'·75N 05°43'·71W; same twr; vis:142°-182° over Russel and Highland Rks; 992351110.
N Hunter Rock ⚓ Q; 54°53'·04N 05°45'·13W.
S Hunter Rock ⚓ VQ(6) + LFl 10s; 54°52'·69N 05°45'·22W; 992351007.
Larne No. 1 ▲ QG; 54°51'·68N 05°47'·67W; *AIS.*
Chaine Twr ⚡ Iso WR 5s 23m **W16M**, R12M; Gy twr; vis: 232°-W-240°-R-000° (shore); 54°51'·27N 05°47'·90W.
Larne Ldg lts 184°, No. 11 Front, 54°49'·60N 05°47'·81W Oc 4s 6m 12M; W 2 with R stripe on R pile structure; vis: 179°-189°. No. 12 Rear, 610m from front, Oc 4s 14m 12M; W 2 with R stripe on R ☐ twr; synch with front, vis: 5° either side of Ldg line.

CARRICKFERGUS
Black Hd ☆ 54°46'·02N 05°41'·34W Fl 3s 45m **27M**; W 8-sided twr; 992351002.
Cloghan Jetty ▲ 54°44'·10N 05°41'·58W QG. Marina Ent Appr ⚡ Dir Oc WRG 3s 5m 3M; vis: G308°- W317·5°- R322·5°-332°; 54°42'·58N 05°48'·78W.

BELFAST LOUGH and BANGOR
Mew Is ☆ NE end 54°41'·92N 05°30'·82W Fl (4) 30s 35m **18M**; B twr, W band; *Racon (O) 14M.* Fog Det Lt VQ; 992351109.
Belfast Fairway ⚓ Iso 4s; 54°42'·32N 05°42'·30W; *Racon (G); AIS.*
Kilroot ▲ Fl QG (sync); 54°42'·85N 05°42'·85W.
Helen's Bay ⚓ QR (sync); 54°41'·86N 05°42'·85W.
No 1 ▲ Fl G 2s (sync); 54°41'·67N 05°46'·51W.
No 2 ⚓ Fl R 2s (sync); 54°41'·55N 05°46'·43W.
Briggs ⚓ Fl (2)R 10s; 54°41'·52N 05°53'·70W; 992351133.

DONAGHADEE
Merandoragh ▲ Fl (2) G 6s (sync); 54°39'·84N 05°32'·24W.
Foreland Spit ⚓ Fl (2) R 6s (sync); 54°39'·64N 05°32'·31W.
Governor's ▲ Fl G 3s (sync); 54°39'·36N 05°31'·99W; 992351134.
Riggs Bank ☉ V–AIS PHM; 54°38'·63N 05°27'·10W; 992356011.
Donaghadee ☆, S Pier Hd 54°38'·71N 05°31'·86W Iso WR 4s 17m **W18M**, R14M; twr; vis: shore-W-326°-R-shore; 002320792.

BALLYWATER and PORTAVOGIE

Ballywalter, Bkwtr Hd Fl WRG 3s 4m 9M; vis: 240°-G-267°-W-277°-R-314°; 54°32'·68N 05°28'·83W.

Skulmartin ⚓ L Fl 10s; 54°31'·82N 05°24'·80W; 992351022.
Portavogie Bkwtr Hd ⚡ Iso WRG 5s 9m W10M, R8M, G8M; ☐ twr; vis: shore-G-258°-W-275°-R-348°; 54°27'·44N 05°26'·14W.
Plough Rock ⬮ 54°27'·40N 05°25'·12W Fl R 3s.
South Rock ⚓ Fl (3) R 30s 7M 54°24'·49N 05°22'·02W; 992351004.

STRANGFORD LOUGH

Strangford ⚓ L Fl 10s; 54°18'·63N 05°28'·69W; 992351028.
Bar Pladdy ⚓ Q (6) + L Fl 15s; 54°19'·34N 05°30'·51W; 992351135.
Dogtail Pt Ldg lts 341°. Front, Iso 4s 2m 5M; G Bn; 54°20'·79N 05°31'·84W. Rear, Gowlands Rk, 0·8M fm front, Iso 4s 6m 5M; W twr, G top, B base.

Swan Is ⚓ Q (3) 5m; BYB brick twr; 54°22'·38N 05°33'·16W.

Strangford East Ldg lts 256°. Front, Oc WRG 5s 6m W9M, R6M, G6M: vis: 190°-R-244°-G-252°-W- 260°-R-294°; 54°22'·29N 05°33'·27W. Rear, 46m from front, Oc R 5s 10m 6M; vis: 250°-264°.

Portaferry Pier Hd ⚡ Oc WR 10s 9m W9M, R6M; Or mast; vis: W335°- R005°- W017°-128°; 54°22'·82N 05°33'·03W.

ARDGLASS

Inner Pier Hd ⚡ Iso WRG 4s 10m W8M, R7M, G5M; twr; vis: shore-G-308°-W-314°-R-shore; 54°15'·79N 05°36'·33W.

DUNDRUM BAY

St John's Point ☆ 54°13'·61N 05°39'·61W Q (2) 7·5s 37m **25M**; B twr, Y bands. **Auxiliary Light** ☆ Fl WR 3s 14m **W15M**, R11M; same twr, vis: 064°-W-078°-R-shore; Fog Det lt VQ 14m vis: 270°; 992351005.
DZ East ⚲ Fl (2) Y 10s; 54°12'·05N 05°45'·30W.
DZ West ⚲ Fl Y 5s; 54°11'·84N 05°50'·82W.

ANNALONG

E Bkwtr Hd ⚡ Oc WRG 5s 8m 9M; twr; vis: 204°-G-249°-W-309°-R-024°; 54°06'·51N 05°53'·73W.

KILKEEL

Pier Hd ⚡ Fl WR 2s 8m 8M; vis: R296°-W313°-017°; 54°03'·46N 05°59'·30W; 992356010 [emergency AIS 992356008].

CARLINGFORD LOUGH and NEWRY RIVER

Hellyhunter ⚓ Q(6) + LFl 15s; 54°00'·35N 06°02'·10W; 992351009.
Haulbowline ⚡ 54°01'·19N 06°04'·74W Fl (3) 10s 32m 10M; Grey twr; 992501235.
Ldg lts 310·4° Front, Oc 3s 7m 11M; R △ on twr; vis: 295°-325°; 54°01'·80N 06°05'·43W; 992501237. Rear, 457m from front, Oc 3s 12m 11M; R ▽ on twr; vis: 295°-325°; both H24.
Newry River Ldg lts 310·4°. Front, 54°06'·37N 06°16'·51W. Rear, 274m from front. Both Iso 4s 5/15m 2M; stone cols.

DUNDALK

Imogene ⬮ Fl (2) R 10s; 53°57'·41N 06°07'·02W; 992501238.
Pile Light ⚡ 53°58'·56N 06°17'·71W Fl WR 15s 10m 10M; W Ho.
Dunany ⬮ Fl R 3s; 53°53'·56N 06°09'·47W; 992501243.

SKERRIES BAY

Skerries Pier Hd ⚡ Oc R 6s 7m 7M; W col; vis: 103°-R-154°; 53°35'·09N 06°06'·49W

DROGHEDA to LAMBAY ISLAND

Port Approach Dir lt 53°43'·30N 06°14'·73W WRG 10m **W19M, R15M**, G15M; vis: 268°-FG-269°-Al WG-269·5°-FW-270·5°-Al WR-271°-FR-272°; H24.
Balbriggan ⚡ Fl (3) WRG 10s 12m W13M, R10M, G10M; W twr; vis: 159°-G-193°-W-288°-R-305°; 53°36'·76N 06°10'·80W.
Rockabill ☆ 53°35'·81N 06°00'·30W Fl WR 12s 45m **W17M, R13M**; W twr, B band; vis: 178°-W-329°-R-178°; 992501246.
Skerries Bay Pier Hd ⚡ Oc R 6s 7m 7M; W col; vis: 103°-154°; 53°35'·09N 06°06'·49W.

9.13.5 PASSAGE INFORMATION

More passage information is threaded between the harbours in this area. Admiralty Leisure Folio 5612 covers Carlingford Lough to Lough Foyle.The latest editions of the Irish Cruising Club Sailing Directions are strongly recommended for all Irish waters (www. irishcruisingclub.com). Particularly useful on the N and W coast where other information is scarce they are published in two volumes, *E and N coasts of Ireland* and *S and W coasts of Ireland*. (Both available from Imray in UK or Todd Chart Services, Bangor, Co Down, ☎028 9146 6640 (www.toddchart.com).

CROSSING THE IRISH SEA

(AC 1123, 1121, 1411) Passages across the Irish Sea range from the fairly long haul from Land's End to Cork (140M), to the relatively short hop from Mull of Kintyre to Torr Pt (11M). Such distances are deceptive, because the average cruising yacht needs to depart from and arrive at a reasonably secure hbr. See 9.0.6 for distances across the Irish Sea. Submarine exercise areas are at 9.9.18.

▶*Strong tidal streams in North Channel can cause heavy overfalls.*◀

Many yachts use the Land's End/Cork route on their way to and from the delightful cruising ground along the S coast of Ireland. Penzance Bay or one of the Scilly Islands' ⚓s are convenient places from which to leave, with good lights to assist departure. Although the Celtic Sea is exposed to the Atlantic, there are no dangers on passage and the tidal streams are weak.

A landfall between Ballycotton and Old Hd of Kinsale (both have good lights) presents no offlying dangers, and in poor vis decreasing soundings indicate approach to land. Outward bound under sail, it is likely that the boat will be on the wind – a possible benefit on the return passage. If however the wind serves, and if it is intended to cruise along the southern coast, a landfall at the Fastnet with arrival at (say) Baltimore will place the yacht more to windward, for a little extra distance (170M).

From Land's End another likely destination is Dun Laoghaire. A stop at Milford Haven enables the skipper to select the best time

for passing the Smalls or Bishops (see 9.11.5) and roughly divides the total passage into two equal parts of about 100M. From S Bishop onwards there are the options of making the short crossing to Tuskar Rk and going N inside the banks (theoretically a good idea in strong W winds) or of keeping to seaward.

▶*But in bad weather the area off Tuskar is best avoided; apart from the Traffic Separation Scheme, the tide is strong at Sp and the sea can be very rough.*◀

The ferry route Holyhead/Dun Laoghaire is another typical crossing (56M) and with relatively straightforward landfalls either end. Beware the TSS off The Skerries and Kish Bank wind farm.

▶*The tide runs hard round Anglesey at Sp, so departure just before slack water minimises the set N or S.*◀

Harbours on the Isle of Man provide convenient staging points for cruising in the Irish Sea, wherever bound.

CROSSING TO/FROM SCOTLAND

(AC 2198, 2199, 2724) Between Scotland and Ireland there are several possible routes, but much depends on weather and tide. Time of departure must take full advantage of the stream, and avoid tide races and overfalls (see 9.9.5). Conditions can change quickly, so a flexible plan is needed. The distance to Mull of Kintyre from Belfast Lough entrance is about 35M.

▶ *A departure at local HW (also HW Dover) provides at least 6hrs of N-going tides, and fair winds make it possible to get past the Mull or Sanda Is on one tide. To be more confident of reaching Gigha or Port Ellen, a departure from Glenarm Marina or Red Bay at HW makes a shorter passage with better stream advantage. The inshore side of the TSS coincides with the outer limit of the race S-SW off the Mull of Kintyre; this runs from HW Dover +0430 until +0610 when a local S-going stream opposes the main N-going stream.*◀

LISCANNOR BAY TO SLYNE HEAD

(AC 2173) The coast NE of Liscannor Bay is devoid of shelter. O'Brien's Tower is conspicuous just N of the 199m high Cliffs of

Moher. Eastward from Black Head there are many bays and inlets, often poorly marked, but providing shelter and exploration.

Galway is a commercial port, with a small marina. 3M SE, New Harbour (AC 1984) is a pleasant ⚓ with moorings off Galway Bay SC. The N side of Galway Bay is exposed, with no shelter. Normal approach to Galway Bay is through South Snd or North Snd. South Snd is 3M wide, with no dangers except Finnis Rock (0·4m) 5ca SE of Inisheer. North Snd is 4M wide from Eagle Rock and other dangers off Lettermullan shore, to banks on S side which break in strong winds. The other channels are Gregory Sound, 1M wide between Inishmore and Inishmaan, and Foul Sound between Inishmaan and Inisheer. The latter has one danger, Pipe Rock and the reef inshore of it, extending 3ca NW of Inisheer.

In good visibility the Connemara coast (AC 2709, 2096) and Aran Islands (AC 3339) give excellent cruising. But there are many rocks and few navigational marks. Between Slyne Hd and Roundstone Bay are many offlying dangers. If coasting, keep well seaward of Skerd Rocks. A conspicuous tower (24m) on Golan Head is a key feature. Going E the better harbours are Roundstone Bay, Cashel Bay, Killeany Bay, Greatman Bay and Cashla Bay. Kilronan (Killeany Bay) on Inishmore is the only reasonable harbour in the Aran Islands, but is exposed in E winds. Disused lighthouse on centre of island is conspicuous.

9.13.6 SPECIAL NOTES FOR IRELAND: See 9.12.6

9.13.7 GALWAY BAY

Galway 53°12'N 09°08'W ※※※※♠♠♠♠♠

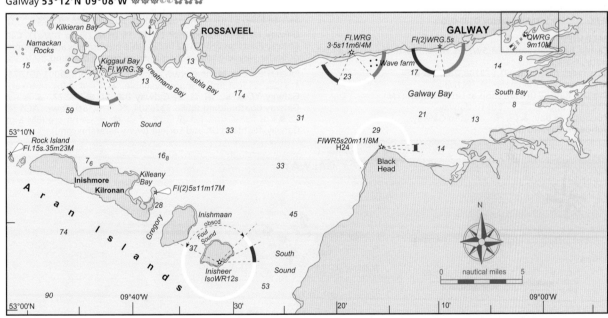

CHARTS AC 2173, 3339, 1984,1904; Imray C55, C54

TIDES –0555 Dover; ML 2·9; Duration 0620

Standard Port GALWAY (→)

Times				Height (metres)			
High Water		Low Water		MHWS	MHWN	MLWN	MLWS
0600	1100	0000	0700	5·1	3·9	2·0	0·8
1800	2300	1200	1900				
Differences KILKIERAN COVE							
+0005	+0005	+0016	+0016	–0·3	–0·2	–0·1	–0·2
ROUNDSTONE BAY							
+0003	+0003	+0008	+0008	–0·7	–0·5	–0·3	–0·3
KILLEANY BAY (Aran Islands)							
0000	0000	+0001	–0001	–0·1	+0·1	+0·1	+0·2
LISCANNOR							
–0003	–0007	+0006	+0002	–0·4	–0·3	ND	ND

SHELTER Galway Bay is sheltered from large swells by Aran Is, but seas get up in the 20M between the Aran Is and Galway. Beware salmon drift nets in the apps to many bays. On the S Shore:

Bays between Black Hd and Galway have rocks and shoals, but give excellent shelter. Kinvarra B, Aughinish B, South B and Ballyvaghan B are the main ones. Enter Kinvarra B with caution on the flood; beware rks. Berth in small drying hbr. Beware fish farm in South B. Ballvaghan B, entered either side of Illaunloo Rk, leads to two piers (both dry) or ⚓ close NE in pool (3m). Best access HW±2.

On the N coast there is no safe hbr from Galway to Cashla (20M).
Cashla Bay: Easiest all weather entry on this coast. To the East side:
Rossaveel is a sheltered but busy fishing and ferry hbr, with small dedicated marina. To the West side: 8 Y ⚓s off Struthan Quay.

The better ⚓s in outer Galway Bay northward to Slyne Head are:
Greatman Bay: Beware English Rk (dries 1·2m), Keeraun Shoal (1·6m, seas break), Arkeena Rk, Trabaan Rk, Rin and Chapel Rks.

⚓ off Natawny Quay (E side), or on 4 Y ⚓s off Maumeen Quay (dries), ⌷ possible.

Kiggaul Bay: Easy ent, H24; ⚓ close W /NW of lt Fl WRG 3s 5/3M. Depth 3 to 4m; exposed to S/SE winds.

Kilkieran Bay: Easy ent abm Golam Tr (conspic). 12 Y ⚓s off Kilkieran. Many ⚓s in 14M long, sheltered bay.

Roundstone Bay: AC2709 (Off chartlet at 53°23'N 09°54'·5W). Safe shelter/access, except in SE'ly. 4 Y ⚓s are 5ca SSE of Roundstone, or ⚓ in 2m off N quay. Other ⚓s E in Bertraghboy and Cashel Bays.

Bunowen Bay: (53°24'·6N 10°06'·9W). Sheltered in W-NE winds; unsafe in S'ly. Easy appr with Bunowen House brg 000°, leaving Mullauncarrickscoltia Rk (1·1m) to port. ⚓ in 3-4m below conspic Doon Hill.

NAVIGATION Enter the Bay by one of four Sounds:
- North Sound between Inishmore and Golam Tr (conspic), 4½M wide, is easiest but beware Brocklinmore Bank in heavy weather.
- Gregory Sound between Inishmore and Inishmaan, is free of dangers, but give Straw Island a berth of 2-3ca.
- Foul Sound between Inishmaan and Inisheer; only danger is Pipe Rock (dries) at end of reef extending 3ca NW of Inisheer.
- South Sound between Inisheer and mainland. Finnis Rock (dries 0·4m) 4½ca SE of E point of Inisheer (marked by ECM buoy Q (3) 10s). From S, beware Kilstiffin Rocks off Liscanor Bay.

LIGHTS AND MARKS Cashla Bay: Killeen Pt Fl (3) WR 10s 6/3M; Lion Pt Dir lt 010°, Iso WRG 4s 8/6M, 008·5°-W-011·5°. Rossaveel ldg lts 116°, Oc 3s. Black Head lt Fl WR 5s 20m 11/8M H24, vis 045°-W-268°-R (covers Illanloo Rk)-276°.

Kiggaul Bay: Fl WR 3s 5m 5/3M, 329°-W-359°-R-059°.

Roundstone Bay: Croaghnakeela Is Fl 3·7s 7m 5M. Inishnee lt Fl (2) WRG 10s 9m 5/3M, W sector 017°-030°.

FACILITIES

Cashla Bay: Carraroe (1M SW) 🛒 Hotel. **Costelloe** (1½M E) ✉ Hotel, Garage.
Rossaveel: Marina ☎091 572108, 28⌷ 🛢 🗼 ⚒.
Greatman Bay: Maumeen 🗼 ☎(1M) ⛽.
Kiggaul Bay: ⛽ (no ☎), shop at Lettermullen (1M).
Kilkieran Bay: ☎ 🛒 ⛽ 🗼 Bus to Galway.
Roundstone Bay: 🗼 ✉ 🗼 ⛽ Bus to Galway.
Bunowen Bay: No facilities.

ARAN ISLANDS Shelter at **Inishmore** in Killeany Bay with harbour at Kilronan, HM ☎099 20946. 8 Y ⚓s available or ⚓ S of Kilronan pier or ⚓ E of Killeany Pt, but exposed to E/NE wind. In good weather ⚓ at Portmurvy. Facilities: 🛢 🗼 ✉ 🗼; Ferry to Rossaveel, **Lights and marks:** Inishmore: Eeragh Island (Rk Is) Fl 15s 35m 23M, W tr, B bands. Killeany Bay: Straw Is, Fl (2) 5s 11m 17M. Ldg lts 192° Oc 5s for Killeany hbr. **Inishmaan** Ldg lts 199°, Oc 6s 8M, lead into small harbour with 3·5m and good shelter. Ferries have priority for ⌷, but do not stay overnight. Inisheer: Iso WR 12s 34m 20/16M, Racon, vis 225°-W-245° (partially obscd 225°-231° beyond 7M), R sector covers Finnis Rk.

GALWAY HARBOUR
53°16'·07N 09°02'·74W ⚓⚓⚓☆☆☆☆☆

SHELTER Very good in Galway hbr, protected from SW'lies by Mutton Island. Dock gates open HW–2 to HW. The approach channel is dredged to 3.4m. Enter Galway Dock and secure in SW corner of basin where there is a 31-berth marina, or ask HM for waiting berth on lead-in pier. Reclamation work at Rinmore Pt has improved shelter at the 'Layby' (a dredged cut NE of New pier) where an additional 8 berths are available on a 60m pontoon. New Harbour (2·5M ESE and home of Galway Bay SC) is nearest safe ⚓ to Galway. 5 Y ⚓s in lee of Mutton Island causeway in 2-4m, may be used while waiting for the dock gates to open.

NAVIGATION Galway WPT 53°14'·80N 09°03'·40W, 061°/1·1M to Leverets lt. Shipping information see www.aislive.com.

LIGHTS AND MARKS See 9.13.4 and chartlet. Leverets Q WRG 9m 10M; B tr, W bands; Rinmore Iso WRG 4s 7m 5M; W □ tr. Appr chan 325° is defined by a Dir lt WRG 7m 3M on New Pier.

COMMUNICATIONS (Code 091) MRCC (066) 9476109; Coast Rescue Service (099) 61107; Police 538000; Dr 562453. HM 561874.

Galway Harbour Radio VHF Ch **12** 16 (HW–2½ to HW+1). Call Galway Pilots on VHF Ch **12** in case ship movements are imminent or under way and follow their instructions on berthing.

FACILITIES **City Marina:** www.galwayharbour.com ☎091 561874, ⚓ 31⌷ €15(<13m) if space available, 🗼 🗼 🗼 🗼(by r/tanker). **Galway YC** 🗼 🗼 🗼 🗼 🗼; **Galway Bay SC** ☎794527, 🗼 🗼 ⛽. **Galway Boatbuilding** ☎086 3379304, ⚒ 🗼 🗼(18t). **Town** ACA ☎ 🗼 🗼 Kos ⚒ 🗼 🗼 🗼 ✉ Ⓑ 🗼 🗼 ✕ 🗼 ⇌ ✈ Carnmore (6M E of Galway, flights to UK and Lorient (www.aerarann.com)).

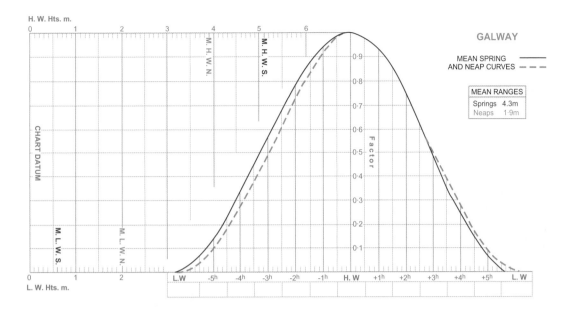

H. W. Hts. m.

GALWAY

MEAN SPRING
AND NEAP CURVES

MEAN RANGES

| Springs | 4.3m |
| Neaps | 1.9m |

SLYNE HEAD TO EAGLE ISLAND

(AC 2420) This coast has many inlets, some sheltered, but few lights.

▶ *Streams are weak offshore.* ◀

Slyne Hd Lt Ho marks SW end of the rocks and islets stretching 2M WSW from coast.

▶ *Here the stream turns N at HW Galway −0320, and S at HW Galway +0300. It runs 3kn at sp and in bad weather causes a dangerous race; keep 2M offshore.* ◀

Seas break on Barret Shoals, 3M NW of Slyne Hd. Cleggan B is moderate ⚓, open to NW but easy access. High Is Sound is usual coasting route, not Friar Is Sound or Aughrus Passage. Clifden Bay (AC 2708) has offlying dangers with breakers; enter 3ca S of Carrickrana Bn and ⚓ off Drinagh Pt, in Clifden Hbr or Ardbear Bay.

Ballynakill Hbr (AC 2706), easily entered either side of Freaghillaun South, has excellent shelter; Tully mountain is conspic to N. Beware Mullaghadrina and Ship Rk in N chan. ⚓ in Fahy, Derryinver or Barnaderg B. There is ⚓ on S side of Inishbofin (lt), but difficult access/exit in strong SW wind or swell. Rks and breakers exist E of Inishbofin and S of Inishshark; see AC 2707 for clearing lines. Lecky Rks lie 1M SSE of Davillaun. Carrickmahoy is a very dangerous rock (1·9m) between Inishbofin and Cleggan Pt.

Killary B (AC 2706) and Little Killary both have good ⚓s in magnificent scenery. Consult sailing directions, and only approach in reasonable weather and good vis.

Clare Is has Two Fathom Rk (3·4m) 5ca off NW coast, and Calliaghcrom Rk 5ca to the N; ⚓ on NE side. In Clew Bay Newport and Westport (AC 2667, 2057) need detailed pilotage directions. S of Clare Is beware Meemore Shoal 1·5M W of Roonagh Hd. 2M further W is the isolated rk Mweelaun. The islands of Caher, Ballybeg, Inishturk (with ⚓ on E side) and Inishdalla have few hidden dangers, but the coast to the E must be given a berth of 1·5M even in calm weather; in strong winds seas break much further offshore.

Achill Sound (ACs 2667/2704) is entered close N of Achillbeg Is and extends N for about 8M to Bull's Mouth. Passage is possible at HW for shallow draught craft (<2m) via a swing bridge, which is rarely opened. ⚓s at each end of Sound, but the stream runs strongly. Rough water is likely off Achill Hd (highest cliffs in the British Islands).

Blacksod B (AC 2704) has easy ent (possible at night) and good shelter. In the appr Black Rk (lt) has rks up to 1·25M SW. From N, in

good weather, there is chan between Duvillaun Beg and Gaghta Is, but in W gales beware breakers 1M SE of Duvillaun More.

Keep 5ca off Erris Hd, and further in bad weather. Unless calm, keep seaward of Eagle Is (lt) where there is race to N. Frenchport (AC 2703) is good temp ⚓ except in strong W winds. Inishkea Is (AC 2704) can be visited in good weather; ⚓ N or S of Rusheen Is. On passage keep 5ca W of Inishkea to avoid bad seas if wind over tide. The sound off Mullett Peninsula is clear, but for Pluddany Rk 6ca E of Inishkea N.

EAGLE ISLAND TO BLOODY FORELAND

Broad Haven (AC 2703) is good ⚓ and refuge, but in N/NW gales sea can break in ent. In approaches beware Slugga Rk on E side with offlier, and Monastery Rk (0·3m) on S side.

The coast E of Broad Haven is inhospitable. Only Portacloy and Belderg give a little shelter. The Stags, 1M N of Buddagh, are steep-to and high. Killala B has temp ⚓ 1M S of Kilcummin Hd, on W side. Proceeding to Killala beware St Patrick's Rks. Ent has ldg lts and marks, but bar is dangerous in strong NE winds.

In Donegal B (AC 2702) beware uncharted rks W of Teelin, a good natural hbr but exposed to S/SW swell. Killybegs has better shelter and is always accessible. Good shelter with fair access in Donegal Hbr (AC 2715). Good ⚓ or ⚓ via YC at Mullaghmore in fair weather; sea state is calm with winds from SE through S to NW. Inishmurray is worth a visit in good weather, ⚓ off S side. There are shoals close E and NE of the Is, and Bomore Rks 1·5M to N. Keep well clear of coast N from Sligo in onshore winds, and beware of lobster pots.

The coast and islands 15M NE from Aran Is give good cruising (AC 1883). An inshore passage avoids offlying dangers: Buniver and Brinlack shoals, which can break; Bullogconnell 1M NW of Gola Is; and Stag Rks 2M NNW of Owey Is. ⚓s include Bunbeg and Gweedore Hbr, and Cruit B which has easier access. Behind Aran Is are several good ⚓s. Use N ent, since S one is shallow (AC 2792). Rutland N Chan is main appr to Burtonport.

Boylagh B has shoals and rks N of Roaninish Is. Connell Rk (0·3m) is 1M N of Church Pool, a good ⚓, best approached from Dawros Hd 4·5M to W. On S side of Glen B a temp ⚓ (but not in W or NW winds) is just E of Rinmeasa Pt. Rathlin O'Birne Is has steps E side; ⚓ SE of them 100m offshore. Sound is 5ca wide; hold Is side to clear rks off Malin Beg Hd.

Off low-lying Bloody Foreland (lt) there is often heavy swell.

STANDARD TIME (UT)
For Summer Time add ONE hour in **non-shaded areas**

GALWAY LAT 53°16'N LONG 9°03'W
TIMES AND HEIGHTS OF HIGH AND LOW WATERS

Dates in red are **SPRINGS**
Dates in blue are **NEAPS**

YEAR 2016

JANUARY

Time	m	Time	m
1 0307 / 0936 / F 1544 / 2209	1.9 / 4.4 / 1.8 / 4.2	**16** 0303 / 0935 / SA 1532 / ☽ 2210	1.3 / 4.8 / 1.3 / 4.5
2 0412 / 1026 / SA 1653 / ☽ 2301	2.1 / 4.2 / 2.0 / 4.1	**17** 0402 / 1035 / SU 1633 / 2315	1.6 / 4.6 / 1.5 / 4.4
3 0531 / 1121 / SU 1804	2.2 / 4.0 / 2.1	**18** 0511 / 1147 / M 1745	1.8 / 4.4 / 1.7
4 0000 / 0636 / M 1226 / 1904	4.0 / 2.2 / 4.0 / 2.0	**19** 0034 / 0636 / TU 1310 / 1915	4.3 / 1.8 / 4.3 / 1.7
5 0113 / 0733 / TU 1348 / 1958	4.1 / 2.1 / 4.1 / 1.9	**20** 0150 / 0806 / W 1421 / 2030	4.4 / 1.6 / 4.5 / 1.6
6 0219 / 0824 / W 1447 / 2046	4.3 / 1.9 / 4.3 / 1.7	**21** 0251 / 0907 / TH 1518 / 2123	4.7 / 1.4 / 4.7 / 1.4
7 0306 / 0910 / TH 1532 / 2129	4.5 / 1.6 / 4.5 / 1.5	**22** 0342 / 0955 / F 1608 / 2207	4.9 / 1.1 / 4.9 / 1.2
8 0348 / 0951 / F 1613 / 2208	4.8 / 1.3 / 4.7 / 1.3	**23** 0428 / 1036 / SA 1653 / 2245	5.2 / 0.9 / 5.1 / 1.1
9 0427 / 1029 / SA 1653 / 2244	5.0 / 1.1 / 4.9 / 1.1	**24** 0511 / 1111 / SU 1735 / ○ 2319	5.3 / 0.8 / 5.2 / 1.0
10 0507 / 1106 / SU 1733 / ● 2320	5.2 / 1.0 / 5.1 / 1.0	**25** 0551 / 1143 / M 1815 / 2353	5.3 / 0.8 / 5.2 / 1.0
11 0547 / 1143 / M 1814 / 2358	5.3 / 0.7 / 5.1 / 0.9	**26** 0629 / 1216 / TU 1854	5.3 / 0.8 / 5.1
12 0627 / 1223 / TU 1855	5.4 / 0.6 / 5.2	**27** 0028 / 0706 / W 1252 / 1930	1.1 / 5.2 / 0.9 / 5.0
13 0040 / 0709 / W 1305 / 1938	0.9 / 5.4 / 0.7 / 5.1	**28** 0105 / 0742 / TH 1328 / 2007	1.2 / 5.0 / 1.1 / 4.8
14 0124 / 0754 / TH 1350 / 2024	0.9 / 5.2 / 0.8 / 4.9	**29** 0141 / 0821 / F 1405 / 2047	1.3 / 4.8 / 1.4 / 4.6
15 0212 / 0842 / F 1438 / 2114	1.1 / 5.1 / 1.0 / 4.7	**30** 0219 / 0902 / SA 1443 / 2130	1.6 / 4.6 / 1.6 / 4.4
		31 0258 / 0948 / SU 1524 / 2217	1.9 / 4.3 / 1.9 / 4.2

FEBRUARY

Time	m	Time	m
1 0344 / 1038 / M 1615 / ☽ 2308	2.1 / 4.1 / 2.1 / 4.1	**16** 0446 / 1124 / TU 1720	1.7 / 4.2 / 1.8
2 0453 / 1133 / TU 1823	2.3 / 3.9 / 2.2	**17** 0008 / 0615 / W 1255 / 1856	4.2 / 1.8 / 4.1 / 1.9
3 0006 / 0700 / W 1238 / 1929	4.0 / 2.2 / 3.9 / 2.1	**18** 0135 / 0756 / TH 1411 / 2020	4.2 / 1.7 / 4.2 / 1.8
4 0114 / 0758 / TH 1356 / 2023	4.1 / 2.0 / 4.0 / 1.9	**19** 0240 / 0902 / F 1509 / 2116	4.4 / 1.4 / 4.5 / 1.5
5 0225 / 0849 / F 1501 / 2110	4.3 / 1.7 / 4.3 / 1.6	**20** 0332 / 0949 / SA 1557 / 2158	4.7 / 1.2 / 4.8 / 1.3
6 0319 / 0933 / SA 1548 / 2152	4.6 / 1.4 / 4.6 / 1.3	**21** 0416 / 1027 / SU 1639 / 2232	5.0 / 1.0 / 5.0 / 1.1
7 0403 / 1014 / SU 1631 / 2230	4.9 / 1.0 / 4.9 / 1.0	**22** 0457 / 1057 / M 1718 / ○ 2301	5.1 / 0.7 / 5.1 / 1.0
8 0445 / 1051 / M 1712 / ● 2306	5.2 / 0.7 / 5.2 / 0.8	**23** 0534 / 1121 / TU 1755 / 2330	5.2 / 0.8 / 5.1 / 0.9
9 0526 / 1128 / TU 1753 / 2344	5.5 / 0.5 / 5.3 / 0.6	**24** 0610 / 1150 / W 1830	5.2 / 0.7 / 5.1
10 0609 / 1206 / W 1836	5.6 / 0.3 / 5.4	**25** 0003 / 0644 / TH 1223 / 1904	0.9 / 5.2 / 0.8 / 5.0
11 0024 / 0652 / TH 1247 / 1918	0.5 / 5.6 / 0.3 / 5.3	**26** 0037 / 0718 / F 1256 / 1938	1.0 / 5.0 / 0.9 / 4.9
12 0106 / 0736 / F 1330 / 2003	0.6 / 5.5 / 0.5 / 5.2	**27** 0110 / 0753 / SA 1330 / 2015	1.1 / 4.8 / 1.1 / 4.7
13 0152 / 0823 / SA 1416 / 2051	0.8 / 5.2 / 0.8 / 4.9	**28** 0145 / 0831 / SU 1404 / 2054	1.3 / 4.6 / 1.4 / 4.5
14 0241 / 0914 / SU 1507 / 2143	1.1 / 4.9 / 1.1 / 4.6	**29** 0222 / 0913 / M 1443 / 2138	1.6 / 4.3 / 1.7 / 4.3
15 0337 / 1012 / M 1606 / ☽ 2245	1.4 / 4.5 / 1.5 / 4.3		

MARCH

Time	m	Time	m
1 0304 / 1000 / TU 1527 / ☽ 2227	1.8 / 4.1 / 1.9 / 4.1	**16** 0427 / 1104 / W 1702 / 2343	1.6 / 4.1 / 1.9 / 4.0
2 0353 / 1054 / W 1621 / 2324	2.1 / 3.9 / 2.2 / 3.9	**17** 0557 / 1241 / TH 1834	1.8 / 4.0 / 2.0
3 0457 / 1156 / TH 1900	2.2 / 3.8 / 2.3	**18** 0119 / 0734 / F 1358 / 1958	4.1 / 1.7 / 4.1 / 1.9
4 0026 / 0732 / F 1306 / 2000	3.9 / 2.1 / 3.9 / 2.0	**19** 0225 / 0845 / SA 1454 / 2057	4.3 / 1.5 / 4.3 / 1.6
5 0138 / 0826 / SA 1426 / 2049	4.1 / 1.7 / 4.2 / 1.6	**20** 0316 / 0931 / SU 1540 / 2139	4.5 / 1.2 / 4.6 / 1.3
6 0247 / 0912 / SU 1522 / 2132	4.5 / 1.3 / 4.6 / 1.2	**21** 0359 / 1007 / M 1620 / 2211	4.7 / 1.0 / 4.8 / 1.1
7 0337 / 0954 / M 1606 / 2210	4.9 / 0.8 / 4.9 / 0.8	**22** 0437 / 1033 / TU 1657 / 2238	4.9 / 0.9 / 5.0 / 0.9
8 0421 / 1031 / TU 1648 / 2247	5.3 / 0.5 / 5.3 / 0.5	**23** 0513 / 1055 / W 1731 / ○ 2306	5.0 / 0.8 / 5.0 / 0.8
9 0504 / 1108 / W 1730 / ● 2324	5.6 / 0.2 / 5.5 / 0.3	**24** 0547 / 1123 / TH 1804 / 2337	5.0 / 0.7 / 5.1 / 0.8
10 0548 / 1145 / TH 1813	5.7 / 0.1 / 5.6	**25** 0621 / 1154 / F 1838	5.0 / 0.8 / 5.0
11 0004 / 0632 / F 1226 / 1856	0.2 / 5.7 / 0.1 / 5.5	**26** 0010 / 0654 / SA 1227 / 1910	0.8 / 4.9 / 0.9 / 4.9
12 0046 / 0717 / SA 1308 / 1941	0.3 / 5.6 / 0.3 / 5.3	**27** 0044 / 0728 / SU 1300 / 1944	0.9 / 4.8 / 1.0 / 4.7
13 0131 / 0805 / SU 1353 / 2028	0.5 / 5.3 / 0.7 / 5.0	**28** 0118 / 0803 / M 1335 / 2020	1.1 / 4.6 / 1.2 / 4.5
14 0220 / 0855 / M 1443 / 2119	0.9 / 4.9 / 1.1 / 4.7	**29** 0155 / 0841 / TU 1413 / 2101	1.3 / 4.3 / 1.5 / 4.3
15 0316 / 0953 / TU 1544 / ☽ 2218	1.3 / 4.5 / 1.6 / 4.3	**30** 0236 / 0927 / W 1457 / 2150	1.6 / 4.1 / 1.8 / 4.1
		31 0324 / 1021 / TH 1549 / ☽ 2247	1.8 / 3.9 / 2.1 / 3.9

APRIL

Time	m	Time	m
1 0422 / 1122 / F 1656 / 2349	2.0 / 3.8 / 2.2 / 3.9	**16** 0054 / 0651 / SA 1335 / 1919	4.0 / 1.7 / 4.0 / 1.9
2 0657 / 1230 / SA 1932	2.0 / 3.9 / 2.0	**17** 0201 / 0805 / SU 1430 / 2021	4.1 / 1.5 / 4.2 / 1.7
3 0058 / 0758 / SU 1348 / 2024	4.1 / 1.6 / 4.1 / 1.6	**18** 0252 / 0856 / M 1516 / 2106	4.3 / 1.3 / 4.5 / 1.4
4 0211 / 0846 / M 1452 / 2107	4.4 / 1.2 / 4.5 / 1.2	**19** 0335 / 0845 / TU 1555 / 2141	4.5 / 1.5 / 4.6 / 1.2
5 0308 / 0928 / TU 1540 / 2146	4.8 / 0.7 / 5.0 / 0.7	**20** 0413 / 0959 / W 1630 / 2211	4.7 / 1.0 / 4.8 / 1.0
6 0356 / 1007 / W 1623 / 2224	5.2 / 0.4 / 5.3 / 0.4	**21** 0447 / 1026 / TH 1703 / 2241	4.8 / 0.9 / 4.9 / 0.9
7 0441 / 1044 / TH 1706 / ● 2303	5.5 / 0.1 / 5.6 / 0.2	**22** 0521 / 1056 / F 1737 / ○ 2313	4.8 / 0.8 / 5.0 / 0.8
8 0527 / 1123 / F 1750 / 2344	5.7 / 0.1 / 5.6 / 0.1	**23** 0556 / 1127 / SA 1811 / 2346	4.8 / 0.8 / 4.9 / 0.8
9 0613 / 1203 / SA 1835	5.7 / 0.1 / 5.6	**24** 0631 / 1200 / SU 1844	4.8 / 0.9 / 4.8
10 0026 / 0659 / SU 1246 / 1920	0.2 / 5.5 / 0.4 / 5.4	**25** 0019 / 0704 / M 1234 / 1917	0.9 / 4.7 / 1.0 / 4.7
11 0112 / 0747 / M 1332 / 2007	0.4 / 5.2 / 0.8 / 5.1	**26** 0055 / 0738 / TU 1310 / 1950	1.0 / 4.5 / 1.2 / 4.6
12 0202 / 0838 / TU 1424 / 2058	0.8 / 4.8 / 1.2 / 4.7	**27** 0134 / 0815 / W 1350 / 2029	1.2 / 4.4 / 1.4 / 4.4
13 0259 / 0935 / W 1526 / 2155	1.2 / 4.4 / 1.6 / 4.3	**28** 0215 / 0859 / TH 1434 / 2118	1.4 / 4.2 / 1.7 / 4.2
14 0411 / 1044 / TH 1645 / ☽ 2311	1.5 / 4.1 / 1.9 / 4.0	**29** 0303 / 0953 / F 1527 / 2215	1.6 / 4.0 / 1.9 / 4.1
15 0532 / 1220 / F 1806	1.7 / 3.9 / 2.0	**30** 0401 / 1053 / SA 1633 / ☽ 2317	1.7 / 3.9 / 2.0 / 4.0

Chart Datum: 0·20 metres below Ordnance Datum (Dublin). HAT is 5·9 metres above Chart Datum.

STANDARD TIME (UT)
For Summer Time add ONE hour in **non-shaded areas**

GALWAY   LAT 53°16'N   LONG 9°03'W

TIMES AND HEIGHTS OF HIGH AND LOW WATERS

Dates in red are **SPRINGS**
Dates in blue are **NEAPS**

YEAR 2016

MAY

Time	m		Time	m
1 0515	1.8	**16** 0125	4.0
1159	4.0	0709	1.7
SU 1820	2.0	M 1357	4.1
		1935	1.8
2 0024	4.1	**17** 0220	4.1
0715	1.5	0803	1.5
M 1313	4.2	TU 1444	4.3
1949	1.6	2025	1.6
3 0136	4.4	**18** 0305	4.3
0813	1.2	0847	1.4
TU 1420	4.6	W 1524	4.5
2037	1.2	2106	1.4
4 0240	4.8	**19** 0343	4.4
0858	0.8	0924	1.2
W 1512	5.0	TH 1600	4.6
2120	0.7	2142	1.2
5 0332	5.2	**20** 0419	4.5
0940	0.5	0958	1.1
TH 1559	5.3	F 1634	4.8
2201	0.4	2217	1.1
6 0420	5.4	**21** 0455	4.7
1020	0.3	1031	1.0
F 1644	5.5	SA 1709	4.9
● 2243	0.2	○ 2251	1.0
7 0507	5.6	**22** 0531	4.7
1101	0.2	1104	1.0
SA 1729	5.6	SU 1744	4.9
2325	0.2	2324	0.9
8 0555	5.6	**23** 0607	4.7
1143	0.3	1137	1.0
SU 1815	5.5	M 1819	4.9
		2359	0.9
9 0010	0.3	**24** 0643	4.7
0642	5.4	1213	1.1
M 1228	0.6	TU 1852	4.8
1901	5.4		
10 0057	0.5	**25** 0037	1.0
0731	5.1	0717	4.6
TU 1315	0.9	W 1251	1.2
1948	5.1	1926	4.7
11 0147	0.8	**26** 0117	1.0
0822	4.8	0754	4.5
W 1407	1.3	TH 1333	1.3
2038	4.8	2005	4.6
12 0243	1.1	**27** 0200	1.2
0916	4.5	0838	4.4
TH 1508	1.6	F 1419	1.5
2132	4.4	2053	4.4
13 0350	1.4	**28** 0248	1.3
1018	4.2	0929	4.2
F 1620	1.9	SA 1511	1.7
◑ 2234	4.1	2148	4.3
14 0500	1.6	**29** 0344	1.5
1139	4.0	1027	4.1
SA 1732	2.0	SU 1614	1.8
		◑ 2249	4.2
15 0009	4.0	**30** 0448	1.5
0607	1.7	1131	4.1
SU 1258	4.0	M 1727	1.8
1836	1.9	2354	4.3
		31 0604	1.5
		1241	4.3
		TU 1853	1.6

JUNE

Time	m		Time	m
1 0106	4.4	**16** 0227	4.1
0726	1.2	0808	1.7
W 1350	4.6	TH 1447	4.3
2002	1.3	2032	1.6
2 0214	4.7	**17** 0310	4.2
0827	1.0	0853	1.5
TH 1448	4.9	F 1527	4.5
2054	0.9	2115	1.5
3 0311	5.0	**18** 0349	4.4
0916	0.7	0933	1.4
F 1538	5.2	SA 1604	4.7
2141	0.6	2155	1.3
4 0403	5.2	**19** 0427	4.5
1001	0.6	1011	1.3
SA 1626	5.4	SU 1641	4.8
2227	0.5	2232	1.1
5 0452	5.4	**20** 0506	4.6
1045	0.5	1046	1.2
SU 1712	5.5	M 1718	4.9
● 2311	0.4	○ 2307	1.0
6 0541	5.4	**21** 0544	4.7
1129	0.6	1120	1.1
M 1759	5.5	TU 1755	5.0
2357	0.4	2343	0.9
7 0628	5.3	**22** 0621	4.8
1213	0.7	1156	1.1
TU 1845	5.4	W 1831	5.0
8 0042	0.6	**23** 0021	0.8
0716	5.1	0658	4.7
W 1259	1.0	TH 1236	1.1
1931	5.1	1907	4.9
9 0130	0.8	**24** 0102	0.9
0804	4.9	0736	4.7
TH 1347	1.3	F 1318	1.2
2018	4.9	1947	4.9
10 0221	1.1	**25** 0146	0.9
0853	4.6	0819	4.6
F 1440	1.6	SA 1404	1.3
2106	4.6	2033	4.7
11 0317	1.4	**26** 0233	1.0
0944	4.3	0907	4.5
SA 1542	1.8	SU 1455	1.4
2157	4.3	2126	4.6
12 0419	1.6	**27** 0325	1.2
1041	4.1	1002	4.4
SU 1650	2.0	M 1553	1.6
◑ 2254	4.1	◑ 2224	4.5
13 0522	1.7	**28** 0424	1.3
1150	4.0	1103	4.3
M 1754	2.0	TU 1658	1.6
		2328	4.4
14 0012	3.9	**29** 0530	1.4
0622	1.8	1212	4.3
TU 1305	4.0	W 1812	1.6
1852	2.0		
15 0133	3.9	**30** 0041	4.4
0717	1.8	0645	1.4
W 1402	4.1	TH 1326	4.5
1945	1.8	1932	1.4

JULY

Time	m		Time	m
1 0155	4.5	**16** 0231	4.0
0801	1.3	0825	1.7
F 1430	4.8	SA 1452	4.3
2037	1.2	2051	1.6
2 0257	4.8	**17** 0319	4.2
0900	1.1	0911	1.6
SA 1524	5.0	SU 1535	4.5
2130	0.9	2134	1.4
3 0351	5.0	**18** 0401	4.4
0949	0.8	0953	1.4
SU 1614	5.2	M 1615	4.8
2217	0.7	2215	1.2
4 0441	5.2	**19** 0441	4.4
1034	0.8	1030	1.2
M 1701	5.4	TU 1653	4.9
● 2302	0.6	○ 2252	0.9
5 0529	5.2	**20** 0520	4.8
1117	0.8	1105	1.0
TU 1746	5.4	W 1731	5.1
2344	0.6	2328	0.8
6 0614	5.2	**21** 0559	4.9
1158	0.9	1141	0.9
W 1830	5.3	TH 1810	5.2
7 0026	0.7	**22** 0005	0.6
0658	5.1	0637	5.0
TH 1239	1.0	F 1221	0.8
1912	5.2	1848	5.2
8 0107	0.8	**23** 0046	0.6
0742	4.9	0717	5.0
F 1321	1.2	SA 1302	0.9
1955	5.0	1930	5.1
9 0150	1.0	**24** 0128	0.6
0825	4.7	0759	4.9
SA 1405	1.4	SU 1347	1.0
2037	4.7	2015	5.0
10 0235	1.3	**25** 0214	0.8
0908	4.5	0846	4.7
SU 1453	1.7	M 1436	1.1
2122	4.5	2105	4.8
11 0326	1.6	**26** 0304	1.0
0954	4.3	0938	4.6
M 1551	1.9	TU 1530	1.4
2209	4.2	◑ 2201	4.6
12 0426	1.8	**27** 0359	1.3
1043	4.1	1037	4.4
TU 1702	2.1	W 1633	1.6
◑ 2301	4.0	2305	4.4
13 0534	1.9	**28** 0503	1.5
1138	4.0	1147	4.3
W 1810	2.1	TH 1746	1.7
14 0000	3.9	**29** 0023	4.3
0637	2.0	0619	1.6
TH 1246	4.0	F 1309	4.3
1909	2.0	1916	1.6
15 0122	3.9	**30** 0144	4.3
0734	1.9	0748	1.5
F 1359	4.1	SA 1418	4.4
2002	1.9	2034	1.3
		31 0248	4.5
		0854	1.3
		SU 1515	4.8
		2128	1.1

AUGUST

Time	m		Time	m
1 0343	4.8	**16** 0336	4.4
0944	1.1	0934	1.4
M 1604	5.1	TU 1549	4.7
2214	0.8	2156	1.1
2 0431	5.0	**17** 0416	4.7
1026	1.0	1013	1.1
TU 1650	5.3	W 1628	5.0
● 2253	0.7	2234	0.8
3 0516	5.1	**18** 0455	4.9
1104	0.9	1048	0.8
W 1733	5.3	TH 1707	5.2
2329	0.6	○ 2310	0.5
4 0558	5.2	**19** 0534	5.1
1139	0.9	1124	0.6
TH 1813	5.3	F 1747	5.4
		2346	0.4
5 0003	0.7	**20** 0614	5.2
0638	5.1	1202	0.5
F 1214	0.9	SA 1828	5.5
1851	5.2		
6 0038	0.8	**21** 0025	0.3
0716	5.0	0654	5.2
SA 1250	1.1	SU 1243	0.5
1929	5.0	1911	5.4
7 0115	0.9	**22** 0107	0.4
0753	4.8	0737	5.1
SU 1328	1.3	M 1327	0.7
2007	4.8	1956	5.2
8 0153	1.2	**23** 0152	0.6
0832	4.6	0823	4.9
M 1406	1.5	TU 1414	0.9
2048	4.5	2045	4.9
9 0232	1.5	**24** 0240	0.9
0914	4.4	0914	4.7
TU 1447	1.8	W 1507	1.2
2132	4.3	2141	4.6
10 0315	1.7	**25** 0335	1.3
1000	4.2	1012	4.4
W 1535	2.0	TH 1611	1.5
◑ 2220	4.0	◑ 2247	4.3
11 0406	2.0	**26** 0442	1.6
1050	4.0	1126	4.2
TH 1716	2.2	F 1730	1.7
2314	3.9		
12 0556	2.1	**27** 0012	4.1
1146	3.9	0606	1.8
F 1837	2.2	SA 1257	4.2
		1915	1.7
13 0014	3.8	**28** 0137	4.1
0705	2.1	0745	1.7
SA 1251	4.0	SU 1410	4.4
1936	2.0	2033	1.4
14 0133	3.9	**29** 0242	4.4
0801	1.9	0850	1.5
SU 1409	4.2	M 1506	4.7
2028	1.7	2125	1.1
15 0249	4.1	**30** 0334	4.7
0850	1.7	0937	1.2
M 1506	4.4	TU 1553	4.9
2114	1.4	2206	0.8
		31 0418	4.9
		1015	1.0
		W 1636	5.1
		2240	0.7

Chart Datum: 0·20 metres below Ordnance Datum (Dublin). HAT is 5·9 metres above Chart Datum.

N Ireland

STANDARD TIME (UT)
For Summer Time add ONE hour in **non-shaded areas**

GALWAY   LAT 53°16'N   LONG 9°03'W
TIMES AND HEIGHTS OF HIGH AND LOW WATERS

Dates in red are **SPRINGS**
Dates in blue are **NEAPS**

YEAR 2016

SEPTEMBER

Time	m		Time	m
1 0459	5.0	**16** 0428	5.0
1047	0.9	1027	0.7
TH 1715	5.2	F 1642	5.4
● 2308	0.6	○ 2247	0.3
2 0538	5.1	**17** 0508	5.3
1116	0.8	1103	0.4
F 1752	5.2	SA 1723	5.6
2335	0.6	2323	0.2
3 0614	5.1	**18** 0549	5.4
1147	0.8	1141	0.3
SA 1828	5.1	SU 1807	5.6
4 0007	0.7	**19** 0002	0.1
0648	5.0	0631	5.4
SU 1220	0.9	M 1222	0.3
1902	5.0	1851	5.5
5 0041	0.9	**20** 0044	0.3
0722	4.9	0715	5.3
M 1255	1.1	TU 1306	0.5
1937	4.8	1938	5.3
6 0115	1.1	**21** 0128	0.6
0759	4.7	0801	5.1
TU 1330	1.3	W 1353	0.8
2016	4.6	2028	4.9
7 0151	1.3	**22** 0217	1.0
0838	4.5	0852	4.7
W 1407	1.6	TH 1448	1.2
2058	4.3	2124	4.5
8 0228	1.6	**23** 0314	1.4
0922	4.3	0950	4.4
TH 1447	1.8	F 1554	1.5
2145	4.1	◑ 2232	4.2
9 0310	1.9	**24** 0426	1.8
1011	4.1	1105	4.1
F 1536	2.1	SA 1721	1.7
◗ 2237	3.8		
10 0402	2.2	**25** 0003	4.0
1105	3.9	0557	1.9
SA 1802	2.2	SU 1245	4.1
2336	3.7	1903	1.7
11 0637	2.2	**26** 0129	4.1
1206	3.9	0730	1.8
SU 1911	2.1	M 1358	4.3
		2020	1.4
12 0043	3.8	**27** 0231	4.3
0738	2.0	0835	1.6
M 1316	4.0	TU 1452	4.6
2005	1.7	2109	1.2
13 0213	4.0	**28** 0319	4.6
0829	1.7	0920	1.3
TU 1431	4.3	W 1537	4.8
2052	1.3	2148	1.0
14 0309	4.4	**29** 0401	4.8
0913	1.3	0956	1.1
W 1520	4.7	TH 1617	5.0
2134	0.9	2218	0.8
15 0350	4.7	**30** 0438	5.0
0952	1.0	1024	1.0
TH 1601	5.1	F 1654	5.1
2212	0.6	2241	0.7

OCTOBER

Time	m		Time	m
1 0513	5.0	**16** 0442	5.4
1050	0.9	1040	0.3
SA 1729	5.1	SU 1701	5.6
● 2306	0.7	○ 2259	0.1
2 0547	5.1	**17** 0525	5.6
1120	0.8	1119	0.2
SU 1802	5.0	M 1746	5.7
2337	0.8	2339	0.2
3 0620	5.0	**18** 0609	5.6
1153	0.9	1202	0.3
M 1836	4.9	TU 1833	5.6
4 0010	0.9	**19** 0022	0.4
0653	4.9	0654	5.4
TU 1227	1.0	W 1247	0.5
1911	4.8	1921	5.3
5 0044	1.1	**20** 0108	0.7
0728	4.8	0742	5.2
W 1302	1.2	TH 1336	0.8
1947	4.6	2012	4.9
6 0119	1.3	**21** 0158	1.1
0805	4.6	0833	4.8
TH 1338	1.4	F 1432	1.2
2027	4.3	2109	4.5
7 0156	1.6	**22** 0257	1.5
0847	4.3	0930	4.5
F 1418	1.7	SA 1540	1.5
2112	4.1	◑ 2216	4.2
8 0238	1.9	**23** 0412	1.9
0934	4.1	1041	4.2
SA 1504	1.9	SU 1705	1.7
2204	3.9	2346	4.0
9 0327	2.1	**24** 0538	2.0
1028	4.0	1221	4.1
SU 1602	2.1	M 1829	1.7
◗ 2302	3.8		
10 0433	2.3	**25** 0108	4.1
1128	3.9	0656	1.9
M 1840	2.1	TU 1335	4.2
		1945	1.5
11 0007	3.8	**26** 0208	4.3
0711	2.1	0803	1.7
TU 1234	4.0	W 1430	4.4
1937	1.7	2038	1.3
12 0123	4.0	**27** 0256	4.5
0835	1.8	0852	1.5
W 1346	4.4	TH 1515	4.6
2026	1.3	2117	1.2
13 0234	4.4	**28** 0337	4.7
0847	1.4	0929	1.3
TH 1446	4.7	F 1554	4.8
2108	0.9	2147	1.0
14 0319	4.8	**29** 0413	4.8
0926	0.9	0958	1.1
F 1533	5.1	SA 1630	4.9
2146	0.5	2212	1.0
15 0400	5.2	**30** 0447	4.9
1003	0.6	1026	1.0
SA 1616	5.5	SU 1704	4.9
2222	0.3	● 2240	0.9
		31 0520	5.0
		1057	0.9
		M 1738	4.9
		2312	0.9

NOVEMBER

Time	m		Time	m
1 0553	5.0	**16** 0550	5.7
1130	0.9	1146	0.4
TU 1813	4.9	W 1817	5.5
2344	1.0		
2 0627	5.0	**17** 0004	0.6
1204	1.0	0636	5.5
W 1848	4.8	TH 1233	0.5
		1906	5.3
3 0018	1.1	**18** 0052	0.8
0701	4.8	0724	5.3
TH 1239	1.1	F 1322	0.8
1924	4.6	1957	5.0
4 0054	1.3	**19** 0142	1.2
0736	4.7	0815	5.0
F 1316	1.3	SA 1417	1.1
2001	4.4	2052	4.7
5 0131	1.5	**20** 0240	1.6
0814	4.5	0909	4.7
SA 1356	1.5	SU 1521	1.4
2043	4.2	2153	4.4
6 0214	1.8	**21** 0350	1.9
0859	4.3	1010	4.4
SU 1442	1.7	M 1635	1.6
2133	4.0	◑ 2309	4.1
7 0303	2.0	**22** 0505	2.0
0953	4.1	1136	4.2
M 1537	1.9	TU 1747	1.7
◗ 2231	3.9		
8 0405	2.2	**23** 0030	4.1
1053	4.1	0614	2.0
TU 1650	2.0	W 1259	4.1
2333	3.9	1852	1.7
9 0549	2.2	**24** 0135	4.2
1157	4.1	0717	1.9
W 1857	1.8	TH 1359	4.2
		1950	1.6
10 0042	4.1	**25** 0226	4.4
0728	1.9	0811	1.8
TH 1306	4.4	F 1447	4.4
1952	1.4	2035	1.5
11 0153	4.4	**26** 0309	4.5
0816	1.5	0854	1.6
F 1412	4.7	SA 1528	4.5
2037	1.0	2112	1.4
12 0248	4.8	**27** 0346	4.7
0859	1.0	0930	1.4
SA 1506	5.1	SU 1604	4.7
2118	0.7	2145	1.2
13 0334	5.2	**28** 0420	4.8
0939	0.7	1004	1.2
SU 1554	5.4	M 1639	4.8
2157	0.4	2218	1.2
14 0419	5.5	**29** 0454	5.0
1019	0.4	1038	1.1
M 1641	5.6	TU 1715	4.8
○ 2238	0.3	● 2251	1.1
15 0504	5.6	**30** 0529	5.0
1102	0.3	1112	1.1
TU 1729	5.6	W 1752	4.8
2320	0.4	2325	1.1

DECEMBER

Time	m		Time	m
1 0605	5.0	**16** 0623	5.6
1146	1.1	1221	0.6
TH 1828	4.8	F 1853	5.4
2359	1.2		
2 0639	4.9	**17** 0038	0.9
1221	1.1	0709	5.4
F 1904	4.7	SA 1308	0.8
		1941	5.1
3 0035	1.3	**18** 0125	1.2
0712	4.8	0756	5.2
SA 1259	1.2	SU 1357	1.0
1939	4.6	2031	4.9
4 0114	1.5	**19** 0216	1.5
0748	4.7	0845	4.9
SU 1340	1.3	M 1451	1.3
2019	4.4	2122	4.6
5 0157	1.6	**20** 0314	1.8
0831	4.6	0936	4.6
M 1425	1.5	TU 1552	1.6
2105	4.3	2218	4.3
6 0245	1.8	**21** 0421	2.0
0922	4.4	1032	4.3
TU 1516	1.6	W 1659	1.8
2200	4.2	◗ 2324	4.1
7 0342	2.0	**22** 0529	2.1
1020	4.3	1145	4.1
W 1617	1.7	TH 1802	1.9
◗ 2300	4.1		
8 0451	2.0	**23** 0041	4.1
1122	4.3	0631	2.1
TH 1729	1.7	F 1312	4.1
		1900	1.9
9 0005	4.2	**24** 0145	4.2
0613	1.9	0728	2.0
F 1231	4.4	SA 1412	4.1
1854	1.5	1953	1.8
10 0116	4.5	**25** 0235	4.3
0735	1.6	0819	1.8
SA 1341	4.7	SU 1459	4.3
2001	1.3	2040	1.7
11 0219	4.8	**26** 0317	4.5
0831	1.2	0903	1.6
SU 1443	5.0	M 1539	4.5
2052	1.0	2121	1.5
12 0312	5.1	**27** 0354	4.7
0919	0.9	0944	1.4
M 1536	5.2	TU 1617	4.6
2138	0.8	2200	1.4
13 0401	5.4	**28** 0431	4.9
1004	0.7	1022	1.3
TU 1627	5.5	W 1654	4.7
2223	0.6	2236	1.3
14 0448	5.6	**29** 0508	5.0
1050	0.5	1058	1.1
W 1716	5.6	TH 1732	4.8
○ 2307	0.6	● 2311	1.2
15 0536	5.6	**30** 0544	5.1
1135	0.5	1132	1.1
TH 1804	5.5	F 1809	4.9
2352	0.7	2345	1.2
		31 0620	5.1
		1207	1.0
		SA 1845	4.9

Chart Datum: 0·20 metres below Ordnance Datum (Dublin). HAT is 5·9 metres above Chart Datum.

9.13.8 WESTPORT (CLEW BAY)
Mayo 53°47'·85N 09°35'·40W ❄❄♤♤❁❁❁

CHARTS AC 2667, 2057; Imray C54

TIDES –0545 Dover; ML 2·5; Duration 0610

Standard Port GALWAY (←)

Times				Height (metres)			
High Water		Low Water		MHWS	MHWN	MLWN	MLWS
0600	1100	0000	0700	5·1	3·9	2·0	0·8
1800	2300	1200	1900				
Differences BROAD HAVEN							
+0035	+0035	+0040	+0040	–1·4	–1·0	–0·6	–0·2
BLACKSOD QUAY							
+0025	+0035	+0040	+0040	–1·2	–1·0	–0·6	–0·4
CLARE ISLAND							
+0015	+0021	+0039	+0027	–0·6	–0·4	–0·1	0·0
INISHGORT/CLEW BAY							
+0035	+0045	+0115	+0100	–0·7	–0·5	–0·2	0·0
KILLARY HARBOUR							
+0021	+0015	+0035	+0029	–1·0	–0·8	–0·4	–0·3
INISHBOFIN HARBOUR							
+0013	+0009	+0021	+0017	–1·0	–0·8	–0·4	–0·3
CLIFDEN BAY							
+0005	+0005	+0016	+0016	–0·7	–0·5	ND	ND

SHELTER Secure ⚓s amongst the islands at all times, as follows: E of Inishlyre in 2m; 2ca NE of Dorinish More, good holding in lee of Dorinish Bar (dries); 6 ⚲s in Rosmoney Hbr 53°49'·75N 09°37'·34W; Westport Quay HW±1½, to dry out on S side; Newport Hbr (dries) can be reached above mid-flood with careful pilotage (AC 2667); dry out against N quay or ⚓ at E end of Rabbit Is.

NAVIGATION WPT 53°49'·20N 09°42'·10W, 071°/1·2M to Inishgort lt ho. Approaches to Westport Channel between Monkellys Rks and Inishmweela are well marked by lit PHMs and SHMs. Beware fish farms E of Clare I and in Newport Bay. Do not ⚓ or ground in the Carricknacally Shellfish Bed Area.

LIGHTS AND MARKS Westport Bay ent via Inishgort lt, L Fl 10s11m 10M (H24), and a series of ldg lines, but not advised at night. Final appr line 080° towards lt bn, Fl 3s. but beware of drying patches on this lead inwards from Pigeon PHM. Chan from Westport Bay to Westport Quay, 1½M approx, is marked by bns.

COMMUNICATIONS (Code 098) MRCC (01) 6620922/3; ⊖ (094) 9021131; Police 9025555; Ⓗ (094) 9021733. No VHF.

FACILITIES **Westport Quays** ⚓ ⚓ ⌂ Kos Ⓟ Ⓔ ☳ ✕ ☐; **Town** Ⓣ Ⓣ ⊠ Ⓑ ⇌ ✈ (Galway/Knock). **Mayo SC Rosmoney** ☎9026160, ⚓ ☐; **Glénans Irish SC** ☎9026046 (Collanmore Is).

OTHER ANCHORAGES FROM SLYNE HEAD TO EAGLE ISLAND
CLIFDEN BAY, Galway, 53°29'·40N 10°05'·90W. AC 1820, 2708. HW –0600 on Dover; duration 0610. Tides at 9.13.8.
Identify conspic W bn on Carrickrana Rks prior to entry; ldg marks: W bn at Fishing Pt on 080° with Clifden Castle (ruin). Avoid Coghan's and Doolick Rks to the E and beware bar 2·4m off Fishing Pt. 8 Y ⚲s are 3ca S of Castle ruins, or ⚓ NE of Drinagh Pt. Beware ruined training wall In the drying creek to Clifden; dry out against the quay or enter Ardbear Bay to ⚓ SE of Yellow Slate Rks in 3·4m. Keep clear of fish farms. Facilities: **Town** ⚓ Ⓣ Ⓣ Kos ⚓ ⊠ Ⓑ ☳ ✕ ☐ Dr Ⓗ bus to Galway.

BALLYNAKILL, Galway, 53°34'·95N 10°03'·00W. AC 2706. Tides as Inishbofin/Killary. Easy appr between Cleggan Pt, Fl (3) WRG 15s and Rinvyle Pt. Then pass N of Freaghillaun South Is, E of which is good passage ⚓ in 7m. Further E, Carrigeen and Ardagh Rks lie in mid-chan. Keep N for ⚓ in Derryinver B. S chan leads to ⚓s: off Ross Pt; S of Roeillaun; in Fahy Bay with 8 Y ⚲s sheltered by bar dries 0·2m. No facilities.

INISHBOFIN, Galway, 53°36'·60N 10°13'·20W. AC 1820, 2707. HW –0555 on Dover; ML 1·9m. Very safe inside narrow ent. Daymarks 2 conspic W trs lead 032° and Dir RWG lt leads 021° at night to increase separation from Gun Rock, Fl (2) 6s 8m 4M, vis 296°-253°. ⚓ between new pier and Port Is. Old pier to the E now refurbished and dredged may provide ⌂. ⚓ ⚓(on both piers) ☳ ✕ ☐ / B&B: ☎(095) 45829.

KILLARY HARBOUR, Mayo/Galway, 53°37'·83N 09°54'·00W, 4ca W of Doonee Is. AC 2706. Tides 9.13.8. A 7M long inlet, narrow and deep. Caution fish farms, some with Fl Y lts. Appr in good vis to identify Doonee Is and Inishbarna bns, both LFl G, ldg 099° to ent. ⚓s off Dernasliggaun, Bundorragha and, at head of inlet, Leenaun with 8 Y ⚲s. (Village:⚓ ☳ ☐ hotel.) Enter **Little Killary Bay** 4ca S of Doonee Is; drying rks at ent. Good ⚓ in 3m at head of bay.

INISHTURK, Mayo, 53°42'·3N 10°05'·2W. 8 Y ⚲s Garranty Hbr. sheltered SW to NNW winds. ⚓ quay ⊠ ✕ ☐ 6⚲s on E side.

CLARE ISLAND, Mayo, 53°48'·1N 09°56'·7W. 6⚲s on SE side untenable in E winds. Hotel ☐ some shops.

BLACKSOD BAY, Mayo, 54°05'·00N 10°02'·00W. AC 2704. HW –0525 on Dover; ML 2·2m; Duration 0610. See 9.13.8. Easy appr, accessible by night. Safe ⚓s depending on wind: NW of Blacksod Pt 6 Y ⚲s in 3m; at Elly B in 1·8m; Elly Hbr 6 Y ⚲s; Saleen Bay (ldg lts 319°, both Oc 4s, lead to pier); N of Claggan Pt. Beware drying rk 3·5ca SSE of Ardmore Pt. Lts: Blacksod Pt, Fl (2) WR 7·5s 13m 9M, see 9.13.4. ECM buoy Q (3) 10s. Ldg lts 181°, Oc W 5s, lead to Blacksod pier hd. Few facilities; nearest town is Belmullet.

ACHILL SOUND ⚓s at each end sheltered by Achill Is. Mayo Co Council (+353 (0)97 81004) requires 2 days' notice to open bridge.

PORTNAFRANKAGH Mayo, 54°14'·95N 10°06'·00W. AC 2703. Tides approx as Broadhaven. A safe passage ⚓, close to the coastal route, but swell enters in all winds. Appr toward Port Pt, thence keep to the N side for better water; middle and S side break in onshore winds. ⚓ in 4–5m on S side, close inshore.⚓ ⚓ at new pier. Unlit, but Eagle Is lt, Fl (3) 15s 67m 26M H24, W tr, is 2M N. No facilities; Belmullet 4M, ☳.

BROAD HAVEN, Mayo, 54°16'·00N 09°53'·20W. AC 2703. A safe refuge except in N'lies. Easy appr across Broadhaven Bay to the ent between Brandy Pt and Gubacashel Pt, Iso WR 4s27m 12/9M, W tr. 7ca S of this lt is Ballyglas Fl G 3s on W side. ⚓ close N or S of Ballyglas which has pier (2m) and 8 Y ⚲s. In E'lies ⚓ 3ca S of Inver Pt out of the tide. Further S off Barrett Pt, the inlet narrows and turns W to Belmullet. Facilities: ⊠ ☳.

N Ireland

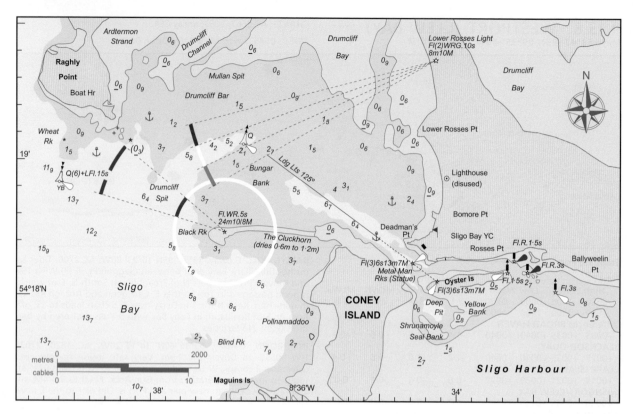

9.13.9 SLIGO

Sligo 54°18'·30N 08°34'·70W ✲✲♒♒♤♤

CHARTS AC 2767, 2852; Imray C54

TIDES –0511 Dover; ML 2·3; Duration 0620

Standard Port GALWAY (←)

Times				Height (metres)			
High Water		Low Water		MHWS	MHWN	MLWN	MLWS
0600	1100	0000	0700	5·1	3·9	2·0	0·8
1800	2300	1200	1900				
Differences SLIGO HARBOUR (Oyster Is)							
+0043	+0055	+0042	+0054	–1·0	–0·9	–0·5	–0·3
MULLAGHMORE							
+0036	+0048	+0047	+0059	–1·4	–1·0	–0·4	–0·4
BALLYSADARE BAY (Culleenamore)							
+0059	+0111	+0111	+0123	–1·2	–0·9	ND	ND
KILLALA BAY (Inishcrone)							
+0035	+0055	+0030	+0050	–1·3	–1·2	–0·7	–0·4

SHELTER The lower hbr is fairly exposed; ⚓ along N side of Oyster Island, or proceed 4M (not at night) up to the shelter of Sligo town; 2nd berth below bridge for yachts.

NAVIGATION WPT 54°19'·15N 08°36'·77W, 125°/1·6M to front ldg lt.
• The passage between Oyster Island and Coney Island is marked 'Dangerous'.
• Pass N of Oyster Is leaving Blennick Rks to port. Passage up to Sligo town between training walls. Appr quays at HW–3 or higher.
• Some perches are in bad repair. Pilots at Raghly Pt and Rosses Pt. *Tide Tables* (based on Inishraher/Inishgort, published by Westport Hbr Comm.) available locally.

LIGHTS AND MARKS See 9.13.4 and chartlet. Channel to Sligo Deepwater Quay is well marked by lit bcns.

COMMUNICATIONS (Code 071) MRCC (01) 6620922/3; ⊜9161064; Police 9157000; Dr 9142886; Ⓗ9171111. Hbr Office 9153819, mob 086 0870767.

Pilots VHF Ch 12 16. HM Ch 16.

FACILITIES 60m visitors pontoon, 20-30ft €19.00; ⚓ ⌂(15t) ⚓; Sligo YC ☎9177168, ⚓s ⚓ ⚓. Services ⚓ ⚓ ⚒ ⚓ ⚓. Town ⚓ ⚓ ⚓ Ⓑ ⚓ ✗ ⚓ ⚓ Irish Rail ☎9169888, Bus ☎9160066, ✈ Londonderry/Derry or Knock, Co Mayo (60-70M).

MINOR HARBOUR/ANCHORAGE TO THE WEST
KILLALA BAY, Sligo/Mayo, 54°13'·02N 09°12'·80W. AC 2715. Tides, see 9.13.19. The bay, open to the N-NE, is entered between Lenadoon Pt and Kilcummin Hd, 6M to the W. 1·1M W of Kilcummin Hd are 8 Y ⚓s off Kilcummin pier, well sheltered from W'lies. Carrickpatrick ECM lt buoy, in mid-bay marks St Patrick's Rks to the W. Thence, if bound for Killala hbr, make good Killala SHM lt buoy, 7½ca to the SSW. The Round Tr and cathedral spire at Killala are conspic. Four sets of ldg bns/lts lead via a narrow chan between sand dunes and over the bar (0·3m) as follows:

• 230°, Rinnaun Pt lts Oc 10s 7/12m 5M, ☐ concrete trs.
• 215°, Inch Is, ☐ concrete trs; the rear has Dir lt Fl WRG 2s.
• 196°, Kilroe lts Oc 4s 5/10m 2M, W ☐ trs, which lead to ⚓ in Bartragh Pool, 6ca NE of Killala hbr; thence
• 236°, Pier lts Iso 2s 5/7m 2M, W ◇ daymarks, lead via narrow, dredged chan to pier where ⚓ is possible in about 1·5m.

Facilities: ⚓ ⚓ ⚓ ⚒ in town ½M.
Other hbrs in the bay: R Moy leading to Ballina should not be attempted without pilot/local knowledge. Inishcrone in the SE has a pier and lt; see 9.13.4.

MINOR HARBOUR ON S SIDE OF DONEGAL BAY
MULLAGHMORE 54°27'·90N 08°26'·80W. AC 2702. Tides see 9.13.9. Pierhead lit Fl G 3s 5m 3M. Fair weather ⚓ in 2-3m off hbr ent, sheltered by Mullaghmore Head, except from N/NE winds. For ⚓s near hbr ent, call Rodney Lomax, ☎071 9166124, mobile 0872 727538. Keep close to N pier to avoid shingle bank drying 1m, ⅔ of the way across the ent toward the S pier. Take the ground or dry out against the piers inside hbr berth on pontoon on bkwtr (min 1m at MLWS). Access approx HW±2 when least depth is 2m. VHF Ch 16, 8, 6. Facilities: ⚓ at S Pier and on bkwtr, ⚓ ⚒ ⚓ ⚓ ✗ ⚓.

9.13.10 KILLYBEGS
Donegal 54°36'·90N 08°26'·80W ❀❀❀❄☖☖☖❀❀❀

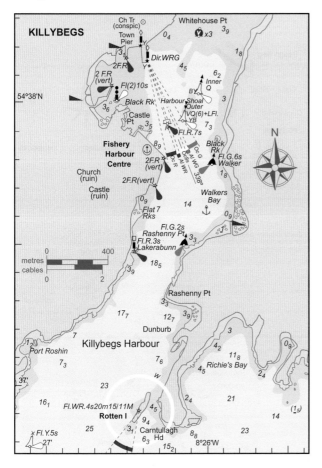

KILLYBEGS

CHARTS AC 2702, 2792; Imray C53

TIDES –0520 Dover; ML 2·2; Duration 0620

Standard Port GALWAY (←—)

Times				Height (metres)			
High Water		Low Water		MHWS	MHWN	MLWN	MLWS
0600	1100	0000	0700	5·1	3·9	2·0	0·8
1800	2300	1200	1900				
Differences KILLYBEGS							
+0040	+0050	+0055	+0035	–1·0	–0·9	–0·5	–0·2
GWEEDORE HARBOUR							
+0048	+0100	+0055	+0107	–1·3	–1·0	–0·5	–0·3
BURTONPORT							
+0042	+0055	+0115	+0055	–1·2	–1·0	–0·6	–0·3
DONEGAL HARBOUR (SALTHILL QUAY)							
+0038	+0050	+0052	+0104	–1·2	–0·9	ND	ND

SHELTER Secure natural hbr, but some swell in SSW winds. A busy major FV port, H24 access.60m pontoon orientated NNE/SSW is 100m N of Black Rk Pier, 3 ⚓ about 2½ca NE of Town Pier, or contact HM and berth at pier. **Bruckless Hbr**, about 2M E at the head of McSwyne's Bay, is a pleasant ⚓ in 1·8m but is severely restricted due to fish cages, sheltered from all except SW winds. Ent on 038° between rks; ICC SDs are essential.

NAVIGATION WPT 54°36'·00N 08°27'·00W, 022°/0·94M to Rotten Is lt. From W, beware Manister Rk (covers at HW; dries at LW) off Fintragh B. Keep mid chan until off new Killybegs Fishery Centre, then follow the Dir lt or Y ◊ ldg marks 338° into hbr.

LIGHTS AND MARKS Rotten Is lt, Fl WR 4s 20m 15/11M, W tr, vis W255°-008°, R008°-039°, W039°-208°. Dir lt 338°, Oc WRG 6s 17m; W sector 336°-340° (see 9.13.4). Harbour Shoal (2·3m) is marked by a SCM and NCM lt buoy.

COMMUNICATIONS (Code 07497) MRCC (01) 6620922/3; Bundoran Inshore Rescue ☎(071) 9841713; ⊜31070; Police 31002; Dr 31148 (Surgery). HM 31032.

HM Ch **14**; essential to request a berth.

FACILITIES **Town Pier** ☎31032, ⚓ ━ ⌂(free) ♁ ♙ ☏ Kos ✖ ⚒ ▣ Ⓔ ⌂(12t) ⛽ ↧ ✖ ⌂. **Black Rock Pier** ⚓━⌂♙. **Town** ✉ Ⓑ ⇌ (bus to Sligo).

OTHER HARBOURS AND ANCHORAGES IN DONEGAL

TEELIN HARBOUR, Donegal, **54°37'·50N 08°37'·87W**. AC 2792. Tides as 9.13.10. A possible passage ⚓, mid-way between Rathlin O'Birne and Killybegs. But Hbr is open to S'ly swell and prone to squalls in NW winds. Ent, 1ca wide, is close E of Teelin Pt lt, Fl R 10s, which is hard to see by day. 4 Y ⚓s and ⚓ on E side in 3m to N of pier, or ⚓ on E side near pier. Many moorings and, in the NE, mussel rafts. Facilities: ♁ ♙ ⛬ at Carrick, 3M inland.

INISHKEEL, Donegal, **54°50'·82N 08°26'·50W**. AC 2792. 6 Y ⚓s in Church Pool, 3ca E of Inishkeel. Open to N/NE'lies.

BURTONPORT, Donegal, **54°58'·93N 08°26'·60W**. AC 1879, 2792. HW –0525 on Dover; ML 2·0m; Duration 0605. See 9.13.10. Only appr via N Chan. Ent safe in all weathers except NW gales. Hbr very full, no space to ⚓; berth on local boat at pier or go to Rutland Hbr or Aran Roads: 6 Y ⚓s 250m NE of Black Rks, Fl R 3s. Ldg marks/lts: N Chan ldg lts on Inishcoo 119·3°, both Iso 6s 6/11m 1M; front W bn, B band; rear B bn, Y band. Rutland Is ldg lts 138°, both Oc 6s 8/14m 1M; front W bn, B band; rear B bn, Y band. Burtonport ldg lts 068°, both FG 17/23m 1M; front Gy bn, W band; rear Gy bn, Y band. HM ☎(075) 42155 (43170 home); VHF Ch 06, 12, **14**, 16. Facilities: ♙ (just inside pier), ♁ (root of pier), ☏ (½M inland). **Village** Kos ✉ ⛬ ✖ ⌂.

GWEEDORE HBR/BUNBEG, Donegal, **55°03'·75N 08°18'·87W**. AC 1883. Tides see 9.13.10. Gweedore hbr, the estuary of the R Gweedore, has sheltered ⚓s or temp ⌂ at Bunbeg Quay, usually full of FVs. Apprs via Gola N or S Sounds are not simple especially in poor visibility. N Sound is easier with ldg lts 171°, both Oc 3s 9/13m 2M, B/W bns, on the SE tip of Gola Is. (There are also ⚓s on the S and E sides of Gola Is.) E of Bo Is the bar 0·4m has a Fl

G 3s and the chan, lying E of Inishinny Is and W of Inishcoole, is marked by a QG and 3 QR. A QG marks ent to Bunbeg. Night ent not advised. Facilities: ♁ ♙ at quay; ✉ Ⓑ⛬ ⌂ at village ½M.

SHEEP HAVEN, Donegal, **55°11'·00N 07°51'·00W**. AC 2699. HW –0515 on Dover. See 9.13.11. Bay is 2M wide with many ⚓s, easily accessible in daylight, but exposed to N winds. Beware rks off Rinnafaghla Pt, and further S: Black Rk (6m) and Wherryman Rks, which dry, 1ca off E shore. ⚓ or 8 Y ⚓s in Downies Bay to SE of pier; in Pollcormick inlet close W in 3m; in Ards Bay for excellent shelter, but beware the bar in strong winds. Lts: Breaghy Hd LFl 10s 3m 6M, Portnablahy ldg lts 125°, both Oc 6s 7/12m 2M, B col, W bands; Downies pier hd, Fl R 3s 5m 2M, R post. Facilities: (Downies) EC Wed; ♁ ☏ ☏(300m) ⛬ ✖ ⌂; (Portnablahy Bay) ☏ ☏⛬ ✖ ⌂.

MULROY BAY, Donegal, **55°15'·30N 07°46'·30W**. AC 2699; HW (bar) –0455 on Dover. See 9.13.11. Beware Limeburner Rk, (NCM buoy, Q) 3M N of ent & the bar which is dangerous in swell and onshore winds. Obtain local advice as the depths on the bar change considerably. Ent at half flood (not HW); chan lies between Bar Rks (Low Bar Rk Fl G 2s) and Sessiagh Rks, it is rep'd to run along the alignment of the charted drying spit thence through First, Second (under bridge 19m) and Third Narrows to Broad Water. HW at the head of the lough is 2¼ hrs later than at the bar. ⚓s: Close SW of Ravedy Is (Fl 3s 9m 3M); Fanny's Bay (2m), excellent; Rosnakill Bay (3·5m) in SE side; Cranford Bay; Milford Port (3 to 4m). Beware power cable 6m, over Moross chan, barring North Water to masted boats. Facilities: **Milford Port**, pier derelict, ⚓ ⛬; **Fanny's Bay** ✉ Shop at Downings village (1M), hotel at Rosepenna (¾M).

9.13.11 LOUGH SWILLY

Donegal 55°17'N 07°34'W ✇✇✇♨♨❀❀❀

CHARTS AC 2697; Imray C53

TIDES –0500 Dover; ML 2·3; Duration 0605

Standard Port GALWAY (←—)

Times				Height (metres)			
High Water		Low Water		MHWS	MHWN	MLWN	MLWS
0200	0900	0200	0800	5·1	3·9	2·0	0·8
1400	2100	1400	2000				
Differences INISHTRAHULL							
+0100	+0100	+0115	+0200	–1·8	–1·4	–0·4	–0·4
TRAWBREAGA BAY							
+0115	+0059	+0109	+0125	–1·1	–0·8	ND	ND
FANAD HEAD							
+0115	+0040	+0125	+0120	–1·1	–0·9	–0·5	–0·3
RATHMULLAN							
+0125	+0050	+0126	+0118	–0·8	–0·7	–0·1	–0·3
MULROY BAY (BAR)							
+0108	+0052	+0102	+0118	–1·2	–1·0	ND	ND
SHEEP HAVEN (DOWNIES BAY)							
+0057	+0043	+0053	+0107	–1·1	–0·9	ND	ND

SHELTER Good, but beware downdrafts on E side. ⚓s N of 55°05'N may suffer from swell. ⚓s from seaward: Ballymastocker Bay, also 8 ♀s, but exposed to E'lies; inside Macamish Pt, sheltered from SE to N; Rathmullan Roads N of pier, where yacht pontoon lies N/S in 3·6m MLWS; Fahan Creek and marina ent at HW–1.

NAVIGATION WPT 55°17'·00N 07°34'·20W, 180°/5M to 5ca W of Dunree Hd lt. Entrance is easy and dangers are well marked. Off W shore: Swilly More Rocks, Kinnegar Spit and Strand; off E shore: Colpagh Rocks, White Strand Rocks and Inch Spit/Flats. Fahan Creek is buoyed; there is a shallow bar S of Inch Spit PHM and Inch Flats dry. Entry should not be attempted before LW+2. There are many marine farms and shellfish beds S from Scraggy Bay (☎074 50172/59071 for permission to use pier).

LIGHTS AND MARKS Fanad Hd lt, Fl (5) WR 20s 38m 18/14M, touching Dunree Hd lt, Fl (2) WR 5s 46m 12/9M, leads 151° into the Lough. Thence Ballygreen Point and Hawk's Nest in line 202°.

COMMUNICATIONS (Codes: both sides of lough 074) Malin Head MRSC 9370103; Police 9153114 (Milford) and 074 9320540 (Buncrana); Dr 9158416 (Rathmullan), 1850 400911 (out of hours), and 9363611 (Buncrana); Rathmullan Pontoon 9158131, 9158315 (eves). No VHF.

FACILITIES Rathmullan Pontoon (Easter–October) ☎087 2480132 €20/craft, ⚓ ⚓ Kos ⊠ ∰ Hotel ✕ ⌂.

Ramelton ⚲(drying) ⊼ ☎9151082, ⚓ 🗂 🗂 Kos ⊠ Ⓒ ✕ ⌂.
Buncrana ⚲(congested) ⬚ 🗂 🗂 Kos ⊠ Ⓑ ∰ Hotel ✕ ⌂ @ (library).

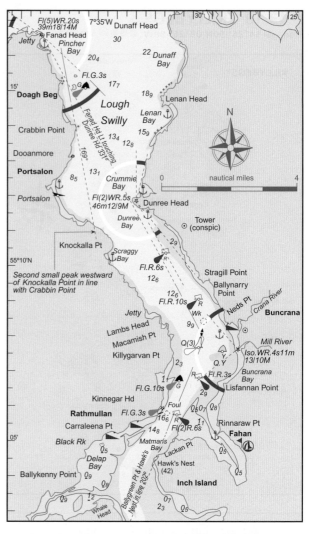

Lough Swilly Marina www.loughswillymarina.com ☎086 108211 200 ⚲ some Ⓥ €3/m, ⚓ ⬚ (development ongoing).

Fahan ⚓ ✕ ⌂ **Lough Swilly YC** www.loughswillyyc.com.

Services (1M SE): 🗂 🗂 ∰ ≋ ✈ (Londonderry 10M bus/taxi).

BLOODY FORELAND TO FAIR HEAD

(AC 2723, 5612) This is a good cruising area, under the lee of land in SW'lies, but very exposed to NW or N. Beware fishing boats and nets in many places and the North Channel TSS.

(AC 2752) Between Bloody Foreland and Horn Head are three low-lying islands: Inishbofin, Inishdooey and Inishbeg. The first is almost part of the mainland; it has a temporary ⚓ on S side and a more sheltered ⚓ on NE side in Toberglassan Bay. 6M offshore is Tory Is (Fl 4 30s) with rocks for 5ca off SW side. The harbour in Camusmore B provides the best shelter on this stretch of coast (details in ICC's SDs).

▶ *In Tory Sound the stream runs W from HW Galway +0230, and E from HW Galway –0530, sp rates 2kn.* ◀

Sheep Haven Bay is easy to enter between Horn Hd and Rinnafaghla Pt, and has good ⚓s except in strong NW or N winds. Beware Wherryman Rks, dry 1·5m, 1ca off E shore.

Between Sheep Haven and Mulroy Bay there is an inshore passage between Guill Rks & Carnabollion and S of Frenchman's Rock. Safe in good weather, otherwise keep 1M offshore. E to Lough Swilly the coast is very foul. Beware Limeburner Rk (2m),

6·8M WNW of Fanad Hd. Mulroy Bay has good ⚓s but needs accurate pilotage (see ICC SDs). Ent to L Swilly is clear except for Swilly Rks off the W shore, SSE of Fanad Hd. From Dunaff Hd at ent to Lough Swilly, keep 5ca offshore to Malin Hd. Trawbreaga Lough (AC 2697) gives shelter, but is shallow, and sea can break on bar; only approach when no swell, and at half flood. **At Malin Hd the direction of buoyage changes to W.**

Garvan Is, Tor Rks and Inishtrahull lie E and NE of Malin Hd. In bad weather it is best to pass at least 3M N of Tor Rks. Inishtrahull is lit and about 1M long; rocks extend N about 3ca into Tor Sound.

▶ *W of Malin Hd a W-going eddy starts at HW Galway +0400, and an E-going one at HW. Inishtrahull Sound, between Inishtrahull and Garvan Is, is exposed; tidal streams up to 4kn sp can raise a dangerous sea with no warning. Stream also sets hard through Garvan Isles, S of which Garvan Sound can be passed safely in daylight avoiding two sunken rks, one 1½ca NE of Rossnabarton, and the other 5ca NW. The main stream runs W for only 3 hrs, from HW Galway –0500 to –0200. W of Malin Hd a W-going eddy starts at HW Galway +0400, and an E-going one at HW Galway –0300.* ◀

Enter Lough Foyle by either the North Chan W of The Tuns, or S chan passing 2ca N of Magilligan Pt and allowing for set towards The Tuns on the ebb (up to 3·5kn).

Proceeding to Portrush, use Skerries Sound in good weather.

▶ A fair tide is essential through Rathlin Sound, as sp rates reach 6kn, with dangerous overfalls. The main stream sets W from HW Dover +0030 for 5 hrs, and E from HW Dover –0530 for 5 hrs. The worst overfalls are S of Rue Pt (Slough-na-more) from HW Dover +0130 to +0230, and it is best to enter W-bound at the end of this period, on the last of fair tide. E-bound enter the Sound at HW Dover –0500. Close inshore between Fair Hd and Carrick-mannanon Rk a counter eddy runs W from HW Dover –0030, and an E-going eddy runs from HW Dover +0200 to +0300. ◀

Pass outside Carrickmannanon Rk (0·3m) and Sheep Is.

There are small hbrs in Church Bay (Rathlin Is) and at Ballycastle.

9.13.12 LOUGH FOYLE

Londonderry/Donegal 55°14'N 06°54'W ✿✿✿⚓⚓✿✿

CHARTS AC 2723, 2798, 2511, 2510, 5612; Imray C53, C64

TIDES
Warren Point: –0430 Dover
Moville: –0350 Dover
Londonderry: –0255 Dover
ML 1·6; Duration 0615

Standard Port RIVER FOYLE (LISAHALLY) (→)

Times				Height (metres)			
High Water		Low Water		MHWS	MHWN	MLWN	MLWS
0100	0800	0200	0700	2·6	1·9	0·9	0·4
1300	2000	1400	1900				
Differences WARREN LIGHTHOUSE							
–0055	–0115	–0155	–0117	–0·3	0·0	ND	ND
MOVILLE							
–0042	–0057	–0127	–0058	–0·3	0·0	+0·1	0·0
QUIGLEY'S POINT							
–0020	–0027	–0040	–0027	–0·3	–0·1	0·0	–0·1
LONDONDERRY							
+0033	+0035	+0032	+0032	+0·1	+0·2	+0·3	+0·2

SHELTER The SE side of the Lough is low lying and shallow. The NW rises steeply and has several village hbrs between the ent and Londonderry (often referred to as Derry).

White Bay: close N of Dunagree Point is a good ⚓ in settled weather on passage.

Greencastle: a busy commercial fishing hbr, open to swell in winds SW to E. No formalised ⚓. Only advised for yachts in emergency.

Moville: a quaint resort on the western shore. A wooden pier, with 1·5m at the end, is near the village (shops closed all day Wed), but is much used by FVs. ⚓ outside hbr is exposed; inside hbr for shoal draught only. 8 Y ⚓s are about 600m upstream.

Carrickarory: pier/quay is used entirely by FVs. ⚓ in bay is sheltered in winds from SW to NNW. Plans for possible new marina.

Culmore Bay: good shelter. ⚓ 1½ca W of Culmore Pt in pleasant cove, 4M from Londonderry.

NAVIGATION WPT Tuns PHM buoy, Fl R 3s, 55°14'·00N 06°53'·49W, 235°/2·5M to Warren Pt lt, 2·5M. The Tuns bank lies 3M NE of Magilligan Pt and may dry. The main or N Chan, ¾M wide, runs NW of The Tuns; a lesser chan, min depth 4m, runs 3ca off shore around NE side of Magilligan Pt. **Beware commercial traffic**. Transiting the Loch the main channel passes NW of McKinney's Bank and the commercial ⚓ off Moville. The maintained channel leads close to the NW shore to the N of North Middle Bank and into the West Channel towards the commercial port of Lisahally. The narrows at Culmore Pt, marked by a GW Twr and medieval castle on the N shore, experience strong currents as the Loch gives way to the river proper. Upstream at Gransha the Foyle Bridge has 31m clearance before the river winds round Madams Bank following the Rosses Bay Channel, which is well marked by leading beacons. The buoyed channel leads to the city itself where further passage is limited by the Peace Bridge (clearance 3·7m) and the Craigavon Bridge (clearance 1·2m). In June and July the channel is at times obstructed by salmon nets at night. In North Chan tidal streams reach 3½kn, and upriver the ebb runs up to 6kn.

LIGHTS AND MARKS Inishowen Fl (2) WRG 10s 28m 18/14M; W tr, 2 B bands; vis 197°–G–211°–W–249°–R–000°. Warren Pt Fl 1·5s 9m 10M; W tr, G abutment; vis 232°–061°. Magilligan Pt QR 7m 4M; ✗ structure. The main chan up to Londonderry is very well lit. Foyle Bridge centre FW each side; VQ G on W pier; VQ R on E.

COMMUNICATIONS (Code 028) CGOC 9146 3933; ⊖ 7126 1937 or 9035 8250; Police 101; Dr 7126 4868; Ⓗ 7034 5171. HM (at Lisahally) 7186 0555.

Yachts are to call *Harbour Radio* on CH 14 for berthing directions at Foyle Marina. When secure ☎7186 0313 for gate code and facilities. VHF Ch **14** 12 16 (H24). Traffic and nav info Ch 14.

FACILITIES **Culdaff Bay** (10M W of Foyle SWM buoy) 6 Y ⚓s at 55°18'N 07°09'·1W, off Bunnagee Port.

Lisahally is the commercial port (55°02'·6N 07°15'·6W).

Londonderry Foyle Marina (pontoon) www.londonderryport.com billmccann@londonderryport.com, ☎71860555, mob 078 41580590 ⬡ in 5–7m, £15–£21/night, ⚓ ⬡£3, adjacent to city centre. All payments can be made at the coffee shop just outside the gate to the 'Northern' pontoon.

Prehen Boat Club (above Craigavon Bridge) ☎7034 3405.

City ▢ ▢ ⚒ ⚓ ⊠ Ⓑ ☐ ✗ ☐ ⇌ ✈.

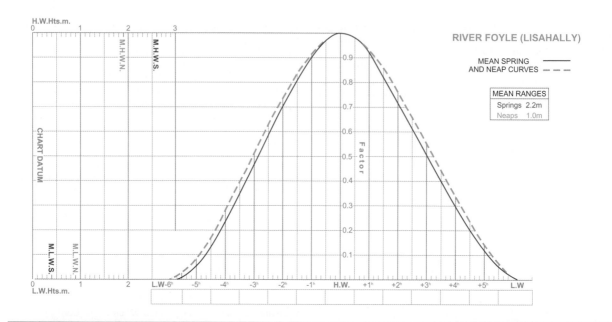

RIVER FOYLE (LISAHALLY)

MEAN SPRING ——————
AND NEAP CURVES — — — —

MEAN RANGES
Springs 2.2m
Neaps 1.0m

STANDARD TIME (UT)
For Summer Time add ONE hour in **non-shaded areas**

RIVER FOYLE LAT 55°03′N LONG 7°16′W
TIMES AND HEIGHTS OF HIGH AND LOW WATERS

Dates in red are **SPRINGS**
Dates in blue are **NEAPS**

YEAR 2016

JANUARY

Day	Time	m	Time	m	Time	m	Time	m
1 F	0537	0.8	1216	2.1	1906	1.0		
2 SA	0052	1.9	0630	0.9	1349	2.0	2011	1.0
3 SU	0209	1.9	0733	1.0	1503	2.0	2118	1.0
4 M	0318	1.9	0842	1.0	1601	2.0	2217	0.9
5 TU	0418	2.0	0951	0.9	1649	2.1	2309	0.8
6 W	0508	2.2	1055	0.9	1729	2.2	2354	0.7
7 TH	0550	2.4	1147	0.8	1804	2.3		
8 F	0035	0.6	0627	2.5	1235	0.7	1840	2.4
9 SA	0114	0.5	0704	2.7	1319	0.6	1918	2.5
10 SU ●	0153	0.5	0742	2.8	1404	0.6	1958	2.5
11 M	0231	0.4	0821	2.9	1448	0.5	2040	2.6
12 TU	0309	0.4	0902	2.9	1533	0.5	2122	2.6
13 W	0348	0.4	0945	2.9	1617	0.5	2205	2.4
14 TH	0428	0.4	1030	2.8	1702	0.5	2250	2.4
15 F	0511	0.5	1119	2.6	1750	0.6	2339	2.2
16 SA ◐	0600	0.7	1219	2.5	1845	0.8		
17 SU	0041	2.1	0701	0.8	1335	2.4	1949	0.8
18 M	0208	2.0	0821	0.9	1455	2.3	2102	0.9
19 TU	0335	2.0	0952	0.9	1608	2.3	2219	0.8
20 W	0451	2.2	1107	0.8	1712	2.4	2324	0.7
21 TH	0547	2.4	1205	0.6	1805	2.5		
22 F	0015	0.6	0631	2.5	1255	0.6	1850	2.5
23 SA	0100	0.5	0710	2.6	1340	0.5	1930	2.6
24 SU ○	0140	0.5	0747	2.7	1423	0.5	2007	2.6
25 M	0217	0.4	0822	2.7	1502	0.5	2042	2.5
26 TU	0252	0.4	0856	2.7	1538	0.6	2115	2.5
27 W	0323	0.4	0930	2.6	1612	0.6	2149	2.4
28 TH	0355	0.5	1003	2.5	1645	0.7	2223	2.3
29 F	0429	0.6	1037	2.3	1721	0.8	2302	2.1
30 SA	0508	0.7	1115	2.1	1805	0.9	2349	2.0
31 SU	0554	0.8	1204	1.9	1859	0.9		

FEBRUARY

Day	Time	m	Time	m	Time	m	Time	m
1 M ◐	0055	1.9	0652	0.9	1329	1.8	2003	1.0
2 TU	0220	1.8	0757	1.0	1507	1.9	2116	0.9
3 W	0333	1.9	0905	1.0	1615	1.9	2228	0.8
4 TH	0434	2.1	1017	0.9	1709	2.0	2325	0.7
5 F	0524	2.2	1125	0.8	1752	2.1		
6 SA	0013	0.6	0607	2.4	1221	0.7	1831	2.3
7 SU	0057	0.5	0647	2.6	1309	0.5	1909	2.4
8 M ●	0138	0.4	0726	2.8	1355	0.4	1948	2.5
9 TU	0218	0.3	0805	2.9	1438	0.3	2027	2.6
10 W	0257	0.2	0845	3.0	1520	0.3	2106	2.6
11 TH	0336	0.2	0927	3.0	1600	0.3	2145	2.4
12 F	0415	0.3	1010	2.9	1641	0.4	2226	2.5
13 SA	0456	0.4	1057	2.7	1723	0.5	2310	2.3
14 SU	0543	0.5	1150	2.5	1811	0.7		
15 M ◐	0003	2.1	0640	0.7	1304	2.2	1910	0.9
16 TU	0129	1.9	0801	0.9	1435	2.1	2026	0.9
17 W	0316	1.9	0945	0.9	1600	2.1	2155	0.9
18 TH	0444	2.1	1102	0.7	1709	2.2	2306	0.8
19 F	0541	2.3	1158	0.6	1801	2.3	2358	0.7
20 SA	0623	2.4	1245	0.5	1844	2.4		
21 SU	0043	0.5	0659	2.5	1327	0.4	1919	2.4
22 M ○	0123	0.4	0733	2.6	1406	0.4	1952	2.4
23 TU	0159	0.4	0804	2.6	1442	0.4	2021	2.5
24 W	0232	0.3	0834	2.6	1514	0.4	2049	2.5
25 TH	0303	0.3	0903	2.6	1543	0.5	2119	2.4
26 F	0332	0.4	0932	2.4	1610	0.5	2151	2.3
27 SA	0405	0.4	1004	2.3	1640	0.6	2227	2.2
28 SU	0442	0.5	1039	2.1	1717	0.7	2308	2.1
29 M	0525	0.7	1120	1.9	1804	0.8		

MARCH

Day	Time	m	Time	m	Time	m	Time	m
1 TU ◐	0001	1.9	0619	0.8	1217	1.7	1904	0.9
2 W	0121	1.8	0723	0.9	1357	1.7	2016	0.9
3 TH	0251	1.8	0832	0.9	1540	1.7	2138	0.9
4 F	0400	1.8	0947	0.8	1648	1.8	2255	0.8
5 SA	0456	2.1	1106	0.7	1738	2.0	2348	0.6
6 SU	0543	2.3	1206	0.5	1819	2.2		
7 M	0034	0.5	0624	2.6	1255	0.4	1855	2.3
8 TU	0117	0.3	0705	2.8	1339	0.2	1931	2.5
9 W ●	0158	0.2	0744	2.9	1421	0.1	2008	2.6
10 TH	0238	0.1	0825	3.0	1500	0.1	2045	2.6
11 F	0318	0.1	0907	3.0	1539	0.1	2123	2.6
12 SA	0358	0.1	0950	2.9	1616	0.2	2203	2.5
13 SU	0440	0.3	1035	2.7	1655	0.4	2246	2.3
14 M	0527	0.5	1128	2.4	1739	0.6	2336	2.1
15 TU ◐	0626	0.7	1239	2.1	1834	0.8		
16 W	0104	1.9	0753	0.8	1416	1.9	1950	0.9
17 TH	0301	1.9	0934	0.8	1543	1.9	2126	0.9
18 F	0424	2.0	1046	0.7	1653	2.0	2239	0.8
19 SA	0522	2.2	1140	0.5	1745	2.1	2333	0.6
20 SU	0605	2.3	1225	0.4	1826	2.2		
21 M	0018	0.5	0641	2.4	1305	0.4	1900	2.3
22 TU	0059	0.4	0713	2.4	1342	0.3	1930	2.3
23 W ○	0136	0.3	0741	2.4	1417	0.3	1956	2.4
24 TH	0209	0.3	0808	2.4	1447	0.3	2022	2.4
25 F	0239	0.3	0835	2.4	1514	0.4	2051	2.4
26 SA	0308	0.3	0905	2.3	1539	0.4	2123	2.4
27 SU	0341	0.4	0937	2.2	1606	0.5	2157	2.3
28 M	0417	0.5	1012	2.1	1638	0.6	2236	2.1
29 TU	0500	0.6	1052	1.9	1718	0.7	2322	2.0
30 W	0551	0.7	1141	1.7	1812	0.8		
31 TH ◐	0029	1.8	0653	0.8	1259	1.6	1923	0.9

APRIL

Day	Time	m	Time	m	Time	m	Time	m
1 F	0206	1.8	0803	0.8	1454	1.6	2046	0.9
2 SA	0322	1.9	0918	0.7	1614	1.7	2212	0.8
3 SU	0421	2.1	1038	0.6	1711	1.9	2315	0.6
4 M	0512	2.3	1141	0.4	1755	2.1		
5 TU	0005	0.4	0557	2.5	1232	0.3	1833	2.3
6 W	0050	0.2	0640	2.7	1316	0.1	1909	2.5
7 TH ●	0133	0.1	0722	2.9	1358	0.0	1946	2.6
8 F	0215	0.0	0803	3.0	1437	0.0	2023	2.7
9 SA	0257	0.0	0846	2.9	1515	0.1	2102	2.6
10 SU	0340	0.1	0931	2.8	1552	0.2	2143	2.5
11 M	0425	0.2	1018	2.6	1631	0.4	2227	2.4
12 TU	0515	0.4	1111	2.4	1713	0.5	2321	2.1
13 W	0617	0.6	1220	2.0	1804	0.7		
14 TH ◑	0053	2.0	0743	0.7	1349	1.8	1915	0.9
15 F	0235	2.0	0908	0.7	1509	1.8	2046	0.9
16 SA	0350	2.0	1014	0.6	1619	1.8	2202	0.8
17 SU	0449	2.1	1108	0.5	1715	2.0	2259	0.6
18 M	0536	2.2	1154	0.4	1758	2.1	2347	0.5
19 TU	0615	2.3	1235	0.4	1834	2.2		
20 W	0029	0.4	0647	2.3	1313	0.4	1903	2.3
21 TH	0107	0.4	0715	2.3	1347	0.3	1929	2.3
22 F ○	0141	0.3	0740	2.3	1419	0.3	1956	2.4
23 SA	0212	0.3	0808	2.3	1446	0.3	2025	2.5
24 SU	0243	0.3	0840	2.3	1512	0.3	2058	2.4
25 M	0317	0.4	0915	2.2	1539	0.4	2134	2.4
26 TU	0356	0.5	0953	2.1	1611	0.5	2212	2.2
27 W	0439	0.5	1035	2.0	1648	0.6	2258	2.1
28 TH	0530	0.6	1124	1.8	1735	0.7	2357	2.0
29 F	0630	0.7	1230	1.7	1839	0.8		
30 SA ◑	0122	1.9	0738	0.7	1403	1.6	2000	0.8

Chart Datum: 1·37 metres below Ordnance Datum (Belfast). HAT is 3·1 metres above Chart Datum.

N Ireland

》》 FREE monthly updates. Register at 《
www.reedsnauticalalmanac.co.uk

625

STANDARD TIME (UT)
For Summer Time add ONE hour in **non-shaded areas**

RIVER FOYLE LAT 55°03′N LONG 7°16′W
TIMES AND HEIGHTS OF HIGH AND LOW WATERS

Dates in red are **SPRINGS**
Dates in blue are **NEAPS**

YEAR **2016**

MAY

Date	Day	Time	m		Date	Day	Time	m
1	SU	0242 / 0848 / 1527 / 2124	2.0 / 0.6 / 1.7 / 0.7		16	M	0409 / 1027 / 1633 / 2219	2.0 / 0.6 / 1.8 / 0.7
2	M	0344 / 1001 / 1633 / 2235	2.1 / 0.5 / 1.9 / 0.6		17	TU	0501 / 1116 / 1723 / 2311	2.1 / 0.5 / 1.9 / 0.6
3	TU	0439 / 1107 / 1725 / 2332	2.3 / 0.4 / 2.1 / 0.4		18	W	0544 / 1200 / 1803 / 2356	2.1 / 0.4 / 2.1 / 0.5
4	W	0529 / 1202 / 1808	2.5 / 0.2 / 2.3		19	TH	0619 / 1240 / 1835	2.1 / 0.4 / 2.2
5	TH	0021 / 0615 / 1250 / 1847	0.2 / 2.7 / 0.1 / 2.4		20	F	0035 / 0647 / 1318 / 1903	0.5 / 2.2 / 0.3 / 2.3
6	F ●	0108 / 0700 / 1333 / 1925	0.1 / 2.8 / 0.1 / 2.6		21	SA ○	0112 / 0714 / 1351 / 1933	0.4 / 2.2 / 0.3 / 2.4
7	SA	0153 / 0744 / 1414 / 2004	0.1 / 2.9 / 0.0 / 2.7		22	SU	0146 / 0744 / 1422 / 2004	0.4 / 2.2 / 0.3 / 2.5
8	SU	0238 / 0829 / 1453 / 2045	0.1 / 2.8 / 0.1 / 2.6		23	M	0220 / 0819 / 1450 / 2039	0.4 / 2.2 / 0.3 / 2.5
9	M	0325 / 0916 / 1532 / 2128	0.2 / 2.7 / 0.2 / 2.6		24	TU	0258 / 0857 / 1520 / 2116	0.4 / 2.2 / 0.4 / 2.4
10	TU	0413 / 1004 / 1611 / 2215	0.3 / 2.5 / 0.3 / 2.4		25	W	0339 / 0939 / 1554 / 2157	0.5 / 2.2 / 0.4 / 2.3
11	W	0506 / 1057 / 1652 / 2311	0.4 / 2.2 / 0.5 / 2.2		26	TH	0425 / 1023 / 1631 / 2243	0.5 / 2.1 / 0.5 / 2.2
12	TH	0607 / 1158 / 1739	0.6 / 2.0 / 0.6		27	F	0515 / 1112 / 1716 / 2338	0.5 / 1.9 / 0.6 / 2.1
13	F ◑	0033 / 0720 / 1312 / 1839	2.1 / 0.7 / 1.8 / 0.8		28	SA	0612 / 1209 / 1811	0.6 / 1.8 / 0.7
14	SA	0200 / 0831 / 1424 / 1957	2.0 / 0.7 / 1.7 / 0.8		29	SU ◑	0049 / 0714 / 1322 / 1924	2.1 / 0.6 / 1.7 / 0.7
15	SU	0309 / 0933 / 1532 / 2115	2.0 / 0.6 / 1.8 / 0.8		30	M	0204 / 0819 / 1441 / 2043	2.1 / 0.5 / 1.8 / 0.7
					31	TU	0309 / 0925 / 1552 / 2158	2.2 / 0.5 / 1.9 / 0.6

JUNE

Date	Day	Time	m		Date	Day	Time	m
1	W	0408 / 1032 / 1654 / 2302	2.3 / 0.4 / 2.0 / 0.4		16	TH	0510 / 1124 / 1730 / 2321	2.0 / 0.5 / 2.0 / 0.6
2	TH	0504 / 1132 / 1744 / 2358	2.5 / 0.3 / 2.2 / 0.3		17	F	0549 / 1208 / 1808	2.0 / 0.5 / 2.1
3	F	0555 / 1224 / 1828	2.6 / 0.2 / 2.4		18	SA	0005 / 0621 / 1249 / 1841	0.6 / 2.1 / 0.4 / 2.3
4	SA	0048 / 0643 / 1311 / 1909	0.2 / 2.7 / 0.2 / 2.5		19	SU	0045 / 0652 / 1327 / 1914	0.5 / 2.1 / 0.4 / 2.4
5	SU ●	0137 / 0730 / 1355 / 1951	0.2 / 2.7 / 0.1 / 2.6		20	M ○	0124 / 0726 / 1402 / 1948	0.5 / 2.2 / 0.3 / 2.5
6	M	0225 / 0816 / 1436 / 2033	0.2 / 2.7 / 0.2 / 2.6		21	TU	0204 / 0803 / 1436 / 2024	0.5 / 2.2 / 0.3 / 2.5
7	TU	0314 / 0903 / 1515 / 2117	0.2 / 2.6 / 0.2 / 2.6		22	W	0246 / 0844 / 1509 / 2103	0.5 / 2.2 / 0.3 / 2.5
8	W	0404 / 0951 / 1554 / 2203	0.3 / 2.4 / 0.3 / 2.5		23	TH	0330 / 0927 / 1544 / 2145	0.4 / 2.2 / 0.3 / 2.5
9	TH	0454 / 1040 / 1633 / 2255	0.4 / 2.2 / 0.4 / 2.3		24	F	0416 / 1011 / 1622 / 2230	0.4 / 2.2 / 0.4 / 2.4
10	F	0548 / 1131 / 1715 / 2358	0.6 / 2.0 / 0.5 / 2.1		25	SA	0503 / 1057 / 1704 / 2320	0.4 / 2.1 / 0.5 / 2.3
11	SA	0647 / 1228 / 1803	0.6 / 1.9 / 0.6		26	SU	0554 / 1147 / 1753	0.5 / 2.0 / 0.5
12	SU ◑	0115 / 0748 / 1332 / 1902	2.0 / 0.7 / 1.8 / 0.7		27	M ◑	0021 / 0719 / 1247 / 1855	2.3 / 0.5 / 1.9 / 0.6
13	M	0225 / 0846 / 1438 / 2015	2.0 / 0.7 / 1.7 / 0.8		28	TU	0131 / 0750 / 1400 / 2010	2.2 / 0.5 / 1.8 / 0.7
14	TU	0327 / 0942 / 1543 / 2128	2.0 / 0.6 / 1.8 / 0.8		29	W	0240 / 0853 / 1517 / 2129	2.2 / 0.5 / 1.9 / 0.6
15	W	0422 / 1035 / 1642 / 2230	2.0 / 0.6 / 1.9 / 0.7		30	TH	0344 / 1001 / 1628 / 2242	2.3 / 0.5 / 2.0 / 0.5

JULY

Date	Day	Time	m		Date	Day	Time	m
1	F	0445 / 1107 / 1728 / 2344	2.4 / 0.4 / 2.1 / 0.4		16	SA	0522 / 1138 / 1743 / 2338	1.9 / 0.5 / 2.1 / 0.7
2	SA	0542 / 1205 / 1817	2.4 / 0.4 / 2.3		17	SU	0601 / 1224 / 1822	2.0 / 0.5 / 2.2
3	SU	0037 / 0633 / 1255 / 1900	0.3 / 2.5 / 0.3 / 2.5		18	M	0025 / 0636 / 1306 / 1857	0.6 / 2.1 / 0.4 / 2.4
4	M ●	0129 / 0721 / 1340 / 1942	0.3 / 2.5 / 0.2 / 2.6		19	TU ○	0110 / 0713 / 1345 / 1933	0.5 / 2.1 / 0.3 / 2.5
5	TU	0218 / 0807 / 1422 / 2023	0.3 / 2.5 / 0.2 / 2.6		20	W	0154 / 0751 / 1423 / 2010	0.5 / 2.2 / 0.3 / 2.6
6	W	0306 / 0851 / 1501 / 2104	0.3 / 2.5 / 0.2 / 2.6		21	TH	0238 / 0831 / 1459 / 2049	0.4 / 2.3 / 0.3 / 2.7
7	TH	0352 / 0934 / 1537 / 2146	0.4 / 2.4 / 0.2 / 2.5		22	F	0321 / 0912 / 1535 / 2129	0.4 / 2.3 / 0.2 / 2.7
8	F	0436 / 1016 / 1613 / 2230	0.4 / 2.3 / 0.3 / 2.4		23	SA	0404 / 0953 / 1612 / 2212	0.3 / 2.3 / 0.3 / 2.6
9	SA	0520 / 1057 / 1649 / 2317	0.5 / 2.1 / 0.4 / 2.2		24	SU	0447 / 1035 / 1651 / 2259	0.3 / 2.2 / 0.3 / 2.5
10	SU	0607 / 1142 / 1728	0.6 / 2.0 / 0.6		25	M	0533 / 1120 / 1736 / 2353	0.4 / 2.1 / 0.4 / 2.4
11	M	0015 / 0657 / 1235 / 1817	2.0 / 0.7 / 1.8 / 0.7		26	TU ◑	0623 / 1212 / 1831	0.5 / 2.0 / 0.6
12	TU ◑	0130 / 0753 / 1340 / 1918	1.9 / 0.7 / 1.7 / 0.8		27	W	0100 / 0719 / 1322 / 1943	2.3 / 0.6 / 1.9 / 0.7
13	W	0240 / 0852 / 1451 / 2028	1.9 / 0.7 / 1.7 / 0.8		28	TH	0216 / 0824 / 1449 / 2110	2.2 / 0.6 / 1.8 / 0.7
14	TH	0342 / 0951 / 1559 / 2140	1.9 / 0.7 / 1.8 / 0.8		29	F	0329 / 0936 / 1614 / 2235	2.2 / 0.6 / 1.9 / 0.7
15	F	0436 / 1048 / 1657 / 2245	1.9 / 0.6 / 1.9 / 0.8		30	SA	0437 / 1050 / 1724 / 2339	2.2 / 0.6 / 2.1 / 0.5
					31	SU	0538 / 1151 / 1814	2.3 / 0.5 / 2.3

AUGUST

Date	Day	Time	m		Date	Day	Time	m
1	M	0033 / 0629 / 1242 / 1855	0.4 / 2.4 / 0.4 / 2.5		16	TU	0010 / 0623 / 1246 / 1838	0.6 / 2.1 / 0.4 / 2.4
2	TU ●	0122 / 0715 / 1326 / 1934	0.4 / 2.4 / 0.3 / 2.6		17	W	0057 / 0659 / 1327 / 1914	0.5 / 2.2 / 0.3 / 2.6
3	W	0209 / 0756 / 1407 / 2011	0.3 / 2.4 / 0.2 / 2.6		18	TH ○	0141 / 0736 / 1405 / 1951	0.4 / 2.3 / 0.2 / 2.7
4	TH	0253 / 0834 / 1444 / 2047	0.3 / 2.4 / 0.2 / 2.6		19	F	0224 / 0813 / 1442 / 2029	0.3 / 2.4 / 0.2 / 2.8
5	F	0333 / 0910 / 1518 / 2123	0.4 / 2.4 / 0.2 / 2.6		20	SA	0305 / 0851 / 1519 / 2109	0.3 / 2.5 / 0.1 / 2.9
6	SA	0411 / 0945 / 1549 / 2159	0.4 / 2.3 / 0.3 / 2.4		21	SU	0345 / 0929 / 1556 / 2150	0.2 / 2.5 / 0.2 / 2.9
7	SU	0446 / 1021 / 1622 / 2236	0.5 / 2.2 / 0.4 / 2.3		22	M	0425 / 1009 / 1635 / 2235	0.3 / 2.4 / 0.3 / 2.7
8	M	0523 / 1059 / 1657 / 2316	0.6 / 2.1 / 0.5 / 2.1		23	TU	0506 / 1051 / 1718 / 2326	0.4 / 2.3 / 0.4 / 2.5
9	TU	0604 / 1144 / 1741	0.7 / 1.9 / 0.7		24	W	0551 / 1140 / 1810	0.5 / 2.1 / 0.6
10	W ◑	0011 / 0654 / 1244 / 1835	1.9 / 0.8 / 1.8 / 0.8		25	TH ◑	0031 / 0646 / 1249 / 1924	2.3 / 0.7 / 1.9 / 0.8
11	TH	0140 / 0755 / 1404 / 1942	1.8 / 0.8 / 1.7 / 0.9		26	F	0158 / 0754 / 1433 / 2106	2.1 / 0.8 / 1.9 / 0.8
12	F	0301 / 0904 / 1521 / 2054	1.7 / 0.8 / 1.8 / 0.9		27	SA	0321 / 0916 / 1610 / 2232	2.1 / 0.8 / 2.0 / 0.7
13	SA	0405 / 1013 / 1626 / 2211	1.8 / 0.7 / 1.9 / 0.8		28	SU	0434 / 1035 / 1719 / 2333	2.1 / 0.7 / 2.2 / 0.6
14	SU	0500 / 1112 / 1719 / 2318	1.9 / 0.6 / 2.1 / 0.7		29	M	0534 / 1135 / 1806	2.2 / 0.6 / 2.4
15	M	0545 / 1201 / 1801	1.9 / 0.5 / 2.2		30	TU	0024 / 0621 / 1225 / 1844	0.4 / 2.3 / 0.4 / 2.5
					31	W	0110 / 0702 / 1308 / 1919	0.4 / 2.4 / 0.3 / 2.6

Chart Datum: 1·37 metres below Ordnance Datum (Belfast). HAT is 3·1 metres above Chart Datum.

STANDARD TIME (UT)
For Summer Time add ONE hour in **non-shaded areas**

RIVER FOYLE   LAT 55°03'N   LONG 7°16'W
TIMES AND HEIGHTS OF HIGH AND LOW WATERS

Dates in red are **SPRINGS**
Dates in blue are **NEAPS**

YEAR 2016

SEPTEMBER

Time	m		Time	m
1 0152	0.3	**16** 0121	0.3
0738	2.4	0714	2.4
TH 1347	0.3	F 1341	0.2
● 1952	2.6	○ 1926	2.9
2 0231	0.3	**17** 0202	0.3
0811	2.5	0749	2.5
F 1422	0.2	SA 1419	0.1
2024	2.6	2005	3.0
3 0307	0.4	**18** 0241	0.2
0842	2.4	0826	2.6
SA 1454	0.2	SU 1457	0.1
2055	2.6	2045	3.0
4 0339	0.4	**19** 0320	0.2
0912	2.4	0904	2.7
SU 1524	0.3	M 1536	0.1
2126	2.5	2127	3.0
5 0408	0.5	**20** 0358	0.3
0945	2.3	0943	2.6
M 1554	0.4	TU 1616	0.3
2157	2.3	2212	2.8
6 0439	0.6	**21** 0437	0.4
1020	2.2	1025	2.5
TU 1628	0.5	W 1701	0.4
2232	2.1	2302	2.5
7 0514	0.7	**22** 0520	0.6
1100	2.1	1114	2.3
W 1709	0.7	TH 1756	0.7
2313	1.9		
8 0559	0.8	**23** 0008	2.2
1152	1.9	0612	0.8
TH 1800	0.8	F 1225	2.1
		◐ 1915	0.8
9 0014	1.7	**24** 0143	2.1
0656	0.9	0721	0.9
F 1314	1.8	SA 1424	2.0
◐ 1904	0.9	2103	0.9
10 0214	1.7	**25** 0309	2.0
0806	0.9	0852	0.9
SA 1445	1.8	SU 1555	2.1
2017	1.0	2219	0.7
11 0334	1.7	**26** 0420	2.1
0930	0.8	1011	0.8
SU 1553	1.9	M 1700	2.3
2138	0.9	2317	0.6
12 0434	1.8	**27** 0517	2.2
1042	0.7	1111	0.7
M 1648	2.1	TU 1746	2.4
2256	0.8		
13 0523	2.0	**28** 0005	0.5
1135	0.6	0602	2.3
TU 1733	2.3	W 1201	0.5
2351	0.6	1824	2.5
14 0603	2.1	**29** 0047	0.4
1219	0.5	0641	2.4
W 1812	2.5	TH 1243	0.4
		1858	2.6
15 0037	0.5	**30** 0126	0.4
0639	2.3	0714	2.5
TH 1301	0.4	F 1322	0.4
1849	2.7	1928	2.6

OCTOBER

Time	m		Time	m
1 0203	0.4	**16** 0134	0.2
0744	2.5	0725	2.7
SA 1357	0.3	SU 1353	0.2
● 1957	2.6	○ 1940	3.1
2 0236	0.4	**17** 0214	0.2
0812	2.5	0802	2.8
SU 1427	0.3	M 1434	0.2
2024	2.5	2022	3.1
3 0305	0.5	**18** 0253	0.2
0840	2.5	0841	2.8
M 1456	0.4	TU 1516	0.2
2053	2.5	2106	3.0
4 0331	0.5	**19** 0332	0.3
0912	2.5	0922	2.8
TU 1526	0.5	W 1600	0.3
2123	2.4	2152	2.8
5 0358	0.6	**20** 0411	0.4
0946	2.4	1006	2.6
W 1601	0.6	TH 1648	0.5
2156	2.2	2244	2.5
6 0429	0.7	**21** 0453	0.6
1023	2.3	1056	2.4
TH 1641	0.7	F 1746	0.7
2234	2.0	2349	2.2
7 0508	0.8	**22** 0543	0.8
1107	2.1	1211	2.2
F 1729	0.9	SA 1908	0.9
2322	1.8	◐	
8 0558	0.9	**23** 0119	2.1
1211	1.9	0647	1.0
SA 1831	1.0	SU 1402	2.2
		2043	0.9
9 0048	1.7	**24** 0242	2.0
0704	1.0	0814	1.0
SU 1358	1.9	M 1523	2.2
◐ 1943	1.0	2153	0.8
10 0249	1.7	**25** 0351	2.1
0823	1.0	0936	0.9
M 1512	2.0	TU 1627	2.3
2101	0.9	2249	0.7
11 0357	1.8	**26** 0449	2.2
0952	0.9	1038	0.8
TU 1608	2.2	W 1717	2.4
2220	0.8	2336	0.6
12 0450	2.0	**27** 0536	2.3
1056	0.8	1129	0.7
W 1657	2.4	TH 1758	2.5
2320	0.6		
13 0534	2.2	**28** 0018	0.5
1145	0.6	0614	2.4
TH 1740	2.6	F 1214	0.6
		1832	2.5
14 0009	0.6	**29** 0056	0.5
0612	2.4	0647	2.5
F 1230	0.4	SA 1253	0.5
1820	2.8	1902	2.5
15 0053	0.3	**30** 0132	0.5
0648	2.5	0716	2.6
SA 1312	0.3	SU 1328	0.5
1900	3.0	● 1929	2.5
		31 0204	0.5
		0744	2.6
		M 1359	0.5
		1955	2.5

NOVEMBER

Time	m		Time	m
1 0233	0.5	**16** 0230	0.3
0813	2.7	0823	2.9
TU 1429	0.5	W 1459	0.3
2023	2.5	2049	3.0
2 0259	0.5	**17** 0310	0.3
0844	2.6	0906	2.9
W 1500	0.6	TH 1547	0.4
2054	2.4	2136	2.8
3 0325	0.6	**18** 0351	0.5
0918	2.6	0952	2.8
TH 1536	0.7	F 1637	0.6
2129	2.3	2228	2.5
4 0355	0.7	**19** 0432	0.6
0955	2.5	1043	2.6
F 1617	0.8	SA 1735	0.8
2208	2.1	2328	2.3
5 0432	0.8	**20** 0518	0.8
1037	2.3	1152	2.4
SA 1704	0.9	SU 1847	0.9
2254	2.0		
6 0516	0.9	**21** 0043	2.1
1129	2.2	0612	0.9
SU 1801	0.9	M 1326	2.3
2357	1.8	◐ 2008	0.9
7 0612	1.0	**22** 0200	2.0
1250	2.1	0722	1.0
M 1910	1.0	TU 1444	2.3
●		2116	0.9
8 0140	1.8	**23** 0309	2.0
0724	1.0	0844	1.0
TU 1421	2.1	W 1549	2.3
2022	0.9	2213	0.8
9 0306	1.9	**24** 0410	2.1
0845	1.0	0956	1.0
W 1525	2.3	TH 1644	2.4
2134	0.8	2302	0.7
10 0408	2.0	**25** 0503	2.2
1003	0.9	1054	0.9
TH 1619	2.5	F 1729	2.4
2240	0.6	2345	0.7
11 0500	2.2	**26** 0546	2.3
1105	0.7	1141	0.8
F 1707	2.7	SA 1807	2.4
2335	0.5		
12 0544	2.4	**27** 0024	0.6
1157	0.5	0620	2.4
SA 1753	2.9	SU 1223	0.7
		1837	2.4
13 0023	0.4	**28** 0101	0.6
0624	2.6	0651	2.6
SU 1244	0.4	M 1259	0.7
1836	3.0	1903	2.4
14 0107	0.3	**29** 0135	0.6
0703	2.8	0720	2.6
M 1329	0.3	TU 1332	0.7
○ 1919	3.1	● 1929	2.5
15 0149	0.2	**30** 0206	0.5
0742	2.9	0750	2.7
TU 1414	0.3	W 1404	0.7
2003	3.1	1959	2.5

DECEMBER

Time	m		Time	m
1 0234	0.6	**16** 0254	0.4
0823	2.7	0854	2.9
TH 1438	0.7	F 1537	0.5
2033	2.4	2123	2.7
2 0303	0.6	**17** 0334	0.4
0859	2.7	0940	2.8
F 1516	0.7	SA 1626	0.6
2111	2.4	2211	2.5
3 0335	0.6	**18** 0414	0.5
0937	2.6	1028	2.7
SA 1558	0.7	SU 1717	0.8
2152	2.3	2301	2.3
4 0411	0.7	**19** 0455	0.7
1018	2.5	1123	2.5
SU 1645	0.8	M 1814	0.9
2237	2.2	2357	2.1
5 0452	0.8	**20** 0539	0.8
1106	2.4	1235	2.3
M 1738	0.9	TU 1918	1.0
2330	2.0		
6 0541	0.9	**21** 0104	2.0
1207	2.3	0632	0.9
TU 1839	0.9	W 1356	2.2
		◐ 2026	1.0
7 0040	1.9	**22** 0215	1.9
0643	0.9	0738	1.0
W 1327	2.3	TH 1506	2.2
● 1945	0.9	2127	0.9
8 0208	1.9	**23** 0322	1.9
0757	1.0	0855	1.0
TH 1441	2.4	F 1609	2.2
2052	0.8	2222	0.9
9 0322	2.0	**24** 0423	2.0
0915	0.9	1010	1.0
F 1543	2.5	SA 1701	2.2
2200	0.7	2310	0.8
10 0425	2.2	**25** 0514	2.2
1028	0.8	1107	0.9
SA 1639	2.6	SU 1744	2.3
2302	0.6	2353	0.7
11 0517	2.4	**26** 0554	2.3
1130	0.6	1153	0.8
SU 1731	2.8	M 1817	2.3
2357	0.5		
12 0604	2.6	**27** 0033	0.6
1223	0.5	0628	2.5
M 1819	2.9	TU 1234	0.8
		1844	2.3
13 0045	0.4	**28** 0111	0.6
0646	2.8	0700	2.6
TU 1313	0.4	W 1311	0.7
1905	3.0	1911	2.4
14 0130	0.3	**29** 0145	0.6
0728	2.9	0733	2.7
W 1401	0.4	TH 1347	0.7
○ 1950	3.0	● 1943	2.4
15 0213	0.3	**30** 0218	0.5
0811	2.9	0807	2.7
TH 1449	0.4	F 1425	0.7
2037	2.9	2019	2.4
		31 0250	0.5
		0844	2.8
		SA 1504	0.7
		2058	2.4

Chart Datum: 1·37 metres below Ordnance Datum (Belfast). HAT is 3·1 metres above Chart Datum.

FREE monthly updates. Register at
www.reedsnauticalalmanac.co.uk

627

N Ireland

9.13.13 RIVER BANN AND COLERAINE

Londonderry/Antrim 55°10'·32N 06°46'·35W ✿✿◊◊✿✿

CHARTS AC 2723, 2798, 2494, 5612; Imray C53, C64

TIDES –0345 Dover (Coleraine); ML 1·1; Duration 0540

Standard Port RIVER FOYLE (←─)

Times				Height (metres)			
High Water		Low Water		MHWS	MHWN	MLWN	MLWS
0100	0800	0200	0700	2·6	1·9	0·9	0·4
1300	2000	1400	1900				
Differences COLERAINE							
–0004	–0106	–0109	–0005	–0.4	–0.1	0.0	0.0

SHELTER Good, once inside the river ent (The Barmouth) between 2 training walls, extending 2ca N from the beaches. Do not try to enter in strong onshore winds or when swell breaks on the pierheads. If in doubt call Coleraine Hbr Radio or ring HM. ⊥ upstream of old CG stn, or berth at Seaton's or Coleraine marinas, 3½M & 4½M from ent, on NE bank.

NAVIGATION WPT 55°11'·00N, 06°46'·65W, 165°/0·72M to ent. Appr from E of N to bring ldg lts into line at bkwtr ends. The sand bar is constantly moving but ent is dredged to approx 3·5m. Beware commercial traffic.

LIGHTS AND MARKS Ldg lts 165°, both Oc 5s 6/14m 2M, front on W pyramidal metal tr; rear W □ tr. Portstewart Pt, Oc R 10s, is 2M ENE.

COMMUNICATIONS (Code 028) CGOC 91463933; Police 101; ⊞ 70344177; Dr 70344831; HM 70342012; Rly Bridge 70325400. Coleraine Hbr Radio Ch 12 (Mon-Fri: HO and when vessel due). Coleraine Marina Ch M.

FACILITIES Coleraine Hbr ☎70342012, ⚓(35t).
Coleraine Marina ☎70344768, ▬ 90+5🅥 £21, ♠(H24) ⚓ ✕ ⊞ ⌂ ⚓(12½t).
Seatons Marina ☎07718 883099, ▬ (£3) ⚓(12t).
Coleraine YC ☎70344503, ⚓ ⌂.
Town ⌂ ⌂ ♠ ⚓ Ⓔ ⚓ ⚓ ⚓ ✉ Ⓑ ⚓ ✕ ⇌ ✈ (Belfast).

MINOR HARBOUR TO THE EAST
Portstewart, Antrim, **55°11'·21N 06°43'·21W**. AC 2494, 5612. Tides as for Portrush. A tiny hbr 1·1ca S of Portstewart Pt lt, Oc R 10s 21m 5M, vis 040°–220°, obscd in final appr. A temp, fair weather berth (£10) at S end of inner basin in 0·8–1·7m; the very narrow ent is open to SW wind and swell. Visitors should obtain fuel at Portrush before proceeding to Portstewart. Beware rocks close to S bkwtr. Facilities: ▬⚓♠⚓ ✕ ⌂ @ at library.

9.13.14 PORTRUSH

Antrim **55°12'·34N 06°39'·49W** ❀❀☼☼☼❀❀❀

CHARTS AC 2798, 2494, 5612; Imray C53, C64

TIDES –0400 Dover; ML 1·1; Duration 0610

Standard Port RIVER FOYLE (←——)

Times				Height (metres)			
High Water		Low Water		MHWS	MHWN	MLWN	MLWS
0100	0800	0200	0700	2·6	1·9	0·9	0·4
1300	2000	1400	1900				
Differences PORTRUSH							
–0046	–0052	–0117	–0057	–0.5	–0.3	+0·1	+0·1

SHELTER Good in hbr, except in strong NW/N winds. Berth on N pier or on pontoon at NE end of it and see HM. A ⚓ may be available, but very congested in season. ⚓ on E side of Ramore Hd in Skerries Roads 1ca S of Large Skerrie gives good shelter in most conditions, but open to NW sea/swell.

NAVIGATION WPT 55°13'·00N 06°41'·00W, 128° to N pier lt, 1·1M.

Ent with onshore winds >F 4 is difficult. Beware submerged bkwtr projecting 20m SW from N pier. Depth is 2·8m in hbr entrance.

LIGHTS AND MARKS Ldg lts 028° (occas, for LB use) both FR 6/8m 1M; R △ on metal bn and metal mast. N pier Fl R 3s 6m 3M; vis 220°-160°. S pier Fl G 3s 6m 3M; vis 220°-100°.

COMMUNICATIONS (Code 028) CGOC 9146 3933; Police 101; Dr 7082 3767; ⊞ 7034 4177. HM 7082 2307.

VHF Ch 12 16 (0900-1700LT, Mon-Fri; extended evening hrs June-Sept; Sat-Sun: 0900–1700, June-Sept only).

FACILITIES Hbr ⚓ ⟋(£3) ⬡ £14, ☒ ⊹ ⛴(0900-1800) ⬡ ⬚ Ⓔ; Portrush YC ☎7082 3932, ⬚; **Town** ⬚⬚⬚⊠Ⓑ⬚⬚✕⬚@ Library ⇌ ✈ (Belfast). Giant's Causeway is 10M ENE.

FAIR HEAD TO LAMBAY ISLAND

(AC 2199/8, 2093, 44) Fair Hd is a bold 190m headland, steep-to all round, but with extensive overfalls in Rathlin Sound. The coast is fairly steep-to except in larger bays, particularly Dundalk.

▶Streams in North Channel can reach more than 4½kn. S of Belfast Lough they are weaker, but run up to 2·5kn S of Rockabill.◀

⚓ in offshore winds in Cushendun Bay, 5M NNW of Garron Pt, and in Red Bay 4M further S. They provide useful ⚓s on passage to/from the Clyde or Western Is. Glenarm Marina provides good shelter for small craft and is useful for on passage to the W coasts of Scotland or Ireland. (AC 2198) 4M N of Larne are the Maidens, two dangerous groups of rks extending 2·5M N/S; E Maiden is lit. Hunter Rk (0·8m), S of the Maidens, is marked by N & S cardinals.

N of Belfast Lough, Muck Island is steep-to on its E side, as is Black Hd. Three routes lead into Belfast Lough from the south:

a. E of Mew Is.

▶Beware Ram Race (N'thly on the ebb and S'thly on the flood.◀

b. Copeland Sound, between Mew Is and Copeland Is, is passable but not recommended.

▶Here the stream runs SSE from HW Belfast +0420 and NW from HW Belfast –0115.◀

c. Donaghadee Sound is buoyed and a good short cut for small craft (see AC 3709).

▶Here the stream runs SSE from HW Belfast +0530 and NW from HW Belfast –0030, 4·5kn max. An eddy extends S to Ballyferris Pt, and about 1M offshore.◀

From Belfast Lough to Strangford Lough, the coast has numerous rocky dangers close inshore, but is reasonably clear if keeping about 2M offshore. In fair weather and good vis, there is a passage between North Rocks and South Rock, otherwise pass to the E using South Rock superbuoy as a guide. Pass E of Butter Pladdy and 5ca off Ballyquintin Point.

▶The tidal cycle is approx 3 hours later than in the N Channel. Flood runs for 6 hours from HW Belfast –0330 (HW Dover –0345), with a maximum rate of 7.5kn at Rue Point. The strong flow flattens the sea in onshore winds and entrance can be made in strong winds. The ebb runs for 6 hours from HW Belfast +0230 (HW Dover +0215), max rate 7.5kn, E of Angus Rk. If entering against ebb use West Channel with care. Smoothest water near Bar Pladdy Buoy when leaving.◀

5M SSW of the entrance to Strangford Lough there is a marina at Ardglass.

There are no offshore dangers SW of Strangford to Carlingford Lough. Skerries Islands (Colt, Shenick's and St Patrick's) are 1M E and SE of Red Island, to E of Skerries hbr. Shenick's Island is connected to shore at LW. Pass between Colt and St Patrick's Islands, but the latter has offliers 3ca to S. Rockabill, two steep-to rks with lt ho, is 2·5M E of St Patrick's Island. Lambay Island is private, and steep-to except on W side, where there can be overfalls.

N Ireland

9.13.15 TIDAL STREAMS AROUND RATHLIN ISLAND

North Ireland 9.13.3   Off Mull of Kintyre 9.9.11   South Ireland 9.12.3
North Irish Sea 9.10.3   SW Scotland 9.9.3

5 Hours before HW Dover (0605 after HW Greenock)

3 Hours before HW Dover (0420 before HW Greenock)

1½ Hours before HW Dover (0250 before HW Greenock)

HW Dover (0120 before HW Greenock)

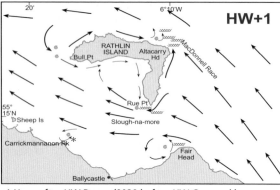

1 Hour after HW Dover (0020 before HW Greenock)

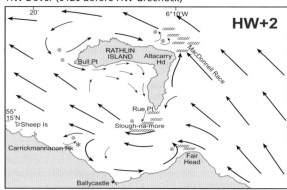

2 Hours after HW Dover (0040 after HW Greenock)

3 Hours after HW Dover (0140 after HW Greenock)

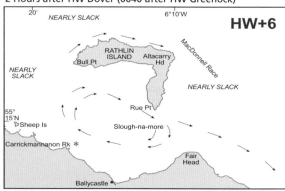

6 Hours after HW Dover (0440 after HW Greenock)

HARBOURS & ANCHORAGES BETWEEN PORTRUSH & LARNE

RATHLIN ISLAND, Antrim, **55°17'·47N 06°11'·78W**. AC 2798, 2494, 5612. HW sp –0445, nps –0200 on Dover. Small harbour in NE corner of Church Bay, sheltered from winds NW through E to SSE. Ferries run from the the Inner Harbour to Ballycastle. Beware sp streams up to 6kn in Rathlin Sound (9.13.15) and the North Channel TSS, just 2M NE of the island. Pass N or E of a wreck 6ca SW of hbr, marked on its SE side by a SCM lt buoy. When clear, appr on NNE to the W and S piers which form an outer hbr. The white sector of Manor House lt, Oc WRG 4s, leads 024·5° to Inner Harbour (2m) ent via channel dredged to 3·5m (2013). ⌓ on yacht pontoons just S of Inner Harbour. ⌓ in outer hbr on NW side in about 1.2m clear of ferry ramp. Outside hbr, close to W pier ⌓ possible in about 5m, but holding poor in rock/weed. Facilities: Church Bay ⌂ ⌗ ⌂ ✕ ferry to Ballycastle.

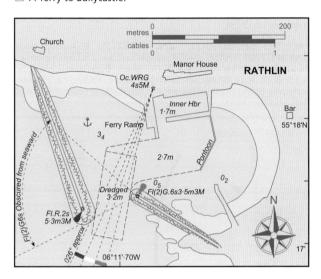

BALLYCASTLE, Antrim, **55°12'·47N 06°14'·27W**. AC 2798, 2494, 5612. Tides as for Rathlin Is; ML 0·8m; –0320 on Dover. Ferries to Rathlin Island berth inside the N breakwater; Ro-Ro disused. The marina is in the S part of the harbour with 2·3m. Outside the harbour is a fair weather ⌓ clear of strong tidal streams, but liable to sudden swell and exposed to onshore winds. Lights: see chartlet and 9.13.4. CG ☎20762226; Ⓗ 20762666. **Marina** ☎(028) 20768525, mob 07803 505084 74 ⌓£2·10, short stay £4/hr. ⌫ ⬧ ⬧ ⬧ ⬧ ⬧ ⬧ ⬧. **Town** ⌂ Ⓑ ⌗ ✕ ⌂.

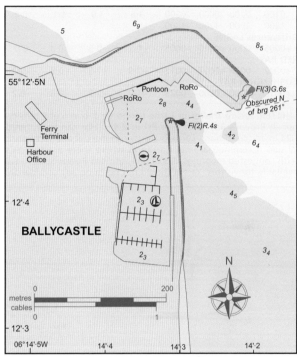

RED BAY, Antrim, **55°03'·91N 06°03'·13W**. AC 2199, 5612. HW +0010 on Dover; ML 1·1m; Duration 0625. See 9.13.16. Good holding, but open to E winds. Beware rocks, 2 ruined piers W of Garron Pt and on S side of bay, approx ½M from shore, fish farms, marked by lit buoys. Glenariff pier is lit Fl 3s 10m 5M. In S and E winds ⌓ 2ca off W stone arch near head of bay in approx 3·5m; in N or W winds ⌓ S of small pier in 2–5m, ½M NE of Glenariff (Waterfoot) village. **Waterfoot** ✕ ⌂ ⌂.
Cushendall (1M N of pier) ⌂ ⌂ ⌂ ⌫ ⬧ ⬧ ⌂ ✕ ⌗ ⬧.

CARNLOUGH HARBOUR, Antrim, **54°59'·87N 05°59'·20W**. AC 2198, 5612. HW +0006 on Dover, +0005 on Belfast; HW –1·6m on Belfast; ML no data; Duration 0625. Good shelter, except in SE gales; do not enter in fresh/strong onshore winds. Entrance shoaling due to build up of kelp, minimum depth 1m (2004). Ldg marks 310°, Y ▽s on B/W posts. N pier lt, Fl G 3s 4m 5M; S pier Fl R 3s 6m 5M, both lts on B/W columns. Beware fish farms in the bay marked by lt buoys (unreliable); and rocks which cover at HW on either side of ent. Small harbour used by yachts and small FVs; visitors welcome. HM mob 07703 606763.
Quay ⌓ £1, ⬧ (see HM) ⬧ (by arrangement) ⬧ ⌫.
Town ⌂ ⌂ ⬧ ⌫ ✕ ⌗ ⌂ ⬧.

GLENARM, Antrim, **54°58'N 05°57'W**. AC 2198, 2199, 5612. HW +0006 on Dover, +0005 on Belfast; HW –1·6m on Belfast; ML no data; Duration 0625. Good shelter in marina, which reports 3m at LW. Hbr ent lts G.3s 3M and R.3s 3M. Beware fish farm about 0·5M NE of entrance. HM / Marina ☎028 2884 1285, Mob 07703 606763.
Marina 40 ⌓ inc 10 Ⓥ in 4-6m, £2.20, ⬧ ⬧ ⬧ ⬧. **Village** ⌗ ⌂ ⌂.

9.13.16 LARNE

Antrim 54°51'·20N 05°47'·50W ✱✱✱✱♨♨✿✿

CHARTS AC 2198, 1237, 5612; Imray C62, C64, C69

TIDES +0005 Dover; ML 1·6; Duration 0620

Standard Port BELFAST (→)

Times				Height (metres)			
High Water		Low Water		MHWS	MHWN	MLWN	MLWS
0100	0700	0000	0600	3·5	3·0	1·1	0·4
1300	1900	1200	1800				
Differences LARNE							
+0005	0000	+0010	−0005	−0·7	−0·5	−0·3	0·0
RED BAY							
+0022	−0010	+0007	−0017	−1·9	−1·5	−0·8	−0·2
CUSHENDUN BAY							
+0010	−0030	0000	−0025	−1·7	−1·5	−0·6	−0·2

SHELTER Secure shelter in Larne Lough or ⚓ overnight outside hbr in Brown's Bay (E of Barr Pt) in 2-4m. Hbr can be entered H24 in any conditions. Larne is a busy commercial and ferry port; W side is commercial until Curran Pt where there are two YCs with congested moorings. ⚓ S of Ballylumford Power Stn. No ⌂ available for visitors. Yachts should not berth on any commercial quays, inc Castle Quay, without HM's permission. Boat Hbr (0·6m) 2ca S of Ferris Pt only for shoal draft craft.

NAVIGATION WPT 54°51'·70N 05°47'·47W, 184°/2·1M to front ldg lt. Beware Hunter Rk 2M NE of hbr ent. Magnetic anomalies exist near Hunter Rk and between it and the mainland. Inside the narrow ent, the recommended chan is close to the E shore. Tide in the ent runs at up to 3½kn.

LIGHTS AND MARKS No1 SHM Q G 1½ca off chartlet N edge; Ldg lts 184°, Oc 2.5s 6m / 5s14m 12M, synch and vis 179°-189°; W ◇ with R stripes. Chaine Tr and fairway lts as chartlet. Note: Many shore lts on W side of ent may be mistaken for nav lts.

COMMUNICATIONS (Code 02828) CGOC 9146 3933; Pilot 273785; Dr 275331; Police 101. HM 872100.

VHF Ch 14 11 16 *Larne Port Control*. Traffic, weather and tidal info available on Ch 14.

FACILITIES Pier ☎279221, ⚓⛽⚓ ⏚(33t); E Antrim Boat Club ☎277204, Visitors should pre-contact Sec'y for advice on ⚓; ⚓⛽⚓🛒🚻; Services 🛢⚓🔧🎣. Town 🛢(delivered, tidal) 🅿 🏧⊠Ⓑ🛒✕🚻⊒ Ferries Cairnryan; 8/day; 1¾ hrs; P&O (www. poferries.co.uk). Troon; 17/week; 1¾ hrs; P&O. ✈ (Belfast City and Belfast/Aldergrove).

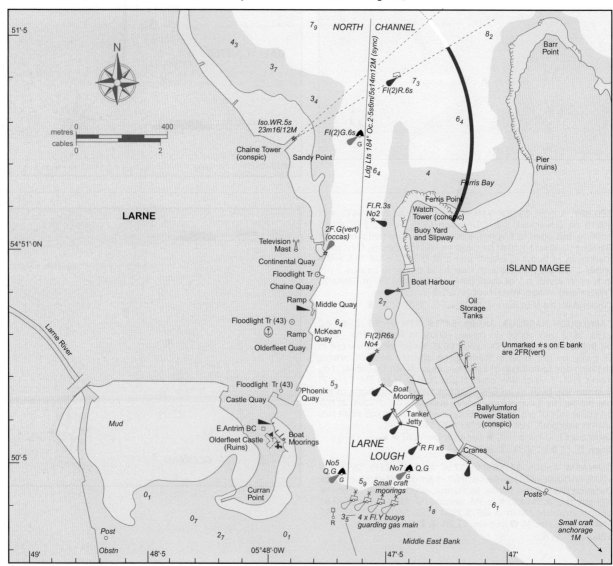

9.13.17 BELFAST LOUGH

County Down and County Antrim **54°42'N 05°45'W**
Bangor ✿✿✿◍◍◍◍✿✿; Carrickfergus ✿✿✿◍◍◍◍✿✿

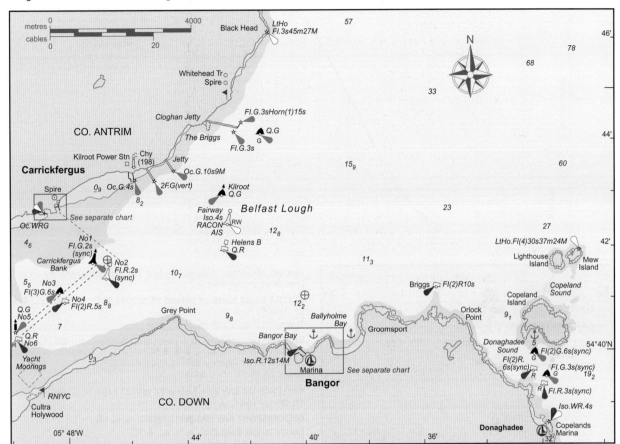

CHARTS AC 2198, 1753, 1752, 5612; Imray C62, C64, C69

TIDES +0007 Dover; ML Belfast 2·0, Carrickfergus 1·8; Duration 0620

Standard Port BELFAST (⟶)

Times				Height (metres)			
High Water		Low Water		MHWS	MHWN	MLWN	MLWS
0100	0700	0000	0600	3·5	3·0	1·1	0·4
1300	1900	1200	1800				
Differences CARRICKFERGUS							
+0005	+0005	+0005	+0005	−0·3	−0·3	−0·2	−0·1
DONAGHADEE							
+0020	+0020	+0023	+0023	+0·5	+0·4	0·0	+0·1

SHELTER Main sailing centres, clockwise around Belfast Lough, are at Bangor, Cultra and Carrickfergus.

Donaghadee: small marina at SE ent to the Lough.

Ballyholme Bay: good ⚓ in offshore winds.

Bangor: exposed to N winds, but well sheltered in marina (depths 2·9m to 2·2m). Speed limits: 4kn in marina, 8kn between Luke's Pt and Wilson's Pt.

Cultra: in offshore winds good ⚓ & moorings off RNoIYC.

Belfast Harbour is a major commercial port, but there are 40 ⌐ in Abercorn Basin, just NE of Lagan Bridge, in the River Lagan.

Carrickfergus: very good in marina, depths 1·8m to 2·3m (2015). The former commercial hbr has 10 yacht berths on the W quay. A stub bkwtr, marked by 2 PHM bns, extends NNE into the hbr from the W pier. The ent and SW part of the hbr are dredged 2·9m (2015); the NE part of the hbr dries 0·7m. Good ⚓ SSE of Carrickfergus Pier, except in E winds.

NAVIGATION Belfast Fairway SWM buoy, 54°42'·32N 05°42'·30W, Traffic passes inbound to N between SWM and Kilroot SHM,

outbound between SWM and Helens Bay PHM. The Lough is well marked. WPT Carrickfergus 54°41'·71N 05°46'·16W, marks start of the buoyed chan to Belfast Hbr. It bears 121°/301° from/to Carrickfergus Marina, 1.7M. Carrickfergus Bk liable to shift. Deep draught yachts should not enter/leave before LW±2. Beware HSS ferry operation. WPT Bangor 54°41'·00N 05°40'·00W, 190°/1M to bkwtr lt. Rounding Orlock Pt beware Briggs Rks extending ¾M offshore.

LIGHTS AND MARKS Chan to Belfast Hbr is well marked/lit by buoys and bns. Beyond No 12 PHM bn it is dangerous to leave the chan. Carrickfergus is easily recognised by conspic castle to E of Marina. On Marina Bkwtr, 30m W of the ✩ QR 7m 3M, is a Dir lt 320°, Oc WRG 3s 5m 3M, (H24).

COMMUNICATIONS (Codes: Belfast 028) CGOC 91463933; 🛟 (08494) 22339; Police 101; Dr 91468521. HM Bangor 91453297; HM Belfast 90553012; Belfast VTS 9055 3504. 0845 600 8000 Bangor Marina Ch **80** M (H24); Bangor HM Ch 11 (H24). Royal N of Ireland YC (at Cultra) Ch **16**; 11 (H24) 80. *Belfast Port Control* (at Milewater Basin) Ch **12** 16 (H24); VTS provides info on request to vessels in port area. The greater part of Belfast Lough is under radar surveillance. Carrickfergus Marina Ch M, 80, M2.

FACILITIES Anti-clockwise around the Lough:

CARRICKFERGUS Marina www.carrickfergus.org/site/marina ☎93366666; entry HW±2; 300 inc 🅥 £2.50 (2nd night free), short stay (max 6 hrs) £1.00 (min charge £7.35) 🔧 ⚓ ⋀ 🚁 Ⓔ. **Carrick SC** ☎351402, ⚓ ⬆ ⊼ ⌐ ⛟. **Hbr:** 10⌐ ⬆ 🚽(45t). **Town** ✉ Ⓑ 🛒 ✕ ⇌ ✈ (Belfast).

BELFAST Abercorn Basin www.belfast-harbour.co.uk 40⌐ £15<10m>£25<12m LOA; short stay (5 hrs) £8; ⚓ 🚁 🚾. **City** all facilities, ⇌ ✈.

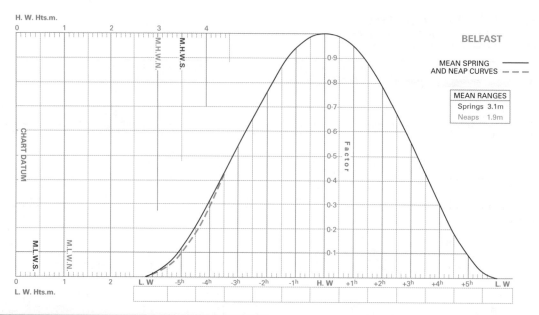

BELFAST

MEAN SPRING
AND NEAP CURVES - - -

MEAN RANGES	
Springs	3.1m
Neaps	1.9m

Ferries: Stranraer; 6/week; 2½ hrs; Stena (www.stenaline.co.uk). Liverpool; 2/day; 8hrs.

CULTRA Royal North of Ireland YC ☎9042 8041, ⚓🏴 ⬛🏊🏪 🏠(½M)✕ 🔧 🅓✕ 🗄.
Town 🏠🏢🏦⬛🅑🛒 ✕🗄🚂✈.

BANGOR Bangor Marina ⚓ www.quaymarinas.com ☎9145 3297, 550🛟+40🆅 £2·55, ⬛🖤🏠🅿✕🔧 🅓🄴⚓🪑🏴(40t) 🔲 Diving 🄛. **Todd Chart Agency** ☎9146 6640, ACA 🄛.
Town ⬛🅑🛒✕🗄🚂✈ (Belfast).

BALLYHOLME BAY Ballyholme YC ☎9127 1467, ✕🗄; **Royal Ulster YC** ☎9127 0568, ⚓✕🗄.

GROOMSPORT BAY HM ☎9127 8040, ⚓🏴⚓.
Cockle Island Boat Club ⚓🏴🏴⚓✕.

HARBOURS BETWEEN BELFAST AND STRANGFORD LOUGHS

DONAGHADEE, Down, **54°38′·71N 05°31′·85W.** 🌸🌸⚓🌸🌸🌸🌸. AC 1753, 5612. HW +0025 on Dover; see 9.13.17; ML no data; Duration 0615. Excellent shelter and basic facilities in tiny marina, access HW±4 over sill; covers approx 1·1m at half tide. Appr on about 275° on ldg marks, orange △s to tricky ent with sharp 90° port turn into marina (pilots available). Appr in strong winds or at night not advised. 3ca to the N, the Old Hbr is small and very full; scend often sets in. Beware rocky reef with less than 2m extends 1·5ca ENE from S pier hd. Max depth in hbr is approx 2·5m, dries at SW end; best berth alongside SE quay. S pier lt, Iso WR 4s 17m 18/14M, W shore–326°, R326°–shore. HM ☎9188 2377. Police 101. Facilities: **Copelands Marina** ☎9188 2184, mob 07802 363382; VHF Ch **16** 11 80; 🛟 for 6 🆅🖤🔲 (20t). **Old Hbr** 🛟 £4. **Town** 🗄🏢🏦 🖤⬛✕🛒🅑🚂 (Bangor), ✈ (Belfast).

PORTAVOGIE, Down, **54°27′·45N 05°26′·08W.** 🌸🌸⚓🌸🌸🌸. AC 2156, 5612. HW +0016 on Dover; ML 2·6m; Duration 0620. See 9.13.18. Good shelter, but hbr so full of FVs as to risk damage; best only for overnight or emergency. Entrance dangerous in strong onshore winds. Beware Plough Rks to SE marked by PHM buoy, Fl (2) R 10s, and McCammon Rks to NE of ent. Keep in W sector of outer bkwtr lt, Iso WRG 5s 9m 9M, shore-G-258°-W-275°-R-348°. Inner bkwtr 2 FG (vert) 6m 4M. Monitor VHF Ch **14** on entry/exit. (Mon-Fri: 0900-1700LT). HM ☎(028) 4277 1470. Facilities: ⚓⚓ (on central quay) ✕🔧 🅓. **Town** 🖤🄛🏢🏦⬛🛒✕.

STANDARD TIME (UT)
For Summer Time add ONE hour in **non-shaded areas**

BELFAST   LAT 54°36'N   LONG 5°55'W
TIMES AND HEIGHTS OF HIGH AND LOW WATERS

Dates in red are **SPRINGS**
Dates in blue are **NEAPS**

YEAR 2016

JANUARY

Day	Time	m	Time	m	Time	m	Time	m
1 F	0336	3.1	0925	1.1	1553	3.4	2150	0.9
16 SA	0320	3.3	0917	0.7	1536	3.6	2159	0.6
2 SA	0426	3.0	1019	1.2	1644	3.3	2248	1.0
17 SU	0420	3.2	1016	0.9	1637	3.4	2306	0.7
3 SU	0519	3.0	1126	1.3	1738	3.2	2356	1.1
18 M	0526	3.2	1125	0.9	1749	3.3		
4 M	0616	3.0	1238	1.3	1836	3.1		
19 TU	0021	0.8	0640	3.1	1244	1.0	1913	3.2
5 TU	0101	1.1	0716	3.0	1340	1.2	1938	3.1
20 W	0130	0.8	0751	3.2	1356	0.9	2025	3.2
6 W	0158	1.0	0820	3.1	1434	1.1	2039	3.2
21 TH	0232	0.7	0853	3.3	1500	0.8	2125	3.3
7 TH	0247	0.9	0915	3.2	1521	0.9	2131	3.3
22 F	0327	0.7	0947	3.4	1554	0.6	2216	3.4
8 F	0330	0.8	0959	3.4	1603	0.8	2214	3.4
23 SA	0416	0.6	1035	3.6	1641	0.5	2303	3.4
9 SA	0411	0.7	1037	3.5	1643	0.7	2251	3.4
24 SU	0459	0.6	1119	3.7	1723	0.5	2346	3.3
10 SU	0451	0.7	1108	3.6	1722	0.6	2326	3.4
25 M	0540	0.6	1200	3.7	1802	0.5		
11 M	0530	0.6	1141	3.6	1802	0.5		
26 TU	0026	3.3	0617	0.7	1239	3.7	1837	0.5
12 TU	0004	3.5	0610	0.6	1220	3.7	1842	0.4
27 W	0103	3.2	0652	0.7	1317	3.7	1911	0.6
13 W	0047	3.4	0652	0.6	1304	3.7	1925	0.4
28 TH	0139	3.2	0726	0.8	1355	3.6	1945	0.6
14 TH	0135	3.4	0737	0.6	1351	3.7	2011	0.4
29 F	0218	3.2	0802	0.8	1435	3.5	2021	0.7
15 F	0226	3.4	0842	0.7	1442	3.7	2102	0.5
30 SA	0259	3.1	0842	0.9	1518	3.4	2103	0.8
31 SU	0345	3.1	0927	1.0	1604	3.3	2149	0.9

FEBRUARY

Day	Time	m	Time	m	Time	m	Time	m
1 M	0434	3.0	1020	1.1	1655	3.1	2245	1.0
16 TU	0458	3.1	1102	0.8	1730	3.2	2359	0.9
2 TU	0529	2.9	1131	1.2	1752	3.0		
17 W	0613	3.1	1226	0.9	1859	3.0		
3 W	0001	1.1	0627	2.9	1258	1.2	1854	2.9
18 TH	0113	0.9	0733	3.1	1343	0.8	2015	3.0
4 TH	0121	1.1	0731	2.9	1403	1.1	1958	3.0
19 F	0219	0.8	0839	3.2	1451	0.7	2115	3.1
5 F	0219	1.0	0835	3.1	1456	0.9	2100	3.1
20 SA	0318	0.7	0933	3.3	1547	0.6	2205	3.2
6 SA	0309	0.8	0928	3.2	1542	0.7	2148	3.2
21 SU	0406	0.7	1020	3.5	1631	0.5	2249	3.2
7 SU	0352	0.7	1008	3.4	1624	0.5	2228	3.3
22 M	0447	0.6	1103	3.6	1707	0.4	2330	3.2
8 M	0433	0.5	1044	3.5	1704	0.4	2306	3.4
23 TU	0522	0.6	1142	3.6	1740	0.4		
9 TU	0512	0.5	1121	3.6	1743	0.3	2346	3.4
24 W	0006	3.2	0555	0.6	1218	3.6	1811	0.5
10 W	0552	0.4	1203	3.7	1824	0.2		
25 TH	0039	3.2	0626	0.6	1252	3.5	1841	0.5
11 TH	0030	3.4	0633	0.4	1249	3.7	1906	0.2
26 F	0110	3.2	0656	0.6	1326	3.5	1912	0.5
12 F	0118	3.4	0718	0.4	1337	3.7	1951	0.3
27 SA	0143	3.2	0729	0.6	1403	3.4	1946	0.6
13 SA	0208	3.4	0804	0.4	1427	3.7	2040	0.4
28 SU	0220	3.2	0806	0.7	1441	3.3	2024	0.6
14 SU	0300	3.3	0855	0.5	1520	3.5	2134	0.5
29 M	0258	3.1	0848	0.8	1521	3.2	2107	0.8
15 M	0356	3.2	0952	0.7	1619	3.4	2240	0.7

MARCH

Day	Time	m	Time	m	Time	m	Time	m
1 TU	0341	3.0	0935	0.9	1610	3.1	2157	0.9
16 W	0433	3.2	1043	0.7	1716	3.0	2338	0.9
2 W	0435	2.9	1034	1.1	1710	2.9	2259	1.1
17 TH	0545	3.0	1210	0.8	1846	2.9		
3 TH	0539	2.8	1213	1.2	1815	2.8		
18 F	0053	1.0	0710	3.0	1327	0.8	2000	2.9
4 F	0039	1.1	0645	2.8	1333	1.0	1921	2.9
19 SA	0203	0.9	0818	3.1	1436	0.7	2058	3.0
5 SA	0152	1.0	0751	2.9	1430	0.8	2025	3.0
20 SU	0303	0.8	0913	3.3	1532	0.5	2147	3.1
6 SU	0245	0.8	0850	3.1	1518	0.5	2118	3.2
21 M	0352	0.7	0959	3.4	1613	0.5	2229	3.2
7 M	0331	0.6	0936	3.3	1601	0.3	2202	3.3
22 TU	0430	0.6	1041	3.5	1645	0.4	2307	3.2
8 TU	0412	0.5	1018	3.5	1642	0.2	2243	3.4
23 W	0501	0.6	1119	3.5	1714	0.5	2342	3.2
9 W	0451	0.3	1100	3.6	1721	0.1	2327	3.5
24 TH	0531	0.6	1153	3.4	1743	0.5		
10 TH	0531	0.3	1145	3.7	1801	0.1		
25 F	0012	3.2	0600	0.6	1225	3.4	1811	0.5
11 F	0012	3.5	0612	0.2	1233	3.7	1843	0.1
26 SA	0039	3.2	0629	0.6	1256	3.4	1842	0.5
12 SA	0101	3.5	0657	0.2	1324	3.7	1929	0.2
27 SU	0108	3.2	0701	0.6	1329	3.4	1915	0.6
13 SU	0151	3.5	0744	0.3	1415	3.6	2018	0.3
28 M	0140	3.2	0737	0.6	1404	3.3	1952	0.6
14 M	0242	3.4	0835	0.4	1509	3.5	2112	0.5
29 TU	0214	3.2	0816	0.7	1443	3.2	2034	0.7
15 TU	0334	3.3	0932	0.5	1606	3.3	2217	0.8
30 W	0253	3.1	0901	0.8	1531	3.1	2121	0.9
31 TH	0341	3.0	0955	0.9	1632	2.9	2219	1.1

APRIL

Day	Time	m	Time	m	Time	m	Time	m
1 F	0447	2.9	1110	1.0	1741	2.8	2334	1.1
16 SA	0026	1.0	0638	3.0	1301	0.7	1937	2.9
2 SA	0602	2.8	1259	0.9	1848	2.9		
17 SU	0134	1.0	0749	3.1	1407	0.7	2034	3.0
3 SU	0117	1.1	0711	2.9	1400	0.7	1952	3.0
18 M	0236	0.9	0845	3.2	1501	0.6	2121	3.1
4 M	0217	0.9	0813	3.1	1451	0.4	2048	3.2
19 TU	0326	0.8	0932	3.3	1542	0.5	2203	3.1
5 TU	0305	0.6	0906	3.3	1535	0.2	2135	3.4
20 W	0405	0.7	1014	3.4	1614	0.5	2240	3.2
6 W	0348	0.4	0953	3.5	1616	0.1	2221	3.5
21 TH	0437	0.6	1051	3.4	1644	0.5	2314	3.2
7 TH	0428	0.3	1039	3.7	1656	0.0	2306	3.6
22 F	0506	0.6	1125	3.4	1714	0.6	2344	3.2
8 F	0509	0.2	1127	3.7	1738	0.0	2354	3.6
23 SA	0536	0.6	1155	3.3	1744	0.6		
9 SA	0552	0.2	1217	3.7	1822	0.1		
24 SU	0010	3.3	0606	0.6	1225	3.3	1815	0.6
10 SU	0043	3.6	0637	0.2	1309	3.7	1908	0.2
25 M	0038	3.3	0637	0.6	1257	3.3	1849	0.6
11 M	0134	3.6	0726	0.3	1402	3.6	1958	0.4
26 TU	0109	3.3	0713	0.6	1333	3.3	1927	0.7
12 TU	0224	3.5	0817	0.3	1455	3.4	2052	0.6
27 W	0145	3.3	0752	0.6	1415	3.2	2009	0.7
13 W	0315	3.4	0915	0.5	1552	3.2	2157	0.8
28 TH	0225	3.3	0837	0.7	1504	3.1	2056	0.8
14 TH	0410	3.3	1026	0.6	1700	3.0	2313	1.0
29 F	0312	3.2	0929	0.8	1603	3.0	2152	1.0
15 F	0515	3.1	1148	0.7	1826	2.9		
30 SA	0410	3.0	1036	0.9	1710	2.9	2258	1.1

N Ireland

Chart Datum: 2·01 metres below Ordnance Datum (Belfast). HAT is 3·9 metres above Chart Datum.

》》 **FREE** monthly updates. Register at 《
www.reedsnauticalalmanac.co.uk

635

STANDARD TIME (UT)
For Summer Time add ONE hour in **non-shaded areas**

BELFAST LAT 54°36'N LONG 5°55'W
TIMES AND HEIGHTS OF HIGH AND LOW WATERS

Dates in red are SPRINGS
Dates in blue are NEAPS

YEAR 2016

MAY

Day	Time	m		Day	Time	m
1 SU	0521 / 1213 / 1818	3.0 / 0.8 / 2.9		**16** M	0054 / 0706 / 1325 / 2000	1.1 / 3.1 / 0.7 / 2.9
2 M	0022 / 0634 / 1324 / 1921	1.1 / 3.0 / 0.6 / 3.1		**17** TU	0154 / 0815 / 1419 / 2048	1.0 / 3.1 / 0.7 / 3.0
3 TU	0138 / 0740 / 1419 / 2019	0.9 / 3.2 / 0.4 / 3.2		**18** W	0248 / 0858 / 1503 / 2131	0.9 / 3.2 / 0.7 / 3.1
4 W	0233 / 0839 / 1506 / 2110	0.7 / 3.4 / 0.2 / 3.4		**19** TH	0332 / 0942 / 1540 / 2209	0.8 / 3.3 / 0.6 / 3.2
5 TH	0321 / 0931 / 1550 / 2159	0.5 / 3.5 / 0.1 / 3.5		**20** F	0409 / 1020 / 1613 / 2243	0.7 / 3.3 / 0.6 / 3.3
6 F ●	0406 / 1020 / 1633 / 2247	0.4 / 3.7 / 0.1 / 3.6		**21** SA ○	0441 / 1056 / 1646 / 2316	0.7 / 3.3 / 0.6 / 3.3
7 SA	0450 / 1110 / 1717 / 2336	0.3 / 3.7 / 0.1 / 3.7		**22** SU	0513 / 1128 / 1719 / 2345	0.6 / 3.3 / 0.6 / 3.4
8 SU	0535 / 1201 / 1803	0.2 / 3.7 / 0.2		**23** M	0547 / 1158 / 1754	0.6 / 3.3 / 0.7
9 M	0025 / 0622 / 1253 / 1851	3.7 / 0.2 / 3.6 / 0.4		**24** TU	0011 / 0621 / 1230 / 1829	3.4 / 0.6 / 3.3 / 0.7
10 TU	0116 / 0711 / 1346 / 1941	3.7 / 0.2 / 3.5 / 0.5		**25** W	0044 / 0657 / 1308 / 1908	3.4 / 0.6 / 3.3 / 0.7
11 W	0205 / 0803 / 1439 / 2036	3.6 / 0.3 / 3.3 / 0.7		**26** TH	0122 / 0736 / 1352 / 1950	3.4 / 0.6 / 3.2 / 0.8
12 TH	0254 / 0859 / 1534 / 2136	3.5 / 0.5 / 3.2 / 0.9		**27** F	0204 / 0819 / 1441 / 2038	3.4 / 0.6 / 3.2 / 0.8
13 F ◐	0345 / 1005 / 1635 / 2244	3.4 / 0.6 / 3.0 / 1.0		**28** SA	0252 / 0911 / 1538 / 2131	3.3 / 0.7 / 3.1 / 0.9
14 SA	0442 / 1117 / 1750 / 2351	3.2 / 0.7 / 2.9 / 1.1		**29** SU ◐	0346 / 1011 / 1641 / 2232	3.2 / 0.7 / 3.0 / 1.0
15 SU	0549 / 1224 / 1902	3.1 / 0.7 / 2.9		**30** M	0448 / 1125 / 1748 / 2340	3.2 / 0.7 / 3.0 / 1.0
				31 TU	0559 / 1243 / 1852	3.2 / 0.6 / 3.1

JUNE

Day	Time	m		Day	Time	m
1 W	0054 / 0710 / 1345 / 1952	0.9 / 3.3 / 0.5 / 3.2		**16** TH	0205 / 0811 / 1422 / 2051	1.0 / 3.1 / 0.8 / 3.1
2 TH	0201 / 0815 / 1438 / 2048	0.8 / 3.4 / 0.4 / 3.4		**17** F	0255 / 0904 / 1505 / 2134	0.9 / 3.1 / 0.8 / 3.2
3 F	0257 / 0912 / 1528 / 2140	0.6 / 3.5 / 0.3 / 3.5		**18** SA	0339 / 0948 / 1544 / 2213	0.8 / 3.2 / 0.7 / 3.3
4 SA	0349 / 1005 / 1615 / 2230	0.4 / 3.6 / 0.3 / 3.6		**19** SU	0416 / 1028 / 1620 / 2249	0.7 / 3.3 / 0.7 / 3.3
5 SU ●	0437 / 1056 / 1701 / 2319	0.3 / 3.6 / 0.3 / 3.7		**20** M ○	0452 / 1103 / 1657 / 2321	0.7 / 3.3 / 0.7 / 3.4
6 M	0524 / 1146 / 1749	0.3 / 3.6 / 0.4		**21** TU	0528 / 1134 / 1734 / 2347	0.6 / 3.3 / 0.7 / 3.4
7 TU	0008 / 0611 / 1237 / 1836	3.7 / 0.3 / 3.5 / 0.5		**22** W	0605 / 1206 / 1813	0.6 / 3.3 / 0.7
8 W	0057 / 0659 / 1328 / 1926	3.7 / 0.3 / 3.4 / 0.6		**23** TH	0020 / 0642 / 1245 / 1852	3.5 / 0.5 / 3.3 / 0.7
9 TH	0144 / 0747 / 1418 / 2017	3.7 / 0.4 / 3.3 / 0.7		**24** F	0101 / 0722 / 1330 / 1935	3.5 / 0.5 / 3.3 / 0.7
10 F	0231 / 0838 / 1508 / 2110	3.6 / 0.5 / 3.1 / 0.9		**25** SA	0145 / 0805 / 1419 / 2021	3.5 / 0.5 / 3.2 / 0.7
11 SA	0319 / 0933 / 1600 / 2207	3.5 / 0.6 / 3.0 / 1.0		**26** SU	0232 / 0853 / 1513 / 2111	3.4 / 0.5 / 3.2 / 0.8
12 SU ◐	0408 / 1036 / 1656 / 2308	3.4 / 0.7 / 2.9 / 1.1		**27** M ◐	0324 / 0948 / 1612 / 2207	3.4 / 0.6 / 3.1 / 0.8
13 M	0502 / 1139 / 1756	3.2 / 0.8 / 2.9		**28** TU	0421 / 1053 / 1716 / 2310	3.3 / 0.6 / 3.1 / 0.9
14 TU	0009 / 0600 / 1239 / 1902	1.1 / 3.1 / 0.8 / 2.9		**29** W	0527 / 1206 / 1822	3.3 / 0.6 / 3.1
15 W	0108 / 0706 / 1334 / 2001	1.1 / 3.1 / 0.8 / 3.0		**30** TH	0021 / 0641 / 1316 / 1927	0.9 / 3.3 / 0.6 / 3.2

JULY

Day	Time	m		Day	Time	m
1 F	0135 / 0755 / 1417 / 2029	0.8 / 3.3 / 0.5 / 3.3		**16** SA	0218 / 0819 / 1432 / 2054	1.0 / 3.0 / 0.9 / 3.1
2 SA	0240 / 0859 / 1511 / 2125	0.7 / 3.4 / 0.5 / 3.5		**17** SU	0308 / 0916 / 1516 / 2142	0.9 / 3.1 / 0.8 / 3.2
3 SU	0337 / 0955 / 1602 / 2217	0.5 / 3.4 / 0.4 / 3.6		**18** M	0351 / 1002 / 1557 / 2221	0.8 / 3.2 / 0.7 / 3.3
4 M ●	0428 / 1046 / 1650 / 2305	0.4 / 3.5 / 0.5 / 3.7		**19** TU ○	0431 / 1039 / 1636 / 2253	0.7 / 3.3 / 0.6 / 3.4
5 TU	0516 / 1134 / 1736 / 2352	0.3 / 3.4 / 0.5 / 3.7		**20** W	0509 / 1111 / 1715 / 2322	0.5 / 3.3 / 0.6 / 3.5
6 W	0601 / 1221 / 1822	0.3 / 3.4 / 0.6		**21** TH	0546 / 1144 / 1754 / 2357	0.5 / 3.3 / 0.6 / 3.5
7 TH	0038 / 0644 / 1308 / 1906	3.7 / 0.4 / 3.3 / 0.7		**22** F	0625 / 1223 / 1834	0.4 / 3.3 / 0.6
8 F	0122 / 0727 / 1353 / 1950	3.7 / 0.4 / 3.2 / 0.8		**23** SA	0038 / 0704 / 1308 / 1915	3.6 / 0.4 / 3.3 / 0.6
9 SA	0206 / 0808 / 1437 / 2034	3.6 / 0.5 / 3.1 / 0.8		**24** SU	0124 / 0746 / 1357 / 2000	3.6 / 0.4 / 3.3 / 0.6
10 SU	0250 / 0851 / 1523 / 2119	3.5 / 0.6 / 3.1 / 0.9		**25** M	0212 / 0832 / 1449 / 2049	3.6 / 0.4 / 3.3 / 0.6
11 M	0335 / 0937 / 1610 / 2209	3.4 / 0.7 / 3.0 / 1.0		**26** TU	0303 / 0924 / 1546 / 2143	3.5 / 0.5 / 3.2 / 0.7
12 TU ◐	0423 / 1032 / 1700 / 2310	3.3 / 0.9 / 2.9 / 1.1		**27** W	0359 / 1024 / 1647 / 2244	3.4 / 0.6 / 3.2 / 0.8
13 W	0515 / 1138 / 1753	3.1 / 0.9 / 2.9		**28** TH	0502 / 1137 / 1754 / 2358	3.3 / 0.7 / 3.1 / 0.9
14 TH	0017 / 0611 / 1244 / 1851	1.1 / 3.0 / 1.0 / 2.9		**29** F	0620 / 1255 / 1907	3.2 / 0.7 / 3.2
15 F	0121 / 0713 / 1341 / 1953	1.1 / 3.0 / 1.0 / 3.0		**30** SA	0119 / 0744 / 1401 / 2016	0.9 / 3.1 / 0.7 / 3.3
				31 SU	0229 / 0853 / 1459 / 2115	0.7 / 3.2 / 0.6 / 3.4

AUGUST

Day	Time	m		Day	Time	m
1 M	0329 / 0949 / 1551 / 2206	0.6 / 3.3 / 0.6 / 3.5		**16** TU	0326 / 0934 / 1534 / 2149	0.7 / 3.2 / 0.7 / 3.3
2 TU ●	0420 / 1038 / 1638 / 2253	0.5 / 3.3 / 0.6 / 3.6		**17** W	0408 / 1013 / 1614 / 2223	0.5 / 3.3 / 0.6 / 3.5
3 W	0505 / 1123 / 1722 / 2336	0.4 / 3.3 / 0.6 / 3.7		**18** TH ○	0446 / 1047 / 1653 / 2256	0.4 / 3.4 / 0.6 / 3.6
4 TH	0546 / 1205 / 1803	0.4 / 3.3 / 0.6		**19** F	0524 / 1122 / 1731 / 2334	0.3 / 3.4 / 0.5 / 3.6
5 F	0018 / 0624 / 1246 / 1840	3.7 / 0.4 / 3.2 / 0.7		**20** SA	0602 / 1202 / 1810	0.3 / 3.4 / 0.5
6 SA	0058 / 0658 / 1324 / 1916	3.6 / 0.5 / 3.2 / 0.7		**21** SU	0017 / 0641 / 1248 / 1852	3.7 / 0.2 / 3.4 / 0.5
7 SU	0138 / 0732 / 1404 / 1952	3.6 / 0.5 / 3.1 / 0.8		**22** M	0104 / 0723 / 1336 / 1937	3.7 / 0.3 / 3.4 / 0.5
8 M	0218 / 0807 / 1445 / 2030	3.5 / 0.6 / 3.1 / 0.8		**23** TU	0153 / 0808 / 1428 / 2026	3.7 / 0.4 / 3.4 / 0.6
9 TU	0300 / 0846 / 1529 / 2113	3.4 / 0.7 / 3.1 / 0.9		**24** W	0245 / 0859 / 1523 / 2120	3.6 / 0.5 / 3.3 / 0.7
10 W ◑	0345 / 0930 / 1617 / 2203	3.3 / 0.8 / 3.0 / 1.1		**25** TH ◑	0341 / 0958 / 1623 / 2222	3.4 / 0.7 / 3.2 / 0.8
11 TH	0434 / 1023 / 1709 / 2309	3.1 / 1.0 / 3.0 / 1.2		**26** F	0446 / 1114 / 1731 / 2343	3.1 / 0.9 / 3.1 / 0.9
12 F	0529 / 1134 / 1805	3.0 / 1.1 / 2.9		**27** SA	0609 / 1239 / 1851	3.0 / 0.9 / 3.1
13 SA	0037 / 0630 / 1300 / 1905	1.2 / 2.9 / 1.1 / 2.9		**28** SU	0109 / 0738 / 1349 / 2004	0.9 / 3.0 / 0.8 / 3.2
14 SU	0144 / 0735 / 1401 / 2009	1.1 / 2.9 / 1.0 / 3.0		**29** M	0221 / 0846 / 1450 / 2103	0.8 / 3.1 / 0.8 / 3.4
15 M	0239 / 0842 / 1451 / 2106	0.9 / 3.0 / 0.9 / 3.2		**30** TU	0322 / 0940 / 1542 / 2153	0.6 / 3.2 / 0.7 / 3.5
				31 W	0411 / 1026 / 1626 / 2238	0.5 / 3.3 / 0.7 / 3.6

Chart Datum: 2·01 metres below Ordnance Datum (Belfast). HAT is 3·9 metres above Chart Datum.

STANDARD TIME (UT)
For Summer Time add ONE hour in **non-shaded areas**

BELFAST LAT 54°36'N LONG 5°55'W
TIMES AND HEIGHTS OF HIGH AND LOW WATERS

Dates in red are **SPRINGS**
Dates in blue are **NEAPS**

YEAR **2016**

SEPTEMBER

Time m	Time m
1 0451 0.4 / 1108 3.3 / TH 1705 0.7 / ● 2318 3.6	**16** 0421 0.3 / 1022 3.5 / F 1628 0.5 / ○ 2232 3.7
2 0526 0.4 / 1146 3.2 / F 1740 0.7 / 2356 3.6	**17** 0458 0.2 / 1101 3.5 / SA 1706 0.5 / 2314 3.7
3 0557 0.5 / 1221 3.2 / SA 1812 0.7	**18** 0536 0.2 / 1143 3.6 / SU 1746 0.4 / 2359 3.8
4 0032 3.6 / 0626 0.6 / SU 1254 3.2 / 1842 0.7	**19** 0616 0.2 / 1229 3.6 / M 1829 0.4
5 0108 3.5 / 0656 0.6 / M 1328 3.2 / 1914 0.8	**20** 0048 3.8 / 0659 0.3 / TU 1319 3.6 / 1915 0.4
6 0145 3.5 / 0728 0.6 / TU 1406 3.2 / 1950 0.8	**21** 0140 3.7 / 0746 0.4 / W 1410 3.5 / 2005 0.5
7 0224 3.4 / 0805 0.7 / W 1447 3.2 / 2031 0.9	**22** 0233 3.6 / 0837 0.6 / TH 1504 3.4 / 2100 0.6
8 0305 3.3 / 0847 0.8 / TH 1532 3.1 / 2117 1.0	**23** 0331 3.4 / 0936 0.8 / F 1603 3.3 / ◗ 2205 0.8
9 0352 3.1 / 0934 1.0 / F 1623 3.0 / ◖ 2213 1.1	**24** 0437 3.2 / 1055 1.0 / SA 1711 3.2 / 2331 0.9
10 0449 3.0 / 1033 1.2 / SA 1722 2.9 / 2342 1.2	**25** 0603 3.0 / 1222 1.1 / SU 1832 3.2
11 0552 2.9 / 1205 1.3 / SU 1824 2.9	**26** 0056 0.9 / 0728 3.0 / M 1335 1.1 / 1946 3.3
12 0111 1.1 / 0658 2.9 / M 1330 1.2 / 1927 3.0	**27** 0209 0.8 / 0832 3.1 / TU 1438 1.0 / 2045 3.4
13 0210 0.9 / 0804 3.0 / TU 1424 1.0 / 2026 3.2	**28** 0310 0.6 / 0923 3.2 / W 1530 0.8 / 2134 3.5
14 0259 0.7 / 0900 3.2 / W 1510 0.8 / 2113 3.4	**29** 0356 0.6 / 1007 3.3 / TH 1611 0.8 / 2217 3.6
15 0342 0.5 / 0943 3.3 / TH 1550 0.6 / 2153 3.5	**30** 0431 0.5 / 1047 3.3 / F 1645 0.7 / 2257 3.6

OCTOBER

Time m	Time m
1 0459 0.6 / 1123 3.3 / SA 1715 0.7 / ● 2333 3.6	**16** 0431 0.2 / 1040 3.7 / SU 1641 0.4 / ○ 2257 3.8
2 0527 0.6 / 1155 3.3 / SU 1744 0.8	**17** 0510 0.2 / 1125 3.7 / M 1723 0.4 / 2345 3.8
3 0005 3.5 / 0554 0.7 / M 1224 3.3 / 1812 0.8	**18** 0552 0.3 / 1213 3.7 / TU 1808 0.4
4 0037 3.5 / 0623 0.7 / TU 1254 3.4 / 1844 0.8	**19** 0036 3.8 / 0637 0.4 / W 1303 3.7 / 1856 0.4
5 0111 3.5 / 0656 0.7 / W 1328 3.4 / 1919 0.8	**20** 0130 3.7 / 0726 0.5 / TH 1355 3.7 / 1947 0.5
6 0148 3.4 / 0732 0.8 / TH 1404 3.4 / 1958 0.9	**21** 0224 3.6 / 0819 0.7 / F 1448 3.6 / 2044 0.6
7 0227 3.3 / 0813 0.9 / F 1443 3.3 / 2043 1.0	**22** 0322 3.4 / 0919 1.0 / SA 1544 3.5 / ◗ 2150 0.8
8 0312 3.2 / 0859 1.0 / SA 1530 3.2 / 2135 1.1	**23** 0427 3.2 / 1035 1.1 / SU 1649 3.3 / 2314 0.9
9 0410 3.0 / 0954 1.2 / SU 1633 3.0 / ◖ 2243 1.2	**24** 0548 3.0 / 1158 1.2 / M 1806 3.3
10 0517 2.9 / 1104 1.3 / M 1742 3.0	**25** 0033 0.9 / 0707 3.0 / TU 1309 1.2 / 1919 3.3
11 0032 1.2 / 0624 2.9 / TU 1249 1.3 / 1847 3.0	**26** 0143 0.8 / 0809 3.1 / W 1413 1.1 / 2019 3.4
12 0136 1.0 / 0729 3.0 / W 1352 1.1 / 1948 3.2	**27** 0243 0.6 / 0859 3.2 / TH 1507 1.0 / 2109 3.5
13 0228 0.7 / 0826 3.2 / TH 1441 0.9 / 2041 3.4	**28** 0328 0.7 / 0943 3.3 / F 1548 0.9 / 2153 3.5
14 0312 0.5 / 0914 3.4 / F 1523 0.7 / 2127 3.6	**29** 0401 0.7 / 1022 3.4 / SA 1622 0.8 / 2233 3.5
15 0352 0.3 / 0957 3.6 / SA 1602 0.5 / 2211 3.7	**30** 0429 0.7 / 1058 3.4 / SU 1651 0.8 / ● 2308 3.5
	31 0458 0.7 / 1129 3.4 / M 1720 0.8 / 2340 3.5

NOVEMBER

Time m	Time m
1 0527 0.8 / 1157 3.5 / TU 1750 0.8	**16** 0534 0.4 / 1158 3.9 / W 1754 0.4
2 0010 3.5 / 0557 0.8 / W 1225 3.5 / 1821 0.8	**17** 0024 3.8 / 0621 0.5 / TH 1249 3.9 / 1842 0.4
3 0043 3.5 / 0630 0.8 / TH 1257 3.5 / 1856 0.8	**18** 0118 3.7 / 0711 0.7 / F 1339 3.8 / 1934 0.5
4 0118 3.4 / 0707 0.8 / F 1331 3.5 / 1934 0.8	**19** 0212 3.5 / 0804 0.8 / SA 1431 3.7 / 2029 0.6
5 0157 3.3 / 0747 0.9 / SA 1409 3.4 / 2017 0.9	**20** 0308 3.4 / 0902 1.0 / SU 1524 3.6 / 2132 0.7
6 0242 3.2 / 0833 1.0 / SU 1453 3.3 / 2107 1.0	**21** 0408 3.2 / 1010 1.2 / M 1621 3.5 / ◗ 2245 0.9
7 0337 3.1 / 0926 1.2 / M 1548 3.2 / ◖ 2208 1.1	**22** 0519 3.0 / 1123 1.2 / TU 1727 3.3 / 2357 0.9
8 0443 3.0 / 1029 1.3 / TU 1656 3.1 / 2328 1.1	**23** 0633 3.0 / 1230 1.2 / W 1840 3.3
9 0551 3.0 / 1145 1.3 / W 1807 3.2	**24** 0101 0.9 / 0735 3.1 / TH 1333 1.2 / 1944 3.3
10 0052 1.0 / 0655 3.1 / TH 1306 1.2 / 1912 3.3	**25** 0159 0.9 / 0828 3.2 / F 1429 1.1 / 2038 3.4
11 0151 0.7 / 0754 3.3 / F 1404 1.0 / 2011 3.5	**26** 0248 0.9 / 0914 3.3 / SA 1517 1.0 / 2125 3.4
12 0239 0.6 / 0846 3.5 / SA 1453 0.8 / 2103 3.6	**27** 0326 0.8 / 0955 3.4 / SU 1555 0.9 / 2206 3.5
13 0323 0.4 / 0934 3.6 / SU 1538 0.6 / 2153 3.8	**28** 0400 0.8 / 1032 3.4 / M 1628 0.9 / 2244 3.5
14 0406 0.3 / 1021 3.8 / M 1622 0.5 / ○ 2242 3.8	**29** 0432 0.8 / 1106 3.5 / TU 1700 0.8 / ● 2318 3.5
15 0449 0.3 / 1109 3.8 / TU 1707 0.4 / 2332 3.8	**30** 0504 0.8 / 1137 3.5 / W 1732 0.8 / 2349 3.4

DECEMBER

Time m	Time m
1 0538 0.8 / 1205 3.6 / TH 1806 0.8	**16** 0012 3.7 / 0609 0.6 / F 1234 3.9 / 1833 0.4
2 0019 3.4 / 0612 0.8 / F 1233 3.6 / 1841 0.8	**17** 0104 3.6 / 0658 0.7 / SA 1323 3.9 / 1922 0.4
3 0053 3.4 / 0649 0.9 / SA 1307 3.6 / 1918 0.8	**18** 0155 3.5 / 0748 0.8 / SU 1411 3.8 / 2013 0.5
4 0132 3.4 / 0729 0.9 / SU 1346 3.5 / 2000 0.8	**19** 0247 3.3 / 0840 0.9 / M 1500 3.7 / 2107 0.7
5 0217 3.3 / 0813 1.0 / M 1430 3.5 / 2046 0.9	**20** 0340 3.2 / 0937 1.1 / TU 1550 3.6 / 2207 0.8
6 0308 3.2 / 0903 1.1 / TU 1519 3.4 / 2141 0.9	**21** 0435 3.1 / 1040 1.2 / W 1644 3.4 / ◗ 2311 0.9
7 0407 3.1 / 1000 1.1 / W 1617 3.3 / ◖ 2245 0.9	**22** 0536 3.0 / 1144 1.2 / TH 1742 3.3
8 0513 3.1 / 1104 1.2 / TH 1724 3.3 / 2359 0.9	**23** 0014 1.0 / 0643 3.0 / F 1246 1.2 / 1848 3.3
9 0620 3.2 / 1215 1.1 / F 1835 3.3	**24** 0112 1.0 / 0747 3.0 / SA 1345 1.2 / 1956 3.2
10 0109 0.8 / 0722 3.3 / SA 1326 1.0 / 1942 3.4	**25** 0205 1.0 / 0841 3.1 / SU 1440 1.1 / 2053 3.2
11 0208 0.6 / 0821 3.4 / SU 1426 0.8 / 2043 3.6	**26** 0252 0.9 / 0927 3.3 / M 1526 1.0 / 2140 3.3
12 0300 0.5 / 0914 3.6 / M 1520 0.6 / 2138 3.7	**27** 0332 0.9 / 1008 3.4 / TU 1606 0.9 / 2221 3.4
13 0348 0.5 / 1005 3.7 / TU 1609 0.5 / 2230 3.7	**28** 0409 0.8 / 1045 3.5 / W 1641 0.8 / 2259 3.4
14 0435 0.5 / 1055 3.8 / W 1657 0.4 / ○ 2321 3.7	**29** 0445 0.8 / 1120 3.5 / TH 1716 0.7 / ● 2333 3.4
15 0522 0.5 / 1144 3.9 / TH 1745 0.4	**30** 0521 0.8 / 1149 3.6 / F 1751 0.7
	31 0001 3.4 / 0557 0.8 / SA 1213 3.6 / 1827 0.7

Chart Datum: 2·01 metres below Ordnance Datum (Belfast). HAT is 3·9 metres above Chart Datum.

》》 FREE monthly updates. Register at 《
www.reedsnauticalalmanac.co.uk

637

N Ireland

STRANGFORD LOUGH

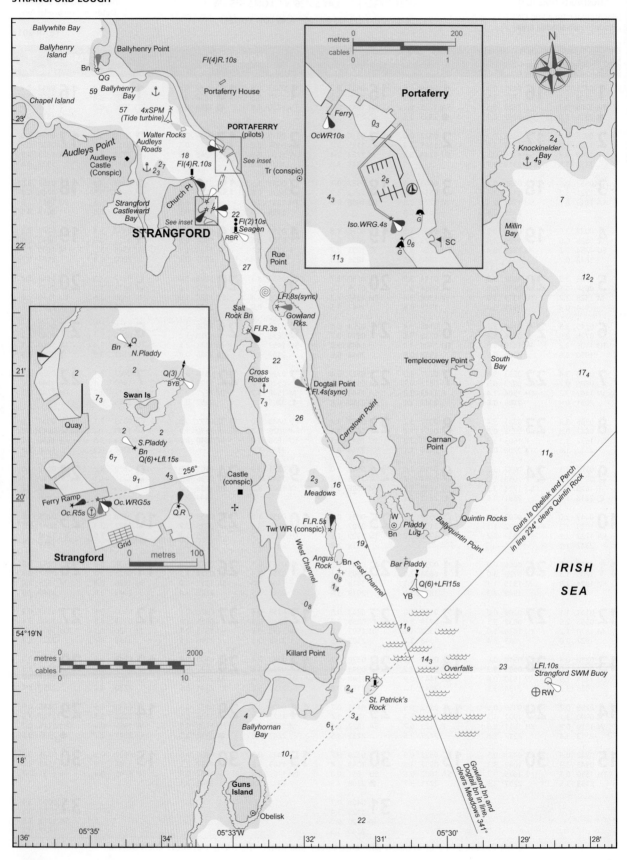

9.13.18 STRANGFORD LOUGH

Down **54°19'·33N 05°30'·85W** (Narrows) ✿✿✿⚓⚓✿✿✿

CHARTS AC 2156, 2159, 5612; Imray C62, C69

TIDES Killard Pt 0000, Strangford Quay +0200 Dover; ML 2·0; Duration 0610

Standard Port BELFAST (←—)

Times				Height (metres)			
High Water		Low Water		MHWS	MHWN	MLWN	MLWS
0100	0700	0000	0600	3·5	3·0	1·1	0·4
1300	1900	1200	1800				
Differences STRANGFORD (The Narrows)							
+0147	+0157	+0148	+0208	+0·1	+0·1	–0·2	0·0
KILLARD POINT (Entr)							
+0011	+0021	+0005	+0025	+1·0	+0·8	+0·1	+0·1
QUOILE BARRIER							
+0150	+0200	+0150	+0300	+0·2	+0·2	–0·3	–0·1
KILLYLEAGH							
+0157	+0207	+0211	+0231	+0·3	+0·3	ND	ND
SOUTH ROCK							
+0023	+0023	+0025	+0025	+1·0	+0·8	+0·1	+0·1
PORTAVOGIE							
+0010	+0020	+0010	+0020	+1·2	+0·9	+0·3	+0·2

SHELTER Excellent; largest inlet on E coast. ⚓ in the Narrows at Cross Roads, off Strangford clear of moorings, in Audley Roads (in about 15m) and in Ballyhenry Bay. Limited ⛵ at Strangford; moorings may be available in Audley Roads (check with Sailing Club). Small marina at Portaferry, where stream runs fast; best to enter near slack water. Many good ⚓ (NB Marine Nature Reserve) and some ⚓s up the lough, by villages and YCs.

NAVIGATION WPT: SWM buoy, L Fl 10s, 54°18'·63N 05°28'·69W, 306°/2·04M to Angus Rk lt twr Fl R 5s 5M. Visitors should use E Chan. Disregard W Chan ≠ 320·5° shown on old editions of AC 2159. Beware St Patricks Rk, Bar Pladdy and Pladdy Lug; overfalls occur in the SE apprs and at the bar, which can be dangerous when ebb meets the NW set outside. Strong tidal streams (7kn at sp) in the Narrows. The bar is passable during the flood; best to enter on young flood or at slack water. The flood starts in the Narrows at HW Belfast –3½ and runs for about 6 hrs; the ebb starts at HW Belfast +2½. Beware car ferry between Strangford, S and E of Swan Is, and Portaferry. Swan Island, seen as a grassy mound at HW, is edged with rks and a reef extends 32m E ending at W bn, Q (3). Further up the lough, despite improved marks and lights, unmarked drying shoals (pladdies) abound; AC 2156 or equivalent is essential.

LIGHTS AND MARKS Off Strangford *Seagen* tide turbine consists of pile with cross beam 50m wide, 3.1m deep at LAT when submerged. Marked as IDM, Fl(2)10s it forms a 43m wide obstruction when raised for maintenance. Vessels should keep 50m clear – visit www.marineturbines.com. Further turbine 1¾Ca WNW of Walter Rks marked by 2 pair Y SPM; outer pr Fl(2) 10s, inner pr FlY 5s.

COMMUNICATIONS (Code 028) CGOC 91463933; Police 101; Medical Clinic 4488 1209. HM at Strangford Ferry Terminal 4488 1637. HM Ch 16 (Mon-Fri 0900-1700). Some YCs Ch **80** M.

FACILITIES Strangford, Portaferry and Killyleagh are easy to access and offer good shops, pubs and restaurants.
STRANGFORD ⛵ 25m ⓥ pontoon, ⚓ ⚓; Cuan Inn ⚓ ⚓.
Islander Marine ☎4488 1449, ⚓ ⚓ ⚓ ⚓ ⚓ ⚓.
PORTAFERRY Marina portaferrymarina.co.uk, mob 7703 209780, ⛵(limited) £1·50, ⚓ ⚓ ⚓. Village ⚓ ⚓ ⚓ Ⓑ ⚓ ⚓ ⚓.
Cook St Pier (2ca S), limited ⛵, ⚓.
QUOILE RIVER: Quoile YC ☎4461 2266, ⚓ ⚓ ⛵ ⚓.
KILLYLEAGH ⚓ ⚓ ⚓ ⚓ ⚓ Kos ⚓ ⚓ ⚓ (Irish Spars & Rigging ☎9751 2830) ⚓ ⚓ ⚓ Ⓑ ⚓ ⚓.
Killyleagh YC ☎07801 291410, N of village.
East Down YC ☎4482 8375, ⛵ (about 1M NE of town).
RINGHADDY QUAY CC, ⚓ ⛵ drying, ⚓ ⚓.
SKETRICK ISLAND In White Rk B,
Strangford Lough YC ☎9754 1883, ⚓ ⛵ ⚓ ⚓ ⚓ ⚓ ⚓.
Down CC, (Old Lightship to NW in Dorn) ⚓ ⚓ ⚓.
KIRCUBBIN (tidal) ⚓ ⚓ ⚓ ⚓ Ⓑ ⚓ ⚓ ⚓.

9.13.19 ARDGLASS

Down **54°15'·63N 05°35'·96W** ✿✿✿⚓⚓✿✿✿

CHARTS AC 2093, 633, 5612; Imray C62

TIDES HW +0025 Dover; ML 3·0; Duration 0620

Standard Port BELFAST (←—)

Times				Height (metres)			
High Water		Low Water		MHWS	MHWN	MLWN	MLWS
0100	0700	0000	0600	3·5	3·0	1·1	0·4
1300	1900	1200	1800				
Differences KILKEEL							
+0040	+0030	+0010	+0010	+1·2	+1·1	+0·4	+0·4
NEWCASTLE							
+0025	+0035	+0020	+0040	+1·6	+1·1	+0·4	+0·1
KILLOUGH HARBOUR							
0000	+0020	ND	ND	+1·8	+1·6	ND	ND
ARDGLASS							
+0010	+0015	+0005	+0010	+1·7	+1·2	+0·6	+0·3

SHELTER Good, except in strong winds from E to S. It is the only all-weather, all-tide shelter between Howth and Bangor. Phennick Cove/Ardglass marina is on W side of hbr, with depths 1·0m to 2·8m. Visitors should check depth of berth offered by marina against depth drawn. The busy fishing port is in South Hbr, with quays (2·1m) on inside of extended S pier. At NW end of hbr, old drying N Dock is also used by FVs.

NAVIGATION WPT 54°15'·30N 05°35'·32W, 131°/5ca to hbr ent. Appr 311° in W sector of WRG Dir lt. Depth in chan 2·4m. Off the tip of the inner breakwater the buoyed marina channel commences in a depth of 1.3m. It is marked by lit PHM and SHM (not shown on chartlet) and turns sharply to port to approach the marina on a heading of 230°. Do not cross the drying SW portion of inner bkwtr, marked by two unlit SCM perches.

LIGHTS AND MARKS Dir lt, 311°, conspic W tr at inner hbr, Iso WRG 4s 10m 8/7/5M, 308°-W-314°; reported hard to see against shore lts. S bkwtr Fl R 3s 10m 5M. W roof of shed on S bkwtr is conspic. If entering S Hbr, avoid Churn Rk, unlit SCM bn. Entrance to marina is buoyed. Castle, spire and water tr (off plan) are conspic to W.

COMMUNICATIONS (Code 028) CGOC 91463933; Police 101; Dr 9084 1242. HM 4484 1291, mob 07790 648274.

Monitor VHF Ch **12** on entry/exit. Marina Ch ⚓ 80.

FACILITIES
Ardglass Marina www.ardglassmarina.co.uk ☎4484 2332, ⚓ (£8·00) 55 inc 22ⓥ £2·40+⚓£1/day, ⚓ ⚓ ⚓ ⚓.
Town ⚓ ⚓ ⚓ ⚓ ⚓ ⚓ ⚓.

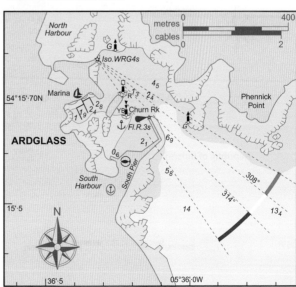

MINOR HARBOURS SOUTH OF ARDGLASS

DUNDRUM BAY, Down. AC 44, 5612. Tides see 9.13.19. This 8M wide bay to the W of St John's Pt is shoal to 5ca offshore and unsafe in onshore winds. The small drying hbr at **Newcastle** (54°11'·8N 05°53'·0W) is for occas use in fair weather. HM ☎(02843) 722804, mob 07803 832515. **Dundrum Hbr (54°14'·2N 05°49'·4W)** No longer used by commercial tfc and only to be used with caution, provides ⚓ in 2m for shoal draught; the bar carries about 0·3m. A steep sea can run on the bar in onshore winds. HW Dundrum is approx that of HW Liverpool; see also 9.13.19 Newcastle. *The Irish Cruising Club's Sailing Directions* are essential for the 1M long, buoyed approach channel. 2 lit DZ buoys offshore are part of the Ballykinler firing range, 2M E of hbr; R flag/lts indicate range active.

ANNALONG HBR, Down, 54°06'·50N 05°53'·65W. AC 44, 5621, 5612. Tides as Kilkeel, see 9.13.8. Slight silting rep't 2006. Entry HW ±2. Excellent shelter in small drying hbr, dredged to 1·5m

80m beyond pier with 30m pontoon (£12/craft), ⚓. Appr in W sector of S breakwater light, Oc WRG 5s 8m 9M, vis 204°-G-249°-W-309°-R-024°. Hug N side of the breakwater to avoid rocky shore to starboard. Surge gate at harbour entrance may be closed in SE winds (3R lts vert shown). IPTS shown at entrance. HM mob 07739 527036. Facilities: 🛢 🚰 ✉.

KILKEEL, Down, 54°03'·47N 05°59'·26W. AC 44, 2800, 5612. HW +0015 on Dover; ML 2·9m; Duration 0620. See 9.13.19. Do not appr in E-SW winds >F4. Inner basin is sheltered, but crowded by FVs; depth off quays approx 1m. Secure in inner basin and see HM. There are drying banks both sides of ent chan and SW gales form a sandbank across ent. This is dredged or slowly eroded in E winds. S bkwtr lt Fl WR 2s 8m 8M, 296°-R-313°-W-017°, storm sigs. Meeney's pier (N bkwtr) Fl G 3s 6m 2M. Monitor VHF Ch **12** on entry/exit. HM ☎(028 417) 62287. Facilities: ⚓ on quay, 🍴 (between fish market and dock). ⚒ 🔧 🖭. **Town** (¾M) 🛢 ✉ 🛒 🔲 ✕ 🚰.

9.13.20 CARLINGFORD LOUGH

Louth/Down **54°01'·25N 06°04'·30W** ✿✿✿✿✿✿

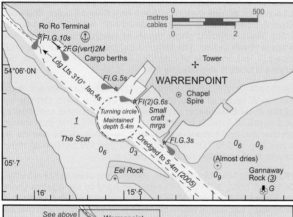

CHARTS AC 44, 2800, 5621, 5612; Imray C62

TIDES Cranfield Pt +0025 and Warrenpoint +0035 Dover; ML 2·9; Duration Cranfield Pt 0615, Warrenpoint 0540

Standard Port DUBLIN (NORTH WALL – AREA 12) ⟵

Times				Height (metres)			
High Water		Low Water		MHWS	MHWN	MLWN	MLWS
0000	0700	0000	0500	4·1	3·4	1·5	0·7
1200	1900	1200	1700				
Differences CRANFIELD POINT							
–0027	–0011	+0005	–0010	+0·7	+0·9	+0·3	+0·2
WARRENPOINT							
–0020	–0010	+0025	+0035	+1·0	+0·7	+0·2	0·0
NEWRY (VICTORIA LOCK)							
+0005	+0015	+0045	Dries	+1·2	+0·9	+0·1	DR
DUNDALK (SOLDIERS POINT)							
–0010	–0010	0000	+0045	+1·0	+0·8	+0·1	–0·1
DUNANY POINT							
–0028	–0018	–0008	–0006	+0·7	+0·9	ND	ND
RIVER BOYNE BAR							
–0005	0000	+0020	+0030	+0·4	+0·3	–0·1	–0·2
BALBRIGGAN							
–0021	–0015	+0010	+0002	+0·3	+0·2	ND	ND

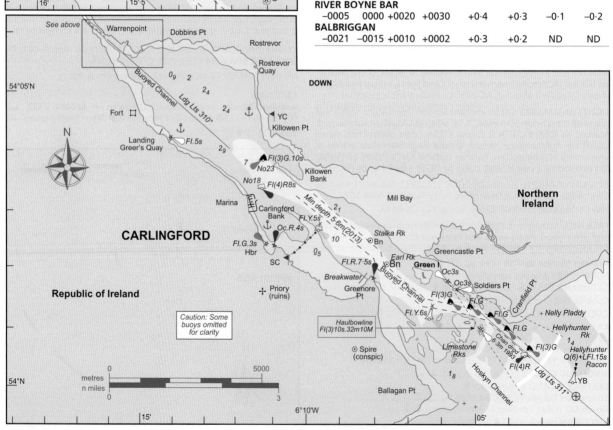

SHELTER

Carlingford Marina is protected on S side by sunken barge; depths 1·5m at LW (check with Marina). Appr from N with Nos 18 and 23 buoys in transit astern 012°, to clear the tail of Carlingford Bank (dries) and beware of marine farms in vicinity.

Carlingford Hbr (dries 2·2m), ⌣ at piers.

Warrenpoint has pontoons on NW side of breakwater (Fl G 3s); access dredged 1·1m. ⌄s clockwise from ent include: off Greenore Pt, between SW end of quay and bkwtr, in 3m clear of commercial traffic; off Greer's Quay in 2m; off Rostrevor Quay, Killowen Pt (YC) and off derelict pier at Greencastle Pt (beware rks). Limited space may be available for short-stay visiting craft in the mussel fishing complex (approx £15/night). Call at least 24 hours before arrival.

NAVIGATION WPT 54°00'·09N 06°02'·00W, (2½ ca S of Hellyhunter SCM lt buoy) 311°/1M to first chan buoys. The main chan is Carlingford Cut (6·3m), about 3ca SW of Cranfield Pt, and passing 2ca NE of Haulbowline lt ho. Drying rocks and shoals obstruct most of the entrance. Small craft should at all times keep clear of commercial shipping in the narrow dredged channel. The ent is impassable in strong on-shore winds. The NE bank is **Northern Ireland**, SW bank is **Republic of Ireland**.

The lough becomes choppy in S'ly winds and, due to the funnelling effect of the mountains, NW winds can cause a higher sea state within the lough than outside it.

Tides run up to 5kn off Greenore Pt through a channel with a minimum depth of 5.6m. Entry is impracticable against the ebb. Beware sudden squalls and waterspouts.

Extensive shellfish beds exist along the NE side of the lough; beware fishing vessels on both sides of the channel, and give them a wide berth.

LIGHTS AND MARKS Haulbowline Fl (3) 10s 32m 10M; granite tower. Ldg lts 310°26': both Oc 3s 7/12m 11M, vis 295°-325°; R △ front, ▽ rear, on framework trs. Greenore Pier Fl R 7·5s 10m 5M. Newry R: Ldg lts 310°, both Iso 4s 5/15m 2M, stone columns. Channel from Green Is to Warren Pt marked by lit lateral buoys.

COMMUNICATIONS (Code Greenore/Carlingford 042; Warrenpoint 028). MRCC (01) 6620922/3 or (02891) 463933; Irish ⊖: Dundalk (042) 34114; Dr (042) 73110; Ⓗ Newry (028) 3026 5511, Dundalk (042) 34701; Police Newry 101, Dundalk (042) 9373102.

Greenore (*Ferry Greenore*) Ch 12 16 (HJ). Carlingford Marina Ch M 16. Warrenpoint Ch 12 16 (H24); call Ch 12 at By No. 23 to enter dredged chan. Dundalk Ch 14 16 (HW±3).

FACILITIES **Carlingford Marina** www.carlingfordmarina.com ☎0429373072, mob 0872 321567, ⎯ 290 inc Ⓥ €4.00 (min€20), short stay <5hrs €12.50, ▣ ▮(0900-2100) ◖(50t) ⌕ ✕ ✕ ▱ @.

Services ⚓ ☎0872 301319, ⚓ ☎(048)44 828882, Divers,.

Carlingford YC ☎(041) 685 1951, ⚓ ⎯ ⚓ ▱.

Carlingford Hbr ⎯ ⌣ ⚓ Kos ✕ ⚓ ▦.

Village ▤ ▤ ✉ Ⓑ ⚲ ✕ ▱.

Warrenpoint HM ☎(02841) 752878; ⚓ (but no access at LW) ⌣ ⚓ ⚓ ▮ ✉ also at Rostrevor, Ⓑ also Dundalk, ⇌ (Dundalk, Newry), ✈ (Dublin).

MINOR HARBOURS AND ANCHORAGES BETWEEN CARLINGFORD LOUGH AND LAMBAY ISLAND

PORT ORIEL, Louth, **53°47'·94N 06°13'·37W**. AC 44. Tides see 9.13.20 (Dunany Pt). Good shelter in small new harbour except in strong ENE winds when a dangerous scend can enter the harbour. Entry reported to be difficult in strong NE winds. Approach from NE on leading marks 174·9°, front and rear Fl 3s (sync) avoiding Wk 1·6m 2.7ca from entrance; Pier Head Light Mo(A) R 9s. ⌣ on pier in 4m, €25.00/craft, or dry out in inner harbour. HM ☎087 2628777. No facilities; ▮ by tanker, ▦ in Clogher Head village (1·2M).

BALBRIGGAN, Dublin, **53°36'·76N 06°10'·84W**. AC 44, 1468. Tides see 9.13.20. Good shelter in small harbour which dries about 1m. Harbour often crowded with FVs and is not recommended for yachts. Approach from E to open the outer harbour; enter from NE but beware severe shoaling on W side of harbour mouth and on both sides of outer harbour. Thence to ⌣ on SE quay in inner

harbour. Lt, Fl (3) WRG 10s 12m 13/10M, conspic W tr on East breakwater head, vis 159°-G193°-W-288°-R-305°. Facilities: ⚓ ▮ (by tanker from Skerries) ⎯ ▮ (HW±2) ▦ ✕ ▱.

SKERRIES, Dublin, **53°35'·09N 06°06'·49W**. AC 633. Tides as Balbriggan, 9.13.20. E & SE of Red Island (a peninsula) lie Colt, Shenick's and St Patrick's Islands, the latter foul to S and SW. Inshore passage between Colt and St Patrick's Is uses transit/brg as on chart to clear dangers. Good shelter and holding in 3m at Skerries Bay, W of Red Is. Approach from E or NE outside PHM buoy, Fl R 10s, off Red Is. ⌄ WNW of pier, Oc R 6s 7m 7M, vis 103°-154°; clear of moorings. Most facilities; Skerries SC ☎1-849 1233, ⎯ (HW±3). Rockabill Lt, Fl WR 12s 45m 17/13M, is conspic 2·4M ENE of St Patrick's Is.

West Denmark

Skagen to Rømø

9.14.1   Contents 643

9.14.2   Area map and Distance table 644

9.14.3   Tidal stream charts see Area 15 Tidal stream charts

9.14.4   Lights, buoys and waypoints 645

9.14.5   Passage information 647
Skagen to Hanstholm – 647 • Hanstholm to Thyborøn – 650 •
Thyborøn to Hvide Sande – 650 • Hvide Sande to Esbjerg – 652 •
Esbjerg to Rømø – 652 • Crossing the North Sea – 652

9.14.6   Special notes for West Denmark 647

9.14.7   Skagen 648

9.14.8   Hirtshals 649

9.14.9   Hanstholm 649

9.14.10 Thyborøn (Western Entrance to Limfjord) 650

9.14.11 Thorsminde 651

9.14.12 Hvide Sande 652
Horns Rev • Grådyb

9.14.13 Esbjerg, tide tables/curves 654

9.14.14 Nordby (Fanø Island) 658

9.14.15 Havneby (Rømø Island) 658

W Denmark

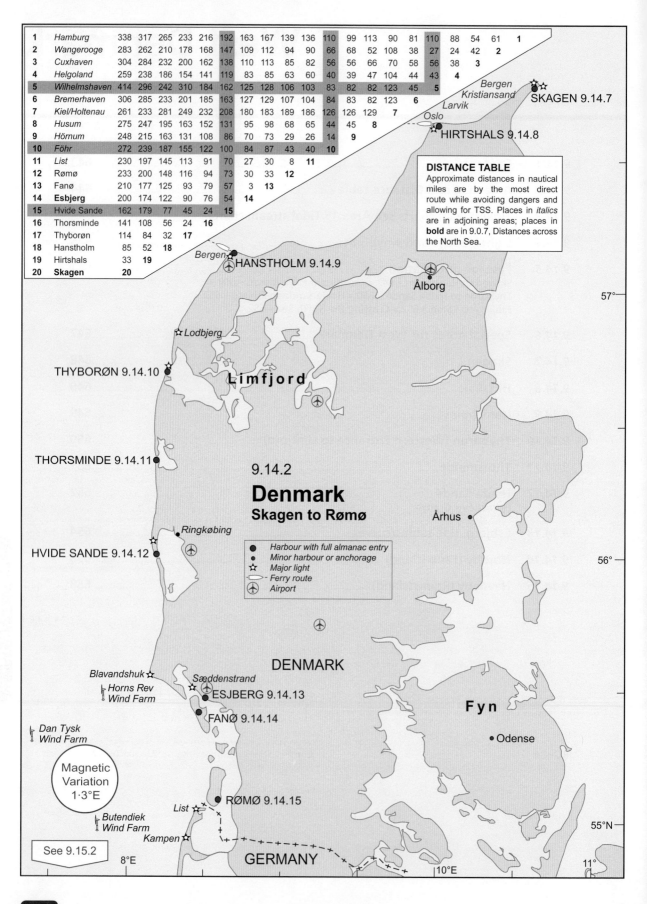

1	*Hamburg*	338	317	265	233	216	192	163	167	139	136	110	99	113	90	81	110	88	54	61	**1**
2	*Wangerooge*	283	262	210	178	168	147	109	112	94	90	66	68	52	108	38	27	24	42	**2**
3	*Cuxhaven*	304	284	232	200	162	138	110	113	85	82	56	56	66	70	58	56	38	**3**
4	*Helgoland*	259	238	186	154	141	119	83	85	63	60	40	39	47	104	44	43	**4**
5	*Wilhelmshaven*	414	296	242	310	184	162	125	128	106	103	83	82	82	123	45	**5**
6	*Bremerhaven*	306	285	233	201	185	163	127	129	107	104	84	83	82	123	**6**
7	*Kiel/Holtenau*	261	233	281	249	232	208	180	183	189	186	126	126	129	**7**
8	*Husum*	275	247	195	163	152	131	95	98	68	65	44	45	**8**
9	*Hörnum*	248	215	163	131	108	86	70	73	29	26	14	**9**
10	*Föhr*	272	239	187	155	122	100	84	87	43	40	**10**
11	*List*	230	197	145	113	91	70	27	30	8	**11**
12	*Rømø*	233	200	148	116	94	73	30	33	**12**
13	*Fanø*	210	177	125	93	79	57	3	**13**
14	**Esbjerg**	200	174	122	90	76	54	**14**
15	*Hvide Sande*	162	179	77	45	24	**15**
16	Thorsminde	141	108	56	24	**16**
17	Thyborøn	114	84	32	**17**
18	Hanstholm	85	52	**18**
19	Hirtshals	33	**19**
20	**Skagen**	**20**

DISTANCE TABLE

Approximate distances in nautical miles are by the most direct route while avoiding dangers and allowing for TSS. Places in *italics* are in adjoining areas; places in **bold** are in 9.0.7, Distances across the North Sea.

SKAGEN 9.14.7

HIRTSHALS 9.14.8

HANSTHOLM 9.14.9

Ålborg

Lodbjerg

THYBORØN 9.14.10

Limfjord

THORSMINDE 9.14.11

9.14.2

Denmark
Skagen to Rømø

Århus

Ringkøbing

HVIDE SANDE 9.14.12

● Harbour with full almanac entry
• Minor harbour or anchorage
☆ Major light
⬭ Ferry route
✈ Airport

DENMARK

Fyn

Blavandshuk ☆

Horns Rev
Wind Farm

Sæddenstrand

ESJBERG 9.14.13

FANØ 9.14.14

Odense

Dan Tysk
Wind Farm

Magnetic
Variation
1·3°E

List ☆

RØMØ 9.14.15

Butendiek
Wind Farm

Kampen ☆

See 9.15.2

GERMANY

9.14.3 TIDAL STREAM CHARTS SEE 9.15.3
9.14.4 LIGHTS, BUOYS AND WAYPOINTS

Bold print = light with a nominal range of 15M or more. CAPITALS = place or feature. *CAPITAL ITALICS* = light-vessel, light float or Lanby. *Italics* = Fog signal. ***Bold italics*** = Racon. Many marks/buoys are fitted with AIS (MMSI No); see relevant charts.

SKAGEN TO THYBORØN

SKAGEN
Skagen W ☆ Fl (3) WR 10s 31m **W17M**/R12M; 053°-W-248°-R-323°; W ○ twr; 57°44'·94N 10°35·70E.
Skagen ☆ Fl 4s 44m **23M**; Gy ○ twr; ***Racon G, 20M***; 57°44'·14N 10°37'·81E.
Skagen No 1A ↓ L Fl 10s; ***Racon N***; 57°43'·46N 10°53'·55E. (Route T)
Skagen No. 2 ↓ L Fl 10s; 57°37'·61N 11°05'·51E. (Route T)
Skagens Rev ↓ Q; 57°45'·97N 10°43'·74E.
↓ Q (3) 10s; 57°43'·87N 10°42'·31E.
Skagen Harbour ↓, Fl(4)Y10s; 57°42'·55N 10°35'·80E.
E bkwtr ≰ FG 8m 5M; G twr; *Horn (2) 30s*; 57°42'·88N 10°35'·66E.

HIRTSHALS
Hirtshals ☆ F Fl 30s 57m **F 18M**; **Fl 25M**; W ○ twr, approx 1M SSW of hbr ent; 57°35'·10N 09°56'·55E.
Ldg lts 166°, both Iso R 2s 10m 11M; R △ on twr; 57°35'·69N 09°57'·64E; marina ent is close N of this lt. Rear, Iso R 4s 18m 11M; R ▽ on twr; 330m from front.
Approach chan ▲ Fl (3) G 10s; 57°36'·18N 09°56·94E.
↘ Fl R 3s; 57°36'·44N 09°57'·67E.
↙ Fl (5) Y 20s; 57°36'·43N 09°57'·72E.
▲ Fl G 5s; 57°36'·11N 09°57'·14E.
Outer W mole ≰ Fl G 3s 14m 6M; G mast; *Horn 15s*; 57°35'·97N 09°57'·37E.
W mole spur ≰ Fl G 5s 9m 4M; G mast; 57°35'·77N 09°57'·51E.
E mole ≰ Fl R 5s 9m 6M; G mast; 57°35'·85N 09°57'·61E.

LØKKEN
Lee bkwtr ≰ Fl 5s 5m 5M; 57°22'·39N 09°42'·11E.

TRANUM STRAND (Firing ranges)
Tranum No. 1 ☆ Al Fl WR 4s 19m W7M, R7M; twr; by day Q 4M; shown when firing in progress; 57°12'·43N 09°30'·34E.
Tranum No. 2 ☆ light as per No.1; 57°10'·57N 09°26'·22E.

LILD STRAND
Bragerne ▲ Fl G 5s; 57°10'·67N 08°56'·35E.
Ldg lts 138°, three ≰: F 12/22m 7/8M; 127°-149°; 3 masts. Front 57°09'·22N 08°57'·75E.

HANSTHOLM
Hanstholm ☆ Fl (3) 20s 65m **26M**; shown by day in poor vis; W 8-sided twr; 57°06'·77N 08°35'·92E, approx 1M S of the hbr ent.
Hanstholm ↓ LFl 10s; 57°08'·06N 08°34'·87E.
Ldg lts 142·6°, both Iso 2s 37/45m 13M; synch; 127·6°-157·6°; R △ on mast. Front, 57°07'·12N 08°36'·15E. Rear, 170m from front, R ▽ on mast.
Note: The 4 outer/inner mole heads are floodlit.
W outer mole ≰ Fl G 3s 11m 9M; G pillar; 57°07'·56N 08°35'·45E. E outer mole ≰ Fl R 3s 11m 9M; R pillar; 57°07'·61N 08°35'·57E.
Roshage ≰ Fl 5s 7m 5M; 57°07'·75N 08°37'·25E (1M E of hbr).

NØRRE VORUPØR
Mole ≰ Fl G 5s 6m 5M; 56°57'·75N 08°21'·679E.
Ldg lts, both Iso R 4s 20/30m 9M, indicate safest landing place for FVs; vis 22·5° either side of ldg line; synch. Front 57°57'·46N 08°22'·10E. Rear, 80m from front.
Lodbjerg ☆ Fl (2) 20s 48m **23M**; ○ twr; 56°49'·40N 08°15'·76E.

THYBORØN
Landfall ↓ L Fl 10s; ***Racon T, 10m***; 56°42'·54N 08°08'·69E.
Agger Tange ldg lts 082°: Front, Iso WRG 4s 8m W11M, R/G8M; 074·5°-G-079·5°-W-084·5°-R-089·5°; R △ on bcn; 56°42'·97N

08°14'·14E. Rear, Iso 4s 17m 11M; 075°-089°; synch; R ▽ on Gy twr; 804m from front.
Off Havmolen ↘ 56°43'·25N 08°12'·52E.
Approach ≰ Fl (3) 10s 24m 12M; intens 023·5°-203·5°; also lit by day in poor vis; lattice twr; 56°42'·49N 08°12'·91E (S side of ent).
Langholm ldg lts 120°, both Iso 2s 7/13m 11M; synch; 113°-127°. Front, R △ on R hut, 56°42'·45N 08°14'·54E. Rear, R ▽ on Gy twr.
Thyborøn Havn ≰ Oc (2) WRG 12s 6m W11M, R/G8M; 122·5°-G-146·5°-W-150°-R-211·3°-G-337·5°-W-340°-R-344°; W twr R band; 56°42'·35N 08°13'·39E (680m N of ent to Yderhavn and Basins).
Yderhavn approach ≰ Dir Iso WRG 2s 21m 5M (0.5M by day); 213°-G-218°-W-223°-R-228°; 56°41'·85N 08°13'·32E
N mole ≰ Fl G 3s 6m 4M; G pedestal; 56°42'·02N 08°13'·52E.
S mole ≰ Fl R 3s 6m 4M; R pedestal; 56°41'·97N 08°13'·53E.

LIMFJORD (Limited coverage eastward, only to 08°42'E)

Sælhundeholm Løb, into Nissum Bredning ('Broad').
No. 1 ▲ Fl (2) G 5s; 56°41'·14N 08°14'·17E.
No. 3 ▲ Fl G 3s; 56°40'·75N 08°13'·79E.
No. 7 ▲ Fl G 3s; 56°40'·31N 08°13'·52E.
No. 11 ↓ Fl G 3s; 56°39'·83N 08°13'·59E.
No. 16 ◢ Fl R 3s; 56°38'·51N 08°13'·89E.
No. 18 ◻ Fl (2) R 5s; 56°38'·20N 08°14'·34E.
No. 21 ▲ Fl G 3s; 56°38'·98N 08°14'·71E.
No. 26 ◻ Fl R 3s; 56°38'·71N 08°15'·38E.
No. 29 ▲ Fl G 3s; 56°38'·40N 08°15'·94E.

APPROACHES TO LEMVIG (Marina and Havn)
Toftum Dir ≰ 120°, Iso WRG 4s 24m, W12M, R/G 8M; 110°-G-120°-W-137°-R-144°; hut; 56°33'·09N 08°18'·33E.
Rønnen ▲ Fl G 3s; 56°36'·71N 08°21'·74E.
Rønnen ↓ L Fl 10s; 56°35'·58N 08°21'·62E.
Søgard Mark ldg lts 243·5°: both FR 20/30m 5M; vis 90° either side of ldg line. Front, ↓ R △ on W bcn; 56°34'·39N 08°17'·29E. Rear, R ▽ on W post; 235m from front.
Chan ▲ 56°33'·27N 08°18'·31E.
Ldg lts, W of hbr 177·7°, both FR 8/20m 5M; 153°-203°. Front ↓ R △ on twr; 56°33'·04N 08°18'·16E. Rear, R ▽ on W post; 184m from front.
Vinkel Hage Marina, N mole ≰ FG 3m 5M; 56°33'·09N 08°18'·33E.
S mole, FR 3m 4M. Marina is 9ca N of the Havn on the W side.
Ostre Havn ↓ 56°33'·17N 08°18'·40E. Havn ent, FG/FR 4m 5M.

THISTED HAVN (Marina)
Outer W mole ≰ Fl R 3s 4m 2M; 56°57'·10N 08°41'·90E.
Outer E mole ≰ Fl G 3s 4m 2M; 56°57'·10N 08°41'·95E.
Thisted Bredning ≰ Aero 3 Fl R 1·5s (vert; 45m apart) 183m 10M; TV mast; 56°58'·52N 08°41'·15E, 1·55M NNW of hbr ent.

THYBORØN TO BLÅVANDS HUK

Bovbjerg ☆ Fl (2) 15s 62m 16M; 56°30'·79N 08°07'·18E.
Wave recorder buoy ↙ Fl (5) Y 20s; 56°28'·49N 08°03'·35E.

THORSMINDE HAVN
(All Lat/Longs for this harbour are approximate)
Lt ho ≰ F 30m 13M; Gy twr; 56°22'·34N 08°06'·99E.
Groyne, N ≰ Fl 5s 8m 5M; Gy hut; 56°22'·46N 08°06'·82E.
N mole ≰ Iso R 2s 9m 4M; R hut; 56°22'·36N 08°06'·62E.
S mole ≰ Iso G 2s 9m 4M; G hut; 56°22'·26N 08°06'·92E.
West hbr, W mole ≰ FG 5m 2M; Gy post; 56°22'·36N 08°07'·12E.
E mole ≰ FR 5m 2M; Gy post; 56°22'·26N 08°07'·18E.
NW dolphin ≰ FG 5m 4M; Gy post; 56°22'·26N 08°07'·22E.
SE dolphin ≰ FG 5m 4M; Gy post; 56°22'·26N 08°07'·26E.
Lock, E side ≰ Iso 4s 12m 4M; Gy mast; 020°-160°; 56°22'·36N 08°07'·22E.
Road bridge ≰ Iso 4s 5m 4M; 200°-340°; 56°22'·36N 08°07'·26E.

HVIDE SANDE
Lyngvig ☆ Fl 5s 53m **23M** (may have Y tint); W ○ twr; 56°02'·99N 08°06'·22E.
N outer bkwtr ≰ Fl R 3s 7m 8M; R hut; 55°59'·94N 08°06'·55E.

N mole ⚲ Fl R 5s 10m 6·5M; R floodlit edifice; 55°59'·98N 08°06'·84E.
S mole ⚲ Fl G 5s 10m 6M; 55°59'·96N 08°06'·98E.
Lt ho ⚲ F 27m 14M; Gy twr; 56°00'·00N 08°07'·35E.
Nordhavn, E pier ⚲ FG 4m 4M; 56°00'·16N 08°07'·42E.
W pier ⚲ 2 FR 3m 2M; W posts; 060°–035°; 56°00'·07N 08°07'·12E.
Sydhavn, W pier ⚲ FG 4m 2M; 56°00'·06N 08°07'·52E.
E Pier ⚲ FR 4m 2M; Gy post; 56°00'·06N 08°07'·53E.
Lock entrance ldg lts 293·5°: both FR 11/14m 2M; 201·6°-021·6°. Front, R △ on Gy tr, 56°00'·0N 08°07'·8E. Rear, R ▽ on Gy twr, 72m from front.
Fjordhavn ldg lts 246·6° (both vis 7·5° either side of ldg line): Front, Iso G 2s 4m 4M; Or △ on mast; 56°00'·5N 08°07'·9E. Rear, Iso G 4s 6m 4M, Or ▽ on mast, 128m from front.

HORNS REV (marks westward from coast)
Oskbøl firing range. Two lights (4M apart) both AIFl WR 4s 35m **16M**, R13M, (by day Q 10M) are shown when firing is in progress: North ☆ 55°37'·24N 08°07'·04E. South ☆ 55°33'·53N 08°04'·67E.
Range safety buoys: ⬦ Fl Y 5s; 55°42'·32N 08°06'·92E.⬦ Fl Y 3s; 55°38'·63N 07°50'·91E.⬦ Fl Y 3s; 55°37'·35N 07°56'·98E.⬦ 55°36'·02N 08°02'·49E.

Blåvands Huk ☆ Fl (3) 20s 55m **23M;** W □ twr; 55°33'·46N 08°04'·95E.

Horns Rev is encircled by:
Tuxen ⚐ Q; 55°34'·22N 07°41'·92E on the N side.
Vyl ⚐ Q (6) + L Fl 15s; 55°26'·22N 07°49'·99E on the S side.
No. 2 ⚑ L Fl 10s; 55°28'·74N 07°36'·49E on the SW side.
Horns Rev W ⚐ Q (9) 15s; 55°34'·47N 07°26'·05E, off the W end.

Slugen Channel (crosses Horns Rev ESE/WNW)
▲ L Fl G 10s; 55°33'·99N 07°49'·38E.
▲ Fl G 3s; 55°32'·26N 07°53'·65E.
⚑ Fl (2) R 5s; 55°31'·46N 07°52'·88E.
▲ Fl (2) G 5s; 55°30'·52N 07°59'·20E.
⚑ Fl (3) R 10s; 55°29'·42N 08°02'·56E.

Søren Bovbjergs Dyb (unlit N/S side channel off Slugen Chan)
⚑ 55°33'·57N 07°55'·55E.
⚑ 55°32'·80N 07°55'·30E.
⚑ 55°32'·19N 07°56'·29E.
⚑ 55°31'·24N 07°57'·46E.

Horns Rev 1 wind farm, 2.7M x 2.5M, centred on 55°29'·22N 07° 50'·21E: 80 turbines all R lts, 12 perimeter turbines Fl (3) Y 10s. NE turbine, *Racon (U);* 55°30'·28N 07°52'·63E. Platform 2 x Mo (U) 15s 15m 5M; 55°30'·52N 07°52'·52E. SW turbine *Racon (U)*, 55° 28'·11N 07°48'·26E.

Horns Rev 2 wind farm, 2M x 6M, centred on 55° 36 08N 07° 34 78E: 91 turbines, all R lts, 14 perimeter turbines Fl (3) Y 10s. NW turbine *Racon (U);* 55° 38 88N 07° 35 55E. Platform 2 x Mo (U) 15s 5M; 55° 35 99N 07° 37 37E.

BLÅVANDS HUK TO RØMØ
APPROACHES TO ESBJERG
Grådyb ⚐ L Fl 10s; *Racon G, 10M;* 55°24'·63N 08°11'·59E.
Sædding Strand 053·8° triple ldg lts: valid up to Nos 7/8 buoys; H24: **Front** Iso 2s 13m **21M**; 052°–056°; R bldg; 55°29'·74N 08°23'·87E. **Middle** Iso 4s 26m **21M**; 051°–057°; R twr, W bands; 55°29'·94N 08°24'·33E, 630m from front. **Rear** F 37m **18M**; 052°–056°; R twr; 55°30'·18N 08°24'·92E, 0·75M from front.
No. 1 ⚐ Q; 55°25'·49N 08°13'·89E.
No. 2 ⚑ Fl (3) R 10s; 55°25'·62N 08°13'·73E.
No. 3 ▲ Fl G 3s; 55°25'·93N 08°14'·84E.
No. 4 ⚑ Fl R 3s; 55°26'·02N 08°14'·72E.
Tide Gauge ⚲ Fl (5) Y 20s 8m 4M; 55°26'·05N 08°15'·93E.
No. 5 ▲ Fl G 5s; 55°26'·32N 08°15'·83E.
No. 6 ⚑ Fl R 5s; 55°26'·44N 08°15'·70E.
No. 7 ⚐ Q; 55°26'·76N 08°16'·91E.
No. 8 ⚑ Fl (2) R 5s; 55°26'·89N 08°16'·81E.

Ldg lts 067°, valid up to Nos 9/10 buoys. Both FG 10/25m **16M**, H24. Front, Gy tripod; rear, Gy twr, 55°28'·76N 08°24'·70E.
No. 9 ▲ Fl (2) G 10s; 55°27'·04N 08°18'·21E.
No.10 ⚑ Fl (2) R 10s; 55°27'·20N 08°18'·04E.
Ldg lts 049°, valid up to No 16 buoy/Jerg. Both FR 16/27m **16M**, H24. Front, W twr; rear, Gy twr, 55°27'·92N 08°23'·75E.
No. 11 ▲ Fl G 3s; 55°27'·71N 08°19'·59E.
No. 12 ⚑ Fl R 3s; 55°27'·88N 08°19'·39E.
No. 13 ▲ Fl G 5s; 55°28'·40N 08°20'·99E.
No. 14 ⚑ Fl R 5s; 55°28'·57N 08°20'·73E.
Jerg ⚲ Fl G 3s 7m 5M; G twr, Y base; 55°28'·88N 08°22'·00E.
No. 16 ⚐ Q (6) + L Fl 15s; 55°29'·04N 08°21'·80E.
No. 15A ▲ Fl (2) G 5s; 55°29'·01N 08°22'·47E.
No 18 ⚑ Fl (2) R 5s; 55°29'·19N 08°22'·71E.
Fovrfelt N ⚲ Oc (2) WRG 6s 7m W6M, R/G4M; 066·5°-G-073°-W-077°-R-085·5°; 327°-G-331°- W-333·5°-R-342°; Y twr; 55°29'·29N 08°23'·79E.
Fovrfelt ⚲ Fl (2) R 10s 11m 6M; R twr; 55°29'·03N 08°23'·75E.
No 15B ▲ Fl (2) G 10s; 55°28'·83N 08°23'·65E.

ESBJERG HAVN
Strandby, shelter mole, NW corner ⚲ Oc WRG 5s 6m W13M, R/G9M; 101·7°-G-105·5°-W-109·5°-R-111·7°; W bldg, R band; 55°28'·76N 08°24'·63E.
Industrifiskerihavn, W mole ⚲ Fl R 5s 6m 4M; R structure; 55°28'·57N 08°24'·89E.
E mole ⚲ Fl G 5s 7m 4M; G structure; 55°28'·52N 08°25'·03E.
Nordsøkai ⚐ Q (9) 15s 5m; Y twr, B band; 55°28'·45N 08°25'·04E.
Konsumfiskerihavn, W mole ⚲ FlR 3s 6m; 203°-119°; R twr; 55°28'·31N 08°25'·33E. Yacht hbr in SSE part of Basin II.
E mole ⚲ Fl G 3s 6m; 023°-256°; G tr; 55°28·31N 08°25·40E.
Trafikhavn, NW corner, ⚲ Oc (2) WRG 12s 6m W11M, R/G8M; 118°-G-124·5°-W-129°-R-131°; W bldg, R band; 55°28·22N 08°25·43E.
N mole ⚲ FR 8m 5M; 232°-135°; R bldg; 55°28'·13N 08°25'·46E.
S mole ⚲ FG 8m 4M; 045°-276°; G bldg; 55°28'·08N 08°25'·51E.
No. 22 ⚑ Fl (2) R 5s; 55°27'·62N 08°25'·50E.
Sønderhavn (industrial) W mole ⚲ FR 9m 4M; 55°27'·49N 08°26'·12E. E mole ⚲ FG 9m 4M; 55°27'·43N 08°26'·31E.
Aero ⚲ 3 x Fl 1·5s (vert, 82m apart) 251m 12M, H24; on chimney; 55°27'·27N 08°27'·32E (1200m ESE of Sønderhavn ent).

FANØ
Slunden outer ldg lts 242°, both Iso 2s 5/8m 3M; 227°-257°. Front, twr; 55°27'·20N 08°24'·53E. Rear, twr, 106m from front.
E shore, reciprocal ldg lts 062°, both FR 10/13m 3M; 047°-077°. Front, twr; 55°27'·65N 08°26'·01E. Rear, twr, 140m from front.
No. 1 ⚲ Fl (2) G l0s 5m 2M; G pile; 55°27'·45N 08°25'·28E.
No. 2 ⚲ Fl (2+1) Y 5s 5m 2M; Y pile; 55°27'·42N 08°25'·32E.
No. 3 ⚲ Fl G 3s 5m 2M; G pile; 55°27'·39N 08°25'·07E.
No. 4 ⚲ Fl R 3s 5m 2M; R pile; 55°27'·35N 08°25'·10E.
Nordby ldg lts 214°, both FR 7/9m 4M; 123·7°-303·7°. Front, W mast;55°26'·94N 08°24'·44E. Rear, Gy twr, 84m from front.
Kremer Sand ⚲ FG 5m 3M; G dolphin; 55°27'·3N 08°24'·9E.
Næs Søjord ⚲ FR 5m 3M; R pile; 55°27'·27N 08°24'·86E.
Nordby marina 55°26'·65N 08°24'·53E.

KNUDEDYB
G ⚐ 55°20'·50N 08°24'·28E.
K ⚐ 55°18'·92N 08°20'·06E.
No. 2 ⚑ 55°18'·82N 08°21'·33E.
No. 4 ⚑ 55°18'·81N 08°22'·21E.
No. 6 ⚑ 55°18'·38N 08°24'·63E.
No. 10 ⚑ 55°18'·68N 08°28'·50E.
Knoben ⚑ 55°18'·71N 08°30'·60E.

JUVRE DYB
No. 4 ⚑ 55°13'·76N 08°24'·73E.
No. 6 ⚑ 55°13'·41N 08°26'·60E.
No. 8 ⚑ 55°12'·69N 08°26'·83E.
No. 10 ⚑ 55°12'·55N 08°28'·72E.
Rejsby Stjært ⚐ 55°13'·15N 08°30'·55E.

OUTER APPROACH (Lister Tief) TO RØMØ
See also Area 15 for details of lights on Sylt.
Rode Klit Sand, Recording stn ⚓ Fl (3) Y 15s, 55°08'·40N 08°07'·25E.
Lister Tief ⚓ Iso 8s, *Whis;* 55°05'·32N 08°16'·80E.
No. 1 ⚓ 55°05'·48N 08°18'·50E.
No. 3 ⚓ Fl G 4s; 55°05'·36E 08°20'·14E.
No. 2 ⚓ Fl (3) R 10s; 55°04'·23N 08°22'·32E.
No. 9 ⚓ Fl (2) G 9s; 55°03'·76N 08°23'·05E.
Lister Landtief No 5 ⚓ 55°03'·68N 08°24'·73E.
No. 4 ⚓ FL (2) R 5s; 55°03'·84N 08°25'·29E.
G1 ⚓ Fl Y 4s; 55°03'·27N 08°28'·32E.

RØMØ DYB and HAVN
No. 1 ⚓ Fl (2) G 10s; 55°03'·23N 08°30'·30E.
No. 10 ⚓ Fl (2) R 10s 5m 3M; R pole; 55°03'·50N 08°31'·10E.
No. 14 ⚓ Fl R 3s 6m 2M; R pole; 55°03'·85N 08°32'·55E.
No. 20 ⚓ Fl R 5s 5m 2M; R pole; 55°04'·79N 08°34'·13E.
No. 9 ⚓ 55°04'·79N 08°34'·63E.
No. 11 ⚓ 55°05'·17N 08°34'·69E.
Rømø Hbr:
S mole ⚓ Fl R 3s 7m 2M; Gy twr; 55°05'·19N 08°34'·31E.
N mole ⚓ Fl G 3s 7m 2M; Gy twr; 55°05'·23N 08°34'·30E.
Inner S mole ⚓ FR 4m 1M; 55°05'·2N 08°34'·2E.

9.14.5   PASSAGE INFORMATION

More Passage Information is threaded between successive harbours in this Area. **Bibliography:** *North Sea (East) Pilot* (NP 55); *Cruising Guide to Germany and Denmark* (Imray/Navin).

SKAGEN TO HANSTHOLM

(AC 1402) From/to **Skagen** keep outside the cardinal buoys marking Skagen Rev. In Jammerbugt (The Bay of Woe: Hirtshals to Hanstholm) a firing practice area extends 7M off Tranum Strand, 57°10'N 09°25'E.

Jutland's low, sandy W coast is not a notable cruising ground and lacks genuine harbours of refuge in strong onshore winds. Even **Esbjerg** is made difficult by a bar on which seas break heavily in strong W or SW winds. **Thyborøn** should not be approached on the ebb with onshore winds >F5. But in settled weather, these and

other ports may break the long passage to the Skagerrak or round **Skagen** (The Skaw) into the Kattegat.

Coastal dangers include: Horns Rev, a large shoal (3m-5m) extending about 22M W of Blåvands Huk, with Horns Rev 1 and 2 wind farms located on it. The main buoyed Slugen channel lies some 5M offshore; further inshore, extensive sand ridges lie parallel to the coast up to 5ca from the shore, with depths of 1m to 5.5m. There are many coastal nature reserves, as charted, into which entry is either restricted or prohibited; but they do not deny access to hbrs.

▶*Tidal streams are weak as far N as Thyborøn and imperceptible further N, but expect wind-driven currents such as the N-going Stryget.*◀ The general direction of buoyage is northerly. Navigational marks are excellent. There is much fishing activity, including pair-trawling.

9.14.6 SPECIAL NOTES FOR W DENMARK

VOCABULARY A few Danish words which may be useful:

Ashore	Paa Lan	Nautical	
Hello	Goodag	Harbourmaster	Havnefoged
Goodbye	Farvel	Customs	Told
Please	Vaer so gud	Hbr Police	Havnepoliti
Thank you	Tak	Boat name	Bådenavn
Yes	Ja	Length	Længde
No	Nej	Beam	Bredde
Sorry	Undskyld	Draught	Dybgang
How much ..?	Hvor meget koster ..?	Water, fresh	Vand
		Electricity	Electricitet
Where is ..?	Hvor er ..?	Petrol	Benzin
Left	Venstre	Shower	Bruser
Right	Højre	Red	Rød
Doctor	Læge	Green	Grøn
Dentist	Tandlæge	White	Hvid
Hospital	Hospitalet	Anchorage	Ankerplads
Chemist	Apotek	Prohibited	Forbudt
Laundry	Vaskeri	High water	Højvande
Post office	Postkontoret	Low water	Lavvande
Bus	Bussen	Chart datum	Kortdatum
Train	Toget	Forecast	Vejrudsigt
Ticket, single	enkeltbillet	Low	Lavtryk
Open	Åben	High	Høj
Closed	Lukket	Gale	Hård kuling
Market	Marked	Squall	Byge
Milk	Mœlk	Rough sea	Grov sø
Bread	Brød	Rain	Regn
Meat	Kød	Fog	Tåge

LANGUAGES English and German are widely spoken.

AMTSKOMMUNERS are given in lieu of 'counties' in the UK.

TIME ZONE is −0100 (subtract 1 hour for UT). DST is kept from the last Sunday in March until the last Sunday in October.

CHARTS Danish charts are issued by Kort-og Matrikelstyrelsen, Rentemestervej 8, DK-2400 København. Those prefixed 'DK' are standard Danish nautical charts. Leisure charts for W Jutland are not available. British Admiralty charts are mostly small scale.

LIGHTS Navigational marks, buoys and lights are excellent.

HARBOURS Most yacht hbrs are owned by the state or community and run by a local Yacht Club (as in Germany).

TRAFFIC SEPARATION SCHEMES There are none off West Jutland, but off Skagen heavy traffic between the Skaggerak and Kattegat follows Transit Routes T or B; see chartlet overleaf.

SIGNALS Local port, tidal, distress and traffic signals are given where possible. When motor sailing, always display a motoring ▼.

TELEPHONES To call UK from Denmark: dial 00 44, then the UK area code, minus the initial zero; followed by the number required. To call Denmark, which has no area codes, from the UK: dial 00 (or +) 45, then the required 8 digit number.

EMERGENCIES Police, Fire and Ambulance: dial 112.

PUBLIC HOLIDAYS New Year's Day, Maundy Thurs (Thu before Easter) thru' to Easter Monday, Common Prayer Day (4th Fri after Easter) Ascension Day, Whit Monday, Constitution Day (5 Jun) Christmas Eve to Boxing Day. Local festivals are common.

CURRENCY The Danish krone (Dkr) = 100 øre. Rate of exchange April 2015: £1.00 = approx Dkr 10.3; 1 Dkr = approx £0.1. VAT (MOMS) is 25%, on everything. Banks with ATMs are as common as in the UK. All major credit cards are accepted.

GAS Propane is the norm. Camping Gaz (butane) is available, but Calor gas, whether butane or propane, is not.

USEFUL ADDRESSES Danish Tourist Board, 55 Sloane St, London SW1X 9SY; ☎020 7259 5959; dtb.london@dt.dk. The Embassy is at the same address.

Denmark: The main Tourist Office is Vesterbrogade 6D, 1620 København V; ☎33 11 14 15; dt@dt.dk; www.dt.dk. There are excellent and very helpful TOs in most fair-sized towns.

W Denmark

British Embassy, Kastelsvej 36-40, DK-2100 København Ø; ☎35 44 52 00.

For other addresses see Chapter 1.

TRANSPORT Ferries run from Hirtshals to Oslo, Larvik and Kristiansand or Hanstholm to Bergen, the Faroe Is and Iceland.

UK scheduled and low-cost flights go to Billund (35M NE of Esbjerg) and Copenhagen. 'Norwegian' have regular budget flights to/from Aalborg (107M NE of Thyboron) to Gatwick and many other domestic and international destinations. Internal flights are limited as Denmark is compact with reliable and fast trains, (DSB) and a broad fare structure. Buses are also widespread and reliable. Bicycles are very popular and easy to hire.

SHOPPING Most towns have a market day on either Wed or Sat and launderettes (*montvaskeri*) coin or token-operated.

9.14.7 SKAGEN

Nordjylland 57°42'·86N 10°35'·63E ❀❀❀♦♦♦❁

CHARTS AC 1402, 2107 with 1:12,500 hbr inset

TIDES Variations in heights are minimal and largely generated by wind. For all practical purposes Hirtshals differences are adequate. HW heights are: MHWS 0·2m, MHWN 0·1m.

TIDAL STREAMS are perceptible only in calm conditions. Wind influence usually creates a NE-going current for 75% of the year.

SHELTER is good, access at all tides/weather. Yachts berth in 4m between Old Pier and Pier 1, NE of the central pontoon. FVs berth either side of Pier 1. There are also quieter berths at the YC pontoons in 3·5m on S side of Bundgarns Bassin.

NAVIGATION From the W, the WPT is 57°45'·98N 10°43'·74E, a NCM buoy Q, off the NE tip of Skagens Rev. Thence 200°/2·2M to an ECM buoy, Q(3) 10s, at 57°43'·87N 10°42'·35E; thence 258°/3·6M to hbr ent. Buoyed routes B and T are for transit shipping. Avoid restricted area in hbr approach.

LIGHTS AND MARKS As chartlet and 9.14.4. The R sector of Skagen West lt ho shows over Skagens Rev. Skagen lt ho is 1·7M NE of the hbr ent, with Racon G, 20M. The flat peninsula is fringed by sand dunes.

COMMUNICATIONS Coast Guard 89 43 30 99; 🛟 Station 98 45 07 57; Ⓗ 70 15 03 00; Police, 96 22 14 48, (emergency 112); Dr 70 15 03 00, (emergency 112); HM 98 44 13 46 (H24) VHF Ch 16, 12; Taxi 98 43 34 34; Tourist Info 98 44 13 77.

FACILITIES ⚓ Ⓥ <11.9m 200Dkr/night, 12·1m-14·9m, 280Dkr/night. ⛽ ⏏ ⚒ ✕ ♨ 🅿 🚻. **YC Skagen Sejl Klub** ☎23 23 38 45/25 12 19 97, call ahead, Ⓥ max LOA <12m 200Dkr/night, ⏏ 🅿 🚻. **Town** ✕ 🍴 🛒 ✉ ⇌ ✈ (Aalborg) ☎98 17 33 11.

9.14.8 HIRTSHALS

Nordjylland 57°35'·90N 09°57'·54E ❀❀♢♢✿

CHARTS AC 1402 with 1:12,500 hbr inset; DK92; D 63

TIDES +0600 Dover, ML 0·0, Duration 0625
Standard Port ESBJERG (→)

Times				Height (metres)			
High Water		Low Water		MHWS	MHWN	MLWN	MLWS
0300	0700	0100	0800	2·0	1·6	0·6	0·2
1500	1900	1300	2000				
Differences HIRTSHALS							
+0055	+0320	+0340	+0100	−1·7	−1·4	−0·5	−0·2

SHELTER Good. Yachts berth in a small marina to stbd and close S of the hbr ent. For possible berth on pontoon in Vestbassin, mostly locals, call HM. Hirtshals is a commercial and fishing hbr which in theory foreign yachts may use only with HM's prior approval. It is not advisable to attempt entrance in strong onshore winds from N to NW due to heavy seas and strong E-going current. In E gales water levels are lowered by up to 1·0m.

NAVIGATION WPT 57°36'·39N 09°57'·32E, 166°/7ca to marina ent in front of rear 166° ldg lt. The outer PHM and SHM buoys funnel vessels into the entrance. Inbound vessels have precedence over out-bound. Turn to stbd before reaching the Inner Hbr lt FG. Caution: FVs and fast ferries to Norway. Speed limit 3kn.

LIGHTS AND MARKS As chartlet and 9.14.4. 166° ldg bcns: front, R ▲ on frame twr; rear, R ▼ on frame twr. The white lt ho is approx 1M SW of the hbr ent.

COMMUNICATIONS Coast Guard 98 92 22 22; Police 96 23 14 48, (emergency 112); ⊞ 70 15 03 00; ⊞ (Hjørring) 98 92 72 44; Dr 98 94 11 77; HM 98 94 14 22, VHF Ch 16, 12, 13, HX; Taxi 98 92 47 00.

FACILITIES ⌒ <10m, 140Dkr, >10m, 160Dkr, ⚓ WC ⚐ ⚒ ☎98 94 12 27.

Town ✗ ⌑ 🛒 ✉, ⇌ ✈ (Aalborg) ☎98 17 33 11. Ferries to Bergen, Oslo, Larvik and Kristiansand.

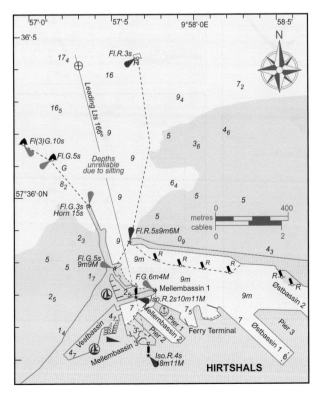

9.14.9 HANSTHOLM

Viborg 57°07'·57N 08°35'·51 ❀❀❀♢♢✿

CHARTS AC 1402 with 1:12,500 hbr inset; DK92, 104; D 82

TIDES +0600 Dover, ML 0.0, Duration 0630
Standard Port ESBJERG (→)

Times				Height (metres)			
High Water		Low Water		MHWS	MHWN	MLWN	MLWS
0300	0700	0100	0800	2·0	1·6	0·6	0·2
1500	1900	1300	2000				
Differences HANSTHOLM							
+0100	+0340	+0340	+0130	−1·7	−1·3	−0·5	−0·2

SHELTER Good. The port is a commercial and fishing hbr which in theory foreign yachts may use only with the HM's prior approval. Yachts berth at the SW corner of Bassin 6 or as directed. S of the FG/FR inner mole heads speed limit 3kn.

NAVIGATION WPT 57°08'·10N 08°34'·94E (SWM buoy, L Fl 10s) 142° about 6ca to the hbr entrance. The approach is clear. Ldg marks 142·6° front, R ▲ on mast; rear, R ▼ on mast, lead between the outer moles.

- Charted depths are unreliable due to silting and particularly after E gales when levels may be lowered by up to 1·3m.
- Check on VHF that the entrances are clear of departing ships which may be hidden behind the moles. In-bound vessels have precedence over out-bound.

LIGHTS AND MARKS As chartlet and 9.14.4. Hanstholm lt ho, a W 8-sided twr with R top, is approx 7ca S of the hbr ent and visible by day from seaward, rising above the flat terrain. The outer and inner mole heads are floodlit.

COMMUNICATIONS Coast Guard 97 92 22 22; Police (Emergency) 112; ⊞ (Thisted) 97 92 44 00, (Skagen) 70 15 03 00; HM 97 96 18 33/96 55 07 10, VHF Ch 16; 12 13, HX; Taxi 97 96 17 11.

FACILITIES ⌒ ⚓ ▮ ⚑ Diver ☎97 96 27 00. **Town** ✗ 🛒 ⇌ (Thisted) ✈ (Thisted) ☎97 96 51 66. Ferries to Bergen, the Faroe Is and Iceland.

HANSTHOLM TO THYBORØN

(AC 1404) Off **Hanstholm** S/SW winds may cause currents up to 4kn in rough weather, but tidal streams are imperceptible for 15-20M offshore. It is best to round Hanstholm outside the SWM buoy. Underwater obstructions and the slim possibility of former WW2 mines may make anchoring dangerous.

Thyborøn is the western entrance to the **Limfjord** (AC 426 to 430) which gives access to the Baltic via Aalborg and Hals/Egense (91 track miles). It is a large area of sheltered water with many harbours, of which **Lemvig** and Thisted are the most W'ly. Some of the marinas are only suitable for drafts <1·5m, but there are plenty of anchorages and it is a justifiably popular cruising

ground. The Danish word 'Bredning', meaning a Broad (as in Norfolk), aptly describes the stretches of water linked by narrow sounds; least depths are 4m.

THYBORØN TO HVIDE SANDE

(AC 1404) 'Dangerous anchoring' areas fringe the coast to 1M offshore. Bovbjerg lt ho, a red tower on a 38m high sand dune, is conspic. **Thorsminde** is a FV hbr between the sea and Nissum Fjord. S of here coastal dunes are high enough to obscure the houses behind them. ▶*Tidal streams (such as they are) divide off Lyngvig lt ho, ie to the S the in-going stream sets S and the out-going stream sets N; the reverse occurs N of Lyngvig.*◀

9.14.10 THYBORØN
Ringkøbing 56°42'·87N 08°12'·81E ✲✲◊◊◊❀❀

THYBORØN *continued*

CHARTS AC 426 (inc 2 hbr plans); DK108S; D 81

TIDES +0530 Dover; ML 0·0; Duration 0620
Standard Port ESBJERG (→)

Times				Heights (metres)			
High Water		Low Water		MHWS	MHWN	MLWN	MLWS
0300	0700	0100	0800	2·0	1·6	0·6	0·2
1500	1900	1300	2000				
Differences THYBORØN							
+0120	+0230	+0410	+0210	−1·6	−1·3	−0·5	−0·2

SHELTER inside is good. Yachts berth in the N'most basin, Nordre Inderhavn, (akaThyborøn Marina) in 3·5m or as directed. Just over 1M south of hbr Thyborøn Lystbådehavn has limited space on two 45m long pontoons in 2m (see chartlet inset).

NAVIGATION Entry can be dangerous in near gale force onshore winds which cause breaking seas and violent eddies. Least depth is usually about 6m on appr and 5m-8m within the bar.
WPT 56°42'·55N 08°08'·70E (SWM buoy L Fl 10s, Racon T) 082°/2·5M to abeam unlit SHM buoy at ent. Here turn stbd SSE, past a SHM buoy Fl (2) G 5s, into the main fairway to the Yderhavn ent. Inside the ent turn 90° stbd to Nordre Inderhavn 4 cables N.

LIGHTS AND MARKS As on chartlet and 9.14.4. The Appr lt, Fl (3) 10s, is an all-round lt ho; in poor vis shown by day. The E mole lt, Oc (2) WRG 12s, has a W sector covering a 148·2° approach from NNW. Yderhavn ent, W sector, Dir Iso WRG 2s.

Outer 082° ldg marks: front, Agger Tange R ▲ on bn; rear, R ▼ on grey frame twr. Langholm 120° ldg marks: front, R ▲ on R hut; rear, R ▼ on grey frame twr. Caution: light sectors and buoys in the approach chans are moved to reflect changes in depths.

COMMUNICATIONS Coast Guard 97 82 13 22; Ⓗ (Thisted) 97 92 44 00; HM 97 83 10 50, VHF Ch 16; 12 13, H24; Taxi 97 83 25 11.

FACILITIES Ⓥ ⌒ or bow/stern on piles and quay<9.4m 100Dkr, 9.4m to <12.4m 150Dkr, >12.4m 170Dkr, inc ⌑ ⓗ WC ⌫ ✂ ☎97 83 11 26, ⌂ ☎97 83 11 25.

Town ✕ ⛽ ⌂ ⊠ ⇌ ✈(Thisted) ☎97 96 51 66.

Limfjord cuts across Jutland via Aalborg to the E ent at Hals in the Kattegat, approx 91M; saving about 90M over the offshore route via Skagen. Thyborøn is the W ent to Limfjord, a notable cruising ground in its own right.

Lemvig (see chartlet opposite) is 13M SSE of Thyborøn. Note: WIP 2014 to enlarge hbr. **Lemvig Marina** is at Vinkel Hage, 56°33'·93N 08°17'·86E, (W bank) 9ca N of Lemvig; www.lemvig-marina.dk

9.14.11 THORSMINDE (TORSMINDE)

Ringkøbing 56°22'·32N 08°06'·82E ❀❀◊◊◊❀

CHARTS AC 1404; DK93; D 82

TIDES +0425 Dover; ML 0·1; Duration 0605
Standard Port ESBJERG (→)

Times				Height (metres)			
High Water		Low Water		MHWS	MHWN	MLWN	MLWS
0300	0700	0100	0800	2·0	1·6	0·6	0·02
1500	1900	1300	2000				
Differences THORSMINDE							
+0045	+0050	+0040	+0010	−1·4	−1·1	−0·5	−0·2

SHELTER Shelter is good inside, but see the notes below. Yachts berth in the Vesthavn between piers 3 and 4. There is no access into Nissum Fjord due to drainage sluices which close off the hbr immediately W of the road bridge.

NAVIGATION WPT 56°22'·34N 08°05'·49E, approx 090°/1M to hbr ent, which is dredged 4m except during onshore gales. Approach directly from seaward and turn stbd into Vesthavn.

Even moderate winds can cause breakers over the bar which normally has less than 2·7m below MSL and the Hbr may have less water than charted due to silting.

WATER LEVEL AND CURRENT

Strong W'lies can raise the water level by up to 3·0m and E'lies lower it by up to 1·8m. Levels are said to be signalled by a complex array of R & W lts, but before approaching, check conditions with the HM and request a berth.

Sluicing causes currents into and out of the hbr; warning signals, as in Hvide Sande, are shown from masts by the sluices.

LIGHTS AND MARKS Lt ho 56°22'·5N 08°07'·2E, F 30m 14M, grey frame tr. Other lts as chartlet and 9.14.4.

TELEPHONE Coast Guard 97 41 22 22; Ⓗ (Holstebro) 97 41 42 00; Police (Emergency) 112; HM 97 49 72 44, VHF Ch 13 16, 0300-1300 1400-2400 LT; Taxi 97 82 14 15.

FACILITIES ⌒ 100Dkr all LOA, ⚓ ⓗ ⌑ ✂. **Town** ✕ ⌂ ⛽ ⇌ (Ramme, Holstebro, Ringkøbing) ✈(Karup) ☎97 10 12 18.

9.14.12 HVIDE SANDE

Ringkøbing 55°59'·89N 08°06'·56E ❄❄🐚🐚🐚🐚🌸🌸

CHARTS AC 1404; DK93, 99; D 82

TIDES +0345 Dover, ML 0·2, Duration 0550. Mean range is about 0·75m.

Standard Port ESBJERG (→)

Times				Height (metres)			
High Water		Low Water		MHWS	MHWN	MLWN	MLWS
0300	0700	0100	0800	2·0	1·6	0·6	0·2
1500	1900	1300	2000				
Differences HVIDE SANDE							
0000	+0010	–0015	–0025	–1·2	–0·9	–0·4	–0·2

SHELTER *Strong W winds can cause breakers over the bar rendering the entrance dangerous.* Good shelter inside, but see notes on Water level and Current. No ⚓ in the hbr channels.

NAVIGATION WPT 55°59'·80N 08°05'·49E, approx 072°/1M Hbr ent. In the approach chan there is usually 2·5m below MSL over two sand ridges, the outer of which is about 3ca offshore. An underwater power cable crosses the ent chan and pipelines cross E of the lock. Hbr speed limit is 3kn.

WATER LEVEL AND CURRENT

- Westerly storms can raise the water level by up to 3·5m, E'lies can lower it up to 2·0m. Levels are not signalled.

- When drainage sluicing from Ringkøbing Fjord takes place, a strong current runs through the outer hbr. Navigation is prohib within areas 250m W of, and 500m E of the sluices.

- Warning signals are hoisted/shown from a mast on the N side of the sluice:

In-going current: ▲ by day; ● over ○ by night.
Out-going current: ▼ by day; ○ over ● by night.
If in doubt, check with the HM by VHF or telephone.

LIGHTS AND MARKS The lt ho, grey frame twr, is on the stbd side of the fairway. Lyngvig lt ho, W round twr, is 3M N of the hbr ent. Other lts as chartlet and 9.14.4.

COMMUNICATIONS Police (Emergency) 112; Customs 72 22 18 18; Ⓗ (Herning) 78 43 00 00; Doctor 97 31 10 23; HM 97 31 16 33 (H24) VHF Ch 16, 12 (0700-1700); Lock 97 31 10 93; Taxi 97 35 33 77.

FACILITIES While the harbour is mainly a fishing port, it also gives access to Ringkøbing Fjord, which is drained by sluices at the E end of the Outer Hbr, and is a popular tourist destination.

Yachts berth on a pontoon in the Sydhavn (first turning to stbd) or as directed by the HM. Or lock through to the Østhavn to enter appr chan (1·8m) and berth in Fjordhavn marina (1·6m). 🛒🗜️📶 🛟 by bowser, contact HM, 📶 🚾 🔧 🛢️ 🗑️ ☎97 31 15 22.

Town ✕ 🏨 🛒 ✉ ⚡ (Ringkøbing) ✈ (Esbjerg).

HVIDE SANDE TO ESBJERG

(AC 1404, 3768, Danish 99) Firing practice areas extend 9M W and 8M NNW from a pos 4M N of Blåvands Huk lt ho; and 7-10M to seaward of Nyminde Gab 55°49'N 08°12'E. Charts warn of the dangers of anchoring and fishing due to wrecks and underwater explosives.

Entrance to **Hvide Sande** can be difficult when sluicing causes strong currents in the entrance. Thence Ringkøbing Fjord, a lagoon 15M long by 6M wide, is entered via a lock. There are several attractive hbrs/marinas (Ringkøbing, Bork, Skaven and Stavning) and ⚓s, but also shallows and conservation areas with prohibited access – see Danish AC 99.

Off Blåvands Huk (AC 3766) Slugen is the buoyed/lit and deep inshore channel through the shallow banks of **Horns Rev**. See Fig 14(1). The general direction of buoyage is NW.

Close SW of Slugen Horns Rev 1 wind farm (80 turbines, all marked by R air obstruction lts) is centred on 55°29'·20N 07°50'·30E; there are Racons (U) at the NE and SW corners.

NW of this wind farm a 12M x 3·5M rectangular area is dangerous for ⚓ing, fishing and seabed ops. It was formerly mined.

Horns Rev 2 wind farm is 11M NW of Horns Rev 1, and has 91 turbines, 14 perimeter turbines Fl (3) Y 10s, and two associated platforms Mo(U)W 15s.

ESBJERG TO RØMØ

(AC 1404, 3768) Grådyb, the ent channel to **Esbjerg**, is well buoyed. See Fig 14(2). ▶*In Esbjerg Roads the stream sets ESE from HW Helgoland –0305 and WSW from +0315. Tidal streams in the Fanø Bugt are shown in tables on AC 3768. In the open sea the flood is SE-going and the ebb NW-going; there is no slack water, just a turning of the stream, usually anti-clockwise. The rate is 1½kn at the change; the ebb is often stronger than the flood.*

Abeam Rømø Havn the ingoing stream begins at HW Helgoland –0215 and runs NNE across the hbr entrance; the outgoing stream begins at +0345 and runs SSW across the entrance; max rate in each direction is about 2kn.◀

From Grådyb SWM buoy/Racon to Lister Tief SWM is 171°/19·5M. Approach Rømø via the buoyed channel.

A firing practice range extends 12M seaward of Rømø.

CROSSING THE NORTH SEA

(AC 2182A, 2182B, 2182C) Important planning considerations to be taken into account include:

- Distances are considerable, eg Harwich to Esbjerg is 330M and Harwich to Skagen is 483M, so experienced crew, used to night passages, and a well found yacht are essential. See 9.0 for distances across the North Sea.

- The many oil/gas rigs must be identified and avoided.

- The extensive Dutch and German TSS south of 54°12'N, and west of Helgoland, must be avoided or carefully negotiated.

From the English S coast it may be prudent to coast-hop via the Netherlands and the Frisian Islands, then make for Helgoland, dependent on weather. A suitable departure point from the Thames estuary is S Galloper SCM, and from Lowestoft, Smiths Knoll SCM. From further N, ie Hartlepool to the Firth of Forth, a track to Esbjerg avoids Dogger Bank and the densest oil/gas rig concentrations. See also Special notes in Areas 15 and 16.

Fig 14(1) Horns Rev, wind farms, and Slugen Channel west of Blåvands Huk

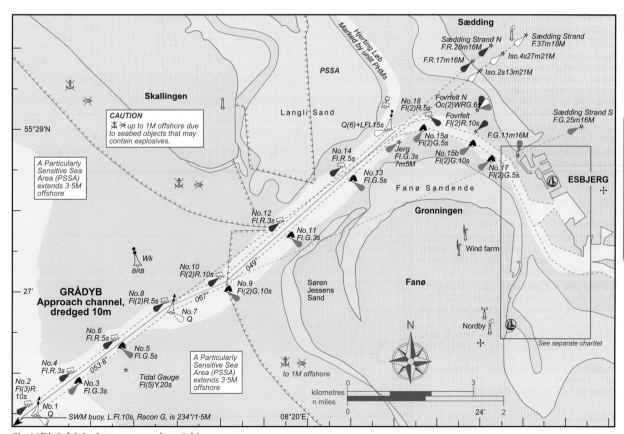

Fig 14(2) Grådyb: Outer approach to Esbjerg

9.14.13 ESBJERG

Ribe **55°28'·29N 08°25'·36E** (Ent to Konsumfiskerihavn Basin II) ✿✿♎♎♍♎♎

CHARTS AC 3766, 420; DK94/5; D 83, 109

TIDES ESBJERG is a Standard Port (➝) +0340 Dover; ML 0·7; Duration 0605; Zone –0100. See differences at Nordby.

SHELTER All tide access. Yachts berth on 2 pontoons in the SE part of FV basin No II as shown. Or at Nordby marina on Fanø.

NAVIGATION Approaching from the N, either pass to seaward of Horns Rev, which extends over 21M W from Blåvands Huk to Horns Rev West WCM buoy, 55°34'·49N 07°26'·08E. See Fig 14(1).

Or take the Slugen chan, 9 to 4M W of Blåvands Huk. This is marked by 2 PHM & 3 SHM lt buoys. The general direction of buoyage through the Slugen is NW. From the E'most PHM buoy, Fl (3) R 10s, to Grådyb SWM is 132°/7·1M.

WPT 55°24'·63N 08°11'·61E, Grådyb SWM buoy L Fl 10s, Racon G. See Fig 14(2). Thence 3 sets of ldg lts (R, G and White) at Sædding Strand lead through Grådyb (ldg lines on the chartlet are colour-coded R, G, B for clarity). At the Y-junction marked by No 16 SCM buoy, turn E/SE into the Hbr proper:

- Outer triple ldg lts (Iso 2s, Iso 4s and F) lead 053·8°/3·7M (black line) to No 7 and No 8 buoys. There jink stbd 13° onto:
- Middle (Sædding S) ldg lts, both FG, 067°/8ca (green line) to Nos 9 & 10 buoys. Thence jink 16° port onto:
- Inner (Sædding N) ldg lts, both FR, lead 049°/2·8M (red line) to the Y-junction at No 16 SCM buoy.

Here follow the buoyed chan to stbd round onto SSE; this turn is covered by 4 directional lights, may be obscured during ongoing works on outer side of N'ly quay. Esbjerg basins lie to port, or deviate stbd at No 21 buoy for Nordby marina.

LIGHTS AND MARKS Blåvands Huk lt ho, W ☐ twr, is 9·6M NNW of Gradyb SWM buoy.

On SW side of the Slugen chan extensive, well lit windfarms have been built in shoal waters; keep well clear. See Chapter 3 and Fig 14(1) – previous page.

Appr daymarks. Outer 053·8° ldg line: Front, R wooden bldg; Middle, R metal twr W bands; Rear, R framework twr.

Hbr daymark: The conspic power stn chimney (250m high) is 7·5 cables E of No 25 SHM lt buoy; see chartlet. Other light details are on the chartlets or in 9.14.4.

COMMUNICATIONS Coast Guard 75 12 82 99; Police (Emergency) 112; Customs 75 12 25 00; ⊞ 75 18 19 00; HM 75 12 41 44, VHF Ch 16: 12 13 14, H24; Taxi 75 12 69 88.

FACILITIES Marina, 100Dkr any LOA, ♪ ⚓ ⚒ ⚓ ⚓ ⚒ ⚒, at Esbjerg Søsport (YC) ☎75 14 47 47; ⚒ A-Z Snedkeriet ☎75 12 65 99; ⚓ Vase As ☎75 12 33 33; Diver, Bjarnes Dykkerservice ☎75 12 04 44.

Town ✕ ☐ ⚒ ⊠ ⇌ ✈ (8km NE).

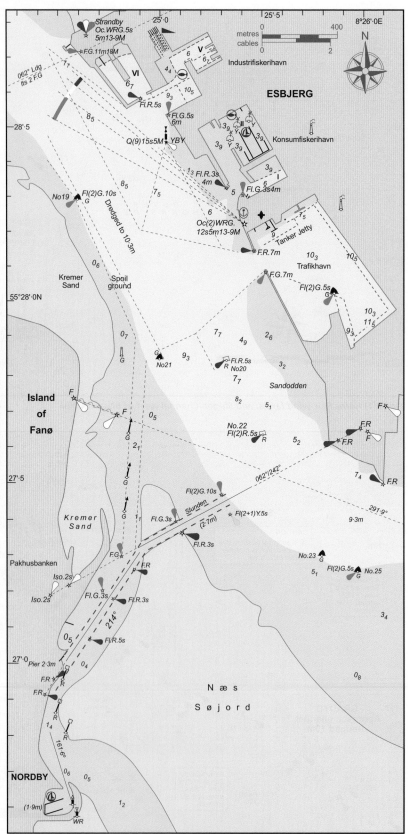

STANDARD TIME UT –01
Subtract 1 hour for UT
For Danish Summer Time add
ONE hour in **non-shaded areas**

ESBJERG LAT 55°28′N LONG 8°27′E
TIMES AND HEIGHTS OF HIGH AND LOW WATERS

Dates in red are **SPRINGS**
Dates in blue are **NEAPS**

YEAR 2016

JANUARY

Time	m		Time	m
1 0038	0.4	**16** 0037	0.2
0658	1.9		0702	2.0
F 1313	0.4	SA 1312	0.2
1929	1.7		1937	1.8
2 0121	0.5	**17** 0127	0.2
0742	1.9		0755	2.0
SA 1359	0.5	SU 1405	0.2
☽ 2018	1.7	☽ 2034	1.7
3 0209	0.5	**18** 0223	0.3
0835	1.9		0857	1.9
SU 1450	0.5	M 1504	0.3
2115	1.7		2139	1.7
4 0304	0.6	**19** 0327	0.4
0936	1.8		1009	1.9
M 1548	0.5	TU 1611	0.4
2220	1.7		2250	1.7
5 0409	0.6	**20** 0439	0.4
1042	1.8		1123	1.8
TU 1654	0.5	W 1721	0.4
2325	1.7		2358	1.8
6 0520	0.6	**21** 0551	0.3
1148	1.8		1231	1.8
W 1757	0.5	TH 1827	0.3
7 0026	1.8	**22** 0100	1.8
0624	0.5		0657	0.2
TH 1248	1.8	F 1333	1.8
1851	0.4		1923	0.3
8 0121	1.9	**23** 0156	1.9
0718	0.4		0754	0.2
F 1343	1.8	SA 1427	1.8
1938	0.4		2013	0.2
9 0209	1.9	**24** 0246	2.0
0804	0.4		0844	0.1
SA 1433	1.9	SU 1515	1.8
2021	0.3	○ 2059	0.2
10 0254	1.9	**25** 0331	2.0
0848	0.3		0929	0.1
SU 1518	1.9	M 1558	1.8
● 2102	0.3		2141	0.2
11 0335	2.0	**26** 0412	2.0
0929	0.2		1011	0.1
M 1600	1.9	TU 1636	1.7
2142	0.2		2219	0.2
12 0414	2.0	**27** 0448	2.0
1010	0.2		1049	0.2
TU 1641	1.9	W 1709	1.7
2224	0.2		2256	0.2
13 0452	2.0	**28** 0520	1.9
1053	0.1		1124	0.2
W 1721	1.8	TH 1740	1.7
2306	0.2		2330	0.2
14 0532	2.0	**29** 0551	1.9
1136	0.1		1159	0.2
TH 1803	1.8	F 1810	1.7
2350	0.2			
15 0615	2.0	**30** 0006	0.3
1222	0.1		0622	1.9
F 1848	1.8	SA 1234	0.3
			1843	1.7
		31 0043	0.3
			0658	1.8
		SU 1312	0.3
			1923	1.7

FEBRUARY

Time	m		Time	m
1 0125	0.3	**16** 0200	0.2
0742	1.8		0839	1.8
M 1355	0.4	TU 1436	0.3
☽ 2012	1.7		2111	1.6
2 0212	0.4	**17** 0303	0.2
0835	1.7		0951	1.7
TU 1445	0.4	W 1544	0.3
2111	1.6		2224	1.6
3 0309	0.4	**18** 0419	0.3
0939	1.7		1107	1.7
W 1545	0.5	TH 1700	0.4
2221	1.6		2336	1.7
4 0418	0.5	**19** 0538	0.2
1054	1.7		1217	1.7
TH 1657	0.5	F 1809	0.3
2335	1.7			
5 0535	0.4	**20** 0041	1.7
1208	1.7		0645	0.2
F 1807	0.4	SA 1318	1.7
			1908	0.2
6 0041	1.8	**21** 0139	1.8
0642	0.4		0741	0.1
SA 1312	1.7	SU 1412	1.7
1906	0.3		1957	0.2
7 0138	1.8	**22** 0229	1.9
0738	0.3		0829	0.0
SU 1407	1.7	M 1459	1.7
1956	0.2	○ 2042	0.1
8 0228	1.9	**23** 0314	1.9
0825	0.1		0912	0.0
M 1457	1.8	TU 1539	1.7
● 2041	0.2		2122	0.1
9 0313	1.9	**24** 0353	1.9
0909	0.1		0951	0.1
TU 1542	1.8	W 1614	1.7
2124	0.1		2159	0.1
10 0356	2.0	**25** 0427	1.9
0952	0.0		1025	0.1
W 1624	1.8	TH 1645	1.7
2206	0.0		2232	0.1
11 0437	2.0	**26** 0456	1.8
1035	0.0		1057	0.1
TH 1705	1.8	F 1712	1.6
2248	0.0		2304	0.1
12 0518	2.0	**27** 0523	1.8
1118	0.0		1127	0.2
F 1745	1.7	SA 1737	1.6
2332	0.0		2336	0.1
13 0600	2.0	**28** 0550	1.8
1203	0.0		1158	0.2
SA 1828	1.7	SU 1805	1.7
14 0018	0.0	**29** 0010	0.1
0647	1.9		0621	1.7
SU 1249	0.1	M 1233	0.2
1915	1.7		1840	1.7
15 0106	0.1			
0739	1.9			
M 1339	0.2			
☽ 2008	1.7			

MARCH

Time	m		Time	m
1 0049	0.2	**16** 0139	0.1
0702	1.7		0821	1.6
TU 1313	0.2	W 1410	0.2
1924	1.6		2045	1.6
2 0134	0.2	**17** 0243	0.2
0751	1.7		0933	1.6
W 1400	0.3	TH 1518	0.3
☽ 2017	1.6		2158	1.6
3 0227	0.3	**18** 0402	0.2
0851	1.6		1049	1.5
TH 1456	0.3	F 1637	0.3
2121	1.6		2312	1.6
4 0331	0.3	**19** 0523	0.2
1006	1.6		1157	1.6
F 1606	0.4	SA 1749	0.3
2240	1.6			
5 0450	0.3	**20** 0018	1.7
1130	1.6		0628	0.1
SA 1724	0.4	SU 1257	1.6
2359	1.6		1848	0.2
6 0608	0.3	**21** 0116	1.8
1241	1.6		0722	0.0
SU 1833	0.3	M 1350	1.7
			1937	0.1
7 0104	1.7	**22** 0207	1.8
0709	0.1		0809	0.0
M 1340	1.7	TU 1436	1.7
1929	0.2		2021	0.0
8 0159	1.8	**23** 0251	1.8
0800	0.0		0850	0.0
TU 1433	1.7	W 1515	1.7
2017	0.1	○ 2100	0.0
9 0248	1.8	**24** 0330	1.8
0847	–0.1		0927	0.0
W 1519	1.8	TH 1550	1.7
● 2102	0.0		2136	0.0
10 0334	1.9	**25** 0403	1.8
0930	–0.2		0958	0.0
TH 1603	1.8	F 1619	1.6
2145	–0.1		2208	0.0
11 0418	1.9	**26** 0432	1.7
1013	–0.2		1027	0.1
F 1644	1.8	SA 1645	1.6
2228	–0.2		2238	0.0
12 0500	1.9	**27** 0457	1.7
1056	–0.2		1055	0.1
SA 1724	1.7	SU 1709	1.6
2312	–0.2		2309	0.0
13 0544	1.9	**28** 0522	1.7
1140	–0.1		1126	0.1
SU 1806	1.7	M 1735	1.6
2357	–0.1		2342	0.0
14 0630	1.8	**29** 0553	1.6
1225	0.0		1200	0.2
M 1851	1.7	TU 1808	1.6
15 0046	0.0	**30** 0021	0.0
0721	1.7		0631	1.6
TU 1315	0.1	W 1241	0.1
☽ 1942	1.6		1849	1.6
		31 0106	0.1
			0719	1.6
		TH 1327	0.2
		☽ 1939	1.6

APRIL

Time	m		Time	m
1 0157	0.1	**16** 0342	0.2
0817	1.5		1024	1.4
F 1421	0.2	SA 1608	0.3
2039	1.6		2242	1.6
2 0259	0.2	**17** 0500	0.1
0929	1.5		1130	1.5
SA 1528	0.3	SU 1721	0.3
2153	1.6		2348	1.6
3 0415	0.2	**18** 0603	0.1
1054	1.5		1229	1.5
SU 1646	0.3	M 1821	0.2
2318	1.6			
4 0534	0.1	**19** 0046	1.7
1210	1.5		0657	0.0
M 1800	0.2	TU 1321	1.6
			1912	0.1
5 0030	1.7	**20** 0138	1.7
0639	0.0		0743	0.0
TU 1312	1.6	W 1407	1.6
1859	0.1		1957	0.0
6 0130	1.7	**21** 0224	1.8
0733	–0.1		0824	–0.1
W 1406	1.7	TH 1448	1.7
1951	0.0		2036	0.0
7 0222	1.8	**22** 0303	1.7
0821	–0.2		0859	0.0
TH 1454	1.7	F 1524	1.7
● 2038	–0.1	○ 2112	0.0
8 0311	1.9	**23** 0338	1.7
0906	–0.2		0930	0.0
F 1539	1.7	SA 1554	1.6
2124	–0.2		2143	0.0
9 0357	1.9	**24** 0408	1.6
0951	–0.2		0959	0.0
SA 1621	1.7	SU 1621	1.6
2208	–0.2		2213	0.0
10 0441	1.8	**25** 0434	1.6
1033	–0.2		1028	0.0
SU 1702	1.7	M 1646	1.6
2253	–0.2		2245	0.0
11 0526	1.8	**26** 0501	1.6
1117	–0.1		1100	0.0
M 1743	1.7	TU 1713	1.6
2339	–0.2		2320	0.0
12 0612	1.7	**27** 0533	1.6
1203	0.0		1136	0.0
TU 1828	1.7	W 1745	1.7
13 0027	–0.1	**28** 0000	0.0
0703	1.6		0611	1.6
W 1251	0.1	TH 1216	0.1
1918	1.7		1824	1.7
14 0121	0.0	**29** 0045	0.0
0803	1.5		0657	1.5
TH 1345	0.2	F 1303	0.1
☽ 2018	1.6		1912	1.6
15 0224	0.1	**30** 0136	0.0
0912	1.5		0754	1.5
F 1450	0.3	SA 1357	0.2
2130	1.6	☽ 2009	1.6

W Denmark

Chart Datum: 0·69 metres below Dansk Normal Null. HAT is 2·2 metres above Chart Datum.

》》 **FREE** monthly updates. Register at 《
www.reedsnauticalalmanac.co.uk

655

STANDARD TIME UT –01
Subtract 1 hour for UT
For Danish Summer Time add
ONE hour in **non-shaded areas**

ESBJERG   LAT 55°28'N   LONG 8°27'E
TIMES AND HEIGHTS OF HIGH AND LOW WATERS

Dates in red are **SPRINGS**
Dates in blue are **NEAPS**

YEAR 2016

MAY

Day	Time m	Time m		Day	Time m	Time m
1 SU	0236 0.1 / 0903 1.5	1500 0.2 / 2119 1.6		**16** M	0425 0.2 / 1052 1.4	1642 0.3 / 2309 1.6
2 M	0347 0.1 / 1023 1.4	1613 0.2 / 2241 1.6		**17** TU	0530 0.1 / 1151 1.5	1745 0.2
3 TU	0502 0.1 / 1138 1.5	1725 0.2 / 2356 1.7		**18** W	0009 1.7 / 0624 0.1	1245 1.6 / 1840 0.1
4 W	0608 0.0 / 1242 1.6	1829 0.1		**19** TH	0103 1.7 / 0712 0.0	1333 1.6 / 1927 0.1
5 TH	0100 1.7 / 0705 -0.1	1338 1.6 / 1924 0.0		**20** F	0151 1.7 / 0754 0.0	1417 1.7 / 2009 0.1
6 F	0157 1.8 / 0756 -0.2	1429 1.7 / ● 2015 -0.1		**21** SA	0233 1.7 / 0830 0.0	1456 1.7 / ○ 2046 0.1
7 SA	0248 1.8 / 0843 -0.2	1515 1.7 / 2103 -0.2		**22** SU	0312 1.6 / 0903 0.0	1530 1.7 / 2120 0.1
8 SU	0336 1.8 / 0928 -0.2	1559 1.7 / 2149 -0.2		**23** M	0345 1.6 / 0933 0.1	1601 1.7 / 2152 0.0
9 M	0423 1.8 / 1012 -0.1	1641 1.7 / 2235 -0.2		**24** TU	0416 1.6 / 1005 0.1	1629 1.7 / 2226 0.0
10 TU	0509 1.7 / 1056 -0.1	1710 1.7 / 2322 -0.2		**25** W	0447 1.6 / 1039 0.1	1658 1.7 / 2303 0.0
11 W	0555 1.6 / 1141 0.0	1808 1.7		**26** TH	0521 1.6 / 1117 0.0	1730 1.7 / 2344 0.0
12 TH	0011 -0.1 / 0645 1.5	1228 0.1 / 1856 1.7		**27** F	0559 1.5 / 1158 0.1	1809 1.7
13 F	0103 0.0 / 0739 1.5	◐ 1319 0.2 / 1951 1.6		**28** SA	0029 0.0 / 0644 1.5	1245 0.1 / 1855 1.7
14 SA	0203 0.1 / 0841 1.4	1418 0.3 / 2056 1.6		**29** SU	0120 0.0 / 0738 1.5	1337 0.1 / ◑ 1950 1.7
15 SU	0312 0.2 / 0948 1.4	1529 0.3 / 2204 1.6		**30** M	0217 0.0 / 0841 1.5	1436 0.2 / 2055 1.7
				31 TU	0321 0.1 / 0953 1.5	1543 0.2 / 2210 1.7

JUNE

Day	Time m	Time m		Day	Time m	Time m
1 W	0430 0.1 / 1106 1.5	1653 0.2 / 2326 1.7		**16** TH	0541 0.2 / 1200 1.6	1800 0.3
2 TH	0537 0.0 / 1212 1.6	1800 0.1		**17** F	0021 1.6 / 0633 0.2	1254 1.6 / 1853 0.2
3 F	0033 1.7 / 0637 0.0	1312 1.6 / 1900 0.0		**18** SA	0113 1.7 / 0719 0.2	1342 1.7 / 1939 0.2
4 SA	0134 1.7 / 0732 -0.1	1405 1.7 / 1954 -0.1		**19** SU	0201 1.7 / 0800 0.1	1426 1.7 / 2021 0.1
5 SU	0229 1.8 / 0821 -0.1	1454 1.7 / ● 2045 -0.1		**20** M	0245 1.7 / 0836 0.1	1506 1.7 / ○ 2057 0.1
6 M	0320 1.7 / 0909 -0.1	1540 1.8 / 2133 -0.2		**21** TU	0324 1.6 / 0910 0.1	1542 1.7 / 2133 0.1
7 TU	0408 1.7 / 0954 0.0	1624 1.7 / 2221 -0.1		**22** W	0400 1.6 / 0945 0.1	1615 1.7 / 2210 0.1
8 W	0454 1.6 / 1037 0.0	1706 1.8 / 2307 -0.1		**23** TH	0436 1.6 / 1022 0.1	1647 1.8 / 2249 0.0
9 TH	0538 1.6 / 1121 0.1	1749 1.8 / 2354 0.0		**24** F	0512 1.6 / 1101 0.1	1722 1.8 / 2330 0.0
10 F	0623 1.5 / 1206 0.1	1833 1.7		**25** SA	0551 1.6 / 1143 0.1	1800 1.8
11 SA	0042 0.1 / 0710 1.5	1253 0.2 / 1921 1.7		**26** SU	0015 0.0 / 0634 1.6	1229 0.1 / 1845 1.8
12 SU	0134 0.1 / 0802 1.4	◐ 1344 0.2 / 2015 1.7		**27** M	0103 0.0 / 0723 1.6	1319 0.1 / ◑ 1936 1.8
13 M	0230 0.2 / 0900 1.4	1442 0.3 / 2117 1.6		**28** TU	0157 0.0 / 0819 1.6	1414 0.1 / 2036 1.8
14 TU	0334 0.2 / 1002 1.5	1548 0.3 / 2221 1.6		**29** W	0255 0.1 / 0924 1.5	1515 0.2 / 2145 1.7
15 W	0440 0.2 / 1103 1.5	1657 0.3 / 2323 1.6		**30** TH	0400 0.1 / 1035 1.5	1624 0.2 / 2300 1.7

JULY

Day	Time m	Time m		Day	Time m	Time m
1 F	0508 0.2 / 1145 1.6	1734 0.2		**16** SA	0542 0.4 / 1208 1.6	1811 0.4
2 SA	0012 1.7 / 0613 0.1	1248 1.7 / 1839 0.1		**17** SU	0033 1.7 / 0639 0.3	1304 1.7 / 1906 0.3
3 SU	0116 1.7 / 0712 0.1	1345 1.7 / 1939 0.0		**18** M	0127 1.7 / 0727 0.3	1354 1.8 / 1953 0.3
4 M	0215 1.8 / 0804 0.1	1438 1.8 / ● 2032 0.0		**19** TU	0218 1.7 / 0809 0.2	1439 1.8 / ○ 2035 0.2
5 TU	0307 1.7 / 0852 0.1	1525 1.8 / 2121 0.1		**20** W	0303 1.7 / 0848 0.2	1520 1.8 / 2115 0.1
6 W	0355 1.7 / 0937 0.1	1609 1.9 / 2207 0.0		**21** TH	0344 1.7 / 0927 0.2	1558 1.9 / 2154 0.1
7 TH	0439 1.7 / 1021 0.1	1651 1.9 / 2251 0.0		**22** F	0423 1.7 / 1006 0.2	1634 1.9 / 2233 0.0
8 F	0519 1.6 / 1102 0.1	1730 1.8 / 2334 0.1		**23** SA	0500 1.7 / 1045 0.1	1711 1.9 / 2315 0.0
9 SA	0558 1.6 / 1143 0.1	1809 1.8		**24** SU	0539 1.7 / 1127 0.1	1751 1.9 / 2358 0.0
10 SU	0017 0.1 / 0637 1.6	1224 0.2 / 1849 1.8		**25** M	0621 1.7 / 1212 0.1	1833 1.9
11 M	0100 0.2 / 0718 1.5	1307 0.2 / 1933 1.8		**26** TU	0044 0.1 / 0706 1.7	1300 0.1 / 1922 1.9
12 TU	0145 0.3 / 0805 1.5	◐ 1354 0.3 / 2022 1.7		**27** W	0134 0.1 / 0757 1.7	1351 0.2 / ◑ 2018 1.9
13 W	0234 0.3 / 0859 1.5	1448 0.4 / 2120 1.7		**28** TH	0229 0.2 / 0857 1.6	1450 0.2 / 2126 1.8
14 TH	0331 0.4 / 1001 1.5	1551 0.4 / 2226 1.6		**29** F	0331 0.3 / 1006 1.6	1558 0.3 / 2242 1.8
15 F	0436 0.4 / 1106 1.6	1703 0.4 / 2331 1.6		**30** SA	0442 0.3 / 1120 1.7	1713 0.3 / 2357 1.8
				31 SU	0552 0.3 / 1228 1.7	1825 0.2

AUGUST

Day	Time m	Time m		Day	Time m	Time m
1 M	0103 1.8 / 0655 0.3	1329 1.8 / 1927 0.2		**16** TU	0053 1.7 / 0651 0.4	1319 1.8 / 1923 0.4
2 TU	0203 1.8 / 0749 0.2	● 1423 1.9 / 2020 0.1		**17** W	0148 1.8 / 0740 0.3	1410 1.9 / 2009 0.3
3 W	0254 1.8 / 0838 0.2	1511 1.9 / 2109 0.1		**18** TH	0238 1.8 / 0824 0.3	1455 2.0 / ○ 2052 0.2
4 TH	0341 1.8 / 0921 0.2	1554 2.0 / 2152 0.1		**19** F	0323 1.8 / 0906 0.2	1537 2.0 / 2133 0.1
5 F	0421 1.8 / 1003 0.2	1633 2.0 / 2233 0.1		**20** SA	0404 1.9 / 0946 0.1	1617 2.1 / 2214 0.1
6 SA	0457 1.7 / 1042 0.2	1709 1.9 / 2311 0.2		**21** SU	0443 1.9 / 1027 0.1	1656 2.1 / 2255 0.0
7 SU	0530 1.7 / 1118 0.2	1742 1.9 / 2347 0.2		**22** M	0522 1.9 / 1109 0.1	1736 2.1 / 2338 0.1
8 M	0602 1.7 / 1154 0.2	1815 1.9		**23** TU	0603 1.9 / 1153 0.1	1820 2.0
9 TU	0023 0.3 / 0635 1.7	1232 0.3 / 1850 1.8		**24** W	0023 0.1 / 0645 1.8	1239 0.1 / 1908 2.0
10 W	0100 0.3 / 0712 1.7	◐ 1312 0.3 / 1931 1.8		**25** TH	0111 0.2 / 0733 1.8	1330 0.2 / ◑ 2003 1.9
11 TH	0140 0.4 / 0756 1.7	1357 0.4 / 2020 1.8		**26** F	0203 0.3 / 0831 1.8	1428 0.3 / 2111 1.8
12 F	0227 0.5 / 0851 1.7	1450 0.4 / 2121 1.7		**27** SA	0305 0.4 / 0942 1.7	1538 0.4 / 2229 1.8
13 SA	0324 0.5 / 0959 1.7	1556 0.5 / 2234 1.7		**28** SU	0418 0.5 / 1059 1.8	1700 0.4 / 2344 1.8
14 SU	0434 0.5 / 1113 1.7	1716 0.5 / 2348 1.7		**29** M	0535 0.5 / 1209 1.8	1814 0.3
15 M	0549 0.5 / 1221 1.7	1827 0.5		**30** TU	0051 1.8 / 0639 0.4	1312 1.9 / 1915 0.2
				31 W	0148 1.9 / 0733 0.3	1406 2.0 / 2006 0.2

Chart Datum: 0·69 metres below Dansk Normal Null. HAT is 2·2 metres above Chart Datum.

⟫ **FREE** monthly updates. Register at ⟨
www.reedsnauticalalmanac.co.uk

STANDARD TIME UT –01
Subtract 1 hour for UT
For Danish Summer Time add
ONE hour in **non-shaded areas**

ESBJERG   LAT 55°28'N   LONG 8°27'E
TIMES AND HEIGHTS OF HIGH AND LOW WATERS

Dates in red are SPRINGS
Dates in blue are NEAPS

YEAR 2016

SEPTEMBER

Time	m		Time	m
1 0239	1.9	**16**	0211	1.9
0821	0.3		0757	0.3
TH 1454	2.1		F 1427	2.1
● 2052	0.1		○ 2027	0.2
2 0322	1.9	**17**	0258	2.0
0904	0.2		0842	0.2
F 1536	2.1		SA 1512	2.1
2133	0.2		2109	0.1
3 0400	1.9	**18**	0341	2.0
0943	0.2		0924	0.2
SA 1614	2.1		SU 1556	2.2
2210	0.2		2151	0.1
4 0433	1.9	**19**	0422	2.0
1019	0.2		1006	0.1
SU 1646	2.0		M 1638	2.2
2245	0.3		2233	0.1
5 0503	1.8	**20**	0502	2.0
1053	0.3		1050	0.1
M 1715	2.0		TU 1720	2.1
2316	0.3		2316	0.1
6 0530	1.8	**21**	0542	2.0
1126	0.3		1134	0.1
TU 1743	1.9		W 1804	2.1
2347	0.4			
7 0557	1.8	**22**	0000	0.2
1200	0.3		0624	1.9
W 1812	1.9		TH 1221	0.2
			1853	2.0
8 0020	0.4	**23**	0048	0.3
0628	1.8		0712	1.9
TH 1236	0.4		F 1312	0.3
1849	1.9		◑ 1949	1.9
9 0057	0.4	**24**	0140	0.5
0708	1.8		0809	1.9
F 1318	0.4		SA 1411	0.4
◐ 1934	1.8		2057	1.8
10 0141	0.5	**25**	0242	0.6
0757	1.8		0920	1.8
SA 1406	0.5		SU 1523	0.5
2030	1.8		2215	1.8
11 0233	0.6	**26**	0357	0.6
0857	1.8		1037	1.9
SU 1506	0.5		M 1646	0.5
2139	1.7		2328	1.8
12 0337	0.6	**27**	0515	0.6
1012	1.8		1148	1.9
M 1623	0.6		TU 1759	0.4
2303	1.7			
13 0457	0.6	**28**	0032	1.9
1132	1.8		0620	0.5
TU 1745	0.5		W 1250	2.0
			1857	0.3
14 0017	1.8	**29**	0127	1.9
0610	0.5		0713	0.4
W 1240	1.9		TH 1345	2.1
1849	0.4		1946	0.2
15 0118	1.9	**30**	0216	2.0
0708	0.4		0800	0.3
TH 1337	2.0		F 1432	2.1
1940	0.3		2030	0.2

OCTOBER

Time	m		Time	m
1 0259	2.0	**16**	0230	2.0
0843	0.3		0815	0.3
SA 1515	2.1		SU 1447	2.2
● 2110	0.2		○ 2044	0.2
2 0336	2.0	**17**	0316	2.1
0921	0.3		0901	0.2
SU 1551	2.1		M 1533	2.2
2145	0.3		2127	0.1
3 0409	2.0	**18**	0359	2.1
0956	0.3		0945	0.1
M 1622	2.0		TU 1618	2.2
2216	0.4		2211	0.2
4 0436	1.9	**19**	0440	2.1
1027	0.4		1030	0.1
TU 1650	2.0		W 1703	2.1
2245	0.4		2254	0.2
5 0501	1.9	**20**	0521	2.1
1058	0.4		1116	0.2
W 1715	1.9		TH 1749	2.1
2315	0.4		2339	0.3
6 0525	1.9	**21**	0605	2.0
1131	0.4		1203	0.2
TH 1742	1.9		F 1839	2.0
2347	0.4			
7 0554	2.0	**22**	0026	0.4
1206	0.4		0653	2.0
F 1817	1.9		SA 1256	0.3
			◑ 1935	1.9
8 0024	0.5	**23**	0118	0.5
0632	2.0		0749	2.0
SA 1248	0.4		SU 1355	0.4
1900	1.9		2041	1.8
9 0106	0.5	**24**	0218	0.6
0718	1.9		0857	1.9
SU 1336	0.5		M 1506	0.5
◐ 1954	1.8		2154	1.8
10 0157	0.6	**25**	0331	0.7
0814	1.9		1012	2.0
M 1433	0.6		TU 1626	0.5
2100	1.8		2303	1.8
11 0258	0.6	**26**	0448	0.6
0921	1.9		1121	2.0
TU 1544	0.6		W 1735	0.4
2221	1.8			
12 0413	0.7	**27**	0005	1.9
1043	1.9		0553	0.5
W 1703	0.5		TH 1222	2.1
2340	1.8		1832	0.4
13 0530	0.6	**28**	0100	2.0
1159	2.0		0648	0.5
TH 1812	0.4		F 1316	2.1
			1921	0.4
14 0045	1.9	**29**	0148	2.0
0633	0.5		0736	0.4
F 1302	2.0		SA 1405	2.1
1909	0.3		2006	0.3
15 0141	2.0	**30**	0232	2.0
0727	0.4		0819	0.3
SA 1357	2.1		SU 1448	2.1
1958	0.2		● 2045	0.3
		31	0310	2.0
			0858	0.3
			M 1526	2.1
			2118	0.4

NOVEMBER

Time	m		Time	m
1 0344	2.0	**16**	0336	2.1
0933	0.4		0927	0.2
TU 1558	2.0		W 1600	2.1
2149	0.4		2150	0.2
2 0412	2.0	**17**	0421	2.1
1004	0.4		1013	0.2
W 1627	2.0		TH 1648	2.1
2218	0.4		2234	0.3
3 0438	2.0	**18**	0503	2.1
1035	0.4		1100	0.2
TH 1652	1.9		F 1735	2.0
2247	0.4		2320	0.3
4 0503	2.0	**19**	0548	2.1
1108	0.4		1149	0.2
F 1721	1.9		SA 1824	1.9
2320	0.4			
5 0531	2.0	**20**	0006	0.4
1144	0.4		0635	2.0
SA 1754	1.9		SU 1240	0.3
2358	0.5		1917	1.8
6 0607	2.0	**21**	0057	0.5
1226	0.4		0729	2.0
SU 1836	1.9		M 1337	0.4
			◑ 2017	1.8
7 0042	0.5	**22**	0153	0.6
0651	2.0		0830	2.0
M 1314	0.4		TU 1442	0.5
◑ 1928	1.9		2122	1.8
8 0131	0.5	**23**	0258	0.6
0743	2.0		0938	2.0
TU 1409	0.5		W 1554	0.5
2030	1.8		2227	1.8
9 0230	0.6	**24**	0410	0.6
0846	2.0		1045	2.0
W 1514	0.5		TH 1702	0.5
2145	1.8		2328	1.8
10 0338	0.6	**25**	0518	0.6
1001	2.0		1146	2.0
TH 1627	0.5		F 1800	0.4
2302	1.8			
11 0451	0.6	**26**	0024	1.9
1119	2.0		0617	0.5
F 1736	0.4		SA 1242	2.0
			1851	0.4
12 0011	1.9	**27**	0115	2.0
0558	0.5		0709	0.4
SA 1228	2.1		SU 1333	2.1
1836	0.3		1937	0.4
13 0110	2.0	**28**	0200	2.0
0657	0.4		0754	0.4
SU 1327	2.1		M 1419	2.0
1930	0.3		2017	0.4
14 0203	2.0	**29**	0242	2.0
0750	0.3		0835	0.4
M 1422	2.1		TU 1500	2.0
○ 2018	0.2		● 2052	0.4
15 0251	2.1	**30**	0319	2.0
0839	0.2		0911	0.4
TU 1512	2.1		W 1535	2.0
2105	0.2		2124	0.4

DECEMBER

Time	m		Time	m
1 0351	2.0	**16**	0404	2.1
0943	0.4		0959	0.2
TH 1607	1.9		F 1633	2.0
2154	0.4		2217	0.3
2 0420	2.0	**17**	0448	2.1
1015	0.4		1047	0.2
F 1636	1.9		SA 1720	1.9
2226	0.4		2302	0.3
3 0447	2.0	**18**	0532	2.1
1049	0.4		1134	0.2
SA 1706	1.9		SU 1806	1.8
2300	0.4		2347	0.4
4 0516	2.0	**19**	0617	2.0
1127	0.4		1223	0.3
SU 1741	1.9		M 1853	1.8
2339	0.4			
5 0551	2.0	**20**	0034	0.4
1209	0.3		0704	2.0
M 1821	1.9		TU 1313	0.3
			1944	1.7
6 0023	0.4	**21**	0124	0.5
0633	2.1		0757	2.0
TU 1256	0.3		W 1408	0.4
1910	1.8		◑ 2039	1.7
7 0112	0.4	**22**	0219	0.5
0722	2.0		0856	1.9
W 1348	0.3		TH 1509	0.5
◑ 2006	1.8		2139	1.7
8 0206	0.5	**23**	0323	0.6
0820	2.0		1000	1.9
TH 1448	0.4		F 1615	0.5
2112	1.8		2241	1.8
9 0308	0.5	**24**	0432	0.6
0928	2.0		1103	1.9
F 1553	0.5		SA 1719	0.5
2225	1.8		2340	1.8
10 0415	0.5	**25**	0539	0.5
1045	2.0		1202	1.9
SA 1701	0.4		SU 1815	0.4
2336	1.8			
11 0525	0.5	**26**	0036	1.9
1157	2.0		0636	0.5
SU 1806	0.3		M 1257	1.9
			1905	0.4
12 0039	1.9	**27**	0126	1.9
0629	0.4		0727	0.4
M 1303	2.0		TU 1347	1.9
1903	0.4		1948	0.4
13 0137	2.0	**28**	0212	1.9
0727	0.3		0811	0.4
TU 1401	2.0		W 1432	1.9
1956	0.3		2027	0.4
14 0230	2.0	**29**	0254	2.0
0821	0.2		0849	0.4
W 1455	2.0		TH 1512	1.9
○ 2045	0.3		● 2100	0.4
15 0318	2.0	**30**	0330	2.0
0911	0.2		0924	0.4
TH 1546	2.0		F 1549	1.8
2132	0.2		2133	0.4
		31	0403	2.0
			0958	0.3
			SA 1623	1.8
			2208	0.3

Chart Datum: 0·69 metres below Dansk Normal Null. HAT is 2·2 metres above Chart Datum.

W Denmark

》》 **FREE** monthly updates. Register at 《
www.reedsnauticalalmanac.co.uk 《

657

ESBJERG
MEAN SPRING AND NEAP CURVES

MEAN RANGES	
Springs	1.8m
Neaps	1.0m

9.14.14 NORDBY (FANØ ISLAND)

Ribe 55°27'·37N 08°25'·09E ✿✿✿⬧⬧✿✿

CHARTS AC 3766, 420

TIDES +0340 Dover; ML 0·7; Duration 0605

Standard Port ESBJERG (→)

Times				Height (metres)			
High Water		Low Water		MHWS	MHWN	MLWN	MLWS
0300	0700	0100	0800	2·0	1·6	0·6	0·2
1500	1900	1300	2000				
Differences BLÅVANDS HUK (55°33'·0N 08°05'·0E)							
−0120	−0110	−0050	−0100	−0·2	−0·2	−0·3	−0·2
GRÅDYB BAR (55°26'·0N 08°15'·0E)							
−0130	−0115	ND	ND	−0·5	−0·4	−0·3	−0·2

SHELTER Good at Nordby Marina in 1·8m; access at all tides.

NAVIGATION As for Esbjerg until nearing No 21 buoy then 154° for 4 cables until ldg lts 242° Iso 2s are abeam to stbd, take Slunden chan (dredged 2·0m) for 3ca thence turn to 214° for FR ldg lts at Nordby. The final appr is on 161·6° transit of two WR bns in the marina.

LIGHTS AND MARKS As Esbjerg and see 9.14.4 for lt details.

COMMUNICATIONS CG 75 16 36 66; Police (Emergency) 112; Customs (Esbjerg) 75 12 25 00; Ⓗ (Esbjerg) 75 18 19 00; HM 75 16 31 00.

FACILITIES Pontoons with ⚓ ⛴ ₩ℂ at SC or tie up on jetty wall 100m S of ferry terminal.

Town: limited ⛏ at Nordby, ✕ ⌂. Full services at Esbjerg via regular ferries.

9.14.15 HAVNEBY (RØMØ ISLAND)

Sønderjylland 55°05'·21N 08°34'·31E ✿✿✿⬧⬧✿✿

CHARTS AC 3766, 3767; DK 94; BSH 3013/2 & 3

TIDES +0315 Dover; ML 0·9; Duration 0620

Standard Port ESBJERG (→)

Times				Height (metres)			
High Water		Low Water		MHWS	MHWN	MLWN	MLWS
0300	0700	0100	0800	2·0	1·6	0·6	0·2
1500	1900	1300	2000				
Difference RØMØ HAVN							
−0040	−0005	0000	−0020	−0·1	0·0	−0·3	−0·3
HOJER (54°58'·0N 08°40'·0E)							
−0020	+0015	ND	ND	+0·4	+0·5	−0·2	−0·2

SHELTER Good, access at all tides. Havneby hbr is on the SE tip of Rømø. ⚓ outside the hbr, close N of the entrance.

NAVIGATION WPT 55°05'·33N 08°16'·83E, Lister Tief SWM buoy, 107°/7M via buoyed channel to G1 Pillar buoy Fl Y 4s. Here enter Rømø Dyb (deep water) marked by 3 PHM lt dolphins and by No 1 SHM lt buoy and 5 unlit SHM buoys to the hbr ent.

LIGHTS AND MARKS Rømø has no major lights. On the N end of Sylt: List West (Ellenbogen) Oc WRG 6s, provides a narrow W sector to mark the outer part of Rømø Dyb. The inner part is marked by a narrow W sector from List Land lt Oc WRG 3s, W mast, R band. List Ost, Iso WRG 6s, W twr, R band, R lantern is the 3rd major light. See Area 15 for lt sectors. Hbr lts as on chartlet.

COMMUNICATIONS Coast Guard 74 75 22 22; Police (Emergency) 112; Ⓗ; (Esbjerg) 75 18 19 00; HM 74 75 55 92, VHF Ch 10, 12, 13, 16 HX; Taxi 74 75 11 00, 74 75 53 94.

RØMØ HAVN

FACILITIES Essentially an isolated fishing and ferry port. Yachts berth at 3 pontoons in about 4m on S side of inner hbr. ⚓ in FV hbr, ⚓ ⚒ ✕ ⛽ ⛴ ₩ℂ at SC on N pier.

Havneby town ⛏. Ferry, every 2 hrs by day, 45 mins to List. Rail at Ribe (mainland, 32km by road). ✈ (Esbjerg).

Germany

Sylt to Emden

9.15.1	Contents	659
9.15.2	Area map, Distance table and Inland waterways	660
9.15.3	Tidal stream charts	662
9.15.4	Lights, buoys and waypoints	664
9.15.5	Passage information	666

North Frisian Islands – 666 • River Eider and Schleswig-Holstein – 666 •
Helgoland to the River Elbe – 666 • From the German Bight to the UK – 670 •
River Weser – 683 • River Jade – 684 • The East Frisians, River Jade to River Ems – 684 •
East Frisian Islands Passage Information – 689 • River Ems – 694

9.15.6	Special notes for Germany	666
9.15.7	North Frisian Islands	667

Sylt (List & Hörnum) • Föhr (Wyk) • Amrum-Hafen • Pellworm • Husum • Tönning

9.15.8	Büsum	669
9.15.9	Helgoland, tide tables/curves	669
9.15.10	River Elbe	674
9.15.11	Cuxhaven (R Elbe), tide tables/curves	676

Otterndorf

9.15.12	Brunsbüttel (R Elbe)	680
9.15.13	Nord-Ostsee Kanal (Kiel Canal)	680
9.15.14	Hamburg (R Elbe)	681
9.15.15	Bremerhaven (R Weser)	682
9.15.16	Hooksiel (R Jade)	683
9.15.17	Wilhelmshaven (R Jade), tide tables/curves	685
9.15.18	East Frisian Islands	689
9.15.19	Wangerooge	690

Harlesiel

9.15.20	Spiekeroog	691

Neuharlingersiel • Bensersiel

9.15.21	Accumersiel	691
9.15.22	Langeoog	692

Baltrum • Nessmersiel

9.15.23	Norderney	692

Norddeich • Juist • Greetsiel

9.15.24	Borkum	693
9.15.25	Emden (R Ems)	694

See 9.14.2

DISTANCE TABLE
Approximate distances in nautical miles are by the most direct route while avoiding dangers and allowing for TSS. Places in *italics* are in adjoining areas; places in **bold** are in 9.0.7. Distances across the North Sea.

- Harbour with full almanac entry
- Minor harbour or anchorage
☆ Major light
RG Emergency RDF station
⛴ Ferry
✈ Airport

DENMARK

GERMANY

Nord-Ostsee Kanal

Rendsburg • (9.15.13)

Husum •

Nordstrand •

Dagebüll •

Wyk •

Föhr

Pellworm •

Eiderdamm

Tönning •

River Eider

Linnenplatte •

Rømø

Kampen

List ☆

Hörnum •

SYLT 9.15.7

Amrun

Wittdün

North Frisian Islands

Westerheversand

River Hever

HELGOLAND 9.15.9

Amrumbank West Wind Farm

Nordsee Ost Wind Farm

Meerwind Süd Ost Wind Farm

Global Tech 1 Wind Farm

9.15.2

Germany
Danish Border to Emden

Magnetic Variation 1·3°E

55°N

30'

30'

7° E

30'

8° E

30'

9° E

1	**Den Helder**	378	322	187	213	192	198	180	229	200	245	192	175	180	159	150	153	148	138	130	132	115	125	95	115	**1**
2	**Delfzijl**	327	222	155	140	119	127	107	159	130	173	120	105	100	89	83	81	65	54	56	58	41	10	22	**2**
3	Borkum	305	200	133	128	97	105	94	104	75	163	110	95	88	80	71	67	55	45	46	48	31	32	**3**
4	Emden	337	232	165	142	129	137	111	174	145	190	137	120	115	106	97	85	80	71	63	65	47	**4**
5	Norderney	291	186	123	110	77	85	70	81	52	137	84	69	62	53	44	44	39	29	23	18	20	**5**
6	Accumersiel	310	206	121	99	74	79	63	116	85	132	79	62	49	44	36	36	37	23	15	2	**6**
7	Langeoog	308	204	119	98	72	77	61	114	85	130	77	60	47	43	43	34	23	21	13	**7**
8	Spiekeroog	301	188	119	98	70	61	53	101	72	119	64	52	47	37	29	32	9	9	**8**
9	Wangerooge	283	178	109	88	60	52	47	61	32	108	55	42	38	27	19	24	9	**9**
10	**Helgoland**	259	154	83	64	38	47	34	88	59	104	51	38	44	43	35	**10**
11	Hooksiel	405	301	116	96	69	73	46	101	72	117	64	47	36	9	**11**
12	**Wilhelmshaven**	414	310	125	104	78	82	54	110	81	123	70	56	45	**12**
13	**Bremerhaven**	306	201	127	107	80	82	55	81	52	131	78	58	**13**
14	Cuxhaven	304	200	110	82	63	66	25	54	25	70	17	**14**
15	**Brunsbüttel**	301	196	126	97	75	76	35	37	8	53	**15**
16	Kiel/Holtenau	231	249	179	153	128	129	85	90	61	**16**
17	Glückstadt	309	204	134	111	83	84	47	29	**17**
18	Hamburg	338	233	163	138	112	113	73	**18**
19	Busum	286	180	97	72	48	43	**19**
20	Husum	268	163	95	70	48	**20**
21	Hörnum Lt (Sylt)	225	120	47	26	**21**
22	List	220	110	36	**22**
23	*Esbjerg*	195	90	**23**
24	*Thyborøn*	104	**24**
25	*Skagen*	**25**

660

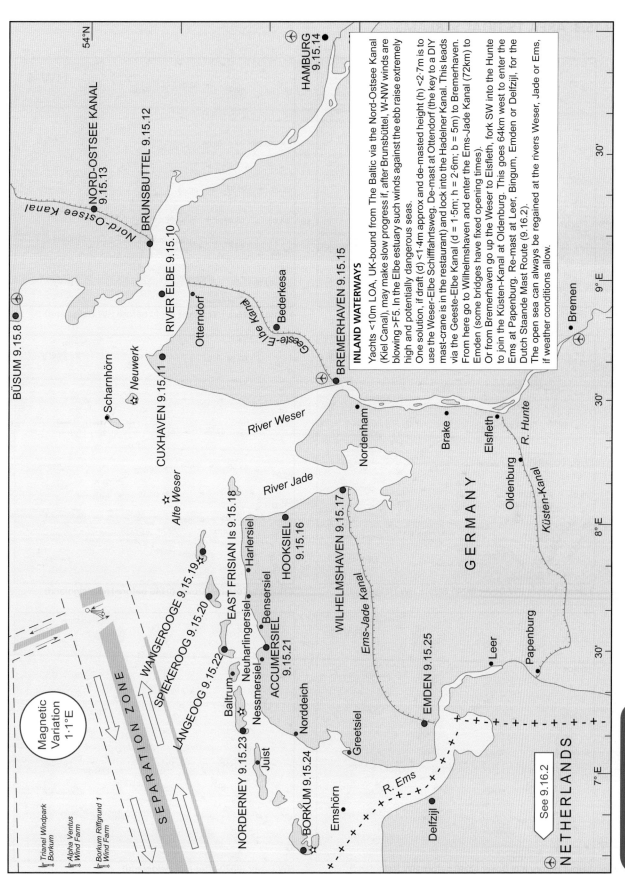

INLAND WATERWAYS

Yachts <10m LOA, UK-bound from The Baltic via the Nord-Ostsee Kanal (Kiel Canal), may make slow progress if, after Brunsbüttel, W-NW winds are blowing >F5. In the Elbe estuary such winds against the ebb raise extremely high and potentially dangerous seas.

One solution, if draft (d) <1·4m approx and de-masted height (h) <2·7m is to use the Weser-Elbe Schifffahrtsweg. De-mast at Ottendorf (the key to a DIY mast-crane is in the restaurant) and lock into the Hadelner Kanal. This leads via the Geeste-Elbe Kanal (d = 1·5m; h = 2·6m; b = 5m) to Bremerhaven. From here go to Wilhelmshaven and enter the Ems-Jade Kanal (72km) to Emden (some bridges have fixed opening times).

Or from Bremerhaven go up the Weser to Elsfleth, fork SW into the Hunte to join the Küsten-Kanal at Oldenburg. This goes 64km west to enter the Ems at Papenburg. Re-mast at Leer, Bingum, Emden or Delfzijl, for the Dutch Staande Mast Route (9.16.2).

The open sea can always be regained at the rivers Weser, Jade or Ems, if weather conditions allow.

Magnetic Variation 1·1°E

Trianel Windpark Borkum

Alpha Ventus Wind Farm

Borkum Riffgrund 1 Wind Farm

SEPARATION ZONE

HAMBURG 9.15.14

NORD-OSTSEE KANAL 9.15.13

Nord-Ostsee Kanal

BRUNSBÜTTEL 9.15.12

RIVER ELBE 9.15.10

Otterndorf

Geeste-Elbe Kanal

Bederkesa

BREMERHAVEN 9.15.15

BÜSUM 9.15.8

Scharnhörn

Neuwerk

CUXHAVEN 9.15.11

River Weser

Nordenham

River Jade

Brake

Elsfleth

R. Hunte

GERMANY

Oldenburg

Küsten-Kanal

Bremen

Alte Weser

EAST FRISIAN Is 9.15.18

Harlersiel

Bensersiel

HOOKSIEL 9.15.16

WANGEROOGE 9.15.19

SPIEKEROOG 9.15.20

Neuharlingersiel

ACCUMERSIEL 9.15.21

LANGEOOG 9.15.22

Baltrum

Nessmersiel

WILHELMSHAVEN 9.15.17

Ems-Jade Kanal

Norddeich

NORDERNEY 9.15.23

Juist

BORKUM 9.15.24

Emshörn

Greetsiel

EMDEN 9.15.25

Leer

Papenburg

R. Ems

Delfzijl

NETHERLANDS

See 9.16.2

54°N

9°E

30'

8°E

30'

7°E

Germany

661

9.15.3 AREA 15 TIDAL STREAMS

5 Hours before HW Dover (0440 before HW Helgoland)

4 Hours before HW Dover (0340 before HW Helgoland)

3 Hours before HW Dover (0240 before HW Helgoland)

2 Hours before HW Dover (0140 before HW Helgoland)

1 Hour before HW Dover (0040 before HW Helgoland)

HW Dover (0020 after HW Helgoland)

South-westward 9.16.3

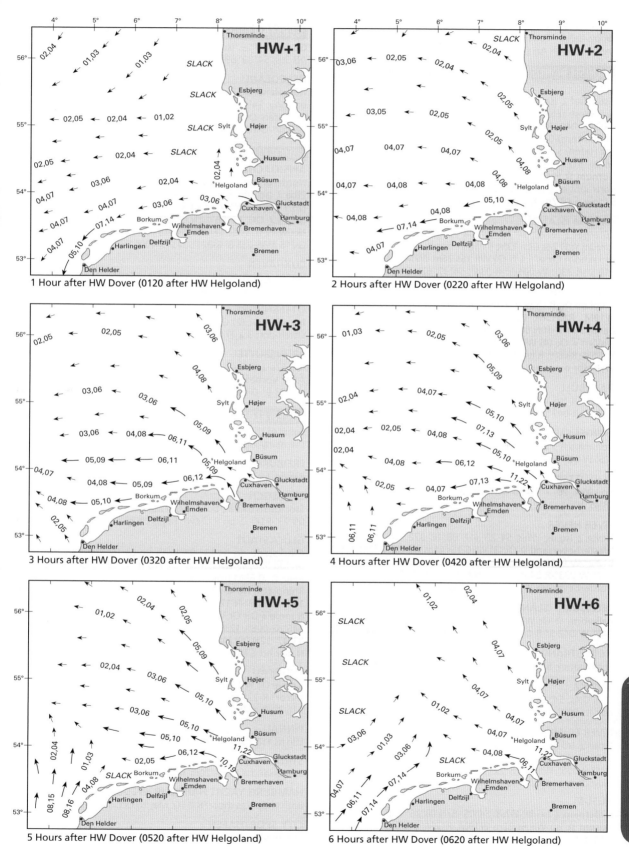

1 Hour after HW Dover (0120 after HW Helgoland)

2 Hours after HW Dover (0220 after HW Helgoland)

3 Hours after HW Dover (0320 after HW Helgoland)

4 Hours after HW Dover (0420 after HW Helgoland)

5 Hours after HW Dover (0520 after HW Helgoland)

6 Hours after HW Dover (0620 after HW Helgoland)

9.15.4 LIGHTS, BUOYS AND WAYPOINTS

Bold print = light with a nominal range of 15M or more. CAPITALS = place or feature. *CAPITAL ITALICS* = light-vessel, light float or Lanby. *Italics* = Fog signal. ***Bold italics*** = Racon. Many marks/buoys are fitted with AIS (MMSI No); see relevant charts.

DANISH BORDER TO BÜSUM

SYLT

Lister Tief ⚓ Iso 8s; *Whis;* 55°05'·33N 08°16'·79E.
List West (Ellenbogen) ⚓ Oc WRG 6s 19m W14M, R11M, G10M; 040°-R-133°-W-227°-R-266·4°-W-268°-G-285°-W-310°-W(unintens)-040°; W twr, R lantern; 55°03'·15N 08°24'·00E.
List Hafen, S mole ⚓ FR 5m 4M; 218°-353°; R mast; 55°01'·01N 08°26'·51E.
Hörnum ☆ Fl (2) 9s 48m **20M**; 54°45'·23N 08°17'·47E.
Vortrapptief ⚓ Iso 4s; 54°34'·80N 08°14'·36E, toward Hörnum.
S mole ⚓ FR 7m 4M; 54°45'·57N 08°17'·97E.

AMRUM ISLAND

Rütergat ⚓ Iso 8s; 54°30'·08N 08°12'·32E.
Amrum ☆ Fl 7·5s 63m **23M**; R twr, W bands; 54°37'·84N 08°21'·23E.
Wriakhorn Cross ⚓ L Fl (2) WR 15s 26m W9M, R7M; 297·5°-W-319·5°-R-330°-W-005·5°-R-034°; 54°37'·62N 08°21'·22E.

FÖHR ISLAND

Nieblum Dir lt showing over Rütergat. ☆ Oc (2) WRG 10s 11m **W19M**, **R/G15M**; 028°-G-031°-W-032·5°-R-035·5°; R twr, W band; 54°41'·10N 08°29'·20E.
Ohlörn ⚓ Oc (4) WR 15s 10m 13/10M, 208°-W-237·5°- R-298°-W-353°- R-080°; R twr, Gy lantern; 54°40'·85N 08°34'·00E (SE Föhr).
Wyk Hbr outer ent, FR 54°41'·55N 08°34'·69E; and FG.

DAGEBÜLL

Dagebüll Iso WRG 8s 23m **W18M**, **R/G15M**; 042°-G-043°-W-044·5°- R-047°; 54°43'·82N 08°41'·43E. FW lts on ent moles.

LANGENESS ISLAND

Nordmarsch ⚓ L Fl (3) WR 20s 13m W14M, R11M; 268°-W-279°-R-306°-W-045°-R-070°-W-218°; dark brown twr; 54°37'·58N 08°31'·85E.

SCHLÜTTSIEL

No. 2/SA 26 ⚓ 54°36'·84N 08°38'·95E. Hbr ent ▲ 54°40'·89N 08°45'·10E.

RIVER HEVER

Hever ⚓ Iso 4s; *Whis;* 54°20'·41N 08°18'·82E.
Westerheversand ☆ Oc (3) WRG 15s 41m **W21M, R17M, G16M**; 012·2°-W-069°-G-079·5°-W-080·5° (ldg sector for Hever)-R-107°-W-233°-R-248°; R twr, W bands; 54°22'·37N 08°38'·36E.
Norderhever No. 1 ⚓ Fl (2+1) R 15s; 54°22'·46N 08°30'·83E.

PELLWORM ISLAND

Pellworm ☆ Oc WRG 5s 38m **20M, R 16M, G 15M**; 037·5°-G-040°-W-042·5°-R-045°; R twr, W band; 54°29'·78N 08°39'·98E.

NORDSTRAND

Suderhafen, S mole ⚓ 54°28'·07N 08°55'·62E.

HUSUM

⚓ Fl G 4s, 54°28'·80N 08°58'·60E, start of access chan.
Ldg lts 090°, both Iso G 8s 7/9m 3M.

RIVER EIDER

Eider ⚓ Iso 4s; 54°15'·70N 08° 29'·20E.
St Peter ☆ L Fl (2) WR 15s 23m **W15M**, R12M; 271°-R-280·5°-W-035°-R-055°-W-068°-R-091°-W-120°; 54°17'·24N 08°39'·10E.
Eiderdamm lock, N mole, W end ⚓ Oc (2) R 12s 8m 5M; W twr.

BÜSUM

Süderpiep ⚓ Iso 8s; *Whis;* 54°05'·82N 08°25'·70E.
Büsum ☆ Iso WR 6s 22m **W19M**, R12M; 248°-W-317°-R-024°-W-148°; 54°07'·60N 08°51'·48E.

E mole ⚓ Oc (3) G 12s 10m 4M; vis 260°-168°; G twr; 54°07'·19N 08°51'·65E.

GERMAN BIGHT TO RIVER ELBE

GB Light V 🚢 Iso 8s 12m **17M**; R hull marked G-B; *Horn Mo (R) 30s;* 54°09'·89N 06°20'·84E.

HELGOLAND

Helgoland ☆ Fl 5s 82m **28M**; brown □ twr, B lantern, W balcony; 54°10'·91N 07°52'·93E.
Vorhafen. Ostmole, S elbow ⚓ Oc WG 6s 5m W6M, G3M; 203°-W-250°-G-109°; G post; fog det lt; 54°10'·31N 07°53'·94E.

RIVER ELBE (LOWER)

APPROACHES

Nordergründe N ⚓ VQ; 53°57'·06N 08°00'·12E.
Elbe ⚓ Iso 10s; ***Racon T, 8M;*** 53°59'·95N 08°06'·49E.
No.1 ⚓ QG; 53°59'·21N 08°13'·20E.
No.25 ⚓ QG; 53°56'·62N 08°38'·25E.
Neuwerk ☆, S side, L Fl (3) WRG 20s 38m **W16M**, R12M, G11M; 165·3°-G-215·3°-W-238·8°-R-321°; 343°-R-100°; 53°54'·92N 08°29'·73E.

CUXHAVEN

No. 31 ⚓ Fl G 4s; 53°53'·93N 08°41'·21E.
Yacht hbr ent, F WR and F WG lts; 53°52'·43N 08°42'·49E.

OTTERNDORF

No. 43 ⚓ Oc (2) G 9s; 53°50'·23N 08°52'·24E.
Otterndorf 3 ▲ Fl (3) G 12s; 53°50'·22N 08°53'·94E.

BRUNSBÜTTEL

No. 57a ⚓ Fl G 4s; 53°52'·62N 09°07'·93E, 030°/ 8 cables to lock.
Alter Vorhafen ent, mole 1 ⚓ F WG 14m W10M, G6M; 266·3°-W-273·9°-G-088·8°, floodlit; 53°53'·27N 09°08'·59E.

RIVER ELBE (BRUNSBÜTTEL TO HAMBURG)

Some of the more accessible yacht hbrs are listed below, but not the many ldg lts and dir lts which define the main river fairway.

FREIBURG

Reede 1 ⚓ Oc (2) Y 9s; 53°50'·41N 09°19'·45E, off ent.

STÖRLOCH

Stör ldg lts 093·8°, both Fl 3s 7/12m 6M; synch. Front, 53°49'·29N 09°23'·91E; △ on R ○ twr. Rear, 200m east; ▽ on white mast.

GLÜCKSTADT

Glückstadt ldg lts 131·8°, both Iso 8s 15/30m **19/21M**; intens on ldg line; W twrs, R bands. **Front** ☆, 53°48'·31N 09°24'·24E.
Rear ☆, 0·68M from front.
Rhinplatte Nord ⚓ Oc WRG 6s 11m W6M, R4M, G3M; 122°-G-144°-W-150°-R-177°- W-122°; 53°48'·09N 09°23'·36E.
N mole ⚓ Dir Oc WRG 6s 9m W9M R7M, G6M; 330°-R-343°-W-346°-G-008°; 123°-G-145°-W-150°-R-170°; W twr with gallery; 53°47'·12N 09°24'·53E.
N pier hd ⚓ FR 5m 5M; 53°47'·10N 09°24'·50E. (S mole hd, FG).

KRUCKAU

S mole Oc WRG 6s 8m, W6M, R4M, G3M; 116·3°-W-120·7° (ldg sector)-R-225°-G-315°-R-331·9°-W-335·4° (ldg sector)-335·4°-G-116·3°; B dolphin; 53°42'·85N 09°30'·72E.

PINNAU

Ldg lts 112·7°, both Iso 4s8/13m 6M. Front, 53°40'·08N 09°34'·02E.

STADE

Stadersand ⚓ Iso 8s 20m 14M; 53°37'·69N 09°31'·64E (at ent).

HAMBURG YACHT HARBOUR, WEDEL

Both entrances show FR & FG, May to Oct; 53°34'·25N 09°40'·77E.

CITY SPORTHAFEN

Brandenburger Hafen Ent ⚓ Iso Or 2s; 53°32'·52N 09°58'·81E.

Plot waypoints on chart before use

WESER ESTUARY

ALTE WESER
Schlüsseltonne ℓ Iso 8s; 53°56'·25N 07°54'·76E.
Alte Weser ☆ F WRG 33m **W23M, R19M, G18M**; 288°-W-352°-R-003°-W-017° (ldg sector for Alte Weser)- G-045°-W-074°-G-118°- W-123° (ldg sector for Alte Weser)- R-140°-G-175°-W-183°-R-196°-W-238°; R ○ twr, 2 W bands, B base; Fog det lt; *Horn Mo (AL) 60s*; 53°51'·79N 08°07'·65E.
16/A15 ⚓ Fl (2+1) R 15s; 53°49'·66N 08°06'·44E (junction with Neuwe Weser).

NEUE WESER
Tegeler Plate, N end ☆ Oc (3) WRG 12s 21m **W21M, R17M, G16M**; 329°-W-340°-R-014°-W-100°- G-116°- W-119° (ldg sector for Neue Weser)-R-123°-G-144°-W-147° (ldg sector for Alte Weser)-R-264°; R ○ twr, gallery, W lantern, R roof; Fog det lt; 53°47'·87N 08°11'·45E.

BREMERHAVEN
No. 61 ℓ QG; 53°32'·26N 08°33'·93E (Km 66·0).
Vorhafen N pier ⚡ FR 15m 5M; 245°-166°; F in fog; 53°32'·15N 08°34'·50E.

RIVER JADE

1b/Jade1ℓ Oc G 4s; 53°52'·40N 07°44'·02E.
Mellumplate ☆ FW 28m **24M**; 116·1°-116·4° (ldg sector for outer part of Wangerooger Fahrwasser); R ☐ twr, W band; 53°46'·28N 08°05'·51E.
No.19 ℓ QG; 53°47'·10N 08°01'·83E.
No. 31/P-Reede/W Siel 1 ℓ Oc (3) G 12s; 53°41'·59N 08°04'·51E.

HOOKSIEL
No. 37/Hooksiel 1 ℓ IQ G 13s; 53°39'·37N 08°06'·58E.
Vorhafen ent ⚡ L Fl R 6s 9m 3M; 53°38'·63N 08°05'·25E.

WILHELMSHAVEN
Fluthafen N mole ⚡ F WG 9m,W6M, G3M; 216°-W-280°- G-010°-W-020°-G-130°; G twr; 53°30'·86N 08°09'·32E.

EAST FRISIAN ISLANDS

NORTH EDGE OF INSHORE TRAFFIC ZONE
TG9/Weser 2 ℓ Fl (2+1) G 15s; 53°55'·0N 07°44'·60E.
TG13 ℓ Oc (3) G 12s; 53°50'·85N 07°15'·43E.
TG7 ℓ Fl (2) G 9s; 53°47'·24N 06°49'·65E.
TG1 ℓ Oc (3) G 12s; 53°44'·90N 06°33'·20E.

WANGEROOGE
Harle ℓ Iso 8s; 53°49'·25N 07°50'·47E.
Wangerooge, W end ☆ Fl R 5s 60m **23M**; R ○ twr, 2 W bands; 53°47'·40N 07°51'·37E.
Dir lt 145·5°, Dir WRG 24m W22M G18M R17M; 119·4°-G-138·8°-W-152·2°-R-159·9°.
Buhne H ℓ VQ (9) 10s; 53°46'·86N 07°49'·65E.
W2 ⚓ ; 53°46'·25N 07°51'·89E, off hbr ent.

SPIEKEROOG
Otzumer Balje ℚ Iso 4s; 53°48'·29N 07°38'·00E (often moved).
Spiekeroog ⚡ FR 6m 4M; 197°-114°; R mast; 53°45'·0N 07°41'·3E.

LANGEOOG
Accumer Ee ℓ Iso 8s; 53°46·92N 07°25·92E (frequently moved).
W mole head ⚡ Oc WRG 6s 8m W7M, R5M, G4M; 064°-G-070°-W-074°-R-326°-W-330°-G-335°-R-064°; R basket on R mast; *Horn Mo (L) 30s* (0730-1800LT); 53°43'·42N 07°30'·13E.

BALTRUM
Groyne hd ⚡ Oc WRG 6s 6m; W6M, R4M, G3M; 074·5°-G-090°-W-095°-R-074·5°; 53°43'·3N 07°21'·7E.

NORDERNEY
Norderney ☆ Fl (3) 12s 59m **23M**; unintens 067°-077° and 270°-280°; R 8-sided twr; 53°42'·54N 07°13'·79E.
Dovetief ℓ Iso 4s; 53°45'·47N 07°12'·70E.

Schluchter ℓ Iso 8s; 53°44'·80N 07°06'·00E.
W mole head ⚡ Oc (2) R 9s 13m 4M; 53°41'·9N 07°09'·9E.

JUIST
Juist-N ℓ VQ; 53°43'·82N 06°55'·42E.
Training wall, S end ⚡ Oc (2) R 9s 7m 3M; 53°39'·65N 06°59'·81E.

MAINLAND HARBOURS: RIVER JADE TO RIVER EMS

HARLESIEL
Carolinensieler Balje, Leitdamm ⚡ L Fl 8s 7m 6M; G mast; 53°44'·13N 07°50'·10E.
Ldg lts 138°, both Iso 6s 12/18m 9M, intens on ldg line; Front 53°40'·70N 07°34'·50E. Rear 167m from front.
N mole head ⚡ Iso R 4s 6m 7M; 53°42'·58N 07°48'·64E.

NEUHARLINGERSIEL
Training wall head ⚡ Oc 6s 6m 5M; 53°43'·22N 07°42'·30E.

BENSERSIEL
E training wall head ⚡ Oc WRG 6s 6m W5M, R3M, G2M; 110°-G-119°-W-121°-R-110°; R post & platform; 53° 41'·80N 07°32'·84E.

DORNUMER-ACCUMERSIEL
W bkwtr head, approx 53°41'·04N 07°29'·30E.

NESSMERSIEL
N mole head ⚡ Oc 4s 6m 5M; G mast; 53°41'·9N 07°21'·7E.

NORDDEICH
W trng wall head ⚡ FG 8m 4M; 021°-327°; G framework twr; 53°38'·62N 07°09'·0E.

GREETSIEL
Meßstation lt bcn, Fl Y 4s 15m 3M; 53°32'·94N 07°02'·18E.

RIVER EMS

APPROACHES
GW/EMS ⬡ Iso 8s 14m **17M** (H24); Emergency Lt, FR, occas. 2 FR vert; R Lt vessel, floodlit; *Horn Mo (R) 30s (H24)*; 54°09'·98N 06°20'· 85E.
Borkumriff ℓ Oc 4s; 53°47'·48N 06°22'·27E.
Osterems ℓ Iso 4s; 53°41'·91N 06°36'·17E.
Riffgat ℓ Iso 8s; 53°38'·96N 06°27'·10E.
Westerems ℓ Iso 4s; 53°36'·93N 06°19'·39E.
H1ℓ 53°34'·91N 06°17'·97E.

BORKUM
Borkum Grosser ☆ Fl (2) 12s 63m **24M**; brown ○ twr; 53°35'·32N 06°39'·64E. Same twr, ⚡ F WRG 46m **W19M, R/G15M;** 107·4°-G-109°-W-111·2°- R-112·6°.
Fischerbalje ⚡ Oc (2) 16s 15m **3M**; 260°-W-123°. Fog det lt; R/W ○ twr on tripod; 53°33'·18N 06°42'·90E.
Schutzhafen, E mole hd ⚡ FG10m 4M; 53°33'·48N 06°45'·02E. W mole hd, FR 8m 4M.

RIVER EMS (LOWER)
No. 27 ℓ Fl G 4s; 53°30'·24N 06°47'·52E.
No. 35 ℓ Fl G 4s; 53°27'·03N 06°52'·80E.
No. 37 ℓ QG; 53°26'·01N 06°54'·85E.
No. 41 ℓ Fl (2) G 9s; 53°24'·25N 06°56'·68E.
No. 49 ℓ Oc (2) G 9s; 53°19'·95N 06°59'·71E.

KNOCK
Knock ⚡ F WRG 28m W12M,R9M, G8M; 270°-W-299°-R-008·3°-G-023°-W-026·8°-R-039°-W-073°-R-119°-W-154°; fog det lt; Gy twr, white conical radar antenna; 53°20'·32N 07°01'·40E.

EMDEN
No. 59 ℓ Oc (2) G 9s; 53°19'·41N 07°03'·86E.
No. 65 ℓ Fl (2) G 9s; 53°19'·82N 07°06'·56E.
Outer hbr, W pier ⚡ FR 10m 4M; R 8-sided twr; *Horn Mo (ED) 30s*; 53°20'·06N 07°10'·49E.
E pier ⚡ FG 7m 5M; G mast on pedestal; 53°20'·05N 07°10'·84E.

9.15.5 PASSAGE INFORMATION

More Passage Information is threaded between successive harbours in this Area. **Bibliography**: NP 55 *North Sea (East)*; 3rd ed 2006 *Cruising Guide to Germany and Denmark* (Imray/Navin). The *Hafenhandbuch, Nord See* (DSV-Verlag) is a well illustrated annual guide, covering Calais to Trondheim and the Lofoten Islands.

NORTH FRISIAN ISLANDS

All the channels from the open sea on this coast are challenging for yachts when the wind is W/NW 4 and start becoming dangerous at F5 and above. From seaward, Butendiek, wind farm centred on 55°02'·31N 07°45'·71E, with surrounding restricted area marked by 10 lit cardinal buoys lies 22M W of List West ☆, NW Sylt.

Lister Tief is the chan between the Danish island of Rømø and the N end of Sylt; it gives access to **List Roads** and hbr as well as to Danish hbrs (AC 1875, 3767). Lister Tief is well buoyed; least depth over the bar is about 4m. Relatively it is the safest chan on this coast (after Süderpiep) available for yachts seeking anch under the lee of Sylt in strong W winds (when however there would be a big swell over the bar on the ebb). Beware buoyed obstructions (ODAS) 18M WSW of List West lt ho.

Sylt is the largest of the N Frisian Islands, almost 20M long from S to N. It has a straight seaward coast, and on its E side is connected by the Hindenburgdamm to the mainland. Vortrapptief is the NNE chan inward between Amrum and Sylt, leading to **Hörnum** Hafen. It has a depth > 10m (but subject to frequent change, and may be < 4m on the bar) and is buoyed and lit. *The area should not be approached in strong W winds.* ▶*The flood (ESE-going) stream begins at HW Helgoland −0350, and the ebb (WNW-going) at HW Helgoland +0110, sp rates 2·5kn.*◀

Rütergat and Schmaltief give access to **Amrum** (Wittdün hbr on SE side) to Wyk on the SE corner of **Fohr** and to Dagebüll (ferry hbr). The R. Hever consists of several chans on the N side of Eiderstedt Peninsula, and S of the North Frisian Islands of **Süderoogsand** and **Pellworm**. Mittelhever is the most important of the three buoyed chans through the outer grounds, all of which meet SE of Süderoogsand. Here they separate into Norderhever, which runs NE between Pellworm and Nordstrand into a number of chans, with Heverstrom leading to Husum.

RIVER EIDER AND SCHLESWIG-HOLSTEIN

Approaching the R. Eider from seaward, find the Eider SWM lt buoy, about 1·2M W of the buoyed ent chan. *The ent can be rough in W winds, and dangerous in onshore gales.* St Peter lt ho can be seen above the dyke. Here the estuary winds up to the Eiderdamm, a storm barrage with a lock (H24) and sluices. **Tönning** is about 5M upriver. The upper Eider parallels the Nord-Ostsee Kanal which it joins at Gieselau.

The W coast of Schleswig-Holstein is flat and marshy, with partly-drying banks extending 5–10M offshore. Between the banks and islands, the chans shift frequently. Süderpiep and Norderpiep join allowing access to **Büsum** and Meldorfer Hafen (marina). Norderpiep has a bar (depth 3m) and is unlit; Süderpiep is deeper and preferable in W'lies, but beware of strong ground seas. Landmarks from seaward are Tertius bn and Blaurtsand bn, and a conspic silo and tall building at Büsum.

HELGOLAND TO THE RIVER ELBE

Helgoland is a useful staging post for yachts entering the Elbe bound for **Cuxhaven** (38M) or direct **Brunsbüttel** (51M) at the ent to the **Nord-Ostsee Kanal** (AC 2469). From the E end of the TSS the Elbe channel is well marked by buoys and beacon trs. Commercial traffic is very heavy. For further details see River Elbe.

▶*At Elbe SWM buoy the E-going (flood) stream begins at HW Helgoland −0500, and the W-going (ebb) stream at HW Helgoland +0050, sp rates 2kn. The stream runs harder N of Scharhörn, up to 4kn on the ebb, when the Elbe estuary is dangerously rough in strong W or NW winds and must not be attempted.*◀

At the mouth of the Elbe estuary (AC 3619, 3625, 3267, 3268) Scharhörn and Neuwerk are islets. The latter features in *Riddle of the Sands*, Erskine Childer's classic sailing and spy thriller.

9.15.6 SPECIAL NOTES FOR GERMANY

LANDS are given in lieu of 'counties' in the UK.

TIME ZONE is −0100 (subtract 1 hour for UT). DST is kept from the last Sunday in March to the last Sunday in October.

CHARTS: German charts are issued by the Bundesamt für Seeschiffahrt und Hydrographie (BSH) Hamburg. In this Almanac German *seekarten* charts are prefixed 'D'.

The 3000 Series of leisure-craft folios are prefixed BSH. 3010-3015 cover the N Sea coast and 3009 the Kiel Canal. They are a handy size (42 x 59cm) stowed in a large clear polythene envelope; each set has up to 16 sheets at medium/large scale.

All charts are referred to WGS 84 and are corrected by *Nachrichten für Seefahrer* (NfS) = Notices to Mariners. Where possible the AC, Imray and Dutch chart numbers are also quoted.

TIDE TABLES: Admiralty charts of German coastal waters are now referred to LAT (not MLWS) as are German charts. Depths in the shallow estuaries, the North and East Frisian islands vary annually, and during the year with wind direction and strength. Treat depth predictions with caution and obtain the latest chart corrections from BSH.

LIGHTS: In coastal waters considerable use is made of sectored lights. A leading or directional sector (usually W, and often intens) may be flanked by warning sectors to show the side to which the vessel has deviated: If to port, a FR lt or a Fl W lt with an even number of flashes; if to starboard, a FG lt or a Fl W lt with an odd number of flashes. Crossing lts with R, W and G sectors indicate (eg) the limits of roadsteads, bends in chans etc.

SIGNALS: IPTS are widely used in major ports. Local port, tidal, distress and traffic signals are given where possible. Visual storm warning signals are not used. When motoring under sail, always display a black motoring cone (▼). Rule 25e is strictly enforced and non-compliance will incur a fine.

Lights at bridges and locks

●●	=	Passage/entry prohib.
●	=	Prepare to pass or enter; exit from lock prohib.
●	=	Exit from lock permitted.
○ ●●	=	Passage or entry permitted if there is sufficient clearance; beware of oncoming traffic which has right of way.
○○ ●●	=	Lifting bridge is at first lift position; and may be passed by vessels for which the vertical clearance is safe.
●●	=	Passage/entry permitted; oncoming traffic stopped.
○ ●●	=	Passage or entry permitted, but beware oncoming traffic which may have right of way.
● ●	=	Bridge, lock or flood barrage closed to navigation.

Signals hoisted at masts

By day	By night	Meaning
R cylinder	○ ● ○	Reduce speed to minimise wash.
● ● ▼	● ● ●	Unusual obstruction to navigation.
● ▼ ▲	● ● ○	Long term closure of waterway.

HARBOURS: Most marinas are owned by the state or community and run by a local YC (as in Belgium and the Netherlands).

TRAFFIC SEPARATION SCHEMES: The extensive TSS (off Terschelling, in the German Bight and in the approaches to the Rivers Jade and Elbe) link up with the various Dutch TSS. Information broadcasts by VTS centres and their associated communications are given in Areas 15 and 16.

Caution: Although the TSS may be crossed in accordance with the IRPCS, vessels in the ITZ should not approach the S edge of the E-bound lane of the Terschelling-German Bight TSS (well marked by buoys TG1 to TG7) closer than 1M, unless in emergency or due to stress of weather. The German Marine Police can impose on-the-spot fines or confiscate equipment to the value of €1000 pending payment of the fine.

Navigators are strongly advised to:

- Be mindful of the value of GPS navigation in these waters.

- Log positions, times, courses, speeds, wind, weather, seastate, other traffic and use of sail or engine. Such records can be important in any dispute or litigation and also keep the skipper alert to any consequences. In addition, they can demonstrate that the boat is being responsibly navigated.

- Always monitor the appropriate VTS channel: Ch 80 in E part of Inner German Bight; Ch 79 in W part. If a deviation from the rules is necessary, inform the VTS traffic centre.

- Beware of deep draught vessels S-bound from the German Bight lt vessel in the Jade approach TSS and crossing E-bound traffic in the Terschelling-German Bight TSS. The former have right of way over E-bound traffic. Yachts either E- or W-bound must take this into account.

HIGH SPEED CRAFT: High speed craft ply between Emden and Borkum, from Hamburg to Stadersand, Cuxhaven and Helgoland, and from an increasing number of ports. Keep a good lookout.

TELEPHONES: To call UK from Germany: dial 00 44; then the UK area code, minus the initial zero; then the number required. To call Germany from UK: dial 00 49; then area code minus initial zero; then desired number. Mobile numbers begin with 015, 016, 017; number of digits varies depending on Service provider.

EMERGENCIES: Police: dial 110. Fire, Ambulance: dial 112.

PUBLIC HOLIDAYS: New Year's Day, Good Friday, Easter Monday, Labour Day (1 May) Ascension Day, Whit Monday, Day of German Unity (3 Oct) Christmas Day and Boxing Day.

NATIONAL WATER PARKS exist in the Wadden Sea areas (tidal mud flats) of Lower Saxony (excluding the Jade and Weser rivers and the Ems-Dollart estuary) and along the west coast of Schleswig-Holstein to the Germany/Denmark border. *These conservation areas have the status of Particularly Sensitive Sea Areas (PSSA) in or near which vessels should take the utmost care to avoid damaging the marine environment and marine organisms living therein. See Chapter 9 Introduction on Environmental Guidance.* The Parks (shown on AC & BSH charts) are divided into 3 zones with certain rules:

Zone 1: comprises the most sensitive areas (about 30%) where yachts must keep to buoyed chans, except HW±3. Speed limit 8kn.

Zone 2: is a buffer zone.

Zone 3: is the remainder. No special constraints exist in Zones 2 and 3.

CUSTOMS: Ports of entry are: Borkum, Norderney, Norddeich, Wilhelmshaven, Bremerhaven, Cuxhaven; but not Helgoland.

GAS: Calor gas bottles can sometimes be re-filled. German 'Flussig' gas can be used with Calor regulators.

CURRENCY is the Euro (€). VAT (MWST) is 19%.

VISITOR'S TAX (KURTAXE) is used to fund local amenities in tourist areas. It is approx €2–€3 per person per day and may be payable in addition to berthing fees.

9.15.7 NORTH FRISIAN ISLANDS

Schleswig-Holstein 54°45'·51N 08°17'·83E (Hörnum hbr)

CHARTS AC 3767; D 107, D 108; BSH 3013.2/3 List, 3013.4/5 Hörnum
Up-to-date charts are essential for these waters.

TIDES +0110 Dover; ML 1·8; Duration 0540

Standard Port HELGOLAND (→)

Times				Height (metres)			
High Water		Low Water		MHWS	MHWN	MLWN	MLWS
0100	0600	0100	0800	3·2	2·8	0·9	0·5
1300	1800	1300	2000				
Differences LIST							
+0252	+0240	+0201	+0210	−0·7	−0·6	−0·2	0·0
HÖRNUM							
+0223	+0218	+0131	+0137	−0·5	−0·3	−0·2	0·0
AMRUM-HAFEN							
+0138	+0137	+0128	+0134	+0·2	+0·3	0·0	0·0
DAGEBÜLL							
+0226	+0217	+0211	+0225	+0·7	+0·7	+0·1	+0·1
SUDEROOGSAND							
+0116	+0102	+0038	+0122	+0·5	+0·5	+0·1	+0·1
HUSUM							
+0205	+0152	+0118	+0200	+1·1	+1·1	+0·1	0·0

Die Nordfriesischen Küste comprises four main islands (Sylt, Amrum, Föhr, Pellworm) many small islets and the adjacent mainland.

SYLT Westerland (the capital) is central, where the Hindenburgdamm links to the mainland. Most of it is nature reserve (landing prohib). Two main hbrs: List and Hörnum, two other hbrs which both dry (soft mud) are Munkmarsch and Rantum; accessible only towards the top of the tide.

FACILITIES WESTERLAND: (04651) CG 85199; Police/Marine police 70470. ⊕ 04651 840. **Town** ✕ ⌂ 🛒 🏬 🏦 ⬛ ⊖ 8356333, ⊠ Ⓑ; ⇌ ✈.

LIST (N END OF SYLT ISLAND)

Sylt **55°00'·99N 08°26'·45E** ❀❀❀❀❀❀

SHELTER The small hbr/marina is sheltered except from NE/E winds.

NAVIGATION WPT 55°05'·32N 08°16'·67E Lister Tief SWM buoy, 111°/7·0M to No 13 SHM buoy. *Strong W/NW winds cause a heavy swell over the bar (depth 3m) on the ebb, which sets on to Salzsand.* Thence 214°/2·0M to the hbr via buoyed chan and W sector of List Land dir lt. Access by day at all tides, but beware strong cross streams. Ent is 25m wide. Enter hbr from SE, keeping List Land lt ho between the mole heads.

LIGHTS AND MARKS Hbr lts as on the chartlet. See 9.15.4 for lts at List West, List Ost, List Land (overlooks List hbr) and Kampen, 5.4M SW. Rømø church, 7M NNE, is conspic.

COMMUNICATIONS (Code 04651 for the whole of Sylt)

CG 870365; Police/Marine police (Westerland) 70470; ⊖ 870413, Dr 870350; HM 870374, VHF Ch **11** (0800-1200, 1600-1800).

FACILITIES Hbr Marina at S end carries 1·1m-2·6m. NE quay is open construction, unsuitable for yachts. ⊊ €1·20, ⚓⊖✚ SC. **Village** ✕⊡🛒 ⓣ🄫. Ferry to Rømø (Denmark).

HÖRNUM (S END OF SYLT ISLAND)

Sylt **54°45'·51N 08°17'·83E** ❀❀❀❀❀❀

SHELTER Good in the small hbr, approx 370m x 90m. Yacht haven at the N end. Good ⚓ in Hörnum Reede in W and N winds and in Hörnumtief in E and S winds.

NAVIGATION WPT: 54°34'·87N 08°12'·97E, Vortrapptief SWM buoy, 016°/10M to S tip of Sylt. Hörnum lt bearing 012° leads through buoyed, lit Vortrapptief between drying banks. In strong W winds the sea breaks on off-lying banks and in the chan. Access at all tides but not at night without radar. Beware strong cross-streams.

LIGHTS AND MARKS Hörnum lt is 3½ca SSW of hbr ent. Other lts as chartlet. Conspic radio mast (5 Fl R lts) 3M N of hbr.

COMMUNICATIONS (Code 04651) MRCC (Bremen) 04215 36870, mobile 124124; CG 881256 (volunteer svc); Police/Marine police (Westerland) 70470; Dr 881016; Brit Consul (030) 204570; HM 881027, VHF Ch **67**, Mon-Thu 0700-1600; Fri 0700-1230, all LT.

FACILITIES Sylter YC ☎880274, www.sylter-yachtclub.de ⊊€1·75 + €3·00 pp/day tax, ⓣ📡🅆🄲 ⊡🄫. **Town** ✕⊡🛒⚒✉ⓑ.

THE HALLIGEN

There are ten German Hallig islands, and one Danish one, dotted around *Die Nordfriesischen Küste*. Halligen are small islands, without the protection of coastal dykes. The Halligen lack harbours and are of little interest to yachtsmen.

OTHER HARBOURS IN THE NORTH FRISIAN ISLANDS

WYK AUF FÖHR, 54°41'·66N 08°34'·72E. AC 3767; D 107, BSH 3013.6. HW +0107 on Dover (UT); +0212 on Helgoland (zone –0100); ML 2·8. Good yacht hbr (1·2-1·7m) to N of ent, sheltered in all winds; access H24; visitors berth on pontoon 1. Inner hbr berths are for locals. Ferry hbr (4m). Lts see 9.15.4: Olhörn (SE point of Föhr) R twr. Mole heads, FR, R mast and FG, G mast. Port VHF Ch **11** 16. Yacht hbr HM ☎04681 3030; Commercial HM ☎500430; Marine Police ☎1280; ⊖ ☎2594; Dr ☎8558; Facilities: **Hbr road** (500m) ⓣ ⓣ🄫 🄲; **Yacht Hbr** €1·75. ✕🛒 🄫(15t); **W Quay** ⛟.

AMRUM-HAFEN, 54°37'·90N 08°23'·04E. AC 3767; D 107, BSH 3013.6 & 7. HW +0107 on Dover (UT); +0137 on Helgoland (zone –0100); ML 2·7m; Duration 0540. See N Frisian Islands. Appr via Rütergat and Norderaue. Make good RGR buoy No 26/AH1, Fl(2+1) R 15s, at 54°37'·76N 08°24'·80E where the hbr chan starts; it carries 2·5m and is marked by PHM withies ⸰. Yachts berth S of stone quay, ⊊ pontoon. Good shelter except in E winds. Ferry quay is 800m to the E. Lt ho Fl 7·5s, R tr, W bands is same location as rear ldg lt. Wriakhörn, Cross lt, is 1M WSW of hbr, see 9.15.4. It is not advisable to enter at night. HM ☎04682 2294; ⊖ ☎2026; Dr☎531; Facilities: ⊊ €1·50 + pp/day tax, ⓣ🄫 ⓣ🄫 (at Nebel); **Amrum YC** ☎2054.

PELLWORM, 54°31'·27N 08°41'·16E. AC 3767; D 106, BSH 3013.11 & 12C. Tides: see Suderoogsand/Husum. Outer appr via Norderhever to NH18 RGR buoy, Fl (2+1) R 15s, at the mouth of the ent chan which has a limiting depth of 0·2m over datum and is marked by withies. VHF Ch 11, 0700-1700. Small yacht hbr dries to soft mud, if you berth on the wall you will need a fender board; 2·8m at MHW. ⊊ €1·00. Village 600m N/W.

MAINLAND HARBOUR 12 MILES EAST OF PELLWORM

HUSUM, Schleswig-Holstein, **54°28'·74N 08°59'·92E.** AC 3767; ⛟.106, BSH 3013.12. HW + 0036 on Dover (UT); ML 1·9m; Duration 0555. See N Frisian Islands. SHM bcn, Fl G 4s, 54°28'·80N 08°58'·60E, marks start of access chan; also marked by two PHM bcns, both Fl (2) R 9s. A sluice/lock, 7ca W of rly bridge, is shut when level > 0·3m over MHW, as shown by ●. Beware the canal effect when meeting big ships in the narrow appr chan. In the outer hbr yachts should turn stbd just before rly bridge for pontoons on S bank, or pass through bridge (clnce 5m when shut) to inner hbr by SC Nordsee; both dry. VHF call *Husum Port* Ch 11. Tfc reports broadcast on Ch 11 every H +00 from HW –4 to HW +2. See 9.15.4 for hbr lts. Night entry not advised. ⊖ ☎61759; HM ☎04841 667218; Sluice ☎2565; **Husum YC** ☎65670; **SC Nordsee** ☎3436, ⊊ €1·00, ⛟⛟ (at hbr).

MAINLAND (RIVER EIDER) HARBOUR 9 MILES NORTH OF BÜSUM
TÖNNING, Schleswig-Holstein, **54°19'·00N 08°57'·00E**. AC 3767 (only appr from Eider SWM buoy); D.104, BSH 3013.14, 3014.8. Eiderdamm lock, 5·5M down-river, opens H24; VHF Ch 14. Lts, see 9.15.4. Above the dam, beware strong currents when sluices are operated HW +3 to HW+½. ⌑ on S side of drying marina (3m at MHW) or on less-comfortable river quays if marina full. Upriver, bridge (clnce 5·6m) opens on request Mon-Sat 0600-SS. HM ☎(04861) 1400; Ⓗ ☎706; Dr ☎389; Facilities: **Quay** ⌑ €1·00, ⚓🏳️🔧🛒🛠(5t) ⚓; **YC** ☎754 (all welcome). **Town** Ⓑ ✉ ⇌ ⚓. Tönning is linked to the Kiel Canal and Rendsburg by the R Eider and the Gieselau Kanal (BSH 3009.7/.6/.5/.2).

9.15.8 BÜSUM

Schleswig-Holstein **54°07'·16N 08°51'·52E** 🌸🌸⊛🌸♎♎♎🌸🌸

CHARTS AC 1875, 3767; D 105, BSH 3014.4

TIDES HW +0036 on Dover (UT); ML 1·9m; Duration 0625
Standard Port HELGOLAND (→)

Times				Height (metres)			
High Water		Low Water		MHWS	MHWN	MLWN	MLWS
0100	0600	0100	0800	3·2	2·8	0·9	0·5
1300	1800	1300	2000				
Differences BÜSUM							
+0054	+0049	−0001	+0027	+0·8	+0·8	0·0	−0·1
LINNENPLATE 54°13'·0N 08°40'·0E							
+0047	+0046	+0034	+0046	+0·6	+0·5	0·0	−0·1
SÜDERHÖFT 54°16'·0N 08°41'·8E							
+0103	+0056	+0051	+0112	+0·7	+0·6	−0·2	0·0
EIDERSPERRWERK 54°16'·0N 08°50'·7E							
+0120	+0115	+0130	+0155	+0·6	+0·5	−0·1	−0·1

SHELTER May be uncomfortable in strong W'ly winds. Beware strong cross-tides and sudden winds over moles. Barrage closes when level > MHW, but craft < 30m LOA can enter via lock. For opening (H24) call VHF or sound 1 long blast in appr chan; 2 long blasts at lock if not already open; obey R/G lts. Turn 90° stbd into marina (1·8m LW) Basin IV. Isolated Meldorf marina is 4M ESE, 2-6m. Appr via Kronenloch (1·1m); enter via barrage/lock.

NAVIGATION WPT 54°05'·82N 08°25'·70E, Süderpiep SWM buoy Iso 8s, 096°/1·0M to No 1 SHM buoy, Fl G 4s. Chan is lit/well-buoyed. Norderpiep is buoyed, but unlit; beware bar if wind-over-tide. Both are approx 15M long; night appr not advised.

LIGHTS AND MARKS Lt ho, ldg lts 355·1° and hbr lts as chartlet. A 22 storey bldg is conspic 8ca NW of hbr; silo is conspic in hbr.

COMMUNICATIONS (Code 04834) Coast Guard 2246; Customs 2376; Doctor 2088; HM 2183, Call *Büsum Port* Ch 11 16 for lock opening; YC 2997.

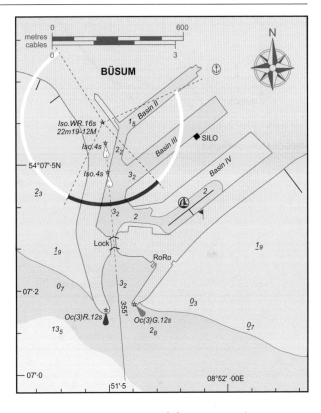

metres / cables 0 — 600, 0 — 3

BÜSUM

Iso.WR.16s 22m19-12M · Iso.4s · Iso.4s · 54°07'·5N · Basin II · Basin III · SILO · Basin IV · Lock · RoRo · Oc(3)R.12s · Oc(3)G.12s · 355° · 08°52'·00E · 07'·2 · 07'·0 · 51'·5

FACILITIES Yacht hbr ⌑ €1·30, ⚓🏳️🔧🛒Ⓒ🔧🛠🛒. **Town** ✕ 🛒 🛍 ✉ Ⓑ ⇌ Ⓗ ✈ (Hamburg).

9.15.9 HELGOLAND

Schleswig-Holstein **54°10'·28N 07°53'·93E** 🌸🌸⊛🌸♎♎♎🌸🌸

CHARTS AC 1635, 1875; Imray C26; D 3, 88; BSH 3014.2/3

TIDES −0030 Dover; ML 1·4; Duration 0540
HELGOLAND is a Standard Port (→).

SHELTER Good in safe artificial hbr (5m); yachts berth on pontoons on the NE side of the Südhafen which gets very crowded, rafting 12 or more deep. Ostkaje now finished, with pontoons and coin operated power. Or ⚓ in the SW part of the Vorhafen, sheltered except in SE'lies; ask permission from Port Control Ch 67.

NAVIGATION WPT 54°08'·96N 07°53'·49E (Helgoland-O ECM buoy, Q (3) 10s, Whis) 020°/1·3M to abeam No 3 buoy. Beware: the Hogstean shoal (1m) 4ca S of Sudmole head lt, and other rky shoals either side of chan; lobster pots around Düne and fast ferries using the S appr chan and ent. Passage between Düne and Helgoland is inadvisable at night or in heavy weather.

- *Between Helgoland and Düne the general direction of buoyage changes at approx 54°11'N; see chartlet overleaf.*

- Entry is prohib at all times into Nature Reserves close to either side of the appr chans from SSW and NW.

> Reserves extend about 2·9M NE and SW of Helgoland; limits marked by 7 card buoys, actively policed. Smaller reserve, lying to the N, E and S of Düne, is a prohib ⚓, with other restrictions. See BSH 3014.2, AC 1635 and 1875 (plan) for limits.

LIGHTS AND MARKS Helgoland is a 60m high red sandstone plateau; in contrast Düne is a low sandy islet. The lt ho is a brown □ twr, B lantern, W balcony. Close S a R/W radio mast and Sig stn are conspic. Düne rear ldg lt is on a red twr with W band.

COMMUNICATIONS (Code 04725) Coast Guard 210; SAR 811305; Met 811007; Police 607; ⊖ 1537 or 304; Brit Consul (030) 204570; Ⓗ 8030; Ambulance 7723; Dr 7345; HM 81593583, or dial 81593-0 and ask for extension; *Helgoland Port Radio* Ch 67, 16 (**1/5–31/8**: Mon-Thur 0700-1200, 1300-2000; Fri/Sat 0700-2000; Sun 0700-1200LT. **1/9–30/4**: Mon-Thur 0700-1200, 1300-1600; Fri 0700-1200).

FACILITIES *Yachts must not enter the Nordost hafen and Dünen hafen nor land on Düne island.*

NOTE: Helgoland is not a port of entry into Germany. Customs in Helgoland are for passport control. Duty free goods are obtainable; hence many day-trippers by ferry. €2.75/pp/day tax. Useful webcam www.rickmers-online.de/webcam.html.

Vorhafen ⚓ ⚓. **Südhafen** ⌑ €1·32. Met offffice above HM. 🏳️ ⚓ on Westkaje, limit 70 litres, €0·50, ⚡ (coins) ✕ 🛒 🔧🛒 🛠(mobile 12t) .

Nordkaje ⚡ €1·00 (coins).

Binnenhafen closed except for fuel, no ⚓ at fuel station.

Wasser Sport Club ☎7422, Berthing <10m with permission; 🏳️⚓ ⚓✕🛒🔧🛒🛠🔧⚓🛠(12t).

Town ✕ 🛒 🛒 ⚓ ✉ Ⓑ ✈ to Bremen, Hamburg, Cuxhaven.

HELGOLAND

MEAN SPRING —————
AND NEAP CURVES – – – –

MEAN RANGES

Springs 2.7m
Neaps 1.9m

HELGOLAND *continued*

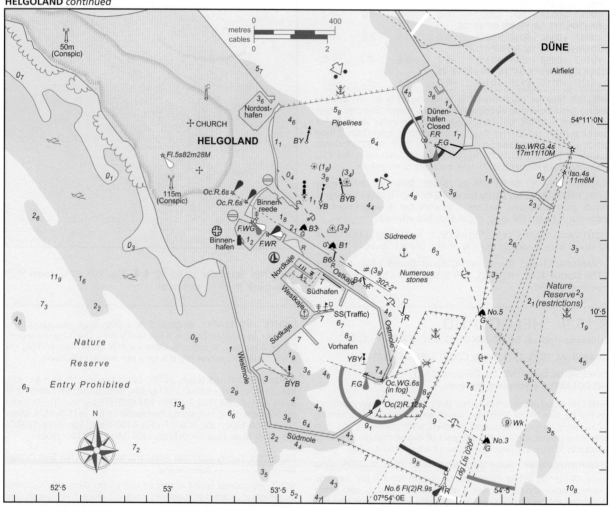

FROM THE GERMAN BIGHT TO THE UK

(AC 2182A, 1423, 1405, 1635) From the Elbe, Weser and Jade estuaries, skirt the E ends of the German Bight western approaches TSS to depart from the vicinity of Helgoland or the GB Lt Vessel. ▶*West of German Bight streams run approx E/W, spring rates <1kn.*◀ Thence parallel the N side of TSS, towards Botney Ground (BG2 buoy) if heading north of Flamborough Head. A WNW track towards the Firth of Forth avoids the areas most densely obstructed by offshore oil and gas installations.

If bound for East Anglia and the Thames Estuary, follow the ITZ westwards until S of Den Helder and Off Texel TSS in about Lat 52° 50'N; thence head west, avoiding areas of offshore industrial activity and the heavy traffic concentrations off the Maas and Westerschelde.

For the English Channel it is best to stay inshore until Cap Gris Nez. For North Sea distances, see Chapter 9, Introduction.

Note: When planning consider that Helgoland is the only all weather harbour of refuge on this coast.

STANDARD TIME UT –01
Subtract 1 hour for UT
For German Summer Time add
ONE hour in **non-shaded areas**

HELGOLAND LAT 54°11'N LONG 7°53'E
TIMES AND HEIGHTS OF HIGH AND LOW WATERS

Dates in red are **SPRINGS**
Dates in blue are NEAPS

YEAR 2016

JANUARY

Time	m	Time	m
1 0354	3.1	**16** 0353	3.2
1041	0.8	1048	0.5
F 1624	2.8	SA 1624	2.9
2249	0.9	2301	0.6
2 0434	3.0	**17** 0444	3.2
1120	0.9	1137	0.6
SA 1707	2.7	SU 1716	2.9
☽ 2333	0.9	☽ 2353	0.7
3 0521	2.9	**18** 0540	3.1
1206	0.9	1231	0.7
SU 1757	2.7	M 1815	2.8
		1902	2.7
4 0029	1.0	**19** 0057	0.8
0620	2.9	0646	3.0
M 1306	0.9	TU 1337	0.7
1900	2.7	1925	2.8
5 0139	1.0	**20** 0216	0.8
0730	2.8	0803	2.9
TU 1417	0.9	W 1455	0.8
2010	2.8	2043	2.9
6 0252	1.0	**21** 0337	0.7
0840	2.9	0921	2.9
W 1526	0.9	TH 1611	0.7
2114	2.9	2154	3.0
7 0359	0.9	**22** 0449	0.6
0942	2.9	1028	2.9
TH 1625	0.8	F 1716	0.6
2210	2.9	2254	3.0
8 0455	0.8	**23** 0547	0.5
1035	3.0	1123	2.9
F 1715	0.7	SA 1807	0.5
2258	3.0	2344	3.1
9 0543	0.7	**24** 0635	0.4
1121	3.0	1210	3.0
SA 1800	0.7	SU 1850	0.5
2341	3.1	○	
10 0627	0.6	**25** 0026	3.2
1205	3.1	0717	0.4
SU 1843	0.6	M 1250	3.0
●		1928	0.5
11 0022	3.2	**26** 0106	3.3
0710	0.6	0756	0.5
M 1248	3.1	TU 1329	3.0
1927	0.6	2006	0.5
12 0103	3.2	**27** 0144	3.3
0753	0.5	0833	0.6
TU 1331	3.1	W 1406	3.0
2008	0.5	2041	0.6
13 0143	3.3	**28** 0220	3.2
0834	0.4	0906	0.6
W 1411	3.0	TH 1440	3.0
2048	0.5	2112	0.6
14 0222	3.3	**29** 0252	3.2
0915	0.4	0938	0.6
TH 1450	3.0	F 1514	2.9
2128	0.5	2144	0.6
15 0305	3.3	**30** 0324	3.1
1000	0.4	1009	0.7
F 1535	2.9	SA 1548	2.9
2213	0.5	2216	0.7
		31 0358	3.0
		1040	0.7
		SU 1622	2.8
		2249	0.8

FEBRUARY

Time	m	Time	m
1 0433	3.0	**16** 0518	3.0
1111	0.8	1202	0.6
M 1659	2.7	TU 1744	2.8
☽ 2330	0.8		
2 0518	2.8	**17** 0032	0.6
1158	0.9	0620	2.8
TU 1751	2.7	W 1306	0.7
		1855	2.8
3 0031	0.9	**18** 0152	0.7
0621	2.7	0741	2.8
W 1307	0.9	TH 1431	0.8
1902	2.7	2020	2.8
4 0150	0.9	**19** 0321	0.6
0741	2.7	0907	2.8
TH 1429	0.9	F 1556	0.7
2021	2.7	2141	2.9
5 0312	0.9	**20** 0440	0.6
0859	2.8	1020	2.8
F 1544	0.8	SA 1705	0.6
2132	2.8	2244	3.0
6 0422	0.7	**21** 0537	0.4
1004	2.8	1114	2.9
SA 1647	0.7	SU 1754	0.5
2230	3.0	2331	3.1
7 0519	0.6	**22** 0621	0.4
1059	2.9	1156	2.9
SU 1739	0.6	M 1835	0.5
2318	3.1	○	
8 0608	0.5	**23** 0010	3.1
1146	3.0	0700	0.4
M 1827	0.5	TU 1233	2.9
●		1912	0.5
9 0002	3.2	**24** 0046	3.1
0654	0.4	0735	0.4
TU 1230	3.0	W 1307	2.9
1912	0.5	1946	0.5
10 0044	3.2	**25** 0121	3.2
0739	0.3	0808	0.5
W 1313	3.0	TH 1340	3.0
1955	0.4	2018	0.4
11 0127	3.3	**26** 0154	3.1
0822	0.2	0839	0.5
TH 1356	3.0	F 1412	3.0
2037	0.3	2048	0.4
12 0209	3.3	**27** 0225	3.1
0905	0.2	0909	0.5
F 1437	3.0	SA 1444	2.9
2119	0.3	2119	0.5
13 0253	3.2	**28** 0257	3.1
0949	0.3	0939	0.5
SA 1520	3.0	SU 1517	2.9
2202	0.4	2150	0.5
14 0339	3.2	**29** 0328	3.0
1033	0.3	1007	0.6
SU 1605	3.0	M 1547	2.9
2247	0.5	2218	0.5
15 0427	3.1		
1116	0.5		
M 1651	2.9		
☽ 2333	0.5		

MARCH

Time	m	Time	m
1 0357	2.9	**16** 0457	2.9
1031	0.6	1136	0.6
TU 1615	2.8	W 1719	2.9
2248	0.6		
2 0430	2.8	**17** 0011	0.5
1104	0.7	0559	2.7
W 1655	2.7	TH 1240	0.7
☽ 2336	0.7	1830	2.8
3 0525	2.6	**18** 0130	0.6
1206	0.8	0721	2.6
TH 1802	2.6	F 1407	0.8
		1958	2.8
4 0054	0.8	**19** 0304	0.6
0645	2.6	0851	2.6
F 1333	0.9	SA 1538	0.7
1928	2.6	2124	2.9
5 0225	0.8	**20** 0425	0.5
0815	2.6	1006	2.7
SA 1503	0.8	SU 1649	0.6
2052	2.8	2227	3.0
6 0348	0.6	**21** 0521	0.4
0933	2.7	1057	2.8
SU 1617	0.6	M 1735	0.5
2159	2.9	2309	3.0
7 0453	0.5	**22** 0559	0.3
1033	2.9	1134	2.9
M 1716	0.5	TU 1812	0.4
2252	3.0	2346	3.0
8 0546	0.3	**23** 0634	0.3
1123	3.0	1208	2.9
TU 1806	0.4	W 1848	0.4
2337	3.2	○	
9 0634	0.3	**24** 0022	3.1
1207	3.0	0708	0.3
W 1852	0.3	TH 1241	2.9
●		1923	0.3
10 0021	3.2	**25** 0056	3.0
0719	0.2	0741	0.4
TH 1251	3.0	F 1313	2.9
1937	0.2	1955	0.3
11 0106	3.2	**26** 0128	3.0
0804	0.1	0811	0.4
F 1336	3.0	SA 1345	2.9
2021	0.2	2025	0.3
12 0152	3.2	**27** 0159	3.0
0849	0.2	0841	0.4
SA 1419	3.0	SU 1416	2.9
2105	0.2	2055	0.3
13 0237	3.2	**28** 0230	3.0
0931	0.2	0911	0.4
SU 1502	3.0	M 1448	2.9
2148	0.2	2126	0.4
14 0322	3.1	**29** 0301	2.9
1012	0.3	0938	0.4
M 1544	3.0	TU 1517	2.9
2230	0.3	2153	0.4
15 0408	3.0	**30** 0330	2.8
1051	0.5	1001	0.5
TU 1627	3.0	W 1544	2.8
☽ 2314	0.4	2220	0.5
		31 0401	2.7
		1030	0.6
		TH 1619	2.7
		☽ 2302	0.6

APRIL

Time	m	Time	m
1 0448	2.6	**16** 0107	0.6
1125	0.8	0658	2.5
F 1718	2.6	SA 1339	0.7
		1929	2.8
2 0012	0.7	**17** 0235	0.5
0604	2.5	0823	2.5
SA 1249	0.8	SU 1507	0.7
1843	2.7	2052	2.8
3 0144	0.7	**18** 0354	0.4
0734	2.6	0936	2.6
SU 1423	0.8	M 1618	0.5
2010	2.8	2155	2.9
4 0312	0.5	**19** 0449	0.3
0858	2.7	1026	2.7
M 1543	0.6	TU 1704	0.4
2123	2.9	2238	2.9
5 0422	0.4	**20** 0526	0.3
1003	2.8	1102	2.8
TU 1646	0.4	W 1741	0.4
2220	3.0	2315	3.0
6 0518	0.2	**21** 0600	0.3
1056	2.9	1137	2.9
W 1740	0.3	TH 1819	0.4
2310	3.1	2353	3.0
7 0609	0.2	**22** 0637	0.3
1143	3.0	1213	2.9
TH 1829	0.2	F 1857	0.3
● 2357	3.2	○	
8 0657	0.1	**23** 0030	3.0
1228	3.0	0712	0.3
F 1916	0.2	SA 1247	2.9
		1931	0.3
9 0044	3.2	**24** 0102	3.0
0743	0.1	0744	0.3
SA 1313	3.0	SU 1319	2.9
2002	0.1	2002	0.3
10 0132	3.2	**25** 0134	2.9
0827	0.1	0814	0.3
SU 1357	3.1	M 1350	2.9
2047	0.1	2033	0.3
11 0219	3.1	**26** 0206	2.9
0910	0.2	0843	0.3
M 1441	3.1	TU 1421	2.9
2131	0.2	2104	0.3
12 0304	3.0	**27** 0238	2.9
0950	0.3	0912	0.4
TU 1523	3.0	W 1453	2.9
2213	0.2	2134	0.3
13 0350	2.9	**28** 0311	2.8
1029	0.4	0941	0.5
W 1607	3.0	TH 1524	2.9
2258	0.4	2205	0.4
14 0441	2.8	**29** 0346	2.7
1115	0.6	1015	0.6
TH 1659	2.9	F 1601	2.9
☽ 2354	0.5	2246	0.5
15 0541	2.6	**30** 0432	2.7
1217	0.7	1051	0.7
F 1807	2.8	SA 1655	2.8
		☽ 2349	0.6

Chart Datum: 1·68 metres below Normal Null (German reference level). HAT is 3·0 metres above Chart Datum.

>> FREE monthly updates. Register at <<
www.reedsnauticalalmanac.co.uk

671

Germany

STANDARD TIME UT –01
Subtract 1 hour for UT
For German Summer Time add
ONE hour in **non-shaded areas**

HELGOLAND   LAT 54°11'N LONG 7°53'E
TIMES AND HEIGHTS OF HIGH AND LOW WATERS

Dates in red are SPRINGS
Dates in blue are NEAPS

YEAR **2016**

MAY

Date	Day	Times / m
1	SU	0538 2.6 · 1219 0.7 · 1809 2.8
2	M	0111 0.6 · 0700 2.6 · 1346 0.7 · 1931 2.8
3	TU	0235 0.5 · 0821 2.7 · 1506 0.6 · 2045 3.0
4	W	0346 0.3 · 0929 2.8 · 1612 0.5 · 2147 3.0
5	TH	0446 0.2 · 1027 2.9 · 1711 0.3 · 2243 3.1
6 ●	F	0542 0.2 · 1119 3.0 · 1807 0.2 · 2335 3.2
7	SA	0634 0.1 · 1206 3.0 · 1856 0.2
8	SU	0024 3.1 · 0720 0.1 · 1250 3.1 · 1943 0.1
9	M	0112 3.1 · 0804 0.2 · 1335 3.1 · 2029 0.1
10	TU	0200 3.0 · 0848 0.2 · 1421 3.1 · 2114 0.2
11	W	0249 3.0 · 0929 0.3 · 1506 3.1 · 2158 0.3
12	TH	0336 2.9 · 1011 0.4 · 1551 3.0 · 2244 0.4
13 ◐	F	0426 2.8 · 1057 0.6 · 1642 3.0 · 2336 0.5
14	SA	0522 2.7 · 1153 0.7 · 1740 2.9
15	SU	0038 0.6 · 0626 2.6 · 1302 0.7 · 1849 2.8
16	M	0151 0.5 · 0738 2.5 · 1419 0.6 · 2003 2.8
17	TU	0303 0.5 · 0848 2.6 · 1528 0.5 · 2108 2.8
18	W	0401 0.4 · 0942 2.7 · 1622 0.5 · 2158 2.9
19	TH	0445 0.3 · 1024 2.8 · 1705 0.4 · 2241 2.9
20	F	0524 0.4 · 1103 2.9 · 1747 0.4 · 2322 2.9
21 ○	SA	0604 0.4 · 1144 3.0 · 1828 0.4
22	SU	0002 3.0 · 0643 0.4 · 1221 3.0 · 1905 0.3
23	M	0036 3.0 · 0716 0.3 · 1254 3.0 · 1939 0.3
24	TU	0110 2.9 · 0749 0.3 · 1326 3.0 · 2012 0.3
25	W	0145 2.9 · 0821 0.4 · 1400 3.0 · 2045 0.4
26	TH	0221 2.9 · 0853 0.4 · 1434 3.0 · 2119 0.4
27	F	0258 2.9 · 0928 0.5 · 1511 3.0 · 2157 0.4
28	SA	0339 2.8 · 1009 0.6 · 1554 3.0 · 2243 0.5
29 ◑	SU	0426 2.8 · 1058 0.6 · 1644 3.0 · 2339 0.5
30	M	0523 2.7 · 1200 0.7 · 1746 2.9
31	TU	0047 0.5 · 0632 2.7 · 1314 0.7 · 1858 3.0

JUNE

Date	Day	Times / m
1	W	0201 0.4 · 0746 2.7 · 1430 0.6 · 2011 3.0
2	TH	0311 0.4 · 0856 2.8 · 1540 0.5 · 2118 3.1
3	F	0416 0.3 · 0958 2.9 · 1645 0.4 · 2219 3.1
4	SA	0517 0.2 · 1055 3.0 · 1745 0.3 · 2317 3.1
5 ●	SU	0613 0.2 · 1146 3.0 · 1838 0.2
6	M	0008 3.1 · 0700 0.2 · 1231 3.1 · 1925 0.1
7	TU	0056 3.0 · 0744 0.2 · 1316 3.1 · 2012 0.2
8	W	0145 3.0 · 0829 0.3 · 1404 3.2 · 2059 0.3
9	TH	0234 3.0 · 0912 0.4 · 1450 3.2 · 2143 0.3
10	F	0321 2.9 · 0954 0.5 · 1534 3.1 · 2227 0.4
11	SA	0406 2.8 · 1036 0.6 · 1619 3.1 · 2312 0.6
12 ◐	SU	0454 2.7 · 1122 0.7 · 1707 3.0
13	M	0000 0.6 · 0545 2.7 · 1216 0.7 · 1801 2.9
14	TU	0055 0.6 · 0643 2.6 · 1318 0.7 · 1904 2.8
15	W	0158 0.6 · 0746 2.6 · 1426 0.6 · 2010 2.8
16	TH	0302 0.6 · 0848 2.7 · 1530 0.6 · 2112 2.9
17	F	0358 0.5 · 0942 2.8 · 1625 0.6 · 2204 2.9
18	SA	0447 0.5 · 1029 2.9 · 1713 0.5 · 2251 2.9
19	SU	0532 0.5 · 1113 3.0 · 1758 0.5 · 2334 3.0
20 ○	M	0613 0.5 · 1153 3.1 · 1839 0.4
21	TU	0013 3.0 · 0652 0.4 · 1230 3.1 · 1917 0.4
22	W	0051 3.0 · 0729 0.4 · 1307 3.1 · 1954 0.4
23	TH	0129 3.0 · 0805 0.5 · 1343 3.2 · 2031 0.4
24	F	0207 3.0 · 0841 0.5 · 1419 3.2 · 2108 0.4
25	SA	0245 2.9 · 0919 0.5 · 1459 3.2 · 2150 0.4
26	SU	0329 2.9 · 1003 0.5 · 1546 3.1 · 2239 0.4
27 ◑	M	0418 2.8 · 1053 0.6 · 1636 3.1 · 2330 0.5
28	TU	0511 2.8 · 1146 0.6 · 1730 3.1
29	W	0026 0.6 · 0609 2.8 · 1248 0.6 · 1833 3.0
30	TH	0130 0.5 · 0715 2.8 · 1400 0.7 · 1943 3.0

JULY

Date	Day	Times / m
1	F	0241 0.5 · 0825 2.9 · 1515 0.6 · 2055 3.0
2	SA	0352 0.5 · 0933 2.9 · 1625 0.5 · 2202 3.0
3	SU	0457 0.4 · 1035 3.0 · 1729 0.3 · 2303 3.0
4 ●	M	0555 0.3 · 1130 3.0 · 1824 0.3 · 2357 3.0
5	TU	0644 0.3 · 1217 3.1 · 1912 0.2
6	W	0045 3.0 · 0728 0.4 · 1302 3.2 · 1957 0.3
7	TH	0131 3.0 · 0812 0.4 · 1348 3.3 · 2042 0.4
8	F	0217 3.0 · 0854 0.5 · 1432 3.3 · 2124 0.5
9	SA	0259 3.0 · 0933 0.5 · 1511 3.2 · 2202 0.5
10	SU	0339 2.9 · 1009 0.6 · 1550 3.1 · 2240 0.6
11	M	0419 2.8 · 1047 0.7 · 1630 3.1 · 2318 0.7
12 ◐	TU	0501 2.8 · 1128 0.7 · 1713 3.0 · 2358 0.7
13	W	0546 2.7 · 1217 0.8 · 1804 2.9
14	TH	0050 0.8 · 0641 2.7 · 1319 0.8 · 1908 2.8
15	F	0155 0.8 · 0747 2.7 · 1431 0.8 · 2018 2.8
16	SA	0305 0.7 · 0853 2.8 · 1541 0.7 · 2124 2.9
17	SU	0408 0.7 · 0952 2.9 · 1640 0.7 · 2220 2.9
18	M	0500 0.6 · 1043 3.0 · 1730 0.6 · 2308 3.0
19 ○	TU	0547 0.6 · 1128 3.1 · 1814 0.5 · 2352 3.0
20	W	0630 0.6 · 1209 3.2 · 1857 0.5
21	TH	0033 3.1 · 0712 0.5 · 1248 3.2 · 1938 0.5
22	F	0113 3.1 · 0752 0.5 · 1326 3.3 · 2017 0.4
23	SA	0151 3.0 · 0829 0.5 · 1404 3.3 · 2056 0.4
24	SU	0230 3.0 · 0908 0.4 · 1446 3.2 · 2139 0.4
25	M	0315 2.9 · 0953 0.5 · 1533 3.2 · 2228 0.4
26	TU	0404 2.9 · 1041 0.5 · 1624 3.2 · 2316 0.5
27 ◑	W	0454 2.9 · 1130 0.6 · 1714 3.1
28	TH	0003 0.6 · 0545 2.9 · 1225 0.7 · 1811 3.0
29	F	0101 0.7 · 0647 2.9 · 1336 0.7 · 1922 3.0
30	SA	0214 0.7 · 0801 2.9 · 1457 0.7 · 2041 3.0
31	SU	0334 0.7 · 0917 3.0 · 1614 0.6 · 2154 3.0

AUGUST

Date	Day	Times / m
1	M	0444 0.6 · 1024 3.0 · 1719 0.4 · 2256 3.0
2 ●	TU	0542 0.5 · 1119 3.1 · 1812 0.4 · 2348 3.0
3	W	0630 0.5 · 1206 3.2 · 1859 0.4
4	TH	0033 3.0 · 0713 0.5 · 1248 3.3 · 1941 0.4
5	F	0114 3.0 · 0754 0.5 · 1329 3.3 · 2021 0.5
6	SA	0154 3.1 · 0832 0.5 · 1408 3.3 · 2057 0.5
7	SU	0231 3.0 · 0906 0.6 · 1444 3.2 · 2131 0.6
8	M	0306 3.0 · 0939 0.6 · 1518 3.2 · 2204 0.6
9	TU	0342 2.9 · 1012 0.7 · 1554 3.1 · 2236 0.7
10 ◐	W	0418 2.9 · 1046 0.7 · 1630 3.0 · 2308 0.8
11	TH	0453 2.8 · 1123 0.8 · 1710 2.9 · 2347 0.9
12	F	0538 2.7 · 1216 0.9 · 1807 2.8
13	SA	0048 0.9 · 0643 2.7 · 1330 0.9 · 1921 2.7
14	SU	0207 0.9 · 0801 2.8 · 1452 0.9 · 2041 2.8
15	M	0326 0.8 · 0914 2.9 · 1605 0.8 · 2149 2.8
16	TU	0430 0.7 · 1013 3.0 · 1702 0.7 · 2244 2.9
17	W	0522 0.7 · 1102 3.1 · 1750 0.6 · 2330 3.0
18 ○	TH	0609 0.6 · 1145 3.2 · 1835 0.5
19	F	0012 3.1 · 0653 0.5 · 1225 3.3 · 1918 0.4
20	SA	0053 3.1 · 0734 0.5 · 1306 3.3 · 2000 0.4
21	SU	0133 3.1 · 0814 0.4 · 1347 3.3 · 2041 0.4
22	M	0213 3.1 · 0854 0.4 · 1430 3.3 · 2124 0.4
23	TU	0256 3.0 · 0938 0.4 · 1516 3.2 · 2209 0.5
24	W	0342 3.0 · 1024 0.5 · 1605 3.2 · 2253 0.6
25 ◑	TH	0429 3.0 · 1110 0.6 · 1655 3.1 · 2337 0.7
26	F	0519 0.7 · 1203 0.7 · 1751 3.0
27	SA	0034 0.8 · 0622 2.9 · 1316 0.8 · 1905 2.9
28	SU	0153 0.9 · 0744 2.9 · 1445 0.8 · 2032 2.9
29	M	0321 0.9 · 0909 3.0 · 1609 0.7 · 2151 2.9
30	TU	0437 0.7 · 1018 3.1 · 1713 0.5 · 2250 3.0
31	W	0531 0.6 · 1108 3.2 · 1800 0.5 · 2335 3.0

Chart Datum: 1·68 metres below Normal Null (German reference level). HAT is 3·0 metres above Chart Datum.

STANDARD TIME UT −01
Subtract 1 hour for UT
For German Summer Time add
ONE hour in **non-shaded areas**

HELGOLAND LAT 54°11'N LONG 7°53'E
TIMES AND HEIGHTS OF HIGH AND LOW WATERS

Dates in red are **SPRINGS**
Dates in blue are **NEAPS**

YEAR 2016

SEPTEMBER

Day	Time	m	Time	m	Time	m	Time	m
1 TH ●	0614	0.6	1150	3.2	1840	0.4		
2 F	0014	3.0	0653	0.6	1229	3.2	1919	0.5
3 SA	0052	3.1	0731	0.6	1306	3.2	1955	0.5
4 SU	0127	3.1	0806	0.5	1342	3.2	2027	0.6
5 M	0201	3.1	0838	0.6	1415	3.2	2058	0.6
6 TU	0233	3.0	0908	0.6	1447	3.1	2128	0.6
7 W	0306	3.0	0940	0.6	1519	3.1	2158	0.7
8 TH	0338	2.9	1010	0.7	1551	3.0	2224	0.8
9 F ☾	0409	2.9	1039	0.8	1625	2.8	2255	0.9
10 SA	0446	2.8	1123	0.9	1714	2.7	2349	1.0
11 SU	0545	2.7	1233	1.0	1828	2.6		
12 M	0110	1.1	0708	2.7	1403	1.0	1956	2.7
13 TU	0241	1.0	0832	2.8	1527	0.9	2115	2.8
14 W	0357	0.9	0940	3.0	1633	0.7	2216	2.9
15 TH	0455	0.7	1032	3.1	1723	0.6	2305	3.0
16 F ○	0544	0.6	1117	3.2	1809	0.5	2348	3.1
17 SA	0629	0.5	1159	3.3	1854	0.4		
18 SU	0030	3.1	0713	0.5	1242	3.3	1938	0.4
19 M	0112	3.1	0756	0.4	1327	3.3	2021	0.4
20 TU	0154	3.1	0838	0.4	1412	3.3	2104	0.5
21 W	0236	3.2	0921	0.4	1457	3.3	2145	0.6
22 TH	0318	3.1	1004	0.5	1544	3.1	2226	0.7
23 F ☽	0403	3.1	1049	0.6	1634	3.0	2311	0.8
24 SA	0455	3.0	1144	0.7	1733	2.9		
25 SU	0010	1.0	0602	2.9	1258	0.8	1851	2.8
26 M	0133	1.0	0728	2.9	1431	0.8	2021	2.8
27 TU	0306	1.0	0856	3.0	1559	0.8	2142	2.9
28 W	0423	0.9	1005	3.1	1701	0.6	2238	2.9
29 TH	0515	0.7	1051	3.2	1740	0.5	2315	3.0
30 F	0552	0.6	1127	3.2	1814	0.5	2349	3.0

OCTOBER

Day	Time	m	Time	m	Time	m	Time	m
1 SA ●	0628	0.6	1204	3.2	1851	0.5		
2 SU	0025	3.0	0706	0.6	1242	3.2	1926	0.5
3 M	0100	3.1	0740	0.5	1316	3.1	1957	0.6
4 TU	0132	3.1	0811	0.5	1347	3.1	2027	0.6
5 W	0203	3.1	0840	0.6	1417	3.1	2055	0.6
6 TH	0233	3.0	0910	0.6	1447	3.0	2122	0.6
7 F	0303	3.0	0939	0.7	1518	2.9	2147	0.8
8 SA	0332	2.9	1007	0.8	1551	2.8	2217	0.9
9 SU ☾	0407	2.9	1045	0.9	1636	2.7	2305	1.1
10 M	0501	2.8	1149	1.0	1744	2.6		
11 TU	0022	1.1	0619	2.8	1316	1.0	1911	2.7
12 W	0154	1.1	0747	2.9	1445	1.0	2035	2.8
13 TH	0317	0.9	0901	3.0	1556	0.7	2142	2.9
14 F	0421	0.8	0958	3.1	1652	0.6	2235	3.0
15 SA	0514	0.7	1047	3.2	1741	0.5	2322	3.1
16 SU ○	0603	0.6	1133	3.3	1828	0.4		
17 M	0006	3.1	0650	0.5	1219	3.3	1914	0.4
18 TU	0049	3.2	0735	0.4	1305	3.3	1958	0.4
19 W	0132	3.2	0820	0.4	1352	3.3	2041	0.5
20 TH	0215	3.2	0904	0.4	1439	3.2	2123	0.6
21 F	0257	3.2	0947	0.5	1525	3.0	2203	0.7
22 SA ☽	0342	3.1	1033	0.6	1616	2.9	2249	0.9
23 SU	0435	3.1	1128	0.8	1717	2.8	2349	1.0
24 M	0541	3.0	1239	0.9	1831	2.7		
25 TU	0107	1.1	0702	2.9	1404	0.9	1956	2.7
26 W	0236	1.0	0827	3.0	1529	0.8	2114	2.8
27 TH	0354	0.9	0936	3.0	1632	0.7	2211	2.9
28 F	0447	0.8	1018	3.1	1711	0.6	2248	2.9
29 SA	0524	0.7	1100	3.1	1744	0.6	2321	3.0
30 SU ●	0600	0.6	1138	3.2	1820	0.6	2358	3.0
31 M	0639	0.6	1216	3.1	1856	0.6		

NOVEMBER

Day	Time	m	Time	m	Time	m	Time	m
1 TU	0033	3.1	0714	0.5	1250	3.1	1929	0.6
2 W	0105	3.1	0745	0.5	1320	3.1	1958	0.6
3 TH	0135	3.1	0815	0.5	1351	3.0	2026	0.6
4 F	0205	3.0	0845	0.6	1422	3.0	2053	0.6
5 SA	0235	3.0	0914	0.6	1454	2.9	2121	0.8
6 SU	0306	3.0	0945	0.7	1529	2.8	2154	0.9
7 M ☾	0343	3.0	1023	0.9	1613	2.8	2240	1.0
8 TU	0432	2.9	1120	1.0	1712	2.7	2346	1.1
9 W	0539	2.9	1236	0.9	1829	2.7		
10 TH	0110	1.1	0700	2.9	1400	0.9	1951	2.7
11 F	0233	1.0	0818	2.9	1515	0.7	2103	2.9
12 SA	0343	0.8	0922	3.1	1617	0.6	2202	3.0
13 SU	0443	0.7	1018	3.2	1712	0.5	2255	3.1
14 M ○	0538	0.6	1110	3.2	1804	0.4	2343	3.1
15 TU	0629	0.5	1158	3.3	1851	0.4		
16 W	0027	3.2	0715	0.4	1245	3.2	1936	0.4
17 TH	0111	3.1	0802	0.4	1334	3.2	2021	0.5
18 F	0157	3.2	0850	0.4	1424	3.1	2105	0.6
19 SA	0242	3.3	0935	0.5	1512	3.0	2147	0.6
20 SU	0327	3.2	1020	0.6	1602	2.9	2233	0.8
21 M ☽	0417	3.1	1111	0.7	1657	2.8	2327	1.0
22 TU	0515	3.1	1211	0.8	1801	2.7		
23 W	0032	1.0	0623	3.0	1322	0.8	1912	2.6
24 TH	0148	1.0	0738	2.9	1436	0.8	2025	2.7
25 F	0302	0.9	0849	2.9	1542	0.7	2127	2.8
26 SA	0404	0.8	0946	3.0	1632	0.7	2213	2.9
27 SU	0450	0.8	1030	3.1	1712	0.7	2251	3.0
28 M	0532	0.7	1111	3.1	1751	0.7	2330	3.0
29 TU ●	0612	0.6	1150	3.0	1828	0.6		
30 W	0007	3.1	0649	0.6	1225	3.0	1902	0.6

DECEMBER

Day	Time	m	Time	m	Time	m	Time	m
1 TH	0040	3.1	0722	0.5	1258	3.0	1933	0.6
2 F	0112	3.1	0754	0.6	1331	3.0	2004	0.6
3 SA	0144	3.1	0826	0.6	1404	3.0	2034	0.7
4 SU	0215	3.1	0857	0.7	1438	3.0	2106	0.7
5 M	0249	3.1	0932	0.7	1515	2.9	2142	0.8
6 TU	0328	3.1	1013	0.8	1558	2.8	2227	0.9
7 W ☾	0414	3.1	1103	0.8	1650	2.8	2322	1.0
8 TH	0510	3.0	1205	0.8	1754	2.7		
9 F	0031	1.0	0620	3.0	1317	0.8	1908	2.8
10 SA	0149	0.9	0735	3.0	1432	0.8	2022	2.9
11 SU	0304	0.9	0847	3.1	1542	0.7	2128	2.9
12 M	0413	0.7	0951	3.1	1645	0.6	2228	3.0
13 TU	0515	0.6	1050	3.1	1743	0.4	2321	3.1
14 W ○	0611	0.4	1142	3.1	1832	0.4		
15 TH	0008	3.2	0700	0.3	1231	3.1	1918	0.4
16 F	0054	3.2	0748	0.4	1321	3.1	2004	0.5
17 SA	0142	3.3	0837	0.4	1412	3.1	2051	0.6
18 SU	0230	3.3	0924	0.5	1500	3.0	2134	0.6
19 M	0313	3.3	1007	0.6	1545	2.9	2215	0.7
20 TU	0356	3.2	1050	0.7	1631	2.8	2300	0.8
21 W ☽	0443	3.1	1136	0.8	1721	2.7	2350	0.9
22 TH	0536	3.0	1228	0.8	1817	2.7		
23 F	0049	0.9	0638	2.9	1329	0.8	1921	2.6
24 SA	0157	0.9	0747	2.9	1437	0.8	2028	2.7
25 SU	0307	0.9	0856	2.9	1542	0.8	2128	2.8
26 M	0410	0.8	0954	3.0	1636	0.8	2213	2.9
27 TU	0502	0.8	1042	3.0	1722	0.7	2303	3.0
28 W	0546	0.7	1125	3.0	1802	0.7	2343	3.1
29 TH ●	0625	0.6	1203	3.0	1839	0.6		
30 F	0019	3.1	0702	0.6	1239	3.0	1914	0.6
31 SA	0053	3.1	0737	0.6	1315	3.0	1949	0.6

Chart Datum: 1·68 metres below Normal Null (German reference level). HAT is 3·0 metres above Chart Datum.

Germany

》 **FREE** monthly updates. Register at 《
www.reedsnauticalalmanac.co.uk

673

9.15.10 RIVER ELBE
Niedersachsen/Schleswig-Holstein

CHARTS AC 3619, 3625, 3267, 3268; D 44, 46, 47, 48; BSH 3010.1-13

TIDES

Standard Port CUXHAVEN (→)

Times				Height (metres)			
High Water		Low Water		MHWS	MHWN	MLWN	MLWS
0200	0800	0200	0900	3·8	3·3	0·9	0·5
1400	2000	1400	2100				

Differences SCHARHÖRN (11M NW of Cuxhaven)

−0045	−0047	−0101	−0103	+0·1	0·0	+0·1	+0·1

GROßER VOGELSAND

−0044	−0046	−0101	−0103	+0·1	0·0	+0·1	+0·1

OTTERNDORF

+0029	+0029	+0027	+0027	−0·1	−0·1	0·0	+0·1

GLÜCKSTADT

+0205	+0214	+0220	+0213	0·0	0·0	0·0	+0·2

STADERSAND

+0241	+0245	+0300	+0254	+0·1	+0·2	−0·1	+0·1

SCHULAU

+0304	+0315	+0337	+0321	+0·3	+0·3	−0·2	0·0

SEEMANNSHÖFT (53°32'·5N 09°52'·7E)

+0324	+0332	+0403	+0347	+0·4	+0·4	−0·3	−0·1

HAMBURG

+0338	+0346	+0422	+0406	+0·5	+0·5	−0·3	−0·2

NAVIGATION Distances: Elbe SWM buoy to Cuxhaven = 24M; Cuxhaven to Wedel Yacht Haven = 45M; Wedel to City Sport Hafen = 11·5M. The river is 13m deep up to Hamburg and tidal to Geesthacht, 24M above Hamburg.

> Strong W winds can raise the level by as much as 4m. Do not try to enter in strong W/NW winds against the ebb. It is a very busy waterway and at night the many lights can be confusing. Yachts should keep to stbd, just outside the marked chan.

SHELTER AND FACILITIES Several hbrs are listed in the RH col, detailed on BSH 3010 sheet Nos. The area is worth exploring.

FREIBURG: 7M above Brunsbüttel on the SW bank. Enter at HW, but mainly commercial at Freiburg Reede; dries, little space for small craft. ML 2·2m. HM ☎(04779) 8314. Facilities: ⚓ 🅿 ✕ 🛠 ⚓ 🛥; **Jugendheim & YC Klubheim** ✕ 🍴 🛒 (Sheet 6).

STÖRLOCH/BORSFLETH: Ent approx 600m above locks on E bank. ML 2·8m. HM ☎(04124) 71437. Facilities: ⚓ YC. Stör Bridge VHF Ch 09, 16 (Sheet 6).

STÖR/BEIDENFLETH: ML 2·8m. **Langes Rack YC** (Sheet 6)

GLÜCKSTADT: E bank, mouth of R Rhin (see inset). Good shelter: 1 marina in outer hbr, 5·8m at MHW. Or via entry gate, opens HW −2 to +½, into inner hbr for 3 marinas in 3-4·5m. VHF Ch 11. HM ☎(04124) 2087; ⊖ ☎2171. 🛥 ⚓ ✕ 🍴 🛒 🛢 YC 🛥 (Sheet 6).

WISCHHAFEN: HM☎(04770) 334; ⊖ 3014 🛥 ⚓ ⚓ 🛥 ML 2·7m. (Sheet 6). Dries.

RUTHENSTROM: 197° ldg bns 2 Oc G 3s with △ topmarks into hbr. ML 2·6m. HM ☎(04143) 5282, ⚓ 🛥 ⚓ 🅿 🛠 🛥 (Sheet 7). Partially dries, hard ground.

KRÜCKAUMÜNDUNG JACHTHAFEN: Entrance via lock gate; ☎(04125) 1521 🛥 ⚓. ML 2·7m. ⊖ ☎20551 (Sheet 7).

PINNAUMÜNDUNG JACHTHAFEN: Ent via lock gate on N bank after passing through main locks. ML 2·5m. HM ☎(04101) 22447 ⚓ 🛥 ⚓ YC, 🛥 (Sheets 7/8).

PINNAU-NEUENDEICH: Drying marina 1½M up the Pinnau from main locks and another at approx 2M (Sheets 7/8).

ABBENFLETH: Ent marked by two bns in line 221° with △ topmarks. (Sheets 7/8). Dries.

HASELDORF Hafen: HM ☎(04129) 268, ⚓ 🛒 🛥 YC (Sheets 8/9). Dries.

R SCHWINGE on SW bank, 12M up-river from Glückstadt: **BRUNSHAUSEN** ☎(04141) 3085. ML 1·1; Duration 0510 (Sheets 8/9).

HÖRNE/WÖHRDEN ML 2·8m.

STADE HM ☎(04141) 101275 ⊖ ☎3014, 🛒 YC, 🛥 . Very good shelter in scenic town (Sheets 8/9).

LÜHE: Hafen in town of Lühe (Sheet 8).

RIVER ELBE, VESSEL TRAFFIC SERVICE

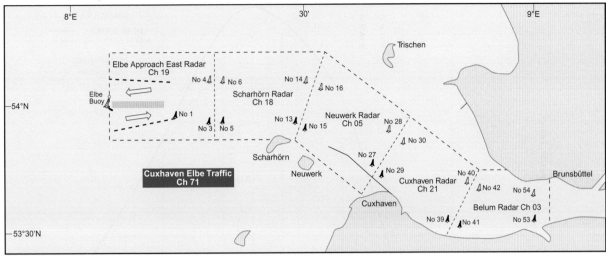

Fig 9.15.10A. Elbe SWM buoy to Brunsbüttel

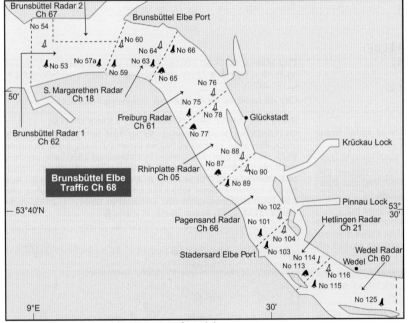

Fig 9.15.10B. Brunsbüttel to 2M east of Wedel

Elbe VTS is not mandatory for yachts, but it is sensible to monitor the 3 Traffic Centres (callsigns and channels below) on your way upriver.

Cuxhaven Elbe Traffic	Ch 71
Brunsbüttel Elbe Traffic	Ch 68
Hamburg Port Traffic	Ch 14/74

Broadcasts of navigational, weather and traffic info are detailed in each port entry.

If radar guidance is required in English, request it from the relevant Traffic Centre.

The radar stations, their areas of coverage and VHF channels are stated for sectors:

Elbe Lt Float-buoy 53	See Cuxhaven
Buoys 51-125	See Brunsbüttel

Then continue with *Hamburg Radar:*

Buoys 125-129	Ch 19
Buoys 129-Seemanshöft (09°53'E)	Ch 03
Seemanshöft-Vorhafen (09°57'E)	Ch 63
09°57'E-Norderelbe bridges	Ch 05

Fig 9.15.10C. 2M east of Wedel to Hamburg City centre

Germany

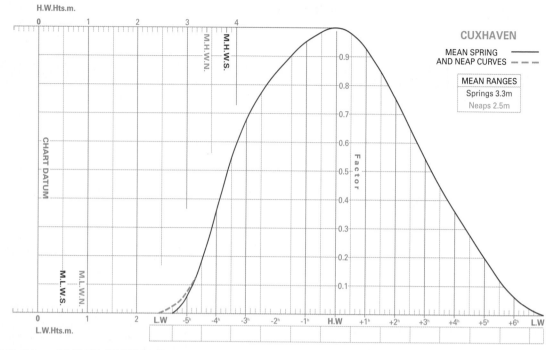

Tidal curve diagram. Axis labels: H.W.Hts.m. (0, 1, 2, 3, 4); L.W.Hts.m. (0, 1, 2); markings M.H.W.N., M.H.W.S., M.L.W.S., M.L.W.N.; CHART DATUM; Factor (0.1–0.9); time axis L.W, -5h, -4h, -3h, -2h, -1h, H.W, +1h, +2h, +3h, +4h, +5h, +6h, L.W.

CUXHAVEN
MEAN SPRING AND NEAP CURVES
MEAN RANGES
Springs 3.3m
Neaps 2.5m

9.15.11 CUXHAVEN (RIVER ELBE)

Niedersachsen 53°52'·43N 08°42'·49E (Marina ent)

CHARTS AC 3619; Imray C26; D 44, BSH 3010.1/2

TIDES +0103 Dover; ML 1·7; Duration 0535

CUXHAVEN is a Standard Port (→)

SHELTER Good in both marinas: YC marina N of radar tower (2·5–3·5m) is open H24 Apr-Oct, direct access from the river.

City Marina, close S of radio mast, is entered from Alter Hafen via bridge which opens (0500-2300) H and H+30, on request Ch 69 *Cuxhaven Lock*, not at night nor in winds > F7. Yachts > 30m LOA berth in the Alter Hafen or Alter Fischereihafen.

NAVIGATION WPT 53°53'·92N 08°41'·20E [No 31 SHM buoy, Fl G 4s], 152°/1·7M to marina ent. From No 27 SHM buoy, appr is in W sector (149°-154°) of Baumrönne ldg lt Iso 4s. Chan well marked/lit, see Lights, buoys & waypoints. *The ebb runs up to 5kn off Cuxhaven. There is much commercial shipping.*

LIGHTS AND MARKS Ent to YC Marina is close NW of conspic radar tr, and N of disused lt ho (dark R with copper cupola). YC marina ent, N side FWG; S side FWR, shown 1 Apr-31 Oct.

COMMUNICATIONS (Code 04721) CG 38011/12, VHF Ch 16; LB 34622, VHF Ch 16; **Cuxhaven Elbe Traffic** (VTS) 567380, broadcasts nav/tfc/weather info for sector Cuxhaven-Brunsbüttel on Ch 71 in German & English every H + 35, yachts should monitor Ch **71**; 681757; Police & Ambulance 110; Waterway Police 745930; 21085; Brit Consul (030) 204570; Dr 112; Port Authority 500150; Cuxhaven Port/lock 500120, Ch 69. *Cuxhaven Elbe Port* (Sig stn/Distress) Ch **12** 16 (H24). *German Bight Traffic* (VTS) Ch 80; see Fig 9.15.16A; Boatyard/Crane 399000; British Cruising Ass'n and Little Ship Club 57270, or 51339. Radar advice for the outer Elbe is provided, in English on request, as follows:

Elbe East Radar	Ch 19	Elbe lt buoy – buoy 5.
Scharhörn Radar	Ch 18	Buoys 3–15.
Neuwerk Radar	Ch 05	Buoys 13–29.
Cuxhaven Radar	Ch 21	Buoys 27–41.
Belum Radar	Ch 03	Buoys 39–53.

FACILITIES YC Marina (Segler-Vereinigung) 34111 (Apr-Oct) €1·50, (cr/card only) (from town) (16A/220V / 32A/380V €0·50/kilowatt) 663650, chart agent (10t) courtesy bikes available.

City Marina 175 90 20015, open all year, 90 berths inc approx 45 for €1·50. (from town) chart agent (25t).

City All facilities, (Bremen, Hamburg, Hanover).

HARBOUR 7M ESE OF CUXHAVEN, ON S BANK OF ELBE

OTTERNDORF, Niedersachsen, 53°50'·10N 08°53'·84E. AC 3619, 3625; BSH 3014.11. HW +0100 on Dover (UT); ML 2·9m; Duration 0525; see River Elbe. Ent marked by Otterndorf 3 SHM and perches. Appr chan (0·8m) from Elbe divides: yachts can take the W branch to Kutterhafen (0·9m) and the R. Medem; or to E, through lock for the Hadelner Kanal, which leads to the Geeste-Elbe Kanal, and turn hard stbd into yacht hbr. Lockmaster (04751) 2190; Yacht hbr 13131, . **Town** (3km) .

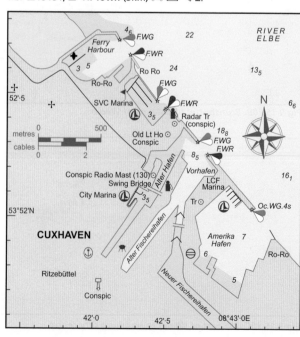

Chart of Cuxhaven harbour showing Ferry Harbour, Ro Ro, SVC Marina, Radar Tr (conspic), Old Lt Ho Conspic, Alter Hafen, Vorhafen, LCF Marina, Amerika Hafen, Conspic Radio Mast (130), Swing Bridge, City Marina, Alter Fischereihafen, Neuer Fischereihafen, Ritzebüttel, CUXHAVEN, RIVER ELBE. Lights: FWG, FWR, Oc.WG.4s. Coordinates 52'·5, 53°52'N, 42'·0, 42'·5, 08°43'·0E.

STANDARD TIME UT –01
Subtract 1 hour for UT
For German Summer Time add ONE hour in **non-shaded areas**

CUXHAVEN   LAT 53°52'N   LONG 8°43'E

TIMES AND HEIGHTS OF HIGH AND LOW WATERS

Dates in red are SPRINGS
Dates in blue are NEAPS

YEAR 2016

JANUARY

Time	m		Time	m
1 0508	3.6	**16** 0506	3.8
1202	0.7		1211	0.5
F 1739	3.3	SA 1741	3.4
2 0007	0.8	**17** 0022	0.6
0547	3.5		0557	3.7
SA 1236	0.8	SU 1258	0.6
◔ 1821	3.2	◑ 1832	3.4
3 0048	0.9	**18** 0112	0.6
0633	3.4		0652	3.6
SU 1320	0.8	M 1349	0.6
1910	3.1		1930	3.3
4 0140	0.9	**19** 0214	0.7
0730	3.4		0758	3.5
M 1419	0.9	TU 1454	0.7
2013	3.2		2040	3.3
5 0249	1.0	**20** 0331	0.8
0840	3.4		0915	3.5
TU 1531	0.9	W 1613	0.7
2123	3.3		2157	3.4
6 0406	0.9	**21** 0454	0.7
0950	3.4		1033	3.5
W 1642	0.8	TH 1730	0.7
2228	3.4		2308	3.5
7 0516	0.9	**22** 0610	0.6
1053	3.5		1141	3.5
TH 1743	0.7	F 1837	0.5
2324	3.5			
8 0616	0.7	**23** 0009	3.6
1147	3.5		0712	0.4
F 1836	0.7	SA 1239	3.5
			1930	0.5
9 0012	3.6	**24** 0058	0.4
0706	0.6		0801	0.4
SA 1236	3.6	SU 1327	3.5
1923	0.6	○ 2013	0.5
10 0055	3.7	**25** 0140	3.8
0752	0.6		0842	0.4
SU 1321	3.6	M 1407	3.5
● 2008	0.6		2051	0.5
11 0137	3.7	**26** 0219	3.8
0836	0.5		0921	0.5
M 1405	3.6	TU 1445	3.5
2053	0.5		2129	0.5
12 0218	3.8	**27** 0257	3.9
0920	0.5		0959	0.6
TU 1448	3.6	W 1521	3.5
2135	0.5		2204	0.6
13 0258	3.8	**28** 0332	3.8
1002	0.4		1032	0.6
W 1529	3.6	TH 1556	3.5
2212	0.5		2235	0.6
14 0336	3.8	**29** 0405	3.7
1042	0.4		1102	0.6
TH 1609	3.5	F 1629	3.5
2251	0.5		2305	0.6
15 0418	3.8	**30** 0438	3.7
1125	0.4		1131	0.6
F 1653	3.5	SA 1703	3.4
2334	0.5		2337	0.7
		31 0512	3.6
			1159	0.7
		SU 1737	3.3

FEBRUARY

Time	m		Time	m
1 0007	0.7	**16** 0051	0.5
0546	3.5		0630	3.5
M 1227	0.7	TU 1320	0.6
◑ 1813	3.2		1859	3.4
2 0043	0.8	**17** 0146	0.6
0629	3.4		0732	3.4
TU 1309	0.8	W 1420	0.7
1904	3.2		2008	3.3
3 0139	0.9	**18** 0303	0.6
0731	3.3		0852	3.3
W 1416	0.9	TH 1544	0.7
2014	3.2		2132	3.3
4 0258	0.9	**19** 0434	0.6
0849	3.2		1019	3.3
TH 1538	0.9	F 1712	0.7
2132	3.3		2252	3.4
5 0424	0.8	**20** 0557	0.5
1006	3.3		1134	3.4
F 1658	0.7	SA 1824	0.6
2243	3.4		2356	3.6
6 0539	0.7	**21** 0700	0.5
1114	3.4		1230	3.4
SA 1805	0.6	SU 1916	0.5
2342	3.5			
7 0640	0.6	**22** 0044	3.7
1212	3.5		0746	0.4
SU 1902	0.6	M 1313	3.5
			○ 1956	0.5
8 0031	3.7	**23** 0123	3.7
0732	0.5		0824	0.4
M 1302	3.6	TU 1349	3.5
● 1952	0.5		2033	0.5
9 0115	3.7	**24** 0159	3.8
0819	0.4		0859	0.4
TU 1348	3.6	W 1422	3.5
2038	0.4		2108	0.4
10 0157	3.8	**25** 0234	3.8
0905	0.3		0903	0.3
W 1431	3.6	TH 1453	3.5
2121	0.3		2140	0.4
11 0239	3.9	**26** 0307	3.7
0949	0.3		1003	0.5
TH 1513	3.6	F 1526	3.5
2202	0.3		2210	0.4
12 0321	3.9	**27** 0338	3.7
1031	0.2		1033	0.6
F 1555	3.6	SA 1558	3.5
2242	0.3		2240	0.5
13 0406	3.9	**28** 0410	3.6
1113	0.2		1102	0.5
SA 1638	3.6	SU 1631	3.5
2325	0.4		2311	0.5
14 0453	3.8	**29** 0441	3.6
1156	0.2		1127	0.5
SU 1723	3.5	M 1700	3.4
			2337	0.5
15 0008	0.4			
0540	3.7			
M 1237	0.5			
◐ 1808	3.5			

MARCH

Time	m		Time	m
1 0510	3.4	**16** 0032	0.4
1147	0.6		0611	3.4
TU 1728	3.3	W 1253	0.6
			1833	3.4
2 0002	0.6	**17** 0122	0.5
0542	3.3		0711	3.2
W 1215	0.7	TH 1351	0.7
◑ 1808	3.2		1942	3.3
3 0044	0.7	**18** 0238	0.6
0635	3.1		0833	3.1
TH 1311	0.8	F 1516	0.7
1913	3.1		2109	3.3
4 0158	0.7	**19** 0413	0.6
0753	3.1		1003	3.2
F 1438	0.8	SA 1650	0.6
2038	3.2		2234	3.4
5 0333	0.7	**20** 0540	0.5
0922	3.2		1120	3.3
SA 1612	0.7	SU 1805	0.5
2201	3.3		2338	3.5
6 0501	0.6	**21** 0640	0.3
1041	3.3		1212	3.4
SU 1733	0.6	M 1855	0.4
2310	3.5			
7 0612	0.4	**22** 0021	3.6
1146	3.4		0721	0.3
M 1838	0.5	TU 1249	3.4
			1932	0.4
8 0004	3.6	**23** 0058	3.6
0709	0.3		0756	0.3
TU 1239	3.5	W 1322	3.5
1931	0.4	○ 2008	0.4
9 0051	3.7	**24** 0134	3.6
0758	0.2		0830	0.3
W 1326	3.6	TH 1354	3.5
● 2017	0.3		2044	0.3
10 0134	3.8	**25** 0208	3.6
0844	0.2		0903	0.3
TH 1410	3.6	F 1426	3.5
2101	0.2		2116	0.3
11 0218	3.8	**26** 0240	3.6
0929	0.1		0934	0.3
F 1453	3.6	SA 1457	3.5
2145	0.2		2146	0.3
12 0304	3.8	**27** 0312	3.6
1012	0.2		1003	0.4
SA 1536	3.6	SU 1529	3.5
2229	0.2		2217	0.3
13 0351	3.8	**28** 0343	3.5
1054	0.2		1032	0.4
SU 1619	3.6	M 1601	3.5
2311	0.2		2247	0.3
14 0437	3.7	**29** 0415	3.5
1135	0.2		1057	0.4
M 1701	3.6	TU 1630	3.4
2352	0.3		2313	0.3
15 0523	3.6	**30** 0443	3.3
1212	0.4		1118	0.4
TU 1743	3.5	W 1657	3.4
●			2336	0.4
		31 0513	3.2
			1142	0.6
		TH 1732	3.3
		◔	

APRIL

Time	m		Time	m
1 0012	0.5	**16** 0215	0.5
0600	3.1		0811	3.1
F 1231	0.7	SA 1446	0.7
1831	3.2		2040	3.3
2 0118	0.6	**17** 0343	0.5
0714	3.0		0936	3.1
SA 1354	0.8	SU 1616	0.6
1953	3.2		2202	3.4
3 0251	0.6	**18** 0507	0.4
0844	3.1		1050	3.2
SU 1531	0.7	M 1731	0.5
2121	3.3		2306	3.5
4 0424	0.5	**19** 0606	0.3
1008	3.2		1140	3.3
M 1657	0.5	TU 1822	0.4
2235	3.5		2349	3.5
5 0540	0.3	**20** 0645	0.2
1116	3.3		1216	3.4
TU 1806	0.4	W 1859	0.3
2333	3.6			
6 0639	0.2	**21** 0026	3.5
1212	3.5		0720	0.3
W 1903	0.3	TH 1250	3.5
			1938	0.3
7 0024	3.7	**22** 0104	3.5
0732	0.1		0758	0.3
TH 1302	3.5	F 1326	3.5
● 1953	0.2	○ 2017	0.3
8 0111	3.8	**23** 0141	3.5
0820	0.1		0833	0.3
F 1346	3.6	SA 1400	3.5
2039	0.1		2052	0.3
9 0158	3.8	**24** 0215	3.5
0905	0.1		0905	0.3
SA 1430	3.7	SU 1431	3.5
2125	0.1		2123	0.2
10 0246	3.8	**25** 0246	3.5
0949	0.1		0934	0.3
SU 1514	3.7	M 1502	3.5
2211	0.1		2154	0.2
11 0334	3.7	**26** 0319	3.4
1032	0.2		1003	0.3
M 1557	3.7	TU 1534	3.5
2254	0.2		2225	0.2
12 0420	3.6	**27** 0352	3.4
1111	0.3		1031	0.4
TU 1639	3.6	W 1605	3.5
2335	0.2		2255	0.3
13 0506	3.5	**28** 0425	3.3
1149	0.4		1058	0.4
W 1722	3.5	TH 1638	3.5
			2324	0.4
14 0015	0.3	**29** 0500	3.2
0555	3.3		1129	0.5
TH 1231	0.5	F 1716	3.4
◑ 1813	3.5			
15 0105	0.5	**30** 0001	0.5
0655	3.2		0546	3.2
F 1327	0.7	SA 1216	0.6
1918	3.4	◑ 1809	3.3

Chart Datum: 2·06 metres below Normal Null (German reference level). HAT is 4·1 metres above Chart Datum.

》》 **FREE** monthly updates. Register at 《
www.reedsnauticalalmanac.co.uk

677

Germany

STANDARD TIME UT –01
Subtract 1 hour for UT
For German Summer Time add
ONE hour in **non-shaded areas**

CUXHAVEN LAT 53°52'N LONG 8°43'E
TIMES AND HEIGHTS OF HIGH AND LOW WATERS

Dates in red are **SPRINGS**
Dates in blue are **NEAPS**

YEAR **2016**

MAY

	Time	m		Time	m
1 SU	0100 0651 1327 1922	0.5 3.1 0.7 3.3	**16** M	0259 0852 1526 2115	0.5 3.0 0.6 3.3
2 M	0221 0814 1455 2044	0.5 3.1 0.6 3.4	**17** TU	0414 1001 1639 2220	0.4 3.1 0.5 3.4
3 TU	0349 0935 1619 2159	0.4 3.2 0.5 3.5	**18** W	0516 1056 1736 2310	0.3 3.2 0.4 3.4
4 W	0504 1044 1730 2301	0.3 3.3 0.4 3.6	**19** TH	0601 1138 1821 2352	0.3 3.4 0.4 3.5
5 TH	0606 1143 1833 2357	0.2 3.4 0.3 3.7	**20** F	0642 1217 1905	0.3 3.5 0.3
6 F	0703 1236 1929	0.1 3.5 0.2	**21** SA	0034 0724 1257 1948	3.5 0.3 3.5 0.3
7 SA	0050 0756 1323 2019	3.7 0.1 3.6 0.1	**22** SU	0114 0803 1334 2026	3.5 0.3 3.6 0.3
8 SU	0139 0842 1407 2105	3.7 0.1 3.7 0.1	**23** M	0149 0837 1407 2100	3.5 0.3 3.6 0.3
9 M	0227 0925 1451 2151	3.7 0.1 3.7 0.1	**24** TU	0224 0909 1439 2133	3.5 0.3 3.6 0.3
10 TU	0316 1008 1536 2237	3.6 0.2 3.7 0.2	**25** W	0300 0941 1513 2207	3.4 0.3 3.6 0.3
11 W	0404 1050 1619 2320	3.5 0.3 3.7 0.3	**26** TH	0336 1013 1547 2241	3.4 0.4 3.6 0.3
12 TH	0451 1130 1704	3.4 0.4 3.6	**27** F	0413 1047 1625 2318	3.4 0.5 3.6 0.4
13 F	0002 0540 1213 1754	0.4 3.3 0.5 3.5	**28** SA	0454 1126 1708	3.3 0.6 3.6
14 SA	0050 0636 1304 1852	0.5 3.2 0.6 3.5	**29** SU	0001 0541 1214 1759	0.5 3.3 0.6 3.5
15 SU	0148 0740 1409 2001	0.5 3.1 0.7 3.4	**30** M	0054 0639 1313 1901	0.5 3.2 0.6 3.5
			31 TU	0201 0749 1427 2013	0.5 3.2 0.6 3.5

JUNE

	Time	m		Time	m
1 W	0315 0903 1544 2126	0.4 3.3 0.6 3.6	**16** TH	0414 1002 1641 2224	0.5 3.2 0.6 3.4
2 TH	0428 1012 1657 2232	0.4 3.4 0.5 3.6	**17** F	0512 1056 1739 2316	0.5 3.3 0.5 3.5
3 F	0534 1114 1804 2333	0.3 3.5 0.4 3.6	**18** SA	0603 1143 1831	0.4 3.5 0.5
4 SA	0637 1211 1907	0.3 3.5 0.2	**19** SU	0003 0649 1227 1918	3.5 0.4 3.6 0.4
5 SU	0031 0733 1302 2000	3.7 0.1 3.6 0.1	**20** M	0046 0732 1307 2000	3.5 0.4 3.6 0.4
6 M	0123 0821 1347 2047	3.6 0.1 3.7 0.1	**21** TU	0126 0811 1345 2038	3.5 0.4 3.7 0.4
7 TU	0212 0905 1432 2133	3.6 0.2 3.7 0.2	**22** W	0205 0849 1421 2117	3.6 0.4 3.7 0.4
8 W	0301 0948 1518 2221	3.5 0.3 3.8 0.3	**23** TH	0245 0926 1457 2154	3.5 0.4 3.8 0.4
9 TH	0350 1031 1602 2305	0.4 0.4 3.8 0.4	**24** F	0324 1001 1533 2231	3.5 0.4 3.7 0.4
10 F	0436 1112 1646 2347	3.4 0.5 3.7 0.4	**25** SA	0403 1038 1613 2313	3.5 0.5 3.7 0.4
11 SA	0522 1153 1731	3.4 0.5 3.6	**26** SU	0446 1122 1659	3.4 0.5 3.7
12 SU	0029 0609 1237 1820	0.5 3.3 0.6 3.5	**27** M	0000 0535 1211 1750	0.4 3.4 0.6 3.7
13 M	0113 0700 1326 1914	0.6 3.2 0.7 3.4	**28** TU	0049 0628 1303 1845	0.5 3.3 0.6 3.7
14 TU	0206 0757 1426 2017	0.6 3.1 0.6 3.4	**29** W	0143 0726 1403 1948	0.5 3.3 0.6 3.6
15 W	0308 0900 1534 2124	0.5 3.1 0.6 3.4	**30** TH	0245 0833 1514 2059	0.5 3.4 0.6 3.6

JULY

	Time	m		Time	m
1 F	0356 0943 1629 2209	0.5 3.4 0.6 3.6	**16** SA	0416 1008 1652 2237	0.7 3.3 0.7 3.4
2 SA	0508 1050 1742 2316	0.4 3.5 0.4 3.6	**17** SU	0521 1107 1755 2332	0.6 3.5 0.6 3.5
3 SU	0615 1151 1848	0.3 3.5 0.3	**18** M	0616 1157 1848	0.5 3.6 0.5
4 M	0017 0715 1245 1945	3.6 0.2 3.6 0.2	**19** TU	0021 0705 1242 1935	3.5 0.5 3.7 0.5
5 TU	0112 0805 1333 2034	3.6 0.2 3.7 0.2	**20** W	0106 0750 1323 2019	3.6 0.5 3.8 0.5
6 W	0201 0849 1417 2119	3.5 0.3 3.8 0.3	**21** TH	0149 0833 1402 2101	3.6 0.5 3.8 0.4
7 TH	0248 0932 1502 2204	3.6 0.4 3.9 0.4	**22** F	0230 0914 1440 2141	3.6 0.4 3.9 0.4
8 F	0333 1014 1544 2246	3.6 0.5 3.9 0.5	**23** SA	0310 0951 1517 2220	3.6 0.4 3.9 0.4
9 SA	0415 1052 1624 2324	3.6 0.5 3.8 0.5	**24** SU	0349 1028 1558 2302	3.5 0.4 3.8 0.4
10 SU	0455 1128 1703 2359	3.4 0.6 3.7 0.6	**25** M	0432 1112 1645 2349	3.5 0.5 3.8 0.4
11 M	0535 1205 1744	3.4 0.6 3.6	**26** TU	0521 1201 1737	3.5 0.5 3.8
12 TU	0034 0616 1243 1827	0.6 3.3 0.7 3.5	**27** W	0036 0610 1249 1829	0.5 3.5 0.6 3.7
13 W	0112 0701 1327 1918	0.7 3.2 0.8 3.4	**28** TH	0122 0702 1342 1926	0.6 3.4 0.6 3.6
14 TH	0200 0756 1426 2022	0.7 3.2 0.8 3.4	**29** F	0217 0804 1449 2037	0.6 3.4 0.6 3.5
15 F	0305 0901 1539 2132	0.7 3.2 0.8 3.4	**30** SA	0329 0919 1610 2155	0.7 3.4 0.7 3.5
			31 SU	0449 1034 1730 2308	0.6 3.5 0.5 3.5

AUGUST

	Time	m		Time	m
1 M	0602 1140 1838	0.5 3.6 0.4	**16** TU	0544 1127 1820 2357	0.7 3.5 0.6 3.5
2 TU	0011 0702 1234 1934	3.5 0.4 3.7 0.3	**17** W	0641 1216 1911	0.6 3.7 0.5
3 W	0104 0751 1321 2021	3.5 0.4 3.8 0.3	**18** TH	0045 0730 1300 1958	3.6 0.6 3.8 0.5
4 TH	0150 0834 1403 2103	3.6 0.5 3.9 0.4	**19** F	0130 0816 1340 2042	3.7 0.5 3.9 0.4
5 F	0232 0914 1443 2144	3.6 0.5 3.9 0.5	**20** SA	0211 0857 1419 2124	3.7 0.4 3.9 0.4
6 SA	0310 0953 1522 2221	3.6 0.5 3.9 0.5	**21** SU	0251 0936 1500 2204	3.6 0.4 3.9 0.3
7 SU	0347 1027 1557 2253	3.6 0.5 3.8 0.6	**22** M	0331 1015 1542 2246	3.6 0.4 3.9 0.4
8 M	0422 1059 1632 2325	3.5 0.6 3.7 0.6	**23** TU	0414 1058 1629 2330	3.6 0.4 3.8 0.5
9 TU	0458 1132 1707 2355	3.5 0.7 3.7 0.7	**24** W	0500 1144 1718	3.6 0.5 3.8
10 W	0533 1204 1743	3.4 0.7 3.6	**25** TH	0014 0546 1230 1808	0.6 3.6 0.5 3.7
11 TH	0023 0608 1236 1824	0.8 3.3 0.8 3.4	**26** F	0057 0635 1320 1906	0.7 3.5 0.6 3.5
12 F	0058 0653 1324 1920	0.8 3.3 0.8 3.3	**27** SA	0150 0738 1429 2020	0.8 3.4 0.7 3.4
13 SA	0156 0757 1436 2035	0.9 3.2 0.9 3.2	**28** SU	0306 0859 1557 2147	0.8 3.4 0.6 3.4
14 SU	0315 0914 1600 2153	0.9 3.3 0.8 3.3	**29** M	0436 1024 1725 2307	0.8 3.5 0.7 3.5
15 M	0435 1027 1718 2301	0.8 3.4 0.7 3.4	**30** TU	0555 1133 1833	0.7 3.6 0.5
			31 W	0008 0652 1224 1922	3.5 0.6 3.7 0.4

Chart Datum: 2·06 metres below Normal Null (German reference level). HAT is 4·1 metres above Chart Datum.

〉〉 **FREE** monthly updates. Register at 〈
www.reedsnauticalalmanac.co.uk 〈

STANDARD TIME UT –01
Subtract 1 hour for UT
For German Summer Time add
ONE hour in **non-shaded areas**

CUXHAVEN LAT 53°52'N LONG 8°43'E

TIMES AND HEIGHTS OF HIGH AND LOW WATERS

Dates in red are **SPRINGS**
Dates in blue are **NEAPS**

YEAR 2016

SEPTEMBER

Time	m		Time	m
1 0053	3.6	**16**	0022	3.6
0735	0.5		0707	0.6
TH 1305	3.8	F	1232	3.7
● 2003	0.4	○	1933	0.4
2 0132	3.6	**17**	0107	3.6
0814	0.5		0753	0.5
F 1344	3.8	SA	1315	3.8
2041	0.5		2019	0.4
3 0209	3.6	**18**	0149	3.7
0853	0.5		0837	0.4
SA 1421	3.8	SU	1357	3.9
2117	0.5		2102	0.4
4 0243	3.6	**19**	0230	3.7
0928	0.5		0919	0.4
SU 1456	3.8	M	1441	3.9
2151	0.6		2145	0.4
5 0317	3.6	**20**	0311	3.7
0959	0.5		1001	0.4
M 1529	3.8	TU	1526	3.9
2221	0.6		2226	0.4
6 0349	3.6	**21**	0354	3.7
1029	0.6		1043	0.4
TU 1601	3.7	W	1612	3.8
2250	0.6		2308	0.5
7 0422	3.5	**22**	0436	3.7
1100	0.6		1126	0.5
W 1632	3.6	TH	1658	3.7
2317	0.7		2348	0.6
8 0453	3.5	**23**	0521	3.6
1129	0.7		1209	0.6
TH 1704	3.5	F	1748	3.5
2340	0.7	☽		
9 0522	3.4	**24**	0030	0.8
1156	0.8		0610	3.5
F 1737	3.3	SA	1300	0.7
☾			1847	3.4
10 0007	0.8	**25**	0125	0.9
0600	3.3		0715	3.5
SA 1233	0.9	SU	1410	0.8
1826	3.2		2005	3.3
11 0056	1.0	**26**	0244	1.0
0659	3.2		0840	3.5
SU 1339	0.9	M	1543	0.8
1940	3.1		2136	3.3
12 0216	1.0	**27**	0419	0.9
0820	3.2		1009	3.6
M 1509	0.9	TU	1714	0.7
2108	3.2		2258	3.4
13 0349	1.0	**28**	0541	0.8
0944	3.4		1119	3.7
TU 1639	0.8	W	1821	0.6
2227	3.3		2356	3.5
14 0511	0.8	**29**	0636	0.7
1054	3.5		1206	3.7
W 1750	0.6	TH	1903	0.5
2330	3.4			
15 0615	0.6	**30**	0034	3.5
1147	3.6		0713	0.6
TH 1845	0.5	F	1243	3.7
			1937	0.5

OCTOBER

Time	m		Time	m
1 0107	3.6	**16**	0041	3.6
0749	0.6		0729	0.5
SA 1319	3.7	SU	1250	3.8
● 2013	0.5	○	1954	0.4
2 0141	3.6	**17**	0125	3.7
0827	0.6		0815	0.5
SU 1356	3.7	M	1335	3.9
2049	0.5		2039	0.4
3 0215	3.6	**18**	0207	3.8
0903	0.5		0900	0.4
M 1430	3.7	TU	1421	3.9
2121	0.5		2123	0.4
4 0247	3.6	**19**	0249	3.8
0934	0.5		0945	0.4
TU 1501	3.7	W	1509	3.9
2150	0.6		2205	0.5
5 0318	3.6	**20**	0332	3.8
1003	0.5		1037	0.4
W 1531	3.6	TH	1555	3.7
2216	0.6		2246	0.6
6 0348	3.6	**21**	0415	3.7
1032	0.6		1110	0.5
TH 1601	3.5	F	1641	3.6
2241	0.6		2326	0.7
7 0417	3.5	**22**	0459	3.7
1059	0.6		1153	0.6
F 1631	3.4	SA	1731	3.4
2305	0.7	☽		
8 0445	3.5	**23**	0009	0.8
1126	0.7		0549	3.6
SA 1703	3.3	SU	1244	0.7
2331	0.9		1831	3.3
9 0521	3.4	**24**	0104	1.0
1159	0.8		0653	3.5
SU 1747	3.2	M	1351	0.8
☾			1945	3.2
10 0015	1.0	**25**	0219	1.0
0614	3.3		0812	3.5
M 1258	0.9	TU	1517	0.8
1856	3.1		2110	3.2
11 0130	1.1	**26**	0348	1.0
0732	3.3		0937	3.5
TU 1425	0.9	W	1645	0.7
2023	3.1		2240	3.3
12 0304	1.0	**27**	0510	0.9
0859	3.4		1049	3.6
W 1558	0.9	TH	1752	0.6
2149	3.3		2328	3.4
13 0433	0.9	**28**	0607	0.7
1015	3.5		1138	3.6
TH 1715	0.7	F	1834	0.5
2258	3.4			
14 0542	0.7	**29**	0005	3.5
1114	3.6		0645	0.7
F 1814	0.5	SA	1215	3.7
2353	3.5		1907	0.5
15 0639	0.6	**30**	0037	3.5
1203	3.7		0722	0.6
SA 1905	0.4	SU	1252	3.7
		●	1943	0.6
		31	0113	3.6
			0801	0.6
		M	1330	3.6
			2020	0.5

NOVEMBER

Time	m		Time	m
1 0148	3.6	**16**	0144	3.8
0838	0.5		0841	0.4
TU 1404	3.6	W	1402	3.8
2053	0.5		2101	0.4
2 0220	3.6	**17**	0227	3.8
0910	0.5		0928	0.4
W 1435	3.6	TH	1451	3.7
2121	0.5		2145	0.5
3 0250	3.6	**18**	0313	3.8
0939	0.5		1015	0.4
TH 1506	3.5	F	1541	3.6
2148	0.5		2228	0.6
4 0319	3.6	**19**	0358	3.8
1008	0.6		1059	0.5
F 1536	3.4	SA	1629	3.5
2214	0.6		2310	0.7
5 0349	3.6	**20**	0442	3.8
1037	0.6		1142	0.6
SA 1608	3.4	SU	1718	3.4
2240	0.7		2353	0.8
6 0420	3.6	**21**	0531	3.7
1106	0.7		1230	0.7
SU 1642	3.3	M	1812	3.3
2311	0.9	☽		
7 0457	3.5	**22**	0043	0.9
1142	0.8		0627	3.6
M 1725	3.2	TU	1326	0.8
● 2353	1.0		1915	3.2
8 0545	3.4	**23**	0145	1.0
1235	0.9		0733	3.5
TU 1824	3.2	W	1435	0.8
			2026	3.1
9 0058	1.0	**24**	0259	0.9
0653	3.4		0847	3.4
W 1349	0.9	TH	1551	0.7
1943	3.2		2139	3.2
10 0222	1.0	**25**	0416	0.9
0814	3.4		0959	3.5
TH 1516	0.8	F	1701	0.7
2107	3.2		2241	3.3
11 0349	0.9	**26**	0521	0.8
0932	3.5		1057	3.5
F 1634	0.7	SA	1753	0.7
2220	3.4		2328	3.4
12 0504	0.8	**27**	0610	0.7
1038	3.6		1143	3.6
SA 1739	0.6	SU	1835	0.6
2321	3.5			
13 0608	0.6	**28**	0007	3.5
1134	3.7		0654	0.7
SU 1836	0.5	M	1225	3.6
			1914	0.6
14 0014	3.6	**29**	0045	3.6
0704	0.6		0736	0.6
M 1226	3.8	TU	1304	3.6
○ 1929	0.4	●	1952	0.6
15 0101	3.7	**30**	0122	3.6
0755	0.5		0814	0.5
TU 1315	3.8	W	1340	3.5
2017	0.4		2026	0.5

DECEMBER

Time	m		Time	m
1 0155	3.6	**16**	0209	3.8
0847	0.5		0913	0.4
TH 1413	3.5	F	1437	3.6
2057	0.5		2128	0.5
2 0226	3.6	**17**	0256	3.9
0919	0.5		1003	0.5
F 1446	3.5	SA	1528	3.6
2128	0.6		2214	0.6
3 0258	3.7	**18**	0343	3.9
0951	0.5		1050	0.5
SA 1520	3.5	SU	1616	3.5
2157	0.6		2257	0.6
4 0330	3.7	**19**	0427	3.8
1023	0.6		1132	0.6
SU 1553	3.4	M	1701	3.4
2227	0.7		2337	0.7
5 0403	3.7	**20**	0510	3.7
1056	0.7		1212	0.6
M 1629	3.4	TU	1747	3.3
2302	0.8			
6 0442	3.6	**21**	0018	0.8
1136	0.7		0556	3.6
TU 1712	3.3	W	1254	0.7
2344	0.9	☽	1836	3.2
7 0527	3.6	**22**	0105	0.8
1223	0.8		0648	3.5
W 1803	3.3	TH	1343	0.7
●			1931	3.1
8 0038	0.9	**23**	0200	0.9
0624	3.5		0748	3.4
TH 1323	0.8	F	1443	0.7
1908	3.2		2034	3.1
9 0146	0.9	**24**	0307	0.9
0733	3.4		0856	3.4
F 1435	0.8	SA	1553	0.8
2023	3.2		2141	3.2
10 0305	0.9	**25**	0420	0.9
0849	3.6		1005	3.4
SA 1551	0.7	SU	1700	0.8
2138	3.4		2242	3.4
11 0424	0.8	**26**	0527	0.8
1001	3.6		1105	3.5
SU 1703	0.5	M	1757	0.7
2245	3.5		2333	3.5
12 0536	0.7	**27**	0623	0.7
1106	3.7		1155	3.5
M 1808	0.4	TU	1843	0.7
2345	3.6			
13 0641	0.5	**28**	0017	3.6
1204	3.7		0710	0.6
TU 1907	0.4	W	1238	3.5
			1924	0.6
14 0038	3.6	**29**	0056	3.6
0737	0.6		0750	0.6
W 1258	3.7	TH	1317	3.5
○ 1958	0.4	●	2002	0.6
15 0124	3.7	**30**	0132	3.7
0826	0.3		0826	0.6
TH 1347	3.6	F	1354	3.6
2043	0.4		2038	0.6
		31	0207	3.7
			0903	0.5
		SA	1431	3.5
			2114	0.6

Chart Datum: 2·06 metres below Normal Null (German reference level). HAT is 4·1 metres above Chart Datum.

Germany

》》 **FREE** monthly updates. Register at 《
www.reedsnauticalalmanac.co.uk

679

9.15.12 BRUNSBÜTTEL (RIVER ELBE)

Schleswig-Holstein 53°53'·27N 09°08'·48E ✦✦✦♤♤♧♧

CHARTS AC 2469, 3625; Imray C26; D 42; BSH 3010.3/4, 3009.1/2

TIDES +0203 Dover; ML 1·4; Duration 0520

Standard Port CUXHAVEN (←)

Times				Height (metres)			
High Water		Low Water		MHWS	MHWN	MLWN	MLWS
0200	0800	0200	0900	3·8	3·3	0·9	0·5
1400	2000	1400	2100				
Differences BRUNSBÜTTEL							
+0057	+0105	+0121	+0112	–0·1	–0·2	–0·1	+0·1

SHELTER Good in the Alter Hafen, which is a drying yacht hbr, W of the locks. Waiting area is close E of FWG ☆ at ent to Alter Vorhafen. No smoking in locks area. Once locked through, good shelter, but noisy and crowded in Kanal-Yachthafen, close NW of the larger Neue locks. 4ca upstream, same bank, is a pier with limited AB.

NAVIGATION WPT 53°53'·01N 09°08'·14E, 039°/6ca to Alte Schleusen (the smaller SE locks, usually for yachts). No navigational dangers in the near apprs, but commercial traffic is heavy. The stream sets strongly across access chan to locks.
Lock Entry Sigs: see Nord-Ostsee Kanal; await White signal or loudspeaker.

LIGHTS AND MARKS Lights as chartlet. Ldg lts 065·5° direct large ships from the main fairway to the lock Approach areas.

COMMUNICATIONS (Code 04852) **VTS**, see River Elbe. *Brunsbüttel Elbe Traffic* broadcasts weather info every H + 05 on Ch 68 in English and German for the inner Elbe. Monitor Ch 68 when underway between buoys 53 and 125; ☃ 36400; Police 112; ⊖ 83000; Brit Consul (030) 204570; Ⓗ 9800; WSA (hbr info) 8850 request *Kiel Kanal I* on Ch 13 for a locking allocation 15 mins before arrival; see Nord-Ostsee Kanal. Radar cover of the inner Elbe is as follows:

Brunsbüttel Radar I	Ch 62	Buoys 51–59;
Brunsbüttel Radar II	Ch 67	Buoys 60–Elbehafen;
(call if entering locks in dense fog)		
S Margarethen Radar	Ch 18	Buoys 57a–65;
Freiburg Radar	Ch 61	Buoys 63–77;
Rhinplate Radar	Ch 05	Buoys 75–89;
Pagensand Radar	Ch 66	Buoys 87–89;
Hetlingen Radar	Ch 21	Buoys 101–115;
Wedel Radar	Ch 60	Buoys 113–125.

FACILITIES Alter Hafen ☎01522 785 6319, €0·95, ✗ ⬩SC; YC WSV ☎51508, www.wsv-bru.de; YC SVB ☎3348, www.svb-bru.de
Town ⬩▯⬩ (as shown on chartlet, 0800-1800 ☎2001) ✗ ▯ 🛒 ▯ ⚒✗⬩ ✉ Ⓑ Ⓗ ⓁⒷ ⇌ ✈ (Hamburg).

9.15.13 NORD-OSTSEE KANAL (KIEL CANAL) SCHLESWIG-HOLSTEIN

CHARTS AC 2469; Kieler Förde, inner part 2344, outer part 2341; D42; BSH 3009.1-4

SIGNALS Two sets of Occulting lights must be obeyed: a. those to move from the Waiting Area (Brunsbüttel) and enter the Approach Area, shown from mast at the lock island; and b. those to enter the lock, shown from a mast at the lock:

○　=　No entry.　　○　=　Prepare to enter.
　　　　　　　　　　●

○　=　Ships exempt from pilotage enter; secure on
●　　central island.

○ Oc　=　**Yachts should enter outer hbr and lock** (berth on low, narrow pontoons).

In the canal Sidings, the following traffic sigs apply to yachts:

● QR　　　　　　=　Entry into a Siding is prohib.
●●● (Oc R, vert)　=　Exit from a Siding is prohib.

COMMUNICATIONS Before entering, yachts should request a lock clearance: **at Brunsbüttel** from *Kiel Kanal I* on Ch 13, or **at Holtenau** from *Kiel Kanal IV* on Ch 12. **In the Canal.** Maintain listening watch as follows:

Kiel Kanal I	Ch 13	Brunsbüttel approaches and lock area.
Kiel Kanal II	Ch 02	Brunsbüttel to Breiholz.
Kiel Kanal III	Ch 03	Breiholz to Holtenau.
Kiel Kanal IV	Ch 12	Holtenau approaches and lock area.

Info broadcasts by *Kiel Kanal II* on Ch 02 at H+15, H+45; and by *Kiel Kanal III* on Ch 03 at H+20, H+50. Vessels should monitor these broadcasts and avoid calling the station if possible. Non AIS-fitted yachts >20m are required to carry an AIS Class B transponder, or a portable Class A transponder which can be hired from the lock master.

NAVIGATION The Canal runs 53·3NM (98·7 km) from Brunsbüttel to Kiel-Holtenau; it is 103-162m wide and 11m deep. *Navigation Rules Kiel Canal* in English/German can be downloaded from www.kiel-canal.org/english. Speed limit 8kn. All 8 bridges have 40m clearance. Yachts may only use the Canal and its ents in official daylight hours (next para) and in good visibility (unless going to Brunsbüttel's Alter or Inner Hbrs, or Holtenau's yacht hbr). Motoring or motor-sailing (hoist a B ▼, or a B pennant) is the norm. Pure sailing (no engine) is prohib.

Daylight hours (LT) in the Kiel Canal are: 1-15 May, 0430-2200; 16-31 May, 0400-2230; 1 June-15 July, 0330-2300; 16-31 July, 0400-2230; 1-15 Aug, 0430-2200; 16-31 Aug 0500-2130; 1-15 Sep, 0530-2100; 16-30 Sep, 0600-2030. WSV.de (Guidance for pleasure craft).

Canal dues (approx €18 <12m) E or W-bound, are paid at Kiel-Holtenau to the Lockmaster if using the Old lock; or at the newspaper stand if using the New lock. Credit cards not accepted.

SHELTER As well as Brunsbüttel and Holtenau, 5 sidings or *weichen* (berthing places, see below) must be reached by yachts during **daylight hours** (see above). Other sidings may not be used; they are likely to be less sheltered and exposed to wash.

- *Dückerswisch siding (km 20·5); W side in turning area.
- *Oldenbüttel siding (km 40·5) off lock to Gieselau Canal (links Kiel Canal to R Eider, thence to Tönning, (Sylt)).
- Rendsburg (ent at km 66). All facilities and 3 marinas in Lake Obereider; the W'most is nearest town centre.
- The Borgstedter See, entrances at km 67.5 and 70.
- Flemhudersee (km 85.4, E of Grossnordsee siding). *Usually for one night only.

KIEL-HOLTENAU (km 98·7) is a major port with two pairs of locks leading into Kieler Förde which is practically tideless and at almost the same level as the canal. Monitor Ch 68 *Kiel Traffic* for VTS info whilst in Kieler Förde. Yachts usually use the smaller Old locks (No smoking); signals and pontoons as at Brunsbüttel.

On the N bank (E of Old locks) is a small Yacht Hbr (max 4 days) no ⚡ no ⚓. HM ☎(0431) 363830. A BP fuel barge is close E of the locks. At Wik (S side of locks) **Nautischer Dienst** (Kapt Stegmann & Co) ☎0431 331772 are ACA and BSH Agents. Kiel city is S of Wik.

The **British Kiel YC** 54°23'·22N 10°09'·75E is 1·5M NNE of the locks and 5ca NNW of Stickenhörn unlit ECM buoy. It is part of a military training ctr, no commercial facilities; but ⚓ (about €15·60) possible for visitors (friendly welcome). Call *Sailtrain* VHF Ch 6, ☎ (0431)398833 or www.bkyc.de/. Kiel 30 min bus ride.

9.15.14　HAMBURG (RIVER ELBE)

Schleswig-Holstein 53°32'·52N 09°58'·84E ✳✳✳⊗▵▵▵❀❀

CHARTS AC 3625, 3267, 3268; D 48; BSH 3010.9-12 (see below)

TIDES See River Elbe. ML 1·3; Duration 0435

SHELTER A busy commercial port with several marinas. Visitors berth at Wedel in Hamburger Yachthafen, 12M W of Hamburg, or in the city centre at City Sporthafen, especially for visitors.

HAMBURGER YACHTHAFEN (53°34'·28N 09°40'·53E, N bank). A conspic 50m high radar twr is at SW corner of hbr. Very good shelter, access H24 via 2 ents painted W. 2100 ⚓ €1·14. ▸◗ (shut Tues) ✕ ⬜🛢🅿 ⊞🄴 ✎ SC, ╼ ▵ ♒ ✗ ▱(16t) ⛟(15t) ⛵(masting). HM ☎(4103) 4438 (office); HM (east) ☎0172 6423227; HM (west) ☎ 0171 2766300; ⊖ ☎0172 6423228; www.hamburger-yachthafen.de (Sheets 9/10).

SCHULAU.(Closed for reconstruction, 2015) Good shelter, 7ca E of Wedel. Beware strong tidal streams across ent, marked by FR and FG lts. Partially dries. HM☎ 2422, ▸◗ ▵ ✎ ⊞🅿 ╼ ⛵ SC (Sheets 9/10).

MÜHLENBERG. 53°33'·19N 09°49'·38E, N bank. 250 ⚓ ╼ ⚓ SC. Yachts < 9m; partially dries (Sheet 10).

TEUFELSBRÜCK. 53°32'·82N 09°52'·06E, N bank, E of ferry pier. Good shelter in small hbr; partially dries. ⬜ SC (Sheets 10/11).

RÜSCHKANAL. 53°32'·57N 09°51'·34E, S bank, narrow ent. ⛵🅿⊞🄳 ✎ ╼ ✗; 4 SCs, pontoons NE side in 3·6m (Sheet 11).

CITY SPORTHAFEN. 53°32'·52N 09°58'·84E HM ☎364297, mob 170 805 2004, out of hrs 227 7651. www.city-sporthafen-hamburg.de hafen@city-sporthafen-hamburg.de 80-120 ⚓ ⓥ €1·50, 1 Apr-31 Oct, €0·80, 1 Nov-31 Mar, (⚡🅳 charged according to use, ⮕ €1·00) pay 6 days, 7th free, 🚻 🚿. Most berths <20m LOA; max draft 2·5-4m, max LOA 50m (Sheet 12).

COMMUNICATIONS (Code 40) Met 3190 8826/7; Auto 0190 116456; Brit Consul (030) 204570; *Hamburg Port Traffic* and Hbr Control vessel Ch 13 14 **73** (H24) all monitor Ch 74. UKHO Chart Agent 374 8110.

HAMBURGER YACHTHAFEN, WEDEL

HAMBURG: CITY SPORTHAFEN

9.15.15 BREMERHAVEN (RIVER WESER)

Federal State of Bremen **53°32'·12N 08°34'·50E** ✿✿◊◊✿✿

CHARTS AC 3617, 3621 plans A & B; Imray C26; D 4; BSH 3011.7/8

TIDES +0051 Dover; ML 2·0; Duration 0600

Standard Port WILHELMSHAVEN (→)

Times				Height (metres)			
High Water		Low Water		MHWS	MHWN	MLWN	MLWS
0200	0800	0200	0900	4·8	4·2	1·1	0·5
1400	2000	1400	2100				
Differences ALTE WESER LT HO (53°51'·8N 08°07'·6E)							
−0055	−0048	−0015	−0029	−1·1	−0·9	−0·1	0·0
DWARSGAT (53°43'·0N 08°18'·0E)							
−0015	+0002	−0006	−0001	−0·6	−0·5	−0·1	0·0
ROBBENSUDSTEERT (53°38'·0N 08°27'·0E)							
−+0009	+0026	+0015	+0020	−0·3	−0·2	−0·1	0·0
BREMERHAVEN							
+0029	+0046	+0033	+0038	−0·1	0·0	−0·1	0·0
NORDENHAM							
+0051	+0109	+0055	+0058	0·0	+0·1	−0·2	−0·1
BRAKE							
+0120	+0119	+0143	+0155	−0·2	−0·1	−0·4	−0·2
ELSFLETH							
+0137	+0137	+0206	+0216	−0·2	−0·1	−0·5	−0·2
VEGESACK							
+0208	+0204	+0250	+0254	−0·2	−0·1	−0·5	−0·2
BREMEN							
+0216	+0211	+0311	+0314	0·0	0·0	−0·6	−0·3

SHELTER There are 6 options for visitors. The first three are most suitable and nearest the city centre, on the upper chartlet:

- Berth in R Geeste downstream of Kennedy fixed bridge (clearance 5·5m) on pontoons, but short stay only as it is noisy and exposed to swell/wash. If able to transit the bridge (flood barrage is normally open) berth on quay upstream.

- Lock into the Schleusenhafen via the smaller W lock. Operational H24, but call *Fischereihafenschleuse* Ch 10 for current opening times; ●●◌ (vert) = enter. Floating pontoons in lock. Berth at Fischereihafen II West Quay. This gives access to 2 other YC/marinas well to the S.

- The Weser YC marina in Hauptkanal (E of Handelshafen); excellent shelter, central location and close to the lock.

- Bremerhaven Marina, 1·3M S of lock on W side (3m).

- Wulsdorf Marina, 0·5M further south in Luneorthafen.

- Lloyd Marina in Neuer Hafen, 0·6M NNW of main hbr ent. Access via Neuer Hafen lock Ch 69, 0600–2200, turn 90° port.

Up the R Weser (AC 3622/3) Nordenham, Rodenkirchen, Rechtenfleth (sand/mud) and Sandstedt (sand) have drying hbrs. Many other non-tidal hbrs available include Brake on the Weser and Elsfleth and Oldenburg on the R Hunte.

NAVIGATION WPT 53°32'·25N 08°33'·94E [SHM buoy 61, QG], 111°/650m to cross into the Vorhafen. Beware much commercial shipping; also ferries using R Geeste. From buoy 51 to 61 (opp container terminal) sailing yachts may not tack in fairway, where commercial traffic has right of way. The Geeste and Hadeler Kanals (1·5m depth; air draft 2·7m) link Bremerhaven to Otterndorf (32M). For R Weser (very well marked) see Passage Information.

LIGHTS AND MARKS See upper chartlet and 9.15.4. Ldg lts lead 150.8° up main fairway (R Weser). Vorhafen ent is close SW of conspic 112m radio twr.

COMMUNICATIONS (Code 0471) Police 94667; British Consul (030) 204570; Ⓗ 2991; HM 59613416, VHF Ch 12 (H24); Neuer Hafen lock 471 941 2840 VHF Ch 69; Fischereihafen lock 59603440, VHF Ch 10 (H24). Weser VTS, see diagram overleaf with radar coverage. Up-river *Bremen-Weser Traffic* broadcasts at H+30 on Ch 19 78 81. *Bremen Port Radio* Ch 03 16 (H24). *Hunte Traffic* broadcasts at H+30 on Ch 63.

FACILITIES R Geeste ☎1428690, 20 ⊖ + 440 Ⓥ.

Lloyd Marina ☎1428690, 180 ⊖ €1·19. VHF Ch 69. www.im-jaich.de ✕.

Weser YC ☎23531, 150 ⊖ + 20 Ⓥ, €1·29, ✕ ⬜ ⬡. www.weser-yacht-club.de

Fischereihafen II W Quay ☎76189, ⛺ ⛾ ⚒.

Bremerhaven Marina ☎77555, 280 ⊖ + 30 Ⓥ 🕯 ⬡(13t) �45.

Wulsdorf Marina ☎73268. 150 ⊖ Ⓥ on request, ⬡ 🔌 �45.

City All facilities: ACA 🕯 ≈ ✈.

RIVER WESER

(AC 1875, 3617, 3622, 3623) The R Weser is a major waterway leading to **Bremerhaven, Nordenham, Brake,** and **Bremen** where it connects with the inland waterways. The upper reaches, *Unterweser,* run from Bremen to Bremerhaven and then the *Aussenweser* flows into a wide estuary (split by two main chans: *Neue Weser* and *Alte Weser*). The position and extent of sandbanks vary: on the W side they tend to be steep-to, but on the E side there are extensive drying shoals (eg Tegeler Plate).

The Jade-Weser SWM lt buoy marks the approach from NW to Neue Weser (the main fairway, recommended) and Alte Weser (not well lit) which are separated by Roter Sand and Roter Grund, marked by the disused Roter Sand lt tr (conspic). Both chans are well marked and converge about 3M S of Alte Weser lt tr (conspic). From this junction *Hohewegrinne* (buoyed) leads inward in a SE direction past Tegeler Plate lt tr (conspic) on the NE side and Hohe Weg lt tr (conspic) to the SSE. From Robbenplate to Wremer Tief (part way towards Bremerhaven) it is constrained by training walls, which may cause tidal accelerations. ▶*In the Aussenweser the stream (> 3kn at sp), often sets towards the banks and the branch chans which traverse them.*◀

The Weser-Elbe Wattfahrwasser is a demanding, extremely shallow inshore passage between R Weser and R Elbe. It leads NNE from the Wurster Arm (E of Hohe Weg) keeping about 3M offshore; SE of Neuwerk and around Cuxhaven training wall. It is tricky, normally requires two or three spring tides for a vessel drawing up to 1·2m, and is not recommended for visitors.

9.15.16 HOOKSIEL (RIVER JADE)

Niedersachsen 53°38'·61N 08°05'·19E ✿✿✿🌊🌊✿✿

CHARTS AC 3618; Imray C26; D 7; BSH 3011.3, 3015.10

TIDES +0034 Dover; ML no data; Duration 0605

Standard Port WILHELMSHAVEN (→)

Times				Height (metres)			
High Water		Low Water		MHWS	MHWN	MLWN	MLWS
0200	0800	0200	0900	4·8	4·2	1·1	0·5
1400	2000	1400	2100				
Differences HOOKSIEL							
−0023	−0022	−0008	−0012	−0.5	−0.4	−0.1	0·0
SCHILLIG							
−0031	−0025	−0006	−0014	−0.7	−0.6	−0.1	0·0

SHELTER Temp ⌒ in the Vorhafen (approx 1m at MLWS) but it is very commercial and uncomfortable in E winds. Beyond the lock there is complete shelter in the Binnentief, 2M long and 2·0-3·5m deep. Best for visitors is Alter Hafen Yacht Hbr in the town; max draft 2m. Larger yachts go to YCs; see lockmaster. Do not enter Watersports Area due to water-ski cables.

NAVIGATION WPT 53°39'·33N 08°06'·52E [No 37/Hooksiel 1 SHM buoy, IQ G 13s], 217°/9ca to H3 SHM buoy, Fl G 4s, approx 350m E of the ent. Caution: aquaculture N of H3, seasonal HS ferry, strong cross tide, transhipment pier and restricted area to SE of ent.

Appr via chan 2·1m deep to Vorhafen (beware shoaling) enclosed by two moles. Inner ldg daymarks 276·5°, two RW bcns, lead thru' ent, but are obsc'd when ferry berthed on N pier.

Lock and opening bridge see www. marinahooksiel.de/ Schleusenzeiten for times. Secure well in lock; climb up to control office to pay €1·00/ metre (each way).

LIGHTS AND MARKS Ldg lts 164·5°, as chartlet. Conspic chys of chemical works and oil refinery SSE of lock. ☆ L Fl R 6s on dayglo R pile on S mole and street lamp on N mole. R/G tfc lts at lock.

COMMUNICATIONS (Code 04425) Coast Guard (0421) 5550555; ⚓ (0190) 116047; Police 269; ⊖ 1302; Brit Consul (030) 204570; Ⓗ (04421) 2080; Dr 1080; HM/Lockmaster, no VHF, ☎430. Jade VTS VHF Ch 63.

FACILITIES Werft BY ☎95850, ▐(25t) ⚓. Wilhelmshaven YC ☎ 04421 22983, 50 Ⓥ berths (⌒) €1·00, ⌂. Alter Hafen Ⓥ berth on quays, first night free then €1·00. Marina Hooksiel ☎958050 hafenmeister@wangerland.de www.marinahooksiel.de. €1·20. Town 🛒🛠🗜✕⌂🍺🔥🔧⛽Ⓑ✉⇌✈ (Wilhelmshaven and Bremen). Wangersiel, 3·4M NNW (53°41'·0N 8°01'·5E) has small marina ☎(04463) 1515, ✕⌂🔩.

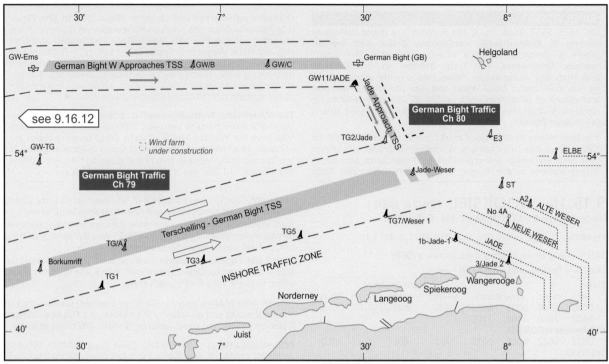

Fig 9.15.16A. Terschelling-German Bight TSS, German Bight Western Approaches TSS and Jade Approach TSS. German Bight VTS: *German Bight Traffic* broadcasts safety and weather info Ch 80 every hour in German/English.

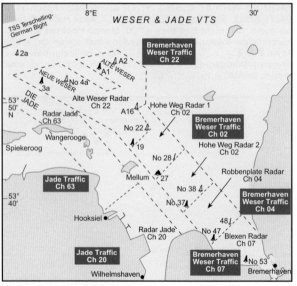

Fig 9.15.16B. Weser and Jade VTS

WESER VTS *Bremerhaven Weser Traffic* broadcasts info in German (English on request) at H+20 on all the chans below. Radar cover in the lower Weser is as Fig 9.15.15B above.

Alte Weser Radar	Ch 22	Neue Weser: buoys 3a–19H.
ditto	ditto	Alte Weser: buoys A1–16a.
Hohe Weg Radar 1 & 2	Ch 02	Buoys 21–37.
Robbenplate Rdr	Ch 04	Buoys 37–47.
Blexen Radar	Ch 07	Buoys 47–63.

JADE VTS *Jade Traffic* broadcasts info in German and English on request and every H+10, Ch 20 63. Radar coverage is as follows:

Jade Radar I	Ch 63	Buoys 1–33;
Jade Radar II	Ch 20	Buoys 33–58.

RIVER JADE

(AC 3617, 3618) The estuaries of the River Weser and River Jade flow seaward between Cuxhaven and Wangerooge.

▶*The outer parts of the rivers Weser and Jade can become very rough with wind against tide, and are dangerous on the ebb in strong NW winds.◀*

Enter from 1b/Jade 1 buoy via Wangerooger Fahrwasser (buoyed fairway) which initially leads ESE to the N of Wangerooge, then S past Minsener Oog and Hooksiel to Wilhelmshaven. S of Wilhelmshaven, the Jadebusen is a large, mostly drying area through which channels run to the small harbours of Dangast and Vareler. Much of it is designated as nature reserves.

Note: Major dredging operations underway for a container term on W bank of the R Jade between Hooksiel and Wilhelmshaven. Beware of hard to see Y measuring buoys outside fairway.

THE EAST FRISIANS, RIVER JADE TO RIVER EMS

(AC 1635. Note: UKHO chart coverage of the East Frisians is too small scale for pilotage. BSH 3015 folio is essential.) Offshore route, passing north of the East Frisian Islands, is about 50 miles long from the E end of Wangerooge to abeam Borkum. This is via the ITZ, south of the Terschellinger-German Bight TSS, whose E-going lane is marked on its S side by SHM buoys TG9/Weser 2 to TG1. These buoys 5M spacing. *Take care to avoid straying inadvertently into the TSS and incurring possible penalties* – it is far better to stay 1M south of these buoys.

4-5M S of this line of buoys the S side of the ITZ is punctuated by landfall buoys marking the approaches to the seven seegaten between the East Frisians. There are major lts on Wangerooge, Norderney and Borkum. But the islands are very low lying with bare dunes and are often difficult to identify with any certainty despite the existence of some landmarks. Even radar identification is none too reliable, best to enter from E or W and traverse well protected from the outside weather.

▶*Near the TSS the E-going stream begins at HW Helgoland −5, and the W-going at HW Helgoland +1, sp rates 1·2kn. Inshore the stream is influenced by the flow through the seegaten.◀*

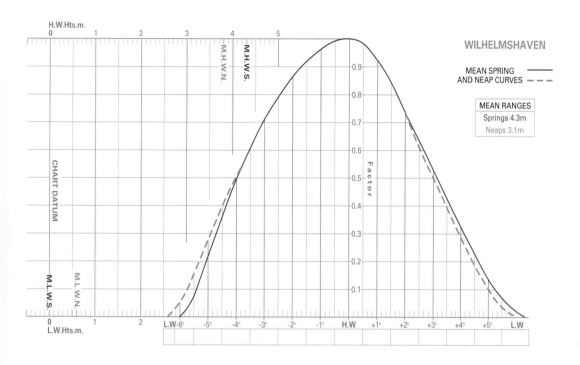

WILHELMSHAVEN

MEAN SPRING ———
AND NEAP CURVES − − −

MEAN RANGES
Springs 4.3m
Neaps 3.1m

9.15.17 WILHELMSHAVEN (R. JADE)

Niedersachsen 53°30'·80N 08°09'·17E ✲✲✲◊◊◊❀❀

CHARTS AC 3617, 3618; Imray C26; D 7; BSH 3011.3/4, 3015.12

WILHELMSHAVEN

Water Tr (conspic)

Church (conspic)

Ems-Jade Kanal

Grosser Hafen

No entry

Kaiser Wilhelm Bridge

Ausrustungs-hafen

Nordhafen

Naval Base

Neuer Vorhafen

RIVER JADE

Ldg Lts 207·8°

Training Wall

TIDES +0050 Dover; ML 2·3; Duration 0615
WILHELMSHAVEN is a Standard Port. (→).

SHELTER Good in yacht hbrs at Nordhafen (rather remote) and Großer Hafen. Access via sea lock which opens H24 at 1 hr's notice. No yacht berths in Neuer Vorhafen (naval port). Nassauhafen in the tidal hbr (*Fluthafen*) has yacht pontoons in about 1·7m, access H24.

NAVIGATION WPT 'A' 53°35'·71N 08°09'·98E (No 45 SHM buoy, QG) 175°/4.4M to WPT 'B' 53°31'·34N 08°10'·57E whence the Fluthafen ent bears 235°/9ca. The fairway is deep, wide and well marked.

Ems-Jade Canal: 39M from Großer Hafen to Emden, can be used by de-masted yachts; max draft 1·7m, min air clearance 3·75m. It has 6 locks. Max speed 4kn.

LIGHTS AND MARKS Principal lts on R Jade, see 9.15.4. Other lts on chartlet.

COMMUNICATIONS (Code 04421) ☎ Bremerhaven 72220; Water Police 942358; Customs 480723; Brit Consul (030) 204570; HM 154580, VHF Ch **11** 16 (H24); Naval base/Port Captain 684923 (operator); VTS 4891281, see Fig. 9.15.15B above for VTS broadcast info, radar assistance; Sealock 7557620, VHF Ch **13**; Bridge opening VHF Ch **11**.

FACILITIES Berthing fee at all marinas is €1·70/m. **Nassauhafen Marina** ☎41439, 28 + 100✔ 🛢 ⚓ ✕ 🍴 ◄ ⚓ ⚓. **Wiking Sportsboothafen** ☎41301, 30 ◯ ✕ 🗑 🖳 🔧 🖂 🛢 ⚓ ⚓ ⚒. **Hochsee YC Germania** ☎44121. There are 16 YCs in the area. **City** 🛒🍴🗑⚓⚓🖳🖂🏧🅱🚂✈.

STANDARD TIME UT −01
Subtract 1 hour for UT
For German Summer Time add ONE hour in **non-shaded areas**

WILHELMSHAVEN LAT 53°31'N LONG 8°09'E
TIMES AND HEIGHTS OF HIGH AND LOW WATERS

Dates in red are **SPRINGS**
Dates in blue are **NEAPS**

YEAR 2016

JANUARY

Day	Time	m	Time	m		Day	Time	m	Time	m
1	0450	4.7	1108	0.9		16	0454	4.8	1113	0.6
F	1718	4.3	2314	1.1		SA	1727	4.4	2323	0.8
2	0529	4.6	1143	1.0		17	0544	4.8	1200	0.7
SA	1758	4.1	◑2352	1.2		SU	1816	4.4	●	
3	0614	4.4	1225	1.1		18	0012	0.8	0637	4.6
SU	1848	4.1				M	1251	0.8	1912	4.3
4	0043	1.3	0712	4.3		19	0112	1.0	0741	4.5
M	1322	1.2	1951	4.1		TU	1354	1.0	2020	4.3
5	0150	1.3	0821	4.3		20	0228	1.0	0858	4.5
TU	1432	1.2	2102	4.2		W	1512	1.0	2137	4.4
6	0307	1.3	0931	4.4		21	0350	1.0	1017	4.5
W	1543	1.1	2208	4.4		TH	1630	0.9	2250	4.5
7	0418	1.2	1034	4.4		22	0506	0.8	1127	4.5
TH	1645	1.0	2305	4.5		F	1737	0.8	2353	4.6
8	0518	1.0	1128	4.5		23	0609	0.6	1226	4.5
F	1739	0.9	2355	4.6		SA	1832	0.6		
9	0609	0.8	1217	4.5		24	0044	4.7	0701	0.5
SA	1829	0.8				SU	1314	4.5	○1918	0.6
10	0039	4.7	0656	0.7		25	0127	4.8	0745	0.5
SU	1303	4.6	●1915	0.7		M	1355	4.5	1959	0.6
11	0121	4.8	0741	0.7		26	0206	4.9	0825	0.6
M	1348	4.6	2000	0.7		TU	1432	4.5	2037	0.7
12	0203	4.9	0825	0.6		27	0244	4.9	0904	0.6
TU	1433	4.6	2043	0.6		W	1508	4.6	2113	0.7
13	0243	4.9	0907	0.5		28	0319	4.9	0938	0.7
W	1514	4.6	2121	0.6		TH	1541	4.5	2144	0.7
14	0323	4.9	0946	0.5		29	0351	4.8	1009	0.8
TH	1555	4.5	2156	0.6		F	1613	4.4	2214	0.8
15	0406	4.9	1027	0.5		30	0423	4.7	1039	0.8
F	1640	4.5	2237	0.7		SA	1645	4.4	2244	0.8
						31	0456	4.6	1108	0.9
						SU	1716	4.3	2313	1.0

FEBRUARY

Day	Time	m	Time	m		Day	Time	m	Time	m
1	0530	4.5	1136	1.0		16	0616	4.6	1223	0.8
M	1751	4.2	◑2347	1.1		TU	1844	4.4		
2	0613	4.3	1215	1.1		17	0043	0.8	0717	4.4
TU	1843	4.1				W	1322	1.0	1951	4.3
3	0042	1.2	0715	4.2		18	0158	0.9	0837	4.3
W	1318	1.2	1955	4.1		TH	1443	1.0	2115	4.3
4	0200	1.3	0833	4.2		19	0330	0.9	1005	4.3
TH	1438	1.2	2115	4.2		F	1611	1.0	2237	4.5
5	0324	1.2	0950	4.3		20	0454	0.8	1121	4.4
F	1557	1.0	2227	4.4		SA	1724	0.8	2343	4.6
6	0438	1.0	1058	4.4		21	0558	0.6	1217	4.4
SA	1705	0.9	2327	4.5		SU	1818	0.7		
7	0539	0.8	1155	4.5		22	0032	4.7	0647	0.5
SU	1805	0.8				M	1301	4.5	○1902	0.6
8	0017	4.7	0632	0.7		23	0113	4.8	0727	0.5
M	1246	4.6	●1857	0.7		TU	1338	4.5	1941	0.6
9	0102	4.8	0721	0.5		24	0149	4.8	0804	0.5
TU	1334	4.6	1944	0.6		W	1412	4.5	2016	0.6
10	0146	4.9	0808	0.4		25	0224	4.8	0838	0.5
W	1420	4.6	2028	0.5		TH	1444	4.5	2048	0.5
11	0230	4.9	0851	0.3		26	0257	4.7	0910	0.6
TH	1503	4.6	2108	0.5		F	1515	4.5	2118	0.5
12	0313	4.9	0933	0.3		27	0328	4.7	0940	0.6
F	1545	4.6	2146	0.4		SA	1545	4.5	2148	0.6
13	0357	4.8	1016	0.4		28	0359	4.7	1009	0.7
SA	1628	4.5	2226	0.4		SU	1615	4.4	2218	0.7
14	0443	4.8	1059	0.5		29	0430	4.6	1036	0.7
SU	1712	4.5	2309	0.6		M	1643	4.4	2243	0.7
15	0529	4.8	1140	0.6						
M	1755	4.5	◑2351	0.7						

MARCH

Day	Time	m	Time	m		Day	Time	m	Time	m
1	0457	4.5	1057	0.8		16	0558	4.4	1158	0.7
TU	1709	4.3	2307	0.8		W	1820	4.4		
2	0529	4.3	1122	0.9		17	0021	0.7	0658	4.2
W	1749	4.1	◑2347	1.0		TH	1254	1.0	1927	4.3
3	0621	4.1	1214	1.1		18	0135	0.9	0820	4.1
TH	1856	4.1				F	1417	1.1	2055	4.3
4	0059	1.1	0739	4.0		19	0309	0.9	0951	4.1
F	1337	1.2	2022	4.1		SA	1550	1.0	2222	4.4
5	0231	1.1	0908	4.1		20	0438	0.7	1108	4.3
SA	1510	1.0	2147	4.3		SU	1706	0.8	2328	4.6
6	0358	0.9	1027	4.3		21	0540	0.5	1201	4.4
SU	1631	0.8	2256	4.5		M	1757	0.6		
7	0508	0.7	1133	4.4		22	0013	4.6	0621	0.4
M	1738	0.7	2353	4.6		TU	1239	4.4	1836	0.6
8	0606	0.5	1228	4.5		23	0050	4.7	0658	0.4
TU	1834	0.6				W	1314	4.5	○1914	0.5
9	0041	4.8	0657	0.4		24	0126	4.7	0735	0.4
W	1316	4.6	●1922	0.4		TH	1348	4.5	1950	0.5
10	0127	4.9	0745	0.2		25	0201	4.7	0809	0.5
TH	1402	4.6	2007	0.3		F	1419	4.5	2023	0.4
11	0213	4.9	0830	0.2		26	0234	4.7	0840	0.5
F	1447	4.6	2051	0.2		SA	1449	4.5	2053	0.4
12	0259	4.9	0915	0.2		27	0305	4.6	0910	0.5
SA	1531	4.6	2133	0.2		SU	1519	4.5	2123	0.4
13	0345	4.9	0959	0.3		28	0335	4.6	0939	0.5
SU	1612	4.6	2213	0.3		M	1548	4.5	2152	0.5
14	0430	4.8	1040	0.4		29	0405	4.5	1005	0.5
M	1653	4.6	2252	0.4		TU	1615	4.5	2217	0.5
15	0513	4.6	1117	0.5		30	0432	4.4	1026	0.6
TU	1733	4.5	◑2331	0.5		W	1641	4.4	2240	0.6
						31	0501	4.2	1050	0.8
						TH	1715	4.2	◑2314	0.8

APRIL

Day	Time	m	Time	m		Day	Time	m	Time	m
1	0546	4.1	1135	1.0		16	0114	0.8	0758	4.0
F	1815	4.1				SA	1348	1.0	2028	4.3
2	0018	0.9	0700	4.0		17	0241	0.8	0924	4.0
SA	1253	1.1	1939	4.1		SU	1516	0.9	2153	4.4
3	0148	0.9	0831	4.1		18	0405	0.6	1039	4.2
SU	1428	1.0	2108	4.3		M	1631	0.7	2258	4.5
4	0319	0.8	0956	4.2		19	0505	0.4	1135	4.4
M	1555	0.8	2223	4.5		TU	1722	0.6	2342	4.5
5	0434	0.5	1106	4.4		20	0545	0.4	1207	4.4
TU	1706	0.6	2324	4.6		W	1800	0.5		
6	0535	0.4	1204	4.5		21	0019	4.6	0622	0.4
W	1805	0.5				TH	1243	4.5	1841	0.5
7	0017	4.8	0630	0.2		22	0058	4.6	0702	0.4
TH	1255	4.6	●1858	0.4		F	1319	4.5	○1921	0.4
8	0107	4.9	0720	0.1		23	0135	4.6	0739	0.4
F	1342	4.7	1945	0.2		SA	1353	4.6	1957	0.4
9	0154	4.9	0808	0.1		24	0209	4.6	0811	0.4
SA	1427	4.7	2031	0.2		SU	1424	4.6	2028	0.3
10	0242	4.9	0853	0.2		25	0241	4.6	0841	0.4
SU	1511	4.7	2115	0.2		M	1454	4.6	2059	0.3
11	0329	4.8	0937	0.2		26	0312	4.5	0910	0.4
M	1553	4.7	2156	0.2		TU	1523	4.5	2129	0.4
12	0414	4.7	1017	0.3		27	0343	4.4	0938	0.5
TU	1633	4.6	2234	0.3		W	1553	4.5	2158	0.4
13	0457	4.5	1055	0.5		28	0414	4.4	1006	0.5
W	1713	4.6	2315	0.5		TH	1623	4.5	2227	0.5
14	0544	4.3	1136	0.7		29	0447	4.3	1037	0.6
TH	1801	4.5	◑			F	1700	4.4	2304	0.6
15	0005	0.6	0642	4.1		30	0531	4.1	1121	0.9
F	1232	0.9	1905	4.4		SA	1753	4.3		

Chart Datum: 2·70 metres below Normal Null (German reference level). HAT is 5·1 metres above Chart Datum.

STANDARD TIME UT –01
Subtract 1 hour for UT
For German Summer Time add
ONE hour in **non-shaded areas**

WILHELMSHAVEN LAT 53°31'N LONG 8°09'E
TIMES AND HEIGHTS OF HIGH AND LOW WATERS

Dates in red are SPRINGS
Dates in blue are NEAPS

YEAR **2016**

MAY

	Time	m		Time	m
1 SU	0001 0637 1229 1908	0.8 4.1 1.0 4.3	**16** M	0159 0839 1427 2104	0.7 4.0 0.9 4.4
2 M	0119 0800 1354 2033	0.8 4.1 0.9 4.4	**17** TU	0313 0950 1538 2211	0.6 4.1 0.7 4.4
3 TU	0244 0923 1519 2149	0.6 4.2 0.8 4.5	**18** W	0416 1046 1635 2302	0.5 4.3 0.6 4.5
4 W	0359 1034 1631 2253	0.5 4.3 0.6 4.7	**19** TH	0503 1128 1722 2344	0.4 4.4 0.6 4.5
5 TH	0503 1135 1735 2352	0.3 4.6 0.5 4.8	**20** F	0545 1208 1807	0.5 4.5 0.5
6 F	0602 1231 1833	0.2 4.6 0.3	**21** SA	0026 0629 1249 1853	4.6 0.5 4.6 0.5
7 SA	0046 0657 1320 1924	4.8 0.2 4.7 0.2	**22** SU	0106 0710 1326 1932	4.6 0.4 4.6 0.4
8 SU	0136 0746 1404 2009	4.8 0.1 4.8 0.1	**23** M	0142 0744 1400 2006	4.6 0.4 4.7 0.4
9 M	0224 0830 1448 2054	4.8 0.2 4.8 0.1	**24** TU	0217 0816 1432 2039	4.6 0.4 4.7 0.4
10 TU	0312 0914 1532 2138	4.7 0.3 4.7 0.2	**25** W	0252 0849 1503 2112	4.5 0.4 4.7 0.4
11 W	0359 0956 1614 2220	4.6 0.4 4.7 0.3	**26** TH	0326 0921 1536 2145	4.4 0.5 4.6 0.4
12 TH	0443 1036 1656 2302	4.5 0.5 4.7 0.4	**27** F	0401 0956 1612 2222	4.4 0.6 4.6 0.5
13 F	0529 1119 1743 2351	4.3 0.7 4.6 0.6	**28** SA	0441 1035 1654 2305	4.3 0.7 4.6 0.6
14 SA	0622 1210 1840	4.2 0.9 4.5	**29** SU	0527 1120 1745 2357	4.3 0.8 4.5 0.6
15 SU	0049 0726 1313 1949	0.7 4.2 0.9 4.4	**30** M	0625 1217 1848	4.2 0.8 4.5
			31 TU	0101 0735 1328 2002	0.6 4.2 0.8 4.5

JUNE

	Time	m		Time	m
1 W	0214 0850 1445 2115	0.6 4.3 0.8 4.6	**16** TH	0317 0949 1542 2213	0.7 4.3 0.8 4.4
2 TH	0327 1001 1558 2223	0.5 4.4 0.7 4.7	**17** F	0417 1044 1642 2306	0.7 4.4 0.7 4.5
3 F	0434 1105 1706 2327	0.4 4.5 0.6 4.7	**18** SA	0509 1132 1735 2353	0.6 4.5 0.7 4.5
4 SA	0538 1205 1809	0.3 4.6 0.4	**19** SU	0557 1217 1824	0.6 4.6 0.6
5 SU	0027 0636 1258 1904	4.8 0.2 4.7 0.2	**20** M	0037 0641 1259 1907	4.6 0.5 4.7 0.5
6 M	0120 0726 1343 1951	4.7 0.2 4.8 0.1	**21** TU	0117 0721 1336 1946	4.6 0.5 4.7 0.5
7 TU	0209 0811 1427 2036	4.7 0.2 4.8 0.2	**22** W	0157 0759 1412 2025	4.6 0.5 4.8 0.5
8 W	0256 0856 1513 2124	4.6 0.3 4.8 0.3	**23** TH	0236 0837 1447 2102	4.6 0.5 4.8 0.4
9 TH	0343 0940 1556 2208	4.5 0.4 4.8 0.4	**24** F	0314 0913 1523 2138	4.5 0.5 4.8 0.5
10 F	0427 1021 1638 2250	4.5 0.6 4.8 0.5	**25** SA	0351 0948 1603 2218	4.4 0.6 4.8 0.5
11 SA	0510 1101 1721 2333	4.4 0.7 4.7 0.6	**26** SU	0435 1030 1649 2304	4.4 0.6 4.7 0.5
12 SU	0555 1144 1807	4.2 0.8 4.6	**27** M	0523 1118 1739 2354	4.4 0.7 4.7 0.6
13 M	0018 0644 1232 1901	0.7 4.1 0.9 4.5	**28** TU	0614 1208 1833	4.3 0.7 4.7
14 TU	0110 0742 1329 2004	0.8 4.0 0.9 4.4	**29** W	0046 0712 1307 1935	0.6 4.3 0.8 4.6
15 W	0211 0846 1435 2111	0.8 4.1 0.9 4.4	**30** TH	0148 0818 1417 2046	0.6 4.3 0.8 4.7

JULY

	Time	m		Time	m
1 F	0259 0929 1531 2159	0.7 4.4 0.8 4.7	**16** SA	0324 0954 1558 2223	0.9 4.3 0.9 4.4
2 SA	0411 1038 1644 2308	0.6 4.5 0.6 4.7	**17** SU	0428 1054 1702 2320	0.8 4.4 0.8 4.4
3 SU	0518 1143 1750	0.5 4.6 0.4	**18** M	0524 1146 1755	0.7 4.6 0.7
4 M	0012 0619 1239 1849	0.4 4.7 0.3 0.3	**19** TU	0009 0615 1231 1842	4.5 0.6 4.7 0.6
5 TU	0108 0711 1327 1939	4.6 0.3 4.8 0.2	**20** W	0055 0702 1313 1928	4.6 0.6 4.8 0.6
6 W	0156 0757 1411 2025	4.6 0.4 4.9 0.3	**21** TH	0139 0746 1353 2011	4.7 0.6 4.9 0.5
7 TH	0242 0842 1455 2111	4.6 0.5 4.9 0.4	**22** F	0221 0827 1432 2050	4.7 0.5 4.9 0.4
8 F	0326 0924 1537 2154	4.6 0.5 4.9 0.5	**23** SA	0301 0904 1510 2128	4.6 0.5 4.9 0.4
9 SA	0406 1003 1615 2231	4.6 0.6 4.9 0.6	**24** SU	0340 0939 1551 2209	4.6 0.5 4.9 0.4
10 SU	0444 1038 1652 2307	4.4 0.7 4.8 0.7	**25** M	0423 1020 1638 2255	4.5 0.5 4.8 0.5
11 M	0521 1114 1731 2344	4.3 0.8 4.7 0.8	**26** TU	0511 1108 1727 2343	4.5 0.6 4.8 0.6
12 TU	0600 1151 1813	4.2 0.9 4.5	**27** W	0558 1156 1816	4.5 0.7 4.8
13 W	0022 0643 1234 1903	0.9 4.2 1.0 4.4	**28** TH	0029 0647 1246 1912	0.7 4.4 0.8 4.6
14 TH	0109 0738 1332 2007	0.9 4.1 1.0 4.3	**29** F	0123 0748 1352 2022	0.7 4.4 0.9 4.6
15 F	0213 0845 1445 2118	1.0 4.1 1.0 4.3	**30** SA	0234 0902 1512 2142	0.9 4.4 0.9 4.5
			31 SU	0354 1019 1632 2258	0.8 4.5 0.7 4.6

AUGUST

	Time	m		Time	m
1 M	0508 1128 1741	0.7 4.6 0.6	**16** TU	0452 1114 1726 2343	0.9 4.5 0.8 4.5
2 TU	0003 0609 1225 1840	4.6 0.5 4.7 0.4	**17** W	0550 1204 1818	0.8 4.7 0.7
3 W	0057 0701 1313 1929	4.6 0.5 4.8 0.4	**18** TH	0033 0642 1249 1906	4.6 0.7 4.8 0.5
4 TH	0143 0746 1355 2013	4.6 0.5 4.9 0.4	**19** F	0119 0728 1331 1951	4.7 0.6 4.9 0.5
5 F	0224 0827 1436 2054	4.6 0.6 4.9 0.5	**20** SA	0203 0811 1412 2033	4.7 0.5 5.0 0.4
6 SA	0303 0906 1514 2132	4.6 0.6 4.9 0.6	**21** SU	0244 0850 1454 2113	4.7 0.4 5.0 0.4
7 SU	0339 0940 1548 2205	4.6 0.6 4.9 0.7	**22** M	0325 0927 1537 2155	4.6 0.4 4.9 0.4
8 M	0412 1011 1622 2236	4.5 0.7 4.8 0.8	**23** TU	0407 1007 1623 2239	4.6 0.5 4.9 0.5
9 TU	0444 1043 1656 2308	4.4 0.8 4.7 0.9	**24** W	0451 1052 1709 2323	4.6 0.6 4.8 0.7
10 W	0516 1114 1730 2338	4.4 0.9 4.6 0.9	**25** TH	0535 1136 1756	4.5 0.7 4.7
11 TH	0549 1146 1808	4.3 1.0 4.4	**26** F	0005 0620 1225 1850	0.8 4.5 0.8 4.5
12 F	0012 0633 1233 1904	1.0 4.2 1.1 4.3	**27** SA	0058 0720 1331 2004	1.0 4.4 1.0 4.4
13 SA	0106 0739 1343 2018	1.2 4.2 1.2 4.2	**28** SU	0212 0841 1459 2132	1.1 4.4 1.0 4.4
14 SU	0222 0858 1507 2137	1.2 4.2 1.1 4.2	**29** M	0342 1007 1628 2254	1.1 4.5 0.9 4.4
15 M	0343 1013 1624 2246	1.0 4.4 0.9 4.4	**30** TU	0502 1118 1738 2356	0.9 4.6 0.7 4.5
			31 W	0601 1212 1830	0.7 4.7 0.6

Chart Datum: 2·70 metres below Normal Null (German reference level). HAT is 5·1 metres above Chart Datum.

Germany

STANDARD TIME UT –01
Subtract 1 hour for UT
For German Summer Time add ONE hour in **non-shaded areas**

WILHELMSHAVEN   LAT 53°31′N   LONG 8°09′E
TIMES AND HEIGHTS OF HIGH AND LOW WATERS

Dates in red are **SPRINGS**
Dates in blue are NEAPS

YEAR 2016

SEPTEMBER

Day	Time m	Time m	Time m	Time m		Day	Time m	Time m	Time m	Time m
1 TH	0043 4.6	0647 0.7	1255 4.8	● 1913 0.5		16 F	0007 4.5	0617 0.7	1221 4.8	○ 1839 0.5
2 F	0124 4.6	0728 0.6	1335 4.8	1952 0.5		17 SA	0055 4.6	0705 0.6	1305 4.9	1926 0.4
3 SA	0201 4.6	0807 0.6	1412 4.9	2029 0.6		18 SU	0140 4.7	0749 0.5	1350 5.0	2011 0.4
4 SU	0236 4.6	0842 0.6	1447 4.8	2103 0.6		19 M	0223 4.7	0832 0.5	1435 5.0	2054 0.4
5 M	0308 4.6	0913 0.6	1520 4.8	2134 0.7		20 TU	0305 4.7	0913 0.4	1520 5.0	2137 0.5
6 TU	0338 4.6	0943 0.7	1551 4.7	2203 0.8		21 W	0346 4.7	0953 0.5	1604 4.8	2218 0.6
7 W	0408 4.5	1013 0.8	1622 4.6	2231 0.8		22 TH	0427 4.7	1032 0.6	1649 4.7	2258 0.7
8 TH	0436 4.5	1040 0.8	1652 4.5	2255 0.9		23 F	0509 4.6	1114 0.7	1735 4.5	◑ 2339 0.7
9 F	0504 4.4	1106 0.9	1722 4.3	◑ 2321 1.1		24 SA	0555 4.5	1203 0.9	1832 4.3	
10 SA	0539 4.2	1142 1.1	1809 4.2			25 SU	0033 1.2	0657 4.4	1313 1.0	1948 4.2
11 SU	0006 1.2	0639 4.1	1246 1.2	1921 4.1		26 M	0151 1.3	0821 4.4	1444 1.1	2119 4.2
12 M	0123 1.3	0802 4.1	1415 1.2	2048 4.1		27 TU	0325 1.3	0951 4.5	1618 1.0	2243 4.3
13 TU	0255 1.2	0927 4.3	1543 1.1	2209 4.1		28 W	0448 1.0	1104 4.7	1727 0.8	2342 4.5
14 W	0417 1.0	1038 4.5	1654 0.8	2314 4.4		29 TH	0545 0.8	1153 4.7	1810 0.6	
15 TH	0522 0.9	1133 4.6	1749 0.7			30 F	0021 4.5	0623 0.7	1231 4.7	1845 0.6

OCTOBER

Day	Time m	Time m	Time m	Time m		Day	Time m	Time m	Time m	Time m
1 SA	0056 4.6	0701 0.7	1308 4.7	● 1922 0.6		16 SU	0028 4.6	0638 0.7	1239 4.9	○ 1859 0.5
2 SU	0132 4.6	0739 0.7	1345 4.7	1959 0.6		17 M	0115 4.7	0726 0.6	1326 4.9	1946 0.4
3 M	0206 4.6	0814 0.6	1420 4.7	2033 0.6		18 TU	0159 4.8	0811 0.5	1413 5.0	2032 0.4
4 TU	0237 4.6	0846 0.6	1451 4.7	2103 0.7		19 W	0242 4.7	0854 0.5	1501 4.9	2116 0.5
5 W	0306 4.6	0915 0.6	1521 4.6	2129 0.7		20 TH	0324 4.8	0936 0.5	1546 4.8	2157 0.6
6 TH	0334 4.6	0943 0.7	1550 4.6	2154 0.8		21 F	0404 4.7	1014 0.5	1631 4.6	2235 0.8
7 F	0401 4.5	1008 0.8	1618 4.4	2217 0.9		22 SA	0446 4.7	1055 0.7	1718 4.4	◑ 2317 1.0
8 SA	0427 4.4	1033 0.9	1647 4.3	2243 1.1		23 SU	0534 4.6	1146 0.9	1815 4.2	
9 SU	0500 4.3	1107 1.1	1728 4.1	◑ 2325 1.3		24 M	0011 1.2	0635 4.5	1253 1.1	1927 4.1
10 M	0553 4.2	1204 1.2	1835 4.0			25 TU	0125 1.4	0754 4.5	1417 1.1	2052 4.1
11 TU	0035 1.4	0711 4.1	1328 1.3	2001 4.1		26 W	0252 1.3	0920 4.5	1546 1.0	2213 4.2
12 W	0208 1.4	0840 4.3	1459 1.1	2128 4.2		27 TH	0413 1.1	1032 4.6	1654 0.8	2311 4.4
13 TH	0336 1.2	0958 4.5	1616 0.9	2239 4.3		28 F	0511 0.9	1122 4.7	1737 0.7	2350 4.5
14 F	0447 0.9	1058 4.6	1715 0.7	2337 4.5		29 SA	0551 0.8	1200 4.7	1812 0.7	
15 SA	0545 0.8	1150 4.8	1808 0.6			30 SU	0024 4.5	0629 0.8	1238 4.7	● 1850 0.7
						31 M	0101 4.6	0710 0.7	1317 4.6	1929 0.7

NOVEMBER

Day	Time m	Time m	Time m	Time m		Day	Time m	Time m	Time m	Time m
1 TU	0136 4.6	0747 0.6	1351 4.6	2002 0.6		16 W	0134 4.8	0748 0.5	1353 4.8	2008 0.5
2 W	0207 4.6	0819 0.6	1423 4.6	2031 0.6		17 TH	0218 4.8	0833 0.5	1442 4.8	2054 0.5
3 TH	0236 4.6	0848 0.6	1453 4.5	2058 0.6		18 F	0302 4.9	0919 0.5	1531 4.7	2137 0.6
4 F	0304 4.6	0916 0.7	1522 4.5	2124 0.7		19 SA	0346 4.8	1002 0.5	1617 4.5	2218 0.8
5 SA	0332 4.6	0944 0.8	1552 4.4	2151 0.9		20 SU	0429 4.8	1044 0.7	1703 4.3	2300 1.0
6 SU	0401 4.5	1013 0.9	1624 4.3	2222 1.1		21 M	0515 4.7	1131 0.9	1755 4.3	◑ 2349 1.2
7 M	0436 4.5	1048 1.0	1705 4.2	◑ 2302 1.2		22 TU	0610 4.6	1228 1.0	1856 4.1	
8 TU	0524 4.4	1139 1.2	1803 4.1			23 W	0050 1.3	0715 4.5	1335 1.1	2007 4.0
9 W	0003 1.4	0632 4.3	1251 1.2	1921 4.1		24 TH	0201 1.3	0830 4.4	1450 1.0	2120 4.1
10 TH	0124 1.4	0754 4.4	1415 1.1	2044 4.2		25 F	0316 1.2	0942 4.5	1600 0.9	2223 4.3
11 F	0250 1.2	0914 4.5	1533 0.9	2159 4.3		26 SA	0421 1.0	1040 4.6	1655 0.8	2310 4.4
12 SA	0406 1.1	1021 4.6	1639 0.8	2302 4.5		27 SU	0512 1.0	1126 4.6	1737 0.8	2350 4.5
13 SU	0511 0.9	1119 4.7	1738 0.6	2359 4.6		28 M	0557 0.9	1208 4.6	1819 0.8	
14 M	0610 0.7	1214 4.8	1833 0.5	○		29 TU	0030 4.6	0641 0.8	1248 4.6	● 1859 0.7
15 TU	0049 4.7	0702 0.6	1304 4.9	1923 0.4		30 W	0107 4.6	0720 0.7	1324 4.6	1933 0.7

DECEMBER

Day	Time m	Time m	Time m	Time m		Day	Time m	Time m	Time m	Time m
1 TH	0140 4.6	0754 0.6	1357 4.5	2005 0.7		16 F	0158 4.8	0816 0.5	1427 4.7	2036 0.6
2 F	0212 4.7	0826 0.6	1431 4.5	2037 0.7		17 SA	0244 4.9	0905 0.5	1517 4.6	2122 0.7
3 SA	0242 4.7	0858 0.7	1503 4.5	2107 0.8		18 SU	0330 4.9	0952 0.6	1603 4.6	2205 0.7
4 SU	0312 4.7	0929 0.7	1535 4.4	2137 0.9		19 M	0413 4.9	1033 0.6	1646 4.4	2244 0.8
5 M	0345 4.7	1002 0.8	1611 4.4	2211 1.0		20 TU	0455 4.8	1113 0.8	1730 4.3	2324 1.0
6 TU	0423 4.6	1040 0.9	1653 4.3	2252 1.1		21 W	0540 4.7	1156 0.9	1817 4.2	◑
7 W	0509 4.6	1126 1.0	1744 4.2	◑ 2342 1.2		22 TH	0009 1.1	0630 4.5	1245 1.0	1911 4.1
8 TH	0605 4.5	1224 1.0	1848 4.2			23 F	0102 1.2	0730 4.4	1344 1.0	2013 4.1
9 F	0047 1.2	0715 4.5	1334 1.0	2002 4.2		24 SA	0207 1.2	0839 4.4	1453 1.1	2121 4.2
10 SA	0205 1.2	0830 4.6	1449 1.0	2117 4.3		25 SU	0320 1.2	0947 4.4	1601 1.1	2223 4.4
11 SU	0323 1.1	0943 4.7	1602 0.9	2226 4.5		26 M	0427 1.1	1047 4.5	1659 1.0	2315 4.5
12 M	0436 0.9	1050 4.7	1708 0.7	2329 4.6		27 TU	0524 1.0	1137 4.5	1747 0.9	
13 TU	0541 0.7	1152 4.7	1809 0.6			28 W	0000 4.6	0612 0.9	1220 4.5	1830 0.8
14 W	0024 4.7	0639 0.6	1247 4.7	○ 1902 0.5		29 TH	0040 4.6	0654 0.8	1259 4.5	● 1909 0.7
15 TH	0112 4.8	0728 0.5	1337 4.7	1949 0.5		30 F	0117 4.7	0732 0.7	1337 4.6	1946 0.7
						31 SA	0151 4.7	0810 0.7	1415 4.5	2023 0.7

Chart Datum: 2·70 metres below Normal Null (German reference level). HAT is 5·1 metres above Chart Datum.

FREE monthly updates. Register at
www.reedsnauticalalmanac.co.uk

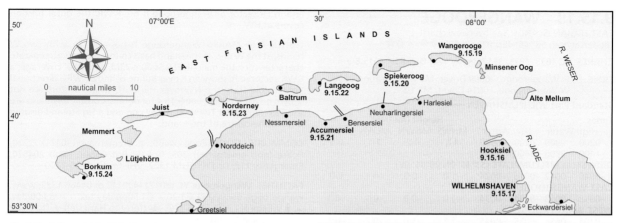

9.15.18   EAST FRISIAN ISLANDS

CHARTS AC 1635, 1875 and 3631 cover the E Frisians at small to medium scales. BSH 3015, sheets 4 to 9, give excellent cover at 1:50,000 with many larger scale insets of the harbours. The *Hafenhandbuch Nordsee* is the most detailed harbour guide with excellent chartlets and aerial shots. It also covers from Calais to Trondheim, Norway.

GEOGRAPHY The East Frisians (*Die Ostfriesischen Küste*) lie from 3 to 11M off the German coast and were once the N coast of the German mainland. Shoals extend seaward for 2 or more miles in places. Groynes and sea defences are on their W and NW sides. Their bare sand dunes are not easy to identify, so the few conspic landmarks must be carefully selected.

The gaps between the islands are called ***Seegats***. They vary in position and depth, and all of them are dangerous on the ebb tide, even in a moderate onshore wind, or in winds >F4 with any N in them. The area between the low-lying islands and the mainland coast is called the ***Wattenmeer***, *wattfahrenwassern* (abbrev: *watt*) being the generic name for tidal channels. Where the *watts* meet (having swept both ways round the island, usually about ⅓ of the way from the E end) are called *Wattenhochs*.

Cruising between the islands and the mainland is generally only for shoal draft boats able to take the ground. Vessels drawing more than 1·4m are likely to be constrained by their draft (see www.wattenschipper.de/Aktuelles.htm). For navigating inside the islands in the shoal or partly drying *watts*, the large scale German leisure charts (BSH) are essential.

TIDES The general strategy, whether E or W-bound, is to leave as early as possible on the flood (dependent on distance) so as to reach your destination or negotiate the shallowest part of the *watts* near HW. Springs occur around midday and give approx 0·5m more water than neaps.

▶*Streams are very slack over the watts but very strong in the seegats, especially on a sp ebb.*◀

Persistent strong W/NW winds may raise the sea level by > 0·75m (exceptionally by 3m); against the ebb they cause steep dangerous seas over the bars near the entrance to the seegats. Strong E/SE winds can lower sea levels, the difference between predicted and expected heights are in German weather forecasts.

WITHIES (*pricken*) are used to mark the *watts*, usually these are PHMs, unbound withies (⚲, twigs pointing up) often marked with red reflective tape; bound withies (⚲, twigs pointing down) are less common (except in ent chans) SHMs.

The **conventional direction of buoyage is always E'ward**, though in places this may conflict with the actual direction of the flood.

NATURE RESERVES inshore of most of the islands must not be entered. Seal and Bird reserves are closed as shown on the charts, but channels are excluded. The *Nationalpark Neidersachsisches Wattenmeer* extends from the mainland to the seaward margin of the drying flats.

National Parks are marked on the charts. Within them Zones 1 are closed from HW +3 to HW −3; landing is prohib year round.

PASSAGE INFORMATION *(SEE ALSO INDIVIDUAL ENTRIES)*

Generally, the Seegats between the islands should be avoided. Vessels small enough to use the Wattenmeer are best advised to enter at the E or W ends and stay inside, enjoying the protection the islands afford from the sea outside.

Alte Mellum, 3015.9, 6M SE of Wangerooge, where the Weser and Jade meet, is part of a bird sanctuary; landing prohib.

Minsener Oog, 3015.9, 1-2·5M SE of Wangerooge, is two small islets linked by a causeway and with many groynes and a conspic radar twr. Beware overfalls in strong NW'lies. Blaue Balje, between Minsener Oog and Wangerooge, is marked by buoys (prefixed 'B' and moved as necessary) but has strong tides and is dangerous in strong W/N winds. Seal (RSG) and Bird (VSG) reserves off the E end of Wangerooge are prohib areas (1/5–1/10 and 1/4–1/10 respectively).

Between **Spiekeroog** and **Langeoog**, 3015.8, the Otzumer Balje (buoys OB1-28/AH2) and unmarked Westerbalje chans lead inward. The former is usually deeper (min 3·5m) but both chans may silt up and unmarked channels are not in any case advised, even in good weather and on a rising tide.

Acummer Ee, 3015.7, leads east of **Baltrum** to **Langeoog** and to **Bensersiel** and **Dornumer-Accumersiel** on the mainland. It is buoyed (A1-A32, moved as required). Shoals extend 2M offshore, chan depths vary greatly, prone to silting. *In onshore winds the sea breaks on the bar (0·8m) between A1 and A1a.*

Norderneyer Seegat, 3015.6, lies between **Norderney** and **Juist**, but is approached across dangerous offshore shoals (Norder-gründe and Nordwestgründe) via two shallow buoyed channels, Dovetief and Schluchter. Dovetief, from the NNE (buoys D1-44) is the main chan, but Schluchter from the WNW (buoys S1-D5/S8) is better protected in NE'lies. *Depths vary considerably and at times the channels silt up and buoys are lifted. Neither channel should be used in strong winds.* Further inshore, Dovetief, now 14–11m deep, hugs the W end of the island and round to Norderney harbour. Beware groynes along the shore.

SW of Norderney, 3015.5, Busetief (buoys B1-24) is deeper than Dovetief and Schluchter. It leads generally S'ward to **Norddeich**, a tidal mainland hbr (2·0m).

▶*The flood begins at HW Helgoland – 0605, and the ebb at HW Helgoland – 0040, sp rates 1kn.*◀

Juist, 3015.4, is the last of the chain of similar, long, narrow and low-lying islands. It has a small drying marina in the hbr mid-way along its southern side.

Memmert and **Kachelot**, S of Juist's W tip, and Lütje Horn, an islet E of Borkum, are bird sanctuaries; landing is prohibited.

Borkum (AC 3631 and 3015.3/4) lies between Osterems and Westerems channels, with high dunes each end so that at a distance it looks like two separate islands. Round the W end of Borkum are unmarked groynes, some extending 3ca offshore. Conspic landmarks, all at the W end of the island, include: Grosse and Neue beacons, a water tower, Borkum Grosser lt ho and two disused lt houses.

9.15.19 WANGEROOGE

EAST FRISIAN ISLANDS. See overview chartlet
Niedersachsen **53° 46'·46N 07° 52'·10E** ❀❀⚓⚓✿✿

CHARTS AC 1635, 1875, 3617; Imray C26; D 2; BSH 3015.8/9

TIDES   E Wangerooge, –0009 Dover; ML 1·9. Duration 0600
W Wangerooge, –0014 Dover; ML 1·5

Standard Port WILHELMSHAVEN (←—)

Times				Height (metres)			
High Water		Low Water		MHWS	MHWN	MLWN	MLWS
0200	0800	0200	0900	4·8	4·2	1·1	0·5
1400	2000	1400	2100				
Differences MELLUMPLATE (53°46'·0N 08°06'·0E)							
–0046	–0042	–0016	–0024	–1·0	–0·8	–0·1	0·0
EAST WANGEROOGE							
–0058	–0053	–0024	–0034	–1·0	–0·9	–0·1	–0·1
WEST WANGEROOGE							
–0101	–0058	–0035	–0045	–1·1	–1·0	–0·2	–0·1

SHELTER Good in most winds, but SW gales, and S'lies at HW, cause heavy swell and difficult conditions in hbr when **Harlesiel**, 5M SSW via Carolinensieler Balje, will be safer. Yachts on marina pontoons at E side (HM) of hbr have 1·3 to 1·8m depth; rafting three or four deep, beware shallower spots. The W jetty (YC) is mainly for ferries but yachts >12m may request a berth.

NAVIGATION WPT 53°49'·25N 07°50'·47E Harle SWM buoy, Iso 8s, 122°/8½ca to H2 buoy; 207°/1·4M to H10 buoy; 154°/8ca to Buhne H WCM buoy, VQ (9) 10s; lastly 141°/1M to H7/OB42 buoy and to hbr ent via D4. The Harle chan has least depth 0·4m near H4 buoy; it shifts and buoys are moved accordingly. Beware Buhne (groyne) H, extending 7½ca, marked by WCM lt buoy

and in FR sector of Wangerooge lt ho, F WR; the outer portion covers at HW.

LIGHTS AND MARKS Wangerooge lt ho, a R twr with two W bands, at the W end of the island has 3 co-located lts with separate characteristics: the main lt, Fl R 5s 60m 23M, with 360° coverage; a F WR sectored lt, showing R over Buhne H; and a F WRG directional light visible from the NW over Harleriff, whose sectors are not shown on the chartlet. See 9.15.4 for details. Conspic daymarks are the 52m high West Tower at the W end and a Sig Stn and disused lt ho in the centre of the island.

COMMUNICATIONS (Code 04469) ♨ Bremerhaven (0471) 72220; Police (non emergency) 205; ⊖ 519; Brit Consul (030) 204570; Emergency 112; Dr 1700; HM 1322, VHF Ch 17, 0700-1700.

FACILITIES **Wangerooge YC** ☎01711439517 or 04469 1322; www.wangerooge.de/wyc/ Ⓥ €1·20 + €1·50pp/day tax; ⛽ €3·00/day. ▲ ▮ **Village** ✕ ⌂ 🛒 🏧 🅟 📮 Ⓑ ⇌ (ferry to Harlesiel) ✈ (to Harle and Helgoland). Ferry: Hamburg-Harwich.

MAINLAND HARBOUR 5M SSW OF WANGEROOGE
HARLESIEL, Niedersachsen, **53°42'·60N 07°48'·63E**. AC 1875, 3617; BSH 3015.8. HW –0100 on Dover (UT); HW (zone –0100) –0005 and ht +0·5m on Helgoland. Appr via Carolinensieler Balje which dries S of the training wall lt, L Fl 8s. Excellent shelter S of lock which has set opening times. 120 pontoon berths to W by village or at ⛏ to E after passing through lock on W side of dyke. Call *Harle Lock*, VHF Ch 17, 0800-2000 or ☎(04464) 472 (Lock & HM). Berthing fees similar to Wangerooge, lock costs depend on how many vessels in lock, max, €13·30, min, €5·15 ⊖ ☎(04462) 6154; YC ☎1473; Facilities: ▮ ▲ ⚓ ◣ ⛽ ⛏ ⌂. **Town** ✕ ⌂ 🛒 🏧 ⚒ ▮ ✉ ✈ and ferry to Wangerooge.

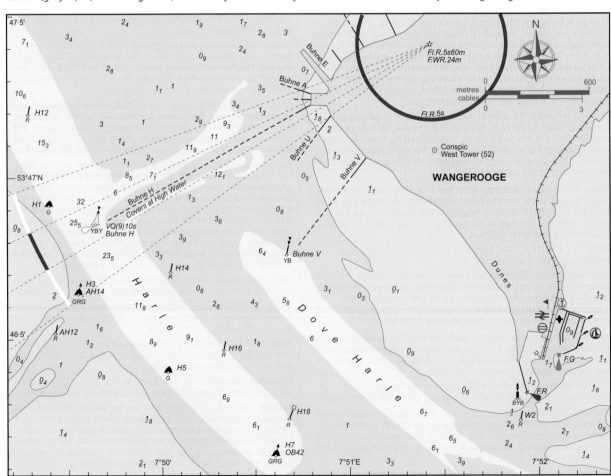

9.15.20 SPIEKEROOG

EAST FRISIAN ISLANDS

Niedersachsen **53°45'·00N 07°41'·27E** (ent to inner chan) ✹⚓⚓✿✿

CHARTS AC 1635, 1875; Imray C26; D 2 89; BSH 3015.8

TIDES HW –0008 on Dover; ML 1·3m; Duration 0555

Standard Port HELGOLAND (←) See differences Langeoog.

SHELTER Good, *but S/SW winds cause heavy swell.* 3 yacht pontoons to the E of ferry berth almost dry to soft mud. Yachts >12m LOA should request a berth on the W jetty.

NAVIGATION WPT 53°48'·29N 07°38'·00E (Otzumer Balje SWM buoy, Iso 4s; often moved) 180°/0·2M to OB1 SHM buoy. Least depth on the bar (between OB2 and OB4) is about 3·5m. Otzumer Balje is well buoyed, but only 4 buoys are lit. The chan skirts the SW end of the island into Schillbalje and to ☆ FR, R mast (lat/long below title) at the ent to inner chan, marked by withies, which leads 020°/1M to small ferry hbr (0·8m). Westerbalje, a shallow, unbuoyed secondary chan, is not advised.

Spiekeroog can also be approached from E and W via the Harlesieler and Neuharlingersieler Wattfahrwassern respectively.

LIGHTS AND MARKS Sand dunes near the W end of the island are up to 24m high, but the E end is low-lying (3m). Two bcns with spherical topmarks are near the centre and E end of island.

COMMUNICATIONS (Code 04976) British Consul (030) 204570; HM 178 879 7741 Mob (0800-1800).

FACILITIES **Hbr** is run by YC: www.ssc-spiekeroog.de 130✅ €1·50 multihulls + 100% extra, <4 hrs €5·00, ⬦ €3·00/day; ▸ 🆆🅲 €2·00/day, ✕ no wi-fi, ⌧(15t). **Town** ½M from hbr. Neuharlingersiel by ferry.

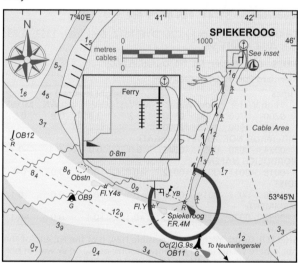

MAINLAND HARBOURS S AND SW OF SPIEKEROOG

NEUHARLINGERSIEL, Niedersachsen, **53°42'·10N 07°42'·29E**. AC 1635, 1875; BSH 3015.8. HW 0000 on Dover (UT); see differences under Langeoog. Appr via Otzumer Balje and Schillbalje to N end of trng wall, Oc 6s. Training wall, 1M long, covers at HW and is marked by SHM withies ⚲. Beware strong tidal streams across the ent. Neuharlinger YC hbr, yachts lie in NE corner of hbr, limited ✅ €1·00/m + €2·00 per person tourist tax, pay at ferry terminal café. Inner berths <1·2m draught. HM ☎(04974) 289. Facilities: **Quay** ⚓. **Village** (picturesque) ✕ ⬜ 🛒. Ferry to Spiekeroog (35 mins).

BENSERSIEL, Niedersachsen, **53°40'·81N 07°34'·20E**. AC 1635, 1875; BSH 3015.7E. HW –0024 on Dover (UT); +0005 on Helgoland (zone –0100); HW ht +0·4m on Helgoland. Appr as for Langeoog thence lit/buoyed Rute chan to training walls with depth of 1·4m to ferry berth, but covering at HW. Very good shelter in the yacht hbr (dries) to stbd just before inner hbr ent (1·0m); also berths on the SW side of main hbr. Lts as 9.15.4. HM VHF Ch 17, ☎(04971) 2502; Facilities: ⚓ ⌧(8t) 🅿(on E pier) 🚽 ⬜ ▸ ✎ ✕. **Town** ⓑ ✕ ⬜ ✉ ⚤ 🛒 🛢. Ferry to Langeoog.

9.15.21 ACCUMERSIEL

Niedersachsen **53°40'·90N 07°29'·25E** ✹✹⚓⚓✿✿

CHARTS AC 1635, 1875; Imray C26; D 89; BSH 3015.7

TIDES –0040 Dover; ML 1·4; Duration 0600

Standard Port HELGOLAND (←) Use differences Langeoog.

SHELTER The marina provides complete shelter in all winds; LW depth, 0.5m-dries in places, very soft mud. Ent is narrow; keep to W of chan on entering. Marina is fenced; obtain a gate key on arrival.

NAVIGATION Appr through Accumer Ee (see Langeoog) leading into Accumersieler Balje and to AB3 SHM buoy, IQ G 13s, 53°41'·44N 07°29'·27E. From here leave all withies ⚲ and posts to stbd. Note warnings on German chart D89. Least depth 0·3m. Locals advise HW, approx 2·0m-2·5m in chan and marina.

LIGHTS AND MARKS None.

COMMUNICATIONS (Code 04933) 🆄🅱 (04972) 247; ⚓ 0190 116048; Police 2218; Customs (04971) 7184; Brit Consul (030) 204570; 🅷 04941-940; HM 2440; Deputy HM 441; *German Bight Tfc* (VTS) broadcasts traffic and weather VHF Ch 80 every H in German/ English.

FACILITIES **Dornumer Yacht Haven, YC** ☎2440, mob 0160 2293046. www.ycacc.de All facilities: 250 ⚓ ✅ €1·10, ▸ 🆆🅲 €2·00/day, 🅿 ✕ ✎ 🅿 ✕ ▬ ⌧(15t); pontoons are lifted out of season. **Town** ✕ ⬜ 🛒 ✉ ⓑ ⚤ (Harlesiel) ✈ (Bremen).

9.15.22 LANGEOOG

EAST FRISIAN ISLANDS
Niedersachsen 53°43'·38N 07°30'·12E ✿✿✿✿✿✿✿

CHARTS AC 1635, 1875; Imray C26; D 89; BSH 3015.7

TIDES –0010 Dover; ML No data; Duration 0600

Standard Port HELGOLAND (←)

Times				Height (metres)			
High Water		Low Water		MHWS	MHWN	MLWN	MLWS
0200	0700	0200	0800	3·2	2·8	0·9	0·5
1400	1900	1400	2000				
Differences LANGEOOG							
+0003	–0001	–0034	–0018	+0·3	+0·2	0·0	–0·1
SPIEKEROOG							
+0003	–0003	–0031	–0012	+0·3	+0·3	0·0	–0·1
NEUHARLINGERSIEL							
+0014	+0008	–0024	–0013	+0·4	+0·4	0·0	–0·1

SHELTER Hbr is well sheltered by 20m high sand dunes, but is open to the S. Marina dredged 1m; E part of hbr dries 1·5m; chan to ferry pier and marina and is marked by several (8+) SHM withies ⌐.

NAVIGATION WPT 53°46'·92N 07°25'·92E [Accumer Ee SWM buoy, Iso 8s; moves frequently, position advised April 2013], approx 180°/5ca to A1–2 chan buoys; thence via by'd chan to A9/B26 GRG by, Fl (2+1) G 15s. W sector of W mole lt leads 072°/1·4M to hbr ent. Ice protectors extend about 40m E of E mole, awash at HW, marked by card buoys. *Buoys shift frequently, call vess Berhard Gruben (Ch 16) for latest info.*

LIGHTS AND MARKS Lt twr on W mole shows all-round red, and WG sectors over the WSW approach and over the SE chan to Bensersiel; see 9.15.4 for details. Daymarks: Langeoog church and water twr 1·5M NNW of hbr.

COMMUNICATIONS (Code 04972) ▨/CG 247; ☏ 0221 3459918; Police 810; Customs 275; British Consul (030) 204570; Dr 589; HM 301, VHF Ch 17, 0700-1700.

FACILITIES Langeoog marina (Segelverein Langeoog) ☎0173 883 2567, 70 + 130❿ €1·40, ⏚ ✗ ⌂ ◔(12t) ▣ ━ .

Village (1½ M by foot or train; motor vehicles are prohibited) ✗ ⌂ 🛒 ⊠ Ⓑ. ⇌ Local train connects with ferry to Bensersiel. ✈ Bremen.

ISLAND HARBOUR 3M NORTH OF NESSMERSIEL
BALTRUM, Niedersachsen, **53°43'·26N 07°21'·80E**. AC 1635, 1875; BSH 3015.6/7. HW –0040 on Dover (UT); Use differences Langeoog. The island is about 3 sq miles, 2·5M long, very low in the E and rising in the W to dunes about 15m high. Shelter is good except in strong SW'lies. Approach at HW via the Norderneyer Wattfahrwasser (which also gives access to Neßmersiel) running W-E, S of Norderney. The appr from seaward via the Wichter Ee is dangerous and is shunned even by locals. Yacht pontoons in 1·5m in the Bootshafen at the E end of the tiny hbr. HM ☎(04939) 448; ⊖ ☎(04939) 668.

Facilities: Hbr ⌇ ⏚; Baltrumer Bootsclub YC. **Village** (¼M NNE) ✗ ⌂ 🛒. No fuel. Ferry from Nessmersiel.

MAINLAND HARBOUR 5M WEST OF DORNUMER-ACCUMER
NESSMERSIEL, Niedersachsen, **53°41'·14N 07°21'·68E**. AC 1635; D89; BSH 3015.6; HW –0040 on Dover (UT); –0020 on Helgoland (zone –0100); HW ht –0·2m on Helgoland. Appr as for Baltrum above, thence via the buoyed Nessmersieler Balje, chan marked by N3–N11 SHM buoys leading to end of trng wall, covers at HW. Here, a lt bn, Oc 4s 5M, and SHM ⌐s and PHM ⌐s mark the 6 cables long trng wall. Yachthafen with pontoons is on W side, beyond ferry quay. Good shelter in all weathers. Hbr dries (2m at HW). HM ☎(04933) 1780.

Facilities: **Nordsee YC Nessmersiel**; ⏚ but no supplies. **Village** (1M to S) has only limited facilities. Ferry to Baltrum.

9.15.23 NORDERNEY

EAST FRISIAN ISLANDS
Niedersachsen 53°41'·90N 07°09'·90E ✿✿✿✿✿✿✿

CHARTS AC 1635; Imray C26; DYC 1812; Zeekarten 1353; D 89; BSH 3012.1, 3015.5/6

TIDES –0042 Dover; ML 1·4; Duration 0605

Standard Port HELGOLAND (←)

Times				Height (metres)			
High Water		Low Water		MHWS	MHWN	MLWN	MLWS
0200	0700	0200	0800	3·2	2·8	0·9	0·5
1400	1900	1400	2000				
Differences NORDERNEY (RIFFGAT)							
–0024	–0030	–0056	–0045	0·0	0·0	–0·1	–0·1
NORDDEICH HAFEN							
–0018	–0017	–0029	–0012	+0·1	+0·1	0·0	0·0
JUIST							
–0026	–0032	–0019	–0008	+0·1	+0·1	0·0	–0·1
MEMMERT							
–0032	–0038	–0114	–0103	+0·1	+0·1	0·0	–0·1

SHELTER Good. Yacht hbr, accessible H24, is at the NE end of the hbr, where yachts lie on finger pontoons; or ⌇ on W wall of hbr, as very crowded in Jul/Aug. Hbr speed limit 3kn.

NAVIGATION Tidal streams set across, not along, the Dovetief and Schluter. **From N/E**, WPT 53°45'·48N 07°12'·70E (Dovetief SWM buoy, Iso 4s) 227°/2·7M to D8 PHM buoy, Oc (2) R 9s. Ent through the Dovetief (see Passage Information) well buoyed but the bar (3·1m) *can be dangerous in onshore winds and following seas which break on it, esp. on the ebb.*

From W, WPT 53°44'·80N 07°06'·00E (Schluter SWM buoy, Iso 8s) 121°/2·1M to S1 SHM buoy, Oc (2) G 9s. *Ent at night is dangerous.* Both chans meet at D5/S8 GRG buoy, Fl (2+1) G 15s.

LIGHTS AND MARKS Lts, see 9.15.4. Land marks: lt ho, 8-sided R brick twr, near centre of island. Conspic water twr in town.

COMMUNICATIONS (Code 04932) Coast Guard 2293; ☏ 549; Police 788; Customs 2386; British Consul (030) 204570; ⊞ 477 and 416; HM 83545, 0175 2785687, VHF Ch 17, Wed–Sun 0700 (0900 Tue)-1200; 1230–1900 (1730 Mon).

FACILITIES Yacht Hbr ⛽ ⌇ ⌇ ◔(10t) ⌂ ⌂ ━ ⚓ ▣ ✂; Yacht Club ⌂. Town ✗ ⌂ 🛒 ⊠ Ⓑ. ⇌ (ferry to Norddeich) ✈ (to Bremen). Ferry Hamburg-Harwich.

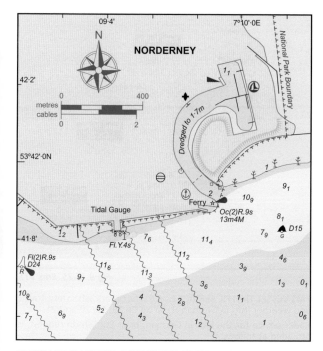

MAINLAND HARBOUR 4·5M SOUTH OF NORDERNEY

NORDDEICH, Niedersachsen, **53°38'·71N 07°08'·88E** (trng wall). AC 1635, BSH 3012.1, 3015.5. HW −0030 on Dover (UT); See Norderney. Appr as for Norderney, then via buoyed/lit Busetief to two training walls (Lat/Long as above) which cover at HW and are marked by Ⅰs. The chan is 100m wide, 2·0m deep and 1·2M long. Much ferry activity can make things interesting at LW. Very good shelter in hbr, divided by central mole. Yacht pontoons and ferry in W hbr; a few yacht berths in Osthafen; FVs in E hbr. Lts as in 9.15.4. HM VHF Ch 17, ☎(04931) 81317; ⊖ ☎(04931) 8435; YC ☎3560; Facilities: ◯ ♨ ☎2721, ♒ ▬ ⚓ BSH agent ◪(5t). **Town** Ⓑ ✕ ◻ 🍴 ♨ Dr, ✉ ⇌. Ferries: Juist and Norderney.

ISLAND HARBOUR 6M WEST OF NORDERNEY

JUIST, Niedersachsen, **53°39'·66N 06°59'83E.** AC 3631; D90; BSH 3012.1, 3015.4. HW −0105 on Dover (UT); HW −0035 and +0·1 on Helgoland; or use differences Memmert (see Norderney). Off the W end of Juist the unmarked, shallow and twisting Haaksgat leads into Juister Balje which then runs E for about 6M to Juist hbr.

The safer approach is via Norderney passing the E tip of Juist and into Memmert and Juister Wattfahrwasseren. Enter hbr via a narrow chan running N into the centre of the S side of island, marked by a red post, Oc (2) R 9s, and by ECM perches to port.

Conspic marks are: West bn, on Haakdünen at W end of island, Juist water twr in centre and East bn, 1M from E end of island. Aero lt Fl 5s at the airfield, see 9.15.4.

There is a small, drying yacht hbr, or yachts can berth in 0.6m alongside quay in ferry hbr. There are extensive No ⚓ areas and underwater power cables, but many drying moorings close E of the hbr. HM ☎(04935) 724; ⊖ ☎1321. Facilities: ♒ ▬ ◪; villages of Ostdorf, Westdorf and Loog in centre of island have limited facilities, ✕ ◻ 🍴 ♨. Ferry to Norddeich.

MAINLAND HARBOUR 14M SOUTH EAST OF BORKUM

GREETSIEL, Niedersachsen, **53°32'·86N 07°02'·07E.** AC 1635, 3631; BSH 89, 3012.1, 3015.2 & .5 (plan). HW −0400 on Dover (UT) −0010 on Helgoland; ML 2·6m. WPT 53°41'·91N 06°36'·17E (Osterems SWM buoy, Iso 4s) 118°/1·2M to chan buoys (O1/O2) passing between Memmert and Borkum. Follow the Osterems chan for 17.5M to O30/L2 buoy; here turn ENE into the Ley chan for 2½M to L8 buoy. Turn 90° stbd to track 165°/7ca, between trng walls to lock which opens HW -4 to +3 (from 0500 to 2200) call *Leysiel Lock* on Ch 17. Thence 4M chan in 3m depths to Greetsiel, FV hbr and ◯ on marina pontoons (2m). HM ☎(04926) 990371; ⊖ ☎(04931) 2764. **Village** YC ✕ ◻ 🍴 ⇌.

EAST FRISIAN ISLANDS
Niedersachsen **53°33'·48N 06°45'·00E** ❋❋❋◊◊♧♧

CHARTS AC 1635, 3631; Imray C26; D 90; BSH 3015.2/3/4; DYC 1812.5; ANWB A

TIDES −0105 Dover; ML 1·4; Duration 0610

Standard Port HELGOLAND (←—)

Times				Height (metres)			
High Water		Low Water		MHWS	MHWN	MLWN	MLWS
0200	0700	0200	0800	3·2	2·8	0·9	0·5
1400	1900	1400	2000				
Differences BORKUM (FISCHERBALJE)							
−0048	−0052	−0124	−0105	0·0	0·0	0·0	0·0

SHELTER Good except in strong winds from S; access H24. Yacht Marina is reported to be silting up, so berth in Burkana Hafen on quays 4 and 5 opposite the ent in 4·1m. Yachts are not allowed to use the ferry berth (*Fährhafen*). The YC in the Schutzhafen has no Ⓥ.

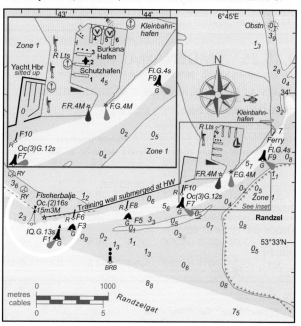

NAVIGATION From the N, WPT 53°38'·96N 06°27'·10E (Riffgat SWM buoy, Iso 8s) about 122°/11M to Fischerbalje lt. See also N Netherlands TSS. **From the W** appr via the Westerems SWM buoy, Iso 4s, Racon (T) (53°36'·93N 06°19'·39E) to join Riffgat which is covered by the 127° W sector of Campen lt, Dir F. (Hubertgat is still marked (unlit R/W buoys H1-H5) least depths around 9m but prone to silting.) Many groynes extend 400m off W end of Borkum. Make for Fischerbalje lt (see 9.15.4 for sectors) at the end of the trng wall which covers at HW. Use the dredged ferry channel and do not cut corners. Beware strong SW'lies and strong currents across the Fischerbalje chan (F1–9 buoys/partly lit). Speed limit in hbrs is 5kn.

LIGHTS AND MARKS Daymarks all WNW of hbr: Water twr, Grosser lt ho (brown brick twr) and disused Kleiner lt ho, R twr with W bands. 3 wind turbines and café immed. NNE of Yacht Hbr ent.

COMMUNICATIONS (Code 4922) Police 91860; Customs 2287; British Consul (030) 204570; Ⓗ 813; Port HM 912828, *Borkum Port* VHF Ch 14 (In season: Mon-Fri 0700-2200; Sat 0800-2100; Sun 0700-2000); *Burkana Hafen* VHF Ch 17.

FACILITIES WSVB Burkana Hafen ☎7877, 80 ◯ (max 12m LOA) €1·20+ ♒ ⚡ ▷ 🅿 WC €2·50 pp/day ⓣ (via ☎7877) ◎.
Yacht Hbr (currently not recommended due to silting) ☎7773, 50 + 200 Ⓥ ◯ €1·20, ♒ ⓣ ♨ ✕ ◻ 🍴 ◎ ▬ ◪ ✎.
Town (7km WNW) ✕ ◻ 🍴 ♨ ♨ ⚓ 🏪 ✉ Ⓑ. Ferry to Emden ⇌ ✈ (& Bremen). S part of island is a Nature Reserve.

RIVER EMS

(AC 3631, 3632) The Ems forms the Dutch-German border and is a major waterway leading to Eemshaven, Delfzijl, Emden, Leer and Papenburg. Approach via Hubertgat or Westerems, which meet W of Borkum; the former is unlit and sparsely buoyed. ▶*Both can be dangerous with a strong W/NW wind against the ebb. The flood begins at HW Helgoland +0530, and the ebb at HWH −0030, sp rates 1·5kn.*◀

Close SW of Borkum the chan divides: Randzel Gat to the NE side and Alte Ems parallel to it on the SW side – as far as Eemshaven on the S bank; here the stream runs 2-3kn. Keep in Ostfriesisches Gatje, the main chan, for Delfzijl and beyond.

Osterems is the unlit E branch of the Ems estuary, passing between Borkum and Memmert. It is shallower than Westerems and leads to Greetsiel via the Ley channel.

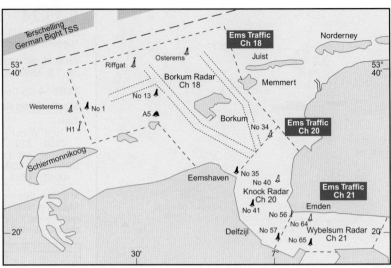

Fig 9.15.25. Ems VTS

9.15.25 EMDEN (RIVER EMS)

Niedersachsen 53°20'·06N 07°10'·69E Hbr ent ✻✻✻♨♨♨✿✿

CHARTS AC 3631, 3632; Imray C26; D 90, 91; BSH 3012.5/6

TIDES HW +0022 on Dover (UT); ML 1·9m

Standard Port HELGOLAND (◄—)

Times				Height (metres)			
High Water		Low Water		MHWS	MHWN	MLWN	MLWS
0200	0700	0200	0800	3·2	2·8	0·9	0·5
1400	1900	1400	2000				
Differences EMSHÖRN (53°29'·6N 06°50'·5E)							
−0037	−0041	−0108	−0047	+0·1	+0·1	−0·1	−0·1
DUKEGAT (53°26'·0N 06°56'·0E)							
−0019	−0023	−0047	−0028	+0·3	+0·3	0·0	0·0
KNOCK (53°19'·6N 07°00'·8E)							
+0018	+0005	−0028	+0004	+0·5	+0·6	0·0	−0·1
EMDEN							
+0041	+0028	−0011	+0022	+0·8	+0·8	−0·1	−0·1

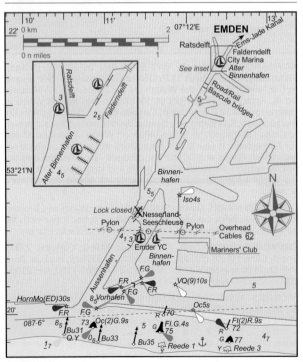

EMS VESSEL TRAFFIC SERVICE *Ems Traffic* broadcasts safety, traffic and weather info in German H+50 on Ch 15, 18, 20 and 21.

Radar advice for the Ems in German, or in English on request, is provided as follows:

Borkum Radar	Ch 18	Buoys 1–35;
Knock Radar	Ch 20	Buoys 35–57;
Wybelsum Radar	Ch 21	Buoy 57–Emden hbr.

SHELTER Good, although the Emder YC marina may be exposed to SSW winds.

NAVIGATION See Borkum and Delfzijl for outer approaches via well buoyed/lit R Ems to No 53 buoy 53°19'·34N 07°01'·15E, south abeam Knock lt ho. Thence 6M east to the hbr ent via Emder Fahrwasser (dredged 8·6m) with drying banks and training walls close outboard of chan buoys. Beware of fast catamaran ferries (*Hochgeschwindigkeitsfahrzeuge*).

Nesserland lock will be closed until 2016 due to major works. Enter Binnenhafen via *Borsummer Schleuse* ☎897265 (Ch 13) which opens on request for small craft between 0800-1230 and 1330-1730.

LIGHTS AND MARKS Fahrwasser outer ldg lts 075°, both Oc (2) 12s 16/28m 12M (at Logum, off chartlet). Inner ldg lts 088°, both Oc 5s 14/30m 12M, lead to hbr ent, marked by FR and FG lts and between Nos 71 and 73 buoys. HT cables on 3 conspic pylons (100m) cross the Außenhafen and Binnenhafen.

COMMUNICATIONS (Code 04921) Police 110; Fire/Ambulance 112; Customs 9279-0; Brit Consul (030) 204570; HM 897260 (H24) Emden Hbr and Locks VHF Ch 13 (H24); Nesserlander lock ☎897270; Ems VTS see above; Tourist Info 97400; No VHF at individual marinas.

FACILITIES Berthing options from seaward:
Emder YC marina (3m) in the Außenhafen, just before Nesserland sealock. HM ☎0175 7528918, 12 ❶ (May–Oct)€0·90, 3 pontoons, most facilities, ⌂.
Through the large Vorhafen sealock to:

Mariners Club, small marina ☎0175 7112027, €0·51; no ⌐.

City Marina, access via two road/rail bridges opening 0650, 0855, 0955, 1140, 1335, 1555, 1735, 1820 and 2140LT on request. For opening sound M (− −). Marina, ☎8907211 or mob 0160 3624744, comprises Alter Binnenhafen (4·5m) Falderndelft (3m) Ratsdelft (3m); all near the city centre. The first two are best for yachts <12m. €0·80, ⌂ ♯. ♠ barge, for location in the hbr call ☎0171 324 4337. **City** all amenities. Ferry to Borkum.

CANALS The Ems-Jade canal, 39M to Wilhelmshaven, is usable by yachts with lowering masts and max draft 1·7m. Access at Emden via the Falderndelft. Min bridge clearance 3·75m. It has 6 locks. Speed limit 4kn.

Netherlands & Belgium

Delfzijl to Nieuwpoort

9.16.1	Contents	**695**
9.16.2	Area map. 9.16.2A Staande Mastroute (Distance table p 736).................	696
9.16.3	Tidal stream chartlets	698
9.16.4	Lights, buoys and waypoints	700
9.16.5	Passage information	704
	Charts, PSSA and TSS – 704 • Crossing the North Sea to the UK – 704 • The West Frisian Islands – 707 • Delfzijl to Ameland – 707 • Terschelling to Texel and Den Helder – 708 • Den Helder to Scheveningen – 718 • Hoek van Holland to the Westerschelde – 718 • Hoek van Holland to the South East UK – 722 • The Belgian Coast – 734 • Crossing the North Sea from Belgium – 740	
9.16.6	Special notes for the Netherlands	704
9.16.7	TSS off the Northern Netherlands	705
9.16.8	Delfzijl, Termunterzijl • Eemshaven	706
9.16.9	Lauwersoog, Zoutkamp • Oostmahorn • Schiermonnikoog • Nes (Ameland)	706
9.16.10	Het Vlie (Zeegat van Terschelling)	708
9.16.11	West Terschelling	709
9.16.12	Vlieland	709
9.16.13	Harlingen	710
9.16.14	Oudeschild	710
9.16.15	Den Helder	711
9.16.16	IJsselmeer	712
9.16.17	IJmuiden	714
9.16.18	Noordzeekanaal	715
9.16.19	Amsterdam	715
9.16.20	Scheveningen	717
9.16.21	TSS and VTS off Southern Netherlands and Belgium	718
9.16.22	Hoek van Holland tides and Maas TSS	719
9.16.23	Hoek van Holland and Nieuwe Waterweg VTS	723
9.16.24	Rotterdam	724
9.16.25	Haringvliet (Stellendam and Hellevoetsluis)	725
9.16.26	Roompotsluis (Oosterschelde)	726
9.16.27	Westerschelde	727
	Ellewoutsdijk • Hoedekenskerke • Hansweert • Walsoorden • Paal • Doel • Lillo	
9.16.28	Vlissingen, tide tables/curves	728
9.16.29	Breskens	733
9.16.30	Terneuzen	733
9.16.31	Special notes for Belgium	734
9.16.32	Antwerpen	734
9.16.33	Zeebrugge, tide tables/curves	735
	Brugge	
	Area distance table	736
9.16.34	Blankenberge	740
9.16.35	Oostende	740
9.16.36	Nieuwpoort	742

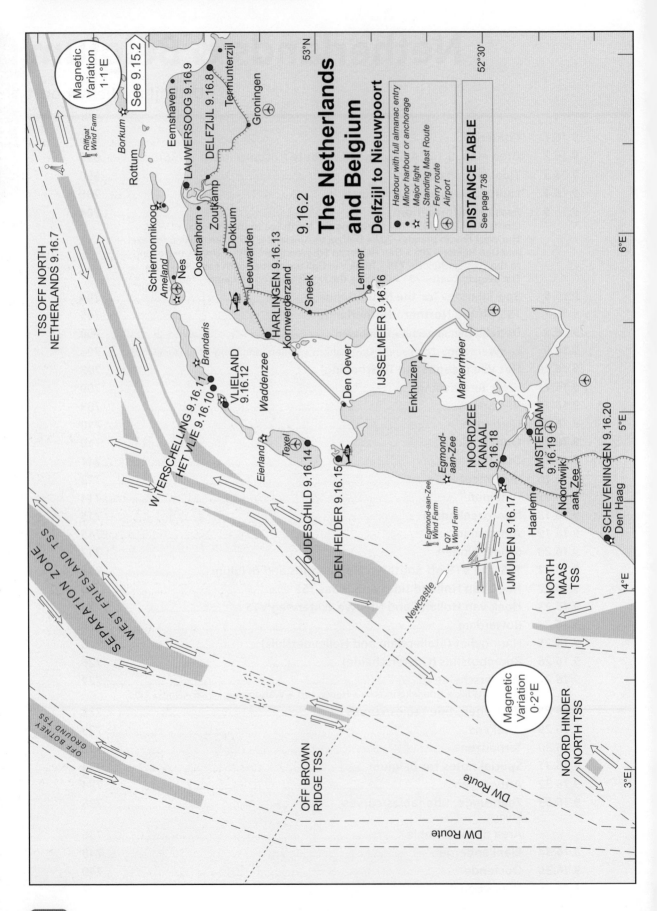

The Netherlands and Belgium

Delfzijl to Nieuwpoort

9.16.2

Magnetic Variation 1·1°E

See 9.15.2

Harbour with full almanac entry
☆ Minor harbour or anchorage
☆ Major light
⚓ Standing Mast Route
⛴ Ferry route
✈ Airport

DISTANCE TABLE
See page 736

TSS OFF NORTH NETHERLANDS 9.16.7

Riffgat Wind Farm

Borkum
Rottum
Eemshaven
LAUWERSOOG 9.16.9
DELFZIJL 9.16.8
Termunterzijl
Groningen
Zoutkamp
Oostmahorn
Dokkum
Schiermonnikoog
Ameland
Nes
Leeuwarden
HARLINGEN 9.16.13
Kornwerderzand
Sneek
Lemmer
IJSSELMEER 9.16.16

Brandaris
VLIELAND 9.16.12
Waddenzee
HET VLIE 9.16.10
W TERSCHELLING 9.16.11
Eierland
Texel
OUDESCHILD 9.16.14
DEN HELDER 9.16.15
Den Oever
Enkhuizen
Markermeer
NOORDZEE KANAAL 9.16.18
AMSTERDAM 9.16.19
Haarlem
Egmond-aan-Zee
Egmond-aan-Zee Wind Farm
Q7 Wind Farm
IJMUIDEN 9.16.17
Noordwijk aan Zee
SCHEVENINGEN 9.16.20
Den Haag
Newcastle

SEPARATION ZONE
WEST FRIESLAND TSS
OFF BOTNEY GROUND TSS
OFF BROWN RIDGE TSS
DW Route
DW Route
NORTH MAAS TSS
NOORD HINDER NORTH TSS
Magnetic Variation 0·2°E

53°N
52°30'
6°E
5°E
4°E
3°E

9.16.2A   THE STAANDE MASTROUTE

The Standing Mast Route (SMR) which does not require masts to be lowered as there are no fixed bridges, can be a useful means of progressing, especially to windward, through the Dutch inland waterways when conditions at sea make this difficult or impossible. Inland the wind may be one or two Forces less and there is usually insufficient fetch to raise a bad sea. The open sea can always be regained at many points. It is a rewarding route and the added interest of progressing through the Dutch countryside and towns is a special bonus.

Pre-planning. To minimise delays at bridges and locks and to have a fair idea of your next night-stop, buy the special SMR chart (wire-bound booklet) published by ANWB at €19.95; it contains detailed information in Dutch but the gist is clear. Michelin road maps give a broader 1: 200,000 perspective.

See IJsselmeer for the ANWB Wateralmanak, Vols 1 and 2 in Dutch.

An International Certificate of Competence (ICC) is required endorsed for Inland Waters, ie CEVNI.

The Route falls broadly into a northern and a southern part, divided by the IJsselmeer/Markermeer. From Delfzijl the Eemskanaal leads to Groningen, thence via the Reitdiep (river) to Zoutkamp and across the Lauwersmeer. Leave via Dokkumer Ee to Leeuwarden where the sea can be regained at Harlingen. If not, continue S past Sneek to enter the IJsselmeer at Lemmer. Cross via Enkhuizen to Amsterdam.

Getting through Amsterdam is complicated by severe restrictions on when bridges may open (due to heavy road/rail traffic). One option is a night convoy through the centre and suburbs. From Het IJ (9.16.19) enter the Westerkanaal near IJ2 buoy. Head SSW past Schiphol airport, Gouda and Rotterdam, thence to Dordrecht.

A simpler daytime option is to detour west via the Noordzeekanaal, turn south at Zijkanaal C through Haarlem and/or Leiden to resume the main route north of Gouda; thence via Rotterdam to Dordrecht.

S of Dordrecht enter Hollandsdiep at which point the SMR has virtually ended. Continue past Willemstad into Volkerak, thence across the Oosterschelde into Veerse Meer and down the Walcheren canal to Vlissingen.

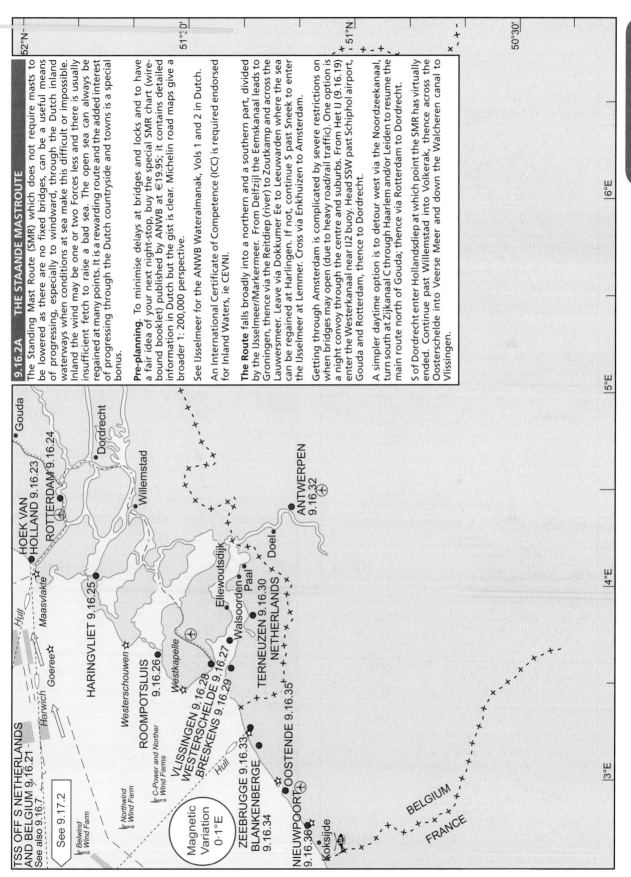

9.16.3 AREA 16 TIDAL STREAMS

5 Hours before HW Dover (0420 after HW Hoek van Holland)

4 Hours before HW Dover (0520 after HW Hoek van Holland)

3 Hours before HW Dover (0605 before HW Hoek van Holland)

2 Hours before HW Dover (0505 before HW Hoek van Holland)

1 Hour before HW Dover (0405 before HW Hoek van Holland)

HW Dover (0305 before HW Hoek van Holland)

South-westward 9.17.3   North-westward 9.4.3   North-eastward 9.15.3

1 Hour after HW Dover (0205 before HW Hoek van Holland) — **HW+1**

2 Hours after HW Dover (0105 before HW Hoek van Holland) — **HW+2**

3 Hours after HW Dover (0005 after HW Hoek van Holland) — **HW+3**

4 Hours after HW Dover (0055 after HW Hoek van Holland) — **HW+4**

5 Hours after HW Dover (0155 after HW Hoek van Holland) — **HW+5**

6 Hours after HW Dover (0255 after HW Hoek van Holland) — **HW+6**

9.16.4 LIGHTS, BUOYS AND WAYPOINTS

Bold print = light with a nominal range of 15M or more. CAPITALS = place or feature. *CAPITAL ITALICS* = light-vessel, light float or Lanby. *Italics* = Fog signal. ***Bold italics*** = Racon. Many marks/buoys are fitted with AIS (<u>MMSI No</u>); see relevant charts.

TSS OFF NORTHERN NETHERLANDS

TERSCHELLING-GERMAN BIGHT TSS
TG1/Ems ↯ IQ G 13s; 53°43'·33N 06°22'·24E.
TE5 ↯ Fl (3) G 10s; 53°37'·79N 05°53'·69E.
TE1 ↯ Fl (3) G 10s; 53°29'·58N 05°11'·31E.

OFF VLIELAND TSS
VL-CENTER ⌑ Fl 5s 12M; ***Racon C, 12–15M;*** 53°26'·93N 04°39'·88E. VL7 ↯ L Fl G 10s; 53°26'·40N 04°57'·60E.
VL1 ↯ Fl (2) G 10s; 53°10'·96N 04°35'·31E.

DELFZIJL TO HARLINGEN

DELFZIJL
PS3/BW26 ↯ Fl (2+1) G 12s; 53°19'·25N 07°00'·32E.
W mole ∮ FG; 53°19'·01N 07°00'·27E.
Ldg lts 203° both Iso 4s. Front, 53°18'·63N 07°00'·17E.

SCHIERMONNIKOOG AND LAUWERSOOG
WG (Westgat) ⌑ Iso 8s; ***Racon N;*** 53°32'·55N 06°00'·70E.
WRG ↯ Q; 53°32'·87N 06°03'·24E.
AM ↯ VQ; 53°30'·95N 05°44'·72E.
Schiermonnikoog ☆ Fl (4) 20s 43m **28M;** dark R ○ twr. Same twr: F WR 29m **W15M,** R12M; 210°-W-221°-R-230°. 53°29'·20N 06°08'·79E.
Lauwersoog W mole ∮ FG; *Horn (2) 30s;* 53°24'·68N 06°12'·00E.

ZEEGAT VAN AMELAND
BR ↯ Q; 53°30'·66N 05°33'·52E.
TS ↯ VQ; 53°28'·15N 05°21'·53E.
WA ⩗ 53°28'·35N 05°28'·64E. (Westgat buoys are all unlit)
Ameland, W end ☆ Fl (3) 15s 57m **30M;** 53°26'·89N 05°37'·42E.

NES
VA2-R1 ↯ VQ(6) + L Fl 10s; 53°25'·71N 05°45'·88E.
Reegeul R3 ∮ Iso G 4s; 53°25'·80N 05°45'·93E.
R7 ∮ QG; 53°25'·91N 05°46'·21E.

HET VLIE (ZEEGAT VAN TERSCHELLING)
ZS (Zuider Stortemelk) ⌑ Iso 4s; ***Racon T;*** 53°19'·56N 04°55'·80E.
ZS1 ↯ VQ G; 53°19'·29N 04°57'·19E.
ZS3 ↯ Fl G 5s; 53° 19'·38N 04° 57'·19E
ZS5 ↯ L Fl G 8s; 53°18'·55N 05°00'·86E.
ZS11-VS2 ↯ Q (9) 15s; 53°18'·66N 05°05'·95E.

NOORD MEEP/SLENK TO WEST TERSCHELLING
WM3 ↯ Iso G 2s; 53°17'·45N 05°12'·28E.
NM 4-S 21 ↯ VQ (3) 5s; 53°19'·02N 05°15'·47E.
SG 15-S 2 ↯ Q (9) 15s; 53°20'·44N 05°11'·70E.
Brandaris Twr ☆ Fl 5s 54m **29M;** Y □ twr partly obscured by dunes; 53°21'·62N 05°12'·86E.
W Terschelling W hbr mole ∮ FR 5m 5M; R post, W bands; *Horn 15s;* 53°21'·26N 05°13'·09E. E pier hd ∮ FG 4m 4M.

VLIELAND
VS3 (Vliesloot) ↯ VQ G; 53°18'·27N 05°06'·25E.
VS14 ⌑ Iso R 4s; 53°17'·58N 05°05'62E.
VS16-VB1 ⌑ Fl (2+1) R 10s; 53°17'·62N 05°05'·19E.
E/W mole hds ∮ FG and ∮ FR; 53°17'·68N 05°05'·51E.
Ldg lts 282° ∮ Iso 4s 10/16m 1M, synch; 53°17'·75N 05°04'·47W (100m apart); mainly for the ferry terminal at E Vlieland.
E Vlieland (Vuurduin) ☆ Iso 4s 54m **20M;** 53°17'·75N 05°03'·49E.

APPROACHES TO HARLINGEN (selected marks):
VLIESTROOM buoys are frequently moved.
VL1 ↯ QG; 53°18'·99N 05°08'·82E. VL 2 ⌑ QR; 53°19'·40N 05°09'·19E.
VL 11 ↯ Iso G 8s; 53°16'·42N 05°09'·70E.

BLAUWE SLENK
BS1-IN2 ↯ VQ (3) 5s; 53°15'·41N 05°09'·41E.
BS13 ↯ QG; 53°13'·31N 05°17'·13E. BS19 ↯ VQ G; 53°11'·90N 05°18'·28E.
BS23 ↯ L Fl G 8s; 53°11'·42N 05°19'·60E.

POLLENDAM
Ldg lts 112°, both Iso 6s 8/19m 13M (H24); B masts, W bands. Front, 53°10'·52N 05°24'·19E. Use only between P2 and P6.
P2 ↧; 53°11'·47N 05°20'·38E on the training wall.
P4, Iso R 8s; P6, Iso R 4s; P8, Iso R 8s; and P10, Iso R 2s.
Yachts should keep outboard of P1 thru 7 SHM buoys. P1 ↯ Iso G 2s; 53°11'·39N 05°20'·32E. P7 ↯ VQ G; 53°10'·68N 05°23'·45E.

HARLINGEN
S mole hd ∮ FG 9m; *Horn (3) 30s;* 53°10'·56N 05°24'·18E.
N mole hd ∮ FR 9m 4M; R/W pedestal; 53°10'·59N 05°24'·32E.

TEXEL AND THE WADDENZEE

TEXEL (from NW and W)
Eierland ☆ Fl (2) 10s 52m **29M;** R ○ twr; 53°10'·93N 04°51'·31E.
TX ↯ Q(9) 15s; 53°06'·81N 04°41'·19E

MOLENGAT (from the N)
MG-A,Y ⌑ ; 53°02'·54N 04°41'·88E.
MG-B, Y ↧; 53°02'·00N 04°42'·11E
MG-D, Y ↧ ; 53°00'·90N 04°42'·04E;
MG-G, Y ↧; 52°59'·30N 04°42'·56E;
MG-H, Y ⌑ ; 52°58'·91N 04°43'·10E.

OUDESCHILD
T12 ⌑ Iso R 8s; 53°02'·23N 04°51'·52E.
Oudeschild Dir ∮ Oc 6s; intens 291°; 53°02'·40N 04°50'·94E; leads 291° into hbr between N mole head FG 6m; and S mole head ∮ FR 6m; 53°02'·33N 04°51'·17E.

APPROACHES TO KORNWERDERZAND SEALOCK

DOOVE BALG (From Texelstroom eastward)
T23 ↯ VQ G; 53°03'·60N 04°55'·85E, 066°/3M from Oudeschild.
T29 ↯ 53°03'·25N 05°00'·05E.
D1 ⌑ Iso G 4s; 53°02'·18N 05°03'·42E.
D21 ⌑ Iso G 8s; 53°03'·55N 05°15'·71E.
BO2-WG1 ↯ Q (6) + L Fl 10s; 53°05'·00N 05°17'·91E.

KORNWERDERZAND SEALOCK
W mole ∮ FG 9m 7M; *Horn Mo(N) 30s;* 53°04'·78N 05°20'·03E.
E mole ∮ FR 9m 7M; 53°04'·70N 05°20'·08E.
W mole elbow ∮ Iso G 6s 6m 7M; 53°04'·61N 05°19'·90E.

APPROACHES TO DEN OEVER SEALOCK

MALZWIN and VISJAGERSGAATJE CHANS TO DEN OEVER
MH4-M1 ↯ VQ (9) 10s; 52°58'·14N 04°47'·49E, close N Den Helder.
M15 ↯ QG; 52°59'·39N 04°55'·48E (hence use DYC 1811.3).
VG1-W2 ↯ Fl (2+1) G 10s; 52°59'·00N 04°56'·85E.
O9 ↯ Iso G 8s; 52°56'·63N 05°02'·38E.

DEN OEVER SEALOCK
Ldg lts 131°, both Oc 10s 6m 7M; 127°-137°. Front, 52°56'·32N 05°02'·98E. Rear, 280m from front.
E end of swing bridge, ∮ Iso WRG 5s 14m 10/7M; 226°-G-231°-W-235°-R-290°-G-327°-W-335°-R-345°; 52°56'·12N 05°02'·52E.

ZEEGAT VAN TEXEL AND DEN HELDER

OFFSHORE MARKS W and SW OF DEN HELDER
NH (Noorderhaaks) ⚓ VQ; 53°00'·24N 04°35'·37E.
MR ⚓ Q (9) 15s; 52°56'·77N 04°33'·82E.
ZH (Zuiderhaaks) ⚓ VQ (6) + L Fl 10s; 52°54'·65N 04°34'·72E.
Vinca G wreck ⚓ Q (9) 15s; *Racon D*; 52°45'·93N 04°12'·31E.

SCHULPENGAT (from the SSW)
Schulpengat Dir ☆ 026·5°, Dir WRG, Al WR, Al WG, **W22M R/G18M**, church spire; 025.1°–FG–025.6°–AlWG–026.3°–FW–026.7°–Al WR–027.4°–F–027.9°; shown H24.
Schilbolsnol ⚮ F WRG 27m W8-3M, R8M, G5M; 333°-W-026°-G-033°-W (Ldg sector for Schulpengat)- 038°-R-070°; Obsc 070°-333°; Post; 53°00'·50N 04°45'·70E (on Texel).
SG ⚓ Mo (A) 8s; *Racon Z*; 52°52'·90N 04°37'·90E.
S1 ⬥ Iso G 4s; 52°53'·53N 04°38'·82E.
S7 ⬥ QG; 52°56'·25N 04°40'·92E. S6A ⚮ QR; 52°56'·52N 04°40'·51E.
S10 ⚮ Iso R 8s; 52°57'·59N 04°41'·57E. S14-MG17 ⚓, see Molengat.
S11 ⬥ Iso G 8s; 52°57'·55N 04°43'·25E.
Huisduinen ⚮ F WR 26m W14M, R11M; 070°-W-113°-R-158°-W-208°; □ twr; 52°57'·14N 04°43'·30E (abeam S10 PHM buoy).
Kijkduin ☆ Fl (4) 20s 56m **30M**; vis 360°, except where obsc'd by dunes on Texel; brown twr; 52°57'·33N 04°43'·58E (mainland).

MARSDIEP and DEN HELDER
T1 ⬥ Fl (3) G 10s; 52°57'·99N 04°44'·62E.
T3 ⬥ Iso G 8s; 52°58'·07N 04°46'·42E.
Den Helder ldg lts 191°, both Oc G 5s 16/25m 3M, synch. Front, vis 183·5°-198·5°; B ▽ on bldg; 52°57'·37N 04°47'·08E.
Marinehaven, W bkwtr head ⚮ QG 11m 8M; *Horn 20s;* 52°57'·95N 04°47'·07E (Harssens Island).
W side, ⚮ Fl G 5s 9m 4M (H24); 180°-067°; 52°57'·78N 04°47'·08E.
Yacht hbr (KMYC) ent ⚮ FR & FG; 165m SW of ⚮ Fl G 5s, above.
E side, MH6 ⚮ Iso R 4s; 52°57'·99N 04°47'·41E.
Ent E side, ⚮ QR 9m 4M (H24); 52°57'·77N 04°47'·37E.

DEN HELDER TO AMSTERDAM
Zanddijk Grote Kaap ⚮ Oc WRG 10s 30m 8M; 041°-G-088°-W-094°-R-120°; brown twr; 52°52'·86N 04°42'·88E.
Petten ⚮ VQ (9) 10s; 52°47'·33N 04°36'·78E (Power stn outfall).
Egmond-aan-Zee ☆ Iso WR 10s 36m **W18M**, R14M; 010°-W-175°-R-188°; W ○ twr; 52°36'·99N 04°37'·16E.
Wind farm approx 6·4M W of Egmond-aan-Zee is marked by: a Meteomast, Mo (U) 15s 11m 10M; 52°36'·36N 04°23'·41E; and by L Fl Y 15s; Horn Mo (U) 30s on 5 of the peripheral wind turbines.

IJMUIDEN
Baloeran ⚓ Q (9) 15s; 52°29'·21N 04°32'·00E.
IJmuiden ⚓ Mo (A) 8s; *Racon Y, 10M;* 52°28'·45N 04°23'·92E.
Ldg lts 100·5° (FW 5M by day; 090·5°-110·5°). **Front** ☆ F WR 30m **W16M**, R13M; 050°-W-122°-R-145°-W-160°; (Tidal and traffic sigs); dark R ○ twrs; 52°27'·70N 04°34'·47E. **Rear** ☆ Fl 5s 52m **29M**; 019°-199° (FW 5M by day; 090·5°-110·5°); 560m from front.
S bkwtr hd ⚮ FG 14m 10M (in fog Fl 3s); W twr, G bands; 52°27'·82N 04°31'·93E.
N bkwtr hd ⚮ FR 15m 10M; 52°28'·05N 04°32'·55E.
IJM 1 ⬥ Iso G 4s; 52°27'·75N 04°33'·59E.
S outer chan ⚮ Iso G 6s, 52°27'·75N 04°33'·81E. ⚮ Iso R 6s, 52°27'·84N 04°34'·39E (Forteiland). Kleine Sluis 52°27'·84N 04°35'·43E.

AMSTERDAM
IJ8 ⚮ Iso R 8s (for Sixhaven marina); 52°22'·86N 04°54'·37E.
Oranjesluizen, N lock 52°22'·93N 04°57'·60E (for IJsselmeer).

AMSTERDAM TO ROTTERDAM
Noordwijk-aan-Zee ☆ Oc (3) 20s 32m **18M**; W□twr; 52°14'·88N 04°26'·02E.

SCHEVENINGEN
Lighthouse ☆ Fl (2) 10s 48m **29M**; 014°-244°; brown twr; 52°06'·23N 04°16'·13E, 5ca E of hbr ent.

Ldg lts 156°, both Iso 4s 14M, H24; synch; Gy masts. Front 52°05'·99N 04°15'·44E; rear 250m from front. Intens at night.
SCH ⚮ Iso 4s; 52°07'·76N 04°14'·12E.
KNS ⚮ Q (9)15s; 52°06'·41N 04°15'·32E.
W mole ⚮ FG 12m 9M; G twr, W bands; 52°06'·23N 04°15'·16E.
E mole ⚮, FR 12m 9M; R twr, W bands; 52°06'·24N 04°15'·37E.
Inner ldg lts 131°: both Iso G 4s synch; Gy posts. Front 52°05'·81N 04°15'·89E. Rear, 34m from front.

NOORD HINDER N & S TSS and JUNCTION
NHR-N ⚓ L Fl 8s; *Racon K, 10M;* 52°10'·91N 03°04'·76E.
Noord Hinder ⚮ Fl (2) 10s; *Horn (2) 30s; Racon T, 12-15M;* 52°00'·10N 02° 51'·11E.
NHR-S ⚮ Fl Y 10s; 51°51'·37N 02°28'·72E.
NHR-SE ⬥ Fl G 5s; 51°45'·42N 02°39'·96E.
Birkenfels ⚓ Q (9) 15s; 51°38'·96N 02°32'·03E.
Twin ⚮ Fl (3) Y 9s; 51°32'·05N 02°22'·62E.
Garden City ⚓ Q (9) 15s; 51°29'·12N 02°17'·92E.

APPROACHES TO HOEK VAN HOLLAND
Europlatform ⬦ Mo (U) 15s; W structure, R bands; helicopter platform; *Horn Mo(U) 30s;* 51°59'·89N 03°16'·46E.
Goeree ☆ Fl (4) 20s 32m **28M**; R/W chequered twr on platform; helicopter platform; *Horn (4) 30s; Racon T, 12-15M;* 51°55'·42N 03°40'·03E.
Maasvlakte ☆ Fl (5) 20s 67m **28M**, H24; 340°-267°; W twr, B bands; 51°58'·20N 04°00'·84E, 1·5M SSW of Maas ent.
Maas Center ⚮ Iso 4s; *Racon M, 10M;* 52°00'·92N 03°48'·79E.
MO ⚮ Mo (A) 8s; 52°01'·10N 03°58'·19E.
MN3 ⬥ Fl (3) G 10s; 52°07'·04N 04°00'·00E.
MN1 ⬥ Fl G 5s; 52°02'·23N 04°01'·91E.

HOEK VAN HOLLAND
Maasmond ldg lts 112° (for deep draught vessels): both Iso 4s 30/46m **21M**; 101°-123°, synch; W twr, B bands. **Front**, 51°58'·88N 04°04'·88E (NW end of Splitsingsdam). **Rear**, 0·6M from front.
Indusbank N ⚓ VQ; 52°02'·89N 04°03'·57E.
MVN ⚓ VQ; 51°59'·61N 04°00'·23E.
MV ⚓ Q (9) 15s; 51°58'·40N 03°56'·58E.
Maas 1 ⬥ L Fl G 5s; 51°59'·35N 04°01'·68E.
Nieuwe Waterweg ldg lts 107°: both Iso R 6s 29/43m **18M**; 099.5°-114.5°; R twr, W bands. Front, 51°58'·55N 04°07'·52E. Rear, 450m from front.
Noorderdam Head ⚮ FR 25m 10M (In fog Al Fl WR 6s; 278°-255°); R twr, W bands; 51°59'·67N 04°02'·80E.
Nieuwe Zuiderdam ⚮ FG 25m 10M, 330°-307°; (In fog Al Fl WG 6s); G twr, W bands; 51°59'·14N 04°02'·49E.

ROTTERDAM
Maassluis ⚮ FG 6m; 51°54'·94N 04°14'·81E; and FR.
Vlaardingen ⚮ FG; 51°53'·99N 04°20'·95E; and FR.
Spuihaven, W ent ⚮ FR; 51°53'·98N 04°23'·97E.
Veerhaven, E ent ⚮ FG; 51°54'·42N 04°28'·75E; and FR.
City marina ent, 51°54'·64N 04°29'·76E.

APPROACHES TO HARINGVLIET
Buitenbank , Iso 4s; 51°51'·16N 03°25'·71E.
Hinder ⚓ Q (9) 15s; 51°54'·55N 03°55'·42E.
SH ⚓ VQ (9) 10s; 51°49'·49N 03°45'·79E.
Westhoofd ☆ Fl (3) 15s 55m **30M**; R □ tr; 51°48'·79N 03°51'·85E.
Ooster ⚓ Q (9) 15s; 51°47'·90N 03°41'·27E.

SLIJKGAT
SG ⚮ Iso 4s; 51°51'·95N 03°51'·42E.
SG 2 ⚮ Iso R 4s; 51°51'·71N 03°53'·45E.
SG 5 ⬥ Iso G 4s; 51°50'·00N 03°55'·56E.
SG 11 ⬥ Iso G 4s; 51°50'·81N 03°58'·52E.
P1 ⬥ Iso G 4s; 51°51'·30N 04°01'·12E.
P3 ⬥ Iso G 8s; 51°51'·12N 04°01'·45E.
P9 ⬥ Iso G 8s; 51°49'·98N 04°02'·15E.

STELLENDAM
N mole ⚡ FG; *Horn (2) 15s;* 51°49'·88N 04°02'·03E.
Buitenhaven ⚡ Oc 6s; 51°49'·73N 04°01'·75E.

APPROACHES TO OOSTERSCHELDE

OUTER APPROACHES
Schouwenbank ◉ Mo (A) 8s; *Racon O, 10M;* 51°44'·94N 03°14'·32E.
Middelbank ◉ Iso 8s; 51°40'·86N 03°18'·20E.
MW ⚓ Q (9) 15s; 51°44'·55N 03°24'·04E (Schouwendiep).
MD 3 ▲ Fl G 5s; 51°42'·70N 03°26'·98E.
SW Thornton ◉ Iso 8s; 51°30'·98N 02°50'·90E.
Wind farm, Thornton Bank in □ 5.5M x 2.7M centred on 51°32'·58N 2°57'·56E: 54 turbines all R lts, the 11 perimeter turbines are lit Mo U Y 15s, Horn Mo U 30s.
Rabsbank ◉ Iso 4s; 51°38'·25N 03°09'·93E.
Westpit ◉ Iso 8s; 51°33'·65N 03°09'·92E.
ZSB ⚓ VQ (9) 10s; 51°36'·57N 03°15'·62E.
OG1 ▲ QG; 51°36'·14N 03°20'·08E.

WESTGAT, OUDE ROOMPOT and ROOMPOTSLUIS
West Schouwen ☆ Fl (2+1)15s 57m **30M;** Gy twr, R diagonals on upper part; 51°42'·52N 03°41'·50E, 5·8M N of Roompotsluis.
OG-WG ⚓ VQ (9) 10s; 51°37'·18N 03°23'·82E.
WG1 ▲ Iso G 8s; 51°38'·00N 03°26'·24E.
WG4 ⚡ L Fl R 8s; 51°38'·62N 03°28'·78E.
WG7 ▲ Iso G 4s 51°39'·40N 03°32'·67E.
WG-GB (Geul van de Banjaard) ⚓ 51°39'·72N 03°32'·69E.
OR1 ⚡ 51°39'·15N 03°33'·59E.
OR5 ▲ Iso G 8s; 51°38'·71N 03°35'·53E.
OR11 ▲ Iso G 4s; 51°36'·98N 03°38'·40E.
OR12 ◤ Iso R 4s; 51°37'·27N 03°39'·25E.
OR-R ⚓ VQ (3) 5s; 51°36'·41N 03°38'·96E.
Roompotsluis ldg lts 073·5°, both Oc G 5s; synch. Front, 51°37'·33N 03°40'·75E. Rear, 280m from front.
N bkwtr ⚡ FR 7m; 51°37'·31N 03°40'·09E.

WESTKAPELLE TO VLISSINGEN

OOSTGAT
Ldg lts 149·5°: Front, Noorderhoofd Oc WRG 10s 20m; W13M, R/G10M; 353°-R-008°-G-029°-W-169°; R ○ twr, W band; 51°32'·40N 03°26'·21E, 0·73M from rear (Westkapelle).
Westkapelle ☆, rear, Fl 3s 50m **28M;** obsc'd by land on certain brgs; □ twr, R top; 51°31'·75N 03°26'·83E.
Kaloo ◉ Iso 8s; 51°35'·55N 03°23'·24E. Chan is well buoyed/lit.
OG5 ▲ Iso G 8s; 51°33'·95N 03°25'·92E.
OG-GR ⚓ VQ (3) 5s; 51°32'·74N 03°24'·71E.
Molenhoofd ⚡ Oc WRG 6s 10m; 306°-R-329°-W-349°-R-008°-G-034·5°-W-036·5°-G-144°-W-169°-R-198°; W mast R bands; 51°31'·58N 03°26'·05E.
Zoutelande FR 21m 12M; 321°-352°; R □ twr; 51°30'·28N 03°28'·41E.
Kaapduinen, ldg lts 130°: both Oc 5s 25/34m 13M; synch; Y □ twrs, R bands. Front, 115°-145°; 51°28'·47N 03°30'·99E. Rear, 107·5°-152·5°; 220m from front.
Fort de Nolle ⚡ Oc WRG 9s 11m W6M, R/G4M; 293°-R-309°-W-324·5°-G-336·5°-R-014°-G-064°-R-099·5°-W-110·5°-G-117°-R-130°; W col, R bands; 51°26'·94N 03°33'·12E.
Ldg lts 117°: Front, Leugenaar, Oc R 5s 6m 7M; intens 108°-126°; W&R pile; 51°26'·43N 03°34'·14E.
Rear, Sardijngeul Oc WR 5s 8m W12M, R9M, synch; 245°-R-272°-W-123.5°-R-147°; R △, W bands on R & W mast; 550m from front; 51°26'·30N 03°34'·56E.

OFFSHORE: W HINDER TSS TO SCHEUR CHANNEL

West Hinder ☆ Fl (4) 30s 23m 13M; *Horn Mo (U) 30s;* *Racon W;* 51°23'·30N 02°26'·27E.
WH Zuid ⚓ Q (6) + L Fl 15s; 51°22'·78N 02°26'·25E.

OostdyckBk ⚓ Q; 51°21'·38N 02°31'·12E.
BerguesBk-N ⚓ Q; 51°19'·95N 02°24'·50E.
Oost-Dyck West ⚓ Q (9) 15s; 51°17'·15N 02°26'·32E.
Oostdyck radar twr; ⚡ Mo (U) 15s 15m 12M on 4 corners; *Horn Mo (U) 30s;* *Racon O.* R twr, 3 W bands, with adjacent red twr/helipad; 51°16'·49N 02°26'·85E.
A-N ◤ Fl (4) R 20s; 51°23'·45N 02°36'·90E.
AS ▲ Fl (3) G 10s; 51°21'·15N 02°36'·92E.
KB2 ⚓ VQ; 51°21'·04N 02°42'·22E.
KB ⚓ Q; *Racon K;* 51°21'·03N 02°42'·83E.
MBN ⚓ Q; 51°20'·82N 02°46'·29E.
Akkaert-SW ⚓ Q (9) 15s; 51°22'·28N 02°46'·34E.
VG ⚓ Q ; 51°23'·38N 02°46'·21E, Vaargeul 1.
VG1 ▲ VQ G; 51°25'·03N 02°49'·04E.
VG2 ◤ Q (6) + L Fl R 15s; *Racon V;* 51°25'·96N 02°48'·16E.
VG3 ▲ QG; 51°25'·05N 02°52'·92E.
VG5 ▲ Fl G 5s; 51°24'·63N 02°57'·92E.
SVG ⚓ Q ; 51°24'·53N 02°59'·92E.
Goote Bank ⚓ Q (3) 10s; 51°26'·95N 02°52'·72E.
A1 ⚓ Iso 8s; 51°22'·36N 02°53'·33E.
A1bis ⚓ L Fl 10s; 51°21'·68N 02°58'·02E.

WESTERSCHELDE APPROACHES

SCHEUR CHANNEL
S1 ▲ Fl G 5s; 51°23'·14N 03°00'·12E.
S3 ⚓ Q; 51°24'·30N 03°02'·92E.
MOW 0 ⊙ Fl (5) Y 20s; *Racon S, 10M;* 51°23'·67N 03°02'·75E.
S5 ▲ Fl G 5s; 51°23'·70N 03°06'·30E.
S7 ▲ Fl G 5s; 51°23'·98N 03°10'·42E.
S9 ▲ QG; 51°24'·42N 03°14'·99E.
S12 ◤ Fl (4) R 10s; 51°24'·67N 03°18'·22E.
S-W ⚓ Q; 51°24'·13N 03°18'·22E, here Wielingen chan merges.
S14 ◤ Fl R 5s; 51°24'·58N 03°19'·67E.

WIELINGEN CHANNEL
BVH ◤ Q (6) + L Fl R 15s; 51°23'·15N 03°12'·04E.
MOW3 tide gauge ⚡ Fl (5) Y 20s; *Racon H, 10M;* 51°23'·38N 03°11'·92E.
W ▲ Fl (3) G 15s; 51°23'·27N 03°14'·92E.
W1 ▲ Fl G 5s; 51°23'·48N 03°18'·22E.
Fort Maisonneuve ⚓ VQ (9) 10s; wreck; 51°24'·20N 03°21'·50E.
W3 ▲ Iso G 8s; 51°23'·96N 03°21'·49E.
W5 ▲ Iso G 4s; 51°24'·31N 03°24'·50E.
W7 ▲ Iso G 8s; 51°24'·60N 03°27'·27E.
W9 ▲ Iso G 4s; 51°24'·96N 03°30'·43E.
Nieuwe Sluis ⚡ Oc WRG 10s 26m W14M, R11M, G10M; 055°-R-089°-W-093°-G-105°-R-134°-W-136·5°-G-156·5°-W-236·5°-G-243°-W-254°-R-292°-W-055°; B 8-sided twr, W bands; 51°24'·41N 03°31'·29E.
Songa ▲ QG; 51°25'·26N 03°33'·66E.
W10 ◤ QR; 51°25'·85N 03°33'·28E.

VLISSINGEN
Koopmanshaven, W mole root, ⚡ Iso WRG 3s 15m W12M, R10M, G9M; 253°-R-277°-W-284°-R-297°- W-306·5°-G-013°-W-024°-G-033°-W-035°-G-039°-W-055°-G-084·5°-R-092°-G-111°-W-114°; R pylon; 51°26'·37N 03°34'·52E.
Sardijngeul Oc WR 5s; 51°26'·30N 03°34'·56E: see OOSTGAT last 3 lines. E mole head, ⚡ FG 7m; W mast; 51°26'·32N 03°34'·67E.
Buitenhaven ent, W side ⚡ FR 10m 5M; also Iso WRG 4s: W073°-324°, G324°-352°, W352°-017°, G017°-042°, W042°-056°, R056°-073°; W post, R bands; tfc sigs; 51°26'·38N 03°36'·06E.
Buitenhaven ent, E side ⚡ FG 7m 4M; 51°26'·41N 03°36'·38E.
Schone Waardin ⚡ Oc WRG 9s 10m W13M, R10M, G9M; 235°-R-271°-W-288°-G-335°-R-341°-G-026°-W-079°-R-091°; R mast, W bands; 51°26'·54N 03°37'·91E (1M E of Buitenhaven ent).

BRESKENS

ARV-VH ↨ Q; 51°24'·71N 03°33'·89E.
VH2 (Vaarwaterlangs Hoofdplaat) ⌐ 51°24'·34N 03°33'·90E.
Yacht hbr, W mole ≰ FG 7m; in fog FY; Gy post; 51°24'·03N 03°34'·06E. E mole ≰ FR 6m; Gy mast; 51°23'·95N 03°34'·09E.

WESTERSCHELDE: TERNEUZEN TO PAAL

TERNEUZEN

Nieuw Neuzenpolder ldg lts 125°, both Oc 5s 6/16m 9/13M; intens 117°-133°; synch. Front, W col, B bands; 51°20'·97N 03°47'·24E. Rear, B & W twr; 365m from front.
Oost Buitenhaven E mole ≰ FR 5M; 51°20'·56N 03°49'·19E.
Former ferry hbr (W part) & marinas (E part) W mole head ≰ FG, Gy mast; 51°20'·57N 03°49'·64E. E mole, FR.
W mole ≰ Oc WRG 5s 15m W9M, R7M, G6M; 090°-R-115°-W-120°-G-130°-W-245°-G-249°-W-279°-R-004°; B & W post; 51°20'·54N 03°49'·58E, close SW of ≰ FG.

HANSWEERT

W mole ≰ Oc WRG 10s 9m W6M, R6M, G6M; (in fog FY); 288°-R-311°-G-320°-W-332·5°-G-348·5°-W-042·5°-R-061·5°-W-078°-G-099°-W-114·5°-R-127·5°-W-288°; R twr, W bands; 51°26'·42N 04°00'·53E.

BELGIUM
ZANDVLIET TO ANTWERPEN

ZANDVLIET

Dir ≰ 118·3°,WRG 20m W4M, R/ G3M; 116·63°-Oc G-117·17°- FG-117·58°-Alt GW-118·63°-F-118·63°-Alt RW-119·18°-FR-119·58°-Oc R-120·13°; 51°20'·61N 04°16'·47E, near Zandvliet locks.

ANTWERPEN

No 107 ♠ Iso G 8s, 51°14'·12N 04°23'·80E (Kattendijksluis for Willemdok ④).
Royerssluis, ldg lts 091°, both FR. Ent FR/FG.
No. 109 ♠ Iso G 8s; 51°13'·88N 04°23'·87E, (off Linkeroever ④).
Linkeroever marina ↨ F WR 9m W3M, R2M; shore-W-283°- R-shore; B ⊙, R lantern; 51°13'·91N 04°23'·70E. Marina ent, FR/FG.

COASTAL MARKS

AW2 ⌐ Fl (4) R 20s; 51°21'·95N 03°00'·94E; Wandelaar.
AW1 ♠ QG; 51°21'·50N 03°02'·59E; Wandelaar.
Oostende Bank N ↨ Q; 51°21'·20N 02°52'·93E.
Wenduine Bank E ⌐ QR; 51°18'·83N 03°01'·64E.
Wenduine Bank W ↨ Q (9) 15s; 51°17'·23N 02°52'·76E.
Nautica Ena wreck ↨ Q; 51°18'·08N 02°52'·79E.
Oostendebank E ⌐ Fl (4) R 20s; 51°17'·35N 02°51'·91E.
Oostendebank W ↨ Q (9)15s; 51°16'·20N 02°44'·74E.
LST 420 ↨ Q (9)15s; 51°15'·45N 02°40'·67E.
MBN ↨ Q; 51°20'·82N 02°46'·29E.
Middelkerke Bank ♠ Fl G 5s; 51°18'·19N 02°42'·75E.
Middelkerke Bank S ⌐ Q (9) R 15s; 51°14'·73N 02°41'·89E.
D1 ↨ Q (3) 10s; 51°13'·95N 02°38'·59E.
BT Ratel ⌐ Fl (4) R 15s; 51°11'·63N 02°27'·92E; Buiten Ratel.

ZEEBRUGGE TO THE FRENCH BORDER

ZEEBRUGGE

AW ↨ Iso 8s; 51°22'·41N 03°07'·05E.
Ldg lts 136°, both Oc 5s 22/45m 8M; 131°-141°; H24, synch; W cols, R bands. Front, 51°20'·71N 03°13'·11E. Rear, 890m SE.
SZ ↨ Q (3) 10s; 51°23'·30N 03°08'·65E (Scheur Channel).
Z ♠ QG; 51°22'·48N 03°09'·95E.
WZ ↨ Q (9) 15s; 51°22'·57N 03°10'·72E.
W outer mole ≰ Oc G 7s 31m 7M; G vert strip lts visible from seaward; 057°-267°; Horn (3) 30s; IPTS; 51°21'·74N 03°11'·17E.

E outer mole ≰ Oc R 7s 31m 7M; R vert strip lts visible from seaward; 087°-281°; Bell 25s; 51°21'·78N 03°11'·86E.
Ldg lts 154°: Front, Oc WR 6s 20m 3M, 135°-W-160°-R-169°; W pylon, R bands; 51°20'·33N 03°12'·89E. Rear, Oc 6s 38m 3M, H24, synch; 520m from front.
Leopold II mole ☆ Oc WR 15s 22m, **W20M, R18M**; 068°-W-145°-R-212°-W-296°; IPTS; Horn (3+1) 90s; 51°20'·85N 03°12'·17E.
Entrance to Marina and FV hbr 51°19'·88N 03°11'·85E.

BLANKENBERGE

Promenade pier Fl (3) Y 20s, 8m 4M; 51°19'·28N 03°08'·18E.
Lt ho ☆ Fl (2) 8s 30m **20M**; 065°-245°; W twr, B top; 51°18'·75N 03°06'·85E.
Ldg lts 134°, both FR 5/9m 3/10M, R cross (X) topmarks on masts; front 51°18'·70N 03°08'·82E; rear 81m from front.
E pier ≰ FR 12m 11M; R290°-245°(315°); W ○ twr; 51°18'·91N 03°06'·56E.
W pier ≰ FG 14m 11M; intens 065°-290°, unintens 290°-335°; W ○ twr; 51°18'·89N 03°06'·42E.
OBST 4 – OBST 14 are eleven ↨s Q approx 3ca offshore, marking Spoil Ground between Blankenberge and Oostende.

OOSTENDE

Oostendebank East ⌐ Fl (4) R 20s; 51°17'·35N 02°51'·91E.
Wenduinebank West ↨ Q (9) 15s; 51°17'·23N 02°52'·76E.
Buitenstroombank ↨ Q; 51°15'·17N 02°51'·71E.
Binnenstroombank ↨ Q (3) 10s; 51°14'·47N 02°53'·65E.
Ldg lts 143°: both Iso 4s (triple vert) 36/46m 4M, 068°-218°; X on metal mast, R/W bands. Front, 51°13'·80N 02°55'·89E.
Oostende lt ho ☆ Fl (3) 10s 65m **27M**; obsc 069·5°-071°; Gy twr, 2 sinusoidal Bu bands; 51°14'·18N 02°55'·84E.
IPTS at E bkwtr Hd, Radar tr, G,R top, 51° 14'·51 02° 55'·19E, plus QY when chan closed.
Inner W pier ≰ FG 12m 10M; G057°-327°(270°); W ○ twr 51°14'·31N 02°55'·03E.

NIEUWPOORT

Zuidstroombank ⌐ Fl R 5s; 51°12'·28N 02°47'·37E.
Weststroombank ⌐ Fl (4) R 20s; 51°11'·34N 02°43'·03E.
Wreck 4 ↨ Q (6) + L Fl 15s; 51°10'·98N 02°40'·10E.
Nieuwpoort Bank ↨ Q (9) 15s; 51°10'·16N 02°36'·09E.
Whitley ↨ Q; 51°09'·15N 02°39'·44E.
Lt ho ☆ Fl (2) R 14s 28m **16M**; R/W twr; 51°09'·27N 02°43'·79E.
E pier ≰ FR 10m 10M; R025°-250°(225°) R307°-347°(40°); W ○ twr; 51°09'·41N 02°43'·08E.
W pier ≰ FG 10m 9M; G025°-250°(225°) G284°-324°(40°); W ○ twr; 51°09'·35N 02°43'·00E.
≰ QG 51°08'·65N 02°44'·31E marks the Y-junction where the channel forks stbd for KYCN and port for WSKLM and VVW-N.

WESTDIEP and PASSE DE ZUYDCOOTE

Den Oever wreck 2 ↨ Q; 51°08'·11N 02°37'·43E.
Wreck 1 ↨ Q; 51°08'·32N 02°35'·03E (adjacent to ⌐ next line).
Wave recorder ⌐ Fl (5) Y 20s; 51°08'·25N 02°34'·98E.
Trapegeer ♠ Fl G 10s; 51°08'·41N 02°34'·36E.
E12 ↨ VQ (6) + L Fl 10s; 51°07'·92N 02°30'·61E.
French waters, for continuity (see also Area 17):
CME ↨ Q (3) 10s; 51°07'·30N 02°30'·00E.
E11 ↨ Fl G 4s; 51°06'·85N 02°30'·83E.
E10 ↨ Fl (2) R 6s; 51°06'·26N 02°30'·40E.
E9 ↨ Fl (2) G 6s; 51°05'·61N 02°29'·60E.
E8 ↨ Fl (3) R 12s; 51°05'·12N 02°28'·60E.

9.16.5 PASSAGE INFORMATION

More Passage Information is threaded between harbours in this Area. **Bibliography**: *N Sea Passage Pilot* (Imray/Navin). *Cruising Guide to the Netherlands* (Imray/Navin). NP 55 *N Sea (East) Pilot*. NP 28 *Dover Strait Pilot*. *Hafenhandbuch Nordsee* (DSV-Verlag).

CHARTS, PSSA AND TSS

While AC 2182A, 1405/06/08, 1630/31/32/33, 1872 suffice for coastal passages and entry to the main ports, larger scale **Dutch yacht charts** (1800 series) are essential for exploring the cruising grounds along this coast or entering the smaller hbrs. Inland, the ANWB booklet-style chart (€19.95) of the *Staande-Mast Route* (Fixed Mast Route; see Stande Maastroute) is very detailed with copious, but intelligible, notes in Dutch.

From the Ems estuary west to Den Helder a **Particularly Sensitive Sea Area** (PSSA) extends 3M seaward from the West Frisian Islands. Yachts should carefully avoid damaging the maritime environment and marine organisms living in it.

The Terschelling-German Bight TSS, Off Vlieland TSS and Off Texel TSS lie between 5 and 10M to seaward of the West Frisian Islands. Cruising yachts are advised to navigate within this relatively narrow ITZ. Further offshore, and particularly in and near the Off Vlieland TSS, West Friesland TSS and Botney Ground TSS, navigation is further complicated by the many oil and gas fields. For general notes on North Sea oil & gas installations, see Chapter 3 and Area 5. Up to date charts are essential, particularly AC 1406,1408 and 1423.

CROSSING THE NORTH SEA TO THE UK

From ports S of Hoek van Holland make for NHR-SE, to cross the TSS for destinations between Harwich and Great Yarmouth (AC 1406, 1408, 1872, 2449, 3371). From ports N of Hoek van Holland passages can be more problematic. For example, a route from IJmuiden to the Humber crosses two DW routes, N and NW of Brown Ridge, and then runs into extensive offshore Gas Fields. These might cause you to opt for two shorter legs, stopping a night at Great Yarmouth. Similar thinking might apply if coming out of Den Helder, even if a stop at Great Yarmouth might incur some southing. From east of Den Helder, make ground west via the ITZ before taking departure.

9.16.6 SPECIAL NOTES: NETHERLANDS

PROVINCES are given in lieu of UK 'counties'.

CHARTS The following types of chart are available from agents. The Chart catalogue (HP7) is downloadable from www.hydro.nl:

- Zeekaarten (equivalent to AC) are issued by the Royal Netherlands Navy Hydrographer and corrected by Notices to Mariners (*Berichten aan Zeevarenden or BaZ*).
- 1800 series *voor Kust-en Binnenwateren* (coastal and inland waters) are yacht charts (DYC) issued annually in March by the Hydrographer in 8 folios (1801-1812, excluding 1802/4/6 & 8); about 9 loose double-sided sheets (54 x 38cm) per folio.

TIME ZONE is –0100, but add 1 hr for DST in the summer months.

TIDES HP 33 *Waterstanden & Stromen* (Tide tables and tidal streams in **English** and Dutch, £16·90 www.bookharbour.com) is useful if cruising Dutch waters for any length of time. It contains tide tables for 15 Dutch and 2 Belgian coastal ports; and 8 tidal stream atlases, including Westerschelde, Oosterschelde and the Maas.

REGULATIONS Discharge of toilet waste from recreational boats is forbidden in all Dutch waters, including inland waterways, lakes and the Waddenzee. Carry evidence (receipts) that any red diesel in your tanks is duty paid.

MARINAS Most marinas are private YCs or Watersport Associations (WSV or WV): *Gemeentelijke (Gem)* = municipal. Marinas with >50 berths must have a pump-out unit ⚓. Sometimes (in Belgium also) berth-holders show a green tally if a berth is free, or a red tally if returning same day, but check with HM. Duty-free fuel (coloured red) is not available for leisure craft and may only be carried in the tank, NOT in cans. A tourist tax (Touristenbelasting) of €0.55–€1.82/head/night is often levied. VAT (BTW) is 21%. A useful website covering other marinas is www.allejachthavens.nl

CUSTOMS Ports of entry are: Delfzijl, Lauwersoog, W Terschelling, Vlieland*, Harlingen, Kornwerderzand, Den Helder, IJmuiden, Scheveningen, Hoek van Holland, Maassluis, Schiedam, Vlaardingen, Rotterdam, Roompot*, Vlissingen, Terneuzen and Breskens. *Summer only. Den Oever and Stellendam are *not* Ports of entry.

FERRIES TO THE UK IJmuiden-Newcastle; Hoek van Holland-Harwich; Rotterdam (Europoort)-Hull.

BUOYAGE Buoys are often named by the abbreviations of the banks or chans which they mark (eg VL = Vliestroom). A division buoy has the abbreviations of both chans meeting there, eg VL2-SG2 = as above, plus Schuitengat.

Some minor, tidal channels are marked by withies: SHM bound ⚲; PHM unbound ⚳. On tidal flats (eg Friesland) where the direction of main flood stream is uncertain, bound withies are on the S side of a chan and unbound on the N side; the banks thus marked are steep-to. In minor chans the buoyage may be moved without notice to accommodate changes.

The SIGNI buoyage system is used in the IJsselmeer, but not in the Eems, Waddenzee and Westerschelde.

SIGNALS When **motor-sailing** yachts must by law hoist a ▼ and when **at anchor** a black ball ●; these laws are rigidly enforced.

IPTS are widely used at coastal ports. Local signals, if any, are given where possible.

Sluicing signals may be shown by day: A blue board, with the word 'SPUIEN' on it; by night 3 ● in a △; sometimes both at once.

Storm warning signals, Its only, are shown by day & night at West Terschelling, Den Helder and IJmuiden; see Chapter 6.

R/T In emergency call *Den Helder Rescue* Ch 16 for Netherlands CG (see Chapter 7); or the working channel of a VTS sector or nearest lock or bridge. Monitor TSS info broadcasts and VTS sector channels. Ch 30/31 is for Dutch marinas (UK VHF sets need to be modified). Note: Do not use Ch M in Dutch waters, where it is a salvage frequency. Ch 13 is for commercial ship-ship calling. English is the second language and widely spoken.

TELEPHONE To call UK from the Netherlands, dial 00-44; then the UK area code minus the prefix 0, followed by the number required. To call the Netherlands from UK dial 00-31 then the area code minus the prefix 0 followed by two or three digits, followed by a 7 or 6 digits subscriber no. Mobile phone Nos start 06.

Emergency: Fire, Police, Ambulance, dial 112 (free); Non-emergency 0900 8844 (local tariff).

PUBLIC HOLIDAYS New Year's Day, Easter Sun and Mon, King's Birthday (27 April) Liberation Day (5 May) Ascension Day, Whit Mon, Christmas and Boxing Days.

BRITISH CONSULS Contact British Consulate-General, Koningslaan 44, 1075 AE Amsterdam; ☎020 676 4343. Or British Embassy, Lange Voorhout 10, 2514 ED The Hague; ☎070 4270 427.

INLAND WATERWAYS The sealocks at Kornwerderzand, Den Oever (IJsselmeer) IJmuiden, Stellendam and Roompotsluis are fully covered. The Staandemast (mast-up) route (Area map) and the IJsselmeer are outlined. Lack of space precludes detailed coverage of other harbours in the very extensive and enjoyable inland seas, waterways and canals.

Regulations. All craft must carry a copy of waterway regulations, *Binnenvaartpolitiereglement (BPR)*, as given in Dutch in the annual ANWB *Wateralmanak Vol 1* or available separately. Vol 2, also in Dutch, is essential reading; it gives pictograph details of marinas and the opening hours of bridges and locks.

Qualifications. Craft >15m LOA or capable of more than 20kph (11kn) must be skippered by the holder of an RYA Coastal Skipper's Certificate or higher qualification; *plus* an International Certificate of Competence (ICC) endorsed for Inland waterways, ie CEVNI.

Bridges and locks mostly work VHF Ch 18, 20 or 22, but CEVNI light signals (shown up/down-stream) largely negate the need for R/T. The most commonly seen signals include:
● = Bridge closed (opens on request).
To request bridges to open, call on VHF low power (1 watt) or sound 'K' (—·—).

● over ● = Bridge about to open.　　● = Bridge open.
● over ● = No service.

9.16.7 TSS OFF THE NORTHERN NETHERLANDS

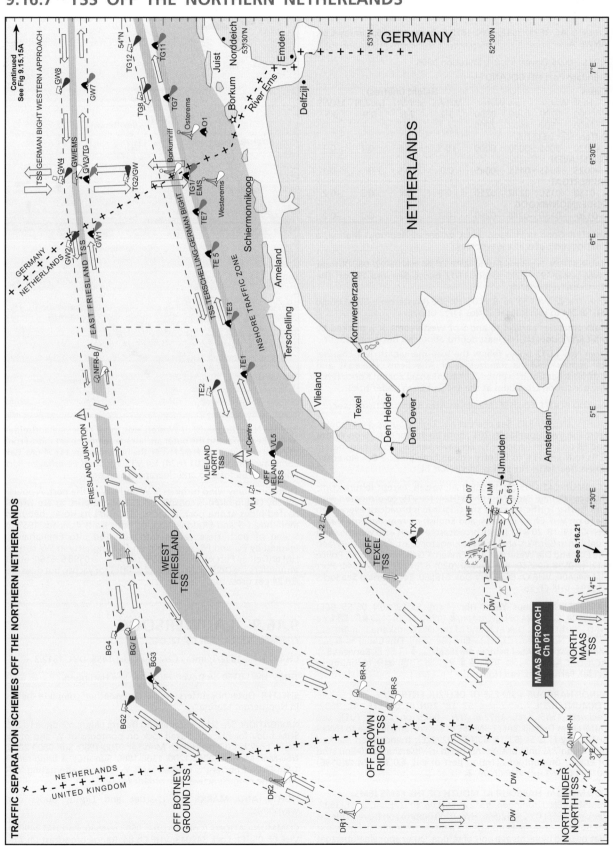

9.16.8 DELFZIJL

Groningen 53°18'·99N 07°00'·45E ✿✿✿◊◊✿✿

CHARTS AC 3631, 3632; DYC 1812.6; Zeekaart 1555; Imray C26; ANWB A

TIDES −0025 Dover; ML 2·1; Duration 0605

Standard Port HELGOLAND (←—)

Times				Height (metres)			
High Water		Low Water		MHWS	MHWN	MLWN	MLWS
0200	0700	0200	0800	3·2	2·8	0·9	0·5
1400	1900	1400	2000				
Differences DELFZIJL							
+0020	−0005	−0040	0000	+0·5	+0·6	−0·1	−0·1
EEMSHAVEN							
−0025	−0045	−0115	−0045	0·0	+0·1	−0·1	−0·1
HUIBERTGAT							
−0150	−0150	−0210	−0210	−0·5	−0·4	−0·2	−0·2
SCHIERMONNIKOOG							
−0120	−0130	−0240	−0220	−0·3	−0·2	−0·2	−0·2

SHELTER Good in Handelshaven, where at Neptunus Marina, a floating jetty, acts as a wavebreak.

NAVIGATION From the E or N, WPT 53°38'·96N 06°27'·06E (Riffgat SWM buoy) 121°/4·2M to Nos 11/12 buoys. See also TSS off the Northern Netherlands.

From the W, WPT 53°36'·93N 06°19'·39E (Westereems SWM buoy) 091°/8·2M to join Riffgat at Nos 11/12 buoys.

Huibertgat lies parallel to and S of Westereems. It is marked by unlit R/W buoys H1-H5, least depths about 9m but prone to silt.

From Nos 11/12 buoys follow the well marked/lit river channel for approx 25M via Randzelgat or Alte Eems, Dukegat and Ostfriesisches Gatje to enter Zeehavenkanaal abeam Knock lt ho. Beware strong cross tides at the ent (Lat/Long under title).

INLAND ROUTE TO IJSSELMEER See Area Map and IJsselmeer.

LIGHTS AND MARKS See chartlet and Lights, buoys & waypoints. From the river, appr ldg lts 203°, both Iso 4s. Hbr ent, FG on W arm and FR on E arm (in fog, FY). Zeehavenkanaal, 2·5M long, has Fl G lts to N and Fl R to S. Entry sigs on both piers: 2 ● = No entry, unless cleared by Hbr office on VHF Ch 14.

COMMUNICATIONS (Code 0596) All vessels, except leisure craft, must call *Delfzijl Radar* VHF Ch 03 (H24) for VTS, co-ordinated with Ems Traffic; Traffic, weather and tidal info is broadcast every even H+10 on VHF Ch 14 in Dutch and English on request; *Ems Traffic* VHF Ch 15, 18, 20 & 21 (H24) broadcasts every H + 50 weather and tidal info in German, and gale warnings for coastal waters between Die Ems and Die Weser, see also Emden; CG (Police) 613831; Police 112; ⊜ 0598 696560; Brit Consul (020) 6764343; Ⓗ 644444; Port HM 640400, VHF Ch 66; HM 't Dok 616560; Eemskanaal Sea locks 613293, VHF Ch 26.

FACILITIES Neptunus Yacht Hbr (4·4m, 53°19'·80N 06°55'·86E) ☎615004, ⌣€0·24 per gross ton, ⬧ (0900-1700) ✕ ☐ ⬧🖻 Ⅶ free wi-fi . **Yacht Hbr 't Dok** at N end of the Old Eemskanaal in 4m, ⌣ €0·41, ⬧. Note: Eems (Dutch) = Ems (German). **Ems Canal** ⬔⚓✕⌣. **Motor Boat Club Abel Tasman** ☎616560 ⌣⬧☐🖤🖻. **Services** ⬧🖻 ACA, DYC Agent, ✎ 🖂. **Town** ⬧🏠✕☐🖤🖂Ⓑ⇌✈ (Groningen/ Eelde). Ferry: Hoek van Holland.

MINOR HARBOUR 1·3M ESE OF DELFZIJL ENTRANCE
TERMUNTERZIJL Groningen, **53°18'·20N 07°02'·21E.** AC 3632; Zeekaart 1555; DYC 1812.6. HW −0025 on Dover (UT); use Differences Delfzijl. Ent (1·3M ESE of Delfzijl ent) is close to BW13 SHM buoy Fl G 5s (53°18'·63N 07°02'·32E); thence chan marked by 7 R and 7 G unlit bns. Yachts berth in Vissershaven (0·9m) stbd of ent, or on pontoons (1m) to port of ent, €0·36. HM ☎(0596) 601891 (Apr-Sept) VHF Ch 09, ⚓ 🚲 ✕ ☐.

COMMERCIAL HARBOUR AT MOUTH OF THE EEMS (EMS)
EEMSHAVEN, Groningen, **53°27'·66N 06°50'·27E.** AC 3631, 3632, Zeekaart 1555, DYC 1812.5/6. HW −0100 (approx) on Dover. Tides as for Borkum Area 15. Eemshaven is a modern commercial port, but may be used by yachts as a port of refuge. Outer appr via Hubertgat

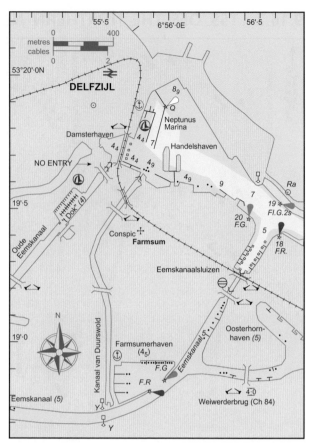

(see TSS N Netherlands) or Westereems; inner appr via Randzelgat or Alte Eems, passing the outer anchorage for merchant ships. From the nearest buoy, A16 PHM Fl R 4s, the hbr ent bears 137°/1·6M. Call *Eemshaven Radar* Ch 01 (H24) for VTS info; and *Eemshaven Port Control* Ch 66 for info and clearance to enter.

There are many wind turbines S, W and NW of the port. A power stn chy (128m high) is conspic 2M ESE of the port; as are white-roofed bldgs at the port. Enter on 175°, ldg lts Iso 4s, between mole hds, FG and FR. Inner ldg lts, both Iso R 4s, leads 195° to S end of port. Here yachts turn 90° stbd into Emmahaven, marked by FG and FR, and berth on floating jetty in SW corner. HM ☎(Delfzijl Port Authority) (0596) 640400; ⊜ 0598 696560; other ☎numbers see Delfzijl. No special facilities for yachts, but dues are €0·24 per gross ton.

9.16.9 LAUWERSOOG

Friesland, **53°24'·68N 06°12'·04E**

CHARTS AC 1632/3; Imray C26; Zeekaart 1458; DYC 1812.3

TIDES HW −0150 on Dover; ML 1·7m. See Harlingen.

SHELTER Outer hbr suffers swell in bad weather; complete shelter in Noordergat Marina.

NAVIGATION See Schiermonnikoog (facing page); continue to Z15 SHM buoy for hbr ent. Await lock on pontoons at W end of FV basin. Lock hrs (LT) **April-Oct**: Mon-Sat 0700-1900; Sun 0800-2000. **Nov-Mar**: Mon-Sat 0700-1200, 1300-1800; Sun shut. A firing range 1·5m ENE of hbr ent is marked by Fl Y 10s bcns, alternating W/R when the range is active; info broadcasts on Ch 71.

LIGHTS AND MARKS See chartlet and Lights, buoys and waypoints.

COMMUNICATIONS (Code 0519) ⊜ 0598 696560; Port HM 349023, VHF Ch 05, 11; Lock 349043, VHF Ch 84; Range broadcast Ch 71.

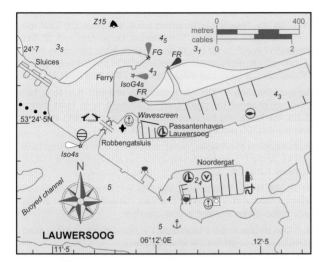

LAUWERSOOG

FACILITIES Noordergat Marina, ☎349040; www.noordergat.nl (2·4m-2·8m). Ⓥs berth on first pontoon, €1·00 + €1·15 pp/day tax. 🛢 (E end of hbr); 🛢 (ferry terminal, W end of hbr) ✕ ⌂ ▦ ⓞ ♨ 🛢 🛒 🅿(30t) ⚓ YC. **Passantenhaven Lauwersoog, ☎**349023. 60 berths (visitors only) €1.50, 🛢 ▦ 🛒. The Staande Mastroute (Area map, Delfzijl-Harlingen) can be entered from the Lauwersmeer at Dokkum.

OTHER HARBOURS IN THE LAUWERSMEER
ZOUTKAMP, Groningen, **53°20'·38N 06°17'·55E**. AC 2593, Zeekaarten 1458, DYC 1812·4; non-tidal. Appr down the Zoutkamperril (2·6-4·5m); approx 2ca before lock/bridge, Hunzegat marina (1·5-2·1m) is to port. ☎(0595) 402875, SC ⚓. Beyond the lock (FR lts) and close to port is Oude Binnenhaven marina (2m) €1·25 + €1·15 pp/day tax. 🔧 ♨ ✕ 🛒(20t) 🅿.

Town 🛢 🛢 ✕ ⌂ 🛒♨⌂ 🛢 ✉ Dr.

OOSTMAHORN, Friesland, **53°22'·94N 06°09'·63E**. AC 2593, Zeekaarten 1458, DYC 1812·4. Non-tidal hbr on W side of Lauwersmeer; lock in as for Lauwersoog. Floating bns with Y flags mark fishing areas. Main hbr with FR and FG lts at ent has marina (2·2-3·0m). VHF Ch 10. HM ☎(0519) 321445; €0·91, 🛢 🛢 ✕ ⌂ 🛒 🛢 ♨ ♨ 🅿(15T) 🔧. Approx 450m to the SSE, next to the Voorm Veerhaven (ferry hbr) is a tiny marina with 1·5-2m.

MINOR HARBOURS IN THE WEST FRISIAN ISLANDS
SCHIERMONNIKOOG, Friesland, **53°28'·07N 06°10'·05E**. AC 3761; Zeekaart 1458; DYC 1812.3. HW –0150 on Dover. See Delfzijl. WPT 53°32'·35N 06°00'·70E, (WG SWM buoy, Iso 8s, Racon N) 149°/1·0M to WG1 SHM buoy, VQ G, at ent to Westgat; buoyed/lit, but in bad weather dangerous due to shoals (2·6m) at seaward end, shifting chan, frequent buoy movements. Call Schiermonnikoog ☆, VHF Ch 5 for update. Follow Westgat chan into Zoutkamperlaag either to Lauwersoog or Schiermonnikoog; leave at Z4 & Z6-GVS bys to enter buoyed Gat van Schiermonnikoog (GVS).

Follow GVS bys to ent of Groote Siege (GS bys), turn port at GS6 into Reegeul chan, marked by withies, and eastwards to the small yacht hbr of Schiermonnikoog, (1·3-1·5m) 1·5m max at HW in apprs. Picturesque, busy in season, but reptd silting 2014; €2·00/m. HM VHF Ch 31; ☎(0519) 531544 (May-Sept). Facilities: 🔧 🗲 ✕.

Note: Lt ho Fl (4) 20s, R twr, conspic 1.35M NNW of yacht hbr. CG at lt ho, 48M radius radar surveillance of the Terschelling/German Bight TSS and coordinates local SAR operations. VHF Ch 05, 10, 16, 67 and 37 (all H24); ☎0519 531247.

NES, Ameland, Friesland, **53°26'·22N 05°46'·53E**. AC 2593, Zeekaart 1458, DYC 1811.10 & .2. HW –0055 on Dover; ML 1·6m; Duration 0625. See Harlingen. Shelter in all but E/S winds. Appr from WA SWM buoy via Westgat, Borndiep and Molengat to VA2-R1 SCM lt perch, VQ(6) + L Fl 10s, beware sandbanks; dangerous in >F6 from SW-N-NE. Lts: see Lights, buoys and waypoints. **Yacht hbr** ('t Leije Gat) HM ☎(0519) 542159. 140 berths, €0·66/m² + €1·18 pp/ day tax. L Fl R 8s and L Fl G 8s piles at ent. Pontoons N end hbr (0·8m) beyond ferry; W side dries. Facilities: 🛢 YC.

THE WEST FRISIAN ISLANDS
This chain of islands stretches from the River Ems estuary W and SSW for some 85M along the Dutch coast to Den Helder (AC 1632, 1633). The islands have similar characteristics – being long, low and narrow, with the major axis parallel to the coast. Texel is the largest and, with Vlieland and Terschelling, lies furthest offshore.

Between the islands, narrow channels (*zeegat* in Dutch, *Seegat* in German) give access to/from the North Sea. Most of these channels are shallow for at least part of their length, and in these shoal areas a dangerous sea builds up in a strong onshore wind against the outgoing (ebb) tide. The Westerems and the Zeegaten van Terschelling and van Texel are safe for yachts in W–NE winds up to force 6. All the others are unsafe in strong onshore winds.

▶*The flood stream along this coast is E-going, so it starts to run in through the zeegaten progressively from W to E. Where the tide meets behind each island, as it flows in first at the W end and a little later at the E end, a bank is formed, called a wantij (Dutch) or Wattenhoch (German).◀*

These banks between the islands and the coast are major obstacles to E/W progress inside the islands. The chans are narrow and winding, marked by buoys and/or withies (🚩 🚩) in the shallower parts, and they mostly dry; so that it is essential to time the tide correctly.

This is an area most suited to shallow-draft yachts that can take the ground easily. Whilst the zeegaten are described briefly below, the many channels inside the islands and across the Waddenzee are mentioned only for general orientation.

DELFZIJL TO AMELAND
The Ems estuary (AC 1633, 3631) flows seaward past the SW side of the German island of Borkum. It gives access to **Delfzijl**, **Termunterzijl** and **Emden**. Hubertgat, which runs parallel to and S of the main Westerems chan, is slightly more direct when bound to/from the W, but in both these chans there is a dangerous sea in strong NW winds over the ebb. Hubertgat is now sparsely buoyed and unlit, but is quite acceptable to yachts. ▶*The E-going (flood) stream begins at HW Helgoland +0530, and the W-going (ebb) stream begins at HW Helgoland –0030, sp rates 1·5kn.◀*

Friesche Zeegat (DYC 1812.3 & .9) between Schiermonnikoog and Ameland, contains two channels: Westgat marked by 'WG' buoys, but increasing shoaling and buoy shifting reported, and Plaatgat, an unmarked channel, which is considered to be dangerous. In strong winds the sea breaks across the whole estuary. Westgat trends SE past Wierumergronden and N7-FA-1 platform, the S past Het Rif and Engelsmanplaat (a prominent sandbank) into Zoutkamperlaag, the main channel (buoys prefixed 'Z') to **Lauwersoog**. Here locks give access to the Lauwersmeer and inland waterways.

Zeegat van Ameland (DYC 1811.6) between Ameland and Terschelling, is fronted by the sandbank of Bornrif about 3M offshore. The main entrance is also called Westgat, with buoys prefixed by letters 'WA', all unlit except WA12 QR. The channel runs ESE close N of Terschelling, and divides into Boschgat (BG) at the E end of West Terschelling and Borndiep (BB) the wider, deeper channel. This skirts the W end of Ameland and at WA22-WG1 buoy gives access via the Molengat to the small ferry port and yacht hbr at Nes.

▶*In Westgat the flood stream begins at HW Helgoland +0425, and the ebb stream at HW Helgoland –0150, sp rates 2kn.◀* A dangerous sea develops in strong onshore winds.

TERSCHELLING TO TEXEL AND DEN HELDER

Zeegat Het Vlie, aka Zeegat van Terschelling (AC 112 and DYC 1811.4, .5 and .9) between Terschelling and Vlieland, gives access to the hbrs of **Vlieland, West Terschelling and Harlingen**; it is also a northern approach to the sealock at Kornwerderzand. Shallow banks extend more than 5M seaward narrowing the ITZ to only 4M wide.

▶*The E-going (flood) stream begins at HW Helgoland +0325, while the W-going (ebb) stream begins at HW Helgoland −0230, sp rates 2·5kn.*◀

The main chan (buoyed/lit) through the banks is Zuider Stortemelk (buoys prefixed 'ZS') running ESE close N of Vlieland into Vliesloot (VS) which leads to **Oost Vlieland** hbr. The Stortmelk (SM) forks eastward at SM1-ZS10 buoy towards the deeper wider channel of Vliestroom.

Approach **West Terschelling** via West Meep and Slenk, **not** by the shorter Schuitengat Zuid which is badly silted. From Zuider Stortemelk the Vliestroom, a deep well buoyed chan (buoys prefixed 'VL') runs S about 4M until its junction with Blauwe Slenk (BS) and Inschot (IN). Blauwe Slenk runs ESE to **Harlingen**; from Harlingen a S-going channel 'Boontjes' leads via the locks at Kornwerderzand into the IJsselmeer.

Eierlandsche Gat (DYC 1811.7) between Vlieland and Texel, consists of dangerous shoals, shallow and unmarked chans used only by fishermen, *and should not be attempted by strangers.*

Zeegat van Texel (AC 1546/DYC 1811.2) lies between Texel and Den Helder, and gives access to the **Waddenzee** and IJsselmeer via locks at Den Oever or Kornwerderzand. Haaksgronden shoals extend 5M seaward. The 2 appr channels are: Molengat from the N along the Texel shore (see next para) and from the SSW the well marked/lit main chan Schulpengat, buoys prefixed 'S'; but strong SW winds cause rough seas against ▶*the SW-going (ebb) stream which begins at HW Helgoland −0330, while the NE-going (flood) stream begins at HW Helgoland +0325, sp rates 1·5kn.*◀

Serious shoaling in the Molengat has reduced this to a fair weather channel for small craft in daylight only, least depth 3.5m, marked along E side by Y spar buoys and Y SPM's at N and S entrances. Do not attempt in fresh NE wind or swell.

▶*In Molengat the N-going (ebb) stream begins at HW Helgoland −0145, and the S-going (flood) stream at HW Helgoland +0425, sp rates 1·25kn.*◀

From the N the Schulpengat involves a southerly deviation of approx 15M and leads W of the very dangerous Zuider Haaks which should be avoided in bad weather.

E of **Den Helder** and the Marsdiep, the flood makes in three main directions through the SW Waddenzee:

- to E and SE through Malzwin and Visjagersgaatje (keeping to the ctr of the channel) to **Den Oever** (where the lock into IJsselmeer is only available during daylight hours); thence NE along the Afsluitdijk, and then N towards Harlingen;

- to NE and E through Texelstroom and Doove Balg towards Kornwerderzand; and

- from Texelstroom, NE and N through Scheurrak, Omdraai and Inschot, where it meets the flood stream from Zeegat van Terschelling. The ebb runs in reverse. The **Kornwerderzand** locks (available H24), near NE end of Afsluitdijk, also give access to the IJsselmeer.

9.16.10 HET VLIE (ZEEGAT VAN TERSCHELLING)

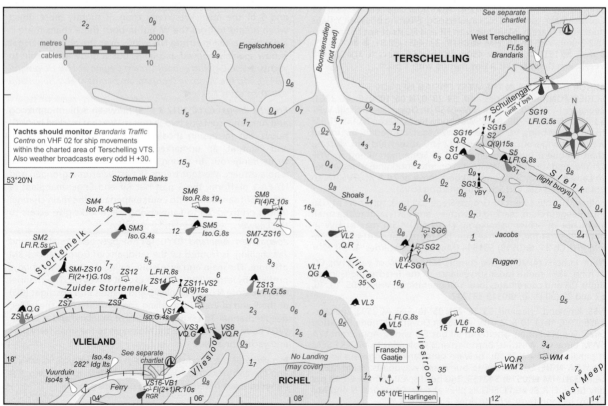

9.16.11 WEST TERSCHELLING

WEST FRISIAN ISLANDS 53°21'·26N 05°13'·13E ✵✵✵❁♨♨♨♨✿✿

CHARTS AC 1633, 112; Zeekaart 1456; DYC 1811.4/.5; Imray C25

TIDES –0300 Dover; ML 1·4; Duration No data

Standard Port HELGOLAND (◀––)

Times				Height (metres)			
High Water		Low Water		MHWS	MHWN	MLWN	MLWS
0200	0700	0200	0800	3·2	2·8	0·9	0·5
1400	1900	1400	2000				
Differences WEST TERSCHELLING							
–0220	–0250	–0335	–0310	–0·8	–0·6	–0·3	–0·2
VLIELAND-HAVEN							
–0250	–0320	–0355	–0330	–0·8	–0·6	–0·3	–0·1

SHELTER Good in the marina; very crowded in season.

NAVIGATION WPT 53°19'·58N 04°55'·77E (ZS buoy) 099°/6·2M to ZS11-VS2 buoy via Zuider Stortemelk which is well marked/lit. In strong W/NW'lies beware dangerous seas from ZS to ZS5/6 buoys. Stortemelk, the deeper N'ly route, is no shorter and is more exposed. From Vlieree (Vlieland Roads) make for the West Meep Channel. Call *Brandaris* Ch 02 or 04; see Communications.

> **Note:** Schuitengat-Zuid, a former approach channel, **is no longer an option for the foreseeable future.**
> West Meep/Slenk is long, but wide, safe and 3·2m deep; hence easier for visitors. Route: WM2, 4 & 6 buoys, NM4–S21 buoy to enter Slenk buoyed/lit chan leading NW into the inner Schuitengat (11m) at SG15/S2 WCM buoy. Expect much commercial traffic and ferries in Slenk.

Ameland can be reached if draft <1.5m by the inshore channels across the Waddenzee (Meep, Noorder Balgen and Oosterom; DYC 1811.4 & .6). Leave at HW –3 to arrive at HW.

LIGHTS AND MARKS Hbr lts as chartlet and Lights, buoys and waypoints. The square, yellow Brandaris light tower dominates the town and hbr.

COMMUNICATIONS (Code 0562) Coast Guard 442341; *Brandaris* (VTS) Ch 02, 04 broadcasts weather, visibility, traffic, tidal data at H+30 in Dutch and English. Yachts must monitor Ch 02; ⊖ 442884; Dr 442181; Port HM 443337 (H24) VHF Ch 12; Marina Ch 31 HO.

FACILITIES **Marina Stichting Passantenhaven** ☎443337 (H24 Apr–

TERSCHELLING

Oct) info@jachthaventerschelling.nl. www.waddenhavens.nl. Notices on pontoons A-G indicate berths by LOA and type of boat (yacht, motor cruiser or traditional craft). 500 ⌷ €0·70/m² + €1·30 pp/day tax, inc ⌁ ⚓ 🔧 WC 🚿; 🛢 at marina ent, @ ▭ 🛒 🛢 🔧 ⚓ chart agent 🔧.

Village is an easy walk: ✗ 🗗 🍽 🏧 ✉ Ⓑ ⛽ (5km). ✈ Amsterdam. Ferry to Harlingen, thence ⇌ for Hook-Harwich.

9.16.12 VLIELAND

WEST FRISIAN ISLANDS

Friesland 53°17'·68N 05°05'·49E ✵✵✵❁♨♨✿✿

CHARTS AC 1633, 112; Zeekaart 1456; DYC 1811.4/.5; Imray C26

TIDES –0300 Dover; ML 1·4; Duration 0610. See West Terschelling. The Dutch use Harlingen as a Standard Port.

SHELTER Good in yacht hbr. Two ⚓s, both affected by swell in strong SE-SW'lies near HW: (a) in up to 3m ½M S of the hbr or (b) in 4-9m 1M W of the hbr (beyond the ferry pier). Do not ⚓ in buoyed chan from hbr ent to ferry pier (no ⌷).

NAVIGATION From ZS buoy (WPT) follow W Terschelling Navigation, lines 1-3. At ZS11-VS2 and VS1 buoys turn S into Vliesloot (narrow and in places only 2·5m MLWS). Keep in mid-chan past VS3-VS7 buoys to hbr ent (strong current across); tight berths difficult in SW wind.

LIGHTS AND MARKS A dk R lt ho (W lantern, R top) stands on Vuurduin, 1·2M W of hbr. Hbr lts as chartlet. Tfc sigs: R Flag or ●● at ent = hbr closed (full). The 282° ldg lts (Lights, bys & waypoints & W Terschelling) lead toward the ferry terminal in E Vlieland.

COMMUNICATIONS (Code 0562) CG 442341; All vessels in Zeegat van Terschelling and N Waddenzee must monitor Ch 02 for *Brandaris Traffic Centre VTS;* Police 451312; ⊖ 058 2949488 (Leeuwarden); Brit Consul (020) 6764343; Dr 451307; HM 451729, mob 06 201 33086, VHF Ch 12.

FACILITIES **Yacht Hbr** ☎451729. www.waddenhavens.nl 300 inc Ⓥ; finger berths on all pontoons in 2m. €0·70 per m² + €1·35 pp/day tax; ▭ 🗑 🛒 ⛽ (10 mins); 🛢 at Harlingen and Terschelling.

Village ✗ 🗗 🍽 🛢 ⛽ @ ✉ Ⓑ ✈(Amsterdam). Ferry as West Terschelling.

VLIELAND

9.16.13 HARLINGEN

Friesland 53°10'·58N 05°24'·23E ❀❀⚓⚓⚓❀❀

CHARTS AC 1633, 112; Zeekaart 1454, 1456; DYC 1811·5; ANWB B; Imray C26

TIDES −0210 Dover; ML 1·2; Duration 0520

Standard Port HELGOLAND (←)

Times				Height (metres)			
High Water		Low Water		MHWS	MHWN	MLWN	MLWS
0200	0700	0200	0800	3·2	2·8	0·9	0·5
1400	1900	1400	2000				
Differences LAUWERSOOG							
−0130	−0145	−0235	−0220	−0·3	−0·2	−0·2	0·0
HARLINGEN							
−0155	−0245	−0210	−0130	−0·8	−0·7	−0·4	−0·1
NES (AMELAND)							
−0135	−0150	−0245	−0225	−0·3	−0·2	−0·2	0·0

SHELTER Very good in Noorderhaven. Beware strong flood stream across outer hbr ent; it can be rough at HW in W/NW'lies.

NAVIGATION WPT 53°11'·48N 05°19'·63E (abeam BS23 buoy) 110°/2·8M to hbr ent. See West Terschelling for appr, thence to the WPT via buoyed Vliestroom and Blauwe Slenk chans; the latter is narrow for the last 2½M. A small craft chan, at least 1·8m, lies parallel to Pollendam trng wall and outboard of the main chan SHM buoys; it is marked by SPM buoys A–N in season. Beware ferries passing very close. Caution: When Pollendam is covered, tidal streams sweep across it.

LIGHTS AND MARKS See chartlet and Lights, buoys and waypoints. Ldg lts 111°, both Iso 6s, H24; B masts, W bands. 2 ch spires are conspic almost on ldg line.

COMMUNICATIONS (Code 0517) Coast Guard 0223 542300; Brandaris Traffic Ctre (0562) 442341, VHF Ch 02; Police 0900 8844/112; ⊜ 088 151444; British Consul (020) 6764343; Ⓗ 058 2866666; Doctor 412544; Harlingen Port Control 412512, VHF Ch

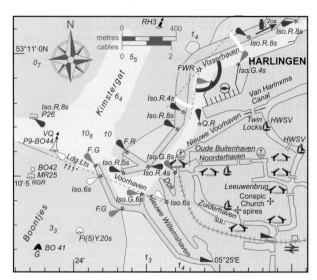

11; HM 492300, VHF Ch 11 (not on Sun); Harinxma Canal locks VHF Ch 22.

FACILITIES Lifting bridges across the Oude Buitenhaven and Noorderhaven open in unison 2 x per hr 0600-2200 in season (on request in winter) but are shut at times of boat trains/ferries and at a tidal height exceeding LAT + 2·54m. Fuel available in the NE corner of the Nieuwe Industriehaven.

Noorderhaven Yacht Hbr ☎415666, ✕ ⬜ 🍴 🚿 🛢 🏬 🏧 ⚒ ⚓ Ⓔ Diving/salvage (by arrangement).

Harlinger Watersportvereningen (HWSV) ☎416898 info@ harlingerwatersportvereningen.nl, useful small marina just inside the locks, 100+ ⚓ inc Ⓥ €1·20 + €0·75 pp/day tax, inc 🔌 ⚓ ⛽ 🚾, 🚿 🏧 @. **Town** ✕ ⬜ 🍴 ✉ Ⓑ ⇌ ✈ (Amsterdam). Ferry see Hoek van Holland.

9.16.14 OUDESCHILD

Texel, 53°02'·35N 04°51'·18E ❀❀❀⚓⚓⚓❀❀

CHARTS AC 1631, 1546; Zeekaart 1546, 1454; DYC 1811·3

TIDES −0355 Dover; ML 1·1m; Duration 0625

Standard Port HELGOLAND (←)

Times				Height (metres)			
High Water		Low Water		MHWS	MHWN	MLWN	MLWS
0200	0700	0200	0800	3·2	2·8	0·9	0·5
1400	1900	1400	2000				
Differences OUDESCHILD							
−0310	−0420	−0445	−0400	−1·4	−1·1	−0·4	−0·2

SHELTER Very good in marina (2·4m) in far NE basin. ⚓ prohib. Visitors are always welcome.

NAVIGATION From Waddenzee appr via Doove Balg and Texelstroom. From seaward, app via Schulpengat or Molengat into Marsdiep (Den Helder). Thence NE via Texelstroom for 3·5M to WPT 53°02'·26N 04°51'·55E (abeam T12 PHM buoy) 291°/400m to hbr ent, keeping the dir lt, Oc 6s, midway between FR and FG mole hd lts. Speed limit 5kn.

LIGHTS AND MARKS See chartlet and Lights, buoys and waypoints. Dir lt Oc 6s, on G mast, is vis only on brg 291°. FR/FG lts are on R/G masts with W bands.

COMMUNICATIONS (Code 0222) Coast Guard 316270; Police/ambulance 0900 8844 or 112; ⊜ via HM; Port HM 312710, mobile 06 1502 8380, VHF Ch 12 0800-2000LT; Marina VHF Ch 31; See Den Helder for VTS and radar assistance/info on request.

FACILITIES

Waddenhaven Texel Marina 250 berths ☎321227. info@ waddenhaventexel.nl €0·70/m² + €1·60pp/day tax 🚿 ⛽ ✕ ⬜ mini 🛒 in season 🏧 🛢 @ ♿ bike hire ⚓.

YC WSV Texel, members only. Dry dock (FVs only).

Village (walking distance) ✕ ⬜ 🍴 ✉ ⚓ ⚒ 🛢 Ⓑ. Ferry from 't Horntje to Den Helder. UK ferries from Hook/Rotterdam. ✈ Amsterdam.

9.16.15 DEN HELDER
Noord Holland **52°57'·94N 04°47'·24E**

CHARTS AC 1408, 1631, 1546, 126; Zeekaart 1454, 1546; DYC 1811.2, 1801.10; ANWB F; Imray C25

TIDES –0430 Dover; ML 1·1; Duration 0540

Standard Port HELGOLAND (◄——)

Times				Height (metres)			
High Water		Low Water		MHWS	MHWN	MLWN	MLWS
0200	0700	0200	0800	3·2	2·8	0·9	0·5
1400	1900	1400	2000				
Differences DEN HELDER							
–0410	–0520	–0520	–0430	–1·3	–1·1	–0·4	–0·2
K13A PLATFORM (53°13'·0N 03°13'·1E; 58M WNW of Den Helder)							
–0420	–0430	–0520	–0530	–1·4	–1·4	–0·3	–0·3

SHELTER Good in all yacht hbrs (see Facilities). Hbr speed limit 5kn. Den Helder is the main base of the Royal Netherlands Navy which manages the Marinehaven Willemsoord and KMYC.

NAVIGATION Caution: fast ferries to/from Texel, many FVs and offshore service vessels, strong tidal streams across the hbr ent. Schulpengat leads into Marsdiep for the hbr ent. It is well marked/lit. WPT 52°52'·90N 04°37'·90E (SG (SWM) buoy, Mo (A) 8s, Racon Z) 026·5°/6M towards S14/MG17 SCM lt buoy.

Although closed as a shipping chan in Apr 2013 due to silting, the Molengat remains a viable fair weather chan for smaller craft, least depth 3.5m, with Y ⚓ marking E side. Do not attempt at night or in fresh NW-N wind or swell.

LIGHTS AND MARKS See chartlet and Lights, buoys and waypoints. Schulpengat: Dir lt Al WRG and F WRG. Kijkduin lt ho, R twr, is conspic 2·3M WSW of hbr. Hbr 191° ldg lts; front B ▲ on bldg; rear B ▼ on B lattice twr. A 60m radar twr is conspic on E side of hbr.

Entry sigs, from Hbr Control twr (W side of ent): ring of R lts = No entry/exit, except for the vessel indicated by Hbr Control Ch 62.

Bridges: Moorman bridge operates H24 7/7, giving access to Nieuwe Diep, Koopvaarder Lock and N Holland Canal.

Van Kinsbergen bridge is operated by HM, on request, Mon-Fri. All bridges are closed 0705-0810, 1200-1215, 1245-1300, 1600-1700, Mon-Fri. All LT.

COMMUNICATIONS (Code 0223) VTS 657522, monitor VHF Ch 62 *Den Helder Traffic* (VTS) in the Schulpengat and Marsdiep; info on request; Police 112/09008844; Water Police 616767; ⊖ 0255 566707; Immigration 08000543; British Consul (020) 676 4343; Ⓗ 696969; HM 613955, VHF Ch 14 (H24) also remote control of van Kinsbergen bridge; *Koopvaarders Lock* VHF Ch 22 (H24) also remote control of Burgemeester Vissersbrug; Moorman bridge VHF Ch 18 (H24).

FACILITIES KMJC is hard to stbd on entering Naval Hbr. ☎652645,

mob 0651007528, VHF Ch 31, www.kmjc.eu €2·10 + €1·24 pp/day tax draught <4m, ⚓ ✕ ⬜ Ⓞ ⬧.

Willemsoord Marina through Zeedoksluis (Ch 14) ☎616100, www. willemsoord.nl €1·65 + €1·24pp/day tax, ⚓ ✕.

In or near **Binnenhaven**, YCs/marinas (MWV, HWN and Breewijd) can only be reached via the Rijkshaven, Moorman bridge, Nieuwe Diep and lock: **MWV YC** ☎652173, www.marinewatersport.nl €1·40 + €1·24pp/day tax. ⚓ Ⓣ ◡. **HWN YC** ☎624422, www. wsvhwn.nl €1·00 + €1·24pp/day tax. **YC WSOV Breewijd** ☎615500 €0·75 + €1·20pp/day tax. **Council harbour** ☎613955 €1·30 + €1·24pp/day tax. **Nauticadam (Den Helder marina)** ☎637444, €0·75 + €1·20pp/day tax Ch 31, ◡ ✕ ⬜ ⬡ ⚓ ⬚ Ⓒ ✕ ⬛ ⬛.

Services: ⬛ ⚓ Floating dock ⚓. Boatyard: Jachtwerf Den Helder ☎636964, ⬛(40t) ✕.

Town ⚓ ✕ ◡ ⬛ ⊠ Ⓑ ⬛. ⇌ ✈ (Amsterdam). Ferry Hook-Harwich; Rotterdam-Hull, IJmuiden-Newcastle.

9.16.16 IJSSELMEER

CHARTS Zeekaart 1351, 1454; In-date DYC 1810, sheets 3-5, are essential to avoid live firing ranges, fishing areas and other hazards; 1810 also has many hbr chartlets.

TIDES The IJsselmeer is non-tidal. Tides seaward of Den Oever and Kornwerderzand locks: –0230 Dover; ML 1·2

Standard Port HELGOLAND (←—)

Times				Height (metres)			
High Water		Low Water		MHWS	MHWN	MLWN	MLWS
0200	0700	0200	0800	3·2	2·8	0·9	0·5
1400	1900	1400	2000				
Differences DEN OEVER							
–0245	–0410	–0400	–0305	–1·2	–1·0	–0·5	–0·2
KORNWERDERZAND							
–0210	–0315	–0300	–0215	–1·0	–0·8	–0·4	–0·2

SHELTER Excellent in the marinas, but in the IJsselmeer strong winds can get up very quickly and often raise short, steep seas.

NAVIGATION Enter from sea by Den Oever or Kornwerderzand locks; also from IJmuiden via the Noordzeekanaal and Amsterdam.

Speed limits: 10·5kn in buoyed chans and <250m from shore. Hbr limits vary; see Special notes for the ANWB *Wateralmanak*.

A firing range, operational Tues to Thurs 1000–1900LT, extends S from Breezanddijk (53°01'·0N 05°12'·5E) to 3M NE of Medemblik then NNW to Den Oever (DYC 1811.3 & 1810.3). Call *Schietterrein Breezanddijk* (range control) Ch 71. Firing times are broadcast on VHF Ch 01 by Centrale Meldpost at H +15 on the day of firing, after weather forecasts.

R/T Nav info VHF Ch 01. Sealocks: Kornwerderzand 18; Den Oever 20. Enkhuizen naviduct 22; Lelystad 20; Oranjesluizen 18.

SEALOCKS Standard signals, see Special notes.

DEN OEVER SEALOCK AND MARINA Approach from seaward via the buoyed/lit Malzwin and Wierbalg channels. Enter on 131° ldg line, then 220° towards Buitenhaven. 2 swing bridges (52°56'·05N 05°02'·50E) and lock open in unison HO. If NW-bound, await lock opening in Binnenhaven. Marina (2-3m) HM ☎(0227) 511789, ▲♦⬚◙⬚(30t).

KORNWERDERZAND SEALOCK From seaward approach from the W via Doove Balg; from NW via shallow Inschot chan; or from N via Boontjes. Entrance to Buitenhaven has FR/G and Iso G 6s lts. When two road bridges (53°04'·44N 05°20'·09E) have opened, yachts enter the smaller Lorentz E lock, operates H24. N-bound, follow 348° ldg lts, Iso 4s, to FR/G at ent. Await lock in Binnenhaven (3·6m). HM ☎(0517) 578170. ⊜ 058 2949444.

FACILITIES clockwise from Kornwerderzand. Overnight fees range from approx €0·80 – €1·80/metre LOA. Most berths are bows on to a pontoon, stern lines to piles; there are few ⚓s.

IJSSELMEER EAST

MAKKUM: About 1M SE of Kornwerderzand, MA5 SHM buoy, Iso G4s, marks the buoyed chan. FR/FG lts lead 090·5° into Makkum. To stbd, **Marina Makkum** (2·5-2·8m) ☎(0515) 232828, ⚓ ⬤ ✕ ▢ 🛒 ▣ ⚙ △. 5 other marinas/YCs are further E; ♨ ⚓ ▣(40t) ⛵(30t). **Town** ✕ 🛒 ▢ Ⓑ Dr.

WORKUM: 5M S of Makkum; FW ldg lts 081° along buoyed chan to **It Soal Marina** ☎(0515) 542937, ⚓ ⬤ ▣ ♨ ⚓ ⚙ ▣ ⛵.

HINDELOOPEN: WV Hylper Haven HM ☎(0514) 522009, ⚓ ⬤ ⚙ ▣ ✕. **Old Hbr** ⚓ ⬤ ⚙ ▣ ✕. **Jachthaven Hindeloopen** (500) HM ☎(0514) 524554, ⚓ ⬤ ✕ ▢ ▣ ⚙ ▣(65t).

STAVOREN: Ent Buitenhaven 048° on Dir lt Iso 4s between FR/G. Marina, €1·00, E of ent in Oudehaven; ☎(0514) 681216, VHF Ch 74. Or, 1km S, ent Nieuwe Voorhaven (FR/G & Fl 5s Dir lt) then via lock to **Marina Stavoren** (3m) ☎(0514) 684684, ⚓ ⬤ ▣ ♨ ⚓ ⛵ △ ▣(65t); **Outer Marina** (3·5m) close S of Nieuwe Voorhaven. Also 3 other marinas. **Town** ✕ ▢ 🛒 Ⓑ Dr ⚲.

LEMMER: Appr on ldg lts 038°, Iso 8s, to KL5/VL2 By; then ldg lts 083°, Iso 8s, through Lemstergeul. Lastly FG and Iso G 4s lead 065° into town and some 14 marinas. **Jachthaven Friese Hoek** (275) HM ☎(0514) 568135. Services: ▢ ⬤ ♨ ⚓ △ ▣(30t).

URK: Hbr ent ½M SE of lt ho. Dir lt Iso G 4s to hbr ent, FR/G. Hbr has 4 basins; keep NNW to berth in Nieuwe Haven, Westhaven or Oosthaven (3·3m). HM ☎(0527) 689970. **Westhaven** ▣ ▣ △; **Oosthaven** ▣ ⚓ ▣ ✕ Ⓔ. **Town** EC Tues; ✉ Ⓑ Dr.

LELYSTAD: **Flevo** is 2M NNE of lock. (550) ☎(0320) 279800, ⚓ ⬤ ✕ ▢ 🛒 ▣ ♨ ⚓ ⛵ ▣(50t) ⛵. **Deko Marina** is close N of the lock; ☎(0320) 269000, ✕ △ ⛵. **WV Lelystad, Houtribhaven** (560) HM ☎(0320) 260198, ⚓ ✕ ▢ 🛒 ▣ ▢ ▣(20t). S of lock **Bataviahaven** (150) Ch 14, HM ☎06 511 77049. **Marina Lelystadhaven** ☎(0320) 260326, ✕.

MARKERMEER is divided from the IJsselmeer by the Houtribdijk 13M long. Access from sea via IJmuiden, Noordzeekanaal, Amsterdam/Oranjesluizen. Hbrs clockwise from Lelystad to Enkhuizen:

DE BLOCQ VAN KUFFELER: (8·5M SW Lelystad) ☎06 2751 2497, ✕.

MUIDERZAND: Ent 1M N of Hollandsebrug (12·7m cl'nce at SE corner of Markermeer) at buoys IJM5 & IJM7/JH2. **Marina** ☎(036) 5369151, ⚓ ⬤ ✕ ▢ ♨ ⛵.

MUIDEN: Ldg lts Q 181° into **KNZ & RV Marina** (2·6m) W of ent; home of Royal Netherlands YC (150 berths). HM ☎(0294) 261450, www.knzrv.nl €1·95, ✕ ▢ ▣ ▢ ▣. On E bank **Stichting Jachthaven** (70) ☎(0294) 261223, €1·60, ⚓ ⬤.

DURGERDAM: ½M E of overhead power lines (70m); convenient for Oranjesluizen. Keep strictly in buoyed chan. **WV Durgerdam** Berth to stbd of ent (1.8m). ☎06-14750510.

UITDAM Appr from MIJ5 buoy Fl Y 5s via buoyed chan to FR/FG at ent. ⚓ ⬤ 🛒 ⬤ ▣(9t). HM (020) 4031433.

MARKEN (picturesque show piece): Ent Gouwzee from N, abeam Volendam, thence via buoyed chan; Dir FW lt 116° between FR/G at hbr (2·2m). Lt ho, conspic, Oc 8s 16m 9M, on E end of island. HM ☎(0299) 601253, free ♥ ⛵ on quay, ✕ 🛒 ▣ ▢ Ⓑ.

MONNICKENDAM: Appr as for Marken, then W to MO10 Iso R 8s and 236° ldg lts FR. **Hemmeland Marina** ☎(0299) 655555, (2·0m) C. **Waterland Marina** ☎652000, (2·5m) ▣ ⛵(15t) www. jachthavenwaterland.nl. **Marina Monnickendam** ☎652595, (2·0m) C. **Zeilhoek Marina** ☎651463, (2·5m) ♨.

VOLENDAM: Fork stbd to old hbr (2·4m). Dir lt Fl 5s 313°. FR/FG at ent. ☎06 51337494, ⚓ ⬤ ⚓ △ ⛵. Fork port via unlit buoyed chan to **Marina Volendam** (2·7m) HM (0299) 320262. ✕ 🛒 Ⓑ △ ⛵(16t).

EDAM: Appr via unlit chan keeping Iso W 8s between FG /R at narrow ent; beware commercial traffic. **Jachthaven WSV De Zeevang** (2·2m) S side of outer hbr. ☎(0299) 350174, ⚓ ▣ ✕. Or lock into the Oorgat and via S canal to Nieuwe Haven. **Town** ✕ ▢ ▢ ⬤ ✉ Ⓑ.

HOORN: Radio twr (80m) 1·5M ENE of hbr. Iso 4s 15m 10M and FR/G at W Hbr ent. Four options: To port **Grashaven** (700) HM ☎(0229) 215208, 🛒 ▢ ▣ ⚓ ▣ ✕ ⛵. To stbd ⚓ in **Buitenhaven** (1.6m). Ahead & to stbd **WSV Hoorn**, (100) ☎213540. Ahead to **Binnenhaven** (2·5m) via narrow lock (open H24) ⛵ ⚓. **Town** ✕ ▢ 🛒 ▣ ▣ ▢ ✉ Ⓑ Dr ⚲.

ENKHUIZEN: 036° ldg lts Iso 4s; 230° ldg lts Iso 8s. Yachts transit the Houtribdijk via a 'Naviduct', ie a twin-chambered lock above the road tunnel below (see chartlet) saving long delays for boat & road traffic; call Ch 22. Krabbersgat lock & bridge, the former bottleneck for both road and sea traffic, can still be used by boats with <6m air clearance.
Compagnieshaven (500) HM ☎(0228) 313353, ⚓ ⬤ ▣ ⬤ ▣(12); **Buyshaven** (195) ☎315660, ⚓ ▢; **Buitenhaven** ☎312444.
Town EC Mon; Market Wed; ✕ ▢ 🛒 ▣ ▢ ▣ ▢ Ⓑ Dr △ ⚓ ▢ ✕ ⛵ ⚲.

BROEKERHAVEN: Small yacht hbr 1·5M SW of Enkhuizen. Appr from KG15-BR2 buoy. WSV de Broekerhaven ☎(0228) 518798.

IJSSELMEER WEST

ANDIJK: Visitors use **Stichting Jachthaven Andijk,** the first of 2 marinas; (600) ☎(0228) 593075, narrow ent with tight turn between FR/G lts. ⚓ ⬤ ▣ ⬤ ⚓ △ ⛵(20t).

MEDEMBLIK: Ent on 232°, Dir lt Oc 5s between FR/G. Thence via Oosterhaven ⚓ into Middenhaven (short stay) and via bridge to Westerhaven. **Pekelharinghaven** (120) HM ☎(0227) 542175; ent is to port by Castle, ✕ ▢ ▣. **Middenhaven** HM ☎541686, ⚓ **Stichting Jachthaven** HM ☎541681 in Westerhaven, ▢ ⛵. **Town** ✕ ▢ 🛒 ▢ ▣ ⚓ ▢ △ ✕ ✉ Ⓑ Dr. **Regatta Centre** 0·5M S of hbr entr. HM ☎(0227) 547781, ⛵ 450, ✕ ▣ ⛵(20t) ⛵.

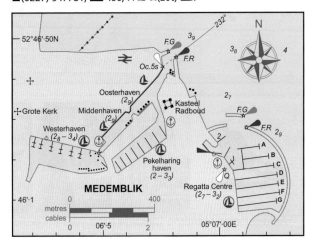

9.16.17 IJMUIDEN

N Holland 52°27'·94N 04°32'·39E (Hbr ent) SPM ⚓⚓♓♓♓❀❀

CHARTS AC 1631, 124; Zeekaart 1450, 1543, 1035, 1350; DYC 1801.8; Imray C25

TIDES +0400 Dover; ML 1·0. Noordzeekanaal level may be above or below sea level

Standard Port VLISSINGEN (→)

Times				Height (metres)			
High Water		Low Water		MHWS	MHWN	MLWN	MLWS
0300	0900	0400	1000	4·9	4·1	1·0	0·5
1500	2100	1600	2200				
Differences IJMUIDEN							
+0145	+0140	+0305	+0325	−2·8	−2·4	−0·7	−0·3
PETTEN (SOUTH) 18M N of IJmuiden							
+0210	+0215	+0345	+0500	−2·8	−2·3	−0·5	−0·2

SHELTER Very good at Seaport (SPM) marina (2·9m-4·8m).

NAVIGATION WPT 52°28'·10N 04°30'·94E, 100°/0·9M to ent. Beware strong tidal streams across hbr ent, scend inside Buitenhaven and commercial tfc – do not impede. Keep a good lookout, especially astern. 3 knots max speed in marina.

Marina access chan is buoyed/lit as per the chartlet.

LOCKS In IJmuiden the 4 locks are North, Middle, South and Small. Yachts normally use the S'most Small lock (Kleine Sluis) which opens H24 on request Ch 22; wait W or E of it.
Lock signals shown W and E from each lock are standard, ie:

- ●
- ● = Lock not in use; no entry.   ● = No entry/exit.
- ●
- ● = Prepare to enter/exit.   ● = Enter or exit.

Tidal & sluicing sigs may be shown from or near the conspic Hbr Ops Centre (HOC) bldg, near the front 100·5° ldg lt:

Tidal signals: ● over Ⓦ = rising tide;   Ⓦ over ● = falling tide.

Sluicing sigs are a △ of 3 horiz lts and 1 lt at the apex. The sluices are N of the N lock, so unlikely to affect yachts.

LIGHTS AND MARKS Both ldg lt ho's 100·5° for Zuider Buitenkanaal are dark R twrs. The HOC bldg is conspic next to front lt; so are 6 chimneys (138-166m high) 7ca N and ENE of Small lock.

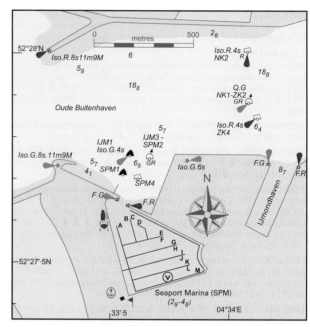

COMMUNICATIONS (Code 0255) Coast Guard 537644; Tfc Centre IJmuiden 564500, VHF Ch 07, see also VTS overleaf; Police 0900 8844; ⊖ 020 5813614; British Consul (020) 6764343; Ⓗ (023) 5453200; Hbr Ops 523934; Seaport Marina call SPM VHF Ch 74; Pilot 564503.

FACILITIES **Seaport Marina** (SPM) ☎560300. www.marinaseaport. nl info@ marinaseaport.nl ⚓ 600 inc Ⓥ €2·20 inc ⚓ ⚓ WC ⚓. Berth as directed, or on M pontoon in a slot with green tally. ⚓ ⚓ ✕ ⚓ ⚓ ⚓ ⚓ ⚓ ⚓ ✕ ⚓ ⚓ Ⓔ ⚓(70t) ⚓. Buses to Amsterdam and Haarlem in summer.

Town ✕ ⚓ ⚓ ⚓ @ ⚓ Ⓑ ⚓ (bus to Beverwijk) ✈ Amsterdam. Ferry IJmuiden-Newcastle.

9.16.18   NOORDZEEKANAAL

VTS Monitor the following VTS stns (H24) in sequence to approach, enter and transit the canal (see diagram opposite):

• *IJmuiden Traffic Centre*	Ch 07	W of IJM C buoy
• *IJmuiden Port Control*	Ch 61	IJM buoy to locks
• *IJmuiden Locks*	Ch 22	at the locks
• *Noordzeekanaal Traffic Centre*	Ch 03	Locks to km 11
• *Amsterdam Port Control*	Ch 68	Km 11 – Amsterdam
• *Sector Schellingwoude*	Ch 60	City centre (04°55'E) to Buiten IJ

Radar surveillance is available on request Ch 07. Visibility reports are broadcast on all VHF chans every H+00 when vis <1000m.

NAVIGATION Canal speed limit is 9 knots. The 13·5M transit is simple, but there is much commercial traffic. Keep as far to stbd as possible. Keep a good lookout for vessels emerging from dock basins and side canals. The banks and ents to basins and side canals are well lit, so night navigation need be no problem apart from the risk of not being seen by other vessels against shore lights.

FACILITIES There are small marinas at IJmond beyond lifting bridge (Ch 18) in Zijkanaal ⌐ km 10 (S bank); and at Nauerna (Zijkanaal ⌐ km 12) (N bank).

WV IJmond ⚓⚓⚓⛟⛟⚘. Ch 31, HM ☎(023) 5375003. ⊂ €1.10. ⌁ ⌐ ⌐(20t) ⌐ ⌐ ✕.

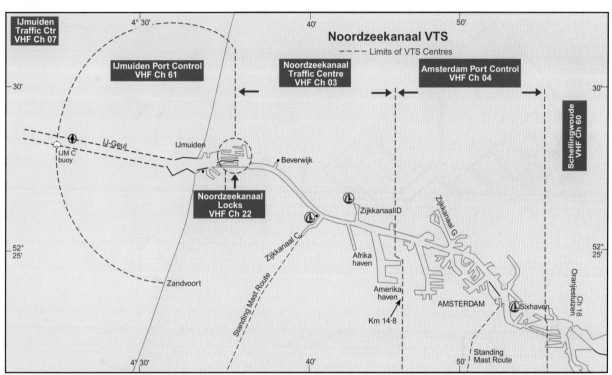

Fig 9.16.18A Noordzeekanaal VTS

9.16.19   AMSTERDAM

Noord Holland **52°22'·97N 04°53'·92E**   Marinas ⚓⚓⚓⛟⛟⚘⚘⚘

CHARTS AC 124; DYC 1801.8, 1810.2; ANWB G, I

NAVIGATION Het IJ (pronounced eye) is the well lit/buoyed chan through the city centre. From IJ 10/11 to IJ 14/15 buoys yachts must keep out of Het IJ, using a yacht chan between the lateral buoys and RW (N side) or GW (S side) buoys further outboard. A second recreational fairway, 100m wide and 5.5m deep (NAP) runs between IJ2 and IJ6.

Two yacht crossing points by YM 14-15 lead to Oranjesluizen, use the N lock, marked 'SPORT'; Ch 18, H24. Waiting piers/jetties either side. The Schellingwoude bridge (9m) opens on the hour and ±20 min, except Mon-Fri 0600-0900 and 1600-1800; VHF Ch 18.

COMMUNICATIONS (Code 020) Emergency 112; Police 0900 8844; ⊖ 5867511; Brit Consul 6764343; Dr 0880 030600; Traffic Info VHF Ch 68; Port Control (East) 6221515; Port Control (West) 0255 514457; Port Info VHF Ch 14; Westerdoks-brug 6241457, VHF Ch 22,69; Oranjesluizen VHF Ch 18; Amsterdam Marina VHF Ch 31.

FACILITIES Complete shelter in all yacht hbrs/marinas.

Amsterdam Marina (52°24'·12N 04°53'·10E) VHF Ch31, ☎6310767. info@marinaamsterdam.com www.marinaamsterdam.com 250+100❶, max LOA 30m. €2·50 inc ⌁; ✕ ⌐ ⌐ ⌐.

Marina Het Realeneiland 52°23'·16N 04°53'·26E, located in the Westerdok, there are two bridges which open simultaneously on request. Booking is required. No VHF. ☎6238855. hetrealeneiland@ upcmail.nl 50 inc❶ max LOA 20m. €2·50. Marina only has ⌐ ⌐ but all other facilities within easy walking distance (<8 mins). Reports welcome.

WVDS Sixhaven 52°22'·90N 04°54'·40E, close NE of IJ8 buoy and NE of conspic Central ⇌ Stn, is small and popular, so in season is often full and log-jammed 1800-1100. ☎6329429. www.sixhaven. nl 100 +❶ max LOA 15m. €1·50 inc ⌁ ⌐(w/e) ⌐ ⌐. WIP, due to end 2017, on adjacent new Metro N-S line detracts only slightly. Free ferries to N bank (ditto for next entry WV Aeolus).

WV Aeolus 52°22'·91N 04°55'·17E. Few ❶ berths and rafting is strictly controlled. ☎6360791. Often full in season, no pre-booking, berth as directed. 45 + 8❶ €1·10, ⌁ ⌐ ⌐ YC.

Aquadam Twellegea 52°23'·04N 04°56'·63E, ent to Zijkanaal K. ☎6320616. €2·50, ⌁ ✕ ⌐(30t). www.aquadam.nl Full ⌐ facilities. Bus to city centre.

WV Zuiderzee 52°22'·95N 04°57'·77E, 140m NE of Oranjesluizen. ☎6946884, www.wsv-dezuiderzee.nl €1·10, YC.

City All needs; ⌐ Ⓔ ACA, DYC Agent, Ⓑ @, ⌐ ⇌ ✈. Amsterdam gives access to the N Holland and Rijn canals.

AMSTERDAM *continued*

9.16.20 SCHEVENINGEN

Zuid Holland **52°06'·24N 04°15'·26E** ❀❀✦✦✦✦❀❀

CHARTS AC 1630, 125, 122; Zeekaart 1035, 1349, 1350, 1449; DYC 1801.7; ANWB H/J; Imray C25

TIDES +0320 Dover; ML 0·9; Duration 0445

Standard Port VLISSINGEN (→)

Times				Height (metres)			
High Water		Low Water		MHWS	MHWN	MLWN	MLWS
0300	0900	0400	1000	4·9	4·1	1·0	0·5
1500	2100	1600	2200				
Differences SCHEVENINGEN							
+0105	+0100	+0220	+0245	−2·7	−2·3	−0·7	−0·2

SHELTER Very good in marina. Ent difficult in SW-N F6 winds, and dangerous in NW F6-8 which cause scend in outer hbr.

NAVIGATION WPT 52°07'·75N 04°14'·12E (SCH buoy) 156°/1·6M to ent; access H24. Close west of the 156° leading line and 1M from the hbr ent, outfalls are marked by E and W cardinal light buoys and by 2 SPM buoys (off chartlet). The promenade pier, Iso 5s, is 1·2M NE of the hbr ent. Caution: Strong tidal streams set across the ent. Slack water is about HW Scheveningen −2 and +3. Beware large ships and FVs entering/leaving.

LIGHTS AND MARKS Daymarks include: the reddish-brown lt ho (5ca E of hbr ent); twr bldgs in Scheveningen and Den Haag.

Traffic signals (from signal mast, N side of ent to Voorhaven):

● over ○ = No entry. ○ over ● = No exit.
Fl ◐ = One or more large vessels are leaving the port.

Tide sigs (same mast): ● over ○ = tide rising. ○ over ● = tide falling.

Shown from SE end of narrow chan between 1st and 2nd Hbrs:
● = vessels must not leave the 2nd Hbr. The chan is generally blind, so go slowly and sound horn.

COMMUNICATIONS (Code 070) Call *Traffic Centre Scheveningen* Ch 21 (H24) prior to entry/departure to avoid FVs; Police 0900 8844; ⊖ 020 5813614; Brit Consul (020) 6764343; Dr 3450650; Port HM 3527711; Radar (info only) on request VHF Ch 21; When in 2nd Hbr call *Yacht Club Scheveningen* VHF Ch 31 for a berth.

FACILITIES YC Scheveningen ☎3520017, mob 0653 293137. www.jachtclubscheveningen.com info@jachtclubscheveningen.com 223 + 100Ⓥ €1·66 + tax €1·27 per adult. Visitors *must* turn to berth/raft bows NE, for fire safety reasons. ✗ ⎕ ☎3520308, ▣.

2nd Harbour Fuel barge GEO (0630-1600) ⬤ only. ⚓ DYC agent; ▯ ⬒⬔ ⏣ ◺ ⚒ ◭.

Town ⬤ ✗ ⎕ ⛽ ⬜ ⊠ Ⓑ ⇄ ✈ Rotterdam and Amsterdam. The town is effectively merged with Den Haag (The Hague) seat of Dutch government and well worth a visit by bus.

DEN HELDER TO SCHEVENINGEN

South of Den Helder the coast is low, and, like most of the Dutch coast, not readily visible from seaward. Conspic landmarks include: chimneys of nuclear power station 1·5M NNE of Petten; Egmond aan Zee lighthouse; chimneys of steelworks N of IJmuiden; two light houses at IJmuiden; big hotels at Zandvoort; Noordwijk aan Zee light house; and Scheveningen light house and big hotels.

A wind farm centred on 52°36'·2N 04°26'·4E (about 6·7M W of Egmond aan Zee) contains a No Entry area marked by 5 peripheral turbines, each L Fl Y 15s 15m 5M, Horn Mo (U) 30s. A second No Entry wind farm is 8M further W, centred on 52°35'·4N 04°13'·2E is marked by 12 peripheral turbines all Mo (U) Y 15s 14m 5M; see AC 1631.

▶*3M W of IJmuiden the N-going stream begins at HW Hoek van Holland –0120, and the S-going at HW Hoek van Holland +0430, sp rates about 1·5kn. Off ent to IJmuiden the stream turns about 1h earlier and is stronger, and in heavy weather there may be a dangerous sea.*◀ From **IJmuiden** the **Noordzeekanaal** leads east to **Amsterdam** and the **IJsselmeer**.

HOEK VAN HOLLAND TO THE WESTERSCHELDE

Off Hoek van Holland at the ent to Europoort and **Rotterdam** shipping is very dense and fast-moving. Yachts must go upriver via the Nieuwe Waterweg which near Vlaardingen becomes the Nieuwe Maas.

The Slijkgat (lit) is the approach chan to **Stellendam** and entry to the Haringvliet. The Schaar, Schouwenbank, Middelbank and Steenbanken lie parallel to the coast off the approaches to Oosterschelde. From the N, Geul van de Banjaard (partly lit) joins Westgat and leads to Oude Roompot, which with Roompot (unlit) are the main, well marked channels to the **Roompotsluis**,

in S half of the barrage. Here the Oosterschelde (AC 192) is entered.

Rounding Walcheren via the Oostgat, close inshore, Westkapelle lt ho is conspic with two smaller lts nearby: Molenhoofd 5ca WSW and Noorderhoofd 7ca NNW. The Oostgat is the inshore link between Oosterschelde and Westerschelde and also the N approach to the latter.

Deurloo and Spleet are unlit secondary channels parallel to Oostgat and a mile or so to seaward. By day they keep yachts clear of commercial traffic in the Oostgat.

All channels converge between **Vlissingen** and **Breskens**. This bottleneck is declared a Precautionary Area in which yachts have no rights of way over other vessels and therefore must keep an above average lookout, staying clear of all traffic. Vessels <20m must give way to larger craft; and yachts <12m should stay just outside the main buoyed chans.

The main approach chan to the Westerschelde from the W is the Scheur, which yachts may follow just outside the fairway. From **Zeebrugge** and the SW use the Wielingen chan, keeping close to S side of estuary until past **Breskens**.

If crossing the Scheur there are two recommended routes. The W route runs to the W of Songa with a dogleg at W10 towards SG-W. The E route from Breskens, head up channel from ARV-VH, staying S of the G bys, turning N to cross the mini-TSS which runs E/W close off Vlissingen. Ensure you cross at right angles. The tide runs hard in the estuary, causing a bad sea in chans and overfalls on some banks in strong winds. There is shoaling.

The passage upriver to **Antwerpen** is best made in one leg, starting from Breskens, Vlissingen or **Terneuzen** in order to get the tidal timing right.

9.16.21 TSS AND VTS OFF SOUTH NETHERLANDS AND BELGIUM

STANDARD TIME UT –01
Subtract 1 hour for UT
For Dutch Summer Time add
ONE hour in **non-shaded areas**

HOEK VAN HOLLAND   LAT 51°59'N   LONG 4°07'E
TIMES AND HEIGHTS OF HIGH AND LOW WATERS

Dates in red are **SPRINGS**
Dates in blue are **NEAPS**

YEAR 2016

Netherlands

JANUARY

Day	Time m	Time m	Time m	Time m		Day	Time m	Time m	Time m	Time m
1 F	0104 0.6	0715 1.9	1234 0.2	1934 2.0		16 SA	0214 0.5	0705 2.0	1214 0.1	1936 2.2
2 SA ☽	0150 0.5	0754 1.9	1329 0.3	2036 1.9		17 SU ●	0217 0.5	0806 2.0	1329 0.1	2040 2.1
3 SU	0250 0.5	0856 1.8	1425 0.3	2135 1.9		18 M	0235 0.5	0909 1.9	1434 0.2	2150 2.0
4 M	0410 0.5	1006 1.7	1650 0.4	2234 1.8		19 TU	0324 0.5	1019 1.9	1556 0.3	2254 2.0
5 TU	0516 0.5	1115 1.7	1750 0.4	2350 1.9		20 W	0434 0.5	1129 1.9	1653 0.3	
6 W	0555 0.5	1216 1.8	1850 0.4			21 TH	0004 2.0	0546 0.5	1236 2.0	1753
7 TH	0045 2.0	0644 0.4	1306 2.0	2034 0.5		22 F	0109 2.1	0926 0.4	1327 2.1	2127 0.4
8 F	0125 2.1	0945 0.4	1345 2.1	1914 0.5		23 SA	0158 2.1	1005 0.4	1416 2.2	2225 0.4
9 SA	0205 2.1	0729 0.4	1420 2.2	1944 0.5		24 SU ○	0244 2.1	1050 0.3	1457 2.3	2305 0.5
10 SU ●	0245 2.2	0806 0.3	1501 2.3	2019 0.5		25 M	0328 2.1	0825 0.3	1538 2.3	2355 0.5
11 M	0325 2.2	0835 0.3	1541 2.3	2315 0.5		26 TU	0404 2.1	0905 0.2	1618 2.3	
12 TU	0405 2.2	0911 0.2	1621 2.4			27 W	0046 0.5	0445 2.1	0946 0.2	1659 2.3
13 W	0004 0.5	0446 2.2	0952 0.2	1700 2.4		28 TH	0114 0.5	0525 2.1	1026 0.2	1739 2.2
14 TH	0105 0.5	0527 2.1	1038 0.1	1745 2.3		29 F	0205 0.5	0559 2.1	1103 0.2	1819 2.2
15 F	0140 0.5	0615 2.1	1126 0.1	1835 2.3		30 SA	0130 0.5	0635 2.0	1206 0.2	1900 2.1
						31 SU	0110 0.5	0715 2.0	1244 0.2	1946 2.0

FEBRUARY

Day	Time m	Time m	Time m	Time m		Day	Time m	Time m	Time m	Time m
1 M ☽	0135 0.4	0805 1.9	1340 0.2	2040 1.9		16 TU	0205 0.4	0846 2.0	1436 0.2	2126 1.9
2 TU	0224 0.4	0910 1.8	1440 0.3	2140 1.8		17 W	0310 0.4	0956 1.9	1546 0.3	2234 1.8
3 W	0427 0.4	1015 1.8	1710 0.4	2250 1.8		18 TH	0426 0.4	1116 1.9	1644 0.3	2354 1.8
4 TH	0534 0.4	1130 1.8	1817 0.4	2354 1.8		19 F	0519 0.4	1221 2.0	2045 0.3	
5 F	0617 0.4	1229 1.9	2010 0.4			20 SA	0105 1.9	0904 0.3	1314 2.1	2140 0.3
6 SA	0100 1.9	0650 0.4	1319 2.0	2136 0.4		21 SU	0155 2.0	1006 0.3	1405 2.2	2226 0.4
7 SU	0146 2.0	0705 0.3	1406 2.2	2225 0.4		22 M ○	0246 2.1	1034 0.3	1445 2.2	2256 0.4
8 M ●	0225 2.1	0746 0.3	1441 2.3	1955 0.5		23 TU	0314 2.1	0809 0.3	1525 2.3	2336 0.4
9 TU	0306 2.2	0816 0.2	1521 2.4	2029 0.5		24 W	0356 2.1	0846 0.2	1558 2.3	
10 W	0345 2.2	0845 0.1	1600 2.4	2355 0.5		25 TH	0015 0.4	0425 2.1	0920 0.2	1635 2.2
11 TH	0427 2.2	0930 0.1	1646 2.4			26 F	0056 0.4	0455 2.1	1304 0.2	1711 2.2
12 F	0034 0.4	0509 2.2	1016 0.1	1727 2.3		27 SA	0125 0.4	0529 2.1	1036 0.2	1748 2.1
13 SA	0126 0.4	0552 2.1	1058 0.1	1815 2.3		28 SU	0130 0.4	0606 2.1	1114 0.2	1815 2.1
14 SU	0205 0.4	0641 2.1	1154 0.1	1909 2.1		29 M	0150 0.4	0631 2.1	1205 0.2	1849 2.0
15 M ☽	0200 0.5	0735 2.0	1314 0.1	2009 2.0						

MARCH

Day	Time m	Time m	Time m	Time m		Day	Time m	Time m	Time m	Time m
1 TU	0050 0.3	0704 2.0	1305 0.2	1936 1.9		16 W	0140 0.3	0826 2.0	1405 0.2	2054 1.8
2 W ☽	0135 0.3	0759 1.9	1405 0.3	2034 1.8		17 TH	0256 0.3	0946 1.9	1525 0.3	2236 1.7
3 TH	0244 0.3	0930 1.8	1504 0.4	2154 1.7		18 F	0354 0.3	1105 1.9	1634 0.3	2349 1.7
4 F	0515 0.4	1045 1.8	1740 0.4	2315 1.7		19 SA	0505 0.3	1216 2.0	2024 0.3	
5 SA	0600 0.3	1154 1.9	1937 0.4			20 SU	0056 1.9	0844 0.3	1304 2.1	2124 0.3
6 SU	0024 1.8	0605 0.3	1255 2.0	2116 0.4		21 M	0146 2.0	0935 0.2	1356 2.1	2205 0.3
7 M	0115 1.9	0635 0.3	1339 2.2	2205 0.4		22 TU	0225 2.0	1030 0.2	1425 2.2	2246 0.4
8 TU	0201 2.0	0710 0.2	1417 2.3	2245 0.4		23 W ○	0259 2.0	0755 0.3	1505 2.2	2310 0.4
9 W ●	0240 2.1	0745 0.1	1457 2.4	2005 0.4		24 TH	0325 2.1	0825 0.2	1536 2.2	2340 0.4
10 TH	0322 2.2	0821 0.1	1541 2.4	2045 0.4		25 F	0359 2.1	1206 0.2	1608 2.2	
11 F	0405 2.2	0905 0.0	1623 2.4			26 SA	0016 0.3	0429 2.1	1224 0.2	1645 2.1
12 SA	0015 0.4	0446 2.2	0949 0.1	1706 2.3		27 SU	0040 0.3	0459 2.1	1245 0.2	1715 2.1
13 SU	0116 0.4	0531 2.2	1039 0.1	1755 2.2		28 M	0116 0.3	0529 2.1	1315 0.3	1746 2.0
14 M	0200 0.3	0615 2.2	1134 0.1	1845 2.1		29 TU	0145 0.3	0555 2.1	1115 0.3	1815 2.0
15 TU ☽	0004 0.4	0709 2.1	1254 0.1	1945 1.9		30 W	0000 0.3	0631 2.1	1220 0.3	1856 2.0
						31 TH ☽	0050 0.2	0720 2.0	1337 0.3	1950 1.8

APRIL

Day	Time m	Time m	Time m	Time m		Day	Time m	Time m	Time m	Time m
1 F	0204 0.2	0824 1.8	1445 0.3	2126 1.7		16 SA	0334 0.2	1040 1.9	1805 0.3	2325 1.7
2 SA	0304 0.3	1010 1.8	1610 0.4	2245 1.7		17 SU	0435 0.2	1155 2.0	1940 0.3	
3 SU	0425 0.3	1126 1.9	1920 0.4	2356 1.7		18 M	0025 1.8	0535 0.2	1245 2.1	2045 0.3
4 M	0525 0.3	1225 2.0	2045 0.3			19 TU	0113 1.9	0905 0.2	1325 2.1	2134 0.3
5 TU	0049 1.9	0559 0.2	1309 2.2	2146 0.3		20 W	0154 1.9	1000 0.2	1405 2.1	2214 0.3
6 W	0136 2.0	0639 0.1	1356 2.3	2220 0.4		21 TH	0229 2.0	0745 0.3	1435 2.1	2235 0.3
7 TH ●	0217 2.1	0720 0.1	1435 2.4	1941 0.4		22 F ○	0259 2.0	0804 0.3	1508 2.2	2310 0.3
8 F	0259 2.2	0800 0.1	1516 2.4	2022 0.3		23 SA	0329 2.1	1050 0.3	1545 2.2	2330 0.3
9 SA	0342 2.2	0842 0.1	1602 2.3	2105 0.3		24 SU	0401 2.1	1125 0.3	1615 2.1	
10 SU	0425 2.3	0928 0.1	1646 2.2	2149 0.3		25 M	0016 0.2	0436 2.1	1220 0.3	1650 2.1
11 M	0507 2.3	1019 0.2	1735 2.1			26 TU	0045 0.2	0505 2.1	1306 0.3	1715 2.0
12 TU	0155	0556 2.2	1125 0.2	1825 2.0		27 W	0125 0.2	0535 2.1	1334 0.3	1752 2.0
13 W	0000 0.3	0645 2.1	1255 0.2	1925 1.8		28 TH	0205 0.2	0609 2.1	1416 0.3	1835 1.9
14 TH ☽	0104 0.2	0756 2.0	1354 0.2	2045 1.7		29 F	0004 0.2	0656 2.0	1347 0.3	1926 1.8
15 F	0220 0.2	0915 1.9	1505 0.3	2216 1.6		30 SA ☽	0136 0.2	0755 1.9	1435 0.4	2056 1.7

Chart Datum is 0·92 metres below NAP Datum. HAT is 2·5 metres above Chart Datum.

»» FREE monthly updates. Register at «
www.reedsnauticalalmanac.co.uk

719

STANDARD TIME UT –01
Subtract 1 hour for UT
For Dutch Summer Time add
ONE hour in **non-shaded areas**

HOEK VAN HOLLAND   LAT 51°59'N   LONG 4°07'E
TIMES AND HEIGHTS OF HIGH AND LOW WATERS

Dates in red are **SPRINGS**
Dates in blue are **NEAPS**

YEAR 2016

MAY

Time	m		Time	m
1 0240	0.2	**16** 0420	0.2
0936	1.9		1120	1.9
SU 1535	0.4	M 1844	0.3
2205	1.7		2356	1.7
2 0335	0.2	**17** 0505	0.2
1050	2.0		1216	2.0
M 1837	0.4	TU 2006	0.3
2315	1.7			
3 0440	0.2	**18** 0039	1.8
1144	2.1		0555	0.2
TU 2005	0.3	W 1300	2.0
			2106	0.3
4 0015	1.8	**19** 0126	1.9
0524	0.2		0644	0.3
W 1241	2.2	TH 1335	2.1
2116	0.3		2156	0.3
5 0108	2.0	**20** 0154	1.9
0616	0.1		0730	0.3
TH 1327	2.3	F 1409	2.1
2156	0.3		2214	0.3
6 0151	2.1	**21** 0224	2.0
0655	0.1		0830	0.3
F 1413	2.3	SA 1445	2.1
● 1919	0.3	○ 2020	0.3
7 0235	2.2	**22** 0301	2.1
0740	0.1		1010	0.3
SA 1458	2.3	SU 1519	2.1
1959	0.3		2300	0.2
8 0319	2.3	**23** 0336	2.1
0821	0.1		1105	0.3
SU 1542	2.3	M 1555	2.1
2045	0.3		2337	0.2
9 0405	2.3	**24** 0409	2.1
0909	0.2		1145	0.3
M 1627	2.2	TU 1626	2.1
2129	0.2			
10 0447	2.3	**25** 0025	0.2
1325	0.2		0441	2.1
TU 1715	2.1	W 1235	0.3
			1659	2.0
11 0145	0.2	**26** 0105	0.1
0535	2.2		0515	2.1
W 1415	0.3	TH 1326	0.3
1805	1.9		1735	2.0
12 0230	0.2	**27** 0155	0.1
0625	2.2		0556	2.1
TH 1454	0.3	F 1406	0.3
1905	1.8		1815	1.9
13 0034	0.1	**28** 0235	0.1
0729	2.1		0639	2.1
F 1345	0.3	SA 1440	0.4
◑ 2004	1.7		1905	1.8
14 0144	0.1	**29** 0050	0.1
0845	2.0		0734	2.0
SA 1434	0.3	SU 1510	0.4
2135	1.6		◑ 2015	1.7
15 0310	0.1	**30** 0154	0.1
1005	1.9		0906	2.0
SU 1746	0.4	M 1520	0.4
2245	1.6		2136	1.7
		31 0305	0.1
			1005	2.0
		TU 1605	0.4
			2246	1.8

JUNE

Time	m		Time	m
1 0359	0.1	**16** 0007	1.7
1118	2.1		0535	0.2
W 1930	0.4	TH 1226	1.9
2346	1.9		1810	0.3
2 0500	0.1	**17** 0050	1.8
1215	2.2		0645	0.3
TH 2034	0.4	F 1306	2.0
			1844	0.3
3 0042	2.0	**18** 0126	1.9
0549	0.1		0720	0.3
F 1308	2.2	SA 1345	2.0
1819	0.3		1936	0.3
4 0128	2.1	**19** 0206	2.0
0640	0.1		0810	0.4
SA 1353	2.2	SU 1419	2.1
1906	0.3		1945	0.3
5 0215	2.2	**20** 0235	2.1
0725	0.2		0805	0.4
SU 1440	2.2	M 1456	2.1
● 1945	0.3	○ 2014	0.2
6 0300	2.3	**21** 0311	2.1
0804	0.3		1025	0.4
M 1527	2.2	TU 1528	2.1
2024	0.2		2050	0.2
7 0345	2.3	**22** 0347	2.1
1210	0.3		1115	0.4
TU 1615	2.1	W 1606	2.1
2115	0.2		2354	0.2
8 0435	2.3	**23** 0420	2.2
1315	0.3		1204	0.4
W 1701	2.0	TH 1645	2.0
9 0125	0.1	**24** 0055	0.1
0518	2.3		0502	2.2
TH 1405	0.3	F 1255	0.4
1748	2.0		1726	2.0
10 0215	0.1	**25** 0137	0.1
0609	2.2		0541	2.2
F 1451	0.3	SA 1346	0.4
1834	1.9		1806	1.9
11 0000	0.1	**26** 0215	0.1
0701	2.1		0628	2.2
SA 1536	0.4	SU 1425	0.4
1934	1.8		1855	1.9
12 0054	0.1	**27** 0015	0.1
0806	2.0		0726	2.1
SU 1404	0.4	M 1510	0.4
◑ 2035	1.7		◑ 2001	1.8
13 0220	0.1	**28** 0119	0.1
0905	1.9		0830	2.1
M 1515	0.4	TU 1500	0.4
2144	1.7		2106	1.8
14 0335	0.2	**29** 0224	0.1
1014	1.9		0940	2.1
TU 1614	0.4	W 1525	0.4
2305	1.7		2216	1.8
15 0444	0.2	**30** 0336	0.1
1125	1.9		1046	2.1
W 1726	0.4	TH 1636	0.4
			2315	1.9

JULY

Time	m		Time	m
1 0435	0.2	**16** 0620	0.4
1152	2.1		1230	1.9
F 1720	0.4	SA 1836	0.3
2 0018	2.0	**17** 0056	1.9
0535	0.2		0700	0.4
SA 1247	2.1	SU 1326	2.0
1804	0.4		1904	0.3
3 0108	2.1	**18** 0135	2.0
0624	0.3		0730	0.4
SU 1341	2.1	M 1358	2.0
1856	0.3		1930	0.3
4 0157	2.2	**19** 0211	2.1
0715	0.3		0734	0.5
M 1431	2.1	TU 1436	2.1
● 1940	0.3	○ 1955	0.3
5 0247	2.3	**20** 0248	2.2
0759	0.4		0805	0.5
TU 1517	2.1	W 1508	2.1
2020	0.2		2026	0.2
6 0330	2.3	**21** 0325	2.3
1156	0.4		0839	0.5
W 1606	2.1	TH 1547	2.1
2059	0.2		2055	0.2
7 0417	2.3	**22** 0406	2.3
1245	0.4		1144	0.5
TH 1645	2.1	F 1628	2.1
2145	0.2		2135	0.1
8 0501	2.3	**23** 0446	2.3
1335	0.4		1235	0.4
F 1735	2.0	SA 1708	2.1
2229	0.1		2216	0.1
9 0546	2.2	**24** 0526	2.3
1426	0.4		1315	0.4
SA 1811	2.0	SU 1749	2.0
2326	0.1		2255	0.1
10 0636	2.2	**25** 0608	2.2
1505	0.4		1405	0.4
SU 1855	1.9	M 1835	2.0
			2345	0.1
11 0019	0.1	**26** 0658	2.2
0726	2.1		1440	0.5
M 1340	0.5	TU 1930	2.0
1945	1.9			
12 0115	0.2	**27** 0045	0.1
0815	2.0		0805	2.1
TU 1430	0.4	W 1440	0.5
◑ 2040	1.8		◑ 2035	1.9
13 0220	0.2	**28** 0204	0.1
0909	1.9		0904	2.0
W 1540	0.4	TH 1504	0.5
2146	1.7		2139	1.9
14 0414	0.3	**29** 0325	0.2
1015	1.8		1014	2.0
TH 1656	0.4	F 1605	0.4
2244	1.7		2256	1.9
15 0526	0.3	**30** 0414	0.3
1135	1.8		1136	2.0
F 1734	0.3	SA 1704	0.4
2355	1.8		2359	2.0
		31 0525	0.3
			1239	2.0
		SU 1754	0.4

AUGUST

Time	m		Time	m
1 0059	2.1	**16** 0110	2.0
0624	0.4		0850	0.5
M 1335	2.1	TU 1329	2.0
1845	0.3		1854	0.3
2 0148	2.2	**17** 0149	2.1
0947	0.4		0714	0.5
TU 1421	2.1	W 1409	2.1
● 1925	0.3		1926	0.3
3 0235	2.3	**18** 0228	2.3
1040	0.5		0739	0.5
W 1509	2.1	TH 1446	2.2
2005	0.3		○ 1956	0.2
4 0317	2.3	**19** 0305	2.4
1130	0.5		0815	0.5
TH 1545	2.1	F 1525	2.2
2046	0.2		2029	0.2
5 0358	2.3	**20** 0342	2.4
1214	0.5		0844	0.5
F 1625	2.1	SA 1606	2.2
2119	0.2		2108	0.1
6 0441	2.3	**21** 0425	2.4
1304	0.5		1215	0.5
SA 1708	2.1	SU 1647	2.2
2205	0.2		2148	0.1
7 0521	2.2	**22** 0505	2.4
1355	0.5		1255	0.5
SU 1745	2.1	M 1725	2.2
2245	0.2		2231	0.1
8 0606	2.2	**23** 0548	2.3
1436	0.5		1334	0.5
M 1819	2.0	TU 1811	2.1
2335	0.2		2321	0.2
9 0646	2.1	**24** 0640	2.2
1310	0.5		1420	0.5
TU 1906	2.0	W 1906	2.1
10 0030	0.2	**25** 0024	0.2
0736	2.0		0736	2.1
W 1330	0.5	TH 1400	0.5
◐ 1950	1.9		◐ 2006	2.0
11 0124	0.3	**26** 0154	0.2
0820	1.9		0846	2.0
TH 1420	0.4	F 1445	0.5
2050	1.8		2116	2.0
12 0225	0.3	**27** 0255	0.3
0926	1.8		0955	1.9
F 1620	0.4	SA 1535	0.5
2144	1.7		2236	1.9
13 0454	0.4	**28** 0415	0.4
1014	1.8		1119	1.9
SA 1705	0.4	SU 1655	0.4
2310	1.8		2349	2.0
14 0555	0.4	**29** 0747	0.4
1145	1.8		1235	1.9
SU 1816	0.4	M 1744	0.4
15 0013	1.9	**30** 0056	2.1
0650	0.5		0916	0.4
M 1245	1.9	TU 1324	2.0
1835	0.4		2124	0.4
		31 0145	2.2
			0954	0.5
		W 1416	2.1
			1909	0.3

Chart Datum is 0·92 metres below NAP Datum. HAT is 2·5 metres above Chart Datum.

STANDARD TIME UT –01
Subtract 1 hour for UT
For Dutch Summer Time add
ONE hour in **non-shaded areas**

HOEK VAN HOLLAND   LAT 51°59'N   LONG 4°07'E
TIMES AND HEIGHTS OF HIGH AND LOW WATERS

Dates in red are **SPRINGS**
Dates in blue are **NEAPS**

YEAR 2016

Netherlands

SEPTEMBER

Time	m		Time	m
1 0225	2.3	**16**	0158	2.3
1030	0.5		0715	0.5
TH 1455	2.1	F	1421	2.2
● 1950	0.3	○	1925	0.2
2 0305	2.3	**17**	0237	2.4
1110	0.6		0745	0.5
F 1535	2.2	SA	1502	2.3
2019	0.3		2001	0.2
3 0341	2.3	**18**	0319	2.5
1156	0.6		0825	0.5
SA 1606	2.2	SU	1542	2.3
2055	0.3		2046	0.2
4 0419	2.3	**19**	0401	2.5
1241	0.5		0905	0.5
SU 1638	2.2	M	1623	2.3
2135	0.3		2125	0.2
5 0455	2.3	**20**	0443	2.4
1326	0.5		0945	0.5
M 1715	2.2	TU	1706	2.3
2216	0.3		2209	0.2
6 0531	2.2	**21**	0527	2.3
1350	0.5		1029	0.5
TU 1748	2.1	W	1748	2.3
2300	0.3		2301	0.3
7 0608	2.1	**22**	0615	2.2
1119	0.5		1123	0.5
W 1819	2.1	TH	1838	2.2
2334	0.4			
8 0640	2.1	**23**	0005	0.3
1220	0.5		0716	2.1
TH 1856	2.0	F	1245	0.5
		☽	1935	2.1
9 0040	0.4	**24**	0135	0.3
0726	2.0		0814	1.9
F 1320	0.4	SA	1416	0.4
☽ 1933	1.9		2055	2.0
10 0134	0.4	**25**	0244	0.4
0804	1.9		0950	1.8
SA 1414	0.4	SU	1520	0.4
2106	1.8		2215	2.0
11 0245	0.5	**26**	0400	0.5
0935	1.8		1115	1.8
SU 1655	0.4	M	1625	0.4
2226	1.8		2346	2.0
12 0536	0.5	**27**	0735	0.5
1050	1.8		1219	1.9
M 1735	0.4	TU	2016	0.4
2340	1.9			
13 0630	0.5	**28**	0046	2.2
1216	1.8		0906	0.4
TU 1820	0.4	W	1316	2.0
			2126	0.3
14 0035	2.0	**29**	0129	2.3
0846	0.5		0944	0.4
W 1259	2.0	TH	1359	2.1
1824	0.4		2155	0.4
15 0126	2.2	**30**	0205	2.3
0934	0.5		1014	0.5
TH 1345	2.1	F	1435	2.2
1856	0.3		1936	0.4

OCTOBER

Time	m		Time	m
1 0245	2.3	**16**	0216	2.5
1035	0.6		0725	0.5
SA 1509	2.2	SU	1435	2.3
● 2005	0.4	○	1935	0.2
2 0317	2.3	**17**	0256	2.5
1117	0.6		0759	0.5
SU 1538	2.2	M	1518	2.4
2035	0.4		2018	0.2
3 0356	2.3	**18**	0336	2.5
1205	0.5		0839	0.5
M 1611	2.3	TU	1559	2.4
2109	0.4		2101	0.2
4 0429	2.3	**19**	0422	2.4
1246	0.5		0926	0.5
TU 1645	2.2	W	1646	2.4
2146	0.4		2149	0.3
5 0501	2.2	**20**	0506	2.3
1304	0.5		1009	0.6
W 1718	2.2	TH	1727	2.45
2226	0.5		2245	0.4
6 0529	2.1	**21**	0555	2.0
1046	0.5		1110	0.4
TH 1746	2.2	F	1817	2.3
2305	0.5		2355	0.4
7 0559	2.1	**22**	0655	2.0
1130	0.4		1200	0.4
F 1818	2.2	SA	1915	2.2
2350	0.5	☽		
8 0635	2.1	**23**	0130	0.4
1226	0.4		0759	1.9
SA 1855	2.1	SU	1346	0.4
			2035	2.1
9 0110	0.5	**24**	0230	0.5
0714	2.0		0936	1.8
SU 1335	0.4	M	1445	0.4
☽ 1950	2.0		2155	2.0
10 0215	0.5	**25**	0600	0.5
0851	1.8		1056	1.9
M 1434	0.4	TU	1616	0.4
2146	1.9		2326	2.1
11 0420	0.6	**26**	0714	0.5
1005	1.8		1155	1.9
TU 1717	0.4	W	1936	0.4
2255	1.9			
12 0610	0.6	**27**	0014	2.2
1126	1.8		0830	0.4
W 1740	0.4	TH	1245	2.0
			2035	0.4
13 0005	2.1	**28**	0110	2.2
0816	0.5		0914	0.4
TH 1226	2.0	F	1336	2.1
1734	0.4		2146	0.4
14 0049	2.3	**29**	0145	2.3
0915	0.5		0954	0.5
F 1316	2.1	SA	1404	2.1
1825	0.3		1913	0.4
15 0135	2.4	**30**	0218	2.3
0954	0.5		1036	0.5
SA 1355	2.2	SU	1445	2.2
1900	0.2	●	1955	0.4
		31	0258	2.3
			0804	0.5
		M	1515	2.2
			2025	0.5

NOVEMBER

Time	m		Time	m
1 0329	2.3	**16**	0319	2.4
0840	0.5		0825	0.4
TU 1545	2.3	W	1539	2.5
2049	0.5		2045	0.3
2 0406	2.3	**17**	0405	2.3
1147	0.4		0906	0.4
W 1615	2.3	TH	1625	2.5
2350	0.5		2131	0.4
3 0435	2.2	**18**	0450	2.2
1236	0.4		0955	0.5
TH 1649	2.3	F	1709	2.4
			2224	0.5
4 0025	0.5	**19**	0538	2.1
0505	2.2		1049	0.5
F 1019	0.4	SA	1758	2.3
1719	2.2			
5 0105	0.5	**20**	0246	0.5
0536	2.1		0635	2.0
SA 1053	0.4	SU	1155	0.3
1755	2.2		1900	2.2
6 0150	0.6	**21**	0055	0.5
0609	2.1		0733	1.9
SU 1145	0.3	M	1305	0.3
1831	2.2	☽	2010	2.1
7 0014	0.6	**22**	0205	0.5
0658	2.0		0845	1.8
M 1244	0.3	TU	1404	0.3
☽ 1925	2.1		2115	2.0
8 0204	0.6	**23**	0524	0.6
0759	1.9		1015	1.8
TU 1406	0.3	W	1524	0.4
2044	2.0		2255	2.0
9 0304	0.6	**24**	0636	0.5
0930	1.8		1119	1.8
W 1505	0.4	TH	1634	0.4
2216	2.0		2344	2.1
10 0410	0.6	**25**	0746	0.5
1039	1.8		1226	1.9
TH 1610	0.4	F	1740	0.4
2315	2.1			
11 0746	0.6	**26**	0040	2.1
1150	2.0		0846	0.5
F 1659	0.3	SA	1306	2.0
			1824	0.4
12 0015	2.3	**27**	0125	2.2
0851	0.5		0935	0.5
SA 1239	2.1	SU	1338	2.1
1749	0.3		1914	0.4
13 0105	2.4	**28**	0155	2.2
0945	0.5		1016	0.5
SU 1327	2.2	M	1420	2.1
1836	0.3		2000	0.5
14 0148	2.5	**29**	0235	2.2
0654	0.5		0810	0.5
M 1413	2.3	TU	1445	2.2
○ 1916	0.3	●	2024	0.5
15 0236	2.5	**30**	0308	2.2
0739	0.4		0830	0.4
TU 1456	2.4	W	1521	2.3
1959	0.4		2236	0.5

DECEMBER

Time	m		Time	m
1 0341	2.2	**16**	0350	2.3
0854	0.4		0856	0.3
TH 1555	2.3	F	1607	2.4
2315	0.5		2114	0.5
2 0411	2.2	**17**	0437	2.2
1155	0.4		0935	0.3
F 1625	2.3	SA	1655	2.4
3 0004	0.5	**18**	0130	0.5
0445	2.2		0525	2.1
SA 1235	0.3	SU	1029	0.2
1701	2.3		1746	2.4
4 0055	0.5	**19**	0226	0.5
0517	2.1		0615	2.1
SU 1034	0.3	M	1125	0.2
1738	2.2		1835	2.3
5 0125	0.5	**20**	0316	0.5
0556	2.1		0704	2.0
M 1119	0.3	TU	1236	0.2
1815	2.2		1936	2.2
6 0205	0.6	**21**	0134	0.5
0639	2.0		0810	1.9
TU 1205	0.2	W	1329	0.2
1905	2.2	☽	2035	2.1
7 0230	0.6	**22**	0246	0.5
0734	1.9		0909	1.8
W 1314	0.2	TH	1450	0.3
☽ 2016	2.1		2145	2.0
8 0235	0.6	**23**	0345	0.5
0856	1.9		1030	1.8
TH 1414	0.3	F	1605	0.4
2135	2.1		2310	1.9
9 0325	0.6	**24**	0435	0.5
1006	1.9		1140	1.8
F 1530	0.3	SA	1705	0.4
2246	2.1			
10 0415	0.6	**25**	0005	2.0
1116	2.0		0545	0.5
SA 1623	0.3	SU	1230	1.9
2346	2.2		1814	0.4
11 0509	0.5	**26**	0056	2.0
1212	2.1		0635	0.5
SU 1726	0.3	M	1316	2.0
			1910	0.4
12 0039	2.3	**27**	0140	2.1
0559	0.5		0945	0.4
M 1305	2.2	TU	1355	2.0
1815	0.3		2000	0.5
13 0128	2.3	**28**	0215	2.1
0646	0.5		0750	0.4
TU 1353	2.3	W	1425	2.1
1905	0.3		2054	0.5
14 0217	2.3	**29**	0245	2.1
0726	0.4		0814	0.4
W 1438	2.4	TH	1506	2.2
○ 1950	0.4	●	2200	0.5
15 0305	2.3	**30**	0326	2.2
0805	0.4		0834	0.3
TH 1523	2.4	F	1536	2.3
2031	0.4		2306	0.5
		31	0356	2.2
			0909	0.3
		SA	1611	2.3
			2350	0.5

Chart Datum is 0·92 metres below NAP Datum. HAT is 2·5 metres above Chart Datum.

⟩⟩ FREE monthly updates. Register at ⟨⟨
www.reedsnauticalalmanac.co.uk

721

HOEK VAN HOLLAND

MEAN SPRING —————
AND NEAP CURVES – – – –

MEAN RANGES
Springs 1.9m
Neaps 1.5m

9.16.22 MAAS TSS

Large ships bound up-Channel for Europoort route via the Dover Strait and the Noord Hinder South TSS to Noord Hinder Junction (a large pentagon roughly centred on 52°N 02°51'·1E). Here ships enter the MAAS TRAFFIC SEPARATION SCHEME.

Maas West Outer TSS starts some 33M offshore and funnels large ships 083°/27M via the TSS lanes or Eurogeul (DW route dredged 24·5m) towards a second pentagon 'Maas Precautionary Area' clear of Maas West Inner TSS and 10M offshore. When abeam Maas Centre SWM buoy, ships enter the Maasgeul (Maas narrows) and track inbound 5·5M on the 112° leading lights to the Maasmond (Maas mouth, ie the harbour entrance).

Yachts should avoid the whole complex by using the ITZ and/or crossing the Maasgeul via the recommended track (see Hoek van Holland and www.maasvlakte2.com/uploads/crossing_route.pdf). Knowing the above is to be aware of what you are avoiding.

HOEK VAN HOLLAND TO THE SOUTH EAST UK

Yachts out of Rotterdam/Hoek van Holland bound for the southern Thames Estuary should route 254°/52M from MV-N buoy to NHR-SE buoy, passing MW5, Goeree tower, MW3 and MW1 buoys. Cross the Noord Hinder South TSS between NHR-SE and NHR-S buoys. Thence depending on destination, set course 242°/44M to Outer Tongue for the River Thames; or 228°/46M to NE Goodwin for Ramsgate or points S.

If bound for the River Orwell or adjacent rivers, it may be best to pass N (rather than S) of the large and busy Sunk Outer and Inner Precautionary Areas and their associated TSS. From NHR-S buoy route via West Gabbard, North Inner Gabbard, S Shipwash and Rough buoys to pick up the recommended yacht track at Cork Sand Yacht beacon; it is a less direct route than going S via Galloper, but it avoids the worst of the commercial traffic.

9.16.22A MAAS TSS, PRECAUTIONARY AREA AND MAASGEUL

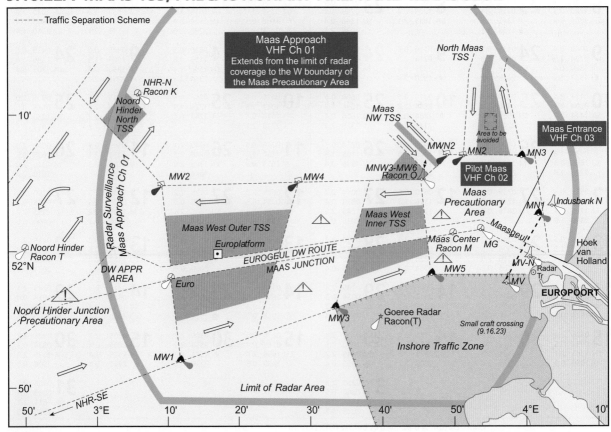

9.16.23 HOEK VAN HOLLAND AND NIEUWE WATERWEG VTS

CROSSING THE MAASGEUL To cross the entrance (ie seaward of the bkwtrs) coasting yachts should advise *Maas Ent* Ch 03 of position/course/speed, then monitor Ch 03. Cross on the recommended track, 046° (226°) under power; see chartlet. Beware strong cross-tides. Keep a keen lookout for fast moving merchantmen; Rules 18 d (ii) and 28.

HOEK VAN HOLLAND

CHARTS AC 122, 132; Zeekaart 1540, 1349, 1350, 1449; DYC 1801.6, 1801.7; Imray C30, Y5

TIDES +0251 Dover; ML 0·9; Duration 0505. HOEK VAN HOLLAND is a Standard Port (◄—), but has no Secondary ports; see also Rotterdam. Double LWs occur, more obviously at sp; in effect a LW stand. The 1st LW is about 5½ hrs after HW and the 2nd LW about 4¼ hrs before the next HW. Predictions are for the *lower* LW. Prolonged NW gales can raise levels by up to 3m.

SHELTER Entry safe, but in strong onshore winds heavy seas/swell develop. The first adequate shelter is 6M up river at Maassluis (3m). Complete shelter 10-19M further E at Rotterdam.

NAVIGATION From N, WPT 52°02'·89N 04°03'·57E (Indusbank NCM lt buoy, Q) 190°/3·2M to Noorderdam lt, FR.

From S, WPT 51°59'·61N 04°00'·20E (MV-N buoy) 101°/3·0M to ent to Nieuwe Waterweg. *Yachts must on no account enter the Calandkanaal or Europoort. Stay in the Nieuwe Waterweg.*

LIGHTS AND MARKS See chartlet and Lights, buoys and waypoints for details. Outer ldg lts 112° to ent; then 107° R (ie keep to port) ldg lts into Nieuwe Waterweg, both R trs, W bands.

Hbr patrol vessels show a Fl Bu lt; additionally a Fl R lt = 'Stop'.

COMMUNICATIONS (Code 0174) TC/Port Authority (010) 2522801; Police (0900) 8844; Brit Consul (020) 6764343; ⊞ 4112800.

FACILITIES Berghaven, use only in emergencies, HM 638850.

Hoek van Holland ⚓ ⛴ ✕ 🛒 🍴 ✉ Ⓑ ⇌ ✈ (Rotterdam). Ferries Hook-Harwich; Rotterdam (Vlaardingen)-Hull.

Maassluis Binnenhaven, bridge, (HM 5912852 Ch 68) ⚓ quay, picturesque town centre, only charge €6 for bridge transit.

NIEUWE WATERWEG VTS

Arrival procedure for yachts First report to *Maas Approach* or *Pilot Maas*, depending on distance offshore (or *Maas Ent* if within 4M of hbr ent) stating name/type of vessel, position and destination. Obey any instructions, monitoring the relevant Radar Ch's (limits as shown by W ☐ signboards on the river banks; Km signs are similar).

Rules for yachts in Nieuwe Waterweg/Nieuwe Maas:
Monitor VTS channels (see below); transmit only in emergency or if obliged to deviate from the usual traffic flow. Keep to extreme stbd limit of buoyed line, avoiding debris between buoys and bank. No tacking/beating; no ⚓. Engine ready for instant start. Able to motor at 3·24kn (6km/hr). Hoist a radar reflector, esp in poor vis or at night. Cross chan quickly at 90°. All docks are prohib to yachts, except to access a marina. *Keep a good lookout, especially astern.*

The 3 **Traffic Centres (TC)** oversee their Radar surveillance stations and sub-sectors (*italics*), on dedicated VHF chans below:

- **TC Hoek van Holland (VCH)** Ch 11. (See diagram above)

Maas Approach	Ch 01	38 – 11M W of Hoek;
Pilot Maas	Ch 02	11 – 4M W of Hoek;
Maas Entrance	Ch 03	4M – km 1031.

  English is the primary language on Ch 01, 02 and 03.

Rozenburg	Ch 65	km 1031 – 1028;
Maasluis	Ch 80	km 1028 – 1017.

- **Botlek Information & Tracking system** Ch 14.

Botlek	Ch 61	km 1017 – 1011;
Eemhaven	Ch 63	km 1011 – 1007.

- **TC Rotterdam (VCR)** Ch 11.

Waalhaven	Ch 60	km 1007 – 1003·5;
Maasbruggen	Ch 81	km 1003·5 – 998;
Brienenoord	Ch 81	km 998 – 993.

Harbour Coordination Centre (HCC) administers and controls Rotterdam port, Ch 19 (H24).

MSI broadcasts by TCs Ch 11 and on request by Radar stns.

Nieuwe Waterweg VTS and Nieuwe Maas into central Rotterdam *continued*

9.16.24 ROTTERDAM

Zuid Holland **51°54′·00N 04°28′·00E** (610m S of Euromast)
❀❀❀🔱🔱🔱❀❀❀

CHARTS AC 122, 132, 133; Zeekaart 1540/1/2; DYC 1809.4, 1809.5

TIDES +0414 Dover; ML 0·9; Duration 0440

Standard Port VLISSINGEN (→)

Times				Height (metres)			
High Water		Low Water		MHWS	MHWN	MLWN	MLWS
0300	0900	0400	1000	4·9	4·1	1·0	0·5
1500	2100	1600	2200				
Differences EUROPLATFORM (30M W of Hoek van Holland)							
+0005	−0005	−0030	−0055	−2·7	−2·3	−0·5	−0·1
MAASSLUIS (Km 1019)							
+0155	+0115	+0100	+0310	−2·9	−2·4	−0·8	−0·3
VLAARDINGEN (Km 1011)							
+0150	+0120	+0130	+0330	−2·8	−2·3	−0·7	−0·2

NOTE: Double LWs occur. Maasluis and Vlaardingen are referenced to Vlissingen, in UK ATT as above. The Dutch HP33 *Tidal heights and streams in Dutch coastal waters* shows the time differences below, relative to HW and the first LW at Hoek van Holland:

HOEK VAN HOLLAND	HW	1st LW
Maasluis	+0102	+0308
Vlaardingen	+0103	+0333
Rotterdam	+0111	+0341

SHELTER Good in the yacht hbrs where visitors are welcome (see Facilities). There is always a considerable chop/swell in the river due to constant heavy traffic to/from Europoort and Rotterdam.

NAVIGATION See 9.16.23 for WPTs (to enter Nieuwe Waterweg) and Yacht Rules. From the hbr ent to the conspic Euromast (51°54′·33N 04°28′·00E) is about 18M (33km). 6M W of the Euromast, the very busy Oude Maas joins at km 1013, giving access to Dordrecht and the Delta network of canals/lakes.

COMMUNICATIONS (Code 010) See chartlet above for VTS in central Rotterdam. In emergency call Rotterdam Tfc Centre Ch 11; Police (0900) 8844; ⊖ 2442266; Brit Consul (020) 676 4343; ⊞ 4112800; Hbr Coordination Centre (HCC) 2522601, also Emergency; see Facilities for VHF Chans at locks/bridges, English is the second language.

FACILITIES Marinas/yacht harbours from seaward:

Vlaardingen YC, 51°53′·98N 04°20′·93E, 400m E of Delta Hotel in Buitenhaven.⌬ (3·6-4·4m) or lock (Ch 20) into Oude Haven (2·7m). HM ☎2484333, 🛢🛥🍴🏖🔧⚓⛽.

Spuihaven, 51°54′·00N 04°24′·00E, immediately E of the ent to Wilhelmina Haven. €1.25/m. No lock/bridge to transit. Schiedam YC mob 0644326722, ⌬ (1·6-2·8m) 🛥🔩🔧🖥🗡⛽.

Coolhaven Yacht Hbr, 51°54′·12N 04°28′·00E; next to Euromast 185m. Access through Parkhaven via lock (Ch 22). ☎4764146, ⌬ (2·7m) 🛢🛥🗡🖥🛒🔩🔧🖥🗡🔑⚓.

Veerhaven. 51°54′·39N 04°28′·76E, 5ca E of Euromast. ☎ 4365446, mob 0653536107, www.veerhavenrotterdam.nl info@ veerhavenrotterdam.nl Centre for traditional sea-going vessels. 🅥 welcome, €1.80. ⌬ (3·9m) 🔩🔧🖥⚓🔩, 350m N,🔩🖥⚓🔩, water taxi to S bank. **Royal Maas YC** ☎4137681 (clubhouse, members only).

City Marina, 51°54′·64N 04°29′·76E. Go under Erasmus bridge (Ch 18) via lifting section at SE end which only opens 1000, 1100, 1330, 1530 & 1900LT or on request Ch 18 2030-0700 one hour before req'd (11m clearance under fixed span). Ldg lts 056·7°, both Iso 2s; then 2nd ent to stbd, via lifting bridge. ☎4854096, mob 0622 215761, www.citymarinarotterdam.nl info@citymarinarotterdam. nl 130 ⌬ +40 🅥 in 4m. €1·85 night for 2 nights; €1·50 3rd-7th nights; €1·25 7 nights. YC, water taxi to N bank, @ 🍽.

WSV IJsselmonde, 51°54′·24N 04°33′·34E, on S bank at km 994, 800m E of Brienenoordbrug, Ch 20 (off chartlet). ☎4828333, ⌬ (1·3-2·1m).

City all facilities, ACA, DYC Agent. Ferry: Rotterdam (Vlaardingen) -Hull; Hook-Harwich.

9.16.25 HARINGVLIET (Stellendam and Hellevoetsluis)

Zuid Holland 51°49'·46N 04°02'·29E (Goereesesluis). Stellendam ✿✿✿✿✿✿

CHARTS AC 1630, 110; Zeekaart 1447, 1448; DYC 1801.6, 1807.6; Imray C30

TIDES +0300 Dover; ML 1·2; Duration 0510

Standard Port VLISSINGEN (→)

Times				Height (metres)			
High Water		Low Water		MHWS	MHWN	MLWN	MLWS
0300	0900	0400	1000	4·9	4·1	1·0	0·5
1500	2100	1600	2200				
Differences HARINGVLIETSLUIZEN							
+0015	+0015	+0015	−0020	−1·9	−1·9	−0·6	−0·2

NOTE: Double LWs occur. The rise after the 1st LW is called the Agger. Water levels on this coast are much affected by weather. Prolonged NW gales can raise levels by up to 3m. Water levels in the Haringvliet can drop by 0·5m an hour when the sluices are open.

SHELTER Good in all the marinas. The entrance to the Slijkgat can be rough in W/NW winds stronger than Force 6, and also with wind against tide. When there is heavy swell, entry into the Slijkgat is not advisable. Call Post Ouddorp, VHF Ch 25 if in doubt. (No service Sunday.)

NAVIGATION WPT 51°51'·95N 03°51'·42E (SG SWM buoy, Iso 4s) 110°/2·7M to SG5/6 chan buoys. Note 1·4m shoal close N bys SG5/8, 7/8 and 9/10. Call Post Ouddorp VHF (see Comms) for latest silting info. Thence 2·4M via the well buoyed/lit Slijkgat chan to SG15/16 bns. Follow Pampus chan P1,3,5,7,9 SHM buoys SSE for 2·2M to the Buitenhaven, avoiding the no-entry sluicing area, marked by 4 SPM buoys.

Access is via a lock which operates 24/7 and has recessed bollards. 2 bridges lift in sequence to minimise road tfc delays. Note: the W bridge has 14m vertical clearance when down (with digital clearance gauge outside) the E bridge only 5·4m.

Hellevoetsluis is 2·6M E of Stellendam marina on the N bank. Three hbrs from W: Heliushaven, Het Groote Dok and Koopvaardijhaven, a LtHo (see below) marks ent to Groote Dok.

LIGHTS AND MARKS Haringvlietsluizen: 3 ● in △ are shown from pier heads on dam when sluicing in progress. Groote Dok LtHo: Iso WRG 10s 16m W11M, R8M, G7M; W tower, R top; 266°-G-275°-W-294°-R-316°-W036°-G-058°-W-095°-R-140°.

COMMUNICATIONS Post Ouddorp (lighthouse) VHF Ch 25 for weather, advice & assistance (H+30 except Sun); ⊖ Rotterdam (010) 4298088 or Vlissingen (0118) 484600; Emergencies 112; Brit Consul (020) 676 4343; Stellendam Port HM 0187 491000; Lock (Walcheren Canal) (0118) 412840; Lock 0187 497350, *Goereese Sluis* VHF Ch 20 for bridges and lock; Hellevoetsluis HM (and locks) VHF Ch 74.

FACILITIES Stellendam Marina ☎0187 493769, VHF Ch 31. www.marinastellendam.nl info@marinastellendam.nl 200 ⌓ inc💲 €2·00 inc tourist tax 🏴 📶. Berth as directed in 3m; all pontoons have fingers. 🛢🔧✕🏪🗑🛒🔨⚓🏗(20t) ☎0187 492344, bike hire.

Town (4½ km) ✕ 🏪 🗑 🛒 ⊠. ✈ (Rotterdam). Bus to Vlissingen & Spijkenisse. Ferry Hoek of Holland-Harwich; Rotterdam-Hull.

Heliushaven On the E side is **Marina Cape Helius** ☎0181 310056, mob 0613181974. www.marina-capehelius.nl havenkantoor@marina-capehelius.nl €1·50 + €0·70 pp/day tax; 🔧. Reports welcome.

There are also 3 YCs situated on the W side of Heliushaven which may be able to help with berthing.

YC Helius ☎0181 316563, mob 0616 756515; €1·40 + €0·70 pp/day tax; 🔧.

YC Hellevoetsluis ☎0181 315868, €1·40 + €0·70 pp/day tax.

YC Haringvliet ☎0181 314748, €1·35 + €0·70 pp/day tax.

Groote Dok. 💲 ⌓ before the bridge, 🔌🛒🏴. In summer the bridge opens every hour. After the bridge is **Marina Hellevoetsluis** ☎0181 312166. www.marina-capehelius.nl info@marina-capehelius.nl €1·50 + €0·70 pp/day tax; 🔧. Reports welcome.

Koopvaardijhaven. 💲 berths with 🔌 on the W side, though there can also be a lot of commercial traffic here. Reports welcome.

Yagra BY (E bank) ☎0181 314748, mob 0629403177 www.jachtservice-reparatie.nl Reports welcome.

Town ✕ 🏪 🛒 ⊠.

9.16.26 ROOMPOTSLUIS (for Oosterschelde)

Zeeland **51°37'·11N 03°41'·08E** Roompot Lock

CHARTS AC 1630, 110; Zeekaart 1448; DYCs 1805.8, 1801.5; Imray C30

TIDES +0230 Dover

Standard Port VLISSINGEN (→)

Times				Height (metres)			
High Water		Low Water		MHWS	MHWN	MLWN	MLWS
0300	0900	0400	1000	4·9	4·1	1·0	0·5
1500	2100	1600	2200				
Differences ROOMPOT BUITEN							
−0015	+0005	+0005	−0020	−1·2	−1·1	−0·3	−0·1

SHELTER An approach in strong SW through NW winds is not advised. The Buitenhaven's inner part is sheltered.

NAVIGATION WPT 51°39'·35N 03°33'·65E (abeam OR1 SHM buoy) 122°/4·2M via Oude Roompot chan to the 073·5° ldg line, both Oc G 5s, into the Buitenhaven. The flood sets E from HW Vlissingen −3 to +1. See Passage information for the several offshore banks.

Alternatively Roompot chan, buoyed but unlit, is further south and closer inshore.

Keep clear of the buoyed areas, W and E of the storm-surge barrier, which are very dangerous due to strong tidal streams and many obstructions.

Roompotsluis lock, ☎0111659265, (H24), beware tide in approach, waiting pontoons (no overnight stays) W and E of lock with intercom to control centre. Custom clearance at lock in season.

The fixed bridge has 18·2m least clearance. At LW clearance is approx 21m. Check with tide gauges or lock-keeper if in doubt. Small bollards, fixed at vertical intervals, are recessed into the lock walls. Ent to/exit from the lock is controlled by R/G traffic lights.

LIGHTS AND MARKS See Lights, buoys and waypoints and chartlet.

COMMUNICATIONS Lock *Roompotsluis* Ch 84. Roompot Marina Ch 31. Monitor Ch 68 which broadcasts local forecasts at H+15.

FACILITIES Roompot Marina Good shelter, 1·5M SE of lock. ☎(0113) 374125, marina@roompot.nl. 250 + 80 💲, €2·00 + €1·60pp/day tax, 🛢🔧✕🏪🛒🗑🔨Ⓑ⚓ Dr ☎372565; ⊖ 0118 484600.

Alternatively, follow G3-G17 buoys into **Betonhaven** about 1M NE of the lock. ⚓ in 3-6m, good holding in sticky black mud. Easy walk to the Delta Expo.

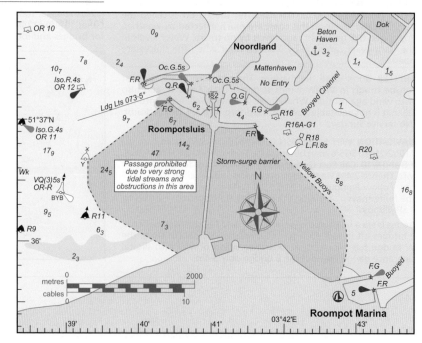

9.16.27 WESTERSCHELDE

Zeeland mostly, but Belgium for the last 12M to Antwerpen

CHARTS AC 1874, 120, 128; Zeekaart 1443; BE 103, 104; DYC 1803; Imray C30

TIDES +0200 Dover; ML Hansweert 2·7, Westkapelle 2·0, Bath 2·8; Duration 0555

Standard Port VLISSINGEN (→)

Times				Height (metres)			
High Water		Low Water		MHWS	MHWN	MLWN	MLWS
0300	0900	0400	1000	4·9	4·1	1·0	0·5
1500	2100	1600	2200				
Differences WESTKAPELLE (8M NW of Vlissingen)							
−0025	−0015	−0010	−0025	−0·6	−0·6	−0·1	−0·1
HANSWEERT							
+0100	+0050	+0040	+0100	+0·7	+0·6	+0·1	0·0
BATH							
+0125	+0115	+0115	+0140	+1·2	+1·0	+0·2	+0·1

SHELTER In strong winds a bad sea can be met in the estuary mouth. Conditions become easier further upriver. See Facilities.

NAVIGATION Westerschelde is the waterway to Antwerpen (and via canal to Gent) very full of ships and barges, especially in the last 15M. The main channel winds through a mass of well marked sand-banks. It is essential to work the tides, which average 2½kn, more at springs. Best timing is most easily achieved by starting from Vlissingen, Breskens or Terneuzen. Yachts should keep to the edge of main chan. Use alternative chans with caution; Vaarwater langs Hoofdplat and de Paulinapolder/Thomasgeul are buoyed shortcuts popular with yachtsmen, but note that channels shift regularly.

Commercial shipping: Yachts should keep just outside the busy shipping chans, ie Wielingen from the SW, Scheur from the W, and Oostgat from the NW. A listening watch on Ch 10 (ship-to-ship) is compulsory above buoy 100 (near Antwerpen).

Be aware of large ship anchorages: Wielingen Noord and Zuid, either side of the fairway, as defined by buoys W6-Trawl-WN6 and W9-Songa. Further E, Flushing Roads anchorage is defined by Songa, SS1, SS5 and ARV-VH buoys. Ocean-going ships often manoeuvre off Vlissingen to transfer pilots.

S of Vlissingen (see Breskens) an E-W TSS (part of a Precautionary area) is best avoided by yachts which should cross the fairway via the recommended N-S track between Buitenhaven and ARV3 SPM lt buoy. Fast ferries, which have right of way, ply half-hourly from Vlissingen Buitenhaven to Breskens. *Keep a sharp lookout.*

Recommended small craft routes: **From the SW** there are few dangers. After Zeebrugge, keep S of the Wielingen chan buoys (W1-9). Off Breskens avoid fast ferries to/from Vlissingen; continue E to SS1 buoy, then cross to Vlissingen on a N'ly track.

From the W, keep clear of the Scheur chan by crossing to the S of Wielingen as soon as practicable.

From the N, the narrow, busy Oostgat can be used with caution, keeping just outside the SW edge of the buoyed/lit chan.

Three lesser, unlit day-only N'ly routes which avoid Oostgat are:
- From Kaloo or DR1 buoys, follow the Geul van de Rassen, Deurloo & Spleet chans to SP4 buoy. Thence E via WN6 buoy.
- Or continue down Deurloo from DL5 by to join Oostgat at OG19 by; thence cross to the N shore when Oostgat tfc permits.
- Another route, slightly further offshore, is to skirt the NW side of Kaloo bank to Botkil-W buoy, thence SE via Geul van de Walvischstaart (PHM buoys only) to Trawl SCM buoy.

LIGHTS AND MARKS The apprs are well lit by lighthouses: on the S shore at Nieuwe Sluis, and on the N shore at Westkapelle; see Lights, buoys and waypoints. The main fairways are, for the most part, defined by ldg lts and by the W sectors of the many Dir lts.

SCHELDEMOND VTS (www.vts-scheldt.net) covers from the North Sea outer approaches upriver to Antwerpen; see the diagram below and TSS off S Netherlands. Yachts should monitor at all times the VHF Ch for the area in which they are, so as to be aware of other shipping and to be contactable if required. Do not transmit, unless called. 7 **Traffic Centres** control the Areas below, within which Radar stations provide radar, weather and hbr info, as shown below:

Outer approaches: (*Traffic Centre* is the callsign prefix)		
Wandelaar	Ch 65	*Zeebrugge Radar* Ch 04.
Zeebrugge	Ch 69	*Radar* as in line above.
Steenbank	Ch 64	*Radar* also on Ch 64.
In the Westerschelde: (*Centrale* is the callsign prefix)		
Vlissingen	Ch 14	*Radar* Ch 21. Vlissingen to E2A/PvN SPR buoys (51°24′N 03°44′E).
Terneuzen	Ch 03	*Radar* Ch 03. Thence to Nos 32/ 35 buoys (51°23′N 03°57′E). Ch 11 Terneuzen-Gent Canal.
Hansweert	Ch 65	*Radar* Ch 65. Thence to Nos 46/ 55 buoys (51°24′N 04°02′E).
Zandvliet	Ch 12	Thence to Antwerpen. *Radar Waarde* 19; *Saeftinge* 21; *Zandvliet* 04; *Kruisschans* 66.

In **emergency**, call initially on the working channel in use; state yacht's name, position and the nature of the problem. You may then be switched to Ch **67** (Emergency) or another VHF channel.

Broadcasts of visibility, Met, tidal data and ship movements are made in Dutch at: H +00 by *Terneuzen* Ch 11; H +10 by *Zeebrugge* Ch 69; H +30 by *Zandvliet* Ch 12; H +50 by *Centrale Vlissingen* Ch 14; **H +55 by Radar Vlissingen Ch 21 in English**.

FACILITIES Some minor hbrs from Terneuzen to Antwerpen (38M) are listed below. They may be useful in emergency or offer shelter, but most dry. Yachts usually go non-stop to Antwerpen.

ELLEWOUTSDIJK, 51°23′·10N 03°49′·05E. BE 103; DYC 1803.2. HW +2 and +0·3m on Vlissingen; ML 2·6m. Small, unlit hbr, 27 ⌒ voluntary donation. 1·5m at MLWS. Strong cross eddy on ebb. HM ☎(0113) 548431/06 251154766, ⚓ @ ⌕(10t) YC.

HOEDEKENSKERKE, 51°25′·11N 03°54′·90E. BE 103; DYC 1803.3. Disused ferry hbr (dries) abeam MG13 SHM buoy (Iso.G.4s).

€1·00/m (min €8·00/night) inc ⟨symbol⟩ ⚓. YC WV Hoedekenskerke ☎(0113) 639278, mob 0653794069. 33 + 6ⓥ ⚓ 🛢 ⚓ ⟨symbol⟩ at campsite 400m. Town ✉ Ⓑ ⇌ (Goes).

HANSWEERT, 51°26'·37N 04°00'·66E. BE 103; DYC 1803.3. Tidal differences above. Temporary stop, but it is the busy ent to Zuid Beveland canal, lock Ch 22. Lt Oc WRG 10s, R lattice tr, W band, at ent. Waiting berths outside lock on E side. **No smoking or naked flames in, or near lock.** Services: 🛢 ⚓ ✗ 🛢 ⚓ ⟨symbol⟩(17t). Town ✉ Ⓑ ⇌ (Kruiningen-Yerseke).

WALSOORDEN, 51°22'·93N 04°02'·10E. BE 103; DYC 1803.3. HW is +0110 and +0·7m on Vlissingen; ML 2·6m. Prone to swell. SHM buoy 57, Iso G 8s, is 500m N of ent where 16 silos are conspic. Ldg lts 220° both Oc 3s. Hbr ent FG & FR. Unmarked stone pier just outside E hbr pier, partly dries at LW. Yacht basin dead ahead on ent to hbr, depths 2 to 2·8m. 4/5 ⟨symbol⟩ €0·30/m, max stay H24. *Zandvliet Radio* VHF Ch 12. WSV d'Ouwe Haven ☎(0114) 681235, ⚓ 🛢 🛢 ⚓ ⟨symbols⟩(16t) ⟨symbol⟩. Town ✗ □ ✉.

PAAL, 51°21'·25N 04°06'·65E. BE 103; DYC 1803.3. HW +0120 and +0·8m on Vlissingen; ML 2·7m. Appr via No. 63 SHM buoy and tide gauge, across drying Speelmansgat. Unlit, drying yacht hbr on S bank at river mouth, ent marked by withy. HM ☎(0114) 314974, 0611 028174. *Zandvliet Radio* Ch 12. **Jachthaven** 150 ⟨symbol⟩ ⟨symbol⟩ €0·50, 🛢 ⚓ Ⓓ ✗ 🍺 @ in YC. ⟨symbol⟩ ⟨symbol⟩ (10 mins walk). **Yachtclub** ☎(0114) 315548; www.wv-saeftinghe.nl

DOEL, 51°18'·67N 04°16'·11E. BE 103; DYC 1803.5. HW +0100 and +0·7m on Vlissingen. Small drying hbr next to 'ghost town' on W bank. Ldg lts 188·5°: front Fl WR 3s on N pier hd; rear Fl 3s, synch. HM ☎(03) 5758103; **YC de Noord** ☎7733669, open to visitors, ⚓ ✗ □.

LILLO, 51°18'·16N 04°17'·30E. BE 103; DYC 1803.5. 1M SE of Doel on opp bank; small drying hbr for shoal-draft only; Customs base. T-jetty has Oc WRG 10s. HM (035) 686456; Small YC, Scaldis, is named after local restaurant original members used for meetings.

9.16.28   VLISSINGEN (FLUSHING)

Zeeland **51°26'·31N 03°34'·61E** Koopmanshaven ⟨symbols⟩

CHARTS AC 1872, 1874, 120; Zeekaart 1442, 1443, 1533; BE 103; DYC 1803.2, 1801.4; Imray C30

TIDES +0210 Dover; ML 2·3; Duration 0555. Note: Vlissingen is a Standard Port (→)

SHELTER Very good in both yacht hbrs.

NAVIGATION WPT 51°25'·16N 03°33'·66E (Songa SHM buoy, QG) 027°/1·29M to Koopmanshaven ent. Keep to W pier side, strong crossflow until piers. Study Westerschelde.

LIGHTS AND MARKS From North West, Oostgat 117° ldg lts: Front RW pile; rear Sardijngeul, Oc WR 5s, R/W banded mast, R △. Note:

Conspic radar twr (close NW of Koopmanshaven) shows a Fl Y lt to warn when ships are approaching in the blind NW arc from Oostgat-Sardijngeul.

A conspic, floodlit R/W metal framework tr (50m) near de Ruyter Marina indicates the appr. Buitenhaven traffic signals from mole W side of ent: R flag or extra ● near FR on W mole hd = No entry.

COMMUNICATIONS (Code 0118) Emergency 112; Schelde Traffic Coordination Centre 424760; Police 0900-8844; ⊖ 484600; Brit Consul (020) 676 4343; Dr 412233; Port HM 0115 647400, VHF Ch 14; Sealock VHF Ch 18; Canal bridges VHF Ch 22, see also Westerschelde; Michiel de Ruyter and Schelde Marinas: nil VHF; use mobile.

FACILITIES

Michiel de Ruyter Marina (2·9m) ☎414498, *Pre-booking in season is strongly advised.* 100 + 40🅥 €2·00/m + €1·60 pp/day tax. ✕ 🗋 🗍 @ free wi-fi; info@brasserieevertsen.nl www. brasserieevertsen.nl

Ent is 6m narrow, over a sill with 1·2m water at MLWS. Access HW±4. Check depth gauge on outer wall or with HM. 2 small footbridges (R/G tfc lts) are swiftly swung open by HM 0800-2200LT. They stay open 2200–0800LT, but only for yachts to leave; ●● (vert) tfc lts prohibit arrival from sea. Storm barrier is open 1/4-1/11. Pilot boats use the adjacent inlet at high speed with scant regard for safety.

VVW Schelde (2·8-4·1m) ☎465912, HM mob 0615538515. www. vvwschelde.nl 90 + 50🅥 €1·60 + €1·30 pp/day tax. 🕬 €1·00/4kWh. 500m by road from ferry. 🛢 ✕ 🗋 🗍 ♂ ⚓ (12t max) @ bike hire.

At ent to the Buitenhaven beware ferries; keep to port, pass S of the ferry terminal for the most N'ly and smallest sealock, which operates H24. Waiting possible on piles to SE.

Town ✕ 🗋 🚟 🏤 🗑 🏦 ✉ Ⓑ ⇌ ✈ (Antwerpen). Foot ferry to Breskens; road tunnel to Terneuzen. At Middelburg: 🏛 🔧 ⚔ ⚓.

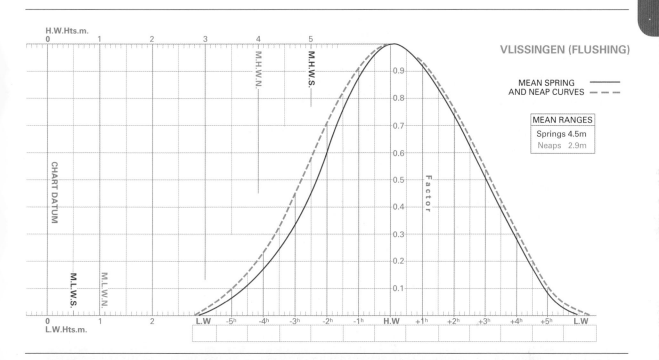

VLISSINGEN (FLUSHING)

MEAN SPRING ———
AND NEAP CURVES - - - -

MEAN RANGES
Springs 4.5m
Neaps 2.9m

STANDARD TIME UT –01
Subtract 1 hour for UT
For Dutch Summer Time add
ONE hour in **non-shaded areas**

VLISSINGEN   LAT 51°27'N   LONG 3°36'E
TIMES AND HEIGHTS OF HIGH AND LOW WATERS

Dates in red are SPRINGS
Dates in blue are NEAPS

YEAR 2016

JANUARY

Day	Time	m	Day	Time	m
1 F	0010 / 0622 / 1256 / 1849	1.1 / 4.4 / 0.9 / 4.4	16 SA	0016 / 0616 / 1256 / 1850	0.9 / 4.7 / 0.5 / 4.7
2 SA ◑	0055 / 0708 / 1335 / 1939	1.2 / 4.3 / 1.0 / 4.2	17 SU ●	0106 / 0816 / 1346 / 1956	1.0 / 4.5 / 0.6 / 4.5
3 SU	0200 / 0801 / 1446 / 2045	1.3 / 4.1 / 1.1 / 4.0	18 M	0206 / 0819 / 1445 / 2102	1.1 / 4.4 / 0.8 / 4.4
4 M	0316 / 0916 / 1540 / 2144	1.4 / 3.9 / 1.2 / 4.0	19 TU	0309 / 0930 / 1556 / 2216	1.1 / 4.3 / 0.9 / 4.4
5 TU	0415 / 1025 / 1645 / 2259	1.3 / 4.0 / 1.2 / 4.1	20 W	0435 / 1039 / 1721 / 2325	1.1 / 4.4 / 0.9 / 4.4
6 W	0515 / 1125 / 1745 / 2356	1.2 / 4.2 / 1.1 / 4.4	21 TH	0556 / 1145 / 1820	1.0 / 4.5 / 0.8
7 TH	0611 / 1215 / 1830	1.1 / 4.4 / 1.0	22 F	0026 / 0650 / 1245 / 1912	4.6 / 0.8 / 4.7 / 0.7
8 F	0035 / 0656 / 1256 / 1916	4.6 / 0.9 / 4.6 / 0.9	23 SA	0115 / 0740 / 1336 / 1955	4.7 / 0.6 / 4.9 / 0.7
9 SA	0119 / 0738 / 1335 / 1956	4.8 / 0.8 / 4.8 / 0.8	24 SU ○	0201 / 0826 / 1418 / 2035	4.9 / 0.5 / 5.0 / 0.7
10 SU ●	0157 / 0818 / 1413 / 2036	4.9 / 0.6 / 5.0 / 0.7	25 M	0240 / 0905 / 1459 / 2116	4.9 / 0.5 / 5.0 / 0.8
11 M	0235 / 0906 / 1453 / 2118	5.0 / 0.5 / 5.1 / 0.7	26 TU	0317 / 0948 / 1537 / 2152	4.9 / 0.5 / 5.0 / 0.8
12 TU	0316 / 0945 / 1533 / 2200	5.0 / 0.4 / 5.1 / 0.7	27 W	0357 / 1021 / 1615 / 2231	4.9 / 0.5 / 4.9 / 0.8
13 W	0356 / 1036 / 1615 / 2246	5.0 / 0.4 / 5.1 / 0.7	28 TH	0435 / 1106 / 1652 / 2306	4.8 / 0.6 / 4.8 / 0.9
14 TH	0436 / 1118 / 1659 / 2330	4.9 / 0.4 / 5.0 / 0.8	29 F	0508 / 1135 / 1728 / 2338	4.7 / 0.7 / 4.7 / 1.0
15 F	0525 / 1205 / 1749	4.8 / 0.4 / 4.9	30 SA	0545 / 1206 / 1808	4.6 / 0.8 / 4.5
			31 SU	0016 / 0626 / 1235 / 1849	1.0 / 4.4 / 0.9 / 4.3

FEBRUARY

Day	Time	m	Day	Time	m
1 M ◑	0046 / 0716 / 1320 / 1946	1.1 / 4.2 / 1.0 / 4.1	16 TU	0140 / 0756 / 1416 / 2036	0.9 / 4.5 / 0.8 / 4.3
2 TU	0135 / 0816 / 1414 / 2046	1.2 / 4.1 / 1.2 / 4.0	17 W	0245 / 0906 / 1524 / 2149	1.1 / 4.3 / 1.0 / 4.1
3 W	0305 / 0919 / 1556 / 2155	1.3 / 4.0 / 1.2 / 4.0	18 TH	0415 / 1025 / 1653 / 2309	1.1 / 4.2 / 1.0 / 4.2
4 TH	0430 / 1035 / 1659 / 2316	1.3 / 4.0 / 1.2 / 4.1	19 F	0535 / 1139 / 1810	1.0 / 4.4 / 0.9
5 F	0525 / 1145 / 1806	1.1 / 4.2 / 1.0	20 SA	0015 / 0640 / 1245 / 1859	4.4 / 0.8 / 4.6 / 0.8
6 SA	0016 / 0625 / 1236 / 1850	4.4 / 0.9 / 4.5 / 0.9	21 SU	0106 / 0729 / 1329 / 1946	4.6 / 0.6 / 4.8 / 0.8
7 SU	0056 / 0715 / 1315 / 1936	4.6 / 0.7 / 4.8 / 0.8	22 M	0148 / 0812 / 1405 / 2019	4.8 / 0.5 / 4.9 / 0.7
8 M	0137 / 0759 / 1356 / 2015	4.8 / 0.6 / 5.0 / 0.7	23 TU	0225 / 0849 / 1446 / 2058	4.8 / 0.5 / 4.9 / 0.7
9 TU	0215 / 0845 / 1435 / 2100	5.0 / 0.4 / 5.2 / 0.6	24 W	0257 / 0928 / 1517 / 2129	4.9 / 0.5 / 5.0 / 0.7
10 W	0256 / 0932 / 1515 / 2146	5.1 / 0.3 / 5.2 / 0.6	25 TH	0336 / 1002 / 1550 / 2205	4.9 / 0.5 / 4.9 / 0.7
11 TH	0336 / 1015 / 1557 / 2228	5.1 / 0.2 / 5.2 / 0.6	26 F	0407 / 1036 / 1625 / 2235	4.9 / 0.5 / 4.9 / 0.8
12 F	0418 / 1102 / 1643 / 2316	5.1 / 0.2 / 5.1 / 0.6	27 SA	0438 / 1106 / 1655 / 2306	4.8 / 0.6 / 4.7 / 0.8
13 SA	0505 / 1145 / 1728 / 2356	5.0 / 0.3 / 5.0 / 0.7	28 SU	0511 / 1130 / 1727 / 2336	4.7 / 0.7 / 4.6 / 0.9
14 SU	0553 / 1229 / 1825	4.9 / 0.4 / 4.8	29 M	0541 / 1156 / 1759	4.6 / 0.8 / 4.5
15 M ◑	0045 / 0649 / 1320 / 1925	0.8 / 4.7 / 0.6 / 4.5			

MARCH

Day	Time	m	Day	Time	m
1 TU	0006 / 0615 / 1229 / 1846	0.9 / 4.4 / 0.9 / 4.3	16 W	0115 / 0729 / 1356 / 2012	0.8 / 4.5 / 0.8 / 4.2
2 W	0050 / 0710 / 1326 / 1950	1.0 / 4.2 / 1.0 / 4.1	17 TH	0225 / 0846 / 1505 / 2135	1.0 / 4.2 / 1.0 / 4.0
3 TH ◑	0150 / 0824 / 1434 / 2110	1.2 / 4.0 / 1.2 / 3.9	18 F	0356 / 1009 / 1635 / 2255	1.0 / 4.2 / 1.1 / 4.1
4 F	0346 / 0952 / 1614 / 2236	1.2 / 4.0 / 1.2 / 4.0	19 SA	0526 / 1130 / 1755 / 2359	0.9 / 4.4 / 0.9 / 4.3
5 SA	0506 / 1105 / 1736 / 2345	1.1 / 4.2 / 1.1 / 4.2	20 SU	0628 / 1225 / 1850	0.7 / 4.6 / 0.8
6 SU	0600 / 1208 / 1825	0.9 / 4.5 / 0.9	21 M	0048 / 0715 / 1311 / 1925	4.5 / 0.6 / 4.8 / 0.8
7 M	0032 / 0656 / 1251 / 1909	4.5 / 0.7 / 4.8 / 0.8	22 TU ○	0131 / 0756 / 1349 / 1959	4.7 / 0.5 / 4.8 / 0.7
8 TU	0115 / 0742 / 1332 / 1955	4.8 / 0.5 / 5.0 / 0.6	23 W ○	0208 / 0825 / 1426 / 2035	4.8 / 0.5 / 4.9 / 0.7
9 W ●	0155 / 0825 / 1413 / 2040	5.0 / 0.3 / 5.2 / 0.5	24 TH	0237 / 0900 / 1455 / 2108	4.8 / 0.5 / 4.9 / 0.7
10 TH	0236 / 0910 / 1455 / 2126	5.1 / 0.2 / 5.3 / 0.5	25 F	0307 / 0936 / 1526 / 2142	4.9 / 0.5 / 4.9 / 0.6
11 F	0316 / 0955 / 1535 / 2208	5.2 / 0.1 / 5.3 / 0.5	26 SA	0339 / 1006 / 1556 / 2216	4.9 / 0.5 / 4.9 / 0.7
12 SA	0358 / 1040 / 1621 / 2255	5.2 / 0.1 / 5.2 / 0.5	27 SU	0408 / 1035 / 1627 / 2239	4.8 / 0.6 / 4.8 / 0.7
13 SU	0446 / 1126 / 1706 / 2338	5.1 / 0.2 / 5.0 / 0.6	28 M	0439 / 1059 / 1656 / 2305	4.8 / 0.7 / 4.7 / 0.7
14 M	0531 / 1208 / 1801	5.0 / 0.2 / 4.7	29 TU	0509 / 1128 / 1726 / 2341	4.7 / 0.8 / 4.5 / 0.8
15 TU ◑	0026 / 0626 / 1251 / 1901	0.7 / 4.7 / 0.6 / 4.5	30 W	0542 / 1202 / 1805	4.6 / 0.8 / 4.4
			31 TH ◑	0020 / 0625 / 1250 / 1854	0.8 / 4.4 / 1.0 / 4.2

APRIL

Day	Time	m	Day	Time	m
1 F	0126 / 0746 / 1355 / 2031	1.0 / 4.1 / 1.2 / 4.0	16 SA	0330 / 0950 / 1620 / 2229	0.9 / 4.2 / 1.1 / 4.0
2 SA	0250 / 0916 / 1534 / 2155	1.1 / 4.0 / 1.2 / 3.9	17 SU	0500 / 1106 / 1735 / 2335	0.9 / 4.3 / 1.0 / 4.2
3 SU	0427 / 1036 / 1656 / 2305	1.0 / 4.2 / 1.1 / 4.2	18 M	0559 / 1201 / 1825	0.7 / 4.6 / 0.9
4 M	0530 / 1135 / 1801	0.8 / 4.5 / 0.9	19 TU	0025 / 0650 / 1248 / 1906	4.4 / 0.6 / 4.7 / 0.8
5 TU	0002 / 0631 / 1225 / 1848	4.5 / 0.6 / 4.8 / 0.7	20 W	0106 / 0726 / 1326 / 1935	4.6 / 0.6 / 4.8 / 0.8
6 W	0045 / 0716 / 1306 / 1936	4.8 / 0.4 / 5.1 / 0.6	21 TH	0139 / 0800 / 1356 / 2008	4.7 / 0.6 / 4.8 / 0.7
7 TH ●	0126 / 0806 / 1348 / 2021	5.0 / 0.3 / 5.2 / 0.5	22 F ○	0207 / 0832 / 1425 / 2042	4.7 / 0.5 / 4.8 / 0.6
8 F	0210 / 0847 / 1433 / 2103	5.2 / 0.2 / 5.3 / 0.4	23 SA	0239 / 0905 / 1455 / 2115	4.8 / 0.5 / 4.9 / 0.6
9 SA	0254 / 0932 / 1515 / 2150	5.3 / 0.1 / 5.3 / 0.4	24 SU	0311 / 0936 / 1527 / 2150	4.9 / 0.5 / 4.9 / 0.6
10 SU	0335 / 1017 / 1602 / 2233	5.2 / 0.2 / 5.1 / 0.4	25 M	0342 / 1005 / 1559 / 2219	4.8 / 0.6 / 4.8 / 0.6
11 M	0421 / 1100 / 1649 / 2321	5.2 / 0.3 / 4.9 / 0.5	26 TU	0415 / 1041 / 1629 / 2249	4.8 / 0.7 / 4.7 / 0.7
12 TU	0511 / 1146 / 1741	5.0 / 0.5 / 4.7	27 W	0446 / 1111 / 1706 / 2326	4.7 / 0.8 / 4.6 / 0.7
13 W	0005 / 0606 / 1230 / 1839	0.6 / 4.8 / 0.7 / 4.4	28 TH	0521 / 1146 / 1742	4.6 / 0.9 / 4.5
14 TH ◑	0054 / 0716 / 1325 / 1945	0.7 / 4.5 / 0.9 / 4.2	29 F	0006 / 0608 / 1229 / 1838	0.8 / 4.5 / 1.0 / 4.2
15 F	0203 / 0825 / 1446 / 2110	0.9 / 4.3 / 1.1 / 4.0	30 SA ◑	0105 / 0716 / 1336 / 2001	0.8 / 4.2 / 1.1 / 4.1

Chart Datum is 2·56 metres below NAP Datum. HAT is 5·4 metres above Chart Datum.

FREE monthly updates. Register at
www.reedsnauticalalmanac.co.uk

STANDARD TIME UT –01
Subtract 1 hour for UT
For Dutch Summer Time add
ONE hour in **non-shaded areas**

VLISSINGEN   LAT 51°27'N   LONG 3°36'E
TIMES AND HEIGHTS OF HIGH AND LOW WATERS

Dates in red are SPRINGS
Dates in blue are NEAPS

YEAR **2016**

Netherlands

MAY

	Time	m		Time	m
1 SU	0226	0.9	16 M	0405	0.9
	0842	4.2		1030	4.3
	1500	1.1		1656	1.1
	2116	4.0		2258	4.1
2 M	0339	0.8	17 TU	0519	0.8
	0955	4.3		1128	4.4
	1613	1.1		1746	1.0
	2226	4.2		2350	4.3
3 TU	0456	0.7	18 W	0616	0.7
	1101	4.6		1216	4.5
	1726	0.9		1830	0.9
	2328	4.5			
4 W	0601	0.6	19 TH	0031	4.4
	1157	4.8		0656	0.7
	1826	0.7		1255	4.6
				1905	0.8
5 TH	0016	4.7	20 F	0108	4.5
	0652	0.4		0726	0.7
	1246	5.0		1328	4.7
	1909	0.6		1940	0.7
6 F ●	0106	5.0	21 SA ○	0139	4.7
	0740	0.3		0800	0.6
	1328	5.2		1359	4.8
	1958	0.5		2016	0.6
7 SA	0147	5.1	22 SU	0212	4.8
	0825	0.2		0831	0.6
	1415	5.2		1431	4.8
	2046	0.4		2049	0.6
8 SU	0233	5.2	23 M	0247	4.8
	0910	0.2		0908	0.6
	1458	5.2		1505	4.9
	2130	0.4		2125	0.6
9 M	0318	5.2	24 TU	0318	4.8
	0955	0.3		0946	0.7
	1545	5.1		1537	4.8
	2217	0.4		2205	0.6
10 TU	0405	5.1	25 W	0352	4.8
	1036	0.4		1021	0.7
	1632	4.9		1610	4.7
	2306	0.4		2240	0.6
11 W	0456	5.0	26 TH	0426	4.7
	1119	0.6		1056	0.8
	1726	4.7		1647	4.6
	2350	0.5		2314	0.6
12 TH	0549	4.8	27 F	0507	4.7
	1208	0.7		1136	0.9
	1819	4.5		1729	4.5
				2359	0.6
13 F ◑	0045	0.6	28 SA	0552	4.6
	0649	4.6		1225	1.0
	1305	0.9		1825	4.3
	1919	4.2			
14 SA	0151	0.7	29 SU ◑	0055	0.7
	0800	4.4		0656	4.4
	1404	1.1		1320	1.0
	2035	4.0		1936	4.2
15 SU	0255	0.8	30 M	0154	0.7
	0916	4.2		0817	4.4
	1524	1.1		1423	1.1
	2152	4.0		2045	4.2
			31 TU	0304	0.7
				0926	4.4
				1534	1.0
				2156	4.3

JUNE

	Time	m		Time	m
1 W	0415	0.7	16 TH	0519	0.9
	1030	4.6		1135	4.3
	1656	0.9		1734	1.0
	2255	4.5		2351	4.3
2 TH	0531	0.6	17 F	0616	0.9
	1127	4.8		1218	4.5
	1756	0.8		1830	0.9
	2352	4.7			
3 F	0626	0.5	18 SA	0036	4.4
	1221	5.0		0649	0.8
	1850	0.6		1255	4.6
				1910	0.8
4 SA	0043	4.9	19 SU	0116	4.6
	0718	0.4		0730	0.8
	1309	5.1		1331	4.7
	1940	0.5		1950	0.7
5 SU ●	0128	5.1	20 M ○	0147	4.7
	0806	0.3		0805	0.7
	1356	5.1		1406	4.8
	2028	0.4		2028	0.6
6 M	0216	5.1	21 TU	0222	4.8
	0850	0.4		0839	0.7
	1443	5.1		1446	4.8
	2116	0.4		2106	0.5
7 TU	0301	5.1	22 W	0259	4.9
	0935	0.4		0919	0.7
	1530	5.0		1518	4.8
	2202	0.4		2145	0.5
8 W	0350	5.1	23 TH	0335	4.9
	1017	0.6		1002	0.7
	1617	4.9		1557	4.8
	2246	0.4		2230	0.5
9 TH	0437	5.0	24 F	0415	4.8
	1059	0.7		1039	0.8
	1706	4.7		1635	4.7
	2336	0.5		2316	0.5
10 F	0528	4.8	25 SA	0455	4.8
	1146	0.8		1126	0.9
	1756	4.6		1717	4.6
				2354	0.5
11 SA	0019	0.5	26 SU	0541	4.7
	0625	4.6		1215	0.9
	1236	1.0		1806	4.5
	1848	4.4			
12 SU ◑	0111	0.7	27 M ◐	0051	0.5
	0726	4.4		0636	4.6
	1330	1.1		1306	1.0
	1945	4.2		1908	4.4
13 M	0209	0.8	28 TU	0146	0.6
	0826	4.2		0746	4.5
	1436	1.2		1359	1.0
	2044	4.0		2016	4.4
14 TU	0316	0.9	29 W	0246	0.6
	0935	4.2		0851	4.5
	1545	1.2		1506	1.0
	2206	4.0		2122	4.4
15 W	0420	0.9	30 TH	0346	0.7
	1046	4.2		1000	4.6
	1645	1.1		1615	1.0
	2305	4.1		2227	4.5

JULY

	Time	m		Time	m
1 F	0506	0.7	16 SA	0536	1.0
	1105	4.6		1142	4.3
	1736	0.9		1749	1.0
	2327	4.6			
2 SA	0606	0.6	17 SU	0002	4.3
	1205	4.8		0621	1.0
	1832	0.7		1230	4.5
				1839	0.9
3 SU	0025	4.8	18 M	0046	4.5
	0701	0.5		0700	0.9
	1255	4.9		1309	4.6
	1926	0.6		1925	0.8
4 M ●	0116	5.0	19 TU ○	0126	4.7
	0745	0.5		0740	0.7
	1345	5.0		1345	4.8
	2015	0.4		2005	0.6
5 TU	0205	5.0	20 W	0206	4.8
	0831	0.5		0818	0.7
	1432	5.0		1422	4.9
	2102	0.4		2046	0.5
6 W	0250	5.1	21 TH	0241	5.0
	0915	0.6		0900	0.7
	1515	5.0		1501	4.9
	2145	0.4		2130	0.4
7 TH	0336	5.1	22 F	0316	5.0
	0956	0.7		0946	0.7
	1559	4.9		1536	4.9
	2232	0.4		2216	0.4
8 F	0420	5.0	23 SA	0356	5.0
	1038	0.8		1025	0.7
	1645	4.8		1619	4.9
	2311	0.5		2259	0.4
9 SA	0508	4.9	24 SU	0438	5.0
	1120	0.9		1109	0.8
	1728	4.7		1700	4.8
	2355	0.5		2346	0.4
10 SU	0551	4.7	25 M	0526	4.9
	1159	1.0		1155	0.8
	1809	4.6		1747	4.7
11 M	0035	0.7	26 TU	0029	0.5
	0640	4.5		0618	4.8
	1245	1.1		1242	0.9
	1856	4.4		1841	4.6
12 TU ◐	0126	0.8	27 W ◐	0121	0.5
	0725	4.3		0720	4.6
	1346	1.2		1336	1.0
	1945	4.2		1948	4.5
13 W	0220	0.9	28 TH	0212	0.6
	0826	4.1		0826	4.5
	1443	1.2		1440	1.0
	2044	4.0		2055	4.4
14 TH	0314	1.0	29 F	0321	0.7
	0936	4.0		0936	4.4
	1549	1.2		1550	1.1
	2206	4.0		2205	4.4
15 F	0436	1.1	30 SA	0436	0.8
	1046	4.1		1045	4.4
	1655	1.1		1710	1.0
	2316	4.1		2312	4.5
			31 SU	0551	0.8
				1151	4.6
				1820	0.8

AUGUST

	Time	m		Time	m
1 M	0016	4.7	16 TU	0019	4.5
	0646	0.7		0635	1.0
	1247	4.8		1246	4.6
	1916	0.6		1859	0.8
2 TU ●	0108	4.9	17 W	0101	4.7
	0735	0.7		0718	0.9
	1335	4.9		1325	4.8
	2003	0.5		1946	0.7
3 W	0157	5.0	18 TH ○	0141	4.9
	0818	0.7		0756	0.8
	1421	5.0		1401	5.0
	2048	0.4		2025	0.5
4 TH	0237	5.1	19 F	0218	5.1
	0858	0.7		0838	0.7
	1500	5.0		1436	5.1
	2129	0.4		2108	0.4
5 F	0321	5.1	20 SA	0256	5.2
	0938	0.8		0926	0.7
	1539	5.0		1517	5.1
	2210	0.4		2153	0.3
6 SA	0359	5.0	21 SU	0336	5.2
	1015	0.8		1005	0.7
	1617	4.9		1557	5.1
	2248	0.5		2238	0.3
7 SU	0438	4.9	22 M	0418	5.2
	1049	0.9		1052	0.7
	1657	4.8		1639	5.1
	2325	0.6		2326	0.4
8 M	0517	4.8	23 TU	0505	5.0
	1131	0.9		1132	0.8
	1732	4.7		1726	4.9
9 TU	0000	0.7	24 W	0008	0.4
	0556	4.6		0555	4.9
	1202	1.0		1219	0.8
	1811	4.5		1816	4.8
10 W ◐	0036	0.9	25 TH ◐	0056	0.6
	0640	4.4		0652	4.6
	1245	1.1		1316	0.9
	1901	4.3		1918	4.6
11 TH	0105	1.0	26 F	0148	0.8
	0731	4.2		0800	4.4
	1324	1.2		1415	1.1
	1956	4.1		2032	4.4
12 F	0204	1.2	27 SA	0244	0.9
	0826	4.0		0912	4.3
	1454	1.3		1535	1.1
	2100	4.0		2145	4.3
13 SA	0346	1.3	28 SU	0425	1.0
	0935	4.0		1036	4.3
	1616	1.3		1706	1.1
	2226	4.0		2301	4.4
14 SU	0445	1.2	29 M	0536	1.0
	1100	4.1		1145	4.5
	1720	1.2		1816	0.8
	2330	4.2			
15 M	0546	1.1	30 TU	0008	4.7
	1200	4.3		0636	0.9
	1816	1.0		1238	4.7
				1905	0.6
			31 W	0101	4.9
				0722	0.8
				1325	4.8
				1952	0.5

Chart Datum is 2·56 metres below NAP Datum. HAT is 5·4 metres above Chart Datum.

STANDARD TIME UT −01
Subtract 1 hour for UT
For Dutch Summer Time add
ONE hour in **non-shaded areas**

VLISSINGEN   LAT 51°27′N   LONG 3°36′E
TIMES AND HEIGHTS OF HIGH AND LOW WATERS

Dates in red are **SPRINGS**
Dates in blue are **NEAPS**

YEAR 2016

SEPTEMBER

Time	m		Time	m
1 0146	5.0	**16** 0115	5.0
0802	0.8	0735	0.8
TH 1406	4.9	F 1335	5.0
● 2032	0.5	○ 2002	0.5
2 0226	5.1	**17** 0153	5.2
0840	0.8	0815	0.7
F 1440	5.0	SA 1413	5.2
2108	0.5	2047	0.4
3 0258	5.1	**18** 0233	5.4
0916	0.8	0903	0.6
SA 1517	5.0	SU 1453	5.3
2146	0.5	2130	0.3
4 0335	5.1	**19** 0314	5.4
0950	0.8	0946	0.6
SU 1551	5.0	M 1536	5.3
2220	0.6	2215	0.3
5 0412	5.0	**20** 0357	5.3
1026	0.9	1030	0.6
M 1625	4.9	TU 1615	5.2
2252	0.7	2257	0.4
6 0445	4.8	**21** 0442	5.1
1056	0.9	1113	0.7
TU 1659	4.8	W 1703	5.1
2326	0.8	2342	0.5
7 0516	4.7	**22** 0530	4.9
1126	1.0	1201	0.8
W 1731	4.7	TH 1755	4.9
2345	0.9		
8 0549	4.5	**23** 0026	0.7
1152	1.0	0627	4.6
TH 1805	4.5	F 1250	0.9
		◑ 1855	4.6
9 0015	1.0	**24** 0119	0.9
0636	4.3	0736	4.4
F 1229	1.1	SA 1356	1.1
◑ 1855	4.3	2009	4.4
10 0059	1.2	**25** 0224	1.1
0725	4.1	0852	4.2
SA 1325	1.3	SU 1516	1.1
2016	4.0	2129	4.3
11 0204	1.4	**26** 0355	1.2
0846	4.0	1020	4.2
SU 1530	1.4	M 1650	1.1
2124	4.0	2256	4.4
12 0416	1.4	**27** 0531	1.1
1006	4.0	1129	4.4
M 1640	1.2	TU 1800	0.8
2249	4.1	2359	4.7
13 0504	1.2	**28** 0625	1.0
1122	4.2	1225	4.6
TU 1746	1.0	W 1849	0.7
2356	4.5		
14 0605	1.1	**29** 0049	4.9
1216	4.5	0706	0.9
W 1836	0.8	TH 1305	4.8
		1931	0.6
15 0035	4.8	**30** 0129	5.0
0652	0.9	0746	0.8
TH 1257	4.8	F 1346	4.9
1918	0.6	2010	0.6

OCTOBER

Time	m		Time	m
1 0206	5.0	**16** 0126	5.3
0818	0.8	0752	0.7
SA 1417	5.0	SU 1345	5.3
● 2041	0.6	○ 2021	0.4
2 0235	5.0	**17** 0208	5.4
0852	0.8	0838	0.6
SU 1451	5.0	M 1428	5.4
2115	0.6	2106	0.3
3 0308	5.0	**18** 0252	5.4
0926	0.8	0926	0.6
M 1526	5.0	TU 1512	5.4
2150	0.7	2150	0.4
4 0341	5.0	**19** 0336	5.3
0958	0.8	1008	0.6
TU 1555	5.0	W 1556	5.3
2219	0.8	2233	0.5
5 0415	4.9	**20** 0422	5.1
1030	0.9	1056	0.6
W 1627	4.9	TH 1643	5.2
2245	0.9	2318	0.6
6 0442	4.7	**21** 0512	4.9
1056	0.9	1139	0.7
TH 1657	4.7	F 1733	4.9
2316	1.0		
7 0515	4.6	**22** 0005	0.8
1120	1.0	0608	4.6
F 1727	4.6	SA 1232	0.8
2342	1.0	◑ 1838	4.7
8 0548	4.5	**23** 0056	1.0
1155	1.0	0715	4.4
SA 1808	4.4	SU 1335	1.0
		1956	4.5
9 0026	1.2	**24** 0206	1.2
0631	4.3	0830	4.2
SU 1250	1.2	M 1444	1.1
◑ 1903	4.2	2116	4.3
10 0126	1.4	**25** 0334	1.3
0801	4.0	0956	4.1
M 1404	1.3	TU 1626	1.0
2045	4.1	2229	4.4
11 0315	1.4	**26** 0506	1.2
0920	4.0	1105	4.3
TU 1555	1.2	W 1735	0.9
2206	4.2	2336	4.6
12 0436	1.3	**27** 0605	1.1
1035	4.2	1159	4.5
W 1706	0.9	TH 1831	0.7
2311	4.5		
13 0535	1.1	**28** 0025	4.8
1135	4.5	0645	1.0
TH 1759	0.8	F 1246	4.7
		1905	0.7
14 0006	4.8	**29** 0107	4.9
0626	0.9	0720	0.9
F 1223	4.8	SA 1318	4.8
1849	0.6	1946	0.7
15 0045	5.1	**30** 0138	4.9
0708	0.8	0752	0.9
SA 1306	5.1	SU 1352	4.9
1938	0.5	● 2016	0.7
		31 0211	5.0
		0826	0.8
		M 1426	4.9
		2048	0.7

NOVEMBER

Time	m		Time	m
1 0242	5.0	**16** 0233	5.3
0859	0.8	0906	0.5
TU 1457	5.0	W 1454	5.4
2120	0.7	2128	0.4
2 0316	5.0	**17** 0319	5.2
0936	0.8	0952	0.5
W 1527	5.0	TH 1538	5.3
2156	0.8	2213	0.5
3 0346	4.9	**18** 0405	5.1
1006	0.8	1038	0.6
TH 1602	4.9	F 1626	5.2
2220	0.9	2256	0.7
4 0417	4.8	**19** 0457	4.9
1035	0.9	1126	0.6
F 1631	4.8	SA 1717	5.0
2245	1.0	2342	0.9
5 0447	4.7	**20** 0549	4.7
1106	0.9	1215	0.7
SA 1702	4.7	SU 1819	4.8
2320	1.1		
6 0521	4.6	**21** 0036	1.0
1140	0.9	0649	4.4
SU 1741	4.6	M 1316	0.9
		◑ 1928	4.5
7 0000	1.1	**22** 0129	1.2
0610	4.4	0756	4.2
M 1230	1.0	TU 1414	1.0
◑ 1835	4.4	2035	4.4
8 0100	1.3	**23** 0256	1.3
0716	4.2	0909	4.1
TU 1333	1.1	W 1534	1.0
2006	4.2	2156	4.3
9 0215	1.4	**24** 0405	1.3
0840	4.1	1025	4.2
W 1516	1.1	TH 1700	1.0
2122	4.3	2302	4.4
10 0346	1.3	**25** 0525	1.2
0955	4.2	1125	4.3
TH 1614	1.0	F 1755	0.9
2232	4.5	2355	4.6
11 0444	1.2	**26** 0615	1.1
1056	4.5	1211	4.5
F 1726	0.8	SA 1841	0.9
2329	4.8		
12 0545	1.0	**27** 0038	4.7
1151	4.8	0650	1.0
SA 1822	0.6	SU 1251	4.6
		1912	0.8
13 0020	5.1	**28** 0112	4.8
0646	0.8	0726	0.9
SU 1238	5.0	M 1327	4.7
1912	0.5	1945	0.8
14 0106	5.3	**29** 0147	4.8
0729	0.7	0759	0.8
M 1326	5.2	TU 1358	4.8
○ 1958	0.4	● 2021	0.8
15 0148	5.3	**30** 0218	4.9
0818	0.6	0838	0.8
TU 1407	5.3	W 1436	4.9
2046	0.4	2056	0.8

DECEMBER

Time	m		Time	m
1 0251	4.9	**16** 0305	5.1
0916	0.7	0938	0.5
TH 1506	4.9	F 1525	5.3
2125	0.8	2153	0.6
2 0325	4.9	**17** 0353	5.0
0945	0.7	1022	0.5
F 1537	4.9	SA 1616	5.2
2158	0.9	2237	0.7
3 0357	4.8	**18** 0441	4.9
1026	0.8	1110	0.5
SA 1612	4.8	SU 1702	5.0
2230	1.0	2320	0.9
4 0432	4.7	**19** 0528	4.8
1056	0.8	1156	0.6
SU 1647	4.8	M 1755	4.8
2308	1.0		
5 0507	4.6	**20** 0005	1.0
1136	0.8	0622	4.6
M 1727	4.7	TU 1246	0.7
2345	1.1	1849	4.6
6 0551	4.5	**21** 0055	1.1
1220	0.8	0716	4.4
TU 1815	4.6	W 1335	0.9
		◑ 1952	4.4
7 0040	1.2	**22** 0155	1.3
0650	4.4	0815	4.2
W 1326	0.9	TH 1440	1.0
◑ 1930	4.4	2105	4.2
8 0139	1.2	**23** 0305	1.3
0806	4.3	0925	4.1
TH 1426	0.9	F 1550	1.1
2046	4.4	2216	4.2
9 0256	1.2	**24** 0420	1.3
0916	4.3	1040	4.1
F 1535	0.9	SA 1655	1.1
2152	4.5	2315	4.3
10 0406	1.2	**25** 0525	1.2
1019	4.5	1136	4.2
SA 1646	0.8	SU 1755	1.0
2256	4.7		
11 0516	1.0	**26** 0005	1.1
1119	4.7	0615	1.1
SU 1749	0.7	M 1221	4.4
2352	4.9	1840	1.0
12 0615	0.9	**27** 0050	4.5
1216	4.9	0654	1.0
M 1851	0.6	TU 1301	4.5
		1918	0.9
13 0046	5.1	**28** 0126	4.6
0712	0.7	0741	0.9
TU 1303	5.1	W 1339	4.7
1941	0.5	1956	0.9
14 0132	5.3	**29** 0155	4.7
0803	0.6	0815	0.9
W 1350	5.2	TH 1415	4.8
○ 2026	0.5	● 2031	0.8
15 0219	5.2	**30** 0236	4.8
0850	0.5	0856	0.7
TH 1438	5.3	F 1447	4.9
2110	0.5	2106	0.8
		31 0305	4.9
		0932	0.6
		SA 1521	4.9
		2146	0.8

Chart Datum is 2·56 metres below NAP Datum. HAT is 5·4 metres above Chart Datum.

9.16.29 BRESKENS

Zeeland **51°24'·00N 03°34'·08E** ❀❀❀◊◊◊◊❀❀❀

CHARTS AC 1874, 1872, 120; Zeekaart 120, 101; BE 103; DYC 1801.4, 1803.2; Imray C30

TIDES +0210 Dover; ML no data; Duration 0600

Standard Port VLISSINGEN (←—) Use Vlissingen data.

SHELTER Good in all winds except N/NW. In fine weather ⌕ off Plaat van Breskens or in peaceful Vaarwater langs Hoofdplaat; no ⌕ in commercial/fishing hbr.

NAVIGATION WPT 51°24'·71N 03°33'·90E (ARV-VH NCM buoy, Q), 170°/7ca to hbr ent. Beware fast ferries and strong tides across the ent. Do not confuse the ent with the ferry port ent, 0·7M WNW, where yachts are prohib.

LIGHTS AND MARKS Large bldg/silo on centre pier in hbr and three apartment blocks (30m) SE of marina are conspic. See chartlet and Lights, buoys and waypoints. Nieuwe Sluis disused lt ho, 28m B/W banded 8-sided twr, is 1·8M W of marina.

COMMUNICATIONS (Code 0117); Police 0900 8844; ⊖ (0118) 484614/484624; British Consul (020) 676 4343; Ⓗ (0117) 459000; Dr 384010, at night/weekends (0115) 643000; Marina VHF Ch 31.

FACILITIES Marina jachthavenbreskens@zonnet.nl ☎381902. www.jachthavenbreskens.nl 580 ⌆ inc Ⓥ €1·90 + €1·15 pp/day tax. Enter marina between two wavebreak barges; access H24, 5m at ent. Berth on 1st pontoon where HM assigns a berth via an intercom ☎. ⚓ ⛽ Ⓛ ACA ⚒ Ⓑ Ⓔ ✗ ⚓ ⚓ ⚑(70t) ⚓(30t) ⚓. **YC Breskens** ☎383278, ✗ ⌸ (book early) @.

Town ✗ ⌸ 🛒 ✉ Ⓑ ⚓ ✈ (Oostende, Antwerp or Brussels). Pedestrian & bike ferry or car tunnel to Vlissingen for ⇌.

9.16.30 TERNEUZEN

Zeeland **51°20'·58N 03°49'·69E** ❀❀❀◊◊◊◊❀❀

CHARTS AC 120; Zeekaart 1443; BE 103, 105; DYC 1803.2; Imray C30

TIDES +0230 Dover; ML 2·5; Duration 0555

Standard Port VLISSINGEN (←—)

Times				Height (metres)			
High Water		Low Water		MHWS	MHWN	MLWN	MLWS
0300	0900	0400	1000	4·9	4·1	1·0	0·5
1500	2100	1600	2200				
Differences TERNEUZEN							
+0020	+0020	+0020	+0030	+0·4	+0·3	+0·1	0·0

SHELTER Very good except in strong N'lies. Exposed ⌕ on N side of fairway between buoys WPT4, WPT6, PvT-ZE & ZE5. Strong currents.

NAVIGATION WPT 51°20'·82N 03°47'·96E (25B SHM buoy), 101°/1M hugging S bank to hbr ent. Fairway only 500m wide; big ships pass very close; lookout E/W and call *Traffic Centre Terneuzen*. Ch 03, when leaving. Very busy traffic from/to the locks; call *Port Control Terneuzen*, Ch 11. If S-bound via Gent Canal transit the E lock to berth in Zijkanaal A. Strong eddy during ebb at Veerhaven ent.

LIGHTS AND MARKS Dow Chemical works and storage tanks are conspic 2M W of hbr. Lt Oc WRG 5s, B/W post, on W mole is a conspic mark for the Veerhaven.

When entry to E Buitenhaven is prohib, a second ● is shown below FR on E mole. For E lock: ● = no entry; ●● (vert) = get ready; ● = go. No yachts in W Buitenhaven, Middle and W locks.

COMMUNICATIONS (Code 0115) Police 0900 8844; ⊖ (0118) 484600; British Consul (020) 676 4343; Ⓗ 688000; Dr 616262; Port HM 612161, call *Port Control* Ch 11 (H24) for locks and Gent canal; also info broadcasts every H+00; East lock VHF Ch 18; No marina VHF.

FACILITIES Two marinas in SE corner of former Veerhaven: **WV Honte Marina** where Ⓥ berths more likely. ☎697089, mob 0651 168987. www.jachthaventerneuzen.nl 130 ⌆ €1.40. 🔌 (230v) €1·25, ⛽ ⚓(6t).
WV Neusen Marina is close ESE (and in Zijkanaal A) ☎696331; www.wvneusen.nl 100 ⌆ €1·00, ⚓ ⛽ no wi-fi. Boatyards (full services): **Aricom** www.aricom.nl ☎614577, €1·00, ⛽ ⚓ Ⓑ ✗ ⚓(50t). **Vermeulen's Yachtwerf** ☎612716, ⌆ €0.75, ⚓(50t).
Town 🏨 🏨 ✗ ⌸ 🛒 Ⓛ ✉ Ⓑ ✈ (Antwerpen). Foot ferry Breskens to Vlissingen. Cars by Schelde tunnel Terneuzen-Ellewoutsdijk.

9.16.31 SPECIAL NOTES FOR BELGIUM

PROVINCES are given in lieu of UK 'counties'.

LANGUAGES Flemish, ie Dutch as spoken in Flanders, is the 1st language along the coast and N of Brussels. English is also widely spoken. French, the 2nd language, is spoken in and S of Brussels.

CURRENCY is the Euro €. VAT (BTW/TVA) is 21%.

CHARTS 'Vlaamse Banken' (BE) issued by the Hydrografische Dienst der Kust are most widely used. Dutch yacht charts (DYC 1800 series) cover all Belgian ports. Imray C30 is also popular.

TIME ZONE is –0100, which is allowed for in tidal predictions but no provision is made for daylight saving schemes.

HARBOURS Ports of entry are Nieuwpoort, Oostende and Zeebrugge; Blankenberge, early April – late Sep. Berths and moorings are administered by local YCs and municipal authorities.

SIGNALS At Oostende and Zeebrugge IPTS are used, and small craft wind warnings (also at Nieuwpoort and Blankenberge). They apply only to craft <6m LOA and indicate onshore wind >F3; offshore wind >F4. Day, 2 ▼s, points together; night, Fl Bu lt.

TELEPHONE To call UK from Belgium, dial 00 44 then the UK area code minus the prefix 0, followed by the number called. To call Belgium from UK, dial 00 32 then the code and 6/7 digit number.

Emergency: ☎101 Police; ☎100 Fire, Ambulance and Marine. ☎112 (EU emergency number) is also operational.

CONSUL British Embassy, Brussels, (consular) 02 287 6248.

MRCC Oostende coordinates SAR operations (5.14). If no contact, call *Oostende Radio* VHF Ch 16 or ☎100. For medical advice call *Radiomédical Oostende* on Ch 16.

PUBLIC HOLIDAYS New Year's Day, Easter Mon, Labour Day (1 May) Ascension Day, Whit Mon, National Day (21 July) Feast of the Assumption (15 Aug) All Saints' Day (1 Nov) Armistice Day (11 Nov) King's Birthday (15 Nov) Christmas Day.

RULES A ▼ when motor-sailing and a black ball ● at ⚓ are strictly enforced. Navigation within 200m of shore (MLWS) and transit through restricted areas around wind farms is prohibited. Red diesel in pleasure craft is illegal in Belgium, see 2.3.2. Ship Radio Licences must be up-to-date and cover all relevant equipment including radar.

INLAND WATERWAYS At www.mobilit.fgov.be download CEVNI-based regulations in French or Dutch. Licence plates (*Immatriculatieplaat*) are required on Belgian waterways. Helmsman's Competence criteria are as for the Netherlands.

THE BELGIAN COAST

Long shoals lie roughly parallel to this 36M long coast (AC 1872). Mostly the deeper, buoyed channels run within 3M of shore, and the outer shoals can give some protection from strong W or SW winds. Strong W to NE winds can be dangerous especially with wind against tide. When approaching, obtain a good fix to correctly identify the required channel.

▶*Off the Belgian coast the E-going stream begins at HW Vlissingen –0320 (HW Dover –0120), and the W-going at HW Vlissingen +0240 (HW Dover +0440), sp rates 2kn. Mostly the streams run parallel with the coast.*◀

From **Zeebrugge** stay a mile offshore inside Wenduine Bank to pass **Oostende**, thence via Kleine Rede or Grote Reede into West Diep off **Nieuwpoort**. At the French border West Diep becomes the narrower, buoyed Passe de Zuydcoote. Thence the very well buoyed route runs close inshore for 25M to Dyck PHM buoy.

Conversely, E-bound from the Thames, if bound for Oostende or the Westerchelde, identify W Hinder lt. From further N, route via NHR-S and NHR-SE buoys or the N Hinder lt buoy. Enter the buoyed channels at Dyck.

Leave Oostende about ▶*HW Vlissingen –0300 on a fair tide*◀. Keep 2M off **Blankenberge**, and 1M or more off Zeebrugge's huge claw-like breakwaters, staying S of the Scheur channel to avoid much commercial traffic. Beware the strong tidal stream and possibly dangerous seas off Zeebrugge.

9.16.32 ANTWERPEN

Belgium, Antwerpen 51°13'·66N 04°23'·79E ❄❄⌂⌂♠♠♠♠♠

CHARTS AC 128; Zeekaart 1443; BE 103, 104; DYC 1803.5

TIDES +0342 Dover; ML 2·9; Duration 0605

Standard Port VLISSINGEN (←—)

Times				Height (metres)			
High Water		Low Water		MHWS	MHWN	MLWN	MLWS
0300	0900	0400	1000	4·9	4·1	1·0	0·5
1500	2100	1600	2200				
Differences ANTWERPEN							
+0128	+0116	+0121	+0144	+1·2	+1·0	+0·1	+0·1

SHELTER Excellent in both marinas. ⚓ in the river is not advised.

NAVIGATION See Westerschelde. Best to check off the buoys coming upriver. There is a gap of 1·4M between No 116 PHM buoy and No 107 SHM buoy. Entrance to the Willemdok marina via Royersluis and Siberia bridge is permitted, but not recommended. Use the Kattendijksluis (abeam buoy 107). VHF 69, operating if tide between TAW 2.2–5.0m, or approximately 3±HW.

COMMUNICATIONS (Code 03) VTS, Zandvliet Centre VHF Ch 12, lock Ch 79; Radar VHF Ch 04, 66 ⊜ 2292004; Police 5460730; Ⓗ 2852000 (W Bank) 234111 (E Bank); Antwerp Port Ops Ch 74 (H24); Royersluis VHF Ch 22; Access bridges VHF Ch 62; Willemdok VHF Ch 23; Linkeroever VHF Ch 09 (HW±1).

FACILITIES from seaward: **Willemdok Marina** 51°13'·78N 04°24'·43E. ☎2315066, mob 0495 535455. www.jachthaven-antwerpen.be jaw@pandora.be. Wait on inshore side of T-shaped ferry pontoon S of No. 107 buoy. H24 access via Royersluis, or from canals/docks. Siberia and Londen bridges open 0630, 0830, 1000, 1130, 1245, 1415, 1515, 1615*, 1730, 1845, 2015, 2145, 2245LT.*Sat/Sun/Public hols. To enter the Dock areas and marina, pre-arrange an FD number via ☎, e-mail or fax to Willemdok (0500-2300). 200 + Ⓥ, €1·50, €8·00 week. ♣ 🛒 🅞 Ⓓ.

Linkeroever Marina (W bank); lat/long under title. Access by gate HW ±1, 0800–2200 (1800 in winter). Y waiting buoy is off the ent. ☎2190895, mob 0475 643957. www.jachthaven-antwerpen.be jachthaven_linkeroever@skynet.be 200+Ⓥ, €1·50, €8·00 week. ♣ ✕ 🛒 🍴 🅞 Ⓓ ⚒ 🏴(38t) ⚓(1·5t) ━.

Royal YC van België ☎2195231. www.rycb.be rycb@rycb.be. ♣ 👤 ✕ 🔲 🅞 🍴 ━.

Kon. Liberty YC ☎2191147. **Services** ⚒ 🏴 ACA, DYC Agent.

City centre is ¾M from Linkeroever via St. Annatunnel (pedestrian) 51°13'·22N. All facilities, @ ⇌ ✈.

9.16.33 ZEEBRUGGE

Belgium, West Flanders **51°21'·83N 03°11'·39E** ❀❀❀♦♦♦❀❀

CHARTS AC 2449, 3371, 1872, 1874; Zeekaart 1441; BE 104; DYC 1801.3, 1803; Imray C30

TIDES +0135 Dover; ML 2·4; Duration 0535. Note: Zeebrugge is a Standard Port (⟶)

SHELTER Very good in the marina, access H24.

NAVIGATION WPT 51°22'·48N 03°09'·95E (Z SHM buoy, QG), 129°/1·1M to ent (lat/long under title). Lts on outer bkwtr heads show high-vis strip lts to seaward. Beware strong currents in hbr apprs (up to 4kn at HW −1). Caution on ent/dep due to limited vis; give all jetties a wide berth. A busy commercial, naval and fishing port and a ferry terminal. An inner WPT 51°20'·85N 03°12'·29E off Leopold II Dam head is useful to avoid getting 'lost' inside the vast outer hbr, especially in poor vis; see chartlet.

LIGHTS AND MARKS Big ship ldg marks (hard to see by day) and lts (as chartlet) lead in sequence to Vissershaven and the marina:

- **136°**: Both W cols, R bands.
- **154°**: Front, W pylon, R bands; rear, W bldg, R bands.
- **220°**: Both W cols, B bands. • **193°**: Both W cols, R bands.

IPTS are shown from E and W outer bkwtrs and Leopold II Dam. Extra sigs: 3 Fl Y lts (vert) + IPTS No 2 or 5 = LNG tanker entering/leaving. No exit/entry without specific permission.

COMMUNICATIONS (Code 050) Coast Guard 545072; Sea Saving Service 544007; Police 544148; ⊕ 545455; British Consul 02 287 6248; Dr 544590; Port HM 543241; Port Control 546867/550801, VHF Ch 71 (H24); Lock Mr 543231, VHF Ch 68; Marina, no VHF.

FACILITIES Royal Belgian SC (VZW) ☎544903, mob 0496 789053. havenmeester@rbsc.be www.rbsc.be 100 + 100Ⓥ €2·64. ⊡ (on coin meter) ⚓ ⚒ ⬛ ⛽ (0830-1200, 1300-1800) ⚔ ⬜ ⚒ ⊞ ⊡ (Clubhouse Alberta) ☎544197.

BZYC Tijdokhaven Marina, www.bzyc.be 94 inc Ⓥ (max 20m) €2·68 inc ⚓ ⊡ ⊳ wi-fi. Reports welcome.

Town ⛽ ⛽ ⚔ ⬜ 🛒 ⬛ ⊠ Ⓑ ⚔ ⚆ 15 mins to Brugge. Tram to Oostende for ✈. Ferries to Hull and Rosyth (Edinburgh).

BRUGGE Masted yachts can transit Zeebrugge's eastern sealock to go 6M S via the Boudewijnkanaal (Ch 68) & 2 lifting bridges to Brugge docks. **Brugse Zeil en Yacht Club (BZYC)** is on the E bank: ☎(0478) 711301. www.bzyc.be info@bzyc.be €1·00/m; ⚓ & ⊡ €2·00/day. 15 mins walk to city centre. Unmasted boats can go clockwise on the Oostende-Gent Canal (Ch 18) to **Flandria Marina** (S of city): ☎380866, mob 0477 384456. www.yachtclubflandria.be 10 mins walk to city centre; close to ⚆ (quick and easy by train).

ZEEBRUGGE

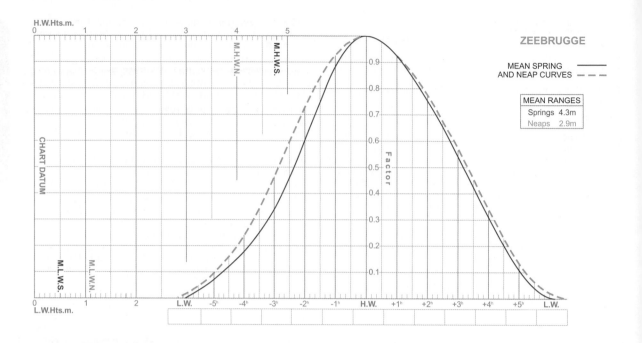

ZEEBRUGGE

MEAN SPRING ───
AND NEAP CURVES ───

MEAN RANGES	
Springs	4.3m
Neaps	2.9m

DISTANCE TABLE
Approximate distances in nautical miles are by the most direct route while avoiding dangers and allowing for TSS. Places in *italics* are in adjoining areas; places in **bold** are in 9.0.7. Distances across the North Sea.

1	*Borkum*	**1**																										
2	**Delfzijl**	22	**2**																									
3	Lauwersoog	30	46	**3**																								
4	Terschelling	65	85	43	**4**																							
5	Vlieland	72	92	52	7	**5**																						
6	Harlingen	80	102	67	19	18	**6**																					
7	Kornwerderzand	86	106	70	24	19	7	**7**																				
8	Den Oever	90	110	88	34	33	21	13	**8**																			
9	Oudeschild	94	114	79	38	34	25	18	10	**9**																		
10	**Den Helder**	95	115	81	39	33	30	24	11	5	**10**																	
11	Medemblik	104	124	99	39	40	27	20	12	22	24	**11**																
12	Enkhuizen	112	132	106	47	41	28	23	18	28	29	10	**12**															
13	Amsterdam	139	159	123	83	84	81	47	62	52	51	38	26	**13**														
14	IJmuiden	126	146	109	70	71	68	57	49	38	38	58	37	13	**14**													
15	Scheveningen	151	171	132	95	96	93	79	74	60	63	80	62	38	25	**15**												
16	Rotterdam	185	205	159	129	130	127	113	108	94	97	114	96	72	59	34	**16**											
17	Hook of Holland	165	185	146	109	110	107	93	88	74	77	94	76	52	39	14	20	**17**										
18	Stellendam	181	201	157	125	126	123	109	104	90	93	110	92	68	55	30	36	16	**18**									
19	**Roompotsluis**	193	204	177	128	124	129	118	109	110	97	130	112	90	63	50	49	29	27	**19**								
20	Vlissingen	211	222	190	146	143	147	137	128	121	115	141	123	108	82	61	67	47	45	21	**20**								
21	Terneuzen	220	240	202	164	165	162	152	143	133	132	153	135	111	98	73	79	59	57	36	12	**21**							
22	Antwerpen	249	269	221	193	194	191	181	172	162	161	182	164	140	127	102	108	88	86	65	41	31	**22**						
23	Breskens	202	230	192	154	155	152	142	133	123	122	143	125	101	88	63	69	49	47	21	3	12	45	**23**					
24	**Zeebrugge**	219	239	194	163	164	161	147	142	128	131	148	130	106	93	68	74	54	50	25	16	28	57	13	**24**				
25	Blankenberge	224	244	196	168	169	166	152	147	133	136	153	135	111	98	73	79	59	55	33	21	33	62	18	5	**25**			
26	Oostende	233	239	205	163	164	161	160	142	141	131	161	143	110	106	81	87	67	72	36	27	41	70	26	13	9	**26**		
27	Nieuwpoort	242	262	215	186	187	184	170	165	151	154	171	153	129	116	91	97	77	83	46	39	51	80	36	23	18	9	**27**	
28	*Dunkerque*	257	277	228	201	202	199	185	180	166	169	186	168	144	131	106	112	92	90	57	50	67	96	52	40	35	26	15	**28**

STANDARD TIME UT –01
Subtract 1 hour for UT
For Belgian Summer Time add
ONE hour in **non-shaded areas**

ZEEBRUGGE LAT 51°21'N LONG 3°12'E
TIMES AND HEIGHTS OF HIGH AND LOW WATERS

Dates in red are **SPRINGS**
Dates in blue are **NEAPS**

YEAR 2016

Belgium

JANUARY

Time	m		Time	m
1 0539	4.2	**16** 0551	4.8
1149	0.9	1230	0.2
F 1811	4.2	SA 1825	4.9
2 0002	1.3	**17** 0051	0.6
0630	4.0	0651	4.7
SA 1245	1.1	SU 1324	0.3
◑ 1908	4.0	● 1928	4.7
3 0106	1.5	**18** 0149	0.7
0732	3.8	0756	4.5
SU 1412	1.2	M 1424	0.5
2015	3.8	2034	4.5
4 0252	1.6	**19** 0254	0.8
0845	3.6	0904	4.4
M 1532	1.3	TU 1533	0.6
2134	3.7	2143	4.4
5 0410	1.5	**20** 0413	0.8
1006	3.6	1014	4.4
TU 1635	1.2	W 1648	0.7
2248	3.8	2250	4.4
6 0511	1.3	**21** 0525	0.8
1113	3.8	1119	4.5
W 1730	1.2	TH 1751	0.7
2343	4.0	2349	4.5
7 0603	1.1	**22** 0621	0.7
1203	4.0	1214	4.6
TH 1816	1.1	F 1842	0.7
8 0026	4.2	**23** 0038	4.6
0645	1.0	0707	0.6
F 1242	4.2	SA 1300	4.7
1856	1.0	1926	0.8
9 0102	4.4	**24** 0121	4.7
0723	0.9	0749	0.6
SA 1316	4.4	SU 1341	4.8
1933	0.9	○ 2004	0.8
10 0134	4.6	**25** 0159	4.7
0800	0.7	0826	0.6
SU 1350	4.7	M 1419	4.8
● 2011	0.8	2039	0.9
11 0207	4.7	**26** 0236	4.7
0839	0.6	0901	0.6
M 1425	4.8	TU 1457	4.8
2051	0.7	2110	0.9
12 0244	4.9	**27** 0313	4.7
0920	0.5	0934	0.5
TU 1505	5.0	W 1535	4.8
2134	0.6	2140	0.9
13 0324	4.9	**28** 0350	4.6
1004	0.3	1006	0.5
W 1548	5.1	TH 1614	4.7
2220	0.6	2213	0.9
14 0409	5.0	**29** 0428	4.5
1050	0.3	1041	0.6
TH 1635	5.1	F 1654	4.5
2308	0.5	2250	0.9
15 0457	4.9	**30** 0507	4.4
1139	0.2	1120	0.7
F 1727	5.0	SA 1735	4.3
2358	0.5	2330	1.1
		31 0549	4.1
		1203	0.9
		SU 1821	4.1

FEBRUARY

Time	m		Time	m
1 0017	1.3	**16** 0118	0.6
0637	3.9	0724	4.5
M 1302	1.2	TU 1354	0.6
◑ 1915	3.8	2002	4.3
2 0129	1.6	**17** 0225	0.9
0739	3.6	0835	4.2
TU 1443	1.4	W 1507	0.8
2028	3.5	2116	4.1
3 0329	1.6	**18** 0352	1.0
0904	3.4	0952	4.1
W 1558	1.4	TH 1629	0.9
2201	3.5	2232	4.1
4 0437	1.5	**19** 0507	0.9
1035	3.6	1105	4.2
TH 1657	1.3	F 1735	0.9
2312	3.8	2335	4.2
5 0534	1.3	**20** 0605	0.8
1134	3.9	1201	4.4
F 1748	1.1	SA 1828	0.9
6 0000	4.1	**21** 0024	4.4
0621	1.0	0652	0.7
SA 1217	4.2	SU 1246	4.6
1830	1.0	1910	0.9
7 0037	4.4	**22** 0104	4.5
0659	0.8	0733	0.6
SU 1253	4.5	M 1324	4.7
1909	0.8	○ 1947	0.9
8 0110	4.6	**23** 0140	4.6
0737	0.5	0808	0.6
M 1328	4.8	TU 1400	4.7
● 1948	0.6	2019	0.9
9 0145	4.9	**24** 0215	4.6
0817	0.3	0841	0.5
TU 1406	5.1	W 1436	4.8
2030	0.4	2049	0.8
10 0224	5.1	**25** 0251	4.7
0859	0.1	0912	0.5
W 1447	5.3	TH 1512	4.8
2115	0.3	2119	0.8
11 0306	5.2	**26** 0326	4.7
0944	0.0	0944	0.5
TH 1530	5.3	F 1549	4.7
2201	0.2	2152	0.8
12 0350	5.3	**27** 0401	4.6
1030	-0.1	1018	0.5
F 1616	5.3	SA 1625	4.6
2247	0.2	2227	0.8
13 0437	5.2	**28** 0436	4.4
1116	-0.1	1053	0.7
SA 1705	5.1	SU 1700	4.4
2335	0.3	2302	1.0
14 0527	5.0	**29** 0511	4.2
1204	0.1	1128	0.9
SU 1758	4.9	M 1737	4.1
		2338	1.2
15 0024	0.4		
0622	4.8		
M 1256	0.3		
◑ 1857	4.6		

MARCH

Time	m		Time	m
1 0550	4.0	**16** 0049	0.7
1207	1.1	0655	4.5
TU 1820	3.8	W 1324	0.8
		1932	4.2
2 0021	1.5	**17** 0156	1.0
0640	3.7	0806	4.1
W 1312	1.4	TH 1442	1.1
◑ 1920	3.5	2048	3.9
3 0241	1.7	**18** 0327	1.1
0756	3.5	0928	4.0
TH 1520	1.3	F 1606	1.2
2100	3.4	2209	3.9
4 0400	1.5	**19** 0443	1.0
0945	3.5	1045	4.1
F 1622	1.4	SA 1714	1.1
2231	3.6	2315	4.0
5 0459	1.3	**20** 0543	0.9
1059	3.9	1142	4.3
SA 1716	1.2	SU 1808	1.0
2326	4.0		
6 0549	1.0	**21** 0004	4.2
1146	4.3	0631	0.8
SU 1802	0.9	M 1226	4.4
		1851	1.0
7 0006	4.4	**22** 0043	4.4
0631	0.6	0712	0.7
M 1226	4.7	TU 1303	4.6
1843	0.7	1926	0.9
8 0044	4.8	**23** 0119	4.5
0711	0.3	0746	0.6
TU 1304	5.0	W 1338	4.7
1924	0.4	○ 1958	0.8
9 0122	5.1	**24** 0153	4.6
0753	0.1	0818	0.6
W 1344	5.3	TH 1413	4.7
● 2008	0.2	2028	0.8
10 0202	5.3	**25** 0228	4.6
0836	-0.1	0850	0.5
TH 1425	5.4	F 1449	4.7
2053	0.0	2100	0.7
11 0244	5.5	**26** 0302	4.6
0921	-0.2	0923	0.5
F 1509	5.5	SA 1524	4.6
2139	0.0	2134	0.7
12 0329	5.5	**27** 0335	4.5
1007	-0.2	0956	0.6
SA 1554	5.4	SU 1556	4.5
2225	0.2	2207	0.8
13 0415	5.4	**28** 0407	4.4
1052	-0.1	1028	0.8
SU 1642	5.2	M 1627	4.3
2311	0.1	2238	1.0
14 0503	5.1	**29** 0439	4.3
1139	0.1	1057	1.0
M 1732	4.9	TU 1700	4.1
2358	0.4	2307	1.2
15 0556	4.8	**30** 0515	4.1
1227	0.4	1128	1.2
TU 1828	4.5	W 1739	3.9
◑		2342	1.3
		31 0602	3.9
		1213	1.4
		TH 1834	3.7
		◑	

APRIL

Time	m		Time	m
1 0047	1.5	**16** 0255	1.2
0711	3.7	0857	3.9
F 1432	1.5	SA 1535	1.4
2007	3.5	2136	3.8
2 0318	1.4	**17** 0410	1.1
0859	3.7	1016	4.0
SA 1544	1.3	SU 1644	1.3
2146	3.7	2246	3.9
3 0418	1.1	**18** 0513	1.0
1019	4.0	1116	4.2
SU 1639	1.1	M 1741	1.1
2247	4.1	2338	4.1
4 0511	0.8	**19** 0604	0.8
1113	4.4	1201	4.3
M 1729	0.8	TU 1826	1.0
2334	4.5		
5 0559	0.5	**20** 0019	4.3
1157	4.8	0645	0.7
TU 1816	0.5	W 1240	4.5
		1902	0.9
6 0016	4.9	**21** 0056	4.4
0643	0.2	0721	0.7
W 1239	5.2	TH 1316	4.6
1900	0.3	1935	0.8
7 0057	5.2	**22** 0131	4.5
0728	-0.1	0754	0.6
TH 1320	5.4	F 1351	4.6
● 1945	0.1	○ 2008	0.8
8 0139	5.4	**23** 0206	4.5
0813	-0.2	0828	0.6
F 1403	5.5	SA 1426	4.6
2031	-0.1	2042	0.7
9 0223	5.5	**24** 0240	4.5
0858	-0.3	0902	0.7
SA 1447	5.5	SU 1500	4.5
2117	-0.1	2117	0.8
10 0307	5.5	**25** 0312	4.4
0943	-0.2	0936	0.8
SU 1533	5.3	M 1531	4.4
2202	0.0	2150	0.9
11 0353	5.4	**26** 0342	4.4
1028	0.0	1007	0.9
M 1619	5.1	TU 1600	4.3
2247	0.2	2220	1.0
12 0441	5.1	**27** 0414	4.3
1113	0.3	1034	1.1
TU 1708	4.8	W 1633	4.2
2333	0.4	2248	1.1
13 0532	4.8	**28** 0452	4.2
1200	0.6	1105	1.2
W 1801	4.5	TH 1714	4.0
		2325	1.2
14 0022	0.7	**29** 0541	4.1
0629	4.5	1154	1.3
TH 1254	1.0	F 1810	3.9
◑ 1901	4.1		
15 0126	1.0	**30** 0040	1.3
0737	4.1	0650	4.0
F 1409	1.3	SA 1340	1.3
2014	3.8	◑ 1936	3.8

Chart Datum is 0·23 metres below TAW Datum. HAT is 5·6 metres above Chart Datum.

STANDARD TIME UT –01
Subtract 1 hour for UT
For Belgian Summer Time add
ONE hour in **non-shaded areas**

ZEEBRUGGE LAT 51°21'N LONG 3°12'E
TIMES AND HEIGHTS OF HIGH AND LOW WATERS

Dates in red are **SPRINGS**
Dates in blue are **NEAPS**

YEAR 2016

MAY

Time	m		Time	m
1 0232	1.1	**16** 0328	1.1
0824	4.0	0934	3.9
SU 1501	1.2	M 1603	1.4
2104	3.9	2207	3.8
2 0336	0.9	**17** 0433	1.0
0940	4.2	1042	4.0
M 1601	0.9	TU 1705	1.2
2209	4.2	2306	3.9
3 0432	0.6	**18** 0529	0.9
1039	4.6	1133	4.2
TU 1656	0.7	W 1756	1.1
2302	4.6	2352	4.1
4 0526	0.3	**19** 0615	0.8
1128	4.9	1215	4.3
W 1749	0.4	TH 1838	0.9
2349	4.9		
5 0617	0.1	**20** 0032	4.2
1214	5.1	0654	0.8
TH 1838	0.2	F 1254	4.4
		1914	0.9
6 0033	5.2	**21** 0110	4.3
0704	0.0	0730	0.8
F 1258	5.3	SA 1331	4.5
● 1925	0.1	○ 1950	0.8
7 0118	5.4	**22** 0146	4.4
0750	-0.1	0806	0.8
SA 1343	5.4	SU 1406	4.5
2011	0.0	2026	0.8
8 0203	5.4	**23** 0219	4.4
0836	-0.1	0841	0.8
SU 1427	5.3	M 1439	4.4
2056	0.0	2102	0.9
9 0248	5.4	**24** 0251	4.3
0920	0.0	0916	0.9
M 1513	5.2	TU 1509	4.3
2141	0.1	2137	0.9
10 0334	5.3	**25** 0323	4.3
1004	0.3	0950	1.0
TU 1559	5.0	W 1540	4.3
2225	0.2	2211	1.0
11 0422	5.0	**26** 0357	4.4
1048	0.5	1022	1.1
W 1646	4.7	TH 1616	4.2
2309	0.5	2246	1.0
12 0511	4.8	**27** 0439	4.4
1131	0.8	1101	1.1
TH 1735	4.4	F 1700	4.2
2355	0.7	2333	1.0
13 0604	4.5	**28** 0530	4.3
1219	1.1	1157	1.1
F 1830	4.1	SA 1757	4.1
◐			
14 0052	0.9	**29** 0043	0.9
0704	4.2	0636	4.3
SA 1323	1.3	SU 1313	1.0
1934	3.8	◑ 1912	4.0
15 0212	1.1	**30** 0155	0.8
0815	3.9	0755	4.3
SU 1450	1.4	M 1422	0.9
2050	3.7	2028	4.1
		31 0257	0.6
		0905	4.4
		TU 1524	0.7
		2133	4.3

JUNE

Time	m		Time	m
1 0356	0.4	**16** 0448	1.0
1006	4.6	1100	3.9
W 1625	0.6	TH 1722	1.2
2231	4.6	2324	3.8
2 0456	0.3	**17** 0542	1.0
1101	4.8	1150	4.1
TH 1725	0.4	F 1813	1.0
2324	4.8		
3 0553	0.2	**18** 0010	4.0
1152	5.0	0627	0.9
F 1819	0.3	SA 1233	4.2
		1854	0.9
4 0013	5.0	**19** 0050	4.1
0644	0.1	0707	0.9
SA 1240	5.1	SU 1311	4.3
1907	0.2	1933	0.9
5 0100	5.2	**20** 0127	4.2
0731	0.1	0744	0.9
SU 1326	5.1	M 1346	4.3
● 1953	0.2	○ 2009	0.8
6 0147	5.2	**21** 0200	4.3
0816	0.2	0821	0.9
M 1411	5.1	TU 1418	4.4
2038	0.2	2046	0.8
7 0232	5.2	**22** 0233	4.4
0859	0.3	0857	0.9
TU 1455	5.0	W 1450	4.4
2121	0.2	2124	0.8
8 0318	5.1	**23** 0307	4.4
0942	0.5	0935	0.9
W 1539	4.8	TH 1524	4.4
2204	0.3	2203	0.8
9 0403	4.9	**24** 0344	4.5
1022	0.7	1014	0.9
TH 1624	4.6	F 1603	4.4
2245	0.5	2246	0.7
10 0449	4.7	**25** 0428	4.6
1102	0.9	1100	0.9
F 1709	4.4	SA 1649	4.4
2327	0.6	2334	0.6
11 0537	4.5	**26** 0518	4.6
1142	1.1	1151	0.8
SA 1758	4.2	SU 1742	4.4
12 0014	0.8	**27** 0027	0.5
0629	4.2	0618	4.5
SU 1231	1.3	M 1249	0.7
◑ 1852	4.0	◑ 1846	4.4
13 0117	1.0	**28** 0124	0.4
0729	4.0	0725	4.5
M 1342	1.4	TU 1348	0.7
1956	3.7	1954	4.4
14 0235	1.1	**29** 0223	0.4
0839	3.8	0832	4.5
TU 1509	1.4	W 1450	0.6
2112	3.6	2059	4.4
15 0346	1.1	**30** 0325	0.3
0956	3.8	0935	4.6
W 1621	1.3	TH 1557	0.6
2226	3.7	2202	4.5

JULY

Time	m		Time	m
1 0431	0.3	**16** 0509	1.2
1036	4.6	1124	3.8
F 1705	0.5	SA 1747	1.2
2302	4.6	2346	3.8
2 0534	0.3	**17** 0601	1.1
1133	4.7	1211	4.0
SA 1804	0.4	SU 1835	1.0
2357	4.8		
3 0628	0.3	**18** 0029	4.0
1225	4.8	0644	1.0
SU 1855	0.4	M 1251	4.2
		1914	0.9
4 0047	4.9	**19** 0106	4.2
0716	0.4	0722	1.0
M 1312	4.8	TU 1324	4.3
● 1940	0.3	○ 1950	0.8
5 0134	5.0	**20** 0139	4.4
0759	0.5	0758	0.9
TU 1356	4.8	W 1356	4.5
2022	0.3	2026	0.7
6 0218	5.0	**21** 0213	4.6
0841	0.6	0836	0.8
W 1438	4.8	TH 1430	4.6
2103	0.4	2105	0.6
7 0300	5.0	**22** 0249	4.7
0919	0.7	0917	0.7
TH 1519	4.7	F 1506	4.7
2143	0.4	2147	0.4
8 0342	4.9	**23** 0329	4.8
0956	0.8	1000	0.6
F 1600	4.6	SA 1547	4.8
2221	0.4	2231	0.3
9 0425	4.7	**24** 0412	4.9
1031	0.9	1046	0.6
SA 1641	4.5	SU 1632	4.8
2258	0.5	2317	0.2
10 0508	4.5	**25** 0501	4.9
1107	1.0	1134	0.5
SU 1724	4.3	M 1722	4.7
2338	0.6		
11 0553	4.3	**26** 0005	0.2
1148	1.1	0555	4.8
M 1811	4.1	TU 1224	0.5
		1819	4.6
12 0026	0.8	**27** 0057	0.2
0643	4.1	0656	4.7
TU 1241	1.2	W 1319	0.6
◑ 1904	3.9	◑ 1922	4.5
13 0132	1.0	**28** 0153	0.3
0743	3.8	0800	4.5
W 1402	1.4	TH 1420	0.6
2010	3.6	2028	4.4
14 0256	1.2	**29** 0257	0.4
0858	3.6	0906	4.4
TH 1536	1.4	F 1533	0.7
2134	3.5	2137	4.4
15 0407	1.2	**30** 0410	0.5
1021	3.7	1015	4.4
F 1647	1.3	SA 1649	0.7
2251	3.6	2245	4.4
		31 0519	0.6
		1120	4.5
		SU 1751	0.6
		2346	4.6

AUGUST

Time	m		Time	m
1 0615	0.6	**16** 0004	4.0
1214	4.6	0619	1.1
M 1843	0.5	TU 1225	4.2
		1849	0.9
2 0037	4.7	**17** 0041	4.3
0702	0.6	0657	0.9
TU 1300	4.7	W 1258	4.4
● 1927	0.5	1925	0.7
3 0121	4.8	**18** 0115	4.6
0744	0.7	0734	0.8
W 1341	4.7	TH 1331	4.7
2007	0.5	○ 2002	0.5
4 0202	4.9	**19** 0150	4.6
0822	0.8	0813	0.6
TH 1419	4.7	F 1406	4.9
2045	0.4	2042	0.3
5 0241	4.9	**20** 0228	5.1
0857	0.8	0855	0.4
F 1457	4.7	SA 1445	5.1
2121	0.4	2124	0.1
6 0319	4.8	**21** 0308	5.2
0930	0.8	0939	0.3
SA 1535	4.7	SU 1527	5.1
2155	0.4	2208	0.0
7 0358	4.8	**22** 0352	5.2
1002	0.8	1025	0.3
SU 1613	4.6	M 1612	5.1
2229	0.5	2254	0.0
8 0438	4.6	**23** 0439	5.1
1036	0.8	1111	0.3
M 1652	4.5	TU 1700	5.0
2306	0.6	2340	0.1
9 0518	4.4	**24** 0530	5.0
1115	0.9	1159	0.4
TU 1732	4.3	W 1753	4.9
2346	0.8		
10 0601	4.2	**25** 0030	0.2
1158	1.1	0627	4.7
W 1817	4.0	TH 1251	0.5
◐		◑ 1853	4.7
11 0036	1.0	**26** 0125	0.4
0650	3.9	0731	4.5
TH 1255	1.4	F 1353	0.7
1911	3.7	2001	4.4
12 0159	1.3	**27** 0233	0.7
0755	3.6	0840	4.3
F 1452	1.6	SA 1513	0.9
2029	3.4	2114	4.3
13 0332	1.4	**28** 0352	0.8
0932	3.4	0955	4.2
SA 1615	1.5	SU 1632	0.9
2213	3.4	2230	4.3
14 0438	1.4	**29** 0503	0.8
1053	3.6	1105	4.3
SU 1718	1.3	M 1735	0.7
2319	3.7	2334	4.5
15 0534	1.3	**30** 0600	0.8
1145	3.9	1200	4.5
M 1810	1.1	TU 1827	0.6
		31 0023	4.7
		0647	0.8
		W 1244	4.6
		1911	0.6

Chart Datum is 0·23 metres below TAW Datum. HAT is 5·6 metres above Chart Datum.

Belgium

STANDARD TIME UT –01
Subtract 1 hour for UT
For Belgian Summer Time add ONE hour in **non-shaded areas**

ZEEBRUGGE LAT 51°21′N LONG 3°12′E
TIMES AND HEIGHTS OF HIGH AND LOW WATERS

Dates in red are SPRINGS
Dates in blue are NEAPS

YEAR 2016

SEPTEMBER

Time	m		Time	m
1 0105	4.8	**16**	0048	4.9
0727	0.8		0707	0.6
TH 1322	4.7	F 1304	5.0
● 1949	0.6	○ 1934	0.3
2 0142	4.8	**17** 0125	5.2
0802	0.9	0748	0.4
F 1358	4.7	SA 1342	5.2
2024	0.5	2016	0.1
3 0219	4.9	**18** 0204	5.4
0834	0.8	0832	0.2
SA 1434	4.8	SU 1422	5.4
2057	0.5	2059	-0.1
4 0255	4.9	**19** 0246	5.4
0905	0.8	0917	0.1
SU 1510	4.8	M 1505	5.5
2129	0.5	2144	-0.1
5 0332	4.8	**20** 0330	5.4
0937	0.8	1002	0.1
M 1546	4.7	TU 1550	5.4
2202	0.5	2229	0.0
6 0408	4.7	**21** 0416	5.3
1011	0.8	1048	0.2
TU 1621	4.6	W 1637	5.3
2236	0.7	2315	0.1
7 0444	4.5	**22** 0506	5.0
1046	0.9	1135	0.4
W 1657	4.4	TH 1729	5.0
2312	0.9	
8 0521	4.2	**23** 0004	0.4
1123	1.2	0600	4.7
TH 1734	4.1	F 1226	0.6
2350	1.2	◐ 1828	4.7
9 0600	3.9	**24** 0059	0.7
1204	1.4	0703	4.4
F 1819	3.8	SA 1328	0.9
◐		1935	4.4
10 0043	1.5	**25** 0209	1.0
0652	3.6	0814	4.2
SA 1327	1.7	SU 1450	1.0
1924	3.5	2052	4.2
11 0256	1.6	**26** 0331	1.1
0831	3.4	0933	4.1
SU 1542	1.6	M 1609	1.0
2125	3.4	2212	4.3
12 0406	1.5	**27** 0442	1.1
1014	3.5	1046	4.2
M 1644	1.4	TU 1713	0.9
2243	3.7	2316	4.4
13 0503	1.3	**28** 0541	1.0
1111	3.9	1141	4.4
TU 1736	1.1	W 1806	0.8
2333	4.1	
14 0549	1.1	**29** 0004	4.6
1153	4.3	0628	1.0
W 1818	0.8	TH 1224	4.5
		1850	0.7
15 0012	4.5	**30** 0044	4.7
0629	0.9	0707	0.9
TH 1229	4.6	F 1301	4.6
1855	0.5	1928	0.7

OCTOBER

Time	m		Time	m
1 0121	4.8	**16** 0100	5.3
0741	0.9	0724	0.3
SA 1336	4.7	SU 1318	5.4
● 2001	0.6	○ 1950	0.0
2 0156	4.9	**17** 0141	5.5
0812	0.8	0809	0.1
SU 1411	4.8	M 1400	5.6
2033	0.6	2035	-0.1
3 0232	4.9	**18** 0224	5.6
0843	0.8	0854	0.0
M 1446	4.8	TU 1444	5.6
2105	0.6	2120	-0.1
4 0307	4.8	**19** 0309	5.5
0916	0.8	0940	0.1
TU 1520	4.7	W 1529	5.5
2138	0.7	2205	0.1
5 0341	4.6	**20** 0355	5.3
0949	0.9	1025	0.2
W 1553	4.6	TH 1617	5.3
2210	0.9	2251	0.3
6 0413	4.4	**21** 0444	5.0
1022	1.0	1112	0.4
TH 1625	4.4	F 1708	5.1
2241	1.1	2338	0.6
7 0445	4.2	**22** 0537	4.7
1052	1.2	1202	0.7
F 1700	4.2	SA 1806	4.8
2312	1.3	◐
8 0521	4.0	**23** 0032	0.9
1125	1.4	0637	4.4
SA 1741	4.0	SU 1302	0.9
2350	1.5	1910	4.4
9 0607	3.7	**24** 0140	1.2
1215	1.6	0745	4.1
SU 1840	3.7	M 1421	1.1
◐		2024	4.2
10 0156	1.7	**25** 0301	1.4
0729	3.5	0902	4.0
M 1502	1.6	TU 1537	1.1
2035	3.6	2144	4.2
11 0330	1.6	**26** 0412	1.3
0927	3.6	1018	4.1
TU 1604	1.4	W 1643	1.0
2201	3.9	2251	4.3
12 0426	1.3	**27** 0514	1.2
1031	4.0	1116	4.2
W 1656	1.0	TH 1739	0.9
2256	4.3	2341	4.5
13 0515	1.0	**28** 0605	1.1
1117	4.4	1200	4.4
TH 1743	0.7	F 1826	0.8
2340	4.7	
14 0559	0.8	**29** 0022	4.6
1158	4.8	0645	1.0
F 1825	0.4	SA 1239	4.5
		1904	0.8
15 0020	5.1	**30** 0059	4.7
0641	0.5	0720	0.9
SA 1237	5.1	SU 1315	4.6
1907	0.2	● 1938	0.8
		31 0135	4.8
		0753	0.9
		M 1350	4.7
		2011	0.8

NOVEMBER

Time	m		Time	m
1 0210	4.8	**16** 0205	5.5
0826	0.9	0833	0.1
TU 1424	4.7	W 1425	5.6
2044	0.8	2058	0.1
2 0245	4.7	**17** 0250	5.4
0900	0.9	0919	0.1
W 1458	4.6	TH 1512	5.5
2117	0.9	2143	0.3
3 0317	4.5	**18** 0337	5.2
0933	1.0	1004	0.2
TH 1530	4.5	F 1600	5.3
2149	1.1	2228	0.5
4 0347	4.4	**19** 0425	5.0
1004	1.2	1050	0.4
F 1601	4.4	SA 1650	5.1
2217	1.2	2313	0.8
5 0418	4.2	**20** 0515	4.7
1032	1.3	1138	0.6
SA 1635	4.3	SU 1743	4.8
2246	1.4	
6 0454	4.1	**21** 0002	1.1
1104	1.4	0609	4.4
SU 1718	4.2	M 1233	0.9
2325	1.5	◑ 1841	4.5
7 0541	3.9	**22** 0101	1.3
1158	1.5	0709	4.1
M 1817	4.0	TU 1342	1.1
●		1946	4.2
8 0045	1.6	**23** 0216	1.5
0655	3.7	0819	3.9
TU 1409	1.4	W 1456	1.1
1954	3.9	2102	4.1
9 0241	1.5	**24** 0331	1.5
0838	3.8	0937	3.9
W 1518	1.2	TH 1604	1.1
2117	4.1	2216	4.1
10 0343	1.2	**25** 0439	1.4
0947	4.1	1044	4.0
TH 1614	0.9	F 1705	1.1
2217	4.5	2313	4.3
11 0438	1.0	**26** 0537	1.2
1040	4.5	1135	4.2
F 1706	0.6	SA 1758	1.0
2307	4.8	2358	4.4
12 0529	0.7	**27** 0624	1.1
1127	4.8	1217	4.3
SA 1755	0.3	SU 1841	0.9
2352	5.1	
13 0616	0.4	**28** 0038	4.5
1211	5.2	0702	1.0
SU 1842	0.1	M 1255	4.4
		1917	0.9
14 0036	5.3	**29** 0116	4.6
0702	0.2	0738	0.9
M 1255	5.4	TU 1332	4.5
○ 1927	0.0	● 1952	0.9
15 0120	5.5	**30** 0152	4.6
0748	0.1	0812	0.9
TU 1340	5.6	W 1407	4.5
2012	0.0	2026	1.0

DECEMBER

Time	m		Time	m
1 0226	4.5	**16** 0234	5.2
0847	1.0	0901	0.2
TH 1440	4.5	F 1457	5.4
2100	1.1	2122	0.5
2 0258	4.4	**17** 0320	5.1
0922	1.0	0945	0.3
F 1511	4.4	SA 1543	5.2
2133	1.2	2205	0.6
3 0328	4.4	**18** 0405	4.9
0955	1.1	1029	0.4
SA 1544	4.4	SU 1630	5.0
2204	1.3	2247	0.8
4 0400	4.3	**19** 0451	4.7
1028	1.2	1113	0.6
SU 1620	4.4	M 1718	4.8
2237	1.3	2329	1.0
5 0438	4.2	**20** 0538	4.5
1107	1.2	1159	0.7
M 1705	4.3	TU 1808	4.5
2322	1.3	
6 0527	4.1	**21** 0015	1.2
1206	1.1	0629	4.2
TU 1802	4.3	W 1254	0.9
		◑ 1902	4.2
7 0031	1.3	**22** 0114	1.4
0632	4.0	0727	4.0
W 1322	1.0	TH 1402	1.1
◑ 1920	4.2	2007	4.0
8 0150	1.2	**23** 0232	1.5
0754	4.1	0838	3.8
TH 1430	0.9	F 1515	1.2
2035	4.3	2126	3.9
9 0257	1.1	**24** 0352	1.5
0904	4.2	1000	3.8
F 1531	0.7	SA 1624	1.2
2138	4.5	2239	3.9
10 0400	0.9	**25** 0503	1.4
1003	4.5	1106	3.9
SA 1631	0.5	SU 1726	1.1
2235	4.8	2334	4.1
11 0500	0.7	**26** 0602	1.2
1058	4.8	1156	4.0
SU 1728	0.3	M 1817	1.1
2327	5.0	
12 0555	0.5	**27** 0019	4.2
1149	5.0	0647	1.1
M 1820	0.2	TU 1238	4.2
		1859	1.1
13 0016	5.1	**28** 0059	4.4
0644	0.3	0725	1.0
TU 1237	5.2	W 1316	4.3
1907	0.2	1935	1.0
14 0103	4.4	**29** 0135	4.4
0731	0.2	0801	0.9
W 1324	5.4	TH 1350	4.4
○ 1953	0.2	● 2009	1.1
15 0149	5.3	**30** 0209	4.4
0816	0.2	0836	0.9
TH 1411	5.4	F 1423	4.4
2038	0.3	2044	1.1
		31 0240	4.4
		0911	0.9
		SA 1456	4.5
		2119	1.1

Chart Datum is 0·23 metres below TAW Datum. HAT is 5·6 metres above Chart Datum.

9.16.34 BLANKENBERGE

Belgium, West Flanders **51°18'·90N 03°06'·48E** ❄❄❄♦♦♦♦✿✿

CHARTS AC 2449, 1872, 1874; DYC 1801.3; Imray C30

TIDES +0130 Dover; ML 2·5; Duration 0535
Standard Port ZEEBRUGGE (←—)

Times				Height (metres)			
High Water		Low Water		MHWS	MHWN	MLWN	MLWS
0300	0900	0300	0900	4·8	3·9	1·1	0·5
1500	2100	1500	2100				
Differences BLANKENBERGE							
–0007		–0002		+0·1	+0·1	–0·1	–0·1

SHELTER Good, but entry is dangerous in NW'lies F6 and above.

NAVIGATION WPT 51°19'·60N 03°05'·31E, 134°/1M to pierheads. Caution: 4 unlit Y SPM buoys and one Fl (3) Y 20s lie E and W of hbr ent about 400m offshore. Beware strong tides (& fishing lines) across ent. Said to be dredged 2·5m during the season; but out of season it silts badly between piers and dries at LW. Do not try to ent/exit around low water, especially at springs and from Oct-end May, unless depths have been pre-checked. Call dredger Ch 10 for clearance to pass.

LIGHTS AND MARKS Conspic high-rise blocks E of lt ho, W twr B top. Ldg lts 134°, both FR, (Red X on front mast) show the best water. FG on W pier is more visible than FR (E pier). A water twr is conspic on E side of new Hbr. See also Lights, buoys and waypoints and chartlet.

COMMUNICATIONS (Code (0)50) *Blankenberge Rescue* Ch 08 or relay via Zeebrugge Traffic Centre VHF Ch 69; ⊖ 544223; Police 429842; Dr 333668; ℍ 413701; Brit Consul (02) 287 6248; VNZ, SYCB and VVW marinas VHF Ch 23; Dredger VHF Ch 10.

FACILITIES A Port of Entry, with Customs early Apr–late Sept. Keep to port, past the FV hbr, for the old Yacht Hbr (1·8m); VNZ pontoons are on N side and SYCB to the E. Or turn 90° stbd into new Hbr/marina (2·5m); 12 pontoons, run by VNZ, SYCB & VVW.

VNZ (YC Vrije Noordzeezeilers) ☎425292, mob 0497 565565. www.vnzblankenberge.be vnz@skynet.be 120 inc ❶⚲, €2·50 inc ⚓ ⛽ wi-fi, multihulls €3·80. ✕ ⬜.

SYCB (Scarphout Royal YC) ☎411420, mob 0476971692 www.scarphout.be scarphout@skynet.be 250 inc ❶⚲, €2·40 inc ⚓ wi-fi ⛽ ☇ coin meters, multihulls +50%. ✕ ⬜ ⬛ ▬ ⤑(10·5t).

VVW Marina ☎417536, mob 0495527536. www.vvwblankenberge.be info@vvweblankenberge.be 350 inc ❶⚲, €1·90 inc ⚓ ☇ wi-fi, ⛽ €2·00 p/d, multihulls +50%, ⬛ ⬛ ⛴ ⛽ ⬛ ⚒ ⬛ ⚓ ☇ ⤑(20t).

Town ✕ ⬜ ⬛ ⬛ ⬛ ⊠ Ⓑ Ⓒ ⇌ ✈Oostende.

9.16.35 OOSTENDE

Belgium, West Flanders **51°14'·35N 02°55'·09E** ❄❄❄♦♦♦♦✿✿✿

CHARTS AC 2449, 1872, 1874, 1873; SHOM 7214; Navi 1010; DYC 1801.2; Imray C30; BE D11

TIDES +0115 Dover; ML 2·79; Duration 0530
Standard Port ZEEBRUGGE (←—)

Times				Height (metres)			
High Water		Low Water		MHWS	MHWN	MLWN	MLWS
0300	0900	0300	0900	4·8	4·0	1·1	0·5
1500	2100	1500	2100				
Differences OOSTENDE							
–0019	–0019	–0008	–0008	+0·4	+0·2	+0·2	+0·1

SHELTER Wash can enter RNSYC (2·2m). Very good in Mercator Marina (5·0m). RYCO (2·7m) may be exposed in strong W/NW'lies.

NAVIGATION WPT 51° 15'·2N 02° 54'·23E, 143°/1M to ent. Avoid the offshore banks esp in bad weather: From the NE stay to seaward of Wenduinebank, except in calm weather. From the NW keep S of W Hinder TSS and approach from MBN buoy. From the SW appr via West Diep and Kleine Rede, inside Stroombank. Note size of Montgomerydok entrance reduced to decrease wash.

LIGHTS AND MARKS Europa Centrum bldg (105m; twice as high as other bldgs) is conspic 4ca SSW of ent. W lt ho, 5ca ESE of the ent, is conspic with 2 sinusoidal blue bands. 143° ldg marks are lattice masts with R/W bands and X topmarks. See chartlet and Lights, buoys and waypoints for lt details.

IPTS from E pier, plus QY = hbr closed, ship moving. At blind exit from Montgomerydok, QY lts = no exit, ship moving.

COMMUNICATIONS (Code 059) ⚓ VHF Ch 27 at 0820 & 1720 UT (English and Dutch); Police 701111; ⊖ 242070; Brit Consul (02) 287 6248; ℍ 552000; HM 321687, Ch 09 (H24); Mercator lock 705762, VHF Ch 14; Marina VHF Ch 14 (lock hrs); Demey lock (VHF Ch 84) gives access to Oostende-Brugge canal.

FACILITIES from seaward: **Royal North Sea YC** (RNSYC) ☎430694 HM 474688278. www.rnsyc.be simon@rnsyc.be No VHF. 150+50❶ much rafting, €2·70/m inc ⛽ ☇ (M/Cruisers €3·25/m). ✕ ⬜ ⬛ ⚒ ⬛ ⚔ ⬛ ⬛ wi-fi, grid.

Mercator lock Must pre-call Ch 14 for lock-in time and berth. Waiting pontoon N of lock. No freeflow. R/G tfc lts. Hrs (times in brackets are Fri-Sun & public hols): May/Jun 0800-2000 (2100); Jul/Aug 0700-2200, 7/7; Sep 0800-1800 (2000). Oct–Apr, see web.

Mercator Marina ☎705762.www.mercatormarina.be 300+50❶ €2·20, (vessels >13·75m +€1·10/m p/d surcharge), ⬛ €1·20 ⚓ €1·00 ⛽ €1·00/1kw; pay cash/card before entering lock outbound.

Royal YC Oostende (RYCO) ☎32477 676391(Dockmaster). ryco@skynet.be www.ryco.be No VHF. 160 + 35❶ €2·60 inc ☇; ⛽ €3·50/day. ✕ ⬜ ▬ (10t) grid ⤑(½t).

Town All needs, ⬛ ⬛ only, ⇌ ✈.

O5
Q.G

Ldg Lts 143°

Q.R
O6

OH
Fl.G.5s

OH2
Q.R

OH1
Fl.G.5s

OH6
Q.R

7 9

F.G.14m10M

OH8
Q.R

Fort Napoleon

N

Radar tower, IPTS,
Storm & Tidal sigs
Q.Y

Lt Ho
Fl(3)10s
65m27M

14'·5

8

Royal North Sea YC

MRCC

Vissenjidok

2 7

C
B A

F E

D

Q.Y

Conspic
Centrum Bldg
(105m)

Montgomerydok

Q.Y

2 2

·5

Tijdok

F.R

RYCO

14'

3 7

8

F.G

Naval Basin
Zeewezendok

G

K

J

H

Mercator
Lock

F

Cathedral
(twin spires)

OOSTENDE

Mercator
Marina

5

ISO4s
38m4M

I

5

Voorhaven

8

5

ISO4s
48m4M

8

5 2

0   metres   600

0   cables   4

8

2 7

RYCO

51°13'·5N

Demey Lock

4 2   Canal to Brugge

See inset

VVW

55'

02°56'E

Royal
North
Sea YC

Waiting
pontoon

Q.Y

MERCATOR MARINA

N

Montgomerydok

Dredged to 2·7m

Waiting
pontoon

I

H   G   F   E

A

A

A

A

5

Mercator STS

D

5

B

1–27

Swing
bridge

B

B

B

C 35–1

9.16.36 NIEUWPOORT

Belgium, W Flanders 51°09'·38N 02°43'·04E ❀❀❀♦♦♦♦❁❁❁

CHARTS AC 2449, 1872, 1873; Belgian 101, D11; DYC 1801.2; SHOM 7214; Navicarte 1010; Imray C30

TIDES +0105 Dover; ML 2·96; Duration 0515

Standard Port ZEEBRUGGE (←—)

Times				Height (metres)			
High Water		Low Water		MHWS	MHWN	MLWN	MLWS
0300	0900	0300	0900	4·8	4·0	1·1	0·5
1500	2100	1500	2100				
Differences NIEUWPOORT							
−0031	−0031	−0010	−0010	+0·7	+0·5	+0·3	+0·1

SHELTER Good except in strong NW'lies. Access H24.

NAVIGATION WPT 51°10'·08N 02°41'·90E, 134°/1M to ent. The bar (1·5m) is liable to silting but there is usually sufficient water for all but the deepest yachts. At sp the stream reaches 2kn across the ent. WIP 2015, inner approach chan, has reduced the navigable width to 35m; (see chartlet). The 1M long access chan to the 3 marinas is marked by white-topped piles and coloured arrows for guidance to each of them (WSKLuM–blue; VVW-N–yellow; KYCN–grey). The chan is dredged 3·1m. Speed limit 5kts in hbr.

LIGHTS AND MARKS The high-rise blocks W of the entrance are in stark contrast to the sand dunes and flatlands to the E. The lt ho is a conspic 28m R twr, W bands; see chartlet and Lights, buoys ,waypoints. The central, pyramidal white HM bldg at VVW-N is conspic.

COMMUNICATIONS (Code 058) MRCC Oostende VHF Ch 16,67; (059) 701000, emergency direct line, SAR support ship 243936; VTS Wandelaar appr VHF Ch 60; Coast Guard/Marine Police 224030; Police 101, emergency 112; Brit Consul (02) 287 6248; Dr 070/24.65.55; Port VHF Ch 60 (H24); Canal locks 233050; *Air Force Yacht Club* (WSKLuM): VHF Ch 72. KYCN: VHF Ch 23. VVW-N: VHF Ch 08.

FACILITIES The 3 marinas are: **WSKLuM**, N & E of Novus Portus. ☎233641, www.wsklum.be info@wsklm.be 453 inc ❷ on pontoon A, €2·20. 🄳 €2·60p/d, 🔧⚓ ♨ wi-fi, 🗜 🖭 🛠 ✈ (at fuel berth, below) ▣(25t) ⛴. At Lombardsijde, 800m NE, ✕ 🛒 ⛽ ice, bike hire.

VVW-N, S & W of Novus Portus. ☎235232, www.vvwnieuwport. be info@vvwnieuwport.be ④ 1000 + ❷ €1·90. 🄳 €2·50p/d. 🔧 ✕ 🛒 🖭 Ⓔ ⚓ ⛴ 🗜 ✈ (at fuel berth) ✕ 🛠 🖭 🛒 ▣(50t) free bikes.

KYCN (Royal Nieuwpoort YC). info@kycn.be www.kycn.be ☎234413. 440 + 75❷ €2·58 Jul/Aug, 🄳 €2·14/day. ⚓ ▣(25t) ⛴. 🐟 pontoon N. ✕ 🛒 🖭 🛠 ⚓ 🛠 ✕.

Fuel 🐟 🐟 available alongside S of VVW-N on the E bank of the river, see inset, 🐟 at KYCN. The river accesses the N Belgian canals.

Town 🐟 🐟 ✕ 🛒 🖭 🐟 ⛽ ✉ Ⓑ ⇌ ✈ (Oostende).

North France

Dunkerque to Cap de la Hague

9.17.1	Contents	743
9.17.2	Area map and Distance table	744
9.17.3	Tidal stream charts	746
9.17.4	Lights, buoys and waypoints	748
9.17.5	Passage information	751

North East France – 751 • Dunkerque to Boulogne – 751 • Boulogne to Dieppe – 751 • Dieppe to Le Havre – 763 • The Seine Estuary – 769 • Deauville to Grandcamp – 777 • Grandcamp to Pointe de Barfleur – 781 • Pointe de Barfleur to Cap de la Hague – 781

9.17.6	Special notes for France	751
9.17.7	Dunkerque, tide tables/curves, Port Est	756
9.17.8	Gravelines	757
9.17.9	Calais, tidal coefficients	759
9.17.10	Boulogne-sur-Mer	760
9.17.11	Étaples	760
	Le Touquet	
9.17.12	St Valéry-sur-Somme	761
	Le Hourdel • Le Crotoy	
9.17.13	Le Tréport	762
9.17.14	Dieppe, tide tables/curves	762
9.17.15	St Valéry-en-Caux	767
9.17.16	Fécamp	768
9.17.17	Le Havre, tide tables/curves	768
9.17.18	Honfleur	773
9.17.19	River Seine	774
	Rouen	
9.17.20	Paris	775
9.17.21	Deauville/Trouville	776
9.17.22	Dives-sur-Mer	777
9.17.23	Ouistreham	778
	Caen	
9.17.24	Courseulles-sur-Mer	779
	Arromanches	
9.17.25	Port-en-Bessin	780
9.17.26	Grandcamp	781
9.17.27	Carentan	782
	Isigny • Iles St Marcouf	
9.17.28	St Vaast-la-Hougue	783
9.17.29	Barfleur	783
9.17.30	Cherbourg, tide tables/curves, tidal coefficients	784
	Port de Lévi • Port du Becquet	
9.17.31	Omonville-la-Rogue	790
	Anse de St Martin	

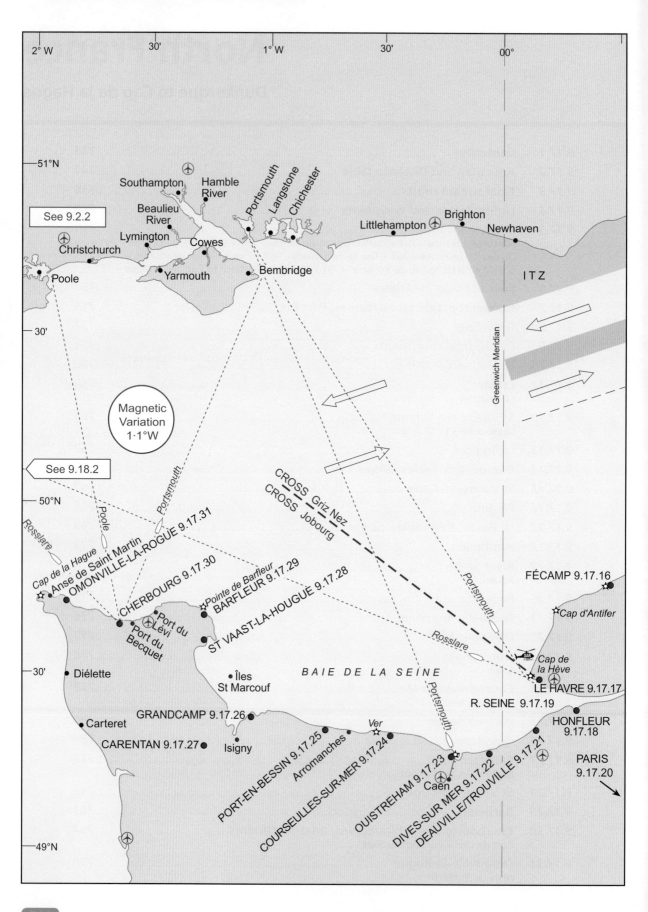

Southampton
Hamble River
Beaulieu River
Lymington
Christchurch
Poole
Cowes
Yarmouth
Bembridge
Portsmouth
Langstone
Chichester
Littlehampton
Brighton
Newhaven

See 9.2.2

I T Z

Greenwich Meridian

Magnetic Variation 1·1°W

See 9.18.2

Rosslare
Poole
Portsmouth

50°N

CROSS Griz Nez
CROSS Jobourg

Portsmouth

FÉCAMP 9.17.16

Cap de la Hague
Anse de Saint Martin
OMONVILLE-LA-ROGUE 9.17.31
CHERBOURG 9.17.30
Port du Lévi
Port du Becquet
Pointe de Barfleur
BARFLEUR 9.17.29
ST VAAST-LA-HOUGUE 9.17.28
Rosslare

Cap d'Antifer

Cap de la Hève
LE HAVRE 9.17.17

30'
Diélette
Îles St Marcouf
BAIE DE LA SEINE

Portsmouth

R. SEINE 9.17.19

Carteret
GRANDCAMP 9.17.26
CARENTAN 9.17.27
Isigny
PORT-EN-BESSIN 9.17.25
Arromanches
Ver
COURSEULLES-SUR-MER 9.17.24
OUISTREHAM 9.17.23
Caen
DIVES-SUR MER 9.17.22
DEAUVILLE/TROUVILLE 9.17.21

HONFLEUR 9.17.18

PARIS 9.17.20

49°N

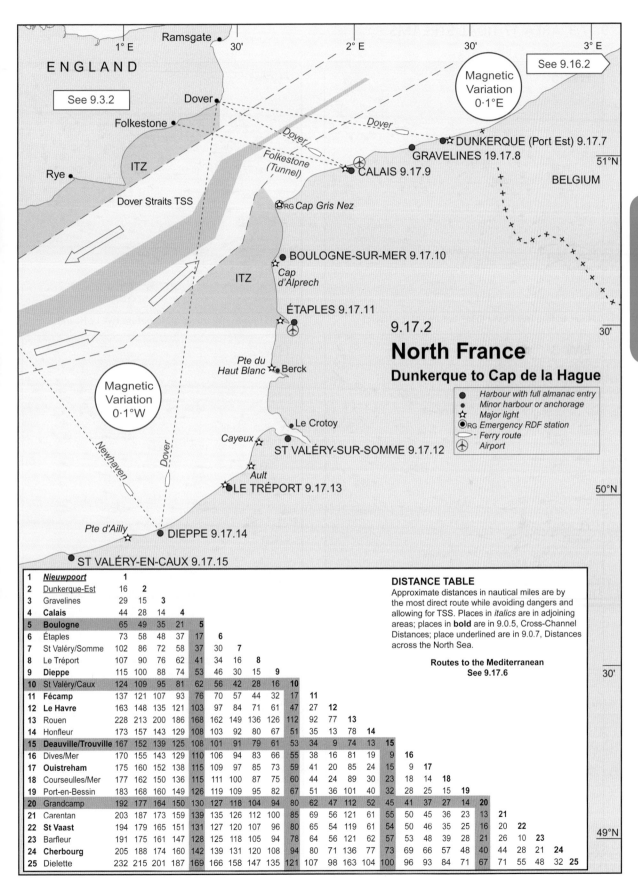

See 9.3.2

ENGLAND

Ramsgate

1° E   30'   2° E   30'   3° E

See 9.16.2

Dover

Folkestone

Rye

ITZ

Dover Straits TSS

Magnetic
Variation
0·1°E

Dover

☆☆ DUNKERQUE (Port Est) 9.17.7
GRAVELINES 19.17.8

Dover/
Folkestone
(Tunnel)

CALAIS 9.17.9

51°N

BELGIUM

⊙RG Cap Gris Nez

ITZ

BOULOGNE-SUR-MER 9.17.10

Cap
d'Alprech

ÉTAPLES 9.17.11

9.17.2

North France

Dunkerque to Cap de la Hague

Pte du
Haut Blanc ☆ Berck

Magnetic
Variation
0·1°W

Newhaven

Dover

Cayeux ☆

Ault

Le Crotoy

ST VALÉRY-SUR-SOMME 9.17.12

LE TRÉPORT 9.17.13

Pte d'Ailly
☆

DIEPPE 9.17.14

● Harbour with full almanac entry
● Minor harbour or anchorage
☆ Major light
⊙RG Emergency RDF station
▭ Ferry route
✈ Airport

30'

50°N

● ST VALÉRY-EN-CAUX 9.17.15

1	*Nieuwpoort*	**1**																						
2	Dunkerque-Est	16	**2**																					
3	Gravelines	29	15	**3**																				
4	**Calais**	44	28	14	**4**																			
5	**Boulogne**	65	49	35	21	**5**																		
6	Étaples	73	58	48	37	17	**6**																	
7	St Valéry/Somme	102	86	72	58	37	30	**7**																
8	Le Tréport	107	90	76	62	41	34	16	**8**															
9	**Dieppe**	115	100	88	74	53	46	30	15	**9**														
10	St Valéry/Caux	124	109	95	81	62	56	42	28	16	**10**													
11	**Fécamp**	137	121	107	93	76	70	57	44	32	17	**11**												
12	**Le Havre**	163	148	135	121	103	97	84	71	61	47	27	**12**											
13	Rouen	228	213	200	186	168	162	149	136	126	112	92	77	**13**										
14	Honfleur	173	157	143	129	108	103	92	80	67	51	35	13	78	**14**									
15	**Deauville/Trouville**	167	152	139	125	108	101	91	79	61	53	34	9	74	13	**15**								
16	Dives/Mer	170	155	143	129	110	106	94	83	66	55	38	16	81	19	9	**16**							
17	**Ouistreham**	175	160	152	138	115	109	97	85	73	59	41	20	85	24	15	9	**17**						
18	Courseulles/Mer	177	162	150	136	115	111	100	87	75	60	44	24	89	30	23	18	14	**18**					
19	Port-en-Bessin	183	168	160	149	126	119	109	95	82	67	51	36	101	40	32	28	25	15	**19**				
20	Grandcamp	192	177	164	150	130	127	118	104	94	80	62	47	112	52	45	41	37	27	14	**20**			
21	Carentan	203	187	173	159	139	135	126	112	100	85	69	56	121	61	55	50	45	36	23	13	**21**		
22	**St Vaast**	194	179	165	151	131	127	120	107	96	80	65	54	119	61	54	50	46	35	25	16	20	**22**	
23	Barfleur	191	175	161	147	128	125	118	105	94	78	64	56	121	62	57	53	48	39	28	21	26	10	**23**
24	**Cherbourg**	205	188	174	160	142	139	131	120	108	94	80	71	136	77	73	69	66	57	48	40	44	28	21	**24**
25	Dielette	232	215	201	187	169	166	158	147	135	121	107	98	163	104	100	96	93	84	71	67	71	55	48	32	**25**

DISTANCE TABLE
Approximate distances in nautical miles are by
the most direct route while avoiding dangers and
allowing for TSS. Places in *italics* are in adjoining
areas; places in **bold** are in 9.0.5, Cross-Channel
Distances; place underlined are in 9.0.7, Distances
across the North Sea.

**Routes to the Mediterranean
See 9.17.6**

30'

49°N

N France

9.17.3 AREA 17 TIDAL STREAMS

HW-5

5 Hours before HW Dover (0345 before HW Le Havre)

HW-4

4 Hours before HW Dover (0245 before HW Le Havre)

HW-3

3 Hours before HW Dover (0145 before HW Le Havre)

HW-2

2 Hours before HW Dover (0045 before HW Le Havre)

HW-1

1 Hour before HW Dover (0015 after HW Le Havre)

HW

HW Dover (0115 after HW Le Havre)

Westward 9.18.3   Channel Is 9.19.3   Northward 9.2.3   North-eastward 9.3.3   Eastward 9.16.3

1 Hour after HW Dover (0215 after HW Le Havre)

2 Hours after HW Dover (0315 after HW Le Havre)

3 Hours after HW Dover (0415 after HW Le Havre)

4 Hours after HW Dover (0515 after HW Le Havre)

5 Hours after HW Dover (0615 after HW Le Havre)

6 Hours after HW Dover (0510 before HW Le Havre)

9.17.4 LIGHTS, BUOYS AND WAYPOINTS

Bold print = light with a nominal range of 15M or more. CAPITALS = place or feature. *CAPITAL ITALICS* = light-vessel, light float or Lanby. *Italics* = Fog signal. ***Bold italics*** = Racon. Many marks/buoys are fitted with AIS (MMSI No); see relevant charts.

OFFSHORE MARKS: W Hinder to Dover Strait
Fairy South ⚓ VQ (6) + L Fl 10s; 51°21'·20N 02°17'·31E.
Fairy West ⚓ VQ (9) 10s 6M; 51°23'·89N 02°09'·27E.
Hinder 1 ⚓ Fl (2) 6s; 51°20'·80N 02°10'·93E.
Bergues ⚓ Fl G 4s 7m 4M; 51°17'·14N 02°18'·63E.
Bergues S ⚓ Q (6) + L Fl 15s; 51°15'·09N 02°19'·42E.
Ruytingen E ⚓ VQ; 51°14'·55N 02°17'·93E.
Ruytingen N ⚓ VQ 4M; 51°13'·16N 02°10'·28E.
Ruytingen SE ⚓ VQ (3) 15s; 51°09'·20N 02°08'·94E.
Ruytingen NW ⚓ Fl G 4s 3M; 51°09'·11N 01°57'·30E.
Ruytingen W ⚓ VQ 4M; 51°06'·93N 01°50'·45E.
Ruytingen SW ⚓ Fl (3) 12s 3M; 51°04'·33N 01°45'·84E.
Sandettié N ⚓ VQ 6M; 51°18'·35N 02°04'·73E.
Sandettié E ⚓ Fl R 4s 7m 3M; 51°14'·88N 02°02'·65E.

DUNKERQUE TO BOULOGNE
PASSE DE ZUYDCOOTE
E12 ⚓ VQ (6) + L Fl 10s; 51°07'·92N 02°30'·61E.(Belgium).
CME ⚓ Q (3) 10s; 51°07'·30N 02°30'·00E.
E11 ⚓ Fl G 4s; 51°06'·85N 02°30'·83E.
E10 ⚓ Fl (2) R 6s; 51°06'·26N 02°30'·40E.
E9 ⚓ Fl (2) G 6s; 51°05'·61N 02°29'·60E.
E8 ⚓ Fl (3) R 12s; 51°05'·12N 02°28'·60E. (E7 does not exist)

PASSE DE L'EST
E6 ⚓ QR; 51°04'·86N 02°27'·08E.
E4 ⚓ Fl R 4s; 51°04'·58N 02°24'·52E.
E1 ⚓ Fl (2) G 6s; 51°04'·12N 02°23'·04E.
E2 ⚓ Fl (2) R 6s; 51°04'·35N 02°22'·31E.
⚓ Q (6) + L Fl 15s; 51°04'·28N 02°21'·73E.

DUNKERQUE PORT EST
E jetty ☆ Fl (2) R 6s 12m 10M; R □, W pylon; 51°03'·59N 02°21'·20E.
W jetty ☆ Fl (2) G 6s 35m 11M; W twr, brown top; 51°03'·63N 02°20'·95E.
Ldg lts 137·5°, front Q7m11M 51°02'·98N 02°22'·05E, rear, Q10m11M, 114m from front, both W cols, R tops, synched.
Inner W jetty ⚓ Q 11m 9M; 51°03'·33N 02°21'·43E.
Dunkerque lt ho ☆ Fl (2) 10s 59m **26M**; 51°02'·93N 02°21'·86E.

DUNKERQUE INTERMEDIATE CHANNEL
DW30 ⚓ QR; 51°04'·14N 02°20'·16E.
DW29 ⚓ QG; 51°03'·85N 02°20'·21E.
DW16 ⚓ Fl (2) R 6s; 51°03'·52N 02°08'·62E.
DKB ⚓ VQ (9) 10s; 51°02'·95N 02°09'·26E.
DW12 ⚓ Fl (3) R 12s; 51°03'·38N 02°05'·36E.
DW11 ⚓ Fl (3) G 12s; 51°02'·79N 02°05'·53E. (DW10-7 omitted)

GRAVELINES
W jetty ⚡ Fl (2) WG 6s 9m W/G4M; 085°-W-224°-G-085°; Y ○ twr, G top; 51°00'·94N 02°05'·49E.

PASSE DE L'OUEST
DW6 ⚓ VQ R; 51°02'·60N 02°01'·01E.
DW5 ⚓ QG; 51°02'·20N 02°00'·92E.
DKA ⚓ L Fl 10s; 51°02'·55N 01°56'·96E.
RCE (Ridens de Calais East) ⚓ Iso G 4s; 51°02'·44N 01°53'·24E.
Dyck ⚓ Fl 3s; ***Racon B***; 51°02'·99N 01°51'·79E.
RCA (Ridens de Calais Approach) ⚓ Q; 51°01'·00N 01°48'·53E.
992271117.

CALAIS
E jetty ☆ Fl (2) R 6s 12m **17M**; (in fog two Fl (2) 6s (vert) on request); Gy twr, R top; *Horn (2) 40s;* 50°58'·39N 01°50'·45E.
W jetty ⚡ Iso G 3s 12m 9M; (in fog Fl 5s on request); W twr, G top; *Bell 5s;* 50°58'·24N 01°50'·40E.

Calais ☆ Fl (4) 15s 59m **22M**; vis 073°-260°; W 8-sided twr, B top; 50°57'·68N 01°51'·21E (440m E of marina entry gate).

CALAIS, WESTERN APPROACH
CA6 ⚓ Fl (3) R 12s; 50°58'·63N 01°49'·92E.
CA4 ⚓ Fl (2)R 6s; 50°58'·38N 01°48'·65E.
CA1 ⚓ Fl G 4s; 50°57'·64N 01°46'·14E (0·5M NNW of Sangatte).
Sangatte ⚡ Oc WG 4s 13m W8M, G5M; 065°-G-089°-W-152°-G-245°; W pylon, B top; 50°57'·19N 01°46'·50E.
Les Quénocs ⚓ VQ; 50°56'·85N 01°41'·12E.
Abbeville wreck ⚓ VQ (9) 10s; 50°56'·08N 01°37'·58E.
Cap Gris-Nez ☆ Fl 5s 72m **29M**; 005°-232°; W twr, B top; 50°52'·09N 01°34'·96E.

OFFSHORE MARKS: DOVER STRAIT TSS, French side
Colbart N ⚓ Q 6M; 50°57'·46N 01°23'·29E.
Colbart SW⚓ VQ (6) + L Fl 10s 8m; 50°48'·87N 01°16'·32E.
ZC2 (Zone Côtière) ⚓ Fl (2+1) Y 15s 5M; 50°53'·54N 01°30'·89E.
ZC1 ⚓ Fl (4) Y 15s 4MN; 50°44'·99N 01°27'·21E.
Ridens SE ⚓ VQ (3) 5s 6M; 50°43'·48N 01°18'·87E.
Bassurelle ⚓ Fl (4) R 15s 6M; ***Racon B, 5-8M;*** 50°32'·74N 00°57'·69E.
Vergoyer N ⚓ VQ 5M; 50°39'·67N 01°22'·21E.
Vergoyer NW ⚓ Fl (2) G 6s 4M; 50°37'·16N 01°17'·85E.
Vergoyer E ⚓ VQ (3) 5s 6M; 50°35'·76N 01°19'·70E.
Vergoyer W ⚓ Fl G 4s 4M; 50°34'·66N 01°13'·57E.
Vergoyer SW ⚓ VQ (9) 10s 6M; 50°27'·01N 01°00'·03E.

BOULOGNE TO DIEPPE
BOULOGNE
Bassure de Baas ⚓ VQ; 50°48'·53N 01°33'·05E.
Approches Boulogne ⚓ VQ (6) + L Fl 10s 8m 6M; 50°45'·31N 01°31'·07E.
Digue N (detached) ⚡ Fl (2) R 6s 10m 7M; 50°44'·71N 01°34'·18E.
Digue S (Carnot) ☆ Fl (2+1) 15s 25m **19M**; W twr, G top; 50°44'·44N 01°34'·05E.
Clearing brg 122·4° : Front, FG in a neon ▽ 4m 5M; 50°43'·71N 01°35'·66E. Rear, FR 44m 11M; intens 113°-133°; 560m from front.
Inner NE jetty ⚡ FR 11m 7M; 50°43'·91N 01°35'·24E. Inner SW jetty ⚡ FG 17m 5M; W col, G top; *Horn 30s;* 50°43'·90N 01°35'·11E.
Cap d'Alprech ☆ Fl (3) 15s 62m **23M**; W twr, B top; 50°41'·91N 01°33'·75E, 2·5M S of hbr ent.

LE TOUQUET/ÉTAPLES
Pointe de Lornel ⚓ VQ (9) 10s 6m 3M; 50°33'·24N 01°35'·12E.
Mérida wreck ⚓ 50°32'·85N 01°33'·44E.
Camiers lt ho ⚡ Oc (2) WRG 6s 17m W10M, R/G7M; 015°-G-090°-W-105°-R-141°; R pylon; 50°32'·86N 01°36'·28E.
Canche Est groyne ⚓ Fl R 4s 8m; 50°32'·57N 01°35'·66E.
Le Touquet ☆ Fl (2) 10s 54m **25M**; Or twr, brown band, W&G top; 50°31'·43N 01°35'·52E.

Pointe du Haut-Blanc ☆ Fl 5s 44m **23M**; W twr, R bands, G top; 50°23'·90N 01°33'·67E (Berck).

BAIE DE LA SOMME
ATSO ⚓ Mo (A) 12s; 50°14'·00N 01°28'·08E (shifts frequently).
Pte du Hourdel ⚡ Oc (3) WG 12s 19m, W12M, G9M; 053°-W-248°-G-323°; tidal sigs; *Horn (3) 30s;* W twr, G top; 50°12'·90N 01°33'·98E.
Cayeux-sur-Mer ☆ Fl R 5s 32m **22M**; W twr, R top; 50°11'·65N 01°30'·72E.
Le Crotoy ⚡ Oc (2) R 6s 19m 8M; 285°-135°; W pylon; 50°12'·91N 01°37'·40E. Marina ⚡ Fl R & Fl G 2s 4m 2M; 50°12'·98N 01°38'·20E.

ST VALÉRY-SUR-SOMME
Trng wall head, ⚓ Fl G 2.5s 2m 1M; 50°12'·25N 01°35'·85E,.
Embankment head ⚡ Iso G 4s 9m 9M; 347°-222°; W pylon, G top; 50°12'·25N 01°36'·02E.
La Ferté môle ⚡ Fl R 4s 9m 9M; 000°-250°; W pylon, R top; 50°11'·18N 01°38'·14E (ent to marina inlet).

LE TRÉPORT
Ault ☆ Oc (3) WR 12s 95m **W15M**, R11M; 040°-W-175°-R-220°;

W twr, R top; 50°06'·28N 01°27'·23E (4M NE of Le Tréport).
W jetty ☆ Fl (2) G 10s 15m **20M**; W twr, G top; *Horn (2) 30s;* 50°03'·88N 01°22'·14E.

DIEPPE

W jetty ⚟ Iso G 4s 11m 8M; W twr, G top; *Horn 30s;* 49°56'·27N 01°04'·97E.

Quai de la Marne �👁 QR 12m 3M; 49°55'·93N 01°05'·20E, E quay.

Pointe d'Ailly ☆ Fl (3) 20s 95m **31M**; W ☐ twr, G top; *Horn (3) 60s;* 49°54'·96N 00°57'·50E.

DIEPPE TO LE HAVRE

SAINT VALÉRY-EN-CAUX

W jetty ⚟ Fl (2) G 6s 13m 11M; W twr, G top; 49°52'·40N 00°42'·54E. Paluel power station 👁 Q; 49°52'·22N 00°38'·03E.

FÉCAMP

N jetty ☆ Fl (2) 10s 15m **16M**; Gy twr, R top; 49°45'·94N 00°21'·80E.

Fécamp NW, Met Mast FECO1-YO2 ⚟ Mo(U) W 15s 13m 10M, Aero F R 36m, Aero UQ R 60m, Aero UQ W 60m(day) Gy pylon, Y platform; Horn Mo(U) 30s; 49°50'·85N 00°13'·14E. 227123456.

PORT D'ANTIFER

Cap d'Antifer ☆ Fl 20s 128m **29M**; 021°-222°; Gy 8-sided twr, G top, on 90m cliffs; 49°41'·01N 00°09'·93E.

A17 👁 Iso G 4s; 49°41'·53N 00°01·75E.

A18 👁 QR; 49°42'·02N 00°02'·18E. Cross the chan W of A17/18.

Ldg lts 127·5°, both Dir Oc 4s 113/135m **22M**; 127°-128°; by day F **33M** 126·5°-128·5° occas. **Front** ☆, 49°38'·32N 00°09'·12E.

LE HAVRE, APPROACH CHANNEL

Cap de la Hève ☆ Fl 5s 123m **24M**; 225°-196°; W 8-sided twr, R top; 49°30'·74N 00°04'·16E.

LHA ⬯ Mo (A) 12s 10m 6M; R&W; *Racon, 8-10M* (a series of 8 dots, or 8 groups of dots; distance between each dot or group represents 0·3M); 49°31'·38N 00°09'·86W. Reserve lt Mo (A).

Ldg lts 106·8°, both Dir F 36/78m **25M** (H24); intens 106°-108°; Gy twrs, G tops. Front, 49°28'·91N 00°06'·50E; rear, 0·73M from front.

LH3 👁 QG; 49°30'·84N 00°04'·02W. (LH1 & 2 buoys do not exist).

LH4 👁 QR; 49°31'·11N 00°03'·90W.

FVs and yachts <19.8m LOA may cross the appr chan west of LH 7/8 buoys, keeping clear of all other shipping, even if under sail:

LH7 👁 Iso G 4s; 49°30'·25N 00°00'·82W.

LH8 👁 Fl (2) R 6s; 49°30'·44N 00°00'·70W.

Note the W-E longitude change. (LH9 buoy does not exist).

LH13 👁 Fl G 4s; 49°29'·32N 00°03'·62E (Ent to Port 2000).

LH14 👁 Fl R 4s; 49°29'·67N 00°03'·43E; 1m shoal depth close W.

LH16 👁 Fl (2) R 6s; 49°29'·45N 00°04'·28E.

LH 2000 👁 VQ (9) 10s; 49°29'·14N 00°04'·78E (Ent to Port 2000).

LE HAVRE

Digue N ☆ Fl R 5s 15m **21M**; IPTS; W ◯ twr, R top; *Horn 15s;* 49°29'·19N 00°05'·44E.

Digue S ⚟ VQ (3) G 2s 15m 11M; W twr, G top; 49°29'·05N 00°05'·38E.

Marina ent, W spur ⚟ Fl (2) R 6s 3M; 49°29'·22N 00°05'·53E.

THE SEINE ESTUARY UP TO HONFLEUR

CHENAL DE ROUEN

Nord du Mouillage 👁 Fl (4) Y 15s; 49°28'·80N 00°00'·22E.

No. 2 👁 QR; *Racon T;* 49°27'·70N 00°00'·60E.

No. 4 👁 Fl R 2·5s; 49°27'·19N 00°01'·97E. Yachts keep N of chan.

Amfard SW 👁 QR; 49°26'·30N 00°04'·82E.

No. 10 👁 Fl R 4s; 49°26'·10N 00°06'·39E.

Digue du Ratier 👁 VQ 10m 4M; 49°25'·94N 00°06'·59E.

Falaise des Fonds ☆ Fl (3) WRG 12s 15m, **W17M**, R/G13M; 040°-G-080°-R-084°-G-100°- W-109°-R-162°-G-260°; W twr, G top; 49°25'·47N 00°12'·85E.

No. 20 👁 QR; 49°25'·85N 00°13'·71E. (Cross here to Honfleur)

HONFLEUR

Digue Ouest ⚟ QG 10m 6M; 49°25'·67N 00°13'·83E.

Digue Est ⚟ Q 9m 8M; *Horn (5) 40s;* 49°25'·67N 00°13'·95E.

Inner E jetty, Oc (2) R 6s 12m 6M; W twr, R top; 49°25'·38N 00°14'·12E.

No. 22 👁 Fl R 4s; 49°25'·85N 00°15'·37E.

TROUVILLE TO COURSEULLES

CHENAL DE ROUEN TO DEAUVILLE and TROUVILLE

Ratelets 👁 Q (9) 15s; 49°25'·29N 00°01'·71E.

Semoy 👁 VQ (3) 5s; 49°24'·15N 00°02'·35E, close to 148° ldg line.

Trouville SW 👁 VQ (9) 10s; 49°22'·54N 00°02'·56E.

DEAUVILLE and TROUVILLE

Ldg lts 148°, both Oc R 4s 11/17m 12/10M: Front, East inner jetty (*estacade*); 330°-150°; W twr, R top; 49°22'·03N 00°04'·48E. Rear, Pte de la Cahotte; synch; 120°-170°; 49°21'·93N 00°04'·58E.

W trng wall ⚟ Fl WG 4s 10m W9M, G6M; 005°-W-176°-G-005°; B pylon, G top; 49°22'·37N 00°04'·11E. Also 4 unlit SHM bcns.

E trng wall ⚟ Fl (4) WR 12s 8m W7M, R4M; 131°-W-175°-R-131°; W pylon, R top; 49°22'·22N 00°04'·33E. Also 3 unlit PHM bcns.

W outer bkwtr ⚟ Iso G 4s 9m 5M; 49°22'·11N 00°04'·33E.

West inner jetty (*estacade*) ⚟ QG 11m 9M; 49°22'·03N 00°04'·43E.

DIVES-SUR-MER *Note the E-W longitude change.*

DI 👁 Iso 4s; 49°19'·17N 00°05'·86W.

No. 1 ⚑ VQ G; 49°18'·50N 00°05'·67W. Ch bys freq moved.

No. 2 ⚑ VQ R; 49°18'·51N 00°05'·56W.

No. 3 👁 QG 7m 4M; W pylon, G top; 49°18'·30N 00°05'·55W.

No. 5 👁 Fl G 4s 8m 4M; W pylon, G top; 49°18'·09N 00°05'·50W.

Bcns 3 & 5, if damaged, may be temporarily replaced by buoys.

No. 7 ⚑ Fl G 4s; 49°17'·65N 00°05'·31W.

OUISTREHAM and CAEN

Merville 👁 VQ; 49°19'·65N 00°13'·39W; spoil ground buoy.

Ouistreham 👁 VQ (3) 5s; wreck buoy; 49°20'·42N 00°14'·81W.

Ldg lts 185°, both Dir Oc (3+1) R 12s 10/30m **17M**; intens 183·5°-186·5°, synch. **Front** ☆, E jetty, W mast, R top, 49°16'·99N 00°14'·81W. **Rear** ☆, 610m from front, tripod, R top.

No. 1 👁 QG; 49°19'·19N 00°14'·67W.

No. 2 👁 QR; 49°19'·17N 00°14'·43W.

Barnabé ⚟ QG 7m 5M; W pylon, G top; 49°18'·02N 00°14'·76W.

St-Médard ⚟ QR 7m 5M; 49°18'·02N 00°14'·62W.

Riva ⚟ Fl G 4s 9m 3M; W pylon, G top; 49°17'·73N 00°14'·79W.

Quilbé ⚟ Fl R 4s 9m 3M; W pylon, R top; 49°17'·72N 00°14'·67W.

Ouistreham lt ho ☆ Oc WR 4s 37m **W17M**, R13M; 115°-R-151°-W-115°; W twr, R top; 49°16'·79N 00°14'·87W.

COURSEULLES-SUR-MER

Courseulles 👁 Iso 4s; 49°21'·28N 00°27'·68W.

W jetty ⚟ Iso WG 4s 7m; W9M, G6M; 135°-W-235°-G-135°; brown pylon on dolphin, G top; 49°20'·41N 00°27'·37W.

E jetty ⚟ Oc (2) R 6s 9m 7M; 49°20'·26N 00°27'·39W.

COURSEULLES TO ST VAAST

Ver ☆ Fl (3)15s 42m **26M**; obsc'd by cliffs of St Aubin when brg >275°; conspic lt ho, W twr, Gy top; 49°20'·41N 00°31'·13W.

ARROMANCHES

Ent buoys: ⚑ 49°21'·35N 00°37'·26W; ⚑ 49°21'·25N 00°37'·30W.

Bombardons 👁 wreck buoys; 49°21'·66N 00°38'·97W.

PORT-EN-BESSIN

Ldg lts 204°, both Oc (3) 12s 25/42m 10/11M; synch. Front, 069°-339°, W pylon, G top; 49°20'·96N 00°45'·53W. Rear; 114°-294°, W and Gy ho; 93m from front.

E mole ⚟ Oc R 4s 14m 7M; R pylon, 49°21'·12N 00°45'·38W.

W mole ⚟ Fl WG 4s 14m, W10M, G7M; G065°-114·5°, W114·5°-065°; G pylon; 49°21'·17N 00°45'·43W.

COASTAL MARKS

Omaha Beach, 1M off : 👁 49°22'·66N 00°50'·28W; 👁 49°23'·17N 00°51'·93W; 👁 49°23'·66N 00°53'·74W.

Broadsword ⚓ Q (3) 10s, wreck buoy; 49°25'·34N 00°52'·96W.
Est du Cardonnet ⚓ VQ (3) 5s; 49°26'·83N 01°01'·10W.

GRANDCAMP

Les Roches de Grandcamp: No. 1 ⚓ 49°24'·72N 01°01'·75W;
No. 3 ⚓ 49°24'·92N 01°03'·70W; No. 5 ⚓ 49°24'·78N 01°04'·98W.
Ldg lts 146°, both Dir Q 9/12m **15M**, 144·5°-147·5°. **Front** ☆,
49°23'·42N 01°02'·90W. **Rear** ☆,102m from front.
Jetée Est ⚓ Oc (2) R 6s 9m 9M; *Horn Mo(N) 30s;* 49°23'·53N
01°02'·96W.
Jetée Ouest ⚓ Fl G 4s 9m 6M; 49°23'·47N 01°02'·96W.

ISIGNY-SUR-MER

IS, small B/Y ⚓, no topmark (⚓ on AC 2135); 49°24'·28N
01°06'·37W.
No 1 ⚓, Fl R 4s; 49°23'·60N 01°07·27W
Dir lts 173°, both Dir Q WRG 7m 9M; 49°19'·57N 01°06·78W.
Training wall heads ⚓ Fl G 4s; 49°21'·40N 01°07'·20W; ⚓ Fl R
4s, 49°21'·40N 01°07'·10W, off Pte du Grouin.

CARENTAN

C-I ⚓ Iso 4s; 49°25'·44N 01°07'·08W; 210°/1·76M to 1 & 2 bys.
No. 1 ⚓ Fl G 2·5s; 49°23'·93N 01°08'·52W.
No. 2 ⚓ Fl R 2·5s; 49°23'·88N 01°08'·37W.
Trng wall ⚓ Fl (4) G 15s; G △ on G bcn;49°21'·96N 01°09'·95W.
Trng wall ⚓ Fl (4) R 15s; R □ on R bcn;49°21'·93N 01°09'·878W.
Ldg lts 209·5°. **Front** ☆ Dir Oc (3) R 12s 6m **18M**; intens 208·2°-
210·7°; W mast, R top; 49°20'·47N 01°11'·17W. Rear, Dir Oc (3) 12s
14m 10M; vis 120°-005°; W gantry, G top; 723m from front.

ÎLES SAINT-MARCOUF

Iles St-Marcouf ⚓ VQ (3) 5s 18m 8M; □ Gy twr, G top; 49°29'·86N
01°08'·81W.
Ouest-Saint-Marcouf ⚓ Q (9) 15s; 49°29'·73N 01°11'·97W.
Saint Floxel ⚓ 49°30'·64N 01°13'·94W.
Quineville ⚓ Q (9) 10s, wreck buoy; 49°31'·79N 01°12'·38W.

ST VAAST TO POINTE DE BARFLEUR

ST VAAST-LA-HOUGUE

Ldg lts 267°: Front, La Hougue Oc 4s 9m 10M; W pylon, G top;
49°34'·25N 01°16'·37W. Rear, Morsalines Oc (4) WRG 12s 90m,
W11M, R/G8M; 171°-W-316°-G-321°-R-342°-W-355°; W 8-sided
twr, G top; 49°34'·16N 01°19'·10W, 1·8M from front.
Le Manquet ⚓ 49°34'·26N 01°15'·56W.
Le Bout du Roc ⚓ 49°34'·68N 01°15'·27W.
La Dent ⚓ 49°34'·57N 01°14'·20W.
Le Gavendest ⚓ Q (6) + L Fl 15s; 49°34'·36N 01°13'·89W.
Jetty ⚓ Dir Oc (2) WRG 6s 12m W10M, R/G7M; 219°-R-237°-G-
310°-W-350°-R-040°; W 8-sided twr, R top; *Siren Mo(N) 30s;*
49°35'·17N 01°15'·41W.
Pte de Saire ⚓ Oc (2+1) 10s 11m 10M; squat W twr, G top;
49°36'·36N 01°13'·78W.

BARFLEUR

Ldg lts 219·5°, both Oc (3) 12s 7/13m 10M; synch. Front, W □
twr;49°40'·18N 01°15'·61W. Rear, 085°-355°; Gy and W □ twr,
G top; 288m from front.
La Grotte ⚓ 49°41'·06N 01°14'·86W.
Roche-à-l'Anglais ⚓ 49°40'·78N 01°14'·93W.
La Vimberge ⚓ 49°40'·54N 01°15'·25W.
W jetty ⚓ Fl G 4s 8m 6M; 49°40'·32N 01°15'·57W.
E jetty ⚓ Oc R 4s 5m 6M; 49°40'·31N 01°15'·47W.
La Jamette ⚓ 49°41'·87N 01°15'·59W.
Pte de Barfleur ☆ Fl (2) 10s 72m **25M**; obsc when brg less than
088°; Gy twr, B top; 49°41'·78N 01°15'·96W.

POINTE DE BARFLEUR TO CAP DE LA HAGUE

Les Équets ⚓ Q 8m 3M; 49°43'·62N 01°18'·36W.
Basse du Rénier ⚓ VQ 8m 4M; 49°44'·84N 01°22'·09W.

Les Trois Pierres ⚓ 49°42'·90N 01°21'·80W.
Anse de Vicq, 158° ldg lts; both Iso R 4s 8/14m 6M; front
49°42'·20N 01°23'·95W.
La Pierre Noire ⚓ Q (9) 15s 8m 4M;49°43'·54N 01°29'·07W.

PORT DU LÉVI

Cap Lévi ☆ Fl R 5s 36m **17M**; Gy □ twr; 49°41'·75N 01°28'·38W.
Port Lévi ⚓ Oc (2) WRG 6s 7m 7M; 055°-G-083°-W-105°-R-163°;
W & Gy hut, W lantern; 49°41'·24N 01°28'·34W.

PORT DU BECQUET

Ldg lts 186·3°, both intens 183°-190°; synch. Front, Dir Oc
(3) 12s 8m 10M; W 8-sided twr; 49°39'·23N 01°32'·84W. Rear,
Dir Oc (3) R 12s 13m 7M. W 8-sided twr, R top; 49m from
front.

CHERBOURG, EASTERN ENTRANCES

Passe Collignon ⚓ Fl (2) R 6s 5m 4M; 49°39'·59N 01°34'·24W.
Passe de l'Est, Jetée des Flamands ldg lts 189°, both Q 9/16m
13M. Front, 49°39'·33N 01°35'·94W . Rear, 516m from front.
Roches du Nord-Ouest ⚓ Fl R 2·5s; 49°40'·64N 01°35'·28W.
La Truite ⚓ Fl (4) R 15s; 49°40'·33N 01°35'·49W.
Fort d'Île Pelée ⚓ Oc (2) WR 6s 19m; W10M, R7M; 055°-W-120°-
R-055°; W & R pedestal; 49°40'·21N 01°35'·08W.
Fort de l'Est ⚓ 49°40'·28N 01°35'·92W, Iso G 4s 19m 9M.
Fort Central ⚓ VQ (6) + L Fl 10s 5m 4M; 322°-032°; 49°40'·40N
01°37'·04W.

CHERBOURG, PASSE DE L'OUEST

CH1 ⊙; V-AIS; 49°43'·24N 01°42'·09W.
Passe de l'Ouest outer ldg lts 141·2°. **Front,** Dir Q (2 horiz, 63m
apart) 5m **17M**; intens 137·3°-143·3° & 139·2°-145·2°; W △ on
bcn; 49°39'·55N 01°37'·95W. **Rear**, Dir Q 35m **19M**; intens 140°-
142·5°; W △ on Gy pylon.
Fort de l'Ouest ☆ Fl (3) WR 15s 19m **W24M, R20M**; 122°-W-
355°-R-122°; Gy twr, R top; 49°40'·45N 01°38'·87W.
Fort de l'Ouest ⚓ Fl R 4s; 49°40'·39N 01°38'·89W.
⚓ Q (6) + L Fl 15s; 49°40'·34N 01°38'·65W.
Digue de Querqueville ⚓ Fl (2) G 6s 8m 6M; W col, G top;
49°40'·30N 01°39'·80W.
Inner ldg lts 124·3°; both intens 114·3°-134·3°: Front, Digue du
Homet head, QG 10m 6M; 49°39'·48N 01°36'·96W. Rear, Dir Iso
G 4s 16m 13M; W col, B bands, 397m from front.
La Ténarde ⚓ VQ 8m 4M; 49°39'·74N 01°37'·75W.

CHERBOURG, PETITE RADE and MARINA

Entrance, W side, Digue du Homet ⚓ QG 10m 8M; intens
114·3°-134·3°; W pylon, G top; 49°39'·48N 01°36'·96W.
E side, ⚓ Q R, off Jetée des Flamands; 49°39'·44N 01°36'·60W.
Marina ent, E side, ⚓ Fl (3) R 12s 6m 6M; W col, R lantern;
49°38'·91N 01°37'·08W.
W mole ⚓ Fl (3) G 12s 7m 6M; G pylon; 49°38'·87N 01°37'·15W.
E quay⚓ Fl (4) R 15s 3m 3M; R bcn; 49°38'·79N 01°37'·12W.
Wavescreen pontoon, N end ⚓ Fl (4) G 15s 4m 2M; W post, G top.

CHERBOURG TO CAP DE LA HAGUE

Raz de Bannes ⚓ 49°41'·32N 01°44'·53W.
Omonville Dir lt 257°: Iso WRG 4s13m; W10M, R/G7M; 180°-G-
252°-W-262°-R-287°; W pylon, G top; 49°42'·24N 01°50'·15W.
L'Étonnard ⚓ 49°42'·33N 01°49'·84W.
Basse Bréfort ⚓ VQ 8m 4M; 49°43'·90N 01°51'·15W.
Jobourg Nuclear plant chimney, R lts; 49°40'·80N 01°52'·91W.
La Plate ⚓ Fl (2+1) WR 10s 11m; W9M, R6M; 115°-W-272°-R-
115°; Y 8-sided twr, with B top; 49°43'·97N 01°55'·74W.
Cap de la Hague (Gros du Raz) ☆ Fl 5s 48m **23M**; Gy twr, W
top; *Horn 30s;* 49°43'·31N 01° 57'·26W.
La Foraine ⚓ VQ (9) 10s, 12m 6M; 49°42'·90N 01°58'·31W.

9.17.5 PASSAGE INFORMATION

More Passage Information is threaded between harbours in this Area. **Bibliography**: *Shell Channel Pilot* (Imray/Cunliffe). *Dover Strait NP28* and *Channel NP27 Pilots*. *Channel Havens* (ACN/ Endean).

NORTH EAST FRANCE

The coasts of Picardy and Normandy are hardly more than an overnight passage from harbours on the UK's S coast – eg, Brighton to Fécamp is only 65M. However some of the harbours dry, so that a boat which can take the ground is an advantage. For a chart of the Dover Strait TSS, see Area 3. Notes on the English Channel and on cross-Channel passages appear below and in Area 3. See Chapter 9, Introduction for cross-Channel distances. For a French glossary, see Chapter 1.

DUNKERQUE TO BOULOGNE

Offshore a series of banks lies roughly parallel with the coast (AC 323, 1892, 2451): Sandettié bank, Outer Ruytingen midway between Sandettié and the coast, and the Dyck banks which extend NE'wards for 30M from a point 5M NE of Calais. There are well-buoyed channels between some of these banks, but great care is needed in poor visibility. In general the banks are steep-to on the inshore side, and slope to seaward. In bad weather the sea breaks on the shallower parts.

Dunkerque Port Est, the old port, has several good yacht marinas and all facilities. Dunkerque Port Ouest is a commercial/ferry port which yachts should not enter. About 3M SW is the drying hbr of **Gravelines**, which should not be entered in strong onshore winds. If E-bound, the Passe de Zuydcoote (4·4m) is a popular inshore link to West Diep and the Belgian ports.

The sea breaks heavily in bad weather on Ridens de Calais, 3M N of **Calais** (AC 1351) and also on Ridens de la Rade immediately N and NE of the harbour where the least charted depth is 0·1m.

The NE-bound traffic lane of the Dover Strait TSS is only 3M off C Gris Nez. Keep a sharp lookout not only for coastal traffic in the ITZ, but also cross-Channel ferries. ▶*1M NW of C Gris Nez the NE-going stream begins at HW Dieppe −0150, and the SW-going at HW Dieppe +0355, sp rates 4kn.*◀ Between C Gris Nez (lt, fog sig) and Boulogne the coastal bank dries about 4ca offshore.

BOULOGNE TO DIEPPE

In the approaches to Pas de Calais a number of shoals lie offshore (AC 2451): The Ridge (or Le Colbart) Les Ridens, Bassurelle, Vergoyer, Bassure de Baas and Battur. In bad weather, and particularly with wind against tide, the sea breaks heavily on all these shoals.

From **Boulogne** (AC 438) to Pte de Lornel the coast dries up to 5ca offshore. ▶*Off Boulogne's Digue Carnot the N-going stream begins HW Dieppe −0130, and the S-going at HW Dieppe +0350, sp rates 1·75kn.*◀

Le Touquet and **Étaples** lie in the Embouchure de la Canche, entered between Pte de Lornel and Pte du Touquet, and with a drying bank extending 1M seaward of a line joining these two points. Le Touquet-Paris-Plage lt is shown from a conspic twr, 1M S of Pte du Touquet.

▶*Off the estuary entrance the N-going stream begins about HW Dieppe −0335, sp rate 1·75kn; and the S-going stream begins about HW Dieppe +0240, sp rate 1·75kn.*◀

From Pte du Touquet the coast runs 17M S to Pte de St Quentin, with a shallow coastal bank which dries to about 5ca offshore, except in the approaches to the dangerous and constantly changing Embouchure de l'Authie (Pte du Haut-Blanc, about 10M S) where it dries up to 2M offshore.

Baie de la Somme, between Pte de St Quentin and Pte du Hourdel (lt) is a shallow, drying area of shifting sands. The access channel to **St Valéry-sur-Somme** and **Le Crotoy**, which runs close to Pte du Hourdel, is well buoyed, but the whole estuary dries out 3M to seaward, and should not be approached in strong W/NW winds.

Offshore there are two shoals, Bassurelle de la Somme and Quémer, on parts of which the sea breaks in bad weather. ▶*4·5M NW of Cayeux-sur-Mer (lt) the stream is rotatory anti-clockwise. The E-going stream begins about HW Dieppe −0200, and sets 070° 2·5kn at sp: the W-going stream begins about HW Dieppe +0600, and sets 240° 1·5kn at sp.*◀

S of Ault (lt) the coast changes from low sand dunes to medium-height cliffs. Between **Le Tréport** and Dieppe, 14M SW, rky banks, drying in places, extend 5ca offshore. Banc Franc-Marqué (3·6m) lies 2M offshore, and some 3M N of Le Tréport. About 3M NW of Le Tréport, Ridens du Tréport (5·1m) should be avoided in bad weather. A prohibited area extends 6ca off Penly nuclear power station and is marked by lt buoys.

9.17.6 SPECIAL NOTES FOR FRANCE

Instead of 'County' the 'Département' is given. For details of documentation apply to the French Tourist Office, 300 High Holborn, London, WC1V 7JH; ☎0906 824 4123; info.uk@ franceguide.com www.franceguide.com

STANDARD TIME is −0100, to which DST should be added.

VOCABULARY See Chapter 1.

AFFAIRES MARITIMES is a part of the central government Ministry of Transport. It oversees all maritime activities (commercial, fishing, pleasure) and helps them develop harmoniously. Information on navigation and other maritime issues can be supplied by a representative whose ☎ is given under each port; he may be based at the nearest commercial or fishing port. The Head Office is: Ministère des Transports et de la Mer, Bureau de la Navigation de Plaisance, 3 Place de Fontenoy, 75700 Paris, ☎01·44·49·80·00. www.mer.gouv.fr

CHARTS Official charts are issued by SHOM (*Service Hydrographique et Oceanographique de la Marine*), the Navy's Hydrographer. (See Chapter 1 for contact details.) SHOM charts are excellent for the smaller French ports not covered by UKHO. A free, annual chart catalogue *Le petit catalogue* is downloadable from www.shom.fr. It includes chart agents and mail order specialists. In this Almanac under 'Facilities', SHOM means a chart agent. Any new edition of a SHOM chart receives a new, different chart number.

TIDAL LEVELS SHOM charts refer elevations of lights to MHWS. Heights of other features are referred to Mean Sea Level (*Niveau moyen*) or as indicated on the chart; it is important to check the chart's title block. Clearance under bridges etc is referred to MSL, MHWS or HAT depending on the recency of the chart. Useful French tidal terms:

LAT	Plus basse mer astronomique (PBMA)
MHWS	Pleine mer moyenne de VE (PMVE)
MHWN	Pleine mer moyenne de ME (PMME)
MLWN	Basse mer moyenne de ME (BMME)
MLWS	Basse mer moyenne de VE (BMVE)
ML/MSL	Niveau moyen (NM)
HW	Vive-eau (VE)
LW	Morte-eau (ME)

TIDAL COEFFICIENTS Popular and much used in France, these are included for a selection of strategic ports, Calais, Cherbourg, St Malo, Brest and Port Bloc, together with explanatory notes.

PUBLIC HOLIDAYS New Year's Day, Easter Sunday and Monday, Labour Day (1 May), Ascension Day, Armistice Day 1945 (8 May), Whit Sunday and Monday, National (Bastille) Day (14 July), Feast of the Assumption (15 Aug), All Saints' Day (1 Nov), Remembrance Day (11 Nov), Christmas Day.

REGULATIONS Flares must be in-date. Carry original ship's registration papers. Yachts built since 2007 must have black water tank(s) or a treatment system. Mandatory boat equipment in French waters is listed at www.developpement-durable.gouv. fr/spip.php?page=article&id_article=5476.

FACILITIES The cost of a visitor's overnight berth is based on high season rates, mostly quoted as €/metre LOA, although sometimes beam may also be taken into account. Low season rates and concessions can be a big saving. The fee usually includes electricity, but rarely showers which range between €1·20 and €2. VAT (TVA) is 20%. Port à Sec offers savings for longer stays if not living aboard. **Fuel** can be obtained H24 in many marinas using modern credit cards to activate/pay at the pumps. Lacking a suitable card, cash/card payments can be made during office hours at the Capitainerie.

N France

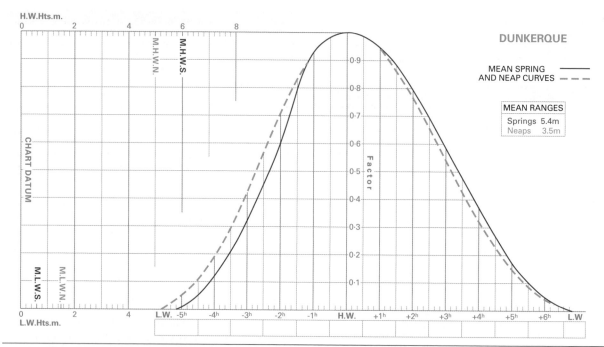

DUNKERQUE

MEAN SPRING ————
AND NEAP CURVES - - - -

MEAN RANGES
Springs 5.4m
Neaps 3.5m

TELEPHONES ☎ numbers contain 10 digits, the first 2 digits being Zone codes (01 to 05) as appropriate to location. Mobile numbers are prefixed 06. Info Nos, eg the recorded weather (Auto) are prefixed 08. The ringing tone is long, equal on/off tones (slower than UK engaged tone). Rapid pips mean the call is being connected. Engaged tone is like that in UK.

To call France from UK, dial + (or 00) 33, then the 9 digit number, omitting the 0 from the Zone code prefix. To call UK from France dial + (or 00) 44 followed by the Area Code (omitting the first 0) and the number. Phonecards for public phones may be bought at the PTT, many Hbr offices, cafés and tabacs. Cheap rates are 2130-0800 Mon-Fri 1330-0800 Sat–Mon.

⚠ **Emergencies, dial:**

15	Ambulance (SAMU), better than a doctor;
17	Police, better than a gendarmerie;
18	Fire (also divers)
112	CROSS & general emergencies

BRITISH CONSULS IN FRANCE

The British Consular service assists UK nationals who are travelling abroad and have encountered difficulties or distress ranging from loss of passport or money to the death of or serious injury to a crew member or fellow traveller.

British Consuls at harbours on the Channel and Atlantic coasts are *the first port of call when help is needed*. The relevant Tel No is given under each hbr in Areas 17-22.

SIGNALS International Port Traffic Signals (Chapter 5) and Storm warning signals (Chapter 6) are used in many French ports. Tidal signals are now rarely seen in French ports.

INLAND WATERWAYS

TOLLS are due on waterways managed by Voies Navigable de France (VNF). Licence discs (*vignettes*) must be visibly displayed stbd side forward. They are obtainable:

- Online from www.vnf.fr/vignettesVNF/ in English. Pay on line by Visa, Mastercard or Eurocard.
- By post from VNF offices listed on the website. Allow 2 weeks.
- Over the counter (cash only) from 35 listed VNF offices.

Licence fees, 2015, were based on LOA and length of stay, with validity for either 1 year, (Jan-Dec) 1 month (30 days) or 3 days. Note: All time periods are consecutive. 1 day licences are also available on request.

	Multiplier	<8m	8–11m	11–14m	>14m
1 year	€8·00	192·00	189·10	365·80	477·60
30 days	€7·20	26·40	38·50	50·60	64·60
3 days	€3·60	14·60	22·20	29·50	36·70

Multiplier x LOA + value from appropriate column, e.g. 10m yacht for 30 days – (€7·20 x 10) + €38·50 = €110·50.

See: www.french-waterways.com for additional information.

CLOSURES

Dates of closures (*chomages*) plus lock hours, water levels and other practicalities, can be obtained from www.vnf.fr or the French National Tourist Office in London (see overleaf).

VNF Head Office, 175 Rue Ludovic Boutleux, BP 820, 62400 Bethune. ☎03·21·63·24·54. www.vnf.fr

QUALIFICATIONS

Helmsmen of craft <15m LOA and not capable of >20 kph (11kn) must have an International Certificate of Competence validated for inland waterways, plus a copy of the CEVNI rules. For larger, faster craft the requirements are under review.

INLAND WATERWAYS TO THE MEDITERRANEAN

For the River Seine, Rouen and Paris, see River Seine and Paris.

The quickest route is via Paris to St Mammès – Canal du Loing – Canal de Briare – Canal Latéral à la Loire – Canal du Centre – Saône – Rhône. Approx 1318km (824M) 182 locks; allow 4 weeks. Max dimensions: LOA 38·5m, beam 5m, draft 1·8m, air draft 3·5m.

Alternatives above St Mammès: R Yonne-Laroche, then either Canal de Bourgogne – Saône; or Canal du Nivernais – Canal du Centre – Saône. Both are slowed by many locks.

There are other routes via R Oise/Aisne or R Marne, both continuing via Canal de la Marne à la Saône.

The Saône can also be accessed via the canals of N France, entered at Calais, Dunkerque or Belgian and Dutch ports.

CANALS IN WESTERN FRANCE

See Area 18 for inland waterways, especially the Ille et Rance Canal/Vilaine River from St Malo to Arzal/Camoel in S Brittany.

See Area 22 for the R Gironde, Canal Latéral à la Garonne and Canal du Midi. Also the *West France Cruising Companion* (NDL/ Featherstone 2010) for an outline of this route.

STANDARD TIME UT –01
Subtract 1 hour for UT
For French Summer Time add ONE hour in **non-shaded areas**

DUNKERQUE   LAT 51°03'N   LONG 2°22'E
TIMES AND HEIGHTS OF HIGH AND LOW WATERS

Dates in red are **SPRINGS**
Dates in blue are NEAPS

YEAR 2016

N France

JANUARY

Time	m		Time	m
1 0452	5.3	**16**	0446	5.7
1149	1.2		1155	0.8
F 1721	5.2	SA	1724	5.7
2 0007	1.6	**17**	0016	1.1
0541	5.1		0544	5.6
SA 1236	1.4	SU	1252	0.9
1814	5.0		1826	5.4
3 0100	1.7	**18**	0119	1.3
0639	4.9		0650	5.4
SU 1337	1.6	M	1401	1.1
1918	4.8		1939	5.3
4 0210	1.8	**19**	0232	1.4
0750	4.8		0808	5.2
M 1448	1.6	TU	1515	1.2
2030	4.8		2101	5.2
5 0323	1.8	**20**	0347	1.4
0903	4.8		0929	5.3
TU 1555	1.6	W	1630	1.1
2136	4.9		2211	5.3
6 0427	1.6	**21**	0503	1.2
1005	5.0		1035	5.5
W 1654	1.4	TH	1741	1.0
2232	5.1		2310	5.5
7 0523	1.4	**22**	0605	1.0
1056	5.3		1130	5.7
TH 1745	1.2	F	1835	0.9
2318	5.4		2359	5.7
8 0610	1.2	**23**	0655	0.8
1139	5.5		1217	5.8
F 1829	1.0	SA	1920	0.8
2358	5.6			
9 0652	1.0	**24**	0041	5.8
1217	5.7		0739	0.6
SA 1910	0.9	SU	1259	5.9
		○	2000	0.8
10 0034	5.8	**25**	0120	5.9
0733	0.8		0819	0.5
SU 1254	5.9	M	1339	6.0
● 1951	0.7		2038	0.8
11 0111	5.9	**26**	0157	5.9
0814	0.6		0857	0.5
M 1333	6.1	TU	1418	6.0
2033	0.7		2113	0.8
12 0149	6.0	**27**	0234	5.9
0857	0.5		0933	0.6
TU 1413	6.1	W	1454	5.9
2115	0.6		2147	0.9
13 0230	6.0	**28**	0307	5.8
0939	0.5		1007	0.7
W 1456	6.1	TH	1528	5.8
2157	0.7		2220	1.0
14 0312	6.0	**29**	0340	5.7
1022	0.6		1039	0.8
TH 1540	6.0	F	1602	5.6
2240	0.8		2252	1.1
15 0356	5.9	**30**	0416	5.5
1106	0.6		1112	1.0
F 1629	5.8	SA	1639	5.4
2325	0.9		2325	1.3
		31	0456	5.3
			1148	1.2
		SU	1723	5.2

FEBRUARY

Time	m		Time	m
1 0005	1.5	**16**	0053	1.2
0544	5.1		0624	5.4
M 1233	1.4	TU	1335	1.1
◑ 1818	4.9		1913	5.1
2 0056	1.7	**17**	0206	1.4
0646	4.8		0745	5.1
TU 1335	1.6	W	1452	1.3
1924	4.7		2040	5.0
3 0210	1.9	**18**	0327	1.4
0759	4.7		0913	5.1
W 1455	1.7	TH	1617	1.3
2038	4.7		2157	5.1
4 0333	1.8	**19**	0450	1.3
0916	4.8		1025	5.3
TH 1609	1.6	F	1728	1.1
2150	4.9		2300	5.3
5 0444	1.6	**20**	0553	1.0
1022	5.1		1122	5.5
F 1713	1.3	SA	1819	1.0
2249	5.2		2350	5.5
6 0542	1.2	**21**	0642	0.8
1114	5.4		1208	5.7
SA 1805	1.1	SU	1904	0.8
2336	5.5			
7 0631	0.9	**22**	0029	5.7
1158	5.7		0723	0.6
SU 1851	0.8	M	1246	5.8
		○	1942	0.8
8 0016	5.7	**23**	0103	5.8
0715	0.6		0801	0.5
M 1237	5.9	TU	1322	5.9
● 1934	0.6		2017	0.7
9 0054	6.0	**24**	0138	5.9
0758	0.4		0836	0.6
TU 1316	6.1	W	1356	5.9
2016	0.5		2049	0.7
10 0132	6.1	**25**	0210	5.9
0840	0.3		0908	0.5
W 1356	6.2	TH	1428	5.9
2059	0.4		2121	0.7
11 0212	6.1	**26**	0240	5.9
0923	0.2		0940	0.6
TH 1438	6.2	F	1458	5.8
2141	0.4		2152	0.8
12 0253	6.2	**27**	0310	5.8
1007	0.2		1010	0.7
F 1522	6.1	SA	1528	5.7
2224	0.5		2221	0.9
13 0337	6.1	**28**	0341	5.7
1050	0.3		1039	0.8
SA 1609	6.0	SU	1601	5.5
2307	0.7		2251	1.1
14 0425	5.9	**29**	0415	5.5
1136	0.6		1110	1.0
SU 1701	5.8	M	1636	5.3
2355	0.9		2325	1.3
15 0520	5.7			
1229	0.8			
M 1800	5.4			
◑				

MARCH

Time	m		Time	m
1 0454	5.2	**16**	0031	1.1
1149	1.3		0604	5.4
TU 1722	5.0	W	1312	1.2
			1849	5.0
2 0008	1.5	**17**	0142	1.4
0550	4.9		0725	5.1
W 1240	1.5	TH	1431	1.5
◑ 1832	4.7		2018	4.8
3 0107	1.7	**18**	0307	1.4
0709	4.7		0855	5.0
TH 1353	1.7	F	1559	1.5
1949	4.6		2138	4.9
4 0233	1.8	**19**	0431	1.3
0830	4.7		1009	5.1
F 1522	1.7	SA	1710	1.3
2108	4.7		2241	5.2
5 0402	1.6	**20**	0534	1.0
0947	4.9		1106	5.4
SA 1638	1.4	SU	1801	1.0
2219	5.0		2330	5.4
6 0512	1.2	**21**	0622	0.8
1048	5.3		1150	5.6
SU 1739	1.1	M	1842	0.9
2312	5.4			
7 0607	0.8	**22**	0007	5.6
1135	5.7		0702	0.7
M 1829	0.8	TU	1225	5.7
2354	5.7		1920	0.8
8 0653	0.5	**23**	0041	5.7
1216	6.0		0737	0.6
TU 1913	0.5	W	1259	5.8
		○	1951	0.7
9 0032	6.0	**24**	0114	5.8
0737	0.3		0809	0.5
W 1255	6.2	TH	1331	5.9
● 1956	0.4		2023	0.7
10 0110	6.2	**25**	0144	5.9
0820	0.1		0840	0.5
TH 1335	6.3	F	1401	5.9
2038	0.3		2054	0.7
11 0151	6.3	**26**	0212	5.9
0903	0.1		0911	0.6
F 1418	6.3	SA	1428	5.8
2121	0.3		2125	0.7
12 0233	6.3	**27**	0241	5.8
0947	0.1		0941	0.7
SA 1502	6.2	SU	1458	5.8
2204	0.4		2154	0.8
13 0318	6.2	**28**	0312	5.7
1031	0.2		1010	0.8
SU 1549	6.0	M	1528	5.6
2248	0.5		2223	0.9
14 0406	6.1	**29**	0342	5.6
1117	0.5		1040	1.0
M 1640	5.8	TU	1558	5.4
2335	0.8		2256	1.1
15 0500	5.8	**30**	0414	5.3
1208	0.9		1117	1.2
TU 1738	5.4	W	1635	5.1
◑			2337	1.3
		31	0501	5.0
			1205	1.4
		TH	1740	4.8
		◑		

APRIL

Time	m		Time	m
1 0033	1.5	**16**	0240	1.4
0628	4.8		0829	4.9
F 1312	1.6	SA	1529	1.5
1909	4.7		2108	4.9
2 0149	1.6	**17**	0400	1.3
0750	4.8		0941	5.1
SA 1438	1.6	SU	1639	1.4
2027	4.7		2210	5.1
3 0320	1.5	**18**	0505	1.1
0910	5.0		1037	5.3
SU 1602	1.4	M	1733	1.1
2144	5.0		2258	5.3
4 0437	1.2	**19**	0554	0.9
1017	5.4		1121	5.5
M 1709	1.0	TU	1815	1.0
2242	5.4		2339	5.5
5 0538	0.8	**20**	0634	0.8
1108	5.7		1159	5.6
TU 1802	0.7	W	1852	0.9
2327	5.7			
6 0628	0.5	**21**	0014	5.7
1151	6.0		0707	0.7
W 1849	0.5	TH	1233	5.7
			1923	0.8
7 0006	6.0	**22**	0048	5.7
0714	0.2		0738	0.7
TH 1231	6.2	F	1304	5.8
● 1933	0.4	○	1954	0.7
8 0046	6.2	**23**	0118	5.8
0758	0.1		0810	0.6
F 1313	6.3	SA	1333	5.8
2016	0.3		2027	0.7
9 0129	6.3	**24**	0146	5.8
0842	0.1		0843	0.6
SA 1356	6.3	SU	1401	5.8
2100	0.2		2100	0.7
10 0213	6.4	**25**	0216	5.8
0926	0.1		0915	0.7
SU 1443	6.2	M	1432	5.8
2145	0.3		2131	0.8
11 0300	6.3	**26**	0248	5.7
1011	0.3		0946	0.8
M 1531	6.0	TU	1503	5.6
2230	0.4		2203	0.9
12 0350	6.1	**27**	0321	5.6
1058	0.6		1019	0.9
TU 1622	5.7	W	1536	5.4
2318	0.7		2237	1.0
13 0444	5.8	**28**	0355	5.4
1148	0.9		1057	1.1
W 1718	5.4	TH	1614	5.2
			2320	1.2
14 0012	1.0	**29**	0444	5.2
0546	5.4		1145	1.3
TH 1249	1.3	F	1715	5.0
◑ 1825	5.0			
15 0120	1.3	**30**	0013	1.3
0703	5.0		0602	5.0
F 1405	1.5	SA	1246	1.5
1950	4.8	◑	1836	4.8

Chart Datum is 2·69 metres below IGN Datum. HAT is 6·4 metres above Chart Datum.

》 **FREE** monthly updates. Register at 《
www.reedsnauticalalmanac.co.uk

753

STANDARD TIME UT –01
Subtract 1 hour for UT
For French Summer Time add
ONE hour in **non-shaded areas**

DUNKERQUE LAT 51°03′N LONG 2°22′E
TIMES AND HEIGHTS OF HIGH AND LOW WATERS

Dates in red are **SPRINGS**
Dates in blue are **NEAPS**

YEAR **2016**

	MAY				JUNE				JULY				AUGUST		
	Time	m	Time	m	Time	m	Time	m	Time	m	Time	m	Time	m	Time	m

MAY

Day	Time	m		Day	Time	m
1 SU	0121 / 0716 / 1404 / 1949	1.4 / 5.0 / 1.5 / 4.9		16 M	0315 / 0901 / 1555 / 2129	1.3 / 5.0 / 1.5 / 5.0
2 M	0244 / 0832 / 1526 / 2104	1.3 / 5.1 / 1.3 / 5.1		17 TU	0421 / 0958 / 1653 / 2220	1.2 / 5.1 / 1.3 / 5.2
3 TU	0402 / 0943 / 1635 / 2207	1.0 / 5.4 / 1.0 / 5.4		18 W	0515 / 1046 / 1741 / 2305	1.0 / 5.3 / 1.1 / 5.4
4 W	0506 / 1039 / 1733 / 2257	0.7 / 5.7 / 0.8 / 5.7		19 TH	0559 / 1128 / 1820 / 2345	0.9 / 5.5 / 1.0 / 5.5
5 TH	0601 / 1125 / 1824 / 2341	0.5 / 6.0 / 0.6 / 6.0		20 F	0635 / 1205 / 1853	0.9 / 5.6 / 0.9
6 F ●	0650 / 1208 / 1910	0.3 / 6.2 / 0.4		21 SA ○	0021 / 0708 / 1237 / 1927	5.6 / 0.8 / 5.7 / 0.8
7 SA	0024 / 0736 / 1251 / 1956	6.2 / 0.2 / 6.3 / 0.3		22 SU	0053 / 0742 / 1307 / 2002	5.7 / 0.8 / 5.7 / 0.8
8 SU	0109 / 0821 / 1338 / 2042	6.3 / 0.2 / 6.2 / 0.3		23 M	0123 / 0818 / 1338 / 2038	5.7 / 0.7 / 5.8 / 0.7
9 M	0156 / 0907 / 1426 / 2127	6.3 / 0.3 / 6.1 / 0.3		24 TU	0155 / 0853 / 1412 / 2113	5.8 / 0.8 / 5.8 / 0.7
10 TU	0245 / 0953 / 1516 / 2214	6.2 / 0.4 / 6.0 / 0.4		25 W	0232 / 0929 / 1448 / 2149	5.7 / 0.8 / 5.7 / 0.8
11 W	0337 / 1040 / 1606 / 2301	6.0 / 0.7 / 5.7 / 0.6		26 TH	0310 / 1005 / 1526 / 2227	5.6 / 0.9 / 5.5 / 0.9
12 TH	0430 / 1128 / 1659 / 2353	5.8 / 1.0 / 5.4 / 0.9		27 F	0351 / 1045 / 1609 / 2309	5.5 / 1.0 / 5.3 / 1.0
13 F ◐	0527 / 1223 / 1758	5.4 / 1.3 / 5.1		28 SA	0441 / 1131 / 1704 / 2359	5.4 / 1.2 / 5.2 / 1.1
14 SA	0052 / 0633 / 1329 / 1911	1.2 / 5.1 / 1.5 / 4.9		29 SU ◑	0542 / 1227 / 1808	1.3 / 5.2 / 5.1
15 SU	0202 / 0752 / 1443 / 2027	1.3 / 4.9 / 1.6 / 4.9		30 M	0100 / 0646 / 1335 / 1914	1.1 / 5.2 / 1.3 / 5.1
				31 TU	0214 / 0755 / 1452 / 2024	1.1 / 5.3 / 1.2 / 5.2

JUNE

Day	Time	m		Day	Time	m
1 W	0329 / 0908 / 1602 / 2133	1.0 / 5.4 / 1.1 / 5.4		16 TH	0425 / 1006 / 1657 / 2229	1.2 / 5.1 / 1.3 / 5.2
2 TH	0435 / 1010 / 1705 / 2230	0.7 / 5.7 / 0.8 / 5.7		17 F	0517 / 1053 / 1744 / 2314	1.1 / 5.3 / 1.2 / 5.3
3 F	0535 / 1103 / 1801 / 2320	0.6 / 5.9 / 0.7 / 5.9		18 SA	0601 / 1135 / 1825 / 2354	1.0 / 5.4 / 1.0 / 5.5
4 SA	0629 / 1150 / 1852	0.5 / 6.0 / 0.5		19 SU	0640 / 1212 / 1902	1.0 / 5.5 / 0.9
5 SU ●	0008 / 0718 / 1236 / 1940	6.1 / 0.4 / 6.1 / 0.4		20 M ○	0030 / 0717 / 1245 / 1940	5.6 / 0.9 / 5.6 / 0.8
6 M	0055 / 0805 / 1324 / 2027	6.2 / 0.4 / 6.1 / 0.3		21 TU	0103 / 0756 / 1319 / 2019	5.7 / 0.8 / 5.7 / 0.7
7 TU	0144 / 0851 / 1413 / 2113	6.2 / 0.5 / 6.1 / 0.3		22 W	0139 / 0835 / 1355 / 2058	5.8 / 0.8 / 5.8 / 0.6
8 W	0234 / 0937 / 1502 / 2159	6.1 / 0.6 / 5.9 / 0.4		23 TH	0218 / 0914 / 1435 / 2137	5.8 / 0.8 / 5.8 / 0.6
9 TH	0324 / 1021 / 1549 / 2244	6.0 / 0.8 / 5.7 / 0.6		24 F	0259 / 0954 / 1516 / 2217	5.8 / 0.8 / 5.7 / 0.7
10 F	0413 / 1106 / 1638 / 2329	5.8 / 1.0 / 5.5 / 0.8		25 SA	0342 / 1034 / 1558 / 2259	5.7 / 0.9 / 5.6 / 0.7
11 SA	0502 / 1152 / 1725	5.5 / 1.2 / 5.3		26 SU	0428 / 1118 / 1646 / 2346	5.6 / 1.0 / 5.5 / 0.8
12 SU	0019 / 0555 / 1245 / 1822	1.0 / 5.2 / 1.4 / 5.1		27 M ◑	0521 / 1208 / 1742	5.5 / 1.1 / 5.4
13 M	0116 / 0659 / 1348 / 1930	1.2 / 5.0 / 1.6 / 4.9		28 TU	0040 / 0619 / 1309 / 1843	0.9 / 5.4 / 1.2 / 5.3
14 TU	0220 / 0809 / 1455 / 2038	1.4 / 4.9 / 1.6 / 4.9		29 W	0147 / 0724 / 1421 / 1951	1.0 / 5.4 / 1.2 / 5.3
15 W	0325 / 0911 / 1559 / 2137	1.3 / 5.0 / 1.5 / 5.0		30 TH	0259 / 0838 / 1533 / 2105	1.0 / 5.4 / 1.1 / 5.4

JULY

Day	Time	m		Day	Time	m
1 F	0408 / 0949 / 1641 / 2212	0.9 / 5.5 / 1.0 / 5.6		16 SA	0434 / 1015 / 1707 / 2242	1.4 / 5.0 / 1.4 / 5.2
2 SA	0514 / 1049 / 1744 / 2309	0.8 / 5.7 / 0.8 / 5.8		17 SU	0528 / 1105 / 1757 / 2329	1.2 / 5.2 / 1.2 / 5.4
3 SU	0614 / 1141 / 1839	0.7 / 5.8 / 0.6		18 M	0615 / 1148 / 1840	1.1 / 5.4 / 1.0
4 M ●	0000 / 0706 / 1229 / 1929	5.9 / 0.6 / 5.9 / 0.5		19 TU ○	0008 / 0656 / 1225 / 1920	5.6 / 0.9 / 5.6 / 0.8
5 TU	0047 / 0753 / 1315 / 2015	6.1 / 0.6 / 6.0 / 0.4		20 W	0045 / 0736 / 1301 / 2000	5.7 / 0.8 / 5.8 / 0.6
6 W	0135 / 0837 / 1401 / 2059	6.1 / 0.6 / 6.0 / 0.4		21 TH	0122 / 0817 / 1338 / 2041	5.9 / 0.7 / 5.9 / 0.5
7 TH	0222 / 0920 / 1445 / 2142	6.1 / 0.7 / 5.9 / 0.4		22 F	0201 / 0858 / 1416 / 2122	6.0 / 0.7 / 5.9 / 0.5
8 F	0306 / 1000 / 1527 / 2222	6.0 / 0.8 / 5.8 / 0.6		23 SA	0241 / 0939 / 1456 / 2203	6.0 / 0.7 / 5.9 / 0.5
9 SA	0350 / 1040 / 1607 / 2302	5.8 / 1.0 / 5.7 / 0.7		24 SU	0322 / 1020 / 1537 / 2245	5.9 / 0.7 / 5.8 / 0.5
10 SU	0431 / 1119 / 1648 / 2342	5.6 / 1.1 / 5.5 / 0.9		25 M	0407 / 1102 / 1623 / 2329	5.8 / 0.8 / 5.7 / 0.6
11 M	0513 / 1201 / 1732	5.4 / 1.3 / 5.3		26 TU	0457 / 1148 / 1716	5.7 / 1.0 / 5.6
12 TU ◑	0027 / 0602 / 1250 / 1826	1.1 / 5.1 / 1.5 / 5.0		27 W ◑	0020 / 0553 / 1244 / 1816	0.8 / 5.6 / 1.1 / 5.5
13 W	0121 / 0700 / 1353 / 1931	1.3 / 4.9 / 1.6 / 4.9		28 TH	0122 / 0658 / 1354 / 1926	1.0 / 5.4 / 1.3 / 5.3
14 TH	0226 / 0808 / 1502 / 2043	1.5 / 4.8 / 1.7 / 4.8		29 F	0234 / 0815 / 1509 / 2047	1.1 / 5.3 / 1.3 / 5.3
15 F	0332 / 0916 / 1608 / 2147	1.5 / 4.9 / 1.6 / 5.0		30 SA	0348 / 0935 / 1624 / 2202	1.1 / 5.3 / 1.2 / 5.5
				31 SU	0503 / 1041 / 1734 / 2304	1.0 / 5.5 / 0.9 / 5.7

AUGUST

Day	Time	m		Day	Time	m
1 M	0606 / 1136 / 1830 / 2356	0.9 / 5.7 / 0.7 / 5.8		16 TU	0550 / 1124 / 1817 / 2346	1.2 / 5.4 / 1.0 / 5.6
2 TU ●	0657 / 1223 / 1918	0.8 / 5.8 / 0.6		17 W	0635 / 1204 / 1859	1.0 / 5.7 / 0.8
3 W	0041 / 0740 / 1304 / 2001	6.0 / 0.7 / 5.9 / 0.5		18 TH ○	0024 / 0716 / 1240 / 1940	5.8 / 0.8 / 5.9 / 0.5
4 TH	0122 / 0821 / 1344 / 2042	6.0 / 0.7 / 6.0 / 0.4		19 F	0101 / 0757 / 1316 / 2021	6.0 / 0.7 / 6.0 / 0.4
5 F	0203 / 0900 / 1422 / 2120	6.0 / 0.7 / 5.9 / 0.5		20 SA	0138 / 0838 / 1353 / 2103	6.1 / 0.6 / 6.1 / 0.3
6 SA	0243 / 0936 / 1459 / 2157	6.0 / 0.8 / 5.9 / 0.6		21 SU	0217 / 0919 / 1432 / 2144	6.2 / 0.5 / 6.1 / 0.3
7 SU	0320 / 1011 / 1534 / 2231	5.9 / 0.9 / 5.8 / 0.7		22 M	0259 / 1001 / 1513 / 2226	6.1 / 0.6 / 6.1 / 0.4
8 M	0355 / 1045 / 1609 / 2305	5.7 / 1.0 / 5.6 / 0.9		23 TU	0343 / 1043 / 1559 / 2310	6.0 / 0.7 / 6.0 / 0.5
9 TU	0431 / 1119 / 1648 / 2341	5.5 / 1.2 / 5.4 / 1.1		24 W	0433 / 1128 / 1652 / 2359	5.8 / 0.9 / 5.8 / 0.8
10 W ◐	0512 / 1158 / 1734	5.3 / 1.4 / 5.2		25 TH ◐	0529 / 1221 / 1753	5.6 / 1.1 / 5.6
11 TH	0024 / 0603 / 1248 / 1833	1.3 / 5.0 / 1.6 / 4.9		26 F	0059 / 0636 / 1331 / 1906	1.1 / 5.3 / 1.3 / 5.3
12 F	0121 / 0705 / 1357 / 1943	1.6 / 4.8 / 1.8 / 4.7		27 SA	0214 / 0758 / 1451 / 2035	1.3 / 5.1 / 1.4 / 5.2
13 SA	0236 / 0819 / 1518 / 2100	1.7 / 4.7 / 1.8 / 4.8		28 SU	0336 / 0924 / 1614 / 2155	1.4 / 5.1 / 1.3 / 5.3
14 SU	0350 / 0933 / 1629 / 2208	1.7 / 4.8 / 1.6 / 5.0		29 M	0457 / 1033 / 1725 / 2257	1.2 / 5.4 / 1.0 / 5.6
15 M	0456 / 1035 / 1729 / 2302	1.5 / 5.1 / 1.3 / 5.3		30 TU	0555 / 1127 / 1819 / 2347	1.0 / 5.6 / 0.8 / 5.7
				31 W	0642 / 1211 / 1903	0.9 / 5.8 / 0.6

Chart Datum is 2·69 metres below IGN Datum. HAT is 6·4 metres above Chart Datum.

STANDARD TIME UT –01
Subtract 1 hour for UT
For French Summer Time add ONE hour in **non-shaded areas**

DUNKERQUE   LAT 51°03′N   LONG 2°22′E
TIMES AND HEIGHTS OF HIGH AND LOW WATERS

Dates in red are **SPRINGS**
Dates in blue are NEAPS

YEAR 2016

N France

SEPTEMBER

Time	m		Time	m
1 TH ●	0028 0724 1247 1943	5.9 0.8 5.9 0.5	**16** F ○	0000 0653 1214 1917	6.0 0.8 6.0 0.5
2 F	0105 0800 1321 2020	6.0 0.8 6.0 0.5	**17** SA	0036 0734 1249 1958	6.2 0.6 6.2 0.3
3 SA	0140 0835 1356 2055	6.0 0.8 6.0 0.5	**18** SU	0112 0815 1327 2040	6.3 0.5 6.3 0.2
4 SU	0215 0908 1428 2128	6.0 0.8 6.0 0.6	**19** M	0152 0857 1407 2122	6.3 0.5 6.3 0.3
5 M	0247 0941 1500 2159	5.9 0.9 5.9 0.7	**20** TU	0235 0939 1451 2205	6.3 0.5 6.3 0.4
6 TU	0317 1012 1531 2229	5.8 1.0 5.7 0.9	**21** W	0320 1023 1538 2250	6.1 0.7 6.1 0.6
7 W	0350 1042 1606 2301	5.6 1.2 5.5 1.1	**22** TH	0410 1109 1632 2339	5.9 0.9 5.9 0.9
8 TH	0426 1115 1646 2337	5.4 1.4 5.3 1.4	**23** F ◑	0508 1202 1735	5.6 1.1 5.6
9 F ◐	0511 1156 1742	5.1 1.6 5.0	**24** SA	0040 0615 1312 1850	1.3 5.3 1.4 5.3
10 SA	0025 0615 1253 1855	1.6 4.8 1.9 4.7	**25** SU	0157 0741 1435 2023	1.5 5.0 1.5 5.1
11 SU	0133 0730 1417 2013	1.9 4.6 2.0 4.7	**26** M	0325 0909 1601 2142	1.6 5.0 1.4 5.3
12 M	0301 0848 1547 2131	1.9 4.7 1.8 4.9	**27** TU	0442 1017 1710 2244	1.4 5.3 1.1 5.5
13 TU	0421 1001 1657 2233	1.6 5.0 1.4 5.3	**28** W	0539 1110 1801 2331	1.1 5.5 0.8 5.7
14 W	0522 1056 1750 2320	1.3 5.4 1.0 5.7	**29** TH	0622 1150 1844	0.9 5.7 0.7
15 TH	0610 1138 1835	1.0 5.7 0.7	**30** F	0009 0702 1224 1921	5.9 0.9 5.9 0.6

OCTOBER

Time	m		Time	m
1 SA ●	0042 0736 1256 1954	6.0 0.9 6.0 0.6	**16** SU ○	0008 0709 1222 1934	6.3 0.6 6.3 0.3
2 SU	0115 0808 1328 2026	6.0 0.9 6.0 0.6	**17** M	0047 0752 1303 2017	6.4 0.5 6.4 0.3
3 M	0145 0840 1358 2057	6.0 0.9 6.0 0.7	**18** TU	0129 0835 1345 2101	6.4 0.5 6.5 0.3
4 TU	0214 0911 1427 2128	6.0 0.9 5.9 0.8	**19** W	0213 0919 1431 2145	6.4 0.5 6.4 0.4
5 W	0243 0942 1458 2157	5.9 1.0 5.8 1.0	**20** TH	0300 1004 1521 2231	6.2 0.6 6.2 0.7
6 TH	0314 1011 1530 2227	5.7 1.1 5.6 1.2	**21** F	0352 1052 1616 2322	5.9 0.8 5.9 1.0
7 F	0346 1042 1604 2302	5.5 1.3 5.4 1.4	**22** SA ◑	0449 1146 1718	5.6 1.1 5.6
8 SA	0421 1121 1647 2346	5.2 1.5 5.1 1.6	**23** SU	0021 0554 1253 1832	1.4 5.2 1.4 5.3
9 SU ◐	0517 1212 1812	4.9 1.8 4.8	**24** M	0136 0717 1413 2001	1.6 5.0 1.5 5.1
10 M	0047 0647 1324 1931	1.9 4.7 1.9 4.7	**25** TU	0301 0842 1535 2118	1.7 5.0 1.4 5.2
11 TU	0209 0804 1457 2049	1.9 4.7 1.8 4.9	**26** W	0416 0949 1644 2218	1.5 5.2 1.2 5.4
12 W	0339 0920 1618 2158	1.7 5.0 1.4 5.3	**27** TH	0514 1041 1737 2305	1.3 5.4 0.9 5.6
13 TH	0448 1022 1718 2250	1.3 5.4 1.0 5.7	**28** F	0558 1122 1819 2343	1.1 5.6 0.8 5.7
14 F	0541 1107 1807 2331	1.0 5.7 0.7 6.0	**29** SA	0637 1157 1855	1.0 5.7 0.8
15 SA	0626 1145 1851	0.8 6.0 0.5	**30** SU ●	0016 0710 1231 1926	5.9 1.0 5.9 0.8
			31 M	0049 0741 1303 1957	5.9 0.9 5.9 0.8

NOVEMBER

Time	m		Time	m
1 TU	0118 0813 1331 2029	5.9 0.9 5.9 0.8	**16** W	0109 0816 1328 2042	6.4 0.5 6.5 0.4
2 W	0145 0845 1400 2101	5.9 0.9 5.9 0.9	**17** TH	0156 0902 1417 2128	6.3 0.5 6.4 0.5
3 TH	0214 0917 1432 2131	5.9 1.0 5.8 1.0	**18** F	0245 0949 1508 2215	6.2 0.6 6.2 0.8
4 F	0247 0947 1505 2202	5.8 1.1 5.7 1.2	**19** SA	0336 1038 1602 2304	5.9 0.8 6.0 1.1
5 SA	0319 1020 1539 2237	5.6 1.2 5.5 1.4	**20** SU	0429 1129 1700 2359	5.6 1.0 5.6 1.4
6 SU	0354 1058 1620 2321	5.3 1.4 5.2 1.6	**21** M ◑	0528 1228 1805	5.3 1.3 5.3
7 M ●	0440 1147 1732	5.1 1.6 5.0	**22** TU	0103 0639 1338 1925	1.6 5.0 1.4 5.1
8 TU	0017 0604 1250 1851	1.7 4.9 1.7 4.9	**23** W	0217 0801 1453 2039	1.8 5.0 1.5 5.1
9 W	0128 0720 1410 2003	1.8 4.8 1.7 5.0	**24** TH	0332 0908 1603 2139	1.7 5.1 1.3 5.2
10 TH	0252 0832 1534 2116	1.7 5.0 1.4 5.3	**25** F	0435 1002 1701 2229	1.5 5.2 1.1 5.4
11 F	0407 0940 1641 2214	1.4 5.3 1.0 5.7	**26** SA	0526 1049 1748 2312	1.3 5.4 1.1 5.6
12 SA	0507 1032 1736 2301	1.1 5.7 0.7 6.0	**27** SU	0609 1130 1826 2350	1.2 5.6 1.0 5.7
13 SU	0559 1116 1825 2343	0.8 6.0 0.5 6.2	**28** M	0644 1207 1858	1.1 5.7 0.9
14 M ○	0645 1159 1911	0.6 6.3 0.4	**29** TU ●	0024 0716 1240 1930	5.8 1.0 5.8 0.9
15 TU	0025 0731 1242 1956	6.4 0.5 6.4 0.4	**30** W	0054 0749 1310 2004	5.8 0.9 5.8 0.9

DECEMBER

Time	m		Time	m
1 TH	0123 0823 1340 2039	5.9 0.9 5.9 0.9	**16** F	0145 0850 1407 2114	6.2 0.4 6.3 0.6
2 F	0154 0858 1414 2113	5.9 0.9 5.9 1.0	**17** SA	0233 0937 1457 2200	6.1 0.5 6.2 0.8
3 SA	0229 0931 1450 2146	5.8 1.0 5.8 1.1	**18** SU	0321 1023 1547 2245	5.9 0.6 6.0 1.0
4 SU	0304 1006 1527 2223	5.7 1.1 5.6 1.2	**19** M	0408 1109 1636 2331	5.7 0.8 5.7 1.3
5 M	0341 1045 1608 2304	5.5 1.2 5.4 1.4	**20** TU	0457 1158 1729	5.5 1.1 5.4
6 TU	0424 1130 1704 2354	5.3 1.3 5.3 1.5	**21** W ◑	0021 0551 1252 1831	1.5 5.2 1.3 5.1
7 W ●	0525 1225 1811	5.1 1.4 5.2	**22** TH	0120 0659 1356 1944	1.7 5.0 1.5 5.0
8 TH	0055 0636 1332 1919	1.6 5.1 1.4 5.2	**23** F	0228 0813 1503 2051	1.8 4.9 1.5 4.9
9 F	0209 0745 1451 2030	1.6 5.1 1.3 5.3	**24** SA	0336 0917 1608 2148	1.7 5.0 1.4 5.1
10 SA	0326 0855 1604 2138	1.4 5.3 1.1 5.6	**25** SU	0442 1012 1707 2239	1.6 5.1 1.3 5.2
11 SU	0433 0959 1706 2235	1.2 5.6 0.8 5.8	**26** M	0535 1100 1754 2323	1.4 5.3 1.2 5.4
12 M	0532 1053 1802 2324	0.9 5.9 0.6 6.0	**27** TU	0618 1143 1833	1.2 5.5 1.1
13 TU	0626 1142 1853	0.7 6.1 0.5	**28** W	0002 0653 1220 1908	5.5 1.1 5.6 1.0
14 W ○	0011 0715 1229 1941	6.2 0.6 6.3 0.5	**29** TH ●	0035 0729 1252 1944	5.7 1.0 5.7 0.9
15 TH	0058 0803 1317 2028	6.2 0.5 6.4 0.5	**30** F	0106 0805 1325 2020	5.8 0.9 5.8 0.9
			31 SA	0139 0842 1400 2057	5.8 0.8 5.9 0.9

Chart Datum is 2·69 metres below IGN Datum. HAT is 6·4 metres above Chart Datum.

FREE monthly updates. Register at
www.reedsnauticalalmanac.co.uk

755

9.17.7 DUNKERQUE (PORT EST)

Nord **51°03′·62N 02°21′·09E** ※※※◊◊◊✿✿

CHARTS AC 323, 1872, 1350, 5605.11; SHOM 6651, 7057; Navi 1010; Imray C30

TIDES Dunkerque is a Standard Port (⟵). +0050 Dover; ML 3·2; Duration 0530. See Gravelines for Dunkerque Port Ouest.

SHELTER Good; access at all tides/weather, but fresh NW–NE winds cause heavy seas at ent and scend in hbr.

NAVIGATION Yachts must not use the W Port. From the east, WPT 51°04′·29N 02°21′·72E (SCM lt buoy) 210°/0·8M to hbr ent. From Nieuwpoort Bank WCM lt buoy, the E1-12 buoys (Lights, buoys and waypoints) lead S of Banc Hills but shore lights make navlights and shipping hard to see. Avoid very large ships manoeuvring into/out of Charles de Gaulle lock.

From the west, WPT 51°03′·83N 02°20′·25E (DW29 lt buoy) 111°/0·6M to hbr ent. Fetch Dyck PHM buoy, 5M N of Calais, thence to DKA SWM buoy, and via DW channel lt buoys past W Port to WPT. Streams reach about 3½kn.

LIGHTS AND MARKS A chimney (116m) 1M WSW of hbr ent and other industrial structures are conspic. The W jetty lt ho is a distinctive, brown brick twr, green top. The E jetty lt pylon is small, slender and less obvious; the jetty covers at HW but is lit by 9 bright sodiums. The ldg lts, W cols, R tops, and the tall lt ho, W tr with B top, lead 137°/1M toward the marinas. (The F Vi(olet) 179° and 185° ldg lts define big ship turning circles.)

IPTS (full) E jetty head and seaward of Port du Grand Large.

Lock Enter Trystram lock (5m) for marinas in Bassin du Commerce and Bassin de la Marine via 3 bridges: (a) at S end of the lock; (b) at SE side of Darse No 1; and (c) at ent to Bassin du Commerce.

The daily schedule (1 Apr–3 Sep, 7/7, LT) for lock and bridges is:

Outward
Bridges (c) and (b) both open 0815, 1045, 1415, 1615, 1830.
Bridge (a) and lock open 0850, 1105, 1435, 1635, 1900.

Inward
Lock opens 1020, 1530, 1830, 2100. Bridge (a) opens 30 mins later. Bridges (b) and (c) open 20 mins after (a); the 1110 opening is only on Sat/Sun.

Lock/bridge, light and sound signals: ● ● = No entry
2 long blasts = Request open lock or bridge
● + Fl ● = Standby to secure on the side of the Fl ●
● + Fl ● = Enter lock and secure on side of the Fl ●

COMMUNICATIONS Monitor *Dunkerque VTS* Ch **73** (H24) for traffic info (English spoken);
CROSS 03·21·87·21·87, VHF Ch 16; ⚓ 03·28·66·45·25;
Auto 08·92·68·08·59;
Police 03·28·59·13·22;
⊖ 03·28·29·25·50;
Brit Consul 01·44·51·31·00;
Ⓗ 03·28·66·70·01;
HM 03·28·28·75·96.
www.portdedunkerque.fr
Port Control 03·28·28·78·78;
Aff Mar 03·28·63·13·93;
Locks VHF Ch 14. Marinas VHF Ch 09.

FACILITIES from seaward: The 3 *marinas are municipal and Ⓥ berths are at different rates; all on the same website www.dunkerque-marina.com

*Port du Grand Large H24 access (3m). ☎03.28.63.23.00. ⚓ 315+25Ⓥ €2·20. ⛽ ⬛ (0800-1200, 1400-1830 daily) 🏗(30t) ✕ 🍴 ⬚ ⬌ YC.

YC Mer du Nord (YCMN) ☎ 03·28·66·79·90. 250+50Ⓥ €2·40 plus €0·20 pp/day tax, ⛽ €1·00. H24 access (3m). secretariat@ycmn.com www.ycmn.com ⬚ ⚒ ◭ ⬛⬛ 0800-1200, 1400-1800 daily, ✕ 🍴 ⬚ wi-fi 🏗(7t).

*Bassin de la Marine 170 incⓋ €2·40. ☎03·28·24·58·80. Access to/from canals via lock at SW end of Darse No 1.

*Bassin du Commerce 140+30 Ⓥ €1·90. Quai des Hollandais (SE bank pontoon) €1·15. YC de Dunkerque (YCD) marina (5m) ☎03.28.21.13.77. ⬛ ⚓ ⬚.

Town ⚓⬛✕🍴🛒 ⬛SHOM ⚒ 🗣 Ⓔ ✕⬚ ⬚✉Ⓑ⇄✈ (Lille). Vehicle ferry (no foot passengers) to Dover.

DUNKERQUE (PORT EST)

9.17.8 GRAVELINES

Nord **51°00'·94N 02°05'·55E** ⚙✿⚓⚓✿✿

CHARTS AC 323, 5605.11, 1350; SHOM 6651, 7057; Navi 1010

TIDES HW +0045 on Dover (UT); ML 3·3m; Duration 0520

Standard Port DUNKERQUE (←)

Times			Height (metres)			
High Water		Low Water		MHWS	MHWN	MLWN	MLWS
0200	0800	0200	0900	6·0	5·0	1·5	0·6
1400	2000	1400	2100				

Differences **GRAVELINES** (3M SW of Dunkerque Ouest)

–0010	–0015	–0010	–0005	+0·5	+0·3	+1·0	0·0

DUNKERQUE PORT OUEST (3M ENE of Gravelines)

–0010	–0015	–0010	–0005	+0·4	+0·3	0·0	0·0

SHELTER Good, but do not enter in strong winds. Appr chan has only 0·5m at LWS. Access approx HW ±3 for shoal draft, ±2 if draft >1·5m; max draft 2·5m. For a first visit HW is recommended.

In strong NE or SW winds the ancient, manually operated bridge over the entry gate cannot be swung, stopping entry and exit to/from the basin.

NAVIGATION WPT 51°01'·30N 02°05'·13E, 142°/0·43M to bkwtr heads; old lt ho (unlit, conspic B/W spiral) in transit with wine-glass shaped water twr 142°. Beware strong E-going stream across ent near HW.

There is often a dredger working at entrance. Do not cut between the E bkwtr head and a NCM lt bcn close N of it.

The approach chan dries to soft mud, all but a trickle; it is about 15m wide and marked by lateral dolphins. Keep to the W on entry and to the E when inside. 650m before the ent to Bassin Vauban the chan turns 60° stbd.

Use pontoon (waiting only) on port hand side of approach to ent, the 9·8m wide entry gate into the marina stays open approx HW ±3 in season, HW ±1½ out of season. A digital gauge shows depth over the sill which dries 0·6m. Ⓥ berth to stbd.

The River Aa, pronounced Ar-Ar, flows through the Bassin via sluice gates upstream which are usually shut. Access to the river/ French canals is not by lock, but through the sluice gates, a 4.3m wide gap, opened briefly at HW for free flow. It is only suitable for small craft; 48 hours, (working days) notice is also required.

LIGHTS AND MARKS Lts as chartlet and Lights, buoys and waypoints. A nuclear power stn with 6 towers (63m) is 1M NE. 1·8M further NE is the ent to Dunkerque Ouest, commercial & ferry port.

The disued lt ho, with continuous black and white spiral band, is very distinctive. The W bkwtr lt shows an approach lt across a sector of arc 139° (085°-224°) G landards.

COMMUNICATIONS

☏ 08·92·68·02·59; Police 03·28·23·15·17/03·28·24·56·45; Brit Consul 01·44·51·31·00; Aff Mar 03·28·26·73·00. Entry gate 03·28·23·13·42, VHF Ch 09 (HO); Marina VHF Ch 09 (HO); Tourist Office 03·28·51·94·00.

FACILITIES

Bassin Vauban General Council Harbour Office, ☎03·59·73·42·42. dvi.portdegravelines@cgn.fr M-F, 0800-1200, 1345-1715. 430+20 Ⓥ, €1·90. Max LOA 15m, max beam 9·8m for entry gate. port. de.plaisance.gravelines@wanadoo.fr www.plaisance-opale.com ✗ 🖵 bike hire. Large well equipped 🍴 ⛴(10t, 3t) 🛥(7t) 🔌 🕭 🔧 🔨.

Berth on Ⓥ pontoon No 1, first to stbd, or as directed. The marina almost dries, but fin keelers sink upright into soft, glorious mud. It may be advisable to close the engine cooling water seacock after shutting down and to clean the water strainer thereafter.

Town 🏤 🏨 (H24 Super-U 600m) ✗ 🖵 🍴 🔲 ✉ Ⓑ ⇌.

9.17.9 CALAIS

Pas de Calais 50°58'·33N 01°50'·42E ✿✿✿◊◊✿✿✿

CHARTS AC 1892, 323, 1351, 5605.10; SHOM 7323, 6651, 7258; Navi 1010; Imray C8

TIDES +0048 Dover; ML 4·0; Duration 0525

Standard Port DUNKERQUE (←)

Times				Height (metres)			
High Water		Low Water		MHWS	MHWN	MLWN	MLWS
0200	0800	0200	0900	6·0	5·0	1·5	0·6
1400	2000	1400	2100				
Differences CALAIS							
–0020	–0030	–0015	–0005	+1·3	+1·0	+0·5	+0·2
SANDETTIE BANK (11M N of Calais)							
–0010	–0015	–0015	–0010	–0·6	–0·6	–0·2	–0·1
WISSANT (8M WSW of Calais)							
–0035	–0050	–0030	–0010	+2·0	+1·5	+0·9	+0·4

SHELTER Very good in the marina (Bassin de l'Ouest, 3-6m). In strong NW-NE winds hbr ent is very rough with heavy swell; access H24. 22 W waiting buoys outside tidal gate (times below).

NAVIGATION WPT 50°57'·64N 01°46'·14E (CA 1 buoy) 077°/2·7M to Jetée Ouest hd, keeping clear to the S of ferry track 083°/263°.

Beware Ridens de la Rade, about 4ca N of ent, a sandbank (0·1m) on which seas break. From the E it is better to keep N of this bank until able to round CA4 PHM lt buoy and appr from the W. But, with sufficient rise of tide, Ridens can be crossed on a SW'ly track.

Byelaws require yachts to have engine running (even if sailing) and to keep clear of ferries/commercial vessels. Speed limits: Avant-Port & Bassin Est 10kn; Arrière-Port 8kn; elsewhere 5kn.

LIGHTS AND MARKS Conspic daymarks: Cap Blanc-Nez and Dover Patrol monument, 5·5M WSW of hbr ent; the lt ho(W, B top) in town centre; two silos on N side of Bassin Est; R/W chimney (78m) 700m SE of Bassin Est; Hbr Control in pyramidal bldg at ent to Arrière Port. Lts as chartlet and Lights, buoys & waypoints.

IPTS (full code) shown from Ⓐ (Jetée Est) Ⓑ and Ⓒ as on chartlet, must be strictly obeyed; no VHF chatter should be necessary. On arrival, first comply with Ⓐ, then Ⓑ to enter Arrière Port.

If no sigs are shown (ie no ferries under way) yachts may enter/exit. Or follow a ferry entering, keeping to the stbd side of the fairway and not impeding said ferry. Keep a very sharp lookout, especially for ferries closing fast from astern.
One ● (alongside lowest IPTS lt) = dredger working.

On leaving, first comply with Ⓒ, then Ⓑ, Ⓐ for seaward. Yachts are often cleared outward as a convoy; monitor Ch 17.

Bassin de l'Ouest (marina). 17m wide entry gate (sill 2m above CD) stays open HW –2 to +2½. Gate signals:

●	=	All movements prohib.
◐	=	10 mins before gate opens.
●	=	Movement authorised.
4 blasts	=	Request permission to enter.

The co-located Hénon swing bridge opens at HW –2, –1, HW, +1¼, +2½. Times are usually strictly observed. The monthly schedule is posted on www.plaisance-opale.com. The lock gates to Bassin de l'Ouest, giving access to the marina, are reported as having been removed until further notice. A gauge shows depth of water over the sill.

Bassin Carnot: If bound for the canals, lock gates open HW –1½ to HW +¾. Lock sigs are as per the CEVNI code. Call VHF Ch 17 or sound 2 blasts to request permission to enter.

COMMUNICATIONS Whilst underway in apprs and hbr, monitor VHF Ch **17** 16 (H24) *Calais Port Control* for VTS, marina and Carnot lock (occas); Channel Navigation Info Service (CNIS): *Gris Nez Traffic* Ch **13** 79 16 (H24). Info broadcasts in English & French Ch 79, at H + 10, also at H + 25 when vis is < 2M; CROSS 03·21·87·21·87, *CROSS Gris Nez* VHF Ch 15, 67, **68**, 73; Ⓛ 03·21·96·31·20; ☎ 03·21·33·24·25; Auto 08·92·68·02·62; Police 03·21·34·86·00; ⊖ 03·21·34·75·40; Brit Consul 01·44·51·31·00; Ⓗ 03·21·46·33·33; HM (Port) 03·21·96·31·20; Aff Mar 03·28·59·19·33.

FACILITIES Marina ☎03·21·34·55·23 www.calais-port.com calais-marina@calais.cci.fr 250 + 30 Ⓥ €2·36 Berth/raft on pontoon A, 1st to stbd, or on fingers S side of B pontoon. ▴ Ⓛ ⏦ Ⓛ ◫(9t). **YC** ☎03·21·82·28·05, www.calais-voile.fr, ◫. For mast unstepping call Calais Nautic, ☎03·21·96·07·57.

Town ✕ ◫ 🛒 🛒 ▮ ☉ ✕ ⚓ Ⓛ ✉ Ⓑ ⇌ ✈ ferries to Dover.

CALAIS-NORD

CALAIS TIDAL COEFFICIENTS 2016

Date	Jan am	Jan pm	Feb am	Feb pm	Mar am	Mar pm	Apr am	Apr pm	May am	May pm	June am	June pm	July am	July pm	Aug am	Aug pm	Sept am	Sept pm	Oct am	Oct pm	Nov am	Nov pm	Dec am	Dec pm
1	57	52	46	41	53	47	39	36	43	46	63	69	68	73	78		87	90	88	90	85	84	80	80
2	47	43	37	35	42	37	37	40	50	56	75	81	77	82	82	86	92	93	90	90	82	81	79	78
3	40	37	34	36	34	33	46	53	63	71	87	92	86		90	92	93	92	89	87	78	76	76	74
4	37	38	39	44	35	40	61	70	78	86	96		90	93	93	94	90	88	85	82	72	69	71	68
5	40	43	50	57	46	53	79	88	93		100	102	95	96	93	92	85	82	78	75	64	60	64	61
6	48	53	64	71	61	70	96		100	105	103	103	96	96	90	87	77	73	70	65	55	51	57	54
7	58	63	78	84	78	87	103	109	109	112	102	100	94	92	83	79	68	62	60	55	46	42	51	49
8	68	74	90		94		114	117	113	112	97	92	88	84	75	70	57	51	49	44	39	38	48	49
9	78		96	100	101	107	118	117	110	106	87	82	80	75	64	59	45	40	39	35	40	43	52	55
10	83	87	104	107	112	115	115	111	101	95	76	70	70	65	53	48	35	32	32	32	48	55	61	67
11	90	93	108	108	116	116	105	99	88	81	64	59	59	54	43	39	31	32	35	40	63	71	73	80
12	95	95	106	103	114	111	91	82	74	66	54	49	49	45	35	34	36	41	47	55	79	86	86	92
13	96	95	99	94	105	99	73	65	59	53	46	44	42	39	34	36	48	56	64	72	94	100	97	101
14	93	91	88	81	91	83	57	50	48	45	43	43	38	39	40	46	64	72	81	88	105	109	104	
15	87	83	73	66	74	66	45	43	43	44	45	47	41	44	52	58	79		96		111		106	106
16	78	74	59	53	57	50	43	45	46	48	50	53	48	52	64		87	93	102	108	112	111	105	103
17	68	64	50	49	46	44	49	54	52	56	57	61	57	62	71	77	99	104	111	114	109	105	99	95
18	60	57	50	53	45	48	59	64	60	63	64		67		83	88	108	110	114	113	100	94	90	84
19	56	56	58	64	53	58	68		67		68	71	71	75	93	97	111	111	111	106	87	79	78	71
20	59	62	69	74	64	69	72	76	70	73	74	77	79	83	100	102	108	105	101	94	72	64	65	59
21	67	72	79		75	79	79	81	75	77	79	80	86	88	103	102	100	93	86	77	58	52	53	49
22	76	81	83	87	82		83	84	79	80	82	82	90	91	101	98	86	78	69	61	48	46	45	43
23	85		89	91	85	87	85	85	81	81	83	82	91	91	94	89	70	61	54	48	45	46	42	43
24	88	90	92	92	89	89	85	84	81	80	81	80	90	87	83	77	54	49	45	45	49	52	45	47
25	92	92	91	90	89	89	82	80	79	77	78	75	85	81	70	63	46	46	47	51	56	60	51	55
26	92	91	87	85	87	86	78	75	75	72	73	70	77	73	58	53	49	54	55	60	63	67	58	62
27	89	87	81	78	83	80	71	67	69	65	67	64	68	64	50	50	59	65	65	70	70	73	66	69
28	84	80	73	68	77	73	63	58	62	58	62	60	60	58	53	57	70	76	74		75	77	72	
29	76	71	63	58	68	63	53	49	56	54	60	60	57	57	62	68	80		77	80	78		75	77
30	67	61			58	52	46	44	53	53	62	65	60	63	73		84	86	82	84	79	80	79	81
31	56	51			47	42			55	59			68	73	79	83			84	85			82	82

Tidal coefficients indicate the magnitude of the tide on any particular day, without having to look up and calculate the range, and thus determine whether it is springs, neaps or somewhere in between. This table is valid for all areas covered by this Almanac. Typical values are:

120	Very big spring tide
95	**Mean spring tide**
70	Average tide
45	**Mean neap tide**
20	Very small neap tide

See also coefficients for Cherbourg (p 786), St Malo (p 805), Brest (p 873) and Port Bloc (p 936).

9.17.10 BOULOGNE-SUR-MER

Pas de Calais 50°44'·58N 01°34'·10E (hbr ent) ✿✿✿✿✿

CHARTS AC 1892, 2451, 438; SHOM 6824, 7247, 7416,7323; Navi 1010, 1011; Imray C8, C31

TIDES 0000 Dover; ML 4·9; Duration 0515

Standard Port DUNKERQUE (←)

Times				Height (metres)			
High Water		Low Water		MHWS	MHWN	MLWN	MLWS
0200	0800	0200	0900	6·0	5·0	1·5	0·6
1400	2000	1400	2100				
Differences Boulogne							
–0045	–0100	–0045	–0025	+2·8	+2·2	+1·1	+0·5

SHELTER Good, except in strong NW'lies. Entry H24 at all tides and in most weather. Exposed ⚓ in N part of outer hbr; no ⚓ SW of pecked magenta line.

NAVIGATION WPT 50°44'·71N 01°33'·00E, 100°/0·71M to hbr ent. No navigational dangers, but keep W and S of the free-standing head of Digue Nord, as outer part of bkwtr covers at HW. After rounding Digue Carnot, head S for 325m to a white □ mark (often faded) on bkwtr, thence track 123° to inner ent; caution drying banks on either side. Obey IPTS; keep clear of ferries & FVs. Yachts must call Port Ctrl, VHF Ch 12, on entry and departure.

LIGHTS AND MARKS 'Approches Boulogne' SCM buoy is 295°/2·1M from Digue Carnot lt. Cap d'Alprech lt ho is 2·5M S of hbr ent. Monument is conspic 2M E of hbr ent. Cathedral dome is conspic, 0·62M E of marina. There are no ldg lts/marks.

IPTS are shown from SW jetty hd (FG ☆) and from Quai Gambetta, opposite ☆ 3FG ▽, visible from marina. A Y lt ● by top IPTS = dredger working, but does not prohibit movements.

COMMUNICATIONS CROSS 03·21·87·21·87; ☎ 08·92·68·02·62; Auto 08·92·65·08·08; Forecasts by CROSS Gris Nez VHF Ch 79 at H+10 (078-1910LT); Maritime Police 03·21·30·87·09; Emergency 17; ☎ 03·21·80·89·90; Brit Consul 01·44·51·31·00; Ⓗ 03·21·99·33·33; Aff Mar 03·21·30·53·23; Control twr 03·21·10·35·47, VHF Ch 12 (H24); Lock into Bassin Napoléon VHF Ch 12 (H24); Marina Ch 09.

FACILITIES Visitors berth in marina (2·7m) on W side of river. FVs berth opposite at Quai Gambetta. Beware turbulence when R Liane is sluicing; indicated at night by 2 Bu lts (hor) S of marina.

Marina ☎03·21·99·66·50 mob 06·76·98·74·98 www. portboulogne.commarina@portboulogne.com114 inc 80Ⓥ €2·26 plus €0·20pp/day tax. ⚓ HW±3, ⛽(card access) Ⓔ ⚒ 🖪 ⛽ Divers ⛴ 🏗(20t). YC Boulonnais ☎03·21·31·80·67, ✕ 🍴 🍺.

Bassin Napoléon pontoons are for locals and long-stay Ⓥ (week or month; 12-25m LOA) by arrangement. Sanson lock (Ch 12) works HW–3¼ to HW+4½ with ent/exits approx every 35 mins, and freeflow HW –2 to +¼. Daily schedules from HM or website.

Town ✕ 🍴 🛒 🚂 🏧 ⚓ SHOM ✉ Ⓑ ⚓ ✈ Le Touquet.

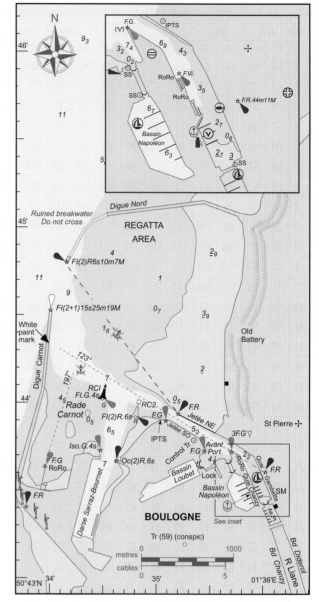

9.17.11 ÉTAPLES

Pas de Calais 50°31'·00N 01°38'·00E ✿✿✿✿✿

CHARTS AC 2451; SHOM 7416; Navi 1011; Imray C31

TIDES –0010 Dover; ML 5·3; Duration 0520

Standard Port DIEPPE (→)

Times				Height (metres)			
High Water		Low Water		MHWS	MHWN	MLWN	MLWS
0100	0600	0100	0700	9·3	7·4	2·5	0·8
1300	1800	1300	1900				
Differences LE TOUQUET (ÉTAPLES)							
+0010	+0015	+0030	+0030	+0·2	+0·2	+0·3	+0·4
BERCK							
+0005	+0015	+0025	+0030	+0·4	+0·5	+0·3	+0·4

SHELTER Good, except in strong W/NW'lies.

NAVIGATION WPT 50°32'·88N 01°32'·36E is on the boundary between W and G sectors of Camiers lt; 090°/0·7M to Mérida unlit

ÉTAPLES *continued*

WCM buoy (50°32'·88N 01°33'·47E) marking a drying wreck 2M NW of Le Touquet lt ho. Estuary dries 1M offshore; do not attempt entry in even moderate SW/W winds, when seas break heavily a long way out. Careful identification of marks is essential.

The outer part of the buoyed chan constantly shifts, such that the chan ent may lie N or S of the ⊕. Nos 1 & 2 buoys (QG and QR) should be visible from the WPT. Bcn, Fl R 4s, 4ca SW of Camiers lt, marks NE end of covering training wall (also marked by posts).

LIGHTS AND MARKS See Lights, buoys and waypoints. Le Touquet is at the S end of 175m high hills, visible for 25M. Le Touquet lt ho, orange with brown band, is conspic; Camiers' R pylon is hard to see by day.

COMMUNICATIONS CROSS 03·21·87·21·87; Auto 08·36·68·08·62; ⊖ 03·21·05·01·72; Brit Consul 01·44·51·31·00; Dr 03·21·05·14·42; Aff Mar Étaples 03·21·94·61·50; HM VHF Ch 09.

FACILITIES **Marina** ☎03·21·84·54·33. cn.canche@wanadoo.fr 182 + 18Ⓥ €2·04. (1·2m) close downriver of low bridge; beware up to 5kn current. Best to berth on first pontoon hammerhead near HW slack. ⬛ ⬛ Ⓗ ⚓ ⚒ ⚔(8t); Quay ⬤(130t). **YC** ☎03.21.94.74.26, ⬛.

Town ⬔ ⬔ ✕ ⬜ ⬚ ⊠ Ⓑ ⇌ ✈ Le Touquet.

Le Touquet Drying moorings to stbd off Le Touquet YC. VHF Ch 09, 77. **Cercle Nautique du Touquet** ☎03·21·05·12·77, ✕ ⬜ ⚓ ⚓ ⚓. **Services** ⬛ ⬛ Ⓗ ⚒ ⚔ ⚠. **Town** ⬔ ⬔ ✕ ⬜ ⬚ ⬛ ⊠ Ⓑ ⇌ ✈.

9.17.12 ST VALÉRY-SUR-SOMME
Somme **50°11'·34N 01°38'·90E** ✲✲⬠⬠✿✿✿

CHARTS AC 2451; SHOM 7416; Navi 1011; Imray C31

TIDES LE HOURDEL –0010 Dover; ML —.

Standard Port DIEPPE (→)

Times				Height (metres)			
High Water		Low Water		MHWS	MHWN	MLWN	MLWS
0100	0600	0100	0700	9·3	7·4	2·5	0·8
1300	1800	1300	1900				
Differences ST VALÉRY-SUR-SOMME							
+0035	+0035	ND	ND	+0·9	+0·7	ND	ND
LE HOURDEL							
+0020	+0020	ND	ND	+0·8	+0·6	ND	ND
CAYEUX							
+0000	+0005	+0015	+0010	+0·5	+0·6	+0·4	+0·4

SHELTER The 3 hbrs (Le Hourdel, Le Crotoy and St Valéry-sur-Somme) are well sheltered.

NAVIGATION WPT 50°14'·00N 01°28'·08E ('ATSO' SWM buoy, Mo (A) 12s). The estuary dries up to 3M seaward of Le Hourdel. It is open to the W and can be dangerous in onshore winds >F6. Silting is a major problem. The sands build up in ridges offshore, but inside Pte du Hourdel are mostly flat, except where R Somme and minor streams scour their way to the sea. If Dover tidal range >4·4m, there is enough water at HW ±1 for vessels <1·5m draft.

Leave 'ATSO' buoy at HW St Valéry –2, then head SE twds R/W lt ho on dunes near Cayeux-sur-Mer to beginning of buoyed chan (S1) Follow chan buoys S1 to S50 (some lit) in strict order (no corner-cutting!) to small unlit WCM BIFurcation buoy, E of Pte du Hourdel. Here fork stbd to St Valéry and its covering training wall with four light beacons; as chartlet and Lights, buoys & waypoints. Continue between the tree-lined promenade to stbd and a covering dyke marked by PHM bns. Bias your track towards the town side, but comply with any buoys. At W lt bcn on head of Digue du Nord turn stbd past town quay into marina, dredged 2m. Enter St

Valéry Marina (max draft 2·0m) at HW ±1; usually met and given a berth on pontoons 4–6. If N-bound leave St Valéry HW–2 for a fair tide.

Abbeville Canal (max draft 2·0m) is navigable, by day only, to Abbeville. Lock 03·22·60·80·23; pre-call Amiens ☎06·74·83·60·69.

LIGHTS AND MARKS Cayeux-sur-Mer lt ho (W + R top, Fl R 5s) and Le Crotoy water twr are conspic. Pte du Hourdel lt ho is partly obscured by dunes until close to. See also Lights, buoys and waypoints.

COMMUNICATIONS CROSS 03·21·87·21·87; Auto 08·92·68·08·80; Police 17 or 03·22·60·82·08; ⊖ 03·22·24·04·75; British Consul 01·44·51·31·00; Dr 03·22·26·92·25; Aff Mar 03·21·30·87·19 (Boulogne); HM VHF Ch 09 (St Valéry HW ±2); Le Crotoy YC (Jul/ Aug only).

FACILITIES **Marina/YC Sport Nautique Valéricain** ☎03·22·60·24·80. snval@wanadoo.fr www.portsaintvalery.com ⚓ Access HW ±2 for 1·5m draft. 250 + 30Ⓥ ⬭ €2·35, multihulls €2·80. ⬜ ⓞ(€5·00) ⊞ free wi-fi, ⬛ ⚒ Ⓗ Ⓔ ⬛(10t).

Town ⬔ ⬔ (500m) ✕ ⬜ ⬚ ⬛ Ⓗ ⊠ Ⓑ ⇌ and ✈ (Abbeville). Ferry see Dieppe, Boulogne or Calais.

ADJACENT HARBOURS

LE HOURDEL A basic FV hbr. Access HW±1½ for 1·5m draft. After lt ho, head SW via unmarked side chan to hug the end of shingle spit. Dry out on hard sand. Impossible to leave before HW –2.

LE CROTOY Somme **50°13'·07N 01°37'·99E** ✲✲⬠✿✿
From the BIF buoy follow the chan E, marked by buoys C1-C10. Enter hbr very close to FV stages (port-side). The badly silted marina can be entered by max draft 0·8m when tidal range at Dover >4·4m (approx Coeff 85). Secure at last FV stage and ask YC for berth on the two drying pontoons. Departure impossible before HW –2. **Marina** (280) ☎03·22·27·83·11 (office in YC bldg), ⬛(6t) ⬛. **YC Nautique de la Baie de Somme** ☎03·22·27·83·11, ⬜. **Town** ✕ ⬜ ⬚ ⬔ ⬔ ⊠ Ⓑ ⇌.

9.17.13   LE TRÉPORT

Seine Maritime **50°03'·89N 01°22'·20E** ※※✿✿✿✿✿✿

CHARTS AC 2451, 1354; SHOM 7207, 7416, 7417; Navi 1011; Imray C31

TIDES –0025 Dover; ML 5·0; Duration 0530

Standard Port DIEPPE (→)

Times				Height (metres)			
High Water		Low Water		MHWS	MHWN	MLWN	MLWS
0100	0600	0100	0700	9·3	7·4	2·5	0·8
1300	1800	1300	1900				
Differences LE TRÉPORT							
+0005	+0000	+0005	+0015	+0·3	+0·2	+0·1	+0·1

SHELTER Good in marina, E of FVs at S side of S Basin (3·7m).

NAVIGATION WPT 50°04'·27N 01°21'·90E, 146°/0·4M to ent. Coast dries to approx 300m off the pier hds. Shingle spit extends NW from E pier hd; keep well to the W. Entry difficult in strong on-shore winds which cause scend in Avant Port. Strong cross-current at entrance; 5kn speed limit. Ent chan and most of Avant Port dry; S chan is dredged 1·5m, but prone to silting; avoid grounding on the triangle between the two channels. Lock opens HW±4; yachts can wait against Quai Francois I with sufficient rise of tide. Berth, if possible, on smooth S wall of lock; vert wires rigged for warps. N wall is corrugated steel shuttering on which fenders can snag. Beware current when leaving.

LIGHTS AND MARKS Le Tréport lies between high chalk cliffs, with crucifix (floodlit) 101m above town and conspic church S of hbr. IPTS at the root of the E Jetty. There are 150° ldg marks, but they are hard to see and not strictly necessary.

Lock sigs shown from head of piled ent chan: 3 ● (vert) = enter; 3 ● (vert) = no entry.

COMMUNICATIONS Ⓛ 02·35·86·30·27; CROSS 03·21·87·21·87; Auto 06·70·63·43·40; Police 02·35·86·12·11; ⊖ 02·35·86·15·34; Brit Consul 01·44·51·31·00; Dr 02·35·86·16·23; Port HM 06·87·70·30·90; *Capitainerie Le Tréport* VHF Ch 12 (HW ±4) for lock/marina; Aff Mar 02·35·06·96·70.

FACILITIES Marina/lock ☎02·35·50·63·06.   www.treport.cci.fr accueilcci@treport.cci.fr 110+15Ⓥ €2·13, ▲ ▮ ⚓ Ⓛ Ⓑ Ⓔ. **Port de Commerce** is an overflow; half-price rates for yachts. ⚓(10t). **YC de la Bresle** ☎02·35·50·17·60, ⬜. **Town** ⓘ ⓘ ✕ ⬜ ⛱ ▮ ⊠ Ⓑ ⇌, ✈ Dieppe.

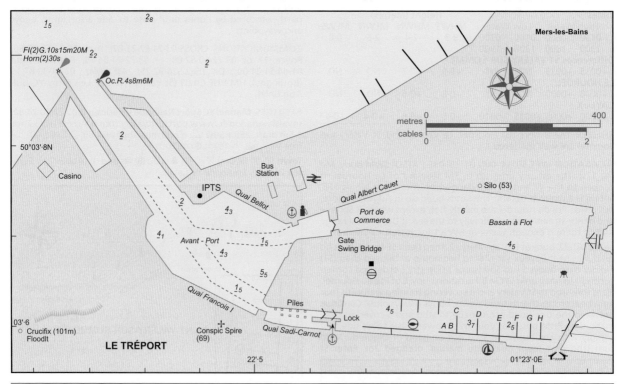

9.17.14   DIEPPE

Seine Maritime **49°56'·31N 01°05'·00E** ※※※✿✿✿✿✿✿✿

CHARTS AC 2451, 2148, 1355; SHOM 7417, 7317; Navi 1011, 1012; Imray C31

TIDES –0011 Dover; ML 4·9; Duration 0535; Standard Port (→)

SHELTER Very good. Access all tides, but ent exposed to NW-NE winds, which cause a heavy scend. A fixed wavebreak at the E end of the marina is effective. 3kn max speed in marina.

NAVIGATION WPT 49°56'·66N 01°04'·83E (off chartlet) 167°/4ca to W jetty head. Ent to hbr is simple, but beware strong stream across ent. Ent chan and Avant Port are dredged 4·5m. Be alert for ferry at the outer hbr, and small coasters or FVs in the narrow chan between outer hbr and marina.

Due to restricted visibility, monitor Ch 12 during arr/dep. It is vital to comply precisely with IPTS (below) both entering and leaving. Anchoring is prohibited in a triangular area 7ca offshore, as charted.

LIGHTS AND MARKS Dieppe lies in a gap between high chalk cliffs, identified by a castle with pinnacle roofs at the W side and a conspic ch spire on the E cliff. DI, ECM wreck buoy, VQ (3) 5s, bears 288°/2·5M from the hbr ent.

IPTS (full code) shown from root of W jetty and S end of marina wavebreak. Additional sigs may be combined with IPTS:

● ● to right   =   Dredger in channel.
Ⓦ to left   =   Bassin Duquesne entry gate open.

COMMUNICATIONS CROSS 03·21·87·21·87; Ⓛ 02·35·84·8·55; ☼ 03·21·31·52·23; Auto 08·36·68·08·76; Police 02·35·06·96·74; ⊖ 02·35·82·24·47; Brit Consul 01·44·51·31·00; Ⓗ 02·35·06·76·76; Port HM 02·35·84·10·55, *Dieppe Port* Ch **12** (HO); Aff Mar 02·35·06·96·70; Marina Ch 09 for a berth. Marina monitors Ch 12 and opens 1/7-24/8 0700-2200.

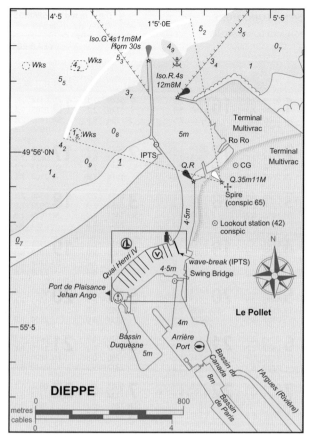

DIEPPE

FACILITIES

Marina Jehan Ango ☎02·35·84·10·55. smpd-plaisance@portde dieppe.fr www.portdedieppe.fr 450 ⬚ inc 50Ⓥ €3·04, on pontoons 10, 9, 8 in about 4·5m but not on hammerheads. If 10 & 9 are full, late arrivals ⬚/raft on E side of pontoon 8; Pontoon 1 W side finger is for boats >14m LOA. Use gangway intercom for shore access code number (if arriving out of hrs, for code call Ch 12). ⬚⬚ cr/card H24, call Ch 09 if unmanned. ⬚ ⬚ Ⓔ ⬚ ⬚ ⬚(12t).

Cercle de la Voile de Dieppe ☎02·35·84·32·99, ⬚ ⬚ ⬚ ⬚(code access).

Bassin Duquesne (70 ⬚) access via lifting bridge/entry gate, **and Bassin de Paris** (70 ⬚) are only for local yachts.

Town ⬚ ⬚ ⬚ ⬚ ⬚ ⬚ ⬚ @ ⬚ Ⓑ ⬚ ✈. Ferry Dieppe-Newhaven; www. transmancheferries.co.uk

Port de Plaisance Jehan Ango

From **Dieppe** to Pte d'Ailly light house the coast (AC 2451) is fringed by a bank, drying in places, up to 4ca offshore.

▶*Off Dieppe the ENE-going stream begins about HW Dieppe –0505, and the WSW-going at about HW Dieppe +0030, sp rates 2kn. E of Pte d'Ailly an eddy runs W close inshore on the first half of the E-going stream.*◀

Drying rks extend 5ca off Pte d'Ailly, including La Galère, a rk which dries 6·8m, about 3ca N of the Pointe (AC 2147). Dangerous wrecks lie between about 1·5M NNW and 1·2M WNW of the lighthouse, the former marked by a NCM light buoy. About 6M N of Pte d'Ailly, Grands and Petits Écamias are banks with depths of 11m, dangerous in a heavy sea.

Between Pte d'Ailly and **St Valéry-en-Caux** (lighthouse) there are drying rks 4ca offshore in places. About 1·5M E of Pte de Sotteville a rocky bank (4·2m) extends about 1M NNW; a strong eddy causes a race, Raz de Saint-Michel, over this bank.

There are rky ledges, extending 4ca offshore in places. Immediately E of St Valéry-en-Caux shallow sandbanks, Les Ridens, with a least depth of 0·6m, extend about 6ca offshore.

▶*At St Valéry-en-Caux entrance the E-going stream begins about HW Dieppe –0550, and the W-going stream begins about HW Dieppe –0015, sp rates 2·75kn. E of the ent a small eddy runs W on the E-going stream.*◀

From St Valéry-en-Caux to **Fécamp**, 15M WSW, the coast consists of high, sheer chalk cliffs broken by valleys. The nuclear power station (prohibited area marked by a NCM lt buoy) at Paluel, 3M W of St Valéry-en-Caux, is conspicuous.

Close NE of Fécamp off Pte Fagnet, Les Charpentiers are unmarked drying rocks extending almost 2 cables offshore.

▶*At Fécamp pierheads the NNE-going stream begins about HW Le Havre –0340, up to 2·9kn sp; the SSW-going about HW Le Havre +0220, much weaker. Diamond A on AC 1354 gives tidal stream data about 9ca WNW of the entrance.*◀

From Fécamp to Cap d'Antifer light house drying rocks extend up to 2½ca offshore. L'Aiguille, a prominent (51m) pinnacle rock is just off Étretat,1·7M NNE of Cap d'Antifer. A large offshore windfarm (WIP 2015/16) is being built 7·5M NW of Fécamp.

Port du Havre-Antifer, a VLCC harbour lies, 1·5M S of Cap d'Antifer (AC 2146, 2990). A 2M long breakwater extends about 1·5M seaward, and should be given a berth of about 1·5M, or more in heavy weather with wind against tide when there may be a race.

The buoyed approach channel starts 11M NW of the harbour. Commercial vessels in it have priority. Yachts should cross it 3·8M to the NW at A17/A18 lt buoys; at 90° to its axis; as quickly as possible, keeping well clear of ships in the channel. Crossing vessels should contact Le Havre d'Antifer and any priority vessel on VHF Ch 22.

▶*Off Cap d'Antifer the NE-going stream begins about HW Le Havre –0430, and the SW-going at about HW Le Havre +0140. There are eddies close inshore E of Cap d'Antifer on both streams.*◀

Le Havre is a large commercial port, as well as a yachting centre. From the NW, the most useful mark is the Le Havre Lanby 'LHA' 9M W of C de la Hève (lt). The well-buoyed/lit access channel runs 6M ESE from LH 3/4 buoys to the hbr ent.

Strong W'lies cause rough water over shoal patches either side of the channel. Coming from the N or NE, there is deep water close off C de la Hève, but from here steer S to join the main ent chan. Beware Banc de l'Éclat (1·0m) which lies on N side of main channel near LH 14 buoy about 1·5M from harbour entrance.

W and S of Digue Sud yachts should not impede vessels entering/ leaving Port 2000, a container terminal. If inbound to the marina it is best to cross early to the N side of the channel.

STANDARD TIME UT −01
Subtract 1 hour for UT
For French Summer Time add ONE hour in **non-shaded areas**

DIEPPE   LAT 49°56'N   LONG 1°05'E
TIMES AND HEIGHTS OF HIGH AND LOW WATERS

Dates in red are **SPRINGS**
Dates in blue are NEAPS

YEAR 2016

JANUARY

	Time	m		Time	m
1	0348	8.0	**16**	0349	8.7
	1030	2.3		1043	1.5
F	1604	7.7	SA	1613	8.5
	2250	2.3		2306	1.5
2	0431	7.6	**17**	0441	8.3
	1114	2.6		1135	1.8
SA	1651	7.3	SU	1712	8.1
◗	2338	2.7	◐		
3	0524	7.2	**18**	0001	1.9
	1209	2.9		0545	8.0
SU	1751	7.0	M	1239	2.1
				1823	7.8
4	0037	2.9	**19**	0111	2.1
	0633	7.1		0701	7.8
M	1317	3.0	TU	1358	2.1
	1907	6.9		1941	7.8
5	0150	2.9	**20**	0232	2.1
	0748	7.2		0818	8.0
TU	1432	2.8	W	1516	1.9
	2020	7.2		2056	8.0
6	0304	2.6	**21**	0346	1.9
	0851	7.6		0927	8.3
W	1538	2.4	TH	1625	1.6
	2118	7.6		2200	8.4
7	0405	2.2	**22**	0451	1.6
	0942	8.1		1025	8.7
TH	1633	1.9	F	1724	1.3
	2207	8.1		2253	8.7
8	0456	1.8	**23**	0546	1.3
	1028	8.5		1114	9.0
F	1722	1.5	SA	1815	1.0
	2251	8.5		2339	9.0
9	0542	1.5	**24**	0633	1.1
	1110	8.8		1156	9.2
SA	1807	1.2	SU	1859	0.9
	2333	8.8	○		
10	0626	1.3	**25**	0020	9.2
	1150	9.1		0714	1.0
SU	1851	1.0	M	1235	9.3
●				1938	0.8
11	0013	9.1	**26**	0057	9.2
	0709	1.1		0751	1.0
M	1231	9.3	TU	1312	9.2
	1934	0.8		2013	0.9
12	0055	9.3	**27**	0132	9.1
	0752	0.9		0826	1.1
TU	1313	9.4	W	1346	9.1
	2017	0.7		2045	1.0
13	0136	9.3	**28**	0206	9.0
	0834	0.9		0856	1.3
W	1355	9.4	TH	1420	8.9
	2058	0.7		2114	1.3
14	0219	9.2	**29**	0239	8.7
	0916	1.0		0924	1.6
TH	1439	9.2	F	1452	8.5
	2138	0.9		2140	1.6
15	0303	9.0	**30**	0310	8.4
	0958	1.2		0952	1.9
F	1524	8.9	SA	1524	8.1
	2220	1.2		2210	1.9
			31	0343	7.9
				1027	2.2
			SU	1559	7.7
				2247	2.3

FEBRUARY

	Time	m		Time	m
1	0422	7.5	**16**	0515	8.0
	1111	2.6		1212	2.1
M	1644	7.2	TU	1756	7.7
◗	2334	2.7			
2	0515	7.1	**17**	0042	2.3
	1208	2.9		0632	7.6
TU	1749	6.8	W	1330	2.3
				1920	7.4
3	0037	3.0	**18**	0207	2.5
	0634	6.9		0800	7.5
W	1321	3.0	TH	1457	2.2
	1921	6.8		2045	7.6
4	0159	3.0	**19**	0331	2.2
	0803	7.1		0917	7.9
TH	1446	2.7	F	1613	1.8
	2040	7.2		2151	8.1
5	0321	2.6	**20**	0441	1.8
	0909	7.6		1015	8.4
F	1557	2.2	SA	1715	1.4
	2138	7.8		2242	8.6
6	0424	2.0	**21**	0536	1.5
	1001	8.2		1102	8.8
SA	1654	1.6	SU	1803	1.1
	2228	8.4		2325	8.9
7	0518	1.5	**22**	0620	1.2
	1048	8.7		1142	9.0
SU	1746	1.2	M	1843	0.9
	2313	8.9	○		
8	0608	1.1	**23**	0002	9.1
	1133	9.2		0657	1.0
M	1835	0.8	TU	1217	9.2
●	2357	9.2		1918	0.8
9	0656	0.8	**24**	0036	9.2
	1216	9.5		0730	1.0
TU	1922	0.5	W	1250	9.2
				1950	0.8
10	0040	9.5	**25**	0108	9.2
	0742	0.6		0801	1.0
W	1259	9.7	TH	1322	9.2
	2006	0.3		2020	0.9
11	0122	9.7	**26**	0139	9.1
	0825	0.5		0830	1.1
TH	1342	9.8	F	1353	9.0
	2047	0.3		2046	1.1
12	0205	9.7	**27**	0208	8.9
	0906	0.5		0856	1.3
F	1425	9.6	SA	1422	8.8
	2127	0.4		2111	1.3
13	0247	9.4	**28**	0236	8.6
	0946	0.8		0922	1.5
SA	1508	9.3	SU	1450	8.4
	2205	0.8		2137	1.7
14	0330	8.9	**29**	0305	8.3
	1027	1.1		0952	1.9
SU	1554	8.8	M	1521	8.0
	2247	1.3		2209	2.1
15	0417	8.5			
	1114	1.6			
M	1647	8.2			
◗	2336	1.8			

MARCH

	Time	m		Time	m
1	0338	7.8	**16**	0450	7.9
	1030	2.3		1150	2.0
TU	1600	7.5	W	1732	7.5
	2250	2.5			
2	0423	7.3	**17**	0018	2.4
	1119	2.7		0605	7.3
W	1654	7.0	TH	1305	2.4
◗	2347	2.9		1859	7.2
3	0528	6.9	**18**	0143	2.6
	1226	2.9		0739	7.2
TH	1816	6.7	F	1434	2.4
				2028	7.4
4	0103	3.1	**19**	0311	2.4
	0706	6.8		0900	7.6
F	1354	2.8	SA	1554	2.0
	1957	7.0		2134	7.9
5	0235	2.8	**20**	0423	1.9
	0832	7.3		0957	8.1
SA	1519	2.3	SU	1655	1.5
	2107	7.6		2223	8.4
6	0351	2.1	**21**	0516	1.5
	0933	8.0		1042	8.6
SU	1625	1.6	M	1741	1.2
	2202	8.3		2303	8.8
7	0451	1.5	**22**	0558	1.3
	1025	8.7		1120	8.8
M	1722	1.1	TU	1819	1.0
	2251	8.9		2338	9.0
8	0546	1.0	**23**	0632	1.1
	1112	9.2		1154	9.0
TU	1815	0.6	W	1852	0.9
	2336	9.4	○		
9	0638	0.6	**24**	0010	9.1
	1157	9.7		0704	1.0
W	1904	0.2	TH	1225	9.1
●				1922	0.9
10	0020	9.7	**25**	0040	9.1
	0726	0.3		0734	1.0
TH	1241	9.9	F	1255	9.1
	1950	0.1		1952	0.9
11	0103	9.9	**26**	0110	9.1
	0810	0.2		0803	1.0
F	1324	10.0	SA	1325	9.0
	2031	0.0		2019	1.0
12	0145	9.9	**27**	0139	9.0
	0851	0.2		0830	1.1
SA	1407	9.9	SU	1354	8.8
	2110	0.3		2044	1.3
13	0227	9.7	**28**	0207	8.7
	0931	0.5		0856	1.4
SU	1450	9.5	M	1423	8.6
	2148	0.7		2111	1.6
14	0309	9.0	**29**	0236	8.4
	1010	0.9		0925	1.7
M	1535	8.9	TU	1455	8.2
	2227	1.2		2141	1.9
15	0355	8.6	**30**	0309	8.0
	1054	1.5		1001	2.0
TU	1626	8.2	W	1532	7.7
◗	2314	1.9		2220	2.4
			31	0351	7.6
				1047	2.4
			TH	1622	7.2
			◗	2313	2.7

APRIL

	Time	m		Time	m
1	0449	7.1	**16**	0114	2.7
	1149	2.7		0705	7.1
F	1734	6.9	SA	1400	2.4
				1956	7.3
2	0025	2.9	**17**	0235	2.5
	0617	6.9		0828	7.4
SA	1312	2.7	SU	1517	2.2
	1914	7.0		2104	7.8
3	0155	2.7	**18**	0346	2.1
	0752	7.3		0928	7.9
SU	1441	2.3	M	1619	1.8
	2032	7.6		2153	8.2
4	0316	2.1	**19**	0441	1.7
	0900	8.0		1013	8.3
M	1552	1.6	TU	1706	1.5
	2132	8.4		2233	8.6
5	0421	1.4	**20**	0524	1.4
	0956	8.7		1051	8.5
TU	1653	1.0	W	1745	1.2
	2224	9.0		2308	8.8
6	0520	0.9	**21**	0601	1.3
	1047	9.3		1125	8.8
W	1750	0.5	TH	1820	1.1
	2312	9.5		2340	8.9
7	0615	0.5	**22**	0634	1.1
	1134	9.7		1157	8.9
TH	1841	0.2	F	1852	1.1
●	2357	9.8	○		
8	0705	0.2	**23**	0011	9.0
	1220	10.0		0707	1.1
F	1928	0.1	SA	1229	8.9
				1923	1.1
9	0042	10.0	**24**	0042	9.0
	0751	0.1		0737	1.1
SA	1305	10.0	SU	1300	8.9
	2011	0.1		1953	1.1
10	0125	9.9	**25**	0112	8.9
	0833	0.2		0806	1.1
SU	1348	9.9	M	1330	8.8
	2051	0.3		2021	1.3
11	0207	9.7	**26**	0142	8.8
	0914	0.4		0836	1.3
M	1432	9.5	TU	1402	8.6
	2130	0.7		2050	1.5
12	0250	9.2	**27**	0214	8.5
	0953	0.9		0908	1.5
TU	1517	8.9	W	1436	8.3
	2209	1.3		2123	1.8
13	0335	8.6	**28**	0250	8.2
	1037	1.4		0944	1.8
W	1608	8.2	TH	1515	7.9
	2256	1.9		2202	2.2
14	0429	7.9	**29**	0332	7.8
	1130	2.0		1028	2.2
TH	1710	7.6	F	1604	7.5
◗	2357	2.4		2253	2.5
15	0538	7.3	**30**	0427	7.4
	1239	2.4		1127	2.4
F	1829	7.2	SA	1710	7.2
			◗	2359	2.6

Chart Datum is 4·43 metres below IGN Datum. HAT is 10·1 metres above Chart Datum.

STANDARD TIME UT –01
Subtract 1 hour for UT
For French Summer Time add
ONE hour in **non-shaded areas**

DIEPPE   LAT 49°56'N   LONG 1°05'E

TIMES AND HEIGHTS OF HIGH AND LOW WATERS

Dates in red are **SPRINGS**
Dates in blue are NEAPS

YEAR 2016

N France

MAY

Day	Time	m	Day	Time	m
1 SU	0544 / 1240 / 1836	7.2 / 2.4 / 7.3	**16** M	0147 / 0736 / 1423 / 2016	2.6 / 7.2 / 2.3 / 7.5
2 M	0120 / 0712 / 1404 / 1955	2.5 / 7.5 / 2.1 / 7.8	**17** TU	0253 / 0843 / 1526 / 2111	2.3 / 7.5 / 2.1 / 7.9
3 TU	0241 / 0825 / 1518 / 2059	2.0 / 8.0 / 1.6 / 8.4	**18** W	0352 / 0935 / 1620 / 2156	2.0 / 7.9 / 1.8 / 8.2
4 W	0350 / 0926 / 1622 / 2155	1.4 / 8.6 / 1.0 / 9.0	**19** TH	0441 / 1017 / 1705 / 2234	1.7 / 8.2 / 1.5 / 8.5
5 TH	0451 / 1020 / 1721 / 2246	0.9 / 9.2 / 0.7 / 9.4	**20** F	0524 / 1054 / 1745 / 2310	1.5 / 8.5 / 1.4 / 8.7
6 F ●	0549 / 1111 / 1815 / 2334	0.6 / 9.6 / 0.4 / 9.7	**21** SA ○	0603 / 1129 / 1822 / 2343	1.3 / 8.6 / 1.3 / 8.8
7 SA	0641 / 1159 / 1905	0.4 / 9.8 / 0.3	**22** SU	0639 / 1203 / 1856	1.2 / 8.7 / 1.2
8 SU	0020 / 0730 / 1245 / 1950	9.8 / 0.3 / 9.8 / 0.3	**23** M	0017 / 0713 / 1237 / 1929	8.9 / 1.2 / 8.8 / 1.2
9 M	0104 / 0814 / 1330 / 2032	9.8 / 0.3 / 9.7 / 0.5	**24** TU	0049 / 0746 / 1311 / 2002	8.9 / 1.1 / 8.8 / 1.3
10 TU	0148 / 0857 / 1415 / 2113	9.6 / 0.5 / 9.4 / 0.9	**25** W	0123 / 0820 / 1346 / 2036	8.8 / 1.2 / 8.7 / 1.4
11 W	0232 / 0938 / 1501 / 2153	9.2 / 0.9 / 8.9 / 1.4	**26** TH	0159 / 0856 / 1424 / 2112	8.6 / 1.3 / 8.5 / 1.6
12 TH	0318 / 1020 / 1549 / 2238	8.6 / 1.4 / 8.3 / 1.9	**27** F	0238 / 0934 / 1505 / 2153	8.4 / 1.6 / 8.2 / 1.9
13 F ◗	0408 / 1109 / 1645 / 2332	8.0 / 1.9 / 7.8 / 2.3	**28** SA	0322 / 1018 / 1553 / 2241	8.1 / 1.8 / 7.9 / 2.1
14 SA	0507 / 1207 / 1750	7.5 / 2.3 / 7.4	**29** SU ◗	0414 / 1111 / 1652 / 2340	7.8 / 2.0 / 7.7 / 2.3
15 SU	0037 / 0618 / 1314 / 1904	2.6 / 7.2 / 2.4 / 7.3	**30** M	0520 / 1215 / 1804	7.6 / 2.1 / 7.7
			31 TU	0051 / 0637 / 1330 / 1918	2.2 / 7.7 / 1.9 / 8.0

JUNE

Day	Time	m	Day	Time	m
1 W	0208 / 0751 / 1444 / 2026	1.9 / 8.1 / 1.6 / 8.4	**16** TH	0300 / 0847 / 1530 / 2113	2.3 / 7.5 / 2.2 / 7.8
2 TH	0319 / 0856 / 1551 / 2126	1.5 / 8.5 / 1.2 / 8.8	**17** F	0357 / 0939 / 1624 / 2159	2.0 / 7.8 / 1.9 / 8.1
3 F	0423 / 0956 / 1652 / 2221	1.1 / 8.9 / 0.9 / 9.2	**18** SA	0447 / 1023 / 1711 / 2240	1.8 / 8.1 / 1.7 / 8.4
4 SA	0523 / 1050 / 1749 / 2312	0.8 / 9.3 / 0.7 / 9.4	**19** SU	0532 / 1103 / 1753 / 2318	1.5 / 8.4 / 1.5 / 8.6
5 SU ●	0618 / 1140 / 1842	0.6 / 9.5 / 0.6	**20** M ○	0613 / 1141 / 1831 / 2355	1.4 / 8.6 / 1.4 / 8.8
6 M	0001 / 0709 / 1228 / 1930	9.6 / 0.5 / 9.5 / 0.6	**21** TU	0652 / 1218 / 1909	1.2 / 8.7 / 1.3
7 TU	0047 / 0757 / 1315 / 2014	9.6 / 0.5 / 9.5 / 0.7	**22** W	0031 / 0730 / 1255 / 1946	8.9 / 1.1 / 8.8 / 1.2
8 W	0132 / 0840 / 1359 / 2056	9.4 / 0.6 / 9.3 / 1.0	**23** TH	0109 / 0809 / 1334 / 2025	8.9 / 1.1 / 8.9 / 1.3
9 TH	0215 / 0922 / 1444 / 2136	9.1 / 0.9 / 8.9 / 1.3	**24** F	0148 / 0848 / 1414 / 2104	8.9 / 1.1 / 8.8 / 1.4
10 F	0259 / 1002 / 1528 / 2216	8.7 / 1.3 / 8.5 / 1.7	**25** SA	0229 / 0928 / 1456 / 2145	8.7 / 1.2 / 8.6 / 1.5
11 SA	0344 / 1043 / 1615 / 2300	8.2 / 1.7 / 8.0 / 2.1	**26** SU	0313 / 1010 / 1541 / 2231	8.5 / 1.4 / 8.4 / 1.7
12 SU ◗	0433 / 1129 / 1706 / 2352	7.7 / 2.1 / 7.6 / 2.4	**27** M ◗	0401 / 1057 / 1633 / 2322	8.2 / 1.6 / 8.2 / 1.9
13 M	0529 / 1223 / 1806	7.3 / 2.4 / 7.3	**28** TU	0458 / 1152 / 1734	8.0 / 1.8 / 8.0
14 TU	0052 / 0634 / 1325 / 1913	2.6 / 7.1 / 2.5 / 7.3	**29** W	0024 / 0607 / 1258 / 1845	2.0 / 7.9 / 1.8 / 8.0
15 W	0156 / 0744 / 1429 / 2019	2.5 / 7.2 / 2.4 / 7.5	**30** TH	0137 / 0721 / 1413 / 1957	1.9 / 8.0 / 1.7 / 8.2

JULY

Day	Time	m	Day	Time	m
1 F	0252 / 0832 / 1525 / 2103	1.7 / 8.3 / 1.5 / 8.5	**16** SA	0313 / 0900 / 1545 / 2124	2.4 / 7.4 / 2.3 / 7.8
2 SA	0400 / 0937 / 1629 / 2203	1.4 / 8.6 / 1.3 / 8.8	**17** SU	0412 / 0953 / 1639 / 2212	2.1 / 7.8 / 2.0 / 8.2
3 SU	0502 / 1035 / 1729 / 2257	1.1 / 8.9 / 1.1 / 9.1	**18** M	0503 / 1038 / 1726 / 2255	1.7 / 8.2 / 1.6 / 8.5
4 M ●	0600 / 1127 / 1823 / 2347	0.9 / 9.2 / 0.9 / 9.3	**19** TU ○	0549 / 1120 / 1810 / 2335	1.4 / 8.6 / 1.4 / 8.8
5 TU	0653 / 1215 / 1913	0.7 / 9.3 / 0.9	**20** W	0633 / 1200 / 1852	1.2 / 8.8 / 1.2
6 W	0033 / 0741 / 1300 / 1957	9.4 / 0.7 / 9.3 / 0.9	**21** TH	0015 / 0716 / 1240 / 1934	9.0 / 1.0 / 9.0 / 1.1
7 TH	0116 / 0823 / 1342 / 2038	9.3 / 0.7 / 9.2 / 1.0	**22** F	0055 / 0758 / 1320 / 2015	9.2 / 0.8 / 9.1 / 1.0
8 F	0157 / 0902 / 1423 / 2115	9.1 / 0.9 / 9.0 / 1.4	**23** SA	0136 / 0839 / 1401 / 2056	9.2 / 0.8 / 9.2 / 1.0
9 SA	0237 / 0938 / 1502 / 2149	8.8 / 1.2 / 8.7 / 1.6	**24** SU	0217 / 0919 / 1442 / 2136	9.1 / 0.8 / 9.1 / 1.1
10 SU	0316 / 1011 / 1541 / 2224	8.5 / 1.6 / 8.3 / 1.9	**25** M	0300 / 0959 / 1526 / 2218	8.9 / 1.0 / 8.8 / 1.3
11 M	0355 / 1046 / 1622 / 2303	8.0 / 1.9 / 7.9 / 2.3	**26** TU	0345 / 1041 / 1612 / 2305	8.7 / 1.3 / 8.5 / 1.6
12 TU ◗	0439 / 1129 / 1709 / 2352	7.6 / 2.3 / 7.5 / 2.6	**27** W ◗	0437 / 1130 / 1707	8.3 / 1.6 / 8.2
13 W	0532 / 1222 / 1808	7.2 / 2.6 / 7.2	**28** TH	0000 / 0540 / 1231 / 1816	1.9 / 8.0 / 1.9 / 7.9
14 TH	0053 / 0640 / 1328 / 1920	2.7 / 7.0 / 2.8 / 7.1	**29** F	0110 / 0657 / 1348 / 1934	2.0 / 7.8 / 2.0 / 7.9
15 F	0204 / 0756 / 1440 / 2028	2.7 / 7.0 / 2.6 / 7.4	**30** SA	0231 / 0815 / 1506 / 2048	1.9 / 7.9 / 1.9 / 8.2
			31 SU	0345 / 0927 / 1615 / 2153	1.6 / 8.3 / 1.6 / 8.5

AUGUST

Day	Time	m	Day	Time	m
1 M	0451 / 1027 / 1717 / 2249	1.3 / 8.7 / 1.3 / 8.9	**16** TU	0434 / 1013 / 1659 / 2232	1.8 / 8.2 / 1.7 / 8.5
2 TU ●	0549 / 1118 / 1811 / 2336	1.0 / 9.0 / 1.1 / 9.1	**17** W	0525 / 1058 / 1748 / 2315	1.4 / 8.6 / 1.3 / 8.9
3 W	0640 / 1203 / 1858	0.9 / 9.2 / 1.0	**18** TH ○	0613 / 1140 / 1834 / 2357	1.0 / 9.0 / 1.0 / 9.2
4 TH	0019 / 0724 / 1244 / 1939	9.3 / 0.8 / 9.3 / 1.0	**19** F	0700 / 1221 / 1919	0.7 / 9.3 / 0.8
5 F	0058 / 0803 / 1322 / 2016	9.3 / 0.8 / 9.3 / 1.0	**20** SA	0038 / 0744 / 1302 / 2002	9.5 / 0.6 / 9.5 / 0.7
6 SA	0135 / 0838 / 1357 / 2049	9.2 / 0.9 / 9.1 / 1.2	**21** SU	0119 / 0825 / 1343 / 2043	9.6 / 0.5 / 9.5 / 0.7
7 SU	0210 / 0909 / 1432 / 2118	9.0 / 1.1 / 8.9 / 1.4	**22** M	0201 / 0905 / 1424 / 2123	9.5 / 0.6 / 9.4 / 0.8
8 M	0244 / 0937 / 1505 / 2146	8.7 / 1.5 / 8.5 / 1.7	**23** TU	0243 / 0943 / 1506 / 2203	9.3 / 0.8 / 9.2 / 1.1
9 TU	0317 / 1005 / 1538 / 2218	8.3 / 1.8 / 8.1 / 2.1	**24** W	0327 / 1023 / 1551 / 2246	8.9 / 1.2 / 8.7 / 1.5
10 W ◗	0351 / 1038 / 1614 / 2258	7.8 / 2.2 / 7.6 / 2.5	**25** TH ◗	0416 / 1109 / 1643 / 2339	8.4 / 1.7 / 8.2 / 1.9
11 TH	0432 / 1123 / 1702 / 2351	7.3 / 2.7 / 7.2 / 2.8	**26** F	0518 / 1209 / 1752	7.9 / 2.1 / 7.8
12 F	0531 / 1223 / 1812	6.9 / 3.0 / 6.9	**27** SA	0050 / 0639 / 1329 / 1917	2.2 / 7.5 / 2.4 / 7.6
13 SA	0100 / 0657 / 1342 / 1941	3.0 / 6.7 / 3.1 / 7.0	**28** SU	0217 / 0806 / 1454 / 2040	2.2 / 7.6 / 2.2 / 7.9
14 SU	0225 / 0822 / 1504 / 2051	2.8 / 7.0 / 2.7 / 7.4	**29** M	0336 / 0922 / 1608 / 2147	1.9 / 8.1 / 1.8 / 8.3
15 M	0337 / 0923 / 1607 / 2145	2.4 / 7.6 / 2.2 / 8.0	**30** TU	0444 / 1019 / 1709 / 2239	1.5 / 8.6 / 1.5 / 8.8
			31 W	0540 / 1106 / 1759 / 2323	1.1 / 9.0 / 1.2 / 9.1

Chart Datum is 4·43 metres below IGN Datum. HAT is 10·1 metres above Chart Datum.

》》 **FREE** monthly updates. Register at 《
www.reedsnauticalalmanac.co.uk

STANDARD TIME UT –01
Subtract 1 hour for UT
For French Summer Time add
ONE hour in **non-shaded areas**

DIEPPE   LAT 49°56'N   LONG 1°05'E
TIMES AND HEIGHTS OF HIGH AND LOW WATERS

Dates in red are **SPRINGS**
Dates in blue are NEAPS

YEAR 2016

SEPTEMBER

Day	Time m	Day	Time m
1	0625 0.9 / 1147 9.2 / TH 1841 1.1 / ●	**16**	0550 0.9 / 1116 9.3 / F 1813 0.9 / ○ 2334 9.5
2	0001 9.2 / 0704 0.9 / F 1223 9.3 / 1917 1.0	**17**	0638 0.6 / 1159 9.6 / SA 1900 0.6
3	0036 9.2 / 0738 0.9 / SA 1256 9.3 / 1950 1.0	**18**	0017 9.7 / 0724 0.4 / SU 1240 9.8 / 1944 0.5
4	0109 9.2 / 0809 1.0 / SU 1328 9.2 / 2020 1.1	**19**	0059 9.9 / 0807 0.3 / M 1322 9.8 / 2026 0.5
5	0141 9.0 / 0837 1.2 / M 1359 9.0 / 2047 1.4	**20**	0142 9.8 / 0846 0.4 / TU 1404 9.7 / 2106 0.6
6	0212 8.8 / 0903 1.4 / TU 1429 8.7 / 2112 1.6	**21**	0224 9.5 / 0925 0.8 / W 1446 9.3 / 2146 1.0
7	0241 8.4 / 0927 1.8 / W 1457 8.3 / 2140 2.0	**22**	0308 9.0 / 1005 1.2 / TH 1530 8.8 / 2229 1.5
8	0310 8.0 / 0957 2.2 / TH 1529 7.8 / 2215 2.4	**23**	0358 8.4 / 1051 1.8 / F 1623 8.2 / ◑ 2322 2.0
9	0346 7.5 / 1036 2.7 / F 1609 7.3 / ◐ 2301 2.8	**24**	0501 7.8 / 1152 2.4 / SA 1733 7.6
10	0436 7.0 / 1130 3.1 / SA 1710 6.9	**25**	0033 2.4 / 0624 7.4 / SU 1315 2.6 / 1902 7.4
11	0005 3.1 / 0553 6.6 / SU 1245 3.3 / 1844 6.7	**26**	0203 2.4 / 0755 7.5 / M 1442 2.4 / 2028 7.7
12	0132 3.1 / 0739 6.8 / M 1419 3.0 / 2014 7.2	**27**	0324 2.0 / 0908 8.0 / TU 1556 2.0 / 2132 8.2
13	0259 2.6 / 0851 7.5 / TU 1534 2.4 / 2115 7.8	**28**	0429 1.6 / 1002 8.6 / W 1654 1.5 / 2221 8.7
14	0404 1.9 / 0945 8.2 / W 1631 1.7 / 2205 8.5	**29**	0521 1.2 / 1046 8.9 / TH 1740 1.3 / 2303 9.0
15	0458 1.3 / 1032 8.8 / TH 1723 1.2 / 2251 9.0	**30**	0602 1.1 / 1123 9.1 / F 1817 1.1 / 2338 9.1

OCTOBER

Day	Time m	Day	Time m
1	0637 1.0 / 1157 9.2 / SA 1850 1.1 / ●	**16**	0613 0.5 / 1133 9.7 / SU 1837 0.5 / ○ 2354 9.9
2	0011 9.2 / 0708 1.0 / SU 1228 9.2 / 1921 1.1	**17**	0701 0.4 / 1217 9.9 / M 1924 0.4
3	0041 9.1 / 0738 1.1 / M 1258 9.2 / 1950 1.2	**18**	0038 10.0 / 0745 0.3 / TU 1300 10.0 / 2007 0.3
4	0112 9.1 / 0806 1.2 / TU 1327 9.0 / 2018 1.3	**19**	0122 9.9 / 0827 0.5 / W 1343 9.8 / 2049 0.6
5	0141 8.8 / 0832 1.5 / W 1356 8.8 / 2043 1.6	**20**	0206 9.6 / 0907 0.8 / TH 1427 9.4 / 2130 1.0
6	0210 8.5 / 0857 1.8 / TH 1424 8.4 / 2110 1.9	**21**	0252 9.1 / 0948 1.3 / F 1513 8.8 / 2214 1.5
7	0239 8.2 / 0926 2.2 / F 1455 8.0 / 2143 2.3	**22**	0343 8.4 / 1035 1.9 / SA 1606 8.2 / ◑ 2306 2.0
8	0314 7.7 / 1002 2.6 / SA 1534 7.5 / 2226 2.7	**23**	0444 7.8 / 1135 2.5 / SU 1713 7.6
9	0401 7.2 / 1053 3.0 / SU 1628 7.0 / ◐ 2325 3.0	**24**	0014 2.5 / 0601 7.4 / M 1253 2.7 / 1835 7.3
10	0508 6.8 / 1202 3.3 / M 1751 6.8	**25**	0137 2.5 / 0727 7.4 / TU 1416 2.6 / 2000 7.5
11	0044 3.1 / 0649 6.9 / TU 1333 3.1 / 1930 7.1	**26**	0255 2.2 / 0840 7.9 / W 1527 2.1 / 2105 8.0
12	0216 2.7 / 0812 7.5 / W 1457 2.5 / 2039 7.8	**27**	0359 1.8 / 0934 8.4 / TH 1624 1.7 / 2155 8.4
13	0329 2.0 / 0911 8.2 / TH 1600 1.8 / 2134 8.5	**28**	0449 1.5 / 1018 8.7 / F 1709 1.5 / 2236 8.7
14	0427 1.3 / 1001 8.9 / F 1655 1.2 / 2223 9.1	**29**	0530 1.3 / 1055 8.9 / SA 1747 1.3 / 2311 8.9
15	0522 0.9 / 1048 9.4 / SA 1747 0.8 / 2309 9.6	**30**	0605 1.2 / 1128 9.1 / SU 1820 1.2 / ● 2343 9.0
		31	0637 1.2 / 1159 9.1 / M 1852 1.2

NOVEMBER

Day	Time m	Day	Time m
1	0014 9.0 / 0709 1.2 / TU 1229 9.1 / 1924 1.2	**16**	0019 9.9 / 0724 0.5 / W 1240 9.9 / 1949 0.4
2	0045 8.9 / 0739 1.3 / W 1259 9.0 / 1953 1.3	**17**	0105 9.8 / 0809 0.6 / TH 1325 9.8 / 2034 0.6
3	0115 8.8 / 0807 1.5 / TH 1329 8.8 / 2021 1.5	**18**	0151 9.6 / 0852 0.9 / F 1411 9.4 / 2117 0.9
4	0146 8.6 / 0834 1.8 / F 1359 8.5 / 2050 1.8	**19**	0238 9.2 / 0934 1.4 / SA 1458 8.9 / 2200 1.4
5	0218 8.3 / 0905 2.1 / SA 1432 8.2 / 2123 2.1	**20**	0327 8.6 / 1020 1.9 / SU 1548 8.3 / 2249 1.9
6	0254 7.9 / 0941 2.4 / SU 1511 7.8 / 2204 2.4	**21**	0422 8.0 / 1114 2.3 / M 1647 7.7 / ◑ 2346 2.3
7	0339 7.5 / 1028 2.8 / M 1602 7.3 / ◐ 2257 2.7	**22**	0526 7.6 / 1219 2.6 / TU 1755 7.4
8	0439 7.2 / 1131 3.0 / TU 1713 7.1	**23**	0054 2.6 / 0639 7.4 / W 1330 2.7 / 1911 7.3
9	0007 2.8 / 0602 7.1 / W 1250 2.9 / 1842 7.2	**24**	0205 2.5 / 0754 7.6 / TH 1439 2.4 / 2023 7.5
10	0131 2.6 / 0727 7.5 / TH 1415 2.5 / 1958 7.8	**25**	0311 2.3 / 0855 7.9 / F 1540 2.1 / 2119 7.9
11	0249 2.0 / 0833 8.2 / F 1525 1.8 / 2059 8.4	**26**	0406 2.0 / 0943 8.3 / SA 1630 1.8 / 2203 8.3
12	0354 1.4 / 0929 8.8 / SA 1625 1.3 / 2153 9.0	**27**	0452 1.7 / 1023 8.6 / SU 1713 1.6 / 2241 8.5
13	0452 1.0 / 1020 9.4 / SU 1720 0.8 / 2244 9.5	**28**	0532 1.5 / 1058 8.8 / M 1751 1.4 / 2316 8.7
14	0546 0.7 / 1108 9.7 / M 1813 0.6 / ○ 2332 9.8	**29**	0608 1.4 / 1132 8.9 / TU 1827 1.3 / ● 2350 8.8
15	0636 0.5 / 1155 9.9 / TU 1903 0.4	**30**	0643 1.4 / 1204 8.9 / W 1900 1.3

DECEMBER

Day	Time m	Day	Time m
1	0023 8.8 / 0715 1.4 / TH 1237 8.9 / 1933 1.3	**16**	0051 9.7 / 0753 0.7 / F 1311 9.7 / 2020 0.6
2	0055 8.8 / 0747 1.5 / F 1309 8.9 / 2004 1.4	**17**	0137 9.5 / 0838 0.9 / SA 1356 9.4 / 2103 0.8
3	0128 8.7 / 0818 1.6 / SA 1342 8.7 / 2037 1.5	**18**	0223 9.2 / 0920 1.2 / SU 1441 9.1 / 2145 1.2
4	0203 8.5 / 0852 1.8 / SU 1418 8.4 / 2112 1.8	**19**	0308 8.8 / 1002 1.6 / M 1527 8.6 / 2226 1.6
5	0241 8.3 / 0929 2.1 / M 1458 8.1 / 2152 2.0	**20**	0355 8.3 / 1045 2.1 / TU 1615 8.0 / 2310 2.1
6	0325 8.0 / 1014 2.4 / TU 1546 7.8 / 2239 2.2	**21**	0445 7.8 / 1134 2.4 / W 1708 7.5 / ◑
7	0417 7.7 / 1108 2.5 / W 1645 7.5 / ◐ 2338 2.4	**22**	0001 2.5 / 0543 7.4 / TH 1232 2.7 / 1811 7.2
8	0523 7.5 / 1214 2.6 / TH 1758 7.5	**23**	0102 2.7 / 0650 7.3 / F 1338 2.7 / 1923 7.1
9	0048 2.4 / 0641 7.7 / F 1331 2.4 / 1916 7.8	**24**	0209 2.7 / 0801 7.4 / SA 1446 2.6 / 2031 7.3
10	0207 2.1 / 0753 8.1 / SA 1447 1.9 / 2025 8.3	**25**	0315 2.5 / 0901 7.7 / SU 1547 2.3 / 2126 7.7
11	0319 1.6 / 0857 8.6 / SU 1555 1.4 / 2126 8.8	**26**	0412 2.2 / 0948 8.1 / M 1639 1.9 / 2211 8.1
12	0423 1.2 / 0954 9.1 / M 1655 1.0 / 2221 9.2	**27**	0500 1.9 / 1030 8.4 / TU 1723 1.6 / 2251 8.4
13	0520 0.9 / 1046 9.5 / TU 1751 0.7 / 2313 9.5	**28**	0542 1.7 / 1108 8.6 / W 1803 1.4 / 2328 8.7
14	0615 0.7 / 1136 9.7 / W 1844 0.5 / ○	**29**	0620 1.5 / 1144 8.8 / TH 1841 1.3 / ●
15	0003 9.7 / 0706 0.6 / TH 1224 9.8 / 1933 0.5	**30**	0004 8.7 / 0656 1.4 / F 1219 8.9 / 1917 1.2
		31	0040 8.9 / 0731 1.4 / SA 1254 9.0 / 1952 1.2

Chart Datum is 4·43 metres below IGN Datum. HAT is 10·1 metres above Chart Datum.

FREE monthly updates. Register at
www.reedsnauticalalmanac.co.uk

DIEPPE

MEAN SPRING ——————
AND NEAP CURVES – – – –

MEAN RANGES
Springs 8.5m
Neaps 4.9m

9.17.15 ST VALÉRY-EN-CAUX

Seine Maritime **49°52'·48N 00°42'·64E** ❁❁❁♦♦♦❁❁❁

CHARTS AC 2451; SHOM 7417; Navi 1012; Imray C31

TIDES −0044 Dover; ML 4·6; Duration 0530

Standard Port DIEPPE (←——)

Times				Height (metres)			
High Water		Low Water		MHWS	MHWN	MLWN	MLWS
0100	0600	0100	0700	9·3	7·4	2·5	0·8
1300	1800	1300	1900				
Differences ST VALÉRY-EN-CAUX							
−0005	−0005	−0015	−0015	−0.5	−0.4	−0.1	−0.1

SHELTER Good, but confused seas break at entrance in fresh W-NE'lies.

NAVIGATION WPT 49°52'·97N 00°42'·14E, 155°/6ca to W pier. Ent is easy, but tidal streams across ent are strong. Shingle builds up against W wall; hug the E side. Inside the pier hds, wave-breaks (marked by posts) reduce the surge. Avant Port dries <u>3</u>m. 5 W waiting buoys are in the pool NE of gate.

Gate is open HW ±2½ if coeff >79; if coeff < 41 gate is open HW −2½ to HW+1¾ (bridge opens H & H+30 in these periods). When entry gate first opens, strong current may complicate berthing on Ⓥ pontoon W side; best to wait >15 mins (i.e. second bridge opening). Ⓥ also berth on 'A' pontoon. Traffic sigs at the bridge/gate are coordinated to seaward/inland: ● = ent/exit; ● = no ent/exit; ● + ● = no movements.

LIGHTS AND MARKS From N or W Paluel nuclear power stn is conspic 3M W of ent. From E a TV mast is 1M E of ent. Hbr is hard to see between high chalk cliffs, but lt ho (W twr, G top) on W pierhead is fairly conspic. Lts as on chartlet and LBW.

COMMUNICATIONS CROSS 03·21·87·28·76; Police 02·35·97·00·17; British Consul 01·44·51·31·00; Ⓗ 02·35·97·00·11; Doctor 02·35·97·05·99; Aff Mar 02·35·10·34·34; HM VHF Ch 09 (French only).

FACILITIES **Marina** ☎02·35·97·01·30, mob 06·07·31·56·95. www.ville-saint-valery-en-caux.fr 575 + 20 Ⓥ €2·20; max LOA 14m. Ⓛ ⒷⒾ Ⓔ ⚒ ⚒ ⛽(5/10t). **CN Valeriquais** ☎02.35.97.25.49, ⏢ ⌧. **Town** ✕ ⏢ 🛒 🎦 🎦 ♨ ✉ Ⓑ bus to ⇌ at Yvetot; ✈ and ferry Dieppe.

ST VALÉRY-EN-CAUX

9.17.16   FÉCAMP

Seine Maritime **49°45'·92N 00°21'·80E** ✳❀♻♻♻✿✿

CHARTS AC 2451, 2148, 1354; SHOM 6857, 7417, 7207; Navi 1012; Imray C31

TIDES –0044 Dover; ML 4·9; Duration 0550

Standard Port DIEPPE (←—)

Times				Height (metres)			
High Water		Low Water		MHWS	MHWN	MLWN	MLWS
0100	0600	0100	0700	9·3	7·4	2·5	0·8
1300	1800	1300	1900				
Differences FÉCAMP							
–0015	–0010	–0030	–0040	–0·8	–0·5	+0·3	+0·4
ETRETAT 49°42'·6N 00°12'·0E							
–0020	–0015	–0050	–0050	–1·1	–0·8	+0·3	+0·4

SHELTER Excellent in basins, but in even moderate W/NW winds a considerable surf runs off the ent and scend can make the Avant Port uncomfortable. Access difficult in onshore winds >F5; do not attempt >F7. Bassin Bérigny for long stay; see HM.

NAVIGATION WPT 49°46'·01N 00°21'·36E, 107°/3ca to hbr ent. Beware the Charpentier Rks off Pte Fagnet and strong cross currents, depending on tides. Ent chan (1·5m) is prone to silting; best water is close to N jetty.

LIGHTS AND MARKS Ent lies SW of conspic ✳, sig stn, TV mast and 5 wind turbines (2 Q) on Pte Fagnet cliff, 142m high. See chartlet and Lights, buoys & waypoints for lts. Pilots use the QR aligned with QG as a 082° ldg line.

IPTS on column at root of S jetty; a W lt is shown next to top lt when Bassin Bérigny gate is open.

COMMUNICATIONS CROSS 03·21·87·21·87; ⒧02·35·28·00·91; Auto 08·92·68·08·76; Police 02·35·28·16·69; ⊖ 02·35·28·19·40; British Consul 01·44·51·31·00; Ⓗ 02·35·28·05·13; HM 02·35·28·25·53, VHF Ch **12** 16 (HW –3 to +1); Marina VHF Ch 09 & Écluse Bérigny.

9.17.17   LE HAVRE

Seine Maritime **49°29'·12N 00°05'·43E** ✳❀❀♻♻♻✿✿

CHARTS AC 2613, 2146, 2990; SHOM 7418, 6683; Navi 526, 1012; Imray C31

TIDES –0103 Dover; ML 4·9; Duration 0543. Le Havre is a Standard Port (←—). There is a stand of about 3 hrs at HW.

SHELTER Port of refuge H24. Excellent in marina (see overleaf).

NAVIGATION WPT 49°29'·86N 00°02'·33E (LH12) 108°/2·1M to Digue Nord. Yachts may only cross frwy E of buoys LH7/LH8. Banc de l'Éclat (1m) is close N of appr chan. Avoid vess ent/dep commercial Port SE of fairway. Beware LW depth in marina ent, give spur Fl(2)R 6s wide berth; at W end Digue Normand deepest water on PH side chan.

LIGHTS AND MARKS 2 chys, RW conspic, on 106·8° ldg line. Twr of St Joseph's ✳ is conspic. Obey IPTS at end of Digue Nord.

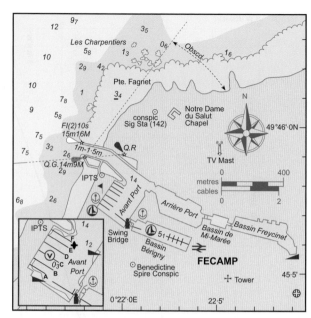

FACILITIES **Marina** plaisance@fecamp-bolbec.cci.fr www.fecamp-bolbec.cci.fr ☎02·35·28·13·58. 725 + 75 ❶ fingers at pontoon C, €2·40. 🛢 (H24 with any cr/card) 🚿♻♻♻♻━ ♁⚓ Ⓔ ✕🖤(mob 20t). **Société des Régates de Fécamp** ☎02·35·28·08·44; 🗐.

Bassin Bérigny is entered via a gate & swing bridge, HW –2½ to HW +¾; pontoons at E end; max LOA for visitors is 45m. ☎02.35.28.23.76, 🔧━ 🖤(30t).

Town ✕🗐🍴🛢✉Ⓑ⇌. ✈ Le Havre.

COMMUNICATIONS CROSS 02·33·52·16·16; ♻ 02·35·42·21·06; Auto 08·92·68·08·76; ⊖ 02·35·41·33·51; Brit Consul 01·44·51·31·00; Dr 02·35·41·23·61; HM 02·35·21·23·95; Port Control & Radar VHF Ch **12**, 20; Aff Mar 02·35·19·29·99; Marina VHF Ch 09; Sas Quinette (lock) VHF Ch 88.

FACILITIES **Marina** ☎02·35·21·23·95. www.lehavreplaisance.com capitainerie@lehavre-plaisance.fr 973 + 60❶ pontoons O or A, €3·00, 🛢♻✕🍴🛢♻━ 🖭Ⓔ ✕ SHOM ACA 🗐🖤(16t) de-mast 🖤(6t) **Port Vauban**, berthing by arrangement with HM, 3+days min. stay, see website for opening times of bridge & lock (twice per day) €3·20, comfortable berth in centre of main shopping centre/facilities.

Sport Nautique du Havre (SNH) ☎02·35·21·01·41, 🗐.

Sté Régates du Havre (SRH) ☎02·35·42·41·21, ✕🗐(shut Aug).

City all facilities.

N France

THE SEINE ESTUARY

The Seine estuary is entered between Le Havre and Deauville, and is encumbered by shallow and shifting banks which extend seawards to Banc de Seine, 15M W of Le Havre. With wind against tide there is a heavy sea on this bank. N and E of it there are 4 big ship waiting anchorages. The estuary is carefully regulated by the port authorities at Le Havre and Rouen. Given the large amount of commercial traffic to/from both these ports, it is sensible for yachtsmen to study AC 2146, 2990 carefully prior to entering the area, so as to gain a sound working knowledge of their routes and the general layout. **Note that the Greenwich Meridian lies about 2·7M W of Cap de la Hève and Deauville; extra caution is necessary in this area to ensure you are using the correct longitude sign (E/W) for your navigation.**

Yachts leaving **Le Havre** Marina near LW should give the inner spur, Fl (2) R 6s, a wide berth due to shoal ground. If bound for **Honfleur, Rouen** or **Paris**, do not try a short cut east of PHM No 2 (Q R pillar buoy at 49°27'·70N 00°00'·60E); to do so risks grounding on Spoil Ground (0·4m). ▶*The SW-going stream begins at HW Le Havre +0400, and the NE-going at HW Le Havre –0300, sp rates 1·5kn.*◀ Between Deauville and Le Havre the sea can be rough in W winds. The inshore **Passage des Pêcheurs**, over the Digue du Ratier, between Honfleur and Deauville is only good HW±2 and requires local knowledge.

Chenal du Rouen is the main chan into R. Seine, and carries much commercial tfc. The S side of the chan is defined by Digue du Ratier, a training wall which extends E to Honfleur. Yachts should stay on the N side of the channel, just outside the PHM buoys, whether east or west bound. This rule applies between No 2 buoy and Tancarville bridge.

With 24 hours access/exit, **Honfleur** is a useful starting port when bound upriver. See routes to the Mediterranean (Special notes), River Seine, Rouen and Paris.

STANDARD TIME UT −01
Subtract 1 hour for UT
For French Summer Time add
ONE hour in **non-shaded areas**

LE HAVRE   LAT 49°29′N   LONG 0°07′E
TIMES AND HEIGHTS OF HIGH AND LOW WATERS

Dates in red are SPRINGS
Dates in blue are NEAPS

YEAR 2016

JANUARY

Day	Time	m	Day	Time	m
1 F	0256 / 0944 / 1504 / 2205	7.0 / 2.7 / 6.9 / 2.7	**16** SA	0258 / 0950 / 1517 / 2212	7.5 / 2.0 / 7.5 / 1.9
2 SA	0344 / 1026 / 1557 / ◑ 2250	6.8 / 3.0 / 6.6 / 3.0	**17** SU	0350 / 1040 / 1615 / ◐ 2308	7.3 / 2.3 / 7.2 / 2.3
3 SU	0445 / 1117 / 1707 / 2344	6.6 / 3.3 / 6.4 / 3.2	**18** M	0455 / 1147 / 1730	7.1 / 2.5 / 7.0
4 M	0550 / 1218 / 1817	6.5 / 3.3 / 6.4	**19** TU	0022 / 0614 / 1308 / 1852	2.5 / 7.0 / 2.6 / 7.0
5 TU	0051 / 0654 / 1334 / 1924	3.2 / 6.6 / 3.2 / 6.5	**20** W	0142 / 0728 / 1424 / 2003	2.5 / 7.2 / 2.4 / 7.2
6 W	0212 / 0755 / 1452 / 2024	3.0 / 6.9 / 2.8 / 6.8	**21** TH	0254 / 0830 / 1534 / 2103	2.3 / 7.4 / 2.0 / 7.4
7 TH	0318 / 0847 / 1547 / 2114	2.6 / 7.2 / 2.4 / 7.1	**22** F	0401 / 0924 / 1638 / 2154	2.0 / 7.6 / 1.7 / 7.6
8 F	0409 / 0931 / 1634 / 2157	2.3 / 7.4 / 2.0 / 7.4	**23** SA	0500 / 1011 / 1731 / 2239	1.8 / 7.7 / 1.4 / 7.7
9 SA	0454 / 1012 / 1718 / 2238	1.9 / 7.7 / 1.7 / 7.6	**24** SU	0549 / 1053 / 1815 / ○ 2321	1.6 / 7.9 / 1.3 / 7.8
10 SU	0537 / 1052 / 1801 / ● 2318	1.7 / 7.9 / 1.4 / 7.8	**25** M	0628 / 1132 / 1852 / 2359	1.5 / 7.9 / 1.2 / 7.8
11 M	0620 / 1132 / 1844 / 2359	1.5 / 8.0 / 1.2 / 7.9	**26** TU	0704 / 1208 / 1926	1.5 / 7.9 / 1.3
12 TU	0703 / 1214 / 1926	1.4 / 8.1 / 1.1	**27** W	0034 / 0738 / 1243 / 1958	7.8 / 1.5 / 7.8 / 1.4
13 W	0042 / 0745 / 1257 / 2007	8.0 / 1.4 / 8.1 / 1.1	**28** TH	0109 / 0809 / 1317 / 2028	7.7 / 1.7 / 7.6 / 1.7
14 TH	0126 / 0826 / 1342 / 2047	7.9 / 1.5 / 7.9 / 1.3	**29** F	0142 / 0839 / 1349 / 2056	7.5 / 2.0 / 7.4 / 2.0
15 F	0211 / 0907 / 1428 / 2127	7.8 / 1.7 / 7.7 / 1.6	**30** SA	0214 / 0907 / 1420 / 2124	7.2 / 2.3 / 7.2 / 2.4
			31 SU	0246 / 0938 / 1455 / 2157	7.0 / 2.7 / 6.8 / 2.7

FEBRUARY

Day	Time	m	Day	Time	m
1 M	0326 / 1019 / 1545 / ◑ 2244	6.7 / 3.0 / 6.5 / 3.1	**16** TU	0421 / 1118 / 1703 / 2352	7.0 / 2.5 / 6.8 / 2.7
2 TU	0428 / 1114 / 1701 / 2344	6.5 / 3.3 / 6.3 / 3.3	**17** W	0544 / 1239 / 1834	6.8 / 2.7 / 6.7
3 W	0550 / 1222 / 1828	6.4 / 3.3 / 6.3	**18** TH	0116 / 0708 / 1403 / 1952	2.8 / 6.8 / 2.6 / 6.9
4 TH	0101 / 0707 / 1351 / 1947	3.3 / 6.5 / 3.1 / 6.5	**19** F	0237 / 0818 / 1524 / 2055	2.6 / 7.0 / 2.3 / 7.1
5 F	0233 / 0815 / 1509 / 2048	3.0 / 6.9 / 2.6 / 6.9	**20** SA	0354 / 0913 / 1632 / 2144	2.3 / 7.3 / 1.9 / 7.4
6 SA	0338 / 0907 / 1606 / 2136	2.5 / 7.3 / 2.1 / 7.3	**21** SU	0454 / 0959 / 1721 / 2226	1.9 / 7.5 / 1.5 / 7.6
7 SU	0430 / 0952 / 1657 / 2220	2.0 / 7.6 / 1.6 / 7.7	**22** M	0537 / 1038 / 1759 / ○ 2303	1.7 / 7.7 / 1.4 / 7.7
8 M	0520 / 1035 / 1746 / ● 2301	1.6 / 7.9 / 1.2 / 7.9	**23** TU	0612 / 1113 / 1832 / 2336	1.5 / 7.8 / 1.2 / 7.8
9 TU	0607 / 1117 / 1833 / 2344	1.2 / 8.1 / 0.9 / 8.1	**24** W	0644 / 1146 / 1903	1.5 / 7.9 / 1.2
10 W	0653 / 1159 / 1917	1.0 / 8.3 / 0.7	**25** TH	0008 / 0715 / 1218 / 1932	7.8 / 1.4 / 7.8 / 1.3
11 TH	0026 / 0736 / 1243 / 1958	8.2 / 0.9 / 8.3 / 0.7	**26** F	0039 / 0744 / 1249 / 2000	7.7 / 1.5 / 7.7 / 1.5
12 F	0109 / 0817 / 1326 / 2037	8.2 / 1.0 / 8.2 / 0.9	**27** SA	0110 / 0810 / 1319 / 2026	7.6 / 1.7 / 7.6 / 1.8
13 SA	0152 / 0856 / 1410 / 2115	8.0 / 1.2 / 8.0 / 1.3	**28** SU	0138 / 0836 / 1348 / 2051	7.4 / 2.0 / 7.3 / 2.1
14 SU	0236 / 0935 / 1456 / 2154	7.7 / 1.6 / 7.6 / 1.7	**29** M	0206 / 0903 / 1420 / 2119	7.2 / 2.3 / 7.0 / 2.5
15 M	0323 / 1019 / 1549 / ◑ 2243	7.4 / 2.1 / 7.2 / 2.3			

MARCH

Day	Time	m	Day	Time	m
1 TU	0241 / 0937 / 1501 / 2158	6.9 / 2.7 / 6.7 / 2.9	**16** W	0353 / 1056 / 1643 / 2329	7.0 / 2.5 / 6.7 / 2.8
2 W	0330 / 1025 / 1601 / ◑ 2255	6.6 / 3.1 / 6.3 / 3.3	**17** TH	0517 / 1213 / 1816	6.6 / 2.8 / 6.6
3 TH	0443 / 1131 / 1731	6.3 / 3.3 / 6.2	**18** F	0051 / 0646 / 1339 / 1937	3.0 / 6.6 / 2.8 / 6.7
4 F	0009 / 0615 / 1256 / 1907	3.4 / 6.4 / 3.2 / 6.4	**19** SA	0217 / 0801 / 1507 / 2040	2.8 / 6.8 / 2.4 / 7.0
5 SA	0149 / 0738 / 1431 / 2020	3.1 / 6.7 / 2.7 / 6.8	**20** SU	0340 / 0857 / 1613 / 2126	2.4 / 7.1 / 2.0 / 7.3
6 SU	0306 / 0840 / 1536 / 2112	2.5 / 7.1 / 2.1 / 7.3	**21** M	0435 / 0940 / 1658 / 2205	2.0 / 7.4 / 1.7 / 7.5
7 M	0405 / 0929 / 1633 / 2157	1.9 / 7.6 / 1.5 / 7.7	**22** TU	0515 / 1017 / 1734 / 2239	1.7 / 7.6 / 1.5 / 7.7
8 TU	0459 / 1013 / 1726 / 2240	1.4 / 7.9 / 1.1 / 8.0	**23** W	0547 / 1050 / 1805 / ○ 2310	1.5 / 7.7 / 1.3 / 7.7
9 W	0550 / 1057 / 1815 / ● 2323	1.0 / 8.2 / 0.7 / 8.2	**24** TH	0619 / 1121 / 1836 / 2340	1.4 / 7.8 / 1.3 / 7.8
10 TH	0637 / 1141 / 1901	0.7 / 8.4 / 0.5	**25** F	0649 / 1152 / 1905	1.4 / 7.7 / 1.3
11 F	0006 / 0721 / 1225 / 1943	8.4 / 0.6 / 8.5 / 0.5	**26** SA	0010 / 0717 / 1223 / 1932	7.7 / 1.4 / 7.7 / 1.5
12 SA	0049 / 0802 / 1308 / 2022	8.4 / 0.6 / 8.4 / 0.7	**27** SU	0040 / 0743 / 1253 / 1958	7.6 / 1.6 / 7.6 / 1.7
13 SU	0132 / 0841 / 1352 / 2059	8.2 / 0.9 / 8.1 / 1.1	**28** M	0108 / 0809 / 1323 / 2024	7.5 / 1.8 / 7.4 / 2.0
14 M	0214 / 0920 / 1437 / 2137	7.8 / 1.4 / 7.6 / 1.7	**29** TU	0138 / 0838 / 1356 / 2054	7.3 / 2.1 / 7.1 / 2.4
15 TU	0258 / 1001 / 1529 / ◑ 2222	7.4 / 2.0 / 7.2 / 2.3	**30** W	0214 / 0910 / 1437 / 2130	7.0 / 2.4 / 6.8 / 2.8
			31 TH	0259 / 0953 / 1532 / ◑ 2221	6.7 / 2.8 / 6.5 / 3.1

APRIL

Day	Time	m	Day	Time	m
1 F	0403 / 1053 / 1650 / 2331	6.4 / 3.1 / 6.3 / 3.3	**16** SA	0022 / 0614 / 1303 / 1907	3.1 / 6.5 / 2.8 / 6.7
2 SA	0529 / 1214 / 1827	6.4 / 3.1 / 6.4	**17** SU	0140 / 0730 / 1423 / 2013	2.9 / 6.6 / 2.6 / 6.9
3 SU	0110 / 0658 / 1353 / 1946	3.1 / 6.6 / 2.7 / 6.9	**18** M	0257 / 0830 / 1531 / 2059	2.6 / 6.9 / 2.3 / 7.2
4 M	0233 / 0807 / 1503 / 2043	2.5 / 7.1 / 2.1 / 7.4	**19** TU	0356 / 0914 / 1619 / 2137	2.2 / 7.2 / 1.9 / 7.4
5 TU	0335 / 0901 / 1603 / 2131	1.9 / 7.5 / 1.5 / 7.8	**20** W	0439 / 0951 / 1658 / 2210	1.9 / 7.4 / 1.7 / 7.6
6 W	0432 / 0949 / 1700 / 2216	1.3 / 7.9 / 1.0 / 8.1	**21** TH	0516 / 1023 / 1733 / 2240	1.7 / 7.5 / 1.5 / 7.7
7 TH	0527 / 1035 / 1753 / ● 2300	0.9 / 8.2 / 0.7 / 8.3	**22** F	0550 / 1054 / 1806 / ○ 2311	1.5 / 7.6 / 1.5 / 7.7
8 F	0617 / 1120 / 1840 / 2344	0.6 / 8.4 / 0.5 / 8.4	**23** SA	0621 / 1126 / 1836 / 2342	1.5 / 7.6 / 1.5 / 7.7
9 SA	0702 / 1205 / 1924	0.5 / 8.4 / 0.5	**24** SU	0650 / 1158 / 1904	1.5 / 7.6 / 1.6
10 SU	0028 / 0745 / 1250 / 2004	8.4 / 0.6 / 8.3 / 0.7	**25** M	0013 / 0718 / 1230 / 1933	7.6 / 1.5 / 7.5 / 1.7
11 M	0111 / 0826 / 1334 / 2043	8.2 / 0.9 / 8.0 / 1.2	**26** TU	0044 / 0747 / 1303 / 2003	7.5 / 1.7 / 7.4 / 2.0
12 TU	0153 / 0905 / 1420 / 2122	7.8 / 1.3 / 7.6 / 1.8	**27** W	0118 / 0819 / 1341 / 2036	7.4 / 1.9 / 7.2 / 2.3
13 W	0237 / 0947 / 1512 / 2207	7.4 / 1.9 / 7.1 / 2.4	**28** TH	0156 / 0853 / 1424 / 2113	7.1 / 2.2 / 6.9 / 2.6
14 TH	0330 / 1039 / 1623 / ◐ 2308	7.0 / 2.5 / 6.7 / 2.9	**29** F	0242 / 0934 / 1516 / 2201	6.8 / 2.5 / 6.7 / 2.9
15 F	0448 / 1147 / 1749	6.6 / 2.8 / 6.6	**30** SA	0340 / 1028 / 1624 / ◐ 2305	6.6 / 2.8 / 6.5 / 3.1

Chart Datum is 4·38 metres below IGN Datum. HAT is 8·4 metres above Chart Datum.

FREE monthly updates. Register at
www.reedsnauticalalmanac.co.uk

STANDARD TIME UT –01
Subtract 1 hour for UT
For French Summer Time add ONE hour in **non-shaded areas**

LE HAVRE LAT 49°29′N LONG 0°07′E
TIMES AND HEIGHTS OF HIGH AND LOW WATERS

Dates in red are **SPRINGS**
Dates in blue are NEAPS

YEAR **2016**

N France

MAY

Day	Time	m	Day	Time	m
1 SU	0455	6.5	16 M	0053	3.0
	1142	2.8		0639	6.5
	1749	6.6		1326	2.8
				1926	6.8
2 M	0034	2.9	17 TU	0200	2.8
	0618	6.7		0746	6.7
	1316	2.5		1432	2.6
	1909	7.0		2020	7.0
3 TU	0158	2.5	18 W	0303	2.5
	0731	7.1		0838	6.9
	1429	2.0		1529	2.3
	2011	7.4		2102	7.2
4 W	0303	1.9	19 TH	0355	2.2
	0831	7.5		0918	7.1
	1531	1.5		1617	2.0
	2103	7.8		2137	7.4
5 TH	0403	1.4	20 F	0439	1.9
	0923	7.8		0954	7.3
	1631	1.1		1659	1.9
	2151	8.0		2210	7.5
6 F ●	0500	1.0	21 SA	0518	1.8
	1012	8.1		1028	7.4
	1727	0.8		1735	1.8
	●2237	8.2		○2244	7.6
7 SA	0554	0.7	22 SU	0553	1.7
	1059	8.2		1103	7.5
	1817	0.7		1808	1.7
	2322	8.3		2317	7.6
8 SU	0643	0.6	23 M	0624	1.6
	1146	8.3		1137	7.5
	1903	0.7		1839	1.7
				2350	7.6
9 M	0007	8.2	24 TU	0656	1.6
	0727	0.7		1211	7.5
	1232	8.2		1912	1.8
	1946	1.0			
10 TU	0051	8.1	25 W	0024	7.6
	0809	0.9		0730	1.6
	1318	7.9		1248	7.4
	2027	1.4		1947	1.9
11 W	0134	7.8	26 TH	0102	7.5
	0851	1.3		0805	1.8
	1405	7.6		1329	7.3
	2107	1.9		2023	2.1
12 TH	0219	7.4	27 F	0144	7.3
	0932	1.8		0842	2.0
	1455	7.2		1414	7.1
	2151	2.4		2102	2.4
13 F	0309	7.0	28 SA	0230	7.1
	1020	2.3		0922	2.2
	1557	6.9		1504	7.0
	◑2245	2.8		2148	2.6
14 SA	0414	6.7	29 SU	0323	6.9
	1116	2.7		1012	2.4
	1708	6.6		1603	6.8
	2347	3.0		◐2246	2.7
15 SU	0529	6.5	30 M	0427	6.8
	1219	2.8		1116	2.5
	1818	6.6		1714	6.8
			31 TU	0002	2.7
				0541	6.9
				1239	2.4
				1830	7.0

JUNE

Day	Time	m	Day	Time	m
1 W	0123	2.4	16 TH	0210	2.8
	0656	7.1		0748	6.7
	1356	2.1		1439	2.7
	1937	7.4		2018	7.0
2 TH	0232	2.0	17 F	0311	2.5
	0802	7.4		0841	6.9
	1501	1.7		1536	2.4
	2035	7.7		2102	7.2
3 F	0334	1.6	18 SA	0402	2.2
	0859	7.7		0925	7.1
	1602	1.4		1623	2.2
	2127	7.9		2141	7.4
4 SA	0434	1.2	19 SU	0445	2.0
	0952	7.9		1004	7.2
	1701	1.2		1704	2.0
	2215	8.0		2219	7.5
5 SU ●	0531	1.0	20 M	0525	1.8
	1041	8.0		1042	7.4
	1755	1.1		1742	1.9
	●2302	8.1		○2255	7.6
6 M	0623	0.9	21 TU	0601	1.7
	1129	8.0		1119	7.5
	1844	1.1		1819	1.8
	2347	8.1		2331	7.7
7 TU	0710	0.9	22 W	0638	1.6
	1216	8.0		1156	7.5
	1928	1.2		1856	1.7
8 W	0032	8.0	23 TH	0009	7.7
	0753	1.1		0716	1.5
	1302	7.8		1236	7.5
	2010	1.8		1934	1.8
9 TH	0115	7.8	24 F	0049	7.7
	0834	1.4		0755	1.5
	1347	7.6		1318	7.5
	2050	1.8		2014	1.8
10 F	0158	7.5	25 SA	0132	7.6
	0914	1.8		0833	1.7
	1433	7.3		1402	7.4
	2130	2.2		2054	2.0
11 SA	0243	7.2	26 SU	0218	7.4
	0954	2.2		0913	1.9
	1523	7.0		1449	7.3
	2214	2.6		2137	2.2
12 SU	0335	6.8	27 M	0307	7.2
	1038	2.6		0958	2.1
	1620	6.7		1541	7.1
	◐2303	2.9		◑2227	2.4
13 M	0437	6.6	28 TU	0402	7.1
	1130	2.8		1053	2.3
	1721	6.6		1642	7.0
				2331	2.5
14 TU	0000	3.0	29 W	0508	7.0
	0542	6.5		1204	2.3
	1228	2.9		1754	7.1
	1823	6.6			
15 W	0102	3.0	30 TH	0049	2.4
	0645	6.5		0625	7.0
	1333	2.9		1324	2.2
	1924	6.8		1907	7.2

JULY

Day	Time	m	Day	Time	m
1 F	0204	2.2	16 SA	0222	2.9
	0738	7.2		0800	6.6
	1434	2.0		1454	2.8
	2011	7.5		2026	6.9
2 SA	0310	1.8	17 SU	0323	2.6
	0841	7.5		0856	6.9
	1538	1.8		1548	2.5
	2107	7.7		2114	7.2
3 SU	0412	1.5	18 M	0413	2.2
	0937	7.6		0941	7.2
	1639	1.6		1635	2.2
	2158	7.8		2156	7.4
4 M ●	0513	1.3	19 TU	0458	1.9
	1027	7.8		1022	7.4
	1737	1.4		1719	1.9
	●2245	7.9		○2235	7.6
5 TU	0607	1.2	20 W	0541	1.6
	1115	7.8		1101	7.5
	1828	1.4		1801	1.7
	2330	8.0		2314	7.8
6 W	0655	1.1	21 TH	0623	1.4
	1200	7.9		1141	7.7
	1911	1.4		1843	1.6
				2354	7.9
7 TH	0013	7.9	22 F	0705	1.3
	0736	1.2		1221	7.8
	1243	7.8		1924	1.5
	1951	1.5			
8 F	0054	7.8	23 SA	0036	7.9
	0813	1.4		0746	1.2
	1324	7.6		1303	7.8
	2027	1.8		2004	1.5
9 SA	0134	7.6	24 SU	0118	7.9
	0849	1.7		0825	1.3
	1404	7.4		1347	7.7
	2102	2.1		2044	1.6
10 SU	0213	7.3	25 M	0202	7.7
	0922	2.0		0903	1.5
	1444	7.2		1431	7.6
	2137	2.4		2124	1.8
11 M	0253	7.0	26 TU	0248	7.5
	0956	2.4		0943	1.8
	1527	6.9		1517	7.4
	2215	2.7		2208	2.1
12 TU	0339	6.7	27 W	0338	7.3
	1035	2.8		1030	2.1
	1619	6.7		1611	7.2
	◐2301	3.0		◑2303	2.4
13 W	0439	6.4	28 TH	0440	7.0
	1124	3.0		1133	2.4
	1721	6.5		1721	7.0
	2357	3.2			
14 TH	0546	6.3	29 F	0019	2.5
	1225	3.2		0601	6.9
	1825	6.5		1256	2.5
				1843	7.0
15 F	0105	3.2	30 SA	0141	2.4
	0653	6.4		0722	7.0
	1340	3.1		1413	2.4
	1928	6.7		1954	7.2
			31 SU	0251	2.1
				0830	7.3
				1520	2.1
				2054	7.5

AUGUST

Day	Time	m	Day	Time	m
1 M	0357	1.8	16 TU	0342	2.3
	0926	7.5		0917	7.1
	1625	1.9		1608	2.3
	2145	7.7		2132	7.4
2 TU	0502	1.6	17 W	0433	1.9
	1016	7.6		1000	7.5
	1726	1.7		1657	1.9
	●2231	7.8		2214	7.7
3 W	0555	1.4	18 TH	0521	1.5
	1100	7.8		1041	7.7
	1814	1.6		1743	1.6
	2313	7.9		○2255	8.0
4 TH	0638	1.3	19 F	0607	1.2
	1142	7.8		1121	7.9
	1853	1.5		1828	1.3
	2353	8.0		2336	8.1
5 F	0715	1.3	20 SA	0651	1.0
	1220	7.8		1203	8.0
	1928	1.5		1910	1.2
6 SA	0029	7.9	21 SU	0018	8.2
	0747	1.4		0732	0.9
	1256	7.7		1245	8.1
	2000	1.7		1951	1.1
7 SU	0105	7.7	22 M	0101	8.2
	0818	1.6		0811	1.0
	1330	7.5		1327	8.0
	2030	1.9		2030	1.3
8 M	0139	7.5	23 TU	0144	8.0
	0847	1.9		0848	1.3
	1404	7.3		1410	7.8
	2058	2.2		2108	1.6
9 TU	0211	7.3	24 W	0229	7.7
	0913	2.3		0925	1.7
	1436	7.1		1454	7.6
	2127	2.6		2149	2.0
10 W	0245	7.0	25 TH	0317	7.4
	0942	2.7		1008	2.1
	1512	6.8		1545	7.2
	◐2202	2.9		◑2240	2.4
11 TH	0328	6.6	26 F	0418	7.0
	1022	3.0		1108	2.6
	1604	6.6		1655	6.9
	2253	3.2		2356	2.7
12 F	0434	6.3	27 SA	0546	6.8
	1120	3.4		1235	2.8
	1721	6.4		1825	6.9
	2359	3.4			
13 SA	0559	6.2	28 SU	0123	2.7
	1235	3.4		0712	6.9
	1839	6.4		1358	2.7
				1942	7.0
14 SU	0124	3.2	29 M	0239	2.4
	0718	6.4		0821	7.1
	1408	3.2		1511	2.4
	1950	6.7		2043	7.3
15 M	0245	2.8	30 TU	0352	2.0
	0826	6.7		0917	7.4
	1516	2.7		1621	2.1
	2047	7.1		2133	7.6
			31 W	0454	1.7
				1002	7.6
				1716	1.8
				2216	7.8

Chart Datum is 4·38 metres below IGN Datum. HAT is 8·4 metres above Chart Datum.

》》FREE monthly updates. Register at **《**
www.reedsnauticalalmanac.co.uk

STANDARD TIME UT –01
Subtract 1 hour for UT
For French Summer Time add
ONE hour in **non-shaded areas**

LE HAVRE   LAT 49°29'N   LONG 0°07'E
TIMES AND HEIGHTS OF HIGH AND LOW WATERS

Dates in red are SPRINGS
Dates in blue are NEAPS

YEAR 2016

SEPTEMBER

Time	m		Time	m
1 0540	1.5	**16** 0456	1.4
1042	7.7	1016	7.9
TH 1756	1.6	F 1721	1.4
● 2254	7.9	○ 2232	8.1
2 0616	1.4	**17** 0545	1.0
1118	7.8	1058	8.1
F 1829	1.5	SA 1808	1.1
2329	8.0	2315	8.3
3 0648	1.3	**18** 0631	0.8
1152	7.8	1141	8.3
SA 1901	1.5	SU 1853	0.9
		2358	8.4
4 0002	7.9	**19** 0714	0.7
0718	1.4	1223	8.3
SU 1224	7.8	M 1934	0.9
1930	1.6		
5 0034	7.8	**20** 0042	8.4
0746	1.6	0754	0.9
M 1256	7.6	TU 1306	8.2
1958	1.8	2014	1.1
6 0106	7.6	**21** 0126	8.2
0812	1.9	0831	1.2
TU 1325	7.5	W 1348	8.0
2023	2.1	2052	1.4
7 0135	7.4	**22** 0210	7.8
0836	2.2	0908	1.7
W 1353	7.3	TH 1432	7.6
2048	2.4	2132	1.9
8 0205	7.1	**23** 0259	7.4
0901	2.6	0950	2.3
TH 1425	7.0	F 1523	7.3
2118	2.8	◗ 2223	2.4
9 0244	6.8	**24** 0403	7.0
0935	3.0	1050	2.8
F 1509	6.7	SA 1635	6.9
◗ 2202	3.1	2339	2.8
10 0338	6.4	**25** 0535	6.7
1028	3.4	1219	3.1
SA 1616	6.4	SU 1809	6.7
2306	3.4		
11 0502	6.2	**26** 0108	2.8
1141	3.6	0700	6.8
SU 1748	6.3	M 1345	2.9
		1927	6.9
12 0029	3.4	**27** 0228	2.5
0638	6.3	0809	7.1
M 1320	3.4	TU 1503	2.5
1913	6.6	2029	7.2
13 0206	3.0	**28** 0339	2.1
0755	6.7	0901	7.4
TU 1443	2.9	W 1607	2.1
2017	7.0	2117	7.5
14 0311	2.4	**29** 0433	1.8
0849	7.2	0943	7.6
W 1540	2.3	TH 1654	1.8
2106	7.5	2156	7.7
15 0405	1.8	**30** 0514	1.6
0934	7.6	1019	7.8
TH 1632	1.7	F 1730	1.7
2150	7.9	2231	7.9

OCTOBER

Time	m		Time	m
1 0547	1.5	**16** 0520	1.0
1052	7.9	1034	8.3
SA 1801	1.6	SU 1746	0.9
● 2303	7.9	○ 2253	8.4
2 0617	1.4	**17** 0608	0.8
1123	7.9	1117	8.4
SU 1832	1.5	M 1832	0.8
2334	7.9	2338	8.5
3 0648	1.5	**18** 0653	0.7
1153	7.8	1201	8.4
M 1901	1.6	TU 1916	0.8
4 0006	7.8	**19** 0023	8.4
0716	1.6	0735	0.9
TU 1223	7.7	W 1245	8.3
1928	1.7	1958	1.0
5 0036	7.7	**20** 0109	8.2
0742	1.9	0816	1.2
W 1252	7.6	TH 1329	8.0
1954	2.0	2039	1.4
6 0106	7.5	**21** 0155	7.8
0807	2.2	0855	1.8
TH 1320	7.4	F 1414	7.7
2020	2.2	2121	1.9
7 0137	7.2	**22** 0245	7.4
0834	2.5	0939	2.3
F 1354	7.1	SA 1505	7.3
2051	2.6	◗ 2212	2.4
8 0217	6.9	**23** 0350	7.0
0907	2.9	1039	2.9
SA 1437	6.8	SU 1616	6.9
2130	3.0	2322	2.8
9 0308	6.6	**24** 0516	6.8
0955	3.3	1158	3.1
SU 1537	6.5	M 1744	6.7
◗ 2227	3.3		
10 0421	6.4	**25** 0042	2.9
1101	3.5	0635	6.8
M 1701	6.4	TU 1319	3.0
2344	3.3	1900	6.8
11 0557	6.4	**26** 0159	2.6
1235	3.4	0743	7.0
TU 1832	6.6	W 1433	2.7
		2005	7.0
12 0124	3.0	**27** 0305	2.3
0719	6.8	0836	7.3
W 1408	2.9	TH 1534	2.3
1942	7.0	2053	7.3
13 0238	2.4	**28** 0357	2.0
0817	7.3	0917	7.6
TH 1509	2.2	F 1619	2.0
2036	7.5	2133	7.6
14 0335	1.8	**29** 0438	1.8
0906	7.7	0952	7.7
F 1604	1.7	SA 1657	1.8
2121	7.9	2207	7.7
15 0429	1.3	**30** 0514	1.7
0950	8.1	1024	7.8
SA 1656	1.2	SU 1732	1.6
2209	8.2	● 2239	7.8
		31 0547	1.6
		1054	7.8
		M 1804	1.6
		2310	7.8

NOVEMBER

Time	m		Time	m
1 0619	1.6	**16** 0634	0.8
1125	7.8	1141	8.4
TU 1835	1.6	W 1900	0.8
2341	7.7		
2 0648	1.7	**17** 0007	8.4
1155	7.8	0719	1.0
W 1903	1.7	TH 1227	8.3
		1944	0.9
3 0013	7.7	**18** 0054	8.2
0716	1.9	0802	1.3
TH 1225	7.7	F 1312	8.1
1931	1.9	2028	1.3
4 0045	7.5	**19** 0142	7.9
0745	2.1	0845	1.8
F 1257	7.5	SA 1358	7.7
2001	2.1	2112	1.7
5 0120	7.3	**20** 0232	7.5
0816	2.4	0930	2.3
SA 1334	7.2	SU 1448	7.3
2034	2.4	2200	2.2
6 0201	7.0	**21** 0332	7.1
0851	2.8	1023	2.7
SU 1418	7.0	M 1550	6.9
2112	2.7	◗ 2256	2.6
7 0251	6.8	**22** 0442	6.9
0935	3.1	1126	3.0
M 1513	6.7	TU 1705	6.7
◗ 2202	3.0		
8 0355	6.6	**23** 0000	2.9
1034	3.3	0552	6.8
TU 1625	6.6	W 1234	3.1
2309	3.1	1817	6.7
9 0516	6.6	**24** 0109	2.8
1154	3.3	0701	6.9
W 1748	6.7	TH 1344	2.9
		1925	6.8
10 0038	2.9	**25** 0215	2.6
0638	6.9	0800	7.1
TH 1327	2.9	F 1447	2.6
1903	7.0	2022	7.1
11 0200	2.4	**26** 0312	2.4
0742	7.3	0846	7.4
F 1436	2.3	SA 1540	2.3
2004	7.5	2106	7.3
12 0303	1.8	**27** 0400	2.1
0835	7.8	0924	7.6
SA 1534	1.7	SU 1624	2.0
2057	7.9	2143	7.5
13 0400	1.4	**28** 0442	1.9
0924	8.1	0958	7.7
SU 1630	1.4	M 1704	1.8
2146	8.2	2216	7.6
14 0454	1.0	**29** 0520	1.8
1010	8.3	1030	7.7
M 1723	0.9	TU 1739	1.7
○ 2233	8.4	● 2250	7.6
15 0546	0.9	**30** 0554	1.8
1056	8.4	1102	7.8
TU 1813	0.8	W 1812	1.7
2320	8.4	2323	7.6

DECEMBER

Time	m		Time	m
1 0626	1.8	**16** 0705	1.1
1134	7.8	1211	8.3
TH 1843	1.7	F 1933	0.9
2356	7.6		
2 0657	1.9	**17** 0041	8.1
1207	7.7	0750	1.3
F 1915	1.7	SA 1256	8.1
		2016	1.2
3 0030	7.5	**18** 0127	7.9
0730	2.0	0833	1.6
SA 1242	7.6	SU 1341	7.8
1948	1.9	2058	1.5
4 0108	7.4	**19** 0214	7.6
0804	2.2	0914	2.0
SU 1321	7.4	M 1427	7.5
2024	2.1	2139	2.0
5 0150	7.3	**20** 0303	7.3
0841	2.5	0957	2.5
M 1405	7.2	TU 1517	7.1
2102	2.3	2222	2.4
6 0237	7.1	**21** 0358	7.0
0924	2.7	1044	2.8
TU 1455	7.0	W 1616	6.8
2147	2.6	◗ 2311	2.8
7 0333	6.9	**22** 0459	6.8
1015	2.9	1139	3.1
W 1555	6.8	TH 1722	6.6
◗ 2243	2.7		
8 0439	6.8	**23** 0008	3.0
1120	3.0	0603	6.7
TH 1706	6.8	F 1242	3.1
2355	2.7	1829	6.6
9 0554	7.0	**24** 0114	3.0
1242	2.8	0708	6.8
F 1823	7.0	SA 1355	3.0
		1938	6.7
10 0119	2.4	**25** 0225	2.8
0705	7.3	0808	7.0
SA 1400	2.4	SU 1459	2.7
1932	7.3	2035	6.9
11 0231	2.0	**26** 0323	2.5
0806	7.6	0854	7.2
SU 1506	1.9	M 1552	2.5
2033	7.7	2119	7.2
12 0333	1.6	**27** 0412	2.3
0900	7.9	0933	7.4
M 1606	1.4	TU 1637	2.1
2127	8.0	2157	7.4
13 0431	1.3	**28** 0454	2.0
0950	8.2	1009	7.6
TU 1703	1.1	W 1716	1.8
2217	8.2	2233	7.5
14 0526	1.1	**29** 0532	1.9
1038	8.3	1044	7.7
W 1756	0.9	TH 1753	1.7
○ 2306	8.3	● 2308	7.6
15 0618	1.0	**30** 0608	1.8
1125	8.3	1119	7.8
TH 1846	0.8	F 1827	1.6
2354	8.3	2343	7.6
		31 0643	1.8
		1154	7.8
		SA 1903	1.5

Chart Datum is 4·38 metres below IGN Datum. HAT is 8·4 metres above Chart Datum.

>> **FREE** monthly updates. Register at
www.reedsnauticalalmanac.co.uk ‹‹

9.17.18 HONFLEUR

Calvados **49° 25'·69N 00° 13'·90E** ✿✿✿✿♨♨♨✿✿✿

CHARTS AC 2146, 2990, 2879, 1349; SHOM 7418, 6683, 7420; Navi 1012; Imray C31

TIDES –0135 Dover; ML 5·0; Duration 0540

Standard Port LE HAVRE (◄—)

Times				Height (metres)			
High Water		Low Water		MHWS	MHWN	MLWN	MLWS
0000	0500	0000	0700	8·0	6·7	2·9	1·2
1200	1700	1200	1900				
LE HAVRE-ANTIFER (49°39'·5N 00°06'·8E)							
+0025	+0015	+0005	–0005	+0·1	0·0	0·0	0·0
CHENAL DE ROUEN (49°26'N 00°07'E)							
0000	0000	0000	+0015	0·0	–0·1	0·0	–0·1
HONFLEUR							
–0150	–0135	+0025	+0040	0·0	0·0	+0·1	+0·4

In the Seine a HW stand lasts about 2hrs 50 mins. The HW time differences refer to the beginning of the stand.

SHELTER Excellent in the Vieux Bassin and Bassin de l'Est.

NAVIGATION WPT No 2 PHM buoy, QR, 49°27'·70N 00°00'·60E, 9M

HONFLEUR
metres 0 ... 400
cables
13'·5 0 ... 2

to Honfleur. Begin appr between LW+1½ to HW–1½ to avoid foul tide. From WPT keep to the N of the Chenal de Rouen between PHM buoys and posts marking Digue basse du Nord; there is no room for E-bound yachts to navigate S of the chan. Beware risk of collision in low vis with W-bound yachts. This rule applies from No 2 buoy to Tancarville bridge.

After No 20 PHM buoy cross to the ent. Caution: strong stream across ent, shoal patches reported at 'shoulders' of lock chamber.

Rouen is 60M and Paris 192M upriver.

LIGHTS AND MARKS See chartlet and Lights, buoys and waypoints. The radar tower by the lock and the Pont de Normandie 1·6M upriver are highly conspic; slightly less so are the wooded hills SW of Honfleur.

COMMUNICATIONS CROSS 02·33·52·16·16; [LB] 02·31·89·20·17; Honfleur Radar Ch **73** (H24); Auto (local) 08·92·68·08·14; Auto (regional) 08·92·68·08·76; Auto (offshore) 08·92·68·08·08; Police 02·31·14·44·45; ⊖ 02·31·14·44·30; Brit Consul 01·44·51·31·00; Dr 02·31·89·34·05; Ⓗ 02·31·89·89·89; Port HM 02·31·14·61·09, VHF Ch **17** 73; Aff Mar 02·31·53·66·50; Lock VHF **17** (H24); Vieux Bassin VHF Ch 09.

THE LOCK (☎02·31·98·72·82) operates H24 and opens at H, every hour for arrivals; and every H +30 for departures. It is 23m wide and has recessed floating bollards, no pontoons. **IPTS** in force. Freeflow operates from HW Le Havre –1½ to HW.

THE ROAD BRIDGE at the Vieux Bassin lifts at times (LT) below. Inbound yachts can wait on E side of the jetty immediately outside the Vieux Bassin. Departing yachts take priority over arrivals; be ready 5 mins in advance.

1 May to 31 Aug
Every day: 0830, 0930, 1030, 1130. 1630, 1730, 1830, 1930.

1 Jan to 30 Apr and 1 Sep to 31 Dec
Every day: 0830, *0930*, *1030*, 1130. 1430, *1530*, *1630*, 1730.
Times in **bold italic** must be requested, Mon-Fri.

FACILITIES **The Avant Port's** W wall, close N of the waiting jetty, can be a temp'y (max 24 hrs) berth, useful for locking out before the Vieux Bassin's 0830 bridge lift or to await the next lock time.

Vieux Bassin ☎02·31·98·87·13 (CNH). info@ville-honfleur.fr 120 +30Ⓥ max LOA 20m, €2·60; €2·30 Avant-Port; multihulls and large motor yachts +50%. Ⓥ raft on pontoons on NW side; or in a CNH berth as assigned by CNH dory which meets and directs all arrivals.

Cercle Nautique d'Honfleur (CNH) www.cnh-honfleur.net controls all yacht berthing. Berthing fees and 🔌 tokens (€2·00) paid for here. ☎02·31·98·87·13. cnh14@wanadoo.fr ⚓ **Bassin de l'Est** can be used by yachts >20m LOA; see HM. 🚿 ✕ Ⓔ 🏗(10t and for masting) 🔧.

Town ✕ 🏠 🛒 🖥 💊 🏦 🛒 🚿 @ ✉ Ⓑ 🚂 ✈ Deauville.

9.17.19 RIVER SEINE

MARINAS

Km		
245	Rouen	€1·20
121	Port de l'Ilon	€1·30
81	Port St Louis	
24	Port Van Gogh Asnières	
0	Paris-Arsenal	€4·00

HALTE NAUTIQUE (HN)

Km	HN
243	Rouen
218	St Aubin-lès-Elbeuf
202	Poses/Amfreville
184	Venables
150	Vernonnet
110	Limay
93	Meulan
70	Coflans St Honorine
49	Bougival
44	Rueil Malmaison

LOCKS

Km	Lock	VHF
202	Amfreville	18
161	Notre Dame de la Garenne	22
121	Méricourt	18
73	Andrésy	22
49	Bougival	22
	or	
44.5	Chatou	18
17	Suresnes	22

TIDES Standard Port LE HAVRE (⬅)

Times				Height (metres)			
High Water		Low Water		MHWS	MHWN	MLWN	MLWS
0000	0500	0000	0700	7·9	6·6	2·8	1·2
1200	1700	1200	1900				

Differences TANCARVILLE* *(Km 338. 49°28'·35N 00°27'·88E)*

−0135	−0120	+0030	+0145	+0·1	+0·1	+0·3	+1·0

QUILLEBEUF* *(Km 332. 49°28'·25N 00°32'·00E)*

−0055	−0110	+0105	+0210	+0·1	+0·1	+0·5	+1·4

VATTEVILLE* *(Km 317. 49°28'·89N 00°40'·01E)*

−0015	−0040	+0205	+0240	−0·1	−0·1	+0·9	+2·2

CAUDEBEC* *(Km 309. 49°31'·35N 00°43'·66E)*

−0005	−0030	+0220	+0300	−0·1	−0·1	+1·0	+2·3

HEURTEAUVILLE* *(Km 295. 49°25'·70N 00°48'·17E)*

+0055	+0005	+0250	+0330	−0·2	−0·1	+1·2	+2·6

DUCLAIR* *(Km 278. 49°28'·73N 00°52'·38E)*

+0210	+0145	+0350	+0410	0·0		+1·5	+3·2

ROUEN *(Km 245. 49°26'·50N 01°03'·13E)*

+0305	+0240	+0505	+0515	−0·1	+0·2	+1·6	+3·3

*HW time differences refer to the start of a stand lasting about 2¾ hrs up to Vatteville, decreasing to about 1¾ hrs at Duclair.

SHELTER Marinas/Haltes Nautiques (HN)/fuel, see above. Masts can be lowered at Dives, Deauville, Le Havre, Honfleur and Rouen.

TIDAL STREAMS require careful study; see *N France and Belgium Cruising Companion* (Featherstone/Fernhurst) for tidal stream graphs and a detailed analysis, plus Rouen and Paris. ▶At Ratier NW SHM buoy the flood starts at LW Le Havre +1 and at La Roque (Km342) at LW+2, and progressively later further up-river. So even a 5kn boat which left Le Havre at LW (or Honfleur at LW+1) and passed the W end of Digue du Ratier at LW+15, should carry the flood to Rouen. If delayed, try another day rather than stopping en route.

Going downriver on the ebb a boat will meet the flood. In a fast boat it is worth continuing, rather than ⚓ for about 4 hrs, because the ebb starts sooner the further downriver one gets.◀

NAVIGATION The tidal section to Rouen should be completed in 1 day, because the strong stream and ships' wash make berthing alongside dangerous and there are no ⚓s. Yacht navigation is prohibited at night. A radar reflector and VHF aerial are needed, even with mast down.

Enter the buoyed Chenal de Rouen abm No 2 buoy. Care is needed to avoid the shoal banks NE of No 2 buoy, especially in morning fog; keep W of Rade de la Carosse WCM buoy. The approach can be rough in a strong W'ly over ebb tide. From No 2 buoy to Tancarville bridge yachts must keep outboard of the PHM chan buoys, in either direction. Pont de Normandie (52m) is conspic 1·7M E of Honfleur.

Entry via the Canal de Tancarville is possible if sea conditions are bad, but expect delays due to 2 locks and 8 bridges. At Le Havre transit the Bassins Bellot, Vétillart, Despujols and the Canal du Havre, leading via the Canal de Tancarville to the lock (Km338).

From Amfreville lock to Paris current is about 1kn in summer, more after winter floods. 6 of 7 locks must be transited; foc 0700-1900.

COMMUNICATIONS Ch 73 *(Rouen Port Control)* gives nav, vis, weather and water level info on request. Monitor Ch 73 for tfc info. Ch 82 broadcasts water levels from Le Havre to Caudebec every 5 mins.

ROUEN, Seine Maritime, **49°26'·07N 01°06'·24E**. AC 2879; SHOM 6796, 6117. Yacht navigation prohib SS+½ to SR–½. ⌂ in the Halte Nautique, NE side of Île Lacroix, Km241 (lat/long above) or **Marina Bassin St Gervais**, N bank, Km245, see below. A one-way system in the chan NE of Île Lacroix only allows navigation *against* the stream. For example, if leaving on the young flood to go upriver, you must first go downriver to the NW tip of the island, then turn upriver via the SW chan. *Rouen Port* VHF Ch **73** 68. HM 02·35·52·54·56; Aff Mar 02·35·98·53·98; ⊖ 02·35·98·27·60; ☎ 02·35·80·11·44.

Facilities: **Marina Bassin St Gervais**, ☎02·35·08·30·59, www.la-crea.fr/port-de-plaisance-de-la-crea.html; €1·20. Mast down/up at pontoon SE side; (25t,3t).

Halte Nautique ☎02·35·07·33·94, www.rouen.fr/adresse/halte-nautique 50 ⌂ 🔧 ⛽ ⬇ ⚓ 🔩 ⚒ Ⓔ (3t). 🔋 from barge at Km239.

ROUEN

Pont Corneille

Km 242 — VNF

One way system:
Only against the flood or ebb

Halte Nautique

Île Lacroix

RIVE GAUCHE

0 — metres — 400

Pont Mathilde

9.17.20 PARIS

Île-de-France, **48°50'·84N 02°22'·02E** (Paris-Arsenal marina).

CHARTS AC & SHOM charts do not cover the Seine E of Rouen, but Fluviacarte No 1 covers Le Havre to Paris in detail.

SHELTER Complete in Paris-Arsenal Marina.

NAVIGATION Between Eiffel Tower and Notre Dame the wash of many *vedettes* can raise quite a chop.

A traffic system (chartlet) around Île de la Cité and Île St Louis must be understood and obeyed. Going upriver, the simpler route is: Hug the S bank in the narrowish one-way Bras de la Monnaie, leaving Notre Dame to port. Or take the wider, two-way N bank chan, noting that this route is only available H to H+20; is regulated by R/G tfc lts on Pont au Change; and requires two cross-overs where shown below.

Going downriver, make good a window H+35 to H+50, obey R/G tfc lts on Pont de Sully, then leave Île St Louis to stbd and Notre Dame to port. Cross-over twice where indicated.

By 2020 Paris should be joined to Cambrai in N France by the 66-mile long Seine-Nord Europe canal, which will be open to leisure craft.

LIGHTS AND MARKS Eiffel Tower and Notre Dame. Check off other landmarks and all bridges carefully; obey cross-overs.

COMMUNICATIONS Brit Consul 01·44·51·31·00; Monitor VHF Ch 10 on the river (mandatory); At the lock to Arsenal talk direct to HM via intercom, CCTV monitors; Marina VHF Ch 09; Canal St Martin VHF Ch 20; Taxi 01·45·85·85·85.

FACILITIES

Paris-Arsenal Marina 11 Blvd de la Bastille, 75012 Paris, ☎01·43·41·39·32. www.fayollemarine.fr No reservations 1/6-30/9, but contact marina several days before arrival. Enter 0800-1900 (2000: w/e, and Jul/Aug) via small lock, with waiting pontoon

outside; obey R/G tfc lts signed 'Plaisance'. Air clearance in lock is 5·2m, least depth 1·9m. Inside, HM and office are to stbd; berth as directed. 176 ⌷ €4·00; max LOA 25m. ⌷ ⌷ (plus on marina trolley, 750m SE) ✕ ⌷ ⌷(7t). Metro stns: Quai de la Rapée (Line 5) at S end of marina and Bastille (Lines 1, 5 & 8) at the N end. Numerous buses from Bastille. Tourist Office at Gare de Lyon (550m).

9.17.21 DEAUVILLE/TROUVILLE

Calvados 49° 22'·38N 00° 04'·15E ❄❄⚓♦♦♦♦❀❀❀

CHARTS AC 2613, 2146, 1349; SHOM 7418, 7420; Navi 526; Imray C32

TIDES –0130 Dover; ML 5·1; Duration 0510

Standard Port LE HAVRE (←—)

Times				Height (metres)			
High Water		Low Water		MHWS	MHWN	MLWN	MLWS
0000	0500	0000	0700	8·0	6·7	2·9	1·2
1200	1700	1200	1900				
Differences TROUVILLE							
–0100	–0010	0000	+0005	+0·4	+0·3	+0·3	+0·1

NOTE: There is a double HW at Trouville. The HW time differences, when referred to the time of HW Le Havre, give the times of the first HW.

SHELTER Access difficult in NW/N winds >F5. Best shelter in Bassin Morny (3m) Port Deauville outer basin (2·8m) can be uncomfortable during freeflow, again in strong winds from the N quadrant. Only FVs berth at Trouville.

NAVIGATION WPT 49°22'·94N 00°03'·62E, 148°/0·65M to abeam W training wall lt, Fl WG 4s, beware drying shoal extending E from the training wall. Do not appr from E of N due to Les Ratelets drying shoal and Banc de Trouville. Trouville SW WCM buoy is 1M WNW of ent. Chan and river dry 2·4m; no access LW±2½, for 2m draft, less water than charted reported in Avant Port.

LIGHTS AND MARKS Trouville's Casino and Deauville's Royal Hotel/Casino (off chartlet to the SW) are conspic. Front 148° ldg lt not visible when brg >150°. Bassin Morny ent gate: **IPTS**, sigs 2-4.

COMMUNICATIONS [LB] 02·31·88·31·70; CROSS 02·33·52·16·16; Auto (local) 08·92·68·08·14; Auto (regional) 08·92·68·08·76; Auto (offshore) 08·92·68·08·08; Police 02·31·88·13·07; ⊖ 02·31·88·35·29; Brit Consul 01·44·51·31·00; Dr & [H] 02·31·14·33·33; Port Deauville and Bassin Morny VHF Ch 09; Aff Mar 02·31·88·36·21; Deauville Tourist office 02·31·14·40·00.

FACILITIES from seaward: **Port Deauville** Lock ☎02·31·98·30·01 free flow approx HW–2 to +2. Marina ☎02·31·98·30·01. port-deauville-sa@wanadoo.fr www.deauville.org 700 + 150 ✔ €3·05 (Jul, Aug). ⚓ €2·00, ⛽ (H24 with cr/card) ✗ 🖳 ⚒ 🔧 ☒ 🔌 ⚓ 🚲(45t).

Port Morny No ✔. Entry gate (☎02·31·98·50·40) portdeplaisance@deauville.fr G IPTS signals when open, about HW ±3. Visit www.dyc14.com portdeplaisance@deauville.fr for actual local times.

Marina ☎02·31·98·50·40 (0830-1200, 1330-1800) www.deauville.org 320 ⚓ no ✔ (but see Bassin Morny below) ⚓ 🖳 🚲 (1km). Water taxi to Trouville runs frequently from the LW berth in the river or HW berth when the gate is open.

Bassin Morny ☎02·31·98·50·40, 80 ✔ ⚓ €3·00, rafting along eastern quay; S end,residents only. From 0700-2300 when entry gate is open, footbridge opens for 30 mins every 50 mins, and remains open 2300-0700 when gate is open. Waiting pontoon, Port Morny, S end pontoon C, by bridge. For cheap petrol (30% less) delivered to boat, (min 500 ltrs) call *Taupin* ☎02·31·79·21·02.

Deauville YC ☎02·31·88·38·19. deauville.yacht.club@wanadoo.fr www.dyc14.com; 🖳 🔧 ☒ 🔧 ⚓(6t). Hotel Ibis @.

Both towns ✗ 🖳 🚊 🖳 ⚓ ⚒ ☒ Ⓑ ⚓ 🔧 🔧 ☒ 🚉 Ⓔ ⇄ ✈ Deauville. Ferry: See Ouistreham.

FI.WG.4s 10m9-6M
0₇
22'·3
Ldg Lts 148°
1₉
1₅
1₂
1₆
2
0₈
Trouville FI(4)WR.12s8m
0₅
Iso.G.4s9m5M
2₈
2₇
Avant Port
4₂
1₇
3
1₅
SS
Q.G.11m9M
3₂
Oc.R.4s12M Horn(2)30s
0₆
49°22'N
1₁
Oc.R.4s17m
1₆
TROUVILLE
Casino (conspic)
Quai Albert 1er
3₅
2₇
5
2₅
1₇
IPTS
Port Morny
Quai Nord
Quai Est
4₆
2₈
Quai des Pêcheurs
4₄
2₁
Residents Only
C
B
A
3
PORT DEAUVILLE
3
Footbridge (lifting)
Port d'échouage
4₉
2₉
Bassin Morny
La Touques
Club House DYC
Quai de la Gare
Residents Only
DEAUVILLE
3₈
N
00°5'·0E
4'·5
1₈
2₆
1₅
2₃
2₇
1₆

metres 0 ... 200
cables 0 ... 1

9.17.22 DIVES-SUR-MER

Calvados 49°17'·68N 00°05'·63W (entry gate) ❀❀❀♦♦♦✿✿

CHARTS AC 2146, 1349; SHOM 7418, 7420 (the preferred chart); Navi 526; Imray C32

TIDES –0135 Dover; ML 5·1m

Standard Port LE HAVRE (←)

Times				Height (metres)			
High Water		Low Water		MHWS	MHWN	MLWN	MLWS
0000	0500	0000	0700	8·0	6·7	2·9	1·2
1200	1700	1200	1900				
Differences DIVES-SUR-MER							
–0100	–0010	0000	0000	+0·3	+0·2	+0·2	+0·1
OUISTREHAM							
–0045	–0010	–0005	0000	–0·3	–0·3	–0·2	–0·3

NOTE: There is a double HW at Dives. HW time differences, referred to the time of HW Le Havre, give times of the first HW.

SHELTER Excellent inside. Appr can be difficult in NW/NE >F5 when a bad sea can build quickly.

NAVIGATION WPT WPT 49°19'·18N 00°05'·85W, SWM buoy 'DI' (off chartlet) Iso 4s, 168°/7ca to Nos 1/2 buoys; thence follow the buoyed chan between sandbanks (drying 4·2m). Appr dries to 1M offshore. See FACILITIES for access times.

4 PHM and SHM buoy/bns are moved to suit shifting chan; Nos 1 (SHM) and 2 (PHM) are the only lit lateral buoys; Nos 3 and 5 SHMs are fixed lt bns which should be given a wide berth. Caution: Buoys/bns may lack numbers.

After last chan buoys (7 & 8) keep to port for the marina ent (400m W of lt ho). Or fork stbd to follow drying Dives river to YC's pontoon (2 ❷ berths) E of footbridge.

LIGHTS AND MARKS See Lights, buoys and waypoints and the chartlet. Wooded hills E of Dives and Houlgate church spire help locate the marina.

COMMUNICATIONS CROSS/⒧⒝ 02·31·88·11·13; ⚓ via HM; Auto 08·92·68·08·14; Police 02·31·28·23·00; Brit Consul 01·44·51·31·00; Dr 02·31·91·60·17; HM VHF Ch 09 (H24).

FACILITIES Marina ☎02·31·24·48·00. sncport guillaume@cegetel. net www.calvados-nautisme.com Access via entry gate HW±3; HW±2½ for draft >1·5m; HW±1½ in bad conditions. Note: unless recently dredged, craft with draft >1·8m should only attempt entry HW±2. Sill below gate is 2·5m above CD; gate opens when tide 4·5m above CD and at springs may create a 4–5kn current

for the first 45 mins. 550+50 ❷ €2·85, max LOA 15m, draft 1·9m. Office hrs: H24 summer; 0900-1200, 1400-1800 winter. ♠♠ H24, ⬚ ⎯ ⌇⚓ ⬚⚓ ✂ ▣(30t) demasting ⬚(1·5t). No @ or wi-fi.

Cabourg YC ☎02·31·91·23·55. cabourg.yacht.club@wanadoo.fr ⬚. **Town** ✗ ⬚ 🍴 ⬚ ⊠ Ⓑ ✈ (Deauville). Ferry Ouistreham.

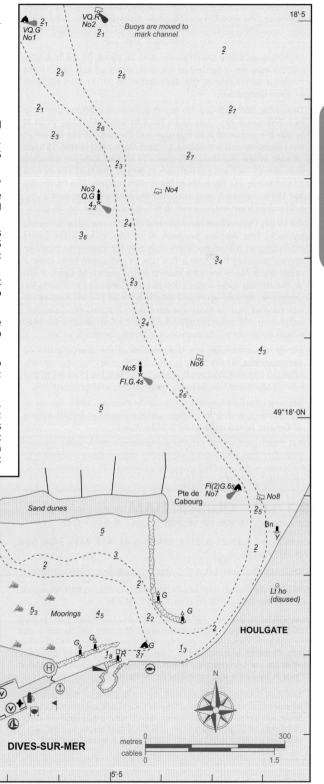

DEAUVILLE TO GRANDCAMP

(AC 2146, 2136). Since access to all the hbrs along this coast is to some extent tidally limited, it is best to cruise this area from W to E, so as to take advantage of the main Channel flood. ▶*Thus you can leave one harbour at the start of the tidal window and reach the next hbr before the window has closed; in the opposite direction this is difficult and may be impossible unless you have a fast boat. Between Le Havre and Ouistreham there is a useful stand at HW of between 2-3hrs.*◀

Le Havre, possibly Ouistreham, and St Vaast (in all but S/SE winds) may be regarded as the only ports of refuge. Strong onshore winds render the approaches to other hbrs difficult or even dangerous.

Deauville, 8M S of Cap de la Hève, is an important yachting hbr. The sands dry more than 5ca offshore; in strong W or N winds the entrance is dangerous and the sea breaks between the jetties. In such conditions it is best attempted within 15 mins of HW, when the stream is slack. To the NE beware Banc de Trouville (1·7m) and Les Ratelets (0·9m) to the N at the mouth of the Seine. 7M SW is **River Dives** with marina. The banks dry for 1M to seaward, and entry is only possible from HW ± 2½, and not in fresh onshore winds when the sea can be rough.

6M W lies **Ouistreham** ferry hbr and marina, the latter accessed by lock. The waiting pontoon N of the locks is for boats intending to transit the locks. It is not for yachts on passage merely pausing to await a fair tide or get some rest. Here a canal leads 7M inland to a marina in the centre of Caen. 2-6M W the drying rocky ledges, Roches de Lion and Les Essarts de Langrune, extend up to 2·25M offshore in places. Roches de Ver (drying rks) lie near the centre of Plateau du Calvados, extending 1M offshore about 2M W of **Courseulles-sur-Mer** which is dangerous to approach in strong onshore winds.

5M W of Courseulles are the remains of the wartime hbr of **Arromanches**, where there is a fair weather anchorage, usually affected by swell. Rochers du Calvados (1·6m) lies close NE of the ruined Mulberry hbr. Beware many wrecks and obstructions.

Between **Port-en-Bessin** and Pte de la Percée a bank lies up to 3ca offshore, with drying ledges and the wrecks and obstructions of Omaha Beach marked by 3 NCM buoys.

▶*A small race forms off Pte de la Percée with wind against tide. Off Port-en-Bessin the E-going stream begins about HW Le Havre −0500, and the W-going at about HW Le Havre +0050, sp rates 1·25kn.*◀

9.17.23 OUISTREHAM

Calvados **49°16'·82N 00°14'·89W** (E lock) ⚓⚓🌸🌸🌸🏵🏵

CHARTS AC 2613, 2146, 2136, 1349; SHOM 7418, 7421, 7420; Navi 526; Imray C32

TIDES −0118 Dover; ML 4·6; Duration 0525. See Dives-sur-Mer.

SHELTER Approach is difficult/dangerous in strong NW-NE winds, and the waiting pontoon is exposed to the N.

NAVIGATION WPT 49° 19'·18N 00° 14'·54W (abm Nos 1 & 2 buoys) 184·5°/2·1M to front ldg lt. In the outer channel, the harbour and the canal to Caen, ships and ferries have priority over all other craft. Outer chan is marked by lt buoys 1-6 and 3 pairs of lt bns; see half-scale inset on chartlet. The W training wall has been extended N'wards by 220m and raised by 2m in height.

> The lock waiting pontoon on E side of Avant Port dries at LWS. Beware current when securing in lock. Lock opens around 5 times each HW. Detailed timings, which vary, posted on waiting pontoon, at Ouistreham Marina (Ch 12) and www.ouistreham-plaisance.com. Yachts usually enter canal via smaller E lock.

LIGHTS AND MARKS See chartlet and Lights, buoys and waypoints. The 184·5° synch ldg lts are easily seen even in poor vis or bright sun. The huge square block of Caen hospital is conspic from afar; ditto the main lt ho, W + R top.

IPTS (full code) shown from lock control twr. A Ⓦ port or stbd of lowest main tfc lt, = E or W lock for yachts. Two Iso 4s, two QW and Fl R 4s define the ferry turning area.

COMMUNICATIONS CROSS 02·33·52·16·16; ⓁⒷ 02·31·97·50·49; ♨ 02·31·26·68·11; Auto 08·92·68·08·77; Police 02·31·97·13·15; ⊖ 02·31·96·89·10; British Consul 01·44·51·31·00; Ⓗ (Caen) 02·31·27·27·27; Dr 02·31·97·18·45; Port HM & lock 02·31·36·22·00; Call *Ouistreham Port* VHF Ch **74**, 68; Lock VHF Ch **12**, 68 (HW –2 to +3); Aff Mar 02·31·53·66·50; Marina VHF Ch 09, 74; Sté des Régates Ch 09 (office hours); Taxi 02·31·97·35·67; Ferry terminal 02·31·96·80·80.

FACILITIES
Marina ☎02·31·96·91·37. www.ouistreham-plaisance.com ⚓ 600 + 65 Ⓥ €2·85. Ⓥ raft on pontoon D, in 3·5m. ⚓ ⚓ (H24) ⚓ ⚓ ⚓ ⚓ Ⓔ SHOM free wi-fi ⚓ ⚓ ⚓ ⚓ ⓯(14t).

Société des Régates de Caen-Ouistreham (SRCO) www.ycf-club. fr ☎02·31·97·13·05. ⚓. **Town** ✕ ⚓ ⚓ ⚓ Ⓑ ⚓ ⚓ (bus to Caen) ✈ (Caen). Ferry to Portsmouth.

HARBOUR AT SOUTH END OF OUISTREHAM CANAL
CAEN 49°10'·95N 00°21'·17W. Charts: AC 1349, SHOM 7420.

NAVIGATION The 8M canal passage is simple and takes approx 1½hrs. Depths 2·5m to 9·8m; max speed 7kn; no overtaking. The 3 opening bridges are: Bénouville (Pegasus) (2½M S of Ouistreham) Colombelles (5M) and La Fonderie, Caen (8M). Tfc lts are not used; N-bound vessels have right of way at bridges.

Bridges open free for boats which transit at the scheduled times, 1/4-30/9, ie S-bound: Dep Bénouville bridge** 1010, 1330, 1630. N-bound dep La Fonderie bridge: 0845, 1200, 1500. Times can vary; check with marina or the website below. *At other times fees are due. **Allow ½hr from Ouistreham lock/marina.

R/T Monitor Ch 68 in the canal. *Caen Port* Ch **74**, 68.

Marina ☎02·31·95·24·47. www.caen-plaisance.com After La Fonderie bridge turn 90° stbd for marina (4m) at Bassin St Pierre in city centre. Ⓥ berths on pontoon E or as directed. 76+16Ⓥ €2·27, max LOA 25m, draft 3.8m. ⚓ ⚓ ⚓ ⚓ ⚓.

City ✕ ⚓ ⚓ ⚓ ⚓ Ⓗ ⚓ Ⓑ @ ⚓ ✈ Caen-Carpiquet.

CAEN MARINA

9.17.24 COURSEULLES-SUR-MER
Calvados 49°20'·40N 00°27'·32W ❀❀◑◐◐❀❀

CHARTS AC 2613, 2136, 1349; SHOM 7421, 7420; Navi 526; Imray C32

TIDES –0145 Dover; ML 4·6; Duration No data. See Port-en-Bessin.

SHELTER Good in Bassin Joinville (3m) but appr becomes difficult in strong N to NE winds. Exposed offshore ⚓ (3-5m) at L'Anneau de la Marguerite, 134°/0·4M from SWM buoy.

NAVIGATION WPT 49° 21'·28N 00° 27'·69W (SWM buoy Iso 4s) 134°/7ca, then 220° to E jetty hd to clear sandbank in approach. Head SE round the bank and then 235° once in the G sector of Iso WG 4s 6M. Plateau du Calvados extends 2M seaward and banks dry for >0·6M offshore, also beware oyster beds 3·5M W of ent. Outer ldg marks are: front, Bernières ⊕ tower ≠ 134° rear, twin spires of Douvres-La-Délivrande (partly obsc'd by trees and below the skyline when close-in). There are 2 other spires in the vicinity. At night the two bkwtr lts ≠ 186° form a useful ldg line. Beware rks awash at CD either side of ldg line.

Avant Port dries 2·5m; best ent at HW –1, keep to the E side. Entry gate into Bassin Joinville stays open approx HW ±2. The Control Twr opens the swing bridge above it on request Ch 09 or when an arriving yacht is spotted, usually quite quickly. **IPTS** in use. There is space at the SW end of the Bassin to turn a boat round.

LIGHTS AND MARKS See chartlet and Lights, buoys and waypoints. Both drying bkwtrs and their perches are easily seen, as is a conspic crucifix at root of the W bkwtr. A Cross of Lorraine is 4ca, and Pte de Ver lt ho is 2·5M W of the hbr ent.

COMMUNICATIONS CROSS 02·33·52·16·16; ⓁⒷ 02·31·37·45·47; ♨ 08·92·68·08·14; Auto 08·92·68·32·50; Police 02·31·29·55·20; ⊖ 02·31·21·71·09 @ Port-en-Bessin; Brit Consul 01·44·51·31·00 (Paris); Ⓗ 02·31·27·27·27 (Caen); Dr 02·31·37·45·24; Ambulance 02·31·44·88·88 or 15; Fire 18; Control twr 02·31·37·46·03, VHF Ch 09 HW±2; Aff Mar 02·31·15·26·15; Tourist Info 02·31·37·46·80.

FACILITIES Bassin Plaisance local shoal draft boats in 1m; no Ⓥ.

Bassin Joinville ☎02·31·37·51·69. Marina hours: Apr-Jun, 7/7 0900-1200. Jul/Aug: 7/7 0900-1200 and at tide times (approx HW ±3). Sep-Mar, M-Sat 0900-1200. Other hours exist; 0900-1200 is a constant. 730+20Ⓥ berth as directed by HM, or if absent, on pontoon X on E Quay. €3·10 and €0·20pp/day tax. 11m max LOA. ⚓ ⚓ ⚓ (600m) ⚓ ⚓ ⚓ ⚓ Ⓔ ⚓ ⚓ ⓯(25t).

Société des Régates de Courseulles (SRC) ☎02.31.37.47.42, ⚓.

Town ✕ ⚓ ⚓ ⚓ ⚓ ⚓ @ ⚓ Ⓑ ⚓ (via bus to Caen) ✈ (Caen-Carpiquet). Taxi ☎02·31·37·90·37. Ferry Ouistreham.

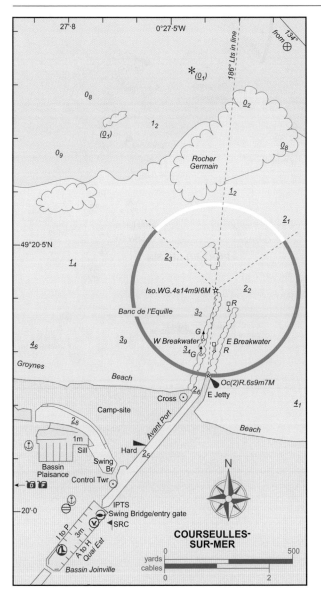

COURSEULLES-SUR-MER

N France

MULBERRY ANCHORAGE 6·5M WEST OF COURSEULLES

ARROMANCHES, Calvados, **49°21'·36N 00°37'·25W**. AC 2613, 2136; SHOM 7421; Navi 527. Tides, see Port-en-Bessin. Strictly a fair weather anchorage, with limited shelter from the ruined WW II Mulberry caissons which are conspicuous especially at LW. Swell intrudes and can make anchoring uncomfortable. Rochers du Calvados, 7ca ENE of the entrance, dries 1·6m but can be crossed near HW.

There are many obstructions within the harbour and offshore: 3 to the W and N are marked by a WCM and two ECM buoys. From

the more N'ly ECM buoy (Roseberry, 49°23'·10N 00°36'·49W) the entrance (lat/long as line 1) bears approx 196°/1·8M; it is marked by a small unlit SHM and PHM buoy.

Enter on about 245° for charted anchorage in 3-5m to S of the caissons, or sound closer inshore with caution. Entry from E or W is not advised due to uncharted obstructions. Caution: rocky plateau (3·4m) and further obstructions W of the dinghy landing area.

Facilities: YC Port Winston ☎02·31·22·31·01, 2 dinghy slips. Worth going ashore for D-Day museums and the panoramic view from the clifftops, but very crowded in season.

9.17.25 PORT-EN-BESSIN

Calvados **49°21'·16N 00°45'·40W** (Hbr ent) ❀⚓❀

CHARTS AC 2613, 2136, 1349; SHOM 7421, 7420; Navi 527; Imray C32

TIDES −0215 Dover; ML 4·4; Duration 0520

Standard Port LE HAVRE (←)

Times				Height (metres)			
High Water		Low Water		MHWS	MHWN	MLWN	MLWS
0000	0500	0000	0700	8·0	6·7	2·9	1·2
1200	1700	1200	1900				
Differences COURSEULLES-SUR-MER							
−0100	−0055	−0015	−0025	−0·6	−0·6	−0·3	−0·1
PORT-EN-BESSIN							
−0055	−0030	−0030	−0035	−0·7	−0·7	−0·2	−0·1
ARROMANCHES							
−0055	−0025	−0025	−0035	−0·6	−0·6	−0·2	−0·2

SHELTER Good. Ent difficult in strong N/NE'lies and dangerous in F8+. No ⚓ in outer hbr nor 1ca either side of 204° ldg line.

NAVIGATION WPT 49°21'·62N 00°45'·09W, 204°/0·5M to hbr ent. Avoid G sector of W mole hd Fl WG 4s which covers the inshore dangers off Omaha Beach, 3 to 6M to the WNW. In the outer hbr beware submerged jetty (marked by R bcn) to E of ent chan.

LIGHTS AND MARKS Conspic marks: a water twr (104m) 1·7M ESE of the hbr; a Sig Stn 7 cables W of hbr. Ldg lts 204°: Front, W pylon, G top, at bottom of hill; rear, W house/grey roof near top of the hillside. See chartlet and Lights, buoys & waypoints.

IPTS control entry to the basins. Entry gate opens HW ±2. While gate stays open, swing bridge opens whenever possible to suit yachts and FVs; there is little road traffic. When shut the bridge shows FR in the middle.

COMMUNICATIONS CROSS 02·33·52·16·16; Sig Stn/Ⓛ 02·31·21·81·51; Auto 08·92·68·08·14; Police 02·31·21·70·10; ⊜02·31·21·71·09; Brit Consul 01·44·51·31·00; Ⓗ 02·31·51·51·51; Dr 02·31·21·74·26; HM 02·31·21·70·49, VHF Ch 18 (HW ±2) for gate opening H24; Aff Mar 02·31·21·71·52; Mairie 02·31·21·72·12.

FACILITIES A major FV port. **Outer Hbr** dries up to 3·7m; ▃ L. Drying � on E side Quai de L'Epi, no charge.

Bassin 1: Inside the entry gate, tight 90° stbd, ❶ berth/raft alongside 25m pontoon in NNW corner. €0·77 + €0·20pp/day tax, ⓘ Ⓦ €3·00 pp/day, code access (free, children <10 yrs) at Tourist Office opp on E quay. Yachts may stay only 24-48 hrs, due to limited space. No berthing on sloping E wall. Call Gate-master ☎02·31·21·71·77 (HW ±2) to pre-book gate time if leaving at early hour.

Bassin 2, no entry. **Services** ▃ ⚓ 🔧 ⊞ Ⓔ ⚒ ✕ ⌐(4t).

Town ✕ 🍴 🛒 ✉ Ⓑ 🏪 🏪 🔌 ⇌ (bus to Bayeux) ✈ (Caen, ☎ 02·31·26·58·00). Ferry see Ouistreham.

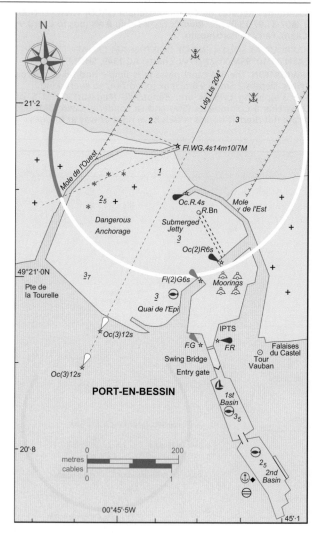

9.17.26 GRANDCAMP

Calvados **49°23'·51N 01°02'·99W** (Pierheads) ✦✦♦♦✿✿✿

CHARTS AC 2613, 2135, 1349; SHOM 7422, 7420; Navi 527; Imray C32

TIDES −0220 Dover; ML Rade de la Capelle 4·4; Duration 0510

Standard Port CHERBOURG (→)

Times				Height (metres)			
High Water		Low Water		MHWS	MHWN	MLWN	MLWS
0300	1000	0400	1000	6·4	5·1	2·6	1·1
1500	2200	1600	2200				
Differences RADE DE LA CAPELLE (3M to NW)							
+0125	+0050	+0125	+0110	+0·7	+0·7	0·0	+0·1
ILES SAINT MARCOUF							
+0130	+0055	+0120	+0110	+0·5	+0·5	0·0	0·0

SHELTER Good, but approach is difficult in NW-NE winds >F6.

NAVIGATION WPT 49°24'·82N 01°04'·33W, 146°/1·6M to E pier head. Nos 1, 3 & 5 NCM buoys mark the seaward limit of Les Roches de Grandcamp, a large plateau of flat rock extending about 1½M offshore and drying approx 1·5m; heavy kelp can cause echosounders to under-read. It can be crossed from most directions, given adequate rise of tide.

Watch out for FVs in the narrow access chan. 14·3m wide gate into wet basin stays open approx HW ±2½; entry is controlled by IPTS. HW −2½ is the same as LW Dunkerque and HW +2½ is the same as HW Dunkerque.

LIGHTS AND MARKS Nos 1, 3 and 5 unlit NCM buoys are 1·4-1·7M NE to NW of the hbr. The 221° ldg lts are primarily for FVs. Maisy's modern church twr is conspic 6ca SSW of hbr ent. See chartlet and Lights, buoys and waypoints and Carentan for other data.

COMMUNICATIONS CROSS 02·33·52·16·16; LB 02·31·22·67·12; ⚓ 02·21·33·25·26; Auto 08·92·68·08·14; Police 02·31·22·00·18; Brit Consul 01·44·51·31·00; Dr 02·31·22·60·44; Hbr/marina VHF Ch 09; Aff Mar 02·31·22·60·65.

FACILITIES Marina ☎02·31·22·63·16. 240 + 10♥ €2·00 (no cr/ cards). ♥ berths in 2·5m at E end of pontoon C, either side; or as directed. ⚓ ⊡ ✗ ⛽ ⚡ ▮(5t) ⛴ YC ☎02.31.22.14.35. **Town** ✗ ⊡ 💈 🏪 🏪 ✉ Ⓑ 🚂 (Carentan) ✈ (Caen). Ferry: Cherbourg, Ouistreham.

GRANDCAMP

GRANDCAMP TO POINTE DE BARFLEUR

(AC 2135) On E side of B du Grand Vey, Roches de Grandcamp (dry) extend more than 1M offshore, N and W of **Grandcamp**, but they are flat and can be crossed from the N in normal conditions HW±1½. Three NCM buoys mark the N edge of this ledge. Heavy kelp can give false echo-sounder readings.

At the head of B du Grand Vey, about 10M S of Îles St Marcouf, are the (very) tidal hbrs of Carentan and Isigny; entry is only possible near HW and should not be attempted in strong onshore winds. The **Carentan** chan is well buoyed and adequately lit. It trends SSW across sandbanks for about 4M, beyond which it runs between two training walls leading to a lock, and thence into the canal to Carentan's popular marina. The **Isigny** chan is slightly deeper, but the hbr dries to steepish mudbanks.

Îles St Marcouf lie 7M SE of St Vaast-la-Hougue, about 4M offshore, and consist of Île du Large (lt) and Île de Terre about ¼M apart. The Banc du Cardonnet, with depths of 5·2m and many wks, extends for about 5M ESE from the islands. Banc de St Marcouf, with depths of 2·4m and many wks, extends 4·5M NW from the isles, and the sea breaks on it in strong N/ NE winds.

Approach **St Vaast-la-Hougue** S of Île de Tatihou, but beware drying rks: La Tourelle (unmarked) Le Gavendest and La Dent, which lie near the approaches and are buoyed. 2M N is Pte de Saire with rks and shoals up to 1M offshore. From Pte de Saire the coast runs 4M NNW to drying **Barfleur**.

Raz de Barfleur, in which the sea breaks heavily, must be avoided in bad weather, particularly with NW or SE winds against the tide, by passing 5-7M seaward of Pte de Barfleur.

▶*In calmer weather it can be taken at slack water (HW Cherbourg −4½ and +2) via an inshore passage, 3ca NE of La Jamette ECM beacon. Expect some turbulence.*◀

Pte de Barfleur marks the W end of Baie de Seine, 56M west of Cap d'Antifer. There are no obstructions on a direct course across the Bay, but a transhipment area for large tankers is centred about 10M ESE of Pte de Barfleur. Be advised: The W end of the Baie de Seine has at times been affected by floating Sargassum weed; in harbours from Grandcamp to St Vaast propellers may be fouled.

POINTE DE BARFLEUR TO CAP DE LA HAGUE

(AC 1114) The N coast of the Cotentin peninsula runs E/W for 26M, mostly bordered by rks which extend 2·5M offshore from Pte de Barfleur to C. Lévi, and 1M offshore between Pte de Jardeheu and C. de la Hague. ▶*Tidal streams reach 5kn at sp, and raise a steep, stopping sea with wind against tide.*◀ Between Pte de Barfleur and Cherbourg, yachts commonly route via Les Equets and Basse du Rénier NCM buoys and La Pierre Noire WCM buoy; be alert for opposite direction traffic using the same Wpts.

The inshore passage between Pte de Barfleur and C. Lévi, well S of the above 3 cardinal buoys, is not advised without local knowledge except in fair weather and good visibility when the transits shown on AC 1114 can be positively identified. Port Lévi and Port de Becquet are two small drying hbrs E of Cherbourg.

▶*Tidal streams run strongly with considerable local variations.*

Off C. Lévi a race develops with wind against tide, and extends nearly 2M N. Off Cherbourg the stream is E-going from about HW −0430 and W-going from HW +0230.◀

Cherbourg is a port of refuge, hugely popular with British yachtsmen. A gentle reminder: If not properly catered for, the tidal streams are strong enough to set you downtide of Cherbourg. Do your sums and bias the approach to be uptide of the harbour. At night positively identify the leading lights through either entrance, *before* entering.

▶*Close inshore between Cherbourg and C. de la Hague a back eddy runs W.*◀ West of **Omonville** and about 2M E of C. de la Hague there is an anchorage off Port Racine in Anse de St Martin, guarded by at least 3 groups of rocks. It is open to the N, but useful to await a fair tide through the Alderney Race.

N France

9.17.27 CARENTAN

Manche **49°19'·09N 01°13'·53W** ✲✲◊◊◊◊✿✿✿

CHARTS AC 2613, 2135; SHOM 7056, 7422; Navi 527; Imray C32

TIDES –0225 Dover; ML Rade de la Capelle 4·4; Duration 0510

Standard Port CHERBOURG (⟶) Use differences RADE DE LA CAPELLE (Grandcamp). HW Carentan = HW Cherbourg +0110.

SHELTER Complete shelter in the land-locked marina (3kn max) but do not attempt to approach in onshore winds >F5.

NAVIGATION WPT 49°25'·45N 01°07'·07W (C-I SWM buoy) 210°/1·8M to Nos 1/2 buoys, not to be confused with the unlit Isigny chan bys 1/2 to the ESE. Start from WPT at HW –2 to –1½. Graph inset gives approx availability. Shallowest water is typically in the area between bys 7-10, but channels in the bay are broad and fairly flat. Once inside trng walls chan deepens.

Appr protected from prevailing S to W winds. Drying out on the hard sands of the estuary is not advised; nor is an early approach, as a bore (*mascaret*) may occur in the river chans between HW –3 and HW –2½ especially at springs. Chan is liable to vary in both depth and direction; buoys are moved accordingly. All buoys have R or G reflective panels; 6 are lit as on chartlet.

3·8M SW a small pool is formed where R Taute and Douve meet. Ahead, the little lock (room for about 6 boats) opens HW –2 to HW +3. Lock sigs: ● = open, ● = shut (use phone at lock office if lock-keeper absent). Small waiting pontoons are on E side, down- and up-stream of the lock. A magnetic anomaly has been reported in the vicinity of the N13 underpass. Electronic autopilots should not be relied upon for steering in this area.

LIGHTS AND MARKS See chartlet and Lights, buoys and waypoints. Maisy Ch twr is a good landmark initially. Ldg lts 209·5° are only valid from the training wall bcns to the front lt, ie <u>not</u> in the buoyed chan.

COMMUNICATIONS CROSS 02·33·52·16·16; Auto 08·92·68·08·50; Police 02·33·42·00·17; ⊖ 02·31·21·71·09; Brit Consul 01·44·51·31·00; ⊞ 02·33·42·14·12; Dr 02·33·42·33·21; Lockmaster 02·33·71·10·85; *Écluse de Carentan* VHF Ch 09 (0800-1200, 1400-1600LT & lock opening hrs); Aff Mar 02·33·44·00·13.

FACILITIES **Marina** ☎02·33·42·24·44, mob 06·89·13·15·56. port-carentan@orange.fr www.cc-carentan-cotentin.fr. 310 + 50♥ fees based on m², so approx fee €2·07/m. Max/min depths 3·5/3m. ♥ berths normally on pontoon K. ⬤♦ (daily: 0900-1000, 1500-1600) ◀ ⬤(35t) ◨(50t) ⚓ ⬀ ⚒ ⬥ ⬥ ✕ ⬜. Additional space for wintering ashore at Top Nautique ☎02·33·42·23·87.

YC Croiseurs Côtiers de Carentan ☎02·33·42·06·61, ⬜.

Town ✕ ☷ ⬜ ⬚ Ⓑ @ in ⬚ ⇌ ✈ and ferry Cherbourg.

Isigny 49°19'·31N 01°06'·24W ✲✲◊◊✿

SHELTER Good in town, but the drying river bed shelves steeply there, soft mud. Cautions in outer approaches as for Carentan, though the Isigny chan is somewhat deeper.

NAVIGATION Leave C-I WPT at HW –2½ tracking 158°/1·25M to IS, a very small unlit NCM buoy at 49°24'·29N 01°06'·37W, near ent to unlit buoyed chan.

From IS buoy track about 220°/0·9M to first unlit chan buoys. Pairs of buoys are spaced at approx 3ca intervals, but between Nos 9/10 and 11/12 the interval is 6ca, with an IDM bn midway.

From Pte du Grouin, ldg lts lead 172·5°/1·9M (as chartlet) between training walls to a Y-junction. Here turn port into R l'Aure for 4ca to town quays on SW bank.

COMMUNICATIONS Ch 09 (0900-1200 & 1400-1800LT).

FACILITIES HM ☎02·31·51·24·01; **Quay** (dries) ⬸ 40+5 ♥. Drying pontoons on SW bank; berth bows-in/stern to buoys. Pay fees at town-hall; ⬚ ⬚ ◀ ⬥ ⚒ ⬀ ⚒ ◨(8t). **Town** ✕ ☷ ⬜ ⬚ Ⓑ.

ADJACENT ANCHORAGE

ILES ST MARCOUF, Manche, **49°29'·78N 01°08'·92W**, the charted ⚓. AC 2613, 2135; SHOM 7422. Tides, Grandcamp. A fair weather ⚓ only, no shelter in S'lies. The two islands, Île du Large and SW of it, Île de Terre, look from afar like ships at ⚓. The former is a fort with a small dinghy hbr on its SW side; the latter is a bird sanctuary, closed to the public. ⚓ SW or SE of Île du Large or SE or NE of Île de Terre. Holding is poor on pebbles and kelp. Île du Large lt, VQ (3) 5s 18m 8M, on top of the fort. Both islands are surrounded by drying rks, uninhabited and have no facilities.

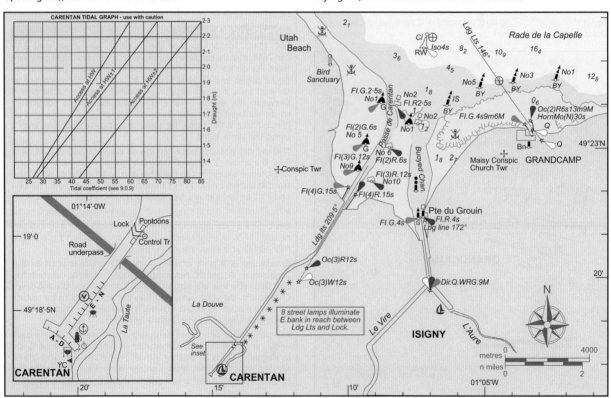

9.17.28 ST VAAST-LA-HOUGUE

Manche **49°35'·17N 01°15'·41W** (Jetty head) ❀❀◍◍◔◔◔❁❁

CHARTS AC 2613, 2135; SHOM 7422, 7120, 7090; Navi 527, 528; Imray C32

TIDES –0240 Dover; ML 4·1; Duration 0530

Standard Port CHERBOURG (⟶)

Times				Height (metres)			
High Water		Low Water		MHWS	MHWN	MLWN	MLWS
0300	1000	0400	1000	6·4	5·1	2·6	1·1
1500	2200	1600	2200				
Differences ST VAAST-LA-HOUGUE							
+0115	+0045	+0120	+0110	+0·3	+0·3	–0·2	–0·1

SHELTER Excellent in marina especially in W'lies. If full, or awaiting entry, ⚓ off between brgs of 330° and 350° on jetty lt, but this ⚓ becomes untenable in strong E-S winds.

NAVIGATION WPT 49°34'·36N 01°13'·88W (Le Gavendest buoy) 310°/1·3M to main jetty lt, Oc (2) WRG 6s; appr in its wide, well marked W sector, dries 1·8m. Beware boats at ⚓ and cross-currents. 'Le Run' appr is only feasible (over extensive oyster beds) if draft <1·2m and when the entry gate is open, HW –2¼ to HW +3½. If barometer is high and Coeff 40-50, gate may close 30 mins early; and 1-1¼ hrs early if Coeff <40. Gate times are on the website.

LIGHTS AND MARKS By day do not confuse similar twrs on Ile de Tatihou and Fort de la Hougue. From N, Pte de Saire is a squat white lt ho/G top. From E, ldg lts 267°: front La Hougue; rear, Morsalines in W sector. Main jetty hd lt is a conspic white twr/R top; W sector covers the bay. See chartlet, Lights, buoys and waypoints.

COMMUNICATIONS CROSS 02·33·52·72·13; ☎ 08·99·71·08·08; Police 17 or 02·33·23·31·90; ⊖ 02·33·23·34·01; British Consul 01·44·51·31·00; Ambulance (SAMU) 15; Aff Mar 02·33·54·51·74; Semaphone stn (Fort de la Hougue) VHF Ch 16; Marina VHF Ch 09; Tourist Office 02·33·23·19·32.

FACILITIES Marina ☎02·33·23·61·00. www.saint-vaast-reville.com port-st-vaast@saint-vaast-reville.com Season 0900-1200, 1400-1700. 604 + 100 ❶ in 2·3m, €2·83 + €0·22pp/day tax, multihulls +50%. ❶ berths by LOA: B >14m; C 12-14m; D no ❶; E (outer part) <12m. 🛁🚿 (pontoon A) ⬚🚻🛢🔧🏪✕⚓🛟(45t).

YC de St Vaast ☎02·33·44·07·87, ✕ 🍴 🛒(3t). **Town** ✕ (closed Mondays) 🍴🛒🛢🔘✉Ⓑ⚡ (bus to Valognes) Cherbourg ✈ ferry.

9.17.29 BARFLEUR

Manche **49°40'·34N 01°15'·48W** ❀❀◍◔◔❁❁❁

CHARTS AC 2613, 2135, 1114; SHOM 7422, 7120; Navi 528; Imray C32

TIDES –0208 Dover; ML 3·9; Duration 0550

Standard Port CHERBOURG (⟶)

Times				Height (metres)			
High Water		Low Water		MHWS	MHWN	MLWN	MLWS
0300	1000	0400	1000	6·4	5·1	2·6	1·1
1500	2200	1600	2200				
Differences BARFLEUR							
+0115	+0050	+0045	+0050	+0·2	+0·1	0·0	+0·1

SHELTER Excellent, except in fresh NE/E winds when ent is difficult and berths uncomfortable. Hbr dries. Rks/shoals in SE of hbr. Safe to ⚓ outside hbr in offshore winds.

NAVIGATION WPT 49°41'·13N 01°14'·41W, 219·5°/1·0M to hbr ent. Positively identify Roche à l'Anglais SHM and Le Hintar PHM buoys before closing the hbr any further. Beware cross-currents.

In rough weather, esp wind against tide, keep 5M off Pte de Barfleur to clear the Race; see Passage information. If using the inshore passage, identify La Jamette ECM bn and La Grotte SHM buoy. From the S, keep seaward of Pte Dranguet and Le Moulard, both ECM bns.

LIGHTS AND MARKS Pte de Barfleur lt ho is conspic 1½M NNW of hbr, a 72m high, grey twr/B top. The squat church twr is a conspic daymark. Ldg lts 219·5°: both W ☐ twrs, not easy to see by day. Light details as chartlet and Lights, buoys and waypoints.

COMMUNICATIONS CROSS 02·33·52·72·13; 📻 02·33·23·10·10; ☎ 02·33·10·16·10; Auto 08·92·68·02·50; Police 17; ⊖ 02·33·23·34·01; Brit Consul 01·44·51·31·00; SAMU 15; Dr 02·33·43·73·55; Dentist 02·33·54·49·31; HM 02·33·54·08·29; Aff Mar 02·33·23·36·00.

FACILITIES

NW Quay Yachts dry out on firm, level sand/mud at SW end of quay, but space may be restricted by FVs. 20❶ 🛟 , 6x<15m, <10m rafted; €1·40. 🔌 (long cable needed) 🚿 at La Blanche Nef (berthing payment receipt req'd) 🛢🛟 .

SC ☎02·33·54·79·08.

Town www.ville-barfleur.fr 🛢🔧 (Montfarville: 1km S) ✕ 🍴🛒✉ Ⓑ no wi-fi, bus to Cherbourg for ⚡ ✈ ferry.

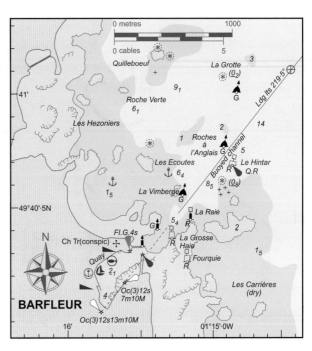

9.17.30 CHERBOURG
Manche **49°38'·94N 01°37'·11W** (Marina ent) ❀❀❀◊◊◊❀❀

CHARTS AC 2656, 2669, 1114, 1112; SHOM 7120, 7092, 7086; Navi 528, 1014; Imray C32, C33A

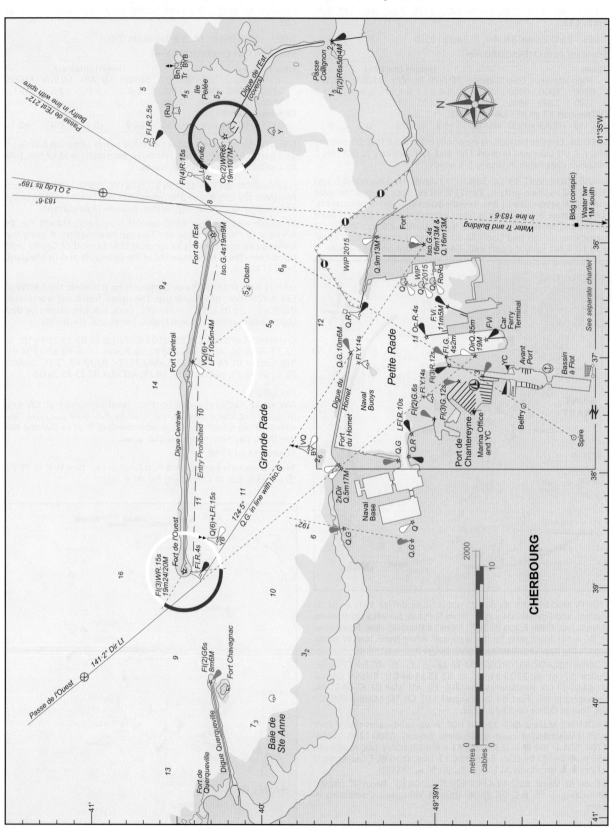

TIDES Cherbourg is a Standard Port (→). –0308 Dover; ML 3·8; Duration 0535

SHELTER Excellent; a port of refuge available H24 in all tides and weather. Small craft ⚓ N of marina bkwtr in 2·3m, keeping clear of the charted limits of the military port, close N and W.

NAVIGATION For coastal features from Pte de Barfleur to Cap de la Hague see Passage Information. Enter the Grande Rade via:

- **Passe de l'Ouest** (W ent). WPT 49°41'·05N 01°39'·81W, 141°/0·85M to abeam Fort de l'Ouest. Rks, marked by a PHM buoy, Fl R 4s, off Fort de l'Ouest, extend about 80m off each bkwtr. From W, the white sector of Fort de l'Ouest lt (or by day bearing >122°) keeps clear of offlying dangers E of Cap de la Hague. From CH1 SWM buoy, L Fl 10s, Fort de l'Ouest bears 145°/3·5M.

- **Passe de l'Est** (E ent, least depth 8m). WPT 49°40'·91N 01°35'·55W, 189°/0·65M to abeam Fort de l'Est. Keep to W side of chan (but at least 80m off Fort de l'Est) to avoid dangers W and NW of Ile Pelée marked by two PHM lt buoys. N/NE of Ile Pelée an extensive drying area is marked by 2 unlit bn trs.

- **Passe Collignon** is a shallow (2m) chan, 93m wide, through Digue de l'Est (covers) near the shore. Tidal stream reaches 3·4 knots. Only advised in good conditions, near HW.

No anchoring in the Passe de L'Ouest and Passe de L'Est. Speed limits: Grande Rade 14kn; Petite Rade 8kn; Marina 3kn. No entry to: the area S of the Digue Centrale due to fish farms, nor to the area east of the Port Militaire. Keep clear of ferries and large prohib area N and E of Port de Commerce, (WIP 2015, land reclamation).

LIGHTS AND MARKS Daymarks include the chimney of Jobourg nuclear plant, 10M W; the cranes of the naval base; and higher ground S of the city. Five powerful lights assist a good cross-Channel landfall:

- 13M E, Pte de Barfleur, Fl (2) 10s 29M
- 5M ENE, Cap Lévi, Fl R 5s 22M
- 12M WNW, Cap de la Hague, Fl 5s 23M
- 28M W, Casquets, Fl (5) 30s 24M

See also Lights, buoys and waypoints (LBW) and Area 19.

Passe de l'Ouest ldg lts 141·2° are a Dir Q rear ldg lt aligned between two Q front ldg lts (horizontally disposed) at base of Digue du Homet. Other lts as on chartlet and LBW.

On entering the Petite Rade make good 200° for marina ent, Fl (3) R 12s and Fl (3) G 12s. Shore lights may confuse and mask navigational lights.

COMMUNICATIONS CROSS 02·33·52·16·16 or 112; *Jobourg Traffic* Ch **13** 80 (H24) provides radar surveillance of the Casquets TSS and ITZ and Mont St Michel to Cap d'Antifer. Radar assistance available on request, Ch 80, to vessels in the 270° sector from S clockwise to E, radius 40M from CROSS Jobourg at 49°41'·1N 01°54'·5W. Jobourg broadcasts nav, weather and traffic info in English and French Ch 80 at H+20 & H+50; ☎ 08·92·68·32·50; Police 02·33·92·70·00; ⊜ 02·33·23·34·02; Brit Consul 01·44·51·31·00; 田 02·33·20·70·00; Dr 02·33·01·58·58; HM (Port & Lock) 02·33·20·41·25; Aff Mar 02·33·23·36·00; Marina: *Chantereyne* VHF Ch 09; lock into Bassin du Commerce (long stay/>25m) VHF Ch 06.

FACILITIES Port Chantereyne Marina ☎ 02·33·87·65·70. www.portchantereyne.fr portchantereyne@ville-cherbourg.fr Access H24. 1310 + 250 Ⓥ €2·77 on pontoons J,N,P, Q, N side of detached pontoon H and W side

of K; max LOA 25m; max draught 3m: 🏂 🛢 (HO) or (H24) cr/debit card. max fill €120·00. 🗑 🔧 @ 🔦 🔧 🔲 Ⓔ ✖ ⚓ SHOM, ━ 🏳(30t). .

YC de Cherbourg ☎02·33·53·02·83, ✖ 🗋.

Bassin du Commerce is for permanent residents only/>25m LOA.

City ✖ 🗋 🛒 🛢 ✉ 🗑 Ⓑ ⇌. Cherbourg (Maupertus) ✈ ☎ 02·33·88·57·60. Ferry: Portsmouth, Poole.

MINOR HARBOURS EAST OF CHERBOURG
PORT DE LÉVI, Manche, **49°41'·23 N 01°28'·46W.** AC 1114; SHOM 7120, 5609, 7092; HW –0310 on Dover (UT); +0024 on Cherbourg. HW ht +0·2m on Cherbourg. Shelter good except in SW to N winds. By day appr on 094° keeping the white wall and lt between the white marks on each pier hd. Lt is Oc (2) WRG 6s; see Lights, buoys and waypoints. Keep in W sector (083°-105°) but night entry not advised due to many pot floats. Berth bows on to NE side below white wall, amongst small FVs. Dries to clean sand. Facilities: Peace. Fermanville (1·5M) has ✖ 🗋 🛒.

PORT DU BECQUET, Manche, **49°39'·304N 01°32'·81W.** AC 1114; SHOM 7120, 7092. Tides as Cherbourg. Ldg lts/marks 186·3°: Both Dir Oc (3) 12s (front lt White; rear lt Red) W 8-sided twrs; see LBW. Large PHM bcn twr, La Tounette, marks rks close E. Shelter good except from N to E when a strong scend enters. Dries, clean sand. Secure to S of the E/W jetty (little space). Facilities: very few; all facilities at Cherbourg 2·5M.

N France

CHERBOURG TIDAL COEFFICIENTS 2016

Date	Jan am	Jan pm	Feb am	Feb pm	Mar am	Mar pm	Apr am	Apr pm	May am	May pm	June am	June pm	July am	July pm	Aug am	Aug pm	Sept am	Sept pm	Oct am	Oct pm	Nov am	Nov pm	Dec am	Dec pm
1	57	52	46	41	53	47	39	36	43	46	63	69	68	73	78	82	90	92	90	90	84	82	80	79
2	47	43	37	35	42	37	37	40	50	56	75	81	77	82	86	90	93	93	90	89	81	78	78	76
3	40	37	34	36	34	33	46	53	63	71	87	92	86	90	92	93	92	90	87	85	76	72	74	71
4	37	38	39	44	35	40	61	70	78	86	96	100	93	95	94	93	88	85	82	78	69	64	68	64
5	40	43	50	57	46	53	79	88	93	100	102	103	96	96	92	90	82		75		60	55	61	
6	48	53	64	71	61	70	96	103	105	109	103	102	96	94	87		77	73	70	65	51		57	54
7	58	63	78	84	78	87	109	114	112	113	100	97	92		83	79	68	62	60	55	46	42	51	49
8	68	74	90	96	94	101	117	118	112	110	92		88	84	75	70	57	51	49	44	39	38	48	49
9	78	83	100	104	107	112	117	115	106	101	87	82	80	75	64	59	45	40	39	35	40	43	52	55
10	87	90	107	108	115	116	111		95		76	70	70	65	53	48	35	32	32	32	48	55	61	67
11	93	95	108	106	116	114	105	99	88	81	64	59	59	54	43	39	31	32	35	40	63	71	73	80
12	95	96	103	99	111	105	91	82	74	66	54	49	49	45	35	34	36	41	47	55	79	86	86	92
13	95	93	94		99		73	65	59	53	46	44	42	39	34	36	48	56	64	72	94	100	97	101
14	91		88	81	91	83	57	50	48	45	43	43	38	39	40	46	64	72	81	88	105	109	104	106
15	87	83	73	66	74	66	45	43	43	44	45	47	41	44	52	58	79	87	96	102	111	112	106	105
16	78	74	59	53	57	50	43	45	46	48	50	53	48	52	64	71	93	99	108	111	111	109	103	99
17	68	64	50	49	46	44	49	54	52	56	57	61	57	62	77	83	104	108	114	114	105	100	95	90
18	60	57	50	53	45	48	59	64	60	63	64	68	67	71	88	93	110	111	113	111	94	87	84	
19	56	56	58	64	53	58	68	72	67	70	71	74	75	79	97	100	111	108	106		79		78	71
20	59	62	69	74	64	69	76	79	73	75	77	79	83	86	102	103	105		101	94	72	64	65	59
21	67	72	79	83	75	79	81	83	77	79	80	82	88	90	102		100	93	86	77	58	52	53	49
22	76	81	87	89	82	85	84	85	80	81	82	83	91	91	101	98	86	78	69	61	48	46	45	43
23	85	88	91	92	87	89	85	85	81	81	82	81	91		94	89	70	61	54	48	45	46	42	43
24	90	92	92	91	89	89	84	82	80	79	80		90	87	83	77	54	49	45	45	49	52	45	47
25	92	92	90	87	89	87	80	78	77		78	75	85	81	70	63	46	46	47	51	56	60	51	55
26	91	89	85	81	86	83	75		75	72	73	70	77	73	58	53	49	54	55	60	63	67	58	62
27	87	84	78	73	80		71	67	69	65	67	64	68	64	50	50	59	65	65	70	70	73	66	69
28	80	76	68		77	73	63	58	62	58	62	60	60	58	53	57	70	76	74	77	75	77	72	75
29	71		63	58	68	63	53	49	56	54	60	60	57	57	62	68	80	84	80	82	78	79	77	79
30	67	61			58	52	46	44	53	53	62	65	60	63	73	79	86	88	84	84	80	80	81	82
31	56	51			47	42			55	59			68	73	83	87			85	85			82	82

Tidal coefficients indicate the magnitude of the tide on any particular day, without having to look up and calculate the range, and thus determine whether it is springs, neaps or somewhere in between. This table is valid for all areas covered by this Almanac. Typical values are:

120	Very big spring tide
95	**Mean spring tide**
70	Average tide
45	**Mean neap tide**
20	Very small neap tide

See also coefficients for Calais (p 759), St Malo (p 805), Brest (p 873) and Port Bloc (p 936).

STANDARD TIME UT –01
Subtract 1 hour for UT
For French Summer Time add
ONE hour in **non-shaded areas**

CHERBOURG   LAT 49°39′N   LONG 1°38′W
TIMES AND HEIGHTS OF HIGH AND LOW WATERS

Dates in red are **SPRINGS**
Dates in blue are NEAPS

YEAR 2016

N France

JANUARY

Day	Time m	Time m	Time m	Time m
1 F	0056 5.4	0735 2.4	1305 5.4	2000 2.3
16 SA	0054 5.9	0743 1.8	1314 5.9	2010 1.7
2 SA ☽	0138 5.2	0822 2.7	1352 5.1	2049 2.6
17 SU ☽	0147 5.6	0839 2.1	1411 5.6	2110 2.0
3 SU	0233 5.0	0920 2.9	1453 4.9	2152 2.8
18 M	0253 5.4	0948 2.3	1528 5.4	2222 2.2
4 M	0342 4.9	1034 2.9	1606 4.9	2307 2.8
19 TU	0414 5.4	1107 2.3	1651 5.4	2340 2.2
5 TU	0457 5.0	1147 2.7	1724 5.0	
20 W	0528 5.5	1224 2.1	1803 5.5	
6 W	0014 2.6	0600 5.3	1246 2.5	1825 5.3
21 TH	0052 2.0	0632 5.7	1330 1.8	1907 5.7
7 TH	0109 2.3	0650 5.6	1336 2.1	1915 5.6
22 F	0152 1.8	0728 6.0	1426 1.5	2002 6.0
8 F	0156 2.0	0735 5.9	1421 1.8	1959 5.8
23 SA	0245 1.6	0817 6.2	1515 1.3	2049 6.1
9 SA	0239 1.8	0816 6.2	1504 1.5	2040 6.1
24 SU ○	0330 1.4	0900 6.4	1557 1.2	2129 6.2
10 SU ●	0321 1.5	0855 6.4	1546 1.3	2120 6.2
25 M	0410 1.3	0938 6.4	1635 1.1	2205 6.3
11 M	0403 1.4	0936 6.5	1627 1.1	2200 6.3
26 TU	0446 1.3	1014 6.4	1710 1.1	2240 6.2
12 TU	0445 1.3	1016 6.4	1709 1.0	2241 6.4
27 W	0520 1.4	1049 6.3	1743 1.3	2314 6.1
13 W	0527 1.3	1057 6.6	1751 1.0	2324 6.3
28 TH	0553 1.6	1122 6.2	1814 1.5	2345 5.9
14 TH	0609 1.4	1140 6.5	1834 1.2	
29 F	0624 1.8	1154 5.9	1845 1.8	
15 F	0008 6.1	0654 1.6	1225 6.2	1919 1.4
30 SA	0015 5.7	0657 1.9	1224 5.6	1918 2.1
31 SU	0049 5.4	0734 2.4	1259 5.3	1956 2.4

FEBRUARY

Day	Time m	Time m	Time m	Time m
1 M ☽	0129 5.1	0819 2.7	1346 5.0	2044 2.7
16 TU	0218 5.4	0920 2.2	1459 5.2	2154 2.4
2 TU	0224 4.9	0919 2.9	1452 4.8	2152 2.9
17 W	0340 5.2	1042 2.4	1630 5.1	2320 2.5
3 W	0339 4.8	1040 2.9	1619 4.7	2318 2.8
18 TH	0506 5.2	1209 2.3	1753 5.2	
4 TH	0511 5.0	1201 2.7	1747 5.0	
19 F	0039 2.5	0620 5.4	1319 2.0	1901 5.5
5 F	0031 2.6	0618 5.3	1303 2.3	1847 5.3
20 SA	0143 2.0	0718 5.7	1416 1.7	1953 5.8
6 SA	0128 2.2	0709 5.7	1355 1.9	1936 5.7
21 SU	0234 1.7	0806 6.0	1501 1.4	2035 6.0
7 SU	0218 1.8	0755 6.1	1443 1.4	2021 6.1
22 M ○	0316 1.5	0845 6.2	1540 1.2	2111 6.2
8 M ●	0304 1.4	0838 6.4	1529 1.1	2105 6.3
23 TU	0352 1.4	0920 6.3	1614 1.1	2144 6.2
9 TU	0349 1.2	0921 6.6	1613 0.8	2148 6.5
24 W	0425 1.3	0953 6.4	1646 1.1	2215 6.2
10 W	0432 1.0	1004 6.8	1656 0.7	2230 6.6
25 TH	0457 1.3	1025 6.3	1716 1.2	2245 6.2
11 TH	0514 0.9	1047 6.8	1737 0.7	2311 6.6
26 F	0526 1.4	1055 6.2	1745 1.4	2313 6.0
12 F	0557 0.9	1128 6.7	1819 0.8	2353 6.4
27 SA	0555 1.6	1123 6.0	1813 1.6	2341 5.8
13 SA	0640 1.2	1210 6.4	1902 1.1	
28 SU	0625 1.8	1151 5.8	1842 1.9	
14 SU	0035 6.1	0725 1.5	1253 6.1	1949 1.6
29 M	0010 5.6	0656 2.1	1222 5.5	1914 2.2
15 M ☽	0120 5.8	0817 1.9	1344 5.6	2044 2.0

MARCH

Day	Time m	Time m	Time m	Time m
1 TU	0043 5.3	0734 2.4	1301 5.1	1955 2.6
16 W	0151 5.4	0857 2.2	1436 5.1	2130 2.5
2 W ☽	0128 5.0	0825 2.7	1357 4.8	2054 2.8
17 TH	0308 5.1	1019 2.5	1611 4.9	2259 2.7
3 TH	0236 4.8	0939 2.8	1522 4.7	2222 2.9
18 F	0441 5.0	1149 2.4	1740 5.1	
4 F	0413 4.8	1112 2.7	1707 4.8	2353 2.7
19 SA	0023 2.5	0601 5.2	1301 2.1	1846 5.4
5 SA	0542 5.1	1228 2.3	1819 5.2	
20 SU	0126 2.1	0700 5.6	1355 1.8	1933 5.7
6 SU	0100 2.3	0640 5.5	1327 1.8	1911 5.7
21 M	0215 1.8	0745 5.9	1438 1.5	2012 5.9
7 M	0154 1.8	0730 6.0	1419 1.4	1959 6.1
22 TU	0254 1.6	0823 6.1	1515 1.3	2046 6.1
8 TU	0243 1.3	0817 6.4	1507 0.9	2045 6.4
23 W ○	0329 1.4	0857 6.2	1547 1.2	2117 6.2
9 W ●	0330 1.0	0903 6.7	1553 0.6	2129 6.7
24 TH	0400 1.4	0929 6.3	1618 1.2	2147 6.2
10 TH	0415 0.7	0947 6.9	1637 0.4	2212 6.8
25 F	0430 1.3	0959 6.3	1647 1.2	2216 6.2
11 F	0458 0.6	1031 7.0	1720 0.5	2254 6.8
26 SA	0459 1.3	1028 6.2	1715 1.4	2243 6.1
12 SA	0541 0.7	1114 6.8	1801 0.7	2334 6.6
27 SU	0528 1.4	1056 6.0	1743 1.6	2311 6.0
13 SU	0623 0.9	1154 6.5	1844 1.1	
28 M	0557 1.6	1125 5.8	1811 1.8	2340 5.8
14 M	0014 6.3	0707 1.3	1237 6.1	1929 1.6
29 TU	0628 1.9	1156 5.6	1843 2.1	
15 TU ☽	0057 5.8	0757 1.8	1326 5.6	2021 2.1
30 W	0013 5.5	0704 2.2	1233 5.3	1922 2.4
31 TH ☽	0054 5.2	0751 2.5	1324 4.9	2017 2.7

APRIL

Day	Time m	Time m	Time m	Time m
1 F	0154 4.9	0858 2.7	1442 4.7	2140 2.9
16 SA	0406 4.9	1117 2.4	1710 5.0	2351 2.6
2 SA	0324 4.8	1027 2.6	1627 4.8	2315 2.7
17 SU	0527 5.1	1226 2.2	1814 5.3	
3 SU	0501 5.0	1150 2.3	1747 5.2	
18 M	0054 2.3	0628 5.4	1321 1.9	1901 5.6
4 M	0028 2.3	0608 5.5	1255 1.8	1843 5.7
19 TU	0143 2.0	0714 5.6	1405 1.7	1940 5.8
5 TU	0126 1.8	0701 6.0	1350 1.3	1932 6.1
20 W	0224 1.7	0753 5.9	1442 1.5	2015 6.2
6 W	0218 1.3	0751 6.4	1441 0.9	2020 6.5
21 TH	0259 1.6	0829 6.0	1516 1.4	2048 6.1
7 TH ●	0307 0.9	0840 6.7	1530 0.6	2106 6.8
22 F ○	0332 1.4	0902 6.1	1548 1.4	2119 6.2
8 F	0354 0.6	0927 6.9	1616 0.4	2151 6.9
23 SA	0403 1.4	0933 6.1	1618 1.4	2148 6.2
9 SA	0439 0.5	1013 6.9	1700 0.5	2234 6.8
24 SU	0434 1.4	1002 6.1	1648 1.5	2216 6.1
10 SU	0523 0.6	1057 6.8	1743 0.7	2315 6.6
25 M	0504 1.4	1032 6.0	1718 1.6	2246 6.0
11 M	0607 0.8	1139 6.4	1826 1.1	2355 6.3
26 TU	0535 1.6	1104 5.9	1748 1.8	2319 5.9
12 TU	0651 1.2	1222 6.0	1911 1.7	
27 W	0607 1.7	1140 5.6	1822 2.1	2355 5.6
13 W	0038 5.9	0739 1.7	1311 5.5	2002 2.2
28 TH	0645 2.0	1219 5.4	1903 2.3	
14 TH ☽	0129 5.4	0837 2.2	1417 5.1	2107 2.6
29 F	0037 5.4	0731 2.2	1308 5.1	1958 2.6
15 F	0239 5.1	0952 2.4	1543 4.9	2231 2.7
30 SA ☽	0133 5.1	0833 2.4	1417 4.9	2112 2.7

Chart Datum is 3·33 metres below IGN Datum. HAT is 7·0 metres above Chart Datum.

〉〉 **FREE** monthly updates. Register at 〈
www.reedsnauticalalmanac.co.uk

CHERBOURG LAT 49°39'N LONG 1°38'W
TIMES AND HEIGHTS OF HIGH AND LOW WATERS

Dates in red are SPRINGS
Dates in blue are NEAPS

YEAR 2016

MAY

Day	Time	m	Time	m	Time	m	Time	m	Day	Time	m	Time	m	Time	m
1 SU	0249	5.0	0952	2.4	1549	5.0	2238	2.6	16 M	0436	5.0	1137	2.4	1726	5.1
2 M	0423	5.1	1113	2.2	1711	5.3	2353	2.2	17 TU	0008	2.5	0541	5.2	1234	2.2	1818	5.4
3 TU	0532	5.5	1221	1.8	1811	5.7			18 W	0101	2.2	0634	5.4	1323	2.0	1902	5.6
4 W	0055	1.8	0630	5.9	1320	1.4	1903	6.1	19 TH	0146	2.0	0718	5.6	1405	1.8	1941	5.8
5 TH	0150	1.3	0724	6.3	1414	1.0	1953	6.5	20 F	0226	1.8	0758	5.8	1443	1.7	2017	6.0
6 F	0243	1.0	0816	6.6	1505	0.8	2041	6.7	21 SA	0302	1.6	0835	5.9	1517	1.6	2051	6.1
7 SA	0333	0.7	0906	6.8	1554	0.7	2128	6.8	22 SU	0336	1.5	0908	6.0	1551	1.6	2123	6.1
8 SU	0420	0.6	0954	6.8	1640	0.7	2213	6.8	23 M	0409	1.4	0940	6.0	1624	1.6	2153	6.1
9 M	0506	0.7	1040	6.6	1725	0.9	2256	6.6	24 TU	0443	1.4	1013	6.0	1657	1.7	2227	6.1
10 TU	0551	0.9	1125	6.3	1808	1.3	2338	6.3	25 W	0518	1.5	1049	5.9	1732	1.8	2304	6.0
11 W	0635	1.2	1209	6.0	1853	1.7			26 TH	0553	1.6	1127	5.8	1809	1.9	2343	5.8
12 TH	0021	5.9	0722	1.7	1257	5.5	1942	2.2	27 F	0633	1.8	1209	5.6	1852	2.1
13 F	0109	5.5	0814	2.1	1353	5.2	2039	2.5	28 SA	0027	5.6	0719	1.9	1258	5.4	1945	2.3
14 SA	0209	5.2	0916	2.4	1502	5.0	2150	2.7	29 SU	0119	5.4	0815	2.1	1357	5.2	2050	2.5
15 SU	0321	5.0	1029	2.5	1618	5.0	2304	2.7	30 M	0223	5.3	0923	2.2	1512	5.2	2204	2.4
									31 TU	0341	5.3	1037	2.1	1632	5.4	2318	2.2

JUNE

Day	Time	m	Time	m	Time	m	Time	m	Day	Time	m	Time	m	Time	m	Time	m
1 W	0457	5.5	1147	1.8	1739	5.7			16 TH	0014	2.5	0548	5.1	1238	2.3	1821	5.4
2 TH	0024	1.9	0601	5.8	1251	1.5	1835	6.0	17 F	0106	2.3	0641	5.3	1327	2.1	1906	5.6
3 F	0124	1.5	0659	6.1	1349	1.3	1928	6.3	18 SA	0151	2.0	0727	5.5	1410	2.0	1947	5.8
4 SA	0220	1.2	0755	6.3	1444	1.1	2019	6.5	19 SU	0232	1.8	0808	5.7	1449	1.8	2025	6.0
5 SU	0314	1.0	0847	6.5	1535	1.0	2108	6.6	20 M	0310	1.6	0846	5.8	1526	1.7	2100	6.1
6 M	0404	0.8	0937	6.5	1623	1.0	2154	6.6	21 TU	0348	1.5	0922	5.9	1604	1.6	2136	6.2
7 TU	0451	0.9	1025	6.4	1708	1.2	2238	6.5	22 W	0425	1.4	0957	6.0	1641	1.6	2211	6.2
8 W	0536	1.0	1109	6.2	1751	1.4	2320	6.3	23 TH	0503	1.4	1035	6.0	1719	1.6	2250	6.2
9 TH	0619	1.2	1153	6.0	1834	1.7			24 F	0542	1.4	1116	5.9	1759	1.7	2332	6.1
10 F	0003	6.0	0701	1.6	1236	5.7	1917	2.0	25 SA	0622	1.5	1158	5.8	1842	1.9
11 SA	0046	5.7	0745	1.9	1322	5.4	2005	2.4	26 SU	0016	5.9	0707	1.6	1244	5.7	1931	2.0
12 SU	0134	5.3	0833	2.2	1415	5.1	2100	2.6	27 M	0104	5.7	0757	1.8	1336	5.5	2028	2.2
13 M	0230	5.1	0931	2.5	1516	5.0	2205	2.7	28 TU	0159	5.6	0856	2.0	1438	5.4	2134	2.3
14 TU	0335	4.9	1037	2.5	1623	5.0	2313	2.7	29 W	0306	5.4	1005	2.0	1554	5.4	2247	2.2
15 W	0444	5.0	1141	2.5	1727	5.2			30 TH	0426	5.5	1118	2.0	1708	5.6	2358	2.0

JULY

Day	Time	m	Time	m	Time	m	Time	m	Day	Time	m	Time	m	Time	m	Time	m
1 F	0537	5.6	1226	1.8	1812	5.8			16 SA	0024	2.5	0601	5.0	1248	2.5	1831	5.4
2 SA	0103	1.7	0641	5.8	1329	1.6	1909	6.1	17 SU	0117	2.3	0656	5.5	1338	2.2	1918	5.6
3 SU	0204	1.4	0740	6.1	1427	1.4	2002	6.3	18 M	0203	2.0	0742	5.6	1423	2.0	2000	5.9
4 M	0259	1.2	0834	6.2	1520	1.3	2052	6.4	19 TU	0246	1.7	0824	5.8	1505	1.8	2040	6.1
5 TU	0350	1.0	0924	6.3	1608	1.2	2137	6.5	20 W	0327	1.5	0904	6.0	1546	1.6	2118	6.3
6 W	0436	1.0	1009	6.3	1651	1.3	2220	6.4	21 TH	0408	1.3	0942	6.1	1626	1.5	2158	6.4
7 TH	0518	1.1	1051	6.2	1732	1.4	2300	6.3	22 F	0449	1.2	1022	6.2	1706	1.4	2237	6.4
8 F	0557	1.2	1130	6.0	1810	1.6	2339	6.2	23 SA	0529	1.1	1102	6.2	1747	1.3	2319	6.4
9 SA	0634	1.5	1208	5.8	1848	1.9			24 SU	0609	1.2	1144	6.1	1829	1.5
10 SU	0017	5.8	0711	1.8	1245	5.6	1926	2.2	25 M	0001	6.3	0651	1.3	1227	6.0	1914	1.7
11 M	0056	5.5	0749	2.1	1325	5.3	2009	2.5	26 TU	0046	6.0	0738	1.6	1313	5.8	2005	1.9
12 TU	0138	5.2	0833	2.4	1411	5.1	2101	2.7	27 W	0136	5.8	0831	1.9	1408	5.5	2106	2.2
13 W	0230	5.0	0928	2.6	1510	4.9	2207	2.8	28 TH	0237	5.5	0936	2.1	1520	5.4	2220	2.3
14 TH	0335	4.8	1037	2.7	1623	4.9	2320	2.7	29 F	0401	5.3	1053	2.2	1643	5.4	2339	2.2
15 F	0451	4.9	1149	2.7	1735	5.1			30 SA	0521	5.4	1210	2.1	1755	5.6
									31 SU	0051	1.9	0630	5.6	1317	1.9	1856	5.9

AUGUST

Day	Time	m	Time	m	Time	m	Time	m	Day	Time	m	Time	m	Time	m	Time	m
1 M	0153	1.6	0731	5.9	1416	1.7	1951	6.1	16 TU	0135	2.1	0716	5.5	1358	2.0	1935	5.9
2 TU	0248	1.4	0825	6.1	1508	1.5	2039	6.3	17 W	0221	1.7	0801	5.9	1443	1.7	2018	6.2
3 W	0336	1.2	0911	6.3	1552	1.4	2121	6.4	18 TH	0306	1.3	0843	6.2	1526	1.4	2059	6.3
4 TH	0418	1.1	0951	6.3	1632	1.3	2200	6.5	19 F	0348	1.1	0924	6.4	1608	1.2	2140	6.7
5 F	0456	1.1	1027	6.2	1709	1.4	2236	6.4	20 SA	0430	0.9	1005	6.5	1650	1.1	2222	6.7
6 SA	0531	1.2	1102	6.1	1742	1.5	2311	6.2	21 SU	0511	0.8	1046	6.5	1731	1.1	2302	6.7
7 SU	0603	1.4	1135	6.0	1815	1.7	2344	6.0	22 M	0552	0.9	1126	6.4	1812	1.2	2344	6.5
8 M	0634	1.7	1206	5.8	1847	2.0			23 TU	0633	1.1	1207	6.2	1856	1.5
9 TU	0016	5.7	0706	2.0	1238	5.5	1922	2.3	24 W	0027	6.2	0718	1.5	1251	6.0	1945	1.8
10 W	0049	5.4	0741	2.3	1315	5.2	2004	2.6	25 TH	0115	5.8	0809	1.9	1342	5.6	2044	2.2
11 TH	0131	5.1	0825	2.6	1403	5.0	2100	2.8	26 F	0214	5.4	0913	2.3	1452	5.3	2200	2.4
12 F	0229	4.8	0927	2.9	1510	4.8	2217	2.9	27 SA	0344	5.2	1036	2.6	1623	5.2	2327	2.4
13 SA	0348	4.7	1052	2.9	1641	4.9	2340	2.8	28 SU	0512	5.2	1200	2.4	1743	5.5
14 SU	0520	4.9	1211	2.7	1756	5.2			29 M	0043	2.1	0625	5.5	1310	2.1	1847	5.7
15 M	0043	2.4	0626	5.2	1309	2.4	1850	5.5	30 TU	0145	1.8	0724	5.8	1407	1.8	1940	6.1
									31 W	0235	1.5	0812	6.1	1453	1.6	2024	6.3

Chart Datum is 3·33 metres below IGN Datum. HAT is 7·0 metres above Chart Datum.

STANDARD TIME UT –01
Subtract 1 hour for UT
For French Summer Time add
ONE hour in **non-shaded areas**

CHERBOURG   LAT 49°39′N   LONG 1°38′W
TIMES AND HEIGHTS OF HIGH AND LOW WATERS

Dates in red are **SPRINGS**
Dates in blue are NEAPS

YEAR 2016

N France

SEPTEMBER

Day	Time m	Time m	Time m	Time m	Day	Time m	Time m	Time m	Time m
1 TH	0318 1.3	0852 6.2	1534 1.4	● 2102 6.4	16 F	0240 1.2	0818 6.4	1503 1.3	○ 2036 6.7
2 F	0356 1.2	0927 6.3	1609 1.4	2136 6.5	17 SA	0325 0.9	0901 6.6	1547 1.0	2119 6.9
3 SA	0430 1.2	0959 6.3	1642 1.4	2209 6.4	18 SU	0408 0.7	0944 6.8	1630 0.9	2203 7.0
4 SU	0501 1.3	1030 6.3	1712 1.5	2241 6.3	19 M	0451 0.7	1025 6.8	1712 0.9	2245 6.9
5 M	0530 1.4	1100 6.1	1742 1.6	2310 6.1	20 TU	0533 0.8	1106 6.7	1754 1.0	2327 6.7
6 TU	0559 1.7	1127 5.9	1811 1.9	2339 5.9	21 W	0615 1.1	1147 6.4	1839 1.3	
7 W	0627 2.0	1157 5.7	1842 2.2		22 TH	0009 6.3	0659 1.6	1230 6.1	1927 1.8
8 TH	0009 5.6	0658 2.3	1228 5.4	1918 2.5	23 F	0055 5.8	0750 2.1	1320 5.7	◑ 2026 2.2
9 F	0045 5.2	0736 2.6	1310 5.1	◑ 2006 2.8	24 SA	0200 5.4	0855 2.5	1431 5.3	2144 2.5
10 SA	0138 4.9	0830 2.9	1413 4.9	2116 3.0	25 SU	0331 5.1	1022 2.7	1604 5.2	2315 2.5
11 SU	0256 4.7	0954 3.1	1544 4.8	2250 2.9	26 M	0502 5.2	1150 2.6	1729 5.3	
12 M	0438 4.8	1131 2.9	1719 5.1		27 TU	0030 2.2	0614 5.5	1257 2.3	1833 5.7
13 TU	0008 2.6	0556 5.2	1239 2.5	1819 5.5	28 W	0128 1.9	0707 5.8	1350 1.9	1922 6.0
14 W	0105 2.1	0649 5.6	1332 2.1	1908 5.9	29 TH	0215 1.6	0749 6.1	1433 1.7	2002 6.2
15 TH	0154 1.6	0735 6.0	1419 1.6	1953 6.3	30 F	0254 1.4	0826 6.2	1510 1.5	2038 6.4

OCTOBER

Day	Time m	Time m	Time m	Time m	Day	Time m	Time m	Time m	Time m
1 SA	0328 1.3	0858 6.3	1543 1.4	● 2111 6.4	16 SU	0300 0.9	0836 6.8	1525 0.9	○ 2057 7.0
2 SU	0400 1.3	0929 6.4	1614 1.4	2142 6.4	17 M	0345 0.7	0921 6.9	1610 0.8	2143 7.0
3 M	0430 1.4	0959 6.3	1643 1.5	2211 6.3	18 TU	0430 0.7	1004 6.9	1654 0.8	2227 6.9
4 TU	0459 1.5	1027 6.2	1712 1.6	2240 6.2	19 W	0514 0.9	1046 6.8	1738 0.9	2311 6.7
5 W	0527 1.7	1054 6.1	1741 1.8	2308 6.0	20 TH	0558 1.2	1129 6.5	1824 1.3	2356 6.3
6 TH	0555 2.0	1123 5.9	1811 2.0	2339 5.7	21 F	0643 1.6	1213 6.1	1913 1.7	
7 F	0625 2.3	1155 5.6	1845 2.3		22 SA	0045 5.8	0735 2.2	1303 5.7	◑ 2010 2.2
8 SA	0015 5.4	0701 2.6	1234 5.3	1929 2.6	23 SU	0148 5.4	0838 2.6	1411 5.3	2124 2.5
9 SU	0104 5.1	0751 2.9	1332 5.0	◑ 2032 2.9	24 M	0311 5.1	1001 2.8	1537 5.2	2251 2.5
10 M	0217 4.8	0909 3.1	1457 4.9	2201 2.9	25 TU	0438 5.2	1126 2.7	1700 5.3	
11 TU	0357 4.9	1048 3.0	1636 5.1	2327 2.6	26 W	0003 2.3	0547 5.4	1231 2.4	1804 5.5
12 W	0522 5.2	1205 2.6	1745 5.5		27 TH	0059 2.0	0638 5.7	1322 2.1	1854 5.8
13 TH	0031 2.1	0618 5.7	1301 2.1	1837 6.0	28 F	0144 1.8	0719 6.0	1405 1.8	1935 6.1
14 F	0124 1.6	0705 6.1	1351 1.6	1925 6.4	29 SA	0223 1.6	0756 6.2	1442 1.7	2011 6.2
15 SA	0213 1.2	0751 6.5	1438 1.2	2011 6.8	30 SU	0258 1.5	0829 6.3	1515 1.6	● 2045 6.3
					31 M	0330 1.5	0901 6.3	1547 1.5	2117 6.3

NOVEMBER

Day	Time m	Time m	Time m	Time m	Day	Time m	Time m	Time m	Time m
1 TU	0401 1.5	0931 6.3	1618 1.5	2146 6.3	16 W	0411 0.8	0944 7.0	1639 0.8	2213 6.9
2 W	0432 1.6	0959 6.3	1648 1.6	2215 6.2	17 TH	0457 1.0	1029 6.8	1725 0.9	2259 6.6
3 TH	0501 1.8	1028 6.2	1719 1.7	2246 6.0	18 F	0543 1.3	1114 6.6	1812 1.2	2346 6.3
4 F	0531 2.0	1100 6.0	1750 1.9	2320 5.8	19 SA	0629 1.7	1159 6.2	1900 1.6	
5 SA	0603 2.2	1134 5.8	1825 2.2	2358 5.5	20 SU	0035 5.9	0719 2.1	1248 5.8	1952 2.0
6 SU	0640 2.5	1215 5.5	1908 2.4		21 M	0132 5.5	0816 2.5	1347 5.5	◑ 2054 2.4
7 M	0045 5.2	0730 2.8	1307 5.2	● 2005 2.6	22 TU	0238 5.2	0924 2.7	1458 5.2	2206 2.5
8 TU	0150 5.0	0839 2.9	1420 5.1	2120 2.7	23 W	0352 5.1	1042 2.8	1613 5.2	2318 2.5
9 W	0316 5.0	1005 2.9	1549 5.1	2243 2.5	24 TH	0502 5.3	1150 2.6	1722 5.3	
10 TH	0441 5.3	1125 2.6	1705 5.5	2353 2.1	25 F	0017 2.3	0559 5.5	1245 2.3	1818 5.5
11 F	0544 5.7	1228 2.1	1804 5.9		26 SA	0107 2.1	0645 5.8	1332 2.1	1903 5.8
12 SA	0052 1.7	0635 6.1	1322 1.6	1857 6.3	27 SU	0150 1.9	0725 6.0	1412 1.9	1944 5.9
13 SU	0145 1.3	0724 6.5	1414 1.2	1947 6.7	28 M	0228 1.8	0802 6.1	1449 1.7	2021 6.1
14 M	0235 1.0	0812 6.8	1503 0.9	○ 2037 6.9	29 TU	0303 1.7	0836 6.2	1523 1.6	● 2055 6.1
15 TU	0324 0.8	0858 6.9	1552 0.8	2125 6.9	30 W	0337 1.7	0908 6.3	1556 1.6	2127 6.1

DECEMBER

Day	Time m	Time m	Time m	Time m	Day	Time m	Time m	Time m	Time m
1 TH	0409 1.7	0939 6.3	1629 1.6	2158 6.1	16 F	0445 1.1	1016 6.8	1714 0.9	2249 6.5
2 F	0442 1.8	1010 6.2	1703 1.6	2231 6.0	17 SA	0530 1.3	1101 6.6	1759 1.1	2334 6.3
3 SA	0515 1.9	1044 6.1	1737 1.7	2307 5.9	18 SU	0615 1.5	1145 6.3	1844 1.4	
4 SU	0550 2.0	1122 6.0	1813 1.9	2347 5.7	19 M	0019 6.0	0659 1.9	1229 6.0	1928 1.8
5 M	0629 2.2	1203 5.8	1855 2.1		20 TU	0105 5.6	0746 2.3	1317 5.6	2016 2.1
6 TU	0032 5.5	0716 2.5	1251 5.5	1946 2.3	21 W	0156 5.3	0838 2.6	1411 5.3	◑ 2110 2.4
7 W	0127 5.3	0815 2.6	1350 5.3	● 2048 2.4	22 TH	0255 5.1	0941 2.8	1514 5.1	2216 2.6
8 TH	0235 5.2	0926 2.7	1503 5.3	2201 2.3	23 F	0402 5.1	1053 2.8	1626 5.0	2324 2.6
9 F	0355 5.3	1042 2.5	1625 5.4	2315 2.1	24 SA	0510 5.2	1159 2.6	1734 5.1	
10 SA	0508 5.6	1153 2.2	1732 5.7		25 SU	0024 2.4	0608 5.4	1255 2.4	1831 5.4
11 SU	0021 1.8	0607 6.0	1255 1.8	1832 6.1	26 M	0116 2.2	0655 5.6	1342 2.1	1917 5.6
12 M	0120 1.5	0701 6.3	1352 1.4	1927 6.4	27 TU	0200 2.0	0736 5.9	1424 1.9	1959 5.8
13 TU	0215 1.2	0752 6.6	1446 1.1	2021 6.6	28 W	0240 1.9	0815 6.1	1502 1.7	2037 5.9
14 W	0307 1.0	0841 6.8	1538 0.9	○ 2112 6.7	29 TH	0316 1.8	0850 6.2	1538 1.6	● 2112 6.0
15 TH	0357 1.0	0929 6.8	1627 0.8	2201 6.7	30 F	0352 1.7	0924 6.3	1614 1.5	2145 6.1
					31 SA	0428 1.6	0957 6.3	1650 1.4	2220 6.1

Chart Datum is 3·33 metres below IGN Datum. HAT is 7·0 metres above Chart Datum.

⟩⟩ **FREE** monthly updates. Register at ⟨
www.reedsnauticalalmanac.co.uk

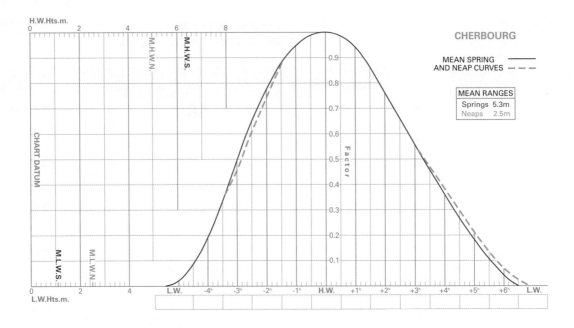

CHERBOURG

MEAN SPRING ———
AND NEAP CURVES - - - -

MEAN RANGES	
Springs	5.3m
Neaps	2.5m

9.17.31 OMONVILLE-LA-ROGUE

Manche 49°42'·29N 01°49'·84W ✿✿✿💧💧✿✿

CHARTS AC 2669, 1114; SHOM 7120, 7158, 5636; Navi 528, 1014; Imray C33A

TIDES −0330 Dover; ML 3·8; Duration 0545

Standard Port CHERBOURG (←—)

Times				Height (metres)			
High Water		Low Water		MHWS	MHWN	MLWN	MLWS
0300	1000	0400	1000	6·4	5·1	2·6	1·1
1500	2200	1600	2200				
Differences OMONVILLE							
−0015	−0010	−0015	−0015	−0·1	−0·1	0·0	0·0
GOURY							
−0100	−0040	−0105	−0120	+1·2	+1·3	+0·7	0·0

OMONVILLE-LA-ROGUE

SHELTER Good, except in moderate/fresh N to SE winds when it may become untenable. Swell tends to work in.

NAVIGATION WPT 49°42'·47N 01°48'·66W, 257°/1·0M to Omonville dir lt. From W or N, keep clear of Basse Bréfort (depth 1m, marked by NCM buoy, VQ) 0·6M N of Pte Jardeheu. Appr on 195° transit (below) passing 100m E of L'Étonnard and into W sector of lt before turning stbd 290° for ⚓s.

From E, appr on 257° transit with church spire and in W sector of lt, until S of L'Étonnard.

Harbour entrance is 100m wide: rocks extend N from Omonville Fort, and ESE from bkwtr to L'Étonnard, G bcn twr.

LIGHTS AND MARKS Omonville lt, on W framework tr with R top, Iso WRG 4s 13m 10/7M, vis 180°-G-252°-W-262°-R-287°. Lt in transit 257° with ✠ steeple (hard to see) 650m beyond, leads S of L'Étonnard, 200m SE of which is a large unlit mooring buoy. From N, L'Étonnard in transit 195° with fort just clears Les Tataquets rocks. Street lights adequately illuminate the hbr area.

COMMUNICATIONS CROSS 02·33·52·16·16; ⚓ 02·33·10·16·10; Auto 08·92·68·02·50; Police (Beaumont Hague) 02·33·01·89·30 or 17; Brit Consul 01·44·51·31·02; Samu (emergency medical) 15; Doctor 02·33·52·78·07/02·33·01·87·00; Fire (pompiers, inc. divers) 18; Aff Mar 02·33·23·36·00.

FACILITIES Harbour There are 6 large W conical ⚓s or ⚓ inside bkwtr; beware rocks off outer end. ⚓🔧.

Village ✕◻🛒🏪 at l'Association du Camping in village centre and at Watersports Centre; 🏨🏤🛢 (at Beaumont-Hague 5km) ✉Ⓑ⇌ (bus to Cherbourg) ✈. Free historical tour www.visiteomonvillelarogue.blogspot.com Ferry see Cherbourg.

MINOR HARBOUR 2M EAST OF CAP DE LA HAGUE

ANSE DE ST MARTIN, Manche, 49°42'·72N 01°53'·78W. AC 1114, 3653; SHOM 7120, 5636. Tides as Omonville-la-Rogue. From N, appr with conspic chy (279m) at Jobourg nuclear plant brg 175°; or from NE via Basse Bréfort NCM buoy, VQ, on with Danneville spire brg 237°. Both lines clear La Parmentière rks awash in centre of bay and Les Herbeuses and Le Grun Rk (1·9m) to W and SE respectively. Care needed, especially in poor visibility.

This bay, 2M E of Cap de la Hague, has ⚓s sheltered from all but onshore winds. Port Racine (said to be the smallest hbr in France) is in the SW corner of Anse de St Martin. ⚓ or moor off the hbr which is obstructed by lines; landing by dinghy.

Central North France

Cap de la Hague to St Quay-Portrieux

9.18.1	**Contents**	791
9.18.2	**Area map and Distance table**	792
9.18.3	**Tidal stream charts**	794
9.18.4	**Lights, buoys and waypoints**	796
9.18.5	**Passage information**	798
	The West Coast of the Cotentin Peninsula – 798 • Cap de la Hague to Mont St Michel – 798 • Mont St Michel to St Malo – 802 • St Malo to L'Ost Pic and Beyond – 802	
9.18.6	**Special notes for France, see Area 17**	
9.18.7	**Diélette**	798
	Goury	
9.18.8	**Carteret**	799
9.18.9	**Portbail**	800
9.18.10	**Granville**	800
	Regnéville	
9.18.11	**Îles Chausey**	801
9.18.12	**St Malo, tidal coefficients, tide tables/curves**	802
	Cancale • Rotheneuf • Dinard	
9.18.13	**River Rance**	809
	Plouër • Lyvet • Dinan	
9.18.14	**Inland waterways of Brittany**	810
9.18.15	**St Cast Le Guildo**	811
	Saint Briac • Erquy	
9.18.16	**Dahouet**	811
9.18.17	**Saint-Brieuc Le Légué**	812
9.18.18	**Binic**	813
9.18.19	**St Quay-Portrieux**	814

Central N France

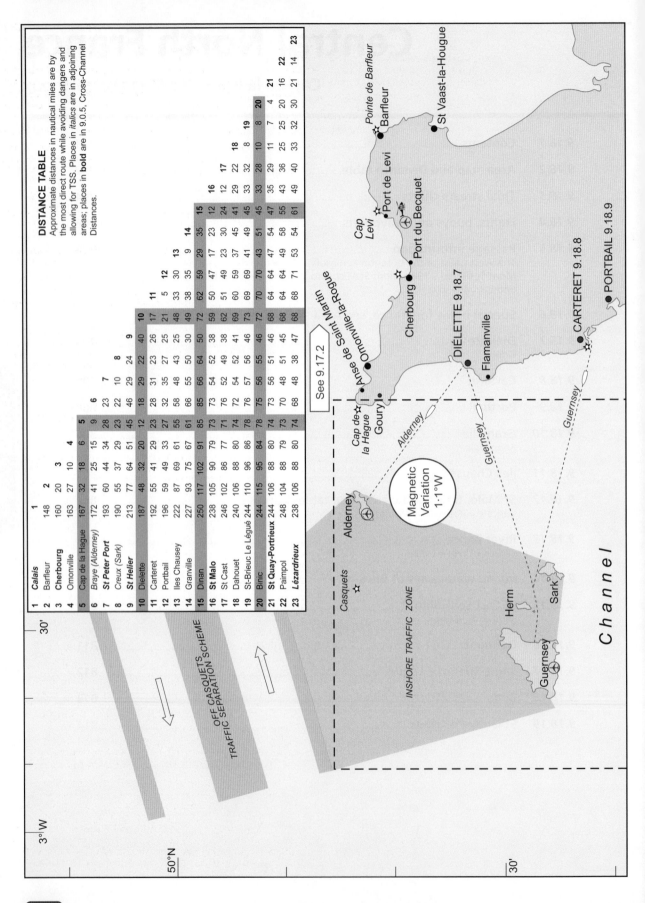

DISTANCE TABLE

Approximate distances in nautical miles are by the most direct route while avoiding dangers and allowing for TSS. Places in *italics* are in adjoining areas; places in **bold** are in 9.0.5, Cross-Channel Distances.

	1	2	3	4	5	6	7	8	9	10	11	12	13	14	15	16	17	18	19	20	21	22	23
1 *Calais*	1																						
2 Barfleur	148	2																					
3 **Cherbourg**	160	20	3																				
4 Omonville	163	27	10	4																			
5 Cap de la Hague	167	32	18	6	5																		
6 *Braye (Alderney)*	172	41	25	15	9	6																	
7 **St Peter Port**	193	60	44	34	28	23	7																
8 *Creux (Sark)*	190	55	37	29	23	22	10	8															
9 **St Helier**	213	77	64	51	45	46	29	24	9														
10 Dielette	187	48	32	20	12	18	29	22	40	10													
11 Carteret	192	55	41	33	23	28	31	23	22	17	11												
12 Portbail	196	59	49	33	27	35	27	25	21	5	12												
13 Iles Chausey	222	87	69	61	55	58	48	43	25	33	30	35	13										
14 Granville	227	93	75	67	61	66	55	50	30	49	48	35	9	14									
15 **Dinan**	250	117	102	91	85	85	66	64	50	72	62	59	38	29	35	15							
16 **St Malo**	238	105	90	79	73	73	54	52	38	59	50	47	17	23	12	16							
17 St Cast	246	102	86	77	71	76	52	49	38	62	51	49	23	30	24	12	17						
18 Dahouet	240	106	88	80	74	76	54	52	41	60	60	59	30	45	41	29	22	18					
19 St-Brieuc Le Légué	244	110	96	86	78	76	57	56	46	73	69	69	37	49	45	33	32	8	19				
20 **Binic**	244	115	95	84	78	75	56	55	46	72	70	70	43	51	45	47	33	10	8	20			
21 **St Quay-Portrieux**	244	106	88	80	74	73	56	51	46	68	64	64	47	54	47	35	29	11	7	4	21		
22 Paimpol	248	104	88	79	73	70	48	51	45	68	64	64	49	58	55	43	36	25	25	20	16	22	
23 *Lézardrieux*	238	106	106	80	74	68	47	38	47	68	68	71	53	54	61	49	40	33	32	30	21	14	23

See 9.17.2

Magnetic Variation 1·1°W

OFF CASQUETS SCHEME
TRAFFIC SEPARATION SCHEME

INSHORE TRAFFIC ZONE

Casquets

Alderney

Herm

Sark

Guernsey

C h a n n e l

Cap de la Hague
Goury
Anse de Saint Martin
Omonville-la-Rogue

Cap Levi
Port de Levi
Port du Becquet
Cherbourg
Port du Becquet

DIÉLETTE 9.18.7
Flamanville
CARTERET 9.18.8
PORTBAIL 9.18.9

Pointe de Barfleur
Barfleur
St Vaast-la-Hougue

Alderney
Guernsey
Guernsey

30'
3°W
50°N
30'

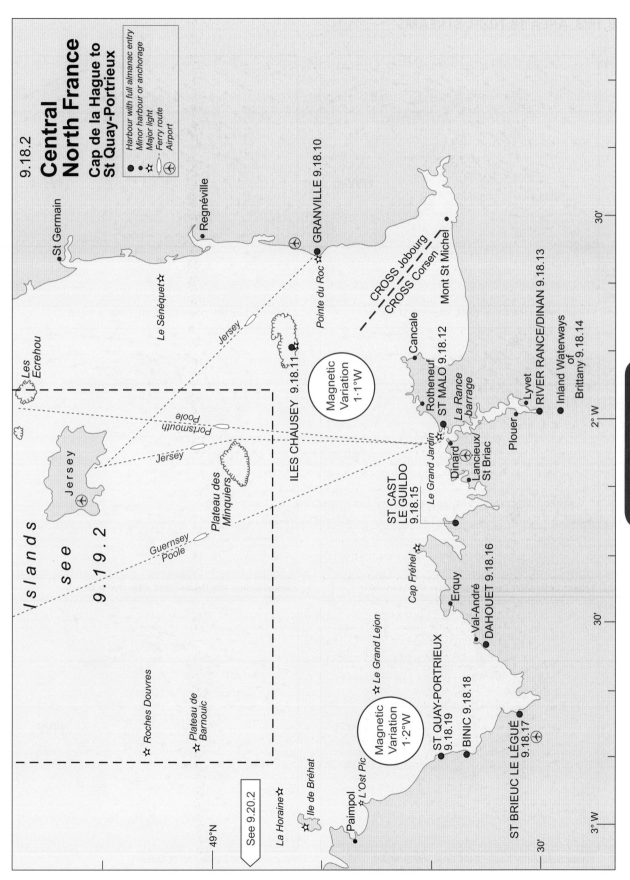

9.18.2

Central North France

Cap de la Hague to
St Quay-Portrieux

Harbour with full almanac entry
Minor harbour or anchorage
Major light
Ferry route
Airport

St Germain

Regnéville

Le Sénéquet ☆

GRANVILLE 9.18.10

Pointe du Roc

CROSS Jobourg
CROSS Corsen

Mont St Michel

Cancale

Rotheneuf
ST MALO 9.18.12

La Rance
barrage

Les Écrehou

Jersey

Portsmouth
Poole

Jersey

ILES CHAUSEY 9.18.11

Magnetic
Variation
1·1'W

Le Grand Jardin

Dinard
Lancieux
St Briac

Plouer

Lyvet
RIVER RANCE/DINAN 9.18.13

Inland Waterways
of
Brittany 9.18.14

2° W

Guernsey
Poole

Plateau des
Minquiers

ST CAST
LE GUILDO
9.18.15

Cap Fréhel ☆

Erquy

Val-André

DAHOUET 9.18.16

30'

Islands
see
9.19.2

Jersey

Roches Douvres ☆

Plateau de
Barnouic ☆

Le Grand Lejon ☆

L'Ost Pic ☆

ST QUAY-PORTRIEUX
9.18.19

BINIC 9.18.18

ST BRIEUC LE LÉGUÉ
9.18.17

Magnetic
Variation
1·2'W

See 9.20.2

Paimpol

Ile de Bréhat

La Horaine ☆

49°N

3° W

30'

30'

9.18.3 AREA 18 TIDAL STREAMS

5 Hours before HW Dover (0006 after HW St Malo)

4 Hours before HW Dover (0106 after HW St Malo)

3 Hours before HW Dover (0206 after HW St Malo)

2 Hours before HW Dover (0306 after HW St Malo)

1 Hour before HW Dover (0406 after HW St Malo)

HW Dover (0506 after HW St Malo)

Westward 9.20.3   Channel Islands 9.19.3   Eastward 9.17.3   Northward 9.2.3

1 Hour after HW Dover (0554 before HW St Malo)

2 Hours after HW Dover (0454 before HW St Malo)

3 Hours after HW Dover (0354 before HW St Malo)

4 Hours after HW Dover (0254 before HW St Malo)

5 Hours after HW Dover (0154 before HW St Malo)

6 Hours after HW Dover (0054 before HW St Malo)

9.18.4 LIGHTS, BUOYS AND WAYPOINTS

Bold print = light with a nominal range of 15M or more. CAPITALS = place or feature. *CAPITAL ITALICS* = light-vessel, light float or Lanby. *Italics* = Fog signal. ***Bold italics*** = Racon. Many marks/buoys are fitted with AIS (<u>MMSI No</u>); see relevant charts.

CAP DE LA HAGUE TO ST MALO

GOURY
La Foraine ⚓ VQ (9) 10s 12m 6M; 49°42'·90 N 01°58'·32W.
Ldg lts 065·2°: Front, QR 5m 7M; R □ in W □ on pier, 49°42'·89N 01°56'·70W. Rear, 116m from front, Q 11m 7M; W pylon/hut. Hervieu ⚓ 49°42'·77N 01°56'·92W.

DIELETTE
W bkwtr Dir lt 140°, Iso WRG 4s 12m W10M, R/G7M; 070°-G-135°-W-145°-R-180°; W twr, G top ;49°33'·18N 01°51'·81W.
E bkwtr ≮ Fl R 4s 6m 2M; 49°33'·21N 01°51'·78W.
Inner N jetty, Fl (2) R 6s. Inner S jetty, Fl (2) G 6s; bth 6m 1M.
Banc des Dious ⚓ Q (9) 15s; 49°32'·58N 01°54'·02W.

CARTERET
Cap de Carteret ☆ Fl (2+1) 15s 81m **26M**; Gy twr, G top; 49°22'·40N 01°48'·41W.
W bkwtr ≮ Oc R 4s 7m 7M; W post, R top; 49°22'·07N 01°47'·32W.
E training wall ≮ Fl G 2·5s 4m 2M; W post, G top; 49°22'·17N 01°47'·30W.
Channel bend ≮ Fl (2) R 6s 5m 1M; R pylon; 49°22'·58N 01°47'·23W. Inside the bend: ≮ Fl (2) G 6s 5m 1M; G pylon; 49°22'·55N 01°47'·20W.
Marina entry sill: ≮ Fl (3) R 12s and Fl (3) G 12s; R & G pylons.

PORTBAIL
PB ⚓ 49°18'·37N 01°44'·75W.
Ldg lts 042°: Front, Q 14m 10M; W pylon, R top, 49°19'·75N 01°42'·50W. Rear, 870m frm front, Oc 4s 20m 10M; ch belfry. ⚓ 49°19'·32N 01°43'·16W.
▲ 49°19'·20N 01°43'·00W.
Training wall head ⚓ Fl (2) R 5s 5m 4M; W mast, R top; 49°19'·43N 01°42'·99W.

REGNÉVILLE
La Catheue ⚓ Q (6) + L Fl 15s, 48°57'·67N 01°42'·23W.
Le Ronquet ⚓ Fl (2) WR 6s, W6M, R4M; 100°-R-293°-W-100°; 49°00'·11N 01°38'·07W.
Pte d'Agon ≮ Oc (2) WR 6s 12m, W10M, R7M; 063°-R-110°-W-063°; W twr, R top, W dwelling; 49°00'·18N 01°34'·63W.
Dir lt 028°, Oc WRG 4s 9m, W12M, R/G9M; 024°-G-027°-W-029°-R-033°; house; 49°00'·63N 01°33'·36W.

PASSAGE DE LA DÉROUTE
Les Trois-Grunes ⚓ Q (9) 15s, 49°21'·84N 01°55'·21W.
Écrevière ⚓ Q (6) + L Fl 15s; *Bell;* 49°15'·27N 01°52'·16W.
Basse Jourdan ⚓ Q (3) 10s; 49°06'·85N 01°43'·96W.
Le Boeuf ⚓ 49°06'·56N 01°47'·17W.
Les Boeuftins ⚓ 49°07'·03N 01°45'·96W.
La Basse du Sénéquet ⚓ 49°05'·96N 01°41'·11W.
Le Sénéquet ≮ Fl (3) WR 12s 18m W13M, R10M; 083·5°-R-116·5°-W-083·5°; W twr; 49°05'·48N 01°39'·73W.
Les Nattes ⚓ 49°03'·46N 01°41'·81W.
International F ⚓ 49°02'·16N 01°42'·98W.
International E ⚓ 49°02'·08N 01°47'·21W.
Basse le Marié ⚓ Q (9) 15s; 49°01'·79N 01°48'·57W.
NE Minquiers ⚓ VQ (3) 5s; *Bell;* 49°00'·85N 01°55'·30W.
Les Ardentes ⚓ Q (3) 10s; 48°57'·89N 01°51'·53W.
SE Minquiers ⚓ Q (3) 10s; *Bell;* 48°53'·42N 02°00'·09W.
S Minquiers ⚓ Q (6) + L Fl 15s; 48°53'·09N 02°10'·10W.

ÎLES CHAUSEY
La Pointue ⚓ 48°54'·44N 01°50'·92W.

L'Enseigne, W twr, B top; 48°53'·67N 01°50'·37W.
L'Etat, BW ⚓, 48°54'·67N 01°46'·21W.
Anvers wreck ⚓ 48°53'·91N 01°41'·07W.
Le Founet ⚓ Q (3) 10s; 48°53'·25N 01°42'·34W.
Le Pignon ≮ Fl (2) WR 6s 10m, W9M, R6M; 005°-R-150°-W-005°; B twr, W band; 48°53'·49N 01°43'·36 W.
La Haute Foraine ⚓ 48°52'·89N 01°43'·66W.
Grande Île ☆ Fl 5s 39m **23M**; Gy □ twr, G top; *Horn 30s;* 48°52'·17N 01°49'·34W.
Channel ⚓ Fl G 2s; 48°52'·07N 01°49'·08W.
La Crabière Est ⚓ Dir Oc(3)WRG 12s 5m, W9M, R/G6M;079°-W-291°-G-329°-W-335°-R-079°; B beacon, Y top; 48°52'·46N 01°49'·39W.
La Cancalaise ⚓ 48°51'·90N 01°51'·11W.

GRANVILLE
Le Videcoq ⚓ VQ (9) 10s; 48°49'·66N 01°42'·06W.
La Fourchie ⚓ 48°50'·15N 01°37'·00W.
Pointe du Roc ☆ Fl (4) 15s 49m **23M;** 48°50'·06N 01°36'·78W.
Le Loup ⚓ Fl (2) 6s 8m 11M; 48°49'·57N 01°36'·24W.
Avant Port, E jetty ≮ Fl G 2·5s 11m 4M; 48°49'·93N 01°36'·19W.
W jetty ≮ Fl R 2·5s 12m 4M; 48°49'·86N 01°36'·23W.
Marina S bkwtr ≮ Fl (2) R 6s 12m 5M; W post, R top; *Horn (2) 40s;* 48°49'·89N 01°35'·90W.
N bkwtr ≮ Fl (2) G 6s 4m 5M; 48°49'·93N 01°35'·90W.
Sill, W side, QR 2M, E side QG 2M; Gy pylons topped R (W) and G (E).

CANCALE
La Fille ⚓ 48°44'·16N 01°48'·46W.
Pierre-de-Herpin ☆ Oc (2) 6s 20m 13M; W twr, B top and base; 48°43'·77N 01°48'·92W.
Ruet ⚓ *Bell;* 48°43'·44N 01°50'·11W.
Grande Bunouze ⚓ 48°43'·17N 01°50'·95W.
Barbe Brûlée ⚓ 48°42'·11N 01°50'·57W.
Jetty hd ≮ Oc (3) G 12s 9m 3M; W pylon B top; 48°40'·10N 01°51'·11W.

ST MALO AND RIVER RANCE

CHENAL DE LA BIGNE
Basse Rochefort (aka Basse aux Chiens) ⚓ 48°42'·69N 01°57'·31W.
La Petite Bigne ⚓ 48°41'·66N 01°58'·72W.
La Crolante ⚓; 48°41'·01N 01°59'·51W, off Pte de la Varde.
Les Létruns ⚓; *Bell;* 48°40'·71N 02°00'·61W.
Roches-aux-Anglais ⚓ Fl G 4s; 48°39'·65N 02°02'·27W.
Les Crapauds-du-Bey ⚓ Fl R 4s; 48°39'·37N 02°02'·57W.

CHENAL DES PETITS POINTUS
Dinard ch spire (74m) brg 203° to right of Le Petit Bé rocks.
La Saint-Servantine ⚓ Fl G 2·5s; *Bell;* 48°41'·93N 02°00'·94W.
Les Petits Pontus ⚓ 48°41'·30N 02°00'·86W.

CHENAL DE LA GRANDE CONCHÉE
Villa Brisemoulin 181·5°, on with LH edge of Le Petit Bé rocks.
La Plate ⚓ Q WRG 11m, W10M, R/G7M; 140°-W-203°-R-210°-W-225°-G-140°; 48°40'·78N 02°01'·91W.
Le Bouton ⚓ 48°40'·59N 02°01'·85W.

CHENAL DU BUNEL
Dinard water twr lt 158·2° on with St Énogat lts.
St Énogat ldg lts 158·2°; both Iso 4s 3/85m 6/8M, synch. Front 48°38'·29N 02°04'·11W, vis 126°-236°. Rear, 1·4M from front, on water twr; vis 143°-210°.
Bunel ⚓ Q (9) 15s; 48°40'·84N 02°05'·38W.

CHENAL DE LA PETITE PORTE
Outer ldg lts 129·7°: **Front, Le Grand Jardin** ☆ Fl (2) R 10s 24m **15M**, 48°40'·20N 02°04'·97W. <u>992271314</u>. Rear, **La Balue** ☆ FG 20m **22M**; 3·1M from front; intens 128°-129·5°; Gy □ twr; 48°38'·16N 02°01'·30W.
Vieux-Banc E ⚓ Q; 48°42'·38N 02°09'·12W.

Vieux-Banc W ⚓ VQ (9) 10s; 48°41'·84N 02°10'·20W.
St Malo Atterrisage (Fairway) ⚓ Iso 4s; 48°41'·39N 02°07'·28W.
Les Courtis ⚑ Fl G 4s 14m 7M; 48°40'·46N 02°05'·80W.

Nearing Le Grand Jardin, move stbd briefly onto Ch du Bunel 152·8° ldg lts to pick up: Inner ldg lts 128·6°, both Dir FG 20/69m **22/25M**; H24. Front, **Les Bas Sablons** ☆, intens 127·2°-130·2°; W □ twr, B top; 48°38'·16N 02°01'·30W. <u>992271312</u>.
Rear, **La Balue** ☆, 0·9M from front; intens 128°-129·5°; Gy □ twr; 48°37'·60N 02°00'·24W. <u>992271313</u>.
Basse du Nord No. 5 ⚑ 48°39'·98N 02°05'·04W.
Les Pierres-Garnier No. 8 ⚑ 48°39'·98N 02°04'·41W.
Les Patouillets ⚑ Fl (3) G 12s, 48°39'·68N 02°04'·30W.
Clef d'Aval No. 10 ⚑ 48°39'·72N 02°03'·91W.
Basse du Buron No. 12 ⚑ Fl (4) R 15s, 48°39'·42N 02°03'·51W.
Le Buron ⚑ Fl (4) G 15s 15m 7M; G twr; 48°39'·32N 02°03'·66W.
Les Grelots ⚓ VQ (6) + L Fl 10s, 48°39'·16N 02°03'·03W.

CHENAL DE LA GRANDE PORTE
Banchenou ⚓ VQ, 48°40'·44N 02°11'·48W.
Outer ldg lts 089·1°: **Front, Le Grand Jardin** ☆ Fl (2) R 10s 24m **15M**, 48°40'·20N 02°04'·97W.

Rear, **Rochebonne** ☆ Dir FR 40m **24M**; intens 088·2°-089·7°; Gy □ twr, R top, 4·2M from front; 48°40'·26N 01°58'·71W. <u>992271311</u>.
Buharats W No. 2 ⚓ Fl R 2·5s, 48°40'·22N 02°07'·50W.
Buharats E No. 4 ⚓ *Bell;* 48°40'·24N 02°07'·20W.
Bas du Boujaron No. 1 ⚑ Fl (2) G 6s, 48°40'·17N 02°05'·97W.
Le Sou ⚓ VQ (3) 5s; 48°40'·11N 02°05'·30W.
Continue on inner 128·6° ldg line: see Chenal de la Petite Porte.

RADE DE ST MALO
Plateau Rance Nord ⚓ VQ, 48°38'·64N 02°02'·35W.
Plateau Rance Sud ⚓ Q (6) + L Fl 15s, 48°38'·43N 02°02'·28W.
Crapaud de la Cité ⚓ QG, 48°38'·34N 02°02'·01W.

ST MALO
Môle des Noires hd ⚑ VQ R 11m 6M; W twr, R top; *Horn (2) 20s*; 48°38'·52N 02°01'·91W.
Écluse du Naye ldg lts 070·4°, both FR 7/23m 3/7M. Front, 48°38'·58N 02°01'·48W. Rear, 030°-120°.
Ferry jetty hd, ⚑ VQ G 6M; 48°38'·44N 02°01'·82W (128·6° ldg line). Ferry jetty, ⚑ Fl R 4s 3m 1M; 260°-080°; 48°38'·44N 02°01'·76W.
Bas-Sablons marina, mole head ⚑ Fl G 4s 7m 5M; Gy mast; 48°38'·42N 02°01'·70W.

LA RANCE BARRAGE
La Jument ⚑ Fl G 4s 6m 4M; G twr, 48°37'·44N 02°01'·76W.
ZI 12 ⚓ Fl R 4s, 48°37'·47N 02°01'·62W.
NE dolphin ⚑ Fl (2) R 6s 6m 5M; 040°-200°, 48°37'·09N 02°01'·71W.
Barrage lock, NW wall ⚑ Fl (2) G 6s 6m 5M, 191°-291°; G pylon, 48°37'·06N 02°01'·73W.
Barrage lock, SW wall, ⚑ Fl (3) G 12s 6m 5M, 48°37'·00N 02°01'·70W.
SE dolphin ⚑ Fl (3) R 12s, 48°36'·97N 02°01'·66W.
ZI 24 ⚓ Fl (2) R 6s, 48°36'·63N 02°01'·33W.

ST MALO TO ST QUAY-PORTRIEUX
ST BRIAC
R. Frémur mouth. Dir lt 125° Iso WRG 4s 10m, W13M, R/G 11M; 121·5°-G-124·5°-W-125·5°-R-129·5°; W mast on hut, 48°37'·07N 02°08'·20W.

ST CAST
Les Bourdinots ⚓ 48°39'·01N 02°13'·48W.
St Cast môle ⚑ Iso WG 4s 12m, W9M, G6M; 180°-G-206°-W-217°-G-235°-W-245°-G-340°; G&W structure; 48°38'·41N 02°14'·61W.
Laplace ⚓, 48°39'·73N 2°16'·45W; 5ca SE of Pte de la Latte.
Cap Fréhel ☆ Fl (2) 10s 85m **29M**; Gy □ twr, G lantern; 48°41'·05N 02°19'·13W. Reserve lt range **15M**.

CHENAL and PORT D'ERQUY
Les Justières ⚓ Q (6) + L Fl 15s; 48°40'·56N 02°26'·48W.
Basses du Courant ⚓ VQ (6) + L Fl 10s; 48°39'·21N 02°29'·16W.
L'Evette ⚓ 48°38'·51N 02°31'·45W.
S môle ⚑ Fl (2) WRG 6s 11m W10M, R/G7M; 055°-R-081°-W-094°-G-111°-W-120°-R-134°; W twr; 48°38'·06N 02°28'·68W.
Inner jetty ⚑ Fl (3) R 12s 10m 3M; R/W twr; 48°38'·09N 02°28'·39W.

DAHOUET
Petit Bignon ⚓ 48°36'·82N 02°35'·06W.
Le Dahouet ⚓ 48°35'·15N 02°35'·43W.
La Petite Muette ⚑ Fl WRG 4s 10m W9M, R/G6M; 055°-G-114°-W-146°-R-196°; W twr, G band; 48°34'·82N 02°34'·30W.
Entry chan, Fl (2) G 6s 5m 1M; 156°-286°; 48°34'·71N 02°34'·19W.

BAIE DE ST BRIEUC
Grand Léjon ☆ Fl (5) WR 20s 17m **W18M**, R14M; 015°-R-058°-W-283°-R-350°-W-015°; R twr, W bands; 48°44'·91N 02°39'·87W. <u>992271325</u>.
Petit Léjon ⚓ ; 48°41'·80N 02°37'·55W.
Les Landas ⚓ Q, 48°41'·43N 02°31'·29W.
Le Rohein ⚓ Q (9) WRG 15s 13m, W8M, R/G5M; 072°-R-105°-W-180°-G-193°-W-237°-G-282°-W-301°-G-330°-W-072°; Y twr, B band; 48°38'·80N 02°37'·77W.

SAINT-BRIEUC LE LÉGUÉ
Tra-Hillion ⚓ 48°33'·38N 02°38'·50W.
Le Légué ⚓ Mo (A)10s; 48°34'·32N 02°41'·15W.
No. 1 ⚑ Fl G 2·5s; 48°32'·42N 02°42'·51W.
No. 2 ⚑ Fl R 2·5s; 48°32'·37N 02°42'·40W.
No. 3 ⚑ Fl (2) G 6s; 48°32'·27N 02°42'·78W.
No. 4 ⚑ Fl (2) R 6s; 48°32'·23N 02°42'·70W.
No. 5 ⚑ Fl (3) G 12s; 48°32'·18N 02°42'·92W.
No. 6 ⚑ Fl (3) R 12s; 48°32'·14N 02°42'·90W.
NE jetty ⚑ VQ R 4M; 48°32'·12N 02°42'·88W.
Pte à l'Aigle jetty ⚑ VQ G 13m 8M; 160°-070°; W twr, G top; 48°32'·12N 02°43'·11W.
No. 7 ⚑ Fl (4) G 15s; 48°32'·11N 02°43'·08W.
No. 8 ⚑ Fl R 2·5s; 48°32'·01N 02°43'·16W.
No. 9 ⚑ Fl (2) G 6s; 48°31'·96N 02°43'·29W.
No. 10 ⚑ Fl (2) R 6s; 48°31'·91N 02°43'·31W.
Jetty ⚑ Iso G 4s 6m 2M; W cols, G top; 48°31'·90N 02°43'·43W.
No. 11 ⚑ Fl (3) G 12s; 48°31'·90N 02°43'·43W.
No. 13 ⚑ Fl (4) G 15s; 48°31'·76N 02°43'·56W.
No. 14 ⚑ Fl (4) R 15s; 48°31'·70N 02°43'·61W.

BINIC
N môle ⚑ Oc (3) 12s 12m 11M; unintens 020°-110°; W twr, G lantern; 48°36'·07N 02°48'·92W.

ST QUAY-PORTRIEUX
Les Hors ⚓ 48°39'·60N 02°44'·04W.
Caffa ⚓ Q (3) 10s; 48°37'·82N 02°43'·08W.
La Longue ⚓ 48°37'·88N 02°44'·68W.
La Roselière ⚓ VQ (6) + L Fl 10s; 48°37'·31N 02°46'·19W.
Herflux ⚑ Dir ⚑ 130°, Fl (2) WRG 6s 10m, W 8M, R/G 6M; 115°-G-125°-W-135°-R-145°; 48°39'·07N 02°47'·95W.
Île Harbour (Roches de Saint-Quay) ⚑ Fl WRG 4s 16m, W9M, R/G 6M; 011°-R-133°-G-270°-R-306°-G-358°-W-011°; W twr & dwelling, R top; 48°39'·99N 02°48'·49W. <u>992271321</u>.
Madeux ⚓ 48°40'·41N 02°48'·81W.
Grandes Moulières de St Quay ⚓ 48°39'·76N 02°49'·91W.
Moulières de Portrieux ⚓ 48°39'·26N 02°49'·21W.
Les Noirs ⚓ 48°39'·09N 02°48·46W.
Marina, **NE mole elbow**, Dir lt 318·2°: Iso WRG 4s 16m **W15M**, R/G11M; W159°-179°, G179°-316°, W316-320·5°, R320·5° -159°; Reserve lt ranges 11/8M; 48°38'·99N 02°49'·09W. <u>992271322</u>.
NE môle hd ⚑ Fl (3) G 12s 10m 2M; 48°38'·84N 02°48'·91W.
S môle hd ⚑ Fl (3) R 12s 10m 2M; 48°38'·83N 02°49'·03W.
Old hbr ent: N side, Fl G 2.5s 11m 2M; 48°38'·71N 02°49'·36W.
S side, Fl R 2.5s 8m 2M; 48°38'·67N 02°49'·35W.

9.18.5 PASSAGE INFORMATION

More Passage Information is threaded between successive harbours in this Area. **Bibliography**: *Shell Channel Pilot* (Imray/Cunliffe); *North Brittany and Channel Islands CC* (Fernhurst/Cumberlidge); *North Brittany* (Imray/RCC/Lawson); *Channel Pilot* (NP 27); *Channel Havens* (ACN/Endean).

THE WEST COAST OF THE COTENTIN PENINSULA

This coast is exposed and often a lee shore (AC 3653, 3655, 3656, 3659). North of Carteret it is mostly rocky. Southward to Mont St Michel and W to St Malo the coast changes to extensive offshore shoals, sand dunes studded with rocks and a series of drying hbrs; there is little depth of water, so that a nasty sea can build.

▶*The seas around this coast are dominated by powerful tidal streams rotating anti-clockwise; and a very large tidal range. In the Alderney Race the main English Channel tidal streams are rectilinear NE/SW. Tidal streams need to be worked carefully; neaps are best, particularly for a first visit. The Admiralty tidal stream atlas NP 264 covers the French coast and Channel Islands. The equivalent SHOM 562-UJA gives more details.*◀

CAP DE LA HAGUE TO MONT ST MICHEL

Cap de La Hague is low-lying, but the ground rises steeply to the south. Between the coast and Gros du Raz lt ho a difficult and narrow passage leads S to Goury. It is sheltered when the Alderney Race is fully exposed, but is not for the inexperienced. The pylons and Jobourg nuclear plant are conspic 3·7M SE. CROSS Jobourg is adjacent.

5M S of Cap de la Hague beware Les Huquets de Jobourg, an extensive unmarked bank of drying (2·1m) and submerged rks, and Les Huquets de Vauville (5·4m) close SE of them.

The drying hbrs at **Goury, Carteret and Portbail** are more readily accessible if cruising from S to N on the flood. The marinas at **Diélette** and Carteret are non-tidal. The former has H24 access for 1·5m draught if coefficient is <80.

Déroute de Terre and Passage de la Déroute are coastal chans, poorly marked, shallow in places and not advised at night.

Déroute de Terre (not shown on AC 2669 and 3655) passes W of Plateau des Trois Grunes; Basses de Portbail; La Basse du Sénéquet WCM buoy; Les Nattes WCM buoy; Roches d'Agon; E of La Catheue SCM lt buoy; E of Iles Chausey; and W of Pte du Roc (Granville). Between Granville and Îles Chausey it carries <1m in places, but with adequate rise of tide this is not a problem for small craft; the *Channel Pilot* gives directions.

DIÉLETTE

Passage de la Déroute is used by ferries from the UK to St Malo. It passes W of Les Trois Grunes; between Basses de Taillepied and Les Écrehou; Chaussée des Boeufs and Plateau de l'Arconie; E of Les Minquiers and Les Ardentes; thence NW of Îles Chausey via Entrée de la Déroute, a side channel.

Granville has an excellent but very crowded half-tide marina. **Iles Chausey**, 8·3M WNW of Granville, is a popular and attractive archipelago. The Sound can be entered from the S or N with careful pilotage and sufficient rise of tide.

The drying expanse of Baie du Mont St Michel should not be entered by sea, but is an enjoyable visit from the mainland.

9.18.6 SPECIAL NOTES FOR FRANCE: see Area 17

MINOR HARBOUR CLOSE SOUTH OF CAP DE LA HAGUE

GOURY, Manche, **49°42'·85N 01°56'·83W**. AC 1114, 3653, 5604.2; SHOM 7158, 5636, 7133 (essential). HW –0410 on Dover (UT); ML 5·1m. See Omonville-la-Rogue. For visitors, appr at slack water nps with no swell and good vis; a fair weather hbr only, dries to flattish shingle. At tidal diamond M, 1·3M W, the N-S stream reaches 9·7kn at sp, 5·8kn at nps.

Cap de la Hague (Gros du Raz) lt ho is 0·5M NW of hbr; La Foraine WCM lt bn is 1·0M to the W. 065° ldg lts, by day: Front W patch with R ■ at end of bkwtr; rear, W pylon, lead between Diotret to S and Les Grios to N. ⚓ W of the 2 🛟 slips in 1·7m or dry out on shingle banks SE of the bkwtr.

☎02·33·52·85·92. Facilities: ✗ 🖵 at Auderville (0·5M).

9.18.7 DIÉLETTE

Manche, **49°33'·20N 01°51'·79W** ❀❀⌕♦♦♦❁❁

CHARTS AC 2669, 3653; SHOM 7158, 7133; Navi 528, 1014; Imray C33A

TIDES HW –0430 on Dover (UT); ML 5·4m

Standard Port ST MALO (→)

Times				Height (metres)			
High Water		Low Water		MHWS	MHWN	MLWN	MLWS
0100	0800	0300	0800	12·2	9·3	4·2	1·5
1300	2000	1500	2000				
Differences DIÉLETTE							
+0045	+0035	+0020	+0035	–2·5	–1·9	–0·7	–0·3

SHELTER Good in marina, but when retaining wall covers, boats surge fore and aft; rig good springs. No entry in strong W'lies.

NAVIGATION WPT 49°33'·45N 01°52'·16W, 140°/650m to W bkwtr lt. Appr is exposed to W'ly winds/swell; 2kn cross tide at hbr ent. Caution: drying rks 3ca E of ⊕, marked by unlit WCM buoy. 1·5M WSW, keep seaward of WCM lt buoy off Flamanville power stn. Outer hbr ent dredged CD +0.5m; with Coefficient <55 accessible for 2m draft. W side of outer hbr dries approx 5m.

LIGHTS AND MARKS The power stn is prominent 1·2M to SW, with pylons on the skyline conspic from afar. W bkwtr Lt ho, W twr/G top. Lts as chartlet.

COMMUNICATIONS CROSS 02·33·52·16·16; 🛟 02·33·04·93·17; ⚓ 08·92·68·08·50; Emergency 15; ⊖ 02·33·23·34·00; Brit Consul 01·44·51·31·00; Dr 02·33·52·99·00; HM VHF Ch 09, summer 0800-1300, 1330-2000LT, winter 0800-1200, 1330-1700LT; Aff Mar 02·33·23·36·00; YC 02·33·04·14·78; Tourist office ☎02·33·52·81·60.

FACILITIES Marina ☎02·33·53·68·78. portdielette@cc-lespieux. com www.cc-lespieux.fr 420+70 ❽ €3·00. Enter about HW±3 for 1·5m draft, tide gate over a sill drops when tide reaches CD+4.5m, tide gauge by sill; waiting pontoon outside. ❽ berths pontoon A and the E ends of B (S side only) and C (N side only). Rig stout springs to minimise surging around HW. ▲♠⚓🅿 ━ free wi-fi ⚑(40t). Ferry to CI.
Village, basic foods, mkt Sun am, ✗ 🖵. **Flamanville**, where third 1,750MW reactor is being built, 15 mins walk away, ▥ ✉.
Les Pieux ✗ 🖵 🛒 ✉ Ⓑ. Free daytime shuttle bus, Marina/ Flamanville/ Les Pieux, Jul/Aug, every 2 or 2½ hrs.

9.18.8 CARTERET

Manche **49°22'·08N 01°47'·33W** ❄❄♻♻♻♻✿✿

CHARTS AC 2669, 3655; SHOM 7157, 7158, 7133; Navi 1014; Imray C33A

TIDES –0440 Dover; ML 5·9; Duration 0545

Standard Port ST MALO (→)

Times				Height (metres)			
High Water		Low Water		MHWS	MHWN	MLWN	MLWS
0100	0800	0300	0800	12·2	9·3	4·2	1·5
1300	2000	1500	2000				
Differences CARTERET							
+0030	+0020	+0015	+0030	–1·6	–1·2	–0·7	–0·3
PORTBAIL							
+0030	+0025	+0025	+0030	–0·8	–0·6	–0·2	–0·1
ST GERMAIN-SUR-AY 49°13'·6N 01°39'·3W							
+0025	+0025	+0035	+0035	–0·7	–0·5	0·0	+0·1
LE SÉNÉQUET 49°05'·9N 01°41'·0W							
+0015	+0015	+0025	+0025	–0·3	–0·3	+0·1	+0·1

SHELTER Good in marina, but very crowded at weekends and in season. No safe ⚓s offshore.

NAVIGATION WPT 49°21'·86N 01°47'·36W, 006°/420m to W bkwtr lt. From N/NW, keep well off shore on appr to avoid rks 1M N of Cap de Carteret extending about 7ca from coast. From W beware Trois Grune Rks (dry 1·6m) about 4M offshore, marked by WCM lt buoy. Appr dries ½M offshore and is exposed to fresh W/SW winds which can make ent rough. Caution: strong cross-streams on the flood.

Best appr at HW–1 to avoid max tidal stream, 4½kn sp. The outer end of W bkwtr covers at big springs. Bar, at right angles to W bkwtr, dries 4m; a SHM buoy east of the W bkwtr marks a shifting shoal patch. Best water is to port of a mid-channel course. The

chan dries progressively to firm sand, and is dredged to 4m and 4·5m just W of the marina.

LIGHTS AND MARKS Cap de Carteret, grey lt twr, G top, and conspic Sig stn are 8ca 295° from the ent. Breakwater and channel lts as chartlet. Marina sill is marked by PHM/SHM lt bcns, and Y poles on the retaining wall.

COMMUNICATIONS CROSS/ⓁⒷ 02·33·52·16·16; ♨ 02·33·22·91·77; Auto 08·92·68·02·50; Police 02·33·53·80·17; ⊖ 02·33·04·90·08; Brit Consul 01·44·51·31·00; Ⓗ (Valognes) 02·33·40·14·39; Signal station VHF 09, 16; Marina VHF Ch 09; Tourist Office 02·33·04·90·58.

FACILITIES

Marina ☎02·33·04·70·84. www.barneville-carteret.net barneville-carteret @wanadoo.fr 311 + 60Ⓥ €2·95 (Apr-Sep; discounts possible Mon-Thu). Access HW ±2½ for 1·5m draft over sill 5m; lifting flapgate retains 2·3m within. IPTS N side of marina ent control ent/exit (sigs 2 & 4). Cross the sill squarely, ie heading NE, to clear the concrete bases each side. Ⓥ ⌐/rafted on E side only of pontoon F (no need to stop at accueil pontoon).

🛶⚓ at accueil pontoon (limited hrs, displayed on the pumps) ⛽ 🔧 ▣(35t) 🏗(24t), 🏴 ▬ ♨ 🚿.

West quay Possible waiting berth clear of ferry, if too late for the marina. A 1·5m draft boat can stay afloat for 6 hrs np, 9 hrs sp. ⌐ free for 6 hrs then at 50% of marina rates. ⚓ ▬.

Other options: Dry out in the tiny Port des Américains and basin close W of marina (up to 5m at HW); or on fine sand SW of marina.

Town ✕🏠🛒 🏪🏫⚓🔧✉ Ⓑ ⇌ (Valognes) ✈ (Cherbourg). Ferry: Cherbourg, Jersey.

9.18.9 PORTBAIL

Manche **49°19'·72N 01°42'·42W** (Basin ent) ✿✿⚓⚓✿✿✿

CHARTS AC 2669, 3655; SHOM 7157, 7133; Navi 1014; Imray C33A

TIDES See Carteret. HW –0440 on Dover; ML 6·3m; Duration 0545

SHELTER Good in all winds. Occasional swell after SW gales.

NAVIGATION WPT 49°18'·25N 01°44'·58W (off chartlet, abeam unlit 'PB' SWM buoy) 042°/1·5M to training wall lt bcn. 7ca offshore the 042° ldg line crosses banks drying 5.3m, thence between a pair of unlit PHM/SHM buoys, least depth 8.1m. Beware very strong tidal stream, esp during sp flood (4-5k) at basin ent. When base of training wall bcn is covered, there is > 2·5m in the chan, dredged 5·2m. Frequent inundations of sand reported, Y buoy marks the worst bank, but following a local vessel (of similar draught) is recommended.

LIGHTS AND MARKS A water twr (38m) is conspic 6ca NNW of ent. Ldg marks/lts 042°: Front (La Caillourie) W pylon, R top; rear, church belfry in town. Trng wall, which covers near HW, head: W mast/R top, followed by 2 R perches. See chartlet and Lights, buoys and waypoints.

COMMUNICATIONS CROSS 02·33·52·16·16; ⊖ 02·33·04·90·08 (Carteret); Brit Consul 01·44·51·31·00; Port/marina VHF Ch 09.

FACILITIES Yacht basin dries 7·0m, frequent dredging needed. Enlarged Ⓥ pontoon parallel to road on NW side of basin; also Ⓥ on the E side of adjacent stone FV jetty; or as directed. At NE side of basin pontoon with fingers is for locals <7.5m LOA.

HM ☎02·33·04·83·48 (15 Jun-31 Aug, tide times; 0830-1200 & 1400-1700). portbail@wanadoo.fr www.portbail.com 160 + 30Ⓥ €1.41. ◣ ✕ ⚒ ᐰ 🅑 ᐅ(5t). **YC** ☎02·33·04·86·15, ✕ 🗆 ⌂.

Town (½M by causeway) ✕ 🗆 🛒 📱 📶 ✉ Ⓑ ⇌ (Valognes).

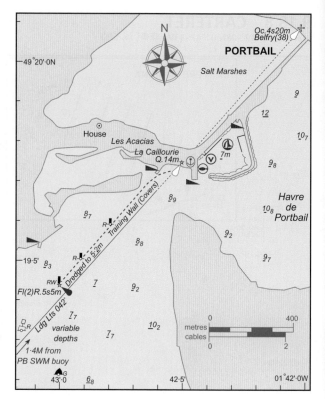

9.18.10 GRANVILLE

Manche **48°49'·94N 01°35'·93W** (Marina ent) ✿✿⚓⚓✿✿

See chartlet opposite

CHARTS AC 3656, 3659; SHOM 7156, 7341; Navi 534, 535; Imray C33B

TIDES –0510 Dover; ML 7·1; Duration 0525

Standard Port ST MALO (→)

Times				Height (metres)			
High Water		Low Water		MHWS	MHWN	MLWN	MLWS
0100	0800	0300	0800	12·2	9·3	4·2	1·5
1300	2000	1500	2000				
Differences REGNÉVILLE-SUR-MER							
+0010	+0010	+0030	+0020	+0·5	+0·4	+0·2	0·0
GRANVILLE							
+0010	+0005	+0025	+0015	+0·7	+0·4	+0·3	0·0
CANCALE							
0000	0000	+0010	+0010	+0·8	+0·6	+0·3	+0·1

SHELTER Good in the marina. Appr is rough in strong W winds. ⚓ in 2m about 4ca WSW of Le Loup to await the tide.

NAVIGATION WPT 48°49'·60N 01°36'·30W (abm Le Loup IDM) 042°/0·40M to S bkwtr lt. Beware rks seaward of La Fourchie WCM bcn twr, pot markers off Pte du Roc and 0·4m patches on Banc de Tombelaine, 1M SSW of Le Loup lt. At night best to keep at least 8ca W of Pte du Roc to avoid these dangers.

At bkwtr hd turn port wide into marina ent to avoid yachts leaving; cross the sill between R/G piles. Ent/exit under power; speed limit 5kn in the near approach, 3kn in hbr. The 5 R piles, Fl Bu 4s, mark the covering wall of a windsurfing/dinghy area to stbd.

LIGHTS AND MARKS Pte du Roc is a conspic headland with large old bldgs and lt ho, grey twr, R top; the Sig stn and church spire are also obvious. The twin domes of St Paul's church in transit 034° with Le Loup lead towards the marina ent. No ldg lts, but S bkwtr hd on with TV mast leads 057° to ent; hbr lts are hard to see against town lts. 3·5M W of Pte du Roc, Le Videcoq rock

drying 0·8m, is marked by a WCM lt buoy. See chartlet and Lights, buoys and waypoints for lt details.

COMMUNICATIONS CROSS 02·33·52·16·16; 📞 02·33·61·26·51; Auto 08·92·68·08·50; Police 02·33·91·27·50; ⊖ 02·33·50·19·90; Brit Consul 01·44·51·31·00; 🏥 02·33·91·50·00; Dr 02·33·50·00·07; Port HM 02·33·50·20·06, VHF Ch 09, 16 (HW±1½); Aff Mar 02·33·91·31·40; Marina VHF Ch 09, 0800-2200 Apr-Oct.

FACILITIES

Hérel Marina ☎02·33·50·20·06. www.granville.cci.fr/herel herel@granville.cci.fr. Depth over sill shown on digital display atop S bkwtr: eg 76 = 7·6m; 00 = no entry. 850+150 Ⓥ Jun-Sep, €3·34 in least depth 2.5m as indicated on SW side pontoon G & NW side pontoon H; in 1.5m pontoon V; (vessels >15m +€4·05 p/d, per additional metre); multihulls +50%. ◣ ᐱ ⚒ 🅟 ⛽ 🅟 ᐰ 🄱 ✕ 🔧 🅹 Ⓔ SHOM, free wi-fi, 🔧(12t) ᐅ(1·5t) ◣

YC de Granville ☎02·33·50·04·25, ✕.🗆 Port á Sec, St Nicolas ☎02·33·61·81·53; STL Nautisme ☎02·33·69·22·75.

Town ✕ 🗆 🛒 📱 📶 ⛽ ✉ Ⓑ 🔧 ⇌ ✈ (Dinard). Ferry: UK via Jersey or Cherbourg.

MINOR HARBOUR 10M NORTH OF GRANVILLE
REGNÉVILLE, Manche, **48°59'·72N 01°34'·05W** (SHM buoy abeam Pte d'Agon). AC 2669, 3656; SHOM 7156, 7133. HW –0500 on Dover (UT); ML 7·0m; Duration 0535. See Granville. A seriously drying hbr on the estuary of R. La Sienne; few yachts visit, a magnet for hardy adventurers seeking sand and solitude.

From 48°57'·65N 01°38'·86W (2·2M E of La Catheue SCM buoy) identify Pte d'Agon lt ho, W twr/R top, and Regnéville's 028° dir lt, both 4M NE at the river mouth (see Lights, buoys & waypoints). Thence track 056°/3·8M across the drying estuary to the SHM buoy (Lat/Long in line 1) marking a drying mole extending SW from the sandspit with landing stages to stbd; a small pontoon at Regneville is 1·2M NNE. Drying heights are from 9m-12·7m off Pte d'Agon.

Access HW –1 to +2 for 1·3m draft. Approx 80 moorings inc Ⓥs. YC ☎02·33·46·36·76 cnregnevillais@free.fr. Facilities: Quay, ᐅ(25t).

Town: ✕ 🗆 🛒 📱 📶✉ Ⓑ Tourist office ☎02·33·45·88·71.

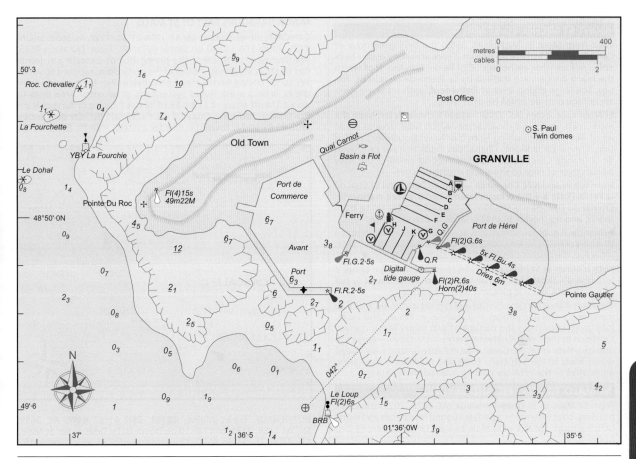

9.18.11   ÎLES CHAUSEY

Manche **48°52′·08N 01°49′·08W** SHM By, S ent ❀❀⚓🏠🏠🏠

CHARTS AC 3656, 3659; SHOM 7156, 7155, 7161, 7134; Navi 534, 535; Imray C33B

TIDES −0500 Dover; ML 7·4; Duration 0530

Standard Port ST MALO (→)

Times				Height (metres)			
High Water		Low Water		MHWS	MHWN	MLWN	MLWS
0100	0800	0300	0800	12·2	9·3	4·2	1·5
1300	2000	1500	2000				
Differences ÎLES CHAUSEY (Grande Île)							
+0005	+0005	+0015	+0015	+0·8	+0·7	+0·6	+0·4
LES ARDENTES (48°58′N 01°52′W, 6M NNW of Grande Île)							
+0010	+0010	+0020	+0010	0·0	−0·1	0·0	−0·1

SHELTER Good except in strong NW or SE winds. Grande Île is privately owned, but may be visited; it is not a Port of Entry.

NAVIGATION WPT 48°51′·43N 01°48′·55W, 332°/1·1M to Crabière lt. In transit with L'Enseigne, W bn tr, B top, leads 332° to Sound. 1 SHM lt buoy, thence between unlit cardinal bns.

The N chan needs adequate ht of tide (max drying ht 5·3m) SHOM 7134, a good Pilot and/or local knowledge, plus careful pilotage. L'Enseigne ≠ Grande Île lt ho leads 156° to the N ent; thence follow charted dogleg. No access 1/4–30/6 to a bird sanctuary, ie all E of line from Grande Île lt ho to L'Enseigne bcn twr (except Aneret).

LIGHTS AND MARKS Grande Île lt ho is conspic. La Crabière lt is on blackish stilts with Y top. See chartlet and Lights, bys & w'points.

COMMUNICATIONS CROSS 02·33·52·16·16; 🛟 02·33·50·28·33; Auto 08·92·68·08·50; Police 02·33·52·72·02.

FACILITIES Moor fore-and-aft to W ⚓s, free, in about 1·3m; some dry at sp. Very crowded w/ends in season, especially as drying out in Port Homard (W side of Grande Île) is discouraged. Tidal streams are moderate. **Village** ⚓ ✕ 🖂 (limited)🛒 🔋.

MONT ST MICHEL TO ST MALO

West from Mont St Michel (AC 3659) the drying harbour of **Cancale** lies 4M S of Pte du Grouin. There are many oyster beds and a fair weather anchorage SE of Île des Rimains, N of which the many dangers are marked by La Pierre-de-Herpin (lt, fog sig). The large drying inlet of **Rothéneuf**, with a fair weather anchorage outside, is 4M NE of St Malo.

St Malo approaches (AC 2700) include 6 channels between the many islets, rocks and shoals across which tidal streams set at up to 4kn sp. Approaching from E or N, with sufficient rise of tide, Chenal de la Bigne, Chenal des Petits Pointus or Chenal de la Grande Conchée can be used in good visibility, but they all pass over or near to shoal and drying patches.

From the W and NW, Chenal de la Grande Porte and Chenal de la Petite Porte are the principal routes and are well marked/lit. Chenal du Décollé is a shallow, ill-marked inshore route from the W, and no shorter than Chenal de la Grande Porte.

▶*Here the E-going stream begins at HW St Helier –0510, and the W-going at HW St Helier +0030, max sp rates 3·75kn. The streams set across the Chenal between Grand Jardin and Buron.*◀

Both marinas at St Malo have tidal restrictions due to either a lock or a sill. **Dinard** has a small basin 2·0m which is usually full of local boats, but it is pleasant to visit by ferry.

The **River Rance** barrage (SHOM 4233) gives access via a lock to many delightful anchorages and marinas at Plouër and Lyvet. It helps to know how the barrage affects water levels upriver. Continue to the lock at Châtelier, thence on to Dinan – subject to draught. Here enter the Canal d'Ille et Rance (Inland waterways) which leads to the S Brittany coast and provides a fascinating alternative to the offshore passage around NW Brittany.

ST MALO TO L'OST PIC AND BEYOND

From **St Malo to St Quay-Portrieux** (AC 2028, 2029) the coast has deep bays (often drying) a few rugged headlands and many offlying rocks. 6M NW of St Malo beware Le Vieux-Banc (1·2m) buoyed.

▶*Here the ESE-going stream begins at HW St Helier –0555, and the NW-going at HW St Helier –0015, sp rates 2·5kn. There are W-going eddies very close inshore on the E-going stream.*◀

Between St Malo and Cap Fréhel there are anchorages S of Île Agot in the apprs to the drying hbr of St Briac and in Baie de l'Arguenon, SE of **St Cast's** excellent all-tide marina. More anchorages may be found in Baie de la Fresnaye and off Pte de la Latte. From Cap Fréhel, a bold steep headland, to Cap d'Erquy there are plenty of inshore rocks but no worthwhile hbrs.

Between Cap d'Erquy and the various rocky patches close to seaward, Chenal d'Erquy runs WSW/ENE into the E side of the Baie de St Brieuc. **Erquy** is a pleasant, drying hbr much used by fishing boats. There are several rocky shoals within the Baie itself, some extending nearly 2M offshore.

Baie de St Brieuc is a 20M wide bight between Cap d'Erquy and L'Ost Pic. Principal features are: to the N, Grand Léjon a rocky shoal extending 2½ca W and 8ca NNE of its lt ho. Petit Léjon (3·1m) lies 3·5M SSE. Rohein lt bcn, 3M further S, is in the centre with rocky plateaux to E and W. A 3M wide chan leads W of these features and S into the shallow, partly-drying S end of the Bay and the interesting hbrs of **Dahouet, St-Brieuc Le Légué** and **Binic**.

On the W side of Bay de St-Brieuc, Roches de St Quay and offlying rocky patches extend 4M east from **St Quay-Portrieux** which can only be approached from NNW or SE. St Quay Marina is a good all-tide base from which to explore the drying hbrs around Baie de St Brieuc.

The next 2 paras are covered in greater detail in Area 20:

To the N and NE of L'Ost-Pic lt ho extensive offshore shoals guard the approaches to Paimpol, Ile de Bréhat and the Trieux river. Further N, the Plateau de Barnouic and Plateau des Roches Douvres (lt, fog sig) must be given a wide berth.

▶*Here the E-going stream begins at about HW St Malo –0400 and the W-going at about HW St Malo +0100 with Spring rates exceeding 4kn.*◀

MINOR HARBOURS EAST OF ST MALO

CANCALE, Ille-et-Vilaine, **48°40'·10N 01°51'·11W**. AC 3659; SHOM 7155, 7131. HW –0510 on Dover (UT); ML 7·2m; Duration 0535; see Granville. La Houle is the drying hbr of Cancale just inside Bay of Mont St Michel, 2·6M S of Pte du Grouin; exposed to winds SW to SE. Drying berths may be available. Pier hd lt, obsc'd from the N, is on a mast. Avoid oyster beds at all costs. Area dries to about 1M off-shore. ⚓ in 4m SE of Île des Rimains or further N off Pte Chatry. Facilities: **Quay** ⊞ ⚓ ⚒ ✕ ⛴(1·5t). **CN de Cancale** ☎02·99·89·90·22. **Town** (famous for oysters) ✕ ⌂ ⛐ ⛽ ⛽ ✉ Ⓑ.

ROTHENEUF, Ille-et-Vilaine, **48°41'·38N 01°57'·64W**. AC 3659; SHOM 7155, 7131. HW –0510 on Dover (UT); Tides as for St Malo; ML 7·0m; Duration 0540. Complete shelter in quiet, delightful sandy hbr which dries up to 8·6m. ⚓ outside in 4m just N of SHM bn marking ent. Rks on both sides of ent which is barely 230m wide. Safest to enter when rks uncovered. Ldg line at 163°, W side of Pte Benard and old windmill (white twr). There are no lts. Facilities: ⚓ ⚓. **Village** ✕ ⌂ ⛐ ⛽ ⛽.

9.18.12 ST MALO

Ille et Vilaine 48°38'·48N 02°01'·91W
Bassin Vauban and Bas Sablons Marinas ❁❁⚓⚓✿✿✿

CHARTS AC 2669, 3659, 2700; SHOM 7155, 7156, 7130; Navi 535; Imray C33B

TIDES –0506 Dover; ML 6·8; Duration 0535

SHELTER Good at Bas Sablons Marina, but in fresh W/NW'lies ♥ berths at outer ends of A & B pontoons are very uncomfortable. Excellent shelter in Bassin Vauban near the walled city.

NAVIGATION Care is needed due to strong tidal streams and many dangerous rocks close to the appr Chans, which from W to E (see chartlet and Lights, buoys and waypoints) are:

1. Grande Porte, from the W, lit. WPT 48°40'·17N 02°08'·49W.
2. Petite Porte, the main lit chan from NW. WPT 48°42'·44N 02°09'·04W (abm Vieux-Banc Est NCM buoy, Q).
3. Bunel, from NNW, lit. WPT 48°42'·29N 02°06'·51W.

The above chans converge at or near Le Grand Jardin lt ho, 48°40'·20N 02°04'·97W. Thence continue in the main fairway 129°/2·8M to Môle des Noires.

4. Grande Conchée, from N, unlit. WPT 48°42'·15N 02°02'·12W.
5. Petits Pointus, from NNE, unlit. WPT 48°42'·34N 02°00'·38W.
6. La Bigne, from NE, unlit. WPT 48°42'·67N 01°57'·25W.

The above chans converge at or near Roches aux Anglais SHM buoy, 48°39'·66N 02°02'·22W. Thence continue 221°/8ca to enter the main 129° fairway.

Central N France

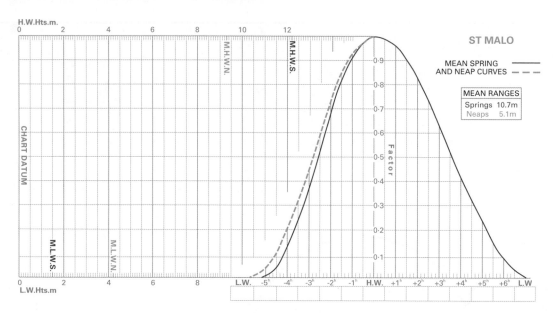

LIGHTS AND MARKS See chartlets/Lights, buoys and waypoints for ldg lts/buoys. Conspic daymarks: Île de Cézembre, Le Grand Jardin lt ho, Le Buron SHM twr, Petit and Grand Bé islets, the W lt twr/R top on Mole des Noires head and St Malo cathedral spire.

COMMUNICATIONS CROSS 02·98·89·31·31; ⚓ 02·99·46·10·46; Auto 08·92·68·08·35; Police 02·99·81·52·30; ⊖ 02·99·81·65·90; Brit Consul 01·44·51·31·00; Ⓗ 02·99·56·56·19; Port HM 02·99·20·63·01, *St Malo Port* Ch **12**; Aff Mar 02·99·56·87·00; Marinas VHF Ch 09.

FACILITIES Les Bas-Sablons Marina, sill 2m. Access for 1·5m draft approx HW −3½ to +4½ sp; H24 at nps. 2 W waiting bys outside. A gauge on S bkwtr head shows depths <3m over sill. A large digital gauge on N side of marina is visible from all berths. ☎02·99·81·71·34. (0900-1200, 1400-1730) port.plaisance@saint-malo.fr www.ville-saint-malo.fr 1216 + 64Ⓥ €3·15. Ⓥ berths on pontoon A: 32-66 (E side) and 43-75 (W side) and on pontoon B: 92-102 and 91-101. ⛽ ⛟ on pontoon I; use French cr/card or pay at office. ✕ ⚓ 🔌 Ⓔ 🔧 ⛴ SHOM ⚒ ⛟ ⚓ ⚓ ☰(20t) ⚓(2·5t). **YC** ⚓. **St Malo/St Servan** ✕ ⚓ 🍴 ⚓ ✉ Ⓑ ⚞. Ferry: Portsmouth, Poole, Weymouth.

Bassin Vauban (6m) is entered by lock; help given with warps. Outside the lock 3 waiting buoys are N of appr chan; keep clear of vedette and ferry berths.
Lock times may vary with traffic & tides. Lock is scheduled to operate 5 or 6 times in each direction, ie:

	−2½	−1½	−½	+½	+1½	+2½
Inward	−2½	−1½	−½	+½	+1½	+2½
Outward		−2	−1	HW	+1	+2

Lock sigs are IPTS Nos 2, 3 and 5. In addition:

🔵 next to the top lt = both lock gates are open, ie freeflow, beware current initially. Main message is the same. Freeflow is rare due to busy road traffic over retracting roller bridges.

🔴 🔴 over 🔵 = all movements prohib, big ship is outbound.

Port Vauban ☎02·99·56·51·91 portplaisancevauban@saint-malo.

cci.fr www.saint-malo.cci.fr 175 + 50Ⓥ ⚓ on S pontoon, €4.60 inc locking, ⚓ ⛴ ⛟ 🚻, multihulls +50%.C (1t). ⚓ in basins; speed limit 3kn. Complete redesign/rebuild/ of marina with new pontoons and fingers, spring 2015, has greatly improved berthing access. No ⚓ in basins; 3kn speed limit. ⚓(1t). **YC** ☎02·99·40·84·42, ⚓ (Ⓥ welcome). Bassin Duguay-Trouin is better for long stay.

DINARD Access by 1m marked chan to yacht basin 2m; see inset prev page, local boats only, no Ⓥ berthing. **HM** ☎02·99·46·65·55, ⚓ ⚓ ⛟ 🔧 Ⓔ ✕⚓. **YC** ☎ 02·99·46·14·32, ✕ ⚓. **Town** ✕ ⚓ 🍴 🛒 ⚓ ⚓ ✉ Ⓑ Ⓗ 02·99·46·18·68. ⚞ ✈. Ferries to St Malo.

ST MALO TIDAL COEFFICIENTS 2016

Date	Jan am	Jan pm	Feb am	Feb pm	Mar am	Mar pm	Apr am	Apr pm	May am	May pm	June am	June pm	July am	July pm	Aug am	Aug pm	Sept am	Sept pm	Oct am	Oct pm	Nov am	Nov pm	Dec am	Dec pm
1	52	47	41		47	42	39	36	43	46	63	69	68	73	78	82	90	92	90	90	84	82	80	79
2	43		37	35	37		37	40	50	56	75	81	77	82	86	90	93	93	90	89	81	78	78	76
3	40	37	34	36	34	33	46	53	63	71	87	92	86	90	92	93	92	90	87	85	76	72	74	71
4	37	38	39	44	35	40	61	70	78	86	96	100	93	95	94	93	88	85	82	78	69	64	68	64
5	40	43	50	57	46	53	79	88	93	100	102	103	96	96	92	90	82	77	75	70	60	55	61	57
6	48	53	64	71	61	70	96	103	105	109	103	102	96	94	87	83	73	68	65	60	51	46	54	51
7	58	63	78	84	78	87	109	114	112	113	100	97	92	88	79	75	62	57	55	49	42	39	49	
8	68	74	90	96	94	101	117	118	112	110	92	87	84	80	70	64	51	45	44	39	38		48	49
9	78	83	100	104	107	112	117	115	106	101	82	76	75	70	59	53	40			35	40	43	52	55
10	87	90	107	108	115	116	111	105	95	88	70	64	65		48		35	32	32	32	48	55	61	67
11	93	95	108	106	116	114	99	91	81	74	59		59	54	43	39	31	32	35	40	63	71	73	80
12	95	96	103	99	111	105	82	73	66		54	49	49	45	35	34	36	41	47	55	79	86	86	92
13	95	93	94	88	99	91	65		59	53	46	44	42	39	34	36	48	56	64	72	94	100	97	101
14	91	87	81	73	83	74	57	50	48	45	43	43	38	39	40	46	64	72	81	88	105	109	104	106
15	83	78	66		66	57	45	43	43	44	45	47	41	44	52	58	79	87	96	102	111	112	106	105
16	74	68	59	53	50		43	45	46	48	50	53	48	52	64	71	93	99	108	111	111	109	103	99
17	64		50	49	46	44	49	54	52	56	57	61	57	62	77	83	104	108	114	114	105	100	95	90
18	60	57	50	53	45	48	59	64	60	63	64	68	67	71	88	93	110	111	113	111	94	87	84	78
19	56	56	58	64	53	58	68	72	67	70	71	74	75	79	97	100	111	108	106	101	79	72	71	65
20	59	62	69	74	64	69	76	79	73	75	77	79	83	86	102	103	105	100	94	86	64	58	59	53
21	67	72	79	83	75	79	81	83	77	79	80	82	88	90	102	101	93	86	77	69	52		49	
22	76	81	87	89	82	85	84	85	80	81	82	83	91	91	98	94	78		61		48	46	45	43
23	85	88	91	92	87	89	85	85	81	81	82	81	91	90	89	83	70	61	54	48	45	46	42	43
24	90	92	92	91	89	89	84	82	80	79	80	78	87	85	77		54	49	45	45	49	52	45	47
25	92	92	90	87	89	87	80	78	77	75	75	73	81	77	70	63	46	46	47	51	56	60	51	55
26	91	89	85	81	86	83	75	71	72	69	70		73		58	53	49	54	55	60	63	67	58	62
27	87	84	78	73	80	77	67	63	65	62	67	64	68	64	50	50	59	65	65	70	70	73	66	69
28	80	76	68	63	73	68	58	53	58		62	60	60	58	53	57	70	76	74	77	75	77	72	75
29	71	67	58	53	63	58	49		56	54	60	60	57	57	62	68	80	84	80	82	78	79	77	79
30	61	56			52	47	46	44	53	53	62	65	60	63	73	79	86	88	84	84	80	80	81	82
31	51	46			42				55	59			68	73	83	87			85	85			82	82

Tidal coefficients indicate the magnitude of the tide on any particular day, without having to look up and calculate the range, and thus determine whether it is springs, neaps or somewhere in between. This table is valid for all areas covered by this Almanac. Typical values are:

120	Very big spring tide
95	**Mean spring tide**
70	Average tide
45	**Mean neap tide**
20	Very small neap tide

See also coefficients for Calais (p 759), Cherbourg (p 786), Brest (p 873) and Port Bloc (p 936).

Central N France

STANDARD TIME UT −01
Subtract 1 hour for UT
For French Summer Time add
ONE hour in **non-shaded areas**

ST MALO LAT 48°38′N LONG 2°02′W
TIMES AND HEIGHTS OF HIGH AND LOW WATERS

Dates in red are **SPRINGS**
Dates in blue are **NEAPS**

YEAR 2016

JANUARY

Time	m	Time	m
1 0539	3.8	**16** 0555	2.6
1109	9.9	1124	11.0
F 1801	3.9	SA 1823	2.7
2337	9.5	2351	10.5
2 0617	4.3	**17** 0644	3.2
1153	9.3	1219	10.3
SA 1843	4.4	SU 1917	3.3
3 0028	9.0	**18** 0051	10.0
0709	4.7	0746	3.7
SU 1253	8.9	M 1328	9.8
1941	4.7	2024	3.6
4 0141	8.7	**19** 0207	9.7
0820	4.9	0903	3.8
M 1411	8.8	TU 1451	9.7
2100	4.7	2142	3.6
5 0301	8.9	**20** 0329	9.9
0939	4.6	1025	3.5
TU 1527	9.1	W 1609	10.0
2215	4.3	2259	3.3
6 0405	9.5	**21** 0440	10.4
1045	4.0	1138	3.0
W 1628	9.6	TH 1715	10.5
2315	3.7		
7 0457	10.1	**22** 0006	2.7
1140	3.4	0539	11.0
TH 1719	10.2	F 1240	2.4
		1810	11.1
8 0007	3.1	**23** 0102	2.3
0543	10.8	0629	11.5
F 1231	2.8	SA 1332	2.0
1805	10.8	1857	11.5
9 0055	2.6	**24** 0150	1.9
0627	11.3	0713	11.9
SA 1318	2.2	SU 1417	1.7
1848	11.3	○ 1940	11.8
10 0141	2.1	**25** 0231	1.7
0708	11.8	0752	12.1
SU 1405	1.8	M 1456	1.6
● 1930	11.7	2018	11.9
11 0225	1.8	**26** 0308	1.7
0749	12.1	0829	12.1
M 1450	1.5	TU 1531	1.6
2012	12.0	2052	11.8
12 0308	1.6	**27** 0341	1.8
0830	12.3	0902	12.0
TU 1533	1.3	W 1603	1.9
2053	12.1	2125	11.5
13 0351	1.5	**28** 0412	2.1
0911	12.4	0934	11.6
W 1615	1.4	TH 1631	2.3
2135	12.0	2155	11.1
14 0431	1.7	**29** 0439	2.6
0953	12.1	1004	11.1
TH 1656	1.6	F 1657	2.8
2217	11.6	2224	10.6
15 0512	2.1	**30** 0505	3.2
1037	11.6	1033	10.5
F 1738	2.1	SA 1722	3.4
2302	11.1	2254	10.0
		31 0533	3.8
		1105	9.8
		SU 1752	4.0
		2328	9.4

FEBRUARY

Time	m	Time	m
1 0610	4.3	**16** 0019	10.0
1145	9.2	0717	3.7
M 1834	4.5	TU 1257	9.6
◑		1953	3.9
2 0018	8.8	**17** 0133	9.4
0705	4.8	0834	4.1
TU 1249	8.7	W 1427	9.2
1938	4.9	2116	4.1
3 0142	8.6	**18** 0307	9.3
0825	4.9	1006	4.0
W 1424	8.6	TH 1557	9.4
2110	4.8	2243	3.8
4 0315	8.5	**19** 0427	9.9
0955	4.5	1126	3.4
TH 1548	9.0	F 1707	10.1
2233	4.3	2353	3.1
5 0424	9.5	**20** 0526	10.6
1105	3.8	1227	2.7
F 1651	9.8	SA 1758	10.8
2335	3.5		
6 0518	10.4	**21** 0048	2.5
1204	3.0	0614	11.3
SA 1743	10.6	SU 1317	2.1
		1842	11.3
7 0031	2.7	**22** 0133	2.1
0607	11.2	0656	11.7
SU 1258	2.2	M 1359	1.8
1831	11.3	○ 1921	11.7
8 0123	2.0	**23** 0212	1.8
0652	11.9	0734	12.0
M 1350	1.5	TU 1436	1.6
● 1916	11.9	1957	11.9
9 0212	1.4	**24** 0247	1.6
0736	12.5	0808	12.1
TU 1438	1.0	W 1508	1.6
1959	12.4	2029	11.9
10 0259	1.0	**25** 0318	1.7
0818	12.9	0839	12.1
W 1524	0.7	TH 1538	1.7
2042	12.7	2058	11.8
11 0342	0.8	**26** 0346	1.9
0900	13.0	0908	11.8
TH 1606	0.6	F 1604	2.0
2123	12.7	2126	11.5
12 0423	0.9	**27** 0411	2.2
0941	12.8	0935	11.4
F 1645	1.0	SA 1627	2.5
2203	12.3	2152	11.0
13 0501	1.4	**28** 0435	2.7
1022	12.2	1002	10.9
SA 1724	1.6	SU 1649	3.0
2243	11.7	2218	10.5
14 0540	2.1	**29** 0500	3.3
1105	11.4	1028	10.2
SU 1803	2.4	M 1715	3.6
2326	10.8	2245	9.9
15 0622	2.9		
1153	10.5		
M 1850	3.2		
◑			

MARCH

Time	m	Time	m
1 0531	3.9	**16** 0653	3.7
1100	9.5	1232	9.4
TU 1750	4.2	W 1925	4.1
2322	9.2		
2 0616	4.5	**17** 0104	9.2
1148	8.9	0808	4.3
W 1842	4.8	TH 1405	8.9
◐		2051	4.5
3 0026	8.6	**18** 0244	9.0
0723	4.9	0945	4.3
TH 1317	8.4	F 1542	9.2
2005	5.0	2222	4.1
4 0217	8.5	**19** 0407	9.6
0902	4.7	1107	3.6
F 1507	8.7	SA 1648	9.9
2148	4.6	2332	3.4
5 0348	9.2	**20** 0506	10.3
1029	4.0	1206	2.9
SA 1622	9.5	SU 1737	10.6
2304	3.7		
6 0451	10.1	**21** 0025	2.7
1135	3.0	0552	11.0
SU 1720	10.5	M 1253	2.3
		1818	11.2
7 0006	2.8	**22** 0109	2.3
0544	11.1	0632	11.5
M 1235	2.1	TU 1333	2.0
1810	11.4	1856	11.6
8 0102	1.9	**23** 0147	1.9
0632	12.0	0709	11.8
TU 1330	1.3	W 1409	1.8
1857	12.2	○ 1930	11.8
9 0155	1.1	**24** 0221	1.8
0718	12.8	0742	12.0
W 1421	0.6	TH 1440	1.7
● 1941	12.8	2001	11.9
10 0243	0.6	**25** 0251	1.7
0802	13.2	0813	12.0
TH 1507	0.3	F 1509	1.8
2024	13.1	2030	11.8
11 0327	0.4	**26** 0319	1.8
0844	13.4	0841	11.8
F 1550	0.2	SA 1535	2.0
2105	13.1	2057	11.7
12 0408	0.5	**27** 0345	2.1
0925	13.1	0908	11.5
SA 1629	0.6	SU 1559	2.3
2144	12.7	2123	11.3
13 0446	1.0	**28** 0410	2.5
1005	12.5	0934	11.1
SU 1706	1.4	M 1622	2.8
2223	12.0	2148	10.8
14 0523	1.8	**29** 0435	3.0
1045	11.5	1001	10.5
M 1743	2.3	TU 1649	3.4
2304	11.0	2216	10.2
15 0603	2.8	**30** 0506	3.6
1131	10.4	1033	9.8
TU 1826	3.3	W 1722	4.0
◐ 2353	10.0	2251	9.6
		31 0547	4.1
		1117	9.2
		TH 1810	4.5
		◐ 2346	9.0

APRIL

Time	m	Time	m
1 0648	4.6	**16** 0210	9.0
1232	8.6	0912	4.4
F 1923	4.9	SA 1509	9.0
		2148	4.3
2 0125	8.7	**17** 0334	9.3
0818	4.6	1031	3.9
SA 1426	8.7	SU 1618	9.6
2106	4.6	2257	3.7
3 0308	9.1	**18** 0434	10.0
0952	4.0	1130	3.3
SU 1549	9.5	M 1707	10.3
2230	3.8	2350	3.1
4 0419	10.1	**19** 0521	10.6
1104	3.0	1217	2.7
M 1651	10.5	TU 1748	10.9
2336	2.8		
5 0516	11.1	**20** 0035	2.6
1208	2.0	0602	11.1
TU 1744	11.5	W 1258	2.4
		1825	11.3
6 0036	1.8	**21** 0114	2.3
0607	12.1	0639	11.4
W 1306	1.2	TH 1335	2.1
1833	12.3	1900	11.6
7 0132	1.0	**22** 0150	2.1
0655	12.8	0713	11.6
TH 1358	0.6	F 1408	2.0
● 1918	12.9	○ 1932	11.7
8 0222	0.5	**23** 0222	2.0
0740	13.2	0745	11.7
F 1446	0.2	SA 1439	2.0
2002	13.2	2001	11.8
9 0308	0.3	**24** 0252	2.0
0824	13.3	0814	11.6
SA 1529	0.3	SU 1507	2.1
2044	13.2	2029	11.7
10 0350	0.5	**25** 0321	2.1
0906	13.1	0843	11.5
SU 1609	0.7	M 1535	2.3
2124	12.8	2057	11.4
11 0429	1.0	**26** 0349	2.3
0947	12.4	0912	11.2
M 1646	1.4	TU 1602	2.7
2203	12.0	2126	11.1
12 0506	1.8	**27** 0418	2.7
1028	11.5	0943	10.7
TU 1723	2.4	W 1631	3.1
2244	11.1	2158	10.6
13 0546	2.7	**28** 0452	3.2
1113	10.4	1019	10.1
W 1806	3.4	TH 1706	3.7
2332	10.1	2237	10.0
14 0634	3.7	**29** 0533	3.7
1211	9.4	1105	9.5
TH 1901	4.2	F 1753	4.2
◐		2331	9.4
15 0038	9.3	**30** 0629	4.1
0742	4.3	1213	9.0
F 1336	8.9	SA 1900	4.5
2021	4.6		

Chart Datum is 6·29 metres below IGN Datum. HAT is 13·6 metres above Chart Datum.

》》 **FREE** monthly updates. Register at 《
www.reedsnauticalalmanac.co.uk 《

STANDARD TIME UT –01
Subtract 1 hour for UT
For French Summer Time add
ONE hour in **non-shaded areas**

ST MALO LAT 48°38′N LONG 2°02′W
TIMES AND HEIGHTS OF HIGH AND LOW WATERS

Dates in red are **SPRINGS**
Dates in blue are NEAPS

YEAR 2016

Central N France

MAY

Time	m		Time	m
1 0053	9.1	**16** 0244	9.1
0748	4.2	0937	4.2
SU 1347	9.1	M 1530	9.3
2031	4.4	2207	4.1
2 0226	9.4	**17** 0350	9.5
0915	3.8	1040	3.8
M 1511	9.6	TU 1625	9.9
2154	3.7	2304	3.6
3 0341	10.1	**18** 0442	10.1
1030	3.0	1132	3.3
TU 1617	10.5	W 1710	10.4
2304	2.8	2353	3.1
4 0443	11.0	**19** 0526	10.5
1136	2.1	1217	2.9
W 1714	11.4	TH 1751	10.9
5 0007	2.0	**20** 0036	2.7
0539	11.9	0606	10.9
TH 1237	1.4	F 1258	2.6
1805	12.2	1827	11.2
6 0105	1.3	**21** 0115	2.5
0630	12.5	0643	11.2
F 1332	0.9	SA 1335	2.4
● 1853	12.7	○ 1902	11.5
7 0158	0.8	**22** 0152	2.3
0718	12.9	0718	11.3
SA 1422	0.6	SU 1410	2.3
1939	13.0	1934	11.6
8 0246	0.6	**23** 0226	2.2
0804	12.9	0750	11.4
SU 1507	0.6	M 1442	2.3
2022	13.0	2006	11.6
9 0331	0.7	**24** 0300	2.1
0848	12.7	0823	11.4
M 1548	1.0	TU 1515	2.3
2104	12.6	2038	11.5
10 0411	1.1	**25** 0333	2.2
0931	12.1	0856	11.2
TU 1627	1.6	W 1547	2.5
2145	12.0	2111	11.3
11 0451	1.8	**26** 0407	2.5
1013	11.4	0932	10.9
W 1705	2.4	TH 1621	2.8
2227	11.2	2148	11.0
12 0531	2.7	**27** 0444	2.8
1057	10.5	1012	10.5
TH 1746	3.3	F 1659	3.3
2313	10.3	2231	10.5
13 0615	3.5	**28** 0527	3.2
1149	9.7	1100	10.0
F 1836	4.0	SA 1746	3.7
◐		2323	10.0
14 0009	9.6	**29** 0619	3.6
0711	4.1	1159	9.6
SA 1256	9.1	SU 1845	4.0
1941	4.5	◑	
15 0124	9.1	**30** 0030	9.7
0823	4.4	0725	3.7
SU 1418	9.0	M 1313	9.5
2058	4.5	2000	4.0
		31 0149	9.7
		0841	3.5
		TU 1431	9.8
		2119	3.6

JUNE

Time	m		Time	m
1 0304	10.2	**16** 0355	9.4
0956	3.0	1042	3.9
W 1541	10.4	TH 1628	9.8
2231	3.0	2308	3.7
2 0411	10.8	**17** 0447	9.9
1105	2.4	1134	3.5
TH 1643	11.1	F 1714	10.3
2337	2.3	2357	3.2
3 0511	11.4	**18** 0532	10.4
1209	1.9	1221	3.1
F 1739	11.8	SA 1755	10.8
4 0039	1.7	**19** 0042	2.8
0607	12.0	0614	10.7
SA 1307	1.4	SU 1304	2.8
1831	12.3	1834	11.2
5 0135	1.3	**20** 0124	2.5
0658	12.3	0653	11.0
SU 1359	1.2	M 1344	2.5
● 1919	12.5	○ 1911	11.4
6 0226	1.1	**21** 0204	2.3
0747	12.4	0730	11.2
M 1446	1.2	TU 1422	2.3
2004	12.6	1947	11.6
7 0313	1.1	**22** 0243	2.1
0833	12.3	0807	11.4
TU 1529	1.4	W 1500	2.2
2047	12.4	2024	11.7
8 0355	1.4	**23** 0322	2.0
0916	11.9	0845	11.4
W 1609	1.8	TH 1537	2.2
2129	12.0	2101	11.7
9 0435	1.8	**24** 0400	2.1
0957	11.4	0925	11.3
TH 1647	2.4	F 1615	2.4
2209	11.4	2141	11.5
10 0513	2.5	**25** 0440	2.3
1038	10.7	1006	11.1
F 1726	3.0	SA 1655	2.7
2251	10.7	2224	11.1
11 0552	3.2	**26** 0522	2.6
1121	10.1	1051	10.7
SA 1806	3.7	SU 1738	3.0
2336	10.0	2312	10.7
12 0634	3.8	**27** 0608	3.0
1211	9.5	1142	10.3
SU 1855	4.2	M 1829	3.4
◐		◑	
13 0030	9.4	**28** 0008	10.3
0727	4.3	0703	3.3
M 1312	9.1	TU 1242	10.0
1957	4.5	1931	3.7
14 0138	9.1	**29** 0115	10.0
0832	4.4	0810	3.4
TU 1425	9.0	W 1353	9.9
2106	4.5	2045	3.6
15 0251	9.1	**30** 0230	10.1
0940	4.3	0924	3.3
W 1533	9.3	TH 1507	10.2
2212	4.2	2202	3.3

JULY

Time	m		Time	m
1 0343	10.4	**16** 0406	9.3
1037	2.9	1052	4.1
F 1617	10.7	SA 1637	9.8
2313	2.8	2320	3.7
2 0450	10.9	**17** 0500	9.9
1145	2.5	1146	3.5
SA 1719	11.2	SU 1726	10.4
3 0019	2.2	**18** 0011	3.1
0550	11.3	0548	10.4
SU 1246	2.0	M 1235	3.0
1814	11.7	1810	11.0
4 0118	1.8	**19** 0058	2.6
0645	11.7	0632	10.9
M 1341	1.7	TU 1321	2.6
● 1904	12.1	○ 1851	11.4
5 0210	1.5	**20** 0144	2.2
0734	11.9	0713	11.3
TU 1429	1.6	W 1405	2.2
1949	12.3	1931	11.8
6 0257	1.4	**21** 0228	1.9
0818	12.0	0754	11.6
W 1512	1.6	TH 1447	2.0
2032	12.3	2011	12.1
7 0338	1.5	**22** 0311	1.7
0859	11.9	0834	11.9
TH 1551	1.8	F 1529	1.8
2111	12.0	2051	12.2
8 0416	1.8	**23** 0353	1.6
0938	11.6	0914	11.9
F 1627	2.2	SA 1609	1.8
2148	11.6	2131	12.1
9 0450	2.3	**24** 0433	1.7
1014	11.1	0955	11.8
SA 1700	2.7	SU 1648	2.0
2224	11.0	2213	11.8
10 0523	2.9	**25** 0512	2.0
1049	10.5	1036	11.4
SU 1732	3.4	M 1727	2.5
2300	10.4	2256	11.3
11 0554	3.5	**26** 0554	2.5
1126	9.9	1121	10.8
M 1807	4.0	TU 1811	3.0
2340	9.7	2345	10.7
12 0631	4.1	**27** 0641	3.1
1210	9.3	1213	10.3
TU 1852	4.5	W 1905	3.5
◐		◑	
13 0030	9.1	**28** 0045	10.1
0720	4.5	0741	3.5
W 1309	8.9	TH 1319	9.9
1954	4.8	2016	3.8
14 0139	8.8	**29** 0201	9.8
0829	4.7	0855	3.7
TH 1426	8.9	F 1440	9.8
2111	4.7	2139	3.7
15 0258	8.9	**30** 0324	9.9
0947	4.6	1015	3.5
F 1540	9.2	SA 1559	10.2
2222	4.3	2258	3.3
		31 0439	10.3
		1129	3.0
		SU 1706	10.8

AUGUST

Time	m		Time	m
1 0007	2.6	**16** 0524	10.3
0541	10.9	1208	3.2
M 1232	2.5	TU 1747	10.9
1802	11.4		
2 0105	2.1	**17** 0034	2.6
0634	11.4	0611	11.0
TU 1327	2.1	W 1259	2.5
● 1851	11.9	1831	11.6
3 0156	1.8	**18** 0124	2.0
0720	11.8	0655	11.6
W 1414	1.8	TH 1347	2.0
1934	12.2	○ 1914	12.1
4 0240	1.6	**19** 0213	1.5
0801	11.9	0737	12.1
TH 1454	1.7	F 1433	1.5
2013	12.3	1955	12.6
5 0319	1.6	**20** 0258	1.2
0839	11.9	0818	12.4
F 1530	1.8	SA 1517	1.3
2049	12.2	2036	12.8
6 0353	1.8	**21** 0340	1.0
0913	11.8	0858	12.5
SA 1602	2.0	SU 1557	1.3
2123	11.9	2116	12.7
7 0423	2.2	**22** 0420	1.2
0945	11.4	0938	12.4
SU 1631	2.5	M 1635	1.5
2154	11.4	2156	12.4
8 0449	2.7	**23** 0458	1.6
1015	10.9	1018	11.9
M 1656	3.1	TU 1713	2.1
2224	10.7	2237	11.7
9 0513	3.3	**24** 0536	2.3
1045	10.3	1059	11.2
TU 1722	3.7	W 1753	2.8
2255	10.0	2322	10.8
10 0540	3.9	**25** 0619	3.1
1117	9.7	1147	10.4
W 1755	4.3	TH 1842	3.5
◖ 2331	9.4	◗	
11 0617	4.5	**26** 0019	10.0
1201	9.1	0715	3.8
TH 1844	4.8	F 1251	9.7
		1952	4.1
12 0027	8.8	**27** 0140	9.4
0713	4.9	0833	4.2
F 1312	8.6	SA 1420	9.4
1958	5.1	2123	4.1
13 0156	8.5	**28** 0314	9.4
0840	5.1	1002	4.0
SA 1448	8.7	SU 1548	9.8
2132	4.8	2250	3.6
14 0326	8.8	**29** 0432	10.0
1010	4.6	1119	3.4
SU 1603	9.3	M 1656	10.5
2245	4.1	2357	2.8
15 0431	9.5	**30** 0531	10.7
1114	3.9	1220	2.7
M 1659	10.1	TU 1749	11.3
2342	3.4		
		31 0052	2.2
		0619	11.3
		W 1311	2.2
		1834	11.8

Chart Datum is 6·29 metres below IGN Datum. HAT is 13·6 metres above Chart Datum.

》》 FREE monthly updates. Register at 《
www.reedsnauticalalmanac.co.uk

807

St Malo tides – River Rance

STANDARD TIME UT –01
Subtract 1 hour for UT
For French Summer Time add
ONE hour in **non-shaded areas**

ST MALO LAT 48°38'N LONG 2°02'W
TIMES AND HEIGHTS OF HIGH AND LOW WATERS

Dates in red are **SPRINGS**
Dates in blue are **NEAPS**

YEAR **2016**

SEPTEMBER

Time	m	Time	m
1 0139	1.9	**16** 0102	1.8
0701	11.8	0632	11.9
TH 1354	1.9	F 1326	1.7
● 1915	12.1	○ 1852	12.5
2 0219	1.7	**17** 0152	1.2
0739	12.0	0715	12.5
F 1432	1.8	SA 1414	1.2
1951	12.3	1935	13.0
3 0254	1.7	**18** 0239	0.8
0813	12.0	0757	12.9
SA 1504	1.8	SU 1500	0.9
2024	12.2	2017	13.2
4 0324	1.8	**19** 0322	0.6
0844	11.9	0838	13.0
SU 1534	2.0	M 1541	0.9
2054	12.0	2058	13.1
5 0351	2.1	**20** 0402	0.9
0913	11.6	0918	12.8
M 1559	2.3	TU 1620	1.2
2122	11.6	2138	12.6
6 0415	2.6	**21** 0440	1.4
0940	11.2	0957	12.2
TU 1623	2.8	W 1657	1.8
2149	11.0	2219	11.8
7 0436	3.1	**22** 0517	2.2
1006	10.6	1038	11.4
W 1646	3.4	TH 1737	2.7
2215	10.3	2303	10.8
8 0500	3.7	**23** 0559	3.2
1032	10.0	1125	10.4
TH 1714	4.0	F 1824	3.6
2245	9.6	◑ 2359	9.8
9 0531	4.3	**24** 0654	4.0
1106	9.3	1229	9.6
F 1755	4.6	SA 1934	4.3
◑ 2327	8.9		
10 0618	4.9	**25** 0124	9.1
1202	8.7	0815	4.5
SA 1857	5.1	SU 1403	9.2
		2110	4.3
11 0050	8.4	**26** 0303	9.2
0735	5.3	0948	4.3
SU 1352	8.5	M 1534	9.6
2035	5.1	2237	3.8
12 0246	8.6	**27** 0421	9.9
0923	5.0	1104	3.6
M 1527	9.0	TU 1639	10.4
2208	4.4	2340	3.0
13 0402	9.3	**28** 0513	10.6
1041	4.2	1201	2.9
TU 1630	9.9	W 1729	11.1
2312	3.5		
14 0458	10.2	**29** 0030	2.4
1140	3.2	0557	11.3
W 1721	10.9	TH 1248	2.3
		1812	11.7
15 0008	2.5	**30** 0113	2.0
0546	11.2	0636	11.7
TH 1234	2.4	F 1328	2.0
1808	11.8	1850	12.0

OCTOBER

Time	m	Time	m
1 0151	1.9	**16** 0128	1.0
0712	11.9	0650	12.8
SA 1404	1.9	SU 1352	1.0
● 1925	12.1	○ 1913	13.1
2 0224	1.8	**17** 0216	0.6
0745	12.0	0734	13.2
SU 1435	1.9	M 1439	0.7
1956	12.1	1956	13.3
3 0253	1.9	**18** 0301	0.5
0814	11.9	0816	13.2
M 1504	2.0	TU 1523	0.7
2025	11.9	2039	13.2
4 0320	2.1	**19** 0343	0.8
0841	11.7	0857	13.0
TU 1530	2.3	W 1604	1.1
2052	11.6	2121	12.7
5 0344	2.5	**20** 0422	1.4
0907	11.4	0938	12.3
W 1554	2.7	TH 1643	1.7
2118	11.2	2203	11.8
6 0407	3.0	**21** 0501	2.3
0933	10.9	1020	11.5
TH 1619	3.2	F 1724	2.6
2145	10.6	2248	10.8
7 0430	3.5	**22** 0543	3.2
0959	10.3	1108	10.5
F 1647	3.7	SA 1811	3.5
2214	9.9	◑ 2344	9.8
8 0501	4.1	**23** 0637	4.1
1031	9.7	1210	9.6
SA 1725	4.3	SU 1916	4.2
2253	9.2		
9 0544	4.7	**24** 0103	9.1
1119	9.0	0752	4.5
SU 1820	4.8	M 1337	9.2
◑		2044	4.4
10 0001	8.6	**25** 0237	9.1
0652	5.1	0921	4.4
M 1254	8.6	TU 1506	9.4
1946	5.0	2207	4.0
11 0201	8.6	**26** 0354	9.7
0834	5.0	1034	3.8
TU 1445	9.0	W 1611	10.1
2125	4.4	2309	3.3
12 0327	9.3	**27** 0445	10.4
1003	4.3	1130	3.2
W 1555	9.8	TH 1701	10.7
2238	3.5	2358	2.8
13 0427	10.3	**28** 0528	11.0
1109	3.3	1216	2.7
TH 1650	10.9	F 1743	11.3
2339	2.5		
14 0518	11.3	**29** 0040	2.4
1206	2.3	0607	11.5
F 1740	11.8	SA 1256	2.3
		1822	11.6
15 0035	1.6	**30** 0118	2.2
0605	12.1	0643	11.7
SA 1300	1.5	SU 1333	2.1
1827	12.6	● 1857	11.8
		31 0152	2.1
		0715	11.8
		M 1406	2.1
		1929	11.8

NOVEMBER

Time	m	Time	m
1 0222	2.1	**16** 0240	0.7
0745	11.8	0756	13.1
TU 1436	2.1	W 1505	0.8
1959	11.7	2022	12.9
2 0251	2.2	**17** 0324	0.9
0813	11.7	0840	12.9
W 1505	2.3	TH 1549	1.0
2027	11.5	2107	12.5
3 0318	2.5	**18** 0406	1.4
0841	11.5	0923	12.4
TH 1533	2.5	F 1631	1.6
2055	11.2	2151	11.8
4 0345	2.8	**19** 0447	2.2
0909	11.1	1006	11.6
F 1601	2.9	SA 1713	2.4
2125	10.8	2236	10.9
5 0412	3.3	**20** 0529	3.0
0939	10.6	1053	10.7
SA 1632	3.4	SU 1758	3.2
2157	10.2	2327	10.0
6 0445	3.8	**21** 0618	3.8
1014	10.1	1147	9.9
SU 1710	3.9	M 1852	3.9
2238	9.6	◑	
7 0527	4.3	**22** 0030	9.4
1101	9.4	0719	4.4
M 1801	4.3	TU 1257	9.3
◑ 2339	9.0	2000	4.3
8 0627	4.7	**23** 0150	9.1
1217	9.0	0834	4.5
TU 1912	4.5	W 1418	9.2
		2116	4.2
9 0113	8.9	**24** 0307	9.3
0752	4.8	0946	4.2
W 1356	9.1	TH 1530	9.6
2041	4.2	2222	3.9
10 0243	9.4	**25** 0407	9.9
0921	4.2	1047	3.7
TH 1514	9.8	F 1625	10.1
2200	3.5	2316	3.4
11 0350	10.2	**26** 0454	10.4
1033	3.3	1137	3.2
F 1616	10.7	SA 1711	10.6
2305	2.6		
12 0446	11.2	**27** 0002	2.9
1135	2.4	0536	10.9
SA 1710	11.6	SU 1221	2.8
		1753	11.0
13 0005	1.8	**28** 0043	2.6
0537	12.0	0613	11.3
SU 1233	1.6	M 1301	2.5
1801	12.4	1830	11.3
14 0101	1.1	**29** 0121	2.4
0625	12.7	0648	11.5
M 1328	1.1	TU 1338	2.3
○ 1850	12.9	● 1904	11.4
15 0153	0.8	**30** 0155	2.3
0711	13.0	0721	11.6
TU 1418	0.8	W 1412	2.2
1937	13.1	1937	11.5

DECEMBER

Time	m	Time	m
1 0228	2.3	**16** 0309	1.1
0752	11.6	0827	12.7
TH 1445	2.2	F 1536	1.0
2008	11.4	2056	12.3
2 0300	2.4	**17** 0352	1.4
0823	11.6	0910	12.4
F 1518	2.3	SA 1619	1.4
2040	11.3	2139	11.9
3 0331	2.6	**18** 0433	1.9
0855	11.4	0953	11.8
SA 1551	2.6	SU 1700	2.0
2114	11.0	2221	11.2
4 0403	2.9	**19** 0513	2.6
0929	11.0	1035	11.1
SU 1625	2.9	M 1740	2.8
2151	10.6	2304	10.5
5 0438	3.3	**20** 0553	3.3
1007	10.6	1119	10.3
M 1704	3.3	TU 1821	3.5
2232	10.1	2351	9.8
6 0519	3.7	**21** 0639	4.0
1053	10.1	1210	9.6
TU 1750	3.7	W 1909	4.1
2325	9.7	◑	
7 0612	4.1	**22** 0048	9.2
1153	9.6	0735	4.4
W 1849	3.9	TH 1314	9.2
●		2010	4.4
8 0034	9.4	**23** 0201	9.0
0719	4.3	0843	4.6
TH 1310	9.5	F 1431	9.0
2002	3.9	2120	4.4
9 0154	9.5	**24** 0316	9.2
0839	4.0	0953	4.3
F 1429	9.8	SA 1541	9.3
2120	3.5	2226	4.1
10 0308	10.1	**25** 0415	9.7
0956	3.4	1054	3.9
SA 1540	10.4	SU 1637	9.8
2231	2.8	2322	3.6
11 0413	10.8	**26** 0504	10.2
1104	2.7	1146	3.4
SU 1642	11.2	M 1723	10.3
2337	2.1		
12 0510	11.6	**27** 0010	3.2
1208	2.0	0546	10.7
M 1739	11.8	TU 1232	2.9
		1805	10.7
13 0037	1.6	**28** 0053	2.8
0604	12.2	0625	11.1
TU 1306	1.4	W 1313	2.6
1832	12.3	1844	11.0
14 0132	1.2	**29** 0132	2.5
0654	12.6	0701	11.4
W 1401	1.0	TH 1352	2.3
○ 1923	12.6	● 1920	11.3
15 0223	1.0	**30** 0210	2.3
0741	12.8	0736	11.6
TH 1450	0.9	F 1430	2.1
2010	12.6	1955	11.4
		31 0246	2.2
		0811	11.7
		SA 1507	2.1
		2031	11.4

Chart Datum is 6·29 metres below IGN Datum. HAT is 13·6 metres above Chart Datum.

9.18.13 RIVER RANCE

Ille-et-Vilaine 48°37′·04N 02°01′·71W (Barrage) ❀❀◊◊☆☆☆

CHARTS AC 3659, 2700; SHOM 7130, 4233; Imray C33B

TIDES Standard Port ST MALO (◄─)
Water levels upriver of the Rance hydro-electric tidal barrage are strongly affected by the operation of the sluice gates and occasional use of the turbines as pumps. On most days from 0700–2100LT, 4m above CD is maintained. There is generally 8·5m above CD for a period of 4 hours between 0700–2000LT. A French language pamphlet, issued by Électricité de France (EDF) should be obtained from HMs at St Malo or Bas Sablons, or from the office at the barrage lock. It gives forecasts for the summer months of when heights of 4m and 8·5m above CD will occur in the period 0700–2000LT. The local daily paper *Ouest France* gives a forecast for the next day of HW and LW upstream of the barrage, under the heading *Usine Marémotrice de la Rance*.

SHELTER Good shelter upriver dependent on wind direction. The principal ⚓s/moorings on the E bank are at St Suliac and Mordreuc, and at La Richardais, La Jouvente, Le Minihic and La Pommeraie on the W bank. Marinas at Plouër, Lyvet (E bank, beyond Chatelier lock) and Dinan: see next column.

NAVIGATION From St Malo/Dinard, appr the lock (at the W end of the barrage) between Pte de la Jument to stbd and a prohib sluicing zone to port, marked by PHM buoys linked by wire cables. 3 white waiting buoys are on the E side of the appr chan, close to the prohibited zone; a similar buoy is upstream.

> **Barrage lock** opens day/night (on request 2030–0430) on the hour, every hour provided the level is at least 4m above CD on both sides of the barrage see: *www.plouer-sur-rance.fr/ fr/horairesdesmarees*. Yachts should arrive at H –20 mins. An illuminated display board gives access times in French/English. Lock entry sigs are **IPTS** sigs 2, 3 and 5. Masted boats entering from sea should berth at the S end of the lock so that the bridge can close astern of them. Vertical wires assist berthing/warp-handling. The lifting road-bridge across the lock opens only between H and H +15. Upstream of the lock a further prohib area to port is marked as above.

The chan upriver is marked by perches; binos needed. The 3M chan to St Suliac has min depth of 2m. The next 6M to the Chatelier lock partially dries and is buoyed; keep to the outside of bends. The suspension bridge and road bridge at Port St Hubert have 20m clearance. A viaduct 1M beyond Mordreuc has 19m clearance. Allow 2-3 hours from the barrage to Chatelier.

Chatelier lock and swing bridge operate 0600-2100LT, provided there is at least 8·5m rise of tide. HW Chatelier is 2-3 hours after HW St Malo depending on the barrage. Entry is controlled by CEVNI sigs (variations on a R and G theme).

The final 3M to Dinan has a published min depth in the marked chan of 1·4m; check with lock-keeper. Dinan gives access to Ille et Rance Canal and River Vilaine to Biscay (see Inland waterways).

LIGHTS AND MARKS Approaching the barrage from seaward: Pte de la Jument bn tr, Fl G 4s 6m 4M; PHM buoy opposite, Fl R 4s (prohib zone). The lock control twr is conspic. NW side of lock, Fl (2) G 6s, with G ▲ on W □. First dolphin, Fl (2) R 6s, with R ■ on W □.

Approaching from Dinan: PHM buoy, Fl (2) R 6s, at S end of prohib zone; leave to stbd. First dolphin, Fl (3) R 12s, with R ■ on W □. SW side of lock, Fl (3) G 12s, G ▲ on W □.

COMMUNICATIONS ☎ 02·99·46·10·46; Auto 08·92·68·08·35; Police 02·99·81·52·30; Brit Consul 01·44·51·31·00; Aff Mar 02·96·39·56·44; Water levels/navigation 02·99·46·14·46; Barrage/lock info 02·99·46·21·87, *Barrage de la Rance* (lock) VHF Ch 13; Chatelier lock 02·96·39·55·66, VHF Ch 14.

FACILITIES St Suliac ✕ ⌂ 🛒 ⚓ ➤ Divers (Convoimer). **Mordreuc** ✕ ⌂ ➤ ⚓ ✦.

La Richardais HM ☎02·99·46·24·20, ✕ ⌂ 🛒 🅱 ⌂ ➤ ✕.
La Jouvente ⌂ ✕ ⌂. **Le Minihic** ⚓ ➤ ✦ ✕.

La Cale de Plouër ⚓ ➤ SC ✕.

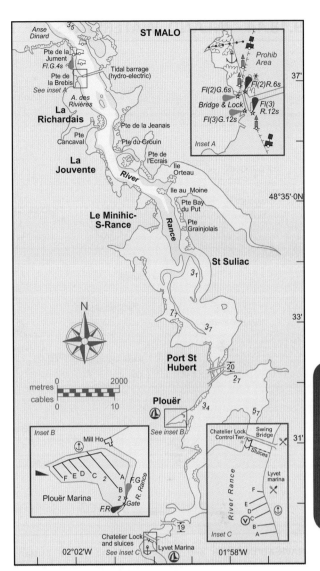

MARINAS ON THE RIVER RANCE
(NB: Berthing space limited, visitors are advised to call ahead.)
PLOUËR, Côtes d'Armor, **48°31′·54N 01°58′·95W**. ❀❀◊◊☆☆. On W bank of R Rance, 6M S of the barrage and 2M N of Chatelier lock. Tide 8m above CD gives 1·5m water over rising gate. Appr with Plouër ✠ spire brg 285°. Off S end bkwtr unlit PHM and SHM perches are 30m from narrow ent between 2 metal lattice pylons marking sill; access when FG/FR lts lit, day/night. Floodlit depth gauge (hard to read). **Marina** plouer.portplaisance@wanadoo.fr www.plouer-sur-rance.fr/fr/ portdeplaisance ☎02·96·86·83·15; 📠02·96·89·11·00. Ch 09. HM, Mlle Liliane Faustin is very helpful. 230 ⌂ +10 ♥ pontoon B, €2·53. ✕ ⌂ ⌂ ➤ ✕ ➤ elec bike hire ▣(14t) ⌂. **Village** ⚊ 1M.

LYVET, **48°29′·40N 02°00′·00W**, on E bank just S of Chatelier lock. ❀❀◊◊☆☆. **La Vicomté Marina** ☎02·96·83·35·57. 210+250♥ €2·00. Berth in 2m on pontoon D, or as directed. ✕⌂ limited 🗑. Draught restricted trot moorings at La Hisse (opp Lyvet), contact La Mairie at Sampson-sur-Rance.

DINAN ❀❀◊◊☆☆☆. **Marina** ☎02·96·39·56·44. ⌂ line the W bank of the river, with fingers (€1·50) near to the Port in about 1·5m. ⚓ ⚓ ✕ ⌂ ➤ for de-masting. Low bridge beyond Port has 2·5m headroom, giving access to the Ille et Rance canal. **Town** (75m above river) ✕ 🗑 ⌂ ⌧ 🅱 ⇌ ✈ (Dinard). Also ⌂ on Quai du Tallard, Lanvally (opp Dinan).

Ille et Rance Canal (opposite): Tolls (see Special notes) are not levied on Breton canals. A CEVNI is required, unless <5m LOA *and not capable* of more than 20kph/11kn.

9.18.14 INLAND WATERWAYS OF BRITTANY

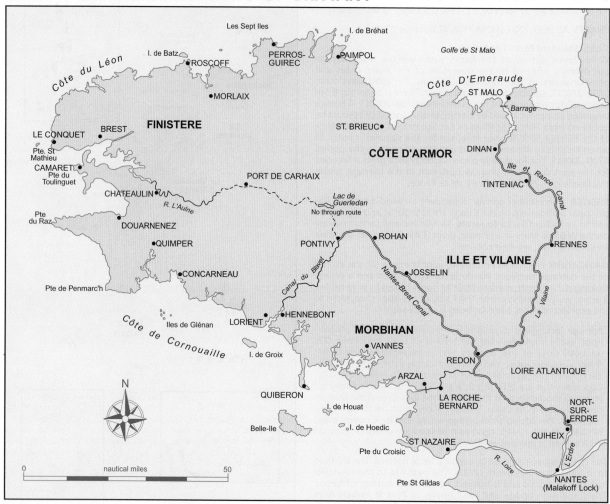

NAVIGATION Canals and rivers across and within Brittany enable boats of limited water and air draughts to go from the English Channel to the Bay of Biscay avoiding the passage around Finistère. Distances, number of locks, boat size and speed limits are summarised below. Dinan-Arzal takes about 5 days. Despite the many locks it can be a thoroughly enjoyable trip through unspoiled countryside and some interesting towns. Most overnight stops are free or at nominal cost.

LOCKS From Apr to Sept locks are worked 7 days a week 0800–1930LT, closing for lunch 1230-1330 approx. All locks are attended, but a fair measure of self-help is the order of the day. In Jul/Aug, in order to conserve water, locks may open on the hour only (and at H+30 if traffic demands).

ACCESS For prior estimate of max possible draught, contact: Equipement, Ille et Vilaine, 1 Avenue de Mail, 35000 Rennes, ☎02·99·59·20·60; or obtain recorded information update on ☎02·99·59·11·12. For latest info on the Ille et Rance Canal/R Vilaine, contact: Rennes ☎02·99·59·20·60 or Redon ☎02·99·71·03·78. For the Lorient-Nantes Canal, contact: Lorient ☎02·97·21·21·54; Hennebont ☎02·97·85·15·15; Pontivy ☎02·97·25·55·21; Nantes ☎02·40·71·02·00. Closures *(Chômages)* for maintenance are scheduled every Weds from approx first week in Nov to last week in March.

INFORMATION *Cruising French Waterways* (McKnight/ACN) *Through the French Canals* (Jefferson/ACN) and *Navicarte Guide No 12 (Bretagne).*

TOLLS may be due on the R Loire only; see Area 17, Special Notes for France, for rates.

SUMMARY	Length km	No of locks	Max draft m	Max air draft m	Max LOA m	Max beam m	Speed limit kn
St MALO–ARZAL (Ille et Rance Canal and La Vilaine river)							
R Rance-Dinan	29·0	1	1·3	19	25	–	5·4
Ille et Rance Canal							
Dinan-Rennes	79·0	48	1·2	2·5	25	4·5	4·3
Rennes-Redon	89·0	13	1·2	3·2/2·6*	25	4·5	4·3
Redon-Arzal	42·0	1	1·3				
*Depending on water level							
LORIENT–NANTES							
Canal du Blavet (See Lorient)							
Lorient-Pontivy	70	28	1·4	2·6	25	4·6	4·3
Nantes-Brest Canal							
	184·3	106	–	3	25	4·6	4·3
Pontivy-Rohan			0·8 (possible closure)				
Rohan-Josselin			1·0				
Josselin-Redon			1·4				
Redon-Quiheix			1·1				
L'Erdre River	27·6	1	1·4	3·8	400	6·2	13·5
R Loire, above Nantes (St Nazaire) may be navigable to Angers.							
R L'AULNE (See Brest)							
Brest-Chateaulin	42	1	3·0	N/A	25	–	
Chateaulin-Carhaix							
	72	33	1·1	2·5	25	4·6	4·3

MINOR HARBOUR BETWEEN ST MALO AND ST CAST LE GUILDO
ST BRIAC 48°37′·31N 02°08′·71W. AC 3659, 2700; SHOM 7155, 7130, 7129. HW –0515 on Dover (UT); ML 6·8m;–Duration 0550. Tides as Dahouet; use Ile des Hébihens. Drying hbr (6m) open to SW-NW. A Dir Iso WRG 4s lt leads 125° between offlying drying and beaconed rocks/islets. The last 6ca are marked by 4 PHM and 3 SHM perches. 10 🅐s in Le Bechet cove or ⚓ on sand W of Ile du Perron in 3-5m. HM ☎02·99·88·01·75, ⚓ ☎. YC de St Briac ☎02·99·88·31·45. **Town** ✕ ⌂ 🍴.

9.18.15 ST CAST LE GUILDO

Côtes d'Armor **48°38′·39N 02°14′·76W** ✿✿✿⚓⚓⚓⚓✿✿✿

CHARTS AC 2669, 3659; SHOM 71, 7129, 7155; Navi 535/6; Imray C33B

TIDES –0520 Dover; ML 6·3; Duration 0550

Standard Port ST MALO (←)

Times				Height (metres)			
High Water		Low Water		MHWS	MHWN	MLWN	MLWS
0100	0800	0300	0800	12·2	9·3	4·2	1·5
1300	2000	1500	2000				
Differences SAINT CAST							
0000	0000	–0005	–0005	–0·2	–0·2	–0·1	–0·1
ERQUY							
0000	–0005	–0020	–0010	–0·8	–0·6	–0·2	–0·2

SHELTER Good shelter in marina, access all states of tide, but exposed shoaling approach in strong NE/E/SE conditions. No anchoring.

NAVIGATION. WPT 48°39′·00N 02°13′·25W E of ECM Les Bourdinots (¾M NE of Pte de St Cast, marks 2m patch) then 225° for 1M to marina ent chan, R & G bys. Beware west mole, submerged HW.

LIGHTS AND MARKS Iso WG 4s 11/8M (appr on W sectors to clear Les Bourdinots) Bright shore lights confusing until within 2M.

COMMUNICATIONS CROSS Corsen 02·98·89·31·31; Ⓑ 02·96·41·76·61; Police 02·99·81·52·30; British Consul 01·44·51·31·00; Ⓗ St Malo 02·99·56·56·19; HM 02·96·70·88·30, VHF Ch 09, 16; Semaphore 02·96·41·85·30; Affaires Maritime 02·96·81·04·43.

FACILITIES Port d'Armor Marina ☎02·96·81·04·43 www. cotesdarmor.cci.fr stcast.plaisance@cotesdarmor.cci.fr; 760+40 Ⓥ €3·10 + €0·20pp/day tax; pontoon D, berth as indicated for LOA, or as instructed. ▶🅷 (H24 cr/card only) ✕ 🗐 🔧(25t) 🚡(10t,12t). **YC** ☎02·96·41·71·71, 🗞. **Town** (10 mins) ✕ ⌂ 🍴 🏦 🖂 🔧 ✕ ✉ Ⓑ ⚡ (Lamballe) ✈ (St Brieuc). Ferry: St Malo.

MINOR HARBOUR BETWEEN ST CAST LE GUILDO AND DAHOUET
ERQUY (pronounced Erky) **48°38′·04N 02°28′·69W.** AC 2669, 2029; SHOM 7154, 7310. HW –0515 on Dover (UT); ML 6·5m; Duration 0550; see St Cast Le Guildo. Sheltered from E, but open to SW/W'lies. An active FV hbr, but yachts can dry out E of the inner jetty hd Fl (3) R 12s; or ⚓ on sand 3ca SW of the outer mole. Beware Plateau des Portes d'Erquy (dry) about 2M to W. Beware rks off Pte de Lahoussaye. Mole hd lt Fl (2) WRG 6s 11m W10M, R/G7M; approach in either W sector (see Lights, buoys and waypoints). HM & ⚓ ☎02·96·72·19·32. **YC** ☎02·96·72·32·40. Facilities: Ⓥ (drying) €0·56, **Quay** 🚡(3·5t) ⚓ ▶🅷. **Town** ✕ ⌂ 🍴 🏦 Ⓑ 🔧 ✕.

9.18.16 DAHOUET

Côtes d'Armor **48°34′·79N 02°34′·39W** ✿✿⚓⚓✿

CHARTS AC 2669, 2029; SHOM 7154, 7310; Navi 536; Imray C33B, C34

TIDES –0520 Dover; ML 6·3; Duration 0550

Standard Port ST MALO (←)

Times				Height (metres)			
High Water		Low Water		MHWS	MHWN	MLWN	MLWS
0100	0800	0300	0800	12·2	9·3	4·2	1·5
1300	2000	1500	2000				
Differences ÎLE DES HÉBIHENS (7M W of St Malo)							
0000	0000	–0005	–0005	–0·2	–0·2	–0·1	–0·1
DAHOUET							
–0005	0000	–0025	–0010	–0·8	–0·7	–0·2	–0·1

SHELTER Good, but ent (dries 4m) has strong currents and is unsafe in fresh/strong NW'lies, when a bar may form.

NAVIGATION WPT 48°35′·15N 02°35′·44W, unlit NCM buoy, 117°/0·8M to La Petite Muette (LPM) lt tr. Appr in W sector, crossing into the R until LPM bears 160°. Enter the narrow break in the cliffs on that track leaving LPM to stbd and 2 W poles to port (they are not ldg marks). It is dangerous to enter S of LPM, due to rocks. At SHM bcn the chan turns E, then SE to the marina sill marked by PHM/SHM perches. Night entry not advised without prior day experience.

LIGHTS AND MARKS La Petite Muette (see Lights, buoys and waypoints) is the key feature. Conspic daymarks: the wide beach NE at Val André and, in the hbr ent, a pagoda-like shrine. See also St Brieuc.

DAHOUET *continued*

COMMUNICATIONS CROSS 02·96·70·42·18; ☎ 02·36·65·02·22; Auto 08·92·68·02·22; Police 02·96·72·22·18; ⊖ 02·96·74·75·32; British Consul 01·44·51·31·00; Ⓗ 02·96·45·23·28; HM VHF Ch 09, 16; Affaires Maritimes 02·96·72·31·42.

9.18.17   SAINT-BRIEUC LE LÉGUÉ

Côtes d'Armor 48°31′·68N 02°43′·97W (lock) ❀❀♦♦♦❀

CHARTS AC 2669, 2029; SHOM 7154, 7128; Navi 536; Imray C34, C33B

TIDES –0520 Dover; ML 6·5; Duration 0550

Standard Port ST MALO (←—)

Times				Height (metres)			
High Water		Low Water		MHWS	MHWN	MLWN	MLWS
0100	0800	0300	0800	12·2	9·3	4·2	1·5
1300	2000	1500	2000				
Differences LE LÉGUÉ (SWM buoy)							
–0010	–0005	–0020	–0015	–0·8	–0·5	–0·2	–0·1

SHELTER Very good in the marina (Bassin No 2, min 3·2m) near the viaduct. Le Légué is also the commercial port for St Brieuc.

NAVIGATION WPT 48°34′·33N 02°41′·15W, Le Légué SWM buoy, Fl Mo (A) 10s, 210°/2·6M to abm Pte à l'Aigle lt. The bay dries E/SE of conspic Pte du Roselier. Access via well buoyed/lit chan, dredged 5m above CD; but not advised in strong N/NE winds. The outer hbr is entered between Pte à l'Aigle on the NW bank of Le Gouet river and the SE jetty head.

Lock opens, relative to height of tide at St Malo, *at local HW:*

FACILITIES Outer hbr (FVs) dries 5·5m. **Marina**, min depth 2m, is entered over 5·5m sill, with depth gauge. Very crowded in season. ☎02·96·72·82·85, 318+25 ❻ berths on 'O' pontoon in about 1·6-2·0m, €2·25. ⌁ ⏃ ⏃ ⏃(8t mobile) ⏃ ⏃ ⏃ ⏃ **YC du Val-André** ☎02·96·72·21·68, ⏃.
Town, ✕ ⏃ ⏃ ⇌ (Lamballe) ✈ (St Brieuc). Ferry: St Malo.

(MHWN 9·3m)	<9·4m	±1
	9·4-10·3m	±1¼
	10·3-10·8m	±1½
(MHWS 12·2m)	>10·8m	−2 to +1½

The lock is 85m x 14m; sill dries 5·0m. Lock staff help with warps. Commercial ships take priority. A low swing bridge at the ent to Bassin No 2 opens when the lock is in use. A sliding footbridge gives access further upstream (no ❻).

LIGHTS AND MARKS Conspic marks: Le Verdelet Is (42m) from E; Rohein twr, from N; Pte du Roselier in near appr; St Brieuc.

COMMUNICATIONS CROSS 02·98·89·31·31; ⒧Ⓑ 02·96·88·35·47; ☎ 02·99·46·10·46; Auto 08·92·68·08·22; Police 02·96·61·22·61; ⊖ 02·96·74·75·32; Brit Consul 01·44·51·31·00; Dr 02·96·61·49·07; Port HM 02·96·33·35·41, *Le Légué Port* VHF Ch 12, 16 (approx HW–2 to +1½); Aff Mar 02·96·68·30·70.

FACILITIES Marina ☎02·96·77·49·85, mob 06·75·91·67·63, legue. plaisance@cotesdarmor.cci.fr/ www.cotesdarmor.cci.fr 250 + 20❻ €2·10. Berth by HM's office on S quay for directions. ✕ ⏃ ⏃ ⏃ ⏃ ⏃ ⏃ ⏃Ⓔ Ⓒ YC ⏃(20t) ⏃(2t for de-masting).

Ship Repair Yard ☎06·76·33·56·66, ⏃(350t).

Town ⏃ ⏃ ⊠ Ⓑ @ ⇌. Ferry St Malo, Roscoff.

9.18.18 BINIC

Côtes d'Armor 48°36´·06N 02°48´·99W ✵✵⚓♦♦✿✿✿

CHARTS AC 2648, 2669, 2029; SHOM 7154, 7128; Navi 536; Imray C33B, C34

TIDES –0525 Dover; ML 6·3; Duration 0550

Standard Port ST MALO (←)

Times				Height (metres)			
High Water		Low Water		MHWS	MHWN	MLWN	MLWS
0100	0800	0300	0800	12·2	9·3	4·2	1·5
1300	2000	1500	2000				
Differences BINIC							
–0005	–0005	–0025	–0015	–0·8	–0·6	–0·2	–0·1

SHELTER Good in marina. Appr not advised in strong E/NE'lies.

NAVIGATION WPT 48°35´·86N 02°46´·68W, 276·5°/1·5M to ent. 2M E of hbr beware oyster beds marked by unlit W bys. Appr dries 7ca offshore. From E, appr via Caffa ECM buoy, whence ent bears 246°/4·2M; or from N through Rade de St Quay-Portrieux.

Avant Port dries 5·5m, is easy day/night, except in E'lies. IPTS, sigs 2 & 3, control entry H24 to the marina via hydraulic flapgate and retracting bridge (office manned only when tide is sufficient). Gate opens when height of tide >8·5m. Times vary (see www. ville-binic.fr) with coefficient/barometer:

Coeff 40	<8·5m	=	No entry; no exit, ie neaped
40-50	8·5m	=	HW –1 to HW +1
50-60	9·15m	=	HW –1½ to HW + 1½
60-80	9·5m	=	HW –2 to HW + 2
>80	10·6m	=	HW –2¼ to HW + 2¼

Outbound vessels have priority. Beware strong current (up to 5kn) in the ent for 20 mins after the gate first opens.

LIGHTS AND MARKS Ldg line 276·5°: N mole lt, W twr, G gallery, ≠ church spire. A water twr is conspic 8ca N of hbr.

COMMUNICATIONS CROSS 02·98·89·31·31; ⊖ 02·96·74·75·32; ⚓ 08·92·68·08·08; Ⓗ 02·96·94·31·71; SAMU 15; Fire 18; Call Port/marina VHF Ch 09 to check tide height and to advise of boat size and length of stay; Aff Mar 02·96·68·30·70.

FACILITIES Marina ☎02·96·73·61·86. English spoken. port@ville-binic.fr www.ville-binic.fr 350 + 50Ⓥ €2·05. Ⓥ berth/raft on pontoon K, to stbd on entry (1·5-2·5m). ⚓ ⊘ ⚓ SHOM ⊾ Ⓔ SC ⚓(20t mobile), hire car available. At St Quay-Portrieux 🏂 🏂 and nearby ⚓ ⚒ ⊡ ✗.

Town ✗ ⌂ 🛒 ⊙ 🛢 @ ✉ Ⓑ. Bus to St Brieuc ⇌. Ferry: St Malo.

Central N France

9.18.19 ST QUAY-PORTRIEUX

Côtes d'Armor **48°38′·84N 02°48′·97W** ✿✿✿❀❀△△✿✿

CHARTS AC 2648, 2669, 3674; SHOM 7154, 7128; Navi 536, 537; Imray C33B, C34

TIDES –0520 Dover; ML 6·3; Duration 0550

Standard Port ST MALO (←)

Times				Height (metres)			
High Water		Low Water		MHWS	MHWN	MLWN	MLWS
0100	0800	0300	0800	12·2	9·3	4·2	1·5
1300	2000	1500	2000				
Differences ST QUAY-PORTRIEUXi							
–0010	–0010	–0030	–0010	–0·9	–0·7	–0·2	–0·1

SHELTER Excellent in marina. ⚓ in the Rade de St Quay-Portrieux, sheltered by Roches de St Quay, but open to NW & SE winds.

NAVIGATION From E and SE appr via Caffa ECM lt buoy to WPT 48°37′·08N 02°46′·18W (2ca S of La Roselière SCM buoy VQ (6) + L Fl 10s); thence 314°/2·5M to NE mole head (works underway).

From the N, WPT 48°40′·94N 02°49′·69W, 169°/2·0M to NE mole elbow. At night, to avoid rocks/shoals, follow the white sectors of 4 Dir lts (see chartlet and tinted box) tracking 169°, 130°, 185° (astern) and 318° in sequence to the marina ent:

- NE mole elbow lt, **W159°-179°**. White concrete twr.
- Herflux lt, **W125°-135°**. A white bcn twr on drying rock.
- Ile Harbour lt, **W358°-011°**. Short W twr/R top on an islet.
- NE mole elbow lt (as first • above) **W316°-320·5°**.

LIGHTS AND MARKS See chartlet and Lights, buoys & waypoints. Signal stn at Pointe de St Quay, NNW of the marina, is conspic. N of the marina Moulières de Portrieux, ECM bcn twr is unlit. E of the marina an unlit WCM buoy marks Les Noirs rks (2·4m).

COMMUNICATIONS CROSS 02·98·89·31·31; ♨02·96·76·76·80; Auto 08·92·68·08·22; Police 02·96·70·61·24; ⊖02·96·33·33·03; Brit Consul 01·44·51·31·00; Dr 02·96·70·41·31; HM (Old Hbr) 02·96·70·95·31; Aff Mar 02·96·68·30·73; Marina VHF Ch 09 (H24).

FACILITIES Marina ☎02·96·70·81·30. welcome@port-armor.com www.port-armor.com ⏀ 900+100Ⓥ Jul/Aug, €3·00, less other months. Ⓥ berths on No 7 pontoon (no fingers NW side). Jul/Aug pre-call for a berth if LOA >12m. 🛢🚿✕🗑🏪🔧⚓ Ⓔ ✕ 🏴(40t) 🅿(20t mobile) 🛗(5t).

SN de St Quay-Portrieux ☎02·96·70·93·34.

Cercle de la Voile de Portrieux ☎02·96·70·41·76.

Old Hbr ☎02·96·70·95·31, dries approx 5·3m. 500+8 ⚓ €1·50. ⚓ ━ 🏴(1·5t).

Town ✕🗑🛒🏪 ⊠ Ⓑ 🛎. Bus to St Brieuc ⇌ ✈. Ferry: St Malo-Poole, Portsmouth.

Channel Islands

Alderney to Jersey

9.19.1	Contents	815
9.19.2	Area map and Distance table	816
9.19.3	Tidal stream charts	818
9.19.4	Lights, buoys and waypoints	820
9.19.5	Passage information	822

The Channel Islands – 822 • The Casquets and Ortac Channel – 822 • Alderney and the Swinge – 822 • The Alderney Race – 822 • Approaches to Guernsey – 825 • Jersey – 834

9.19.6	Special notes for The Channel Islands	823
9.19.7	Alderney	823

Saye Bay • Corblets Bay • Longy Bay • La Tchue • Telegraph Bay • Hannaine Bay • Platte Saline Bay • Burhou

9.19.8	Braye Harbour (Alderney)	824
9.19.9	Little Russel Channel (Guernsey/Herm)	825

Minor harbours and anchorages around Guernsey:
Bordeaux Harbour • St Sampson • Havelet Bay • Soldier's Bay • Fermain Bay • Petit Port • Moulin Huet • Saints Bay • Icart Bay • Jaonnet • Petit Bôt Bay • Portelet • Portelet Harbour • Lihou Island • Perelle Bay • Vazon Bay • Cobo Bay • Grande Havre • L'Ancresse Bay • Fontenelle Bay

9.19.10	Beaucette (Guernsey)	826
9.19.11	St Peter Port (Guernsey), tide tables/curves	827
9.19.12	Herm	832
9.19.13	Sark	833

Saignie Bay • Port à la Jument • Havre Gosselin • Port és Saies • La Grande Grève • Port Gorey • Rouge Terrier • Dixcart Bay • Derrible Bay • Grève de la Ville • Les Fontaines

9.19.14	St Helier (Jersey)	834

Minor harbours and anchorages around Jersey:
St Aubin • Belcroute Bay • Portelet Bay • St Brelade • Grève au Lancon • Grève de Lecq • Bonne Nuit Bay • Bouley Bay • Rozel Bay • St Catherine • La Rocque • Les Écrehou • Plateau des Minquiers

9.19.15	Gorey (Jersey)	836
	St Helier (Jersey), tide tables/curves	837

Channel Islands

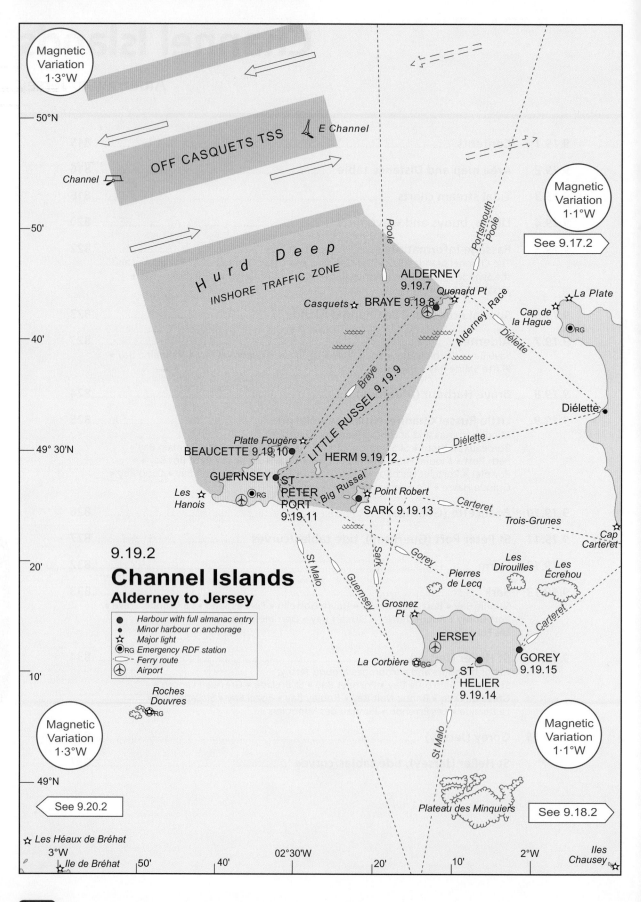

Magnetic Variation 1·3°W

50°N

E Channel

OFF CASQUETS TSS

Channel

Magnetic Variation 1·1°W

See 9.17.2

50'

Poole

H u r d   D e e p

INSHORE TRAFFIC ZONE

ALDERNEY 9.19.7

Quenard Pt

La Plate

Casquets ☆ BRAYE 9.19.8

Cap de la Hague

●RG

40'

Alderney Race

Diélette

Diélette

Braye

Diélette

Dièlette

Diélette

Carteret

LITTLE RUSSEL 9.19.9

49° 30'N

Platte Fougère ☆

BEAUCETTE 9.19.10

HERM 9.19.12

GUERNSEY

Les ☆
Hanois

●RG

ST
PETER
PORT
9.19.11

Big Russel

Point Robert

SARK 9.19.13

Carteret

Trois-Grunes

Cap
Carteret

20'

St Malo

Sark

Gorey

Les
Dirouilles

Les
Écrehou

Guernsey

Pierres
de Lecq

9.19.2
Channel Islands
Alderney to Jersey

● Harbour with full almanac entry
• Minor harbour or anchorage
☆ Major light
●RG Emergency RDF station
⊹ Ferry route
✈ Airport

Grosnez
Pt ☆

JERSEY

GOREY
9.19.15

La Corbière ●RG

ST
HELIER
9.19.14

10'

Roches
Douvres

●RG

Magnetic
Variation
1·3°W

Magnetic
Variation
1·1°W

49°N

See 9.20.2

Plateau des Minquiers

See 9.18.2

☆ Les Héaux de Bréhat

3°W   50'   40'   02°30'W   20'   10'   2°W

Iles
Chausey ☆

Ile de Bréhat

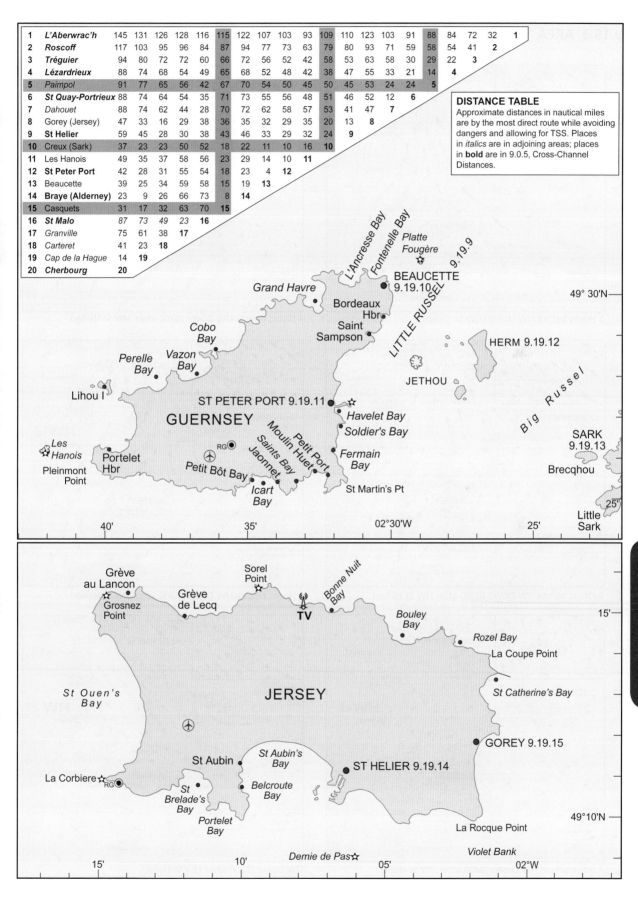

1	*L'Aberwrac'h*	145	131	126	128	116	115	122	107	103	93	109	110	123	103	91	88	84	72	32	1
2	*Roscoff*	117	103	95	96	84	87	94	77	73	63	79	80	93	71	59	58	54	41	2
3	**Tréguier**	94	80	72	72	60	66	72	56	52	42	58	53	63	58	30	29	22	3
4	**Lézardrieux**	88	74	68	54	49	65	68	52	48	42	38	47	55	33	21	14	4
5	*Paimpol*	91	77	65	56	42	67	70	54	50	45	50	45	53	24	24	5
6	**St Quay-Portrieux**	88	74	64	54	35	71	73	55	56	48	51	46	52	12	6
7	*Dahouet*	88	74	62	44	28	70	72	62	58	57	53	41	47	7
8	**Gorey (Jersey)**	47	33	16	29	38	36	35	32	29	35	20	13	8
9	**St Helier**	59	45	28	30	38	43	46	33	29	32	24	9
10	Creux (Sark)	37	23	23	50	52	18	22	11	10	16	10
11	Les Hanois	49	35	37	58	56	23	29	14	10	11
12	**St Peter Port**	42	28	31	55	54	18	23	4	12
13	Beaucette	39	25	34	59	58	15	19	13
14	**Braye (Alderney)**	23	9	26	66	73	8	14
15	Casquets	31	17	32	63	70	15
16	*St Malo*	87	73	49	23	16
17	*Granville*	75	61	38	17
18	*Carteret*	41	23	18
19	*Cap de la Hague*	14	19
20	*Cherbourg*	20

DISTANCE TABLE
Approximate distances in nautical miles are by the most direct route while avoiding dangers and allowing for TSS. Places in *italics* are in adjoining areas; places in **bold** are in 9.0.5, Cross-Channel Distances.

49° 30'N

BEAUCETTE
9.19.10

L'Ancresse Bay
Fontenelle Bay
Platte
Fougère
9.19.9

Grand Havre
Bordeaux
Hbr
Saint
Sampson

LITTLE RUSSEL

HERM 9.19.12

Cobo
Bay

JETHOU

Big Russel

Perelle
Bay
Vazon
Bay

Lihou I

ST PETER PORT 9.19.11

GUERNSEY

Havelet Bay
Soldier's Bay

SARK
9.19.13

Les
Hanois

Portelet
Hbr

Petit Port
Moulin Huet
Saints Bay
Jaonnet

Fermain
Bay

Brecqhou

Pleinmont
Point

Petit Bôt Bay

Icart
Bay

St Martin's Pt

Little
Sark

25'

40'   35'   02°30'W   25'

Grève
au Lancon

Sorel
Point

Bonne Nuit
Bay

15'

Grosnez
Point

Grève
de Lecq

TV

Bouley
Bay

Rozel Bay
La Coupe Point

St Ouen's
Bay

JERSEY

St Catherine's Bay

GOREY 9.19.15

St Aubin
St Aubin's
Bay

ST HELIER 9.19.14

La Corbiere

St
Brelade's
Bay

Belcroute
Bay

La Rocque Point

49° 10'N

Portelet
Bay

Demie de Pas

Violet Bank

15'   10'   05'   02°W

Channel Islands

817

9.19.3 AREA 19 TIDAL STREAMS

5 Hours before HW Dover (HW St Helier)

4 Hours before HW Dover (0100 after HW St Helier)

3 Hours before HW Dover (0200 after HW St Helier)

2 Hours before HW Dover (0300 after HW St Helier)

1 Hour before HW Dover (0400 after HW St Helier)

HW Dover (0500 after HW St Helier)

Westward 9.20.3   Southward 9.18.3   Northward 9.2.3   Eastward 9.17.3

1 Hour after HW Dover (0600 after HW St Helier)

2 Hours after HW Dover (0530 before HW St Helier)

3 Hours after HW Dover (0430 before HW St Helier)

4 Hours after HW Dover (0330 before HW St Helier)

5 Hours after HW Dover (0230 before HW St Helier)

6 Hours after HW Dover (0130 before HW St Helier)

Channel Islands

9.19.4 LIGHTS, BUOYS AND WAYPOINTS

Bold print = light with a nominal range of 15M or more. CAPITALS = place or feature. *CAPITAL ITALICS* = light-vessel, light float or Lanby. *Italics* = Fog signal. ***Bold italics*** = Racon. Many marks/buoys are fitted with AIS (MMSI No); see relevant charts.

MID-CHANNEL MARKS

CHANNEL LT VESSEL ⌑ Fl 15s 12m **15M**; R hull with lt twr amidships; *Horn (20s)*; ***Racon O, 15M***; 49°54'·46N 02°53'·74W; 992351028.
E Channel ⌀ Fl Y 5s 6M; ***Racon T, 10M***; 49°58'·67N 02°29'·01W.

COTENTIN PENINSULA (NW COAST)

La Plate ⌀ Fl (2+1) WR 10s 11m; W9M, R6M; 115°-W-272°-R-115°; Y 8-sided twr, with B top; 49°43'·98N 01°55'·76W.
Cap de la Hague (Gros du Raz) ☆ Fl 5s 48m **23M**; Gy twr, W top; *Horn 30s*; 49°43'·31N 01° 57'·27W.
La Foraine ⌀ VQ (9) 10s 12m 6M; 49°42'·90 N 01°58'·32W.
Cap de Carteret ☆ Fl (2+1) 15s 81m **26M**; Gy twr, G top; 49°22'·41N 01°48'·41W.

THE CASQUETS AND ALDERNEY

Casquets ☆ Fl (5) 30s 37m **18M**, H24; W twr, 2 R bands; NW'most of three; ***Racon T, 25M***; 49°43'·32N 02°22'·63W. 992351121.
Ortac rock, unlit; 49°43'·40N 02°17'·44W.
Pierre au Vraic, unmarked rk 1·2m ⊛; 49°41'·61N 02°16'·94W.
Quenard Pt (Alderney) ⌀ Fl (4) 15s 37m 12M; 085°-027°; W ○ twr, B band; 49°43'·75N 02°09'·86W.
Château à L'Étoc Pt ⌀ Iso WR 4s 20m W10M, R7M; 071·1°-R-111·1°-W-151·1°; 49°43'·94N 02°10'·63W.

BRAYE

Ldg bns 142° (to clear the submerged Adm'ty bkwtr). Front, W ⌀, 49°43'·90N 02°10'·97W. Rear, BW ⌀; 720m from front.
Ldg lts 215°: both Q 8/17m 9/12M, synch; 210°-220°; orange △s. Front, old pier elbow, 49°43'·40N 02°11'·91W. Rear, 215°/335m.
Admiralty bkwtr head ⌀ L Fl 10s 7m 5M; 49°43'·82N 02°11'·67W.
FairwayNo. 1 ⌀ QG; 49°43'·72N 02°11'·72W.
No. 2 ⌀ QR; 49°43'·60N 02°11'·75W.
Inner fairway ⌀ Q (2) G 5s; 49°43'·58N 02°11'·98W.
Braye quay ⌀ 2 FR (vert) 8m 5M; 49°43'·53N 02°12'·00W.
Little Crabby hbr ent ⌀ FG & FR 5m 2M; 49°43'·45N 02°12'·12W.

GUERNSEY, NORTHERN APPROACHES

LITTLE RUSSEL CHANNEL

Grande Amfroque, two unlit bcn twrs: larger, BW-banded; smaller, white; 49°30'·57N 02°24'·62W.
Tautenay ⌀ Q (3) WR 6s 7m W7M, R6M; 050°-W-215°- R-050°; B & W striped bcn 49°30'·11N 02°26'·84W.
Platte Fougère ☆ Fl WR 10s 15m **16M**; 155°-W-085°-R-155°; W 8-sided twr, B band; *Horn 45s*; ***Racon P***; 49°30'·83N 02°29'·14W.
Corbette d'Amont ⌀ Y bcn twr, topmark; 49°29'·64N 02°29'·38W.
Roustel ⌀ Q 8m 7M; BW chequered base, W framework col; 49°29'·23N 02°28'·79W.
Rousse, Y bcn twr, topmark ⌗; 49°28'·98N 02°28'·36W.
Platte ⌀, Fl WR 3s 6m, W7M, R5M; 024°-R-219°-W-024°; G conical twr; 49°29'·08N 02°29'·57W.
Vivian bcn twr, BW bands; 49°28'·45N 02°30'·66W.
Brehon ⌀ Iso 4s 19m 9M; bcn on ○ twr, 49°28'·28N 02°29'·28W.
Demie Flieroque, Y bcn twr, topmark F; 49°28'·13N 02°31'·38W.

BIG RUSSEL

Noire Pute ⌀ Fl (2) WR 15s 8m 6M; 220°-W-040°-R-220°; on 2m high rock; 49°28'·21N 02°25'·02W; unreliable.
Fourquies ⌀ Q; 49°27'·34N 02°26'·47W.
Lower Heads ⌀ Q (6) + L Fl 15s; *Bell*; 49°25'·85N 02°28'·55W.

GUERNSEY, HERM AND SARK

BEAUCETTE MARINA

Petite Canupe ⌀ Q (6) + L Fl 15s; 49°30'·20N 02°29'·14W.
Ldg lts 277°: Both FR. Front, W ⬜, R stripe; 49°30'·19N 02°30'·23W. Rear, R ⬜, W stripe; 185m from front.
Appr chan buoys: SWM L Fl 10s 49°30'·15N 02°29'·66W. SHM Fl G 5s, PHM Fl R 5s & Fl (3) R 5s. NCM perch Q 49°30'·165N 02°30'·06W. SHM perch Q (3) G 5s at ent; 49°30'·18N 02°30'·24W.

ST SAMPSON

Ldg lts 286°: Front, FR 3m 5M; 230°-340°; tfc sigs; 49°28'·90N 02°30'·74W. Rear, FG 13m; clock twr, 390m from front.
N Pier ⌀ FG 3m 5M; 230°-340°; 49°28'·92N 02°30'·71W.
Crocq pier ⌀ FR 11m 5M; 250°-340°; 49°28'·98N 02°31'·00W.

ST PETER PORT

Outer ldg lts 220°: **Front**, Castle bkwtr, Al WR 10s 14m **16M**; 187°-007°; dark ○ twr, W on NE side; *Horn 15s*; 49°27'·31N 02°31'·45W.
Rear 220° ldg lt, Belvedere, Oc 10s 61m 14M; 179°-269°; intens 217°-223°; W ⬜ on W twr; 980m from front. The lts are synchronised so that the rear lt is on when the front is W, and the off when the front is R.
Queen Elizabeth II marina, 270° Dir Oc ⌀ WRG 3s 5m 6M; 258°-G-268°-W-272°-R-282°; 49°27'·73N 02°31'·87W.
Reffée ⌀ Q (6) + L Fl 15s; 49°27'·74N 02°31'·27W.
Appr buoys: outer pair ⌀ QG; 49°27'·83N 02°31'·54W. ⌀ QR; 49°27'·71N 02°31'·52W. Inner pair: ⌀ QG; 49°27'·76N 02°31'·74W. ⌀ QR; 49°27'·72N 02°31'·74W.
The Pool ldg lts 265° (*not* into moorings): Front, Victoria marina, S Pier, ⌀ Oc R 5s 10m 14M; 49°27'·32N 02°32'·03W. Rear, ⌀ Iso R 2s 22m 3M; 260°-270°; on Creasey's bldg, 160m from front.
White Rock pier ⌀ Oc G 5s 11m 14M; intens 174°-354°; ○ twr; tfc sigs; 49°27'·38N 02°31'·59W.
S Fairway ⌀ QG; 49°27'·30N 02°31'·76W; pontoon W of buoy. ⌀ Fl R; 49°27'·28N 02°31'·74W. 49°27'·27N 02°31'·80W. 49°27'·27N 02°31'·86W.

HAVELET BAY

Oyster Rock ⌀ Y bcn, topmark 'O'; 49°27'·09N 02°31'·46W.
Oyster Rock ⌀ QG; 49°27'·04N 02°31'·47W.
Moulinet ⌀ QR; 49°26'·97N 02°31'·54W.
Moulinet ⌀ Y bcn, topmark 'M'; 49°26'·95N 02°31'·58W.

GUERNSEY, SOUTH-EAST and SOUTH COASTS

Anfré, Y bcn, topmark 'A'; 49°26'·45N 02°31'·48W.
Longue Pierre, Y bcn, topmark 'LP'; 49°25'·36N 02°31'·48W.
St Martin's Pt ⌀ Fl (3) WR 10s 15m 14M; 185°-R-191°-W-011°-R-061·5°; flat-topped, W bldg. *Horn (3) 30s*; 49°25'·30N 02°31'·70W.

GUERNSEY, NORTH-WEST COAST

Les Hanois ☆ Fl (2) 13s 33m **20M**; 294°-237°; Gy ○ twr, B lantern, helicopter platform; *Horn (2) 60s*. 49°26'·10N 02°42'·15W.
4 FR on mast 1·3M ESE of Les Hanois lt ho.
Portelet Hbr, bkwtr bcn, 49°26'·16N 02°39'·84W.
Cobo Bay, Grosse Rock, B bcn 11m; 49°29'·02N 02°36'·19W.
Grand Havre, Rousse Point bkwtr, B bcn; 49°29'·92N 02°33'·05W.

HERM

Corbette de la Mare, Wh disc (W side) and R disc (E side) on Y pole, 49°28'·48N 02°28'·72W.
Petit Creux ⌀ QR; red 'C' on red pole; 49°28'·09N 02°28'·72W.
Alligande ⌀ Fl (3) G 5s; B pole, Or 'A'; 49°27'·86N 02°28'·78W.
Épec ⌀ Fl G 3s; black 'E' on G mast; 49°27'·98N 02°27'·89W.
Vermerette ⌀ Fl (2) Y 5s; Or 'V' on bcn; 49°28'·12N 02°27'·75W.
Gate Rock (Percée Pass) ⌀ Q (9) 15s; 49°27'·88N 02°27'·54W.
Hbr ldg lts 078°: White drums. ⌀ 2F occas; 49°28'·25N 02°27'·11W.
Hbr pier, N end ⌀ 2 FG; G ⬜ on G bcn; 49°28'·21N 02°27'·26W.

SARK

Courbée du Nez ⌀ Fl (4) WR 15s 14m 8M; 057°-W-230°-R-057°; W structure on rock; 49°27'·09N 02°22'·17W.

Point Robert ☆ Fl 15s 65m **20M**; vis 138°-353°; W 8-sided twr; *Horn (2) 30s;* 49°26'·19N 02°20'·75W.
Founiais ⌁, topmark 'F'; 49°26'·02N 02°20'·36W.
Blanchard ⌁ Q (3) 10s; *Bell;* 49°25'·36N 02°17'·42W.
Pilcher monument (070° appr brg); 49°25'·71N 02°22'·44W.

JERSEY, WEST AND SOUTH COASTS

Desormes ⌁ Q (9) 15s; 49°18'·94N 02°17'·98W.
Grosnez Point ☆ Fl (2) WR 15s 50m **W19M, R17M**; 081°-W-188°-R-241°; W hut; 49°15'·50N 02°14'·80W.
L a Rocco twr (conspic) 15m; 49°11'·90N 02°14'·05W.
La Frouquie ⌁ (seasonal); 49°11'·30N 02°15'·38W.
La Corbière ☆ Iso WR 10s 36m **W18M, R16M**; shore-W-294°-R-328°-W-148°-R-shore; W ○ twr; *Horn Mo (C) 60s;* 49°10'·79N 02°15'·01W.
Pt Corbière ⌁ FR; R □, W stripe; 49°10'·87N 02°14'·38W.

WESTERN PASSAGE

Ldg lts 082°. Front, La Gréve d'Azette Oc 5s 23m 14M; 034°-129°; 49°10'·16N 02°05'·09W. Rear, Mont Ubé, Oc R 5s 46m 12M; 250°-095°; 1M from front.
Passage Rock ⌁ VQ; 49°09'·54N 02°12'·26W.
Les Fours ⌁ Q; 49°09'·59N 02°10'·16W.
Noirmont Pt ⌁ Fl (4) 12s 18m 10M; B twr, W band; 49°09'·91N 02°10'·08W.
Pignonet ⌁ 49°09'·88N 02°09'·70W.
Ruaudière Rock ⌁ Fl G 3s; *Bell;* 49°09'·74N 02°08'·62W.

ST AUBIN'S BAY and HARBOUR

Les Grunes du Port ⌁ 49°10'·02N 02°09'·14W.
Diamond Rock ⌁ Fl (2) R 6s; 49°10'·12N 02°08'·64W.
Castle pier ⌁ Fl R 4s 8m 1M; 49°11'·13N 02°09'·634W.
North pier, Dir lt 254° ⌁ F WRG 5m, 248°-G-253°-W-255°-R-260°; 49°11'·22N 02°10'·03W. Same col, Iso R 4s 12m 10M.
Beach Rock ⌁ 49°11'·29N 02°08'·42W, (Apr-Oct).
Rocquemin ⌁ 49°10'·64N 02°07'·95W.
Baleine ⌁ 49°10'·41N 02°08'·23W.

ST HELIER

Elizabeth marina, west appr: Dir ⌁ 106°: F WRG 4m 1M; 096°-G-104°-W-108°-R-119°; R dayglo □, B stripe; 49°10'·76N 02°07'·12W.
La Vrachiére ⌁ 49°10'·90N 02°07'·59W, Fl (2) 5s 1M.
Fort Charles North ⌁ 49°10'·81N 02°07'·50W.
Marina ent ⌁ Oc G & Oc R, both 4s 2M; 49°10'·83N 02°07'·13W.
Red & Green Passage, ldg lts 022·7° on dayglo R dolphins: Front, Elizabeth E berth Dn ⌁ Oc G 5s 10m 11M; 49°10'·63N 02°06'·94W. Rear, Albert Pier root ⌁ Oc R 5s 18m 12M; synch; 230m SSW.
East Rock ⌁ QG; 49°09'·96N 02°07'·28W.
Dog's Nest Rk ⌁ Fl Y 3s 3M; Y cross on bcn; 49°09'·99N 02°06'·95W.
Oyster Rocks, R/W bcn, topmark 'O'; 49°10'·10N 02°07'·49W.
Platte Rock ⌁ Fl R 1·5s 6m 5M; R col; 49°10'·16N 02°07'·34W.
Small Road No. 2 ⌁ QR; 49°10'·39N 02°07'·24W.
No. 4 ⌁ QR; 49°10'·53N 02°07'·12W.
Elizabeth marina, S appr: E1 ⌁ Fl G 3s; 49°10'·59N 02°07'·09W.
E2 ⌁ Fl R 2s; 49°10'·58N 02°07'·13W.
E5 ⌁ Fl G 5s; 49°10'·70N 02°07'·17W.
E6 ⌁ Fl R 2s; 49°10'·69N 02°07'·21W.
Fort Charles East ⌁ Q (3) 5s 2m 1M; 49°10'·74N 02°07'·26W.
La Collette basin ⌁ QR; 49°10'·54N 02°06'·91W. ⌁ 49°10'·52N 02°06'·90W.
St Helier Hbr, ldg lts 078°, both FG on W cols. Front, 49°10'·62N 02°06'·66W. Rear, 80m from front.
Victoria pier hd, Port control twr; IPTS; 49°10'·57N 02°06'·88W.

JERSEY, SOUTH-EAST AND EAST COASTS

Hinguette ⌁ QR; 49°09'·33N 02°07'·32W.
Hettich ⌁ Fl Y 5s; 49°08'·10N 02°09'·00W.
South Pier Marine ⌁ Fl Y 5s; 49°09'·10N 02°06'·30W.

Demie de Pas ⌁ Mo (D) WR 12s 11m, W14M, R10M; 130°-R-303°-W-130°; *Horn (3) 60s;* **Racon T, 10M;** B bn twr, Y top; 49°09'·01N 02°06'·15W. Icho Tower (conspic, 14m) 49°08'·89N 02°02'·90W.
Canger Rock ⌁ Q (9) 15s; 49°07'·35N 02°00'·38W.
La Conchière ⌁ Q (6) + L Fl 15s 2M; 49°08'·22N 02°00'·17W.
Frouquier Aubert ⌁ Q (6) + L Fl 15s; 49°06'·08N 01°58'·84W.
Violet ⌁ L Fl 10s; 49°07'·81N 01°57'·14W.
Petite Anquette; SCM ⌁; 49°08'·47N 01°56'·29W.
Grande Anquette ⌁ VQ(9) 10s; 49°08'·32N 01°55'·20W.
Le Cochon ⌁ 49°09'·77N 01°58'·80W.
La Noire ⌁ 49°10'·13N 01°59'·23W. Le Giffard ⌁ 49°10'·59N 01°59'·00W.

GOREY

Pier Hd Dir lt 298°, ⌁ WRG, 6m 8M; 293·5°-G-296·5°-W-299·5°-R-302·5°; ⌁ Oc RG 5s 6m 8M; 304°-R-353°-G-304°; W twr on pierhead; 49°11'·80N 02°01'·34W.
Horn Rock ⌁, topmark 'H'; 49°10'·96N 01°59'·85W.
Les Burons, R bcn, topmark 'B'; 49°11'·33N 02°00'·81W.
Fairway ⌁ QG; 49°11'·50N 02°00'·34W.
Écureuil Rock ⌁ 49°11'·67N 02°00'·78W.
Equerrière Rk, bcn 'fishtail' topmark; 49°11'·80N 02°00'·67W.
Les Arch ⌁, BW bcn, 'A' topmark; 49°12'·02N 02°00'·60W.

ST CATHERINE BAY

St Catherine Bay, Le Fara ⌁ Q (3) 10s 3M; 49°12'·85N 02°00'·48W.
Archirondel Tower (conspic, 16m) 49°12'·72N 02°01'·42W.
Verclut bkwtr ⌁ Fl 1·5s 18m 13M; 49°13'·34N 02°00'·64W. In line 315° with unlit turret, 49°13'·96N 02°01'·57W, on La Coupe Pt.

JERSEY, NORTH COAST

Rozel Bay Dir lt 245°, F WRG 11m 5M; 240°-G-244°-W-246°-R-250°; W col; 49°14'·21N 02°02'·76W.
Bonne Nuit Bay ldg lts 223°: both FG 7/34m 6M. Front, Pier 49°15'·10N 02°07'·17W. Rear, 170m from front.
Demie Rock ⌁ 49°15'·56N 02°07'·36W.
Sorel Point ☆ L Fl WR 7·5s 50m **15M**; 095°-W-112°-R-173°-W-230°-R-269°-W-273°; W ○ twr, only 3m high; 49°15'·60N 02°09'·54W.

OFFLYING ISLANDS

LES ÉCREHOU
Écrevière ⌁ Q (6) + L Fl 15s; 49°15'·26N 01°52'·15W.
Mâitre Ile ⌁ 49°17'·08N 01°55'·60W.

PLATEAU DES MINQUIERS
N Minquiers ⌁ Q; 49°01'·64N 02°00'·58W.
NE Minquiers ⌁ VQ (3) 5s; *Bell;* 49°00'·85N 01°55'·30W.
SE Minquiers ⌁ Q (3) 10s; *Bell;* 48°53'·42N 02°00'·09W.
S Minquiers ⌁ Q (6) + L Fl 15s; 48°53'·09N 02°10'·10W.
SW Minquiers ⌁ Q (9) 15s 5M; *Whis;* 48°54'·35N 02°19'·41W.
NW Minquiers ⌁ Q 5M; *Whis;* 48°59'·63N 02°20'·59W.
Refuge ⌁; B/W bcn; 49°00'·13N 02°10'·16W.
Demie de Vascelin ⌁ 49°00'·81N 02°05'·17W.
Grand Vascelin, BW bcn ⌁ 48°59'·97N 02°07'·26W.
Maitresse Ile, Puffin B&W bcn twr ⌁ 48°58'·33N 02°03'·65W.
Le Coq Reef, ⌁ Q (3) 10s 6m 3M; 48°57'·88N 02°01'·29W.

FRENCH MARKS NORTH OF ILE DE BRÉHAT

Roches Douvres ☆ Fl 5s 60m **24M**; pink twr on dwelling with G roof; 49°06'·30N 02°48'·87W. 992271319. (16M NNE of Ile de Bréhat).
Barnouic ⌁ VQ (3) 5s 15m 7M; 49°01'·64N 02°48'·41W.
Roche Gautier ⌁ VQ (9) 10s; 49°02'·013N 02°54'·36W.
Les Héaux de Bréhat ☆ Fl (4) WRG 15s 48m, **W15M**, R/G11M; 227°-R-247°-W-270°-G-302°-W-227°; Gy twr; 48°54'·50N 03°05'·18W (4·7M WNW of Ile de Bréhat).

Channel Islands

9.19.5 PASSAGE INFORMATION

More Passage Information is threaded between successive harbours in this Area. **Bibliography:** *The Channel Islands* (Imray/RCC/Carnegie); *N Brittany & Channel Islands CC* (Fernhurst/Cumberlidge); *Channel Pilot* (NP27). *Channel Havens* (ACN/Endean); *Shell Channel Pilot* (Imray/Cunliffe).

THE CHANNEL ISLANDS

In an otherwise delightful cruising area (AC 2669) the main problems include fog and poor visibility, the very big tidal range, strong tidal streams, overfalls and steep seas which get up very quickly. The shoreline is generally rugged with sandy bays and many offlying rks. It is important to use large scale AC, and recognised leading marks (of which there are plenty) when entering or leaving many of the harbours and anchs. Several passages are marked by beacons and perches identified by an alphabetical letter(s) in lieu of topmark. Beware high speed ferries.

From the N, note the Casquets TSS and ITZ. Soundings of Hurd Deep can help navigation. The powerful lights at the Casquets, Alderney, Cap de la Hague, Cap Levi, Barfleur and Les Hanois greatly assist a night or dawn landfall. By day Alderney is relatively high and conspicuous. Sark is often seen before Guernsey which slopes down from S to N. Jersey is low-lying in the SE. The islands are fringed by many rky dangers. In poor visibility stay in harbour.

▶*Over a 12 hour period tidal streams broadly rotate anti-clockwise around the Islands, particularly in open water and in wider chans. The E-going (flood) stream is of less duration than the W-going, but is stronger. The islands lie across the main direction of the streams, so eddies are common along the shores. The range of tide is greatest in Jersey (9·6m sp, 4·1m np), and least in Alderney (5·3m sp, 2·2m np). Streams run hard through the chans and around headlands and need to be worked carefully; neaps are easier, particularly for a first visit. Strong W'lies cause a heavy sea, usually worst HW ±3.*◀

As well as the main harbours, described in some detail, there are also many attractive minor harbours and anchorages. In the very nature of islands a lee can usually be found somewhere. Boats which can take the ground can better explore the quieter hbrs. Avoid lobster pots and oyster beds.

THE CASQUETS AND ORTAC CHANNEL

Casquets lt ho is conspic on the largest island of this group of rks 5·5M W of Alderney (AC 60). Off-lying dangers extend 4ca W and WSW (The Ledge and Noire Roque) and 4ca E (Pt Colote). The tide runs very hard around and between these obstructions. A shallow bank, on which are situated Fourquie and l'Equêt rks (dry) lies from 5ca to 1M E of Casquets, and should not be approached. Ortac Rk (24m) is 3·5M E of Casquets. Ortac Chan runs NNW/SSE 5ca W of Ortac.

▶*Here the stream starts to run NE at HW St Helier – 2½, and SW at HW St Helier +0355; sp rates up to 5½kn (7kn reported).*◀

Ortac Channel should not be used in bad weather due to tremendous overfalls; these and violent eddies also occur over Eight Fathom Ledge (8½ca W of Casquets) and over the SW, SSW and SSE Casquet Banks where least depths of 7·3m occur.

ALDERNEY AND THE SWINGE

Braye Harbour (AC 2845) together with pleasant bays and anchs around the island, offer shelter from different wind/sea directions. The sunken NE extremity of Admiralty Breakwater should only be crossed in calm conditions, outside LW±2 and about 50m off the head of the Breakwater where there is 2·3m.

The **Swinge** lies between Burhou with its bordering rks, and the NW coast of Alderney. It can be a dangerous chan, and should only be used in reasonable vis and fair weather. On the N side of the

Swinge the main dangers are Boues des Kaines, almost awash at LW about 7½ca ESE of Ortac, and North Rk 2½ca SE of Burhou. On S side of the Swinge beware Corbet Rk (0·5m high) with drying outliers, 5ca N of Fort Clonque, and Barsier Rk (0·9m) 3½ca NNW of Fort Clonque.

▶*The SW-going stream begins at HW St Helier +0340, and the NE stream at HW St Helier –0245, sp rates 7-8kn. On the NE-going stream, beware the very strong northerly set in vicinity of Ortac.*◀

The tide runs very hard, and in strong or gale force winds from S or W there are very heavy overfalls on the SW-going stream between Ortac and Les Etacs (off W end of Alderney).

In strong E winds, on the NE-going stream, overfalls occur between **Burhou** and Braye breakwater. These overfalls can mostly be avoided by choosing the best time and route (see below) but due to the uneven bottom and strong tides broken water may be met even in calm conditions.

▶*It is best to pass SW through the Swinge is about HW St Helier +4, when the SW-going stream starts; hold to the SE side of the chan since the stronger stream runs on the Burhou side. But after HW St Helier +5, to clear the worst of the overfalls keep close to Burhou and Ortac, avoiding North Rk and Boues des Kaines.*◀

Pierre au Vraic (dries 1·2m) is an unmarked pinnacle rock at 49°41'·61N 02°16'·94W,1·8M S of Ortac and 1·8M WSW of Les Étacs, almost in the fairway to/from the Swinge. Arriving on a fair tide from Guernsey it will be well covered, but it is a serious hazard if leaving the Swinge on a SW-going spring tide close to local LW. Carefully study the clearing bearings shown on AC 60.

Heading NE at about HW St Helier –2, Great Nannel in transit with E end of Burhou clears Pierre au Vraic to the E, but passes close W of Les Etacs. On this transit, when Roque Tourgis fort is abeam, alter slightly to stbd to pass 1ca NW of Corbet Rk; keep near SE side of the channel.

THE ALDERNEY RACE

This race, characterised by very strong tidal streams, runs SW/NE between Cap de la Hague and Alderney, but its influence extends at least as far SW as 02° 20'W (AC 3653). The fairway, approx 4M wide, is bounded by Race Rk and Alderney S Banks to the NW, and to the SE by rky banks 3M E of Race Rk, and by Milieu and Banc de la Schôle (least depth 2·7m). These dangers which cause breaking seas and heavy overfalls should be carefully avoided. In bad weather and strong wind-against-tide the seas break in all parts of the Race and passage is not advised. Conditions are worst at springs.

▶*In mid-chan the SW-going stream starts at HW St Helier +4½ (HW Dover) and the NE-going stream at HW St Helier –2¼ (HW Dover +5½), sp rates both 5·5kn. The times at which the stream turns do not vary much in various places, but the rates do; for example, 1M W of Cap de la Hague sp rates reach 7-8kn.*◀

For optimum conditions, timing is of the essence. As a rule of thumb the Race should be entered on the first of the fair tide so as to avoid peak tidal streams with attendant overfalls/seas.

▶*Thus SW-bound, arrrive off Cap de la Hague at around HW St Helier +4½ (HW Dover –½ to HW Dover) when the stream will be slack, whilst just starting to run SW off Alderney. A yacht leaving Cherbourg at HW Dover –3 will achieve the above timing by utilising the inshore W-going tidal eddy.*

Conversely, NE-bound, leave St Peter Port, say, at approx local HW St Helier –4½ (HWD+3) with a foul tide so as to pass Banc de la Schôle as the fair tide starts to make.◀ A later departure should achieve a faster passage, but with potentially less favourable conditions in the Race. On the NE-going stream the worst overfalls are on the French side.

9.19.6 SPECIAL NOTES: CHANNEL ISLANDS

The Channel Islands (Alderney, Guernsey, Sark, Jersey and other smaller islands) lie, not in the Channel, but in the Bay of St Malo. Alderney, Herm, Jethou, Sark and Brecqhou are all part of the Bailiwick of Guernsey and the States (Parliament) of Alderney have seats in the States of Guernsey.

The Islands are self-governing with their own laws, judiciary and Customs, but conduct their foreign affairs mostly through the English crown. They are not part of either the UK or the EU. The French call the CI Les Îles Anglo-Normandes (Aurigny, Guernsey, Sercq et Jersey). Standard Time is UT (Zone 0).

Charts Admiralty Leisure folio SC5604 (£44·30, 7th ed. published Feb 2013) contains 17 sheets of the CIs These and other charts are listed under individual ports.

Ports of entry Braye, Beaucette, St Sampson, St Peter Port, St Helier and Gorey.

Customs British yachts entering the CI, except Jersey, *must* complete and deposit the local Customs declaration form and may have to show the vessel's registration documents; they are no longer expected to fly the Q flag arriving from the UK or France. On arrival back in UK they are subject to Customs formalities. Yachts going to France need usual documents (passports etc).

Speed limit, Jersey, 5 kts in all hbrs, and within 200m of water's edge in any bay, inside 10m depth contour, Les Écrehou and Les Dirouilles, inside 5m depth contour, Les Minquiers.

Medical Jersey has reciprocal medical arrangements with the UK, but the other CI do not and visitors will have to pay for costly inpatient / outpatient care. Medical insurance recommended.

Weather forecasts prepared by Jersey Met Office for the CI area and adjacent coasts of Normandy and Brittany are available ☎0900 669 0022 (premium rate) throughout the CI and UK. From France replace the first 0 by 00 44. More details, Chapter 6. **SAR**

operations are co-ordinated by Guernsey Coastguard in the N area and Jersey Coastguard in the S (see Chapter 5). Major incidents are co-ordinated with CROSS Jobourg and Falmouth CGOC. There are ⒧ⒷⓈ at Braye, St Peter Port, St Helier and St Catheriness (Jersey).

Telephones CI phones are integrated with the UK, but charges, especially on mobiles, may vary. Toll-free and premium numbers may be blocked when roaming with non-CI mobiles. For an on-line telephone directory visit www.theguernseydirectory.com

Courtesy flags Many yachts fly a courtesy flag in CI ports as a mark of politeness but it is not essential. The local flags are: *Jersey* R ensign with the Jersey Royal Arms (three lions passant with gold crown above); see Fig 5(9).

Guernsey R ensign with Duke William's cross in the fly. Vessels owned by Guernsey residents may wear this ensign; see Fig 5(9).

Sark The English (St George's) flag with the Normandy arms in the canton.

Alderney The English (St George's) flag and in the centre a green disc charged with a gold lion.

Herm The English (St George's) flag, and in the canton the Arms of Herm (three cowled monks on a gold diagonal stripe between blue triangles containing a silver dolphin).

Cars can be hired in Jersey, Guernsey and Alderney. In Sark cars are prohib, but bikes and horse-drawn carriages can be hired.

Animals may only be landed from boats from the UK, Ireland, Isle of Man or other CI, but not if the boat has visited France. No vessel with an animal on board may lie alongside a pontoon or quay without a Revenue Officer's permission.

UK currency can be used in the Islands but not vice versa. Postage stamps, issued by Jersey, Guernsey and Alderney must be used in the appropriate Bailiwick; Guernsey and Alderney stamps are interchangeable. There is only one class of post.

Gas Some UK Gas retailers may not accept Channel Island bottles.

9.19.7 ALDERNEY

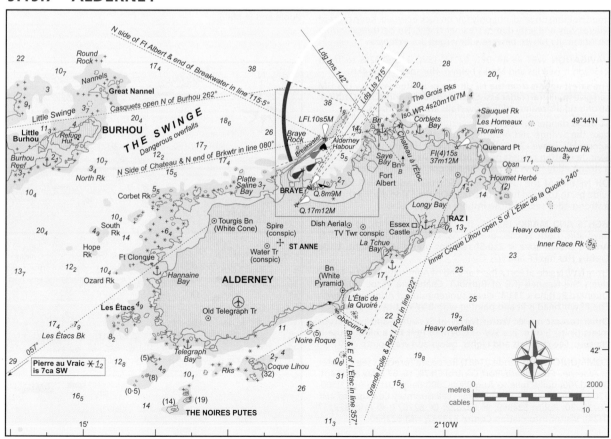

ANCHORAGES AROUND ALDERNEY There are several ⚓s, all picturesque but only safe in settled weather and offshore winds; most are very small and many have offlying rocks. Entry to these anchorages is not advised without local knowledge, advice from the HM, a good Pilot book plus AC 60 and/or preferably 2845. None provide any facilities. Clockwise from Braye they are:

Saye Bay 49°43′·85N 02°10′·87W. Small sandy bay 4ca E of ent to Braye Hbr. Appr on transit 142° of ldg bns; move E to open up the ent. Ent is 100m wide between Homet des Pies and Homet des Agneaux. Exposed to N'lies. Château a l'Étoc ☆ is 300m ENE. Speed limit 4kn.

Longy Bay 49°43′·06N 02°10′·29W. Wide drying bay with good holding on sand. Speed limit 4kn. Appr on N between Queslingue (a 14m high stack) and a rk 0·6m to stbd. ⚓ in 3·5m closer to Essex Castle than to Raz Island to await fair tide in the Race. Whilst the stream is setting SW offshore, a NE-going eddy develops along the island's southern shore and this would assist a straightforward arrival.

La Tchue 49°42′·78N 02°10′·75W. Good holding in small bay surrounded by cliffs. La Rocque Pendante to the E and smoking rubbish tip to the NW are both conspic.

Telegraph Bay 49°42′·01N 02°13′·51W. Pleasant sandy bay on SW tip of the island but ringed by rks. Appr 035° between Noires Putes and Coupé. Old Telegraph Twr (93m) is conspic until obsc'd close inshore, when a pillar on the cliff edge offers brgs, as does the astern transit of Coupé (7m) and and Orbouée (0.5m).

Hannaine Bay 49°42′·73N 02°13′·98W. A good place to await the flood tide. Appr Tourgis Bn △ ≠ 057° SE side of Fort Clonque. Beware rks either side. ⚓ on sand in 3m, 100m S of Fort.

Platte Saline Bay 49°43′·37N 02°12′·65W. Good shelter from E'lies. Les Jumelles and Outer Fourchie guard the ent. Beach is unsafe for swimming.

Burhou 49°43′·64N 02°15′·49W. Temp'y ⚓ in bay SW of Burhou, only on SW stream; exposed at HW on the NE-going stream. Appr on 010° for the gap between Burhou and Little Burhou.

9.19.8 BRAYE HARBOUR

Alderney 49°43'·77N 02°11'·51W ❀❀❀⬥⬥⬥✿✿✿

CHARTS AC 2669, 3653, 60, 2845, 5604.7/.8; SHOM 7158, 6934; Navi 1014; Imray C33A, 2500

TIDES –0400 Dover; ML 3·5; Duration 0545

Standard Port ST HELIER (→)

Times				Height (metres)			
High Water		Low Water		MHWS	MHWN	MLWN	MLWS
0300	0900	0200	0900	11·0	8·1	4·0	1·4
1500	2100	1400	2100				
Differences BRAYE							
+0050	+0040	+0025	+0105	–4·8	–3·4	–1·5	–0·5

SHELTER Good, but exposed to strong N/NE'lies. ⚓s as charted; good holding on sand, but only fair on rock or weed. Keep clear of the fairway and jetty due to steamer traffic. No berthing/landing on Admiralty bkwtr; beware submerged 3ca extension.

NAVIGATION WPT 49°44′·08N 02°11′·18W, 215°/0.82M to front 215° ldg lt (see Lts and Marks below). Hbr speed limit 4kn.

The safest appr is from the NE. In fog or limited visibility the HM can give small craft radar assistance and RDF bearings HO. Caution: Strong tidal streams, eddies, overfalls and many rocks.

To avoid dangerous overfalls in certain wind and tide conditions take the Swinge and Alderney Race at/near slack water; this is approx HW Braye +2½ & LW +2½ (or HW St Helier +3½ and –3). The Swinge is often calmest near Corbet Rk. At mid-flood (NE-going) a strong eddy flows SW past the hbr ent.

Off NE end of the island give Brinchetais Ledge and Race Rk a wide berth to avoid heavy overfalls. On the S side during the ebb, a strong eddy sets NE close inshore of Coque Lihou.

LIGHTS AND MARKS Hbr ldg lts 215°, both Q synch; co-located daymarks, both W cols, orange △s. Old Hbr white bcn and St Anne's church spire ≠ 210·5° remains useful in daylight. Little Crabby Hbr has FR and FG.

From NW, N side of Fort Albert and hd of Admiralty Bkwtr ≠ 115·5° clears the Nannels (NE of Burhou). Château à l'Etoc lt ho and Quenard Pt lt ho ≠ 111·1° clears sunken ruins of Admiralty Bkwtr. The Bkwtr head is lit and painted with B/W vertical stripes.

Other conspic marks: Water twr W of St Anne's and two lattice masts (R lts) E of it. 2ca SW of Quenard Pt lt ho a blockhouse is conspic. See chartlets and Lights, buoys and waypoints.

COMMUNICATIONS (Code 01481) Recorded forecasts see Chapter 6; *Alderney Coastguard* /Port Control Ch **74**, 16, 67 (May and Sept, 0800-1700, daily; June to Aug 0800-2000 daily; Oct to Apr, 0800-1700, Mon-Fri: all LT) at other times call *Guernsey Coastguard* VHF Ch 16, 20; Police 824999; Ⓗ 822822; Dr 822077; ♥ Info 823737 (H24); Visitor Information Centre 823737; HM 822620.

FACILITIES 70 Y ⚓s (£15·00) with pick-up lines lie parallel to the Admiralty bkwtr and in blocks E of Braye Jetty and SW of Toulouse Rock. Orange buoys are for locals only. If >12m LOA, check suitability of ⚓ with HM. ⚓s (£5·00 inc ♖) as on chartlet.

HM (aka CG & ⊖): ☎822620. mark.gaudion@gov.gg www.sailalderney.com/harbour.html www.alderney.gov.gg

Commercial Quay ⚓ ♖ ⬚ ▬ ℗ Ⓔ ✕ ⬙(25t).

Little Crabby Hbr Access at the top of the tide only for **Mainbrayce**, VHF Ch 37 *Mainbrayce*, ☎822772, mob 07781 415420: ⬙ ⚓ ◨ 🔧 ACA ⬙ 🗉. Water taxi call *Mainbrayce* Ch M (0800-2359) ☎ as above, £1·50 each way till 2130, £2·50 after 2130.

Alderney SC ☎822758, ⬚ visitors welcome.

Town (St Anne) ✕ ⬚ 🛒 ✉ Ⓑ. ✈ direct to Guernsey, Southampton; other connections via Guernsey. Ferries: via Guernsey to Jersey, Poole and St Malo.

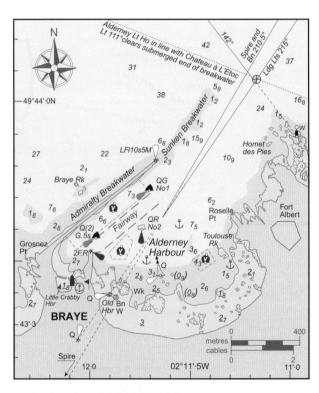

APPROACHES TO GUERNSEY

From the N/NE, The Little Russel Channel (AC 3654, 808) gives the most convenient access to Beaucette Marina and St Peter Port (AC 3140). But it needs care, due to rks which fringe the chan and appr, and the strong tide which sets across the ent.

▶In mid chan, S of Platte and NW of Bréhon, the NE-going stream begins at HW St Peter Port –0245, and the SW stream at HW St Peter Port +0330, sp rates both 5·25kn which can raise a very steep sea with wind against tide.◀

With lts on Platte Fougère, Tautenay, Roustel, Platte and Bréhon, plus the 220° ldg lts for St Peter Port, the Little Russel can be navigated day or night in reasonable vis, even at LW.

The Big Russel is wider and easier. In bad weather or poor vis it may be a better approach to St Peter Port, via Lower Heads SCM lt buoy, though may be lumpy on a rising sp tide. From the NW, Doyle Passage, which is aligned 146°/326° off Beaucette, can be used but only by day with local knowledge. From the S or W, the natural route is around St Martin's Pt, keeping 1·5ca ENE of Longue Pierre bcn (LP) and a similar distance off Anfré bcn (A).

In onshore winds keep well clear of the W coast, where in bad weather the sea breaks on dangers up to 4M offshore. But in settled weather it is worth exploring the rock strewn bays.

9.19.9 LITTLE RUSSEL CHANNEL

See notes above and AC 5604.9/10, 807, 808

MINOR HARBOURS AND ANCHORAGES AROUND GUERNSEY

Hbrs and ⚓ages are listed clockwise from the north. All have buses to St Peter Port (from cliff-top level on the E and S coasts).

EAST COAST

Bordeaux Hbr 49°29'·31N 02°30'·37W. Small drying hbr full of moorings. Rky appr and strong cross tide. Open to E. Café.

St Sampson's Hbr 49°28'·91N 02°30'·73W. Dries 3·6m. Port of entry. Good shelter but predominantly a working hbr. ♥ only when visiting for work (prior arrangement). Entry over lifting flapgate HW -1 to +2. Marina tfc sigs SW cnr Abraham's Bosom: steady ● = ent/exit; Fl ◐ or steady ● = No ent/exit. WPT 49°28'·70N 02°29'·65W (intersection of St Peter Port and St Sampson ldg lts). 2 chimneys NW of hbr are conspic. Ldg lts 286°: Front on S Pier, FR; rear on clocktower, FG. Other lts as chartlet and Lights, buoys and waypoints. On S Pier a 360° Fl R above FR (shown landward) = large commercial vessel under way in the hbr; no ent/exit for other craft without Dockmaster's approval on Ch 12 (H24) or ☎01481 720229. Speed limit 6kn. No ⚓ in hbr. Services: ⚓ (Marine & General 01481 243048, VHF Ch 17) ⬛(75t) ⚒ ⬚ ⚓ ⚒ ⬚ ⚓. ▲ by bowser (2 companies) at Abraham's Bosom (E side) must pre-book ☎01481 200800/245858.

Havelet Bay 49°27'·06N 02°31'·68W. Enter between SHM buoy QG and PHM buoy QR, marking Oyster Rk, bn 'O', and Moulinet Rk, bn M. SHM and PHM lt buoys about 100m closer inshore. Crowded ⚓ in summer but sheltered in SW gales; no ⚓s. Landing slip close W of Castle Cornet. Power cables on chartlet overleaf are normally buried to 1m but should be avoided. Beware swimmers far out.

Soldier's Bay 49°27'·06N 02°31'·68W. Good holding on sand; open to E. Beware Boue Sablon and rks off Les Terres Point. Anfré Rk (3₃) bn A, is 4ca offshore. Steps to cliff path.

Fermain Bay 49°26'·05N 02°31'·95W. Good holding on sand; open to E. Beware Gold Fisher Rk (3·6) and Gabrielle Rk (2·1). Popular tourist beach. ✕ hotel ⬚.

SOUTH COAST
In centre of first bay W of St Martin's Pt, beware Mouillière (8·5) 49°25'·28N 02°32'·86W. S'lies can bring swell above half tide. In this bay are:

Petit Port Good holding on sand. Steep steps lead up to The Auberge, a popular fine dining restaurant and bar with a spectacular view.

Moulin Huet Good holding on sand. Tea garden and hotel.

Saints Bay Good holding on sand. Below half tide beware uncharted rock in middle and unburied telephone cable on E side. ⚓ outside moorings with trip line. Café, hotel, bar.

Icart Bay Beware Fourquie de la Moye (3·1) 49°25'·28N 02°32'·86W between Icart Pt and Pte de la Moye. Within Icart Bay are:

Jaonnet. Good holding on sand off small beach or further E off rky shore. Exposed to S. No facilities. Cliff path inland.

Petit Bôt Bay Beware drying reef on E side. A short swell often works in. ✕ up hill to hotel, ⬚ ✈ ✉ ⬛. Better ⚓ close W at:

Portelet 49°25'·11N 02°35'·28W. Good holding on sand. Sheltered from N and W. No access inland. Facilities via dinghy/Petit Bôt.

WEST COAST
Good visibility and AC 807 essential; an E'ly wind is desirable. Pass outside Les Hanois, unless bound for:

Portelet Harbour ⚓ at 49°26'·40N 02°40'·26W, outside moorings. Good holding on sand. Exposed to W. Rky appr; local knowledge advised. Avoid small drying stone quay. ✕ hotel ⬚.

Lihou Island ⚓ 49°27'·88N 02°39'·46W, off NE corner. Sheltered from E. Between Lihou and Guernsey is a drying causeway. Tide runs fast. Avoid bird sanctuaries on off-lying islets. No facilities.

Perelle Bay 49°28'·23N 02°38'·89W. Good holding on sand. Exposed to W; rky appr. Beware Colombelle Rk (1₅) NE of brg 128° on cement mill. ⚓ outside moorings. ⬚ hotel ⬛ ⬚.

Vazon Bay 49°28'·29N 02°37'·09W. Wide sandy beach for settled conditions, but open to the W. Beware Boue Vazon (3) in appr, many lobster pots, surfers and bathers. Long surf line. Hotel, ✕ ⬚.

Cobo Bay 49°28'·66N 02°36'·33W. Beware lobster pots and Boue Vazon (3) in appr. Rky appr from S of Moulière. Good holding on sand; ⚓ outside moorings. Hotel, ✕ ⬚ ✉ ⬛ ⬚ ⬚.

Grande Havre 49°30'·11N 02°33'·09W. Very popular, many local moorings. Rky appr 171° as charted; exposed to NW, sheltered from S'lies. ⚓ to W of Hommet de Grève. Stone slip, busy in summer; lying alongside not advised. Facilities: Hotel ⬚.

L'Ancresse Bay 49°30'·45N 02°31'·38W. Good holding on sand. Exposed to the N, but good shelter from S/SW. Hotel ⬚ café.

Fontenelle Bay 49°30'·48N 02°30'·69W. Good holding on sand. Exposed to the N. Beware drying rks on E of ent. No facilities.

9.19.10 BEAUCETTE
Guernsey 49°30'·19N 02°30'·20W ✲✲◉⚓⚓⚓⚓✿✿✿

CHARTS AC 3654, 807, 808, 5604.9; SHOM 7159, 6904, 6903; Navi 1014; Imray C33A, 2500

TIDES −0450 Dover; ML 5·0; Duration 0550. See Facilities and St Peter Port for access times.

SHELTER Excellent in marina. Entry not advised in strong N–SE onshore winds.

NAVIGATION WPT 49°30'·15N 02°28'·85W, 277°/8½ ca to ent. Appr from Little Russel to mid-way between Platte Fougère lt tr (W with B band, 25m) and Roustel lt tr. Pick up the ldg marks/lts, SWM L Fl 10s, a SHM Fl G 5s, 2 PHM buoys Fl R 5s & Fl (3) R 5s, a NCM perch Q and a SHM perch Q (3) G 5s at the ent.

Beware stream setting across the appr chan and drying rocks either side.

LIGHTS AND MARKS See chartlet and Lights, buoys and waypoints. Petite Canupe is a very spindly SCM lt bcn. 277° ldg lts: both FR. Front, R arrow on W background, stbd side of ent; rear, W arrow on R background, on roof of W bldg, with windsock.

COMMUNICATIONS (Code 01481); Police 725111; ⊜ 245000; St John Ambulance 725211; HM & water taxi VHF Ch 80 (0700-2200).

FACILITIES Marina sill dries 2·37m. 6 Y waiting buoys lie outside and a tide gauge is inside the 8m wide ent channel. For entry sequence and a berth it is essential to pre-call HM with boat details: ☎245000, mob 07781102302. www.beaucettemarina.com info@beaucettemarina.com ④

115 inc approx 45▼ £2·25, min charge £15·75/boat; £0·75 for <4 hrs. ▲ on pontoon B, ▲✕⬚⬛✉▲ ⚓ ⬚(16t) ⬚(12t) ☎247066, bike hire, ♿ access ramp is steep @ LW.

Village Basic food shop. Good coastal walks. For further information see St Peter Port; 20 mins by bus 6/6A (but bus stop is nearly a mile away).

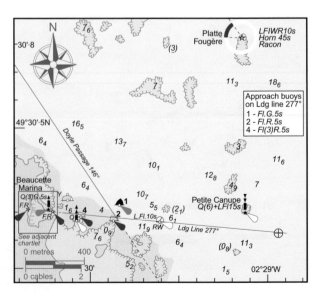

9.19.11 ST PETER PORT
Guernsey ✿✿✿✿◊◊◊✿✿✿

N appr WPT: 49°30'·23N 02°27'·68W (Little Russel)
S appr WPT: 49°25'·00N 02°31'·17W (St Martin's Point)

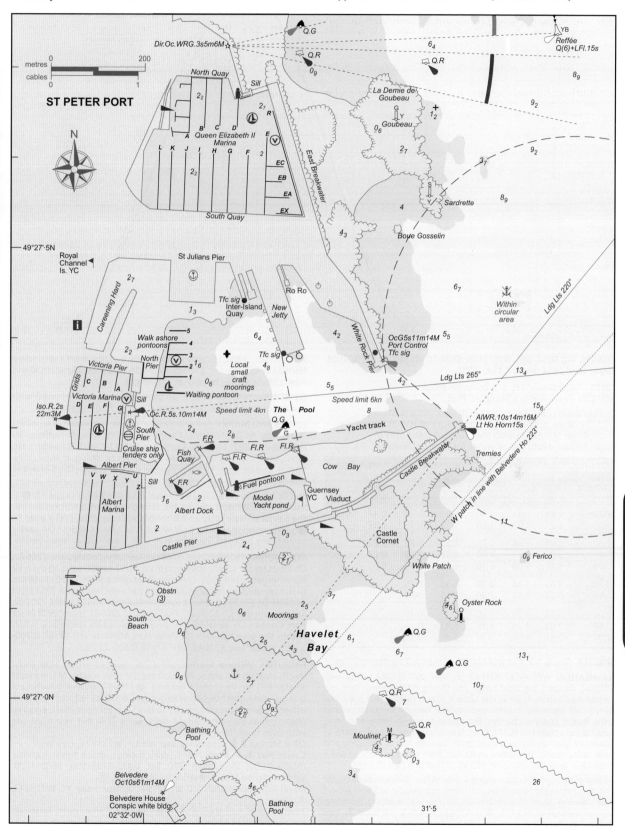

ST PETER PORT

Channel Islands

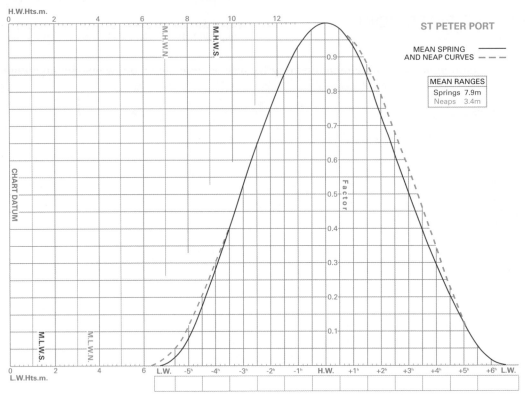

ST PETER PORT

MEAN SPRING ———
AND NEAP CURVES -- -- --

MEAN RANGES
Springs 7.9m
Neaps 3.4m

CHARTS AC 3654, 808, 807, 3140, 5604.10/11; SHOM 7159, 6903, 6904; Navi 1014; Imray C33A, 2500

TIDES –0439 Dover; ML 5·2; Duration 0550
NOTE: St Peter Port is a Standard Port.

To find depth of water over the sill into Victoria Marina:
1. Look up predicted time and height of HW St Peter Port.
2. Enter table below on the line for height of HW.
3. Extract depth (m) of water for time before/after HW.

Ht (m) of HW St Peter Port	Depth of water in metres over the sill (dries 4·2 m)						
	HW	±1hr	±2hrs	±2½hrs	±3hrs	±3½hrs	±4hrs
6·20	1·85	1·67	1·30	1·03	0·75	0·47	0·20
·60	2·25	2·00	1·50	1·13	0·75	0·37	–
7·00	2·65	2·34	1·70	1·23	0·75	0·27	–
·40	3·05	2·67	1·90	1·33	0·75	0·27	–
·80	3·45	3·00	2·10	1·43	0·75	0·07	–
8·20	3·85	3·34	2·30	1·53	0·75	–	–
·60	4·25	3·67	2·50	1·63	0·75	–	–
9·00	4·65	4·00	2·70	1·73	0·75	–	–
·40	5·05	4·34	2·90	1·83	0·75	–	–
·80	5·45	4·67	3·10	1·93	0·75	–	–

SHELTER Good, but Victoria Marina is exposed to strong E'lies.

NAVIGATION WPT 49°27'·82N 02°30'·78W, 227°/0·68M to hbr ent. Offlying dangers, big tidal range and strong tidal streams demand careful navigation. From N the Little Russel is most direct appr, but needs care especially in poor visibility; see Passage Information and Little Russel Channel chartlet. Big Russel between Herm and Sark, passing S of Lower Hds SCM lt buoy, is an easier appr. From W and S of Guernsey, give Les Hanois a wide berth. Beware ferries, shipping and unlit racing marks. Hbr speed limits: 6kn from outer pier hds to line S from New Jetty to Castle pier; 4kn W of that line (see chartlet).

Access via buoyed/lit chan along S side of hbr. Beware busy cruise ship tenders using same chan and landing pontoon at E end Albert Pier. Marina boat directs arrivals to marina, waiting pontoon (colour coded) or, N of it, ♥ pontoons nos 1–5. All have ⚓, nos 1–4 are walk-ashore (seasonal only). Pontoons for tenders each

side of marina ent. Local moorings, centre of hbr, with a secondary fairway N of them. ⚓ prohib. ♥ berths in Queen Elizabeth II and Albert Marinas only by prior arrangement.

GY RDF beacon, 304·50kHz, on Castle Bkwtr is synchronised with the co-located horn (15s) to give distance finding; see LBW.

LIGHTS AND MARKS See chartlet and Lights, buoys and waypoints. Outer ldg lts 220°: Front, Castle bkwtr hd; rear, Belvedere. By day, white patch at Castle Cornet in line 223° with Belvedere Ho (conspic). Inner ldg lts 265° are for ferries berthing at New Jetty. Yachts should appr Victoria marina via the buoyed/lit S channel (dashed line).

Traffic Signals When a large vessel is under way a single FR ● is shown from White Rock pier hd, facing landward and/or seaward; the S ends of New Jetty and Inter-Island Quay show Q Fl Y over FR ● = 'No vessel may enter or leave hbr', but boats <15m LOA under power may proceed, keeping well clear commercial vessel berths.

COMMUNICATIONS (Code 01481) *Guernsey Coastguard* 720672, Ch 20 (H24) for safety traffic, VHF Ch 16/67 for DF brgs, VHF Ch 62 for emergency link calls; Guernsey Met office 0906 713 0111 (from Guernsey only, lowest premium rate) for weather; Police 725111; ⊖ 741410; Dr 711237 (H24); St John Ambulance 725211; HM 720229; *St Sampson* VHF Ch 12 (H24); Port Control 720481, monitor *St Peter Port Control* Ch 12 (H24) call only if absolutely necessary, when within the pilotage area; *St Peter Port Marinas* VHF Ch 80 (0700-2300); Water taxi 424042, VHF Ch 10 (0800-2359LT).

FACILITIES Victoria Marina, sill 4·2m; gauges either side show depth over sill (on entry, PH accurate, SH over-reads). Marina staff and/or R/G tfc lts control ent/exit. ☎725987. guernsey.harbour@ gov.gg www.guernseyharbours.gov.gg 400♥ £2·60, Jul/Aug, less other months. Max LOA/draft, 12·8m/1·8m (deeper draft, enquire ahead); max stay normally 14 days; ⚓ but you must use your own hose; free wi-fi.
QE II Marina 🛥🚶 times vary, esp Winter, ☎726071, free wi-fi.
Castle Pier ⚓🛥 ☎726071 (dries springs, check tides) Summer (01/05–01/09) M–F, 0800–1730. Sat, 0830–1730 Sun, 0830–1230. Winter, closed Sun. ⛽ 🛥 ACA, 🔧 ⚓ 🔩 Ⓔ ⓦ.
Royal Channel Islands YC ☎723154, 🍴. **Guernsey YC** ☎725342, ✕🍴.
Town ✕🍴🛒🏧⛽🅿@✉Ⓑ. Ferries to Poole, Portsmouth, Jersey (St Helier/Gorey), St Malo, Diélette, Carteret, Sark, Herm; ✈.

STANDARD TIME (UT)
For Summer Time add ONE hour in **non-shaded areas**

ST PETER PORT LAT 49°27'N LONG 2°32'W
TIMES AND HEIGHTS OF HIGH AND LOW WATERS

Dates in red are **SPRINGS**
Dates in blue are **NEAPS**

YEAR 2016

JANUARY

Day	Time	m	Day	Time	m
1 F	0440 / 1039 / 1708 / 2306	3.2 / 7.7 / 3.2 / 7.3	16 SA	0446 / 1051 / 1717 / ☽2318	2.3 / 8.4 / 2.3 / 8.0
2 SA	0524 / 1125 / 1755 / ☽2359	3.6 / 7.2 / 3.6 / 7.0	17 SU	0540 / 1148 / 1815	2.7 / 8.0 / 2.7
3 SU	0622 / 1224 / 1856	3.9 / 6.9 / 3.8	18 M	0021 / 0649 / 1300 / 1928	7.6 / 3.1 / 7.6 / 3.0
4 M	0109 / 0732 / 1338 / 2006	6.8 / 4.0 / 6.9 / 3.8	19 TU	0140 / 0815 / 1424 / 2052	7.4 / 3.2 / 7.5 / 3.1
5 TU	0224 / 0843 / 1450 / 2113	7.0 / 3.8 / 7.0 / 3.5	20 W	0303 / 0938 / 1540 / 2206	7.6 / 3.0 / 7.7 / 2.8
6 W	0329 / 0946 / 1551 / 2212	7.3 / 3.4 / 7.4 / 3.1	21 TH	0411 / 1043 / 1643 / 2306	8.0 / 2.5 / 8.1 / 2.4
7 TH	0421 / 1040 / 1642 / 2303	7.8 / 2.9 / 7.8 / 2.7	22 F	0506 / 1138 / 1735 / 2358	8.5 / 2.0 / 8.5 / 1.9
8 F	0508 / 1129 / 1730 / 2349	8.3 / 2.4 / 8.3 / 2.4	23 SA	0555 / 1226 / 1822	8.9 / 1.6 / 8.8
9 SA	0552 / 1214 / 1814	8.7 / 1.9 / 8.6	24 SU	0043 / 0638 / 1311 / ○1905	1.6 / 9.2 / 1.4 / 9.0
10 SU	0033 / 0634 / 1259 / ●1857	1.9 / 9.1 / 1.6 / 8.9	25 M	0126 / 0719 / 1351 / 1944	1.4 / 9.4 / 1.3 / 9.1
11 M	0116 / 0715 / 1342 / 1938	1.6 / 9.3 / 1.3 / 9.1	26 TU	0203 / 0756 / 1427 / 2019	1.4 / 9.3 / 1.3 / 9.0
12 TU	0158 / 0756 / 1423 / 2019	1.4 / 9.4 / 1.2 / 9.2	27 W	0237 / 0830 / 1500 / 2052	1.5 / 9.2 / 1.6 / 8.8
13 W	0238 / 0837 / 1504 / 2100	1.4 / 9.4 / 1.3 / 9.1	28 TH	0308 / 0902 / 1529 / 2122	1.8 / 8.9 / 1.9 / 8.5
14 TH	0318 / 0919 / 1545 / 2142	1.6 / 9.2 / 1.5 / 8.9	29 F	0336 / 0932 / 1556 / 2151	2.2 / 8.5 / 2.3 / 8.1
15 F	0400 / 1003 / 1629 / 2227	1.9 / 8.9 / 1.9 / 8.5	30 SA	0404 / 1003 / 1624 / 2223	2.6 / 8.0 / 2.8 / 7.7
			31 SU	0436 / 1038 / 1657 / 2300	3.1 / 7.5 / 3.2 / 7.3

FEBRUARY

Day	Time	m	Day	Time	m
1 M	0516 / 1121 / 1742 / ☽2350	3.5 / 7.1 / 3.6 / 6.9	16 TU	0618 / 1229 / 1854	3.1 / 7.3 / 3.2
2 TU	0618 / 1222 / 1853	3.9 / 6.8 / 3.9	17 W	0106 / 0745 / 1402 / 2027	7.2 / 3.4 / 7.0 / 3.4
3 W	0106 / 0742 / 1348 / 2017	6.7 / 3.9 / 6.7 / 3.8	18 TH	0242 / 0923 / 1529 / 2153	7.2 / 3.3 / 7.2 / 3.2
4 TH	0236 / 0901 / 1509 / 2132	6.9 / 3.6 / 7.0 / 3.5	19 F	0358 / 1033 / 1632 / 2254	7.6 / 2.8 / 7.7 / 2.7
5 F	0346 / 1008 / 1614 / 2234	7.4 / 3.1 / 7.5 / 2.9	20 SA	0453 / 1126 / 1722 / 2344	8.1 / 2.3 / 8.2 / 2.2
6 SA	0442 / 1104 / 1708 / 2328	7.9 / 2.5 / 8.0 / 2.3	21 SU	0540 / 1211 / 1806	8.6 / 1.8 / 8.6
7 SU	0531 / 1156 / 1757	8.5 / 1.9 / 8.6	22 M	0027 / 0621 / 1252 / ○1846	1.7 / 9.0 / 1.4 / 8.9
8 M	0017 / 0617 / 1244 / ●1842	1.7 / 9.1 / 1.3 / 9.1	23 TU	0107 / 0659 / 1330 / 1922	1.4 / 9.2 / 1.3 / 9.1
9 TU	0103 / 0701 / 1329 / 1925	1.3 / 9.5 / 0.9 / 9.4	24 W	0142 / 0735 / 1403 / 1955	1.3 / 9.3 / 1.2 / 9.1
10 W	0146 / 0743 / 1411 / 2006	0.9 / 9.8 / 0.6 / 9.6	25 TH	0214 / 0807 / 1433 / 2026	1.3 / 9.2 / 1.4 / 9.0
11 TH	0228 / 0825 / 1452 / 2047	0.8 / 9.9 / 0.6 / 9.6	26 F	0242 / 0836 / 1500 / 2053	1.5 / 9.0 / 1.7 / 8.7
12 F	0308 / 0906 / 1531 / 2127	0.9 / 9.7 / 0.9 / 9.3	27 SA	0308 / 0904 / 1524 / 2120	1.8 / 8.7 / 2.0 / 8.4
13 SA	0347 / 0947 / 1611 / 2208	1.2 / 9.3 / 1.3 / 8.8	28 SU	0333 / 0932 / 1548 / 2147	2.2 / 8.2 / 2.5 / 8.0
14 SU	0429 / 1031 / 1655 / 2253	1.8 / 8.7 / 2.0 / 8.3	29 M	0400 / 1001 / 1616 / 2217	2.7 / 7.8 / 2.9 / 7.5
15 M	0516 / 1122 / 1746 / ☽2348	2.4 / 8.0 / 2.7 / 7.6			

MARCH

Day	Time	m	Day	Time	m
1 TU	0433 / 1037 / 1652 / ☽2258	3.2 / 7.3 / 3.4 / 7.1	16 W	0554 / 1204 / 1826	3.0 / 7.2 / 3.4
2 W	0519 / 1128 / 1748 / 2359	3.6 / 6.9 / 3.8 / 6.8	17 TH	0035 / 0719 / 1341 / 2002	7.0 / 3.5 / 6.8 / 3.7
3 TH	0637 / 1245 / 1919	3.9 / 6.6 / 4.0	18 F	0219 / 0905 / 1512 / 2134	6.9 / 3.4 / 7.0 / 3.4
4 F	0132 / 0817 / 1426 / 2053	6.7 / 3.7 / 6.8 / 3.7	19 SA	0338 / 1014 / 1614 / 2235	7.3 / 2.9 / 7.5 / 2.9
5 SA	0308 / 0936 / 1545 / 2206	7.1 / 3.2 / 7.3 / 3.1	20 SU	0433 / 1105 / 1701 / 2322	7.8 / 2.4 / 8.0 / 2.3
6 SU	0414 / 1039 / 1644 / 2305	7.8 / 2.5 / 7.9 / 2.3	21 M	0518 / 1147 / 1743	8.4 / 1.9 / 8.5
7 M	0508 / 1134 / 1736 / 2357	8.5 / 1.7 / 8.6 / 1.6	22 TU	0003 / 0557 / 1227 / 1820	1.9 / 8.8 / 1.6 / 8.8
8 TU	0557 / 1223 / 1822	9.2 / 1.1 / 9.2	23 W	0041 / 0634 / 1303 / ○1855	1.5 / 9.1 / 1.4 / 9.1
9 W	0044 / 0642 / 1309 / ●1906	1.0 / 9.7 / 0.5 / 9.7	24 TH	0116 / 0708 / 1335 / 1928	1.4 / 9.2 / 1.3 / 9.1
10 TH	0129 / 0726 / 1353 / 1948	0.5 / 10.1 / 0.2 / 9.9	25 F	0147 / 0740 / 1404 / 1958	1.3 / 9.2 / 1.4 / 9.1
11 F	0212 / 0808 / 1434 / 2029	0.3 / 10.2 / 0.2 / 9.9	26 SA	0215 / 0810 / 1430 / 2025	1.4 / 9.0 / 1.6 / 8.8
12 SA	0252 / 0849 / 1513 / 2108	0.4 / 10.0 / 0.5 / 9.6	27 SU	0241 / 0838 / 1454 / 2051	1.7 / 8.7 / 1.9 / 8.5
13 SU	0331 / 0930 / 1552 / 2148	0.8 / 9.5 / 1.1 / 9.1	28 M	0306 / 0905 / 1519 / 2118	2.0 / 8.3 / 2.3 / 8.2
14 M	0411 / 1012 / 1633 / 2230	1.5 / 8.8 / 1.9 / 8.4	29 TU	0333 / 0933 / 1546 / 2147	2.5 / 7.9 / 2.8 / 7.8
15 TU	0456 / 1100 / 1721 / ☽2321	2.3 / 7.9 / 2.7 / 7.6	30 W	0404 / 1009 / 1621 / 2226	2.9 / 7.5 / 3.2 / 7.4
			31 TH	0447 / 1057 / 1712 / ☽2323	3.3 / 7.1 / 3.6 / 7.0

APRIL

Day	Time	m	Day	Time	m
1 F	0555 / 1208 / 1831	3.6 / 6.8 / 3.9	16 SA	0142 / 0830 / 1440 / 2101	6.9 / 3.5 / 6.9 / 3.6
2 SA	0045 / 0735 / 1343 / 2014	6.8 / 3.6 / 6.8 / 3.7	17 SU	0304 / 0941 / 1543 / 2203	7.1 / 3.1 / 7.3 / 3.1
3 SU	0224 / 0902 / 1511 / 2135	7.1 / 3.1 / 7.3 / 3.1	18 M	0401 / 1033 / 1631 / 2251	7.6 / 2.7 / 7.8 / 2.6
4 M	0341 / 1010 / 1616 / 2238	7.7 / 2.4 / 8.0 / 2.3	19 TU	0447 / 1116 / 1712 / 2332	8.1 / 2.2 / 8.3 / 2.2
5 TU	0440 / 1107 / 1709 / 2332	8.5 / 1.7 / 8.7 / 1.5	20 W	0527 / 1154 / 1749	8.5 / 1.9 / 8.6
6 W	0532 / 1158 / 1758	9.2 / 1.0 / 9.3	21 TH	0010 / 0604 / 1230 / 1824	1.8 / 8.8 / 1.6 / 8.9
7 TH	0021 / 0619 / 1246 / ●1843	0.9 / 9.7 / 0.5 / 9.8	22 F	0045 / 0639 / 1303 / ○1858	1.6 / 8.9 / 1.5 / 9.0
8 F	0108 / 0705 / 1331 / 1927	0.4 / 10.1 / 0.2 / 10.1	23 SA	0117 / 0712 / 1333 / 1929	1.5 / 9.0 / 1.6 / 9.0
9 SA	0152 / 0749 / 1413 / 2008	0.2 / 10.2 / 0.2 / 10.0	24 SU	0147 / 0744 / 1401 / 1959	1.5 / 8.9 / 1.7 / 8.8
10 SU	0234 / 0831 / 1454 / 2049	0.3 / 9.9 / 0.5 / 9.7	25 M	0216 / 0813 / 1428 / 2026	1.7 / 8.7 / 1.9 / 8.6
11 M	0314 / 0912 / 1534 / 2129	0.7 / 9.4 / 1.1 / 9.2	26 TU	0244 / 0842 / 1456 / 2055	2.0 / 8.4 / 2.3 / 8.3
12 TU	0355 / 0955 / 1615 / 2211	1.4 / 8.7 / 1.9 / 8.5	27 W	0313 / 0914 / 1526 / 2128	2.3 / 8.0 / 2.6 / 8.0
13 W	0440 / 1043 / 1702 / 2300	2.2 / 7.9 / 2.7 / 7.7	28 TH	0347 / 0951 / 1603 / 2209	2.7 / 7.7 / 3.0 / 7.8
14 TH	0536 / 1143 / 1803 / ☽	2.9 / 7.2 / 3.4	29 F	0431 / 1041 / 1653 / 2304	3.0 / 7.3 / 3.4 / 7.3
15 F	0007 / 0653 / 1311 / 1929	7.1 / 3.5 / 6.8 / 3.8	30 SA	0534 / 1146 / 1805 / ○	3.3 / 7.0 / 3.6

Chart Datum is 5·06 metres below Ordnance Datum (Local). HAT is 10·3 metres above Chart Datum.

Channel Islands

》FREE monthly updates. Register at 《
www.reedsnauticalalmanac.co.uk

829

STANDARD TIME (UT)
For Summer Time add ONE hour in **non-shaded areas**

ST PETER PORT LAT 49°27'N LONG 2°32'W
TIMES AND HEIGHTS OF HIGH AND LOW WATERS

Dates in red are SPRINGS
Dates in blue are NEAPS

YEAR 2016

MAY

Day	Time	m		Day	Time	m
1 SU	0017	7.2		16 M	0212	7.0
	0700	3.3			0851	3.3
	1309	7.1			1457	7.1
	1937	3.5			2116	3.4
2 M	0144	7.3		17 TU	0316	7.3
	0826	3.0			0948	3.0
	1433	7.4			1549	7.5
	2100	3.0			2209	3.0
3 TU	0304	7.8		18 W	0407	7.7
	0937	2.4			1035	2.7
	1542	8.0			1634	7.9
	2207	2.3			2254	2.6
4 W	0408	8.4		19 TH	0450	8.0
	1037	1.7			1116	2.3
	1639	8.7			1714	8.3
	2304	1.6			2333	2.2
5 TH	0504	9.0		20 F	0530	8.3
	1131	1.1			1154	2.1
	1731	9.3			1751	8.6
	2357	1.0				
6 F	0555	9.5		21 SA	0010	2.0
	1221	0.7			0608	8.6
	1819	9.7			1230	1.9
●				○	1827	8.8
7 SA	0045	0.6		22 SU	0046	1.8
	0643	9.8			0644	8.7
	1308	0.5			1303	1.8
	1905	9.9			1902	8.8
8 SU	0132	0.4		23 M	0121	1.7
	0730	9.9			0719	8.7
	1353	0.5			1336	1.8
	1949	9.9			1935	8.8
9 M	0217	0.5		24 TU	0154	1.8
	0814	9.7			0752	8.6
	1436	0.8			1408	2.0
	2031	9.7			2006	8.7
10 TU	0259	0.9		25 W	0227	1.9
	0857	9.3			0826	8.4
	1518	1.3			1440	2.2
	2112	9.2			2039	8.5
11 W	0342	1.4		26 TH	0301	2.1
	0941	8.7			0901	8.2
	1600	1.9			1515	2.4
	2155	8.5			2116	8.3
12 TH	0427	2.1		27 F	0339	2.4
	1027	8.0			0942	7.9
	1645	2.7			1555	2.7
	2241	7.9			2159	8.0
13 F	0518	2.8		28 SA	0424	2.7
	1121	7.4			1030	7.7
	1739	3.3			1644	3.0
◐	2338	7.3			2252	7.7
14 SA	0621	3.3		29 SU	0521	2.9
	1230	7.0			1129	7.5
	1847	3.7			1746	3.2
				◑	2357	7.6
15 SU	0051	7.0		30 M	0632	3.0
	0737	3.5			1239	7.4
	1349	6.9			1904	3.2
	2007	3.7				
				31 TU	0111	7.6
					0750	2.8
					1356	7.6
					2025	2.9

JUNE

Day	Time	m		Day	Time	m
1 W	0228	7.9		16 TH	0316	7.2
	0903	2.4			0944	3.2
	1508	8.0			1549	7.5
	2136	2.4			2207	3.0
2 TH	0338	8.3		17 F	0409	7.6
	1008	2.0			1033	2.8
	1610	8.5			1636	7.9
	2238	1.9			2254	2.7
3 F	0438	8.7		18 SA	0455	7.9
	1105	1.5			1117	2.5
	1706	9.0			1719	8.2
	2334	1.4			2337	2.3
4 SA	0534	9.1		19 SU	0538	8.2
	1159	1.1			1157	2.2
	1758	9.4			1759	8.5
5 SU	0026	1.0		20 M	0018	2.0
	0625	9.4			0618	8.4
	1249	0.9			1237	2.0
●	1846	9.6		○	1838	8.7
6 M	0115	0.8		21 TU	0058	1.8
	0713	9.5			0658	8.6
	1336	0.9			1316	1.9
	1932	9.7			1915	8.8
7 TU	0202	0.8		22 W	0137	1.7
	0800	9.4			0736	8.6
	1421	1.1			1353	1.9
	2015	9.5			1951	8.8
8 W	0246	1.0		23 TH	0215	1.7
	0844	9.1			0814	8.6
	1503	1.4			1430	1.9
	2057	9.2			2028	8.8
9 TH	0329	1.4		24 F	0253	1.8
	0926	8.7			0852	8.5
	1544	1.9			1508	2.1
	2138	8.7			2108	8.7
10 F	0411	2.0		25 SA	0332	1.9
	1009	8.2			0933	8.4
	1626	2.5			1548	2.3
	2220	8.1			2151	8.5
11 SA	0455	2.5		26 SU	0416	2.2
	1053	7.7			1019	8.1
	1710	3.0			1634	2.5
	2305	7.6			2239	8.2
12 SU	0544	3.0		27 M	0506	2.4
	1144	7.2			1111	7.9
	1802	3.4			1728	2.8
◐	2359	7.2		◑	2336	7.9
13 M	0640	3.4		28 TU	0606	2.6
	1245	7.0			1212	7.7
	1903	3.7			1835	3.0
14 TU	0104	7.0		29 W	0042	7.8
	0743	3.5			0716	2.7
	1352	7.0			1323	7.7
	2010	3.6			1952	2.9
15 W	0214	7.0		30 TH	0158	7.8
	0847	3.4			0832	2.6
	1456	7.2			1438	7.9
	2113	3.4			2109	2.7

JULY

Day	Time	m		Day	Time	m
1 F	0313	8.0		16 SA	0326	7.2
	0943	2.4			0949	3.3
	1547	8.2			1559	7.5
	2217	2.3			2216	3.0
2 SA	0419	8.3		17 SU	0422	7.5
	1046	2.0			1042	2.9
	1647	8.6			1649	7.9
	2317	1.8			2307	2.6
3 SU	0518	8.7		18 M	0511	7.9
	1142	1.7			1130	2.5
	1741	9.0			1734	8.3
					2354	2.2
4 M	0012	1.4		19 TU	0557	8.3
	0611	9.0			1215	2.1
	1235	1.4			1817	8.7
●	1831	9.3		○		
5 TU	0102	1.1		20 W	0038	1.8
	0700	9.2			0640	8.6
	1322	1.2			1259	1.8
	1917	9.5			1858	9.0
6 W	0149	1.0		21 TH	0122	1.5
	0746	9.2			0721	8.9
	1407	1.2			1340	1.6
	2000	9.4			1938	9.2
7 TH	0232	1.1		22 F	0203	1.3
	0828	9.1			0802	9.0
	1447	1.4			1420	1.5
	2040	9.2			2017	9.2
8 F	0312	1.4		23 SA	0243	1.3
	0908	8.8			0841	9.0
	1525	1.8			1459	1.6
	2118	8.9			2057	9.2
9 SA	0349	1.8		24 SU	0323	1.4
	0944	8.4			0921	8.9
	1601	2.2			1539	1.7
	2153	8.4			2139	9.0
10 SU	0425	2.3		25 M	0403	1.7
	1020	8.0			1004	8.6
	1636	2.7			1621	2.1
	2230	7.9			2223	8.6
11 M	0502	2.8		26 TU	0448	2.1
	1059	7.5			1050	8.2
	1715	3.2			1708	2.5
	2311	7.4		◑	2314	8.2
12 TU	0544	3.3		27 W	0541	2.5
	1145	7.2			1145	7.9
	1804	3.5			1807	2.8
◐						
13 W	0001	7.1		28 TH	0016	7.8
	0637	3.6			0646	2.9
	1244	6.9			1254	7.6
	1904	3.8			1923	3.1
14 TH	0107	6.8		29 F	0134	7.5
	0741	3.7			0806	3.0
	1353	6.9			1415	7.5
	2013	3.7			2050	3.0
15 F	0220	6.9		30 SA	0258	7.6
	0848	3.6			0927	2.8
	1501	7.1			1532	7.8
	2118	3.5			2207	2.7
				31 SU	0410	7.9
					1035	2.5
					1636	8.3
					2308	2.2

AUGUST

Day	Time	m		Day	Time	m
1 M	0509	8.4		16 TU	0448	7.8
	1132	2.0			1107	2.6
	1730	8.7			1711	8.3
					2333	2.2
2 TU	0001	1.7		17 W	0536	8.3
	0600	8.8			1155	2.1
	1222	1.6			1757	8.8
●	1817	9.1				
3 W	0050	1.3		18 TH	0020	1.6
	0646	9.0			0621	8.8
	1308	1.4			1242	1.6
	1901	9.4		○	1840	9.3
4 TH	0133	1.2		19 F	0105	1.2
	0729	9.2			0704	9.2
	1350	1.3			1325	1.2
	1942	9.4			1922	9.6
5 F	0213	1.2		20 SA	0147	0.9
	0808	9.2			0745	9.4
	1427	1.4			1406	1.0
	2019	9.3			2002	9.7
6 SA	0249	1.3		21 SU	0228	0.8
	0843	9.0			0825	9.5
	1501	1.6			1446	1.0
	2052	9.0			2042	9.7
7 SU	0321	1.7		22 M	0307	1.0
	0915	8.7			0904	9.3
	1532	2.0			1524	1.3
	2123	8.6			2122	9.4
8 M	0350	2.1		23 TU	0346	1.3
	0945	8.3			0945	9.0
	1600	2.4			1604	1.7
	2154	8.2			2205	8.9
9 TU	0418	2.6		24 W	0428	1.9
	1016	7.8			1028	8.5
	1630	2.9			1648	2.3
	2227	7.7			2252	8.3
10 W	0448	3.1		25 TH	0517	2.5
	1052	7.4			1120	7.9
	1706	3.4			1744	2.9
◐	2306	7.2		◑	2352	7.6
11 TH	0530	3.6		26 F	0621	3.1
	1137	7.0			1229	7.4
	1759	3.8			1901	3.3
12 F	0000	6.8		27 SA	0116	7.2
	0635	3.9			0748	3.4
	1245	6.7			1400	7.2
	1916	4.0			2041	3.4
13 SA	0120	6.7		28 SU	0252	7.3
	0756	3.9			0919	3.2
	1411	6.8			1524	7.6
	2035	3.8			2201	2.9
14 SU	0245	6.8		29 M	0404	7.7
	0910	3.7			1027	2.8
	1524	7.2			1626	8.1
	2144	3.3			2300	2.4
15 M	0353	7.3		30 TU	0459	8.2
	1013	3.2			1120	2.2
	1622	7.7			1717	8.6
	2242	2.8			2348	1.9
				31 W	0546	8.7
					1207	1.8
					1801	9.0

Chart Datum is 5·06 metres below Ordnance Datum (Local). HAT is 10·3 metres above Chart Datum.

》》FREE monthly updates. Register at 《
www.reedsnauticalalmanac.co.uk 《

STANDARD TIME (UT)
For Summer Time add ONE hour in **non-shaded areas**

ST PETER PORT LAT 49°27'N LONG 2°32'W

TIMES AND HEIGHTS OF HIGH AND LOW WATERS

Dates in red are SPRINGS
Dates in blue are NEAPS

YEAR **2016**

SEPTEMBER

Time m	Time m
1 0032 1.5 / 0627 9.0 / TH 1249 1.5 / ● 1841 9.3	**16** 0559 9.1 / 1220 1.4 / F 1818 9.5 / ○
2 0112 1.3 / 0706 9.2 / F 1327 1.3 / 1919 9.4	**17** 0043 1.0 / 0642 9.5 / SA 1305 1.0 / 1901 9.9
3 0148 1.2 / 0741 9.3 / SA 1402 1.3 / 1953 9.4	**18** 0127 0.6 / 0724 9.8 / SU 1347 0.7 / 1943 10.1
4 0221 1.4 / 0814 9.1 / SU 1433 1.5 / 2024 9.1	**19** 0208 0.5 / 0805 9.9 / M 1428 0.7 / 2024 10.0
5 0249 1.7 / 0843 8.9 / M 1500 1.9 / 2052 8.8	**20** 0248 0.7 / 0844 9.7 / TU 1507 1.0 / 2104 9.6
6 0314 2.1 / 0910 8.5 / TU 1525 2.3 / 2120 8.4	**21** 0327 1.2 / 0925 9.3 / W 1547 1.5 / 2146 9.0
7 0337 2.5 / 0937 8.1 / W 1550 2.8 / 2148 7.9	**22** 0408 1.8 / 1007 8.6 / TH 1631 2.2 / 2233 8.3
8 0402 3.0 / 1007 7.6 / TH 1620 3.3 / 2221 7.4	**23** 0456 2.6 / 1058 7.9 / F 1726 3.0 / ◑ 2333 7.5
9 0435 3.5 / 1045 7.1 / F 1702 3.7 / ◐ 2307 6.9	**24** 0600 3.3 / 1208 7.3 / SA 1846 3.5
10 0527 4.0 / 1142 6.8 / SA 1813 4.0	**25** 0104 7.0 / 0733 3.7 / SU 1346 7.1 / 2032 3.5
11 0019 6.6 / 0701 4.2 / SU 1314 6.7 / 1953 4.0	**26** 0244 7.1 / 0908 3.5 / M 1511 7.4 / 2149 3.1
12 0203 6.7 / 0834 3.9 / M 1448 7.0 / 2113 3.5	**27** 0351 7.6 / 1012 3.0 / TU 1610 8.0 / 2243 2.5
13 0325 7.2 / 0945 3.4 / TU 1554 7.6 / 2216 2.9	**28** 0441 8.1 / 1102 2.4 / W 1657 8.5 / 2328 2.0
14 0423 7.8 / 1043 2.7 / W 1646 8.3 / 2309 2.1	**29** 0524 8.6 / 1145 2.0 / TH 1738 8.9
15 0513 8.5 / 1133 2.0 / TH 1734 9.0 / 2358 1.5	**30** 0008 1.7 / 0602 9.0 / F 1224 1.6 / 1816 9.2

OCTOBER

Time m	Time m
1 0045 1.5 / 0638 9.2 / SA 1301 1.5 / ● 1852 9.3	**16** 0019 0.9 / 0617 9.7 / SU 1243 0.9 / ○ 1839 10.0
2 0119 1.4 / 0712 9.3 / SU 1333 1.5 / 1925 9.3	**17** 0104 0.6 / 0701 10.0 / M 1327 0.6 / 1923 10.2
3 0149 1.5 / 0743 9.2 / M 1403 1.6 / 1955 9.1	**18** 0147 0.5 / 0743 10.1 / TU 1410 0.6 / 2005 10.1
4 0216 1.7 / 0811 9.0 / TU 1429 1.9 / 2023 8.9	**19** 0229 0.7 / 0825 9.9 / W 1451 0.9 / 2048 9.7
5 0240 2.1 / 0838 8.6 / W 1455 2.2 / 2050 8.5	**20** 0310 1.2 / 0906 9.4 / TH 1533 1.4 / 2131 9.0
6 0304 2.5 / 0904 8.2 / TH 1520 2.7 / 2117 8.0	**21** 0352 1.9 / 0950 8.8 / F 1618 2.2 / 2218 8.2
7 0330 3.0 / 0933 7.8 / F 1549 3.1 / 2149 7.6	**22** 0440 2.7 / 1040 8.0 / SA 1713 2.9 / ◑ 2317 7.5
8 0401 3.4 / 1008 7.4 / SA 1628 3.6 / 2232 7.1	**23** 0542 3.4 / 1147 7.4 / SU 1828 3.5
9 0447 3.9 / 1102 7.0 / SU 1728 3.9 / ◐ 2339 6.8	**24** 0043 7.0 / 0708 3.8 / M 1319 7.1 / 2005 3.6
10 0605 4.2 / 1222 6.8 / M 1909 4.0	**25** 0218 7.1 / 0841 3.7 / TU 1443 7.3 / 2121 3.3
11 0116 6.7 / 0754 4.1 / TU 1404 7.0 / 2038 3.6	**26** 0325 7.5 / 0946 3.2 / W 1542 7.8 / 2215 2.8
12 0250 7.2 / 0913 3.5 / W 1519 7.6 / 2146 2.9	**27** 0414 8.0 / 1035 2.7 / TH 1629 8.2 / 2259 2.4
13 0353 7.9 / 1015 2.7 / TH 1616 8.3 / 2242 2.1	**28** 0456 8.4 / 1117 2.3 / F 1710 8.6 / 2338 2.0
14 0445 8.6 / 1107 2.0 / F 1706 9.0 / 2332 1.5	**29** 0533 8.8 / 1155 2.0 / SA 1748 8.9
15 0532 9.2 / 1156 1.4 / SA 1754 9.6	**30** 0014 1.8 / 0608 9.1 / SU 1231 1.8 / ● 1823 9.1
	31 0048 1.7 / 0642 9.2 / M 1304 1.7 / 1856 9.1

NOVEMBER

Time m	Time m
1 0118 1.7 / 0713 9.2 / TU 1334 1.7 / 1928 9.0	**16** 0128 0.7 / 0724 10.1 / W 1353 0.7 / 1949 9.9
2 0146 1.9 / 0743 9.0 / W 1402 1.9 / 1957 8.8	**17** 0212 0.9 / 0808 9.9 / TH 1438 0.9 / 2034 9.6
3 0213 2.1 / 0811 8.7 / TH 1430 2.2 / 2026 8.5	**18** 0255 1.3 / 0851 9.5 / F 1522 1.4 / 2118 9.0
4 0240 2.5 / 0839 8.4 / F 1459 2.6 / 2056 8.2	**19** 0339 1.9 / 0936 8.9 / SA 1608 2.0 / 2205 8.4
5 0308 2.9 / 0910 8.1 / SA 1530 2.9 / 2130 7.8	**20** 0425 2.6 / 1024 8.2 / SU 1659 2.7 / 2258 7.7
6 0342 3.3 / 0947 7.7 / SU 1610 3.3 / 2213 7.4	**21** 0520 3.2 / 1121 7.6 / M 1800 3.3 / ◑
7 0427 3.7 / 1038 7.3 / M 1705 3.6 / ◐ 2313 7.0	**22** 0006 7.2 / 0629 3.7 / TU 1234 7.2 / 1915 3.6
8 0533 4.0 / 1149 7.1 / TU 1826 3.7	**23** 0129 7.0 / 0751 3.8 / W 1355 7.2 / 2033 3.5
9 0034 7.0 / 0708 3.9 / W 1317 7.2 / 1956 3.5	**24** 0242 7.2 / 0903 3.6 / TH 1501 7.4 / 2134 3.2
10 0205 7.3 / 0835 3.5 / TH 1438 7.7 / 2110 2.9	**25** 0337 7.6 / 0958 3.2 / F 1553 7.8 / 2223 2.9
11 0316 7.9 / 0942 2.8 / F 1542 8.3 / 2210 2.2	**26** 0422 8.0 / 1043 2.8 / SA 1637 8.1 / 2304 2.5
12 0414 8.6 / 1039 2.1 / SA 1637 8.9 / 2304 1.6	**27** 0501 8.4 / 1123 2.4 / SU 1717 8.4 / 2342 2.3
13 0505 9.2 / 1131 1.4 / SU 1729 9.4 / 2354 1.1	**28** 0538 8.7 / 1200 2.2 / M 1754 8.7
14 0553 9.7 / 1220 1.0 / M 1817 9.8 / ○	**29** 0017 2.1 / 0613 8.9 / TU 1235 2.0 / ● 1830 8.8
15 0042 0.8 / 0639 10.0 / TU 1308 0.7 / 1904 10.0	**30** 0050 2.0 / 0648 9.0 / W 1309 1.9 / 1904 8.8

DECEMBER

Time m	Time m
1 0121 2.0 / 0720 9.0 / TH 1341 2.0 / 1938 8.7	**16** 0159 1.0 / 0755 9.8 / F 1427 0.9 / 2022 9.5
2 0153 2.1 / 0752 8.8 / F 1414 2.1 / 2010 8.5	**17** 0243 1.3 / 0838 9.6 / SA 1511 1.2 / 2106 9.1
3 0224 2.3 / 0823 8.6 / SA 1446 2.3 / 2042 8.3	**18** 0326 1.7 / 0921 9.1 / SU 1554 1.7 / 2149 8.6
4 0256 2.6 / 0857 8.3 / SU 1521 2.6 / 2119 8.0	**19** 0408 2.3 / 1004 8.6 / M 1638 2.3 / 2233 8.0
5 0332 2.9 / 0936 8.0 / M 1601 2.9 / 2202 7.7	**20** 0453 2.8 / 1050 8.0 / TU 1725 2.9 / 2322 7.5
6 0415 3.2 / 1024 7.8 / TU 1650 3.1 / 2255 7.5	**21** 0543 3.4 / 1142 7.4 / W 1818 3.4 / ◑
7 0511 3.5 / 1124 7.5 / W 1753 3.3 / ◐	**22** 0021 7.1 / 0643 3.7 / TH 1246 7.1 / 1921 3.6
8 0001 7.3 / 0624 3.6 / TH 1236 7.5 / 1910 3.2	**23** 0133 7.0 / 0754 3.8 / F 1359 7.0 / 2030 3.7
9 0119 7.4 / 0750 3.4 / F 1355 7.7 / 2028 2.9	**24** 0244 7.1 / 0904 3.7 / SA 1505 7.2 / 2134 3.4
10 0237 7.8 / 0906 2.9 / SA 1508 8.1 / 2137 2.5	**25** 0341 7.5 / 1001 3.3 / SU 1559 7.5 / 2225 3.1
11 0343 8.3 / 1011 2.4 / SU 1611 8.6 / 2238 2.0	**26** 0427 7.9 / 1048 2.9 / M 1645 7.9 / 2309 2.7
12 0440 8.9 / 1108 1.8 / M 1707 9.0 / 2333 1.5	**27** 0509 8.3 / 1130 2.5 / TU 1728 8.2 / 2348 2.4
13 0532 9.4 / 1202 1.3 / TU 1800 9.4	**28** 0549 8.6 / 1210 2.2 / W 1808 8.5
14 0024 1.1 / 0621 9.7 / W 1253 1.0 / ○ 1849 9.6	**29** 0026 2.2 / 0626 8.8 / TH 1248 2.0 / ● 1846 8.6
15 0113 1.0 / 0709 9.9 / TH 1341 0.8 / 1937 9.6	**30** 0103 2.0 / 0703 8.9 / F 1325 1.9 / 1923 8.7
	31 0138 2.0 / 0738 8.9 / SA 1402 1.8 / 1958 8.7

Chart Datum is 5·06 metres below Ordnance Datum (Local). HAT is 10·3 metres above Chart Datum.

Channel Islands

9.19.12 HERM

Herm 49°28'·22N 02°27'·26W ✤✤◈◊✿✿✿

CHARTS and TIDES As for St Peter Port.

SHELTER Good shelter E or W of island depending on winds.

NAVIGATION Access is not difficult, if properly planned. Herm is reached from the Little Russel Channel via any of 7 passages all of which require reasonable visibility and care with tidal streams. AC 807 & 808 list the marks and bearings of all the passages.

The Alligande Passage is the simplest and most direct from St Peter Port. Approach with the Vermerette bcn (topmark 'V') in transit 074° with W patch on Hbr quay. Leave Alligande 'A', Godfrey 'GB' and Epec 'E' bcns to starboard; skirt close N of Vermerette. When its base is awash, there is 1m at hbr ent. The tide will be setting N. Sand build-up W of Vermerette affects craft NW-bound in the Percée Passage.

The appr from the Big Russel is more open and leads easily to pleasant ⚓s at Belvoir Bay and Shell Bay.

LIGHTS AND MARKS Ldg lts, both FW (occas) and W drums at 078°. 2FG (vert) on quay hd. Night appr not advised for visitors.

COMMUNICATIONS (Code 01481) Island Admin ☎750000 for overnight stay in hbr.

FACILITIES

Herm Harbour, options:

• Lie to mooring lines secured to N and S walls inside hbr.
• Dry out on the beach to the E, moored fore/aft to chains.
• 5 ⚓s to N and 6 to W of hbr dry out; 6 more to NW: rky bottom to the N; isolated boulders/sand to W.

www.herm.com for navigation details and webcams of hbr & buoys. No fees; donations welcome. ⚓ ⚑ ✕ ▭ limited ⛟. Can be busy at summer weekends, better mid-week.

Rosière Steps Easiest appr is from the Big Russel via Percée passage; avoid Fourquies (2₃, NCM lt buoy) Meulettes (1₇) and Tinker (2₇). ⚓ 49°28'·22N 02°27'·22W NW of Rosière steps; good holding on sand, but exposed to S and SW. Access for landing only; do not linger alongside. Buoys are for ferries and Herm-owned boats only; hourly ferries by day. The bottom step is about 1.5m above the seabed which may be inconvenient at LWS. Caution: From just before HW to HW+2 tide sets hard onto the steps.

Belvoir Bay 49°28'·40N 02°26'·34W and **Shell Bay** are good ⚓s on sand, sheltered from W. Easy access from E; from S keep 400m offshore. Beach café or walk 800m to Harbour village. Note: Jethou, Crevichon and Grande Fauconnière are private. No landing.

9.19.13 SARK

Sark 49°25'·81N 02°20'·45W Creux ✤✤◈◊✿✿✿

CHARTS AC 808, 5604.12; SHOM 7159, 6904; Navi 1014; Imray C33A, 2500

TIDES −0450 Dover; ML 5·3; Duration 0550
Standard Port ST HELIER (➝)

Times				Height (metres)			
High Water		Low Water		MHWS	MHWN	MLWN	MLWS
0300	0900	0200	0900	11·0	8·1	4·0	1·4
1500	2100	1400	2100				
Differences SARK (MASELINE PIER)							
+0005	+0015	+0005	+0010	−2·1	−1·5	−0·6	−0·3

Tidal streams *Beware large tidal range, strong streams and many lobster pots. In Gouliot (W coast) and Goulet Passages the streams reach 6-7kn at springs. Note that at about half-tide the streams are slack around Sark. At HW the stream sets hard to the N, ie onto Little Sark. At LW the stream sets hard to the S, ie onto Bec du Nez (N tip). If bound from Guernsey to Sark's E coast, go N-about at HW and S-about at LW; conversely on the return.*

SHELTER Sark is fringed by rocks, but the centres of the bays are mainly clear of dangers. A safe ⚓age or ⚓s can usually be found sheltered from offshore winds. But, depending on wind and tide, they may be uncomfortable, except in settled weather; see also Facilities.

NAVIGATION From the West, WPT 49°25'·27N 02°24'·30W, 070°/1·29M towards the Pilcher monument (Sark mill is obsc'd by trees) for Havre Gosselin or La Grande Grève.
From the N or after rounding Bec du Nez, the WPT is 49°27'·30N 02°21'·42W, on the 153° charted transit (aka the outside passage) towards Grève de la Ville and Maseline. Noire Pierre rk is unlit but marked with a Y post & radar reflector. The inside passage, W of drying Pécheresse, is used by locals but ill advised for visitors.

LIGHTS AND MARKS Point Robert lt ho and Courbée du Nez are the only navigational lights on Sark; see chartlet and Lights, buoys and waypoints.

COMMUNICATIONS (Code 01481) Police (Guernsey) 725111; ⊖ (Guernsey) 726911; Dr 832045; HM 832323, VHF Ch 10; Maseline Hbr VHF Ch 13, season only; Tourist Office 832345.

ANCHORAGES, ⚓s, HARBOURS AND FACILITIES
Anti-clockwise from Bec du Nez, the ⚓s below (all unlit) are safe in offshore winds; in other conditions they can be exposed and sometimes dangerous. Some are only suitable around LW.

⚓s at Havre Gosselin and Grève de la Ville are free (courtesy Sark Moorings ☎832260, 07781 106065). Donations welcome in local boxes or c/o Le Grand Fort, Sark, Channel Islands, GY9 0SF. Water taxi Ch 10. All other moorings are private; use only in emergency.

WEST COAST (all ⚓s are exposed to W'lies)
Saignie Bay 49°26'·53N 02°22'·10W. Sand and shingle with fair holding. Picturesque rock formations.

Port à la Jument 49°26'·17N 02°22'·42W. Sand and shingle with fair holding. Difficult shore access.

Brecqhou Island is strictly private; landing prohibited.

Havre Gosselin 49°25'·77N 02°22'·68W. Popular deep (4-9m) ⚓. 20 Y ⚓'s (see above). Beware of drying rk at extreme NW of bay. Crowded in summer. 299 steps to cliff top and panoramic views.

Port és Saies 49°25'·41N 02°22'·28W. Sandy inlet, steep cliff path.

La Grande Grève 49°25'·38N 02°22'·59W. Wide sandy bay, subject to swell, but popular day ⚓age. Beware two rks (drying 0·3m and ⊛) in the appr. Temporary steps to cliff-top panoramic views.

LITTLE SARK
Port Gorey 49°24'·60N 02°22'·72W. ⚓ or pick up Y ⚓ (free) in centre of deep, weedy bay over LW only; heavy swell begins near half-flood. Rocky appr from just NW of Grande Bretagne (18m high) then 045° into bay. Rocks must be positively identified. Remains of quay with ladder. Cliff walk past silver mine ruins to hotel.

Rouge Terrier 49°24'·78N 02°21'·87W. Sandy with some local moorings under high cliffs. Exposed to E. Landing with cliff path to hotel. Also ⚓ 49°25'·09N 02°21'·70W, 4ca NNE in Baleine Bay.

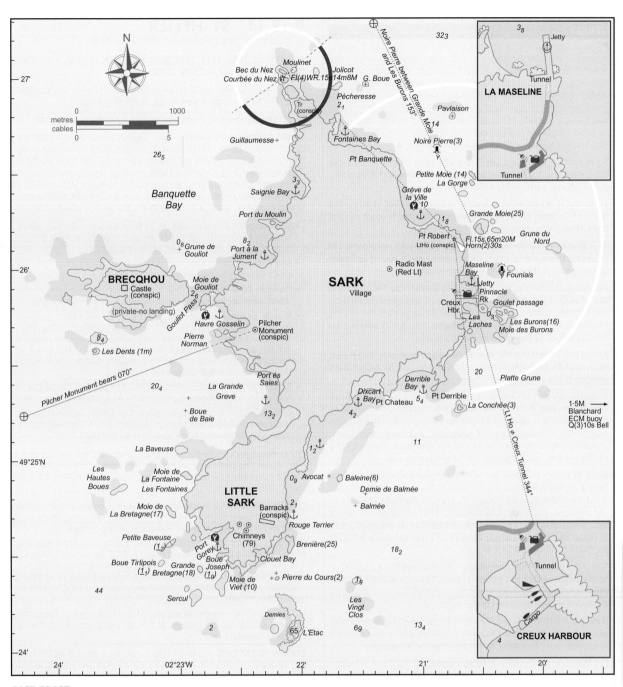

EAST COAST

Dixcart Bay 49°25´.33N 02°21´.48W. Popular sandy bay with good holding, but open to S'lies. Drying rocks either side of the bay but not in the centre. Cliff path and walk to hotels. Beware rockslides.

Derrible Bay 49°25´.41N 02°21´.07W. Sandy bay with good holding. Open to the S. No dangers in the bay, but keep off SW tip of Derrible Pt. Picturesque caves and steep climb ashore.

Creux Harbour 49°25´.80N 02°20´.61W dries completely, fair weather only, prone to surge/swell. Local transits: Pinnacle Rk on with E edge of Grand Moie 001°; or Creux tunnel (white arch) on with Pt Robert lt ho 344°. The lt ho dips behind the cliffs when close inshore. In the Goulet Passage (E of Creux) S-bound navigation is prohibited to commercial vessels (except with HM's permission) but not to yachts in either direction.

Dry out bow to NE wall, stern to ⚓. ⌑ possible only at inner part of the S pier, clear of steps and cargo berth at outer end. ☎832025

(kiosk) 🅟(free) ⚓ ⤵ ⤴ 🅆 🅟 via HM; walk or tractor up steep hill to village ✕ ⌂ 🛒 🛈 🅆 🅟 ✉ Ⓑ. Bikes and horse-drawn carriages for hire.

Maseline Harbour 49°25´.96N 02°20´.60W is prone to surge/swell. Busy ferry hbr with no yacht berths (except to land people). Call HM Ch 13 for approval to ⚓. Moorings are private. ☎832070 (kiosk) ⌑(3t). Condor catamaran to Jersey/St Malo.

Grève de la Ville 49°26´.35N 02°21´.06W. ⚓ close in out of tide. Fair holding sand/shingle. Open to E. Seven Y 🅐s (free, see p 832). Easy walk ashore.

Les Fontaines 49°26´.78N 02°21´.62W. Sand and shingle with fair holding. Reef drying 4·5m extends 1ca N from shore. ⚓ between reef and Eperquerie headland. Exposed to the E.

With grateful acknowledgements to John Frankland, author of *Sark, Round the Island*.

▶The rotatory pattern of tidal streams affecting the Channel Islands as a whole dictates that when streams are slack on the N and S coasts of Jersey, they are running strongly on the E and W coasts; and vice versa. If approaching Jersey from Guernsey/ Alderney at HW St Helier +4, a fair tide can be carried for at least 6 hrs down the W coast and along the S coast to St Helier. From the S, leave St Malo at about HW, keeping E of the Minquiers, in order to carry a fair tide for 6 hrs to St Helier. Follow similar tidal tactics when coasting around the island.◀

To N and NE of Jersey, Les Pierres de Lecq (Paternosters) Les Dirouilles and Les Écrehou are groups of islets and drying rks, 2–4M offshore (AC *3655*, 1136, 1137, 1138).

From the N, a convenient landfall is Desormes WCM buoy, 4M NNW of Grosnez Pt (conspic lookout tr). On the N coast several bays offer anchs sheltered in offshore winds.

In St Ouen Bay on the W coast, which has drying rks almost 1M offshore, there are no good anchorages except NW of La Rocco tower which is sheltered in offshore winds.

Rounding the SW tip, to clear offlying dangers by 1M, keep the top of La Corbière lt ho (conspic) level with or below the FR lt on the clifftops behind. The inshore Boat Passage over the drying causeway between Jersey and Corbière lt ho is not advised. La Frouquie rock (9·8m) is no longer marked.

Along the S coast the **NW and W Passages** (buoyed) lead E past Noirmont Pt towards St Helier. St Brelade and St Aubin Bays provide some shelter from W'lies.

From the SW and S St Helier can be approached via **Danger Rock Passage** or **Red & Green Passage**. Both require good visibility to identify the transit marks and care to maintain the transits exactly. Only the Red & Green Passage is lit, but it needs sufficient water to pass over Fairway Rk (1·2m). Although charted, the marks for the South Passage 341° are hard to see.

SE of St Helier the drying, rocky Violet Bank extends 1M S to Demie de Pas lt bcn, thence E past Icho Twr (conspic). It extends 1·7M S and 2M SE of La Rocque Pt. Further rky plateaux extend 1M offshore. The **Violet Channel** (AC 1138) although buoyed, is best avoided in bad weather, wind-over-tide or poor vis. From St Helier make good Canger Rk WCM lt buoy, thence track 078° for 2·2M to Violet SWM lt buoy.

Turn N to pick up the charted ldg lines towards Gorey (dries) or to St Catherine Bay, both popular hbrs. The safe width of Violet Chan is only 5ca in places. The E coast of Jersey is well sheltered from W'lies, but requires careful pilotage.

If bound for the adjacent French coast, proceed NE from Violet buoy via the **Anquette Channel**, between Petite and Grande Anquette bcns. See Area 18 for Passage de la Déroute.

Les Minquiers, an extensive rocky plateau 10–18M S of St Helier, can be left to port, ie via NW and SW Minquiers buoys, if making for St Malo; or to stbd via NE and SE Minquiers buoys. A more direct route via N and SE Minquiers buoys requires sufficient height of tide to clear rocks drying 2m in the northern part. For a first visit the plateau should only be entered in settled weather, with exquisite care and a good Pilot book – not to mention *The Wreck of the Mary Deare* by Hammond Innes. The anchorage off Maîtresse Île is the principal attraction. From Jersey approach via the Demie de Vascelin SHM buoy with bcns in transit 167°. See St Helier.

9.19.14 ST HELIER

Jersey 49°10'·57N 02°06'·98W ✿✿❀❀⚓⚓⚓❀❀

CHARTS AC 3655, 1137, 3278, 5604.14/15; SHOM 7160, 7161, 6938; Navi 534, 1014; Imray C33B, 2500

TIDES –0455 Dover; ML 6·1; Duration 0545. St Helier is a Standard Port. The tidal range is very large, 9·6m at springs.

SHELTER Excellent in 2 marinas and La Collette basin (no ❶).

NAVIGATION WPT (where all apprs converge) 49°09'·95N 02°07'·38W, 023°/0·74M to front 023° ldg lt. WPT is near the centre of a Precautionary Area extending up-hbr to the ferry terminal. Caution: Rule 9 (narrow channels) applies; vessels under sail not fitted with a working engine must notify St Helier VTS VHF Ch 14 before entering Precautionary area. ⚓ in Small Road discouraged. Speed limit 5kn N of La Collette Yacht Basin. Approach channels:

- **W Passage** 082° ldg lts and Dog's Nest Rock bcn, Fl Y 3s 3M (Y cross on top) lead N of Les Fours and Ruaudière Rock SHM lt buoys, to WPT. Beware race off Noirmont Pt.
- **NW Passage** (095°/110°, much used by yachts) passes 6ca S of La Corbière lt ho to join W Passage abeam Noirmont Pt.
- **Danger Rock Passage** (044°) unlit; past rky, drying shoals; needs care and good vis. Beware overfalls S of Danger Rk.
- **Red & Green Passage** Ldg line 022·7° passes over Fairway Rk 1·2m, 8ca SSW of the WPT, with drying rocks either side; thence via Small Road to the ferry terminal, where daymarks are red dayglow patches on front dolphin and rear lt twr. The fairway's W side is beaconed and buoyed. Outer pier hds and dolphin are painted white and floodlit.
- **S Passage** (341°) unlit marks are hard to see. Demie de Pas in transit 350° with floodlit power stn chy is easier to see D/N.
- **E Passage** 290° from Canger Rock WCM lt buoy passes close SW of Demie de Pas, B tr/Y top (at night stay in W sector); thence 314° past Hinguette buoy and 341° to the WPT.

Other Passages Middle Passage, unlit, leads 339° towards St Aubin Hbr. Violet Channel skirts round SE tip of Jersey to/from the N and would normally link to the E Passage to/from St Helier. Caution: many offlying reefs.

LIGHTS AND MARKS See chartlet and Lights, buoys and waypoints for lights. Power station chy (95m, floodlit) Sig mast, W concave roofs of Fort Regent and Elizabeth Castle are conspic.

IPTS Sigs 1-4 are shown from the VTS tower and are easily seen from Small Road and the Main Hbr. An Oc Y 4s, shown above Sigs 1-4, exempts power-driven craft <25m LOA from the main signals. All leisure craft should: *Keep to starboard, well clear of shipping; maintain a sharp all-round lookout and monitor VHF Ch 14 during arrival/departure.*

Sig 2 is also shown when vessels depart the Tanker Basin. Departing tankers, which can be hidden at LW, sound a long blast if small craft are approaching.

COMMUNICATIONS (Code 01534) *Jersey Coastguard* 447705; Ch 82, 16 H24; monitor *St Helier VTS* Ch 14 H24 for ferry/shipping movements; *St Helier Pierheads* on Ch 18 broadcasts recorded wind direction and speed; Police 612612; ⊖ 448000; Ⓗ 622000. No marina VHF, but call *VTS* if essential; Marine Leisure Centre (enquiries) 447708.

FACILITIES Port Control ☎447708 or 447788, ⚓ 🛥(32t) ⚓ Grids. www.jerseymarinas.je marinas@gov.je

Fuel 🛢🛢 at the ent to La Collette (HW ±4; 07700 347313); a fuel barge by ent to Old Hbr (H24, self service, cr/debit card; ☎525247 07829 876543); Elizabeth marina (HW ±3; ☎07797 723090) self service, cr/debit card.

La Collette Basin, good shelter in 1·8m. Access H24. Caution: Ent narrow at LWS, least depth 1·1m; keep close to W side; PHM buoys mark shoal on E side. 130 ⌇ locals only, no ❶ berths, access H24. 🛢🛢 (at top of tide) 🏳(65t and 25t, ☎447773) ⚓.

St Helier YC ☎721307, ✕ 🍴.

Continued on page 836

ST HELIER

metres 0 400
cables 0 2

Causeway

La Vrachière
Fl(2)5s
BRB

Dir Bn in line with R Tr 106°

Fort Charles
North

Dries

Fort Charles
East
Q(3)5s

North Flagstaff

Fort Charles

Elizabeth
Castle

Crow
Rk

Hermitage Rk

SE Rock
(4) 7·4

Fish
Storage
Boxes

Crapaud of
the Castle
Q.R
No2

Fish
Storage
Boxes

B/W vert
stripes

Fl.R.1·5s6m5M
Platte Rk

O 4·5
RWBn
Oyster Rk

R and G Passage Ldg Lts 023°

Q.G
G East Rock
0·1

Ldg Lts: Front Oc.5s
Rear Oc.R.5s

W. Passage Ldg Lts 082°

PRECAUTIONARY
AREA

Danger Rk Passage 044°
Hermitage Bkwtr end Ø
Fort Regent Signal Mast
(off chartlet)

S. Passage B/W Marks Ø 341°
with Gros du Château

N

Bn YBY
Roquemin

Gros du Château
(2) (3)
Fort Elizabeth
FS(31m)

Les Bûts YB 6·4

Rear Mark=B/W vert stripes on sea wall

Hermitage Breakwater

E6
Fl.R.2s
E5
Fl.G.5s
2·5

E4

E2
Fl.R2s
0
E3
E1
Fl.G.3s

3xOc.Bu.8s
FWRG
Oc.G.4s
Bn R
Oc.R.4s
Sill
IPTS

Sill

Ro
Ro

Ro
Ro

Red Daymark

Oc.R.5s18m12M

Elizabeth
Marina

J 2·1m
H G
F
E
D 3.5m
C
K 2.1m
B
A

St Helier
Marina
G Grid
F V
E 2
D
C
B 2
A
3 IPTS
Sill (3·6)

New North Quay

Old Harbour (dries)

Holding
Pontoon

Red Daymark

Albert Pier

Maintained
depth 2.0m

South Pier
St Helier
YC F.G
F.G

VTS
IPTS
Victoria
Pier

Q.R No4
Oc.G.5s
10m11M
3
Q.R No4
078°
2·5

Dn
Q.R
1·8

La Collette Yacht
Basin (no V)
D C B A
Fish Quay

Chy conspic
(95m)

Tanker
Basin

La Collette Harbour

Oil
Tanks

2·5

Land Reclamation

Rock breakwater

Quereme Rk (2m)

4

Fl.Y.3s
Dog's Nest Rk

7

3

2·5

5·4

Nipple Rk
9·7

Small
Road

5·5

2·5

6·4

0·8

0·1

7·6

Corbière Lt Ho

Noirmont

Dog's Nest

Platte

Demie de Pas

Channel Islands

St Helier Marina ④ ☎447708. Sill dries 3·6m; hinged gate rises 1·4m above sill to retain 5m. Electronic tide gauge flashes when more than 2·5m depth over sill. Adjacent IPTS then control ent/exit and also shows R when depth 2m or less over sill, when tide gauge will read zero. A 150m long waiting pontoon is SW of marina ent, with ⚓ ⬚; rafting and overnight stays are permitted.

Ⓥ berths: E, F and G in 2m; yachts >14m or >2·1m draught call 447730/447708, £2·85. 🛢 🔌 🗑 🛁 🔧 📞 Ⓔ 🛒 ⚠ Grid, www. portsofjersey.je marinas@gov.je

Elizabeth Marina Preferred appr, when height of tide >7m, is from the W on 106°. Ldg marks, just S of ent, are orange □s with B vert line; at night stay in the W sector of Dir lt at front daymark. Pass S of La Vrachière IDM lt bn, cross the causeway 5·3m, turn ENE then NNE into the ent.

The marina is mainly for locals, or large visiting yachts by prior arrangement. Access over sill/flapgate; max LOA 20m, drafts 2·1–3·5m. At ent a digital gauge reads depth over sill. IPTS (sigs 2 & 3) control one-way ent/exit, usually 10 mins in each direction. ☎447708, 564🔌📞 (H24) ⚓.

The secondary appr chan 338° from Small Road (ie N of No 4 PHM buoy) is marked by 3 pairs of PHM and SHM buoys, 1st and 3rd pairs are lit. 5 Y can waiting buoys in about 1·0m are outboard of the lateral buoys (3 to the W, 2 to the E).

Town ✕ 🗑 ✉ 🛒 🅿 Ⓑ ✈. Fast ferries (Mar-Nov): Poole, St Malo, Granville, Sark, Guernsey, Carteret, Ro Ro Ferry: Portsmouth.

OTHER HARBOURS AND ANCHORAGES AROUND JERSEY
The main hbrs and ⚓s, clockwise from St Helier, are:

SOUTH COAST
St Aubin 49°11'·21N 02°10'·02W. Dries 6·8m. From seaward Middle Passage (339°, Mon Plaisir Ho ≠ twr on St Aubin Fort) leads towards St Aubin Fort. Final appr with N pier head bearing 254° lies N of St Aubin Fort; at night in W sector of the pier head lt – see chartlet and Lights, buoys and waypoints. Castle pier head is also lit. Beware very strong tidal streams in ent during full flood. Dry out alongside N pier or on wide grid. With offshore winds, quiet ⚓ in bay (partly dries). Hbr is administered by St Helier. Facilities: ⚓£14·50, ⚓ ◣ (1t 5t) Grid. **Royal Channel Islands YC** ☎741023, ✕ 🗑 ⚓ ◣. **Town** ✕ 🗑 🛒 🍴 📷 📞 , bus to St Helier.

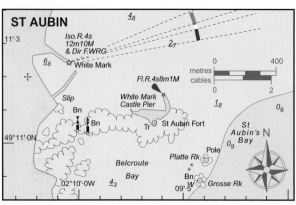

Belcroute Bay 49°10'·43N 02°09'·60W. Beware shellfish beds. Excellent shelter from W/SW'lies, but dries 3·5m and many moorings; ⚓ off Pt de Bût in 2·7m. Platte and Grosse Rks are marked by poles.

Portelet Bay 49°10'·13N 02°10'·61W. W of Noirmont Pt. Good ⚓ in N'lies, either side of Janvrin Twr.

St Brelade's Bay 49°10'·64N 02°11'·84W. Good shelter from N and W, but open to SW'ly swell. Beware many drying rks, especially Fournier Rk (0·9m) and Fourché (3·4m) in centre. A quiet ⚓ is in Beau Port. Bouilly Port in NW corner has local moorings and a small stone jetty. Very popular sandy tourist beach with hotels.

NORTH COAST
Grève au Lançon 49°15'·50N 02°14'·07W. Between Grosnez Pt's squat lt ho and Plemont Pt (conspic hotel) a wide, part-drying sandy bay, suitable for short stay on calm days. Open to swell.

Grève de Lecq 49°14'·98N 02°12'·12W. Ldg line 202°: W Martello tr on with W hotel with grey roof. ⚓ in 5m N of pier and W of submarine cables. Exposed to swell. ✕ 🗑, and bus to St Helier.

Bonne Nuit Bay 49°15'·19N 02°07'·09W. To the NNW beware Demie Rk (5·2m; SHM buoy); and to the E Les Sambues 5·5m. Conspic TV mast (232m) is ½M W of the bay. Ldg lts 223°. ⚓ in 5m NNE of pier and W of Chevel Rk. Hbr dries to sand/shingle. Many local moorings. ✕ 🗑.

Bouley Bay 49°14'·50N 02°04'·70W. Good ⚓ on sand in 2·5m SE of pier (rocky footings). Exposed to NE, see Gorey for tides. Local moorings. ✕ 🗑.

Rozel Bay 49°14'·25N 02°02'·43W. Appr with pierhead brg 245°, in W sector of Dir lt on shore. Conspic bldg close N on Nez du Guet. Pass between pierhead and WCM bn. Hbr, dries 1·5m to sand/shingle, and is full of moorings; ⚓ outside in 4-5m S of appr. Shops ✕ 🗑, bus to St Helier.

EAST COAST
St Catherine's Bay 49°13'·30N 02°00'·90W. Local moorings off W end of Verclut bkwtr. Clear the bkwtr hd by at least 50m due to strong tidal streams and the new rock armour extending 15m seaward. ⚓ S of bkwtr in 3–7m, but not in S/SE winds. Land on slip at root of bkwtr. See Gorey for tides, beware rocky St Catherine Bank, 3·1m (ECM bcn) and submarine cables to the S. Dinghy SC, RNLI 🅸 ✕.

La Rocque 49°09'·80N 02°01'·88W. Small hbr dries 6·5m–9·6m; bkwtr, local moorings, sandy beach, slip. Appr 330° across Violet Bank requires detailed local knowledge. No facilities.

OFFLYING ISLANDS
Les Écrehou 49°17'·39N 01°55'·58W (States ⚓). AC 3655, 5604.15; SHOM 6937 (1:25,000). 5M NE of Rozel Bay, has about 12 cottages. ML 6·2m. Arrive at about ½ tide ebbing; see Gorey. Appr with Bigorne Rk on 022°; when SE of Maître Ile alter to 330° for FS on Marmotière Is. Beware of strong and eddying tidal streams 4–8kn. No lts. Pick up ⚓ close SE of Marmotière or ⚓ in a pool 3ca WSW of Marmotière (with houses); other islands are Maître Ile (one house) Blanche and 5 other small islets. Local knowledge or a detailed pilotage book is essential.

Plateau des Minquiers 48°58'·04N 02°03'·74W. AC 5604.17 (1:25,000); 3656 and SHOM 7161 are both 1:50,000. About 12M S of Jersey, encircled by six cardinal light buoys. See Gorey for tides. ML 6·4m. Beware of strong and eddying tidal streams. Appr by day in good vis from Demie de Vascelin SHM buoy, 49°00'·81N 02°05'·15W, on 161°: Jetée des Fontaines RW bn in transit with FS on Maîtresse Ile. Further transits skirt the W side of the islet to ⚓ due S of it; safe only in settled weather and light winds. Maîtresse Ile has about a dozen cottages. Land at the slipway NW of the States of Jersey ⚓. Without local knowledge a detailed pilotage book is essential.

9.19.15 GOREY
Jersey 49°11'·78N 02°01'·37W ※※⚓⚓☆☆☆

CHARTS AC 3655, 5604./16, 1138; SHOM 7157, 7160, 6939; Navi 534, 1014; Imray C33B, 2500

TIDES –0454 Dover; ML 6·0; Duration 0545 Note the very large tidal range.

Standard Port ST HELIER (→)

Times				Height (metres)			
High Water		Low Water		MHWS	MHWN	MLWN	MLWS
0300	0900	0200	0900	11·0	8·1	4·0	1·4
1500	2100	1400	2100				
Differences ST CATHERINE BAY							
0000	+0010	+0010	+0010	0·0	−0·1	0·0	+0·1
BOULEY BAY							
+0002	+0002	+0004	+0004	−0·3	−0·3	−0·1	−0·1
LES ECREHOU							
+0005	+0009	+0011	+0009	−0·2	+0·1	−0·2	0·0
LES MINQUIERS							
−0014	−0018	−0001	−0008	+0·5	+0·6	+0·1	+0·1

SHELTER Good in the hbr, except in S/SE winds. ⚓ about 2ca E of pier hd or in deeper water in the Roads.

ST HELIER

MEAN SPRING ——————
AND NEAP CURVES – – – – –

MEAN RANGES	
Springs	9.6m
Neaps	4.1m

NAVIGATION WPT 49°11′·18N 01°59′·59W, 298°/1·3M to pier hd/front ldg lt. On appr, keep well outside all local bcns until the ldg marks are identified, but beware Banc du Chateau (0·4m least depth) 1M offshore NE of 298° ldg line; and Azicot Rk (dries 2·2m) just S of 298° ldg line, 2ca from ent. There are at least 3 approaches:

- Dir lt 298°: Gorey pierhead, W framework twr; rear ldg mark House (west gable end). Best for visitors.
- Pierhead ≠ ✠ spire 304° (line between R & G sectors) leads close to Road Rk (3·3m) and over Azicot Rk (2·2m).
- Pierhead ≠ white house/R roof 250° leads close to Les Arch bn (B/W with A topmark) and Pacquet Rk (0·3m).

See Passage information for the Violet Chan to St Helier. The Gutters and Boat Passage across Violet Bank are not advised.

LIGHTS AND MARKS See chartlet and Lights, buoys and waypoints. Mont Orgueil Castle (67m) is highly conspic.

COMMUNICATIONS (Code 01534) Marinecall 09068 969656; ⊖ 833833; for Dr contact Port Control St Helier 447788; *Gorey Hbr* Ch 74 (not permanently manned); Info *Jersey Coastguard* Ch 82.

FACILITIES Hbr dries to 6·9m. 12 drying ⚓s, 150m W of pier hd, 4 drying ⬡ against pier hd, all £14·50. Port Control ☎447708. 🅗 🅟 HW ±3 (HO) by hose at pierhead mob07797 742384, ⚓ ♠ ⚒ 🅑 🅘(7t).

Town ✕ 🏠 🛒 🅗 ✉ Ⓑ bus to St Helier. Ferries to Guernsey, Carteret.

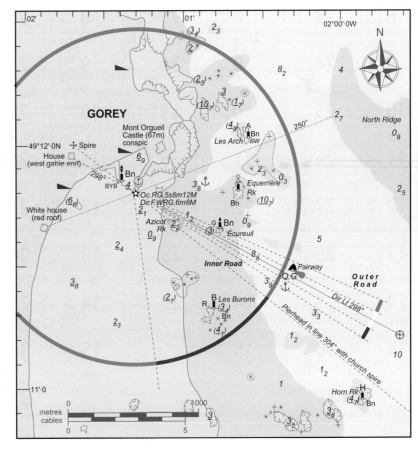

STANDARD TIME (UT)
For Summer Time add ONE hour in **non-shaded areas**

ST HELIER   LAT 49°11'N   LONG 2°07'W
TIMES AND HEIGHTS OF HIGH AND LOW WATERS

Dates in red are **SPRINGS**
Dates in blue are **NEAPS**

YEAR 2016

JANUARY

Time	m	Time	m
1 0443	3.4	**16** 0455	2.4
1025	9.0	1041	10.0
F 1707	3.6	SA 1724	2.4
2255	8.5	☽ 2308	9.5
2 0525	3.9	**17** 0547	2.9
1111	8.4	1135	9.3
SA 1753	4.0	SU 1819	2.9
☽ 2350	8.1		
3 0620	4.3	**18** 0006	9.0
1213	8.0	0650	3.3
SU 1855	4.3	M 1243	8.8
		1928	3.3
4 0105	7.9	**19** 0121	8.7
0732	4.4	0809	3.5
M 1334	7.9	TU 1405	8.7
2012	4.3	2048	3.3
5 0223	8.1	**20** 0243	8.8
0850	4.2	0930	3.2
TU 1447	8.2	W 1525	8.9
2124	3.9	2203	3.0
6 0324	8.5	**21** 0356	9.3
0955	3.7	1042	2.7
W 1545	8.6	TH 1632	9.4
2223	3.4	2309	2.5
7 0414	9.1	**22** 0456	9.9
1049	3.1	1143	2.2
TH 1634	9.2	F 1727	10.0
2313	2.8		
8 0458	9.7	**23** 0005	2.0
1138	2.5	0545	10.4
F 1719	9.8	SA 1236	1.8
		1814	10.4
9 0000	2.3	**24** 0053	1.7
0541	10.2	0629	10.8
SA 1224	2.0	SU 1321	1.5
1802	10.2	○ 1856	10.6
10 0045	2.0	**25** 0135	1.6
0623	10.7	0709	10.9
SU 1309	1.6	M 1400	1.5
● 1845	10.6	1934	10.7
11 0127	1.7	**26** 0212	1.6
0705	11.0	0745	10.9
M 1351	1.4	TU 1434	1.5
1928	10.9	2008	10.6
12 0208	1.5	**27** 0244	1.7
0747	11.2	0819	10.8
TU 1432	1.2	W 1504	1.8
2010	10.9	2040	10.4
13 0248	1.5	**28** 0313	2.0
0829	11.2	0850	10.4
W 1513	1.2	TH 1532	2.1
2052	10.8	2110	10.0
14 0329	1.6	**29** 0341	2.4
0911	11.0	0919	10.0
TH 1554	1.5	F 1559	2.5
2135	10.5	2139	9.5
15 0411	1.9	**30** 0409	2.8
0955	10.5	0949	9.4
F 1637	1.9	SA 1627	3.0
2219	10.0	2210	9.0
		31 0441	3.4
		1022	8.8
		SU 1700	3.6
		2246	8.4

FEBRUARY

Time	m	Time	m
1 0521	3.9	**16** 0621	3.3
1103	8.2	1211	8.6
M 1745	4.1	TU 1856	3.5
☽ 2337	7.9		
2 0619	4.3	**17** 0046	8.4
1206	7.8	0740	3.7
TU 1852	4.4	W 1339	8.2
		2022	3.8
3 0058	7.7	**18** 0219	8.3
0739	4.5	0912	3.6
W 1337	7.7	TH 1512	8.4
2021	4.4	2149	3.4
4 0230	7.9	**19** 0342	8.8
0905	4.1	1032	3.0
TH 1501	8.1	F 1622	9.0
2141	3.9	2258	2.8
5 0337	8.6	**20** 0443	9.5
1014	3.4	1133	2.4
F 1603	8.8	SA 1715	9.7
2243	3.1	2352	2.2
6 0431	9.3	**21** 0532	10.1
1111	2.6	1222	1.9
SA 1656	9.5	SU 1759	10.2
2337	2.4		
7 0520	10.1	**22** 0038	1.8
1203	1.9	0613	10.5
SU 1745	10.2	M 1304	1.6
		○ 1838	10.5
8 0027	1.8	**23** 0117	1.6
0606	10.7	0650	10.8
M 1253	1.4	TU 1340	1.5
● 1830	10.8	1913	10.7
9 0114	1.3	**24** 0151	1.5
0651	11.3	0724	10.9
TU 1339	0.9	W 1411	1.5
1915	11.2	1945	10.7
10 0158	1.0	**25** 0220	1.5
0735	11.6	0755	10.8
W 1422	0.7	TH 1438	1.6
1958	11.5	2014	10.6
11 0239	0.8	**26** 0247	1.7
0817	11.7	0824	10.6
TH 1503	0.6	F 1504	1.8
2039	11.4	2041	10.3
12 0319	0.9	**27** 0314	1.9
0859	11.5	0851	10.3
F 1542	0.8	SA 1529	2.1
2120	11.1	2107	9.9
13 0359	1.2	**28** 0341	2.4
0940	11.0	0917	9.8
SA 1622	1.3	SU 1555	2.6
2200	10.5	2132	9.4
14 0439	1.8	**29** 0409	2.9
1022	10.3	0944	9.2
SU 1703	2.1	M 1623	3.2
2243	9.8	2200	8.8
15 0524	2.6		
1109	9.4		
M 1751	2.9		
☽ 2335	9.0		

MARCH

Time	m	Time	m
1 0441	3.5	**16** 0558	3.3
1016	8.5	1147	8.4
TU 1657	3.8	W 1829	3.7
☽ 2237	8.2		
2 0526	4.1	**17** 0017	8.2
1104	8.0	0717	3.9
W 1750	4.3	TH 1318	7.9
2341	7.7	1959	4.0
3 0637	4.4	**18** 0156	8.1
1228	7.6	0854	3.8
TH 1918	4.5	F 1455	8.2
		2130	3.7
4 0126	7.7	**19** 0323	8.5
0812	4.3	1013	3.2
F 1415	7.8	SA 1604	8.8
2057	4.1	2239	3.0
5 0259	8.2	**20** 0423	9.3
0937	3.6	1112	2.5
SA 1533	8.6	SU 1655	9.5
2212	3.3	2331	2.4
6 0403	9.1	**21** 0511	9.9
1043	2.7	1159	2.0
SU 1632	9.5	M 1737	10.1
2312	2.4		
7 0457	10.0	**22** 0015	2.0
1140	1.8	0551	10.3
M 1724	10.3	TU 1239	1.7
		1814	10.4
8 0005	1.6	**23** 0052	1.7
0546	10.9	0626	10.6
TU 1232	1.1	W 1313	1.6
1812	11.0	○ 1847	10.6
9 0055	1.0	**24** 0124	1.6
0633	11.5	0659	10.7
W 1320	0.6	TH 1343	1.5
● 1857	11.5	1918	10.7
10 0141	0.6	**25** 0153	1.5
0718	11.9	0729	10.8
TH 1405	0.3	F 1410	1.6
1940	11.8	1946	10.7
11 0224	0.4	**26** 0221	1.6
0801	12.0	0757	10.6
F 1447	0.2	SA 1436	1.7
2021	11.8	2013	10.5
12 0305	0.5	**27** 0248	1.8
0842	11.8	0824	10.4
SA 1526	0.5	SU 1502	2.0
2101	11.4	2038	10.2
13 0344	0.9	**28** 0316	2.1
0923	11.2	0850	9.9
SU 1604	1.2	M 1528	2.4
2140	10.7	2103	9.7
14 0423	1.6	**29** 0344	2.6
1003	10.3	0916	9.4
M 1643	2.0	TU 1556	3.0
2220	9.9	2129	9.2
15 0505	2.5	**30** 0415	3.2
1048	9.3	0947	8.8
TU 1728	3.0	W 1628	3.6
☽ 2308	9.0	2204	8.6
		31 0455	3.7
		1032	8.2
		TH 1715	4.1
		☽ 2300	8.0

APRIL

Time	m	Time	m
1 0558	4.1	**16** 0124	8.0
1147	7.8	0823	3.9
F 1833	4.4	SA 1424	8.1
		2058	3.9
2 0036	7.8	**17** 0250	8.4
0728	4.1	0940	3.4
SA 1333	7.9	SU 1532	8.7
2014	4.1	2206	3.3
3 0219	8.2	**18** 0352	9.0
0859	3.5	1038	2.8
SU 1500	8.6	M 1624	9.3
2137	3.4	2258	2.7
4 0332	9.1	**19** 0440	9.6
1010	2.7	1124	2.4
M 1604	9.5	TU 1706	9.8
2242	2.4	2342	2.3
5 0430	10.1	**20** 0521	10.0
1111	1.8	1204	2.1
TU 1659	10.4	W 1744	10.2
2339	1.6		
6 0522	10.9	**21** 0019	2.0
1206	1.1	0557	10.3
W 1748	11.1	TH 1240	1.9
		1817	10.4
7 0031	0.9	**22** 0054	1.8
0611	11.6	0630	10.4
TH 1257	0.5	F 1311	1.8
● 1835	11.7	○ 1848	10.6
8 0120	0.5	**23** 0125	1.7
0657	11.9	0701	10.5
F 1343	0.3	SA 1341	1.7
1919	11.9	1917	10.6
9 0205	0.3	**24** 0155	1.7
0741	12.0	0730	10.5
SA 1427	0.3	SU 1410	1.8
2001	11.9	1946	10.5
10 0247	0.4	**25** 0225	1.8
0824	11.7	0759	10.3
SU 1507	0.7	M 1438	2.0
2041	11.5	2013	10.3
11 0328	0.8	**26** 0255	2.0
0905	11.1	0828	10.0
M 1546	1.3	TU 1507	2.3
2121	10.8	2042	9.9
12 0408	1.6	**27** 0325	2.4
0947	10.2	0859	9.6
TU 1626	2.1	W 1537	2.8
2201	9.9	2112	9.5
13 0450	2.4	**28** 0359	2.9
1032	9.2	0933	9.1
W 1710	3.0	TH 1612	3.3
2248	9.0	2150	8.9
14 0541	3.3	**29** 0440	3.5
1129	8.4	1020	8.5
TH 1808	3.8	F 1659	3.8
☽ 2353	8.3	2245	8.4
15 0654	3.8	**30** 0538	3.7
1252	7.9	1129	8.2
F 1931	4.1	SA 1808	4.0

Chart Datum is 5·88 metres below Ordnance Datum (Local). HAT is 12·2 metres above Chart Datum.

》》**FREE** monthly updates. Register at 《
www.reedsnauticalalmanac.co.uk 《

STANDARD TIME (UT)
For Summer Time add ONE hour in **non-shaded areas**

ST HELIER LAT 49°11'N LONG 2°07'W
TIMES AND HEIGHTS OF HIGH AND LOW WATERS

Dates in red are SPRINGS
Dates in blue are NEAPS

YEAR **2016**

MAY

Day	Time	m	Day	Time	m
1 SU	0007 / 0656 / 1259 / 1937	8.2 / 3.7 / 8.2 / 3.9	16 M	0201 / 0849 / 1446 / 2118	8.3 / 3.7 / 8.5 / 3.6
2 M	0141 / 0821 / 1424 / 2101	8.5 / 3.3 / 8.7 / 3.3	17 TU	0307 / 0950 / 1542 / 2214	8.7 / 3.3 / 9.0 / 3.2
3 TU	0257 / 0935 / 1532 / 2209	9.2 / 2.6 / 9.5 / 2.5	18 W	0400 / 1040 / 1628 / 2301	9.1 / 2.9 / 9.4 / 2.8
4 W	0400 / 1039 / 1630 / 2309	10.0 / 1.9 / 10.3 / 1.7	19 TH	0445 / 1123 / 1709 / 2342	9.5 / 2.5 / 9.8 / 2.4
5 TH	0456 / 1136 / 1722	10.7 / 1.2 / 11.0	20 F	0524 / 1202 / 1745	9.8 / 2.3 / 10.1
6 F ●	0004 / 0547 / 1230 / 1810	1.1 / 11.3 / 0.8 / 11.5	21 SA ○	0020 / 0559 / 1239 / 1818	2.1 / 10.1 / 2.1 / 10.3
7 SA	0056 / 0635 / 1320 / 1857	0.7 / 11.6 / 0.6 / 11.7	22 SU	0056 / 0633 / 1314 / 1850	1.9 / 10.2 / 2.0 / 10.4
8 SU	0144 / 0722 / 1406 / 1941	0.5 / 11.7 / 0.6 / 11.7	23 M	0131 / 0706 / 1347 / 1922	1.8 / 10.3 / 2.0 / 10.5
9 M	0229 / 0806 / 1449 / 2023	0.6 / 11.4 / 0.9 / 11.3	24 TU	0205 / 0739 / 1419 / 1955	1.8 / 10.3 / 2.0 / 10.4
10 TU	0312 / 0849 / 1530 / 2104	1.0 / 10.9 / 1.5 / 10.7	25 W	0238 / 0813 / 1451 / 2029	1.9 / 10.1 / 2.2 / 10.2
11 W	0354 / 0932 / 1611 / 2145	1.6 / 10.1 / 2.2 / 10.0	26 TH	0312 / 0849 / 1526 / 2105	2.2 / 9.8 / 2.5 / 9.8
12 TH	0437 / 1016 / 1654 / 2230	2.4 / 9.3 / 3.0 / 9.2	27 F	0350 / 0929 / 1605 / 2147	2.5 / 9.4 / 2.9 / 9.4
13 F ◑	0524 / 1108 / 1745 / 2326	3.1 / 8.6 / 3.6 / 8.5	28 SA	0433 / 1017 / 1652 / 2240	2.8 / 9.0 / 3.3 / 9.0
14 SA	0623 / 1216 / 1852	3.7 / 8.1 / 4.0	29 SU ◑	0526 / 1116 / 1752 / 2347	3.1 / 8.7 / 3.5 / 8.7
15 SU	0041 / 0736 / 1336 / 2009	8.2 / 3.9 / 8.1 / 4.0	30 M	0632 / 1229 / 1906	3.3 / 8.6 / 3.5
			31 TU	0105 / 0747 / 1347 / 2025	8.8 / 3.1 / 8.9 / 3.2

JUNE

Day	Time	m	Day	Time	m
1 W	0221 / 0901 / 1458 / 2136	9.2 / 2.7 / 9.4 / 2.6	16 TH	0313 / 0949 / 1545 / 2215	8.6 / 3.4 / 8.9 / 3.3
2 TH	0329 / 1007 / 1600 / 2239	9.8 / 2.1 / 10.0 / 2.0	17 F	0405 / 1040 / 1631 / 2303	9.0 / 3.1 / 9.3 / 2.8
3 F	0429 / 1108 / 1657 / 2339	10.3 / 1.7 / 10.6 / 1.5	18 SA	0449 / 1126 / 1712 / 2348	9.4 / 2.7 / 9.7 / 2.5
4 SA	0525 / 1205 / 1749	10.8 / 1.3 / 11.1	19 SU	0530 / 1208 / 1750	9.7 / 2.4 / 10.1
5 SU ●	0034 / 0616 / 1259 / 1837	1.1 / 11.1 / 1.1 / 11.3	20 M ○	0029 / 0608 / 1249 / 1827	2.1 / 10.0 / 2.2 / 10.3
6 M	0126 / 0705 / 1347 / 1923	0.9 / 11.2 / 1.0 / 11.4	21 TU	0110 / 0646 / 1327 / 1904	1.9 / 10.2 / 2.0 / 10.5
7 TU	0214 / 0751 / 1433 / 2006	1.0 / 11.1 / 1.2 / 11.2	22 W	0148 / 0724 / 1404 / 1941	1.8 / 10.3 / 2.0 / 10.6
8 W	0259 / 0834 / 1515 / 2048	1.2 / 10.7 / 1.6 / 10.7	23 TH	0226 / 0803 / 1440 / 2020	1.7 / 10.3 / 2.0 / 10.6
9 TH	0340 / 0916 / 1555 / 2128	1.7 / 10.2 / 2.1 / 10.1	24 F	0304 / 0843 / 1518 / 2100	1.7 / 10.2 / 2.1 / 10.3
10 F	0420 / 0957 / 1634 / 2208	2.3 / 9.5 / 2.8 / 9.5	25 SA	0343 / 0925 / 1559 / 2143	2.0 / 10.0 / 2.3 / 10.0
11 SA	0500 / 1040 / 1715 / 2254	2.9 / 8.9 / 3.3 / 8.9	26 SU	0426 / 1010 / 1644 / 2231	2.2 / 9.6 / 2.7 / 9.6
12 SU ◑	0544 / 1131 / 1804 / 2350	3.4 / 8.4 / 3.8 / 8.4	27 M ◔	0514 / 1101 / 1736 / 2326	2.6 / 9.3 / 3.0 / 9.3
13 M	0638 / 1236 / 1906	3.8 / 8.2 / 4.0	28 TU	0609 / 1201 / 1839	2.8 / 9.0 / 3.2
14 TU	0100 / 0742 / 1347 / 2015	8.2 / 3.9 / 8.2 / 4.0	29 W	0033 / 0715 / 1311 / 1952	9.0 / 3.0 / 8.9 / 3.2
15 W	0212 / 0850 / 1452 / 2120	8.3 / 3.8 / 8.5 / 3.7	30 TH	0148 / 0829 / 1425 / 2107	9.0 / 2.9 / 9.2 / 2.9

JULY

Day	Time	m	Day	Time	m
1 F	0302 / 0940 / 1535 / 2216	9.3 / 2.6 / 9.6 / 2.5	16 SA	0324 / 0958 / 1555 / 2227	8.4 / 3.6 / 8.8 / 3.3
2 SA	0409 / 1046 / 1637 / 2320	9.8 / 2.2 / 10.1 / 2.0	17 SU	0417 / 1052 / 1642 / 2318	8.9 / 3.1 / 9.4 / 2.7
3 SU	0509 / 1147 / 1732	10.2 / 1.8 / 10.6	18 M	0503 / 1141 / 1725	9.4 / 2.6 / 9.9
4 M ●	0019 / 0602 / 1243 / 1821	1.6 / 10.6 / 1.5 / 10.9	19 TU ○	0005 / 0546 / 1227 / 1806	2.2 / 9.9 / 2.2 / 10.3
5 TU	0113 / 0651 / 1332 / 1907	1.3 / 10.8 / 1.4 / 11.1	20 W	0051 / 0628 / 1310 / 1847	1.9 / 10.3 / 1.9 / 10.7
6 W	0201 / 0736 / 1417 / 1950	1.2 / 10.9 / 1.4 / 11.1	21 TH	0134 / 0710 / 1351 / 1928	1.6 / 10.6 / 1.7 / 10.9
7 TH	0244 / 0817 / 1458 / 2030	1.3 / 10.7 / 1.6 / 10.8	22 F	0215 / 0751 / 1431 / 2009	1.4 / 10.7 / 1.6 / 11.0
8 F	0322 / 0856 / 1534 / 2106	1.6 / 10.3 / 2.0 / 10.4	23 SA	0254 / 0832 / 1510 / 2051	1.3 / 10.8 / 1.6 / 10.9
9 SA	0357 / 0932 / 1607 / 2142	2.0 / 9.9 / 2.5 / 9.9	24 SU	0334 / 0914 / 1550 / 2133	1.4 / 10.6 / 1.7 / 10.6
10 SU	0429 / 1007 / 1640 / 2218	2.6 / 9.3 / 3.0 / 9.3	25 M	0415 / 0956 / 1632 / 2216	1.7 / 10.2 / 2.1 / 10.2
11 M	0502 / 1046 / 1717 / 2259	3.1 / 8.8 / 3.5 / 8.7	26 TU ◔	0458 / 1041 / 1719 / 2305	2.1 / 9.7 / 2.6 / 9.6
12 TU ◔	0541 / 1133 / 1804 / 2353	3.6 / 8.3 / 3.9 / 8.2	27 W	0547 / 1133 / 1814	2.6 / 9.2 / 3.1
13 W	0632 / 1237 / 1905	4.0 / 8.0 / 4.2	28 TH	0004 / 0647 / 1239 / 1924	9.0 / 3.1 / 8.8 / 3.4
14 TH	0104 / 0740 / 1353 / 2019	8.0 / 4.2 / 8.0 / 4.1	29 F	0120 / 0802 / 1358 / 2045	8.7 / 3.3 / 8.8 / 3.3
15 F	0221 / 0854 / 1500 / 2129	8.0 / 4.0 / 8.4 / 3.8	30 SA	0243 / 0921 / 1518 / 2203	8.8 / 3.1 / 9.1 / 2.9
			31 SU	0358 / 1033 / 1625 / 2311	9.2 / 2.7 / 9.7 / 2.3

AUGUST

Day	Time	m	Day	Time	m
1 M	0500 / 1136 / 1720	9.8 / 2.2 / 10.3	16 TU	0439 / 1116 / 1702 / 2342	9.3 / 2.7 / 9.8 / 2.2
2 TU ●	0010 / 0551 / 1231 / 1808	1.8 / 10.3 / 1.8 / 10.7	17 W	0526 / 1205 / 1747	10.0 / 2.1 / 10.5
3 W	0101 / 0637 / 1318 / 1852	1.5 / 10.6 / 1.5 / 11.0	18 TH ○	0030 / 0610 / 1252 / 1830	1.7 / 10.5 / 1.7 / 11.0
4 TH	0146 / 0718 / 1400 / 1931	1.3 / 10.8 / 1.5 / 11.0	19 F	0117 / 0653 / 1336 / 1912	1.3 / 11.0 / 1.3 / 11.4
5 F	0224 / 0756 / 1436 / 2007	1.4 / 10.8 / 1.6 / 10.9	20 SA	0200 / 0735 / 1417 / 1954	1.0 / 11.2 / 1.1 / 11.6
6 SA	0258 / 0830 / 1507 / 2040	1.5 / 10.5 / 1.8 / 10.6	21 SU	0241 / 0816 / 1457 / 2036	0.8 / 11.3 / 1.1 / 11.5
7 SU	0327 / 0902 / 1536 / 2111	1.9 / 10.2 / 2.2 / 10.2	22 M	0320 / 0857 / 1537 / 2117	0.9 / 11.1 / 1.3 / 11.1
8 M	0354 / 0931 / 1604 / 2141	2.3 / 9.7 / 2.7 / 9.6	23 TU	0359 / 0937 / 1617 / 2158	1.3 / 10.7 / 1.7 / 10.5
9 TU	0421 / 1002 / 1635 / 2213	2.9 / 9.2 / 3.2 / 9.0	24 W	0440 / 1019 / 1700 / 2243	1.9 / 10.0 / 2.4 / 9.7
10 W	0452 / 1036 / 1712 / 2252	3.4 / 8.6 / 3.8 / 8.4	25 TH ◔	0525 / 1107 / 1752 / 2339	2.7 / 9.3 / 3.1 / 8.9
11 TH	0533 / 1123 / 1803 / 2349	4.0 / 8.1 / 4.2 / 7.8	26 F	0623 / 1210 / 1904	3.4 / 8.6 / 3.6
12 F	0632 / 1238 / 1916	4.4 / 7.7 / 4.5	27 SA	0058 / 0743 / 1339 / 2033	8.3 / 3.7 / 8.4 / 3.7
13 SA	0117 / 0755 / 1411 / 2041	7.6 / 4.5 / 7.9 / 4.2	28 SU	0235 / 0911 / 1509 / 2157	8.4 / 3.5 / 8.8 / 3.2
14 SU	0244 / 0918 / 1521 / 2153	7.9 / 4.1 / 8.4 / 3.6	29 M	0353 / 1026 / 1616 / 2303	8.9 / 3.0 / 9.4 / 2.5
15 M	0347 / 1022 / 1614 / 2250	8.6 / 3.4 / 9.1 / 2.9	30 TU	0451 / 1126 / 1709 / 2358	9.6 / 2.3 / 10.1 / 1.9
			31 W	0538 / 1216 / 1754	10.2 / 1.9 / 10.6

Channel Islands

Chart Datum is 5·88 metres below Ordnance Datum (Local). HAT is 12·2 metres above Chart Datum.

STANDARD TIME (UT)
For Summer Time add ONE hour in **non-shaded areas**

ST HELIER   LAT 49°11′N   LONG 2°07′W
TIMES AND HEIGHTS OF HIGH AND LOW WATERS

Dates in red are **SPRINGS**
Dates in blue are **NEAPS**

YEAR 2016

SEPTEMBER

Time m	Time m
1 0044 1.6 / 0619 10.6 / TH 1259 1.6 / ● 1833 10.9	**16** 0006 1.5 / 0548 10.8 / F 1230 1.4 / ○ 1809 11.3
2 0125 1.4 / 0657 10.8 / F 1337 1.5 / 1909 11.0	**17** 0055 1.0 / 0632 11.3 / SA 1316 1.0 / 1853 11.7
3 0159 1.4 / 0730 10.8 / SA 1409 1.6 / 1941 11.0	**18** 0140 0.7 / 0715 11.7 / SU 1359 0.8 / 1935 11.9
4 0228 1.6 / 0801 10.7 / SU 1437 1.7 / 2012 10.8	**19** 0222 0.5 / 0756 11.7 / M 1440 0.7 / 2017 11.8
5 0254 1.8 / 0830 10.5 / M 1504 2.0 / 2040 10.4	**20** 0302 0.7 / 0837 11.5 / TU 1520 1.0 / 2058 11.3
6 0319 2.2 / 0856 10.1 / TU 1530 2.4 / 2106 9.9	**21** 0341 1.2 / 0917 10.9 / W 1600 1.6 / 2140 10.6
7 0345 2.7 / 0922 9.5 / W 1558 3.0 / 2133 9.3	**22** 0421 2.0 / 0957 10.1 / TH 1643 2.4 / 2224 9.6
8 0412 3.3 / 0949 8.9 / TH 1630 3.6 / 2203 8.6	**23** 0506 2.8 / 1044 9.3 / F 1735 3.2 / ◑ 2320 8.7
9 0445 3.9 / 1024 8.3 / F 1713 4.2 / ◑ 2247 8.0	**24** 0604 3.6 / 1148 8.5 / SA 1849 3.8
10 0536 4.5 / 1122 7.7 / SA 1821 4.6	**25** 0044 8.1 / 0728 4.1 / SU 1323 8.2 / 2023 3.9
11 0008 7.5 / 0658 4.7 / SU 1310 7.6 / 1953 4.5	**26** 0226 8.2 / 0900 3.8 / M 1457 8.6 / 2145 3.3
12 0201 7.7 / 0836 4.4 / M 1445 8.1 / 2117 3.9	**27** 0340 8.8 / 1012 3.2 / TU 1601 9.3 / 2247 2.6
13 0318 8.4 / 0951 3.7 / TU 1546 9.0 / 2221 3.0	**28** 0434 9.6 / 1108 2.5 / W 1650 10.0 / 2337 2.1
14 0413 9.3 / 1049 2.6 / W 1637 9.9 / 2316 2.2	**29** 0518 10.2 / 1154 2.0 / TH 1732 10.5
15 0502 10.1 / 1140 2.0 / TH 1724 10.7	**30** 0020 1.8 / 0556 10.6 / F 1234 1.8 / 1809 10.8

OCTOBER

Time m	Time m
1 0057 1.6 / 0631 10.8 / SA 1308 1.7 / ● 1843 10.9	**16** 0028 0.9 / 0608 11.5 / SU 1251 0.9 / ○ 1830 11.8
2 0128 1.6 / 0702 10.8 / SU 1338 1.7 / 1914 10.9	**17** 0116 0.6 / 0652 11.9 / M 1337 0.7 / 1915 12.0
3 0155 1.7 / 0731 10.8 / M 1406 1.8 / 1942 10.8	**18** 0200 0.5 / 0735 11.9 / TU 1421 0.7 / 1958 11.8
4 0221 1.9 / 0759 10.6 / TU 1433 1.9 / 2010 10.5	**19** 0242 0.8 / 0817 11.6 / W 1503 1.0 / 2041 11.3
5 0247 2.2 / 0824 10.3 / W 1501 2.3 / 2036 10.0	**20** 0323 1.3 / 0858 11.0 / TH 1545 1.6 / 2124 10.5
6 0314 2.6 / 0849 9.8 / TH 1529 2.8 / 2102 9.5	**21** 0405 2.1 / 0940 10.2 / F 1630 2.4 / 2209 9.6
7 0341 3.2 / 0915 9.2 / F 1600 3.4 / 2131 8.8	**22** 0450 3.0 / 1026 9.3 / SA 1722 3.2 / ◑ 2305 8.7
8 0412 3.8 / 0946 8.6 / SA 1639 4.0 / 2210 8.2	**23** 0548 3.7 / 1127 8.6 / SU 1832 3.8
9 0457 4.4 / 1034 8.0 / SU 1739 4.4 / ◑ 2319 7.7	**24** 0024 8.1 / 0707 4.2 / M 1257 8.2 / 2000 3.9
10 0611 4.7 / 1209 7.7 / M 1907 4.5	**25** 0200 8.2 / 0835 4.0 / TU 1428 8.5 / 2118 3.5
11 0111 7.7 / 0751 4.5 / TU 1401 8.0 / 2038 4.0	**26** 0313 8.7 / 0945 3.4 / W 1533 9.1 / 2218 2.9
12 0242 8.3 / 0915 3.8 / W 1513 8.9 / 2148 3.1	**27** 0406 9.4 / 1039 2.8 / TH 1623 9.7 / 2307 2.4
13 0343 9.2 / 1018 2.9 / TH 1608 9.8 / 2246 2.2	**28** 0450 10.0 / 1124 2.4 / F 1705 10.2 / 2348 2.1
14 0435 10.2 / 1112 2.0 / F 1658 10.7 / 2338 1.5	**29** 0528 10.3 / 1203 2.1 / SA 1742 10.5
15 0523 10.9 / 1203 1.4 / SA 1745 11.4	**30** 0023 2.0 / 0602 10.6 / SU 1237 1.9 / ● 1815 10.6
	31 0055 1.9 / 0633 10.7 / M 1308 1.9 / 1846 10.7

NOVEMBER

Time m	Time m
1 0124 1.9 / 0703 10.7 / TU 1338 1.9 / 1916 10.6	**16** 0139 0.8 / 0716 11.8 / W 1403 0.8 / 1942 11.6
2 0153 2.0 / 0731 10.6 / W 1408 2.0 / 1944 10.4	**17** 0224 1.0 / 0759 11.6 / TH 1449 1.0 / 2026 11.2
3 0221 2.2 / 0758 10.4 / TH 1437 2.2 / 2013 10.1	**18** 0308 1.4 / 0842 11.1 / F 1533 1.6 / 2111 10.5
4 0250 2.5 / 0826 10.0 / F 1508 2.6 / 2042 9.7	**19** 0351 2.1 / 0925 10.4 / SA 1618 2.3 / 2156 9.7
5 0319 3.0 / 0854 9.5 / SA 1540 3.1 / 2114 9.1	**20** 0436 2.8 / 1010 9.6 / SU 1707 3.0 / 2247 8.9
6 0352 3.5 / 0928 9.0 / SU 1619 3.6 / 2154 8.6	**21** 0527 3.6 / 1104 8.8 / M 1806 3.6 / ◑ 2350 8.3
7 0435 4.0 / 1015 8.5 / M 1712 4.0 / ◑ 2255 8.1	**22** 0633 4.0 / 1214 8.4 / TU 1918 3.9
8 0538 4.4 / 1131 8.1 / TU 1827 4.2	**23** 0110 8.2 / 0749 4.1 / W 1337 8.3 / 2031 3.8
9 0024 8.0 / 0706 4.4 / W 1309 8.2 / 1954 3.9	**24** 0226 8.4 / 0900 3.8 / TH 1449 8.7 / 2135 3.4
10 0156 8.4 / 0833 3.8 / TH 1431 8.8 / 2109 3.2	**25** 0326 8.9 / 0959 3.3 / F 1545 9.1 / 2227 3.0
11 0306 9.2 / 0942 3.0 / F 1534 9.7 / 2212 2.4	**26** 0415 9.4 / 1047 2.9 / SA 1631 9.6 / 2311 2.7
12 0403 10.0 / 1041 2.2 / SA 1629 10.5 / 2308 1.7	**27** 0456 9.9 / 1129 2.6 / SU 1711 9.9 / 2349 2.4
13 0455 10.8 / 1135 1.5 / SU 1720 11.2	**28** 0532 10.2 / 1206 2.3 / M 1748 10.1
14 0001 1.1 / 0544 11.4 / M 1227 1.0 / ○ 1808 11.6	**29** 0024 2.2 / 0606 10.4 / TU 1242 2.1 / ● 1821 10.3
15 0051 0.8 / 0630 11.7 / TU 1316 0.8 / 1856 11.8	**30** 0058 2.1 / 0638 10.5 / W 1315 2.0 / 1853 10.4

DECEMBER

Time m	Time m
1 0131 2.1 / 0709 10.5 / TH 1348 2.0 / 1925 10.3	**16** 0210 1.1 / 0746 11.5 / F 1437 1.0 / 2014 11.1
2 0202 2.2 / 0740 10.5 / F 1421 2.1 / 1957 10.2	**17** 0255 1.4 / 0829 11.2 / SA 1522 1.4 / 2057 10.5
3 0233 2.4 / 0812 10.2 / SA 1454 2.4 / 2031 9.9	**18** 0337 1.9 / 0911 10.6 / SU 1604 2.0 / 2139 10.0
4 0306 2.7 / 0845 9.9 / SU 1528 2.7 / 2107 9.5	**19** 0419 2.5 / 0952 10.0 / M 1646 2.6 / 2222 9.3
5 0341 3.1 / 0922 9.5 / M 1608 3.0 / 2148 9.1	**20** 0500 3.1 / 1035 9.3 / TU 1729 3.3 / 2309 8.7
6 0423 3.5 / 1007 9.1 / TU 1655 3.4 / 2240 8.7	**21** 0547 3.7 / 1126 8.7 / W 1820 3.8 / ◑
7 0516 3.8 / 1107 8.7 / W 1756 3.6 / ◑ 2348 8.5	**22** 0008 8.3 / 0645 4.1 / TH 1232 8.3 / 1924 4.0
8 0627 3.9 / 1223 8.6 / TH 1910 3.6	**23** 0121 8.1 / 0756 4.2 / F 1348 8.2 / 2034 4.0
9 0107 8.6 / 0748 3.9 / F 1344 8.8 / 2027 3.2	**24** 0233 8.3 / 0905 4.0 / SA 1458 8.4 / 2137 3.7
10 0224 9.0 / 0904 3.2 / SA 1457 9.4 / 2137 2.7	**25** 0333 8.8 / 1004 3.5 / SU 1554 8.8 / 2231 3.3
11 0330 9.7 / 1010 2.5 / SU 1600 10.0 / 2239 2.1	**26** 0422 9.2 / 1054 3.1 / M 1641 9.3 / 2316 2.9
12 0429 10.4 / 1110 1.9 / M 1657 10.7 / 2337 1.5	**27** 0504 9.7 / 1137 2.7 / TU 1722 9.6 / 2357 2.6
13 0522 11.0 / 1206 1.4 / TU 1751 11.1	**28** 0542 10.0 / 1218 2.3 / W 1759 9.9
14 0031 1.2 / 0612 11.4 / W 1300 1.0 / ○ 1841 11.4	**29** 0036 2.3 / 0617 10.3 / TH 1257 2.1 / ● 1835 10.2
15 0122 1.0 / 0700 11.6 / TH 1350 0.9 / 1928 11.4	**30** 0113 2.1 / 0652 10.5 / F 1334 1.9 / 1910 10.3
	31 0149 2.1 / 0727 10.6 / SA 1409 1.9 / 1946 10.3

Chart Datum is 5·88 metres below Ordnance Datum (Local). HAT is 12·2 metres above Chart Datum.

North Brittany

Paimpol to Pointe de Penmarc'h

9.20.1	Contents	841
9.20.2	Area map, Distance table and TSS off Ouessant	842
9.20.3	Tidal stream chartlets	844
9.20.4	Lights, buoys and waypoints	846
9.20.5	Passage information	850
	North Brittany – 849 • Paimpol to Ploumanac'h – 850 • Les Sept Îles to Baie de Morlaix – 850 • Île de Batz to Le Four – 850 • Ouessant (Ushant) – 850 • Chenal du Four – 866 • Chenal de la Helle – 866 • Approaches to Brest – 866 • L'Iroise and Baie de Douarnenez – 866 • Raz de Sein to Pointe de Penmarc'h – 866	
9.20.6	Special notes for France; see Area 17	
9.20.7	Paimpol	851
9.20.8	Île de Bréhat	851
	Passe de la Gaine	
9.20.9	Lézardrieux	853
9.20.10	Pontrieux	854
9.20.11	Tréguier	854
9.20.12	Port Blanc	855
9.20.13	Perros-Guirec	855
	Les Sept Îles	
9.20.14	Ploumanac'h	857
	Trégastel-Plage	
9.20.15	Trébeurden	858
9.20.16	Léguer River (Lannion)	858
	Locquémeau • Locquirec	
9.20.17	Primel-Trégastel	859
9.20.18	Morlaix	860
	Penzé River	
9.20.19	Roscoff	862
	Île de Batz • Moguériec • Brignogan (Pontusval)	
9.20.20	L'Aber Wrac'h	863
9.20.21	L'Aber Benoit	864
	Portsall • Argenton	
9.20.22	L'Aber-Ildut	864
9.20.23	Le Conquet	865
	Anse de Bertheaume	
9.20.24	Chenal du Four	867
9.20.25	Île d'Ouessant	868
	Île de Molène	
9.20.26	Rade de Brest, Brest tide tables/curves	869
	R L'Elorn • R L'Aulne	
9.20.27	Brest, tidal coefficients	874
9.20.28	Camaret	876
9.20.29	Morgat	876
9.20.30	Douarnenez	877
9.20.31/32	Raz de Sein and Île de Sein	878
9.20.33	Audierne	879
	Ste Evette • St Guénolé • Le Guilvinec • Lesconil	

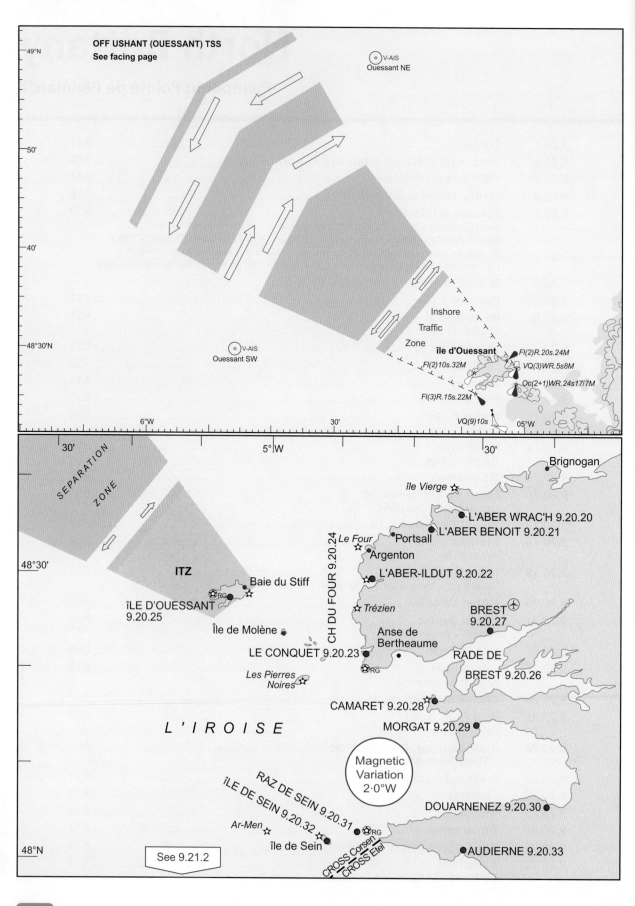

OFF USHANT (OUESSANT) TSS
See facing page

V-AIS
Ouessant NE

49°N

50'

40'

Inshore

Traffic

Zone

48°30'N

V-AIS
Ouessant SW

île d'Ouessant

Fl(2)R.20s.24M

VQ(3)WR.5s8M

Fl(2)10s.32M

Oc(2+1)WR.24s17/7M

Fl(3)R.15s.22M

VQ(9)10s

05°W

6°W

30'

SEPARATION

ZONE

Brignogan

île Vierge ☆

L'ABER WRAC'H 9.20.20

L'ABER BENOIT 9.20.21

Le Four ☆

Portsall

Argenton

ITZ

L'ABER-ILDUT 9.20.22

48°30'

Baie du Stiff

☆ *Trézien*

BREST
9.20.27

îLE D'OUESSANT
9.20.25

Île de Molène

RG

Anse de
Bertheaume

RADE DE

Le Conquet 9.20.23

BREST 9.20.26

Les Pierres
Noires

RG

CAMARET 9.20.28

L' I R O I S E

MORGAT 9.20.29

Magnetic
Variation
2·0°W

RAZ DE SEIN 9.20.31

DOUARNENEZ 9.20.30

îLE DE SEIN 9.20.32

Ar-Men ☆

RG

île de Sein

CROSS Corsen
CROSS Etel

See 9.21.2

AUDIERNE 9.20.33

48°N

30'

5°W

30'

CH DU FOUR 9.20.24

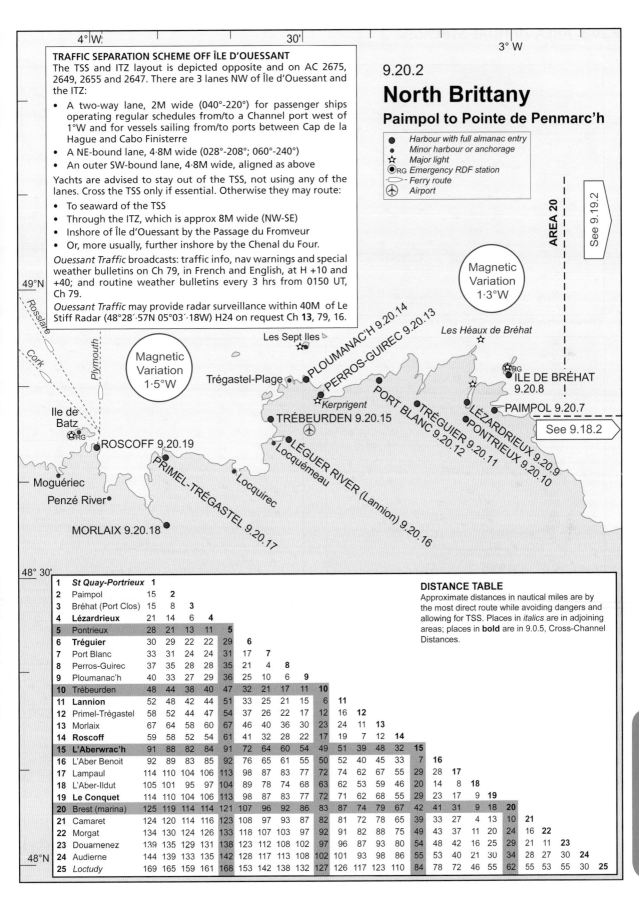

See 9.19.2

TRAFFIC SEPARATION SCHEME OFF ÎLE D'OUESSANT

The TSS and ITZ layout is depicted opposite and on AC 2675, 2649, 2655 and 2647. There are 3 lanes NW of Île d'Ouessant and the ITZ:

- A two-way lane, 2M wide (040°-220°) for passenger ships operating regular schedules from/to a Channel port west of 1°W and for vessels sailing from/to ports between Cap de la Hague and Cabo Finisterre
- A NE-bound lane, 4·8M wide (028°-208°; 060°-240°)
- An outer SW-bound lane, 4·8M wide, aligned as above

Yachts are advised to stay out of the TSS, not using any of the lanes. Cross the TSS only if essential. Otherwise they may route:

- To seaward of the TSS
- Through the ITZ, which is approx 8M wide (NW-SE)
- Inshore of Île d'Ouessant by the Passage du Fromveur
- Or, more usually, further inshore by the Chenal du Four.

Ouessant Traffic broadcasts: traffic info, nav warnings and special weather bulletins on Ch 79, in French and English, at H +10 and +40; and routine weather bulletins every 3 hrs from 0150 UT, Ch 79.

Ouessant Traffic may provide radar surveillance within 40M of Le Stiff Radar (48°28'·57N 05°03'·18W) H24 on request Ch **13**, 79, 16.

9.20.2
North Brittany
Paimpol to Pointe de Penmarc'h

- ● Harbour with full almanac entry
- ● Minor harbour or anchorage
- ☆ Major light
- ◉RG Emergency RDF station
- ⊏⊐ Ferry route
- ⊕ Airport

Magnetic Variation 1·3°W

AREA 20

Magnetic Variation 1·5°W

Rosslare
Cork
Plymouth

Les Héaux de Bréhat ☆

Les Sept Iles ▷
☆☆ PLOUMANAC'H 9.20.14
PERROS-GUIREC 9.20.13
◉RG ÎLE DE BRÉHAT 9.20.8
Trégastel-Plage
☆ Kerprigent
PORT BLANC 9.20.12
TRÉGUIER 9.20.11
LÉZARDRIEUX 9.20.9
PAIMPOL 9.20.7
☆ PONTRIEUX 9.20.10
TRÉBEURDEN 9.20.15
Ile de Batz
◉RG
ROSCOFF 9.20.19
LÉGUER RIVER (Lannion) 9.20.16
Locquémeau
Moguériec
PRIMEL-TRÉGASTEL 9.20.17
Locquirec
Penzé River
MORLAIX 9.20.18

See 9.18.2

DISTANCE TABLE
Approximate distances in nautical miles are by the most direct route while avoiding dangers and allowing for TSS. Places in *italics* are in adjoining areas; places in **bold** are in 9.0.5, Cross-Channel Distances.

		1	2	3	4	5	6	7	8	9	10	11	12	13	14	15	16	17	18	19	20	21	22	23	24	25
1	*St Quay-Portrieux*	**1**																								
2	Paimpol	15	**2**																							
3	Bréhat (Port Clos)	15	8	**3**																						
4	**Lézardrieux**	21	14	6	**4**																					
5	Pontrieux	28	21	13	11	**5**																				
6	**Tréguier**	30	29	22	22	29	**6**																			
7	Port Blanc	33	31	24	24	31	17	**7**																		
8	Perros-Guirec	37	35	28	28	35	21	4	**8**																	
9	Ploumanac'h	40	33	27	29	36	25	10	6	**9**																
10	Trébeurden	48	44	38	40	47	32	21	17	11	**10**															
11	**Lannion**	52	48	42	44	51	33	25	21	15	6	**11**														
12	Primel-Trégastel	58	52	44	47	54	37	26	22	17	12	16	**12**													
13	Morlaix	67	64	58	60	67	46	40	36	30	23	24	11	**13**												
14	**Roscoff**	59	58	52	54	61	41	32	28	22	17	19	7	12	**14**											
15	**L'Aberwrac'h**	91	88	82	84	91	72	64	60	54	49	51	39	48	32	**15**										
16	L'Aber Benoit	92	89	83	85	92	76	65	61	55	50	52	40	45	33	7	**16**									
17	Lampaul	114	110	104	106	113	98	87	83	77	72	74	62	67	55	29	28	**17**								
18	L'Aber-Ildut	105	101	95	97	104	89	78	74	68	63	62	53	59	46	20	14	8	**18**							
19	**Le Conquet**	114	110	104	106	113	98	87	83	77	72	71	62	68	55	29	23	17	9	**19**						
20	Brest (marina)	125	119	114	114	121	107	96	92	86	83	87	74	79	67	42	41	31	9	18	**20**					
21	Camaret	124	120	114	116	123	108	97	93	87	82	81	72	78	65	39	33	27	4	13	10	**21**				
22	Morgat	134	130	124	126	133	118	107	103	97	92	91	82	88	75	49	43	37	11	20	24	16	**22**			
23	Douarnenez	139	135	129	131	138	123	112	108	102	97	96	87	93	80	54	48	42	16	25	29	21	11	**23**		
24	Audierne	144	139	133	135	142	128	117	113	108	102	101	93	98	86	55	53	40	21	30	34	28	27	30	**24**	
25	*Loctudy*	169	165	159	161	168	153	142	138	132	127	126	117	123	110	84	78	72	46	55	62	55	53	55	30	**25**

N Brittany

9.20.3 AREA 20 TIDAL STREAMS

5 Hours before HW Dover (0215 after HW Brest)

4 Hours before HW Dover (0315 after HW Brest)

3 Hours before HW Dover (0415 after HW Brest)

2 Hours before HW Dover (0515 after HW Brest)

1 Hour before HW Dover (0615 after HW Brest)

HW Dover (0510 before HW Brest)

Southward 9.21.3   Eastward 9.18.3   Northward 9.1.3

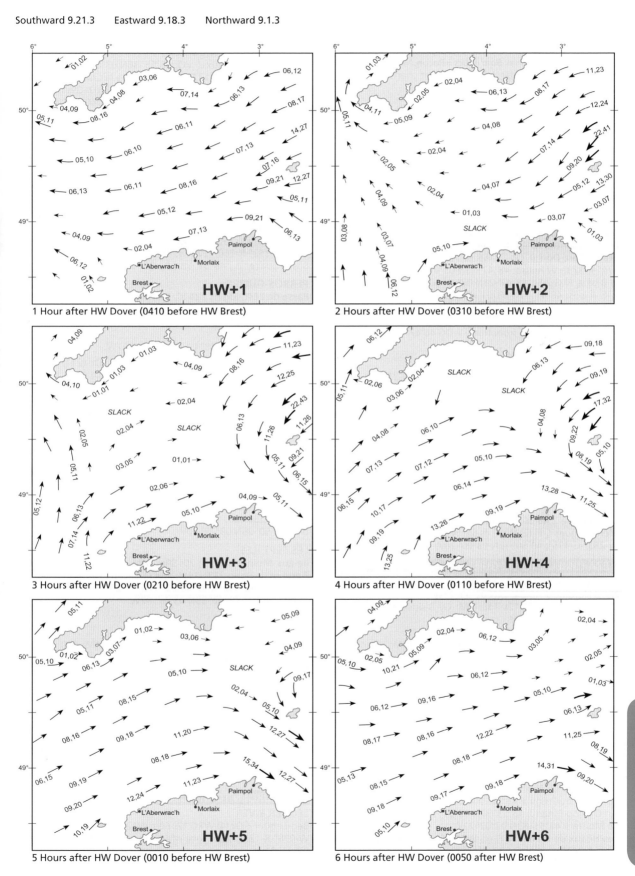

1 Hour after HW Dover (0410 before HW Brest)

2 Hours after HW Dover (0310 before HW Brest)

3 Hours after HW Dover (0210 before HW Brest)

4 Hours after HW Dover (0110 before HW Brest)

5 Hours after HW Dover (0010 before HW Brest)

6 Hours after HW Dover (0050 after HW Brest)

N Brittany

9.20.4 LIGHTS, BUOYS AND WAYPOINTS

Bold print = light with a nominal range of 15M or more. CAPITALS = place or feature. *CAPITAL ITALICS* = light-vessel, light float or Lanby. *Italics* = Fog signal. ***Bold italics*** = Racon. Many marks/buoys are fitted with AIS (<u>MMSI No</u>); see relevant charts.

OFFSHORE MARKS

Roches Douvres ☆ Fl 5s 60m **24M**; pink twr on dwelling with G roof; 49°06'·30N 02°48'·87W. <u>992271319</u>.
Barnouic ⌀ VQ (3) 5s 15m 7M; 49°01'·63N 02°48'·41W.
Roche Gautier ⌀ VQ (9) 10s; 49°02'·013N 02°54'·36W.

PAIMPOL TO ÎLE DE BRÉHAT

PAIMPOL
Les Calemarguiers ⌀ 48°46'·98N 02°54'·84W.
L'Ost Pic ⚲ Fl(4)WR 15s 20m, W9M, R6M; 105°-W-116°-R-221°-W-253°- R-291°-W-329°; obsc by islets near Bréhat when brg < 162°; W twr/turret, R top; 48°46'·77N 02°56'·42W.
Les Charpentiers ⌀ 48°47'·89N 02°56'·01W.
Pte de Porz-Don ☆ Oc (2) WR 6s 13m **W15M**, R11M; 269°-W-272°-R-279°; W house; 48°47'·48N 03°01'·55W. <u>992271327</u>.
El Bras ⌀ Fl G 2·5s; 48°47'·21N 03°01'·50W.
⌀ Fl R 2·5s; 48°47'·17N 03°01'·49W.
Ldg lts 262·2°, both QR 5/12m 7/10M. Front, Kernoa jetty; W & R hut; 48°47'·09N 03°02'·44W. Rear, Dir QR, intens 260·2°-264·2°; W pylon, R top; 360m from front. ⚲ QG 48°47'·12N 03°02'·47W.

CHENAL DU FERLAS (277°-257°-271°)
Lel Ar Serive ⌀ 48°49'·98N 02°58'·76W.
Loguivy Dir lt 257°: Q WRG 12m 10/8M; 254°-G-257°-W-257·7°-R-260·7°; Gy twr; 48°49'·37N 03°03'·67W.
Les Piliers ⌀ 48°49'·77N 02°59'·99W.
Kermouster Dir ⚲ 271°: Fl WRG 2s 16m, W 10M, R/G 8M; 267°-G-270°-W-272°- R-274°; W col; 48°49'·55N 03°05'·19W (R. Trieux).

ÎLE DE BRÉHAT
Le Paon ⚲ Oc WRG 4s 22m W11M, R/G8M; 033°-W-078°-G-181°-W-196°- R-307°-W-316°-R-348°; Y twr; 48°51'·92N 02°59'·15W.
Roche Guarine ⌀ 48°51'·63N 02°57'·63W, Chenal de Bréhat.
Rosédo ☆ Fl 5s 29m **20M**; W twr; vis 006°-340° (334°); 48°51'·45N 03°00'·29W.
La Chambre ⌀ 48°50'·16N 02°59'·58W.
Men-Joliguet ⌀ Fl (2) WRG 4s 6m W11M, R/G8M; 255°-R-279°-W-283°-G-175°; 48°50'·12N 03°00'·20W.

LÉZARDRIEUX TO TRÉGUIER

LE TRIEUX RIVER to LÉZARDRIEUX
Nord Horaine ⌀ 48°54'·54N 02°55'·23W.
La Horaine ⌀ Fl (3) 12s 13m 7M; Gy 8-sided twr on B hut; 48°53'·50N 02°55'·22W.
Men-Marc'h ⌀ 48°53'·17N 02°51'·82W.
Ldg lts 224·8°: Front, **La Croix** ☆ Q 15m **18M**; intens 215°-235°; two Gy ○ twrs joined, W on NE side, R tops; 48°50'·23N 03°03'·25W. Rear **Bodic** ☆ Dir Q 55m **22M**; intens 221°-229°; W ho with G gable; 2·1M from front.
Les Sirlots ⌀ *Whis;* 48°52'·95N 02°59'·58W.
Men-Grenn ⌀ Q (9) 15s 7m 7M; 48°51'·22N 03°03'·89W.
Coatmer ldg lts 218·7°. Front, Q RG 16m R/G7M; 200°-R-250°-G-053°; W gable; 48°48'·26N 03°05'·75W. Rear, QR 50m 7M; vis 197°-242°; W gable; 660m from front.
Les Perdrix ⌀ Fl (2) WG 6s 5m, W6M, G3M; 165°-G-197°-G-202·5°-G-040°; G twr; 48°47'·74N 03°05'·79W.

CHENAL DE LA MOISIE (339·4°); PASSE DE LA GAINE (241·5°)
La Vieille du Tréou ⌀ 48°52'·00N 03°01'·09W.
An Ogejou Bihan ⌀ 48°53'·37N 03°01'·91W.
La Moisie ⌀ 48°53'·83N 03°02'·22W.
Les Héaux de Bréhat ☆ Fl (4) WRG 15s 48m, **W15M**, R/G11M; 227°-R-247°-W-270°-G-302°-W-227°; Gy ○ twr; 48°54'·50N 03°05'·17W. 992271320.

Basse des Héaux ⌀ 48°54'·07N 03°05'·28W.
Pont de la Gaine ⌀ 48°53'·12N 03°07'·40W.

JAUDY RIVER TO TRÉGUIER
La Jument des Héaux ⌀ VQ; 48°55'·36N 03°08'·04W.
Grande Passe ldg lts 137°. Front, Port de la Chaine, Oc 4s 12m 11M; 042°-232°; W house; 48°51'·55N 03°07'·89W. Rear, **St Antoine** ☆ Dir Oc R 4s 34m **15M**; intens 134°-140°; R & W house; 0·75M from front. (Both marks are hard to see by day.)
Basse Crublent ⌀ QR; *Whis;* 48°54'·29N 03°11'·16W.
Le Corbeau ⌀ Fl R 4s; 48°53'·35N 03°10'·26W.
Pierre à l'Anglais ⌀ Fl G 4s; 48°53'·21N 03°10'·46W.
Petit Pen ar Guézec ⌀ Fl (2) G 6s; 48°52'·52N 03°09'·44W.
La Corne ⚲ Fl (3) WRG 12s 14m W8M, R/G6M; 052°-W-059°-R-173°-G-213°-W-220°-R-052°; W twr, R base; 48°51'·35N 03°10'·62W.

TRÉGUIER TO TRÉBEURDEN

PORT BLANC
Le Voleur Dir ⚲ 150°: Fl WRG 4s 17m, W14M, R/G 11M; 140°-G-148°-W-152°-R-160°; W twr; 48°50'·20N 03°18'·52W.
Basse Guazer ⌀; 48°51'·58N 03°20'·96W.

PERROS-GUIREC
Passe de l'Est, ldg lts 224·8°. Front, Le Colombier ⚲ Dir Q 28m 14M; intens 214·5°-234·5°; W house; 48°47'·87N 03°26'·66W. Rear, **Kerprigent** ☆, Dir Q 79m **21M**; intens 221°-228°; W twr, 1·5M from front.
Pierre du Chenal ⌀ 48°49'·29N 03°24'·67W.

Passe de l'Ouest. Kerjean ☆ Dir lt 143·6°, Oc (2+1) WRG 12s 78m, **W15M**, R/G12M; 133·7°-G-143·2°-W-144·8°- R-154·3°; W twr, B top; 48°47'·79N 03°23'·39W.
Les Couillons de Tomé ⌀ 48°50'·88N 03°25'·68W.
La Horaine ⌀ 48°49'·89N 03°27'·26W.
Roche Bernard ⌀ 48°49'·43N 03°25'·46W.
Gommonénou ⌀ VQ R 1M; 48°48'·27N 03°25'·83W.
Jetée du Linkin ⚲ Fl (2) G 6s 4m 6M; W pile, G top; 48°48'·20N 03°26'·31W.

PLOUMANAC'H
Mean-Ruz ⚲ Oc WR 4s 26m W12M, R9M; 226°-W-242°-R-226°; obsc by Pte de Trégastel when brg <080°; partly obsc by Les Sept-Îles 156°-207° and partly by Île Tomé 264°-278°; pink □ twr; 48°50'·25N 03°29'·00W.

LES SEPT ÎLES
Île-aux-Moines ☆ Fl (3) 20s 59m **24M**; obsc by Îliot Rouzic and E end of Île Bono 237°-241°, and in Baie de Lannion when brg <039°; Gy twr/dwelling; 48°52'·72N 03°29'·40W. <u>992271326</u>.
Les Dervinis ⌀ 48°52'·35N 03°27'·32W.

TRÉGASTEL-PLAGE
Île Dhu ⌀ 48°50'·37N 03°31'·24W.
Les Triagoz ⚲ Fl (2) WR 6s 31m W14M, R11M; 010°-W-339°-R-010°; obsc in places 258°-268° by Les Sept-Îles; Gy □ twr, R lantern; 48°52'·28N 03°38'·80W. <u>992271324</u>.
Bar-ar-Gall ⌀ VQ (9) 10s; 48°49'·79N 03°36'·22W.
Le Crapaud ⌀ Q (9) 15s; 48°46'·67N 03°40'·59W.

TRÉBEURDEN
Pte de Lan Kerellec ⚲ Iso WRG 4s; W8M, R/G5M; 058°-G-064°-W-069°-R-130°; Gy twr; 48°46'·74N 03°35'·06W.
Ar Gouredec ⌀ VQ (6) + L Fl 10s; 48°46'·41N 03°36'·59W.
An Ervennou ⌀ Fl (2) R 6s; 48°46'·48N 03°35'·96W.
NW bkwtr ⚲ Fl G 2·5s 8m 2M; IPTS; 48°46'·33N 03°35'·21W.

TRÉBEURDEN TO MORLAIX

LÉGUER RIVER
Beg-Léguer ⚲ Oc (4) WRG 12s 60m W12M, R/G9M; 007°-G-084°-W-098°-R-129°; west face of W house, R lantern; 48°44'·31N 03°32'·91W.
Kinierbel ⌀; *Bell;* 48°44'·15N 03°35'·18W.

LOCQUÉMEAU
Ldg lts 121°: Front, ⚡ QR 21m 7M; 068°-228°; W pylon, R top; 48°43'·41N 03°34'·44W. Rear, ⚡ QR 39m 7M; 016°-232°; W gabled house; 484m from front.
Locquémeau ⚓ *Whis*; 48°43'·86N 03°35'·93W.

LOCQUIREC
Gouliat ⓘ 48°42'·57N 03°38'·97W.

PRIMEL-TRÉGASTEL
Méloine ⓘ 48°45'·56N 03°50'·68W.
Ldg lts 152°, both ⚡ QR 35/56m 7M. Front, 134°-168°; W □, R stripe, on pylon; 48°42'·45N 03°49'·19W. Rear, R vert stripe on W wall; 172m from front.
W bkwtr ⚡ Fl G 4s 6m 7M; 48°42'·77N 03°49'·52W.

BAIE DE MORLAIX
Chenal de Tréguier ldg lts 190·5°: Front, ⚡ Île Noire Oc (2) WRG 6s 15m, W11M, R/G8M; 051°-G-135°-R-211°-W-051°; obsc in places; W □ twr, R top; 48°40'·35N 03°52'·54W. Common Rear, **La Lande** ☆ Fl 5s 85m **23M**; obsc by Pte Annelouesten when brg >204°; W □ twr, B top; 48°38'·20N 03°53'·14W.
La Pierre Noire ⓘ 48°42'·56N 03°52'·20W.
La Chambre ⓘ 48°40'·74N 03°52'·51W.

Grande Chenal ldg lts 176·4°: Front, **Île Louet** ☆ Oc (3) WG 12s 17m **W15M**,G10M; 305°-W (except where obsc by islands)-244°-G-305°; W □ twr, B top; 48°40'·41N 03°53'·34W. Common Rear, **La Lande** as above.
Pot de Fer ⓘ *Bell*; 48°44'·23N 03°54'·02W.
Stolvezen ⓘ 48°42'·64N 03°53'·41W.
Ricard ⓘ 48°41'·54N 03°53'·51W.
Vieille ⓘ 48°42'·60N 03°54'·11W. (Chenal Ouest de Ricard 188·8°)
La Noire ⓘ 48°41'·65N 03°54'·06W.
Corbeau ⓘ 48°40'·63N 03°53'·33W.

MORLAIX RIVER
Barre de-Flot No. 1 ⓘ 48°40'·18N 03°52'·95W.
No. 2 ⚡ Fl R 2s; 48°39'·88N 03°52'·53W.
No. 3 ⓘ Fl G 2s; 48°39'·25N 03°52'·32W.
No. 4 ⚡ Fl R 2s; 48°38'·62N 03°51'·62W.
No. 5 ⚡ Fl G 2s; 48°38'·04N 03°51'·34W.
No. 7 ⓘ 48°37'·68N 03°51'·03W.

PENZÉ RIVER
Cordonnier ⓘ 48°42'·93N 03°56'·59W.
Guerhéon ⓘ 48°42'·73N 03°57'·18W.
Trousken ⓘ 48°42'·26N 03°56'·54W.
Pte Fourche ⓘ 48°42'·14N 03°56'·76W.
Ar Tourtu ⓘ 48°41'·99N 03°56'·57W.
An Nehou (Caspari) ⓘ 48°41'·56N 03°56'·45W.
Le Figuier ⓘ 48°40'·46N 03°56'·16W.

ROSCOFF TO ÎLE VIERGE
BLOSCON/ROSCOFF
Astan ⓘ VQ (3) 5s 9m 6M; 48°44'·91N 03°57'·66W.
Le Menk ⓘ Q (9) WR 15s 6m W5M, R3M; 160°-W-188°-R-160°; 48°43'·28N 03°56'·71W.
Basse de Bloscon ⓘ VQ; 48°43'·71N 03°57'·55W.
Bloscon pier ⚡ Fl WG 4s 9m W10M, G7M (in fog Fl 2s); 200°-W-210°-G-200°; W twr, G top; 48°43'·21N 03°57'·69W.
Roscoff Marina ent, E side, ⚡ Fl (2) R 6s 4m; 48°43'·10N 03°57'·90W.
Ar Pourven ⓘ Q; 48°43'·04N 03°57'·70W.
Ar-Chaden ⓘ Q (6) + L Fl WR 15s 14m, W8M, R6M; 262°-W-293°-R-326°- W-110°; YB twr; 48°43'·94N 03°58'·26W.
Men-Guen-Bras ⓘ Q WRG 14m, W9M, R/ G6M; 068°-W-073°-R-197°-W-257°-G-068°; BY twr; 48°43'·76N 03°58'·07W.
Roscoff ldg lts 209°: Front, N môle ⚡ Oc (3) G 12s 7m 7M; synch; 078°-318°; W col, G top; 48°43'·56N 03°58'·67W. **Rear** ☆ Oc (3) 12s

24m **15M**; 062°-242°; Gy □ twr, W on NE side; 430m from front.

CANAL DE L'ÎLE DE BATZ
Roc'h Zu ⓘ 48°43'·88N 03°58'·59W.
Jetty hd (LW landing) ⚡ Q 5m 1M; BY col; 48°43'·92N 03°58'·97W.
Run Oan ⓘ 48°44'·10N 03°59'·30W.
Perroch ⓘ 48°44'·10N 03°59'·71W.
Tec'hit Bihan ⓘ 48°44'·02N 04°00'·88W.
La Croix ⓘ 48°44'·19N 04°01'·30W.
L'Oignon ⓘ 48°44'·04N 04°01'·35W.
Basse Plate ⓘ 48°44'·25N 04°02'·53W.

ÎLE DE BATZ
Île aux Moutons ⓘ VQ (6)+ L Fl 10s 3m 7M; 48°44'·25N 04°00'·52W; landing stage, S end and ent to hbr.
Malvoch ⓘ 48°44'·26N 04°00'·67W, W side of ent.
Lt ho ☆ Fl (4) 25s 69m **23M**; Gy twr; 48°44'·71N 04°01'·62W. Same twr, auxiliary lt, FR 65m 7M; 024°-059°. 992271315.

MOGUÉRIEC
Ldg lts 162°: Front ⚡ Iso WG 4s 9m W11M, G6M; 158°-W-166°-G-158°; W twr, G top; jetty 48°41'·34N 04°04'·47W. Rear ⚡ Iso G 4s 22m 7M, synch; 142°-182°; W col, G top; 440m from front.

PONTUSVAL
Pointe de Pontusval ⓘ 48°41'·42N 04°19'·32W.
Ar Peich ⓘ 48°40'·90N 04°19'·16W.
Pte de Beg-Pol ⚡ Oc (3) WR 12s 16m W10M, R7M; Shore-W-056°-R-096°-W-shore; W twr, B top, W dwelling; 48°40'·67N 04°20'·76W. QY and FR lts on towers 2·4M S.
Aman-ar-Ross ⓘ Q 7M; 48°41'·88N 04°27'·03W.
Lizen Ven Ouest ⓘ VQ (9) 10s 5M; 48°40'·53N 04°33'·63W.
Île-Vierge ☆ Fl 5s 77m **27M**; 337°-325°; Gy twr; *Horn 60s*; 48°38'·33N 04°34'·05W. 992271317.

ÎLE VIERGE TO L'ABER-ILDUT
L'ABER WRAC'H
Libenter ⓘ Q (9) 15s 6M; 48°37'·45N 04°38'·46W.
Outer ldg lts 100·1°: Front, Île Wrac'h ⚡ QR 20m 7M; W □ twr, Or top; 48°36'·88N 04°34'·56W. Rear, Lanvaon ⓘ Q 55m 12M; intens 090°-110°; W □ twr, Or △ on top; 1·63M from front.
Grand Pot de Beurre ⓘ 48°37'·21N 04°36'·48W.
Petit Pot de Beurre ⓘ 48°37'·12N 04°36'·23W.
Basse de la Croix ⚡ Fl G 2·5s; 48°36'·92N 04°35'·99W.
Breac'h Ver ⚡ Fl (2) G 6s 6m 3M; △ on twr; 48°36'·63N 04°35'·38W.
Dir ⚡ 128°, Oc (2) WRG 6s 5m W13M, R/G11M; 125·7°-G-127·2°-W-128·7°-R-130·2°; 48°35'·89N 04°33'·82W, root of W bkwtr. Marina ent: QG 3m 2M; vis 186°-167° (341°); 48°35'·97N 04°33'·64W. QR 3m 2M; vis 244°-226° (342°).

L'ABER BENOÎT
Petite Fourche ⓘ 48°36'·99N 04°38'·75W.
Rusven Est ⓘ 48°36'·30N 04°38'·63W.
Rusven Ouest ⓘ *Bell*; 48°36'·07N 04°39'·43W.
Basse du Chenal ⓘ 48°35'·81N 04°38'·53W.
Poul Orvil ⓘ 48°35'·52N 04°38'·29W.
La Jument ⓘ 48°35'·10N 04°37'·41W.
Ar Gazel ⓘ 48°34'·90N 04°37'·27W.
Le Chien ⓘ 48°34'·67N 04°36'·88W.

ROCHES DE PORTSALL
Le Relec ⓘ 48°35'·99N 04°40'·84W.
Corn-Carhai ⚡ Fl (3) 12s 19m 9M; W 8-sided twr, B top; 48°35'·19N 04°43'·94W.
Grande Basse de Portsall ⓘ VQ (9) 10s 9m 4M; 48°36'·70N 04°46'·13W.
Basse Paupian ⓘ 48°35'·31N 04°46'·27W.
Bosven Aval ⓘ 48°33'·82N 04°44'·27W.
Men ar Pic ⓘ 48°33'·65N 04°44'·03W.

Portsall ⚓ Oc (4) WRG 12s 9m W13M, R/G10M; 058°-G-084°-W-088°-R-058°; W col, R top; 48°33'·84N 04°42'·26W.

ARGENTON to L'ABER-ILDUT
Le Taureau ⚓ 48°31'·46N 04°47'·34W.
Argenton, Île Dolvez, front ldg bcn 086° ⚓ 48°31'·25N 04°46'·23W.
Le Four ☆ Fl (5) 15s 28m **22M**; Gy ○ twr; 48°31'·38N 04°48'·32W.
L'Aber-Ildut ☆ Dir Oc (2) WR 6s 12m **W25M, R20M**; 081°-W-085°-R-087°; W bldg; 48°28'·26N 04°45'·56W.
Le Lieu ⚓Fl R 2.5s 4M; 48°28'·25N 04°46'·66W.

CHENAL DU FOUR AND CHENAL DE LA HELLE
Ldg lts 158·5°. Front, **Kermorvan** ☆ Fl 5s 20m **22M**; obsc'd by Pte de St Mathieu when brg <341°; W ☐ twr; *Horn 60s*; 48°21'·72N 04°47'·40W. 992271316.
Rear, **Pte de St Mathieu** ☆ Fl 15s 56m **24M**; W twr, R top; 48°19'·79N 04°46'·26W. 992271318. Same twr: Dir F 54m **28M**; intens 157·5°-159·5°.
⚓ Q WRG 26m, W12M, R/G 8M; 085°-G-107°-W-116°-R-134°; W twr 54 m WNW of previous entry; 48°19'·80N 04°46'·30W.
Plâtresses N ⚓ Fl G 2·5s; 48°26'·46N 04°50'·71W.
Valbelle ⚓ Fl R 2·5s; 48°26'·43N 04°50'·03W.
Plâtresses SE ⚓ Fl (2) G 6s; 48°25'·96N 04°50'·52W.
Saint Paul ⚓ Fl (2) R 6s; 48°24'·82N 04°49'·16W.
Pte de Corsen ⚓ Dir Q WRG 33m W12M, R/G8M; 008°-R-012°-W-015° (ldg sector)- G-021°; W hut; 48°24'·89N 04°47'·61W.

CHENAL DE LA HELLE
Ldg lts 137·9°: Front, **Kermorvan** ☆ see above. Rear, **Lochrist** ☆ Dir Oc (3) 12s 49m **22M**; intens 135°-140°; W 8-sided twr, R top; 48°20'·55N 04°45'·82W. 992271310.
Luronne ⚓; 48°26'·61N 04°53'·78W.
Ldg lts 293·5° astern (to join Ch du Four S of St Paul ⚓): Front, Le Faix ⚓ VQ 16m 8M; 48°25'·73N 04°53'·91W. Rear, **Le Stiff** ☆ (below at Ouessant); 6·9M from front.
Ldg line 142·5°, optional day only (to join Ch du Four SE of St Pierre ⚓): Front, **Kermorvan** ☆ see above. Rear, two W gables (Les Pignons de Kéravel, 48m) 48°20'·10N 04°45'·55W
Pourceaux ⚓ Q; 48°24'·00N 04°51'·31W.
Saint-Pierre ⚓ 48°23'·09N 04°49'·09W.

Rouget ⚓ Fl G 4s; 48°22'·04N 04°48'·88W (Ch du Four).
Grande Vinotière ⚓ Fl R 4s 12m 5M; R 8-sided twr; 48°21'·94N 04°48'·42W.

Le Conquet, Môle Ste Barbe ⚓ Fl G 2·5s 5m 6M; 48°21'·57N 04°46'·98W.

Les Renards ⚓ 48°21'·00N 04°47'·48W.
Tournant et Lochrist ⚓ Fl (2) R 6s; 48°20'·64N 04°48'·12W,
Ar Christian Braz ⚓ 48°20'·68N 04°50'·14W, E of Ile de Béniguet.
Ldg line 325°: Front, Grand Courleau ⚓; rear, La Faix (above).
Ldg lts 007°: Front, **Kermorvan** ☆ above. Rear, **Trézien** ☆ Dir Oc (2) 6s 84m **20M**; intens 003°-011°; Gy twr, W on S side; 48°25'·41N 04°46'·74W. 992271309.
Les Vieux-Moines ⚓ Fl R 4s 16m 5M; 280°-133°; R 8-sided twr; 48°19'·33N 04°46'·63W, 5ca SSW of Pte de St Mathieu.
La Fourmi ⚓ Fl G 4s; 48°19'·25N 04°47'·96W.

OUESSANT AND ÎLE MOLÉNE
OFF USHANT TSS
Ouessant NE ⊙; V-AIS; 48°59'·51N 05°24'·00W.
Ouessant SW ⊙; V-AIS 48°30'·00N 05°45'·00W.

ÎLE D'OUESSANT
Le Stiff ☆ Fl (2) R 20s 85m **24M**; two adjoining W twrs; 48°28'·47N 05°03'·41W. Radar twr, conspic, 340m NE, ⚓ Q (day); ⚓ FR (night).
Port du Stiff, E môle ⚓ Dir Q WRG 11m W10M, R/G7M; 251°-G-254°-W-264°-R-267°; W twr, G top; 48°28'·12N 05°03'·25W.

Baie du Stiff: Gorle Vihan ⚓ 48°28'·32N 05°02'·60W.
Men-Korn ⚓ VQ (3) WR 5s 21m 8M; 145°-W-040°-R-145°; BYB twr; 48°27'·96N 05°01'·32W.
Men ar Froud ⚓ 48°26'·62N 05°03'·67W.
La Jument ⚓ Fl (3) R 15s 36m 10M; 241°-199°; Gy 8-sided twr, R top; 48°25'·33N 05°08'·03W.
Nividic ⚓ VQ (9) 10s 28m 10M; 290°-225°; Gy 8-sided twr, helicopter platform; 48°26'·74N 05°09'·06W.
Créac'h ☆ Fl (2) 10s 70m **32M**; obsc 247°-255°; *Horn (2) 120s*; W twr, B bands; 48°27'·55N 05°07'·76W.

ÎLE MOLÈNE and ARCHIPELAGO
Kéréon ☆ Oc (2+1) WR 24s 38m **W18M,** R14M; 019°-W-248°-R-019°; Gy twr; 48°26'·24N 05°01'·54W; unreliable.
Les Trois-Pierres ⚓ Iso WRG 4s 15m W9M, R/G6M; 070°-G-147°-W-185°-R-191°-G-197°-R-213°-R-070°; W col; 48°24'·70N 04°56'·84W.
Molène, Old môle Dir ⚓ 191°, Fl (3) WRG 12s 6m W9M, R/G7M; 183°-G-190°-W-192°-R-203°; 48°23'·85N 04°57'·29W.
Same structure: Chenal des Laz , Dir ⚓ 261°: Fl (2) WRG 6s 9m W9M, R/G7M ; 252·5°-G-259·5°-W-262·5°-R-269·5°.

Pierres-Vertes ⚓ VQ (9) 10s 9m 5M; 48°22'·19N 05°04'·76W.
Pierres Noires ⚓; 48°18'·47N 04°58'·15W.
Les Pierres Noires ☆ Fl R 5s 27m **19M**; W twr, R top; 48°18'·67N 04°54'·88W.

BREST AND APPROACHES
Basse Royale ⚓ Q (6) + L Fl 15s; 48°17'·45N 04°49'·61W.
Vandrée ⚓ VQ (9) 10s; *Whis*; 48°15'·21N 04°48'·24W.
Goëmant ⚓; 48°15'·12N 04°46'·34W.
La Parquette ⚓ Fl RG 4s 17m R6M, G6M; 244°-R-285°-G-244°; W 8-sided twr, B diagonal stripes; 48°15'·90N 04°44'·29W.
Coq Iroise ⚓ 48°19'·07N 04°43'·99W.
Charles Martel ⚓ Fl (4) R 15s; 48°18'·95N 04°41'·92W.
Trépied ⚓ 48°16'·73N 04°41'·50W.
Swansea Vale ⚓ Fl (2) 6s; *Whis*; 48°18'·27N 04°38'·85W.

GOULET DE BREST
Le Chat ⚓ 48°20'·29N 04°41'·72W (Anse de Bertheaume).
Le Trez Hir ⚓ 48°20'·78N 04°41'·96W.
Pte du Petit-Minou ☆ Fl (2) WR 6s 32m **W19M, R15M**; Shore-R-252°-W-260°-R-307°-W(unintens)-015°-W-065·5°; 036·5°-R(intens)-039·5°; 70·5°-W-shore; Gy twr, R top; 48°20'·19N 04°36'·86W. 992271300.
Ldg lts 068°, both Dir Q 30/54m **23/22M. Front, Pte du Petit-Minou**, intens 067·3°-068·8°. **Rear, Pte du Portzic ☆**, Aux Dir Q 54m **22M**; intens 065°-071° (see below).
Fillettes ⚓ VQ (9) 10s; 48°19'·75N 04°35'·66W.
Kerviniou ⚓ Fl R 2·5s; 48°19'·77N 04°35'·25W.
Basse Goudron ⚓ Fl (2) R 6s; 48°20'·03N 04°34'·86W.
Mengam ⚓ Fl (3) WR 12s 10m W11M, R8M; 034°-R-054°-W-034°; R twr, B bands; 48°20'·32N 04°34'·56W.
Pte du Portzic ☆ Oc (2) WR 12s 56m **W19M, R15M**; 219°-R-259°-W-338°-R-000°-W-065·5° (vis 041°-069° when west of Goulet)-W-219°; Gy twr; 48°21'·49N 04°32'·05W. Same lt ho, Aux Dir Q, rear 068° ldg lt. Aux ☆ Dir Q (6) + L Fl 15s 54m **23M**; intens 045°-050°.

BREST
Pénoupèle ⚓ Fl R 2·5s; 48°21'·45N 04°30'·53W.
S jetty ⚓ QR 10m 7M; 094°-048°; W/R twr; 48°22'·10N 04°29'·46W.
E jetty ⚓ QG 10m 7M; W/G twr; 48°22'·15N 04°29'·22W.
Ldg lts 343.5°: Front ⚓ Dir Q WRG 24m W9M, R/G6M; 334°-G-342°-W-346°-R-024°; 48°22'·80N 04°29'·62W. Rear ⚓ Dir Q 32m 9M; intens 342°-346°; 115m from front.
Le Château Marina, S Jetty hd, ⚓ Fl G 2·5s 8m 4M; 48°22'·66N 04°29'·49W.
Le Château Marina, N Jetty hd, ⚓ Fl R 2·5s 8m 4M; 48°22'·67N 04°29'·45W.

Port de Commerce, E ent: N jetty ⚓ Oc (2) G 6s 8m 7M; W/G pylon; 48°22'·76N 04°28'·53W. S jetty ⚓ Oc (2) R 6s 8m 5M; W pylon, R top; 48°22'·69N 04°28'·48W.
R2 ⚓ Fl (2) R 6s; 48°22'·01N 04°28'·75W.
R1 ⚓ Fl G 6s; 48°21'·83N 04°28'·29W.
R4 ⚓ Fl (3) R 12s; 48°22'·22N 04°28'·06W.
R3 ⚓ Q (6) + L Fl 15s; 48°22'·48N 04°27'·36W.

LE MOULIN BLANC MARINA

Moulin Blanc ⚓ Fl (2) R 6s; 48°22'·80N 04°25'·91W
MBA ⚓ Q(3) 10s 3m 2M; pontoon elbow; 48°23'·54N 04°25'·74W.

CAMARET

N môle ⚓ Iso WG 4s 7m W12M, G9M; 135°-W-182°-G-027°; W pylon, G top; 48°16'·85N 04°35'·31W.
Port Vauban wavebreak, S end, ⚓ Fl (2) G 6s 2m 2M; 48°16'·74N 04° 35'·33W.
S môle ⚓ Fl (2) R 6s 9m 5M; R pylon; 48°16'·63N 04°35'·33W.

POINTE DU TOULINGUET TO RAZ DE SEIN

Pte du Toulinguet ☆ Oc (3) WR 12s 49m **W15M**, R11M; Shore-W-028°-R-090°-W-shore; W ☐ twr; 48°16'·81N 04°37'·72W.
La Louve ⚓ 48°16'·76N 04°38'·03W.
Mendufa ⚓ 48°16'·05N 04°39'·43W.
Basse du Lis ⚓ Q (6) + L Fl 15s 9m 6M; 48°12'·99N 04°44'·53W.
Le Chevreau △ 48°13'·30N 04°36'·98W.
Le Bouc ⚓ Q (9) 15s; 48°11'·51N 04°37'·37W.
Basse Vieille ⚓ Fl (2) 6s 8m 7M; 48°08'·22N 04°35'·75W.

MORGAT

Pointe de Morgat ☆ Oc (4) WRG 12s 77m **W15M**, R11M, G10M; Shore-W-281°-G-301°-W-021°-R-043°; W ☐ twr, R top, W dwelling; 48°13'·17N 04°29'·80W. ⚓ Fl R 4s; 48°13'·60N 04°29'·55W.

DOUARNENEZ to RAZ DE SEIN

Île Tristan ⚓ Fl (3) WR 12s 35m, W6M, R6M; Shore-W-138°-R-153°-W-shore; Gy twr, W band, B top; 48°06'·14N 04°20'·24W.
Pointe du Millier ☆ Oc (2) WRG 6s 34m **W16M**, R12M, G11M; 080°-G-087°-W-113°-R-120°-W-129°-G-148°-W-251°-R-258°; W house; 48°05'·92N 04°27'·93W. <u>992271015</u>.
Basse Jaune ⚓ 48°04'·70N 04°42'·45W.

RAZ DE SEIN

Tévennec ⚓ Q WR 28m W9M R6M; 090°-W-345°-R-090°; W ☐ twr and dwelling; 48°04'·28N 04°47'·73W. Same twr, Dir ⚓ Fl 4s 24m 12M; intens 324°-332° (through Raz de Sein).
La Vieille ☆ Oc (2+1) WRG 12s 33m **W18M**, R13M, G14M; 290°-W-298°-R-325°-W-355°-G-017°-W-035°-G-105°-W-123°-R-158°-W-205°; Gy ☐ twr; 48°02'·44N 04°45'·40W.

La Plate ⚓ Q (9) 15s 19m 6M; 48°02'·37N 04°45'·58W.
Le Chat ⚓ Fl (2) WRG 6s 27m, W9M, R/ G6M; 096°-G-215°-W-230°-R-271°-G-286°-R-096°; YB twr; 48°01'·43N 04°48'·86W.

CHAUSSÉE DE SEIN and ÎLE DE SEIN

Cornoc-An-Ar-Braden ⚓ Fl G 4s; 48°03'·23N 04°50'·84W.
Men-Brial ⚓ Oc (2) WRG 6s 16m, W12M, R9M, G7M; 149°-G-186°-W-192°-R-221°-W-227°-G-254°; G&W twr; 48°02'·28N 04°50'·97W.
Île de Sein Fl (4) 25s 49m **29M**; W twr, B top; 48°02'·62N 04°52'·02W. Same twr, Dir Q WRG, W8M, R/G 6M; 267°-G-269°-W-271°-R-275°.
Ar-Men Fl (3) 20s 29m **23M**; W twr, B top; 48°03'·01N 04°59'·87W.
Chaussée de Sein ⚓ VQ (9) 10s 9m 6M; *Whis*; *Racon O (3cm), 10M*; 48°03'·74N 05°07'·75W.

AUDIERNE

Pointe de Lervily ⚓ Fl WR 4s 20m W14M, R11M; 236°-W-269°-R-294°-W-087°-R-109°; W twr, R top; 48°00'·04N 04°33'·94W.
Gamelle W ⚓ VQ (9) 10s; 47°59'·46N 04°32'·85W.
Jetée de Ste-Évette ⚓ Fl (2) R 6s 2m 7M; R lantern; 090°-000° (270°); 48°00'·31N 04°33'·07W.
Passe de l'Est ldg lts 331°: Front, Jetée de Raoulic ⚓ Fl (3) RG 12s 11m, R/G6M; 085°-R-034°-G-085°-R-085°; 48°00'·55N 04°32'·46W.
Kergadec ⚓ 006°: Dir Q WRG 43m W12M, R/G9M; 000°-G-005·3°-W-006·7°-R-017°; W 8-sided twr, R top; 48°00'·96N 04°32'·78W.
Gamelle E ⚓; 47°59'·46N 04°32'·05W.

POINTE DE PENMARC'H

Eckmühl ☆ Fl 5s 60m **23M**; Gy 8-sided twr; 47°47'·89N 04°22'·36W.
Men-Hir ⚓ Fl (2) WG 6s 19m, W7M, G4M; G135°-315°, W315°-135°; W twr, B band; 47°47'·74N 04°23'·99W.
Cap Caval ⚓ Q (9) 15s; 47°46'·47N 04°22'·68W.
Locarec ⚓ Iso WRG 4s 11m, W9M, R/G6M; 063°-G-068°-R-271°-W-285°-G-340°-R-063°; iron col on rk; 47°47'·29N 04°20'·31W.

LE GUILVINEC

Névez ⚓ Fl G 2·5s; 47°45'·84N 04°20'·08W.
Ldg lts (triple Q) 053°, all synch: **Front**, ⚓ Q 7m 8M; 233°-066°; W pylon; Môle de Léchiagat, spur; 47°47'·43N 04°17'·07W.
Middle, ⚓ Q WG 12m W14M, G11M; 006°-W-293°-G-006°; synch; R ☐ on R col; Rocher Le Faoutés, 210m from front. Rear, ⚓ Q 26m 8M, 051·5°-054·5°, R ☐ on W twr; 0·58M from front.
Spineg ⚓ Q (6) + L Fl 15s; 47°45'·19N 04°18'·90W.

LESCONIL

Karek Greis ⚓ Q (3) 10s; 47°46'·03N 04°11'·36W.
⚓ Q (6) + LFl 15s; 47°46'·53N 04°10'·63W
Men-ar-Groas ⚓ Fl (3) WRG 12s 14m, W10M, R/G7M; 268°-G-313°-W-333°-R-050°; W lt ho, G top; 47°47'·79N 04°12'·68W.

9.20.5 PASSAGE INFORMATION

More Passage Information is threaded between successive harbours in this Area. **Bibliography**: *North Brittany* (Imray/RCC/Lawson) *North Brittany and Channel Islands CC* (Fernhurst/Cumberlidge); *Shell Channel Pilot* (Imray/Cunliffe); *Channel Pilot* (NP 27); *Channel Havens* (ACN/Endean); *Secret anchorages of Brittany* (Imray/Cumberlidge).

NORTH BRITTANY

This ever-popular cruising ground (AC 2648, 2647, 2643, 2028) is not one to trifle with, but conversely it is rewarding to navigate into a remote anchorage or small fishing harbour. Here the unique character and appeal of N Brittany is all around you.

Good landfall marks must be carefully identified, as a back-up to GPS, before closing the rock-strewn coast. In rough weather, poor visibility (fog and summer haze are frequent) or if uncertain of position, it may be prudent to lie off and wait for conditions

to improve; there are few safe havens. Closer inshore the tidal streams are variable with associated overfalls. Thorough careful planning remains the key to safe pilotage.

In the E of the area the outer approaches to **Paimpol, Île de Bréhat, Lézardrieux** and **Tréguier** may be complicated by strong tidal streams. At times there may be a proliferation of landmarks which must be selectively identified. Concentrate on those which affect your pilotage and discard those which are non-essential and may even distract you from the task.

Once W of **Roscoff** there are few safe hbrs and anchorages until **L'Aberwrac'h** where many British yachts tend to pause and await good conditions for negotiating the Chenal du Four. This is rarely as difficult as it is imagined to be. The Atlantic swell can be a new experience for some, but in moderate winds it is rarely a hazard. It can however much reduce the range at which objects, especially floating marks and other small craft, are seen.

N Brittany

PAIMPOL TO PLOUMANAC'H

In the offing, 11-18M NNE of Île de Bréhat and on a direct track from/to Guernsey, are Plateau de Barnouic (lit) and **Plateau des Roches Douvres** (light house); both with drying and submerged rks, to be given a wide berth in poor vis (AC 2027, 2028, 3673).

Approaching L'Ost-Pic from the SE, keep to seaward of the three ECM marks E of it or enter Baie de **Paimpol** from a point about 1M E of the most N'ly ECM (Les Charpentiers); but it is safe to pass about 300m east of L'Ost Pic in moderate weather.

The Ferlas channel (AC 3673) lies E-W south of **Ile de Bréhat**, and is useful if entering/leaving R. Trieux from/to the E. It is easiest to beacon-hop past the 5 SCM bcns and 1 IDM bcn marking the N side of the chan; the S side is less well marked.

For the many yachts approaching from Guernsey, Les Héaux-de-Bréhat light house is a conspic landfall day/night for either Tréguier or Lézardrieux. Closer in or from the E, La Horaine (lt beacon) is a better landfall for the latter. It marks the Plateaux de la Horaine and des Échaudés and other rocks to the SE. In poor visibility it should be closed with caution and left at least 7ca to the SE, as the flood stream sets strongly onto it.

The Grand Chenal 224·7° is the main, lit channel into the R. de Trieux for **Lézardrieux** and further upriver to **Pontrieux**. From NW the unlit Chenal de La Moisie leads 159° to join the Grand Chenal off Ile de Bréhat.

The **Passe de la Gaine** 241·5° is a useful inshore route between Lézardrieux and Tréguier, avoiding a detour round Les Héaux. If taken at above half tide, it presents no problem in fair weather. But it is unlit and from the NE the distant ldg marks are hard to see in poor visibility or against a low sun. However two SHM beacons to the S and SW of Les Héaux-de-Bréhat lighthouse can be used as a handrail towards Pont de la Gaine.

The Grande Passe 137·3° into **Rivière de Tréguier** is well lit, but ldg marks are hard to see by day. Use the NE Passage with caution.

Between Basse Crublent lt buoy and Port Blanc unmarked rocks extend 2M offshore. **Port Blanc** can be difficult to identify by day until it opens on 150°. **Perros-Guirec** is approached either side of Ile Tomé from NE or NW via well lit/marked channels; take care not to be neaped. **Ploumanac'h** can only be entered by day. It boasts some of the most spectacular pink granite along the North Brittany coast.

LES SEPT ÎLES TO BAIE DE MORLAIX

(AC 2026). **Les Sept Îles** (AC 2027) consist of five main islands and several islets, through which the tide runs strongly. Île aux Moines is lit, and all the islands are bird sanctuaries. Further W, Plateau des Triagoz has offlying dangers WSW and NE of the lt, where the sea breaks heavily.

▶*Here the stream turns ENE at HW Brest –0230, and WSW at HW Brest +0330, sp rates both 3·8kn.*◀

Trégastel-Plage is a small anchorage W of Ploumanac'h. To the SW the coast as far as **Trébeurden** is not easily approached due to many offlying rks. The radome NE of Trébeurden is conspic. Further S in the Baie de Lannion is Locquémeau and anchorages near the mouth of the drying R. Léguer up to **Lannion**. Locquirec is a pleasant drying harbour 3·5M SW of Locquemeau. **Primel-Trégastel**, at the E ent to Baie de Morlaix, is a useful inlet to await the tide up to Morlaix. To the N Plateau de la Méloine dries.

The Baie de Morlaix (AC 2745) is bestrewn with drying rks and shoals, all marked. Careful pilotage and adequate visibility are needed to negotiate any of the chans which are narrow in parts. Chenal de Tréguier 190·5° and the Grand Chenal 176·4° are both lit. The former should only be used HW ±3 due to shoals at the S end. Grand Chenal passes close E of Île Ricard with Île Louet and La Lande lights in transit; abeam Calhic bn tr alter to port to pass between Château du Taureau (conspic) and Île Louet. Continue SSE and up-river to lock into **Morlaix** Marina in complete shelter. The ⌔age NE of Carantec is reached from Chenal Ouest de Ricard.

ÎLE DE BATZ TO LE FOUR

▶*2M N of Île de Batz the E-going stream begins at HW Brest –0400, and the W-going stream at HW Brest +0200, sp rates 3·8kn. The flood and ebb streams deflected N of Ile de Batz may be >5kn, in opposition and violent. Against any existing sea state, the resultant breaking seas can endanger small craft which are therefore strongly advised to pass at least 2-2½M north of Batz. See the French tidal stream atlas 563-UJA.*◀

Approaching **Roscoff** from the NE, leave Astan ECM lt buoy to stbd tracking with Men Guen Bras lt bcn in transit 213° with white Chapelle St Barbe, to round Ar Chaden lt bcn for Roscoff's drying harbour. Bound for Roscoff Marina pass E of No1 pillar by and Basse de Bloscon NCM, then S towards Bloscons prominent ferry terminal pier. (AC 2026, 2025, 2745)

S of the island **Canal de L'Île de Batz**, day only and above half tide, requires careful pilotage at its narrower eastern end. From near Ar Chaden steer 275° for the Q bn at end of the conspic LW ferry pier. Pass 30m W of this bn, then alter to 300° for Run Oan SCM. Thence steer 283°, leaving Perroch NCM bcn twrs 100m to port. When clear of this rky, drying shoal alter to 270°, leaving Porz Kernok hbr bkwtrs well to stbd then midway between L'Oignon NCM and La Croix SCM bcns. With these abeam steer 281° for Basse Plate NCM bn; thence West into open waters.

Proceeding W from Île de Batz towards Le Four there are many offlying dangers, in places 3M offshore. Swell may break on shoals even further to seaward. With strong tides, in poor vis or bad weather it is a coast to avoid. But in fair weather enjoy its delightful harbours such as Moguériec, Brignogan, L'Aberwrac'h, L'Aberbenoit, Portsall and Argenton. N of L'Aberwrac'h is the conspic Île Vierge lt ho, reputedly the tallest in the world.

▶*N of Le Libenter buoy, off L'Aberwrac'h, the NE-going stream starts at HW Brest –0500, sp rate 3·8kn, and the WSW-going stream at HW Brest +0100.*◀

L'Aberwrac'h is accessible at all tides and strategically placed to catch the tide S through Chenal du Four. The marina is also well protected from NW'lies.

W of L'Aberwrac'h a daytime only/good vis chan leads inshore of Roches de Portsall to Le Four lt ho. The distant marks are not easy to identify and it is only 1M shorter than the offshore route via Basse Paupian WCM buoy. AC 1432 or SHOM 7094 and a detailed pilot book essential.

OUESSANT (USHANT)

(AC 2356) Île d'Ouessant lies 10M off NW Brittany, surrounded by five lt ho's. It is a rky island, with dangers extending 5ca to NE, 7½ca to SE, 1·5M to SW and 1M to NW; here Chaussée de Keller is a dangerous chain of unmarked drying and submerged rks running 1M W of Île de Keller and into the ITZ. There are anchorages and moorings at **Lampaul** and **Baie du Stiff**, the former exposed to the SW and the latter to the NE.

Ouessant is an important landfall, but avoid in thick weather. Ideally, stay in harbour until the vis improves. In fair weather and reasonable visibility the pilotage in the well marked chans between it and the mainland is not very demanding. But the tide runs hard in places, causing overfalls when against wind >F5.

▶*Tidal streams are strong around the island, and in the chans between it and the mainland. Off Baie du Stiff (lt, radar twr) the stream turns NW at HW Brest +0400, and SE at HW Brest –0200, max rates 2¾kn.*◀

The routes outside Ouessant TSS or via the ITZ add much to the distance and are exposed to sea, swell and shipping. Yachts usually take the inshore chans, ie: Chenal du Four, most direct and popular; Chenal de la Helle, an alternative to N part of Chenal du Four (also gives access to Île Molène) is less direct but better in bad weather. Passage du Fromveur, SE of Ouessant, is easiest but longer and can be extremely rough. ▶*The NE-going stream starts HW Brest –0515, max rates 8-9kn; the SW-going stream starts around HW Brest, max rates 7-8kn.*◀

9.20.6 **Special Notes for France:** See Area 17.

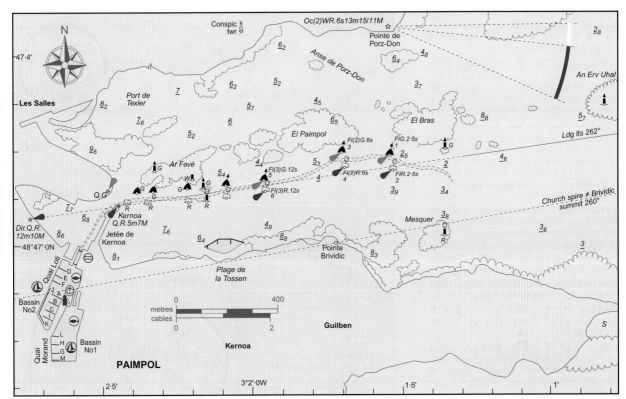

9.20.7 PAIMPOL

Côtes d'Armor **48°47´·00N 03°02´·56W** ❄❄❄◊◊◊❀❀

CHARTS AC 2648, 2027, 2028, 3673; SHOM 7152, 7154, 7127; Navi 537; Imray C34

TIDES Dover –0525; ML 6·1; Duration 0600

Standard Port ST MALO (←──)

Times				Height (metres)			
High Water		Low Water		MHWS	MHWN	MLWN	MLWS
0100	0800	0300	0800	12·2	9·3	4·2	1·5
1300	2000	1500	2000				
Differences PAIMPOL							
–0010	–0005	–0035	–0025	–1·4	–0·9	–0·4	–0·1

SHELTER Good shelter from all winds in hbr, but few ⚓s as most of the Anse de Paimpol dries, including the appr chan to hbr.

NAVIGATION Bearing in mind the very large tidal range, there is enough water in the bay for most craft from half-flood.

WPT 48°47´·82N 02°54´·58W, 262°/5·2M to Kernoa jetty hd. Chenal de la Jument 260° is the outer appr. After La Jument PHM bn tr, alter onto 262·2° inner ldg line; or ⚓ in 5m to await the tide. Small buoys/bns in final 1M of chan (dries 6-7m) may be hard to see against a low evening sun. The drying rks (El Paimpol, El Bras and Ar Fav) are close N of the ldg line.

An alternative appr from Île de Bréhat lies E of Les Piliers NCM bn tr, thence S past Pte de la Trinité. Or appr via Cadenenou NCM and Chenal du Denou 193°. SHOM 7127 is essential for these and other inshore passages which should only be attempted towards HW.

LIGHTS AND MARKS From the S L'Ost-Pic lt is a conspic □ W tr 4M E of Paimpol. Pte de Porz-Don, lt ho on a white house, is 7ca ENE of hbr ent; its W sector leads 270° to intercept the inner 262·2° ldg lts at La Jument bcn twr. A conspic twr (52m) is 3ca W of Porz-Don.

Chenal de la Jument, outer ldg marks 260°: Paimpol ✠ spire (the N'ly of two) on with the ill-defined summit of Pte Brividic. Inner ldg lts, both QR, 262·2°: front, Jetée de Kernoa.

COMMUNICATIONS CROSS 02·98·89·31·31; Auto 08·92·68·08·22; Police 02·96·20·80·17; ⊖ 02·96·20·81·87; Brit Consul 01·44·51·31·00; Ⓗ 02·96·55·60·00; Dr 02·96·55·15·15; Port Mgr 02·96·20·80·77; HM and Lock VHF Ch 09 (0800-1200LT and lock opening hrs); Aff Mar 02·96·55·35·00.

FACILITIES Lock opens HW ±2½, 60m x 12m, ☎02·96·20·90·02. **Marina** ☎02·96·20·47·65. port-plaisance.paimpol@wanadoo.fr www.ville-paimpol.fr **Basin No 2** 306+24 Ⓥ, May-Sep €2·62; Ⓥ berths, pontoon A, min depth 3·8m. ⛽ (hose) ⛴ @ ⓢ. **Basin No 1** Yachts up to 40m LOA. ⛴(25t, 6t, 4t). **E Quays** ⚒ ⚒ ⊕⛽ⓛ. SHOM Ⓔ. **Town** ╳ ⌂ ⬚ ⬚ 🛒 🏨 ⧓ ⊠ Ⓑ Ⓗ ⇌ ✈ Dinard, Brest, Rennes. Ferry: Roscoff, St Malo.

9.20.8 ÎLE DE BRÉHAT

Côtes d'Armor **48°51´·00N 03°00´·00W** ❄❄❄◊◊◊❀❀❀

CHARTS AC 2648, 2027, 2028, 3673; SHOM 7152, 7154; Navi 537; Imray C34

TIDES –0525 Dover; ML 5·8; Duration 0605

Standard Port ST MALO (←──)

Times				Height (metres)			
High Water		Low Water		MHWS	MHWN	MLWN	MLWS
0100	0800	0300	0800	12·2	9·3	4·2	1·5
1300	2000	1500	2000				
Differences LES HEAUX DE BRÉHAT							
–0015	–0010	–0100	–0045	–2·3	–1·6	–0·6	–0·1
ÎLE DE BRÉHAT							
–0015	–0005	–0040	–0030	–1·6	–1·2	–0·5	0·0

SHELTER Good in Port Clos and at Port de la Corderie, except in W-NW'lies. See Facilities for other ⚓s depending on wind direction.

NAVIGATION From the NE use R. Trieux WPT (Lézardrieux) and 224·8° ldg line. From the N use Ch de Bréhat 167·4° (distant ldg marks) then enter Ferlas Chan from the E via WPT 48°49´·39N 02°55´·08W, 277°/3·4M to abm Port Clos. For the R Trieux, follow Ferlas Chan transits Ⓒ – Ⓕ, as shown on AC 3673.

LIGHTS AND MARKS Chapelle St Michel, Amer du Rosedo and nearby Sig stn are all conspic. For the 3 main lts, see chartlet and Lights, buoys and waypoints.

COMMUNICATIONS CROSS 02·98·89·31·31; ⌂ 02·96·20·00·14; Auto 08·92·68·08·22; Police 02·96·20·80·17; ⊖ 02·96·20·81·87; ⊞ 02·96·55·60·00; Dr 02·96·20·09·51; Signal station VHF Ch 16, 10, day only.

FACILITIES Port Clos the main hbr (dries) is busy with vedettes. No ⊂; ⌷ clear of fairway. SW of Port Clos ⌷ is prohib due to cables/pipe across the Ferlas Chan. So moor/⌷ in **Le Kerpont** W of Bréhat. **Port de la Corderie** dries; some ⌷s but get well out of strong tidal streams.

E of Le Bourg there are free drying private buoys near ⌷, but some reported unsafe due to lack of maintenance.

La Chambre ⌷ in upper reaches just S of the ⚓ area. ⌐ can be floodlit by pressing button on lamp post at top of slip.

Guerzido off the Chenal de Ferlas has good holding, out of the strong tides.

Hbrs (no HM) full access at HW, ⚓⛵ from fuel barge at Port Clos, ⌐; CN de Bréhat, ⛵⌂⚒.

Village ✕⌂🛒⏱🚰✉ Ⓑ. Ferry to Pte de l'Arcouest, bus to Paimpol thence ⇌ to Paris, Brest, Roscoff and St Malo. ✈ Dinard, Brest, Rennes to London. No cars on island.

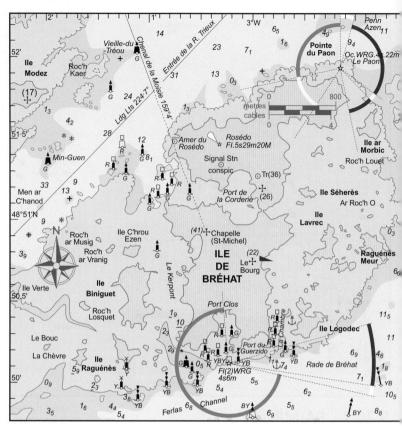

OUTER APPROACHES TO LÉZARDRIEUX AND TRÉGUIER WITH PASSE DE LA GAINE BETWEEN

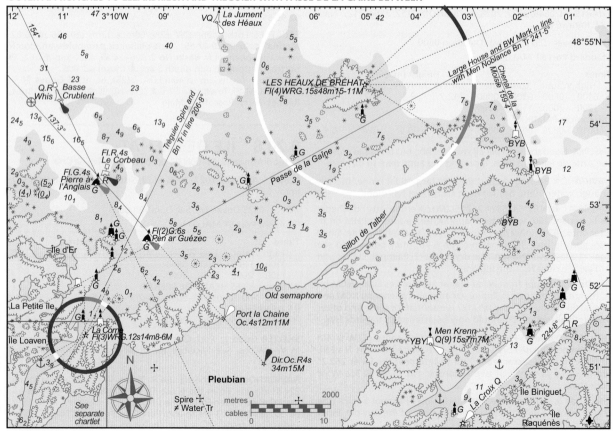

9.20.9 LÉZARDRIEUX

Côtes d'Armor **48°47′·34N 03°05′·89W** ✿✿✿✿♠♠♠✿✿

CHARTS AC 2648, 2027, 2028, 3673; SHOM 7152/3, 7126/7; Navi 537; Imray C34

TIDES –0510 Dover; ML 5·9; Duration 0610

Standard Port ST MALO (←)

Times				Height (metres)			
High Water		Low Water		MHWS	MHWN	MLWN	MLWS
0100	0800	0300	0800	12·2	9·3	4·2	1·5
300	2000	1500	2000				
Differences LÉZARDRIEUX							
–0020	–0015	–0055	–0045	–1·7	–1·3	–0·5	–0·2

SHELTER Very good in all weathers. Strong stream at half tide.

NAVIGATION WPT 48°54′·98N 02°56′·07W, 224·7°/6·7M to La Croix, front 224·8° ldg lt. Offshore dangers include Roches Douvres, Barnouic, Plateau de la Horaine and rky shoals to the W in the outer apps. Off the river ent allow for strong cross-streams.

The 3 well-marked approach channels are:

- **Grand Chenal**, main lit chan from NE, best for strangers. Ldg lts 224·7°: Front, La Croix, two double-barrelled trs, W on NE side with R tops. Rear, Bodic (2·1M from front) high up amongst the trees.
- **Ch de la Moisie**, unlit from the NW: Amer du Rosédo, W obelisk on 159·4° with St Michael's chapel (both conspic on Île de Bréhat). It also connects with Passe de la Gaine, short cut from/to Tréguier; see chartlet on previous page.
- **Ferlas Chan** (lit) from E or W, passing S of Île de Bréhat. By day follow the 5 SCM bcns on the N side of the chan. At night: W sector (281°) of Men Joliguet, WCM bcn twr.
  W sector (257°) of Roche Quinonec at Loguivy.
  W sector (271°) of Kermouster joins Coatmer 218·7° ldg line.

Within the Trieux river:

- Coatmer ldg lts 218·7°: front, low down amongst trees; rear, 660m from front, high up almost obsc'd by trees.
- W sector (200°) of Les Perdrix, a stout green bcn twr. Speed limit 5kn from Perdrix to the bridge.

LIGHTS AND MARKS See Lights, buoys and waypoints for offshore lts at Roches Douvres lt ho, Barnouic ECM bn tr, Les Héaux de Bréhat lt ho and, on Île de Bréhat, Pte du Paon and Rosédo. Outer and inner approach marks are described above. There are no navigational lights S of Perdrix.

Beware, at night, the unlit, 16m high Roc'h Donan 400m S of Perdrix. At the tidal marina floodlights on shore may dazzle yachts arriving.

COMMUNICATIONS CROSS 02·98·89·31·31; Auto 08·92·68·08·22; Police 02·96·20·18·17; ⊖ at Paimpol 02·96·20·81·87; Brit Consul 01·44·51·31·00; Dr 02·96·20·18·30; Aff Mar (Paimpol) 02·96·55·35·00; Marina VHF Ch 09 (0730-2200 Jul/Aug; 0800-1200 and 1400-1800 rest of year).

FACILITIES Tidal marina ☎02·96·20·14·22. http://port.3douest. com/lezardrieux/ port-lezardrieux@wanadoo.fr Access H24. 380 + 50 ♥. Berth on the first pontoon (No 3) or as directed by HM. ⌐€2·39. ⚑ €2·00. Multihulls and boats >12·5m LOA should moor on ⚓s €1·67 or ♥ pontoon. ▮▯✕◻◻▮▮⚒ ⊞ Ⓔ ⚠ ✖ Divers, ⊫⊳(50t).

YC de Trieux ☎02·96·20·10·39.

Non-tidal marina (247 ⌐) has some ♥ berths (2·4m inside). The retaining wall is marked by 5 unlit Y SPM perches, the sill by PHM/SHM perches. Access over sill 4·9m with automatic flap. As sill covers on the flood, at CD +6·15m, flap drops to give 1·1m clearance. A depth gauge shows water over sill. The waiting pontoon is as on chartlet. IPTS sigs 2 & 4 control entry/exit.

Town ✕◻▤⚱▮⊠Ⓑ ⇌ (occas bus to Paimpol & Lannion) ✈ Lannion. Ferry: Roscoff.

Yachts can continue about 12km upriver (via bridge, clearance 17m) to lock in at Pontrieux in complete shelter.

9.20.10 PONTRIEUX

Côtes d'Armor 48°42'·74N 03°08'·98W ❀❀♠♦♦♦❀❀

CHARTS Navi 537; AC 3673 & SHOM 7126 end S of Lézardrieux. A useful river guide/map is available from HM Pontrieux.

TIDES HW at Pontrieux lock is also HW St Malo (Area 18).

SHELTER Complete.

NAVIGATION See Lézardrieux for the river up to it. Not before HW –3, proceed via suspension bridge (17m clearance at MHWS) for 6M upriver, keeping to high, rky bank on bends. Between river bends the best water is indicated by the alignment of reflective posts (1-2m high) near the bends. Allow 1½hrs for the passage, and aim to reach the lock at HW –1, to avoid sand coasters (*sabliers*) which use the lock occas at HW. Below Château de la Roche Jagu a waiting buoy is accessible HW±3.

Lock (48°42'·78N 03°08'·31W) ☎02·96·95·60·70; opens HW –2¼ to +2¼, all year round. 2 half-tide waiting buoys.

LIGHTS AND MARKS River is unlit; few marks.

COMMUNICATIONS Auto 08·36·68·08·22; Lock and port VHF Ch 12, (lock hrs); ☎ link to HM Pontrieux; Public ☎ at Ch Roche Jagu.

FACILITIES Berth in 1·5-3m at quay (SE bank) about 1km WSW of lock. HM ☎02·96·95·34·87, 06·73·87·07·95. www.letrieux.com 80+40♥ ⌘ €1·88, ⚓ ⊕🅿️🅾️✕🅰️.
Town ✕ 🃏 🍴 🛒 🏧 🏤 @ ⊠ Ⓑ ⚓ ⇌ Paimpol/Guingamp, ✈ Brest, Rennes, Dinard & Paris. Taxi 02·96·95·60·43.

9.20.11 TRÉGUIER

Côtes d'Armor 48°47'·21N 03°13'·27W ❀❀❀♦♦♦❀❀❀

CHARTS AC 2648, 2027, 2028; SHOM 7152, 7126; Navi 537; Imray C34

TIDES –0540 Dover; ML 5·7; Duration 0600

Standard Port ST MALO (←)

Times				Height (metres)			
High Water		Low Water		MHWS	MHWN	MLWN	MLWS
0100	0800	0300	0800	12·2	9·3	4·2	1·5
1300	2000	1500	2000				
Differences TRÉGUIER							
–0020	–0020	–0100	–0045	–2·3	–1·6	–0·6	–0·2
PORT-BÉNI (5ca SSE of La Corne lt twr)							
–0020	–0020	–0105	–0050	–2·4	–1·7	–0·6	–0·2

SHELTER Good. Possible ⚓s, keeping clear of fairway and fish farms. 7ca SW of La Corne lt tr, but exposed to N'lies; N and S of La Roche Jaune village; and in well sheltered pool (6m) 1ca NE of No 10 buoy (8ca N of marina).

NAVIGATION WPT 48°54'·13N 03°11'·50W (abeam Basse Crublent PHM buoy) 137°/2·1M to Pen ar Guézec SHM light buoy, a key turning point. Caution: strong tidal streams across the well lit chan. The three approach channels, with marks, are:

- **Grande Passe** 137·3°. Ldg marks (hard to see by day): Front, Port de la Chaine, white ho; rear, St Antoine, RW ho. At Pen ar Guézec SHM buoy alter 216° in La Corne's W sector. A clearer transit from Basse Crublent is: Pleubian spire (charted) and adj water twr (on skyline) ≠ 154° to pass between Pierre à l'Anglais and Le Corbeau bys (both lit).

- **Passe de la Gaine** Unlit short cut to/from Lézardrieux. See chartlet below Île De Bréhat. Adequately marked, navigable by day with care/sufficient tide, good vis; but marks hard to see from afar in low sun/ poor vis. 241·5° ldg marks: Front, Men Noblance bcn twr, W with B band, ≠ rear mark (W wall with B vert stripe) below the skyline, right of conspic Plougrescant ✠ spire. Stay on transit to hold narrow chan. See Passage info.

- **Passe du Nord-Est** 205°/207°: unlit, dangerous with W-NW winds as sea breaks across the shallowest part of chan. On 205° keep W of La Jument NCM buoy and adjacent rky shoals; jink port onto 207° transit of Tréguier spire and Skeiviec bcn twr, for direct appr to La Corne.

In river, heed buoys and bns, eg keep E of Taureau SHM light buoy, 300m SW of La Corne lt ho, to clear adjacent drying mudbank. Speed limit 5kn south of No 3 SHM lt buoy. Less water reported from No 11 lt by to marina, approx 5M from La Corne lt hse.

LIGHTS AND MARKS See Lights, buoys and waypoints/chartlet. Important marks: La Corne, a stumpy WR lt tr; 6ca to the N is Men Noblance black/white bcn tr (for Passe de la Gaine); and 4ca SW is Skeiviec W bn tr. The spire of Tréguier cathedral is 4·6M SSW of La Corne, but obsc. by high wooded banks entering the estuary.

COMMUNICATIONS CROSS 02·96·89·31·31; Auto 08·92·68·08·22; Police 02·96·20·84·30; ⊖ 02·96·20·81·37; Brit Consul 01·44·51·31·00; Ⓗ 02·96·05·71·11; Dr 02·96·92·32·14; Aff Mar 02·96·92·30·38 (Paimpol); Marina VHF Ch 09, office hrs in season, 7/7: 0700-2145; Winter, M-F 0800-1200 & 1330-1730 (Sat 1630): Sun shut.

FACILITIES A 30m waiting pontoon, least depth 1·2m LWS, 500m N of marina, boats >15m LOA & multihulls can use overnight. **Marina** *Arr/dep at slack water*, tide sets hard N/S diagonally through pontoons. Berth bows to tide on outer part of pontoons C, D, E. F pontoon is exclusively for vessels >14m, max LOA/Wt, 30m/100t, depth 3m LAT on outside, 2.8m inside. HM ☎02·96·92·42·37, mob 06·72·70·70·20 portplaisance.treguier@wanadoo.fr www.ville-treguier.fr/culture-sports-loisirs/port-plaisance.php 260 + 70♥ €2·80 > 20m, €57.00 + €3.00 per/m extra. 🅾️ wi-fi @, 🔧 ⊕ 🅶🅻 ⚓ ✕ 🏗(40t) ⚓ 🅁 ☎02·96·92·15·15/93·55.

Bar des Plaisanciers ☎02·96·92·41·56, 365/365. **CN de Tréguier** ☎02·96·92·37·49, opposite marina, 🃏 Sat evenings only. **Town** ✕ 🃏 🍴 (small s/market near cathedral delivers) 🏧 🏤 ⊠ Ⓑ. Market Wed. ⇌ via bus to Paimpol, Guingamp, Lannion. ✈ Brest, Rennes, Dinard. Ferry: Roscoff, St Malo.

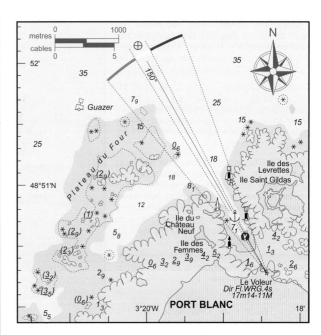

150° towards Le Voleur Dir lt; the rear ldg mark is hard to see amongst trees. From the SW approach via Basse Guazer PHM buoy, 7ca SW of the WPT. Beware ebb tide initially setting hard toward drying Plateau du Four.

LIGHTS AND MARKS 150° ldg marks: Front, Le Voleur Dir lt, low down amongst trees (between white cottage and Grand Hotel, conspic white block); rear, 5ca from front, La Comtesse Mill, unlit and obscured by trees on skyline. A white obelisk (16m) on Île du Château Neuf is conspic to stbd of the fairway. A smaller obelisk to port on Île St Gildas is less obvious.

COMMUNICATIONS CROSS 02·33·52·72·13; Auto 08·92·68·02·22; ⊖ 02·96·20·81·87; Brit Consul 01·44·51·31·00; *Port Blanc* VHF Ch 09, 16; Aff Mar 02·96·92·30·38;.

FACILITIES 5 W 🛟s, marked VISIT in approx 5m. Safe ⚓ and good holding, where charted in 7m. Or dry out alongside quays, 1·3m; best to pre-recce. HM ☎02·96·92·89·11 in sailing school. 🛟 or 🛟 €7·00 all LOA, ⚓ 🍴 🛒 🔧 ✂ ✗ ⚓(16t) ⚓.

Town ✕ 🏠 🍺(basic). Taxi 02·96·92·64·00, 🚆 Lannion. ✈ Brest, Rennes, Dinard. Ferry: Roscoff.

9.20.13 PERROS-GUIREC

Côtes d'Armor **48° 48'·17N 03° 26'·21W** (Jetty hd)
❄❄⚓⚓⚓⚓❀❀

CHARTS AC 2648, 2027; SHOM 7152, 7125; Navi 537, 538; Imray C34

TIDES –0550 Dover; ML 5·4; Duration 0605

Standard Port ST MALO (◄—)

Times				Height (metres)			
High Water		Low Water		MHWS	MHWN	MLWN	MLWS
0100	0800	0300	0800	12·2	9·3	4·2	1·5
1300	2000	1500	2000				
Differences PERROS-GUIREC							
–0035	–0035	–0115	–0100	–2·8	–2·0	–0·7	–0·1

SHELTER Very good in marina. Off Pointe du Chateau safe ⚓, except in NE'lies, in approx 3m good holding; plus 5 small W 🛟s.

NAVIGATION From the E, WPT 48°49'·86N 03°23'·66W, 224·8°/2·4M to Jetée Linkin. From the W, WPT 48°50'·53N 03°26'·45W, 143·5°/1·8M to join 224·8° ldg line. Caution: Rocks extend 7ca W and 6ca NE of Île Tomé. Local moorings E of Jetée du Linkin are marked by 2 SHM buoys.

9.20.12 PORT BLANC

Côtes d'Armor **48°50'·54N 03°18'·89W** ❄❄⚓⚓⚓❀❀

CHARTS AC 2648, 2027; SHOM 7152, 7125/6; Navi 537, 538; Imray C34

TIDES –0545 Dover; ML 5·3; Duration 0600. Interpolate between Les Héaux de Bréhat and Perros-Guirec. HW –0040 and ht –2·0m on St Malo. MHWS 9·0m; MHWN 6·8m.

SHELTER Good in natural hbr (aka Port Bago) but very uncomfortable in winds with any N in them.

NAVIGATION WPT 48°52'·24N 03°20'·16W, 150°/1·9M to first 🛟. The fairway does not open up until at the WPT. Appr on ldg line

LIGHTS AND MARKS 225° ldg marks are hard to see by day, but bcns/buoys in the appr are adequate. Passe de l'Ouest: Kerjean Dir lt 143·6° also hard to see, but old lt ho (gable end) on the foreshore is clearer.

COMMUNICATIONS CROSS 02·98·89·31·31; ⌨02·96·91·40·10; Auto 08·92·68·02·22; Police 02·96·23·20·17; ⊖02·96·20·81·87; Brit Consul 01·44·51·31·00; Dr 02·96·23·20·01; Entry gate 02·96·23·19·03; Aff Mar 02·96·91·21·28; Marina VHF Ch 09, 16.

FACILITIES Access: 6m wide gate (no lock) opens when rise of tide reaches 7m. Sill under gate dries 3·5m, giving 3·5m water inside gateway on first opening, less water reported near fuel pontoon. Caution: strong current initially. Gate sigs: IPTS. No ⚓ in basin. R & W poles mark retaining wall; do not try to cross it near HW.

Gate opens depending on tidal coefficient (see Chapter 9) may open > 30 mins earlier and close slightly later:

Coeff <40	=	Shut, 4 days max
40-50	=	HW –½ to HW
50-60	=	HW –1 to +½
60-70	=	HW ±1
>70	=	approx HW ±1½

Marina ☎02·96·49·80·50. www.port-perros-guirec.com 720+80⓿ €3·30. ⓿ berth on N'most 2 pontoons in 2·5m. 🚿🚽 ⌕@🔧Ⓔ SHOM △⛽🛒⚒⛵🏪(7t). **YC** ☎02·96·91·12·65. **Town** ✕🍴🍺✉Ⓑ. ⇌ Lannion. ✈ Brest. Ferry: Roscoff.

LES SEPT ÎLES *See notes on facing page*

OFFSHORE ISLANDS

LES SEPT ÎLES, Côte d'Armor, **48° 52´·72N 03°29´·16W**. AC 2026, 2027; SHOM 7152 (large scale 1:20,000). See chartlet on previous page. HW-0550 on Dover (UT); +0005 on Brest. HW ht −1·8m on Brest. ML 5·2m. Use Ploumanac'h differences.

The whole archipelago, outlined in magenta, is a no-entry bird sanctuary. Île aux Moines is the only one of the 7 islands/islets on which landing is permitted; no shore facilities. It is safe to sail NW of Les Dervinis (SCM buoy) and Les Noires de Rouzic (always visible) a steep-to rocky chain SE of the main group.

Straightforward approach to Île aux Moines, allowing for strong stream. Avoid a rock (0·6m) 300m SE of the island. Best ⚓ (Lat/ Long as on line 1) is due E of jetty at E end of Île aux Moines and S of W end of Île Bono; overnight only in calm settled weather. Vedettes use a mooring buoy here. More exposed ⚓s are S of the Old Fort (W end of island) or close S of Île Bono; both require care due to rocks, clearly shown on SHOM 7152. The most dangerous is at 48° 52´·45N 03°28´·68W; it dries 0·1, is rarely seen but often encountered. Only slightly less dangerous is the rock drying 1·4 at 48° 52´·64N 03°28´·28W.

Île aux Moines lt ho is a very conspic grey twr whose 24M beam is obscured by Bono and Rouzic in a 4° arc (237°-241°) and when brg < 039°, ie if tucked into the Baie de Lannion.

9.20.14 PLOUMANAC'H

Côtes d'Armor **48° 49´·78N 03° 29´·51W** (Sill) ❋❋⚓🛥⚓🌸🌸🌸

CHARTS AC 2648, 2027 (limited scale 1:48,700); SHOM 7125 (1:10,000 inset); Navi 537, 538; Imray C34

TIDES −0550 Dover; ML 5·5; Duration 0605
Standard Port ST MALO (⟵)

Times				Height (metres)			
High Water		Low Water		MHWS	MHWN	MLWN	MLWS
0100	0800	0300	0800	12·2	9·3	4·2	1·5
1300	2000	1500	2000				
Differences PLOUMANAC'H							
−0035	−0040	−0120	−0100	−2·9	−2·0	−0·7	−0·2

SHELTER Good inside. Appr is uncomfortable against the ebb in strong N'lys, favour the PHMs; dangerous during spring ebb with strong NE winds.

NAVIGATION WPT 48°51´·44N 03°29´·09W, 188°/1·2M to ent between Mean Ruz lt ho and Château Costaérès (conspic). At the entrance beware An Dreuzinier, unmarked rocks (drying 1·4m) 100m N and NE of No 1 SHM perch; not a problem with sufficient rise of tide, but near LW stick to a line 225° through the Nos 2 and 4 PHM perches. 11 unlit perches mark the chan.

Entry is over a sill, drying 2·55m, which retains 1·2m to 2·3m within. If concrete base of No 5 perch (the 3rd SHM) is covered, depth over sill is >1·4m. No 12 PHM perch right next to sill has depth gauge; note: gauge not visible when >3m over sill.

LIGHTS AND MARKS See chartlet and Lights, buoys and waypoints. Mean Ruz lt ho is a reddish square twr. Signal stn, 8ca SSE, is conspic from the E. Night entry not advised.

COMMUNICATIONS ⒧Ⓑ 02·96·20·00·45; Auto 08·92·68·08·22; Brit Consul 01·44·51·31·00; Dr 02·96·91·42·00; HM VHF Ch 09.

FACILITIES Between Nos 8 and 8bis perches on the E side of chan are 3 waiting buoys marked VPG in 1·5m least depth.

Port de Plaisance Inside the sill, are moorings for FVs and yachts >12m in a trot to stbd of ent in 3m, with 2 drying trots of small craft outboard. For best water keep to port and approach the yacht moorings from NW. ⚓s are 1st trot (A) of dumbell buoys; max LOA 12m. Yachts >12m may be able to tie up to the fishing vessel trots perpendicular to the sill. SE & SW sides of hbr are very shallow.

HM (part time) ☎02·96·91·44·31 or 02·96·49·80·50 (Perros-Guirec). www.port-perros-guirec.com 230 + 20 ⚓s, €3·30. The full time Perros HM also oversees Ploumanac'h and Anse de Trestraou, a pleasant anchorage on N side of Perros town, well sheltered from S and W winds; exposed to NE'lies.

Quai Bellevue ⚓🛥 🏠🅿. 🔧 ☎02·96·23·05·89. **YC de Perros-Guirec** Bus to Lannion and Perros. Ferry see Roscoff.

ANCHORAGE 1M WEST OF PLOUMANAC'H

TRÉGASTEL-PLAGE, Côtes d'Armor, **48°50´·04N 03°31´·29W**, AC 2026, 2027; SHOM 7152, 7125. HW −0550 on Dover (UT); +0005 and −1·8m on Brest HW; ML 5·1m; Duration 0605. Use Ploumanac'h differences.

A large W radome, conspic 3·2M S of the ent, gives general orientation. Enter between PHM bcn on Île Dhu (just W of Île Dé, with dice-shaped Pierre Pendue) and SHM bcn (or buoy if bcn destroyed) off Le Taureau, rk drying 4·5m. A conspic house with □ turret, brg approx 165° leads between Île Dhu and Le Taureau. But it is easier to track 183° leaving the 3 PHM bcns about 60m to port. SHOM 7125 (1: 20,000) is strongly advised.

Beyond SHM bcn, ⚓ or pick up westernmost ⚓s €10·00/night, S of Île Ronde (nearest the ent). Good ⚓ in 2m, but exposed W to N, with swell near HW springs; more comfortable at neaps.

Facilities ⚓ in E part of hbr. **CN de Trégastel** ☎02·96·23·45·05.

Town (Ste Anne, 0·5M inland) ✕ ⌂ 🛒 🏦 🔧 ✉ Ⓑ; Dr 02·96·23·88·08/02·96·15·96·49.

9.20.15 TRÉBEURDEN

Côtes d'Armor **48°46′·29N 03°35′·15W** ❀❀◊◊◊❁❁

TIDES –0605 Dover; ML 5·5; Duration 0605

Standard Port BREST (→)

Times				Height (metres)			
High Water		Low Water		MHWS	MHWN	MLWN	MLWS
0000	0600	0000	0600	7·0	5·5	2·7	1·1
1200	1800	1200	1800				
Differences TRÉBEURDEN							
+0100	+0110	+0120	+0100	+2·2	+1·8	+0·8	+0·3

CHARTS AC 2026; SHOM 7151/2, 7125, 7124; Navi 537, 538; Imray C34

SHELTER Good in marina (2–2·6m). ⚓ off NE side of Île Milliau is exposed to W'lies.

NAVIGATION WPT 48°45′·34N 03°40′·00W, 067°/3M to first PHM buoy in white sector of Lan Kerellec Dir lt. From WSW, go direct to ⊕; thence enter the buoyed chan (105°/088°) to marina.

From E and N, round Bar-ar-Gall and Le Crapaud WCM buoys; continue S for 1·4M, before altering 067° towards the ⊕. If properly planned, the short-cut 183° between Le Crapaud reef and reefs N and S of Île Losket is safe in offshore winds.

LIGHTS AND MARKS See chartlet and Lights, buoys and waypoints. Lan Kerellec Dir lt leads 067° to the start of buoyed chan, aligned approx 100° with ✠ spire (conspic, 115m).

COMMUNICATIONS CROSS 02·98·89·31·31; Police 02·96·23·51·96 (Jul/Aug); ⊖ 02·96·48·45·32; Brit Consul 01·44·51·31·00; *Port Trébeurden* VHF Ch 09, 16; Aff Mar 02·96·91·21·28.

FACILITIES 10 Y waiting buoys lie S of the fairway and some 600m W of marina ent. Access times based on Coefficients are:

95 (sp)	HW –3½ to +4
70	HW –3¾ to +4¼
5 (np)	HW –4¾ to +5¼

An automatic flap (2·5m) above the sill starts to open when ht of tide is 4·0m and is fully open at 4·3m (conversely on the ebb). Best to calculate tidal heights.

IPTS (sigs 2 and 4) on bkwtr and at gate show when access is possible. Do not ent/exit within 20 mins of the flap moving, due to fast, turbulent water in the entrance.

Enter between gate posts in retaining wall, which dries 3·5m and is marked by 4 Y SPM bns. A floodlit tide gauge is port side of the 15m wide gate.

Marina ☎02·96·23·64·00. portrebeurden@ wanadoo.fr www.port-trebeurden.com 625 + 70 Ⓥ on pontoon G (NE side <11m, SW side >11m) July/Aug €3·70, inc ⊕ @, ⟜€1·00. ⚓⛽⛴⚙⚓⚓⚒⚓⚓(16t mobile) **YC** ☎02·96·15·45·97 (July-Aug).

Town ✕ ⌂ ☰ ✉ Ⓑ. ✈ Brest, Rennes, Dinard. Ferry: Roscoff or St Malo.

9.20.16 LÉGUER RIVER (Lannion)

Côtes d'Armor **48°43′·76N 03°33′·28W (2nd G bcn)** ❀❀◊❁❁❁

CHARTS AC 2648, 2026; SHOM 7124; Navi 537, 538; Imray C34

TIDES –0605 Dover; ML 5·4; Duration: no data

Standard Port BREST (→)

Times				Height (metres)			
High Water		Low Water		MHWS	MHWN	MLWN	MLWS
0000	0600	0000	0600	7·0	5·5	2·7	1·1
1200	1800	1200	1800				
Differences LOCQUIREC							
+0100	+0110	+0125	+0100	+2·2	+1·8	+0·8	+0·4

SHELTER Good, except in strong W/NW winds. No access at LW sp, esp in strong NW'lies when seas break on the drying bar. ⚓ in non-drying pools off Le Yaudet and up to the 2nd Y perch, but very little space due to silting and many local moorings; a vacant mooring is possible or ⚓ outside and recce by dinghy.

NAVIGATION WPT 48°44′·38N 03°36′·87W, 091°/2·6M to Pte Beg-Léguer lt (also on Locquémeau ldg lts 121°). Leave Kinierbel SHM buoy close to stbd; stand on until Trébeurden spire bears 004° (to clear a drying patch) then alter 135° for the ent, passing close to two G bcn twrs. Le Yaudet pool is 8ca further E.

Chan upriver to Lannion is narrow, steep-to, dries and is unmarked by bcns; best seen at LW before attempting. Just below Lannion town, opp Quai de Loguivy, drying out on the N bank may be feasible; a recce by dinghy or on foot is advised.

LIGHTS AND MARKS See chartlet and Lights, buoys and waypoints. Pte de Beg-Léguer lt ho is a squat twr, R top, built onto a house and on high cliffs. Le Yaudet spire helps to pinpoint the river ent.

COMMUNICATIONS CROSS 02·98·89·31·31; Auto 08·92·68·08·22; Police 02·96·46·69·50; ⊖ 02·96·48·45·32; Brit Consul 01·44·51·31·00; Dr 02·96·47·23·15; Aff Mar 02·96·55·35·00.

FACILITIES HM ☎02·96·47·29·64, ⎯ (dries) ⚓ ⌂(1t) ⚒ ⌂ only in emergency. ✕ ⌂ in Le Yaudet, but few supplies.

Lannion, ✕ ⌂ 🛒 ⬧ ⚒ ⬧ ⌂ ✕ SHOM Ⓔ ✉ Ⓑ.

ADJACENT HARBOURS IN BAIE DE LANNION

LOCQUÉMEAU, Côtes d'Armor, **48°43′·60N 03°34′·96W** (1st SHM perch). AC 2648, 3669; SHOM 7124. HW –0600 on Dover (UT); +0110 on Brest; HW ht +1·5m on Brest; ML 5·3m. SHOM 7124 is desirable as it shows the various perches. A small drying hbr by ent to Lannion River (above); use the same WPT. Approach leaving Locquémeau SHM buoy close to stbd. There are two quays: the outer is accessible at LW, but open to W winds. Yachts can dry out at inner quay, on S side. Ldg lts 121° to outer quay: front, W pylon + R top; rear, hard-to-see W gable and R gallery. Facilities: ⚒ ⌂ ✕. **Town** ✕ ⌂.

LOCQUIREC, Finistère, **48°41′·41N 03°38′·79W**. AC 3669, 2648; SHOM 7124. Tidal data see Lannion. Small drying hbr mouth of Le Douron R, with good shelter from W'lies. The whole estuary dries to extensive, shifting sandbanks (2·9m to 6·1m). No lights, but Gouliat NCM buoy is 7ca N of Pte du Château and a SHM bcn twr (6ca S of the hbr, marking Roche Rouge) is conspic. E of Pte du Château, off Île Vert, there are 25 W moorings +2 🅜, €6·50p/n, in up to 4m. Beyond, Locquirec harbour and bay has 304 W drying moorings, + 5 🅜, also €6·50, pleasant temp ⚓, overnight if able to take the ground, or dry ⌂ on stone quay, ⚓ ⍢. Land at slip. A small, laid-back resort, inc some fine trad. hotel-restaurants and ⊙. Mairie (Port Authority) ☎02·98·67·42·20.Good LW cockling.

9.20.17 PRIMEL-TRÉGASTEL

Finistère **48°42′·79N 03°49′·46W** ❀⚓⚓⚓⚓

CHARTS AC 2026, 2745; SHOM 7151, 7124, 7095; Navi 538; Imray C34, C35

TIDES See Morlaix for differences ANSE DE PRIMEL. –0610 Dover; ML 5·3; Duration 0600.

SHELTER Access at all tides, but exposed to N/NW'lies; if strong, seas break across ent. Drying upper reaches are well sheltered.

NAVIGATION WPT 48°43′·59N 03°50′·10W, 152°/0·9M to the hbr bkwtr. Beware drying rks NE of Pte de Primel. Enter exactly on 152° ldg marks, past a SHM bcn marking Ar Zammeguez (3m high). The ent, between a PHM and 2nd SHM bcns, is narrow. A useful hbr to await the tide up to Morlaix.

LIGHTS AND MARKS Pte de Primel is a conspic 42m high rky islet NNE of the hbr. 152° ldg marks/lights are clear but precise: three whited walls with R vert stripe in serried ranks up the hillside surrounded by small pine trees; ldg lts, both QR, are on the front and rear marks.

COMMUNICATIONS Ⓛ 06·14·38·26·81; Dr 02·98·67·30·20; Brit Consul 01·44·51·31·00; HM 06·81·79·67·49 (English speaker); VHF Ch 09, 16 (season); Rolland Marine B'yard 02·98·72·30·06.

FACILITIES Le Diben (Plougasnou) on W side is a FV hbr. FVs occupy most of the bkwtr; rafting possible. In summer a line is strung along the quay for yachts to tie on to, the E end is non-drying; not safe in N'lies. A small ⚓age and 16 W 🅜s in about 1m (13 are drying) are SE of the bkwtr head. Or ⚓ to SE or SW of bkwtr hd in 2·9m. ⎯ ⬧ Ⓔ ⌂ ✕ Ⓔ ⌂(45t). **Primel-Trégastel** ⌂ ⌂ ✕ ⌂ 🛒 ⌂ ⬧ ✉ Ⓑ. Ferry: Roscoff.

9.20.18 MORLAIX

Finistère **48°35′·31N 03°50′·30W** (Lock) ✿✿✿◊◊◊✿✿✿

CHARTS AC 2026, 2745; SHOM 7151, 7095; Navi 538; Imray C34, C35

TIDES –0610 Dover; ML 5·3; Duration 0610

Standard Port BREST (→)

Times				Height (metres)			
High Water		Low Water		MHWS	MHWN	MLWN	MLWS
0000	0600	0000	0600	7·0	5·5	2·7	1·1
1200	1800	1200	1800				

Differences MORLAIX (CHÂTEAU DU TAUREAU)
+0055 +0105 +0115 +0055 +1·9 +1·6 +0·7 +0·2
ANSE DE PRIMEL
+0100 +0110 +0120 +0100 +2·0 +1·6 +0·7 +0·2

SHELTER Good in the bay, except in onshore winds. Strong N'lies can raise a steep sea even in the estuary. ⚓ NE of Carantec is exposed, especially to NW; landing stage dries about 4·6m. Complete shelter in the marina.

NAVIGATION The three appr channels have rocky dangers and require careful pilotage; they converge near No 1 buoy, Barre de Flot. La Lande is the rear ldg lt/mark for the first two channels.

- **Chenal de Tréguier** is best at night, but almost dries: WPT 48°42′·55N 03°51′·93W (abm La Pierre Noire SHM bcn) 190·5°/1·84M to La Chambre bcn twr. Ldg lts 190·5°: Front mark, Île Noire lt bcn.

- **Grand Chenal** is shallower but lit: WPT 48°42′·65N 03°53′·53W, (abm Stolvezen PHM buoy) 176°/2·23M to Île Louet lt, passing E of Ile Ricard. Ldg lts 176·4°: Front mark, Île Louet lt ho.

- **Chenal Ouest de Ricard** is the deepest, but unlit. It deviates 188·8° from Grand Chenal, passing W of Ile Ricard. 188·8° ldg marks: Pierres de Carantec and Kergrist, both painted white.

The Morlaix River is buoyed/bcn'd but unlit and prone to silting. ⚓ off Pen al Lann and Dourduff (dries) clear of extensive oyster beds marked by small orange buoys/stakes.

Morlaix lock, marina and town are 5·5M upriver from No 1 SHM buoy.

LIGHTS AND MARKS See Lights, buoys and waypoints and chartlet. Château du Taureau fort is conspic 4ca N of No 1 buoy. Some marks omitted for clarity.

Morlaix

Tobacco Factory

Town centre

COMMUNICATIONS CROSS 02·98·89·31·31; ⓁⒷ 02·98·72·35·10; Auto 08·92·68·02·29; Police 02·98·88·17·17; ⊖ 02·98·69·70·84; British Consul 01·44·51·31·00; Ⓗ 02·98·62·62·45; Port and Marina VHF Ch 09; Aff Mar 02·98·62·10·47.

FACILITIES Morlaix lock ☎02·98·88·15·10/06·77·50·15·90 opens by day only (SR-SS) at HW −1½, HW and HW+1; wait alongside W quay.

Marina ☎02·98·62·13·14 www.plaisancebaiedemorlaix.com plaisance.morlaix@morlaix.cci.fr 150+32Ⓥ Jun-Sep €2·50; multihulls x150%. Ⓥ pontoon (fingers) on E bank, parallel to road. In lock hours a movable footbridge across the marina is usually open. 🐟 🗠 ⚓ 🛠 ⊠(8t) ◣.

Town ✕ ⌂ 🛒 ⊙ 🛏 💈 🍷 ⚓ △ Ⓔ SHOM ⊠ Ⓑ. ✈ Brest, ferry Roscoff.

ANCHORAGES W AND NW OF MORLAIX

PENZÉ RIVER, Finistère, Ent **48°42′·24N 03°56′·66W**, AC 2745; SHOM 7095. HW −0610 on Dover (UT) +0105 on Brest; HW ht +1·2m Brest; ML 5·0m; Duration 0605.

The Penzé river lies W of Île Callot. From NNW, appr between Cordonnier and Guerhéon bcn twrs at mid-flood. Or, for deeper water, from the ENE pass between Les Bizeyer reef and Le Paradis bcn twr; thence S via beaconed channel.

Passe aux Moutons, between Carantec and Île Callot, is a drying (6·2m) short cut from the E into the Penzé, with adequate rise.

S of Le Figuier IDM bcn the chan narrows and is scantily marked by small oyster withies, then by mooring trots. Best shelter is SW of Pte de Lingos: moor in 1·2m off St Yves where the old ferry slips provide landing places.

S of Pont de la Corde (10m clearance) the river is buoyed and navigable on the tide for 3M to Penzé. No access 1 Mar–31 Aug to a Nature Reserve at the head of the river.

Facilities at Carantec 🐟 ⚓ 🗠 ⚓ 🛠 ✕ ⚓ Ⓔ. **Town** ✕ ⌂ 🛒 ⊠ Ⓑ.

Penzé ⌂ (drying) limited ✕ ⌂ 🛒.

9.20.19 ROSCOFF

Finistère **48°43'·54N 03°58'·59W** ✱❀✿⚓⚓✿✿✿

CHARTS AC 2026, 2745; SHOM 7151, 7095; Navi 538; Imray C35

TIDES –0605 Dover; ML 5·2; Duration 0600

Standard Port BREST (→)

Times				Height (metres)			
High Water		Low Water		MHWS	MHWN	MLWN	MLWS
0000	0600	0000	0600	7·0	5·5	2·7	1·1
1200	1800	1200	1800				
Differences ROSCOFF							
+0055	+0105	+0115	+0055	+1·9	+1·6	+0·7	+0·2
ÎLE DE BATZ							
+0045	+0100	+0105	+0055	+1·9	+1·5	+0·8	+0·3
BRIGNOGAN (48°40'·6N 04°19'·2W)							
+0040	+0045	+0100	+0040	+1·4	+1·1	+0·5	+0·1

SHELTER Good in drying Vieux Port except in strong N/E'lies. 12 W ⚓s, €8·50 all LOA, in 4-5m close W of Ar Chaden are exposed. Shelter very good in large modern marina.

NAVIGATION Roscoff Marina (S of ferry port): Appr H24 on 205° in W sector of mole hd lt. Avoid ferries; do not attempt to enter marina from south. Strong tide in marina, close to E bkwtr. **Vieux Port**: From the N, WPT 48°45'·88N 03°56'·00W, 213°/2·5M to Men Guen Bras lt. From the E approach between Plateau de la Méloine and Plateau des Duons; or pass S of the latter and N of Le Menk bcn lt twr in Ar Chaden's W sector (289·5°-293°). From the W, WPT 48°44'·27N 04°04'·06W, 090°/1M to Basse Plate bcn twr; thence via well-marked Canal de l'Île de Batz; see Passage information. Approach near HW over many large drying rks on 209° ldg line: see Lights, buoys and waypoints. N mole hd has B/W vert stripes.

LIGHTS AND MARKS See chartlet and Lights, buoys and waypoints. Île de Batz lt ho is conspic through 360°. E ent to Canal de l'Île de Batz flanked by bcn twrs: Ar Chaden YB and Men-Guen-Bras BY.

COMMUNICATIONS CROSS 02·98·89·31·31; ⑬ 06·82·18·01·34; Auto 08·92·68·08·29; Police 02·98·69·00·48; ⊖ (Bloscon) 02·98·69·70·84; ⊖ (Roscoff) 02·98·69·70·15; Brit Consul 01·44·51·31·00; Ⓗ 02·98·88·40·22; Dr 02·98·69·71·18; Roscoff HM 02·98·69·76·37, mob 06·70·50·96·68, VHF Ch 09; *Bloscon* VHF Ch 09, (0700-2200) all vessels must call for authorisation to enter marina, only authorised vessels to proceed on IPTS GWG vert, no vessels to leave, IPTS 3R vert. Aff Mar 02·98·69·70·15.

FACILITIES
Vieux Port, S Basin 5m. Dry out with fender board against rough inner jetty or quay. 280+20❂ €0·73. ⓛ �🚿 ⚒ ⒷⒹ ✕. N Basin for FVs, ⛴(5t). **CN de Roscoff** ☎02·98·69·72·79, ⌸.

Bloscon ferry terminal. HM ☎02·98·61·27·84, www.roscoff.fr. Adjacent ⚓age may be used only with HM's approval.

Roscoff Marina ☎02·98·79·79·49, www.plaisancebaiedemorlaix. com plaisance.roscoff@morlaix.cci.fr 625 + 40❂Jun-Sep, €2·80, on S side pontoon B, N side pontoon D. Convenient lift from pontoons to quayside level. 🛢🚿 (H24) ✕ ⌸ 🛒 ⌾ ⚒ ⛴ ⓛ ◪(50t) ⛴(2t). Shuttle bus to town, June-Aug, 0700–1700, or 10 min walk.

Roscoff Town ✕⌸🛒ⓛ🛢⊠Ⓑ✎ⒷⒹ✕. ⇌✈ Brest, Morlaix. Ferries to Plymouth, Rosslare, Cork.

OTHER HARBOURS AND ANCHORAGES WEST OF ROSCOFF

ÎLE DE BATZ, Finistère, **48°44'·26N 04°00'·60W** (Porz-Kernok ent). AC 2026, 2745; SHOM 7151, 7095; HW +0610 Dover (UT); ML 5·2m. See Roscoff. Good shelter but hbr dries up to 5m; ⚓ where space permits. E landing stage is for ferries only. Charted ⚓ in the chan E of hbr ent, but poor holding and may be exposed. ⚓ prohib between SE tip of Batz and Roscoff LW landing jetty, due to cables. Île de Batz lt ho (gy twr, B top) the CG Stn and church spire are all conspic. Facilities: R, basic shops, bike hire.

MOGUÉRIEC, Finistère, **48°41'·35N 04°04'·41W**, AC 2648, 2026; SHOM 7151. Tides approx as Île de Batz. A small drying fishing harbourr, 3M SSW of W entrance to Canal de l'Île de Batz, open to NW swell. Ldg lts 162°, both W towers/G tops: Front on jetty, Iso WG 4s 9m, W158°-166°; rear Iso G 4s 22m. Beware Méan Névez rk, dries 3·3m, to W of approach. ⚓ close SW of Île de Siec, or 3ca

WSW of Ar Skeul WCM bn tr in 4m, or 2ca NNE of Moguériec's drying jetty. ⚓ against jetty, but clear of FVs. ⚓ also possible against jetty at Île de Siec's drying hbr, but rky bottom. Facilities: ✕⌸🛒.

BRIGNOGAN (Pontusval), Finistère, **48°40'·59N 04°19'·17W**, about 10M ENE of Île Vierge lt ho. AC 2025; SHOM 7150. HW +0605 on Dover (UT); ML 4·7m; Duration 0600; see Roscoff.

Pontusval is the port, Brignogan a small resort. App from ECM buoy (48°41'·42N 04°19'·33W) via 178° ldg marks, W bn on with Plounéour ch spire 1M S. Ent between Ar Neudenn R bn tr to E and 3 white-topped rks to W. ⚓ here in approx 4m or dry out closer in. Hbr is open to N winds and often full of FVs. Entry at night prohib as it is unlit. Pte de Beg Pol, Oc (3) WR 12s, 1M to the west. Facilities: ⚓✕⌸🛒 5 ⚓'s (free).

9.20.20 L'ABER WRAC'H

Finistère **48°35'·97N 04°33'·62W** (marina ent) 🌸🌸⚓🌸🌸🌸🌸

CHARTS AC 2025, 1432; SHOM 7150, 7094; Navi 539; Imray C35

TIDES +0547 Dover; ML 4·5; Duration 0600

Standard Port BREST (→)

Times				Height (metres)			
High Water		Low Water		MHWS	MHWN	MLWN	MLWS
0000	0600	0000	0600	7·0	5·5	2·7	1·1
1200	1800	1200	1800				
Differences L'ABER WRAC'H, ÎLE CÉZON							
+0030	+0030	+0040	+0035	+0·7	+0·6	+0·1	−0·1

SHELTER Good in marina, but ⚓s uncomfortable in strong NW wind. No ⚓ off the marina. Paluden, 1·5M upriver (unlit/unbuoyed; beware oyster beds) is well sheltered in all winds.

NAVIGATION WPT (Grand Chenal) 48°37'·35N 04°38'·51W, abm Libenter WCM buoy, 100°/1·5M to Petit Pot de Beurre ECM bcn twr. Here pick up the inner appr chan 128°/1·7M, in W sector of Dir Lt Oc (2) WRG 6s, passing a SHM lt buoy, SHM bcn lt twr and a PHM lt buoy, to the last SHM bcn twr; thence visual pilotage for 700m to the pool/pontoon at L'Aber Wrac'h. Caution: large unlit mooring buoys.

Chenal de la Malouine is a narrow short cut from/to N & E, only by day and in good weather: 176·2° transit of Petit Pot de Beurre with Petite Île W obelisk; precise tracking is required. Beware breakers and cross-tides.

Chenal de la Pendante 135·7° is rarely used as it crosses drying rks, saves few miles and the unlit marks are hard to see. Front, B/W 'smiley' on Île Cézon; rear, W obelisk/orange top, 650m S of hbr.

Caution: SE of marina entrance, rocks drying 0·5m are marked by 3 Y SPM buoys outside and 5 inside the wavebreak. Inside the marina 2 PHM perches define the associated No Entry area. Speed limits 5kn in chan, 3kn in marina. Crowded at peak times.

LIGHTS AND MARKS See chartlets and Lights, buoys and waypoints. Grand Chenal 100·1° ldg marks/lts: front on Île Wrac' h, W □ twr orange top, QR; rear, 1·63M E at Lanvaon, W □ twr orange △ top, Q. A Dir lt Oc (2) WRG 6s 128° and two hard-to-see ldg twrs, both W with R tops, mark the inner chan. QG and QR lts mark the marina ent.

COMMUNICATIONS CROSS 02·98·89·31·31; Auto 08·92·68·02·29; Police 02·98·04·00·18 or just 17; ⊜ 02·98·85·07·40; Brit Consul 01·44·51·31·00; Ⓗ 02·98·22·33·33; Dr 02·98·04·91·87; HM 02·98·04·91·62, VHF Ch 09, 16 (0800-2000LT in season); Aff Mar 02·98·04·90·13.

FACILITIES

Marina 70 Ⓥ ⌷ on inside of wavebreak (and outside in good weather) (see inset) Jun-Sep, €2·76; 20 numbered W ⚓s €2·17 (+ €0.22pp/day tax) multihulls +150%. www.port-aberwrach.com aberwrach@port.cci-brest.fr. 🅽 🅐 (H24 with cr/card) ⚓ 🔌 🔧 🅱 ✗ Ⓔ 🛢(15t) 🛢(1t) ⌐. **YC des Abers** ☎02·98·04·92·60, 🗆 🏴 ⚒, Ⓞ.

L'Aberwrac'h ✗ 🗆 🛢.

Landéda (Jul/Aug bus €1·00 rtn: M-Sat 11 times/day; Sun 3 times/day; or 20 mins up-hill walk): 🛒🅿🛈✉ Ⓑ a.m. Tues-Fri. Bus to Brest ⇌ & ✈. Ferry Roscoff.

Paluden. HM ☎02·98·04·63·12. 20 dumbell ⚓s in 5m ✗ 🗆 🏴.

9.20.21   L'ABER BENOIT

Finistère **48°34'·64N 04°36'·89W** ❄⊛♨♨❁❁❁

CHARTS AC 2025, 1432; SHOM 7150, 7094; Navi 539, 540; Imray C35

TIDES +0535 Dover; ML 4·7; Duration 0555

Standard Port BREST (→)

Times				Height (metres)			
High Water		Low Water		MHWS	MHWN	MLWN	MLWS
0000	0600	0000	0600	7·0	5·5	2·7	1·1
1200	1800	1200	1800				
Differences L'ABER BENOIT							
+0022	+0025	+0035	+0020	+0·9	+0·8	+0·3	+0·1
PORTSALL							
+0015	+0020	+0025	+0015	+0·5	+0·4	0·0	−0·1

SHELTER Excellent, but do not enter at night, in poor vis nor in strong WNW winds; best near LW when dangers can be seen.

NAVIGATION WPT 48°37'·00N 04°38'·84W, close W of Petite Fourche WCM buoy; thence follow the tracks shown on chartlet. Beware oyster beds. Tidal streams up to 3kn at half-tide. River is navigable on the tide to Tréglonou bridge 3M upstream.

LIGHTS AND MARKS Unlit. Vital to identify correctly buoys and beacons in the approach chan.

COMMUNICATIONS CROSS/ⓁⒷ 02·98·89·31·31; Auto 08·92·68·08·29; Police 02·98·48·10·10 or 17; ⊖ 02·98·85·07·40; British Consul 01·44·51·31·00; HM VHF Ch 09, 06·19·87·75·39 (0830–1230, 1500–1830 Jul–Aug).

FACILITIES Many moorings in lower reach, call HM who will allocate a vacant buoy, Jul–Aug, €1·50, otherwise half price. ⚓ poss upriver nps. ►♠⚓⚒✕(☎02·98·89·86·55) ✎ Ⓑ. **Village** Ⓟ ⓉⒹ(☎02·98·89·87·06) ⬛✉ Ⓑ. Bus to Brest ⇌ ✈. Ferry: Roscoff.

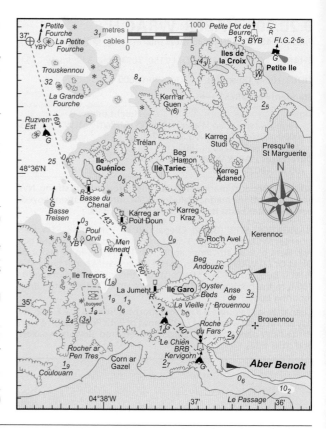

THE PASSAGE INSHORE OF ROCHES DE PORTSALL AND ROCHES D'ARGENTON

This 8·5M passage is only 1M shorter than the offshore passage from L'Aber Wrac'h towards Le Four lt ho. It requires about 6M visibility in daylight (to see the more distant marks) and is navigationally challenging because of strongish tidal streams and the risk of mis-identifying the marks. A well illustrated Pilot and detailed study of AC 1432, or SHOM 7094, are essential.

The passage from NE to SW is summarised in 3 main sections:

- 7 ca SW of Libenter WCM buoy (see L'Aber Wrac'h) enter Ch du Relec 218·5°: Pte de Landunvez W bcn twr, R top (rear mark, 6M distant) just open left of Petit Men Louet, W bcn twr 4M distant. Leave Le Relec ECM buoy (1·5M distant) close to stbd.

- With Corn Carhai lt ho almost abeam to stbd, alter stbd into Ch du Raous: Bosven Kreiz W bcn twr in transit 249° with rks S of Le Gremm, a 12m high rk. Hold this for 7 ca then alter port 228°: Le Four lt ho open left of Bosven Aval W bcn twr. 7ca later jink 7° to port to clear Bosven Aval; then pick up:

- The astern transit 036° of Bosven Aval with Bosven Kreiz. After 1·5M (Pte de Landunvez to port) intercept Chenal Méridional de Portsall, another astern transit 049°: the saddle of Le Yurc'h (7m high rk) on with Grand Men Louet W bcn twr. After 2·2M pass SE of Le Four lt ho into the Ch du Four.

HARBOURS ADJACENT TO THE INSHORE PASSAGE

PORTSALL, Finistère, **48°33'·79N 04°43'·04W**. AC 3688, 1432; SHOM 7150, 7094. HW +0535 on Dover (UT); ML 4·4m; Duration 0600. See L'Aber Benoit.

Small drying hbr at head of bay. Good shelter except in strong NW'lies. Outer 109° ldg marks: Le Yurc'h rk (7m) on with Ploudalmézeau spire. Inner 085° ldg marks (W cols; rear with R top). Front col has Oc (4) WRG 12s, W sector 084°–088°. Appr marked by 3 W bn trs to port, 1 SHM and 1 NCM bcn twr. ⚓ in >10m to W of ent, or enter inner hbr to berth on quay. **Tel/Facilities:** ⓁⒷ☎02·98·48·69·17; Police 02·98·48·10·10; Dr ☎02·98·48·75·52; Aff Mar ☎02·98·48·66·54; Taxi ☎02·98·48·68·48. **Quay** ◔(0·5t) ⬛♠⚓►; CN Portsall Kersaint ☎02·98·48·77·49; **Ancre an Eor** ☎02·98·48·73·19, Ⓛ. **Town** ✕Ⓓ🍴✉.

ARGENTON, Finistère, **48°31'·26N 04°46'·34W**. AC 1432, 3345; SHOM 7150, 7122. HW +0535 on Dover (UT); ML 4·6m; Duration 0600; use PORTSALL diffs L'Aber Benoit.

Small hbr drying approx 5·5m; good shelter except in W winds when swell enters. WPT 48°31'·14N 04°48'·29W (2½ca S of Le Four lt ho) 084·6°/1·4M to first of 2 W bcn twrs and a white wall with R vert stripe, all on Île Dolvez and in transit. A conspic water twr is slightly S of this ldg line and 1M inland. The N side of chan is marked by 3 PHM bcn twrs. Beware strong NE-SW tidal streams in the approach.

⚓ W of Île Dolvez in about 2m; or skirt round its N side to enter the inner hbr and dry out against the stone quay/slip on N side. 10 🅾s reported but not seen. Facilities: ⚓ ♠ on quay. **Nautisme en Pays d'Iroise** ☎02·98·89·54·04, Ⓚ. **Village** Ⓓ✕🍴.

9.20.22   L'ABER-ILDUT

Finistère **48°28'·24N 04°45'·75W** (1st PHM bcn) ❄⊛♨♨❁❁❁

CHARTS AC 2647, 2356, 3345; SHOM 7149, 7122; Navi 540; Imray C36

TIDES +0520 on Dover (UT); ML 4·2m

Standard Port BREST (→)

Times				Height (metres)			
High Water		Low Water		MHWS	MHWN	MLWN	MLWS
0000	0600	0000	0600	7·0	5·5	2·7	1·1
1200	1800	1200	1800				
Differences L'ABER-ILDUT							
+0010	+0010	+0020	+0010	+0·3	+0·2	−0·1	−0·1

SHELTER Good, but in the appr a high swell runs in strong W'lies; strong river current inside. In fair weather ⚓ outside, as chartlet.

NAVIGATION WPT 48°28'·07N 04°47'·95W, 083°/1·6M to Dir lt which is on conspic white gable end of house. Beware: Les Liniou rks 1·5M NNW of the WPT, Plateau des Fourches 1·2M S and strong cross tides in the approach. Drying rks are marked by Pierre de l'Aber SHM bcn and Le Lieu PHM bcn twr 5m; the latter is easier to see. At the narrow ent (2m) leave Men Tassin PHM bn and Rocher du Crapaud, a large rounded rock, 30m to port to clear drying spit on S side, marked by SHM buoy. Drop sails before entering under power alone. Care needed at LWS (max draft 2m) visit first in daylight. All quays dry at LW. A good hbr to await the tide in the Ch du Four; see Passage information, Chenal du Four.

LIGHTS AND MARKS See chartlet and Lights, buoys and waypoints. Glizit water twr is conspic 6·5ca N of the ent. Dir lt 083° is a good daymark (see above); its 2° R sector and Le Lieu bcn, Fl R 2·5s 4M, mark dangers on N side of chan, but do not take liberties with the S side. An inland uncharted water twr is almost on 083°. Lanildut and Brélès spires, in transit 078·5°, are small, very hard to see and obsc'd by trees when close in.

COMMUNICATIONS CROSS 02·98·89·31·31; 02·98·89·30·31; Meteo 08·92·68·08·08; Auto 08·36·68·08·08; Police 02·98·48·10·10 (or emergency 17); 02·98·44·35·20; Dr 02·98·04·33·08; HM VHF Ch 09, Jun-Sep in office hours; Aff Mar 02·98·48·66·54; Tourism 02·98·48·12·88.

FACILITIES HM ☎02·98·04·36·40 (season) mob 06·31·93·58·71. www.lanildut.fr port.aber-ildut@wanadoo.fr **Hbr** 410 M, inc 16 s; €2·00 (average) multihull +50%; max LOA 11m, draft 2m.1 x 13m. Raft on dumbells (V trot) just before 2nd PHM bcn or beyond FV quay (landing) on outer dumbells (G trot). No ⌀ inside hbr. ask at , ice R ⊠. at short stay (1hr max) on yacht pontoon close N of FV jetty. 1hr stay also poss at Porscave (opp bank) on E side of pontoon.

Village ⚓. Bus to Brest 25km. Ferries: Ouessant, Molène.

9.20.23 LE CONQUET

Finistère **48°21'·60N 04°47'·00W** (Hbr ent) ❀❀◊◊✿✿

CHARTS AC 2356, 3345 with 1:10,000 inset; SHOM 7149, 7148, 7122; Navi 540; Imray C36

TIDES +0535 Dover; ML 3·9; Duration 0600

Standard Port BREST (→)

Times				Height (metres)			
High Water		Low Water		MHWS	MHWN	MLWN	MLWS
0000	0600	0000	0600	7·0	5·5	2·7	1·1
1200	1800	1200	1800				
Differences LE CONQUET							
0000	0000	+0005	+0005	−0·2	−0·2	−0·1	0·0
LE TREZ HIR (Anse de Bertheaume)							
−0005	−0005	−0015	−0010	−0·4	−0·4	−0·2	−0·1

SHELTER Good except in strong W'lies. Nearest ⌀ is in Anse des Blancs Sablons, sheltered from W/SW but fair weather only.

NAVIGATION WPT 48°21'·68N 04°48'·50W, 096°/1M to Mole Ste Barbe lt. From the NW, beware Grande and Petite Vinotière rks. From the SSW keep seaward of Les Renards IDM buoy and R bcn twr closer inshore. In hbr ent keep closer to Mole Ste Barbe lt, to avoid La Basse du Filet (1·4m) on N side.

LIGHTS AND MARKS See chartlet and Lights, buoys and waypoints. Mole Ste-Barbe lt in line 096° with church spire leads to hbr ent. 079° transit of La Louve R bn tr and end of Mole St Christophe (inner mole) is only valid outside the hbr.

COMMUNICATIONS CROSS 02·98·89·31·31 and 1616; ☎ 02·98·84·60·64; Auto 08·92·68·08·29; Police 02·98·89·00·13; British Consul 01·44·51·31·00; Dr 02·98·89·01·86; *Le Conquet Port* (HM) VHF 08, 16; Aff Mar 02·98·89·00·05.

FACILITIES The Avant Port is reserved for FVs & workboats; ferries use Mole Ste Barbe; ⌀ prohib. Call the HM for a vacant mooring or dry out further ENE on one of 12 s; all LOA €7·00 per day. **HM** ☎02·98·89·18·58, mob 06·71·11·67·41 yann.lagadec@leconquet.fr. **Town** ⚓ Ⓑ. Bus to Brest. Ferries to: Molène, Brest and Ouessant.

ANCHORAGE EAST OF PTE DE ST MATHIEU
ANSE DE BERTHEAUME, 48°20'·44N 04°41'·84W. AC 2350, 3427. SHOM 7401. A useful passage ⌀ to await the tide E to Brest, N into Ch du Four, or S toward Raz de Sein. ⌀ in 3-5m between slip NNW of the Fort and an ECM bcn 3ca further N. Good shelter in W'lies. 1ca NE of Fort de Bertheaume (32m, conspic) Le Chat rk (6·8m) is marked by an ECM bcn. No lights.

CHENAL DU FOUR

The Chenal du Four (and de la Helle; AC 3345, 2356) are unhappily split between the small scale Admiralty tidal stream atlases for the English Channel (NP 250) and W France (NP 265). Both channels and the Raz de Sein are better covered at large scale by the French atlas (560-UJA: NW Brittany, Brignogan to Penmarc'h).

Using tides to advantage it is possible for a 5½kn boat to pass through the Chenal du Four and Raz de Sein in one leg from L'Aberwrac'h to Audierne or vice versa. Anse de Bertheaume (anchorage 4M east of Pte de Ste Mathieu) Camaret or Brest are alternative interim stops.

▶*1M W of Le Four the S-going stream begins around HW Brest (arrive a little early); the N-going stream at HW Brest –5¾, sp rates 3·6kn. Further S, off Pte de Corsen, the stream is weaker, max 2·3kn at springs.*

The tide runs strongest at S end of Chenal du Four, off Le Conquet. Here the S-going stream starts at HW Brest +¼, max 5kn; the N-going stream begins at HW Brest –5¾, 5·2kn max at sp. Wind-over-tide may raise many short, steep seas.◀

Pilotage. On the mainland coast are 6 major lights, either directional or leading. The charted transits are: St Mathieu directional lt and Kermorvan ≠ 158·5°; Pte de Corsen directional lt bearing 012° (in its white sector and on your port quarter). Thence pass between Rouget SHM lt buoy and Grande Vinotière lt bn. Maintain the 325·5° astern transit of Grand Courleau unlit NCM bcn twr ≠ Le Faix NCM lt bcn twr. Track 145·5° to pass midway between La Fourmi SHM lt buoy and Vieux-Moines lt bcn twr.

Alternatively a simple buoy-hopping sequence (below) can be used by day or night. The 3 legs are short and the speed over the ground is likely to be quite high, but other buoys between the waypoints provide a useful check on progress and any deviation from the planned track.

Leg 1:	From ⊕ 48°31'·5N 04°49'·0W (5ca W of Le Four) track 188°/5M (leaving Les Liniou well to port) to Valbelle PHM buoy.
Leg 2:	Thence track 168°/4·6M to abeam Grande Vinotière lt bn.
Leg 3:	Finally track 172°/2·7M to La Fourmi SHM lt buoy (for Raz de Sein); or
Leg 3A:	156°/2·9M to Vieux-Moines lt bcn (for Brest/Camaret).

Homeward-bound, or along the N Brittany coast, enter the S end of Chenal du Four at LW Brest; a fair tide can then be carried through the channel and NE past Île Vierge. The reverse sequence of pilotage is followed. Two hours is par for the 12M passage.

L'Aber-Ildut, 3·5M SSE of Le Four lt ho, and **Le Conquet**, 3ca SE of Pte de Kermorvan, are the only ports on the mainland coast: the former is safe if there is no swell or W'lies; the latter has limited facilities for yachts but in offshore winds there are anchorages in Anse de Porsmoguer and Anse des Blancs-Sablons, both between Corsen and Kermorvan.

CHENAL DE LA HELLE

▶*At N end of Chenal de la Helle the ENE-going stream starts at HW Brest –0520 (sp rate 2·8kn), and the SW-going stream at HW Brest –0045 (sp rate 3·8kn).*◀

The Chenal de la Helle converges at a 20° angle with the Chenal du Four and the two meet WSW of Pte de Corsen or just N of Grande Vinotière, depending on which of two track variants you take (AC 3345, 2356).

A simple GPS-based buoy-hopping sequence is as follows:

From close abeam Luronne WCM buoy track 137°/4·7M to St Pierre SHM buoy, passing Le Faix lt twr and Le Pourceau NCM lt buoy. Thence 169°/3·9M to La Fourmi SHM buoy, passing Rouget SHM lt buoy, Grande Vinotière and Tournant et Lochrist PHM lt buoy en route.

APPROACHES TO BREST

The outer approaches lie between Chaussée des Pierres Noires and Pte St Mathieu to the N and Pte du Toulinguet to the S (AC 2350, 3427, 3428). From the W maintain the 068° transit of Petit-Minou and Portzic ldg lts on the N shore; or keep slightly north of this transit to avoid large ships. From the S steer NNE towards Pte du Petit-Minou to pick up the transit, but beware rks 7M W and SW of Pte du Toulinguet. Yachts <25m LOA are exempt from VTS, but should monitor VHF Ch 08 or 16.

Abeam Petit-Minou lt ho the Goulet (Straits) de Brest narrows to 1M. Drying rocks almost in mid-stream may be left to port or starboard; they are marked by a WCM lt buoy, 2 PHM lt buoys and Roche Mengam IDM lt bn. A course of 075° through the Passe Nord leaves the latter 2ca to stbd.

▶*In Passe Sud a useful back-eddy close inshore sets ENE during the ebb. Tidal streams reach 4·5kn in the Goulet.*◀

Beyond Pte du Portzic a buoyed chan leads ENE to **Brest** and the marinas at Le Château and Moulin Blanc. In the **Rade de Brest** (AC 3429) around Île Longue (a peninsula) are DG ranges, no-anchoring and no-entry areas, but there is room to anchor off Roscanvel in the lee of the Quélern peninsula. Further east the beautiful River Aulne can be explored well inland to Port Launay and as far as **Chateaulin**.

L'IROISE AND BAIE DE DOUARNENEZ

L'Iroise is the sea area SW of Chaussée des Pierres Noires and N of Chaussée de Sein (AC 2350, 2349). On the NE side of L'Iroise (AC 3427) a chain of rks extends 7M W from Pte du Toulinguet. There are several chans through these rocks, of which the simplest for Brest and Camaret is the 3ca wide **Chenal du Toulinguet** which runs NNW/SSE between La Louve WCM bcn twr (1ca W of Pte du Toulinguet) on E side and Le Pohen rk on the W side.

▶*Here the N-going stream begins at HW Brest –0550, and the S-going at HW Brest +0015, sp rates 2·75kn.*◀

3·5M SSE of Pte du Toulinguet is Le Chevreau (WCM bcn; dries 5·9m) with La Chèvre 5ca NE of it. 1·9M S of Le Chevreau is Le Bouc (WCM lt buoy; dries 7·4m). 5·7M SW of Pte du Toulinguet lies Basse du Lis, rky shoals with depth of 2·7m and WCM lt buoy, and the last of a string of underwater rocks extending seaward from Les Tas de Pois (Pile of Peas). Close E of Les Tas de Pois, the Anse de Pen-Hir is a useful anch in NE'lies.

The Baie de Douarnenez is entered between C. de la Chèvre and Pte du Van. Off C. de la Chèvre various dangers, on which the sea breaks, extend SW for 2·25M to Basse Vieille (dries 0·8m) lt buoy. Basse Laye (dries 0·7m) is unmarked 7ca SSE of the CG stn on C. de la Chèvre and a hazard if rounding the headland close inshore. **Morgat** lies 4M NNE of C. de la Chèvre. 2·5M ESE of Morgat beware Les Verrès (12m high rk) and a group of drying rks, including La Pierre-Profonde and Le Taureau.

Approaching **Douarnenez** beware Basse Veur and Basse Neuve (2·2m). The S shore of the B is clear of dangers more than 2ca offshore, except for Duellou Rk (7m high) 5ca offshore, and other rks 1M eastward. Further W Basse Jaune is an isolated rk (dries 1·4m; IDM buoy) about 1M N of Pte du Van. This rocky headland with offliers is 2·25M NE of Pte du Raz and La Vieille lt twr and is the NE corner of the Raz de Sein.

RAZ DE SEIN TO POINTE DE PENMARC'H

Chaussée de Sein (AC 2350, 2819, 2348) is a chain of islands, rks and shoals extending 15M W of the Pointe du Raz. A WCM lt buoy marks the seaward end. See section on **Raz de Sein** for directions through it. **Audierne** is 10M ESE where **Ste Evette** is a good passage ⚓ before or after tackling the Raz.

Off Pte de Penmarc'h (light house, fog signal) depths shoal very rapidly from >60m to <5m in the space of 7ca; dangers extend 1M to NW, W and S, and 3M to SE, with breaking seas in onshore winds; keep at least 3M off the light house. Le Guilvinec, a busy fishing harbour is well sheltered, but has a difficult entrance. If weather permits, Lesconil, a former fishing port, now welcomes leisure craft Jun-Sept, with much improved facilities.

9.20.24 CHENAL DU FOUR

AC 3345. Simple buoy-hopping/GPS track: _ _ _ _ _ _ _ _   *See Passage information for notes on this and Chenal de la Helle.*

9.20.25 ÎLE D'OUESSANT

Finistère **48°26'·64N 05°07'·49W** (B de Lampaul) ✸✸♦✿✿✿

CHARTS AC 2356; SHOM 7149, 7123; Navi 540; Imray C36; Stan 2

TIDES +0522 Dover; ML 3·9; Duration 0555
Standard Port BREST (→)

Times				Height (metres)			
High Water		Low Water		MHWS	MHWN	MLWN	MLWS
0000	0600	0000	0600	7·0	5·5	2·7	1·1
1200	1800	1200	1800				
Differences BAIE DE LAMPAUL							
+0010	+0010	−0005	+0005	−0·1	−0·1	−0·1	−0·1
ÎLE DE MOLENE							
+0015	+0010	+0020	+0020	+0·3	+0·3	+0·1	+0·1

SHELTER Good in Baie de Lampaul in N-E winds, and in Baie du Stiff in S-NW winds; fresh/strong winds from the opposite directions render the bays untenable. For a first visit settled weather, no swell and good vis (often poor in July) are ideal.

NAVIGATION See Area map for notes on Ouessant TSS.

Lampaul WPT 48°25'·38N 05°09'·96W, 054·4°/2·4M to LH edge of Youc'h Korz (a 28m high rock which may be passed on either side) in transit with Le Stiff lt ho. Do not attempt Chenal de la Fourche, a short cut inside La Jument lt ho.

Baie du Stiff WPT 48°28'·38N 05°01'·32W, 259°/1·3M to Dir lt.

LIGHTS AND MARKS The radar twr (132m) at Le Stiff is highly conspic. See Lights, buoys and waypoints for La Jument, Nividic, Creac'h and Le Stiff lt ho's; also Gorle Vihan IDM bcn twr in Baie du Stiff, where a Dir lt Q WRG marks the 259° appr for ferries; Men-Korn bcn twr (lit) is 1.3M E. Kéréon lt ho marks the S side of Passage du Fromveur, although has been reported unreliable.

COMMUNICATIONS CROSS 02·98·89·31·31 & 1616; ℻ 02·98·48·84·33; Meteo 08·92·68·08·08; Auto 08·92·68·08·29; Police (summer only) 02·98·48·81·61; Brit Consul 01·44·51·31·00; Dr 02·98·48·83·22; HM 02·98·48·80·06; Aff Mar 02·98·89·00·05; Tourist Office 02·98·48·85·83.

FACILITIES Lampaul has room to ‡ in 6-10m and 29 free ⚓s SE of the PHM/SHM bcn twrs off the narrow drying hbr (not usually accessible to visitors). Ferries no longer use Lampaul. ⏚ ⏛. **Village** ✕ ⌂ ⊡ ▮ ⊠ Ⓑ @ bike hire ✈.

Baie du Stiff has 4 W ⚓s (free) in 5m on S side at 48°28'·03N 05°03'·11W, but there is little space to ‡ and holding is poor. S of the lt ho, Porz Liboudou is for ferries to Le Conquet and Brest, berthing on either jetty; keep clear.

ADJACENT ISLAND HARBOUR

ÎLE MOLÈNE, Finistère, **48°24'·07N 04°57'·30W** (🅛🅑 mooring).

CHARTS AC 2356, 3345; SHOM 7149, 7148, 7122, 7123; Navi 540; Imray C36

TIDES See Île d'Ouessant; +0520 Dover; ML 4·6

SHELTER Good, except in strong N/NE'lies; streams are strong. ‡ near 🅛🅑 in about 1·2m. Appr is easier in settled weather, good vis and nps.

NAVIGATION The simpler appr is from the NE, near Luronne WCM buoy, to WPT 48°25'·00N 04°56'·97W; thence 191°/8·5ca to the 🅛🅑 mooring (a good ref point in the hbr) passing between PHM buoy QR and SHM buoy Iso G; see inset on Chenal du Four.

From the E (Chenal des Laz) WPT 48°24'·35N 04°51'·30W (650m N of Pourceaux NCM lt buoy) 264°/2·7M to 48°24'·07N 04°55'·41W. Here make good 315°/1·1M to skirt round Les Trois Pierres lt twr and continue via the N ⊕. This appr is used by mainland ferries.

Day transits define both appr chans: from the N, Moulin Sud and white mark on N jetty ≠ 190°. From the E, Moulin Nord (W bcn, orange top) and white twr ≠ 264°.

LIGHTS AND MARKS The church spire and old CG twr are both conspic from afar. At old S pier the two lts (on same column) are: Fl (3) WRG 12s whose W sector covers the 191° appr; and Fl (2) WRG 6s whose W sector covers the Chenal des Laz 261°.

COMMUNICATIONS HM, via Mairie 02·98·07·39·05.

FACILITIES 12 white ⚓s, €12·00 all LOA. Ferries berth on N jetty. ⏚ ✕ ⌂ 🛒 ⚓ ⊠; ⚑ near museum. www.molene.fr

9.20.26 RADE DE BREST

Finistère **48°20′·75N 04°32′·10W** (Goulet de Brest, E end)

CHARTS AC 3427, 3428, 3429; SHOM 7401, 7400, 7397, 7398, 7399; Navi 542; Imray C36. Beyond the N and E edges of AC 3429 & SHOM 7400, the Rivers Elorn and Aulne are usefully covered on IGN land maps 0516, 0517 & 0518 (1:25,000). These maps show some depths, Lat/Long values, much valuable topographical data and are readily available.

TIDES
Standard Port BREST (→)

Times				Height (metres)			
High Water		Low Water		MHWS	MHWN	MLWN	MLWS
0000	0600	0000	0600	7·0	5·5	2·7	1·1
1200	1800	1200	1800				

Differences TINDUFF (48°20′·2N 04°21′·9W)
| +0005 | +0005 | +0005 | +0005 | +0·3 | +0·2 | +0·1 | +0·1 |

PONT DE TERENEZ (48°16′·1N 04°15′·7W)
| +0005 | +0005 | +0010 | +0005 | +0·4 | +0·3 | +0·2 | +0·2 |

In most of the bays in the Rade de Brest tidal streams are weak, but in the main rivers they can exceed 2kn at sp.

SHELTER The Rade (50 sq miles) is a sheltered cruising ground when the weather offshore is bad. There are attractive ⚓s in the SE part of the Rade and up the Rivers Elorn and Aulne.

NAVIGATION WPT 48°18′·20N 04°44′·26W, 068°/5·3M to front ldg lt (Pte du Petit Minou); thence through the Goulet de Brest. See Lights, buoys & waypoints for light details. Tidal streams run hard, max 4·6kn on the spring ebb and 3·5kn on the flood.

Pass either side of Plateau des Fillettes, rks well marked by 3 lt buoys and Mengam IDM lt bcn twr. Keep well inshore to avoid warships and commercial vessels and to cheat any foul tide.

The area S and SW from the Moulin Blanc Marina is very shoal, and yachts should head WSW from the marina to R1 buoy and Basse du Renard WCM buoy or Pte de l'Armorique.

In the SW part of Rade de Brest are various naval sites with no ⚓, prohib zones and DG ranges around Île Longue. There are however ⚓s at Le Fret (SE of Île Longue) whence a ferry runs to Brest; and off Roscanvel in the lee of the Quélern Peninsula. There are 2 free ⚓s at both Roscanvel and Quélern and 3 white ⚓s, €10·00, at Lanveoc, marked 'Accuil'.

E of Pte de l'Armorique there are ⚓s and W ⚓s in 3m in Anse de L'Auberlac'h, Baie de Daoulas and its inlets/creeks (very shoal) at the mouth of the River de l'Hôpital and up the drying/buoyed River de Faou. S of here the much larger River Aulne (see below) can easily be explored 18M upriver from Landévennec to Châteaulin.

RIVERS FLOWING INTO THE RADE DE BREST

RIVER ÉLORN, entered 1·2M E of Moulin Blanc Marina, is navigable to Landerneau, 6M upriver. From Pont Albert-Louppe (29m) and Pont de l'Iroise (25m clearance) the river is buoyed/lit for 2M. Further up it is marked by smaller unlit buoys and perches, some of which may well be some distance outboard of navigable water. Good ⚓s are at Le Passage, 8ca above the bridges, or further NE in Anse St Nicolas.

The drying port of Landerneau can be reached on the tide via a lifting bridge, ☎06·11·03·31·20. Dry out against the N quay. Few nautical facilities, but a pleasant town.

RIVER AULNE To the SE of Rade de Brest, L'Aulne is a lovely river with steep wooded banks. There are ⚓s off Landévennec, below both Térénez bridges (27m) and also ⚓s 1½M above the E bridge. Charted depths are optimistic.

At Guily-Glaz, 14M above Landévennec, a lock (operating HW Brest −2 to +1½, but not between 2100 and 0730; ☎02·98·86·03·21) gives access to the canalised river (2·7m) and Port Launay on a U-bend. Port Launay is a popular place to winter against the attractive stone quays in complete shelter. Facilities: ⌷ on quay, ⬡ (apply to Town Hall, weekdays only before 1700 ☎02·98·86·09·63) ⚓ ✕. Easy access to the N165 motorway.

2·25M further on is Châteaulin, a charming country town and head of the navigable river. Beware shallow spots mid-river. Facilities: small **Ⓥ** pontoon, no ⬡ ☎02·98·86·10·05; ⚓ ⛟ 100m from quay.

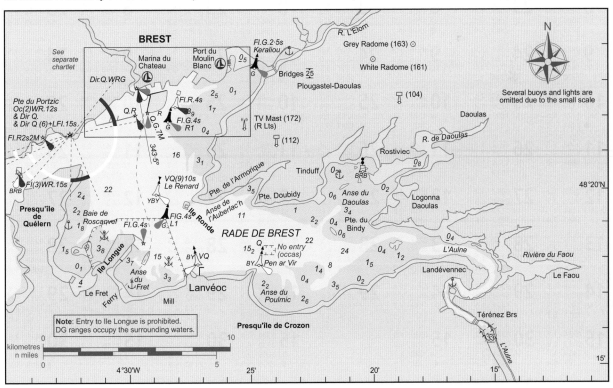

STANDARD TIME UT –01
Subtract 1 hour for UT
For French Summer Time add
ONE hour in **non-shaded areas**

BREST   LAT 48°23'N   LONG 4°30'W

TIMES AND HEIGHTS OF HIGH AND LOW WATERS

Dates in red are **SPRINGS**
Dates in blue are **NEAPS**

YEAR 2016

JANUARY

Day	Time m	Time m	Day	Time m	Time m
1 F	0321 2.5	0916 5.8	16 SA	0324 1.8	0927 6.5
	1551 2.6	2148 5.6		1553 1.8	2156 6.1
2 SA	0408 2.8	1007 5.5	17 SU	0421 2.1	1026 6.2
	1644 2.8	2246 5.3		1653 2.1	2301 5.9
3 SU	0506 3.0	1112 5.3	18 M	0527 2.3	1136 5.9
	1746 2.9	2354 5.3		1803 2.3	
4 M	0614 3.0	1224 5.3	19 TU	0016 5.8	0642 2.4
	1853 2.9			1254 5.9	1918 2.3
5 TU	0102 5.4	0723 2.9	20 W	0132 6.0	0757 2.2
	1331 5.5	1956 2.7		1408 6.1	2028 2.1
6 W	0203 5.7	0823 2.6	21 TH	0239 6.2	0903 1.9
	1428 5.8	2049 2.4		1510 6.3	2127 1.8
7 TH	0254 6.0	0914 2.3	22 F	0335 6.6	0959 1.6
	1517 6.1	2136 2.1		1601 6.6	2218 1.6
8 F	0339 6.4	0959 1.9	23 SA	0423 6.8	1047 1.3
	1600 6.4	2219 1.8		1647 6.8	2303 1.4
9 SA	0421 6.7	1041 1.6	24 SU	0506 7.0	1130 1.2
	1641 6.6	2300 1.5		1727 6.9	2344 1.3
10 SU	0502 6.9	1122 1.3	25 M	0545 7.1	1209 1.2
	1721 6.8	2341 1.3		1805 6.9	
11 M	0542 7.1	1203 1.1	26 TU	0022 1.3	0621 7.1
	1801 7.0			1246 1.2	1839 6.8
12 TU	0022 1.3	0623 7.2	27 W	0057 1.4	0654 6.9
	1244 1.0	1843 7.0		1320 1.4	1912 6.6
13 W	0104 1.2	0705 7.2	28 TH	0132 1.6	0727 6.7
	1327 1.1	1926 6.9		1354 1.7	1945 6.4
14 TH	0148 1.3	0749 7.1	29 F	0206 1.8	0759 6.4
	1411 1.2	2011 6.7		1428 2.0	2019 6.1
15 F	0234 1.5	0835 6.8	30 SA	0241 2.2	0833 6.1
	1459 1.5	2101 6.4		1504 2.3	2056 5.8
			31 SU	0320 2.5	0912 5.7
				1545 2.6	2142 5.5

FEBRUARY

Day	Time m	Time m	Day	Time m	Time m
1 M	0406 2.8	1002 5.4	16 TU	0502 2.3	1109 5.8
	1637 2.9	2243 5.3		1737 2.5	2349 5.7
2 TU	0505 3.0	1111 5.2	17 W	0618 2.5	1234 5.6
	1745 3.0	2358 5.2		1857 2.6	
3 W	0620 3.0	1233 5.2	18 TH	0114 5.7	0742 2.4
	1902 3.0			1356 5.7	2014 2.4
4 TH	0114 5.4	0736 2.8	19 F	0227 6.0	0853 2.1
	1348 5.4	2011 2.7		1500 6.1	2115 2.0
5 F	0219 5.7	0841 2.4	20 SA	0323 6.3	0948 1.8
	1448 5.8	2107 2.3		1550 6.4	2205 1.7
6 SA	0313 6.2	0933 2.0	21 SU	0410 6.6	1034 1.5
	1537 6.2	2155 1.8		1632 6.6	2248 1.5
7 SU	0359 6.6	1019 1.5	22 M	0450 6.8	1114 1.3
	1622 6.6	2240 1.4		1709 6.8	2325 1.3
8 M	0443 7.0	1103 1.1	23 TU	0526 7.0	1149 1.2
	1704 7.0	2323 1.1		1743 6.9	
9 TU	0526 7.3	1146 0.8	24 W	0000 1.3	0558 7.0
	1746 7.2			1221 1.2	1814 6.9
10 W	0006 0.9	0608 7.5	25 TH	0032 1.3	0628 6.9
	1228 0.6	1828 7.3		1252 1.3	1843 6.8
11 TH	0049 0.8	0650 7.6	26 F	0104 1.4	0658 6.8
	1311 0.6	1910 7.3		1323 1.5	1913 6.6
12 F	0132 0.8	0733 7.5	27 SA	0135 1.6	0728 6.6
	1354 0.8	1953 7.1		1353 1.8	1944 6.3
13 SA	0217 1.1	0817 7.1	28 SU	0207 1.9	0758 6.3
	1440 1.2	2039 6.7		1425 2.1	2016 6.0
14 SU	0305 1.4	0905 6.7	29 M	0242 2.2	0832 5.9
	1530 1.7	2131 6.3		1500 2.4	2054 5.7
15 M	0359 1.9	1000 6.2			
	1628 2.1	2232 5.9			

MARCH

Day	Time m	Time m	Day	Time m	Time m
1 TU	0322 2.5	0913 5.6	16 W	0440 2.3	1047 5.6
	1544 2.8	2145 5.4		1713 2.6	2325 5.6
2 W	0414 2.8	1011 5.3	17 TH	0557 2.6	1213 5.6
	1644 3.0	2258 5.2		1836 2.7	
3 TH	0522 3.0	1135 5.1	18 F	0053 5.5	0723 2.6
	1805 3.1			1339 5.5	1956 2.5
4 F	0024 5.3	0646 2.9	19 SA	0208 5.8	0835 2.3
	1304 5.3	1929 2.8		1443 5.9	2057 2.2
5 SA	0141 5.6	0803 2.5	20 SU	0304 6.1	0929 1.9
	1416 5.7	2036 2.4		1531 6.2	2146 1.9
6 SU	0243 6.1	0903 2.0	21 M	0349 6.4	1013 1.6
	1512 6.2	2129 1.8		1611 6.5	2226 1.6
7 M	0334 6.6	0954 1.4	22 TU	0428 6.7	1050 1.4
	1559 6.7	2217 1.3		1646 6.7	2302 1.4
8 TU	0421 7.1	1041 0.9	23 W	0501 6.8	1124 1.3
	1644 7.2	2302 0.9		1717 6.8	2334 1.4
9 W	0505 7.5	1125 0.6	24 TH	0532 6.9	1154 1.3
	1727 7.5	2346 0.6		1746 6.9	
10 TH	0548 7.8	1209 0.4	25 F	0005 1.3	0601 6.9
	1809 7.6			1224 1.3	1815 6.8
11 F	0030 0.5	0631 7.8	26 SA	0036 1.4	0630 6.8
	1252 0.4	1851 7.5		1253 1.5	1844 6.7
12 SA	0114 0.5	0714 7.6	27 SU	0107 1.5	0659 6.6
	1335 0.7	1933 7.3		1323 1.7	1914 6.5
13 SU	0159 0.8	0758 7.2	28 M	0138 1.8	0729 6.4
	1421 1.1	2018 6.8		1353 2.0	1945 6.2
14 M	0246 1.3	0845 6.7	29 TU	0212 2.0	0802 6.1
	1509 1.6	2108 6.4		1428 2.3	2022 5.9
15 TU	0339 1.8	0939 6.2	30 W	0252 2.3	0842 5.7
	1605 2.2	2208 5.9		1510 2.6	2110 5.6
			31 TH	0341 2.6	0936 5.4
				1606 2.9	2217 5.3

APRIL

Day	Time m	Time m	Day	Time m	Time m
1 F	0444 2.8	1053 5.2	16 SA	0022 5.5	0653 2.6
	1721 3.0	2341 5.3		1307 5.4	1925 2.7
2 SA	0603 2.8	1222 5.3	17 SU	0137 5.7	0804 2.4
	1848 2.8			1412 5.7	2028 2.4
3 SU	0102 5.6	0724 2.5	18 M	0234 5.9	0859 2.1
	1341 5.7	2002 2.4		1501 6.1	2117 2.1
4 M	0209 6.1	0831 2.0	19 TU	0320 6.2	0943 1.8
	1442 6.2	2100 1.8		1542 6.3	2158 1.8
5 TU	0305 6.6	0926 1.4	20 W	0358 6.4	1020 1.7
	1533 6.8	2151 1.2		1617 6.5	2234 1.6
6 W	0355 7.1	1015 0.9	21 TH	0432 6.6	1054 1.5
	1620 7.2	2239 0.8		1649 6.7	2307 1.5
7 TH	0442 7.5	1102 0.5	22 F	0504 6.7	1125 1.5
	1704 7.5	2325 0.5		1719 6.8	2338 1.5
8 F	0527 7.8	1147 0.4	23 SA	0534 6.7	1156 1.5
	1748 7.7			1749 6.8	
9 SA	0010 0.4	0611 7.8	24 SU	0010 1.5	0605 6.7
	1232 0.4	1831 7.6		1226 1.6	1819 6.7
10 SU	0055 0.5	0655 7.6	25 M	0042 1.6	0635 6.6
	1316 0.7	1914 7.3		1257 1.7	1850 6.6
11 M	0141 0.8	0740 7.2	26 TU	0115 1.7	0707 6.4
	1401 1.1	1959 6.9		1329 1.9	1923 6.4
12 TU	0229 1.3	0828 6.6	27 W	0151 1.9	0742 6.1
	1449 1.7	2048 6.4		1406 2.2	2002 6.1
13 W	0321 1.8	0921 6.1	28 TH	0231 2.1	0823 5.8
	1544 2.2	2146 5.9		1449 2.4	2050 5.8
14 TH	0421 2.3	1026 5.6	29 F	0320 2.4	0917 5.6
	1649 2.6	2259 5.6		1543 2.7	2152 5.6
15 F	0533 2.6	1146 5.4	30 SA	0419 2.6	1027 5.4
	1807 2.8			1652 2.8	2308 5.5

Chart Datum is 3·64 metres below IGN Datum. HAT is 7·9 metres above Chart Datum.

STANDARD TIME UT –01
Subtract 1 hour for UT
For French Summer Time add
ONE hour in **non-shaded areas**

BREST   LAT 48°23'N   LONG 4°30'W
TIMES AND HEIGHTS OF HIGH AND LOW WATERS

Dates in red are **SPRINGS**
Dates in blue are NEAPS

YEAR 2016

MAY

Time	m		Time	m
1 0531	2.6		**16** 0050	5.5
1147	5.4		0718	2.5
SU 1812	2.7		M 1327	5.5
			1945	2.6
2 0025	5.7		**17** 0151	5.7
0648	2.3		0817	2.3
M 1305	5.7		TU 1421	5.8
1927	2.3		2039	2.3
3 0134	6.1		**18** 0241	5.9
0757	1.9		0904	2.1
TU 1410	6.2		W 1506	6.1
2029	1.8		2123	2.1
4 0234	6.6		**19** 0323	6.1
0856	1.4		0945	1.9
W 1505	6.7		TH 1544	6.3
2124	1.3		2202	1.9
5 0328	7.0		**20** 0400	6.3
0949	1.0		1022	1.8
TH 1555	7.1		F 1619	6.5
2215	0.9		2238	1.7
6 0418	7.4		**21** 0435	6.4
1039	0.7		1056	1.7
F 1642	7.4		SA 1653	6.6
● 2303	0.6		○ 2313	1.6
7 0506	7.6		**22** 0509	6.5
1126	0.6		1130	1.6
SA 1727	7.6		SU 1726	6.7
2351	0.5		2347	1.6
8 0552	7.5		**23** 0542	6.5
1212	0.6		1203	1.6
SU 1812	7.5		M 1759	6.7
9 0037	0.6		**24** 0021	1.6
0638	7.3		0615	6.5
M 1258	0.9		TU 1236	1.7
1857	7.3		1832	6.6
10 0124	0.9		**25** 0056	1.6
0724	7.0		0650	6.4
TU 1343	1.2		W 1312	1.8
1942	6.9		1909	6.5
11 0212	1.3		**26** 0134	1.7
0811	6.5		0728	6.2
W 1431	1.7		TH 1351	2.0
2030	6.3		1949	6.3
12 0303	1.8		**27** 0216	1.9
0902	6.1		0812	6.0
TH 1522	2.2		F 1435	2.2
2124	6.0		2037	6.1
13 0358	2.2		**28** 0304	2.1
1000	5.7		0903	5.8
F 1620	2.5		SA 1527	2.4
◑ 2228	5.7		2134	5.9
14 0501	2.5		**29** 0400	2.2
1108	5.4		1006	5.7
SA 1728	2.8		SU 1629	2.5
2339	5.5		◑ 2241	5.8
15 0611	2.6		**30** 0504	2.3
1220	5.4		1116	5.6
SU 1840	2.7		M 1740	2.5
			2351	5.9
			31 0615	2.2
			1229	5.8
			TU 1853	2.2

JUNE

Time	m		Time	m
1 0101	6.1		**16** 0155	5.6
0724	1.9		0820	2.4
W 1337	6.2		TH 1424	5.8
1959	1.9		2044	2.4
2 0205	6.4		**17** 0245	5.8
0828	1.6		0907	2.2
TH 1437	6.6		F 1510	6.0
2059	1.5		2129	2.1
3 0304	6.8		**18** 0329	6.0
0925	1.2		0950	2.0
F 1532	6.9		SA 1550	6.2
2153	1.1		2210	1.9
4 0357	7.1		**19** 0408	6.2
1018	1.0		1029	1.9
SA 1622	7.2		SU 1628	6.4
2245	0.9		2248	1.7
5 0448	7.2		**20** 0446	6.4
1107	0.9		1106	1.7
SU 1710	7.3		M 1705	6.6
● 2334	0.8		○ 2326	1.6
6 0536	7.2		**21** 0522	6.5
1155	0.9		1143	1.6
M 1756	7.3		TU 1741	6.7
7 0022	0.8		**22** 0003	1.5
0623	7.1		0559	6.5
TU 1241	1.1		W 1219	1.6
1841	7.1		1818	6.7
8 0108	1.0		**23** 0041	1.5
0708	6.8		0637	6.5
W 1326	1.3		TH 1258	1.6
1925	6.9		1857	6.7
9 0154	1.3		**24** 0120	1.5
0753	6.5		0717	6.4
TH 1410	1.7		F 1338	1.7
2009	6.5		1939	6.6
10 0241	1.7		**25** 0203	1.6
0838	6.1		0800	6.3
F 1457	2.0		SA 1423	1.8
2056	6.1		2024	6.4
11 0329	2.1		**26** 0249	1.7
0927	5.8		0849	6.1
SA 1547	2.4		SU 1512	2.0
2148	5.8		2116	6.2
12 0422	2.4		**27** 0341	1.9
1022	5.5		0944	5.9
SU 1642	2.6		M 1608	2.2
◑ 2247	5.5		◑ 2215	6.1
13 0520	2.6		**28** 0439	2.1
1124	5.4		1048	5.8
M 1745	2.8		TU 1712	2.3
2352	5.4		2320	6.0
14 0622	2.7		**29** 0546	2.1
1229	5.4		1157	5.9
TU 1850	2.8		W 1822	2.2
15 0057	5.4		**30** 0031	6.0
0724	2.6		0656	2.0
W 1330	5.5		TH 1308	6.0
1951	2.6		1933	2.0

JULY

Time	m		Time	m
1 0141	6.2		**16** 0205	5.5
0803	1.8		0829	2.5
F 1415	6.3		SA 1434	5.7
2038	1.7		2055	2.4
2 0245	6.4		**17** 0258	5.8
0905	1.6		0918	2.3
SA 1514	6.6		SU 1522	6.0
2138	1.4		2142	2.1
3 0343	6.7		**18** 0343	6.0
1001	1.4		1002	2.0
SU 1607	6.9		M 1604	6.3
2231	1.2		2225	1.8
4 0434	6.9		**19** 0424	6.3
1052	1.2		1043	1.8
M 1656	7.1		TU 1645	6.6
● 2321	1.0		○ 2305	1.5
5 0522	6.9		**20** 0504	6.5
1139	1.1		1123	1.5
TU 1742	7.1		W 1724	6.8
			2344	1.3
6 0007	1.0		**21** 0543	6.7
0607	6.9		1203	1.4
W 1224	1.2		TH 1803	7.0
1825	7.1			
7 0051	1.1		**22** 0024	1.2
0649	6.8		0622	6.8
TH 1306	1.3		F 1243	1.3
1905	6.9		1843	7.0
8 0133	1.3		**23** 0105	1.1
0729	6.6		0703	6.8
F 1347	1.6		SA 1324	1.3
1944	6.6		1924	7.0
9 0214	1.6		**24** 0147	1.2
0808	6.3		0745	6.6
SA 1427	1.9		SU 1407	1.4
2023	6.3		2008	6.8
10 0254	1.9		**25** 0231	1.4
0848	6.0		0831	6.5
SU 1509	2.2		M 1454	1.7
2104	6.0		2056	6.5
11 0337	2.3		**26** 0320	1.7
0932	5.7		0922	6.2
M 1554	2.5		TU 1547	1.9
2151	5.6		2150	6.2
12 0425	2.6		**27** 0416	2.0
1024	5.4		1020	6.0
TU 1647	2.8		W 1647	2.2
◑ 2249	5.4		◑ 2253	6.0
13 0522	2.8		**28** 0520	2.2
1127	5.3		1129	5.8
W 1750	2.9		TH 1757	2.3
2355	5.2			
14 0625	2.8		**29** 0006	5.8
1234	5.3		0632	2.3
TH 1857	2.9		F 1246	5.8
			1913	2.2
15 0104	5.3		**30** 0125	5.9
0730	2.8		0746	2.2
F 1339	5.5		SA 1400	6.1
2001	2.7		2026	2.0
			31 0235	6.1
			0853	1.9
			SU 1503	6.4
			2128	1.7

AUGUST

Time	m		Time	m
1 0333	6.4		**16** 0318	6.0
0950	1.6		0936	2.1
M 1557	6.7		TU 1540	6.4
2221	1.4		2200	1.8
2 0423	6.7		**17** 0402	6.4
1039	1.4		1020	1.7
TU 1644	6.9		W 1623	6.7
● 2309	1.2		2243	1.4
3 0508	6.8		**18** 0444	6.7
1124	1.3		1102	1.4
W 1727	7.0		TH 1704	7.1
2351	1.1		○ 2324	1.1
4 0549	6.9		**19** 0524	6.9
1205	1.2		1143	1.1
TH 1805	7.0		F 1744	7.3
5 0031	1.1		**20** 0005	0.9
0626	6.8		0604	7.1
F 1243	1.3		SA 1224	1.0
1841	6.9		1825	7.4
6 0107	1.3		**21** 0046	0.8
0700	6.7		0645	7.1
SA 1319	1.5		SU 1306	1.0
1914	6.7		1906	7.3
7 0142	1.6		**22** 0128	0.9
0733	6.4		0726	7.0
SU 1354	1.7		M 1350	1.1
1947	6.5		1949	7.1
8 0217	1.8		**23** 0212	1.2
0807	6.2		0810	6.8
M 1430	2.0		TU 1436	1.4
2021	6.1		2035	6.8
9 0252	2.2		**24** 0300	1.5
0843	5.9		0859	6.4
TU 1508	2.4		W 1527	1.8
2059	5.8		2128	6.4
10 0332	2.5		**25** 0354	2.0
0926	5.5		0956	6.0
W 1552	2.7		TH 1626	2.2
◑ 2146	5.4		◑ 2231	5.9
11 0420	2.8		**26** 0458	2.4
1022	5.3		1107	5.7
TH 1649	3.0		F 1738	2.4
2250	5.2		2350	5.6
12 0523	3.0		**27** 0615	2.5
1135	5.2		1231	5.7
F 1800	3.0		SA 1901	2.5
13 0008	5.1		**28** 0116	5.7
0638	3.0		0735	2.4
SA 1252	5.3		SU 1351	5.9
1915	2.9		2018	2.2
14 0125	5.3		**29** 0228	6.0
0750	2.8		0844	2.1
SU 1359	5.6		M 1454	6.3
2021	2.6		2119	1.8
15 0228	5.6		**30** 0323	6.3
0848	2.5		0939	1.8
M 1453	5.9		TU 1545	6.6
2114	2.2		2209	1.5
			31 0410	6.6
			1025	1.5
			W 1629	6.9
			2253	1.3

Chart Datum is 3·64 metres below IGN Datum. HAT is 7·9 metres above Chart Datum.

N Brittany

FREE monthly updates. Register at
www.reedsnauticalalmanac.co.uk

871

STANDARD TIME UT –01
Subtract 1 hour for UT
For French Summer Time add
ONE hour in **non-shaded areas**

BREST   LAT 48°23'N   LONG 4°30'W
TIMES AND HEIGHTS OF HIGH AND LOW WATERS

Dates in **red** are **SPRINGS**
Dates in blue are NEAPS

YEAR 2016

SEPTEMBER
Time m

1 0450 6.8 / 1106 1.4 / TH 1707 7.0 / ● 2331 1.2	**16** 0420 6.9 / 1039 1.2 / F 1640 7.3 / ○ 2301 0.9		
2 0526 6.9 / 1143 1.3 / F 1742 7.0	**17** 0502 7.3 / 1121 0.9 / SA 1722 7.6 / 2343 0.6		
3 0006 1.2 / 0559 6.9 / SA 1217 1.3 / 1813 7.0	**18** 0543 7.4 / 1204 0.7 / SU 1804 7.7		
4 0038 1.3 / 0630 6.8 / SU 1249 1.5 / 1843 6.8	**19** 0026 0.6 / 0624 7.5 / M 1247 0.7 / 1846 7.6		
5 0110 1.5 / 0700 6.6 / M 1321 1.7 / 1913 6.6	**20** 0109 0.7 / 0706 7.3 / TU 1331 0.9 / 1930 7.3		
6 0140 1.8 / 0730 6.4 / TU 1354 1.9 / 1944 6.3	**21** 0153 1.1 / 0750 7.0 / W 1418 1.3 / 2017 6.8		
7 0212 2.1 / 0802 6.1 / W 1428 2.3 / 2017 5.9	**22** 0241 1.5 / 0839 6.5 / TH 1509 1.8 / 2110 6.3		
8 0246 2.4 / 0839 5.7 / TH 1507 2.6 / 2056 5.6	**23** 0334 2.1 / 0937 6.1 / F 1610 2.2 / ◑ 2215 5.8		
9 0328 2.8 / 0927 5.4 / F 1557 2.9 / ◑ 2152 5.2	**24** 0439 2.5 / 1050 5.7 / SA 1724 2.6 / 2338 5.5		
10 0424 3.1 / 1037 5.2 / SA 1704 3.1 / 2313 5.1	**25** 0559 2.7 / 1219 5.6 / SU 1850 2.6		
11 0543 3.2 / 1204 5.2 / SU 1827 3.0	**26** 0105 5.6 / 0722 2.6 / M 1338 5.9 / 2005 2.3		
12 0043 5.2 / 0708 3.0 / M 1322 5.5 / 1944 2.7	**27** 0214 5.9 / 0829 2.3 / TU 1439 6.2 / 2103 1.9		
13 0156 5.5 / 0816 2.6 / TU 1422 5.9 / 2043 2.2	**28** 0307 6.3 / 0922 1.9 / W 1527 6.6 / 2150 1.6		
14 0251 6.0 / 0908 2.1 / W 1512 6.4 / 2133 1.7	**29** 0350 6.6 / 1006 1.6 / TH 1608 6.8 / 2231 1.4		
15 0337 6.5 / 0955 1.6 / TH 1557 6.9 / 2218 1.2	**30** 0427 6.8 / 1044 1.5 / F 1643 6.9 / 2306 1.3		

OCTOBER
Time m

1 0501 6.9 / 1118 1.4 / SA 1715 7.0 / ● 2339 1.3	**16** 0439 7.4 / 1059 0.7 / SU 1700 7.7 / ○ 2321 0.6
2 0531 6.9 / 1150 1.4 / SU 1745 6.9	**17** 0522 7.6 / 1143 0.6 / M 1744 7.8
3 0009 1.4 / 0600 6.9 / M 1221 1.5 / 1814 6.8	**18** 0005 0.5 / 0604 7.7 / TU 1228 0.6 / 1828 7.7
4 0039 1.6 / 0630 6.7 / TU 1252 1.7 / 1844 6.6	**19** 0050 0.7 / 0648 7.5 / W 1314 0.8 / 1914 7.3
5 0109 1.8 / 0659 6.5 / W 1324 1.9 / 1913 6.4	**20** 0135 1.1 / 0733 7.1 / TH 1402 1.2 / 2002 6.8
6 0139 2.1 / 0730 6.3 / TH 1357 2.2 / 1945 6.1	**21** 0224 1.6 / 0823 6.6 / F 1454 1.7 / 2055 6.3
7 0212 2.4 / 0805 5.9 / F 1435 2.5 / 2023 5.7	**22** 0317 2.1 / 0920 6.2 / SA 1554 2.2 / ◑ 2200 5.8
8 0252 2.7 / 0849 5.6 / SA 1522 2.8 / 2114 5.4	**23** 0421 2.6 / 1032 5.8 / SU 1706 2.6 / 2318 5.5
9 0344 3.0 / 0953 5.4 / SU 1623 3.0 / ◑ 2229 5.2	**24** 0537 2.8 / 1155 5.7 / M 1827 2.6
10 0456 3.2 / 1119 5.3 / M 1742 3.0	**25** 0040 5.6 / 0657 2.7 / TU 1312 5.8 / 1939 2.4
11 0000 5.2 / 0624 3.0 / TU 1241 5.5 / 1903 2.7	**26** 0148 5.8 / 0804 2.4 / W 1413 6.1 / 2037 2.1
12 0119 5.6 / 0740 2.6 / W 1347 6.0 / 2009 2.2	**27** 0240 6.2 / 0856 2.1 / TH 1501 6.4 / 2123 1.8
13 0219 6.1 / 0837 2.1 / TH 1442 6.5 / 2102 1.7	**28** 0323 6.4 / 0940 1.8 / F 1541 6.6 / 2203 1.7
14 0309 6.6 / 0927 1.5 / F 1530 7.0 / 2150 1.2	**29** 0401 6.7 / 1018 1.7 / SA 1616 6.7 / 2238 1.5
15 0355 7.1 / 1013 1.1 / SA 1615 7.4 / 2236 0.8	**30** 0434 6.8 / 1052 1.6 / SU 1648 6.8 / ● 2311 1.5
	31 0505 6.9 / 1124 1.5 / M 1719 6.8 / 2342 1.5

NOVEMBER
Time m

1 0535 6.9 / 1156 1.6 / TU 1749 6.8	**16** 0548 7.7 / 1212 0.6 / W 1813 7.6
2 0012 1.6 / 0606 6.8 / W 1228 1.7 / 1820 6.6	**17** 0033 0.8 / 0633 7.5 / TH 1259 0.8 / 1900 7.3
3 0043 1.8 / 0636 6.6 / TH 1300 1.8 / 1851 6.4	**18** 0120 1.1 / 0719 7.2 / F 1348 1.2 / 1948 6.8
4 0115 2.0 / 0708 6.4 / F 1334 2.1 / 1924 6.2	**19** 0208 1.6 / 0808 6.8 / SA 1439 1.7 / 2040 6.4
5 0149 2.3 / 0743 6.2 / SA 1413 2.3 / 2002 5.9	**20** 0300 2.0 / 0902 6.3 / SU 1535 2.1 / 2139 5.9
6 0229 2.6 / 0827 5.9 / SU 1458 2.6 / 2052 5.6	**21** 0357 2.5 / 1004 5.9 / M 1639 2.5 / ◑ 2246 5.6
7 0319 2.8 / 0925 5.6 / M 1555 2.8 / ◑ 2159 5.4	**22** 0504 2.7 / 1117 5.7 / TU 1749 2.6 / 2358 5.5
8 0423 3.0 / 1040 5.5 / TU 1704 2.8 / 2320 5.4	**23** 0616 2.8 / 1230 5.7 / W 1858 2.6
9 0542 2.9 / 1159 5.6 / W 1821 2.6	**24** 0106 5.7 / 0725 2.6 / TH 1334 5.8 / 1959 2.4
10 0038 5.6 / 0659 2.6 / TH 1309 6.0 / 1931 2.2	**25** 0204 5.9 / 0822 2.4 / F 1427 6.1 / 2049 2.2
11 0143 6.1 / 0803 2.1 / F 1409 6.5 / 2030 1.7	**26** 0251 6.2 / 0909 2.1 / SA 1511 6.3 / 2132 2.0
12 0239 6.6 / 0858 1.6 / SA 1502 7.0 / 2123 1.2	**27** 0331 6.4 / 0950 1.9 / SU 1549 6.5 / 2210 1.8
13 0329 7.1 / 0949 1.1 / SU 1552 7.4 / 2212 0.9	**28** 0408 6.6 / 1027 1.8 / M 1623 6.6 / 2244 1.7
14 0416 7.4 / 1037 0.8 / M 1639 7.6 / ○ 2300 0.7	**29** 0441 6.6 / 1101 1.7 / TU 1657 6.6 / ● 2318 1.7
15 0502 7.6 / 1125 0.6 / TU 1726 7.7 / 2347 0.6	**30** 0514 6.8 / 1135 1.6 / W 1729 6.7 / 2351 1.7

DECEMBER
Time m

1 0547 6.8 / 1209 1.7 / TH 1802 6.6	**16** 0019 0.9 / 0621 7.5 / F 1247 0.8 / 1847 7.2
2 0024 1.8 / 0620 6.7 / F 1243 1.7 / 1835 6.5	**17** 0105 1.1 / 0706 7.3 / SA 1334 1.1 / 1933 6.9
3 0058 1.9 / 0654 6.6 / SA 1318 1.9 / 1910 6.3	**18** 0151 1.4 / 0752 6.9 / SU 1421 1.5 / 2020 6.5
4 0134 2.1 / 0731 6.4 / SU 1357 2.0 / 1950 6.1	**19** 0238 1.8 / 0838 6.5 / M 1510 1.9 / 2108 6.1
5 0215 2.3 / 0814 6.2 / M 1441 2.2 / 2037 5.9	**20** 0327 2.2 / 0929 6.1 / TU 1602 2.3 / 2202 5.7
6 0302 2.5 / 0905 6.0 / TU 1532 2.4 / 2134 5.7	**21** 0422 2.6 / 1026 5.8 / W 1700 2.6 / ◑ 2304 5.5
7 0358 2.6 / 1007 5.8 / W 1632 2.5 / ◑ 2243 5.6	**22** 0523 2.8 / 1132 5.6 / TH 1804 2.7
8 0505 2.7 / 1118 5.8 / TH 1742 2.5 / 2356 5.7	**23** 0010 5.5 / 0630 2.8 / F 1240 5.5 / 1909 2.6
9 0618 2.5 / 1229 6.0 / F 1853 2.2	**24** 0115 5.6 / 0736 2.7 / SA 1344 5.6 / 2008 2.6
10 0106 6.0 / 0728 2.2 / SA 1336 6.3 / 1959 1.9	**25** 0213 5.8 / 0833 2.5 / SU 1437 5.9 / 2058 2.3
11 0209 6.4 / 0830 1.8 / SU 1436 6.7 / 2057 1.5	**26** 0301 6.1 / 0921 2.3 / M 1521 6.1 / 2141 2.1
12 0305 6.9 / 0927 1.3 / M 1532 7.1 / 2151 1.1	**27** 0342 6.3 / 1002 2.0 / TU 1601 6.3 / 2220 1.9
13 0357 7.2 / 1020 1.0 / TU 1623 7.3 / 2242 0.9	**28** 0420 6.5 / 1040 1.8 / W 1637 6.5 / 2257 1.8
14 0447 7.5 / 1110 0.8 / W 1713 7.4 / ○ 2331 0.8	**29** 0456 6.7 / 1116 1.7 / TH 1712 6.6 / ● 2333 1.7
15 0534 7.6 / 1159 0.7 / TH 1800 7.4	**30** 0531 6.8 / 1152 1.6 / F 1747 6.6
	31 0008 1.6 / 0607 6.8 / SA 1227 1.5 / 1823 6.6

Chart Datum is 3·64 metres below IGN Datum. HAT is 7·9 metres above Chart Datum.

》》**FREE** monthly updates. Register at 《
www.reedsnauticalalmanac.co.uk 《

BREST TIDAL COEFFICIENTS 2016

Date	Jan am	Jan pm	Feb am	Feb pm	Mar am	Mar pm	Apr am	Apr pm	May am	May pm	June am	June pm	July am	July pm	Aug am	Aug pm	Sept am	Sept pm	Oct am	Oct pm	Nov am	Nov pm	Dec am	Dec pm
1	52	47	41	37	47	42	36		43	46	63	69	68	73	78	82	90	92	90	90	84	82	80	79
2	43	40	35	34	37	34	37	40	50	56	75	81	77	82	86	90	93	93	90	89	81	78	78	76
3	37	37		36	33		46	53	63	71	87	92	86	90	92	93	92	90	87	85	76	72	74	71
4		38	39	44	35	40	61	70	78	86	96	100	93	95	94	93	88	85	82	78	69	64	68	64
5	40	43	50	57	46	53	79	88	93	100	102	103	96	96	92	90	82	77	75	70	60	55	61	57
6	48	53	64	71	61	70	96	103	105	109	103	102	96	94	87	83	73	68	65	60	51	46	54	51
7	58	63	78	84	78	87	109	114	112	113	100	97	92	88	79	75	62	57	55	49	42	39	49	48
8	68	74	90	96	94	101	117	118	112	110	92	87	84	80	70	64	51	45	44	39	38	40	49	52
9	78	83	100	104	107	112	117	115	106	101	82	76	75	70	59	53	40	35	35	32	43			55
10	87	90	107	108	115	116	111	105	95	88	70	64	65	59	48	43	32			32	48	55	61	67
11	93	95	108	106	116	114	99	91	81	74	59	54	54	49	39	35	31	32	35	40	63	71	73	80
12	95	96	103	99	111	105	82	73	66	59	49	46	45	42		34	36	41	47	55	79	86	86	92
13	95	93	94	88	99	91	65	57	53	48		44		39	34	36	48	56	64	72	94	100	97	101
14	91	87	81	73	83	74	50	45		45	43	43	38	39	40	46	64	72	81	88	105	109	104	106
15	83	78	66	59	66	57		43	43	44	45	47	41	44	52	58	79	87	96	102	111	112	106	105
16	74	68	53	50	50	46	43	45	46	48	50	53	48	52	64	71	93	99	108	111	111	109	103	99
17	64	60		49		44	49	54	52	56	57	61	57	62	77	83	104	108	114	114	105	100	95	90
18	57		50	53	45	48	59	64	60	63	64	68	67	71	88	93	110	111	113	111	94	87	84	78
19	56	56	58	64	53	58	68	72	67	70	71	74	75	79	97	100	111	108	106	101	79	72	71	65
20	59	62	69	74	64	69	76	79	73	75	77	79	83	86	102	103	105	100	94	86	64	58	59	53
21	67	72	79	83	75	79	81	83	77	79	80	82	88	90	102	101	93	86	77	69	52	48	49	45
22	76	81	87	89	82	85	84	85	80	81	82	83	91	91	98	94	78	70	61	54	46	45	43	
23	85	88	91	92	87	89	85	85	81	81	82	81	91	90	89	83	61	54	48			46	42	43
24	90	92	92	91	89	89	84	82	80	79	80	78	87	85	77	70	49		45	45	49	52	45	47
25	92	92	90	87	89	87	80	78	77	75	75	73	81	77	63	58	46	46	47	51	56	60	51	55
26	91	89	85	81	86	83	75	71	72	69	70	67	73	68		53	49	54	55	60	63	67	58	62
27	87	84	78	73	80	77	67	63	65	62	64	62	64	60	50	50	59	65	65	70	70	73	66	69
28	80	76	68	63	73	68	58	53	58	56	60		58		53	57	70	76	74	77	75	77	72	75
29	71	67	58	53	63	58	49	46	54	53	60	60	57	57	62	68	80	84	80	82	78	79	77	79
30	61	56			52	47	44			53	62	65	60	63	73	79	86	88	84	84	80	80	81	82
31	51	46			42	39			55	59			68	73	83	87			85	85			82	82

Tidal coefficients indicate the magnitude of the tide on any particular day, without having to look up and calculate the range, and thus determine whether it is springs, neaps or somewhere in between. This table is valid for all areas covered by this Almanac. Typical values are:

120	Very big spring tide
95	**Mean spring tide**
70	Average tide
45	**Mean neap tide**
20	Very small neap tide

See also coefficients for Calais (p 759), Cherbourg (p 786), St Malo (p 805) and Port Bloc (p 936).

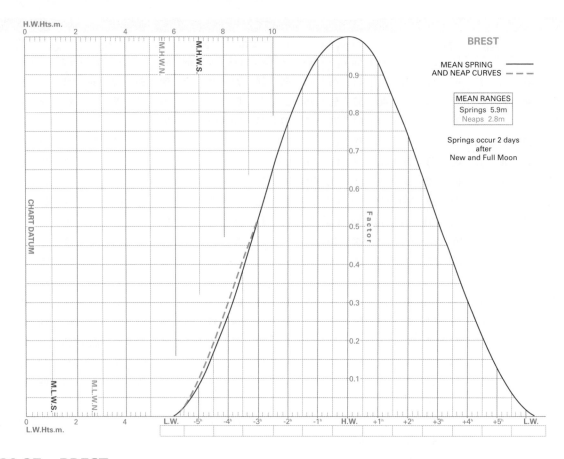

BREST

MEAN SPRING ————
AND NEAP CURVES − − −

| MEAN RANGES |
| Springs 5.9m |
| Neaps 2.8m |

Springs occur 2 days
after
New and Full Moon

9.20.27 BREST

Finistère 48°22′·00N 04°28′·28W (Port centre) ✿✿✿✿✿✿✿✿

CHARTS See under Rade de Brest.

TIDES +0520 Dover; ML 4·0; Duration 0605. Note: Brest is a Standard Port (←). Tidal coefficients are in Chapter 9.

SHELTER Excellent in both Moulin Blanc marina and Marina du Château. Access all tides H24. Brest is a busy naval, commercial and fishing port.

NAVIGATION See Rade de Brest for the outer approaches. Ldg Its 068° at Pte du Petit Minou and Pte du Portzic. Navigation within the Naval Port and the River Penfeld is strictly prohibited; see AC 3428 and chartlet. Keep clear of all warships.

For **Marina du Château** from S WPT 48°21′·70N 04°29′·15W, 344°/1M to marina ent via Passe Sud, leaving the Naval Base to port; or from E via S part of the Port de Commerce using Passe de l'Est and Passe de la Santé. Le Château and the American Monument are conspic.

For **Port du Mouin Blanc** marina WPT 48°22′·80N 04°25′·88W 007°/7ca via MB1/MB2 chan buoys to Q (3) 10s in centre of Port du Moulin Blanc marina, thence as directed into N or basins. The Océanopolis bldg, W roof, is conspic except near LW.

COMMUNICATIONS CROSS 02·98·89·31·31; Monitor *Brest Port* (at Pte du Portzic) VHF Ch **08** (controls apprs to Brest); ⚓02·98·32·55·55; Auto 08·92·68·08·29; Police 02·98·43·77·77; ⊖ 02·98·44·35·20; Brit Consul 01·44·51·31·00; Ⓗ 02·98·22·33·33; Dr 02·98·44·38·70; HM Brest 02·98·33·41·41; Aff Mar 02·98·80·62·25; Marinas VHF Ch 09 (HO).

FACILITIES Sté des Régates de Brest ☎02·98·02·53·36, ✕. **YC Rade de Brest ☎**02·98·44·63·32. **CN Municipal ☎**02·98·34·64·64.

Marina du Château ent 48°22′·68N 04°29′·48W. ☎02·98·33·12·50 (0800-2200 Jul-Aug) www.marinasbrest.fr chateau@marinasbrest. fr 650 ⌓ + 120 Ⓥ €2·86, multihulls +150%, pontoon A, and S side of pontoon PC. English spoken. 🛓 🛓 (H24) ✕ 🗔 🛒. ▣ ⛟ ⬛ 🛓 🚿 ice, free wi-fi. ⚓(25t) 5 mins to city centre by bus/tram.

Port du Moulin Blanc ④ ☎02·98·02·20·02. www.marinasbrest. fr moulinblanc@marinasbrest.fr. Excellent facilities, and very helpful bi-lingual staff. 1340 ⌓ + 120Ⓥ €2·40, no surcharge for multihulls, 🛓 🛓 H24 (French cr/card or pre-arrange with HM) ✕ 🗔 🛒 ▣ 🛓 ⛟ ⬛ 🚿. SHOM ⬍ ⬚ Ⓔ ✕ ⚓(14t,40t) ⟱(18t) ⬛. No 3 bus, 20 mins to city centre.

City all facilities: ✉ Ⓑ ⇌ Guipavas ✈. Ferry: Roscoff. Brest Festival of the Sea, 2nd wk July every 4 years, 2016 is a festival year!

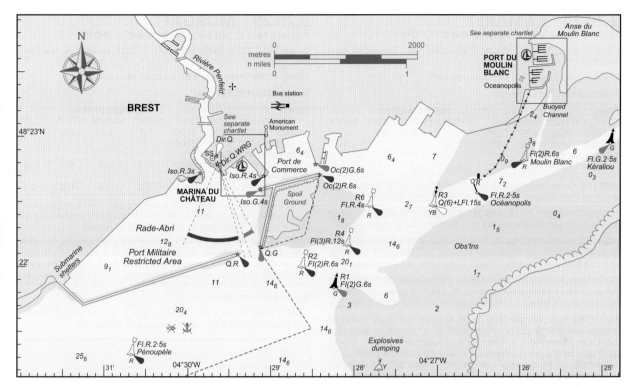

BREST

48°23'N

metres 0   2000
n miles 0   1

Riviére Penfeld

See separate chartlet

Bus station

American Monument

See separate chartlet

Dir.Q.

Dir.Q.WRG

SS

Iso.R.3s

Iso.R.4s

MARINA DU CHÂTEAU

Iso.G.4s

Port de Commerce

Oc(2)G.6s

Oc(2)R.6s

Spoil Ground

6₄

6₄   7

R6
Fl.R.4s

R4
Fl(3)R.12s

Q.G

R2
Fl(2)R.6s

R1
Fl(2)G.6s

Q.R

11

Rade-Abri

12₈

Port Militaire
Restricted Area

Submarine shelters

9₁

11

14₆

20₄

14₆

25₆

Fl.R.2·5s
Pénoupèle

31'

04°30'W

29'

14₆

1₈

2₇

14₆

20₁

6

3

14₆

Explosives
dumping

28'

PORT DU
MOULIN
BLANC

Oceanopolis

Anse du
Moulin Blanc

See separate chartlet

Buoyed Channel

2₄

3₈

0₉

R

Fl(2)R.6s
Moulin Blanc

6

Fl.G.2·5s
Kéraliou

0₃

R3
Q(6)+LFl.15s
YB

R

Fl.R.2·5s
Océanopolis

7₂

1₅

0₄

Obs'tns

1₇

2

04°27'W

04°26'W

25'

Y

Le Château

MARINA DU
CHÂTEAU

metres 0   200
cables   1

N

Bassin
No1

Bassin
No2

5₃

4₂

5₇

Port de
Commerce

4₆

D
E
J
I
C
B
A
F
G
H
N
M
PC
V
V
V

Iso.R.4s

Fl.R.2·5s

Fl.G.2·5s

Iso.G.4s

22'·8

343·5°

7₈

3₄

48°22'·6N

6₉

Military Port
(Prohib)

29'·4

29'·2

4°29'·0W

PORT DU
MOULIN
BLANC

N

metres 0   200
cables   1

3

1₁

Line of Y buoys

M
L
K
J
I
H
G

Bassin Nord

2

1₅

1₈

Wavebreak

1₃

V

V

Q(3)10s
MBA

Fl(2)G.2·5s
MB5

1₅

F
E
D
C
B
A

Bassin Sud

2₁

4

Fl.R.2·5s
MB4

1₅

Fl.G.2·5s
MB3

0₆

48°23'·5N

Oceanopolis

4₆

1₇

5₂

23'·3

Fl(3)R.12s
MB2

Fl(3)G.12s
MB1

04°26'·0W

25'·6

9.20.28 CAMARET

Finistère **48°16'·85N 04°35'·31W** ✿✿✿✿♦♦♦✿✿

CHARTS AC 2350, 3427; SHOM 7149, 7148, 7401; Imray C36; Navi 540, 542

TIDES +0500 Dover; ML 3·8; Duration 0610

Standard Port BREST (◄—)

Times				Height (metres)			
High Water		Low Water		MHWS	MHWN	MLWN	MLWS
0000	0600	0000	0600	7·0	5·5	2·7	1·1
1200	1800	1200	1800				
Differences CAMARET							
−0010	−0010	−0015	−0010	−0·4	−0·4	−0·3	−0·1

SHELTER Good, except in strong E'lies. Access H24. Good ⚓ and ⚓s SE of Port Vauban. No ⚓ within fairway and hbr.

NAVIGATION WPT 48°17'·44N 04°36'·09W, 139°/0·8M to N mole lt and in its W sector. Beware rks W & N of Pte du Grand Gouin.

LIGHTS AND MARKS The W SHM bcn twr at the E end of the N mole is very conspic; ditto Tour Vauban and chapel. Fish farms, one 3ca ENE and the other 3ca SE of N mole, unlit Y buoys.

COMMUNICATIONS CROSS 02·98·89·31·31; Auto 08·92·68·08·29; Police 02·98·27·84·94 (Jul/Aug); ⊜02·98·33·97·11 (Brest); Brit Consul 01·44·51·31·00; SAMU 15; HM VHF Ch 09; Aff Mar 02·98·27·93·28.

FACILITIES Port Vauban (5m) inside N mole; enter via S end of wavebreak. 200 ♥ ▭ rafted on outside of bkwtr or pontoons 2,3,4,5 and S end of 6; Jun-Sep €2·46; multihulls +50%; ⚓ ▮€2·00; all finger berths are residents only; ☎02·98·27·95·99 (0830-1200; 1600-2000 July/Aug, varies other months) mob ☎06·43·09·26·83, in emergency. ⚓ ⚓ local wi-fi €5·00 p/w. 20 W ⚓s marina rate −50%.

Port du Notic, in town has 200 ▭ and 50♥ price as above, outer side E pontoon. Also ▭ rafted on outside of A pontoon in 2·0m for yachts <13m LOA (1·0m at W end). HM☎ 02·98·27·89·31. www. camaret-sur-mer.com   capitainerie.camaret@wanadoo.fr 🅿 ♿ 🄿 🄿 ⚓ SHOM @ ⚒ 🄷 Ⓔ ⚠ ⚒ ▭(12t) careening area.

Port du Styvel is strictly for local boats; no ♥.

Town ✕ ▭ 🛒 ▮ 🄿 ✉ Ⓑ. ⇌ (Brest) ✈ (Brest or Quimper).

9.20.29 MORGAT

Finistère **48°13'·56N 04°29'·76W** ✿✿✿✿♦♦ ✿✿✿

CHARTS AC 2350, 2349; SHOM 7172, 7121; Imray C36; Navi 541, 542

TIDES +0500 Dover; ML 3·8; Duration No data

Standard Port BREST (◄—)

Times				Height (metres)			
High Water		Low Water		MHWS	MHWN	MLWN	MLWS
0000	0600	0000	0600	7·0	5·5	2·7	1·1
1200	1800	1200	1800				
Differences MORGAT							
−0005	−0010	−0020	−0005	−0·5	−0·4	−0·2	0·0
DOUARNENEZ							
−0010	−0010	−0020	−0010	−0·4	−0·4	−0·2	−0·1

SHELTER Good, except in W-N winds. Access H24. ⚓ in the bay in 2m on sand. Pleasant day ⚓s, sheltered from the W, are S of Morgat in the bays of St Hernot, St Norgard and St Nicolas.

NAVIGATION WPT 48°13'·60N 04°29'·15W, 270°/500m to first PHM buoy. Chan is nominally dredged 1·5m, but is prone to silting; it shoals abruptly on its unmarked N edge. Rocks lurk close under the cliffs S of Pte de Morgat. Les Verrès, 2–2·5M ESE of ent, are in the G sector of Pte de Morgat lt. See Passage information for dangers off Cap de la Chèvre.

LIGHTS AND MARKS Pte de Morgat lt is on the clifftop among trees. See chartlet and Lights, buoys and waypoints.

COMMUNICATIONS CROSS 02·98·89·31·31; ⦿ 02·98·10·51·41; Auto 08·92·68·08·29; Police 02·98·27·00·22; ⊜02·98·33·97·11 (Brest); Brit Consul 01·44·51·31·00; 🄷 02·98·27·05·33; Aff Mar 02·98·27·09·95; Marina VHF Ch 09; Tourist Office 02·98·27·29·49 (Crozon).

FACILITIES Marina ☎02·98·27·01·97   port-de-morgat@crozon.fr www.crozon.com 646 + 35♥ €1·87. ♥ on small pontoon parallel to fuel dock: fingers on E side or ▭ on W side for larger yachts. 40 unmarked ⚓s, €0·93, ▭ ⚓ ▮ ♿ 🄷 ⚒ Ⓔ 🅿 ♿ ▭(6t).

YC ☎02·98·16·00·01. **Town** (Crozon) ✕ ▭ 🛒 ▮ ✉ Ⓑ ⇌ ✈ (Brest or Quimper). Ferry: Roscoff.

9.20.30 DOUARNENEZ

Finistère **48°05'·98N 04°20'·35W** ✿❀♦♦♦✿✿✿

CHARTS AC 2349; SHOM 7121; Imray C36; Navi 542.

TIDES Differences Morgat. +0500 Dover; ML 3·7; Duration 0615

SHELTER In strong NW'lies due to seas/swell the outer Grande Passe ❶ pontoons can be uncomfortable. Rade de Guet ⚓ areas in 1·5-3m are exposed to N/NE'lies. Good shelter in Port Rhu.

NAVIGATION WPT 48°06'·71N 04°20'·83W, 157°/0·8M to abeam outer ❶ pontoon. Basse Veur (4·8m) and Basse Neuve (2·2m) are 8 and 4ca NNW of Île Tristan lt and in its R sector. 5ca NW of Tréboul jetty, QG, keep well NE of Rochers Le Coulinec and La Tête de Pierre, a rocky ledge, both off chartlet.

LIGHTS AND MARKS Daymarks: Two conspic spires and Île Tristan lt twr are in transit 147°. Île Tristan is a cliffy, wooded island. See chartlet and Lights, buoys and waypoints for details.

COMMUNICATIONS CROSS 02·98·89·31·31; ⚓ 02·98·84·60·64; Auto 08·92·68·08·29; ⊖ 02·98·33·97·11 (Brest); Brit Consul 01·44·51·31·00; Ⓗ 02·98·92·25·00; Port HM 02·98·92·14·85; Port Rhu VHF Ch 09; FV Port VHF Ch 12; Aff Mar 02·98·75·31·30; Marina VHF Ch 09.

FACILITIES from seaward: Grande Passe 60❶ Apr-Sep, €2·50; the southerly pontoons project slightly into the fairway. ⬚ ⚓ access H24. Other facilities as per Tréboul.

Tréboul Marina (1·5m) is full of local boats. ☎02·98·74·02·56. www. inet-bretagne.fr 463 ⬚ Apr-Sep, €2·50; ▣ N of ⛴ ⛴ berth, ⬚ ⚒ ⚠ Ⓔ ⛽ ⬚ ✕ ⬚(12t) ⬚(6t) ⚓. YC ☎02·98·92·02·03.

Port Rhu ☎02·98·92·00·67. 30❶ Apr-Sep, €2·50; If conditions are bad in Grande Passe pre-arrange a berth at Port Rhu's non-tidal basin (3·2m); or at pontoons S of the bridge (16·5m clnce at HW, coeff 80) in 2·2m. Enter over a sill (1·1m) via 10m wide gate which opens approx HW ±2 when Coeff >70.

Port Neuf is for FVs only. **Port de Rosmeur** is full of moorings, but ⚓ S of it, as charted, in about 1·5m; convenient for town.

Town ✕ ⬚ 🛒 ⛽ @ ✉ Ⓑ ⇌ ✈ (Quimper). Ferry Roscoff.

9.20.31 THE RAZ DE SEIN

GEOGRAPHY The Raz (pronounced *Raah*) is bounded by Le Chat bcn twr to the W (at E end of Chaussée de Sein) and to the E by La Vieille lt ho which with La Plate lt tr marks the dangers extending 8ca off Pte du Raz. Plateau de Tévennec, 2·9M NNE of Le Chat, consists of islets, drying rks and shoals which extend 5ca in all directions from the lt ho thereon. Other dangers on the N side of the Raz are rks and shoals extending nearly 1M W and WSW from Pte du Van, and Basse Jaune (1·4m) 1M to N.

On the S side the main dangers, all 1·5M off La Vieille, are: to the SW, Kornog Bras, a 3·7m deep rk; to the S, Masklou Greiz, rky shoals (7·6m) on which sea can break heavily; and to the SE, Roche Moulleg (5·2m).

TIDAL STREAMS ►In the middle (48°02'·88N 04°46'·76W) of the Raz the NE-going (flood) stream begins at HW Brest +0550, sp rate 6·5kn; the SW-going (ebb) stream begins at HW Brest −½, sp rate 5·5kn. Slack water, as the N-going flood stream expires, occurs between about HW Brest −1 and −½; if S-bound aim for the middle of that window, ±15 mins. If N-bound, slack water is about HW Brest +5½. Study tidal diamond 'D' on chart 2348. There are eddies near La Vieille on both streams.

*As a rule the Raz should always be taken at slack water to ensure the least uncomfortable conditions. Precise timing is vital: even an hour early or late on slack water can greatly affect the sea state. In good weather, near nps, wind and tide together, it is not difficult. But in moderate to strong winds it **must** be taken at slack water. In strong winds-against-tide which raise steep breaking seas and overfalls, the chan **must not** be used.◄*

NAVIGATION By day, WPT 'D' 48°02'·37N 04°45'·88W, 2ca W of La Plate, allows visual pilotage in normal sea states. By night use WPT 'N' 48°02'·91N 04°46'·61W, 308°/9ca from La Plate, and at the intersection of the white sectors of Tévennec, La Vieille and Le Chat. Take care to correct minor track deviations before they become major. Ensure accurate timing. Take all the usual safety precautions. Sig stn at Pte du Raz monitors Ch 16.

MARKS Tévennec lt ho is a white twr perched on a house, atop a substantial and distinctive rock. La Vieille is a stone twr, B top, with a derrick on its E side. La Plate is a smaller YBY bcn twr. Le Chat is a B twr, Y top marking rocks on its W, N and E sides; keep at least 0·5M E of it. See chartlet/Lights, buoys and waypoints for lt details.

9.20.32 ÎLE DE SEIN

Finistere 48°02'·34N 04°50'·89W ❀⚓✿

CHARTS AC 2348*; SHOM 7147, 7148, 7423*; Imray C36, C37; Navi 541. *Both have 1:10,000 insets of the island.

TIDES ML 3·6

Standard Port BREST (←—)

Times				Height (metres)			
High Water		Low Water		MHWS	MHWN	MLWN	MLWS
0000	0600	0000	0600	7·0	5·5	2·7	1·1
1200	1800	1200	1800				
Differences ÎLE DE SEIN							
−0005	−0005	−0015	−0010	−0·9	−0·7	−0·4	−0·2

SHELTER The non-drying N part of the hbr is open to N and E winds; the drying S part is protected from E and S by two large bkwtrs. The island itself affords some shelter from W'lies. ⚓ near the Ⓛ, but clear of 3 pairs of small lateral buoys (Apr-Oct) marking the fairway between N mole and S quay, which is used at HW by ferries. Île de Sein is worth visiting in fair weather, good visibility and preferably near nps.

NAVIGATION WPT 48°03'·67N 04°50'·71W, 187°/1·4M to Men-Brial lt ho ≠ B vert stripe on W house, (3rd from left, close S of lt ho and within its W sector (186°-192°)). This N appr is advised except in strong N'lies. Cornoc-An-Ar-Braden SHM lt buoy, where the tide sets across the chan, is 4ca S of the ⊕.

Chenal d'Ar Vas Du is aligned 224°: bcn on An Nerroth ≠ Men-Brial lt ho W sector (221°-227°). Chenal Oriental is aligned 270° on Sein lt ho and in the W sector (269°-271°) of its Dir Q WRG lt. By day appr with front bcn (close N of An Nerroth) ≠ 265° bcn close SSW of Sein lt ho. All bcns are reported hard-to-identify.

LIGHTS AND MARKS Men-Brial lt ho, G/W bands, has two white sectors leading towards the hbr. Cornoc-an-ar-Braden SHM lt buoy marks a rock on the W side of the N'ly white sector. Sein lt ho, W twr/B top, and two lattice masts are conspic at the W end of Île de Sein (which first appears as a low rounded hummock). See Lights, buoys and waypoints for lts further W on Chaussée de Sein.

COMMUNICATIONS Mairie 02·98·70·90·35; Sig stn at Pte du Raz VHF Ch 16.

FACILITIES ✗ ⌷ limited 🛒, ⌂, ✉ ✗. www.enezsun.com

9.20.33 AUDIERNE

Finistere 48°00′·55N 04°32′·42W ✳✳✳🅐⛵❀❀❀

CHARTS AC 2819 inc inset; SHOM 7147, 7148; Imray C36, C37; Navi 541

TIDES +0440 Dover; ML 3·1; Duration 0605

Standard Port BREST (◄—)

Times				Height (metres)			
High Water		Low Water		MHWS	MHWN	MLWN	MLWS
0000	0600	0000	0600	7·0	5·5	2·7	1·1
1200	1800	1200	1800				
Differences AUDIERNE							
−0035	−0030	−0035	−0030	−1·8	−1·4	−0·7	−0·3
ANSE DE FEUNTEUN AOD (6M WNW Audierne)							
−0030	−0040	−0035	−0025	−1·4	−1·2	−0·6	−0·2
LE GUILVINEC							
−0010	−0025	−0025	−0015	−1·9	−1·5	−0·7	−0·2
LESCONIL							
−0010	−0030	−0030	−0020	−2·0	−1·5	−0·7	−0·2

SHELTER Good in marina; see also Ste Evette below. There is limited room to ⚓ in the bight N of La Petite Gamelle SCM bcn.

NAVIGATION From the W, WPT 47°59′·47N 04°33′·01W (abeam Gamelle Ouest WCM buoy) 006°/1·5M to Kergadec Dir lt. Appr between La Gamelle, rks drying 0·9m in the middle of the bay, and Le Sillon de Galets rks to the W. Appr is difficult in strong S'lies, when seas break at the ent.

From the SE, WPT 47°59′·46N 04°31′·56W (3ca E of Gamelle Est SCM buoy) 331°/1·2M to bkwtr lt. This track leaves La Gamelle reef 300m to port. If bound for Ste Evette, do not turn WNW until well clear N of La Gamelle and Pte de Lervily's R sector.

Bar at ent; inside, dredged chan initially lies about 25m off the bkwtr. Pick up the 359° ldg marks, vert R/W chevrons, to avoid rocks at foot of bkwtr and banks to stbd drying 1·1m to 2·1m.

At root of bkwtr pick up 2nd set of R/W chevrons on corner of fish market bldg ≠ 045°; leave the fish market close (15m) to stbd, then alter 90° port towards the marina.

Leave the FV quays and yacht pontoons close to port to clear the bank to stbd, drying 1·9m and more. On departure there are 225° ldg chevrons by the roadside at root of W bkwtr.

LIGHTS AND MARKS See chartlet and Lights, buoys and waypoints. By day from the W WPT, Kergadec, W 8-sided lt tr, R top, ≠ old lt ho (hard to see) leads 006°. At night stay in the W sector (005°-007°) of Kergadec. From SE WPT, 331° ldg marks (no longer lit) are: Front, conspic white twr on Jetée du Raoulic (long W bkwtr); rear Kergadec. Fish market is a conspic B/W bldg.

COMMUNICATIONS CROSS 02·98·89·31·31; ⒷⒷ 02·98·70·07·54; Auto 08·92·68·08·29; Police 02·98·70·04·38; ⊜ 02·98·33·97·11 (Brest); Brit Consul 01·44·51·31·00; Ⓗ 02·98·75·10·10; Aff Mar 02·98·70·03·33; HM VHF Ch 09.

FACILITIES Marina Access HW −2 to +1 for 2m draft. Quays reserved for FVs. Berth initially on any hammerhead. Pontoons A-C are dredged 2m at the outer end, but 1·6m will be found between dredgings. Pontoons D-G are dredged 2m; avoid F & G as the

flood sets strongly towards the bridge. HM ☎02·98·75·04·93, mob 06·72·91·70·52. www.audierne.info 100 + 30Ⓥ €2·50. **Poulgoazec** ⌂(15t) ⬛s.

Town ✕ ◻🍴 🍴 🅟 Ⓛ 🔧 🖂 ✕ Ⓔ 🛢 🖂 Ⓑ. Bus to Quimper for ⇌ ✈. Ferry Roscoff.

SAINTE EVETTE The long bkwtr gives good shelter, but in strong SE-SW winds it is very uncomfortable and swell intrudes. Access H24. 30 W ⚓s (€1·29) in 1-2m, some very close together. Or ⚓ free in 3-4m close to seaward of ⚓s in moderate holding on sand. Avoid the S side of N pier/slip used by vedettes.

HM ☎02·98·70·00·28. 🛢 by hose from N side of N jetty (approx HW ±3) 🍴 🅞 ▷ ✕ ◻ limited victuals.

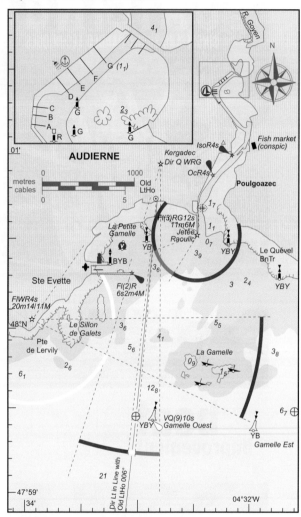

COMMUNICATIONS

on W pylon; rear, 0·58M from front, R ◻ on W pylon with R stripe. VHF Ch 12. HM (Plaisance) ☎02·98·58·14·47; www.leguilvinec.com Facilities: 4 ⚓s, 4 ◯ at NE end of hbr, €1·56, ⊜ ▲ 🍴 🖙 🔧 Ⓛ Ⓔ 🖙.

LESCONIL, Finistere, 47°47′·70N 04°12′·64W. AC 2820, 3640; SHOM 7146, 6646; Navi 543. Tides, see Bénodet; ML 3m. Former fishing port 3M SW of Loctudy. Appr is difficult in strong S'lys. Appr from Karek Greis ECM buoy, on ldg line 325°, church spire just open W of Men ar Groas lt ho, W twr/G top; at night in W sector 313°-333°. Bkwtr lts are QG and Oc R 4s.

Yachts welcomed Apr/May-Sept only, either on 6+ ⚓s (1·5-3·0m) or up to 22 Ⓥ ◯ on seasonal pontoon and the old fish quay, (max LOA 15m/draught 2·5m); call ahead. HM ☎06·72·04·55·80/ 06·47·82·77·12 (0800-1000; 1700-2000) www.plobannalec-lesconil. com. Facilities: ⚓ ✕ ◻ limited 🍴 🅞 🖙 Ⓛ Ⓒ.

FISHING HARBOURS N AND E OF THE POINTE DE PENMARC'H

ST GUÉNOLÉ, Finistere, 47°48′·68N 04°22′·93W. AC 2819, 2820; SHOM 7146/7, 6645; Navi 543. Strictly a fishing port; yachts only in emergency/force majeure. Narrow ent difficult in fresh W'lies, impossible in heavy weather. 3 sets of ldg marks/lts (see SHOM 6645). Pilot book essential. HM ☎02·98·58·60·43/ 06·78·85·38·85.

LE GUILVINEC, Finistere, 47°47′·46N 04°17′·17W. AC 2819, 2820, 3640; SHOM 7146/7, 6646. HW + 0447 on Dover (UT); ML 3·0m. See Bénodet. Good shelter and useful passage port (3m) but total priority to FVs; no ent/exit 1600-1830. Access H24 for <2·5m draft. Start from Névez SHM buoy. Appr is easy if able to see 053° triple ldg lts or the large, conspic dayglo red daymarks. Close in, beware Lost Moan rocks, SE of ent, marked by RW bcn lt twr. Good ⚓ SE of hbr ent in 7·7m staying clear of fairway. Three 053° ldg lts, all Q synch (see Area 21): front, W pylon; middle, 210m from front, R ◻

South Brittany

Loctudy to St Nazaire

9.21.1	Contents	881
9.21.2	Area map and Distance table	882
9.21.3	Tidal stream chartlets	884
9.21.4	Lights, buoys and waypoints	886
9.21.5	Passage information	890

South Brittany – 890 • Loctudy to Concarneau – 890 • Concarneau to Lorient – 894 • Île de Groix, Lorient and Belle Île – 896 • Baie de Quiberon – 903 • Crouesty to Le Croisic – 905 • Le Croisic to the River Loire and to Nantes – 908

9.21.6	Special notes for France, see Area 17	
9.21.7	Loctudy	890
9.21.8	Bénodet	891
9.21.9	Port-la-Forêt	891
9.21.10	Îles de Glénan	892
9.21.11	Concarneau	893
9.21.12	Minor harbours between Concarneau and Lorient	894

Aven and Bélon Rivers • Brigneau • Merrien • Doëlan • Le Pouldu

9.21.13	Port Tudy (Île de Groix)	895

Loc Maria

9.21.14	Lorient	896
9.21.15	River Étel	898
9.21.16	Sauzon (Belle Île)	898

Ster Wenn

9.21.17	Le Palais (Belle Île)	899
9.21.18	Passage de la Teignouse	899
9.21.19	Houat and Hoëdic	899
9.21.20	Port Haliguen	901

Port Maria

9.21.21	La Trinité-sur-Mer	901
9.21.22	Golfe du Morbihan	902
9.21.23	Vannes	904
9.21.24	Crouesty	905

Pénerf

9.21.25	Vilaine River	906

Arzal • La Roche Bernard

9.21.26	Piriac	907
9.21.27	La Turballe	907
9.21.28	Le Croisic	908
9.21.29	Le Pouliguen	909
9.21.30	Pornichet	909
9.21.31	Saint-Nazaire (R. Loire)	910

Nantes • Port de Comberge • Port de la Gravette • St Gildas

Area map – Distance table

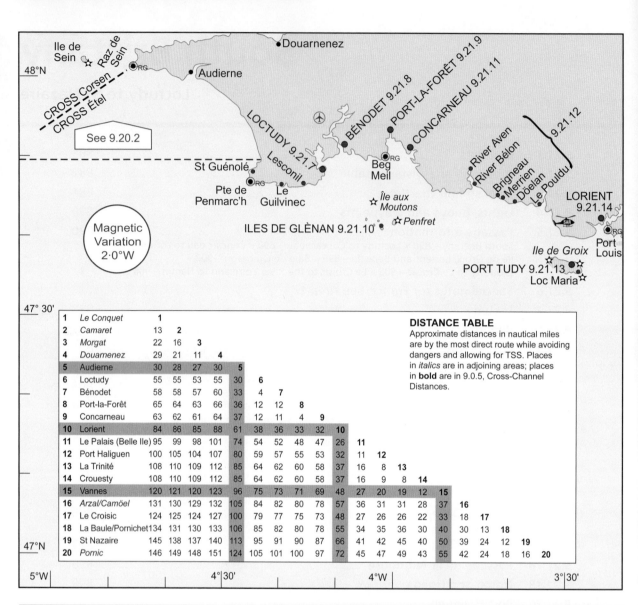

Map labels

Ile de Sein ☆
Raz de Sein
CROSS Corsen
CROSS Étel
48°N
•Douarnenez
•Audierne
LOCTUDY 9.21.7
BÉNODET 9.21.8
PORT-LA-FORÊT 9.21.9
CONCARNEAU 9.21.11
9.21.12
River Aven
River Bélon
Brigneau
Merrien
Döelan
Le Pouldu
LORIENT 9.21.14
See 9.20.2
St Guénolé
Lesconil
Beg Meil
Pte de Penmarc'h
Le Guilvinec
☆ Île aux Moutons
☆ Penfret
Magnetic Variation 2·0°W
ILES DE GLÉNAN 9.21.10
Port Louis
Ile de Groix
PORT TUDY 9.21.13
Loc Maria ☆
47° 30'

DISTANCE TABLE

Approximate distances in nautical miles are by the most direct route while avoiding dangers and allowing for TSS. Places in *italics* are in adjoining areas; places in **bold** are in 9.0.5, Cross-Channel Distances.

1	*Le Conquet*	**1**																		
2	*Camaret*	13	**2**																	
3	*Morgat*	22	16	**3**																
4	*Douarnenez*	29	21	11	**4**															
5	Audierne	30	28	27	30	**5**														
6	Loctudy	55	55	53	55	30	**6**													
7	Bénodet	58	58	57	60	33	4	**7**												
8	Port-la-Forêt	65	64	63	66	36	12	12	**8**											
9	Concarneau	63	62	61	64	37	12	11	4	**9**										
10	Lorient	84	86	85	88	61	38	36	33	32	**10**									
11	Le Palais (Belle Ile)	95	99	98	101	74	54	52	48	47	26	**11**								
12	Port Haliguen	100	105	104	107	80	59	57	55	53	32	11	**12**							
13	La Trinité	108	110	109	112	85	64	62	60	58	37	16	8	**13**						
14	Crouesty	108	110	109	112	85	64	62	60	58	37	16	9	8	**14**					
15	Vannes	120	121	120	123	96	75	73	71	69	48	27	20	19	12	**15**				
16	*Arzal/Camöel*	131	130	129	132	105	84	82	80	78	57	36	31	31	28	37	**16**			
17	*Le Croisic*	124	125	124	127	100	79	77	75	73	48	27	26	26	22	33	18	**17**		
18	*La Baule/Pornichet*	134	131	130	133	106	85	82	80	78	55	34	35	36	30	40	30	13	**18**	
19	*St Nazaire*	145	138	137	140	113	95	91	90	87	66	41	42	45	40	50	39	24	12	**19**
20	*Pornic*	146	149	148	151	124	105	101	100	97	72	45	47	49	43	55	42	24	18	16	**20**

47°N

5°W   4°|30'   4°W   3°|30'

TABLE OF BUDGET AIRLINE ROUTES

The table below shows direct flights by 5 budget airlines from 6 coastal airports (Nantes and Bordeaux are only 45 minutes by train to/from the coast) to no less than 10 UK regional and 3 London airports.

FRANCE: N – S \ UK: A – Z	Birmingham	Bristol	East Midlands	Edinburgh	Exeter	Gatwick	Glasgow	Leeds/Bradford	Liverpool	Luton	Manchester	Southampton	Stansted
Brest	FB			FB		EJ	FB	FB			FB	FB	
Nantes				FB	FB	EJ	FB		EJ		FB		
La Rochelle	FB	EJ	RY	FB		EJ	FB				FB	FB	RY
Bordeaux		EJ		RY		BA/EJ			EJ	EJ	FB		
Biarritz													RY
Rennes				FB	FB		FB	FB			FB	FB	

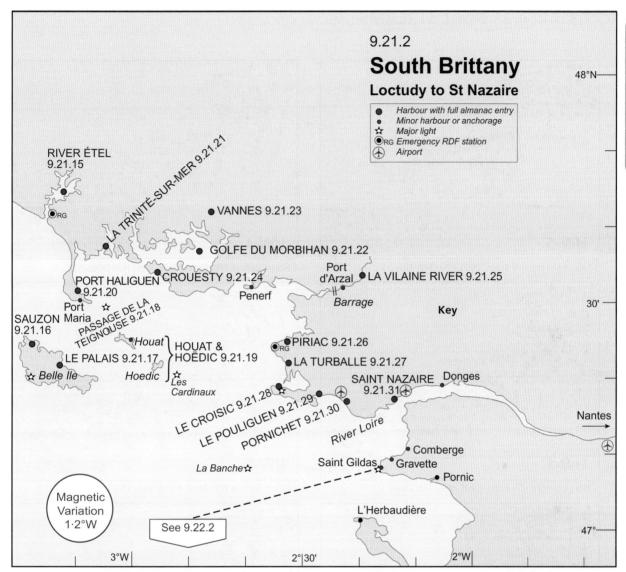

9.21.2
South Brittany
Loctudy to St Nazaire

48°N

●	Harbour with full almanac entry
•	Minor harbour or anchorage
☆	Major light
⊚RG	Emergency RDF station
✈	Airport

RIVER ÉTEL
9.21.15

LA TRINITÉ-SUR-MER 9.21.21

● VANNES 9.21.23

● GOLFE DU MORBIHAN 9.21.22

Port
d'Arzal ● LA VILAINE RIVER 9.21.25

PORT HALIGUEN CROUESTY 9.21.24
9.21.20

Key

Penerf

Barrage

30'

Port
SAUZON Maria ☆
9.21.16 PASSAGE DE LA
TEIGNOUSE 9.21.18

● *Houat*

⊚RG ● PIRIAC 9.21.26

LE PALAIS 9.21.17 HOUAT &
HOËDIC 9.21.19

● LA TURBALLE 9.21.27

☆ *Belle Ile* *Hoedic* ☆
*Les
Cardinaux*

SAINT NAZAIRE ● Donges
9.21.31

LE CROISIC 9.21.28 ✈

Nantes
→

LE POULIGUEN 9.21.29

PORNICHET 9.21.30

River Loire

✈

● Comberge

Saint Gildas ● Gravette

La Banche ☆

● Pornic

Magnetic
Variation
1·2°W

See 9.22.2

L'Herbaudière

47°

3°W 2°|30' 2°W

KEY AND CONTACT DETAILS OF BUDGET AIRLINES

Key	Airline	Website	UK	France
BA	British Airways	www.ba.com	0844 493 0787	08·25·82·54·00
EJ	EasyJet	www.easyjet.com	0913 232 3636	08·99·70·00·41
FB	Flybe	www.flybe.com	0871 700 2000	+44 1392 268529
RY	Ryanair	www.ryanair.com	0871 246 0000	08·92·56·21·50*

* French only

9.21.3 AREA 21 TIDAL STREAMS

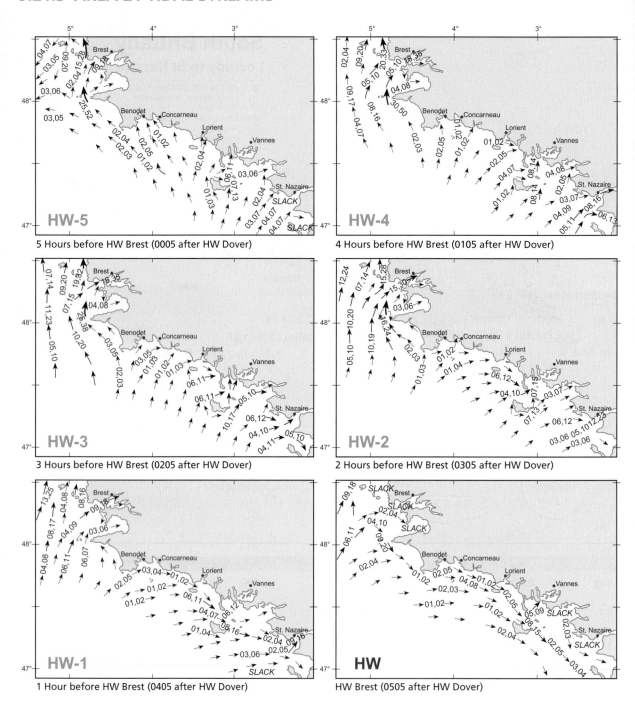

HW-5
5 Hours before HW Brest (0005 after HW Dover)

HW-4
4 Hours before HW Brest (0105 after HW Dover)

HW-3
3 Hours before HW Brest (0205 after HW Dover)

HW-2
2 Hours before HW Brest (0305 after HW Dover)

HW-1
1 Hour before HW Brest (0405 after HW Dover)

HW
HW Brest (0505 after HW Dover)

CAUTION: Due to the very strong rates of the tidal streams in some of the areas covered by these chartlets, many eddies and overfalls may occur. Where possible some indication of these eddies has been included. In many areas there is either insufficient information or the eddies are unstable. Tidal streams are generally weak offshore.

Strong winds may also have a great effect on the rate and direction of the tidal streams in all these chartlets.

Note: These chartlets are based on the Admiralty tidal stream atlas for France, W coast (NP 265, 2nd edition 2005) which uses data from actual observations out to 20-35M offshore.

The comparable French atlas (558-UJA; 2005) gives data for further offshore, based on computer predictions. It covers from Audierne to Le Croisic and contains large scale chartlets of the areas from Loctudy to Concarneau inc Iles de Glénan; Lorient and Ile de Groix; Belle Ile and the west part of Quiberon Bay.

560-UJA (1994) is recommended for large scale coverage of the Chenal du Four and Raz de Sein.

Northward 9.20.3   Southward 9.22.3

1 Hour after HW Brest (0605 after HW Dover)

2 Hours after HW Brest (0515 before HW Dover)

3 Hours after HW Brest (0415 before HW Dover)

4 Hours after HW Brest (0315 before HW Dover)

5 Hours after HW Brest (0215 before HW Dover)

6 Hours after HW Brest (0115 before HW Dover)

CAUTION: Due to the very strong rates of the tidal streams in some of the areas covered by these chartlets, many eddies and overfalls may occur. Where possible some indication of these eddies has been included. In many areas there is either insufficient information or the eddies are unstable. Tidal streams are generally weak offshore.

Strong winds may also have a great effect on the rate and direction of the tidal streams in all these chartlets.

Note: These chartlets are based on the Admiralty tidal stream atlas for France, W coast (NP 265, 2nd edition 2005) which uses data from actual observations out to 20-35M offshore.

The comparable French atlas (558-UJA; 2005) gives data for further offshore, based on computer predictions. It covers from Audierne to Le Croisic and contains large scale chartlets of the areas from Loctudy to Concarneau inc Iles de Glénan; Lorient and Ile de Groix; Belle Ile and the west part of Quiberon Bay.

560-UJA (1994) is recommended for large scale coverage of the Chenal du Four and Raz de Sein.

S Brittany

9.21.4 LIGHTS, BUOYS AND WAYPOINTS

Bold print = light with a nominal range of 15M or more. CAPITALS = place or feature. *CAPITAL ITALICS* = light-vessel, light float or Lanby. *Italics* = Fog signal. ***Bold italics*** = Racon. Many marks/buoys are fitted with AIS (<u>MMSI No</u>); see relevant charts.

LOCTUDY TO CONCARNEAU

Rostolou ⤬ 47°46'·64N 04°07'·29W.
Boulanger ⤬ VQ (6) + L Fl 10s;47°47'·38N 04°09'·13W.
Chenal de Bénodet ⤬ 47°48'·53N 04°07'·04W.
Bilien ⤬ VQ (3) 5s; 47°49'·10N 04°08'·09W.

LOCTUDY

Pte de Langoz ☆ Fl (4) WRG 12s 12m, **W15M**, R/G11M; 115°-W-257°-G-284°-W-295°-R-318°-W-328°-R-025°; W twr, R top; 47°49'·88N 04°09'·55W.
Karek-Saoz ⤬ Fl R 2·5s 3m 1M; R twr; 47°50'·02N 04°09'·38W.
Men Audierne ⤬ 47°50'·34N 04°08'·94W.
No. 2 (Karek-Croisic) ⤬ Fl (2) R 6s; 47°50'·19N 04°09'·49W.
No. 1 ⤬ Fl (2) G 6s; 47°50'·22N 04°09'·73W.
No. 3 ⤬ Fl (3) G 12s; 47°50'·20N 04°09'·99W.
Groyne head ⤬ Q 3m 10M; 47°50'·21N 04°10'·34W.
Le Blas ⤬ Fl (4) G 15s 5m 1M; truncated col; 47°50'·28N 04°10'·23W.

BENODET

Ldg lts 345·5°: Front, Pte du Coq ⤬ Dir Oc (3) G 12s 11m 10M; intens 345°-347°; W ○ twr, G stripe; 47°52'·31N 04°06'·70W.
Pyramide ⤬ Oc (3) 12s 48m 14M; 338°-016°, synch; W twr, G top; common rear, 348m from front.
Lts in line 000·5°: Front, Pte de Combrit ⤬ Oc (2) WR 6s 19m, W12M, R9M; 325°-W-017°-R-325°; W ☐ twr, Gy corners 47°51'·86N 04°06'·78W. Common rear, Pyramide; see above.
Rousse ⤬ 47°51'·55N 04°06'·47W. La Potée ⤬ 47°51'·78N 04°06'·55W.
Pte du Toulgoët ⤬ Fl R 2·5s 2m 1M; 47°52'·30N 04°06'·86W.
Le Taro ⤬ 47°50'·51N 04°04'·83W.
La Voleuse ⤬ Q (6) + L Fl 15s; 47°48'·76N 04°02'·49W.

ILE AUX MOUTONS and LES POURCEAUX

Île-aux-Moutons ☆ Iso WRG 2s 18m, **W15M**, R/G11M; 035°-W-050°-G-063°-W-081°-R-141°-W-292°-R-035°; W ☐ twr and dwelling; 47°46'·48N 04°01'·66W. <u>992271013</u>.
Rouge de Glénan ⤬ VQ (9) 10s 8m 8M; 47°45'·48N 04°03'·95W.
Grand Pourceaux ⤬ Q; 47°45'·98N 04°00'·80W.

ÎLES DE GLÉNAN

Penfret ☆ Fl R 5s 36m **21M**; W ☐ twr, R top; 47°43'·26N 03°57'·18W. Same twr: auxiliary ⤬ Dir Q 34m 12M; 295°-315°.
La Pie ⤬ Fl (2) 6s 9m 3M; 47°43'·75N 03°59'·76W.
Pte de la Baleine ⤬ VQ (3) 5s 2M; 47°43'·26N 03°59'·20W.
Les Bluiniers ⤬ 47°43'·35N 04°03'·81W.
Offlying marks, anticlockwise from the west:
Basse Pérennès ⤬ Q (9) 15s 8m 5M; 47°41'·06N 04°06'·14W.
Jument de Glénan ⤬ Q (6) + L Fl 15s 10m 4M; 47°38'·76N 04°01'·41W.
Laoennou ⤬ 47°39'·65N 03°54'·70W.
Jaune de Glénan ⤬ Q (3) 10s; 47°42'·56N 03°49'·83W.

PORT-LA-FORÊT

Linuen ⤬ Q (3) 10s; 47°50'·76N 03°57'·31W.
Laouen Pod ⤬ 47°51'·23N 03°58'·00W.
Cap Coz mole ⤬ Fl (2) WRG 6s 5m, W7M, R/G5M; Shore-R-335°-G-340°-W-346°-R-shore; 47°53'·49N 03°58'·28W.
Le Scoré ⤬ 47°52'·75N 03°57'·56W.
Les Ormeaux ⤬ 47°53'·27N 03°58'·34W.
Entry channel ⤬ Fl G 2·5s; 47°53'·39N 03°58'·12W.
⤬ Fl R 2·5s; 47°53'·39N 03°58'·22W.
Kerleven mole ⤬ Fl G 4s 8m 6M; 47°53'·60N 03°58'·37W.

CONCARNEAU

Ldg lts 028·5°: Front, La Croix ⤬ Q 14m 13M; 006·5°-093°; R&W twr; 47°52'·16N 03°55'·08W. **Rear, Beuzec** ☆ Dir Q 87m **23M**;

synch, intens 026·5°-030·5°; spire, 1·34M from front.
⤬ QR; 47°51'·40N 03°55'·75W.
Le Cochon ⤬ Fl (3) WRG 12s 5m W9M, R/G6M; 230°-R-352°-W-048°-G-207°-Obsc'd-230°; G twr; 47°51'·47N 03°55'·55W.
Basse du Chenal ⤬ Fl R 4s; 47°51'·54N 03°55'·66W.
Men Fall ⤬ Fl (2) G 6s; 47°51'·76N 03°55'·28W.
Kersos ⤬ 47°51'·80N 03°54'·93W (Anse de Kersos).
Lanriec ⤬ QG 13m 8M; 063°-078°; G window on W gable end; 47°52'·01N 03°54'·64W.
La Medée ⤬ Fl (3) R 12s 9m 4M; 47°52'·07N 03°54'·80W.
Ville-Close ⤬ Q WR, W9M, R6M; 209°-R-354°-W-007°-R-018°; R turret; 47°52'·27N 03°54'·76W.
No. 1 ⤬ Fl (4) G 15s 4m 5M; G turret; 47°52'·16N 03°54'·72W.
Marina wavescreen ⤬ Fl (4) R 15s 3m 1M; 47°52'·23N 03°54'·79W.
No. 2 ⤬ Fl R 4s 4m 5M; G turret; 47°52'·31N 03°54'·71W.
Ville-Close (NE end) ⤬ Fl (2) R 6s; R twr; 47°52'·37N 03°54'·68W.
Ent to FV basin ⤬ Fl (2) G 6s; G twr; 47°52'·36N 03°54'·63W.
Pouldohan ⤬ Fl G 4s 6m 8M; 053°-065°; W ☐ twr, G top; 47°50'·97N 03°53'·69W.
Roché Tudy ⤬ 47°50'·52N 03°54'·49W.

CONCARNEAU TO ÎLE DE GROIX

PTE DE TRÉVIGNON TO PORT MANEC'H

Les Soldats ⤬ VQ (9) 10s; 47°47'·86N 03°53'·42W.
Trévignon mole ⤬ Fl G 4s 5m 6M; 47°47'·69N 03°51'·30W.
Trévignon bkwtr root ⤬ Fl (2) WRG 6s 11m, W10/7M, R/G7M; 004°-W-051°-G-085°-W-092°-R-127°; 322°-R-351°; W ☐ twr, G top; 47°47'·59N 03°51'·33W.
Men Du ⤬ 47°46'·41N 03°50'·50W.
Men an Tréas ⤬ 47°45'·77N 03°49'·66W.

PORT MANEC'H (Aven and Bélon rivers)

Port Manech ⤬ Oc (4) WRG 12s 38m, W10M, R/G7M; obscd by Pte de Beg-Morg when brg <299°; 050°-W (unintens)-140°-W-296°-G-303°-W-311°-R (over Les Verrès)-328°-W-050°; W & R twr; 47°47'·99N 03°44'·34W.
Les Verrès ⤬ 47°46'·65N 03°42'·71W.

BRIGNEAU

Brigneau ⤬ 47°46'·11N 03°40'·10W.
W mole, ⤬ Oc (2) WRG 6s 7m, W12M, R/G9M; R col, W top; 280°-G-329°-W-339°-R-034°; 47°46'·86N 03°40'·18W. In line 331° with W ☐ daymark behind, Hard to see, rear beacon unlit.

MERRIEN

⤬ Dir QR 26m 7M; 004°-009°; W ☐ twr, R top; 47°47'·04N 03°38'·97W.
Ent, W side ⤬. E side ⤬ 47°46'·45N 03°38'·89W.

DOËLAN

Ldg lts 013·8°: Front ⤬ Oc (3) WG 12s 20m, W13M, G10M; W shore-305°, G305°-314°, W314°-shore; W lt ho, G band and top; 47°46'·32N 03°36'·51W. Rear ⤬ Oc (3)R 12s 27m 9M; W lt ho, R band & top; 326m from front.

LE POULDU

Ent ⤬ 47°45'·74N 03°32'·25W.
Grand Cochon ⤬ 47°43'·03N 03°30'·81W.
Pte de Kerroc'h ⤬ Oc (2) WRG 6s 22m W11M, R/G8M; 096·5°-R-112°·5-G-132°-R-302°-W-096·5°; W twr, R top; 47°41'·97N 03°27'·66W. <u>992271009</u>.

ÎLE DE GROIX

Pen Men ☆ Fl (4) 25s 60m **29M**; 309°-275°; W ☐ twr, B top; 47°38'·84N 03°30'·56W.
Speerbrecker, ⤬ 47°39'·10N 03°26'·33W.
Port Tudy, N môle ⤬ Iso G 4s 12m 6M; W twr, G top; 47°38'·72N 03°26'·74W.
E môle ⤬ Fl (2) R 6s 11m 6M; 112°-226°; W twr R top; 47°38'·68N 03°26'·78W.
Basse Melité ⤬ 47°38'·84N 03°25'·54W.

Pte de la Croix ⚹ Oc WR 4s 16m, W12M, R9M; 169°-W-336°-R-345°-W-353°; W pedestal, R lantern; 47°38'·05N 03°25'·02W.
Edouard de Cougy ⚐ 47°37'·93N 03°23'·90W.
Pointe des Chats ☆ Fl R 5s 16m **19M**; W ☐ twr and dwelling; 47°37'·24N 03°25'·30W.
Les Chats ⚐ Q (6) + L Fl 15s; 47°35'·70N 03°23'·57W.

LORIENT AND RIVER ÉTEL
LORIENT, PASSE DE L'OUEST
Passe de l'Ouest ldg lts 057°: both Dir Q 11/22m 13/**18M**. Front, Les Sœurs; intens 042·5°-058·5°, 058·5°-042·5°, 4M range only; R twr, W bands; 47°42'·13N 03°21'·84W. Rear **Port Louis** ☆, 740m from front (see below).
Banc des Truies ⚐ Q (9) 15s; 47°40'·76N 03°24'·48W.
Loméner, Anse de Stole, Dir ⚹ 357·2°: Q WRG 13m, W10M, R/G8M; 349·2°-G-355·2°-W-359·2°-R-005·2°; W twr, R top; 47°42'·30N 03°25'·54W.
A2 ⚐ Fl R 2·5s; 47°40'·94N 03°24'·98W.
Les Trois Pierres ⚐ Q RG 11m R/G6M; 060°-G-196°-R-002°; B twr, W bands; 47°41'·53N 03°22'·47W.
A8 ⚐ Fl R 2·5s; 47°41'·90N 03°22'·52W.

PASSE DU SUD
Bastresses Sud ⚐ QG; 47°40'·77N 03°22'·09W.
Les Errants ⚐ Fl (2) R 6s; 47°41'·10N 03°22'·38W.
Bastresses Nord ⚐ Fl (2) G 6s; 47°41'·11N 03°22'·20W.
Locmalo ⚐ Fl (3) G 12s; 47°41'·67N 03°22'·13W.

SOUTH OF PORT LOUIS
La Paix ⚓; 47°41'·97N 03°21'·84W.
Île aux Souris ⚹ Dir Q WG 6m, W3M, G2M; 041·5°-W-043·5°-G-041·5°; G twr; 47°42'·15N 03°21'·52W.
⚓ Fl G 2·5s; 47°42'·13N 03°21'·21W.
Ban-Gâvres FV/yacht hbr, W jetty ⚹ Fl (2) G 6s 3M; 47°42'·06N 03°21'·11W. E jetty ⚹ Fl (2) R 6s 3M; 47°42'·06N 03°21'·06W.

ENTRANCE CHANNEL
Ldg lts 016·5°, both Dir QG 8/14m 13M; intens 014·5°-017·5°; synch; W twrs, G tops. Front ☆, 47°43'·47N 03°21'·63W (Île St Michel). Rear ☆, 306m from front.
La Citadelle ⚓ Oc G 4s 6m 6M; 012°-192°; 47°42'·59N 03°21'·94W.
Secondary yacht chan, ▭ RGR; 47°42'·49N 03°22'·12W.
La Petite Jument ⚐ Oc R 4s 5m 6M; 182°-024°; R twr; 47°42'·58N 03°22'·07W.
Le Cochon ⚐ Fl R 4s 5m 5M; RGR twr; 47°42'·80N 03°22'·00W.
No. 1 ⚐ 47°42'·80N 03°21'·84W.

PORT LOUIS (Port de la Pointe)
D 1 ⚐ 47°42'·76N 03°21'·52W.
Jetty ⚹ Fl G 2·5s 7m 6M; W twr, G top; 47°42'·71N 03°21'·37W.
Anse du Driasker Stage Head Fl R 2·5s 2M, metal mast.

KERNEVEL
Kéroman ldg lts 350°, both Dir Oc (2) R 6s 25/31m **15M**; synch; intens 349°-351°: **Front** ☆, 47°43'·60N 03°22'·02W, R ho, W bands. **Rear** ☆ R&W topmark on Gy pylon, R top; 91m from front.
Banc du Turc ⚐ Fl G 2·5s; 47°43'·33N 03°21'·85W.
Kernével marina, entrance: ⚹ QR 1M; 47°43'·39N 03°22'·10W.
Ldg lts 217°, both Dir QR 10/18m **15M**; intens 215°-219°; synch: **Front, Kernével** ☆ R&W twr; 47°43'·02N 03°22'·32W. **Rear** ☆ W ☐ twr, R top; 290m from front.
⚐ Q (3) 10s; marks wreck; 47°43'·49N 03°22'·05W.

NORTHERN PART OF LORIENT HARBOUR
Grand Bassin (FV hbr) E side of ent, ⚹ Fl RG 4s 7m 6M; 000°-G-235°-R-000°; W twr, G top; 47°43'·63N 03°21'·87W.
Ste Catherine marina ent ⚹ QG 5m 3M; 47°43'·52N 03°21'·12W.
N side of marina ent ⚹ QR 2m 2M 47°43'·53N 03°21'·10W.
Pengarne ⚐ Fl G 2·5s 3m 3M; G twr; 47°43'·88N 03°21'·23W.
No. 9 ⚐ Fl (2) G 6s; 47°43'·95N 03°21'·16W.
No. 11 ⚐ Fl (3) G 12s; 47°44'·04N 03°21'·01W.

Pen-Mané marina, bkwtr elbow ⚹ Fl (2) G 6s 4M; 47°44'·11N 03°20'·86W.
Blavet River, No. 1 ⚐ Fl (2+1)G 10s; 47°44'·20N 03°20'·78W.
No. 2 ⚐ Fl R 2·5s; 47°44'·26N 03°20'·74W.
Pointe de l'Espérance, Dir ⚹ 037°, Q WRG 8m W10M, R/G8M; 034·2°-G-036·7°-W-037·2°-R-047·2°; W twr, G top; 47°44'·51N 03°20'·66W.
Ro-Ro jetty ⚹ QR 7m 2M; 47°44'·42N 03°20'·96W.
No. 8 ⚐ Fl R 2·5s; 47°44'·56N 03°20'·98W, ent to Lorient marina.

RIVIÈRE D'ÉTEL
Roheu ⚓ 47°38'·54N 03°14'·70W.
Épi de Plouhinec ⚹ Fl R 2·5s 7m 2M; 47°38'·59N 03°12'·86W.
W side ent ⚹ Oc (2) WRG 6s 13m W9M, R/G6M; 022°-W-064°-R-123°-W-330°-G-022°; R twr; 47°38'·69N 03°12'·83W.
Conspic R/W radio mast (CROSS Étel) 47°39'·73N 03°12'·11W.
Les Pierres Noires ⚓ 47°35'·53N 03°13'·29W.

BELLE ÎLE AND QUIBERON BAY (W OF 2° 50'W)
PLATEAU DES BIRVIDEAUX (6M NNW of Belle Île)
⚐ Fl (2) 6s 24m 10M; BRB twr; 47°29'·15N 03°17'·46W.

BELLE ÎLE
Pte des Poulains ☆ Fl 5s 34m **23M**; 023°-291°; W ☐ twr and dwelling; 47°23'·31N 03°15'·11W.
Les Poulains ⚐ 47°23'·44N 03°16'·68W.
N Poulains ⚐ 47°23'·68N 03°14'·88W.
Sauzon, NW jetty ⚹ Fl G 4s 8m 8M; 47°22'·49N 03°13'·04W.
SE jetty ⚹ Fl R 4s 8m 8M; 315°-272°; W twr, R top; 47°22'·45N 03°13'·00W. Inner hbr, W jetty ⚹, QG 9m 5M; 194°-045°; W twr, G top; 47°22'·37N 03°13'·13W.
Le Palais, N jetty ⚹ QG 11m 7M; obsc 298°-170° by Ptes de Kerdonis and Taillefer; W twr, G top; 47°20'·84N 03°09'·04W.
No 1 ⚐ Fl (2) G 6s; 47°20'·82N 03°08'·96W;
No 2 ⚐ Fl R 4s; 47°20'·71N 03°08'·89W;
No 4 ⚐ Fl (2) R 6s; 47°20'·78N 03°09'·01W.
S jetty ⚹ QR 11m 7M; obsc'd 298°-170° (see N jetty); W twr, R lantern; 47°20'·82N 03°09'·07W.
Pointe de Kerdonis ☆ Fl (3) R 15s 49m **15M**; obsc'd 225°-320° (095°) by vegetation, 004°-155° (151°) by land. W ☐ twr, R top and W dwelling; 47°18'·60N 03°03'·58W.
Les Galères ⚐ 47°18'·77N 03°02'·76W.
SW coast of Belle Île: La Truie ⚓ 47°17'·11N 03°11'·79W.
Goulphar ☆ Fl (2) 10s 87m **27M**; Gy twr; 47°18'·65N 03°13'·63W.

PORT MARIA (Quiberon ferry port)
Light ho ⚹ Q WRG 28m W14M, R/ G10M; 246°-W-252°-W-297°-G-340°-W-017°-R-051°-W-081°-G-098°-W-143°; W twr, G lantern; 47°28'·78N 03°07'·45W.
Le Pouilloux ⚐ 47°27'·88N 03°08'·01W.
Ldg lts 006·5°, both Dir QG 5/13m **16/17M**; intens 005°-008°; W twrs, B bands: **Front**, 47°28'·63N 03°07'·18W. **Rear**, 230m north; 992271030.
Les Deux Frères ⚐ Fl R 2·5s; 175°-047°; 47°28'·34N 03°07'·29W.
S bkwtr ⚹ Oc (2) R 6s 9m 7M; W twr, R top; 47°28'·54N 03°07'·31W.

CHAUSSÉE and PASSAGE DE LA TEIGNOUSE
Le Four ⚓ 47°27'·78N 03°06'·54W.
Les Trois Pierres ⚓ 47°27'·46N 03°05'·24W.
Roc er Vy ⚓ 47°27'·70N 03°04'·40W; Chenal en Toull Bras.
Roc er Vy ⚓ 47°27'·86N 03°04'·35W.
Basse Cariou ⚐ 47°26'·94N 03°06'·38W.
Basse du Chenal ⚐ 47°26'·66N 03°05'·77W.
Goué Vaz S ⚐ Q (6) + L Fl 15s; 47°25'·76N 03°04'·93W.
La Teignouse ☆ Fl WR 4s 20m **W15M**, R11M; 033°-W-039°-R-033°; W ○ twr, R top; 47°27'·44N 03°02'·75W. 992271007.
Basse du Milieu ⚐ Fl (2) G 6s 9m 2M; 47°25'·91N 03°04'·12W.
Goué Vaz E ⚐ Fl (3) R 12s; 47°26'·23N 03°04'·28W.
NE Teignouse ⚐ Fl (3) G 12s; 47°26'·55N 03°01'·87W.
Basse Nouvelle ⚐ Fl R 2·5s; 47°26'·98N 03°01'·99W.

CHAUSSÉE and PASSAGE DU BÉNIGUET
Les Esclassiers ⚓ 47°25'·68N 03°03'·05W.
Le Grand Coin ⚓ 47°24'·45N 03°00'·26W.
Bonen Bras ⚓ 47°24'·26N 02°59'·88W.

ÎLE DE HOUAT
Le Rouleau ⚓ 47°23'·67N 03°00'·31W.
Port St-Gildas N môle ⚟ Fl (2) WG 6s 8m W9M, G6M; 168°-W-198°-G-210°-W-240°-G-168°; W twr, G top; 47°23'·57N 02°57'·34W.
Mussel beds, 1 -1.5M NNE of above ⚟ and in its G sector (198°-210°) marked by 2 ⌰ at S end, and Ɫ VQ (3) 5s & Ɫ VQ 5s at N end.
Bcns SE and S of Houat: Men Groise ⚓ 47°22'·77N 02°55'·00W.
Er Spernec Bras ⚓ 47°22'·09N 02°55'·23W.
Men er Houteliguet ⚓ 47°22'·54N 02°56'·38W.

ÎLE DE HOËDIC
Les Sœurs ⚓ 47°21'·14N 02°54'·74W.
La Chèvre ⌰ 47°21'·08N 02°52'·55W.
Port de l'Argol bkwtr ⚟ Fl WG 4s 10m W9M, G6M; 143°-W-163°-G-183°-W-194°-G-143°; W twr, G top; 47°20'·66N 02°52'·48W.
Er Gurranic'h ⚓ 47°20'·51N 02°50'·44W.
Cohfournik ⚓ 47°19'·49N 02°49'·69W.
Grands Cardinaux ⚟ Fl (4) 15s 28m 13M; R and W twr; 47°19'·27N 02°50'·10W. 992271025; 2·1M SE of Argol.
Le Chariot ⌰ 47°18'·87N 02°52'·95W.
Er Palaire ⚓ 47°20'·16N 02°55'·01W.

PORT HALIGUEN
Banc de Quiberon S ⌰ Q (6) + L Fl 15s; 47°28'·04N 03°02'·34W.
Banc de Quiberon N ⌰ 47°29'·65N 03°02'·58W.
E bkwtr hd ⚟ Oc (2) WR 6s 10m, W11M, R8M; 233°-W-240·5°-R-299°-W-306°-R-233°; W twr, R top; 47°29'·36N 03°05'·94W.
Port Haliguen ⌰ 47°29'·38N 03°05'·55W; marks 1·7m patch.
NW bkwtr hd ⚟ Fl G 2·5s 9m 6M; 47°29'·34N 03°06'·02W.
Bugalet wreck ⌰ 47°31'·19N 03°05'·45W.
Men er Roué ⌰ 47°32'·25N 03°06'·06W.

LA TRINITÉ-SUR-MER
Rivière de Crac'h ⚟ Q WRG 11m W10M, R/G7M; 321°-G-345°-W-013·5°-R-080°; W twr, G top; 47°34'·09N 03°00'·37W.
Buissons de Méaban ⌰ 47°31'·66N 02°58'·49W.
Le Petit Buisson ⌰ 47°32'·14N 02°58'·57W.
Roche Révision ⌰ 47°32'·63N 02°59'·36W.
Petit Trého ⌰ Fl R 2·5s; 47°33'·47N 03°00'·71W.
R. de Crac'h Dir ⚟ 347°: Oc WRG 4s 9m W13M, R/G 11M; 345°-G-346°-W-348°-R-349°; W twr; 47°35'·03N 03°00'·98W.
S pier ⚟ Q WR 6s 6m, W9M, R6M; 090°-R-293·5°-W-300·5°-R-329°; W twr, R top; 47°35'·09N 03°01'·50W.
Marina wavebreak/pier ⚟ Iso R 4s 8m 5M; 47°35'·28N 03°01'·46W.

GOLFE DU MORBIHAN
Méaban ⌰ 47°30'·77N 02°56'·23W.
Outer ldg marks 001°: Front, Petit Vezid ⌂; W obelisk; 47°34'·17N 02°55'·23W. Rear, Baden ch spire (83m) 47°37'·20N 02°55'·14W.
Pointe de Port-Navalo ☆ Oc (3) WRG 12s 32m, **W15M**, R/G11M; 155°-W-220°; 317°-G-359°-W-015°-R-105°; W twr and dwelling; 47°32'·87N 02°55'·11W.
Ldg marks 359°: Front, Grégan ⚓ Q (6) + L Fl 15s 3m 8M; 47°33'·91N 02°55'·05W. Rear, Baden ch spire (5 lines above).
Auray river: Catis ⚓ 47°36'·14N 02°57'·23W.
César ⚓ 47°38'·36N 02°58'·22W.
No. 13 ⌰ 47°39'·48N 02°58'·64W (last bcn before 14m bridge).
Morbihan: Grand Mouton ⌰ QG; 47°33'·71N 02°54'·85W.
Gavrinis ⚓ 47°34'·21N 02°54'·05W.
Jument ⚓ 47°34'·29N 02°53'·43W.
Creizic S ⌰ 47°34'·63N 02°52'·84W.
Creizic N ⌰ 47°34'·93N 02°52'·20W.
Les Rechauds, two ⌰ 47°36'·18N 02°51'·30W.
Truie d'Arradon ⚓ 47°36'·58N 02°50'·27W.

Logoden ⚓ 47°36'·70N 02°49'·91W.
Drenec ⚓ 47°36'·84N 02°48'·39W.
Bœdic ⌰ 47°36'·85N 02°47'·59W.
Roguédas ⚓ Fl G 2·5s 4m 4M; G twr; 47°37'·12N 02°47'·28W.

CROUESTY
Ldg lts 058°, both Dir Q 10/27m 8M; intens 056·5°-059·5°: Front ⚟ R panel, W stripe; 47°32'·53N 02°53'·95W. Rear ⚟, W twr; 315m from front.
No. 1 ⚓ QG; 47°32'·26N 02°54'·76W.
No. 2 ⚓ 47°32'·26N 02°54'·76W.
N jetty ⚟ Fl R 4s 9m 7M; R&W □ twr; 47°32'·47N 02°54'·14W.
S jetty ⚟ Fl G 4s 9m 7M; G&W □ twr; 47°32'·45N 02°54'·10W.

PLATEAU DU GRAND MONT
Chimère ⌰ 47°28'·83N 02°53'·98W.
L'Epieu ⌰ 47°29'·51N 02°52'·91W.
St Gildas ⌰ 47°29'·78N 02°52'·87W.
Grand Mont ⌰ 47°28'·98N 02°51'·12W.

QUIBERON BAY (E OF 2° 50'W) TO PTE DU CROISIC
PLATEAU DE SAINT JACQUES
Le Bauzec ⚓ 47°28'·89N 02°49'·40W.
St Jacques ⌰ 47°28'·17N 02°47'·52W.
Port St Jacques, jetty ⚟ Oc (2) R 6s 5m 6M; W 8-sided twr, R top; 47°29'·24N 02°47'·41W.

PLATEAU DE LA RECHERCHE
Recherche ⌰ Q (9) 15s; 47°25'·56N 02°50'·39W.
Locmariaquer ⌰ 47°25'·82N 02°47'·36W.

PÉNERF
⌰ Fl (5) Y 20s; 47°27'·58N 02°39'·56W.
Penvins ⌰ 47°28'·93N 02°40'·09W. Borenis ⌰ 47°29'·21N 02°38'·35W.
Tour des Anglais ⌂ W bcn; 47°30'·21N 02°37'·94W; appr brg 031·4°.
Pignon ⚓ Fl (3) WR 12s 6m W9M, R6M; 028·5°-R-167°-W-175°-R-349·5°-W-028·5°; R twr; 47°30'·03N 02°38'·89W.

VILAINE RIVER
Pte de Penlan ☆ Oc (2) WRG 6s 26m, **W15M**, R/G11M; 292·5°-R-025°-G-052°-W-060°-R-138°-G-180°; W sector defines Passe de la Grande Accroche; W twr, R bands; 47°30'·97N 02°30'·11W.
Bertrand ⚓ Iso WG 4s 6m, W9M, G6M; 040°-W-054°-G-227°-W-234°-G-040°; G twr; 47°31'·06N 02°30'·72W.
Basse de Kervoyal ⚓ Dir Q WR W8M, R5M; 269°-W-271°-R-269°; W sector leads to/from Nos 1 & 2 buoys; 47°30'·37N 02°32'·62W.
No. 1 ⚓ Fl G 2·5s; 47°30'·32N 02°28'·73W.
No. 2 ⚓ Fl R 2·5s; 47°30'·41N 02°28'·71W.
Pointe du Scal ⚟ QG 8m 4M; 47°29'·67N 02°26'·87W.

ÎLE DUMET
Fort ⚟ Fl (3) WRG 15s 14m, W7M, R/G4M; 090°-G-272°-W-285°-R-335°-W-090°; W col, G top; 47°24'·69N 02°37'·21W.
Basse-Est Île Dumet ⌰ Q (3) 10s; 47°25'·20N 02°34'·93W.
Île Dumet ⌰ Q; 47°25'·90N 02°36'·05W.

MESQUER
Laronesse ⌰ 47°25'·96N 02°29'·50W.
Basse Normande ⌰ 47°25'·47N 02°29'·80W.
Jetty ⚟ Oc WRG 4s 7m W10M, R/G7M; 067°-W-072°-R-102°-W-118°-R-293°-W-325°-G-067°; W col & bldg; 47°25'·31N 02°28'·05W.

PIRIAC-SUR-MER
Grand Norven ⌰ Q; 47°23'·55N 02°32'·89W.
Inner mole ⚟ Oc (2) WRG 6s 8m, W10M, R/G7M; 066°-R-148°-G-194°-W-201°-R-221°; W col; 47°22'·93N 02°32'·71W. *Siren 120s (occas), 35m SW.* E bkwtr ⚟ Fl R 4s 4m 5M; W pylon, R top; 47°23'·00N 02°32'·67W.
Les Bayonnelles ⌰ Q (9) 15s; 47°22'·56N 02°35'·23W.

LA TURBALLE
Ldg lts 006·5°, both Dir Iso R 4s 11/19m 3M; intens 004°-009°: Front, 47°20'·70N 02°30'·87W. Rear, 275m N of the front lt.
Ouest Jetée de Garlahy ⚓ Fl (4) WR 12s 13m, W10M, R7M; 060°-R-315°-W-060°; W pylon, R top; 47°20'·70N 02°30'·93W.
Digue Tourlandroux ⚓ Fl G 4s 4m 2M; G post, E side of access chan; 47°20'·75N 02°30'·88W.

LE CROISIC
Basse Hergo ⚓ Fl G 2·5s 5m 3M; 47°18'·62N 02°31'·69W.
Jetée du Tréhic ⚓ Iso WG 4s 12m W14M, G11M; 042°-G-093°-W-137°-G-345°; Gy twr, G top; 47°18'·49N 02°31'·42W. F Bu fog det lt, 100m SE.
Outer ldg lts 155·5°, both Dir Q 10/14m 13M; intens 154°-158°: Front ☆ 47°17'·96N 02°30'·99W . Rear ☆, 116m from front.
Middle ldg lts 174°, both QG 5/8m 11M; 170·5°-177·5°: Front 47°18'·06N 02°31'·07W. Rear, 48m from front.
Le Grand Mabon ⚓ Fl (3) R 12s 6m 2M; 47°18'·05N 02°31'·03W.
Inner ldg lts 134·7°, both QR 6/10m 8M; intens 132·5°-143·5°; synch.
Basse Castouillet ⚓ Q (9) 15s; 47°18'·12N 02°34'·37W.
Basse Lovre ⚓ 47°15'·87N 02°29'·58W, Chenal du Nord.

PLATEAU DU FOUR/BANC DE GUÉRANDE
Bonen du Four ⚓ Q; 47°18'·53N 02°39'·29W.
Le Four ☆ Fl 5s 23m **18M**; W twr, B diagonal stripes; 47°17'·87N 02°38'·05W. 992271016.
Ouest Basse Capella ⚓ Q (9) 15s; *Whis;* 47°15'·66N 02°42'·78W.
Goué-Vas-du-Four ⚓ Q (6) + L Fl 15s; 47°14'·91N 02°38'·20W.
Sud Banc Guérande ⚓ VQ (6) + L Fl 10s; 47°08'·80N 02°42'·81W.
Restricted Area centred on 47°14'·37N 2°46'·65W, limits marked by 4 ⚓ Fl Y 4s, enclosing 2 ⚓ Fl (5) Y 20s ODAS bys.

POINTE DU CROISIC TO POINTE DE ST GILDAS
PLATEAU DE LA BANCHE
NW Banche ⚓ Q 8m 4M; 47°12'·86N 02°31'·02W.
W Banche ⚓ Q (9) 10s; 47°11'·64N 02°32'·41W.
La Banche ☆ Fl (2) WR 6s 22m **W15M**, R11M; 266°-R-280°-W-266°; B twr, W bands; 47°10'·62N 02°28'·08W. 992271020.
SE Banche ⚓ 47°10'·40N 02°26'·10W.

PLATEAU DE LA LAMBARDE
NW Lambarde ⚓ 47°10'·84N 02°22'·93W.
SE Lambarde ⚓ Q (6) + L Fl 15s; 47°10'·04N 02°20·80W.

BAIE DU POULIGUEN (or Baie de la Baule)
Penchateau ⚓ Fl R 2·5s; 47°15'·24N 02°24'·35W.
Les Guérandaises ⚓ Fl G 2·5s; 47°15'·03N 02°24'·29W.
⚓ 47°15'·24N 02°23'·50W.
Les Evens ⚓ 47°14'·32N 02°22'·55W.
Les Troves ⚓ 47°14'·23N 02°22'·41W.
NNW Pierre Percée ⚓ 47°13'·63N 02°20'·63W.
La Vieille ⚓ 47°14'·03N 02°19'·51W.
Sud de la Vieille ⚓ 47°13'·75N 02°19'·54W.
Le Caillou ⚓ 47°13'·65N 02°19'·18W.
Le Petit Charpentier ⚓ 47°13'·34N 02°18'·95W.
Le Grand Charpentier ⚓ Q WRG 22m, W14M, R/G10M; 020°-G-054°-W-062°-R-092°-W-111°-R-310°-W-020°; Gy lt ho, G lantern; 47°12'·83N 02°19'·13W. 992271026.

LE POULIGUEN
Basse Martineau ⚓ 47°15'·54N 02°24'·34W.
Petits Impairs ⚓ Fl (2) G 6s 6m 2M; 47°15'·99N 02°24'·61W.
SW jetty ⚓ QR 13m 9M; 171°-081°; W col; 47°16'·39N 02°25'·39W.

PORNICHET (La Baule)
S bkwtr ⚓ Iso WRG 4s 11m, W10M, R/G7M; 303°-G-081°-W-084°-R-180°; W twr, G top; 47°15'·49N 02°21'·14W.
Ent, W side ⚓ QG 3m 1M; B perch, G top; 47°15'·51N 02°21'·12W.
E side ⚓ QR 4m 1M; B perch, R top; 47°15'·50N 02°21'·10W.

ST NAZAIRE APPROACH (Chenal du Sud)
S-N1 ⚓ L Fl 10s 8m 5M; *Racon Z (3cm), 3–8M;* 47°00'·07N 02°39'·84W.
S-N2 ⚓ Iso 4s 8m 5M; 47°02'·08N 02°33'·47W.

Les Chevaux ⚓ Fl G 2·5s; 47°03'·53N 02°26'·37W.
Thérèsia ⚓ Fl R 2·5s; 47°04'·84N 02°27'·27W.
La Couronnée ⚓ Fl (2) G 6s; *Racon T, 3–5M;* 47°07'·60N 02°20'·05W.
Lancastria ⚓ Fl (2) R 6s; 47°08'·88N 02°20'·37W.

PASSE DES CHARPENTIERS
Portcé ☆ ldg lts 025·5°, both Dir Q 6/36m **22/24M**: **Front**, intens 024·7°-026·2°; synch; W col; 47°14'·57N 02°15'·44W. 992271021.
Rear ☆ (H24); intens 024°-027°; W twr; 0·75M from front.
Wreck (Y) ⚓ Fl (3) R 12s; 47°09'·87N 02°19'·46W.
No. 1 ⚓ VQ G; 47°09'·94N 02°18'·40W.
No. 2 ⚓ VQ R; 47°10'·06N 02°18'·72W.
No. 6 ⚓ Fl (3) R 12s; 47°12'·05N 02°17'·31W.
No. 5 ⚓ Fl (4) G 15s; 47°12'·65N 02°16'·57W.
No. 8 ⚓ Fl (4) R 15s; 47°12'·75N 02°16'·86W.
No. 7 ⚓ VQ G; 47°13'·30N 02°16'·11W.
No. 10 ⚓ VQ R; 47°13'·62N 02°16'·16W. (Buoys 9 - 18 not listed).
Pointe d'Aiguillon ☆ Oc (3) WR 12s 27m, W13M, R10M; 207°-R-233°-W-293°; 297°-W-300°-R-327°-W-023°; 027°-R-089°; W lt ho, conspic; 47°14'·54N 02°15'·79W. 992271018.
Ville-ès-Martin jetty ⚓ Fl (2) 6s 10m 10M; W twr, R top; 47°15'·33N 02°13'·65W.
Morées ⚓ Fl (3) WR 12s 12m, W6M, R4M; W058°-224°, R300°-058°; G twr; 47°15'·00N 02°13'·02W.

SAINT NAZAIRE
W jetty ⚓ Oc (4) R 12s 11m 8M; W twr, R top; 47°15'·97N 02°12'·25W.
E jetty ⚓ Oc (4) G 12s 11m 11M; W twr, G top; 47°15'·99N 02°12'·14W. (Ent lock for big ships into Bassin St Nazaire)
No. 20 ⚓ QR; 47°16'·19N 02°11'·37W.
Sud Basse Nazaire ⚓ Q (6) + L Fl 15s; 47°16'·21N 02°11'·51W.
Basse Nazaire ⚓ Fl R 2·5s; 47°16'·24N 02°11'·63W.
SE Old mole ⚓ Fl R 2·5s; 47°16'·24N 02°11'·73W.
Old Môle ⚓ Q (3) 10s 18m 11M; 153·5°-063·5°; W twr, R top; weather signals; 47°16'·27N 02°11'·82W.
⚓ Fl (2) R 6s 5m 1M; R dolphin ; 47°16'·37N 02°11'·80W.
Entrée Est ⚓ Fl (3) R 12s 9m 9M; R pylon; 47°16'·47N 02°11'·85W.
Ldg lts 280° (into E lock) 2 VQ Vi; front 47°16'·49N 02°11'·92W.

EAST SIDE OF RIVER, SOUTH TO PTE DE ST GILDAS
Le Pointeau, Digue S ⚓ Fl WG 4s 4m, W10M, G7M; 050°-G-074°-W-149°-G-345°-W-050°; G&W ○ hut; 47°14'·02N 02°10'·96W.
Port de Comberge La Truie ⚓ 47°12'·06N 00°13'·36W.
S jetty ⚓ Oc WG 4s 7m W9M, G6M; 123°-W-140°-G-123°; W twr, G top; 47°10'·70N 02°09'·90W.
Port de la Gravette La Gravette ⚓ 47°09'·81N 02°13'·06W.
Bkwtr hd ⚓ Fl (3) WG 12s 7m, W8M, G5M; 183°-W-188°-G-124°-W-138°-G-183°; W structure, G top; 47°09'·64N 02°12'·70W.
Anse du Boucau bkwtr ⚓ Fl (2) G 6s 3M; 47°08'·41N 02°14'·80W.
Pte de Saint Gildas ⚓ Q WRG 20m, W14M, R/G10M; 264°-R-308°-G-078°-W-088°-R-174°-W-180°-G-264°; col on W house, G top; 47°08'·02N 02°14'·74W.
Nord Couronnée ⚓ Q; 47°07'·34N 02°17'·78W.

RIVER LOIRE TO NANTES
Well buoyed/lit to Île de Bois, thence G lts, S side; R lts N side.
No. 21 ⚓ VQ G; 47°16'·96N 02°10'·25W.
Suspension bridge, mid-span ⚓ Iso 4s 55m, vis up/down stream; 47°17'·10N 02°10'·24W; between 2 conspic R/W twrs.
MA ⚓ Fl (2) G 6s; 47°17'·50N 02°09'·22W.
MB ⚓ Fl (3) G 12s; 47°17'·95N 02°07'·53W.
Fernais-25 ⚓ Fl (4) G 15s; 47°18'·10N 02°06'·60W.
Paimboeuf, môle root ⚓ Oc (3) WG 12s 9m, W10M, G7M; Shore-G-123°-W-shore; W twr, G top; 47°17'·42N 02°01'·95W.
Île de Bois (upstream) Fl (2) G 6s; 47°12'·79N 01°48'·27W.
Ldg lts 064° Rear Dir VQ R; 47°12'·00N 01°34'·35W, Y-junction with Bras de la Madeleine.

9.21.5 PASSAGE INFORMATION

More Passage information is threaded between ports. **Bibliography:** *Atlantic France* (Imray/RCC/Parkinson); *West France Cruising Companion* (Fernhurst/Featherstone); NP 22 *Bay of Biscay Pilot*.

SOUTH BRITTANY

▶*Tidal streams are weak offshore, but can be strong in estuaries, channels and around headlands, especially in the north of the Area. The Admiralty tidal stream atlas for W France uses data from actual observations. The French equivalent gives much more computer-based data.*◀

Mist and haze are quite common in the summer, fog less so. Winds are predominantly from SW to NW, often light and variable in summer, but in early and late season N or NE winds are common. Summer gales are rare and usually related to passing fronts; the sea is often calm or slight, but swell, usually from W or NW, can severely affect exposed anchorages.

In summer the sea/land breeze cycle, known locally as the *vent solaire*, can disrupt moorings and anchorages open to the NE: After a quiet forenoon, a W'ly sea breeze sets in about midday, blowing onshore. It slowly veers to the NW, almost parallel to the coast, reaching Force 4 by late afternoon; it then veers further to the N, expiring at dusk. Around midnight a land breeze may pipe up from the NE and freshen sufficiently to kick up rough seas. By morning the wind has abated.

9.21.7 LOCTUDY

Finistere 47°50´·34N 04°10´·55W ✲❀◊◊◊◊❀❀

CHARTS AC 2820, 3641; SHOM 7146, 6649; Navi 543; Imray C37, 38

TIDES +0505 Dover (UT); ML 3·0; Duration 0615. See Bénodet.

SHELTER Excellent in marina and hbr, except in strong ESE'lies. Little space to ⚓ due many ⚓s/moorings in the river.

NAVIGATION WPT 47°50´·21N 04°09´·06W, 270°/0·8M to abeam the SHM bcn Fl (4) G 15s; thence alter stbd 305° to marina. Stay on the N side of buoyed chan (2 SHM lt buoys). Karek Croisic PHM lt buoy marks a 1·1m rock. Beware shoals and drying patches from W to ENE of Men Audierne SHM unlit spar buoy. From S and W, appr via Bilien ECM buoy to WPT. Sp ebb runs at 3½kn.

The following Breton words have navigational significance:

Aber: estuary. *Aven*: river, stream. *Bann*: hill. *Bian*: small. *Bras*: great. *Du*: black. *Enez*, *Inis*: island. *Garo*: rough, hard. *Glas*: green. *Goban*: shoal. *Gwenn*: white. *Karreg*: rock. *Ker*: house. *Men*, *mein*: rock, stone. *Morlenn*: creek. *Penn*: strait. *Porz*: harbour. *Raz*: tide race. *Ruz*: red. *Trez*: sand.

LOCTUDY TO CONCARNEAU

Loctudy is well sheltered from W/SW, but be wary of shallow patches in the outer approaches (AC 2646, 2820). Further east the larger ports are **Bénodet, Port-la-Forêt** and **Concarneau**, all with marinas. Anse de Bénodet has rocky shoals on both sides but is clear in the middle. The coast from Pte de Mousterlin to Beg Meil is fringed by rocks, many of which dry, extending 1M offshore. Chaussée de Beg Meil extends 8½ca SE, where Linuen rock (dries) is marked by bn. From Concarneau to Pte de Trévignon rocks extend nearly 1·5M offshore in places.

Îles de Glénan lie to seaward of Loctudy and Concarneau. With offlying dangers they stretch 5M from W to E and 4M from N to S. The islands are interesting to explore, but anchorages are rather exposed. Between Îles de Glénan and Bénodet lie Les Pourceaux, reefs which dry, and Île aux Moutons which has dangers extending SW and NW.

9.21.6 Special Notes for France: See Area 17.

Yachts must enter/leave hbr under power. Keep clear of FVs, esp 1630-1830LT daily when yachts should NOT enter/exit hbr.

Pont l'Abbé (3M upriver; 2m) is accessible to shoal draft boats on the flood; chan marked by perches.

LIGHTS AND MARKS Perdrix twr, N of appr chan, has conspic B/W chequers.

COMMUNICATIONS CROSS 02·97·55·35·35; ⓛⒷ 02·98·87·41·12; ⓦ 08·36·68·08·29; Auto 08·92·68·08·29; ⊖02·98·58·28·80; Ⓗ (6km) 02·98·82·40·40; Dr 02·98·87·41·80; Port VHF Ch 12; Aff Mar 02·98·87·41·79; Marina Ch 09 (Office hrs).

FACILITIES

Marina ☎02·98·87·51·36. www. loctudy.fr plaisance.loctudy@wanadoo.fr 592 + 65 ♥⌂ pontoon A, €2·52; 50 ⚓s €1·90. ⚓ ⚓ ⚓ Ⓞ ⚐ ⚒ ✕ ⚒ ⚒ ⚒(9t). **Town** ✕ ⚐ ⚒Ⓑ.

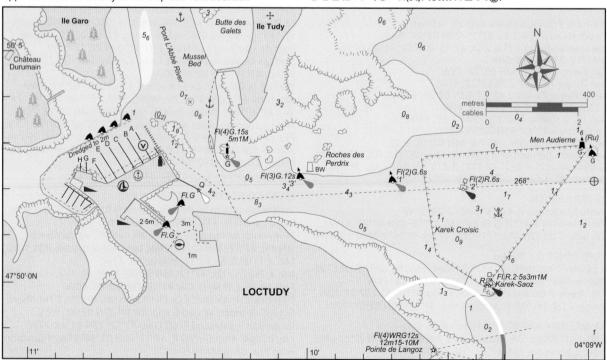

47°50´·0N LOCTUDY 04°09´W

9.21.8 BÉNODET

Finistere 47°51'·56N 04°06'·42W ※※※◊◊◊◊❀❀❀

CHARTS AC 2820, 3641; SHOM 6649, 6679; Navi 543; Imray C37

TIDES +0450 Dover; ML 3·1; Duration 0610

Standard Port BREST (←)

Times				Height (metres)			
High Water		Low Water		MHWS	MHWN	MLWN	MLWS
0000	0600	0000	0600	7·0	5·5	2·7	1·1
1200	1800	1200	1800				
Differences BÉNODET							
–0010	–0025	–0030	–0020	–1·7	–1·4	–0·6	–0·1
PORT DU CORNIGUEL							
+0015	+0010	–0015	–0010	–2·1	–1·7	–1·1	–0·8
LOCTUDY							
–0010	–0030	–0030	–0020	–2·1	–1·7	–0·9	–0·4

SHELTER Good in Sainte-Marine (W bank) and Penfoul (E bank) marinas; see Facilities for note about strong tidal streams. ⚓ in Anse du Trez in offshore winds or N of Cornouaille bridge. Between these two points ⚓ in the fairway is prohib.

NAVIGATION WPT 47°51'·25N 04°06'·30W, 345·5°/1·1M to front ldg lt. The centre of the bay is clear for small craft, but beware Roches de Mousterlin at the SE end of the bay (La Voleuse SCM lt buoy) and various rks off Loctudy to the SW.

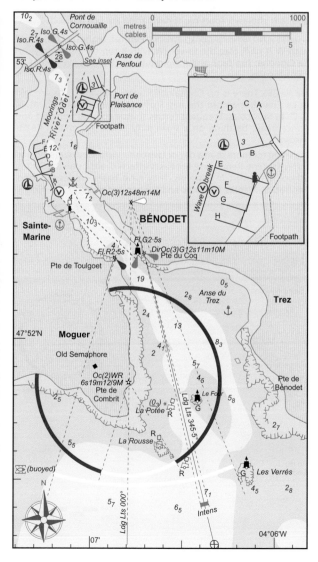

The 000° ldg line has no special merit and must be vacated early to avoid grounding on Pte de Combrit. Beware drying ledges around Les Verrès, La Rousse and Le Four unlit bcn twrs. Speed limit 3kn in hbr.

LIGHTS AND MARKS 345·5° ldg lts, synch/daymarks: Front, W ○ tr, G vert stripe, G top (hard to see until close); rear, conspic W tr, G top.

COMMUNICATIONS CROSS 02·97·55·35·35; 🖩 02·98·57·02·00; ⚓ 02·98·94·03·43; ⊖ 02·98·55·04·19 (Quimper); Brit Consul 01·44·51·31·00; Dr 02·98·57·22·21; Aff Mar 02·98·58·13·13; Marinas Ch 09 (0800-2000LT in season).

FACILITIES Both marinas accessible H24 at any tide, but it is best to arr/dep near slack water to avoid the strong stream, esp 4kn ebb, through the pontoons with risk of damage. At Ste-Marine the hammerheads and at Penfoul berths on both inside and outside of the wavebreak are best avoided except by larger vessels.

SAINTE-MARINE: **Marina** ☎02·98·56·38·72. port.plaisance@combrit-saintemarine.fr www.combrit-saintemarine.fr 670+100Ⓥ €2·64; or €1·87 for ⚓ ▶ €2·00. ⚓. **Town** ✕ 🗔 🛒 🍴 Ⓔ ✉ Ⓑ. Pedestrian ferry to Bénodet, only in July and August.

BÉNODET: **Penfoul Marina** ☎02·98·57·05·78. www.benodet.fr 510+40Ⓥ ⌂ €2·47; also 175 buoys +15 ⚓ €1·80. ⛽ ▮(0600-2300 cr/card) ✕ 🛒 🖲 ⚓ free ▶ wi-fi ⚒ ♺ ⊞ Ⓔ ✕ △ Divers 🏗(10t).

Town All facilities, ⚓ ✉ Ⓑ. Bus to Quimper for ⇌ ✈.

> Above the Cornouaille Bridge (28m) the charming R Odet is navigable on the tide for some 7M. Beware vedettes. There are pleasant ⚓s at Anse de Combrit, Anse de Kérautren, Porz Keraign, Porz Meilou, Anse de Toulven (a hidden creek) and SW of Lanroz. Beware marine farms. SHOM 6679 is advised.
>
> N of Lanroz the river shoals progressively to 0·5m in places and is less interesting. Quimper is the limit of navigation for Bridge (5·8m clearance) 0·5M below the city.

9.21.9 PORT-LA-FORÊT

Finistere 47°53'·49N 03°58'·24W ※※※◊◊◊◊❀❀

CHARTS AC 2820, 3641; SHOM 7146, 6650; Navi 543, 544; Imray C38

TIDES +0450 Dover; ML 2·9; Duration 0615
Use Differences CONCARNEAU.

SHELTER Very good in marina (2m). Restricted ⚓ inside Cap Coz; or to seaward on the W side of the Bay N'ward from Beg-Meil.

NAVIGATION WPT 47°52'·53N 03°57'·85W, 343°/1·0M to Cap Coz lt. Beware: Basse Rouge 0·8m, 6·5ca S of Cap Coz; Le Scoré (unlit SCM perch) and buoyed oyster farms in apprs. At sp a shoal patch 0·8m just S of the ent denies access around LW; but there are 3 W waiting buoys close SSW of ent. Shoaling reported in the inner chan, dredged 1·2m, to marina.

LIGHTS AND MARKS Cap Coz, Dir lt, leads 343° into chan marked with buoys and bns; the first pair of chan buoys are lit, Fl R 2·5s and Fl G 2·5s. The only conspic landmarks are Beg-Meil sig stn, 2·2M S of Cap Coz; the bldgs of Concarneau to the E and a forest of masts at PLF marina.

COMMUNICATIONS CROSS-Etel 02·97·55·35·35; 🖩 02·98·94· 99·91 (Beg Meil); Auto 08·92·68·08·29; Police 02·98·50·15·00 (Concarneau); ⊖ and Ⓗ see Concarneau; British Consul 01·44·51·31·00; HM VHF Ch 09; Aff Mar 02·98·60·51·42.

FACILITIES **Marina** ☎02·98·56·98·45. capitainerie@port-la-foret.fr www.port-la-foret.fr. 1069 inc Ⓥ €3·30 (multis LOA x1.5), plus 61 ⚓. Ⓥ pontoon opp inner end E bkwtr on N side of largest boat hoist basin. 240m pontoon berthing for larger vessels, dredged 5m, lies along the inner W side of ent. ⛽ ▮(H24/cr/card) ✕ 🗔 🛒(delivers 02·98·56·97·44) 🖲 defib Ⓒ ⚓ ⚒▮⚓ ✕ △ ▮(70t,25t) 🏗(3t) ⚓ YC ANPLF ☎02·98·56·84·13.

Town 🗔 ✕ 🛒 ⚓ ✉ Ⓑ. ⇌ ✈ Quimper. Ferry: Roscoff.

See chartlet overleaf

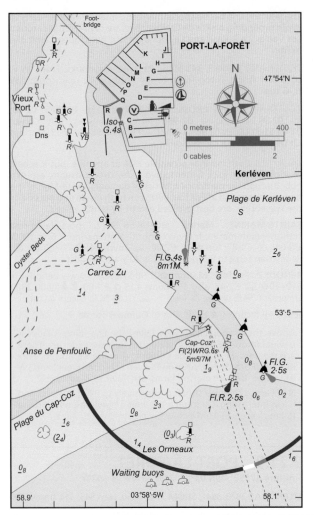

9.21.10 ÎLES DE GLÉNAN

Finistere **47°42′·98N 03°59′·60W** (Île Cignone) ✹✹⌂✿✿✿

CHARTS AC 2820, 3640; SHOM 7146, 6648; Navi 243; Imray C38

TIDES ML 3·0. See Concarneau for differences.

SHELTER Visit in settled weather as ⌂s can be exposed, especially near HW. The more popular are close W of N tip of Île de Penfret and on its E side; in La Chambre; close E of Île Cigogne; N of Île du Loc'h; and NW of Île de Bananec. Other ⌂s can be found.

NAVIGATION Cardinal buoys mark the S and SW limits of the islands; and Basse Jaune, a large, offlying shoal to the SE. Les Pourceaux (rocks) and Île aux Moutons are close N.

WPT 47°44′·15N 03°57′·04W, 186°/0·9M to Penfret lt ho, is the easiest entry point to Glénan. There is enough water HW±3 for most boats, but below half-tide careful pilotage is needed and a constant awareness of the state of the tide. Boats of the Glénan Sailing School abound and are best avoided. Speed limit is 8kn mid-Jun to mid-Sept.

Other approaches: Also from N, Cigogne twr in transit 182° with chimney on Île du Loc'h leads close E of La Pie IDM lt bcn; avoiding the drying Pierres-Noires further E.

From the W, broad Chenal des Bluiniers, 095°/088°, used by Bénodet-St-Nicolas ferries, passes S of Les Bluiniers SCM, and close N of Broc'h NCM, (NW of Île Drénec) drying 0·7m between Îles Drénec and St-Nicolas. S apprs/night navigation are not advised.

LIGHTS AND MARKS Conspic marks: Penfret, highest (17m) island with W lt ho/R top; the tower (W with B top on E side only) on Île Cigogne; fish tank and a wind turbine (5 FR) at SW end of Île Saint-Nicolas.

FACILITIES There is no Capitainerie. www.glenans.fr ⌂s are laid in La Chambre and in the bay S of La Pie; fees €14·50/boat. ✗ ⌂ ⅋ in St-Nicolas. Frequent ferries shuttle to/from Bénodet and Concarneau, berthing on a small jetty at the SW end of Île St-Nicolas.

If visiting for more than a day or two come victualled, watered and self-sufficient. Repairs are not easily effected, although the Sailing School may assist in extremis; they have bases on Îles Cignone, Penfret, St-Nicolas, Bananec and Drénec and monitor VHF Ch 16.

9.21.11 CONCARNEAU

Finistere **47°52´·15N 03°54´·77W** ✿✿✿✸◊◊◊❀❀❀

CHARTS AC 2820, 3641; SHOM 7146, 6650; Navi 543/4; Imray C38

TIDES +0455 Dover; ML 3·0; Duration 0615

Standard Port BREST (←—)

Times				Height (metres)			
High Water		Low Water		MHWS	MHWN	MLWN	MLWS
0000	0600	0000	0600	7·0	5·5	2·7	1·1
1200	1800	1200	1800				
Differences CONCARNEAU							
−0010	−0030	−0030	−0020	−1·9	−1·5	−0·7	−0·2
ÎLE DE PENFRET (Îles de Glénan)							
−0005	−0030	−0030	−0020	−2·0	−1·6	−0·8	−0·3

SHELTER Good, except in strong S'lies; not ideal for wintering.

NAVIGATION WPT 47°50´·62N 03°56´·33W, 028·5°/1·0M to abm Le Cochon SHM bcn twr. Beware large FVs and rks around Men Cren and Le Cochon. Speed limits: 5kn between le Cochon and Men Fall; 4kn inshore of Men Fall SHM lt buoy.

LIGHTS AND MARKS See chartlet and Lights, buoys and waypoints. Ldg lts 028·5°: front, small RW twr, very hard to see against bldgs behind; rear, Beuzec spire on skyline, 1·34M from front. Pass between Le Cochon G bcn twr (easier to see than the ldg marks) and Basse du Chenal PHM buoy, QR. After Men Fall SHM buoy,

track 070° toward Lanriec lt, in a G window on W gable end. It is better to lower sail here as there is little room to do so further up-chan. Note the 0·5m depth close SSW of La Médée PHM bcn twr.

COMMUNICATIONS CROSS 02·97·55·35·35; Police 17; ⊖ 02·98·97· 01·73; Brit Consul 01·44·51·31·00; ⊞ 02·98·52·60·60; Dr 02·98·97·85·95; Fire 18; HM (Port) 02·98·97·57·96, VHF Ch 12 (H24); Aff Mar 02·98·60·55·56; Marina Ch 09 (0700-2100LT in season).

FACILITIES

Marina ☎02·98·60·51·10 port-de-plaisance@concarneau.fr is very crowded in high season. 290 + 52♥ €3.25 (first two hrs free) on both sides of pontoon D (2m). Avoid arriving Fridays when D and E pontoons are usually busy with sailing school boats changing crews etc. Yachts can also berth on the inside of the anti-wash barrier. ▲ ⚓ berth in about 1m is only accessible near HW, depending on draft. ◻ ⚒ ⚓ ⊡ ⚑ SHOM ACA ⚓ ⛴(17t) ⚓.

Yacht pontoon on NE corner of La Ville Close is only for locals. The Arrière Port is solely for FVs. It has good heavy maintenance/repair facilities; or go to Port-la-Forêt or Bénodet.

⚓ing in Anse de Kersos may now be subject to charges.

Town ✗ ◻ 🛒 ⬤ Ⓔ ✉ Ⓑ bus to Quimper ⇌ and ✈. Ferry: via Roscoff.

CONCARNEAU TO LORIENT

Between Pte de Trévignon and Lorient are rocky cliffs and several interesting minor hbrs and anchs, delightful in fair weather; but most dry and are dangerous to approach in strong onshore winds. The Aven and Bélon rivers, Brigneau, Merrien, Doëlan and Le Pouldu (Rivière de Quimperlé), are described below.

Hazards SE and E of Pte de Trévignon include: Men Du, a rk 0·3m

high, marked by IDM bn, about 1·25M SE of the same Pte. Corn Vas, depth 1·8m, and Men ar Tréas, a rk which dries, are close S, both buoyed. Île Verte lies 6ca S of Île de Raguénès, with foul ground another 2ca to S. The approaches to Aven and Bélon Rivers are clear, except for Le Cochon and Les Verrés (IDM bn) to the SE. Between Le Pouldu and Lorient, Grand Cochon (SCM buoy) and Petit Cochon lie about 1M offshore.

9.21.12 MINOR HARBOURS BETWEEN CONCARNEAU AND LORIENT

The six places listed below are all in Finistere and share the same charts: AC 2821, SHOM 7138 (essential) and Navi 544.

AVEN and BELON RIVERS 47°47′·94N 03°43′·87W, HW +0450 on Dover (UT) –0030 on Brest. HW ht –1·9m on Brest; ML 2·8m; Duration 0608.

1·6M SW of ent beware Les Cochons de Rospico (0·5m) and 1·6M to SE Le Cochon (0·8m) and Les Verrès (IDM bcn). The latter are in the R sector of Port Manec'h's lt ho, Oc (4) WRG 12s. Night entry not advised to either river. Both rivers have bars and are shallow/dry in their upper reaches. Seas rarely break on the Aven bar, but the Belon bar is impassable in bad weather. SHOM 7138 and a good pilot book are essential.

The Aven Port Manec'h has Øs (possible risk of grounding) and a good ‡ in 2·5m outside the bar which is crossed by a buoyed chan dredged at least 0·6m. Beyond, the river has patches drying up to 1·6m. Very good shelter at Rosbraz in a crowded pool with up to 1·6m depth but no Øs. Pont-Aven, 3·6M upriver, is only accessible for shoal craft or a dinghy exped. Moorings in 2·5m; or ◻ at the quay dries 2·5m.

Facilities HM ☎02·98·71·08·65 (also for Brigneau, Belon, and Merrien); **YC de l'Aven** (Port Manec'h). **Town** ✕ ◻ ♣ 🗑 🏧 ⚒ ➤ ⛽.

The Bélon Cross the unmarked bar (0·6m) HW –1 in mid-river. After about 700m the chan S-bends; take care to avoid drying banks. 1M upriver off NW bank poss ‡ in pool up to 12·4m deep, unreliable holding. Beware 3kn streams. ✕ ◻ ➤ ⛽.

BRIGNEAU, 47°46′·87N 03°40′·13W. HW –0020 on Brest; ML 2·8m. Small drying, fair weather hbr. In strong onshore winds ent is dangerous and the hbr untenable due to swell. WPT 47°46′·25N 03°39′·67W, 331°/7ca to front leading mark ≰Oc (2) WRG 6s 331° ldg line: bkwtr Dir lt, W twr/R top, ≠ rear W panel (hard to see); at night the white sector (329°-339°) leads to ent; see Lights, buoys and waypoints for details. Some Øs (€1·10/m) or ‡ E side of ent or ◻ on drying W quay. ♣ ✕ ◻ ☕ 🍴 ⛽.

MERRIEN, 47°46′·77N 03°39′·02W. Tides as Brigneau. Small drying inlet, rky ledges either side of appr; sandbar 0·1m. Appr with W ◻ lt twr/R top (on skyline above trees) brg 005°; at night (not advised) QR, vis 004°-009°. ‡ or moor to 2 W Øs outside ent (€1·10/m) or ◻ on drying quay SE side; keep clear of oyster beds beyond. ♣ ✕ ◻.

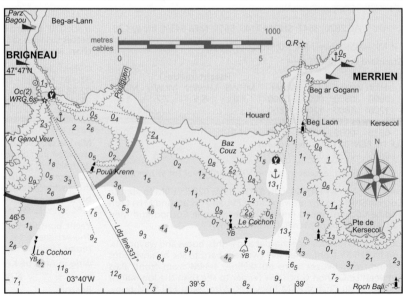

DOËLAN, 47°46′·21N 03°36′·55W. HW +0450 on Dover (UT) –0035 on Brest; HW ht –2·2m on Brest; ML 3·1m; Duration 0607. Fair weather only, open to onshore winds. 6 large W round buoys (€1·10/m) just N and S of bkwtr are for temp/short stay. 4 small dayglo red 🅐s (€1·10/m) afloat in outer hbr. Drying 🛏 at W quay or raft on FV up-river. Daymarks: Two outsize lt ho's in transit 013·8°: front (*aval*) W twr/G band & top; rear (*amont*) W twr/R band & top. Conspic factory chy E of front ldg lt. Facilities: HM ☎02·98·71·53·98. 🄻 🄴 🕭 ⚒. **Village** 🅟 🈁 ⚓ ✕ 🍴 Dr ✉.
LE POULDU, 47°45′·70N 03°32′·23W. Tides as Brigneau. Strictly

fair weather HW appr, close to W side of the wide Quimperlé estuary, marked by low cliffs and 2 unlit PHM bcns; chan shifts. In onshore winds the bar is dangerous on the ebb, up to 6kn sp. Guidel marina (E bank): adequate shelter for boats <9·5m/1·5m draft, 99+3 🆅, €1·70. HM ☎02·97·05·99·92/06·07·18·11·54. Ch 09. www.sellor.com ⚓ ✕ 🍴 🛒 🚌 Bus. Otherwise ⚓ 300m NW in 2·8m pool.

9.21.13 PORT TUDY (Île de Groix)
Morbihan 47°38′·72N 03°26′·73W ❀❀❀🌊🌊🌸🌸

CHARTS AC 2821, 2822; SHOM 7031, 7032, 7139; Navi 544; Imray C38

TIDES +0505 Dover; ML 3·1; Duration 0610

Standard Port BREST (⟵)

Times				Height (metres)			
High Water		Low Water		MHWS	MHWN	MLWN	MLWS
0000	0600	0000	0600	7·0	5·5	2·7	1·1
1200	1800	1200	1800				
Differences PORT TUDY (Île de Groix)							
0000	–0025	–0025	–0015	–1·8	–1·4	–0·6	–0·2

SHELTER Good, but in strong N/NE'lies swell enters outer hbr.

NAVIGATION WPT 47°39′·11N 03°26′·29W (100m E of Speerbrecker ECM buoy) 218°/0·5M to N mole hd lt, just in R sector of E mole hd lt. Beware rks SE of appr, and ferries. From E or SE pass N of Basse Mélite NCM buoy.

Use SHOM 7139 (1:20,000) for exploring the island, including Locmaria (below) Port St Nicolas and Port Lay.

LIGHTS AND MARKS The three principal lts on Groix are: Pen-Men lt ho at NW end of island, W □ tr, B top; Pte de la Croix lt at NE end; and Pte des Chats at SE end. There are no ldg lts, although the mole head lts in transit 219° clear offlying rks to the E. By day the N mole lt ho and conspic water twr (1M SW of hbr) ≠ 220° lead close W of Speerbrecker to ent.

COMMUNICATIONS CROSS 02·97·55·35·35; Auto 08·92·68·08·56; Police 02·97·86·81·17; ⊖ 02·97·86·80·93; Brit Consul 01·44·51·31·00; Ⓗ (Lorient) 02·97·83·04·02; HM VHF Ch 09 during opening hours for gate into wet basin; Aff Mar 02·97·37·16·22.

FACILITIES HM ☎02·97·86·54·62. Access H24.

Outer hbr, moor F&A; 120 🅐s in 2-4m, <10M €11·00, >10M €14·50. No ⚓ in hbr, often full in season. Caution: Ferry wash/noise.

Marina 43 🆅 🛏 €2·53, on first 2 pontoons inside E mole, dredged 2·0-2·5m.

Wet basin (104 🛏 mostly locals) access via gate HW±2 (0630–2200) but less at small neaps. No opening if swell/backwash.

Quay 🈁 🈁 (0800–1200 & 1400–1900, ☎02·97·86·80·96) ⚒ 🄴 ⚒ 🅱(3t) 🄻. **Town** ✕ 🍴 🛒 🄾 ✉. Ferry to Lorient.

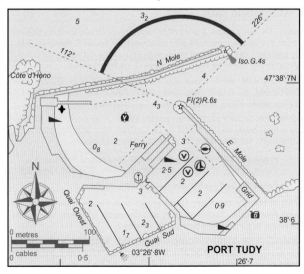

DRYING HARBOUR NEAR SE END OF ÎLE DE GROIX

LOCMARIA, Morbihan, **47°37'·40N 03°26'·42W**. AC 2820/1, SHOM 7139; –0020 Brest; ML 2·8m. Appr is open to the S, but in offshore winds tiny drying hbr gives shelter. Steer N initially for G bcn twr; then pick up 350° ldg line, W bcn ≠ conspic Ho. Enter between PHM/SHM bcns; a SCM and NCM bcn drying reef. Limited space inside to dry out; or ‡ outside the hbr either side of the ldg line. Facilities in village, ✕ ☲. Le Bourg is approx 1M walk.

ÎLE DE GROIX, LORIENT AND BELLE ÎLE

(AC 2352, 2353) Île de Groix lies 4M SW of Lorient. Its main offlying dangers are to the E and SE: shoals off Pte de la Croix; Les Chats which extend 1M SE from Pte des Chats; and shoals extending 7½ca S of Loc Maria. Several live firing exercise stations exist in this area, refer to Navtex and VHF for information. **Port Tudy** is the main hbr with easy access and good shelter except from NE'lies.

Lorient is a port of refuge, sheltered apprs and 5 marinas. 7M SE of Lorient, **River Étel** has a pleasant marina but access is via a potentially difficult ent, shifting bar; only approach in good weather and on the last of the flood. Further S do not approach the isthmus of the Quiberon peninsula closely due to rky shoals. 6M W of Quiberon lies Plateau des Birvideaux (lt) a rky bank (depth 4·6m) on which the sea breaks in bad weather.

Belle Île has no dangers more than 2½ca offshore, apart from buoyed rks which extend 7½ca W of Pte des Poulains, and La Truie rk marked by IDM bn tr 5ca off the S coast. The S coast is much indented and exposed to swell from W. In good settled weather (only) and in absence of swell there is an attractive ‡ in Port du Vieux Château (Ster Wenn) 1M S of Pte des Poulains; see *North Biscay Pilot*. On the NE coast lie **Sauzon**, which partly dries but has good anch off and is sheltered from S and W, and **Le Palais**.

9.21.14 LORIENT

Morbihan **47°42'·60N 03°22'·00W** ❄❄❄♠♠♠♠✿✿

CHARTS AC 2821, 304; SHOM 7031, 7032, 7139, 7140; Navi 544, 545; Imray C38

TIDES +0455 Dover; ML 3·1; Duration 0620

Standard Port BREST (←)

Times				Height (metres)			
High Water		Low Water		MHWS	MHWN	MLWN	MLWS
0000	0600	0000	0600	7·0	5·5	2·7	1·1
1200	1800	1200	1800				
Differences LORIENT and PORT LOUIS							
+0005	−0020	−0020	−0010	−1·9	−1·5	−0·7	−0·3
HENNEBONT (Blavet River)							
+0015	−0015	+0005	+0005	−2·0	−1·6	−0·9	−0·3
PORT D'ETEL							
+0020	−0010	+0030	+0010	−2·1	−1·4	−0·5	+0·4

SHELTER Very good, a port of refuge. Île de Groix shelters the ent from SW'lies. Access H24 all tides/weather to all 5 marinas. No ‡ in Lorient hbr/chans, but moorings ENE of La Citadelle and ‡ for shoal draft in Petite Mer de Gâvres (E of Port Louis).

NAVIGATION WPT Passe de l'Ouest, 47°40'·80N 03°24'·85W, 057°/2·0M to 016·5° ldg line. WPT Passe du Sud 47°40'·47N 03°22'·46W, 008·5°/2·15M to abm La Citadelle.

Here a secondary yacht chan passes W of La Jument. It is marked by a RGR buoy to the S and by Le Cochon RGR bcn twr to the N. Ldg/dir lts for the Passe de l'Ouest is shown below:

- **Passe de l'Ouest**: ldg lts 057°, both Dir Q near Port Louis. Front, R twr, W bands. Rear, W daymark, R bands on bldg.
- **Passe du Sud** no longer has leading lights.

Within the hbr there are few navigational dangers and chans are well marked/lit. Yachts do not usually need the inner ldg lines. Beware large multihulls moving at speed.

LIGHTS AND MARKS Conspic daymarks: Water twr 8ca W of Kernével; La Citadelle stbd of ent; Port-Louis spire further E, submarine pens at Pointe de Kéroman and 2 silos N of Île St Michel.

COMMUNICATIONS CROSS 02·97·55·35·35; Ⓘ 02·97·86·29·62; ⊖ 02·97·37·34·66; Auto 08·92·68·08·56; Police 02·97·64·27·17; Sea Police 02·97·12·12·12; Brit Consul 01·44·51·31·00; Ⓗ 02·97·64·81·58; Port HM 02·97·37·11·86; VHF Ch 11 *Lorient Port* (H24); Aff Mar 02·97·37·16·22; Marinas VHF Ch 09 (in season: 0800-1230, 1330-2000, out of season: 0830-1200, 1400-1800).

FACILITIES Lorient Marina, inset on the main chartlet. ☎02·97·21·10·14. www.ports-paysdelorient.fr port-lorient@sellor.com ⌘ in Avant Port 2·5-3m, crowded in season. 320+50Ⓥ €3·40 Jul/Aug; €1·80 Apr/May/Jun/Sep. Wet Basin 2·5m. ▯ ⚒ Ⓒ Ⓔ ✕ ⚓ SHOM ACA ⚓ ⚓ ▱(25t) ⌘(1½t). The only ▮ ▮ in Lorient is at S end of Kernével Marina; call Ch 09. Water bus to/from city. **YC de Lorient**, ▱. **City** All facilities, ✉ Ⓑ ⇌ ✈.

Kernével ☎02·97·65·48·25. www.ports-paysdelorient.fr port-kernevel@sellor.com. Enter at N end of wave-break, Ⓥ berths at S end, 3m. 1000+70Ⓥ pricing same as Lorient, ▮ ▮ ▯ ▱ YC.

Ban-Gâvres ☎02·97·65·48·25. Tiny, part-drying harbour. Strong tides need care on the approach which is from the SW, passing W and N of Île aux Souris; recce by foot ferry from/to Port Louis. 57+8 Ⓥ berths (2m) on W end of yacht pontoon, €1·75. Water bus to/from city.

Port Louis ☎02·97·82·59·55. www.portspaysdelorient.fr port-portlouis@sellor.com 450+50Ⓥ €2·80, dredged 2m, ▱(50t). Exposed to strong N'lies. A few FVs; basic facilities, but friendly; interesting town. Water bus to/from city.

Locmiquélic ☎02·97·33·59·51. www.locmiquelic.com locmiquelic@sagemor.fr 447+30Ⓥ €3·60 Jul/Aug; €2·30 Apr/May/Jun/Sep, in 1·5-3m. No Ⓥ pontoon: pre-call Ch 09 or berth as available and check in. No allocated ⚓s, but call Ch 09 for availability. The town is 500m walk. ▯ ⚓ (limited) ▱. **Pen Mané Marina** has no Ⓥ ⌘ but some ⚓s, beware strong cross-tide during full flood/ebb.

River Blavet is navigable on the flood for 6·3M to Hennebont. Use AC 304 plan Ⓐ. Leave Lorient at about HW–2. The river is attractive once clear of industrial areas in the buoyed lower reaches. The 3 bridges have least clearance of 21m. 0·3m patches occur beyond the 1st bridge and before and after the 2nd.

At Hennebont the sheltered 14m Ⓥ pontoon is 1st to stbd; the 2nd is for vedettes only. No ⏃ ⚓; 12 hrs max stay. Good facilities in town. See also Area 18 for the Canal du Blavet to Pontivy.

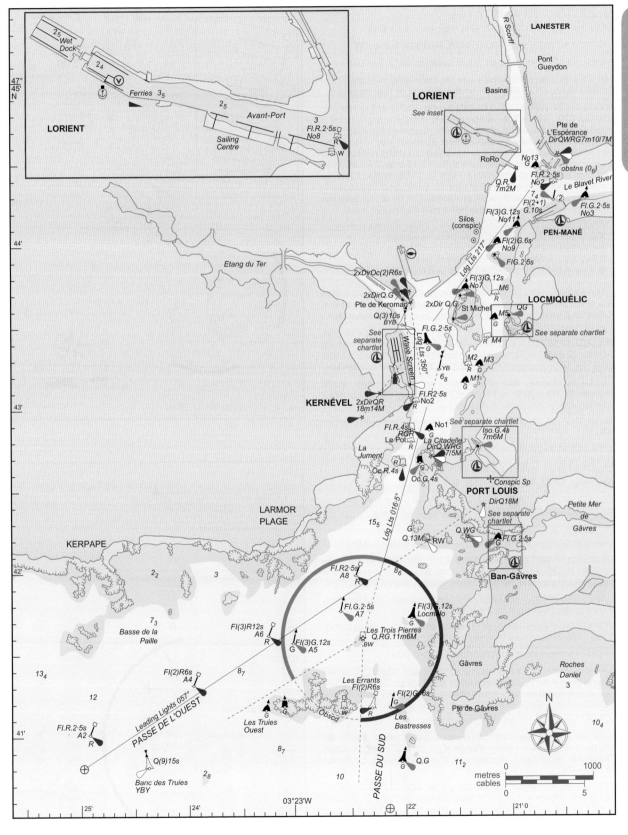

9.21.15 RIVER ÉTEL

Morbihan **47°38'·65N 03°12'·69W** (Hbr ent) ✵⚓⚓✿✿

CHARTS AC 2822; SHOM 7032, 7138; Navi 545; Imray C38

TIDES +0505 Dover (UT); ML 3·2m; Duration 0617. See Lorient.

SHELTER Excellent in marina inside the town quay. Possible ⚓s S of conspic 🏠 ho, off Le Magouër on the W bank or N of marina.

NAVIGATION WPT 47°37'·31N 03°13'·44W, 020°/1·4M to hbr ent, aligned 020° with a R/W radiomast (off chartlet). Night entry is strongly discouraged by Hbr Authority; daytime, approach only in good vis and settled weather; best at about HW –1½ on the last of the flood. The chan is well buoyed but shifts radically; bar dries approx 0·4m. Call *Semaphore d'Etel* Ch 13 or ☎02·97·55·35·59 HW-3 to HW+2 (HW-4 to HW+2 15/06 to 31/08) 0600-2200LT, or daylight hours if less, for clearance to enter, and directions in simple French. If no VHF radio, pre-notify ETA by ☎, for pilotage signals to be shown by movable red arrow on Fenoux mast:

> Arrow vertical (also stowed position) = Enter, maintain track.
> Arrow deflected L/R = turn in direction indicated.
> Arrow horiz = no vessels to ent/exit, conditions dangerous.
> R flag = no ent/exit, insufficient depth over bar.
> ● = no ent/exit for undecked boats and craft <8m LOA.

LIGHTS AND MARKS Dir lt W side of ent, Oc (2) WRG 6s; no ⚓ within 5ca of it. Épi de Plouhinic bn, Fl R 2·5s 7m 2M, marks groyne at ent. Other lts & marks as chartlet and Lights, buoys and waypoints.

COMMUNICATIONS CROSS 02·97·55·35·35; Auto 08·92·65·02·56; Police 02·97·55·32·11; ⊖ 02·97·36·20·76; Brit Consul 01·44·51·31·00; Aff Mar 02·97·55·30·32; Semaphore Stn 02·97·55·35·59; Marina VHF Ch 9, 13, 16 (in season 7/7 0830-1230/1330-1900 and tide times).

FACILITIES Marina ☎02·97·55·46·62. www.ria-etel.com etel@ sagemor.fr 383 + 20 Ⓥ in 1·5-2m, €2·90 Jul/Aug; €2·30 Apr/May/ Jun/Sep. **Quay** ✕ ⬚ 🛢 ⛽ ⚒ 🔧 ⚓(6t) ⚓.

Town ✕ ⬚ 🛒 🗑 🏤 📮 🏦. Bus to ⇌ (Auray 15km) ✈ (Lorient 32km).

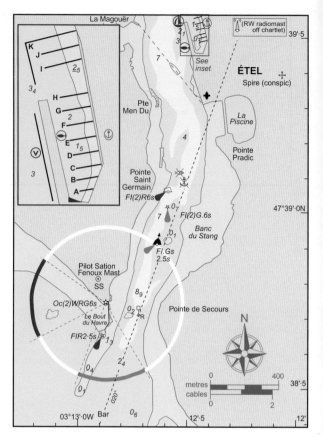

9.21.16 SAUZON (Belle Île)

Morbihan **47°22'·47N 03°13'·01W** ✵✵⚓⚓✿✿✿

CHARTS AC 2822; SHOM 7032, 7142 (1:7,500 plan); Navi 545

TIDES +0450 Dover; ML 3·0; Duration 0615. Use differences for Le Palais.

SHELTER Good except in NE'lies. Drying creek extends 0·5M S.

NAVIGATION WPT 47°22'·88N 03°12'·73W, 205°/685m to NW Jetée lt. Main lt, QG, and NW Jetée, Fl G 4s, form a 205° transit. No navigational dangers. No ⚓ between Sauzon and Le Palais.

LIGHTS AND MARKS Le Gareau SHM bcn is conspic 0·9M N of hbr; ditto Pte des Poulains lt ho 1·6M NW.

COMMUNICATIONS HM 02·97·31·63·40, VHF Ch 09, Jul/Aug: 0900-1230, 1700-2000; closed Sun; see also Le Palais.

FACILITIES www.sauzon.fr/Leport.htm Yachts may:

> • ⚓ ENE of hbr ent • Pick up one of 22 W ⚓s (<15m LOA) N of the N jetty • Raft up between 20 W ⚓s in about 2m on W side of Avant Port (FVs moor E side of Avant Port) • Dry out on the E side of the inner hbr on 50 R ⚓s, fore and aft; firm, level mud/sand. Inner hbr dries about 1·6–2m.

Note: ⚓s are R in avant port, W outside; locals' buoys are G. ⚓s €2·10, ⚓ €1·05. ⚓ on quay. **Village** ✕ ⬚ 🛒 🗑 📮 Ⓑ Ⓒ; market Thurs. Bus to Le Palais. Repairs best at Port Haliguen, La Trinité or Crouesty.

STER WENN, Morbihan, **47°22'·30N 03°15'·24W**. AC 2822; SHOM 7032, 7142. Tides as Sauzon. A very popular W coast anchorage 1·1M S of the N tip of Belle Île (recce from Sauzon, 3km by bike). From the NW enter Ster-Vraz, the outer inlet. When abeam the 5th hole of the golf course to the N, Ster Wenn opens to stbd. ⚓ where space permits, stern line ashore. Shelter is good in all winds. Facilities: None, other than relative peace.

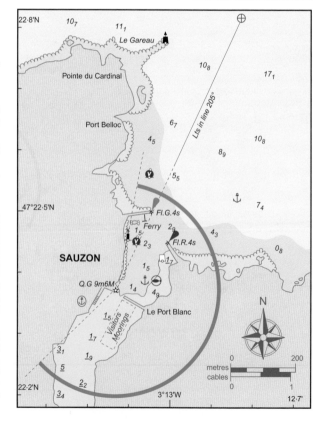

9.21.17　LE PALAIS (Belle Île)

Morbihan 47°20'·84N 03°08'·98W ❋❋❋⚓⚓♜♜♜

CHARTS AC 2822, 2823; Imray C39; SHOM 7032, 7142; Navi 545

TIDES +0458 Dover; ML 3·1; Duration 0615

Standard Port BREST (←—)

Times				Height (metres)			
High Water		Low Water		MHWS	MHWN	MLWN	MLWS
0000	0600	0000	0600	7·0	5·5	2·7	1·1
1200	1800	1200	1800				
Differences LE PALAIS							
−0005	−0025	−0025	−0010	−1·9	−1·4	−0·8	−0·3

SHELTER Good, but strong E'lies cause heavy swell in Avant Port. No ⚓ between Le Palais and Sauzon, nor in hbr.

NAVIGATION WPT 47°21'·20N 03°08'·39W, 231°/6ca to N Pier lt. Obey 3 lateral lt buoys off hbr ent; give way to fast ferries.

LIGHTS AND MARKS Lts as chartlet. La Citadelle is conspic.
COMMUNICATIONS CROSS 02·97·55·35·35; Auto 08·92·68·08·56; Police 02·97·31·80·22; ⊖ 02·97·31·85·95; Brit Consul 01·44·51·31·00; Ⓗ 02·97·31·48·48; Dr 02·97·31·40·90; HM 02·97·31·42·90, VHF Ch 09 (Season: 0800-1200, 1400-2000. Out of season 0800-1200, 1500-1800); Aff Mar 02·97·31·47·08.

FACILITIES www.belle-ile.com **Avant Port** 70 ⚓s (€2·00) in 3 trots inside Mole Bourdelle for yachts <18m LOA; shallow draft on ⚓s to port. Multihulls +150%, +200% Jul/Aug. ⛟ ⛵ ⚓ ◗(10t,5t) ⚓. Packed in season. **Inner hbr** dries. Poss berthing on stbd quay.
Bassin à Flot Enter via a gate/small lifting bridge open HW −1½ to +1 (0600-2200LT) for rafted berthing on either side in 2·5m. ⛵ (90) €2·50, multihulls +150%, +200% Jul/Aug; max LOA 30m. ⚓ ⊞ ⚒.
Marina (Bassin de la Saline, 1·7–2m) prior arrangement only (24hrs notice) if finger berths vacant; max LOA 13m. Enter via lifting bridge beyond Bassin à Flot. 8 ♥ €2·50. **YC** ☎02·97·31·55·85.
Town ✕ ⊡ ⛻ ⊙ ⛟ ⚖ ⊠ Ⓑ. Ferry to Quiberon ⇌ ✈. UK ferry: via Roscoff.

9.21.18　PASSAGE DE LA TEIGNOUSE　*See the upper chartlet on the next page*

This is the easiest route through the Chaussée de La Teignouse being deep, wide, well buoyed and lit. See AC 2357; SHOM 7141. From WPT 47°25'·35N 03°05'·00W appr with La Teignouse lt ho brg 036° (by night in its W sector 033°-039°). The chan is marked by Goué Vas Sud SCM, Basse du Milieu SHM and Goué Vas Est PHM lt buoys. Abeam Goué Vas Est, turn stbd 068°, passing between Basse Nouvelle PHM and NE Teignouse SHM lt buoys, into open waters. Minor variations on this route can be used.　Further SE, **Passage**

du Béniguet 044° lies NW of Houat and **Passage des Soeurs** 016° between Houat and Hoëdic, both unlit.
The **shortest passage**, saving 3M if to/from Port Haliguen, lies between Le Four SCM and Les Trois Pierres NCM bcn twrs with La Teignouse lt ho brg 094°. Turn N as the rocky chan opens between Iniz en Toull Bras to stbd and a SCM and 2 ECM perches to port; see pecked line. Only for experienced navigators by day, exercising caution at LW; pre-study SHOM 7141 carefully; expect strong cross-streams.

9.21.19　HOUAT and HOËDIC

❋❋❋⚓♜♜♜

CHARTS AC 2823, 2835; Imray C39; SHOM 7033, 7143; Navi 545

TIDES +0505 Dover; ML 3·1; Duration 0605

Standard Port BREST (←—)

Times				Height (metres)			
High Water		Low Water		MHWS	MHWN	MLWN	MLWS
0000	0600	0000	0600	7·0	5·5	2·7	1·1
1200	1800	1200	1800				
Differences HOUAT							
+0005	−0025	−0025	−0010	−1·8	−1·4	−0·7	−0·4
HOËDIC							
+0010	−0035	−0025	−0020	−1·9	−1·5	−0·8	−0·4

HOUAT 47°23'·57N 02°57'·34W (Port St Gildas). Appr from N or NE passing abeam La Vieille rk (conspic 17m) NNE of which mussel beds are marked by 4 buoys: NCM VQ, ECM VQ(3) 5s and 2 unlit SPM. The only lt on the Island is on the hbr bkwtr head, Fl (2) WG 6s; the G sector covers La Vieille and buoyed mussel beds. At night appr in W sectors on either side; details in Lts, bys and waypts.

Good shelter, except from N/NE'lies, at Port St Gildas, near E end of the N coast. Moor fore & aft on two dumbbell ⚓ trots in centre hbr and along N wall; no ⚓ in hbr. S part of hbr dries. Keep clear of ferries on W quay and FVs on N quay. E of the hbr are 15 W ⚓s and space to ⚓.
Tréach er Gourhed, aka Rade de Houat, is a 7 cables wide, crab-shaped bay at the ESE end of Houat, designated a No-anch area due to cables. Other ⚓s at Tréach er Béniguet, Portz Ler and Portz Navallo, all at NW end of Houat. No ent (15/4-31/8) to Nature reserves at Er Yoc'h and Ile Guric and Séniz at NW end. Facilities: Mairie 02·97·30·68·04, ✕ ⊡ ⛻ ⊙ ⛟ ⚖ ⊠ Dr. Ferries to Quiberon, Lorient, La Turballe, Le Croisic.

HOËDIC 47°20'·67N 02°52'·45W (Port de L'Argol). Appr from NNW or NNE passing abeam La Chèvre rk (unlit IDM bcn) and W sectors of bkwtr lt, Fl WG 4s; its G sector (163°-183°) covers La Chèvre.
Good shelter, except in N/NE'lies, at Port de L'Argol, on N coast. Ferries berth at the W end hbr wall; a 72m walk ashore pontoon with finger berths lies inside the E end, a trot of 4 large ⚓ in hbr centre, or ⚓ outside to NNE. On the S coast, Port de la Croix, dries but ⚓ outside in 2-4m. Other ⚓s in the island's lee. 1·5M SE of Hoëdic, Grands Cardinaux lt ho, R twr/W band Fl (4) 15s, marks the outer reefs. Facilities: ✕ ⊡ limited ⛻. Ferries as for Houat.

9.21.20 PORT HALIGUEN

Morbihan **47°29'·35N 03°05'·97W** ❀❀❀⚓⚓⚓🌸🌸

CHARTS AC 2357; Imray C38, 39; SHOM 7141, 7032, 7033; Navi 545

TIDES +0500 Dover; ML 3·1; Duration 0615

Standard Port BREST (⟵)

Times				Height (metres)			
High Water		Low Water		MHWS	MHWN	MLWN	MLWS
0000	0600	0000	0600	7·0	5·5	2·7	1·1
1200	1800	1200	1800				
Differences PORT HALIGUEN							
+0010	−0020	−0015	−0010	−1·7	−1·3	−0·7	−0·3
PORT MARIA							
+0010	−0025	−0025	−0015	−1·7	−1·4	−0·7	−0·2
LA TRINITÉ							
+0025	−0020	−0015	−0010	−1·7	−1·2	−0·6	−0·3

SHELTER Good, but uncomfortable in strong NW to NE winds.

NAVIGATION WPT 47°29'·80N 03°05'·00W, 237°/7ca to bkwtr lt. From W or S, appr via Passage de la Teignouse.

LIGHTS AND MARKS Hazards: Banc de Quiberon (2·7m at S end) marked by an unlit NCM and a lit SCM buoy. The W sector (299°-306°) of bkwtr lt, Oc (2) WR 6s, clears its S end. A 1·7m shoal marked by an unlit SCM buoy 'Port Haliguen', is avoided by approaching in the other W sector (233°-240·5°) of the bkwtr lt.

COMMUNICATIONS CROSS 02·97·55· 35·35; 🕽 02·97·64·34·86; Auto 08·92·68·08·56; Police 02·97·50·07·39; Dr 02·97·50·13·94; Marina Ch 09.

FACILITIES **Marina** Access H24. ☎02·97·50·20·56. www.sagemor. com port-haliguen@sagemor.fr 970 + 100**Ⓥ** €3·60. **Ⓥ** berths alongside/rafted both sides of pontoon V, with 10 finger berths on the NW end arm. 🛥🛢✕🖥🛒🔧⚓🛞 Ice 🔲(13t) 🛞(2t) 🛒. **Town** (Quiberon) ✕🖥🛒🛢✉Ⓑ➘✈.

PORT MARIA, Morbihan, **47°28'·54N 03°07'·31W**. AC 2357; SHOM 7032, 7141. See Port Haliguen. Shelter good in all winds, but access dangerous in strong SE–SW winds. It is a very busy ferry/FV port, only feasible for yachts in emergency. ⚓ in SW of hbr in approx 2m. HM ☎02·97·50·08·71. Facilities: ⚓ at E quay, 🖥🔧🛞 SHOM 🛞(6t).

9.21.21 LA TRINITÉ-SUR-MER

Morbihan **47°35'·28N 03°01'·46W** ❀❀❀⚓⚓⚓🌸🌸

CHARTS AC 2371, 2357; Imray C39; SHOM 7033, 7141,7034; Navi 545, 546

TIDES +0455 Dover; ML 3·2; Duration 0610

Standard Port BREST (⟵). Differences see Port Haliguen.

SHELTER Very good, except in strong SE/S winds near HW when La Vaneresse sandbank is covered. No ⚓/fishing in river.

NAVIGATION WPT 47°33'·38N 03°00'·42W, 347·2°/1·0M to No 2 PHM buoy. No navigational dangers; the Rivière de Crac'h is well marked by buoys and perches. Best water close to E bank. Beware many oyster beds, marked with perches. The channel is buoyed on both W sectors. Speed limit 5kn.

LIGHTS AND MARKS Daymarks include: Mousker, a 4·5m rk, off-white paint; and ✠ spire at La Trinité. See chartlet and Lights, buoys & waypoints. Rivière de Crac'h Q WRG, W twr/G top, but amid trees. 1M up-river: Dir lt 347°, Oc WRG 4s, W twr. S Pier, Q WR 6s 9/6M, W twr/R top, W 293·5°-300·5°.

COMMUNICATIONS CROSS 02·97·55·35·35; 🔲 02·97·55·01·15; 🕽 02·97·64·34·86; Auto 08·92·68·08·56; Police 02·97·55·71·62; ⊖ 02·97·55·73·46; Brit Consul 01·44·51·31·00; Dr 02·97·55·74·03; Marina VHF Ch 09; Aff Mar 02·97·24·01·43.

FACILITIES **Marina** ☎02·97·55·71·49. www.baie-de-quiberon.com access H24 at all tides. Marina boat will meet. **Ⓥ** pontoon (fingers) is first N of ✫ Iso R 4s. 1030 +120**Ⓥ** €3·60. 🛥🛢🔧🖥🛞 SHOM, Ⓔ 🛞 🔧 Ice 🔲(36t) Grid.

CN ☎02·97·55·73·48.

Town ✕🖥🛒🛢✉Ⓑ. ➘ Auray. ✈ Lorient, Rennes. UK Ferry: via Roscoff, St Malo.

9.21.22 GOLFE DU MORBIHAN

Morbihan **47°32'·87N 02°55'·28W** (Ent) ❄❄💧💧💧✿✿✿

CHARTS AC 2823, 2371; SHOM 7033, 6992, 7034; Imray C39; Navi 546. Road map N° 0921 (IGN TOP 25 series, 1:25,000) is most useful in the Morbihan as a shoreside complement to charts.

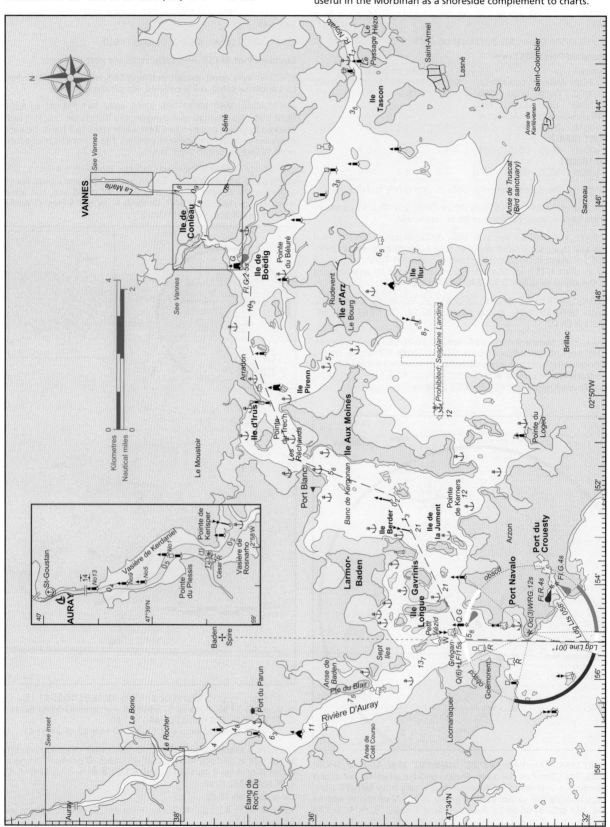

TIDES +0515 Dover; ML 3·0

Standard Port BREST (◀——)

Times				Height (metres)			
High Water		Low Water		MHWS	MHWN	MLWN	MLWS
0000	0600	0000	0600	7·0	5·5	2·7	1·1
1200	1800	1200	1800				
Differences PORT NAVALO							
+0030	−0005	−0010	−0005	−2·1	−1·6	−0·9	−0·4
AURAY							
+0035	0000	+0015	−0005	−2·3	−1·9	−1·2	−0·5
ARRADON							
+0135	+0145	+0140	+0115	−3·9	−3·1	−1·9	−0·8
VANNES							
+0200	+0150	+0140	+0120	−3·8	−3·0	−2·1	−0·9
LE PASSAGE (47°35′·4N 02°43′·1W)							
+0200	+0200	+0210	+0135	−3·8	−3·0	−2·0	−0·9
LE LOGEO							
+0140	+0140	+0145	+0115	−4·1	−3·2	−2·1	−0·9

SHELTER The Golfe du Morbihan is a large, inland tidal sea with deep apprs and entrance. A significant sea state is rare, except for tide rips in the ent. Shelter can usually be found from all winds, but the many ⚓s (see chartlet and Facilities) are increasingly restricted by extensive local moorings. It is essential to use chain when ⚓ing, unless well out of the tide. Île aux Moines and Île d'Arz are the only public islands. Vannes is the only marina.

NAVIGATION WPT 47°31′·97N= 02°55′·29W (as for Crouesty) 001°/1·7M to abeam Grand Mouton lt bcn. Pilotage is not difficult and the strong streams are not inherently dangerous, it's worth pre-plotting the desired tracks/distances within the channels. Marks can then be more readily identified and track adjusted with ease, especially if beating. Night navigation is not advised.

Caution: Many vedettes in season. Frequent ferries cross from Port Blanc to Île aux Moines, at times with scant regard for IRPCS; keep well clear of them. Avoid numerous bird sanctuaries in SE, and oyster beds marked by withies. Much of the E & SE dries.

TIDAL STREAMS The very strong tidal streams which reach 9·1kn between Île Berder and Île de la Jument dictate ent/exit times.

On AC 2371 study tidal diamonds ◇ B (off Port Navalo lt ho), ◇ D at Grand/Petit Mouton and ◇ G at the S tip of Ile Berder. These give a good idea of when the flood and ebb should start, hence earliest times to ent/exit. At ◇ B the flood starts at about HW Brest −4½. At ◇ G the ebb starts at about HWB +2¼. Their rates however may be conservative. At ◇ D for example the max rate is 5·2kn whereas the tidal vectors here show 6·9 to 8kn.

For a first visit avoid springs; entry at night with a tide running is hazardous. Off Grand Mouton sp rates reach 8kn and in the ent can exceed 4kn. Flood starts HW Brest – 0400 and turns at HW Brest + 0200. Here the flood divides: the weaker flow enters the R. Auray; whilst the major stream sets strongly toward Petit & Grand Mouton rks before setting into the main chan. Maintain 001° until Grand Mouton lt bcn is abaft the beam, but beware shoals E of Goémorent R bcn twr.

The streams ease in the upper reaches except between Pte de Toulindag and Port Blanc where ◇ K indicates >5kn. HW times become later further E, eg HW Vannes is 2 hrs after HW Pt Navalo. HW Auray is only 25 mins after HW Port Navalo.

LIGHTS AND MARKS The 001° appr ldg daymarks are: front, Petit Vézid W obelisk (from afar looks like a yacht sail); rear, Baden ⊹ spire (3·3M). Chans and dangers are well marked. The only lts are: Port Navalo lt ho at the ent. Inside ent: Grand Mouton SHM bcn and Le Grégan, a squat SCM bcn. Roguédas, a large SHM bcn is near the appr chan to Vannes.

FACILITIES, clockwise from **AURAY RIVER**

LOCMARIAQUER 47°34′·19N 02°55′·81W. Two drying quays 500m apart: To S, ferries use the buoyed chan to Cale du Guilvin, 🛒 (S side only, max 30 mins) ⚓ fuel 06.03.09.15.06; drying ⚓ to W. To N, Cale de Locmariaquer (used by ferries) slip. Village: ✕ 🍴 🛒 © ✉ bike hire, buses.

SEPT ÎLES 47°35′·09N 02°56′·18W: Small, quiet ⚓. Enter Anse de Baden, mostly dries, between mainland and W end of island.

PORT DU PARUN 47°36′·76N 02°56′·96W: Drying inlet: no moorings but 🍴 with quay, slip & all facilities. ☎02·97·57·00·15.

LE ROCHER 47°38′·17N 02°57′·75W: Sheltered, but full of moorings.

LE BONO 47°38′·43N 02°57′·21W: Moor or ⚓ (rky bottom) off Banc de la Sarcelle. 🛒 in drying basin, ☎02·97·57·88·98. Village: ✕ 🍴 © ✎ 🔋 ✉.

AURAY 47°39′·84N 02°58′·73W: The river shoals to 0·2m, but St Goustan is accessible at mid-flood via a bridge with 14m MSL and 0·4m depth; gauge shows air clearance. Calculate tides carefully for a safe passage. 12 dumbbell ⚓s in a pool upriver; drying 🛒 at St Goustan; or ⚓s S of 14m bridge. HM ☎02.97.56.29.08. ⚓ ✎ 🖥 ✕. Town ✕ 🍴 🛒 © ✉ ⚓.

NW AND CENTRAL AREAS

ÎLE LONGUE 47°34′·17N 02°54′·44W: ⚓ out of the stream. No landing.

LARMOR BADEN 47°34′·82N 02°53′·62W: good ⚓s to S and further W, but many moorings, HM ☎02·97·57·20·86; 06.14.19.12.76. Village: ✕ 🍴 🛒 ✉ ⚓ 🛢(4t). Ferry to Île Gavrinis all year.

ÎLE BERDER 47°34′·81N 02°53′·05W: pleasant ⚓ E of the island.

PORT BLANC 47°36′·32N 02°51′·66W: HM ☎02·97·26·30·57, mob 06·11·07·56·89. Ch 09, 0845-1230; many moorings €2·40. Quay, ⚓ ✕ 🍴 🔧 🖥 © ✉ free seasonal water taxi, Ch 09 *Service de Rade*. Ferry ½hrly to Île aux Moines.

ÎLE AUX MOINES: a popular public island. The narrows between Pte de Toulindag and Port Blanc can be rough; beware Les Réchauds drying rks. ⚓ 47°36′·22N 02°51′·11W; or ⓥ pontoon (🖥 🔧) the W'ly of two; land at small marina. Free seasonal water taxi, VHF Ch 09 *Service de Rade* or sound horn. HM ☎02·97·26·30·57, 0845-1230, 🔧. Quieter ⚓s S & E of Pte du Trec'h and SE tip of island.

ARRADON: limited ⚓ 47°36′·84N 02°50′·24W, exposed to S. HM ☎ 02·97·44·01·23. Ch 09. Apr-Sep 1400-1700. Many local moorings, €2·40. ⚓ 🔧. Free seasonal water taxi, Ch 09 *Service de Rade*. Pier, depth gauge/disabled access, max 15 mins stay on W side only.

ÎLE PIRENN: exposed ⚓ 47°35′·79N 02°49′·42W in tidal stream.

ÎLE D'ARZ, a public island. ⚓ 47°36′·45N 02°47′·43W NE of Pte du Béluré (ferry landing); E of Le Bourg (good shelter); or to the W, depending on winds. Rudevent village: ✎ 🖥 ✕.

ÎLE DE BOËDIG: sheltered ⚓ 47°37′·00N 02°46′·53W, if moorings permit, in Chenal de Badel N of the island.

ÎLE DE CONLEAU 47°37′·62N 02°46′·73W: ⚓ or moor in bight just S of village, as space permits. ✕ ✎ 🖥 ✕ in village.

EAST AND SOUTH AREAS

SÉNÉ 47°37′·19N 02°44′·49W: dry out in traditional hbr at head of inlet; ✎. Easy bike recce 2M SSE from Vannes.

LE PASSAGE: ⚓ 47°35′·43N 02°42′·93W off Pte du Passage (depths upriver not charted). Village: (St Armel): 🍴. Seasonal ferry.

KERNERS: ⚓ 47°33′·65N 02°52′·44W in 3-6m off Anse de Kerners or further SE off Anse de Pen Castel. Ferry to Île aux Moines.

ÎLE DE LA JUMENT 47°34′·20N 02°53′·15W: good shelter E of island out of the tide; handy for leaving the gulf on the tide.

PORT NAVALO 47°33′·00N 02°54′·86W: ⚓ in bay, but full of moorings, and open to W/NW winds. Convenient to await the tide. HM ☎02·97·53·82·12. All facilities at Crouesty (½M by road).

BAIE DE QUIBERON

Open to the E and SE, this is an important and attractive cruising area with centres at **Port Haliguen**, **La Trinité**, the **Morbihan** and **Crouesty** (AC 2823). The S side of the bay is enclosed by a long chain of islands, islets, rocks and shoals from Presqu'île de Quiberon to Les Grands Cardinaux 13M SE. The islands of **Houat** and **Hoëdic** are well worth visiting, ideally mid-week.

From W or S, enter the Bay via Passage de la Teignouse in W sector (033°-039°) of La Teignouse lt ho; thence 068° between Basse Nouvelle lt buoy and NE Teignouse lt buoy.

▶*In this channel the NE-going flood starts at HW Brest −6¼, and the SW-going at HW Brest, sp rates 3·75kn; in strong winds it is best to pass at slack water.*◀

Passage du Béniguet and Passage des Soeurs are alternatives.

9.21.23 VANNES

Morbihan 47°38'·45N 02°45'·62W ✺✺◊◊◊✿✿✿

CHARTS AC 2823, 2371; Imray C39; SHOM 7034; Navi 546

TIDES See GOLFE DU MORBIHAN. ML 2·0m

SHELTER Excellent in all winds.

ACCESS by day only. Whilst the entry gate stays open (HW±2½) Kérino **swing bridge** (road tfc) opens only in daylight at approx H and H+30 in season (15 Jun–15 Sep) and at weekends; but only at H out of season. During the first and final ½ hour periods when the entry gate is open, the bridge will open on request VHF Ch 09. Outbound craft have priority over arrivals. Waiting pontoons (partly drying) are up-/downstream; intercom to HM on downstream pontoon. Bridgemaster may occasionally speak by loudhailer.

> Traffic signals (3 vert lts) on the bridge's western pier are:
> F ●●● = no passage; F ●●● = proceed;
> Oc ●●● = No passage unless committed, but
> unmasted boats may transit.

An **entry gate** (not a lock) approx 250m N of the bridge and remotely-controlled by the HM, stays open HW±2½. HW Vannes –2½, when the gate opens, is the same as HW Port Tudy (a French Standard Port). The sill retains 2·1m in the wet basin. F ●●●●● = NO GO; F ●●● = GO.

Guide pratique du Port (a free annual schedule of gate hrs etc) is available from Port de Plaisance, La Rabine, 56019 Vannes or download from www.mairie-vannes.fr/getFile.aspx?FILEID=8400

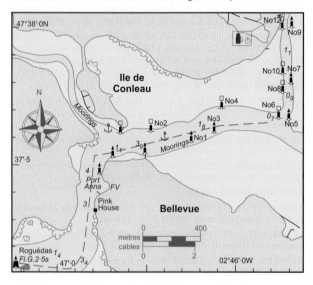

NAVIGATION WPT: see Golfe du Morbihan. After Roguédas SHM lt bcn do not cut the shallow corner to port; if anything head ESE until the chan opens up. Turn N'wards when a pink house on its E bank bears 020°. Passage upstream past Île de Conleau is easy and well marked. Beacon'd appr chan to Vannes is narrow, least charted depth 0·7m, but dredged 2·1m from No 6 bcn to the bridge.

LIGHTS AND MARKS No navigation lts N of Roguédas, Fl G 2.5s.

COMMUNICATIONS CROSS 02·97·55·35·35; ⚓ 02·97·42·49·49; Police 02·97·47·19·20; ⊖ 02·97·01·36·00; Brit Consul 01·44·51·31·00; Ⓗ 02·97·01·41·41; HM VHF Ch 09 (0800-1200, 1500-1900 Jul-Aug); Aff Mar 02·97·63·40·95.

FACILITIES **Marina** ☎02·97·01·55·20. www.mairie-vannes.fr port. plaisance@mairie-vannes.fr 214+70Ⓥ €2·64, multihulls +50%. Marina launch may allot a finger berth. If not, visitors berth/raft N/S on pontoons D and G, just S of mobile foot-bridge (*passerelle*).

🅙 Ⓒ 🛠 🏷 ⛽ 🅛 ⚒ ⚓ ⛲ Ⓔ ◭ ⚓ 🏗(10t); nearest fuel by hose is at Crouesty.

City ✕ 🍴 🖂 SHOM ⊠ Ⓑ ⇌ ✈.

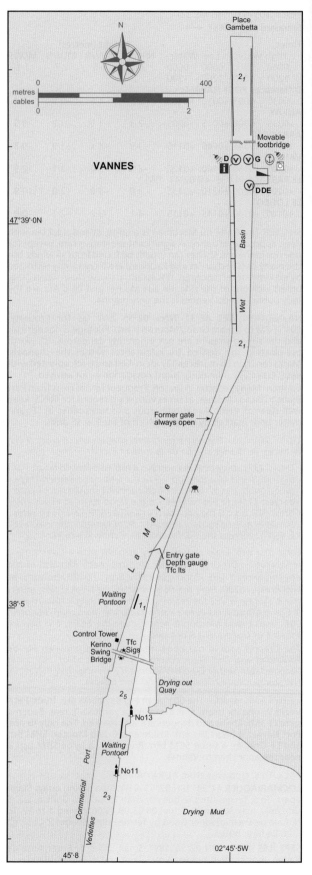

CROUESTY TO LE CROISIC

East of the Morbihan entrance, dangers extend 1M seaward of Pte de St Jacques (where a small hbr dries). 3M offshore Plateau de la Recherche has depths of 1·8m (AC 2823). SE of **Pénerf**, which provides good anchorage, Plateau des Mats is a rocky bank, drying in places, extending up to 1·75M offshore.

Across the ent to the **Vilaine River** beware La Grande Accroche, a large shoal with least depth 1m. The main lit chan leads NW of La Grande Accroche, to the bar on N side.

▶*Here the flood begins at HW Brest –0515, and the ebb at HW Brest +0035, sp rates 2·5kn.*◀

In SW winds against tide the sea breaks heavily; the Passe de la Varlingue, 5ca W of Pte du Halguen, is then better, but beware La Varlingue (dries 0·3). At **Arzal/Camoël** yachts can lock into the non-tidal Vilaine River for **La Roche Bernard** (6km) and **Redon** (29km) to access the **Ille et Rance Canal** to Dinan/St Malo.

South of Pte du Halguen in a little frequented corner of Quiberon

Bay, other dangers include the rocky shoals, both 0·6m, of Basse de Loscolo and Basse du Bile, close inshore. Off Pte du Castelli, the Plateau de Piriac extends about 1·75M NW with depths of 2·3m and drying rocks closer inshore. The small hbr and marina at **Piriac** lies on the N side of Pointe du Castelli and Les Bayonelles (dry) extend 5ca W. A chan runs between Plateau de Piriac and Île Dumet (lt) which is fringed by drying rocks and shoals particularly on N and E sides.

In the Rade du Croisic are the fishing harbours of **La Turballe** and **Le Croisic**, the former with a marina. Off Pointe du Croisic dangers extend 1M to N and W. Plateau du Four, a dangerous drying bank of rocks, lies about 4M W and WSW of Pointe du Croisic, marked by buoys and lt ho near N end. Approx 3M SW of Ouest Basse Capella WCM, a Restricted Area centred on 47° 14'·37N 02° 46'·65W, with limits marked by 4 SPM Fl Y 4s, encloses 2 ODAS bys, Fl (5) Y 20s. 2·8M ESE of Pointe du Croisic, Basse Lovre is a rocky shoal with depths of 0.7m, 5ca offshore, marked by a SCM buoy.

9.21.24 CROUESTY

Morbihan **47°32'·46N 02°54'·13W** ❀❀❀♦♦♦❀❀

CHARTS AC 2823, 2371; Imray C39; SHOM 6992, 7034, 7033; Navi 546

TIDES +0505 Dover; ML 3·0; Duration 0555

Standard Port BREST (←—)

Times				Height (metres)			
High Water		Low Water		MHWS	MHWN	MLWN	MLWS
0000	0600	0000	0600	7·0	5·5	2·7	1·1
1200	1800	1200	1800				
Differences CROUESTY							
+0010	–0025	–0015	–0010	–1·6	–1·2	–0·7	–0·3

SHELTER Very good, but strong onshore winds can raise dangerous seas in the ent. Quiberon peninsula gives some shelter from W'lies.

NAVIGATION WPT 47°31'·97N 02°55'·30W, 058°/0·9M to hbr ent. Access chan is dredged 1·8m and buoyed. Caution: drying shoal close NW of ldg line between Nos 6 and 8 buoys; stay in the chan especially near LW.

LIGHTS AND MARKS Ldg marks/lts 058°: Front W vert stripe on R panel; rear the tall grey lt ho. Only No.1 SHM lead-in buoy is lit, QG. See chartlet and Lights, buoys and waypoints for details.

COMMUNICATIONS CROSS 02·97·55·35·35; Ⓛ 02·97·41·35·35; Auto 08·92·68·08·56; Police 02·97·53·71·65; ⊜ 02·97·01·36·00; Brit Consul 01·44·51·31·00; Dr 02·97·53·71·61; HM VHF Ch 09; Aff Mar 02·97·41·84·10.

FACILITIES The enormous **Port du Crouesty Marina** has 6 large separate basins; see large scale chartlet below. HM ☎02·97·53·73·33. www.baie-de-quiberon.com 1432+130❶ €3·60. On pontoons V1, V2 and V3 in ❶ basin on S side of fairway; boats >12m LOA can berth on the wall between V1 and V2. For free water taxi call VHF Ch 09. ♣ ♠ (H24 with cr/card) Ⓛ Ⓔ ⚓🔧 🕭 ✕ ► ➶(45t) ⏴(10t).

Town (Arzon) ✕ ⬚ 🛒 ⚓✉ Ⓑ. ≋ Vannes. ✈ Vannes, Lorient, St Nazaire, Rennes.

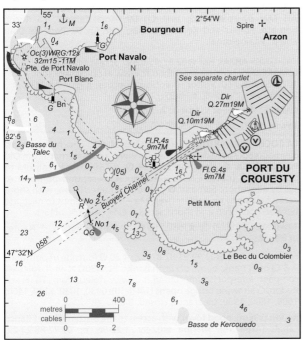

MINOR HARBOUR, approx 13 track miles east of Crouesty

PÉNERF RIVER, Morbihan, **47°30'·03N 02°38'·89W**, AC 2823; SHOM 7033, 7135 (advised, plus good Pilot). HW +0515 on Dover (UT); Duration 0610; ML 3·3m. See Vilaine River. Shelter good, except in fresh W'lies. The 3 ents require careful planning & pilotage:

Passe de l'Ouest is shoal and ill marked. **Passe du Centre** is the widest and easiest, least depth 0·7m. Appr on 359° ldg line, Le Pignon red lt bcn ≠ Le Tour du Parc spire, passing between Penvins PHM buoy and Borénis SHM buoy. Leave a SHM bcn (marking a drying reef) 150m to the E; thence pass 40m E of Le Pignon, whose white sector covers the appr, but night entry is not advised for a first visit.

Passe de l'Est has 4m, but is narrower and rks are close to stbd. It joins Passe du Centre E of Le Pignon bcn. Once in the river, head ENE for 1M to ⚓ off Pénerf quay. Beware oyster beds. Facilities: ► **Village** ✕ ⬚ 🛒 (at Damgan).

9.21.25 VILAINE RIVER

Morbihan 47°30'·38N 02°28'·71W (1st lt buoys) ⚙⚙◊◊◊🌸🌸✿

CHARTS AC 2823; SHOM 7033, 7144; Navi 546; Imray C39. Note: 2823 ends at the river mouth. 7144 continues to La Roche-Bernard.

TIDES +0500 Dover; ML (Pénerf) 3·3; Duration 0610

Standard Port BREST (⟵)

Times				Height (metres)			
High Water		Low Water		MHWS	MHWN	MLWN	MLWS
0000	0600	0000	0600	7·0	5·5	2·7	1·1
1200	1800	1200	1800				
Differences TRÉHIGUIER							
+0035	−0020	−0005	−0010	−1·5	−1·1	−0·6	−0·4
PÉNERF RIVER							
+0015	−0025	−0015	−0015	−1·6	−1·2	−0·7	−0·4

SHELTER Some shelter on 4 blue ⚓s to E of Tréhiguier on S side of river. Total shelter in marinas at Arzal, La Roche Bernard and Foleux. Avoid anchoring in centre of river, which is dredged day and night.

NAVIGATION WPT 47°28'·95N 02°33'·96W, 052°/3·3M to Penlan lt ho; see chartlet. The 3 appr's to the river mouth are:

- Passe de la Grande Accroche: Penlan lt ho ≠ Abbey de Prières 052°. Marks are reportedly conspic by day.
- Penlan lt ≠ Billiers ch tr 025·5°, leaving La Varlingue Rk (0·3m) close to stbd.
- Passe de la Varlingue (unlit): W wall ≠ Abbey Tr 023°, passing close to WCM bcn and oyster poles off Le Logui.

In strong onshore winds, especially at sp ebb, seas break on La Vilaine bar (min 0·5m). Best to enter/leave HW. River is well buoyed to Tréhiguier, adequately beyond; keep to byd chan, river silts.

Arzal lock usually opens hourly, up to 9 times per day, 0700–2100 (LT) Jul/Aug; rest of year, 0800, 0900, 1100,1200, 1400, 1600, 1800, 2000 (LT) but times can vary: call HM Ch 11, or ☎09·69·32·22·97 for recorded info. Keep to the byd chan to avoid the prohib areas (Y buoys) below/above the barrage. Below Arzal lock there is a lge W waiting and 200m waiting pontoon with shore access on N bank. Inside lock secure to vertical chains or raft alongside other vessels and anticipate surge when upper gates open. Above lock, 2 lge W waiting buoys SW Arzal.

LIGHTS AND MARKS See chartlet. Two Dir lts in approaches are:

- Basse Bertrand, 14m G twr. The W sector (040°-054°) overlaps the W sector of Penlan in Passe de la Grande Accroche.
- Pte de Penlan, 18m W twr, R bands; W sector 052°-060°.

At river ent Petit Sécé W bcn twr is easier to see than Nos 1 & 2 buoys. On S side of lock the control twr is conspic.

COMMUNICATIONS CROSS 02·97·55·35·35; Auto 08·92·68·08·56; ⊜ 02·97·63·18·71 (Vannes); Brit Consul 01·44·51·31·00; ⊞ 02·99·90·61·20; Dr 02·97·45·01·21; Pont de Cran 02·99·90·11·31; Aff Mar 02·99·90·32·62; VHF Ch 09: Arzal-Camoël, La Roche Bernard, Foleux and Redon marinas; VHF Ch 10: Pont de Cran and R. Vilaine/canal.

FACILITIES ARZAL HM ☎02·97·45·02·97 arzal-cameol@ compagniedesportsdumorbihan.fr www.arzal-cameol.com 1078+54🅥 €3·60. Visitor reception, pontoon A (🍴) ⚓ 🛠 H24. ✕ 🖥️🅿️@🅿️⚓⚒🛥🛒Ⓔ✕🏴(35t).

CAMOËL 🏴. Check in first at Arzal.

Towns (both 3km) ✕ 🏬🛒🛍 Ⓑ ✉.

LA ROCHE BERNARD (6km): HM ☎02·99·90·62·17. roche-bernard@ compagniedesportsdumorbihan.fr www. roche-bernard.com **Old Port** 96 🅥 108 ⚓. **New Port** 335+56 🅥, €2·90, best to call HM to book ahead, (note: 3 hour lunch break) reception pontoon A, 🍴 by allocation. Summer 🅥 pontoon an alternative, with shore access by dinghy on endless fixed line – do not attempt to navigate between pontoon and shore. ⚓ wi-fi 🅿️✕ 🛥Ⓔ✕🏴(13t).

Town ✕ 🏬🛒🖥🛍🛒 ⚓ Ice ✉ Ⓑ ⛽. ⇌ Vannes/Redon. ✈ Nantes or Rennes.

Foleux (12·5km) ☎02·99·91·87·87. 321 + 20 🅥 (outside of pontoons A & C) on N bank; buoys off both banks. ✕ 🏴(30t) ✕ wi-fi. foleux@ compagniedesportsdumorbihan.fr www.foleux.com To reach Redon, Cran swing bridge opens (1/4-1/10): 0930, 1030, 1130, 1400, 1600, 1800, 1900. Clearance, closed, 5·8m.

Redon (29km): good town, with access to the Brittany/ Ille et Rance canals (Area 18). **Marina** ☎02·99·71·22·96, or 07·77·88·23·22, cepredon@yahoo.fr. 128 +10🅥 €2·20, ⚓ mast 🏴(500kg).

9.21.26 PIRIAC

Loire Atlantique 47°23'·02N 02°32'·67W ✿✿✿✿✿✿

CHARTS AC 2823; Imray C39, 40; SHOM 7033, 7136; Navi 546

TIDES Interpolate between Vilaine River (Penerf) and Le Croisic. HW +0505 on Dover (UT); ML 3·3m; Duration 0605. Locals use predictions for Le Croisic.

SHELTER Good in marina, near HW can be exposed to N'lies.

NAVIGATION WPT 47°23'·69N 02°32'·38W, 197°/7ca to hbr ent in the W sector (194°-201°) of Piriac lt, Oc (2) WRG 6s.

Plateau de Piriac extends about 1M W and N from the hbr; to the SW it is marked by a WCM buoy, Q (9) 15s, and to the NNW by Le Rohtrès NCM bcn twr and by Grand Norven NCM lt bcn, Q. Due to shoals and drying patches do not attempt to pass S of these bcns even near HW.

At night the W sectors of Île Dumet lt (272°-285°) and of Pte de Mesquer lt (067°-072°) help to position within the W sector of Piriac lt, as well as bns, lit QG to N and QR to NE of hbr ent.

LIGHTS AND MARKS Piriac church belfry is a conspic daymark, aligned approx 197° with chan. The Sig stn at Pte de Castelli, 8ca SW of the hbr, is a conspic white, prow-shaped structure.

COMMUNICATIONS CROSS 02·97·55·35·35; ⊞ 02·40·23·55·74 (Jul-Aug only); Auto 08·92·68·08·44; ⊜ 02·40·23·32·51 (Turballe); Brit Consul 01·44·51·31·00; HM VHF Ch 09; Aff Mar 02·40·23·33·35.

FACILITIES Marina ☎02·40·23·52·32. portpiriacsurmer@sainte-nazaire.cci.fr. Flap gate over sill drops once tide reaches CD +2·2m giving access HW±4 for draft <1.5m, more at nps, beware strong currents when entering. Retaining wall marked by Y SPM perches, flap gate by a PHM & SHM perch. IPTS, sigs 2 or 4; the latter shows when depth over sill is >1·5m. 750+70❂ on pontoon E, €3·40. ⛽ ⓉⒺ (on quay) ⬌ Ⓔ ⊞⊟ ⚒ ⚓ ⛴(12t).

Town The attractive small town has ✕ ⬜ ⚒ ⊠ Ⓑ market.

PIRIAC-SUR-MER

9.21.27 LA TURBALLE

Loire Atlantique 47°20'·70N 02°30'·89W ✿✿✿✿✿✿✿

CHARTS AC 2823; Imray C39, 40; SHOM 6826, 7033; Navi 546

TIDES Approx as for Le Croisic

SHELTER Good in all winds, but in strong SSW'lies heavy swell can close the hbr. In N to E winds it is possible to ⚓ S of the hbr.

NAVIGATION WPT 47°20'·19N 02°30'·99W, 006·5°/5ca to W bkwtr lt. The ent is not easily seen until S of the hbr. Appr in W sector (315°-060°) of W bkwtr lt to pick up the 006·5° ldg lts. A number of small unlit R and G bys close to the ldg transit may be ignored. Jun-Sep a buoyed No ⚓ area lies between 060° and 090° 8 cables to the west bkwtr lt. From the SW avoid Plateau du Four and shoals off Le Croisic. There is an active fleet of 80 FVs.

LIGHTS AND MARKS By day Trescalan ⊕ and water tr (conspic) 1M ENE of hbr, lead 070° to just S of ent. R bcn twr is 80m off W bkwtr.

COMMUNICATIONS CROSS 02·97·55·35·35; Auto 08·92·68·08·44; ⊜ 02·40·23·37·51; HM call VHF Ch 09 before entering. If in Jul/Aug the marina is full and closed to new arrivals, it is announced on Ch 09; Aff Mar 02·40·23·33·35.

FACILITIES Marina ☎02·40·23·41·65. www.port-peche-turballe.fr plaisance.turballe@wanadoo.fr 315+30❂ €2·10.

Access H24. Inside ent, turn 110° stbd round a blind corner into the marina (1·5-2m). There are shoal patches between A–G and B–F pontoons, therefore craft >1·5m draught should not enter LW springs ±2 to avoid grounding.

Visitors are rafted tight into the rectangular box off B pontoon; if staying a few days a vacant finger berth may be allotted on request. Call before entering; see R/T.

⛽(min 10 ltrs); pay by French cr/card or at Coopérative Maritime ☎02·40·62·80·05 (facing hbr); ⓉⒺ at garage 500m. ⬌ ⚒ ⊞ Ⓔ ⚓ ⛴(32t) ☎02·40·62·80·40, ⚓ ⊟ in Capitainerie, ⚓ ⚒ YC SHOM Ice @ in Capitainerie and the Municipal library.

Town ✕ ⚒ ⬜ ⊠ Ⓑ Dr. Buses to St Nazaire and Nantes.

LA TURBALLE

9.21.28 LE CROISIC

Loire Atlantique 47°18'·52N 02°31'·35W ※※◊◊◊◊✿✿✿

CHARTS AC 2646, 2986, 2823; Imray C39; SHOM 7033, 7395, 7145 (the preferred large scale chart); Navi 546, 547

TIDES +0450 Dover; ML 3·3; Duration 0605

Standard Port BREST (←)

Times				Height (metres)			
High Water		Low Water		MHWS	MHWN	MLWN	MLWS
0000	0600	0000	0600	7·0	5·5	2·7	1·1
1200	1800	1200	1800				
Differences LE CROISIC							
+0015	−0040	−0020	−0015	−1·6	−1·2	−0·7	−0·4

SHELTER Good. No ⚓ in Le Poul; full of moorings, very strong ebb.

NAVIGATION WPT 47°18'·97N 02°31'·65W, 155·5°/5ca to Tréhic bkwtr hd. Beware rks and shoals close in to Hergo SHM bcn twr in Tréhic's W sector (093°-137°). In its SW G sector dangers extend 2M W to Basse Castouillet WCM buoy. Keep strictly on the ldg lines as appr and hbr dry extensively; chan is dredged 2m at first, then 1·6m and 1·2m. Sp tides reach 4kn. 4 drying basins (*Chambres*) are formed by islets (*Jonchères*).

LIGHTS AND MARKS A ✠ dome, almost on the 155·5° ldg line, and hospital at Pte de Pen Bron are conspic. 3 sets of ldg lines:

Outer 155·5°: both Dir Q; Or □s on W pylons, rear has G top

Middle 173·4°: both Dir QG; Y □s with G stripe on G & W pylons, almost obsc'd by trees.

Inner 134·7°: both Dir QR; R/W chequered □s on fish market roof. At the marina the BY is a large, white conspic shed.

COMMUNICATIONS CROSS 02·97·55·35·35; ⚓ 08·92·68·02·44; Police 02·40·23·00·19; ⊖ 02·40·23·05·38; Brit Consul 01·44·51·31·00; Ⓗ 02·40·23·01·12; Dr 02·40·23·01·70; Aff Mar 02·40·23·06·56; Marina VHF Ch 09 (0800-1200; 1330-2000 in season).

FACILITIES **Marina** (*Chambre 4*). ☎09·81·12·75·92, 06·65·93·34·05. le-croisic-plaisance.peche-plaisance44.fr/   plaisance.lecroisic@ lapp44.fr Visitors dry out against the marina's outer wall or bows-in to inner wall. 360+27◊ €1·90, plus €0·20pp/day tax. ⬛ ⚓ ▣ ✕ ◗ Ⓔ YC Divers ▣ ▢(180t) ▷(8t).

Town ✕ ◻ ⬛ ▤ ▤ ▤ ▮ ⊠ Ⓑ ⇌ ✈ (St Nazaire).

9.21.29   LE POULIGUEN

Loire Atlantique **47°16´·42N 02°25´·33W** ✿✿✿✿✿✿✿✿

CHARTS AC 2646, 2986; Imray C39; SHOM 7395, 7145; Navi 547

TIDES Sp +0435 Dover, Nps +0530 Dover; ML 3·3; Duration Sp 0530, Nps 0645

Standard Port BREST (◀—)

Times				Height (metres)			
High Water		Low Water		MHWS	MHWN	MLWN	MLWS
0000	0600	0000	0600	7·0	5·5	2·7	1·1
1200	1800	1200	1800				
Differences LE POULIGUEN							
+0020	−0025	−0020	−0025	−1·6	−1·2	−0·7	−0·4

SHELTER Very good, except in SE winds. In strong S'lies, beware swell and breakers in the approach chan. Pornichet, 3M to the E, has a much easier approach and ent.

NAVIGATION WPT 47°15´·54N 02°24´·34W, Basse Martineau unlit PHM buoy, 321°/1.1M to hbr ent. Reefs running 4M ESE towards Grand Charpentier lt ho, Q WRG, form a barrier across the Baie du Pouliguen which may be entered via any of 4 passes; see AC 2986 or the larger scale SHOM 7145.

Best appr at HW −1 from W/SW between Pte de Penchâteau and Les Evens (drying reef). From Basse Martineau, leave La Vieille SHM perch and Petits Impairs bcn twr, Fl (2) G 6s, well to stbd, and 3 PHM bns closer to port. The inner access chan shifts, dries approx 1·5m and is marked by one PHM and 5 SHM poles.

LIGHTS AND MARKS See chartlet/Lights, buoys and waypoints for lt details. W jetty, QR, is a slim, conspic white column, R top. The final SHM pole marks the narrowing chan and E training wall which covers. Navigational lights are very hard to see against shore lts of La Baule and night appr is not advised without a prior day visit.

COMMUNICATIONS LB 02·40·61·03·20; CROSS 02·97·55·35·35; Auto 08·92·68·08·44; Police 02·40·24·48·17; ⊖ 02·40·61·32·04; Brit Consul 01·44·51·31·00; Dr 08·36·69·12·34; HM VHF Ch 09, 0900-2000 in season; Aff Mar 02·40·23·06·56.

FACILITIES 30 ♥ berths on pontoon A, to stbd at ent. Here yachts up to 2m draft can stay afloat but further N will be aground. First fixed bridge has 2·5m clearance. Beware strong ebb tide.

9.21.30   PORNICHET

Loire Atlantique **47°15´·51N 02°21´·11W** ✿✿✿✿✿✿✿✿

CHARTS AC 2986, 2989; Imray C39; SHOM 7395, 6797, 7145; Navi 547

TIDES Sp +0435 Dover, Nps +0530 Dover; ML 3·3; Duration Sp 0530, Nps 0645

Standard Port BREST (◀—)

Times				Height (metres)			
High Water		Low Water		MHWS	MHWN	MLWN	MLWS
0000	0600	0000	0600	7·0	5·5	2·7	1·1
1200	1800	1200	1800				
Differences PORNICHET							
+0020	−0045	−0022	−0022	−1·5	−1·1	−0·6	−0·3
LE GRAND CHARPENTIER							
+0015	−0045	−0025	−0020	−1·6	−1·2	−0·7	−0·4

SHELTER Very good in large man-made marina; full facilities.

NAVIGATION WPT 47°15´·41N 02°22´·15W, 082·5°/0·7M to N-facing ent in W sector (081°-084°) of bkwtr lt. From SW, track 037°/1·5M between Les Evens PHM and Les Troves SHM unlit buoys. From SSE track 333°/3M from Le Grand Charpentier lt ho.

LIGHTS AND MARKS Navigational lts are very hard to see against shore lts of La Baule. Masts in the marina are conspicuous.

COMMUNICATIONS CROSS 02·97·55·35·35; ☎ 02·40·90·08·80; Auto 08·92·68·08·44; ⊖ 02·40·45·88·78; Brit Consul 01·44·51·31·00;

HM (E Quay) ☎02·40·11·97·97. port.labaulelepouliguen@saint-nazaire.cci.fr www.lepouliguen.fr 720+30♥ max LOA 11·5m, €3·20. ⚓🛠️⛽🔧🏗️ Ⓔ 🛢️ Divers 🔺 ⛵(17t) ⛵(Mast 500kg) ▃ YC ☎02·40·60·57·87.

Town (W bank) ✕ 🍴 🛒 ⛽ ⊠ Ⓑ ⇌. ✈ St Nazaire.

Ⓗ 02·40·90·60·60; Dr 02·40·60·17·20; HM VHF Ch 09; Aff Mar 02·40·23·06·56.

FACILITIES Marina ☎02·40·61·61·06. portechouagepornichet@nantesstnazaire.cci.fr Access H24 for <2·5m draft. Best to pre-call, or tempy berth on hammerheads A, B, D, F, G, L, M and N. 1000+150♥ €1·70. ▃ ⚓🛠️✕🍴🔧 SHOM Ⓔ 🛢️🔧⛽ 🔘(25t).

Town ✕ 🍴 🛒 🅿️ Dr, Ⓑ ⊠ ⇌. ✈ St Nazaire (Montoir).

9.21.31 SAINT-NAZAIRE

Loire Atlantique **East lock 47°16′·51N 02°11′·93W** ✪ ◉ ⚓ ✿.

CHARTS AC 2986, 2989, 2985; Imray C40; SHOM 7395, 6797, 7396; Navi 248, 547

TIDES Saint-Nazaire: Sp +0445 Dover, Nps −0540 Dover; ML 3·6; Duration Sp 0640, Nps 0445.

Standard Port BREST (←——)

Times				Height (metres)			
High Water		Low Water		MHWS	MHWN	MLWN	MLWS
0000	0600	0000	0600	7·0	5·5	2·7	1·1
1200	1800	1200	1800				
Differences SAINT- NAZAIRE							
+0030	−0040	−0010	−0010	−1·2	−0·9	−0·5	−0·3
NANTES							
+0135	+0055	+0125	+0215	−0·7	−0·4	−0·9	−0·2

SHELTER Complete in Bassin de St-Nazaire or at S end of Bassin Penhoët. Commercial ops may require yachts to move. Nearest ⚓ is close inshore 5ca NNE of Pte d'Aiguillon.

NAVIGATION From all directions make good a WPT 47°13′·94N 02°16′·01W: SCM buoy, the first of 6 SCM lt buoys (all Q (6) + L Fl 15s) marking a secondary channel parallel to and NW of the PHM buoys of the main chan (Chenal de Bonne-Anse). Track 053°/3·7M until 1 ca before Sud de la Basse Nazaire buoy; thence alter 344° between two lateral buoys, Fl R and G 2·5s, for the small E lock & swing bridge which open (0800–2000 every 2 hrs ±10mins). Waiting, s'ward, is poss against the piled quay. Deep draught vessels may not be admitted near LWS.

LIGHTS AND MARKS Ldg lts VQ Vi 280° into E lock. The Loire suspension bridge (R/W twrs, R lts) is conspic.

COMMUNICATIONS CROSS 02·97·55·35·35; Monitor *Loire Ports Control* VHF Ch 06, **14**, 16, 67, 69 (H24). Tidal info is broadcast VHF Ch 23 at H, +15, +30, +45; ⚓ 02·40·90·00·80; Auto 08·92·68·08·44; ⊜ 02·40·45·88·70; Brit Consul 01·44·51·31·00; ⊞ 02·72·27·80·00; Dr 02·40·22·15·32; Port HM 02·40·45·39·00; pre-call lock on VHF Ch 14 *Loire Port Control* or ☎02·40·66·09·26 for entry into lock/Bassins.

FACILITIES HM ☎02·40·00·45·89, once berthed call at Capitainerie SE of Bassin St-Nazaire to register and leave mobile phone number in case your vessel needs moving. Berth/raft on rough stone quays W side of Bassin St-Nazaire, avoiding ferry berth (Y stripes); <10m, €3·00, no ⚓ or ⌁ ⚓ ⚒ ⊠ ✕ Ⓔ ⚓. www.mairie-saintnazaire.fr

City All facilities; ⇌ ✈. Re-generation WW2 sub pens ongoing.

Passage to Nantes: Les Portes de L'Atlantique boatyard, ☎ 02·40·27·10·10, www.placedeport.com nicholas@placedeport.fr 7.5M upstream of St-Nazaire bridge, S bank, E of Paimboeuf, 60m

pontoon, ❶ welcome, ⊇ €3·00, ⚓⚓ (H24) ⚓ ⌁ ▢ ⚓ wi-fi, ⚒ ⚓ ▣ (50t) courtesy cars/bicycles to access **Town** 2M ✕ ▢ ⚒ ⊠.

NANTES AC 2985 ⒹⒹ, SHOM 6797, 7396. Ponton des Chantiers (SE bank of Bras de la Madeleine) mainly for ferries; €2·30, 1st night free, (but no yachts 15/06–31/08); ⚓ ⌁ no ⚓. Ponton Anne de Bretagne (NW bank nearest bridge) 20 ❶ ⊇; €2·60, 1st night free; ⚓ ⌁. Call Nantes-Port VHF ch 14, ☎02·40·37·04·62 for les3ports@ nge-nantes.fr. Both pontoons are exposed to ferry wash and tide (4+ kts). Trentemoult marina is badly silted; no visitors.

MINOR HARBOURS FROM ST-NAZAIRE TO PTE ST GILDAS
Three small drying hbrs (below, from N to S) are NE of Pte de St-Gildas, on the E side of the R. Loire. Flanked by shellfish beds on rky ledges drying up to 4ca offshore, bkwtrs shelter from W'lies but the hbrs are exposed to N/NE.

Charts: AC 2986, 2981, 2989; SHOM 7395, 6797; Navi 547.

Tides: Interpolate Le Grand Charpentier, St-Nazaire and Pornic.

PORT DE COMBERGE, Loire Atlantique, **47°10′·52N 02°09′·96W**. 2·7M NW is La Truie rk, 3m marked by unlit IDM bn; 1·5M W of it are rocks drying 1·2m. Appr with bkwtr lt in transit with the old lt ho beyond, or night,136° in the W sector. Beware Les Moutons, rk drying 0·7m, 7½ca NW of the bkwtr lt near approach track. Narrow ent; tiny hbr dries 2m, access half-flood. HM ☎02·40·27·82·85. Facilities: ▲ ⚓ YC ⚓ ▢ ⚓ and at St Michel-Chef-Chef.

PORT DE LA GRAVETTE, Loire Atlantique, **47°09′·64N 02°12′·71W**, 2·2M NE of Pte de St Gildas. Daymarks, bkwtr lt in transit 130° with La Treille water tr, 2M inland. Unlit NCM bns mark shellfish beds, beyond 600m bkwtr, turn stbd between lateral buoys; the N part of the hbr has about 1·2m water, but dries inshore.

ST GILDAS (Anse du Boucau) Loire Atlantique, **47°08′·47N 02°14′·74W**. The hbr is 5ca N of Pte de St Gildas lt ho, Q WRG. The hbr bkwtr extends 3ca N, with a large tide gauge and ✩ at its N end. An unlit SHM bn, and a SHM buoy, Fl G 2·5s (May-Sept) lie 150m and 300m NNW of bkwtr hd. Appr from 1M N of Pte de St Gildas on brg 177°, or night, in the W sector 174°-180° of St Gildas lt ho. L'Ilot rky ledge is marked by a PHM bn. Use a mooring in 1·5m N part of the hbr or dry out further S. HM ☎02·40·21·60·07. VHF Ch 09. Facilities: ⚓ ⚓ YC ◫(5t) ⚒ at Préfailles 1M to E.

South Biscay

River Loire to the Spanish border

9.22.1	Contents	911
9.22.2	Area map and Distance table	912
9.22.3	Tidal stream charts	914
9.22.4	Lights, buoys and waypoints	916
9.22.5	Passage information	919
	The Bay of Biscay and West Coast of France – 919 • Baie de Bourgneuf – 920 • Ile d'Yeu to Pertuis Breton – 924 • Pertuis Breton – 924 • Pertuis d'Antioche – 924 • Île d'Oléron – 928 • Pointe de Grave to Arcachon – 940 • Arcachon to the Spanish Border – 941	
9.22.6	Special notes for France. See Area 17	
9.22.7	Pornic	919
9.22.8	L'Herbaudière (Île de Noirmoutier)	920
	Noirmoutier-en-l'Ile • Fromentine	
9.22.9	Port Joinville (Île d'Yeu)	921
	Port de la Meule	
9.22.10	St Gilles-Croix-de-Vie	922
9.22.11	Les Sables d'Olonne	923
9.22.12	Bourgenay	924
	Jard-sur-Mer • L'Aiguillon/La Faute-sur-Mer • Marans • La Flotte, Ile de Ré	
9.22.13	Ars-en-Ré	926
9.22.14	St Martin-de-Ré	926
9.22.15	La Rochelle	927
9.22.16	Île d'Oléron: Port St Denis, Le Douhet, Boyardville	928
	Île d'Aix (La Charente River)	
9.22.17	Rochefort	930
	La Seudre River (Marennes, La Tremblade)	
	Pointe de Grave, tidal coefficients, tide tables/curves	931
9.22.18	The Gironde Estuary	934
9.22.19	Royan	935
9.22.20	Port Bloc	937
9.22.21	Port Médoc	937
9.22.22	The River Gironde	938
	Meschers • Mortagne • Bourg	
9.22.23	Pauillac	938
9.22.24	Bordeaux	939
9.22.25	Canals to the Mediterranean	939
9.22.26	Landes range	940
9.22.27	Bassin d'Arcachon	941
	La Vigne • Andernos • Fontainevieille • Audenge	
9.22.28	Capbreton	943
9.22.29	Anglet/Bayonne	943
9.22.30	St Jean de Luz	944
9.22.31	Hendaye	945

S Biscay

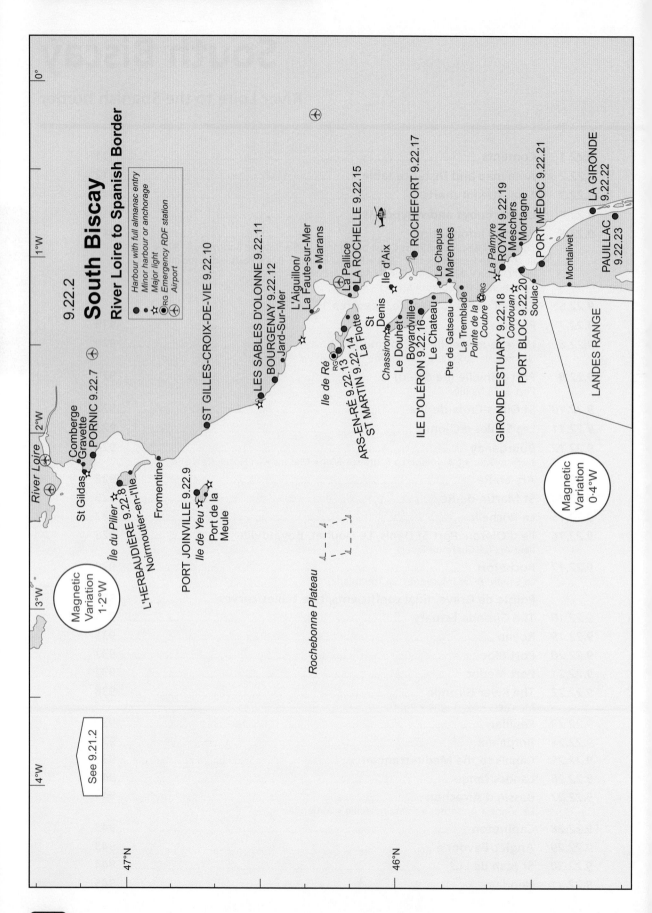

9.22.2

South Biscay

River Loire to Spanish Border

Legend:
- Harbour with full almanac entry
- Minor harbour or anchorage
- ☆ Major light
- ⊙ RG Emergency RDF station
- ⊕ Airport

River Loire

Magnetic Variation 1·2°W

St Gildas ☆
Comberge Gravette
PORNIC 9.22.7 ⊕

Île du Pilier ☆
L'HERBAUDIÈRE 9.22.8
Noirmoutier-en-l'Île
Fromentine

PORT JOINVILLE 9.22.9
Île de Yeu ☆
Port de la Meule

Rochebonne Plateau

ST GILLES-CROIX-DE-VIE 9.22.10

LES SABLES D'OLONNE 9.22.11 ☆
BOURGENAY 9.22.12
Jard-Sur-Mer

L'Aiguillon/
La Faute-sur-Mer
• Marans

Île de Ré ☆
RG
ARS-EN-RÉ 9.22.13
ST MARTIN 9.22.14
La Flotte

La Pallice
La Rochelle 9.22.15
Île d'Aix
St Denis
Chassiron ☆
Le Douhet
Boyardville
Le Chateau
Pte de Gateau

ROCHEFORT 9.22.17

Le Chapus
Marennes

ILE D'OLÉRON 9.22.16
La Tremblade
Pointe de la Coubre ☆
RG

La Palmyre
ROYAN 9.22.19
Meschers
Mortagne

GIRONDE ESTUARY 9.22.18
Cordouan ☆
PORT BLOC 9.22.20
Soulac
• Montalivet

PORT MÉDOC 9.22.21

LA GIRONDE 9.22.22

PAUILLAC 9.22.23

LANDES RANGE

Magnetic Variation 0·4°W

See 9.21.2

Magnetic Variation 1·2°W

47°N
46°N

4°W   3°W   2°W   1°W   0°

DISTANCE TABLE

Approximate distances in nautical miles are by the most direct route while avoiding dangers and allowing for TSS. Places in *italics* are in adjoining areas; places in **bold** are in 9.0.5. Cross-Channel Distances.

	1	2	3	4	5	6	7	8	9	10	11	12	13	14	15	16	17	18	19	20	21	22	23	24	25
1	*Le Conquet*																								
2	St Nazaire	145	2																						
3	Pornic	146	16	3																					
4	L'Herbaudière	144	16	10	4																				
5	Port Joinville	147	35	30	20	5																			
6	St Gilles-C-de-Vie	166	45	40	29	18	6																		
7	Sables d'Olonne	181	65	55	40	31	16	7																	
8	Bourgenay	190	74	64	49	40	25	9	8																
9	Ars-en-Ré	204	88	71	63	51	41	23	17	9															
10	St Martin (I de Ré)	208	92	75	67	55	44	27	20	8	10														
11	La Rochelle	218	101	92	76	66	51	36	29	17	12	11													
12	Boyardville	227	110	101	85	66	55	40	37	25	20	15	12												
13	Rochefort	242	126	116	102	84	75	61	54	40	36	26	21	13											
14	R La Seudre	242	125	116	100	89	71	58	52	49	33	24	36	30	14										
15	Port St Denis	220	103	94	78	59	48	33	30	22	21	13	7	26	22	15									
16	Port Bloc/Royan	234	122	115	107	97	85	71	60	56	51	51	5	7	7	44	16								
17	Pauillac	266	154	147	139	129	117	103	92	88	84	81	65	100	59	74	32	17							
18	Bordeaux	289	177	170	162	152	140	126	115	115	111	107	104	123	82	97	55	23	18						
19	Cap Ferret	274	168	160	153	138	130	113	110	111	102	98	95	114	75	88	70	100	123	19					
20	Capbreton	316	220	215	211	192	186	169	166	168	165	156	152	172	131	145	124	156	179	58	20				
21	Anglet/Bayonne	328	232	223	223	200	195	181	178	176	177	168	164	184	143	157	133	164	187	70	12	21			
22	St Jean de Luz	338	242	233	233	210	205	191	188	183	187	178	177	194	153	167	143	174	197	80	22	31	22		
23	Hendaye	341	245	236	236	213	208	194	191	185	190	181	177	197	156	170	145	177	200	83	25	34	8	23	
24	Santander	300	243	237	229	212	210	204	204	207	206	202	199	218	184	192	179	212	235	133	106	103	88	94	24
25	*Cabo Finisterre*	382	404	399	390	377	395	393	394	407	406	407	404	423	399	397	403	433	456	376	370	373	354	360	274	25

See 9.23.2

Magnetic Variation 0·3°W

9.22.3 AREA 22 TIDAL STREAMS

5 Hours before HW Brest
(0005 after HW Dover)
(0510 before HW Pointe de Grave)

4 Hours before HW Brest
(0105 after HW Dover)
(0410 before HW Pointe de Grave)

3 Hours before HW Brest
(0205 after HW Dover)
(0310 before HW Pointe de Grave)

2 Hours before HW Brest
(0305 after HW Dover)
(0310 before HW Pointe de Grave)

1 Hour before HW Brest
(0405 after HW Dover)
(0110 before HW Pointe de Grave)

HW Brest
(0505 after HW Dover)
(0010 before HW Pointe de Grave)

CAUTION: Due to the very strong rates of the tidal streams in some of the areas covered by these chartlets, many eddies and overfalls may occur. Where possible some indication of these eddies has been included. In many areas there is either insufficient information or the eddies are unstable. Tidal streams are generally weak offshore.

Strong winds may also have a great effect on the rate and direction of the tidal streams in all these chartlets.

Northward 9.21.3

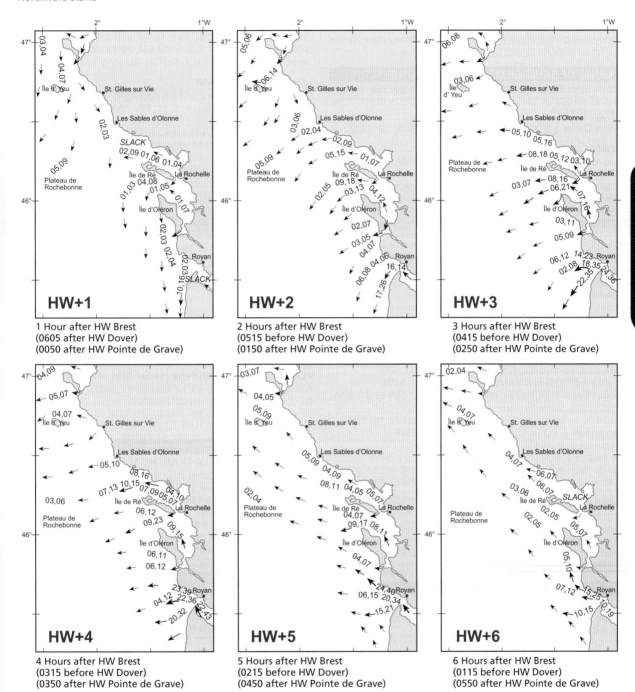

1 Hour after HW Brest
(0605 after HW Dover)
(0050 after HW Pointe de Grave)

2 Hours after HW Brest
(0515 before HW Dover)
(0150 after HW Pointe de Grave)

3 Hours after HW Brest
(0415 before HW Dover)
(0250 after HW Pointe de Grave)

4 Hours after HW Brest
(0315 before HW Dover)
(0350 after HW Pointe de Grave)

5 Hours after HW Brest
(0215 before HW Dover)
(0450 after HW Pointe de Grave)

6 Hours after HW Brest
(0115 before HW Dover)
(0550 after HW Pointe de Grave)

Note: These chartlets are based on the Admiralty tidal stream atlas for France, W coast (NP 265, 2nd edition 2005) which uses data from actual observations out to 20-35M offshore.

The comparable French atlas (559-UJA; 2001) gives data for further offshore, based on computer predictions. It covers at large scale the approaches to St Nazaire; Les Sables d'Olonne to Pertuis de Maumusson; and the mouth of the Gironde.

No tidal stream atlas is published by either the French or UK Hydrographic Offices for the waters southward to the Spanish border.

9.22.4 LIGHTS, BUOYS AND WAYPOINTS

Bold print = light with a nominal range of 15M or more. CAPITALS = place or feature. *CAPITAL ITALICS* = light-vessel, light float or Lanby. *Italics* = Fog signal. ***Bold italics*** = Racon. Many marks/buoys are fitted with AIS (<u>MMSI No</u>); see relevant charts.

POINTE DE SAINT GILDAS TO FROMENTINE

Pte de Saint Gildas ⚡ Q WRG 20m, W14M, R/G10M; 264°-R-308°-G-078°-W-088°-R-174°-W-180°-G-264°; 47°08'·02N 02°14'·75W. <u>992271017</u>.

Notre Dame ⚓ VQ (9) 10s 7m 3M; 47°05'·42N 02°08'·26W.

PORNIC

Approaches ⚓ Fl G 4s; 47°06'·39N 02°06'·60W; ⚓ Fl R 4s; 47°06'·40N 02°06'·64W

Marina ent, S side ⚡ Fl (2) R 6s 4m 2M; 47°06'·47N 02°06'·67W.
Pte de Noëveillard ⚡ Oc (4) WRG 12s 22m W13M, R/G9M; Shore-G-051°-W-079°-R-shore; W □ twr, G top, W house; 47°06'·62N 02°06'·92W.

ÎLE DE NOIRMOUTIER anti-clockwise from the East

Pte des Dames ☆ Oc (3) WRG 12s 34m, **W19M, R/G15M**; 016·5°-G-057°-R-124°-G-165°-W-191°-R191°-267°-W-357°-R-016·5°; W □ twr; 47°00'·67N 02°13'·26W. La Chaise ⚓ 47°01'·21N 02°12'·64W.

L'Herbaudière, ldg lts 187·5°, both Q 5/21m 7M, Gy masts. Front, 47°01'·59N 02°17'·84W. Rear, 310m from front.
Martroger ⚓, Q WRG 11m W9M, R/G6M; 033°-G-055°-W-060°-R-095°-G-124°-W-153°-R-201°-W-240°-R-033°; 47°02'·61N 02°17'·11W.
W jetty ⚡ Oc (2+1) WG 12s 9m W10M, G7M; 187·5°-W-190°-G-187·5°; W col and hut, G top; 47°01'·63N 02°17'·84W.
E jetty ⚡ Fl (2) R 6s 8m 4M; 47°01'·61N 02°17'·81W.
Île du Pilier ☆ Fl (3) 20s 33m **29M**; Gy twr; 47°02'·55N 02°21'·60W. <u>992271019</u>. Same twr, aux ⚡ QR 10m 11M, 321°-034°.
Passe de la Grise ⚓ Q (6) + L Fl 15s; 47°01'·66N 02°19'·97W.
Les Boeufs ⚓ VQ (9) 10s; 46°55'·04N 02°27'·99W.
Réaumur ⚓ Q (9) 15s; 46°57'·46N 02°24'·22W.
Le Bavard ⚓ VQ (6) + L Fl 10s; 46°56'·79N 02°23'·34W.
Pte du Devin (Morin) ⚡ Oc (4) WRG 12s 10m W11M, R/G8M; 314°-G-028°-W-035°-R-134°; W col & hut, G top; 46°59'·15N 02°17'·60W.

GOULET DE FROMENTINE

Fromentine ⚲ L Fl (3) 10s 3m 6M; 46°53'·10N 02°11'·40W
Milieu ⚓ Fl (4) R 15 s 6m 5M; 46°53'·59N 02°09'·63W.
Pte de Notre Dame-de-Monts ⚡ Dir Oc (2) WRG 6s 21m, W13M, R/G10M; 000°-G-043°-W-063°-R-073°-W-094°-G-113°-W-116°-R-175°-G-196°-R-230°; W twr, B top; 46°53'·33N 02°08'·54W.

ÎLE D'YEU TO BOURGENAY

ÎLE D'YEU

Petite Foule (main lt) ☆ Fl 5s 56m **24M**; W □ twr, G lantern; 46°43'·04N 02°22'·91W. <u>992271031</u>.
Les Chiens Perrins ⚓ Q (9) WG 15s 16m W7M, G4M; 330°-G-350°-W-200°; 46°43'·60N 02°24'·58W.
Pont d'Yeu ⚓ 46°45'·81N 02°13'·81W.
La Sablaire ⚓ Q (6) + L Fl 15s; 46°43'·62N 02°19'·49W.
Port Joinville ldg lts 219°, both QR 11/16m 6M, 169°-269°: Front, Quai du Canada 46°43'·61N 02°20'·94W. Rear, 85m from front.
NW jetty ⚡ Oc (3) WG 12s 7m, W11M, G8M; Shore-G-150°-W-232°-G-279°-W-285°-G-shore; W 8-sided twr, G top; 46°43'·77N 02°20'·82W. La Galiote ⚏ 46°43'·73N 02°20'·71W.

Pte des Corbeaux ☆ Fl (2+1) R 15s 25m **20M**; 083°-143° obsc by Île de Yeu; W □ twr, R top; 46°41'·41N 02°17'·08W.
Port de la Meule ⚡ Oc WRG 4s 9m, W9M, R/G6M; 007·5°-G-018°-W-027·5°-R-041·5°; Gy twr, R top; 46°41'·62N 02°20'·72W.

SAINT GILLES-CROIX-DE-VIE

Pte de Grosse Terre ☆ Fl (4) WR 12s 25m, **W18M, R15M**; 290°-R-339°-W-125°-R-145°; W truncated twr; 46°41'·54N 01°57'·92W.

Ldg lts 043·7°, both Q 7/28m **15M**; 033·5°-053·5°; synch; W □ twrs, R tops: Front, 46°41'·86N 01°56'·75W. <u>992271027</u>. Rear, 260m NE.
Pilours ⚓ Q (6) + L Fl 15s; *Bell;* 46°40'·99N 01°58'·09W.
Jetée de la Garenne ⚡ Fl G 4s 8m 6M; 46°41'·46N 01°57'·25W.
Jetée de Boisvinet ⚡ Fl R 4s 8m 6M; 46°41'·62N 01°57'·16W.

LES SABLES D'OLONNE

Les Barges ⚡ Fl (2) R 10s 25m 13M; Gy twr; 46°29'·70N 01°50'·50W.
Petite Barge ⚓ Q (6) + L Fl 15s 3M; *Whis;* 46°28'·90N 01°50'·61W.
L'Armandèche ☆ Fl (2+1) 15s 42m **24M**; 295°-130°; W 6-sided twr, R top; 46°29'·39N 01°48'·29W. <u>992271035</u>.
Nouch Sud ⚓ Q (6) + L Fl 15s; 46°28'·55N 01°47'·41W.
Nouch Nord ⚓ 46°28'·88N 01°47'·36W.
SW Pass, ldg lts 032·5°, both Iso 4s 12/33m **16M**, H24: **Front** ☆, 46°29'·42N 01°46'·37W. <u>992271024</u>. Mast. **Rear** ☆, **La Potence**, W □ twr.
SE Pass, ldg lts 320°: Front, E jetty ⚡ QG 11m 8M; W twr, G top; 46°29'·44N 01°47'·51W. <u>992271032</u>. Rear, Tour d'Arundel ⚡ Q 33m 13M, synch; large Gy □ twr; 46°29'·63N 01°47'·74W. <u>992271022</u>.
Jetée St Nicolas (W jetty) ⚡ QR 1·2s 16m 8M; 143°-094°; W twr, R top; 46°29'·23N 01°47'·52W.

BOURGENAY

Ldg lts 040°, both QG 9/19m 7M. Front, 020°-060°; on S bkwtr; 46°26'·34N 01°40'·62W. Rear, 010°-070°; 162m from front.
Landfall ⚓ L Fl 10s; 46°25'·26N 01°41'·93W.
Ent ⚡ Fl R 4s 8m 9M & ⚡ Iso G 4s 6m 5M; 46°26'·29N 01°40'·75W.

PLATEAU DE ROCHEBONNE (Shoal 32M offshore)

NW ⚓ Q (9) 15s; *Whis;* 46°12'·92N 02°31'·58W.
NE ⚓ Iso G 4s; 46°12'·73N 02°24'·83W.
SE ⚓ Q (3) 10s; *Bell;* 46°09'·20N 02°21'·13W.
SW ⚓ Fl (2) R 6s; 46°10'·11N 02°27'·04W.

PERTUIS BRETON

JARD-SUR-MER and LA TRANCHE-SUR-MER

Ldg marks 036°, two unlit W bcns; 46°24'·47N 01°34'·16W.
Pte du Grouin du Cou ☆ Fl WRG 5s 29m, **W20M, R/G16M**; 034°-R-061°-W-117°-G-138°-W-034°; W 8-sided twr, B top; 46°20'·67N 01°27'·83W.
La Tranche pier ⚡ Fl (2) R 6s 6m 6M; R col; 46°20'·55N 01°25'·63W.
Mussel farm, No entry, marked by: ⚓ Q, 46°17'·17N 01°22'·19W. ⚓ Q (3) 10s, 46°15'·73N 01°18'·42W. ⚓ Q (6) + L Fl 15s, 46°15'·16N 01°19'·98W. ⚓ Q (9) 15s, 46°16'·28N 01°22'·88W.

L'AIGUILLON and LA FAUTE-SUR-MER

Le Lay ⚓ Q (6) + L Fl 15s; 46°16'·10N 01°16'·53W.
No. 1 ⚓ 46°16'·65N 01°16'·25W. Many mussel beds.

ANSE DE L'AIGUILLON (for La Sèvre Niortaise & Marans)

ATT de L'Aiguillon ⚓ L Fl 10s; 46°15'·33N 01°11'·50W.
Inner fairway ⚓ 46°17'·19N 01°09'·65W.
Port du Pavé ⚡ Fl G 4s 9m 7M; W col, G top; 46°18'·15N 01°08'·01W.

ÎLE DE RÉ

Les Baleineaux ⚡ VQ 23m 7M; pink twr; R top; 46°15'·81N 01°35'·22W.
Les Baleines ☆ Fl (4) 15s 53m **27M**; conspic Gy 8-sided twr, R lantern; 46°14'·64N 01°33'·69W.

ARS-EN-RÉ

Dir ⚡ 268° Oc WRG 4s, W10M, R/G 7M, 275·5°-G-267·5°-W-268·5°-R-274·5°; W hut with W □ topmark; 46°14'·05N 01°28'·60W.
Les Islattes ⚓ Q 13m 3M; NCM bcn twr 46°14'·03N 01°23'·32W.
Bûcheron No 1 ⚓ Fl G 2·5s; 46°14'·21N 01°25'·98W.
Bûcheron No 3 ⚓ Fl (2) G 6s; 46°14'·14N 01°26'·94W.

Le Fier d'Ars, inner ldg lts 232·5°, both Q 5/13m 9/11M: Front ⚡ R/W frame on W col; 46°12'·75N 01°30'·59W. Rear ⚡ 142°-322°; B vert rectangle on W mast, 370m from front.

ST MARTIN DE RÉ

Lt ho, E of ent ⚡ Oc (2) WR 6s 18m W10M, R7M; Shore-W-245°-R-281°-W-shore; W twr, R top; 46°12'·44N 01°21'·89W.
W mole ⚡ Fl G 2·5s 10m 5M; obsc'd by Pte du Grouin when brg <124°; W post, G top; 46°12'·49N 01°21'·91W.

LA FLOTTE

N bkwtr ⚡ Fl WG 4s 10m W12M, G9M; 130°-G-205°-W-220°-G-257°; W ○ twr, G top; 46°11'·33N 01°19'·30W.
Dir lt 212·5°, Moiré effect, is next to the main lt twr.
ÎLE DE RÉ (South coast, from the W, inc bridge)
Chanchardon ⚡ Fl WR 4s 15m W11M, R8M; 118°-R-290°-W-118°; B 8-sided twr, W base; 46°09'·71N 01°28'·41W.
Chauveau ☆ Oc (3) WR 12s 27m **W15M**, R11M; 057°-W-094°-R-104°-W-342°-R-057°; W ○ twr, R top; 46°08'·03N 01°16'·42W.
Bridge span (30m cl'nce) for SE-bound vessels, ⚡ Iso 4s 34m 8M; 46°10'·18N 01°14'·75W . Span (30m cl'nce) for NW-bound vessels, ⚡ Iso 4s 34m 8M; 46°10'·26N 01°14'·52W.
La Pallice, W mole, ⚡ Fl G 4s 5m 6M; 46°09'·78N 01°14'·45W.
S end, ⚡ Q (6) + L Fl 15s 6m 9M; 46°09'·36N 01°14'·52W.

LA ROCHELLE AND LA CHARENTE TO ROCHEFORT

LA ROCHELLE

Chauveau ⚓ VQ (6) + L Fl 10s; 46°06'·56N 01°16'·06W.
Roche du Sud ⚓ Q (9) 15s; 46°06'·37N 01°15'·22W.
La Rochelle ldg lts 059°, both Dir Q 15/25m 13/14M; synch; by day Fl 4s. Front; intens 056°-062°; R ○ twr, W bands; 46°09'·35N 01°09'·16W. Rear, 044°-074°, obsc'd 061°-065°; W 8-sided twr, G top, 235m from front.
Lavardin ⚓ Fl (2) 6s 14m 7M; BRB IDM twr; 46°08'·09N 01°14'·52W.
Les Minimes ⚓ Q (9) 15s; 46°08'·01N 01°11'·60W.
Pte des Minimes ⚡ Fl (3) WG 12s 8m; W8M, G5M; 059°-W-213°; 313°-G-059°; 8-sided twr on piles; 46°08'·27N 01°10'·76W.
Chan buoy ⚓ QG; 46°08'·53N 01°10'·87W; marking hbr limit.
Tour Richelieu ⚓ Fl R 4s 10m 9M; R twr; 46°08'·90N 01°10'·34W.
Port des Minimes, W bkwtr ⚡ Fl (2) G 6s 9m 7M; 46°08'·82N 01°10'·15W.

LA CHARENTE

Fort Boyard twr ⚡ Q (9) 15s; 45°59'·96N 01°12'·85W.
Ldg lts 115°, both Dir QR 8/21m **19/20M**; intens 113°-117°; W□ twr, R top: **Front** ☆, 45°57'·96N 01°04'·38W. **Rear, Soumard** ☆, 600m from front. Same twr; ⚡ QR 21m 8M; vis 322°-067°.
Île d'Aix ☆ Fl WR 5s 24m **W24M, R20M**; 103°-R-118°-W-103°; two conspic W ○ twrs, R top; 46°00'·60N 01°10'·67W.
Pte Ste Catherine, jetty hd ⚓ Q (6) + L Fl 15s 5m 4M; 45°59'·05N 01°07'·85W. Les Palles ⚓ Q; 45°59'·54N 01°09'·56W.
Fouras, Port Sud bkwtr ⚡ Fl WR 4s 6m 9/6M; 115°-R-177°-R-115°; 45°58'·97N 01°05'·72W. Rear 75m from front.

ROCHEFORT

Night passage by yachts is not advised.
⚡ Fl (2) G 6s; bcn 45°55'·21N 00°56'·98W, 1·4M S of Rochefort.

ÎLE D'OLÉRON

Pte de Chassiron ☆ Fl 10s 50m, **28M**; conspic W twr, B bands; 46°02'·80N 01°24'·62W, NW tip of island.
Antioche ⚓ Q 20m 11M; 46°03'·94N 01°23'·71W.

ST DENIS

Dir ⚡ 205°, Iso WRG 4s 14m, W11M, R/G8M; 190°-G-204°-W-206°-R-220°; 46°01'·61N 01°21'·92W.
E jetty ⚡ Fl (2) WG 6s 6m, W9M, G6M; 205°-G-277°-W-292°-G-165°; □ hut; 46°02'·10N 01°22'·06W.

PORT DU DOUHET and PASSAGE DE L'OUEST

Seaweed farm ⚓ VQ; 46°00'·65N 01°17'·38W.
N ent ⚓ 46°00'·10N 01°19'·20W.
Fishfarm ⚓ Q; 46°00'·24N 01°15'·34W. ⚓ Q (3) 10s; 45°59'·84N 01°14'·79W.

BOYARDVILLE

Mole ⚡ Fl (2) R 6s 8m 5M; obsc'd by Pte des Saumonards when brg <150°; W twr, R top; 45°58'·24N 01°13'·83W.

COUREAU D'OLÉRON and LE CHÂTEAU D'OLÉRON

Juliar ⚓ Q (3) WG 10s 12m, W11M; G8M; 147°-W-336°-G-147°; 45°54'·11N 01°09'·48W.
Chateau d'Oléron, ldg lts 318·5°, both QR 11/24m 7M; synch. Front, 191°-087°; R line on W twr; 45°53'·03N 01°11'·45W. Rear, W twr, R top; 240m from front.

LA SEUDRE RIVER

Pte du Mus de Loup ⚡ Oc G 4s 8m 6M; 118°-147°; 45°47'·70N 01°08'·60W.
Pertuis de Maumusson. Depths & buoys subject to change.
ATT Maumusson ⚓ L Fl 10s; 45°46'·46N 01°18'·13W.
La Barre ⚓ 45°46'·59N 01°16'·25W.
Tabouret ⚓ 45°46'·93N 01°16'·15W.
Mattes ⚓ 45°46'·93N 01°15'·29W.
Gatseau ⚓ 45°47'·36N 01°14'·70W.

LA COTINIÈRE (SW side of island)

Ent ldg lts 339°, both ⚡ Dir Oc (2) 6s 6/14m 13/12M; synch: Front, 329°-349°; W twr, R top; 45°54'·72N 01°19'·79W. Rear, intens 329°-349°; W twr, R bands; 425m from front on W bkwtr.
W bkwtr head, Fl R 4s 10m; 45°54'·63N 01°19'·70W.

GIRONDE APPROACHES AND TO BORDEAUX

GRANDE PASSE DE L'OUEST

Pte de la Coubre ☆ Fl (2) 10s 64m **28M**; W twr, R top; 45°41'·78N 01°13'·99W. Same twr, Q R 42m10M; 030°-R-110°.
BXA ⚓ Iso 4s 8m 7M; *Racon B*; 45°34'·10N 01°26'·63W. 992271005.
La Palmyre Wtr Twr Dir ☆ 056·7°, WRG 54m **W25M R/G20M**; 055·25°-FG-055·95°-AlWG-056·58°-FW-057·42°-AlWR-058·02-FR-058·72°; Gy Wtr Twr; 45°41'·65N 01°10'·18W.
Duc d'Albe ⚡ Fl(2) 6s 10m 2M; W pylon on dolphin; 45°39'·56N 01°08'·77W.
No. 1 ⚓ QG; 45°36'·17N 01°21'·70W; 992271036 .
No. 6 ⚓ VQ (6) + L Fl 10s; 45°38'·27N 01°17'·91W.
No. 7a ⚓ Fl G 4s; 45°38'·96N 01°14'·85W.
No. 9 ⚓ Q; 45°39'·43N 01°12'·79W.
No. 11 ⚓ Iso G 4s; 45°39'·06N 01°10'·60W.
No. 11A ⚓ Fl G 2·5s; 45°38'·37N 01°07'·90W.
No. 13 ⚓ Fl (2) G 6s; 45°37'·34N 01°06'·41W.
No. 13A ⚓ Fl (3) G 12s; 45°35'·68N 01°04'·19W.
No. 13B ⚓ QG; 45°34'·65N 01°02'·99W.
Ldg lts 327° (down-river): Front, Terre-Nègre ⚡, Oc (3) WRG 12s 39m W14M, R/G11M; 304°-R-319°-W-327°-G-000°-W-004°-G-097°-W-104°-R-116°; W twr, R top on W side; 45°38'·77N 01°06'·39W. Rear, **La Palmyre** Oc (3) R 12s 57m 17M; W radar Twr; 45°39'·72N 01°07'·25W. 1.1M from front (Terre Nègre above).
Cordouan ☆ Oc (2+1) WRG 12s 60m, **W19M, R16M, G15M**; 014°-W-126°-G-178·5°-W-250°-W (unintens) -267°-R (unintens)-294·5°-R-014°; obsc'd in estuary when brg >285°; W twr, Gy band; 45°35'·18N 01°10'·40W.

PASSE SUD (or DE GRAVE)

Ldg lts 063°: **Front, St Nicolas** ☆ Dir QG 22m **16M**; intens 061·5°-064·5°; W □ twr; 45°33'·73N 01°05'·01W.
Rear, Pte de Grave ☆ Oc WRG 4s 26m, **W17M**, R/G13M; 033°-W(unintens)-054°-W-233·5°-R-303°-W-312°-G-330°-W-341°-W(unintens)-025°; W □ twr, B corners and top, 0·84M from front. G ⚓ 45°30'·33N 01°15'·55W. G3 ⚓ 45°32'·79N 01°07'·72W.

Ldg lts 041°, both Dir QR 33/61m **18M**. **Front, Le Chay** ☆ 45°37'·31N 01°02'·43W. **Rear, St Pierre** ☆, 0·97M from front. G4 �𝑙 45°34'·70N 01°05'·80W. G6 �𝑙 45°34'·88N 01°04'·80W.

ROYAN
R1 ⟊ Iso G 4s; 45°36'·56N 01°01'·96W.
S jetty ⚡ Fl (2) R 10s 11m 12M; 199°-116°; 45°37'·00N 01°01'·81W.
Hbr ent, W jetty ⚡ Fl (3) R 12s 8m 6M; 45°37'·12N 01°01'·53W.

PORT BLOC and PORT-MÉDOC
Pte de Grave, jetty ⟊ Q 6m 2M; 45°34'·42N 01°03'·68W.
Port Bloc, ent N side ⚡ Fl G 4s 9m 3M; 45°34'·14N 01°03'·74W.
Port-Médoc, N bkwtr head ⚡ QG 4M; 45°33'·40N 01°03'·39W.

MESCHERS, MORTAGNE, PORT MAUBERT, PAUILLAC and BLAYE
Meschers ldg lts 354·4°: QG (privately maintained). Front 45°33'·20N 00°56'·63W, on E bkwtr head. Rear, 130m from front.
Mortagne ent ⟊ VQ (9) 10s; 45°28'·24N 00°49'·02W.
Port Maubert, ldg lts 024·5°, both QR 5/9m 5M; 45°25'·56N 00°45'·47W.
Pauillac, No. 43 ⟊ Fl (2) G 6s; 45°12'·44N 00°44'·27W.
NE elbow ⚡ Fl G 4s 7m 5M; 45°11'·95N 00°44'·59W.
Ent E side ⚡ QG 7m 4M; 45°11'·83N 00°44'·59W.
Blaye, No. S9 ▲ 45°09'·35N 00°40'·27W; 1·8M N of Blaye.
N quay ⚡ Q (3) R 5s 6m 3M; 45°07'·49N 00°39'·99W.
D6 ⟊ Fl (2) R 6s; 45°06'·91N 00°39'·93W, 0.6M S of Blaye.
Bec d'Ambés ⟊ QG 5m 5M; 45°02'·53N 00°36'·47W (here Rivers Dordogne and Garonne flow into River Gironde).

BORDEAUX
Pont d'Aquitaine ⚡ 4 F Vi; 44°52'·82N 00°32'·28W.
No 73 ⟊ Iso G 4s; 44°52'·72N 00°32'·24W.
Lormont Iso R 4s, S end of ❶ pontoon; 44°52'·62N 00°32'·12W.
Lock ent to Bassins Nos 1 and 2, 44°51'·72N 00°33'·07W.
Pont de Pierre, 44°50'·30N 00°33'·78W, Km 0.

POINTE DE GRAVE TO THE SPANISH BORDER
Hourtin ☆ Fl 5s 55m **23M**; R □ twr; 45°08'·51N 01°09'·63W.
Cap Ferret ☆ Fl R 5s 53m **22M**; W ○ twr, R top; 44°38'·72N 01°14'·90W. Same twr, ⚡ Oc (3) 12s 46m 11M; 045°-135°.

ARCACHON, PASSE NORD
Buoys may be moved, or added, as the chan shifts. Positions shown here are approximate and should be used with caution. Nos 10 &12 buoys do not exist; 11, 13, 14 lie btwn nos 9N &15.
ATT-ARC SWM ⟊ L Fl 10s 8m 5M; 44°34'·046N 01°18'·75W.
Outer/entrance channel inward to Cap Ferret:

1N	⟊	44°34'·571N 01°17'·81W.
2N	⟊	44°34'·681N 01°17'·84W.
3N	⟊	44°34'·49N 01°16'·43W.
4N	⟊	44°34'·61N 01°16'·57W.
5N	⟊	44°34'·78N 01°15'·92W.
7N	⟊	44°35'·12N 01°15'·33W.
6N	⟊	44°35'·87N 01°14'·66W.
8N	⟊	44°36'·53N 01°14'·39W.
9N	⟊	44°36'·96N 01°14'·34W.
15	⟊	44°39'·85N 01°12'·04W.

Marina W bkwtr ⚡ QG 6m 6M; 44°39'·77N 01°09'·14W.
E bkwtr ⚡ QR 6m 6M; 44°39'·76N 01°09'·07W.
Port de La Vigne ⚡ Iso R 4s 7m 5M, occas; 44°40'·43N 01°14'·36W.
La Salie ⟊ Fl (2) 6s; 44°30'·43N 01°17'·75W.
Contis ☆ Fl (4) 25s 50m **23M**; W ○ twr, B diagonal stripes; 44°05'·59N 01°18'·99W.

CAPBRETON
Digue Nord ⚡ Fl (2) R 6s 13m 12M; W ○ twr, R top; 43°39'·36N 01°26'·90W.
Estacade Sud ⚡ Fl (2) G 6s 9m 8M; 43°39'·27N 01°26'·83W.

ANGLET and BAYONNE
BA ⟊ L Fl 10s; 43°32'·59N 01°32'·85W, 144°/0·93M to hbr ent.
Outer ldg lts 090°, both Dir Q 9/15m **19M**; intens 086·5°-093·5°; synch; W twrs, R tops. Front, **Boucau** ☆ 43°31'·81N 01°31'·23W. Rear, 250m from front.
Outer N bkwtr ⟊ QR 11m 8M; W/R twr; 43°31'·88N 01°31'·99W.
Outer S bkwtr ⟊ Q (9) 15s 15m 6M; 43°31'·53N 01°31'·76W.
Inner ldg lts 111·5°, both Dir QG 6/10m 14M; intens 109°-114° Front, W/G hut; 43°31'·63N 01°30'·94W. Rear,149m from front.
Inner N jetty ⚡ Iso R 4s 12m 8M; 43°31'·82N 01°31'·42W.
Inner S jetty ⚡ Iso G 4s 9m 10M; W □ twr, G top; 43°31'·73N 01°31'·48W. IPTS from twr 260m ESE.
N training wall ⚡ Fl (2) R 6s 9m 8M; 43°31'·76N 01°31'·25W.
N training wall ⚡ Fl (3) R 12s 9m 3M, 43°31'·75N 01°31'·18W.
N training wall ⚡ Fl (4) R 15s 9m 3M, 43°31'·73N 01°30'·97W.
S training wall ⚡ Fl (3) G 12s 9m 3M, 43°31'·67N 01°31'·21W.
S training wall ⚡ Fl (4) G 15s 9m 3M, 43°31'·64N 01°30'·96W.
S training wall ⚡ QG 7m 8M; 43°31'·63N 01°30'·90W.
Anglet marina ent ⚡Fl G 2·5s 6m 2M; 43°31'·57N 01°30'·51W.

Upriver toward Bayonne fixed bridge, 5·2m clearance:
Digue Basse ⚡ Fl G 4s 3m 4M, 43°31'·54N 01°30'·08W.
Ldg lts 322·4°: Front, Forges QR 14m 8M; vis 188°-098°; 43°31'·61N 01°29'·92W. Rear, Dir Iso R 4s 18m 13M; intens 312·4°-332·4°.
No. 1 ⟊ Iso G 4s; 43°31'·34N 01°29'·72W.
No. 2 ⟊ QR; 43°30'·89N 01°29'·46W.
Pte de Blanc-Pignon ⚡ Fl (2) G 6s 8m 9M; 43°30'·92N 01°29'·61W.
No. 10 ⟊ VQR; 43°30'·11N 01°29'·39W.
Pont de l'Aveugle ⚡ Dir Q WRG W10M, R/G7M; 165°-G-170°-W-175°-R-195°; 43°30'·05N 01°29'·53W.

BIARRITZ
Pointe Saint-Martin ☆ Fl (2) 10s 73m **27M**; W twr, B top; 43°29'·62N 01°33'·24W.

ST JEAN DE LUZ
Outer ldg lts 138·5° (Passe d'Illarguita): Front, Socoa ⚡ Q WR 36m W12M, R8M; shore-W-264°-R-282°-W-shore; W □ twr, B stripe; 43°23'·70N 01°41'·20W.

Rear, Bordagain, Dir Q 67m **20M**; intens 134·5°-141·5°; B/W panel, B/W pylon; 43°23'·12N 01°40'·50W.

Middle ldg lts 101°, both Oc (4) R 12s 30/47m **18M**; intens 095°-107°; synch. **Front, Ste Barbe** ☆, W △ on W hut; 43°23'·96N 01°39'·88W. **Rear** ☆, B △ on W □ twr, 340m from front.

Inner ldg lts 150·7°, both Dir QG 18/27m **16M**; intens 149·5°-152·2°. **Front, E jetty** ☆ W □ twr, R stripe; 43°23'·25N 01°40'·15W. **Rear** ☆, W □ twr, G stripe; 410m from front, at S corner of marina.

Digue des Criquas (W bkwtr) ⚡ Iso G 4s 11m 6M; G □ twr; 43°23'·84N 01°40'·67W.

East Groyne ⚡ Fl R 4s 7m 2M; 43°23'·29N 01°40'·18W.

HENDAYE
Cabo Higuer ☆ Fl (2) 10s 63m **23M**; 072°-340°; □ twr, R cupola; 43°23'·52N 01°47'·52W (in Spain).
Refuge port ent, Fl (2) G 7s 7m 5M; 43°23'·13N 01°47'·35W.
W training wall ⚡ Fl (3) G 9s 9m 5M; 43°22'·82N 01°47'·34W.
E training wall hd ⚡ L Fl R 10s 8m 5M; 43°22'·67N 01°47'·22W.
E training wall root, Pte des Dunes ⚡ Fl R 2·5s 6m 4M; 43°22'·38N 01°47'·34W.
◣ Fl G 5s; 43°22'·33N 01°47'·48W.
⟊ VQ (3) G 5s 3m 3M; G bent mast; 43°22'·10N 01°47'·27W.
Marina, W bkwtr, elbow ⚡ Fl (2) R 6s 6m 2M; 294°-114°; 43°22'·08N 01°47'·16W.
S jetty ⚡ Fl Y 4s 5m 4M; Y col; 43°22'·05N 01°47'·05W.
⟊ L Fl G 10s 3m 4M; G mast; 43°21'·99N 01°47'·07W.

9.22.5 PASSAGE INFORMATION

More Passage Information is threaded between successive harbours in this Area. **Bibliography**: *Atlantic France* (Imray/RCC/Parkinson); *West France CC* (Fernhurst/Featherstone); *South Biscay* (Imray/RCC/Pickard); The *Bay of Biscay Pilot* NP22. AC 1104, 20, 2664, 1102; carry also large scale French charts for inshore waters.

THE BAY OF BISCAY AND WEST COAST OF FRANCE

Despite its reputation, weather in the southern part of the Bay is often warm and settled in summer when the Azores high and Spanish heat low are the dominant weather features.

NE'lies prevail in sea area Fitzroy in summer and gales may occur twice monthly, although forecast more frequently. Atlantic lows can bring W'ly spells at any time together with long swells which are dangerous inshore. SE or S winds are rare, but wind direction and speed often vary from day to day. Sea and land breezes can be well developed in the summer. Off N Spain *Galernas* are dangerous squally NW winds which blow with little warning. Visit www.franksingleton. clara.net/biscay.html for useful Met and routeing information.

Rainfall is moderate, increasing in the SE, where thunder is more frequent. Sea fog occurs May-Oct, but is less common in winter. Japanese seaweed has been reported as a nuisance.

▶*Tidal streams are weak offshore, but can be strong in estuaries and channels, and around headlands. The tidal stream chartlets at 9.22.3 are based on NP 265 (Admiralty tidal stream atlas for France, W Coast) which uses data from actual observations out to 15-25M offshore. The equivalent French Atlas gives more data, based on computer predictions.*◀

The general direction and rate of the surface current much depends on wind: in summer it is SE, towards the SE corner of B of Biscay, where it swings W along the N coast of Spain. In winter with W gales, the current runs E along N coast of Spain, sometimes at 3kn or more. When crossing the Bay of Biscay, allow for a likely set to the E, particularly after strong W winds.

The Continental Shelf with mean depths of 130m is a major feature of the Bay of Biscay. From 75M SW of Pointe du Raz it closes the French coast near Capbreton. AC 1104 shows depths plummeting to 4000m in the space of 20M. This steep slope allied to Atlantic swell can cause severe sea states.

9.22.6 Special Notes for France See Area 17.

9.22.7 PORNIC

Loire Atlantique **47°06′·54N 02°06′·59W** ✴✴✴✴♒♒♒❀❀❀

CHARTS AC 2646, 2986, 2981; Imray C40; SHOM 7395, 7394; Navi 547, 549

TIDES +0515 Dover; ML 3·6; Duration 0540

Standard Port BREST (←)

Times				Height (metres)			
High Water		Low Water		MHWS	MHWN	MLWN	MLWS
0500	1100	0500	1100	7·0	5·5	2·7	1·1
1700	2300	1700	2300				
Differences PORNIC							
−0050	+0030	−0010	−0010	−1·2	−0·9	−0·5	−0·3
POINTE DE SAINT GILDAS							
−0030	+0025	−0005	−0020	−1·4	−1·0	−0·6	−0·3

SHELTER Very good in the large marina.

NAVIGATION WPT 47°05′·98N 02°08′·86W, 073°/1·6M to hbr ent, is in the W sector of Pte de Noëveillard lt ho. A SPM buoy, Fl Y 2·5s, at 47°06′·17N 02°07′·40W marks the seaward end of an outfall.

Beware Banc de Kerouars 4-6M WSW of harbour, least depth 0·7m, on which seas break; it is unmarked, but the W sector of Pte de Noëveillard lt clears it by night.

From the NW pass between this bank and the mainland. All other hazards are well marked. The S end of B de Bourgneuf is full of oyster beds, and many obstructions.

Do not attempt entry at LWS. Approach to marina reported silting, best depths close to west side of channel. Two sets of PHM and SHM beacons avoid rocky spurs close to E and S bkwtrs.

LIGHTS AND MARKS See chartlet and Lights, buoys and waypoints. The W sector (051°-079°) of Pte de Noëveillard lt ho lies between Notre-Dame WCM bcn twr VQ (9) 10s, and the E end of Banc de Kerouars 0·7m. The lt ho and trees behind are not easily seen.

COMMUNICATIONS CROSS 02·97·55·35·35; 🔲09·65·22·36·72; Auto 08·92·68·08·44; Police 02·40·82·00·29; Brit Consul 01·44·51·31·00; Dr 02·40·82·01·80; HM VHF Ch 09 (H24); Taxi 02·40·82·87·62.

FACILITIES Marina ☎02·51·74·15·65 accueil.portdepornic@gmail.com www. portdepornic.com. Depth 2m, no access LW±2 for draft >1·7m when Coeff >75, nor in SE–SW winds >F7. Call ahead VHF Ch 09, ❤ berths are at S ends of pontoons P3 (1st to stbd, LOA <8m) P2 (<10m) or where directed. ☎02·40·82·05·40. 754+40❤ €3·00. 🛢🛢 on small pontoon with drying ledge inshore, ▣🚿 artisan bread/ croissants sold on quay daily (am) ✎ ▣ ✗ Ⓔ ⚓ 🏳(50t).

CN de Pornic ☎02·40·82·34·72.

Old hbr dries 2·5m; access via marked, lit drying channel. Locals only, 🏳(6t).

Town Market Thurs & Sun am. ✗ 🏠 🛒 🛢 ✉ Ⓑ ⚓ ✈ (Nantes). Ferry: via Roscoff/St Malo.

S Biscay

9.22.8 L'HERBAUDIÈRE Loire Atlantique 47°06'·54N 02°06'·59W ✳✳✳✿◊◊◊❀❀

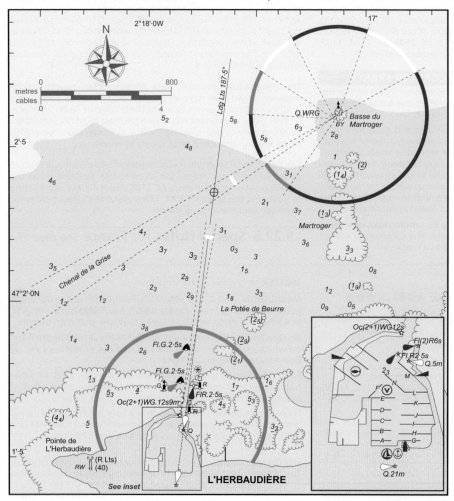

is 2·7M WNW of hbr. Other lts/marks as chartlet and Lights, buoys and waypoints.

At night initial positioning is assisted by any of the 3 W sectors (see Lights, buoys and waypoints) of Basse de Martroger, NCM bcn twr. Ldg lts 187·5°, both hard-to-see-by-day grey masts, Q, lead over Banc de la Blanche (1·5m) approx 2M N of hbr.

COMMUNICATIONS
🛟 02·51·39·33·90;
CROSS 02·97·55·35·35;
⚓ 02·40·84·80·19;
Auto 08·92·68·08·85;
Police 02·51·39·04·36;
✉ 02·51·39·06·80;
British Consul 01·44·51·31·00;
Dr 02·51·39·05·64;
HM VHF Ch 09 (HO);
Aff Mar 02·51·39·94·03.

FACILITIES
Marina ☎02·51·39·05·05
herbaudiere.port@wanadoo.fr
www.portdeplaisance-herbaudiere.com. Dredged 1·5m.
❶s berth on pontoon F or as directed. 572 + 50❶ €3·06. ⚓
🚿(Jul/Aug) ⬛⚒🔧⚓🏪🅴🔨
⬛(30t) 🔌 SC.

Town ✕🛒🍴✉🅑⚡✈ (Nantes).
Ferry: via Roscoff/St Malo.

CHARTS AC 2646, 2986, 2981; Imray C40; SHOM 7395, 7394; Navi 547, 549

TIDES +0500 Dover; ML 3·4

Standard Port BREST (←)

Times				Height (metres)			
High Water		Low Water		MHWS	MHWN	MLWN	MLWS
0500	1100	0500	1100	7·0	5·5	2·7	1·1
1700	2300	1700	2300				
Differences L'HERBAUDIÈRE							
−0045	+0025	−0020	−0020	−1·5	−1·1	−0·6	−0·3
FROMENTINE							
−0045	+0020	−0015	+0005	−1·8	−1·4	−0·9	−0·2

SHELTER Good in marina, except in N to E winds. The popular ⚓ off Pte des Dames lt, 4+M to the E, is exposed to N and E winds.

NAVIGATION WPT 47°02'·32N 02°17'·70W, 187·5°/7ca to hbr ent. There are rocks and banks to the SW, NW and NE of Pte de l'Herbaudière. Ldg lts and white sector (187·5°-190°) of the W jetty lt both lead into ent chan, dredged to 1·2m and passing close W of two 0·3m patches; care is needed at LWS. Two SHM buoys, both Fl G 2·5s, and a PHM buoy, Fl R 2·5s, mark the last 2ca of the chan. W bkwtr obscures vessels leaving. FVs leave hbr (W side) early morning.

LIGHTS AND MARKS Visibility in summer is often poor. Conspic daymarks are R/W radio mast 500m W of hbr and water tr about 1M SE. Ile du Pilier lt ho, with two distinctive twrs,

---

BAIE DE BOURGNEUF

The Baie de Bourgneuf (AC 2646, 2981) lies between Pointe de St Gildas and Pte de l'Herbaudière and is sheltered except in W winds, which can raise a heavy sea on the ebb. The E and S sides of the Baie are full of shoals, rocks and oyster or mussel fisheries. The only marinas are at Pornic and L'Herbaudière.

From the NW (AC 2986) the approach is simple, but beware La Couronnée (dries 2·2m) a buoyed rocky bank 2M WSW of Pte de St Gildas. Banc de Kerouars (least depth 1m; breaks) extends 3M further E. From the W approach **Pornic** either N or S of La Couronnée and Banc de Kerouars; thence pass NW of Notre Dame WCM beacon tower marking a line of rocks extending ESE to the coast.

To approach **L'Herbaudière** Marina from the SW, skirt round Chaussée des Boeufs, buoyed rocks, some drying, on to which the tide sets; and which extend 6M to seaward of the island. Thence enter Chenal de la Grise (3m) between Île du Pilier and Pte de l'Herbaudière and in the 055°-060° W sector of Martroger NCM bcn.

If heading NE to Pornic, pass N of Martroger, and clear of Roches des Pères about 1M ENE.

The S ent to the Bay via Goulet de Fromentine (SHOM 7394; ECM 549) is difficult due to a shifting bar and 8 hrs of W-going stream; the conspic bridge has 24m clearance. The chan between the SWM lt buoy and Milieu beacon tower shifts and is buoyed accordingly. Inside, Route du Gois, a causeway drying 3m, limits further progress NNE to shoal draft at sp HW±1.

MINOR HARBOUR ON NE CORNER OF ÎLE DE NOIRMOUTIER

NOIRMOUTIER-EN-L'ÎLE **46°59´·33N 02°13´·17W**. AC 2981; SHOM 7394. Tides as opposite. Good shelter and ⌒ 4M SE of L'Herbaudière, but this FV hbr dries up to 3·6m on mud/gravel. Appr across drying rock ledges to the N and E of ent. The E'ly of 2 chans runs SSW from approx 47°00´·70N 02°11´·10W, then doglegs WNW to the ent. It is marked by 5 unlit PHM bns. An inshore chan runs S from off Pte des Dames lt ho, Oc (3) WRG12s. It is more easily followed, keeping about 400m off 5 unlit SHM bns along the shore. Both chans meet at the ent where the flood tide sets S. S jetty hd has ☆ Oc (2) R 6s. Follow the jetty on N side for about 1M to hbr. A recce is advised (3M overland from L'Herbaudière).

HM ☎02·51·39·08·39. **Quay** ⌒ ⊥ ⚓ ⚒ 🅴 ⚒ △ ⌒(4t).

Town ✕ 🛒 ⚒ ⚒ ⚒ ♨ ⊠ 🅑 ⚒ ✈ at Nantes, via ferry to Pornic and bus. Ferry: via Roscoff or St Malo.

MINOR HARBOUR AT SOUTH END OF ÎLE DE NOIRMOUTIER

FROMENTINE, Vendée, **46°53´·60N 02°08´·60W**. AC 2981, SHOM 7394. HW +0550 on Dover (UT); ML 3·2m; Duration 0540. Tides L'Herbaudière. Do not appr from Baie de Bourgneuf as there is a road causeway (Le Gois) dries 3m, from Île de Noirmoutier to the mainland.

From Fromentine SWM lt buoy, L Fl 10s, 46°53´·10N 02°11´·40W, track 072°/1·4M via Goulet de Fromentine to Tourelle Milieu, Fl (4) R 15s, R twr, and Boisvinet W bcn twr. The chan is buoyed, moved as necessary, but is very shallow, so dangerous in bad weather. At sp the ebb can reach 8kn, the flood 5kn. Do not attempt night entry. Pass under the bridge (clearance 24m, two x Iso 4s 32m 18M). ⚓ W of pier near PHM buoy Fl R 2s. At Pte du Notre Dames-de-Monts,

there is a Dir lt, Oc (2) WRG 6s; see Lights, buoys and waypoints for sector details.

Facilities: Quay ⊥ ⚓ ⚒ ⚒ ✕ ⌒ ⌒(3t).

9.22.9 PORT JOINVILLE (Île d'Yeu)

Vendée **46°43´·75N 02°20´·77W** ❄❄⚓⚓⚓✿✿✿

Continued overleaf

PORT JOINVILLE *continued*

CHARTS AC 2663, 2997, 3640; Imray C40; SHOM 7402, 7410; Navi 549

TIDES +0550 Dover; ML 3·1; Duration 0600

Standard Port BREST (←)

Times				Height (metres)			
High Water		Low Water		MHWS	MHWN	MLWN	MLWS
0500	1100	0500	1100	7·0	5·5	2·7	1·1
1700	2300	1700	2300				
Differences *PORT JOINVILLE							
−0040	+0015	−0030	−0035	−2·0	−1·5	−0·8	−0·4

*Local information gives a little more water at all states.

SHELTER Good in marina, but swell enters outer hbr in N/NE winds. Visitors' dedicated ‡age is E of the marina as per chartlet; 4-5m on sand, but only in suitable conditions. No ‡ in outer hbr.

NAVIGATION WPT 46°44′·30N 02°20′·13W (7ca SW of Basse Mayence NCM buoy) 219°/7ca to abeam bkwtr lt. Approach with caution in strong N-E winds; if F8 do not attempt entry. Beware Basse du Bouet (dries 0·6m) 3ca NW, La Sablaire shoal to the E and rks along the coast both sides of hbr ent. Outer chan dredged 1·2m; appr with care at LW, avoiding bank (dries 2·3m) filling the W half of outer hbr. Keep clear of ferries.

LIGHTS AND MARKS Daymarks: Very conspic water twr leads 224° to hbr. Two green-topped white twrs (front, Quai du Canada Iso G 4s; rear, Old lt ho) ≠ 203° also lead into hbr. Conspic chimney W of hbr ent. Ldg lts 219°, both QR. Other lts: see Lts, buoys and waypoints.

COMMUNICATIONS CROSS 02·97·55·35·35; [LB] 02·51·58·32·01; ⚓ 02·51·36·10·78; Auto 08·92·68·08·85; Police 02·51·58·30·05;

⊜ 02·51·39·06·80; Brit Consul 01·44·51·31·00; [H] 02·51·68·30·23; Dr 02·51·59·30·00/02·51·58·30·58; Aff Mar 02·51·59·42·60; Marina VHF Ch 09, 16 (HO).

FACILITIES
Marina Very crowded in season; best to pre-book as it is the only secure hbr on Yeu. Ⓥ at E end of pontoons A & B. ☎02·51·58·38·11. plaisance.iledyeu@vendee.cci.fr 330+170Ⓥ €3·30. (15t). **CN Ile d'Yeu** ☎02·51·58·31·50.

The wet basin (3·7m) access HW±1½ via entry gate 02·51·58·37·01, is mainly for FVs, but possible overflow for yachts in high season; no pontoons; R/G tfc lts at the gate.

Town ✕ ⌂ ⛽ 🛒 🗑 ⊠ Ⓑ ⇌ (St Gilles-Croix-de-Vie) ✈ (Nantes). Helicopter from hbr to mainland, La Barre de Monts. Flights from airfield 2M west of hbr to Nantes and (summers only) to Les Sables d'Olonne. Ferry: via Roscoff or St Malo, then Fromentine, St Gilles and Les Sables d'Olonne.

ANCHORAGES ON S COAST OF ILE D'YEU
PORT DE LA MEULE, 46°41′·68N 02°20′·75W. AC 2663; SHOM 7410. −0050 sp and −0020 nps on Brest; ML 3·0m. Small, drying fishing hbr on S side of Ile d'Yeu, only safe in settled offshore weather; untenable in S winds. Many little FVs inside hbr; best to ‡ outside and W of the ent.

Between Pte de la Père to the W and Pte de la Tranche to the SE, appr on 023° towards W square patch on Gy □ lt tr, R top, Oc WRG 4s; at night in W sector (018°-027·5°) but night ent not advised. Within 1ca of ent beware rks, first to port, then to stbd. Few facilities: ⚓s, 1½ miles to ✕ ⌂ 🛒.

ANSE DES VIEILLES, 46°41′·53N 02°18′·70W; 1M W of Pte des Corbeaux. Use SHOM 7410. Good shelter from N'lies. ‡ clear of bathing area in 8-9m N or NE of Ours des Vieilles rks drying 3·5m.

9.22.10   ST GILLES-CROIX-DE-VIE

Vendée **46°41′·47N 01°57′·29W** (abm bkwtr hd) ❄❄❄🌢🌢🌢✿✿

CHARTS AC 2663, 2997; Imray C40; SHOM 7402; Navi 1022, 549

TIDES +0500 Dover; ML 3·2; Duration 0600

Standard Port BREST (←—)

Times				Height (metres)			
High Water		Low Water		MHWS	MHWN	MLWN	MLWS
0500	1100	0500	1100	7·0	5·5	2·7	1·1
1700	2300	1700	2300				
Differences ST GILLES-CROIX-DE-VIE							
−0030	+0015	−0030	−0030	−1·9	−1·4	−0·7	−0·4

SHELTER Good in marina and easy access except in strong SW'lies or swell when breakers form off ent. Waiting ⚓ off ent, NW or SE of ldg line in 2·0–3·5m.

NAVIGATION WPT 46°40′·94N 01°58′·02W, on the 043·7° ldg line and abeam Pilours SCM lt buoy, 043·7°/0·72M to Jetée de la Garenne lt. From W & NW, beware Rocher Pill'Hours (2·8m) and drying reefs extending 1ca SE.

Ent chan (⚓ prohib) is dredged 1·0m, but narrow and very shallow near bkwtr hds due to silting which persists despite dredging; depths may be further reduced. Keep well off the first two chan buoys which are laid outboard of drying rks; and, tfc permitting, keep slightly W of 043·7° ldg line.

Best arr/dep HW −2 to HW to avoid strong ebb, up to 6kn in ent at springs which may complicate berthing on outer ends of pontoons. Do not arr/dep LW±2 if >1·5m draft.

LIGHTS AND MARKS Daymarks are Pte de Grosse-Terre (rky hdland) with lt ho, W truncated conical tr; and two spires NE of the marina. The front ldg mark is initially hard to see against bldgs behind.

COMMUNICATIONS CROSS 02·97·55·35·35; Ⓛ 02·51·55·01·19; ♒ 02·51·36·10·78; Auto 08·92·68·08·85; Police 02·51·55·01·19; ⊖ 02·51·55·10·18; Brit Consul 01·44·51·31·00; Dr 02·51·55·11·93; HM VHF Ch 09 (season 0600-2200; out of season 0800-1200, 1400-1800LT); Aff Mar 02·51·55·10·58.

FACILITIES

Marina ☎02·51·55·30·83. port.la.vie@wanadoo.fr www.semvie.com Dredged 1·3–1·5m. Ⓥ berth/raft on pontoon No. 8 or as directed, or pick up ⚓ S of marina. 900+100Ⓥ €2·67. ▲ ⚓ ✕ 🖙 🛒 🗑 🛢 ▲ SHOM ⚒ 🖛 ⚓ ✕ Ⓔ 🅿(35t) ⛴(2·2t) ◥. **CN** ☎02·51·54·09·31.

Town ✕ 🖙 🛒 ▲ ⊠ Ⓑ ⇌. A ferry runs to Ile d'Yeu. Ferry: Roscoff or St Malo.

9.22.11 LES SABLES D'OLONNE

Vendée 46°29′·42N 01°47′·54W 🌸🌸⚓⚓⚓🌸🌸

CHARTS AC 2663, 2998, 3638; Imray C40, C41; SHOM 7402, 7403, 7411; Navi 1022

TIDES +0530 Dover; ML 3·2; Duration 0640

Standard Port BREST (←—)

Times				Height (metres)			
High Water		Low Water		MHWS	MHWN	MLWN	MLWS
0500	1100	0500	1100	7·0	5·5	2·7	1·1
1700	2300	1700	2300				
Differences LES SABLES D'OLONNE							
−0030	+0015	−0035	−0035	−1·8	−1·3	−0·7	−0·4

SHELTER Very good in Quai Garnier and Port Olona Marinas.

NAVIGATION WPT 46°28′·49N 01°47′·23W (abeam Nouch Sud SCM buoy, Q (6)+L Fl 15s) 032·5°/5ca to intersection with 320° ldg line; thence 7ca to the hbr ent. To the W, beware Les Barges d'Olonne, extending 1·3M W of Pte de Aiguille.

The 2 appr chans are: the SW chan with La Potence ldg lts 032·5°, which lead into SE chan on ldg line 320°. In bad weather use the SE chan. Le Noura (1·3m) and Le Nouch (0·7m) are isolated rks on shoals S of Jetée St Nicolas. Further SE, Barre Marine breaks, even in moderate weather. A buoyed wk (dries) lies off hbr ent, to E of 320° ldg line.

Access at all tides; entry is easy except in strong winds from SE to SW when apprs get very rough. Sailing is prohib in the narrow dredged chan (2m). Best water initially is on the E side, then mid-chan. Caution: oncoming vessels passing very close.

LIGHTS AND MARKS See Lights, buoys and waypoints and chartlet for lt details. Les Barges lt ho, 2M W of ent. L'Armandèche lt ho, conspic 6ca W of ent. SW Chan ldg lts 032·5°: both Iso 4s H24. SE Chan ldg lts 320°: front QG on E bkwtr; rear, Q on Tour d'Arundel a large grey twr. St Nicolas jetty hd, QR. Inner ldg lts 328·1°, both Iso R 4s; R/W vert stripes, hard to see by day but not essential for yachts.

COMMUNICATIONS CROSS 02·97·55·35·35; ♒ 02·51·36·10·78; Auto 08·92·68·08·85; Police 02·51·33·69·91; ⊖ 02·51·23·58·00; Brit Consul 01·44·51·31·00; Ⓗ 02·51·21·06·33; Dr 02·51·95·14·47; Port HM 02·51·95·11·79, VHF Ch 12 (0800-1800); Aff Mar 02·51·28·81·91; Marinas VHF Ch 09 *Plaisance Quai Garnier* or *Port Olona* (0600-2400LT in season).

Continued overleaf

LES SABLES D'OLONNE *continued*

FACILITIES Marina Quai Garnier ☎06·78·07·15·15/ 06·73·76·94·65. www.vendee.cci.fr 119 ⌷ inc Ⓥ on 10 pontoons (J-A); berthing prohibited on hammerheads; €3·00; Occas wash from FVs. N of pontoon J at the gate into the Commercial basin an Oc Y lt shows when a vessel is entering/leaving (HW ±2); keep clear. Readers report good facilities and helpful staff. Very close to town centre, shops and many restaurants.

Port Olona Marina ☎02·51·32·51·16. www.portolona.fr portolona@wanadoo.fr (1·5–3·5m). Access H24. Visitors check in at accueil/fuel pontoon port side, by Capitainerie. Distant pontoon ⌦, at NE end, is for Ⓥ and multihulls, but on request a berth may be found in pontoons A-C. 990+110Ⓥ €2·95. ⌦ ⌷ ⌷ (0800-2000) ⚓ ⚒ ⌷ ⚓ Ⓔ SHOM ⌷ Divers ⌷(28t).

Town ⌷ ⌷ × ⌷ ⌷ Ⓑ ⇌ ✈. Ferry: Roscoff or St Malo.

9.22.12 BOURGENAY

Vendée 46° 26'·27N 01° 40'·70W ✿❀♦♦♦✿✿

CHARTS AC 2663, 2998; Imray C41; SHOM 7069, 7403; Navi 1022

TIDES +0600 Dover; ML 3·1; Duration 0640. Use Differences LES SABLES D'OLONNE, 5·5M NW

SHELTER Good in marina (2m). Even in moderate weather, and especially with SW'lies, a big swell can break at the ent; care needed and/or divert to Les Sables d'Olonne.

NAVIGATION WPT 46°25'·27N 01°41'·93W (SWM buoy, L Fl 10s) 040°/1·32M to pier head. 600m ENE of WPT, beware Roches du Joanne (2·9m; dangerous in bad weather) and shoal/drying patches to E of ent, marked by unlit SHM buoy and bcn.

West pierhead is painted white. Ent chan, dredged 1·0m, makes a pronounced, blind S-bend marked by reflective chevrons; 3kn speed limit. May be less water than charted.

LIGHTS AND MARKS Ldg lts 040° QG; front W hut; rear W pylon, both with G □ & W border. Large white bldg is conspic near ldg lts. The Iso G 4s and Fl (2) R 6s are obsc'd until inside hbr ent.

COMMUNICATIONS CROSS 02·97·55·35·35; Auto 08·92·68·08·85; Police 02·51·90·60·07; ⊖ 02·51·95·11·33; Brit Consul 01·44·51·31·00; Ⓗ 02·51·21·85·85; Dr 02·51·90·62·68; HM VHF Ch 09 16 (office hrs: in season 0800-2100LT, out of season 0900-1200, 1400-1800); Aff Mar 02·51·21·81·81.

FACILITIES Marina ☎02·51·22·20·36 portbourgenay@wanadoo.fr 610+50Ⓥ €2·67. Ⓥ ⌷/raft on long pontoon (no fingers) N side of E bkwtr or as directed. ⚓ ⚓ ⌷ ⌦ ⚓ ⌷ ⌷(20t). YC ☎08·74·50·57·57.

Town × ⌷ ⌷ ⌷ Ⓑ. ⇌ (Les Sables d'Olonne). ✈(La Lande, Chateau d'Olonne). Ferry: via Roscoff or St Malo.

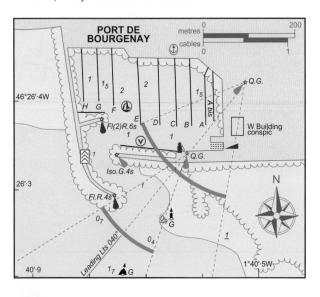

Île d'Yeu, 30m high, has its main lt ho near the NW end where Les Chiens Perrins lt bcn marks offliers. On the NE coast a very conspicuous water twr gives a good lead into **Port Joinville**, crowded in season. The passage along the NE of the island carries 6-7m nearer to the island. Pte des Courbeaux lt ho is at the low SE end of the island. The passage along the NE of the island carries 6-7m nearer to the island. ▶*Here the E-going stream starts at HW Brest –5¾ and the W-going at HWB –¼.*◀ The SW coast is steep-to and rky, with a tiny drying hbr at Port de la Meule (best to anch outside) and, further E, anch at Anse des Vieilles, both only tenable in settled conditions.

Les Marguerites, rky shoals, lie SSW of Goulet de Fromentine, with the part-drying reef, Pont d'Yeu (SCM buoy) extending midway between the mainland and the Île d'Yeu; here anch is prohibited in a 2·2M wide strip due to underwater cables (AC 2663, 2997, 2998).

The low-lying, wooded Côte de la Vendée continues 40M SE to Pte du Grouin-du-Cou with few dangers more than 1·5M offshore, except near Les Sables d'Olonne. 17M E of Port Joinville is **St Gilles-Croix-de-Vie**. Thence 17M SSE is **Les Sables d'Olonne**, with Les Barges drying reef (lt) extending 2·5M W of the entrance. **Bourgenay** is 6M further SE. The approaches to these secure hbrs, all with marinas, are exposed to onshore winds from SE to NW, and susceptible to swell. **Jard-sur-Mer** is a small drying harbour midway between Bourgenay and Pte du Grouin-du-Cou.

PERTUIS BRETON

Pertuis Breton is entered between Pte du Grouin-du-Cou and Pointe des Baleines (both lit) on **Île de Ré**, which is surrounded by shallows and drying areas (AC 2999). Beware rocky ledges (dry) extending 2·5M NW from Les Baleines. Pertuis Breton gives access to the harbours of **Ars-en-Ré**, **St Martin** and **La Flotte** on the N shore of Île de Ré. From St Martin to Pte de Sablanceaux (SE tip) there are extensive oyster beds.

On the mainland side, in fresh NW winds against tide a bad sea builds on the bank which extends 8M W of Pointe du Grouin du Cou. 1M SSE of the Pointe is the unmarked Roche de l'Aunis (depth 0·8m). SE of the Pte the coast is fronted by sand dunes and mussel beds, whose seaward limits are marked by unlit buoys. 8M ESE is the drying ent to Rivière Le Lay, with a bar (dries 1m) dangerous in bad weather. Thence the channel to L'Aiguillon and La Faute-sur-Mer is marked by bcns and buoys. Offshore a large mussel farm (3M x 1M) is marked by 4 cardinal and 2 SPM light buoys.

5M further E is the entrance to Anse de l'Aiguillon, in which there are extensive mussel beds. Here the Sèvre Niortaise estuary, after 3·5M, leads via a canal into the port of **Marans**. Further S the main route continues SE and E, passing under the fine road bridge (30m clearance) between the mainland and Île de Ré and into Rade de la Pallice; thence to La Rochelle or the Pertuis d'Antioche.

PERTUIS D'ANTIOCHE

The low-lying shores of the Pertuis d'Antioche (AC 2999) converge eastwards to **La Rochelle**, **Ile d'Aix**, La Charente River and **Rochefort**. Île de Ré forms the N side, fringed by rocky ledges extending 2·5M SE from Pte de Chanchardon (lt) and nearly 1M from Pte de Chauveau (marked by lt twr and two bcns).

Île d'Oléron lies SE of Pertuis d'Antioche. At its N tip, off Pte de Chassiron (light house and signal station), reefs extend 5ca W, 1·5M N to Rocher d'Antioche (lit) and 1·5M E; here there is often a nasty sea.

Well offshore, 43M WNW of Pointe de Chassiron, The Plateau de Rochebonne is a chain of rocks on which the sea breaks dangerously. It has least depth 3·3m, is marked by 2 cardinal and 2 lateral light buoys and is enclosed by a circular 'Area to be avoided', radius 7M centred on 46°10'·00N 02°25'·60W – although the rocks only occupy the NNE half of the circle.

MINOR HARBOURS ON THE N SHORE OF PERTUIS BRETON

JARD-SUR-MER, Vendée, **46°24´·43N 01°34´·78W**. ✿❁❁✿. AC 2663, 2998 (1:10,000 inset); SHOM 7403. HW +0600 on Dover (UT); HW−0010 & ht −2·0m on Brest; ML 3·1m; Duration 0640. 2 W bcns 4ca E of hbr ent lead 036° between the drying Roches de l'Islatte and Roches de la Brunette, marked by buoys (May-Sep). Inner RW bcns at W end of hbr lead 293° into small drying unlit hbr, 4·5m max at HW.

Moorings inside bkwtr, inc 7 Y ⚓s. HM's office with blue roof and adjacent bldgs are conspic. HM (occas) ☎02·51·33·90·61; ☎02·51·95·11·33. Facilities: Jetty ⚓ 🏪 Divers ⛴(5t).

L'AIGUILLON/LA-FAUTE-SUR-MER, Vendée, **46°19´·95N 01°18´·78W**. ✿❁❁✿✿. AC 2663, 2999; SHOM 7404. HW +0535 on Dover (UT) HW −0030, ht +0·6m on Pte de Grave (Zone −0100); ML 3·4m. The bar dries and is dangerous in bad weather; avoid in strong S or W winds. The ent is only safe in fine weather with offshore winds. A transformer at 46°18´·20N 01°14´·90W is conspic on a hill NE of Pte d'Arcay. Beware mussel beds with steel piles which cover at HW; also oyster beds.

Access HW±3, max 1·5m draft; not suitable for keel boats. Appr from Le Lay SCM buoy, 46°16´·15N 01°16´·49W, with transformer brg 033°. The drying berths at both hbrs are small enclosures made of rough timber framing; no facilities for visitors. Upriver the low bridge has 3m clearance. HM L'Aiguillon ☎05·51·97·06·57; HM La Faute ☎05·51·56·45·02, ✂ 🏪.

MARANS, Vendée, approx **46°18´·70N 01°00´·00W**. ✿❁❁❁✿✿. AC 2663, 2999; SHOM 7404; Navi 551. Tides: see L'Aiguillon above; HW at Brault lock = HW La Rochelle + 0020. From SWM buoy,

L Fl 10s, (46° 15´·45N 01°11´·39W) appx 2·5M c. 035° to beginning of the buoyed chan (dries 1·0m) which leads NE past Pavé jetty, Fl G 4s. In the estuary and lower chan beware mussel beds with many posts which cover at HW.

Waiting buoys below Brault bridge, which lifts at between HW±1 when the lock opens, day only. Pre-check times ☎06·64·49·29·60 before entering the river. The vast lock (104 x 45m wide, with sloping sides) contains a waiting pontoon and small swing bridge at the far end.

A straight 3M long canal (4·7m) leads to good shelter in non-tidal hbr, max LOA 16m. Beyond new finger pontoons raft on other yachts. HM ☎05·46·01·02·99/06·25·79·16·35, portdemarans@orange.fr; ⊂ (40+10 Ⓥ) €1·20, ⚓ 🗓 🗓 🗓 ⨂ ✕ (wood) ⚠ 🏪(8t). Town ✕ 🗓 🕮 Market (Tues & Sat) . 🖂 ☎05·46·01·53·33, 0·5M below Brault bridge, has 🖂 for larger yachts.

MINOR HARBOUR ON ÎLE DE RÉ

LA FLOTTE, Charente Maritime, **46°11´·34N 01°19´·30W**. ✿❁❁❁✿✿. AC 2999; SHOM 7404, 7412. HW +0535 on Dover (UT). From NW keep clear of Le Couronneau; from E, keep N of bcn off Pte des Barres. Appr on 212·5° in W sector (205°-220°) of La Flotte lt ho, W tr + G top, Fl WG 4s. Alongside it a Moiré indicator shows vert B line when on track 212·5°, or chevrons pointing in the direction to regain track. 5 waiting buoys outside; or ⚓ off in 3m, sheltered from S & W.

Outer hbr is sheltered by mole and dries 2·0m. 3 pontoons (6 Ⓥ berths, max LOA 10m, on mole, €2·60) in outer hbr: pre-booking advised. Inner hbr dries 2·8m. HM ☎05·46·09·67·66. **Quay** ━ ⚓ Grid 🗓 🗓 🏪 ✂ ⚠ 🗓. **CN de la Flotte-en-Ré** ☎05·46·09·97·34, (Jul-15 Sep) 🗓.

ILE DE RÉ (9.22.13 and 9.22.14) *See overleaf*

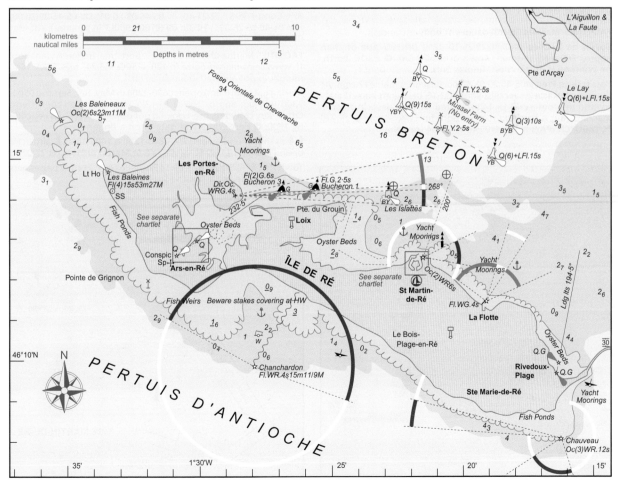

9.22.13 ARS-EN-RÉ

Charente Maritime 46°12'·64N 01°30'·71W ✿✿✿✿✿✿✿✿

CHARTS AC 2998, 2999; Imray C41; SHOM 7404, 7412; Navi 1022

TIDES +0540 Dover; ML 3·7m; See St Martin-de-Ré.

SHELTER ⚓ clear of channel in a pool (2m) close S of Pte du Fier. Both small marinas are sheltered, but in the confined space with strong NE'lies manoeuvring may be difficult for LOA >13m.

NAVIGATION WPT 46°14'·18N 01°23'·25W, 268°/1·8M to first SHM chan buoy, Bûcheron 1. Beware shoal ground 0·8m to 2·5m both sides of the outer appr chan. Continue to the inner, buoyed and beaconed 232·5° ldg line, where buoys are moved to mark new sand/mudbanks. Port d'Ars is at the head of a dog-legged chan in the SW corner of the bay, Le Fier d'Ars, which dries to salt pans & oyster beds. Night appr not advised without prior day visit. HW –1 is best time to arr/dep the basins. Times vary with draft (max 1·8m) coefficient, wind and barometer.

LIGHTS AND MARKS Les Islattes NCM bcn lt twr is conspic S of the ⊕. Loix water twr is conspic S of the 268° appr chan.

Dir ⚡, Oc WRG 4s (hard to see by day) covers the 268° outer appr chan. To the right of it and 3M west, Les Baleines lt ho may be seen.

Inner 232·5° ldg lts, hard to see by day, lead across Fiers d'Ars into Port d'Ars: front, W □ with R lantern; rear, B □ on W framework tr, G top. See Lights, buoys and waypoints and chartlet for light details. In the town the ✷ spire, white with black top, is conspic about 330m SSW of the rear 232·5° ldg lt.

COMMUNICATIONS CROSS 02·97·55·35·35; Auto 08·92·68·08·17; Police 05·46·84·32·67; ⊖ 05·46·41·11·73; Brit Consul 05·57·22·21·10; Dr 05·46·29·44·19; Aff Mar 05·46·09·68·89 (at La Flotte); Marinas VHF Ch 09.

FACILITIES Marinas (depth gauges at both entrances):

Bassin de la Criée ☎05·46·29·25·10. (2m) on NW side of chan approx 600m NE of town. Access over sill 2·5m. Ⓥ €3·16, berth, on pontoon H, first to port, fingers only.

Bassin de la Prée ☎05·46·29.08.52 portarsenre@mairie17.com. At head of chan access over sill 2·9m. Ⓥ €3·16, berth/raft immediately to stbd on pontoon D in 1·8m. Rafted up to 5 deep in season; manoeuvring & berthing may be tight for LOA >13m. ➘ ⚒ ⊞ ⚓ ⛟(6t). **CN d'Ars-en-Ré** ☎05·46·29·23·04 (Apr-Nov).

9.22.14 ST MARTIN-DE-RÉ

Charente Maritime 46°12'·50N 01°21'·93W ✿✿✿✿✿✿✿✿

CHARTS AC 2998, 2999; Imray C41; SHOM 7404, 7412; Navi 551, 1022

TIDES +0535 Dover; ML 3·7

Standard Port POINTE DE GRAVE (→)

Times				Height (metres)			
High Water		Low Water		MHWS	MHWN	MLWN	MLWS
0000	0600	0000	0500	5·3	4·4	2·1	1·0
1200	1800	1200	1700				
Differences ST MARTIN, Ile de Ré							
+0005	–0030	–0025	–0030	+0·6	+0·3	+0·2	–0·1

SHELTER 4 W waiting buoys or ⚓ off ent which is difficult in strong NW-NE winds. These winds also render untenable a waiting pontoon (in season; prev dredged 2m, but now dries, very soft mud) inside NW mole. Complete shelter in marina (depth 3m).

NAVIGATION WPT 46°14'·68N 01°20'·74W, 200°/2·3M to St Martin mole hd. From the NW, keep N and E of Banc du Rocha (1·2m) extending 2½M ENE from Pte du Grouin. From SE, pass well N of unlit NCM bcn, about ¾M NE of ent, marking Le Couronneau drying ledge in R sector (245°-281°) of St Martin lt ho. By day appr 210·5° lt ho ≠ □ ✷ tr; or 202° mole hd ≠ ✷ tr; the lt ho is far easier to see than the mole head.

LIGHTS AND MARKS Conspic daymarks: La Citadelle 3ca E of hbr ent; Lt ho, Oc (2) WR 6s, W twr/R top, on ramparts SE of ent; and the □ ✷ tr and nearby ruins. Mole hd, Fl G 2·5s, is obsc'd by Pte du Grouin when brg <124°. See chartlet and Lights, buoys and waypoints for details.

COMMUNICATIONS CROSS 02·97·55·35·35; ⚓ 05·46·41·29·14; Auto 08·92·68·08·17; Police 05·46·09·21·17; ⊖ 05·46·09·21·78; Brit Consul 05·57·22·21·10; Ⓗ 05·46·09·20·01; Dr 05·46·09·20·08; HM 05·46·09·26·69, VHF Ch 09 (0700-1900LT in summer); Aff Mar 05·46·09·68·89.

FACILITIES Marina ☎05·46·09·26·69. portstmartindere@wanadoo. fr www.saint-martin-de-re.fr 170 + 50Ⓥ €3·24; max LOA 16m, multihulls not allowed in Bassin à Flot.

Marina gate (sill 0·8m) opens 0630-2200 in May, Jun, Sep and 0500-2300 Jul/Aug; approx HW–2 to +2½, depending on coefficient. Often very crowded, pre-booking not possible. Get ⏰ ♿ key code by 1900. Ⓥ berth/raft W of gate or on a finger.

⛽ 🛢 as shown on chartlet; 7/7, 0800-1200 & 1400-1800. 🛢 also in FV basin (1·5m) ⚒ Ⓔ ⊞ ⚓ ⛟(10t). **YC St Martin** ☎05·46·09·22·07.

Town ✕ ⛴ 🛒 🛍 🏦. SHOM ⛽ ⊠ Ⓑ. La Rochelle ⇌ ✈. Ferries: passengers only, La Rochelle, La Tranche sur Mer.

9.22.15 LA ROCHELLE

Charente Maritime **46°08'·83N 01°10'·11W** ❋❋❋❋♒♒♒❀❀❀

CHARTS AC 2999, 3000, 2743; Imray C41; SHOM 7404, 7413; Navi 551, 1022

TIDES +0515 Dover; ML 3·8

Standard Port POINTE DE GRAVE (→)

Times				Height (metres)			
High Water		Low Water		MHWS	MHWN	MLWN	MLWS
0000	0600	0000	0500	5·3	4·4	2·1	1·0
1200	1800	1200	1700				
Differences LA ROCHELLE and LA PALLICE							
+0015	−0030	−0020	−0025	+0·7	+0·5	+0·3	−0·1

SHELTER Very good in Port des Minimes, excellent in Vieux Port.

NAVIGATION WPT 46°08'·41N 01°11'·40W, 059°/0·9M to abm Tour Richelieu (with tide gauge) marking a drying rky spit N of the chan. Appr from N or S of Plateau du Lavardin to pick up 059° ldg line. Pte des Minimes lt bcn (aka Bout du Monde) and a WCM lt buoy mark drying rks extending 4ca SW of the marina.

Shallow (0·5m) appr chan needs care at MLWS; speed limit 5kn. Ent to Port des Minimes, 1ca past Tour Richelieu, is marked by WCM VQ(9) 10s and 2 PHM buoys. Speed limit 3kn in all basins.

For Vieux Port stay on 059° ldg line in chan (35m narrow) leaving 4 small PHM buoys well to port. Caution: Vedettes run at all hours. The unlit chan is not advised at night.

LIGHTS AND MARKS See chartlet/Lts, buoys and waypoints. The rear 059° ldg lt, Q, is briefly obscured by Tour St Nicolas between 061°-065°, ie when approaching between La Pallice and Lavardin plateau. Tour St Nicolas, Tour de la Chaine and Tour de la Lanterne, near the Vieux Port, are conspic by day and floodlit at night.

COMMUNICATIONS CROSS 02·97·55·35·35; ⚓ 05·46·50·86·00; Auto 08·92·68·08·17; Police 05·46·00·50·99; ⊖ 05·46·41·11·73; Brit Consul 05·57·22·21·10; Ⓗ 05·46·27·33·33; Dr 05·46·42·19·22; Port des Minimes VHF Ch 09 (H24); Vieux Port VHF Ch 09; Aff Mar 05·46·28·07·28.

FACILITIES Port des Minimes Marina ☎05·46·44·41·20. www.portlarochelle.com capitainerie@portlarochelle.com. 4,500, inc 400 Ⓥ on pontoon extending S from the accueil pontoon to the Capitainerie, max LOA 25m. €3·00 (pay 2 nights, 3rd free). ⛽ ⚓ (H24 with cr/card) ✕ ⬜ Ⓞ ⬚ ⬚ SHOM Ice ⊠ Ⓑ ⚒ ✕ Ⓔ 🅿 ☗(50t) ⬚(10t). Water bus to the town (Bassin d'Echouage) every H, 1000-2000, except 1300; Jul/Aug H and H+30, 0900-2330, except 1300. The major extension NW'wards was completed Summer 2014; it is now the largest marina on the Atlantic coast. Reports welcome.

Vieux Port comprises 3 separate basins, call ☎05·46·41·32·05, mob ☎06·03·54·00·57. From S to N:

Bassin des Chalutiers (5m). Access by prior arrangement. Entry gate ☎ 05·46·52·14·56, mob 06·16·46·60·26. Mainly used for events and by long-stay and mega-yachts.

Bassin d'Echouage (part-dredged 1·3m) is entered between the twin twrs of St Nicolas and La Chaine. It is an atmospheric tidal basin, but can be noisy at night. vieuxport@portlarochelle.com 75 + 40Ⓥ €2·40. First 2 pontoons for Ⓥ.

Bassin des Yachts (3m). Access HW −2 to +½ via entry gate with sill 1·2m; gate ☎05·46·52·14·56, mob 06·16·46·60·26. Night ent by prior arrangement. 100 ⬭ €2·40; 15m max LOA. ⬚ ⚓ ⚒ ✕ ☗(150t) ⬚(10t).

Town ✕ ⬜ 🛒 ⬚ ⬚ @ ⊠ Ⓑ ⇌ Laleu ✈ direct flts to London, Birmingham, So'ton etc; see Area 21. Ferries: via Roscoff, St Malo.

La Pallice, 3M W, is a commercial/FV port with no yacht facilities.

> **Note:** The adjacent bridge to Ile de Ré is lit and buoyed for big ships: N-bound between piers Nos 13 and 14; S-bound between piers Nos 10 and 11; 30m air clearance. Yachts may transit other spans, subject to air and water clearances.

ÎLE D'OLÉRON

On the NE coast of Ile d'Oléron (AC 2663, 2999, 3000) there are marinas at **Port St Denis** and **Le Douhet** at the N end; further S is a yacht basin at **Boyardville** and a fishing harbour at Le Château. All are sheltered from the prevailing W'lies.

From Pertuis d'Antioche, Grande Rade des Trousses is entered via either Passage de l'Est close to Île d'Aix or Passage de l'Ouest, which run each side of La Longe and Le Boyard, an extensive sandbank on which stands Fort Boyard with light tower. From Grande Rade, where good anch is found except in fresh NW winds, the narrow and shallow Coureau d'Oléron winds between ledges, oyster beds and constantly shifting shoals, with buoys moved to conform. About 2M SE of Le Chateau it is crossed by a bridge, clearance 18m; the bridge arch for the navigable chan is marked at road level by W □ boards, with ▲ or ■ superimposed, illuminated at night. Just N of bridge is Fort-Louvois (32m) connected to mainland by causeway. SHOM 7414 is needed.

▶*S-going stream starts at HW Pte de Grave –2½; N-going at HW Pte de Grave +5, sp rates 2kn.*◀ Up the Seudre River there are anchs and yacht facilities at **Marennes** and **La Tremblade**.

The W coast, from Pointe de Chassiron 15M SSE to Pte de Gatseau, is bounded by drying rks and shoals. In bad weather the sea breaks 4 or 5M offshore. La Cotinière, the only hbr, is almost exclusively a fishing port, exposed to the Atlantic.

▶*Tidal streams are weak, sp rate 1kn, starting NW at HW Pte de Grave +0300 and SE at HW Pte de Grave –0505, but often overcome by current due to prevailing wind. The rate, however, increases towards Pte de Gatseau.*◀

Here **Pertuis de Maumusson** separates the island from the mainland. A SWM lt buoy 'ATT Maumusson' about 3½M WSW of Pointe de Gatseau marks the approach. Banc de Gatseau and Banc des Mattes, both of which dry up to 3.9m, lie N and S of the chan; the sand bar usually has a depth of about 1·5m. Depth and position vary, and unlit lateral buoys are moved accordingly. Any swell speedily forms breakers, and the chan is very dangerous then or in any onshore winds, especially on the ebb (sp rate 4kn). In calm weather with no swell, a stout craft and reliable engine, and having gained local knowledge, consider entering about HW –1; ideally following a local FV with deeper draught.

9.22.16 ILE D'OLÉRON

CHARTS AC 2999, 3000; Imray C41; SHOM 7404, 7405, 7414, 7415; Navi 552

TIDES +0545 Dover; ML 3·9; Duration 0540

Standard Port POINTE DE GRAVE (→)

Times				Height (metres)			
High Water		Low Water		MHWS	MHWN	MLWN	MLWS
0000	0600	0000	0500	5·3	4·4	2·1	1·0
1200	1800	1200	1700				
Differences ILE D'AIX							
+0015	–0040	–0025	–0030	+0.8	+0.5	+0.3	–0.1
LE CHAPUS (Bridge to mainland)							
+0015	–0040	–0015	–0025	+0.7	+0.6	+0.4	+0.2
POINTE DE GATSEAU (S tip of Île d'Oléron)							
+0015	0000	–0020	–0015	0.0	–0.1	+0.2	+0.1

OVERVIEW Marinas at St Denis, Le Douhet and Boyardville. La Cotinière is a FV hbr prone to frequent swell; yachts only in emergency.

ST DENIS D'OLÉRON 46°02'·07N 01°22'·06W ✸❀♨♨♨♨♨

SHELTER Very good in marina. www.st-denis-oleron.com

NAVIGATION WPT 46°03'·26N 01°20'·82W, 205°/1·35M to chan ent. Appr chan (about 1·1m) is buoyed. Daymarks: SHM perch ≠ ✠ spire leads 257° to ent. By night 2 Dir lts lead 205° and 284° in sequence. Beware fishing nets with very small floats.

LIGHTS AND MARKS Pte de Chassiron, W lt ho + B bands, is conspic 1·9M WNW at N tip of island. Antioche lt bcn marks dangerous reef 2·2M NNW. Dir lt, ½M S of hbr ent, leads 205° in W sector (204°-206°) to intercept W sector (277°-292°) of second Dir lt on N

pier; or use this sector if coming from the SE.

COMMUNICATIONS CROSS 02·97·55·35·35; Auto 08·92·68·08·17; Ⓗ (12km) 05·46·47·00·86; HM VHF Ch 09; Aff Mar 05·46· 85·14·33.

FACILITIES Marina ☎05·46·47·97·97 port.stdenis@wanadoo.fr 3 W waiting bys about 700m E of ent in 0·7–1m. Access over sill 1·5m approx HW±2½ for 2m draft. Tide gauge, E bkwtr by sill, best depth by short stay, walk ashore (drying) pontoon inside E bkwtr end. 600+70♥ €2·47. ♥ pontoon A, max depth 2·5m: N side if <10m, S side >10m LOA. ⛽🚿🚻♿ wi-fi €2·00p/d ⚓ ⚓(10t).

LE DOUHET 46°00'·09N 01°19'·20W ⊛⊛⊛♨♨🌸🌸

SHELTER Very good in marina, SE part of hbr; FVs use NW part.

NAVIGATION WPT 46°00'·65N 01°17'·38W, NCM buoy VQ, 246°/1·4M to ent. Unlit, buoyed appr chan dries about 1·6m to approx 0·35M offshore. Ent difficult in fresh NE'lies against ebb; beware swell, oyster beds and overfalls.

LIGHTS AND MARKS The WPT buoy is 1·4M W of another NCM lt buoy Q; the latter plus an ECM buoy, Q (3) 10s, mark a fish farm in Passage de l'Ouest (toward Boyardville). La Longe and Le Boyard, rocky/sandy shoals, are marked by an unlit WCM buoy and by Fort Boyard, conspic twr 29m, Q (9) 15s. The Capitainerie and yacht masts are the only conspic features; no lts at marina.

COMMUNICATIONS Ch 09; port-plaisancedudouhet@wanadoo.fr

FACILITIES Marina ☎05·46·76·71·13. Access over a sill marked by tide gauge, 1·8m CD. Caution: sandbanks encroach on approach channel, reduce access times and need frequent dredging, sometimes closing hbr completely. Not advised for draft >1.5m. 326 + 30🅥 €1·96; max LOA 15m. ⛽🛒 at St Georges d'Oléron and La Brée.

BOYARDVILLE 45°58'·25N 01°13'·83W ⊛⊛♨♨🌸🌸

TIDES HW +0545 on Dover (UT); use Ile d'Aix.

SHELTER Very good in non-tidal marina (2·2m). 3 W waiting buoys are 7ca N of bkwtr hd, or ⚓ there in 3m.

NAVIGATION WPT 45°58'·44N 01°13'·36W, La Perrotine SHM buoy, 239°/3ca between sandbanks to S bkwtr head; in the ent chan hug the S bkwtr for best water. Strong river current, 2kn @ sp. Stand on beyond the ent to avoid a bank to stbd, then turn stbd onto about 350° for the gate; allow departures to leave first.

LIGHTS AND MARKS Le Boyard, a sandy shoal, is marked by Fort Boyard, conspic twr 29m, Q (9) 15s. A fish farm in the Passage de l'Ouest is marked by an ECM buoy, Q (3) 10s and NCM buoy Q.

COMMUNICATIONS Auto 08·92·68·08·17; Brit Consul 05·57·22·21·10; Gatemaster 05·46·47·23·71; HM VHF Ch 09; Aff Mar 05·46·47·00·18.

FACILITIES Marina ☎05·46·76·48·56. port-plaisance.boyardville@wanadoo.fr Drying appr chan. Waiting pontoon on NW bank is 50m SW of automatic gate which opens H24, when level reaches CD+4.3M, approx HW±1½ nps and HW±3 sp. ● = ent gate is shut. 🅥 berths/rafting against NE quay to stbd. 190+60🅥 €1·96. on SE bank ☎05·46·47·01·36; 0800-1230, 1400-1830. (14t). YCB ☎05·46·47·10·28.

LA CHARENTE RIVER

CHARTS AC 3000, 2747; SHOM 7414, 7415

TIDES See Île d'Oléron/Rochefort; interpolate between Ile d'Aix and Rochefort. Ile d'Aix, HW + 0545 (UT) on Dover; ML 3.9m.

NAVIGATION See notes under Rochefort. Fort Boyard, 1·6M WSW of Pte St Catherine's, is conspic and lit Q (9) 15s 29m; ringed by a No entry area, radius 200m. Upriver expect to meet sea-going cargo ships which have priority. Night navigation by yachts is prohibited because the banks cannot be seen.

MINOR HARBOURS & ANCHORAGES

ILE D'AIX Charente Maritime, 46°00'·60N 01°10'·45W. Fair weather ⚓s are charted SW, S and NE of the island. On W side of island, near LW keep 3ca off the two WCM perches. St Catherine's Pt (S tip of d'Aix) Fl WR 5s, two conspic white lt twrs, R tops; 103°-R-118°-W-103° (345°). R sector (15°) covers rks off NW tip of Ile d'Oléron.

Off Pte St Catherine are 25 drying 🅥s (W with dayglo G band, €10); buoys with R band = locals only; Blu band = with HM's approval. HM 06·17·80·38·07.

Facilities: ⚓ ✕ 🛒 📮 🛢 ⛽. Ferry to Fouras. **CNIA** ☎05·46·84·69·89.

Continued overleaf

FOURAS 45°58'·99N 01°05'·69W (S Hbr). A small town on N bank of the Charente from which ferries cross to Ile d'Aix; good beaches and many mussel/oyster beds. The 3 drying hbrs are: South, North and Port de la Fumée. A chequered twr near S Hbr and the ch spire are both conspic.

South Harbour Access HW ±2 for max draft 1·5m. 55 + 20 Ⓥ on drying pontoons. Facilities: ⚓ 🚿 ⛴ ⛽(3t) 🏠. Lights: Fouras Sud ≰ Fl WR 4s. Ldg lts 042·5°, both ≰ Oc (2) R 6s; see Lights, buoys & waypoints for details. HM ☎05·46·84·23·10 (1 Jun-30 Sep); other months, Mairie ☎05·46·84·66·09. **YC** Cercle Nautique de Fouras.

North Harbour No 3 ⚓ is free but dries. ⚓ ⛴ 🚿. Fouras Nord pier head ≰ Oc (3+1) WRG 12s. Mairie 05·46·84·60·11.

Port de la Fumée Mairie ☎05·46·84·60·11. Access H24, except near LWS. Jetty is for ferries. No 53 ⚓ is free but dries, 🚿. SHM bcn, ≰ QG.

PORT DES BARQUE approx **45°57'N 01°04'·W**. 2 free ⚓s (afloat) Apr-Oct; shift to afloat pontoon overnight (when ferry stops running). HM 05·46·84·51·24/ 06·78·90·45·10, ⚓ ⛴ 🚿 🏠 👤.

PORT NEUF 45°57'·02N 00°59'·80W. 2 free ⚓s (afloat); also a 64m long pontoon at the drying line. **CN Rochefortais** ☎05·46·87·34·61 (on site), 🚿 ⚓.

SOUBISE 45°55'·75N 01°00'·32W, 8 berths (€17, pay at Capitanerie) on a 48m afloat pontoon, ⛴ SW bank. Beware big ship wash. Mairie 05·46·84·92·04. Facilities: 🏠 village shops.

MARTROU (Echillais), 45°54'·95N 00°57'·85W approx; 10·5M from Fouras S. This is a No ⚓ Area. Mairie 05·46·83·03·74. Village shops at Martrou.

9.22.17 ROCHEFORT

Charente Maritime 45°56'·55N 00°57'·29W ✴✴♦♦♦♦ ✿✿✿

CHARTS AC 3000, 2747; Imray C41; SHOM 7414, 7415; Navi 552

TIDES +0610 Dover

Standard Port POINTE DE GRAVE (→)

Times				Height (metres)			
High Water		Low Water		MHWS	MHWN	MLWN	MLWS
0000	0600	0000	0500	5·3	4·4	2·1	1·0
1200	1800	1200	1700				
Differences ROCHEFORT							
+0035	−0015	+0125	+0030	+1·2	+0·9	+0·1	−0·2
LA CAYENNE (R. Seudre)							
+0030	−0015	−0005	−0010	+0·3	+0·2	+0·3	0·0

SHELTER Excellent in the marina, popular for over-wintering.

NAVIGATION WPT 45°59'·63N 01°09'·51W (N-abeam Les Palles NCM buoy, Q) 115°/2·1M to next ldg line 134·5°. Beware drying wreck just S of WPT. The bar, 1M S of Fouras, carries least depth 0·9m. Beware very strong currents here, except at HW or LW. Stream in river runs about 2kn (4kn in narrows) and at sp there is a small bore. Leave Fort Boyard/Ile d'Aix approx HW −3 to reach Rochefort (about 13M from the WPT) near HW while the ent gate is open. The river is navigable 3·5M on to Tonnay-Charente.

Progress is straightforward and swift on the flood. An elegant viaduct, 32m clearance, and 400m east an aerial transporter bridge (to which yachts should give way if it is operating) are 2M downstream of Rochefort.

LIGHTS AND MARKS Ldg lts at river mouth (for big ships): First 115°, both QR, W ☐ trs/R tops; Soumard lt ho is the conspic rear mark. Second, abeam Fontenelles ◢: Port-des-Barques ldg lts 134·5°, both Iso G 4s, W ☐ trs; rear has a B band on its west side. 19 pairs of unlit white ldg bcns (lettered TT to AA; no S) are mainly for big ships; ditto 10 Y SPM buoys, Fl Y 4s.

COMMUNICATIONS CROSS 02·97·55·35·35; ⚓05·46·41·11·11; Auto 08.92.68.08.17; Police 05·46·87·26·12; ⊜ 05·46·99·03·90; Brit Consul 05·57·22·21·10; Dr 05·46·99·61·11; Commercial port VHF Ch 12; Aff Mar 05·46·84·22·67; Marina VHF Ch 09 (HW±1).

FACILITIES Marina ☎05·46·83·99·96, mob 06·86·01·64·29, port. plaisance@ville-rochefort.fr www.ville-rochefort.fr 300+40 Ⓥ in Bassins 1 & 2, €1·98 + €0·20 pp/day tax. Marina Entry gate opens on average HW La Rochelle −½ to +1 0530-2230 in season; pre-check with HM. A waiting pontoon, just outside gate, dries to soft mud. Road bridge between Bassins Nos 1 and 2 lifts when gate open; berth as directed by HM. 🅿 ⚓ ⛴ ⚒ Ⓔ ♿, ⛟ 🏗(16t, 7t) A 36m pontoon, 120m S of ent to Bassin 1, is short stay €2·10/m + ⛴ ⚓.

Town ✕ 🍴 🛒 🏦 📷 📮 🅱 ⚕ ✈ (Rochefort, winter only, and La Rochelle). Ferry: via Roscoff or St Malo.

MINOR HARBOURS ON THE RIVER SEUDRE
AC 3000; SHOM 7405, 7414 (essential). Tides, see Rochefort. +0545 Dover; Duration Sp 0545, Np 0700. Good shelter in both hbrs.

NAVIGATION Warning: Passage through Pertuis de Maumusson (S of Ile d'Oléron) is not advised even in good weather. The N appr applies to both hbrs. WPT 45°56'·00N 01°08'·56W, Chenal Est-Nord ⨑, thence Chenal Est, Coureau d'Oléron, Viaduc d'Oléron (18m) and Ch de la Soumaille (dries initially about 0·7m). Power cables (16m cl'nce) cross Canal de Marennes and de la Tremblade. Marennes spire (88m) is conspic. Lts see Lights, buoys and waypoints.

FACILITIES There are ⚓s off La Cayenne village (very crowded) La Grève (½M upstream) and at the ent to Chenal de la Tremblade.

MARENNES, 45°49'·17N 01°06'·67W. Yacht basin ☎05·46·85·02·68 (2·5m). Access via automatic ent gate, opens about HW±2 sp, HW±1 np. 🚿 ⛴ ⚒ ⚓ 🏗(6t).
Town ✕ 🍴 🛒 📷 📷 ✈ (La Rochelle).

LA TREMBLADE 45°46'·07N 01°08'·17W. HM ☎05·46·76·47·97. Ch 09 Small hbr (3m) in the town centre; yachts dry on soft mud against stone quay. 🅿 ⚓ ⚒ ⚓ 🚿 🏗(5t).
Town 🚿 ✕ 🍴 🛒 📷 📷 🏦 📮 🅱 ⚓.

STANDARD TIME UT –01
Subtract 1 hour for UT
For French Summer Time add
ONE hour in **non-shaded areas**

POINTE DE GRAVE   LAT 45°34'N   LONG 1°04'W
TIMES AND HEIGHTS OF HIGH AND LOW WATERS

Dates in red are **SPRINGS**
Dates in blue are NEAPS

YEAR 2016

S Biscay

JANUARY

Time	m	Time	m
1 0308	1.9	**16** 0309	1.5
0935	4.5	0947	5.0
F 1541	2.0	SA 1537	1.5
2208	4.3	2223	4.7
2 0359	2.2	**17** 0407	1.7
1033	4.3	1053	4.8
SA 1637	2.2	SU 1639	1.7
◑ 2313	4.2	◐ 2337	4.6
3 0501	2.3	**18** 0514	1.8
1141	4.2	1209	4.7
SU 1742	2.3	M 1750	1.8
4 0023	4.2	**19** 0053	4.6
0609	2.3	0629	1.8
M 1253	4.2	TU 1325	4.7
1848	2.2	1905	1.8
5 0129	4.3	**20** 0201	4.7
0713	2.2	0743	1.7
TU 1358	4.2	W 1432	4.8
1948	2.1	2015	1.7
6 0224	4.5	**21** 0300	4.9
0810	2.1	0849	1.6
W 1451	4.5	TH 1530	4.9
2039	1.9	2114	1.5
7 0310	4.7	**22** 0352	5.1
0900	1.8	0945	1.4
TH 1535	4.7	F 1619	5.0
2124	1.7	2204	1.3
8 0351	4.9	**23** 0437	5.2
0945	1.6	1034	1.2
F 1615	4.9	SA 1702	5.1
2206	1.5	2249	1.2
9 0431	5.1	**24** 0518	5.3
1028	1.4	1118	1.1
SA 1655	5.1	SU 1741	5.2
2247	1.4	○ 2330	1.2
10 0511	5.3	**25** 0557	5.3
1110	1.2	1158	1.1
SU 1734	5.2	M 1818	5.2
● 2329	1.2		
11 0551	5.4	**26** 0009	1.2
1152	1.1	0632	5.3
M 1814	5.2	TU 1235	1.2
		1851	5.1
12 0010	1.2	**27** 0044	1.2
0633	5.5	0705	5.2
TU 1234	1.1	W 1310	1.3
1856	5.2	1922	5.0
13 0052	1.2	**28** 0119	1.4
0716	5.5	0736	5.0
W 1316	1.1	TH 1343	1.4
1940	5.1	1953	4.8
14 0135	1.2	**29** 0153	1.6
0802	5.4	0809	4.9
TH 1359	1.2	F 1417	1.6
2027	5.0	2028	4.6
15 0220	1.3	**30** 0229	1.7
0851	5.2	0846	4.6
F 1445	1.3	SA 1454	1.8
2120	4.8	2109	4.4
		31 0309	2.0
		0931	4.4
		SU 1536	2.1
		2203	4.3

FEBRUARY

Time	m	Time	m
1 0358	2.2	**16** 0449	1.8
1030	4.2	1148	4.5
M 1632	2.3	TU 1724	1.9
◑ 2312	4.1		
2 0504	2.3	**17** 0032	4.4
1145	4.1	0609	1.9
TU 1747	2.4	W 1312	4.4
		1846	2.0
3 0031	4.1	**18** 0149	4.5
0621	2.3	0731	1.9
W 1308	4.1	TH 1425	4.5
1901	2.3	2002	1.9
4 0143	4.3	**19** 0252	4.7
0730	2.2	0840	1.7
TH 1416	4.3	F 1522	4.7
2004	2.1	2102	1.7
5 0240	4.6	**20** 0343	4.9
0829	1.9	0935	1.4
F 1509	4.6	SA 1608	4.9
2057	1.8	2152	1.4
6 0327	4.8	**21** 0425	5.1
0921	1.6	1020	1.3
SA 1554	4.8	SU 1647	5.0
2145	1.5	2234	1.3
7 0410	5.1	**22** 0502	5.2
1008	1.4	1101	1.2
SU 1637	5.1	M 1721	5.1
2229	1.3	○ 2313	1.2
8 0453	5.4	**23** 0536	5.3
1053	1.1	1138	1.1
M 1718	5.3	TU 1753	5.1
● 2313	1.1	2348	1.2
9 0535	5.6	**24** 0607	5.3
1136	0.9	1211	1.1
TU 1759	5.4	W 1822	5.1
2356	0.9		
10 0617	5.7	**25** 0021	1.2
1219	0.8	0636	5.2
W 1841	5.4	TH 1243	1.2
		1850	5.1
11 0038	0.9	**26** 0052	1.3
0701	5.7	0705	5.1
TH 1301	0.8	F 1313	1.3
1924	5.4	1919	5.0
12 0120	0.9	**27** 0122	1.4
0746	5.6	0734	4.9
F 1343	0.9	SA 1342	1.5
2009	5.2	1951	4.8
13 0203	1.0	**28** 0153	1.6
0833	5.3	0807	4.7
SA 1426	1.1	SU 1413	1.7
2057	5.0	2026	4.6
14 0250	1.2	**29** 0227	1.8
0925	5.0	0844	4.5
SU 1515	1.4	M 1448	1.9
2154	4.7	2108	4.4
15 0343	1.5		
1028	4.7		
M 1613	1.7		
◑ 2307	4.5		

MARCH

Time	m	Time	m
1 0308	2.0	**16** 0426	1.8
0932	4.3	1129	4.3
TU 1532	2.2	W 1659	2.0
2207	4.2		
2 0402	2.2	**17** 0009	4.3
1041	4.1	0548	2.0
W 1636	2.4	TH 1257	4.3
◑ 2328	4.1	1823	2.1
3 0522	2.3	**18** 0131	4.4
1212	4.0	0714	1.9
TH 1807	2.4	F 1410	4.4
		1942	2.0
4 0055	4.1	**19** 0237	4.6
0645	2.2	0822	1.7
F 1338	4.2	SA 1506	4.6
1925	2.2	2043	1.8
5 0204	4.5	**20** 0327	4.8
0755	2.0	0915	1.5
SA 1439	4.5	SU 1549	4.8
2027	1.9	2131	1.5
6 0258	4.8	**21** 0406	5.0
0853	1.6	0958	1.4
SU 1529	4.8	M 1624	4.9
2120	1.6	2213	1.4
7 0346	5.1	**22** 0440	5.1
0944	1.3	1037	1.2
M 1614	5.1	TU 1655	5.0
2207	1.2	2250	1.3
8 0431	5.4	**23** 0510	5.2
1031	1.0	1112	1.2
TU 1657	5.4	W 1724	5.1
2253	0.9	○ 2323	1.2
9 0514	5.7	**24** 0539	5.2
1116	0.8	1143	1.2
W 1740	5.5	TH 1753	5.1
● 2337	0.7	2354	1.2
10 0558	5.8	**25** 0607	5.2
1159	0.7	1213	1.2
TH 1822	5.6	F 1821	5.1
11 0020	0.7	**26** 0024	1.3
0642	5.8	0636	5.1
F 1241	0.7	SA 1241	1.3
1905	5.5	1850	5.0
12 0102	0.7	**27** 0053	1.3
0727	5.6	0705	5.0
SA 1323	0.8	SU 1310	1.5
1949	5.4	1920	4.9
13 0145	0.9	**28** 0123	1.5
0814	5.4	0736	4.8
SU 1406	1.1	M 1339	1.6
2036	5.1	1954	4.8
14 0230	1.1	**29** 0156	1.6
0905	5.0	0811	4.6
M 1453	1.4	TU 1413	1.8
2129	4.8	2034	4.6
15 0322	1.5	**30** 0235	1.8
1007	4.6	0856	4.4
TU 1548	1.7	W 1455	2.0
◑ 2239	4.5	2127	4.4
		31 0324	2.0
		0959	4.2
		TH 1552	2.2
		◑ 2240	4.2

APRIL

Time	m	Time	m
1 0434	2.2	**16** 0100	4.3
1126	4.1	0643	2.0
F 1716	2.3	SA 1341	4.3
		1910	2.1
2 0006	4.3	**17** 0207	4.5
0600	2.2	0752	1.8
SA 1257	4.2	SU 1437	4.5
1842	2.2	2013	1.9
3 0124	4.5	**18** 0258	4.6
0716	1.9	0845	1.6
SU 1406	4.4	M 1519	4.7
1950	1.9	2102	1.7
4 0225	4.8	**19** 0337	4.8
0819	1.6	0929	1.5
M 1500	4.9	TU 1554	4.8
2048	1.5	2144	1.5
5 0317	5.2	**20** 0411	4.9
0915	1.3	1007	1.4
TU 1548	5.2	W 1625	4.9
2140	1.2	2221	1.4
6 0406	5.5	**21** 0441	5.0
1005	0.9	1041	1.3
W 1633	5.4	TH 1654	5.0
2229	0.9	2254	1.3
7 0452	5.7	**22** 0511	5.1
1052	0.7	1112	1.3
TH 1717	5.6	F 1724	5.1
● 2315	0.7	○ 2326	1.3
8 0538	5.8	**23** 0541	5.1
1137	0.6	1142	1.3
F 1801	5.7	SA 1755	5.1
2359	0.6	2357	1.3
9 0623	5.8	**24** 0611	5.0
1220	0.7	1212	1.3
SA 1845	5.6	SU 1826	5.1
10 0043	0.6	**25** 0028	1.3
0709	5.6	0641	4.9
SU 1302	0.8	M 1242	1.4
1930	5.4	1857	5.0
11 0126	0.8	**26** 0100	1.4
0757	5.3	0713	4.8
M 1345	1.1	TU 1314	1.6
2016	5.1	1932	4.9
12 0211	1.1	**27** 0134	1.5
0847	4.9	0750	4.6
TU 1431	1.4	W 1350	1.7
2108	4.8	2014	4.7
13 0302	1.4	**28** 0214	1.7
0948	4.6	0836	4.5
W 1525	1.7	TH 1433	1.9
2214	4.5	2106	4.5
14 0403	1.8	**29** 0302	1.8
1106	4.3	0936	4.3
TH 1632	2.0	F 1527	2.1
◐ 2338	4.3	2211	4.4
15 0521	2.0	**30** 0404	2.0
1230	4.2	1054	4.2
F 1752	2.1	SA 1641	2.2
		◐ 2328	4.4

Chart Datum is 2·83 metres below IGN Datum. HAT is 6·1 metres above Chart Datum.

⟩⟩ **FREE** monthly updates. Register at ⟨
www.reedsnauticalalmanac.co.uk ⟨

931

POINTE DE GRAVE

LAT 45°34'N LONG 1°04'W

TIMES AND HEIGHTS OF HIGH AND LOW WATERS

STANDARD TIME UT –01
Subtract 1 hour for UT
For French Summer Time add
ONE hour in **non-shaded areas**

Dates in red are **SPRINGS**
Dates in blue are **NEAPS**

YEAR **2016**

MAY

Time	m	Time	m
1 0521	2.0	**16** 0118	4.3
1219	4.3	0707	1.9
SU 1801	2.1	M 1352	4.3
		1930	2.0
2 0044	4.6	**17** 0215	4.4
0636	1.8	0805	1.8
M 1330	4.6	TU 1439	4.5
1911	1.8	2024	1.8
3 0151	4.8	**18** 0259	4.6
0743	1.5	0852	1.7
TU 1429	4.9	W 1517	4.7
2014	1.5	2109	1.7
4 0248	5.1	**19** 0337	4.7
0843	1.3	0932	1.5
W 1520	5.2	TH 1552	4.8
2111	1.2	2148	1.6
5 0340	5.4	**20** 0411	4.8
0937	1.0	1008	1.5
TH 1608	5.4	F 1625	4.9
2203	0.9	2224	1.5
6 0430	5.6	**21** 0445	4.9
1027	0.8	1041	1.4
F 1655	5.5	SA 1659	5.0
● 2252	0.7	○ 2258	1.4
7 0518	5.7	**22** 0518	4.9
1114	0.7	1114	1.4
SA 1741	5.6	SU 1732	5.1
2339	0.7	2332	1.3
8 0606	5.6	**23** 0551	4.9
1158	0.8	1147	1.4
SU 1827	5.5	M 1806	5.1
9 0024	0.7	**24** 0007	1.3
0653	5.5	0624	4.9
M 1242	0.9	TU 1221	1.4
1913	5.4	1841	5.0
10 0109	0.9	**25** 0042	1.3
0741	5.2	0659	4.8
TU 1326	1.1	W 1257	1.5
2000	5.1	1919	4.9
11 0154	1.1	**26** 0119	1.4
0830	4.9	0738	4.7
W 1411	1.4	TH 1335	1.6
2049	4.9	2002	4.8
12 0243	1.4	**27** 0200	1.5
0925	4.6	0824	4.6
TH 1502	1.7	F 1419	1.7
2147	4.6	2052	4.7
13 0338	1.7	**28** 0247	1.6
1031	4.3	0920	4.4
F 1601	1.9	SA 1511	1.8
◗ 2256	4.4	2151	4.6
14 0445	1.9	**29** 0343	1.7
1143	4.2	1028	4.4
SA 1712	2.1	SU 1615	1.9
		◑ 2258	4.6
15 0010	4.3	**30** 0450	1.8
0558	2.0	1144	4.4
SU 1253	4.2	M 1727	1.9
1825	2.1		
		31 0010	4.7
		0601	1.7
		TU 1257	4.6
		1836	1.9

JUNE

Time	m	Time	m
1 0119	4.8	**16** 0214	4.3
0709	1.5	0808	1.9
W 1359	4.8	TH 1438	4.5
1942	1.5	2029	1.9
2 0221	5.0	**17** 0301	4.5
0813	1.3	0854	1.7
TH 1454	5.0	F 1520	4.6
2043	1.3	2114	1.7
3 0318	5.2	**18** 0342	4.6
0911	1.1	0935	1.6
F 1546	5.2	SA 1559	4.8
2139	1.0	2154	1.6
4 0411	5.3	**19** 0420	4.7
1004	1.0	1014	1.5
SA 1635	5.4	SU 1636	4.9
2232	0.9	2233	1.4
5 0502	5.4	**20** 0457	4.8
1053	0.9	1051	1.4
SU 1723	5.5	M 1712	5.0
● 2321	0.8	○ 2311	1.3
6 0550	5.4	**21** 0533	4.9
1140	0.9	1128	1.3
M 1810	5.4	TU 1749	5.1
		2350	1.2
7 0008	0.8	**22** 0609	4.9
0638	5.3	1206	1.3
TU 1224	1.0	W 1827	5.1
1856	5.3		
8 0053	0.9	**23** 0028	1.2
0724	5.1	0647	4.9
W 1307	1.1	TH 1244	1.3
1942	5.1	1907	5.1
9 0137	1.1	**24** 0108	1.2
0809	4.8	0728	4.8
TH 1351	1.3	F 1324	1.4
2027	4.9	1950	5.0
10 0222	1.3	**25** 0149	1.3
0855	4.6	0812	4.7
F 1437	1.6	SA 1408	1.5
2114	4.7	2038	4.9
11 0310	1.6	**26** 0233	1.4
0945	4.4	0904	4.6
SA 1528	1.8	SU 1456	1.6
2207	4.5	2131	4.8
12 0404	1.8	**27** 0324	1.5
1043	4.2	1003	4.5
SU 1626	2.0	M 1552	1.7
◗ 2308	4.3	◗ 2232	4.7
13 0505	2.0	**28** 0423	1.6
1148	4.1	1113	4.5
M 1732	2.1	TU 1657	1.7
		2341	4.7
14 0013	4.2	**29** 0530	1.6
0611	2.0	1226	4.5
TU 1252	4.2	W 1807	1.7
1837	2.1		
15 0117	4.2	**30** 0053	4.7
0714	2.0	0640	1.6
W 1350	4.3	TH 1334	4.7
1937	2.0	1916	1.6

JULY

Time	m	Time	m
1 0201	4.8	**16** 0225	4.3
0748	1.5	0818	1.9
F 1435	4.9	SA 1450	4.5
2022	1.4	2041	1.8
2 0302	5.0	**17** 0314	4.5
0851	1.3	0906	1.7
SA 1529	5.1	SU 1533	4.7
2122	1.2	2127	1.6
3 0358	5.1	**18** 0357	4.6
0947	1.1	0949	1.6
SU 1620	5.2	M 1613	4.9
2217	1.0	2211	1.4
4 0449	5.2	**19** 0437	4.8
1038	1.0	1030	1.4
M 1708	5.3	TU 1652	5.1
● 2308	0.9	○ 2252	1.3
5 0536	5.2	**20** 0516	4.9
1124	1.0	1111	1.3
TU 1755	5.3	W 1731	5.2
2354	0.9	2333	1.1
6 0621	5.1	**21** 0554	5.0
1208	1.0	1151	1.2
W 1838	5.3	TH 1811	5.3
7 0038	1.0	**22** 0014	1.0
0703	5.0	0633	5.0
TH 1249	1.1	F 1231	1.1
1920	5.1	1852	5.3
8 0119	1.1	**23** 0054	1.0
0742	4.9	0714	5.0
F 1329	1.3	SA 1312	1.1
1958	5.0	1935	5.2
9 0158	1.3	**24** 0135	1.1
0819	4.7	0757	4.9
SA 1410	1.4	SU 1354	1.2
2036	4.7	2021	5.1
10 0239	1.5	**25** 0217	1.2
0857	4.5	0845	4.8
SU 1452	1.7	M 1439	1.3
2118	4.5	2111	5.0
11 0322	1.7	**26** 0304	1.3
0943	4.3	0939	4.7
M 1540	1.9	TU 1530	1.5
2208	4.3	2209	4.8
12 0412	1.9	**27** 0359	1.5
1041	4.2	1045	4.6
TU 1637	2.1	W 1632	1.6
◗ 2308	4.2	◗ 2318	4.6
13 0512	2.1	**28** 0504	1.7
1148	4.1	1202	4.5
W 1742	2.2	TH 1743	1.7
14 0016	4.1	**29** 0035	4.6
0618	2.1	0617	1.7
TH 1258	4.1	F 1318	4.5
1848	2.1	1859	1.7
15 0125	4.1	**30** 0149	4.6
0722	2.1	0731	1.7
F 1359	4.3	SA 1423	4.7
1948	2.0	2011	1.5
		31 0253	4.7
		0839	1.5
		SU 1520	4.9
		2114	1.3

AUGUST

Time	m	Time	m
1 0349	4.9	**16** 0333	4.6
0936	1.3	0926	1.6
M 1610	5.1	TU 1550	4.9
2208	1.2	2148	1.4
2 0437	5.0	**17** 0415	4.8
1026	1.2	1010	1.4
TU 1655	5.2	W 1630	5.2
● 2256	1.0	2232	1.2
3 0521	5.1	**18** 0455	5.0
1110	1.1	1052	1.2
W 1737	5.3	TH 1711	5.4
2339	1.0	○ 2314	1.0
4 0601	5.1	**19** 0535	5.2
1151	1.1	1133	1.0
TH 1816	5.2	F 1752	5.5
		2356	0.9
5 0019	1.0	**20** 0615	5.3
0637	5.0	1215	0.9
F 1229	1.1	SA 1833	5.5
1851	5.2		
6 0056	1.1	**21** 0037	0.9
0709	4.9	0657	5.3
SA 1305	1.2	SU 1255	0.9
1924	5.0	1917	5.5
7 0131	1.2	**22** 0117	0.9
0740	4.8	0739	5.2
SU 1340	1.4	M 1337	1.0
1956	4.8	2002	5.3
8 0205	1.4	**23** 0159	1.1
0813	4.6	0825	5.0
M 1416	1.6	TU 1421	1.2
2032	4.6	2052	5.1
9 0241	1.6	**24** 0244	1.3
0851	4.4	0917	4.8
TU 1455	1.8	W 1510	1.4
2114	4.4	2150	4.9
10 0321	1.9	**25** 0337	1.5
0940	4.2	1019	4.6
W 1541	2.0	TH 1611	1.7
◗ 2208	4.2	◗ 2302	4.5
11 0410	2.1	**26** 0441	1.8
1045	4.1	1145	4.4
TH 1642	2.2	F 1726	1.8
2318	4.0		
12 0519	2.3	**27** 0027	4.4
1204	4.0	0600	1.9
F 1758	2.3	SA 1308	4.5
		1850	1.8
13 0037	4.0	**28** 0144	4.5
0635	2.3	0721	1.9
SA 1320	4.2	SU 1417	4.7
1909	2.2	2005	1.4
14 0150	4.1	**29** 0248	4.6
0742	2.1	0830	1.7
SU 1419	4.4	M 1513	4.9
2010	2.0	2106	1.4
15 0247	4.4	**30** 0340	4.8
0838	1.9	0925	1.4
M 1507	4.7	TU 1600	5.1
2102	1.7	2156	1.2
		31 0423	5.0
		1012	1.3
		W 1640	5.2
		2240	1.1

Chart Datum is 2·83 metres below IGN Datum. HAT is 6·1 metres above Chart Datum.

FREE monthly updates. Register at
www.reedsnauticalalmanac.co.uk

STANDARD TIME UT –01
Subtract 1 hour for UT
For French Summer Time add
ONE hour in **non-shaded areas**

POINTE DE GRAVE   LAT 45°34'N   LONG 1°04'W
TIMES AND HEIGHTS OF HIGH AND LOW WATERS

Dates in red are **SPRINGS**
Dates in blue are NEAPS

YEAR 2016

S Biscay

SEPTEMBER

Time	m		Time	m
1 0501	5.1	**16**	0432	5.2
1053	1.2		1029	1.1
TH 1717	5.2	F 1648		5.5
● 2319	1.1	○ 2251		0.9
2 0535	5.1	**17**	0513	5.4
1131	1.1		1112	1.1
F 1750	5.2	SA 1730		5.7
2355	1.1	2334		0.8
3 0606	5.1	**18**	0554	5.5
1206	1.2		1155	0.8
SA 1820	5.2	SU 1814		5.7
4 0029	1.2	**19**	0016	0.8
0635	5.0		0637	5.5
SU 1238	1.2	M 1236		0.8
1849	5.1	1858		5.6
5 0100	1.3	**20**	0058	0.9
0704	4.9		0720	5.4
M 1309	1.4	TU 1319		0.9
1919	4.9	1945		5.4
6 0130	1.5	**21**	0140	1.1
0735	4.8		0806	5.1
TU 1341	1.6	W 1403		1.1
1951	4.7	2035		5.1
7 0201	1.7	**22**	0225	1.3
0809	4.6		0858	4.9
W 1414	1.8	TH 1452		1.4
2028	4.5	2135		4.7
8 0235	1.9	**23**	0317	1.6
0851	4.4		1003	4.6
TH 1453	2.0	F 1553		1.7
2115	4.2	◑ 2252		4.3
9 0316	2.1	**24**	0422	1.9
0947	4.2		1132	4.4
F 1544	2.2	SA 1711		2.0
◐ 2221	4.0			
10 0414	2.3	**25**	0021	4.4
1107	4.1		0544	2.1
SA 1701	2.4	SU 1258		4.5
2350	4.0	1839		1.9
11 0543	2.4	**26**	0136	4.5
1235	4.1		0707	2.0
SU 1828	2.3	M 1407		4.7
		1953		1.8
12 0115	4.1	**27**	0237	4.6
0704	2.3		0814	1.8
M 1344	4.4	TU 1502		4.9
1936	2.1	2050		1.5
13 0217	4.4	**28**	0325	4.8
0806	2.0		0907	1.6
TU 1437	4.7	W 1544		5.0
2032	1.7	2136		1.3
14 0306	4.7	**29**	0403	5.0
0858	1.7		0952	1.4
W 1523	5.0	TH 1620		5.1
2122	1.4	2217		1.2
15 0350	5.0	**30**	0436	5.1
0945	1.4		1031	1.3
TH 1606	5.3	F 1652		5.2
2207	1.1	2254		1.2

OCTOBER

Time	m		Time	m
1 0507	5.1	**16**	0450	5.6
1107	1.2		1048	0.9
SA 1722	5.2	SU 1710		5.8
● 2327	1.2	○ 2310		0.8
2 0536	5.1	**17**	0533	5.6
1140	1.3		1133	0.8
SU 1751	5.2	M 1755		5.8
2358	1.3	2354		0.8
3 0605	5.1	**18**	0617	5.6
1210	1.3		1217	0.8
M 1819	5.1	TU 1841		5.7
4 0028	1.4	**19**	0037	0.9
0634	5.0		0703	5.5
TU 1240	1.4	W 1301		0.9
1848	5.0	1930		5.4
5 0057	1.5	**20**	0121	1.1
0704	4.9		0750	5.3
W 1310	1.6	TH 1347		1.1
1919	4.8	2022		5.1
6 0127	1.7	**21**	0207	1.4
0737	4.7		0843	5.0
TH 1342	1.7	F 1437		1.4
1954	4.6	2122		4.7
7 0159	1.9	**22**	0259	1.7
0816	4.5		0949	4.7
F 1419	1.9	SA 1536		1.8
2036	4.3	◑ 2240		4.5
8 0238	2.1	**23**	0402	2.0
0906	4.3		1114	4.5
SA 1505	2.2	SU 1651		2.0
2138	4.1			
9 0331	2.3	**24**	0003	4.4
1018	4.2		0520	2.2
SU 1611	2.3	M 1237		4.5
◐ 2306	4.0	1815		2.0
10 0450	2.5	**25**	0116	4.4
1146	4.2		0641	2.1
M 1741	2.3	TU 1345		4.6
		1927		1.9
11 0036	4.2	**26**	0214	4.6
0620	2.4		0748	1.9
TU 1303	4.4	W 1439		4.8
1857	2.1	2024		1.7
12 0143	4.4	**27**	0300	4.8
0728	2.1		0842	1.7
W 1403	4.8	TH 1521		4.9
1957	1.8	2110		1.5
13 0236	4.8	**28**	0337	4.9
0824	1.7		0926	1.6
TH 1453	5.1	F 1556		5.0
2050	1.4	2150		1.4
14 0322	5.1	**29**	0409	5.0
0915	1.4		1005	1.5
F 1540	5.4	SA 1627		5.1
2139	1.1	2225		1.2
15 0406	5.4	**30**	0439	5.1
1002	1.1		1040	1.4
SA 1625	5.6	SU 1656		5.2
2226	0.9	● 2258		1.3
		31	0509	5.2
			1112	1.4
		M 1726		5.2
		2329		1.4

NOVEMBER

Time	m		Time	m
1 0540	5.2	**16**	0601	5.7
1144	1.4		1200	0.8
TU 1756	5.1	W 1828		5.6
2359	1.4			
2 0611	5.1	**17**	0019	0.9
1215	1.5		0648	5.6
W 1826	5.0	TH 1246		0.9
		1917		5.4
3 0029	1.6	**18**	0104	1.1
0642	5.0		0737	5.4
TH 1246	1.6	F 1333		1.1
1857	4.8	2009		5.1
4 0101	1.7	**19**	0150	1.3
0716	4.9		0830	5.1
F 1320	1.7	SA 1422		1.4
1932	4.7	2106		4.8
5 0135	1.9	**20**	0240	1.7
0755	4.7		0930	4.8
SA 1357	1.8	SU 1517		1.7
2015	4.5	2213		4.5
6 0215	2.0	**21**	0338	1.9
0844	4.6		1041	4.6
SU 1442	2.0	M 1622		1.9
2112	4.3	◗ 2326		4.4
7 0305	2.2	**22**	0446	2.1
0947	4.4		1156	4.5
M 1540	2.2	TU 1735		2.0
◗ 2229	4.2			
8 0413	2.3	**23**	0037	4.4
1103	4.4		0600	2.2
TU 1656	2.2	W 1306		4.5
2354	4.3	1847		2.0
9 0534	2.3	**24**	0138	4.5
1220	4.5		0710	2.1
W 1812	2.1	TH 1405		4.6
		1948		1.9
10 0106	4.5	**25**	0227	4.6
0645	2.1		0808	1.9
TH 1327	4.8	F 1450		4.7
1918	1.8	2037		1.7
11 0203	4.8	**26**	0307	4.8
0747	1.8		0855	1.8
F 1423	5.1	SA 1528		4.8
2017	1.5	2119		1.6
12 0254	5.1	**27**	0341	4.9
0843	1.4		0936	1.7
SA 1515	5.4	SU 1602		4.9
2110	1.2	2156		1.5
13 0342	5.4	**28**	0414	5.0
0935	1.2		1013	1.6
SU 1604	5.6	M 1635		5.0
2200	1.0	2230		1.5
14 0428	5.6	**29**	0447	5.1
1025	0.9		1047	1.5
M 1652	5.7	TU 1707		5.1
○ 2248	0.8	● 2303		1.5
15 0514	5.7	**30**	0520	5.2
1113	0.8		1121	1.5
TU 1739	5.7	W 1739		5.0
2334	0.8	2336		1.5

DECEMBER

Time	m		Time	m
1 0554	5.2	**16**	0004	1.0
1154	1.4		0636	5.6
TH 1811	5.0	F 1233		0.9
		1904		5.4
2 0009	1.5	**17**	0049	1.1
0627	5.1		0724	5.4
F 1229	1.5	SA 1319		1.1
1844	4.9	1952		5.1
3 0043	1.6	**18**	0134	1.3
0703	5.0		0812	5.2
SA 1304	1.6	SU 1405		1.3
1920	4.8	2041		4.9
4 0120	1.7	**19**	0220	1.5
0742	4.9		0902	4.9
SU 1343	1.7	M 1453		1.5
2002	4.6	2133		4.6
5 0200	1.9	**20**	0310	1.8
0828	4.8		0956	4.7
M 1426	1.8	TU 1546		1.8
2054	4.5	2231		4.4
6 0248	2.0	**21**	0406	2.0
0923	4.7		1057	4.5
TU 1517	1.9	W 1646		2.0
2158	4.4	◗ 2336		4.3
7 0346	2.1	**22**	0511	2.2
1028	4.6		1205	4.4
W 1620	2.0	TH 1753		2.1
◗ 2313	4.4			
8 0454	2.1	**23**	0043	4.3
1140	4.6		0619	2.2
TH 1731	2.0	F 1313		4.4
		1900		2.1
9 0027	4.5	**24**	0144	4.4
0605	2.0		0724	2.1
F 1251	4.8	SA 1412		4.4
1840	1.8	1958		2.0
10 0132	4.8	**25**	0233	4.6
0711	1.8		0820	2.0
SA 1355	5.0	SU 1459		4.6
1944	1.6	2047		1.9
11 0229	5.0	**26**	0315	4.7
0813	1.5		0907	1.8
SU 1453	5.2	M 1539		4.7
2043	1.3	2128		1.7
12 0321	5.3	**27**	0353	4.9
0912	1.3		0947	1.7
M 1547	5.4	TU 1616		4.8
2138	1.1	2206		1.6
13 0411	5.5	**28**	0429	5.0
1006	1.0		1025	1.6
TU 1638	5.6	W 1651		4.9
2229	1.0	2242		1.5
14 0459	5.6	**29**	0504	5.1
1058	0.9		1102	1.5
W 1727	5.6	TH 1725		5.0
○ 2317	0.9	● 2318		1.5
15 0548	5.6	**30**	0539	5.2
1146	0.8		1138	1.4
TH 1816	5.5	F 1759		5.0
		2354		1.4
		31	0614	5.2
			1215	1.4
		SA 1834		5.0

Chart Datum is 2·83 metres below IGN Datum. HAT is 6·1 metres above Chart Datum.

》 **FREE** monthly updates. Register at 《
www.reedsnauticalalmanac.co.uk

933

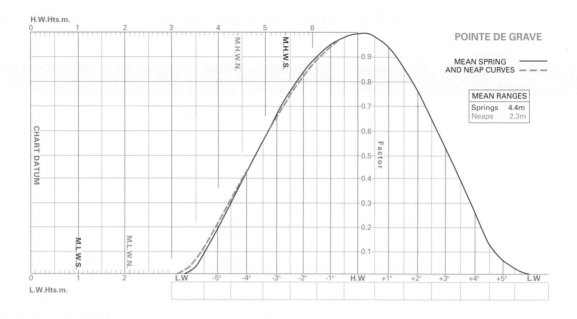

POINTE DE GRAVE

MEAN SPRING
AND NEAP CURVES

MEAN RANGES	
Springs	4.4m
Neaps	2.3m

9.22.18 THE GIRONDE ESTUARY

CHARTS AC 3057/8, 3068; SHOM 7425/6/7: all these are essential.

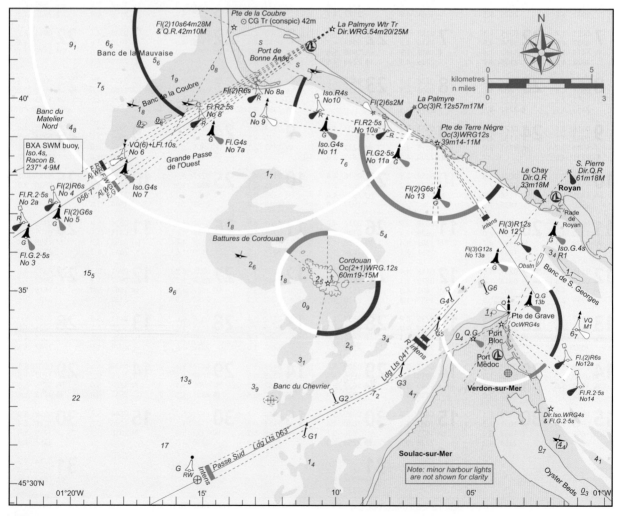

TIDES
Standard Port POINTE DE GRAVE (←)

Times				Height (metres)			
High Water		Low Water		MHWS	MHWN	MLWN	MLWS
0000	0600	0000	0500	5·3	4·4	2·1	1·0
1200	1800	1200	1700				
Differences CORDOUAN							
–0010	–0010	–0025	–0015	–0·4	–0·4	–0·1	–0·2
RICHARD							
+0020	+0020	+0035	+0030	0·0	–0·1	–0·4	–0·5
LAMENA							
+0035	+0045	+0125	+0100	+0·3	+0·1	–0·5	–0·3
LA REUILLE							
+0135	+0145	+0305	+0230	–0·1	–0·3	–1·3	–0·7
LE MARQUIS							
+0145	+0150	+0320	+0245	–0·2	–0·4	–1·5	–0·9
LIBOURNE (La Dordogne)							
+0250	+0305	+0540	+0525	–0·6	–0·9	–2·0	–0·4

In the Gironde and the R Dordogne the sp flood reaches 3kn and the ebb 4kn, continuing for 1½hrs after predicted LW time. In the Garonne the sp flood starts with a small bore and runs at 3kn, ebb reaches 5kn. *The Annuaire des Marées Estuaire de la Gironde*, downloadable from www.bordeaux-port.fr, is full of useful tidal information for the Gironde and also Arcachon.

▶*Off Terre-Nègre lt the SE-going stream begins at HW –0500 (sp 1·5kn), and the NW-going at HW+0130 (sp 2·5kn).*◀

Between Pte de Grave and Bordeaux the height of water at Verdon, Richard, Lamena, Pauillac, Fort Médoc, Bec d'Ambes, Le Marquis and Bordeaux (see R Gironde chartlet) is automatically broadcast H24 on Ch 17 every 5 mins. From these read-outs it is possible to deduce, for example, when the flood starts to make.

SHELTER See Royan to Bordeaux.

NAVIGATION La Gironde is a substantial waterway. The mouth of the estuary is 9M wide between Pte de la Coubre and Pte de Grave and narrows from 6M wide off Royan to 2.5M at Pauillac. The Garonne and Dordogne flow into the Gironde at Bec d'Ambes (45°02'N). Be alert for ferries and shipping.

The outer apprs can be dangerous due to Atlantic swell, very strong tidal streams and currents, extensive shoals and shifting sandbanks. Swell, strong W'lies and an ebb tide raise dangerous, breaking seas some 5m high; in such conditions do not attempt entry. Nevertheless at the right time and in the right weather both entrances are straightforward by day and night.

The two approach channels are:

- **Grande Passe de l'Ouest** WPT BXA SWM buoy, Iso 4s, 45°37'·53N 01°28'·68W, 057°/4·0M to Nos 1 & 2 buoys. The chan is deep and well marked/lit. Keep well clear of La Mauvaise and Matelier Nord banks (overfalls) the drying Banc de la Coubre and Battures de Cordouan.

  Leave No 1 buoy at LW on 057° Dir lt WRG, La Palmyre Wtr Tr. Hold to stbd side of buoyed channel, gradually swinging eastward to No 7a SHM, Fl G 4s, and once past No 9 NCM, Q, head south eastwards in W sector (097°-104°) of Tèrre Negre lt. Using stern transit 327° (Tèrre Negre Oc W / La Palmyre Oc R) enter R Gironde past No 13b SHM QG, off Pointe de Grave.

- **Passe du Sud** is a useful short cut if arriving from the S. It carries approx 5m through shoals, but Platin de Grave, N of G4 and G6 buoys, has only 1·4m or less. Not advised in poor vis or heavy swell. The six lateral buoys are unlit.

  WPT 45°30'·00N 01°15'·33W, abeam 'G' unlit SWM buoy, 063°/5·5M (063° ldg lts: Front, QG; rear, Oc WRG 4s) to G3 buoy. Here pick up the 041° ldg lts (both QR) to G4 buoy where alter stbd to G6 buoy passing S of Platin de Grave. Thence track direct to Royan or via No 13b buoy to Port-Médoc.

LIGHTS AND MARKS See chartlet and Lts, buoys and waypoints for details of lts mentioned above. **Other major lts in estuary**: La Coubre lt ho, W twr, top third R. 2M NNE is a R/W CG twr. Cordouan lt ho is an elegant, W sculpted twr on a drying plateau in mid-estuary.

COMMUNICATIONS *Bordeaux Port Control* Ch 12 (H24) provides radar surveillance from BXA buoy to Bordeaux, plus MSI on request.

9.22.19 ROYAN
Charente Maritime 45°37'·12N 01°01'·50W ✿✿✿◊◊◊◊✿✿✿

CHARTS AC 3057, 3058; Imray C42; SHOM 7426, 7425; Navi 553, 554

TIDES +0530 Dover; ML 3·2; Duration Sp 0615, Np 0655

Standard Port POINTE DE GRAVE (←)

Times				Height (metres)			
High Water		Low Water		MHWS	MHWN	MLWN	MLWS
0000	0600	0000	0500	5·3	4·4	2·1	1·0
1200	1800	1200	1700				
Differences ROYAN							
0000	–0005	0000	0000	–0·1	–0·1	0·0	0·0

SHELTER Good in marina. Access difficult in strong W/NW'lies.

NAVIGATION WPT 45°36'·56N 01°01'·96W (R1 SHM buoy, Iso G 4s; marks wk (3m) 260m E) 030°/0·64M to hbr ent. The Gironde apprs can be dangerous; see The Gironde Estuary for details; allow about 2 hrs from Pte de la Coubre to Royan. Beware Banc de St Georges (0·4m) 1M S of WPT. Off hbr ent, there is an eddy, running S at about 1kn on the flood and 3kn on the ebb. Caution: FVs and fast ferries.

LIGHTS AND MARKS Cathedral spire is conspic modern 'spike'. See chartlet and Lights, buoys and waypoints for lt details.

COMMUNICATIONS CROSS 02·97·55·35·35; Auto 08.92.68.08.17; ☎ 05·56·34·20·11; Police 05·46·38·34·22; ⊖ 05·46·23·10·24; Brit Consul 05·57·22·21·10; Ⓗ 05·46·38·01·77; HM VHF Ch 09, 16 (season 0800-2000; otherwise 0900-1800LT); Aff Mar 05·46·39·26·30.

FACILITIES Marina Care needed near LW sp. Ent chan, least depth 0·1m, silts but is regularly dredged to at least 0·8m; it narrows markedly off New Jetty hd. Accueil pontoon is dead ahead, W of the ent.

Marina ☎05·46·38·72·22 (Jul/Aug 7/7 0800-2000) www.port-royan.com port.royan@wanadoo.fr 900 + 100Ⓥ €2·12. ♠ ♠ (ⓐ 05·46·38·53·39. 0830-1200, 1430-1830, 7/7 in season) ⛽ SHOM ⚒ Ⓗ ⚒ Grid Ice Ⓔ ⚓ de-mast ⌂(1·5t) ⚑(26t) ⚓. **Base Nautique (YC)** ☎05·46·05·44·13.

Town ✕ ⌂ ☲ ⛐ ⬤ ⊠ Ⓑ ⇌ ✈ (Bordeaux). Ferry via Roscoff or St Malo; fast ferries to Port Bloc.

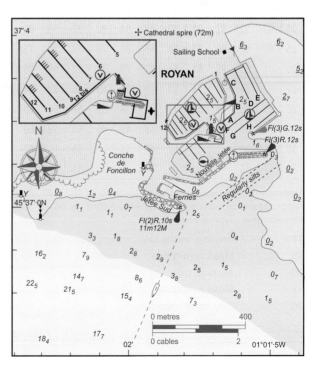

POINTE DE GRAVE TIDAL COEFFICIENTS 2016

Date	Jan am	Jan pm	Feb am	Feb pm	Mar am	Mar pm	Apr am	Apr pm	May am	May pm	June am	June pm	July am	July pm	Aug am	Aug pm	Sept am	Sept pm	Oct am	Oct pm	Nov am	Nov pm	Dec am	Dec pm
1	52	47	41	37	47	42	36		43	46	63	69	68	73	78	82	90	92	90	90	84	82	80	79
2	43	40	35		37	34	37	40	50	56	75	81	77	82	86	90	93	93	90	89	81	78	78	76
3	37		34	36	33		46	53	63	71	87	92	86	90	92	93	92	90	87	85	76	72	74	71
4	37	38	39	44	35	40	61	70	78	86	96	100	93	95	94	93	88	85	82	78	69	64	68	64
5	40	43	50	57	46	53	79	88	93	100	102	103	96	96	92	90	82	77	75	70	60	55	61	57
6	48	53	64	71	61	70	96	103	105	109	103	102	96	94	87	83	73	68	65	60	51	46	54	51
7	58	63	78	84	78	87	109	114	112	113	100	97	92	88	79	75	62	57	55	49	42	39	49	48
8	68	74	90	96	94	101	117	118	112	110	92	87	84	80	70	64	51	45	44	39	38	40	49	
9	78	83	100	104	107	112	117	115	106	101	82	76	75	70	59	53	40	35	35		43		52	55
10	87	90	107	108	115	116	111	105	95	88	70	64	65	59	48	43	32		32	32	48	55	61	67
11	93	95	108	106	116	114	99	91	81	74	59	54	54	49	39		31	32	35	40	63	71	73	80
12	95	96	103	99	111	105	82	73	66	59	49		45		35	34	36	41	47	55	79	86	86	92
13	95	93	94	88	99	91	65	57	53	48	46	44	42	39	34	36	48	56	64	72	94	100	97	101
14	91	87	81	73	83	74	50		45		43	43	38	39	40	46	64	72	81	88	105	109	104	106
15	83	78	66	59	66	57	45	43	43	44	45	47	41	44	52	58	79	87	96	102	111	112	106	105
16	74	68	53		50		43	45	46	48	50	53	48	52	64	71	93	99	108	111	111	109	103	99
17	64	60	50	49	46	44	49	54	52	56	57	61	57	62	77	83	104	108	114	114	105	100	95	90
18	57		50	53	45	48	59	64	60	63	64	68	67	71	88	93	110	111	113	111	94	87	84	78
19	56	56	58	64	53	58	68	72	67	70	71	74	75	79	97	100	111	108	106	101	79	72	71	65
20	59	62	69	74	64	69	76	79	73	75	77	79	83	86	102	103	105	100	94	86	64	58	59	53
21	67	72	79	83	75	79	81	83	77	79	80	82	88	90	102	101	93	86	77	69	52	48	49	45
22	76	81	87	89	82	85	84	85	80	81	82	83	91	91	98	94	78	70	61	54	46		43	
23	85	88	91	92	87	89	85	85	81	81	82	81	91	90	89	83	61	54	48		45	46	42	43
24	90	92	92	91	89	89	84	82	80	79	80	78	87	85	77	70	49		45	45	49	52	45	47
25	92	92	90	87	89	87	80	78	77	75	75	73	81	77	63		46	46	47	51	56	60	51	55
26	91	89	85	81	86	83	75	71	72	69	70	67	73	68	58	53	49	54	55	60	63	67	58	62
27	87	84	78	73	80	77	67	63	65	62	64	62	64		50	50	59	65	65	70	70	73	66	69
28	80	76	68	63	73	68	58	53	58	56	60		60	58	53	57	70	76	74	77	75	77	72	75
29	71	67	58	53	63	58	49	46	54	53	60	60	57	57	62	68	80	84	80	82	78	79	77	79
30	61	56			52	47	44		53		62	65	60	63	73	79	86	88	84	84	80	80	81	82
31	51	46			42	39			55	59			68	73	83	87			85	85			82	82

Tidal coefficients indicate the magnitude of the tide on any particular day, without having to look up and calculate the range, and thus determine whether it is springs, neaps or somewhere in between. This table is valid for all areas covered by this Almanac. Typical values are:

120	Very big spring tide
95	**Mean spring tide**
70	Average tide
45	**Mean neap tide**
20	Very small neap tide

See also coefficients for Calais (p 759), Cherbourg (p 786), St Malo (p 805) and Brest (p 873).

9.22.20 PORT BLOC

Gironde 45°34'·12N 01°03'·72W ✦✦✦✦✦

CHARTS AC 3057, 3058; Imray C42; Navi 553, 554; SHOM 7426, 7425

TIDES Use predictions for Pointe de Grave (⟵) 4ca N. Caution strong tidal streams.

SHELTER Good, but primarily a ferry harbour (caution: wash) and Service station for Bordeaux port authority and pilot boats, plus some fishing vessels. Swell enters in bad weather.

NAVIGATION WPT 45°34'·66N 01°02'·98W (13b SHM buoy, QG) 224°/7ca to harbour ent. Ent is 30m wide and is effectively fully used by ferries which have priority. Pauillac 25M; Bordeaux 53M.

LIGHTS AND MARKS Pte de Grave lt ho, W □ twr, B corners and top. Woods behind the hbr. See chartlet and Lights, buoys and waypoints.

COMMUNICATIONS CROSS 02·97·55·35·35; Radar Verdon VHF Ch 12, 11 (H24); ⚓ 05·56·34·20·11; Auto 08·92·68·08·17; Police 05·56·09·80·29; ⊖ 05·56·09·65·14; Brit Consul 05·57·22·21·10; Dr 05·56·09·60·04; HM 05·56·09·63·91; Aff Mar 05·56·09·60·23.

FACILITIES Space for yachts is very limited; possible ⌒ on pontoons W side of the hbr dredged 3m. Visitor berthing requires prior approval from AUPB (Port Bloc Users' Association). No 🏴 or 🚻's. Royan (3·4M NNE) or Port-Médoc 7ca S are far better options.

AUPB ☎05·56·09·72·64. **Moto YC de la Pointe de Grave** ☎05·56·09·84·02, ⚓ good ⛽🔧 🕔 ⚒ ⛴(10t).

Town ✗ 🍴 🏪 🛒 ⛽ ⊠ Ⓑ (Verdon) ⇌ ✈ Bordeaux. Fast local ferries to Royan every 45 mins.

9.22.21 PORT MÉDOC

Gironde 45°33'·40N 01°03'·46W ✦✦✦✦✦✦✦

CHARTS AC 3057, 3058; SHOM 7426, 7425; Imray C42; Navi 553, 554

TIDES Pointe de Grave is a Standard Port whose predictions (⟵) apply to Port Bloc and Port Médoc; no tidal constraints.

SHELTER Good, in depths 2-3m. It is approx 1M S of Pte de Grave and 0·7ca S of Port Bloc on the W bank of the R Gironde.

NAVIGATION WPT 45°34'·66N 01°02'·97W (13b SHM buoy, QG) about 195°/1·3M to hbr ent, dredged 3m. Access H24 at all states of the tide. Le Verdon oil jetty, about 8ca SE, is dangerous and in a prohib area; no landing. Caution strong tidal streams.

LIGHTS AND MARKS Marina's N and S bkwtr heads are lit QG and QR respectively. Pte de Grave lt ho (see Lights, buoys and waypoints) is 8ca NNW of the marina ent. Just N of the lt ho a 51m high red radio mast is marked by R lts. At Le Verdon cranes adj to the oil jetty are conspic. The oil jetty is marked by: Fl (3) G 12s at its N end and Fl G 2·5s at the S end, no landing. 9 ca further S is a Dir lt 171·5°, Iso WRG 4s.

COMMUNICATIONS CROSS 02·97·55·35·35; *Radar de Bordeaux* (VTS) Ch 12 (H24); Auto 08·92·68·02·17; Police 05·56·09·80·29; ⊖ 05·56·09·65·14; Brit Consul 05·57·22·21·10; Dr 05·56·09·60·04; Aff Mar 05·56·09·60·23; Marina VHF Ch 09.

FACILITIES Marina ☎05·56·09·69·75. www.port-medoc.com capitainerie@port-medoc.com ⚓ 740 + 60Ⓥ, max LOA 25m, €2·42. The reception pontoon is at the N end of marina. Work on 400 new berths between N & Z, due to start in 2015. 🏴🏴 ✗ 🛒 some shops, ⛽🔧 🕔 ⚒ 🏴(45t) bike hire. Footpath to town starts near pontoon N. 10m wide free public ⚓ just N of hbr, W of ECM.

Le Verdon (1·3km S) ✗ good 🍴 ⊠ Ⓑ ⇌ ✈ (Bordeaux). Fast ferry Port Bloc to Royan every 45 mins €3·00; info ☎05·56·73·37·73.

Soulac-sur-Mer is 7·5km SW.

9.22.22 THE RIVER GIRONDE

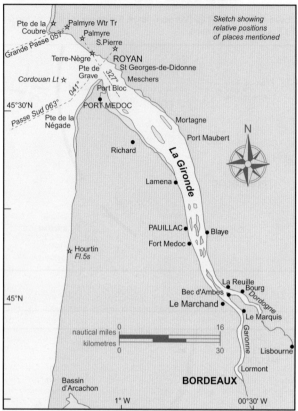

Sketch showing relative positions of places mentioned

MINOR HARBOURS ON THE GIRONDE

MESCHERS-sur-GIRONDE, Charente Maritime, **45°33′·20N 00°56′·63W**. AC 3058, 3057 (with 1:10,000 inset); SHOM 7426; Navi 554. Tides as for ROYAN. Talmont church is conspic 1·8M upriver. Appr close to Pte de Meschers between PHM and SHM unlit perches. Access via narrow, drying chan perched to port and mudbank close to stbd. Ldg marks/lts 354·4°, 2 G panels, W stripe, on G metal pylons with QG lts (privately maintained); just before the front ldg mark turn port 330° to enter hbr.

To stbd drying marina basin, pontoons A-C. Dead ahead an automatic entry gate gives access HW±2½ over sill 2m above CD to a wet basin in 2·5m on pontoons D-F; H24 in season, otherwise 0730-1930; waiting pontoons. Good shelter in all weathers. Both basins are small; max LOA 9m.

HM (M. Jean-Luc Rat, available around HW) ☎05.46.02.56.89; Auto 05.36.68.08.17. **Marinas** 125 ⌷ in drying basin, 123 in wet basin: 18 Ⓥ €1·80. **Town** ✕ ⌷ ☵ 🅟 🅣.

MORTAGNE-sur-GIRONDE, Charente Maritime, **45°28′·21N 00°48′·81W**. AC 3058, 3057; SHOM 7426. Tides, use RICHARD differences (Gironde Estuary). Good shelter in marina on E bank of river, 14M from Royan/40M from Bordeaux (near the 75km mark). Leave the main Gironde chan at No 18 PHM buoy, Fl (2) R 6s; track 094°/4·6M to Mortagne access chan.

Ent is marked by a WCM buoy, VQ (9) 10s, an unlit PHM buoy, PHM and SHM perches and the ruins of a bcn twr. Buoyed chan, 2·5m at half tide, trends 063°/0·9M across drying mudbanks to entry gate which opens HW−1 to HW (give 48 hrs notice out of season). VHF Ch 09. ☎05.46.90.63.15. Facilities: **Marina** 130+20 Ⓥ in 4·5m €1·65, 🚿 🛢 🝙 ☵ 🛒(10t) ☵ Ice Aff Mar, ⊖ and Ⓛ at Royan; Auto 05.36.68.08.17.

BOURG-sur-GIRONDE (Actually on the Dordogne) **45°02′·28N 00°33′·61W**. AC 3068, 3069; SHOM 7030. Tides, use LA REUILLE (Gironde Estuary). 18 ⌷ + 2 Ⓥ <12m LOA, no rafting, on the W'ly of 2 long pontoons (parallel to shore); in strong W'lies very choppy against the ebb. HM ☎05.57.68.40.06. ⚓ ➔ 🝙 ☵(9T) ◣.

9.22.23 PAUILLAC

Gironde 45°11′·83N 00°44′·59W ❀❀⌂⌂⌂❀❀❀

CHARTS AC 3058, 3068 (Inset D; 1: 20,000); SHOM 7427

TIDES HW +0620 on Dover (UT); ML 3·0m

Standard Port POINTE DE GRAVE (←)

Times				Height (metres)			
High Water		Low Water		MHWS	MHWN	MLWN	MLWS
0000	0600	0000	0500	5·3	4·4	2·1	1·0
1200	1800	1200	1700				
Differences PAUILLAC							
+0100	+0100	+0205	+0135	+0·2	0·0	−1·0	−0·5
BORDEAUX							
+0200	+0225	+0405	+0330	0·0	−0·2	−1·7	−1·0

SHELTER Excellent in marina. ⚓ possible close E of marina; or 500m further S clear of a 250m wide No-anch area (due to underwater cables).

NAVIGATION WPT 45°12′·45N 00°44′·27W [No. 43 SHM buoy, Fl (2) G 6s], 196°/0·65M to hbr ent. Depths 2·0-4·5m; at LW hug the E side of entrance, due to silting. Do NOT arrive on the ebb; only manoeuvre in Marina at HW, approx HW Pte de Grave −3 and +2; see tidal diamond 'C' (45°13′·80N, 2M N). Strong current in approach, from S leave channel at No 43 buoy only; from N leave at No 45. Avoid the Airbus delivery ship depot, 1·2-1·7M north.

LIGHTS AND MARKS See chartlet and Lights, buoys and waypoints for lt details. The church twr, NW of the marina ent is conspic. So too is a 4m high bottle of Pauillac on the W side of the ent.

COMMUNICATIONS CROSS 02·97·55·35·35; ⚓ 05·36·68·08·33; ⊖ 05·56·59·04·01; Brit Consul 05·57·22·21·10; Marina VHF Ch 09 (0800–1200 & 1400-1800LT).

FACILITIES Marina ☎05·56·59·12·16. 130+20Ⓥ €2·50; max LOA 15m. Ⓥ berth on W end of pontoons A or B or any hammerhead; see HM for vacant berth. ◣ (haul-out on trailer, max 12m/15t) ✎ 🅑 ✕ de-masting ⌫(15m max) 🅣 🅟(0800-1800) from local garage only.

CN de Pauillac ☎05·56·59·12·58.

Town ✕ ⌷ ☵ 🅾 ✉ Ⓑ.

9.22.24   BORDEAUX

Gironde, **44°52´·79N 00°32´·18W** (Pont d'Aquitaine) ❀❀❀⌂⌂⌂✿✿✿

CHARTS AC 3068, 3069; SHOM 7030

TIDES HW +0715 on Dover (UT); ML 2·4m. Slack water is at approx HW Pte de Grave –2 and +2½; see tidal ◊s C and D.

SHELTER Good. Both Lormont and the Halte Nautique may be exposed to wash and flotsam. Do not berth on city quays.

NAVIGATION The chan is well marked/lit. Beware big ships, strong currents (up to 5kn if river in spate) and possibly large bits of flotsam. At No 62 WCM buoy (close NW of Bec d'Ambès) keep to stbd into the R. Garonne, with 12M to run to Bordeaux. De-mast before the Pont de Pierre (4·2m). Impressive vertical lift bridge, Pont Chaban-Delmas.

LIGHTS AND MARKS Pont d'Aquitaine (49m) and the new Pont Chaban-Delmas (53m) are good landmarks. Lts as on chartlet.

COMMUNICATIONS *Bordeaux Traffic* VHF Ch 12; ⊖ 05·57·81·03·60; Brit Consul 05·57·22·21·10; Bordeaux Port (PAB) 05·56·90·59·57, postoffice@bordeaux-port.fr www.bordeaux-port.fr; Aff Mar 05·56·00·83·00; Bègles marina VHF Ch 09.

FACILITIES from seaward: **Lormont YC** ☎05·56·31·50·10 is almost under Pont d'Aquitaine on E bank: limited space on S pontoon; de-masting crane; Clubhouse. FVs/ferries use the N pontoon. (The marina on the opposite bank is private, no ❶.)
Bassin No 2 (1·5M S of bridge). Pre-arrange with HM ☎05·56·90·59·57 mob 06·13·79·10·75, or Ch 12. A waiting pontoon is just S of the lock; ☎05·56·90·59·85 for latest info. No 🅿 or security. Facilities await development. ⚓ 🅻 SHOM ⚒ 🖅 ✕ Ⓔ ⛽(mobile, 5t de-masting).
Halte Nautique, E bank close SE of Pont de Pierre. 🅿.
Bègles Marina (Port Garonne), W bank, Km66. 4km S of Pont de Pierre. ☎05·56·85·76·04. 60+ 20 ❶ at N end of the wave-break.

9.22.25   CANALS TO THE MEDITERRANEAN

The Canal Latéral à la Garonne and **Canal du Midi** form a popular route to the Mediterranean, despite some 120 locks. The transit can be done in about a week, but 2-3 weeks is more relaxed. Masts can be lowered at Royan, Port-Médoc, Pauillac or Bordeaux.

Leave Bordeaux at LW Pointe de Grave for the 30M leg upriver to the first lock at Castets. Commercial traffic and W-bound boats have right of way. Most of the locks on the Canal Latéral à la Garonne are automatic. On the older and more interesting Canal du Midi there are many hire cruisers in summer. Depths vary with winter rainfall, summer drought and silting/dredging.

🛢 🚰 is available by hose at Bordeaux (Bègles) Mas d'Agenais, Agen, Toulouse (Port Sud), Castelnaudary, Port la Robine and by can elsewhere. 🗑 and ⚓ are readily obtainable. Tolls are listed in Special Notes for France Area 17. *Guide Vagnon No 7* or *Navicarte No 11* are advised. Further info from: Service de la Navigation de Toulouse, 8 Port St Etienne, 31079 Toulouse Cedex, ☎05·61·36·24·38, www.sn-sud-ouest.equipement.gouv.fr/en. See also *The Canal du Midi* (ACN).

SUMMARY Canal	From	To	Km/ Locks	Min Depth(m)	Min Ht(m)
Canal Latéral à la Garonne	Castets	Toulouse	193/53	2·2	3·5
Canal du Midi	Toulouse	Sète	240/65	1·6	3·0
Canal de la Nouvelle	Salleles	Port la Nouvelle	37/14	1·5	3·1

Dimensions: Max LOA 30m; draft 1·5m (varies with season); max beam 5·5m. Headroom of 3·3m is to centre of arch; over a width of 4m, clearance is about 2·4m.

Speed limits: 8km/hr (about 4½kn) but 3km/hr under bridges and over aqueducts.

POINTE DE GRAVE TO ARCACHON

(AC 2664, 1102) From Pte de la Négade to Capbreton, the coast is a featureless stretch of 107M broken only by the estuary which gives access to Arcachon. It is bordered by sand dunes and pine trees, and is often a lee shore with no shelter from W winds. 5M offshore a current usually sets N at about 0·5kn, particularly with a S wind; in winter this may be stronger after W winds. Within 1M of the coast there may be a S'ly counter-current.

A missile range, operated by Centre d'Essais des Landes, lies between Pointe de la Négade and Capbreton and extends up to 45M offshore. For details of boundaries, activity and sources of information, see below. Make every effort to obtain the latest information about range activity.

9.22.26 LANDES RANGE

Centre d'Essais des Landes (CEL: Landes Test Centre) firing range extends from Pointe de la Negade to Capbreton and up to 45M offshore. It is bounded by 45°27'N 01°13'W (inshore N) 45°12'N 02°00'W (NW corner) 44°00'N 02°25'W (SW corner) and 43°41'N 01°30'W (inshore S). The inshore boundary lies parallel to and about 3M off the coast. See chartlet, AC 1104 & 2664. CEL Range Control is at 44°26'N 01°15'W.

Sectors: The range is in two blocks, N and S of an 8M wide, 39M long, safe access corridor to Arcachon, orientated E-W from ATT-ARC buoy to the range boundary at 44°34'·05N 02°12'·91W.
Within the two blocks those sectors which are active, or are planned to be, are referred to by their Lat/Long coordinates in all radio broadcasts, telephone recordings and enquiries.

Range activity: Sectors may be active Mon-Fri from 0800 LT, but are usually inactive Sat/Sun, at night and in August. **However there can be exceptions to the above.** Navigation may be prohibited in active sectors inside the 12M territorial limit. Beyond 12M it *'is strongly discouraged due to the particularly dangerous tests carried out'*. A No-entry box (1·6M x 1M) is centred on 44°22'·70N 01°25'·50W, 13M SSW of ATT-ARC buoy. **You must check range activity before entering the range.**

Information on Range activity is available from:
- ☎+33 (0)5·58·82·22·58 recorded info H24, inc daily activity.
- Semaphore stations at Chassiron (Île d'Oléron) Pte de Grave, Cap Ferret, Messanges and Socoa (St Jean-de-Luz): on request Ch 16 for advance notice of activity. See also Chapter 5 for ☎ Nos. Some English spoken.
- CROSS Etel broadcasts 0703, 0715, 0733, 0745 and 0803LT or ☎+33 (0)2·97·55·35·35.
- CEL, after announcement on Ch 16, broadcasts Ch 06, 10: Mon-Thu, 0815 & 1615, and Fri 0815 & 1615 (when firing); all LT.
- CEL on request Ch 06, Mon-Thu 0800-1700 (1100 Fri).
- CEL ☎+33 (0)5·58·78·18·00; same hrs as previous line.
- Local NMs (AvUrNavs) at HMs' offices and Affaires Maritimes in Arcachon.

Options:
- Keep clear to seaward of the range. For example, a rhumb line track 200°/173M from PA buoy (22M W of La Rochelle) to Bilbao clears the range by at least 6M.

Subject to notified range activity and monitoring Ch 16:
- Navigate within the 3M wide coastal strip.
- Cross the range when it is usually inactive, ie at night, at weekends and in August.

S Biscay

ARCACHON TO THE SPANISH BORDER

Arcachon is accessible from offshore via an E-W corridor (not normally active) through the Landes range. The entrance to Bassin d'Arcachon and the marina is via the well buoyed, but unlit, Passe Nord between extensive, shifting sandbanks on which the sea breaks even in calm conditions. Study the text for optimum timing, swell and weather conditions before attempting entry (AC 1102, 1343).

▶ *The times of HW and LW occur progressively later from the seaward to the inner ends of the access channel. HW at Arcachon np is 10 mins and at sp is 25 mins after HW Pointe de Grave. LW at np is 20 mins after, and at sp the same as LW Pointe de Grave. Tidal streams in the appr can attain in excess of 4kn on the ebb at sp. N of the town, streams can reach in excess of 3.5kn sp.*◀

It is worth inserting in your GPS the Lat/Long of the key buoys as waypoints; see Lights, buoys and waypoints and also get the latest update from Cap Ferret sig station on Ch 13 or ☎05.56.60.60.03. Positively identify each buoy as you pass it; do not cut corners. Passe Sud is closed to navigation and most of the buoys have been lifted.

Strong N or W winds and swell make the narrow approach chan to the large marina at **Capbreton** impassable. Anglet/Bayonne may then be a safer option; or stay at sea. The Fosse (or Gouf) de Capbreton, a submarine canyon, lies at right angles to the coast. The 50m depth contour is 3ca W of Capbreton hbr bkwtr and the 100m line is 4ca further W. In strong W winds a dangerous sea breaks along the N and S edges of it, but within the very deep waters of the Fosse the sea is relatively calm.

Anglet has a marina, but there are few if any facilities for yachts further up the R. Adour at Bayonne.

▶*At L'Adour entrance the flood runs E and SE from HW Pointe de Grave (PdG) –6½, sp rate 2-4kn. The ebb runs W from HW PdG –½, sp rate 3-5kn. Slack water lasts for a mere 15 minutes, after which the rate increases rapidly.*◀

S of Pte St Martin the coast has mostly sandy beaches and rocky cliffs, with offlying rocky shoals and the Pyrenees visible as a backdrop.

St Jean-de-Luz (AC 1343) is best approached for the first time or in bad weather through Passe d'Illarguita (between Illarguita and Belhara Perdun shoals): follow the 138° transit (Le Socoa lt on with Bordagain lt) until the Ste Barbe ldg lts (101°) are in transit; thence enter by Passe de l'Ouest on the 151° transit of the inner hbr ldg lts. In strong W winds the sea breaks on Plateau de St Jean-de-Luz, a chain of rky shoals lying 1-4M offshore.

▶*On the rising tide the incoming stream sets more strongly through the East entrance, then follows the E shore of the Bay to where the River Nivelle enters. In the yacht anchorages in the NW part of the Bay an eddy sets N at about 0.5kn.*◀

During the falling tide the out-going flow from La Nivelle sets through the West entrance; the rate in the yacht anchorages is 1kn. Heavy seas in the entrance can stop the flow from leaving the Bay and it continues clockwise around the eastern part of the Bay.◀

Baie de Fontarabie (Rada de Higuer in Spanish) lies on the border of France and Spain and is entered between Pte Ste Anne and Cabo Higuer (a bare, rugged cape with lt ho) 1.75M WNW. In the middle of the bay a neutral area is marked by beacons as shown on AC 1171. To seaward of this area the boundary line (approximately 01°46'.2W) runs N from a white pyramid on the S shore, about 1M SW of Pte Ste Anne.

Les Briquets (dry) lie 1M N of Pte Ste Anne. Keep to W of Banc Chicharvel and Bajo Iruarri in ent to B. Entry should not be attempted with strong onshore winds or heavy swell. R La Bidassoa (Ría de Fuenterrabía in Spanish) is entered between breakwaters in SW corner of the B, giving access to the marina at **Hendaye-Plage**. See Area 23 for the Spanish part of the harbour and the marina at Fuenterrabía.

9.22.27 BASSIN D'ARCACHON

Gironde **44°39´.77N 01°09´.11W** (marina ent) ❀◊◊◊✿✿✿

CHARTS AC 2664; Imray C42; SHOM 6766, 7428; Navi 255, 1024

TIDES +0620 Dover; ML 2.5

Standard Port POINTE DE GRAVE (◀—)

Times				Height (metres)			
High Water		Low Water		MHWS	MHWN	MLWN	MLWS
0000	0600	0000	0500	5.3	4.4	2.1	1.0
1200	1800	1200	1700				
Differences ARCACHON (7 cables WNW of marina)							
+0005	+0020	+0030	+0005	–1.0	–1.0	–0.8	–0.6
CAP FERRET							
–0005	0000	+0020	–0005	–1.1	–1.1	–0.7	–0.4

SHELTER Good in marina. If full, ⚓ N of it, except in strong N'lies. No access to Bassin d'Arcachon in strong SW-N winds or at night.

NAVIGATION WPT 44°33´.048N 01°18´.714W (ATT-ARC SWM buoy) 088°/0.64M to 1N & 2N buoys in the well buoyed, unlit N Passe.

Before approaching, visitors should make every effort to contact the English-speaking *Cap Ferret Semaphore* Ch 16, 13 (HJ) ☎05.56. 60.60.03 for a buoyage/navigational update.

Buoys may at times be off station or missing and are only relaid annually.

N Passe trends NE then N between Banc d'Arguin and Banc du Toulinguet towards Cap Ferret. The longer S Passe is un-marked and closed to navigation, although some local FVs may still use it. In any wind the sea breaks on the shifting sand banks between Cap Ferret and Wharf de la Salie, but the chan buoys can be seen between the breakers (not as fearsome as it sounds).

Start appr ideally at HW–1 and no later than HW+1. Due to the very strong ebb (6kn sp) the chan bar (mean depth 3.7-4.5m) is impassable from HW+1 until LW, better to wait until LW+3. Best to leave on the last of the flood. When swell is higher than 1m, bar may be dangerous. If in doubt, stay out.

Be aware of any activity in firing ranges between Pte de la Negade and Capbreton, out to 40M offshore; see Landes Range.

LIGHTS AND MARKS ATT-ARC (SWM lt buoy) is moved to show approach to N Passe. See Lights, buoys and waypoints and website Updates for the most recent positions of appr buoys for use as waypoints.

Cap Ferret is low-lying but its white/red lt ho is conspic, as are a water twr close N and the white Sig stn close S. The Dune de Pyla (103m high) on the E shore is very conspic, unless hidden in haze.

COMMUNICATIONS ⒧Ⓑ 05.56.22.36.75; CROSS 02.97.55.35.35; Auto 08.92.68.08.33; Police 05.56.83.04.63; ⊖ 05.56.72.29.24; Brit Consul 05.57.22.21.10; Ⓗ 05.56.83.39.50; Dr 05.56.83.04.72; HM VHF Ch 09, 16 (H24); Aff Mar 05.57.52.57.07; Sig Stn 05.56.60.60.03.

FACILITIES Marina ☎05.56.22.36.75. plaisance@port-arcachon. com www.arcachon.com 2350+250 Ⓥ; €4.22, 2nd night free. Very crowded in season. Berth/raft on accueil pontoon (35m long); 2 days max stay. There are *very few* Ⓥ berths in Jul/Aug and on any public holiday in summer. Waiting list: 9,000+ applicants and 25+ years!

Extensive moorings (Areas U-Z) along the S shore west of the marina are not for daily use. They are patrolled by police launches.

⚓◣ ☎05.56.22.36.74 (0800-1845 Aug; OT 0815-1845) ⚓ Ⓞ ⚓. SHOM ⚓ ⒧Ⓔ ✕⚓ ◣ ▬(45t) ⚓ (20t).
YC du Bassin d'Arcachon ☎05.56.83.22.11, ◣ ✕ ⌂.
Town ✕ ⌂ ☲ ◨ ⊠ Ⓑ ⇌. Electric minibus into town approx every 20 mins. ✈ Bordeaux. Ferry: Roscoff/St Malo.

MINOR HARBOURS IN THE ARCACHON BASIN

Around the Basin are many small hbrs which dry LW ±2 or 3 hours, but are worth exploring by shoal draft boats: To the W, La Vigne*, Le Canon, Piquey and Claouey; to the NE, Port de Lège, Arès, Andernos*, Fontanievieille*, Lanton (Cassy) and Audenge*; and to the S, La Teste and Gujan. *See next page. Secondary chans in the Bassin d'Arcachon are marked by piles lettered A to K, clockwise from the N, plus pile number.

LA VIGNE, 44°40′·44N 01°14′·35W. HW time & ht approx as Cap Ferret above; ML 2·4m. See Bassin d'Arcachon. Good shelter, but crowded; beware strong currents across hbr ent. 2 perches mark the ent and on the SW point, a lt Iso R 4s 7m 5M. A small bkwtr (unlit) protrudes into the ent from the NE side. Aff Mar ☎05·56·60·52·76. Facilities: **Marina** ☎05·56·60·54·36. 268 + 2, max LOA 8·5m. ⚓ 🏠 ⚓ 🅿 🖾(2t).

ANDERNOS, 44°44′·53N 01°06′·54W. HW time & ht approx as Arcachon 9.22.27. Dredged chan to Bétey is marked by lateral poles D0 to D14; thence side chan to Andernos. Ent Fl G 4s and Fl R 4s. Jetty + pontoon for yachts able to dry out, max LOA 12m. Bad silting may occur despite dredging. Also ⚓ on flat drying foreshore. HM ☎05·56·82·00·12. Few facilities.

FONTAINEVIEILLE, 44°43′·30N 01°04′·63W. Tides as Bassin D'Arcachon. Drying marina on NE side of Bassin d'Arcachon, access via Chenal de Mouchtalette. Proceed from E0 PHM pile to E8, where fork left onto NNE for 7ca to hbr ent. No lts. Boats dry out on pontoons. Auto 05·36·65·08·33.

Facilities: **Marina** ☎05·56·82·17·31. 178+ 2, ⚓ 🏠 ⚓ 🔧.

AUDENGE, 44°40′·58N 01°01′·57W. Tides as 9.22.27. Drying marina and oyster port 5·5M E of Arcachon. Appr from G0 PHM pile via drying Chenal d'Audenge to G8 pile, 5ca short of the ent. Mairie ☎05·56·03·81·50. The Old Port (84 berths) is to the N; the New Port has 130 pontoon berths, ⚓ YC 🅿 🖾.

Other drying hbrs: to the NE, Port du Bétey & de Cassy; and E of Arcachon, Ports de La Teste & de La Hume and Port du Teich.

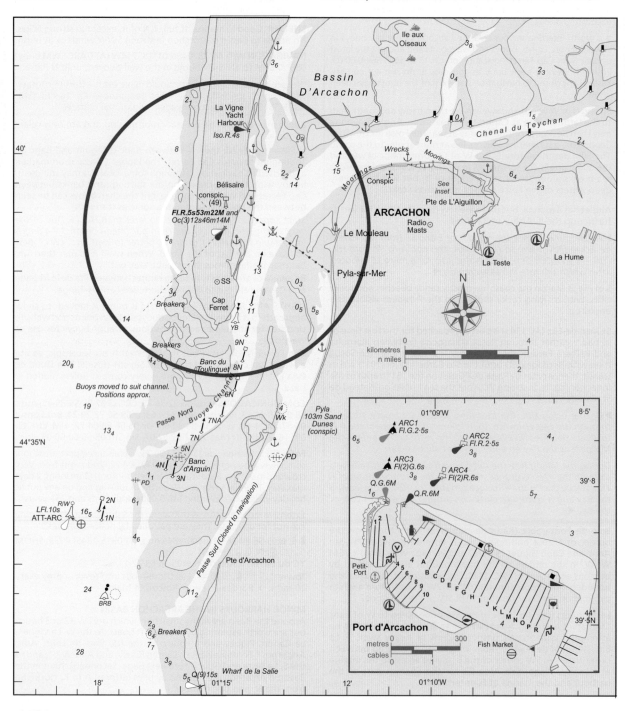

Port d'Arcachon

9.22.28 CAPBRETON

Landes **43°39'·34N 01°27'·00W** ❀❀❀❀❀❀ ❀❀❀

CHARTS AC 1102; SHOM 7440; Navi 555, 1024

TIDES +0450 Dover; ML 2·3. Use differences for St Jean de Luz.

SHELTER Good, but the narrow canalised ent (Canal du Boucarot) is dangerous in strong winds from W to N. Do not enter if swell or seas are seen to be breaking in mid-chan, although they often break either side. No ‡ outside the entrance.

NAVIGATION WPT 43°39'·63N 01°27'·49W, 123°/5ca to N pier lt. Bkwtr lts in line 123° lead to hbr ent. Gouf de Capbreton is a submarine canyon off the hbr ent where in last 3ca depths shoal rapidly from 50m to 3m; see Passage information. Access ±2hr for 2m draft. Silting occurs around head of S jetty, a wooden *estacade* which should be given a wide berth. An obstn 255° S jetty lt 300m is dangerous at LW; take great care.

Inside Canal du Boucarot best water is close to N jetty initially; from abeam small statue of Virgin Mary, move to mid-chan or just S of mid-chan. Marina entrance is via the obvious gap in training wall on SE side of chan, abeam conspic Capitainerie.

LIGHTS AND MARKS A casino and red-roofed sanatorium are conspic S of ent and a water twr 1M ENE. Lts as chartlet and 9.22.4. S jetty lt is on an extension close W of the former lt bcn.

COMMUNICATIONS CROSS 02·97·55·35·35; Auto 08·92·68·08·40; Police 05·59·50·31·52; ⊜ 05·59·46·68·80; Brit Consul 05·57·22·21·10; Ⓗ (Bayonne) 05·59·44·35·35; HM VHF Ch 09 (0700-2100 in season); Aff Mar 05·58·72·10·43.

FACILITIES Marina ☎05·58·72·21·23. www.port-capbreton.com info@port-capbreton.com. The 3 basins are in an inverted Y-shape. Chan and basins dredged 1·5m. 890+60❷ €2·47. ❷ pontoon B (2nd to stbd of marina ent). ⚓ ⚓ (☎05·58·72·15·66, 0830-1300, 1400-2000) ⊕ ⊕ ⚒ ⛵ ⚓ Ⓔ ⚓ ⚓(30t) ⚓(1·5t).

CNCP ☎05·58·72·67·09.

Town: ✕ ⊡ ⚑ ⊠ Ⓑ ⚑ Bayonne (17km); ✈ Biarritz (25km).

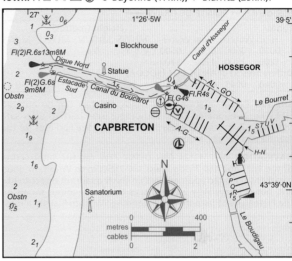

9.22.29 ANGLET/BAYONNE

Pyrénées Atlantique **43°31'·81N 01°31'·95W** ❀❀❀❀❀❀

CHARTS AC 1102, 1175; SHOM 7440, 7430; Navi 555

TIDES +0450 Dover (UT); ML 2·5

Standard Port POINTE DE GRAVE (◄—)

Times				Height (metres)			
High Water		Low Water		MHWS	MHWN	MLWN	MLWS
0000	0600	0000	0500	5·3	4·4	2·1	1·0
1200	1800	1200	1700				
Differences L'ADOUR (BOUCAU)							
−0030	−0035	−0040	−0025	−1·1	−1·1	−0·4	−0·3

SHELTER Very good in Anglet Marina, 7ca E of ent on S bank of R Adour, but in NE winds cement dust from the industrial N bank may be a problem. No ‡ due to commercial ships to which yachts must give way.

NAVIGATION WPT 43°32'·59N 01°32'·76W (BA SWM buoy, L Fl 10s) 142°/0·9M to N bkwtr lt. Easy access except in strong W winds. Strong tidal streams: 1·5-4kn flood, 2·4-5kn at sp ebb.

At Bayonne no passage for masted craft beyond bridge (9·5m clearance); interesting old quarter, 3M upriver on S bank.

LIGHTS AND MARKS Outer ldg lts 090°, both Q, W pylons/R tops. Inner ldg lts 111°, both QG, lit when chan is practicable. 3 more sets of ldg lts upriver to Bayonne. Other lts as chartlet and 9.22.4. Pte St Martin lt ho is 2·45M SSW of hbr.

IPTS (full code) from conspic Sig twr (W, B stripe) S side of ent.

COMMUNICATIONS CROSS 02·97·55·35·35; ⚓ 05·59·23·84·15; Auto 08·92·65·08·64; ⊜ 05·59·59·08·29; Brit Consul 05·57·22·21·10; Ⓗ 05·59·44·35·35; HM Port 05·59·63·11·57; Port/pilots VHF 12, 16 (0800-1200, 1400-1800LT); Aff Mar 05·59·55·06·68; Marina VHF Ch 09.

FACILITIES Marina ☎05·59·63·05·45, 🖷05·59·63·22·12. portdeplaisance@agglo-bab.fr www.agglo-bab.fr (Port du Brise-Lames; 1·7-3·5m) 367+58❷ €2·38. ⚓ ⚓ ⊕ ⊕ SHOM ⚒ ⚓ ⚓ ⛵ Ⓔ ⚓(25t) ⚓(1·3t).

YC Adour Atlantique ☎05·59·63·16·22.

Bayonne ✕ ⊡ ⚑ ⊠ Ⓑ ⚑ ✈ (Biarritz). Ferry: Bilbao.

9.22.30 ST JEAN DE LUZ
(Basque name: Donibane Lohizune)

Pyrénées Atlantique, **43°23′·85N 01°40′·61W** ✲✲🌢🌢🌢🌣🌣🌣

CHARTS AC 1102, 1170; SHOM 7440, 7431; Navi 555

TIDES HW +0435 on Dover (UT); ML 2·5m
Standard Port POINTE DE GRAVE (⟵)

Times				Height (metres)			
High Water		Low Water		MHWS	MHWN	MLWN	MLWS
0000	0600	0000	0500	5·3	4·4	2·1	1·0
1200	1800	1200	1700				
Differences ST JEAN DE LUZ (SOCOA)							
−0040	−0045	−0045	−0030	−1·0	−1·1	−0·5	−0·4

SHELTER Except in strong NW winds, the bay can be entered H24. The 3 designated yacht anchorages (see chartlet) offer safe holding in 2-4m; the SE is less prone to swell. See Facilities for the marina and Socoa harbour.

NAVIGATION WPT 43°24′·09N 01°40′·79W (at the intersection of the 101° and 150·7° ldg lines) 150·7°/500m to W ent. In good weather yachts can approach within the N quadrant direct to hbr ent, but in heavy W'ly weather seas break on various shoals on the Plateau de St Jean-de-Luz, extending SW from Biarritz almost to Hendaye. 3M W of hbr ent and 2M NE of Hendaye, beware Les Briquets rks, drying 0·4m.

The 3 appr chans (outer, middle, inner) are for big ships or smaller craft in bad weather and should be taken in sequence:

> **Outer chan** leads 138° between Illarguita and Belhara Perdun banks. It is the main and only safe approach at night and in bad weather. Ldg Its: Front, Socoa QWR, 12m; W □ twr, B stripes (R sector covers Socoa hbr ent). Rear Q; B ■/W bands on W pylon/B bands, hard to see by day, but the nearby Bordagain twr, 100m, is more readily visible.
>
> **Middle chan** 101° is intercepted at 43°24′·23N 01°41′·90W. Thence, or if arriving direct from the W, it leads past the hbr's W bkwtr, to the WPT. Ldg Its at Ste Barbe: Both Oc (4) R 12s; front, W △ on W bldg; rear, B ▲ on W twr.
>
> **Inner chan** leads 150·7° from the WPT through the W ent into the bay and via a buoyed chan to St Jean de Luz hbr. Ldg Its: Both Dir QG, conspic white twrs: front has a R vert stripe; rear, a G vert stripe.

The unmarked E ent to the bay is used by locals, but is not recommended. Beware: a large unlit mooring buoy S of Digue d'Artha, antipollution booms off the SW and SE beaches and an unmarked submerged jetty in the SE corner of bay.

Speed limits: in the bay 7kn; port area 5kn. Ent/exit the port only under power. FVs tend to ent/exit the FV hbr at speed and with much wash in the narrow access channel.

LIGHTS AND MARKS La Rhune, an 898m high conical mountain bears 160°/5·5M from the W ent; conspic in good visibility. Digue des Criquas (the W bkwtr) is lit, Iso G 4s. Digue d'Artha is a detached unlit bkwtr across the middle of the bay. See chartlet and Lights, buoys and waypoints for details.

COMMUNICATIONS CROSS 02·97·55·35·35; ⓛⓑ 05·59·47·22·98; ⚓ 05·59·22·03·30; Auto 08·92·68·08·64; Police 05·59·26·01·55; Brit Consul 05·57·22·21·10; Aff Mar 05·59·47·14·55; Marina Ch 09, 16; 0700-1230 & 1400-2000 in season.

FACILITIES

ST JEAN DE LUZ Marina ☎05·59·47·26·81, ▤05·59·47·86·11. Small facility (2·5m) 73+8❶ max LOA 16m; €3·03. at Ciboure, in Bassin de Larraldénia, close to rear 150·7° ldg lt and W and SW of the FV hbr. Berth/raft on hammerheads of the first 2 pontoons, or as directed. East of the pontoons beware the long slipway marked by 2 PHM bcns. ⚓ by tanker, ▬ ⚓ Ⓔ ⚒ 🄳 ⚔ 🅑(6t).
Town ✕ 🍴 ▤ 🛒 🅞 ✉ Ⓑ ⇌.

Socoa in the NW corner of the bay is close S of conspic fort. Hbr dries about 0·5m; tide gauge at ent. Here small craft can dry out, but it is little more than a landing place for yachts anchored E & SE of it in designated anchorages. It is over 2km walk to the town centre.
Jetty ⚓ by tanker, ⚓ ▬ 🢅 ⚒ 🄳 ⚔ 🅑(1t).
YC Basque ☎05·59·47·18·31.

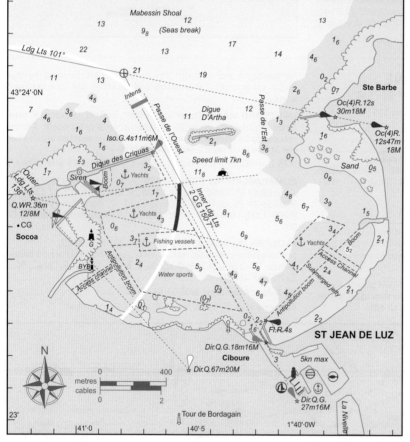

9.22.31 HENDAYE

Pyrénées Atlantique **43°22′·75N 01°47′·28W** ✿✿✿✿✿✿✿ ✿✿✿

CHARTS AC 1102, 1170, 1171; SHOM 7440, 7431; Navi 555

TIDES HW +0450 on Dover (UT); ML 2·3m
Use differences ST JEAN DE LUZ (SOCOA).

SHELTER Excellent in marina. Access H24, but exposed to N/NE and S/SW gales. Possible ⚓ in river or the Baie de Chingoudy but both are overcrowded. A Spanish port of refuge (2-3m) lies S of C Higuer, for FVs rather than for yachts. Note: Fuenterrabía (Hondarrabia, Area 23) is Spanish. See chartlet for neutral zone.

NAVIGATION WPT 43°23′·81N 01°46′·51W, 212°/1·15M to W bkwtr hd. Beware Les Briquets 8ca N of Pte Ste Anne at E end of the Baie; near centre of B, keep clear of Bajo Iruarri. River ent 6·5-1·3m is easy except in heavy N'ly swell; sp ebb is very strong. Inshore of Pte des Dunes, Fl R 2·5s, hug the E training wall for best water. A spit drying 1·3m (SHM bn, VQ (3) G 5s) off Fuenterrabía narrows the chan to about 100m wide before the marina ent opens.

LIGHTS AND MARKS La Rhune, a 900m high conical mountain, is conspic in good vis 8M ESE of hbr. Cabo Higuer lt ho is conspic at W side of bay on a rugged headland.

COMMUNICATIONS CROSS 02·97·55·35·35; ⚓ 05·59·24·58·80; Auto 08·92·68·08·64; Police 05·59·50·31·52; ⊖ 05·59·48·10·68; Brit Consul 05·57·22·21·10; Ⓗ 05·59·20·08·22; Aff Mar 05·59·47·14·55; Marina VHF Ch 09 (H24).

FACILITIES

Marina ☎05·59·48·06·10. www.hendaye-tourisme.fr station. littorale@hendaye.com (2·5-3m) 803 + 50Ⓥ €3·10 (3rd night free) is entered close W of a Fl Y 4s and a conspic RW TV relay mast (40m). Accueil pontoon A in NW corner. 🛠🚿 (H24) ⛽ ⚓ 🔧 📞 Ⓔ 🔨 ⊙ @ ⛽(30t).

Town ✕ 🍴 🍺 🛒 ✉ Ⓑ ⇌ (TGV Paris-Madrid) ✈ Fuenterrabía, or for UK flights Biarritz (☎05·59·43·83·83). Local ferry from Hendaye to Fuenterrabía every 15 mins. UK ferry from Bilbao or Santander.

Fuenterrabía Marina, see Area 23.

N & NW Spain

Fuenterrabía to Bayona

9.23.1	Contents	947
9.23.2	Area map and Distance table	948
9.23.3	Lights, buoys and waypoints	950
9.23.4	Passage information	953
	Bay of Biscay: Wind, Weather and Sea – 953 • Crossing the Bay of Biscay – 953 • French Border to Santander – 953 • Santander to Cabo Ortegal – 960 • Cabo Ortegal to Cabo Finisterre – 965 • Cabo Finisterre to the Portuguese Border – 973	
9.23.5	Special notes for Spain	953
9.23.6	Fuenterrabía	954
9.23.7	Pasajes	954
9.23.8	San Sebastián	955
	Ría de Orio	
9.23.9	Guetaria	955
9.23.10	Zumaya	956
	Motrico	
9.23.11	Lequeitio	956
9.23.12	Elanchove	957
	Bermeo	
9.23.13	Bilbao	957
9.23.14	Castro Urdiales	958
9.23.15	Laredo	959
	Santoña	
9.23.16	Santander	959
	Suances	
9.23.17	San Vicente de la Barquera	961
9.23.18	Ribadesella	961
	Llanes	
9.23.19	Lastres	962
9.23.20	Gijón	962
	Avilés • San Esteban de Pravia	
9.23.21	Cudillero	963
9.23.22	Luarca	963
9.23.23	Ría de Ribadeo	964
9.23.24	Ría de Viveiro	964
9.23.25	Ría de Cedeira	965
	Ría del Barquero • Ensenada de Santa Marta de Ortigueira	
9.23.26	Ferrol	966
9.23.27	Ría de Ares	966
9.23.28	Ría de Betanzos	967
9.23.29	La Coruña, tide tables/curves	968
9.23.30	Ría de Corme y Laxe	972
9.23.31	Ría de Camariñas	972
	Finisterre • Ensenada del Sardineiro • Ría de Corcubión	
9.23.32	Ría de Muros	975
9.23.33	Ría de Arosa	975
9.23.34	Ría de Pontevedra	977
9.23.35	Ría de Vigo	978
9.23.36	Bayona	980

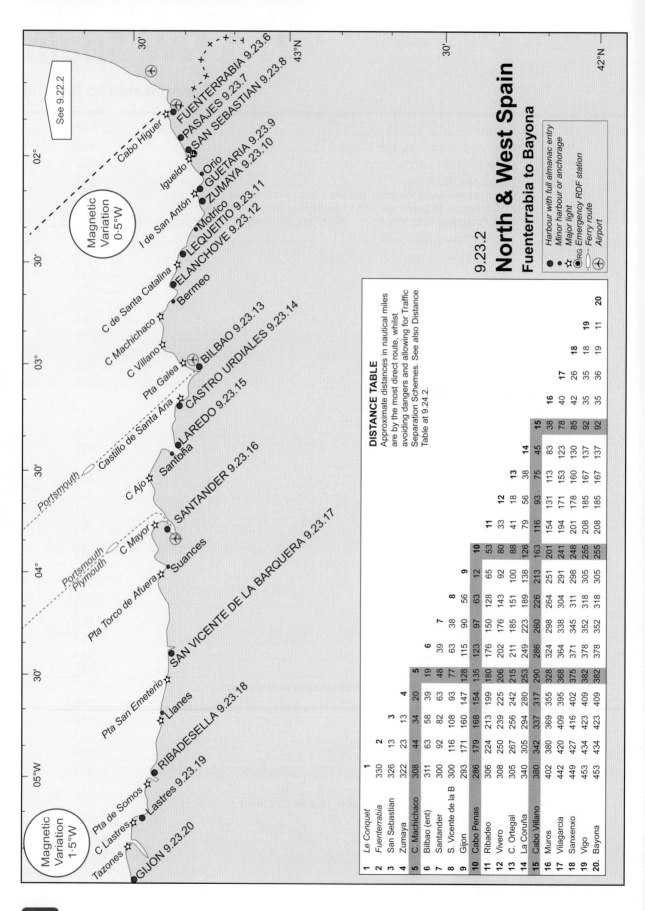

See 9.22.2

Magnetic Variation 0·5°W

Magnetic Variation 1·5°W

9.23.2

North & West Spain
Fuenterrabia to Bayona

- ● Harbour with full almanac entry
- ● Minor harbour or anchorage
- ☆ Major light
- RG Emergency RDF station
- --- Ferry route
- ✈ Airport

Map labels: Cabo Higuer, FUENTERRABIA 9.23.6, PASAJES 9.23.7, SAN SEBASTIAN 9.23.8, Igueldo, Orio, GUETARIA 9.23.9, I de San Antón, ZUMAYA 9.23.10, Motrico, LEQUEITIO 9.23.11, ELANCHOVE 9.23.12, Bermeo, C de Santa Catalina, C Machichaco, BILBAO 9.23.13, C Villano, Pta Galea, CASTRO URDIALES 9.23.14, Castillo de Santa Ana, LAREDO 9.23.15, Santoña, C Ajo, SANTANDER 9.23.16, C Mayor, Suances, Pta Torco de Afuera, SAN VICENTE DE LA BARQUERA 9.23.17, Pta San Emeterio, Llanes, RIBADESELLA 9.23.18, Pta de Somos, C Lastres, Lastres 9.23.19, Tazones, GIJON 9.23.20, Portsmouth, Plymouth, Le Conquet

DISTANCE TABLE

Approximate distances in nautical miles are by the most direct route, whilst avoiding dangers and allowing for Traffic Separation Schemes. See also Distance Table at 9.24.2.

	1	2	3	4	5	6	7	8	9	10	11	12	13	14	15	16	17	18	19	20
1 Le Conquet	**1**																			
2 Fuenterrabia	330	**2**																		
3 San Sebastian	326	13	**3**																	
4 Zumaya	322	23	13	**4**																
5 C. Machichaco	308	44	34	20	**5**															
6 Bilbao (ent)	311	63	58	39	19	**6**														
7 Santander	300	92	82	63	48	39	**7**													
8 S. Vicente de la B	300	116	108	93	77	63	38	**8**												
9 Gijon	293	171	160	147	128	115	90	56	**9**											
10 Cabo Penas	286	179	168	154	135	123	97	63	12	**10**										
11 Ribadeo	306	224	199	180	176	150	128		65	53	**11**									
12 Vivero	308	250	239	225	206	202	176	143	92	80	33	**12**								
13 C. Ortegal	305	267	256	242	215	211	185	151	100	88	41	18	**13**							
14 La Coruña	340	305	294	280	253	249	223	189	138	126	79	56	38	**14**						
15 Cabo Villano	380	342	337	317	290	286	260	226	213	163	116	93	75	45	**15**					
16 Muros	402	380	369	355	328	324	298	264	251	201	154	131	113	83	38	**16**				
17 Vilagarcia	442	420	409	395	368	364	338	304	291	241	194	171	153	123	78	40	**17**			
18 Sanxenxo	449	427	416	402	375	371	345	311	298	248	201	178	160	130	85	42	26	**18**		
19 Vigo	453	434	423	409	378	352	318	305	255	208	185	167	137	92	35	35	35	18	**19**	
20. Bayona	453	434	423	409	378	352	318	305	255	208	185	167	137	92	92	35	36	19	11	**20**

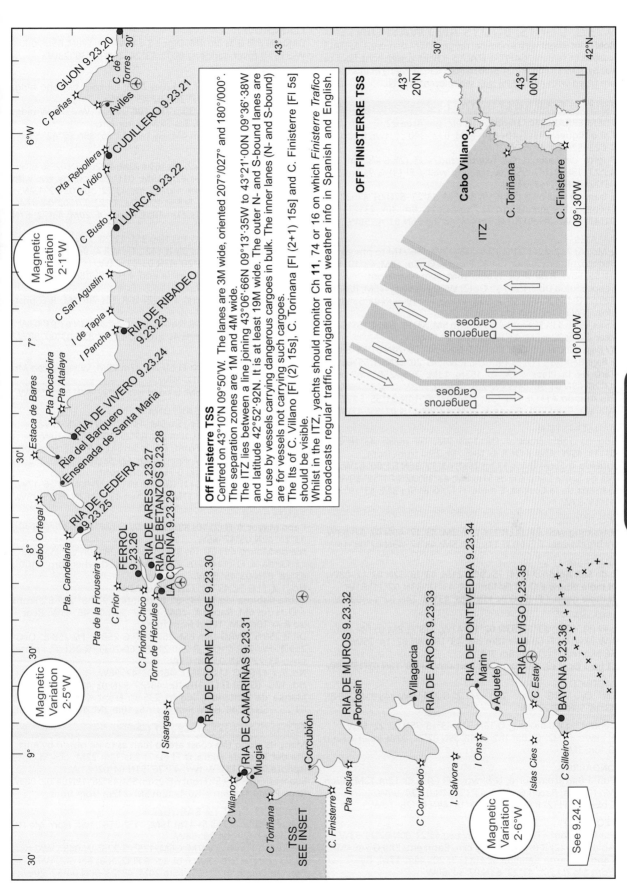

Off Finisterre TSS

Centred on 43°10'N 09°50'W. The lanes are 3M wide, oriented 207°/027° and 180°/000°.
The separation zones are 1M and 4M wide.
The ITZ lies between a line joining 43°06'·66N 09°13'·35W to 43°21'·00N 09°36'·38W and latitude 42°52'·92N. It is at least 19M wide. The outer N- and S-bound lanes are for use by vessels carrying dangerous cargoes in bulk. The inner lanes (N- and S-bound) are for vessels not carrying such cargoes.
The lts of C. Villano [Fl (2) 15s], C. Torinana [Fl (2+1) 15s] and C. Finisterre [Fl 5s] should be visible.
Whilst in the ITZ, yachts should monitor Ch **11**, 74 or 16 on which *Finisterre Trafico* broadcasts regular traffic, navigational and weather info in Spanish and English.

OFF FINISTERRE TSS

N & NW Spain

9.23.3 LIGHTS, BUOYS AND WAYPOINTS

Bold print = light with a nominal range of 15M or more. CAPITALS = place or feature. *CAPITAL ITALICS* = light-vessel, light float or Lanby. *Italics* = Fog signal. ***Bold italics*** = Racon. Many marks/buoys are fitted with AIS (MMSI No); see relevant charts.

FUENTERRABÍA TO SANTANDER

FUENTERRABÍA For Hendaye lts etc, see Area 22.
Cabo Higuer ☆ Fl (2) 10s 63m **23M**; 072°-340°; ☐ twr, R cupola; 43°23'·52N 01°47'·52W.
Puerto de Refugio, NE bkwtr head ⚡ Fl (2) G 7s 7m 5M; 43°23'·13N 01°47'·35W.W training wall ⚡ Fl (3) G 9s 9m 5M; 43°22'·83N 01°47'·34W.
E training wall hd ⚡ L Fl R 10s 8m 5M; 43°22'·67N 01°47'·22W.
Fuenterrabía marina ent, ⚡ Fl (4) G 11s 9m 3m; 43°22'·59N 01°47'·51W. ⚡ Fl (4) R 11s 9m 1M; 43°22'·56N 01°47'56W.

PASAJES

⚓ Mo (A) 7s 7M; 43°21'·19N 01°56'·12W; 166°/1M to hbr ent.
Senocozulúa lts in line 154·8°: **Front** ☆ Q 67m **18M**; 43°19'·89N 01°55'·60W. **Rear** ☆ Oc 3s 86m **18M**, 40m from front.
Senocozulúa Dir ⚡ 155·75°: Oc (2) WRG 12s 50m W6M, R/G3M; 129·5°-G-154·8°-W-157°-R-190°; W twr; 43°19'·90N 01°55'·61W, close NW of the two ldg lts.
Inner ldg lts 114·3°: Front ⚡ Iso 4s 11m 5M; 43°19'·70N 01°55'·31W. Rear ⚡ Q 1s 32m 5M; 43°19'·56N 01°55'·17W.
W Bajo Bancha ⬩ Fl G 5s 14m 7M; G twr; 43°20'·24N 01°55'·88W.
E Bajo Bancha ⬩ Fl R 5s 14m 7M; R twr; 43°20'·22N 01°55'·65W.
Faro de la Plata ⚡ Oc 4s 153m 13M; 285°-250°; W bldg; 43°20'·08N 01°56'·04W (W side of ent).
Pta Arando ⚡ Fl (2) R 7s 12m 11M; 43°20'·14N 01°55'·68W
Atalaya, 96m; IPTS; ***Racon K, 20M***; 43°20'·11N 01°55'·47W.

SAN SEBASTIÁN

Ldg lts 158°: Front ⚡ QR 10m 7M; Gy mast; 43°18'·85N 01°59'·45W. Rear ⚡ Oc R 4s 16m 7M; 25m from front.
Igueldo ☆ Fl (2+1) 15s 132m **26M**; 43°19'·36N 02°00'·61W.
Isla de Santa Clara ⚡ Fl 5s 51m 9M; 43°19'·32N 01°59'·89W.
Dársena de la Concha, E mole, Fl (2) G 8s 9m 3M; 43°19'·34N 01°59'·37W.

RÍO DE ORIO

E training wall ⚡ Fl (4) R 11s 12m 3M; 43°17'·49N 02°07'·93W.
W breakwater ⚡ Fl (4) G 11s 18m 5M; 43°17'·49N 02°08'·13W.

GUETARIA

I. de San Antón ☆ Fl (4) 15s 91m **21M**; 43°18'·62N 02°12'·08W.
N mole elbow ⚡ Q (3) 5s 14m 3M; 43°18'·33N 02°11'·81W.
N mole head ⚡ Fl (3) G 9s 11m 5M; 43°18'·24N 02°11'·891W.

ZUMAYA

⚓ Fl (5) Y 20s; 43°19'·83N 02°13'·47W, 213°/1·8M to hbr ent.
NW bkwtr ⚡ Fl (2) G 7s 18m 5M; 43°18'·38N 02°14'·75W.
SE bkwtr ⚡ Fl (2) R 7s 8m 3M.
Lt ho ⚡ Oc (1+3) 12s 39m 12M; Port sigs; 43°18'·18N 02°15'·08W.

MOTRICO

Dir Oc WRG 12m 3M; 241.2°-G-244·1°-W-245·9°-R-248·8°; 43°18'·50N 02°22'·80W.
Central Tidal Barrage ⚓, Q 18m 3M; 43°18'·75N 02°22'·60W.
N bkwtr ⚡ Fl (2) G 10m 1M; 43°18'·59N 02°22'·80W. Inner ⚡ Fl (3) G 10m 1M.

ONDÁRROA

Punta Barracomuturra, N bkwtr ⚡ Fl (3) G 8s 13m 12M; *Siren Mo (O) 20s;* ***Racon G, 12M;*** 43°19'·54N 02°24'·94W.
S bkwtr ⚡ Fl (2) R 6s 7m 6M; 43°19'·49N 02°25'·10W.

LEQUEITIO

Pta Amandarri ⚡ Fl G 4s 8m 5M; Gy twr; 43°22'·00N 02°29'·93W.
Aislado, Fl (2) R 8s 5m 4M.
Cabo de Santa Catalina ☆ Fl (1+3) 20s 44m **17M**; Gy ○ twr; *Horn Mo (L) 20s;* 43°22'·63N 02°30'·60W.

ELANCHOVE

Digue N ⚡ Fl G 3s 8m 4M. Digue S ⚡ F WR 7m W8M,R5M; 000°-W-315°-R (over dangers)-000°; 43°24'·24N 02°38'·23W.

BERMEO

Cabo Machichaco ☆ Fl 7s 120m **24M**; lookout twr and bldg; *Siren Mo (M) 60s;* 43°27'·30N 02°45'·19W.
Rosape ⚡ Fl (2) WR 10s 36m 7M; 108°-R-204°-W-232°; W bldg; 43°25'·07N 02°42'·76W, 3ca SSW of hbr ent.
N bkwtr ⚡ Fl G 4·5s 16m 4M; Gy twr; 43°25'·39N 02°42'·59W.

BILBAO

C. Villano (Gorliz) ☆ Fl (1+2) 16s 163m **22M**; 43°26'·03N 02°56'·63W.
Punta Galea ☆ Fl (3) 8s 82m **19M**; 011°-227°; stone twr with dwelling, Gy cupola; *Siren Mo (G) 30s;* 43°22'·30N 03°02'·13W.
E bkwtr hd (detached) ⚡ Fl R 5s 19m 7M; 43°22'·77N 03°04'·67W.
W bkwtr head ⚡ Fl G 5s 21m 10M; ***Racon X, 20M;*** 43°22'·67N 03°05'·04W.
Santurce Quay ⚡ Fl (3) G 10s 18m 3M; 43°20'·79N 03°01'·91W.
Contradique de Algorta dir lt 142°, Oc WR 4s 9m 3M; 134°-W-149°-R-174°; Gy col; 43°20'·47N 03°01'·21W.
Contradique de Algorta mole end ⚡ Fl R 5s 18m 3M, W twr; 43°20'·51N 03°01'·67W.
Mole head (cruise liners) ⚡ Fl (2+1) R 10s 7m 1M; R/G post; 43°20'·34N 03°01'·57W.
Getxo marina bkwtr ⚡ QR 5m 1M, R col; 43°20'·23N 03°01'·03W.
Las Arenas marina ⚡ Oc G 4s 1m 1M; 43°19'·83N 03°00'·98W.

CASTRO URDIALES

Castillo de Santa Ana ☆ Fl (4) 24s 47m **20M**; W twr; *Siren Mo (C) 60s;* 43°23'·07N 03°12'·88W.
N bkwtr ⚡ Fl G 3s 12m 6M; 8-sided twr; 43°22'·87N 03°12'·54W.

LAREDO and RIA DE SANTOÑA

Laredo N bkwtr ⚡ Fl R 5s 15m 5M; 43°25'·11N 03°25'·17W.
Laredo S bkwtr ⚡ Fl G 5s 10m 3M; 43°25'·10N 03°25'·10W.
Santoña ldg lts 283·5°: Front, ⚡ Fl 2s 5m 8M; B ▼ on framework; 43°26'·34N 03°27'·60W. Rear, ⚡ Oc (2) 5s 12m 11M; 279·5°-287·5°; ○ on framework; 0·75M from front.
Pta Pescador lt ho ⚡ Fl (3+1) 18s 37m 9M; 43°27'·90N 03°26'·17W.
C. Ajo ☆ Oc (3) 16s 69m **17M**; W ○ twr; 43°30'·67N 03°35'·73W.

SANTANDER

Cabo Mayor ☆ Fl (2) 10s 89m **21M**; W ○ twr; *Horn Mo (M) 40s;* 43°29'·42N 03°47'·45W.
Isla de Mouro ⚡ Fl (3) 16s 37m 11M; 43°28'·40N 03°45'·35W.
La Cerda ⚡ Fl (1+4) 20s 22m 7M; obsc'd when brg <160°; 43°28'·01N 03°45'·83W.
No. 1 ⚓ Fl G 5s; 43°27'·73N 03°46'·22W.
Punta Rabiosa ldg lts 235·8°, both intens 231·8°-239·8°: Front ⚡ Q 1s 7m 6M; ***Racon K, 10M;*** 43°27'·52N 03°46'·43W. Rear ⚡ Iso R 4s 10m 6M, 100m from front.
Dir lt 259·5°, WRG 18m 6M; 257·8°-Fl G-258·3°-FG-258·8°-OcG-259·4°-FW-259·6°-Oc R-260·2°-FR-260·7°-Fl R-261·2°; metal post; 43°27'·33N 03°48'·61W.
No. 7 ⚓ Fl (4) G 11s; 43°27'·46N 03°47'·98W.
No. 17 ⚓ (GRG) Fl (2+1)G 10s; 43°25'·93N 03°48'·25W.
Marina de Santander, ldg lts 235·6°: Front ⚡ Iso 2s 9m 2M; 43°25'·75N 03°48'·84W. Rear ⚡ Oc 5s 10m 2M; 46m from front.

SANTANDER TO CABO PEÑAS

Many lts along this coast are so high as to be obsc'd by mist.
Pta del Torco de Fuera ☆ Fl (1+2) 24s 33m **22M**; obsc'd close inshore 091°-113°; W twr; 43°26'·51N 04°02'·61W.
Suances ldg lts 149·5°: Front ⚡ Fl 2s 8m 5M; 43°26'·19N 04°02'·08W. Rear, Punta Marzán ⚡ Iso 4s 12m 5M; 212m from front.

SAN VICENTE DE LA BARQUERA

Pta de la Silla ⚡ Oc 3·5s 41m 13M; 115°-250°; twr; *Horn Mo (V) 30s;* 43°23'·58N 04°23'·56W.
W bkwtr ⚡ Fl WG 2s 7M; G6M; 175°-G-225°-W-045°; W/G twr; 43°23'·71N 04°23'·11W. E bkwtr ⚡ Fl (2) R 8s 6m 5M; W/R twr.
Pta San Emeterio ☆ Fl 5s 66m **20M**; 43°23'·90N 04°32'·20W.

LLANES
Pta de San Antón ☆ Oc (4) 30s 16m **15M**; W 8-sided twr; 43°25'·21N 04°44'·88W, 150m S of hbr ent.
N bkwtr (Osa) ≮ Fl G 5s 13m 5M; 43°25'·18N 04°44'·90W.
Dock hd ≮ Fl (3) G 9s 6m 1M.

RIBADESELLA
Pta del Caballo ≮ Fl (4) R 11s 10m 5M; 278·4°-212·9°; 43°28'·08N 05°03'·98W. Fl R 5s 7m 3M; R post; 43° 28'·06N 05°03'·92W.
Somos ☆ Fl (2+1) 12s 113m **25M**; twr; 43°28'·36N 05°04'·97W.

LASTRES
E bkwtr ≮ Fl (3) G 9s 13m 4M; 43°30'·89N 05°15'·90W.
C. Lastres ☆ Fl (5) 25s 116m **23M**. W ○ twr; 43°32'·04N 05°18'·02W.
Tazones ☆ Fl 7·5s 127m **20M**; W 8-sided twr; 43°32'·87N 05°23'·94W. Bkwtr head, ≮ Fl (2) G 7s 10m 5M.

GIJÓN
El Musel, NE mole ≮ Fl (2+1) G 10s 22m 7M; 43°34'·26N 05°40'·58W.
Piedra Sacramento ⚓ Fl (2) G 6s 11m 5M; 8-sided twr; 43°32'·83N 05°40'·30W. Marina ent ≮ Fl (2) R 6s 9m 3M; 43°32'·79N 05°40'·21W.
Cabo de Torres ☆ Fl (2) 10s 80m **18M**; 43°34'·31N 05°41'·96W.
Puerto de Candás ☆ Oc (2)10s 38m **15M**; R twr, W ho; *Horn Mo (C) 60s;* 43°35'·66N 05°45'·65W.
Punta del Cuerno ≮ Dir Fl WRG 4s 20m 3M; 284°-R-286°-W-288°-G-290°; R/W col; 43°35'·69N 05°45'·61W
N bkwtr hd ≮ Fl (3) G 9s 11m 3M.
Luanco ldg lts 255°: Front, inner bkwtr ≮ Fl G 5s 3m 5M; 43°36'·91N 05°47'·36W. Rear ≮ Oc G 4s 6m 5M, 200m from front.
Cabo Peñas ☆ Fl (3) 15s 115m **35M**; Gy 8-sided twr; 43°39'·33N 05°50'·92W.

CABO PEÑAS TO PUNTA DE LA ESTACA DE BARES
Avilés, Pta de la Forcada ≮ Fl R 5s 23m 7M; R mast on W truncated twr; 43°35'·82N 05°56'·84W. 80m E: *Siren Mo (A) 30s 3M.*
Pta del Castillo ☆ Oc WR 5s 38m **W20, R17M**; 091·5°-R-113°-W-091·5°; W □ twr; 43°35'·72N 05°56'·72W. ≮ Fl (2) R 7s is 30m SW.
Ent chan, S side ≮ Fl (2) G 7s 12m 5M; 106°-280°; W ○ twr, G band; 43°35'·57N 05°56'·49W. ≮ Fl G 5s 15m 5M is 390m west.
San Estaban de Pravia, W bkwtr ☆ Fl (2) 12s 19m **15M**; W ○ twr, B bands; *Siren Mo (N) 30s;* 43°34'·21N 06°04'·64W.
Ldg lts 206·2°: Front, Q 7m 1M; W twr, R bands; 43°34'·05N 06°04'·71W. Rear, Iso 4s 28m 1M; W pylon, R bands; 206·2°/360m.
Cudillero, Pta Rebollera ☆ Oc (4) 16s 42m **16M**; W 8-sided twr; *Siren Mo (D) 30s;* 43°33'·96N 06°08'·68W. N bkwtr ≮ Fl (3) G 9s.
Cabo Vidio ☆ Fl 5s 99m **25M**; ○ twr; 43°35'·62N 06°14'·61W.
Cabo Busto ☆ Fl (4) 20s 84m **25M**; 43°34'·15N 06°28'·17W.

LUARCA
Punta Altaya ≮ Oc (3) 15s 63m 14M; W □ twr; *Siren Mo (L) 30s;* 43°32'·98N 06°31'·95W.
Ldg lts 170°, both W cols, R bands: Front ≮ Fl 5s 19m 3M; 43°32'·78N 06°32'·11W. Rear ≮ Oc 4s 26m 3M; 41m from front. W Dique ≮ Fl (3) G 9s 6m 3M; 43°32'·96N 06°32'·24W.
Cabo de San Agustín ☆ Oc (2) 12s 70m **25M**; W ○ twr, B bands; 43°33'·81N 06°44'·05W.
C. San Sebastían ☆ Fl (2+1) 19s 22m **18M**; 43°34'·43N 06°56'·75W.

RÍA DE RIBADEO
Punta de la Cruz ≮ Fl (4) R 11s 16m 7M; 43°33'·39N 07°01'·74W.
Isla Pancha ☆ Fl (3+1) 20s 26m **21M**; W ○ twr, B bands; 43°33'·39N 07°02'·52W, W side of entrance.
1st ldg lts 140°, both R ◇s, W twrs. Front, Pta Arroxo ≮ Iso R 18m 5M; 43°32'·81N 07°01'·52W. Rear ≮ Oc R 4s 24m 5M; 228m SE.
2nd ldg lts 205°. Front, Muelle de García ≮ VQ R 8m 3M; R ◇, W twr; 43°32'·49N 07°02'·23W. Rear ≮ Oc R 2s 18m 3M.
Marina, ≮ Fl G 5s 9m 3M; 43°32'·45N 07°02'·16W. ≮ Fl R 5s 9m 1M.
Punta Atalya (San Ciprian) ☆ Fl (5) 20s 39m **20M**; W ○ twr, B band; 43°42'·03N 07°26'·20W. Alúmina Port, N bkwtr Fl (2) WG 8s 17m 4M; 110°-W-180°-G-110°; 43°42'·98N 07°27'·58W.

RÍA DE VIVEIRO
Pta Roncadoira ☆ Fl 7·5s 92m **21M**; W twr; 43°44'·15N 07°31'·58W.
Pta de Faro ≮ Fl R 5s 18m 7M; W twr; 43°42'·74N 07°35'·03W.
Pta Socastro ≮ Fl G 5s 18m 7M; G twr; 43°43'·08N 07°36'·42W.
Cillero, outer bkwtr ≮ Fl (2) R 7s 8m 5M; 43°40'·93N 07°36'·16W.
Marina ent ≮ Fl (2+1) G 15s 7m 1M; 43°40'·15N 07°35'·71W.

RÍA DEL BARQUERO
Isla Coelleira ≮ Fl (4) 16s 87m 7M; 43°45'·52N 07°37'·77W.
Pta de la Barra ≮ Fl WRG 3s 15m 5M; 198°-G-213°-W-240°-R-255°-G-213°; W twr; 43°44'·51N 07°41'·31W.
Pta del Castro ≮ Fl (2) 7s 14m 5M; 43°44'·46N 07°40'·48W.

PTA DE LA ESTACA DE BARES TO CABO VILLANO
Pta de la Estaca de Bares ☆ Fl (2) 7·5s 99m **25M**; obsc'd when brg >291°; 8-sided twr, R roof; 43°47'·17N 07°41'·06W.
Espasante, W bkwtr ≮ Fl R 5s 11m 3M; 43°43'·36N 07°48'·88W.
Cariño, bkwtr ≮ Fl G 9s 12m 5M; 43°44'·05N 07°51'·75W.
Cabo Ortegal (Pta de los Aguillones) ☆ Fl (2+1) 15s 122m **18M**; W ○ twr, R band; 43°46'·26N 07°52'·18W.
Punta Candelaria ☆ Fl (3+1) 24s 87m **21M**; 43°42'·67N 08°02'·95W.

RÍA DE CEDEIRA
Punta del Sarridal ≮ Oc WR 6s 39m 11M; Shore-R -145°-W-172°-R-145°; R ○ twr; 43°39'·65N 08°04'·53W.
Pta Promontorio ≮ Oc (4) 10s 24m 11M; 43°39'·05N 08°04'·20W.
Piedras de Media Mar ⚓ Fl (2) 7s 12m 4M; BRB twr; 43°39'·37N 08°04'·79W.
Bkwtr hd ≮ Fl (2) R 7s 10m 5M; R post; 43°39'·30N 08°04'·20W.
Punta Frouxeira ☆ Fl (5) 15s 73m **20M**; 43°37'·05N 08°11'·25W.
Cabo Prior ☆ Fl (1+2) 15s 105m **22M**; 055·5°-310°; 6-sided twr; 43°34'·06N 08°18'·86W. ODAS buoy Fl (5) Y 20s at 43°35'·41N 08°19'·45W, 1·4M NNW of Cabo Prior.
Punta del Castro ≮ Fl (2) 7s 42m 8M; W 6-sided twr; 43°30'·47N 08°19'·73W.

RÍA DE FERROL
Cabo Prioriño Chico ☆ Fl 5s 35m **23M**; 225°-129·5°; W 8-sided twr; 43°27'·53N 08°20'·40W.
Ent bkwtr head ≮ Fl R 5s 21m 7M; *Racon N 12M*; 43°27'·23N 08°19'·80W. Bkwtr spur on NE side Fl (2) R 7s 12m 3M.
Batería de San Cristóbal ≮ 042° Dir WRG 33m 5M, 038·2°-FG-040°-AlWG-041·5°-W-042·5°-AlWR-044°-FR-045·7°; white □ twr; 43°27'·93N 08°18'·25W.
Ldg lts 085·2°: Front, Pta de San Martín ≮ Fl 1·5s 10m 5M; 43°27'·58N 08°17'·06W. Rear ≮ Oc 4s 5M; 701m from front.
Dársena de Curuxeiras ≮ Fl (4) R 11s 5m 3M; 43°28'·54N 08°14'·67W.

RÍA DE ARES and RÍA DE BETANZOS
Ares bkwtr ≮ Fl (4) R 11s 10m 5M; R twr; 43°25'·35N 08°14'·29W.
Sada N pier ≮ Fl (4) G 11s 11m 5M; G twr; 43°21'·77N 08°14'·52W.

LA CORUÑA
Torre de Hércules ☆ Fl (4) 20s 104m **23M**; □ twr, 8-sided top; 43°23'·15N 08°24'·39W.
Canal del Oeste ldg lts 108·5°, both W 8-sided twrs: Front, Pta de Mera ≮ Oc WR 4s 54m, W8M R3M; 000°-R-023°; 100·5°-R-105·5°-W-114·5°-R-153°; *Racon M, 18M; 020°-196°*; 43°23'·00N 08°21'·27W. Rear ≮ LFl 8s 80m 8M; 357·5°-177·5°; 300m from front.
Canal del Este ldg lts 182°, both R/W chequered □ twrs: Front, Pta Fiaiteira ≮ Iso WRG 2s 28m, W10M, R/G7M; 146·4°-G-180°-W-184°-R-217·6°; *Racon X, 11-21M*; 43°20'·59N 08°22'·25W. Rear ≮ Oc R 4s 52m 5M; 380m from front.
Dique d'Abrigo ≮ Fl G 3s 16m 6M; 43°21'·90N 08°22'·48W.
Dársena de la Marina ≮ Fl (3) G 9s 8m 2M; 43°22'·01N 08°23'·66W.
Islas Sisargas ☆ Fl (3) 15s 108m **23M**; W 8-sided twr and W bldg; *Siren Mo (S) 30s;* 43°21'·60N 08°50'·78W. AIS, 992242125.
Punta Nariga ☆ Fl (3+1) 20s 53m **22M**; W ○ twr; 43°19'·24N 08°54'·69W.

RÍA DE CORME Y LAXE
Pta del Roncudo ≮ Fl 6s 36m10M; 43°16'·50N 08°59'·47W.

Pta de Laxe ☆ Fl (5) 20s 64m **20M**; W twr; 43°13'·84N 09°00'·65W.
Corme, mole ⚓ Fl (2) R 5s 12m 3M; 43°15'·64N 08°57'·80W.
Laxe, N mole ⚓ Fl G 3s 15m 4M; G ○ twr; 43°13'·37N 08°59'·96W.

CABO VILLANO TO THE PORTUGUESE BORDER

RÍA DE CAMARIÑAS and to PUNTA INSUA

C. Villano ☆ Fl (2) 15s 103m **28M**; 031·5°-228·5°; 8-sided twr, Gy cupola; *Racon M, 35M*; 43°09'·63N 09°12'·69W.
Camariñas ldg lts 081°: Front, Pta Villueira ⚓ Fl 5s 13m 9M; W ○ twr; 43°07'·38N 09°11'·56W. Rear, Pta del Castillo, ⚓ Iso 4s 26m 11M; 079·5°-082·5° (3°); W twr, 612m from front.
Camariñas, outer bkwtr ⚓ Fl R 5s 7m 3M; 43°07'·45N 09°10'·69W.
Pta de la Barca ⚓ Oc 4s 11m 10M; 43°06'·79N 09°13'·19W.
Muxia bkwtr hd ⚓ Fl (4) G 11s 14m 5M; ○ twr with G △ topmark; 43°06'·36N 09°12'·73W; obscured 146°-175° over El Carreiro
Pta de Lago ⚓ Oc (2) WRG 6s 13m, W6M, R/G4M; 029·5°-W-093°-G-107·8°-W-109·1°-R-139·3°-W-213·5°; W twr; 43°06'·60N 09°10'·01W.
Cabo Toriñana ☆ Fl (2+1) 15s 63m **24M**; W ○ twr; *(1·7M SE of ☆)*; 43°03'·19N 09°17'·84W.
Cabo Finisterre ☆ Fl 5s 142m **23M**; 8-sided twr; obscd when brg >149°; *Racon O, 35M*; 42°52'·94N 09°16'·31W.

SENO DE CORCUBIÓN

Lobeira Grande ⚓ Fl (3) 15s 16m 9M; 42°52'·84N 09°11'·12W.
Carrumeiro Chico ⚓ Fl (2) 7s 7m 6M; 42°54'·37N 09°10'·76W.
Cabo Cée ⚓ Fl (5) 13s 25m 7M; Gy twr; 42°54'·98N 09°11'·02W.
Corcubión bkwtr ⚓ Fl (2) R 7s 10m 5M; 42°56'·70N 09°11'·39W.
Pta Insúa ☆ Fl (3) WR 9s 26m; **W15M**, R14M; 020°-R-045°-W-070°-R-090°-W-125°-R-152°-W-020°; 42°46'·28N 09°07'·55W.

RÍA DE MUROS

Pta Queixal ⚓ Fl (2+1) 12s 25m 9M; 42°44'·36N 09°04'·73W.
C. Rebordiño ⚓ Fl (2) R 7s 17m 10M; 42°46'·20N 09°02'·90W.
Muros Fl (4) G 11s 9m 3M; G/W bcn; 42°46'·64N 09°03'·29W.
Portosin bkwtr ⚓ Fl (3) G 9s 8m 5M; 42°45'·94N 08°56'·93W.
Pta Cabeiro ⚓ Oc WR 3s 35m W9M, R6M; 050°-R-054·5°-W-058·5°-R-099·5°-W-189·5°; Gy twr; 42°44'·36N 08°59'·44W.
O Son outer bkwtr ⚓ Fl (2) G 7s 5m 5M; G/W banded ○ twr; 42°43'·73N 08°00'·05W.
Pta Focha ⚓ Fl 5s 27m 4M; ○ twr; 42°41'·98N 09°01'·59W.

RÍA DE AROSA (Selected lights only)

C. Corrubedo ☆ Fl (2+3) WR 20s 31m **15M**; dangerous sector 347°-R-040°; *Racon K*; ○ twr, W cupola; 42°34'·57N 09°05'·39W.
Isla Sálvora ☆ Fl (3+1) 20s 38m **21M**; vis 217°-126°; 42°27'·97N 09°00'·80W. Same twr, ⚓ Fl (3) 20s; 126°-160° danger sector.
Santa Uxia marina bkwtr knuckle ⚓ Fl (2+1) G 15s 8m 3M; G ○ twr +R band; 42°33'·76N 08°59'·27W. 1ca NW ⚓ Fl (2) R 7s 8m 3M.
Isla Rúa ⚓ Fl (2) WR 7s 25m 12M; 121·5°-R-211·5°-W-121·5°; *Racon G*; 42°32'·96N 08°56'·38W.
Pobra do Caramiñal Marina bkwtr ⚓ Fl (2+1) R 15s 3M; R post, G band; 42°36'·33N 08°56'·04W.
Puerto Cabo de Cruz S bkwtr head ⚓ Fl (2) G 7s 3M; G post, W band; 42°36'·87N 08°53'·56W
Rianxo NW bkwtr hd ⚓ Fl (3) G 9s 3M; 42°39'·00N 08°49'·54W.
Vilagarcia, Dir lt 070·5° Fl WRG 6s, **W15M** R/G12M; 0608-G-070°-W-071°-R-073°; 42°36'·41N 08°46'·07W.
N mole ⚓ Fl (3) R 9s 7m 3M; ○ twr; 42°36'·14N 08°46'·37W. I Arosa Pta Caballo ⚓ Fl (4) 11s 12m 10M; 42°34'·34N 08°53'·04W.

PENINSULA DEL GROVE

Roca Pombeiriño ⚓ Fl (2) G 12s 13m 8M; WGW bcn twr; 42°28'·93N 08°56'·83W.
Piedras Negras marina, bkwtr hd ⚓ Fl (4) WR 11s 6m, W4M R3M; 305°-W-315°-R-305; R ○ twr; 42°27'·47N 08°55'·09W.
Bajo Piedra Seca ⚓ Fl (3) R 9s 4m 5M; 42°27'·00N 08°54'·62W.

RÍA DE PONTEVEDRA

Bajo Fagilda ⚓ QR 1s 16M; 42°24'·76N 08°53'·77W.
Bajo Picamillo ⚓ Fl G 5s 10m 8M; twr; 42°24'·25N 08°53'·46W.
Bajo Los Camoucos ⚓ Fl (3) R 9s 12m 8M; 42°23'·71N 08°54'·75W.

Isla Ons ☆ Fl (4) 24s 126m **25M**; 8-sided twr; 42°22'·95N 08°56'·19W.
Almacén pier ⚓ Fl R 4s 7m 2M; R ○ twr; 42°22'·63N 08°55'·77W.
Portonovo mole ⚓ Fl(2) R 7s 10m 5M; 42°23'·65N 08°49'·13W.
Sangenjo marina, ⚓ Fl R 5s 5m 3M; 42°23'·81N 08°48'·06W.
NE side ⚓ Fl G 5s 5m 3M; 42°23'·83N 08°48'·03W.
Cabezo de Morranzán ⚓ Iso R 5s; 42°22'·37N 08°47'·04W.
Rajó mole ⚓ Fl (2) R 8s 9m 3M; R ○ twr; 42°24'·10N 08°45'·26W.
Isla Tambo ⚓ Fl (3) 8s 35m 11M; W ○ twr; 42°24'·48N 08°42'·47W.
Combarro S jetty hd ⚓ Fl (3) R 9s 3M; 42°25'·62N 08°42'·22W.
Rio Lérez mouth ⚓ Fl R 5s, ⚓ Fl G 5s; 42°24'·84N 08°41'·22W.
Aguete ⚓ Fl (4) G 11s 3M; 42°22'·66N 08°44'·20W.
Marina, wavebreak, E end ⚓ Fl (2) G 7s 1M; 42°22'·55N 08°44'·16W.
Bueu N bkwtr ⚓ Fl G 5s 9m 5M; G twr; 42°19'·79N 08°47'·02W.
Beluso N bkwtr ⚓ Fl (4) G 11s 7m 5M; 42°20'·02N 08°47'·90W.
Cabeza de la Mourisca ⚓ Fl (2) G 7s 10m 5M; 42°20'·90N 08°49'·17W.
Aldán jetty ⚓ Fl (3) R 9s 7m 5M; R twr; 42°16'·95N 08°49'·38W.
Punta Couso ⚓ Fl (3) WG 10·5s 18m, W10M, G8M; 060°-W-190°-G-000°; 42°18'·56N 08°51'·33W.

ISLAS CÍES

Isla del Norte, Pta de Monte Agudo ⚓ Fl G 5s 24m 10M; 42°14'·60N 08°54'·19W.
Monte Faro ☆ Fl (2) 8s 186m **22M**; obsc'd 315°-016·5° over Bajos de Los Castros and Forcados; ○ twr; 42°12'·85N 08°54'·90W. 992242128.
Isla del Faro, Pta Canabal ⚓ Fl (3) 20s 63m 9M; W twr; 42°12'·73N 08°54'·77W.
Isla de San Martin, Cabo Vicos ⚓ Fl (3) R 9s 93m 10M; obsc'd 108°-210·5°; W twr; 42°11'·51N 08°53'·48W.
Islote Boeiro ⚓ Fl (2) R 8s 22m 5M; W twr; 42°10'·75N 08°54'·61W.

RÍA DE VIGO

N Chan, outer ldg lts 129°, both 090°-180°; W ○ twrs: Front, Cabo del Home ⚓ Fl 3s 36m 9M; 42°15'·15N 08°52'·37W. Rear, Pta Subrido ⚓ Oc 6s 53m 11M; 815m from front,.
Punta Robaleira ⚓ Fl (2) WR 7·5s 25m, W11M, R9M; 300·5°-W-321·5°-R-090°; 115·5°-R-170·5°; R twr; *Racon C*; 42°15'·00N 08°52'·38W (285m S of Cabo del Home ldg lt).
Cangas NE mole ⚓ Fl R 5s 8m 5M; R ○ twr; 42°15'·61N 08°46'·83W.
S Chan ldg lts 069·3°, both R twrs, W bands; vis 066·3°-072·3°. **Front** ☆ Iso 2s 16m **18M**; *Horn Mo (V) 60s*; *Racon B, 22M*; 42°11'·11N 08°48'·88W. **Rear** Oc 4s 48m **18M**, 660m from front.
Bouzas, NE head ⚓ Fl (2+1) G 15s 9m 5M; 42°14'·07N 08°44'·67W.
Marina Dávila Sport ⚓ Fl (2) G 7s 10m 5M; 42°13'·99N 08°44'·42W.
Muelle del Berbés, N head ⚓ Fl (3) G 9s 9m 5M; G twr; 42°14'·43N 08°44'·01W.
Muelle de Transatlánticos, NE head ⚓ Fl (4) G 11s 5M; G post; 42°14'·59N 08°43'·56W.
Vigo marina ent, QG and QR 9m 5M; 42°14'·57N 08°43'·41W.
Pta Lagoa marina, bkwtr hd, ⚓ QG 1s 1m 2M, 42°15'·50N 08°42'·30W.
La Guia ☆ Oc (2+1) 20s 37m **15M**; W twr; 42°15'·57N 08°42'·14W.

BAYONA to the PORTUGUESE BORDER

Las Serralleiras ⚓ Q (9) 15s 4M; 42°09'·23N 08°53'·35W.
Pta Lameda ⚓ Fl (2) G 8s 27m 5M; W twr; 42°09'·39N 08°50'·98W.
Ldg/Dir lts 084°, both W twrs: Front, Cabezo de San Juan ⚓ Fl 6s 8m 10M; 42°08'·13N 08°50'·10W. Rear, **Panjón Dir** ☆ 084°, Oc WRG 4s **W18M, R15M**, G14M; 079°-G-083°-W-085°-R-088°; 42°08'·34N 08°48'·84W.
Dique de Abrigo ⚓ QG 10m 5M; G/W twr; 42°07'·49N 08°50'·58W.
C. Silleiro ☆ Fl (2+1) 15s 84m **24M**; W 8-sided twr, R bands; 42°06'·17N 08°53'·78W. ⚓ Q; 42°07'·29N 08°54'·73W
La Guardia N bkwtr ⚓ Fl (2) R 7s 11m 5M; R twr; *Siren Mo (L) 30s*; 41°54'·07N 08°52'·90W. S bkwtr ⚓ Fl (2) G 7s 11m 5M; G/W twr.

9.23.4 PASSAGE INFORMATION

More information is threaded between successive hbrs in this Area.
Bibliography: *NW Spain & Portugal* (Fernhurst/Jens); *South Biscay* (Imray/RCC/Pickard); *Cruising Galicia* (Imray/Rojas/Bailey); *Atlantic Spain and Portugal* (Imray/RCC/Walker/Buchanan); *Bay of Biscay Pilot* (NP 22); *W Coasts of Spain and Portugal Pilot* (NP 67).

BAY OF BISCAY: WIND, WEATHER AND SEA

NE'lies prevail in sea area Finisterre in summer and gales may occur twice monthly, although forecast more frequently. Atlantic lows can bring W'ly spells at any time. Wind direction and speed often varies daily, but SE or S winds are rare. Sea and land breezes can be well developed in the summer. Off N Spain *Galernas,* dangerous squally NW winds, can blow with little warning. Coastal winds are intensified by the Cordillera Cantábrica (2615m mountains).

Despite its reputation, the S part of the Bay is often warm and settled in summer when the Azores high and Spanish heat low are the dominant weather features. Rainfall is moderate, increasing in the SE, with thunder more frequent. Sea fog, which often precedes strong winds, occurs May-Oct, but is less common in winter.

▶*There are no tidal stream atlases. Streams are weak offshore, but can be strong in narrow channels and off headlands.*◀

Surface current much depends on wind: in summer it sets SE ½–¾kn, towards the SE corner of Bay of Biscay, thence W along N coast of Spain. When crossing the Bay, allow for some set to the E, particularly after strong west winds.

CROSSING THE BAY OF BISCAY

Distances across the Bay are tabulated at 9.0.8. Leaving the English Channel, the track (208°/362M) from Ushant ITZ to Cabo Villano (AC 1104) lies undesirably close to busy shipping lanes, but a track (213°/287M) from, say, harbours between **Loctudy** and **Concarneau**, to Cabo Ortegal is substantially offset to the SE of the shipping route. Cabo Ortegal as a landfall permits onward passage to **La Coruña**. Or, if conditions for rounding Ortegal are bad (as is not unknown) **Ría de Viveiro** is a safe refuge. From Scilly or Ireland the direct track lies in about 7°-8°W.

Within the Bay itself, sailing down the French coast is attractive and allows a 200M passage from **La Rochelle** to **Santander**. Landes missile range between Pointe de Grave and Capbreton extends up to 45M offshore. It may inhibit coastal passages, but is usually inactive at night, w/ends and August.

The Continental Shelf, where depths plummet from 150m to over 4000m in only 30M, can cause dangerous seas in swell and bad weather. From a position about 60M SW of Ile de Sein it trends SE to the Franco/Spanish border as shown on AC 1104.

Atlantic swell runs in mainly from W or NW. It is not a problem well offshore, except over the Continental Shelf in bad weather. Considerable rolling is probable; other yachts or even small ships may be lost to view in the troughs. Closer inshore swell will exacerbate the sea state and render entry to, and exit from, lee shore harbours dangerous. With a 2m swell running, crossing a bar with 4m depth over it and breaking seas would be foolhardy. In winter some harbours may be closed for long periods.

FRENCH BORDER TO SANTANDER

The N coast of Spain is bold and rocky (AC 1290, 1291, 1292). In clear visibility the peaks of the **Cordillera Cantábrica** may be seen from well offshore. The coast is mostly steep-to, but it is best to keep 2-3M offshore (beyond the 50m contour) to avoid short, steep seas breaking on isolated shoals. Many major lights are sited high, and often obscured by low cloud/mist in onshore weather. Most small rivers have bars and none are navigable far inland.

From Hendaye to Santander harbours are closely spaced compared with further west. Ports of refuge in NW gales are: **Pasajes** (no yacht facilities)**, Guetaria, Bermeo, Bilbao, Castro Urdiales** and **Santander**. San Sebastián is an attractive ⚓ but exposed to the NW. **Bilbao** and **Santander** are ferry ports and major cities with marinas. Fishing hbrs with � or ⚓ include: **Guetaria, Motrico, Bermeo, Lequeitio, Elanchove** and **Laredo**.

9.23.5 SPECIAL NOTES FOR SPAIN

Regions/Provinces: Spain is divided into 17 autonomous regions, ie in Area 23, Guipuzcoa (Basque Country) Cantábrica, Asturias and Galicia. Most regions are sub-divided into provinces, eg in Galicia: Lugo, La Coruña, Pontevedra and Orense (inland). The province is shown below the name of each main port.

Language: This Almanac recognises regional differences (eg Basque, Gallego) but in the interests of standardisation uses the spelling of Castilian Spanish where practicable.

Spanish charts (SC). Contact Instituto Hidrográfico de la Marina, Plaza San Severiano 3, 11007 Cádiz, ☎(956) 599 414; www. armada.mde.es/ihm or order from chart agents in Madrid, Bilbao, Santander, Gijon, La Coruña, Villagarcia de Arosa and Vigo.

Currency is the Euro €. VAT (IVA) 21%.

Courtesy ensign: E of Bilbao it may be politic to fly the Basque flag, Fig 5 (9) or none at all. The Spanish ensign may give offence, especially if flown above the autonomous Basque flag.

Fuel: Diesel (*Gasoleo*) is of two types: *Gasoleo A* is taxed and supplied to yachts from marina pumps or by cans from a filling station ashore. *Gasoleo B* is untaxed and usually only available to FVs from a pump on the fish quay.

Standard Time in Spain is UT –1, but in Portugal it is UT.

Spanish secondary ports referenced to Lisboa. Time differences for those Spanish ports which are referenced to Lisboa (all south of Ría de Muros) when applied to the printed times of HW and LW for Lisboa, give HW and LW times in Spanish Standard Time. DST is the only correction required.

Telephone: To call Spain from UK dial 00 34, then the area code, followed by the 6 digit ☎ number. Internally dial the area code plus the ☎ number. To call UK from Spain, dial 07 44, then area code less the initial 0, plus the ☎ number.

Emergencies: ☎900 202 202 for Maritime Rescue. Fire, Police and Ambulance call 112. *Rioja Cruz* (Red Cross) operate ⑬s.

Public Holidays: Jan 1, 6; Good Friday; 1 May (Labour Day); June 11 (Corpus Christi); Aug 15 (Assumption); Oct 12 (National Day); Nov 1 (All Saints Day); Dec 6, 8 (Immaculate Conception) 25.

Representation: Spanish Tourist Office, 64 North Row, London W1K 7DE; ☎020 7317 2011. info.londres@tourspain.es www.spain.info

British Embassy, Paseo de la Castellana 259D, 28046 Madrid; ☎(91) 334 2194. Consular Section ☎(902) 109 356.

Buoyage: Preferred channel buoys [RGR, Fl (2+1) R and GRG, Fl (2+1) G] are quite widely used; see Chapter 3.

Documents: (See Chapter 2 Cruising Formalities) Spain, although an EU member, still checks papers. This can be a time-consuming, repetitive and inescapable process. The only palliatives are courtesy, patience and good humour. Organise your documents to include:

Personal – Passports; crew list, ideally on headed paper with the yacht's rubber stamp, giving DoB, passport nos, where joined/intended departure. Certificate of Competence (Yachtmaster Offshore, ICC/HOCC etc). Radio Operator's certificate. EHIC is advised for medical treatment.

Yacht – Registration certificate, Part 1 or SSR. Proof of VAT status. Marine insurance. Ship's Radio licence. Itinerary and ship's log.

Access: Ferries from UK to Santander and Bilbao. Flights from UK to Bilbao, Santiago de Compostela and Madrid. Internal flights via Madrid to Asturias, La Coruña and Vigo. Buses are mostly good and trains adequate, except in more remote regions.

SAR: A Spanish Government agency SASEMAR is responsible for the all-weather lifeboats, tugs, search helicopters and aircraft. Irrespective of value, assisted vessels <20m LOA are charged €350/hour including towing; 20m-30m LOA €749/hour (2015).

9.23.6 FUENTERRABÍA (Hondarribia)

Guipúzcoa 43°22'·56N 01°47'·52W ✴✴✴✦✦✿✿✿

CHARTS AC 1292, 1170, 1171; SC 391, 3910; SHOM 6556, 6558, 6786; Navi 555. *Spanish* and French names for the same feature: *Rada de Higuer* = Baie de Fontarabie or de Figuier; *Ría de Fuenterrabía* = La Bidassaoa (river).

TIDES HW +0450 on Dover (UT); ML 2·3m
Interpolate between **SOCOA** (St Jean de Luz) and **PASAJES**.

SHELTER Excellent in marina, access H24. S of Cabo Higuer, a FV hbr, 2-3m, is a port of refuge. Good ⚓ in river off Pta Roca.

NAVIGATION WPT 43°23'·87N 01°46'·44W, 212°/1·15M to W bkwtr hd. Beware Les Briquets 8ca N of Pte Ste Anne at E end of the Baie. Near centre of B, keep clear of Bajo Iruarri. River ent (6·5-1·3m) is easy except in heavy N'ly swell; sp ebb is very strong.
The neutral zone between France and Spain is delineated by the alignment of 4 pairs of beacons; see Hendaye (Area 22).

LIGHTS AND MARKS Cabo Higuer lt ho, a square stone twr with R top, is conspic at W side of bay, as is the high dorsal ridge which runs down to the headland. A red-roofed castle is prominent 2 cables SSE of Cabo Higuer lt ho. La Rhune, an 898m high conical peak, bears 120°/8M from the marina.

COMMUNICATIONS (Code 943) MRSC (942) 213030; Emergency 112; Marina VHF Ch 09 (H24); Club Nautico ☎642788.

FACILITIES Marina ☎641711 (3m) 595 + 25 Ⓥ; berth on pontoon G (1st to port). Pontoons A-E are NW of the central pontoon F; G-J are to the SE. €2.50 (30 days pa max); max LOA 16m. Boats > 16m LOA should ⚓ in the Bay or in Baie de Chingoudy. 🏃🚻 (UK cr/cards) Ⓞ ⚒ 🔧(35t) 🔧(3t).

Town ✕ 🏠 🍴 🛢✉Ⓑ ⇌✈ (Fuenterrabía or Biarritz). Ferries, local: Fuenterrabía to Hendaye. UK: Santander, Bilbao.

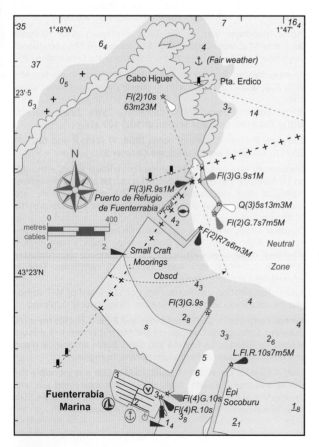

9.23.7 PASAJES (Pasaia)

Guipúzcoa 43°20'·23N 01°55'·82W ✴✴✴✦✦✿

CHARTS AC 1292, 1171, 1157; SC 391/2, 3911; SHOM 6375, 6558

TIDES See San Sebastian.

SHELTER A port of refuge (3·5M E of San Sebastián, 7M from France). Yachts may ⚓ as shown on the chartlet: the bay NE of Dir lt is fair weather only; the other two ⚓s are outside local small craft moorings, as space allows; the N'ly is subject to wash.

NAVIGATION WPT 43°21'·15N 01°56'·12W, SWM buoy [Mo (A) 7s], 161°/1·0M to hbr ent, a 200m wide cleft in spectacular cliffs. Appr on 154·8° transit of Dir lt, Oc (2) WRG 12s (Racon M) with ldg lts: front Q, rear Oc 3s; all on the W cliff. Inner ldg line 144·3° on Iso 4s, Rear Q 1s on E shore.

Banks, 3·4 and 1·5m W and E of ent, are marked by conspic R and G 18m high lt twrs. Inner chan, dredged 10m, is only 100m wide, but well marked/lit; do not impede large vessels/FVs; be alert for departing ships.

LIGHTS AND MARKS Observe and obey IPTS located high on the E cliff at Atalya de Pasajes (Racon K). Sigs 1, 2 and 3 are shown to seaward and inwards; the ent is blind and narrow for big ships.

COMMUNICATIONS (Code 943) MRSC (942) 213030; Emergency 112; Brit Consul (902)109356; HM 351816 ppasajes@sarenet.es; Port VHF Ch 09 (H24).

FACILITIES Pasajes is a busy FV and commercial hbr with no yacht facilities. ✕ 🏠 Ⓔ ⚒. **Town** ✕ 🏠 🍴 ✉ Ⓑ.

9.23.8 SAN SEBASTIÁN (Donostia)

Guipúzcoa **43°19'·48N 01°59'·79W** ❀❀❀⚓♨♨❀❀❀

CHARTS AC 1292, 1171; SC 391, 392, 3910; SHOM 6558, 6786

TIDES

Standard Port POINTE DE GRAVE (←—)

Times				Height (metres)			
High Water		Low Water		MHWS	MHWN	MLWN	MLWS
0000	0600	0000	0500	5·3	4·4	2·1	1·0
1200	1800	1200	1700				
Differences SAN SEBASTIÁN							
–0110	–0030	–0040	–0020	–1·1	–1·2	–0·5	–0·4
PASAJES							
–0050	–0030	–0045	–0015	–1·1	–1·3	–0·5	–0·5

SHELTER Fair; the ⚓ becomes lively or dangerous in any NW/N'lies when heavy swell enters. ⚓ S of Isla Santa Clara or SW of YC, in both cases clear of moorings and with ⚓ buoyed.

NAVIGATION WPT 43°19'·85N 02°00'·00W, 158°/0·4M to hbr ent (between Isla de Santa Clara and Monte Urgull). Avoid La Bancha on which the sea breaks in swell and heavy weather. Open up the bay before standing in.

LIGHTS AND MARKS Monte Urgull (huge statue of Virgin Mary) is prominent from all directions; from E it looks like an island. Monte Igueldo on W side of ent is only slightly less obvious. Isla de Santa Clara is lower and inconspicuous; not seen until the nearer appr. Ldg marks (grey masts) are hard to see, but a large R and W Palladian-styled villa is conspic on 158°. Ldg lts, although intense, may be masked by shore lts.

COMMUNICATIONS (Code 943) MRSC (942) 213030; Emergency 112; Met 274030; Auto 906 365320; Brit Consul (902)109356; Real Club Nautico VHF Ch 09 (H24).

FACILITIES for visiting boats do not match the city's elegance. **Real Club Náutico** ☎423574, www.rcnss.com; ⚓(best to pre-arrange) ✕ ⌂ Ice. YC launch will meet and direct (Ch 09) to a vacant ⚓.

Dársena de la Concha is overcrowded, with limited space on ⓥ pontoon dead ahead of ent; W part shelves quite steeply; inner basins are locals/FVs only.

City ✕ ⌂ 🖳 🛒 🛄 ⊠ Ⓑ Ⓗ. ⇌ & ✈ Fuenterrabía (25km). ✈ also at Bilbao (92km) and ferry to UK.

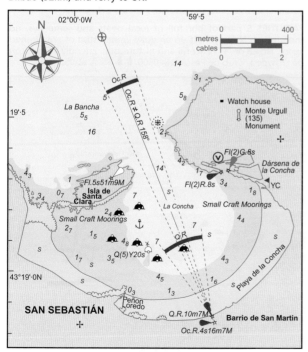

MINOR HARBOUR 3M ESE OF GUETARIA

RÍA DE ORIO Guipúzcoa **43°17'·49N 02°07'·98W**. AC 1292; SC 944, 303A; SHOM 6786, 6379. Tides: MHWS 4·1m, MLWS 0·5m; MHWN 3·2m, MLWN 1·4m. Interpolate between San Sebastián & Guetaria. WPT 43°17'·97N 02°07'·95W, 180°/5ca to W bkwtr between Mte Talai (360m) to the E and Mte Amesti (188m) to the W. Small marina (2-3m) 5ca upriver from the ent, on the east bank and just N of the road bridge, 17·5m clearance. Best water to the E. Good shelter, but exposed to N'lies when swell may work in. Fair weather tempy ⚓ outside to the E off San Juan beach in 5m.

Best appr just before HW. See Lights, buoys and waypoints for the two entrance lts. A first visit at night is not advised despite a total of 17 nav lts.

HM ☎(943) 249400. www.ekpsa.com VHF 09. kirolportua@ekpsa.com 300 �_ €0.71 x LOA x beam (m)/per day (30 days max); max LOA 12m. Facilities: ⚓ ⌂ 🛠 🛢 (H24) ⌂ 🔧(32t) ⚓. **Town** 800m south. ✕ ⌂ 🛒.

9.23.9 GUETARIA (Getaria)

Guipúzcoa **43°18'·23N 02°11'·93W** ❀❀❀⚓⚓❀❀❀

CHARTS AC 1292, 1171; SC 392, 3921; SHOM 6379, 6786

TIDES

Standard Port POINTE DE GRAVE (←—)

Times				Height (metres)			
High Water		Low Water		MHWS	MHWN	MLWN	MLWS
0000	0600	0000	0500	5·3	4·4	2·1	1·0
1200	1800	1200	1700				
Differences GUETARIA							
–0110	–0030	–0040	–0020	–0·9	–1·0	–0·5	–0·4

SHELTER The only good shelter in S-NW winds between Pasajes and Bilbao. No ⚓ in hbr which is generally foul. In fair weather ⚓ on sand or pick up a buoy close S of hbr ent, but exposed to E'lies.

NAVIGATION WPT 43°18'·90N 02°11'·00W, 225°/9½ca to hbr ent. Appr is straightforward day/night in fair visibility.

LIGHTS AND MARKS Lts as chartlet & Lts, bys and wypts. The mouse-like profile of Isla de San Anton is distinctive; lt ho on its N side.

COMMUNICATIONS (Code 943) MRSC (942) 213030; Emergency 112; Met 906 365 320; Auto 906 365365; ⊖ via Marina; Brit Consul (902)109356; HM VHF Ch 09 (H24).

FACILITIES Marina ☎580959 thage@infonegocio.com pontoons A-F, ⓥ welcome but space limited, no rafting allowed, 300 ⌒; €2.40. max LOA 14m. 🛢 🛒 🛠 ⚓ ⚓ 🔧(30t) ⚓(5t) ⚓. FVs use N side of outer hbr and all the inner hbr.

Club Náutico y Pesca ☎140201, ✕ ⌂. **Town** ✕ ⌂ 🖳 🛒 🛢 ⊠ ⇌ ✈ Bilbao (UK ferry).

9.23.10 ZUMAYA (Zumaia)

Guipúzcoa 43°18'·36N 02°14'·92W (Hbr ent) ✺✺✺🤿🤿🤿🏵🏵🏵

CHARTS AC 1292; SC 392, 3921; SHOM 6379, 6786

TIDES As for Guetaria.

SHELTER Good in the marina, but in strong onshore winds the appr could be difficult. In E'lies a fair weather ‡ off Pta Izustarri, clear of rks at sandy beach 3ca NE of hbr ent. No ‡ in hbr.

NAVIGATION WPT 43°19'·85N 02°14'·51W (8ca W of ODAS buoy, Fl (5) Y 20s) 190°/1·5M to hbr ent. Easy appr from NNE, but keep clear of rocks W of Pta Iruarriaundieta and S of Pta Izustarri. The access chan, least depth 0·1m as shown on SC 3921, is said to be dredged 3m. Best water is to stbd; near LW keep 10m off the NW bkwtr and river bank. The Rio Urola flows through the hbr.

LIGHTS AND MARKS Lt ho, Oc (1+3) 12s, is a white 8-sided tower with blue cupola. Lts as chartlet and Lts, buoys and waypoints.

COMMUNICATIONS (Code 943) MRSC (942) 213030; Emergency112; Brit Consul (902)109356; Marina VHF Ch 09 (0900-1300 & 1400-1930 daily)

FACILITIES Marina Urola (2·5m) ☎860938 www.marinaurola.com info@marinaurola.com Reception pontoon in front of marina. 515 inc🅥 €3.00; max LOA 15m. ⚓ ✕ 🅿 ⚒ ⚓ 🖧 🖳(35t) ▬.

Town ✕ 🍴 🛒 🛢 ✉ ⇌ ✈ Bilbao (UK ferry).

MINOR HARBOUR 6M W OF ZUMAYA

MOTRICO (Mutriku) Guipúzcoa **43°18'·5N 02°22'·9W**. Tides: interpolate between Guetaria & Lequeitio. AC 1292; SC 392, 3922; SHOM 6379. Small FV hbr (4m) in rky inlet. Good shelter from NE winds due to a large new outer bkwtr, but swell may still work in. Readers report that berthing is limited, enter the inner hbr and tie up alongside the wall to port. Further reports welcome.

WPT 43°18'·76N 02°21'·99W, on the 245° WRG Oc. dir lt, 7ca past outer bkwtr which must be left to stbd. The ent is easily seen from E, but from W open hbr & town before approaching. Lts as chartlet.

‡ inside the new bkwtr in 5m; or possible ◯ at inner end of SE quay if FVs are away. HM ☎(943) 603204. www.mutriku.net udala@mutriku.net Facilities: (FV quay): 🛢 🔧 🅿 ⚓ Ice 🖳(5t) ▬.
Town ✕ 🍴 🛒 🛒.

9.23.11 LEQUEITIO (Lekeitio)

Vizcaya 43°22'·00N 02°29'·91W ✺✺✺🤿🤿🏵🏵

CHARTS AC 1292, 1171; SC 392, 393; SHOM 6379, 5009

TIDES
Standard Port POINTE DE GRAVE (←)

Times				Height (metres)			
High Water		Low Water		MHWS	MHWN	MLWN	MLWS
0000	0600	0000	0500	5·3	4·4	2·1	1·0
1200	1800	1200	1700				
Differences LEQUEITIO							
−0115	−0035	−0045	−0025	−1·1	−1·2	−0·5	−0·4

SHELTER Good. ‡ in about 3m or pick up a vacant buoy at the S end of basin where there are extensive small craft trots. In settled weather ‡ further out, to E or W of Isla de San Nicolas.

NAVIGATION WPT 43°22'·25N 02°29'·68W, 214°/3ca to abeam Pta Amandarri lt. Avoid rky shoals close E of 213·8° ldg line and ruins of old bkwtr (Fl (2) R 8s) further S.

LIGHTS AND MARKS Conspic daymarks: Cabo de Santa Catalina 0·8M to the NW. Isla de San Nicolas is rky, steep-sided, wooded and linked to mainland by a part-drying causeway. Church dome SSW of hbr. Palacio Uribarren, the rear 213·8° ldg mark.

COMMUNICATIONS (Code 946) MRSC (942) 213030; Emergency 112; Brit Consul (902)109356; HM 243324, VHF Ch 09.

FACILITIES S part of hbr full of local boats and some FVs, no mooring on N quay. Raft up on quay immed stbd of ent, limited space. No 🚾 or 🚿. Peaceful and picturesque. Friendly YC but can be crowded. Club de Pesca ☎840500. 🛢 ⚓ 🍴 🛒 🅿 ▬ ⚓.

9.23.12 ELANCHOVE (Elantxobe)

Vizcaya **43°24'·23N 02°38'·21W** ❀❀◉⚓♣♣♣♣

CHARTS AC 1292, 1171; SC 393; SHOM 6380, 6991

TIDES Interpolate between Lequeitio and Bermeo.
Standard Port POINTE DE GRAVE (←)

Times				Height (metres)			
High Water		Low Water		MHWS	MHWN	MLWN	MLWS
0000	0600	0000	0500	5·3	4·4	2·1	1·0
1200	1800	1200	1700				
Differences BERMEO							
−0055	−0015	−0025	−0005	−0·7	−0·7	−0·5	−0·4

SHELTER Sheltered from W in lee of Cabo Ogoño, but exposed to N/E. In calm weather temporary ⚓ to E of hbr ent in 5m.

NAVIGATION WPT 43°24'·67N 02°37'·61W, 225°/6ca to N mole hd. From the W, hbr is not visible until bearing >205°.

LIGHTS AND MARKS The S Mole, F WR lt shows Red (315°-000°) over inshore dangers to the SE. Cabo Machichaco, Fl 7s, is 6M WNW.

COMMUNICATIONS (Code 944) MRSC (942) 213030; Emergency

112; Brit Consul (902)109356; HM 881323.

FACILITIES Tiny hbr in spectacular setting at foot of steep 100m cliffs. The N part of outer hbr and all the inner hbr dry. At ent turn hard port into S part of outer hbr to pick up buoy in about 3m or ⚓ with tripping line as the bottom may well be foul. ◯ on S mole is feasible, but a 1m wide underwater ledge requires a holding-off line to kedge ⚓.

Village ⚓ ✕ ▯ Basic ▨; Bus to Bilbao for ⇌ ✈. 15km south by road is Guernica, the ancient sacred city of the Basques.

HARBOUR 2·7M SE OF CABO MACHICHACO
BERMEO 43°25'·35N 02°42'·57W. Tides: see Elanchove. AC 1292, 1172; SC 393, 3931; SHOM 6380, 6991. A busy commercial & FV port, well sheltered from the N, but swell enters in E'lies.

Hbr is between the wide drying estuary of Rio Mundaca and conspic Cabo Machichaco to NW. Appr on 218° in W sector (204°-232°) of Rosape Dir lt to clear unlit Isla de Izaro, 1M ENE of hbr and the N mole.

NW corner of outer harbour taken up with fore/aft moorings for locals between buoys and piles, poss ◯ in 3-4m on pier with FG lt on chartlet. Inner hbr, **Marina Bermeo** (2m) mostly locals Ⓥ limited, €0·76 per m². VHF Ch 09,16. HM ☎(946) 18 64 45 or mobile ☎908 87 33 30. www.euskadi.net/portuak/ ⚓ ♠ ♦ ﹏◗(12t) ⚒ ✕ Ice. **Town** ✕ ▯ ﹏ ≞ ▮ ⌂ Ⓔ ✉ Ⓑ. ⇌ to Guernica. Reports welcome.

9.23.13 BILBAO (Bilbo)

Vizcaya **43°22'·73N 03°04'·87W** (outer ent) ❀❀◉⚓♦♦♣♣♣

CHARTS AC 1292, 1174, 1173; SC 394, 394A, 3941; SHOM 6774, 6991

TIDES
Standard Port POINTE DE GRAVE (←); ML 2·4.

Times				Height (metres)			
High Water		Low Water		MHWS	MHWN	MLWN	MLWS
0000	0600	0000	0500	5·3	4·4	2·1	1·0
1200	1800	1200	1700				
Differences ABRA DE BILBAO							
−0125	−0045	−0055	−0035	−1·1	−1·2	−0·5	−0·4
PORTUGALETE (INNER HARBOUR)							
−0100	−0020	−0030	−0010	−0·6	−1·2	−0·2	−0·6

SHELTER Excellent in both marinas. ⚓ in 3-5m between these two havens which are adjacent to a pleasant suburb.

NAVIGATION WPT 43°23'·17N 03°05'·21W, 150°/5ca to outer hbr ent. This is formed by a W bkwtr extending 1·25M NE from Punta Lucero, a 308m high unlit headland on the W side. The submerged E bkwtr extends WNW for almost 2M from Punta Galea, Fl (3) 8s, on the E side of El Abra. It is marked by 3 SPM lt buoys, but can be crossed with caution; approx 7m depth was found close off Pta Galea. A mini-TSS is aligned 150°/330° 1·6M toward the outer hbr ent.

From the outer ent follow the buoyed fairway/TSS 120°/2M and 143°/1M across El Abra (= outer hbr/roads) to Dique de Santurtzi. thence 195°/3 cables to the inner hbr. Yachts should stay just outside the fairway buoys and keep clear of ships using the TSS. After passing Contradique de Algorta, Fl R 5s, round the head,

Fl (2+1) R 10s, of a new jetty (for cruise liners) extending W from Getxo bkwtr. When clear turn SE for Getxo Marina and SSE across the inner hbr to Las Arenas yacht hbr.

LIGHTS AND MARKS 5 wind turbines (inc 3 lit) near the head of W bkwtr are prominent. The Port Control bldg/twr is conspic at the SE end of Dique de Santurtzi. 2 power stn chimneys (200m high) with R/W bands are conspic 1M SW of Port Control. Glare from industrial plants may be seen from afar at night. Caution: Flood lights on the beach at Algorta (43°21'·34N 03°01'·08W) shine seaward and are very distracting when inbound at night, up to 0100LT.

COMMUNICATIONS (Code 944) MRCC 839411; Met via YC; Emergency 112; ⊖ 234700; Brit Consul (902)109356; Ⓗ 903100; Port HM 241416; Port Control VHF Ch 05, **12**, 16; Marinas VHF Ch 09.

FACILITIES **Getxo Marina** ☎912367 www.getxokaia.com getxokaia@getxokaia.com (1·9-2·9m) on S side of Contramuelle de Algorta. 827 ◯ incⓋ €5.00; max LOA 18m, southern pontoon J only, prone to wash, ♠♦ (H24) ⛽⚒ ⌂ Ⓔ ✕⚓ ▯(45t) ◗(5t). Metro line 1 for downtown, board at Gobela/Neguri stations.

Las Arenas Yacht hbr ☎637600; www.rcmarsc.es puerto@rcmarsc.es, near the mouth of R. Bilbao in 1-3m, is run jointly by **Real Club Maritimo del Abra** and **Real Sporting Club**. 250 + 50Ⓥ €3.00 + ⊞ €3.44/day. ♠♠♦ ✕ ﹏ ﹏ ⚒ ✕⚓ ⌂ Ⓔ ▯(35t) ◗(5t). Metro line 1 for downtown, board at Areeta station.

City centre is 10km SE of marinas. There are no yacht berths up-river. Bilbao is a major industrial city, but also a leading cultural and commercial centre. The Guggenheim museum is close to the Old Quarter. Main tourist office ☎910800.

All facilities; ⇌ Sondika ✈ is 10km NE of the city centre. P&O ferry to/from Portsmouth, approx 33 hrs, every 3 days.

BILBAO *continued*

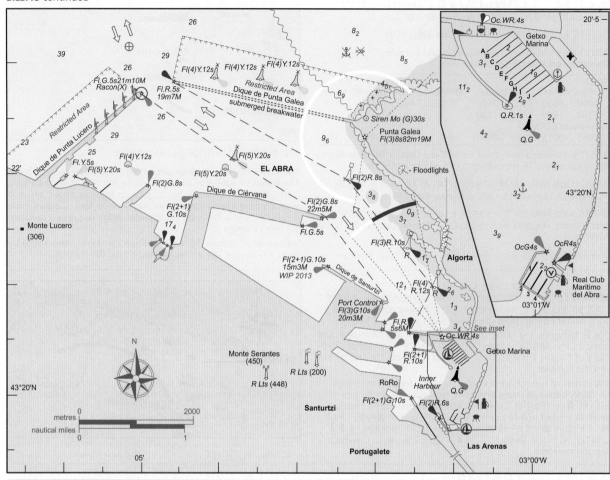

9.23.14 CASTRO URDIALES

Cantabria 43°22'·80N 03°12'·60W ⚓⚓⚓⚓♦♦♦♦♦♦

CHARTS AC 1292, 1174; SC 394, 394A; SHOM 3542, 6991

TIDES Differences, see Laredo facing page.

SHELTER Good, except in N-E winds when swell enters. A few ⚓s or ⚓ in 9m on mud to seaward of the 6 lines of mooring trots which fill the NE part of hbr.

NAVIGATION WPT 43°23'·55N 03°11'·69W, 222°/1·0M to hbr ent, as under title. N mole lt ≠ S mole lt form an approx 222° ldg line. The approach is straightforward with no hazards. In the hbr use the fairway to avoid extensive mooring trots.

LIGHTS AND MARKS From the NW, town and hbr are obsc'd until rounding Pta del Rabanal with its conspic cemetery. The ✠, lt ho and castle are easily identified on Santa Ana which dominates the hbr. Lts as chartlet and Lights, buoys and waypoints.

COMMUNICATIONS (Code 942) MRCC (944) 839411; Emergency 112; CG/🄻🄱 900 202 202; Met via HM (below); ⊜ 861146; Brit Consul (902)109356; HM 861585; YC VHF Ch 09, call *Club Náutico*.

FACILITIES 150+ private moorings in the outer hbr; FVs fill inner hbr. ☎ on N mole possible if calm and no swell. Club ☎861585. www.cncu.es cnautico@cncu.es; ask at club for any available moorings; ⚓ free, but €15.00 YC fee includes use of YC launch *Blancona* (Ch 09) & club facilities. Space for tenders at YC limited; ➤🄑🛠. **YC** ✕🍴 ☎861234 to reserve a table.

Town ✕🛒🍺 ✉ Ⓑ 🏦 📞, ✈ Bilbao.

9.23.15 LAREDO

Cantabria **43°25'·08N 03°25'·25W (ent)** ❀❀♎♎♎✿✿✿

For Santoña 283·5°/1·3M to the narrows N abeam of Pta del Pasaje. The least charted depth on the ldg line is 3·5m, but expect less due to silting. The sandy estuary is exposed to E and S. A shallow chan runs 2·3M S to Colindres.

LIGHTS AND MARKS Santoña 283·5° ldg marks are hard to see. Monte Ganzo, 374m conspic mountain on N side is 1.2M NW of the Wpt. Pta del Pescador lt ho is 1M N of Monte Ganzo.

CHARTS AC 1292, 1171; SC 394, 3942; SHOM 3542

TIDES Standard Port POINTE DE GRAVE (←)

Times				Height (metres)			
High Water		Low Water		MHWS	MHWN	MLWN	MLWS
0000	0600	0000	0500	5·3	4·4	2·1	1·0
1200	1800	1200	1700				
Differences CASTRO URDIALES							
–0040	–0120	–0110	–0020	–1·3	–1·5	–0·6	–0·6
RIA DE SANTOÑA (03°28'W)							
–0005	–0045	–0035	+0015	–0·6	–1·2	–0·3	–0·7

SHELTER Good in marina, also W of Pta del Pasaje (CN de Laredo has ⚓ sheltered from E'lies).

NAVIGATION WPT 43°25'·97N 03°25'·47W. For Laredo, 170°/9ca to ent. Caution due to extensive fish havens N of Laredo ent.

COMMUNICATIONS (Code 942) MRCC (944) 839411; Emergency 112; Brit Consul (902)109356. Marina, VHF Ch 09.

FACILITIES Marina de Laredo, ☎606720. www.marinadelaredo.com info@marinadelaredo.es access 24 hours, all weather and tides, 857 ⌂ inc♥ max LOA 20m, €4.00. Pontoon A at north end, or pontoon G south end by ⛽. ♠ ♣ ✕ 🛒 🚿 🏪 free wi-fi ♨ 🔧(50t).

Laredo YC, ☎605812/16, www.rcnlaredo.es info@rcnlaredo.es W of Pta del Pasaje, a few ⚓s or ⚓ in 2m in area marked by 4 Y buoys, Fl Y 2s. LW landing on jetty/pontoons. ✕ 🛒 ⛽ ⚓ 🔧(6t).

Village: Several 🛒 on outskirts, ✕🛒✉Ⓑ✈ Bilbao and Santander, bus station. Colindres for large 🛒.

Santoña has 2 basins; 5 ♥, max LOA 8·5m, lots of FVs. Reports of yachts being made unwelcome. ♠ ⚓ at the SE corner; ✕🛒🚿.

9.23.16 SANTANDER

Cantabria **43°27'·66N 03°46'·13W (between Nos 1 & 2 chan buoys)** ❀❀❀♎♎♎✿✿✿

CHARTS AC 1291/2, 1145; SC 401, 4011; SHOM 7365, 6991

TIDES Standard Port POINTE DE GRAVE (←); ML 2·5

Times				Height (metres)			
High Water		Low Water		MHWS	MHWN	MLWN	MLWS
0000	0600	0000	0500	5·3	4·4	2·1	1·0
1200	1800	1200	1700				
Differences SANTANDER							
–0020	–0100	–0050	0000	–0·6	–1·2	–0·3	–0·7
RÍA DE SUANCES (04°03'W)							
0000	–0030	–0020	+0020	–1·4	–1·5	–0·6	–0·6

SHELTER Access H24, all weather/tides. Fair shelter off the city but subject to wash; very good shelter up-river at marina.

NAVIGATION From the E, WPT 43°28'·45N 03°44'·54W (E of Isla de Mouro) 236°/1·4M to Nos 1/2 buoys.

From the W, WPT 43°29'·49N 03°46'·57W (4ca E of Cabo Mayor). Thence in fair weather head SE between Isla de Mouro and Peninsula de la Mágdalena (min depth 7·3m). In heavy weather keep NE of Isla de Mouro. Punta Rabiosa ldg lts lead 235·8° into the well-buoyed/lit DW chan, defined by a Dir WRG lt 259·5°.

For Marina Santander (see chartlet inset) follow the buoyed chan

to No 17 small GRG buoy, Fl (2+1) G 10s; here turn WSW for marina ent. Caution: the hard-to-see 235·6° ldg marks to marina lead close to or over a shifting, drying sandbank. Night entry not advised unless already visited by day.

LIGHTS AND MARKS The university, formerly a Royal Palace, is conspic on Peninsula de la Mágdalena. City buildings are prominent. Long oil jetty is conspic N of Marina Santander; 400m westward a glazed pyramid-shaped building is a good daymark.

COMMUNICATIONS (Code 942) MRSC 213030; Emergency 112; Local weather broadcast VHF Ch 11 every 4 hrs, 0245-2245 and via HM; ☺ via HM; HM 223900; Port Authority *Santander Prácticos* VHF Ch **14**, 16; Marinas VHF Ch 09; Airport 202120.

FACILITIES Four options, from seaward:

- **Club Nautico de La Horadada** (CNH) ☎280402. www.horadada.es ⚓. ⚓ for 7–10m LOA craft NW of Nos 1 & 3 buoys, in 5m on rock; bay is almost full of local moorings. Tripping line essential.
- **Pedreña Marina,** 8ca SSE of Pta Rabiosa and the main channel. Access from No 4 buoy via PHM marked/lit chan with least depth 1·4m. Limited ♥ (vessels <10m only) access is tight, reports would be welcomed. www.marinapedrena.com
- **Darsena de Molnedo:** Real Club Maritimo de Santander (RCMS; near city) ☎214050 www.rcmsantander.com. 108 ⌂ max LOA 12m in 1·7m but few if any ♥. €3.10/m. RCMS

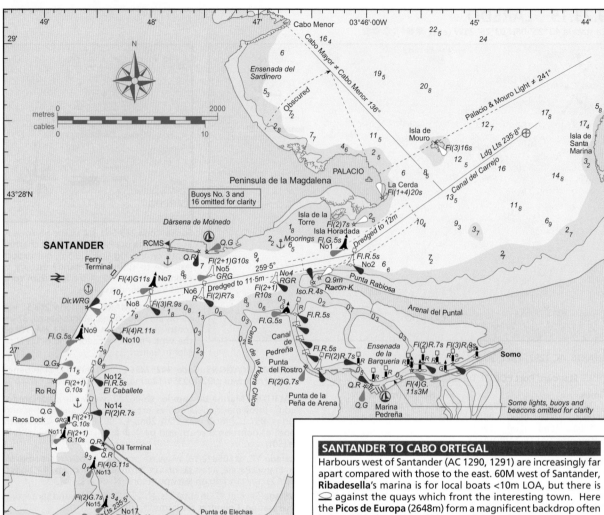

SANTANDER TO CABO ORTEGAL

Harbours west of Santander (AC 1290, 1291) are increasingly far apart compared with those to the east. 60M west of Santander, **Ribadesella**'s marina is for local boats <10m LOA, but there is ⌓ against the quays which front the interesting town. Here the **Picos de Europa** (2648m) form a magnificent backdrop often being visible from 15M offshore whilst cloud obscures their lower slopes.

Further west a heavy NW swell, known as a *Vaga de Mar*, breaks over banks and closes any ports which have bars. It can occur quite suddenly in calm weather, but is usually associated with strong NW'lies. In summer NE winds predominate and fog occurs fairly often especially in late June to early July.

Gijón with a modern marina offers refuge; it lies close SE of, but quite distinct from, El Musel, the very large commercial harbour whose northerly extension, under construction may be a trap for unwary yachts cruising close inshore.

Cudillero and **Luarca** are small fishing hbrs worth visiting, but beware swell especially off Cudillero. **Ría de Ribadeo** (with a much improved marina) is the first of the *Rías altas* (ie upper or northern) sunken estuaries not unlike a Scottish sea loch. **San Ciprian**'s aluminium port (AC 1122) offers within its long bkwtrs a sheltered ⚓ of refuge close to its W shore.

Ría de Viveiro has a well sheltered marina and several anchorages to await fair winds around Cabo Ortegal. **Ría del Barquero** and the **Ensenada de Santa Marta**, to E and W of Pta de la Estaca de Bares, offer many attractive ⚓s exposed only to N/NE winds; but beware S/SW winds off the mountains being accelerated through the valleys. This stretch of steep, rocky coast is deeply indented and rises to over 500m a few miles inland.

▶*From Pta de la Estaca de Bares to Cabo Ortegal (8M) the tidal stream sets E or NE on a rising tide and W or SW when falling, rates up to 2kn inshore and perceptible for a considerable distance offshore.*◀

berths S side, misc N side. 🛥 ⚓ ⌐(27t) ⌓(1·5t). ⚓s and ⚓ S of the dock are exposed to wash.

- **Marina de Santander** ☎369298 is approx 2·7M further upriver, but 20 mins bus/taxi ride to the city. 1400 ⌓ inc ⓥ in 2·2-3·1m, max LOA 23m, €3.50, max 15 days. 🛥 🅿 ✕ 🖥🛒 (taxi req'd) ⌐ ⌂🔧✕ 🅿 Ⓔ ⚓⌐(27t) ⌓. Security adequate to leave a yacht. Little noise from adjacent Parayas airport. marinasantander@marinasantander.com www. marinasantander.com

City: www.puertosantander.com All amenities; 🚆 & bus; Brittany ferry terminal ☎360611 is 6ca W of Darsena de Molnedo; ✈.

SUANCES, Cantabria. **43°25'·78N 04°02'·08W** (marina). AC 1291, 1150; SC 4012; SHOM 2042. Tidal differences overleaf. No access in N'lies or when swell breaking over bar, 1m; not advised at night. On late flood from ⊕ 43°25'·78N 04°02'·08W track 149·5°/8ca on ldg bcns (Q & Iso 4s) to Fl(4)R 11s bcn at hd of trng wall. Best water E side of chan. Marina (Fl (4) R & G 11s) has small craft pontoon. If >8m LOA, ⌓ on piled N or W walls. Basics in village, 🍴✕.

9.23.17
SAN VICENTE DE LA BARQUERA
Cantabria. 43°23'·69N 04°23'·06W ❁❀♦♦♧♧ ✿✿✿

CHARTS AC 1291; SC 402, 4021; SHOM 6381, 6991

TIDES
Standard Port POINTE DE GRAVE (←)

Times				Height (metres)			
High Water		Low Water		MHWS	MHWN	MLWN	MLWS
0000	0600	0000	0500	5·3	4·4	2·1	1·0
1200	1800	1200	1700				
Differences SAN VICENTE DE LA BARQUERA							
−0020	−0100	−0050	0000	−1·4	−1·5	−0·6	−0·6
RÍA DE TINA MAYOR (04°31'W, 6M W of San Vicente)							
−0020	−0100	−0050	0000	−1·3	−1·5	−0·6	−0·6

SHELTER Well protected large shallow estuary, An active FV hbr with virtually no yacht berth availability, the planned 700 berth marina at S end of Puenta Nuevo will not materialise, FVs occupy newly-built pontoons nearby. ⚓ing very difficult due to local moorings. Berthing might be possible on end of the fish quay, which is very busy but may be quieter than by the bridge. Reports welcome.

NAVIGATION WPT 43°24'·39N 04°22'·16W, 223°/1M to hbr ent (abeam Isla Peña Menor). Beware shallow bar and tidal streams 3-4kn. Avoid access near LW if there is any swell at all. Depths are as per SC 4021 but may be more water.

LIGHTS AND MARKS Pta de la Silla lt ho and hbr lts. Beach E of ent, castle & church upriver are conspic.

COMMUNICATIONS (Code 942) MRCC 326050; Emergency 112; Brit Consul (902)109356; HM 710004.

FACILITIES Town About 41km NExE of the Picos de Europa. ⚓✕◻🛒🏪🅿🛰 ⊠ Ⓑ ✈ (Santander).
www.sanvicentedelabarquera.org

HARBOUR 16M WEST OF SAN VICENTE DE LA BARQUERA

LLANES, Asturias. **43°25'·15N 04°44'·92W.** AC 1291; SC 403; SHOM 5009. Tides: Interpolate between San Vicente and Ribadesella. The small FV hbr faces E. Do not enter in strong N to E winds or heavy swell, or below half-tide. The N pierhead, ⚡ Fl G 5s, is conspic due to brilliantly coloured concrete rock armour. The S side of the ent is marked by ⚡ Fl R 5s and San Anton lt ho, easily seen from the E. Approx 150m further W a ⚡ Fl (3) G 9s marks a small basin to stbd; berth/raft in 2m where space permits (very limited due to FVs) on smooth N or E walls or as directed. No room to ⚓. ⚓ ✕ ◻ good shops 🅿 🏪 ⚓ 🛒 ⚒. turismo@ayuntamientodellanes.com www.llanes.com **Marina** ☎985 401421, ent via a lock. 140, €0·15/ m²/day, max LOA <10m, few spaces for visitors. Lock traffic signals: 3 G (vert) for open and 3 R (vert) for closed.

9.23.18 RIBADESELLA
Asturias 43°28'·06N 05°03'·99W ❁❀♦♦♧♧ ✿✿✿

CHARTS AC 1291, 1150; SC 403, 4031; SHOM 6381, 6691

TIDES
Standard Port POINTE DE GRAVE (←)

Times				Height (metres)			
High Water		Low Water		MHWS	MHWN	MLWN	MLWS
0000	0600	0000	0500	5·3	4·4	2·1	1·0
1200	1800	1200	1700				
Differences RIBADESELLA							
+0005	−0020	−0020	+0020	−1·3	−1·3	−0·6	−0·4

SHELTER Good. No space within the hbr to ⚓.

NAVIGATION WPT 43°28'·59N 05°04'·02W, 180°/5ca to Pta del Caballo. Bajo Serropio, 7ca NNE of ent, breaks in heavy seas. Easy appr around the top of the tide, but do not attempt in strong onshore winds when seas and swell break on the bar, reported as 2m, but less after NW gales. Hug the N bank 25m off the promenade to clear drying, buoyed sandbanks to stbd.

LIGHTS AND MARKS E of hbr ent smooth, dark cliffs and a chapel are distinctive. To the W a white beach is obvious. The channel up-river is now well lit and buoyed; see chartlet.

COMMUNICATIONS (Code 985) MRCC 326050; Emergency 112; Brit Consul (902)109356; HM 860038, VHF Ch 09 occasionally.

FACILITIES Berth/raft on the NE/E quays in 2-3m, clear of FVs and well N of the 2m high bridge; pre-rig a fender board. ⚓ ⚒ 🛢 📠 ✕ Ⓔ YC. **Marina** off the W bank in 3m, ent dredged 2·5m, has ❶ pontoon, only charge, small deposit for gate key. No ⚡.

Town ✕ ◻ 🛒 ⊠ Ⓑ 🅿 🏪 ⚓ ✈ Oviedo (78 km).

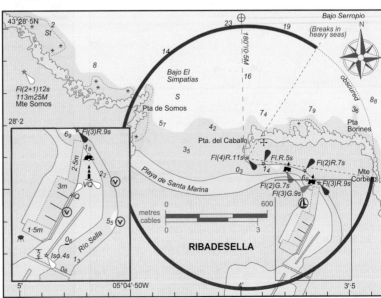

9.23.19 LASTRES

Asturias 43°30'·90N 05°15'·93W ✿✿🌕🌕✿✿

CHARTS AC 1291, 1153, 1150; SC 403, 4041; SHOM 6991, 5009

TIDES Interpolate between Ribadesella and Gijon.

SHELTER Well protected from W'lies by steep hillside, but in strong E'lies it becomes uncomfortable despite a substantial E bkwtr. ⚓ about 1ca S of the hbr in 5m on sand, clear of a fish farm marked by ⬦ Fl Y 2·5s.

NAVIGATION WPT 43°31'·16N 05°15'·35W, 235°/5ca to hbr ent. Access H24. Easy approach in all but strong E'lies, avoiding Bajo de la Plancha (5·7m) 3ca NNE of ent, breaks in heavy seas; also a pinnacle rk (0·3m) close NNE of the E bkwtr.

LIGHTS AND MARKS A 118m high, R/W TV mast (R lts) is conspic close W of the hbr. Cabo Lastres lt ho is 1·9M NW.

COMMUNICATIONS (Code 985) MRCC 326050; Emergency 112; Brit Consul (902)109356; HM 850003; FV President 850075; Tourist office (Colunga) 852200.

FACILITIES Berth/raft at the S end of the E bkwtr in 4-7m. FVs berth on W quays or moor in E-W trots across the centre of the hbr. At the N end of the hbr a pontoon with fingers is for local boats only, ditto the inlet in SW corner. ⚓ ⛟ (FV quay) 🏗(4 t).

Village ✗ ⬚ Basic shops, ⇌ Oviedo (62 km), Asturias ✈ is 5M E of Cudillero. www.infoasturias.com

9.23.20 GIJÓN

Asturias 43°32'·78N 05°40'·26W ✿✿✿🌕🌕🌕✿✿✿

CHARTS AC 1290, 1291, 1153, 1154; SC 404, 404A, 4042; SHOM 6381, 5009

TIDES
Standard Port POINTE DE GRAVE (←); ML 2·3

Times				Height (metres)			
High Water		Low Water		MHWS	MHWN	MLWN	MLWS
0000	0600	0000	0500	5·3	4·4	2·1	1·0
1200	1800	1200	1700				
Differences GIJON							
−0005	−0030	−0030	+0010	−0·9	−1·4	−0·4	−0·7
LUANCO (05°47'W)							
−0010	−0035	−0035	+0005	−1·3	−1·3	−0·6	−0·4
AVILÉS (05°56'W)							
−0100	−0040	−0050	−0015	−1·1	−1·6	−0·5	−0·7
SAN ESTABAN DE PRAVIA (06°05'W)							
−0005	−0030	−0030	+0010	−1·3	−1·3	−0·6	−0·4

SHELTER Excellent in marinas; care needed in strong N'lies.

NAVIGATION WPT 43°33'·78N 05°40'·26W, 180°/1·3M to marina ent, 1M SE of Puerto de El Musel, the vast conspic industrial port. The WPT is at the Y-junction of two mini-TSS chans, which lead in from NE and NNW, primarily for big ships. Banco Las Amosucas, marked by 4 cardinal lt buoys, lies within the Y. From the E yachts can approach more directly, passing N of Peninsula de Sta Catalina. In the near appr leave Sacramento G twr (marks drying reef) close to stbd.

LIGHTS AND MARKS NW of the marina are conspic W tanks in El Musel, many nav/shore lts and Cabo de Torres, Fl (2) 10s.

COMMUNICATIONS (Code 985) MRCC 326050; ⊖ & Met via marina; Puerto de El Musel VHF Ch 10, 11, 12, 14, 16; Marinas VHF Ch 09.

FACILITIES Marina (aka Puerto Local) ☎344543. www.puertodeportivogijon.es info@puertodeportivogijon.es ⚓ 658+88 Ⓥ €0.64/m², inc ⬚ 🏴. Access H24 (1·2-2·5m) in 4 basins. Berth initially dead ahead for check-in/fuel, then shift to Ⓥ finger berths in N Basin (Antepuerto). ⚓ ⛟ 🔧 🖻 ⚒ 🏪 ⚠ Ⓔ no wi-fi 🏗(30t). YC.

Marina Yates (El Musel) ☎687549658. www.marinayates.es info@marinayates.es 156 incⓋ €0.45/m². Access H24 (4·5m). Berth G dead ahead for check/in. ⚓ ⛟ ✗ 🚲 bike hire 🏗(64t) 🏗(5t).

Town ✗ ⬚ 🛒 @ ⊠ Ⓑ ⇌ ✈ Oviedo (42 km).

HARBOUR 6M SOUTH WEST OF CABO PEÑAS
AVILÉS, Asturias. **43°35'·65N 05°56'·77W**. AC 1290, 1142; SC 405A, 4052; SHOM 5009, 7361. Tides: 9.23.20. A port of refuge, except in W/NW'lies >F9, but also heavily industrialised and a FV hbr. The ent, S of Pta del Castillo lt ho and N of conspic white beach, is open to W/NW but well sheltered from N'lies. The well marked/lit ent chan runs ExS for 7ca then turns S for 2M to club pontoons at 43°33'·66N 05°55'·28W, W side of chan; max LOA about 12m, ☎ 985 565479 (for a gate key and to pay). FV supplies only. Caution FV wash. Port VHF Ch 12 16. HM ☎(985) 541 111, paviles@paviles.com. YC at Salinas, 2M NW. Old town is picturesque and close SW. No ⚓; €1.18/m. ✈ Asturias

HARBOUR 3M EAST OF CUDILLERO
SAN ESTEBAN DE PRAVIA, Asturias. **43°34'·21N 06°04'·54W**. AC 1290, 1133; SC 405, 405A, 4041; SHOM 5009. Tides: Gijón. A former coaling port, now only a few FVs berth on E bank at San Juan de la Arena quay, 90m long. The 206·2° ldg lts lead very close to the W bkwtr hd. Appr near HW due to silting in the ent. Possible ⚓ on FV quay or on facing quays on W bank. Or ⚓ in river on sand in 2·2m just S of ent to town quays.

9.23.21 CUDILLERO

Asturias **43°33'·96N 06°08'·80W** (W bkwtr hd) 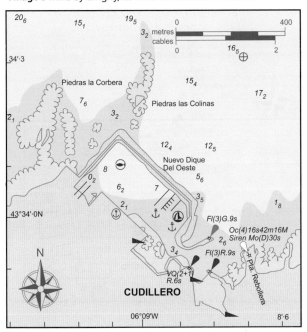 ⚓⚓⚓

CHARTS AC 1290; SC 934, 405, 405A; SHOM 5009

TIDES Use San Estaban de Pravia, 3M W; see Gijón.

SHELTER Good in large modern basin, NW of hbr ent, but N'ly swell can make entry/exit difficult and may work into the basin.

NAVIGATION WPT 43°34'·33N 06°08'·71W, 190°/3½ca to W bkwtr hd. From the W keep at least 5ca offshore until clear of Piedras las Colinas, tall islets/reefs NW of ent. From the E appr with Pta Rebollera lt ho on initial bearing of 200°. The rocky ent opens up only when close to. Narrow chan skirts the W bkwtr into basin. Before leaving look over the bkwtr to assess swell.

LIGHTS AND MARKS Pta Rebollera lt ho is conspic W 8-sided twr.

COMMUNICATIONS (Code 985) MRCC 326050; Emergency 112; Brit Consul (902)109356; HM VHF Ch 27.

FACILITIES Take one of the 12 yellow/red fore-and-aft pick up buoys SW of the pontoons, currently FOC. FVs moor/berth at NW end of the basin. The old drying hbr, S of ent, is exposed to swell, but convenient for dinghies; yachts can dry out on slip. HM ☎591114, ⛽ (gasoleo B) from FV quay, ⌕ ⚒ ⚓ small commercial shipyard.

Village 5 mins by dinghy, 25 mins on foot. ✗ ⌂ ☷ ✉.

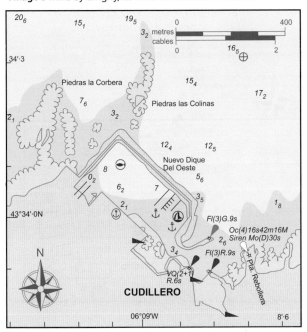

9.23.22 LUARCA

Asturias **43°32'·95N 06°32'·20W** ⚓⚓⚓⚓

CHARTS AC 1290, 1133; SC 933, 934, 4061; SHOM 6381, 5009

TIDES
Standard Port POINTE DE GRAVE (⟵)

Times				Height (metres)			
High Water		Low Water		MHWS	MHWN	MLWN	MLWS
0000	0600	0000	0500	5·3	4·4	2·1	1·0
1200	1800	1200	1700				
Differences LUARCA							
+0010	−0015	−0015	+0025	−1·1	−1·1	−0·5	−0·3

SHELTER The outer hbr is protected from all but N'lies, but swell can intrude (reported to badly affect the ⚓s). No room to ⚓ inside the hbr.

NAVIGATION WPT 43°33'·51N 06°32'·29W, 170°/6ca to E bkwtr hd lt ho which is conspic and almost in transit with the ldg marks. Appr on the ldg lts/marks 170° to clear rky shoals either side.

LIGHTS AND MARKS Cabo Busto is 3M ENE of hbr ent. Pta Blanca (or Focicón) lt ho is W ⬜ twr and house on prominent headland 300m ENE of hbr ent; conspic ✠ close by. 170° ldg lts are on thin W lattice pylons, R bands; not easy to see until close in.

COMMUNICATIONS (Code 985) MRCC 326050; Emergency 112; Brit Consul (902)109356; HM 640176/640083.

FACILITIES Moor fore-and-aft to the E bkwtr, 4 x stern to steel ⚓s which are close together. Underwater obst'ns of 1·3m, close bkwtr reported, so caution at springs, but in calm conditions it is possible to lie alongside the bkwtr.

Narrow chan leads SE to inner hbr dredged 2m (prone to silting). Pontoons in SE corner are full of local craft, little chance of a berth. Temp'y ⚓ on NW side may be possible. ⚓ from FV quay, ⌕ ⚒ ⚓(8t).

Town Friendly YC next to front ldg lt, ✗ ⌂ ☷ ✉ ⑧ ≈.

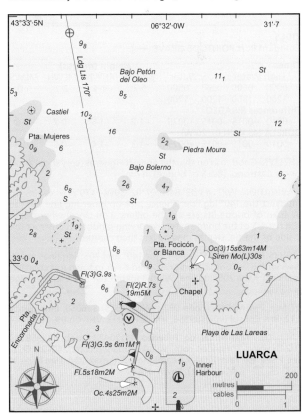

9.23.23 RÍA DE RIBADEO

Lugo **43°32'·44N 07°02'·16W** (Marina ent) ✿✿❀☙☙✿✿✿

CHARTS AC 1290, 1096; SC 932, 4071; SHOM 5009

TIDES
Standard Port POINTE DE GRAVE (←)

Times				Height (metres)			
High Water		Low Water		MHWS	MHWN	MLWN	MLWS
0000	0600	0000	0500	5·3	4·4	2·1	1·0
1200	1800	1200	1700				
Differences RIBADEO							
+0010	−0015	−0015	+0025	−1·2	−1·5	−0·7	−0·8
BURELA (43°40'N 07°20'W)							
+0010	−0015	−0015	+0025	−1·3	−1·4	−0·5	−0·4

SHELTER Good. ⚓ as on chartlet and off Figueras (very shoal appr) or off Castropol (8ca S of bridge) in 4m.

NAVIGATION WPT 43°33'·84N 07°02'·30W, 170°/4½ cables to intersect the 140° ldg line. Thence pick-up inner ldg line 205° to W span of bridge (lts are on the pillars; 32m clnce) where tides run hard. S of the bridge extensive drying sandbanks obstruct the E side of the ría. Depths may be less than charted.

LIGHTS AND MARKS Daymarks for 140° and 205° ldg lines are W twrs with R ◇. Conspic R onion-shaped ⊕ twr in town (off chartlet) is almost ≠ the 205° ldg line. White lookout twr at E end of bridge is conspic.

COMMUNICATIONS (Code 982); MRCC (985) 326050; Emergency 112; Brit Consul (902)109356; HM ☎128005, VHF Ch 09, 16.

FACILITIES Marina/CN de Ribadeo ☎120428, mob 648958143 (H24); www. cnribadeo.com vela@cnribadeo.com; 240+20 Ⓥ⌂ see inset, €0·78m² (LOA x beam) May-Oct, max LOA 18m ⊖ ⚒ ✗ ⌂ⓘ🗑 ⚒ ⚓ ✗ ━ ⌂(8t). **Town** ✗⌂🛒 ⬛ ✉ Ⓑ.

9.23.24 RÍA DE VIVEIRO

Lugo **43°40'·93N 07°36'·16W** (Celeiro bkwtr hd) ✿✿✿❀☙☙✿✿

CHARTS AC 1290, 1122; SC 931, 4082; SHOM 5009, 6383

TIDES
Standard Port POINTE DE GRAVE (←)

Times				Height (metres)			
High Water		Low Water		MHWS	MHWN	MLWN	MLWS
0000	0600	0000	0500	5·3	4·4	2·1	1·0
1200	1800	1200	1700				
Differences RÍA DE VIVEIRO							
+0010	−0015	−0015	+0025	−1·3	−1·3	−0·6	−0·4

SHELTER Good, except in N'lies. ⚓ in E'lies inside Isla Insua d'Area, 1·3M NE of Celeiro. In S-W winds ⚓ at the head of the ria off Playa de Covas in 6-7m.

NAVIGATION Easy ent, 7ca wide, between Isla Gabeira and Pta de Faro. WPT 43°43'·25N 07°35'·30W (mouth of the ria) 195°/2·4M to round Celeiro bkwtr hd. Keep clear of this major FV hbr.

LIGHTS AND MARKS Mte Faro Juances, is a distinctive conical peak (193m) close SE of Pta de Faro.

COMMUNICATIONS (Code 982) MRCC (985) 326050; Emergency 112; Celeiro HM 560074; Celeiro Port VHF Ch 16; Marina VHF Ch 09.

FACILITIES Marina Viveiro ☎mob 690604452; www. northwestmarinas.com/es/marinaviveiro/marina-viveiro.htm marinaviveiro@marinaviveiro.com 243 inc Ⓥ⌂ €2·80, max 18m LOA on fingers, 30m LOA quayside, multihulls +50%. ⚒ 🚶⚒ ✗ Ⓔ ━ wi-fi ⌂(35t) ⌂(8t). **Town** ✗ ⌂ 🛒 ⚒ ✉ Ⓑ ⇌.

MINOR HARBOUR SOUTH OF PUNTA DE LA ESTACA DE BARES
RÍA DEL BARQUERO. 43°45'·00N 07°40'·70W. AC 1290, 1122; SC 931, 4082; SHOM 6383. Tides: Ría de Viveiro. A delightful ría, exposed only to NE'lies. There is an ⚓ at Bares below the steep, high W bank and 2·5M SSW a shallow access channel leads to a tiny hbr at El Barquero in depths 1·5-2·5m.

On the E side of the ría, yachts may ⚓ N of Pta del Castro, clear of an outfall; or a berth may be found in the modern FV port at Vicedo on the S or W quays.

See Lights, buoys and waypoints for lts at: Isla Coelleira (an islet between this ría and Ría de Viveiro); Pta de la Estaca de Barra; Pta de la Barra; Pta del Castro; and Vicedo.

MINOR HARBOURS SOUTH OF CABO ORTEGAL
CARIÑO. 43°44'·21N 07°52'·14W is 2M S of Cabo Ortegal (AC 1122) and well sheltered from all W'lies. After rounding the outer commercial jetty, ⚓ in 5-6m on sand within the bay. Dinghies can be left on an extended pontoon in the NW corner of the hbr; Ashore: shops ⌂ ✕ 🏧, 🗓 📞. In E'lies, ⚓ off **Espasante** FV hbr, 2·3M ESE of Cariño (AC 1111).

RÍA DE SANTA MARTA DE ORTIGUEIRA. 43°42'·78N 07°50'·60W. Tides: +0050 La Coruña (➡); MHWS –0·1; MHWN +0·4; MLWN –0·1; MLWS +0·3; streams reach 5kn sp. SC 4083 (1:20,000) is essential for pilotage. WPT 43°45'·00N 07°50'·60W, 180°/2·4M to the Ría ent (L/L above) at HW –2. ⚓s in 3 pools off Pta Postina (11·4m) Pta Sismundi (11·1m) and Pta Redonda (5·2m).

The access chan crosses the 0·4m bar on a S'ly track with Isla de San Vicente 50m to port. Reports received of silting over the bar (2012) caution recommended. It then turns W, SW and S in three legs each about 1M long and is marked by small lateral buoys and occas perches/withies. The eponymous town will be seen when 7 ca WNW of it. In sequence pass first a quay with blue crane, then a small private jetty, finally the marina ent. Rewarding pilotage to an elegant town.

Facilities: ♥ welcome; 3 YC pontoons (2m max) max LOA 12m. ✕ ⌂ 🍴 🗓 📞 ✉ Ⓑ; coast road & ⇌ to La Coruña ✈.

CABO ORTEGAL TO CABO FINISTERRE
Cabo Ortegal should be rounded at least 2M off due to the offlying needle rocks, Los Aguillones. Most of the major headlands should be given a good offing to avoid fluky winds and, in some cases, pinnacle rocks. The deeply indented coast (AC 1111, 3633, 3764) begins to trend WSW, with 600m high mountains rising only 5M inland. Coastal features are often hard to identify against this background. Here many major lights are sited so high as to be obscured by low cloud/mist in onshore weather.

▶ *Tidal streams set SW on the ebb and NE on the flood. Any current tends to run SW, then S.* ◀

The **Ría de Cedeira** is entered 3M SSW of Pta Candelaria lt ho. This small attractively wooded ría offers refuge by day/night to yachts unable to round Cabo Ortegal in strong NE'lies. Several banks along this stretch break in heavy weather when they should be passed well to seaward.

About 20M further SW, having passed Pta de la Frouxeira, Cabo Prior, Pta de Castro and Cabo Priorño Chico, all lit, is the entrance to **Ría de Ferrol**, a well sheltered commercial and naval port, but with limited yacht facilities. The adjacent **Ría de Ares** and **Ría de Betanzos** are attractive alternatives, both with easy access to their marinas.

La Coruña is a port of refuge, accessible in all weathers and has excellent yacht facilities with two marinas near the city centre and another to the south.

W of the conspicuous Torre de Hércules lt ho a 19M wide bight is generally foul as far as Islas Sisargas, three islets 2M offshore; here the inshore passage is not advised. 8M SW of Pta Nariga, between Pta del Roncudo and Pta de Lage, is the unspoiled **Ría de Corme y Laxe** with two small fishing hbrs in the NNE and SSW corners of the bay.

Cabo Villano, a rocky headland with lighthouse, siren and conspicuous wind turbines, is a possible landfall after crossing Biscay. 4ca NW of it El Bufardo (43°10'·00N 09°13'·28W) is a dangerous pinnacle rock awash at CD.

20M NW of Cabo Villano and 22M NW of Punta de la Barca is the **Off Finisterre TSS**; see Area Map. Yachts have more than adequate searoom within the ITZ whose southern boundary lies due west of Cabo Finisterre. *Finisterre Tráfico* (VTS) on VHF **11**, 74, 16 broadcasts regular information about Navwarnings, weather and traffic in Spanish and English. On request it can also provide the vessel's position, course and speed and the identity of other vessels in the vicinity.

The entrance to **Ría de Camariñas**, the last of the rías altas and a safe refuge, is 2·4M S of Cabo Villano. There is a small, friendly marina at Camariñas, and on the SW side of the ria a new 224 berth marina at Mugia, and a sheltered anchorage just S.

Cabo Toriñana and **Cabo Finisterre** both have charted dangers, mostly pinnacle rocks, lying up to 1M offshore; hence the popular name *Costa del Morte* for this wild, magnificent and sometimes forbidding coast.

9.23.25 RÍA DE CEDEIRA
La Coruña 43°39'·38N 08°03'·86W (⚓) ❀❀❀✿✿✿✿✿

CHARTS AC 1290, 1111, 1122; SC 411; SHOM 5009, 3007

TIDES Interpolate between Santa Marta de Ortigueira (Ría de Viveiro) and Ferrol. ML no data.

SHELTER Very good; handy for awaiting a fair wind to round Cabo Ortegal. ⚓ in 3-4m about 500m E of the modern FV port, excellent holding on sand. Although the whole ría is a marine reserve (see chartlet) there are no constraints on visitors who can ⚓ anywhere for free.

NAVIGATION WPT 43°41'·00N 08°05'·41W, 155°/1·7M to Pta del Sarridal. From N keep > 1M off Pta Candelaria until the ría opens. From the S a similar offing clears rks N/NW of Pta Chirlateira.

155° ldg line Pta Promontorio lt ho ≠ Pta del Sarridal. Abeam Pta Chirlateira with Pta Xian brg 161°, track midway between Pta del Sarridal and Piedras de Media Mar. When the bkwtr hd bears 060° turn toward the anchorage. The E end of the bay shoals rapidly toward the drying mouth of Rio de Cedeira.

LIGHTS AND MARKS Piedras de Media Mar, BRB lt twr conspic in centre of ría, can be passed on either side. A chapel is conspic 4ca N of the hbr. Lts as chartlet and Lights, buoys and waypoints.

COMMUNICATIONS (Code 981) MRSC 209541; Emergency 112; Brit Consul (902)109356; Fishermen Cofradia ☎480389, No VHF, MF *Cedeira Cofradia* 1800kHz, 2182.

FACILITIES Hbr ⊖ 🚢 ⚓ ⚓ 🗜 ✕ 🏴(100t) ☎492212.

Town ✕ ⌂ 🍴 🗓 🗓 📞 Ⓔ ✉ Ⓑ; ⇌ & ✈ La Coruña.

9.23.26   FERROL

La Coruña 43°28'·56N 08°14'·61W (Curuxeiras dock) ✵✵✵◊✿✿

CHARTS AC 1111, 1094, 1118, 1117; SC 412, 412A, 4122, 4123; SHOM 3007

TIDES
Standard Port LA CORUÑA (→); ML 2·2.

Times				Height (metres)			
High Water		Low Water		MHWS	MHWN	MLWN	MLWS
0000	0600	0000	0500	4·2	2·8	1·6	0·3
1200	1800	1200	1700				
Differences FERROL (La Grana)							
–0002	–0002	+0001	+0001	0·0	0·0	0·0	0·0

SHELTER Very good inside ría, but be aware of strong tides and occasional squalls in the narrows.

⚓s, from seaward, N shore of appr chan: Ensenada de Cariño (poor holding, good shelter except in S'lies); inshore of Pereiro PHM buoy in 5-6m. Inside the ría (see chartlet) S shore: Ensenada de el Baño and off Mugardos; on N shore off Pta de Caranza, ESE of chapel and near No 2 PHM buoy (off chartlet).

NAVIGATION WPT 43°26'·59N 08°20'·17W (close to WCM lt buoy and in the white sector (041·5°-042·5°) of Batería de San Cristóbal Dir lt) 042°/1·3M to pick up the 085·2° ldg line. Caution: From near Cabo Prioriño Chico lt ho a substantial bkwtr extends 145°/0·5M, halfway across the ent to the Ria. Chan is deep and adequately buoyed/lit. In bad weather avoid the outer banks (Tarracidos, Cabaleiro and Leixiñas) to the N and W.

LIGHTS AND MARKS C. Prioriño Chico bkwtr head, Fl R 5s 21m 7M; adjacent T-jetty Fl (2) R 7s 10m 3M. Ldg lts 085·2° through the narrows.

COMMUNICATIONS (Code 981) MRSC 209541; Emergency 112; Brit Consul (902)109356; HM 364055.

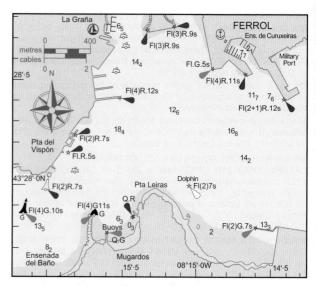

FACILITIES Limited yacht facilities in this naval, commercial and ship-building port. **Marina Terramar** at La Graña: approx 50 ⚓ a few ⚓s or ⚓. ⊖ ⚒ 🅿 🅿 ⊾ ⚓ 🛒 ✕ ⚓(1·5t) YC. Bus every ½hr to the city (20 mins).

Dársena de Curuxeiras is central, but ⚓ in 2m on NE side is against stone walls and prone to ferry wash. The 3 pontoons at the head of this inlet are not available to visitors. ☎321594, 🛒 ⚓ 🅿 🅿 🛒 🅱 ⚓(8t).

Other options: Ría de Ares has yacht pontoons in the small FV hbr. Sada (Ría de Betanzos) has two marinas and good bus links to La Coruña.

City all facilities.

9.23.27   RÍA DE ARES

Marina 43°25'·34N 08°14'·38W ✵✵✵◊✿✿✿

CHARTS AC 1111, 1094; SC 412, 412A; SHOM 3007, 7598, 6665

TIDES Use differences for La Coruña.

SHELTER Good in the marina/fishing hbr at the W end of the 1M wide, sheltered bay (Ensenada de Ares) but the ría is exposed to strong NW'lies and swell. ⚓ NE of the hbr in 3-5m, good holding on muddy sand. The ambience is peaceful and relaxed. The ría extends E past Redes toward Puentedeume, where there is day ⚓ in shoaling waters, clear of mussel beds (*viveiros*).

NAVIGATION WPT 43°24'·88N 08°14'·03W, 340°/0·5M to hbr's E mole hd. The appr is simple, but keep clear of Bajo La Miranda (3·7m; 1·5M WSW of the hbr) and adjacent islets off Pta Miranda. In the near approach avoid Bajo Cagarroso (2·1m) 5 cables SSW of the E mole head. To the W of Ares there are two Measured miles with transit bcns, should you need to check your speedo.

LIGHTS AND MARKS As on the chartlet. Cabo Prioriño Chico is 5M WNW. Sada/Fontan marina/hbr is 4M S; see Ría de Betanzos. A conspic rounded hill (Monte de San Miguel de Breamo) with chapel on top is 2·7M ESE of the hbr.

COMMUNICATIONS (Code 981) MRSC 209541; Emergency 112; ⊖ & Met via Club (below) which is welcoming; Brit Consul (902)109356; Marina/Club VHF Ch 09, English spoken, ask for Marco.

FACILITIES Marina/CN Ría de Ares ☎468787 mob 610737344. secretaria@nauticoares.com www.nauticoares.com 4 pontoons in 2m for yachts; FVs berth inside E mole. 325 + 16♥ €2.79 max LOA 20m on fingers; mostly on outer end of pontoon No 1. 🛒 🅰 ✕ 🖥 🅱 🛒 🖥(35t) ⚓(8t).

Town ✕ 🖥 🛒 🖂 ⊠ 🅱. Bus to Ferrol.

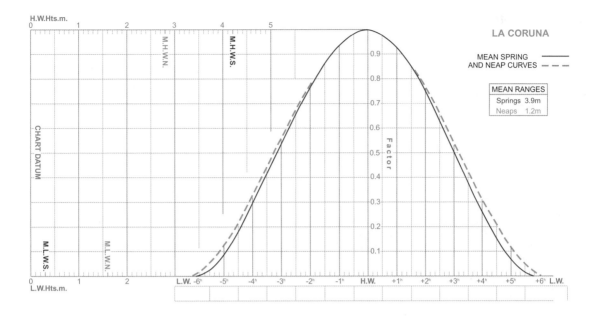

LA CORUNA

MEAN SPRING ———
AND NEAP CURVES - - -

MEAN RANGES	
Springs	3.9m
Neaps	1.2m

9.23.28   RÍA DE BETANZOS

La Coruña **43°21'·73N 08°14'·45W** ✿✿✿✿◊◊◊✿✿✿

CHARTS AC 1111, 1094; SC 412, 412A; SHOM 3007, 7598, 6665

TIDES Use differences for La Coruña.

SHELTER Very good in two marinas in Fontán hbr, Marina Sada, S side of hbr and Club Náutico de Sada, NW side. Keep close to the eastern hammerheads on arrival. Visitors are very welcome at this well-run fully-equipped marina, a safe haven for a longer stay or winter lay-up. Alternatively, ⚓ N or E of the hbr in 4-7m, but beware holding on weedy bottom.

NAVIGATION WPT 43°24'·34N 08°15'·07W, 171°/2·6M to Fontán hbr's N mole hd. Appr the WPT on an ESE track midway between Pta Miranda to the NE and Pta Redonda* to the SW; Bajo La Miranda (3·7m) lies 6ca SW of the former. The Ria is wide and easy to enter, but shoals steadily S of the hbr. There are shellfish beds visible close to the shore from 2M NW of Fontán.

*Note: a second Pta Redonda is 4M ESE of the first mentioned, opposite Fontán.

LIGHTS AND MARKS Hbr lts proliferate, as on the chartlet, mostly marking FV quays on the N and W side of the hbr. 1·4M NE of the hbr Pta de San Pedro lt (off chartlet) is listed as Perbes pierhd, Dir Fl (2) WR 7s. At the head of the ría a bridge across the Río Mandeo is clearly visible.

COMMUNICATIONS (Code 981) MRSC 209541; Emergency 112; Brit Consul (902)109356; HM 620005; Marinas VHF Ch 09.

FACILITIES Marina Sada ☎619015, mob 661356269, (1000-1400, 1600-1900) www. marinasada.com marinasada@marinasada.com 630+ Ⓥ ◡ €2·72, max 20m LOA, max stay 6mths. ⛽ ⛽ (1000-2000) ✕ ⌂ 🛒 ⊟ ◉ ⚓ ◣ ⚒ △ 🕻 Ⓔ 🗜 ⚒ ⚓ 🖋 ⊟(32t). Safe over-wintering.

Club Náutico de Sada ☎619018, mob 653469579, www. nauticodesada.es info@clubnauticodesada.com 313 inc Ⓥ, ◡ €2·20, in 2-3m, max 20m LOA, facilities same as Marina Sada, except ⊟(50t).

Town Facilities at Sada (10 mins walk) are better than Fontán: ✕ ⌂ ✉ Ⓑ Ⓗ. La Coruña by taxi 20 mins, by hourly bus 30 mins.

STANDARD TIME UT –01
Subtract 1 hour for UT
For French Summer Time add ONE hour in **non-shaded areas**

LA CORUÑA   LAT 43°22'N   LONG 8°24'W
TIMES AND HEIGHTS OF HIGH AND LOW WATERS

Dates in red are **SPRINGS**
Dates in blue are **NEAPS**

YEAR 2016

JANUARY

Time	m	Time	m
1 F 0222 1.3 / 0842 3.2 / 1503 1.3 / 2116 2.9		**16** SA 0222 0.9 / 0845 3.6 / 1458 0.9 / 2119 3.3	
2 SA 0316 1.5 / 0936 3.0 / 1601 1.4 / 2218 2.9		**17** SU 0324 1.1 / 0948 3.4 / 1603 1.1 / 2230 3.2	
3 SU 0420 1.6 / 1041 2.9 / 1708 1.5 / 2327 2.9		**18** M 0438 1.2 / 1102 3.3 / 1717 1.1 / 2347 3.2	
4 M 0532 1.6 / 1151 2.9 / 1814 1.5		**19** TU 0558 1.2 / 1219 3.2 / 1832 1.1	
5 TU 0032 3.0 / 0639 1.5 / 1255 3.0 / 1911 1.4		**20** W 0100 3.3 / 0711 1.1 / 1329 3.3 / 1937 1.0	
6 W 0127 3.1 / 0735 1.3 / 1349 3.1 / 1959 1.2		**21** TH 0202 3.5 / 0813 0.9 / 1428 3.4 / 2031 0.9	
7 TH 0214 3.3 / 0823 1.2 / 1434 3.3 / 2041 1.0		**22** F 0254 3.6 / 0905 0.8 / 1518 3.5 / 2118 0.7	
8 F 0255 3.5 / 0905 1.0 / 1515 3.4 / 2120 0.9		**23** SA 0339 3.8 / 0951 0.6 / 1602 3.6 / 2200 0.6	
9 SA 0334 3.7 / 0944 0.8 / 1555 3.6 / 2158 0.7		**24** SU 0420 3.9 / 1032 0.5 / 1642 3.7 / 2239 0.6	
10 SU 0413 3.8 / 1023 0.6 / 1634 3.7 / 2236 0.6		**25** M 0458 3.9 / 1110 0.5 / 1718 3.6 / 2316 0.6	
11 M 0452 3.9 / 1102 0.5 / 1714 3.7 / 2315 0.5		**26** TU 0534 3.9 / 1146 0.6 / 1753 3.6 / 2351 0.7	
12 TU 0532 4.0 / 1143 0.4 / 1755 3.7 / 2356 0.5		**27** W 0609 3.8 / 1221 0.7 / 1827 3.5	
13 W 0615 4.0 / 1226 0.5 / 1839 3.7		**28** TH 0026 0.8 / 0643 3.6 / 1255 0.8 / 1902 3.4	
14 TH 0040 0.6 / 0700 3.9 / 1312 0.6 / 1926 3.6		**29** F 0102 0.9 / 0717 3.5 / 1331 1.0 / 1938 3.2	
15 F 0128 0.7 / 0749 3.7 / 1401 0.7 / 2019 3.4		**30** SA 0139 1.1 / 0755 3.3 / 1409 1.2 / 2019 3.1	
		31 SU 0222 1.3 / 0837 3.1 / 1455 1.4 / 2110 2.9	

FEBRUARY

Time	m	Time	m
1 M 0314 1.5 / 0932 2.9 / 1554 1.5 / 2215 2.8		**16** TU 0415 1.2 / 1040 3.1 / 1651 1.2 / 2324 3.1	
2 TU 0423 1.6 / 1043 2.8 / 1708 1.6 / 2331 2.8		**17** W 0541 1.3 / 1205 3.0 / 1815 1.3	
3 W 0544 1.6 / 1204 2.8 / 1824 1.5		**18** TH 0046 3.1 / 0703 1.2 / 1322 3.1 / 1927 1.2	
4 TH 0043 3.0 / 0657 1.4 / 1314 2.9 / 1926 1.3		**19** F 0152 3.3 / 0807 1.0 / 1422 3.2 / 2022 1.0	
5 F 0142 3.2 / 0755 1.2 / 1408 3.1 / 2016 1.1		**20** SA 0244 3.5 / 0857 0.8 / 1509 3.4 / 2107 0.8	
6 SA 0230 3.4 / 0842 1.0 / 1454 3.3 / 2059 0.9		**21** SU 0327 3.6 / 0938 0.7 / 1548 3.5 / 2146 0.7	
7 SU 0313 3.6 / 0924 0.7 / 1536 3.6 / 2139 0.6		**22** M 0405 3.8 / 1014 0.6 / 1623 3.6 / 2221 0.6	
8 M 0354 3.9 / 1005 0.5 / 1617 3.7 / 2219 0.5		**23** TU 0439 3.8 / 1048 0.5 / 1656 3.6 / 2254 0.6	
9 TU 0434 4.0 / 1045 0.3 / 1657 3.9 / 2259 0.3		**24** W 0511 3.8 / 1120 0.5 / 1727 3.6 / 2327 0.6	
10 W 0515 4.1 / 1125 0.2 / 1738 3.9 / 2341 0.3		**25** TH 0541 3.8 / 1151 0.6 / 1757 3.6 / 2358 0.7	
11 TH 0558 4.2 / 1208 0.2 / 1821 3.9		**26** F 0612 3.7 / 1221 0.7 / 1828 3.5	
12 F 0024 0.2 / 0643 4.1 / 1252 0.3 / 1907 3.8		**27** SA 0030 0.9 / 0643 3.5 / 1252 0.9 / 1900 3.4	
13 SA 0110 0.5 / 0730 3.9 / 1339 0.5 / 1956 3.6		**28** SU 0104 1.0 / 0716 3.4 / 1326 1.1 / 1936 3.2	
14 SU 0202 0.7 / 0823 3.6 / 1432 0.8 / 2053 3.4		**29** M 0141 1.1 / 0754 3.2 / 1404 1.3 / 2019 3.1	
15 M 0301 1.0 / 0925 3.3 / 1535 1.1 / 2202 3.2			

MARCH

Time	m	Time	m
1 TU 0226 1.3 / 0841 3.0 / 1454 1.4 / 2116 2.9		**16** W 0357 1.2 / 1022 3.0 / 1628 1.3 / 2303 3.0	
2 W 0326 1.5 / 0945 2.8 / 1603 1.6 / 2231 2.8		**17** TH 0526 1.3 / 1151 2.9 / 1757 1.4	
3 TH 0449 1.6 / 1112 2.8 / 1732 1.6 / 2356 2.9		**18** F 0028 3.1 / 0649 1.2 / 1309 3.0 / 1911 1.3	
4 F 0616 1.5 / 1237 2.9 / 1849 1.4		**19** SA 0135 3.2 / 0752 1.1 / 1407 3.1 / 2006 1.1	
5 SA 0107 3.1 / 0723 1.2 / 1340 3.1 / 1947 1.2		**20** SU 0226 3.4 / 0839 0.9 / 1450 3.3 / 2048 0.9	
6 SU 0201 3.3 / 0815 0.9 / 1430 3.3 / 2034 0.9		**21** M 0307 3.5 / 0917 0.7 / 1527 3.4 / 2125 0.8	
7 M 0248 3.6 / 0900 0.6 / 1513 3.6 / 2116 0.6		**22** TU 0342 3.6 / 0950 0.6 / 1559 3.6 / 2158 0.7	
8 TU 0331 3.9 / 0941 0.3 / 1555 3.8 / 2158 0.3		**23** W 0414 3.7 / 1021 0.6 / 1629 3.6 / 2230 0.6	
9 W 0413 4.1 / 1023 0.1 / 1636 4.0 / 2239 0.2		**24** TH 0444 3.7 / 1051 0.6 / 1659 3.7 / 2301 0.6	
10 TH 0455 4.3 / 1104 0.1 / 1718 4.1 / 2322 0.1		**25** F 0513 3.7 / 1120 0.6 / 1728 3.7 / 2331 0.7	
11 F 0539 4.3 / 1146 0.1 / 1801 4.1		**26** SA 0543 3.7 / 1149 0.7 / 1758 3.6	
12 SA 0006 0.2 / 0624 4.1 / 1230 0.2 / 1846 3.9		**27** SU 0002 0.8 / 0613 3.6 / 1219 0.9 / 1829 3.5	
13 SU 0053 0.4 / 0712 3.9 / 1317 0.5 / 1935 3.7		**28** M 0035 0.9 / 0646 3.4 / 1252 1.0 / 1904 3.4	
14 M 0144 0.8 / 0804 3.6 / 1409 0.8 / 2031 3.4		**29** TU 0112 1.1 / 0723 3.2 / 1328 1.2 / 1945 3.2	
15 TU 0243 0.9 / 0905 3.3 / 1510 1.1 / 2139 3.2		**30** W 0155 1.2 / 0808 3.1 / 1414 1.4 / 2037 3.1	
		31 TH 0251 1.4 / 0908 2.9 / 1518 1.5 / 2147 2.9	

APRIL

Time	m	Time	m
1 F 0407 1.5 / 1031 2.8 / 1644 1.6 / 2311 3.0		**16** SA 0621 1.2 / 1241 2.9 / 1841 1.3	
2 SA 0534 1.4 / 1159 2.9 / 1809 1.4		**17** SU 0105 3.1 / 0722 1.1 / 1338 3.1 / 1937 1.2	
3 SU 0027 3.1 / 0647 1.2 / 1308 3.1 / 1913 1.2		**18** M 0157 3.2 / 0809 1.0 / 1422 3.2 / 2020 1.0	
4 M 0128 3.4 / 0743 0.9 / 1401 3.4 / 2005 0.9		**19** TU 0238 3.4 / 0847 0.9 / 1458 3.4 / 2058 0.9	
5 TU 0218 3.7 / 0831 0.6 / 1447 3.7 / 2051 0.6		**20** W 0313 3.5 / 0921 0.8 / 1531 3.5 / 2132 0.8	
6 W 0304 3.9 / 0915 0.3 / 1531 3.9 / 2135 0.3		**21** TH 0346 3.6 / 0952 0.7 / 1601 3.6 / 2204 0.7	
7 TH 0349 4.2 / 0958 0.1 / 1614 4.1 / 2218 0.2		**22** F 0416 3.6 / 1022 0.7 / 1631 3.7 / 2236 0.7	
8 F 0434 4.3 / 1041 0.1 / 1657 4.2 / 2303 0.1		**23** SA 0446 3.6 / 1052 0.7 / 1701 3.7 / 2307 0.7	
9 SA 0519 4.3 / 1125 0.1 / 1741 4.2 / 2349 0.2		**24** SU 0517 3.6 / 1121 0.8 / 1732 3.7 / 2339 0.8	
10 SU 0606 4.1 / 1210 0.3 / 1827 4.0		**25** M 0549 3.5 / 1153 0.9 / 1805 3.6	
11 M 0037 0.4 / 0655 3.9 / 1257 0.6 / 1917 3.8		**26** TU 0013 0.9 / 0624 3.4 / 1226 1.0 / 1842 3.5	
12 TU 0129 0.6 / 0748 3.5 / 1348 0.9 / 2012 3.5		**27** W 0051 1.0 / 0702 3.3 / 1305 1.1 / 1923 3.4	
13 W 0228 0.9 / 0848 3.2 / 1448 1.2 / 2117 3.2		**28** TH 0135 1.1 / 0748 3.1 / 1351 1.3 / 2014 3.2	
14 TH 0339 1.2 / 1002 3.0 / 1602 1.4 / 2236 3.1		**29** F 0229 1.3 / 0846 3.0 / 1451 1.4 / 2118 3.1	
15 F 0502 1.3 / 1126 2.9 / 1727 1.4 / 2357 3.0		**30** SA 0338 1.3 / 1000 2.9 / 1608 1.5 / 2233 3.1	

Chart Datum is 1·86 metres below NMMA Datum. HAT is 4·5 metres above Chart Datum.

〉〉 FREE monthly updates. Register at 〈〈
www.reedsnauticalalmanac.co.uk

STANDARD TIME UT –01
Subtract 1 hour for UT
For French Summer Time add
ONE hour in **non-shaded areas**

LA CORUÑA   LAT 43°22'N   LONG 8°24'W
TIMES AND HEIGHTS OF HIGH AND LOW WATERS

Dates in red are **SPRINGS**
Dates in blue are NEAPS

YEAR 2016

N & NW Spain

MAY

	Time	m		Time	m
1 SU	0456 1122 1728 2348	1.3 3.0 1.4 3.2	**16** M	0020 0639 1256 1856	3.0 1.2 3.0 1.3
2 M	0608 1232 1837	1.1 3.2 1.2	**17** TU	0116 0729 1344 1945	3.1 1.1 3.1 1.2
3 TU	0053 0709 1329 1934	3.4 0.9 3.4 0.9	**18** W	0202 0811 1424 2026	3.2 1.0 3.3 1.1
4 W	0148 0801 1420 2024	3.7 0.6 3.7 0.6	**19** TH	0241 0848 1500 2103	3.3 0.9 3.4 0.9
5 TH	0238 0849 1506 2112	3.9 0.4 3.9 0.4	**20** F	0316 0922 1533 2138	3.4 0.9 3.5 0.9
6 F	0327 0935 1552 2159	4.1 0.2 4.1 0.2	**21** SA	0349 0954 1605 2212	3.5 0.8 3.6 0.8
7 SA	0414 1020 1637 2246	4.2 0.2 4.2 0.2	**22** SU	0422 1026 1638 2246	3.5 0.8 3.7 0.8
8 SU	0502 1105 1723 2334	4.2 0.2 4.2 0.2	**23** M	0455 1058 1712 2320	3.5 0.8 3.7 0.8
9 M	0550 1151 1810	4.0 0.4 4.0	**24** TU	0530 1132 1747 2357	3.5 0.8 3.6 0.8
10 TU	0023 0640 1239 1900	0.4 4.0 0.6 3.8	**25** W	0607 1208 1826	3.4 0.9 3.6
11 W	0115 0732 1329 1953	0.6 3.5 0.9 3.6	**26** TH	0037 0648 1249 1909	0.9 3.3 1.0 3.5
12 TH	0211 0829 1425 2053	0.9 3.2 1.1 3.3	**27** F	0121 0735 1336 1958	1.0 3.2 1.1 3.4
13 F	0315 0933 1531 2200	1.1 3.0 0.9 3.1	**28** SA	0213 0829 1432 2056	1.1 3.1 1.2 3.3
14 SA	0426 1046 1644 2313	1.2 2.9 1.4 3.0	**29** SU	0314 0934 1539 2203	1.1 3.0 1.3 3.2
15 SU	0537 1157 1756	1.3 2.9 1.4	**30** M	0422 1047 1652 2313	1.1 3.1 1.3 3.3
			31 TU	0532 1157 1802	1.1 3.2 1.1

JUNE

	Time	m		Time	m
1 W	0020 0636 1259 1905	3.4 0.9 3.4 0.9	**16** TH	0119 0732 1346 1952	3.0 1.2 3.1 1.2
2 TH	0121 0734 1354 2001	3.6 0.7 3.6 0.7	**17** F	0206 0815 1428 2035	3.1 1.1 3.3 1.1
3 F	0217 0826 1445 2053	3.8 0.5 3.8 0.5	**18** SA	0246 0853 1506 2114	3.2 1.0 3.4 1.0
4 SA	0309 0915 1534 2144	3.9 0.4 4.0 0.4	**19** SU	0324 0929 1542 2151	3.3 0.9 3.5 0.8
5 SU	0359 1003 1622 2233	4.0 0.4 4.1 0.3	**20** M	0401 1004 1618 2227	3.4 0.8 3.6 0.8
6 M	0448 1049 1709 2321	4.0 0.4 4.1 0.3	**21** TU	0437 1039 1654 2304	3.5 0.8 3.7 0.7
7 TU	0536 1135 1755	3.8 0.5 4.0	**22** W	0514 1115 1732 2342	3.5 0.8 3.7 0.7
8 W	0010 0624 1221 1843	0.4 3.7 0.7 3.8	**23** TH	0553 1154 1812	3.5 0.8 3.7
9 TH	0058 0712 1308 1931	0.6 3.5 0.8 3.6	**24** F	0023 0635 1235 1855	0.7 3.4 0.8 3.6
10 F	0148 0802 1358 2022	0.8 3.2 1.0 3.4	**25** SA	0107 0720 1321 1942	0.8 3.3 0.9 3.5
11 SA	0241 0856 1452 2118	1.0 3.0 1.2 3.2	**26** SU	0155 0811 1413 2035	0.9 3.3 1.0 3.4
12 SU	0339 0955 1554 2218	1.2 2.9 1.4 3.0	**27** M	0250 0909 1513 2136	1.2 3.2 1.1 3.3
13 M	0442 1100 1700 2323	1.3 2.9 1.4	**28** TU	0352 1016 1621 2243	1.0 3.2 1.2 3.3
14 TU	0545 1203 1805	1.3 2.9 1.4	**29** W	0500 1126 1734 2353	1.0 3.2 1.1 3.3
15 W	0025 0642 1259 1903	3.0 1.3 3.0 1.3	**30** TH	0608 1234 1843	1.0 3.3 1.0

JULY

	Time	m		Time	m
1 F	0101 0712 1336 1945	3.4 0.9 3.5 0.8	**16** SA	0131 0742 1356 2007	3.0 1.2 3.2 1.2
2 SA	0202 0809 1431 2042	3.6 0.7 3.7 0.7	**17** SU	0219 0826 1440 2051	3.1 1.1 3.3 1.0
3 SU	0257 0901 1522 2134	3.7 0.6 3.8 0.5	**18** M	0301 0906 1520 2131	3.2 0.9 3.5 0.8
4 M	0348 0950 1610 2223	3.8 0.5 3.9 0.4	**19** TU	0341 0944 1558 2209	3.4 0.8 3.6 0.7
5 TU	0436 1035 1655 2309	3.8 0.5 4.0 0.4	**20** W	0419 1021 1636 2246	3.5 0.7 3.8 0.6
6 W	0521 1119 1739 2353	3.7 0.5 3.9 0.5	**21** TH	0457 1058 1714 2325	3.6 0.6 3.8 0.5
7 TH	0604 1201 1822	3.6 0.6 3.8	**22** F	0536 1138 1755	3.6 0.5 3.9
8 F	0036 0646 1243 1904	0.6 3.5 0.8 3.6	**23** SA	0005 0618 1219 1837	0.5 3.6 0.6 3.8
9 SA	0118 0729 1326 1946	0.8 3.3 0.9 3.4	**24** SU	0048 0702 1304 1923	0.6 3.5 0.7 3.7
10 SU	0202 0812 1411 2031	1.0 3.1 1.1 3.2	**25** M	0134 0750 1353 2014	0.6 3.4 0.8 3.6
11 M	0249 0901 1501 2121	1.1 3.0 1.3 3.1	**26** TU	0225 0845 1449 2112	0.8 3.3 1.0 3.4
12 TU	0342 0957 1600 2220	1.3 2.9 1.4 2.9	**27** W	0324 0948 1556 2219	1.0 3.2 1.1 3.3
13 W	0443 1102 1708 2327	1.4 2.8 1.5	**28** TH	0433 1101 1713 2335	1.1 3.2 1.2 3.2
14 TH	0549 1207 1816	1.4 2.9 1.5	**29** F	0548 1217 1830	1.1 3.2 1.1
15 F	0033 0650 1306 1916	2.9 1.4 3.0 1.3	**30** SA	0050 0659 1325 1939	3.2 1.1 3.4 0.9
			31 SU	0155 0800 1423 2036	3.4 0.9 3.6 0.8

AUGUST

	Time	m		Time	m
1 M	0251 0852 1514 2127	3.5 0.7 3.7 0.6	**16** TU	0238 0842 1456 2108	3.2 1.0 3.5 0.8
2 TU	0340 0938 1559 2211	3.6 0.6 3.9 0.5	**17** W	0319 0921 1535 2147	3.4 0.7 3.7 0.6
3 W	0423 1020 1640 2252	3.7 0.6 3.9 0.5	**18** TH	0358 1000 1614 2225	3.6 0.6 3.9 0.4
4 TH	0503 1100 1719 2331	3.7 0.5 3.9 0.5	**19** F	0436 1038 1653 2304	3.7 0.4 4.0 0.3
5 F	0540 1138 1756	3.6 0.6 3.8	**20** SA	0516 1118 1734 2344	3.8 0.4 4.1 0.3
6 SA	0008 0616 1215 1832	0.6 3.5 0.7 3.7	**21** SU	0557 1159 1817	3.8 0.4 4.0
7 SU	0044 0652 1251 1908	0.7 3.4 0.8 3.5	**22** M	0026 0640 1244 1903	0.4 3.8 0.5 3.9
8 M	0120 0729 1330 1946	0.9 3.3 0.9 3.3	**23** TU	0111 0728 1333 1953	0.5 3.6 0.7 3.7
9 TU	0159 0810 1412 2028	1.1 3.1 1.2 3.1	**24** W	0201 0821 1429 2051	0.5 3.4 0.9 3.4
10 W	0243 0858 1503 2119	1.3 3.0 1.4 2.9	**25** TH	0300 0926 1538 2202	1.0 3.3 1.1 3.2
11 TH	0339 0959 1608 2226	1.5 2.8 1.5 2.8	**26** F	0412 1044 1701 2325	1.2 3.1 1.2 3.1
12 F	0449 1112 1726 2345	1.5 2.8 1.6 2.8	**27** SA	0535 1207 1826	1.3 3.1 1.2
13 SA	0605 1224 1840	1.5 2.9 1.5	**28** SU	0046 0652 1319 1936	3.1 1.2 3.3 1.0
14 SU	0056 0708 1324 1939	2.9 1.4 3.1 1.3	**29** M	0152 0753 1416 2030	3.2 1.0 3.5 0.8
15 M	0152 0759 1413 2027	3.0 1.2 3.3 1.0	**30** TU	0244 0842 1503 2115	3.4 0.9 3.7 0.7
			31 W	0327 0924 1544 2155	3.5 0.7 3.8 0.5

Chart Datum is 1·86 metres below NMMA Datum. HAT is 4·5 metres above Chart Datum.

》》 FREE monthly updates. Register at 《
www.reedsnauticalalmanac.co.uk

969

LA CORUÑA LAT 43°22'N LONG 8°24'W

TIMES AND HEIGHTS OF HIGH AND LOW WATERS

STANDARD TIME UT −01
Subtract 1 hour for UT
For French Summer Time add
ONE hour in **non-shaded areas**

Dates in red are **SPRINGS**
Dates in blue are **NEAPS**

YEAR 2016

SEPTEMBER

Day	Time m	Time m	Time m	Time m
1 TH ●	0405 3.6	1002 0.6	1620 3.9	2231 0.5
16 F ○	0333 3.8	0936 0.5	1549 4.0	2200 0.3
2 F	0440 3.7	1037 0.6	1655 3.9	2304 0.5
17 SA	0413 3.9	1015 0.3	1630 4.2	2239 0.2
3 SA	0512 3.7	1111 0.6	1727 3.8	2336 0.6
18 SU	0453 4.0	1056 0.3	1712 4.2	2320 0.3
4 SU	0544 3.6	1145 0.7	1759 3.7	
19 M	0535 4.1	1139 0.3	1756 4.2	
5 M	0008 0.7	0616 3.5	1218 0.8	1831 3.6
20 TU	0003 0.3	0619 4.0	1225 0.4	1843 4.0
6 TU	0040 0.9	0649 3.4	1253 1.0	1905 3.4
21 W	0049 0.5	0707 3.8	1315 0.6	1935 3.7
7 W	0114 1.1	0725 3.3	1330 1.2	1942 3.2
22 TH	0139 0.8	0801 3.6	1413 0.9	2035 3.4
8 TH	0153 1.3	0808 3.1	1415 1.4	2029 3.0
23 F ◑	0239 1.1	0907 3.3	1525 1.1	2149 3.1
9 F ◑	0241 1.5	0903 2.9	1514 1.5	2131 2.8
24 SA	0354 1.3	1029 3.2	1652 1.3	2317 3.0
10 SA	0349 1.6	1017 2.9	1636 1.6	2257 2.7
25 SU	0522 1.4	1155 3.2	1818 1.2	
11 SU	0516 1.6	1140 2.9	1801 1.5	
26 M	0039 3.1	0641 1.3	1307 3.3	1925 1.1
12 M	0021 2.8	0633 1.5	1250 3.1	1908 1.3
27 TU	0141 3.2	0740 1.1	1402 3.5	2016 0.9
13 TU	0124 3.0	0730 1.3	1344 3.3	1958 1.1
28 W	0229 3.4	0826 1.0	1446 3.6	2057 0.8
14 W	0212 3.3	0815 1.0	1428 3.6	2041 0.8
29 TH	0308 3.5	0905 0.8	1523 3.7	2132 0.7
15 TH	0254 3.5	0856 0.8	1509 3.8	2121 0.5
30 F	0342 3.6	0940 0.7	1557 3.8	2204 0.6

OCTOBER

Day	Time m	Time m	Time m	Time m
1 SA ●	0413 3.7	1013 0.7	1628 3.8	2235 0.6
16 SU ○	0348 4.1	0953 0.3	1607 4.3	2215 0.2
2 SU	0444 3.7	1045 0.7	1658 3.8	2306 0.7
17 M	0430 4.2	1036 0.2	1652 4.3	2258 0.2
3 M	0514 3.7	1117 0.7	1729 3.7	2335 0.8
18 TU	0514 4.2	1122 0.2	1738 4.2	2343 0.3
4 TU	0544 3.7	1149 0.8	1759 3.6	
19 W	0600 4.1	1209 0.4	1827 4.0	
5 W	0005 0.9	0616 3.5	1222 1.0	1832 3.4
20 TH	0029 0.5	0649 3.9	1301 0.6	1920 3.7
6 TH	0037 1.1	0650 3.4	1258 1.2	1908 3.3
21 F	0121 0.8	0744 3.7	1400 0.9	2021 3.4
7 F	0113 1.3	0730 3.3	1340 1.3	1952 3.1
22 SA ◑	0220 1.1	0850 3.4	1511 1.1	2134 3.1
8 SA	0158 1.5	0821 3.1	1435 1.5	2051 2.9
23 SU	0334 1.4	1008 3.2	1635 1.3	2259 3.0
9 SU ◑	0300 1.6	0930 3.0	1551 1.6	2213 2.8
24 M	0459 1.5	1131 3.2	1756 1.3	
10 M	0425 1.7	1053 3.0	1718 1.5	2341 2.9
25 TU	0017 3.0	0617 1.4	1242 3.3	1901 1.2
11 TU	0551 1.6	1209 3.1	1830 1.4	
26 W	0118 3.2	0716 1.2	1337 3.4	1951 1.0
12 W	0049 3.1	0654 1.3	1308 3.3	1924 1.1
27 TH	0204 3.3	0803 1.1	1421 3.5	2031 0.9
13 TH	0141 3.3	0744 1.1	1357 3.6	2010 0.8
28 F	0242 3.5	0841 1.0	1458 3.6	2105 0.8
14 F	0225 3.6	0828 0.8	1441 3.9	2052 0.5
29 SA	0316 3.6	0916 0.9	1531 3.7	2137 0.8
15 SA	0307 3.9	0911 0.5	1524 4.1	2134 0.3
30 SU ●	0347 3.7	0949 0.8	1602 3.7	2208 0.8
31 M	0417 3.7	1021 0.8	1632 3.7	2237 0.8

NOVEMBER

Day	Time m	Time m	Time m	Time m
1 TU	0447 3.8	1053 0.8	1703 3.7	2307 0.9
16 W	0457 4.3	1108 0.3	1723 4.1	2326 0.4
2 W	0518 3.7	1125 0.9	1734 3.6	2338 1.0
17 TH	0545 4.2	1157 0.4	1813 3.9	
3 TH	0550 3.7	1159 1.0	1807 3.5	
18 F	0013 0.6	0634 4.0	1249 0.6	1906 3.7
4 F	0010 1.1	0625 3.5	1235 1.1	1844 3.3
19 SA	0104 0.8	0728 3.8	1346 0.8	2004 3.4
5 SA	0047 1.2	0705 3.4	1317 1.2	1928 3.2
20 SU	0200 1.1	0828 3.5	1450 1.1	2109 3.1
6 SU	0130 1.4	0753 3.3	1408 1.4	2023 3.0
21 M ◑	0305 1.3	0936 3.3	1602 1.2	2223 3.0
7 M ◑	0226 1.5	0854 3.1	1514 1.5	2135 2.9
22 TU	0420 1.4	1050 3.2	1716 1.3	2337 3.0
8 TU	0340 1.6	1008 3.1	1632 1.5	2257 2.9
23 W	0535 1.5	1201 3.2	1821 1.3	
9 W	0503 1.6	1124 3.2	1746 1.3	
24 TH	0040 3.1	0639 1.4	1300 3.2	1915 1.2
10 TH	0009 3.1	0614 1.4	1229 3.4	1846 1.1
25 F	0131 3.2	0731 1.3	1349 3.3	1959 1.1
11 F	0107 3.4	0711 1.1	1324 3.6	1938 0.8
26 SA	0213 3.3	0814 1.1	1429 3.4	2036 1.0
12 SA	0156 3.7	0800 0.8	1414 3.9	2025 0.6
27 SU	0249 3.5	0852 1.0	1505 3.5	2111 0.9
13 SU	0242 3.9	0847 0.6	1501 4.1	2110 0.4
28 M	0322 3.6	0927 0.9	1538 3.5	2143 0.9
14 M ○	0326 4.1	0933 0.4	1548 4.2	2154 0.3
29 TU ●	0355 3.7	1001 0.9	1610 3.6	2214 0.9
15 TU	0411 4.2	1020 0.3	1635 4.2	2240 0.3
30 W	0426 3.7	1034 0.9	1643 3.6	2246 0.9

DECEMBER

Day	Time m	Time m	Time m	Time m
1 TH	0459 3.7	1108 0.9	1716 3.5	2318 0.9
16 F	0532 4.2	1147 0.4	1801 3.9	2359 0.6
2 F	0532 3.7	1142 0.9	1751 3.5	2352 1.0
17 SA	0620 4.0	1236 0.5	1850 3.7	
3 SA	0608 3.6	1219 1.0	1828 3.4	
18 SU	0046 0.7	0709 3.8	1326 0.7	1940 3.4
4 SU	0029 1.1	0648 3.6	1300 1.1	1911 3.3
19 M	0136 1.0	0800 3.6	1419 1.0	2033 3.2
5 M	0112 1.2	0733 3.4	1348 1.2	2001 3.1
20 TU	0229 1.2	0855 3.3	1516 1.2	2133 3.0
6 TU	0202 1.3	0827 3.3	1444 1.3	2101 3.0
21 W ◑	0329 1.4	0956 3.1	1619 1.3	2239 2.9
7 W ◑	0304 1.4	0930 3.2	1550 1.3	2213 3.0
22 TH	0437 1.5	1103 3.0	1726 1.4	2346 2.9
8 TH	0417 1.4	1040 3.3	1700 1.3	2326 3.1
23 F	0547 1.5	1210 3.0	1828 1.4	
9 F	0531 1.4	1149 3.3	1807 1.1	
24 SA	0047 3.0	0650 1.5	1309 3.1	1922 1.3
10 SA	0031 3.3	0637 1.2	1253 3.5	1907 0.9
25 SU	0138 3.1	0743 1.3	1358 3.2	2006 1.2
11 SU	0129 3.6	0735 0.9	1350 3.7	2000 0.7
26 M	0221 3.3	0827 1.2	1440 3.3	2046 1.1
12 M	0221 3.8	0828 0.7	1443 3.9	2051 0.5
27 TU	0259 3.4	0907 1.0	1517 3.4	2121 1.0
13 TU	0310 4.0	0919 0.5	1534 4.0	2139 0.4
28 W	0334 3.6	0943 0.9	1552 3.4	2155 0.9
14 W ○	0358 4.2	1009 0.3	1623 4.1	2226 0.4
29 TH ●	0409 3.7	1018 0.8	1626 3.5	2229 0.8
15 TH	0445 4.2	1058 0.3	1712 4.0	2312 0.4
30 F	0443 3.7	1053 0.8	1701 3.5	2302 0.8
31 SA	0517 3.8	1128 0.7	1737 3.5	2338 0.8

Chart Datum is 1·86 metres below NMMA Datum. HAT is 4·5 metres above Chart Datum.

9.23.29 LA CORUÑA (A Coruña)

La Coruña **43°22'·01N 08°23'·69W** (Marina) ✳✳✳⚓⚓⚓🏳🏳🏳

CHARTS AC 1111, 1094, 1110; SC 412, 412A, 4126; SHOM 3007, 7598, 6665

TIDES ML 2·2. Note: La Coruña is a Standard Port (←).

SHELTER from seaward: Marina Coruña, close N of Castillo de San Antón, is sheltered from wash/swell by two wavebreaks. Note: ⚓ prohibited SE of wavebreaks and within hbr limits, you will be moved by Maritime Police; no ♀s. Marina Real in 3-4m is well sheltered. Possible ⚓'s in the Ria de la Coruna include NE corner, Ensenada de Mera, in NW/NE winds, and S corner, E end Playa de Santa Christina in W/S winds.

NAVIGATION WPT 43°23'·21N 08°22'·12W, 189°/1·3M to head of Dique de Abrigo; thence 6-9ca west to marina. The WPT is where the 108·5° & 182° ldg lines intersect. Avoid Banco Yacentes, about 1M NW of WPT (off chartlet). Speed limit 3kn in hbr.

LIGHTS AND MARKS Conspic marks: Torre de Hércules lt ho; 5·3M SW of which is power stn chy (218m); twin white twrs (85m, R lts) of Hbr Control on Dique Barrié de la Maza.

COMMUNICATIONS (Code 981) MRSC 209541; Emergency 112; VTS VHF Ch 13; ⊖ & Met via YCs; Brit Consul (902)109356; HM 216161; Port Control VHF Ch 12; Marinas VHF Ch 09.

FACILITIES Marina Coruña ☎920482; www.northwestmarinas.com jesusga@marinacoruna.es or marinacoruna@marinacoruna.es ⚓ 700 +100Ⓥ, ⚲ €2.72, Jun-Sep, max 50m LOA, multihulls + 50%, discounts for longer stays. 🔌 🛢 (note: pontoon, 4m LW); ✕ 🍴 🛒 🗐 🛢 ⬧ 🅿(50t) 🛥(50t). For repairs go to **Marina Seca** 1·5M S at 43°20'·92N 08°22'·97W; ☎ as above.

Marina Real (Real Club Nautico de La Coruna) ☎226880 (ext 3) mob 650053130 www.marinarealcoruna.com marina@rcncoruna. com 350 inc 40 Ⓥ, ⚲ €2.67, Jun-Sep, max 30m LOA, multihulls + 50%. 🗐 🛢 ⬧ ✎ 🛒 Ⓔ ⚒ ⚐, wi-fi 🅿(32t).

City All amenities. Larger quantities duty free 🛢 available for registered (non private) commercially run vessels/charter yachts from Bunoil S.L. ☎136248 (HO), or mob 639186552, in Dársena de Oza, inside outer end bkwtr, opposite Marina Seca. Bus, ≈ ✈.

9.23.30 RÍA DE CORME Y LAXE

La Coruña. Corme lt **43°15'·63N 08°57'·83W** ✿✿✿✦✿✿✿
Laxe, E lt **43°13'·35N 08°59'·95W** ✿✿✿✦✿✿✿

CHARTS AC 3633, 1111, 1113; SC 928; SHOM 3007

TIDES
Standard Port LA CORUÑA (←—)

Times				Height (metres)			
High Water		Low Water		MHWS	MHWN	MLWN	MLWS
0000	0600	0000	0500	4·2	2·8	1·6	0·3
1200	1800	1200	1700				
Differences RÍA DE CORME							
–0005	–0005	–0004	–0004	–0·3	–0·2	–0·1	0·0

SHELTER Corme is well sheltered from N and E winds. Its pier has been extended to enlarge the crowded ⚓age. **Laxe** (Lage) is well sheltered from W and S. Both hbrs may be affected by swell.

NAVIGATION WPT 43°15'·31N 09°00'·63W: 081°/2·1M to Corme pierhd; 166°/2·0M to Lage pierhd. The ría is over 2M wide and the appr is straightforward, but beware rocks off Pta del Roncudo and Pta de Laxe. In bad weather follow the ldg lines to Corme to avoid Bajo de La Averia (5·6m). On SC 4131 many depths in apprs and in the ría may be less than charted on AC 1113.

LIGHTS AND MARKS All lts and ldg lines as on the chartlet.

COMMUNICATIONS (Code 981). MRSC 209541; See Area map and Passage information for the TSS/ITZ off Finisterre; Emergency 112; Brit Consul (902)109356.

FACILITIES Corme, Fisherman Cofradia ☎738043, ⛽(4t) ⬛. **Laxe** Fisherman Cofradia ☎728255 mob 699998686. ✎ ⛽(16t). Space may be available on the piers, raft up on FVs or ⚓ close in between the pierheads, good holding on sand in 5m. ⌂▷🛟. **Both villages** are small: ✕ ⬜ 🗑 ✉ ⑧ @ 🅳 🅟. Buses to La Coruña, ✈ Santiago.

9.23.31 RÍA DE CAMARIÑAS

La Coruña **43°07'·58N 09°10'·94W (CNC)** ✿✿✿✦✦✿✿✿

CHARTS AC 1111, 3633, 1113; SC 928, 927; SHOM 3007

TIDES
Standard Port LA CORUÑA (←—)

Times				Height (metres)			
High Water		Low Water		MHWS	MHWN	MLWN	MLWS
0000	0600	0000	0500	4·2	2·8	1·6	0·3
1200	1800	1200	1700				
Differences RÍA DE CAMARIÑAS							
–0005	–0005	–0005	–0005	–0·2	–0·2	–0·1	0·0

SHELTER Good in **Camariñas**, or ⚓ to N or NW of hbr except in E/NE'lies, **Mugia** (43°06'·36N 09°12'·73W) is well sheltered in all but strong E/SE'lies either in Muxia Marina Cataventos, or ⚓ outside hbr in lee of bkwtr.

NAVIGATION El Bufardo (43°10'·00N 09°13'·28W, 4ca NW of C. Villano) a dangerous pinnacle rk awash at CD. From N pass E of Las Quebrantas shoal (0·6m). From S appr on the 081° ldg lts. WPT 43°07'·27N 09°12'·65W (intersection of the 108° and 081° ldg lines). Thence track 108°/1·1M, then 039°/0·7M to Camariñas bkwtr hd. For Mugia bkwtr 180°/0·9M.

LIGHTS AND MARKS C. Villano, with 23 conspic wind turbines close SE, is 2·2M N of ent to ría. Pta de la Barca lt ho marks the SW side of the ría ent. Pta de Lago dir lt ≠ bcn leads 108° into the ría.

COMMUNICATIONS (Code 981) MRSC 209541; Emergency 112; Brit Consul (902)109356; Marinas VHF Ch 09

FACILITIES CN Camariñas (CNC) ☎737130 www.cncmarinas.com cncamarinas@cncamarinas.com 60 + ♥ €1.50 F&A, on 3 YC pontoons. max 18m LOA. ⊖ 🛟🅟🛠🔧✎ ⬜✕.

Muxia Marina Cataventos mob 673168199/ 666369324; www.marinamuxia.com info@marinamuxia.com 232⛽ inc ♥, €1·82,multihulls +50%, max 22m LOA. 🛟🐟▷✎ ✕⬛ wi-fi 🛢 (from truck). **Town** ✕ ⬜ 🗑 ✉ ⑧. Buses to La Coruña, ✈, Santiago de Compostela. Reports welcomed.

CABO FINISTERRE TO THE PORTUGUESE BORDER

NE of Cabo Finisterre (AC 3674) the Ensenada del Sardineiro and Ría de Corcubión are sheltered ⚓s except in S'lies and despite fish farms. From 5 to 11M south, beware various islets, reefs and shoals N and NW of Pta Insua, notably the unmarked Las Arrosas and Bajo de los Meixidos; the latter is in the red sector (070°-090°) of Pta Insua and can be passed inshore if bound to/from Ría de Muros.

The 40M stretch (AC 3633) from Pta Queixal to Cabo Silleiro makes an impressive cruising ground, encompassing the four major rías bajas (lower or southern) from N to S: Muros, Arosa, Pontevedra and Vigo. All resemble large and scenic Scottish sea lochs with interesting pilotage to fishing hbrs and the many sheltered anchorages. In summer NE winds can blow strongly down all rías, usually without raising any significant sea. The last three rías are sheltered from onshore winds by the coastal islands at their mouth. These Galician islands have protected National Park status, with restrictions on pleasure craft, anchoring, fishing and diving, see: www.iatlanticas.es.

Ría de Muros (AC 1756) is the smallest and easiest to enter. Monte Louro at the N side of the entrance is a distinctive, 239m high, conical hill. The ancient town of Muros itself is worth visiting and the 212 berth **Marina Muros** usually has plenty of space for visitors. The anchorage off the harbour is renowned for poor holding in weed. Most of the bays and inlets are occupied by fish farms, often unlit and unmarked, so night pilotage is not advised. On the eastern shore **Portosin** is a very well run, modern marina with every facility.

Ría de Arosa (AC 1764) the largest ría, runs 14M inland and is up to 6M wide. Isla Sagres and Isla Salvora at the western side of the entrance provide shelter from SW'lies and are within the National Park. **Villagarcia** is the principal hbr and marina, but there are other marinas at **Santa Uxia de Ribeira, Caramiñal, Puerto Cabo de Cruz, Rianxo, Vilanova** and **Piedras Negras**. There are many anchorages, mostly good holding on mud, several active fishing fleets and numerous fish farms and nets. Isla de Arosa and Peninsula del Grove, both in the SE part of the Ria, provide intricate pilotage and secluded anchorages.

Ría de Pontevedra (AC 1732) can be entered from the NW by one of two minor channels (Canals de la Fagilda and de Los Camoucos) between the mainland and Isla Ons. The main entrance, Boca del Sudoeste, is from the SW where the channel is unobstructed and 3·5M wide. On the N shore **Sanxenxo** is a modern marina whilst near the N end (AC 1733) **Combarro** has a marina and is a popular tourist spot. The naval base at Marin has nothing for yachts, but **Aguete** has a small marina and **Beluso** caters for yachts and FVs. Isla Ons and Isla Onzas are protected by National Park rules.

Ría de Vigo (AC 1730) is notable for the large busy port of **Vigo** and, to the S, Bayona's pleasant hbr with 2 marinas. The beautiful Islas Cies, comprising Isla del Norte and the smaller Isla de San Martin, are within the National Park and also provide considerable shelter to the ría against onshore winds. The two approach channels, from NW and SW, have been formalised into 2 mini-TSS which meet in a Precautionary area where the ría narrows down and the docks begin. **Marina Davila Sport** in the Bouzas area has first class BY, 1M SW of the long-established but rather small marina near central Vigo. 1M NE **Punta Lagao** Marina is on the W side of Monte de la Guia which is topped by a conspic chapel. Several lesser marinas are on the N bank and within the Ensenada de San Simón beyond the Rande bridge (38·8m clearance).

Bayona AC 3633 is too small a scale, 1:200,000, for pilotage. AC 1730, 1:25,000, subject to bilateral agreement with Spain, will hopefully include Bayona which is usually approached from seaward on the 084° leading line, passing between Las Serralleiras, rky islets to the N, and to the S Cabo Silleiro and its offshore reefs marked by a NCM buoy. The long-established Monte Real YC marina has been supplemented by Bayona marina close S; there is still space to anchor.

The Río Miño, 15M S of Cabo Silleiro, is the Spanish/Portuguese border. The river ent is difficult and best not attempted.

SENO DE CORCUBIÓN

MINOR HARBOURS IN SENO DE CORCUBIÓN

FINISTERRE. 42°54'·59N 09°15'·36W (hbr ent). AC 1111, 3764; SC 927, 9270. Tides: Ría de Muros, Corcubión. Finisterre (or Fisterra, the end of the Ancient World) is famous and imposingly massive. The touristy town has no yacht facilities. But it is worth walking 1·5M from the hbr to the granite lt ho for the views and ambience. Quickly forming dense fog is frequent on this coast. From the N, Centolo de Finisterre is a 23m high pinnacle rock 9 cables WNW of the lt ho. Beware La Carraca with 2·1m over it (42°53'·93N 09°18'·53W) less obvious therefore more dangerous. El Turdeiro (1·9m) is 300m due S of the cape, with a sprinkling of lesser rocks closer inshore where small FVs cluster; keep well clear. The E side is mostly steep-to.

The hbr (see chartlet) is usually full of small FVs, but in settled weather it is safe to ⚓ outside in 5-10m, either close S of the bkwtr or to the N off the Playa de Llagosteira.

ENSENADA DEL SARDIÑEIRO 42°56'·36N 09°13'·82W, ⚓ off the village of Sardineiro at the NW corner. Charts and Tides as for Ría de Muros. This S-facing inlet is about 6ca wide and lies between Playa de Lagosteira to the W and to the E the peninsula which forms the W bank of Ría de Corcubión. There are no lights.

Beware 0.7m patch midway along west shore. Lovely ⚓ and sandy beach at Playa de Estorde, on E side, or ⚓, in 4-5m off the sandy beach fronting the village of Sardiñeiro. Only basic facilities available ashore.

RÍA DE CORCUBIÓN 42°56'·77N 09°11'·28W (⚓ at head of ria). AC 1111, 3764; SC 927, 9270. Tides: Ría de Muros. A pleasant ría, well sheltered from the summer NE'lies, open only to S'lies. Three features south of the ría which are seen on the approach are: Lobeira Grande, a group of yellowish rocky islets with lt; Carrumeiro Grande, a single flat island with ldg bcn; and Carrumeiro Chico, a small rock with IDM bcn twr and lt. It bears 162°/6ca from Cabo Cée lt ho at the SW side of the ent to ría.

The simplest approach for yachts is to pass between Cabo Cée and Carrumeiro Chico on a NE'ly heading, thence straight up the ría on 342° for 1·5M, with Carrumeiro Grande bearing 162° astern, if this is found necessary. A night appr presents no great difficulties: the ría is 0·5M wide and of the 3 lts at its head, as on the chartlet, keep midway between ⚡ Fl (2) R 7s on the W bank and ⚡ Fl (4) G 11s on the E bank.

Best ⚓ SE of the ⚡ Fl (2) R 7s in 6-7m, clear of the commercial quays. Holding is variable. The 100m jetty at Corcubión is easy to access with 4m LWS, and the town is nearer and more attractive than Cée at the head of the ría; both have the usual facilities and shops.

9.23.32   RÍA DE MUROS

La Coruña **42°46'·56N 09°03'·41W** (Muros) ⚓⚓⚓🌊🌊🌀🌀🌀☆

CHARTS AC 1756; SC 415, 415A; SHOM 3007

TIDES

Standard Port LISBOA (→)

Times				Height (metres)			
High Water		Low Water		MHWS	MHWN	MLWN	MLWS
0500	1000	0300	0800	3·8	3·0	1·5	0·6
1700	2200	1500	2000				
Differences CORCUBION							
+0055	+0110	+0120	+0135	−0·5	−0·4	−0·3	−0·1
MUROS							
+0050	+0105	+0115	+0130	−0·3	−0·3	−0·2	−0·1

SHELTER Excellent in Marina Muros. ⚓ off harbour (but beware poor holding in weed) or at Ensenada de San Francisco to the SSW. Noya, at the head of the ría, is only accessible to shoal draft. CN Portosín on the E shore is sheltered, efficient and welcoming – a real pleasure; see Facilities.

NAVIGATION WPT 42°43'·60N 09°03'·36W, 1·25M SE of Punta Queixal lt and 2·0M NW of Punta Focha lt (off chartlet). From the N, the Canal de los Meixidos (between the mainland and Bajo de los Meixidos and Los Bruyos, where its least width is 1.1M) is usable with care. From the S avoid Bajos la Baya. The ría is mostly deep and clear, but beware mussel rafts (*bateas*) off the N shore.

LIGHTS AND MARKS Mte Louro is a conspic conical hill (239m) N of Pta Queixal. Isla Creba is a good mark NW of Portosín. Principal lts: Pta Queixal, C. Reburdino, Pta Cabeiro, Pta Focha.

COMMUNICATIONS (Code 981) MRCC 767320; Emergency 112; Brit Consul (902) 109356; HM 826005; *Marina Muros* VHF Ch 09, H24. *Club Náutico Portosín* VHF Ch 09.

FACILITIES Marina Muros 42°46'·64N 09°03'·29W ☎827660; www.muport.es info@muport.es 212⊜ inc Ⓥ €25·50 up to 12m, outer hbr, inner hbr for locals/FVs. ✗ ⬜ 🍴 🔧 🚾 (not near LW) ⛽ ✉ Ⓑ wi-fi; ⭢ and ✈ Santiago de Compostela (40 km by bus).

CN Portosín 42°45'·94N 08°56'·92W. ☎766598 info@cnportosin.com www.cnportosin.com ④ 220 inc Ⓥ €2·52 F&A/fingers in 3·5-8m; ⚓ ✗ ⬜ ⛽ 🔧 @ ⚒ 🔧 ✂ ⚓ 🅿(32t) ⚓(3t).

Village 2km: ✗ ⬜ 🍴 ✉ Ⓑ car hire. ✈ Santiago, 1 hr's drive.

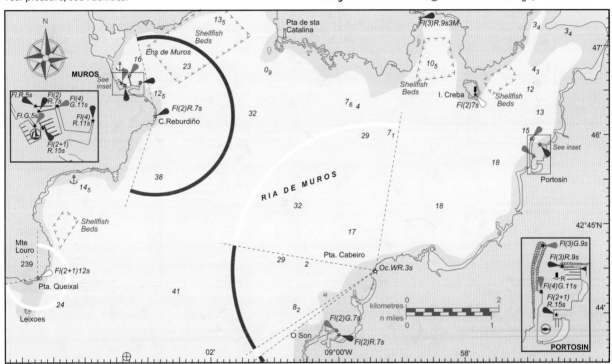

9.23.33   RÍA DE AROSA

La Coruña (NW shore) Pontevedra (SE shore).
42°34'·58N 08°54'·15W (mid-Ría) ⚓⚓⚓🌊🌊🌀🌀🌀☆

CHARTS AC 1734, 1764, 1762, 1755; SC 415, 415B/C

TIDES

Standard Port LISBOA (→); ML 2·05

Times				Height (metres)			
High Water		Low Water		MHWS	MHWN	MLWN	MLWS
0500	1000	0300	0800	3·8	3·0	1·5	0·6
1700	2200	1500	2000				
Differences VILAGARCIA							
+0040	+0100	+0110	+0120	−0·2	−0·3	0·0	−0·4

SHELTER Shelter can be found from most winds. Many ⚓ages await exploration with large scale charts and a Pilot book.

NAVIGATION WPT 42°27'·00N 08°59'·50W, 025°/6·6M to 5ca E of Is Rua (lt) 42°32'·96N 08°56'·38W. Isla Sálvora (lt) is usually left to port; the inshore passage to NW between Isla Sagres and Pta Falcoeiro, Canal de Sagres, is rock bordered but viable in daylight with good weather and AC 1734 or SC 4152/3, essential for cruising Ría de Arosa, the largest ría (approx 14M x 7M). The fairway to Vilagarcia is buoyed/lit, but the many mussel rafts (*bateas*) on either side are either unlit, or marked by Y lt buoys. The low bridge from Isla de Arosa to the mainland prevents passage.

LIGHTS AND MARKS Principal lts/marks are on the chartlet as scale permits. Isla Rúa is a prominent rky islet with lt ho.

COMMUNICATIONS (Codes: 981 La Coruna, NW shore. 986 Pontevedra, SE shore). MRCC (981) 767320; Emergency 112; Brit Consul (902) 109356; Santa Uxia, Caramiñal, Rianxo and Vilagarcia VHF Ch 09, 16.

FACILITIES Marinas (see insets)
Santa Uxia 42°33'·80N 08°59'·34W. ☎(981) 873801. 70 F & A on 4 pontoons in 4m. **Town** all facilities.

Caramiñal 42°36'·27N 08°56'·10W. ☎608174395, www.muport.es info@muport.es 305⊜ in 3·5m inc 30 Ⓥ, €3·00 inc 🔧 ⛽ 📶 🚾 🔧 🅿 by taxi, ⚓.

Marina Cabo de Cruz 42°36'·60N 08°53'·30W, VHF Ch 09, ☎ (622)884846,   info@nauticoboiro.com   www.nauticoboiro.com

N & NW Spain

CARAMIÑAL

PIEDRAS NEGRAS

Fl(4)G.11s4m1M
Fl(4)WR.11s7m4/3M
Q(6)+LFl.15s3M

SANTA UXIA

Fl(2+1)G.15s
Fl(2)G.7s
Fl(2)R.7s
Fl.R.5s
Fl(2+1)R.15s

VILAGARCIA DE AROSA

Fl(2)R.5s
Fl(3)R.9s
Fl(4)R.11s
Fl(2)G.5s
Fl(2)G.7s
Fl(2)R.7s
Fl.G.3s
Fl.G.5s
Fl(2+1)R.14.5s
Fl(2+1)G.14.5s
Maintained depth 9m
Ro Ro

CARAMIÑAL
Fl(3)G.9s
Fl(2+1)R.15s

PIEDRAS NEGRAS

Puerto Cabo de Cruz
Fl(3)R.9s
Fl(2)G.7s3M

Santa Uxia de Ribeira
Q.R
Q.G

A Pobrado Caramiñal
See separate chartlet
Fl(2)WR.7s Escarabote
Fl(2)7s YC
Fl(3)R.9s
Fl(2)G.7s3M

Puerto Cabo de Cruz
See inset

Bodion
Fl(4)G.11s
Fl(4)G.11s
Fl(3)G.9s
Fl(4)G.11s
Q(9)15s
Rianxo
See inset

Fl.G.5s

Fl(3)R.10s
Fl(2)R.7s
Carril
Fl(2)G.7s
Fl(4)R.11s
Fl.Y.5s
Fl.G.5s
F(2)G.7s
See separate chartlet
VILAGARCIA
Vilaxoan

Palmeira

Ens de Corrubedo

I.Rua(22)
Fl(2)WR.7s
Racon G

Fl(4)R.8s
Fl(2+1)R.7s
Fl.G.3s
R
RW
Fl.R.5s

Pta Caballo
Fl(4)11s
Fl(2+1)G.15s
S.Xulian
VQ.W.0·5s
Q.R
Fl(4)G.11s

Vilanova
See inset

ISLA DE AROSA

Fl(2)7s
BRB
Fl(4)G.11s
Fl(3)G.9s
Fl(3)R.9s
Q(9)15s
Pta Fincheira
Piedra Couto
RIANXO
Fl.G.5s
Fl(4)R.11s

PTA FALCOEIRO
Aguiño
Fl.R.3s
Q.R
I.Sagres
Q.G

Santa Uxia de Ribeira
See separate chartlet
Q.R
Q.G

Fl(3)G.9s
WGW

Fl(2)G.7s
GW

Fl(3)G.9s

Fl(2)WR.7s

GRG
Fl(2+1)G.14·5s

Fl.G.5s
Fl(4)Y.12s
Fl.R.5s

Fl(4)Y.12s
Fl(3)R9s
Cambados
Q.G
GW
Fl(2)G.7s
Fl(2)R.7s

San Martin del Grove

National Park
obscd
Fl(3)

ISLA SALVORA

Fl(3+1)20s
Fl(3)20s

Fl(3+1)

Fl(3)G.10s

Fl.G.5s
WGW

PENINSULA DEL GROVE
Fl(4)WR.11s **Piedras Negras**
See separate chartlet
Q(6)+LFl.15s
Fl(3)R.9s
Fl(3)G.9s
Ens de la Lanzada
08°55'W

VILANOVA
Fl.R.5s
Fl.G.5s
Fl(4)R.11s
Fl.R.5s
Fl(2+1)G.15s
Fl(4)G.11s

nautical miles 0 ... 2

RIA DE AROSA
CANAL PRINCIPAL

200&+30 ⓥ, 6m-max 22m LOA in 2m-5m, €2·10, inc 🔌 ⚓ 🚿 🚾 🛟 🚶 ✕ 🛒 📶 wi-fi. **Town** Boiro, 4km/10mins taxi, shops, ✕ 🛒 🛍 ✉.
Rianxo 42°39'·00N 08°49'·51W. ☎866107. 2 pontoons at SE corner in 1·4m; 80 ⛵ or ⚓ just SE of the town off the beach.
Vilagarcia 42°36'·04N 08°46'·21W. ☎986 511175. marinavilagarcia @marinavilagarcia.com 450 ⛵ inc ⓥ €2.00 as directed or on 1st pontoon to stbd in 3-5m; 🔧 🛠 ✕ Ⓔ 🏗(70t) 🏗(50t) YC.

Town 🏨 🅿 nearby (unmanned, Spanish cr/cards). ✕ 🛒 🛍 ✉ Ⓑ 🏧 📞 (service only) 🚂 ✈ Santiago.
Vilanova Marina 42°33'·98N 08°50'·05W. ☎986 554113. €2.11/m. Due to extensive *viveros* the only appr is from the N: WPT at 42°35'·09N 08°50'·87W (⚓ Fl (3) G 10s); thence 179°/1·2M to 0·5M abm Vilanova. 🛟 🛒 🏗(35t) 🔧 in FV basin to the S.

Piedras Negras 42°27'·50N 08°55'·09W. ☎986 738430. www.cnsvicente.org club.nautico.s.v.@wanadoo.es ⊕42°26'·97N 08°54'·28W, 310°/8ca to ent in W sector (305°-315°) of Fl (4) WR 11s. 134 ⌣ inc ❶ €1.55; 2 pontoons + fingers, 12m max LOA; depths 4-6m. 🛢🛢🗜🏁(35t). **CN** ☎738325, ✕🍴 ☎738435.

FV harbours with yacht facilities or ⚓s include:

San Xulian (Isla Arosa, N Bay) 42°33'·94N 08°52'·14W; ⚓ 1ca NW of the southernmost pier, on the W side of which are 4 pontoons for small craft. ⚓ in S Bay at 42°33'·36N 08°52'·57W in 5-8m; restricted by many moorings. ⛵ at head of the bay.

Cambados 42°30'·69N 08°49'·11W, S of Tragove FV hbr. Very shallow, 0·3m; only advised for multihulls and bilge-keelers.

9.23.34 RÍA DE PONTEVEDRA

Pontevedra 42°22'·00N 08°50'·00W (mid-Ría) ❀❀❀❀⚓⚓⚓❀❀❀   CHARTS AC 3633, 1732/3; SC 416A/B; SHOM 3007

Continued overleaf

TIDES
Standard Port LISBOA (➡); ML 1·9

Times				Height (metres)			
High Water		Low Water		MHWS	MHWN	MLWN	MLWS
0500	1000	0300	0800	3·8	2·9	1·5	0·6
1700	2200	1500	2000				
Differences MARIN (42°24'N 08°42'W)							
+0050	+0110	+0120	+0130	–0·4	–0·3	–0·1	0·0

SHELTER Good shelter can be found from any wind. Islas Ons and Onza across the mouth of the ría offer a barrier to wind and swell. Marinas, hbrs and ⚓s are listed under Facilities.

NAVIGATION From WPT 42°18'·00N 08°57'·16W, (which also lies on the 129° ldg lts for Ria de Vigo) track 053° to enter via the main chan, Boca del Sudoeste, into centre of ria between Cabo de Udra and Pta Cabicastro.

From the NW small craft normally transit the Canal de la Figilda on 125° or the Canal de Los Camoucos on 180°; both are approx 3 cables wide and viable in rougher weather with due care.

LIGHTS AND MARKS As chartlet. Isla Ons is a steep, cliffy, rocky island with conspic octagonal lt ho.

COMMUNICATIONS (Code 986). MRCC (981) 767320; Emergency 112; Marin VHF Ch 12, 16; Marinas VHF Ch 09.

FACILITIES (Clockwise from NW)
Porto Novo 42°23'·71N 08°49'·28W. FV hbr with marina (100+ berths) and Real Club Nautico. ☎720802. Ferry to Isla Ons.

Off **Playa de Silgar** (between Porto Novo and Sanxenxo) there is a pleasant ⚓ on sand in 3-5m, exposed only to S'lies.

Sanxenxo Marina 42°23'·83N 08°48'·04W. SHM lt bcn marks drying rks, N side ent. Avoid ferry pier. **CN** ☎720517, nauta@sanxenxo.org www.rcnsanxenxo.com 500 ⌣ 🛢🛢✕🗜🚿Ⓑ🅱🏁⊠🏁(5t) ⛵.

Raxó 42°24'·06N 08°45'·34W. Temp'y ⚓ in about 5m off bkwtr hd.

Combarro 42°25'·38N 08°42'·12W. AC 1733. 287 + 45 ❶ <12m & 2 ❶ max 20m LOA, €2.80/m. Good depths in marina and channel but do not pass north of Isla Tambo; leave mussel rafts to port in final approach. ☎(986) 778415, www.combarromar.com comercial@combarromar.com; 🛢🛢⛽✕🗜🛢🏁(50t)🏁(6t) ⛵.

Pontevedra 42°24'·86N 08°41'·35W, mouth of Rio Lérez. A LW dinghy recce of shallow chan between training walls is advised; least depth 1·3m. Access restricted by overhead cables (19m clearance) and motorway bridge (12m clearance). **Town** The provincial capital has all domestic facilities and good road links.

Marin 42°23'·94N 08°42'·21W, a naval, commercial and FV port, with no yacht facilities (tiny marina in W basin, private). Isla Tambo, NW, is a restricted military area; landing prohib. Buoyed ship ⚓age NW of port.

Aguete 42°22'·55N 08°44'·16W. **Marina** ☎702373. Approx 100 ⌣ inc Ⓥ. N'most pontoon is a wavebreak, many ⚓s. To clear offlying rks, round the SHM buoy before turning S. ⓜ ⚓ ⚓ 🔌 ⚓ ⛽(28t) ⚓(8t).
Real Club de Mar ✕ ⌂. **Village** few shops.

Bueu 42°19'·79N 08°47'·02W. A FV hbr, no yacht berths; ⚓ to W of hbr in about 5m. Caution: mussel rafts (*bateas*) to NNW and NNE. Hbr ent is lit as chartlet. HM ☎320042, ✕ ⌂ 🛒 good shops 🅿 🅟 ⛽. Ferry to Isla Ons.

Beluso 42°20'·01N 08°47'·90W. Mussel beds (*viveiros*) are E/NE of the hbr. 2nd pontoon to stbd is for yachts, F & A berths, but access to shore may be difficult due to coded gates. HM ☎400870 if he can be found. Excellent ✕. 20 mins walk to Bueu.

Ria de Aldán Punta del Con at 42°20'·01N 08°47'·90W, head of Ria. Worth exploring in settled weather; beware mussel rafts on W side. Temp'y ⚓s may be found on E side and at head of ría; uncomfortable in fresh N'lies.

Isla Ons Almost uninhabited, but very popular with tourists in season. At Almacen jetty (Fl R 4s) ferry wash detracts from the ⚓ or possible ⚓s; ✕ ⌂. Fair weather ⚓ in 5m off Melide beach 42°23'·37N 08°55'·50W, but within National Park. No landing on Isla Onza (bird sanctuary).

9.23.35   RÍA DE VIGO

Pontevedra **42°13'·85N 08°48'·26W** (centre of precautionary area) ⚓⚓⚓⚓⚓⚓⚓⚓

CHARTS AC 3633, 1730, 1731; SC 416, 416B; SHOM 7595

TIDES
Standard Port LISBOA (→); ML 1·96

Times				Height (metres)			
High Water		Low Water		MHWS	MHWN	MLWN	MLWS
0500	1000	0300	0800	3·8	3·0	1·5	0·6
1700	2200	1500	2000				
Differences VIGO							
+0040	+0100	+0105	+0125	0·0	−0·3	0·0	−0·2

SHELTER The ría is sheltered by Islas Cies, which have attractive ⚓s on E side; also sheltered ⚓s in Ens de San Simon. 3 marinas near Vigo centre are well sheltered, unless a NW'ly scend enters.

NAVIGATION N Chan (129° ldg lts) WPT 42°16'·09N 08°53'·94W, thence 160°/2M via a northern TSS to intersect the 095° ldg line.

S Chan (069° ldg lts) WPT 42°09'·42N 08°54'·86W, thence via a southern TSS to intersect the 041° ldg line.

A precautionary area is centred on the intersection (see L/L under title) of the 095° and 041° ldg lines. Small craft should not impede big ships in the TSS and/or the Precautionary area.

The buoyed inner fairway is defined by brgs of 068° and 074° on the conspic Hermitage (chapel) tower on Monte de la Guía. Beware extensive unlit mussel rafts (*bateas*) along the N shore.

LIGHTS AND MARKS Islas Cies are easily identified by their high, rugged bare slopes and white beaches on the E side. Ldg lts/marks for the appr chans are as per the chartlet and Lights, buoys and waypoints.

COMMUNICATIONS (Code 986); MRCC 222230; Emergency 112; ⊖ & Met at marinas; Brit Consul (902)109356; Port Control *Vigo Prácticos* VHF Ch 14, 16; Marinas Ch 09; Club Náutico de Rodiera (Cangas) Ch 06; ⚓269414; ✈ 268200; Turismo 224757.

FACILITIES heading NE up the Ria:
NW shore
Cangas 42°15'·66N 08°46'·97W. Rodeira is a small friendly YC marina ☎300165 clubnauticocangas@hotmail.com in the NW corner of Ensenada de Cangas. ⊇ on fingers as directed; approx €2.50. ▮▯🔧(64t) 🔲(8t). **YC** (English spoken) 🔲🔲. Good security, 24 hrs guard. If full, ⚓ ESE of marina clear of ferries and FVs. **Town** Usual facilities, 🔲🔲. Ferry to Vigo ½ hourly.

Moaña Marina 42°16'·66N 08°44'·00W. ☎311140. www.moanamar. es; Appr from S, passing El Con hbr. Avoid *viveros* to NE and E. €1.37. Limited 🔲⊇ on fingers at N end of 1st pontoon (other fingers too short) telephone before arrival. ▬.

Domayo Marina 42°17'·26N 08°40'·60W, 7ca W of suspension bridge between Domayo hbr Fl R 5s and conspic factory quay to NE. Best approach from the SW, close inshore of fish farms (*viveros*).

NE of Rande suspension bridge (38.8m clearance)
Ensenada de San Simón, 42°17'·29N 08°39'·63W (AC 1730) has peaceful ⚓s in 3-4m mud off W, E and S sides, clear of *viveros*. €2.50; ✕ (lunch only).

San Adrián de Cobres 42°18'·08N 08°39'·20W. **Marina** ☎874007, contacto@acimutnorte.com www.acimutnorte.com Skirt a large *vivero* S and E of the marina to appr from the NE. Pierhead lt Fl (4) R 11s is the only lt, close NW of the marina. 204⊇ on fingers inc🔲 max LOA 18m, €1.25 ✕ @.

SE shore (Vigo city)
Marina Dávila Sport 42°13'·97N 08°44'·42W, ☎244612 www. davilasport.es recepcion@davilasport.es. In Darsena de Bouzas, Dutch manager (good English). €3.20. Many 🔲. ▮▯🔧 modern repair facilities for all yacht sizes 🔧(70t). **CN** ☎232442, ✕🔲.

Vigo Marina 42°14'·56N 08°43'·42W. Narrow ent, between PHM/ SHM lt twrs. ☎449694, mob 638037201/ VHF Ch 09 (both H24); www.rcnauticovigo.com puertodeportivo@rcnauticovigo.com 436 F&A, fingers 🔲 €2.65, in west basin (Dársena de A Laxe) max 10m LOA; east basin (Dársena de AS Avenidas) >10m to max 35m LOA and also on dolphins at marina ent. ▮▯ (0900-2100) 🔧🔧🔲Ⓔ 🔧✕ wi-fi security H24, key card access, ▬ 🔧(32t) 🔲(3t). **RCN** clubhouse ☎433588, ✕🔲🔲🔲.

Pta Lagoa Marina 42°15'·51N 08°42'·46W. ☎410096. www. yatesport.es lagoa@yatesport.es VHF 09. 250 + 100🔲 €3.40, max LOA 24m, draft 3·5m. ▮▯🔧🔧 🔲Ⓔ ⚓ 🔧(70t); 2½km N of city.

City of Vigo All amenities, ⚓ ✈ (internal flights).

9.23.36 BAYONA

Pontevedra 42°07'·45N 08°50'·55W (MRCY) ❋❋❋❋♋♋♋❀❀❀

CHARTS AC 3633, 1730; SC 417, 416B; SHOM 7596, 7595

TIDES
Standard Port LISBOA (→)

Times				Height (metres)			
High Water		Low Water		MHWS	MHWN	MLWN	MLWS
0500	1000	0300	0800	3·8	3·0	1·5	0·6
1700	2200	1500	2000				
Differences BAYONA							
+0035	+0050	+0100	+0115	−0·3	−0·3	−0·2	−0·1
LA GUARDIA (41°54'N 08°53'W)							
+0040	+0055	+0105	+0120	−0·5	−0·4	−0·3	−0·2

SHELTER Excellent in both marinas; wavebreak at Bayona Marina 20m NE of pontoons, now has low level Q.Fl.Y lts at each end. ↓ E/NE of marinas, clear of fairway to FV jetty.

NAVIGATION WPT 42°07'·78N 08°55'·06W (see Ría de Vigo chartlet) 084°/2·8M to SPM buoy QY. From S keep 1M off Cabo Silleiro to clear reefs marked by NCM buoy, Q. Las Serralleiras, rks/islets N of ldg line, are marked by SCM & WCM lt buoys and by a lt bcn. The Canal de la Porta, between Monte Ferro and Las Estelas is a useful shortcut to/from Ría de Vigo, but hold close to the island to avoid a 0.9m patch and rock which dries close W of mid-channel.

LIGHTS AND MARKS Ldg/Dir lts 084° as chartlet and Lights, buoys and waypoints, hard-to-see white conical twrs, the front on a tiny rky islet. N of the bay, a prominent white monument is on Monte Ferro. Dique de Abrigo (white wall) has a G/W tower QG. Other conspic daymarks: castle walls (parador) and MRCY flag-mast. FV jetty has white crane.

COMMUNICATIONS (Code 986) MRCC 222230; Emergency112; ⊖ & Met via YC/marina; Brit Consul (902)109356; *Puerto Deportivo de Bayona* VHF Ch 09; *Monte Real Club de Yates* VHF Ch 71.

FACILITIES Monte Real Club de Yates (MRCY) ☎385000. secretaria@ mrcyb.com www.mrcyb.es 200 F&A + some 🛥s in 6m, €2.95, multihulls +50%, 🛠🚿✕ ◻ Ice ▥ ▬ ▯(20t) ⚓(1·5t) ⚒. Clubhouse is very well appointed.

Bayona Marina ☎385107, mob 626299162. puertobaiona@ puertobaiona.com www.puertobaiona.com 297 + 19🅥 €3.00. 🛠🚿 ✕ ◻ 🛒 🚾 key access, Ice ▥ ▯(35t) ⚓(6t) ▬.

Town Most domestic needs inc 🛒 ▣ ⚒ 📷 Ⓔ ▤ ⚠ ✕; ⇌ and buses to Vigo, 21 km (✈, internal flights).

Portugal

Viana do Castelo to Vila Real de Santo António

9.24.1	Contents	981
9.24.2	Area map and Distance table	982
9.24.3	Lights, buoys and waypoints	984
9.24.4	Passage information	986
	Wind and Weather – 986 • The Spanish Border to Cabo Raso – 986 • Cabo Raso to Cape St Vincent – 993 • Southern Portugal: the Algarve – 999	
9.24.5	Special notes for Portugal	986
9.24.6	Viana do Castelo	987
9.24.7	Póvoa de Varzim	987
9.24.8	Leixões	988
	Rio Douro • Porto	
9.24.9	Porto (Oporto)	988
9.24.10	Aveiro	989
9.24.11	Figueira da Foz	989
9.24.12	Nazaré	990
9.24.13	Peniche	990
9.24.14	Cascais	991
9.24.15	Lisboa (Standard port, tide tables/curves)	991
9.24.16	Sesimbra	997
9.24.17	Setúbal	997
	Arrábida	
9.24.18	Sines	998
	Baleeira	
9.24.19	Lagos	998
	Alvor	
9.24.20	Portimão	999
9.24.21	Albufeira	1000
9.24.22	Vilamoura	1000
9.24.23	Faro and Olhão	1001
	Tavira	
9.24.24	Vila Real de Santo António	1002

Portugal

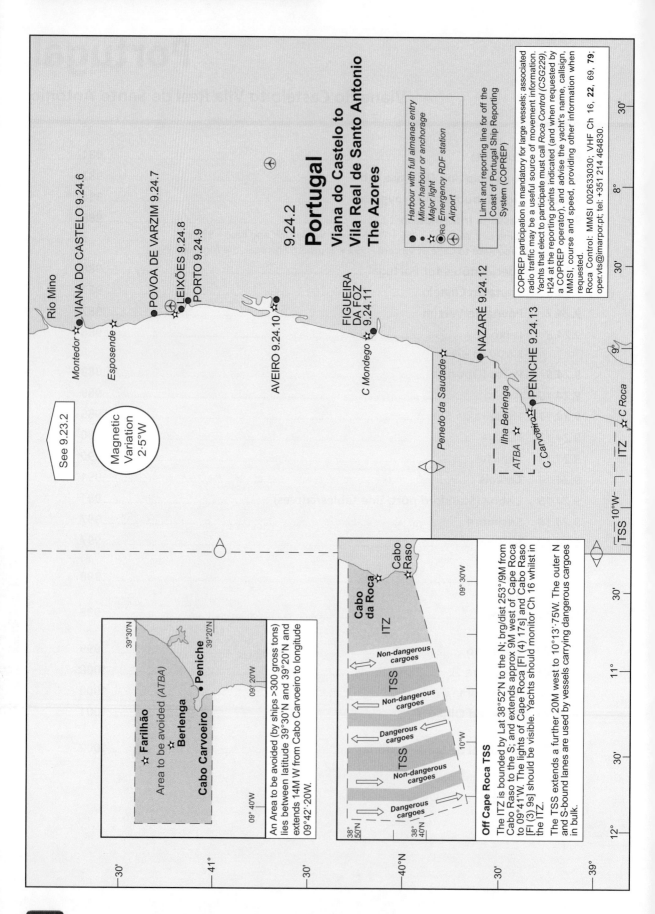

See 9.23.2

Magnetic Variation 2·5°W

Rio Mino

☆ Montedor

● VIANA DO CASTELO 9.24.6

☆ Esposende

● PÓVOA DE VARZIM 9.24.7

☆⊕● LEIXÕES 9.24.8
● PORTO 9.24.9

● AVEIRO 9.24.10

FIGUEIRA DA FOZ 9.24.11
● ☆ C Mondego

9.24.2

Portugal

Viana do Castelo to Vila Real de Santo Antonio
The Azores

● Harbour with full almanac entry
● Minor harbour or anchorage
☆ Major light
⊛ RG Emergency RDF station
⊕ Airport

☐ Limit and reporting line for off the Coast of Portugal Ship Reporting System (COPREP)

COPREP participation is mandatory for large vessels; associated radio traffic may be a useful source of movement information. Yachts that elect to participate must call *Roca Control* (CSG229), H24 at the reporting points indicated (and when requested by a COPREP operator), and advise the yacht's name, callsign, MMSI, course and speed, providing other information when requested.
Roca Control: MMSI 002633030; VHF Ch 16, **22**, 69, **79**; oper.vts@imarpor.pt; tel: +351 214 464830.

● NAZARÉ 9.24.12

☆ Penedo da Saudade

☆ Ilha Berlenga
|ATBA|
☆ C Carvoeiro ● PENICHE 9.24.13
L C Carvoeiro

TSS
ITZ
☆ C Roca

—TSS—10°W—|—30'—|—8°—|—30'—

Inset (upper left): Berlenga area

39°30'N
39°20'N
09°40'W 09°30'W 09°20'W

☆ Farilhão
Area to be avoided (*ATBA*)
Berlenga
☆ Cabo Carvoeiro ● Peniche

An Area to be avoided (by ships >300 gross tons) lies between latitude 39°30'N and 39°20'N and extends 14M W from Cabo Carvoeiro to longitude 09°42'·20W.

Inset (lower): Cabo da Roca TSS

Cabo da Roca
ITZ
☆ Cabo Raso

Non-dangerous cargoes
TSS
Non-dangerous cargoes
Dangerous cargoes
TSS
Non-dangerous cargoes
Dangerous cargoes

38°50'N
38°40'N
09°30'W 10°W

Off Cape Roca TSS

The ITZ is bounded by Lat 38°52'N to the N; brg/dist 253°/9M from Cabo Raso to the S; and extends approx 9M west of Cape Roca to 09°41'W. The lights of Cape Roca [Fl (4) 17s] and Cabo Raso [Fl (3) 9s] should be visible. Yachts should monitor Ch 16 whilst in the ITZ.

The TSS extends a further 20M west to 10°13'·75W. The outer N and S-bound lanes are used by vessels carrying dangerous cargoes in bulk.

9° 30' 11° 30' 12°

30'
41°
30'
40°N
30'
39°

Magnetic Variation 2·1'W

See 9.25.2

DISTANCE TABLE
Approximate distances in nautical miles are by the most direct route, whilst avoiding dangers and allowing for Traffic Separation Schemes. Places in italics are off UK and North West coast of France.

Off Cape St Vincent TSS

The ITZ is bounded by Lat 37°01'N (Cape St Vincent. Fl 5s) to the N; brg/dist 186°/9M from Punta de Sagres, Iso R 2s, to the E; and extends in a segmented arc approx 9M west and south of these lights which should be visible. Yachts should monitor Ch 16 whilst in the ITZ.

The TSS extends a further 25M west and south from the outer boundary of the ITZ. The outer N and S-bound lanes are used by vessels carrying dangerous cargoes in bulk.

SPAIN

LISBOA 9.24.15

SETÚBAL 9.24.17

SINES 9.24.18

Arrábida 9.24.16
C Espichel 9.24.16

CASCAIS 9.24.14

SESIMBRA 9.24.14

C Roca
C Raso

Magnetic Variation 2·5'W

ITZ
TSS

C Sardão

C de São Vicente

Baleeira

Pta da Piedade

ITZ
TSS

LAGOS 9.24.19
PORTIMÃO 9.24.20
ALBUFEIRA 9.24.21
VILAMOURA 9.24.22
FARO/OLHÃO 9.24.23
VILA REAL DE SANTO ANTÓNIO 9.24.24

Tavira
C de Santa Maria
Rompido de Cartaya
Ayamonte

Cabo de S Vicente

Pta de Sagres

36°50'N

36°30'N

ITZ

Non-dangerous cargoes

TSS

Non-dangerous cargoes

TSS

Non-dangerous cargoes

Dangerous cargoes

Dangerous cargoes

Dangerous cargoes

09°00'W
09°15'W
09°30'W

Portugal

		1	2	3	4	5	6	7	8	9	10	11	12	13	14	15	16	17	18	19	20
1	*Longships*	1																			
2	*Ushant (Créac'h)*	100	2																		
3	La Coruña	418	338	3																	
4	Cabo Villano	439	365	43	4																
5	Bayona	510	436	114	71	5															
6	Viana do Castelo	537	468	141	98	32	6														
7	Leixões (Pórto)	565	491	169	126	63	33	7													
8	Nazaré	659	585	263	220	156	127	97	8												
9	Cabo Carvoeiro	670	596	274	231	171	143	114	22	9											
10	Cabo Raso	710	636	314	271	211	183	154	62	40	10										
11	Lisboa (bridge)	686	652	330	287	227	199	170	78	56	16	11									
12	Cabo Espichel	692	658	336	293	233	205	176	84	62	22	23	12								
13	Sines	726	692	370	327	267	239	210	118	96	54	57	34	13							
14	Cabo São Vicente	777	743	421	378	318	290	261	169	147	104	108	85	57	14						
15	Lagos	797	763	441	398	338	310	281	189	167	124	128	105	77	20	15					
16	Vilamoura	820	786	464	421	361	333	304	212	190	147	151	128	100	43	27	16				
17	Cádiz	911	877	555	512	452	424	395	303	281	238	242	219	191	134	120	95	17			
18	Cabo Trafalgar	928	894	572	529	469	441	412	320	298	255	259	236	208	151	139	115	28	18		
19	Tarifa	952	918	596	553	493	465	436	344	322	279	283	260	232	175	163	139	52	24	19	
20	Gibraltar	968	934	612	569	509	481	450	360	338	295	299	276	248	191	179	155	68	40	16	20

9.24.3 LIGHTS, BUOYS AND WAYPOINTS

Bold print = light with a nominal range of 15M or more. CAPITALS = place or feature. *CAPITAL ITALICS* = light-vessel, light float or Lanby. *Italics* = Fog signal. ***Bold italics*** = Racon. Many marks/buoys are fitted with AIS (MMSI No); see relevant charts.

RIO MIÑO (SPANISH BORDER) TO LISBOA

Río Miño ent, Fort Ínsua ⚓ Fl WRG 4s 16m W12M, R8M, G9M; 204°-G-270°-R-357°-W-204°; W col; 41°51'·57N 08°52'·48W.
Montedor ☆ Fl (2) 9·5s 102m **22M**; Twr, R cupola and bldg; 41°45'·07N 08°52'·41W.

VIANA DO CASTELO

Dir lt Barra ⚓ Oc WRG 4s 14m 8/6M; 350°-G-005°-W-010°-R-025°; on Pilots' twr; 41°41'·05N 08°50'·30W.
Outer mole ⚓ Fl R 3s 9M; W col, R bands; 41°40'·46N 08°50'·66W.
No. 2 ⚓ Fl R 3s; 41°40'·53N 08°50'·48W.
E mole ⚓ Fl G 3s 9M; 41°40'·67N 08°50'·25W.
No. 1 ⚓ Fl G 3s; 41°40'·68N 08°50'·29W.
Ldg lts 012·5° (for docks only); R twrs, W stripes: **Front** ☆ Iso R 4s 14m 5M; 41°41'·33N 08° 50'·35W. **Rear** ☆ Oc R 6s 32m 5M.
No. 3 ⚓ Fl (2) G 3s; 41°40'·86N 08°50'·24W.
No. 4 ⚓ Fl (2+1) R 5s; 41°40'·89N 08°50'·36W.
Rio Lima: Nos. 5-13 ⚓s are Fl G 3s. Nos. 6-14 ⚓s are Fl R 3s.
No. 13 ⚓ Fl G 3s; 41°41'·49N 08°49'·27W, SSE of marina ent.
No. 14 ⚓ Fl R 3s; 41°41'·56N 08°49'·33W, SSE of marina ent.
Marina ent, unlit; 41°41'·60N 08°49'·33W.

PÓVOA DE VARZIM

Molhe N ⚓ Fl R 3s 14m 12M; *Siren 40s;* 41°22'·23N 08°46'·28W.
Molhe S ⚓ L Fl G 6s 4M; 41°22'·27N 08°46'·07W.

LEIXÕES

Tanker mooring ⚓ Fl (3) 15s 6M; *Horn (U) 30s;* 41°12'·10N 08°45'·06W. Oil refinery 2M E of this buoy is conspic D/N.
Leça ☆ Fl (3) 14s 56m **28M**; W twr, B bands; 41°12'·08N 08°42'·73W.
Outer N mole, Quebra-Mar ⚓ Fl WR 5s 24m W12M, R9M; 001°-R-180°-W-001°; Gy twr; *Horn 20s;* 41°10'·37N 08°42'·49W.
S mole ⚓ Fl G 4s 17m 7M; 328°-285°; *Horn 30s;* 41°10'·68N 08°42'·34W.
Inner N mole ⚓ Fl R 4s 9m 6M; 173°-353°; 41°10'·70N 08°42'·51W.
No. 2 ⚓ Fl (3) R 8s; 41°10'·77N 08°42'·51W.
No. 4 ⚓ Fl (2) R 5s; 41°10'·86N 08°42'·60W .
Marina bkwtr ⚓ L Fl (2) R 12s 4m 2M; 41°11'·08N 08°42'·25W.

RIO DOURO/PORTO

N mole head Fl R 5s 16m 9M; W twr, R stripes; 41°08'·73N 08°40'·72W.
S mole head Fl G 5s 14m 6M; W col, G stripes; 41°08'·67N 08°40'·53W.
Up-river buoys in sequence:
No. 1 ⚓ Fl G 3s; 41°08'·74N 08°40'·29W.
No. 2 ⚓ Fl R 3s; 41°08'·72N 08°39'·77W.
No. 3 ⚓ Fl (2) G 5s; 41°08'·65N 08°39'·52W.
No. 4 ⚓ Fl (2) R 5s; 41°08'·80N 08°39'·12W.
No. 5 ⚓ Fl (3) G 6s; 41°08'·78N 08°38'·85W.
No. 6A ⚓ Fl (3) R 6s; 41°08'·44N 08°37'·26W.
No. 8 ⚓ Fl R 4s; 41°08'·39N 08°37'·12W; 4 ca W of bridge, 8·8m.

AVEIRO

Boia Aveiro, ⚓ Fl(3)G 7s; 40°37'·86N 08° 46'·98W.
Lt ho Aero ☆ Fl (4) 13s 65m **23M**; R/W twr; 40°38'·57N 08°44'·87W.
Ldg lts 060·6°; both Or posts, G bands: Front ⚓, Oc R 3s 10m 9M (4M day); 40°38'·81N 08°45'·04W. Rear ⚓ Oc R 6s 15m 9M (4M day) 440m from front; vis 05° each side of Ldg line.
Molhe N, SW extension, outer end, Mohle ⚓ Fl Y 1.5s 12m 2M, W col, R bands; 40° 38'·57N 08° 45'·87W.
Molhe N ⚓ Fl R 3s 15m 8M; W twr, R bands; 40°38'·54N 08°45'·92W.

Molhe S ⚓ Fl G 3s 15m 9M; W col, G bands; 40°38'·53N 08°45'·47W
Molhe Central ⚓ L Fl G 5s 8m 3M; 40°38'·64N 08°44'·95W.
Monte Farinha ⚓ Fl (2+1) R 6s 6m 4M; 40°39'·48N 08°43'·51W;
Y-junction of Canal Principal and Canal de San Jacinto, to the N. Canal Principal is well buoyed and defined by 3 sets of ldg lts.
Terminal Sul ⚓ Q (2+1) G 6s 6m 3M; 40°38'·26N 08°41'·33W.
Inner Canal Principal: 2 lateral lt buoys, 1 lt bcn and 7 unlit lateral buoys. Lock, 40°38'·75N 08°39'·74W, into Canal das Piramides for yacht berths.

FIGUEIRA DA FOZ

Cabo Mondego ☆ Fl 5s 96m **28M**; W twr and house; 40°11'·46N 08°54'·31W, 3M NNW of Figueira da Foz.
Hbr ent ldg lts 046·5°, both W strucs, R bands: Front ⚓ Oc R 3s 6m 2M; 40°08'·83N 08°52'·21W. Rear, ⚓ Oc R 3s 15m 6M; vis 039° - 055°.
Ldg lts 081·5°, both W cols, R bands: Front ⚓ Iso R 5s 6m 4M; 40°08'·83N 08°51'·23W. Rear, ⚓ Oc R 6s 10m 4M; 081·5°/239m.
Molhe N ⚓ Fl R 6s 13m 9M; *Horn 35s;* 40°08'·60N 08°52'·70W.
Molhe S ⚓ Fl G 6s 13m 7M; 40°08'·59N 08°52'·41W.
Inner N bkwtr ⚓ Fl R 3s 8m 4M; 40°08'·78N 08°52'·10W.
Inner S bkwtr ⚓ Fl G 3s 8m 4M; 40°08'·69N 08°52'·08W.
Marina ent, Fl (2) G 8s & Fl (2) R 8s, 6m 2M; G & R twrs; 40°08'·81N 08°51'·52W.
Penedo da Saudade ☆ Fl (2) 15s 54m **30M**; ☐ twr, and house; 39°45'·84N 09°01'·86W, 10M N of Nazaré.

NAZARÉ

Pontal da Nazaré ⚓ Oc 3s 49m 14M; twr & bldg; 39°36'·25N 09°05'·15W, 9ca NNW of hbr ent.
Molhe N ⚓ L Fl R 5s 14m 9M; 39°35'·45N 09°04'·59W.
Molhe S ⚓ L Fl G 5s 14m 8M; 39°35'·36N 09°04'·70W.

SÃO MARTINHO DO PORTO

Ponta de Santo António ⚓ Iso R 6s 32m 9M; W col, R bands; *Siren 60s;* 39°30'·61N 09°08'·61W (NE side of ent).
Ldg lts 145·1°, both W cols, R bands: Front ⚓ Iso R 1·5s 9m 9M; 39°30'·14N 09°08'·39W. Rear ⚓ Oc R 6s 11m 9·5M, 129m from front.

LOS FARILHÕES/ILHA DA BERLENGA

Farilhão Grande ⚓ Fl (2) 5s 99m 13M; 39°28'·73N 09°32'·72W.
Ilha da Berlenga ☆ Fl 10s 120m **16M**; W ☐ twr and houses; *Horn 28s;* 39°24'·90N 09°30'·60W (5·7M NW of C. Carvoeiro).

PENICHE

Cabo Carvoeiro ☆ Fl R 6s 56m **15M**; W☐twr; *Horn 35s;* 39°21'·63N 09°24'·47W, on the W tip of the Peniche peninsula.
Molhe W ⚓ Fl R 3s 13m 9M; W twr, R bands; *Siren 120s;* 39°20'·87N 09°22'·53W.
Molhe E ⚓ Fl G 3s 13m 9M; W twr, G bands; 39°20'·98N 09°22'·39W.
Porto das Barcas ⚓ L Fl R 6s 29m 6M; post; 39°13'·81N 09°20'·47W.
Porto Dineiro ⚓ Oc 3s 19m 8M; 39°12'·83N 09°20'·55W.
Assenta ⚓ L Fl 5s 74m 13M; W structure on conical base; 39°03'·53N 09°24'·85W (17M N of Cabo da Roca).
Ericeira ldg lts 083°; Front, N Pier Head ⚓ Fl (3) R 12s 11m 6M; 38°57'·79N 09°25'·41W. Rear, ⚓ Oc R 13s 34m 6M; *Siren 70s;* 38°57'·85N 09°25'·06W.
C. da Roca ☆ Fl (4) 17s 164m **26M**; W twr and bldgs; 38°46'·90N 09°29'·84W. R lt (527m) 5M E at Pena.
Cabo Raso ☆ Fl (3) 9s 22m **15M**; 324°-189°; R twr; 38°42'·57N 09°29'·15W, 3·2M WNW of Cascais.
C2 ⚓ Fl Y 5s; outfall buoy; 38°39'·99N 09°27'·97W.

CASCAIS

Ldg lts 284·7°: Front, **Forte de Santa Marta** ☆ Oc WR 6s 24m **W18M**, R14M; 233°-R-334°-W-098°; W ☐ twr, Bu bands; 38°41'·42N 09°25'·25W. Rear, **Guia** ☆ Iso WR 2s **W19M, R16M**; 326°-W-092°; 278°-R-292°; W twr; 38°41'·74N 09°26'·78W.

MC1, 2 & 3 all ⚓s, VQ (6) + L Fl 10s, mark the SE side of marina.
CC2 ⚓ Fl R 4s; NE of bkwtr hd; 38°41'·63N 09°24'·80W.
No. 2 ⚓ Fl R 10s (May-Oct); 38°41'·69N 09°24'·99W.
Marina Molhe Sul ⚡ Fl (3) R 4s 8m 6M; 38°41'·59N 09°24'·84W.
Molhe Norte ⚡ Fl (2) G 4s 8m 3M; 38°41'·60N 09°24'·90W.

RIO TEJO, N BANK

Marconi ⚡ Iso WR 3s 17m 9M; 048°-R-058°-W-068°; Y twr; 38°40'·86N 09°20'·78W, on Ponta da Rana 3·25M ESE of Cascais.
Forte de São Julião ⚡ Oc R 5s 38m 14M; 38°40'·47N 09°19'·52W.
Oeiras marina, S pier hd ⚡ Fl R 3s 9m 5M; 38°40'·56N 09°19'·00W.
N pier hd ⚡ Fl G 3s 9m 5M; 38°40'·62N 09°18'·99W.
Doca de Paço de Arcos, W pier ⚡ Fl R 5s 9m 4M; 38°41'·43N 09°17'·61W. E pier ⚡ Fl G 5s; 38°41'·46N 09°17'·59W.

LISBOA

Triple ldg lts 047·1°: Front, **Gibalta** ☆ Oc R 6s 30m **21M**; 039·5°-054·5°; twr & cupola, R stripes; 38°41'·95N 09°15'·97W.
Middle, **Esteiro** ☆ Oc R 6s 81m **21M**; 039·5°-054·5°; W twr, R bands; *Racon Q, 15M*; 38°42'·23N 09°15'·59W.
Rear, **Mama** ☆ Oc R 6s 153m **21M**; 045·5°-048·5°; platform; 38°43'·65N 09°13'·63W.
No. 2 ⚓ Fl R 10s; 38°37'·29N 09°23'·28W.
No. 1 ⚓ Fl G 2s; 38°39'·55N 09°18'·79W.
Forte Bugio ☆ Fl G 5s 27m **15M**; ○ twr on fortress; *Horn Mo (B) 30s*; 38°39'·63N 09°17'·93W.
No. 3 ⚓ Fl G 3s; 38°40'·04N 09°18'·26W.
No. 5 ⚓ Fl G 4s; 38°40'·43N 09°17'·66W.
No. 7 ⚓ Fl G 5s; 38°40'·65N 09°16'·90W.
No. 9 ⚓ Fl G 6s; 38°40'·55N 09°14'·57W.
Bridge (Ponte 25 de Abril). Fl (3) G 9s and Fl (3) R 9s on the N (38°41'·64N 09°10'·69W) and S (38°41'·10N 09°10'·56W) pillars; vis up- and downstream, covering main (70m) and inshore spans.
Lisnave ⚡ Fl G 2s 4m 3M; 38°41'·29N 09°08'·63W, S bank. Docks & industrial areas lie S and E of this lt.

Cabo Espichel ☆ Fl 4s 167m **26M**; W 6-sided twr; *Horn 31s*; 38°24'·92N 09°12'·97W.

SESIMBRA

Cavalo Fort ⚡ Oc 5s 34m 14M; R ○ twr; 38°26'·01N 09°07'·01W.
Ldg lts 003·5°, both ⚡ L Fl R 5s 9/21m 7/6M: Front, 38°26'·51N 09°06'·09W. Rear 34m from front. Pierhead ⚡ Fl R 3s 8M.

SETÚBAL

Ldg lts 039·7°, both ☆ Iso Y 6s 12/60m **22M** (by day 5/6M). Front, R structure, W stripes; 38°31'·08N 08°54'·00W (fish dock). Rear, **Azêda**, vis 038·3°-041·3°; R hut on piles; 1·7M from front.
No. 1 ⚓ Fl G 3s 5M; 38°26'·99N 08°58'·17W.
No. 2 ⚓ Fl (2) R 10s 13m 9M; *Racon B, 15M*; 38°27'·22N 08°58'·45W.
No. 4 ⚓ Fl R 4s 13m 4M; 38°27'·92N 08°57'·71W.
No. 3 ⚓ Fl G 5s; 38°28'·33N 08°56'·78W.
No. 5 ⚓ Fl G 4s 13m 4M; 38°29'·20N 08°55'·36W.
Forte de Outão ⚡ Oc R 6s 33m 12M; 38°29'·31N 08°56'·05W.
João Farto ⚓ Q (9) 15s; 38°30'·33N 08°54'·44W.
Forte de Albarquel ⚡ Iso R 2s 16m 6M; 38°30'·65N 08°54'·79W.
Commercial dock (yachts in W half); 38°31'·22N 08°53'·13W.
Pinheiro da Cruz ⚡ Fl 3s 66m 11M; W ○ col, R stripes; 38°15'·47N 08°46'·28W. Firing area extends 6M offshore.

SINES

Cabo de Sines ☆ Fl (2) 15s 55m **26M;** W twr & house; obsc 001°-003° (2°) & 004°- 007° (3°); 37°57'·57N 08°52'·82W.
⚓ Fl (5) Y 20s; 37°55'·36N 08°55'·80W, 2·2M WSW of W mole hd.
W mole hd ⚡ Fl 3s 20m 12M; 37°56'·48N 08°53'·33W.
Sines W ⚓ Fl R 3s 6M; 37°56'·12N 08°53'·25W, 4ca S of W mole hd.
E mole, NW head ⚡ L Fl G 8s 16m 6M; 37°56'·32N 08°51'·95W.
Borboleta ⚓ VQ (3) 5s; 37°55'·18N 08°50'·89W.

Marina mole ⚡ Fl G 4s 4M; W twr, G bands; 37°57'·03N 08°52'·03W.
FV bkwtr ⚡ Fl R 6s 6M; W twr, R bands; 37°57'·06N 08°52'·19W.
Ponta da Gaivota ⚡ L Fl 7s 19m 13M; 37°51'·10N 08°47'·78W.
Rio Mira, ent ⚡ Fl 3s 22m 12M; 37°43'·12N 08°47'·40W.
Cabo Sardão ☆ Fl (3) 15s 67m **23M**; 37°35'·93N 08°49'·01W.

CAPE ST VINCENT TO THE SPANISH BORDER

CAPE ST VINCENT/SAGRES

Cabo de São Vicente ☆ Fl 5s 84m **32M**; W twr and bldg; *Horn Mo (I) 30s;* 37°01'·36N 08°59'·78W.
Ponta de Sagres ⚡ Iso R 2s 51m 11M; 36°59'·67N 08°56'·95W.
Baleeira mole ⚡ Fl WR 4s 12m, W14M, R11M; 254°-W-355°-R-254°; W twr; 37°00'·67N 08°55'·43W.

LAGOS

Pta da Piedade ☆ Fl 7s 50m **20M**; 37°04'·81N 08°40'·14W.
W bkwtr ⚡ Fl R 4s 7M; W col, R bands; 37°05'·93N 08°39'·83W.
E bkwtr ⚡ Fl (2) G 6s 6M; W col, G bands; 37°05'·91N 08°39'·92W.
Alvor ent, ⚡ Fl R/G 4s; W twrs, R/G bands; 37°07'·10N 08°37'·11W.

PORTIMÃO

Ponta do Altar ☆ L Fl 5s 31m **16M**; vis 290°-170°; W twr and bldg; 37°06'·34N 08°31'·17W.
Ldg lts 019·1°: Both ⚡ Iso R 6s 18/30m 6M, W cols, R bands. Front; 37°07'·35N 08°31'·31W. Rear ⚡ 54m from front.
Outer E mole ⚡ Fl G 5s 9m 7M; 37°06'·50N 08°31'·19W.
Outer W mole ⚡ Fl R 5s 9m 7M; 37°06'·52N 08°31'·76W.
Marina S end ⚡ Iso R 2s 6m 3M; 37°06'·90N 08°31'·62W.
No. 2 ⚓ Fl R 4s; 37°07'·04N 08°31'·46W.
Marina ent, S side ⚡ Fl R 6s 3M; 37°07'·10N 08°31'·58W.
N side ⚡ Fl G 6s 3M; 37°07'·14N 08°31'·58W.
Marina N end ⚡ Iso R 4s 8m 3M; 37°07'·33N 08°31'·58W.

Pta de Alfanzina ☆ Fl (2) 14s 62m **29M**; 37°05'·19N 08°26'·54W.
Armacão de Pera ⚡ Oc R 5s 24m 6M; 37°06'·00N 08°21'·36W.

ALBUFEIRA

Ponta da Baleeira ⚡ Oc 6s 30m 11M; 37°04'·81N 08°15'·80W.
N bkwtr ⚡, Fl (2) G 5s 9m 4M, approx 37°04'·81N 08°15'·57W.
S bkwtr ⚡, Fl (2) R 5s 9m 4M, approx 37°04'·76N 08°15'·57W.
Olhos de Água ⚡ L Fl 5s 29m 7M; 37°05'·42N 08°11'·35W.

VILAMOURA

Vilamoura ☆, Fl 5s 17m **19M**; 37°04'·460N 08°07'·37W.
Marina, W mole ⚡ Fl R 4s 13m 5M; 37°04'·08N 08°07'·38W.
E mole ⚡ Fl G 4s 13m 5M; 37°04'·12N 08°07'·30W.

FARO, OLHÃO and TAVIRA

Ent from sea: E mole ⚡ Fl G 4s 9m 6M; 36°57'·79N 07°52'·14W.
W mole ⚡ Fl R 4s 9m 6M; 37°57'·84N 08°52'·25W, appr on 352°.
Access ldg lts 020·9°: Front, Barra Nova ⚡ Oc 4s 8m 6M; 37°58'·22N 07°52'·00W. Rear, **Cabo de Santa Maria** ☆ Fl (4) 17s 49m **25M**; W ○ twr; 36°58'·48N 07°51'·88W.
No. 6 ⚓ Fl R 6s; 36°58'·40N 07°52'·20W (NW to Faro; NE to Olhão).
No. 20 ⚓ Fl R 6s, 37°00'·05N 07°55'·1W (approx 2M to **Faro**).
No. 8 ⚓ Fl R 3s; 36°59'·82N 07°51'·16W; 2·2M N & ENE to **Olhão**.
Tavira ldg lts 325·9°: Front, ⚡ Fl R 3s 9m 4M. Rear, Fl R 3s 10m 5M; sync'd with front.
E mole ⚡ Fl G 2·5s 5m 6M; 37°06'·82N 07°37'·03W.

VILA REAL DE SANTO ANTONIO

Lt ho ☆ Fl 6·5s 51m **26M**; W twr, B bands; 37°11'·20N 07°24'·91W.
Bar, No 1 buoy ⚓ Q (3) G 6s; approx 37°08'·90N 07°23'·44W.
No 2 ⚓ Fl R 4s; 37°08'·83N 07°23'·74W.
W bkwtr ⚡ Fl R 5s 4M; 37°09'·75N 07°24'·03W.
Marina, QR at S end; QR/QG at ent; QR at N end.

Portugal

9.24.4 PASSAGE INFORMATION

More Passage Information is threaded between successive harbours in this Area. **Bibliography**: *SW Spain & Portugal CC* (Fernhurst/ Jens); *Atlantic Spain and Portugal* (Imray/RCC/Walker/Buchanan); *W coasts of Spain and Portugal Pilot* (NP 67).

WIND AND WEATHER

In summer the Portuguese Trades (*Nortada*) are N'ly F4-6 and the Portugal Current runs S at ½-¾kn. Summer gales are rare; coastal fog is common in the mornings. If N-bound, especially if lightly crewed, try making daily passages between about 0400 and 1200 to avoid the stronger afternoon winds, possibly augmented by a fresh onshore sea breeze.

▶*Tidal streams appear to set N on the flood and S on the ebb, but are poorly documented.*◀

PORT CLOSURES Some hbrs, especially on this coast, may be closed for weather reasons, ie strong to gale force onshore winds and/or heavy swell causing dangerous breaking seas at the hbr ent or over the bar. Closures may be promulgated by radio and Navtex. At Leixões, Aveiro and Figueira da Foz the following signals, with minor variations = port closed:

By day: cylinder/ball hoisted close up; at half-mast = enter with caution. By night: ●●●(vertical) steady or flashing.

THE SPANISH BORDER TO CABO RASO

The Río Miño (*Minho* in Portuguese) 15M S of Cabo Silleiro, forms the northern border with Spain. The river ent is difficult and best not attempted.

The 150M long coastline is hilly as far S as **Porto**, then generally low and sandy, backed by pine forests, to Cabo Carvoeiro; there are few prominent features (AC 3633, 3634, 3635). Coasting yachts should keep at least 3M offshore; see **Viana do Castelo**, a commercial and fishing port, with a marina at the NE end. **Povoa de Varzim** has a marina. A restricted area centred on 41°27'·67N 8° 51'·10W (6.45M NNW of Póvoa de Vazim, 3M offshore) surrounds a large 'Windfloat' wind turbine (see below) comprising a yellow, triangular floating base, supporting the 100m high white tower and blades, R lt, 3 Mo (U). The undersea power line to the shore is marked by three SPM buoys, (Y) the outer two Fl Y 2.5s, as shown on AC 3634.

Photo: John Frankland

Seaplane Operating Areas are used for landing, take-off and drawing water to fight forest fires along this, and the rest of the Portuguese coast. They are approx 1.5M long x 0.5M wide and must be avoided when in use, usually in summer.

Leixões is an industrial port 2M N of the mouth of the R. Douro; it may be closed in bad weather due to heavy swell breaking on the bar. There is a seaplane operating area with an associated no entry zone N of Leixões which requires caution. The R. Douro is not easily entered, but two substantial bkwtrs at the river mouth have improved access; there is a new marina 3M upriver. Oporto can easily be visited by road from Leixões to recce the latest situation.

Aveiro (⚓) and **Figueira da Foz** (marina) are exposed to the W and can be closed in bad weather; the latter can be identified from N or S by the higher ground (257m) of Cabo Mondego. 34M further

SSW **Nazaré** is an artificial fishing harbour and port of refuge with easy entrance and a marina.

Cabo Carvoeiro, the W tip of Peniche peninsula, looks like an island from afar (do not confuse with Ilha Berlenga). Ilha da Berlenga and Os Farilhões, both lit, are respectively 5 and 9·5M NW of Cabo Carvoeiro, and within an '**Area to be avoided**' (by big ships) formerly an ITZ; see Area map.

The normal coastal route is between Cabo Carvoeiro and Berlenga; this chan is deep, clear and 5M wide, although it appears narrower until opened up. **Peniche** fishing port/marina on the S side of the peninsula is well sheltered from N'lies, but open to SW swell. The coastline rises steadily to the 527m high ridge of Sintra, inland of Cabo da Roca, S of which it drops steeply to the low headland of Cabo Raso. The **Off Cabo da Roca Traffic Separation Scheme** is 14M to seaward of that cape and extends a further 20M west with extra lanes for dangerous cargoes. The ITZ is 14M wide; see Area map.

9.24.5 SPECIAL NOTES FOR PORTUGAL

Districts/Provinces: Portugal is divided into 18 administrative districts. The names of the former coastal provinces (N-S: Minho, Douro, Beira Litoral, Estremadura, Alentejo and Algarve) are still used and appear below the name of each hbr.

Charts: The 2 folios are: FA, old (*antigo*) only a few charts, with 2 digit chart nos; and F94. F94 charts (5 digit nos) include: 4 offshore charts 23201-04; 6 coastal charts 24201-06; 12 hbr charts 26301-12; 8 hbr appr charts 26401-08; and 3 charts of hbr plans 27501-03. 12 Leisure (*recreio*) charts at 1:150,000 scale, Nos 25R01-12, cover the whole coast N-S. Some long-established marinas are not yet charted at large scale.

Portuguese charts and publications are available from Chart agents, one of the largest : J Garraio & Ca Lda, ave 24 de Julho, 2 - 1st Dto, 1200-478 Lisboa; ☎213 473 081. info@jgarraio.pt www. jgarraio. For Portuguese Hydrographic Office, see Ch 1.

Standard Time is UT (**NOT** UT –1); add 1 hr for DST, in force from the last Sun in Mar until the Sat before the last Sun in Oct.

Representation: Portuguese Tourist Office, 11 Belgrave Square, London SW1X 8PP; ☎020 7201 6666.

Tourist Office, Ave António Augusto de Aguiar 86, P-1000 Lisboa; ☎213 425 231.

British Embassy, Rua de S. Bernardo 33, 1249-082 Lisboa; ☎21 359 4082.

There are Consulates in Lisbon/Portimão & Hon Consuls at Oporto and Ponta Delgada (Azores) ☎808 203537 (all same number).

Telephone: To call Portugal from UK dial +351, the area code, then the ☎ number. To call UK from Portugal, dial +44, the area code less the initial 0, then the ☎ number. The area code is mandatory, even within the same area; ie nine numbers in all. National emergency ☎112; National SAR ☎214 401 919.

R/T: VHF Ch 09 is the recreational craft channel; it is used by all Portuguese marinas, in addition to any Port VHF channels.

Access: There are daily flights in season to/from the UK via Porto, Lisboa and Faro with connections by bus and/or train to other towns. See also Area 25 for Spanish flights.

Currency is the Euro €. VAT (IVA) is 23%. Credit cards are widely accepted; cash dispensers are in all towns.

Public Holidays: Jan 1; Shrove Tues; Good Friday; April 25 (Liberation Day); May 1 (Labour Day); Jun 10 (Camões Day); Aug 15 (Assumption); Dec 8 (Immaculate Conception) 25 (Christmas). Also every town has a local *festa*.

Documents: (See Ch 2 Cruising Formalities) Portugal observes EU regulations but formalities may be lengthy. Papers needed: *Personal* – Passports; crew list, ideally on headed paper with the yacht's rubber stamp, giving DoB, passport nos, where joined/ intended departure. Certificate of Competence (Yachtmaster Offshore, ICC/HOCC etc). Radio Operator's certificate. EHIC.

Yacht – Registration certificate. Proof of VAT status. Marine insurance. Ship's Radio licence. Itinerary, backed up by ship's log.

9.24.6 VIANA DO CASTELO

Minho **41°41'·68N 08°49'·21W** (Marina ent) ❀❀❀❀♦♦❀❀❀

CHARTS AC 3633/4, 3257; PC 24201, 26401; SC 41B

TIDES

Standard Port LISBOA (⟶); ML 2·0

Times				Height (metres)			
High Water		Low Water		MHWS	MHWN	MLWN	MLWS
0400	0900	0400	0900	3·8	3·0	1·5	0·6
1600	2100	1600	2100				
Differences VIANA DO CASTELO							
−0020	0000	+0010	+0015	−0·3	−0·3	−0·2	−0·1
ESPOSENDE (41°32'N)							
−0020	0000	+0010	+0015	−0·6	−0·5	−0·2	−0·1

SHELTER Excellent in marina; possible strong current across its ent. No ⚓ in river. Night entry not advised if any swell.

NAVIGATION WPT 41°40'·18N 08°50'·37W, 007·5°/7ca in the W sector of dir lt Oc WRG 4s to abeam No 3 SHM lt buoy. Ignore the 012·5° ldg line which applies solely to the shipyard/FV area.

The outer mole hd (Fl R 3s) should be given a wide clearance; best water lies E of No 2 buoy. The flood reaches 2kn max; the ebb 3kn, but 6kn if river in spate. Entry to the hbr prohibited during seaplane firefighting operations, 3.3M SSW of Outer Mole.

Caution 1: 13M S of Viana 3 SPM buoys, the 2 outer buoys Fl Y 2.5s, mark a wave-generated power trial. Keep well seaward of the floating equipment which may be hard to see. See opposite.

LIGHTS AND MARKS Montedor lt ho is a R twr 4M NNW. ⊕ dome on Monte Santa Luzia is conspic, almost aligned with 007·5° dir lt. Storm sigs are displayed close SE of dir lt Oc WRG 4s. See Lights, buoys and waypoints for details. Nos 5 to 14 buoys are Fl R 3s or Fl G 3s.

COMMUNICATIONS (Code 258) MRCC (21) 4401919; Police 822345; ⊖ 823346; Brit Consul 808 203 537; Port HM 829096, Call *Porto de Viana* Ch 16; 11 (0900-1200; 1400-1700LT); Yacht Club and Marina Ch 09.

FACILITIES Marina ☎359500. www.apvc.pt apvc@apvc.pt (3m) 200 inc❶ €2·60 up to 12m LOA is abeam Nos 13 & 14 buoys and 200m short of low (4·6m) and noisy road/rail bridge. A small pedestrian bridge across the marina ent lifts on request, if shut berth on waiting pontoon, bridge remains open 0000–0800. Limited ❶ berths F&A on W side of first pontoon D. 🅟🅐⊙🅐🔧 🅚🛆🅐🅔🅖 🍴🔧🛟(20t). YC. No sign of a marina planned in the long FV basin (41°41'·34N 08°50'W).

Town ✕🅟🛒✉🅑🅗🚉 (1km). ✈ Porto (50km).

9.24.7 PÓVOA DE VARZIM

Minho **41°22'·20N 08°46'·00W** (Marina ent) ❀❀❀♦♦❀❀

CHARTS AC 3634 (too small scale, 1:200,000, to be of much practical value); PC 24201

TIDES

Standard Port LISBOA (⟶)

Times				Height (metres)			
High Water		Low Water		MHWS	MHWN	MLWN	MLWS
0400	0900	0400	0900	3·8	3·0	1·5	0·6
1600	2100	1600	2100				
Differences PÓVOA DE VARZIM							
−0020	0000	+0010	+0015	−0·3	−0·3	−0·1	−0·1

SHELTER Good in all winds, but ent is rough in heavy swell and becomes impracticable when wave height >3m. In particular give a wide berth to the head of the N mole where there may be broken water. Reports of larger yachts being denied entry to the marina with bad weather approaching.

NAVIGATION WPT 41°21'·48N 08°46'·73W, 030°/1M to hbr ent. Follow SHM buoys (marking a shoal near the S bkwtr) round to the marina. Tidal streams are weak but there is an appreciable current, mainly S-going, along the coast.

LIGHTS AND MARKS Mole hd lts are on R and G posts with W bands. High rise blocks/hotels and sandy beaches are conspic. Monte São Félix is an isolated rounded hill (209m) 4·5M NNE.

COMMUNICATIONS (Code 252) MRCC (21) 4401919; Police 620026; Brit Consul 808 203 537; 🅗 690600; Hbr VHF Ch 11, 16 (M-F 0900-1200, 1400-1700); Marina VHF Ch 09; Tourist office 298120.

FACILITIES FVs use the N part of hbr; the friendly marina is to the SE. ❶ berth on the N'most reception/fuel pontoon in 2·4 to 3m.

Marina ☎688121. www.clubenavalpovoense.com 241, inc some ❶ €1·50; 18m max LOA. Good facilities, many yachts lay up here: 🅝 (at the FV hbr) ⊙🅐🅐🔧 🅚🅐🔧 🅔🏴(35t).

Town ✕🅟🛒✉🅑🅗🚉 (15 mins walk; trains to Porto, 30km and ✈ 18km).

9.24.8 LEIXÕES

Douro **41°11'·08N 08°42'·24W** (Marina ent) ❀❀❀◍◔❀

CHARTS AC 3634, 3258; PC 23201, 24201, 26402; SC 418A

TIDES

Standard Port LISBOA (⟶); ML 2·0

Times				Height (metres)			
High Water		Low Water		MHWS	MHWN	MLWN	MLWS
0400	0900	0400	0900	3·8	3·0	1·5	0·6
1600	2100	1600	2100				
Differences LEIXÕES							
−0025	−0010	0000	+0010	−0·3	−0·3	−0·2	−0·1
RIO DOURO ENT							
−0010	+0005	+0015	+0025	−0·6	−0·4	−0·1	+0·1
PORTO							
+0002	+0002	+0040	+0040	−0·5	−0·4	−0·2	0·0

SHELTER Good in marina, but scend enters in gales. The port may be closed occas in strong W'lies and/or swell, as shown by tfc sigs (see Passage information) on a mast at the Hbr Office close NE of marina. There have been reports of larger yachts being denied entry to the marina with bad weather approaching. Charted ⚓ outside hbr in 4-6m on sand/mud needs HM's approval.

NAVIGATION WPT 41°10'·09N 08°42'·26W, 350°/6ca to inner mole hds. Leça lt ho, 1·7M N of hbr ent, on a brg of 350·5° leads into inner hbr; thence track 015° to marina. Ent is usually simple. Keep at least 200m off the N mole hd (Quebra-mar) due to a sunken bkwtr (2·5m) and a wreck (8m) 100m SW and S of it.

LIGHTS AND MARKS N of Leça lt are many R/W banded chy's at the oil refinery. From the S, bldgs of Porto are conspic.

COMMUNICATIONS (Code 229) MRCC (21) 4401919; LB 226. 170091; Met 484527 (Porto airport); Police 383649; ⊖ 951476; Brit Consul 808 203 537; Fire 380018; H 391000; HM 953000; Marina *Porto Atlântico* VHF Ch 09 (0900-1230; 1400-1800).

FACILITIES Marina Porto Atlântico ☎964895 www. marinaportoatlantico.net (2·9-1·5m) is rather oily with limited facilities onsite. 200 F&A + 40❷ €1·64; reception W quay. ✕ ⌂ ▢ ⌂ @ 🔧 ⛽ ⚓ ✈ 🗑 🅿 (6t). **Clube de Vela Atlântico** ✕

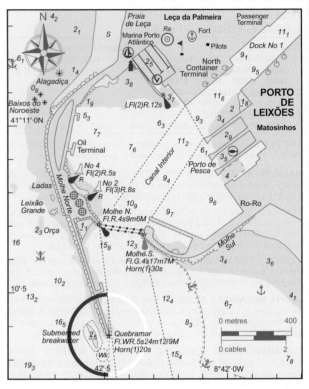

☎963547. **Clube Naval de Leça** Bar ☎951700. **Sport Club do Porto** ☎952225.

Town ✕ ⌂ 🛒 ✉ Ⓑ. Bus & Metro to Porto (5km); ⇌ and ✈.

9.24.9 PORTO (OPORTO)

Douro **41°08'·80N 08°40'·55W** ❀◍◔❀❀

CHARTS AC 3258; PT 26402; 426402

TIDES see above.

SHELTER Excellent at Douro marina. Also possible ☋ at small quay by Cantareira, or at Cais de Estiva between the first and second bridges. A road recce of the Rio Douro is advised.

NAVIGATION Difficult river ent, prone to swell and fast current; dangerous in strong W'lies (see Passage information for port closure signals). New moles at the ent; see AC 3258 inset. N mole is lit by Fl R 5s 16m 9M (W twr, R stripes); S mole by Fl G 5s 15m 6M. Chan has 2 PHM and 2 SHM bys (lit, numbered 1 to 4). When abeam PHM no 4, head 161°/0·3M to Douro Marina. The bar shifts and charted depths are unreliable, the effect of the breakwater works is so far unknown. For buoys up-river see Lights, buoys and waypoints.

FACILITIES

Douro Marina (41°08'·57N 08°39'·02W) ☎351 220 907 300, mob 351 918 501 474 www.douromarina.com info@douromarina.com 1M from the ent to the Rio Douro. VHF Ch 09 call *Douro Marina*. 240 + 60 ❷ €2·46/m + VAT, max LOA <50m. ⛽ ⚓ ⚓ ▢ Ⓒ 🏳(50t) 🗑(3t). Car and bicycle hire also available. Reports welcomed.

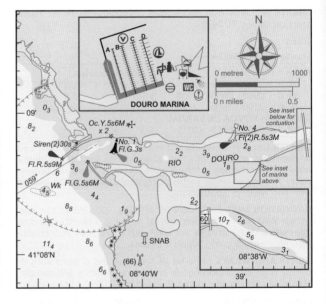

9.24.10 AVEIRO

Beira Litoral **40°38'·51N 08°45'·86W** (Hbr ent) ✿✿✿✿◊◊✿✿✿

CHARTS AC 3634, 3227; PC 24201, 24202, 26403, 59; SC 4219

TIDES

Standard Port LISBOA (⟶); ML 2·0

Times				Height (metres)			
High Water		Low Water		MHWS	MHWN	MLWN	MLWS
0400	0900	0400	0900	3·8	3·0	1·5	0·6
1600	2100	1600	2100				
Differences AVEIRO							
+0005	+0010	+0010	+0015	−0·5	−0·4	−0·1	0·0
SAN JACINTO (40°40'N 08°44'W)							
+0025	+0030	+0030	+0045	−0·7	−0·5	−0·1	+0·2
FIGUEIRA DA FOZ							
−0015	0000	+0010	+0020	−0·3	−0·3	−0·2	−0·1

SHELTER Good, but the bar (c. 6m) becomes impassable in strong onshore winds and the port is often one of the first on this coast to be closed by bad weather or swell, even if there is no wind.

Two small marinas can usually accommodate visitors (see Facilities) or ☊ at the N end of Baia de San Jacinto in about 7m having avoided the large drying sandbank (buoyed) inside the outer end of the bay's N bkwtr.

NAVIGATION WPT (off chartlet) Boia Aveiro buoy, 40°37'·86N 08°46'·98W, 056°/1·1M to hbr ent. Caution: strong tidal streams. From abeam the lt ho the ent to B de San Jacinto is about 9 ca NE. Do not proceed further E.

LIGHTS AND MARKS Hbr lts as chartlet/Lights, buoys and waypoints. The R/W banded lt ho and R/W Aero lt ho twr are conspic on either side of ent. Tfc sigs (Passage information) are shown from a mast close N of the lt ho. The 060·6° ldg line (marks hard to see) leads between the moles. A military airfield with conspic control twr is NE of the hbr ent.

Caution: changes to depths S of main appr reported: an area liable to change exists centred 40°37'·69N 08°46'·00W.

COMMUNICATIONS (Code 234) ⊖ ☇ ☎393300; Brit Consul 808 203 537; HM 393300, www.portodeaveiro.pt geral@portodeaveiro.pt; Port VHF Ch 11, 16 (M-F, 0900-1200 & 1400-1700).

FACILITIES A helpful, English/French-speaking contact in Aveiro is Senhor Cala Barras, Avela YC. A ferry reported to run from near the ☊ in Baia de San Jacinto to Aveiro.

CVCN YC marina,**Ⓥ** via Mira Canal (2m) and road bridge (14m air draught) www.cvcn.pt

Avela YC marina 3/4**Ⓥ** ☎422142 www.avela.pt near electrical cable (21m air draught)

Town The attractive old town, with its *moliceiros* (mussel boats) fine buildings and a network of canals, is reminiscent of a mini-Venice. ✗ ⌂ ☵ ⊠ Ⓑ ⇌ S to Figueira da Foz and Coimbra (worth a visit) and N to Porto. ✈ Pôrto & Lisboa (77km & 256km) by motorway.

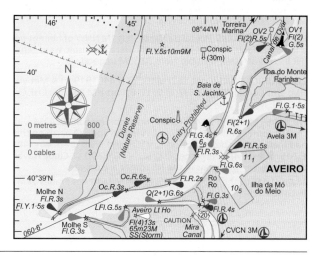

9.24.11 FIGUEIRA DA FOZ

Beira Litoral **40°08'·68N 08°52'·48W** (Hbr ent) ✿✿◊◊✿✿

CHARTS AC 3634, 3635, 3228; PC 24201/2, 26404, 34, 64; SC 42A

TIDES See under Aveiro; ML 2·0

SHELTER Excellent in marina on N bank, 0·75M from hbr ent.

Bar at ent has charted depth of 4·1m, but this cannot be relied upon. It shifts/shoals constantly and can be highly dangerous in swell, especially Nov-Mar with strong W/NW winds. Conditions on the bar are signalled from mast at Forte de Santa Catarina (N of inner moles) and may be broadcast on VHF Ch 11.

NAVIGATION WPT 40°08'·61N 08°53'·12W (off chartlet) 081°/5ca to hbr ent. The bar in bad weather is the only offshore danger.

LIGHTS AND MARKS Cabo Mondego lt ho is 3M NW of hbr ent. Buarcos light is on the beach 1·1M N of hbr ent. Ldg lts lead 081·5° past marina ent. Lts as chartlet and Lights, buoys and waypoints. Extensive beaches N and S of hbr ent and a white suspension bridge (39m cl'nce) 1·5M E of ent are conspic from seaward.

COMMUNICATIONS (Code 233) Maritime Police 916 352629; Port VHF Ch 11, 422955; Marina VHF Ch 08 (call *Capmarfoz*).

FACILITIES Marina ☎402910, ipc.gap@mail.telepac.pt (2-2·8m). Strong ebb tide can affect pontoons nearest marina ent. Reception/fuel berth is dead ahead of ent. 200 inc**Ⓥ** €2·65, plus possible surcharge for beam, multihulls +50%. ⛽ ▣ ⚓ ≜ ✕ ⚒ ⛫.

Town ✗ ⌂ ☵ Ⓑ ⇌ ✈ Pôrto.

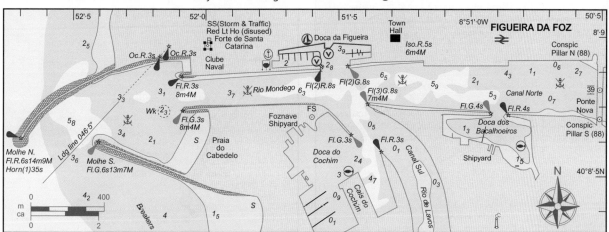

9.24.12 NAZARÉ

Estremadura **39°35'·42N 09°04'·63W** ❄❄❄♨♨❀❀

CHARTS AC 3635; PC 24203, 26302, 34, 65

TIDES

Standard Port LISBOA (⟶); ML 2·0

Times				Height (metres)			
High Water		Low Water		MHWS	MHWN	MLWN	MLWS
0400	0900	0400	0900	3·8	3·0	1·5	0·6
1600	2100	1600	2100				
Differences NAZARÉ (Pederneira)							
−0030	−0015	−0005	+0005	−0·5	−0·4	−0·1	0·0

SHELTER Good with all-weather access. No ⚓ in outer harbour.

NAVIGATION WPT 39°35'·69N 09°05'·95W, 106°/1M to hbr ent. Due to an underwater canyon there are depths of 452m and 152m 5M and 7ca offshore respectively.

LIGHTS AND MARKS Pontal da Nazaré lt is 1M NNW of hbr ent on a low headland; keep at least 2ca off. Here a conspic wind farm (FR lts & W strobes) is visible from 20-30M in good vis.

COMMUNICATIONS (Code 262) MRCC (21) 4401919; ⊖ & Met via HM; Brit Consul 808 203 537; 🏥 561116; Port HM 561255, CHF Ch 11 (HO); Marina VHF Ch 09.

FACILITIES Marina ☎561401 (British manager) mob 968074254 is in the SW corner of the inner hbr, call (VHF) on appr; ❶ berth on the hammerheads or outer pontoons, <15m and 3·5m draft. 40+14❶ €2·15. 🛢🚻 (H24) in NE corner, 🅿Ⓒ🔄🔧✕🖥(80t) 🛒. **Clube Naval de Nazaré** The pontoons in the NE corner are private.

Town (1½M to the N) ✕ 🏪 🍽 ✉ Ⓑ. ⇌ Valado dos Frades 7km. ✈ Lisboa 125km.

9.24.13 PENICHE

Estremadura **39°20'·82N 09°22'·48W** ❄❄❄♨♨❀❀

CHARTS AC 3635; PC 24203, 26405, 36

TIDES

Standard Port LISBOA (⟶); ML 2·0

Times				Height (metres)			
High Water		Low Water		MHWS	MHWN	MLWN	MLWS
0400	0900	0400	0900	3·8	3·0	1·5	0·6
1600	2100	1600	2100				
Differences PENICHE							
−0035	−0015	−0005	0000	−0·3	−0·3	−0·1	−0·1
ERICEIRA (38°58'N, 24M S)							
−0040	−0025	−0010	−0010	−0·4	−0·3	−0·1	0·0

SHELTER Good all-weather hbr, but uncomfortable if SW swell works in; also noisy and prone to FV wash. Possible moorings or ⚓ in SE part of hbr; holding is good, but ground may be foul. The ⚓ in 4m on N side of peninsula is open to swell, even in S'lies.

NAVIGATION WPT 39°20'·42N 09°22'·53W, 000°/5ca to W mole hd. 3kn speed limit in hbr. The coastal route, a 5M wide chan between Cabo Carvoeiro and Ilha da Berlenga, is often rough.

Peniche and the offshore islands are in an 'Area to be avoided', only by ships >300 tons (for environmental reasons); see Area map.

On the SE side of Ilha Berlenga ⚓ below the lt ho. Os Farilhões lies 4·5M further NW. These islands lie within a Nature Reserve.

LIGHTS AND MARKS Peniche looks like an island from afar, not to be confused in poor vis with Berlenga. An isolated square tower block is conspic about 2ca NNE of the harbour. Cabo Carvoeiro is 2M WNW of hbr ent and steep-to apart from Nau dos Carvos, a conspic high rk close-in. Other lts as chartlet and Lights, buoys and waypoints. Storm sigs are shown from the root of W mole.

COMMUNICATIONS (Code 262) MRCC (21) 4401919; 🖥 789629; Brit Consul 808 203 537; Ⓗ 781702; Port HM 781153, VHF Ch 16 (H24); Marina VHF Ch 09.

FACILITIES Berth on outer pontoon (2·4-3·5m) of small marina ☎ 783331. 130+ 20Ⓥ €1·60. ⊖ ⚒ 🛢 🗃 (or tanker) ⚓ ⬛ 🛒 🔧 🗜 Ⓔ 🛟 ◄ YC.

Town ✕ ⛲ 🍴 ✉ Ⓑ ⇌ Obidos (30km) ✈ Lisboa (80km).

9.24.14 CASCAIS

Estremadura **38°41'·59N 09°24'·88W** 🌀🌀🌀⚓⚓⚓🏵🏵🏵

CHARTS AC 3635, 3220; PC 27504, 26303; SC 4310

TIDES
Standard Port LISBOA (⟶); ML 2·0

Times				Height (metres)			
High Water		Low Water		MHWS	MHWN	MLWN	MLWS
0400	0900	0400	0900	3·8	3·0	1·5	0·6
1600	2100	1600	2100				
Differences CASCAIS							
−0040	−0025	−0015	−0010	−0·3	−0·2	0·0	+0·1

SHELTER Good in marina or on outer pontoons. ⚓ N of marina in designated area 3-5m, sheltered from all but E winds.

NAVIGATION WPT 38°41'·28N 09°24'·57W, 334°/4ca to CC2 🔲 and on the 285° ldg line. MC1-3 SCM buoys mark no-entry area.

LIGHTS AND MARKS Guia, rear 285° ldg lt, is 1·2M from the front, Santa Marta. Other lts as chartlet.

COMMUNICATIONS (Code 214) MRCC 401919; Brit Consul 808 203 537; Marina VHF Ch 09 (0900-2000).

FACILITIES Marina ☎824857. www.mymarinacascais.com info@marinacascais.pt (6m) 638 inc 125Ⓥ €3.00, multihulls +50%. ⚓⚓ (0900-2000) 🗃 ⬛ ⚓ Ⓒ Ⓑ 🔧 🛒(70t) ◣(2t). Mail pickup: Casa de S° Bernardo, Marina de Cascais, 2750-800 Cascais.

Town All amenities. ⇌ to Lisboa (½ hr) every ¼ hr, 0530-0230.

9.24.15 LISBOA

Estremadura **38°41'·75N 09°10'·71W** (N span of Ponte 25 de Abril) 🌀🌀🌀⚓⚓⚓🏵🏵🏵

CHARTS AC 3635, 3220, 3221, 3222; PC 26406, 26303, 26304, 26305, 26306, 26307, 24204, 24203, 25R07; SC 4310

TIDES
Standard Port LISBOA (⟶); ML 2·0

TimesHeight (metres)							
High Water		Low Water		MHWS	MHWN	MLWN	MLWS
0400	0900	0400	0900	3·8	3·0	1·5	0·6
1600	2100	1600	2100				
Differences PACO DE ARCOS (5·5M W of Tejo bridge)							
−0020	−0030	−0005	−0005	−0·4	−0·4	−0·2	−0·1
PEDROUCOS (2·2M W of Tejo bridge)							
−0010	−0015	0000	0000	−0·3	−0·2	−0·1	0·0

SHELTER is good in the marinas along the N bank of R. Tejo.

NAVIGATION From N/NW keep at least 5ca off Cabo da Roca (see Area map for TSS) and Cabo Raso. Magnetic anomalies 3M SSE of Cabo Raso can alter local variation by +5° to −3°.

WPT 38°36'·90N 09°22'·90W (5ca SE of No 2 buoy) 047°/5·4M to abeam No 5 buoy. This is the main DW chan, Barra Sul, into R. Tejo proper; shoals lie on either side and break in bad weather. Tidal streams reach 5kn in the river and the ebb sets towards the shoals E of Forte Bugio. Caution: cross-current at the ent to marinas. Speed limit E of Torre de Belém: 10kn, but 5kn within 300m of shore.

Barra Norte, a lesser chan carrying 5·2m, is a short cut for yachts between Cascais and R. Tejo in fair weather; it passes close S of Forte de S. Julião and is orientated 285°/105° on ldg lts at Forte de Sta Marta and Guia (obscured by trees from the east by day).

LIGHTS AND MARKS See Lts, buoys and waypoints, chartlets and hbr text. Conspic features, W-E, include: Torre de Belém, Padrao dos Descobrimentos, Ponte 25 de Abril and statue of Cristo Rei.

COMMUNICATIONS (Code 21) MRCC 4401919; Monitor *Lisboa Port Control* (VTS, applies to yachts >20m LOA) Ch 12, 13 - **74** broadcasts MSI as required; Brit Consul 808 203 537; Port HM **12**, 01, 05, 60; Marinas VHF Ch 09.

FACILITIES
Oeiras Marina 38°40'·59N 09°18'·99W. ☎(21) 440 1510. Ch 09. www.portoderecrio.oeirasviva.pt porto.recreio@oeirasviva.pt 275 ◻ €2.91, multihulls +100%. ⚓⚓ ⬛ (0800-2200) ✕ ⛲ 🗃 Ⓒ @ ⚓ ◄ ◣(8t). Fast trains to Lisboa 16km; see Cascais.

Marinas Ⓐ-Ⓓ are run by Lisboa port authority (www.portodelisboa.pt ☎(3611000). In all these marinas average cost is €2.25.

Ⓐ **Doca de Bom Sucesso** 38°41'·55N 09°12'·70W. ☎3922080. doca.bomsucesso@portodelisboa.pt 163 mostly local yachts, F&A in 3m. ✕ 🗃 🗃 ◄ ⇌ bus to Lisboa.

Ⓑ **Doca de Belém** 38°41'·64N 09°13'·24W. ☎3922203. doca.belem@portodelisboa.pt 194 yachts, v few Ⓥ; F&A in 3m. ⚓ (hose) ✕ ⚓ 🗃 ⬛ ◣. ⇌ bus to Lisboa.

Ⓒ **Doca de Santo Amaro** 38°41'·92N 09°12'·65W, almost below the suspension bridge (noise from road traffic and night life). ☎3922011. doca.stamaro@portodelisboa.pt 287 ◻ (few Ⓥ) in 4m, ✕ ⛲ 🍴. ⇌ bus 2M to city centre.

Ⓓ **Doca de Alcântara** 38°42'·15N 09°09'·42W (best for foreign visitors). Chan doubles back W into marina at end of dock, via a swing bridge usually open; Ch 68. doca.alcantara@portodelisboa.pt ☎3922048, 440 F&A inc Ⓥ in 8m. Secure, 4 floating ✕ ⛲ 🍴. ⇌ bus 2M to city centre.

Ⓔ **Marina Parque das Nações** 38°45'·37N 09°05'·92W 1·35M S of Vasco de Gama bridge. ☎(218) 949066. www.marinaparquedasnacoes.pt info@marinaparquedasnacoes.pt 400 (inc Ⓥ) €2.35, multihulls +50%. Reception pontoon in outer hbr W of sluices. ⚓ ✕ 🗃 🗃 ⚓ ⚓ 🗃 🔧 Ⓒ ⛲ 🛒(35t) ◣(3t). Reports welcome.

Centro Náutico de Algés Boatyard, between Oeiras and Torre de Belem, 38°41'·47N 09°14'·07, ☎213032440; www.cnalges.pt geral@cnalges.pt ⬛ ⚓ 300+ ⛲ 🗃 🔧 ⚒ 🗃 🛒(50t).

Tagus Yacht Centre, Amora, Seixal Bay 38°38'·07N 09°06'·77W. ☎227 6400. www.tagusyachtcenter.com info@tagusyachtcenter.com. 🗃 🛒(70t). Ferries Seixal to Lisboa

City All amenities; ⇌ ✈.

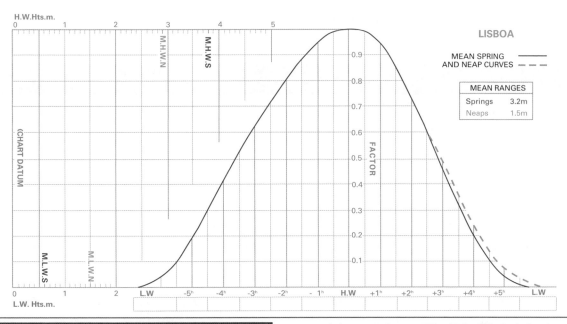

LISBOA

MEAN SPRING ————
AND NEAP CURVES – – – –

MEAN RANGES	
Springs	3.2m
Neaps	1.5m

CABO RASO TO CAPE ST VINCENT

Cascais is an expensive stop for yachts on passage or not wishing to go 12M E to Lisboa. **Oeiras Marina**, 5M E of Cascais, is also a good stop. From Cascais to the Rio Tejo (Tagus) use the Barra Norte (least depth 5m) which joins the main Barra Sul abeam São Julião lt. In **Lisboa** 5 marinas line the N bank of the R Tejo. If S-bound, stand on 1M beyond Forte Bugio before turning for flat-topped Cabo Espichel (AC 3635, 3636).

Eastward the coast rises to 500m high cliffs. **Sesimbra** is well sheltered from N'lies. Near the ent to R Sado, Arrábida's shallow bay offers temp'y ⌁. **Setúbal** has few yacht facilities. Thence unbroken beaches stretch 35M to Cabo de Sines, light house and conspicuous chimneys. **Sines** is a strategically placed commercial port 56M N of Cabo São Vicente; its small marina has limited facilities. A tiny bay (37°17'·51N 08°52'·08W) below the high cliffs of Punta da Arrifana gives shelter from N'lies.

Portugal

993

STANDARD TIME (UT)
For Summer Time add ONE hour in **non-shaded areas**

LISBOA   LAT 38°43'N   LONG 9°07'W
TIMES AND HEIGHTS OF HIGH AND LOW WATERS

Dates in **red** are **SPRINGS**
Dates in blue are NEAPS

YEAR 2016

	JANUARY				FEBRUARY				MARCH				APRIL		
	Time m		Time m		Time m		Time m		Time m		Time m		Time m		Time m
1	0101 1.2 / 0730 3.1 / F 1337 1.2 / 2004 2.9	**16**	0105 0.8 / 0746 3.4 / SA 1339 0.8 / ☾ 2020 3.2	**1**	0156 1.4 / 0823 2.8 / M 1433 1.4 / ◑ 2105 2.8	**16**	0253 1.1 / 0938 3.1 / TU 1531 1.2 / 2218 3.1	**1**	0107 1.3 / 0736 2.9 / TU 1334 1.4 / ◑ 2009 2.9	**16**	0235 1.2 / 0919 3.0 / W 1508 1.4 / 2156 3.1	**1**	0248 1.5 / 0924 2.8 / F 1525 1.6 / 2203 3.0	**16**	0504 1.4 / 1137 3.1 / SA 1727 1.5 / 2359 3.3
2	0155 1.4 / 0823 2.9 / SA 1436 1.3 / ◑ 2107 2.8	**17**	0206 1.0 / 0849 3.3 / SU 1444 1.0 / 2129 3.1	**2**	0307 1.5 / 0935 2.7 / TU 1549 1.4 / 2222 2.8	**17**	0420 1.2 / 1102 3.0 / W 1655 1.3 / 2338 3.2	**2**	0208 1.5 / 0839 2.8 / W 1444 1.5 / 2121 2.8	**17**	0406 1.3 / 1047 3.0 / TH 1639 1.4 / 2320 3.1	**2**	0417 1.4 / 1054 3.0 / SA 1650 1.5 / 2324 3.2	**17**	0604 1.2 / 1233 3.2 / SU 1822 1.3
3	0302 1.5 / 0931 2.8 / SU 1545 1.4 / 2219 2.8	**18**	0318 1.1 / 1002 3.2 / M 1558 1.1 / 2243 3.1	**3**	0428 1.5 / 1055 2.8 / W 1703 1.4 / 2334 2.9	**18**	0544 1.1 / 1217 3.1 / TH 1809 1.1	**3**	0332 1.5 / 1003 2.7 / TH 1611 1.5 / 2246 2.9	**18**	0532 1.3 / 1203 3.1 / F 1755 1.3	**3**	0530 1.2 / 1206 3.2 / SU 1756 1.2	**18**	0051 3.4 / 0650 1.1 / M 1318 3.4 / 1906 1.1
4	0416 1.5 / 1044 2.8 / M 1653 1.3 / 2325 2.9	**19**	0438 1.1 / 1118 3.2 / TU 1713 1.0 / 2354 3.3	**4**	0539 1.5 / 1205 2.9 / TH 1804 1.2	**19**	0044 3.3 / 0649 1.0 / F 1317 3.3 / 1906 1.0	**4**	0457 1.4 / 1129 2.9 / F 1728 1.4 / 2359 3.1	**19**	0027 3.3 / 0634 1.1 / SA 1301 3.3 / 1850 1.1	**4**	0028 3.5 / 0629 0.9 / M 1302 3.5 / 1850 0.9	**19**	0134 3.5 / 0728 1.0 / TU 1356 3.5 / 1943 1.0
5	0523 1.4 / 1148 2.9 / TU 1751 1.2	**20**	0553 1.0 / 1226 3.2 / W 1819 0.9	**5**	0034 3.1 / 0636 1.1 / F 1302 3.1 / 1855 1.0	**20**	0138 3.5 / 0739 0.8 / SA 1406 3.4 / 1952 0.8	**5**	0605 1.2 / 1235 3.1 / SA 1827 1.1	**20**	0120 3.5 / 0720 0.9 / SU 1346 3.4 / 1933 1.0	**5**	0122 3.8 / 0718 0.6 / TU 1351 3.8 / 1938 0.6	**20**	0211 3.6 / 0802 0.9 / W 1430 3.6 / 2017 0.9
6	0020 3.0 / 0618 1.2 / W 1242 3.1 / 1839 1.1	**21**	0056 3.4 / 0656 0.8 / TH 1326 3.4 / 1915 0.8	**6**	0125 3.3 / 0724 0.9 / SA 1351 3.3 / 1940 0.8	**21**	0224 3.6 / 0821 0.7 / SU 1448 3.5 / 2031 0.7	**6**	0058 3.4 / 0658 0.9 / SU 1328 3.4 / 1917 0.8	**21**	0203 3.6 / 0758 0.8 / M 1425 3.5 / 2010 0.9	**6**	0211 4.0 / 0804 0.4 / W 1437 4.0 / 2024 0.4	**21**	0245 3.7 / 0834 0.8 / TH 1503 3.7 / 2049 0.8
7	0107 3.2 / 0704 1.0 / TH 1329 3.2 / 1922 0.9	**22**	0150 3.6 / 0748 0.7 / F 1418 3.5 / 2004 0.7	**7**	0211 3.6 / 0807 0.6 / SU 1437 3.5 / 2023 0.6	**22**	0304 3.7 / 0857 0.6 / M 1525 3.6 / ○ 2107 0.6	**7**	0148 3.7 / 0745 0.6 / M 1416 3.6 / 2002 0.6	**22**	0241 3.7 / 0832 0.7 / TU 1500 3.6 / 2043 0.8	**7**	0258 4.2 / 0848 0.3 / TH 1522 4.2 / ● 2108 0.3	**22**	0317 3.7 / 0905 0.8 / F 1534 3.7 / ○ 2120 0.8
8	0150 3.4 / 0746 0.9 / F 1412 3.3 / 2002 0.8	**23**	0238 3.7 / 0834 0.6 / SA 1504 3.5 / 2047 0.6	**8**	0256 3.8 / 0850 0.4 / M 1521 3.7 / ● 2105 0.4	**23**	0340 3.8 / 0930 0.5 / TU 1559 3.6 / 2140 0.6	**8**	0235 3.9 / 0829 0.4 / TU 1500 3.9 / 2046 0.4	**23**	0314 3.8 / 0903 0.7 / W 1532 3.7 / ○ 2115 0.7	**8**	0343 4.3 / 0931 0.2 / F 1605 4.2 / 2152 0.2	**23**	0349 3.7 / 0935 0.8 / SA 1604 3.7 / 2152 0.8
9	0232 3.6 / 0827 0.7 / SA 1455 3.5 / 2041 0.6	**24**	0322 3.8 / 0915 0.5 / SU 1545 3.6 / ○ 2125 0.6	**9**	0340 4.0 / 0931 0.3 / TU 1604 3.8 / 2147 0.3	**24**	0413 3.7 / 1001 0.6 / W 1630 3.6 / 2211 0.6	**9**	0320 4.1 / 0911 0.2 / W 1544 4.0 / ● 2128 0.3	**24**	0346 3.8 / 0933 0.6 / TH 1602 3.7 / 2145 0.7	**9**	0428 4.3 / 1014 0.3 / SA 1649 4.2 / 2236 0.3	**24**	0420 3.6 / 1006 0.8 / SU 1635 3.7 / 2223 0.9
10	0314 3.7 / 0907 0.5 / SU 1537 3.6 / ● 2121 0.6	**25**	0402 3.8 / 0953 0.5 / M 1623 3.5 / 2202 0.6	**10**	0423 4.1 / 1013 0.2 / W 1646 3.8 / 2229 0.3	**25**	0443 3.7 / 1032 0.6 / TH 1659 3.5 / 2242 0.7	**10**	0404 4.2 / 0953 0.2 / TH 1627 4.1 / 2211 0.2	**25**	0415 3.7 / 1003 0.7 / F 1631 3.7 / 2216 0.7	**10**	0513 4.1 / 1057 0.4 / SU 1734 4.0 / 2322 0.5	**25**	0452 3.5 / 1036 0.9 / M 1708 3.6 / 2256 0.9
11	0356 3.8 / 0947 0.4 / M 1619 3.6 / 2201 0.5	**26**	0437 3.7 / 1027 0.5 / TU 1657 3.5 / 2236 0.6	**11**	0506 4.1 / 1055 0.2 / TH 1729 3.8 / 2311 0.3	**26**	0512 3.6 / 1102 0.7 / F 1728 3.5 / 2313 0.8	**11**	0447 4.2 / 1035 0.2 / F 1709 4.0 / 2254 0.3	**26**	0445 3.7 / 1032 0.8 / SA 1700 3.6 / 2246 0.8	**11**	0600 3.9 / 1142 0.7 / M 1820 3.8	**26**	0526 3.4 / 1109 1.0 / TU 1742 3.5 / 2332 1.1
12	0438 3.9 / 1028 0.4 / TU 1702 3.6 / 2243 0.5	**27**	0510 3.6 / 1100 0.6 / W 1729 3.4 / 2309 0.7	**12**	0549 4.0 / 1138 0.2 / F 1813 3.7 / 2356 0.5	**27**	0541 3.5 / 1133 0.8 / SA 1759 3.3 / 2346 0.9	**12**	0531 4.1 / 1118 0.3 / SA 1753 3.9 / 2338 0.4	**27**	0515 3.6 / 1102 0.9 / SU 1730 3.5 / 2318 0.9	**12**	0011 0.7 / 0650 3.6 / TU 1231 1.0 / 1912 3.6	**27**	0603 3.3 / 1145 1.2 / W 1822 3.4
13	0521 3.8 / 1111 0.4 / W 1746 3.6 / 2326 0.6	**28**	0541 3.5 / 1133 0.7 / TH 1800 3.3 / 2343 0.9	**13**	0635 3.8 / 1223 0.6 / SA 1900 3.5	**28**	0613 3.3 / 1206 1.0 / SU 1833 3.2	**13**	0617 3.9 / 1202 0.6 / SU 1839 3.7	**28**	0546 3.4 / 1134 1.0 / M 1804 3.4 / 2353 1.1	**13**	0108 1.0 / 0748 3.3 / W 1329 1.3 / 2014 3.3	**28**	0015 1.2 / 0647 3.2 / TH 1230 1.3 / 1910 3.2
14	0606 3.8 / 1156 0.5 / TH 1832 3.5	**29**	0612 3.4 / 1208 0.9 / F 1833 3.2	**14**	0045 0.7 / 0725 3.5 / SU 1314 0.8 / 1954 3.3	**29**	0022 1.1 / 0650 3.1 / M 1244 1.2 / 1915 3.0	**14**	0027 0.7 / 0706 3.6 / M 1251 0.9 / 1931 3.5	**29**	0622 3.2 / 1209 1.2 / TU 1843 3.2	**14**	0218 1.3 / 0900 3.0 / TH 1444 1.5 / ◑ 2131 3.2	**29**	0108 1.3 / 0743 3.0 / F 1330 1.5 / 2011 3.1
15	0013 0.7 / 0653 3.6 / F 1244 0.7 / 1922 3.3	**30**	0019 1.0 / 0646 3.2 / SA 1246 1.0 / 1911 3.0	**15**	0142 0.9 / 0824 3.3 / M 1415 1.0 / ◑ 2059 3.1			**15**	0123 1.0 / 0805 3.3 / TU 1350 1.2 / ◑ 2035 3.2	**30**	0034 1.3 / 0706 3.1 / W 1253 1.4 / 1932 3.1	**15**	0343 1.4 / 1024 3.0 / F 1612 1.6 / 2252 3.2	**30**	0218 1.4 / 0857 3.0 / SA 1450 1.5 / ◑ 2129 3.1
		31	0102 1.2 / 0728 3.0 / SU 1333 1.2 / 2001 2.8							**31**	0130 1.4 / 0803 2.9 / TH 1357 1.5 / ◑ 2038 3.0				

HAT is 4·3 metres above Chart Datum.

LISBOA LAT 38°43'N LONG 9°07'W
TIMES AND HEIGHTS OF HIGH AND LOW WATERS

STANDARD TIME (UT)
For Summer Time add ONE hour in **non-shaded areas**

Dates in red are **SPRINGS**
Dates in blue are NEAPS

YEAR 2016

MAY

Time	m	Time	m
1 0340	1.4	**16** 0520	1.3
1021	3.0	1154	3.1
SU 1613	1.4	M 1742	1.4
2249	3.3		
2 0454	1.2	**17** 0012	3.2
1134	3.3	0610	1.2
M 1723	1.2	TU 1241	3.3
2356	3.5	1830	1.3
3 0556	0.9	**18** 0057	3.3
1232	3.5	0652	1.1
TU 1821	0.9	W 1321	3.4
		1911	1.1
4 0054	3.7	**19** 0137	3.4
0650	0.7	0730	1.0
W 1324	3.8	TH 1358	3.5
1913	0.7	1948	1.0
5 0146	4.0	**20** 0214	3.5
0738	0.5	0804	0.9
TH 1412	4.0	F 1433	3.6
2001	0.5	2023	0.9
6 0235	4.1	**21** 0249	3.5
0825	0.4	0837	0.9
F 1459	4.1	SA 1506	3.7
● 2048	0.5	○ 2056	0.9
7 0323	4.2	**22** 0324	3.6
0910	0.3	0910	0.8
SA 1545	4.2	SU 1540	3.7
2134	0.3	2130	0.8
8 0410	4.1	**23** 0358	3.5
0954	0.4	0942	0.9
SU 1631	4.2	M 1614	3.7
2221	0.4	2204	0.8
9 0457	4.0	**24** 0433	3.5
1039	0.6	1016	0.9
M 1717	4.0	TU 1650	3.6
2308	0.5	2240	0.9
10 0544	3.8	**25** 0511	3.4
1124	0.8	1051	1.0
TU 1804	3.8	W 1727	3.6
2357	0.8	2318	0.9
11 0634	3.5	**26** 0550	3.3
1213	1.0	1130	1.1
W 1853	3.6	TH 1809	3.5
12 0051	1.0	**27** 0002	1.0
0728	3.3	0636	3.2
TH 1307	1.3	F 1216	1.2
1949	3.4	1856	3.4
13 0154	1.2	**28** 0053	1.1
0832	3.1	0729	3.1
F 1413	1.5	SA 1312	1.3
◐ 2056	3.2	1953	3.3
14 0305	1.4	**29** 0155	1.2
0945	3.0	0834	3.1
SA 1529	1.6	SU 1421	1.3
2209	3.1	◑ 2101	3.3
15 0417	1.4	**30** 0307	1.2
1055	3.0	0949	3.2
SU 1642	1.5	M 1537	1.3
2316	3.1	2215	3.3
		31 0418	1.1
		1100	3.3
		TU 1649	1.2
		2324	3.4

JUNE

Time	m	Time	m
1 0524	0.9	**16** 0014	3.1
1202	3.5	0612	1.2
W 1752	0.9	TH 1243	3.2
		1836	1.2
2 0026	3.6	**17** 0101	3.2
0621	0.7	0656	1.1
TH 1258	3.7	F 1325	3.3
1849	0.7	1919	1.1
3 0122	3.8	**18** 0143	3.3
0714	0.6	0735	1.0
F 1350	3.9	SA 1403	3.5
1942	0.6	1957	1.0
4 0215	3.9	**19** 0222	3.4
0804	0.5	0811	0.9
SA 1440	4.0	SU 1441	3.6
2032	0.4	2034	0.9
5 0306	3.9	**20** 0301	3.4
0851	0.5	0847	0.8
SU 1528	4.1	M 1518	3.6
● 2120	0.4	○ 2111	0.8
6 0355	3.9	**21** 0339	3.5
0937	0.5	0922	0.8
M 1615	4.1	TU 1556	3.7
2207	0.4	2148	0.7
7 0442	3.8	**22** 0418	3.5
1022	0.6	0959	0.8
TU 1701	4.0	W 1635	3.7
2254	0.5	2226	0.7
8 0529	3.7	**23** 0458	3.5
1107	0.8	1037	0.8
W 1747	3.8	TH 1715	3.7
2341	0.7	2306	0.7
9 0615	3.5	**24** 0539	3.4
1152	1.0	1118	0.9
TH 1832	3.6	F 1758	3.6
		2349	0.8
10 0028	0.9	**25** 0624	3.3
0702	3.3	1203	0.9
F 1240	1.2	SA 1844	3.5
1919	3.4		
11 0119	1.1	**26** 0037	0.9
0753	3.1	0714	3.3
SA 1333	1.3	SU 1255	1.0
2011	3.2	1936	3.4
12 0216	1.3	**27** 0132	1.0
0852	3.0	0811	3.2
SU 1435	1.5	M 1355	1.1
◑ 2112	3.0	◑ 2036	3.3
13 0319	1.3	**28** 0236	1.0
0957	2.9	0918	3.2
M 1543	1.5	TU 1505	1.2
2218	3.0	2145	3.3
14 0423	1.3	**29** 0345	1.0
1101	3.0	1028	3.2
TU 1650	1.5	W 1618	1.1
2320	3.0	2256	3.3
15 0522	1.3	**30** 0454	1.0
1156	3.1	1135	3.4
W 1747	1.3	TH 1727	1.0

JULY

Time	m	Time	m
1 0003	3.4	**16** 0023	3.0
0558	0.9	0622	1.2
F 1236	3.5	SA 1250	3.2
1831	0.8	1850	1.2
2 0104	3.5	**17** 0112	3.1
0656	0.7	0706	1.1
SA 1333	3.7	SU 1335	3.3
1928	0.6	1933	1.0
3 0200	3.7	**18** 0157	3.3
0748	0.6	0746	0.9
SU 1425	3.8	M 1417	3.5
2020	0.5	2013	0.9
4 0253	3.7	**19** 0239	3.4
0837	0.6	0825	0.8
M 1514	3.9	TU 1458	3.7
● 2108	0.5	○ 2052	0.7
5 0341	3.7	**20** 0320	3.5
0923	0.6	0904	0.7
TU 1601	3.9	W 1538	3.8
2154	0.5	2131	0.6
6 0427	3.7	**21** 0402	3.6
1006	0.6	0943	0.6
W 1644	3.9	TH 1620	3.8
2237	0.5	2210	0.5
7 0510	3.6	**22** 0443	3.6
1047	0.7	1023	0.6
TH 1726	3.8	F 1701	3.8
2318	0.6	2251	0.5
8 0550	3.5	**23** 0525	3.6
1127	0.8	1104	0.6
F 1805	3.6	SA 1743	3.8
2358	0.8	2333	0.6
9 0629	3.3	**24** 0608	3.5
1207	1.0	1148	0.7
SA 1843	3.4	SU 1828	3.7
10 0039	1.0	**25** 0018	0.7
0709	3.1	0655	3.4
SU 1250	1.2	M 1236	0.8
1923	3.2	1917	3.6
11 0124	1.1	**26** 0109	0.8
0754	3.0	0747	3.3
M 1340	1.3	TU 1331	1.0
2010	3.0	◑ 2013	3.4
12 0217	1.3	**27** 0207	1.0
0849	2.9	0849	3.2
TU 1440	1.5	W 1437	1.1
◑ 2109	2.9	2119	3.3
13 0320	1.4	**28** 0316	1.1
0955	2.9	1001	3.2
W 1551	1.5	TH 1554	1.1
2218	2.9	2234	3.2
14 0428	1.4	**29** 0431	1.1
1102	2.9	1114	3.3
TH 1700	1.4	F 1712	1.1
2325	2.9	2348	3.3
15 0530	1.3	**30** 0542	1.0
1200	3.0	1222	3.4
F 1800	1.3	SA 1821	0.9
		31 0054	3.4
		0644	0.9
		SU 1321	3.6
		1920	0.8

AUGUST

Time	m	Time	m
1 0151	3.5	**16** 0133	3.3
0738	0.8	0723	1.0
M 1414	3.8	TU 1352	3.6
2011	0.6	1950	0.8
2 0241	3.6	**17** 0217	3.5
0825	0.7	0804	0.8
TU 1501	3.9	W 1436	3.8
● 2056	0.5	2031	0.6
3 0326	3.7	**18** 0300	3.6
0907	0.6	0845	0.6
W 1544	3.9	TH 1519	3.9
2136	0.5	○ 2111	0.5
4 0407	3.7	**19** 0342	3.8
0946	0.6	0925	0.5
TH 1624	3.9	F 1601	4.0
2214	0.6	2151	0.4
5 0445	3.6	**20** 0424	3.8
1023	0.7	1005	0.4
F 1700	3.8	SA 1643	4.1
2249	0.6	2231	0.4
6 0520	3.5	**21** 0506	3.8
1058	0.8	1047	0.5
SA 1733	3.7	SU 1725	4.0
2323	0.8	2313	0.4
7 0553	3.4	**22** 0548	3.8
1133	0.9	1130	0.6
SU 1805	3.5	M 1810	3.9
2358	0.9	2357	0.6
8 0626	3.3	**23** 0634	3.7
1209	1.0	1217	0.7
M 1839	3.3	TU 1857	3.7
9 0036	1.1	**24** 0045	0.8
0702	3.1	0724	3.5
TU 1251	1.2	W 1310	0.9
1918	3.1	1952	3.5
10 0120	1.3	**25** 0142	1.0
0748	3.0	0825	3.3
W 1342	1.4	TH 1416	1.1
◑ 2008	3.0	◑ 2100	3.2
11 0217	1.4	**26** 0252	1.2
0847	2.9	0939	3.2
TH 1449	1.5	F 1538	1.3
2114	2.8	2221	3.1
12 0329	1.5	**27** 0415	1.3
1001	2.8	1100	3.3
F 1609	1.6	SA 1705	1.2
2234	2.8	2341	3.2
13 0444	1.5	**28** 0534	1.2
1114	2.9	1212	3.4
SA 1722	1.5	SU 1817	1.1
2346	2.9		
14 0547	1.4	**29** 0047	3.3
1216	3.1	0637	1.1
SU 1820	1.3	M 1311	3.6
		1913	0.9
15 0044	3.1	**30** 0140	3.5
0638	1.2	0728	0.9
M 1307	3.3	TU 1400	3.8
1908	1.0	1958	0.7
		31 0226	3.6
		0810	0.8
		W 1444	3.9
		2038	0.6

Portugal

HAT is 4·3 metres above Chart Datum.

STANDARD TIME (UT)
For Summer Time add ONE hour in **non-shaded areas**

LISBOA LAT 38°43'N LONG 9°07'W
TIMES AND HEIGHTS OF HIGH AND LOW WATERS

Dates in red are **SPRINGS**
Dates in blue are **NEAPS**

YEAR 2016

SEPTEMBER

Time	m		Time	m
1 0306	3.7	**16** 0237	3.8
0849	0.7	0823	0.6
TH 1523	3.9	F 1456	4.1
● 2113	0.6	○ 2048	0.4
2 0343	3.7	**17** 0319	4.0
0924	0.7	0904	0.4
F 1558	3.9	SA 1539	4.2
2146	0.6	2129	0.3
3 0417	3.7	**18** 0402	4.1
0957	0.7	0946	0.4
SA 1631	3.8	SU 1622	4.3
2218	0.7	2210	0.3
4 0448	3.7	**19** 0444	4.1
1029	0.8	1028	0.4
SU 1701	3.7	M 1706	4.2
2249	0.8	2252	0.4
5 0517	3.6	**20** 0527	4.0
1101	0.9	1112	0.5
M 1731	3.6	TU 1751	4.0
2320	0.9	2336	0.6
6 0548	3.5	**21** 0613	3.8
1134	1.0	1159	0.7
TU 1802	3.4	W 1839	3.8
2354	1.1		
7 0621	3.3	**22** 0024	0.9
1210	1.2	0704	3.6
W 1837	3.2	TH 1253	1.0
		1935	3.5
8 0032	1.3	**23** 0120	1.2
0701	3.1	0805	3.4
TH 1254	1.4	F 1401	1.2
1922	3.0	◐ 2046	3.2
9 0120	1.5	**24** 0234	1.4
0753	3.0	0922	3.3
F 1354	1.6	SA 1529	1.4
◐ 2022	2.9	2211	3.1
10 0228	1.6	**25** 0404	1.5
0903	2.9	1046	3.3
SA 1516	1.6	SU 1658	1.3
2144	2.8	2331	3.2
11 0355	1.6	**26** 0525	1.4
1026	3.0	1158	3.4
SU 1642	1.6	M 1806	1.2
2310	2.9		
12 0511	1.5	**27** 0034	3.4
1140	3.1	0625	1.2
M 1748	1.4	TU 1255	3.6
		1857	1.0
13 0015	3.1	**28** 0123	3.5
0609	1.3	0711	1.0
TU 1237	3.4	W 1341	3.7
1840	1.1	1938	0.9
14 0107	3.4	**29** 0205	3.7
0657	1.0	0751	0.9
W 1327	3.7	TH 1421	3.8
1925	0.8	2014	0.8
15 0153	3.6	**30** 0242	3.8
0741	0.8	0826	0.8
TH 1412	3.9	F 1458	3.9
2007	0.6	2047	0.7

OCTOBER

Time	m		Time	m
1 0316	3.8	**16** 0255	4.1
0859	0.8	0842	0.4
SA 1531	3.9	SU 1516	4.3
● 2118	0.7	○ 2106	0.3
2 0347	3.8	**17** 0339	4.2
0930	0.8	0926	0.3
SU 1602	3.8	M 1601	4.3
2148	0.8	2148	0.3
3 0417	3.8	**18** 0423	4.2
1001	0.8	1010	0.3
M 1631	3.7	TU 1647	4.2
2218	0.8	2231	0.5
4 0446	3.7	**19** 0508	4.1
1032	0.9	1056	0.5
TU 1700	3.6	W 1734	4.0
2248	1.0	2317	0.7
5 0516	3.6	**20** 0555	3.9
1104	1.0	1145	0.7
W 1732	3.5	TH 1824	3.8
2319	1.1		
6 0549	3.4	**21** 0005	1.0
1139	1.2	0646	3.7
TH 1807	3.3	F 1240	1.0
2354	1.3	1921	3.5
7 0627	3.3	**22** 0102	1.3
1219	1.4	0747	3.5
F 1848	3.1	SA 1348	1.2
		◐ 2031	3.2
8 0037	1.5	**23** 0214	1.5
0714	3.1	0901	3.3
SA 1312	1.5	SU 1512	1.4
1944	3.0	2153	3.1
9 0137	1.7	**24** 0342	1.6
0817	3.0	1023	3.3
SU 1428	1.6	M 1636	1.4
◐ 2101	2.9	2310	3.2
10 0304	1.7	**25** 0501	1.5
0940	3.0	1134	3.4
M 1558	1.6	TU 1742	1.3
2232	3.0		
11 0430	1.6	**26** 0010	3.3
1101	3.2	0601	1.3
TU 1711	1.4	W 1230	3.5
2343	3.2	1831	1.1
12 0535	1.4	**27** 0058	3.5
1205	3.4	0647	1.2
W 1808	1.1	TH 1315	3.6
		1911	1.0
13 0038	3.5	**28** 0138	3.6
0628	1.1	0726	1.0
TH 1258	3.7	F 1354	3.7
1856	0.8	1946	0.9
14 0126	3.7	**29** 0214	3.7
0715	0.8	0801	0.9
F 1346	4.0	SA 1430	3.8
1940	0.6	2019	0.8
15 0211	4.0	**30** 0247	3.8
0759	0.6	0834	0.9
SA 1431	4.2	SU 1503	3.8
2023	0.4	● 2050	0.8
		31 0319	3.8
		0906	0.8
		M 1534	3.7
		2121	0.8

NOVEMBER

Time	m		Time	m
1 0349	3.8	**16** 0405	4.2
0937	0.9	0955	0.3
TU 1605	3.7	W 1631	4.1
2151	0.9	2214	0.5
2 0420	3.7	**17** 0452	4.1
1009	0.9	1043	0.4
W 1636	3.6	TH 1720	3.9
2221	1.0	2301	0.7
3 0451	3.6	**18** 0540	4.0
1041	1.0	1132	0.6
TH 1709	3.5	F 1810	3.7
2253	1.1	2349	0.9
4 0525	3.5	**19** 0630	3.7
1116	1.1	1226	0.9
F 1745	3.3	SA 1904	3.4
2327	1.2		
5 0603	3.4	**20** 0043	1.2
1156	1.3	0726	3.5
SA 1826	3.2	SU 1327	1.1
		2006	3.2
6 0009	1.4	**21** 0147	1.4
0648	3.3	0831	3.3
SU 1245	1.4	M 1437	1.3
1918	3.0	◐ 2118	3.0
7 0104	1.5	**22** 0302	1.5
0745	3.1	0944	3.2
M 1351	1.5	TU 1553	1.4
◐ 2027	2.9	2230	3.0
8 0219	1.6	**23** 0418	1.5
0900	3.1	1055	3.2
TU 1512	1.5	W 1700	1.3
2151	3.0	2333	3.1
9 0344	1.5	**24** 0523	1.4
1020	3.2	1154	3.3
W 1628	1.3	TH 1754	1.2
2306	3.2		
10 0456	1.4	**25** 0024	3.3
1129	3.4	0614	1.3
TH 1731	1.1	F 1242	3.3
		1838	1.1
11 0006	3.4	**26** 0107	3.4
0555	1.1	0657	1.1
F 1227	3.7	SA 1324	3.4
1825	0.8	1917	1.0
12 0058	3.7	**27** 0145	3.5
0647	0.8	0735	1.0
SA 1319	3.9	SU 1402	3.5
1913	0.6	1952	0.9
13 0146	3.9	**28** 0220	3.6
0735	0.6	0810	0.9
SU 1408	4.1	M 1437	3.5
1959	0.4	2025	0.8
14 0233	4.1	**29** 0253	3.7
0822	0.4	0844	0.8
M 1456	4.2	TU 1510	3.5
○ 2044	0.4	● 2057	0.8
15 0319	4.2	**30** 0326	3.7
0908	0.3	0917	0.8
TU 1543	4.2	W 1544	3.5
2129	0.4	2129	0.8

DECEMBER

Time	m		Time	m
1 0359	3.7	**16** 0439	4.0
0950	0.8	1031	0.4
TH 1617	3.5	F 1707	3.8
2201	0.9	2246	0.6
2 0433	3.6	**17** 0525	3.9
1024	0.9	1119	0.5
F 1653	3.4	SA 1754	3.6
2234	0.9	2332	0.8
3 0509	3.6	**18** 0612	3.7
1100	0.9	1206	0.7
SA 1730	3.3	SU 1841	3.4
2310	1.0		
4 0547	3.5	**19** 0019	1.0
1140	1.0	0659	3.5
SU 1811	3.2	M 1256	0.9
2352	1.1	1931	3.2
5 0631	3.4	**20** 0110	1.2
1226	1.1	0751	3.3
M 1900	3.1	TU 1351	1.1
		2028	3.0
6 0042	1.3	**21** 0209	1.4
0723	3.3	0850	3.1
TU 1322	1.2	W 1453	1.3
1959	3.0	◐ 2133	2.9
7 0144	1.4	**22** 0318	1.4
0826	3.2	0958	3.0
W 1430	1.2	TH 1601	1.3
◐ 2111	3.0	2240	2.9
8 0259	1.4	**23** 0429	1.4
0939	3.2	1104	3.0
TH 1544	1.2	F 1705	1.3
2226	3.1	2340	3.0
9 0414	1.3	**24** 0533	1.3
1052	3.3	1202	3.0
F 1652	1.0	SA 1800	1.2
2332	3.3		
10 0522	1.1	**25** 0031	3.1
1156	3.5	0626	1.2
SA 1753	0.8	SU 1251	3.1
		1846	1.1
11 0030	3.5	**26** 0115	3.3
0621	0.9	0710	1.1
SU 1254	3.7	M 1334	3.2
1848	0.6	1926	1.0
12 0123	3.8	**27** 0154	3.4
0715	0.6	0749	0.9
M 1348	3.9	TU 1413	3.3
1939	0.5	2002	0.9
13 0214	3.9	**28** 0230	3.5
0806	0.4	0825	0.8
TU 1440	3.9	W 1450	3.4
2027	0.4	2037	0.8
14 0303	4.0	**29** 0306	3.6
0856	0.3	0900	0.8
W 1530	4.0	TH 1526	3.4
○ 2115	0.4	● 2111	0.8
15 0351	4.1	**30** 0342	3.6
0944	0.3	0935	0.7
TH 1619	3.9	F 1603	3.4
2201	0.5	2145	0.7
		31 0418	3.6
		1010	0.7
		SA 1640	3.4
		2221	0.7

HAT is 4·3 metres above Chart Datum.

》FREE monthly updates. Register at 《
www.reedsnauticalalmanac.co.uk

9.24.16 SESIMBRA

Baixo Alentejo **38°26'·30N 09°06'·49W** (Mole hd) ❀❀◊◊❀❀

CHARTS AC 3635, 3636; PC 24204, 26407; SC42B

TIDES Standard Port LISBOA (←); ML 2·0

Times				Height (metres)			
High Water		Low Water		MHWS	MHWN	MLWN	MLWS
0400	0900	0400	0900	3·8	3·0	1·5	0·6
1600	2100	1600	2100				
Differences SESIMBRA							
−0045	−0030	−0020	−0010	−0·4	−0·4	−0·1	0·0

SHELTER Good in marina. Or ⚓ in about 5m to N or E of ent and clear of fairway, local moorings and underwater cables. Strong N'ly gusts blow down from high ground late pm/evening.

NAVIGATION WPT 38°25'·86N 09°06'·15W, 333°/0·5M to break-water head, which is 475m W of the ldg line; and on the 003·5° ldg line.

LIGHTS AND MARKS 003·5° ldg lts/marks (NE of the hbr) are hard to see by day and night. Bkwtr hd lt, W twr, R bands, and Forte do Cavalo lt, R ○ twr are both conspic. See Lights, buoys and waypoints.

COMMUNICATIONS (Code 212) Brit Consul 808 203 537; Port VHF Ch 11; Marina VHF Ch 09.

FACILITIES YC Marina ☎233451 130 ⌣ but often full in summer, €2.60, multihulls +50%. 🐟 (fish market). www.naval-sesimbra.pt

Town: (1½M to the NE) ✕ ⌸ 🍴 ⌧ Ⓑ. Bus into town. ⇌ Seixal. ✈ Lisboa. Tourist amenities. 30 min bus to Lisboa or, to avoid traffic jams, local bus to Cacilhas and ferry across the Tejo.

9.24.17 SETÚBAL

Baixo Alentejo **38°31'·22N 08°53'·14W** (Marina) ❀❀◊◊❀❀

CHARTS AC 3635, 3636, 3259; PC 24204, 26308/09; SC42B

TIDES
Standard Port LISBOA (←); ML 2·0

Times				Height (metres)			
High Water		Low Water		MHWS	MHWN	MLWN	MLWS
0400	0900	0400	0900	3·8	3·0	1·5	0·6
1600	2100	1600	2100				
Differences SETÚBAL							
−0020	−0015	−0005	+0005	−0·3	−0·3	−0·2	−0·1

SHELTER Good in marina, but almost always full. A new marina is at Troia; see Facilities. ⚓ SW of the FV basin in up to 13m.

NAVIGATION WPT 38°27'·06N 08°58'·26W (abm No 1 buoy & No 2 bcn, Racon B) 040°/5·2M to front ldg lt. The 10m bar is dangerous only in strong SW'lies against a spring ebb. R. Sado is shoal either side until N of Forte de Outão; heed the ldg line/buoys/bcns.

LIGHTS AND MARKS As Lights, buoys and waypoints/chartlet.

COMMUNICATIONS (Code 265) VTS *Setúbal Port Control* VHF Ch 73 (mandatory for yachts >15m LOA; MSI broadcasts on request) Brit Consul 808 203 537; Ⓗ 561116; Port VHF Ch 11; Marina VHF Ch 09.

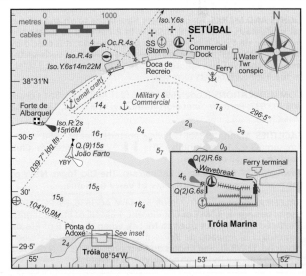

FACILITIES Setúbal's so called marina is crowded with small FVs and local boats and most unlikely to accommodate visiting yachts. ☎542076. 155 ⌣. 🐟⚒⛽⊕♨.

Doca de Recreio at Setúbal is even smaller; local boats only. Industrial quays extend 6M upriver to the ESE.

Town ✕ ⌸ 🍴 ⌧ Ⓑ good train/bus to Lisboa ⇌ ✈.

ADJACENT ANCHORAGE
ARRÁBIDA 38°28'·52N 08°58'·77W, 1·5M NNW of ent to Rio Sado (AC 3259). Popular but shallow day ⚓ in 2·5m at SW end of the bay, sheltered from N by high ground, but also prone to katabatic gusts. Appr from SSW, 1-2ca offshore to avoid Baixo de Alpertuche (0·6m). Caution: rocky reef at NE end of bay.

HARBOUR ACROSS THE RIO SADA
Tróia Marina at 38°29'·66N 08°54'·14W has modern facilities and is part of an up-market development. From Setúbal it bears 207°/1·75M.

From a WPT at 38°29'·88N 08°55'·28W (on the 039·7' leading line) make good 104°/9 cables to the marina ent; lights as chartlet. 187 ⌣ (fingers) inc ❶ in 2·5-4·0m, max LOA 18m.

☎499333, mob 933 389847. May-Sep 0830-2100. VHF Ch 09. marina@troiaresort.pt; reported quite expensive; 🐟🛒✕⌸🍴 4 mins walk, ⚓▷◫@.

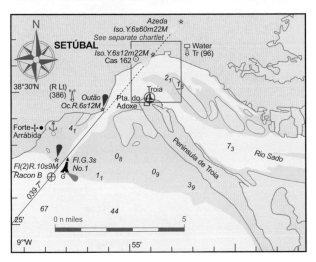

9.24.18 SINES

Baixo Alentejo 37°57'·04N 08°52'·11W (marina ent) ⊕⊛⚓♨♦♦✿✿

CHARTS AC 3636, 3224; PC 24204, 24205, 26408; SC 430A

TIDES

Standard Port LISBOA (←—); ML 2·0

Times				Height (metres)			
High Water		Low Water		MHWS	MHWN	MLWN	MLWS
0400	0900	0400	0900	3·8	3·0	1·5	0·6
1600	2100	1600	2100				
Differences SINES							
−0050	−0030	−0020	−0010	−0·4	−0·4	−0·1	0·0
MILFONTES (37°43'·0N 08°47'·0W)							
−0040	−0030	ND	ND	−0·1	−0·1	0·0	+0·1
ARRIFANA (37°17'·5N 08°52'·0W)							
−0030	−0020	ND	ND	−0·1	0·0	−0·1	+0·1

SHELTER Good in marina protected by E mole. Or ⚓ NW of the marina in 4m; holding is reportedly poor E of 08°52'·00W.

NAVIGATION WPT 37°55'·70N 08°52'·97W, 027°/1·5M to ent to the marina bay. The outer ent, between W and E moles, is 1M wide. From the NW, caution: the derelict outer 400m section of the W mole dries 0·6m. A PHM buoy, Fl R 3s, 680m S of the mole hd ☆ (Fl 3s) *must* be rounded; no short cuts! In strong S'lies swell may cause turbulence off the W mole. Sines is strategically located 51M SSE of Cascais and 57M N of Cape St Vincent. Beware seaplane operating area to SE of hbr.

LIGHTS AND MARKS An oil refinery flare in the town and several tall chy's/masts 3-4M to the E (all with R lts) help to locate the port. Cabo de Sines lt ho is 9ca NW of the marina; it is obsc'd by oil tanks 001°-003° and 004°-007°, leaving a visible 1° wide sector in which the WPT lies. Ignore the 357·3° ldg lts which lead solely to the Refined Products terminal.

COMMUNICATIONS (Code 269) MRCC (21) 4401919; Brit Consul 808 203 537; *Porto de Sines* (Port Authority) VHF Ch 11, 13; Marina VHF Ch 09.

FACILITIES Marina ☎860612. www.portodesines.pt 250 ⚊ in 4m, some♥ €2.11, multihulls +50%. ♠♠⚒🗵; no internet. **Club Náutico** close SE of marina.

Town ✕🖪🛒♠✉Ⓑ⇌✈ Lisboa/Faro. The Old Town, NNW of the marina, is pleasant; birthplace of Vasco da Gama.

BALEEIRA, Algarve 37°00'·72N 08°55'·44W. AC 3636, 89. PC 27502. Useful passage ⚓, 3·5M E of C St Vincent, 1·6M NE of Pta de Sagres. Sheltered from W, but open to NE'lies, when the ⚓s E of Pta de Sagres and in Enseada de Belixe may be better. Enter between bkwtr hd, Fl WR 4s, and Ilhotes do Martinhal, 4ca NE. ⚓ N of the 2 jetties in 6m. See lts in Lights, buoys and waypoints.

9.24.19 LAGOS

Algarve 37°05'·93N 08°39'·94W (Hbr ent) ⊕⊛⊛♨♦♦♦✿✿✿

CHARTS AC 3636, 89; PC 24206, 2705, 25R11

TIDES

Standard Port LISBOA (←—); ML 2·0

Times				Height (metres)			
High Water		Low Water		MHWS	MHWN	MLWN	MLWS
0400	0900	0400	0900	3·8	3·0	1·5	0·6
1600	2100	1600	2100				
Differences LAGOS							
−0100	−0040	−0030	−0025	−0·4	−0·4	−0·1	−0·1
ENSEADA DE BELIXE (Cape St Vincent)							
−0050	−0030	−0020	−0015	+0·3	+0·2	+0·2	+0·2

SHELTER Very good in marina; access H24 all tides. Swell possible at ent in strong SE winds. The ⚓ close NE of the E mole in 3-5m is exposed only to E and S winds and possible SW swell.

NAVIGATION WPT 37°05'·75N 08°39'·06W, 284°/7ca to hbr ent. Fishing nets may extend up to 1M offshore. Beware fish farms SW of Ponta de Piedade, marked by 4 Y SPM bys, Fl(4)Y12s. From Ponta da Piedade keep 5ca offshore to clear rocks; otherwise there are no offshore dangers. Final appr is on 282°, W mole head lt ≠ Santo António church. Shoaling in ent; care needed at LW if draft >1·5m. Note, small craft moor W of chan. 3kn hbr speed limit.

Berth at arrival pontoon stbd side close SE of lifting foot-bridge which opens on request 0800-2200, 1/6-15/9; 0900-1800 16/9-31/5, except for some ¼hr spells for train passengers. Unmasted craft may transit when bridge is down.

LIGHTS AND MARKS Lts as chartlet and Lts, buoys and waypoints. The river chan is unlit, except by shore lts. Pta da Piedade is 1·2M S, with conspic. 20m lattice radio/radar mast close behind Lt Ho. Pta do Altar (Portimão) is 7M E.

COMMUNICATIONS (Code 282) MRCC (21) 4401919; 🖃 461406; Met (218) 447000; 🕮 VHF Ch 12 at 1000 and 1400 LT in season; Police 762930; Brit Consul 808 203 537; 🖽 770100; Fire 770790; HM 769161; *Marina de Lagos* VHF Ch 09 for bridge, pump-out and fuel.

FACILITIES Marina ☎770210 marina@marlagos.pt www. marinadelagos.pt. 462+♥ €4·61,(multihulls +100% high season, +50% med and low). Short stay €8·61/hr. ♠♠✕🖪🛒⚙🗐⊖Ⓒ@♂ ⛽🚿⚒🗵⚒⚓🖵(35t) **Club de Vela** ☎762256. **Town:** all usual amenities, ⇌✈ Faro (75km).

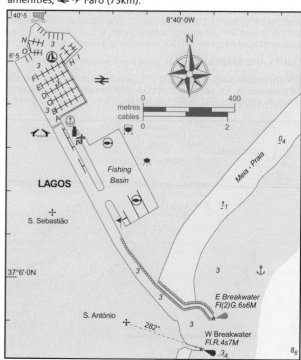

ANCHORAGE 2·6M ENE OF LAGOS

ALVOR 37°07'·06N 08°37'·16W. PC 27502. Beware of an Aquatic Revival Park S of Alvor, centred on 37°05'·75N 08°35'·09W, marked with 4 Y special marks. A peaceful but shallow lagoon entered between moles Fl R 4s & G 4s, but not in strong onshore winds nor at night. ⌗ just inside the ent to port in about 2·2m or off Alvor (1·2M E) in about 1·6m. The channel is buoyed, but the sandbanks frequently move, best to enter on the young flood whilst the drying banks can be seen. While not geared for yachtsmen, it is a lively tourist spot, not devoid of charm.

SOUTHERN PORTUGAL: THE ALGARVE

An ITZ extends 15M W/SW of Cape St Vincent and the TSS 20M further west; see Area map & AC 89, 93. Passage ⌗s, sheltered from N'lies, are at Enseada de Belixe (E of Cabo São Vicente) and at Enseadas de Sagres and da Baleeira (NE of Pta de Sagres).

The direct route to Gibraltar is 110°/175M (Cabo São Vicente to Tarifa). Alternatively make a series of coastal legs using the good marinas at Lagos, Portimão, Albufeira and Vilamoura.

E of Cabo São Vicente the summer climate becomes more Mediterranean, ie hotter, clearer, winds more from NW to SW. Swell may decrease, tidal streams are slight and the current runs SE. Hbrs along the coasts of the Algarve and SW Andalucía are sheltered from all but S'lies (infrequent). The 'Nortada', which comes in most afternoons from the NW, can inc to F5 or 6.

Nets around tuna farms can be a problem NE of Ponta de Sagres and 4M SSW of the entrance to Faro/Olhão at Cabo de Santa Maria. These are charted (AC 93) but dimly lit and hard to see against shore lights. Ideally, stay outside the 50m contour and keep a careful lookout. More nets lie S and E of Ilha de Armona. Numerous lobster pot markers are a hazard all along the Algarve coast.

At Cabo de Santa Maria a gap in the low-lying dunes gives access to the lagoons and chans leading to Faro and Olhão, where a marina has been built (but may not be open to visitors); there are also some peaceful anchorages, notably Culatra Is (ferries to Olhão).

The marina at Vila Real de Santo António is on the Portuguese bank of the Rio Guadiana, which forms the Spanish border.

9.24.20 PORTIMÃO

Algarve **37°06'·51N 08°31'·71W** (Hbr ent) ✵✵✵❀⌗⌗⌗❁❁

CHARTS AC 91, 3636, 89, 83; PC 24206, 26310

TIDES
Standard Port LISBOA (⟵); ML 2·0

Times				Height (metres)			
High Water		Low Water		MHWS	MHWN	MLWN	MLWS
0400	0900	0400	0900	3·8	3·0	1·5	0·6
1600	2100	1600	2100				
Differences PORTIMÃO							
–0100	–0040	–0030	–0025	–0·5	–0·4	–0·1	+0·1
PONTA DO ALTAR (Lt ho E of ent)							
–0100	–0040	–0030	–0025	–0·3	–0·3	–0·1	0·0
ENSEADA DE ALBUFEIRA							
–0035	+0015	–0005	0000	–0·2	–0·2	0·0	+0·1

SHELTER Good in marina on W bank. Good ⌗ off Ferragudo. ⌗ inside E molhe in about 4m is prone to swell and FV wash.

NAVIGATION WPT 37°06'·04N 08°31'·94W, 021°/5ca to hbr ent. Beware an artificial reef W of Portimão, approx 1M SSW of Ponta do Faco (marked by four Y lit bys). Seaplane operating area approx 1.5M SW of hbr ent. Beyond W molhe stay on ldg line until past No 2 by, due to shifting sandbank just inside ent. River is straightforward, but flood & ebb streams run hard; berth bows to the stream. Caution: 2 low (1.4m cl'nce) fixed road/rail bridges N of Portimão centre.

LIGHTS AND MARKS Ponta do Altar lt ho is on low cliffs to E of ent. The R-roofed church at Ferragudo is conspic by day, almost on the 019° ldg line. Ldg lts as chartlet and Lights, buoys and waypoints; both on R/W banded columns, moved when chan changes.

COMMUNICATIONS (Code 282) MRCC (21) 4401919; Brit Consul (located here) 808 203537; Port VHF Ch 11; Marina VHF Ch 09.

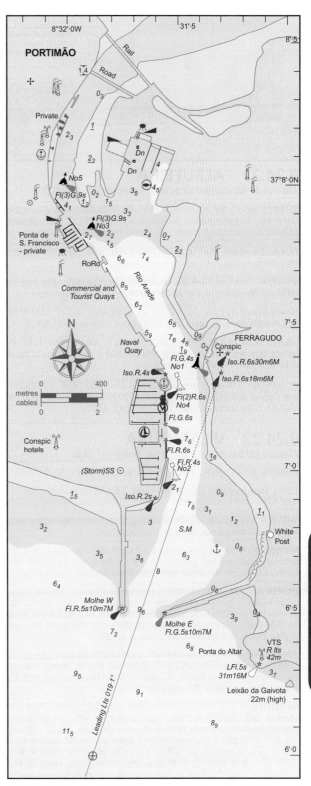

FACILITIES Marina ☎400680. marinaportimao@mail.telepac.pt www.marinadeportimao.com.pt. 620 ⌂; max LOA 30m, max draft 4m, €4.34 multihulls +150%. 🔧🛠️🚻♨️@⛽🚿🔧(2t). Town centre 20 mins by dinghy, long walk, poor bus. Pontoons 2M N of hbr ent are nearer town centre but access is limited.
Fishing Hbr 🚽✕🔧⛽Ⓔ🔧(350t).

Town ✕🛒🍴🛏✉Ⓑ🚆. ✈ Faro (65km).

In the absence of an official large scale chart, this chartlet is drawn from publicity material. It is therefore only a guide and lacks Lat, Long and scale

ALBUFEIRA MARINA

9.24.21 ALBUFEIRA

Algarve **37°04'·90N 08°15'·52W** (N bkwtr lt) ❀❀✿◊◊◊✿✿

CHARTS AC 91, 89; PC 24206, 27503, 90 (1:10,000) only shows the outer hbr, not the marina. No large scale AC (2012). Positions are approx; the Lat/Long above is by hand-held GPS (WGS 84).

TIDES Standard Port LISBOA (←); **ML 2·0.** See differences Portimão.

SHELTER Good, in 2·5–4·0m. In strong E'lies the outer ent may be difficult. No space to ⚓ in outer hbr due to local moorings.

NAVIGATION WPT 37°03'·88N 08°15'·02W, approx 338°/1M to hbr ent. No offshore dangers, but rocks close inshore to W & E.

LIGHTS AND MARKS Ponta de Baleeira lt, Oc 6s 30m 11M, W lattice twr with R top, is near the root of S bkwtr. Bkwtr lts as chartlet, both W cols, R/G bands.

COMMUNICATIONS (Code 289) MRCC (21) 4401919; ⊖ 589363; Brit Consul 808 203 537 (Portimão); Ⓗ 802555 (Faro); Taxi 583230; *Marina de Albufeira* Ch 09 (1/6-15/9: 0900-2100).

FACILITIES Marina ☎289 514282, mob 964249596, 475 +Ⓥ €2·70. info@marinadealbufeira.com www.marinadealbufeira.com ⚓🛥🛥 ✕🗔🖻🛢🛢🛢⚓⚠@⚒🔌Ⓔ✕🍴 mob 962079546, ⛽(70t) ⛏(6t) ━ YC. **Town** Most facilities; Ⓑ ⇌. ✈ Faro 30km.

9.24.22 VILAMOURA

Algarve **37°04'·10N 08°07'·35W** ❀❀❀◊◊◊✿✿

CHARTS AC 91, 89; PC 24206, 27503; SHOM 7300

TIDES Standard Port LISBOA (←); **ML 2·0.** Use differences for Albufeira (6M to the west).

SHELTER Very good, except in strong S'lies when the 100m wide ent can be dangerous. No ⚓ in the outer hbr.

NAVIGATION WPT 37°03'·60N 08°07'·40W, 007°/5ca to hbr ent. No offshore dangers, but groynes to the E. Hbr speed limit 3kn. The fairway through the outer hbr is dredged 4m.

LIGHTS AND MARKS Marina is surrounded by apartment bldgs; large hotel is conspic on E side of ent. Main lt, Fl 5s 17m 19M, yellow twr on HM's bldg; this lt between the W and E mole head lts (R/W and G/W banded twrs) leads 007° into the outer hbr. See Lights, buoys & waypoints for other lt details.

COMMUNICATIONS (Code 289) MRCC (21) 4401919; Maritime Police 310575; Brit Consul 808 203 537; Ⓗ 891100 (Faro); Medic 314243; *Vilamoura Radio* VHF Ch 09, 16 (0830-2130/1830 off season) outside these hrs expect only 1M reception range.

FACILITIES
Marina ☎310560 🖷310580 marinavilamoura@lusort.pt www.marinadevilamoura.com 1000 incⓋ max LOA 43m. €3.62 Jul/Aug, inc ⟨▷. Dredged 3·3m, reducing to 2·0m in the NE corner near pontoon A. Reception pontoon, port side of access chan. 🛥🛥✕ 🗔🛒🖻🛢🛢⚓@⚒🔌Ⓔ🍴✕⚠━⛽(60t)⛏(2t,6t). Local Ⓑ M-F 0830-1445, plus Ⓒs. Free ⚓ by sludge boat comes to your yacht. Mail pick-up, if addressed to: Marina de Vilamoura, 8125-409 Quarteira, Portugal.

Clube Náutico (by reception area) welcomes visitors: lounge, ✕🗔@.

Faro (20km) has most facilities, inc ⇌ ✈ ☎800800.

VILAMOURA

9.24.23 FARO and OLHÃO

Algarve 36°57'·81N 07°52'·19W (Hbr entrance) ❀❀♨♠♣

CHARTS AC 91, 89, 83; PC 24206, 26311; SC 44B

TIDES Standard Port LISBOA (←); ML 2·0

Times

High Water		Low Water		MHWS	MHWN	MLWN	MLWS
0400	0900	0400	0900	3·8	3·0	1·5	0·6
1600	2100	1600	2100				

Differences PORTO DE FARO-OLHÃO

–0050	–0030	–0015	+0005	–0.4	–0.4	–0.1	–0.1

SHELTER Generally good in the lee of coastal dunes. Popular ⚓s in 3-5m are inside Ilha da Culatra, abm lt Fl 5s or off jetty Oc G 4s.

NAVIGATION WPT 36°57'·04N 07°52'·12W, 352°/7ca to narrow ent (dangerous in strong S'lies); extensive shoals to the W. Seaplane operating area 1.4M E of ent. Enter near HW between moles, Fl R 4s & G 4s, and training walls. Abeam Cabo de Santa Maria lt ho, chan forks into salt marshes: WNW to Faro or NE to Olhão.

The **Faro** chan carries 5·4m for the buoyed/lit 4M to the Commercial quay (not for yachts); but beyond depths vary.

The **Olhão** buoyed/lit chan (5M) is narrower and carries 2·1m least depth. From N of No 5 buoy enter the marina by an E-W chan, depth approx 2m, marked by 3 PHM buoys.

LIGHTS AND MARKS Cabo de Santa Maria lt ho near root of long E trng wall and adjacent VTS twr (42m) are conspic. Lt ho is also rear 021° ldg lt to abeam front ldg lt; see chartlet/Lts, buoys and waypoints.

COMMUNICATIONS (Code 289) IPTM (Institute of Ports & Shipping) Olhão 715912; Brit Consul 808 203 537; both ports VHF Ch 11, 09: Faro H24; Olhão, M-F office hrs.

FACILITIES

Faro HM ☎822025. ⚓ in 3m, 3ca SW of Doca de Recreio, small craft only; low bridge across ent, clearance 0.5m W sp; 0.5m depth LW sp. Some noise from airport, but not after 2100LT.

Olhão HM ☎703160. Marina, W end; max LOA 17m, draft 3m, rarely ❶ space, call, if available, keep N of PHM bys, berth on wavebreak. No ⛟ or ⚑ 🛢 ⛽. ⚓ in fairways; ⌂ poss outer wall Puerto deportivo.

Upstream, Marina Formosa ♫☎702162, 🔧 (20t, 2m draft); 🔩⚒⚙ 🔟; secure winter lay up ashore.

Towns: Both are sizeable, all amenities inc @ ⇌ ✈.

MINOR PORT 11M W OF VILA REAL DE SANTO ANTÓNIO

TAVIRA. 37°06'·69N 07°37'·01W. AC 89, 93; PC 24206, 27503; SC 44A/B. Tide, interpolate between Faro/Olhão and Vila Real de San António. The narrow (about 70m) ent between I de Tavira and sand dunes to the E is lit by Fl G and R lts 2·5s and ldg lts 325·9° over the bar, front Fl R 3s, front & rear sync'd. ‡ in about 3m on sand SW of the entrance in the Rio Formosa and SE of the small community of Quatro Aguas (a planned marina is on hold). Show an ‡ lt, due to FVs.

Town is 1M (20 mins walk or dinghy) NW of the ‡ and, apart from a small craft BY and diesel supply, has little to offer a visiting yacht, being essentially a working FV port. Access is not possible for masted boats due to a low road bridge. Facilities: YC ☎281 326 858, ✕ ⌂ ♨ ⊠ Ⓑ Ⓗ ➘ ⇌ ✈ (Faro).

9.24.24
VILA REAL DE SANTO ANTÓNIO

Algarve **37°10'·80N 07°24'·49W** (Marina) ✿✿♨♨✿✿✿

CHARTS AC 91, 89, 93; PC 24206, 26312; SC 440, 440A

TIDES

Standard Port LISBOA (←—); ML 2·0

Times				Height (metres)			
High Water		Low Water		MHWS	MHWN	MLWN	MLWS
0400	0900	0400	0900	3·8	3·0	1·5	0·6
1600	2100	1600	2100				
Differences VILA REAL DE SANTO ANTÓNIO							
−0050	−0015	−0010	0000	−0·4	−0·4	−0·1	+0·1

SHELTER Good in marina, but river ent difficult in strong S'lies.

NAVIGATION WPT 37°09'·00N 07°23'·35W (No 1 SHM buoy) 340°/9ca to hd of W trng wall. Caution: Extensive drying areas E of the trng walls stretch 2M east. The bar (1·0m) is marked by Nos 3/4 P/SHM lt buoys. Hug the W bank between trng walls (E partly covers) and up to marina. Access best at slack water as streams & current are very strong: 2½kn (up to 6kn in winter).

LIGHTS AND MARKS High bldgs, FR lts, are 3M WNW of ent. Vila Real lt ho, W twr/B bands, is conspic. 2M inland a white suspension bridge (18m cl'nce MSL) is visible from seaward.

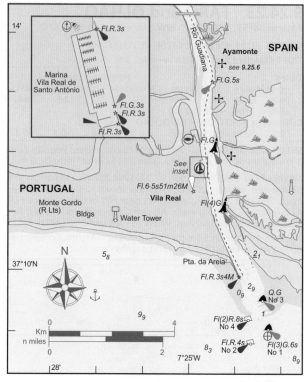

COMMUNICATIONS (Code 281) MRCC (21) 4401919; Brit Consul 808 203 537; *Porto de Vila Real* VHF Ch 11, 16; Marina VHF Ch 09.

FACILITIES Marina ☎513769 anguadiana@mail.telepac.pt (2m) Ⓥ ⌿ €2.43, 25% surcharge for reception pontoon. ⚓ on pontoon just N of marina (cards not accepted); ♨ 1NM downstream ☎542069, marinaguadiana@sapo.pt ⌖(70t).YC ☎511306, ✕ ➘. **Town** ✕ ⌂ ♨ ⊠ Ⓑ.

Upriver better water is off the Spanish bank and Ayamonte. The river Guadiana is navigable for about 20M to ‡ off Alcoutim (pontoons €7.50/day for 1st week) and Pomerão (27M).

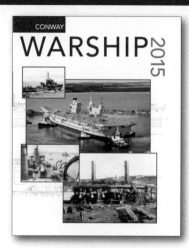

South West Spain, Gibraltar & Morocco

Ayamonte to Europa Point and Ceuta

9.25.1	Contents	1003
9.25.2	Area map	1004
9.25.3	Lights, buoys and waypoints	1004
9.25.4	Passage information	1006
	Portuguese Border to Rio Guadalquivir – 1006 • Chipiona to Cape Trafalgar – 1010 • Strait of Gibraltar – 1015 • Morocco – 1020	
9.25.5	Special notes for SW Spain	1006
9.25.6	Ayamonte	1007
9.25.7	Isla Canela	1007
9.25.8	Isla Cristina	1008
9.25.9	El Rompido (Rio de las Piedras)	1008
9.25.10	Punta Umbria	1009
9.25.11	Mazagón (Huelva River)	1009
9.25.12	Chipiona	1010
9.25.13	Rio Guadalquivir	1010
9.25.14	Sevilla	1011
9.25.15	Bay of Cádiz	1011
	Rota • Puerto Sherry • Puerto Santa Maria • Puerto America	
9.25.16	Sancti-Petri	1013
9.25.17	Barbate	1013
9.25.18	Tarifa	1015
9.25.19	Algeciras	1015

UNITED KINGDOM

9.25.20	Gibraltar, Special notes, tide tables/curves	1016
	La Linea	

MOROCCO

9.25.21	Tangier	1020
9.25.22	Ceuta (to Spain)	1020

SW Spain & Gibraltar

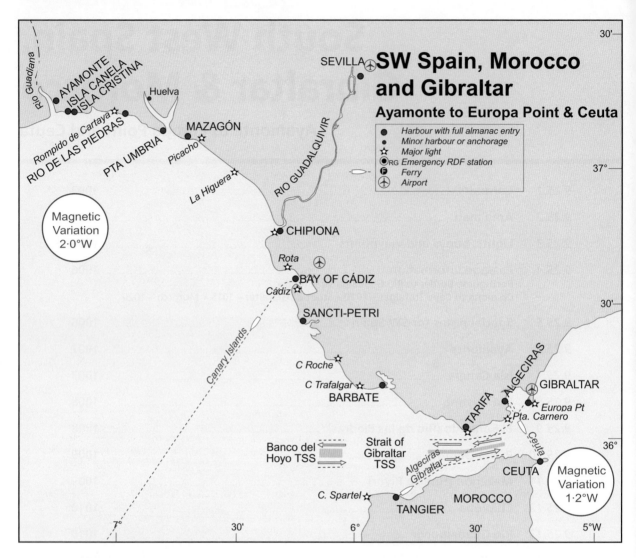

9.25.3 LIGHTS, BUOYS AND WAYPOINTS

Bold print = light with a nominal range of 15M or more. CAPITALS = place or feature. *CAPITAL ITALICS* = light-vessel, light float or Lanby. *Italics* = Fog signal. ***Bold italics*** = Racon. Many marks/buoys are fitted with AIS (<u>MMSI No</u>); see relevant charts.

RIO GUADIANA TO SEVILLA

AYAMONTE (E side of Rio Guadiana)
No. 1 ⯭ Fl (3) G 6s; 37°09'·00N 07°23'·35W.
No. 2 ⯑ Fl R 4s; 37°08'·63N 07°23'·23W.
E trng wall ⚡ Fl(3) G 9s 3M; 37°09'·97N 07°23'·70W.
W trng wall ⚡ Fl R 3s 4M; 37°09'·77N 07°23'·93W (Portugal).
Vila Real de Santo Antònio ☆ Fl 6·5s 51m **26M**; W ○ twr, B bands; 37°11'·20N 07°24'·91W (Portugal).
Marina ent, ⚡ QR and ⚡ QG; 37°12'·70N 07°24'·54W.
Baluarte (ferry landing) Fl (2) G 5s 1M; G ○ twr; 37°12'·78N 07°24'·63W.

ISLA CANELA and ISLA CRISTINA

Appr ⚓ Fl 10s 5M; 37°10'·51N 07°19'·49W.
W mole hd ⚡ VQ (2) R 5s 9m 4M; 37°10'·83N 07°19'·60W.
Ldg lts 312·8°; framework twrs. Front ⚡ Q 8m 5M; 37°11'·50N 07°20'·30W approx. Rear ⚡ Fl 4s 13m 5M, 100m from front.
Isla Canela marina ent, QR and QG; 37°11'·40N 07°20'·26W.
Isla Cristina marina ent, QR and QG; 37°11'·89N 07°19'·65W.

RIO DE LAS PIEDRAS

El Rompido ☆ Fl (2) 10s 41m **24M**; W & R ○ twr; 37°13'·12N 07°07'·72W, 4M WNW of the bar buoy.

RIA DE HUELVA

Punta Umbria, ⚓ L Fl 10s; 37°08'·78N 06°56'·74W.
No. 1 ⚓ Fl (2) G 10s; 37°09'·13N 06°56'·58W. No. 2 ⚓ Fl (2) R 10s.
Bkwtr hd ⚡ VQ (6) + L Fl 10s 8m 5M; 37°09'·85N 06°56'·93W.
No. 3 ⚓ Fl (3) G 15s; 37°09'·70N 06°56'·40W. No. 4 ⚓ Fl (3) R 15s.
No. 5 ⚓ Fl (4) G 15s; 37°10'·10N 06°56'·60W. No. 6 ⚓ Fl (4) R 15s.
No. 7 ⚓ Fl G 5s; 37°10'·50N 06°57'·00W. No. 8 ⚓ Fl R 5s.
Marina anti-pollution boom, S hd, Fl (2) R 10s 1M; 37°10'·81N 06°57'·34W. N head, Fl (3) R 15s 1M; 37°10'·88N 06°57'·44W.
Tanker mooring ⚓ Fl (4) Y 20s 8M; *Siren Mo (E) 30s*; 37°04'·79N 06°55'·59W. Oil pipeline N'ward marked by 3 Y ⚓s, Fl (4) Y 20s.
Huelva, main chan, Dir ⚡ 339·2°, WRG 59m 8M; 337·2°- G-337·8°-FG-338·3°-OcWG-338·9°-FW-339·1°-OcWR-339·7°- FR-340·2°-Fl R-340·7°; W twr; 37°08'·57N 06°50'·67W, 8ca NW of marina ent.
⚓ Q (9) 15s; 37°05'·55N 06°49'·11W.
No. 1 Fl G 5s; 37°06'·34N 06°49'·49W.
No. 2 ⚓ Fl R 5s; 37°06'·22N 06°49'·65W.
Juan Carlos I bkwtr hd ⚡ Fl (3+1) WR 20s 29m, W12M, R9M; 165°-W-100°-R-125°; ***Racon K, 12M***; 37°06'·48N 06°49'·94W.
No. 3 ⚓ Fl (2) G 10s; 37°06'·90N 06°49'·75W.
No. 4 ⚓ Fl (2) R 10s; 37°06'·85N 06°49'·94W.

No. 5 ⟟ Fl (3) G 15s; 37°07'·54N 06°50'·06W.
No. 6 ⟟ Fl (3) R 15s; 37°07'·51N 06°50'·27W.
No. 7 ⟟ Fl (4) G 20s; 37°07'·79N 06°50'·27W.
No. 8 ⟟ Fl (4) R 20s; 37°07'·78N 06°50'·52W.

MAZAGÓN

Picacho lt ho ☆ Fl (2+4) 30s 50m **25M**; 37°08'·11N 06°49'·56W.
Marina, S bkwtr ⚓ QG 7m 2M; 37°07'·89N 06°50'·18W.
N bkwtr ⚓ QR 5m 2M; 37°07'·95N 06°50'·08W.
Inner ent, Fl G 5s 3m 1M and Fl R 5s 3m 1M.

⟟ Fl (2) 10s; 37°04'·30N 06°43'·71W, off Torre del Loro (ruins).
La Higuera ☆ Fl (3) 20s 45m **20M**; 37°00'·51N 06°34'·18W.

CHIPIONA

Bajo Salmedina ⟟ Q (9) 15s 9m 5M; 36°44'·46N 06°28'·35W.
Pta de Chipiona ☆ Fl 10s 67m **25M**; 36°44'·26N 06°26'·52W.
Marina, No. 2 ⟟ Fl (2) R 7s; 36°45'·14N 06°25'·58W.
No. 4 ⟟ Fl (3) R 11s; 36°45'·05N 06°25'·51W.
N bkwtr head ⚓ Fl (2) G 10s 6m 5M; 36°44'·96N 06°25'·69W.
N inner jetty ⚓ Fl (3) G 9s 2m 1M; 36°44'·89N 06°25'·77W.
SE side of ent ⚓ Fl (4) R 11s 3m 3M; 36°44'·89N 06°25'·70W.

RÍO GUADALQUIVIR

No. 1 ⟟ L Fl 10s; *Racon M, 10M*; 36°45'·85N 06°26'·85W.
No. 2 ⟟ Q (9) 15s; 36°47'·35N 06°26'·76W (1·6M N of No.1).
Ldg lts 069°, Y☐ on metal towers. Front ⚓ Q 28m 5M; 36°47'·85N 06°20'·25W. Rear, ⚓ Iso 4s 61m 10M; 5ca from front.
Selected buoys in sequence as far as Bonanza:
No. 3 ⟟ Fl G 5s; 36°46'·20N 06°25'·45W.
No. 4 ⟟ Fl R 5s; 36°46'·27N 06°25'·60W.
No. 7 ⟟ Fl (3) G 10s; 36°46'·58N 06°24'·14W.
No. 8 ⟟ Fl (3) R 10s; 36°46'·68N 06°24'·19W.
No. 11 ⟟ Fl G 5s; 36°46'·95N 06°22'·94W.
No. 12 ⟟ Fl R 5s; 36°47'·05N 06°22'·93W.
No.13 ⟟ Fl (2) G 6s; 36°47'·10N 06°22'·47W.
No. 14 ⟟ Fl (2) R 6s; 36°47'·19N 06°22'·45W.
No.17 ⟟ Fl (4) G 12s; 36°47'·51N 06°21'·18W.
No. 20 ⟟ Fl R 5s; 36°47'·81N 06°20'·70W.
Bonanza lt ho Fl 5s 22m 6M; R twr; 36°48'·17N 06°20'·15W.
Upstream the river is marked by lt bcns, lt buoys and a few ldg lts.

SEVILLA

Esclusa Beacon C ⚓ Fl (2+1) R 14·5s 9m 5M; 37°18'·92N 06°00'·83W, G ○ twr, R band. Fork stbd to city centre via lock, or port for Gelves marina, ent Fl R 5s (37°20'·43N 06°01'·40W) and Fl G 3s.

CADIZ TO CABO TRAFALGAR

ROTA (Clockwise around the Bay of Cadiz)

Bajo El Quemado ⟟ Fl (2) R 9s; 36°35'·88N 06°23'·95W.
Rota Aero ☆ Alt Fl WG 9s 79m **17M**; R/W chequered water twr on 8 cols, conspic; 36°38'·13N 06°20'·834W.
Rota ⚓ Oc 4s 33m 13M; W ○ lt ho, R band; 36°36'·95N 06°21'·43W.
Marina, S pier ⚓ Fl (3) R 10s 8m 3M; 36°36'·86N 06°21'·04W.
N pier ⚓ Q (6) G 12s; G post; 36°36'·91N 06°21'·10W.
Las Cabezuelas ⟟ Fl (4) R 10s; 36°35'·21N 06°19'·95W.

PUERTO SHERRY and PUERTO DE SANTA MARIA

La Galera ⟟ Q; 36°34'·53N 06°17'·64W.
Puerto Sherry marina, S bkwtr ⚓ Oc R 4s 4M; 36°34'·64N 06°15'·25W. N bkwtr ⚓ Oc G 5s 3M. Basin ent, QR and QG.
Santa María ldg lts 040°: Front ⚓ QG 16m 4M; 36°35'·77N 06°13'·35W. Rear ⚓ Iso G 4s 20m 4M; 253m from front.
W trng wall head ⚓ Fl R 5s 10m 3M; 36°34'·35N 06°14'·95W.
W trng wall root ⟟ Fl (2) R 5s 1M; 36°35'·07N 06°14'·12W.
E trng wall ⚓ Fl (2) G 7s 9m 3M; 36°34'·68N 06°14'·39W.
E trng wall ⚓ Q (9) 15s 3M; 36°34'·46N 06°14'·35W.

CÁDIZ CITY and PUERTO AMERICA

Castillo de San Sebastián ☆ Fl (2) 10s 38m **25M**; metal ○ twr; 36°31'·70N 06°18'·97W.
Bajos de San Sebastián ⟟ Q (9) 15s; 36°31'·30N 06°20'·36W.

⚓ L Fl 10s; for Canal Principal; 36°33'·84N 06°19'·90W.
No. 1 ⟟ Fl G 3s; 36°33'·12N 06°19'·07W.
No. 3 ⟟ Fl (2) G 4s; 36°33'·15N 06°18'·31W.
No. 4 ⟟ Fl (2) R 4s; 36°33'·57N 06°17'·88W.
No. 5 ⟟ Fl (3) G 13s; 36°33'·08N 06°17'·31W.
No. 6 ⟟ Fl (3) R 10s; 36°33'·39N 06°17'·19W.
Spur N of S. Felipe ⚓ Fl (4) G 10s 10m 3M; 36°32'·72N 06°16'·79W.
San Felipe mole ⚓ Fl G 3s 10m 5M; G twr; 36°32'·57N 06°16'·77W.
E mole ⚓ Fl R 2s 11m 5M; 36°32'·44N 06°16'·63W.
Puerto America, NE bkwtr, ⚓ Fl (4) G 16s 1M; G twr; 36°32'·47N 06°16'·94W.
Real Club Nautico pier ⚓ FG; 36°32'·35N 06°17'·13W.
International Free Zone hbr: No. 1 ⟟ Fl (3) G 9s; 36°30'·66N 06°15'·45W. **Puerto Elcano** marina ent 36°30'·08N 06°15'·45W.

SANCTI PETRI

Punta del Arrecife ⚓ Q (9) 15s 7m 3M; 36°23'·73N 06°13'·51W.
Sancti Petri castle ⚓ Fl 3s 18m 9M; twr; 36°22'·76N 06°13'·29W.
Ldg lts 050°: Front ⚓ Fl 5s 12m 6M; 36°22'·91N 06°12'·24W.
Rear ⚓ Oc (2) 6s 16m 6M.
No. 1 ⟟ Fl (3) G 9s; 36°22'·50N 06°12'·90W.
No. 2 ⟟ Fl (3) R 9s; 36°22'·50N 06°13'·00W.
No. 3 ⟟ Fl (4) G 11s; 36°22'·64N 06°12'·65W.
No. 4 ⟟ Fl (4) R 11s; 36°22'·70N 06°12'·70W.
Inner ldg lts 346·5°: Front ⚓ Fl 5s 11m 6M. Rear ⚓ Oc (2) 6s 21m.
⟟ Fl G 5s 8m 2M; 36°23'·02N 06°12'·74W; and ⟟ Fl R 5s 8m 2M.

Cabo Roche ☆ Fl (4) 24s 44m **20M**; 36°17'·74N 06°08'·41W.
Cabo Trafalgar ☆ Fl (2+1) 15s 50m **22M**; 36°10'·97N 06°02'·09W.

THE GIBRALTAR STRAIT

BARBATE

Barbate lt ho ⚓ Fl (2) WR 7s 22m, W10M, R7M; 281°-W-015°-R-095°; W ○ twr, R bands; 36°11'·22N 05°55'·42W.
SW mole ⚓ Fl R 4s 11m 5M; 36°10'·79N 05°55'·55W.
Inner NE mole ⚓ Fl G 3s 8m 2M; 36°10'·95N 05°55'·79W.
Marina ent, Fl (2+1) R 21s 2M, 7 Fl (4) Y 20s (anti-oil boom) and Fl R 4s 2M.

△-shaped tunny net, N end, ⟟ Q (Mar-Sep); 36°10'·72N 05°55'·39W (1ca E of hbr ent). Other marks clockwise:
⟟ Q (3) 10s (Mar-Sep); 36°08'·89N 05°55'·39W.
⟟ VQ (6) + L Fl 12s (Jul-Sep); 36°08'·07N 05°55'·07W.
⟟ VQ (6) + L Fl 12s (Mar-Jun); 36°08'·82N 05°57'·12W.
⟟ VQ (9) 10s (Mar-Sep); 36°09'·20N 05°56'·00W, centre of net.
Punta de Gracia ⚓ Oc (2) 5s 74m13M; 36°05'·39N 05°48'·57W.
Pta Paloma ⚓ Oc WR 5s 44m, W10M, R7M; 010°-W-340°-R-010°, over Bajo de Los Cabezos; 36°03'·84N 05°43'·19W.

TARIFA

Tarifa ☆ Fl (3) WR 10s 40m **W26M, R18M**; 113°-W-089°-R-113°; W twr; *Racon C, 20M*; 36°00'·07N 05°36'·58W. 992242147.
Lado E ⚓ Fl R 5s 12m 3M; 36°00'·23N 05°36'·38W.
Outer SE mole ⚓ Fl G 5s 11m 5M; vis 249°-045°; G twr with statue of Virgin Mary; 36°00'·39N 05°36'·21W.
Inner S mole ⚓ Fl (2) R 6s 7m 1M; 36°00'·51N 05°36'·16W.

ALGECIRAS

Pta Carnero ☆ Fl (4) WR 20s 42m, **W16M**, R13M; 018°-W -325°-R-018°; W ○ twr; 36°04'·62N 05°25'·58W.

⟟ Fl (2) 7s; 36°06'·73N 05°24'·76W, 1·2M ESE of marina ent.
Darsena del Saladillo (marina) outer approach buoys: ⟟ Fl(2)G 7s; 36°07'·18N 05°25'·83W. ⟟ Fl (3) G 9s; 36°07'·19N 05°25'·97W.
⟟ Fl (2) R 7s; 36°07'·07N 05°25'·73W.
Marina, off outer S jetty ⟟ Fl (3) R 9s, 36°07'·13N 05°26'·10W.
N jetty ⚓ Fl (4) G 11s 5m 1M; 36°07'·13N 05°26'·22W.
Inner S jetty ⚓ Fl (4) R 11s 4m 1M; 36°07'·14N 05°26'·28W.

LA LÍNEA

⟟ Fl (3) R 6s; 36°09'·56N 05°21'·98W.
⟟ Fl (3) G 6s; 36°09'·52N 05°22'·00W.

Dique de Capitania ⚓ Fl (2) G 7s 5m 3M; G twr; 36°09'·50N 05°21'·60W.
Marina ent jetty ⚓ QR 7m 2M; 36°09'·56N 05°21'·67W.

GIBRALTAR

Aero ☆ Mo (GB) R 10s 405m **30M**; 36°08'·57N 05°20'·59W.
Europa Pt ☆ Iso 10s 49m **19/15M**; W vis 197°-042° and 067°-125°; W ○ twr, R band; 36°06'·58N 05°20'·69W. Also ⚓ FR 44m **15M**; R 042°-067°; Same twr and vis arc, ⚓ Oc R 10s 49m **15M**. *Horn 20s.*
A head ⚓ Fl 2s 18m 5M; *Horn 10s;* 36°08'·03N 05°21'·85W.
Queensway Quay marina, N ent ⚓ FR/FG, 36°08'·16N 05°21'·39W.
Coaling Island, SW corner ⚓ 2 FR (vert); 36°08'·19N 05°21'·46W.
B head ⚓ QR 9m 5M; 36°08'·14N 05°21'·84W.
C head ⚓ QG 10m 5M; 36°08'·54N 05°22'·04W.
D head ⚓ QR 18m 5M; 36°08'·65N 05°21'·95W.
Runway threshold, both sides, VQ occas; 36°09'·01N 05°21'·59W.

MOROCCO (WEST TO EAST)

Cap Spartel ☆ Fl (4) 20s 92m **30M**; Y □ twr; *Dia (4) 90s;* 35°47'·47N 05°55'·43W.

TANGIER

Monte Dirección (Le Charf) ☆ Oc (3) WRG 12s 88m **W16M**, R12M, G11M; 140°-G-174·5°-W-200°-R-225°; 35°45'·98N 05°47'·35W.
N bkwtr ⚓ Fl (3) 12s 20m 14M; 35°47'·47N 05°47'·60W.
SE mole ⚓ Oc R 4s 7m 6M; 35°47'·29N 05°47'·87W.
Jetée des Yachts ⚓ Iso G 4s 6m 6M; 35°47'·23N 05°48'·14W.
Pta Malabata ☆ Fl 5s 77m **22M**. W □ twr; 35°48'·99N 05°44'·92W.
Basses des Almirantes ⚲ VQ; 35° 49'· 72N 5° 45'· 75W.
Ksar es Srhir ⚓ Fl (4) 12s 16m 8M; 35°50'·91N 05°33'·67W.
Pta Círes ☆ Fl (3) 12s 44m **18M**; 060°-330°; 35°54'·42N 05°28'·89W.

CEUTA (Spanish enclave)

Punta Almina ⚓ Fl (2) 10s 148m 22M; 35°53'·90N 05°16'·85W.
W pier ⚓ Fl G 5s 13m 5M; *Racon O, 12M*; 35°53'·75N 05°18'·68W; 992241092.
E pier ⚓ Fl R 5s 13m 5M; vis 245°-210°; 35°53'·73N 05°18'·47W. 05°18'·90W.
Marina ent, N bkwtr ⚓ Fl (4) R 11s 8m 1M; 35°53'·45N 05°18'·90W.
Marina ent, S bkwtr elbow and S bkwtr head: two synchronised ⚓ Fl (4) G 11s 6/5m 1M; 35°53'·42N 05°18'·94W and 65m ESE.

9.25.4 PASSAGE INFORMATION

More Passage Information is threaded between successive harbours in this Area.

Bibliography: *Atlantic Spain and Portugal* (Imray/RCC/Walker/Buchanan); *SW Spain and Portugal Cruising Companion* (Fernhurst/Jens); *W Coasts of Spain and Portugal Pilot* (Admiralty, NP 67) covers Cabo Ortegal to Gibraltar.

The Straits Handbook by Colin Thomas of Straits Sailing (☎+35051372) is very informative and enjoyable reading. It covers Gibraltar and Spanish ports from Chipiona to Benalmadena on the Med coast; also Ceuta and Marina Smir in Morocco.

PORTUGUESE BORDER TO RIO GUADALQUIVIR

(AC 93) On the Spanish bank of the Rio Guadiana **Ayamonte** has a marina, the most W'ly of 16 modern marinas in SW Andalucía; see Special Notes below.

4M to the E, **Isla Canela** W bank and **Isla Cristina** E bank of Ria de la Higuerita both have marinas. 13M further E, Rio de las Piedras requires care on entry due to shifting shoals and bar, but is a peaceful river with a marina at **El Rompido**. S of Huelva there are marinas at **Punta Umbria** and **Mazagón** (AC 73). Keep clear of commercial ships going to/from **Huelva**.

About 30M SE is the mouth of the **Rio Guadalquivir** (AC 85) navigable 54M to **Sevilla**. Caution: NW of the estuary large fish havens in shoal waters extend 5M offshore. **Chipiona** marina is a good place to check the latest info for the passage to and lock times at Sevilla; and to time your departure so as to carry a fair tide and synchronise your arrival with lock and bridge opening hours.

9.25.5 SPECIAL NOTES FOR SW SPAIN

Regions/Provinces: Spain is divided into 17 autonomous regions, eg Andalucía. Most regions are sub-divided into provinces, eg in Atlantic Andalucía: Huelva, Sevilla and Cadiz. The province is shown below the name of each main port.

Language: This Almanac recognises regional differences, but in the interests of standardisation uses the spelling of Castilian Spanish where practicable.

Spanish Charts (SC): Contact Instituto Hidrográfico de la Marina, Plaza San Severiano 3, 11007 Cádiz, ☎(956) 599 414; www.armada.mde.es/ihm Or order from chart agents in Madrid, Huelva, Sevilla, Cadiz and Algeciras, also larger UK chart agents.

Standard Time in Spain is UT –1, but in Portugal it is UT.

Spanish secondary ports referenced to Lisboa: Time differences for these Spanish ports (all down to Barbate) when applied to the printed times of HW and LW for Lisboa (UT) give HW and LW times in Spanish Standard Time. Add DST, when applicable.

Telephone: To call Spain from UK dial 00-34, then the area code, followed by the ☎ number. To call UK from Spain, dial 07-44, then area code, less the initial 0, then the ☎ number.

Emergencies: ☎900 202 202 for Fire, Police and Ambulance. *Rioja Cruz* (Red Cross) operate Ⓛ Ⓢ s.

Public Holidays: Jan 1, 6; Good Friday; 1 May (Labour Day); June 11 (Corpus Christi); Aug 15 (Assumption); Oct 12 (National Day); Nov 1 (All Saints Day); Dec 6, 8 (Immaculate Conception) 25.

Representation: Spanish Tourist Office, 64 North Row, London W1K 7DE; ☎020 7317 2011. info.londres@tourspain.es www.spain.info

British Embassy, Paseo de la Castellana 259D, 28046 Madrid; ☎917 146 301. There are British Consuls at Malaga ☎902 109 356.

Buoyage: IALA Region A system is used. Preferred chan buoys (RGR, Fl (2+1) R and GRG, Fl (2+1) G) are quite widely used.

Documents: (See Chap 2 Cruising Formalities) Spain, although an EU member, still checks papers. This can be a time-consuming, repetitive and inescapable process. The only palliatives are courtesy, patience and good humour. Organise your papers to include:

Personal – Passports; crew list, ideally on headed paper with the yacht's rubber stamp, giving DoB, passport nos, where joined/intended departure. Certificate of Competence (Yachtmaster Offshore, ICC/HOCC etc). Radio Operator's certificate. EHIC (advised for medical treatment).

Yacht – Registration certificate, Part 1 or SSR. Proof of VAT status. Marine insurance. Ship's Radio licence. Itinerary, backed up by ship's log.

Currency is the Euro €. VAT (IVA) 21%.

Marina fees: Marinas run by or linked to the Andalucían Junta (www.eppa.es) ie all but Algeciras, charge on average €2.45/m LOA inc 21% VAT in season (1/6-30/9); low season 50% less.

Fuel: Diesel (*Gasoleo*) is of two types: *Gasoleo A* is taxed and available from marina pumps or by cans from a filling station ashore. *Gasoleo B* is untaxed and usually only available to FVs.

Travel: UK ferries from/to Santander and Bilbao.

Direct UK flights from Jerez, Sevilla, Madrid, Malaga & Gibraltar.

Internal flights via Madrid to Jerez, Sevilla and Malaga.

Buses are usually good and trains adequate, except in the more remote regions.

9.25.6   AYAMONTE

Huelva **37°12'·70N 07°24'·50W** ❋❋⚓⚓❀❀

CHARTS AC 93; SC 440, 440A; PC 24206, 26312

TIDES
Standard Port LISBOA (←); ML 1·8

Times				Height (metres)			
High Water		Low Water		MHWS	MHWN	MLWN	MLWS
0500	1000	0500	1100	3·8	3·0	1·5	0·6
1700	2200	1700	2300				
Differences AYAMONTE							
+0005	+0015	+0025	+0045	−0·7	−0·6	−0·1	−0·2
RÍA DE HUELVA, BAR							
0000	+0015	+0035	+0030	−0·1	−0·6	−0·1	−0·4

SHELTER Good in Vila Real Marina on W bank of Rio Guadiana.

NAVIGATION WPT 37°08'·51N 07°23'·09W (No 1 SHM buoy, Q (3) G 6s) 340°/9ca to hd of W trng wall. Least depth over bar 2·0m, avoid near low water. See Vila Real de Santo Antonio for chartlet, lts and directions upriver. A poorly lit pontoon to port just inside the ent (former fuel berth) is a hazard at night. Sanlucar, 20M up Rio Guadiana, is accessible on the flood. Ayamonte Marina, beware narrow ent chan and increasing silting/reduced depths within; only enter after half flood.

LIGHTS AND MARKS High bldgs, FR lts, are 3M WNW of ent. White suspension bridge 2M inland is visible from seaward. Vila Real lt ho, ☆ Fl 6·5s 51m **26M** W ○ tr, B bands, is 1·5M S. ⚲ Fl G 5s, 350m N of marina, marks ferry berths.

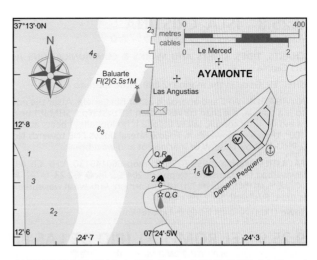

COMMUNICATIONS (Code 959). Huelva MRSC 243000; Brit Consul 902 109356; HM 471407; Marina VHF Ch 09 (H24).

FACILITIES **Marina** (0.6m-1·5m) ☎034498, mob 6001491 is W'most of 13 marinas run by Junta de Andalucía, www.eppa.es. ayamonted@eppa.es 317 inc🅥 €2·23. ⊖ 🛠 🅛 🗝 🕮.
Town ✕ 🛒 🍴 ◎ 🅛 ✉ Ⓑ YC. Ferries to Vila Real de Santo António. Bus Huelva (60km); ✈ Faro (48km).

9.25.7   ISLA CANELA

Huelva **37°11'·24N 07°20'·45W**
❋❋⚓⚓❀❀

CHARTS AC 93; SC 440, 440A

TIDES Use Differences Ayamonte

SHELTER Good in the marina (2·5m) but little room to ⚓. The appr is open to SE'lies.

NAVIGATION Isla Canela and Isla Cristina share a common outer approach into Ría de la Higuerita. Best appr at half-flood, when drying sand banks can be seen.

WPT 37°10'·51N 07°19'·46W (SWM buoy, Fl 10s, off chartlet) 000°/4ca to W mole head, where depth is 1·2m. Pick up the 312·8° ldg lts. After 7ca between trng walls, for Isla Canela marina turn 90° port between QR and QG lights.

See Isla Cristina for continuation.

LIGHTS AND MARKS Ldg lts (front Q 7m 5M; rear Fl 4s 12m 5M) aligned 312·8° between the W mole head VQ (2) R 5s 9m 4M, and the covering trng wall to stbd. The powerful lt Fl 6·5s 51m **26M** at Vila Real de Santo António is 4M to the W.

COMMUNICATIONS (Code 959) Huelva MRSC 243000; Brit Consul 902 109356; Marina VHF Ch 09.

FACILITIES **Marina** ☎4790000 www. marina-islacanela.com marina@islacanela. es 231 ⤓ inc🅥 max LOA 12m, €2.82 ✕ 🛒 🍴 ⊖ 🛠 🅛 ✉ Ⓑ ✕ 🏳(32t) 🗝(2t).
Town Infrequent bus Ayamonte (17km); ✈ Faro (65km).

9.25.8 ISLA CRISTINA
Huelva **37°11'·88N 07°19'·65W** ✳✳✳♨♨ ✿✿

CHARTS, TIDES, LIGHTS AND MARKS & COMMUNICATIONS As Isla Canela.

SHELTER Good in the marina (2·0m). Little space to ⚓.

NAVIGATION WPT and approach as for Isla Canela. After the ent to Isla Canela, the chan curves N then NE past No 1 SHM lt buoy. Stand on until conspic tall W bldg (a lt ho lookalike) bears about 100°, then alter towards it. Follow lateral buoys ENE towards No 2 PHM lt buoy, close W of marina ent and wavebreak.

FACILITIES **Marina** ☎959 077613; mob 600149130. VHF Ch 09. www.eppa.es islacristinad@eppa.es 204 ⌾ inc Ⓥ €2·23. ⊖ ⚲ ♦ ✕ ⌸ ⓞ ⚒ 🖭 ⚒ Ice ▄ 🗗(32t) ⏏(5t). Caution: Anti-wash wavebreak is unlit. FVs berth N of the marina.

Town (1km) ✕ ⌸ 🖳 ⊠ Ⓑ ✈ Faro (65km).

ISLA CRISTINA

Caution: This chartlet is sourced from local plans and aerial photos, due to lack of large scale Spanish and UKHO chart coverage. It is believed to be generally correct.

9.25.9 EL ROMPIDO (RIO DE LAS PIEDRAS)
Huelva **37°13'·00N 07°07'·72W** (Marina) ✳♨♨ ✿✿✿

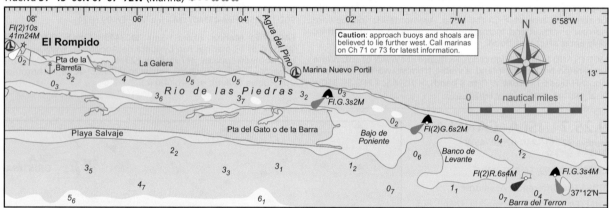

Caution: approach buoys and shoals are believed to lie further west. Call marinas on Ch 71 or 73 for latest information.

CHARTS AC 93; SC 440, 441A

TIDES Ayamonte 9.25.6.

SHELTER Good in lee of sand dunes. Do not enter in strong onshore winds. Tidal streams may exceptionally exceed 2kn.

NAVIGATION WPT No. 1 'Recalada' landfall ↗, 37°11'·65N 07°03'·05W. Note: Bar (0·4m) and shoals shift frequently. No 1 and lateral buoys 2-11 are moved as required. Visit the website below for the latest positions or call HM on VHF/phone before entry. Puerto El Terron 37°13'·50N 07°10'·49W is shallow (<1·2m).

FACILITIES **Marina El Rompido** ☎(959) 399614. www.puerto elrompido.com info@puertoelrompido.com. VHF Ch 71. 331 ⌾ inc Ⓥ €2·12. Free berthing help available on request and recommended in bad conditions. All berths are aligned F & A with the tidal stream. ⋒ ☎399026 ⚒ ⓞ ♦ ♦ ☎6209 22969, 0800-2000 daily in summer.

Marina Nuevo Portil, ☎(359) 504100, adnnportil@yahoo.es www. adnneuvoportil.es; VHF Ch 73, 395 ⌾ (max 16m) inc Ⓥ, €1·83. Free berthing help if required, berths aligned F&A with tidal stream. ⟲ ✕ ⓞ 🖳 wi-fi ♦ ♦ (2M upstream).

Puerto El Terron ☎382225. VHF Ch 09. 32 ⌾ limited Ⓥ. ✕ 🖳 ⊠ YC, 6ca E, ☎399349 ✕ ⌸. **Village** ✕ ⌸. Golf/beach resort is adjacent.

MARINA EL ROMPIDO

Caution: This chartlet is sourced from local plans and aerial photos, due to lack of large scale Spanish and UKHO chart coverage. It is believed to be generally correct.

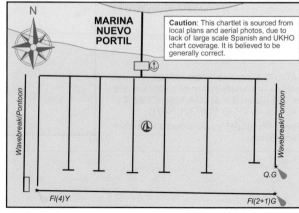

MARINA NUEVO PORTIL

Caution: This chartlet is sourced from local plans and aerial photos, due to lack of large scale Spanish and UKHO chart coverage. It is believed to be generally correct.

9.25.10 PUNTA UMBRIA

Huelva **37°10'·89N 06°57'·48W** ✿✿❄❄❄✿✿

CHARTS As for Mazagón. **TIDES** As for Special Notes. Tidal streams are strong and arrival near slack water will assist berthing.

SHELTER Good. A new wavebreak at Punta Umbria Marina offers better shelter than previously.

NAVIGATION WPT 37°08'·78N 06°56'·74W (off chartlet) ⚓ L Fl 10s, 000°/1M via Nos 1 & 2 buoys to bkwtr hd (1·1m). Thence the chan is marked by Nos 3-9 lateral lt buoys. 2M SE of ent 4 ⚓s, Fl (4) Y 20s, mark a pipeline extending 5M S (entry prohib).

LIGHTS AND MARKS See Lights, buoys and waypoints.

COMMUNICATIONS Marina VHF Ch 06.

FACILITIES from seaward: **Real Club Maritimo** ☎(959)311899, www.remtpu.com remtpu@remtpu.com 200+60 Ⓥ. **Punta Umbria Marina** ☎314304. www.eppa.es puntaumbriad@eppa.es 267 ⌒ inc Ⓥ in 6m, max LOA 12m, €2·06. ⬛⬛⬛⬛⬛. **Punta Umbria RCN Marina** is 5ca further upriver. Privately run, limited space for Ⓥ but more peaceful than downstream. **Town** ✕⬜🛒◎.

9.25.11 MAZAGÓN

Huelva **37°07'·92N 06°50'·13W** (Marina ent) ✿✿❄❄❄✿✿

CHARTS AC 91, 93, 73; SC 441, 4411, 4413; SHOM 6862, 7300

TIDES See Ayamonte, 9.25.6, Ría de Huelva, Bar; ML 1.8 .

SHELTER Good in marina (overleaf) entered hdg ESE downriver.

NAVIGATION WPT 37°05'·55N 06°49'·11W, abm WCM buoy, Q (9) 15s (off chartlet) 339°/2·5M to marina. After the first pair of chan buoys (off chartlet) the straightforward appr is sheltered by a 7M long bkwtr.

At No 7 SHM buoy turn stbd to the marina ent; the deep, well lit, buoyed 339·2° channel diverges WNW into Huelva's industrial complex. Keep clear of large ships.

About 5M W of the WPT an oil pipe line runs N to a refinery. It is marked by an SBM and 3 SPM buoys, all Fl (4) Y 20s; keep 500m clear. ⚓ prohib within 500m either side of the line of buoys.

LIGHTS AND MARKS Dir lt 339·2° (off chartlet); W twr. The bkwtr hd has a W twr, R band. Picacho lt ho, W twr with R corners/R roofed bldg, is 5ca NE of marina (off chartlet). See Lights, buoys and waypoints.

COMMUNICATIONS (Code 959) MRCC (956) 684740; Brit Consul Malaga 902 109356; Huelva Port VHF Ch 06, 11, 12, **14**; Marina VHF Ch 09.

FACILITIES in large, modern **Marina** (4m): ☎536251. mazagon@eppa.es 647 ⌒ incⓋ €2·06. ⬛⬛✕⬜⬛⬛⬛ Ice YC ⬛ ⬛(32t). ⬛ Huelva (24km); ✈ Sevilla (130km).

MAZAGÓN MARINA

MAZAGÓN

CHIPIONA TO CAPE TRAFALGAR

Around **Cadiz Bay** (AC 86, 88) there are marinas at **Rota, Puerto Sherry, Santa Maria** and two in the hbr of Cádiz itself. S of Cádiz keep about 4M offshore on the 30m line to clear inshore banks. **Sancti-Petri** marina lies on a remote and interesting river with a slightly tricky approach.

Artificial reefs, which may reduce charted depths by approx 2·5m, are common from Cadiz to C. Trafalgar; see AC 93 & 142. Cabo Trafalgar may be rounded within 100m of the lt ho, inshore of a tidal race, or about 4M off to clear foul ground/ shoals lying from SW to NW of the cape. Plans to build a wind farm (up to 500 turbines) about 2·5M off Cabo Trafalgar are strongly opposed by local fishermen.

Tunny nets are laid up to 7M offshore, usually marked by cardinal lt buoys. Positions vary each year, but the more established nets are laid:

• Very close off Barbate hbr, extending 2·7M S, Mar–Aug.
• Off Zahara beach (5M SE of Barbate) extending 1·7M SW to two WCM and one SCM lt buoys, Mar–Sep.
• 1M NW of Tarifa, to 1M SW, buoyed as Zahara, Mar–Jul.

9.25.12 CHIPIONA

Cádiz **36°44'·96N 06°25'·69W** (W bkwtr hd) ✦✦✦❋❋❋

CHARTS AC 85; SC 4421, 4422 (Sheets 2 and 3)

TIDES
Standard Port LISBOA (←); ML 1·9; see Rio Guadalquivir.

SHELTER Good in modern marina, but swell enters in NW gales.

NAVIGATION WPT 36°45'·75N 06°26'·82W, adj No 1 SWM buoy, L Fl 10s (Racon M) 125°/1·2M to marina ent. Caution: extensive fish havens with obstructions 2·5m high lie from 3·5M to 7·5M NW of the WPT. Same WPT is on the 068·9° ldg line for buoyed chan into Rio Guadalquivir.

LIGHTS AND MARKS Chipiona (Pta del Perro) lt ho is conspic 1M SW of marina. 1·7M W of this lt ho, the drying Bajo Salmedina (off chartlet) is marked by a WCM bcn twr which must be rounded if coming from S. Inside hbr bkwtr, the Fl (3) G 9s light, near FV berths, is not visible from seaward. See Lights, buoys and waypoints for light details.

COMMUNICATIONS (Code 956) MRCC 684740; Port and Rio Guadalquivir VHF Ch 12; Marina VHF Ch 09.

FACILITIES Marina ☎373844, chipiona@eppa.es 412 ⌣ inc ❤ in 3·5m, €2.20. ❤ berth on fuel/reception hammerhead; only larger yachts in first basin on port side. ⛴▲✕◻⌣♨◻▬◻✎◻✕ ◻(32t). FVs berth against the NW bkwtr.
Town ⛟ ✉ Ⓑ ⇌ and ✈ Jerez (32km).

9.25.13 RIO GUADALQUIVIR

Huelva/Cádiz **36°45'·73N 06°27'·03W** (No 1 SWM buoy)

CHARTS AC 85 or SC 4422 (18 sheets, mostly 1:12,500) corrected up-to-date as buoys/bcns are liable to change.

TIDES
Standard Port LISBOA (←); ML 1·9

Times				Height (metres)			
High Water		Low Water		MHWS	MHWN	MLWN	MLWS
0500	1000	0500	1100	3·8	3·0	1·5	0·6
1700	2200	1700	2300				
Differences RIO GUADALQUIVIR, BAR (36°45'N 06°26'W)							
−0005	+0005	+0020	+0030	−0·7	−0·5	−0·2	−0·1
BONANZA (36°48'N 06°20'W)							
+0025	+0040	+0100	+0120	−0·8	−0·7	−0·5	−0·2
CORTA DE LOS JERONIMOS (37° 08'N 06°05'W, 16M S of Sevilla)							
+0210	+0230	+0255	+0345	−1·2	−0·9	−0·5	−0·1
SEVILLA							
+0400	+0430	+0510	+0545	−1·8	−1·3	−0·7	−0·2

SHELTER There are no recognised stopping places and any ⚓ is vulnerable to passing traffic; monitor VHF Ch 12.

NAVIGATION The river is not difficult to navigate but can be hot and uninspiring, through the flat and almost featureless terrain of the Doñana National Park. From No 1 SWM buoy (shown above and as WPT for Chipiona) the Bonanza ldg lts (both Y ◻, on metal towers) lead 069° for 10M through lit, buoyed access channel.

From No 1 SWM buoy to a lock 3M S of Sevilla is about 49M. At Bonanza (36°48'N 06°20'W) the channel becomes truly riverine, 750m wide; 250m nearer Sevilla. The banks are well marked throughout by lt beacons. Depths are rarely less than 5m, best water usually being on the outside of bends, which are often buoyed in the broader lower reaches. A chart plotter or similar is a great boon in some of the unmarked reaches.

TIDAL STREAMS The flood makes for about 7hrs (3kn sp, 1kn nps) the ebb for 5½ hrs; so it should be possible to reach the lock on one tide, passing Bonanza N-bound at LW −½hr.

Around the Chipiona chart:
10 8₄ Fl(2)G10s ✦
Fl(3)G9s 4 Fl(4)R.11s
LFl10s Racon (M) No1 Bonanza Ldg lts 069° to R.Guadalquivir Nos 17 & 18 buoys for R.Guadalquivir
9₈ RW ⊕ 2₅ CHIPIONA
45'·5 5₁
N 4₁ 3₂ Fl(2)R.7s NO 2
3₆ 1₈ Fl(3)R.11s NO 4
36°45'N Fl(2)G10s
Fl(3)G9s Fl(4)R.11s
0 metres 1000 2₉ See inset
cables 0 5 CHIPIONA
Fl10s 68m25M 06°26'·0W
27'·0 Pta del Perro

9.25.14 SEVILLA

Sevilla **37°19'·86N 05°59'·74W** (Lock) ✹✹✹♤♤♤♤✿✿✿

CHARTS & TIDES AC 85; SC 4422 (sheets 17/18). ML Sevilla 1·3M

SHELTER Good; safe to over-winter. Very hot 40°C+ in summer.

NAVIGATION A new lock, S of Pta del Verde, ☎(954) 454196, opens 0100, 0400, 0700, 0900, 1100, 1300, 1600, 1900 & 2100LT daily; secure to ladders or diagonal rubber strips.

Delicias bridge (10·1m; ☎ 954) 247630) lifts Mon, Wed, Fri 2200; Sat/Sun/Hols 2000. 1 hr before ETA request passage Ch 12. R lts at lock & bridge = no go; Lock, Intens Fl W = enter; Bridge, G = pass.

COMMUNICATIONS (Codes 954, 955) Brit Consul 902 109 356; Lock, Port, Delicias bridge VHF Ch 12; marinas Ch 09.

FACILITIES Gelves 37°20'·43N 06°01'·39W on the river, ie **not** accessed via the lock, therefore tidal; may silt. Overhead cables (16·5m) downriver; approx 3M from the city centre. ☎(955) 761212. www.puertogelves.com 133+❶ €0.93, in about 3m, waiting pontoon. ✕ ⛟ ◻ ◙ ◖ ⚓ ▣(25t).

Sun Sails Marina 37°19'·90N 05°59'·49W, 300m E of the old lock almost below HT cables; a long pontoon in a quiet creek 3M from city; useful stop if waiting to transit Delicias bridge 2M N. ☎(954) 184848 www.sunsails.es. 150 ⌐ inc ❶ in 6m, ◖ ◙ ◖ ⚓ ⚓.

Club Náutico de Sevilla 37°22'·18N 05°59'·60W, excellent facilities close to city centre; also possible ⌐ on E quay. ☎(954)454777. www.nauticosevilla.com 100+❶ in 3m; pre-book €0.52/m²/day. ◣ ✕ ◻ Ice ◙. **City** All amenities, ⇌ ✈ 10km.

9.25.15 BAY OF CÁDIZ

Cádiz **36°33'·91N 06°19'·88W** (SWM buoy) ✹✹✹♤♤♤♤✿✿✿

CHARTS AC 93, 86, 88; SC 443, 443A, 443B, 4430, 4431, 4437

TIDES
Standard Port LISBOA (←—); ML 1·8

Times				Height (metres)			
High Water		Low Water		MHWS	MHWN	MLWN	MLWS
0500	1000	0500	1100	3·8	3·0	1·5	0·5
1700	2200	1700	2300				
Differences ROTA							
–0010	+0010	+0025	+0015	–0.7	–0.7	–0.4	–0.1
PUERTO DE SANTA MARIA							
+0006	+0006	+0027	+0027	–0.6	–0.5	–0.4	–0.1
PUERTO CÁDIZ							
0000	+0020	+0040	+0025	–0.5	–0.6	–0.3	0.0

SHELTER is good in the 4 marinas listed below, clockwise from N.

NAVIGATION Rota WPT 36°35'·97N 06°21'·01W, 000°/9ca to marina ent. Airfield lt (Al Fl WG 9s) and conspic water tank bear 004°/2·1M from WPT. From W/NW keep 1·5M offshore to clear shoals. Naval vessels may be at ⚓ S of Rota Naval base. Significant shoaling reported; keep to stbd (N bkwtr) side of ent chan.

Puerto Sherry WPT 36°33'·98N 06°15'·22W, 000°/7ca to marina entrance. Modern bldgs around the marina are conspic.

Puerto de Santa Maria same WPT as P. Sherry, follow the W sector of Dir Fl WRG 6s to marina in canalised Rio Guadalete, dredged 4·5m. Ldg lts 040°: front, QG; rear Iso G 4s. From W, the N Chan trends ESE, inshore of shoals toward both marinas.

Puerto América (at N tip of Cádiz). WPT 36°32'·75N 06°16'·36W, 238°/0·5M to marina ent passing close S of Dique de San Felipe head. From S, keep to seaward of extensive shoals N & W of Cádiz. The S Chan (Canal del Sur) is not navigable for draft >2·5m. The shore is low-lying, but next to the marina a W bldg on Dique de San Felipe is conspic.

LIGHTS AND MARKS Lts/buoys as chartlets and Lights, buoys and waypoints. Conspic daymarks: Rota lt ho 33m, W twr/R band, at marina. Puerto Sherry, W bkwtr hd ○ twr. Cádiz city, golden-domed cathedral and radio twr (113m) close SE. Further SSE: Two power cable pylons (155m) dockyard cranes and long bridge to mainland.

COMMUNICATIONS (Code 956) Cádiz Trafico VHF Ch **74**; ⊖ & Met via marinas (see below); Brit Consul 902 109 356; Port HM 224011; Marinas VHF Ch 09.

FACILITIES Rota Marina ☎840069, rota@eppa.es 509 ⌐ inc ❶ in 4·5m, €2.06. ◖ ◣ ◖ ⚓ ⚓ ▣(32t) ◛(5t).

Puerto Sherry ☎870103. www.puertosherry.com 753 ⌐ inc❶ €2.31; 3·0-4·5m. ◖ ◣ ◖ ▣(50t) ◖ ◙ ◖ △ Ⓔ ◙ ◖ ⚒ ◙ ✕ ◻ YC ☎ 858751. Arrivals berth as chartlet.

Puerto de Santa Maria (RCN) ☎852527. rcnpuerto@ono.com www.rcnpsm.com Very pleasant Real Club Náutico has 250 ⌐ + a few ❶ berths in 3·0- 8·0m; €2.98, pre-booking is not essential. 10 hammerhead pontoons (A-J) on NW bank plus two mid-stream pontoons. ◖ ◣ ⚒ ◖ ◛(5t) ▣(25t) ◖ ... ✕ ◻. FVs berth upriver. Ferry to Cádiz.

Puerto América ☎223666. puertoamerica@eppa.es 253 ⌐ inc ❶ in 7·5m, €2.06. ◖ ◣ ◖ ⚓ ▣ ◛(10t).

Real Club Náutico de Cádiz A private basin close SW of Puerto America. ☎213262. www.cherrytel.com 160 ⌐ in 3m. ❶ berths unlikely. ◖ good ✕ ◻.

Cadiz City all amenities, ⇌; ✈ Jerez (25km). Spanish Hydrographic Office (IHM) is at 36°31'·38N 06°17'·04W (1·5M from Puerto America): Plaza San Severiano 3, 11007 Cádiz; ☎599414. www.armada.mde.es/ihm

BAY OF CÁDIZ including marinas at: ROTA, PUERTO SHERRY, PUERTO DE SANTA MARIA and PUERTO AMERICA

9.25.16   SANCTI-PETRI

Cádiz **36°23'·80N 06°12'·55W** ✿✿✿✿✿✿

CHARTS AC 93; SC 443B, 4438

TIDES Interpolate between differences Puerto Cádiz and Cabo Trafalgar (Barbate); ML No data

SHELTER Good, except with swell and in S'lies. ⚓ or moor in stream, W of marina.

NAVIGATION WPT 36°22'·32N 06°13'·15W 050°/5ca to intercept 346·5° ldg line which leads between 4 lateral lt buoys. From 2nd pair of buoys track on the 346·5° ldg line for 6ca to the gateway formed by R/G columns (Fl R/G 5s 7m) which are easily seen. A night appr is not advised first-time.

Best appr at half-flood in fair vis; least charted depth 2·2m, but silting occurs. Sp ebb can be > 4kn. Swell and/or strong S'lies render the appr dangerous. El Arrecife, a long drying reef, and other shoals prevent a direct appr from W/NW. Tunny nets may be set about 9M SSE off Conil. Planned marina at Conil has not materialised (2015).

LIGHTS AND MARKS See chartlet and Lights, buoys and waypoints. The castle's 16m square twr, Fl 3s, is conspic. 050° and 346·5° ldg marks are hard-to-see lattice twrs. 4 lateral buoys define the 050° ldg line.

COMMUNICATIONS (Code 956) MRCC ☎684740; Brit Consul 902 109356; Marina *Puerto Sancti-Petri VHF Ch 09*.

FACILITIES Marina ☎496169. sanctipetri@eppa.es 90 ⌷ in 5m, inc 2 ♥ ⌷ on pontoons in 2–5m, €2·06; see HM. ⚓s €0·66. 🛢 limited 🔌 and 🚿 in season ⚓ 🛒. Club Náutico ✕ 🍴.

Village Few facilities in abandoned 'ghost town' at mouth of sandy, peaceful lagoon; ⇌ San Fernando (18km); ✈ Jerez (50km).

9.25.17   BARBATE

Cádiz **36°10'·79N 05°55'·54W** ✿✿✿✿✿✿

CHARTS AC 91, 142; SC 444, 4441

TIDES

Standard Port LISBOA (◄—); ML 1·2

Times				Height (metres)			
High Water		Low Water		MHWS	MHWN	MLWN	MLWS
0500	1000	0500	1100	3·8	3·0	1·5	0·5
1700	2200	1700	2300				
Differences CABO TRAFALGAR							
–0003	–0003	+0026	+0026	–1·4	–1·2	–0·6	–0·1
RIO BARBATE							
+0016	+0016	+0045	+0045	–1·9	–1·6	–0·5	+0·1
PUNTA CAMARINAL (36°05'N 05°48'W)							
–0007	–0007	+0013	+0013	–1·7	–1·5	–0·7	–0·2

SHELTER Good. FVs berth in the large outer basin.

NAVIGATION WPT 36°10'·81N 05°55'·38W, 270°/200m to abeam SW mole hd. From the W there is room to enter between the SW mole hd and the NCM lt buoy (see below) but do not wander S of the buoy, especially at night. From the SE keep 2M offshore to avoid tunny nets NW of Cabo Plata (36°06'·15N 05°49'·50W).

Near the WPT a NCM lt buoy marks the apex of a triangular tunny net, which extends 2·7M S and is marked by 2 SCM lt buoys at its base, an ECM lt buoy on its E side and a WCM lt buoy near its centre; see chartlet overleaf, Lights, buoys and waypoints and AC 142. A guard vessel, showing ●● (vert) and listening Ch 70, may be on station. Tunny nets, Mar-Sep, are a real hazard, especially at night. Nets may be laid E of Pta Camarinal and NW of Tarifa.

LIGHTS AND MARKS Barbate lt ho, Fl (2) WR 7s, W twr + R bands, is 4ca N of hbr ent, on edge of town. Other lights as chartlet and Lights, buoys and waypoints. Orange sodium lts in the hbr make navigational lts hard to discern.

COMMUNICATIONS (Code 956) MRCC ☎684740; Barbate is at W end of Gibraltar Strait VTS, monitor *Tarifa Traffic* VHF Ch 10 on passage; Brit Consul 902 109356; Marina VHF Ch 09.

FACILITIES

Marina ☎431907. barbated@eppa.es in the 3 inner basins (2·8m). A movable anti-pollution boom lies E of the marina ent; its SSE end is lit Fl (4) Y 20s 1M. 314 ⌷ inc ♥ in 3m, €1·93. 🔌 🚿 ✕ 🛢 🔩 🚻 🛢 ⚓ 🛒 🅑 ✕ Ice ⛴(45t). **Club Náutico** ☎433905.

Town is about 2M away, although 🚿 is closer; ✕ 🍴 ✉ ⓑ. Bus to Cádiz (61km). ✈ Jerez, Gibraltar or Malaga.

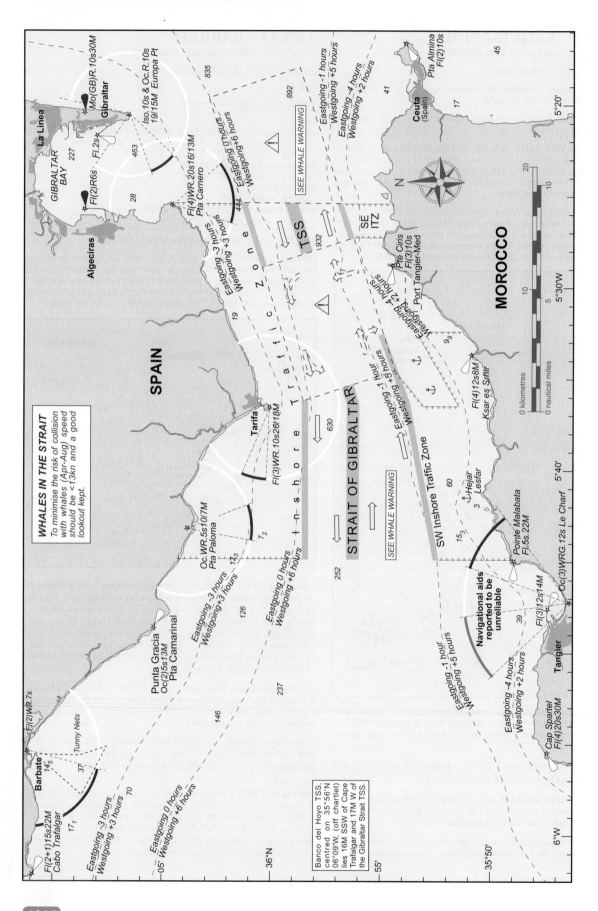

WHALES IN THE STRAIT
To minimise the risk of collision with whales (Apr-Aug) speed should be <13kn and a good lookout kept.

GIBRALTAR BAY

La Linea

Gibraltar

Mo(GB)R.10s30M

Iso.10s & Oc.R.10s 19/15M Europa Pt

Fl.2s

Fl(2)R6s

227

463

28

Fl(4)WR.20s16/13M Pta Carnero

444

Algeciras

835

892

Eastgoing 0 hours
Westgoing +6 hours

SEE WHALE WARNING

Eastgoing -1 hours
Westgoing +5 hours

Eastgoing -4 hours
Westgoing +2 hours

Pta Almina
Fl(2)10s

45

Ceuta (Spain)

17

41

N

Eastgoing -3 hours
Westgoing +3 hours

19

Traffic Zone

TSS

1932

SE ITZ

932

Pte Ciris
Fl(3)10s

Port Tangier-Med

MOROCCO

20

10

0 kilometres

0 nautical miles

10

5

5°30'W

5°20'

SPAIN

In-shore Traffic Zone

630

Tarifa

Fl(3)WR.10s26/18M

Oc.WR.5s10/7M
Pta Paloma

252

1₂

12₃

Eastgoing -4 hours
Westgoing +2 hours

9.9

Fl(4)12s8M
Ksar es Srhir

60

Hejar
Lestar

3

15₃

STRAIT OF GIBRALTAR

SEE WHALE WARNING

Eastgoing -1 hour
Westgoing +5 hours

SW Inshore Traffic Zone

Eastgoing -1 hour
Westgoing +5 hours

Pointe Malabata
Fl.5s.22M

Navigational aids
reported to be
unreliable

Fl(3)12s14M

Oc(3)WRG.12s Le Charf

Punta Gracia
Oc(2)5s13M
Pta Camarinal

Eastgoing -3 hours
Westgoing +3 hours

126

237

146

Eastgoing 0 hours
Westgoing +6 hours

Eastgoing -4 hours
Westgoing +2 hours

39

Tangier

Cap Spartel
Fl(4)20s30M

35°50'

5°40'

Barbate

Fl(2)WR.7s

Tunny Nets

37

14₅

17₁

Fl(2+1)15s22M
Cabo Trafalgar

70

Eastgoing -3 hours
Westgoing +3 hours

Eastgoing 0 hours
Westgoing +6 hours

36°N

05'

55'

6°W

Banco del Hoyo TSS, centred on 35°56'N 06°09'W, (off chartlet) lies 16M SSW of Cape Trafalgar and 17M W of the Gibraltar Strait TSS.

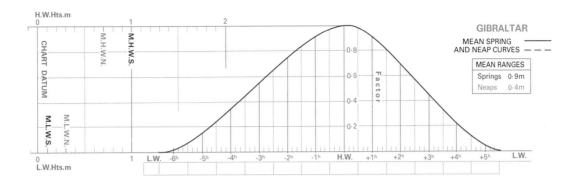

STRAIT OF GIBRALTAR

▶*Tidal streams:* Near the middle of the Strait the stream sets east from HW Gib and west from HWG –6. Closer inshore, streams start progressively earlier as shown on the chartlet opposite. To find when (approx) the tidal stream turns, apply the time intervals on the pecked lines to the time of HW Gib. Max rates, about 3kn in both directions, are usually found close inshore off headlands, decreasing offshore to < 1·5kn.◀

A surface current, as distinct from a tidal stream, always flows east in the centre of the Strait reaching about 2kn at the eastern end. Its speed decreases from the centre of the Strait towards the shore where the tidal stream influence increases.

Surface flow is the combined effect of tidal streams and the constant surface current into the Med. The resultant max surface flow is about 2kn W-going and 5kn E-going (further influenced by recent E or W winds). W-bound yachts should hug the Spanish coast.

Tidal races on the Spanish side form: off Cape Trafalgar, usually up to 2M SW, but 5-12M in heavy weather; near Bajo de Los Cabezos (2-4M S of Pta Paloma); and NW of **Tarifa**. The position and violence of these races depends on wind (especially over tide) springs/neaps and the state of the tide. For more information, see *W Coasts of Spain and Portugal* Pilot.

TSS is aligned 090°/270° and 073°/253°s in mid-strait. Yachts bound to/from **Gibraltar** usually use the northern ITZ. Traffic from/to the Moroccan port of Tangier-Med joins/leaves the TSS north of Punta Ciris.

Tarifa VTS, c/s *Tarifa Trafico* Ch **10**, 16, 67 gives radar surveillance on request and offers a reporting system (voluntary for yachts) between 05°58'W and 05°15'W. MSI is broadcast routinely and on request in Spanish and English Ch 67.

9.25.18 TARIFA

Cádiz **36°00'·35N 05°36'·26W** (Outer hbr) ⚙⚙⚓♨♨

CHARTS AC 142; SC 445, 445B, 4450; SHOM 1619, 7042

TIDES See Gibraltar for differences.

SHELTER A bolt-hole in bad conditions. In W'lies ⚓ off hbr ent in 5m. In a *levanter* ⚓ NW of the causeway as on the chartlet.

NAVIGATION WPT 36°00'·07N 05°35'·91W, 315°/4ca to hbr ent. S of Tarifa lt the ITZ is only 1·5M wide. A SCM lt buoy, 1M NNW of Tarifa lt, marks a tunny net (Mar-Jul) extending WNW to 2 WCM lt buoys thence to shore; care needed if approaching the NW ⚓.

LIGHTS AND MARKS See chartlet and Lts, buoys and waypoints.

COMMUNICATIONS HM 956 585400; no port VHF. See Passage information above for Tarifa VTS.

FACILITIES Only ⚓. No specific ⛴ for yachts; space on the dog-legged SE mole, clear of ferries, may be possible.

Inner basin: Ferries berth at the terminal in the NE corner and on S side. FVs on the S and W sides; *Guardia Civil* vessels and small naval craft use pens on N side.

Town ✕ ⌂ ☷ ✉ Ⓑ ⇌ ✈ Jerez, Gibraltar, Malaga.

9.25.19 ALGECIRAS

Cádiz **36°07'·12N 05°26'·16W** (Marina ent) ⚙⚙⚓♨♨♨

CHARTS AC 3578, 142, 1448, 1455; SC 445, 445A, 4451, 4452

TIDES

Standard Port GIBRALTAR (➡); **ML 0·66**; differences overleaf.

SHELTER Good in marina, except perhaps in strong SE'lies.

NAVIGATION WPT 36°06'·73N 05°24'·76W (Isolated Danger Mark, Fl (2) 7s, off chartlet) 295°/0·97M to first appr buoys. From S, keep 5ca off Pta de San García and beware drying reefs S of hbr ent. From N/E, beware big ships at ⚓, high-speed ferries.

LIGHTS AND MARKS The Hbr Control twr is conspic by day, close WNW of the marina.

COMMUNICATIONS (Code 956) MRCC 684740; Brit Consul 902 109356; HM 572620, VHF Ch 09, 12; Port Authority 585400; Marina VHF Ch 09 16.

FACILITIES **Marina/Real Club Náutico** ☎601402, 666000. Max LOA 16m. ♨ ✕ ⌂ ⊖ ♨♨ ⚓ ⌂ ▣(25t).

Town ✕ ⌂ ☷ ✉ Ⓑ ⇌ ✈ Jerez, Gibraltar, Malaga.

9.25.20 GIBRALTAR

36°08'·96N 05°22'·00W (100m off 'E' Hd) ✱✱✱✿✿✿✿✿✿✿

CHARTS AC 142, 3578, 1448, 144, 45 (Note: these charts do not accurately depict the layout of Marina Bay and Queensway Quay Marinas); SC 445, 445A, 4452.

TIDES Standard Port GIBRALTAR (⟶); ML 0·5

Times				Height (metres)			
High Water		Low Water		MHWS	MHWN	MLWN	MLWS
0000	0700	0100	0600	1·0	0·7	0·3	0·1
1200	1900	1300	1800	Note the very small range			
Differences TARIFA							
–0038	–0038	–0042	–0042	+0·3	+0·3	+0·2	+0·1
PUNTA CARNERO							
–0010	–0010	0000	0000	0·0	+0·1	+0·1	+0·1
ALGECIRAS							
–0010	–0010	–0010	–0010	+0·1	+0·2	+0·2	+0·2
SANDY BAY (E side of the Rock)							
–0011	–0011	–0016	–0016	–0·2	–0·1	0·0	0·0

SHELTER Good in both marinas (best to pre-book) but beware swell and fierce gusts in the E'ly *Levanter*. It is no longer permitted to ⚓ NW of the runway. ⚓ off La Linea in Spanish waters is feasible.

NAVIGATION From the SW, WPT 36°04'·62N 05°24'·34W (1M E of Pta Carnero) 024°/4·7M to 'E' Hd; 030°/4M to 'A' Hd. The Rock is steep-to all round; from the SW, beware shoals to S of Pta Carnero. From the E, clear Europa Pt by 3ca to avoid overfalls.

On arrival yachts with masts >10m high, intending to go to the yacht fuel berth (Waterport) or to Marina Bay Marina, must obtain instructions from *Gibraltar Port* Ch 12, 20 before rounding 'E' Head/North mole. VQ strobe lts and loudspeakers warn of aircraft activity; yachts must then stay out of a Restricted Entry area west of the runway, as shown on the chartlet.

All yachts now clear Customs in/out at their chosen marina.

LIGHTS AND MARKS The Rock (423m) is easily seen, but from the W not until rounding Pta Carnero, 5M to SW. The Aero lt atop the Rock is obsc'd when within 2M W of it. Europa Pt, almost at sea level, has 3 sectored lts; see Lights, buoys & waypoints. At N end of bay, industrial plants are conspic day/night. Shore lts may mask commercial vessels at ⚓ and navigational lights.

COMMUNICATIONS International code +350 (Area code 200) ⊖ 72901; Met 53416; Ambulance/Police 199; Fire 190; Ⓗ 79700; Port Captain (HM) 77254, VHF Ch 06; QHM 55901, VHF Ch 08; Gibraltar Bay VHF Ch 12, 71; Queensway Quay and Marina Bay Marinas: VHF Ch 71; ✈ 73026.

FACILITIES Queensway Quay Marina ☎ 44700, info@queenswayquay.com www.queenswayquay.com 160 inc❤ mostly F&A, and some ⌣; £2·00 summer. At its NW corner the marina entrance (least depth 3·5m) is closed by an anti-wash boom 2100-0830. 🅛 🗖 wi-fi, @ no fuel. Close to town centre.

Marina Bay ☎ 73300. pieroffice@marinabay.gi www.marinabay.gi. 209 ⌣ mostly F&A, in 1·8-3·7m, inc 29 ❤: £1·54 1 Mar-31 Oct; £0·93 1 Nov-28/29 Feb. Close to airport runway but only occas noise. 🛢🛠✕🛒🗖🅛🔌 ACA ⚠ wi-fi Ice.

Ocean Village Marina ☎ 73300 www.oceanvillage.gi some❤ but predominantly for superyachts. Better shelter than at Marina Bay. Phase 1 of this retail development is now open.

Sheppard's chandlery is near Marina Bay; workshops at Coaling Island; ☎ 75148. ⚓ yachtrep@gibraltar.gi 🛠🖂✕🅛 Ⓔ 🏳(6t). www.sheppard.gi admin@sheppard.gi

Yacht fuel is available at Waterport, 36°08'·90N 05°21'·37W from Mobil (Mon-Sat 0900-2300; Sun 0900-1400; also Shell, shorter hrs).

Royal Gibraltar YC ☎ 78897, 48847. www.rgyc.gi. info@rgyc.gi

Town ✕ 🏪 🛒 🖂 Ⓑ Ⓗ, ✈.

SPECIAL NOTES FOR GIBRALTAR

Standard time is UT –1, as in Spain. DST (UT –2) is kept from the last Sun in Mar until the Sat before the last Sun in Oct.

Telephone The international dial code is +350 (this includes calling from Spain) then 200, then the 5 digit tel No.

Customs Visiting RIBs must obtain prior written permission to enter from Collector of Customs, Customs House, Waterport, Gibraltar, ☎78879; VHF Ch 14 – or entry may be refused.

Local winds Levanters are E'lies, common with high pressure to the N and low to the S. A persistent Levanter produces the roughest seas in the Strait, severe squalls (violent in a F8) and eddies in the lee of the Rock. The Poniente is a frequent W'ly, slightly less common in summer. The Vendavale is a strong SW'ly associated with lows, with rain/drizzle near coasts. In light summer conditions, a strong daytime sea breeze (SSW) may affect Europa Point.

MARINA 1M N OF GIBRALTAR

LA LINEA, 36°09'·49N 05°21'·60W, Alcaidesa Marina ☎ (956) 021660 mob 690 339636; marina@alcaidesa.com www.alcaidesamarina.com ⚓ 624 inc 60❤ €1·55, (+ taxes). 🛢🛠🅛🗖(75t)ⵣ⚓🔌🖂✕. La Linea, town, 🏪✕🛒. Reports welcomed.

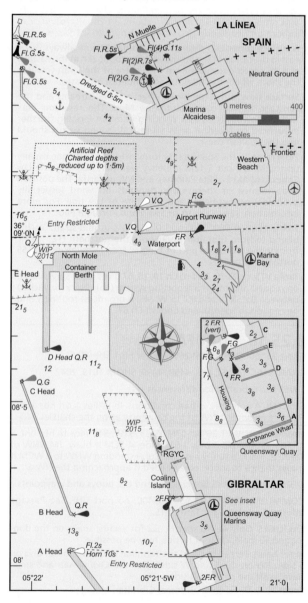

STANDARD TIME UT –01
Subtract 1 hour for UT
For Gibraltar Summer Time add
ONE hour in **non-shaded areas**

GIBRALTAR LAT 36°08'N LONG 5°21'W
TIMES AND HEIGHTS OF HIGH AND LOW WATERS

Dates in **red** are **SPRINGS**
Dates in **blue** are **NEAPS**

YEAR 2016

JANUARY

Time	m	Time	m
1 0051	0.2	**16** 0054	0.2
0752	0.8	0742	0.9
F 1331	0.3	SA 1332	0.2
2011	0.7	2014	0.8
2 0147	0.3	**17** 0159	0.2
0844	0.7	0845	0.8
SA 1436	0.3	SU 1446	0.2
◑ 2105	0.7	◐ 2121	0.7
3 0258	0.3	**18** 0319	0.2
0942	0.7	0956	0.8
SU 1548	0.3	M 1611	0.2
2209	0.7	2242	0.7
4 0418	0.3	**19** 0446	0.2
1047	0.7	1112	0.8
M 1659	0.3	TU 1733	0.2
2322	0.7		
5 0525	0.3	**20** 0000	0.7
1148	0.7	0556	0.2
TU 1756	0.2	W 1220	0.8
		1834	0.1
6 0021	0.7	**21** 0103	0.8
0614	0.3	0650	0.1
W 1238	0.8	TH 1316	0.9
1840	0.2	1923	0.1
7 0108	0.8	**22** 0155	0.8
0654	0.2	0736	0.1
TH 1322	0.8	F 1406	0.9
1919	0.1	2006	0.0
8 0150	0.8	**23** 0241	0.9
0731	0.2	0818	0.1
F 1405	0.9	SA 1452	0.9
1957	0.1	2046	0.0
9 0230	0.8	**24** 0324	0.9
0807	0.1	0858	0.0
SA 1447	0.9	SU 1536	0.9
2035	0.1	○ 2123	0.0
10 0310	0.9	**25** 0403	0.9
0846	0.1	0936	0.0
SU 1529	1.0	M 1616	0.9
● 2113	0.1	2158	0.0
11 0351	1.0	**26** 0440	0.9
0926	0.1	1013	0.0
M 1611	1.0	TU 1654	0.9
2153	0.0	2231	0.1
12 0432	1.0	**27** 0516	0.9
1007	0.1	1051	0.1
TU 1653	1.0	W 1731	0.8
2233	0.0	2304	0.1
13 0514	1.0	**28** 0551	0.8
1050	0.1	1124	0.1
W 1737	1.0	TH 1808	0.8
2314	0.1	2336	0.1
14 0558	1.0	**29** 0627	0.8
1137	0.1	1200	0.1
TH 1824	0.9	F 1845	0.8
15 0000	0.1	**30** 0010	0.2
0647	0.9	0705	0.7
F 1230	0.2	SA 1242	0.2
1915	0.9	1925	0.7
		31 0050	0.2
		0749	0.7
		SU 1331	0.2
		2012	0.7

FEBRUARY

Time	m	Time	m
1 0140	0.3	**16** 0247	0.2
0841	0.7	0932	0.8
M 1438	0.3	TU 1549	0.2
◑ 2107	0.6	2220	0.7
2 0257	0.3	**17** 0427	0.3
0943	0.7	1055	0.7
TU 1605	0.3	W 1727	0.3
2217	0.6	2348	0.7
3 0436	0.3	**18** 0551	0.2
1058	0.7	1212	0.8
W 1725	0.3	TH 1832	0.2
2337	0.6		
4 0546	0.3	**19** 0056	0.7
1205	0.7	0646	0.2
TH 1819	0.2	F 1312	0.8
		1918	0.1
5 0039	0.7	**20** 0148	0.8
0633	0.2	0730	0.1
F 1259	0.8	SA 1400	0.9
1902	0.1	1958	0.0
6 0128	0.8	**21** 0232	0.8
0713	0.2	0808	0.1
SA 1346	0.8	SU 1443	0.9
1941	0.1	2033	0.0
7 0212	0.8	**22** 0310	0.9
0753	0.1	0845	0.0
SU 1431	0.9	M 1522	0.9
2020	0.0	○ 2107	0.0
8 0254	0.9	**23** 0346	0.9
0833	0.1	0920	0.0
M 1515	1.0	TU 1558	0.9
● 2100	0.0	2139	0.0
9 0336	1.0	**24** 0419	0.9
0915	0.0	0953	0.0
TU 1558	1.0	W 1633	0.9
2140	0.0	2209	0.0
10 0418	1.0	**25** 0450	0.9
0957	0.0	1025	0.0
W 1642	1.0	TH 1706	0.9
2221	0.0	2239	0.0
11 0500	1.0	**26** 0520	0.8
1040	0.0	1057	0.1
TH 1725	1.0	F 1738	0.8
2302	0.1	2307	0.1
12 0544	1.0	**27** 0551	0.8
1124	0.0	1128	0.1
F 1811	0.9	SA 1811	0.8
2344	0.0	2337	0.1
13 0631	1.0	**28** 0624	0.8
1212	0.1	1201	0.2
SA 1901	0.9	SU 1847	0.7
14 0032	0.1	**29** 0008	0.2
0723	0.9	0703	0.7
SU 1307	0.1	M 1240	0.2
1956	0.8	1930	0.7
15 0131	0.2		
0822	0.8		
M 1416	0.2		
◑ 2100	0.7		

MARCH

Time	m	Time	m
1 0047	0.2	**16** 0222	0.3
0751	0.7	0912	0.7
TU 1333	0.3	W 1525	0.3
2022	0.6	2200	0.7
2 0145	0.3	**17** 0408	0.3
0851	0.6	1037	0.7
W 1500	0.3	TH 1712	0.3
◑ 2127	0.6	2329	0.7
3 0330	0.3	**18** 0539	0.2
1008	0.6	1200	0.7
TH 1647	0.3	F 1816	0.2
2249	0.6		
4 0513	0.3	**19** 0039	0.7
1131	0.7	0633	0.2
F 1753	0.2	SA 1300	0.8
		1900	0.2
5 0005	0.7	**20** 0130	0.8
0610	0.2	0714	0.1
SA 1235	0.7	SU 1346	0.8
1839	0.1	1936	0.1
6 0102	0.8	**21** 0211	0.8
0654	0.1	0750	0.1
SU 1326	0.8	M 1425	0.8
1920	0.1	2010	0.1
7 0149	0.9	**22** 0247	0.9
0736	0.1	0824	0.1
M 1412	0.9	TU 1501	0.9
2000	0.1	2042	0.1
8 0233	1.0	**23** 0320	0.9
0817	0.0	0857	0.0
TU 1457	1.0	W 1535	0.9
2041	0.0	○ 2113	0.1
9 0317	1.0	**24** 0351	0.9
0900	0.0	0929	0.0
W 1542	1.0	TH 1607	0.9
● 2123	-0.1	2143	0.1
10 0400	1.0	**25** 0421	0.9
0943	-0.1	1000	0.0
TH 1626	1.0	F 1639	0.9
2204	-0.1	2212	0.1
11 0443	1.1	**26** 0450	0.9
1025	-0.1	1030	0.1
F 1710	1.0	SA 1710	0.9
2245	0.0	2240	0.1
12 0527	1.0	**27** 0520	0.8
1108	0.0	1059	0.1
SA 1756	1.0	SU 1742	0.8
2327	0.0	2308	0.1
13 0614	1.0	**28** 0551	0.8
1153	0.1	1130	0.1
SU 1846	0.9	M 1817	0.8
		2339	0.2
14 0012	0.1	**29** 0628	0.7
0704	0.9	1204	0.2
M 1244	0.1	TU 1859	0.7
1940	0.8		
15 0107	0.2	**30** 0015	0.2
0803	0.8	0714	0.7
TU 1350	0.2	W 1251	0.2
◑ 2043	0.7	1950	0.7
		31 0107	0.3
		0814	0.7
		TH 1408	0.3
		◑ 2052	0.7

APRIL

Time	m	Time	m
1 0237	0.3	**16** 0508	0.3
0927	0.6	1132	0.7
F 1559	0.3	SA 1741	0.3
2208	0.7		
2 0430	0.3	**17** 0006	0.8
1054	0.7	0606	0.2
SA 1717	0.2	SU 1233	0.8
2328	0.7	1827	0.2
3 0541	0.2	**18** 0058	0.8
1206	0.7	0647	0.2
SU 1809	0.2	M 1319	0.8
		1904	0.2
4 0030	0.8	**19** 0139	0.8
0630	0.1	0723	0.1
M 1301	0.8	TU 1358	0.8
1853	0.1	1937	0.1
5 0121	0.9	**20** 0215	0.9
0714	0.1	0757	0.1
TU 1349	0.9	W 1433	0.9
1935	0.0	2010	0.1
6 0208	1.0	**21** 0248	0.9
0757	0.0	0830	0.1
W 1436	1.0	TH 1507	0.9
2017	0.0	2042	0.1
7 0253	1.0	**22** 0320	0.9
0841	0.0	0903	0.1
TH 1522	1.0	F 1540	0.9
● 2100	0.0	○ 2114	0.1
8 0338	1.1	**23** 0351	0.9
0925	-0.1	0935	0.1
F 1607	1.0	SA 1612	0.9
2143	0.0	2144	0.1
9 0423	1.1	**24** 0422	0.9
1008	-0.1	1006	0.1
SA 1653	1.0	SU 1643	0.9
2225	0.0	2214	0.1
10 0509	1.0	**25** 0453	0.9
1051	0.0	1036	0.1
SU 1740	1.0	M 1717	0.9
2308	0.0	2244	0.1
11 0556	1.0	**26** 0526	0.8
1135	0.1	1107	0.1
M 1830	0.9	TU 1753	0.8
2354	0.1	2317	0.2
12 0647	0.9	**27** 0604	0.8
1223	0.1	1142	0.2
TU 1925	0.8	W 1836	0.8
		2355	0.3
13 0047	0.2	**28** 0650	0.7
0745	0.8	1227	0.2
W 1325	0.2	TH 1927	0.7
2025	0.8		
14 0158	0.3	**29** 0047	0.3
0851	0.7	0746	0.7
TH 1451	0.3	F 1336	0.3
◑ 2136	0.7	2026	0.7
15 0334	0.3	**30** 0206	0.3
1010	0.7	0856	0.7
F 1632	0.3	SA 1510	0.3
2256	0.7	◑ 2134	0.7

Chart Datum is 0·25 metres below Alicante Datum (ie Mean Sea Level, Alicante). HAT is 1·2 metres above Chart Datum.

FREE monthly updates. Register at
www.reedsnauticalalmanac.co.uk

1017

SW Spain & Gibraltar

STANDARD TIME UT –01
Subtract 1 hour for UT
For Gibraltar Summer Time add
ONE hour in **non-shaded areas**

GIBRALTAR LAT 36°08'N LONG 5°21'W

TIMES AND HEIGHTS OF HIGH AND LOW WATERS

Dates in red are **SPRINGS**
Dates in blue are **NEAPS**

YEAR **2016**

MAY

Time	m		Time	m
1 0342	0.3	**16**	0521	0.3
1016	0.7		1152	0.7
SU 1632	0.2	M	1741	0.3
2250	0.8			
2 0502	0.2	**17**	0014	0.8
1133	0.7		0610	0.2
M 1733	0.2	TU	1243	0.8
2356	0.8		1824	0.2
3 0601	0.1	**18**	0059	0.8
1233	0.8		0650	0.2
TU 1823	0.1	W	1324	0.8
			1902	0.2
4 0050	0.9	**19**	0137	0.8
0650	0.1		0727	0.2
W 1324	0.9	TH	1401	0.8
1908	0.1		1937	0.2
5 0140	1.0	**20**	0213	0.9
0736	0.0		0802	0.1
TH 1412	1.0	F	1437	0.9
1952	0.0		2011	0.2
6 0228	1.0	**21**	0247	0.9
0821	0.0		0837	0.1
F 1500	1.0	SA	1511	0.9
● 2037	0.0	○	2045	0.2
7 0315	1.0	**22**	0321	0.9
0907	0.0		0910	0.1
SA 1547	1.0	SU	1545	0.9
2122	0.0		2118	0.2
8 0403	1.0	**23**	0356	0.9
0951	0.0		0943	0.1
SU 1635	1.0	M	1620	0.9
2206	0.0		2150	0.2
9 0450	1.0	**24**	0430	0.9
1034	0.0		1015	0.1
M 1723	1.0	TU	1655	0.9
2250	0.1		2224	0.2
10 0538	1.0	**25**	0506	0.9
1118	0.1		1049	0.2
TU 1813	0.9	W	1733	0.9
2336	0.1		2300	0.2
11 0629	0.9	**26**	0546	0.8
1204	0.1		1126	0.2
W 1907	0.9	TH	1816	0.8
			2342	0.2
12 0028	0.2	**27**	0631	0.8
0725	0.8		1211	0.2
TH 1300	0.2	F	1905	0.8
2004	0.8			
13 0131	0.3	**28**	0033	0.3
0826	0.8		0725	0.8
F 1410	0.3	SA	1311	0.2
◑ 2106	0.8		2001	0.8
14 0248	0.3	**29**	0141	0.3
0933	0.7		0829	0.7
SA 1530	0.3	SU	1427	0.3
2212	0.8	◑	2104	0.8
15 0411	0.3	**30**	0301	0.3
1046	0.7		0941	0.7
SU 1645	0.3	M	1545	0.3
2319	0.8		2213	0.8
		31	0420	0.2
			1058	0.8
		TU	1654	0.2
			2322	0.8

JUNE

Time	m		Time	m
1 0530	0.2	**16**	0013	0.8
1204	0.8		0615	0.2
W 1753	0.1	TH	1247	0.8
			1826	0.3
2 0021	0.9	**17**	0058	0.8
0627	0.1		0657	0.2
TH 1300	0.9	F	1329	0.8
1843	0.1		1906	0.2
3 0115	1.0	**18**	0137	0.8
0717	0.0		0735	0.2
F 1351	1.0	SA	1407	0.8
1931	0.1		1942	0.2
4 0205	1.0	**19**	0216	0.9
0805	0.0		0811	0.1
SA 1441	1.0	SU	1444	0.9
2018	0.0		2018	0.2
5 0255	1.0	**20**	0254	0.9
0851	0.0		0846	0.1
SU 1530	1.0	M	1521	0.9
● 2104	0.0	○	2054	0.2
6 0344	1.0	**21**	0332	0.9
0936	0.0		0921	0.1
M 1618	1.0	TU	1558	0.9
2150	0.1		2130	0.2
7 0432	1.0	**22**	0411	0.9
1019	0.0		0956	0.1
TU 1705	1.0	W	1635	0.9
2234	0.1		2207	0.2
8 0520	0.9	**23**	0450	0.9
1101	0.1		1032	0.1
W 1754	0.9	TH	1714	0.9
2319	0.1		2246	0.2
9 0609	0.9	**24**	0530	0.9
1144	0.1		1111	0.1
TH 1843	0.9	F	1757	0.9
			2329	0.2
10 0006	0.2	**25**	0615	0.9
0700	0.8		1155	0.2
F 1231	0.2	SA	1844	0.9
1935	0.8			
11 0059	0.2	**26**	0018	0.2
0754	0.8		0706	0.8
SA 1325	0.2	SU	1247	0.2
2028	0.8		1936	0.9
12 0159	0.2	**27**	0118	0.2
0851	0.7		0805	0.8
SU 1427	0.3	M	1351	0.2
◑ 2123	0.8	◑	2035	0.9
13 0307	0.3	**28**	0227	0.2
0952	0.7		0911	0.8
M 1536	0.3	TU	1503	0.2
2222	0.8		2140	0.8
14 0419	0.3	**29**	0344	0.2
1058	0.7		1026	0.8
TU 1644	0.3	W	1619	0.2
2322	0.8		2250	0.8
15 0525	0.3	**30**	0504	0.3
1158	0.7		1139	0.8
W 1741	0.3	TH	1729	0.2
			2356	0.9

JULY

Time	m		Time	m
1 0611	0.1	**16**	0017	0.8
1241	0.8		0628	0.2
F 1827	0.1	SA	1256	0.8
			1838	0.3
2 0054	0.9	**17**	0105	0.8
0705	0.1		0709	0.2
SA 1335	0.9	SU	1338	0.8
1917	0.1		1917	0.2
3 0148	0.9	**18**	0148	0.9
0753	0.0		0746	0.2
SU 1426	1.0	M	1418	0.9
2005	0.1		1955	0.2
4 0239	1.0	**19**	0230	0.9
0839	0.0		0822	0.1
M 1515	1.0	TU	1457	0.9
● 2051	0.1	○	2032	0.2
5 0328	1.0	**20**	0311	0.9
0922	0.0		0859	0.1
TU 1602	1.0	W	1536	1.0
2135	0.1		2111	0.1
6 0415	1.0	**21**	0352	1.0
1002	0.0		0936	0.1
W 1647	1.0	TH	1615	1.0
2217	0.1		2151	0.1
7 0500	0.9	**22**	0433	1.0
1040	0.1		1014	0.1
TH 1730	0.9	F	1655	1.0
2258	0.1		2231	0.1
8 0545	0.9	**23**	0514	1.0
1118	0.1		1053	0.1
F 1814	0.9	SA	1737	1.0
2339	0.1		2314	0.1
9 0630	0.8	**24**	0559	1.0
1157	0.2		1135	0.1
SA 1858	0.9	SU	1822	1.0
10 0022	0.2	**25**	0000	0.2
0716	0.8		0647	0.9
SU 1239	0.2	M	1223	0.2
1943	0.8		1911	0.9
11 0110	0.2	**26**	0054	0.2
0804	0.7		0743	0.9
M 1328	0.3	TU	1320	0.2
2030	0.8		2008	0.9
12 0206	0.3	**27**	0157	0.2
0856	0.7		0846	0.8
TU 1427	0.3	W	1430	0.3
◑ 2121	0.7	◑	2111	0.9
13 0311	0.3	**28**	0315	0.3
0955	0.7		1000	0.8
W 1538	0.3	TH	1552	0.3
2219	0.7		2224	0.8
14 0429	0.3	**29**	0447	0.2
1103	0.7		1121	0.8
TH 1653	0.3	F	1714	0.3
2321	0.7		2338	0.8
15 0538	0.3	**30**	0603	0.2
1205	0.7		1230	0.8
F 1753	0.3	SA	1818	0.2
		31	0043	0.9
			0658	0.1
		SU	1327	0.9
			1909	0.2

AUGUST

Time	m		Time	m
1 0138	0.9	**16**	0123	0.9
0744	0.1		0720	0.2
M 1416	0.9	TU	1352	0.9
1955	0.1		1932	0.2
2 0227	0.9	**17**	0207	0.9
0825	0.1		0757	0.1
TU 1502	1.0	W	1433	1.0
● 2037	0.1		2011	0.1
3 0314	1.0	**18**	0250	1.0
0904	0.1		0835	0.1
W 1545	1.0	TH	1513	1.0
2118	0.1	○	2051	0.1
4 0357	1.0	**19**	0332	1.0
0941	0.1		0913	0.1
TH 1624	1.0	F	1554	1.1
2156	0.1		2132	0.1
5 0437	0.9	**20**	0414	1.1
1015	0.1		0953	0.1
F 1702	1.0	SA	1635	1.1
2233	0.1		2214	0.1
6 0516	0.9	**21**	0456	1.1
1048	0.1		1033	0.1
SA 1739	0.9	SU	1716	1.1
2309	0.1		2256	0.1
7 0554	0.9	**22**	0541	1.0
1121	0.1		1114	0.1
SU 1815	0.9	M	1800	1.1
2345	0.2		2340	0.1
8 0633	0.8	**23**	0629	1.0
1155	0.2		1159	0.2
M 1852	0.8	TU	1849	1.0
9 0023	0.2	**24**	0029	0.2
0714	0.8		0722	0.9
TU 1234	0.3	W	1253	0.2
1932	0.8		1943	0.9
10 0108	0.3	**25**	0130	0.3
0801	0.7		0825	0.8
W 1321	0.3	TH	1402	0.3
◐ 2019	0.8	◐	2047	0.9
11 0205	0.3	**26**	0251	0.3
0855	0.7		0940	0.8
TH 1428	0.4	F	1533	0.3
2114	0.7		2203	0.8
12 0325	0.4	**27**	0437	0.3
1001	0.7		1107	0.8
F 1558	0.4	SA	1706	0.3
2222	0.7		2328	0.8
13 0456	0.3	**28**	0557	0.3
1118	0.7		1222	0.8
SA 1718	0.4	SU	1811	0.3
2336	0.8			
14 0559	0.3	**29**	0037	0.9
1221	0.8		0648	0.2
SU 1812	0.3	M	1318	0.9
			1859	0.2
15 0034	0.8	**30**	0130	0.9
0642	0.3		0730	0.2
M 1309	0.8	TU	1403	1.0
1853	0.3		1941	0.2
		31	0215	0.9
			0806	0.1
		W	1444	1.0
			2019	0.1

Chart Datum is 0·25 metres below Alicante Datum (ie Mean Sea Level, Alicante). HAT is 1·2 metres above Chart Datum.

STANDARD TIME UT –01
Subtract 1 hour for UT
For Gibraltar Summer Time add ONE hour in **non-shaded areas**

GIBRALTAR LAT 36°08'N LONG 5°21'W
TIMES AND HEIGHTS OF HIGH AND LOW WATERS

Dates in **red** are **SPRINGS**
Dates in blue are NEAPS

YEAR 2016

SEPTEMBER

Time	m		Time	m
1 0256	1.0	**16** 0227	1.0
0841	0.1	0810	0.1
TH 1522	1.0	F 1448	1.1
● 2055	0.1	○ 2029	0.1
2 0334	1.0	**17** 0310	1.1
0914	0.1	0849	0.1
F 1557	1.0	SA 1530	1.2
2130	0.1	2111	0.0
3 0410	1.0	**18** 0353	1.1
0945	0.1	0930	0.1
SA 1630	1.0	SU 1613	1.2
2204	0.1	2153	0.0
4 0443	1.0	**19** 0437	1.1
1016	0.1	1011	0.1
SU 1701	1.0	M 1655	1.2
2237	0.1	2235	0.1
5 0516	0.9	**20** 0521	1.1
1046	0.2	1053	0.1
M 1732	0.9	TU 1740	1.1
2308	0.2	2319	0.1
6 0549	0.9	**21** 0609	1.0
1117	0.2	1137	0.2
TU 1803	0.9	W 1828	1.0
2341	0.2		
7 0625	0.8	**22** 0006	0.2
1149	0.3	0702	1.0
W 1839	0.8	TH 1230	0.3
		1922	0.9
8 0018	0.3	**23** 0104	0.3
0708	0.8	0806	0.9
TH 1229	0.3	F 1339	0.4
1922	0.8	◗ 2027	0.9
9 0106	0.3	**24** 0229	0.4
0802	0.7	0921	0.8
F 1325	0.4	SA 1515	0.4
◑ 2018	0.7	2147	0.8
10 0223	0.4	**25** 0422	0.4
0907	0.7	1051	0.8
SA 1501	0.4	SU 1653	0.4
2128	0.7	2319	0.8
11 0413	0.4	**26** 0542	0.3
1026	0.7	1207	0.9
SU 1642	0.4	M 1757	0.3
2253	0.8		
12 0527	0.3	**27** 0028	0.9
1142	0.8	0630	0.3
M 1743	0.3	TU 1301	0.9
		1842	0.3
13 0005	0.8	**28** 0117	0.9
0615	0.3	0708	0.2
TU 1238	0.9	W 1343	1.0
1828	0.3	1919	0.2
14 0058	0.9	**29** 0157	1.0
0654	0.2	0741	0.2
W 1324	1.0	TH 1420	1.0
1908	0.2	1955	0.2
15 0144	1.0	**30** 0234	1.0
0731	0.1	0813	0.2
TH 1406	1.0	F 1454	1.0
1948	0.1	2029	0.1

OCTOBER

Time	m		Time	m
1 0307	1.0	**16** 0248	1.1
0844	0.1	0825	0.1
SA 1526	1.0	SU 1506	1.2
● 2102	0.1	○ 2049	0.0
2 0340	1.0	**17** 0332	1.1
0915	0.1	0907	0.1
SU 1556	1.0	M 1551	1.2
2134	0.1	2132	0.0
3 0411	1.0	**18** 0417	1.1
0945	0.2	0949	0.1
M 1626	1.0	TU 1635	1.2
2206	0.1	2215	0.1
4 0441	1.0	**19** 0502	1.1
1015	0.2	1033	0.1
TU 1655	1.0	W 1721	1.1
2236	0.2	2258	0.1
5 0512	0.9	**20** 0550	1.0
1044	0.2	1116	0.2
W 1725	0.9	TH 1810	1.0
2307	0.2	2345	0.2
6 0546	0.9	**21** 0644	1.0
1116	0.3	1210	0.3
TH 1759	0.9	F 1905	0.9
2340	0.3		
7 0627	0.8	**22** 0041	0.3
1152	0.3	0746	0.9
F 1842	0.8	SA 1319	0.4
		◗ 2009	0.9
8 0021	0.3	**23** 0202	0.4
0720	0.8	0859	0.8
SA 1242	0.4	SU 1451	0.4
1937	0.8	2126	0.8
9 0128	0.4	**24** 0348	0.4
0824	0.8	1023	0.8
SU 1410	0.4	M 1624	0.4
◑ 2047	0.8	2255	0.8
10 0325	0.4	**25** 0510	0.4
0939	0.8	1138	0.9
M 1559	0.4	TU 1729	0.3
2211	0.8		
11 0451	0.4	**26** 0005	0.8
1100	0.4	0600	0.3
TU 1710	0.3	W 1232	0.9
2332	0.8	1815	0.3
12 0544	0.3	**27** 0055	0.9
1203	0.9	0639	0.3
W 1800	0.3	TH 1314	0.9
		1853	0.2
13 0031	0.9	**28** 0133	0.9
0626	0.2	0712	0.2
TH 1253	1.0	F 1350	1.0
1842	0.2	1927	0.2
14 0119	1.0	**29** 0208	1.0
0705	0.1	0744	0.2
F 1338	1.1	SA 1423	1.0
1924	0.1	2001	0.2
15 0203	1.1	**30** 0240	1.0
0744	0.1	0816	0.2
SA 1422	1.1	SU 1455	1.0
2006	0.1	● 2035	0.1
		31 0311	1.0
		0847	0.2
		M 1526	1.0
		2108	0.1

NOVEMBER

Time	m		Time	m
1 0342	1.0	**16** 0359	1.1
0918	0.2	0932	0.1
TU 1556	1.0	W 1617	1.1
2139	0.2	2159	0.1
2 0413	1.0	**17** 0446	1.1
0949	0.2	1017	0.1
W 1627	1.0	TH 1705	1.1
2210	0.2	2243	0.1
3 0445	1.0	**18** 0534	1.0
1020	0.2	1104	0.2
TH 1659	0.9	F 1754	1.0
2241	0.2	2329	0.2
4 0520	0.9	**19** 0627	1.0
1052	0.3	1155	0.3
F 1735	0.9	SA 1848	0.9
2314	0.3		
5 0600	0.9	**20** 0021	0.3
1128	0.3	0725	0.9
SA 1817	0.8	SU 1258	0.3
2353	0.3	1948	0.8
6 0650	0.8	**21** 0129	0.3
1216	0.4	0831	0.8
SU 1910	0.8	M 1416	0.4
		◗ 2055	0.8
7 0051	0.3	**22** 0254	0.4
0750	0.8	0942	0.8
M 1332	0.4	TU 1538	0.4
◗ 2014	0.8	2211	0.8
8 0230	0.4	**23** 0417	0.4
0859	0.8	1054	0.8
TU 1511	0.4	W 1648	0.3
2130	0.8	2326	0.8
9 0404	0.4	**24** 0520	0.3
1015	0.8	1154	0.9
W 1630	0.3	TH 1742	0.3
2253	0.8		
10 0507	0.3	**25** 0022	0.8
1126	0.9	0605	0.3
TH 1728	0.2	F 1240	0.9
		1824	0.2
11 0000	0.9	**26** 0105	0.9
0556	0.2	0643	0.3
F 1222	1.0	SA 1319	0.9
1816	0.2	1902	0.2
12 0053	1.0	**27** 0142	0.9
0639	0.1	0718	0.2
SA 1311	1.0	SU 1354	0.9
1901	0.1	1938	0.2
13 0141	1.0	**28** 0216	0.9
0721	0.1	0752	0.2
SU 1357	1.1	M 1428	1.0
1945	0.0	2013	0.2
14 0227	1.1	**29** 0248	0.9
0804	0.1	0824	0.2
M 1444	1.1	TU 1501	1.0
○ 2030	0.0	● 2047	0.1
15 0313	1.1	**30** 0320	1.0
0848	0.1	0857	0.2
TU 1531	1.1	W 1535	1.0
2115	0.0	2120	0.1

DECEMBER

Time	m		Time	m
1 0353	1.0	**16** 0434	1.0
0929	0.2	1007	0.1
TH 1609	0.9	F 1652	1.0
2152	0.2	2233	0.1
2 0427	0.9	**17** 0521	1.0
1002	0.2	1053	0.1
F 1643	0.9	SA 1740	0.9
2225	0.2	2316	0.1
3 0503	0.9	**18** 0610	0.9
1036	0.2	1140	0.2
SA 1720	0.9	SU 1830	0.9
2259	0.2		
4 0542	0.9	**19** 0002	0.2
1114	0.3	0701	0.9
SU 1801	0.9	M 1234	0.2
2338	0.3	1923	0.8
5 0628	0.9	**20** 0054	0.2
1200	0.3	0757	0.8
M 1849	0.8	TU 1335	0.3
		2019	0.8
6 0028	0.3	**21** 0157	0.3
0722	0.8	0857	0.8
TU 1303	0.3	W 1443	0.3
1947	0.8	◗ 2120	0.7
7 0140	0.3	**22** 0310	0.3
0825	0.8	1000	0.8
W 1425	0.3	TH 1555	0.3
◗ 2054	0.8	2229	0.7
8 0307	0.3	**23** 0426	0.3
0934	0.8	1105	0.8
TH 1545	0.3	F 1702	0.3
2212	0.8	2339	0.7
9 0424	0.3	**24** 0529	0.3
1047	0.9	1201	0.8
F 1655	0.2	SA 1757	0.3
2327	0.2		
10 0526	0.2	**25** 0033	0.8
1151	0.9	0616	0.3
SA 1754	0.2	SU 1248	0.8
		1840	0.2
11 0028	0.9	**26** 0117	0.8
0617	0.1	0656	0.2
SU 1247	1.0	M 1328	0.8
1844	0.1	1919	0.2
12 0121	0.9	**27** 0155	0.8
0704	0.1	0732	0.2
M 1337	1.0	TU 1406	0.9
1932	0.1	1956	0.1
13 0210	1.0	**28** 0230	0.9
0750	0.1	0807	0.2
TU 1426	1.0	W 1443	0.9
2018	0.0	2031	0.1
14 0259	1.0	**29** 0305	0.9
0835	0.0	0841	0.2
W 1515	1.1	TH 1519	0.9
○ 2104	0.0	● 2106	0.1
15 0347	1.0	**30** 0339	0.9
0921	0.1	0915	0.2
TH 1604	1.0	F 1555	0.9
2149	0.0	2140	0.1
		31 0414	0.9
		0949	0.2
		SA 1632	0.9
		2214	0.1

Chart Datum is 0·25 metres below Alicante Datum (ie Mean Sea Level, Alicante). HAT is 1·2 metres above Chart Datum.

》》 FREE monthly updates. Register at 《
www.reedsnauticalalmanac.co.uk

SW Spain & Gibraltar

MOROCCO

From Cap Spartel eastward for some 35M to Punta Almina the coast is generally rugged with mountains rising progressively higher to the east. Jbel Musa (one of the ancient Pillars of Hercules) towers 848m/2782ft high, only 8ca from the coast.

If using the SW and/or SE ITZs, be aware of the large commercial port of Tanger-Med which extends 2·7M SW of Pta Ciris and connects into the Gibraltar TSS.

The Spanish enclave of **Ceuta** (AC 2742) and, 10M further south, Marina Smir are popular day trips from Gib. Be aware of frequent high-speed ferries crossing the Strait and warships.

9.25.21 TANGIER

Morocco 35°47'·42N 05°47'·45W ✻✻✻◊✿✿

CHARTS AC 142, 1912; SC 445, 446, 4461

TIDES Standard Port GIBRALTAR (→); ML 0·5

Times				Height (metres)			
High Water		Low Water		MHWS	MHWN	MLWN	MLWS
0000	0700	0100	0600	1·0	0·7	0·3	0·1
1200	1900	1300	1800	Note the very small tidal range.			
Differences TANGIER*		*Moroccan Zone Time is 0 (UT)					
–0030	–0030	–0020	–0020	+1·3	+1·0	+0·5	+0·3
HEJAR LESFAR*							
–0035	–0035	–0005	–0005	+0·8	+0·6	+0·4	+0·2
PUNTA CIRIS*							
–0110	–0110	–0110	–0110	+0·2	+0·2	+0·2	+0·1
CEUTA*							
–0045	–0045	–0050	–0050	0·0	+0·1	+0·1	+0·1

SHELTER Good, except in N to E winds to which the bay and hbr are exposed. In E winds ⚓ in 6m 4ca SSW of Pte Malabata lt ho in the lee of high ground.

NAVIGATION WPT 35°48'·33N 05°46'·94W, 205°/1M to N mole hd, within the white sector of El Charf Dir lt, passing W of an IDM lt buoy. The approach is straightforward, but from W, beware WIP 2014, harbour under construction.

9.25.22 CEUTA

Spanish enclave 35°53'·75N 05°18'·62W ✻✻✻◊◊✿✿✿

CHARTS AC 3578, 142, 2742; SC 44C, 445, 451, 4511

TIDES Standard Port GIBRALTAR (→); ML 0·5; see Tangier.

SHELTER Good in marina. The harbour ent is exposed to the *levanter* and NE winds. Fresh SE'lies raise heavy seas in the Bay of Ceuta. In strong W'lies heavy squalls blow down from the mountains. No ⚓ to N and NE of the E pier. Ensenada de la

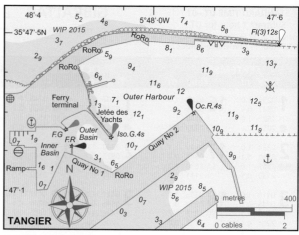

TANGIER

LIGHTS AND MARKS Caution: Locally navigational aids are reported to be unreliable, missing, unlit or out of position.

Brg/dist of lights and conspic landmarks from the N mole hd (whose 16m high W lt twr is conspic) are: Cap Spartel, 270°/6·4M. Conspic W water twr, 265°/1·6M. El Charf (Monte Dirección) 172°/1·5M, a 100m high conical hill with bldg/minaret; below the summit is a Dir lt whose WRG sectors cover the bay. IDM lt buoy, 067°/0·5M. Pte Malabata lt ho 055°/2·6M. Other lights as chartlet and Lights, buoys and waypoints.

Hbr signals may be flown from the HM's office: R flag = Hbr open; Y flag = Hbr closed; Blue flag = Bad weather expected.

COMMUNICATIONS (Code 00 2129) Port 3993 7495. Arabic, French and Spanish spoken, call Ch 16 for a working channel.

FACILITIES Yachts enter the inner hbr and berth in its innermost basin (0·7-3·4m) with FVs and other small craft; completion of a marina is imminent, and a purpose built fishing hbr in 2016. Commercial ships and ferries to Gibraltar and Spain have now been relocated to Tangier Med commercial port. Bureaucratic procedures may detract from Tangier's other merits.

Almadraba (S of the marina isthmus) is a sheltered ⚓ in W or N winds. Beware tunny nets (Jun-Dec) south of Ceuta.

NAVIGATION WPT 35°54'·75N 05°18'·62W, 180°/1·0M to W side of hbr ent. From there make good 222° to the marina ent. From the E give a wide berth to the N side of the Almina Peninsula from which islets and Bajo Isabel (3·6m) extend about 4ca N. Beware high-speed ferries at the hbr ent.

LIGHTS AND MARKS Jbel Musa 848m (one of the ancient Pillars of Hercules) is 4·7M W of the hbr. Monte Hacho fort 204m is 1·1M E of the hbr, with Punta Almina lt ho a further 0.5M east. At the N end of Muelle Espana the Port Control twr is conspic by day; leave it to stbd for the marina ent.

COMMUNICATIONS (Code 956, as if on the Spanish mainland); Port VHF Ch 09, **12**; *Banda Marina* VHF Ch 09; Tourist info 506275.

FACILITIES **Marina Hércules** ☎525001 (0800-1330 and 1700-2000) www.mahersa. es info@mahersa.es 300 ⚓ max LOA 20m. Expect to berth on fingers at pontoon G (furthest east). Depths vary from 4m to 2m. Fees: €0.60/m²/day not inc 🔌 & ⚓🚿🛢️🛒 🚹⊖🚻⚓🪝🛢️🔧. Heliport – Malaga & Algeciras.

Real Club Náutico ✕ 🏠.

Town ✕ 🏠🚆 Good local market, ✉ Ⓑ ⇌. Frequent ferries to Algeciras. ✈ Tangier, Gibraltar, Malaga.

CEUTA

Portuguese Islands

Flores, Faial, Pico, São Jorge, Graciosa, Terceira, Santa Maria and São Miguel; Porto Santo and Madeira

9.26.1	Contents	1021
9.26.2	Area map and Distance table	1022
9.26.3	Special notes for the Portuguese Islands	1023

THE AZORES ARCHIPELAGO

9.26.4	Lights, buoys and waypoints	1024
9.26.5	Porto das Lajes (Flores)	1026
9.26.6	Horta (Faial)	1026
9.26.7	Madalena (Pico)	1027
9.26.8	Cais (São Roque) do Pico (Pico)	1028
9.26.9	Velas (São Jorge)	1028
9.26.10	Calheta (São Jorge)	1028
9.26.11	Vila da Praia (Graciosa)	1029
9.26.12	Angra do Heroísmo (Terceira)	1029
9.26.13	Praia da Vitória (Terceira)	1030
9.26.14	Vila do Porto (Santa Maria)	1030
9.26.15	Ponta Delgada (São Miguel)	1031
9.26.16	Vila Franca do Campo (São Miguel)	1031
	Ponta Delgada (Standard port, tide tables)	1032

THE MADEIRA ARCHIPELAGO

9.26.17	Lights, buoys and waypoints	1035
9.26.18	Porto de Porto Santo	1036
9.26.19	Quinta do Lorde (Madeira)	1036
	Baía de Abra • Caniçal • Machico	
9.26.20	Funchal (Madeira)	1037
9.26.21	Calheta (Madeira)	1037
	Madalena do Mar • Porto Moniz •	
	Casablanca (Standard port, tide tables/curves)	1038

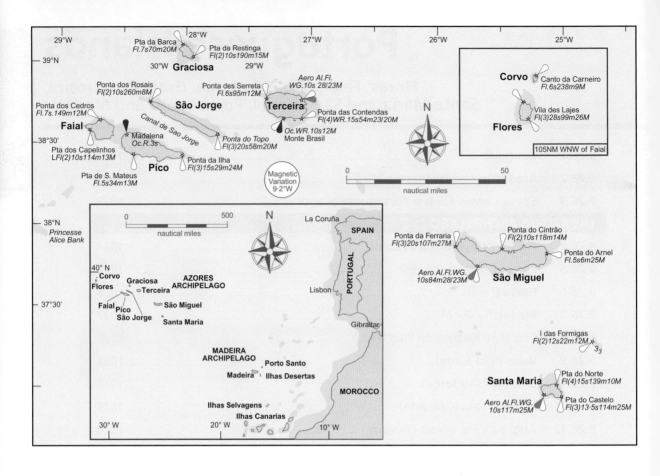

PORTUGUESE ISLANDS DISTANCE TABLE

Approximate rhumb line distances in nautical miles between selected harbours/anchorages in the Madeira and Azores archipelagos. Distances shown are by the most direct route, considering likely dangers and typical currents/weather.

	1	2	3	4	5	6	7	8	9	10	11	12	13	14	15	16	17
1 Porto das Lajes (Flores)	1																
2 Horta (Faial)	131	2															
3 Madalena (Pico)	135	5	3														
4 Cais/São Roque (Pico)	143	16	12	4													
5 Velas (São Jorge)	144	22	15	11	5												
6 Vila da Praia (Graciosa)	152	48	45	42	37	6											
7 Angra de Heroismo (Terceira)	190	69	65	54	52	44	7										
8 Praia do Vitoria (Terceira)	199	81	78	66	63	49	14	8									
9 Ponta Delgada (São Miguel)	277	150	147	135	135	138	93	91	9								
10 Vila do Porto (Santa Maria)	320	192	191	179	179	185	143	142	54	10							
11 Porto (Porto Santo)	817	691	691	677	676	679	636	634	544	500	11						
12 Quinta do Lorde (Madeira)	817	689	689	672	673	674	631	629	540	494	31	12					
13 Funchal (Madeira)	828	699	699	683	684	663	620	618	529	483	41	12	13				
14 Calheta (Madeira)	787	659	659	646	645	648	605	602	514	467	56	27	16	14			
15 Ilha Deserta Grande											38	18	22	36	15		
16 Ilha Selvagem Grande											179	165	162	172	148	16	
17 Ilha Selvagem Pequena											183	168	165	173	152	10	17

Ilhas Selvagens and Ilhas Desertas may not be visited without a permit, only obtainable *in advance* from Madeira/Porto Santo.

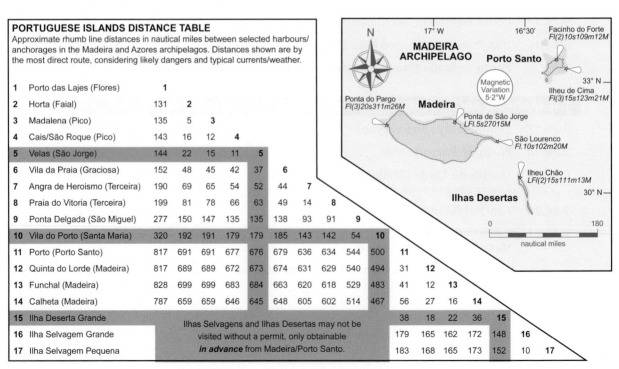

OCEANIC DISTANCE TABLE
Approximate distances in nautical miles are by the most direct route while avoiding dangers and allowing for TSS and common shipping concentrations. Oceanic legs over 60 miles have been calculated as great circles

		1	2	3	4	5	6	7	8	9	10	11	12	13	14	15	16	17
1	English Harbour (Antigua)	1																
2	Road Town (Tortola)	186	2															
3	St Georges (Bermuda)	945	847	3														
4	Cork	3280	3313	2697	4													
5	Falmouth	3368	3409	2819	189	5												
6	Brest	3346	3404	2835	286	136	6											
7	La Coruña	3132	3197	2695	504	440	350	7										
8	Lisboa	3026	3115	2700	836	780	658	327	8									
9	Vilamoura	3065	3178	2771	969	928	853	464	151	9								
10	Porto (Porto Santo)	2630	2738	2423	1180	1144	1068	751	492	472	10							
11	Funchal (Madeira)	2595	2704	2397	1217	1186	1110	793	528	511	41	11						
12	Europa Point (Gibraltar)	3178	3291	2910	1114	1087	960	618	294	155	577	617	12					
13	Casablanca	3061	3174	2835	1135	1114	978	637	313	208	438	472	188	13				
14	Ponta Delgada (São Miguel)	2270	2349	1930	1133	1183	1145	862	771	839	544	529	984	918	14			
15	Horta (Faial)	2165	2235	1791	1169	1235	1202	966	904	979	691	699	1125	1065	150	15		
16	Las Palmas (Gran Canaria)	2662	2785	2552	1458	1410	1334	1016	705	651	301	283	705	527	775	911	16	
17	Mindelo (Cape Verdes)	2109	2263	2342	2242	2218	2145	1840	1553	1505	1078	1042	1563	1386	1252	1314	869	17

9.26.3 PORTUGUESE ISLANDS: SPECIAL NOTES

Geography The Azores comprises 9 islands in 3 groups:

- Western group: Corvo and Flores
- Central group: Faial, Pico, São Jorge, Graciosa and Terceira
- Eastern group: São Miguel and Santa Maria

The Madeira archipelago comprises:

- Porto Santo
- Madeira
- Islas Desertas – three islands: Ilhéu Chão, Ilha Deserta Grande and Ilhéu Bugio
- Islas Selvagens – three small islands: Selvagem Grande, Selvagem Pequena, Ilhéu de Fora, plus a group of rocks, Ilhéus do Norte

Both in the Azores and Madeira the islands are well spread out. See the Distance table.

Harbours
Azores – The largest and most visited harbours are Horta (Faial) Angra do Heroísmo (Terceira) and Ponta Delgada (São Miguel). Lajes on Flores is oft-visited because it is closest to the USA and to the E-bound transatlantic track. At least one harbour suitable for yachts, for each of the rest of the islands, is shown.
Madeira – The largest and most visited yacht harbours are Porto de Porto Santo, which is the closest to mainland Portugal, and Quinta do Lorde (E Madeira). Other yacht harbours include Funchal and Calheta, both on Madeira's south coast.

Standard Time
Azores – is UT +1, ie 1 hour behind UT.
Madeira – is UT, i.e. the same time as the UK.
In both island groups one hour is added in March for summer time and removed in October, the same as the rest of Europe.

Tides
Azores – Ponta Delgada is the Standard Port with a regular, semi-diurnal regime. In summer HWS is usually around 0200 and 1400 LT, and HWN 0700 and 2000 LT. Tidal range is small: 1·4m sp and 0·6m np. Secondary port differences are < 8 minutes and < 0·2m, thus interpolation is not really necessary. The rising tide sets E-NE, the ebb W-SW. Rates are at their greatest at mid-tide springs and are influenced by strong winds.
Madeira – Casablanca is the Standard Port with a regular, semidiurnal regime. Tidal range is < 2·2m; HWS = HW Dover +3, HWN = HW Dover +3·5. Spring rates max 1·25kn (neaps 0·5kn) floods NE/ENE, ebb is reciprocal.

Currents
Azores – < 1 kn and variable, but tending towards S or SE-going; locally influenced by each island's topography.
Madeira – the around 0·5kn ambient S or SW-going current is influenced by an inshore current flowing NW-SW from Pta do Pargo to Funchal, causing eddies offshore.

Navigation
Azores – Jul-Aug is a good time for UK-based yachts to visit, after trans-Atlantic boats have passed through in May-Jun. It is best to get back to the UK by end Aug to avoid autumnal gales.
A DGPS station on Horta at 38°32'N 28°37'W is operational on 308kHz out to a range of 295M. W variation as area map, decreasing 9' annually.
Madeira – June-Sept is a good time for UK-based yachts to visit, before boats heading for a transatlantic circuit have passed through in the autum/winter. W variation as area map, decreasing 8' annually.

Lights, buoys, daymarks All islands are well lit (see area maps) but some higher lts may be obscured by low cloud. In the Azores, Pico (a 2351m/7713ft dormant volcano) is visible from 50M in good visibility. Pico Ruivo de Santana (Madeira) is lower (1862m/6109ft) but will still be raised earlier than Porto Santo (516m/1693ft) on approach from the NE.

The buoyage system is IALA 'A' (reverse of IALA 'B' as used in the Americas). Thus red can buoys are to be left to PORT and green conical buoys to STARBOARD, whilst following the general direction of buoyage. There are few buoys.

Weather Jetstream migrations-permitting, the Azores High usually predominates from June to mid-Sept with light winds/calms and warm settled conditions in the Azores group. If it is not established, windier more variable European conditions may prevail. In summer winds are usually NE'ly over the Central and Eastern groups and visibility is good; fog is rare. Humidity may be high but not uncomfortable. Summers average about 22°C and winters 14°C.

Being SE of the Azores High, for the main part of the year the Madeira archipelago experiences N to NE'ly winds, rarely more than 25kn, often with patches of calm. The temperature remains fairly constant in the summer at about 24°C and about 16°C in winter.

In the winter both island groups may be affected by depressions tracking across the Atlantic and this can produce very strong SW'ly winds.

NAVTEX – São Miguel, Cencomaraçores (F) transmits a 24H forecast in English on 518kHz for sea areas Altair, Açores, Irving, Milne and Marsala every 4 hrs from 0050 UT; see Chapter 6 and Fig 6(14). São Miguel, Cencomaraçores (J) transmits the forecast on 490kHz in Portuguese every four hours from 0130.

Porto Santo – Cencomarmadeira (P) transmits a 24H forecast in English on 518 kHz, for sea areas Josephine, Madeira, Casablanca, Canaries, Meteor, every four hours from 0230, including nav warnings within a 20M radius of Madeira and Porto Santo. Covering the same information, Porto Santo, Cencomarmadeira (M) transmits in Portuguese on 490kHz, every four hours from 0100.

See Chapter 6 for forecasts on VHF Ch 11.

Books – *W Coasts of Spain & Portugal* (Admiralty NP67). *The Atlantic Crossing Guide* (RCC/ACN). The detailed *Atlantic Islands* (Anne Hammick, RCC/Imray) is essential and gives thorough treatment to these Portuguese islands as well as the Canaries and Cape Verdes.

Charts – are listed under each harbour.

Azores – AC 1950 and PC 41101 cover the whole archipelago. PC 43101 covers the western island group, 43102 the central group and 43103 the eastern.

Madeira – AC 4104 and PC 33101 (both 1:350,00) cover the whole archipelago. AC1831 and PC 36201 cover Porto Santo, Madeira and Islas Desertas. PC 36407 covers Ilhas Selvagens.

Formalities
Brit Hon Consul is at Ponta Delgada (São Miguel) or ☎808 203 537 (Lisbon).

Madeira and the Azores are in the EU, but arr/dep procedures can differ for each island group. For E-bound yachts the Azores may be the first port of entry into Europe, so the process is important. Fly flag 'Q' on arrival. Well organised ship's and personal papers, plus courtesy, good humour and patience are a great help. Expect to be processed by Immigration (*Serviço Estrangeiros e Fronteiras (SEF)* Customs (*Alfândega*) and *Policia Maritima*; plus *Guarda Nacional Republicana* (tax authority) and the local HM (*Capitão do Porto*) all of whom may (not) be in the same building.

It is acceptable to fly the Azorean/Madeiran flags on the stbd halyard *below* the Portuguese courtesy ensign. Departure clearance, valid for 24hrs, is also required, even between islands. If you cancel, the process must re-start. In Madeira, all 4 marinas are licensed to process yacht clearance paperwork and may be treated as ports of entry.

Foreign yachts staying in Portuguese waters (or hauled ashore) for more than 183 days may be subject to a 'long stay tax'. Long stay yachts from other EU countries will need to provide evidence that VAT has been paid within the EU. Otherwise VAT will be claimed by the authorities as part of the temporary importation process.

Ilhas Desertas and Ilhas Selvagens – permits must be obtained before visiting these islands. There are officials stationed on them who **will not** allow you to land if you do not have the correct documents. The permits are FOC and can be obtained in Funchal at the National Parks Department, or via any of the four Madeiran marinas (though allow 2 days for them to arrange this). Ship's papers and passports are required to obtain a permit.

Telephone – International dialling code for both groups is 351. Area codes are: 291 for Madeira and Porto Santo; 292 for Flores, Corvo, Faial and Pico; 295 for São Jorge, Graciosa and Terceira; and 296 for São Miguel and Santa Maria.

On all islands the area code <u>must</u> be dialled even within the same island, 9 digits in all. Faxes & e-mail are easy to send/receive.

In both the Azores and Madeira archipelago mobile voice and data coverage is good in populated areas, patchy elsewhere.

Public Holidays – For both groups they are mostly the same as Portugal, see Area 24. Every town has a local festa.

Azores – Local festas are: Ponta Delgada 1 April; Velas 23 April; Porto das Lajes 1 June; Praia da Vítoria 20 June; Horta / Angra de Heroismo / Vila Franca do Campo 24 June; Madalena 22 July; Vila da Praia 12 August; Calheta 25 November.

Madeira – An additional public holiday, Madeira Day, is on 1 July. Porto Santo and Calheta's festa is 24 June and Funchal's is 21 August.

Flights, Azores – Most European airlines, eg TAP, route via Lisboa to São Miguel, Faial and Terceira, but Sata (Air Azores; ☎00 351 296 209720; www.sata.pt) flies direct from Gatwick (London) to São Miguel every Tues. Sata also shuttles between all 9 islands. There are direct flights from/to Boston, USA and Canada.

Madeira – There are two scheduled airlines, TAP and Easyjet. TAP route via Lisboa to Porto Santo & Funchal and Easyjet route via London Gatwick and Bristol to Funchal, and Transavia with regular weekly flights to Paris. Several other airlines fly into Funchal on a less regular basis, these include Thomas Cook, Thompson Flights, Monarch and Jet2. A car ferry runs between Porto Santo and Funchal daily during the summer, see www.portosantoline.pt

Value Added Tax (IVA) – is 22% Madeira and 18% Azores.

AZAB – The Azores and Back race (Falmouth–Ponta Delgada) is held every 4 years. It continues to be a true corinthian event, open to all comers, with limited sponsorship. The next race will be in 2019, see: www.azab.co.uk.

THE AZORES ARCHIPELAGO

9.26.4 AZORES: SELECTED LIGHTS

Bold print = light with a nominal range of 15M or more. CAPITALS = place or feature. *CAPITAL ITALICS* = light-vessel, light float or Lanby. *Italics* = Fog signal. ***Bold italics*** = Racon. All positions are WGS 84 Datum. Sequence is from WNW to ESE.

ILHA DO CORVO

Canto de Carneira ⚡ Fl 6s 238m 9M; W twr/R bands; 39°43'·04N 31°05'·35W.

Pta Negra Fl 5s 22m 6M; col on W building; 39°40'·18N 31°06'·86W.

ILHA DAS FLORES

Ponta do Albarnaz ⚡, Fl 5s 103m **22M**, 035°-258°; W twr, R cupola; 39°31'·10N 31°13'·90W.

Santa Cruz Pedra Açucareiro ⚡, Fl R 5s 13m 4M; 39°27'·27N 31°07'·48W.

PORTO VELHO (SANTA CRUZ), ldg lts 261·6°: Front, L Fl G 5s 13m 4M, 39°27'·24N 31°07'·53W; rear L Fl G 5s 15m 4M, 17m from front.

PORTO DAS POÇAS (SANTA CRUZ), ldg lts 284·6°: Front, Fl R 3s 8m 5M, 39°27'·08N 31°07'·62W; rear Fl R 3s 17m 5M, 40m from front.

PORTO DAS LAJES, ldg lts 270°: Dir Iso WRG 6s 9m W9M R/G6M; 265·5°-G-268·5°-W-271·5°-R-274·5°; R ◊ on W column, R bands; 39°22'·81N 31°10'·27W.

Bkwtr hd ⚡ Fl (2) R 12s 9m 2M; 39°22'·75N 31°09'·94W.

Ponta das Lajes ☆, Fl (3) 28s 98m **26M**, 263°-054°; W twr, R cupola; 39°22'·53N 31°10'·60W, SE corner of island.

Faja Grande ⚡, Fl 5s 13m 4M; W post, R bands; 39°27'·56N 31°15'·78W.

ILHA DO FAIAL

Ponta dos Cedros ⚡, Fl 7s 148m 12M; 38°38'·29N 28°43'·29W.

Ponta da Ribeirinha ⚡, Fl (3) 20s 135m 12M; Post; 38°35'·73N 28°36'·15W, E end of island.

HORTA

Boa Viagem, ldg Lt 285° ⚡ Dir Iso WRG 6s 12m W9M R/G6M; 276°-G-282°-W-288°-R-294°; R ◊ on W post; 38°32'·23N 28°37'·54W.

Bkwtr hd ⚡ Fl R 3s 20m 11M; W structure; 38°32'·03N 28°37'·27W.

Ldg lts 195·3°: both Iso G 2s 13/15m 2M; Red X on W posts, R bands. Front 38°31'·67N 28°37'·51W; rear 12m from front.

ILHA DO PICO

MADALENA
N Bkwtr ⸔ Oc R 3s 11m 10M; W twr, R bands; 38° 32'·17N 28° 32·0W.
W Bkwtr Fl(3) G 8s 5m 5M, W col, G bands; 38° 32'·13N 28°32'·12W
Ldg lts 138·9°: both Fl G 6s 16/20m 5M; W posts, R bands. Front 38°31'·99N 28°31'·84W; rear 128m from front.
Ponta da Ilha ☆ Fl (3) 15s 28m **24M**, 166°-070°; W twr, R cupola and bldg; 38°24'·75N 28°01'·84W, E tip of island.

CAIS DO PICO
Cais do Pico ⸔ Oc R 6s 10m 6M; R lantern on W structure, 5m; 38° 31'·62N 28° 19·20W.
Mole hd ⸔ Fl G 3s 11m 2M; W twr, G bands; 38° 31'·85N 28° 19·15W.
Cabos Galeão ⸔ Fl (2) R 8s 8m 9M; 38°25'·50N 28°25'·38W.
Ponta de São Mateus ⸔ Fl 5s 33m 13M; 284°-118°; W twr; 38°25'·37N 28°26'·91W.

ILHA DE SÃO JORGE
Ponta dos Rosais ⸔ Fl (2) 10s 282m 8M; 38°45'·22N 28°18'·73W.
Ponta do Norte Grande ⸔ Fl 6s 34m 12M; R col, W bands; 38°40'·83N 28°03'·03W.
Ponta do Topo ☆ Fl (3) 20s 60m **20M**, 133°-033°; W twr, R cupola and bldg; 38°32'·92N 27°45'·28W, E tip of island.
Ponta do Junçal Fl 3s 64m 6M; R col, W bands; 38° 35'·66N 27° 58·77W.

CALHETA
Pier ⸔ Fl R 3s 12m 10M; W twr, R bands; 38° 36'·04N 28° 00·61W.
Ponta da Queimada Fl 5s 37m 10M; W column, R bands; 38° 40'·15N 28° 11·58W.

VELAS
Quay Pier Head ⸔ Fl R 5s 14m 7M, W twr, R bands; 38° 40'·68N 28° 12·18W.
Marina Pier Head ⸔ Fl(3)G 9s 8m 3M; G twr, W bands; 38° 40'·78N 28° 12·18W.
Mole Pier Head ⸔ Fl(3) R 9s 4m 3M; R post, W bands; 38° 40'·78N 28° 12·21W.
Cabos Velas ⸔ Fl(2) R 8s 12m 9M; W post, R bands; 38° 40'·71N 28° 12·22W; marks submarine cable area.

ILHA GRACIOSA
Ponta da Barca ☆ Fl 7s 71m **20M**; vis 029°-031°, 035°-251°, 267°-287°; W twr, R cupola & B bands; 39°05'·61N 28°02'·98W.

VILA DA PRAIA
Ldg lts 306·9°: Front, Fl R 2s 8m 6M vis 293°-321°; W twr, R bands; 39°03'·21N 27°58'·16W. Rear, Iso R 4s 13m 6M vis 293°-321°; W column, R bands; 39°03'·24N 27°58'·19W.
Mole hd ⸔ Fl G 3s 15m 9M, W twr, G bands; 39°03'·16N 27°57'·96W.
Cabos Praia ⸔ Fl(2) R 8s 5m 9M; R lantern on building; 39°03'·07N 27°58'·26W; marks submarine cable area.
Ponta da Restinga ☆ Fl (2) 10s 188m **15M**; 165°-098°; R twr & cupola; 39°00'·84N 27°57'·29W.

ILHA TERCEIRA
Ponta da Serreta ⸔ Fl 6s 99m 12M; W twr, R bands; 38°45'·95N 27°22'·45W, NW end of island.
Lajes ☆ Aero Al Fl WG 10s 132m **W28M, G23M**; 38°45'·59N 27°04'·84W.

PRAIA DA VITORIA
N mole ⸔ Fl 5s 19m 10M; W post, Bk bands; 38°43'·59N 27°03'·06W. Marina in NW corner of hbr.
S mole Fl R 3s 22m 8M; W twr, R bands; 38°43'·24N 27°02'·92W.
Pta das Contendas ☆ Fl (4) WR 15s 53m **W23M, R20M**; 220°-W-020°-R-044°-W-072°-R-093°; W twr, R top; 38°38'·62N 27°05'·07W.

ANGRA DO HEROÍSMO
Ldg lts 340·9°: front, Fl R 4s 20m 7M; R post, 38°39'·25N 27°13'·09W. Rear, Oc R 6s 54m 7M; R lantern on Y structure; 505m from front.
Porto Pipas (marina ent, S side) ⸔ Fl G 3s 13m 6M.
N mole, ⸔ Fl (2) R 6s 8m 3M; 38°39'·12N 27°12'·95W.
Monte Brasil ⸔ Oc WR 10s 27m 12M; 191°-R-295°-W-057°; W col, R bands; 38°38'·60N 27°13'·04W.

ILHA DE SAO MIGUEL
Ferraria ☆ Fl (3) 20s 106m **27M**; 339°-174°; W twr, R cupola; 37°51'·21N 25°51'·02W, NW end of island.
Ponta do Arnel ☆ Aeromarine Fl 5s 65m **25M**; 157°-355°; W twr, R cupola; 37°49'·45N 25°08'·14W, E end of island.
Ponta Garça ☆ L Fl WR 5s 100m **W16M**, R13M; 240°-W-080-R-100°; twr and house; 37°42'·85N 25°22'·18W, S side of island.

VILAFRANCA DO CAMPO
Marina N pier head: Fl G 4s 11m 9M; W twr, G bands; 37° 42'·82N 25° 25'·77W.
Marina S pier head: Fl R 4s 11m 9M; W col, R bands; 37°42'·79N 25°25'·71W

PONTA DELGADA
Bkwtr head ⸔ L Fl R 6s 16m 10M; W twr, R bands; 37°44'·18N 25°39'·40W, 450m SSE of marina ent.
Naval Club Dir lt 309°: Fl (2) WRG 5s 24m 10M (H24); 301·5°-G-306·5°-W-311·5°-R-316·5°; 37°44'·45N 25°39'·39W.
Terminal pier E hd ⸔ Fl(2)R 5s 2m 3M; W col, R bands; 37°44'·35N 25°39'·61W.
Terminal pier W hd ⸔ Fl(3)G 15s 2m 3M; W col, G bands; 37°44'·24N 25°39'·83W.
Quebra floating bkwtr S end ⸔ Q(6)+LFl W 15s 3M; 2 ▼ (vert) on Bk bcn, Y top; 37°44'·26N 25°39'·97W.
Marina mole hd ⸔ Fl G 3s 12m 10M; W twr, G bands; 37°44'·32N 25°39'·57W, 450m NNW of bkwtr head.
Forte S. Bras ⸔ Fl (4) WRG 8s 13m 10M (H24); 259°-G-261°-W-263°-R-265°; 37°44'·13N 25°40'·37W.
Santa Clara ☆ L Fl 5s 26m **15M**; 282°-102°; R lantern; 37°44'·00N 25°41'·16W, between airport and Ponta Delgada.
Airport ☆ Aero Al Fl WG 10s 83m **W28M, G23M**; 282°-124°; control twr; 37°44'·62N 25°42'·46W, 2·5M W of Ponta Delgada.
Ilhéus das Formigas ⸔ Fl 5s 22m 9M; W twr; 37°16'·26N 24°46'·83W (islet 20M NE of Ilha de Santa Maria).

ILHA DE SANTA MARIA
Pta do Norte ⸔ Fl (4) 15s 138m 12M; W structure, R bands. 37°00'·77N 25°03'·55W.
Pta do Castelo ☆ Aeromarine Fl (3) 13·5s 113m **25M**; 181°-089°; W twr, R cupola and bldg. 36°55'·77N 25°00'·97W, SE corner of island.

VILA DO PORTO
Control twr mole head ⸔ L Fl R 5s 14m 5M; R trunc twr, W stripes; 36°56'·49N 25°08'·92W.
Naval Club, ⸔ Fl (2) WRG 8s 24m 10M; 346°-G-351°-W-354°-R-359°; 36°56'·80N 25°08'·86W.
Pta do Malmerendo ⸔ Fl (2) 10s 49m 12M; W bldg, R bands; vis 282°-091°; 36°56'·43N 25°09'·42W.
Airport ☆ Aero Al Fl WG 10s 116m **WG25M**; 021°-121°; shown by day in poor vis; control twr. 36°58'·37N 25°09'·90W, W end of Is.

9.26.5 PORTO DAS LAJES (FLORES)

Flores **39°22'·82N 31°09'·96W** (Hbr ent) ✿✿✿✤✤✤✿✿✿

CHARTS AC 1950, 1959; PC 41101, 43101, 46401

TIDES MHWS 1·5M; MHWN 1·2M; MLWN 0·7M; MLWS 0·4M

SHELTER is good, and the new bkwtr gives good protection against NE'lies in the marina, though there may be a scend. ⚓ where indicated on sand; elsewhere patches of gravel and rock offer poor holding and uncomfortable in N-NE'lies.

Lajes has supplanted the much smaller and more difficult hbr at Santa Cruz, 5M NNE.

NAVIGATION WPT 39°22'·81N 31°09'·00W, 270·0°/9ca to hbr ent. The approach is straightforward by day, but Flores is fringed by rocks; keep at least 1M off. The 9M wide passage between Flores and Corvo can be choppy at sp; deepest water is in the N half.

LIGHTS AND MARKS See Azores Selected Lts for details. The bkwtr lt is often destroyed in winter and is not wholly reliable in summer. The 270° ldg lts lead clear of inshore rocks to stbd and the bkwtr to port. By day the church, water tower and lt ho are all conspic; the latter's light is obscured N of a brg of 263°.

COMMUNICATIONS (Code 292). Brit Consul 808 203 537; Medical emergency 112. Ⓗ 592294; HM 593437, Ch 14,16; Tourist Info 592369 (Santa Cruz).

FACILITIES lajes.flores@aptosa.com, €1·25/m 🛢️ 🛢️(up to 500ltrs) ☎ 292 593813, ⚓ 🛒(12t,40t) 🔧 🚿 ⛽. Mail pick-up via Café Beira Mar, Lajes das Flores 9960, Azores. Internet access foc at library.

Clube Naval welcomes/helps visitors: ☎292 593145. info@cnlflores.com www.cnlflores.com

Town ✕ 🍴 🛒 ⊠ Ⓑ Ⓒ 🏧 Ⓗ ✈ at Santa Cruz, 12km.

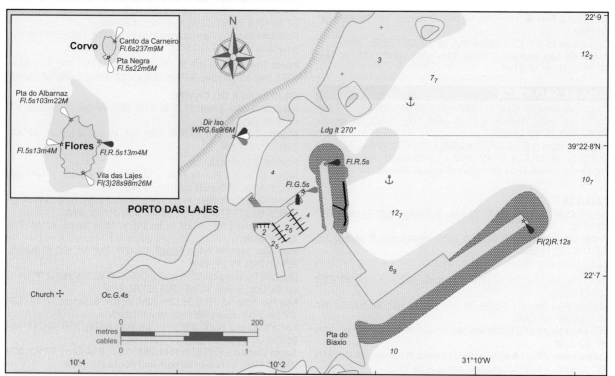

MINOR HARBOUR 9M N OF ILHA DO FLORES
CORVO 39°40'·30N 31°06'·00W. PC 43101, 46401. A small hbr approx 3 ca N of Punta Negra. The hbr is exposed to N/NE and suffers badly from swell much of the time; it is unwise to berth alongside, the ⚓ is unreliable: yachts should not be left unattended.

9.26.6 HORTA (FAIAL)

Faial **38°32'·05N 28°37'·38W** (Hbr ent) ✿✿✿✤✤✤✿✿✿

CHARTS AC 1950, 1956, 1957; PC 43102, 46403; SHOM 7605

TIDES MHWS 1·6M; MHWN 1·2M; MLWN 0·7M; MLWS 0·4M

SHELTER Good, except in strong SW-SE winds which raise a steep, confused sea in the hbr. ⚓ in hbr allowed (patchy holding). No ⚓ at Porto Pim, an inlet S of the hbr.

NAVIGATION WPT 38°32'·04N 28°36'·00W 270°/1M to bkwtr hd. The 195° ldg line is not for yachts. Streams in the 2·6M wide Canal do Faial between Faial and Pico reach 2½kn eddying SSW towards Horta. Beware many ferries, Baixa do Sul rk (7·1m) in mid-chan which breaks in SW gales. The modern cruise ship terminal lies north of the marina.

LIGHTS AND MARKS See Azores Selected Lts for details. Near the W end of Faial, Vale Formoso, L Fl (2) 10s 113m 13M, is reported to have a range of only about 5M. Monte da Guia, 1M S of the harbour, is a steep, rounded extinct volcano with conspic radio mast. SSW of the marina the Observatory and a hotel tower block are conspic.

COMMUNICATIONS (Code 292) Brit Consul 808 203 537; Medical emergency 112; ⊞ 200200; APTO (Port Authority) 208300, call *Capimarhorta* VHF Ch 11, 16; Call *Marina da Horta* VHF Ch 10/16 (0800-2000 daily); Portuguese, English, French & Spanish spoken; MAYS *Mid-Atlantic* VHF Ch 77 (HO); Tourist Info 292237.

FACILITIES Marina ☎292391693, 🖷292 293986. www. marinasazores.com/horta hortamarina@aptosa.com 300 inc 🅥 €1·25/m, multihull +50%. In 2 basins: the long-established N basin (2-3m, for locals and smaller visitors) and a new S basin mainly for mega yachts (but <35m LOA, 6m draft). Reception/fuel pontoon is between the basins. Private yacht moorings lie outside the marina. 🅗 🅗(cash only) ⬚ 🅞 🅗 ⚓ 🅗 via MAYS 🅗 🅗 🅗 🅗 🅗 🅔 🅗 🅗(20t). Mail pick-up via marina, Av'da Gago Sacadura Cabral 7, 9900 Horta, Faial, Açores; or via Café Sport or MAYS.

Long term plans envisage a new bkwtr extending SSE from a point about 250m NNE of Boa Viagem lt, giving better shelter and more room to ⚓.

Mid-Atlantic Yacht Services (MAYS) at N end of marina is the key company for visiting yachts, for all repairs, advice, handling mail, fax and e-mail. Helpful and friendly. ☎292 391616. mays@mail. telepac.pt

Clube Naval da Horta ☎292 391719/ 200680 (S of marina) Ch 16 during office hours. secretariado@cnhorta.org Welcomes visitors for racing. Small ✕ ⬚.

Café Sport ☎292 292327. www.petercafesport.com Welcoming yachtsmen since about 1975, no sailing trip to Horta is considered complete without a stop at this famous bar. There is also a scrimshaw museum upstairs.

Town ✕ ⬚ 🛒 ⊠ Ⓑ Ⓒ ACA ⊞ ✈ (UK direct).

9.26.7 MADALENA (PICO)

Pico 38°32'·15N 28°32'·00W (Ent) ✿✿✿

CHARTS AC 1950,1956,1957; PC 43102, 46403; SHOM 7605

TIDES Use Horta, see Azores Special notes.

SHELTER Reasonable in harbour, though some swell finds its way in and wash from ferries can be troublesome. Alongside space may be at a premium, so ⚓ is usually the best option.

NAVIGATION WPT 38°32'·46N 28°32'·33W thence 139°/4ca to ent. Beware ferry traffic manoeuvring in hbr and WIP 2014.

LIGHTS AND MARKS Madalena N Bkwtr Oc R 3s 11m 10M; W twr, R bands; 38° 32'·2N 28°32'·0W.

W Bkwtr Fl (3) G 8s 5m 5M, W col, G bands; 38°32'·13N °32'·12W.

Ldg lts 138°·9 Front Fl G 6s 15m 5M; W post R bands; 38° 32'·0N 28° 31'·9W.

COMMUNICATIONS (Code 292) Brit Consul 808 203 537; Emergency 112; Medical Emergency 115; Dr 622241; Hbr Auth VHF Ch 11, 16; Tourist Info 623345/623524.

FACILITIES Poss 🅥 ➰ N pier. ☎292 622365/623303. marina@azores. com €1·23/m, no ⛽ ⬚.

Clube Naval do Madalena welcomes visitors: ☎292 623042. cnmpico@hotmail.com.

Town ✕ ⬚ 🛒 🅞 ⊠ Ⓑ Ⓒ. ✈ Cachorro (5km) or via ferry to Horta. Reports welcome.

9.26.8 CAIS (SÃO ROQUE) DO PICO

Pico 38°31'·80N 28°19'·12W (ent) ⚓🔆✿

CHARTS AC 1950, 1956; PC 47501, 46201; SHOM 7605

SHELTER/NAVIGATION Can be rolly: exposed to E and subject to katabatics in SW-NW winds. WPT 38° 32'·04N 28° 18'·90W, then 215°/3ca to ent. ⚓ (deep/rky) clear of bkwtr, or ⬭ on it if space.

LIGHTS AND MARKS See chartlet and Azores Selected Lights.

COMMUNICATIONS (Code 292) Emergency 112; Brit Consul 808 203 537; Medical emergency 115; Dr 672123; HM 642326, VHF Ch 11,16; Tourist Info 679320/623524.

FACILITIES Clube Naval ☎292 642150, ⚓ (and on quay) 🖒.
Town 🖂✕🛒🛍🅿🖂Ⓑ Ⓒ ✈ at Cachorro, 12km.

9.26.9 VELAS (SÃO JORGE)

São Jorge 38°40'·61N 28°12'·14W (bkwtr)⚓🔆🔆✿✿

CHARTS AC 1950, 1956; PC 43102, 178; SHOM 7605

SHELTER/NAVIGATION Well protected, but very small. Keep clear of ferry traffic.

LIGHTS AND MARKS Marina Pier Head Fl(3)G 9s 8m 3M G tr, W bands.Mole Pier Head Fl(3) R 9s 4m 3M R post, W bands.

COMMUNICATIONS (Code 295) Maritime Police 432388; Emergency 112; Brit Consul 808 203 537; Medical emergency 115; Dr 412122; HM 412336, VHF Ch 11; *Velas Marina* VHF Ch 16/10 (0800-2000 mid May-mid Aug); Tourist Info 432395/412440.

FACILITIES Marina ☎295 698900. saojorge@aptosa.com www.marinasazores.com 76 incⓋ €1·25/m, multihull +50%. 🔌🚽 no wi-fi 🖒. **Clube Naval das Velas** ☎295 432169 ✕.
Town ✕🖂🛒🖂Ⓑ Ⓒ Ⓗ.

9.26.10 CALHETA (SÃO JORGE)

São Jorge 38°35'·96N 28°0'·54W (bkwtr) ⚓🔆✿✿

CHARTS AC 1950, 1956; PC 43102, 178; SHOM 7605

NAVIGATION/SHELTER/LIGHTS WPT 38°35'·77N 28°0'·65W, then 025°/2 ca to bkwtr. Keep close in to bkwtr and quay, rocky shallows close E. Bkwtr lt uncertain. Shelter better than it looks.

LIGHTS AND MARKS New bkwtr Fl R 3s 9m 10M; W twr, R bands; 38° 36'·0N 28° 00'·6W.

COMMUNICATIONS (Code 295) Emergency 112; Brit Consul 808 203 537; Medical emergency 115; Dr 416498; Tourist Info 416252/412440.

FACILITIES HM ☎295 416142. €1·25/m, ⚓ only.
Town ✕🖂🛒🖂Ⓑ Ⓒ ✈ at Velas, 16km.

9.26.11 VILA DA PRAIA (GRACIOSA)

Graciosa 39°03'·18N 27°58'·11W ❀⚓✿✿✿

CHARTS AC 1950, 1956, 1957; PC 43102, 113; SHOM 7605

TIDES Santa Cruz MHWS 1·6M; MHWN 1·3M; MLWN 0·7M; MLWS 0·3M, see Azores Special notes.

SHELTER ⚓ can be rolly, afforded some protection by Ilhéu da Praia, but exposed E to S. Fishing hbr, very crowded, ⓥ not encouraged.

NAVIGATION WPT 39°03'·05N 27°57'·88W then 1.5 ca along 306·9° to ⚓.

LIGHTS AND MARKS Ldg lts 306·9° Front: Fl R 2s 8m 6M vis 293°-321°; W twr, R bands; Rear: Iso R 4s 14m 6M vis 293°-321°; W col, R bands. N Mole, Fl G 3s 15m 9M; W mast. Hbr ent, S pier Fl R 3s 10m 6M, N quay, Fl (2) G 6s 5m3M. Submarine cables warning lt, Fl (2) R 8s 13m 9M, vis 264°-292°.

COMMUNICATIONS (Code 295) Emergency 112; Brit Consul 808 203 537; Medical emergency 115; HM 712128, VHF Ch 11,14,16; Tourist Info 712888/712059.

FACILITIES Town ✕ ⌂ 🛒 ⚏ 🅿 🛢 ✉ Ⓑ Ⓒ Ⓗ and small ✈ at Santa Cruz, 7km. Ferries to Faial, Horta.

9.26.12 ANGRA DO HEROÍSMO (TERCEIRA)

Terceira 38°39'·10N 27°12'·93W (Marina ent) ❀❀❀⚓⚓✿✿✿✿

CHARTS AC 1950, 1956, 1957; PC 43102, 46405; SHOM 7605

TIDES MHWS 1·6M; MHWN 1·3M; MLWN 0·7M; MLWS 0·4M

SHELTER Good. Exposed only to strong S/SE winds and, if any S in the wind, swell may enter.

A bkwtr extending 1ca W from Porto Pipas ⚡ Fl G 3s is planned.

⚓ SW of the marina off the YC, but clear of Baixa da Prainha, a rock awash at CD. Private moorings are laid NE of this rock.

NAVIGATION WPT 38°38'·00N 27°12'·54W, 341°/1·1M to abeam marina ent and on 341° ldg line. Beware unlit Ilhéus das Cabras (146m high islets) 3·3M ESE of the hbr and 9ca offshore.

2M SE of these islets is Ilhéus Fradinhos 8m high pinnacle rks. Both are in the R sectors of Pta das Contendas lt ho.

Monte Brasil has a marine reserve extending about 1M off the coast around it, starting from S. Benedicto castle on the E side and continuing round to Ponta de S. Diogo on the W side. No ⚓'ing or fishing permitted in the reserve.

LIGHTS AND MARKS See Azores Selected lights for details. Front ldg lt is on a R mast at a conspic twin-towered church; rear is a yellow structure, R top. Monte Brasil, a 205m high peninsula with wooded slopes, is close SW of the marina. Ponta do Farol lt on its SE flank is on a W col, R bands.

COMMUNICATIONS (Code 295) Maritime Police via HM; Brit Consul 808 203 537; Medical emergency 112; Ⓗ 212122 @ Praia da Vitória; HM 540500, VHF Ch 11, 16; Marina VHF Ch 09, 16 (0800-2000 daily); Taxi 212004; Tourist Info 213393.

FACILITIES Angra do Heroísmo (Angra) is a handsome city, a fine natural hbr and a UNESCO World Heritage Site since 1983.

Marina ☎295 540000. www.marinasazores.com/angra marina. angra@aptg.pt 260 inc 60 ⓥ (18m max LOA) €1·25/m, multihull +50%; ⓥ finger berths usually on adjacent pontoon G. Depths 3·4m, but less at the NW end. From the ent, reception is due N. ⛽ ⛽ (reception pontoon) ✕ ⌂ 🛢 🍴 ⚒ ⚓ Ⓔ ⌿ ⚠ 🏗(50t). Mail pick-up from Marina, Código Postal 9700-154, Angra do Heroísmo, Terceira, Açores, Portugal.

Angra YC (Porto Pipas mole) welcomes visitors, ☎295 216530.

Town ✕ ⌂ 🛒 🛢 ✉ Ⓑ Ⓒ Ⓗ. ✈ Lajes (24km) direct to UK.

9.26.13 PRAIA DA VITÓRIA (TERCEIRA)

Terceira 38°43'·37N 27°02'·91W (hbr ent) ⊛⚓✿

CHARTS AC 1950, 1956, 1957; PC 43102, 46405; SHOM 7605

TIDES MHWS 1.7; MHWN 1.3; MLWN 0.7; MLWS 0.3; see Azores Special notes.

SHELTER Good in marina and if ⚓ in main harbour SE of marina ent, though the ⚓'age is exposed to E/SE and affected by swell.

NAVIGATION WPT 38°43'·37N 27° 02'·27W is in W sector of Dir lt. Then 5ca/090° to hbr ent wpt. Marina ent lies 6ca 313°, beware ships turning in this busy port.

LIGHTS AND MARKS Mohle Head N Fl 5s 11m 10M; W post, B bands. Mohle Head S Fl R 3s 8M; W twr, R bands; Dir lt Fl (2) WRG 5s 24m 10M.

COMMUNICATIONS (Code 295) Emergency 112; Brit Consul 808 203 537; Medical emergency 115; Ⓗ 212122; HM 540500, VHF 11,16; **Call** *Marina* on VHF Ch 16 / 09 (0830-1930 daily); Tourist info 543251.

FACILITIES Port of Entry.

Marina ☎295 512082/540019. €1·25/m, 210 inc Ⓥ; 🛢 (tanker 24hrs notice) 🅿 🚿 (48 hrs notice) ◣ 🏳(35t). Popular for laying up ashore.

Town ✕ 🏪 🛒 Market ✉ Ⓑ Ⓒ. ✈ at Lajes, 4km.

9.26.14 VILA DO PORTO (SANTA MARIA)

Santa Maria 36°56'·5N 25°08'·85W (bkwtr) ⊛⚓✿

CHARTS AC 1950, 1959; PC 43102, 46407; SHOM 7605

TIDES As Ponta Delgada, see Azores Special notes.

SHELTER Excellent in marina. ⚓ in outer hbr allowing turning and berth space for ferry.

NAVIGATION WPT 36° 56'·0N 25° 8'·85W then 0°/5ca to title WPT off main bkwtr. Marina a further 1 ca 325°. Santa Maria now has marine reserves to the N and S of the island, all recreational marine activities are forbidden within them.

LIGHTS AND MARKS Bkwtr hd: LFl R 5s 14m 5M; R truncated tower, W stripes.

COMMUNICATIONS (Code 296). Emergency 112; Brit Consul 808 203 537; Medical emergency 115; Dr 820100; HM 882282, VHF Ch 11,16; Call *Vila do Porto Marina* on VHF Ch 16 / 10 (0800-2000 daily); Tourist Info 886355.

FACILITIES Port of Entry. Port Authority: portovdp@apsm.pt

Vila do Porto Marina ☎296 882282. 📠296 882176 marinavdp@ marinasazores.com www.marinasazores.com 120 inc Ⓥ €1·25/m, multihull +50%. 🛢 🛢🏳(75t) ⚓ ✕ 🅿 🛠.

Clube Naval de Santa Maria (CNSM) very helpful and friendly. cnsm@mail.telpac.pt

Town ✉ Ⓑ Ⓒ 🏪 ✕ 🛒 ✈ at Santa Cruz, 2·5km.

9.26.15 PONTA DELGADA
(SÃO MIGUEL)
São Miguel **37°44'·23N 25°39'·24W** ✻✻✻♨♨♨✿✿✿

CHARTS AC 1950, 1895; PC 43103, 46406; SHOM 7432

TIDES MHWS 1·7M; MHWN 1·3M; MLWN 0·7M; MLWS 0·3M

SHELTER Good inside the high and substantial bkwtr, but in strong E-S winds a heavy swell enters. ⚓ is allowed, but not encouraged, at the W end of the hbr in 8-10m on sand and well clear of any ship operations.

NAVIGATION WPT 37°44'·27N 25°39'·12W, where the W sectors of the 309° and 262° Dir lts intersect; thence 284°/4ca to marina.

LIGHTS AND MARKS See Azores Selected lights for details of directional lts. The 300-1100m high island is easily seen from afar. Conspic marks include the W twr + R bands on bkwtr head and the Solmar bldg, a white 22 storey hotel, close W of the marina.

COMMUNICATIONS (Code 296) MRCC Ponta Delgada 281777 – only accepts safety and distress calls, mrcc.delgada@mail.telepac.pt; ⚓ and nav bulletins VHF Ch 11 in Portuguese; see Chapters 5 & 6 for times and content; Police/Fire/Ambulance 112; Brit Consul 808 203 537; Rua Domingos Rebelo 43A, 9500-234 Ponta Delgada; ⊞ 203000, ask at marina for doctor or dentist; Port HM 205240, VHF 11, 16; *Marina* VHF Ch **9**/16 (0900-1830 M-Fri; 0900-1730 Sat/Sun).

FACILITIES Marina (3·5m-8m) ☎296 281510/11/12. marinapdl@apsm.pt www.marinasazores.com 670 inc ❶ €1·25/m, multihull +50%. Reception pontoon and 🚿 ▲ are N side of ent; best to fill up on arrival. ⬛ ⬥ ▲ via HM ⬛ ⬥ 🖭 Ⓔ ⚠ ♫ Diver ⚓ ◨(25t).

Mail pick-up: c/o Marina de Ponta Delgada, Aptdo 3 – Calheta, 9500 Ponta Delgada, São Miguel, Açores, Portugal.

Clube Naval de Ponta Delgada (E of marina) welcomes visitors: ✗ ⬛ ◳ (3t). ☎296 308380. atendimento@cnpdl.pt www.cnpdl.pt

City ✗ ⬛ ☕ Market ✉ Ⓑ Ⓒ ⊞. ✈ 4km W: flights direct to Gatwick, UK every Tues and daily via Lisboa to other European destinations.

9.26.16 VILA FRANCA DO CAMPO
(SÃO MIGUEL)
São Miguel **37°42'·79N 25°25'·78W** (marina ent) ✻✻♨♨✿✿

CHARTS AC 1950, 1895; PC 43103, 46406, 165; SHOM 7432

TIDES As Ponta Delgada **SHELTER** Excellent in marina.

NAVIGATION Pass inside or outside Ilhéu da Vila to WPT 37° 42'·54N 25° 25'·56W then 325°/3ca to marina ent (title wpt). Caution necessary due to construction works.

LIGHTS AND MARKS Marina Pier Head: Fl G 4s 11m 9M; W twr, G bands; 37° 42'·80N 25° 25'·80W. 2 x Y marks (B pole, Y stripes) indicate works area.

COMMUNICATIONS (Code 296). Emergency 112; Brit Consul 808 203 537; ⊞ 203000; Dr 582117; *APTO* VHF Ch 09/16.

FACILITIES Marina ☎296 581488 (0830-1230 & 1330-1630). marinaem@sapo.pt; 156 limited ❶, €1·70/m. 🚿 ▲ Ⓒ ⬛ ▲ ✕ ⚠ ◳(6t).

Clube Naval de Vila Franca do Campo welcomes visitors ☎296 582333. cnvfc@mail.telepac.pt / cnvfc@clix.pt; ⚓.

Town ✗ ⬛ ☕ ⬛ ✉ Ⓑ Ⓒ ⊞ ✈ 23km W, flights direct to UK.

STANDARD TIME UT +01
Add 1 hour for UT
For Azores Summer Time add
ONE hour in **non-shaded areas**

PONTA DELGADA   LAT 37°44'N   LONG 25°40'W
TIMES AND HEIGHTS OF HIGH AND LOW WATERS

Dates in red are **SPRINGS**
Dates in blue are **NEAPS**

YEAR 2016

JANUARY

Time	m	Time	m
1 0558 / 1221 / F 1830	1.4 / 0.7 / 1.3	**16** 0558 / 1226 / SA 1838 ☽	1.6 / 0.5 / 1.4
2 0035 / 0658 / SA 1323 / ☽ 1939	0.7 / 1.3 / 0.7 / 1.2	**17** 0047 / 0705 / SU 1336 / 1953	0.6 / 1.5 / 0.6 / 1.4
3 0148 / 0806 / SU 1431 / 2052	0.8 / 1.3 / 0.7 / 1.3	**18** 0207 / 0821 / M 1452 / 2110	0.6 / 1.4 / 0.6 / 1.4
4 0303 / 0914 / M 1535 / 2155	0.8 / 1.3 / 0.7 / 1.3	**19** 0327 / 0937 / TU 1603 / 2218	0.6 / 1.4 / 0.6 / 1.5
5 0406 / 1012 / TU 1628 / 2245	0.7 / 1.3 / 0.6 / 1.4	**20** 0437 / 1043 / W 1703 / 2316	0.6 / 1.5 / 0.5 / 1.5
6 0457 / 1100 / W 1712 / 2327	0.6 / 1.4 / 0.6 / 1.5	**21** 0535 / 1139 / TH 1753	0.5 / 1.5 / 0.4
7 0540 / 1143 / TH 1752	0.6 / 1.5 / 0.5	**22** 0005 / 0623 / F 1226 / 1837	1.6 / 0.4 / 1.6 / 0.4
8 0005 / 0620 / F 1222 / 1829	1.5 / 0.5 / 1.5 / 0.4	**23** 0048 / 0706 / SA 1308 / 1916	1.7 / 0.4 / 1.6 / 0.3
9 0042 / 0658 / SA 1301 / 1907	1.6 / 0.4 / 1.6 / 0.4	**24** 0128 / 0745 / SU 1347 / ○ 1953	1.7 / 0.3 / 1.6 / 0.3
10 0120 / 0737 / SU 1340 / ● 1945	1.7 / 0.3 / 1.6 / 0.3	**25** 0205 / 0821 / M 1424 / 2028	1.8 / 0.3 / 1.6 / 0.3
11 0159 / 0817 / M 1421 / 2026	1.8 / 0.3 / 1.6 / 0.3	**26** 0242 / 0856 / TU 1459 / 2102	1.7 / 0.3 / 1.6 / 0.4
12 0240 / 0859 / TU 1504 / 2108	1.8 / 0.3 / 1.6 / 0.3	**27** 0317 / 0930 / W 1533 / 2136	1.7 / 0.4 / 1.5 / 0.4
13 0323 / 0944 / W 1550 / 2154	1.8 / 0.3 / 1.6 / 0.3	**28** 0351 / 1005 / TH 1608 / 2210	1.6 / 0.5 / 1.5 / 0.5
14 0410 / 1032 / TH 1639 / 2243	1.7 / 0.3 / 1.6 / 0.4	**29** 0428 / 1041 / F 1646 / 2248	1.5 / 0.5 / 1.4 / 0.6
15 0501 / 1125 / F 1734 / 2340	1.6 / 0.4 / 1.5 / 0.5	**30** 0507 / 1122 / SA 1729 / 2334	1.4 / 0.6 / 1.3 / 0.7
		31 0554 / 1212 / SU 1824	1.4 / 0.7 / 1.3

FEBRUARY

Time	m	Time	m
1 0034 / 0655 / M 1317 / ☽ 1937	0.7 / 1.3 / 0.7 / 1.2	**16** 0147 / 0802 / TU 1430 / 2050	0.6 / 1.4 / 0.6 / 1.4
2 0154 / 0811 / TU 1434 / 2058	0.8 / 1.3 / 0.7 / 1.3	**17** 0317 / 0928 / W 1551 / 2207	0.6 / 1.3 / 0.6 / 1.4
3 0318 / 0927 / W 1545 / 2206	0.8 / 1.3 / 0.7 / 1.3	**18** 0433 / 1039 / TH 1655 / 2306	0.6 / 1.4 / 0.6 / 1.5
4 0424 / 1029 / TH 1641 / 2258	0.7 / 1.3 / 0.6 / 1.4	**19** 0530 / 1133 / F 1744 / 2354	0.5 / 1.4 / 0.5 / 1.6
5 0515 / 1118 / F 1727 / 2341	0.6 / 1.4 / 0.5 / 1.5	**20** 0614 / 1217 / SA 1825	0.4 / 1.5 / 0.4
6 0559 / 1202 / SA 1809	0.5 / 1.5 / 0.4	**21** 0034 / 0652 / SU 1254 / 1901	1.6 / 0.4 / 1.6 / 0.4
7 0022 / 0640 / SU 1243 / 1849	1.6 / 0.4 / 1.6 / 0.3	**22** 0110 / 0726 / M 1329 / ○ 1934	1.7 / 0.3 / 1.6 / 0.4
8 0102 / 0720 / M 1324 / ● 1930	1.8 / 0.3 / 1.7 / 0.3	**23** 0144 / 0758 / TU 1401 / 2006	1.7 / 0.3 / 1.6 / 0.3
9 0142 / 0800 / TU 1405 / 2011	1.8 / 0.2 / 1.7 / 0.2	**24** 0217 / 0828 / W 1433 / 2036	1.7 / 0.3 / 1.6 / 0.4
10 0224 / 0842 / W 1448 / 2053	1.9 / 0.2 / 1.8 / 0.2	**25** 0248 / 0859 / TH 1503 / 2107	1.7 / 0.4 / 1.6 / 0.4
11 0307 / 0925 / TH 1532 / 2138	1.9 / 0.2 / 1.7 / 0.3	**26** 0320 / 0930 / F 1534 / 2139	1.6 / 0.4 / 1.5 / 0.5
12 0352 / 1011 / F 1619 / 2225	1.8 / 0.3 / 1.7 / 0.3	**27** 0352 / 1001 / SA 1607 / 2212	1.6 / 0.5 / 1.5 / 0.5
13 0441 / 1100 / SA 1711 / 2319	1.7 / 0.4 / 1.6 / 0.4	**28** 0426 / 1036 / SU 1643 / 2251	1.5 / 0.5 / 1.4 / 0.6
14 0535 / 1157 / SU 1811	1.6 / 0.5 / 1.5	**29** 0506 / 1117 / M 1728 / 2340	1.4 / 0.6 / 1.3 / 0.7
15 0024 / 0641 / M 1306 / ☽ 1925	0.6 / 1.4 / 0.6 / 1.4		

MARCH

Time	m	Time	m
1 0557 / 1211 / TU 1829 / ☽	1.3 / 0.7 / 1.3	**16** 0132 / 0746 / W 1408 / 2029	0.6 / 1.3 / 0.7 / 1.4
2 0052 / 0709 / W 1328 / 1954	0.8 / 1.2 / 0.8 / 1.3	**17** 0305 / 0917 / TH 1535 / 2149	0.7 / 1.3 / 0.7 / 1.4
3 0226 / 0839 / TH 1456 / 2119	0.8 / 1.2 / 0.7 / 1.3	**18** 0420 / 1029 / F 1640 / 2249	0.6 / 1.3 / 0.6 / 1.5
4 0348 / 0955 / F 1606 / 2223	0.7 / 1.3 / 0.7 / 1.4	**19** 0514 / 1120 / SA 1727 / 2334	0.5 / 1.4 / 0.6 / 1.5
5 0447 / 1052 / SA 1700 / 2313	0.6 / 1.4 / 0.6 / 1.5	**20** 0555 / 1200 / SU 1806	0.5 / 1.5 / 0.5
6 0534 / 1139 / SU 1746 / 2357	0.5 / 1.5 / 0.4 / 1.7	**21** 0013 / 0629 / M 1234 / 1840	1.6 / 0.4 / 1.5 / 0.4
7 0616 / 1222 / M 1828	0.3 / 1.6 / 0.3	**22** 0047 / 0701 / TU 1306 / 1911	1.6 / 0.4 / 1.6 / 0.4
8 0039 / 0658 / TU 1304 / 1910	1.8 / 0.2 / 1.7 / 0.2	**23** 0119 / 0730 / W 1336 / ○ 1942	1.7 / 0.4 / 1.6 / 0.4
9 0121 / 0739 / W 1346 / ● 1953	1.9 / 0.2 / 1.8 / 0.2	**24** 0150 / 0759 / TH 1406 / 2011	1.7 / 0.3 / 1.6 / 0.4
10 0204 / 0821 / TH 1429 / 2036	1.9 / 0.1 / 1.8 / 0.2	**25** 0220 / 0828 / F 1435 / 2041	1.6 / 0.3 / 1.6 / 0.4
11 0247 / 0904 / F 1513 / 2121	1.9 / 0.2 / 1.8 / 0.2	**26** 0250 / 0858 / SA 1505 / 2112	1.6 / 0.4 / 1.6 / 0.4
12 0333 / 0949 / SA 1559 / 2209	1.8 / 0.2 / 1.7 / 0.3	**27** 0321 / 0928 / SU 1536 / 2145	1.5 / 0.5 / 1.5 / 0.5
13 0422 / 1037 / SU 1649 / 2302	1.7 / 0.4 / 1.6 / 0.4	**28** 0355 / 1001 / M 1611 / 2222	1.5 / 0.5 / 1.5 / 0.6
14 0516 / 1132 / M 1748	1.5 / 0.5 / 1.5	**29** 0433 / 1039 / TU 1652 / 2309	1.4 / 0.6 / 1.4 / 0.7
15 0008 / 0622 / TU 1240 / ☽ 1902	0.6 / 1.4 / 0.6 / 1.4	**30** 0521 / 1128 / W 1747	1.3 / 0.7 / 1.3
		31 0014 / 0628 / TH 1238 / ☽ 1904	0.7 / 1.3 / 0.7 / 1.3

APRIL

Time	m	Time	m
1 0143 / 0757 / F 1409 / 2032	0.7 / 1.2 / 0.7 / 1.3	**16** 0351 / 1002 / SA 1610 / 2218	0.6 / 1.3 / 0.7 / 1.4
2 0309 / 0919 / SA 1529 / 2144	0.7 / 1.3 / 0.7 / 1.4	**17** 0444 / 1053 / SU 1700 / 2305	0.6 / 1.4 / 0.6 / 1.5
3 0413 / 1022 / SU 1629 / 2240	0.6 / 1.4 / 0.7 / 1.5	**18** 0525 / 1133 / M 1739 / 2344	0.5 / 1.5 / 0.6 / 1.5
4 0504 / 1112 / M 1719 / 2328	0.4 / 1.5 / 0.4 / 1.7	**19** 0559 / 1207 / TU 1814	0.5 / 1.5 / 0.5
5 0549 / 1158 / TU 1805	0.3 / 1.7 / 0.3	**20** 0019 / 0631 / W 1239 / 1846	1.6 / 0.4 / 1.6 / 0.4
6 0014 / 0633 / W 1242 / 1849	1.8 / 0.2 / 1.8 / 0.2	**21** 0051 / 0701 / TH 1310 / 1917	1.6 / 0.4 / 1.6 / 0.4
7 0058 / 0716 / TH 1325 / ● 1933	1.9 / 0.1 / 1.9 / 0.2	**22** 0123 / 0731 / F 1340 / ○ 1947	1.6 / 0.4 / 1.6 / 0.4
8 0143 / 0759 / F 1409 / 2018	1.9 / 0.1 / 1.9 / 0.2	**23** 0153 / 0800 / SA 1410 / 2018	1.6 / 0.4 / 1.6 / 0.4
9 0228 / 0843 / SA 1454 / 2105	1.9 / 0.2 / 1.8 / 0.2	**24** 0225 / 0830 / SU 1440 / 2050	1.6 / 0.4 / 1.6 / 0.4
10 0315 / 0928 / SU 1540 / 2154	1.8 / 0.3 / 1.8 / 0.3	**25** 0257 / 0901 / M 1512 / 2125	1.5 / 0.4 / 1.6 / 0.5
11 0405 / 1016 / M 1631 / 2249	1.7 / 0.4 / 1.6 / 0.4	**26** 0332 / 0935 / TU 1548 / 2204	1.5 / 0.5 / 1.5 / 0.6
12 0500 / 1110 / TU 1729 / 2354	1.5 / 0.5 / 1.5 / 0.5	**27** 0411 / 1014 / W 1631 / 2251	1.4 / 0.6 / 1.5 / 0.6
13 0605 / 1217 / W 1839	1.4 / 0.6 / 1.4	**28** 0500 / 1103 / TH 1724 / 2353	1.3 / 0.6 / 1.4 / 0.7
14 0113 / 0726 / TH 1340 / ☽ 2000	0.6 / 1.3 / 0.7 / 1.4	**29** 0604 / 1209 / F 1833	1.3 / 0.7 / 1.4
15 0239 / 0852 / F 1504 / 2117	0.6 / 1.3 / 0.7 / 1.4	**30** 0111 / 0724 / SA 1331 / ☽ 1953	0.7 / 1.3 / 0.7 / 1.4

HAT is 1·9 metres above Chart Datum.

PONTA DELGADA LAT 37°44′N LONG 25°40′W

TIMES AND HEIGHTS OF HIGH AND LOW WATERS

Portuguese Islands

STANDARD TIME UT +01
Add 1 hour for UT
For Azores Summer Time add
ONE hour in **non-shaded areas**

Dates in red are **SPRINGS**
Dates in blue are NEAPS

YEAR 2016

MAY

Time	m		Time	m
1 0230	0.6	**16**	0401	0.6
0844	1.3		1014	1.4
SU 1451	0.6	M 1621	0.6
2106	1.5		2226	1.4
2 0337	0.5	**17**	0446	0.6
0949	1.4		1058	1.4
M 1556	0.6	TU 1706	0.6
2207	1.5		2309	1.5
3 0432	0.4	**18**	0524	0.5
1044	1.5		1136	1.5
TU 1651	0.4	W 1744	0.5
2300	1.6		2348	1.5
4 0521	0.3	**19**	0559	0.5
1133	1.7		1211	1.5
W 1741	0.3	TH 1819	0.5
2349	1.8		
5 0608	0.2	**20**	0023	1.5
1219	1.8		0632	0.4
TH 1828	0.3	F 1243	1.6
			1853	0.5
6 0036	1.8	**21**	0056	1.5
0653	0.2		0704	0.4
F 1304	1.8	SA 1315	1.6
● 1915	0.2	○ 1925	0.4
7 0124	1.8	**22**	0129	1.5
0738	0.2		0735	0.4
SA 1350	1.9	SU 1347	1.6
2002	0.2		1959	0.4
8 0211	1.8	**23**	0203	1.5
0824	0.2		0807	0.4
SU 1436	1.8	M 1419	1.6
2051	0.2		2033	0.4
9 0300	1.7	**24**	0238	1.5
0910	0.3		0841	0.4
M 1524	1.8	TU 1454	1.6
2141	0.3		2110	0.5
10 0350	1.6	**25**	0315	1.5
0958	0.4		0918	0.5
TU 1614	1.7	W 1533	1.6
2236	0.4		2152	0.5
11 0444	1.5	**26**	0357	1.4
1050	0.5		0959	0.5
W 1709	1.6	TH 1617	1.5
2336	0.5		2240	0.5
12 0544	1.4	**27**	0447	1.4
1150	0.6		1049	0.6
TH 1811	1.5	F 1708	1.5
			2337	0.6
13 0044	0.6	**28**	0546	1.4
0654	1.3		1149	0.6
F 1301	0.7	SA 1810	1.5
◐ 1921	1.4		
14 0157	0.6	**29**	0043	0.6
0810	1.3		0656	1.3
SA 1418	0.7	SU 1301	0.6
2033	1.4	◑ 1920	1.4
15 0305	0.6	**30**	0155	0.6
0919	1.3		0809	1.4
SU 1527	0.7	M 1411	0.6
2135	1.4		2031	1.5
		31	0302	0.5
			0917	1.5
		TU 1525	0.6
			2136	1.5

JUNE

Time	m		Time	m
1 0402	0.4	**16**	0446	0.6
1016	1.5		1102	1.4
W 1626	0.5	TH 1712	0.6
2234	1.6		2316	1.4
2 0456	0.4	**17**	0527	0.5
1109	1.6		1142	1.5
TH 1720	0.4	F 1753	0.6
2328	1.7		2355	1.5
3 0546	0.3	**18**	0604	0.5
1159	1.7		1218	1.5
F 1811	0.3	SA 1830	0.5
4 0019	1.7	**19**	0032	1.5
0634	0.3		0639	0.5
SA 1247	1.8	SU 1252	1.6
1901	0.3		1905	0.5
5 0108	1.7	**20**	0108	1.5
0721	0.2		0713	0.4
SU 1334	1.8	M 1326	1.6
● 1950	0.2	○ 1941	0.4
6 0157	1.7	**21**	0144	1.5
0807	0.3		0748	0.4
M 1421	1.8	TU 1401	1.6
2038	0.3		2018	0.4
7 0245	1.7	**22**	0221	1.5
0853	0.3		0824	0.4
TU 1507	1.8	W 1438	1.7
2127	0.3		2057	0.4
8 0333	1.6	**23**	0301	1.5
0939	0.4		0904	0.4
W 1555	1.7	TH 1519	1.6
2216	0.4		2139	0.4
9 0422	1.5	**24**	0344	1.5
1026	0.5		0946	0.4
TH 1644	1.6	F 1603	1.6
2308	0.5		2225	0.4
10 0514	1.4	**25**	0432	1.5
1117	0.6		1035	0.5
F 1737	1.5	SA 1652	1.6
			2318	0.5
11 0003	0.6	**26**	0527	1.4
0611	1.3		1130	0.5
SA 1214	0.7	SU 1748	1.5
1835	1.4		
12 0103	0.6	**27**	0017	0.5
0717	1.3		0629	1.4
SU 1321	0.7	M 1235	0.6
◐ 1938	1.4	◑ 1852	1.5
13 0207	0.6	**28**	0123	0.5
0822	1.3		0738	1.4
M 1430	0.7	TU 1347	0.6
2043	1.4		2001	1.5
14 0307	0.6	**29**	0232	0.5
0925	1.3		0848	1.4
TU 1534	0.7	W 1500	0.6
2141	1.4		2111	1.5
15 0401	0.6	**30**	0337	0.5
1018	1.4		0954	1.5
W 1627	0.7	TH 1607	0.5
2232	1.4		2215	1.5

JULY

Time	m		Time	m
1 0437	0.4	**16**	0456	0.6
1052	1.6		1113	1.4
F 1707	0.4	SA 1727	0.6
2314	1.6		2329	1.4
2 0531	0.4	**17**	0538	0.5
1145	1.7		1152	1.5
SA 1801	0.4	SU 1807	0.5
3 0007	1.6	**18**	0009	1.5
0621	0.3		0616	0.5
SU 1234	1.7	M 1229	1.6
1851	0.3		1845	0.5
4 0057	1.6	**19**	0047	1.5
0708	0.3		0652	0.4
M 1321	1.8	TU 1305	1.6
● 1939	0.3	○ 1922	0.4
5 0144	1.7	**20**	0125	1.6
0752	0.3		0729	0.4
TU 1405	1.8	W 1342	1.7
2024	0.3		2000	0.4
6 0229	1.6	**21**	0204	1.6
0834	0.3		0808	0.3
W 1449	1.8	TH 1421	1.7
2107	0.3		2040	0.3
7 0312	1.6	**22**	0244	1.6
0916	0.4		0848	0.3
TH 1531	1.7	F 1502	1.7
2150	0.4		2122	0.3
8 0354	1.5	**23**	0327	1.6
0957	0.4		0931	0.3
F 1614	1.6	SA 1546	1.7
2233	0.4		2206	0.3
9 0438	1.5	**24**	0413	1.6
1040	0.5		1017	0.4
SA 1658	1.5	SU 1633	1.7
2317	0.5		2255	0.4
10 0523	1.4	**25**	0504	1.5
1126	0.6		1109	0.5
SU 1745	1.4	M 1726	1.6
			2350	0.5
11 0006	0.6	**26**	0603	1.5
0616	1.3		1210	0.5
M 1220	0.7	TU 1827	1.5
◑ 1840	1.4		
12 0102	0.7	**27**	0054	0.5
0717	1.3		0711	1.4
TU 1326	0.7	W 1323	0.6
1943	1.3		1938	1.4
13 0206	0.7	**28**	0207	0.6
0827	1.3		0826	1.4
W 1439	0.7	TH 1443	0.6
2050	1.3		2055	1.4
14 0310	0.7	**29**	0321	0.6
0932	1.3		0939	1.5
TH 1546	0.7	F 1559	0.6
2152	1.3		2207	1.4
15 0407	0.6	**30**	0427	0.5
1027	1.4		1042	1.5
F 1641	0.7	SA 1702	0.5
2244	1.4		2308	1.5
		31	0523	0.4
			1136	1.6
		SU 1756	0.4

AUGUST

Time	m		Time	m
1 0001	1.5	**16**	0552	0.5
0612	0.4		1205	1.6
M 1223	1.7	TU 1822	0.4
1843	0.3		
2 0047	1.6	**17**	0025	1.5
0655	0.3		0630	0.4
TU 1307	1.8	W 1242	1.7
● 1925	0.3		1900	0.3
3 0129	1.6	**18**	0104	1.6
0735	0.3		0709	0.3
W 1347	1.8	TH 1321	1.8
2005	0.3	○ 1939	0.3
4 0208	1.6	**19**	0143	1.7
0813	0.3		0748	0.3
TH 1426	1.8	F 1400	1.8
2042	0.3		2018	0.2
5 0246	1.6	**20**	0224	1.7
0849	0.4		0829	0.3
F 1503	1.7	SA 1442	1.8
2118	0.4		2100	0.3
6 0322	1.6	**21**	0307	1.7
0925	0.4		0912	0.3
SA 1540	1.6	SU 1525	1.8
2154	0.4		2144	0.3
7 0359	1.5	**22**	0352	1.7
1001	0.5		0958	0.3
SU 1617	1.6	M 1612	1.7
2231	0.5		2231	0.3
8 0437	1.4	**23**	0442	1.6
1040	0.6		1049	0.4
M 1657	1.5	TU 1704	1.6
2312	0.6		2325	0.4
9 0520	1.4	**24**	0539	1.5
1125	0.7		1151	0.5
TU 1744	1.4	W 1806	1.5
10 0000	0.7	**25**	0029	0.6
0613	1.3		0648	1.4
W 1223	0.8	TH 1307	0.6
◑ 1842	1.3	◑ 1922	1.4
11 0101	0.7	**26**	0149	0.6
0722	1.3		0810	1.4
TH 1340	0.8	F 1437	0.6
1955	1.3		2048	1.4
12 0215	0.7	**27**	0312	0.6
0841	1.3		0930	1.4
F 1502	0.8	SA 1557	0.6
2111	1.3		2205	1.4
13 0327	0.7	**28**	0422	0.6
0950	1.3		1035	1.5
SA 1609	0.7	SU 1700	0.5
2214	1.3		2305	1.5
14 0424	0.6	**29**	0516	0.5
1043	1.4		1127	1.6
SU 1701	0.6	M 1748	0.4
2303	1.4		2352	1.5
15 0511	0.6	**30**	0601	0.4
0919	—		1210	1.7
M 1744	0.5	TU 1829	0.4
2346	1.5		
		31	0033	1.6
			0640	0.4
		W 1249	1.7
			1906	0.3

HAT is 1·9 metres above Chart Datum.

》》 FREE monthly updates. Register at 《
www.reedsnauticalalmanac.co.uk

1033

STANDARD TIME UT +01
Add 1 hour for UT
For Azores Summer Time add ONE hour in **non-shaded areas**

PONTA DELGADA LAT 37°44'N LONG 25°40'W
TIMES AND HEIGHTS OF HIGH AND LOW WATERS

Dates in red are **SPRINGS**
Dates in blue are NEAPS

YEAR 2016

SEPTEMBER

Day	Time	m	Time	m	Time	m	Time	m
1 TH ●	0110	1.6	0715	0.3	1325	1.7	1940	0.3
16 F ○	0040	1.7	0646	0.3	1256	1.8	1914	0.2
2 F	0144	1.6	0749	0.3	1359	1.7	2013	0.3
17 SA	0121	1.8	0726	0.2	1337	1.9	1955	0.2
3 SA	0217	1.6	0821	0.4	1433	1.7	2044	0.4
18 SU	0202	1.8	0809	0.2	1420	1.9	2037	0.2
4 SU	0250	1.6	0853	0.4	1506	1.6	2116	0.4
19 M	0245	1.8	0853	0.2	1505	1.8	2121	0.2
5 M	0322	1.5	0926	0.5	1539	1.6	2149	0.5
20 TU	0331	1.7	0940	0.3	1553	1.7	2208	0.3
6 TU	0356	1.5	1001	0.5	1614	1.5	2224	0.6
21 W	0421	1.6	1033	0.4	1646	1.6	2302	0.5
7 W	0433	1.4	1040	0.6	1655	1.4	2306	0.6
22 TH	0518	1.5	1136	0.5	1750	1.4		
8 TH	0518	1.3	1131	0.7	1747	1.3		
23 F ◑	0008	0.6	0629	1.4	1257	0.6	1911	1.3
9 F ◐	0000	0.7	0620	1.3	1243	0.8	1859	1.2
24 SA	0133	0.7	0755	1.4	1430	0.6	2043	1.3
10 SA	0116	0.8	0744	1.2	1416	0.8	2027	1.2
25 SU	0301	0.7	0917	1.4	1550	0.6	2159	1.4
11 SU	0242	0.8	0907	1.3	1535	0.7	2141	1.3
26 M	0411	0.6	1021	1.5	1648	0.5	2254	1.4
12 M	0350	0.7	1008	1.4	1631	0.6	2235	1.4
27 TU	0503	0.5	1110	1.6	1732	0.4	2337	1.5
13 TU	0442	0.6	1055	1.5	1716	0.5	2320	1.5
28 W	0544	0.5	1151	1.6	1809	0.4		
14 W	0526	0.5	1137	1.6	1756	0.4		
29 TH	0014	1.5	0620	0.4	1227	1.7	1842	0.4
15 TH	0001	1.6	0606	0.4	1216	1.7	1835	0.3
30 F ●	0047	1.6	0653	0.4	1300	1.7	1913	0.3

OCTOBER

Day	Time	m	Time	m	Time	m	Time	m
1 SA	0119	1.6	0724	0.4	1332	1.7	1943	0.3
16 SU ○	0058	1.8	0706	0.2	1315	1.9	1931	0.1
2 SU	0149	1.6	0755	0.4	1403	1.7	2012	0.4
17 M	0141	1.9	0749	0.2	1400	1.9	2015	0.2
3 M	0220	1.6	0825	0.4	1435	1.6	2042	0.4
18 TU	0225	1.8	0836	0.2	1447	1.8	2100	0.2
4 TU	0250	1.6	0857	0.5	1506	1.5	2113	0.5
19 W	0312	1.6	0925	0.3	1536	1.7	2148	0.3
5 W	0322	1.5	0940	0.5	1540	1.5	2146	0.5
20 TH	0403	1.7	1020	0.4	1631	1.6	2242	0.5
6 TH	0357	1.5	1007	0.6	1618	1.4	2224	0.6
21 F	0500	1.6	1124	0.5	1736	1.4	2348	0.6
7 F	0438	1.4	1054	0.7	1705	1.3	2312	0.7
22 SA ◑	0610	1.5	1243	0.6	1855	1.3		
8 SA	0533	1.3	1159	0.7	1812	1.2		
23 SU	0110	0.7	0731	1.4	1411	0.6	2023	1.3
9 SU ◐	0022	0.8	0650	1.3	1329	0.8	1941	1.2
24 M	0237	0.7	0851	1.4	1527	0.6	2138	1.3
10 M	0152	0.8	0817	1.3	1454	0.7	2102	1.3
25 TU	0347	0.6	0955	1.5	1623	0.5	2232	1.4
11 TU	0311	0.7	0927	1.4	1555	0.6	2203	1.4
26 W	0439	0.6	1045	1.5	1707	0.5	2314	1.5
12 W	0409	0.6	1020	1.5	1644	0.5	2251	1.5
27 TH	0521	0.5	1126	1.6	1743	0.4	2350	1.5
13 TH	0457	0.5	1106	1.6	1726	0.4	2334	1.6
28 F	0557	0.5	1201	1.6	1815	0.4		
14 F	0540	0.3	1149	1.7	1808	0.3		
29 SA	0023	1.6	0630	0.4	1234	1.6	1845	0.4
15 SA	0016	1.7	0623	0.3	1232	1.8	1849	0.2
30 SU ●	0054	1.6	0701	0.4	1306	1.6	1915	0.4
31 M	0124	1.6	0732	0.4	1338	1.6	1944	0.4

NOVEMBER

Day	Time	m	Time	m	Time	m	Time	m
1 TU	0155	1.6	0803	0.4	1409	1.6	2014	0.4
16 W	0209	1.9	0823	0.2	1432	1.8	2043	0.2
2 W	0225	1.6	0834	0.5	1441	1.5	2045	0.4
17 TH	0257	1.8	0914	0.3	1523	1.7	2132	0.3
3 TH	0257	1.5	0908	0.5	1515	1.5	2118	0.5
18 F	0348	1.7	1008	0.3	1617	1.5	2224	0.4
4 F	0331	1.5	0946	0.6	1553	1.4	2155	0.6
19 SA	0442	1.5	1108	0.5	1717	1.4	2323	0.6
5 SA	0412	1.4	1031	0.6	1639	1.3	2241	0.6
20 SU	0544	1.5	1217	0.6	1826	1.3		
6 SU	0502	1.4	1130	0.7	1739	1.3	2343	0.7
21 M ◑	0034	0.7	0655	1.4	1332	0.6	1944	1.3
7 M ◐	0608	1.3	1246	0.7	1857	1.2		
22 TU	0153	0.7	0809	1.4	1445	0.6	2058	1.3
8 TU	0104	0.7	0727	1.4	1407	0.7	2018	1.3
23 W	0306	0.7	0916	1.4	1545	0.6	2157	1.4
9 W	0225	0.7	0842	1.4	1514	0.6	2125	1.4
24 TH	0405	0.6	1010	1.4	1632	0.6	2244	1.4
10 TH	0332	0.6	0943	1.5	1609	0.5	2220	1.5
25 F	0452	0.6	1055	1.5	1712	0.5	2323	1.5
11 F	0426	0.5	1035	1.6	1657	0.4	2308	1.6
26 SA	0531	0.5	1134	1.5	1747	0.5	2359	1.5
12 SA	0515	0.4	1123	1.7	1742	0.3	2353	1.7
27 SU	0607	0.5	1210	1.5	1820	0.4		
13 SU	0602	0.3	1210	1.8	1827	0.2		
28 M	0032	1.6	0640	0.5	1244	1.5	1851	0.4
14 M ○	0038	1.8	0648	0.2	1256	1.8	1911	0.2
29 TU ●	0103	1.6	0713	0.4	1316	1.6	1922	0.4
15 TU	0123	1.9	0735	0.2	1344	1.8	1957	0.2
30 W	0134	1.6	0745	0.4	1349	1.5	1954	0.4

DECEMBER

Day	Time	m	Time	m	Time	m	Time	m
1 TH	0206	1.6	0819	0.4	1422	1.5	2026	0.4
16 F	0244	1.8	0903	0.3	1509	1.6	2116	0.3
2 F	0239	1.6	0854	0.5	1457	1.5	2100	0.5
17 SA	0332	1.8	0953	0.3	1558	1.6	2203	0.4
3 SA	0314	1.6	0932	0.5	1536	1.4	2137	0.5
18 SU	0421	1.7	1044	0.4	1650	1.5	2253	0.5
4 SU	0355	1.5	1015	0.6	1620	1.4	2221	0.6
19 M	0513	1.6	1139	0.5	1745	1.4	2349	0.6
5 M	0441	1.5	1107	0.6	1714	1.3	2316	0.6
20 TU	0610	1.5	1239	0.6	1848	1.3		
6 TU	0538	1.4	1209	0.6	1819	1.3		
21 W ◑	0054	0.7	0714	1.4	1345	0.6	1958	1.3
7 W ◑	0023	0.7	0645	1.4	1320	0.6	1933	1.3
22 TH	0207	0.7	0822	1.4	1450	0.6	2107	1.3
8 TH	0140	0.7	0757	1.4	1431	0.6	2045	1.4
23 F	0317	0.7	0926	1.4	1549	0.6	2206	1.3
9 F	0253	0.6	0905	1.5	1534	0.5	2148	1.5
24 SA	0416	0.7	1020	1.4	1638	0.6	2254	1.4
10 SA	0357	0.5	1006	1.6	1630	0.4	2244	1.6
25 SU	0504	0.6	1106	1.4	1719	0.6	2335	1.5
11 SU	0453	0.4	1101	1.6	1721	0.3	2334	1.7
26 M	0545	0.6	1147	1.5	1757	0.5		
12 M	0546	0.3	1153	1.7	1810	0.3		
27 TU	0011	1.5	0622	0.5	1224	1.5	1831	0.5
13 TU ○	0023	1.8	0636	0.3	1243	1.8	1857	0.2
28 W	0045	1.6	0657	0.5	1259	1.5	1904	0.4
14 W	0110	1.8	0725	0.2	1332	1.8	1943	0.2
29 TH ●	0117	1.6	0731	0.4	1333	1.5	1937	0.4
15 TH	0157	1.9	0814	0.2	1421	1.7	2030	0.3
30 F	0150	1.6	0805	0.4	1407	1.5	2010	0.4
31 SA	0224	1.6	0840	0.4	1443	1.5	2045	0.4

HAT is 1·9 metres above Chart Datum.

>> FREE monthly updates. Register at <<
www.reedsnauticalalmanac.co.uk

THE MADEIRA ARCHIPELAGO

9.26.17 LIGHTS, BUOYS AND WAYPOINTS

Bold print = light with a nominal range of 15M or more. CAPITALS = place or feature. *CAPITAL ITALICS* = light-vessel, light float or Lanby. *Italics* = Fog signal. ***Bold italics*** = Racon. Many marks/buoys are fitted with AIS (<u>MMSI No</u>); see relevant charts.

ILHA DE PORTO SANTO

Ilhéu de Cima ☆ Fl (3) 15s 123m **21M**; aeromarine; W ☐ twr and dwelling; 33°03'·28N 16°16'·75W.
Ilhéu de Ferro ⚓ LFl 15s 129m 13M; W twr, R lantern with building; 33°02'·3N 16°24'·4W.
N Coast ⚓ Fl (2) 10s 109m 12M; vis 064°-243° (179°); W twr with R bands; 33°05'·79N 16°20'·42W.

PORTO DE PORTO SANTO

S mole head ⚓ Fl G 4s 16m 6M; W col, G bands; 33°03'·5N 16°18'·9W.
N mole head ⚓ Fl R 4s 12m 7M; W col, R bands; 33°03'·58N 16°18'·91W.
Cabos Cabeço ⚓ Fl (2) R 8s 9m 9M; W post, R bands; vis 308°-328° (20°) marks submarine cable; 33°02'·0N 16°21'·4W.

ILHA MADEIRA

E Point, **Ilhéu de Fora** ☆ Fl 10s 102m **20M**; W twr, R cupola, W building; 32°43'·82N 16°39'·41W.
W Point, **Ponta do Pargo** ☆ Fl (3) 20s 311m **26M**; W twr, R cupola; 32°48'·6N 17°15'·5W.
N Point, **Porto de São Jorge** ☆ LFl 5s 270m **15M**; twr and dwelling; 32°49'·9N 16°54'·1W.
Porto do Moniz ⚓ Fl WR 5s; twr; 116°-R-127°-W-116° (349°); 32°52'·2N 17°09'·8W.

QUINTA DO LORDE

S mole head⚓ Fl (2) G 5s 8m 6M; W post, G bands; 32°44'·4N 16°42'·9W.
N mole head ⚓ Fl (2) R 5s 8m 3M; W post, R bands; 32°44'·5N 16°42'·8W.

QUINTA DO LORDE TO FUNCHAL

Caniçal Pier Head ⚓ Fl (2) G 5s 15m 9M; W twr, G bands; 32°43'·97N 16°44'14W.
Machico, São Roque ⚓ LFl WR 5s 16m W9M, R7M; W twr, G bands; 32°43'·97N 16°44'14W.

Santa Catarina Airport ⚓ Aero AIFI WG 10s 72m; control twr; 32°41'·62N 16°46'·53W.

FUNCHAL

Funchal mole head ⚓ Fl R 5s 14m 8M; W col, R bands; vis 275°-075° (160°); 32°38'·5N 16°54'·4W.
Funchal marina E mole ⚓ F G 9m 2M; W col, G bands; 32°38'·7N 16°54'·6W.
Funchal marina W mole ⚓ F R 10m 3M; W col, R bands; 32°38'·69N 16°54'·66W.

FUNCHAL TO CALHETA

Cabos Gorgulho ⚓ Fl (2) R 8s 8m 9M; lantern with R bands; vis 330°-025° (55°); marks submarine cable; 32°38'·2N 16°56'·1W.
Camara de Lobos ⚓ Oc R 6s 22m 9M; W structure; 32°38'·8N 16°58'·5W.
Ribeira Brava ⚓ Fl R 5s 6m 6M; W twr, R bands; 32°40'·07N 17°03'·8W.

CALHETA

E pier head ⚓ Fl G 5s 9m 6M; G/W twr; 32°43'·0N 17°10'·3W.
W pier head ⚓ Fl R 5s 9m 6M; R/W twr; 32°42'·99N 17°10'·3W.
E pier root ⚓ Fl (2) G 5s 7m 2M; G/W twr on wall; vis 002°-042° (40°) & vis 075°-125° (50°); 32°43'·0N 17°10'·3W.
W pier root ⚓ Fl (2) R 5s 7m 2M; lantern on wall; vis 130°-320° (190°); 32°43'·0N 17°10'·3W.
Inner bkwtr ⚓ Fl (3) G 5s 7m 2M; lantern on wall; 32°42'·8N 17°10'·1W.

ILHAS DESERTAS

Ilhéu Chão (Farilhão) ⚓ LFl (2) 15s 111m 13M; twr and building; 32°35'·2N 16°32'·5W.
Ilhéu Bugio (Ponta da Agulha) ⚓ Fl 4s 71m 13M; 163°-100°; twr; 32°24'·0N 16°27'·6W.

ILHAS SELVAGENS

Selvagem Grande ⚓ Fl 4s 162m 13M; W col R bands; 30°08'·73N 15°52'·27W. (Can often be extinguished due to storms)
Selvagem Pequena ⚓ Fl (2) 6s 49m 12M; R lantern on col; 30°02'·16N 16°01'·66W. (Can often be extinguished due to storms.)

9.26.18 PORTO DO PORTO SANTO
33°03'·45N 16°18'·80W (Hbr ent) ✿✿✿✿◊◊◊✿✿

CHARTS AC 1689; PC 36401; Imray E3

TIDES

Standard Port CASABLANCA (→)

Times				Height (metres)			
High Water		Low Water		MHWS	MHWN	MLWN	MLWS
ND	ND	ND	ND	3·5	2·8	1·5	0·8
Differences PORTO SANTO							
−0014	ND	−0011	ND	−1·1	−1·0	−0·6	−0·4

SHELTER Good but can be a strong surge.

NAVIGATION Ilha de Porto Santo lies approx 440M SW of Portugal. Porto do Porto Santo is the only destination for yachts on the island. If approaching from the NE head for WPT 33°08'N 16°13'W to avoid Ilhéu de Fora (note there is one of these near Madeira too). Then proceed round the E side of Ilhéu de Cima, looking out for the lighthouse Fl(3) 15s 123m 21M. The passage between this island and Porto Santo island is very rocky. WPT 33°02·80N 16°16·00W, then 285° for 2·6M to reach close SW of the hbr ent. The marina is in the NW corner of the harbour.

LIGHTS AND MARKS See chartlet and Lights, buoys and waypoints.

COMMUNICATIONS Emergency 112; Police 291 208 400; British Consul 808 20 35 37; Hospital 291 705 600; GNR (Customs) 291 982 328; Polícia Maritima 291 980 076; HM 291 980 070, VHF Ch 11; Marina VHF Ch 09; Tourist Office 291 982 361.

FACILITIES Visitors should call ahead on Ch 09 (0800-1800) and then go to the arrivals pontoon on the end of the concrete wall, even if wanting to take one of the moorings.

Marina and Boatyard ☎291 980 080, marinaportosanto@quintadolorde.pt www.quintadolorde.pt 140 (inc ✔) + 40 ⚓; €2·35 (pontoon inc ⚡ and ♣) max LOA 25m and draught <3·5m; multihull +50%; ⚓ €1·35/m discount for 7+ days. ♠ (cash only) in the Commercial hbr (HW only), book through the marina. ▣ ♠ ♨ (basic) Ⓔ ⊞ ⚒ ⚔ Ice ⚑ ◪(25t).

Town ✕ ⬚ ☷ 🏧 Ⓒ ✉. Ferry to Funchal (in Commercial hbr) ✈ 5km (to Funchal).

ANCHORAGE W OF PORTO SANTO MARINA

When the weather is fair it is possible to ⚓ off the beach at Baia do Porto Santo in the small craft ⚓age. However, this is not free, approx €6·00 per vessel <15m which is payable to the Harbour Authority, and no facilities are supplied. There can be some katabatic gusts down the valley from Pico do Facho.

9.26.19 QUINTA DO LORDE
32°44'·43N 16°42'·86W (S mole) ✿✿✿✿◊◊◊✿✿

CHARTS AC 1831, 1684, 1685; PC 36402; Imray E3

TIDES See Funchal.

SHELTER Good in the marina. However, Quinta do Lorde often experiences light winds even if the weather is strong offshore, due to being in the shelter of high cliffs.

NAVIGATION Ilhéu de Fora/São Lourenço ltho is conspic. WPT 32°43'·32N 16°38'·95W is 0·5M SW of Ilhéu de Fora; then 0·5M W and 290°/3·95M to the marina ent. Maintain 0·5M offing when rounding the island to avoid underwater rocks, esp if going twds Baia de Abra ⚓age 3·5M E of Quinta do Lorde.

LIGHTS AND MARKS Ilhéu de Fora lighthouse, Fl.10s 103m 20M, a W twr with red cupola on top of a W building. See also chartlet and Lights, buoys and waypoints.

COMMUNICATIONS Lifeboat (SANAS) 291 230112; Emergency 112; Police 291 208400; British Consul 808 20 35 37; Hospital 291 705 600; Marina VHF 09, 16; Boatyard VHF Ch 09.

FACILITIES Visitors should call ahead. If not met, make your way to the waiting pontoon which is along the bkwtr.

Marina ☎291 969607, marina@quintadolorde.pt www.quintadolorde.pt 264 (inc 150 ✔); €2·20, draught <4·5m; multihulls +50%. ♠ ♨ (extra charge for card payment) ✕ ⬚ ☷ ▣ Ⓒ ♨ ♠ ⊹ ⚓ Ice Car hire. Swimming pool in 5 star resort nearby.

Boatyard (3·7M SW of the marina, beneath the airport runway) ☎291 969 800, www.repmaritima.com Ⓔ ⊞ ⚒ ⚔ ◪(35t).

✈ Madeira International (10km).

ANCHORAGE 3·5M E OF QUINTA DO LORDE MARINA

Baía de Abra 32°44·60N 16°41·60W. Approx 3·5M E of Quinta do Lorde marina. Good protection from N'lies but NE'lies can occasionally make it a bit choppy. Peaceful. No facilities ashore, but some good walking.

MINOR HARBOURS W OF QUINTA DO LORDE MARINA

Caniçal 32°44·00N 16°44·21W. Approx 1·25M SW of Quinta do Lorde Marina. Commercial hbr with no facilities for yachts. The ⚓age is off the small beach W of the hbr, near the Yacht Club. This area is also popular with local fishermen. Most facilities ashore including a BY.

Machico 32°42·94N 16°45·68W. Approx 3M SW of Quinta do Lorde marina. When coming from the E, give a wide berth to Baixa da Cruz, a rock with 1·2M clearance which is very hard to locate visually. There is a small marina here, but only for local boats. ⚓ off the beach. In W to SW winds, ⚓ing below the São Roque light may give better shelter. There may be the possibility of rafting up on the extended E mole. Entry at night is not recommended due to many unlit moorings and difficulty identifying navigation lights.

9.26.20 FUNCHAL

32°38'·30N 16°54'·20W (0·3M SE of bkwtr head) ⊛❀⚓⚓♧♧♧

CHARTS AC 1831, 1685, 1689; PC 36402; Imray E3

TIDES

Standard Port CASABLANCA (→)

Times				Height (metres)			
High Water		Low Water		MHWS	MHWN	MLWN	MLWS
ND	ND	ND	ND	3·5	2·8	1·5	0·8
Differences FUNCHAL							
−0023	ND	−0019	ND	−1·1	−1·0	−0·5	−0·4

SHELTER The hbr is open to the E and S and unlike the rest of the island, Funchal often experiences S to SW'ly winds, dying at night. This can leave an onshore swell overnight which makes the hbr uncomfortable.

The marina is completely sheltered, but is often overcrowded. The permitted ⚓age is a small area E of the marina off the beach which can be very exposed and is partly cordoned off by six lit Y buoys.

Works to rebuild the seafront, including a new cruise ship terminal and pier E of the marina, are ongoing. The future of the small craft anchorage remains uncertain (2015). Yachts are not permitted to ⚓ S of the 10M contour.

NAVIGATION No hazards when appr from E or W. Watch out for high speed vessels when close-in and at night, beware of large unlit mooring buoys in the anchorage.

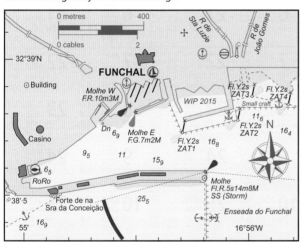

LIGHTS AND MARKS In daylight from the E, Ponta do Garajau, (3·5M E of Funchal) is very visible due to a large floodlit statue of Christ on top. From the W, the oil terminal at Praia Formosa is visible by day and at night the lt at Camara de Lobos, Oc R 6s 22m 9M, acts as a guide. For hbr lts see the chartlet and Lights, buoys and waypoints.

COMMUNICATIONS Lifeboat (SANAS) 291 230112; Guarda Nacional Republicana (Customs) 291 214460; Polícia Maritima 291 213110; Brit Consul 808 20 35 37; HM 291 225281/225288 (Capimarfunchal) VHF Ch 11, 12, 13; Marina VHF Ch 09; Tourist Office 291 211902.

FACILITIES Marina ☎291 232717 geral@marina-funchal.com www.marina-funchal.com 210, €2·50, draught <3m. Full of local boats, very limited space for ❶. The reception dock/fuel berth is immed st'bd as you enter the hbr. ▮▮◨◨ⓔ▦⚒⚠.

Town ✕⌂☴Ⓑ✗.

Ferries Porto Santo. ✈ Madeira International 18km.

9.26.21 CALHETA

32°42'·85N 17°10'·60W (0·3M SW of bkwtr head) ⊛❀⚓♧♧

CHARTS AC 1831; PC 36403; Imray E3

TIDES See Funchal.

SHELTER Despite the addition of outer moles, swell still seems to enter the marina and is more of an issue at the W end.

NAVIGATION Best to avoid entry at night on your first visit. At low tide, deeper draught yachts should keep to mid channel (reports of concrete debris on S side of ent chan).

COMMUNICATIONS Lifeboat (SANAS) 291 230 112; Brit Consul 808 20 35 37; Marina VHF Ch 09, 16.

FACILITIES Marina ☎291 824 003 info@portoderecreiodacalheta. com www.portoderecreiodacalheta.com 339 inc ❶ €2·50, draught <3·5m; multihulls +70% Call ahead on VHF for a berth. Otherwise tie up to the pontoon on the W end of the W bkwtr. ▮▮ (cr/cards accepted) ✕⌂☴Ⓒ◨⚓⚒.

Clube Naval do Calheta ☎291 823675, ✈ Madeira International 40km.

Town (1·25M away) ✕⌂☴✉Ⓑ.

ANCHORAGE 2M E OF CALHETA MARINA
Madalena do Mar 32°41·90N 017°08·30W. Fair weather ⚓age only, with village supplying basic facilities.

STANDARD TIME (UT)
For Moroccan Summer Time
add ONE hour in
non-shaded areas

CASABLANCA LAT 33°36′N LONG 07°37′W
TIMES AND HEIGHTS OF HIGH AND LOW WATERS

Dates in red are **SPRINGS**
Dates in blue are NEAPS

YEAR 2016

JANUARY

	Time	m		Time	m
1 F	0007 0638 1247 1910	1.5 3.1 1.5 2.9	**16** SA ☽	0009 0634 1246 1911	1.2 3.4 1.2 3.2
2 SA ☽	0102 0734 1348 2014	1.7 2.9 1.6 2.8	**17** SU	0114 0739 1356 2023	1.3 3.2 1.3 3.1
3 SU	0211 0840 1458 2122	1.8 2.8 1.7 2.8	**18** M	0234 0855 1516 2140	1.4 3.1 1.4 3.1
4 M	0328 0948 1605 2225	1.7 2.8 1.6 2.9	**19** TU	0357 1013 1632 2252	1.4 3.1 1.3 3.2
5 TU	0435 1049 1701 2318	1.6 2.9 1.5 3.0	**20** W	0510 1123 1736 2353	1.3 3.1 1.2 3.3
6 W	0530 1142 1749	1.5 3.0 1.4	**21** TH	0610 1221 1828	1.1 3.3 1.1
7 TH	0004 0616 1227 1831	3.2 1.3 3.1 1.2	**22** F	0044 0700 1311 1913	3.5 0.9 3.4 0.9
8 F	0046 0658 1309 1910	3.4 1.2 3.3 1.1	**23** SA	0130 0744 1355 1954	3.6 0.8 3.5 0.9
9 SA	0126 0737 1349 1948	3.5 1.0 3.4 1.0	**24** SU ○	0212 0823 1435 2031	3.7 0.8 3.5 0.8
10 SU ●	0205 0816 1429 2026	3.7 0.9 3.5 0.9	**25** M	0251 0859 1512 2106	3.7 0.8 3.5 0.9
11 M	0245 0855 1509 2106	3.8 0.8 3.6 0.8	**26** TU	0327 0933 1547 2139	3.7 0.9 3.4 0.9
12 TU	0325 0935 1550 2146	3.8 0.8 3.6 0.8	**27** W	0402 1007 1622 2213	3.6 1.0 3.4 1.0
13 W	0407 1016 1633 2229	3.8 0.8 3.6 0.9	**28** TH	0438 1041 1658 2249	3.5 1.1 3.3 1.2
14 TH	0452 1100 1719 2315	3.7 0.9 3.5 1.0	**29** F	0514 1117 1736 2327	3.3 1.2 3.1 1.3
15 F	0539 1149 1810	3.6 1.0 3.3	**30** SA	0554 1157 1817	3.1 1.4 3.0
			31 SU	0010 0638 1243 1907	1.5 3.0 1.6 2.8

FEBRUARY

	Time	m		Time	m
1 M ☽	0103 0733 1344 2011	1.6 2.8 1.7 2.7	**16** TU	0212 0835 1451 2118	1.4 2.9 1.5 3.0
2 TU	0217 0844 1501 2125	1.7 2.7 1.7 2.7	**17** W	0343 0959 1617 2237	1.4 2.9 1.4 3.0
3 W	0342 1000 1616 2234	1.7 2.7 1.6 2.9	**18** TH	0501 1114 1725 2341	1.3 3.0 1.3 3.2
4 TH	0453 1106 1716 2332	1.6 2.9 1.5 3.0	**19** F	0601 1212 1817	1.1 3.1 1.2
5 F	0548 1200 1805	1.4 3.0 1.3	**20** SA	0032 0647 1258 1859	3.3 1.0 3.3 1.0
6 SA	0019 0634 1245 1848	3.3 1.1 3.2 1.1	**21** SU	0115 0726 1338 1936	3.5 0.9 3.4 0.9
7 SU	0102 0714 1327 1928	3.5 0.9 3.4 0.9	**22** M ○	0153 0801 1414 2010	3.6 0.8 3.4 0.8
8 M ●	0143 0755 1408 2007	3.7 0.7 3.6 0.7	**23** TU	0228 0834 1448 2042	3.6 0.8 3.5 0.8
9 TU	0224 0834 1449 2048	3.9 0.6 3.7 0.6	**24** W	0301 0905 1520 2114	3.6 0.8 3.5 0.8
10 W	0306 0914 1531 2129	4.0 0.5 3.8 0.6	**25** TH	0334 0936 1552 2146	3.6 0.9 3.4 0.9
11 TH	0349 0956 1614 2212	4.0 0.6 3.7 0.7	**26** F	0407 1008 1626 2219	3.5 1.0 3.3 1.0
12 F	0433 1040 1659 2258	3.9 0.7 3.6 0.8	**27** SA	0441 1041 1700 2254	3.4 1.1 3.2 1.2
13 SA	0521 1127 1749 2349	3.6 0.9 3.4 1.0	**28** SU	0517 1116 1737 2332	3.2 1.3 3.1 1.3
14 SU	0614 1221 1846	3.4 1.1 3.2	**29** M	0556 1155 1819	3.0 1.5 2.9
15 M ☽	0052 0717 1327 1956	1.2 3.1 1.3 3.0			

MARCH

	Time	m		Time	m
1 TU ☽	0016 0643 1244 1912	1.5 2.8 1.6 2.8	**16** W ☽	0155 0817 1428 2055	1.4 2.9 1.5 2.9
2 W	0118 0747 1357 2025	1.6 2.7 1.7 2.7	**17** TH	0326 0944 1556 2217	1.4 2.8 1.5 3.0
3 TH	0246 0910 1526 2146	1.7 2.7 1.7 2.8	**18** F	0445 1058 1706 2322	1.3 2.9 1.4 3.1
4 F	0411 1028 1639 2254	1.6 2.8 1.5 3.0	**19** SA	0542 1154 1757	1.2 3.0 1.3
5 SA	0515 1129 1735 2348	1.3 3.0 1.3 3.2	**20** SU	0012 0625 1238 1838	3.2 1.1 3.2 1.1
6 SU	0604 1218 1822	1.1 3.2 1.1	**21** M	0053 0701 1315 1913	3.4 1.0 3.3 1.0
7 M	0035 0647 1302 1904	3.5 0.9 3.5 0.8	**22** TU	0129 0734 1349 1946	3.5 0.9 3.4 0.9
8 TU	0118 0729 1344 1945	3.7 0.6 3.7 0.6	**23** W ○	0202 0805 1421 2018	3.5 0.8 3.5 0.9
9 W ●	0200 0810 1426 2027	3.9 0.5 3.9 0.5	**24** TH	0234 0835 1452 2049	3.5 0.9 3.5 0.9
10 TH	0244 0851 1509 2110	4.0 0.4 3.9 0.4	**25** F	0305 0906 1524 2121	3.5 0.9 3.5 0.9
11 F	0328 0935 1554 2155	4.0 0.5 3.9 0.5	**26** SA	0338 0937 1557 2154	3.5 1.0 3.4 1.0
12 SA	0415 1019 1640 2242	3.9 0.6 3.7 0.7	**27** SU	0412 1010 1631 2229	3.3 1.1 3.3 1.1
13 SU	0503 1106 1729 2334	3.6 0.9 3.5 0.9	**28** M	0448 1044 1707 2305	3.2 1.2 3.2 1.3
14 M	0557 1159 1825	3.4 1.1 3.3	**29** TU	0525 1121 1746 2347	3.0 1.4 3.0 1.4
15 TU ☽	0036 0700 1304 1933	1.2 3.1 1.4 3.0	**30** W	0610 1207 1835	2.9 1.5 2.9
			31 TH ☽	0043 0709 1312 1940	1.5 2.7 1.6 2.8

APRIL

	Time	m		Time	m
1 F	0203 0829 1440 2100	1.6 2.7 1.7 2.9	**16** SA	0414 1029 1634 2252	1.4 2.9 1.5 3.0
2 SA	0331 0950 1600 2214	1.5 2.8 1.5 3.0	**17** SU	0511 1125 1727 2343	1.3 3.0 1.4 3.1
3 SU	0438 1055 1702 2314	1.3 3.0 1.3 3.2	**18** M	0554 1209 1810	1.2 3.1 1.3
4 M	0532 1148 1753	1.1 3.3 1.1	**19** TU	0024 0630 1246 1846	3.2 1.1 3.2 1.1
5 TU	0005 0618 1234 1838	3.5 0.8 3.5 0.8	**20** W	0100 0703 1320 1920	3.3 1.0 3.3 1.0
6 W	0052 0702 1319 1922	3.7 0.6 3.7 0.6	**21** TH	0134 0734 1352 1953	3.4 1.0 3.4 1.0
7 TH ●	0137 0745 1403 2007	3.9 0.5 3.9 0.4	**22** F ○	0206 0806 1425 2026	3.4 0.9 3.5 0.9
8 F	0223 0829 1448 2053	4.0 0.4 4.0 0.4	**23** SA	0239 0838 1458 2059	3.4 0.9 3.5 0.9
9 SA	0310 0914 1535 2140	4.0 0.5 3.9 0.5	**24** SU	0313 0911 1532 2134	3.4 1.0 3.5 1.0
10 SU	0358 1000 1623 2229	3.8 0.6 3.8 0.7	**25** M	0349 0945 1607 2210	3.3 1.1 3.4 1.1
11 M	0448 1048 1713 2321	3.6 0.9 3.6 0.9	**26** TU	0425 1020 1644 2247	3.2 1.2 3.3 1.2
12 TU	0541 1140 1808	3.3 1.1 3.3	**27** W	0504 1059 1724 2330	3.1 1.3 3.2 1.3
13 W	0021 0643 1242 1912	1.1 3.0 1.4 3.1	**28** TH	0549 1145 1811	3.0 1.5 3.1
14 TH ☽	0134 0756 1359 2028	1.3 2.9 1.5 3.0	**29** F	0023 0645 1245 1911	1.4 2.9 1.6 3.0
15 F	0258 0917 1524 2146	1.4 2.8 1.6 3.0	**30** SA ☽	0132 0757 1404 2024	1.5 2.8 1.6 3.0

HAT is 4·0 metres above Chart Datum.

Portuguese Islands

STANDARD TIME (UT)
For Moroccan Summer Time add ONE hour in **non-shaded areas**

CASABLANCA LAT 33°36'N LONG 07°37'W
TIMES AND HEIGHTS OF HIGH AND LOW WATERS

Dates in red are **SPRINGS**
Dates in blue are NEAPS

YEAR 2016

MAY

Day	Time	m	Day	Time	m
1 SU	0252	1.4	**16** M	0428	1.4
	0914	2.9		1046	2.9
	1523	1.5		1648	1.5
	2137	3.1		2305	3.0
2 M	0404	1.3	**17** TU	0516	1.3
	1021	3.1		1134	3.0
	1629	1.3		1736	1.4
	2241	3.3		2350	3.1
3 TU	0500	1.1	**18** W	0556	1.2
	1118	3.3		1214	3.2
	1724	1.1		1817	1.3
	2337	3.5			
4 W	0551	0.8	**19** TH	0030	3.2
	1208	3.5		0632	1.1
	1814	0.8		1250	3.3
				1854	1.2
5 TH	0028	3.7	**20** F	0106	3.3
	0638	0.7		0706	1.0
	1255	3.7		1324	3.4
	1902	0.6		1930	1.1
6 F ●	0117	3.8	**21** SA ○	0141	3.3
	0724	0.5		0740	1.0
	1343	3.9		1359	3.4
	1950	0.5		2005	1.0
7 SA	0205	3.9	**22** SU	0216	3.4
	0809	0.5		0814	1.0
	1430	3.9		1434	3.5
	2038	0.5		2040	1.0
8 SU	0254	3.9	**23** M	0252	3.4
	0856	0.6		0849	1.0
	1519	3.9		1510	3.5
	2127	0.5		2117	1.0
9 M	0344	3.7	**24** TU	0329	3.3
	0943	0.7		0925	1.1
	1607	3.8		1547	3.5
	2216	0.7		2154	1.0
10 TU	0434	3.5	**25** W	0407	3.3
	1031	0.9		1002	1.1
	1657	3.6		1626	3.4
	2307	0.9		2233	1.1
11 W	0525	3.3	**26** TH	0448	3.2
	1121	1.1		1043	1.2
	1749	3.4		1707	3.3
				2316	1.2
12 TH	0002	1.1	**27** F	0533	3.1
	0622	3.1		1129	1.3
	1216	1.3		1754	3.2
	1847	3.2			
13 F ◐	0104	1.3	**28** SA	0005	1.3
	0726	2.9		0626	3.0
	1322	1.4		1225	1.4
	1952	3.0		1849	3.2
14 SA	0216	1.4	**29** SU ◐	0106	1.3
	0838	2.8		0730	3.0
	1439	1.6		1334	1.5
	2103	3.0		1954	3.1
15 SU	0328	1.5	**30** M	0217	1.3
	0948	2.8		0842	3.0
	1550	1.6		1449	1.4
	2210	3.0		2105	3.1
			31 TU	0329	1.2
				0950	3.1
				1559	1.3
				2212	3.2

JUNE

Day	Time	m	Day	Time	m
1 W	0432	1.1	**16** TH	0520	1.4
	1051	3.3		1139	3.1
	1700	1.1		1747	1.4
	2313	3.4		2359	3.0
2 TH	0527	0.9	**17** F	0602	1.3
	1146	3.5		1220	3.2
	1755	0.9		1829	1.2
3 F	0009	3.5	**18** SA	0040	3.1
	0618	0.8		0641	1.2
	1237	3.6		1258	3.3
	1848	0.7		1908	1.1
4 SA	0102	3.7	**19** SU	0118	3.2
	0707	0.7		0717	1.1
	1327	3.8		1335	3.4
	1938	0.6		1946	1.0
5 SU ●	0153	3.7	**20** M ○	0156	3.3
	0755	0.6		0753	1.0
	1416	3.9		1412	3.5
	2027	0.5		2022	1.0
6 M	0242	3.7	**21** TU	0233	3.3
	0841	0.7		0830	1.0
	1504	3.9		1450	3.6
	2115	0.6		2059	0.9
7 TU	0330	3.6	**22** W	0311	3.4
	0926	0.8		0907	1.0
	1551	3.8		1528	3.6
	2202	0.7		2137	0.9
8 W	0417	3.5	**23** TH	0350	3.4
	1011	0.9		0945	1.0
	1638	3.6		1608	3.6
	2248	0.8		2216	0.9
9 TH	0504	3.3	**24** F	0431	3.3
	1057	1.1		1026	1.1
	1725	3.5		1649	3.5
	2335	1.0		2258	1.0
10 F	0553	3.1	**25** SA	0515	3.3
	1145	1.3		1111	1.2
	1815	3.3		1735	3.4
				2344	1.1
11 SA	0026	1.2	**26** SU	0605	3.2
	0647	3.0		1203	1.2
	1240	1.4		1827	3.3
	1911	3.1			
12 SU ◐	0125	1.4	**27** M ◐	0039	1.2
	0749	2.9		0704	3.1
	1344	1.6		1306	1.3
	2013	3.0		1928	3.2
13 M	0231	1.5	**28** TU	0145	1.2
	0856	2.8		0812	3.1
	1456	1.6		1419	1.4
	2118	2.9		2037	3.1
14 TU	0336	1.5	**29** W	0258	1.3
	0959	2.9		0923	3.1
	1602	1.6		1534	1.3
	2220	2.9		2149	3.1
15 W	0432	1.4	**30** TH	0409	1.2
	1053	2.9		1029	3.2
	1659	1.5		1643	1.2
	2313	2.9		2257	3.2

JULY

Day	Time	m	Day	Time	m
1 F	0511	1.1	**16** SA	0533	1.4
	1130	3.4		1150	3.1
	1745	1.0		1804	1.3
	2358	3.3			
2 SA	0607	0.9	**17** SU	0015	3.0
	1224	3.5		0617	1.2
	1840	0.8		1232	3.2
				1846	1.2
3 SU	0052	3.5	**18** M	0056	3.1
	0657	0.8		0656	1.1
	1315	3.7		1312	3.4
	1930	0.7		1925	1.0
4 M ●	0143	3.6	**19** TU ○	0135	3.3
	0743	0.7		0733	1.0
	1403	3.8		1350	3.5
	2017	0.6		2002	0.9
5 TU	0230	3.6	**20** W	0213	3.4
	0830	0.7		0810	0.9
	1449	3.8		1428	3.6
	2101	0.6		2038	0.8
6 W	0314	3.6	**21** TH	0251	3.5
	0909	0.8		0847	0.8
	1532	3.7		1507	3.7
	2142	0.7		2116	0.7
7 TH	0356	3.5	**22** F	0330	3.5
	0949	0.9		0926	0.8
	1614	3.6		1546	3.7
	2222	0.8		2155	0.7
8 F	0438	3.3	**23** SA	0411	3.5
	1029	1.0		1006	0.9
	1656	3.5		1628	3.7
	2302	1.0		2236	0.8
9 SA	0520	3.2	**24** SU	0454	3.4
	1110	1.2		1050	1.0
	1739	3.3		1713	3.6
	2344	1.2		2321	0.9
10 SU	0604	3.1	**25** M	0542	3.3
	1156	1.3		1140	1.1
	1825	3.1		1804	3.4
11 M	0031	1.3	**26** TU ◐	0012	1.1
	0656	2.9		0637	3.2
	1249	1.5		1240	1.2
	1919	2.9		1903	3.2
12 TU ◐	0128	1.3	**27** W	0115	1.2
	0756	2.8		0744	3.1
	1355	1.6		1355	1.3
	2021	2.8		2015	3.1
13 W	0235	1.6	**28** TH	0232	1.3
	0903	2.8		0900	3.0
	1510	1.6		1518	1.3
	2129	2.8		2133	3.0
14 TH	0343	1.6	**29** F	0352	1.3
	1007	2.8		1014	3.1
	1619	1.6		1635	1.2
	2233	2.8		2248	3.1
15 F	0443	1.5	**30** SA	0502	1.2
	1116	2.9		1120	3.3
	1716	1.5		1740	1.0
	2328	2.9		2351	3.2
			31 SU	0559	1.0
				1216	3.4
				1834	0.9

AUGUST

Day	Time	m	Day	Time	m
1 M	0045	3.3	**16** TU	0032	3.1
	0648	0.9		0632	1.1
	1305	3.6		1246	3.4
	1921	0.7		1900	1.0
2 TU ●	0132	3.4	**17** W	0112	3.3
	0731	0.8		0710	1.0
	1349	3.7		1325	3.6
	2003	0.6		1937	0.8
3 W	0215	3.5	**18** TH ○	0149	3.5
	0811	0.8		0747	0.8
	1431	3.7		1403	3.7
	2041	0.7		2013	0.7
4 TH	0254	3.5	**19** F	0227	3.6
	0848	0.8		0825	0.7
	1510	3.7		1442	3.8
	2117	0.7		2051	0.6
5 F	0331	3.5	**20** SA	0307	3.7
	0923	0.8		0904	0.7
	1546	3.6		1523	3.9
	2152	0.8		2130	0.6
6 SA	0407	3.4	**21** SU	0348	3.7
	0959	1.0		0945	0.7
	1623	3.5		1606	3.8
	2226	1.0		2212	0.7
7 SU	0444	3.3	**22** M	0431	3.6
	1035	1.1		1029	0.8
	1700	3.3		1652	3.6
	2302	1.1		2256	0.8
8 M	0522	3.1	**23** TU	0519	3.4
	1114	1.3		1119	1.0
	1741	3.1		1742	3.4
	2342	1.3		2347	1.0
9 TU	0605	3.0	**24** W	0613	3.3
	1158	1.4		1219	1.2
	1826	2.9		1843	3.2
10 W ◐	0029	1.5	**25** TH ◐	0049	1.3
	0659	2.8		0720	3.1
	1254	1.6		1336	1.3
	1922	2.8		1957	3.0
11 TH	0129	1.6	**26** F	0210	1.4
	0800	2.7		0841	3.0
	1409	1.7		1508	1.4
	2033	2.7		2123	2.9
12 F	0247	1.7	**27** SA	0340	1.4
	0914	2.7		1003	3.1
	1535	1.7		1631	1.3
	2149	2.7		2242	3.0
13 SA	0403	1.6	**28** SU	0454	1.3
	1023	2.8		1111	3.2
	1644	1.5		1734	1.1
	2256	2.8		2345	3.1
14 SU	0503	1.5	**29** M	0551	1.1
	1119	3.0		1206	3.4
	1738	1.4		1824	0.9
	2348	2.9			
15 M	0551	1.3	**30** TU	0035	3.3
	1205	3.2		0636	0.9
	1821	1.2		1252	3.5
				1906	0.8
			31 W	0117	3.4
				0716	0.9
				1333	3.6
				1943	0.8

HAT is 4·0 metres above Chart Datum.

》 **FREE** monthly updates. Register at 《
www.reedsnauticalalmanac.co.uk

1039

STANDARD TIME (UT)
For Moroccan Summer Time add ONE hour in **non-shaded areas**

CASABLANCA LAT 33°36'N LONG 07°37'W
TIMES AND HEIGHTS OF HIGH AND LOW WATERS

Dates in red are **SPRINGS**
Dates in blue are **NEAPS**

YEAR 2016

SEPTEMBER

Time	m	Time	m
1 0155	3.5	**16** 0123	3.6
0751	0.8	0722	0.8
TH 1410	3.7	F 1338	3.8
● 2017	0.7	○ 1947	0.6
2 0231	3.5	**17** 0202	3.8
0825	0.8	0832	0.6
F 1444	3.6	SA 1418	4.0
2049	0.8	2025	0.5
3 0304	3.5	**18** 0243	3.9
0857	0.9	0843	0.6
SA 1517	3.6	SU 1501	4.0
2119	0.9	2106	0.5
4 0336	3.4	**19** 0325	3.9
0929	1.0	0926	0.6
SU 1550	3.5	M 1545	3.9
2151	1.0	2149	0.6
5 0409	3.3	**20** 0410	3.7
1002	1.1	1012	0.7
M 1624	3.3	TU 1633	3.7
2223	1.1	2235	0.8
6 0444	3.2	**21** 0459	3.6
1037	1.2	1103	0.9
TU 1701	3.2	W 1725	3.4
2259	1.3	2326	1.1
7 0522	3.1	**22** 0554	3.3
1117	1.4	1203	1.2
W 1742	3.0	TH 1826	3.1
2339	1.5		
8 0606	2.9	**23** 0028	1.3
1204	1.6	0701	3.1
TH 1830	2.8	F 1322	1.4
		◖ 1943	2.9
9 0029	1.6	**24** 0151	1.5
0701	2.8	0823	3.0
F 1309	1.7	SA 1457	1.4
◖ 1936	2.7	2112	2.9
10 0143	1.7	**25** 0326	1.5
0816	2.7	0948	3.0
SA 1443	1.7	SU 1620	1.3
2101	2.6	2231	2.9
11 0316	1.7	**26** 0441	1.4
0936	2.8	1057	3.2
SU 1607	1.6	M 1721	1.2
2219	2.7	2331	3.1
12 0429	1.6	**27** 0536	1.3
1043	3.0	1150	3.3
M 1705	1.4	TU 1807	1.1
2318	2.9		
13 0522	1.4	**28** 0018	3.2
1135	3.2	0619	1.1
TU 1751	1.2	W 1234	3.5
		1845	1.0
14 0004	3.2	**29** 0057	3.4
0605	1.2	0656	1.0
W 1218	3.4	TH 1312	3.5
1831	1.0	1918	0.9
15 0044	3.4	**30** 0132	3.5
0644	1.0	0729	1.0
TH 1258	3.6	F 1346	3.6
1909	0.8	1950	0.9

OCTOBER

Time	m	Time	m
1 0205	3.5	**16** 0138	3.9
0801	0.9	0741	0.6
SA 1418	3.6	SU 1357	4.0
● 2019	0.9	○ 2003	0.5
2 0236	3.5	**17** 0221	4.0
0832	0.9	0825	0.5
SU 1449	3.5	M 1442	4.0
2049	0.9	2046	0.6
3 0307	3.5	**18** 0306	4.0
0903	1.0	0911	0.6
M 1521	3.5	TU 1529	3.9
2119	1.0	2131	0.7
4 0338	3.4	**19** 0353	3.8
0936	1.1	0959	0.7
TU 1554	3.4	W 1618	3.7
2151	1.1	2218	0.9
5 0412	3.3	**20** 0443	3.7
1010	1.2	1051	0.9
W 1629	3.2	TH 1712	3.4
2225	1.3	2309	1.2
6 0448	3.2	**21** 0538	3.4
1046	1.4	1151	1.2
TH 1708	3.0	F 1812	3.2
2301	1.5		
7 0528	3.0	**22** 0010	1.4
1129	1.5	0642	3.2
F 1752	2.9	SA 1305	1.4
2346	1.6	◖ 1926	3.0
8 0617	2.9	**23** 0127	1.6
1225	1.7	0759	3.1
SA 1851	2.7	SU 1433	1.5
		2050	2.9
9 0049	1.8	**24** 0259	1.6
0722	2.8	0921	3.1
SU 1347	1.7	M 1554	1.4
◖ 2012	2.7	2207	2.9
10 0222	1.8	**25** 0414	1.6
0845	2.8	1031	3.1
M 1520	1.7	TU 1655	1.4
2136	2.8	2307	3.1
11 0346	1.7	**26** 0511	1.4
1000	3.0	1126	3.2
TU 1627	1.5	W 1741	1.2
2241	3.0	2353	3.2
12 0446	1.5	**27** 0555	1.3
1058	3.2	1210	3.3
W 1717	1.4	TH 1818	1.2
2331	3.2		
13 0534	1.2	**28** 0032	3.3
1147	3.5	0632	1.2
TH 1800	1.0	F 1247	3.4
		1851	1.1
14 0015	3.5	**29** 0107	3.4
0617	1.0	0707	1.1
F 1230	3.7	SA 1321	3.5
1841	0.8	1922	1.0
15 0056	3.7	**30** 0139	3.5
0658	0.8	0739	1.1
SA 1313	3.9	SU 1354	3.5
1921	0.6	● 1953	1.0
		31 0210	3.5
		0811	1.0
		M 1425	3.5
		2023	1.0

NOVEMBER

Time	m	Time	m
1 0242	3.5	**16** 0251	4.0
0843	1.1	0900	0.6
TU 1458	3.4	W 1517	3.9
2055	1.1	2117	0.7
2 0314	3.5	**17** 0340	3.9
0916	1.1	0950	0.7
W 1532	3.4	TH 1608	3.7
2127	1.2	2205	0.9
3 0348	3.4	**18** 0430	3.8
0951	1.2	1041	0.9
TH 1607	3.2	F 1700	3.5
2201	1.3	2254	1.1
4 0424	3.3	**19** 0522	3.6
1028	1.3	1136	1.1
F 1645	3.1	SA 1755	3.2
2237	1.5	2349	1.4
5 0503	3.2	**20** 0620	3.3
1108	1.5	1238	1.3
SA 1727	3.0	SU 1859	3.0
2319	1.6		
6 0548	3.1	**21** 0052	1.6
1157	1.6	0727	3.2
SU 1819	2.9	M 1352	1.5
		◗ 2012	2.9
7 0014	1.7	**22** 0212	1.7
0644	3.0	0841	3.1
M 1303	1.7	TU 1509	1.5
◗ 1929	2.8	2126	2.9
8 0131	1.8	**23** 0331	1.7
0757	3.0	0952	3.1
TU 1426	1.6	W 1615	1.5
2049	2.9	2230	3.0
9 0257	1.7	**24** 0435	1.6
0913	3.1	1051	3.1
W 1542	1.5	TH 1706	1.4
2159	3.0	2321	3.1
10 0406	1.5	**25** 0525	1.5
1019	3.2	1139	3.2
TH 1640	1.3	F 1747	1.3
2256	3.3		
11 0502	1.3	**26** 0003	3.2
1114	3.4	0607	1.4
F 1729	1.1	SA 1220	3.3
2345	3.5	1823	1.2
12 0551	1.1	**27** 0040	3.3
1204	3.6	0644	1.3
SA 1815	0.9	SU 1256	3.3
		1857	1.2
13 0032	3.7	**28** 0114	3.4
0638	0.8	0719	1.1
SU 1252	3.8	M 1331	3.4
1900	0.7	1930	1.1
14 0117	3.9	**29** 0148	3.5
0724	0.7	0753	1.1
M 1340	3.9	TU 1406	3.4
○ 1945	0.6	● 2003	1.1
15 0204	4.0	**30** 0221	3.5
0812	0.6	0828	1.1
TU 1428	3.9	W 1440	3.4
2030	0.6	2036	1.1

DECEMBER

Time	m	Time	m
1 0256	3.5	**16** 0328	3.9
0902	1.1	0940	0.7
TH 1516	3.4	F 1555	3.7
2111	1.2	2150	0.9
2 0331	3.5	**17** 0415	3.8
0938	1.2	1027	0.8
F 1552	3.3	SA 1643	3.5
2146	1.2	2236	1.0
3 0408	3.5	**18** 0503	3.6
1014	1.2	1114	1.0
SA 1629	3.2	SU 1731	3.3
2222	1.3	2322	1.2
4 0445	3.4	**19** 0552	3.4
1053	1.3	1204	1.2
SU 1709	3.1	M 1823	3.1
2302	1.4		
5 0527	3.3	**20** 0013	1.4
1137	1.4	0647	3.2
M 1756	3.0	TU 1301	1.4
2350	1.5	1922	2.9
6 0617	3.2	**21** 0113	1.6
1231	1.5	0749	3.0
TU 1854	3.0	W 1407	1.6
		◗ 2030	2.9
7 0052	1.6	**22** 0229	1.7
0718	3.1	0857	2.9
W 1338	1.5	TH 1517	1.6
◗ 2004	3.0	2139	2.9
8 0208	1.6	**23** 0343	1.7
0829	3.1	1004	2.9
TH 1454	1.5	F 1620	1.6
2117	3.1	2240	2.9
9 0325	1.5	**24** 0446	1.6
0940	3.2	1102	3.0
F 1604	1.3	SA 1712	1.5
2222	3.2	2330	3.0
10 0431	1.3	**25** 0538	1.5
1045	3.3	1150	3.0
SA 1702	1.2	SU 1755	1.4
2320	3.4		
11 0529	1.1	**26** 0012	3.2
1143	3.5	0621	1.4
SU 1755	1.0	M 1232	3.1
		1834	1.3
12 0012	3.6	**27** 0050	3.3
0622	0.9	0700	1.3
M 1236	3.7	TU 1310	3.2
1844	0.8	1911	1.2
13 0102	3.8	**28** 0127	3.4
0713	0.7	0737	1.2
TU 1328	3.8	W 1347	3.3
1932	0.7	1946	1.1
14 0151	3.9	**29** 0203	3.4
0803	0.6	0813	1.1
W 1418	3.8	TH 1424	3.4
○ 2019	0.7	● 2021	1.1
15 0240	4.0	**30** 0239	3.6
0852	0.6	0848	1.0
TH 1507	3.8	F 1500	3.4
2105	0.7	2055	1.0
		31 0315	3.6
		0923	1.0
		SA 1536	3.4
		2130	1.1

HAT is 4·0 metres above Chart Datum.

〉〉**FREE** monthly updates. Register at 〈
www.reedsnauticalalmanac.co.uk 〈

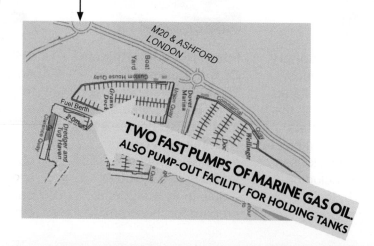

A

Abandon ship 113
Abdominal pain 121
Accidents and injuries 122
Acute illness 121
Admiralty chart symbols 6
Agents 2
Allergies 121
Area (map) 130
Automatic Transmission Identification
    System (ATIS) 65
Automatic Identification System (AIS) 42

B

BBC World Service 12
Beaufort scale 82
Belgium
    Coast Radio Stations 70
    Special notes 734
Bites and stings 122
Bleeding
    Internal 122
    Nosebleed 125
    Open wound 122
Budget flights 878
Buoyage, direction of 39
Burgee 79
Burns and scalds 122

C

Cardinal marks 38
Certificate of Competence, International 22
Channel crossing
    Central part 198
    Eastern part 260
    From SW England 179
Channel Islands
    Coast Radio Stations 70
    SAR operations 821
    Special notes 821
    Weather forecasts 96
Chart
    Admiralty chart symbols 6
    Datum (CD) 54
    Datum (horizontal) 40
    Electronic
        Raster 40
        Vector 40
    Plotters 41
Chest injury 122
Childbirth 125
Children
    Acute illness 121
    Resuscitation 119
Choking 120
Coastguard, HM 114
    MSI broadcasts 89
    Remote stations 91
Coast radio stations 68
    Belgium 100
    Channel Islands 70
    Denmark 71
    Germany 72, 98
    Ireland 70, 96
    Spain 76, 102
Coefficients, tidal 136
Collision regulations 24
Communications (Chapter 5) 63
    Coast radio stations 68
    Flag etiquette 79
    International code of signals 64
    International Port Traffic Signals (IPTS) 78
    Vessel Traffic Services (VTS) 68, 78
Constipation 121
Conversion tables 11
CROSS, VHF broadcasts 101
Cruising formalities 22
Crush injuries 122
Customs, Foreign 23

Customs, HM 22
    Arrival from an EU country 23
    Arrival from a non-EU country 23
    Notice of departure 22
Cuts 122

D

Dead reckoning 35
Declination, sun 47
Denmark
    Coast radio stations 71
    Special notes 647
    Weather forecasts 97
Dental problems 123
Diabetes 121
Diarrhoea 121
Digital Selective Calling 108
Directing signals by aircraft 113
Disabled facilities information 131
Dislocations 123
Distances
    Across the Bay of Biscay 135
    Across the English Channel 132
    Across the Irish Sea 133
    Across the North Sea 134
Distress 108
    Alerting 108
    Mobile telephones 110
    Raising the alarm 114
    Signals and emergences 110
Documents, personal 22
Double lift 113
Drowning 120
Drugs and medication 125
Drug smuggling 23
Dutchman's log 36
Duty free stores 22

E

EasyTide 58
Eclipses 10
EGNOS 39
Emergency, definitions of 108
Emergency resuscitation (ABC) 118
Ensigns 79, 80
Environmental guidance 132
EPIRB, Registration and false alerts 107
Exercise areas, Submarine
    NW Scotland 449
    SW Scotland 482
Eye problems 123

F

Ferries: Scotland, west coast 464
Fever 121
Firing ranges
    Castlemartin Range 544
    Lulworth Range 188
    Manorbier Range 544
    Pembrey Range 544
    Penally Range 544
    Pendine Range 544
First Aid (Chapter 8) 117
    Abdominal pain 121
    Acute (sudden) illness 121
    Allergies 121
    Bites and stings 122
    Bleeding 122
    Books on first aid 126
    Burns and scalds 122
    Chest injury 122
    Childbirth 125
    Children 121
    Choking 120
    Constipation 121
    Crush injuries 122
    Cuts and wounds 122
    Dental problems 123
    Diabetes 121

    Diarrhoea 121
    Drowning 120
    Drugs and medication 125
    Emergency resuscitation 118
    Eye problems 123
    Fever 121
    First Aid kit 125
    Fish hook injury 122
    Fractures and dislocations 123
    Frostbite 124
    Head injury 124
    Heart attack 121
    Hypothermia 120
    Injections 126
    Medical advice by R/T 118
    Medical care abroad 118
    Nail injuries 124
    Nosebleed 125
    Observation form 127
    Poisoning 125
    Seasickness 121
    Shock 120
    Strains and sprains 125
First Aid kit 125
Fish hook injuries 122
Flags
    Code 64
    Ensigns 79, 80
    Etiquette 79
Fractures 123
France
    Special notes 751
Frostbite 124

G

Gale warnings 85
Geostrophic scales 85
German Bight, crossing to UK 670
Germany
    Coast radio stations 72, 98
    Special notes 666
    Weather forecasts 98
Gibraltar
    Special notes 1010
    Weather forecasts 104
GMDSS 107
GPS 39
Gunfacts 68

H

Harbour information 131
Harbours, Coasts and Tides (Chapter 9) 129
Hazards 43
Head injury 124
Health regulations 22
Heart attack 121
Helicopter rescue 113
HF radio 66
HF radio facsimile broadcasts 91
Highest Astronomical Tide (HAT) 54
High seas bulletins 89
HM Coastguard 114
Horizontal datums 40
Hypothermia 120

I

IALA buoyage 38
    Cardinal marks 38
    Isolated danger marks 39
    Lateral marks 38
    Preferred channel marks 38
    Safe water marks 39
    Special marks 39
Immigration 23
Importation, Temporary (TI) 23
Injections 126
Injury
    Chest 122
    Crush 122

Fish hook 122
Head 124
Nail 124
Inland waterways
Belgium 734
Brittany 807
Caledonian Canal...................394, 408, 456
France 752
Netherlands..704
To the Mediterranean 934
Insurance 22
International Certificate of Competence.............22
International code of signals.............................64
International Port Traffic Signals (IPTS) 78
Inter-ship VHF channels 65
Ireland
Coast radio stations 96
Special notes 570
Weather forecasts.................................96

L

Lateral marks 38
Lifeboats 75
Light(s)
Characteristics 35, 36
Lights and shapes 31
Navigation......................................29, 30
Rising or dipping distance.......................49
Sectors, arcs of visibility.........................36
Link calls
MF 68
Paying for 68
VHF 68
LORAN-C, NELS and eLORAN.........................40
Lowest Astronomical Tide (LAT) 54

M

Map of almanac areas.................................130
MAYDAY calls 108, 109
Medical care abroad 118
Medical help by R/T 110
Medical observation form 127
Metmarine 102
MF link calls 68
MF radio 66
MHWS, MHWN, MLWS, MLWN 54
Mines, explosive 43
Moon
Phases 52
Rise/set tables.......................................51
MSI broadcasts by HM Coastguard 89

N

Nail injuries...124
National Coastwatch Institution (NCI) 69, 89, 115
Naval exercise areas, SW England 165
Navareas 43
Navigational tables
Distance off by vertical sextant angle..........46
Distance off when rising or dipping..............49
Distance of horizon 44
Distance, speed and time........................45
Sunrise/set 49
Navigation (Chapter 3) 33
Automatic Identification System (AIS) 42
Chart plotters 41
Dead reckoning 35
Electronic charts....................................40
Hazards...43
IALA buoyage 38
Lights 36, 37, 49
LORAN-C, NELS and eLORAN 40
Moonrise/set 51
Navigational tables.................................44
Offshore energy installations 44
Passage planning 35
Radar 41
Satellite navigation
EGNOS..39
GPS...39

SOLAS V..35
Warnings 43
Waypoints 40
Navigation lights 30
NAVTEX 86
Netherlands
Coastguard (IJmuiden) 100
Weather forecasts.................................99
North Sea, crossing to UK
From Denmark 652
From Netherlands 704
Nosebleed 125
NW European Loran-C System (NELS) 40

O

Offshore energy installations 44
Oil and gas installations..............................336

P

PAN-PAN, Urgency signal.............................109
Passage planning 35
Plotters, chart 41
Poisoning 125
Port operations VHF channels 65
Port Traffic Signals, International (IPTS) 78
Portugal
Special notes 980
Weather forecasts.................................103
Wind and weather.................................980
Portuguese Islands, Special notes.................1017
Prowords (radio)..65
Public holidays, UK 10

R

Racons (radar beacons) 43
Radar
Collision avoidance................................42
Navigation aid 42
Radio
Licences 65
Radio silence, cancelling and relaxing 109
R/T procedures 65
Time signals 12
Weather information 96
Reference Data (Chapter 1) 5
Admiralty chart symbols..........................6
Conversion tables 11
Eclipses 10
E-mail and postal addresses....................9
Public holidays, UK 10
Radio time signals 12
Symbols and abbreviations 7
Time, UT and DST 12
Vocabularies..12
Websites 9
Regulations (Chapter 2) 21
Collision regulations 24
Cruising formalities 22
Customs 22, 23
Lights and shapes 31
Navigation lights 30
Part A: General 24
Part B: Steering and sailing rules.............24
Part C: Lights and shapes 29
Part D: Sound and light signals................29
Red diesel 23
Travelling with pets 23
Resuscitation
Children 119
Emergency (ABC) 118
Problems..120
Problems during; drowning and
hypothermia 120

S

Safety 105
Equipment recommended 107
Signals 108, 109

Satellite communications..............................67
Satellite navigation 39
EGNOS 39
GPS 39
Schengen Treaty 23
Sea areas (GMDSS) 108
Search and Rescue (SAR)
Resources..71
UK 114
Seasickness 121
Sea states 83
Secondary ports.........................54, 55, 56, 58
SÉCURITÉ, Safety signal..............................110
Semaphore (Signal) stations 75
Shipping forecast, BBC Radio 4 94
Shock 120
Signals, international
Code of signals 64
Port Traffic Signals (IPTS) 78
Storm warning signals............................86
Sound signals 32
SOLAS V 35
Spain
Coast radio stations 76, 102
Special notes 947, 1000
Special notes
Belgium 734
Channel Islands.............................820, 821
Denmark 647
France 751
Germany 666
Gibraltar 1010
Ireland 570
Portugal 980
Portuguese Islands...............................1017
Spain 947, 1000
Standard ports...........................54, 55, 56
Storm warnings 61
Storm warning signals, International 86
Strains and sprains 125
Stroke 122
Strong wind warnings...................................85
Subfacts..68
Submarines 43
Exercise areas
NW Scotland 449
SW Scotland.....................................482
Symbols and abbreviations 7

T

Tidal coefficients
Brest..873
Calais 759
Cherbourg 786
Pointe de Grave 936
St Malo 805
Tidal curves
Aberdeen 384
Avonmouth 554
Belfast..634
Brest..874
Cherbourg 790
Chichester Harbour................................241
Christchurch to Selsey, special curves 209
Cobh 588
Cuxhaven 676
Dartmouth 173
Dieppe 767
Dover 262
Dublin 574
Dunkerque 752
Esbjerg 658
Falmouth 152
Galway 615
Gibraltar 1015
Greenock 486
Harwich 316
Helgoland..670
Hoek van Holland..................................722
Holyhead 521
Immingham...340
Invergordon 403

La Coruña......................................967
Le Havre.......................................769
Leith...379
Lerwick...422
Lisboa...993
Liverpool.......................................513
London Bridge...............................297
Lowestoft......................................323
Margate...284
Milford Haven................................540
Oban...460
Plymouth.......................................165
Pointe de Grave.............................934
Poole...204
Portland...185
Portsmouth....................................231
River Foyle.....................................624
River Tees......................................349
River Tyne......................................359
Sheerness......................................287
Shoreham......................................253
Southampton.................................226
St Helier..837
St Malo..804
St Peter Port..................................828
Stornoway......................................440
Tarbert Island................................601
Ullapool...436
Walton-on-the-Naze.......................310
Wick..411
Wilhelmshaven...............................685
Zeebrugge.....................................736
Tidal streams
  Belgium......................................698
  Biscay, S.....................................914
  Brittany, N..................................844
  Brittany, S...................................884
  Channel Islands...........................818
  England, Central S.......................194
  England, E...................................272
  England, NE.................................332
  England, NW................................498
  England, SE..................................248
  England, SW................................140
  France, Central N.........................794
  France, N.....................................746
  Germany......................................662
  Ireland, N....................................608
  Ireland, S....................................566
  Isle of Wight and Solent..............218
  Isles of Scilly...............................140
  Mull of Kintyre............................477
  Netherlands.................................698
  Portland.......................................180
  Rathlin Island..............................630
  Scotland, NE................................392
  Scotland, NW...............................430
  Scotland, SE.................................366
  Scotland, SW...............................468
  Thames Estuary...........................278
  Wales, S and Bristol Channel........528
Tide tables
  Aberdeen.....................................385
  Avonmouth..................................551
  Belfast...635
  Brest..870
  Casablanca.................................1038
  Cherbourg....................................787
  Chichester Harbour.......................242
  Cobh..585
  Cuxhaven.....................................677
  Dartmouth....................................170
  Dieppe...764
  Dover...264
  Dublin..575
  Dunkerque....................................753
  Esbjerg...655
  Falmouth......................................153
  Galway...616
  Gibraltar.....................................1017
  Greenock......................................487
  Harwich..317
  Helgoland.....................................671

Hoek van Holland...........................719
Holyhead.......................................522
Immingham....................................341
Invergordon...................................404
La Coruña......................................968
Le Havre.......................................770
Leith...376
Lerwick...423
Lisboa...994
Liverpool.......................................510
London Bridge...............................298
Lowestoft......................................324
Margate...281
Milford Haven................................541
Oban...457
Plymouth.......................................162
Pointe de Grave.............................931
Ponta Delgada.............................1032
Poole...206
Portland...182
Portsmouth....................................232
River Foyle.....................................625
River Tees......................................350
River Tyne......................................356
Sheerness......................................290
Shoreham......................................254
Southampton.................................223
St Helier..838
St Malo..806
St Peter Port..................................829
Stornoway......................................437
Tarbert Island................................602
Ullapool...444
Vlissingen......................................730
Walton-on-the-Naze.......................311
Wick..412
Wilhelmshaven...............................686
Zeebrugge.....................................737
Tides (Chapter 4)..............................53
  Barometric pressure effects............61
  Calculations............................55, 56
  Clearances under bridges................58
  Coefficients...................................54
  Computation of stream rates.....61, 62
  Computer predictions.....................58
    EasyTide..................................58
    TotalTide.................................59
  Definitions.....................................54
  Duration.......................................54
  Factors..58
  Interval...54
  Meteorological effects....................61
  MHWS, MHWN, MLWS, MLWN.......54
  Range...54
  Rise/Fall.......................................54
  Secondary Ports.............54, 55, 56, 58
  Standard Ports...............................55
  Storm warnings.............................61
  Streams, diamonds & atlases.....59, 60
  Streams in rivers...........................61
  Twelfths rule.................................58
  Wind, effect on tides......................61
Tides, special curves Christchurch to Selsey.........209
Time, UT and DST.............................12
TotalTide..59
Traffic Separation
  Banco del Hoyo.............................998
  Botney Ground, off.........696, 704, 705
  Brown Ridge, off...................696, 705
  Cape Roca, off.......................976, 980
  Cape St Vincent, off...............977, 993
  Casquets, off........139, 790, 814, 820
  Dover Strait..................250, 260, 745
  Fastnet Rock, off...................564, 590
  Finisterre, off........................943, 959
  Friesland, East..............................705
  Friesland, West......................704, 705
  German Bight, western approaches....684, 705
  Gibraltar Strait...................998, 1008
  Harwich approaches...............314, 315
  Hinder, West.................................718
  Holyhead.......................................521
  Humber approaches.......................331

Jade approach.........................667, 684
Land's End, off.............138, 144, 145
Maas...722
Maas, North............................696, 722
Maas, West Inner and Outer............722
Neist Point, off..............................428
Netherlands, North, off.....705, 696, 661
Netherlands, South, and Belgium.......697
Netherlands, South, off...................718
Noord Hinder North.........696, 718, 722
Noord Hinder South........................718
North Channel........................467, 622
Ouessant (Ushant), off.......840, 841, 848
Scilly, South of...............138, 144, 145
Scilly, West of................138, 144, 145
Skerries, off............................607, 612
Smalls, off..............................526, 532
Terschelling-German Bight...667, 684, 694, 705
Texel, off................................704, 705
Tuskar Rock, off............565, 572, 581
Vigo approaches.....................967, 972
Vlieland North................................705
Vlieland, off...................................704
Twelfths rule......................................58

U
Updates..2
Urgency calls.............................108, 109

V
Vent solaire.....................................886
Vessel Traffic Services (VTS)........78, 675
  Elbe, River....................................675
  Ems, River....................................694
  Nieuwe Waterweg..........................723
  Noordzeekanaal.............................715
  Weser, River.................................684
VHF
  CROSS broadcasts....................100, 101
  Inter-ship channels.........................65
  Link calls.......................................68
  Port operations channels.................65
  Radio..65
Visual signals, shore to ships............113
Vocabularies......................................12
  Danish...647
  Breton..886
Volmet...96
Voluntary Separation Scheme (VSS)
  Dover...260

W
Waypoints..40
Weather (Chapter 6)...........................81
  Actual weather reporting stations.....95
  Beaufort scale................................82
  Fax forecasts, HF radio broadcasts....91
  High seas bulletins.........................89
  Inshore waters forecast..................94
  Land forecasts...............................94
  Mobile phone forecasts, Metmarine...102
  MSI broadcasts by HM Coastguard.....89
  NAVTEX..86
  Press forecasts..............................96
  Sea states.....................................83
  Sources abroad
    Channel Islands........................96
    Denmark..................................97
    France...................................100
    Germany..................................98
    Gibraltar................................104
    Ireland....................................96
    Netherlands..............................99
    Portugal & Portuguese Islands...103
    Spain.....................................102
  Telephone forecasts, Météo............102
  Television forecasts........................96
  Terminology...................................85
Wind farms..44
Wireless internet (wi-fi).......................66
Wounds...122

A

Abbenfleth 674
Aberaeron 536
Aberdaron 521
Aberdeen 384
Aberdour 368, 374
Aberdovey 536
Aberporth 536
Abersoch 533
Aberystwyth 536
A'bhodaich 480
Acairseid Mhór 440
Accumersiel 691
Achillbeg Island (South Point) 610
Adrigole 595
Ailsa Craig 471
Albert Bridge 293
Albufeira 985, 1000
Aldeburgh 322
Alde, River 321
Alderney 820, 822, 823, 824
Algeciras 1005, 1015, 1016
Allington Lock 289
Altacarry Head (Rathlin East) 611
Alte Mellum 689
Alte Weser light house 683
Althorne 303
Alt, River 514
Alum Bay 211
Alvor 999, 1026
Amble 360
Ameland 707
Amlwch 521
Amrum-Hafen 667, 668
Amsterdam 715
Andernos 942
Andijk 713
Anglesey 516
Anglet 918, 943
Angra do Heroísmo 1025, 1029
Annalong 640
Annan Waterfoot 494
Anse de Bertheaume 865
Anse de Feunteun Aod 879
Anse de Penfoul 891
Anse de Primel 860
Anse de St Martin 790
Anse des Vieilles 922
Anstey's Cove 174
Anstruther 379
Antwerpen 734
Anvil Point 143
Aran Islands 614
Aranmore light house 611
Arbroath 382
Arcachon 918, 941
Ardglass 639
Ardgroom 596
Ardinamar 461
Ardnakinna Point 569
Ardnamurchan 433
Ardnave Point 473
Ardrishaig 475
Ardrossan 471, 490
Argenton 864
Arinagour 453
Arisaig 450
Arklow 579
Arklow Roadstone Harbour 579
Armadale Bay 448
Ar-Men 849
Arnish Point 432
Arnside 507
Arrábida 997
Arradon 903
Arrifana 998
Arrochar 484
Arromanches 780
Ars-en-Ré 916, 926
Arun, River 252
Ashlett Creek 209, 229
Audenge 942
Audierne 879, 883

Aulne, River 869
Aultbea 447
Ault light house 748
Auray 888, 903
Auskerry 395, 417
Aust 550
Aveiro 984, 989
Aven, River 886
Aveton Gifford 166
Avilés 951, 962, 963
Avoch 407
Avonmouth 556
Avon, River (Devon) 166
Avon, River (Dorset) 210
Ayamonte 1004, 1007
Ayr 471, 490, 491
Azores
    Angra do Heroísmo 1025, 1029
    Cais (São Roque) do Pico 1025, 1028
    Calheta 1025, 1028
    Ferraria 1025
    Horta 1024, 1026
    Madalena 1025, 1027
    Ponta da Barca 1025
    Ponta da Sereta 1025
    Ponta das Lajes (Flores) 1024
    Ponta Delgada 1025, 1031
    Ponta do Albarnaz 1024
    Ponta do Arnel 1025
    Ponta do Castelo 1025
    Ponta do Topo 1025
    Ponta Garça 1025
    Porto das Lajes 1024, 1026
    Praia de Vitória 1025, 1030
    Santa Clara 1025
    Santa Maria 1025
    Velas 1025, 1028
    Vila da Praia 1025
    Vila do Porto 1025, 1030
    Vila Franca do Campo 1025

B

Babbacombe Bay 174
Backaland Bay 417
Badachro 447
Badcall Bay 443
Badentarbat Bay 447
Baginbun Head 582
Baie de Bourgneuf 916, 920, 1035
Baie de Lampaul 868, 869
Baie de Morlaix 850
Baily 568
Balbriggan 641
Ballinskelligs Bay 596
Ballycastle 631
Ballycotton 569
Ballycrovane 596
Ballyholme Bay 633
Ballysadare Bay 620
Balta Sound 396, 421
Baltimore 569, 593
Baltrum 665, 692
Banff 394, 399
Bangor (Belfast Lough) 611, 634
Bangor (Menai) 501, 516, 518
Banjo Pier (Looe) 142, 158
Bann, River 628
Bantham 166
Bantry Bay 569, 595
Barbate 1005, 1013
Bardsey Island 501, 516, 521, 530
Barfleur 750, 783
Barloge Creek 592
Barmouth 535
Barnstaple 559
Barra 440
Barra Head 432, 440
Barrow-in-Furness 500, 504, 507
Barry 548
Bartlett Creek 289
Basta Voe 421
Bath (Westerschelde) 727
Battlesbridge 303

Bay of Biscay 919, 953
Bay of Cádiz 1005, 1011, 1012
Bay of Laig 452
Bayona 952, 980
Bayonne 941, 943
Beachley 550
Beachy Head 250, 251, 260
Beadnell Bay 360
Beaucette 826
Beaulieu, River 196, 215
Beaumaris 518
Bee Ness 289
Beidenfleth 674
Belcroute Bay 836
Belfast Harbour 633
Belfast Lough 612, 633
Belle Île 887, 898, 899
Beluso 978
Belvoir Bay 832
Bembridge 198, 231
Bénodet 890, 886, 891, 894, 866
Bensersiel 665, 691
Berck 760
Berghaven 723
Berkeley 550
Bermeo 950, 957
Berwick-upon-Tweed 335, 361, 362
Beuzec 886
Bideford 559
Bilbao (Bilbo) 950, 957
Binic 797, 813
Bishop Rock 142
Bishops, the 532
Black Ball Harbour 595
Black Nore Point 531
Blackpool 500, 508
Black Rock (Falmouth) 142, 150
Black Rock (North Ireland) 610
Black Rock (Yarmouth, IoW) 196, 212
Blacksod Bay 619
Blacktoft 344
Blackwater, River 275, 307
Blakeney 336
Blankenberge 703, 740
Blavands Huk 646
Blue Mull 421
Blue Mull Sound 422
Blyth 359
Boddam 398
Bodic 846
Bognor Regis 252
Bonanza 1010
Bonawe 460
Bonne Nuit Bay 836
Boontjes 700
Bordeaux (France) 918, 938, 939
Bordeaux (Guernsey) 825
Borkum 693
Borsfleth 674
Boscastle 560, 561
Bosham 240
Boston 334, 339
Bouley Bay 836
Boulmer 360
Boulogne 748, 760
Bound Skerry 395
Bourgenay 916, 924
Bourg-sur-Gironde 938
Bovisand Pier 159
Boyardville 917
Boyne, River Bar 640
Bradwell 307
Brake 682
Brancaster Staithe 334, 337
Brandaris light tower 700, 709
Brandon Bay 570, 599
Braye 820, 824
Brehon 820
Bremen 682
Bremerhaven 665, 682
Brentford Dock Marina 297
Breskens 733
Bressay 395
Brest 866, 874

Bridgwater 557
Bridlington 347
Brightlingsea 275, 308
Brighton 250, 257
Brigneau 886, 894
Brignogan 862
Bristol 554
Bristol Channel 530
Brixham 173
Broad Haven 619
Brodick 479
Broekerhaven 713
Brough Haven 345
Brough of Birsay 395
Brownsea Island 196, 200
Brugge 735
Bruichladdich 473
Brunsbüttel 664, 680
Bryher 146
Buckie 394, 400
Bucklers Hard 215
Bude 560, 561
Bueu 978
Bull Point 531
Bull Rock 569
Bull Sand Fort 344
Bunbeg 621
Burela 964
Burghead 394, 401
Burhou 824
Burnham-on-Sea 557
Burnmouth 368, 371
Burntisland 368, 372
Burnt Islands 479
Burra Voe 422
Burray Ness 418
Burry Port 546
Bursledon 229
Burton Stather 344
Bur Wick 418
Busum 669
Bute 481
Butt of Lewis 432

C

Cabo Ajo 950
Cabo Busto 951
Cabo Carvoeiro 984
Cabo Corrubedo 952
Cabo da Roca 984
Cabo de Cruz 952, 975
Cabo de San Agustín 951
Cabo de Santa Catalina 950
Cabo de Santa Maria 985, 1001
Cabo de São Vicente 985
Cabo de Sines 985
Cabo de Torres 951
Cabo Espichel 985
Cabo Finisterre 952, 973
Cabo Higuer 918, 950
Cabo Lastres 951
Cabo Machichaco 950
Cabo Mayor 950
Cabo Mondego 984
Cabo Ortegal 951, 965
Cabo Peñas 951
Cabo Prior 951
Cabo Prioriño Chico 951
Cabo Raso 984, 986
Cabo Roche 1005
Cabo San Sebastián 951
Cabo Sardão 985
Cabo Silleiro 952
Cabo Spartel 1006
Cabo Trafalgar 1005, 1010, 1013
Cabo Vidio 951
Cabo Villano (Bilbao) 950
Cabo Villano, (Camariñas) 952, 972
Cadgwith 149
Cádiz 1012
Caen 749, 779
Caernarfon 518, 519
Caernarfon Bar 519

Cahersiveen Marina 597
Cais (São Roque) do Pico 1025, 1028
Caister-on-Sea 327
Caladh Harbour 479
Calais 748, 758
Caledonian Canal 394, 408, 434, 456
Calf of Man 500
Calf Sound 505
Calheta (Madeira) 1035, 1037
Calheta (São Jorge) 1025, 1028
Calshot Spit 197
Camaret 849, 876
Cambados 977
Campbeltown 478
Camusnagaul 447
Cancale 796, 800, 802
Cangas 979
Canna 433, 452
Cantick Head 395
Canvey Island 293, 301
Capbreton 918, 943
Cap d'Alprech 748
Cap d'Antifer 763
Cap de Carteret 796, 820
Cap de la Hague 781, 820
Cap de la Hève 749
Cape Cornwall 561
Cape Wrath 394, 416, 432, 435
Cap Ferret 941
Cap Fréhel 797
Cap Gris-Nez 748
Cap Lévi 750
Caramiñal 952, 975
Cardiff 530, 544, 548
Cardigan 537
Cardigan Bay 530
Carentan 750, 782
Cargreen 159
Cariño 951, 965
Carlingford Lough 640
Carloway 441
Carmarthen 545, 546
Carna 568
Carnlough Harbour 611, 631
Carnsore Point 582
Carradale Bay 478
Carreg-y-Chwislen 521
Carrickarory 623
Carrickfergus 612, 633
Carrigaholt Bay 600
Carsaig Bay 454, 474
Carteret 796, 799
Cascais 984, 991
Cashla Bay 614
Casquets 820, 822
Castillo de San Sebastián 1005
Castillo de Santa Ana 950
Castlebay 440, 441
Castlemartin Range 544
Castletown 500, 503
Castletown Bay 503
Castletownbere 595
Castro Urdiales 950, 958
Cat Firth 420
Caudebec 774
Cayeux-sur-Mer 748, 761
Cemaes Bay 521
Ceuta 1006, 1020
Channel Light Vessel 820
Chanonry 394
Chapman's Pool 189
Charlestown 156
Chatham 289
Chauveau 917
Cheek Point 583
Chelsea Harbour 297
Chenal de la Helle 866
Chenal du Four 866, 867
Chepstow 548
Cherbourg 750, 784
Chesil Beach/Cove 178
Chester 513
Chichester Harbour 198, 239
Chipiona 1005, 1010

Christchurch 210
Church Ope Cove 189
Clacton-on-Sea 308
Clare Island 619
Cleanderry 596
Clear Island 593
Cleavel Point 200
Clevedon 556
Clew Bay 619
Clifden Bay 619
Clonakilty Bay 591
Clovelly 559
Clyde, River 485
Clyth Ness 394
Cobo Bay 826
Cockenzie 373
Codling Lanby 568
Coefficients, tidal 759, 786, 805, 873, 936
Colchester 308
Coleraine 628
Colne, River 275, 301, 308
Colonsay 470, 473
Colwyn Bay 513
Combarro 977
Concarneau 886, 893, 894
Coningbeg 568
Connah's Quay 500, 513
Connel 460
Contis 918
Conwy 504, 515
Conyer Creek 286
Copinsay 395
Coquet Island 335, 346
Cordouan 917, 935
Cork 569, 588
Corpach 434, 455
Corran Narrows 454
Corryvreckan 472
Corta de Los Jeronimos 1010
Coryton 301
Cotehele Quay 159
Cotentin Peninsula 798
Courseulles-sur-Mer 749, 779
Courtmacsherry 569, 591
Courtown 579
Coverack 149
Covesea Skerries 394
Cowes/River Medina 196, 221
Crac'h, River 888
Craighouse 473
Craignure 455
Crail 379
Cramond 374
Cranfield Point 640
Craster 360
Créac'h 848
Creux Hbr 833
Criccieth 534
Crinan 472
Crinan Canal 474
Cromarty 407, 409
Cromer 336
Crookhaven 594
Crouch, River 275, 303
Crouesty 903, 905
Crowlin Islands 448
Cuan Sound 461
Cudillero 951, 963
Cullen 399
Cullercoats 355
Culmore Bay 623
Cultra 633
Cumberland Basin 554
Cumbrae 480
Cushendall 631
Cushendun Bay 632
Cuxhaven 664, 676

D

Dagebüll 664, 667
Dahouet 797, 811
Dale Roads 538
Darnett Ness 289

Darrynane 596
Dartmouth 168
Deal 250, 263
Deauville 749, 776
Deben, River 320
De Block van Kuffeler 713
Dee, River 500, 513
Deer Sound 418
Deganwy 504, 515
Delfzijl 706, 707
Dell Quay 240
Den Helder 711, 718
Den Oever 700, 712, 713
Derby Haven 505
Déroute de Terre 798
Derrible Bay 833
Derwent, River 502
Diélette 796, 798
Dieppe 749, 762
Dinan 809
Dinard 802, 804
Dingle Bay 570, 592, 597
Dingwall 407
Dittisham 168
Dives-sur-Mer 749, 777
Dixcart Bay 833
Dodman Point Firing Range 156
Doel 728
Doëlan 886, 895
Donaghadee 633, 634
Donegal Harbour 621
Donostia (San Sebastián) 950, 955
Doove Balg 700
Dornie Bridge 450
Dornoch Firth 394, 409
Douarnenez 849, 866, 876, 877
Douglas Head 500
Dover 250, 263
Dover Strait 267
Drummore 486, 493, 494
Dubh Artach 434
Dublin 568, 574
Duclair 774
Duddon Bar 503
Dunany Point 640
Dunbar 371
Dunbeacon Harbour 594
Duncansby Head 394, 410
Dundalk 640
Dundee 381
Dundrum Bay 640
Dungarvan Bay 588
Dungarvan Harbour 584
Dungeness 251, 260
Dunkerque 748, 756, 757
Dunkerron Harbour 596
Dun Laoghaire 568, 578
Dunmanus Bay 594
Dunmanus Harbour 594
Dunmore East 569, 584
Dunnet Head 394
Dunoon 483
Dunstaffnage 461
Durdle Door 189
Durgerdam 713
Dury Voe 422

E
Eastbourne 250, 259
East Channel light buoy 820
East Frisian Islands 689
East Loch Tarbert 476, 440, 476, 440
East Pier Head 569
East Wangerooge 690
Eckmühl 849
Edam 713
Eday 417
Eddystone 142
Eemshaven 706
Eeragh (Rock Island) 610
Egmond-aan-Zee 701
Eidersperrwerk 669
Eigg Harbour 452

Eilean a' Cheo (Skye) 433, 447, 448, 450
Eilean Glas 432
Elanchove (Elantxobe) 950, 957
Elbe, River 666, 674
Elberry Cove 173
Elie 369, 374
Ellewoutsdijk 703, 727
Élorn, River 869
El Rompido 1004, 1008
Elsfleth 682
Emden 665, 694
Emshorn 694
Ems, River 665
Emsworth 198, 240
English and Welsh Grounds 556
Enkhuizen 713
Enseada da Baleeira 998
Enseada de Belixe 998
Ensenada del Sardiñeiro 974
Ensenada de San Simón 979
Ericeira 990
Erme, River 166
Erquy 797, 811
Esbjerg 646, 654
Esha Ness 396
Esposende 987
Esteiro 985
Étaples 760
Étel, River 887, 896, 898
Etretat 768
Europa Point 1006, 1016
Europlatform 724
Exe, River 175
Exeter, Exmouth Dock 177
Exmouth 175
Eyemouth 368, 370

F
Fair Isle 395, 421, 422
Fairlie Quay 481
Falaise des Fonds 749
Falmouth and River Fal 150
Fambridge 303
Fanad Bay 622
Fanad Head 611
Fanø 658
Farne Islands 361
Faro 985, 1001
Faslane 484
Fastnet 569
Faversham 286
Fawley 228
Fécamp 749, 768
Felixstowe Ferry 320
Felixstowe Pier 320
Fermain Bay 826
Ferraria 1025
Ferrol 951, 966
Ferryside 546
Fiddler's Ferry 513
Fidra 371
Fife Ness 369
Figueira da Foz 984, 989
Filey Bay 347
Findhorn 394, 402
Findochty 394, 400
Finisterre 973
Firth of Clyde 481
Firth of Forth 373
Firths Voe 395
Fischerbalje 665
Fishbourne Channel 240
Fishcombe Cove 173
Fisherrow 374
Fishguard 537
Flamborough Head 335
Flannan Isles 433, 441
Flat Holm 530, 548
Fleetwood 508
Flotta 417
Flowerdale Bay 447
Föhr 667
Folkestone 250, 262

Folly Inn 221
Folly Reach 221
Fontainevieille 942
Fontenelle Bay 826
Formby 509
Fort Belan 518
Fort (Cromwell) Point 570
Fort de l'Ouest 750
Forte de Santa Marta 984
Forth and Clyde Canal 375
Forth, River 368
Fortrose 407
Fort William 434, 455
Foula 396, 421, 422
Fouras 930
Fowey 157
Fraserburgh 394, 398
Freiburg 674
Fremington 559
Freshwater Bay 210, 211
Fromentine 916, 920, 921
Fuenterrabía (Hondarribia) 950, 954
Funchal 1035, 1037

G
Galley Head 569
Gallions Point Marina 296
Galway Bay 610, 613
Galway Harbour 614
Gardenstown 398
Gareloch, Garelochhead 470, 484
Garlieston 493, 494
Garnish Bay 596
Getaria (Guetaria) 950
Gibalta 985
Gibraltar 1006, 1016
Gibraltar Strait 1005, 1015
Gijón 951, 962
Gills Bay 415
Girvan 471, 491
Gladstone Dock 513
Glandore (Unionhall) 591
Glasson Dock 507
Glenarm 631
Glengariff 595
Glengarrisdale Bay 473
Glin 600
Gluckstadt 674
Gluss Island 396
Goeree light tower 701
Goleen 594
Golfe du Morbihan 888, 902
Golspie 409
Gometra 454
Goole 344
Gorey 836
Gorleston 276, 327
Gorran Haven 152
Gosport 236, 237
Gott Bay 452, 453
Goulphar 887
Gourdon 383
Gourock 483
Goury 796, 798
Grådyb 653, 658
Grandcamp 750, 781
Grande Havre 826
Grande Île 796
Grand Léjon 797
Grangemouth 369, 374
Granton 368, 374
Granville 800
Gravelines 748, 757
Gravesend 274, 293
Greatman Bay 614
Great Yarmouth 276, 309, 327, 334
Greenway Quay 168
Greetsiel 665, 693
Grève au Lancon 836
Grève de la Ville 833
Grève de Lecq 836
Greystones 574
Gribbin Head 156, 157

Griminish...432
Grimsby..345
Grosnez Point..821
Grosser Vogelsand................................674
Grunna Voe..420
Gruting Voe...420
Grutness Voe...420
Guernsey.......................................820, 825
Guetaria (Getaria).........................950, 955
Gugh..145
Gull Stream...250

H

Hale Head..513
Hamble, River........................197, 228, 229
Hamburg..674, 681
Hammersmith Bridge............................293
Hamna Voe..420
Hamworthy...200
Hannaine Bay..824
Hanstholm......................................645, 649
Hansweert.....................................727, 728
Haringvliet....................................701, 718
Harlesiel.......................................665, 690, 691
Harlingen.......................................700, 710
Harrington..502
Hartland Point...............................531, 556
Hartlepool....................................335, 346, 353
Harwich...276, 309
Haseldorf..674
Haskeir Island.......................................433
Hastings..250, 259, 261
Haulbowline..612
Havelet Bay...................................820, 826
Havengore...302
Haverfordwest......................................538
Havre Gosselin......................................832
Hayle...531, 561
Hayling Island.......................................240
Hejar Lesfar..1020
Helford River...149
Helgoland......................................664, 669
Helmsdale...410
Hendaye.......................................918, 945
Hengistbury Head.................................210
Hennebont..896
Herm..832
Herne Bay......................................274, 285
Hestan Islet...494
Het Vlie...708
Heugh (Hartlepool)...............................335
Heugh (Holy Island)......................335, 361
Heurteauville..774
Hever, River..664
Heysham..507
Hilbre Island...513
Hillswick..422
Hindeloopen..713
Hinkley Point...557
Hirtshals.......................................645, 649
Hoedekenskerke....................................727
Hojer...658
Holehaven...293
Holliwell Point.......................................303
Holyhead...520
Holy Island....................................335, 361
Holy Loch..483
Hondarribia (Fuenterrabía)...........950, 954
Honfleur..773
Hook Head...568
Hooksiel..665, 683
Hoorn..713
Hope Cove..166
Hopeman.......................................394, 401
Horne..674
Horns Rev...653
Hörnum..664, 667, 668
Horse Sand Fort............................197, 237
Horseshoe Harbour........................578, 593
Horta...1024, 1026
Hourtin...918
Houton Bay...417
Howth...573

Hoy...417
Hugh Town..145
Huibert Gat...706
Hull...344
Hullbridge...303
Humber Bridge.............................335, 340, 344
Hunstanton...337
Hurst Point....................................196, 210
Husum...664, 667, 668
Hvide Sande..................................647, 652
Hyskeir...433
Hythe Range...260
Hythe (Southampton)....................197, 226

I

Icart Bay...826
Igueldo...950
IJmuiden.......................................701, 714
IJsselmeer...712
Île-aux-Moines......................................903
Île-aux-Moines light house..................846
Île-aux-Moutons...................................886
Île Berder...903
Île Cézon..863
Île d'Aix.......................................917, 928, 929
Île d'Arz...903
Île de Batz...................................847, 850, 862
Île de Boëdig..903
Île de Bréhat.................................846, 851
Île de Conleau...............................903, 904
Île de Groix...886
Île de Hoëdic................................888, 899
Île de Houat..................................888, 899
Île de la Jument....................................903
Île de Molène................................848, 868
Île de Noirmoutier.........................916, 921
Île de Penfret................................886, 892, 893
Île de Ré.......................................916, 925
Île de Sein....................................849, 878
Île des Hébihens...................................811
Île d'Oléron...................................917, 928
Île d'Ouessant...............................848, 850, 868
Île Dumet...888
Île du Pilier...916
Île d'Yeu.......................................916, 924
Île Harbour..797
Île Longue...903
Île Louet.......................................847, 860
Île Pirenn...903
Îles Chausey...801
Îles de Glénan...............................886, 892
Îles Saint-Marcouf........................750, 781, 782
Île-Vierge light house...........................847
Ilfracombe....................................531, 558
Ilha da Berlenga...................................984
Ilha de Porto Santo............................1035
Ilha Madeira.......................................1035
Ilhas Desertas.....................................1035
Ilhas Selvagens...................................1035
Immingham..341
Inchcolm...374
Inishbofin...619
Inisheer..610
Inishowen...611
Inishtrahull...611
Inner Dowsing light tower.....................339
Instow...531
Inveraray...475
Inverewe Gardens.................................447
Invergordon...................................394, 407
Inverie Bay..452
Inverkip...483
Inverness.......................................394, 407
Inverness Firth..............................394, 409
Inward Rocks...550
Iona..454
Ipswich...276
Irvine..490
Isigny-sur-Mer...............................750, 782
Isla Canela...................................1004, 1007
Isla Cristina..................................1004, 1008
Isla de San Antón..................................950
Isla de Tapia...951
Isla Ons..952, 978

Isla Pancha...951
Isla Sálvora...952
Islas Cíes.......................................952, 978
Isla Sisargas...951
Islay...470, 472
Isle Martin..443
Isle of Man....................................500, 503, 504
Isle of May....................................369, 379
Isle of Walney......................................500
Isle of Whithorn...........................493, 494
Isle Ristol...447
Isles of Scilly................................142, 144, 145

J

Jade, River..684
Jaonnet...826
Jard-sur-Mer.................................916, 925
Jaudy, River..846
Jersey...821, 834, 836
Johnshaven...383
Juist...665, 692, 693
Jupiter Point...159
Jura...473

K

Kallin..440
Kenmare..596
Kenmare River......................................596
Kéréon..848
Kerjean...846
Kermorvan..848
Kerners...903
Kerprigent..846
Ketelhaven..713
Kettletoft Pier...............................395, 417, 419
Kew Bridge..293
Keyhaven...210
Kiel Canal...680
Kiel-Holtenau..681
Kiel-Holtenau: British Kiel YC...............681
Kiggaul Bay...................................610, 614
Kildalloig Bay..478
Kilkeel...639, 640
Kilkieran Bay...614
Kilkieran Cove.......................................613
Killadysert..600
Killala Bay...620
Killard Point...639
Killary Harbour.....................................619
Killeany Bay..613
Killough Harbour...................................639
Killybegs..621
Killyleagh..639
Kilmakilloge..596
Kilmokea Point......................................583
Kilmore Quay..................................568, 582
Kilrush...570, 601
Kingsbridge...167
King's Lynn....................................334, 338
Kinlochbervie.................................432, 442
Kinnaird Head.......................................394
Kinsale..569, 590
Kippford...494
Kirkaldy..379
Kirkcaldy..369
Kirkcudbright..494
Kirk Sound..417
Kirkwall...395, 418
Kish Bank..568
Knock..665
Knude...646
Kornwerderzand....................................712
Kruckaumundung..................................674
Kyle Akin..450
Kyle of Durness.....................................415
Kyle of Lochalsh....................................450
Kyle of Tongue......................................415

L

La Balue..796
La Banche..889

Labasheeda	600
L'Aber Benoît	847, 864
L'Aber-Ildut	848, 864
L'Aberwrac'h	847, 863
La Cayenne	930
La Chambre	852
La Charente, River	917, 929, 930
La Collette	834
La Corbière	821
La Coruña	951, 971
La Cotinière	917
La Croix	846
La Dordogne, River	935
L'Adour, River	918, 943
La Flotte	917, 925
La Gironde, River	934, 939
Lagos	985, 998
La Grande Grève	832
La Guardia	980
La Guia	952
La Higuera	1005
L'Aiguillon/La-Faute-sur-Mer	916, 925
La Jouvente	809
La Jument	848
La Lande	847
Lambay Island	568, 611, 629
Lamena	935
Lamlash	479
Lampaul	868
Lancaster	507
L'Ancresse Bay	826
Landes Range	940
Land's End	559
Langeoog	665, 692
Langstone Harbour	198, 238
Lannion (Léguer River)	858
La Pallice	927
La Pierre-de-Herpin	796
La Pommeraie	809
La Rance, Barrage and River	796, 797, 809
Laredo	950, 959
La Reuille	935
Largs	481
La Richardais	809
L'Armandèche	916
Larmor Baden	903
Larne	612, 632
La Roche Bernard	906
La Rochelle	917, 927
La Rocque	836
La Seudre, River	930
Lastres	962
La Tchue	824
La Teignouse	887
La Tremblade	930
La Trinité-sur-Mer	888, 901
La Turballe	889, 907
Lauwersoog	700, 706, 710
La Vieille	849
La Vigne	942
Lawrence Cove	595
Laxey	500, 503
Le Bono	903
Leça	984, 1024
Le Chapus	928
Le Colombier	846
Le Conquet	848, 865
Le Croisic	889, 905, 908
Le Crotoy	761
Leeds & Liverpool Canal	513
Lee-on-The-Solent	236
Le Four light house	848, 850
Le Four (Loire Atlantique)	889
Le Grand Jardin	796, 797
Le Guilvinec	849, 879
Le Havre	749, 768, 773
Le Hourdel	761
Leigh-on-Sea	274, 301
Leith	368, 373
Leixões	984, 988
Le Légué	797, 812
Le Logeo	903
Lelystad	713
Le Marquis	935
Le Minihic	809
Lemmer	713
Lemvig	645, 650, 651
Leopold II mole head	703
Le Palais	887, 899
Le Passage	903
Le Pointeau	889
Le Pouldu	895
Le Pouliguen	889, 909
Lequeitio (Lekeitio)	950, 956
Le Rocher	903
Lerwick	422
Les Ardentes	801
Les Baleines	916
Les Bas Sablons	797
Lesconil	849, 879
Les Ecrehou	836
Le Sénéquet	799
Les Fontaines	833
Les Hanois	820
Les Héaux de Bréhat	821, 846, 851
Les Minquiers	836
Les Pierres Noires	848
Les Sables d'Olonne	916, 923
Les Sept Îles	846, 850, 856, 857
Le Stiff	848
Le Touquet	748, 760
Le Tréport	748, 762
Le Trez Hir	865
Lézardrieux	846, 852, 853
L'Herbaudière	916, 920
Libourne	935
Lihou Island	826
Lild Strand	645
Lillo	728
Limehouse Basin	296
Limfjord	645, 651
Linklet Bay	417
Linnenplate	669
L'Iroise	866
Lisboa	985, 991
Liscannor	612, 613
Lismore	434
List	664, 667, 668
Little Bernera	436
Littlehampton	252
Little Haven	537
Little Killary Bay	619
Little Loch Broom	447
Little Russel Channel	825
Little Samphire Island	570
Liverpool	513
Lizard Point	142, 148
Llanddwyn Island	518
Llandudno	513
Llanelli	546
Llanes	951, 961
Llangwm	538
Loch A'bhraige	448
Loch Aline	455
Loch Alsh	450
Loch Beag	463
Loch Bervie	442
Loch Boisdale	432, 440
Loch Carnan	432, 440
Loch Carron	449
Loch Craignish	434, 463
Loch Creran	454
Loch Dunvegan	448
Loch Eatharna	452, 453
Loch Eil	455
Loch Eport	440
Loch Eriboll	415
Loch Ewe	443, 447
Loch Eynort	440
Loch Feochan	461
Loch Gairloch	443
Lochgoilhead	484
Loch Harport	448
Loch Hourn	450
Loch Inver	432, 442, 443
Loch Lathaich	454
Loch Laxford	442, 443
Loch Leven	454
Loch Linnhe	434, 455
Loch Long	485
Loch Maddy	440
Loch Melfort	434, 462
Loch Moidart	452
Loch Na Keal	454
Loch Nedd	443
Loch Nevis	433, 450
Loch Ranza	479
Loch Riddon	479
Lochrist	848
Loch Shell	440
Loch Shieldaig	447
Loch Shuna	463
Loch Skipport	440
Loch Snizort	448
Loch Spelve	454
Loch Sunart	453
Loch Sween	473
Loch Tarbert	473
Loch Thurnaig	447
Loch Torridon	435, 448
Locmaria	896
Locquemeau	859
Locquirec	859
Loctudy	886, 890
Loire, River	908, 910
London Bridge	277, 297
Londonderry	623
Lonehort Harbour	595
Long Hope	417
Longships	142
Longy Bay	824
Looe	158
Looe Channel	237
Loop Head	570
Lopwell	159
Lorient	887, 896
Lossiemouth	394, 400
L'Ost Pic	802
Lostwithiel	157
Loth	417, 419
Lough Foyle	611, 623
Lough Swilly	622
Lowestoft	276, 323
Luanco	962
Luarca	951, 963
Luce Bay Range	486
Luhe	674
Lulworth Cove	185, 189
Lulworth Range	188
Lundy Island	531, 558, 560
Lune, River	500
Lybster	394, 409
Lydd firing ranges	260
Lyme Bay	175
Lyme Regis	178
Lymington	196, 213
Lyness	417
Lyngvig	645
Lynher, River	161, 165
Lynmouth	557, 558
Lynmouth Foreland	531
Lynn of Lorn	434, 454
Lyvet	809

M

Maassluis	723, 724
Maasvlakte light house	701
MacDuff	394, 399
Machrihanish	473
Madalena	1025, 1027
Madeira	
Calheta	1035, 1037
Funchal	1035, 1037
Ilha de Porto Santo	1035
Ilha Madeira	1035
Ilhas Desertas	1035
Ilhas Selvagen	1035
Porto de Porto Santo	1035, 1036
Quinta do Lorde	1035, 1036
Makkum	713
Malahide	568, 572

Maldon...307
Mallaig...433, 452
Mama...985
Manchester Ship Canal...............................513
Marans...925
Marazion...148
Manorbier Range.......................................544
Marchwood..226
Marennes...930
Margate..274, 285
Marin..977
Marino Point..588
Marken...713
Markermeer...713
Marsdiep...701
Martin's Haven..537
Martrou...930
Maryport..500, 502
Maseline Harbour......................................833
Maughold Head...500
Mazagón...1005, 1009
McDermott Base..402
Medemblik...713
Medina, River/Cowes.............................196, 221
Medway...274
Medway, River..288
Meikle Ferry...409
Memmert...692
Menai Bridge..518
Menai Strait...........................501, 516, 518
Merrien...886, 894
Mersey and Dee, Rivers...............................500
Mersey, River..514
Meschers-sur-Gironde.................................938
Methil...379
Mevagissey...156
Mew Island..611
Middlesbrough.....................................348, 353
Mid Yell Voe...421
Milfontes...998
Milford Haven...538
Millport...480
Milton Creek...286
Minehead (Bristol Channel)....................531, 557
Mine Head (South Ireland)...........................569
Minsener Oog..689
Minsmere...322
Mizen Head.......................................569, 590
Mochras...534
Moclett..417
Moelfre..521
Moguériec..862
Monach Islands...441
Monnickendam...713
Monte Dirección (Le Charf).........................1006
Montedor...984
Monte Faro...952
Montrose...369, 382
Moray Firth..397, 407
Mordreuc...809
Morecambe..500, 507
Morgat...849, 876
Morlaix...847, 860
Morocco...1020
Mortagne...938
Mostyn Quay..513
Motrico (Mutriku).................950, 955, 956
Moulin Huet..826
Mousehole..147
Mouth...374
Moville..623
Muck..449
Muckle Flugga...396
Muckle Skerry..418
Mugia (Muxia)...965
Mühlenberg..681
Muiden..713
Muiderzand...713
Mullaghmore...620
Mullion Cove..148
Mull of Galloway..486
Mull of Kintyre....................................470, 475
Mulroy Bay.........................611, 621, 622
Mumbles..530, 547

Mumbles Head..544
Munkmarsch...667
Mupe Bay..185
Muxia (Mugia)..965

N

Naarden...713
Nab Deep Water Channel..............................193
Nab Tower..198
Nairn..394, 402
Narlwood Rocks...550
Nash...530
Nazaré..984, 990
Neath, River...547
Needles...196
Needles: Needles Channel.......................196, 211
Neist Point...433
Nes (Ameland)....................700, 707, 710
Ness...394
Nessmersiel.......................................665, 692
Netley...226
Neuharlingersiel..................665, 691, 692
Neuwerk..666
Newarp..334
Newbiggin..359
Newburgh...381
Newcastle...639
Newcastle-upon-Tyne..................................355
New Grimsby Sound....................................145
Newhaven...250, 258
New Hythe..289
Newlyn..147
Newport...221, 548, 549
Newquay (Cornwall)...................................561
New Quay (Wales)...............................530, 536
New Ross..583
Newry..640
Newton Haven...360
Newtown River..214
Neyland...538
Nieblum...664
Nieuwe Waterweg................................701, 723
Nieuwpoort..703, 742
Noirmoutier-en-l'Île....................................921
No Man's Land Fort.............................230, 237
Noodzeekanaal..715
Noordwijk-aan-Zee.....................................701
Noordzeekanaal...701
Nordby (Fanø)..658
Norddeich....................665, 692, 693
Nordenham...682
Norderney...665, 692
Nord-Ostsee Kanal.....................................680
Norfolk Broads..327
North Bank...568
North Bay...440
North Berwick.....................................368, 371
North Bull..568
North Channel (East Solent).........................197
North Channel (Hurst Point).........................196
North Channel (Poole).................................196
North Fambridge..303
North Foreland...................................251, 260
North Frisian Islands...................................667
North Harbour...593
Northney..240
North Rona...394
North Ronaldsay...395
North Shields..355
North Sunderland (Seahouses)...............335, 360
North Uist..432
Noss Head..394
Noup Head...395
Nuevo Portil..1008

O

Oban..434, 460
Odet, River...891
Oeiras...991
Old Grimsby Sound.....................................145
Old Head of Kinsale.............568, 569, 581
Olhão...985

Olna Firth...420
Omonville-la-Rogue..............................750, 790
Oostende..703, 740
Oosterschelde...726
Oostmahoorn..707
Oporto (Porto)....................................984, 988
Ore, River...321
Orford Ness.................276, 309, 322
Orkney Islands....................................395, 416
Ormond's Harbour......................................596
Ornsay..433
Orsay Island...473
Ortac Rock...822
Orwell, River.....................................276, 316
Osea Island...307
Otterndorf...674, 676
Oudeschild..700, 710
Ouessant SW Lanby....................................848
Ouistreham..749, 778
Outer Hebrides...........433, 435, 440, 441
Out Skerries..421
Owers..250, 251
Oyster Haven..590

P

Paal...728
Paco de Arcos..991
Padstow...561
Pagham...252
Paignton..173
Paimpol.......................846, 850, 851
Papa Westray..417
Par..156
Paris...775
Pasajes (Pasaia)..................................950, 954
Passage de la Déroute..................................798
Passage de la Teignouse.........................887, 899
Passe de la Gaine..852
Pauillac...918, 938
Pedroucos...991
Peel...505
Pegal..417
Pellworm..668
Pembrey Range..544
Penally Range..544
Penarth...531, 548
Pendeen...531
Pendine Range...544
Penedo da Saudade.....................................984
Pénerf, River................888, 905, 906
Peniche...984, 990
Peninnis Head.....................................142, 145
Pen Men (Île de Groix).................................886
Pennan Bay...398
Penryn..150
Pentland Firth..411
Penzance..148
Penzé River...861
Perelle Bay...826
Perranporth...561
Perros-Guirec.....................................846, 855
Perth..381
Pertuis Breton....................................916, 924
Pertuis d'Antioche......................................924
Peterhead...397
Petit Bôt Bay...826
Petite Foule..916
Petit Port...826
Picacho..1005
Piedras Negras...977
Pierowall..417, 419
Pierre de Herpin..796
Pierres Noires..848
Pile light (Dundalk).....................................612
Pinnaumundung...674
Pinnau-Neuendeich.....................................674
Piriac..888, 907
Pittenweem...379
Pladda...471
Plateau de Rochebonne................................916
Plateau des Minquiers..........................821, 836
Platte...820
Platte Fougère...820

Platte Saline Bay 824
Plockton 448, 449
Plouër 809
Ploumanac'h 850, 857
Plymouth 142, 159
Pointe de Barfleur 750, 781
Pointe de Chassiron 917
Pointe de Gatseau 928
Pointe de Grave 917, 918, 931
Pointe de Grosse Terre 916
Pointe de Kerdonis 887
Pointe de la Coubre 917
Pointe de Langoz 886
Pointe de Penlan 888, 906
Pointe de Porz-Don 846
Pointe de Saint-Gildas 910, 916
Pointe des Chats 887
Pointe des Corbeaux 916
Pointe des Dames 916
Pointe des Poulains 887
Pointe de St Mathieu 848
Pointe de Ver light house 749, 779
Pointe du Grouin du Cou 916
Pointe du Haut-Blanc 748
Pointe du Millier 849
Pointe du Petit-Minou 848
Pointe du Portzic 848, 866
Pointe du Roc 796
Pointe du Toulinguet 849
Pointe St Martin 918
Point Lynas 501
Point of Ayre 500
Point of Fethaland 396
Point Robert light house 821
Polduff Pier 579
Polperro 158
Ponta da Barca 1025
Ponta da Sereta 1025
Ponta das Lajes (Flores) 1024
Ponta Delgada 1025, 1031
Ponta do Albarnaz 1024
Ponta do Altar 985, 999
Ponta do Arnel 1025
Ponta do Castelo 1025
Ponta do Topo 1025
Ponta Garça 1025
Pont de Térénez 869
Pontevedra 977
Pontrieux 854
Pontusval 847
Poolbeg 568
Poole Harbour 196, 200
Poolewe Bay 447
Poplar Dock Marina 296
Porlock Weir 557
Pornic 916, 919
Pornichet 889, 909
Portaferry 639
Port à la Jument 832
Port Askaig 473
Portavadie 476
Portavogie 612, 634, 639
Portbail 796, 800
Port Bannatyne 480
Port-Béni 854
Port Blanc (Brittany) 855
Port Blanc (Morbihan) 903
Port Bloc 918, 937
Port Cardigan 537
Portcé 889
Portchester Lake 236
Port Clos 851
Port Crannaich 478
Port d'Antifer 749, 773
Port de Comberge 910
Port de la Corderie 851
Port de la Gravette 910
Port de la Meule 922
Port de Lévi 750, 785
Port des Barques 930
Port des Minimes 927
Port d'Etel 896
Port Dinorwic 518
Port du Becquet 750, 785

Port du Corniguel 891
Port du Douhet 917, 929
Port du Parun 903
Port Edgar 373
Portelet Bay (Jersey) 836
Portelet Harbour (Guernsey) 820, 826
Port Ellen 470, 473
Port-en-Bessin 749, 777, 780
Port Erin 503, 505
Port és Saies 832
Port Glasgow 485
Port Gorey 832
Port Haliguen 888, 901
Porthcawl 530, 547
Porth Conger 145
Porth Cressa 145
Porth Dinllaen 521
Porthgain 537
Porthleven 148
Porthmadog 535
Porth Mellin 148
Porth Trecastell 521
Porth Ysgaden 521
Portimão 985, 999
Port Isaac 561
Portishead 556
Portishead Point 531
Port Joinville 916, 921, 922
Portknockie 399
Port-la-Forêt 886, 891
Portland Bill/Harbour 143, 179
Portland Race 179
Port Louis 887, 896
Portmagee 598
Portmahomack 409
Port Manec'h 886
Port Maria 901
Port-Médoc 918, 937
Portmellon 152
Portmore 623
Port Navalo 888, 903
Port Neuf 930
Porto das Lajes 1024, 1026
Porto do Porto Santo 1035, 1036
Porto Novo 977
Porto (Oporto) 984, 988
Portosín 975
Portpatrick 492, 493
Portreath 561
Portree 448
Portrush 629
Portsall 864
Portscatho 152
Portsmouth 197, 235
Portsoy 399
Port St Mary (Isle of Man) 505
Port Talbot 547
Port Tudy 886, 895
Portugalete (Bilbao) 957
Port William 494
Pottery Pier (Poole) 200
Póvoa de Varzim 984, 987
Praia de Vitória 1025, 1030
Preston 509
Primel-Trégastel 847, 859
Puerto America 1005, 1011, 1012
Puerto Cabo de Cruz 975
Puerto Cádiz 1011, 1012
Puerto de San Esteban 951
Puerto de Santa Maria 1005, 1011, 1012
Puerto Sherry 1005, 1011, 1012
Puilladobhrain 461
Punta Atalya 951
Punta Candelaria 951
Punta Caraminal 1013
Punta Carnero 1005, 1016
Punta Círes 1006, 1020
Punta da Piedade 985
Punta de Alfanzina 985
Punta de la Estaca de Bares 951
Punta de la Frouxeira 951
Punta del Castillo 951
Punta del Cuerno 951
Punta del Torco de Fuera 950

Punta de San Antón 951
Punta Galea 950
Punta Insúa 952
Punta Malabata 1006
Punta Nariga 951
Punta Rebollera 951
Punta Roncadoira 951
Punta San Emeterio 950
Punta Umbria 1004, 1009
Pwllheli 534

Q

Queenborough 287
Quenard Point light house 820
Quiberon 901
Quillebeuf 774
Quimper 891
Quinta do Lorde 1035, 1036
Quiole Barrier 639

R

Rade de Brest 869
Rade de la Capelle 781
Ramsey Sound 532, 537
Ramsgate 251, 268
Ramsholt 320
Rantum 667
Rapness 417
Rathlin Island 631
Rathlin O'Birne, west side 610
Rathmullan 622
Rattray Head 394
Ravenglass 503
Raxó 977
Raz de Sein 849, 878
Red Bay 611, 631
Redbridge 226
Redkirk 494
Regnéville-sur-Mer 796, 800
Renfrew 485
Restronguet 150
Rhinns of Islay 470
Rhu 484
Rhubh' a Mháil 473
Rhubh' a Mháil (Ruvaal) 470
Ría de Aldán 978
Ría de Ares 951, 966
Ría de Arosa 952, 975
Ría de Betanzos 951, 967
Ría de Camariñas 972
Ría de Cedeira 951, 965
Ría de Corcubión 974, 975
Ría de Corme y Laxe 951, 952, 972
Ría de Huelva 1007
Ría de Huelva Bar 1004
Ría del Barquero 951, 965
Ría de Muros 952, 975
Ría de Pontevedra 952, 977
Ría de Ribadeo 951, 964
Ría de Santa Marta de Ortigueira 964, 965
Ría de Suances 959, 960
Ría de Tina Mayor 959
Ría de Vigo 952, 978
Ría de Vivero 951, 964
Rianxo 952, 976
Ribadesella 951, 961
Ribble, River 500, 509
Richard 935
Richborough 268
Richmond Lock 293
Ringaskiddy 588
Ringstead Bay 189
Río de Las Piedras 1004, 1008
Río de Orio 950, 955
Rio Douro entrance 988
Rio Guadalquivir 1005, 1006, 1010
Río Guadiana 1004, 1007
Río Miño 973, 986
Roach, River 302
Roancarrigmore 569
Roaringwater Bay 593

Roberts Cove...588
Rochebonne...797
Rochefort...917, 930
Roches Douvres...............................821, 846
Roche's Point..569
Rochester...289
Rochford...302
Rockabill..611
Rockall..441
Rømø (Havneby)...............................646, 658
Rona..415, 433
Roompotsluis..726
Roscoff...847, 862
Rosédo light house....................................846
Rosehearty..398
Rosière Steps...832
Rosneath...484
Rosslare..568, 581
Rota..1005, 1011
Rotheneuf...802
Rother, River...261
Rothesay..480
Rotten Island...610
Rotterdam..723, 724
Rouen...773, 774
Rouge Terrier...832
Round Island..142
Roundstone Bay..614
Rousay..417
Roustel...820
Royan...918, 935
Rozel Bay...836
Rubha nan Gall..434
Rubha Reidh...432
Rum...452
Runswick Bay...348
Rüschkanal...681
Ruthenstrom..674
Ryde..197, 230
Ryde Sand...230
Rye..250, 261

S

Sagres...985
Saignie Bay...832
Saints Bay..826
Salcombe...167
Salen..453, 455
Saltash..159
Sancti-Petri..1013
Sanday..417
Sandend Bay...399
Sandettie Bank..758
Sandown..231
Sandwich...268
Sandy Bay..1016
San Estaban de Pravia..............951, 962, 963
San Jacinto..989
San Sebastián (Donostia)................950, 955
Santa Clara...1025
Santa Maria (Azores)...............................1025
Santander...950, 959
Santoña...950, 959
San Vicente de la Barquera...........950, 961
Sanxenxo..977
San Xulian...977
Sark...832
Saundersfoot..545
Sauzon...887, 898
Saye Bay...824
Scalasaig..473
Scalloway..396, 422
Scapa Bay..395, 416
Scarborough.......................................335, 347
Scarinish..433
Scharhorn..674
Scheveningen.....................................701, 717, 718
Schiermonnikoog..............................706, 707
Schilling...683
Schleswig-Holstein....................................666
Schulau...674, 681
Schull..569, 594
Schulpengat...701

Schwinge, River..674
Scilly, Isles of........................142, 144, 145
Scolpaig...441
Scoraig...447
Scrabster...415
Seaham...335, 354
Seahouses (North Sunderland)..............360
Sea Mills..554
Seemannshöft..674
Seil Sound..462
Seine, River...769, 774
Sella Ness...396
Selsey Bill...250, 251
Séné...903
Sept Îles...903
Sesimbra..985, 997
Setubal...985, 997
Seven Stones..142
Severn Bridges...550
Severn Estuary...544
Sevilla..1005, 1010, 1011
Shaldon Bridge...174
Shambles Bank...179
Shannon Estuary, River...............570, 599, 600
Shapinsay..417
Sharfleet Creek..289
Sharpness...550
Sheep Haven.......................................621, 622
Sheep's Head..569
Sheerness..274
Shell Bay..832
Shetland Islands................395, 419, 420
Shiant Islands...440
Shieldaig...448
Shingles Bank...209
Shirehampton...554
Shoeburyness...301
Shoreham...250, 253
Silloth...500, 502
Sines...998
Skadan South...395
Skagen..645, 648
Skegness..339
Skeld...421
Skelligs Rock...570
Skerries...501, 641
Skerries Bank.....................................143, 168
Skerryvore...434
Skokholm Island...530
Skomer Island..537
Skroo..395
Skye (Eilean a' Cheo)..........433, 447, 448, 450
Sligo..620
Slyne Head..610
Small Isles..452
Smalls..530
Smerwick Harbour............................598, 599
Sneem...596
Soay...448
Socoa...944
Soldier's Bay...826
Solent...209, 220
   Central...197
   Eastern..197, 237
   Marks..196
   Precautionary Area..............................220
   Western..196
Solva..537
Solway Firth..500, 501
Somos...951
Sorel Point light house............................821
Soubise..930
Sound of Harris..................................432, 440
Sound of Iona..454
Sound of Jura...472
Sound of Mull.....................................434, 454
Sound of Raasay...433
Sound of Sleat...433
Southampton.....................197, 209, 226
South Bishop..530
South Dock Marina.....................................296
Southend (Mull of Kintyre)....................478
Southend-on-Sea..............................274, 301
Southerness Point.......................................494

South Ferriby...345
South Ness..396
Southport..513
South Queensferry.....................................374
South Rock..639
South Ronaldsay...417
Southsea..236
South Stack..501
South Wick..417
Southwold...276, 322
Sovereign Harbour...........................250, 259
Spiekeroog...691, 692
Spit Sand Fort.....................................197, 236
Spurn Head.........................335, 340, 344
St Abb's..371
St Abb's Head...368
Stackpole Quay..545
Stade...664
Stadersand..674
Staffa..454
St Agnes..145
St Alban's Head..................................179, 188
St Andrew's...381
St Ann's Head...530
Stansore Point...215
St Anthony Head.................................142, 150
St Antoine...846
Starcross..175
Start Point (Devon)...............143, 156, 167
Start Point (Orkney)..................................395
St Aubin...821, 836
Stavoren...713
St Bees Head...500
St Brelade Bay...836
St Briac...797, 811
St Brides Bay..537
St Brieuc...797, 812
St Cast Le Guildo................................797, 811
St Catherine Bay (Jersey).......................836
St Catherine's Point (Fowey)..................142
St Catherine's Point (IoW).............198, 209
St Denis d'Oléron..............................917, 928
Steep Holm..548
Stellendam..702
Ster Wenn...898
St Evette..879
St Germain-sur-Ay.......................................799
St German's River...............................159, 165
St Gilles-Croix-de-Vie..............916, 922, 923
St Guénolé..879
St Helen's Fort....................................198, 231
St Helen's Pool...145
St Helier...821, 834
St Hernot Bay..876
St Ives..531, 561
St Jean de Luz....................................918, 944
St Jean de Monts...916
St Katharine Haven....................................296
St Kilda...433, 441
St Malo.......................................796, 798, 802
St Martin de Ré...................................917, 926
St Mary's..142, 147
St Mawes..150
St Michael's Mount....................................148
St Monans...369, 374
St Nazaire..889, 910
St Nicolas Bay (Morgat)...........................876
St Nicolas (Les Sables d'Olonne)...........923
St Nicolas (Pointe de Grave)..................917
St Ninian's Bay...479
St Norgard Bay..876
Stoer Head..432
Stokes Bay...209
Stonehaven..369, 383
Stor..674
Storloch..674
Stornoway..432, 435
Stour, River (Dorset)..................................210
Stour, River (Essex/Suffolk)...........276, 315
Stour, River (Kent)......................................268
St Peter (Eider, River)................................664
St Peter Port..820, 827
St Quay-Portrieux................................797, 814
Straight Point......................................168, 176

Strangford Lough................................639
Stranraer...492
Strathy Point.....................................394
Straw Island......................................610
Stroma......................................394, 415
Stromness.................................395, 418
Stronsay..419
Strumble Head...................................530
St Sampson's Harbour 825
St Suliac..809
St Thomas Head.................................556
St Tudwall's Roads 533
Studland Bay..............................200, 231
St Vaast-la-Hougue....................750, 783
St Valéry-en-Caux......................749, 767
St Valéry-sur-Somme..................748, 761
Sudbrook..550
Süderhöft...669
Suderoogsand....................................667
Sule Skerry................................394, 415
Sullom Voe.................396, 420, 422
Sumburgh..422
Sumburgh Head.................................395
Summer Isles.....................................443
Sunderland.................................335, 354
Sunk Head...307
Sunk Inner Precautionary Area 271
Swale, River...............................274, 286
Swanage.....................................196, 199
Swanwick...229
Swarbacks Minn.................................420
Swinge...822
Sylt..667

T

Tabs Head 339
Tamar, River......................................165
Tancarville Bridge..............................774
Tanera Beg...447
Tanera Mor..447
Tangier....................................1006, 1020
Tarbat Ness.......................................394
Tarbert Island (Shannon) 600
Tarbert (Loch Fyne)...........................476
Tarifa......................1005, 1015, 1016
Tarn Point...503
Tater-du...142
Tautenay..820
Tavira......................................985, 1002
Tavy, River..159
Taw and Torridge, Rivers...........531, 559
Tay, River..381
Tazones...951
Tean Sound..145
Teddington Lock................................295
Tees, River...348
Teignmouth..174
Telegraph Bay....................................824
Tenby...545
Termunterzijl.....................................706
Terneuzen..733
Terre-Nègre.......................................917
Terschelling...............................708, 709
Teüfelsbruck......................................681
Thames Estuary/River 274, 277, 280, 293
Thames Tidal Barrier 275, 295
Thisted Havn......................................645
Thorney Channel........................198, 240
Thorsminde.................................650, 652
Thurso...394
Thyborøn...650
Tidal coefficients 759, 786, 805, 873, 936
Tighnabruaich....................................480
Tilbury...293
Tinduff...869
Tingwall...418
Tinker's Hole.....................................454
Tiumpan Head....................................432
Tobermory...454
Tollesbury..307

Tonning..669
Topsham...175
Torduff Point......................................494
Torquay..174
Torre de Hércules...............................951
Torridge, River...................................559
Tory Island..611
Totland Bay.................................210, 211
Totnes..168
Trawbreaga Bay..................................623
Trearddur Bay....................................521
Trébeurden.................................846, 858
Tréboul..877
Trefor...521
Trégastel-Plage...........................846, 857
Tréguier.......................846, 852, 854
Tréhiguier..906
Tresco.......................................145, 146
Treshnish Isles...................................454
Trévignon...886
Trevose Head.....................................531
Trézien...848
Troon.......................................471, 491
Trouville....................................749, 776
Truro..150
Trwyn Dinmor....................................521
Tuckton Bridge...........................210, 218
Turf Lock...175
Turnberry Point.................................471
Turnchapel..159
Tuskar Rock...............................568, 572
Tyne, River.................................335, 355

U

Ullapool.....................................435, 443
Ulva Sound...454
Ulverston...507
Unionhall (Glandore) 591
Upnor...289
Ura Firth..420
Urk...713
Ushenish..432

V

Vaila Sound.................................420, 421
Val-André...811
Valentia Harbour...............................597
Vannes..903, 904
Vatersay Sound..................................432
Vatteville...774
Vazon Bay..826
Vegesack..682
Velas...1025, 1028
Ventnor..231
Viana do Castello.......................984, 987
Vila da Praia......................................1025
Vila do Porto..............................1025, 1030
Vila Franca do Campo........................1025
Vilagarcia.................952, 975, 976
Vilaine, River.............................888, 906
Vilamoura.................................985, 1000
Vilanova..976
Vila Real de Santo Antònio 985, 1002, 1004
Village Bay (St Kilda).........................441
Vlaardingen.......................................724
Vlieland..709
Vlissingen..728
Volendam...713

W

Wadebridge..561
Wainfleet..339
Waldringfield......................................320
Wallasea...303
Walsoorden..728
Walton Backwaters.....................276, 310
Wangerooge..690
Wareham.....................................200, 202

Warrenpoint (Carlingford Lough).......640
Warren Point (R Exe).........................175
Warren Point (R Yealm).....................166
Warsash......................................228, 229
Wash..334, 337
Watchet..557
Waterfoot...631
Waterford...583
Watermill Cove...................................145
Watermouth..558
Watwick Point....................................530
Welland, River....................................339
Wellhouse Rock..................................550
Wells-next-the-Sea.....................334, 337
Wemyss Bay.......................................483
Weser, River.................664, 665, 683
West Bay (Bridport)...........................178
West Burra Firth Inner.......................396
West Cove...596
Westerheversand.................................664
Westerschelde.............................718, 727
West Frisian Islands............................707
Westhoofd light house 701
Westkapelle................................702, 727
West Kyle...479
West Loch Tarbert......................473, 436
West Mersea.......................................307
Weston...226
Weston-super-Mare.............................556
Westport...619
West Schouwen light house 702
West Stones..337
West Terschelling.......................708, 709
West Wangerooge...............................690
Wexford......................................579, 580
Weymouth...187
Whalsay......................................395, 420
Whitaker Beacon.................................303
Whitby..335, 348
Whitehaven.................................500, 503
White House.......................................550
Whitsand Bay.....................................158
Whitstable...................................274, 285
Wick...394, 410
Wicklow......................................568, 579
Wicklow Head.....................................568
Widewall Bay......................................418
Widnes...513
Wielingen Channel.............................702
Wilhelmshaven...........................665, 685
Winteringham Haven..........................345
Winterton-on-Sea...............................327
Wisbech......................................334, 338
Wischhafen...674
Wolf Rock..142
Woolwich..293
Wootton Creek...........................197, 230
Worbarrow Bay...................................189
Worhden...674
Workington.................................500, 502
Workum..713
Worthing.....................................250, 253
Wouldham..289
Wyk..668
Wyre light house................................508
Wyre Sound (Orkney).........................417

Y

Yare, River..327
Yarmouth (Isle of Wight) 196, 212
Yealm, River.......................................166
Yelland Marsh.....................................559
Yell Sound..................395, 396, 420
Youghal...584

Z

Zeebrugge...735
Zeegat van Terschelling......................708
Zoutkamp...707
Zumaya (Zumaia)........................950, 956

WWW.VETUS.COM

Total Solutions

with VETUS systems

- **Engines**
- **Stern gear**
- **Fuel**
- **Instruments**
- **Fresh water**
- **Waste water**
- **Manoeuvring**
- **Electrical**
- **Steering**
- **Hatches**
- **Ventilation**
- **Anchoring**

Our innovative **systems thinking** offers you a **total solution** for all your on-board equipment. Above or below decks – power or sail – we have more than 2,500 product lines, backed by worldwide warranty and service

Visit our website for the online VETUS catalogue and to find your local distributor

☎ **023 8045 4507**

sales@vetus.co.uk | www.vetus.com

CREATORS OF
BOAT SYSTEMS

vetus 5
5-year warranty on engines*

vetus 3
3-year warranty on equipment*

*Warranty terms and conditions can be found at www.vetus.com

ENGINES | STERN GEAR | FUEL | INSTRUMENTS | FRESH WATER | WASTE WATER | MANOEUVRING | ELECTRICAL | STEERING | HATCHES | VENTILATION | ANCH